Medical Physiology

Principles for Clinical Medicine

FOURTH EDITION

Medical Physiology

Principles for Clinical Medicine

FOURTH EDITION

EDITED BY

Rodney A. Rhoades, Ph.D.

Professor Emeritus
Department of Cellular and Integrative Physiology
Indiana University School of Medicine
Indianapolis, Indiana

David R. Bell, Ph.D.

Associate Professor
Department of Cellular and Integrative Physiology
Indiana University School of Medicine
Fort Wayne, Indiana

Wolters Kluwer | Lippincott Williams & Wilkins
Health

Philadelphia • Baltimore • New York • London
Buenos Aires • Hong Kong • Sydney • Tokyo

Acquisitions Editor: Crystal Taylor
Product Managers: Angela Collins & Catherine Noonan
Development Editor: Kelly Horvath
Marketing Manager: Joy Fisher-Williams
Vendor Manager: Bridgett Dougherty
Manufacturing Manager: Margie Orzech
Design & Art Direction: Doug Smock & Jen Clements
Compositor: SPi Global

Fourth Edition

351 West Camden Street
Baltimore, MD 21201

Two Commerce Square
2001 Market Street
Philadelphia, PA 19103

Printed in China

ISBN 978-1-4511-1039-5

DISCLAIMER
Care has been taken to confirm the accuracy of the information present and to describe generally accepted practices. However, the authors, editors, and publisher are not responsible for errors or omissions or for any consequences from application of the information in this book and make no warranty, expressed or implied, with respect to the currency, completeness, or accuracy of the contents of the publication. Application of this information in a particular situation remains the professional responsibility of the practitioner; the clinical treatments described and recommended may not be considered absolute and universal recommendations.

The authors, editors, and publisher have exerted every effort to ensure that drug selection and dosage set forth in this text are in accordance with the current recommendations and practice at the time of publication. However, in view of ongoing research, changes in government regulations, and the constant flow of information relating to drug therapy and drug reactions, the reader is urged to check the package insert for each drug for any change in indications and dosage and for added warnings and precautions. This is particularly important when the recommended agent is a new or infrequently employed drug.

Some drugs and medical devices presented in this publication have Food and Drug Administration (FDA) clearance for limited use in restricted research settings. It is the responsibility of the health care provider to ascertain the FDA status of each drug or device planned for use in their clinical practice.

To purchase additional copies of this book, call our customer service department at (800) 638-3030 or fax orders to (301) 223-2320. International customers should call (301) 223-2300.

Visit Lippincott Williams & Wilkins on the Internet: http://www.lww.com. Lippincott Williams & Wilkins customer service representatives are available from 8:30 am to 6:00 pm, EST.

9 8 7 6 5 4 3 2 1

CCS1211

Preface

The function of the human body involves intricate and complex processes at the cellular, organ, and systems level. The fourth edition of *Medical Physiology: Principles for Clinical Medicine* explains what is currently known about these integrated processes. Although the emphasis of the fourth edition is on normal physiology, discussion of pathophysiology is also undertaken to show how altered functions are involved in disease processes. This not only reinforces fundamental physiologic principles, but also demonstrates how basic concepts in physiology serve as important principles in clinical medicine.

Our mission for the fourth edition of *Medical Physiology: Principles for Clinical Medicine* is to provide a clear, accurate, and up-to-date introduction to medical physiology for medical students and other students in the health sciences as well as to waste no space in so doing—each element of this textbook presents a learning opportunity; therefore we have attempted to maximize those opportunities to the fullest.

AUDIENCE AND FUNCTION

This book, like the previous edition, is written for medical students as well as for dental, nursing graduate, and veterinary students who are in healthcare professions. This is not an encyclopedic textbook. Rather, the fourth edition focuses on the basic physiologic principles necessary to understand human function, presented from a fundamentally clinical perspective and without diluting important content and explanatory details. Although the book is written primarily with the student in mind, the fourth edition will also be helpful to physicians and other healthcare professionals seeking a physiology refresher.

In the fourth edition, each chapter has been rewritten to minimize the compilation of isolated facts and make the text as lucid, accurate, and up-to-date as possible, with clearly understandable explanations of processes and mechanisms. The chapters are written by medical school faculty members who have had many years of experience teaching physiology and who are experts in their field. They have selected material that is important for medical students to know and have presented this material in a concise, uncomplicated, and understandable fashion. We have purposefully avoided discussion of research laboratory methods, and/or historical material. Although such issues are important in other contexts, most medical students prefer to focus on the essentials. We have also avoided topics that are as yet unsettled, while recognizing that new research constantly provides fresh insights and sometimes challenges old ideas.

CONTENT AND ORGANIZATION

This book begins with a discussion of basic physiologic concepts, such as homeostasis and cell signaling, in Chapter 1. Chapter 2 covers the cell membrane, membrane transport, and the cell membrane potential. Most of the remaining chapters discuss the different organ systems: nervous (Chapters 3–7), muscle (Chapter 8), cardiovascular (Chapters 11–17), respiratory (Chapters 18–21), renal (Chapters 22–23), gastrointestinal (Chapters 25 and 26), endocrine (Chapters 30–35), and reproductive physiology (Chapters 36–38). Special chapters on the blood (Chapter 9), immunology (Chapter 10), and the liver (Chapter 27) are included. The immunology chapter emphasizes physiologic applications of immunology. Chapters on acid–base regulation (Chapter 24), temperature regulation (Chapter 28), and exercise (Chapter 29) discuss these complex, integrated functions. The order of presentation of topics follows that of most United States medical school courses in physiology. After the first two chapters, the other chapters can be read in any order, and some chapters may be skipped if the subjects are taught in other courses (e.g., neurobiology or biochemistry).

An important objective for the fourth edition is to demonstrate to the student that physiology, the study of normal function, is key to understanding pathophysiology and pharmacology, and that basic concepts in physiology serve as important principles in clinical medicine.

KEY CHANGES

As in previous editions, we have continued to emphasize basic concepts and integrated organ function to deepen reader comprehension. Many significant changes have been instituted in this fourth edition to improve the delivery and, thereby, the absorption of this essential content.

Art

Most striking among these important changes is the use of full color to help make the fourth edition not only more visually appealing, but also more instructive, especially regarding the artwork. Rather than applying color arbitrarily, color itself is used with purpose and delivers meaning. Graphs, diagrams, and flow charts, for example, incorporate a coordinated scheme. Red is used to indicate stimulatory, augmented, or increased effects, whereas blue connotes inhibitory, impaired, or decreased effects.

A coordinated color scheme is likewise used throughout to depict transport systems. This key, in which pores and channels are blue, active transporters are red, facilitated transport is purple, cell chemical receptors are green, co- and counter-transporters are orange, and voltage-gated transporters are yellow, adds a level of instructiveness to the figures not seen in other physiology textbooks. In thus differentiating these elements integral to the workings of physiology by their function, the fourth edition artwork provides visual consistency with meaning from one figure to the next.

Artwork was also substantially overhauled to provide a coherent style and point of view. An effort has also been made to incorporate more conceptual illustrations alongside the popular and useful graphs and tables of data. These beautiful new full-color conceptual diagrams guide students to an understanding of the general underpinnings of physiology. Figures now work with text to provide meaningful, comprehensible content. Students will be relieved to find concepts "clicking" like never before.

Text

Another important improvement for the fourth edition is that most chapters were not only substantially revised and updated, but they were also edited to achieve unity of voice as well as to be as concise as possible, both of which approaches considerably enhance clarity.

Features

Finally, we have also revised and improved the features in the book to be as helpful and useful as possible. First, a set of *active* learning objectives at the beginning of each chapter indicate to the student what they should be able to *do* with the material in the chapter once it has been mastered, rather than merely telling them *what* they should master, as in other textbooks. These objectives direct the student to apply the concepts and processes contained in the chapter rather than memorize facts. They urge the student to "explain," "describe," or "predict" rather than "define," "identify," or "list."

Next, chapter subheadings are presented as active concept statements designed to convey to the student the key point(s) of a given section. Unlike typical textbook subheadings that simply title a section, these are given in full sentence form and appear in bold periodically throughout a chapter. Taken together, these revolutionary concept statements add up to another way to neatly summarize the chapter for review.

The clinical focus boxes have once again been updated for the fourth edition. These essays deal with clinical applications of physiology rather than physiology research. In addition, we are reprising the "From Bench to Bedside" essays introduced in the third edition. Because these focus on physiologic applications in medicine that are "just around the corner" for use in medical practice, readers will eagerly anticipate these fresh, new essays published with each successive edition.

Students will appreciate the book's inclusion of such helpful, useful tools as the glossary of text terms, which has been expanded by nearly double for the fourth edition and corresponds to bolded terms within each chapter. Updated lists of common abbreviations in physiology and of normal blood values are also provided in this edition.

As done previously, each chapter includes two online case studies, with questions and answers. In addition, a third, new style of case study has been added in each chapter, designed to integrate concepts between organ function and the various systems. These might require synthesizing material across multiple chapters to prepare students for their future careers and get them thinking like clinicians.

All of the abundant chapter review questions (now numbering over 500) are again online and interactive. They have been updated to United States Medical Licensing Examination (USMLE) format with explanations for right and wrong answers. These questions are analytical in nature and test the student's ability to apply physiologic principles to solving problems rather than test basic fact-based recall. These questions were written by the author of the corresponding chapter and contain explanations of the correct and incorrect answers.

Also, the extensive test bank written by subject matter experts is once again available for instructors using this textbook in their courses.

PEDAGOGY

This fourth edition incorporates many features designed to facilitate learning. Guiding the student along his or her study of physiology are such in-print features as:

- **Active Learning Objectives.** These active statements are supplied to the student to indicate what they should be able to do with chapter material once it has been mastered.
- **Readability.** The text is a pleasure to read, and topics are developed logically. Difficult concepts are explained clearly, in a unified voice, and supported with plentiful illustrations. Minutiae and esoteric topics are avoided.
- **Vibrant Design.** The fourth edition interior has been completely revamped. The new design not only makes navigating the text easier, but also draws the reader in with immense visual appeal and strategic use of color. Likewise, the design highlights the pedagogical features, making them easier to find and use.
- **Key Concept Subheadings.** Second-level topic subheadings are active full-sentence statements. For example, instead of heading a section "Homeostasis," the heading is "Homeostasis is the maintenance of steady states in the body by coordinated physiological mechanisms." In this way, the key idea in a section is immediately obvious. Add them up, and the student has another means of chapter review.
- **Boldfacing.** Key terms are boldfaced upon their first appearance in a chapter. These terms are explained in the text and defined in the glossary for quick reference.
- **Illustrations and Tables.** Abundant full-color figures illustrate important concepts. These illustrations often show interrelationships between different variables or components of a system. Many of the figures are color-coded flow diagrams, so that students can appreciate the sequence of events that follow when a factor changes. Red is used to indicate stimulatory effects, and blue indicates inhibitory effects. All illustrations are now rendered in full color to reinforce concepts and enhance reader comprehension. Review tables provide useful summaries of material explained in more detail in the text.
- **Clinical Focus and Bench to Bedside Boxes.** Each chapter contains two Clinical Focus boxes and one all-new Bench to Bedside box, which illustrate the relevance of

the physiology discussed in the chapter to clinical medicine and help the reader make those connections.

- **Bulleted Chapter Summaries**. These bulleted statements provide a concise summative description of the chapter, and provide a good review of the chapter.
- **Abbreviations and Normal Values.** This third edition includes an appendix of common abbreviations in physiology and a table of normal blood, plasma, or serum values on the inside book covers for convenient access. All abbreviations are defined when first used in the text, but the table of abbreviations in the appendix serves as a useful reminder of abbreviations commonly used in physiology and medicine. Normal values for blood are also embedded in the text, but the table on the inside front and back covers provides a more complete and easily accessible reference.
- **Index.** A comprehensive index allows the student to easily look up material in the text.
- **Glossary.** A glossary of all boldfaced terms in the text is included for quick access to definition of terms.

Ancillary Package

Still more features round out the colossal ancillary package online at thePoint. These bonus offerings provide ample opportunities for self-assessment, additional reading on tangential topics, and animated versions of the artwork to further elucidate the more complex concepts.

- **Case Studies.** Each chapter is associated with two online case studies with questions and answers. These case studies help to reinforce how an understanding of physiology is important in dealing with clinical conditions. A new integrated case study has also been added to each chapter to help the student better understand integrated function.
- **Review Questions and Answers.** Students can use the interactive online chapter review questions to test whether they have mastered the material. These USMLE-style questions should help students prepare for the Step 1 examination. Answers to the questions are also provided online and include complete explanations as to why the choices are correct or incorrect.
- **Suggested Reading.** A short list of recent review articles, monographs, book chapters, classic papers, or websites where students can obtain additional information associated with each chapter is provided online.
- **Animations.** The fourth edition contains online animations illustrating difficult physiology concepts.
- **Image Bank for Instructors.** An image bank containing all of the figures in the book, in both pdf and jpeg formats is available for download from our website at thePoint.

Rodney A. Rhoades, Ph.D.
David R. Bell, Ph.D.

Contributors

DAVID R. BELL, PH.D.
Associate Professor of Cellular and Integrative Physiology
Indiana University School of Medicine
Fort Wayne, Indiana

ROBERT V. CONSIDINE, PH.D.
Associate Professor of Medicine and Physiology
Indiana University School of Medicine
Indianapolis, Indiana

JEFFREY S. ELMENDORF, PH.D.
Associate Professor of Cellular and Integrative Physiology
Physiology
Indiana University School of Medicine
Indianapolis, Indiana

PATRICIA J. GALLAGHER, PH.D.
Associate Professor of Cellular and Integrative Physiology
Indiana University School of Medicine
Indianapolis, Indiana

JOHN C. KINCAID, M.D.
Professor of Neurology and Physiology
Indiana University School of Medicine
Indianapolis, Indiana

RODNEY A. RHOADES, PH.D.
Professor Emeritus
Department of Cellular and Integrative Physiology
Indiana University School of Medicine
Indianapolis, Indiana

GEORGE A. TANNER, PH.D.
Emeritus Professor of Cellular and Integrative Physiology
Indiana University School of Medicine
Indianapolis, Indiana

GABI NINDL WAITE, PH.D.
Associate Professor of Cellular and Integrative Physiology
Indiana University School of Medicine
Terre Haute Center for Medical Education
Terre Haute, Indiana

FRANK A. WITZMANN, PH.D.
Professor of Cellular and Integrative Physiology
Indiana University School of Medicine
Indianapolis, Indiana

JACKIE D. WOOD, PH.D.
Professor of Physiology
Ohio State University College of Medicine
Columbus, Ohio

Acknowledgments

We would like to express our deepest thanks and appreciation to all of the contributing authors. Without their expertise and cooperation, this fourth edition would have not been possible. We also wish to express our appreciation to all of our students and colleagues who have provided helpful comments and criticisms during the revision of this book, particularly, Shloka Anathanarayanan, Robert Banks, Wei Chen, Steve Echtenkamp, Alexandra Golant, Michael Hellman, Jennifer Huang, Kristina Medhus, Ankit Patel, and Yuri Zagvazdin. We would also like to give thanks for a job well done to our editorial staff for their guidance and assistance in significantly improving each edition of this book. A very special thanks goes to our Developmental Editor, Kelly Horvath, who was a delight to work with, and whose patience and editorial talents were essential to the completion of the fourth edition of this book. We are indebted as well to our artist, Kim Battista. Finally, we would like to thank Crystal Taylor, our Acquisitions Editor at Lippincott Williams and Wilkins, for her support, vision, and commitment to this book. We are indebted to her administrative talents and her managing of the staff and material resources for this project.

Lastly, we would like to thank our wives, Pamela Bell and Judy Rhoades, for their love, patience, support, and understanding of our need to devote a great deal of personal time and energy to the development of this book.

Contents

Preface *v*

Contributors *viii*

Acknowledgments *ix*

PART I • CELLULAR PHYSIOLOGY **1**

CHAPTER 1 • Homeostasis and Cellular Signaling 1
Patricia J. Gallagher, Ph.D.
Basis of Physiologic Regulation 1
Communication and Signaling Modes 6
Molecular Basis of Cellular Signaling 9
Second Messengers Roles 15
Mitogenic Signaling Pathways 21

CHAPTER 2 • Plasma Membrane, Membrane Transport, and Resting Membrane Potential **24**
Robert V. Considine, Ph.D.
Plasma Membrane Structure 24
Solute Transport Mechanisms 26
Water Movement Across the Plasma Membrane 37
Resting Membrane Potential 39

PART II • NEUROMUSCULAR PHYSIOLOGY **42**

CHAPTER 3 • Action Potential, Synaptic Transmission, and Maintenance of Nerve Function **42**
John C. Kincaid, M.D.
Neuronal Structure 42
Action Potentials 46
Synaptic Transmission 51
Neurotransmission 54

CHAPTER 4 • Sensory Physiology **61**
David R. Bell, Ph.D., Rodney A. Rhoades, Ph.D.
Sensory System 61
Somatosensory System 67
Visual System 69
Auditory System 76
Vestibular System 82
Gustatory and Olfactory Systems 85

CHAPTER 5 • Motor System **91**
John C. Kincaid, M.D.
Skeleton as Framework for Movement 91
Muscle Function and Body Movement 91
Nervous System Components for the Control of Movement 92
Spinal Cord in the Control of Movement 96
Supraspinal Influences on Motor Control 98

Cerebral Cortex Role in Motor Control 100
Basal Ganglia and Motor Control 103
Cerebellum in the Control of Movement 105

CHAPTER 6 • Autonomic Nervous System 108
John C. Kincaid, M.D.
Overview of the Autonomic Nervous System 108
Sympathetic Nervous System 110
Parasympathetic Nervous System 113
Control of the Autonomic Nervous System 114

CHAPTER 7 • Integrative Functions of the Central Nervous System 119
John C. Kincaid, M.D.
Hypothalamus 119
Brain Electrical Activity 125
Functional Components of the Forebrain 128
Higher Cognitive Skills 134

CHAPTER 8 • Skeletal and Smooth Muscle 138
David R. Bell, Ph.D.
Skeletal Muscle 138
Motor Neurons and Excitation-Contraction Coupling in Skeletal Muscle 143
Mechanics of Skeletal Muscle Contraction 148
Smooth Muscle 158

PART III • BLOOD AND IMMUNOLOGY 166

CHAPTER 9 • Blood Components 166
Gabi Nindl Waite, Ph.D.
Blood Functions 166
Whole Blood 167
Soluble Components of Blood and Their Tests 167
Formed Elements of Blood and Their Tests 170
Red Blood Cells 175
White Blood Cells 178
Blood Cell Formation 180
Blood Clotting 182

CHAPTER 10 • Immunology, Organ Interaction, and Homeostasis 188
Gabi Nindl Waite, Ph.D.
Immune System Components 188
Immune System Activation 189
Immune System Detection 191
Immune System Defenses 191
Cell-Mediated and Humoral Responses 194
Acute and Chronic Inflammation 201
Chronic Inflammation 204
Anti-Inflammatory Drugs 204
Organ Transplantation and Immunology 205
Immunologic Disorders 206
Neuroendoimmunology 209

PART IV • CARDIOVASCULAR PHYSIOLOGY 212

CHAPTER 11 • **Overview of the Cardiovascular System and Hemodynamics** 212
David R. Bell, Ph.D.
Functional Organization 213
Physics of Blood Containment and Movement 216
Physical Dynamics of Blood Flow 218
Distribution of Pressure, Flow, Velocity, and Blood Volume 224

CHAPTER 12 • **Electrical Activity of the Heart** 227
David R. Bell, Ph.D.
Electrophysiology of Cardiac Muscle 228
Electrocardiogram 236

CHAPTER 13 • **Cardiac Muscle Mechanics and the Cardiac Pump** 248
David R. Bell, Ph.D.
Cardiac Excitation-Contraction Coupling 249
Cardiac Cycle 251
Determinants of Myocardial Performance 253
Cardiac Output 260
Cardiac Output Measurement 262
Imaging Techniques for Measuring Cardiac Structures, Volumes, Blood Flow, and Cardiac Output 263

CHAPTER 14 • **Systemic Circulation** 267
David R. Bell, Ph.D.
Determinants of Arterial Pressures 267
Arterial Pressure Measurement 270
Peripheral and Central Blood Volume 271
Coupling of Vascular and Cardiac Function 274

CHAPTER 15 • **Microcirculation and Lymphatic System** 279
David R. Bell, Ph.D.
Components of the Microvasculature 279
Lymphatic System 282
Materials Exchange between the Vasculature and Tissues 283
Regulation of Microvascular Resistance 288

CHAPTER 16 • **Special Circulations** 295
David R. Bell, Ph.D.
Coronary Circulation 295
Cerebral Circulation 297
Small Intestine Circulation 301
Hepatic Circulation 303
Skeletal Muscle Circulation 304
Cutaneous Circulation 305
Fetal and Placental Circulations 306

CHAPTER 17 • **Control Mechanisms in Circulatory Function** 311
David R. Bell, Ph.D.
Autonomic Neural Control of the Circulatory System 311
Hormonal Control of the Cardiovascular System 317
Circulatory Shock 321

PART V • RESPIRATORY PHYSIOLOGY 326

CHAPTER 18 • **Ventilation and the Mechanics of Breathing** 326
Rodney A. Rhoades, Ph.D.
Lung Structural and Functional Relationships 327
Pulmonary Pressures and Airflow During Breathing 328
Spirometry and Lung Volumes 333
Minute Ventilation 336
Lung and Chest Wall Mechanical Properties 341
Airflow and the Work of Breathing 349

CHAPTER 19 • **Gas Transfer and Transport** 356
Rodney A. Rhoades, Ph.D.
Gas Diffusion and Uptake 356
Diffusing Capacity 358
Gas Transport by the Blood 359
Respiratory Causes of Hypoxemia 363

CHAPTER 20 • **Pulmonary Circulation and Ventilation/Perfusion** 369
Rodney A. Rhoades, Ph.D.
Functional Organization 369
Hemodynamic Features 370
Fluid Exchange in Pulmonary Capillaries 374
Blood Flow Distribution in the Lungs 376
Shunts and Venous Admixture 378
Bronchial Circulation 380

CHAPTER 21 • **Control of Ventilation** 382
Rodney A. Rhoades, Ph.D.
Generation of the Breathing Pattern 382
Lung and Chest Wall Reflexes 386
Feedback Control of Breathing 387
Chemoresponses to Altered Oxygen and Carbon Dioxide 390
Control of Breathing During Sleep 392
Control of Breathing in Unusual Environments 394

PART VI • RENAL PHYSIOLOGY AND BODY FLUIDS 399

CHAPTER 22 • **Kidney Function** 399
George A. Tanner, Ph.D.
Overview of Structure and Function 399
Urine Formation 403
Renal Blood Flow 407
Glomerular Filtration 409
Transport in the Proximal Tubule 413
Tubular Transport in the Loop of Henle 417
Tubular Transport in the Distal Nephron 417
Urinary Concentration and Dilution 419
Inherited Defects in Kidney Tubule Epithelial Cells 424

CHAPTER 23 • **Regulation of Fluid and Electrolyte Balance** 427
George A. Tanner, Ph.D.
Fluid Compartments of the Body 427
Fluid Balance 432

Sodium Balance 436
Potassium Balance 442
Calcium Balance 445
Magnesium Balance 446
Phosphate Balance 446
Urinary Tract 447

CHAPTER 24 • Acid–Base Homeostasis 451
George A. Tanner, Ph.D.
Basic Principles of Acid–Base Chemistry 451
Acid Production 453
Blood pH Regulation 454
Intracellular pH Regulation 462
Acid–Base Balance Disturbances 463

PART VII • GASTROINTESTINAL PHYSIOLOGY 471

CHAPTER 25 • Neurogastroenterology and Motility 471
Jackie D. Wood, Ph.D.
Musculature of the Digestive Tract 472
Neural Control of Digestive Functions 475
Synaptic Transmission in the Enteric Nervous System 480
Enteric Motor Neurons 483
Gastrointestinal Motility Patterns 487
Esophageal Motility 490
Gastric Motility 490
Small Intestinal Motility 495
Large Intestinal Motility 499

CHAPTER 26 • Gastrointestinal Secretion, Digestion, and Absorption 505
Rodney A. Rhoades, Ph.D.
Salivary Secretion 505
Gastric Secretion 508
Pancreatic Secretion 511
Biliary Secretion 515
Intestinal Secretion 519
Carbohydrates Digestion and Absorption 520
Lipid Digestion and Absorption 523
Protein Digestion and Absorption 526
Vitamin Absorption 528
Electrolyte and Mineral Absorption 530
Water Absorption 534

CHAPTER 27 • Liver Physiology 536
Rodney A. Rhoades, Ph.D.
Liver Structure and Function 536
Drug Metabolism in the Liver 539
Energy Metabolism in the Liver 540
Protein and Amino Acid Metabolism in the Liver 544
Liver as Storage Organ 545
Endocrine Functions of the Liver 548

PART VIII • TEMPERATURE REGULATION AND EXERCISE PHYSIOLOGY 550

CHAPTER 28 • Regulation of Body Temperature 550
Frank A. Witzmann, Ph.D.

Body Temperature and Heat Transfer 551
Balance between Heat Production and Heat Loss 553
Metabolic Rate and Heat Production at Rest 554
Heat Dissipation 558
Thermoregulatory Control 561
Thermoregulatory Responses During Exercise 564
Heat Acclimatization 565
Responses to Cold 567
Clinical Aspects of Thermoregulation 570

CHAPTER 29 • Exercise Physiology 575
Frank A. Witzmann, Ph.D.

Oxygen Uptake and Exercise 575
Cardiovascular Responses to Exercise 577
Respiratory Responses to Exercise 580
Skeletal Muscle and Bone Responses to Exercise 582
Gastrointestinal, Metabolic, and Endocrine Responses to Exercise 585
Aging and Immune Responses to Exercise 586

PART IX • ENDOCRINE PHYSIOLOGY 589

CHAPTER 30 • Endocrine Control Mechanisms 589
Jeffrey S. Elmendorf, Ph.D.

General Concepts of Endocrine Control 589
Hormone Classes 593
Mechanisms of Hormone Action 600

CHAPTER 31 • Hypothalamus and the Pituitary Gland 604
Robert V. Considine, Ph.D.

Hypothalamic-Pituitary Axis 604
Posterior Pituitary Hormones 606
Anterior Pituitary Hormones 608

CHAPTER 32 • Thyroid Gland 621
Robert V. Considine, Ph.D.

Functional Anatomy 621
Thyroid Hormone Synthesis, Secretion, and Metabolism 622
Thyroid Hormone Mechanism of Action 626
Thyroid Hormone Function 627
Thyroid Function Abnormalities in Adults 630

CHAPTER 33 • Adrenal Gland 633
Robert V. Considine, Ph.D.

Functional Anatomy 633
Metabolism of Adrenal Cortex Hormones 635
Adrenal Medulla Hormones 647

CHAPTER 34 • Endocrine Pancreas 649

Jeffrey S. Elmendorf, Ph.D.

Islets of Langerhans 649

Insulin and Glucagon Influence on Metabolic Fuels 656

Diabetes Mellitus 660

CHAPTER 35 • Endocrine Regulation of Calcium, Phosphate, and Bone Homeostasis 664

Jeffrey S. Elmendorf, Ph.D.

Overview of Calcium and Phosphate in the Body 664

Calcium and Phosphate Metabolism 667

Plasma Calcium and Phosphate Regulation 669

Bone Dysfunction 673

PART X • REPRODUCTIVE PHYSIOLOGY 676

CHAPTER 36 • Male Reproductive System 676

Jeffrey S. Elmendorf, Ph.D.

Endocrine Glands of the Male Reproductive System 676

Testicular Function and Regulation 677

Male Reproductive Organs 679

Spermatogenesis 683

Endocrine Function of the Testis 685

Androgen Action and Male Development 688

Male Reproductive Disorders 690

CHAPTER 37 • Female Reproductive System 693

Robert V. Considine, Ph.D.

Hormonal Regulation of the Female Reproductive System 693

Female Reproductive Organs 695

Folliculogenesis, Steroidogenesis, Atresia, and Meiosis 696

Follicle Selection and Ovulation 701

Menstrual Cycle 703

Estrogen, Progestin, and Androgen Metabolism 708

Infertility 709

CHAPTER 38 • Fertilization, Pregnancy, and Fetal Development 712

Robert V. Considine, Ph.D.

Ovum and Sperm Transport 713

Fertilization and Implantation 714

Pregnancy 717

Fetal Development and Parturition 720

Postpartum and Prepubertal Periods 724

Sexual Development Disorders 729

Appendix: Common Abbreviations in Physiology 732

Glossary 735

Index 795

1 Homeostasis and Cellular Signaling

ACTIVE LEARNING OBJECTIVES

Upon mastering the material in this chapter you should be able to:

- Identify important variables essential for life and discuss how they are altered by external and internal forces. Explain how homeostasis benefits the survival of an organism when such forces alter these essential variables.
- Explain the differences between negative and positive feedback and discuss their relationship to homeostasis.
- Contrast steady and equilibrium states in terms of whether an organism must expend energy to create either state.
- Understand how gap junctions and plasma membrane receptors regulate communications between cells.
- Explain how paracrine, autocrine, and endocrine signaling are different relative to their roles in the control of cell function.
- Understand how second messengers regulate and amplify signal transduction.
- Explain the interrelationship between the control of intracellular calcium concentration or the ways in which calcium is stored in terms of how it is used to transduce cell signals.
- Explain how reactive oxygen species can be both second messengers as well as have pathologic effects.
- Explain how mitogenic signaling regulates cell growth, proliferation, and survival.
- Contrast apoptosis and necrosis in terms of the normal regulation of cell life cycles versus pathologic cell damage and death.

Physiology is the study of processes and functions in living organisms. It is a dynamic and expansive field that encompasses many disciplines, with strong roots in physics, chemistry, and mathematics. Physiologists assume that the same chemical and physical laws that apply to the inanimate world govern processes in the body. They attempt to describe functions in chemical, physical, and engineering terms. For example, the distribution of ions across cell membranes is described in thermodynamic terms, muscle contraction is analyzed in terms of forces and velocities, and regulation in the body is described in terms of control systems theory. Because the functions of a living system are carried out by its component structures, an understanding of its structure from its gross anatomy to the molecular level is relevant to the understanding of physiology.

The scope of physiology ranges from the activities or functions of individual molecules and cells to the interaction of our bodies with the external world. In recent years, we have seen many advances in our understanding of physiologic processes at the molecular and cellular levels. In higher organisms, changes in cell function occur in the context of the whole organism, and different tissues and organs can affect one another. The independent activity of an organism requires the coordination of function at all levels, from molecular and cellular to the whole individual. An important part of physiology is understanding how different cell populations that make up tissues are controlled, how they interact, and how they adapt to changing conditions. For a person to remain healthy, physiologic conditions in the body must be optimal and they are closely regulated. Regulation requires efficient communication between cells and tissues. This chapter discusses several topics related to regulation and communication: the internal environment, homeostasis of extracellular fluid, intracellular homeostasis, negative and positive feedback, feedforward control, compartments, steady state and equilibrium, intercellular and intracellular communication, nervous and endocrine systems control, cell membrane transduction, and other important signal transduction cascades.

BASIS OF PHYSIOLOGIC REGULATION

Our bodies are made up of incredibly complex and delicate materials, and we are constantly subjected to all kinds

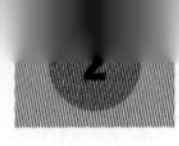

of disturbances, yet we keep going for a lifetime. It is clear that conditions and processes in the body must be closely controlled and regulated—that is, kept within appropriate values. Below we consider, in broad terms, physiologic regulation in the body.

Stable internal environment is essential for normal cell function.

The 19th-century French physiologist Claude Bernard was the first to formulate the concept of the internal environment (*milieu intérieur*). He pointed out that an external environment surrounds multicellular organisms (air or water) and a liquid internal environment (extracellular fluid) surrounds the cells that make up the organism (Fig. 1.1). These cells are not directly exposed to the external world but, rather, interact with it through their surrounding environment, which is continuously renewed by the circulating blood.

For optimal cell, tissue, and organ function in animals, several facets of the internal environment must be maintained within narrow limits. These include but are not limited to (1) oxygen and carbon dioxide tensions; (2) concentrations of glucose and other metabolites; (3) osmotic pressure; (4) concentrations of hydrogen, potassium, calcium, and magnesium ions; and (5) temperature. Departures from optimal conditions may result in dysfunction, disease, or death. Bernard stated, "Stability of the internal environment is the primary condition for a free and independent existence." He recognized that an animal's independence from changing external conditions is related to its capacity to maintain a relatively constant internal environment. A good example is the ability of warm-blooded animals to live in different climates. Over a wide range of external temperatures, core temperature in mammals is maintained constant by both physiologic and behavioral mechanisms. This stability offers great flexibility and has an obvious survival value.

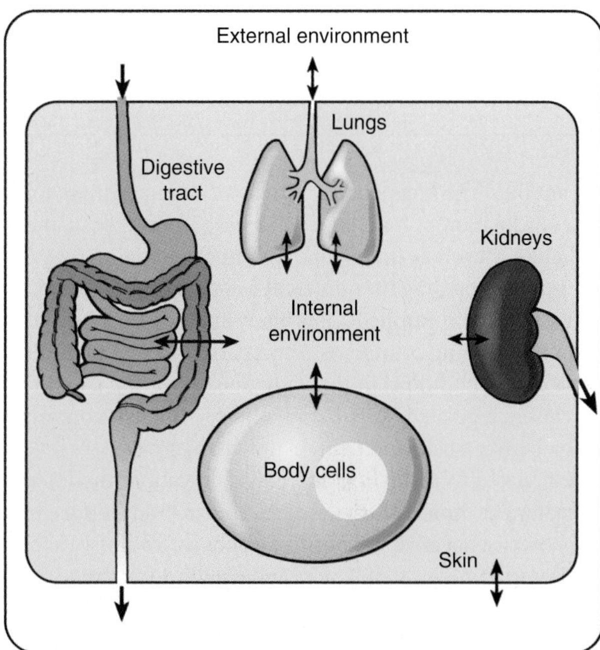

Figure 1.1 The living cells of our body, surrounded by an internal environment (extracellular fluid), communicate with the external world through this medium. Exchanges of matter and energy between the body and the external environment (indicated by *arrows*) occur via the gastrointestinal tract, kidneys, lungs, and skin (including the specialized sensory organs).

Homeostasis is the maintenance of steady states in the body by coordinated physiologic mechanisms.

The key to maintaining the stability of the body's internal environment is the masterful coordination of important regulatory mechanisms in the body. The renowned physiologist Walter B. Cannon captured the spirit of the body's capacity for self-regulation by defining the term **homeostasis** as the maintenance of steady states in the body by coordinated physiologic mechanisms.

Understanding the concept of homeostasis is important for understanding and analyzing normal and pathologic conditions in the body. To function optimally under a variety of conditions, the body must sense departures from normal and then be able to activate mechanisms for restoring physiologic conditions to normal. Deviations from normal conditions may vary between too high and too low, so mechanisms exist for opposing changes in either direction. For example, if blood glucose concentration is too low, the hormone glucagon is released from the alpha cells of the pancreas and epinephrine is released from the adrenal medulla to increase it. If blood glucose concentration is too high, insulin is released from the beta cells of the pancreas to lower it by enhancing the cellular uptake, storage, and metabolism of glucose. Behavioral responses also contribute to the maintenance of homeostasis. For example, a low blood glucose concentration stimulates feeding centers in the brain, driving the animal to seek food.

Homeostatic regulation of a physiologic variable often involves several cooperating mechanisms activated at the same time or in succession. The more important a variable, the more numerous and complicated are the mechanisms that operate to keep it at the desired value. When the body is unable to restore physiologic variables, then disease or death can result. The ability to maintain homeostatic mechanisms varies over a person's lifetime, with some homeostatic mechanisms not being fully developed at birth and others declining with age. For example, a newborn infant cannot concentrate urine as well as an adult and is, therefore, less able to tolerate water deprivation. Older adults are less able to tolerate stresses, such as exercise or changing weather, than are younger adults.

Intracellular homeostasis

The term *homeostasis* traditionally refers to the extracellular fluid that bathes our tissues—but it can also be applied to conditions within cells. In fact, the ultimate goal of maintaining a constant internal environment is to promote intracellular homeostasis, and toward this end, conditions in the cytosol of cells are closely regulated.

The multitude of biochemical reactions characteristic of a cell must be tightly regulated to provide metabolic energy and proper rates of synthesis and breakdown of cellular constituents. Metabolic reactions within cells are catalyzed by enzymes and are therefore subject to several factors that regulate or influence enzyme activity:

- First, the final product of the reactions may inhibit the catalytic activity of enzymes, a process called **end-product inhibition**. End-product inhibition is an example of negative-feedback control (see below).
- Second, intracellular regulatory proteins such as the calcium-binding protein calmodulin may associate with enzymes to control their activity.
- Third, enzymes may be controlled by covalent modification, such as **phosphorylation** or dephosphorylation.
- Fourth, the ionic environment within cells, including hydrogen ion concentration ($[H^+]$), ionic strength, and calcium ion concentration, influences the structure and activity of enzymes.

Hydrogen ion concentration or pH affects the electrical charge of the amino acids that comprise a protein, and this contributes to their structural configuration and binding properties. A measure of acidity or alkalinity, pH affects chemical reactions in cells and the organization of structural proteins. Cells can regulate their pH via mechanisms that buffer intracellular hydrogen ions and by extruding H^+ into the extracellular fluid (see Chapter 2, "Plasma Membrane, Membrane Transport, and Resting Membrane Potential," and Chapter 24, "Acid–Base Homeostasis").

The structure and activity of cellular proteins are also affected by the salt concentration or ionic strength. Cytosolic ionic strength depends on the total number and charge of ions per unit volume of water within cells. Cells can regulate their ionic strength by maintaining the proper mixture of ions and unionized molecules (e.g., organic osmolytes such as sorbitol). Many cells use calcium as an intracellular signal or "messenger" for enzyme activation and, therefore, must possess mechanisms for regulating cytosolic $[Ca^{2+}]$. Such fundamental activities as muscle contraction; the secretion of neurotransmitters, hormones, and digestive enzymes; and the opening or closing of ion channels are mediated by transient changes in cytosolic $[Ca^{2+}]$. Cytosolic $[Ca^{2+}]$ in resting cells is low, about 10^{-7} M, and far below the $[Ca^{2+}]$ in extracellular fluid (about 2.5 mM). Cytosolic $[Ca^{2+}]$ is regulated by the binding of calcium to intracellular proteins, transport is regulated by adenosine triphosphate (ATP)-dependent calcium pumps in mitochondria and other organelles (e.g., sarcoplasmic reticulum in muscle), and the extrusion of calcium is regulated via cell membrane Na^+/Ca^{2+} exchangers and calcium pumps (see Chapter 2, "Plasma Membrane, Membrane Transport, and Resting Membrane Potential"). Toxins or diminished ATP production can lead to an abnormally elevated cytosolic $[Ca^{2+}]$. Abnormal cytosolic $[Ca^{2+}]$ can lead to hyperactivation of calcium-dependent enzyme pathways, and high cytosolic $[Ca^{2+}]$ levels can overwhelm calcium regulatory mechanisms, leading to cell death.

Negative feedback promotes stability, and feedforward control anticipates change.

Engineers have long recognized that stable conditions can be achieved by negative-feedback control systems (Fig. 1.2). Feedback is a flow of information along a closed loop. The components of a simple negative-feedback control system include a regulated variable, sensor (or detector), controller (or comparator), and **effector**. Each component controls the next component. Various disturbances may arise within or outside the system and cause undesired changes in the regulated variable. With **negative feedback**, a regulated variable is sensed, information is fed back to the controller, and the effector acts to oppose change (hence, the term *negative*).

A familiar example of a negative-feedback control system is the thermostatic control of room temperature. Room temperature (regulated variable) is subjected to disturbances. For example, on a cold day, room temperature falls. A thermometer (sensor) in the thermostat (controller) detects the room temperature. The thermostat is set for a certain temperature (set point). The controller compares the actual temperature (feedback signal) with the set point temperature, and an error signal is generated if the room temperature falls below the set temperature. The error signal activates the furnace (effector). The resulting change in room temperature is monitored, and when the temperature rises sufficiently, the furnace is turned off. Such a negative-feedback system allows some fluctuation in room temperature, but the components

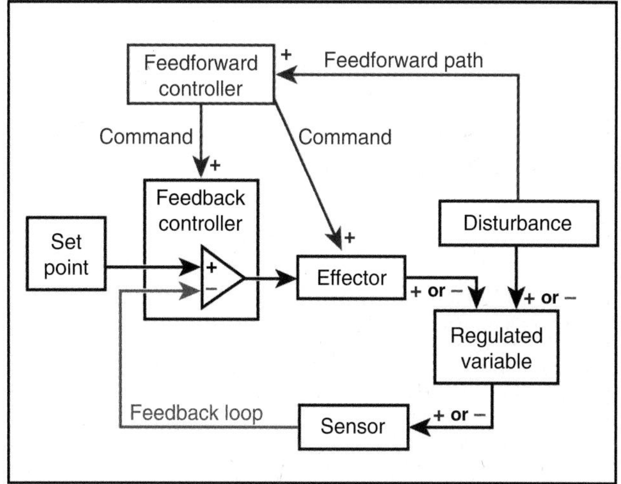

Figure 1.2 Elements of negative-feedback and feedforward control systems. In a negative-feedback control system, information flows along a closed loop. The regulated variable is sensed, and information about its level is fed back to a feedback controller, which compares it with a desired value (set point). If there is a difference, an error signal is generated, which drives the effector to bring the regulated variable closer to the desired value. In this example, the negative sign at the end of the feedback bath signifies that the controller is signaled to move the regulated variable in the opposite direction of the initial disturbance. A feedforward controller generates commands without directly sensing the regulated variable, although it may sense a disturbance. Feedforward controllers often operate through feedback controllers.

act together to maintain the set temperature. Effective communication between the sensor and effector is important in keeping these oscillations to a minimum.

Similar negative-feedback systems exist to maintain homeostasis in the body. For example, the maintenance of water and salts in the body is referred to as **osmoregulation** or fluid balance. During exercise, fluid balance can be altered as a result of water loss from sweating. Loss of water results in an increased concentration of salts in the blood and tissue fluids, which is sensed by the cells in the brain as a negative feedback (see Chapter 23, "Regulation of Fluid and Electrolyte Balance"). The brain responds by telling the kidneys to reduce secretion of water and also by increasing the sensation of being thirsty. Together the reduction in water loss in the kidneys and increased water intake return the blood and tissue fluids to the correct osmotic concentration. This negative-feedback system allows for minor fluctuations in water and salt concentrations in the body but rapidly acts to compensate for disturbances to restore physiologically acceptable osmotic conditions.

Feedforward control is another strategy for regulating systems in the body, particularly when a change with time is desired. In this case, a command signal is generated, which specifies the target or goal. The moment-to-moment operation of the controller is "open loop"; that is, the regulated variable itself is not sensed. Feedforward control mechanisms often sense a disturbance and can, therefore, take corrective action that anticipates change. For example, heart rate and breathing increase even before a person has begun to exercise.

Feedforward control usually acts in combination with negative-feedback systems. One example is picking up a pencil. The movements of the arm, hand, and fingers are directed by the cerebral cortex (feedforward controller); the movements are smooth, and forces are appropriate only in part because of the feedback of visual information and sensory information from receptors in the joints and muscles. Another example of this combination occurs during exercise. Respiratory and cardiovascular adjustments closely match muscular activity, so that arterial blood oxygen and carbon dioxide **tensions** (the **partial pressure** of a gas in a liquid) hardly change during all but exhausting exercise (see Chapter 21, "Control of Ventilation"). One explanation for this remarkable behavior is that exercise simultaneously produces a centrally generated feedforward signal to the active muscles and the respiratory and cardiovascular systems; feedforward control, together with feedback information generated as a consequence of increased movement and muscle activity, adjusts the heart, blood vessels, and respiratory muscles. In addition, control system function can adapt over a period of time. Past experience and learning can change the control system's output so that it behaves more efficiently or appropriately.

Although homeostatic control mechanisms usually act for the good of the body, they are sometimes deficient, inappropriate, or excessive. Many diseases, such as cancer, diabetes, and hypertension, develop because of defects in these control mechanisms. Alternatively, damaged homeostatic mechanisms can also result in autoimmune diseases, in which the immune system attacks the body's own tissue. Formation of a scar is an example of an important homeostatic mechanism for healing wounds, but in many chronic diseases, such as pulmonary fibrosis, hepatic cirrhosis, and renal interstitial disease, scar formation goes awry and becomes excessive.

Positive feedback promotes a change in one direction.

With **positive feedback**, a variable is sensed and action is taken to reinforce a change of the variable. The term positive refers to the response being in the same direction, leading to a cumulative or amplified effect. Positive feedback does not lead to stability or regulation, but to the opposite—a progressive change in one direction. One example of positive feedback in a physiologic process is the sensation of needing to urinate. As the bladder fills, mechanosensors in the bladder are stimulated and the smooth muscle in the bladder wall begins to contract (see Chapter 23, "Regulation of Fluid and Electrolyte Balance"). As the bladder continues to fill and become more distended, the contractions increase and the need to urinate becomes more urgent. In this example, responding to the need to urinate results in a sensation of immediate relief upon emptying the bladder, and this is positive feedback. Another example of positive feedback occurs during the follicular phase of the menstrual cycle. The female sex hormone estrogen stimulates the release of luteinizing hormone, which in turn causes further estrogen synthesis by the ovaries. This positive feedback culminates in ovulation (see Chapter 37, "Female Reproductive System"). A third example is calcium-induced calcium release in cardiac muscle cells that occurs with each heartbeat. Depolarization of the cardiac muscle plasma membrane leads to a small influx of calcium through membrane calcium channels. This leads to an explosive release of calcium from the intracellular organelles, a rapid increase in the cytosolic calcium level, and activation of the contractile machinery (see Chapter 13, "Cardiac Muscle Mechanics and the Cardiac Pump"). Positive feedback, if unchecked, can lead to a vicious cycle and dangerous situations. For example, a heart may be so weakened by disease that it cannot provide adequate blood flow to the muscle tissue of the heart. This leads to a further reduction in cardiac pumping ability, even less coronary blood flow, and further deterioration of cardiac function. The physician's task sometimes is to disrupt detrimental cyclical positive-feedback loops.

Steady state and equilibrium are both stable conditions, but energy is required to maintain a steady state.

Physiology often involves the study of exchanges of matter or energy between different defined spaces or compartments, separated by some type of limiting structure or membrane. Simplistically, the whole body can be divided into two major compartments: intracellular fluid and extracellular fluid, which are separated by cell plasma membranes (Fig. 1.3). The fluid component of the body comprises about 60% of the

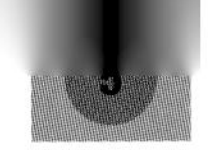

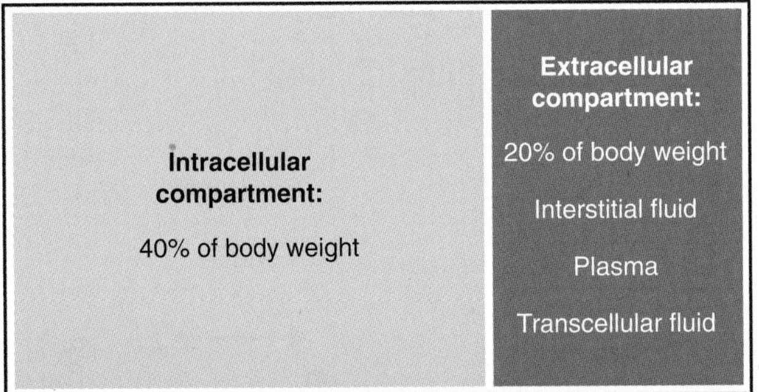

Figure 1.3 Fluid compartments in the body. The body's fluids, which comprise about 60% of the total body weight, can be partitioned into two major compartments: the intracellular compartment and the extracellular compartment. The intracellular compartment, which is about 40% of the body's weight, is primarily a solution of potassium, other ions, and proteins. The extracellular compartment, which is about 20% of the body weight, comprising the interstitial fluids, plasma, and other fluids, such as mucus and digestive juices, is primarily composed of NaCl and $NaHCO_3$.

total body weight. The intracellular fluid compartment comprises about two thirds of the body's water and is primarily composed of potassium and other ions as well as proteins. The extracellular fluid compartment is the remaining one third of the body's water (about 20% of your weight), consists of all the body fluids outside of cells, and includes the interstitial fluid that bathes the cells, lymph, blood plasma, and specialized fluids such as cerebrospinal fluid. It is primarily a sodium chloride (NaCl) and sodium carbonate ($NaHCO_3$) solution that can be divided into three subcompartments: the interstitial fluid (lymph and plasma); plasma that circulates as the extracellular component of blood; and transcellular fluid, which is a set of fluids that are outside of normal compartments, such as cerebrospinal fluid, digestive fluids, and mucus.

When two compartments are in **equilibrium**, *opposing forces are balanced*, and there is no net transfer of a particular substance or energy from one compartment to the other. Equilibrium occurs if sufficient time for exchange has been allowed and if no physical or chemical driving force would favor net movement in one direction or the other. For example, in the lung, oxygen in alveolar spaces diffuses into pulmonary capillary blood until the same oxygen tension is attained in both compartments. Osmotic equilibrium between cells and extracellular fluid is normally present in the body because of the high water permeability of most cell membranes. An equilibrium condition, if undisturbed, remains stable. No energy expenditure is required to maintain an equilibrium state.

Equilibrium and steady state are sometimes confused with each other. A **steady state** is simply a condition that does not change with time. It indicates that the amount or concentration of a substance in a compartment is constant. In a steady state, there is no net gain or net loss of a substance in a compartment. Steady state and equilibrium both suggest stable conditions, but a steady state does not necessarily indicate an equilibrium condition, and energy expenditure may be required to maintain a steady state. For example, in most body cells, there is a steady state for Na^+ ions; the amounts of Na^+ entering and leaving cells per unit time are equal. But intracellular and extracellular Na^+ ion concentrations are far from equilibrium. Extracellular $[Na^+]$ is much higher than intracellular $[Na^+]$, and Na^+ tends to move into cells down concentration and electrical gradients. The cell continuously uses metabolic energy to pump Na^+ out of the cell to maintain the cell in a steady state with respect to Na^+ ions. In living systems, conditions are often displaced from equilibrium by the constant expenditure of metabolic energy.

Figure 1.4 illustrates the distinctions between steady state and equilibrium. In Figure 1.4A, the fluid level in the sink is constant (a steady state) because the rates of inflow and outflow are equal. If we were to increase the rate of inflow (open the tap), the fluid level would rise, and with time, a new steady state might be established at a higher level. In Figure 1.4B, the fluids in compartments X and Y are not in equilibrium (the fluid levels are different), but the system as a whole and each compartment are in a steady state, because

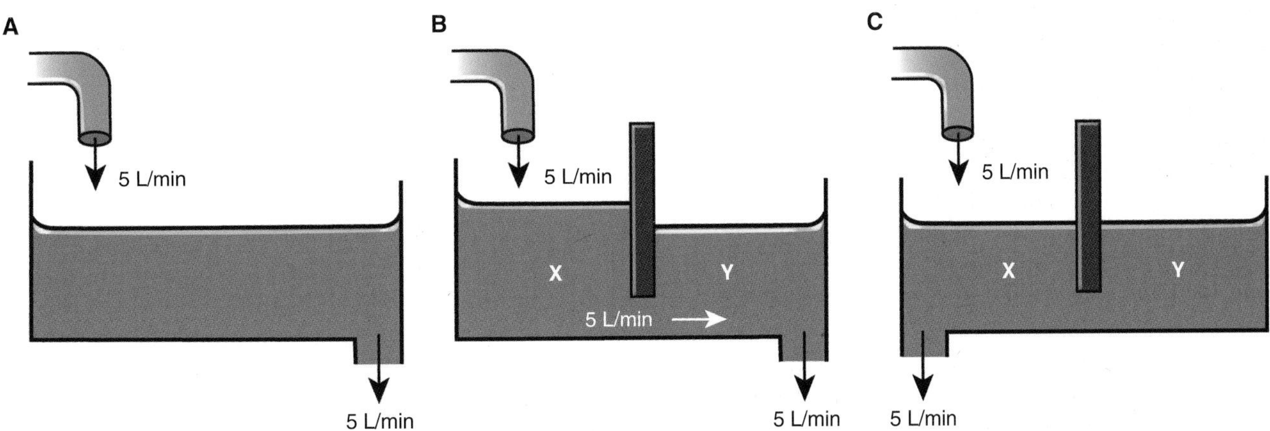

Figure 1.4 Models of the concepts of steady state and equilibrium. Parts **(A–C)** depict a steady state. In **(C)**, compartments X and Y are in equilibrium.

inputs and outputs are equal. In Figure 1.4C, the system is in a steady state and compartments X and Y are in equilibrium. Note that the term *steady state* can apply to a single or several compartments; the term *equilibrium* describes the relation between at least two adjacent compartments that can exchange matter or energy with each other.

Coordinated body activity requires integration of many systems.

Body functions can be analyzed in terms of several systems, such as the nervous, muscular, cardiovascular, respiratory, renal, gastrointestinal, and endocrine systems. These divisions are rather arbitrary, however, and all systems interact and depend on each other. For example, walking involves the activity of many systems besides the muscle and skeletal systems. The nervous system coordinates the movements of the limbs and body, stimulates the muscles to contract, and senses muscle tension and limb position. The cardiovascular system supplies blood to the muscles, providing for nourishment and the removal of metabolic wastes and heat. The respiratory system supplies oxygen and removes carbon dioxide. The renal system maintains an optimal blood composition. The gastrointestinal system supplies energy-yielding metabolites. The endocrine system helps adjust blood flow and the supply of various metabolic substrates to the working muscles. Coordinated body activity demands the integration of many systems.

Recent research demonstrates that many diseases can be explained on the basis of abnormal function at the molecular level. These investigations have led to incredible advances in our knowledge of both normal and abnormal cellular functions. Diseases occur within the context of a whole organism, however, and it is important to understand how all cells, tissues, organs, and organ systems respond to a disturbance (disease process) and interact. The saying, "The whole is more than the sum of its parts," certainly applies to what happens in living organisms. The science of physiology has the unique challenge of trying to make sense of the complex interactions that occur in the body. Understanding the body's processes and functions is clearly fundamental to both biomedical research and medicine.

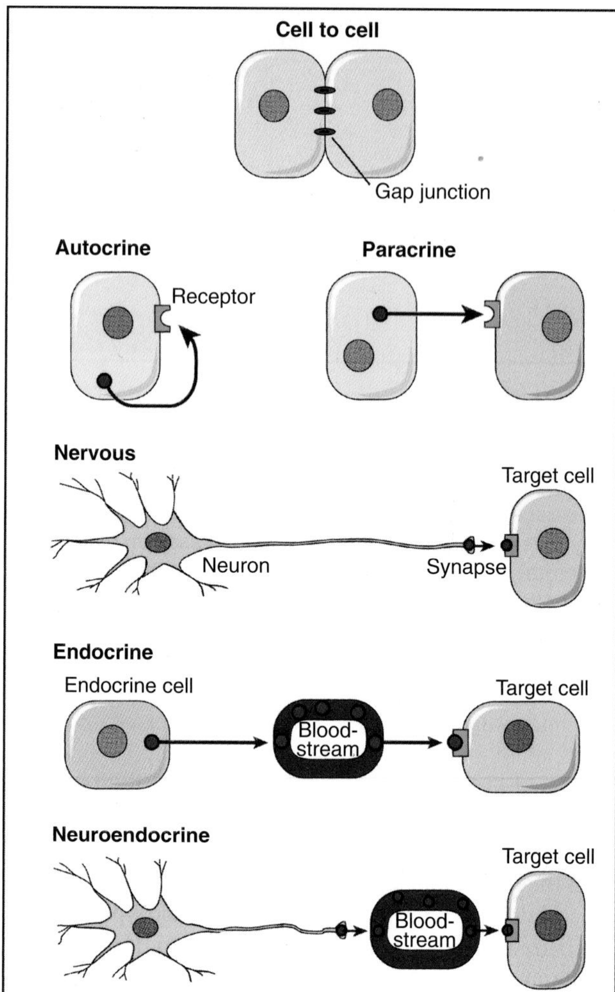

Figure 1.5 Modes of intercellular signaling. Cells may communicate with each other directly via gap junctions or chemical messengers. With autocrine and paracrine signaling, a chemical messenger diffuses a short distance through the extracellular fluid and binds to a receptor on the same cell or a nearby cell. Nervous signaling involves the rapid transmission of action potentials, often over long distances, and the release of a neurotransmitter at a synapse. Endocrine signaling involves the release of a hormone into the bloodstream and the binding of the hormone to specific target cell receptors. Neuroendocrine signaling involves the release of a hormone from a nerve cell and the transport of the hormone by the blood to a distant target cell.

COMMUNICATION AND SIGNALING MODES

The human body has several means of transmitting information between cells. These mechanisms include direct communication between adjacent cells through gap junctions, autocrine and paracrine signaling, and the release of neurotransmitters and **hormones** (chemical substances with regulatory functions) produced by endocrine and nerve cells (Fig. 1.5).

Gap junctions provide a pathway for direct communication between adjacent cells.

Adjacent cells sometimes communicate directly with each other via **gap junctions**, specialized protein channels in the plasma membrane of cells that are made of the protein **connexin** (Fig. 1.6). Six connexins assemble in the plasma membrane of a cell to form a half channel (hemichannel), called a **connexon**. Two connexons aligned between two neighboring cells then join end to end to form an intercellular channel between the plasma membranes of adjacent cells. Gap junctions allow the flow of ions (hence, electrical current) and small molecules between the cytosol of neighboring cells (see Fig. 1.5). Gap junctions are critical to the function of many tissues and allow rapid transmission of electrical signals between neighboring cells in the heart, smooth muscle cells, and some nerve cells. They may also functionally couple adjacent epithelial cells. Gap

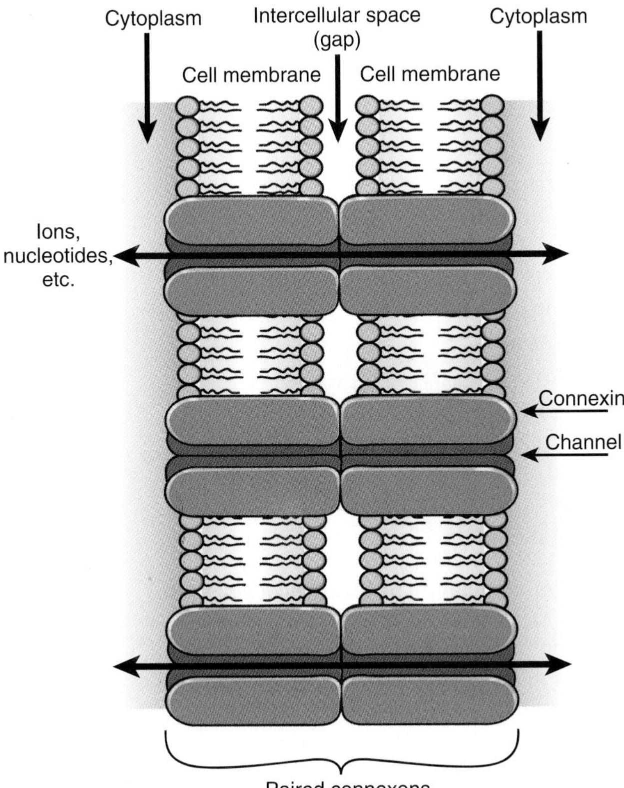

Figure 1.6 The structure of gap junctions. The channel connects the cytosol of adjacent cells. Six molecules of the protein connexin form a half channel called a *connexon*. Ions and small molecules such as nucleotides can flow through the pore formed by the joining of connexons from adjacent cells.

junctions are thought to play a role in the control of cell growth and differentiation by allowing adjacent cells to share a common intracellular environment. Often when a cell is injured, gap junctions close, isolating a damaged cell from its neighbors. This isolation process may result from a rise in calcium or a fall in pH in the cytosol of the damaged cell.

Cells communicate locally by paracrine and autocrine signaling.

Cells may signal to each other via the local release of chemical substances. This means of communication does not depend on a vascular system. In **paracrine signaling**, a chemical is liberated from a cell and diffuses a short distance through the extracellular fluid to act on nearby cells. Paracrine-signaling factors affect only the immediate environment and bind with high specificity to cell receptors on the plasma membrane of the receiving cell. They are also rapidly destroyed by extracellular enzymes or bound to extracellular matrix, thus preventing their widespread diffusion. **Nitric oxide (NO)**, originally called *endothelium-derived relaxing factor* (*EDRF*), is an example of a paracrine-signaling molecule because it has an intrinsically short half-life and thus can affect cells located directly next to the NO-producing cell. Although most cells can produce NO, it has major roles in mediating vascular smooth muscle tone, facilitating central nervous system (CNS) neurotransmission activities, and modulating immune responses (see Chapter 15, "Microcirculation and Lymphatic System," and Chapter 26, "Gastrointestinal Secretion, Digestion, and Absorption"). The production of NO results from the activation of **nitric oxide synthase (NOS)**, which deaminates arginine to citrulline (Fig. 1.7). NO, produced by endothelial cells, regulates vascular tone by diffusing from the endothelial cell to the underlying vascular smooth muscle cell, where it activates its effector target, a cytoplasmic enzyme **guanylyl cyclase (GC)**. The activation of cytoplasmic or soluble GC results in increased intracellular **cyclic guanosine monophosphate (cGMP)** levels and the activation of **cGMP-dependent protein kinase**, also known as **protein kinase G (PKG)**. This enzyme phosphorylates potential target substrates such as calcium pumps in the sarcoplasmic reticulum or sarcolemma, leading to reduced cytoplasmic levels of calcium. In turn, this deactivates the contractile machinery in the vascular smooth muscle cell and produces relaxation or a decrease of tone (see Chapter 8, "Skeletal and Smooth Muscle," and Chapter 15, "Microcirculation and Lymphatic System").

In contrast, during **autocrine signaling**, the cell releases a chemical messenger into the extracellular fluid that binds to a receptor on the surface of the cell that secreted it (see Fig. 1.5). Eicosanoids (e.g., prostaglandins) are examples of signaling molecules that can act in an autocrine manner. These molecules act as local hormones to influence a variety of physiologic processes such as uterine smooth muscle contraction during pregnancy.

Nervous system provides for rapid and targeted communication.

The CNS includes the brain and spinal cord, which links the CNS to the peripheral nervous system (PNS), which is composed of nerves or bundles of neurons. Together the CNS and the PNS integrate and coordinate a vast number of sensory processes and motor responses. The basic functions of the nervous system are to acquire sensory input from both the internal and external environment, integrate the input, and then activate a response to the stimuli. Sensory input to the nervous system can occur in many forms, such as taste, sound, blood pH, hormones, balance or orientation, pressure, or temperature, and these inputs are converted to signals that are sent to the brain or spinal cord. In the sensory centers of the brain and spinal cord, the input signals are rapidly integrated, and then a response is generated. The response is generally a motor output and is a signal that is transmitted to the organs and tissues, where it is converted into an action such as a change in heart rate, sensation of thirst, release of hormones, or a physical movement. The nervous system is also organized for discrete activities; it has an enormous number of "private lines" for sending messages from one distinct locus to another. The conduction of information along nerves occurs via electrical signals, called *action potentials*, and signal transmission between nerves or between nerves and effector structures takes place at a **synapse**. Synaptic transmission is almost always mediated by the release of specific chemicals or **neurotransmitters** from the nerve terminals (see Fig. 1.5).

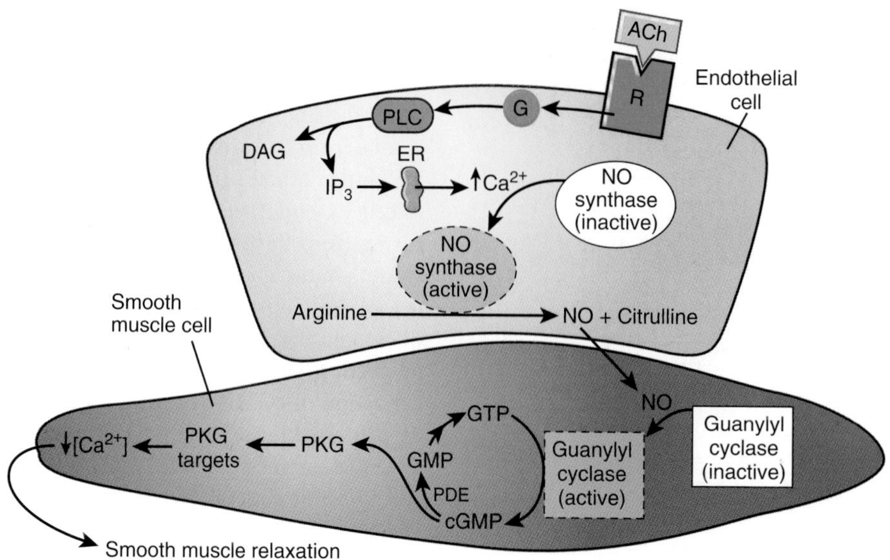

Figure 1.7 Paracrine signaling by nitric oxide (NO) after stimulation of endothelial cells with acetylcholine (ACh). The NO produced diffuses to the underlying vascular smooth muscle cell and activates its effector, cytoplasmic guanylyl cyclase (GC), leading to the production of cyclic guanosine monophosphate (cGMP). Increased cGMP leads to the activation of cGMP-dependent protein kinase, which phosphorylates target substrates, leading to a decrease in cytoplasmic calcium and relaxation. Relaxation can also be mediated by nitroglycerin, a pharmacologic agent that is converted to NO in smooth muscle cells, which can then activate GC. G, G protein; PLC, phospholipase C; DAG, diacylglycerol; IP_3, inositol trisphosphate; GTP, guanosine triphosphate; R, receptor; ER, endoplasmic reticulum.

Innervated cells have specialized protein molecules (receptors) in their cell membranes that selectively bind neurotransmitters. Serious consequences occur when nervous transmission is impaired or defective. For example, in **Parkinson disease**, there is a deficiency in the neurotransmitter dopamine caused by a progressive loss of dopamine-secreting neurons, which results in both the cognitive impairment (e.g., slow reaction times) and behavioral impairment (e.g., tremors) of this devastating disease. Chapter 3 will discuss the actions of various neurotransmitters and how they are synthesized and degraded. Chapters 4 to 6 will discuss the role of the nervous system in coordinating and controlling body functions.

Endocrine system provides for slower and more diffuse communication.

The endocrine system produces hormones in response to a variety of stimuli, and these hormones are instrumental in establishing and maintaining homeostasis in the body. In contrast to the rapid, directed effects resulting from neuronal stimulation, responses to hormones are much slower (seconds to hours) in onset, and the effects often last longer. Hormones are secreted from endocrine glands and tissues and are broadcast to all parts of the body by the bloodstream (see Fig. 1.5). A particular cell can only respond to a hormone if it possesses the appropriate receptor ("receiver") for the hormone. Hormone effects may also be focused. For

Clinical Focus / 1.1 

Dopamine and Parkinson Disease

Parkinson disease (PD) is a degenerative disorder of the central nervous system that gradually worsens, affecting motor skills and speech. PD is characterized by muscle rigidity, tremors, and slowing of physical movements. These symptoms are the result of excessive muscle contraction, which is a result of insufficient **dopamine**, a neurotransmitter produced by the dopaminergic neurons of the brain. The symptoms of PD result from the loss of dopamine-secreting cells in a region of the brain that regulates movement. Loss of dopamine in this region of the brain causes other neurons to fire out of control, resulting in an inability to control or direct movements in a normal manner. There is no cure for PD, but several drugs have been developed to help patients manage their symptoms, although they do not halt the disease. The most commonly used drug is levodopa (L-DOPA), a synthetic precursor of dopamine. L-DOPA is taken up in the brain and changed into dopamine, allowing the patient to regain some control over his or her mobility. Other drugs, such as carbidopa, entacapone, and selegilin, inhibit the degradation of dopamine and are generally taken in combination with L-DOPA. A controversial avenue of research that has potential for providing a cure for this devastating disease involves the use of embryonic stem cells. Embryonic stem cells are undifferentiated cells derived from embryos, and scientists think they may be able to encourage these cells to differentiate into neuronal cells that can replace those lost during the progression of this disease. Other scientific approaches are aimed at understanding the molecular and biochemical mechanisms by which the dopaminergic neurons are lost. Based on a better understanding of these processes, neuroprotective therapies are being designed.

example, arginine vasopressin specifically increases the water permeability of kidney collecting duct cells but does not alter the water permeability of other cells. Hormone effects can also be diffuse, influencing practically every cell in the body. For example, thyroxine has a general stimulatory effect on metabolism. Hormones play a critical role in controlling such body functions as growth, metabolism, and reproduction.

Cells that are not traditional endocrine cells produce a special category of chemical messengers called **tissue growth factors**. These growth factors are protein molecules that influence cell division, differentiation, and cell survival. They may exert effects in an autocrine, paracrine, or endocrine fashion. Many growth factors have been identified, and probably many more will be recognized in years to come. **Nerve growth factor** enhances nerve cell development and stimulates the growth of axons. **Epidermal growth factor (EGF)** stimulates the growth of epithelial cells in the skin and other organs. **Platelet-derived growth factor** stimulates the proliferation of vascular smooth muscle and endothelial cells. **Insulin-like growth factors** stimulate the proliferation of a wide variety of cells and mediate many of the effects of growth hormone. Growth factors appear to be important in the development of multicellular organisms and in the regeneration and repair of damaged tissues.

Nervous and endocrine control systems overlap.

The distinction between nervous and endocrine control systems is not always clear. This is because the nervous system exerts control over endocrine gland function, most if not all endocrine glands are innervated by the PNS, and these nerves can directly control the endocrine function of the gland. In addition, the innervation of endocrine tissues can also regulate blood flow within the gland, which can impact the distribution and thus function of the hormone. On the other hand, hormones can affect the CNS to alter behavior and mood. Adding to this highly integrated relationship is the presence of specialized nerve cells, called **neuroendocrine**, or **neurosecretory cells**, which directly convert a neural signal into a hormonal signal. These cells thus directly convert electrical energy into chemical energy, and activation of a neurosecretory cell results in hormone secretion. Examples are the hypothalamic neurons, which liberate releasing factors that control secretion by the anterior pituitary gland, and the hypothalamic neurons, which secrete arginine vasopressin and oxytocin into the circulation. In addition, many proven or potential neurotransmitters found in nerve terminals are also well-known hormones, including arginine vasopressin, cholecystokinin, enkephalins, norepinephrine, secretin, and vasoactive intestinal peptide. Therefore, it is sometimes difficult to classify a particular molecule as either a hormone or a neurotransmitter.

MOLECULAR BASIS OF CELLULAR SIGNALING

Cells communicate with one another by many complex mechanisms. Even unicellular organisms, such as yeast cells, use small peptides called **pheromones** to coordinate mating events that eventually result in haploid cells with new assortments of genes. The study of intercellular communication has led to the identification of many complex signaling systems that are used by the body to network and coordinate functions. These studies have also shown that these signaling pathways must be tightly regulated to maintain cellular homeostasis. Dysregulation of these signaling pathways can transform normal cellular growth into uncontrolled cellular proliferation or cancer.

Signal transduction refers to the mechanisms by which **first messengers** from transmitting cells can convert its information to a **second messenger** within the receiving cells. Signaling systems consist of **receptors** that reside either in the plasma membrane or within cells and are activated by a variety of extracellular signals or first messengers, including peptides, protein hormones and growth factors, steroids, ions, metabolic products, gases, and various chemical or physical agents (e.g., light). Signaling systems also include **transducers** and effectors, which are involved in conversion of the signal into a physiologic response. The pathway may include additional intracellular messengers, called second messengers (Fig. 1.8). Examples of second messengers are cyclic nucleotides such as **cyclic adenosine monophosphate (cAMP)** and cGMP, lipids such as inositol 1,4,5-trisphosphate (IP_3) and diacylglycerol (DAG), ions such as calcium, and gases such as NO and carbon

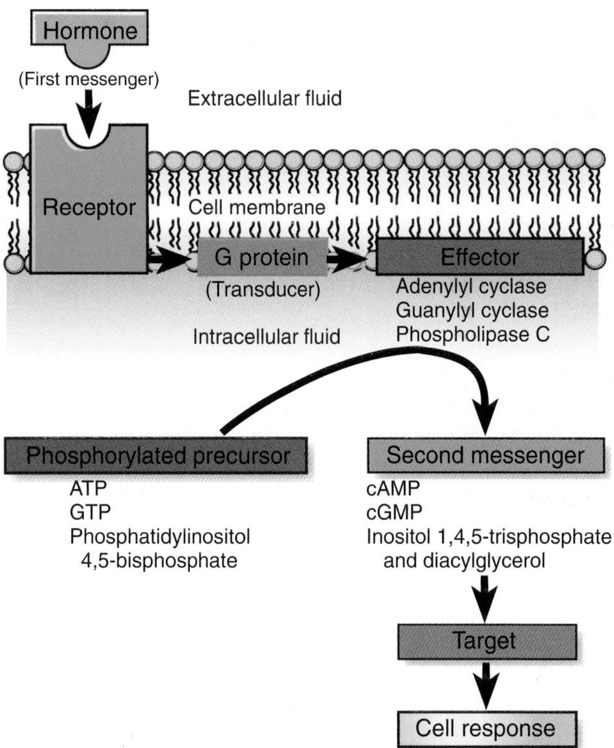

Figure 1.8 Signal transduction blueprints common to second messenger systems. A protein or peptide hormone binds to a plasma membrane receptor, which stimulates or inhibits a membrane-bound effector enzyme via a G protein. The effector catalyzes the production of many second messenger molecules from a phosphorylated precursor (e.g., cyclic adenosine monophosphate [cAMP] from adenosine triphosphate [ATP], cGMP from guanosine triphosphate [GTP], or inositol 1,4,5-trisphosphate and diacylglycerol from phosphatidylinositol 4,5-bisphosphate). The second messengers, in turn, activate protein kinases (targets) or cause other intracellular changes that ultimately lead to the cell response.

monoxide (CO). A general outline for a **signal cascade** is as follows: Signaling is initiated by binding of a first messenger to its appropriate ligand-binding site on the outer surface domain of its relevant membrane receptor. This results in activation of the receptor; the receptor may adopt a new conformation, form aggregates (multimerize), and/or become phosphorylated or dephosphorylated. These changes usually result in association of adapter signaling molecules that couple the activated receptor to downstream molecules that transduce and amplify the signal through the cell by activating specific effector molecules and generating a second messenger. The outcome of the signal transduction cascade is a physiologic response, such as secretion, movement, growth, division, or death. It is important to remember these physiologic responses are the collective result of a multitude of signaling messengers that transmit signals to the cells in various tissues.

Plasma membrane receptors activate signal transduction pathways.

As mentioned above, the molecules that are produced by one cell to act on itself (autocrine signaling) or other cells (paracrine, neural, or endocrine signaling) are ligands or first messengers. Many of these ligands bind directly to receptor proteins that reside within and extend both outside and inside of the plasma membrane. Other ligands cross the plasma membrane and interact with cellular receptors that reside in either the cytoplasm or the nucleus. Thus, cellular receptors are divided into two general types: *cell-surface receptors* and *intracellular receptors*. Three general classes of cell-surface receptors have been identified: *G protein–coupled receptors (GPCRs)*, *ion channel–linked receptors*, and *enzyme-linked receptors*. Intracellular receptors include steroid and thyroid hormone receptors and are discussed in a later section in this chapter. Some but not all of these cell-surface receptors may be found in organized structures that form "microdomains" within the plasma membrane. These specialized microdomains are referred to as **lipid rafts** and are distinct from the rest of the plasma membrane in that they are highly enriched in cholesterol and sphingolipids such as sphingomyelin and have lower levels of phosphatidylcholine than the surrounding bilayer. Lipid rafts can act to compartmentalize and organize assembly of signaling complexes. Their reduced fluidity and tight packing allows them to "float" freely in the membrane bilayer. Examples of membrane receptors that may require lipid rafts for effective signal transduction include EGF receptor, insulin receptor, B-cell antigen receptor, and T-cell antigen receptor. In addition to membrane receptors several ion channels have been linked to a requirement for lipid rafts for efficient function.

G protein–coupled receptors transmit signals through the trimeric G proteins.

GPCRs are the largest family of cell-surface receptors, with more than 1,000 members. These receptors indirectly regulate their effector targets, which can be ion channels or plasma

Clinical Focus / 1.2

Tyrosine Kinase Inhibitors for Chronic Myeloid Leukemia

Cancer may result from defects in critical signaling molecules that regulate many cell properties, including cell proliferation, differentiation, and survival. Normal cellular regulatory proteins or **proto-oncogenes** may become altered by mutation or abnormally expressed during cancer development. **Oncogenes**, the altered proteins that arise from proto-oncogenes, in many cases, are signal transduction proteins that normally function in the regulation of cellular proliferation. Examples of signaling molecules that can become oncogenic span the entire signal transduction pathway and include ligands (e.g., growth factors), receptors, adapter and effector molecules, and transcription factors.

There are many examples of how normal cellular proteins can be converted into oncoproteins. One occurs in **chronic myeloid leukemia (CML)**. This disease is characterized by increased and unregulated clonal proliferation of myeloid cells in the bone marrow. CML results from an inherited chromosomal abnormality that involves a reciprocal translocation or exchange of genetic material between chromosomes 9 and 22 and was the first malignancy to be linked to a genetic abnormality. The translocation is referred to as the *Philadelphia chromosome* and results in the fusion of the *bcr* gene, whose function is unknown, with part of the cellular *abl* (c-*abl*) gene. The c-*abl* gene encodes a protein tyrosine kinase whose normal substrates are unknown. This abnormal Bcr–Abl fusion protein (composed of fused parts of *bcr* and c-*abl*) has unregulated tyrosine kinase activity, and through SH2 and SH3 binding domains in the Abl part of the protein, the mutant protein binds to and phosphorylates the interleukin 3β(c) receptor. This receptor is linked to control of cell proliferation, and the expression of the unregulated Bcr–Abl protein activates signaling pathways that control the cell cycle, which speeds up cell division.

The chromosomal translocation that results in the formation of the Bcr–Abl oncoprotein occurs during the development of hematopoietic stem cells, and the observance of a shorter Philadelphia 22 chromosome is diagnostic of this cancer. The translocation results initially in a CML that is characterized by a progressive leukocytosis (increase in number of circulating white blood cells) and the presence of circulating immature blast cells. However, other secondary mutations may spontaneously occur within the mutant stem cell and can lead to acute leukemia, a rapidly progressing disease that is often fatal.

Historically, CML was treated with chemotherapy, interferon administration, and bone marrow transplantation. More recently, the understanding of the molecules and signaling pathways that result in this devastating cancer have led to targeted therapeutic strategies to attenuate the disease. Toward this end, a pharmacologic agent that inhibits tyrosine kinase activities has been developed. Although treatment of patients with CML with the drug Gleevec® (imatinib mesylate) does not eradicate the disease, it can greatly limit the development of the tumor clone and improve the quality of life and lifespan of the patient.

membrane–bound effector enzymes, through the intermediary activity of a separate membrane-bound adapter protein complex called the *trimeric guanosine triphosphate (GTP)-binding regulatory protein* or *trimeric G protein* (Fig. 1.9). GPCRs mediate cellular responses to numerous types of first messenger signaling molecules, including proteins, small peptides, amino acids, and fatty acid derivatives. Many first messenger ligands can activate several different GPCRs. For example, serotonin can activate at least 15 different GPCRs.

GPCRs are structurally and functionally similar molecules. They have a ligand-binding extracellular domain on one end of the molecule, separated by a seven-pass transmembrane-spanning region from the cytosolic regulatory domain at the other end, where the receptor interacts with the membrane-bound G protein. Binding of ligand or hormone to the extracellular domain results in a conformational change in the receptor that is transmitted to the cytosolic regulatory domain. This conformational change allows an association of the ligand-bound, activated receptor with a trimeric G protein associated with the inner leaflet of the plasma membrane. The interaction between the ligand-bound, activated receptor and the G protein, in turn, activates the G protein, which dissociates from the receptor and transmits the signal to its effector enzyme (e.g., adenylyl cyclase [AC]) or ion channel.

The trimeric G proteins are named for their requirement for **guanosine triphosphate (GTP)** binding and hydrolysis and have been shown to have a broad role in linking various seven-pass transmembrane receptors to membrane-bound effector systems that generate intracellular messengers. G proteins are tethered to the membrane through lipid linkage and are heterotrimeric, that is, composed of three distinct subunits. The subunits of a G protein are an α subunit, which binds and hydrolyzes GTP, and β and γ subunits, which form a stable, tight noncovalent-linked βγ dimer. When the α subunit binds **guanosine diphosphate (GDP)**, it associates with the βγ subunits to form a trimeric complex that can interact with the cytoplasmic domain of the GPCR. The conformational change that occurs upon ligand binding causes the GDP-bound trimeric (αβγ complex) G protein to associate with the ligand-bound receptor. The association of the GDP-bound trimeric complex with the GPCR activates the exchange of GDP for GTP. Displacement of GDP by GTP is favored in cells because GTP is in higher concentration. The displacement of GDP by GTP causes the α subunit to dissociate from the receptor and from the βγ subunits of the G protein. This exposes an effector-binding site on the α subunit, which then associates with an effector enzyme (e.g., AC or phospholipase C [PLC]) to result in the generation of second messengers (e.g., cAMP or IP_3 and DAG). The hydrolysis of GTP to GDP by the α subunit results in the reassociation of the α and βγ subunits, which are then ready to repeat the cycle.

The cycling between inactive (GDP bound) and active forms (GTP bound) places the G proteins in the family of **molecular switches**, which regulate many biochemical events. When the switch is "off," the bound nucleotide is GDP. When the switch is "on," the hydrolytic enzyme (G protein) is bound to GTP, and the cleavage of GTP to GDP will reverse the switch to an "off" state. Although most

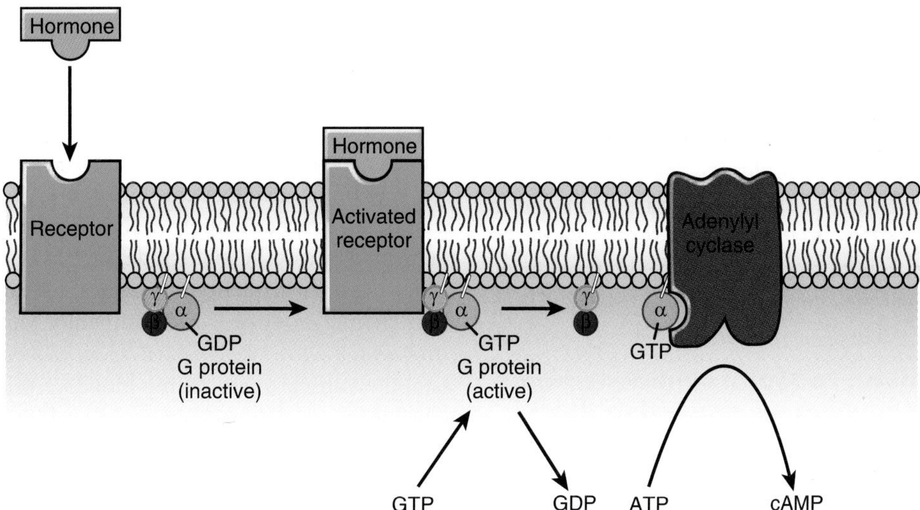

Figure 1.9 Activation of a G protein–coupled receptor and the production of cyclic adenosine monophosphate (cAMP). When bound to guanosine diphosphate (GDP), G proteins are in an inactive state and are not associated with a receptor. Binding of a hormone to the receptor results in association with the inactive, GDP-bound trimeric G protein. The interaction of the GDP-bound trimeric G protein with the activated receptor results in activation of the G protein via the exchange of GDP for guanosine triphosphate (GTP) by the α subunit. The α and βγ subunits of the activated GTP-bound G protein dissociate. The activated, GTP-bound α subunit of the trimeric G protein can then interact with and activate the membrane effector protein adenylyl cyclase to catalyze the conversion of adenosine triphosphate (ATP) to cAMP. The intrinsic GTPase activity in the α subunit of the G protein hydrolyzes the bound GTP to GDP. The GDP-bound α subunit reassociates with the βγ subunit to form an inactive, membrane-bound trimeric G-protein complex.

of the signal transduction produced by G proteins is a result of the activities of the α subunit, a role for βγ subunits in activating effectors during signal transduction is beginning to be appreciated. For example, βγ subunits can activate K^+ channels. Therefore, both α and βγ subunits are involved in regulating physiologic responses.

The catalytic activity of a trimeric G protein, which is the hydrolysis of GTP to GDP, resides in its Gα subunit. Each Gα subunit within this large protein family has an intrinsic rate of GTP hydrolysis. The intrinsic catalytic activity rate of G proteins is an important factor contributing to the amplification of the signal produced by a single molecule of ligand binding to a GPCR. For example, a Gα subunit that remains active longer (slower rate of GTP hydrolysis) will continue to activate its effector for a longer period and result in greater production of second messenger.

The G proteins functionally couple receptors to several different effector molecules. Two major effector molecules that are regulated by G-protein subunits are **adenylyl cyclase (AC)** and PLC. The association of an activated Gα subunit with AC can result in either the stimulation or the inhibition of the production of cAMP. This disparity is a result of the two types of α subunit that can couple AC to cell-surface receptors. Association of an α_s subunit (s for stimulatory) promotes the activation of AC and production of cAMP. The association of an α_i (i for inhibitory) subunit promotes the inhibition of AC and a decrease in cAMP. Thus, bidirectional regulation of AC is achieved by coupling different classes of cell-surface receptors to the enzyme by either G_s or G_i (Fig. 1.10).

In addition to α_s and α_i subunits, other isoforms of G-protein subunits have been described. For example, α_q activates PLC, resulting in the production of the second messengers, DAG and inositol trisphosphate. Another Gα subunit, α_T or **transducin**, is expressed in photoreceptor tissues and has an important role in signaling in light-sensing rod cells in the retina by activation of the effector *cGMP phosphodiesterase* (PDE), which degrades cGMP to 5′GMP (see Chapter 4, "Sensory Physiology"). All three subunits of G proteins belong to large families that are expressed in different combinations in different tissues. This tissue distribution contributes to both the specificity of the transduced signal and the second messenger produced.

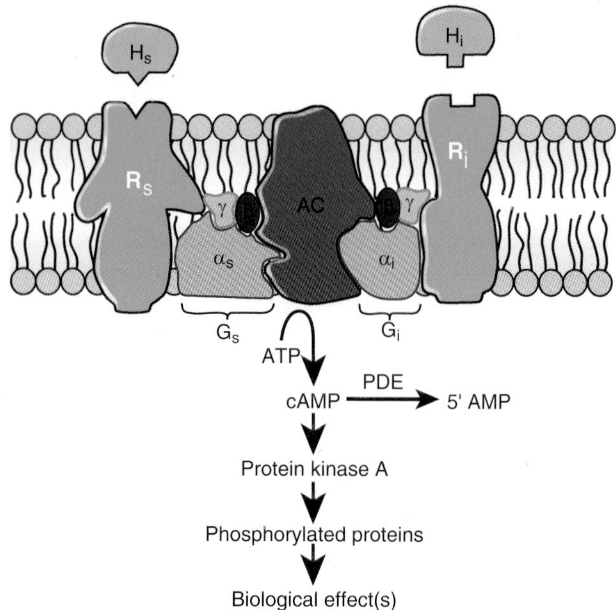

Figure 1.10 Stimulatory and inhibitory coupling of G proteins to adenylyl cyclase (AC). Stimulatory (G_s) and inhibitory (G_i) G proteins couple hormone binding to the receptor with either activation or inhibition of AC. Each G protein is a trimer consisting of Gα, Gβ, and Gγ subunits. The Gα subunits in G_s and G_i are distinct in each and provide the specificity for either AC activation or AC inhibition. Hormones (H_s) that stimulate AC interact with "stimulatory" receptors (R_s) and are coupled to AC through stimulatory G proteins (G_s). Conversely, hormones (H_i) that inhibit AC interact with "inhibitory" receptors (R_i) that are coupled to AC through inhibitory G proteins (G_i). Intracellular levels of cyclic adenosine monophosphate (cAMP) are modulated by the activity of phosphodiesterase (PDE), which converts cAMP to 5′AMP and turns off the signaling pathway by reducing the level of cAMP. ATP, adenosine triphosphate.

Ion channel–linked receptors mediate some forms of cell signaling by regulating the intracellular concentration of specific ions.

Ion channels, found in all cells, are transmembrane proteins that cross the plasma membrane and are involved in regulating the passage of specific ions into and out of cells. Ion channels may be opened or closed by changing the membrane potential or by the binding of ligands, such as neurotransmitters or hormones, to membrane receptors. In some cases, the receptor and ion channel are one and the same molecule. For example, at the neuromuscular junction, the neurotransmitter acetylcholine binds to a muscle membrane nicotinic cholinergic receptor that is also an ion channel. In other cases, the receptor and an ion channel are linked via a G protein, second messengers, and other downstream effector molecules, as in the muscarinic cholinergic receptor on cells innervated by parasympathetic postganglionic nerve fibers. Another possibility is that the ion channel is directly activated by a cyclic nucleotide, such as cGMP or cAMP, produced as a consequence of receptor activation. This mode of ion channel control is predominantly found in the sensory tissues for sight, smell, and hearing as well as others like the smooth muscle surrounding blood vessels. The opening or closing of ion channels plays a key role in signaling between electrically excitable cells, such as nerve and muscle.

Tyrosine kinase receptors signal through adapter proteins to activate the mitogen-activated protein kinase pathway.

Many hormones and growth factors (**mitogens**) signal their target cells by binding to a class of receptors that have tyrosine kinase activity and result in the phosphorylation of tyrosine residues in the receptor and other target proteins. Many of the receptors in this class of plasma membrane receptors have an intrinsic tyrosine kinase domain that is part of the cytoplasmic region of the receptor (Fig. 1.11). Another

From Bench to Bedside / 1.1

Nitric Oxide, Phosphodiesterase, Angina, Pulmonary Hypertension, and Erectile Dysfunction—What is the Link?

A phosphodiesterase (PDE) is an enzyme that hydrolyzes a phosphodiester bond. Cyclic nucleotide PDEs are particularly important in the clinical setting as they control the cellular levels of the second messengers, cyclic adenosine monophosphate (cAMP) and cGMP, and the signal transduction pathways modulated by these molecules. A large family of cyclic nucleotide PDEs have been identified and they are classified according to sequence, regulation, substrate specificity, and tissue distribution. Because some PDEs are expressed in a tissue-specific manner, this presents an opportunity to target a specific PDE with an inhibitory or activating drug.

Therapeutic agents for angina pectoris (severe chest pain resulting from insufficient blood supply to cardiovascular tissues) generally included the administration of nitrates, a commonly used agent that reduces myocardial oxygen demand. Nitrates act as an exogenous source of nitric oxide (NO), which can stimulate soluble GCs and increase cellular levels of cGMP. Formation of cGMP transduces a signal that promotes relaxation of vascular smooth muscle in arteries and veins. Nitrates' salutary effect in treating myocardial ischemia is to dilate veins, which allows blood to translocate from inside the ventricles into the peripheral tissues. This reduces stretch and strain on the heart, which reduces myocardial oxygen demand. Although nitrates provide a relatively easy solution, a common side effect is **tachyphylaxis**, or reduced responsiveness to a chronically used drug. The search for new drugs to treat angina pectoris and other similar cardiovascular diseases led to the discovery of silendafil, which is now marketed under the trade name Viagra. Silendafil is a fairly selective inhibitor of PDE5, and its administration enhances cGMP levels in vascular smooth muscle cells, leading to vasodilation. Unfortunately, the relatively short half-life thwarted the usefulness of this drug as a practical treatment for chronic angina. In addition, several side effects were noted during clinical trials including the ability of sildenafil to augment the vasodilatory effects of nitrates. One other interesting, common side effect noted was penile erection, and subsequent clinical trials validated the use of this drug as an effective therapeutic agent for **erectile dysfunction (ED)**.

There are many causes of ED, including psychological conditions like depression as well as a host of clinical conditions. Common clinical conditions associated with ED include vascular disease; diabetes; neurologic conditions such as spinal cord injury, multiple sclerosis, and Parkinson disease; and numerous inflammatory conditions. During sexual stimulation, the penile cavernosal arteries relax and dilate, allowing increased blood flow. This increase in blood volume and compression of the trabecular muscle result in collapse and obstruction of venous outflow to produce a rigid erection. Because NO is the principal mediator of smooth muscle relaxation, it is essential for an erection to occur. Nitric oxide activates soluble guanylate cyclase causing increased synthesis of cGMP. Cellular levels of cGMP reflect a balance of activities between NO production by NO synthase and degradation of cGMP by a cyclic PDE. Thus, the use of a transient inhibitor of PDE5, the main PDE in the cavernosal arteries and trabecular muscle, provides a rational, temporary vasodilation in those tissues.

Following its wide use as a therapeutic drug for ED, another therapeutic use for sildenafil was discovered, and sildenafil is now considered a promising treatment for pulmonary hypertension for which it is administered under the trade name Revatio. Pulmonary hypertension results from high blood pressure in the pulmonary circulation and is a highly progressive disease with a poor prognosis due to the ensuing right heart dysfunction. It is often fatal. The usefulness of Revatio is based on the findings that in animal models of pulmonary hypertension, the levels of PDE5 increase in the pulmonary aorta and other arteries of the lung, leading to decreased cGMP and increased tone in this vessel. Thus, administration of sildenafil has a beneficial effect by increasing cGMP and relaxation. Certainly with an increased understanding of PDEs, this story will have more chapters as more uses are discovered.

group of related receptors lacks an intrinsic tyrosine kinase but, when activated, becomes associated with a cytoplasmic tyrosine kinase. Both families of tyrosine kinase receptors use similar signal transduction pathways, and they will be discussed together.

Structurally, **tyrosine kinase receptors** consist of a hormone-binding region that is exposed to the extracellular space, a transmembrane region, and a cytoplasmic tail domain. Examples of **agonists** (molecules that bind and activate receptors; **ligand**) for these receptors include hormones (e.g., insulin) or growth factors (e.g., epidermal, fibroblast, and platelet-derived growth factors). The signaling cascades generated by the activation of tyrosine kinase receptors can result in the amplification of gene transcription and *de novo* transcription of genes involved in growth, cellular differentiation, and movements such as crawling or shape changes.

The general scheme for this signaling pathway begins with the agonist binding to the extracellular portion of the receptor (Fig. 1.12). The binding of the agonist causes two of the agonist-bound receptors to associate (**dimerization**), and this, in turn, triggers the built-in or associated tyrosine kinases to become activated. The activated tyrosine kinases then phosphorylate tyrosine residues in the other subunit (cross-phosphorylation) of the dimer to fully activate the receptor. These phosphorylated tyrosine residues in the cytoplasmic domains of the dimerized receptor now serve as "docking sites" for additional signaling molecules or adapter proteins that have a specific sequence called an **SH2 domain**. The SH2-containing adapter proteins may be serine/threonine protein kinases, phosphatases, or other bridging proteins that help in the assembly of the cytoplasmic signaling complexes that transmit the signal from an activated receptor

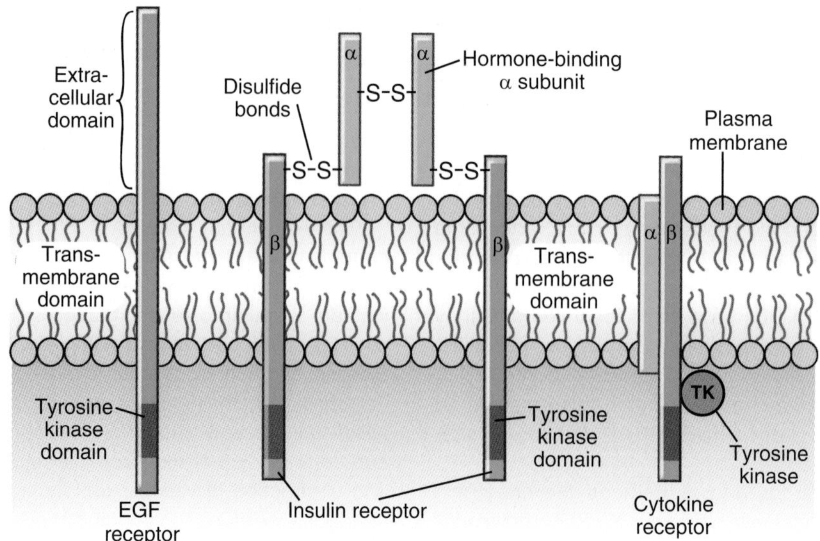

Figure 1.11 General structures of the tyrosine kinase receptor family. Tyrosine kinase receptors have an intrinsic protein tyrosine kinase activity that resides in the cytoplasmic domain of the molecule. Examples are the epidermal growth factor (EGF) and insulin receptors. The EGF receptor is a single-chain transmembrane protein consisting of an extracellular region containing the hormone-binding domain, a transmembrane domain, and an intracellular region that contains the tyrosine kinase domain. The insulin receptor is a heterotetramer consisting of two α and two β subunits held together by disulfide bonds. The α subunits are entirely extracellular and involved in insulin binding. The β subunits are transmembrane proteins and contain the tyrosine kinase activity within the cytoplasmic domain of the subunit. Some receptors become associated with cytoplasmic tyrosine kinases following their activation. Examples can be found in the family of cytokine receptors, some of which consist of an agonist-binding subunit and a signal-transducing subunit that become associated with a cytoplasmic tyrosine kinase.

to many signaling pathways, ultimately leading to a cellular response. A notable difference in signaling pathways activated by tyrosine kinase receptors is that they do not generate second messengers such as cAMP or cGMP.

One signaling pathway associated with activated tyrosine kinase receptors results in activation of another type of GTPase (monomeric) related to the trimeric G proteins described above. Members of the **ras** family of *monomeric G proteins* are activated by many tyrosine kinase receptor growth factor agonists and, in turn, activate an intracellular signaling cascade that involves the phosphorylation and activation of several protein kinases called *mitogen-activated protein kinases* (*MAPKs*). In this pathway, the activated MAPK translocates to the nucleus, where it activates transcription of a cohort of genes needed for proliferation and survival or cell death.

Hormone receptors bind specific hormones to initiate cell signaling in the cells.

Hormone receptors reside either on the cell surface or inside the cell. There are two general kinds of hormones that activate these receptors: the **peptide hormones** and the **steroid hormones**. *Peptide hormone* receptors are usually plasma membrane proteins that belong to the family of GPCR and effect their signaling by generation of second messengers such as cAMP and IP_3 and by the release of calcium from its storage compartments. GPCR signaling has already been described and will not be further discussed here. The second major group of hormones, the *steroid hormones*, binds either to soluble receptor proteins located in the cytosol or nucleus (type I) or to receptors already bound to the gene response elements (promoter) of target genes (type II). Examples of type I cytoplasmic or nuclear steroid hormone receptors include the sex hormone receptors (androgens, estrogen, and progesterone), glucocorticoid receptors (cortisol), and mineralocorticoid receptors (aldosterone). Examples of type II, DNA-bound steroid hormone receptors include vitamin A, vitamin D, retinoid, and thyroid hormone receptors.

Generally, steroid hormone receptors have four recognized domains, including variable, DNA-binding, hinge, and hormone-binding and dimerization domains. The N-terminal *variable domain* is a region with little similarity between these receptors. A centrally located *DNA-binding domain* consists of two globular motifs where zinc is coordinated with cysteine residues (**zinc finger**). This is the domain that controls the target gene that will be activated and may also have sites for phosphorylation by protein kinases that are involved in modifying the transcriptional activity of the receptor. Between the central DNA-binding and the C-terminal hormone-binding domains is located a *hinge domain*, which controls the movement of the receptor to the nucleus. The carboxyl-terminal *hormone-binding and dimerization domain* binds the hormone and then allows the receptor to dimerize, a necessary step for binding to DNA. When

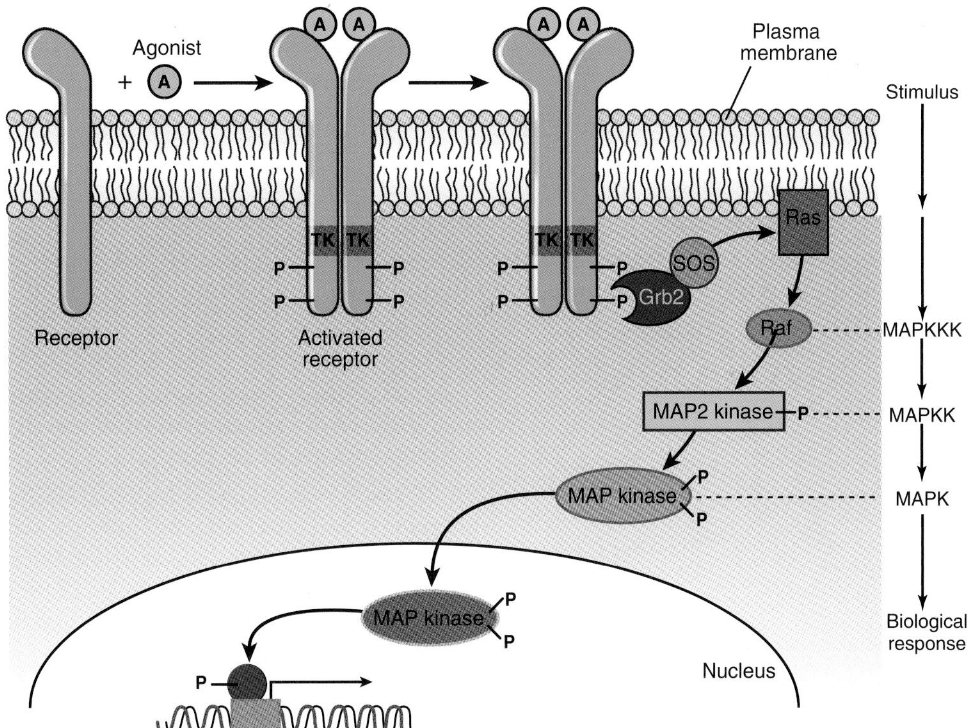

Figure 1.12 A signaling pathway for tyrosine kinase receptors. Binding of agonist to the tyrosine kinase receptor (TK) causes dimerization, activation of the intrinsic tyrosine kinase activity, and phosphorylation of the receptor subunits. The phosphotyrosine residues serve as docking sites for intracellular proteins (P), such as Grb2, which recruits son of sevenless (SOS), a guanine nucleotide exchange factor, to the receptor complex. SOS interacts with and modulates the activity of Ras by promoting the exchange of guanosine diphosphate (GDP) for guanosine triphosphate (GTP). Ras-GTP (active form) activates the serine/threonine kinase Raf, initiating a phosphorylation cascade that results in the activation of mitogen-activated protein kinase (MAPK). MAPK translocates to the nucleus and phosphorylates transcription factors to modulate gene transcription. The right side of the figure illustrates the hierarchical organization of the MAPK signaling cascade. The generic names in this pathway are shown aligned to specific memebers of a typical tyrosine kinase pathway. Proteins with P attached represent phosphorylation at either tyrosine or serine/threonine residues.

steroid hormones bind their receptor, the hormone–receptor complex moves to the nucleus, where it binds to a specific DNA sequence in the gene regulatory (promoter) region of a hormone-responsive gene. The targeted DNA sequence in the promoter is called a **hormone response element (HRE)**. Binding of the hormone–receptor complex to the HRE can either activate or repress transcription. Although most effects involve increased production of specific proteins, repressed production of certain proteins by steroid hormones can also occur. The result of stimulation by steroid hormones is a change in the readout or transcription of the genome. These newly synthesized proteins and/or enzymes will affect cellular metabolism with responses attributable to that particular steroid hormone. The binding of the activated hormone–receptor complex to chromatin results in alterations in RNA polymerase activity that lead to either increased or decreased transcription of specific portions of the genome. As a result, mRNA is produced, leading to the production of new cellular proteins or changes in the rates of synthesis of preexisting proteins. Steroid hormone receptors are also known to undergo phosphorylation/dephosphorylation reactions. The effect of this covalent modification is also an area of active research. The model of steroid hormone action shown in Figure 1.13 is generally applicable to all steroid hormones. In contrast to steroid hormones, the thyroid hormones and retinoic acid bind to receptors that are already associated with the DNA response elements of target genes. Examples of these type II receptor hormones include thyroid hormones, retinoids, vitamin A, and vitamin D. The unoccupied receptors are inactive until the hormone binds, and they serve as repressors in the absence of hormone. These receptors are discussed in Chapter 31, "Hypothalamus and the Pituitary Gland," and Chapter 33, "Adrenal Gland."

SECOND MESSENGER ROLES

The concept of second messengers and their vital roles in signaling began with Earl Sutherland, Jr., who was awarded the Nobel Prize in 1971 "for his discoveries concerning the mechanisms of action of hormones." Sutherland discovered cyclic adenosine monophosphate (cAMP) and showed it was a critical intermediate in cellular responses to hormones. **Second messengers** transmit and amplify signals from

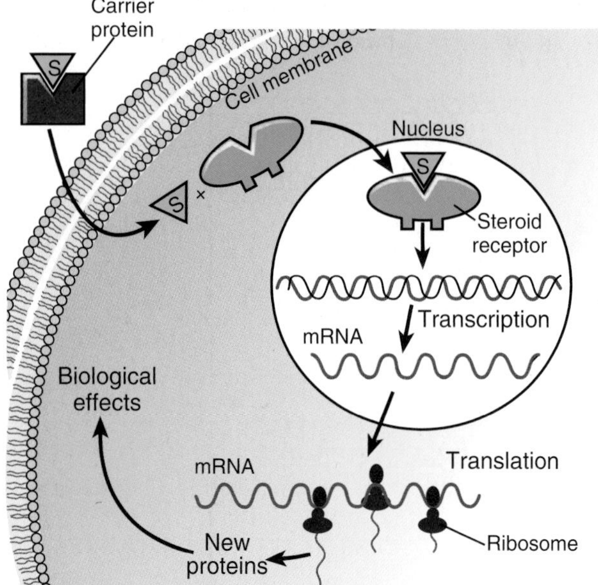

Figure 1.13 The general mechanism of action of steroid hormones. Steroid hormones (S) are lipid soluble and pass through the plasma membrane, where they bind to a cognate receptor in the cytoplasm. The steroid hormone–receptor complex then moves to the nucleus and binds to a HRE in the promoter-regulatory region of specific hormone-responsive genes. Binding of the steroid hormone–receptor complex to the response element initiates transcription of the gene to form messenger RNA (mRNA). The mRNA moves to the cytoplasm, where it is translated into a protein that participates in a cellular response. Thyroid hormones are thought to act by a similar mechanism, although their receptors are already bound to a HRE, repressing gene expression. The thyroid hormone–receptor complex forms directly in the nucleus and results in the activation of transcription from the thyroid hormone–responsive gene.

receptors to downstream target molecules as part of signaling pathways inside the cell. There are three general types of second messengers: hydrophilic, water-soluble messengers, such as IP_3, cAMP, cGMP, or Ca^{2+}, that can readily diffuse throughout the cytosol; hydrophobic water insoluble, lipid messengers, which are generally associated with lipid-rich membranes such as DAG and phosphatidylinositols (e.g., PIP_3); and gases, such as NO, CO, and reactive oxygen species (ROS), which can diffuse both through the cytosol as well as across cell membranes. A critical feature of second messengers is that they are able to be *rapidly synthesized and degraded* by cellular enzymes, rapidly sequestered in a membrane-bound organelle or vesicle or have a restricted distribution within the cell. It is the rapid appearance and disappearance that allow second messengers to amplify and then terminate signaling reactions. *Amplification* of signaling is also a cornerstone of signaling, allowing fine-tuning of the response. For example, when a cell receptor is only briefly stimulated with a ligand, the generation of a second messenger will terminate rapidly. Conversely, when a large amount of ligand persists to stimulate a receptor, the increased levels of second messenger in the cell will be sustained for a longer period of time before termination. Because the second messengers responsible for relaying signals within target cells are limited, and each target cell has a different complement of intracellular signaling pathways, the physiologic responses can vary. Thus, every cell in our body is programmed to respond to specific combinations of first and second messengers, and these same messengers can elicit a distinct physiologic response in different cell types. For example, the neurotransmitter acetylcholine can cause heart muscle to relax, skeletal muscle to contract, and secretory cells to secrete.

cAMP is the predominant second messenger for both hormonal and nonhormonal first messengers in all cells.

As a result of binding to specific GPCRs, many peptide hormones and catecholamines produce an almost immediate increase in the intracellular concentration of cAMP. For these ligands, the receptor is coupled to a stimulatory G protein ($G\alpha_s$), which upon activation and exchange of GDP for GTP can diffuse within the membrane to interact with and activate AC, a large transmembrane protein that converts intracellular ATP to the second messenger cAMP. The second messenger cAMP participates in transducing the signals from a vast array of hormones and receptors mainly through activation of **cAMP-dependent protein kinase** (also called **protein kinase A or PKA**) but also functions to directly activate some calcium channels.

In addition to the hormones that stimulate the production of cAMP through a receptor coupled to $G\alpha_s$, some hormones act to decrease cAMP formation and, therefore, have opposing intracellular effects. These hormones bind to receptors that are coupled to an inhibitory ($G\alpha_i$) rather than a stimulatory ($G\alpha_s$) G protein. cAMP is perhaps the most widely used second messenger and has been shown to mediate numerous cellular responses to both hormonal and nonhormonal stimuli, not only in higher organisms but also in various primitive life forms, including slime molds and yeasts. The intracellular signal provided by cAMP is rapidly terminated by its hydrolysis to 5′AMP by members of a family of enzymes known as **phosphodiesterases (PDEs)**, which, in some cases, are activated by high levels of cyclic nucleotides.

Protein kinase A is the major target mediating the signaling effects of cAMP.

The cyclic nucleotide cAMP activates PKA, which, in turn, catalyzes the phosphorylation of various cellular proteins, ion channels, and transcription factors. This phosphorylation alters the activity or function of the target proteins and ultimately leads to a desired cellular response. PKA is a tetramer that, when inactive, consists of two catalytic and two regulatory subunits, with the protein kinase activity residing in the catalytic subunit. When cAMP concentrations in the cell are low, the two catalytic subunits are bound to the two regulatory subunits, forming an inactive tetramer (Fig. 1.14). When cAMP is formed in response to hormonal stimulation, two molecules of cAMP bind to each of the

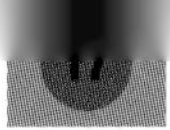

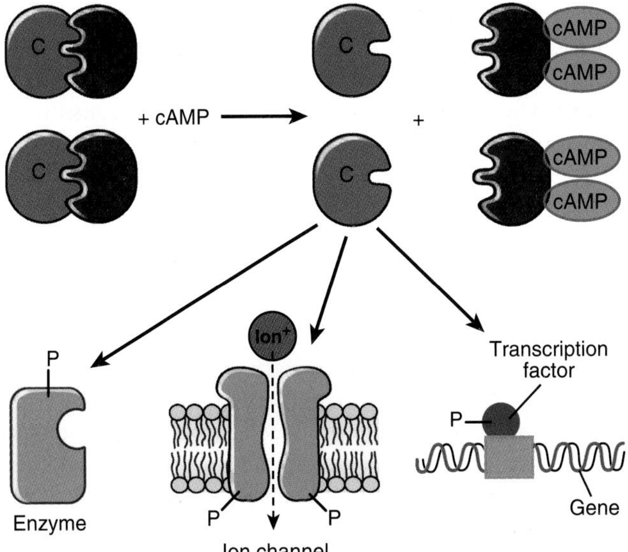

Figure 1.14 Activation and targets of protein kinase A. Inactive protein kinase A consists of two regulatory subunits complexed with two catalytic subunits. Activation of adenylyl cyclase results in increased cytosolic levels of cyclic adenosine monophosphate (cAMP). Two molecules of cAMP bind to each of the regulatory subunits, leading to the release of the active catalytic subunits. These subunits can then phosphorylate target enzymes, ion channels, or transcription factors, resulting in a cellular response. R, regulatory subunit; C, catalytic subunit; P, phosphate group.

regulatory subunits (R), causing them to dissociate from the catalytic subunits. This relieves inhibition of the catalytic subunits (C), thus activating PKA to result in phosphorylation of target substrates and to cause a biologic response to the hormone (see Fig. 1.14).

In addition to activating PKA and phosphorylation of target proteins, in some cell types, cAMP can directly bind and alter the activity of ion channels. Cyclic nucleotide–gated ion channels may be regulated by either cAMP or cGMP and are especially important in the olfactory and visual systems. For example, there are a vast number of odorant receptors that are coupled to G proteins, and like GPCRs when stimulated by a specific odorant, AC is activated and cAMP generated. The cAMP then binds a cAMP-gated ion channel that opens to allow calcium (Ca^{2+}) into the cell causing membrane "depolarization" (influx of positive ions) as part of the sensing of the odor.

cGMP, NO, and CO are important second messengers in smooth muscle and sensory cells.

The second messenger cGMP is generated by the enzyme GC. Although the full role of cGMP as a second messenger is not as well understood, its importance is finally being appreciated with respect to signal transduction in sensory tissues (see Chapter 4, "Sensory Physiology") and smooth muscle tissues (see Chapter 8, "Skeletal and Smooth Muscle," Chapter 9, "Blood Components," and Chapter 15, "Microcirculation and Lymphatic System"). The main target of cGMP is PKG. cGMP can also directly activate several ion channels or ion pumps, which collectively participate in modulating cytoplasmic Ca^{2+} levels not only in smooth muscle but also in sensory tissue. Activation of these ion channels or pumps either directly by cGMP binding or as a result of phosphorylation by PKG can also alter cytoplasmic Ca^{2+} concentration, which among other things mediates contraction and relaxation of smooth muscle cells. The production of cGMP is regulated by the activation of one of two forms of GC, a soluble, cytoplasmic form or a particulate, membrane localized form. GCs are targets of the paracrine-signaling molecule NO that is produced by endothelial as well as other cell types, and this pathway can mediate smooth muscle relaxation (see Fig. 1.7; also see Chapter 8, "Skeletal and Smooth Muscle," and Chapter 15, "Microcirculation and Lymphatic System") and neurotransmission (see Chapter 6, "Autonomic Nervous System") as well as gene regulation and other signaling pathways. Nitric oxide, or nitrogen monoxide, is not to be confused with nitrous oxide or laughing gas, which is used as an anesthetic. NO is a highly reactive, free radical that was initially called **endothelial-derived relaxing factor (EDRF)**. Research showing that EDRF was actually a gas, NO, resulted in a Nobel Prize in 1998 that was awarded to Robert Furchgott, Louis Ignarro, and Ferid Murad. NO is produced through the action of the enzyme NOS in a reaction that converts L-arginine to L-citurilline.

Endothelial cell production of NO is used to transduce a relaxation signal to the neighboring smooth muscle cells (see Fig. 1.7). In this pathway, endothelial cells are stimulated by a number of factors including blood flow, acetylcholine, or cytokines, which results in activation of NOS. NO rapidly diffuses into the smooth muscle cells where it activates *soluble GC* to generate cGMP (see Fig. 1.7). Soluble GC is a heterodimeric protein that also contains two heme (an organic compound consisting of iron bound to a heterocyclic ring called *porphorin*) prosthetic groups, which in their iron bound form can associate with NO. Binding of NO to these heme prosthetic groups activates GC leading to the production of cGMP. The second messenger cGMP activates PKG leading to phosphorylation of a number of proteins including regulators of calcium channels and pumps. These ion channels and pumps collectively cause a reduction in cytoplasmic calcium concentrations in the cell. This reduction in calcium ultimately results in relaxation of the vascular smooth muscle. Degradation of cGMP is mediated by a PDE. Activation of the PDE occurs in response to high levels of cGMP, which binds to PDE. This circuit serves as a negative-feedback loop to modulate intracellular calcium levels in smooth muscle, tone (continuous partial contraction of muscle), and, in part, blood pressure (see Fig. 1.7). This signaling pathway is also terminated by decreases in NO, a highly reactive molecule with a very short half-life. Thus, the production of NO by endothelial cells in blood vessels is an important factor in regulating vascular tone by mediating signal transduction pathways that cause vasodilation and vasocontraction (see Chapter 15, "Microcirculation and Lymphatic System"). As mentioned above, there is another form of GC called *particulate GC*. Particulate GC functions as a transmembrane protein that is a receptor for the atrial natriuretic (ANP) peptide produced by cardiomyocytes in response to increased blood volume. Binding of ANP to particulate GC results in production of cGMP, which leads to reduction of water and sodium concentrations and blood volume in the circulation (see Chapter 17, "Control Mechanisms in Circulatory Function," and Chapter 23, "Regulation of Fluid and Electrolyte Balance").

Less well understood as a second messenger is the gas CO. Like NO, CO binds to the iron at the active site of heme; thus, CO can activate sGC to produce cGMP, although not as potently as can NO. The fact that CO binds to iron at the active site of heme prosthetic groups also explains the toxicity of inhaled CO, which can bind the heme in hemoglobin, thereby displacing oxygen. CO also binds to the heme-containing protein, cytochrome oxidase, a key mitochondrial enzyme needed for ATP production. Inhibition of cytochrome oxidase by CO binding reduces ATP levels. In cells, CO is normally produced as a by-product of the reaction catalyzed by heme oxygenase (HO). HO oxidation of heme results in the production of CO and biliverdin (responsible for the green color of bruises). Although it is a weak activator of sGC, HO-derived CO is thought to have a role in neuronal signaling, including olfactory transmission, vascular tone, and platelet aggregation to name a few physiologic processes. CO can also modulate MAPK activity, and the numerous signaling pathways regulated by these signaling molecules extend CO a role in proliferation, inflammation, and cell death.

Lipids have important second messenger regulatory functions, including immune response mediation.

Because lipids can freely diffuse through plasma and organelle membranes, they cannot be stored in membrane-bound vesicles and must be synthesized on demand in the location where they are needed. Many lipid second messengers are derived from two sources: phosphatidylinositol (PIP_2) and sphingolipid. Other lipid messengers, including steroids, retinoic acid derivatives, prostaglandins, and lysophosphatidic acid, are also important regulators of many cellular functions, derived via various mechanisms. Important PIP_2-derived messengers, such as IP_3 and DAG, have been well studied and are described in the next section. Ceramide is a lipid second messenger that is generated from sphingomyelin through the action of sphingomyelinase, an enzyme localized in the plasma membrane. Activation of sphingomyelinase occurs through binding of the **cytokines** (small, secreted peptides, including **tumor necrosis factor** [TNF] and **interleukin-1**, that mediate immune and inflammatory responses) to their receptors. These activated receptors are then coupled to sphingomyelinase, leading to its activation and generation of ceramide and subsequent activation of the MAPK pathway.

Diacylglycerol and inositol trisphosphate

Some GPCRs are coupled to a different effector enzyme, **phospholipase C (PLC)**, which is localized to the inner leaflet of the plasma membrane. Similar to other GPCRs, binding of a ligand or an agonist to the receptor results in activation of the associated G protein, usually $G\alpha_q$ (or G_q). Depending on the isoform of the G protein associated with the receptor, either the α or the $\beta\gamma$ subunit may stimulate PLC. Stimulation of PLC results in the hydrolysis of the membrane phospholipid PIP_2 into **DAG** and **IP_3**. Both DAG and IP_3 serve as second messengers in the cell (Fig. 1.15).

In its second messenger role, DAG accumulates in the plasma membrane and activates the membrane-bound calcium- and lipid-sensitive enzyme **protein kinase C (PKC)**. When activated, this enzyme catalyzes the phosphorylation of specific proteins, including other enzymes and transcription factors, in the cell to produce appropriate physiologic effects such as cell proliferation. Several tumor-promoting **phorbol esters** that mimic the structure of DAG have been shown to activate PKC. They can, therefore, bypass the receptor by passing through the plasma membrane and directly activating PKC, causing the phosphorylation of downstream targets to result in cellular proliferation.

IP_3 promotes the release of calcium ions into the cytoplasm by activation of endoplasmic or sarcoplasmic reticulum IP_3-gated calcium release channels (see Chapter 8, "Skeletal and Smooth Muscle"). The concentration of free calcium ions in the cytoplasm of most cells is in the range of 10^{-7} M. With appropriate stimulation, the concentration may abruptly increase 1,000 times or more. The resulting increase in free cytoplasmic calcium synergizes with the action of DAG in the activation of some forms of PKC and may also activate many other calcium-dependent processes.

Mechanisms exist to reverse the effects of DAG and IP_3 by rapidly removing them from the cytoplasm. IP_3 is dephosphorylated to inositol, which can be reused for phosphoinositide synthesis. DAG is converted to phosphatidic acid by the addition of a phosphate group to carbon number 3. Like inositol, phosphatidic acid can be used for the resynthesis of membrane inositol phospholipids (see Fig. 1.15). On removal of the IP_3 signal, calcium is quickly pumped back into its storage sites, restoring cytoplasmic calcium concentrations to their low prestimulus levels.

In addition to IP_3, other, perhaps more potent phosphoinositols, such as IP_4 or IP_5, may also be produced in response to stimulation. These are formed by the hydrolysis of appropriate phosphatidylinositol phosphate precursors found in the cell membrane. The precise role of these phosphoinositols is unknown. Evidence suggests that the hydrolysis of other phospholipids such as phosphatidylcholine may play an analogous role in hormone-signaling processes.

Cells use calcium as a second messenger by keeping resting intracellular calcium levels low.

The levels of cytosolic calcium in an unstimulated cell are about 10,000 times lower than in the extracellular fluid (10^{-7} M vs. 10^{-3} M). This large gradient of calcium is maintained by the limited permeability of the plasma membrane to calcium, by calcium transporters in the plasma membrane that extrude calcium, by calcium pumps in the membranes

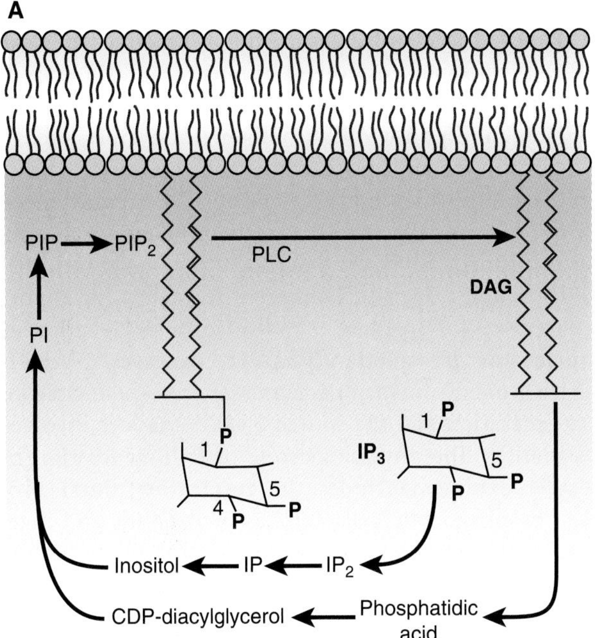

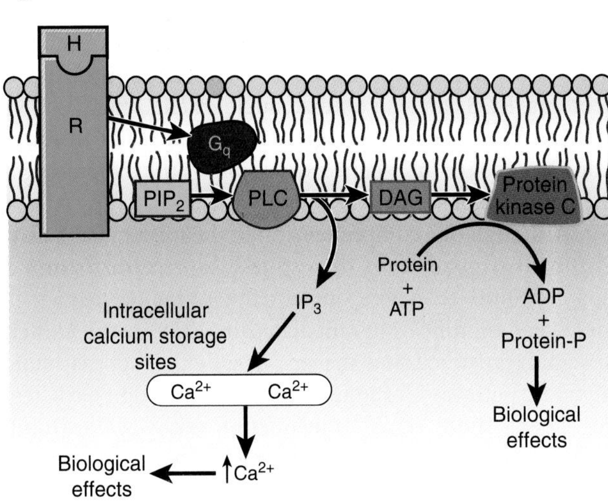

Figure 1.15 The phosphatidylinositol second messenger system. (A) The pathway leading to the generation of inositol trisphosphate and diacylglycerol (DAG). The successive phosphorylation of phosphatidylinositol (PI) leads to the generation of phosphatidylinositol 4,5-bisphosphate (PIP_2). Phospholipase C (PLC) catalyzes the breakdown of PIP_2 to inositol trisphosphate (IP_3) and 1,2-DAG, which are used for signaling and can be recycled to generate phosphatidylinositol. **(B)** The generation of IP_3 and DAG and their intracellular signaling roles. The binding of hormone (H) to a G protein–coupled receptor (R) can lead to the activation of PLC. In this case, the Gα subunit is G_q, a G protein that couples receptors to PLC. The activation of PLC results in the cleavage of PIP_2 to IP_3 and DAG. IP_3 interacts with calcium release channels in the endoplasmic reticulum, causing the release of calcium to the cytoplasm. Increased intracellular calcium can lead to the activation of calcium-dependent enzymes. An accumulation of DAG in the plasma membrane leads to the activation of the calcium- and phospholipid-dependent enzyme protein kinase C and phosphorylation of its downstream targets. Protein-P, phosphorylated protein; ATP, adenosine triphosphate; ADP, adenosine diphosphate.

of intracellular organelles that store calcium, and by cytoplasmic and organellar proteins that bind calcium to buffer its free cytoplasmic concentration. Several plasma membrane ion channels serve to increase cytosolic calcium levels. Either these ion channels are voltage gated and open when the plasma membrane depolarizes or they may be controlled by phosphorylation by PKA or PKC, which is important for regulating the contractile functions of smooth and cardiac muscle (see Chapter 8, "Skeletal and Smooth Muscle," and Chapter 13, "Cardiac Muscle Mechanics and the Cardiac Pump").

In addition to the plasma membrane ion channels, the endoplasmic reticulum, an extensive membrane-bound organelle, has two other main types of ion channels that when activated, release calcium into the cytoplasm, causing an increase in cytoplasmic calcium. The small water-soluble molecule IP_3 activates the IP_3-gated **calcium release channel** in the membrane of the endoplasmic or sarcoplasmic (a specialized type of endoplasmic reticulum in smooth and striated muscle) reticulum. The activated channel opens to allow calcium to flow down a concentration gradient into the cytoplasm. The IP_3-gated channels are structurally similar to the second type of calcium release channel, the **ryanodine receptor**, found in the sarcoplasmic reticulum of muscle cells and neurons. In cardiac and skeletal muscle, ryanodine receptors release calcium to trigger muscle contraction when an action potential invades the transverse tubule system of these cells (see Chapter 8, "Skeletal and Smooth Muscle"). Both types of channels are regulated by positive feedback, in which the released cytosolic calcium can bind to the receptor to enhance further calcium release. This form of positive feedback is referred to as **calcium-induced calcium release** and causes the calcium to be released suddenly in a spike, followed by a wavelike flow of the ion throughout the cytoplasm (see Chapter 8, "Skeletal and Smooth Muscle," and Chapter 13, "Cardiac Muscle Mechanics and the Cardiac Pump").

Increasing cytosolic free calcium activates many different signaling pathways and leads to numerous physiologic events, such as muscle contraction, neurotransmitter secretion, and cytoskeletal polymerization. Calcium acts as a second messenger in two ways:

- It binds directly to an effector target such as PKC to promote in its activation or
- It binds to an intermediary cytosolic calcium-binding protein such as calmodulin.

Calmodulin is a small protein (16 kDa) with four binding sites for calcium. The binding of calcium to calmodulin causes calmodulin to undergo a dramatic conformational change and increases the affinity of this intracellular calcium "receptor" for its effectors (Fig. 1.16). Calcium–calmodulin complexes bind to and activate a variety of cellular proteins, including protein kinases that are important in many physiologic processes, such as smooth muscle contraction (myosin light-chain kinase; see Chapter 8, "Skeletal and Smooth Muscle") and hormone synthesis (aldosterone synthesis; see Chapter 34, "Endocrine Pancreas"), and ultimately result in altered cellular function.

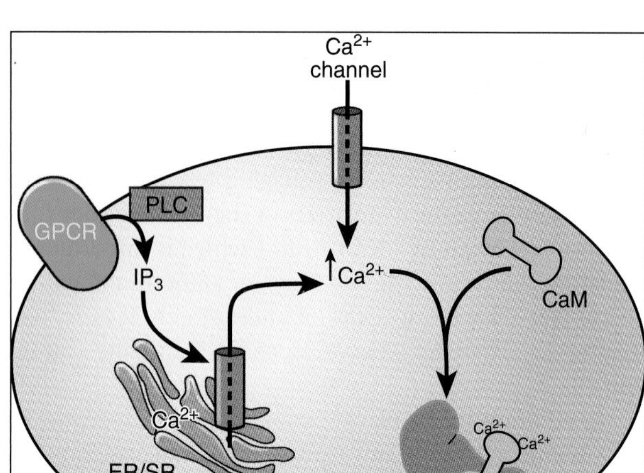

Figure 1.16 The role of calcium in intracellular signaling and activation of calcium–calmodulin-dependent protein kinases. Membrane-bound ion channels that allow the entry of calcium from the extracellular space or release calcium from internal stores (e.g., endoplasmic reticulum, sarcoplasmic reticulum in muscle cells, and mitochondria) regulate levels of intracellular calcium. Calcium can also be released from intracellular stores via the G protein–mediated activation of phospholipase C (PLC) and the generation of inositol trisphosphate (IP_3). IP_3 causes the release of calcium from the endoplasmic or sarcoplasmic reticulum in muscle cells by interaction with calcium ion channels. When intracellular calcium rises, four calcium ions complex with the dumbbell-shaped calmodulin protein (CaM) to induce a conformational change. Ca^{2+}/CaM can then bind to a spectrum of target proteins including Ca^{2+}/CaM-PKs, which then phosphorylate other substrates, leading to a response. IP_3, inositol trisphosphate; PLC, phospholipase C; CaM, calmodulin; Ca^{2+}/CaM-PK, calcium–calmodulin-dependent protein kinases; ER/SR, endoplasmic/sarcoplasmic reticulum; GPCR, G protein–coupled receptor.

Two mechanisms operate to terminate calcium action: The IP_3 generated by the activation of PLC can be dephosphorylated by cellular phosphatases leading to inactivation of this second messenger. In addition, the calcium that enters the cytosol can be rapidly removed. The plasma membrane, endoplasmic reticulum, sarcoplasmic reticulum, and mitochondrial membranes all have ATP-driven calcium pumps such as the **plasma membrane calcium ATPase (PMCA)** that pumps the free calcium out of the cytosol to the extracellular space or into an intracellular organelle. Lowering cytosolic calcium concentrations shifts the equilibrium in favor of the release of calcium from calmodulin. Calmodulin then dissociates from the various proteins that were activated, and the cell returns to its basal state.

Reactive oxygen species can act as second messengers as well as pathologic mediators.

ROS are molecules that include both free radical molecules, such as **superoxide (O_2^-)**, **hydroxyl radical**, and NO, and nonradical molecules such as **hydrogen peroxide (H_2O_2)**. These molecules are highly reactive (they can oxidize amino acids in proteins or nucleic acids in RNA or DNA) because they have an unpaired electron. ROS can be generated in response to environmental activators, such as pollutants in the air, smoke, smog, and exposure to radiation (e.g., ultraviolet light). Under normal circumstances, **oxidoreductases** that are part of the mitochondrial electron transport system generate ROS, but there are other cellular sources, such as xanthine oxidoreductases, lipoxygenases, cyclooxygenases, and **nicotinamide adenine dinucleotide phosphate (NADPH) oxidases**. NADPH oxidase is one of the major enzyme sources responsible for ROS generation and the source of ROS that are involved in signaling. The physiologic role of ROS generation by NADPH oxidases includes the **respiratory burst** produced by phagocytic cells such as neutrophils and macrophages that results in large amounts of ROS production (see Chapter 9, "Blood Components"). The respiratory burst is a critical feature in the host response to infection and leads to the destruction of bacteria or fungi. A second physiologic role of NADPH oxidase–generated ROS arises from their ability to react with amino acid residues in proteins, leading to modifications in their activities, localization, and stability. In addition to direct modification of proteins, ROS can also oxidize nucleic acids, such as RNA and DNA. Oxidative damage to DNA can result in mutations in genes or alter gene expression by the mispairing of the damaged bases.

Although our comprehension of the mechanisms leading to ROS generation in response to receptor stimulation continues to evolve, our understanding of the specific molecular modifications mediated by ROS in the context of signal transduction is meager, despite the fact that many signaling pathways responsive to ROS generation have been described. These signal transduction pathways are quite diverse, including those regulating cell growth, survival, differentiation, and death. This form of signaling in which ROS are generated and serve as second messengers is sometimes referred to as *redox signaling*, and NADPH oxidases are thought to be the major source of ROS for this purpose (Fig. 1.17). The NADPH oxidase (Nox) complex consists of six subunits, including p22-phox, gp91-phox (the catalytic subunit), p67-phox, p47-phox, and a small GTP-binding from the Rho family (Rac1 or Rac2). The gp22-phox1/gp91-phox subunits are transmembrane proteins, which localize the NADPH complex to plasma or organellar membranes. In response to stimulation, the cytosolic regulatory proteins are recruited to the p22/p91 heterodimer in the membrane to form an active NADPH oxidase enzyme complex leading to the generation of O_2^-, which is rapidly converted to H_2O_2 by a scavenger enzyme such as **superoxide dismutase (SOD)**. Many first messengers have been found to stimulate assembly of active NADPH oxidase and generation of ROS, including vasoactive factors, such as angiotensin II and endothelin, and cytokines such as TNF, as well as various growth factors and hormones. In addition to these messengers, mechanical forces, including

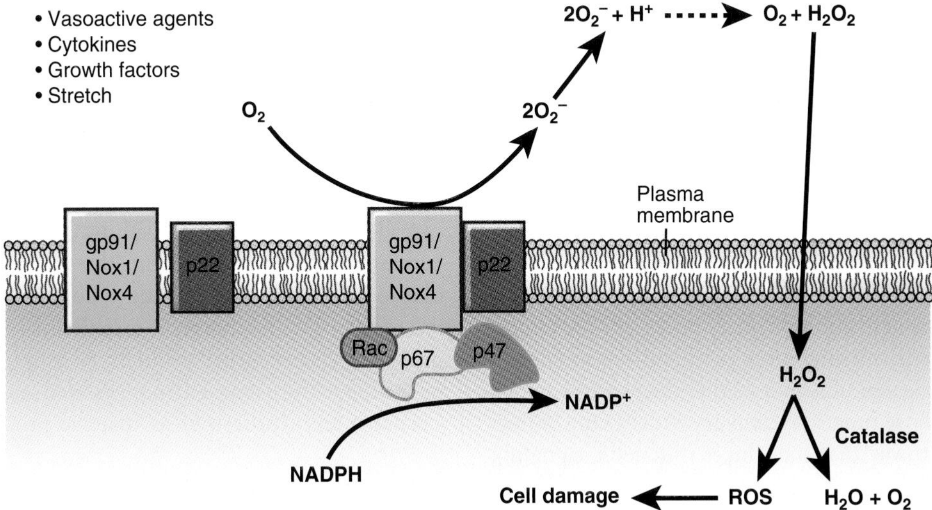

Figure 1.17 Balancing the levels of reactive oxygen species. One of the main sources of superoxide anion (O_2^-) is its synthesis by nicotinamide adenine dinucleotide phosphate oxidase (NADPH oxidase). NADPH oxidase is a multisubunit complex that is comprised of two transmembrane proteins gp91-phox/p22-phox that are localized to the plasma or organellar membranes. In response to various stimuli such as vasoactive factors like angiotensin II, or endothelin, cytokines like tumor necrosis factor, growth factors, hormones, or sheer stress, three cytosolic regulatory proteins p67-phox; p47-phox; and the small guanosine triphosphate (GTP)-binding protein, Rac, are recruited to form active NADPH oxidase. NADPH oxidase converts NADPH to superoxide using nicotinamide adenine dinucleotide phosphate ($NADP^+$), a product of the pentose phosphate pathway. Superoxide, a highly reactive anion, is rapidly converted to hydrogen peroxide (H_2O_2) and oxygen by the antioxidant scavenger enzyme, superoxide dismutase (SOD). Diffusion back into the cell leads to conversion of H_2O_2 into water and oxygen by another scavenger enzyme, catalase. Alternatively, H_2O_2 that escapes destruction, may cause cell damage as a reactive oxygen species (ROS).

shear stress from fluid movements, and stretching forces also activate NADPH oxidases and ROS production. As mentioned above, the various NADPH oxidases have distinct cellular expression, localization, and compositions, which accounts not only for the amount of ROS produced but also for the variety of signal transduction pathways they modulate.

The normal modest levels of ROS production that are efficiently used for signal transduction are thought to oxidize only the most highly reactive, oxidation-sensitive targets. Conversely, higher levels of ROS are likely to oxidize additional more resistant, less reactive targets, and in this context ROS can promote a condition termed **oxidative stress** (an imbalance between production and degradation of ROS). To maintain cellular homeostasis, balance between production of ROS and utilization or destruction must be achieved. Countering the systems for generation of ROS are mechanisms for detoxifying these reactive molecules. First, the half-lives of ROS molecules are relatively short at high concentrations. Second, the ability to diffuse across membranes for some ROS, such as O_2^-, is restricted, and this restriction can be circumvented by using ion channels to move between the outside and inside of the cell or organelles. Third is the presence of cellular antioxidant enzymes, which have a vital role in maintaining homeostasis. Examples of these **antioxidants** include SOD and **catalase**, which reduce O_2^- to water and oxygen. Oxidative stress has been implicated in numerous cardiovascular diseases, such as atherosclerosis and ischemic damage to tissues (e.g., stroke and heart attack), and in neurologic diseases, such as Parkinson disease, **Alzheimer disease**, and amyotrophic lateral sclerosis (also known as *Lou Gehrig disease*). Attempts to counter oxidative stress in patients with these and other diseases using dietary supplements such as vitamin E, diets rich in antioxidants, or by administration of radical scavenging drugs have given mixed results. Based on the lack of solid support for this approach, a more directed effort to better understand ROS targets and effects is warranted, with the goal of better-designed therapeutic agents.

MITOGENIC SIGNALING PATHWAYS

Mammalian cells have numerous signaling pathways that collectively result in a number of different outcomes including cell growth, proliferation, and differentiation. Mitogenic first messengers, for example, fibroblast growth factor (FGF), insulin-like growth factor, and granulocyte colony–stimulating factor to name a few, can act to stimulate the progression through the cell cycle and mitosis. These peptide factors have other functions as well as promoting cell growth and mitosis. For example, FGF can also stimulate mesodermal differentiation and **angiogenesis**.

Some of these pathways ultimately rule a cell's fate, deciding between survival and death. These physiologic outcomes are the results of the cell's interpretation of its environment, and much of this information is received by receptors on the plasma membrane and within the cell. The information from these receptors flows through signaling cascades, where it is passed from molecule to molecule in the form of messengers such as cAMP as well as modifications such as phosphorylation to the individual proteins that link the pathway together. Many of these signal transduction cascades result in the activation of genes necessary for alterations in metabolism, cell migration, proliferation, and death. In this way, a single stimulus may lead to the expression of a group of genes whose functions can vary widely. One important signaling pathway that transduces mitogenic signaling is the MAPK pathway.

MAPK signaling pathways operate without second messengers.

There are three major MAPK pathways, referred to as *MAPK/ERK*, *SAPK/JNK*, and *p38*. These MAPK pathways are downstream of many receptors and transduce a variety of external signals to result in different cellular responses such as mitosis, growth, differentiation, and inflammation. The MAPK signaling pathways are one of the few pathways that operate in the absence of second messengers; instead they rely on a modular cascade consisting of three protein kinases arranged in a hierarchical pathway. The general modular arrangement of these pathways was shown in Figure 1.12. The MAPK pathway can be activated by binding of a ligand, which leads to activation of the apical kinase of cascade, MAP kinase kinase kinase (MAPKKK). Activated MAPKKK then phosphorylates MAP kinase kinase (MAPKK, or MAP2 kinase), which, in turn, phosphorylates MAPK. MKKK (also called *Raf kinase*) is activated by interaction with a member of the Ras family of small G proteins, which are bound to the plasma membrane (see Fig. 1.12). Ras becomes activated (Ras-GTP) in response to growth factor binding to its cognate receptor (i.e., FGF to FGF receptor). Phosphorylation and activation of the last member of the cascade, MAPK, causes its translocation from the cytoplasm to the nucleus, where it phosphorylates proteins including transcription factors that regulate expression of genes important for activation of the cell cycle and mitosis, growth, differentiation, and inflammation. Two examples of genes expressed in response to MAPK signaling are the transcription factors *c-Myc* and *c-Fos*. These transcription factors stimulate the expression of proteins needed to progress through the cell cycle such as cyclin D, which is needed for transition from G_1 to the S phase. When mutant or oncogenic forms of *c-Myc* and *c-Fos* are expressed at high levels in cells, unregulated cell proliferation and cancer may result.

Loss of mitogenic signaling can result in cell death.

When cells are deprived of mitogens and survival signals, infected with viruses, exposed to toxic chemicals, or suffer extensive DNA damage or inflammation, signaling programs promoting cell death are activated. This type of cell death, called **programmed cell death** or **apoptosis**, is an altruistic cell death that does not result in the exposure of surrounding cells to toxic contents of the dead cell. Rather, the apoptosing cell shrinks, the cytoskeleton collapses, chromosomes are fragmented, and the cell breaks down into small membrane-bound structures that are engulfed by neighboring cells or **macrophages** (Fig. 1.18). There are numerous signaling cascades that can result in cell death, but they share some common features and most involve activation of a protease cascade. This protease cascade is composed of several proteases having a cysteine residue in their active sites and these proteases cleave target proteins at aspartic acid residues, hence the name **caspase** (from **c**ysteine–**asp**artic prote**ases**). Caspases are synthesized as inactive procaspases that can

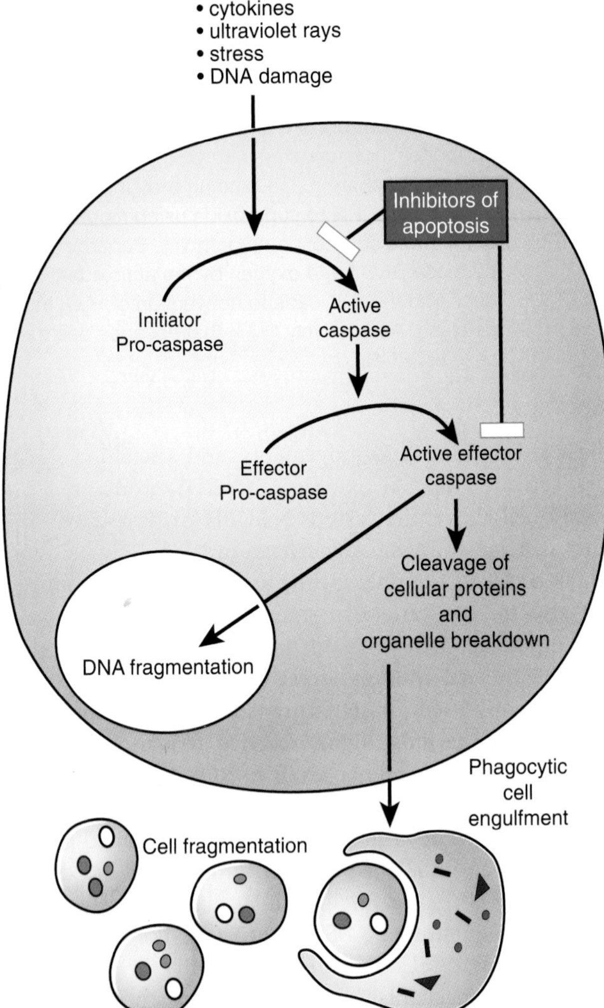

Figure 1.18 Apoptotic cell death. Loss or destruction of factors that promote cell growth and maintenance results in apoptotic cell death. A key component of this cell death is disassembly of the cell structure into small membrane-bound fragments through the action of proteolytic enzymes called *caspases*. The resulting cell fragments are phagocytized, thus preventing the cell components from spilling over into adjacent tissue where they might otherwise initiate broad inflammation-mediated cell damage.

be activated by different mechanisms to result in removal of the prodomain and formation of an active caspase. The caspase cascade commences when initiator caspases cluster and self-activate. These initiator caspases then cleave downstream caspases, called *effector caspases*, to amplify this proteolytic cascade. These activated caspases cleave key cellular proteins, causing a general breakdown in cell structures and organelles. In some cases, proteolytic cleavage of target proteins by a caspase can activate a latent enzymatic activity such as DNA degradation (activation of DNAse). As a result, the cell disassembles, or fragments into small membrane-bound bodies, and neighboring cells or macrophages engulf these cellular remnants. Because caspases are part of the normal complement of cellular proteins, which, when activated, commit the cell to death, there are numerous mechanisms in place to tightly regulate them and suppress this death program. These suppressors include regulating not only aggregation to activate initiator caspases but also the expression of other cellular proteins that block caspase activation, called **inhibitors of apoptosis**. Key to the apoptotic program is absence of release of toxic cellular products into the surrounding tissue space. This distinguishes apoptotic or programmed cell death from **necrosis**. Necrotic cell death usually results from acute injury to cells, and in response the cells rupture and release their contents on neighboring cells, an event that can stimulate an inflammatory response and cause more damage.

Not all cell death is pathologic. Over the course of a day, up to a million cells can die by apoptosis. These cell deaths serve to maintain a homeostatic balance by the elimination of old or unhealthy cells that are replaced with new, healthy cells. Nonpathologic cell death occurs during development, which is essential for sculpting the body (i.e., organs, fingers, and toes). Another example of nonpathologic cell death occurs during brain development and serves to eliminate excess neurons. A final example of necessary cell death is the elimination of immune cells recognizing "self"-antigens. Failure to eliminate these immune cells can result in a number of diseases, such as type 1 diabetes, systemic lupus erythematosus, rheumatoid arthritis, and multiple sclerosis to name a few.

Apoptosis can also be pathologic when too many cells are eliminated. Examples of this are cell death in response to stroke in which brain cells die from lack of blood supply and Parkinson disease, a degenerative disorder of the nervous system. Conversely, evasion of apoptosis is the foundation of such diseases as cancer and leukemia.

Chapter Summary

- Physiology is the study of the functions of living organisms and how they are regulated and integrated.
- A stable internal environment is necessary for normal cell function and survival of the organism.
- Homeostasis is the maintenance of steady states in the body by coordinated physiologic mechanisms.
- Negative and positive feedbacks are used to modulate the body's responses to changes in the environment.
- Steady state and equilibrium are distinct conditions. Steady state is a condition that does not change over time, whereas equilibrium represents a balance between opposing forces.
- Cellular communication is essential to integrate and coordinate the systems of the body so they can participate in different functions.
- Modes of cell communication differ in terms of distance and speed.
- A hallmark of cellular signaling is that it is regulatable, with a variety of mechanisms to both activate and terminate signal transduction.
- Activators of signal transduction pathways are called *first messengers*, and they include ions, gases, small peptides, protein hormones, metabolites, and steroids.
- Receptors are the receivers and transmitters of first messenger signaling molecules; they are located either on the plasma membrane or within the cell.
- Second messengers are important for amplification and flow of the signal received by plasma membrane receptors. Some second messengers such as calcium interact with accessory proteins such as calmodulin to stimulate the signal transduction flow.
- Reactive oxygen species represent a class of second messengers that are highly reactive and transduce signals by oxidizing proteins and nuclei acids. These reactive molecules can be produced in a "redox signaling" pathway involving nicotinamide adenine dinucleotide phosphate oxidases.
- Mitogenic signaling molecules (e.g., growth factors) activate signaling cascades that promote cell growth, proliferation, and differentiation.
- An absence of mitogenic signaling, in addition to cell stress or damage, can activate an intrinsic cell death pathway called *apoptosis*. The hallmark of apoptosis signaling is the activation of a proteolytic cascade involving proteases called *caspases*. Apoptosis differs from necrotic cell death in that cellular contents are engulfed rather than spilled into the extracellular space and resulting in inflammation.

thePoint *Visit http://thePoint.lww.com for chapter review Q&A, case studies, animations, and more!*

2 Plasma Membrane, Membrane Transport, and Resting Membrane Potential

ACTIVE LEARNING OBJECTIVES

Upon mastering the material in this chapter you should be able to:

- Understand how proteins and lipids are assembled to form a selectively permeable barrier known as the plasma membrane.
- Explain how the plasma membrane maintains an internal environment that differs significantly from the extracellular fluid.
- Understand how voltage-gated channels and ligand-gated channels are opened.
- Explain how carrier-mediated transport systems differ from channels.
- Understand the importance of adenosine triphosphate–binding cassette transporters to lipid transport and development of multidrug resistance.
- Explain the difference between primary and secondary active transport.
- Explain how epithelial cells are organized to produce directional movement of solutes and water.
- Explain how many cells can regulate their volume when exposed to osmotic stress.
- Understand why the Goldman equation gives the value of the membrane potential.
- Understand why the resting membrane potential in most cells is close to the Nernst potential for K^+.

The intracellular fluid of living cells, the **cytosol**, has a composition very different from that of the extracellular fluid (ECF). For example, the concentrations of potassium and phosphate ions are higher inside cells than outside, whereas sodium, calcium, and chloride ion concentrations are much lower inside cells than outside. These differences are necessary for the proper function of many intracellular enzymes; for instance, the synthesis of proteins by the ribosomes requires a relatively high potassium concentration. The **plasma membrane** of the cell creates and maintains these differences by establishing a permeability barrier around the cytosol. The ions and cell proteins needed for normal cell function are prevented from leaking out; those not needed by the cell are unable to enter the cell freely. The plasma membrane also keeps metabolic intermediates near where they will be needed for further synthesis or processing and retains metabolically expensive proteins inside the cell.

The plasma membrane is necessarily selectively permeable. Cells must receive nutrients to function, and they must dispose of metabolic waste products. To function in coordination with the rest of the organism, cells receive and send information in the form of chemical signals, such as hormones and neurotransmitters. The plasma membrane has mechanisms that allow specific molecules to cross the barrier around the cell. A selective barrier surrounds not only the cell but also every intracellular organelle that requires an internal milieu different from that of the cytosol. The cell nucleus, mitochondria, endoplasmic reticulum, Golgi apparatus, and lysosomes are delimited by membranes similar in composition to the plasma membrane. This chapter describes the specific types of membrane transport mechanisms for ions and other solutes, their relative contributions to the resting membrane electrical potential, and how their activities are coordinated to achieve directional transport from one side of a cell layer to the other.

PLASMA MEMBRANE STRUCTURE

The first theory of membrane structure proposed that cells are surrounded by a double layer of lipid molecules, a **lipid bilayer**. This theory was based on the known tendency of lipid molecules to form lipid bilayers with low permeability to water-soluble molecules. However, the lipid bilayer theory did not explain the selective movement of certain

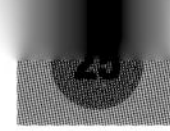

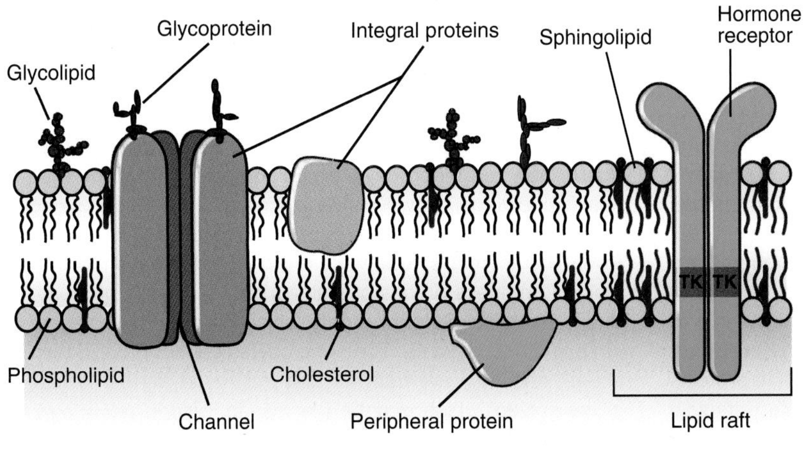

Figure 2.1 The fluid mosaic model of the plasma membrane. Lipids are arranged in a bilayer. Cholesterol provides rigidity to the bilayer. Integral proteins are embedded in the bilayer and often span it. Some membrane-spanning proteins form pores and channels; others are receptors. Peripheral proteins do not penetrate the bilayer. Lipid rafts form stable microdomains composed of sphingolipids and cholesterol.

water-soluble compounds, such as glucose and amino acids, across the plasma membrane. In 1972, Singer and Nicolson proposed the **fluid mosaic model** of the plasma membrane, which described the organization and interaction of proteins with the lipid bilayer (Fig. 2.1). With minor modifications, this model is still accepted as the correct picture of the structure of the plasma membrane.

Plasma membrane consists of different types of membrane lipids with different functions.

Lipids found in cell membranes can be classified into two broad groups: those that contain fatty acids as part of the lipid molecule and those that do not. Phospholipids are an example of the first group, and cholesterol is the most important example of the second group.

Phospholipids

The fatty acids present in **phospholipids** are molecules with a long hydrocarbon chain and a carboxyl terminal group. The hydrocarbon chain can be saturated (no double bonds between the carbon atoms) or unsaturated (one or more double bonds present). The long hydrocarbon chain tends to avoid contact with water and is described as **hydrophobic.** The carboxyl group at the other end is compatible with water and is termed **hydrophilic.** Fatty acids are said to be **amphipathic** because both hydrophobic and hydrophilic regions are present in the same molecule.

Phospholipids are the most abundant complex lipids found in cell membranes. They are amphipathic molecules formed by two fatty acids (normally, one saturated and one unsaturated) and one phosphoric acid group substituted on the backbone of a glycerol or sphingosine molecule. This arrangement produces a hydrophobic area formed by the two fatty acids and a polar hydrophilic head. When phospholipids are arranged in a bilayer, the polar heads are on the outside and the hydrophobic fatty acids on the inside (see Fig. 2.1). It is difficult for water-soluble molecules and ions to pass directly through the hydrophobic interior of the lipid bilayer.

The phospholipids, with a backbone of sphingosine (a long amino alcohol), are usually called *sphingolipids* and are present in all plasma membranes in small amounts. They are especially abundant in brain and nerve cells. Ceramide is a lipid second messenger that is generated from the sphingolipid sphingomyelin (see Chapter 1, "Homeostasis and Cellular Signaling," for a discussion of lipid second-messenger signaling).

Glycolipids are lipid molecules that contain sugars and sugar derivatives (instead of phosphoric acid) in the polar head. They are located mainly in the outer half of the lipid bilayer, with the sugar molecules facing the ECF. Proteins can associate with the plasma membrane by linkage to the extracellular sugar moiety of glycolipids.

Cholesterol

Cholesterol is an important component of mammalian plasma membranes. The proportion of cholesterol in plasma membranes varies from 10% to 50% of total lipids. Cholesterol has a rigid structure that stabilizes the cell membrane and reduces the natural mobility of the complex lipids in the plane of the membrane. Increasing amounts of cholesterol make it more difficult for lipids and proteins to move in the membrane. Some cell functions, such as the response of immune system cells to the presence of an antigen, depend on the ability of membrane proteins to move in the plane of the membrane to bind the antigen. A decrease in membrane fluidity resulting from an increase in cholesterol will impair these functions.

Lipid microdomains

Aggregates of sphingolipids and cholesterol can form stable microdomains termed **lipid rafts** that diffuse laterally in the phospholipid bilayer. The protein **caveolin** is present in a subset of lipid rafts (termed **caveolae**), causing the raft to form a cavelike structure. It is believed that one function of both noncaveolar and caveolar lipid rafts is to facilitate interactions between specific proteins by selectively including (or excluding) these proteins from the raft microdomain.

Clinical Focus / 2.1

Hereditary Spherocytosis

Red blood cells (erythrocytes) contain hemoglobin, which binds oxygen for transport from lungs to other organs and tissues. A normal erythrocyte has a marked biconcave structure, which permits the cell to deform for easier passage through narrow capillaries, especially in the spleen. The unusual shape is maintained by the **cytoskeleton**, formed by both integral and peripheral plasma membrane proteins. A key player is the long filamentous protein known as **spectrin**, a peripheral protein that forms a meshwork on the cytoplasmic surface. Linkages between spectrin and several integral proteins serve to anchor the plasma membrane to the cytoskeleton. **Hereditary spherocytosis (HS)** is a genetic disease that affects proteins in the erythrocyte membrane, and the result is a defective cytoskeleton. The incidence of HS is 1 in 5,000. The most common defect is deficiency of spectrin, and the result is that regions of the membrane break off because they are no longer anchored to the cytoskeleton. The remaining membrane reseals spontaneously, but after several "sheddings," the cell eventually becomes small and spherical. The presence of these **microspherocytes** is the hallmark of HS disease, which can range from asymptomatic to severe hemolytic anemia. **Hemolysis** (cell bursting) is present because the spherocytes are fragile to osmotic stress. Any entry of water will increase intracellular volume, and the cell membrane will break and release the hemoglobin. In contrast, normal biconcave erythrocytes can swell without bursting to accommodate osmotic water entry. Hence the lifespan of a spherocyte is considerably shorter than the 90- to 120-day lifespan of an erythrocyte. The shorter life is often compensated by accelerated production of new red blood cells (**erythropoiesis**) in an otherwise healthy individual. However, the anemia may be severe during an illness involving fever because erythropoiesis is slowed by fever. The anemia is usually the reason that HS patients complain of tiredness or loss of stamina and shortness of breath during exercise. **Splenomegaly** (enlarged spleen) is present in 75% of HS patients. The spherocytes cannot deform their shape, and they are trapped and destroyed (by hemolysis) in the spleen, which may be the reason it increases in size. Removal of spherocytes by the spleen is a major cause of the anemia and the shortened lifespan. Surgery to remove the spleen is often recommended if the anemia is severe and will improve spherocyte survival significantly. This enables most patients with HS to maintain a normal hemoglobin level.

Treatment of patients with HS has produced an appreciation of the role of the spleen in maintaining the integrity of the red blood cell population. Laboratory investigation of HS has helped the understanding of the structure and function of the membrane cytoskeleton, which is present in almost all types of cells, and the point mutations that produce defective spectrin have been identified.

For example, lipid rafts can mediate the assembly of membrane receptors and intracellular signaling proteins as well as the sorting of plasma membrane proteins for internalization.

Proteins are integrally and peripherally associated with the plasma membrane.

Proteins are the second major component of the plasma membrane, present in about equal proportion by weight with the lipids. Two different types of proteins are associated with the plasma membrane. **Integral proteins** (or intrinsic proteins) are embedded in the lipid bilayer; many span it completely, being accessible from the inside and outside of the membrane. The polypeptide chain of these proteins may cross the lipid bilayer once or may make multiple passes across it. The membrane-spanning segments usually contain amino acids with nonpolar side chains and are arranged in an ordered α-helical conformation. **Peripheral proteins** (or extrinsic proteins) do not penetrate the lipid bilayer. They are in contact with the outer side of only one of the lipid layers—either the layer facing the cytoplasm or the layer facing the ECF (see Fig. 2.1). Many membrane proteins have carbohydrate molecules, in the form of specific sugars, attached to the parts of the proteins that are exposed to the ECF. These molecules are known as **glycoproteins**. Some of the integral membrane proteins can move in the plane of the membrane, like small boats floating in the "sea" formed by the lipid bilayer. Other membrane proteins are anchored to the cytoskeleton inside the cell or to proteins of the extracellular matrix.

The proteins in the plasma membrane play a variety of roles. Many peripheral membrane proteins are enzymes, and many membrane-spanning integral proteins are carriers or channels for the movement of water-soluble molecules and ions into and out of the cell. Another important role of membrane proteins is structural; for example, certain membrane proteins in the erythrocyte help maintain the biconcave shape of the cell. Finally, some membrane proteins serve as highly specific receptors on the outside of the cell membrane to which extracellular molecules, such as hormones, can bind. If the receptor is a membrane-spanning protein, it provides a mechanism for converting an extracellular signal into an intracellular response.

SOLUTE TRANSPORT MECHANISMS

All cells must import oxygen, sugars, amino acids, and small ions and export carbon dioxide, metabolic wastes, and secretions. At the same time, specialized cells require mechanisms to transport molecules such as enzymes, hormones, and neurotransmitters. The movement of large molecules is carried out by endocytosis and exocytosis, that is, the transfer of substances into or out of the cell, respectively, by vesicle formation and vesicle fusion with the plasma membrane. Cells also have mechanisms for the rapid movement of ions and solute molecules across the plasma membrane. These mechanisms are of two general types: **passive transport**, which requires no direct expenditure of metabolic energy, and **active transport**, which uses metabolic energy to move solutes across the plasma membrane.

Import of extracellular materials occurs through phagocytosis and endocytosis.

Phagocytosis is the ingestion of large particles or microorganisms, usually occurring only in specialized cells such as macrophages (Fig. 2.2). An important function of macrophages is to remove invading bacteria from the body. The phagocytic vesicle (1–2 μm in diameter) is almost as large as the phagocytic cell itself. Phagocytosis requires a specific stimulus. It occurs only after the extracellular particle has bound to the extracellular surface. The particle is then enveloped by expansion of the cell membrane around it.

Endocytosis is a general term for the process in which a region of the plasma membrane is pinched off to form an endocytic vesicle inside the cell. During vesicle formation, some fluid, dissolved solutes, and particulate material from the extracellular medium are trapped inside the vesicle and internalized by the cell. Endocytosis produces much smaller endocytic vesicles (0.1–0.2 μm in diameter) than phagocytosis. It occurs in almost all cells and is termed a constitutive process, because it occurs continually and specific stimuli are not required. In further contrast to phagocytosis, endocytosis originates with the formation of depressions in the cell membrane. The depressions pinch off within a few minutes after forming and give rise to endocytic vesicles inside the cell.

Two types of endocytosis can be distinguished (see Fig. 2.2). **Fluid-phase endocytosis** is the nonspecific uptake of the ECF and all its dissolved solutes. The material is trapped inside the endocytic vesicle as it is pinched off inside the cell. The amount of extracellular material internalized by this process is directly proportional to its concentration in the extracellular solution. **Receptor-mediated endocytosis** is a more efficient process, which uses receptors on the cell surface to bind specific molecules. These receptors accumulate at specific depressions known as **coated pits**, so named because the cytosolic surface of the membrane at this site is covered with a coat of several proteins. The coated pits pinch off continually to form endocytic vesicles, providing the cell with a mechanism for rapid internalization of a large amount of a specific molecule without the need to endocytose large volumes of ECF. The receptors also aid the cellular uptake of molecules present at low concentrations outside the cell. Receptor-mediated endocytosis is the mechanism by which cells take up a variety of important molecules, including hormones, growth factors, and serum transport proteins such as **transferrin** (an iron carrier). Foreign substances, such as diphtheria toxin and certain viruses, also enter cells by this pathway.

Export of macromolecules occurs through exocytosis.

Many cells synthesize important macromolecules that are destined for **exocytosis** or export from the cell. These molecules are synthesized in the endoplasmic reticulum, modified in the Golgi apparatus, and packed inside transport vesicles. The vesicles move to the cell surface, fuse with the cell membrane, and release their contents outside the cell (see Fig. 2.2).

There are two exocytic pathways—constitutive and regulated. Some proteins are secreted continuously by the cells

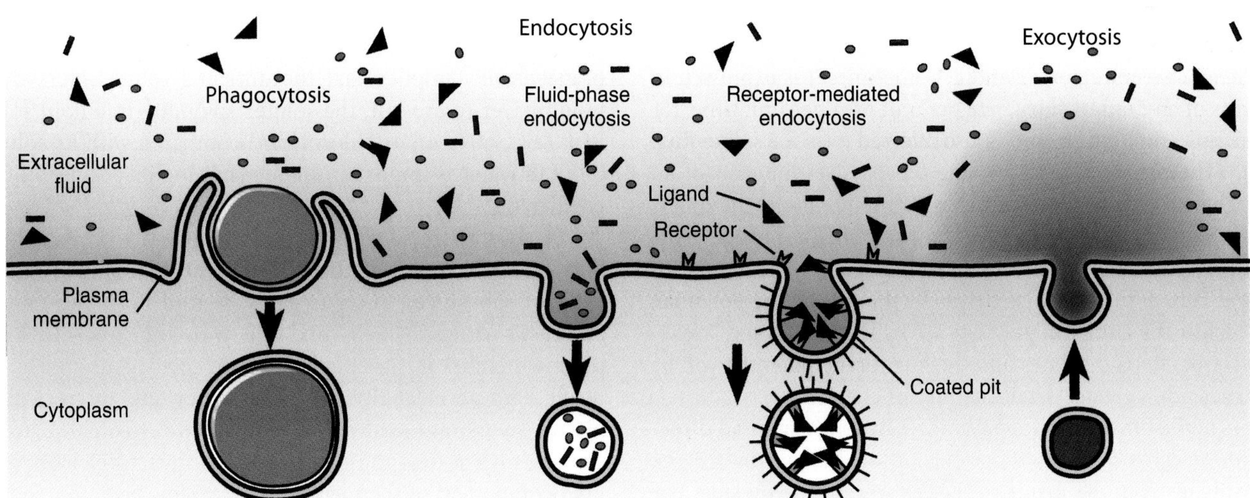

Figure 2.2 The transport of macromolecules across the plasma membrane by the formation of vesicles. Particulate matter in the extracellular fluid (ECF) is engulfed and internalized by phagocytosis. During fluid-phase endocytosis, ECF and dissolved macromolecules enter the cell in endocytic vesicles that pinch off at depressions in the plasma membrane. Receptor-mediated endocytosis uses membrane receptors at coated pits to bind and internalize specific solutes (ligands). Exocytosis is the release of macromolecules destined for export from the cell. These are packed inside secretory vesicles that fuse with the plasma membrane and release their contents outside the cell.

that make them. Secretion of mucus by **goblet cells** in the small intestine is a specific example. In this case, exocytosis follows the *constitutive pathway,* which is present in all cells. In other cells, macromolecules are stored inside the cell in secretory vesicles. These vesicles fuse with the cell membrane and release their contents only when a specific extracellular stimulus arrives at the cell membrane. This pathway, known as the *regulated pathway,* is responsible for the rapid "on-demand" secretion of many specific hormones, neurotransmitters, and digestive enzymes.

Uncharged solutes cross the plasma membrane by passive diffusion.

Any solute will tend to uniformly occupy the entire space available to it. This movement, known as **diffusion**, is a result of the spontaneous Brownian (random) movement that all molecules experience. A drop of ink placed in a glass of water will diffuse and slowly color all the water. The net result of diffusion is the movement of substances from regions of high concentration to regions of low concentration. Diffusion is an effective way for substances to move short distances.

The speed with which the diffusion of a solute in water occurs depends on the difference of concentration, the size of the molecules, and the possible interactions of the diffusible substance with water. These different factors appear in the **Fick law**, which describes the diffusion of any solute in water. In its simplest formulation, the Fick law (also known as *Fick's law*) can be written as

$$J = DA(C_1 - C_2)/\Delta X \quad (1)$$

where J is the flow of solute from region 1 to region 2 in the solution; D is the diffusion coefficient of the solute, which is determined by factors such as solute molecular size and interactions of the solute with water; A is the cross-sectional area through which the flow of solute is measured; C is the concentration of the solute at regions 1 and 2; and ΔX is the distance between regions 1 and 2. Sometimes, J is expressed in units of amount of substance per unit area per unit time, for example, mol/cm^2/h, and is also referred to as the solute **flux**.

The principal force driving the passive diffusion of an uncharged solute across the plasma membrane is the difference of concentration between the inside and the outside of the cell (Fig. 2.3). In the case of an electrically charged solute, such as an ion, diffusion is also driven by the membrane potential, which is the electrical gradient across the membrane. Movement of charged solutes and the membrane potential will be discussed in greater detail later in this chapter.

Diffusion across a membrane has no preferential direction; it can occur from the outside of the cell toward the inside or from the inside of the cell toward the outside. For any substance, it is possible to measure the **permeability coefficient (P)**, which gives the speed of the diffusion across a unit area of plasma membrane for a defined driving force. The Fick law for the diffusion of an uncharged solute across a membrane can be written as

$$J = PA(C_1 - C_2) \quad (2)$$

which is similar to equation 1. P includes the membrane thickness, the diffusion coefficient of the solute within the membrane, and the solubility of the solute in the membrane. Dissolved gases, such as oxygen and carbon dioxide, have high permeability coefficients and diffuse across the plasma membrane rapidly. As a result, gas exchange in the lungs is very effective. Because diffusion across the plasma membrane usually implies that the diffusing solute enters the lipid bilayer to cross it, the solute's solubility in a lipid solvent (e.g., olive oil or chloroform) compared with its solubility in water is important in determining its permeability coefficient.

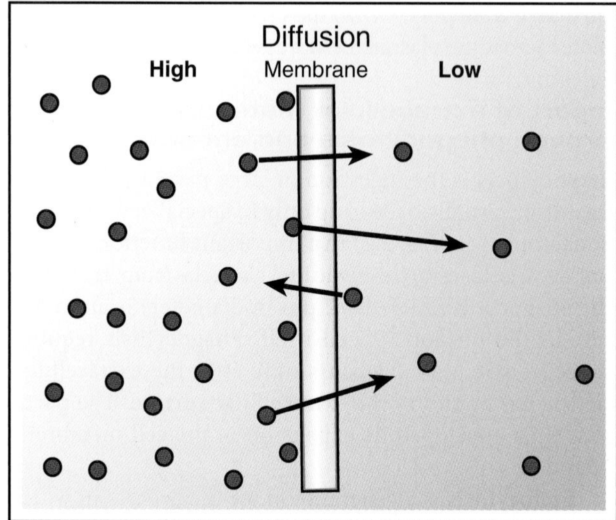

Figure 2.3 The diffusion of gases and lipid-soluble molecules through the lipid bilayer. In this example, the diffusion of a solute across a plasma membrane is driven by the difference in concentration on the two sides of the membrane. The solute molecules move randomly by Brownian movement. Initially, random movement from left to right across the membrane is more frequent than movement in the opposite direction because there are more molecules on the left side. This results in a net movement of solute from left to right across the membrane until the concentration of solute is the same on both sides. At this point, equilibrium (no net movement) is reached because solute movement from left to right is balanced by equal movement from right to left.

A substance's solubility in oil compared with its solubility in water is its **partition coefficient**. Lipophilic (lipid-soluble) substances, such as gases, steroid hormones, and anesthetic drugs, which mix well with the lipids in the plasma membrane, have high partition coefficients and, as a result, high permeability coefficients; they tend to cross the plasma membrane easily. Hydrophilic (water-soluble) substances, such as ions and sugars, do not interact well with the lipid component of the membrane, have low partition coefficients and low permeability coefficients, and diffuse across the membrane more slowly.

Solutes such as oxygen readily diffuse across the lipid part of the plasma membrane by simple diffusion. Thus, the relationship between the rate of movement and the difference in concentration between the two sides of the membrane is linear (Fig. 2.4). The larger the difference in concentration ($C_1 - C_2$), the greater the amount of substance crossing the membrane per unit time.

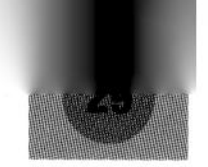

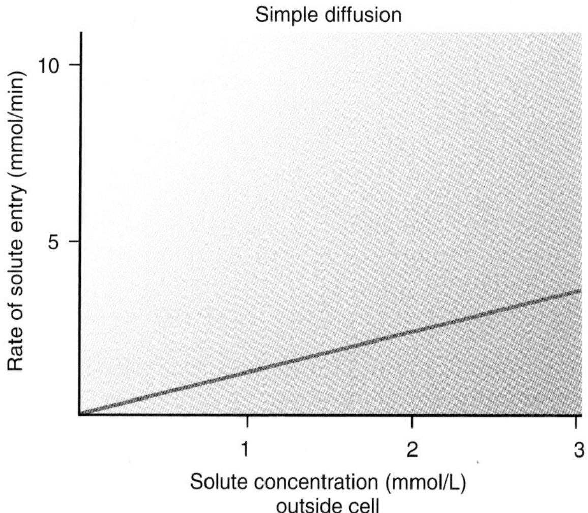

Figure 2.4 A graph of solute transport across a plasma membrane by simple diffusion. The rate of solute entry increases linearly with extracellular concentration of the solute. Assuming no change in intracellular concentration, increasing the extracellular concentration increases the gradient that drives solute entry.

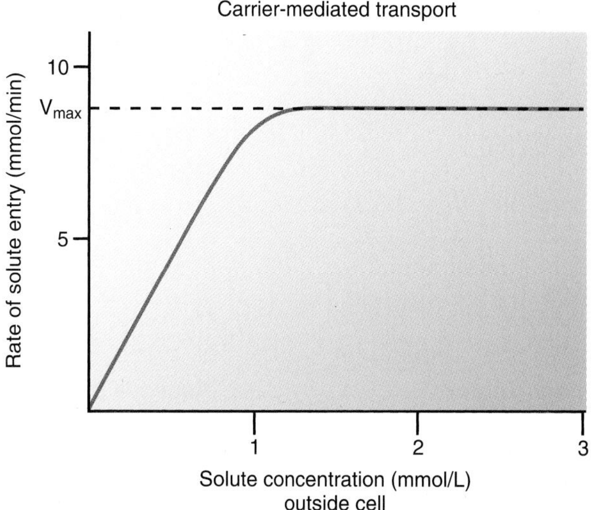

Figure 2.5 A graph of solute transport across a plasma membrane by facilitated diffusion. The rate of transport is much faster than that of simple diffusion (see Fig. 2.4) and increases linearly as the extracellular solute concentration increases. The increase in transport is limited, however, by the availability of channels and carriers. Once all are occupied by solute, further increases in extracellular concentration have no effect on the rate of transport. A maximum rate of transport (V_{max}) is achieved that cannot be exceeded.

Integral membrane proteins facilitate diffusion of solutes across the plasma membrane.

For many solutes of physiologic importance, such as ions, sugars, and amino acids, the rate of transport across the plasma membrane is much faster than expected for simple diffusion through a lipid bilayer. Furthermore, the relationship between transport rate and concentration difference of these hydrophilic substances follows a curve that reaches a plateau (Fig. 2.5). Membrane transport with these characteristics is often called **facilitated diffusion** or **carrier-mediated diffusion**, because an integral membrane protein facilitates (or assists) the movement of a solute through the membrane. Integral membrane proteins can form pores, channels, or carriers, each of which facilitates the transport of specific molecules across the membrane.

There are a limited number of pores, channels, and carriers in any cell membrane; thus, increasing the concentration of the solute initially uses the existing "spare" pores, channels, or carriers to transport the solute at a higher rate than by simple diffusion. As the concentration of the solute increases further and more solute molecules associate with the pore, channel, or carrier, the transport system eventually reaches saturation, when all the pores, channels, and carriers are involved in translocating molecules of solute. At this point, additional increases in solute concentration do not increase the rate of solute transport (see Fig. 2.5).

The types of integral membrane protein transport mechanisms considered here can transport a solute along its concentration gradient only, as in simple diffusion. Net movement stops when the concentration of the solute has the same value on both sides of the membrane. At this point, with reference to equation 2, $C_1 = C_2$ and the value of J is 0. The transport systems function until the solute concentrations have equilibrated. However, equilibrium is attained much faster than with simple diffusion.

Membrane pores

A pore provides a conduit through the lipid bilayer that is always open to both sides of the membrane. **Aquaporins** in the plasma membranes of specific kidney and gastrointestinal tract cells permit the rapid movement of water. Within the **nuclear pore complex**, which regulates movement of molecules into and out of the nucleus, is an aqueous pore that only allows the passive movement of molecules smaller than 45 kDa and excludes molecules larger than 62 kDa. The **mitochondrial permeability transition pore** and **mitochondrial voltage-dependent anion channel (VDAC)**, which cross the inner and outer mitochondrial membranes, promote mitochondrial failure when formed, resulting in the generation of **reactive oxygen species** and cell death.

Gated channels

Small ions, such as Na^+, K^+, Cl^-, and Ca^{2+}, cross the plasma membrane faster than would be expected based on their partition coefficients in the lipid bilayer. The electrical charge of an ion makes it difficult for the ion to move across the lipid bilayer. The rapid movement of ions across the membrane, however, is an aspect of many cell functions. The excitation of nerves, the contraction of muscle, the beating of the heart, and many other physiologic events are possible because of the ability of small ions to enter or leave the cell rapidly. This movement occurs through selective ion channels.

Ion channels are composed of several polypeptide subunits that span the plasma membrane and contain a gate

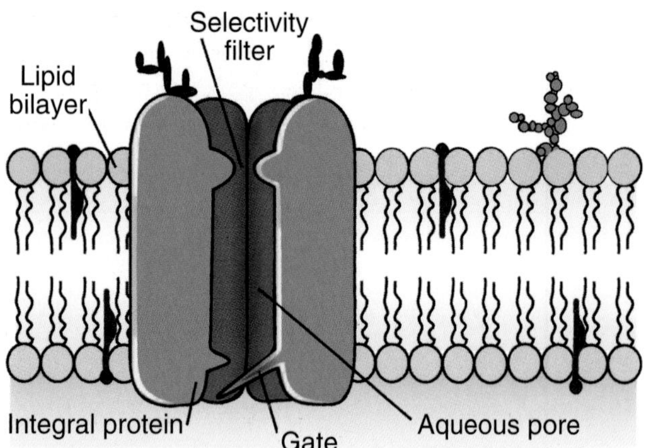

Figure 2.6 An ion channel. Ion channels are formed between the polypeptide subunits of integral proteins that span the plasma membrane, providing an aqueous pore through which ions can cross the membrane. Different types of gating mechanisms are used to open and close channels. Ion channels are often selective for a specific ion.

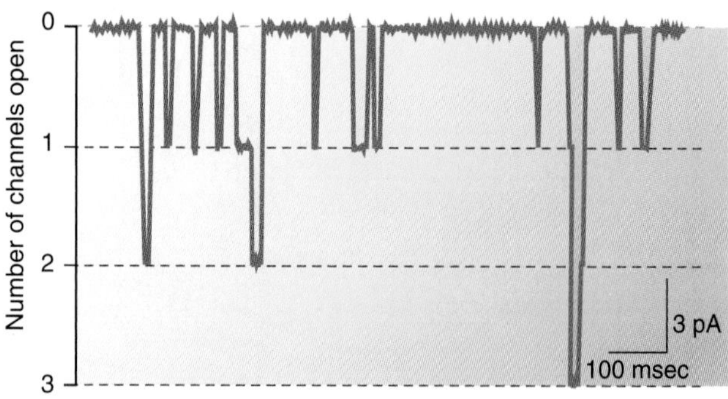

Figure 2.7 A patch clamp recording from a frog muscle fiber. Ions flow through the channel when it opens, generating a current. The current in this experiment is about 3 pA and is detected as a downward deflection in the recording. When more than one channel open, the current and the downward deflection increase in direct proportion to the number of open channels. This record shows that up to three channels are open at any instant.

that determines if the channel is open or closed. Specific stimuli cause a conformational change in the protein subunits to open the gate, creating an aqueous channel through which the ions can move (Fig. 2.6). In this way, ions do not have to enter the lipid bilayer to cross the membrane; they are always in an aqueous medium. When the channels are open, the ions diffuse rapidly from one side of the membrane to the other down the concentration gradient. Specific interactions between the ions and the sides of the channel produce an extremely rapid rate of ion movement; in fact, ion channels permit a much faster rate of solute transport (about 10^8 ions/s) than the carrier-mediated systems discussed below.

Ion channels have a selectivity filter, which regulates the transport of certain classes of ions such as anions or cations or specific ions such as Na^+, K^+, Ca^{2+}, and Cl^- (see Fig 2.6). The amino acid composition of the channel protein does not appear to confer the ion selectivity of the channel.

The **patch clamp** technique has revealed a great deal of information about the characteristic behavior of channels for different ions. The small electrical current caused by ion movement when a channel is open can be detected with this technique, which is so sensitive that the opening and closing of a single ion channel can be observed (Fig. 2.7). In general, ion channels exist either fully open or completely closed, and they open and close very rapidly. The frequency with which a channel opens is variable, and the time the channel remains open (usually a few milliseconds) is also variable. The overall rate of ion transport across a membrane can be controlled by changing the frequency of a channel opening or by changing the time a channel remains open.

Most ion channels usually open in response to a specific stimulus. Ion channels can be classified according to their gating mechanisms, the signals that make them open or close. There are voltage-gated channels and ligand-gated channels. Some ion channels are more like membrane pores in that they are always open; these ion transport proteins are referred to as *nongated channels* (see Chapter 3, "Action Potential, Synaptic Transmission, and Maintenance of Nerve Function").

Voltage-gated ion channels open when the membrane potential changes beyond a certain threshold value. Channels of this type are involved in conducting the excitation signal along nerve axons and include sodium and potassium channels (see Chapter 3, "Action Potential, Synaptic Transmission, and Maintenance of Nerve Function"). Voltage-gated ion channels are found in many cell types. It is thought that some charged amino acids located in a membrane-spanning α-helical segment of the channel protein are sensitive to the transmembrane potential. Changes in the membrane potential cause these amino acids to move and induce a conformational change of the protein that opens the way for the ions.

Ligand-gated ion channels cannot open unless they first bind to a specific agonist. The opening of the gate is produced by a conformational change in the protein induced by the ligand binding. The ligand can be a neurotransmitter arriving from the extracellular medium. It can also be an intracellular second messenger, produced in response to some cell activity or hormone that reaches the ion channel from the inside of the cell. The nicotinic acetylcholine receptor channel found in the postsynaptic neuromuscular junction (see Chapter 3, "Action Potential, Synaptic Transmission, and Maintenance of Nerve Function" and Chapter 9, "Blood Components") is a ligand-gated ion channel that is opened by an extracellular ligand (acetylcholine). Examples of ion channels gated by intracellular messengers also abound in nature. This type of gating mechanism allows the channel to open or close in response to events that occur at other locations in the cell. For example, a sodium channel gated by intracellular cyclic guanosine monophosphate (cGMP) is involved in the process of vision (see Chapter 4, "Sensory Physiology"). This channel is located in the rod cells of the retina and opens in the presence of cGMP. The generalized structure of one subunit of an ion channel gated by cyclic nucleotides is shown in Figure 2.8. There are six membrane-spanning

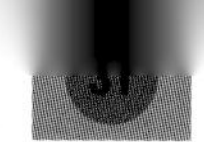

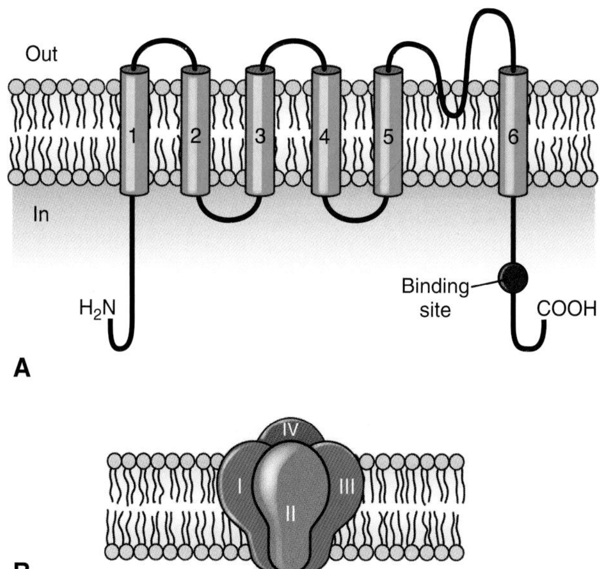

Figure 2.8 Structure of a cyclic nucleotide-gated ion channel. (A) The secondary structure of a single subunit has six membrane-spanning regions and a binding site for cyclic nucleotides on the cytosolic side of the membrane. **(B)** Four identical subunits (I–IV) assemble together to form a functional channel that provides a hydrophilic pathway across the plasma membrane.

regions, and a cyclic nucleotide-binding site is exposed to the cytosol. The functional protein is a tetramer of four identical subunits. Other cell membranes have potassium channels that open when the intracellular concentration of calcium ions increases. Several known channels respond to inositol 1,4,5-trisphosphate, the activated part of G proteins, or **adenosine triphosphate (ATP)**. The epithelial chloride channel that is mutated in cystic fibrosis is normally gated by ATP.

Carrier-mediated transport moves a range of ions and organic solutes passively across membranes.

In contrast to pores and ion channels, integral membrane proteins that form carriers provide a conduit through the membrane that is never open to both sides of the membrane at the same time. This is due to the presence of two gates (Fig. 2.9). During carrier-mediated transport, binding of the solute to one side of the carrier induces a conformational change in the protein, which closes one gate and opens the second gate, allowing the solute to pass through the membrane. As with pores and channels, carriers function until the solute concentrations have equilibrated.

Carrier-mediated transport systems have several characteristics:

- They allow the transport of polar (hydrophilic) molecules at rates much higher than that expected from the partition coefficient of these molecules.
- They eventually reach saturation at high substrate concentration (see Fig 2.5).
- They have structural specificity, meaning each carrier system recognizes and binds specific chemical structures (a carrier for D-glucose will not bind or transport L-glucose).
- They show competitive inhibition by molecules with similar chemical structure. For example, carrier-mediated transport of D-glucose occurs at a slower rate when molecules of D-galactose are also present. This is because galactose, structurally similar to glucose, competes with glucose for the available glucose carrier proteins.

A specific example of carrier-mediated transport is the movement of glucose from the blood to the interior of cells. Most mammalian cells use blood glucose as a major source of cellular energy, and glucose is transported into cells down its concentration gradient. The transport process in many cells, such as erythrocytes and the cells of fat, liver, and muscle tissues, involves a plasma membrane protein called *GLUT1* (glucose transporter-1). The erythrocyte GLUT1 has an affinity for D-glucose that is about 2,000-fold greater than the affinity for L-glucose. It is an integral membrane protein that contains 12 membrane-spanning α-helical segments.

Carrier-mediated transport, like simple diffusion, does not have a directional preference. It functions equally well bringing its specific solutes into or out of the cell, depending on the concentration gradient. Net movement by carrier-mediated transport ceases once the concentrations inside and outside the cell become equal.

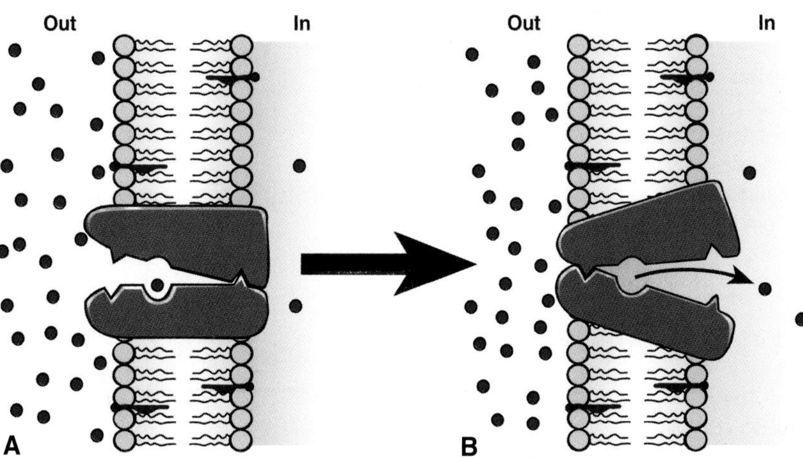

Figure 2.9 The role of a carrier protein in facilitated diffusion of solute molecules across a plasma membrane. In this example, solute transport into the cell is driven by the high solute concentration outside compared with inside. **(A)** Binding of extracellular solute to the carrier, a membrane-spanning integral protein, may trigger a change in protein conformation that exposes the bound solute to the interior of the cell. **(B)** Bound solute readily dissociates from the carrier because of the low intracellular concentration of solute. The release of solute may allow the carrier to revert to its original conformation **(A)** to begin the cycle again.

The **anion exchange protein (AE1)**, the predominant integral protein in the mammalian erythrocyte membrane, provides a good example of the reversibility of transporter action. AE1 is folded into at least 12 transmembrane α helices and normally permits the one-for-one exchange of Cl^- and HCO_3^- ions across the plasma membrane. The direction of ion movement is dependent only on the concentration gradients of the transported ions. AE1 has an important role in transporting CO_2 from the tissues to the lungs. The erythrocytes in systemic capillaries pick up CO_2 from tissues and convert it to HCO_3^-, which exits the cells via AE1. When the erythrocytes enter pulmonary capillaries, the AE1 allows plasma HCO_3^- to enter erythrocytes, where it is converted back to CO_2 for expiration by the lungs (see Chapter 21, "Control of Ventilation").

Active transport systems move solutes against gradients.

All the passive transport mechanisms tend to bring the cell into equilibrium with the ECF. Cells must oppose these equilibrating systems and preserve intracellular concentrations of solutes, in particular ions that are compatible with life.

Primary active transport

Integral membrane proteins that directly use metabolic energy to transport ions against a gradient of concentration or electrical potential are known as **ion pumps**. The direct use of metabolic energy to carry out transport defines a **primary active transport mechanism**. The source of metabolic energy is ATP synthesized by mitochondria, and the different ion pumps hydrolyze ATP to **adenosine diphosphate (ADP)** and use the energy stored in the third phosphate bond to carry out transport. Because of this ability to hydrolyze ATP, ion pumps also are called **ATPases**.

The most abundant ion pump in higher organisms is the sodium–potassium pump or **Na^+/K^+-ATPase**. It is found in the plasma membrane of practically every eukaryotic cell and is responsible for maintaining the low sodium and high potassium concentrations in the cytoplasm by transporting sodium out of the cell and potassium ions in. The sodium–potassium pump is an integral membrane protein consisting of two subunits. The α subunit has 10 transmembrane segments and is the catalytic subunit that mediates active transport. The smaller β subunit has one transmembrane segment and is essential for the proper assembly and membrane targeting of the pump. The Na^+/K^+-ATPase is known as a **P-type ATPase** because the protein is phosphorylated during the transport cycle (Fig. 2.10). The pump counterbalances the tendency of sodium ions to enter the cell passively and the tendency of potassium ions to leave passively. It maintains a high intracellular potassium concentration, which is necessary for protein synthesis. It also plays a role in the resting membrane potential by maintaining ion gradients. The sodium–potassium pump can be inhibited either by metabolic poisons that stop the synthesis and supply of ATP or by specific pump blockers,

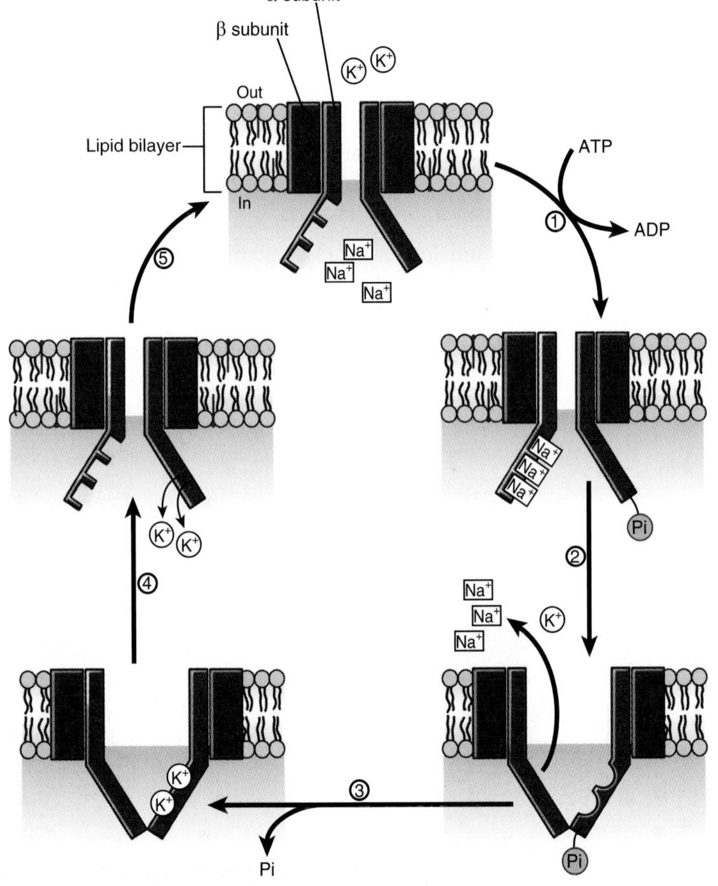

Figure 2.10 Function of the sodium–potassium pump. The pump is composed of two large α subunits that hydrolyze adenosine triphosphate (ATP) and transport the ions. The two smaller β subunits are molecular chaperones that facilitate the correct integration of the α subunits into the membrane. In step 1, three intracellular Na^+ bind to the α subunit, and ATP is hydrolyzed. Phosphorylation (Pi) of the α subunit results in a conformational change, exposing the Na^+ to the extracellular space (step 2). In step 3, the Na^+ diffuses away and two K^+ bind, resulting in dephosphorylation of the α subunit. Dephosphorylation returns the α subunit to an intracellular conformation. The K^+ diffuses away, and ATP is rebound to start the cycle over again (step 6). ADP, adenosine diphosphate.

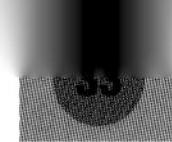

From Bench to Bedside / 2.1

Multidrug Resistance and Cancer

Some studies indicate that up to 40% of human cancers develop resistance to multiple anticancer agents, and it is a major problem when treating patients with malignant disease. A number of cellular mechanisms are known to lead to drug resistance, but it is now recognized that the most common mechanism is the efflux of chemotherapeutic drugs from the tumor cell by adenosine triphosphate–binding cassette (ABC) family drug transporters. The reduced accumulation of drugs in tumor cells was originally thought to result from overexpression of a 170-kDa protein termed **P-glycoprotein** (now known as *MDR1* or *ABCB1*) in the plasma membrane, resulting in expulsion of cytotoxic drugs, thereby reducing the intracellular concentration to below the threshold for cell killing. Since the discovery of P-glycoprotein, a number of additional drug resistance transporters have been identified. These include the 190-kDa **multidrug resistance (MDR)-associated proteins (MRP1)**, or ABCC1, and **MRP2**, or ABCC2, and the **breast cancer–resistance protein (BCRP)**, or ABCG2. Unlike the classical mammalian transport systems, which can be very selective in their substrate preferences, the multidrug transporters are highly promiscuous and can recognize and transport a wide range of substrates. P-glycoprotein and BCRP preferentially exclude large positively charged molecules, whereas the MRP family can exclude both uncharged molecules and water-soluble anionic compounds. Thus, the most logical approach to reversing MDR is to find compounds that can block ABC transporter activity.

First-generation modulators of MDR were the calcium-channel blocker, verapamil, and the immunosuppressant, cyclosporine A. Although these compounds were effective in cell culture and animal models, side effects due to the high dose of drug required to reverse chemotherapeutic drug resistance halted clinical trials. A number of derivatives (structurally similar molecules) for verapamil and cyclosporine A were subsequently tested. Unfortunately, these compounds interfered with drug metabolism and elimination, resulting in overexposure to cytotoxic chemotherapeutic agents. Using structure–activity relationships and combinatorial chemistry, a third generation of MDR modulators are being studied that are effective at nanomolar concentrations and do not affect the pharmacokinetics of chemotherapeutic agents.

The search for modulators of MDR is also currently focused on naturally occurring compounds, with the rationale that such products would be less toxic. Naturally occurring modulators of MDR include curcumin, coumarin, flavonoids, chokeberry, and mulberry leaves. Curcumin, in particular, has been shown to inhibit all three major ABC transporters, MDR1, MRP, and BCRP. Low oral bioavailability and rapid metabolism have prompted investigation into improved methods of delivery such as encapsulation in liposomes.

Finally, the use of transcriptional and translational inhibitors to specifically block synthesis of MDR1, MRP, and BCRP is under investigation. However, the recent discovery of **single-nucleotide polymorphisms** among the ABC drug transporters may complicate this approach. Polymorphisms in MDR1 have been shown to alter expression and function of the protein. Such changes in transporter function could contribute to variation between different individuals and ethnic groups in both chemotherapeutic drug response and the effect of inhibitors on transporter activity.

such as **digoxin**, a **cardiac glycoside** used to treat a variety of cardiac conditions.

As the Na^+/K^+-ATPase specifically moves sodium and potassium ions against their concentration or electrical potential, a number of other pumps move specific substrates across membranes utilizing the energy released by ATP hydrolysis.

- **Calcium pumps** are P-type ATPases located in the plasma membrane and the membrane of intracellular organelles. Plasma membrane Ca^{2+} ATPases pump calcium out of the cell. Calcium pumps in the membrane of the endoplasmic reticulum and in the sarcoplasmic reticulum membrane within muscle cells (termed *SERCAs* for sarcoplasmic and endoplasmic reticulum calcium ATPases) pump calcium into the lumen of these organelles. The organelles store calcium and, as a result, help maintain a low cytosolic concentration of this ion (see Chapter 1, "Homeostasis and Cellular Signaling").
- The **H^+/K^+-ATPase** is another example of a P-type ATPase present in the luminal membrane of the parietal cells in the oxyntic (acid-secreting) glands of the stomach. By pumping protons into the lumen of the stomach in exchange for potassium ions, this pump maintains the low pH in the stomach that is necessary for proper digestion. It is also found in the colon and in the collecting ducts of the kidney. Its role in the kidney is to secrete H^+ ions into the urine, when blood pH falls, and to reabsorb K^+ ions (see Chapter 25, "Neurogastroenterology and Motility").
- **Proton pumps** or **H^+-ATPases** are found in the membranes of the lysosomes and the Golgi apparatus. They pump protons from the cytosol into these organelles, keeping the inside of the organelles more acidic (at a lower pH) than the rest of the cell. These pumps, classified as **V-type ATPases** because they were first discovered in intracellular vacuolar structures, have now been detected in plasma membranes. For example, the proton pump in the plasma membrane of specialized bone and kidney cells is characterized as a V-type ATPase. The secretion of protons by **osteoclasts** helps to solubilize the bone mineral and creates an acidic environment for bone breakdown by enzymes. The proton pump in the kidney is present in the same cells as the H^+/K^+-ATPase and helps to secrete H^+ ions into the urine when blood pH falls.
- **ATP-binding cassette (ABC) transporters** are a super family of transporters composed of two transmembrane domains and two cytosolic nucleotide-binding domains. The transmembrane domains recognize specific solutes and transport them across the membrane using a number

of different mechanisms, including conformational change. The nucleotide-binding domain, or ABC domain, has a highly conserved sequence. ABC transporters are involved in a number of cellular processes, including nutrient uptake, cholesterol and lipid trafficking, resistance to cytotoxic drugs and antibiotics, cellular immune response, and stem cell biology.

- **ABCA1**, a member of the ABC subfamily A, has an important role in effluxing cholesterol, phospholipids, and other metabolites out of cells. ABCA1 transfers lipids and cholesterol to lipid-poor **high-density lipoproteins (HDLs)**. ABCA1 is a unique ABC transporter because it is also a receptor, binding the lipid-poor HDL to facilitate the loading of the cholesterol that the transporter is moving out of the cell.
  - ABC subfamily C transporters play a crucial role in the development of **multidrug resistance (MDR)**. There are a number of different transporters encoded by multiple MDR genes. The MDR1 transporter is widely distributed in the liver, brain, lung, kidney, pancreas, and small intestine and transports a wide range of antibiotics, antivirals, and chemotherapeutic drugs out of the cell. **MDR-associated protein transporters** are a related class of ABCC transporters that also interfere with antibiotic and chemotherapy. The cystic fibrosis transmembrane conductance regulator (ABCC7) is another member of this family.
  - **Organic anion transporting polypeptides (OATPs)** are members of the solute carrier family and are highly expressed in the liver, kidney, and brain. OATPs transport anionic and cationic chemicals, steroid, and peptide backbones generally into cells. **Thyroxine**, **bile acids**, and **bilirubin** are important solutes transported by OATPs. These transporters also import agents such 3-hydroxy-3-methylglutaryl-CoA reductase inhibitors (statins), angiotensin-converting enzyme inhibitors, angiotensin receptor II antagonists, and cardiac glycosides into cells.
  - **F-type ATPases** are located in the inner mitochondrial membrane. This type of proton pump normally functions in reverse. Instead of using the energy stored in ATP molecules to pump protons, its principal function is to synthesize ATP by using the energy stored in a gradient of protons that is crossing the inner mitochondrial membrane down its concentration gradient. The proton gradient is generated by the respiratory chain.

Secondary active transport

The net effect of ion pumps is maintenance of the various environments needed for the proper functioning of organelles, cells, and organs. Metabolic energy is expended by the pumps to create and maintain the differences in ion concentrations. Besides the importance of local ion concentrations for cell function, differences in concentrations represent stored energy. An ion releases potential energy when it moves down an electrochemical gradient, just as a body releases energy when falling to a lower level. This energy can be used to perform work. Cells have developed several carrier mechanisms to transport one solute against its concentration gradient by using the energy stored in the favorable gradient of another solute. In mammals, most of these mechanisms use sodium as the driver solute and use the energy of the sodium gradient to carry out the "uphill" transport of another important solute (Fig. 2.11). Because the sodium gradient is maintained by the action of the Na^+/K^+-ATPase, the function of these transport systems depends on the function of the Na^+/K^+-ATPase. Although they do not directly use metabolic energy for transport, these systems ultimately depend on the proper supply of metabolic energy to the sodium–potassium pump. They are called **secondary active transport mechanisms**. Disabling the pump with metabolic inhibitors or pharmacologic blockers causes these transport systems to stop when the sodium gradient has been dissipated.

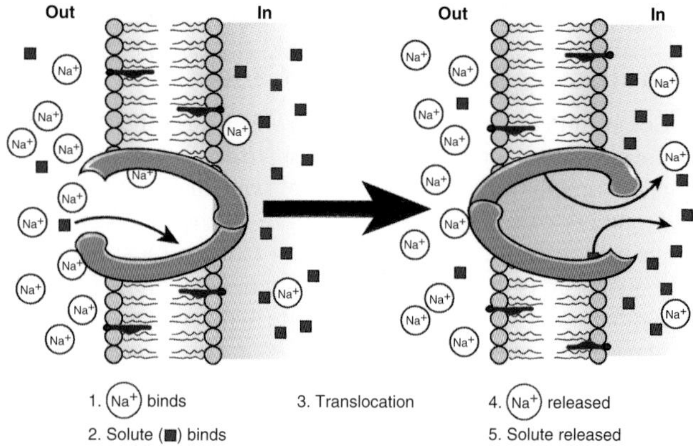

Figure 2.11 Mechanism of secondary active transport. A solute is moved against its concentration gradient by coupling it to Na^+ moving down a favorable gradient. Binding of extracellular Na^+ to the carrier protein (step 1) may increase the affinity of binding sites for solute, so that solute also can bind to the carrier (step 2), even though its extracellular concentration is low. A conformational change in the carrier protein (step 3) exposes the binding sites to the cytosol, where Na^+ readily dissociates because of the low intracellular Na^+ concentration (step 4). The release of Na^+ decreases the affinity of the carrier for solute and forces the release of the solute inside the cell (step 5), where solute concentration is already high. The free carrier then reverts to the conformation required for step 1, and the cycle begins again.

Similar to passive carrier-mediated systems, secondary active transport systems are integral membrane proteins; they have specificity for the solute they transport and show saturation kinetics and competitive inhibition. They differ, however, in two respects. First, they cannot function in the absence of the driver ion, the ion that moves along its electrochemical gradient and supplies energy. Second, they transport the solute against its own concentration or electrochemical gradient. Functionally, the different secondary active transport systems can be classified into two groups: **symport** (cotransport) systems, in which the solute being transported moves in the same direction as the sodium ion; and **antiport** (exchange) systems, in which the sodium ion and the solute move in opposite directions (Fig. 2.12).

Examples of symport mechanisms are the sodium-coupled sugar transport system and the several sodium-coupled amino acid transport systems found in the small intestine and the renal tubule. The symport systems allow efficient absorption of nutrients even when the nutrients are present at low concentrations. The sodium-dependent glucose transporter-1 (SGLT1) in the human intestine contains 664 amino acids in a single polypeptide chain with 14 membrane-spanning segments (Fig. 2.13). One complete cycle or turnover of a single SGLT1 protein, illustrated in Figure 2.11, can occur 1,000 times/s at 37°C. In reality, this cycle probably involves a coordinated trapping–release cycle and/or tilt of membrane-spanning segments rather than the simplistic view presented in Figure 2.11. Another example of a symport system is the family of sodium-coupled phosphate transporters (termed *NaPi*, types I and II) in the intestine and renal proximal tubule. These transporters have six to eight membrane-spanning segments and contain 460 to 690 amino acids. Sodium-coupled chloride transporters in the kidney are targets for inhibition by specific diuretics. The Na^+–Cl^- cotransporter in the distal tubule, known as *NCC*, is inhibited by thiazide diuretics, and the Na^+–K^+–$2Cl^-$ cotransporter in the ascending limb of the loop of Henle, referred to as *NKCC*, is inhibited by bumetanide.

The most important examples of antiporters are the Na^+/H^+ exchange and Na^+/Ca^{2+} exchange systems, found mainly in the plasma membrane of many cells. The first uses the sodium gradient to remove protons from the cell, controlling

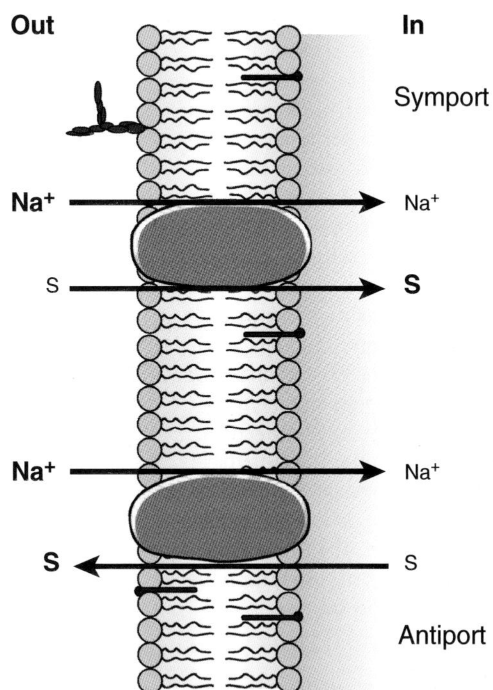

Figure 2.12 Secondary active transport systems. In a symport system (*top*), the transported solute (S) is moved in the same direction as the Na^+ ion. In an antiport system (*bottom*), the solute is moved in the opposite direction to Na^+. *Boldfaced* and *lightfaced* fonts indicate high and low concentrations, respectively, of Na^+ ions and solute.

Clinical Focus / 2.2

Hexose Malabsorption in the Intestine

Malabsorption of hexoses in the intestine can be the indirect result of a number of circumstances, such as an increase in intestinal motility or defects in digestion because of pancreatic insufficiency. Although less common, malabsorption may be a direct result of a specific defect in hexose transport. Regardless of the cause, the symptoms are common and include diarrhea, abdominal pain, and gas. The challenge is to identify the cause so proper treatment can be applied. Some infants develop a copious watery diarrhea when fed milk that contains glucose or galactose or the disaccharides lactose and sucrose. The latter are degraded to glucose, galactose, and fructose by enzymes in the intestine. The dehydration can begin during the first day of life and can lead to rapid death if not corrected. Fortunately, the symptoms disappear when a carbohydrate-free formula fortified with fructose is used instead of milk. This condition is a rare inherited disease known as **glucose–galactose malabsorption** (**GGM**), and about 200 severe cases have been reported worldwide. At least 10% of the general population has glucose or lactose intolerance, however, and it is possible that these people may have milder forms of the disease. A specific defect in absorption of glucose and galactose can be demonstrated by tolerance tests in which oral administration of these monosaccharides produces little or no increase in plasma glucose or galactose. The primary defect lies in the Na^+–glucose cotransporter protein (SGLT, Fig. 2.13), located in the apical plasma membrane of intestinal epithelial cells (Fig. 2.14). Glucose and galactose have very similar structures, and both are substrates for transport by SGLT. Fructose transport is not affected by a defect in SGLT because a specific fructose transporter named **GLUT5** is present in the apical membrane. Human SGLT was cloned in 1989, and almost 30 different mutations have been identified in GGM patients. Many of the mutations produce premature cessation of SGLT protein synthesis or disrupt the trafficking of mature SGLT to the apical plasma membrane. In a few cases, the SGLT reaches the apical membrane but is no longer capable of glucose transport. The result in all cases is that functional SGLT proteins are not present in the apical membrane so glucose and galactose remain in the lumen of the intestine. As these solutes accumulate in the lumen, the osmolality of the fluids increases and retards absorption of water, leading to diarrhea and severe water loss from the body. Identification of specific changes in defective SGLT proteins in patients has provided clues about the specific amino acids that are essential for the normal function of SGLT. At the same time, advances in molecular biology have allowed a better understanding of the genetic defect at the cellular level and how this leads to the clinical symptoms. GGM is an example of how information from a disease can further understanding of physiology and vice versa.

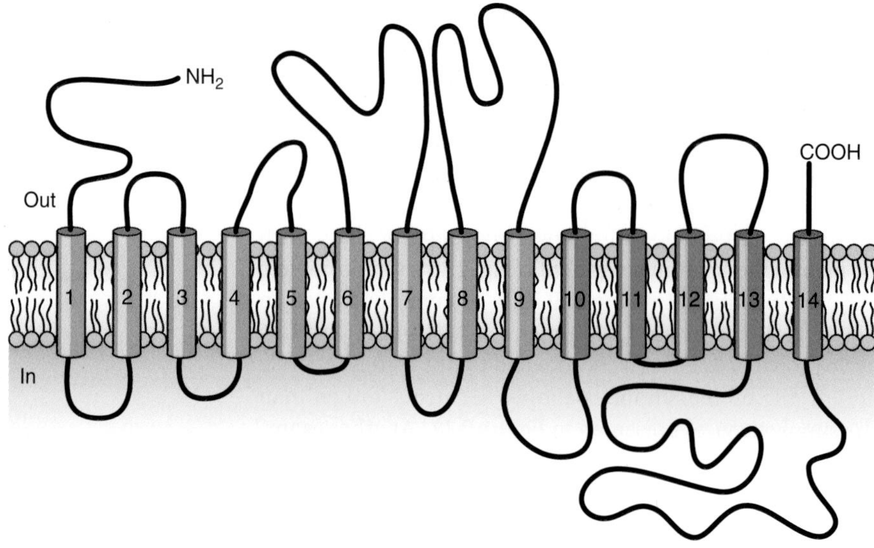

Figure 2.13 A model of the secondary structure of the Na^+-glucose cotransporter protein (SGLT1) in the human intestine. The polypeptide chain of 664 amino acids passes back and forth across the membrane 14 times. Each membrane-spanning segment consists of 21 amino acids arranged in an α-helical conformation. Both the NH_2 and the COOH ends are located on the extracellular side of the plasma membrane. In the functional protein, the membrane-spanning segments are clustered together to provide a hydrophilic pathway across the plasma membrane. The N-terminal portion of the protein, including helices 1 to 9, is required to couple Na^+ binding to glucose transport. The five helices (10–14) at the C terminus may form the transport pathway for glucose.

the intracellular pH and counterbalancing the production of protons in metabolic reactions. It is an **electroneutral system** because there is no net movement of charge. One Na^+ enters the cell for each H^+ that leaves. The second antiporter removes calcium from the cell and, together with the different calcium pumps, helps maintain a low cytosolic calcium concentration. It is an **electrogenic system** because there is a net movement of charge. Three Na^+ enter the cell and one Ca^{2+} ion leaves during each cycle.

The structures of the symport and antiport protein transporters that have been characterized (see Fig. 2.13) share a common property with ion channels (see Fig. 2.9) and equilibrating carriers, namely the presence of multiple membrane-spanning segments within the polypeptide chain. This supports the concept that, regardless of the mechanism, the membrane-spanning regions of a transport protein form a hydrophilic pathway for rapid transport of ions and solutes across the hydrophobic interior of the membrane lipid bilayer.

Transcellular transport

Epithelial cells occur in layers or sheets that allow the directional movement of solutes not only across the plasma membrane but also from one side of the cell layer to the other. Such regulated movement is achieved because the plasma membranes of epithelial cells have two distinct regions with different morphologies and different transport systems. These regions are the **apical membrane**, facing the lumen, and the **basolateral membrane**, facing the blood supply (Fig. 2.14). The specialized or polarized organization of the cells is maintained by the presence of **tight junctions** at the areas of contact between adjacent cells. Tight junctions prevent proteins on the apical membrane from migrating to the basolateral membrane and those on the basolateral membrane from migrating to the apical membrane. Thus, the entry and exit steps for solutes can be localized to opposite sides of the cell. This is the key to **transcellular transport** across epithelial cells.

An example is the absorption of glucose in the small intestine. Glucose enters the intestinal epithelial cells by active transport using the electrogenic Na^+–glucose cotransporter system (SGLT) in the apical membrane. This increases the

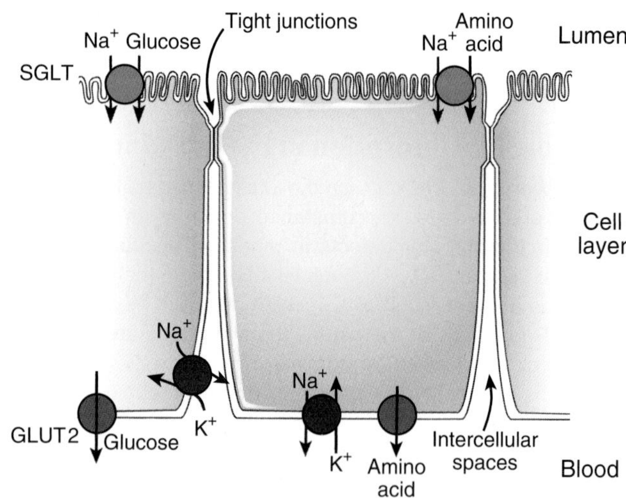

Figure 2.14 The localization of transport systems to different regions of the plasma membrane in epithelial cells of the small intestine. A polarized cell is produced, in which entry and exit of solutes, such as glucose, amino acids, and Na^+, occur at opposite sides of the cell. Active entry of glucose and amino acids is restricted to the apical membrane, and exit requires equilibrating carriers located only in the basolateral membrane. For example, glucose enters on sodium-dependent glucose transporter (SGLT) and exits on glucose transporter-2 (GLUT2). Na^+ that enters via the apical symporters is pumped out by the Na^+/K^+-ATPase on the basolateral membrane. The result is a net movement of solutes from the luminal side of the cell to the basolateral side, ensuring efficient absorption of glucose, amino acids, and Na^+ from the intestinal lumen.

intracellular glucose concentration above the blood glucose concentration, and the glucose molecules move passively out of the cell and into the blood via an equilibrating carrier mechanism (GLUT2) in the basolateral membrane (see Fig. 2.14). The intestinal GLUT2, like the erythrocyte GLUT1, is a sodium-independent transporter that moves glucose down its concentration gradient. Unlike GLUT1, the GLUT2 transporter can accept other sugars, such as galactose and fructose, that are also absorbed in the intestine. The Na^+/K^+-ATPase that is located in the basolateral membrane pumps out the sodium ions that enter the cell with the glucose molecules on SGLT. The polarized organization of the epithelial cells and the integrated functions of the plasma membrane transporters form the basis by which cells accomplish transcellular movement of both glucose and sodium ions.

WATER MOVEMENT ACROSS THE PLASMA MEMBRANE

Water can move rapidly in and out of cells, but the partition coefficient of water into lipids is low; therefore, the permeability of the membrane lipid bilayer for water is also low. Specific membrane proteins that function as water channels explain the rapid movement of water across the plasma membrane. These water channels are small (molecular weight of about 30 kDa), integral membrane proteins known as **aquaporins**. Ten different forms have been discovered so far in mammals. At least six forms are expressed in cells in the kidney and seven forms in the gastrointestinal tract, tissues in which water movement across plasma membranes is particularly rapid.

In the kidney, aquaporin-2 (AQP2) is abundant in the collecting duct and is the target of the hormone **arginine vasopressin**, also known as *antidiuretic hormone*. This hormone increases water transport in the collecting duct by stimulating the insertion of AQP2 proteins into the apical plasma membrane. Several studies have shown that AQP2 has a critical role in inherited and acquired disorders of water reabsorption by the kidney. For example, **diabetes insipidus** is a condition in which the kidney loses its ability to reabsorb water properly, resulting in excessive loss of water and excretion of a large volume of very dilute urine (**polyuria**). Although inherited forms of diabetes insipidus are relatively rare, it can develop in patients receiving chronic lithium therapy for psychiatric disorders, giving rise to the term **lithium-induced polyuria**. Both of these conditions are associated with a decrease in the number of AQP2 proteins in the collecting ducts of the kidney.

Water movement across the plasma membrane is driven by differences in osmotic pressure.

The spontaneous movement of water across a membrane driven by a gradient of water concentration is the process known as **osmosis**. The water moves from an area of high concentration of water to an area of low concentration. Because concentration is defined by the number of particles per unit of volume, a solution with a high concentration of solutes has a low concentration of water, and vice versa. Osmosis can, therefore, be viewed as the movement of water from a solution of high water concentration (low concentration of solute) toward a solution with a lower concentration of water (high solute concentration). Osmosis is a passive transport mechanism that tends to equalize the total solute concentrations of the solutions on both sides of every membrane.

If a cell that is in osmotic equilibrium is transferred to a more dilute solution, water will enter the cell, the cell volume will increase, and the solute concentration of the cytoplasm will be reduced. If the cell is transferred to a more concentrated solution, water will leave the cell, the cell volume will decrease, and the solute concentration of the cytoplasm will increase. As discussed below, many cells have regulatory mechanisms that keep cell volume within a certain range. Other cells such as mammalian erythrocytes do not have volume regulatory mechanisms, and large volume changes occur when the solute concentration of the ECF is changed.

The driving force for the movement of water across the plasma membrane is the difference in water concentration between the two sides of the membrane. For historical reasons, this driving force is not called the chemical gradient of water but the difference in osmotic pressure. The **osmotic pressure** of a solution is defined as the pressure necessary to stop the net movement of water across a selectively permeable membrane that separates the solution from pure water. When a membrane separates two solutions of different osmotic pressure, water will move from the solution with low osmotic pressure (high water and low solute concentrations) to the solution of high osmotic pressure (low water and high solute concentrations). In this context, the term *selectively permeable* means that the membrane is permeable to water but not solutes. In reality, most biologic membranes contain membrane transport proteins that permit solute movement.

The osmotic pressure of a solution depends on the *number* of particles dissolved in it, the total concentration of all solutes, regardless of the type of solutes present. Many solutes, such as salts, acids, and bases, dissociate in water, so the number of particles is greater than the molar concentration. For example, NaCl dissociates in water to give Na^+ and Cl^-, so one molecule of NaCl will produce two osmotically active particles. In the case of $CaCl_2$, there are three particles per molecule. The equation giving the osmotic pressure of a solution is

$$\pi = nRTC \qquad (3)$$

where π is the osmotic pressure of the solution, n is the number of particles produced by the dissociation of one molecule of solute (2 for NaCl, 3 for $CaCl_2$), R is the universal gas constant (0.0821 L·atm/mol·K), T is the absolute temperature, and C is the concentration of the solute in mol/L. Osmotic pressure can be expressed in atmospheres (atm). Solutions with the same osmotic pressure are called **isosmotic**. A solution is **hyperosmotic** with respect to another solution if it has a higher osmotic pressure and **hyposmotic** if it has a lower osmotic pressure.

Equation 3, called the *van't Hoff equation*, is valid only when applied to very dilute solutions, in which the particles of solutes are so far away from each other that no interactions occur between them. Generally, this is not the case at physiologic concentrations. Interactions between dissolved particles, mainly between ions, cause the solution to behave as if the concentration of particles is less than the theoretical value (nC). A correction coefficient, called the *osmotic coefficient* (φ) of the solute, needs to be introduced in the equation. Therefore, the osmotic pressure of a solution can be written more accurately as

$$\pi = nRT\Phi C \quad (4)$$

The osmotic coefficient varies with the specific solute and its concentration. It has values between 0 and 1. For example, the osmotic coefficient of NaCl is 1.00 in an infinitely dilute solution but changes to 0.93 at the physiologic concentration of 0.15 mol/L.

At any given T, because R is constant, equation 4 shows that the osmotic pressure of a solution is directly proportional to the term nφC. This term is known as the **osmolality** or *osmotic concentration* of a solution and is expressed in Osm/kg H_2O. Most physiologic solutions such as blood plasma contain many different solutes, and each contributes to the total osmolality of the solution. The osmolality of a solution containing a complex mixture of solutes is usually measured by freezing point depression. The freezing point of an aqueous solution of solutes is lower than that of pure water and depends on the total number of solute particles. Compared with pure water, which freezes at 0°C, a solution with an osmolality of 1 Osm/kg H_2O will freeze at −1.86°C. The ease with which osmolality can be measured has led to the wide use of this parameter for comparing the osmotic pressure of different solutions. The osmotic pressures of physiologic solutions are not trivial. Consider blood plasma, for example, which usually has an osmolality of 0.28 Osm/kg H_2O, determined by freezing point depression. Equation 4 shows that the osmotic pressure of plasma at 37°C is 7.1 atm, about seven times greater than the atmospheric pressure.

Many cells can regulate their volume.

Cell volume changes can occur in response to changes in the osmolality of ECF in both normal and pathophysiologic situations. Accumulation of solutes also can produce volume changes by increasing the intracellular osmolality. Many cells can correct these volume changes.

Volume regulation is particularly important in the brain where cell swelling can have serious consequences because expansion is strictly limited by the rigid skull.

Tonicity

A solution's osmolality is determined by the total concentration of all the solutes present. In contrast, the solution's **tonicity** is determined by the concentrations of only those solutes that do not enter ("penetrate") the cell. Tonicity determines cell volume, as illustrated in the following examples. Na^+ behaves as a nonpenetrating solute because it is pumped out of cells by the Na^+/K^+-ATPase at the same rate that it enters. A solution of NaCl at 0.2 Osm/kg H_2O is hypo-osmotic compared with cell cytosol at 0.3 Osm/kg H_2O. The NaCl solution is also **hypotonic** because cells will accumulate water and swell when placed in this solution. A solution containing a mixture of NaCl (0.3 Osm/kg H_2O) and urea (0.1 Osm/kg H_2O) has a total osmolality of 0.4 Osm/kg H_2O and will be hyperosmotic compared with cell cytosol. The solution is **isotonic**, however, because it produces no permanent change in cell volume. The reason is that cells shrink initially as a result of loss of water, but urea is a penetrating solute that rapidly enters the cells. Urea entry increases the intracellular osmolality, so water also enters and increases the volume. Entry of water ceases when the urea concentration is the same inside and outside the cells. At this point, the total osmolality both inside and outside the cells will be 0.4 Osm/kg H_2O and the cell volume will be restored to normal. By extension, it can be seen that normal blood plasma is an isotonic solution because Na^+ is the predominant plasma solute and is nonpenetrating. This stabilizes cell volume while other plasma solutes (glucose, amino acids, phosphate, urea, etc.) enter and leave the cells as needed.

Volume regulation mechanisms

When cell volume increases because of extracellular hypotonicity, the response of many cells is rapid activation of transport mechanisms that tend to decrease the cell volume (Fig. 2.15A). Different cells use different **regulatory volume decrease (RVD) mechanisms** to move solutes out of the cell and decrease the number of particles in the cytosol, causing water to leave the cell. Because cells have high intracellular concentrations of potassium, many RVD mechanisms involve an increased efflux of K^+, either by stimulating the opening of potassium channels or by activating symport mechanisms for KCl. Other cells activate the efflux of some amino acids, such as taurine or proline. The net result is a decrease in intracellular solute content and a reduction of cell volume close to its original value (see Fig. 2.15A).

When placed in a **hypertonic** solution, cells rapidly lose water and their volume decreases. In many cells, a decreased volume triggers **regulatory volume increase (RVI) mechanisms**, which increase the number of intracellular particles, bringing water back into the cells. Because Na^+ is the main extracellular ion, many RVI mechanisms involve an influx of sodium into the cell. Na^+–Cl^- symport, Na^+–K^+–$2Cl^-$ symport, and Na^+/H^+ antiport are some of the mechanisms activated to increase the intracellular concentration of Na^+ and increase the cell volume toward its original value (Fig. 2.15B).

Mechanisms based on an increased Na^+ influx are effective for only a short time because, eventually, the sodium pump will increase its activity and reduce intracellular Na^+ to its normal value. Cells that regularly encounter hypertonic ECFs have developed additional mechanisms for maintaining normal volume. These cells can synthesize specific organic solutes, enabling them to increase intracellular osmolality for a long time and avoiding altering the concentrations of ions they must maintain within a narrow range of values. The organic solutes are usually small molecules that do not interfere with normal cell function when they accumulate inside

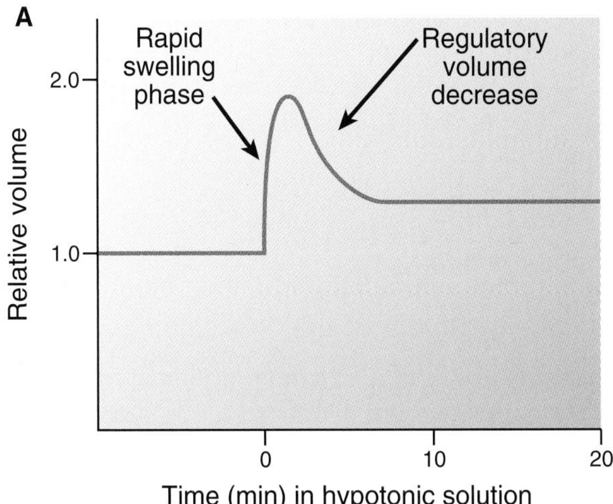

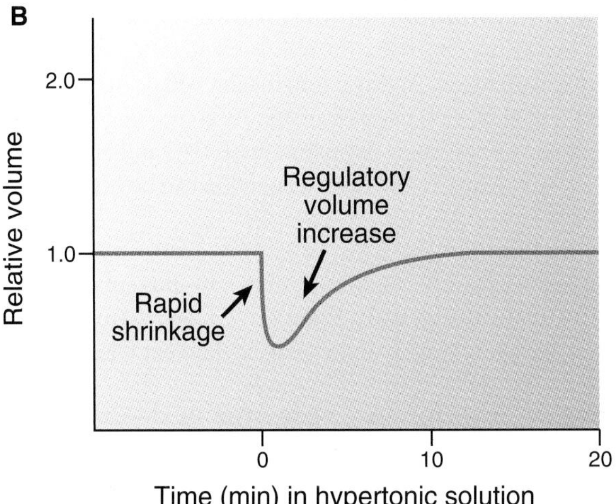

Figure 2.15 The effect of tonicity changes on cell volume. Cell volume changes when a cell is placed in either a hypotonic or a hypertonic solution. **(A)** In a hypotonic solution, the reversal of the initial increase in cell volume is known as a regulatory volume decrease. Transport systems for solute exit are activated, and water follows movement of solute out of the cell. **(B)** In a hypertonic solution, the reversal of the initial decrease in cell volume is a regulatory volume increase. Transport systems for solute entry are activated, and water follows solute into the cell.

the cell. For example, cells of the medulla of the mammalian kidney can increase the level of the enzyme aldose reductase when subjected to elevated extracellular osmolality. This enzyme converts glucose to an osmotically active solute, sorbitol. Brain cells can synthesize and store inositol. Synthesis of sorbitol and synthesis of inositol represent different answers to the problem of increasing the total intracellular osmolality, allowing normal cell volume to be maintained in the presence of hypertonic ECF.

Oral rehydration therapy is driven by solute transport.

Oral administration of rehydration solutions has dramatically reduced the mortality resulting from cholera and other diseases that involve excessive losses of water and solutes from the gastrointestinal tract. The main ingredients of rehydration solutions are glucose, NaCl, and water. The glucose and Na^+ ions are reabsorbed by SGLT1 and other transporters in the epithelial cells lining the lumen of the small intestine (see Fig. 2.14). Deposition of these solutes on the basolateral side of the epithelial cells increases the osmolality in that region compared with the intestinal lumen and drives the osmotic absorption of water. Absorption of glucose, and the obligatory increases in absorption of NaCl and water, helps to compensate for excessive diarrheal losses of salt and water.

RESTING MEMBRANE POTENTIAL

The different passive and active transport systems are coordinated in a living cell to maintain intracellular ions and other solutes at concentrations compatible with life. Consequently, the cell does not equilibrate with the ECF but rather exists in a **steady state** with the extracellular solution. For example, intracellular Na^+ concentration (10 mmol/L in a muscle cell) is much lower than extracellular Na^+ concentration (140 mmol/L), so Na^+ enters the cell by passive transport through nongated (always open) Na^+ channels. The rate of Na^+ entry is matched, however, by the rate of active transport of Na^+ out of the cell via the sodium–potassium pump (Fig. 2.16). The net result is that intracellular Na^+ is maintained constant and at a low level, even though Na^+ continually enters and leaves

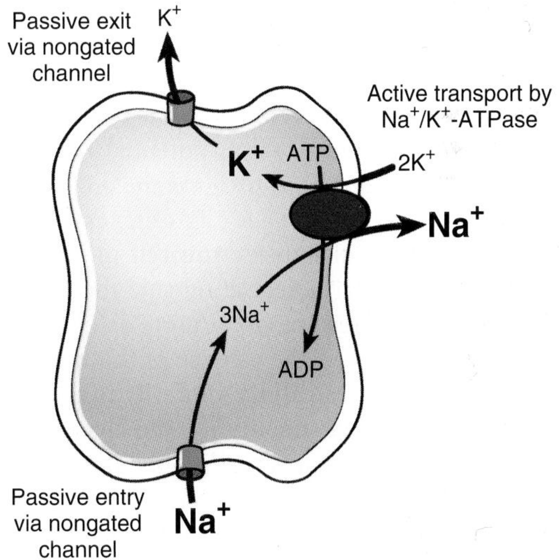

Figure 2.16 The concept of a steady state. Na^+ enters a cell through nongated Na^+ channels, moving passively down the electrochemical gradient. The rate of Na^+ entry is matched by the rate of active transport of Na^+ out of the cell via the Na^+/K^+-ATPase. The intracellular concentration of Na^+ remains low and constant. Similarly, the rate of passive K^+ exit through nongated K^+ channels is matched by the rate of active transport of K^+ into the cell via the pump. The intracellular K^+ concentration remains high and constant. During each cycle of the ATPase, two K^+ are exchanged for three Na^+, and one molecule of adenosine triphosphate (ATP) is hydrolyzed. *Boldfaced* and *lightfaced* fonts indicate high and low ion concentrations, respectively. ADP, adenosine diphosphate.

the cell. The reverse is true for K^+, which is maintained at a high concentration inside the cell relative to the outside. The passive exit of K^+ through nongated K^+ channels is matched by active entry via the pump (see Fig. 2.16). Maintenance of this steady state with ion concentrations inside the cell different from those outside the cell is the basis for the difference in electrical potential across the plasma membrane or the **resting membrane potential**.

Ion movement is driven by the electrochemical potential.

If there are no differences in temperature or hydrostatic pressure between the two sides of a plasma membrane, two forces drive the movement of ions and other solutes across the membrane. One force results from the difference in the concentration of a substance between the inside and the outside of the cell and the tendency of every substance to move from areas of high concentration to areas of low concentration. The other force results from the difference in electrical potential between the two sides of the membrane and applies only to ions and other electrically charged solutes. When a difference in electrical potential exists, positive ions tend to move toward the negative side, whereas negative ions tend to move toward the positive side.

The sum of these two driving forces is called the *gradient* (or difference) of **electrochemical potential** across the membrane for a specific solute. It measures the tendency of that solute to cross the membrane. The expression of this force is given by

$$\Delta\mu = RT \ln C_i/C_0 + zF(E_i - E_0) \quad (5)$$

where μ represents the electrochemical potential (Δμ is the difference in electrochemical potential between two sides of the membrane); C_i and C_o are the concentrations of the solute inside and outside the cell, respectively; E_i is the electrical potential inside the cell measured with respect to the electrical potential outside the cell (E_o); R is the universal gas constant (2 cal/mol·K); T is the absolute temperature (K); z is the valence of the ion; and F is the Faraday constant (23 cal/mV·mol). By inserting these units in equation 5 and simplifying, the electrochemical potential will be expressed in cal/mol, which is the unit of energy. If the solute is not an ion and has no electrical charge, then z = 0 and the last term of the equation becomes zero. In this case, the electrochemical potential is defined only by the different concentrations of the uncharged solute, called the **chemical potential**. The driving force for solute transport becomes solely the difference in chemical potential.

Net ion movement is zero at the equilibrium potential.

Net movement of an ion into or out of a cell continues as long as the driving force exists. Net movement stops and equilibrium is reached only when the driving force of electrochemical potential across the membrane becomes zero. The condition of equilibrium for any permeable ion will be Δμ = 0. Substituting this condition into equation 5, we obtain

$$0 = RT \ln\left(\frac{C_i}{C_o}\right) + zF(E_i - E_o)$$

$$E_i - E_o = -\frac{RT}{zF}\ln\left(\frac{C_i}{C_o}\right) \quad (6)$$

$$E_i - E_o = +\frac{RT}{zF}\ln\left(\frac{C_o}{C_i}\right)$$

Equation 6, known as the **Nernst equation**, gives the value of the electrical potential difference ($E_i - E_o$) necessary for a specific ion to be at equilibrium. This value is known as the **Nernst equilibrium potential** for that particular ion and it is expressed in millivolts (mV). At the equilibrium potential, the tendency of an ion to move in one direction because of the difference in concentrations is exactly balanced by the tendency to move in the opposite direction because of the difference in electrical potential. At this point, the ion will be in equilibrium and there will be no net movement. By converting to $\log_{10}$ and assuming a physiologic temperature of 37°C and a value of +1 for z (for Na^+ or K^+), the Nernst equation can be expressed as

$$E_I - E_O = 61\, LOG_{10}(C_O/C_I) \quad (7)$$

Because Na^+ and K^+ (and other ions) are present at different concentrations inside and outside a cell, it follows from equation 7 that the equilibrium potential will be different for each ion.

Resting membrane potential is determined by the passive movement of several ions.

The resting membrane potential is the electrical potential difference across the plasma membrane of a normal living cell in its unstimulated state. It can be measured directly by the insertion of a microelectrode into the cell with a reference electrode in the ECF. The resting membrane potential is determined by those ions that can cross the membrane and are prevented from attaining equilibrium by active transport systems. Potassium, sodium, and chloride ions can cross the membranes of every living cell, and each of these ions contributes to the resting membrane potential. By contrast, the permeability of the membrane of most cells to divalent ions is so low that it can be ignored in this context.

The **Goldman equation** gives the value of the membrane potential (in mV) when all the permeable ions are accounted for

$$E_i - E_o = \frac{RT}{F}\ln\left(\frac{P_K[K^+]_o + P_{Na}[Na^+]_o + P_{Cl}[Cl^-]_i}{P_K[K^+]_i + P_{Na}[Na^+]_i + P_{Cl}[Cl^-]_o}\right) \quad (8)$$

where P_K, P_{Na}, and P_{Cl} represent the permeability of the membrane to potassium, sodium, and chloride ions, respectively, and brackets indicate the concentration of the ion inside (i) and outside (o) the cell. If a specific cell is not permeable to one of these ions, the contribution of the impermeable ion to the membrane potential will be zero. For a cell that is permeable to an ion other than the three considered in the Goldman equation, that ion will contribute to the membrane potential and must be included in equation 8.

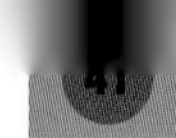

It can be seen from equation 8 that the contribution of any ion to the membrane potential is determined by the membrane's permeability to that particular ion. The higher the permeability of the membrane to one ion relative to the others, the more that ion will contribute to the membrane potential. The plasma membranes of most living cells are much more permeable to potassium ions than to any other ion. Making the assumption that P_{Na} and P_{Cl} are zero relative to P_K, equation 8 can be simplified to

$$E_i - E_o = \frac{RT}{F} \ln\left(\frac{P_K[K^+]_o}{P_K[K^+]_i}\right)$$
$$E_i - E_o = \frac{RT}{F} \ln\left(\frac{[K^+]_o}{[K^+]_i}\right) \qquad (9)$$

which is the Nernst equation for the equilibrium potential for K^+ (see equation 6). This illustrates two important points:

- In most cells, the resting membrane potential is close to the equilibrium potential for K^+.
- The resting membrane potential of most cells is dominated by K^+ because the plasma membrane is more permeable to this ion compared with the others.

As a typical example, the K^+ concentrations outside and inside a muscle cell are 3.5 and 155 mmol/L, respectively. Substituting these values in equation 7 gives a calculated equilibrium potential for K^+ of −100 mV, negative inside the cell relative to the outside. Measurement of the resting membrane potential in a muscle cell yields a value of −90 mV (negative inside). This value is close to, although not the same as, the equilibrium potential for K^+.

The reason the resting membrane potential in the muscle cell is less negative than the equilibrium potential for K^+ is as follows. Under physiologic conditions, there is passive entry of Na^+ ions. This entry of positively charged ions has a small but significant effect on the negative potential inside the cell. Assuming intracellular Na^+ to be 10 mmol/L and extracellular Na^+ to be 140 mmol/L, the Nernst equation gives a value of +70 mV for the Na^+ equilibrium potential (positive inside the cell). This is far from the resting membrane potential of −90 mV. Na^+ makes only a small contribution to the resting membrane potential because membrane permeability to Na^+ is low compared with that of K^+.

The contribution of Cl^- ions need not be considered because the resting membrane potential in the muscle cell is the same as the equilibrium potential for Cl^-. Therefore, there is no net movement of chloride ions.

In most cells, as shown above using a muscle cell as an example, the equilibrium potentials of K^+ and Na^+ are different from the resting membrane potential, which indicates that neither K^+ ions nor Na^+ ions are at equilibrium. Consequently, these ions continue to cross the plasma membrane via specific nongated channels, and these passive ion movements are *directly* responsible for the resting membrane potential.

The Na^+/K^+-ATPase is important *indirectly* for maintaining the resting membrane potential because it sets up the gradients of K^+ and Na^+ that drive passive K^+ exit and Na^+ entry. During each cycle of the pump, two K^+ ions are moved into the cell in exchange for three Na^+, which are moved out (see Fig. 2.16). Because of the unequal exchange mechanism, the pump's activity contributes slightly (about −5 mV) to the negative potential inside the cell.

Chapter Summary

- The plasma membrane consists of proteins in a phospholipid bilayer. Integral proteins are embedded in the bilayer, whereas peripheral proteins are attached to the outer surface.
- Macromolecules cross the plasma membrane by endocytosis and exocytosis.
- Passive movement of a solute across a membrane dissipates the gradient (driving force) and reaches an equilibrium at which point there is no net movement of solute.
- Simple diffusion is the passage of lipid-soluble solutes across the plasma membrane by diffusion through the lipid bilayer.
- Facilitated diffusion is the passage of water-soluble solutes and ions through a hydrophilic pathway created by a membrane-spanning integral protein.
- Facilitated diffusion of small ions is mediated by specific pores and ion channel proteins.
- Active transport uses a metabolic energy source to move solutes against gradients, and the process prevents a state of equilibrium.
- Polarized organization of epithelial cells ensures directional movement of solutes and water across the epithelial layer.
- Water crosses plasma membranes rapidly via channel proteins termed aquaporins. Water movement is a passive process driven by differences in osmotic pressure.
- Cells regulate their volume by moving solutes in or out to drive osmotic entry or exit of water, respectively.
- The driving force for ion transport is the sum of the electrical and chemical gradients, known as the gradient of electrochemical potential across the membrane.
- The resting membrane potential is determined by the passive movements of several ions through nongated channels, which are always open. It is described most accurately by the Goldman equation, which takes into account the differences in membrane permeability of different ions. In a muscle cell, for example, the membrane permeability to Na^+ is low compared with K^+ and the resting membrane potential is a result primarily of passive exit of K^+.

3 Action Potential, Synaptic Transmission, and Maintenance of Nerve Function

ACTIVE LEARNING OBJECTIVES

Upon mastering the material in this chapter you should be able to:

- Relate the design of the unique anatomic and cellular features of neurons to their function.
- Explain the role of the neuronal cytoskeleton in axonal transport, growth, and metabolic maintenance of neurons.
- Explain the process of electronic conduction and why it cannot be used for electrical signaling over long distances in the nervous system.
- Explain the underlying electrical phenomena expressed in the space and time constants that increase or decrease axonal conduction velocity.
- Explain the molecular mechanisms of ionic conductance events that underlie initiation, termination, and propagation of the action potential.
- Explain how combinations of temporal and spatial summation of end plate potentials can move a neuron closer to or further from its threshold for firing an action potential.
- Explain how neurotransmitter release alters postsynaptic membrane potentials.
- Contrast ionotropic and metabotropic neurotransmission.

The nervous system coordinates the activities of the other organs in response to signals from the external and internal environment. Examples of nervous system functions include activation of muscles for movement, control of glandular secretion, the regulation of heart rate and blood pressure, and the maintenance of body temperature. In this chapter, we examine the aspects of neuronal structure necessary for the maintenance of cell function, the properties that permit the generation of an electrical potential across the cell membrane, the process by which electrical impulses are propagated along the length of the cell, and the mechanisms for transmission of activity between cells.

NEURONAL STRUCTURE

Neurons are the key element of the nervous system. The structure and function of neurons permit rapid transmission of information from one cell to another using electrochemical signals.

Special anatomic features of neurons adapt them for communicating information.

The structure of a neuron reflects the specialized function of these types of cells (Fig. 3.1). The cell body of the neuron, also called the **soma** or **perikaryon**, is the location of the genetic material and synthetic activity of the neuron. The **dendrites** are shrub-like extensions from the cell body that receive incoming information from other neurons, which allows their function to be modulated by inputs from many different sources. The axon conducts information collected by the dendrites and cell body away from the cell body by electrical signals called **action potentials**. Axons can be very long and therefore facilitate transmission of information over large distances in the body. Axons are surrounded by **glial cells**—oligodendrocytes in the central nervous system (CNS) or Schwann cells in the peripheral nervous system (PNS). Glia provide structural and biochemical support for neuronal functions. Some axons are enveloped by multiple layers of glial cell membrane producing the **myelin** sheath, which acts

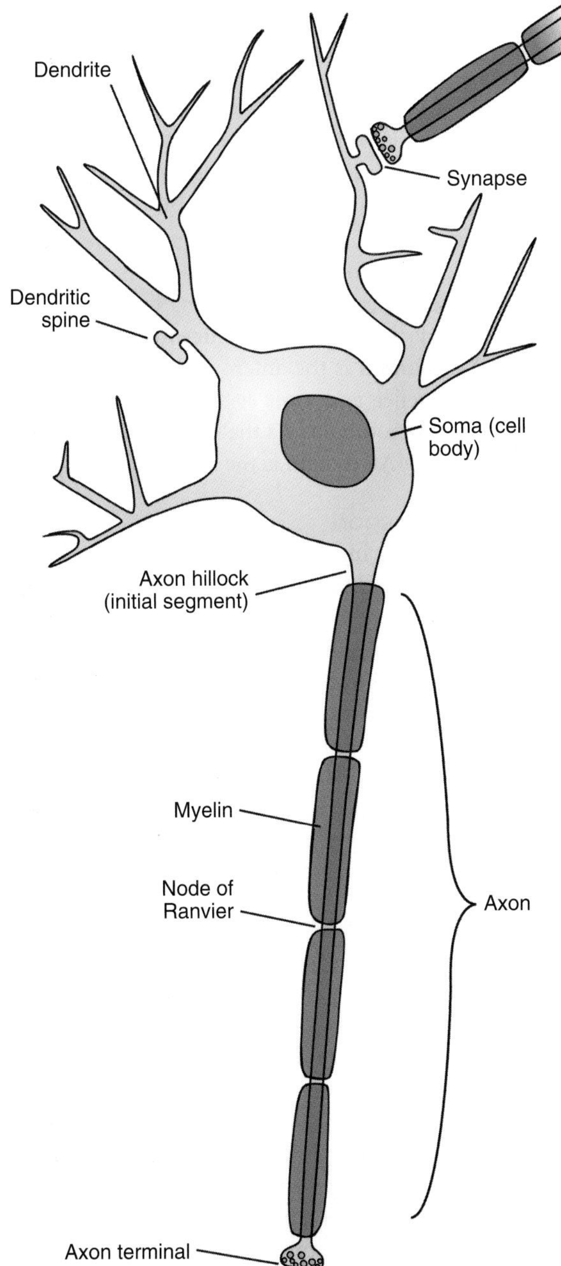

Figure 3.1 Structural components of a neuron with a myelinated axon.

as an insulating medium. Intermittent gaps in the glial sheath along the axon are called **nodes of Ranvier**. The combination of myelin and nodes allows action potentials to be conducted down the axon with much greater velocity than could be produced in an unmyelinated axon. The terminal region of an axon stores chemical neurotransmitters whose release relays information to other neurons or other types of effector cells. Arrival of an action potential at the axon terminal causes the release of neurotransmitter. The neurotransmitter molecules bind to **receptors** located on target cells. The site of interaction between the axon and the target cell is called the **synapse**. Neurotransmitter activation of receptors on postsynaptic cells is the primary means by which the originating signal in the presynaptic neuron is passed on to the postsynaptic cell.

Nerve cells have specialized functional requirements for maintaining viability.

Neurons have unique metabolic needs. The axons of some neurons can exceed 1 m in length. Consider the control of toe movement in a tall person. Neurons in the motor cortex of the brain have axons that synapse with the motor neurons in the lumbar region of the spinal cord. The axons of the motor neurons in the spinal cord synapse with muscles in the foot, which move the toes. In an adult, the axons of both these types of neurons are 60 to 80 cm in length, whereas the cell body might be only 50 μm in diameter. An enormous amount of membrane and intra-axonal material must be synthesized in the soma and delivered to the axon to maintain its structural integrity and function. The soma also supports the structure and function of the dendrites. In some cell types, the surface of the dendritic arborization may be 2 to 5 mm in length.

Protein synthesis

The nucleus of a neuron is large, and a substantial portion of the genetic information it contains is continuously transcribed. Based on hybridization studies, it is estimated that one third of the genome in brain cells is actively transcribed, producing more mRNA than any other kind of cell in the body. Because of the high level of transcriptional activity, the nuclear chromatin is dispersed. In contrast, the chromatin in nonneuronal cells in the brain such as glial cells, is found in clusters on the internal face of the nuclear membrane.

Most of the proteins formed by free ribosomes and polyribosomes remain within the soma, whereas proteins formed by **rough endoplasmic reticulum (RER)** are exported to the dendrites and the axon. Polyribosomes and RER are found predominantly in the soma of neurons. Axons contain no RER and are unable to synthesize proteins. The **Golgi apparatus** in neurons is found only in the soma. As in other types of cells, this structure is engaged in the terminal glycosylation of proteins synthesized in the RER. The Golgi apparatus forms export vesicles for proteins produced in the RER. These vesicles are released into the cytoplasm, and some are transported to the axon terminals and dendrites.

Cytoskeleton forms infrastructure and transport systems within neurons.

The highly specialized shape of the neuron and the cell's ability to transport proteins and other structural components throughout the cell depend on the internal cytoskeleton. Figure 3.2 shows key components of the neuronal cytoskeleton including microfilaments, neurofilaments, and microtubules. **Microfilaments** are composed of actin, which is similar in structure to the contractile protein found in muscle. Actin monomers are arranged into two intertwined strings. The resulting filaments are 4 to 5 nm in diameter and about a micrometer in length. Microfilaments are concentrated in growth cones, dendritic spines, and axonal nerve terminals. Microfilaments are involved in a variety of functions, including growth cone motility, cytoskeletal integrity, distribution and maintenance of specific proteins

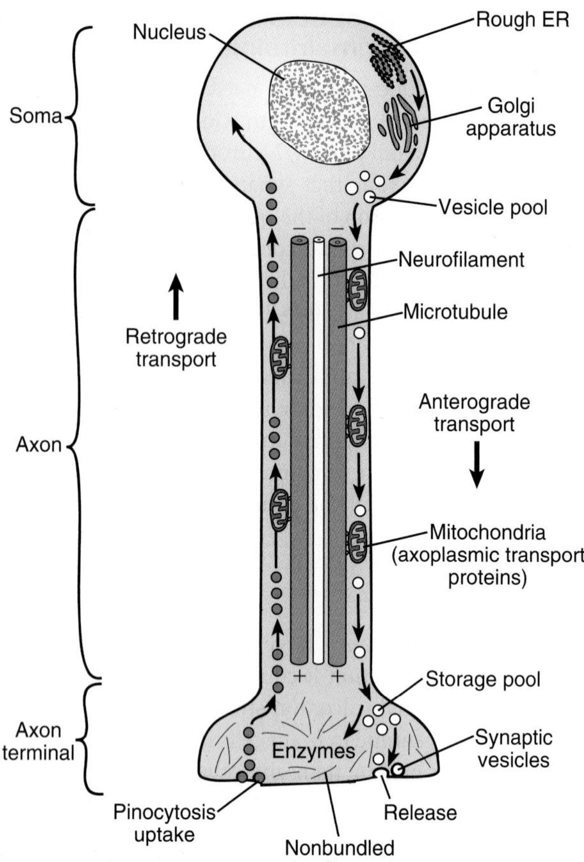

Figure 3.2 Key components of the neuron important for maintenance of the axon. Microtubules are the framework for fast axoplasmic transport. Kinesin is the molecular motor for anterograde (outward) transport, whereas dynein is the motor for retrograde (inward) transport. Mitochondria and vesicles move along the axon by this process. Microfilaments provide the framework of the nerve terminal. ER, endoplasmic reticulum.

and organelles in particular regions of the neuronal membrane, and neurotransmitter release. **Neurofilaments** are the intermediate type filaments of neurons. They are found in axons and are thought to provide structural rigidity. Neurofilaments are about 10 nm in diameter and form rope-like filaments that are several micrometers in length. The core of neurofilaments consists of a 70-kDa protein subunit, similar to intermediate filaments in other cells. Neurofilaments are unique from the intermediate filaments found in other cell types in that they have subunits that extend side arms. These side arms keep the neurofilaments from being as densely packed as intermediate filaments in other cells and aid in the maintenance of the caliber of axons.

Microtubules provide framework for the transport of structural material in axons and dendrites. They are composed of a core polymer of 50-kDa tubulin subunits that align to form protofilaments. Thirteen of these protofilaments assemble to form a tubular structure whose diameter is about 25 nm. Microtubules elongate by addition of additional tubulin at the "plus" end relative to the initial subunit, which is termed the "minus" end. In axons, microtubules are oriented with their plus ends outward and can be quite long (>100 μm). In dendrites, microtubules have mixed orientations with some having the plus end extending outward from the soma and others with the minus end oriented outward.

Neuronal transport mechanisms

Proteins, organelles, and other cellular materials must be transported throughout the neuron for the maintenance of structural integrity and cellular function. The neuron has transport mechanisms for moving cellular components in an **anterograde** direction, away from the soma, or in a **retrograde** direction, toward the soma. The **microtubule-associated protein (MAP) kinesin** is the anterograde molecular motor that moves organelles and vesicles from the minus to the plus ends of the microtubules via the hydrolysis of adenosine triphosphate. The MAP **dynein** is the retrograde motor.

Anterograde transport

Anterograde transport in the axon occurs at both slow and fast rates. The rate of **slow axoplasmic transport** is 1 to 2 mm/d. Structural proteins, such as actin, neurofilaments, and microtubules, are transported at this speed. Slow axonal transport is rate limiting for the regeneration of axons following injury to the neuron. The rate of **fast axoplasmic transport** is about 400 mm/d. Fast transport mechanisms are used for organelles, vesicles, and membrane glycoproteins needed at the axon terminal. In dendrites, anterograde transport occurs at a rate of approximately 0.4 mm/d.

Retrograde transport

Retrograde transport is the process by which material is moved from terminal endings back to the cell body. This process also provides a mechanism for the cell body to sample the environment around its synaptic terminals. In some neurons, maintenance of synaptic connections depends on the **transneuronal transport** of trophic substances such as nerve growth factor, across the synapse from target cells. After retrograde transport to the soma, nerve growth factor activates mechanisms for protein synthesis. Some neurodegenerative disorders may reflect loss of trophic substance delivery as a result of a defect in axonal transport machinery. Some of the toxic effects of cancer chemotherapy occur secondary to impairment of axoplasmic transport.

Electrical potentials across the cell membrane are the basis of signaling by neurons.

Nerve cells have an electric potential across the cellular membrane. This is referred to as the **membrane potential.** It can be recorded by inserting a microelectrode into the nerve cell. The voltage measured inside the cell is compared with the value detected by a reference electrode located in the extracellular fluid. The value at the latter location is considered to be zero. The difference in voltage between the two measurements is the membrane potential (Fig. 3.3). For a neuron in the unstimulated state this value is approximately −60 to −90 mV and is termed the **resting membrane potential.**

The resting membrane potential reflects an equilibrium state between the ability of ions to flow across the cell

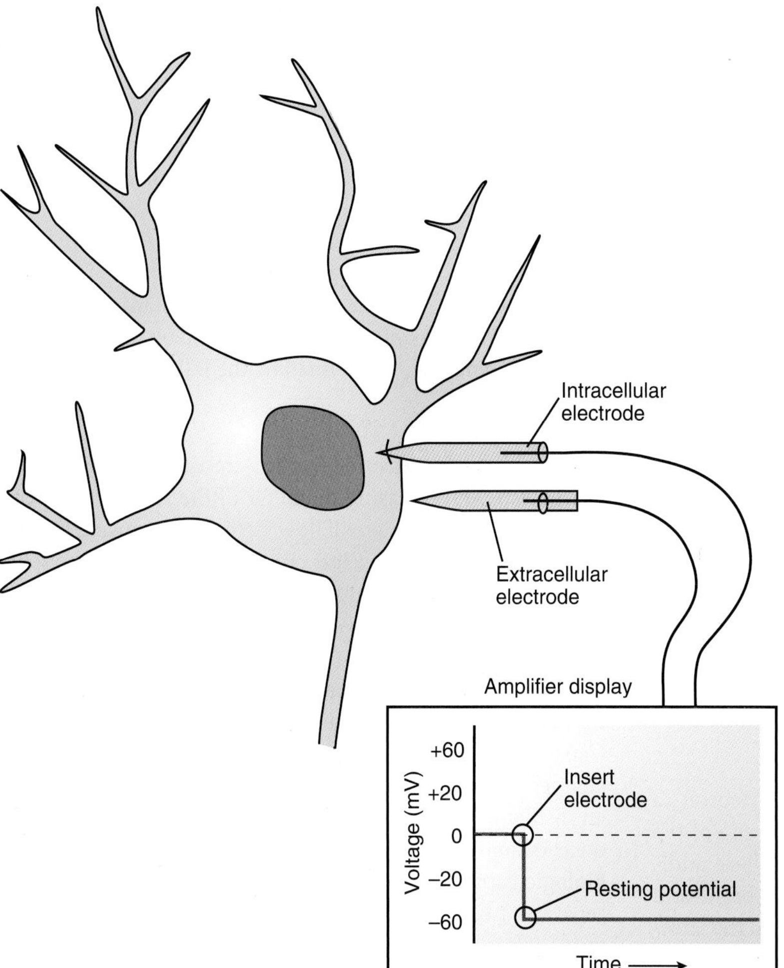

Figure 3.3 Recording the resting potential of a neuron. When both microelectrodes are in the extracelluar fluid, no potential difference is present between them. When one electrode is inserted through cell membrane, an inside negative potential of about 60 mV is recorded relative to the extracellular electrode.

membrane along their chemical gradients and the electrical potential that results from those ion movements. In excitable cells such as nerve or muscle, the intracellular concentration of K^+ is much greater than the extracellular, whereas the opposite is true for Na^+. This concentration difference is largely due to the action of the Na^+/K^+ pump (see Chapter 2, "Plasma Membrane, Membrane Transport, and Resting Membrane Potential"). The permeability of the membrane to potassium (P_K) is much greater than sodium (P_{Na}) in the resting state because of the presence of ion channels that allow K^+ to flow across the membrane along its concentration gradient. This type of channel is termed the *potassium inward rectifier*, or *Kir* (Fig. 3.4). The K^+ that flows out of the cell leaves some negative ions in the form of large nondiffusible

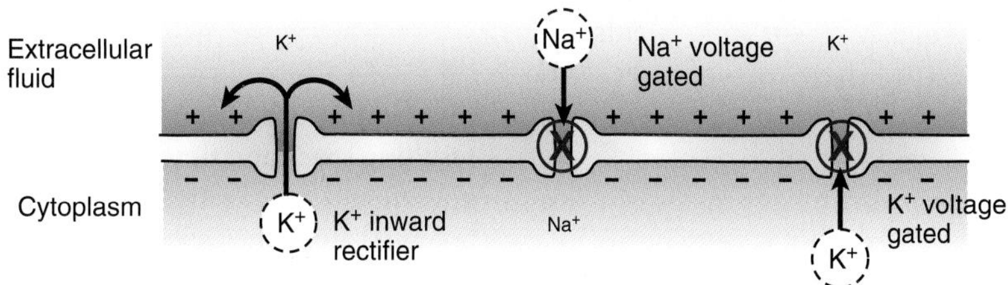

Figure 3.4 Origin of the resting membrane potential. Potassium is high in concentration in the cytoplasm (large K^+). The inward rectifier type of potassium channel allows potassium to diffuse out of the cell along its concentration gradient, but an inside negative charge develops across the cell membrane as potassium leaves. An equilibrium develops between potassium's tendency for outward diffusion and the resulting inside negative charge, which restrains diffusion. The electrical potential across the cell membrane at the equilibrium state is the resting membrane potential. Sodium is high in concentration outside the cell (large Na^+), but the inward rectifier channel does not allow Na^+ to enter. In the resting state, the voltage-gated Na^+ and K^+ channels are closed.

proteins behind inside the cell, and, hence, the inside of the cell has a negative voltage value. The positive and negative charges resulting from ion movement are located on the outer and inner aspects of the cell membrane. Only a miniscule amount of the total amount of potassium actually leaves the cell in the establishment of the membrane potential. The value of the membrane potential can be closely approximated by the **Nernst** and **Goldman equations** (see Chapter 2, "Plasma Membrane, Membrane Transport, and Resting Membrane Potential"). Typical values for equilibrium potentials in neurons are −90 mV for potassium (E_K) and +50 mV for sodium (E_{Na}). Because of the membrane's permeability to K^+, the resting membrane potential is closer to the equilibrium potential for potassium (E_K).

Because sodium is far from its equilibrium potential when the cell is in the resting state, both chemical and electrical forces act on it to enter the negatively charged interior of the cell. As the cell membrane is only minimally permeable to Na^+ in the resting state, the ability of Na^+ to flow across the membrane depends on opening Na^+ permeable channels in the membrane.

Electrical properties of the neuronal membrane affect flow.

When an electrical stimulus is applied to the neuron, a change in the membrane potential occurs at that site of the stimulus due to extra ions accumulating there. These additional ions redistribute themselves along the membrane and produce change in the membrane potential around the stimulus site. These local changes in membrane potential have both length and time components. The electrical properties of the neuronal membrane govern how perturbations of the resting membrane potential spread throughout the neuron.

The ease with which ions flow across the membrane is a measure of the membrane's **conductance**: the greater the conductance, the greater the flow of ions. Conductance is the inverse of resistance and is measured in siemens. For an individual ion channel and a given ionic solution, the conductance is a constant value determined, in part, by factors such as the relative size of the ion with respect to that of the channel and the charge distribution within the channel. The **Ohm law** describes the relationship between a single channel conductance, ionic current, and the membrane potential:

$$I_{ion} = g_{ion}\,(E_m - E_{ion}) \qquad (1)$$

or

$$g_{ion} = I_{ion}/(E_m - E_{ion}) \qquad (2)$$

where I_{ion} is the ion current flow, E_m is the membrane potential, E_{ion} is the equilibrium (Nernst) potential for a specified ion, and g_{ion} is the channel conductance for an ion. Notice that if $E_m = E_{ion}$, there is no net movement of the ion and $I_{ion} = 0$. The conductance for a nerve membrane is the summation of all of its single channel conductances.

A perturbation in membrane potential spreads along the membrane but diminishes in magnitude away from the point of origination. The voltage change at any point along the membrane is a function of current and resistance as defined by the Ohm law. The reason for the decline in voltage change with distance is that some of the ion current leaks back out across the cell membrane, thereby leaving less to perturb more distant sites. This type of signal propagation is termed **electrotonic conduction** (also known as *passive conduction*). The distance along the membrane at which 63% of the total change in the membrane potential that will occur due to the stimulus is defined as the **space constant** (also called *length constant*) or **λ**. The value of the space constant depends on the internal axoplasmic resistance (R_a) and on the transmembrane resistance (R_m) as defined by the following equation:

$$\lambda = \sqrt{R_m/R_a} \qquad (3)$$

R_m is usually measured in ohm-cm and R_a in ohm/cm. R_a decreases with increasing diameter of the axon or dendrite; therefore, more current will flow farther along inside the cell. Similarly, if R_m increases, less current leaks out and the space constant is larger. The larger the space constant, the farther along the membrane a voltage change is observed after a local stimulus is applied to the cell.

Another electrical property of the nerve membrane that influences spread of a stimulus along the cell is **capacitance**. Capacitance is the membrane's ability to store electrical charge. A capacitor consists of two conductors separated by an insulator. Positive charge accumulates on one of the conductors, and negative charge accumulates on the other. The biologic capacitor is the lipid bilayer cell membrane, which separates two conductors, the extracellular and intracellular fluids. Membrane capacitance is measured in units of **farads (F)**. In the resting state, positive charges have accumulated on the extracellular side, whereas negative charges are present on the intracellular side. When an electrical stimulus is applied to a neuron, the amount of charge stored across the membrane will be altered. The speed with which the change occurs depends on the resistance of the circuit. Ions move more rapidly when resistance is low. The time course of change in membrane potential following an applied stimulus is quantified as the **time constant** or **τ**. The time constant is the time required for the membrane potential to change by 63% of the total amount that will occur under the influence of the stimulus. Its relationship to capacitance (C) and resistance (R) is defined by the following equation:

$$\tau = RC \qquad (4)$$

Membrane resistance and capacitance acting through the resulting space and time constants play key roles in the integration and propagation of perturbations of the membrane potential along the cell membrane.

ACTION POTENTIALS

The action potential is a transient depolarization and repolarization of the membrane potential (Fig. 3.5). The key property of the action potential is that it can regenerate itself as it propagates along the cell membrane. In this manner, a signal generated in a neuron soma can travel the entire distance of neuron through the axon to the axon terminus.

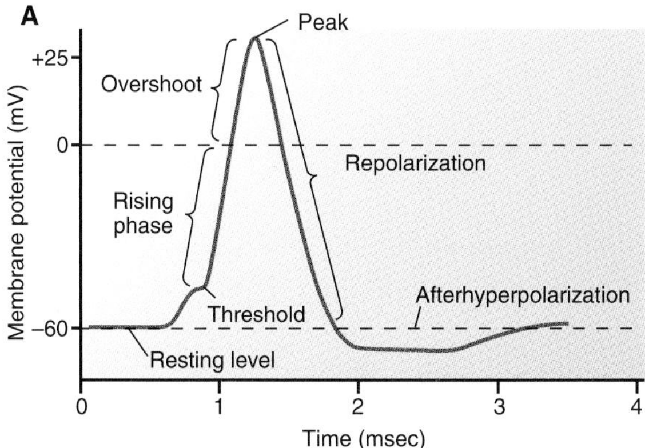

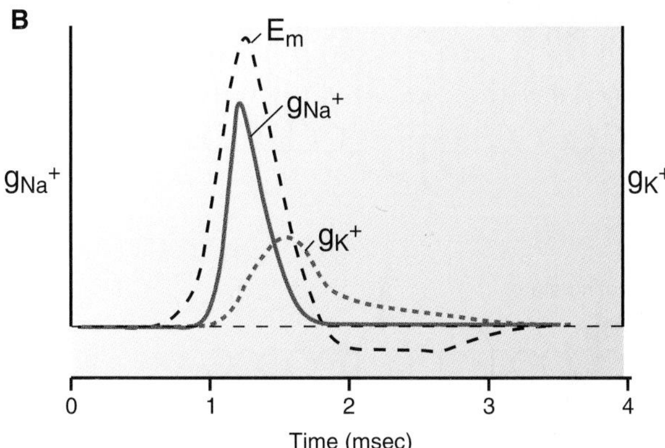

Figure 3.5 Phases of the action potential and the associated changes in sodium and potassium conductances. (A) The components of a nerve action potential. **(B)** Changes in sodium (g_{Na^+}) and potassium (g_{K^+}) conductance during an action potential are shown. Sodium conductance greatly increases during the rising and peak phases and then decreases back to baseline during the repolarization phase. Potassium conductance increases in a delayed fashion after the sodium increase and then continues after sodium conductance has returned to its resting level.

Generation of action potentials allows signal propagation over long distances along the axon.

The passive or electrotonic conduction mechanisms described above propagate changes in membrane potential only over short segments of the cell membrane. Nerve axons are almost always longer than their length constants. Therefore, electrotonic conduction is not useful for transmitting a signal over the distances commonly used in nerve transmission. For changes in the membrane potential to spread over longer distances, such as the entire length of an axon, an additional process is required.

Generating an action potential in a neuron requires the presence of voltage-gated sodium channels. When activated (opened), these channels allow for a transient depolarization of the cell membrane and the creation of a local current away from the site of the depolarization. These types of sodium channels are present only in the axon. Thus, although changes in membrane potential can occur in neuronal dendrites and cell bodies, these changes are only conducted electrotonically (passively); these structures cannot conduct action potentials. In most types of neurons, the region of the axon nearest to the soma is the site of action potential initiation. This region is referred to as the **axon hillock**, also called the *initial segment* or the *trigger zone* (see Fig. 3.1). This portion of the axon has a very high density of voltage-gated sodium channels.

Action potentials are created by time-dependent opening and closing of voltage-gated ion channels.

The action potential's phases occur because of changes in membrane conductance to sodium and potassium that result from activation of voltage-gated Na^+ and K^+ channels. These types of channels contain locations within them that serve as "gates" allowing or blocking the passage of ions through the channel. Sodium channels contain **activation** and **inactivation gates** that are moved by membrane depolarization but with slightly different opening and closing times. The interplay between these two gates allows the channel to assume either an open or closed state.

When a neuron is at rest, only the K^+ channels are open, thus establishing the resting potential. There are numerous different types of K^+ gates in neurons and other excitable tissues, but the simplest form of this gated channel opens when the membrane is depolarized, and closes when it is repolarized toward the cell's resting membrane potential.

Action potentials do not occur in nerves until a certain level, or **threshold**, membrane depolarization is reached. A nerve can be brought to threshold depolarization by the passive (electrotonic) conduction mechanisms described above. When the membrane is depolarized to the threshold for channel activation, the Na^+ channels open the activation gate, thus allowing the Na^+ ions to flow into the cell along the electrical and chemical gradient for sodium (Figs. 3.5 and 3.6).

The influx of Na^+ ions during the rising phase causes the membrane to further depolarize into the positive range. This is the **overshoot** phase. The quantity of ions moving across the membrane during an action potential is very small; $<0.01\%$ of the sodium pool is exchanged. Depolarization of the cell membrane both opens the Na^+ channel activation gate and closes the channel inactivation gate. However, the latter occurs more slowly than the time required to open the activation gate. Thus, during the rising phase of the action potential, the activation gates are open, whereas the inactivation gates have yet to close, and the sodium conductance across the neuronal membrane is high.

During the *peak phase* of the action potential, however, the sodium conductance begins to fall as the inactivation gates of the channels close. The voltage-gated K^+ channels open more slowly in response to depolarization and do not fully activate until after the peak of the action potential is reached. These channels allow positively charged K^+ ions to leave the neuron because the inside of the cell has become less negative. These types of K^+ channels are termed *delayed rectifiers* because they tend to counteract the depolarization of the cell membrane.

The net effect of opening additional K^+ channels while inactivating Na^+ channels is to repolarize the cell membrane back toward its resting value. This phase, then, is called the **repolarization** phase.

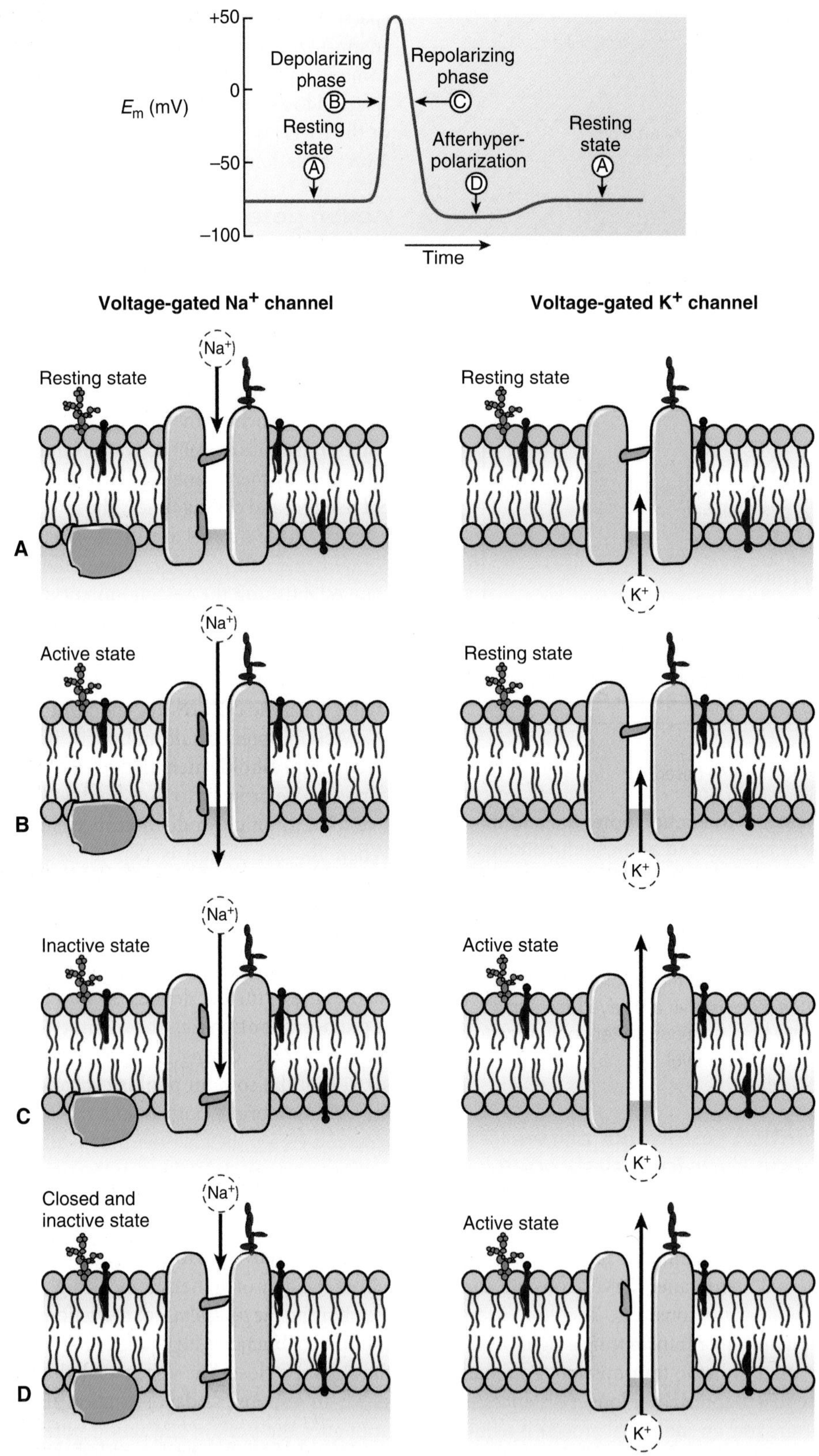

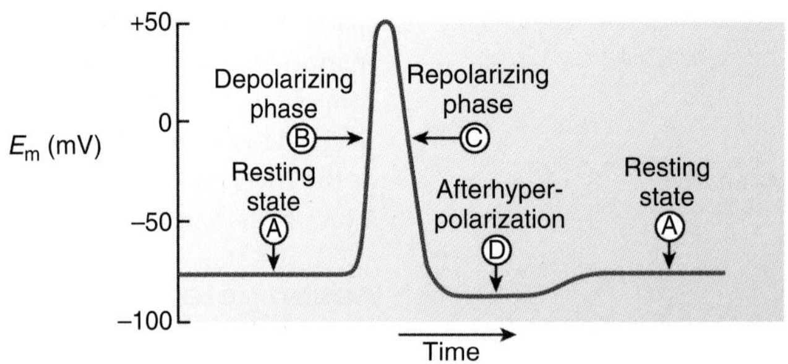

Figure 3.6 Voltage-gated channel activity during the action potential. (A) In the resting state, both sodium- and potassium-gated channels are closed. **(B)** In the depolarizing phase, sodium channels are activated, whereas potassium wchannels remain mostly closed. **(C)** In the repolarizing phase, sodium channels close by inactivation, and potassium channels are activated. **(D)** In the afterhyperpolarization phase, potassium channels remain activated. As this phase ends, sodium channels reset the activation and inactivation gates to their resting positions and are available to mediate the next cycle of opening.

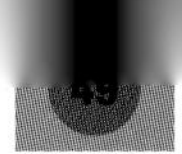

As the membrane continues to repolarize, the membrane potential may become more negative than it was at its resting level. This **afterhyperpolarization** phase is a result of voltage-gated K^+ channels remaining open, allowing further efflux of K^+ ions down their concentration gradient. In this phase of the action potential, the membrane's conductance to K^+ is higher than when the neuron is at rest due to conductance through both the inward rectifier channels, which set the resting membrane potential, and the delayed rectifier channels. Consequently, the membrane potential is driven even more toward the K^+ equilibrium potential.

Inactivation of voltage-gated sodium channels makes the axon temporarily refractory to action potential generation.

During action potential generation, there is a time when all sodium channel activation gates are in the open position; further stimulation of the neuron, therefore, cannot create any more signal. Also, once cell depolarization closes all inactivation gates in the sodium channels, it is impossible for any additional cell depolarizing stimuli to activate sodium channels; the inactivation gate must be reset to its resting (open) position. This resetting requires a finite amount of time, and, until accomplished, another action potential cannot be generated. The timeframe in which sodium channels have not totally reset to their resting state is the **absolute refractory period**. The absolute refractory period limits the rate of firing of action potentials. It also prevents action potentials from traveling backward along the axon, because the region of the axon that has just produced the action potential is "refractory."

After the absolute refractory period, the neuron enters a phase in which it will generate another action potential if a stronger-than-normal stimulus is applied to the neuron. This is called the **relative refractory period**. In this period, some portion of the voltage-gated Na^+ channels have reset and are available for opening. However, these channels have higher depolarization thresholds and are open, whereas K^+ conductance is still higher than it is in the resting state. Therefore, a stronger depolarizing stimulus is required to open the available channels.

Ion channel mutations in disease states

Abnormalities in the proteins that make up the voltage-gated sodium and potassium channels (as well as in voltage-gated calcium and chloride channels) are now known to be the basis of several diseases of nerve and muscle known as **channelopathies** (see Clinical Focus 3.1.) The structure of some of these channels is shown in Figure 3.7.

Clinical Focus / 3.1

Channelopathies

The ability for ions to flow across nerve and muscle cell membranes in a controlled manner is fundamental to normal function of the body. Voltage-gated channels for sodium, potassium, calcium, and chloride underlie the excitability of neurons and muscle cells. Ligand-gated channels in postsynaptic receptors are critical for synaptic transmission. Genetic alterations in the structural components of the channels or the development of antibodies that cross-react with a component of the channel can alter normal functioning. Diseases based on altered ion channel function are called **channelopathies**. Disorders of this nature can affect neurons and skeletal and cardiac muscles.

One type of channelopathy that affects cardiac muscle is the long QT (LQT) syndrome. The QT interval is the time interval on the electrocardiogram between the beginning of ventricular depolarization and the end of ventricular repolarization. In patients with this disorder, the QT interval is abnormally long because of defective membrane repolarization. The defect is caused by either a prolonged inward sodium current or a reduced outward potassium current. Mutations in potassium channels account for two types of LQT syndromes, and a third derives from a sodium channel mutation. These abnormalities can lead to abnormal cardiac rhythm and sudden death.

Another type of channelopathy causes myotonia, a condition characterized by delayed relaxation of skeletal muscle. Difficulty releasing grip during a handshake and quickly opening the eyelids after a sneeze are common manifestations. Repetitive action potential generation in the muscle cell membrane is the underlying abnormality. Several types of myotonia exist. Some are due to genetic abnormalities in the skeletal muscle chloride channel, and others due to abnormal function of the sodium channel.

Abnormalities in voltage-gated calcium channels cause an inherited disorder of coordination termed *spinocerebellar ataxia*. Ataxias are a disruption in walking or finger aiming caused by abnormal function of the neurons in the cerebellum or spinal sensory neurons. Another calcium channel abnormality, *familial hemiplegic migraine*, causes attacks of migraine headache accompanied by temporary paralysis of one side of the body.

Channelopathies can also be caused by abnormal reactivity of the body's immune system. The resulting production of antibodies, termed *autoimmunity*, can cause abnormality when those antibodies cross-react with normal structures. One example of this type of disorder is **Lambert–Eaton myasthenic syndrome (LEMS)**. Fatigable weakness of skeletal muscle is the main manifestation of LEMS. The antibodies impair opening of voltage-gated calcium channels in the axon terminals of motor neurons and reduce the release of acetylcholine during action potentials. Synaptic transmission to muscle cells is impaired. Synaptic transmission in the autonomic nervous system impairs the function of salivary glands and the bladder.

The production of antibodies that cross-react with voltage-gated K^+ channels on neurons in the cerebral areas associated with memory impairs the ability to form new short-term memories. Patients with this condition termed *limbic encephalitis* may have no recollection of what was said to them a few minutes earlier.

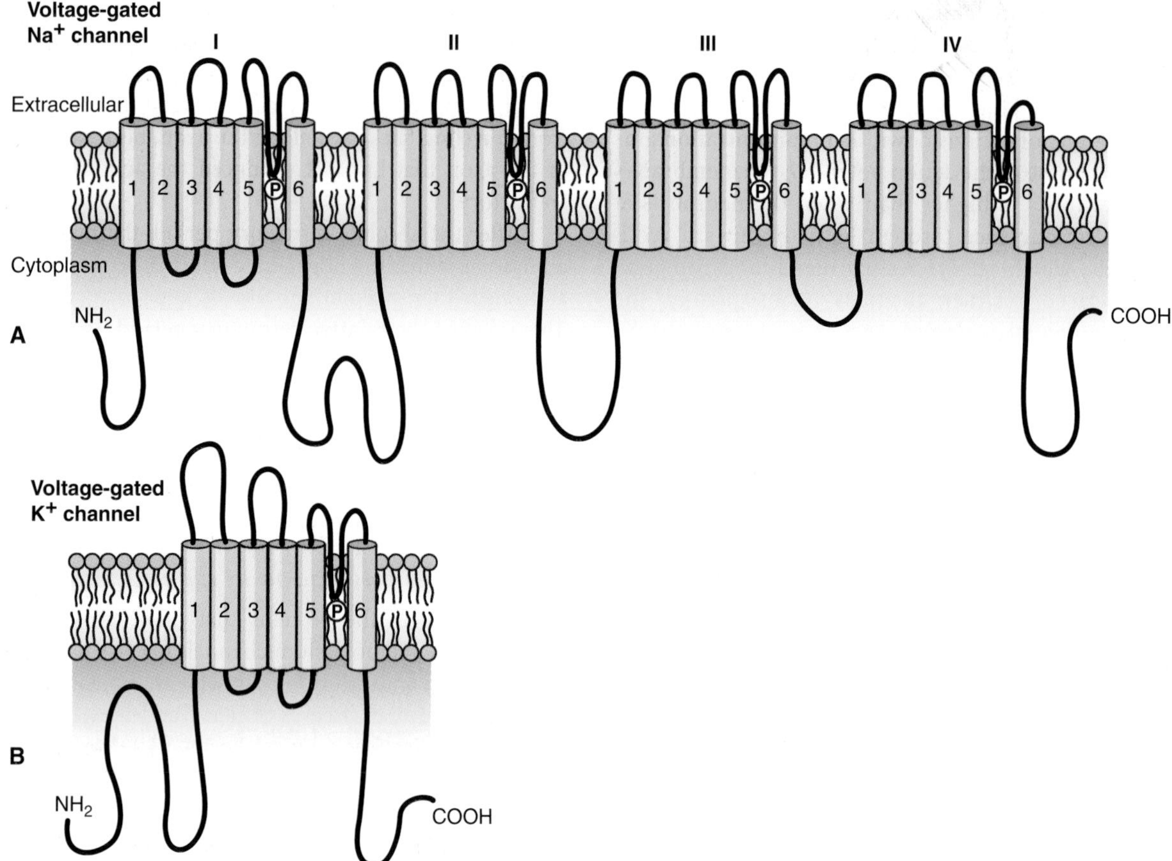

Figure 3.7 Structure of the voltage-gated sodium and potassium channels. The basic building block is a six membrane–spanning region of the polypeptide chain. The sodium channel **(A)** consists of four of these regions, which form into a circular pore to create the ion channel. The potassium channel **(B)** is formed by assembly of four separate segments to form the ion channel. The fourth transmembrane component of each segment is thought to be the sensor of change in the membrane potential as the action potential arrives at the channel.

Speed of propagation of the action potential depends on axon diameter and myelination.

After an action potential is generated, the speed at which it propagates along the axon depends on the diameter of the axon and whether the axon is myelinated. Larger-diameter axons have faster action potential conduction velocities than do smaller-diameter axons. In unmyelinated axons, voltage-gated Na^+ and K^+ channels are distributed uniformly along the length of the axonal membrane. The entry of Na^+ ions into the axon hillock causes the adjacent region of the axon to depolarize as the ions that entered the cell during the action potential flow away from the entry point. This local, passive spread of current depolarizes the adjacent region of the axon to the threshold for opening of voltage-gated Na^+ channels, resulting in the regeneration of the action potential in that region. By sequentially depolarizing adjacent segments of the axon, the action potential moves along the length of the axon from point to point, like a wave (Fig. 3.8A).

Large-diameter axons have less cytoplasmic resistance, just as a large-diameter hose allows a greater flow of water than does a smaller-diameter one, because of decreased internal resistance. By this same principle, large-diameter axons permit easier flow of ions. The improved ion flow in the cytoplasm allows a greater length of the axon to be depolarized by current entering through open Na^+ channels. Recall that the space constant, λ, defines the length along the axon that a voltage change is observed after a local stimulus is applied. In this case, the local stimulus is the inward sodium current that accompanies the action potential. The larger the space constant, the farther along the membrane a change in voltage spreads after a local stimulus is applied. The result is that the speed at which action potentials are conducted, the **conduction velocity**, increases as a function of increasing axon diameter.

Conduction velocity of action potentials can also be increased by myelination of the axon. Schwann cells in the PNS and oligodendrocytes in the central nervous system (CNS) encircle axons with multiple layers of cell membrane to form the myelin sheath. Myelin attenuates the passage of ions through the axonal membrane beneath the myelin (i.e., it reduces the leak of current out through the membrane) as seen in Figure 3.8B. The length of a myelinated segment, termed the *internode*, is determined by the longitudinal span of a Schwann cell or oligodendrocyte. At the end of a myelinated segment, the axon is free of myelin for a short distance. These myelin-free intervals are the nodes of Ranvier.

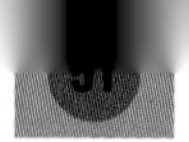

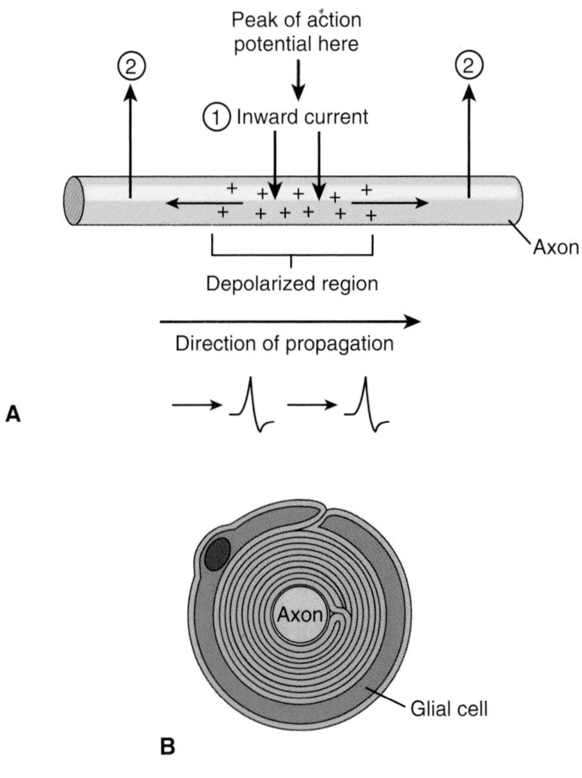

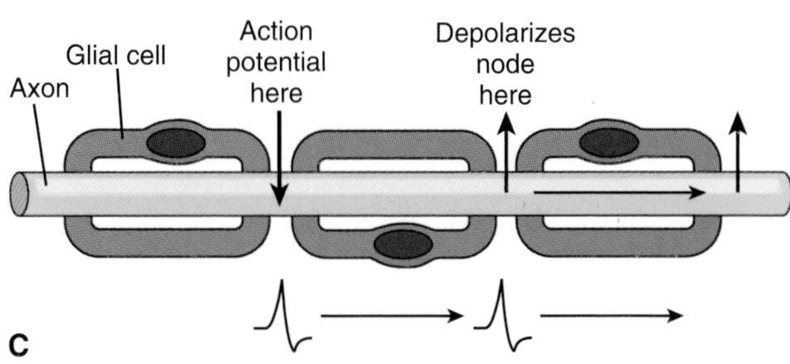

Figure 3.8 Propagation of an action potential along an axon. (A) An action potential travels along the axon as a wave of depolarization. In the region of activated voltage-gated sodium channels, current in the form of sodium ions flow into the axon depolarizing it. The depolarization spreads along the axon both ahead of and behind the region of active sodium channels. Depolarization ahead of the action potential can activate voltage-gated channels in that region and reproduce an action potential there. This process continues along the axon. In an unmyelinated axon, the sodium channels of each continuous segment must be serially activated for the action potential to propagate. **(B)** Myelinated axons are wrapped by overlapping layers of the Schwann or oligodendroglial cell membrane forming the myelin sheath. Myelin limits ion flow through the underlying axon membrane. **(C)** Myelinated axons locate sodium channels in the gaps (nodes of Ranvier) between adjacent myelinated (internodal) segments. Depolarization at one node of Ranvier can spread over the entire internodal segment of the axon due to myelin's ability to limit the decrement of depolarization in that segment. The action potential occurs only at the nodes of Ranvier rather than having to be generated in each adjacent segment of axon, as is required in an unmyelinated axon. Myelination and the resulting longer distance between sodium channels allow the action potential to propagate at much faster velocity.

In general, axons larger than approximately 1 μm in diameter are myelinated, and the thickness of the myelin increases as a function of axon diameter. Myelination increases the resistance of the axonal membrane, R_m, because ions that flow across the axonal membrane must also flow through the layers of myelin before they reach the extracellular fluid. This increase in R_m increases the space constant. The layers of myelin also decrease the effective capacitance of the axonal membrane because the distance between the extracellular and intracellular conducting fluid compartments is increased. Because the capacitance is decreased, the time constant is decreased, thereby increasing the speed at which changes in membrane potential spread along the axon.

The effects of myelination on R_m and capacitance are beneficial for increasing conduction velocity, but, even more importantly, myelination alters the nature of action potential conduction. In myelinated axons, the Na^+ channels are highly concentrated in the nodes of Ranvier and are in low density in the internodal segment. When an action potential is initiated at one node of Ranvier, the influx of Na^+ ions there causes the adjacent node to then depolarize by passive mechanisms and generate an action potential there. Because action potentials are successively generated at neighboring nodes of Ranvier, the action potential in a myelinated axon appears to jump from one node to the next, a process called **saltatory conduction** (see Fig. 3.8C). This process results in much faster conduction velocities for myelinated than for unmyelinated axons. Conduction in myelinated neurons is like taking normal steps across a distance, whereas that in unmyelinated neurons is like traversing that same distance by walking successively "heel to toe." The conduction velocity in mammals ranges from 3 to 120 m/s for myelinated axons and 0.5 to 2.0 m/s for unmyelinated ones.

SYNAPTIC TRANSMISSION

Neurons communicate at synapses. Two types of synapses have been identified: electrical and chemical. At **electrical synapses**, passageways known as **gap junctions** connect the cytoplasm of adjacent neurons and permit the bidirectional passage of ions from one cell to another. Electrical synapses are uncommon in the adult mammalian nervous system. Typically, they are found at dendrodendritic sites of contact; they are thought to synchronize the activity of neuronal populations. Gap junctions are more common in the embryonic nervous system, where they may act to aid the development of appropriate synaptic connections based on synchronous firing of neuronal populations. Gap junctions are also important for cell-to-cell communication in smooth and cardiac muscle (see Chapter 8, "Skeletal and Smooth Muscle").

Synaptic transmission occurs via release of chemical neurotransmitters.

The structure of a **chemical synapse** includes the terminal portion of the axon, called the **axon terminal**, a **synaptic cleft** of about 50 nm width, and a group of receptors in the

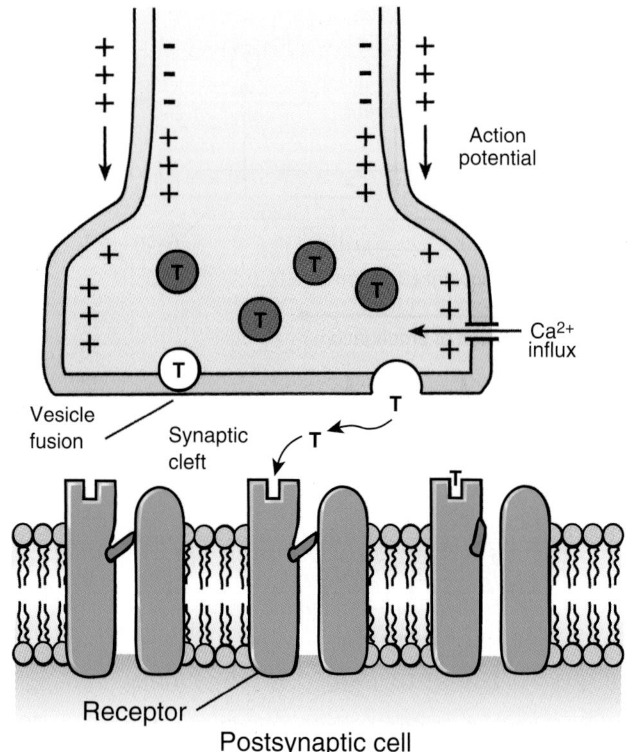

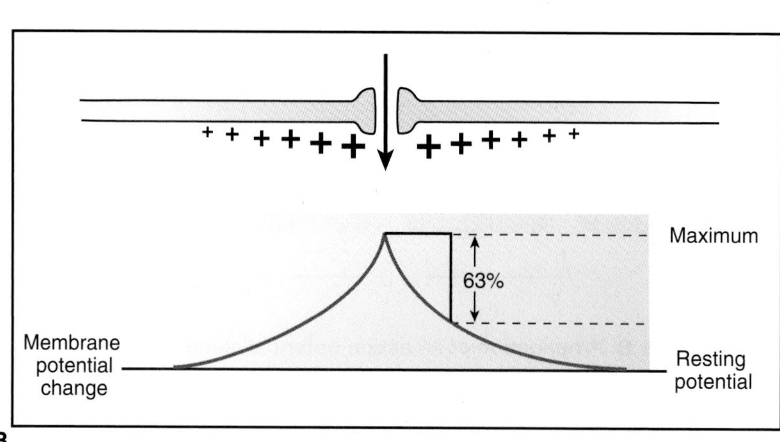

Figure 3.9 Electrical events at a synapse. (A) Depolarization of the nerve terminal by arrival of the action potential opens voltage-gated calcium channels. Influx of Ca^{2+} initiates fusion of synaptic vesicles with the cell membrane and then release of neurotransmitter (T) into the synaptic cleft. The transmitter diffuses across the cleft and binds to receptors. In this example, transmitter binding to the receptor opens ion channels in the postsynaptic membrane. **(B)** Ions entering through open channels at the synapse produce the greatest change in membrane potential at the synapse. The change in membrane potential spreads along the cell membrane by passive (electrotonic) means and decreases in amount further along the membrane. The value of the space constant (λ_m) is defined as the site along the membrane where the potential has changed by 63% of the total that will occur. A cell membrane with a longer space constant allows spread of the change in potential further along the membrane.

postsynaptic cell membrane as shown in Figure 3.9. The axon terminal contains vesicles filled with chemical neurotransmitter molecules. These molecules are released into the synaptic cleft when an action potential enters the axon terminal.

Arrival of an action potential to the nerve terminal depolarizes it and opens voltage-gated Ca^{2+} channels located in the cellular membrane of the axon terminal. These channels allow Ca^{2+} to enter along its electrochemical gradient. The resulting increase in intracellular calcium causes mobilization of the neurotransmitter containing vesicles and fusion of those with the presynaptic membrane. The number of transmitter molecules contained by a vesicle is called a **quantum**, and the total number of quanta released when the synapse is activated is called the **quantum content**. Release of neurotransmitter following entry of Ca^{2+} into the terminals occurs via the process of **regulated exocytosis** and involves a series of protein interactions. Calcium activates a kinase that phosphorylates the cytoskeletal protein, synapsin. Synapsin then releases vesicles from attachment to the cytoskeleton in the axon terminal. The protein Rab3 escorts the mobilized vesicles to sites on the cellular membrane. Two proteins in the vesicle wall, **synaptotagmin** and **synaptobrevin**, interact with complementary proteins snap-25 and syntaxin in the membrane of the axon terminal to locate the vesicles at the proper sites for transmitter release. These locating proteins are called **SNARES**. Those on the vesicles are called *v-SNARES*, whereas those at the target locations on the axon terminal membrane are called *t-SNARES*. The process of locating and then attaching the vesicles to the membrane is termed *docking*.

To complete the process begun by Ca^{2+} entry into the nerve terminal, the docked and bound vesicles must fuse with the membrane and create a pore through which the transmitter is released into the synaptic cleft. After releasing transmitter, the vesicles either fuse with the cellular membrane of the axon terminal for later retrieval out of the membrane into new synaptic vesicles or detach from the membrane back into the vesicle pool. The latter process is termed "kiss and go."

Once released into the synaptic cleft, neurotransmitter molecules bind to one of two types of receptors in the postsynaptic membrane. In one, the macromolecule that contains the receptor also forms an ion channel, which often shows some ion type selectivity. This type of receptor is termed **ionotropic** or *directly ligand gated.* Interaction between the transmitter and this type of receptor leads to opening or closing of specific ion channels, which, in turn, alters the membrane potential or other functions of the postsynaptic cell. For example, when transmitter binds to the receptor of an ion channel, changes in membrane conductance occur. The result is depolarization or hyperpolarization of the postsynaptic cell, depending on the channel type: An increase in Na^+ conductance causes depolarization; an increase in K^+ or Cl^- conductance causes hyperpolarization.

In the other type of postsynaptic receptor, the receptor activates a G protein and second messenger system. These types of receptors are termed **metabotropic** or *indirectly ligand gated.* Second messenger systems include **adenylyl cyclase**, which produces cyclic adenosine monophosphate (or, *cAMP*); **guanylyl cyclase**, which produces cyclic guanosine monophosphate (or, *cGMP*); and **phospholipase C**, which leads to the formation of diacylglycerol (DAG) and inositol trisphosphate (IP_3) (see Chapter 1, "Homeostasis and Cellular Signaling").

Postsynaptic potentials in the dendrites and soma are summated in time and space.

The transmission of information between neurons is mediated by membrane potential changes in the dendrites and soma of postsynaptic cell resulting from the synaptic transmission process. The changes are not action potentials but, rather, small changes in the polarization of the postsynaptic membrane (see Fig. 3.9B). Because membrane depolarization leads to the activation of the neuron, postsynaptic depolarization was termed an **excitatory postsynaptic potential (EPSP)** by the early investigators in neurophysiology. In contrast, membrane hyperpolarization prevents the cell from being activated and was termed an **inhibitory postsynaptic potential (IPSP)** as shown in Figure 3.10.

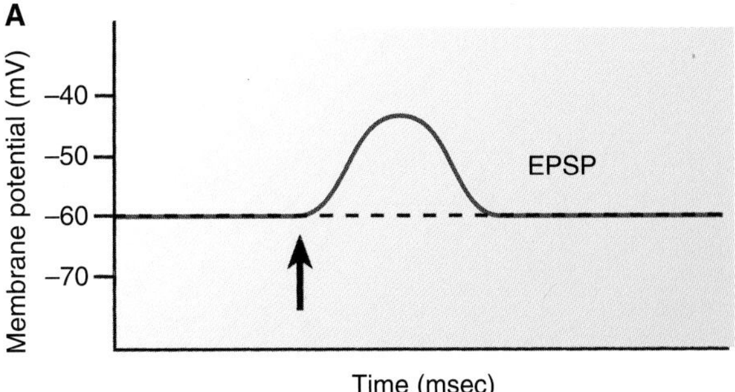

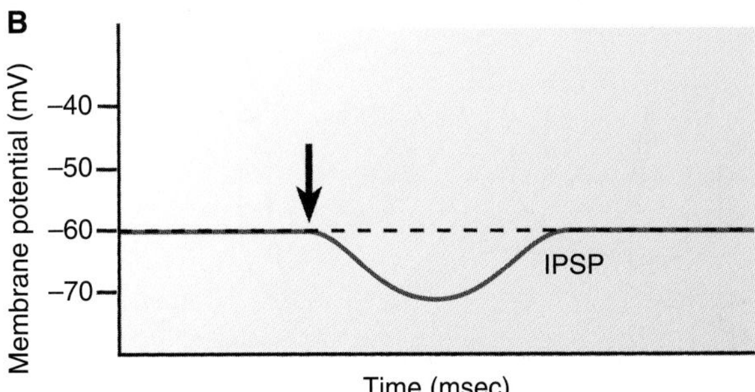

Figure 3.10 Excitatory and inhibitory postsynaptic potentials. (A) Inflow of positive ions at a synapse (*arrow*) makes the postsynaptic membrane potential less negative and increases the likelihood of the postsynaptic cell producing an action potential. This is an excitatory postsynaptic potential (EPSP). **(B)** Inflow of negative ions or outflow of positive ones makes the postsynaptic membrane potential more negative and decreases the likelihood of the cell producing an action potential. This is an inhibitory postsynaptic potential (IPSP).

Postsynaptic potentials are transient. When the ions channels are opened by synaptic transmission, the membrane potential of the postsynaptic cell in the region of the synapse changes from its resting value and then returns to it when the channels close. The rate at which it depolarizes or repolarizes depends on the membrane time constant, τ (Fig. 3.11). In segments of the membrane that have longer time constants, the potential changes more slowly and persists longer. Slower rates of change allow additional time for other postsynaptic potentials to summate. The process in which postsynaptic membrane potentials add together in the time domain is called **temporal summation** (Fig. 3.12). If the summated EPSPs, which propagate by passive, electrotonic conduction are sufficient to lower the membrane potential to the threshold for voltage-gated sodium channel activation at the axon hillock, an action potential will be generated in the postsynaptic neuron.

Summation of postsynaptic potentials can also occur when several synapses located at different sites of contact

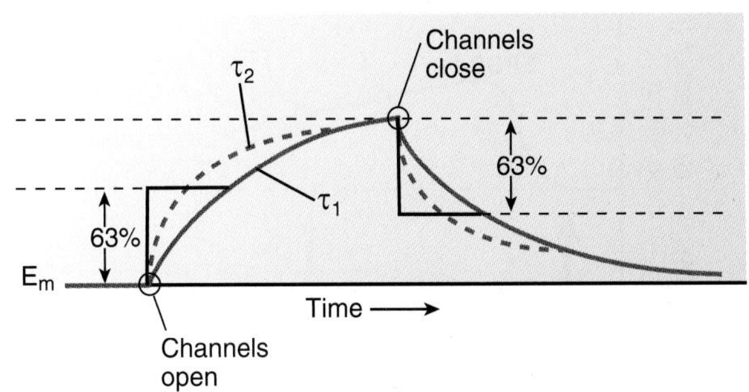

Figure 3.11 Membrane potential change and the time constant. Inflow of ions at a synapse alters the membrane potential. Because the cell membrane is a capacitor, the change in membrane potential follows an exponential time course. When inflow of ions stops, the membrane returns to the resting value in a similar fashion. The time constant is the time required for the membrane potential to change by 63% of the total that will occur. Membranes with a shorter time constant (*dotted line*) change potential quicker.

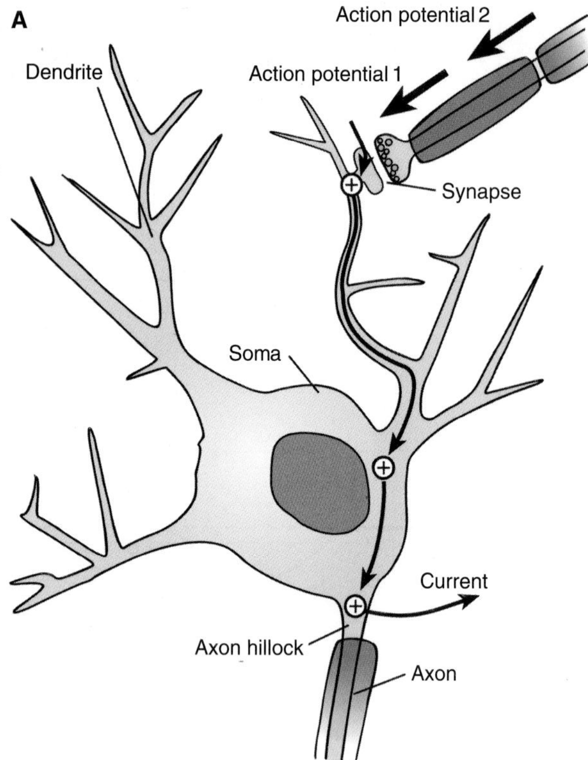

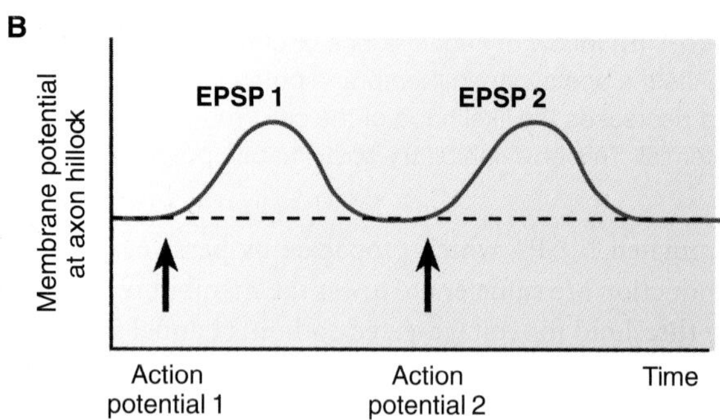

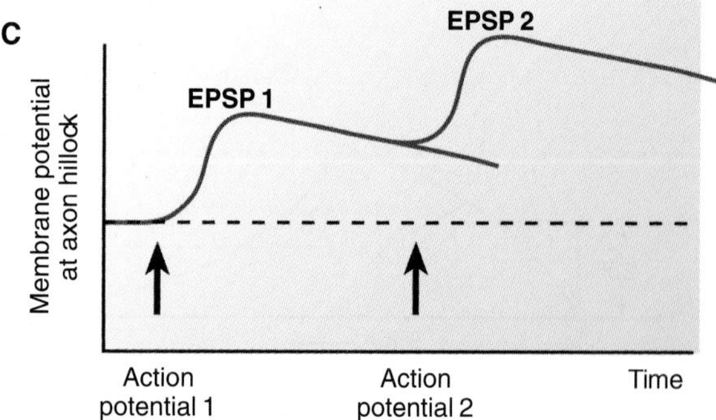

Figure 3.12 The process of temporal summation. **(A)** Excitatory postsynaptic potentials (EPSPs) produced at one synapse by two sequential action potentials are shown. **(B)** In a cell membrane with a shorter time constant, the first EPSP terminates before the second is produced. **(C)** When the membrane has a longer time constant, some of the depolarization from the first EPSP is still present when the second occurs, and the individual depolarizations can summate.

are activated. This process is called **spatial summation.** The postsynaptic potential is maximal at the synapse and then decreases with distance away from the synapse because of the process of passive conduction (Fig. 3.13). The rate of decrease is determined by the length or space constant, λ. The larger the value of the space constant, the greater amount of change in membrane potential is delivered to more distant segments of the cell membrane. The location of the synapse impacts the ability to activate a postsynaptic neuron. For example, axodendritic synapses, located in distal segments of the dendritic tree, are far removed from the axon hillock, and their activation alone has little individual impact on the membrane potential at the axon hillock. In contrast, axosomatic synapses have a greater effect in altering the membrane potential at the axon hillock because of their closer location. Potentials produced by activation of other synapses can summate along the length of the dendrite or soma. As with temporal summation, if the change in membrane potential from spatial summation is sufficient to alter the membrane potential in the region of axon hillock to the threshold for activation of voltage-gated sodium channels, the neuron will generate an action potential.

NEUROTRANSMISSION

Neurons communicate with other cells by the release of chemical neurotransmitters (see Fig. 3.9B). The human nervous system has some 100 billion neurons, each of which communicates with postsynaptic targets via chemical neurotransmission. The first neurotransmitters described were acetylcholine (ACh) and norepinephrine (NE). These were identified at synapses in the PNS. Many others transmitters have been identified since then, but, even counting all the peptides known to act as transmitters, the number is well less than 50. The specific neuronal signaling that allows the enormous complexity of function in the nervous system is largely a result of the specificity of neuronal connections made during development and the distribution of specific classes of neurotransmitter receptors (Table 3.1).

Differences among neurotransmitters contribute to the ability of the nervous system to perform complex functions.

Transmitters fall into three traditional classes: amino acids, monoamines, and polypeptides. Examples of *amino acid* transmitters include glutamate (GLU), aspartate, glycine (GLY), and γ-aminobutyric acid (GABA). The *monoamines* (or biogenic amines) are so named because they are synthesized from a single, readily available amino acid precursor. Examples of monoaminergic neurotransmitters are ACh, derived from choline; the catecholamine transmitters dopamine (DA), NE, and epinephrine (EPI), derived from the amino acid tyrosine; and an indoleamine, serotonin or 5-hydroxytryptamine (5-HT), derived from tryptophan. The *polypeptide transmitters* (or neuropeptides) consist of an amino acid chain, varying in length from three to several dozen. Examples of polypeptide transmitters are the opioids and substance P. More recently, an additional set of

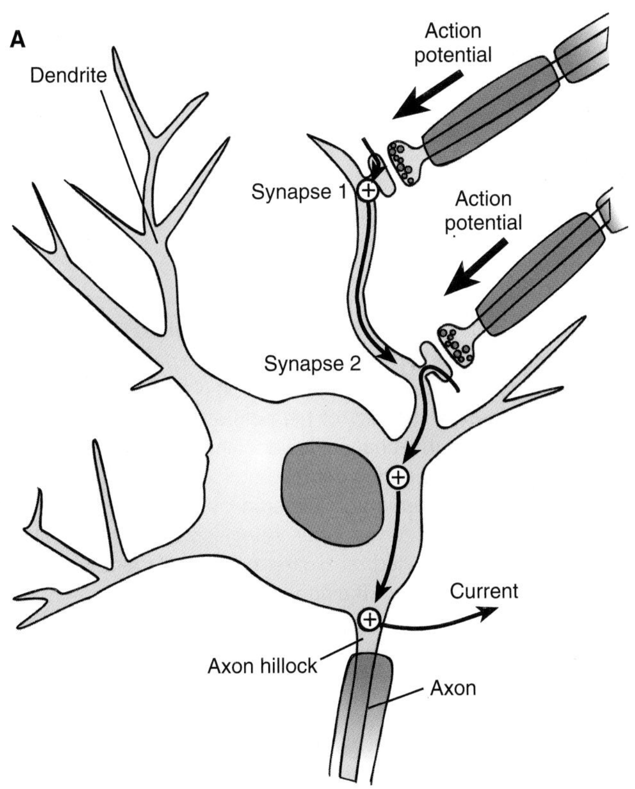

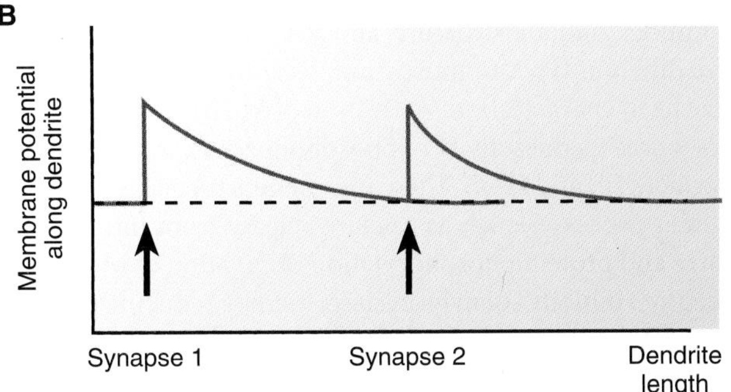

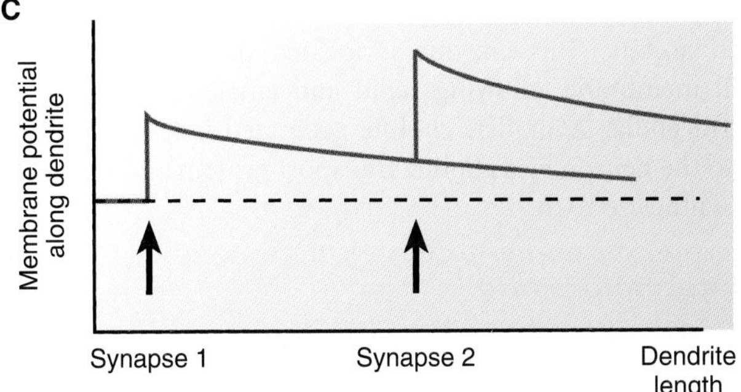

Figure 3.13 The process of spatial summation. **(A)** Action potentials in two neurons produce excitatory postsynaptic potentials (EPSPs) at their synapses, which propagate by passive conduction to the soma and axon hillock. **(B)** When the space constant of the cell membrane in the dendrite is shorter, the EPSP from synapse 1 decays back to the resting membrane potential value before propagating to the site of synapse 2. **(C)** When the cell membrane has a longer space constant, the EPSP from synapse 1 is able to propagate further along the membrane and can summate with the EPSP from synapse 2, resulting in more depolarization to propagate to the soma and axon hillock.

TABLE 3.1 General Functions of Neurotransmitters

Neurotransmitter	Function
Acetylcholine	Control of movement, cognition, autonomic control
Dopamine	Affect, reward, control of movement
γ-Aminobutyric acid	General inhibition
Glutamate	General excitation, sensation
Glycine	General inhibition
Nitric oxide	Vasodilation, metabolic signaling
Norepinephrine	Affect, alertness
Opioid peptides	Control of pain
Serotonin	Mood, arousal, modulation of pain, gut regulation
Substance P	Transmission of pain

membrane-soluble molecules that act as signaling molecules between neurons and their target has been identified. The best known of these are nitric oxide (NO) and arachidonic acid.

Acetylcholine

Neurons that use **acetylcholine (ACh)** as their neurotransmitter are known as **cholinergic** neurons. ACh is synthesized in the cholinergic neuron from choline and acetate, under the influence of the enzyme **choline acetyltransferase**. ACh is synthesized and stored in synaptic vesicles in the axon terminal region of the cholinergic neuron. The storage vesicles and choline acetyltransferase are produced in the soma and are transported to the axon terminals. The rate-limiting step in ACh synthesis in the nerve terminals is the availability of choline, of which specialized mechanisms ensure a continuous supply.

The receptors for ACh, known as *cholinergic receptors*, fall into two categories, based on the drugs that mimic or antagonize the actions of ACh on its target cell types. In classical studies dating to the early 20th century, the drugs muscarine, isolated from poisonous mushrooms, and nicotine, isolated from tobacco, were used to distinguish the two types of receptors. Muscarine stimulated some of the receptors, and nicotine stimulated all the others, hence the designation *muscarinic* or *nicotinic* (Fig. 3.14). The **nicotinic receptor** is composed of five subunits: two α and single β, γ, and δ subunits. The ion channel is in the center of the macromolecule. There are multiple forms of the subunits, giving rise to a family of nicotinic receptors. The two α subunits are binding sites for ACh. When ACh molecules bind to both α subunits, a conformational change occurs, thereby opening the ion channel. An increase in conductance for Na^+ and, to a much lesser extent, K^+, occurs. The strong inward electrical and chemical gradient for Na^+ predominates over the outward

From Bench to Bedside / 3.1

Red Fiery Feet

The investigation of a rare familial disorder that causes patients to experience recurring attacks of severe burning pain in the feet and hands accompanied by red discoloration of the involved body parts has enhanced our understanding of mechanisms of pain. The condition is named **erythromelalgia**. The attacks of pain are often provoked by warming of the limbs. Patients manage the pain attacks by placing their feet and hands in pans of ice water or by directing the air from vigorously blowing fans onto the painful areas. Conventional pain medications provide only modest at most relief from the pain.

Recent genetic studies in several families who showed a dominant pattern of inheritance for this disorder have linked the condition to a gene that codes for the voltage-gated sodium channel $Na^+_v1.7$. This type of channel is found in the distal ends of the sensory nerve fibers that transduce hot stimuli into nerve action potentials. The amino acid sequence of the channel protein is altered from the wild-type sequence in affected families. Electrophysiologic analysis of the channel's function demonstrates that mutant channels allow greater inflow of Na^+ ions during activation than do wild-type channels and can result in a longer train of pain sensation carrying action potentials from a stimulus than would occur in a normal neuron. Activation of the pain-sensing neurons can cause secondary activation of the sympathetic neurons controlling blood flow through cutaneous blood vessels in the region of the pain. Dilation of these vessels causes the redness of the painful areas.

Treatment with medications that inhibit Na^+ channel function can lessen the intensity of the pain attacks. One medication that affects this channel is lidocaine. This drug is the local anesthetic used during dental procedures or suturing of a laceration. Lidocaine is administered by injection, and its action lasts for less than an hour. Administration of an oral analog of lidocaine called *mexiletine* can greatly reduce the frequency and intensity of the attacks.

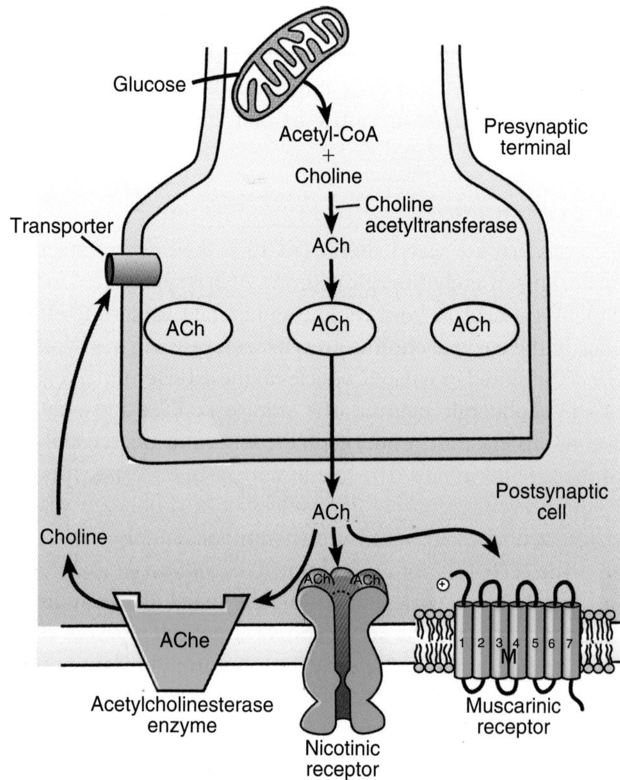

Figure 3.14 Cholinergic neurotransmission. Arrival of the nerve action potential to the presynaptic nerve terminal allows calcium influx and then the release of acetylcholine (ACh) into the synaptic cleft. ACh binds to nicotinic or muscarinic receptors in the postsynaptic membrane. ACh is hydrolyzed by acetylcholinesterase (AChE) to choline and acetate. Choline is taken up into the presynaptic nerve terminal by a transporter and is then resynthesized to ACh.

gradient for K^+ ions. The net influx of positively charged ions depolarizes the postsynaptic cell in the region of the synapse.

The **muscarinic receptors** have a seven membrane–spanning domain structure and act through a G protein. Five different types of muscarinic receptors, M_1 through M_5, have been characterized. Activation of M_1, M_3, and M_5 receptors increases the activity of phospholipase C, which, in turn, produces IP_3 and DAG. These messengers then activate other cellular processes such as calcium release from intracellular stores and protein phosphorylation. Activation of M_2 and M_4 receptors inhibits adenylyl cyclase, thereby reducing the level of cAMP. This process activates a class of potassium channels that hyperpolarizes the cell.

After release of ACh into the synaptic cleft, the enzyme **acetylcholinesterase (AChE)** hydrolyzes ACh to choline and acetate. This enzyme is located in the postsynaptic cell membrane, allowing rapid and efficient hydrolysis of extracellular ACh. The choline generated is taken back up into the neuron by a choline transport protein and is used to resynthesize ACh.

Catecholamines

The *catecholamines* are so named because they consist of a catechol moiety (a phenyl ring with two attached hydroxyl groups) and an ethylamine side chain. The catecholamines **dopamine (DA)**, **norepinephrine (NE)**, and **epinephrine (EPI)** share a common pathway for biosynthesis (Fig. 3.15). Three of the enzymes involved—tyrosine hydroxylase (TH), dopamine β hydroxylase (DBH), and phenylethanolamine *N*-methyltransferase—are unique to catecholamine-secreting cells. **Dopaminergic neurons** express only TH, NE **(noradrenergic) neurons** express both TH and DBH, and EPI-secreting cells (adrenergic) express all three. EPI-secreting cells include a small population of CNS neurons as well as the chromaffin cells of the adrenal medulla, which

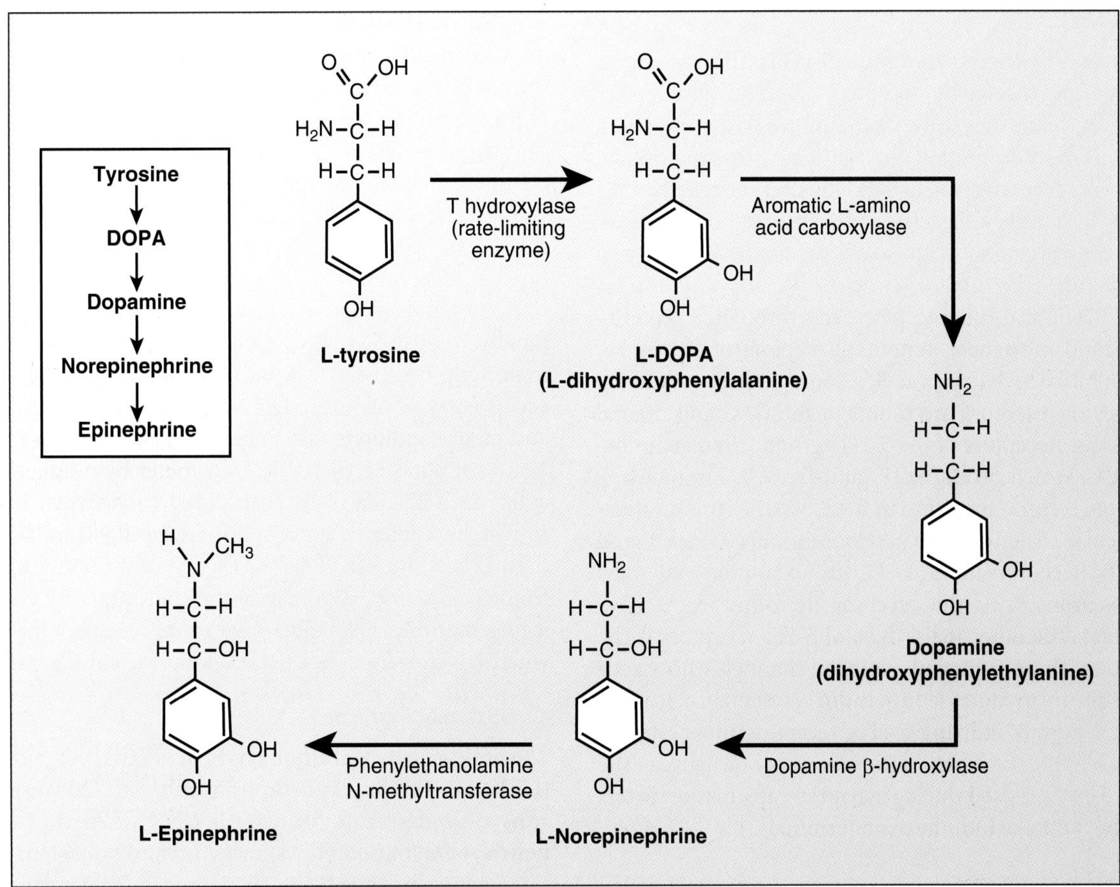

Figure 3.15. Catecholamine synthesis. Synthesis begins with tyrosine. Neurons that use norepinephrine and epinephrine contain the enzymes required for further processing of dopamine into those transmitters.

secrete EPI during generalized activation of the sympathetic component of the autonomic nervous system in the "fight-or-flight response" (see Chapter 6, "Autonomic Nervous System"). The rate-limiting enzyme in catecholamine biosynthesis is **tyrosine hydroxylase**, which converts L-tyrosine to L-3,4-dihydroxyphenylalanine.

The receptors for catecholamines share the seven-transmembrane domain macromolecular structure described above for muscarinic cholinergic receptors. Five subtypes of DA receptors (D_1–D_5) have been described in the CNS, but these can be grouped into two functional classes. The **D_1-like receptors** (D_1 and D_5) are coupled to G proteins (G_s), which stimulate adenylyl cyclase. **D_2-like receptors** (D_2–D_4) are coupled to G proteins (G_i), which inhibit adenylyl cyclase.

Adrenergic receptors, stimulated by NE and EPI, are located on cells throughout the body, including the CNS and the peripheral target organs of the sympathetic nervous system (see Chapter 6, "Autonomic Nervous System"). Adrenergic receptors are classified as either α or β, based on the potency of catecholamines and related analogs in stimulating each type of receptor. The analogs originally used to distinguish α- from β-adrenergic receptors were NE, EPI, and the two synthetic compounds isoproterenol (ISO) and phenylephrine (PE). In 1948, α receptors were designated as those receptors in which EPI was highest in potency and ISO was least potent (EPI > NE >> ISO). β Receptors exhibited a different order of responsiveness with ISO being the most potent and EPI either more potent or equal in potency to NE. Studies with PE also distinguished these two classes of receptors: α Receptors were stimulated by PE, whereas β receptors were not. α Receptors were subsequently further divided into α_1 and α_2 subtypes. Activation of α_1 receptors increases release of Ca^{2+} from intracellular stores and increases protein kinase C activity. Activation of α_2 receptors inhibits adenylyl cyclase, and thereby, reduces cAMP levels. β Receptor activation increases adenylyl cyclase activity.

After release into the synapse, a specific membrane transporter rapidly moves most of the catecholamine back into the presynaptic nerve terminals for repacking into synaptic vesicles.

Two enzymes are involved in degrading the catecholamines. **Monoamine oxidase (MAO)** removes the amine group, and **catechol-*O*-methyltransferase (COMT)** methylates the 3-OH group on the catechol ring. MAO is localized in the mitochondria of presynaptic and postsynaptic cells. COMT is localized in the cytoplasm of postsynaptic cells. At synapses of NE neurons in the PNS, the postsynaptic COMT-containing cells are the smooth muscle and glandular cells that receive sympathetic innervation (see Chapter 6, "Autonomic Nervous System"). In the CNS, most of the COMT is localized in glial cells, especially astrocytes, rather than in postsynaptic target neurons.

Serotonin

Serotonin or **5-hydroxytryptamine (5-HT)** is the transmitter in **serotonergic neurons**. Tryptophan hydroxylase, a marker of serotonergic neurons, converts the amino acid tryptophan, which is derived from the diet, to 5-hydroxytryptophan, which is, in turn, converted to 5-HT by decarboxylation.

5-HT is stored in vesicles and is released by exocytosis upon depolarization of the axon terminals. Two general types of 5-HT receptors are described. One type, comprising 5-HT_1, HT_2, and HT_4 receptors, acts through G protein–linked second messenger systems. Receptors of the 5-HT_1 family inhibit adenylyl cyclase. Serotonergic pathways using these types of receptors are found in the CNS and cranial blood vessels. Receptors of the 5-HT_2 group stimulate phospholipase C, which produces IP_3 and DAG. Neurons using this type of receptor are found in the CNS, the smooth muscle of vascular structures and gastrointestinal (GI) tract, and in platelets. Receptors in the 5-HT_4 group stimulate adenylyl cyclase. Neurons using this receptor are found in the CNS and GI tract. The other major type of 5-HT receptors is the 5-HT_3 group. These are ligand-gated ion channels whose activation results in an increase in sodium conductance, leading to EPSPs. Neurons utilizing 5-HT_3 receptors innervate the brainstem region associated with sensations of nausea. The action of 5-HT released during synaptic transmission is terminated by reuptake into the axon terminals.

Glutamate

Glutamate (GLU) is the principal excitatory transmitter of the CNS. GLU neurons are the predominant type in the transmission of sensory information into the CNS. GLU is an amino acid, which is an important substrate for transamination reactions in all cells, but in certain neurons it also serves as a neurotransmitter. The main source of GLU for neurotransmission is mitochondrial conversion of α ketoglutarate derived from the **Krebs cycle** to which an amino group is added. GLU for use as a neurotransmitter is stored in synaptic vesicles and released by exocytosis. After release, it interacts with specific receptors on the postsynaptic neuron. GLU receptors have one of the two different general structures. One is a ligand-gated ionotropic cation channel that opens during binding of GLU. The other is metabotropic G protein linked. The ionotropic receptors are named for the synthetic analogs that best activate them—kainate, α-amino-3-hydroxy-5-methyl-4-isoxazole propionic acid (AMPA), and *N*-methyl-D-aspartate (NMDA). Activation of the kainate and AMPA receptors produces EPSPs by opening ion channels that increase Na^+ and, to a lesser degree, Ca^{2+} conductance. Activation of the **NMDA receptor** is unique in that GLY must also bind to the site on the NMDA receptor along with GLU for channel opening to occur. Activation of the NMDA receptor increases both Na^+ and Ca^{2+} conductance. The ion channel is blocked by Mg^{2+} when the membrane is in the resting state and early phase of membrane depolarization. As the membrane becomes further depolarized, the Mg^{2+} block is relieved. GLU transmission through the NMDA receptor is important for the development of specific neuronal connections related to learning and memory.

The **metabotropic glutamate receptors (mGluRs)** have a seven-transmembrane structure similar to that of the ACh muscarinic and adrenergic receptors. The mGluRs use a G protein to initiate second messenger cascades that elicit a number of different outcomes, including modulation of ion channels. Activation of all of these types of receptors inhibits voltage-gated Ca^{2+} channels. The function of other ligand-gated channels can be modulated by the activation of these types of receptors.

Two efficient active transport mechanisms remove GLU rapidly from the synapse. Control of free glutamine is very important, because it can be neurotoxic insofar as allowing excess entry of Ca^{2+} through NMDA receptors during stroke and other conditions that impair oxygen supply to the brain. Neuronal uptake recycles the transmitter by restorage in synaptic vesicles. Glial cells (particularly astrocytes) contain a similar, high-affinity, active transport mechanism that ensures the efficient removal of excitatory neurotransmitter molecules from the synapse. Glia recycle the transmitter by converting it to **glutamine**. Glial glutamine readily reenters the neuron, where it is converted back to GLU for use again as a transmitter.

GABA and glycine

The amino acids **γ-aminobutyric acid (GABA)** and **glycine (GLY)** are inhibitory neurotransmitters. **GABAergic neurons** are widespread through the CNS, whereas **glycinergic neurons** are found in the spinal cord and brainstem.

The synthesis of GABA in neurons is by decarboxylation of GLU by glutamic acid decarboxylase, a marker of GABAergic neurons. GABA is stored in synaptic vesicles and released by exocytosis when the axon terminal is depolarized. There are two types of GABA receptors: $GABA_A$ and $GABA_B$. The $GABA_A$ receptor is a ligand-gated channel similar in structure to the inotropic ACh receptor. Activation of this receptor produces IPSPs by allowing influx of Cl^- ions. The $GABA_B$ receptor is metabotropic and uses a G protein. Activation of these types of receptors inhibits the cell by enhancing the K^+ channel or alternatively inhibiting the Ca^{2+} channels.

The action of GABA is terminated by removal from the synaptic cleft into the presynaptic terminal and glial cells (astrocytes) by high-affinity transports. GABA enters the Krebs cycle in both neuronal and glial mitochondria and is converted to succinic semialdehyde by the enzyme GABA-transaminase. This enzyme is also coupled to the conversion of α ketoglutarate into GLU. The GLU produced in the glial cell is converted to glutamine and is transported into the presynaptic terminal, where it is converted back into GLU.

The function of $GABA_A$ receptors is enhanced by **benzodiazepines**, drugs that are widely used to treat anxiety and some of the general anesthetics. Drugs that inhibit GABA transmission cause epileptic seizures, indicating a major role for inhibitory mechanisms in normal brain function.

GLY is produced from serine by the enzyme serine hydroxymethyltransferase. GLY is stored in synaptic vesicles and released by depolarization of the axon terminal. The GLY receptor is similar in structure to the $GABA_A$ receptor. Activation of the receptor opens a channel that allows Cl^- to enter the postsynaptic cell, producing hyperpolarization.

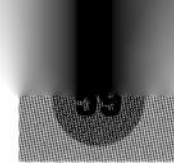

Clinical Focus / 3.2

Inherited Peripheral Neuropathy

Disorders of the PNS are termed *peripheral neuropathies*. This type of disease causes muscle weakness and loss of sensation, in which symptoms are more prominent in the distal portions of the legs and arms. Some types of this disorder are inherited, and great progress has occurred over the last 15 years in understanding their molecular basis. These carry the name **Charcot–Marie–Tooth (CMT) disease**, after the physicians who originally characterized these disorders. Abnormalities in the myelin sheath underlie several of these diseases.

This most common type of this disorder, $CMT1_A$, is now known to be due to abnormalities in the protein PMP-22. This protein plays a role in the location and stabilization of adjacent layers of the myelin sheath. Because of abnormality in the myelin sheath, patients with this disorder have abnormally slow nerve conduction velocities, often as slow as 20 m/s secondary to impairment of saltatory conduction. Normal conduction velocity would be in the range of 50 m/s.

Another disorder related to this same protein causes patients' peripheral nerves to be abnormally sensitive to compression experienced in prolonged sitting or leaning on the arms. Deficits such as foot drop, wrist drop, and weakness of hand muscles can occur. In contrast to the transient numbness and weakness of a hand or foot normal individuals experience from sitting or lying in a position for an extended period of time, these patients may require days or weeks for nerve function to return to normal. This disorder, termed *hereditary neuropathy with liability to pressure palsy*, is also due to abnormality in the PMP-22 protein.

Recall also that GLY also interacts with the glutaminergic NMDA receptor. The action of GLY is terminated by reuptake into the presynaptic axon terminals or into the glia.

Neuropeptides

Peptides can function as neurotransmitters. Many of these substances were discovered in the pituitary and GI systems and later found also to be present in neurons. Peptide transmitters are derived from precursors that are often much longer in length. The precursors are synthesized in cell body and packaged by into vesicles by the Golgi apparatus. As the vesicles are moved down the neuron by anterograde axoplasmic transport, the precursor peptides are modified by proteases that cleave them into smaller peptides. Commonly, vesicles containing neuropeptides are colocalized with vesicles containing another transmitter in the same neuron and both can be released during nerve stimulation. In these instances, release of the peptide-containing vesicles generally occurs at higher action potential frequencies than does release of the vesicles containing nonpeptide neurotransmitters.

The list of candidate peptide transmitters continues to grow; it includes well-known GI hormones, pituitary hormones, and hypothalamic-releasing factors. As a class, the neuropeptides fall into several families of peptides, based on their origins, homologies in amino acid composition, and similarities in the response they elicit at common or related receptors. Table 3.2 lists some members of each of these families.

Peptides interact with specific receptors located on postsynaptic target cells. These receptors most often have the same seven membrane–spanning motif used by G protein–coupled receptors. The most common removal mechanism for synaptically released peptides appears to be diffusion, a slow process that ensures a longer-lasting action of the peptide in the synapse and in the extracellular fluid surrounding it. Peptides are degraded by proteases in the extracellular space; some of this degradation may occur within the synaptic cleft. Unlike with

TABLE 3.2 Some Neuropeptide Neurotransmitters

Neuropeptide	Amino Acid Composition
Opioids	
Met-enkephalin	Tyr-Gly-Gly-Phe-Met-OH
Leu-enkephalin	Tyr-Gly-Gly-Phe-Leu-OH
Dynorphin	Tyr-Gly-Gly-Phe-Leu-Arg-Arg-Ile
β Endorphin	Tyr-Gly-Gly-Phe-Met-Thr-Glu-Lys-Ser-Gln-Thr-Pro-Leu-Val-Thr-Leu-Phe-Lys-Asn-Ala-Ile-Val-Lys-Asn-His-Lys-Gly-Gln-OH
Gastrointestinal peptides	
Cholecystokinin octapeptide	Asp-Tyr-Met-Gly-Trp-Met-Asp-Phe-NH2
Substance P	Arg-Pro-Lys-Pro-Gln-Gln-Phe-Phe-Gly-Leu-Met
Vasoactive intestinal peptides	His-Ser-Asp-Ala-Val-Phe-Thr-Asp-Asn-Tyr-Thr-Arg-Leu-Arg-Lys-Gln-Met-Ala-Val-Lys-Lys-Tyr-Leu-Asn-Ser-Ile-Leu-Asn-NH_2
Hypothalamic and pituitary peptides	
Thyrotropin-releasing hormone	Pyro-Glu-His-Pro-NH_2
Somatostatin	Ala-Gly-Cys-Asn-Phe-Phe-Trp-Lys-Thr-Phe-Thr-Ser-Cys
Luteinizing hormone–releasing hormone	Pyro-Glu-His-Trp-Ser-Tyr-Gly-Leu-Arg-Pro-Gly
Vasopressin	Cys-Tyr-Phe-Gln-Asn-Cys-Pro-Arg-Gly-NH_2
Oxytocin	Cys-Tyr-Ile-Gln-Asn-Cys-Pro-Leu-Gly-NH_2

more classical neurotransmitters, there are no mechanisms for the recycling of peptide transmitters into the axon terminal.

Nitric oxide, arachidonic acid, and endocannabinoids

More recently, a novel type of neurotransmission has been identified. In this case, membrane-soluble molecules diffuse through neuronal membranes and activate "postsynaptic" cells via second messenger pathways. **Nitric oxide (NO)** is a labile free-radical gas that is synthesized on demand from its precursor, L-arginine, by nitric oxide synthase (NOS). Because NOS activity is exquisitely regulated by Ca^{2+}, the release of NO is calcium dependent, even though it is not packaged into synaptic vesicles. NO was identified as the endothelial-derived relaxing factor in blood vessels before it was known to be a neurotransmitter. It is a neurotransmitter in some peripheral autonomic pathways as well as the CNS. The effects of NO are mediated through its activation of second messengers, particularly guanylyl cyclase.

Arachidonic acid is a fatty acid released from phospholipids in the membrane when phospholipase A2 is activated by ligand-gated receptors. The arachidonic acid then diffuses retrograde to affect the presynaptic cell by activating second messenger systems. NO can also act in this retrograde fashion as a signaling molecule.

The **endocannabinoids** are another set of lipid molecules that exhibit synaptic signaling and can work in a retrograde direction to alter function in the presynaptic terminal. These compounds are the endogenous ligands for the receptors that mediate the effects of the psychoactive component of marijuana, Δ^9-tetrahydrocannabinol. There are two cannabinoid receptors, CB_1 and CB_2. The endocannabinoids have effects in multiple neuronal and physiologic systems and have recently become intense targets for drug discovery. The importance of this system is underscored by the fact that CB_1 is the most abundant G protein–coupled receptor in the brain. The details of endocannabinoid signaling are still being worked out, but it is clear that their synthesis increases when intracellular Ca^{2+} increases.

Chapter Summary

- The morphology of neurons mediates complex circuitry and presents a unique metabolic challenge.
- Anterograde axonal transport supplies axon terminals with organelles, metabolic components, synaptic vesicles, and receptors.
- Retrograde axonal transport recycles waste products and brings nutrients and signaling messengers from the terminal environment to the cell body.
- Ion channels establish the resting membrane potential of neurons.
- Voltage-gated ion channels are responsible for the action potential and the release of neurotransmitter.
- Ligand-gated ion channels predominate on the soma and dendritic parts of the cell membrane and cause membrane depolarization or hyperpolarization in response to neurotransmitter attachment.
- Voltage-gated channels are largely restricted to the axon and its terminals.
- Membrane conductance and capacitance affect the spread of local changes in membrane potential and the speed of electrical transmission in neurons.
- Initiation of an action potential occurs when an axon hillock is depolarized to a threshold for activation of voltage-gated sodium channels.
- An action potential is a transient depolarization of the cell membrane that results from a rapid activation of voltage-gated sodium channels followed by an inactivation of those channels and the delayed activation of voltage-gated potassium channels.
- Action potentials are self-regenerating depolarizations that travel down the length of the axon and that result from the depolarizing signal of one action potential bringing the next voltage-gated sodium channel in line to its threshold.
- Nerve conduction velocity increases with an increase in the diameter of the axon and the thickness of its myelin sheath, if present.
- Following an action potential, that axonal region is temporarily unable to generate another because of voltage-gated sodium channel inactivation.
- Voltage-gated calcium channels in the axon terminal allow calcium to enter the terminal and start a cascade of events leading to the release of neurotransmitter.
- Most neurotransmitters are stored in synaptic vesicles and released upon nerve stimulation by a process of calcium-mediated exocytosis.
- Once released, the neurotransmitter binds to and stimulates its receptors on postsynaptic cells briefly before being rapidly removed from the synapse.
- Removal of neurotransmitter from the synapse occurs by active transport into the presynaptic cell (or glia in the central nervous system), diffusion out of the synaptic cleft, or enzymatic degradation in the region of the synaptic cleft.
- Neurotransmitter is synthesized and recycled in the nerve terminal, with the exception of peptide neurotransmitters.

thePoint *Visit http://thePoint.lww.com for chapter review Q&A, case studies, animations, and more!*

Sensory Physiology

ACTIVE LEARNING OBJECTIVES

Upon mastering the material in this chapter you should be able to:

- Explain how the many different forms of environmental energy are transformed by sensory receptors into the same type of electrical signal in sensory nerves.
- Explain why sensory receptors have specificity to sense one type of environmental energy over another.
- Explain why sensory receptors that respond to a selective type of environmental energy can respond to a different type, yet the sensation produced is the same as if the receptor responded to the energy for which it was designed to detect.
- Postulate the value of adaptation of sensory receptors in relation to the selective type of sensation the receptor is designed to detect.
- Explain why action potential frequency is the mode of information transmission in sensory systems and how changes in that frequency are related to graded electrical potential changes in sensory receptors.
- Explain how the processes of adaptation and accommodation permit the processing of a wide range of stimulus intensities.
- Distinguish between rod and cone vision and point out the special features of each type.
- Describe the organization of the retina, relating each layer to its specific and integrated function in the process of vision.
- Diagram the paths of light rays in a myopic eyeball, with and without corrective lenses.
- Compare and contrast the sensory processes in the related senses of smell and taste.
- Distinguish among sensory receptors on the basis of internal versus external, or by sensory modality.
- Distinguish among the different types of pain sensation.
- Explain, in terms of biochemical events, the process of light and dark adaptation in the retina.

SENSORY SYSTEM

The survival of any organism depends on having adequate information about the external environment as well as information about the state of internal bodily processes and functions. Information about our external and internal environments is gathered by the sensory branch of the nervous system. This chapter discusses the functions of the sensory receptors, the organs that permit us to gather this information. The discussion emphasizes somatic sesnsations, that is, those dealing with the external aspect of the body. This chapter does not specifically address the visceral sensations that come from internal organs.

Sensory systems transform physical and chemical signals from the external and internal environments into information in the form of nerve action potentials.

Events in our external and internal worlds must first be translated into signals that our nervous systems can process. Despite the wide range of types of information to be sensed and acted on, a small set of common principles underlies all sensory processes.

Furthermore, although the human body contains a very large number of different sensory receptors, these receptors have many functional features in common. The process of sensation essentially involves sampling selected small amounts of energy from the environment by sensory receptors and using it to control the generation of trains of action potentials in sensory nerves (Fig. 4.1). The pattern of sensory action potentials, along with the specific nature of the sensory receptor and its nerve pathways in the brain, provides an internal representation of a specific component of the external world. The wide variety of specialized sensory functions is a result of structural and physiologic adaptations that fit a particular receptor for its role in the organism. Finally, the process of sensation is a portion of the more complex process of perception, in which sensory information is integrated with previously learned information and other sensory inputs, enabling us to make judgments about the quality, intensity, and relevance of what is being sensed.

Sensory stimuli

A factor in the environment that produces an effective response in a sensory receptor is called a **stimulus**. Stimuli involve exchanges of energy between the environment and the receptors. Typical stimuli include electromagnetic quantities, such as radiant heat or light; mechanical quantities, such as pressure, sound waves, or vibrations; and chemical

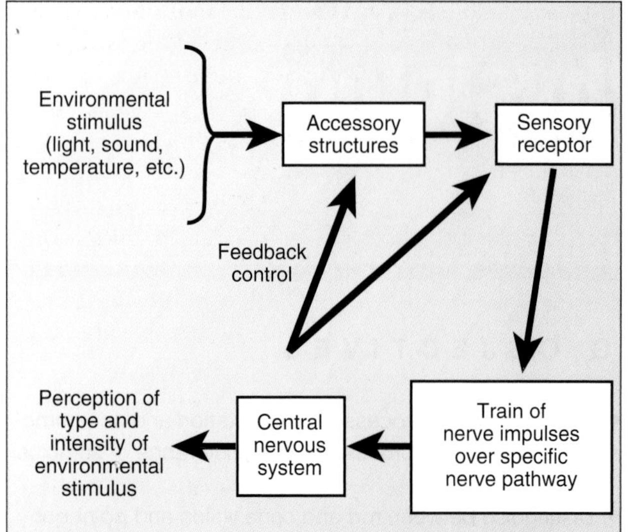

Figure 4.1 A basic model for the translation of an environmental stimulus into a perception. Although the details vary with each type of sensory modality, the overall process is similar.

qualities, such as acidity and molecular shape and size. Regardless of the sensation experienced, all sensory systems transmit information to the brain with the four common attributes of intensity, modality (e.g., temperature, pressure, and chemical type), location, and timing.

A fundamental property of sensory receptors is their ability to respond to different intensities of stimulation with a representative output. *Intensity* is a measure of the energy content available to interact with a sensory receptor. Most individuals are familiar with sensory intensity in phenomena such as the brightness of light, the "loudness" of sound, or the force of physical contact of an object with the body. Chemicals sensed by the body also have intensity. For such stimuli, the *concentration* of a chemical signal is considered analogous to intensity.

Sensory receptors are capable of detecting the modality of a stimulus. **Sensory modality** is the term used to describe the *kind* of sensation present. Common sensory modalities include taste, smell, touch, sight, and hearing (the traditional five senses) as well as more complex sensations, such as slipperiness or wetness. Sensory modality is often used to classify receptors. For example, **photoreceptors** that serve a visual function sense light. **Chemoreceptors** detect chemical signals and serve the senses of taste and smell. **Mechanoreceptors** sense physical deformation and serve such sensory modalities as touch, hearing, and the amount of stress in a tendon or muscle. Heat, or its relative lack, is detected by **thermoreceptors**. Many sensory modalities are a combination of simpler sensations. For example, the sensation of wetness is actually composed of sensations of pressure and temperature. You can demonstrate this phenomenon yourself by placing your hand in a plastic bag and then immersing it in cold water. Even though your hand will remain dry inside the bag, the perception you sense will be that of wetness.

Although sensory receptors are often classified by the modality to which they respond, other sensory receptors are classified instead by their "vantage point" in the body. For example, **exteroceptors** detect stimuli from outside the body; **enteroceptors** detect internal stimuli; and **proprioceptors** (*proprio* is Latin for "one's own") provide information about the positions of joints, muscle activity, and the orientation of the body in space. **Nociceptors** (pain receptors) detect noxious agents, both internally and externally.

It is often difficult to communicate a precise definition of a sensory modality because of the subjective perception or **affect** that accompanies it. This property has to do with the psychologic feeling attached to the stimulus. For example, the sensation of cold or touch can be pleasurable at certain intensities but rise to an impression of discomfort or extreme pain at very high intensities. Stimulus affect, however, is more complex than a combination of intensity with modality. Previous experience and learning play a role in determining the affect of a sensory perception. The affect of identical visual stimuli, for example, can provoke pleasure, anger, fear, or terror depending on the experiences of those perceiving the stimulus and the memories associated with it.

Sensory receptors

Most sensory receptors are optimized to respond preferentially to a single kind of environmental stimulus. The usual stimulus for the eye is light and that for the ear is sound. This specificity is a result of several features that match a receptor to its preferred stimulus. The usual and appropriate stimulus for a type of sensory receptor is called its **adequate stimulus**. This is the stimulus for which the receptor has the lowest **threshold**, that is, the lowest stimulus intensity that can be reliably detected. Threshold, however, can be difficult to quantify because it can vary over time, and it can be altered by either the presence of interfering stimuli or the action of **accessory structures**. These structures are not considered part of the sensory receptor itself. However, in many cases, such as the lens of the eye or the structures of the outer and middle ear, these structures either enhance the specific sensitivity of the receptor or exclude unwanted stimuli. Often these accessory structures are part of a control system that adjusts the sensitivity according to the information being received (see Fig. 4.1).

Most receptors will respond to additional types of stimuli other than the adequate stimulus. However, the threshold for the inappropriate stimuli is much higher. For example, applying pressure to the eye will cause one to "see lights" although light itself is the energy type for which the sensory receptors of the eye have the lowest threshold. The pain perceived during extremes of temperature is another example of one sensory modality activating a receptor that was designed to detect another type of stimulus. Finally, almost all receptors can be stimulated electrically to produce sensations that mimic the one usually associated with that receptor.

Another feature of sensory system specificity resides in the central nervous system (CNS). The CNS pathway over which sensory information travels also plays a role in determining the nature of the perception. Information arriving by way of the optic nerve, for example, is always perceived as light and never as sound. This is known as the concept of the **labeled line**. Details of all neural sensory pathways are beyond the scope of this text.

Sensory transduction changes environmental energy into sensory nerve action potentials.

The key physiologic function of a sensory receptor is to translate nonelectrical forms of environmental energy into electrical events that can be transmitted and processed by the nervous system. These electrical events are nerve action potentials, which are the fundamental units of information in the nervous system. A device that translates one form of energy signal into another is called a **transducer**. Sensory receptors in the body, therefore, can be thought of as biologic transducers. A typical sequence of electrical events in the sensory transduction process is shown in Figure 4.2.

Generator potentials

Figure 4.2 depicts a typical type of sensory receptor system. This example is a representation of a mechanoreceptor, that is, a receptor that translates energy in the form of a physical stress or strain (i.e., pressure, displacement, stretch, etc.) into an electrical signal that will eventually produce action potentials in sensory nerves. In the example in Figure 4.2, the tip of the receptor has been deflected (strained) and held constant (1, 2). This deflection deforms the cell membrane of the receptor, causing a portion of it to become more permeable to positive ions (*shaded region*, 3). The resulting influx of positive charge (mostly from influx of sodium ions) then leads to a localized depolarization, called the **generator potential**. The generator potential creates a current through the axon membrane called a **local excitatory current**. It provides the link between the formation of the generator potential and the excitation of the nerve fiber membrane.

The production of the generator potential is of critical importance in the transduction process because it is the step in which information related to stimulus intensity and duration is transduced. The strength (intensity) of the stimulus applied (in Fig. 4.2, the amount of deflection) determines the size of the generator potential depolarization. Varying the intensity of the stimulation will correspondingly vary the generator potential, although the changes are not always directly proportional to the intensity. This is called a **graded response**, in contrast to the all-or-none response of an action potential, and causes a similar gradation of the strength of the local excitatory currents.

As currents resulting from the depolarization of the generator potential spread down the neuron, they reach a region of the receptor membrane called the **impulse initiation**

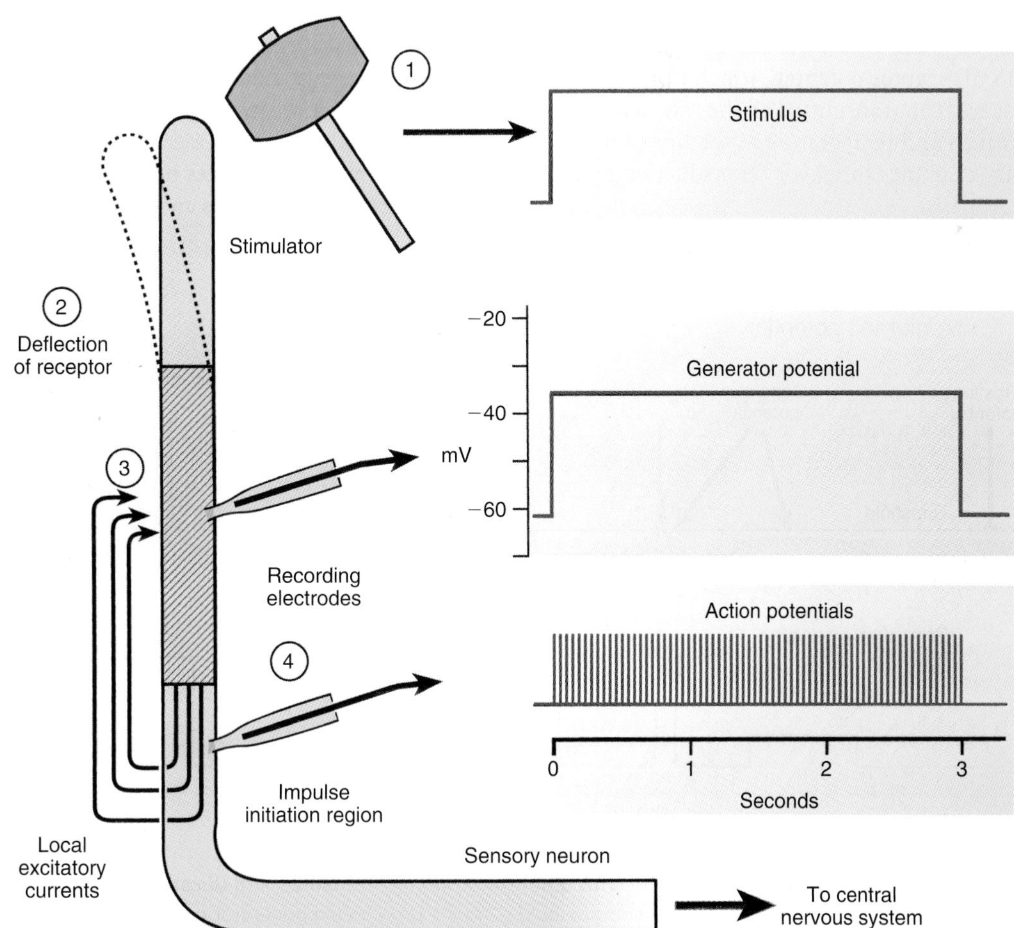

Figure 4.2 The relation between an applied stimulus and the production of sensory nerve action potentials. (See text for details.)

region (4). This region is the location of the formation of action potentials in the neuron and constitutes the next important link in the sensory process. In this region, action potentials are produced at a frequency related to the strength of the current caused by the generator potential (and, thus, related to the original stimulus intensity). This region is, therefore, sometimes called the *coding region*, because here the generator potential and flowing current caused by the initial stimulus to the receptor are translated into a train of action potentials of a frequency proportional (but not always in a linear fashion) to the size of the generator potential. The train of action potentials in the sensory nerve fiber is the electrical signal that is sent to the CNS. In complex sensory organs that contain a great many individual receptors, the generator potential may arise from several sources within the organ. Thus, the train of action potentials formed results from multiple inputs to the sensory receptor. In these instances, the generator potential is often called a **receptor potential**.

Sensory receptor subspecialization

The mechanoreceptor depicted in Figure 4.2 is an example of a sensory receptor subtype called a *long receptor*. In such receptors, the formation of the receptor and action potentials occurs in the same cell, the neuron. This type of sensory receptor-to-action potential arrangement is best suited for transmission of sensory information over long distances. In these cases, the initial physical stimulus creates a local receptor potential, which triggers action potentials that are then transmitted all the way to the CNS. The sense of touch and temperature at the tips of our fingers or toes are examples of sensory transduction by long receptors.

In contrast, other types of sensory transduction in the body occur through *short receptors*. Short receptors are highly specialized cells that, like long receptors, also convert a nonelectrical form of energy into a receptor potential within them. However, in short receptors, the receptor generator potential and neuron action potentials occur in two different cells. In short receptors, the local potential event is coupled to the release of neurotransmitter stored in the receptor cell. This transmitter is then released onto synapses of a postsynaptic primary afferent neuron in which it produces another localized generator potential. It is this second generator potential that triggers trains of action potentials in the afferent neuron that are then sent to the CNS. Short receptors produce either depolarizing or hyperpolarizing generator potentials. In some cases, the same receptor can produce depolarizing and hyperpolarizing generator potentials depending on the stimulus they receive. Positive generator potentials increase the rate of transmitter release, whereas negative potentials decrease this rate. The photoreceptors of the eye and the hair receptors of the cochlea of the inner ear are examples of short sensory receptors (see below).

Sensory nerve impulse frequency is modulated by the magnitude and duration of the generator potential.

Figure 4.3 shows how the magnitude and duration of a generator potential affect subsequent action potential formation in a sensory nerve. This example is typical of what might be seen in a long receptor-type sensory nerve. The threshold (*black line* in the figure) is a critical level of depolarization. Membrane potential changes below this level are caused by the local excitatory currents and vary in proportion to them,

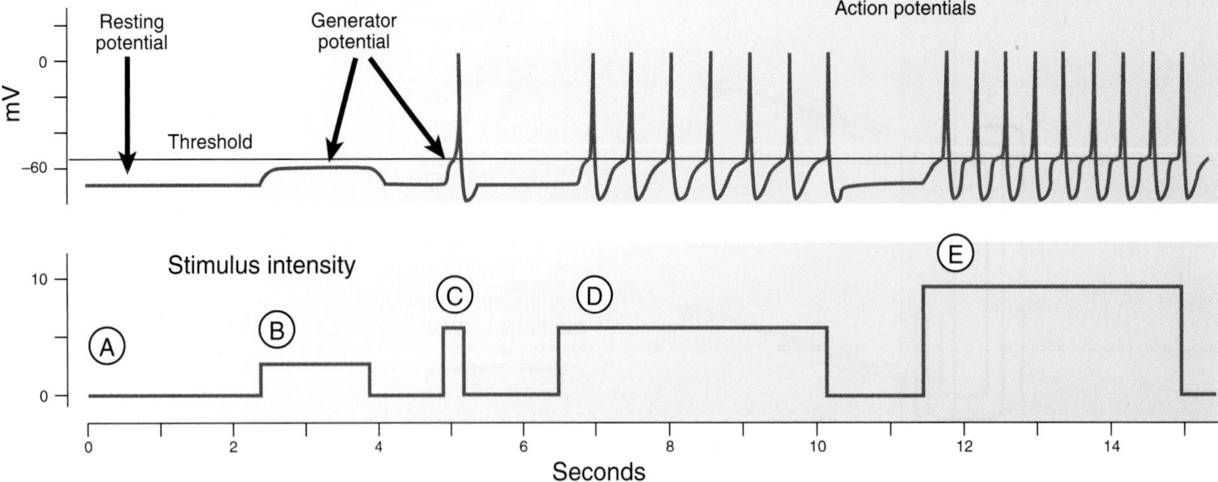

Figure 4.3 Sensory nerve activity with different stimulus intensities and durations. (A) With no stimulus, the membrane is at rest. **(B)** A subthreshold stimulus produces a generator potential too small to cause membrane excitation. **(C)** A brief, but intense, stimulus can cause a single action potential. **(D)** Maintaining this stimulus leads to a train of action potentials. **(E)** Increasing the stimulus intensity leads to an increase in the action potential firing rate.

whereas the membrane activity above the threshold level consists of locally produced **action potentials**. The lower trace shows a series of different stimuli applied to the receptor, and the upper trace shows the resulting electrical events in the impulse initiation region. No stimulus is given at **A**, and the membrane voltage is at the resting potential. At **B**, a small stimulus is applied, producing a generator potential too small to bring the impulse initiation region to threshold; no action potential activity results, and the stimulus causing this level of potential generation would, therefore, not be sensed at all. In contrast, at **C**, a brief stimulus of greater intensity is given, and the resulting generator potential displacement is of sufficient amplitude to trigger a single action potential. As in all excitable and all-or-none nerve membranes, the action potential is immediately followed by repolarization, often to a level that transiently hyperpolarizes the membrane potential because of temporarily high potassium conductance. Because the brief stimulus has been removed by this time, no further action potentials are produced. A longer stimulus of the same intensity **(D)** produces repetitive action potentials because, as the membrane repolarizes from the action potential, local excitatory currents are still flowing. They bring the repolarized membrane to threshold at a rate proportional to their strength. During this time interval, the fast sodium channels of the membrane are being reset, and another action potential is triggered as soon as the membrane potential reaches threshold. As long as the stimulus is maintained, this process will repeat itself at a rate determined by the stimulus intensity. If the intensity of the stimulus is increased **(E)**, the local excitatory currents will be stronger and the threshold will be reached more rapidly. This will result in a reduction of the time between each action potential and, as a consequence, a higher action potential frequency. This change in action potential frequency is critical in communicating the intensity of the stimulus to the CNS.

Sustained sensory receptor stimulation can result in diminished action potential generation over time.

In the presence of a constant stimulus, only some sensory receptors maintain the magnitude of their initial generator potential. In most sensory receptors, the magnitude of the generator potential decays over time, even when the original sensory stimulus does not change. This phenomenon is called **adaptation**. Figure 4.4A shows the output from a receptor in which there is no adaptation. As long as the stimulus is maintained, there is a steady rate of action potential firing. Figure 4.4B shows **slow adaptation**; the generator potential amplitude declines slowly, and the interval between the action potentials increases correspondingly. Figure 4.4C demonstrates **rapid adaptation**; the action potential frequency falls rapidly and then maintains a constant slow rate that does not show further adaptation. Responses in which there is little or no adaptation are called *tonic*, whereas those in which significant adaptation occurs are called *phasic*. In some cases, tonic receptors may be called *intensity receptors* and phasic receptors may be called *velocity receptors*. Many receptors—*muscle spindles*, for example—show a combination of responses. On application of a stimulus, a rapidly adapting phasic response is followed by a steady tonic response. Both of these responses may be graded by the intensity of the stimulus. As a receptor adapts, the sensory input to the CNS is reduced, and the sensation is perceived as less intense.

The phenomenon of adaptation is important in preventing "sensory overload" and allows less important or unchanging environmental stimuli to be partially ignored. When a change occurs, however, the phasic response will occur again, and the sensory input will become temporarily more noticeable. Rapidly adapting receptors are also important in sensory systems that must sense the rate of change of a stimulus, especially when the intensity of the stimulus can vary over a range that would overload a tonic receptor.

Receptor adaptation can occur at several places in the transduction process. As mentioned above, adaptation of the generator potential itself can produce adaptation of the

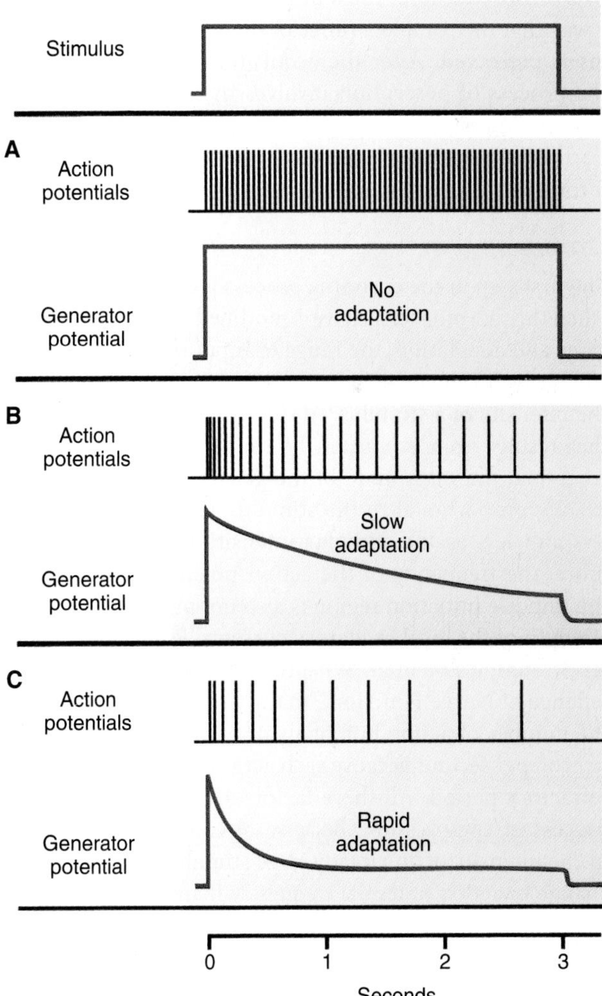

Figure 4.4 Adaptation. Adaptation in a sensory receptor is often related to a decline in the generator potential with time. **(A)** The generator potential is maintained without decline, and the action potential frequency remains constant. **(B)** A slow decline in the generator potential is associated with slow adaptation. **(C)** In a rapidly adapting receptor, the generator potential declines rapidly.

overall sensory response. In some cases, however, the receptor's sensitivity is changed by the action of accessory structures, as in the constriction of the pupil of the eye in the presence of bright light (an example of feedback-controlled adaptation). In the sensory cells of the eye, light-controlled changes in the amounts of the visual pigments also can change the basic sensitivity of the receptors and produce adaptation. Finally, the phenomenon of **accommodation** in the impulse initiation region of the sensory nerve fiber can slow the rate of action potential production even though the generator potential may show no change. Accommodation refers to a gradual increase in threshold caused by prolonged nerve depolarization and results from the inactivation of sodium channels.

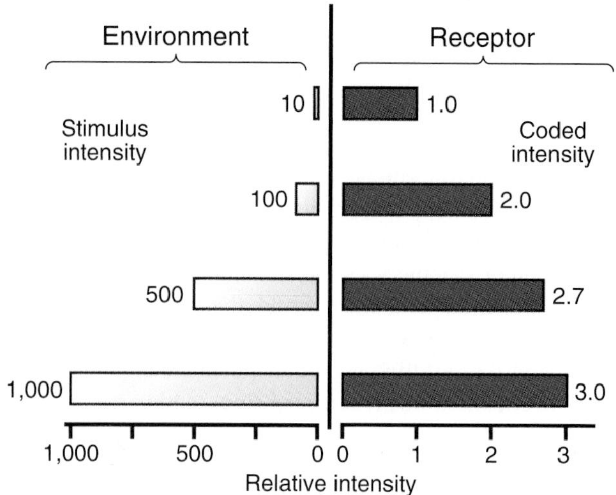

Figure 4.5 Compression in sensory process. By a variety of means, a wide range of input intensities is coded into a much narrower range of responses that can be represented by variations in action potential frequency.

Perception of sensory information requires encoding and decoding electrical signals sent to the central nervous system.

Environmental stimuli that have been partially processed by a sensory receptor must be conveyed to the CNS in such a way that the complete range of the intensity of the stimulus is preserved. After the acquisition of sensory stimuli, the process of perception involves the subsequent encoding and transmission of the sensory signal to the CNS. Further processing, or decoding, yields biologically useful information.

Compression

The first step in the encoding process is **compression**. Even when the receptor sensitivity is modified by accessory structures and adaptation, the range of input intensities is large, as shown in Figure 4.5. At the left is a 100-fold range in the intensity of a stimulus. At the right is an intensity scale that results from events in the sensory receptor. In most receptors, the magnitude of the generator potential is not exactly proportional to the stimulus intensity; it increases less and less as the stimulus intensity increases. Furthermore, the frequency of the action potentials produced in the impulse initiation region is also not proportional to the strength of the local excitatory currents. Recall from Chapter 3, "Action Potential, Synaptic Transmission, and Maintenance of Nerve Function," that there is an upper limit to the number of action potentials that can be generated in a neuron per second because each action potential has a finite refractory period. All these factors are responsible for the process of compression. Because of compression, changes in the intensity of an already large stimulus cause a smaller change in action potential frequency than the same change would cause if the change occurred from an initially small stimulus. As a result, the hundredfold variation in the stimulus is compressed into a threefold range after the receptor has processed the stimulus. Some information is necessarily lost in this process, although integrative processes in the CNS can restore the information or compensate for its absence. Physiologic evidence for compression is based on the observed nonlinear (logarithmic or power function) relation between the actual intensity of a stimulus and its perceived intensity.

Transmission

Sensory information initiated at the receptor level must be transmitted to the CNS in order to be perceived. The encoding processes in the receptors provide some basis for this transfer by producing a series of action potentials related to the stimulus intensity. A special process is necessary for the complete transfer, however, because of the nature of the conduction of action potentials. Recall that action potentials are transmitted down nerve axons without decrement; their duration and amplitude do not change during conduction and, therefore, the parameters of electrical potential amplitude and duration are lost as a means of transmitting different forms of information to the CNS. It follows, then, that the only information that can be conveyed by a single action potential is its presence or absence.

Relationships between and among action potentials, however, can convey large amounts of information, and this is the system found in the sensory transmission process. To review, in physiologic sensory systems, an input signal provided by some physical quantity is continuously measured and converted into an electrical signal (the generator potential), whose amplitude is proportional to the input signal. This signal controls the frequency of action potential generation in the impulse initiation region of a sensory nerve fiber. Because action potentials are of a constant height and duration, the amplitude information of the original input signal is now in the form of a frequency of action potential generation. These trains of action potentials are sent along a nerve pathway to some distant point in the CNS, where they produce an electrical voltage in postsynaptic neurons that is proportional to the frequency of the arriving pulses. This voltage is a replica of the original input voltage. Further processing can produce a graphic record of the input data, which is accompanied by processing and interpretation in the CNS.

Perception

The interpretation of encoded and transmitted information into a perception requires several other factors. For instance, the interpretation of sensory input by the CNS depends on the neural pathway it takes to the brain. For this reason, for example, all information arriving on the optic nerves is interpreted as light, even though the signal may have arisen as a result of pressure applied to the eyeball. The localization of a cutaneous sensation to a particular part of the body also depends on the particular pathway it takes to the CNS. Often a sensation (usually pain) arising in a visceral structure (e.g., heart, gallbladder) is perceived as coming from a portion of the body surface, because developmentally related nerve fibers come from these anatomically different regions and converge on the same spinal neurons. Such a sensation is called **referred pain**.

The remainder of this chapter deals with specific sensory receptors, including those of both the somatosensory system (i.e., touch, temperature, and pain) and the **special senses** (i.e., sight, hearing, taste, and smell).

SOMATOSENSORY SYSTEM

The somatosensory system is a diverse sensory system comprising receptors and processing centers that produce sensory modalities, such as touch (tactile sensation), proprioception (body position), temperature, and pain (nociception). These sensory receptors cover the skin and epithelia, skeletal muscles, bones and joints, and internal organs. The system responds to a number of diverse stimuli using the following receptors: mechanoreceptors (movement/pressure), chemoreceptors, thermoreceptors, and nociceptors. Signals from these receptors pass via sensory nerves through tracts in the spinal cord and into the brain. Neural information is processed primarily in the somatosensory area in the parietal lobe of the cerebral cortex (see Chapter 7, "Integrative Functions of the Nervous System," for details).

Skin receptors provide sensory information from the body surface.

The skin has a rich supply of sensory receptors that provide cutaneous sensations, such as touch (light and deep pressure), temperature (warmth and coldness), and pain. More complicated modalities include itch, tickle, wetness, etc. Cutaneous receptors can be mapped over the skin by using highly localized stimuli of heat, cold, pressure, or vibration. Some areas require a high degree of spatial localization (e.g., fingertips, lips) and, accordingly, have a high density of specific sensory receptors. These areas are correspondingly well represented in the somatosensory area of the cerebral cortex (see Chapter 7, "Integrative Functions of the Nervous System").

Touch

Several types of sensory receptors are involved in the sensation of touch (Fig. 4.6). Three of these receptors (Merkel disk, Meissner corpuscles, and Pacinian corpuscles) are distributed on various areas of the skin but are concentrated in regions void of hair (e.g., palm of the hand). The nerve endings of these receptors respond to pressure and texture; therefore, they are classified as **mechanoreceptors**. All three mechanoreceptor types are stimulated by deformation of the nerve ending. When these receptors are deformed, either due to pressure or release of pressure, they cause receptor potentials to be generated by opening pressure-sensitive sodium ion channels in the axon membrane, which allows an influx of Na^+ and resulting in the eventual formation of an action potential in the sensory neuron.

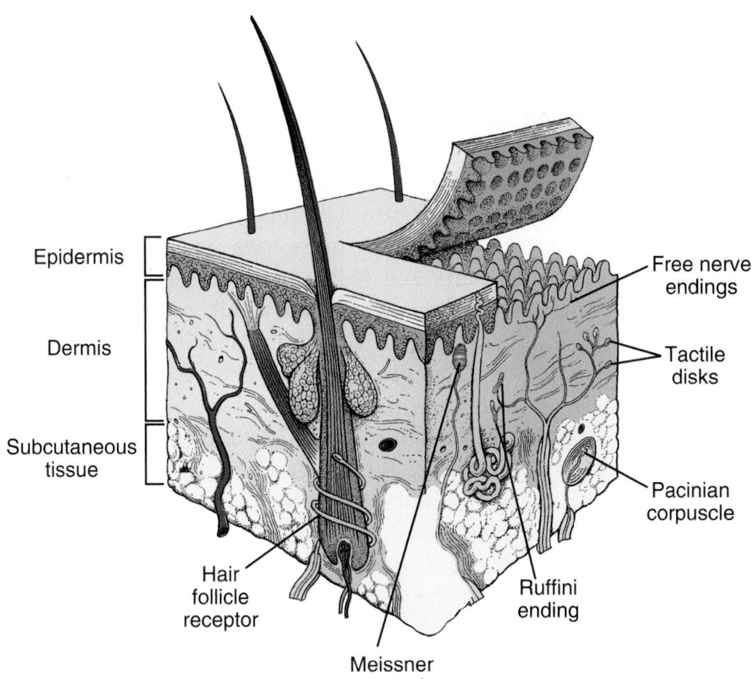

Figure 4.6 Tactile receptors in the skin. Cutaneous receptors providing a sense of touch, pressure, stretch, and vibration are the Merkel disk, Meissner corpuscle, Pacinian corpuscle, Ruffini corpuscle, and the hair follicle receptor.

The **Merkel disks** are receptors (located in the lowest layer of the epidermis) that exhibit slow adaptation and also respond to steady pressure.

The **Meissner corpuscle** is a cutaneous touch receptor that mediates the sensation of light touch. These receptors are distributed on various areas of the skin but are concentrated in areas sensitive to light touch, such as fingertips, lips, and nipples. The Meissner corpuscle is a capsule surrounding the ending of a sensory nerve (or nerves) that wind between stacks of flattened Schwann cells in the capsule interior. These Schwann "cushions" dissipate deformation of the nerve endings caused by sustained points of pressure on the skin. This makes these receptors rapidly adaptive.

The **Pacinian corpuscle** is a very highly adaptive receptor that is responsible for sensitivity to pressure as well as pain. These receptors comprise nerve endings surrounded by multiple gel-like capsules within capsules that act collectively as a mechanical filter. When the capsule is deformed, for example, by a point of pressure, the underlying never ending is deformed and generates an electrical potential. However, the multiple gel capsules almost immediately equalize pressure surrounding the nerve, thus removing any local deformation. Consequently, the local receptor potential ceases to be generated. The nerve ending only deforms again to generate an electrical potential when the initial deforming pressure is suddenly released. Thus, the Pacinian corpuscle best detects the beginning and ending of a rapidly changing local pressure on the skin. This makes these receptors especially suited for detecting fast-changing stimuli such as vibrations.

Thermoreceptors are sensory nerve endings that code for absolute and relative temperatures.

Temperature receptors (thermoreceptors) are classified as nonspecialized naked nerve endings supplied by either thin myelinated fibers (cold receptors) or nonmyelinated fibers (warm receptors) with low conduction velocity. The density of temperature receptors differs at various places on the body surface. They are present in much lower numbers than cutaneous mechanoreceptors, and there are many more cold receptors than warm receptors. The cold-sensitive receptors are preferentially found in moderate density not only in the skin, but also in the cornea, tongue, bladder, and facial skin. The adequate stimulus for a warm thermoreceptor is warming, which results in a concomitant increase in rate of firing of its action potentials. Conversely, cooling results in a decrease in the action potential discharge rate of the warm receptor. Cold-sensitive thermoreceptors give rise to the sensation of cooling or coldness. Their rate of firing increases with cooling and decreases during warming.

The perception of temperature stimuli is closely related to the properties of the receptors. The phasic component of the response is apparent in our adaptation to sudden immersion in, for example, a warm bath. The sensation of warmth, apparent at first, soon fades away, and a less intense impression of the steady temperature may remain. Moving to somewhat cooler water produces an immediate sensation of cold that soon fades away. Over an intermediate temperature range (the "comfort zone"), there is no appreciable temperature sensation. This range is approximately 30°C to 36°C for a small area of skin; the range is narrower when the whole body is exposed. Outside this range, steady temperature sensation depends on the ambient (skin) temperature. Skin temperatures in the range of 30°C up to about 45°C are sensed as harmless warmth, whereas skin temperatures >45°C elicit pain. At skin temperatures <17°C, **cold pain** is sensed, but this sensation arises from pain receptors, not cold receptors. Cold pain can be induced when something cold is swallowed quickly such as ice cream, hence the term "ice-cream headache." Cold pain can be activated under certain abnormal conditions, such as with peripheral neuropathy or from chemotherapy. Severely diabetic patients who suffer from peripheral neuropathy and patients who undergo chemotherapy can exhibit a cold-induced pain response when they touch a cool object. Both result from damage to the receptors. At very high skin temperatures (>45°C), there is a sensation of **paradoxical cold**, caused by activation of a part of the cold receptor population.

Temperature perception is subject to considerable processing by higher centers. Although the perceived sensations reflect the activity of specific receptors, the phasic component of temperature perception may take several minutes to be completed, whereas the adaptation of the receptors is complete within seconds.

Nociceptors are free nerve endings that trigger the sensation of pain in the brain.

Pain is sensed by a population of specific receptors called nociceptors. These receptors can detect mechanical, thermal, or chemical changes and are found in the skin, on internal organ surfaces, and on joint surfaces. They typically have a high threshold for mechanical, chemical, and thermal stimuli (or a combination) of intensity sufficient to cause tissue destruction. All pain receptors are free nerve endings, and the concentration of these receptors varies throughout the body, mostly on the skin and less so in the internal organs. In the skin, these are the free endings of thin myelinated and nonmyelinated fibers with characteristically low conduction velocities. The skin has many more points at which pain can be elicited than it has mechanically or thermally sensitive sites. Because of the high threshold of pain receptors (compared with that of other cutaneous receptors), we are usually unaware of their existence.

Superficial pain may often have two components: an immediate, sharp, and highly localizable *initial pain* and, after a latency of about 1 second, a longer lasting and more diffuse *delayed pain*. These two submodalities appear to be mediated by different nerve fiber endings. In addition to their normally high thresholds, both cutaneous and deep pain receptors show little adaptation, a fact that is unpleasant but biologically necessary. Deep and visceral pain appears to be sensed by similar nerve endings, which may also be stimulated by local metabolic conditions such as an electrolyte imbalance leading to muscle cramps. Referred pain is a term to describe the pain phenomenon that is perceived at a site adjacent to the site of injury. Sometimes this pain is also referred to as **reflective pain**. One of the best examples of referred pain is during an episode of ischemia brought on by a myocardial infarction (heart attack) in which pain is frequently experienced in the neck, left arm, shoulders, and back rather than in the chest, the original site of injury.

VISUAL SYSTEM

All sensory systems are important, but, in primates, the visual system is one of the most important. The visual system is the part of the CNS that enables mammals to process visual details. The eyes are organs that detect light and convert it to electrochemical impulses. Complex neural pathways exist that connect the eye, via the **optic nerve**, to the visual cortex and other areas of the brain, which converts these impulses into visual images.

Eyes comprise three layers of specialized tissue.

The eye is a fluid-filled, spherical organ enclosed by a three-layered structure of specialized tissue (Fig. 4.7). The outermost layer consists of a tough layer of connective tissue, called the *sclera*, and accounts for most of the eye's mechanical strength. The **extraocular muscles** are attached to the sclera and control the direction of the eye. The anterior portion of the sclera consists of the transparent layer, the **cornea**, which allows light rays to pass into the interior of the eye. The middle layer of the eye is known as the *choroid layer*, which is highly pigmented and highly vascularized. The *iris* is housed in the middle layer and is a circular smooth muscle structure that forms the *pupil*. The pupil is the neurally controlled aperture that controls the amount of light admitted to the interior of the eye. The iris also gives the eye its characteristic color. The transparent *lens* is located just behind the iris and is suspended from the *ciliary body* by strands of fibers called *zonule fibers*. It is important to note that the lens is not part of any of the three tissue layers. The iris separates the space between the cornea and the lens into the *anterior* and *posterior chambers*. The anterior chamber is a space filled with a clear watery liquid, called aqueous humor, and is similar in composition to cerebrospinal fluid. The aqueous humor carries nutrients to the cornea and lens, both of which lack a vascular supply. This liquid is continuously secreted by ciliary epithelial cells located behind the iris. As aqueous fluid accumulates, it is drained through the *canal of Schlemm* and into the venous circulation. If the drainage of aqueous humor is impaired, pressure builds up in the anterior chamber, and internal structures are compressed, leading to optical nerve damage. This condition is known as **glaucoma** and can cause blindness. The posterior chamber lies behind the iris and contains **vitreous humor**, a clear gelatinous liquid that helps to maintain the spherical shape of the eye.

The innermost layer of the eye is actually a bilayer. The outer of these two layers is a one-cell sheet of pigment epithelium, the **retina**, which extends from the optic disk all the way to the edge of the pupil. The inner portion of the bilayer forms the photosensitive cells in the retina, called **rods** and **cones**, and functions in the early stage of image processing. Electrical impulses generated from the rods and cones are carried via the optic nerve to the cerebral cortex and other parts of the brain. Slightly off to the nasal side of the retina is the **optic disc**, where the optic nerve leaves the retina. There are no photoreceptor cells here, resulting in a **blind spot** in the field of vision.

Functional optics of the eye resemble those of a camera.

The image that falls on the retina is real and inverted, as in a camera. Neural processing restores the upright appearance

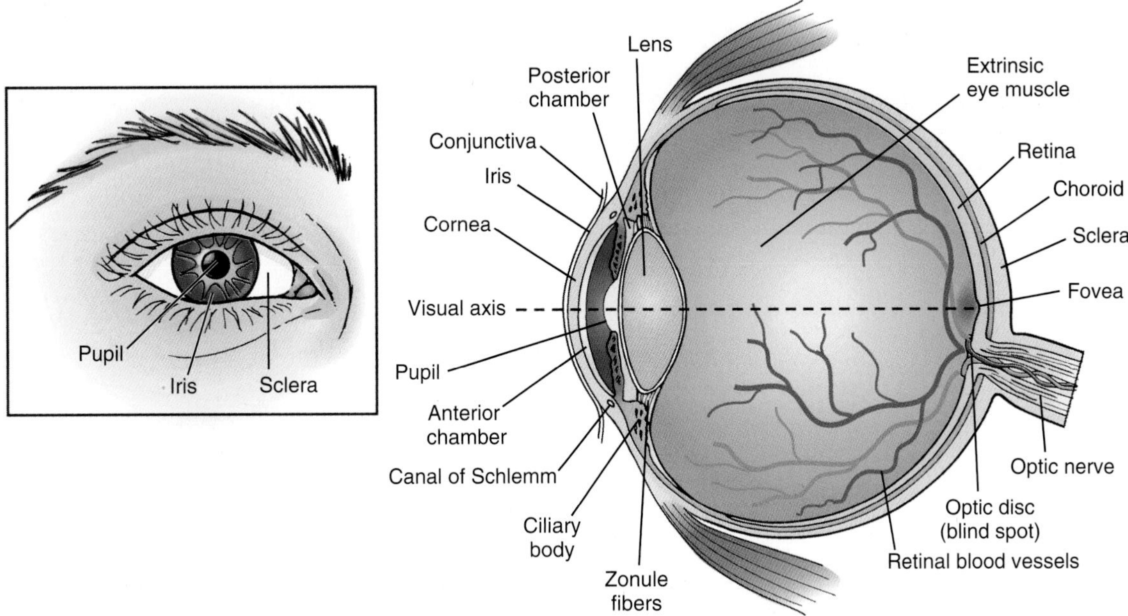

Figure 4.7 The major parts of the human eye. This is a view showing the relative positions of its optical and structural parts.

of the field of view. The image itself can be modified by optical adjustments made by the lens and the iris.

Light is defined as electromagnetic radiation composed of packets of energy called **photons** that travel in the form of waves. The distance between two waves is known as **wavelengths**. Light represents a small part of the electromagnetic spectrum, and the photoreceptors of the eye are sensitive only to wavelengths between 770 nm (red) and 380 nm (violet). The familiar colors of the spectrum all lay between these limits. A wide range of intensities (amplitude or height of the wave), from a single photon to the direct light of the sun, exists in nature. Light rays travel in a straight line in a given medium. Light rays are refracted (bent) as they pass between different media (e.g., air → liquid, or air → glass), and the degree light is bent gives rise to the various refractive indices.

Focusing

Parallel light waves entering the eye must be bent in order to be focused to a common point, called a **focal point**. This is accomplished by the eye's convex lens, which forms a real image. The distance from the lens to this point is its **focal length**, which is expressed in meters (Fig. 4.8). The muscles of the ciliary body can change the focal length by changing the curvature of the lens. When the muscles are fully relaxed, the lens is at its flattest, and the eye and the converging light rays are focused behind the retina. When the ciliary muscle is fully contracted, the lens is at its most curved state, and the eye is focused at its nearest point of vision. The ability to adjust the strength of the lens for close vision is called accommodation. With age, the lens loses its elasticity, and the point of vision moves farther away. This condition is called **presbyopia**, and supplemental refractive power, in the form of external lenses (reading glasses), is required for distinct near vision. Another age-related problem is the loss of transparency in the lens, a condition known as **cataracts**. Normally, the elastic fibers in the lens are completely transparent to visible light. These fibers occasionally become opaque, and light rays have difficulty passing through the opaque lens. Cataracts are treated by surgical removal of the defective lens. An artificial lens may be implanted in its place, or eyeglasses may be used to replace the refractive power of the lens.

Errors of refraction are common (Fig. 4.9). They can be corrected with external lenses (eyeglasses or contact lenses). Farsightedness, or **hyperopia**, is caused by the eyeball being physically too short to focus on distant objects. The natural accommodation mechanism may compensate for distance vision, but the near point will still be too far away. The use of a positive (converging) lens corrects this error. If the eyeball is too long, nearsightedness, or **myopia**, results. In effect, the converging power of the eye is too great; close vision is clear, but the eye cannot focus on distant objects. Concave lenses can correct this defect. If the curvature of the cornea is not symmetric, **astigmatism** results.

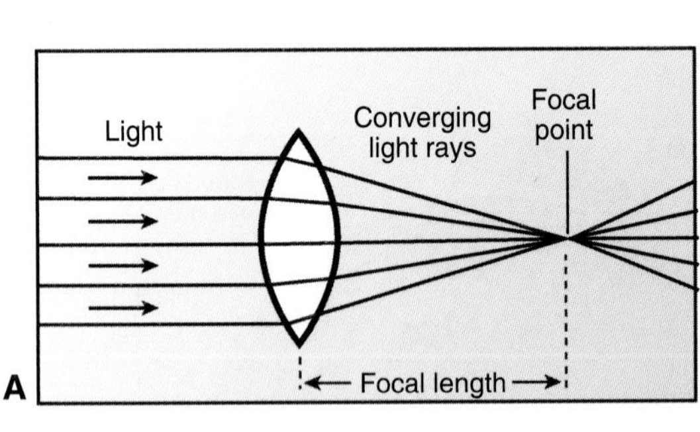

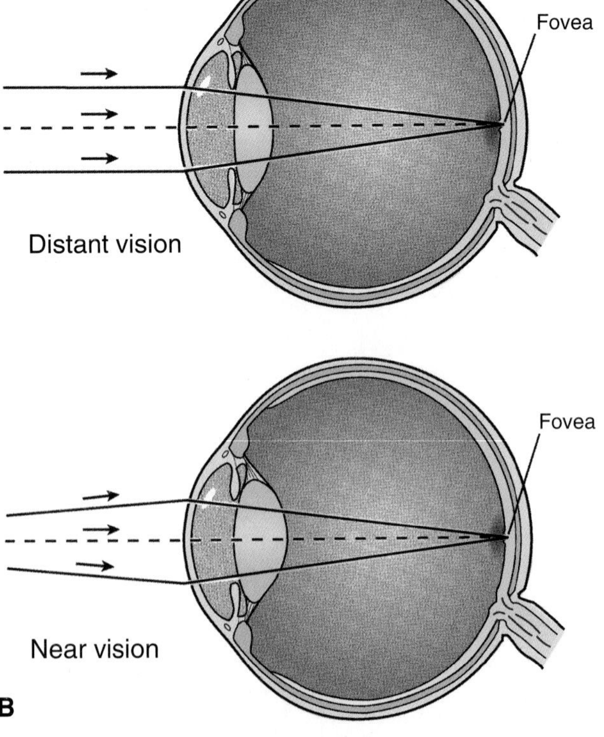

Figure 4.8 The eye as an optical device. (A) The eye converges light to a focal point on the retina by a convex lens. **(B)** During fixation, the center of the image falls on the fovea. For distant vision, the lenses are flattened, and rays from a distant object are brought to a sharp focus. For near vision, the lens curvature increases with accommodation, and rays from a nearby object are focused.

From Bench to Bedside / 4.1

Age-Related Macular Degeneration

The leading cause of impaired vision in older people is a condition known as **age-related macular degeneration (AMD)**. Approximately 30% of persons older than 75 years have some form of this condition, which leads to decrease of the sharpness of central vision, especially in bright light. It is a progressive condition whose causes are obscure and for which no cure is available. Despite this, a number of current and emerging treatments can slow the progress of the disease and provide some improvement in function.

Approximately 10% to 15% of AMD cases are of the *exudative*, or "wet" type. It arises from a proliferation of blood vessels in the choroid layer of the retina behind the macula lutea, which is a small area of densely packed rod and cone cells directly at the fixation point of straight-ahead vision. This area is responsible for our sharpest vision, and its mechanical disruption by the proliferating blood vessels leads to distortion of vision and progressive loss of photoreceptors. The process does not affect areas of the retina peripheral to the macula, so a considerable amount of peripheral vision can be preserved. The remaining AMD cases are of the "dry" type, with changes in the retinal pigment epithelium and "drusen" (see below). This latter form poses less threat to vision, and high levels of zinc and antioxidants can slow its progress.

The retina of an affected eye is characterized by the presence of **drusen**, which are localized deposits of cellular debris whose accumulation further distorts the shape of the retina. An important diagnostic tool, and one used by patients to follow the course of the disease, is the *Amsler grid*, a grid of black lines forming a horizontal and vertical meshwork. The lines, when observed by one eye at a time, will appear "wavy" when the retina is distorted. Thus the patient may detect sudden changes in visual acuity and seek therapy.

The causes of the condition are still obscure. There is a strong suggestion of a hereditary predisposition, and it is most common among white persons from North America and Europe. The primary biologic agent associated with the process of **choroidal neovascularization** is a molecule known as **vascular endothelial growth factor (VEGF)**. This factor is responsible for the growth of blood vessels (angiogenesis) at many sites in the body. Some evidence also implicates an inflammatory process as an etiologic factor. The disease is also associated with cardiovascular illness and the risk factors that it entails. Smoking also is an aggravating factor. Although there are no natural animal models for a human-like AMD, several transgenic mouse models can duplicate features of the disease and are being used in its study. It has recently been found that mice with a congenital lack of photoreceptors can be made sensitive to light by the retinal implantation of stem cell–derived embryonic photoreceptor (rod) cells. This appears to be a promising avenue for eventual treatments of a number of types of blindness including AMD.

For the early stages of AMD there is little specific treatment available. Some help has been obtained by dietary supplementation with high-dose antioxidant vitamins (such as C, E, and β carotene) and zinc. Lutein may provide some benefit, and the effects of corticosteroids are being investigated. For advanced AMD, there are several avenues of therapy, and others are being developed. One of the first such approaches was *laser coagulation* of the invading blood vessels. This is a thermal process, and it damages the overlying retinal tissue, producing blind spots. This limits its use to areas outside the macula. A more recent technique is *verteporfin photodynamic laser treatment*; in this method, a dye is injected into the retinal circulation. When a low-intensity laser pulse activates it, the dye releases free radicals that cause occlusion in the target vessel without damage to the retina. Because of the role of VEGF in the neovascularization process, use of an isoform-specific VEGF inhibitor called *pegaptanib sodium* has achieved some success. This drug is well tolerated, and its effectiveness has been established. However, it must be delivered by a long series of intraocular injections, and forms of sustained-release delivery systems to simplify treatment are under development. The U.S. Food and Drug Administration has recently approved the drug *ranibizumab*, which is a recombinant humanized monoclonal antibody fragment that targets all of the isoforms of VEGF. In patients with the wet form of AMD, this drug (trade name Lucentis) has been shown to improve vision. Finally, a surgical approach involves the implantation of a miniature telescope in the anterior chamber of the eye in the place of the biologic lens. This optical system spreads the light usually delivered to the fovea (5° of the visual field) to a much larger and usually undamaged surface of the more peripheral retina (a 55° field). Such a device, medically suited to only a subgroup of AMD patients, requires relearning of the visual process but can significantly increase the amount of useful vision.

AMD is a serious detriment to the quality of life in older persons and to those who care for them. Current research is aimed at a better understanding of the cause and progression of the disease and at finding specific therapies that are both effective and practical to apply.

Light regulation

The iris, which has both sympathetic and parasympathetic innervation, controls the diameter of the pupil (Fig. 4.10). It is capable of a 30-fold change in area and in the amount of light admitted to the eye. This change is under complex reflex control, and bright light entering just one eye will cause the appropriate constriction response in both eyes. As with a camera, when the pupil is constricted, less light enters, but the image is focused more sharply because the more poorly focused peripheral rays are cut off.

Phototransduction by retinal cells converts light energy into neural electrical-chemical signals.

The major function of the eye is to focus light waves from the external environment and create an image of the visual world on the retina, which serves much the same function as film in a camera. The retina is an extension of the CNS and is a multilayered structure containing the photoreceptor cells and a complex web of several types of nerve cells (Fig. 4.11). Light striking the photosensitive cells (rods and cons) initiates a

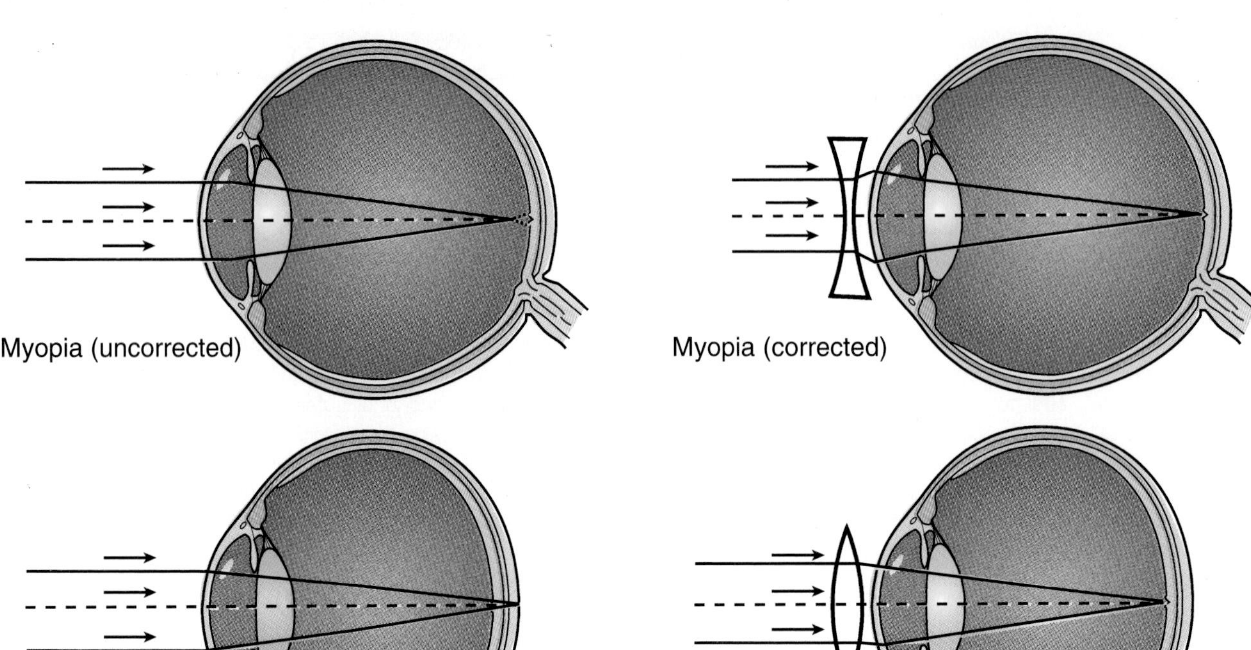

Figure 4.9 The use of external lenses to correct refractive errors. The external optical corrections change the effective focal length of the natural optical components. For myopia (nearsightedness), vision is corrected with concave lenses, which divert light before it reaches the eye. For hyperopia (farsightedness), vision is corrected with convex lenses, which converge light before it reaches the eye.

cascade of chemical and electrical events that ultimately trigger nerve impulses, which are sent to various visual centers in the brain through the fibers of the optic nerve.

The retina consists of a stack of four main neuronal layers: pigment epithelium, the photoreceptor layer, the neural network layer, and, finally, the ganglion cell layer. The pigment layer consists of cells with high **melanin** content and functions to sharpen an image by preventing the scattering of stray light. People with albinism lack this pigment layer and have blurred vision that cannot be corrected effectively with external glasses.

The photoreceptor layer captures the visual image on the rods and cones. The rods and cones are packed tightly side by side, with a density of many thousand per square millimeter.

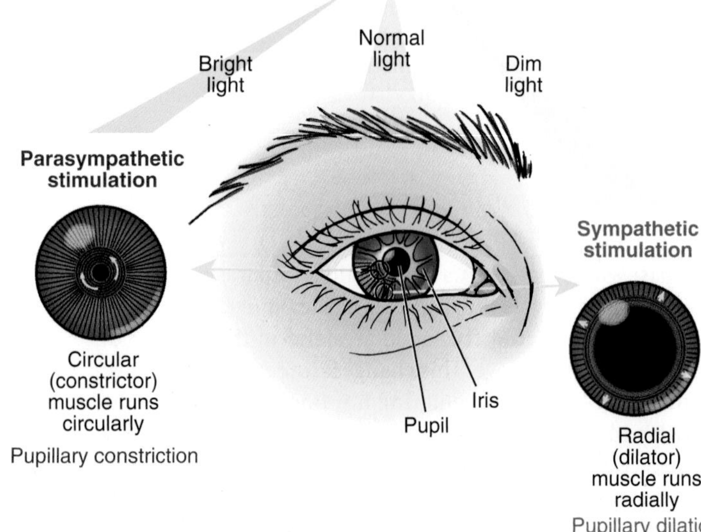

Figure 4.10 Control of pupillary diameter. The pupil controls the amount of light entering the eye by variable contraction of the iris muscles to admit more or less light. The iris muscles run circular (constriction) and radial (dilation).

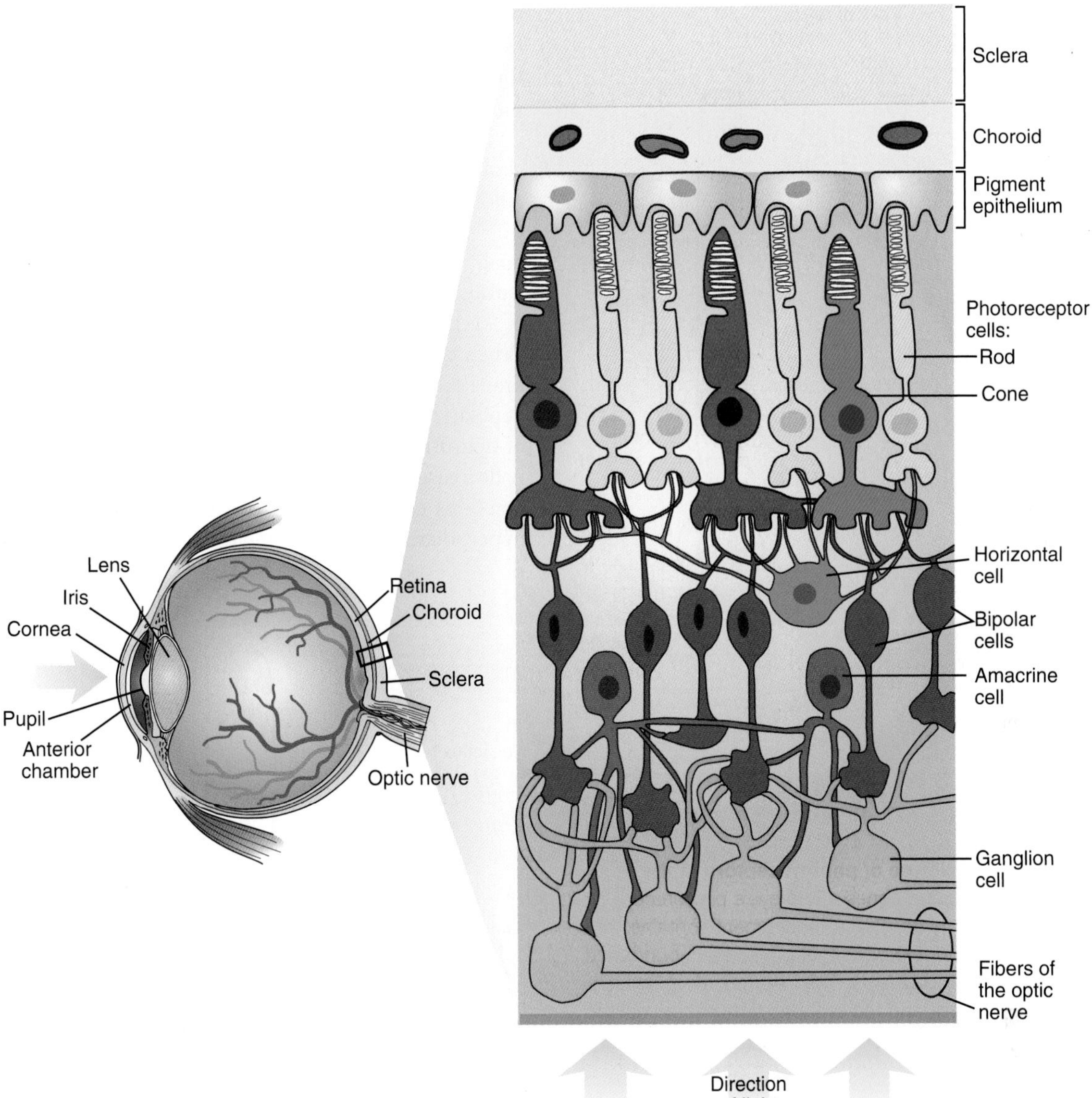

Figure 4.11 Organization of the human retina. The path of the visual retinal layers extends from the rods and cones (photoreceptor cells) to the ganglion cells. The light-sensitive end of the photoreceptor cells faces the choroid layer and away from the incoming light. The amacrine and horizontal cells are involved in retinal processing of visual input.

The photoreceptors are divided into two classes. The cones are responsible for **photopic (daytime) vision**, which is in color (chromatic), and the rods are responsible for **scotopic (nighttime) vision**, which is not in color. Their functions are basically similar, although they have important structural and biochemical differences.

Cone cells have an *outer segment* that tapers to a point (Fig. 4.12). Three different photopigments are associated with cone cells. The pigments differ in the wavelength of light that optimally excites them. The peak spectral sensitivity for the *red-sensitive pigment* is 560 nm; for the *green-sensitive pigment*, it is about 530 nm; and for the *blue-sensitive pigment*, it is about 420 nm. The corresponding photoreceptors are called *red, green*, and *blue cones*, respectively. At wavelengths away from the optimum, the pigments still absorb light but with reduced sensitivity. Because of the interplay between light intensity and wavelength, a retina with only one class of cones would not be able to detect colors unambiguously. The presence of two of the three pigments in each cone removes this uncertainty. Color blind individuals, who have a genetic lack of one or more of the pigments or who lack an associated transduction mechanism, cannot distinguish between the affected colors. Loss of a single color system produces *dichromatic vision*, and lack of two of the systems causes *monochromatic vision*. If all three are lacking, vision is monochromatic and depends only on the rods.

A **rod cell** is long, slender, and cylindrical, and it is larger than a cone cell (see Fig. 4.12). Its outer segment contains numerous photoreceptor disks composed of cellular membrane in which the molecules of the photopigment

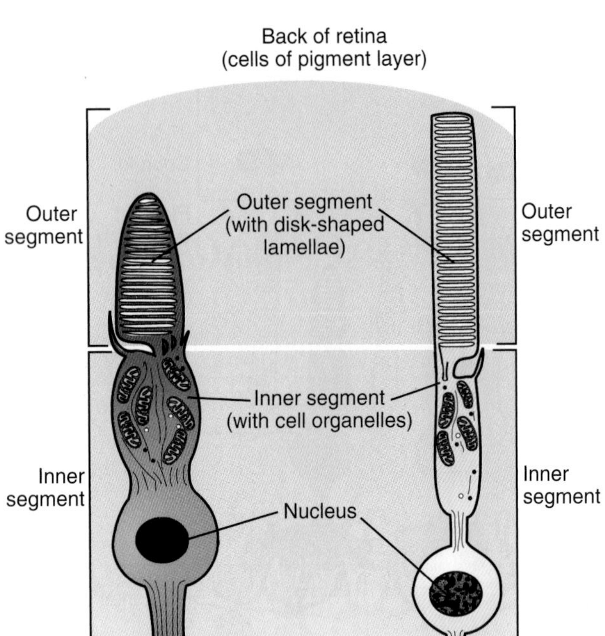

Figure 4.12 Structure of photoreceptors of the human retina. The rod and cones of the eye's photoreceptors contain three segments. The outer segment contains stacked flattened membrane disks that contain an abundance of light-absorbing photopigments. The inner segment houses the metabolic machinery, and the synaptic terminal stores and releases neurotransmitters.

rhodopsin are embedded. The lamellae near the tip are regularly shed and replaced with new membrane synthesized at the opposite end of the outer segment. The *inner segment*, connected to the outer segment by a modified **cilium**, contains the cell nucleus, many mitochondria that provide energy for the phototransduction process, and other cell organelles. At the base of the cell is a synaptic body that makes contact with one or more bipolar nerve cells and liberates a transmitter substance in response to changing light levels.

The visual pigments of the photoreceptor cells convert light to a nerve signal. This process is best understood in the case of the rod cells. In the dark, the pigment rhodopsin (or *visual purple*) consists of a light-trapping **chromophore** (the part of a molecule responsible for its color) called **scotopsin**, which is chemically conjugated with **11-*cis*-retinal**, the aldehyde form of vitamin A1. When struck by light, rhodopsin undergoes a series of rapid chemical transitions, with the final intermediate form metarhodopsin II providing the critical link between this reaction series and the electrical response. The end products of the light-induced transformation are the original scotopsin and an all-*trans* form of retinal, now dissociated from each other. Under conditions of both light and dark, the all-*trans* form of retinal is isomerized back to the 11-*cis* form, and the rhodopsin is reconstituted. All of these reactions take place in the highly folded membranes comprising the outer segment of the rod cell.

The cellular process of visual signal transduction is shown in Figure 4.13. The coupling of the light-induced reactions and the electrical response involves the activation of **transducin**, a G protein; the associated exchange of guanosine triphosphate for guanosine diphosphate activates a **phosphodiesterase**. This, in turn, catalyzes the breakdown of cyclic guanosine monophosphate (cGMP) to 5′-GMP. When cellular cGMP levels are high (as in the dark), membrane sodium channels are kept open, and the cell is relatively depolarized. Under these conditions, there is a tonic release of neurotransmitter from the synaptic body of the rod cell. A decrease in the level of cGMP as a result of light-induced reactions causes the cell to close its sodium channels and hyperpolarize, thus reducing the release of neurotransmitter.

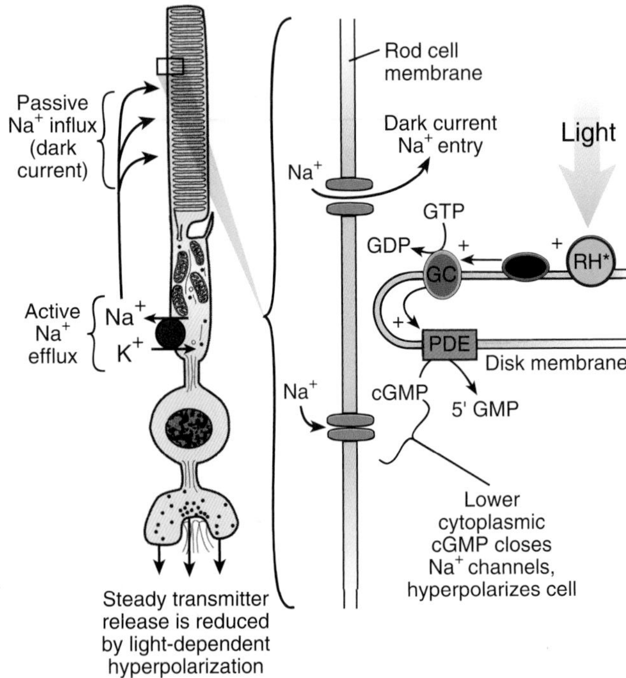

Figure 4.13 Visual signal transduction and photoreceptor activity in the dark and light. *Left*: An active Na^+/K^+ pump maintains the ionic balance of a rod cell, while Na^+ enters passively through channels in the plasma membrane, causing a maintained depolarization and a dark current under conditions of no light. *Right*: The amplifying cascade of reactions (which take place in the disk membrane of a photoreceptor) allows a single activated rhodopsin molecule to control the hydrolysis of 500,000 cyclic guanosine monophosphate (cGMP) molecules. (See text for details of the reaction sequence.) In the presence of light, the reactions lead to a depletion of cGMP, resulting in the closing of cell membrane Na^+ channels and the production of a hyperpolarizing generator potential. The release of neurotransmitter decreases during stimulation by light. RH*, activated rhodopsin; TR, transducin; GC, guanylyl cyclase; PDE, phosphodiesterase; GTP, guanosine triphosphate; GDP, guanosine diphosphate.

This change is the signal that is further processed by the nerve cells of the retina to form the final response in the optic nerve. An active sodium pump maintains the cellular concentration at proper levels. A large amplification of the light response takes place during the coupling steps. One activated rhodopsin molecule will activate approximately 500 transducins, each of which activates the hydrolysis of several thousand cGMP molecules. Under proper conditions, a rod cell can respond to a single photon striking the outer segment. The processes in cone cells are similar, although there are three different opsins (with different spectral sensitivities), and the specific transduction mechanism is also different. In addition, the overall sensitivity of the transduction process is lower.

In the light, much rhodopsin is in its unconjugated form, and the sensitivity of the rod cell is relatively low. During the process of *dark adaptation*, which takes about 40 minutes to complete, the stores of rhodopsin are gradually built up, with a consequent increase in sensitivity (by as much as 25,000 times). Cone cells adapt more quickly than rods, but their final sensitivity is much lower. The reverse process, *light adaptation*, takes about 5 minutes.

Neural network layer

Bipolar cells, **horizontal cells**, and **amacrine cells** comprise the *neural network layer* of the eye (see Fig. 4.11). These cells together are responsible for considerable initial processing of visual information. Because the distances between neurons here are so small, most cellular communication involves the **electrotonic spread** of cell potentials, rather than propagated action potentials. Light stimulation of the photoreceptors produces hyperpolarization that is transmitted to the bipolar cells. Some of these cells respond with a depolarization that is excitatory to the ganglion cells, whereas other cells respond with a hyperpolarization that is inhibitory. The horizontal cells also receive input from rod and cone cells but spread information laterally, causing inhibition of the bipolar cells on which they synapse. Another important aspect of retinal processing is **lateral inhibition.** A strongly stimulated receptor cell can inhibit, via lateral inhibitory pathways, the response of neighboring cells that are less well illuminated. This has the effect of increasing the apparent contrast at the edge of an image. Amacrine cells also send information laterally but synapse on ganglion cells.

Ganglion cell layer

In the *ganglion cell layer* (see Fig. 4.11), the results of retinal processing are finally integrated by the retinal **ganglion cells**, whose axons form the optic nerve. These cells are tonically active, sending action potentials into the optic nerve at an average rate of five per second, even when unstimulated. Input from other cells converging on the ganglion cells modifies this rate up or down.

Many kinds of information regarding color, brightness, contrast, and so on are passed along the optic nerve. The output of individual photoreceptor cells converges on the ganglion cells. In keeping with their role in visual acuity, relatively few cone cells converge on a ganglion cell, especially in the fovea, where the ratio is nearly 1:1. Rod cells, however, are highly convergent, with as many as 300 rods converging on a single ganglion cell. Although this mechanism reduces the sharpness of an image, it allows for a great increase in light sensitivity.

Signals from the retina are modified and separated before reaching the thalamus and visual cortex.

The retina, unlike a camera, does not simply send a picture to the brain. The retina spatially encodes (compresses) the image to fit the limited capacity of the optic nerve. Encoding is necessary because there are 100 times more photoreceptor cells than ganglion cells. The encoding is carried out by bipolar and ganglion cells. Once the image is spatially encoded, the signal is sent out the optical nerve (via the axons of the ganglion cells) through the optic chiasm to the **lateral geniculate nucleus (LGN)**, which is in the thalamus.

Image crossover

Information from the right and left visual fields transmitted to opposite sides of the brain represents another visual modification. The optic nerves, each carrying about 1 million fibers from each retina, enter the rear of the orbit and pass to the underside of the brain to the **optic chiasma** (Fig. 4.14), where about half of the fibers from each eye "cross over" to the other side. Fibers from the temporal side of the retina do

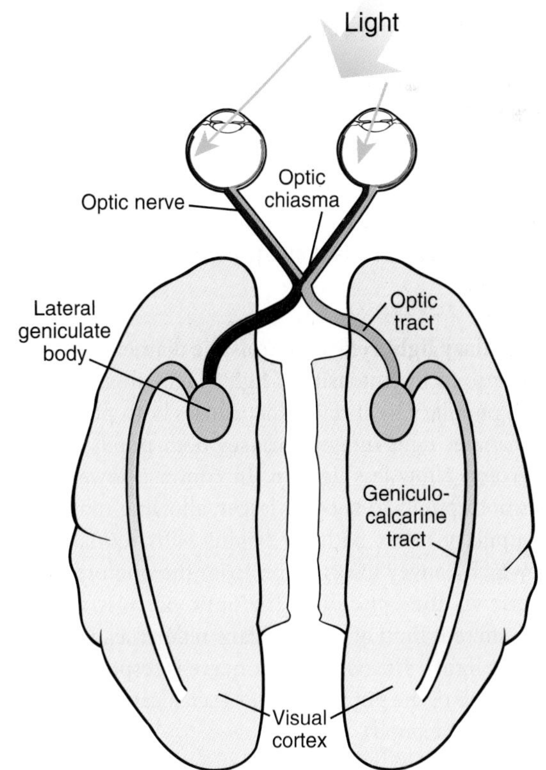

Figure 4.14 The central nervous system pathway for visual information. Fibers from the right visual field will stimulate the left half of each retina, and nerve impulses will be transmitted to the left hemisphere.

not cross the midline but travel in the **optic tract** on the same side of the brain. Fibers originating from the nasal side of the retina cross the optic chiasma and travel in the optic tract to the opposite side of the brain.

The first stop in the brain where information is modified from the visual pathways is the LGN in the thalamus, where the divided output information is separated and relayed to fiber bundles known as **optic radiations** to different zones of the **visual cortex** in the **occipital lobe** of the brain (see Fig. 4.14). Mechanisms in the visual cortex detect and integrate visual information, such as shape, contrast, line, and intensity, into a coherent visual perception.

Information from the optic nerves is also sent to the **suprachiasmatic nucleus** of the hypothalamus, where it participates in the regulation of circadian rhythms; to the **pretectal nuclei**, which are concerned with the control of visual fixation and pupillary reflexes; and to the **superior colliculus**, which coordinates simultaneous bilateral eye movements, such as tracking and convergence.

Depth perception

Depth perception is the visual ability to see the world in three dimensions ("3D") and arises from binocular depth cues that require input from both eyes. Depth perception is lost when one eye is damaged. By using two images of the same scene obtained from slightly different angles, it is possible to triangulate the distance to an object. For example, if the object is far away, the disparity of the image falling on both retinas will be small. If the object is closer, the disparity of that image will be large. This binocular oculomotor cue provides a high degree of accuracy for depth/perception.

Visual reflexes are partially under central nervous system control.

Like a modern camera, the eye has an autofocus for light and an autofocus for distance as well as other reflex mechanisms. These reflexes are mediated, in part, by CNS control of the eye, and they help to adapt and protect the eye from injury.

Pupillary reflex

The **pupillary light reflex** controls the diameter of the pupil in response to the intensity of light. Light shone in one eye elicits a pupillary light reflex that causes both pupils to contract. Greater light intensity causes both pupils to become smaller and allow less light in. In contrast, lower intensity causes both pupils to become larger, allowing more light in. The pupillary reflex pathway begins with retinal ganglion cells, which convey information from the photoreceptors to the optic via the optic disc. The optic nerve is responsible for the afferent limb of the pupillary reflex (i.e., it senses the incoming light). The oculomotor nerve is responsible for the efferent limb of the pupillary reflex (i.e., it causes the muscles to constrict the pupil).

The pupillary light reflex performs several important functions: (1) It regulates the intensity of light that falls on the retina, thereby assisting in adaption to various levels of light and darkness; (2) It aids in retinal sensitivity to light; and (3) It protects the eye from retinal damage from overexposure to light. Aging affects the pupillary light reflex, which becomes most apparent in going from a light to dark room. With aging, it takes longer to adjust to the darkness.

Corneal reflex

The **corneal reflex** is an involuntary blinking of the eyelids elicited by stimulation of the cornea with a foreign object or bright light. Just touching one cornea causes the reflex blinking of both eyes. The corneal reflex protects the eye from foreign material and bright lights. The blink reflex is also initiated with loud sounds (>40 dB). The reflex is mediated through the ophthalmic branch of the trigeminal nerve (fifth cranial nerve) that senses the stimulus on the cornea. The seventh cranial nerve (facial nerve) initiates the motor response.

This reflex is absent in newborns and is diminished with use of contacts lenses. The corneal reflex is often part of a neurologic examination, especially when a patient is evaluated for a coma.

Accommodation reflex

Accommodation is the process whereby the eye changes its optical power to maintain a clear vision on an object as its distance changes. This reflex action is in response to focusing on a near object and then looking at a distant object (and vice versa). When someone accommodates to a near object, two things happen. First, the eyes converge, which is accomplished by the ciliary muscles contracting, thus making the lenses more convex. This shortens the focal length so that the image lands on both retinas. Second, the pupil constricts in order to prevent divergent light rays from hitting the periphery of the retina and resulting in blurred vision. Pupillary constriction also improves the depth of vision. Both things happen automatically as part of the accommodation reflex. Accommodation is another visual reflex that diminishes with age universal occurrence. At about 60 years of age, most people will have noticed a decrease in their ability to focus on close objects.

AUDITORY SYSTEM

The sensory systems of primates are astonishing achievements in design and efficiency. In many ways, the ear stands out among the sensory organs. The ear is the organ that not only receives sound but also plays a major role in the sense of balance and body position. The ear is part of the auditory system and consists of three components: the outer, middle, and inner ear (Fig. 4.15). The outer ear collects sound, and the middle ear amplifies the sound pressure before transmitting it to the fluid-filled inner ear. The inner ear houses two separate sensory systems: the auditory system, which contains the **cochlea** whose receptors convert sound waves into nerve impulses, and the vestibular system, which is involved in balance and special position.

Sound is an oscillating pressure wave composed of frequencies that are transmitted through different media.

Sound is an oscillating pressure wave traveling through a medium (i.e., air, liquid, solid). This process involves both

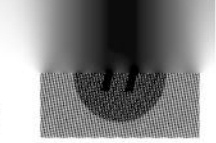

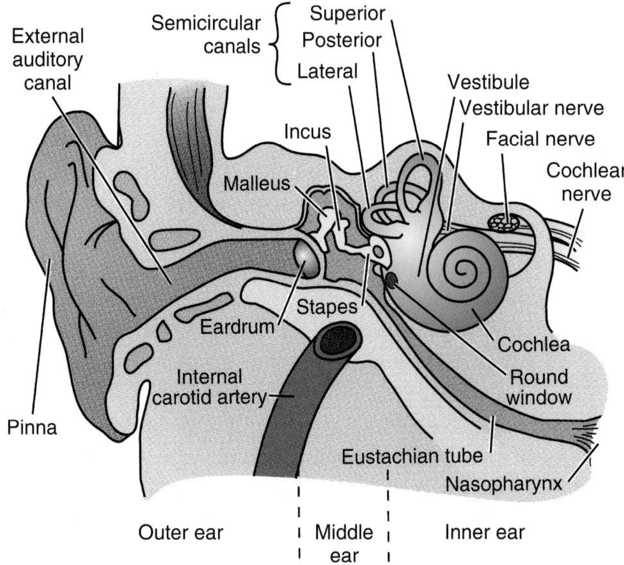

Figure 4.15 Basic structure of the human ear. Each ear consists of three basic parts: the external, middle, and the inner ear. The structures of the middle and inner ear are encased in the temporal bone of the skull. The external and middle portions of the ear transmit airborne sound waves to the fluid-filled inner ear, amplifying sound waves in the process.

compression and **rarefaction** (Fig. 14.16). The distance between the compression peaks is called the wavelength of sound and is inversely related to the frequency. Hearing in humans is normally limited to frequencies between 12 and 20,000 Hz (1 Hz = 1 hertz = 1 cycle/s). Bats, however, can perceive a wider range of frequencies (2–110 kHz). Pitch or tone of sound is determined by frequency. *Sinusoidal sound waves* (those that have regularly repeating oscillations) contain all of their energy at one frequency and are perceived as pure tones. *Complex sound waves,* such as those in speech or music, consist of the addition of several simpler waveforms of different frequencies and amplitudes. Intensity or loudness depends on the amplitude of the sound wave and is expressed as a **decibel (dB)** scale:

$$\text{Decibels (dB)} = 20 \log P/P_0 \qquad (1)$$

where P is sound pressure and P_0 is a reference level pressure (i.e., threshold for normal hearing at the best frequency). The threshold stimulus is set at 0 dB ($\log P/P_0 = 0$). For a sound that is 10 times greater than the reference, the expression becomes

$$\text{dB} = 20 \log (0.002/0.002) = 20 \qquad (2)$$

Thus, any two sounds having a 10-fold difference in intensity have a decibel difference of 20. A 100-fold difference would mean a 40-dB difference, and a 1,000-fold difference would mean a 60-dB difference. Usually the reference value is assumed to be constant and standard, and it is not expressed when measurements are reported. Common sounds range from a faint whisper (20 dB) to a jet taking off (140 dB).

External ear functions in sound localization.

An overall view of the human ear is shown in Figure 4.15. The auricle or *pinna,* the visible portion of the outer ear, collects sound waves and channels them down the external auditory canal. The external auditory canal extends inward through the temporal bone. Wax-secreting glands line the canal, and its inner end is sealed by the tympanic membrane or *eardrum,* a thin, oval, slightly conical, flexible membrane that is anchored around its edges to a ring of bone. An incoming pressure wave traveling down the external auditory canal causes the eardrum to vibrate back and forth in step with the compressions and rarefactions of the sound wave. This is the first mechanical step in the transduction of sound. The overall acoustic effect of the outer ear structures is to produce an amplification of 10 to 15 dB in the frequency range broadly centered around 3,000 Hz.

Middle ear mechanically converts tympanic vibrations to fluid waves in the inner ear.

The middle ear is an air-filled cavity in the temporal bone and functions as the mechanical transformer. The middle ear contains three tiny bones that couple vibration of the eardrum

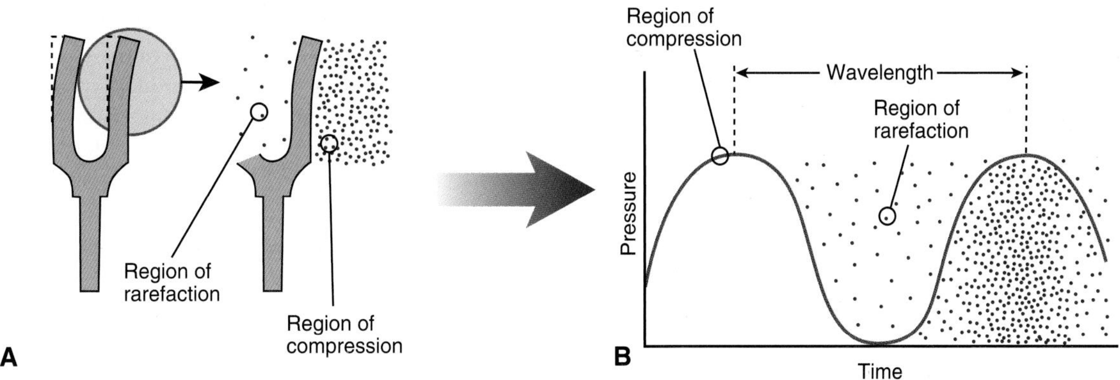

Figure 4.16 Sound wave formations. (A) Sound waves generated from a tuning fork cause molecules ahead of the advancing arm to be compressed and the molecules behind the arm to be rarified. **(B)** Sound waves are propagated as sinusoidal, alternating regions of compression and rarefaction of air molecules. The wavelength of a sinusoidal wave is the spatial period between two peak compression waves.

into waves in the fluid and membranes of the inner ear. The hollow space of the middle ear is called the *tympanic cavity*. The eustachian tube joins the tympanic cavity, and the tube opens briefly during swallowing, allowing equalization of the pressures on either side of the eardrum. During rapid external pressure changes (such as in an elevator ride or during takeoff or descent in an airplane), the unequal forces displace the eardrum. Such physical deformation may cause discomfort or pain and, by restricting the motion of the tympanic membrane, may impair hearing. Blockages of the eustachian tube or fluid accumulation in the middle ear (as a result of an infection) can also lead to difficulties with hearing.

Bridging the gap between the tympanic membrane and the inner ear is a chain of three small bones, the ossicles (Fig. 4.17). The *malleus* (Latin for "hammer") is attached to the eardrum in such a way that the back-and-forth movement of the eardrum causes a rocking movement of the malleus. The *incus* (Latin for "anvil") connects the head of the malleus to the third bone, the *stapes* (Latin for "stirrup"). This last bone, through its oval footplate, connects to the **oval window** of the inner ear and is anchored there by an annular ligament.

Four separate suspensory ligaments hold the ossicles in position in the middle ear cavity. The superior and lateral ligaments lie roughly in the plane of the ossicular chain and anchor the head and shaft of the malleus. The anterior ligament attaches the head of the malleus to the anterior wall of the middle ear cavity, and the posterior ligament runs from the head of the incus to the posterior wall of the cavity. The suspensory ligaments allow the ossicles sufficient freedom to transmit the vibrations of the tympanic membrane to the oval window. This mechanism is especially important because, although the eardrum is suspended in air, the oval window seals off a fluid-filled chamber. When sound waves in air strike liquid, most of the energy (~99%) is reflected off the liquid and lost. The middle ear allows the impedance matching of sound traveling in air to acoustic waves traveling in a system of fluids and membranes in the inner ear. Two principles of mechanical advantage are used in the impedance matching, the hydraulic principle and the lever principle. The shape of the articulated ossicular chain is lever-like and has a lever ratio of about 1.3:1.0, producing a slight gain in force. In addition, the vibrating portion of the tympanic membrane is coupled to the smaller area of the oval window (~ a 17:1 ratio). These conditions result in a pressure gain of around 25 dB, largely compensating for the potential loss. Although the efficiency depends on the frequency, approximately 60% of the sound energy that strikes the eardrum is transmitted to the oval window.

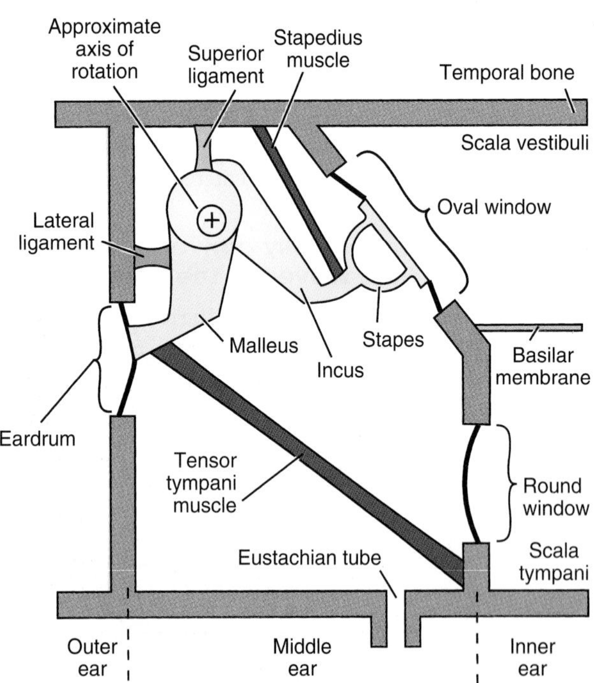

Figure 4.17 Model of the middle ear. Vibrations from the eardrum are transmitted by the lever system formed by the ossicular chain to the oval window of the scala vestibuli. The anterior and posterior ligaments, part of the suspensory system for the ossicles, are not shown. The combination of the four suspensory ligaments produces a virtual pivot point (marked by a *cross*); its position varies with the frequency and intensity of the sound. The stapedius and tensor tympani muscles modify the lever function of the ossicular chain.

The middle ear is also able to substantially dampen sound conduction when faced with very loud sounds by noise-induced reflex contraction of the middle-ear muscles. These small muscles are attached to the ossicular chain and help hold the bones in position and modify their function (see Fig. 4.17). The *tensor tympani muscle* inserts on the malleus (near the center of the eardrum), passes diagonally through the middle ear cavity, and enters the *tensor canal*, in which it is anchored. Contraction of this muscle limits the vibration amplitude of the eardrum and makes sound transmission less efficient. The *stapedius muscle* attaches to the stapes near its connection to the incus and runs posteriorly to the mastoid bone. Its contraction changes the axis of oscillation of the ossicular chain and causes dissipation of excess movement before it reaches the oval window. These muscles are activated by a reflex (simultaneously in both ears) and contract in response to moderate and loud sounds. They act to reduce the transmission of sound to the inner ear, thereby protecting its delicate structures. This is called the **acoustic reflex.**

Because this reflex requires up to 150 milliseconds to operate (depending on the loudness of the stimulus), it cannot provide protection from sharp or sudden bursts of sound.

The process of sound transmission can bypass the ossicular chain entirely. If a vibrating object, such as a tuning fork, is placed against a bone of the skull (typically the mastoid), the vibrations are transmitted mechanically to the fluid of the inner ear, where the normal processes act to complete the hearing process. Bone conduction is used as a means of diagnosing hearing disorders that may arise because of lesions in the ossicular chain. Some hearing aids employ bone conduction to overcome such deficits.

Inner ear transduces sound.

The inner ear is a tortuous system of fluid-filled channels of the bony structure, the cochlea, which contains the receptor cells for both the auditory and vestibular systems. The inner ear is where the actual process of sound transduction takes place.

Cochlea

The cochlea defines the overall structure of the auditory transducer. It propagates the mechanical signals as waves in fluid and membranes and tranduces them to nerve impulses, which are transmitted to the brain.

The auditory structures are located in the cochlea (Fig. 4.18), part of a cavity in the temporal bone called the *bony labyrinth*. The cochlea (Latin for "snail") is a fluid-filled spiral tube that arises from a cavity called the *vestibule*, with which the organs of balance also communicate. It is partitioned longitudinally into three divisions (canals) called the *scala vestibuli* (into which the oval window opens), the *scala tympani* (sealed off from the middle ear by the *round window*), and the *scala media* (in which the sensory cells are located). Arising from the bony center axis of the spiral (the *modiolus*) is a winding shelf called the *osseous spiral lamina*. Opposite to it on the outer wall of the spiral is the *spiral ligament*, and connecting these two structures is a highly flexible connective tissue sheet, the **basilar membrane**, which runs for almost the entire length of the cochlea. The basilar membrane separates the scala tympani (*below*) from the scala media (*above*). The **hair cells**, which are the actual sensory receptors, are located on the upper surface of the basilar membrane (see Fig. 4.18C). They are called "hair cells" because each has a bundle of hair-like cilia at the end that project away from the basilar membrane.

The *Reissner membrane*, a delicate sheet only two cell layers thick, divides the scala media (*below*) from the scala vestibuli (*above*) (see Fig. 4.18B). The scala vestibuli communicates with the scala tympani at the apical (distal) end of the cochlea via the *helicotrema*, a small opening where a portion of the basilar membrane is missing. The scala vestibuli and scala tympani are filled with **perilymph**, a fluid high in sodium and low in potassium. The scala media contains **endolymph**, a fluid high in potassium and low in sodium. The endolymph is secreted by the **stria vascularis**, a layer of fibrous vascular tissue along the outer wall of the scala media. Because the cochlea is filled with incompressible fluid and is encased in hard bone, pressure changes caused by the in-and-out motion at the oval window (driven by the stapes) are relieved by an out-and-in motion of the flexible round window membrane.

Sensory structures

The **organ of Corti** is formed by structures located on the upper surface of the basilar membrane and runs the length of the scala media (see Fig. 4.18C). It contains one row of some 3,000 *inner hair cells*. The *arch of Corti* and other specialized supporting cells separate the inner hair cells from the three or four rows of *outer hair cells* (about 12,000)

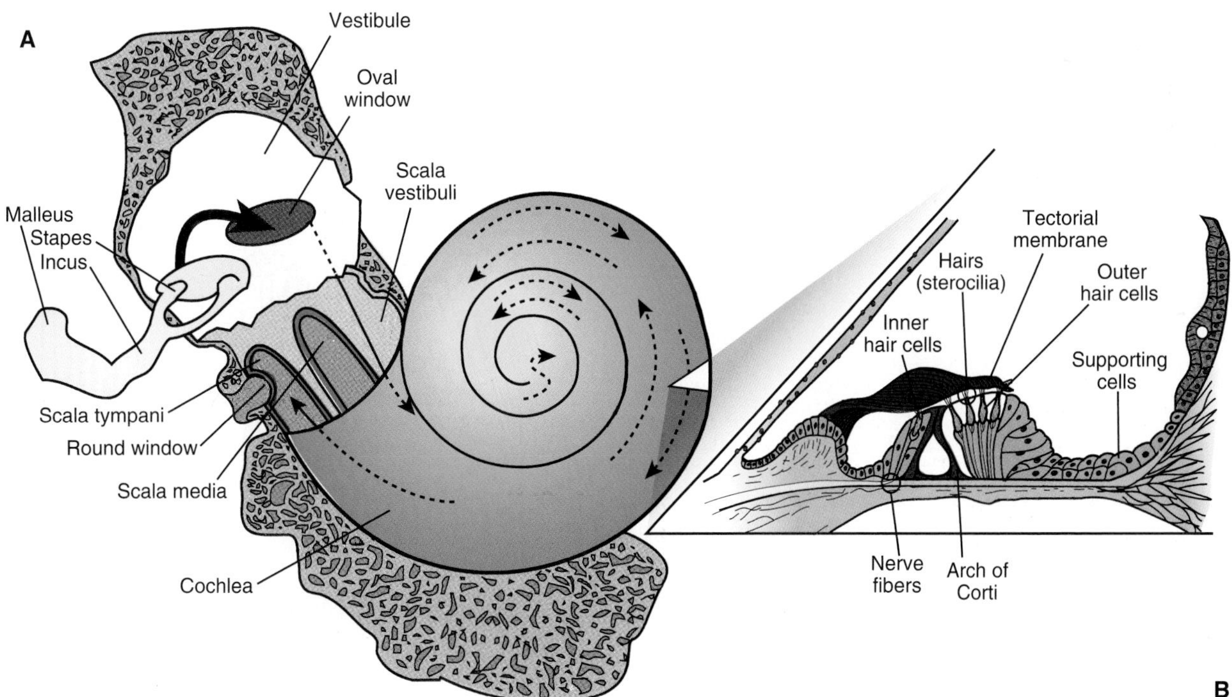

Figure 4.18 The cochlea and the organ of Corti. (A) The pea-sized, snail-shaped cochlea is a coiled system and is the hearing portion of the inner ear. **(B)** An enlargement of a cross section of the organ of Corti, showing the relationships among the hair cells and the membranes. Hair cells in the organ of Corti transduce fluid movement into neural signals.

located on the stria vascularis side. The rows of inner and outer hair cells are inclined slightly toward each other and are covered by the *tectorial membrane*, which arises from the *spiral limbus*, a projection on the upper surface of the osseous spiral lamina.

Nerve fibers from cell bodies located in the *spiral ganglia* form radial bundles on their way to synapse with the inner hair cells. Each nerve fiber makes synaptic connection with only one hair cell, but each hair cell is served by 8 to 30 fibers. Although the inner hair cells comprise only 20% of the hair cell population, they receive 95% of the afferent fibers. In contrast, many outer hair cells are each served by a single external spiral nerve fiber. The collected afferent fibers are bundled in the *cochlear nerve*, which exits the inner ear via the *internal auditory meatus*. Some efferent fibers also innervate the cochlea. They may serve to enhance pitch discrimination and the ability to distinguish sounds in the presence of noise. Recent evidence suggests that efferent fibers to the outer hair cells may cause them to shorten (contract), altering the mechanical properties of the cochlea.

Hair cells

The hair cells of the inner and the outer rows are similar anatomically. Both sets are supported and anchored to the basilar membrane by *Deiters cells* (see Fig. 4.18B) and extend upward into the scala media toward the tectorial membrane. Extensions of the outer hair cells actually touch the tectorial membrane, whereas those of the inner hair cells appear to stop just short of contact. The hair cells make synaptic contact with afferent neurons that run through channels between Deiters cells. A chemical transmitter of unknown identity is contained in synaptic vesicles near the base of the hair cells. As in other synaptic systems, the entry of calcium ions (associated with cell membrane depolarization) is necessary for the migration and fusion of the synaptic vesicles with the cell membrane prior to transmitter release.

At the apical end of each inner hair cell is a projecting bundle of about 50 **stereocilia**, rod-like structures packed in three parallel, slightly curved rows (see Fig. 4.18C). Minute strands link the free ends of the stereocilia together, so the bundle tends to move as a unit. The height of the individual stereocilia increases toward the outer edge of the cell (toward the stria vascularis), giving a sloping appearance to the bundle. Along the cochlea, the inner hair cells remain constant in size, whereas the stereocilia increase in height from about 4 μm at the basal end to 7 μm at the apical end. The outer hair cells are more elongated than the inner cells, and their size increases along the cochlea from base to apex. Their stereocilia (about 100 per hair cell) are also arranged in three rows that form an exaggerated "W" figure. The height of the stereocilia also increases along the length of the cochlea, and they are embedded in the tectorial membrane. The stereocilia of both types of hair cells extend from the *cuticular plate* at the apex of the cell. The diameter of an individual stereocilium is uniform (about 0.2 μm) except at the base, where it decreases significantly. Each stereocilium contains cross-linked and closely packed **actin** filaments, and, near the tip, is a cation-selective transduction channel.

The process of mechanical transduction in hair cells is shown in Figure 4.19. When a hair bundle is deflected slightly (the threshold is <0.5 nm) toward the stria vascularis, minute mechanical forces open the transduction channels, and cations (mostly potassium) enter the cells. The resulting **depolarization**, roughly proportional to the deflection, causes the release of transmitter molecules, generating afferent nerve action potentials. Approximately 15%

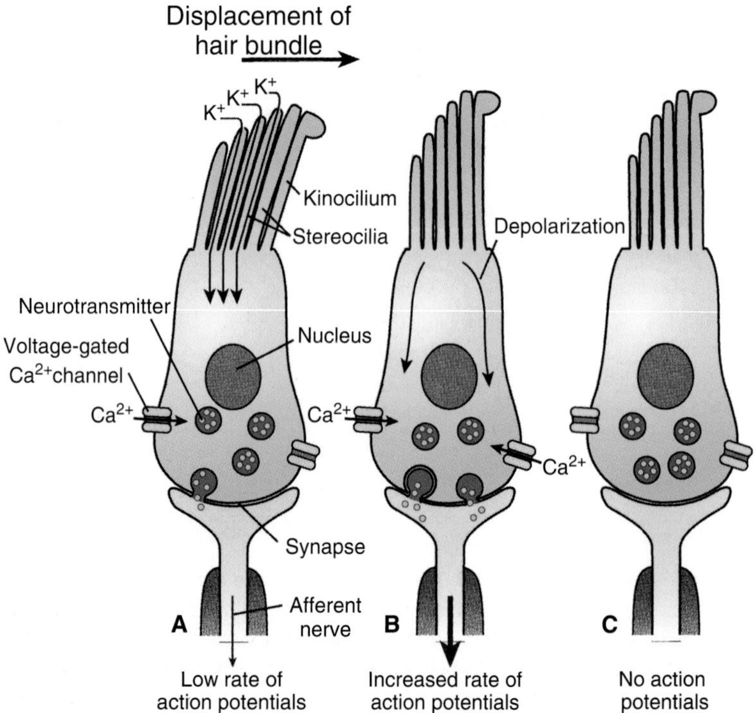

Figure 4.19 Role of stereocilia in sound transduction in the hair cells of the ear. (A) Deflection of the stereocilia opens apical K^+ channels. **(B)** K^+ enters the hair cells causing depolarization. The resulting depolarization opens the voltage-gated Ca^{2+} channels at the basal end of the cell. **(C)** This causes the release of the neurotransmitter, thereby exciting the afferent nerve.

of the transduction channels are open in the absence of any deflection, and bending in the direction of the modiolus of the cochlea results in hyperpolarization, thereby increasing the range of motion that can be sensed. Hair cells are quite insensitive to movements of the stereocilia bundles at right angles to their preferred direction.

The response time of hair cells is remarkable; they can detect repetitive motions of up to 100,000 times per second. They can, therefore, provide information throughout the course of a single cycle of a sound wave. Such rapid response is also necessary for the accurate localization of sound sources. When a sound comes from directly in front of a listener, the waves arrive simultaneously at both ears. If the sound originates off to one side, it reaches one ear sooner than the other and is slightly more intense at the nearer ear. The difference in arrival time is on the order of tenths of a millisecond, and the rapid response of the hair cells allows them to provide temporal input to the auditory cortex. The timing and intensity information are processed in the auditory cortex into an accurate perception of the location of the sound source.

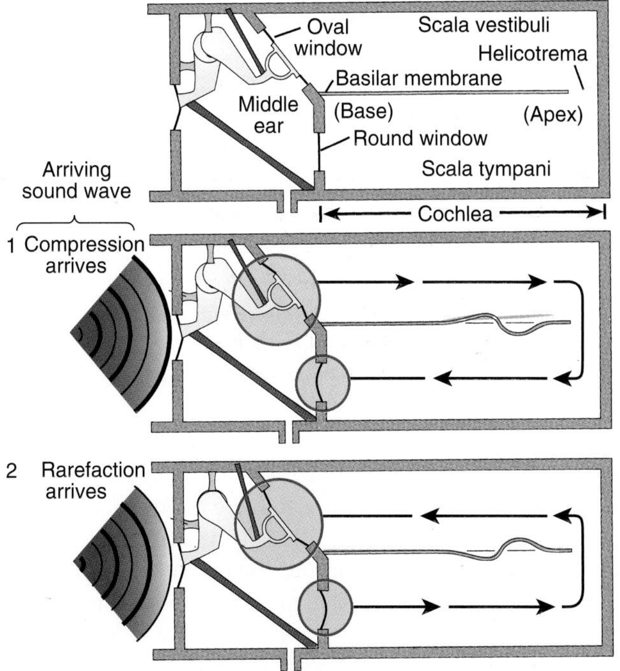

Figure 4.20 The mechanics of the cochlea, showing the action of the structures responsible for pitch discrimination (with only the basilar membrane of the organ of Corti shown). When the compression phase of a sound wave arrives at the eardrum, the ossicles transmit it to the oval window, which is pushed inward. A pressure wave travels up the scala vestibuli and (via the helicotrema) down the scala tympani. To relieve the pressure, the round window membrane bulges outward. Associated with the pressure waves are small eddy currents that cause a traveling wave of displacement to move along the basilar membrane from base to apex. The arrival of the next rarefaction phase reverses these processes. The frequency of the sound wave, interacting with the differences in the mass, width, and stiffness of the basilar membrane along its length, determines the characteristic position at which the membrane displacement is maximal.

Organ of Corti

The actual transduction of sound requires an interaction among the tectorial membrane, the arches of Corti, the hair cells, and the basilar membrane. When a sound wave is transmitted to the oval window by the ossicular chain, a pressure wave travels up the scala vestibuli and down the scala tympani (Fig. 4.20). The canals of the cochlea, being encased in bone, are not deformed, and movements of the round window allow the small volume change needed for the transmission of the pressure wave. Resulting eddy currents in the cochlear fluids produce an undulating distortion in the basilar membrane. Because the stiffness and width of the membrane vary with its length (it is wider and less stiff at the apex than at the base), the membrane deformation takes the form of a "traveling wave," which has its maximal amplitude at a position along the membrane corresponding to the particular frequency of the sound wave (Fig. 4.21). Low-frequency sounds cause a maximal displacement of the membrane near its apical end (near the helicotrema), whereas high-frequency sounds produce their maximal effect at the basal end (near the oval window). As the basilar membrane moves, the arches of Corti transmit the movement to the tectorial membrane, the stereocilia of the outer hair cells (embedded in the tectorial membrane) are subjected to lateral shearing forces that stimulate the cells, and action potentials arise in the afferent neurons.

Because of the tuning effect of the basilar membrane, only hair cells located at a particular place along the membrane are maximally stimulated by a given frequency (pitch). This localization is the essence of the **place theory** of pitch discrimination, and the mapping of specific tones (pitches) to specific areas is called **tonotopic organization.** As the signals from the cochlea ascend through the complex pathways of the auditory system in the brain, the tonotopic organization of the neural elements is at least partially preserved, and pitch can be spatially localized throughout the system.

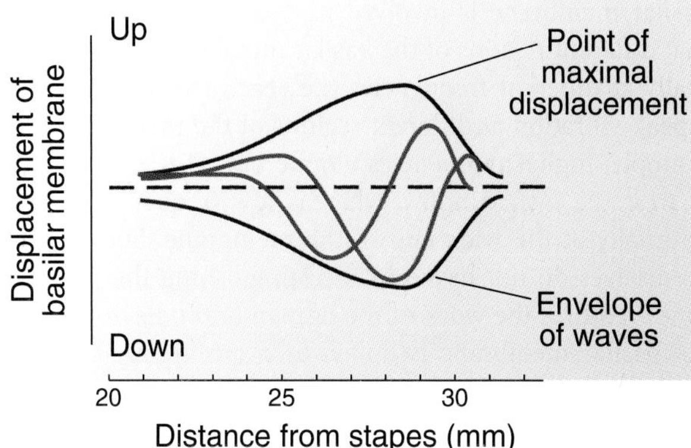

Figure 4.21 Membrane localization of different frequencies. An illustration of a traveling sound wave being displaced along the basilar membrane at two instants. Over time, the peak excursions of many such waves form an envelope of displacement with a maximal value at about 28 mm from the stapes (lower portion). At this position, its stimulating effect on the hair cells will be most intense. Low frequency (<800 Hz) produces maximal effects at the apex of the basilar membrane.

The sense of pitch is further sharpened by the resonant characteristics of the different-length stereocilia along the length of the cochlea and by the frequency–response selectivity of neurons in the auditory pathway. The cochlea acts as both a transducer for sound waves and a frequency analyzer that sorts out the different pitches so that they can be separately distinguished. In the midrange of hearing (around 1,000 Hz), the human auditory system can sense a difference in frequency of as little as 3 Hz. The tonotopic organization of the basilar membrane has facilitated the invention of prosthetic devices whose aim is to provide some replacement of auditory function to people suffering from deafness that arises from severe malfunction of the middle or inner ear.

Central auditory pathways

Nerve fibers from the cochlea enter the spiral ganglion of the organ of Corti. From there, fibers are sent to the *dorsal* and *ventral cochlear nuclei*. The complex pathway that finally ends at the **auditory cortex** in the superior portion of the **temporal lobe** of the brain involves several sets of synapses and considerable crossing over and intermediate processing. As with the eye, there is a spatial correlation between cells in the sensory organ and specific locations in the primary auditory cortex. In this case, the representation is called a *tonotopic map*, with different pitches being represented by different locations, even though the firing rates of the cells no longer correspond to the frequency of sound originally presented to the inner ear.

Pitch discrimination

Pitch represents the perceived frequency of sound and is one of the major auditory attributes to musical tones. Pitch is determined by frequency (cycles per second or hertz) and corresponds closely to the repetition rate of sound waves. Although pitch discrimination (the ability to distinguish between various frequencies) is based on quantified frequencies, it is not an objective physical property but is completely subjective and is based on perceived sound. The basilar membrane is involved in determining pitch. Recall that different regions of the basilar membrane vibrate maximally at different frequencies (i.e., each frequency exhibits a peak vibration at different regions of the membrane). For example, higher frequencies vibrate best at the narrow end nearest the oval window, whereas low frequencies vibrate optimally at the wide end of the membrane. Sound waves themselves do not have pitch; distinguishing the perceived tones requires the work of the human brain. Each region of the basilar membrane is linked to a specific region of the auditory cortex, and tones are perceived in different regions of the primary auditory cortex. Tone deafness is the lack of relative pitch, or the inability to discriminate between musical notes. However, tone-deaf individuals can fully interpret the pitch of human speech.

Loudness

Loudness is another auditory attribute and depends on the amplitude of vibration. Sound waves originating from louder sounds cause the eardrum to vibrate more vigorously. Recall that a greater tympanic membrane deflection is converted into greater amplitude of basilar membrane movement. The CNS interprets this membrane oscillation and hair bending as a louder sound. Although the auditory system is sensitive and can detect very faint sounds, loud sounds can be damaging to the ear. The protective reflexes of the middle ear cannot sufficiently attenuate very loud sounds. Violent vibrations of the basilar membrane that come from sources, such as a jet engine, a siren, or a rock concert, for example, can permanently damage hair cells leading to impaired hearing.

Hearing loss results from a mechanical or a neural problem.

Hearing loss or *deafness* refers to conditions in which individuals are unable to detect or perceive some frequencies of sound. Deafness can be temporary or permanent (partial or complete). Hearing loss is the second most common physical disability in North America, affecting approximately 10% of the population. Hearing impairments are categorized by their types—**conductive deafness** and **sensorineural deafness**. A conductive deafness is an impairment resulting from dysfunction in any of the mechanisms that normally conducts sound waves through the outer ear, the eardrum, or the bones of the middle ear. Common causes include the buildup of earwax that blocks the ear canal, a ruptured eardrum, a middle ear infection, and restriction of ossicular movements of bony structures of the middle ear.

A sensorineural hearing impairment occurs as a result of (1) inner ear dysfunction, especially of the cochlea, where sound vibrations are converted into neural signals, or (2) any part of the brain that subsequently processes these signals. The vast majority of sensorineural deafness is associated with damage to the hair cells of the organ of Corti in the cochlea. This dysfunction may be present from birth from genetic or developmental abnormalities or arise through trauma or disease. Diseases causing deafness includes measles, meningitis, mumps, and venereal disease. Some medications can also cause irreversible damage to the ear. The most important group is the aminoglycosides (primarily gentamicin). Other medications include anti-inflammatory (nonsteroidal) and diuretic drugs. Narcotic painkillers, in particular Vicodin and OxyContin, are another source of medication that causes permanent hearing loss.

One of the most common causes of partial hearing loss is **neural presbycusis**, the progressive loss of the ability to hear high frequency with increasing age as the hair cells "wear out" over time. Long-term exposure to environmental noise damages the hair cells, which increases the onset of neural presbycusis.

VESTIBULAR SYSTEM

Animals need feedback information to keep track of which end is up—that is, how they are oriented to their surrounding environment. The eyes help to keep track of movement with other objects, but the sensory system that also contributes to our balance and sense of spatial orientation is the vestibular system. This system provides dominant input about movement and **equilibrioception**. The **vestibular apparatus** sends signals primarily to the neural structures that control the eye movements and the muscles that keep us upright.

Vestibular system comprises the semicircular canal and the otoliths.

Each ear contains three **semicircular canals** and two **otolithic organs**, the *utricle* and the *saccule* (Fig. 4.22). These structures are located in the bony labyrinth of the temporal bone. As with hearing, the basic sensing elements are hair cells. The canals are arranged three-dimensionally in planes that are at approximately right angles to each other and are called the *horizontal* (lateral), *anterior vertical* (superior), and the *posterior canals* (inferior). The anterior and posterior canals may be collectively called *vertical semicircular canals.* The three pair of canals works in a push–pull fashion—when one canal is stimulated, its corresponding partner on the other side is inhibited and vice versa. The push–pull systems allow the vestibular apparatus to sense all directions of rotation. For example, the right horizontal canal gets stimulated during head rotations to the right, and the left horizontal canal gets stimulate by head rotations to the left. Vertical canals are coupled in a crossed fashion, which allows for stimulations for the anterior canal to also be inhibitory for the contralateral posterior canal and vice versa.

Movement of fluid within the horizontal canal corresponds to rotation of the head around a vertical axis (i.e., the neck). The anterior and posterior canals detect rotations of the head in the sagittal plane (i.e., nodding) and in the frontal plane, as when doing a cartwheel. Both anterior and posterior canals are oriented at approximately 45° between the frontal and sagittal planes. Near its junction with the utricle, each canal has a swollen portion called the **ampulla**. Each ampulla contains the sensory structure for that semicircular canal. It is composed of ciliary bundles encapsulated by a *cupula*, a gelatinous mass (Fig. 4.23). The ciliary bundle consists of a hair cell called a **kinocilium** and a tuft of stereocilia. When the stereocilia are deflected by movement of the endolymph fluid in the canals, the hair bundle pushes on the hair cells in the cupula, which, in turn, converts the mechanical movement into electrical signals. The hair cells synapse with terminal endings of afferent neurons whose axons form the *vestibular nerve.* Depolarization increases the release of neurotransmitters in the hair cells, bringing about a concomitant increased rate of firing in the afferent fibers. Conversely, hyperpolarization inhibits neurotransmitter release from the hair cells and, in turn, decreases the frequency of action potentials in the afferent nerve fibers.

Otolithic organs

Whereas the semicircular canals respond to rotation, the otolithic organs sense linear accelerations. There are two on each side, the utricle and the saccule, are sac-like structures located in the bony chamber, and are housed between semicircular canals and the cochlea. Calcium carbonate crystals, the **otoliths**, are suspended on a viscous gel layer and are heavier than their surroundings (Fig. 4.24). During linear acceleration, the otoliths get displaced, which, in turn, deflects the ciliary bundle of hair cells and produces a sensory signal. Most of the utricular signals elicit eye movements, whereas most of the saccular signals go to muscles that control our posture.

Vestibulo-ocular reflex

The **vestibulo-ocular reflex (VOR)** is a reflex of eye movement that stabilizes images on the retina during head

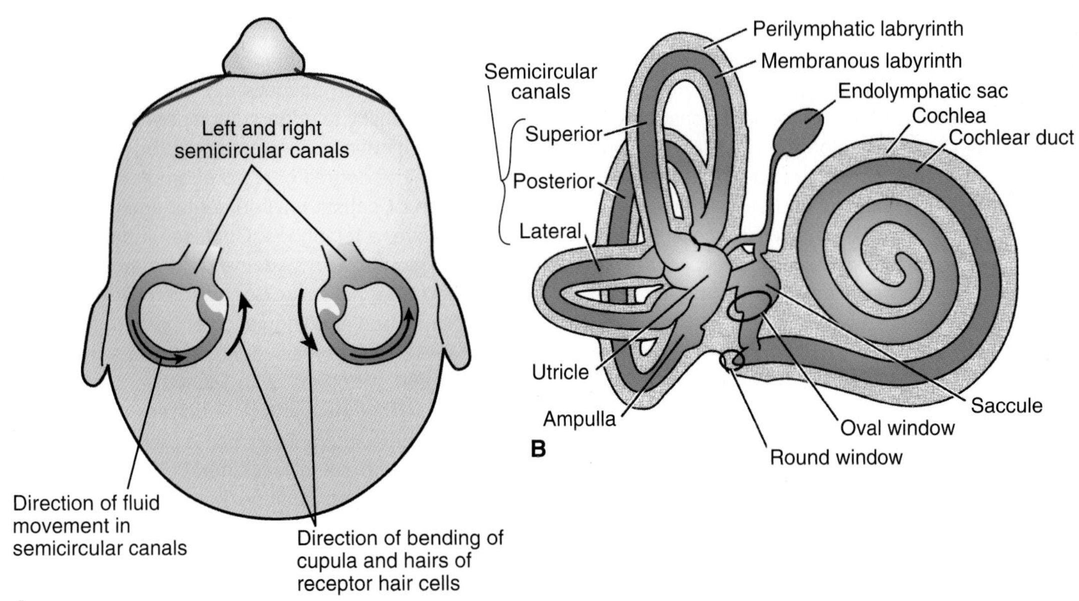

Figure 4.22 The vestibular apparatus in the bony labyrinth of the inner ear. (A) Activation of hair cells in the semicircular canals. **(B)** The semicircular canals sense rotary acceleration and motion, whereas the utricle and saccule sense linear acceleration and static position.

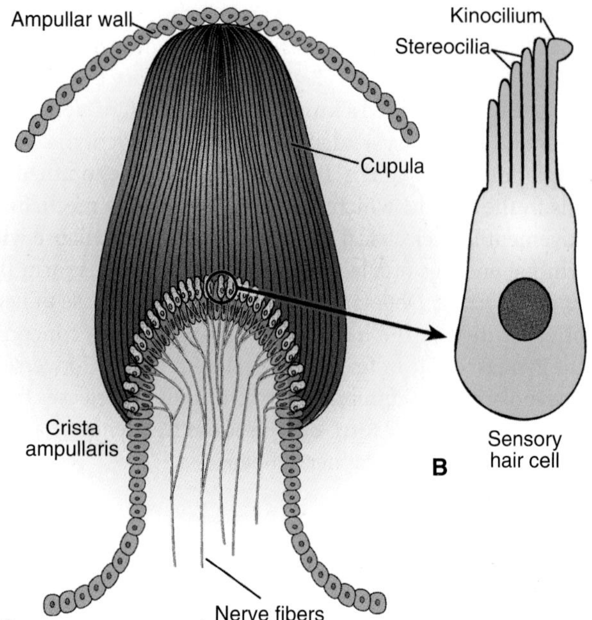

Figure 4.23 The sensory structure of the semicircular canals. (A) The crista ampullaris contains the hair (receptor) cells, and the whole structure is deflected by motion of the endolymph. **(B)** An individual hair cell.

movement (Fig. 4.25). This reflex produces an eye movement in the direction opposite to the head movement, thus preserving the image on the center of the visual field. The reflex is initiated when the rotation of the head is detected, which, in turn, triggers an inhibitory signal to the extraocular muscles on one side and an excitatory signal to the muscle on the other side. This reflex results in a compensatory movement of the eyes. The VOR has both rotational and translational inputs. When the head rotates about any axis (horizontal, vertical, or torsional), distant images are stabilized by rotating the eyes about the same axis, but in the opposite direction. Because slight head movements are happening all of the time, the VOR is very important for stabilizing vision. Patients whose VOR is impaired find it extremely difficult to read, because they cannot stabilize the eyes during small head movements. The VOR is driven by signals from the vestibular apparatus in the inner ear. The semicircular canals detect head rotation and drive the rotational VOR. The otoliths detect head translation and drive the translational VOR.

Vestibular dysfunction may result in immobilization.

The vestibular system is exquisitely designed to maintain equilibrium. However, there are instances wherein the system

Clinical Focus / 4.1

Cochlear Implants

Disorders of hearing are broadly divided into the following categories: **conductive hearing loss**, related to structures of the outer and middle ear; **sensorineural hearing loss** ("nerve deafness"), dealing with the mechanisms of the cochlea and peripheral nerves; and central hearing loss, concerning processes that lie in higher portions of the central nervous system. Damage to the cochlea, especially to the hair cells of the organ of Corti, produces sensorineural hearing loss by several means. Prolonged exposure to loud occupational or recreational noises can lead to hair cell damage, including mechanical disruption of the stereocilia. Such damage is localized in the outer hair cells along the basilar membrane at a position related to the pitch of the sound that produced it. Antibiotics such as streptomycin and certain diuretics can cause rapid and irreversible damage to hair cells similar to that caused by noise, but it occurs over a broad range of frequencies. Diseases such as meningitis, especially in children, can also lead to sensorineural hearing loss. In carefully selected patients, the use of a cochlear implant can restore some function to the profoundly deaf. The device consists of an external microphone, amplifier, and speech processor coupled by a plug-and-socket connection, magnetic induction, or a radio frequency link to a receiver implanted under the skin over the mastoid bone. Stimulating wires then lead to the cochlea. A single extracochlear electrode, applied to the round window, can restore perception of some environmental sounds and aid in lip reading, but it will not restore pitch or speech discrimination. A multielectrode intracochlear implant (with up to 22 active elements spaced along it) can be inserted into the basal turn of the scala tympani. The linear spatial arrangement of the electrodes takes advantage of the tonotopic organization of the cochlea, and some pitch (frequency) discrimination is possible. The external processor separates the speech signal into several frequency bands that contain the most critical speech information, and the multielectrode assembly presents the separated signals to the appropriate locations along the cochlea. In some devices, the signals are presented in rapid sequence, rather than simultaneously, to minimize interference between adjacent areas.

When implanted successfully, such a device can restore much of the ability to understand speech. Considerable training of the patient and fine-tuning of the speech processor are necessary. The degree of restoration of function ranges from recognition of critical environmental sounds to the ability to converse over a telephone. Cochlear implants are most successful in adults who became deaf after having learned to speak and hear naturally. Success in children depends critically on their age and linguistic ability; currently, implants are being used in children as young as 2 years of age.

Infrequent problems with infection, device failure, and natural growth of the auditory structures may limit the usefulness of cochlear implants for some patients. In certain cases, psychologic and social considerations may discourage the advisability of using auditory prosthetic devices in general. From a technical standpoint, however, continual refinements in the design of implantable devices and the processing circuit are extending the range of subjects who may benefit from cochlear implants. Research directed at external stimulation of higher auditory structures may eventually lead to even more effective treatments of profound hearing loss.

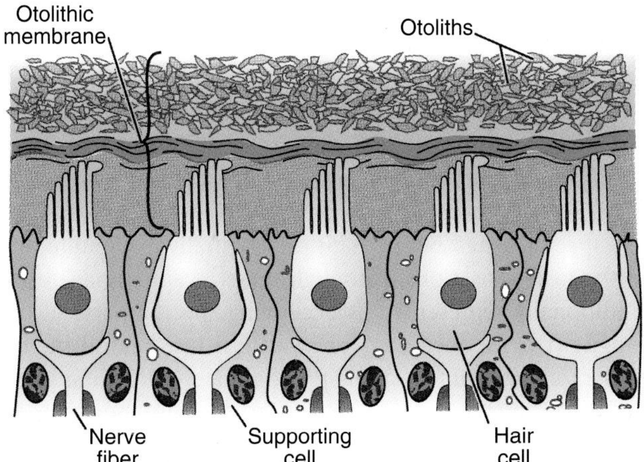

Figure 4.24 The relation of the otoliths to the sensory cells in the macula of the utricle and saccule. The gravity-driven movement of the otoliths stimulates the hair cells.

malfunctions and patients become immobilized. One such condition is **vertigo**, which results in dizziness or spinning movement when the patient is stationary. Vertigo is often associated with nausea and vomiting as well as difficulties standing or walking. Excessive consumption of alcohol notoriously causes symptoms of vertigo. If the otolithic organs are stimulated rhythmically, as by the motion of a ship or automobile, distressing symptoms of **motion sickness** (dizziness and nausea) may result. Over time, these symptoms diminish and disappear. The underlying mechanisms causing vertigo and motion sickness are as yet poorly understood.

GUSTATORY AND OLFACTORY SYSTEMS

Unlike the photoreceptors of the eye and the mechanoreceptors of the ear, the receptors of taste and smell are chemoreceptors. The sensation of taste and smell are two other sensory mechanisms that provide specific information about the external environment. In lower animals, the mechanisms of taste (**gustation**) and smell (**olfaction**) play a major role in finding food, seeking prey, finding directions, bonding with offspring and mates, and avoiding danger/predators. In some species (e.g., dogs) smell is often linked to defecating. In the case of humans, most of these neural signals are associated with food, fragrance, and odors of our surrounding environment. Although our taste and smell are less sensitive than those of other primates, millions of dollars are spent on additives to make our food taste better and on deodorants and perfumes to make us more desirable/attractive and sociable.

Taste buds house the receptor cells for gustatory sensations.

Taste is the major determinant in whether food in the mouth is a dangerous substance or safe to consume. The multicellular receptors for taste are packaged primarily in taste buds (~10,000) located on the upper surface of the soft palate, esophagus, and epiglottis, with the majority

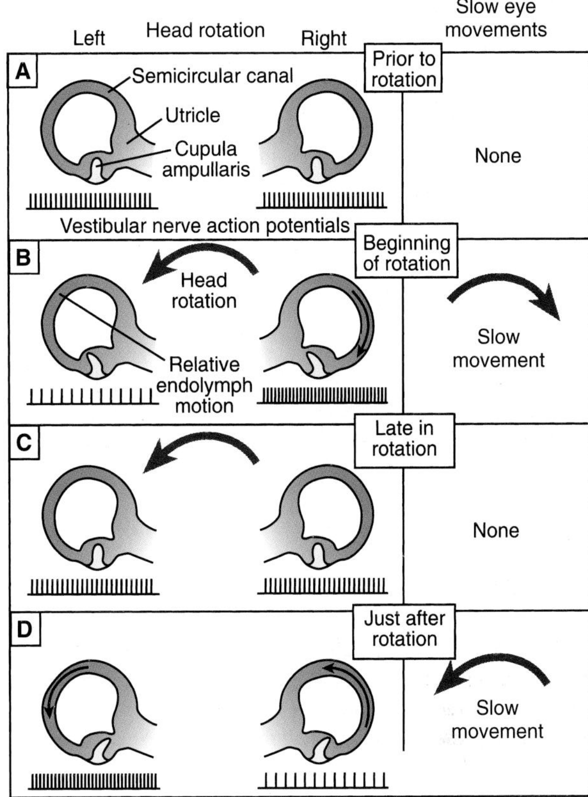

Figure 4.25 The role of the semicircular canals in sensing rotary acceleration. This sensation is linked to compensatory eye movements by the vestibulo-ocular reflex. Only the horizontal canals are considered here. This pair of canals is shown as if one were looking down through the top of a head looking toward the top of the page. Within the ampulla of each canal is the cupula, an extension of the crista ampullaris, the structure that senses motion in the endolymph fluid in the canal. Below each canal is the action potential train recorded from the vestibular nerve. **(A)** The head is still, and equal nerve activity is seen on both sides. There are no associated eye movements (right column). **(B)** The head has begun to rotate to the left. The inertia of the endolymph causes it to lag behind the movement, producing a fluid current that stimulates the cupulae (*arrows* show the direction of the relative movements). Because the two canals are mirror images, the neural effects are opposite on each side (the cupulae are bent in relatively opposite directions). The reflex action causes the eyes to move slowly to the right, opposite to the direction of rotation (right column). They then snap back and begin the slow movement again as rotation continues. The fast movement is called *rotatory nystagmus*. **(C)** As rotation continues, the endolymph "catches up" with the canal because of fluid friction and viscosity, and there is no relative movement to deflect the cupulae. Equal neural output comes from both sides, and the eye movements cease. **(D)** When the rotation stops, the inertia of the endolymph causes a current in the same direction as the preceding rotation, and the cupulae are again deflected, this time in a manner opposite to that shown in **(B)**. The slow eye movements now occur in the same direction as the former rotation.

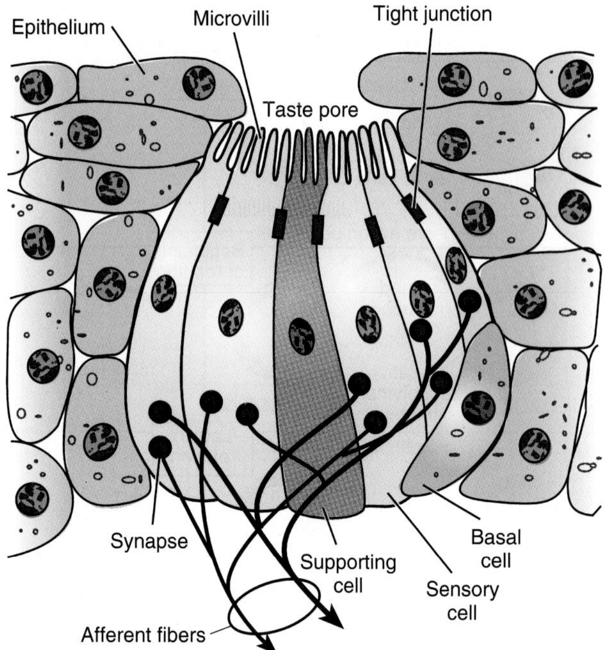

Figure 4.26 The sensory and supporting cells in a taste bud. Taste buds consist of taste cells surrounded by supporting epithelial cells. The afferent nerve synapses with the basal areas of the sensory cells.

located on the dorsal portion of the tongue's surface. Taste buds are a collection of about 50 long, spindled sensory cells. Among the sensory cells are elongated *supporting cells* that do not have synaptic connections. The sensory cells along with the supporting cells are arranged like slices of an orange. The cells lie mostly buried in the surface, and food dissolved in saliva comes in contact with the sensory cells by way of **taste pores** (Fig. 4.26). Taste receptor cells are modified epithelial cells, and their apical ends are connected laterally by tight junctions, which bear microvilli that greatly increase the surface area exposed to oral fluid. Only food that is dissolved in saliva can attach to taste receptor cells and evoke a neural sensation of taste. Stimulating the receptor cell alters the cell's ion channels that produce a depolarizing potential. Similar to other sense receptors, a depolarizing potential opens the voltage-gated Ca^{2+} channels, thereby flooding the cell with an influx of Ca^{2+} and leading to a neurotransmitter release. This neurotransmitter, in turn, triggers an action potential that is sent to cortical gustatory areas of the brain via the seventh, ninth, and tenth cranial nerves.

The sensory cells typically have a lifespan of 10 days. They are continually replenished by new sensory cells formed from the basal cells of the lower part of the taste buds. When a sensory cell is replaced by a maturing basal cell, the old synaptic connections are broken, and new ones must be formed.

Chemoreception distinguishes five primary taste categories.

Among thousands of different taste sensations, humans can discriminate between five specific tastes received by the gustatory receptors. These are salty, sweet, bitter, sour, and ***umami***, which means "savory" or "meaty" in Japanese. There are also two "accessory qualities" of taste sensation, alkaline (soapy) and metallic. Recently, another taste sensation—fatty—has been proposed, because scientists have identified a sensory receptor for long-chain fatty acids. In some Asian countries, piquancy (spiciness) is also considered a basic taste. Sensory receptor cells respond in varying degrees to all five specific tastes, but perhaps preferentially respond to one of the taste sensations. Although these receptors have been traditionally represented as occupying specific parts of the tongue (i.e., the receptors for sweetness just behind the tip of the tongue, sour receptors predominately along the sides, salt at the tip, and bitter across the rear), this "tongue map" theory has been refuted.

Salt receptors

Salt plays a critical role in ion and water homeostasis in animals, especially mammals. NaCl is especially important in the mammalian kidney as an osmotically active molecule that facilitates passive reuptake of water. Because of this, salt elicits a pleasant taste in most mammals, which potentially can lead to excess salt intake in the human diet. The simplest receptor found in the mouth is the NaCl receptor. An ion channel in the taste cell receptor allows Na^+ ions to enter the cell directly to depolarize the cell and open the voltage-regulated Ca^{2+} on its own. This sodium channel is known as the **epithelial sodium channel**, or **ENaC**, and is composed of three subunits. The ENaC protein channel can be blocked by the drug amiloride in many mammals. However, the action of amiloride is much less pronounced on the salt receptors in humans, leading to the speculation that there may be an additional receptor protein besides ENaC.

Sour receptors

Sour taste signals the presence of acidic foods (H^+ ions in solution). The citric taste of lemon is a good example of our experience with the distinctive sour taste. Sour taste can be mildly pleasant in small quantities and is linked to salt flavor. However, in larger quantities it becomes more of an unpleasant taste. This is a protective mechanism and can signal an overripe fruit, rotten meat, or other spoiled foods that can be toxic to the body because of bacteria contamination. Also, sour taste signals acid (H^+ ions) to the brain, which can warn of serious tissue damage.

There are three different receptor proteins involved in the detection of the sour taste. The first is a simple ion channel that allows hydrogen ions to flow directly into the receptor cell. The protein for this ion channel is ENaC, the same protein that is involved in the detection of the salty taste. This overlapping protein may explain the relationship between salt and sour receptors and also why salty taste is suppressed when a sour taste is present. The second type of receptor present is an H^+-gated channel, which blocks K^+ from escaping from inside the cell. The third protein receptor opens to Na^+ ions when an H^+ ion attaches to it. This allows the sodium ions to flow down a concentration gradient into the cell, leading to the opening of a voltage-regulated Ca^+ gate. All three receptor proteins work in concert and lead

Clinical Focus / 4.2 

Vertigo

A common medical complaint is dizziness. This symptom may be a result of several factors, such as cerebral ischemia ("feeling faint"), reactions to medication, disturbances in gait, or disturbances in the function of the vestibular apparatus and its central nervous system (CNS) connections. Such disturbances can produce the phenomenon of vertigo, which may be defined as the illusion of motion (usually rotation) when no motion is actually occurring. Vertigo is often accompanied by autonomic nervous system symptoms of nausea, vomiting, sweating, and pallor. The body uses three integrated systems to establish its place in space: the vestibular system, which senses position and rotation of the head; the visual system, which provides spatial information about the external environment; and the somatosensory system, which provides information from joint, skin, and muscle receptors about limb position. Several forms of vertigo can arise from disturbances in these systems. Physiologic vertigo can result when there is discordant input from the three systems. Seasickness results from the unaccustomed repetitive motion of a ship (sensed via the vestibular system). Rapidly changing visual fields can cause visually induced motion sickness, and space sickness is associated with multiple-input disturbances. Central positional vertigo can arise from lesions in cranial nerve VIII (as may be associated with multiple sclerosis or some tumors), vertebrovascular insufficiency (especially in older adults), or from the impingement of vascular loops on neural structures. It is commonly present with other CNS symptoms. Peripheral vertigo arises from disturbances in the vestibular apparatus itself. The problem may be either unilateral or bilateral. Causes include trauma, physical defects in the labyrinthine system, and pathologic syndromes such as **Ménière disease**. As in the cochlea, aging produces considerable hair cell loss in the cristae and maculae of the vestibular system. Caloric stimulation can be used as an indicator of the degree of vestibular function. The most common form of peripheral vertigo is **benign paroxysmal positional vertigo (BPPV)**. This is a severe vertigo, with incidence increasing with age. Episodes appear rapidly and are limited in duration (from minutes to days). They are usually brought on by assuming a particular position of the head such as one might do when painting a ceiling. BPPV is thought to be a result of the presence of **canaliths**, debris in the lumen of one of the semicircular canals. The offending particles are usually clumps of otoconia (otoliths) that have been shed from the maculae of the saccule and utricle, whose passages are connected to the semicircular canals. These clumps act as gravity-driven pistons in the canals, and their movement causes the endolymph to flow, producing the sensation of rotary motion. Because they are in the lowest position, the posterior canals are the most frequently affected. In addition to the rotating sensation, this input gives rise, via the vestibular system, to a pattern of nystagmus (eye movements) appropriate to the spurious input.

The specific site of the problem can be determined by using the *Dix–Hallpike maneuver*, which is a series of physical maneuvers (changes in head and body position). By observing the resulting pattern of nystagmus and reported symptoms, the location of the defect can be deduced. Another set of maneuvers known as the *canalith repositioning procedure of Epley* can cause gravity to collect the loose canaliths and deposit them away from the lumen of the semicircular canal. This procedure is highly effective in cases of true BPPV, with a cure rate of up to 85% on the first attempt and nearly 100% on a subsequent attempt. Patients can be taught to perform the procedure on themselves if the problem returns.

Ménière disease is a syndrome of uncertain (but peripheral) origin associated with vertigo. Its cause(s) and precipitating factors are not well understood. Typical associated findings include fluctuating hearing loss and tinnitus (ringing in the ears). Episodes involve increased fluid pressure in the labyrinthine system, and symptoms may decrease in response to salt restriction and diuretics. Other cases of peripheral vertigo may be caused by trauma (usually unilateral) or by toxins or drugs (such as some antibiotics); this type is often bilateral.

Central and peripheral vertigo often may be differentiated on the basis of their specific symptoms. Peripheral vertigo is more severe, and its nystagmus shows a delay (latency) in appearing after a position change. Its nystagmus fatigues and can be reduced by visual fixation. Position-sensitive and of finite duration, the condition usually involves a horizontal orientation. Central vertigo, usually less severe, shows a vertically oriented nystagmus without latency and fatigability; it is not suppressed by visual fixation and may be of long duration.

Treatment of vertigo, beyond that mentioned above, can involve bed rest and vestibular inhibiting drugs (such as some antihistamines). However, these treatments are not always effective and may delay the natural compensation that can be aided by physical motion such as walking (unpleasant as that may be). In severe cases that require surgical intervention (labyrinthectomy, etc.), patients can often achieve a workable position sense via the other sensory inputs involved in maintaining equilibrium. Some activities, such as underwater swimming, must be avoided by those with an impaired sense of orientation, because false cues may lead to moving in inappropriate directions and increase the risk of drowning.

to depolarization of the receptor cell and the concomitant neurotransmitter release.

Bitter receptors

The bitter taste is almost completely unpleasant in humans. This is because many nitrogenous organic compounds that have a pharmacologic effect on humans taste bitter. These include caffeine, the stimulant in coffee, and nicotine, the addictive compound in cigarettes. Almost all poisonous/toxic plants taste bitter; the bitter taste is a protective mechanism, and the reaction is a last-line warning system before the compound is swallowed to cause injury or death. However, humans have evolved a very sophisticated sense for bitter substances and can distinguish between many different compounds. As a result, they have overcome their innate aversion to certain bitter tastes, as evidenced by the wide consumption of caffeinated drinks enjoyed around the world.

Cells that detect bitter flavors have >50 receptors, each of which responds to a different bitter flavor. This mechanism allows for many different classes of bitter compounds to be detected, which can be chemically very different. Bitter compounds act through **G protein–coupled receptors (GPCRs)** in the taste cells. **Gustducin** is the G protein that the bitter substance is coupled to, and when the bitter compound activates the GPCR, gustducin is released. The release of gustducin activates a second messenger in the cell, which closes potassium ion channels. Activation of the second messenger also stimulates the endoplasmic reticulum to release Ca^{2+}, which also contributes to depolarization. This leads to a further buildup of potassium ions, depolarization, and neurotransmitter release. The G protein that triggers the second messenger pathway in the taste cell is similar to the visual G protein, transducin.

Sweet receptors

Sweet taste signals the presence of readily usable food. Humans have evolved to seek out the highest caloric intake, because our ancient ancestors were not always sure when the next meal would be available. Carbohydrates have a very high caloric count (therefore, much energy) and, as a result, have been sought out by the human body. Unfortunately, this evolutionary pattern has lead to a craving of sweets. Carbohydrates are used as direct energy (sugars) and as stored energy (glycogen).

Like bitter tastes, sweet taste transduction involves GPCRs. Natural sugars, such as cane sugar, activate the GPCR, which releases gustducin. Gustducin, in turn, activates adenylate cyclase and increases the concentration of cAMP (adenosine 3′,5′-cyclic monophosphate). Elevated cAMP leads to depolarization and neurotransmitter release. There are several noncarbohydrate molecules that also trigger a sweet taste response, which has lead to the industrial development of artificial sweeteners, including saccharin, sucralose, and aspartame. These artificial sweeteners activate different GPCRs, which, ultimately, leads to the blocking of potassium ion channels.

Umami receptors

The umami taste signals the presence of meat, thus encouraging the intake of peptides and proteins. The signal is triggered by amino acids (specifically glutamine), which are used to build such things as muscle, organs, transport molecules (e.g., hemoglobin), and cellular enzymes. Amino acids are critical for the human body to function; therefore, it is important to have a steady supply. Hence, the umami taste signals a pleasurable response to a desirable, nutritionally rich source of protein.

Umami receptors act much the same way as bitter and sweeter receptors and involve GPCRs. Glutamate binds to the GPCRs and activates a second-messenger pathway. However, not much is known about the specifics of the activation pathway. Umami receptors also respond to monosodium glutamate (commonly known as *MSG*), which is used as a food additive and is very popular in Asia, especially Japanese dishes.

Olfactory apparatus serves the sense of smell.

The olfactory system has a much broader function than does the gustatory system. It not only participates in the selection and enjoyment of food but also is involved in detecting smell from the surrounding environment (e.g., fragrance of flowers, other people, and dangerous odors that can be harmful to the body). Appreciation of flavors is especially important and compliments the gustatory system. For example, savoring the aroma of a glass of red wine often surpasses the gustatory system. Many patients who complain about the loss of taste often have an olfactory disorder.

Humans, compared with other mammals, have a relative poor sense of smell. However, the olfactory system is still quite extraordinary insofar as the nose contains >5 million olfactory receptors (10 times more than taste receptors) that can differentiate between 1,000 different odorants. The main olfactory system detects odorants that are inhaled through the nose, where they contact the main olfactory epithelium, which contains various olfactory receptors. The receptor organ for olfaction is located in the olfactory epithelium in the roof of the nasal cavity. Normally, there is little airflow in this region of the nasal tract, but sniffing serves to direct air upward, increasing the likelihood of an odor being detected. In contrast to the taste sensory cells, the olfactory cells are neurons and, as such, are *primary receptors*. These cells are interspersed among supporting (sustentacular) cells and basal cells that bind the cells together at their sensory ends (Fig. 4.27). The olfactory receptors are membrane proteins located in the olfactory epithelium. Rather than binding specific ligands like most receptors, olfactory receptors have a binding affinity for a range of odorant molecules. This allows the olfactory system to discriminate among a wide range of different odor molecules. Olfactory thresholds vary widely from substance to substance; the threshold concentration for the detection of ethyl ether (used as a general anesthetic) is around 5.8 mg/L air, whereas that for methyl mercaptan (the odor of garlic) is approximately 0.5 ng/L. This represents a 10 million–fold difference in sensitivity. The basis for odor discrimination is not well understood.

Signal transduction appears to involve the binding of a molecule of an odorous substance to a GPCR on a sensory cell. This binding causes the production of cAMP, which binds to, and opens, sodium channels in the ciliary membrane. The resulting inward sodium current depolarizes the cell to produce a generator potential, which causes action potentials in the afferent fiber. The frequency of the action potential is dependent on the concentration of the odorant. The sense of smell shows a high degree of adaptation, some of which takes place at the level of the generator potential and some of which may be a result of the action of efferent neurons in the olfactory bulb. Discrimination between odor intensities is not well defined; detectable differences may be about 30%.

The mechanism of the olfactory system can be divided into a peripheral mechanism, one that senses an external stimulus and encodes it as an electrical signal, and a central

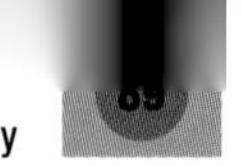

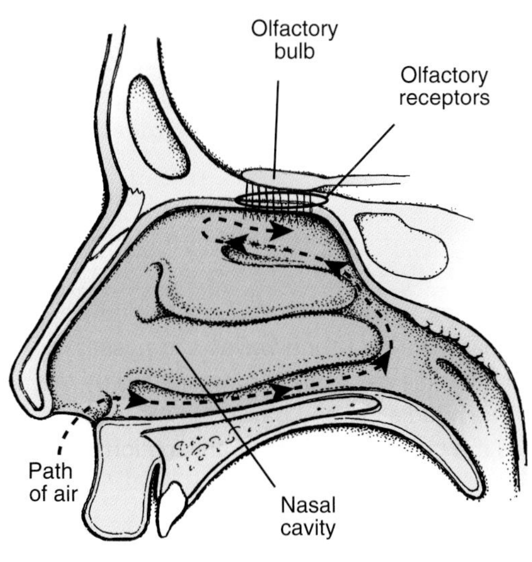

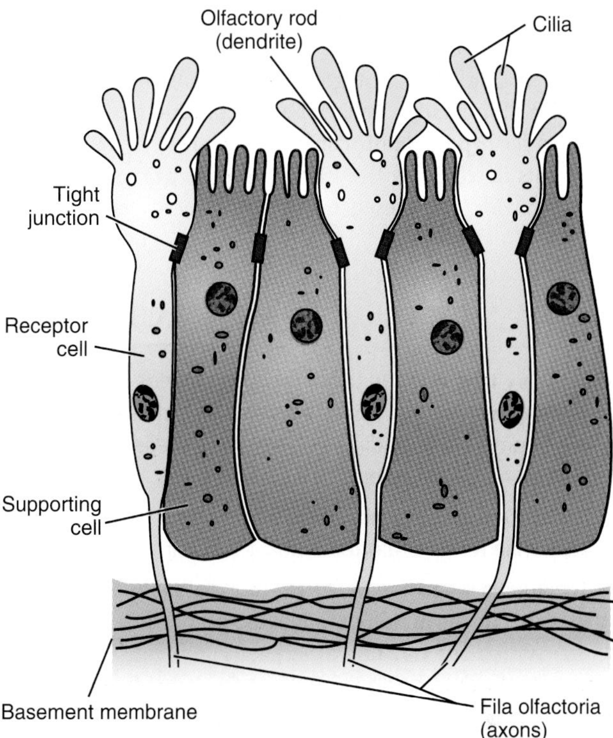

Figure 4.27 The sensory cells in the olfactory mucosa. The fila olfactoria, the axons leading from the receptor cells, are part of the sensory cells. The axons leading from the receptor cells form the olfactory tract, which innervates the limbic system and the orbitofrontal cortex.

one where the signal in the neurons are integrated and processed in the CNS. The signals travel along the olfactory nerves (afferent nerves) that belong to the peripheral nervous system. These nerves terminate and immediately synapse into the **olfactory bulb**, which belongs to the CNS. There are two olfactory bulbs, one on each side and consist of a complex multilayer neural architecture located in the forebrain. Thus, the olfactory bulb serves as a relay station that transmits information from the nose to the brain. The ends of olfactory nerve axons cluster in a ball-like neural junction known as the *olfactory glomeruli*. Each glomerulus receives input primarily from olfactory receptor neurons that express the same olfactory receptor. Thus, the glomeruli serve to categorize the sense of smell. Glomeruli are also permeated by dendrites from neurons termed **mitral cells**, which, in turn, send out information to the olfactory cortex. As a neural circuit, the olfactory bulbs have one source of sensory input (axons from olfactory receptor neurons) and one output (mitral cell axons), which transmit signals to the olfactory cortex.

Sneezing

The olfactory mucosa also contains sensory fibers from the trigeminal (V) cranial nerve. They are sensitive to irritants and certain odorous substances, such as peppermint and chlorine, and play a role in the initiation of reflex responses (e.g., sneezing) that result from irritation of the nasal tract. Sneezing typically occurs when an irritant pass through the nasal hairs to reach the nasal mucosa. This triggers the release of histamine, which, in turn, irritates the nerve cells in the nose, resulting in an electrical signal being sent to the brain via the fifth cranial nerve to initiate the sneeze reflex. The brain then activates the pharyngeal, tracheal, and chest muscles to expel a large volume of air from the lungs through the nose and mouth. Sneezing is also triggered by sinus infection and allergies.

Vomeronasal organ

In addition to olfactory receptors, the nose also contains an auxiliary sensory organ, the **vomeronasal organ (VNO)**. This organ is located at the base of the nasal cavity adjacent to the vomer bone in the nose, hence its name. The sensory neurons of the VNO detect specific chemicals that give rise to animal scent. Some scents act as chemical-communication signals known as **pheromones**. These are chemical messengers that carry information between individuals of the same species, sometimes referred to as the "sixth sense." These pheromones are secreted by the skin, from the reproductive organs, and released in the urine. Pheromones trigger behavior and social responses between members of the same species. They can act as alarm pheromones, sexual pheromones, and scents that can modify other types of behavior. For example, pheromones are thought to be involved in a mother accepting or rejecting offspring and offspring being accepted within a litter.

The VNO has two separate and distinct types of receptors, V1R and V2R, which are coupled to G proteins and allow for the discrimination of different types of scents. Unlike the main olfactory bulb, which sends signals to the olfactory cortex, the VNO sends neuronal signals to the amygdala and hypothalamus. The hypothalamus not only functions as the body's thermostat, but also as the body's neuroendocrine system (see Chapter 7, "Integrative Functions of the Nervous System," for details). Behavior and reproductive physiology are under hypothalamic control, which may explain why some scents detected by the VNO influence aggressive and mating behavior. The presence and function of pheromones in humans are still somewhat controversial. There is evidence that supports that the spontaneous feeling between individuals as good chemistry (such as "good vibes," "love at first sight," and "spiritual attraction") or bad chemistry (such as "bad vibes," aggressive posture, and negative behavior) from another individual just encountered is the work of pheromones.

Chapter Summary

- Sensory transduction takes place in a series of steps, starting with stimuli from the external or internal environment and ending with neural processing in the central nervous system.
- The structure of sensory organs optimizes their response to the preferred types of stimuli.
- A stimulus gives rise to a generator potential, which, in turn, causes action potentials to be produced in the associated sensory nerve.
- The speeds of adaptation of particular sensory receptors are related to their biologic roles.
- Specific sensory receptors for a variety of types of tactile stimulation are located in the skin.
- Somatic pain is associated with the body surface and the musculature; visceral pain is associated with the internal organs.
- The sensory function of the eyeball is determined by structures that form and adjust images and by structures that transform images into neural signals.
- The retina contains a number of layers and several cell types, each with a specific role in the process of visual transduction.
- The rod cells in the retina have a high sensitivity to light but produce indistinct images without color, whereas the cones provide sharp color vision with less sensitivity to light.
- The visual transduction process requires many steps, beginning with the absorption of light and ending with an electrical response.
- The outer ear receives sound waves and passes them to the middle ear. They are transmitted by the bones of the middle ear and passed to the inner ear, where the process of sound transduction takes place.
- The transmission of sound through the middle ear greatly increases the efficiency of its detection, whereas its protective mechanisms guard the inner ear from damage caused by extremely loud sounds. Disturbances in this transmission process can lead to hearing impairments.
- Sound vibrations enter the cochlea through the oval window and travel along the basilar membrane, where their energy is transformed into neural signals in the organ of Corti.
- Displacements of the basilar membrane cause deformation of the hair cells, the ultimate transducers of sound. Different sites along the basilar membrane are sensitive to different frequencies.
- The vestibular apparatus senses the position of the head and its movements by detecting small deflections of its sensory structures.
- Taste is mediated by sensory epithelial cells in the taste buds. There are five fundamental taste sensations: sweet, sour, salty, bitter, and umami.
- Smell is detected by nerve cells in the olfactory mucosa. Thousands of different odors can be detected and distinguished.

thePoint *Visit http://thePoint.lww.com for chapter review Q&A, case studies, animations, and more!*

Motor System

ACTIVE LEARNING OBJECTIVES

Upon mastering the material in this chapter, you should be able to:

- Explain the role of α motor neurons in skeletal muscle control.
- Explain how the actions of the muscle spindle and Golgi tendon organ regulate muscle action.
- Be familiar with the anatomy of motor neuron subgroups in relation to the muscle groups controlled.
- Explain the three classic spinal reflexes and what they reveal about spinal cord circuitry.
- Explain how the brainstem nuclei and their associated descending spinal cord tracts influence motor neuron function.
- Explain the role of the cerebral cortex, the corticospinal tract, and the basal ganglia in the control of movement.
- Explain the influence of the cerebellum in motor control.

The finger movements of a neurosurgeon manipulating microsurgical instruments while repairing a cerebral aneurysm and the eye-hand-body coordination of a professional basketball player making a rimless three-point shot are two examples of the motor control system operating at high skill levels. The coordinated contraction of the hip flexors and ankle extensors to clear a slight pavement irregularity encountered during walking is a familiar example of the motor control system working at a seemingly automatic level. Conversely, the stiff-legged stride of a patient who experienced a stroke and the swaying walk of an intoxicated person are examples of perturbed motor control.

Although our understanding of the physiology of the motor system is still far from complete, a significant fund of knowledge exists. This chapter proceeds through the major components of the motor system, beginning with the skeleton and ending with the brain.

SKELETON AS FRAMEWORK FOR MOVEMENT

Bones are the body's framework as well as a system of levers. They are the elements that move. The way adjacent bones articulate determines the motion and range of movement at a joint. Ligaments hold the bones together across the joint. Movements are described on the basis of the anatomic planes through which the skeleton moves and the physical structure of the joint. Most joints move in only one plane, but some permit movement in multiple anatomic reference planes (Fig. 5.1).

Hinge joints, such as the elbow, are uniaxial, permitting movements in the sagittal plane. The wrist is an example of a biaxial joint. The shoulder is a multiaxial joint; movement can occur in oblique planes as well as the three major planes of that joint. Flexion and extension describe movements in the sagittal plane. **Flexion** movements decrease the angle between the moving body segments. **Extension** describes movement in the opposite direction. **Abduction** moves the body part away from the midline, whereas **adduction** moves the body part toward the midline.

MUSCLE FUNCTION AND BODY MOVEMENT

Muscles span joints and are attached at two or more points to the bony levers of the skeleton. The muscles provide the power that moves the body's levers. Muscles are described in terms of their origin and insertion attachment sites. The *origin* tends to be the more fixed, less mobile location, whereas the *insertion* refers to the skeletal site that is more mobile. Movement occurs when a muscle generates force on its attachment sites and undergoes shortening. This type of action is termed an **isotonic** or **concentric contraction**. Another form of muscular action is a controlled lengthening while still generating force. This is an **eccentric contraction**. A muscle may also generate force but hold its attachment sites static, as in **isometric contraction**.

Because muscle contraction can produce movement in only one direction, at least two muscles opposing each other at a joint are needed to achieve motion in more than one direction. When a muscle produces movement by shortening, it is an **agonist**. The **prime mover** is the muscle that contributes most to the movement. Muscles that oppose the action of the prime mover are **antagonists**. The quadriceps and hamstring muscles are examples of agonist–antagonist pairs in knee extension and flexion. During both simple and light-load skilled movements, the antagonist is relaxed. Contraction of the agonist with concomitant relaxation of the antagonist occurs by the nervous system function of

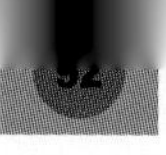

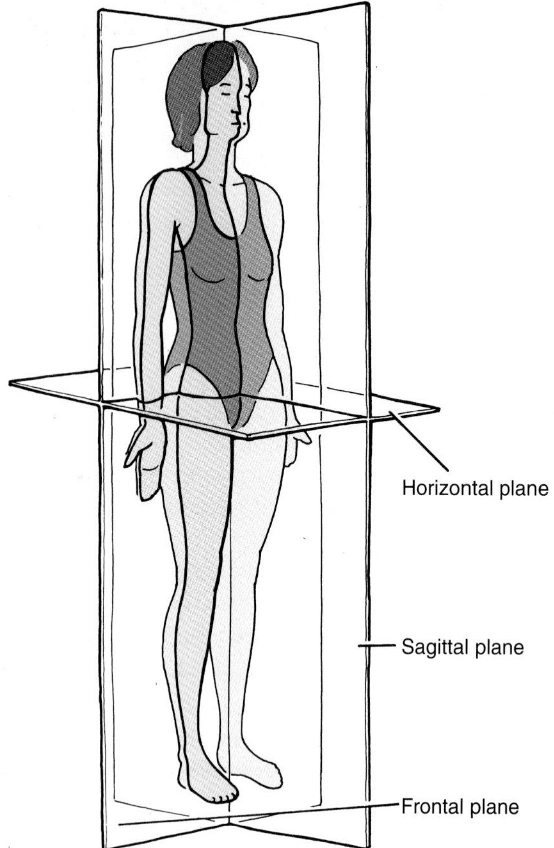

Figure 5.1 Anatomic reference planes. The figure is shown in the standard anatomic position with the associated primary reference planes.

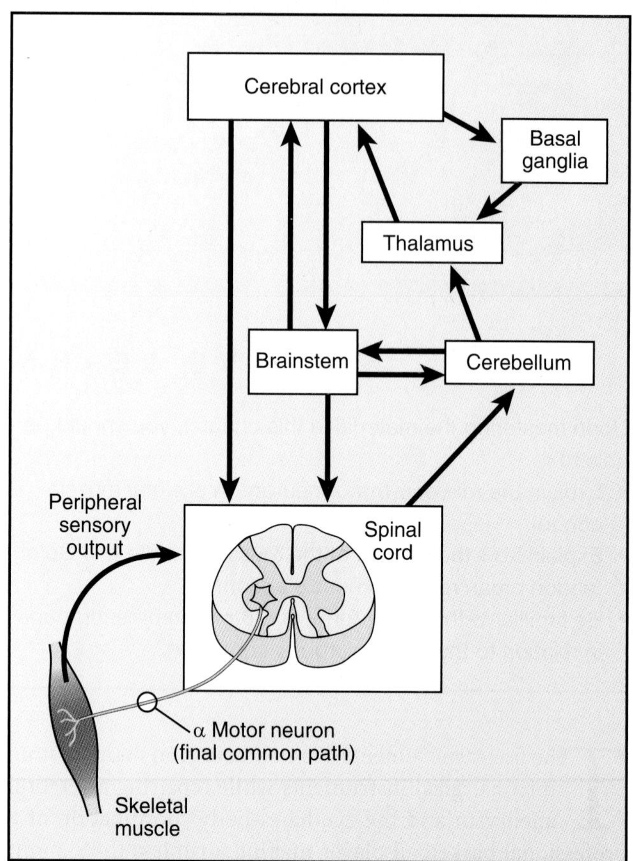

Figure 5.2 Motor control system. α Motor neurons are the final common path for motor control. Peripheral sensory input and spinal cord tracts that descend from the brainstem and the cerebral cortex influence the motor neurons. The cerebellum and basal ganglia contribute to motor control by modifying the brainstem and the cortical activity.

reciprocal inhibition. Cocontraction of agonist and antagonist occurs during movements that require precise control.

A muscle functions as a **synergist** if it contracts at the same time as the agonist while cooperating in producing the movement. Synergistic action can aid in the following: producing a movement (e.g., the activity of both the flexor carpi ulnaris and extensor carpi ulnaris are used in producing ulnar deviation of the wrist); eliminating unwanted movements (e.g., the activity of wrist extensors prevents flexion of the wrist when the finger flexors contract in closing the hand); or stabilizing proximal joints (e.g., isometric contractions of muscles of the forearm, upper arm, shoulder, and trunk accompany a forceful grip of the hand).

NERVOUS SYSTEM COMPONENTS FOR THE CONTROL OF MOVEMENT

Here, we identify the components of the nervous system that are predominantly involved in the control of motor function and discuss the probable roles for each of them. It is important to appreciate that even the simplest reflex or voluntary movement requires the interaction of multiple levels of the nervous system (Fig. 5.2).

The α motor neurons are the **final common path**, the route by which the central nervous system (CNS) controls the skeletal muscles. The motor neurons located in the ventral horns of the spinal gray matter and brainstem nuclei are influenced by both local reflex circuitry and by pathways that descend from the brainstem and cerebral cortex. The brainstem-derived pathways include the rubrospinal, vestibulospinal, and reticulospinal tracts; the cortical pathways are the corticospinal and corticobulbar tracts. Although some of the cortically derived axons terminate directly on motor neurons, most of the axons of the cortical-derived tracts and the brainstem-derived tracts terminate on interneurons, which then influence motor neuron function. The outputs of the basal ganglia of the brain and cerebellum provide fine-tuning of the cortical and brainstem influences on motor neuron functions.

α Motor neurons are the final common path for motor control.

Motor neurons segregate into two major categories, α (alpha) and γ (gamma). **α Motor neurons** innervate the **extrafusal muscle fibers**, which are responsible for force generation. γ **Motor neurons** innervate the **intrafusal muscle fibers**, which are components of the muscle spindle. γ Motor neurons are discussed further in the next section. An α motor neuron controls several muscle fibers, 10 to 1,000, depending on the muscle. The term **motor unit** describes an α motor neuron, its axon, the branches of the axon, the

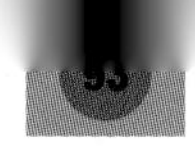

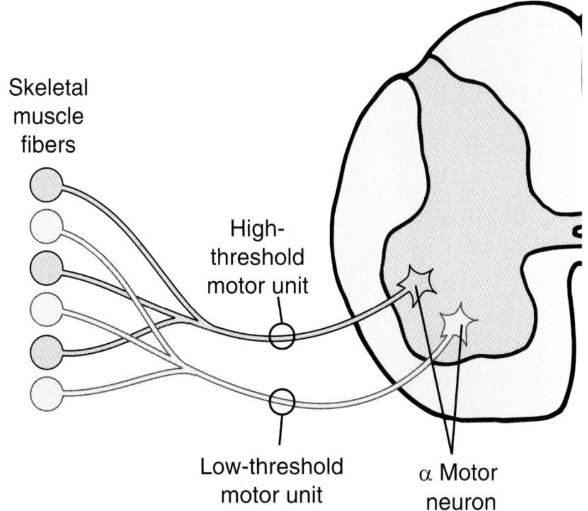

Figure 5.3 Motor unit structure. A motor unit consists of an α motor neuron and a group of extrafusal muscle fibers it innervates. Functional characteristics, such as activation threshold, twitch speed, twitch force, and resistance to fatigue, are determined by the motor neuron. Low-threshold and high-threshold motor units are shown.

neuromuscular junction synapses at the distal end of each axon branch, and all of the extrafusal muscle fibers innervated by that motor neuron (Fig. 5.3). When a motor neuron is activated, all of its muscle fibers are also activated.

α Motor neurons can be separated into two general populations according to their cell body size and axon diameter. The larger cells have a high threshold to synaptic stimulation, have fast action potential conduction velocities, and are active in high-effort force generation. They innervate fast-twitch, high-force but fatigable muscle fibers. The smaller motor neurons have lower thresholds to synaptic stimulation, conduct action potentials at slightly slower velocities, and innervate slow-twitch, low-force, fatigue-resistant muscle fibers (see Chapter 8, "Skeletal and Smooth Muscle"). The muscle fibers of a motor unit are homogeneous, either fast-twitch or slow-twitch. The twitch property of a muscle fiber is determined by the motor neuron, presumably by trophic substances released at the neuromuscular junction. Muscle fibers that are denervated secondary to disease of the axon or nerve cell body change twitch type if reinnervated by an axon sprouted from a different type of motor neuron.

The organization into different motor unit types has important functional consequences for the production of smooth, coordinated contractions. The smallest neurons have the lowest threshold and are, therefore, activated first when synaptic activity is low. These motor units produce sustainable, relatively low-force tonic contractions in slow-twitch, fatigue-resistant muscle fibers. If additional force is required, synaptic drive from higher centers increases the action potential firing rate of the initially activated motor neurons and then activates additional motor units of the same type. If yet higher force levels are needed, the larger motor neurons are recruited, but their contribution is less sustained as a result of their tendency for fatigability. This orderly process of motor unit recruitment obeys what is called the **size principle**—the smaller motor neurons are activated first. A logical corollary of this arrangement is that muscles concerned with endurance, such as antigravity muscles, contain predominantly slow-twitch muscle fibers in accordance with their function of continuous postural support. Muscles that contain predominantly fast-twitch fibers, including many physiologic flexors, are capable of producing high-force but less sustainable contractions.

Afferent muscle innervation provides feedback for motor control.

The muscles, joints, and ligaments are innervated with sensory receptors that inform the CNS about body position and muscle activity. Skeletal muscles contain muscle spindles, **Golgi tendon organs (GTOs)**, free nerve endings, and some pacinian corpuscles. Joints contain Ruffini endings and pacinian corpuscles, joint capsules contain nerve endings, and ligaments contain Golgi tendon-like organs. Together, these are the **proprioceptors**, providing sensation from the deep somatic structures. The output of these receptors provides feedback that is necessary for the control of movements.

Muscle spindles provide information about the muscle length and the velocity at which the muscle is being stretched. GTOs provide information about the force being generated. Spindles are located in the mass of the muscle, in parallel with the extrafusal muscle fibers. GTOs are located at the junction of the muscle and its tendons, in series with the muscle fibers (Fig. 5.4).

Muscle spindles

Muscle spindles are sensory receptors found in almost all of the skeletal muscles. They occur in greatest density in small muscles serving fine movements, such as those of the hand, and in the deep muscles of the neck. Muscle spindles function primarily to detect changes in muscle length. They convey length information to the CNS via sensory neurons, and the brain processes this information to help determine the position. The response of muscle spindles to changes in length also play a key role in regulating muscle contraction, by activating motor neurons via the stretch reflex to resist muscle stretch.

The muscle spindle, named for its long fusiform shape, is attached at both ends to extrafusal muscle fibers. Within the spindle's expanded middle portion is a fluid-filled capsule containing 2 to 12 specialized striated muscle fibers entwined by sensory nerve terminals. These intrafusal muscle fibers, about 300 μm long, have contractile filaments at both ends. The noncontractile midportion contains the cell nuclei (see Fig. 5.4B). γ Motor neurons innervate the contractile elements. There are two types of intrafusal fibers: **nuclear bag fibers**, named for the large number of nuclei packed into the midportion, and **nuclear chain fibers**, in which the nuclei are arranged in a longitudinal row. There are about twice as many nuclear chain fibers as nuclear bag fibers per spindle. The nuclear bag type fibers are further classified as bag_1 and bag_2, based on whether they respond best in the dynamic or static phase of muscle stretch, respectively.

Sensory axons surround both the noncontractile midportion and paracentral region of the contractile ends of the intrafusal fiber. The sensory axons are categorized as

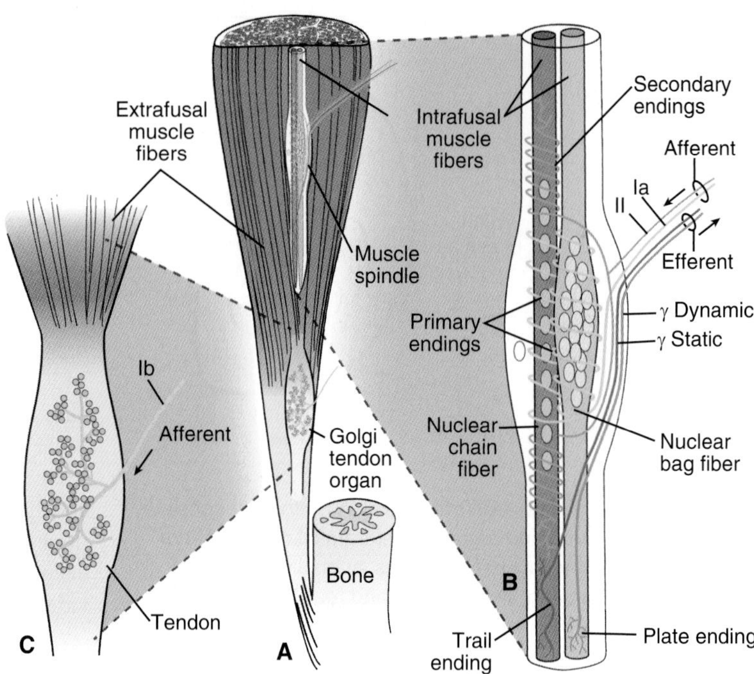

Figure 5.4 Muscle spindle and Golgi tendon organ (GTO) structure. (A) Muscle spindles are arranged parallel to extrafusal muscle fibers; GTOs are in series. **(B)** This enlarged spindle shows the following: nuclear bag and nuclear chain types of intrafusal fibers; afferent innervation by Ia axons, which provide primary endings to both types of fibers; type II axons, which have secondary endings mainly on chain fibers; and motor innervation by the two types of γ motor axons: static and dynamic. **(C)** An enlarged GTO. The sensory receptor endings interdigitate with the collagen fibers of the tendon. The axon is type Ib.

primary (type Ia) and **secondary (type II)**. The axons of both types are myelinated. Type Ia axons are larger in diameter (12–20 μm) than type II axons (6–12 μm) and have faster conduction velocities. Type Ia axons have spiral endings that wrap around the middle of the intrafusal muscle fiber (see Fig. 5.4B). Both nuclear bag and nuclear chain fibers are innervated by type Ia axons. Type II axons innervate mainly nuclear chain fibers and have nerve endings that are located along the contractile components on either side of the type Ia spiral ending. The nerve endings of both primary and secondary sensory axons of the muscle spindles respond to stretch by generating action potentials that convey information to the CNS about changes in the muscle length and the velocity of length change (Fig. 5.5). The primary endings temporarily cease generating action potentials during the release of a muscle stretch (Fig. 5.6).

Golgi tendon organs

GTOs are 1-mm-long, slender receptors encapsulated within the tendons of the skeletal muscles (see Fig. 5.4A,C). The distal pole of a GTO is anchored in collagen fibers of the tendon. The proximal pole is attached to the ends of the extrafusal muscle fibers. This arrangement places the GTO in series with the extrafusal muscle fibers such that contractions of the muscle stretch the GTO.

A large-diameter, myelinated type Ib afferent axon arises from each GTO. These axons are slightly smaller in diameter than the type Ia variety, which innervate the muscle spindle. Muscle contraction stretches the GTO and generates action potentials in type Ib axons. The GTO output provides information to the CNS about the force of the muscle contraction.

Information entering the spinal cord via types Ia and Ib axons is directed to many targets, including the spinal interneurons that give rise to the **spinocerebellar tracts**. These tracts convey information to the cerebellum about the status of muscle length and tension.

γ Motor neurons adjust spindle output during muscle contraction.

α Motor neurons innervate the extrafusal muscle fibers, and γ motor neurons innervate the intrafusal fibers. Cells bodies of both α and γ motor neurons reside in the ventral horns of the spinal cord and in nuclei of the cranial motor nerves. Nearly one third of all motor nerve axons are destined for intrafusal muscle fibers. This high number reflects the complex role of the spindles in motor system control. Intrafusal muscle fibers likewise constitute a significant portion of the total number of muscle cells, yet they contribute little or nothing to the total force generated when the muscle contracts. Rather, the contractions of intrafusal fibers play a modulating role in spindle output as they alter the length and, thereby, the sensitivity of the muscle spindles.

Even when the muscle is at rest, the muscle spindles are slightly stretched, and type Ia afferent nerves exhibit a slow discharge of action potentials. Contraction of the muscle increases the firing rate in type Ib axons from GTOs, whereas type Ia axons temporarily cease or reduce firing because the shortening of the surrounding extrafusal fibers unloads the intrafusal muscle fibers. If a load on the spindle were reinstituted, the Ia nerve endings would resume their sensitivity to stretch. The role of the γ motor neurons is to "reload" the spindle during muscle contraction by activating the contractile elements of the intrafusal fibers. This is accomplished by coordinated activation of the α and γ motor neurons during muscle contraction (see Fig. 5.5).

The γ motor neurons and the intrafusal fibers they innervate are traditionally referred to as the **fusimotor system**. Axons of the γ neurons terminate in one of two types of endings, each located distal to the sensory endings on the striated poles of the spindle's muscle fibers (see Fig. 5.4B). The nerve terminals are either plate endings or trail endings; each

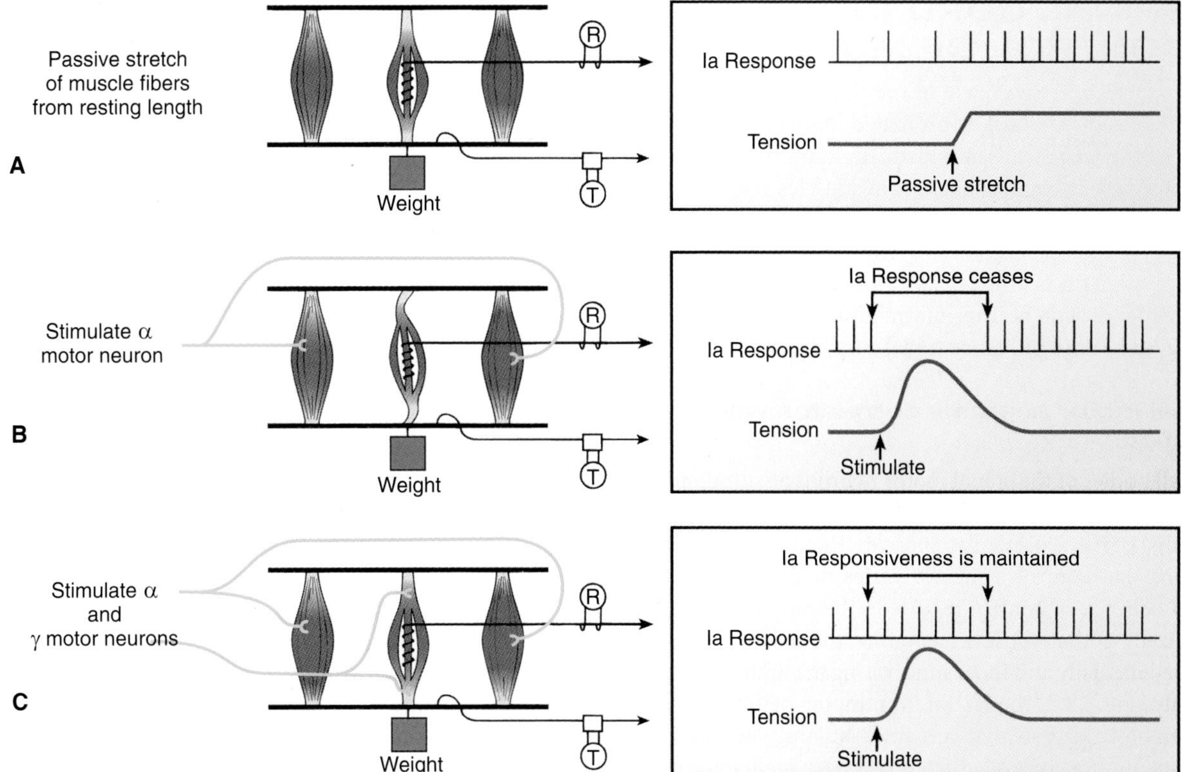

Figure 5.5 Action potentials (R) from type Ia endings versus muscle tension (T). (A) The Ia sensory endings from the muscle spindles discharge at a slow rate when the muscle is at its resting length and show an increased firing rate when the muscle is stretched. **(B)** α Motor neuron activation shortens the muscle and releases tension on the muscle spindle. Ia activity ceases temporarily during the tension release. **(C)** Concurrent α and γ motor neuron activation, as occurs in normal, voluntary muscle contraction, shortens the muscle spindle along with the extrafusal fibers, maintaining the spindle's responsiveness to the stretch.

intrafusal fiber has only one of these two types of endings. *Plate endings* occur predominantly on bag$_1$ fibers (dynamic), whereas *trail endings*, primarily on chain fibers, are also seen on bag$_2$ (static) fibers. This arrangement allows for largely independent control of the nuclear bag and nuclear chain fibers in the spindle.

γ Motor neurons with plate endings are designated *dynamic*, and those with trail endings are designated *static*. This functional distinction is based on experimental findings showing that stimulation of γ neurons with plate endings enhanced the response of type Ia sensory axons to stretch, but only during the dynamic (muscle length–changing) phase of a muscle stretch. During the static phase of the stretch (muscle length increase maintained), stimulation of the γ neurons with trail endings enhanced the response of type II sensory axons. Static γ neurons can affect the responses of both types Ia and II sensory axons; dynamic γ neurons affect the response of only type Ia axons. These differences suggest that the motor system has the ability to monitor muscle length more precisely in some muscles and the speed of contraction in others.

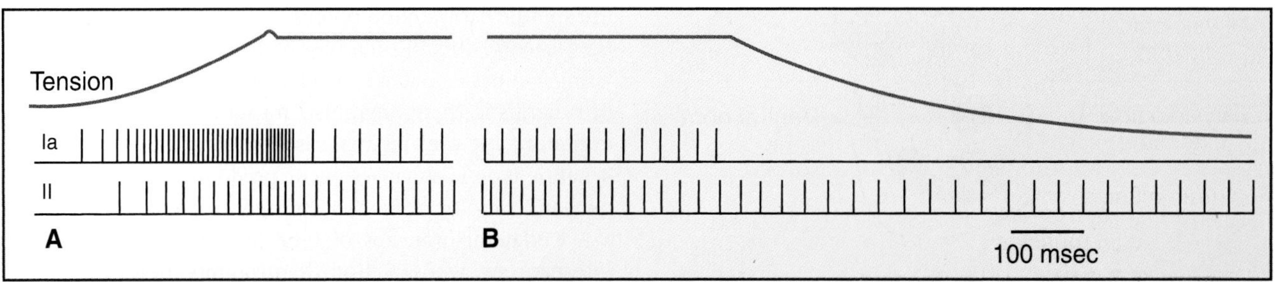

Figure 5.6 Response of types Ia and II sensory endings during muscle stretch. (A) During rapid stretch, type Ia endings show a greater firing rate increase, whereas type II endings show only a modest increase. **(B)** With the release of the stretch, Ia endings cease firing, whereas firing of type II endings slows. Ia endings report both the velocity and the length of muscle stretch; type II endings report the length.

SPINAL CORD IN THE CONTROL OF MOVEMENT

Muscles interact extensively in the maintenance of posture and the production of coordinated movement. The circuitry of the spinal cord automatically controls much of this interaction. Sensory feedback from muscles reaches motor neurons of related muscles and, to a lesser degree, of more distant muscles. In addition to activating local circuits, muscles and joints transmit sensory information up the spinal cord to higher centers. This information is processed and can be relayed back to influence spinal cord circuits.

Spinal motor anatomy correlates with function.

The cell bodies of the spinal motor neurons are grouped into pools in the ventral horns. A **pool** consists of the motor neurons that serve a particular muscle. The number of motor neurons that control a muscle varies in direct proportion to the delicacy of control required. The motor neurons are organized so that those innervating the axial muscles are grouped medially and those innervating the limbs are located laterally (Fig. 5.7). The lateral limb motor neuron areas are further organized so that proximal actions, such as girdle movements, are controlled from relatively medial locations, whereas distal actions, such as finger movements, are located the most laterally. Neurons innervating flexors and extensors are also segregated. A motor neuron pool may extend over several spinal segments, and the axons then emerge in the ventral nerve roots of two or even three adjacent spinal levels. A physiologic advantage to such an arrangement is that injury to a single nerve root, as might occur by herniation of an intervertebral disk, will not completely paralyze a muscle.

Between the spinal cord's dorsal and ventral horns lies the intermediate zone, which contains an extensive network of interneurons that interconnect motor neuron pools (see Fig. 5.7). Some interneurons make connections in their own cord segment; others have longer axon projections that travel in the white matter to terminate in other segments of the spinal cord. These longer interneurons, termed **propriospinal cells**, carry information that aids coordinated movement. The importance of spinal cord interneurons is clear, insofar as they comprise most of the neurons in the spinal cord as well as being the origin of most of the synapses on motor neurons.

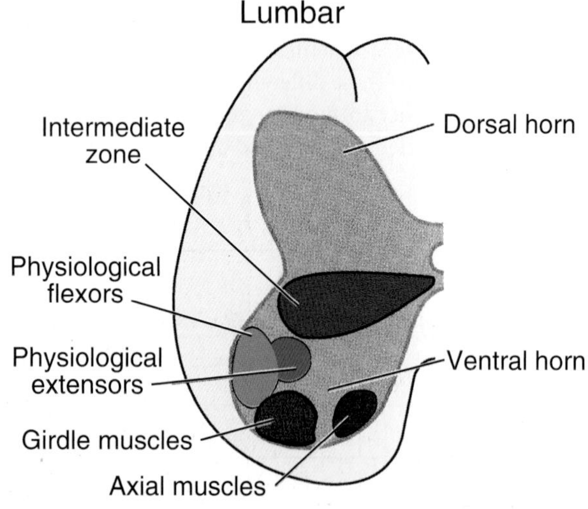

Figure 5.7 Spinal cord motor neuron pools. Motor neurons controlling axial, girdle, and limb muscles are grouped in pools oriented in a medial-to-lateral fashion. Flexor and extensor motor neurons also group into pools.

Spinal cord mediates reflex activity.

The spinal cord contains neural circuitry to generate **reflexes**, stereotypical actions produced in response to a stimulus. One function of a reflex is to generate a rapid response. A familiar example is the rapid, involuntary withdrawal of a hand after touching a dangerously hot object well before the heat or pain is perceived. This type of reflex protects the organism before higher CNS levels identify the problem. Some reflexes are simple, whereas others are much more complex. Even the simplest requires coordinated action in which the agonist muscle contracts while the antagonist relaxes. The functional unit of a reflex consists of a sensor, an afferent pathway, an integrating center, an efferent pathway, and an effector. The sensory receptors for spinal reflexes are the proprioceptors and cutaneous receptors. Impulses initiated in these receptors travel along afferent nerves to the spinal cord, where interneurons and motor neurons constitute the integrating center. The motor neurons are the efferent pathway to the effector organs, the skeletal muscles. The responsiveness of such a functional unit can be modulated by higher motor centers acting through descending pathways to facilitate or inhibit its activation.

Study of the three types of spinal reflexes—the myotatic, the inverse myotatic, and the flexor withdrawal—provides a basis for understanding the general mechanism of reflexes.

Myotatic (muscle stretch) reflex

Stretching or elongating a muscle—such as when the patellar tendon is tapped with a reflex hammer or when a quick change in posture is made—causes it to contract within a short duration. The duration between the onset of a stimulus and the response, the **latency period**, is on the order of 30 msec for a knee-jerk reflex in a human. This response, called the **myotatic** or **muscle stretch reflex**, is a result of monosynaptic circuitry, in which an afferent sensory neuron synapses directly on the efferent motor neuron (Fig. 5.8). The stretch activates muscle spindles. Type Ia sensory axons from the spindle carry action potentials to the spinal cord, where they synapse directly on motor neurons of the same (homonymous) muscle that was stretched and on motor neurons of synergistic (heteronymous) muscles. These synapses are excitatory and use glutamate as the neurotransmitter. Monosynaptic type Ia synapses occur predominantly on α motor neurons; γ motor neurons seemingly lack such connections.

Collateral branches of type Ia axons also synapse on interneurons, whose action then inhibits motor neurons of antagonist muscles (see Fig. 5.8). This synaptic pattern, called *reciprocal inhibition*, serves to coordinate muscles of opposing function around a joint. Secondary (type II) spindle

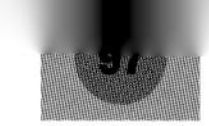

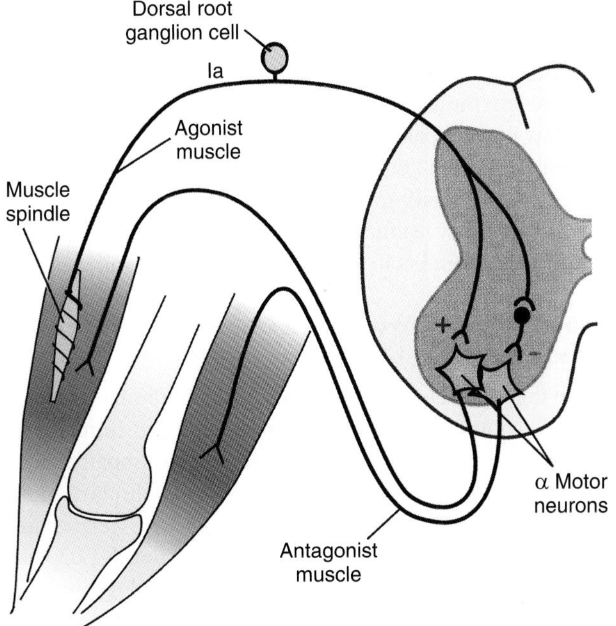

Figure 5.8 Myotatic reflex circuitry. Ia afferent axons from the muscle spindle make excitatory monosynaptic contact with homonymous motor neurons and with inhibitory interneurons that synapse on motor neurons of antagonist muscles. The plus sign indicates excitation; the minus sign indicates inhibition.

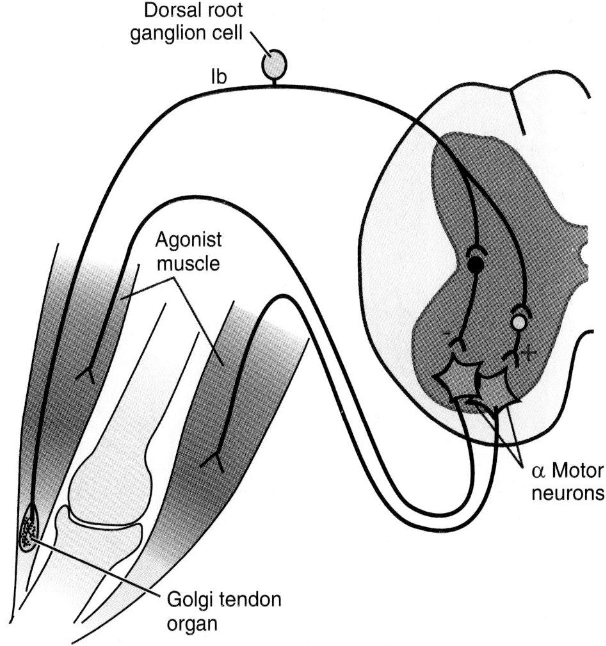

Figure 5.9 Inverse myotatic reflex circuitry. Contraction of the agonist muscle activates the Golgi tendon organ and Ib afferents that synapse on interneurons that inhibit agonist motor neurons and excite the motor neurons of the antagonist muscle.

afferent fibers also synapse with homonymous motor neurons, providing excitatory input through both monosynaptic and polysynaptic pathways. GTO input via type Ib axons has an inhibitory influence on homonymous motor neurons.

The myotatic reflex has two components: a *phasic part*, exemplified by the quick limb movement after tendon tapping with the reflex hammer, and a more sustained *tonic part*, thought to be important for posture maintenance. These two components blend together, but either one may predominate, depending on whether other synaptic activity, such as from cutaneous afferent neurons or pathways descending from higher centers, influences the motor response. Primary spindle afferent fibers probably mediate the tendon jerk, with secondary afferent fibers contributing mainly to the tonic phase of the reflex.

The myotatic reflex performs many functions. At the most general level, it produces rapid corrections of motor output in the moment-to-moment control of movement. It also forms the basis for postural reflexes, which maintain body position despite a varying range of loads on the body.

Inverse myotatic reflex

Active contraction of a muscle also causes reflex inhibition of the contraction. This response is called the **inverse myotatic reflex** because it produces an effect that is opposite to that of the myotatic reflex. Active muscle contraction stimulates GTOs, producing action potentials in the type Ib afferent axons. Those axons synapse on inhibitory interneurons that influence homonymous and heteronymous motor neurons and on excitatory interneurons that influence motor neurons of antagonists (Fig. 5.9).

The function of the inverse myotatic reflex appears to be a tension feedback system that can adjust the strength of contraction during sustained activity. The inverse myotatic reflex does not have the same function as reciprocal inhibition. Reciprocal inhibition acts primarily on the antagonist, whereas the inverse myotatic reflex acts on the agonist.

The inverse myotatic reflex, like the myotatic reflex, has a more potent influence on the physiologic extensor muscles than on the flexor muscles, suggesting that the two reflexes act together to maintain optimal responses in the antigravity muscles during postural adjustments. Another hypothesis about the conjoint function is that both of these reflexes contribute to the smooth generation of tension in muscle by regulating muscle stiffness.

Flexor withdrawal reflex

Cutaneous stimulation—such as touch, pressure, heat, cold, or tissue damage—can elicit a **flexor withdrawal reflex**. This reflex consists of a contraction of flexors and a relaxation of extensors in the stimulated limb. The action may be accompanied by a contraction of the extensors on the contralateral side. The axons of cutaneous sensory receptors synapse on interneurons in the dorsal horn. Those interneurons act ipsilaterally to excite the motor neurons of flexor muscles and inhibit those of extensor muscles. Collaterals of interneurons cross the midline to excite contralateral extensor motor neurons and inhibit flexors (Fig. 5.10).

There are two types of flexor withdrawal reflexes: those that result from innocuous stimuli and those that result from potentially injurious stimulation. The first type produces a localized flexor response accompanied by slight or no limb

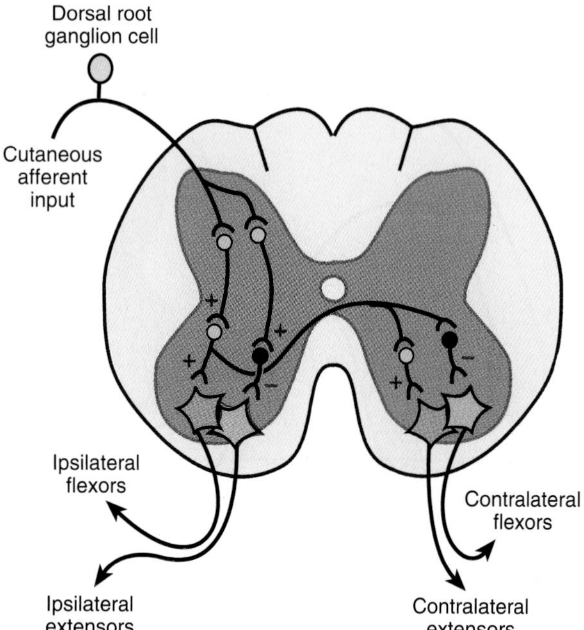

Figure 5.10 Flexor withdrawal reflex circuitry. Stimulation of cutaneous afferents activates ipsilateral flexor muscles via excitatory interneurons. Ipsilateral extensor motor neurons are inhibited. Contralateral extensor motor neuron activation provides postural support for the limb that is flexing.

withdrawal; the second type produces widespread flexor contraction throughout the limb and abrupt withdrawal. The function of the first type of reflex is less obvious but may be a general mechanism for adjusting the movement of a body part when an obstacle is detected by cutaneous sensory input. The function of the second type is protection of the individual. The endangered part is rapidly removed, and postural support of the opposite side is strengthened to support the body.

Collectively, these reflexes provide for stability and postural support (the myotatic and inverse myotatic) and mobility (flexor withdrawal). The reflexes provide a foundation of automatic responses on which more complicated voluntary movements are built.

Spinal cord can produce basic locomotor actions.

For locomotion, muscle action must occur in the limbs, but the posture of the trunk must also be controlled to provide a foundation from which the limb muscles can act. For example, when a human takes a step forward, not only must the advancing leg flex at the hip and knee, the opposite leg and bilateral truncal muscles must also be properly activated to prevent collapse of the body as weight is shifted from one leg to the other. Responsibility for the different functions that come together in successful locomotion is divided among several levels of the CNS.

Studies in experimental animals, mostly cats, have demonstrated that the spinal cord contains the capability for generating basic locomotor movements. This neural circuitry, called a **central pattern generator (CPG)**, can produce the alternating contraction of limb flexors and extensors that is needed for walking. It has been shown experimentally that application of an excitatory amino acid, such as glutamate, to the spinal cord produces rhythmic action potentials in motor neurons. Each limb has its own pattern generator, and the actions of different limbs are then coordinated by interneurons. The normal strategy for generating basic locomotion engages CPGs and uses both sensory feedback and efferent impulses from higher motor control centers for the refinement of control.

Spinal cord injury alters voluntary and reflex motor activity.

When the spinal cord of a human or other mammal is severely injured, voluntary and reflex movements are immediately lost caudal to the level of injury. This acute impairment of function is called **spinal shock**. The loss of voluntary motor control is termed **plegia**, and the loss of reflexes is termed **areflexia**. Spinal shock may last from days to months, depending on the severity of cord injury. Reflexes tend to return, as may some degree of voluntary control. As recovery proceeds, myotatic reflexes become hyperactive, as demonstrated by an excessively vigorous response to tapping the muscle tendon with a reflex hammer. A single tap, or limb repositioning that produces a change in the muscle length, may also provoke **clonus**, a condition characterized by repetitive contraction and relaxation of a muscle in an oscillating fashion every second or so. Flexor withdrawal reflexes may also reappear and be provoked by lesser stimuli than normally would be required. The acute loss and eventual overactivity of each of these reflexes results from the lack of influence of the neural tracts that descend from higher motor control centers to the motor neurons and associated interneuron pools.

SUPRASPINAL INFLUENCES ON MOTOR CONTROL

Descending signals from the cervical spinal cord, brainstem, and cortex can influence the rate of motor neuron firing and the recruitment of additional motor neurons to increase the force of muscle contraction. A dog, whose right and left limbs show alternating contractions during walking and then a pattern of both sides contracting in synchrony during running, illustrates the influence of higher motor control centers.

The brainstem contains the neural circuitry for initiating locomotion and for controlling posture. The maintenance of posture requires coordinated activity of both axial and limb muscles in response to input from proprioceptors and spatial position sensors such as the inner ear. Cerebral cortex input through the corticospinal system is necessary for the control of fine individual movements of the distal limbs and digits. Each higher level of the nervous system acts on lower levels to produce appropriate, more refined movements.

Brainstem is the origin of three descending tracts that influence movement.

Three brainstem nuclear groups give rise to descending motor tracts that influence motor neurons and their

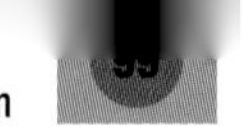

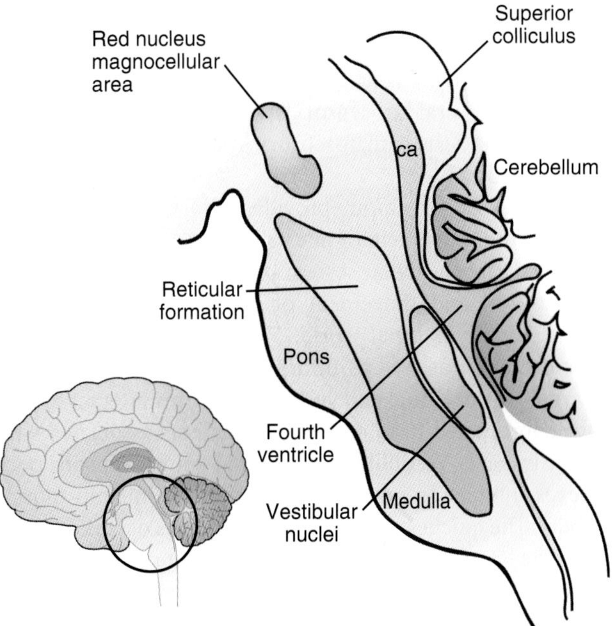

Figure 5.11 Brainstem nuclei of descending motor pathways. The magnocellular portion of the red nucleus is the origin of the rubrospinal tract. The lateral vestibular nucleus is the source of the vestibulospinal tract. The reticular formation is the source of two tracts: one from the pontine portion and one from the medulla. Structures illustrated are from the monkey. ca, cerebral aqueduct.

associated interneurons. These are the **red nucleus**, the **vestibular nuclear complex**, and the **reticular formation** (Fig. 5.11). The other major descending influence on the motor neurons is the corticospinal tract, the only volitional control pathway in the motor system. This tract is discussed later. In most cases, the descending pathways act through synaptic connections on interneurons. The connection is less commonly made directly with motor neurons.

Rubrospinal tract

The red nucleus of the mesencephalon receives major input from both the cerebellum and the cerebral cortical motor areas. Output via the **rubrospinal tract** is directed predominantly to contralateral spinal motor neurons that are involved with movements of the distal limbs. The axons of the rubrospinal tract are located in the lateral spinal white matter. Rubrospinal action enhances the function of motor neurons innervating limb flexor muscles while inhibiting extensors. This tract may also influence γ motor neuron function.

Electrophysiologic studies reveal that many rubrospinal neurons are active during locomotion, with more than one half showing increased activity during the swing phase of stepping, when the flexors are most active. This system appears to be important for the production of movement, especially in the distal limbs. Experimental lesions that interrupt rubrospinal axons produce deficits in distal limb flexion, with little change in more proximal muscles. In higher animals, the corticospinal tract supersedes some of the function of the rubrospinal tract.

Vestibulospinal tract

The vestibular system regulates muscular function for the maintenance of posture in response to changes in the position of the head in space and accelerations of the body. There are four major nuclei in the vestibular complex:the *superior, lateral, medial*, and *inferior vestibular nuclei*. These nuclei, located in the pons and medulla, receive afferent action potentials from the vestibular portion of the ear, which includes the semicircular canals, the utricle, and the saccule (see Chapter 4, "Sensory Physiology"). Information about rotatory and linear motions of the head and body is conveyed by this system. The vestibular nuclei are reciprocally connected with the superior colliculus on the dorsal surface of the mesencephalon. Input from the retina is received there and is used in adjusting eye position during movement of the head. Reciprocal connections to the vestibular nuclei are also made with the cerebellum and reticular formation.

The chief output to the spinal cord is the **vestibulospinal tract**, which originates predominantly from the lateral vestibular nucleus. The tract's axons are located in the anterior spinal white matter and carry excitatory action potentials to ipsilateral extensor motor neuron pools, both α and γ. The extensor motor neurons and their musculature are important in the maintenance of posture. Lesions in the brainstem secondary to stroke or trauma may abnormally enhance the influence of the vestibulospinal tract and produce dramatic clinical manifestations.

Reticulospinal tracts

The reticular formation in the central gray matter core of the brainstem contains many axon bundles interwoven with cells of various shapes and sizes. A prominent characteristic of reticular formation neurons is that their axons project widely in ascending and descending pathways, making multiple synaptic connections throughout the neuraxis. The medial region of the reticular formation contains large neurons that project upward to the thalamus as well as downward to the spinal cord. Afferent input to the reticular formation comes from the spinal cord, vestibular nuclei, cerebellum, lateral hypothalamus, **globus pallidus (GP)**, tectum, and sensorimotor cortex.

Two areas of the reticular formation are important in the control of motor neurons. The descending tracts arise from the **nucleus reticularis pontis oralis** and **nucleus reticularis pontis caudalis** in the pons, and from the **nucleus reticularis gigantocellularis** in the medulla. The pontine reticular area gives rise to the ipsilateral **pontine reticulospinal tract**, whose axons descend in the anterior medial spinal white matter. These axons carry excitatory action potentials to interneurons that influence α and γ motor neuron pools of axial muscles. The medullary area gives rise to the **medullary reticulospinal tract**, whose axons descend in the ipsilateral anterior spinal white matter. These axons have inhibitory influences on interneurons that modulate extensor motor neurons.

Terminations of the brainstem motor tracts correlate with their functions.

The vestibulospinal and reticulospinal tracts terminate in the ventromedial part of the intermediate zone, an area

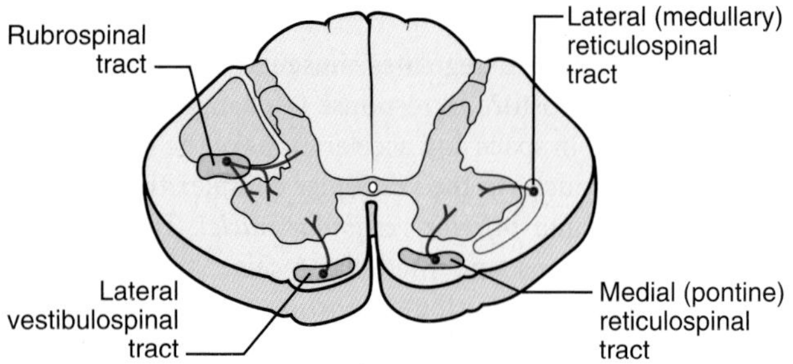

Figure 5.12 Brainstem motor control tracts. The vestibulospinal and reticulospinal tracts influence motor neurons that control axial and proximal limb muscles. The rubrospinal tract influences motor neurons controlling distal limb muscles. Excitatory pathways are shown in *red*.

in the gray matter containing propriospinal interneurons (Fig. 5.12). There are also some direct connections with motor neurons of the neck and back muscles and the proximal limb muscles. These tracts are the main CNS pathways for maintaining posture and head position during movement.

The rubrospinal tract terminates mostly on interneurons in the lateral spinal intermediate zone, but it also has some monosynaptic connections directly on motor neurons to muscles of the distal extremities. This tract supplements the medial descending pathways in postural control and the corticospinal tract for independent movements of the extremities.

In accordance with their medial or lateral distributions to spinal motor neurons, the reticulospinal and vestibulospinal tracts are thought to be most important for the control of axial and proximal limb muscles, whereas the rubrospinal (and corticospinal) tracts are most important for the control of distal limb muscles, particularly the flexors.

Sensory and motor systems work together to control posture.

The maintenance of an upright posture in humans requires active muscular resistance against gravity. For movement to occur, the initial posture must be altered by flexing some body parts against gravity. Balance must be maintained during movement, which is achieved by postural reflexes initiated by several key sensory systems. Vision, the vestibular system, and the somatosensory system are important for postural reflexes.

Somatosensory input provides information about the position and movement of one part of the body with respect to others. The vestibular system provides information about the position and movement of the head and neck with respect to the external world. Vision provides both types of information as well as information about objects in the external world. Visual and vestibular reflexes interact to produce coordinated head and eye movements associated with a shift in gaze. Vestibular reflexes and somatosensory neck reflexes interact to produce reflex changes in limb muscle activity. The quickest of these compensations occurs at about twice the latency of the monosynaptic myotatic reflex. These response types are termed **long-loop reflexes.** The extra time reflects the action of other neurons at different anatomic levels of the nervous system.

CEREBRAL CORTEX ROLE IN MOTOR CONTROL

The cerebral cortical areas concerned with motor function exert the highest level of motor control. It is difficult to formulate an unequivocal definition of a **cortical motor area**, but three criteria may be used. An area is said to have a motor function if:

- Stimulation using low current strengths elicits movements.
- Destruction of the area results in a loss of motor function.
- The area has output connections going relatively directly (i.e., with a minimal number of intermediate connections) to brainstem or spinal motor neurons.

Some cortical areas fulfill all of these criteria and have exclusively motor functions. Other areas fulfill only some of the criteria yet are involved in movement, particularly volitional movement.

Clinical Focus / 5.1

Decerebrate Rigidity

A patient with a history of poorly controlled hypertension is brought to the emergency department because of sudden collapse and subsequent unresponsiveness. A neurologic examination performed about 30 minutes after onset shows no response to verbal stimuli. Spontaneous movements of the limbs are absent. A painful stimulus, compression of the soft tissue of the supraorbital ridge, causes immediate extension of the neck and all of the limbs. This posture relaxes within a few seconds after the stimulation is stopped. After the patient is stabilized medically, he undergoes a magnetic resonance imaging (MRI) study of the brain. The study demonstrates a large area of hemorrhage bilaterally in the upper portion of the brainstem.

The posture this patient demonstrated in response to the noxious stimulus is termed **decerebrate rigidity**. Its occurrence is associated with lesions of the mesencephalon that eliminate the influence of higher brainstem and cortical centers. The abnormal posture is a result of extreme activation of the antigravity extensor muscles by the unopposed action of the lateral vestibular nucleus and the vestibulospinal tract. A model of this condition can be produced in experimental animals by a surgical lesion located between the mesencephalon and pons. It can also be shown in experimental animals that a destructive lesion of the lateral vestibular nucleus relieves the rigidity.

Distinct cortical areas participate in voluntary movement.

The **primary motor cortex (M1)**, Brodmann area 4, fulfills all three criteria for a motor area (Fig. 5.13). The **supplementary motor area (SMA)**, which also fulfills all three criteria, is rostral and medial to M1 in Brodmann area 6. Other areas that fulfill some of the criteria include the rest of Brodmann area 6; areas 1, 2, and 3 of the postcentral gyrus; and areas 5 and 7 of the parietal lobe. Each of these areas contributes fibers to the **corticospinal tract**, the efferent motor pathway from the cortex.

Primary motor cortex

This cortical area corresponds to Brodmann area 4 in the precentral gyrus. Area 4 is structured in six well-defined layers (I–VI), with layer I being closest to the pial surface. Afferent fibers terminate in layers I to V. Thalamic afferent fibers terminate in two layers: those that carry somatosensory information end in layer IV and those from nonspecific nuclei end in layer I. Cerebellar afferents terminate in layer IV. Efferent axons arise in layers V and VI to descend as the corticospinal tract. Body areas are represented in an orderly manner, as **somatotopic maps**, in the motor and sensory cortical areas (Fig. 5.14). Those parts of the body that perform fine movements, such as the digits and the facial muscles, are controlled by a greater number of neurons that occupy more cortical territory than the neurons for the body parts only capable of gross movements.

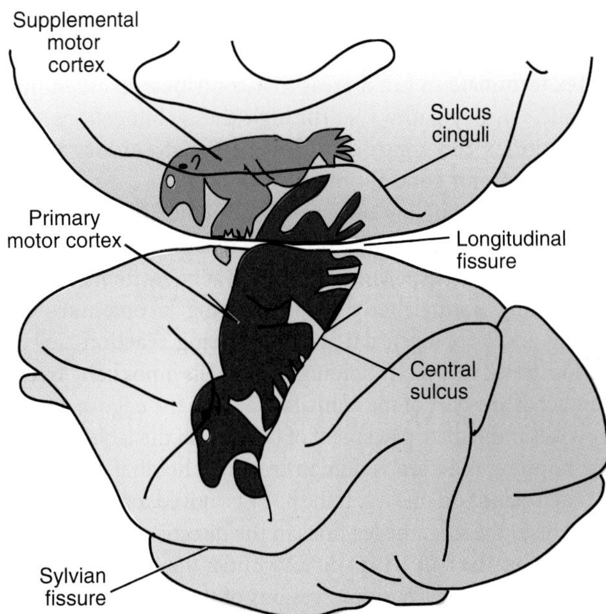

Figure 5.14 A cortical map of motor functions. Primary motor cortex and supplementary motor cortex (SMA) areas in the monkey brain. SMA is on the medial aspect of the hemisphere.

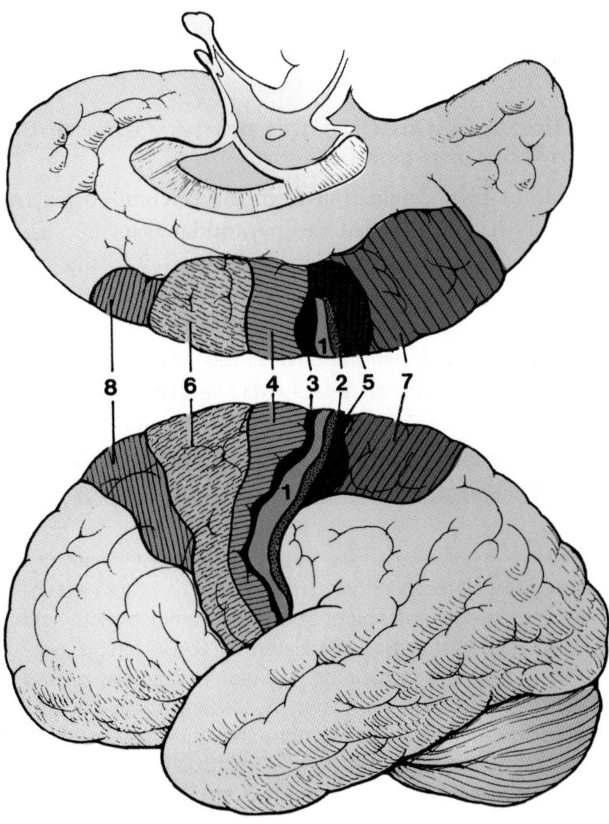

Figure 5.13 Brodmann cytoarchitectural map of the human cerebral cortex. Area 4 is the primary motor cortex; area 6 is the premotor cortex and includes the supplementary motor area on the medial aspect of the hemisphere. Area 8 influences voluntary eye movements. Areas 1, 2, 3, 5, and 7 have sensory functions but also contribute axons to the corticospinal tract. The central sulcus divides areas 4 and 3.

Low-level electrical stimulation of surgically exposed M1 produces twitch-like contraction of a few muscles or, less commonly, a single muscle. Slightly stronger stimuli also produce responses in adjacent muscles. Movements elicited from area 4 have the lowest stimulation thresholds and are the most discrete of any movements elicited by stimulation. Stimulation of M1 limb areas produces contralateral movement, whereas cranial cortical areas may produce bilateral motor responses. Destruction of any part of the primary motor cortex leads to immediate paralysis of the muscles controlled by that area. In humans, some function may return weeks to months later, but the movements lack the fine degree of muscle control of the normal state. For example, after a lesion in the arm area of M1, the use of the hand recovers, but the capacity for discrete finger movements does not.

Neurons in M1 encode the capability to control muscle force, muscle length, joint movement, and position. The area receives somatosensory input, both cutaneous and proprioceptive, via the ventrobasal thalamus. The cerebellum projects to M1 via the red nucleus and ventrolateral thalamus. Other afferent projections come from the nonspecific nuclei of the thalamus, the contralateral motor cortex, and many other ipsilateral cortical areas. There are many axons between the precentral (motor) and postcentral (somatosensory) gyri and many connections to the visual cortical areas. Because of their connections with the somatosensory cortex, the cortical motor neurons can also respond to sensory stimulation. For example, cells innervating a particular muscle may respond to cutaneous stimuli originating in the area of skin that moves when that muscle is active, and they may respond to proprioceptive stimulation from the muscle to which they are related. Many efferent fibers from the primary motor cortex terminate in brain areas that contribute to ascending somatic sensory pathways. Through these connections, the motor cortex can control the flow of somatosensory information to motor control centers.

The close coupling of sensory and motor functions may play a role in two cortically controlled reflexes that were originally described in experimental animals as being important for maintaining normal body support during locomotion—the placing and hopping reactions. The **placing reaction** can be demonstrated in a cat by holding it so that its limbs hang freely. Contact of any part of the animal's foot with the edge of a table provokes immediate placement of the foot on the table surface. The **hopping reaction** is demonstrated by holding an animal so that it stands on one leg. If the body is moved forward, backward, or to the side, the leg hops in the direction of the movement so that the foot is kept directly under the shoulder or hip, stabilizing the body position. Lesions of the contralateral precentral or postcentral gyrus abolish placing. Hopping is abolished by a contralateral lesion of the precentral gyrus.

Supplementary motor area

The SMA is located on the medial surface of the hemispheres, above the cingulate sulcus and rostral to the leg area of the primary motor cortex (see Fig. 5.14). This cortical region within Brodmann area 6 has no clear cytoarchitectural boundaries; that is, the shapes and sizes of cells and their processes are not obviously compartmentalized, as in the layers of M1.

Electrical stimulation of SMA produces movements, but a higher stimulus strength is required than for M1. The movements produced by stimulation are also qualitatively different from M1; they last longer, the postures elicited may remain after the stimulation is over, and the movements are less discrete. Bilateral responses are common. SMA is reciprocally connected with M1 and receives input from other motor cortical areas. Experimental lesions in M1 eliminate the ability of SMA stimulation to produce movements.

Current knowledge is insufficient to adequately describe the unique role of SMA in higher motor functions. SMA is thought to be active in bimanual tasks, in learning and preparing for the execution of skilled movements, and in the control of muscle tone. The mechanisms that underlie the more complex aspects of movement, such as thinking about and performing skilled movements and using complex sensory information to guide movement, remain incompletely understood.

Primary somatosensory cortex

The **primary somatosensory cortex (S1)** (Brodmann areas 1, 2, and 3) lies on the postcentral gyrus (see Fig. 5.13) and has a role in movement. Electrical stimulation here can produce movement, but thresholds are two to three times higher than in M1. S1 is reciprocally interconnected with M1 in a somatotopic pattern—for example, arm areas of sensory cortex project to arm areas of motor cortex. Efferent fibers from areas 1, 2, and 3 travel in the corticospinal tract and terminate in the dorsal horn areas of the spinal cord.

Superior parietal lobe

The **superior parietal lobe** (Brodmann areas 5 and 7) also has important motor functions. In addition to contributing fibers to the corticospinal tract, it is well connected to the motor areas in the frontal lobe. Lesion studies in animals and humans suggest this area is important for the use of complex sensory information in the production of movement.

Corticospinal tract is the primary efferent path from the cortex.

The motor tract originating from the cerebral cortex traditionally has been called the **pyramidal tract** because it traverses the medullary pyramids on its way to the spinal cord (Fig. 5.15). This path is the corticospinal tract. All other descending motor tracts emanating from the brainstem were generally grouped together as the extrapyramidal system. Cells in Brodmann area 4 (M1) contribute 30% of the

From Bench to Bedside / 5.1

Spasticity

Patients who develop lesions, such as an infarction in the internal capsule of the cerebrum or spinal cord trauma affecting the anterior two thirds of the cord, develop a disorder of motor control that is termed **spasticity**. Clinically this is characterized by: (1) weakness, slowness, and clumsiness of movement in body parts below the level of the lesion; (2) abnormally increased muscle stiffness to passive lengthening; and (3) abnormally overactive muscle stretch reflexes in those limbs. The neurophysiology of spasticity is not fully understood, but the spinal interneurons involved with motor control become disinhibited and overactive as a result of impaired inhibitory influence from the brainstem and cortex. Although the increased muscle tone is partly compensatory for the impaired corticospinal tract input, it may become so exaggerated that the affected limbs become even less functional as a result of the exaggerated muscle stiffness.

Several medications are available to treat spasticity. Baclofen is a medication that stimulates the receptors at synapses that release the neurotransmitter γ-aminobutyric acid (GABA). The interneurons stimulated by GABA inhibit overactivity of the α motor neurons. Injection of precisely diluted amounts of the toxin produced by the anaerobic bacteria *Clostridium botulinum* into the abnormal muscles is another method to reduce their overactivity. The toxin impairs release of the neurotransmitter acetylcholine from the motor neuron at the synapse with muscle cells. Proper dosing partially weakens the muscle and reduces the abnormally increased muscle stiffness. Gaining an understanding of the physiology of synaptic vesicle release and discovering the role of the botulinum toxins in disturbing this process have allowed treatment with these agents to be made available to patients with conditions that formerly were difficult to treat.

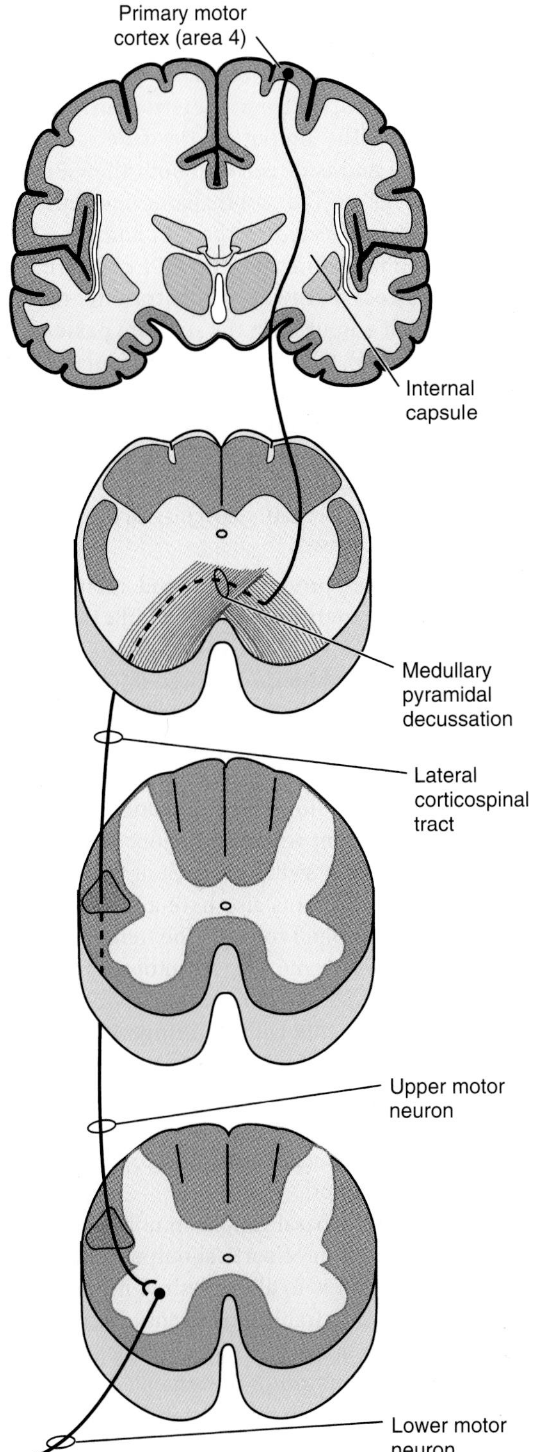

Figure 5.15 The corticospinal tract. Axons arising from cortical motor areas descend through the internal capsule, decussate in the medulla, descend in the lateral funiculus of the spinal cord as the lateral corticospinal tract, and terminate on motor neurons and interneurons in the ventral horn areas of the spinal cord. Note the designations of upper and lower motor neurons.

corticospinal fibers; area 6 (SMA) 30% of the fibers; and the parietal lobe, especially Brodmann areas 1, 2, and 3, the remaining 40%. In primates, 10% to 20% of corticospinal fibers end directly on motor neurons; the others end on interneurons associated with motor neurons.

From the cerebral cortex, the corticospinal tract axons descend through the brain along a path located between the basal ganglia and the thalamus, known as the **internal capsule**. They then continue along the ventral brainstem as the **cerebral peduncles** and on through the pyramids of the medulla. Most of the corticospinal axons cross the midline in the medullary pyramids. Thus, the motor cortex in each hemisphere controls the muscles on the contralateral side of the body. After crossing in the medulla, the corticospinal axons descend in the dorsal lateral columns of the spinal cord posterior to the rubrospinal tract and terminate in lateral motor pools that control distal muscles of the limbs. A smaller group of axons does not cross in the medulla and descends in the anterior spinal columns. These axons terminate in the motor neuron pools and adjacent intermediate zone interneurons that control the axial and proximal musculature. Deficits in corticospinal tract function produce a characteristic type of impairment in motor control.

The corticospinal tract is estimated to contain about 1 million axons at the level of the medullary pyramid. The largest-diameter, heavily myelinated axons are between 9 and 20 μm in diameter, but that population of axons accounts for only a small fraction of the total. Most corticospinal axons are small, 1 to 4 μm in diameter, and one half are unmyelinated.

In addition to the direct corticospinal tract, there are other indirect pathways by which cortical fibers influence motor function. Some cortical efferent fibers project to the reticular formation, then to the spinal cord via the reticulospinal tract; others project to the red nucleus, then to the spinal cord via the rubrospinal tract. Despite the fact that these pathways involve intermediate neurons on the way to the cord, volleys relayed through the reticular formation can reach the spinal cord motor circuitry at the same time as, or earlier than, some volleys along the corticospinal tract.

BASAL GANGLIA AND MOTOR CONTROL

The **basal ganglia** are a group of subcortical nuclei located primarily in the base of the forebrain, with some in the diencephalon and upper brainstem. The striatum, GP, subthalamic nucleus, and substantia nigra comprise the basal ganglia. Input is derived from the cerebral cortex, and output is directed to the cortical and brainstem areas concerned with movement. Basal ganglia action influences the entire motor system and plays a role in the preparation and execution of coordinated movements.

The forebrain (telencephalic) components of the basal ganglia consist of the **striatum**, which is made up of the **caudate nucleus** and the **putamen**, and the GP. The caudate nucleus and putamen are histologically identical but are separated anatomically by fibers of the anterior limb of the internal capsule. The GP has two subdivisions: the *external segment (GPe)*, adjacent to the medial aspect of the putamen, and the *internal segment (GPi)*, medial to the GPe. The other main nuclei of the basal ganglia are the **subthalamic nucleus** in the diencephalon and the **substantia nigra** in the mesencephalon.

Basal ganglia are extensively interconnected.

Although the circuitry of the basal ganglia appears complex at first glance, it can be simplified into input, output, and internal pathways (Fig. 5.16). Input from the cerebral cortex is directed to the striatum and the subthalamic nucleus. The predominant nerve cell type in the striatum is termed the **medium spiny neuron**, based on its cell body size and dendritic structure. This type of neuron receives input from all of the cerebral cortex except for the primary visual and auditory areas. The input is roughly somatotopic and is excitatory via glutamine-containing neurons. The caudate receives input from the prefrontal cortex, whereas the putamen receives most of the cortical input from sensorimotor areas. The subthalamus receives excitatory glutaminergic input from cortical areas concerned with motor function, including eye movement.

Basal ganglia output is from the GPi and one segment of the substantia nigra. The GPi output is directed to ventrolateral and ventral anterior nuclei of the thalamus, which feed back to the cortical motor areas. Output from the GPi is also directed to a region in the upper brainstem termed the **midbrain extrapyramidal area**. This area then projects to the neurons of the reticulospinal tract. The substantia nigra output arises from the **pars reticulata (SNr)**, which is histologically similar to the GPi. The output is directed to the superior colliculus of the mesencephalon, which is involved in eye movement control. The GPi and SNr output is inhibitory via neurons that use γ-aminobutyric acid (GABA) as the neurotransmitter.

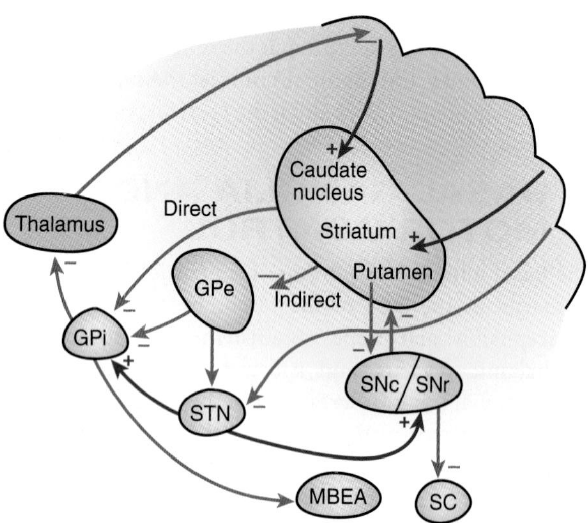

Figure 5.16 Basal ganglia nuclei and circuitry. The neuronal circuit of cerebral cortex to striatum to globus pallidus interna (GPi) to thalamus and back to the cortex is the main pathway for basal ganglia influence on motor control. Note the direct and indirect pathways of the striatum, GPi, globus pallidus externa (GPe), and subthalamic nucleus (SNT). GPi output also flows to the midbrain extrapyramidal area (MBEA). The substantia nigra pars reticulata (SNr) to superior colliculus (SC) pathway is important in eye movements. Excitatory pathways are shown in *red*, and inhibitory pathways are shown in *blue*. SNc, substantia nigra pars compacta.

The GPe, the subthalamic nucleus, and the pars compacta region of the substantia nigra (SNc) form an important internal pathway within the basal ganglia. The GPe receives inhibitory input from the striatum via GABA-releasing neurons. The output of the GPe is inhibitory via GABA release and is directed to both the GPi and the subthalamic nucleus. The subthalamic nucleus output is excitatory and is directed to the GPi and the SNr. The striatum–GPe–subthalamic nucleus–GPi circuit has been termed the **indirect pathway** in contrast to the **direct pathway**, in which output from the striatum passes directly to GPi (see Fig. 5.16). The SNc output modulates cortical influence on the striatum via dopamine-releasing neurons, which are excitatory to the direct pathway and inhibitory to the indirect pathway.

Functions of the basal ganglia are partially revealed by disease.

Basal ganglia diseases produce profound motor dysfunction in humans and experimental animals. The disorders can result in reduced motor activity, **hypokinesis**, or abnormally enhanced activity, **hyperkinesis**. Two well-known neurologic conditions that show histologic abnormality in basal ganglia structures, **Parkinson disease (PD)** and Huntington disease, illustrate the effects of basal ganglia dysfunction. Patients with PD show a general slowness of initiation of movement and paucity of movement when in motion. The latter takes the form of reduced arm swing and lack of truncal swagger when walking. These patients also have a resting tremor of the hands, described as "pill rolling." The tremor stops when the hand goes into active motion. At autopsy, patients with PD show a severe loss of dopamine-containing neurons in the SNc region. Patients with **Huntington disease** have uncontrollable, quick, brief movements of individual limbs. These movements are similar to what a normal individual might show when flicking a fly off a hand or when quickly reaching up to scratch an itchy nose. At autopsy, a severe loss of striatal neurons is found.

The function of the basal ganglia in normal individuals appears to be modulation of cortical output. One theory is that the primary action is to allow desired motions to proceed and to inhibit undesirable movements. Neuronal activity is increased in the appropriate areas of the basal ganglia prior to the actual execution of movement. The basal ganglia act as a brake on undesirable motion by inhibitory output from the GPi back to the thalamus and, ultimately, to the motor cortex. The loss of dopamine-releasing neurons in PD is thought to enhance this braking effect by increasing the activity of the indirect basal ganglia pathway. This effect ultimately increases the inhibitory output of the GPi because the subthalamus is disinhibited. Decreased motor action is the final result. Hyperkinetic disorders such as Huntington disease are thought to result from decreased GPi output resulting from loss of inhibitory influence of the direct pathway.

The braking effect of the basal ganglia may also be active in modulating emotional and intellectual activities of the cortex.

CEREBELLUM IN THE CONTROL OF MOVEMENT

The **cerebellum**, or "little brain," lies caudal to the occipital lobe and is attached to the posterior aspect of the brainstem through three paired fiber tracts: *inferior*, *middle*, and *superior* **cerebellar peduncles**. Input to the cerebellum comes from peripheral sensory receptors, the brainstem, and the cerebral cortex. The inferior, middle, and, to a lesser degree, superior cerebellar peduncles carry the input. The output projections are mainly, if not totally, to other motor control areas of the CNS and are mostly carried in the superior cerebellar peduncle. The cerebellum contains three pairs of intrinsic nuclei: the **fastigial**, **interposed** (or **interpositus**), and **dentate**. In some classification schemes, the interposed nucleus is further divided into the *emboliform* and *globose* nuclei.

Structural divisions of the cerebellum correlate with function.

The cerebellar surface is arranged in multiple, parallel, longitudinal folds termed **folia**. Several deep fissures divide the cerebellum into three main morphologic components—*anterior*, *posterior*, and *flocculonodular lobes*, which also correspond with the functional subdivisions of the cerebellum (Fig. 5.17). The functional divisions are the *vestibulocerebellum*, the *spinocerebellum*, and the *cerebrocerebellum*. These divisions appear in sequence during evolution. The lateral cerebellar hemispheres increase in size along with expansion of the cerebral cortex. The three divisions have similar intrinsic circuitry; thus, the function of each depends on the nature of the output nucleus to which it projects.

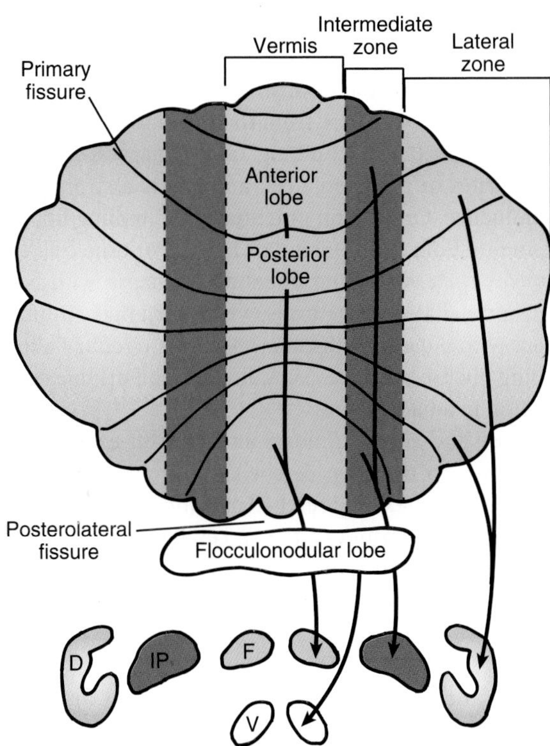

Figure 5.17 The structure of the cerebellum. The three lobes are shown: anterior, posterior, and flocculonodular. The functional divisions are demarcated by color. The vestibulocerebellum (*white*) is the flocculonodular lobe, which projects to the vestibular (V) nuclei. The spinocerebellum includes the vermis (*green*) and the intermediate zone (*violet*), which project to the fastigial (F) and interposed (IP) nuclei, respectively. The cerebrocerebellum (*pink*) projects to the dentate nuclei (D).

The **vestibulocerebellum** is composed of the flocculonodular lobe. It receives input from the vestibular system and visual areas. Output goes to the vestibular nuclei in the brainstem, rather than to nuclei intrinsic to the cerebellum. The vestibulocerebellum functions to control equilibrium and eye movements.

The medially placed **spinocerebellum** consists of the midline **vermis** plus the medial portion of the lateral hemispheres, called the **intermediate zones**. Spinocerebellar pathways carrying somatosensory information terminate in the vermis and intermediate zones in somatotopic arrangements. The auditory, visual, and vestibular systems and sensorimotor cortex also project to this portion of the cerebellum. Output from the vermis is directed to the fastigial nuclei, which project through the inferior cerebellar peduncle to the vestibular nuclei and reticular formation of the pons and medulla. Output from the intermediate zones goes to the interposed nuclei and then to the red nucleus. The ultimate target is the motor cortex via the ventrolateral nucleus of the thalamus. It is believed that both the fastigial and interposed nuclei contain a complete representation of the muscles. The fastigial output system controls antigravity muscles in posture and locomotion, whereas the interposed nuclei, perhaps, act on stretch reflexes and other somatosensory reflexes.

The **cerebrocerebellum** occupies the lateral aspects of the cerebellar hemispheres. Input comes exclusively from the cerebral cortex, relayed through the middle cerebellar peduncles of the pons. The cortical areas that are prominent in motor control are the sources for most of this input. Output is directed to the dentate nuclei and, ultimately, to the motor and premotor areas of the cerebral cortex via the ventrolateral thalamus.

Intrinsic circuitry of the cerebellum is regulated by Purkinje cells.

The cerebellar cortex is composed of five types of neurons arranged into three layers (Fig. 5.18). The *molecular layer* is the outermost and consists mostly of axons and dendrites plus two types of interneurons: **stellate cells** and **basket cells**. The *Purkinje cell layer* contains the **Purkinje cells**, whose dendrites reach upward into the molecular layer in a fanlike array. The Purkinje cells are the efferent neurons of the cerebellar cortex. Their action is inhibitory, with GABA being the neurotransmitter. Deep to the Purkinje cells is the **granular layer**, containing **Golgi cells** and small local circuit neurons, the **granule cells**. The granule cells are numerous; there are more granule cells in the cerebellum than neurons in the entire cerebral cortex!

Afferent axons to the cerebellar cortex are of two types: mossy fibers and climbing fibers. **Mossy fibers** arise from the spinal cord and brainstem neurons, including those of the pons that receive input from the cerebral cortex. Mossy

Clinical Focus / 5.2

Stereotactic Neurosurgery and Deep Brain Stimulation for Parkinson Disease

Parkinson disease is a central nervous system (CNS) disorder producing a generalized slowness of movement and resting tremor of the hands. Loss of dopamine-producing neurons in the substantia nigra pars compacta is the cause of the condition. Treatment with medications that stimulate an increased production of dopamine by the surviving substantia nigra neurons has revolutionized the management of Parkinson disease. Unfortunately, the benefit of the medications tends to lessen after 5 to 10 years of treatment. Increasing difficulty in initiating movement and worsening slowness of movement are features of a declining responsiveness to medication. Improved knowledge of basal ganglia circuitry has enabled neurosurgeons to develop surgical procedures to ameliorate some of the effects of the advancing disease.

Degeneration of the dopamine-releasing cells of the substantia nigra allows inhibitory output of the putamen and the globus pallidus external segment (GPe) to greatly increase via the indirect pathway. This results in decreased inhibitory GPe influence on the subthalamic nucleus, which then acts to excessively stimulate the globus pallidus internal segment (GPi). Enhanced activity of the GPi increases its inhibitory influence on the thalamus and results in decreased excitatory drive back to the cerebral cortex.

Stereotactic neurosurgery is a technique in which a small probe can be precisely placed into a target within the brain. Magnetic resonance imaging (MRI) of the brain defines the three-dimensional location of the GPi. The surgical probe is introduced into the brain through a small hole made in the skull and is guided to the target by the surgeon by using the MRI coordinates. When the target location is reached, the probe is heated to a temperature that destroys a precisely controllable amount of the GPi. The inhibitory outflow of the GPi is reduced and movement improves.

More recently, the use of stimulators whose electrodes can be implanted in specific areas of the basal ganglia has been investigated as an alternative method to improve motor function in patients with Parkinson disease and other types of movement disorders. High-frequency electrical stimulation of the subthalamic nucleus paradoxically reduces its output and, thereby, improves function of the thalamocortical pathway for motor control.

fibers make complex multicontact synapses on granule cells. The axons of the granule cells then ascend to the molecular layer and bifurcate, forming the **parallel fibers**. These travel perpendicular to and synapse with the dendrites of Purkinje cells, providing excitatory input via glutamate. Mossy fibers discharge at high tonic rates, 50 to 100 Hz, which increase further during voluntary movement. When mossy fiber input is of sufficient strength to bring a Purkinje cell to threshold, a single action potential results.

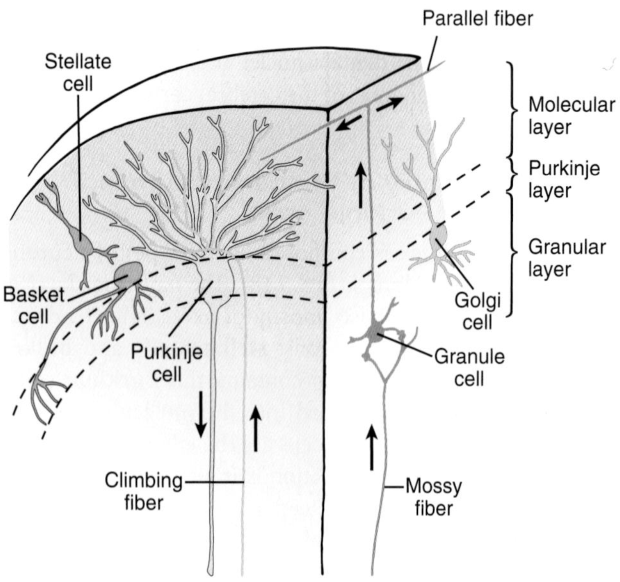

Figure 5.18 Cerebellar circuitry. The cell types and action potential pathways are shown. Mossy fibers are afferent from the spinal cord and the cerebral cortex. Climbing fibers are afferent from the inferior olive nucleus in the medulla and synapse directly on the Purkinje cells. The Purkinje cells are the efferent pathways of the cerebellum.

Climbing fibers arise from the **inferior olive**, a nucleus in the medulla. Each climbing fiber synapses directly on the dendrites of a Purkinje cell and exerts a strong excitatory influence. One action potential in a climbing fiber produces a burst of action potentials in the Purkinje cell, called a *complex spike*. Climbing fibers also synapse with basket, Golgi, and stellate interneurons, which then make inhibitory contact with adjacent Purkinje cells. This circuitry allows a climbing fiber to produce excitation in one Purkinje cell and inhibition in adjacent ones.

Mossy and climbing fibers also give off excitatory collateral axons to the deep cerebellar nuclei before reaching the cerebellar cortex. The cerebellar cortical output (Purkinje cell efferents) is inhibitory to the cerebellar and vestibular nuclei, but the final output of the cerebellum by those nuclei is predominantly excitatory.

Cerebellar function

Lesions of the cerebellum produce impairment in the coordinated action of agonists, antagonists, and synergists. This impairment is clinically known as **ataxia**. The control of limb, axial, and cranial muscles may be impaired depending on the site of the cerebellar lesion. Limb ataxia might manifest as the coarse jerking motions of an arm and hand during reaching for an object instead of the expected, smooth actions. This jerking type of motion is also referred to as **action tremor**. The swaying walk of an intoxicated individual is a vivid example of truncal ataxia.

Cerebellar lesions can also produce a reduction in muscle tone, called **hypotonia**, which manifests as a decrease in the level of resistance to passive joint movement present in a normal relaxed person. Myotatic reflexes produced by tapping a tendon with a reflex hammer reverberate for several cycles (pendular reflexes) because of impaired damping from the reduced muscle tone. The hypotonia is likely a result of impaired processing of cerebellar afferent action potentials from the muscle spindles and GTOs.

Although these lesions help define cerebellar function, it is less clear what the cerebellum does in the normal state. Current research proposes that the cerebellum compares the intended movement with the actual movement and adjusts motor system output in ongoing movements. The cerebellum functions in planning movements and learning new motor skills and also has a role in nonmotor cognitive processes such as word association.

Chapter Summary

- The contraction of skeletal muscle produces movement by acting on the skeleton.
- Motor neurons activate the skeletal muscles.
- Sensory feedback from muscles is important for precise control of contraction.
- The output of sensory receptors, such as the muscle spindle, can be adjusted.
- The spinal cord is the source of reflexes that are important in the initiation and control of movement.
- Spinal cord function is influenced by higher centers in the brainstem.
- The highest level of motor control comes from the cerebral cortex.
- The basal ganglia and the cerebellum provide feedback to the motor control areas of the cerebral cortex and brainstem.

thePoint *Visit http://thePoint.lww.com for chapter review Q&A, case studies, animations, and more!*

Autonomic Nervous System

ACTIVE LEARNING OBJECTIVES

Upon mastering the material in this chapter you should be able to:

- Explain the role of the autonomic nervous system in regulating the involuntary "background operating system" functions of the body.
- Explain the anatomic and physiologic bases for division of the autonomic nervous system into the sympathetic, parasympathetic, and enteric divisions.
- Know the neurotransmitters used by the sympathetic and parasympathetic divisions.
- Explain how the central nervous system regulates autonomic function.

The **autonomic nervous system (ANS)** is the part of the peripheral nervous system that acts to control visceral functions largely below the level of consciousness, such as heart rate, digestion, respiratory rate, salivation, perspiration, pupil diameter, micturition, and sexual arousal. For instance, the sweating sunbather lying quietly in the summer sun or the racing heart and "hair-standing-on-end" sensations experienced by a person suddenly frightened in a horror movie are familiar examples of the body responding automatically to events in the physical or emotional environment. Responses such as these occur as a result of the actions of the autonomic portion of the nervous system and take place without conscious action on the part of the person. The term *autonomic* is derived from the root *auto* (meaning "self") and *nomos* (meaning "law"). Our concept of the autonomic part of the nervous system has evolved over several centuries. The recognition of anatomic differences between the spinal cord and peripheral nerve pathways that control visceral functions and those that control skeletal muscles was a major step. Observations about the effect on heart rate of a substance released by the vagus nerve helped begin the understanding of the biochemical features.

The functions of the ANS fall into three major categories:

- Maintaining homeostatic conditions within the body
- Coordinating the body's responses to exercise and stress
- Assisting the endocrine system to regulate reproduction

The ANS regulates the functions of the involuntary organs, which include the heart, the blood vessels, the exocrine glands, and the visceral organs. In some organs, the actions of the ANS are joined by circulating endocrine hormones and by locally produced chemical mediators to complete the control process.

OVERVIEW OF THE AUTONOMIC NERVOUS SYSTEM

On the basis of anatomic, functional, and neurochemical differences, the ANS is usually subdivided into three divisions: sympathetic, parasympathetic, and enteric. The enteric nervous system is concerned with the regulation of gastrointestinal (GI) function and is covered in more detail in Chapter 26, "Gastrointestinal Secretion, Digestion, and Absorption." The sympathetic and parasympathetic divisions are the primary focus of this chapter.

Coordination of the body's activities by the nervous system was the process of **sympathy** in classical anatomic and physiologic thinking. Regulation of the involuntary organs came to be associated with the portions of the nervous system that were located, at least in part, outside the standard spinal cord and peripheral nerve pathways. The ganglia, located along either side of the spine in the thorax and abdominal regions and somewhat detached from the nerve trunks destined for the limbs, were found to be associated with some of the involuntary bodily functions and, therefore, designated the **sympathetic division**. This collection of structures was also termed the **thoracolumbar division** of the ANS because of the location of the ganglia and the neuron cell bodies that supply axons to the ganglia. Nuclei and their axons that controlled internal functions were also found in the brainstem and associated cranial nerves (CNs) as well as in the most caudal part of the spinal cord. Those pathways were somewhat distinct from the sympathetic system and were designated the **parasympathetic division**. The term **craniosacral** was applied to this portion of the ANS because of the origin of cell bodies and axons.

Neurochemical differences were recognized between these two divisions, leading to the designation of the sympathetic system as **adrenergic**, for the adrenaline-like actions resulting from sympathetic nerve activation. The parasympathetic system was designated as **cholinergic**, for the acetylcholine-like actions when nerves of this division were stimulated.

The functions of the sympathetic and parasympathetic divisions are often simplified into a two-part scheme. The *sympathetic division* is said to preside over the use of metabolic resources and emergency responses of the body. The *parasympathetic division* presides over the restoration of the body's reserves and the elimination of waste products. In reality, most of the organs supplied by the ANS receive both sympathetic and parasympathetic innervation. In many instances, the two divisions are activated in a reciprocal fashion, so that if the firing rate in one division is increased, the rate is decreased in the other. An example is controlling the heart rate: Increased firing in the sympathetic nerves and simultaneous decreased firing in the parasympathetic nerves result in increased heart rate.

In some organs, the two divisions work synergistically. For example, during secretion by exocrine glands of the GI tract, the parasympathetic nerves increase volume and enzyme content at the same time that sympathetic activation contributes mucus to the total secretory product. Some organs, such as the skin and blood vessels, may receive only sympathetic innervation and are regulated by increases or decreases in a baseline firing rate of the nerves.

Autonomic nervous system consists of a two-neuron efferent chain.

In the motor system, there is a single axon path from a motor neuron's cell body, located in either the ventral horn of the spinal cord or a brainstem motor nucleus, to the skeletal muscle cells controlled by that neuron (see Chapter 5, "Motor System"). In the ANS, the efferent path consists of a two-neuron chain with a synapse interposed between the central nervous system (CNS) and the effector cells (Fig. 6.1). The cell bodies of the autonomic motor neurons are located in the spinal cord or specific brainstem nuclei. An efferent fiber emerges as the **preganglionic axon** and then synapses with neurons located in a peripheral ganglion. The neuron in the ganglion then projects a **postganglionic axon** to the autonomic effector cells.

Automatic nervous system releases neurotransmitters that are complementary in function.

In the somatic nervous system, neurotransmitter is released from specialized nerve endings that make intimate contact with the target structure. The mammalian motor endplate, with one nerve terminal to one skeletal muscle fiber, illustrates this principle. This arrangement contrasts with the ANS, where postganglionic axons terminate in varicosities, swellings containing synaptic vesicles, which release the transmitter into the extracellular space surrounding the effector cells (see Fig. 6.1). The response to the ANS output originates in some of the effector cells and then propagates to the remainder via gap junctions.

Acetylcholine

The preganglionic nerve terminals of both the sympathetic and the parasympathetic divisions release **acetylcholine (ACh)** (Fig. 6.2). The synapse at those sites uses a **nicotinic receptor** similar in structure to the receptor at the neuromuscular junction. Parasympathetic postganglionic neurons release ACh at the synapse with the effectors. The postganglionic sympathetic neurons to the sweat glands and to some blood vessels in skeletal muscle also use ACh as the neurotransmitter. The synapse between the postganglionic parasympathetic neuron and the effectors uses a **muscarinic receptor**. This classification of cholinergic synapses is based on the response to the application of the alkaloids nicotine and muscarine, which act as agonists at their respective type of synapse.

The nicotinic receptor is of the direct ligand-gated type, meaning that the receptor and the ion channel are contained in the same structure. The muscarinic receptor is of the indirect ligand-gated type and uses a G protein to link receptor and effector functions (see Chapter 3, "Action Potential,

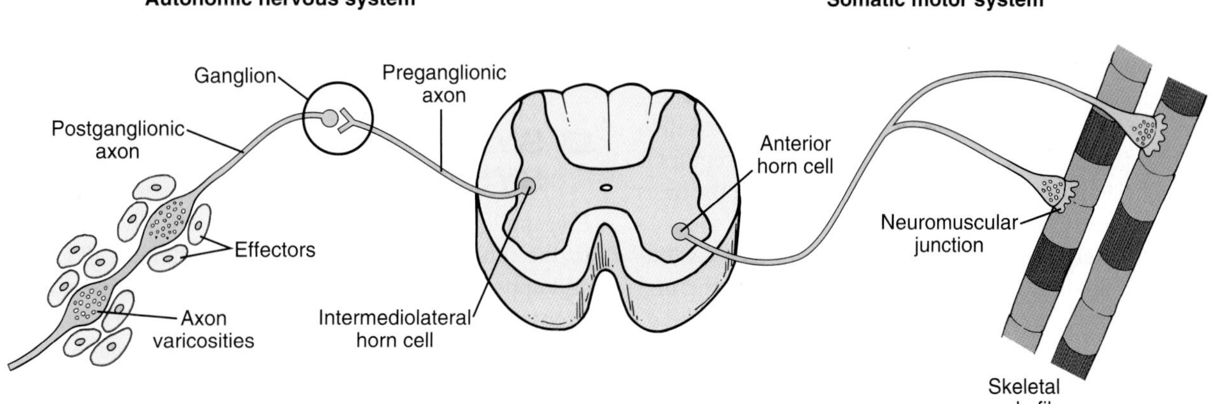

Figure 6.1 The efferent path of the autonomic nervous system (ANS) as contrasted with the somatic motor system. The ANS uses a two-neuron pathway. Note the structural differences between the synapses at autonomic effectors and skeletal muscle cells.

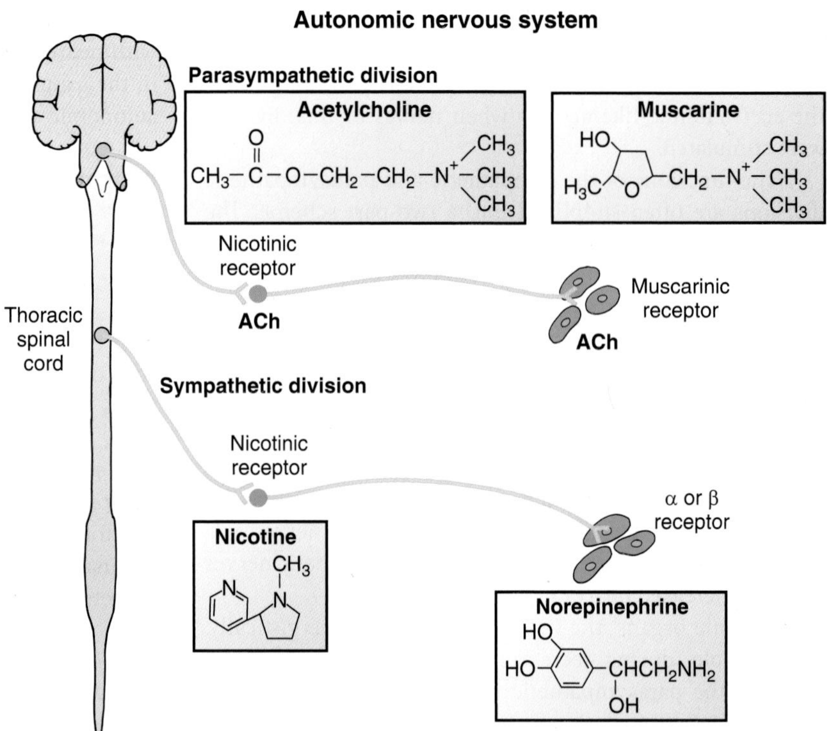

Figure 6.2 The neurochemistry of the autonomic paths. The structures of the neurotransmitters and the agonists for which the synapses were originally named are shown. ACh, acetylcholine.

Synaptic Transmission, and Maintenance of Nerve Function"). The action of ACh is terminated by diffusion of the transmitter away from the receptor and also by action of the enzyme acetylcholinesterase, which breaks ACh down. Choline released by the enzyme action is taken back into the nerve terminal and resynthesized into ACh.

Norepinephrine

The catecholamine **norepinephrine (NE)** is the neurotransmitter released at the postganglionic synapses of the sympathetic division (see Fig. 6.2), with the exception of the sweat glands. The synapses that use NE receptors can also be activated by the closely related compound epinephrine (adrenaline), which is released into the general circulation by the adrenal medulla—hence the original designation of this type of receptors as adrenergic. Adrenergic receptors are classified as either α or β, based on their responses to pharmacologic agents that mimic or block the actions of NE and related compounds. α receptors respond best to epinephrine, less well to NE, and least well to the synthetic compound isoproterenol. β receptors respond best to isoproterenol, less well to epinephrine, and least well to NE. Each class of receptors is further subclassified as α_1 or α_2 and β_1, β_2, or β_3 on the basis of responses to additional pharmacologic agents.

The adrenergic receptors are of the indirect, ligand-gated, G protein–linked type. They share a general structural similarity with the muscarinic type of ACh receptor. The α_1 receptors activate phospholipase C and increase the intracellular concentrations of diacylglycerol and inositol trisphosphate. The α_2 receptors inhibit adenylyl cyclase, whereas the β types stimulate it. The action of NE and epinephrine at a synapse is terminated by diffusion of the molecule away from the synapse and reuptake into the nerve terminal.

Other neurotransmitters

Other neurotransmitters are released by autonomic neurons. Neurally active peptides are often colocalized with small-molecule transmitters and are released simultaneously during nerve stimulation in the CNS. This process is the same in the ANS, especially in the intrinsic plexuses of the gut, where amines, amino acid transmitters, and neurally active peptides are widely distributed. In the ANS, examples of a colocalized amine and peptide are seen in the sympathetic division, where NE and neuropeptide Y are coreleased by vasoconstrictor nerves. **Vasoactive intestinal polypeptide (VIP)** and calcitonin gene-related peptide are released along with ACh from nerve terminals innervating the sweat glands.

Nitric oxide (NO) is another type of neurotransmitter produced by some autonomic nerve endings. The term **nonadrenergic noncholinergic** has been applied to such nerves. NO is a highly diffusible substance important in the regulation of smooth muscle contraction (see Chapter 1, "Homeostasis and Cellular Signaling").

SYMPATHETIC NERVOUS SYSTEM

Preganglionic neurons of the sympathetic division originate in the intermediolateral horn of the thoracic (T1 to T12) and upper lumbar (L1 to L3) spinal cord. The preganglionic axons exit the spinal cord in the ventral nerve roots. Immediately after the ventral and dorsal roots merge to form the spinal nerve, the sympathetic axons leave the spinal nerve via the **white ramus** and enter the **paravertebral sympathetic ganglia** (Fig. 6.3). The paravertebral ganglia form an interconnected chain located on either side of the vertebral column. These ganglia extend above and below the thoracic

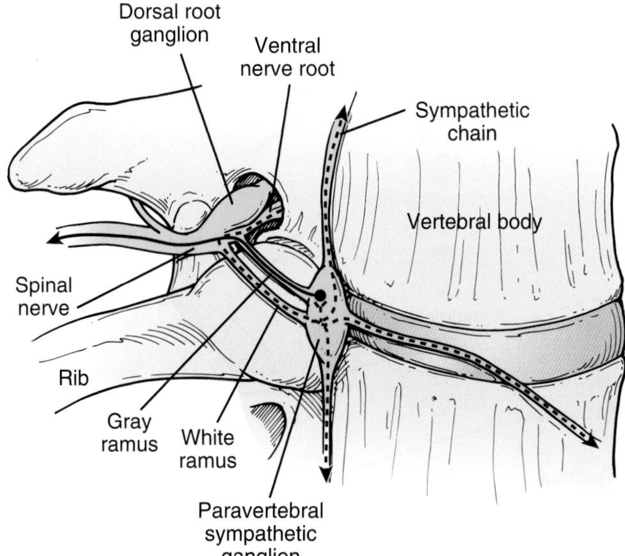

Figure 6.3 Peripheral sympathetic anatomy. The preganglionic axons course through the spinal nerve and white ramus to the paravertebral ganglion. Synapse with the postganglionic neuron may occur at the same spinal level or at levels above or below. Postganglionic axons rejoin the spinal nerve through the gray ramus to innervate structures in the limbs or proceed to organs, such as the lungs or heart, in discrete nerves. Preganglionic axons may also pass to a prevertebral ganglion without synapsing in a paravertebral ganglion.

and lumbar spinal levels, where preganglionic fibers emerge, to provide postganglionic sympathetic axons to the cervical and lumbosacral spinal nerves (Fig. 6.4). The preganglionic axons that ascend to the cervical levels arise from T1 to T5 and form three major ganglia: the **superior**, the **middle**, and the **inferior cervical ganglia**. Preganglionic axons descend below L3, forming two additional lumbar and at least four sacral ganglia. The preganglionic axons may synapse with postganglionic neurons in the paravertebral ganglion at the same level, ascend or descend up to several spinal levels and then synapse, or pass through the paravertebral ganglia en route to a **prevertebral ganglion**.

Postganglionic axons that are destined for somatic structures—such as sweat glands, pilomotor muscles, or blood vessels of the skin and skeletal muscles—leave the paravertebral ganglion in the **gray ramus** and rejoin the spinal nerve for distribution to the target tissues. Postganglionic axons to the head, heart, and lungs originate in the cervical or upper thoracic paravertebral ganglia and make their way to the specific organs as identifiable, separate nerves (e.g., the cardiac nerves), as small-caliber individual nerves that may group together, or as perivascular plexuses of axons that accompany arteries.

The superior cervical ganglion supplies sympathetic axons that innervate the structures of the head. These axons travel superiorly in the perivascular plexus along the carotid arteries. Structures innervated include the following: the radial muscle of the iris, responsible for dilation of the pupil; the Müller muscle, which assists in elevating the eyelid; the salivary glands; and the sweat glands of the face. Lesions that interrupt this pathway result in a group of clinical signs. The middle and inferior cervical ganglia innervate organs of the chest, including the trachea, esophagus, heart, and lungs.

Postsynaptic axons destined for the abdominal and pelvic visceral organs arise from the prevertebral ganglia (see Fig. 6.4). The three major prevertebral ganglia, also called **collateral ganglia**, overlie the celiac, superior mesenteric, and inferior mesenteric arteries at their origin from the aorta and are named accordingly. The **celiac ganglion** provides sympathetic innervation to the stomach, liver, pancreas, gallbladder, small intestine, spleen, and kidneys. Its preganglionic axons originate in the T5 to T12 spinal levels. The **superior mesenteric ganglion** innervates the small and large intestines. Its preganglionic axons originate primarily in T10 to T12. The **inferior mesenteric ganglion** innervates the lower colon and rectum, urinary bladder, and reproductive organs. Its preganglionic axons originate in L1 to L3.

Sympathetic division can produce local or widespread responses.

The sympathetic division exerts a continuous influence on the organs it innervates. This continuous level of control, called **sympathetic tone**, is produced by a persistent discharge of the sympathetic nerves. When the situation dictates, the rate of firing to a particular organ can be increased, such as when the sympathetic neurons supplying the iris dilate the pupil in dim light, or decreased to produce pupillary constriction during drowsiness.

The number of postganglionic axons emerging from the paravertebral ganglia is greater than the number of preganglionic neurons that originate in the spinal cord. It is estimated that postganglionic sympathetic neurons outnumber preganglionic neurons by 100:1 or more. This spread of influence, termed **divergence**, is accomplished by collateral branching of the presynaptic sympathetic axons, which then make synaptic connections with postganglionic neurons both above and below their original level of emergence from the spinal cord. Divergence enables the sympathetic division to produce widespread responses of many effectors when physiologically necessary.

Stimulation of the adrenal medulla mimics sympathetic function.

In addition to the divergence of postganglionic axons, the sympathetic division has a hormonal mechanism to activate target tissues endowed with adrenergic receptors, including those innervated by the sympathetic nerves. The hormone is the catecholamine **epinephrine**, which is secreted along with lesser amounts of NE by the adrenal medulla during generalized response to stress.

The adrenal medulla, a neuroendocrine gland, forms the inner core of the adrenal gland situated on top of each kidney. Cells of the adrenal medulla are innervated by the lesser splanchnic nerve, which contains preganglionic sympathetic axons originating in the lower thoracic spinal cord (see Fig. 6.4). These axons pass through the paravertebral ganglia and the celiac ganglion without synapsing and terminate on

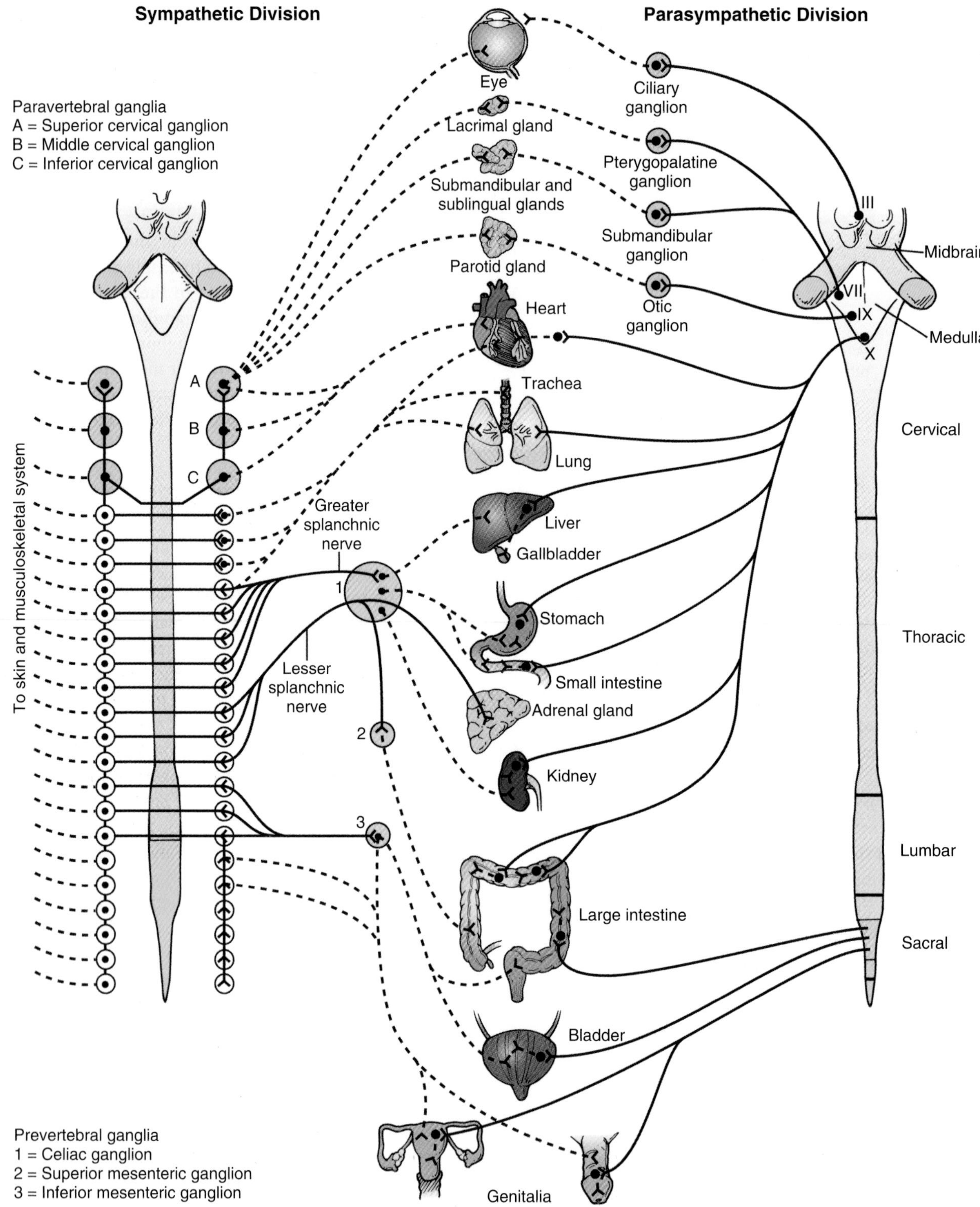

Figure 6.4 The organ-specific arrangement of the autonomic nervous system. Preganglionic axons are indicated by *solid lines*, postganglionic axons by *dashed lines*. Sympathetic axons destined for the skin and musculoskeletal system are shown on the left side of the spinal cord. Note the named paravertebral and prevertebral ganglia.

the **chromaffin cells** of the adrenal medulla (Fig. 6.5). The chromaffin cells are modified ganglion cells that synthesize both epinephrine and NE in a ratio of about 8:1 and store them in secretory vesicles. Unlike neurons, these cells possess neither axons nor dendrites but function as neuroendocrine cells that release hormone directly into the bloodstream in response to preganglionic axon activation.

Circulating epinephrine and NE mimic the actions of sympathetic nerve stimulation at both α-adrenergic and β-adrenergic receptors. Epinephrine can also stimulate

Clinical Focus / 6.1

Horner Syndrome

Lesions of the sympathetic pathway to the head produce abnormalities that are easily detectable on physical examination. The deficits of function occur ipsilateral to the lesion and include the following:

- Partial constriction of the pupil as a result of loss of sympathetic pupillodilator action
- Drooping of the eyelid, termed **ptosis**, as a result of loss of sympathetic activation of the Müller muscle of the eyelid
- Dryness of the face as a result of the lack of sympathetic activation of the facial sweat glands

A pattern of historical or physical examination findings that is consistent from patient to patient is often termed a syndrome. Johann Horner, a 19th-century Swiss ophthalmologist, described this pattern of eye and facial abnormalities in patients, and these are referred to as **Horner syndrome**. Etiologies for Horner syndrome include the following:

- Brainstem lesions, such as produced by strokes, which interrupt the tracts that descend to the sympathetic neurons in the spinal cord
- Upper thoracic nerve root lesions, such as those produced by excessive traction on the arm or from infiltration of the nerve roots by cancer spreading from the lung
- Cervical paravertebral ganglia lesions from accidental or surgical trauma or metastatic cancer
- Arterial injury in the neck, from neck hyperextension, or direct trauma, which interrupt the postganglionic axons traveling in the carotid periarterial plexus

adrenergic receptors on cells that receive little or no direct sympathetic innervation, such as liver and adipose cells for mobilizing glucose and fatty acids, and blood cells, which participate in the clotting and immune responses.

Sympathetic system is widely activated in the fight-or-flight response.

This response is the classic example of the sympathetic nervous system's ability to produce widespread activation of its effectors. The response is activated when an organism's survival is in jeopardy and the animal may have to fight or flee. Some components of the response result from the direct effects of sympathetic neuron activation, although the action of epinephrine secreted by the adrenal medulla also contributes.

Sympathetic stimulation of the heart and blood vessels results in a rise in blood pressure because of increased cardiac output and increased total peripheral resistance. There is also a redistribution of the blood flow so that the muscles and heart receive more blood, whereas the splanchnic territory and the skin receive less. The need for an increased exchange of blood gases is met by acceleration of the respiratory rate and dilation of the bronchiolar tree. The volume of salivary secretion is reduced, but the relative proportion of mucus increases, permitting lubrication of the mouth despite increased ventilation. The potential demand for an enhanced supply of metabolic substrates, such as glucose and fatty acids, is met by the actions of the sympathetic nerves and circulating epinephrine on hepatocytes and adipose cells. **Glycogenolysis** mobilizes stored liver glycogen, increasing plasma levels of glucose. **Lipolysis** in fat cells converts stored triglycerides to free fatty acids that enter the bloodstream.

The skin plays an important role in maintaining body temperature in the face of increased heat production from contracting muscles. The sympathetic innervation of the skin vasculature can adjust blood flow and heat exchange by vasodilation to dissipate heat or by vasoconstriction to protect blood volume. The eccrine sweat glands are important structures that also can be activated to enhance heat loss. Sympathetic nerve stimulation of the sweat glands results in the secretion of a watery fluid, and evaporation then dissipates body heat. Constriction of the skin vasculature, concurrent with sweat gland activation, produces the cold, clammy skin typical of a frightened person. Hair-standing-on-end sensations result from activation of the piloerector muscles associated with hair follicles. In humans, this action is likely a phylogenetic remnant from animals that use hair erection for body temperature preservation or to enhance the apparent body size.

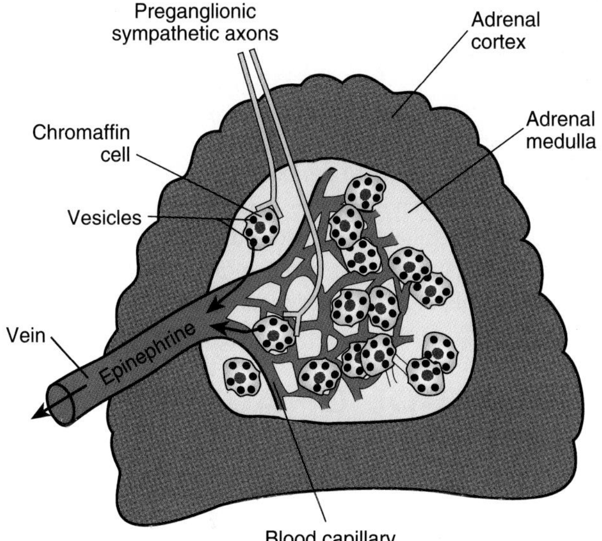

Figure 6.5 Sympathetic innervation of the adrenal medulla. Preganglionic sympathetic axons terminate on the chromaffin cells. When stimulated, the chromaffin cells release epinephrine into the circulation.

PARASYMPATHETIC NERVOUS SYSTEM

The parasympathetic division comprises a *cranial portion*, emanating from the brainstem, and a *sacral portion*, originating in the intermediate gray zone of the sacral spinal cord (see Fig. 6.4). In contrast to the widespread activation pattern of the sympathetic division, the neurons of the parasympathetic division are activated in a more localized fashion. There is also much less tendency for divergence of the presynaptic influence to multiple postsynaptic neurons—on average, one presynaptic parasympathetic neuron synapses with 15 to 20 postsynaptic neurons. An example of localized activation is seen in the vagus nerve, where one portion of its outflow can be activated to slow the heart rate without altering the function of the stomach.

Ganglia in the parasympathetic division are either located close to the organ innervated or embedded within its walls. The organs of the GI system demonstrate the latter pattern. Because of this arrangement, preganglionic axons are much longer than postganglionic axons.

Brainstem parasympathetic neurons innervate structures in the head, chest, and abdomen.

Of the 12 CNs, 4 contain parasympathetic axons—numbers III, VII, IX, and X. The nuclei of these nerves, which occupy areas of the tectum in the midbrain, pons, and medulla, are the centers for the initiation and integration of autonomic reflexes for the organ systems they innervate. Parasympathetic and sympathetic activities are coordinated by these nuclei.

The *oculomotor nerve*, CN III, originates from nuclei in the tectum of the midbrain, where synaptic connections with the axons of the optic nerves provide input for ocular reflexes. The parasympathetic neurons are located in the **Edinger-Westphal nucleus**. The presynaptic axons travel in the superficial aspect of CN III to the **ciliary ganglion**, located inside the orbit, where the synapse occurs. The postganglionic axons enter the eyeball near the optic nerve and travel between the sclera and the choroid. These axons supply the sphincter muscle of the iris, the ciliary muscle, which focuses the lens, and the choroidal blood vessels. About 90% of the axons are destined for the ciliary muscle, whereas only about 3% to 4% innervate the iris sphincter. The parasympathetic presynaptic axons of the *facial nerve*, CN VII, arise from the **superior salivatory nuclei** in the rostral medulla. Presynaptic axons pass from the facial nerve into the greater superficial petrosal nerve and synapse in the **pterygopalatine ganglion**. The postsynaptic axons from that ganglion innervate the lacrimal gland and the glands of the nasal and palatal mucosa. Other facial nerve presynaptic axons travel via the chorda tympani and synapse in the **submandibular ganglion**. These postsynaptic axons stimulate the production of saliva by the submandibular and sublingual glands. Parasympathetic activation can also produce dilation of the vasculature within the areas supplied by the facial nerve.

The parasympathetic presynaptic axons of the *glossopharyngeal nerve*, CN IX, arise from the **inferior salivatory nuclei** of the medulla. The axons follow a circuitous course through the lesser petrosal nerve to reach the **otic ganglion**, where they synapse. From the otic ganglion, the postsynaptic axons join the auriculotemporal branch of CN V and arrive at the parotid gland, where they stimulate secretion of saliva.

The *vagus nerve*, CN X, has an extensive autonomic component, which arises from the **nucleus ambiguus** and the **dorsal motor nuclei** in the medulla. It has been estimated that vagal output comprises up to 75% of total parasympathetic activity. Long preganglionic axons travel in the vagus trunks to ganglia in the heart and lungs and to the intrinsic plexuses of the GI tract.

The right vagus nerve supplies axons to the sinoatrial node of the heart, and the left vagus nerve supplies the atrioventricular node. Vagal activation slows the heart rate and reduces the force of contraction. The vagal efferents to the lung control smooth muscle that constricts bronchioles and also regulate the action of secretory cells. Vagal output to the esophagus and stomach regulates motility and influences secretory function in the stomach. ACh and VIP are the transmitters of the postsynaptic neurons.

There is also vagal innervation to the kidneys, liver, spleen, and pancreas, but its roles in these organs have not been fully established.

Sacral parasympathetic neurons innervate structures in the pelvis.

Preganglionic fibers of the sacral division originate in the intermediate gray matter of the sacral spinal cord, emerging from segments S2, S3, and S4 (see Fig. 6.4). These preganglionic fibers synapse in ganglia in or near the pelvic organs, including the lower portion of the GI tract (the sigmoid colon, rectum, and internal anal sphincter), the urinary bladder, and the reproductive organs. Improved understanding of ANS actions in the male genitalia has allowed development of pharmacologic treatment of **erectile dysfunction**.

Autonomic activity has opposing actions on organ responses.

As noted earlier, most involuntary organs are dually innervated by the sympathetic and parasympathetic divisions, often with opposing actions. A list of these organs and a summary of their responses to sympathetic and parasympathetic stimulation are given in Table 6.1. The type of receptor at the synapse with the effectors is also indicated. The chapters on the specific organ systems provide more detailed discussions of the effects of autonomic nerve activation.

CONTROL OF AUTONOMIC NERVOUS SYSTEM

The ANS uses a hierarchy of reflexes to control the function of autonomic target organs. These reflexes range from local, involving only a part of one neuron, to regional, requiring mediation by the spinal cord and associated autonomic ganglia, to the most complex, requiring action by the brainstem and cerebral centers. In general, the higher the level of complexity, the more likely the reflex will require coordination of

TABLE 6.1 Responses of Effectors to Parasympathetic and Sympathetic Stimulation

Effector	Parasympathetic	Sympathetic
Eye		
Pupil	Constriction	Dilation (α_1)
Ciliary muscle	Contraction	Relaxation (β_2)
Müller muscle	None	Contraction (α_1)
Lacrimal gland	Secretion	None
Nasal glands	Secretion	Inhibition (α_1)
Salivary glands	Secretion	Amylase secretion (β)
Skin		
Sweat glands	None	Secretion (cholinergic muscarinic)
Piloerector muscles	None	Contraction (α_1)
Blood vessels		
Skin (general)	None	Constriction (α)
Facial and neck muscles	Dilation (?)*	Constriction (α) Dilation (?)*
Skeletal muscle	None	Dilation (β_2) Constriction (α)
Viscera	None	Constriction (α_1)
Heart		
Rate	Decrease	Increase (β_1, β_2)
Force	Decrease	Increase (β_1, β_2)
Lungs		
Bronchioles	Constriction	Dilation (β_2)
Glands	Secretion	Decrease (α_1) Increase (β_2)
Gastrointestinal tract		
Wall muscles	Contraction	Relaxation (α, β_2)
Sphincters	Relaxation	Contraction (α_1)
Glands	Secretion	Inhibition
Liver	None	Glycogenolysis and gluconeogenesis (α_1, β_2)
Pancreas (insulin)	None	Decreased secretion (α_2)
Adrenal medulla	None	Secretion of epinephrine (cholinergic nicotinic)
Urinary system		
Ureter	Relaxation	Contraction (α_1)
Detrusor	Contraction	Relaxation (β_2)
Sphincter	Relaxation	Contraction (α_1)
Reproductive system		
Uterus	Variable	Contraction (α_1)
Genitalia	Erection (ACh and NO)	Ejaculation/vaginal contraction (α)
Adipose cells	None	Lipolysis (β)

ACh, acetylcholine; NO, nitric oxide.

*The status of vasodilator activity to the face (as shown in blushing and/or response to cold exposure) is still controversial.

both sympathetic and parasympathetic responses. Somatic motor neurons and the endocrine system may also be involved.

Sensory input contributes to autonomic function.

The ANS is traditionally regarded as an efferent system, and the sensory neurons innervating the involuntary organs are not considered part of the ANS. Sensory input, however, is important for autonomic functioning. Sensory axons that provide input important for autonomic function travel in CN IX. The carotid bodies sense the concentrations of oxygen and carbon dioxide in blood flowing in the carotid arteries and transmit that chemosensory information to the medulla via glossopharyngeal afferents. The carotid sinus, which is located in the proximal internal carotid artery, monitors blood pressure and transmits this baroreceptor information to the **tractus solitarius** in the medulla. The sensory innervation to the visceral organs, blood vessels, and skin forms the afferent limb of autonomic reflexes (Fig. 6.6). Most of the sensory axons from ANS-innervated structures are unmyelinated C fibers.

Sensory information from these pathways may not reach the level of consciousness. Sensations that are perceived may be vaguely localized or may be felt in a somatic structure rather than the organ from which the afferent action potentials originated. The perception of pain in the left arm during a myocardial infarction is an example of **referred pain** from a visceral organ.

Local axon reflexes are paths for autonomic activation.

A sensory neuron may have several peripheral terminal branches that enlarge the receptive area and innervate multiple receptors. As sensory action potentials that originated in one of the terminal branches propagate afferently, or **orthodromically**, they may also enter other branches of that same axon and then conduct efferently, or **antidromically**, for short distances. Propagation of the action potential by this process can result in a more wide-ranging reaction than that produced by the initial discrete stimulus. The distal ends of the sensory axons may release neurotransmitters in response to these antidromic action potentials. Release of neurotransmitter from a neuron that innervates blood vessels or sweat glands can produce reddening of the skin as a result of vasodilation, local sweating as a result of sweat gland activation, or pain as a result of activation of nociceptive sensory nerve endings. This process is called a **local axon reflex** (see Fig. 6.6). It differs from the usual reflex pathway in that a synapse with an efferent neuron in the spinal cord or peripheral ganglion is not required to produce a response. The neurotransmitter producing this local reflex is likely the same as that released at the synapse in the spinal cord—substance P or glutamate for sensory neurons or ACh and NE at the target tissues for autonomic neurons. Local axon reflexes in nociceptive nerve endings that become persistently activated after local trauma can produce dramatic clinical manifestations.

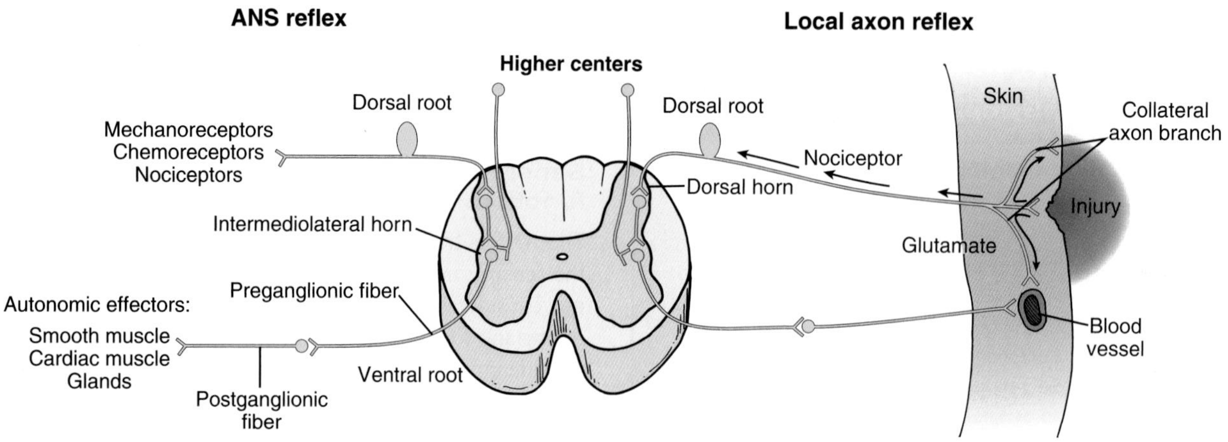

Figure 6.6 Sensory components of autonomic function. *Left,* sensory action potentials from mechanical, chemical, and nociceptive receptors that propagate to the spinal cord can trigger autonomic nervous system (ANS) reflexes. *Right*, local axon reflexes occur when an orthodromic action potential from a sensory nerve ending propagates antidromically into collateral branches of the same neuron. The antidromic action potentials may provoke release of the same neurotransmitters, such as substance P or glutamate, from the nerve endings as would be released at the synapse in the spinal cord. Local axon reflexes may perpetuate pain, activate sweat glands, or cause vasomotor actions.

Autonomic ganglia can modify reflexes.

Although the paravertebral ganglia may serve merely as relay stations for the synapse of preganglionic and postganglionic sympathetic neurons, evidence suggests that synaptic activity in these ganglia may modify efferent activity. Input from other preganglionic neurons provides the modifying influence. Prevertebral ganglia also serve as integrative centers for reflexes in the GI tract. Chemoreceptors and mechanoreceptors located in the gut produce afferent action potentials that pass to the spinal cord and then to the celiac or mesenteric ganglia where changes in motility and secretion may be instituted during digestion. The integrative actions of these ganglia are also responsible for halting motility and secretion in the GI tract during a generalized stress reaction (the fight-or-flight response).

The intrinsic plexuses of the GI visceral wall are reflex integrative centers where input from presynaptic parasympathetic axons, postganglionic sympathetic axons, and the action of intrinsic neurons may all participate in reflexes that influence motility and secretion. The intrinsic plexuses also participate in centrally mediated GI reflexes (see Chapter 26, "Gastrointestinal Secretion, Digestion, and Absorption").

Spinal cord coordinates many autonomic reflexes.

Reflexes coordinated by centers in the lumbar and sacral spinal cord include micturition (emptying the urinary bladder), defecation (emptying the rectum), and sexual response (engorgement of erectile tissue, vaginal lubrication, and ejaculation of semen). Sensory action potentials from receptors in the wall of the bladder or bowel report about degrees of distention. Sympathetic, parasympathetic, and somatic efferent actions require coordination to produce many of these responses.

Higher centers provide facilitating or inhibiting influences to the spinal cord reflex centers. The ability to voluntarily suppress the urge to urinate when the sensation of bladder fullness is perceived is an example of higher CNS centers inhibiting a spinal cord reflex. Following injury to the cervical or upper thoracic spinal cord, micturition may occur involuntarily or be provoked at much lower-than-normal bladder volumes. Episodes of hypertension and piloerection in patients with spinal cord injuries are another example of uninhibited autonomic reflexes arising from the spinal cord.

From Bench to Bedside / 6.1

Saying "NO" to Erectile Dysfunction

Problems of achieving penile erection sufficient for sexual intercourse affect about 40% of middle-aged males. A significant advance in the ability to treat this clinical problem has emerged from the understanding of the role of parasympathetic neurons that release **nitric oxide (NO)** as a neurotransmitter. During sexual arousal, NO is released by a class of parasympathetic neurons that innervate the corpora cavernosa in the penis. The presence of NO causes enhanced synthesis of **cyclic guanosine monophosphate (cGMP)** in vascular smooth muscle cells. Elevated levels of GMP produce local vascular relaxation and resulting engorgement of the penile erectile tissues. cGMP activity is terminated when NO release from neurons stops, and when the enzyme phosphodiesterase is activated. The development of inhibitors of the specific type of this enzyme, phosphodiesterase type 5, which is localized to the vascular structures of the penis, has allowed the naturally occurring elevation of GMP levels to be prolonged and erectile function to be improved. For the medication to be effective, parasympathetic innervation of the corpora cavernosa must be intact.

Brainstem is a major control center for autonomic reflexes.

Areas within all three levels of the brainstem are important in autonomic function (Fig. 6.7). The **periaqueductal gray matter** of the midbrain coordinates autonomic responses to painful stimuli and can modulate the activity of the sensory tracts that transmit pain. The **parabrachial nucleus** of the pons participates in respiratory and cardiovascular control. The medulla contains several key autonomic areas. The **nucleus of the tractus solitarius** receives afferent input from cardiac, respiratory, and GI receptors. The **ventrolateral medullary area** is the major center for control of the preganglionic sympathetic neurons in the spinal cord. Vagal efferent axons also arise from this area. Neurons that control specific functions, such as blood pressure and heart rate, are clustered within this general region. The descending paths for regulation of the preganglionic sympathetic and spinal parasympathetic neurons are not yet fully delineated. The reticulospinal tracts may carry some of these axons. Autonomic reflexes coordinated in the brainstem include pupillary reaction to light, lens accommodation, salivation, tearing, swallowing, vomiting, blood pressure regulation, and cardiac rhythm modulation.

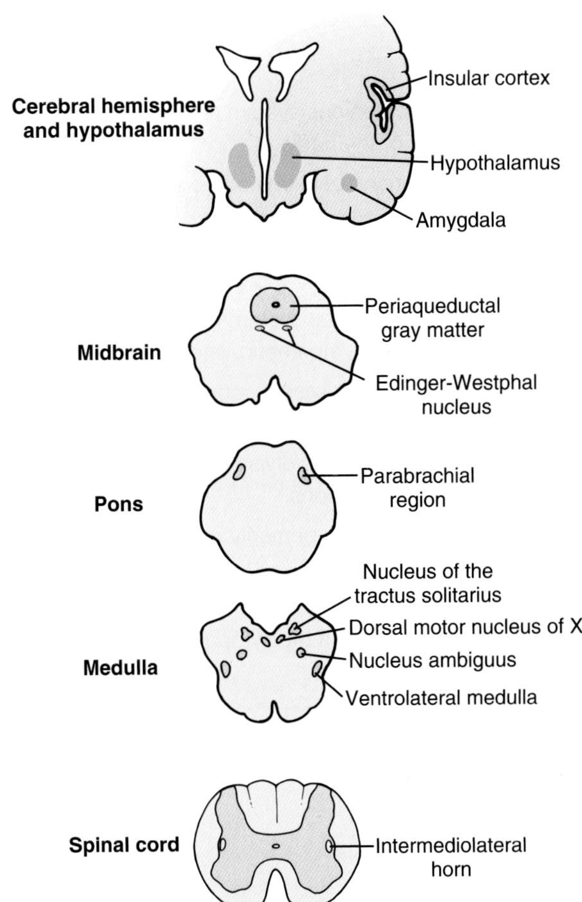

Figure 6.7 The central autonomic network. Note the cerebral, hypothalamic, brainstem, and spinal cord components. A hierarchy of reflexes initiated from these different levels regulates autonomic function.

Hypothalamus and cerebral hemispheres provide the highest levels of autonomic control.

The periventricular, medial, and lateral areas of the **hypothalamus** in the diencephalon control circadian rhythms and homeostatic functions, such as thermoregulation, appetite, and thirst. Because of the major role of the hypothalamus in autonomic function, it has at times been labeled the "head ganglion of the ANS." The **insular** and **medial prefrontal areas** of the cerebral cortex are the respective sensory and motor areas involved with the regulation of autonomic function. The **amygdala** in the temporal lobe coordinates the autonomic components of emotional responses.

The areas of the cerebral hemispheres, diencephalon, brainstem, and central pathways to the spinal cord that are involved in the control of autonomic functions are collectively termed the **central autonomic network** (see Fig. 6.7).

Clinical Focus / 6.2

Reflex Sympathetic Dystrophy

Reflex sympathetic dystrophy (RSD) is a clinical syndrome that includes spontaneous pain, painful hypersensitivity to nonnoxious stimuli such as light touch or moving air, and evidence of autonomic nervous system dysfunction in the form of excessive coldness and sweating of the involved body part. The foot, knee, hand, and forearm are the more common sites of involvement. Local trauma, which may be minor in degree, and surgical procedures on joints or bones are common precipitating events. The term *dystrophy* applies to atrophic changes that may occur in the skin, soft tissue, and bone in the painful areas if the condition goes untreated for many months. A full explanation of the pathogenetic mechanisms is still lacking. Local axon reflexes in traumatized nociceptive neurons and reflex activation of the sympathetic nervous system are thought to be contributors. Blockade of sympathetic neuron action by local anesthetic injection into the paravertebral ganglia serving the involved body part, followed by mobilization of the body part in a physical therapy program, is the mainstay of treatment. If repeated local anesthetic sympathetic blocks do not provide lasting improvement, surgical sympathectomy may be performed. RSD is also called *complex regional pain syndrome type I*.

Chapter Summary

- The autonomic nervous system regulates the involuntary functions of the body.
- One of the primary functions of the autonomic nervous system is to regulate the body's internal environment (e.g., body temperature, fluid/electrolyte balance, pH, respiration, and blood pressure).
- The autonomic nervous system has three divisions: sympathetic, parasympathetic, and enteric.
  1. The sympathetic division coordinates the body's response to stress.
  2. The parasympathetic division regulates several homeostatic functions.
  3. The enteric system primarily regulates gastrointestinal activity.
- Parasympathetic postganglionic fibers release acetylcholine, whereas the sympathetic fibers release norepinephrine.
- The parasympathetic and sympathetic systems dually innervate most internal organs.
- Each autonomic efferent nerve innervates each organ in a two-neuron chain pathway. The adrenal medulla is a modified part of the sympathetic system.
- Many regions of the central nervous system control autonomic function through a hierarchy of reflexes and integrative centers.
  1. Some automatic reflexes (e.g., urination, defecation, and erection) are integrated at the spinal cord level.
  2. The medullary region in the brainstem serves as a control center for regulating cardiovascular, respiratory, and digestive activity via the autonomic nervous system.
  3. The hypothalamus plays an important role in integrating somatic and endocrine responses associated with anger, fear, and stress.

thePoint *Visit http://thePoint.lww.com for chapter review Q&A, case studies, animations, and more!*

7 Integrative Functions of the Central Nervous System

ACTIVE LEARNING OBJECTIVES

Upon mastering the material in this chapter you should be able to:

- Explain how the hypothalamus interfaces among the endocrine, autonomic, and limbic systems.
- Explain the difference between a biologic rhythm and a cycle.
- Explain the role of the brainstem reticular formation in consciousness and arousal.
- Describe the stages of sleep.
- Describe the functional organization of the forebrain.
- Describe the functional components of the limbic system.
- Explain the brain's reward system.
- Explain the components of the limbic system involved in mood disorders and schizophrenia.
- Explain the concept of long-term potentiation and how it relates to learning.
- Describe the areas of the brain that involve memory and learning.
- Describe how the Broca and Wernicke areas serve as functional components of language.

The **central nervous system (CNS)** detects and processes sensory information from both the internal and external environment. The motor system directs information to skeletal muscle for locomotion and fine movement. Both sensory and motor functions do not occur in isolation but are highly coordinated. The CNS can be grouped according to integrative function and comprises the following:

- **I.** Brainstem
  - **A.** Medulla
  - **B.** Pons
  - **C.** Midbrain
- **II.** Cerebellum
- **III.** Forebrain
  - **A.** Diencephalon
    - **1.** Hypothalamus
    - **2.** Thalamus
  - **B.** Cerebrum
    - **1.** Basal nuclei
    - **2.** Cerebral cortex

The brain area involved in the coordinated activities of human behavior is part of the central integrative system. An overview of these components is shown in Figure 7.1 with their respective integrative functions shown in Table 7.1. Functional components of the central integrative system can be classified into several broad categories. *Motivational systems*, for example, are those responsible for the human drive to satisfy basic needs such as hunger and thirst. Another is the group of systems that carry out functions such as *learning* and *memory*. These systems endow us with the ability to acquire and store new information. Still another category involves systems responsible for *communication*, specifically, the form of the written and spoken language. Many of these functions depend on input from the **brainstem** regarding both internal and external environmental stimuli; the brainstem is also critical to the efferent control mediated by higher centers.

This chapter describes the brain area responsible for these integrated functions and also examines how the brain collects and integrates information from its two hemispheres, specifically, how similar functions are located on one side of the brain.

HYPOTHALAMUS

The **hypothalamus** is the portion of the brain that contains a number of small nuclei with a variety of functions. The hypothalamus is located just below the thalamus and just above the brainstem. It forms the ventral part of the **diencephalon**. The diencephalon includes the hypothalamus, thalamus, and subthalamus (see Fig. 7.1). The rostral border of the hypothalamus is at the optic chiasm, and its caudal border is at the mammillary body. On the basal surface of the hypothalamus, exiting the median eminence, the pituitary stalk contains the **hypothalamo-hypophyseal portal blood vessels** (Fig. 7.2). Neurons within specific nuclei of the hypothalamus secrete releasing factors into these portal vessels. The releasing factors are then transported to the anterior pituitary, where they stimulate secretion of trophic hormones to other glands of the endocrine system (see Chapter 31, "Hypothalamus and the Pituitary Gland"). The pituitary stalk also contains the axons of magnocellular neurons whose cell bodies are located in the supraoptic and paraventricular hypothalamic

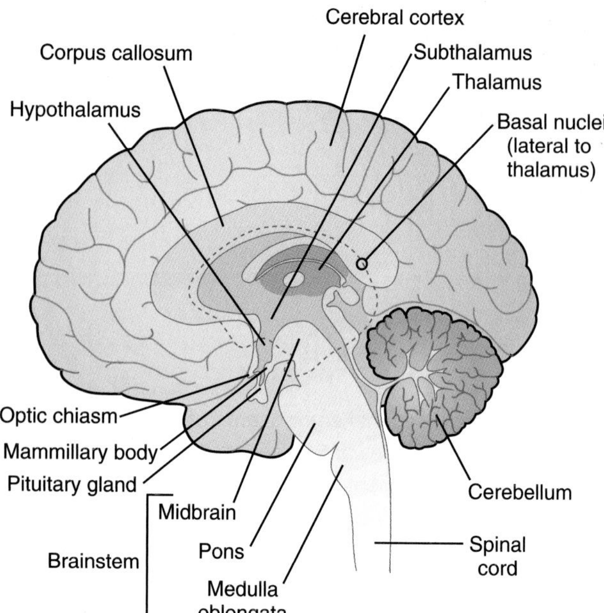

Figure 7.1 A midsagittal section through the brain showing the most prominent components. The cerebrum (*dotted line*) is the largest portion of the human brain.

TABLE 7.1 Major Integrative Functions of the Brain

Component	Major Functions
Cerebral cortex	1. Sensory perception 2. Voluntary movement 3. Language 4. Personality 5. Memory/decision making
Basal nuclei	1. Muscle tone 2. Coordination of slow/sustained movements 3. Inhibition of random movement
Thalamus	1. Relay station for all synaptic input 2. Awareness of sensations 3. Motor control
Hypothalamus	1. Homeostatic regulation for temperature, thirst, food intake, fluid balance, and biological rhythms 2. Relay station between nervous and endocrine systems 3. Sex drive and sexual behavior 4. Emotions and behavior patterns 5. Sleep–awake cycle
Cerebellum	1. Balance 2. Muscle tone 3. Coordination of voluntary muscle activity
Brainstem	1. Origin of most peripheral cranial nerves 2. Control centers for cardiovascular, respiratory, and digestive regulation 3. Regulation of muscle reflexes for posture and equilibrium 4. Some control of sleep–wake cycle 5. Integration of all synaptic input from spinal cord 6. Activation of cerebral cortex

nuclei. These axons form the **hypothalamo-hypophyseal tract** within the pituitary stalk and represent the efferent limbs of neuroendocrine reflexes that lead to the secretion of the hormones **arginine vasopressin** and **oxytocin** into the blood. These hormones are made in the magnocellular neurons and released by their axon terminals next to the blood vessels within the posterior pituitary.

All vertebrate brains contain a hypothalamus, and in the adult human, it is approximately the size of an almond. Despite its small size, it serves as the major headquarters for coordinating many functions. The hypothalamus is responsible for coordinating autonomic reflexes of the brainstem and the spinal cord. It also activates the endocrine and somatic motor systems when responding to signals generated either within the hypothalamus or brainstem or in higher centers, such as the **limbic system**, where the emotion and motivation are generated. The hypothalamus can accomplish this integration by virtue of its unique location at the interface between the limbic system and the endocrine and autonomic nervous systems.

As a major regulator of homeostasis, the hypothalamus receives input about the internal environment of the body via signals in the blood and from visceral afferents relaying through the brainstem. In most of the brain, capillary endothelial cells are connected by tight junctions that prevent substances in the blood from entering the brain. These tight junctions are part of the **blood–brain barrier**. The blood–brain barrier is missing in several small regions of the brain, called **circumventricular organs**, which are adjacent to the fluid-filled ventricular spaces. Several circumventricular organs are in the hypothalamus. Capillaries in these regions, similar to those in other organs, are fenestrated ("leaky"), allowing the cells of hypothalamic nuclei to sample freely, from moment to moment, the composition of the blood. Neurons in the hypothalamus then initiate the mechanisms necessary to maintain levels of constituents at a given set point, fixed within narrow limits by specific hypothalamic nuclei. Homeostatic functions regulated by the hypothalamus include body temperature, water and electrolyte balance, blood glucose levels, and energy balance.

The hypothalamus is connected to the endocrine system via the pituitary and serves as the major regulator of endocrine function. These connections include direct neuronal innervations of the posterior pituitary lobe by specific hypothalamic nuclei and a direct hormonal connection between specific hypothalamic nuclei and the anterior pituitary. *Hypothalamic-releasing hormones* are neurohormones, also designated as **releasing factors**, which reach the anterior pituitary lobe by a portal system of capillaries. Releasing factors then regulate the secretion of most hormones of the endocrine system.

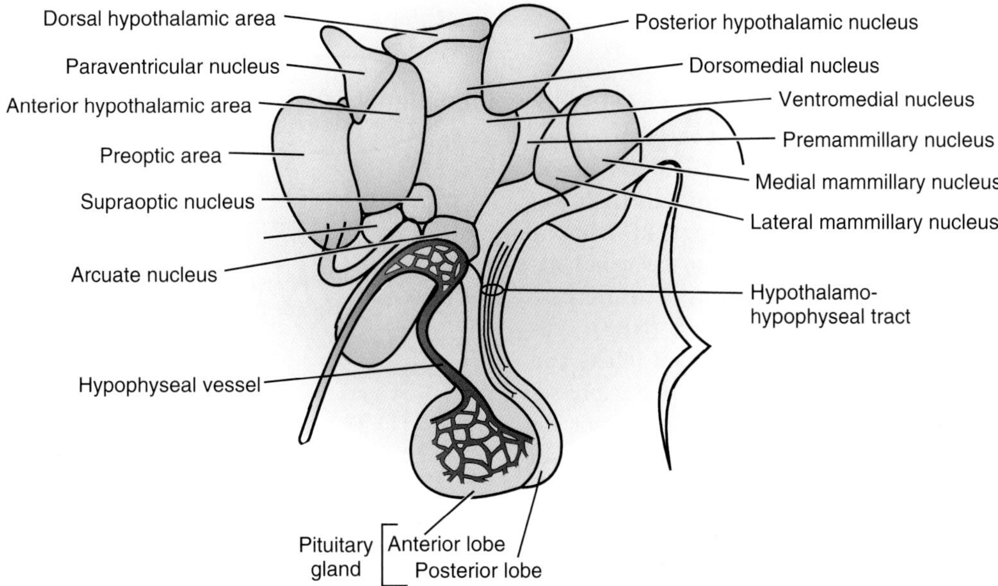

Figure 7.2 The hypothalamus and its nuclei. The hypothalamus comprises several different clusters of cell nuclei that are involved in many of the integrative functions. The connections between the hypothalamus and the pituitary gland are also shown.

Hypothalamus consists of distinct nuclei that interface between the endocrine, autonomic, and limbic systems.

The nuclei of the hypothalamus have ill-defined boundaries, despite their customary depiction (see Fig. 7.2). Many are named according to their anatomic location (e.g., anterior hypothalamic nuclei, ventromedial nucleus) or for the structures they lay next to (e.g., the periventricular nucleus surrounds the third ventricle, the **suprachiasmatic nucleus [SCN]** lies above the optic chiasm).

The hypothalamus receives afferent inputs from all levels of the CNS and makes reciprocal connections with the limbic system via fiber tracts in the **fornix** and **stria terminalis.** The hypothalamus also makes extensive reciprocal connections with the brainstem, including the reticular formation and the medullary centers of cardiovascular, respiratory, and gastrointestinal regulation. Many of these connections travel within the **medial forebrain bundle**, which also connects the brainstem with the cerebral cortex.

Several major connections of the hypothalamus are one-way rather than reciprocal. One of these, the **mammillothalamic tract**, carries information from the mammillary bodies of the hypothalamus to the anterior nucleus of the **thalamus**, from where information is relayed to limbic regions of the cerebral cortex. A second one-way pathway carries visual information from the retina to the SCN of the hypothalamus via the optic nerve. Through this retinal input, the light cues of the day–night cycle entrain or synchronize the "biologic clock" of the brain to the external clock. A third one-way connection is the hypothalamo-hypophyseal tract from the supraoptic and paraventricular nuclei to the posterior pituitary gland. The hypothalamus also projects directly to the spinal cord to activate sympathetic and parasympathetic preganglionic neurons (see Chapter 6, "Autonomic Nervous System").

Hypothalamus regulates energy balance by integrating metabolism and eating behavior.

Energy balance is the homeostasis of energy in living systems. It is measured by the following equation: Energy intake = energy expenditure + storage. *Energy intake* is dependent on diet, which is mainly regulated by hunger and calories consumed. *Energy expenditure* is the sum of internal heat production and external work produced. Internal heat production is, in turn, the **basal metabolic rate (BMR)** and the thermal effect of food. External work is determined by physical activity. The hypothalamus plays a key role in regulating the homeostatic energy balance by controlling both energy intake and energy expenditure. With respect to eating, the ventromedial nucleus of the hypothalamus was thought to serve as a **satiety** center and the lateral hypothalamic area as a feeding center. Together, these areas would then coordinate the processes that govern eating behavior and the subjective perception of satiety. These hypothalamic areas also influence the secretion of hormones, particularly from the thyroid gland, adrenal gland, and pancreatic islet cells, in response to changing metabolic demands. Although crude lesion and stimulation studies supported the satiety versus feeding center model of the regulation of feeding, this simplistic model is no longer tenable.

The regulation of eating behavior is part of a complex pathway that regulates food intake, energy expenditure, and reproductive function in the face of changes in nutritional state. In general, the hypothalamus regulates caloric intake, use, and storage in a manner that tends to maintain the body weight in adulthood. The presumptive set point around which it attempts to stabilize body weight is remarkably constant but can be altered by changes in physical activity, composition of the diet, emotional states, stress, aging, pregnancy, and pharmacologic agents such as atypical antipsychotics.

Several **orexigenic** (stimulating appetite) and **anorexigenic** (causing loss of appetite) peptides and neurotransmitters contribute to the regulation of energy balance. Some, such as **leptin** and **insulin**, act on a long time course to maintain body weight and blood glucose levels, and others, such as **cholecystokinin (CCK)** and **ghrelin**, act on a shorter time course to regulate initiation or cessation of a meal. Many of these signals are detected in the **arcuate nucleus (ARC)** of the hypothalamus, some by directly bathing ARC neurons and others by relaying through vagal inputs to the hypothalamus (Fig. 7.3). Within the ARC, two groups of neurons are critical. One group expresses **neuropeptide Y (NPY)**; activation of these neurons causes increased food intake and decreased energy expenditure. The other group expresses **proopiomelanocortin (POMC)**, which is a precursor for melanocortin peptides; activation of these neurons causes decreased food intake and increased energy expenditure. Outputs from the ARC that mediate alterations in feeding or energy expenditure include projections to the lateral hypothalamus, dorsomedial nucleus, and paraventricular nucleus, as well as a variety of structures in the brainstem and forebrain limbic system involved in motivated behavior. Direct output to parasympathetic and sympathetic preganglionic neurons in the brainstem and spinal cord derives from the paraventricular nucleus and the ARC. The paraventricular nucleus engages the endocrine system through its regulation of the pituitary gland.

A key player in the regulation of body weight is the hormone leptin, which is released by white fat cells (adipocytes). As fat stores increase, plasma leptin levels increase; conversely, as fat stores are depleted, leptin levels decrease. Leptin inhibits the NPY cells and stimulates the POMC cells in the ARC (see Fig. 7.3). Physiologic responses to low leptin levels (starvation) are initiated by the hypothalamus to increase food intake, decrease energy expenditure, decrease reproductive function, decrease body temperature, and increase parasympathetic activity. Physiologic responses to high leptin levels (obesity) are initiated by the hypothalamus to decrease food intake, increase energy expenditure, and increase sympathetic activity.

In addition to long-term regulation of body weight, the hypothalamus also regulates eating behavior more acutely. Factors that limit the amount of food ingested during a single feeding episode originate in the gastrointestinal tract and influence the hypothalamic regulatory centers. These include the following: sensory signals carried by the vagus nerve that signify stomach filling; chemical signals giving rise to the sensation of satiety, including absorbed nutrients (glucose, certain amino acids, and fatty acids); and gastrointestinal hormones, especially **CCK**, produced by endocrine cells in the gut wall, and ghrelin, produced by the epithelia of the stomach and small intestine. CCK is a satiety signal mediated via the vagus, whereas ghrelin stimulates food intake by activating the NPY neurons in the ARC. The success of bariatric gastric bypass surgery as a treatment for morbid obesity is thought to be a result of the decreased secretion of ghrelin.

The complex regulation of energy balance and the evolutionary pressure to conserve energy conspire to make weight loss a difficult proposition. The ready availability of highly palatable foods and the decrease in exercise levels required to procure them are thought to underlie the increased incidence of obesity in the population. Drugs that target increasing energy expenditure by activating the sympathetic nervous system have unacceptable side effects. With rare exceptions, obesity is associated with leptin resistance, rather than a lack of leptin, and so initial hopes that leptin would be therapeutic have not held up. Recently, drug development has focused on blocking a central signaling system that stimulates appetite, the **endocannabinoids**. These compounds are the endogenous ligands for the receptors that mediate the effects of the psychoactive component of marijuana, which has long been

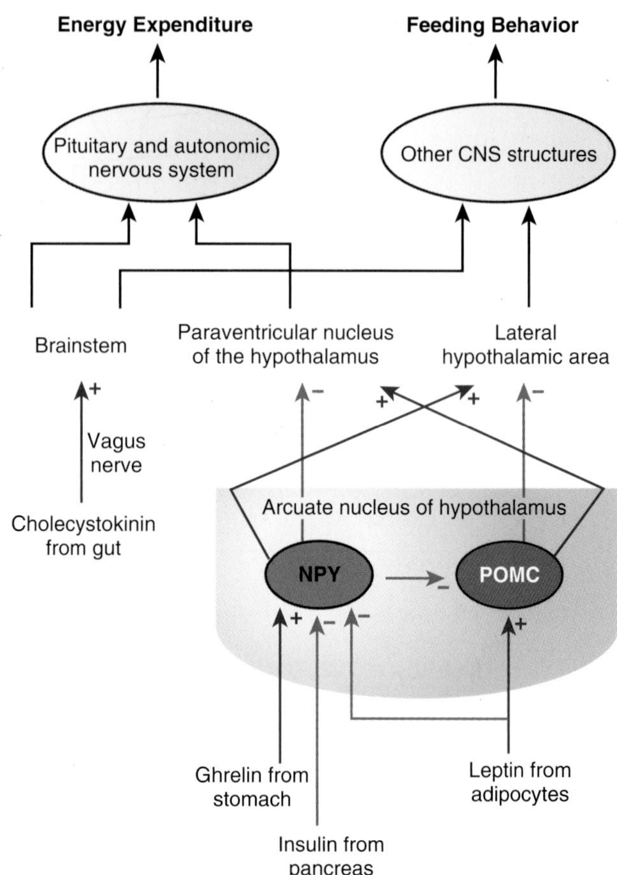

Figure 7.3 Control of energy expenditure and feeding behavior. Cell groups within the arcuate nucleus (ARC) integrate peripheral orexigenic and anorexigenic signals. Activation of neuropeptide Y (NPY) neurons within the ARC leads to a decrease in energy expenditure acting via the paraventricular nucleus of the hypothalamus (PVH), the pituitary, and the autonomic nervous system and an increase in feeding behavior acting via the lateral hypothalamic area (LH) and several other central nervous system (CNS) sites. Activation of proopiomelanocortin (POMC) neurons within the ARC increases energy expenditure and decreases feeding behavior acting via the same areas. Additional peripheral signals activate the brainstem via the vagus nerve; the brainstem then bypasses the ARC to act directly on the PVH and LH. Many central pathways (*not shown*) involved in feeding behavior and/or energy expenditure also project to the ARC, PVH, and LH.

known to stimulate carbohydrate craving and hyperphagia. Endocannabinoids signal both in the ARC and in several forebrain regions associated with motivated behavior. Levels of endocannabinoids in the hypothalamus decrease with leptin administration, consistent with their role as an appetite stimulant.

Sexual drive and sexual behavior are controlled by the hypothalamus.

The anterior and preoptic hypothalamic areas are sites for regulating gonadotropic hormone secretion and sexual behavior. Neurons in the preoptic area secrete **gonadotropin-releasing hormone (GnRH)**, beginning at puberty, in response to signals that are not understood. These neurons contain receptors for gonadal steroid hormones, testosterone and/or estradiol, which regulate GnRH secretion in either a cyclic (female) or a continual (male) pattern following the onset of puberty.

At a critical period in fetal development, circulating testosterone secreted by the testes of a male fetus changes the characteristics of cells in the preoptic area that are destined later in life to secrete GnRH. These cells, which would secrete GnRH cyclically at puberty had they not been exposed to androgens prenatally, are transformed into cells that secrete GnRH continually at a homeostatically regulated level. As a result, males exhibit a steady-state secretion rate for gonadotropic hormones and, consequently, for testosterone (see Chapter 36, "Male Reproductive System").

In the absence of androgens in fetal blood during development, the preoptic area remains unchanged, so that at puberty the GnRH-secreting cells begin to secrete in a cyclic pattern. This pattern is reinforced and synchronized throughout female reproductive life by the cyclic feedback of ovarian steroids, estradiol and progesterone, on secretion of GnRH by the hypothalamus during the menstrual cycle (see Chapter 37, "Female Reproductive System").

Steroid levels during prenatal and postnatal development are known to mediate differentiation of sexually dimorphic regions of the brain of most vertebrate species. Sexually dimorphic brain anatomy, behavior, and susceptibility to neurologic and psychiatric illness are evident in humans. However, with the exception of the GnRH-secreting cells, it has been difficult to definitively show a steroid dependency for sexually dimorphic differentiation in the human brain.

Hypothalamus contains the "biological clock" that controls rhythms and cycles.

Many physiologic events or processes repeat themselves with periodicity. When an event is repeated daily in the body, it is *rhythmic*, and if the event follows a monthly pattern it is *cyclic*. Daily rhythms are referred to as **circadian rhythms**. Circadian rhythms are linked to the light–dark cycle. If the activity occurs during the day, the circadian rhythm is referred to as a **diurnal rhythm**. *Nocturnal* is another type of daily rhythm, and the behavior is characterized by activity during the night. Nocturnal rhythms are most prevalent in other vertebrates. However, people who work a night shift and develop nocturnal habits are often referred to as "night owls." Examples of a diurnal rhythms is body temperature, which peaks during the day and falls at night.

An example of the monthly cycle is the female menstrual cycle, in which a series of physiologic events is repeated every 28 days.

From Bench to Bedside / 7.1

The Endocannabinoid System as a Therapeutic Target for Antiobesity Medications

Marijuana (*Cannabis sativa*) has been a cultivated crop with known psychoactive properties for more than 4,000 years, but its active constituent, Δ-9 tetrahydrocannabinol, was not isolated until 1964. A high-affinity binding site in the brain was not discovered until 1988, and the first cannabinoid receptor (CB_1) was cloned in 1991. Soon thereafter, the first endogenous cannabinoid, anandamide, was discovered. These findings set the stage for intensive efforts to develop pharmacological agonists and antagonists that target the endogenous pathways. A major motivation for this effort is the hyperphagia and carbohydrate craving associated with marijuana intoxication, with the implication that the endocannabinoids are involved in the regulation of feeding behavior. Researchers hope to be able to target this system to increase appetite in cancer patients and to decrease appetite in obese patients. Early results indicate that the antiobesity strategy may be fruitful, although side effects may sideline these drugs as well.

A selective CB_1 receptor blocker, rimonabant, was characterized in 1994 and has been studied in several large phase 3 trials as an appetite suppressant. These and other studies have shown that the endocannabinoids have peripheral and central effects that influence metabolism. Notably, peripheral lipogenesis increases along with appetite when the system is stimulated. CB_1 blockade has been shown to decrease food intake, weight, triglyceride levels, low-density lipoprotein levels, C-reactive protein levels, and insulin resistance while increasing high-density lipoprotein levels. This combined profile is encouraging in that a number of cardiovascular risk factors associated with obesity are targeted apparently in addition to the effect on appetite and ensuing weight loss. Rimonabant is in clinical trials also for tobacco cessation and appears to prevent the weight gain typically associated with cessation of smoking.

Despite early success, CB_1 receptor blockade is not yet proven to be a safe and effective treatment for weight loss. The endocannabinoid system (like most neurotransmitter systems) is widespread in the brain and likely involved in many functional systems whose disruption may be problematic. The biggest concern for rimonabant is the potential for neuropsychiatric effects, especially an observed increased incidence of anxiety and depressive disorders. These are common disorders, and the effect of CB_1 blockade in persons suffering from them is currently unknown.

The hypothalamus is thought to play a major role in regulating all of these biologic rhythms. Furthermore, these rhythms appear to be endogenous (within the body) because they persist even in the absence of time cues, such as day–night cycles for light and dark periods, lunar cycles for monthly rhythms, or changes in temperature and day length for seasonal change. Accordingly, most organisms, including humans, are said to possess an endogenous timekeeper, a so-called **biologic clock** that times the body's regulated functions.

Most homeostatically regulated functions exhibit peaks and valleys of activity that recur approximately daily. The "master clock" that controls daily rhythms of the body is located in the SCN, a center in the hypothalamus that serves as the brain's biologic clock. The SCN, which influences many hypothalamic nuclei via its efferent connections, has the properties of an oscillator whose spontaneous firing patterns change dramatically during a day–night cycle. This diurnal cycle of activity is maintained in vitro and is an internal property of SCN cells. The molecular basis of the cellular rhythm is a series of transcriptional–translational feedback loops. The genes involved in these loops are apparently conserved from prokaryotes to humans. An important pathway influencing the SCN is the afferent **retinohypothalamic tract** of the optic nerve, which originates in the retina, enters the brain through the optic chiasm, and terminates in the SCN. This pathway is the principal means by which light signals from the outside world transmit the day–night rhythm to the brain's internal clock, thereby entraining the endogenous oscillator to the external clock.

Figure 7.4 illustrates some of the circadian rhythms of the body. One of the most vivid is alertness, which peaks in the afternoon and is lowest in the hours preceding and following sleep. Another, body temperature, ranges approximately 1°C (~2°F) throughout the day, with the low point occurring during sleep. Plasma levels of growth hormone increase greatly during sleep, in keeping with this hormone's metabolic role as a glucose-sparing agent during the nocturnal fast. Cortisol, on the other hand, has its highest daily plasma level prior to arising in the morning. The mechanism by which the SCN can regulate diverse functions is related to its control of the production of **melatonin** by the **pineal gland**. Melatonin levels increase with decreasing light as night ensues.

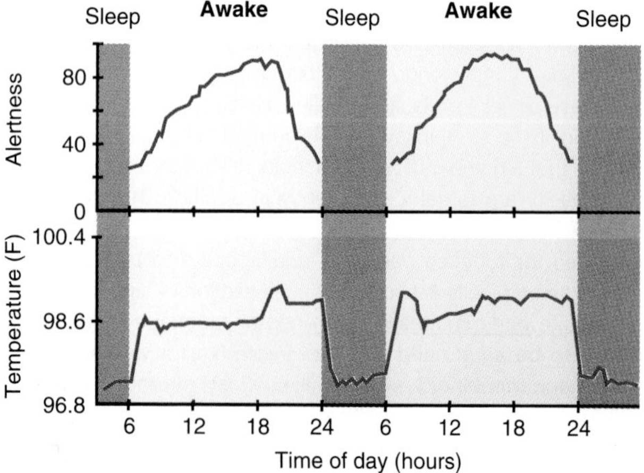

Figure 7.4 Circadian rhythms in some homeostatically regulated functions during two 24-hour periods. Alertness is measured on an arbitrary scale between sleep and most alert. Note that alertness and body temperature parallel each other.

Other homeostatically regulated functions exhibit diurnal patterns as well. When they are all in synchrony, they function harmoniously and impart a feeling of well-being. When there is a disruption in rhythmic pattern, such as by sleep deprivation or when passing too rapidly through several time zones, the period required for reentrainment of the SCN to the new day–night pattern is characterized by a feeling of malaise and physiologic distress. This is commonly experienced as "jet lag" in travelers crossing several time zones or by workers changing from day shift to night shift or from night shift to day shift. In such cases, the hypothalamus requires time to "reset its clock" before the regular rhythms are restored and a feeling of well-being ensues. The SCN uses the new pattern of light–darkness, as perceived in the retina, to entrain its firing rate to a pattern consistent with the external world. Resetting the clock may be facilitated by the judicious use of exogenous melatonin and by altering exposure to light.

In addition to entrainment of the biologic clock by light cycles, recent evidence suggests that the homeostatic mechanisms related to energy balance are intimately associated with the sleep–wake cycle. The link is a set of neurons in the posterolateral hypothalamus that express **hypocretin** (also known as **orexin**). These neurons are sensitive to nutritional status, simulate appetite, and project to arousal areas of the brain. A functional deficit in hypocretin was found to be the cause of **narcolepsy** in dog and mouse models of the disease. The SCN projects to the hypocretin-containing neurons and also receives input from the ARC. However, the SCN is not required for entrainment of a circadian oscillator of food anticipatory behavior by a fixed daily meal. The exact identification of the food-entrainable clock is not yet known.

Reticular formation governs the regulation of muscle tone, pain, and the sleep-awake pattern.

The brainstem contains anatomic groupings of cell bodies clearly identified as the nuclei of cranial sensory and motor nerves or as relay cells of ascending sensory or descending motor systems. The remaining cell groups of the brainstem, located in the central core, constitute a diffuse-appearing system of neurons with widely branching axons, known as the **reticular formation** and is centered roughly in the pons. The reticular formation constitutes the core of the brainstem and runs through the midbrain, pons, and medulla. The ascending reticular activating system connects to areas in the cortex, thalamus, and hypothalamus. A unique characteristic of neurons of the reticular formation is their widespread system of axon collaterals, which make extensive synaptic contacts and, in some cases, travel over long distances in the CNS. A striking example is the demonstration using intracellular labeling of individual cells and their processes, that one axon branch descends all the way into the spinal cord, whereas the collateral branch projects rostrally all the way to the forebrain, making myriad synaptic contacts along both axonal pathways. The reticular formation consists of >100 small neural networks, with varied functions including the regulation of the sleep–wake cycle, modulation of pain, and regulation of muscle tone.

Ascending reticular activating system mediates consciousness and arousal.

Sensory neurons bring peripheral sensory information to the CNS via specific pathways that ascend and synapse with specific nuclei of the thalamus, which, in turn, innervate primary sensory areas of the cerebral cortex. These pathways involve three to four synapses, starting from a receptor that responds to a specific sensory modality—such as touch, hearing, or vision. Each modality has, in addition, a nonspecific form of sensory transmission, in that axons of the ascending fibers send collateral branches to cells of the reticular formation (Fig. 7.5). The latter, in turn, send their axons to the **intralaminar nuclei of the thalamus**, which innervate wide areas of the cerebral cortex and limbic system. In the cerebral cortex and limbic system, the influence of the nonspecific projections from the reticular formation is arousal of the organism. This series of connections from the reticular formation through the intralaminar nuclei of the thalamus and on to the forebrain is termed the **ascending reticular activating system**.

The reticular formation also houses the neuronal systems that regulate sleep–wake cycles and consciousness.

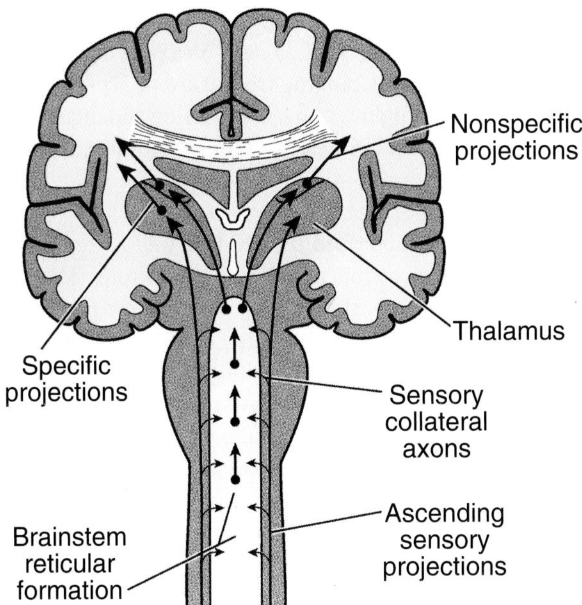

Figure 7.5 The brainstem reticular formation and reticular activating system. Ascending sensory tracts send axon collateral fibers to the reticular formation. Neurons in the reticular formation give rise to fibers that synapse in the intralaminar nuclei of the thalamus. From there, these nonspecific thalamic projections influence widespread areas of the cerebral cortex and limbic system.

So important is the ascending reticular activating system to the state of arousal that a malfunction in the reticular formation, particularly the rostral portion, can lead to a loss of consciousness and coma.

BRAIN ELECTRICAL ACTIVITY

The influence of the ascending reticular activating system on the brain's activity can be monitored via **electroencephalography (EEG)**. EEG is a sensitive recording device for picking up the electrical activity of the brain's surface through electrodes placed on designated sites on the scalp. This noninvasive tool measures simultaneously, via multiple leads, the electrical activity of the major areas of the cerebral cortex. It is also the best diagnostic tool available for detecting abnormalities in electrical activity, such as in epilepsy, and for diagnosing sleep disorders.

The detected electrical activity reflects the extracellular recording of the myriad postsynaptic potentials in cortical neurons underlying the electrode. The summated electrical potentials recorded from moment to moment in each lead are influenced greatly by the input of sensory information from the thalamus via specific and nonspecific projections to the cortical cells, as well as inputs that course laterally from other regions of the cortex.

EEG waves have characteristic patterns corresponding to state of arousal.

An EEG is the recorded electrical activity along the scalp produced by the firing of neurons within the brain. EEG frequency usually ranges from less than 1 to about 30 Hz, and amplitude, or height of the wave, usually ranges from 20 to 100 μV. Because the waves are a summation of activity in a complex network of neuronal processes, they are highly variable. However, during various states of consciousness, EEG waves have certain characteristic patterns. At the highest state of alertness, when sensory input is greatest, the waves are of high frequency and low amplitude, as many units discharge asynchronously. At the opposite end of the alertness scale, when sensory input is at its lowest in deep sleep, a synchronized EEG has the characteristics of low frequency and high amplitude. An absence of EEG activity is the legal criterion for death in the United States.

EEG wave patterns are classified according to their frequency (Fig. 7.6). **Alpha waves**, a rhythm ranging from 8 to 13 Hz, are observed when the person is awake but relaxed with the eyes closed. When the eyes are open, the added visual input to the cortex imparts a faster rhythm to the EEG, ranging from 13 to 30 Hz and designated **beta waves**. The slowest waves recorded occur during sleep: **theta waves** at 4 to 7 Hz and **delta waves** at 0.5 to 4 Hz, in deepest sleep.

Abnormal wave patterns are seen in **epilepsy**, a neurologic disorder of the brain characterized by spontaneous discharges of electrical activity, resulting in abnormalities ranging from momentary lapses of attention, to seizures of varying severity, to loss of consciousness, if both brain hemispheres participate in the electrical abnormality.

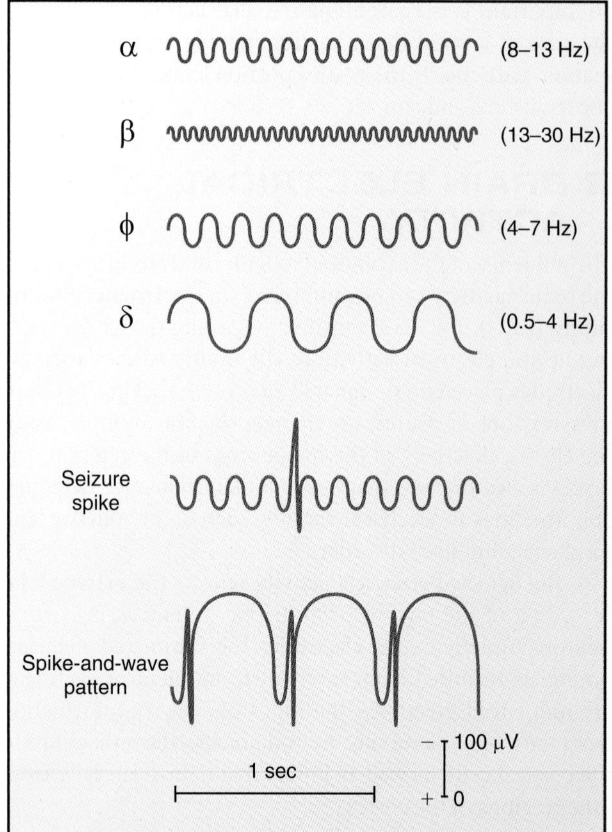

Figure 7.6 Patterns of brain waves recorded on an electroencephalogram. Wave patterns are designated alpha, beta, theta, or delta waves, based on frequency and relative amplitude. In epilepsy, abnormal spikes and large summated waves appear, as many neurons are activated simultaneously.

The characteristic waveform signifying seizure activity is the appearance of spikes or sharp peaks, as abnormally large numbers of units fire simultaneously. Examples of spike activity occurring singly and in a spike-and-wave pattern are shown in Figure 7.6.

Sleep stages are defined by the EEG.

Sleep follows a circadian rhythm and is a naturally recurring altered state of consciousness, in which sensory and motor activity are suspended, as well as nearly all voluntary muscle activity. During sleep, a heightened anabolic state exists, accentuating growth and rejuvenating the immune, nervous, and muscular systems. Sleep is regulated by the biologic clock in the reticular formation. The biologic clock for sleep (an inner time-keeping, temperature-fluctuating, enzyme-controlling device) works in tandem with **adenosine**, a neurotransmitter that inhibits most of the functional processes associated with wakefulness. Adenosine rises over the course of the day, and high levels induce drowsiness. The hormone melatonin is released during drowsy sleep and causes a concomitant gradual decrease in core temperature.

Electrical activity associated with sleep was first discovered in 1937 using EEG. The EEG recorded during sleep reveals a persistently changing pattern of wave amplitudes and frequencies, indicating that the brain remains continually active even in the deepest stages of sleep. The EEG pattern recorded during sleep varies in a cyclic fashion that repeats approximately every 90 minutes, starting from the time of falling asleep to awakening 7 to 8 hours later. Sleep stages and other characteristics of sleep are assessed in a specialized sleep laboratory. Measurements include EEG of brain waves, *electrooculography* of eye movements, and *electromyography* of skeletal muscle activity. Sleep is divided into two broad categories: **rapid eye movement (REM)** and **non-rapid eye movement (NREM) sleep**. Each type has a distinct set of functional, neurologic, and physiologic features. NREM is further divided into three sub stages: N1, N2, and N3. The latter is also called *delta sleep* or **slow-wave sleep (SWS)**. Sleep proceeds in cycles of REM and NREM, the order normally being: awake → N1 → N2 → N3 → N2 → REM. REM sleep is sometimes referred to as *stage 5*. There is a greater amount of deep sleep (stage N3) early in the night, whereas the proportion of REM sleep increases later in the night and just before natural awakening. In humans, each cycle last on the average 90 to 110 minutes. NREM sleep stages are characterized as follows:

- Stage N1 is referred to as *somnolence* or *drowsy sleep stage*. Brian waves in stage N1 transition from alpha waves (awake state with a frequency of 8–13 Hz) to theta waves (frequency of 4–7 HZ). During this transition, sudden twitches, also known as **myoclonus**, are often associated with this stage. Also, muscle tone and most conscious awareness are lost during N1.
- During stage N2, muscular activity decreases and conscious awareness of the external environment completely disappear. This stage occupies 45% to 55% of total sleep in adults.
- Stage N3 is characterized as *deep* or SWS with the presence of delta waves ranging from 0.4 to 4 Hz. This is the stage in which night terrors, bed wetting (nocturnal enuresis), sleepwalking, and sleeptalking (somniloquy) occur.

REM sleep

REM sleep is characterized by REM as well as rapid, low-voltage EEG and accounts for 20% to 25% of total sleep time in most adults. This is the stage in which most memorable dreaming occurs. REM sleep is also known as **paradoxical sleep**, because of the seeming contradictions in its characteristics. First, the EEG exhibits unsynchronized, high-frequency, low-amplitude waves (i.e., a beta rhythm), which is more typical of the awake state than sleep, yet the subject is as difficult to arouse. Second, the autonomic nervous system is in a state of excitation; blood pressure and heart rate are increased and breathing is irregular. In males, autonomic excitation in REM sleep includes penile erection. This reflex is used in diagnosing impotence, to determine whether erectile failure is based on a neurologic or a vascular defect (in which case, erection does not accompany REM sleep). When subjects are awakened during a REM period, they usually report dreaming. Accordingly, it is customary to consider REM sleep as dream sleep. Another curious characteristic of REM sleep is that most voluntary muscles are temporarily paralyzed. Two exceptions, in addition to the muscles of respiration, include the extraocular muscles, which contract rhythmically to produce the REMs, and the muscles of the middle ear, which protect the inner ear (see Chapter 4, "Sensory Physiology"). Muscle paralysis is caused by an active inhibition of motor neurons mediated by a group of neurons located close to the locus coeruleus in the brainstem. Many of us have experienced this muscle paralysis on waking from a bad dream, feeling momentarily incapable of running from danger. In certain sleep disorders in which skeletal muscle contraction is not temporarily paralyzed in REM sleep, subjects act out dream sequences with disturbing results, with no conscious awareness of this happening.

Sleep in humans varies with developmental stage. Newborns sleep approximately 16 hours per day, of which about 50% is spent in REM sleep. Normal adults sleep 7 to 8 hours per day, of which about 25% is spent in REM sleep. The percentage of REM sleep declines further with age, together with a loss of the ability to achieve stage 3, illness, of SWS. **Sleep debt** is not getting enough sleep. Sleep deprivation can lead to a large sleep debt, which can cause emotional, mental, and physical problems. Sleep deprivation leads to increased irritability, cognitive impairment, memory loss, muscle tremors, and daytime yawning. Sleep deprivation has been shown to be a risk factor for weight gain, hypertension, type 2 diabetes, and impaired immune system. There are many reasons for poor sleep that lead to sleep deprivation. Pain, drugs, and stress can be the culprit. Sleep apnea, narcolepsy, insomnia, restless leg syndrome, and peripheral neuropathy are also causes of poor sleep. Peripheral neuropathy is especially prevalent in individuals with severe type II diabetes and in cancer survivors who have undergone chemotherapy.

Abnormal brain waves indicate epilepsy.

Epilepsy (also known as *seizures*) is defined as a transient symptom of abnormal excessive activity in the brain. The outward effect of epileptic seizures is manifested in three different ways: (1) dramatic wild thrashing movements (tonic–clonic seizure), (2) mild loss of awareness, or (3) convulsions. Sometimes the seizure is not accompanied by convulsions, but as "full-body slump," in which the patient loses consciousness and "slumps" to the ground. After a seizure, while the brain is recovering, there can be a transient loss of memory, usually classified as *short term*. Epilepsy can be controlled but not cured with medication. Epilepsy is divided into three major types: **tonic–clonic**, **absence**, and **partial seizures**.

Tonic–clonic seizures

Tonic–clonic seizures (formerly known as *grand mal seizures*) are a type of generalized seizures that affects the entire brain and are the most commonly associated with epilepsy/seizures in general. In the *tonic phase*, the person quickly loses consciousness and the skeletal muscles suddenly become tense, causing extremities to be pulled inward toward the body. Often a loud moan during the tonic stage occurs due to air forcibly being expelled from the lungs. A typical EEG is shown in Figure 7.7 from the cortex during the tonic phase.

During the *clonic phase*, muscles contract and relax rapidly, causing convulsions. The convulsions can range from exaggerated twitches to violent shaking. The eyes typically roll back or close, and the tongue often suffers bruises sustained from jaw contractions. Incontinence sometimes occurs. Confusion and complete amnesia are usually experienced upon regaining consciousness. Due to physical and nervous exhaustion, a long sleep period usually follows a tonic–clonic seizure.

Absence seizures

Absence seizures (formerly known as *petit mal seizures*) are characterized by the lack of jerking or twitching movements of the eyes. The person appears to stare into space with episodes lasting up to 10 seconds. Sometimes they experience, in the absence of seizures, moving from one location to another without any purpose. Some patients report seeing flashing or blinking lights in the corner of one or both eyes shortly before they slip into the seizure. A typical EEG pattern exhibited during absence seizures is also shown in Figure 7.7.

Partial seizure

Partial seizures (formerly known as *focal epilepsy*) are preceded by an isolated disturbance of a cerebral function. For example, a disturbance may include a twitching of part of

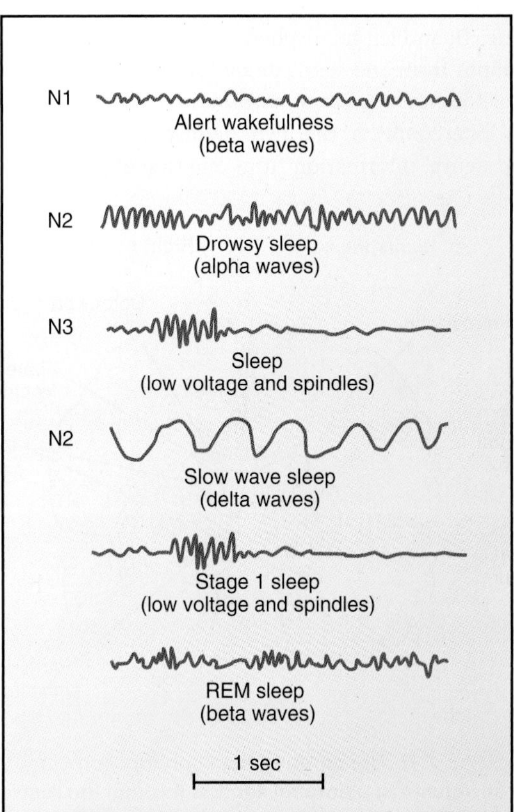

Figure 7.7 The brain wave patterns during a normal sleep cycle. Sleep cycles are divided into REM (rapid eye movement) and NREM (nonrapid eye movement) sleep. NREM is further divided into three stages: N1, N2, and N3. REM sleep is sometimes referred to as *stage 5.*

the body, such as a limb, or a deceptive or illusory sensation. Partial seizures are further classified as (1) simple partial (with no interruption to consciousness) and (2) complex partial (interrupts consciousness). An example of a brain wave exhibited during a simple partial seizure is illustrated in the bottom tracing in Figure 7.7. Most often, partial seizures result from a localized lesion or a functional abnormality. Examples include scar tissue that adheres to adjacent brain tissue, a brain tumor that compresses a discrete area of the brain, or a dysfunctional local circuitry.

FUNCTIONAL COMPONENTS OF THE FOREBRAIN

The forebrain is the rostralmost portion of the brain that contains the thalamus, hypothalamus, and the cerebrum. The forebrain is responsible for controlling body temperature, reproductive functions, eating, sleeping, and emotional affect.

The cerebrum consists of the **cerebral cortex** and the **subcortical structures** rostral to the diencephalon. The cortex, a few-millimeters-thick outer shell of the cerebrum, has a rich, multilayered array of neurons and their processes forming columns perpendicular to the surface. The axons of cortical neurons give rise to descending fiber tracts and intrahemispheric and interhemispheric fiber tracts, which, together with ascending axons coursing toward the cortex, make up the prominent white matter underlying the outer cortical gray matter. A deep sagittal fissure divides the cortex into a right and left hemisphere, each of which receives sensory input from and sends its motor output to the opposite side of the body. A set of **commissures** containing axonal fibers interconnects the two hemispheres, so that processed neural information from one side of the forebrain is transmitted to the opposite hemisphere. The largest of these commissures is the **corpus callosum**, which interconnects the major portion of the hemispheric regions (Fig. 7.8). Among the subcortical structures located in the forebrain are the components of the limbic system, which regulates emotional response, and the **basal ganglia** (caudate, putamen, and globus pallidus), which are essential for coordinating motor activity (see Chapter 5, "Motor System") and thought processes.

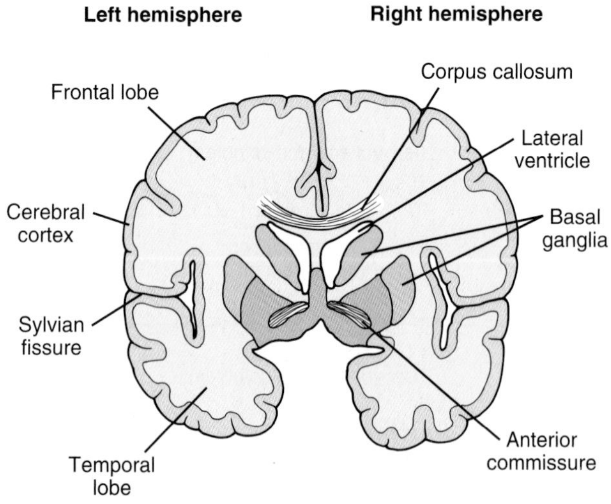

Figure 7.8 The cerebral hemispheres and some deep structures in a coronal section through the rostral forebrain. The corpus callosum is the major commissure that interconnects the right and left hemispheres. The anterior commissure connects rostral components of the right and left temporal lobes. The cortex is an outer rim of gray matter (neuronal cell bodies and dendrites); deep to the cortex is white matter (axonal projections) and then subcortical gray matter.

Cerebral cortex comprises three functional areas: sensory, motor, and associated areas.

In the human brain, the surface of the cerebral cortex is highly convoluted, with gyri (singular, **gyrus**) and sulci (singular, **sulcus**), which are akin to hills and valleys, respectively. Deep sulci are also called *fissures*. Two deep fissures form prominent landmarks on the surface of the cortex. The **central sulcus** divides the **frontal lobe** from the **parietal lobe**, and the **Sylvian fissure** divides the parietal lobe from the **temporal lobe** (Fig. 7.9). The **occipital lobe** has less prominent sulci separating it from the parietal and temporal lobes.

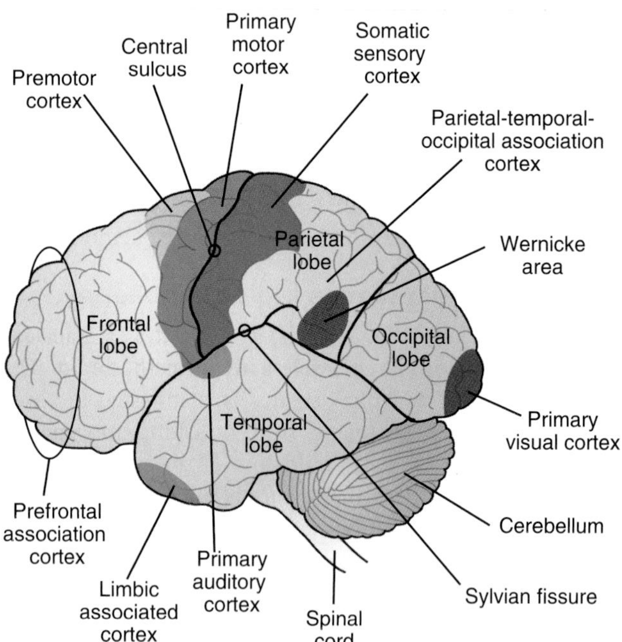

Figure 7.9 The four lobes of the cerebral cortex and their respective functional areas. The central sulcus and Sylvian fissure are prominent landmarks used in defining the cortical lobes. The primary auditory cortex is responsible for hearing. The limbic association cortex is the region for motivation, emotions, and memory. The prefrontal association cortex is responsible for voluntary activity, decision making, and personality traits. The premotor cortex is responsible for coordinating complex motor movements. The primary motor cortex is responsible for voluntary movements. The somatosensory cortex is responsible for sensation and proprioception. The parietal temporal occipital association cortex integrates all sensory input and is important in language. The Wernicke area is responsible for speech comprehension.

Topographically, the cerebral cortex is divided into areas of specialized functions, including the primary sensory areas for vision (occipital cortex), hearing (temporal cortex), somatic sensation (postcentral gyrus), and motor functions (precentral gyrus) (see Chapter 4, "Sensory Physiology," and Chapter 5, "Motor System"). Sensory and motor functions are controlled by cortical structures in the contralateral hemisphere (see Chapter 4, "Sensory Physiology," and Chapter 5, "Motor System"). Particular cognitive functions or components of these functions may also be lateralized to one side of the brain.

As shown in Figure 7.9, these well-defined areas comprise only a small fraction of the surface of the cerebral cortex. The majority of the remaining cortical area is known as **association cortex**, where the processing of neural information is performed at the highest levels of which the organism is capable. Among vertebrates, the human cortex contains the most extensive association areas. The association areas are also sites of long-term memory, and they control such human functions as language acquisition, speech, musical ability, mathematical ability, complex motor skills, abstract thought, symbolic thought, and other cognitive functions.

Association areas interconnect and integrate information from the primary sensory and motor areas via intrahemispheric connections. The **parietal–temporal–occipital association cortex** integrates neural information contributed by visual, auditory, and somatic sensory experiences. The **prefrontal association cortex** is that part of the frontal lobe other than the specific motor and speech regions. The prefrontal area governs the following: **attention**, the monitoring of the external and internal milieu; **intention**, the shaping of behavior in accordance with internal motivation and external context; and **execution**, the orchestration of complex sensorimotor sequences in a seamless path toward a goal. The selection and execution of cognitively driven sensorimotor sequences depend on integration with the basal ganglia in the same way that motor sequences do (see Chapter 5, "Motor System"). The prefrontal cortex is responsible for executive

Clinical Focus / 7.1

The Split Brain

Patients with life-threatening, intractable epileptic seizures were treated in the past by surgical commissurotomy or cutting of the corpus callosum (see Fig. 7.8). This procedure effectively cut off most of the neuronal communication between the left and right hemispheres and vastly improved patient status because seizure activity no longer spread back and forth between the hemispheres.

There was a remarkable absence of overt signs of disability following commissurotomy; patients retained their original motor and sensory functions, learning and memory, personality, talents, emotional responding, and so on. This outcome was not unexpected because each hemisphere has bilateral representation of most known functions; moreover, those ascending (sensory) and descending (motor) neuronal systems that crossed to the opposite side were known to do so at levels lower than the corpus callosum.

Notwithstanding this appearance of normalcy, following commissurotomy, patients were shown to be impaired to the extent that one hemisphere literally did not know what the other was doing. It was further shown that each hemisphere processes neuronal information differently from the other and that some cerebral functions are confined exclusively to one hemisphere.

In an interesting series of studies by Nobel laureate Roger Sperry and colleagues, these patients with a so-called split brain were subjected to psychophysiological testing in which each disconnected hemisphere was examined independently. Their findings confirmed what was already known: sensory and motor functions are controlled by cortical structures in the contralateral hemisphere. For example, visual signals from the left visual field were perceived in the right occipital lobe, and there were contralateral controls for auditory, somatic sensory, and motor functions. (Note that the olfactory system is an exception, as odorant chemicals applied to one nostril are perceived in the olfactory lobe on the same side.) However, the scientists were surprised to find that language ability was controlled almost exclusively by the left hemisphere. Thus, if an object was presented to the left brain via any of the sensory systems, the subject could readily identify it by the spoken word. However, if the object was presented to the right hemisphere, the subject could not find words to identify it. This was not due to an inability of the right hemisphere to perceive the object, as the subject could easily identify it among other choices by nonverbal means, such as feeling it while blindfolded. From these and other tests, it became clear that the right hemisphere was mute; it could not produce language.

In accordance with these findings, anatomic studies show that areas in the temporal lobe concerned with language ability, including the Wernicke area, are anatomically larger in the left hemisphere than in the right in a majority of humans, and this is seen even prenatally. Corroborative evidence of language ability in the left hemisphere is shown in persons who have had a stroke, where aphasias are most severe if the damage is on the left side of the brain. Analysis of people who are deaf who communicated by sign language prior to a stroke has shown that sign language is also a left-hemisphere function. These patients show the same kinds of grammatical and syntactical errors in their signing following a left-hemisphere stroke as do speakers.

In addition to language ability, the left hemisphere excels in mathematical ability, symbolic thinking, and sequential logic. The right hemisphere, on the other hand, excels in visuospatial ability, such as three-dimensional constructions with blocks and drawing maps, and in musical sense, artistic sense, and other higher functions that computers seem less capable of emulating. The right brain exhibits some ability in language and calculation, but at the level of children aged 5 to 7. It has been postulated that both sides of the brain are capable of all of these functions in early childhood, but the larger size of the language area in the left temporal lobe favors development of that side during language acquisition, resulting in nearly total specialization for language on the left side for the rest of one's life.

function (making decisions) and, by virtue of its connections with the limbic system, coordinates emotionally motivated behaviors. The prefrontal cortex receives neural input from the other association areas and regulates motivated behaviors by direct input to the **premotor area**, which serves as the association area of the motor cortex.

Cortical and subcortical structures are part of the architectural design of the limbic system.

The limbic system comprises specific areas of the cortex and subcortical structures interconnected via circuitous pathways that link the cerebrum with the diencephalon and brainstem (Fig. 7.10). The limbic system includes the hippocampus, amygdala, anterior thalamic nuclei, and the limbic cortex. The limbic cortex supports a variety of functions, including long-term memory, behavior, emotions, and olfaction. The limbic system operates by influencing the automatic nervous system and the endocrine system. It is richly innervated by the **nucleus accumbens** (the brain's pleasure center) and plays a role in sexual arousal and the emotional "high" derived from recreational drugs. Many of the psychiatric disorders, including bipolar disorder, depression, schizophrenia, and dementia, involve malfunctions in the limbic system.

Originally, the limbic system was considered to be restricted to a ring of structures surrounding the corpus callosum, including the olfactory system, **cingulate gyrus**, **parahippocampal gyrus**, and **hippocampus**, together with the fiber tracts that interconnect them with the diencephalic components of the limbic system, the hypothalamus, and the anterior thalamus. Current descriptions of the limbic system also include the **amygdala** (deep in the temporal lobe), the nucleus accumbens (the limbic portion of the basal ganglia), the **septal nuclei** (at the base of the forebrain), the **prefrontal cortex** (anterior and inferior components of the frontal lobe), and the **habenula** (in the diencephalon).

Circuitous loops of fiber tracts interconnect the limbic structures. The main circuit links the hippocampus to the mammillary body of the hypothalamus by way of the fornix, the hypothalamus to the anterior thalamic nuclei via the mammillothalamic tract, and the anterior thalamus to the cingulate gyrus by widespread, anterior thalamic projections (Fig. 7.11). To complete the circuit, the cingulate gyrus connects with the hippocampus, to enter the circuit again. Other structures of the limbic system form smaller loops within this major circuit, forming the basis for a wide range of emotional behaviors. Of particular importance is the amygdala, which interconnects with the hypothalamus and basal forebrain region via the stria terminalis and is responsible for fear conditioning and vigilance in learning and memory paradigms.

The fornix also connects the hippocampus to the base of the forebrain, where the septal nuclei and nucleus accumbens reside. The prefrontal cortex and other areas of association cortex provide the limbic system with information based on previous learning and currently perceived needs. Inputs from the brainstem provide visceral and somatic sensory signals, including tactile, pressure, pain, and temperature information from the skin and sexual organs and pain information from the visceral organs.

At the caudal end of the limbic system, the brainstem has reciprocal connections with the hypothalamus (see Fig. 7.11). As noted earlier, all ascending sensory systems in the brainstem send axon collaterals to the reticular formation, which, in turn, innervates the limbic system, particularly via monoaminergic pathways. The reticular formation also forms the ascending reticular activating system, which serves not only to arouse the cortex but also to impart an emotional tone to the sensory information transmitted nonspecifically to the cerebral cortex.

Monoaminergic innervations

Monoaminergic neurons release monoamine neurotransmitters derived from aromatic amino acids (e.g., tryptophan, tyrosine, and phenylalanine). Monoamine neurotransmitters include *histamine*, *catecholamines* (e.g., dopamine,

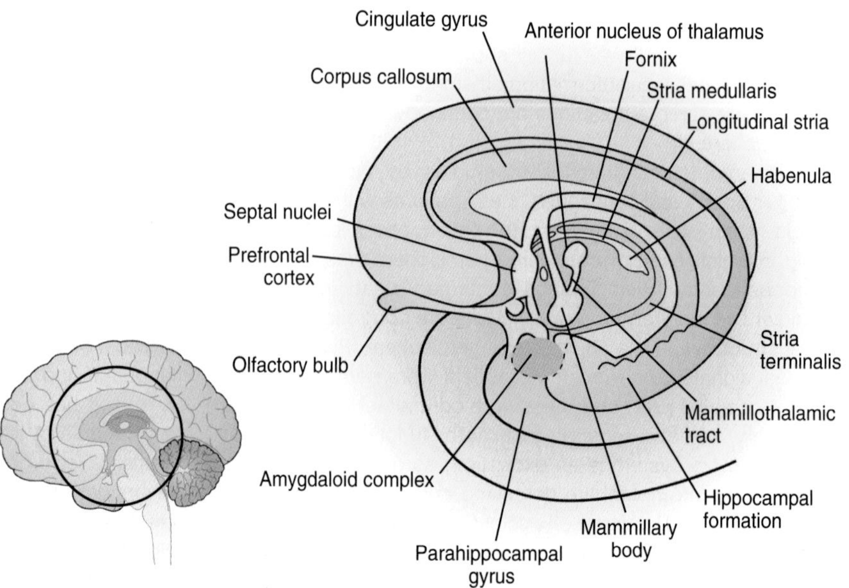

Figure 7.10 The cortical and subcortical structures of the limbic system extending from the cerebral cortex to the diencephalon. The fiber tracts that interconnect the structures of the limbic system are also shown.

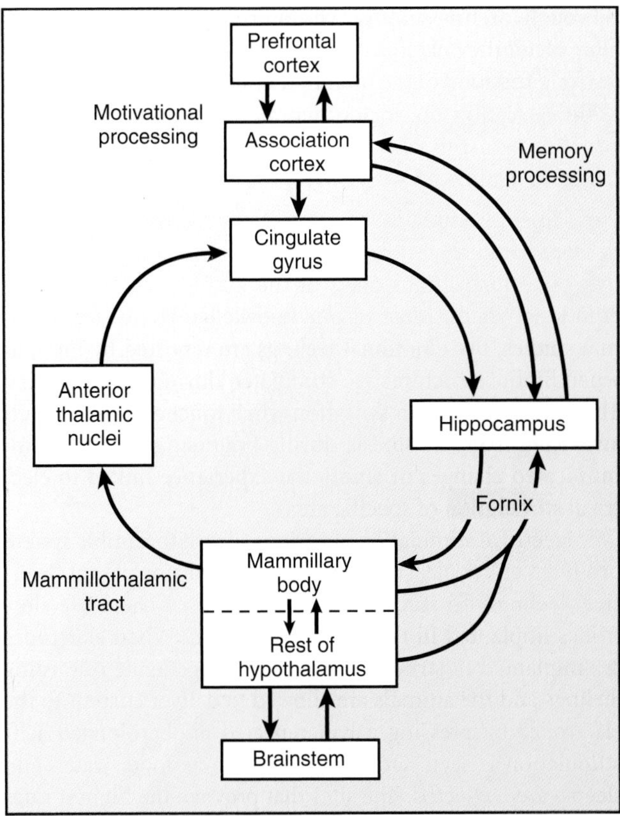

Figure 7.11 The main circuit of the limbic system.

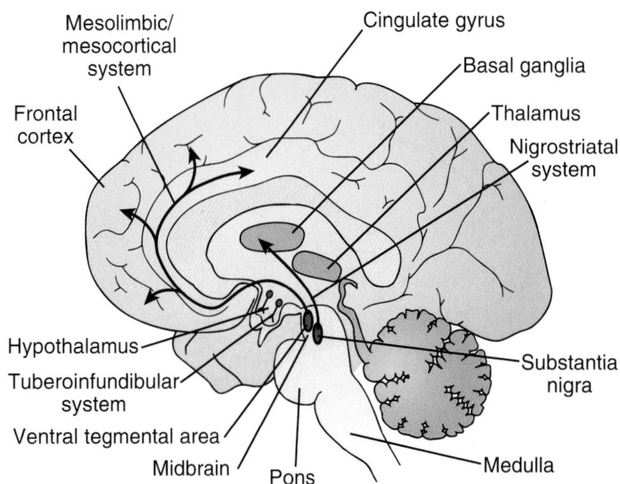

Figure 7.12 The origins and projections of the three major dopaminergic systems and the origins and projections of the serotonergic system.

norepinephrine [NE], and epinephrine), and *tryptamines* (melatonin and serotonin). Monoaminergic neurons innervate all parts of the CNS via widespread, divergent pathways starting from cell groups in the reticular formation. The limbic system and basal ganglia are richly innervated by catecholaminergic (noradrenergic and dopaminergic) and serotonergic nerve terminals emanating from brainstem nuclei that contain relatively few cell bodies compared with their extensive terminal projections. From neurochemical manipulation of monoaminergic neurons in the limbic system, it is apparent that they play a major role in determining emotional state.

Dopaminergic neurons are located in three major pathways originating from cell groups in either the midbrain (the substantia nigra and ventral tegmental area) or the hypothalamus (Fig. 7.12). The **nigrostriatal system** consists of neurons with cell bodies in the substantia nigra (pars compacta) and terminals in the neostriatum (caudate and putamen) located in the basal ganglia. This dopaminergic pathway is essential for maintaining normal muscle tone and initiating voluntary movements (see Chapter 5, "Motor System"). The **tuberoinfundibular system** of dopaminergic neurons is located entirely within the hypothalamus, with cell bodies in the arcuate nucleus and periventricular nuclei and terminals in the median eminence on the ventral surface of the hypothalamus. The tuberoinfundibular system is responsible for the secretion of hypothalamic releasing factors into a portal system that carries them through the pituitary stalk into the anterior pituitary lobe (see Chapter 31, "Hypothalamus and the Pituitary Gland"). The dopaminergic neurons inhibit prolactin release from the pituitary.

Dopamine pathways

A major dopaminergic pathway in the brain is the *mesolimbic pathway*, part of the **mesolimbic/mesocortical system.** This pathway begins in the ventral tegmental area of the midbrain of the brainstem and innervates the limbic system, amygdala, hippocampus, and the medial prefrontal cortex. It functions in modulating behavioral responses, especially motivation and drive, to stimuli through the neurotransmitter dopamine (see Fig. 7.12). Drugs that increase dopaminergic transmission, such as cocaine, which inhibits dopamine reuptake, and amphetamine, which promotes dopamine release and inhibits its reuptake, lead to repeated administration and abuse presumably because they stimulate the brain's reward system. The mesolimbic dopaminergic system is also the site of action of **neuroleptic** drugs, which are used to treat schizophrenia (discussed later) and other psychotic conditions.

The *mesocortical pathway* is another dopaminergic pathway of the mesolimbic/mesocortical system that connects the ventral tegmentum to the cerebral cortex, particularly the frontal lobes. It is one of the four major dopamine pathways in the brain and is essential to the normal cognitive function of the prefrontal cortex. It is thought to be involved in motivation and emotional response.

Noradrenalin pathways

Noradrenergic neurons (containing NE) are located in cell groups in the medulla and pons (Fig. 7.13). The medullary cell groups project to the spinal cord, where they influence cardiovascular regulation and other autonomic functions. Cell groups in the pons include the lateral system, which innervates the basal forebrain and hypothalamus, and the **locus coeruleus**, which sends efferent fibers to nearly all parts of the CNS.

Noradrenergic neurons innervate all parts of the limbic system and the cerebral cortex, where they play a major role

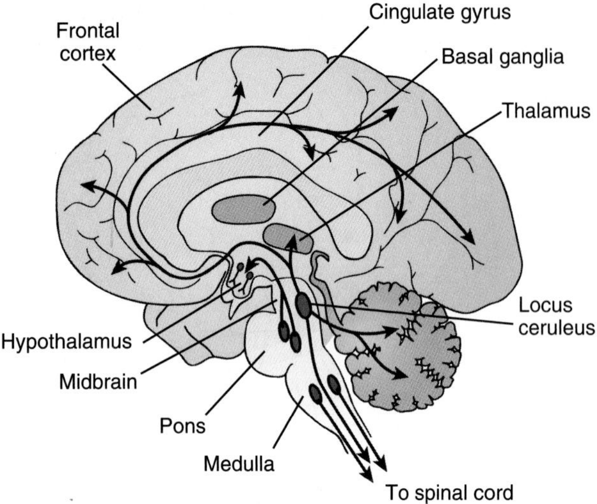

Figure 7.13 The origins and projections of five of seven cell groups of noradrenergic neurons of the brain. The depicted groups originate in the medulla and pons. Among the latter, the locus coeruleus in the dorsal pons innervates most parts of the central nervous system (CNS).

in setting **mood** (sustained emotional state) and **affect** (the emotion itself; e.g., euphoria, depression, anxiety). Drugs that alter noradrenergic transmission have profound effects on mood and affect. For example, reserpine, which depletes brain NE, induces a state of depression. Drugs that enhance NE availability, such as **monoamine oxidase inhibitors (MAOIs)** and inhibitors of reuptake, reverse this depression. Amphetamines and cocaine have effects on boosting noradrenergic transmission similar to those described for dopaminergic transmission; they inhibit reuptake and/or promote the release of NE. Increased noradrenergic transmission results in an elevation of mood, which further contributes to the potential for abusing such drugs, despite the depression that follows when drug levels fall. Some of the unwanted consequences of cocaine or amphetamine-like drugs reflect the increased noradrenergic transmission, in both the periphery and the CNS. This can result in a hypertensive crisis, myocardial infarction, or stroke, in addition to marked swings in affect, starting with euphoria and ending with profound depression.

Serotonin pathways

Serotonergic neurons (neurons that make serotonin) also innervate most parts of the CNS. Cell bodies of these neurons are located at the midline of the brainstem (the raphe system) and in more laterally placed nuclei, extending from the caudal medulla to the midbrain (see Fig. 7.12). Serotonin functions in mood, memory, sleep, and cognitive skills. Decreased serotonin levels are linked to affective disorders (discussed later). Drugs that increase serotonin transmission are effective antidepressant agents. Serotonin also regulates heart rate and breathing, and low levels of serotonin have been linked to **sudden infant death syndrome (SIDS)** or *crib death.* SIDS is marked by the sudden death of an infant that is unexpected by history and cannot be explained following a thorough forensic autopsy. An abnormality in serotonin signaling is not the only link; the frequency of SIDS appears to be a strong function of the infant's sex, age, and ethnicity as well as the education and socioeconomic status of the parents.

Limbic reward system

Experimental studies beginning early in the last century demonstrated that stimulating the limbic system or creating lesions in various parts of the limbic system can alter emotional states. Most of our knowledge comes from animal studies, but emotional feelings are reported by humans when limbic structures are stimulated during brain surgery. The brain has no pain sensation when touched, and subjects awakened from anesthesia during brain surgery have communicated changes in emotional experience linked to electrical stimulation of specific areas.

Electrical stimulation of various sites in the limbic system produces either pleasurable (rewarding) or unpleasant (aversive) feelings. To study these findings, researchers use electrodes implanted in the brains of animals. When electrodes are implanted in structures presumed to generate rewarding feelings and the animals are allowed to deliver current to the electrodes by pressing a bar, repeated and prolonged self-stimulation is seen. Other needs—such as food, water, and sleep—are neglected. The sites that provoke the highest rates of electrical self-stimulation are in the ventral limbic areas, including the septal nuclei and nucleus accumbens. Extensive studies of electrical self-stimulatory behavior indicate that dopaminergic neurons play a major role in mediating reward. The nucleus accumbens is thought to be the site of action of addictive drugs, including opiates, alcohol, nicotine, cocaine, and amphetamine.

Aggressive behavior

A **fight-or-flight response**, including the autonomic components (see Chapter 6, "Autonomic Nervous System") and postures of rage and aggression characteristic of fighting behavior, can be elicited by electrical stimulation of sites in the hypothalamus and amygdala. If the frontal cortical connections to the limbic system are severed, rage postures and aggressiveness become permanent, illustrating the importance of the higher centers in restraining aggression and, presumably, in invoking it at appropriate times. By contrast, bilateral removal of the amygdala results in a placid animal that cannot be provoked.

Sexual arousal

The biologic basis of human sexual activity is poorly understood because of its complexity and because findings derived from nonhuman animal studies cannot be extrapolated. The major reason for this limitation is that the cerebral cortex, uniquely developed in the human brain, plays a more important role in governing human sexual activity than the instinctive or olfactory-driven behaviors in nonhuman primates and lower mammalian species. Nevertheless, several parallels in human and nonhuman sexual activities exist, indicating that the limbic system, in general, coordinates sex drive and mating behavior, with higher centers exerting more or less overriding influences.

Copulation in mammals is coordinated by reflexes of the sacral spinal cord, including male penile erection and ejaculation reflexes and engorgement of female erectile tissues as well as the muscular spasms of the orgasmic response. Copulatory behaviors and postures can be elicited in animals by stimulating parts of the hypothalamus, olfactory system, and other limbic areas, resulting in mounting behavior in males and lordosis (arching the back and raising the tail) in females. Ablation studies have shown that sexual behavior also requires an intact connection of the limbic system with the frontal cortex.

Olfactory cues are important in initiating mating activity in seasonal breeders. Driven by the hypothalamus' endogenous seasonal clock, the anterior and preoptic areas of the hypothalamus initiate hormonal control of the gonads. Hormonal release leads to the secretion of odorants (pheromones) by the female reproductive tract, signaling the onset of estrus and sexual receptivity to the male. The odorant cues are powerful stimulants, acting at extremely low concentrations to initiate mating behavior in males. The olfactory system, by virtue of its direct connections with the limbic system, facilitates the coordination of behavioral, endocrine, and autonomic responses involved in mating.

Although human and nonhuman primates are not seasonal breeders (mating can occur on a continual basis), vestiges of this pattern remain. These include the importance of the olfactory and limbic systems and the role of the hypothalamus in cyclic changes in female ovarian function and the continuous regulation of male testicular function. More important determinants of human sexual activity are the higher cortical functions of learning and memory, which serve to either reinforce or suppress the signals that initiate sexual responding, including the sexual reflexes coordinated by the sacral spinal cord.

Limbic system malfunction leads to psychiatric disorders.

The major psychiatric disorders, including affective disorders and schizophrenia, are disabling diseases with a genetic predisposition and no known cure. The functional basis for these disorders remains obscure. In particular, the role that environmental influences play on individuals with a genetic predisposition to developing a disorder is also unclear. Altered states of the brain's monoaminergic systems have been a major focus as possible underlying factors, based on extensive human studies in which neurochemical imbalances in catecholamines, **acetylcholine (ACh)**, and serotonin have been observed. Another reason for focusing on the monoaminergic systems is that the most effective drugs used in treating psychiatric disorders are agents that alter monoaminergic transmission.

Affective disorders

Illnesses marked by abnormal mood regulation and associated signs and symptoms are called **affective disorders** or **mood disorders**. The word "mood" describes an overall emotional tone that provides the background to one's internal life. Thus, mood is hypothesized to be the underling cause of affective disorders. Mood disorders are broadly classified as **major depression**, which can be so profound as to provoke suicide, and **bipolar disorders** (formerly known as *manic-depressive disorder*), in which periods of profound depression are followed by periods of mania, in a cyclic pattern. Neurophysiologic studies indicate that depressed patients show decreased use of brain NE. In manic periods, NE transmission increases. Whether in depression or in mania, all patients seem to have decreased brain serotonergic transmission, suggesting that serotonin may exert an underlying permissive role in abnormal mood swings, in contrast with NE, whose transmission, in a sense, titrates the mood from highest to lowest extremes. The most effective treatments for depression are antidepressant drugs such as MAOIs and **selective serotonin reuptake inhibitors (SSRIs)**. With electroconvulsive therapy, they have in common the ability to stimulate both noradrenergic and serotonergic neurons serving the limbic system. A therapeutic response to these treatments ensues only after treatment is repeated over time. Similarly, when treatment stops, symptoms may not reappear for several weeks. This time lag in treatment response is presumably a result of alterations in the long-term regulation of receptor and second messenger systems in relevant regions of the brain. The potential to understand depression—and other complex mental illness—has dramatically increased with the advent of both molecular genetics and imaging technologies. An important finding was that a genetic variation in the serotonin reuptake transporter predicts susceptibility to depression subsequent to a number of stressors. Imaging techniques have shown that the amygdala is hyperreactive to emotionally provocative stimuli in individuals with the genetic variant of the serotonin transporter as compared with those who do not express the variant.

The need for new technologies to explain disease mechanisms is highlighted by the other side of the affective spectrum. The most effective long-term treatment for mania is lithium, although antipsychotic (neuroleptic) drugs, which block dopamine receptors, are effective in the acute treatment of mania. The therapeutic actions of lithium remain unknown, but the drug has an important action on a receptor-mediated second messenger system. Lithium interferes with regeneration of phosphatidylinositol in neuronal membranes by blocking the hydrolysis of inositol-1-phosphate. Depletion of phosphatidylinositol in the membrane renders it incapable of responding to receptors that use this second messenger system. Without knowing the specific neuroanatomic underpinnings of the disease, it is difficult to ascertain how these mechanisms might relate to amelioration of symptoms.

Seasonal affective disorder (SAD), also known as *winter depression* and *winter blues*, also falls under mood disorders. Some people have a seasonal pattern, with episodes of SAD coming on in the autumn and/or winter with the depression resolving in the spring. Diagnosis is made on the basis of at least two episodes that have occurred in one winter season with no other times over a 2-year period. The lack of sunlight, especially for individuals that live at higher latitudes, has been hypothesized as the cause of SAD. However, the amount of sunlight reaching the eyes does not appear to be the only underlying cause. SAD is more prevalent in people who are younger and typically affects females more than males.

Schizophrenia involves the uncoupling of thought and emotional processes.

Schizophrenia is the collective name for a group of psychotic disorders that vary greatly in symptoms among individuals. It most commonly manifests as **delusions** (fixed false beliefs) or **hallucinations** (perceptions in the absence of a stimulus) that are coupled to disorganized thinking and speech. Delusions are usually persecutory in nature and tend to follow themes of being harassed, cheated on, conspired against, drugged, or poisoned. Hallucination (most commonly hearing voices) is paranoid in nature and often tied to social dysfunctions. The onset of symptoms typically occurs in young adulthood with approximately 2% reporting lifetime prevalence. No neurologic tests for schizophrenia currently exist, and the diagnosis is based exclusively on the patient's self-reporting experiences and observed behavior.

Although the exact cause of schizophrenia is not known, increased dopamine activity in the mesolimbic pathway of the brain (see Fig. 7.12) is commonly found in patients with schizophrenia. The mainstay of treatment is antipsychotic medication that primarily works on suppressing dopamine activity. Current research is focused on finding the subtype of dopamine receptor that mediates mesocortical/mesolimbic dopaminergic transmission but does not affect the nigrostriatal system, which controls motor function (see Fig. 7.12). So far, neuroleptic drugs that block one pathway almost always block the other as well, leading to unwanted neurologic side effects, including abnormal involuntary movements after long-term treatment or Parkinsonism in the short term. Similarly, some patients with Parkinson disease who receive L-DOPA to augment dopaminergic transmission in the nigrostriatal pathway must be taken off the medication because they develop hallucinations.

Newer, so-called **atypical antipsychotics** ameliorate the symptoms of schizophrenia without targeting the dopamine receptors and thus imply that dopamine may not be the primary culprit in the disease. Evidence is accumulating that there may be a hypofunctioning of glutamate signaling through the *N*-methyl-D-aspartate (NMDA) receptor (see Chapter 3, "Action Potential, Synaptic Transmission, and Maintenance of Nerve Function"). The idea is that glutamate may normally activate pathways that inhibit dopaminergic pathways; and therefore, hypofunction of NMDA receptor signaling would increase dopamine signaling.

HIGHER COGNITIVE SKILLS

Cognitive science is the interdisciplinary study of how information (e.g., concerning perception, language, speech, reasoning, and emotion) is represented and transformed in the brain. Multiple research disciplines (neuroscience, psychology, artificial intelligence, and learning science) are brought together to investigate neural circuitry and modular brain organization in an effort to analyze information from low-level learning and decision-making mechanisms to high-level logic and planning.

Cerebral cortex and the limbic system provide the architectural components for learning and memory systems.

Learning involves synthesizing different types of information in order to acquire: (1) new knowledge, (2) behavior, (3) skills, (4) values, and (5) preferences. Memory allows the brain to store information for later retrieval and consists of both long-term (i.e., days, months, years) and short-term (i.e., minutes, hours) store. Memory and learning are inextricably linked because part of the learning process involves the assimilation of new information and its commitment to memory. The most likely sites of learning in the human brain are the large association areas of the cerebral cortex, in coordination with subcortical structures deep in the temporal lobe, including the hippocampus and amygdala. The association areas draw on sensory information received from the primary visual, auditory, somatic sensory, and olfactory cortices and on emotional feelings transmitted via the limbic system. This information is integrated with previously learned skills and stored memory, which presumably also reside in the association areas.

The learning process itself is poorly understood, but it can be studied experimentally at the synaptic level in isolated slices of mammalian brain or in more simple invertebrate nervous systems. Synapses subjected to repeated presynaptic neuronal stimulation show changes in the excitability of postsynaptic neurons. These changes include the facilitation of neuronal firing, altered patterns of neurotransmitter release, second messenger formation and, in intact organisms, evidence that learning occurred. The phenomenon of increased excitability and altered chemical state on repeated synaptic stimulation is known as **long-term potentiation**, a persistence beyond the cessation of electrical stimulation, as is expected of learning and memory. Ca^{2+} entry through activation of NMDA receptors is critical to the development of long-term potentiation. An early event in long-term potentiation is a series of protein phosphorylations induced by receptor-activated second messengers and leading to activation of a host of intracellular proteins and altered excitability. In addition to biochemical changes in synaptic efficacy associated with learning at the cellular level, structural alterations occur. The number of connections between sets of neurons increases as a result of experience.

Much of our knowledge about human memory formation and retrieval is based on studies of patients in whom stroke, brain injury, or surgery resulted in memory disorders. Such knowledge is then examined in more rigorous experiments in nonhuman primates capable of cognitive functions. From these combined approaches, we know that the prefrontal cortex is essential for coordinating the formation of memory, starting from a learning experience in the cerebral cortex, then processing the information and communicating it to the subcortical limbic structures. The prefrontal cortex receives sensory input from the parietal, occipital, and temporal lobes and emotional input from the limbic system. Drawing on skills such as language and mathematical ability, the prefrontal cortex integrates these inputs in light of previously acquired learning. The prefrontal cortex can thus be considered the site of **working memory**, where new experiences are processed, as opposed to sites that consolidate the memory and store it. The processed information is then transmitted to the hippocampus, where it is consolidated over several hours into a more permanent form that is stored in, and can be retrieved from, the association cortices.

Long-term and short-term memory comprise the brain's memory system.

Memory can be divided into that which can be recalled for only a brief period (seconds to minutes) and that which can be recalled for weeks to years. Newly acquired learning experiences can be readily recalled for only a few minutes or more using **short-term memory**. An example of short-term memory is looking up a telephone number, repeating it mentally until you finish dialing the number, then promptly forgetting it as you focus your attention on starting the conversation. Short-term memory is a product of working memory; the decision to process information further for permanent storage is based on judgment as to its importance or on whether it is associated with a significant event or emotional state. An active process involving the hippocampus must be employed to make a memory more permanent.

The conversion of short-term to **long-term memory** is facilitated by repetition, by adding more than one sensory modality to learn the new experience (e.g., writing down a newly acquired fact at the same time one hears it spoken) and, even more effective, by tying the experience (through the limbic system) to a strong, meaningful emotional context. The role of the hippocampus in consolidating the memory is reinforced by its participation in generating the emotional state with which the new experience is associated.

Declarative memory (sometimes referred to as *explicit memory*) is one of the two subtypes of long-term memory systems. Declarative memory refers to the memory system that consciously recalls facts and events (e.g., birthdays, state capitals, musical event, and personal events). The counterpart to declarative memory is known **nondeclarative** or **procedural memory** (sometimes referred to as *implicit memory*) and refers to unconscious memories of how to do things (e.g., from tying shoes to driving a car). These memories are automatically retrieved and occur without the need for conscious control.

Memory loss

Memory loss can be partial or total. Unfortunately, memory loss is considered normal when it comes with aging. Sudden memory loss can result from brain trauma, strokes, meningitis, and epilepsy. Trauma to the brain is not the only cause of sudden memory loss. It may appear as a side effect from chemotherapy in which cytotoxic drugs are used to treat cancer or from drugs that are prescribed for lowering blood cholesterol. Sudden memory loss can be permanent or temporary. When it is caused by other medical conditions, such as Alzheimer disease or Parkinson disease, memory loss is gradual and tends to be permanent.

Cholinergic innervations play an important role in memory, with ACh as the major neurotransmitter in cognitive function, learning, and memory. The basal forebrain region contains prominent populations of cholinergic neurons that project to the hippocampus and to all regions of the cerebral cortex (Fig. 7.14). These cholinergic neurons are known generically as **basal forebrain nuclei** and include the septal nuclei, the nucleus basalis, and the nucleus accumbens. Another major cholinergic projection derives from a region of the brainstem reticular formation known as the *pedunculopontine nucleus*, which projects to the thalamus, spinal cord, and other regions of the brainstem. Approximately 90% of brainstem inputs to all nuclei of the thalamus are cholinergic.

Cortical cholinergic connections are thought to control selective attention, a function congruent with the cholinergic brainstem projections through the ascending reticular activating system. Loss of cholinergic function is associated with **dementia**, an impairment of memory, memory loss, abstract thinking, and judgment. Other cholinergic neurons include motor neurons and autonomic preganglionic neurons as well as a major interneuronal pool in the striatum.

Language and speech are coordinated in specific areas within the association cortex.

The ability to communicate by language, verbally and in writing, is one of the most difficult cognitive functions to study because only humans are capable of these skills. Thus, our knowledge of language processing in the brain has

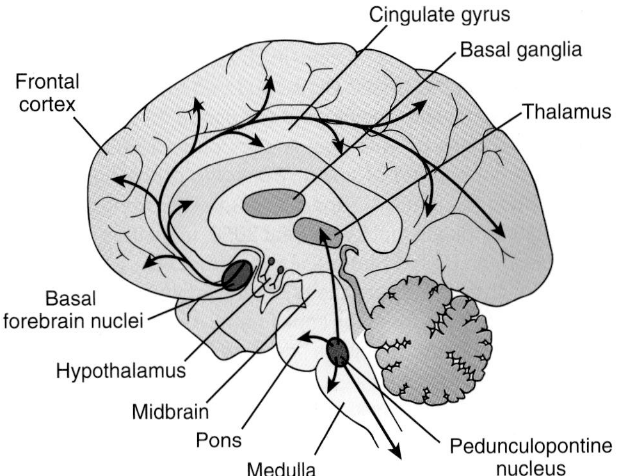

Figure 7.14 The origins and projections of major cholinergic neurons. Cholinergic neurons in the basal forebrain nuclei innervate all regions of the cerebral cortex. Cholinergic neurons in the brainstem's pedunculopontine nucleus provide a major input to the thalamus and also innervate the brainstem and spinal cord. Cholinergic interneurons are found in the basal ganglia. Not shown are peripherally projecting neurons, the somatic motor neurons, and autonomic preganglionic neurons, which also are cholinergic.

been inferred from clinical data by studying patients with **aphasias**—disturbances in producing or understanding the meaning of words—following brain injury, surgery, or other damage to the cerebral cortex.

Two areas appear to play an important role in language and speech: the **Wernicke area**, in the upper temporal lobe, and the **Broca area**, in the frontal lobe (Fig. 7.15). Both of these areas are located in the association cortex, adjacent to cortical areas that are essential in language communication. The Wernicke area is in the parietal–temporal–occipital association cortex, a major association area for processing sensory information from the somatic sensory, visual, and auditory cortices. The Broca area is in the prefrontal association cortex, adjacent to the portion of the motor cortex that regulates movement of the muscles of the mouth, tongue, and throat (i.e., the structures used in the mechanical production of speech). A fiber tract, the arcuate fasciculus, connects the Wernicke area with the Broca area to coordinate aspects of understanding and executing speech and language skills.

Clinical evidence indicates that the Wernicke area is essential for the comprehension, recognition, and construction of words and language, whereas the Broca area is essential for the mechanical production of speech. Patients with a defect in the Broca area show evidence of comprehending a spoken or written word but are not able to say the word. In contrast, patients with damage in the Wernicke area can produce speech, but the words they put together have little meaning.

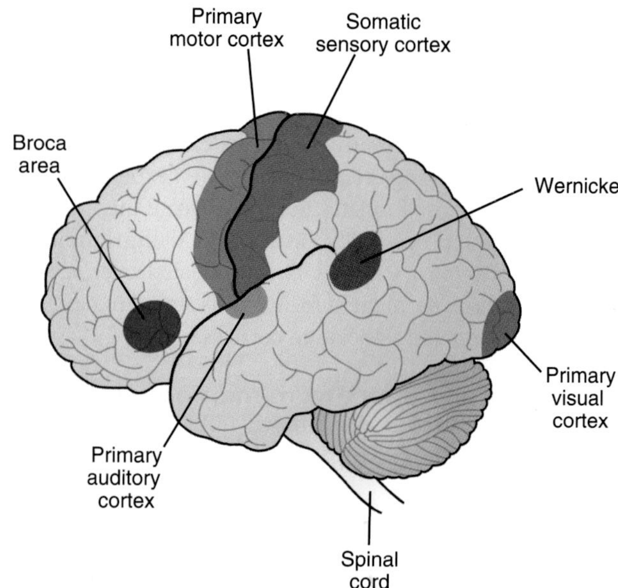

Figure 7.15 Wernicke and Broca areas and the primary motor, visual, auditory, and somatic sensory cortices.

Language is a highly lateralized function of the brain residing in the left hemisphere (see Clinical Focus Box 7.2). This dominance is observed in left-handed as well as right-handed people. Moreover, it is language that is lateralized, not the reception or production of speech. Thus, native signers (people who use sign language) who have been deaf since birth still show left-hemisphere language function.

Clinical Focus / 7.2

Alzheimer Disease

Alzheimer disease (AD) is the most common cause of dementia in older adults. The cause of the disease still is unknown, and there is no cure. In 2003, an estimated 4.5 million people in the United States suffered from AD. Although the disease usually begins after the age of 65 years and the risk of AD goes up with age, it is important to note that AD is not a normal part of aging. The aging of the baby boom generation has made AD one of the fastest growing diseases. Estimates indicate that by the year 2050, more than 13 million people in the United States will suffer from AD.

Cognitive deficits are the primary symptoms of AD. Early on, there is mild memory impairment. As the disease progresses, memory problems increase and difficulties with language are generally observed, including word-finding problems and decreased verbal fluency. Many patients also exhibit difficulty with visuospatial tasks. Personality changes are common, and patients become disoriented as the memory problems worsen. A progressive deterioration of function follows and, at late stages, the patient is bedridden, nearly mute, unresponsive, and incontinent. A definitive diagnosis of AD is not possible until autopsy, but the constellation of symptoms and disease progression allow a reasonably certain diagnosis.

Gross pathology consistent with AD is mild-to-severe cortical atrophy (depending on the age of onset and the age at death). Microscopic pathology indicates two classic signs of the disease even at the earliest stages: the presence of senile plaques (SPs) and neurofibrillary tangles (NFTs). As the disease progresses, synaptic and neuronal loss or atrophy and an increase in SPs and NFTs occur.

Although many neurotransmitter systems are implicated in AD, the most consistent pathology is the loss or atrophy of cholinergic neurons in the basal forebrain. Medications that ameliorate the cognitive symptoms of AD are cholinergic function enhancers. These observations emphasize the importance of cholinergic systems in cognitive function.

Chapter Summary

- The central nervous system receives, integrates, and acts on sensory stimuli from the internal and external environment.
- The hypothalamus is the interface between the endocrine, autonomic, and limbic systems.
- The hypothalamus regulates energy balance by integrating metabolism and eating behavior, it controls the gonads and sexual activity, and it contains the biological clock.
- The brainstem reticular formation is critical for arousal and sleep.
- The ascending reticular activating system mediates consciousness and arousal.
- An electroencephalogram records electrical activity at the brain's surface in waves that have characteristic patterns depending on the state of arousal.
- The forebrain contains the thalamus, hypothalamus, and the cerebrum and processes sensory information, executes behavior, makes decisions, and is capable of learning.
- The cerebral cortex comprises three parts: sensory, motor, and associated areas.
- The limbic system contains cortical and subcortical structures and is the seat of emotions, containing a reward system, mediating aggression, and coordinating sexual activity as well as being implicated in psychiatric disorders.
- Monoaminergic innervation of the forebrain derives from the brainstem.
- Affective disorders are marked by abnormal mood regulation.
- Schizophrenia involves disordered thought processes, delusions, and hallucinations.
- Memory and learning require the cerebral cortex and limbic system.
- Declarative and nondeclarative memory involve different central nervous system structures.
- Cholinergic innervation is associated with cognitive function.
- Language and speech are coordinated in specific areas of the association cortex.

thePoint *Visit http://thePoint.lww.com for chapter review Q&A, case studies, animations, and more!*

Skeletal and Smooth Muscle

ACTIVE LEARNING OBJECTIVES

Upon mastering the material in this chapter you should be able to:

- Explain why impulses from motor neurons are required for contraction of skeletal muscle and how these impulses are linked to the generation of a muscle contraction.
- For each specific step in excitation–contraction coupling explain how failure of that step prevents skeletal muscle contraction.
- Explain why skeletal muscle contraction is either "full on" or "full off."
- Explain how force generation in muscle is enhanced by rapid repeated stimulation of the muscle and describe the value of this mechanism in terms of specific mechanical tasks muscles perform.
- Explain how lack of oxygen to skeletal muscle locks it into a rigor state.
- Distinguish between isotonic and isometric conditions and contractions and give examples of each.
- Explain the molecular basis of why the production of isometric force depends on the length to which muscle is stretched prior to its activation.
- Show how the isometric length–tension curve sets the limit for all isotonic contractions in skeletal muscle.
- Contrast the important steps in the regulation of smooth versus skeletal muscle contraction by calcium.
- Explain why the contraction of a smooth muscle cell can be graded, whereas that of a skeletal muscle fiber cannot.
- Explain the mechanism by which smooth muscle reduces its energy requirements during long-duration, sustained contractions.

Muscles are organs composed of cells that possess the ability to contract and, thus, shorten. Muscles, therefore, have the ability to generate force and movement. Muscle is commonly classified into subtypes based on anatomical location and appearance. *Skeletal muscle* is associated with the bones, or skeleton, of the body. *Cardiac muscle* is found exclusively in the heart, where it is responsible for creating the force needed to move blood throughout the circulatory system. *Smooth muscle*, so named for its unremarkable or "smooth" appearance when viewed microscopically, plays a key role in the functions of the alimentary tract, the genitourinary system, blood vessels, bronchi, and the eye.

Anatomical classifications, however, are of little use for distinguishing the cellular physiology of these three structural subtypes of muscle. Although the ability of all muscle cells to contract and relax is rooted in the chemical and mechanical interaction of two basically similar intracellular proteins, the mechanisms initiating and regulating those protein interactions are markedly different among the three histologic muscle types. Furthermore, the mechanics of contraction of skeletal, cardiac, and smooth muscle differ in significant ways as well. This chapter will explain the cellular physiology and mechanics of skeletal and smooth muscle. The physiology and mechanics of cardiac muscle will be discussed in Chapters 12, "Electrical Activity of the Heart," and 13, "Cardiac Muscle Mechanics and the Cardiac Pump."

SKELETAL MUSCLE

Skeletal muscle is primarily associated with the bones of the skeleton. However, this type of muscle also makes up the muscles of the diaphragm and parts of the esophagus. Skeletal muscles move the eyes as well. Muscles of the skeleton are responsible for large and forceful movements of the body, such as those involved in walking, running, and lifting heavy objects. Large skeletal muscle groups, such as those in the lower back, buttocks, and thigh, are also responsible for stabilizing the body's position against gravity. This function allows us to sit and stand upright without teetering. Skeletal muscle, however, is not restricted to the makeup of large muscle groups and movements. This muscle type is also used for the fine, delicate movements of the hands, such as those used in writing, playing the violin, and manipulating tiny objects. Finally, a secondary function of skeletal muscle is the production of heat. Heat given off in shivering or during general exercise is an important source of heat generation in the body (see Chapters 28, "Regulation of Body Temperature," and 29, "Exercise Physiology").

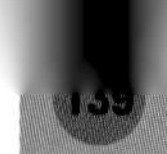

From Bench to Bedside / 8.1

New Pharmacotherapy for Muscle Wasting Disease

Contraction of a skeletal muscle is totally dependent upon impulses from motor neurons innervating muscle fibers. Disease or damage to any part of a motor nerve pathway in a motor unit results in flaccid paralysis of all muscle fibers in that motor unit. This is easy to predict based on basic physiology of skeletal muscle contraction. Skeletal muscle cannot generate its own action potential, which is required to mobilize its calcium stores for contraction. Instead, acetylcholine, released from motor nerve endings, is required to generate local depolarizations that are sufficient to start muscle excitation–contraction coupling. It is now accepted that motor neurons also influence tissue growth and maintenance in skeletal muscle. Motor neurons exert a trophic influence on skeletal muscle. A skeletal muscle group will atrophy significantly if its associated motor neurons are severed, damaged, or destroyed.

Spinobulbar muscular atrophy (also known as *Kennedy disease*) is a type of skeletal muscle disorder that affects predominantly males. It is an X-linked recessive inherited disease associated with a mutation in the androgen receptor (AR) gene. The specific gene defect is an expanded CAG triplet repeat, which results in expression of an altered structure of the receptor protein normally coded for by the AR gene. The male sex steroids, testosterone and dihydrotestosterone (DHT), bind to this abnormal receptor on motor neurons in the brain and spinal cord, initiating events that result in motor neuron damage. Without properly functioning motor neurons, the skeletal muscles innervated by the neurons begin to weaken and atrophy. Motor neuron damage and loss of muscle mass in the arms and legs result in twitching, tremors, muscle cramps, and generalized muscle weakness. Difficulty walking and a tendency to fall characterize the effects of this disease. Muscles in the face and throat (bulbar muscles) are also affected, resulting in difficulty in swallowing and slurred speech.

Currently, there is no cure for spinobulbar muscular atrophy. Treatment for this disease has been largely relegated to supportive physical therapy and rehabilitation. However, a recently completed phase II clinical trial involving the drug dutasteride has provided evidence in favor of a possible effective pharmacologic therapy for this disorder. Dutasteride reduces the production of DHT in the body and is currently recognized as a drug used for the treatment of prostatic hyperplasia (an enlarged prostate). The rationale for using this drug for the treatment of spinobulbar muscular atrophy is that reduced DHT levels in the body could reduce activation of the defective AR responsible for the motor neuron damage that leads to muscle weakness and wasting in this condition.

In a recent limited study, men with spinobulbar muscular atrophy were treated with dutasteride daily over a 24-month trial period and compared against age-matched men with this condition who received only a placebo control. Early results from this small study have indicated that dutasteride slows the progression of muscle weakness and reduces plasma creatine levels (used to indicate muscle damage) in patients with spinobulbar muscular atrophy. There is also some evidence in support of a reduction in the progression of swallowing difficulty associated with spinobulbar muscular atrophy in patients treated with dutasteride as well as a modest improvement in motor neuron function in the drug treated group. Overall, patients receiving dutasteride reported noticeable improvement in their physical quality of life index compared to patients receiving only a placebo. These results should warrant further investigation and indicate that drug treatment for a previously incurable muscle disease may be forthcoming in the near future.

Protein filaments provide the architecture of skeletal muscle and form the molecular basis of its contractile machinery.

All skeletal muscles have a striated (striped) appearance when viewed with a light microscope or an electron microscope. This appearance comes from the highly organized arrangement of protein filaments within each muscle cell. These proteins, their particular cellular arrangement, and other associated proteins play key roles in the ability to initiate and regulate muscle contraction at the cellular level. The proteins that make up the structure of the contractile mechanism also participate in the chemical processes that are responsible for converting chemical energy into mechanical energy and, thus, make the mechanical function of muscle possible.

A whole skeletal muscle is composed of numerous muscle cells, also called **muscle fibers**. A cell can be up to 100 μm in diameter and many centimeters long, especially in larger muscles. The fibers are multinucleate, and the nuclei occupy positions near the periphery of the fiber. Skeletal muscle has an abundant supply of **mitochondria**, which are vital for supplying chemical energy in the form of **adenosine triphosphate (ATP)** to the contractile system. The process of muscle contraction requires an elaborate cellular structure that couples biochemical reactions to physical movements. This involves a cyclic process that can be either activated (to contract the muscle) or inactivated (to allow the muscle to relax).

Thick filaments

The contractile proteins inside a muscle cell are actually polymers that are described generically as *thick* and *thin filaments.* The fundamental unit of a thick filament is **myosin** (molecular weight, ~500,000), a complex protein molecule with several distinct regions (Fig. 8.1A). Most of the length of the molecule consists of a long, straight portion (the "tail" portion) composed of light meromyosin. The remainder of the molecule, heavy meromyosin (HMM), consists of a protein chain that terminates in a globular head portion, called the *S1 region* (or *subfragment 1*). This region is responsible for the enzymatic and chemical activity that results in muscle contraction. It contains a binding site, by which it can interact with another protein filament with the muscle cell called

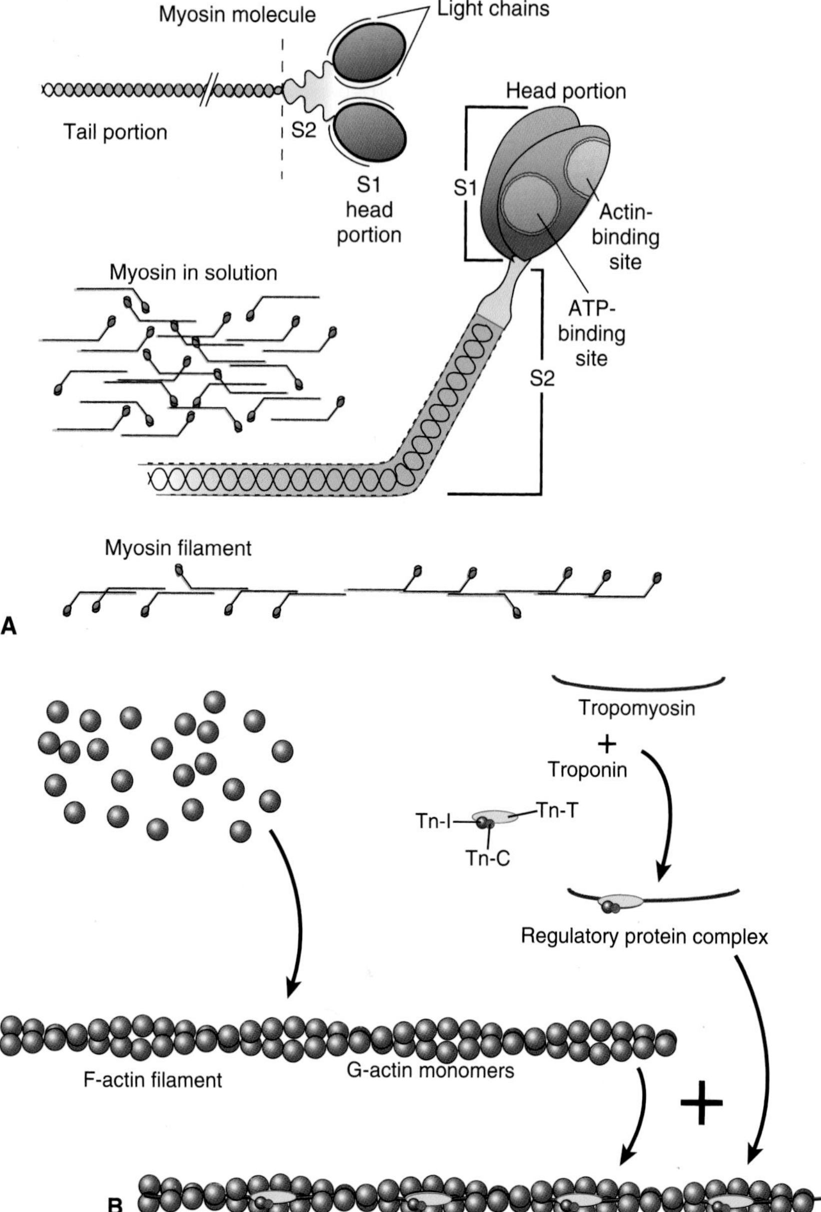

Figure 8.1 Myofilament components of skeletal muscle fibers. **(A)** The myosin molecular structure and its assembly into thick filaments. The S2 region of the molecule is flexible, whereas the tail region is stiff. **(B)** The assembly of the thin (actin) filaments of skeletal muscle from actin monomers, troponin, and tropomyosin. ATP, adenosine triphosphate; Tn-C, troponin-C; Tn-T, troponin-T; Tn-I, troponin-I; G-actin, globular actin; F-actin, filamentous actin.

actin (see below). The head portion of myosin also contains an ATP-binding site that is involved in the supply of energy for the actual process of contraction. The chain portion of HMM, the *S2 region* (or *subfragment 2*), serves as a flexible link between the head and tail regions.

Associated with the S1 region are two loosely attached peptide chains ("light chains") of a much lower molecular weight. One light chain, called the **essential light chain**, is necessary for myosin stability, and the other chain, called the **regulatory light chain**, is phosphorylated during muscle activity and serves to modulate muscle function. Functional myosin molecules are paired; their tail and S2 regions are wound about each other along their lengths, and the two heads, each bearing its two light chains and its own ATP-binding and actin-binding sites, lie adjacent to each other.

The assembly of individual myosin dimers into thick filaments involves close packing of the myosin molecules such that their tail regions form the "backbone" of the thick filament, with the head regions extending outward in a helical fashion. The effect is like that of a bundle of golf clubs bound tightly by the handles, with the heads projecting from the bundle. The myosin molecules are packed so that they are tail-to-tail in the center of the thick filament and extend outward from the center in both directions, creating a bare zone (i.e., no heads protruding) in the middle of the filament (see Fig. 8.1A).

Thin filaments

The other major member of contractile filaments in muscle cells is the thin filament. The backbone of each thin filament (also called an *actin filament*) consists of two strands of macromolecular subunits entwined about each other (see Fig. 8.1B). The strands are composed of repeating subunits (monomers) of the globular protein G-actin (molecular weight, 41,700). These slightly ellipsoid molecules are joined front to back into long chains that wind about each other, forming a helical structure that undergoes a half turn every seven G-actin monomers—this chain is called *F-actin* (or *filamentous actin*). In the groove formed down the length of the helix, there is an end-to-end series of fibrous protein molecules (molecular weight, 50,000) called **tropomyosin**. Each tropomyosin molecule extends a distance of seven G-actin monomers along the F-actin groove. Near one end of each tropomyosin molecule is a protein complex called **troponin**, composed of three attached subunits: troponin-C (Tn-C), troponin-T (Tn-T), and troponin-I (Tn-I). Tropomyosin and troponin are involved in regulating the interaction of actin with myosin heads, which is responsible for muscle contraction and relaxation as described later in this chapter.

Structural muscle filaments

In addition to the proteins directly involved in the process of contraction, there are several other important structural proteins within a muscle fiber. **Titin**, a large filamentous protein, extends from the Z lines to the bare portion of the myosin filaments. Titin may help prevent overextension of the sarcomeres and maintain the central location of the A bands. **Nebulin**, a filamentous protein that extends along the thin filaments, may play a role in stabilizing thin filament length during muscle development. The protein **α actinin**, associated with the Z lines, serves to anchor the thin filaments to the structure of the Z line. **Dystrophin**, which lies just inside the sarcolemma, participates in the transfer of force from the contractile system to the outside of the cells via membrane-spanning proteins called **integrins**. External to the cells, the protein **laminin** forms a link between integrins and the extracellular matrix. These proteins are disrupted in the group of genetic diseases collectively called **muscular dystrophy**.

Interdigitating regions of thick and thin filaments form the fundamental contractile unit of a muscle fiber called the sarcomere.

A single muscle fiber (cell) is divided lengthwise into several hundred to several thousand parallel **myofibrils**. Each myofibril has alternating light and dark bands that repeat at regular intervals, giving the fiber as a whole a striated (striped) appearance. This appearance results from the way thick and thin filaments interdigitate within a myofibril (Fig. 8.2A,B). The dark band of a myofibril is called an **A band**. It is divided at its center by a narrow, lighter-colored region called an **H zone**. In many skeletal muscles, a prominent **M line** is found at the center of the H zone. Low-density **I bands** lie between the A bands. Crossing the center of the I band is a dark structure called a **Z line** (sometimes termed a *Z disk* to emphasize its three-dimensional nature). The filaments of the I band attach to the Z line and extend in both directions into the adjacent A bands. Closer examination shows the A and I bands to be composed of the two kinds of myofilaments. The I band contains only thin (actin)

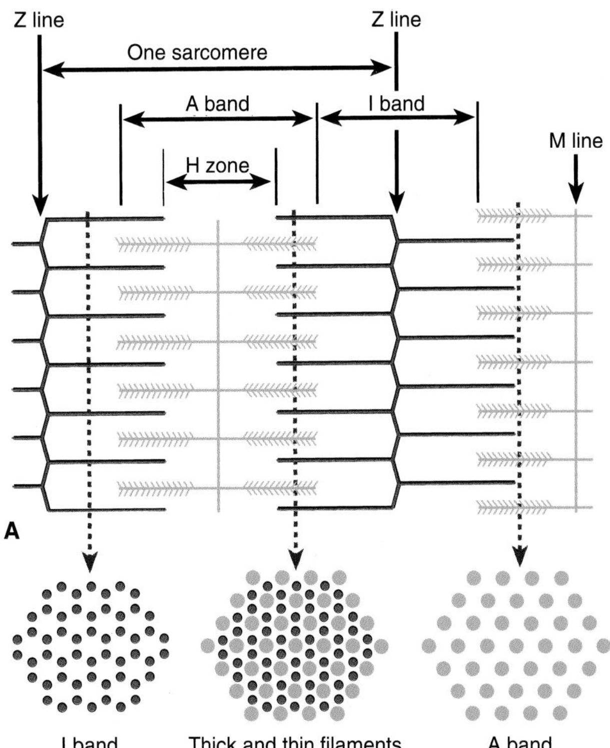

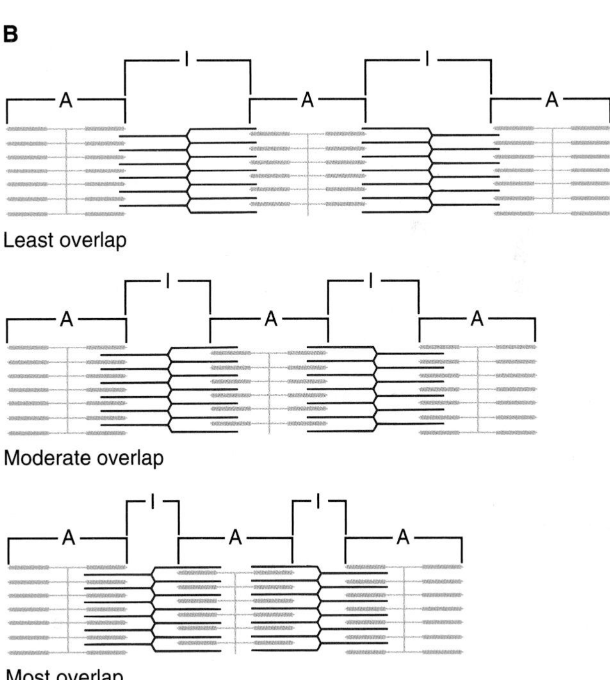

Figure 8.2 Molecular organization of the sarcomere. **(A)** The arrangement of the elements in a sarcomere viewed longitudinally. **(B)** Cross sections through selected regions of the sarcomere, showing the overlap of myofilaments at different parts of the sarcomere. The multiplying effect of sarcomeres placed in series. The overall shortening is the sum of the shortening of the individual sarcomeres.

Clinical Focus / 8.1

Muscular Dystrophy

Muscular dystrophies are a collection of inherited genetic disorders affecting males that result in muscle wasting, weakness, and eventual death from respiratory insufficiency and decompensated cardiac failure. **Duchenne muscular dystrophy (DMD) and Becker muscular dystrophy (BMD)** are the two most prevalent forms of many types of muscular dystrophies. Both are X-linked disorders with the female carriers showing essentially no clinical symptoms of the disease. DMD is the most severe form of this disease. Affected males begin to show symptoms of muscle wasting and weakness in early childhood that can become totally debilitating before the onset of puberty. Muscle weakness first appears in the pelvic girdle and spreads to the shoulders. Death often occurs before the affected individual reaches 30 years of age.

The primary defect in DMD resides in abnormalities in the gene that codes for the skeletal muscle mechanical linking protein, dystrophin. Dystrophin molecules concentrate in the vicinity of the **Z bands** in the sarcomere. These molecules are believed to interact or connect actin to the sarcolemmal membrane and, thus, form a mechanical link between the contractile apparatus of the cell and the cell itself. Patients with DMD have little or no dystrophin within their skeletal muscle fibers. Similar deficiencies of or abnormal dystrophin levels are also present in patients with BMD.

Current research is being directed at gene therapy for the treatment of DMD. However, immunologic reactions to gene products from this therapy have diminished the success of this approach in experimental animals to date.

filaments, whereas the A bands contain the thick (myosin) filaments and the overlapping projections of the thin filaments from the I band.

The fundamental repeating unit of the myofibril bands is called a **sarcomere** and is defined as the space between (and including) two successive Z lines (see Fig. 8.2B).

Contraction of skeletal muscle fibers results from controlled interactions between myosin and actin.

The process of contraction of a muscle cell involves a cyclic interaction between the thick and thin filaments in a sequence of events called the **crossbridge cycle**. This cycle is shown in Figure 8.3. Details of the mechanism of this cycle, its ability to produce force and movement, and its regulation are presented later in this chapter. To describe it briefly here, this cycle involves attachment of thick-filament myosin heads to binding sites on actin monomers (forming the **crossbridge**), sarcomere shortening, detachment, and then reattachment of myosin heads along the thin filaments. This cycling of crossbridges results both in force generation and movement through the whole muscle fiber.

Crossbridge cycling does not occur randomly on its own. Actin and myosin are normally prevented from attaching and forming cycling crossbridges by the presence of allosteric blocking, regulatory proteins associated with F-actin. This is the natural state of resting skeletal muscle, which is functionally flaccid because actin and myosin are not attached to one another. In simple terms, the crossbridge cycle must be initiated by signals to the muscle cell that release the inhibitory effect of the regulatory proteins on F-actin. This initiation starts first by the formation of an action potential in the muscle cell. This action potential causes the release of calcium from the sarcoplasmic reticulum (SR) in the muscle cell, causing the intracellular calcium concentration to rise. This rise in calcium removes the inhibitory action of regulatory proteins on F-actin and, thus, allows myosin to bind to actin, initiating the crossbridge cycle. The link between the muscle fiber action potential and the subsequent contraction of a muscle fiber is called **excitation–contraction coupling**. The details of this process and the resulting crossbridge cycle are described later in this chapter.

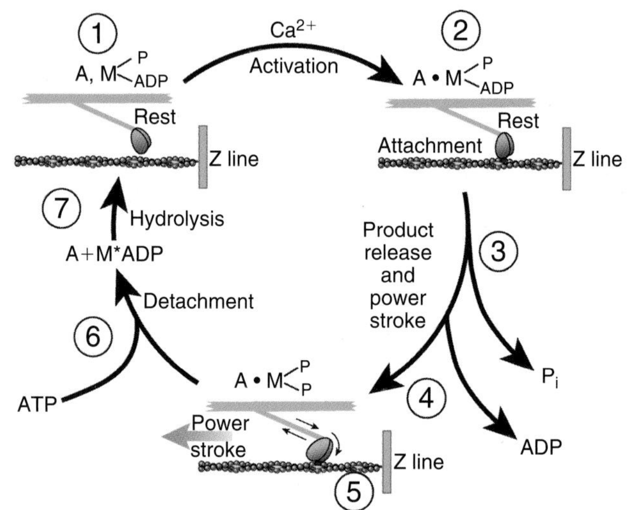

Figure 8.3 Events of the crossbridge cycle in skeletal muscle. (1) At rest, adenosine diphosphate (ADP) and inorganic phosphate are bound to the myosin head, which is in position to interact with actin. The interaction, however, is blocked allosterically by tropomyosin. (2) Inhibition of actin–myosin interaction is removed by binding of calcium to troponin-C; the myosin head binds to actin. (3) The release of ADP and phosphate changes the conformation of the myosin head from 90° to 45°, stretching the myosin S2 region. (4) Recoil of the S2 region creates the power stroke. (5) The rotated and still-attached crossbridge is now in the rigor state. (6) Detachment is possible when a new adenosine triphosphate (ATP) molecule binds to the myosin head and is subsequently hydrolyzed. (7) Energy from ATP hydrolysis resets the myosin head from a 45° conformation back to its original 90° conformation, thereby returning the myosin and actin positions to their original resting state. These cyclic reactions can continue as long as the ATP supply remains and activation via Ca^{2+} maintained. (See text for further details.) A, actin; M, myosin; P_i, inorganic phosphate ion; –, chemical bond.

MOTOR NEURONS AND EXCITATION–CONTRACTION COUPLING IN SKELETAL MUSCLE

Skeletal muscle is considered an "excitable" tissue, which means that its membrane potential can be altered in a manner that results in activation of that cell's function (in this case, contraction). Depolarization of the skeletal muscle cell membrane of sufficient magnitude can lead to the generation of muscle cell action potentials, which, in turn, leads to an increase in intracellular Ca^{2+} concentration, activation of the crossbridge cycle, and contraction of the cell. *Skeletal muscle cells, however, cannot generate their own action potentials.* Instead they must receive a depolarizing chemical signal released from motor neurons that innervate the cell. In other words, all skeletal muscles are externally controlled; these cannot contract without a signal from the somatic nervous system. This system is analogous to the light in a table lamp. The light cannot be turned on unless an electric cord connects the lamp to a source of electricity. Similarly, skeletal muscle cells cannot contract if the motor neurons innervating them are severed or neurochemical transmission to them is blocked. In such situations, the muscle group innervated by the motor neuron is flaccidly paralyzed.

Acetylcholine released from motor neurons initiates action potential formation in muscle fibers.

The contraction of skeletal muscle occurs in response to action potentials that travel down somatic motor nerves originating in the central nervous system (CNS). Upon reaching a muscle cell, the axon of a motor neuron typically branches into several terminals, each of which constitutes the *presynaptic* portion of a special synapse called the **neuromuscular junction (NMJ)**, also called the **myoneural junction** or **motor endplate** (Fig. 8.4). The terminals, which are all covered by a Schwann cell, lie in grooves or "gullies" in the surface of the muscle cell membrane. The transfer of the signal from nerve to muscle takes place across the NMJ. This type of synapse has a close association between the membranes of the nerve and muscle. It possesses many of the features of a nerve–nerve synapse in the CNS. It is somewhat simpler, however, because it is only excitatory and accepts no feedback from the postsynaptic cell.

Electrochemical events at the neuromuscular junction link synaptic transmission to muscle activity.

Within the axoplasm of the motor nerve terminals are located numerous membrane-enclosed vesicles containing **acetylcholine (ACh)**. The postsynaptic portion of the NMJ or **endplate membrane** is formed into postjunctional folds that contain **nicotinic ACh receptors** (see Chapter 3, "Action Potential, Synaptic Transmission, and Maintenance of Nerve Function").

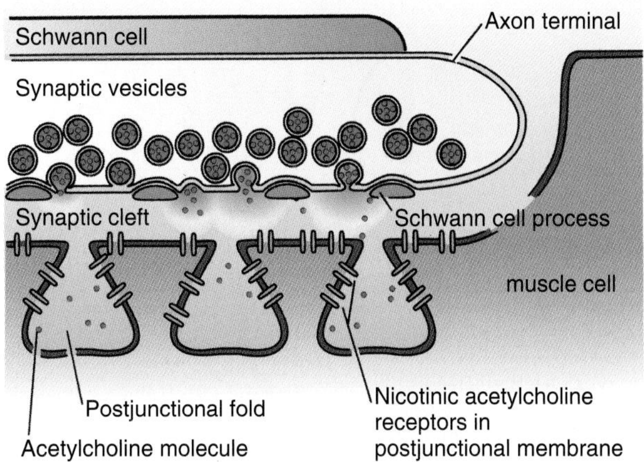

Figure 8.4 Structural features of the neuromuscular junction. Processes of the Schwann cell that overlies the axon terminal wrap around under it and divide the junctional area into active zones. Acetylcholine released from membrane vesicles is released from the axon terminal and diffuses to the underlying membrane where it binds with nicotinic receptors concentrated in invaginations of the sarcolemma.

When action potentials reach the terminal of a motor axon, the resulting depolarization of the terminal causes membrane channels to open, allowing external calcium ions to enter the axon. This causes the axoplasmic vesicles of ACh to migrate to the inner surface of the axon membrane, where they fuse with the membrane and release their contents. The vesicles are all approximately the same size and all release about the same amount, or **quantum**, of neurotransmitter. ACh released from the motor neuron terminal then diffuses across the NMJ and binds to the ACh receptors in the motor endplate. When two ACh molecules are bound to a receptor, it undergoes a configurational change, or "opening," that allows the relatively free passage of sodium and potassium ions down their respective electrochemical gradients. The opening of the channels depends only on the presence of neurotransmitter and not on membrane voltage. Sodium and potassium permeability changes occur simultaneously, and both ions share the same membrane channels. When the channel is open, sodium enters the muscle cell, while potassium simultaneously leaves.

The net result of these altered permeabilities is an inward positive current, known as the **endplate current**, that depolarizes the postsynaptic membrane. This positive voltage change is called the **endplate potential** (Fig. 8.5).

Endplate depolarization is *graded*, that is, its amplitude will vary with the number of receptors bound with ACh. The number of receptors bound is proportional to the local concentration of ACh in the NMJ, which, in turn, is a function both of the ACh released by the motor neuron and the amount of ACh degraded by **acetylcholinesterase (AChE)** in the NMJ. ACh binds loosely to its receptor. It can detach and diffuse away and become hydrolyzed into choline and acetate by AChE. This enzymatic degradation of ACh terminates its function as a transmitter molecule. At low action potential frequencies of the motor neuron, the amount

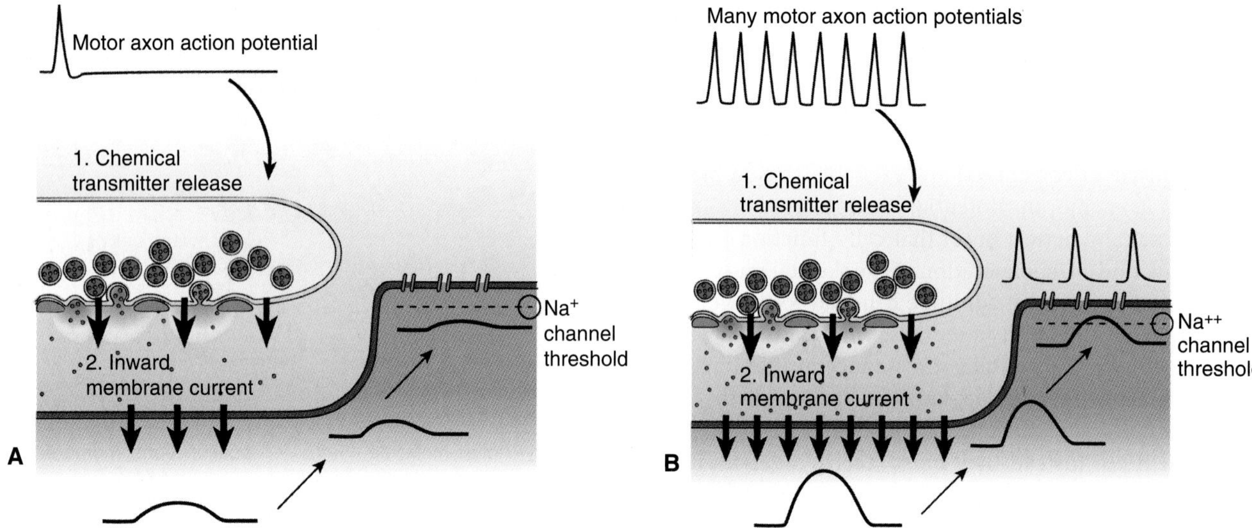

Figure 8.5 Electrical activity at the neuromuscular junction. (A) An action potential from a motor neuron releases a small amount of acetylcholine into the myoneural junction. This causes a small depolarization in the motor end plate that is transmitted electrotonically through the myoplasm. Because electrotonic conduction proceeds with decrement, the amount of depolarization reaching the first available voltage-gated sodium channel in the cell membrane is below threshold for opening the channel, and no muscle action potential is formed. **(B)** With greater frequency of action potentials reaching the motor axon terminal, acetylcholine reaches a higher concentration in the myoneural junction. This creates a larger endplate potential than that seen with lesser motor neuron activation. Although this potential proceeds down the muscle fiber with decrement, the depolarization reaching the first sodium channels is greater than its threshold, and the muscle fiber is able to fire an action potential. Action potentials are self-reinforcing and travel down the muscle fiber without decrement in a manner similar to that seen in unmyelinated nerve fibers.

of ACh entering the synaptic cleft is low (see Fig. 8.5A). Combined with the fact that AChE hydrolyzes a portion of this ACh, the concentration of ACh receptors occupied will be very low, and the amplitude of the resulting endplate depolarization will, therefore, be small. Endplate potentials are simple local depolarizations that create a current that flows outward through the muscle cytoplasm (myoplasm), across the adjacent muscle membrane, and back through the extracellular fluid. As such, an endplate potential is transmitted outward from underneath the motor endplate with decrement.

The motor endplate does not contain voltage-gated sodium channels and cannot produce self-propagating action potentials. However, regions of the sarcolemma away from the motor endplates do contain voltage-gated sodium channels. If these channels can be brought to threshold, the muscle fiber can form self-propagating action potentials. This is not likely to happen with generation of small endplate potentials, because, as these potentials decremate, the voltage finally reaching sarcolemmal sodium channels may be insufficient to bring the sodium channel to its threshold for firing an action potential (see Fig. 8.5A). In contrast, large amounts of ACh released into the NMJ will cause a proportionally larger endplate depolarization (see Fig. 8.5B). Although this endplate potential will also spread out across the muscle membrane with decrement, enough depolarization will reach the region of voltage-gated sodium channels in the muscle membrane to cause them to open and fire an action potential. Once formed, the muscle action potential will then be propagated along the muscle cell membrane by regenerative local currents in a fashion similar to the process of conduction in a nonmyelinated nerve fiber.

Neuromuscular transmission can be altered by toxins, drugs, and trauma.

Under normal circumstances, the endplate potential is much more than sufficient to produce a muscle action potential; this reserve, referred to as a **safety factor**, can help preserve function under abnormal conditions. However, neuromuscular transmission is subject to interference at several steps. Presynaptic blockade can occur if calcium does not enter the presynaptic terminal. Inhibition of choline uptake by the presynaptic terminal caused by the drug hemicholinium results in the depletion of ACh in the neuron. **Botulinum toxin** reduces ACh release. This bacterial toxin is used to treat focal dystonias and, more recently, for the cosmetic treatment of facial wrinkles (see Clinical Focus 8.2).

Postsynaptic blockade can result from drugs that partially mimic the action of ACh. Derivatives of **curare**, originally used as arrow poison in South America, bind tightly to ACh receptors. This binding does not result in opening of the ion channels, however, and the endplate potential is reduced in proportion to the number of receptors occupied by curare, resulting in muscle paralysis. Although the muscle can be directly stimulated electrically, nerve stimulation is ineffective. The drug **succinylcholine** blocks the NMJ in a

Clinical Focus / 8.2

Focal Dystonias and Botulinum Toxin

Focal dystonias are neuromuscular disorders characterized by involuntary and repetitive or sustained skeletal muscle contractions that cause twisting, turning, or squeezing movements in a body part. Abnormal postures and considerable pain, as well as physical impairment, often result. Usually the abnormal contraction is limited to a small and specific region of muscles, hence, the term **focal** (by itself). *Dystonia* means "faulty contraction." *Spasmodic torticollis* and *cervical dystonia* (involving neck and shoulder muscles), *blepharospasm* (eyelid muscles), *strabismus* and **nystagmus** (extraocular muscles), *spasmodic dysphonia* (vocal muscles), *hemifacial spasm* (facial muscles), and *writer's cramp* (finger muscles in the forearm) are common dystonias. Such problems are neurologic, not psychiatric, in origin, and sufferers can have severe impairment of daily social and occupational activities.

The specific cause is located somewhere in the central nervous system, but usually its exact nature is unknown. A genetic predisposition to the disorder may exist in some cases. Centrally acting drugs are of limited effectiveness, and surgical denervation, which carries a significant risk of permanent and irreversible paralysis, may provide only temporary relief.

Botulinum toxin, which produces chemical denervation of motor neurons resulting in flaccid muscle paralysis, has shown potential benefit in the treatment of dystonias. Botulinum toxin is produced from the bacterium *Clostridium botulinum* and is one of the most potent natural toxins known; a lethal dose for a human adult is about 2 to 3 μg. Type A toxin, the complex form most often used therapeutically, is sold under the trade names Botox and Oculinum.

The toxin first binds to the cell membrane of presynaptic nerve terminals in skeletal muscles. The initial binding does not appear to produce flaccid paralysis until the toxin is actively transported into the cell, a process requiring more than an hour. Once inside the cell, the toxin disrupts calcium-mediated acetylcholine release, producing an irreversible transmission block at the neuromuscular junction. The nerve terminals begin to degenerate, and the denervated muscle fibers atrophy. Eventually, new nerve terminals sprout from the axons of affected nerves and make new synaptic contact with the chemically denervated muscle fibers. During the period of denervation, which may be several months, the patient usually experiences considerable relief of symptoms. The relief is temporary, however, and the treatment must be repeated when reinnervation has occurred.

Clinically, highly diluted toxin is injected into the individual muscles involved in the dystonia. Often this is done in conjunction with electrical measurements of muscle activity (electromyography) to pinpoint the muscles involved. Patients typically begin to experience relief in a few days to a week. Depending on the specific disorder, relief may be dramatic and may last for several months or more. The abnormal contractions and associated pain are greatly reduced, speech can become clear again, eyes reopen and cease uncontrolled movements and, often, normal activities can be resumed.

The principal adverse effect is a temporary weakness of the injected muscles. However, this property has been exploited for the cosmetic use of Botox for releasing muscle tension in the face and forehead, which then reduces the formation of wrinkles in the skin. Studies have shown that the toxin's activity is confined to the injected muscles, with no toxic effects noted elsewhere. Long-term effects of the treatment, if any, are unknown.

slightly different way; this molecule binds to the receptors and causes the channels to open. Because it is hydrolyzed slowly by AChE, its action is long lasting and the channels remain open, which depolarizes the muscle. This prevents resetting of the inactivation gates of voltage-gated sodium channels near the endplate region and blocks subsequent action potentials. Drugs that produce extremely long-lasting endplate potentials are referred to as **depolarizing blockers**. Compounds such as **physostigmine** (**eserine**) are potent inhibitors of AChE and produce a depolarizing blockade. In carefully controlled doses, they can temporarily alleviate symptoms of **myasthenia gravis**, an autoimmune condition that results in a loss of postsynaptic ACh receptors. The principal symptom is muscular weakness caused by endplate potentials of insufficient amplitude. Partial inhibition of the enzymatic degradation of ACh allows it to remain in higher concentration in the NMJ and, thus, to compensate for the loss of receptor molecules.

Muscle action potentials release calcium from the sarcoplasmic reticulum to activate the crossbridge cycle.

The outer surface of a skeletal muscle fiber is surrounded by an electrically excitable cell membrane supported by an external meshwork of fine fibrous material. Together these layers form the cell's surface coat, the **sarcolemma.** As noted above the sarcolemma generates and conducts action potentials much like those of nerve cells. Contained wholly within a skeletal muscle cell is another set of membranes called the **sarcoplasmic reticulum (SR)**, which is a specialization of the endoplasmic reticulum (Fig. 8.6). The SR is specially adapted for the uptake, storage, and release of calcium ions, which are critical in controlling the processes of contraction and relaxation. Within each sarcomere, the SR consists of two distinct portions. The longitudinal element forms a system of hollow sheets and tubes that are closely associated with the myofibrils. The ends of the longitudinal elements terminate in a system of **terminal cisternae** (or *lateral sacs*). These contain a protein, calsequestrin, that weakly binds calcium, and most of the stored calcium is located in this region.

Closely associated with both the terminal cisternae and the sarcolemma are the **transverse tubules** (**T tubules**), inward extensions of the cell membrane whose interior is continuous with the extracellular space. Although they traverse the muscle fiber, T tubules do not open into its interior. The association of a T tubule and the two terminal cisternae at its sides is called a **triad**, a structure important in excitation–contraction coupling.

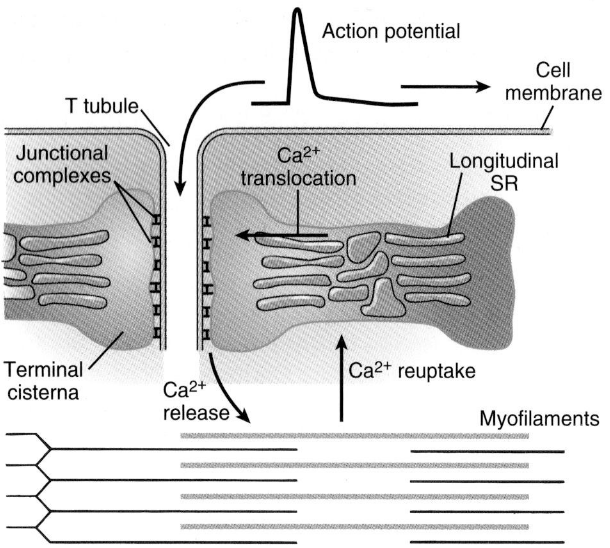

Figure 8.6 Release of activator calcium from the SR by a muscle action potential. Excitation–contraction coupling and the cyclic movement of calcium is shown. (See text for details of the process.) T tubule, transverse tubule; SR, sarcoplasmic reticulum.

The large diameter of skeletal muscle cells places interior myofilaments out of range of the immediate influence of events at the cell surface, but the T tubules, SR, and their associated structures act as a specialized internal communication system that allows the surface action potential signal to penetrate to interior parts of the cell. This process begins in skeletal muscle with the electrical excitation of the surface membrane (see Fig. 8.6). An action potential sweeps rapidly down the length of the fiber. The action potential propagates down the T-tubule membrane. This propagation results in numerous action potentials, one in each T tubule, traveling toward the center of the fiber.

At some point along the T tubule, the action potential reaches the region of a triad where the presence of the action potential is communicated to the terminal cisternae of the SR. Here the T-tubule action potential affects specific protein molecules called **dihydropyridine receptors (DHPRs)**. These molecules, which are embedded in the T-tubule membrane in clusters of four, serve as voltage sensors that respond to the T-tubule action potential. They are located in the region of the triad where the T-tubule and SR membranes are the closest together, and each group of four is located in close proximity to a specific channel protein called a **ryanodine receptor** (**RyR**), which is embedded in the SR membrane. The RyR serves as a controllable channel through which calcium ions can move readily when it is in the open state. These are therefore sometimes termed **calcium release channels.** DHPR and RyR form a functional unit called a **junctional complex** (see Fig. 8.6).

When the muscle is at rest, the RyR is closed; when T-tubule depolarization reaches the DHPR, it causes the RyR to open and release calcium from the SR. This leads to rapid release of calcium ions from the terminal cisternae into the intracellular space surrounding the myofilaments.

Calcium released from the sarcoplasmic reticulum removes inhibition of the crossbridge cycle in resting muscle.

The rapid release of calcium from the SR upon stimulation by action potentials entering the triad serves as the trigger for activating the skeletal muscle crossbridge cycle. Although resting skeletal muscle is not contracted, the chemical processes of its crossbridge cycle are in a state of constant readiness. Contraction due to crossbridge cycling does not occur in resting skeletal muscle because the troponin–tropomyosin complex of the thin myofilaments physically obstructs the actin-binding site, thereby preventing the natural interaction between actin and myosin (Fig. 8.7).

The calcium ion concentration in the region of the myofilaments is low in resting, or nonactivated, skeletal muscle, and the muscle is thus relaxed. When calcium ion concentrations increase, the ions bind to the Tn-C subunit associated with each tropomyosin molecule. Through the action of Tn-I and Tn-T, calcium binding causes the tropomyosin molecule to change its position slightly, uncovering the myosin-binding sites on the actin filaments. The myosin is, thus, allowed to interact with actin, and the events of the crossbridge cycle take place. As long as calcium ions are bound to the Tn-C subunit, the crossbridge cycle will operate. The switching action of the calcium–troponin–tropomyosin complex in skeletal muscle is extended by the structure of the thin filaments, which allows one troponin molecule, via its tropomyosin connection, to control seven

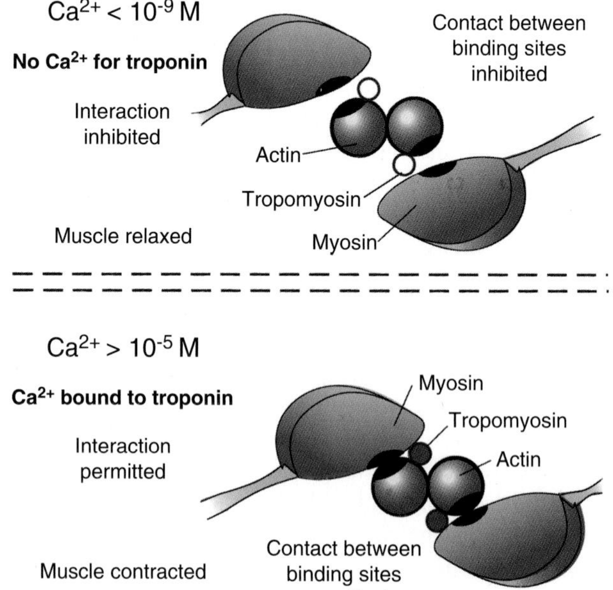

Figure 8.7 The calcium switch for controlling skeletal muscle contraction. Calcium ions, via the troponin–tropomyosin complex, control the blocking of the interaction between the myosin heads (the crossbridges) and the active site on the thin filaments. The thin filaments are seen in cross section.

actin monomers. Because the calcium control in striated muscle is exercised through the thin filaments, it is termed **actin-linked regulation**.

State of total relaxation to maximum contraction in skeletal muscle is tightly linked to its intracellular calcium concentration.

The relationship between the relative force developed and the calcium concentration in the region of the myofilaments is steep. At a calcium concentration of 1×10^{-8} M, the interaction between actin and myosin is negligible, whereas an increase in the calcium concentration to 1×10^{-5} M produces essentially full force development (see Fig. 8.7). This process is saturable, so that further increases in calcium concentration lead to little increase in force. In skeletal muscle, an excess of calcium ions is usually present during activation, and the contractile system is nearly saturated.

Even during calcium release from the terminal cisternae, Ca^{2+} ATPases in the SR membranes take up free calcium ions from the myofilament space and actively pump them into the interior of the SR. The rapid release process quickly stops: There is only one burst of calcium ion release for each action potential, and the continuous **calcium pump** in the SR membrane reduces calcium in the region of the myofilaments to a low level (1×10^{-8} M). With calcium ions no longer available, the contractile activity ceases and relaxation begins. The resequestered calcium ions are moved along the longitudinal elements to storage sites in the terminal cisternae, and the system is ready to be activated again. This entire process takes place in a few tenths of a millisecond and may be repeated many times each second.

Cyclic interaction of actin and myosin is the molecular engine driving muscle contraction.

At rest, the myosin heads in the thick filaments are bound with **adenosine diphosphate (ADP)** and inorganic phosphate (P_i) and situated 90° to the axis of the thin filaments. Myosin does not bind to actin in this state because of the inhibitory action of tropomyosin on crossbridge attachments (see Fig. 8.3, step 1). When the inhibition of tropomyosin on crossbridge attachments is removed by calcium released from the SR, actin and myosin bind strongly, and the crossbridges become firmly attached (see Fig. 8.3, step 2). When myosin is attached to actin, it loses its affinity for ADP and P_i, which then fall free into the cytoplasm (see Fig. 8.3, step 3). The release of ADP and P_i from myosin causes it to undergo a conformational change (see Fig. 8.3, step 4). This occurs because the tip of the myosin head is still attached to actin so that the portion of the head attached to the flexible S2 region pitches outward by 45°, thereby stretching this flexible chain region in a manner similar to stretching a spring. This "spring," however, snaps back as soon as it is stretched, and, because myosin is still attached to actin, the snapping back pulls the actin filaments past the myosin filaments in a movement called the **power stroke** (see Fig. 8.3, step 5).

Following this movement (which results in a relative filament displacement of around 10 nm), the actin–myosin binding is still strong, and the crossbridge cannot detach. At this point in the cycle, the state of actin–myosin binding is termed a **rigor crossbridge** (see Fig. 8.3, step 5). If ATP is present, it binds to the rigor crossbridge at the myosin head, resulting in detachment of myosin from actin (see Fig. 8.3, step 6). The ATPase property of myosin then partially hydrolyzes ATP to ADP and P_i, which remain bound to the myosin head. The energy released from this hydrolysis is used to reset the conformation of the myosin head to its original 90° orientation, and the newly recharged myosin head can begin the cycle of attachment, power stroke, and detachment again (see Fig. 8.3, step 1). This cycle can repeat as long as the muscle is activated, a sufficient supply of ATP is available, and the physical limit to shortening has not been reached. If cellular energy stores are depleted, as happens after death, the crossbridges cannot detach because of the lack of ATP, and the cycle stops in an attached state (see Fig. 8.3, step 5). This produces an overall stiffness of the muscle, which is observed as the **rigor mortis** that sets in shortly after death.

Cellular structure of skeletal muscle transforms crossbridge cycling into mechanical motion.

The amounts of movement (about 10 nM) and force (about 4 pN) produced by a single crossbridge interaction are exceedingly small, and it is the organization of the myofilaments into sarcomeres that both allows useful work to be done and creates the chemically and mechanically cyclic nature of the crossbridge reactions. In a shortening muscle, the power stroke moves the two sets of filaments past one another, and the crossbridge that is formed following one "turn" of the cycle involves myosin attaching to a new actin monomer further down the thin filament.

The width of the A bands (thick-filament areas) in striated muscle remains constant, regardless of the length of the entire muscle fiber, whereas the width of the I bands (thin-filament-only areas) varies directly with the length of the fiber. The spacing between Z lines also depends directly on the length of the fiber. The lengths of the thin and thick myofilaments themselves remain constant despite changes in fiber length. The **sliding filament theory** proposes that changes in overall fiber length are directly associated with changes in the overlap between the thick and thin filaments. During contraction, this is accomplished by the interaction of the globular heads of the myosin molecules with binding sites on the actin filaments.

The total shortening of each sarcomere is only about 1 μm, but a muscle contains many thousands of sarcomeres placed end to end (in series). This arrangement has the effect of adding up all the small sarcomere length changes into a large overall shortening of the muscle. Similarly, the amount of force exerted by a single sarcomere is small (a few hundred micronewtons), but, again, there are thousands of sarcomeres side by side (in parallel), resulting in the production of considerable force.

MECHANICS OF SKELETAL MUSCLE CONTRACTION

Muscle is a biologic motor. In common with other motors, it produces physical work by transforming the energy (a chemical fuel) into the generation of mechanical force and movement. In spite of the remarkable variety of controlled muscular movements that humans can make, the fundamental mechanical events of the contraction process can be described by a relatively small set of specially defined functions that result from particular muscle capabilities.

Contraction of large muscle groups results from temporal and spatial summation of single contractions in muscle units.

A single muscle action potential leads to a single brief contraction of the muscle called a **twitch** (Fig. 8.8A). Though the contractile machinery (i.e., the crossbridge cycle mechanism) may be fully activated (or nearly so) during a twitch, some of this active contraction must be used first to stretch a series of elastic elements in the muscle fibers before full contractile force can be transmitted to the muscle fiber. Thus, the amount of force produced by a single twitch is relatively low. The duration of the action potential in a skeletal muscle fiber, however, is shorter (about 5 ms) than the duration of a twitch (tens or hundreds of milliseconds). For this reason, the muscle fiber can be activated again long before the muscle has relaxed. Figure 8.8B shows the result of stimulating a muscle that is already active as a result of a prior stimulus. In this example, the second stimulus is given before the fiber has relaxed completely from the first twitch. At this point, the fiber is well outside its electrical refractory period caused by the first stimulus, but it is not fully relaxed. Thus, the force of the second stimulus is added to the remaining force from the first stimulus, and significant additional force is developed. In a similar manner, when the second stimulus closely follows the first (even before force has begun to decline), the resulting peak force is greater than that of a single twitch for that fiber (see Fig. 8.8C). The examples above demonstrate that individual twitch contractions can be summed temporally to enhance the contractile force generated by a muscle fiber.

As an extension of this phenomenon, if many stimuli are given repeatedly and rapidly to the muscle fiber, the result is an amplified, *sustained* contraction called a **tetanus** (Fig. 8.9). The amount of force produced in a tetanus is typically several times that of a twitch, and the disparity between the two is expressed as the **tetanus/twitch ratio.** When the contractions occur so close together that no fluctuations in force are observed, a **fused tetanus** results. The repetition rate at which this occurs is the **tetanic fusion frequency**, typically 20 to 60 stimuli per second, with the higher rates found in muscles that contract and relax rapidly. Figure 8.9 shows these effects in a special situation, in which the interval between successive stimuli is steadily reduced and the muscle responds at first with a series of twitches that become fused into a smooth tetanus at the highest stimulus frequency. Because it involves events that occur close together in time, a tetanus is a form of **temporal summation.**

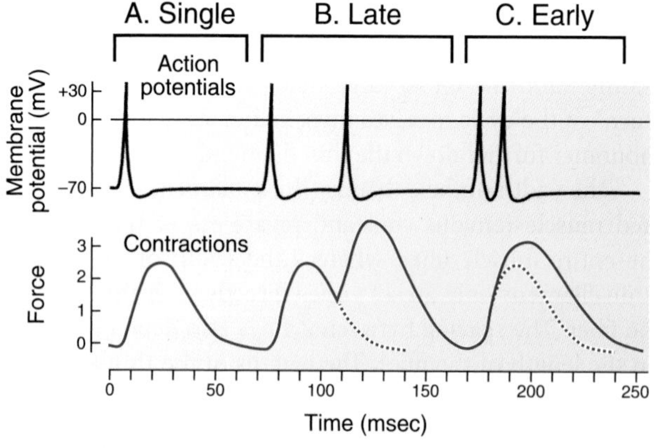

Figure 8.8 Temporal summation of muscle twitches. **(A)** The first contraction is in response to a single action potential. **(B)** The next contraction shows the summed response to a second stimulus given during relaxation; the two individual responses are evident, but the contraction produced during the second response is added onto residual contraction from the first response. **(C)** The last contraction is the result of two stimuli in quick succession so that no relaxation occurs between the contractions. The *dashed line* in the *lower graph* shows the force generated by one twitch, and the *solid line* represents the actual summed tension of two activations of the muscle in rapid succession.

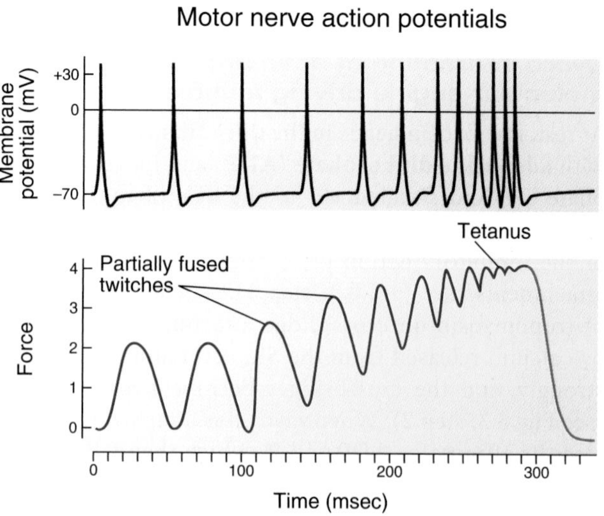

Figure 8.9 Fusion of twitches into a smooth tetanus. The muscle is activated in rapid succession. When the interval between successive activations allows twitches to partially relax in between contractions, peak tension increases but oscillates. This is called *partial tetanus*. As the interval between successive stimuli steadily decreases, twitches fuse on top of one another, and a sustained, smooth higher force is created. This is called *tetanus.*

Intracellular Ca^{2+} levels are different in muscle cells during fused tetanus than during partially fused twitches (partial tetanus). Action potentials do not vary significantly in muscle cells, and each action potential releases essentially the same amount of Ca^{2+} from the SR. This amount of calcium is sufficient to cause a twitch. During partial tetanus, the calcium increase with each action potential is very transient because SR Ca^{2+} ATPases rapidly pump the calcium back into the SR. Calcium levels peak and return to control levels repeatedly in this situation, and, any increase, or summation, of twitch force is due to incomplete mechanical relaxation of a twitch before the muscle fiber is activated again, adding its force to that remaining from the prior twitch. During complete, or fused, tetanus, the muscle fiber is activated at a frequency that adds calcium into the intracellular space faster than it can be pumped into the SR. In this case, total tension and intracellular calcium concentration is increased and maximized, with the high calcium levels continually activating muscle contraction as long as the muscle is stimulated at a high enough rate.

Muscles are organized into functional neuromuscular units that allow partial activation of a whole muscle.

Skeletal muscles are made up of many muscle fibers. These fibers require innervation from a motor neuron in order to be functional, but each fiber is not innervated by its own single neuron. Instead, a typical motor axon branches as it courses through the muscle, with each of its terminal branches innervating a single muscle fiber. In this manner, one neuron can innervate many muscle fibers. In some cases, a neuron may innervate just a few fibers, whereas in other cases a single neuron, through its branching, can innervate tens of thousands of muscle fibers. A group of muscle fibers innervated by branches of the same motor neuron is called a **motor unit**. Any large muscle group (e.g., the biceps, quadriceps, and gastrocnemius) contains many motor units. This arrangement allows for activation of only a portion of the muscle at any one time and represents a spatial control of muscle contraction. The pattern of activation is determined by the CNS and distribution of the motor axons among the muscle fibers.

A motor unit functions as a single mechanical entity; all the fibers in the unit will contract together when an action potential travels from the CNS through the motor neuron that innervates the muscle fibers contained within the unit. Independent contraction of only some of the fibers in a motor unit is impossible, so the motor unit is normally the smallest functional unit of a muscle. In muscles adapted for fine and precise control, only a few muscle fibers are associated with a given motor axon. In muscles in which high force is more important, a single motor axon controls many thousands of muscle fibers.

The number of motor units active at any one time determines the total force produced by a muscle. As more motor units are brought into play, the force increases. This is called **motor unit summation** or **spatial summation** and is illustrated in Figure 8.10. This type of modulation of muscle function endows a large muscle group with the ability to deal with light loads with an economy of function. When light loads need to be moved or supported with the muscles of an arm, for example, the CNS activates only enough motor units to accomplish the task of moving such a load, rather than modulating the contractile force of all the muscle fibers (and, thus, all the motor units) of all the muscle groups in the arm. This becomes particularly important when a limb needs to support a weight that is far less than the maximum weight that could be supported by all the muscle groups in the limb. Indeed, it is impossible, for example, to hold a few pieces of paper steady in your hand against gravity by activating all the muscle fibers in the arm at less than their composite maximum contractile force generation capability. Holding any weight against gravity requires fused, or tetanic, contractions of the muscle groups involved in supporting the weight. The total tetanic force generated by the arm muscles results in far more force generation than that needed to support a few pieces of paper. If the CNS were to simply stimulate all the muscle fibers in the arm (and, thus, all the motor units) to just hold pieces of paper, it would be

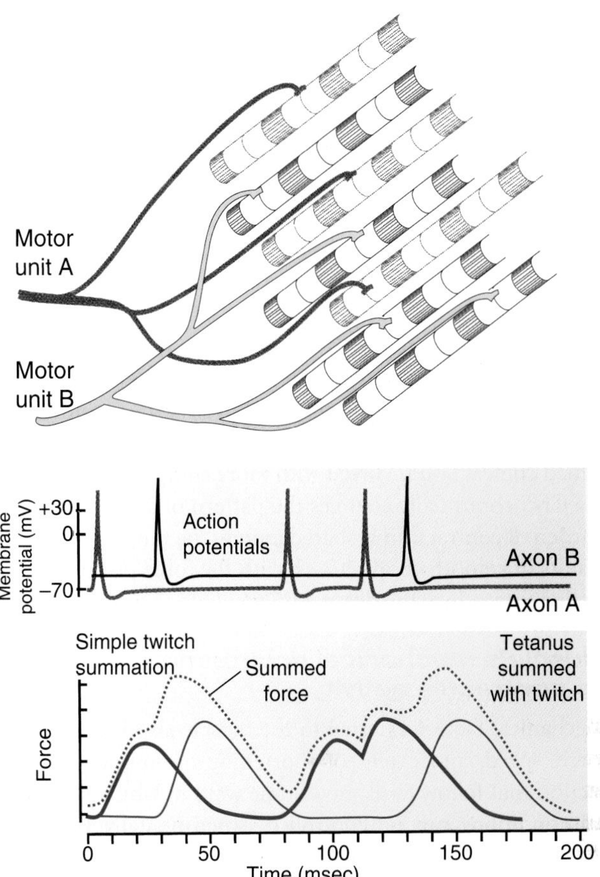

Figure 8.10 Motor unit summation. Two motor units, A and B, are shown above with their motor nerve action potentials and muscle twitches below. The summed force represents the total force generated in the whole muscle due to activation of both motor units. In the first example, fibers from motor units A and B are activated in close succession. Total force is the simple spatial summation of twitches from the two motor units; in the second, a brief tetanus in motor unit A (temporal summation) is spatially added to a twitch from motor unit B to create an even larger generation of force by the whole muscle.

expending far more energy than necessary to accomplish the task. Furthermore, if the CNS, instead, were to simply reduce the frequency of motor neuron action potential generation to all the arm muscles in order to reduce the total force generated by those muscles so that only the force needed to hold the papers was produced, the muscles would be unable to form the fused contraction needed to just hold the papers steady against gravity. The muscle contraction, instead, would oscillate between larger and smaller forces (see Fig. 8.9). Instead of using either of these modes of muscle activation, the CNS stimulates to tetany (or fuses) only the number of muscle fibers (units) needed to hold the weight of the paper steady against gravity; other motor units within the muscle groups are not activated. During a sustained contraction involved with more complex movements, the CNS continually changes the pattern of activity, and the burden of contraction is shared among the motor units. This results in smooth contractions, with the force precisely controlled to produce the desired movement.

Muscle's mechanical environment modifies its contractile activity.

Mechanical factors external to the muscle also influence the force, speed, and extent of shortening of contraction. The section that follows will show some ways in which mechanical conditions can be controlled experimentally to aid in the analysis of the mechanical function of muscle. Generally, these arrangements represent "artificial" conditions that are better controlled (and hence less complex) than those encountered in daily activities. Nevertheless, these types of analyses demonstrate how certain mechanical variables alter the contractile performance of muscle. Such relationships are not applicable only to understanding skeletal muscle as a mechanical engine. The relationships described in the following sections of this chapter are critical components that control the mechanical performance of the heart as well. An understanding of such mechanical relationships is critical to understanding the function of the heart in health and disease in the clinical setting.

Isometric muscle contraction occurs when muscle contracts against a load that is too heavy to move.

If a muscle is attached on its ends to a permanent fixture so that it cannot move when activated, the muscle will express its contractile activity by developing force without shortening. This type of contraction is termed **isometric contraction** (meaning "same length") and is demonstrated when we try to lift a weight far in excess of what our muscles can lift, pull on an object permanently affixed to a stationary support, or push against a wall. We feel our muscles contract, tense, and "harden" in these situations, although no muscle or object is actually moved.

The forces developed during an isometric contraction can be studied by attaching a dissected muscle to an apparatus similar to that shown in Figure 8.11. This arrangement provides for setting the length of the muscle and tracing a record of force versus time. In this example, the muscle is

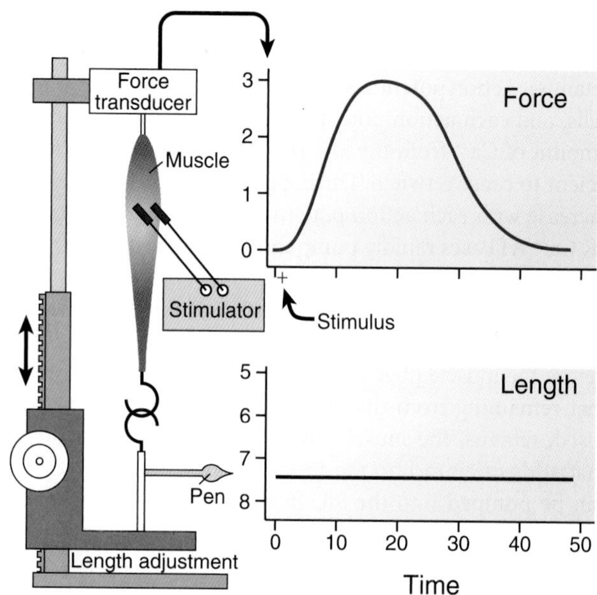

Figure 8.11 A simple apparatus for recording isometric contractions. The length of the muscle (marked on the graph by the pen attached near its lower end) is adjustable at rest and is called the *preload* on the muscle. In this apparatus, the muscle is prevented from shortening following its activation. The force transducer provides a record of the force produced by the muscle in response to a single stimulus at a fixed length (preload). This force is isometric by definition. (Force, length, and time units in the figure are arbitrary.)

stimulated only once but with sufficient strength to activate all its motor units. This produces a single twitch in which isometric force develops relatively rapidly (subsequent isometric relaxation is somewhat slower). The durations of both contraction time and relaxation time are related to the rate at which calcium ions can be delivered to and removed from the region of the crossbridges, the actual sites of force development. During an isometric contraction, no actual physical work is done on the external environment because no movement takes place while the force is developed. The muscle, however, still consumes energy to fuel the processes that generate and maintain force.

Isotonic contractions in muscle result in movement of a load because the muscle generates a force greater than the load.

When conditions are arranged so that a muscle can generate a force larger than the load (weight) to which it is attached, the muscle will have the ability to shorten and, thus, move the mass. To accomplish load movement, the muscle first develops force enough to equal the weight of that mass to which it is attached, then begins to move the mass with whatever force-generating capability remains for that muscle contraction. This type of contraction is called **isotonic contraction** (meaning "same force"). In the simplest conditions, this constant force is the load a muscle moves. This load is called an **afterload**, because its magnitude and presence are not apparent to the muscle until *after* it has begun to shorten.

Recording an isotonic contraction requires modification of the apparatus used to study isometric contraction (Fig. 8.12). Here the muscle is allowed to shorten while lifting an afterload, which is provided by the attached weight, chosen to be somewhat less than the peak force capability of the muscle. When the muscle is stimulated, it will begin to develop force without shortening, because it takes some time to build up enough force to begin to lift the weight. This means that early on, the contraction is isometric (see Fig. 8.12, phase 1). After sufficient force has been generated, the muscle will begin to shorten and lift the load (see Fig. 8.12, phase 2). The contraction then becomes isotonic because the force exerted by the muscle exactly matches that of the weight, and the mass of the weight does not vary. Therefore, the solid line in the upper tracing in Figure 8.12 shows a flat line representing constant force during this phase, whereas the muscle length (lower tracing) is free to change. As relaxation begins (see Fig. 8.12, phase 3), the muscle lengthens at constant force because it is still supporting the load. This phase of relaxation is isotonic, and the muscle is re-extended by the weight. When the muscle has been extended sufficiently to return to its original length, conditions again become isometric (see Fig. 8.12, phase 4), and the remaining force in the muscle declines as it would in a purely isometric twitch. In almost all situations encountered in daily life, isotonic contraction is preceded by isometric force development; such contractions are called **mixed contractions** (isometric–isotonic–isometric).

The duration of the early isometric portion of the contraction varies, depending on the afterload. At low afterloads, the muscle requires little time to develop sufficient force to begin to shorten, and conditions will be isotonic for a longer time. Figure 8.13 presents a series of three twitches. At the lowest afterload (weight A only), the isometric phase is the briefest and the isotonic phase is the longest with the lowest force. With the addition of weight B, the afterload is doubled and the isometric phase is longer, whereas the isotonic phase is shorter with twice the force. If weight C is added, the combined afterload represents more force than the muscle can exert, and the contraction is isometric for its entire duration. Also note from this figure that both the velocity and extent of shortening depend on the afterload; an isotonic contraction with a heavy afterload cannot move that load as fast or as far as it can a lighter afterload. This unique relationship between afterload and the extent and

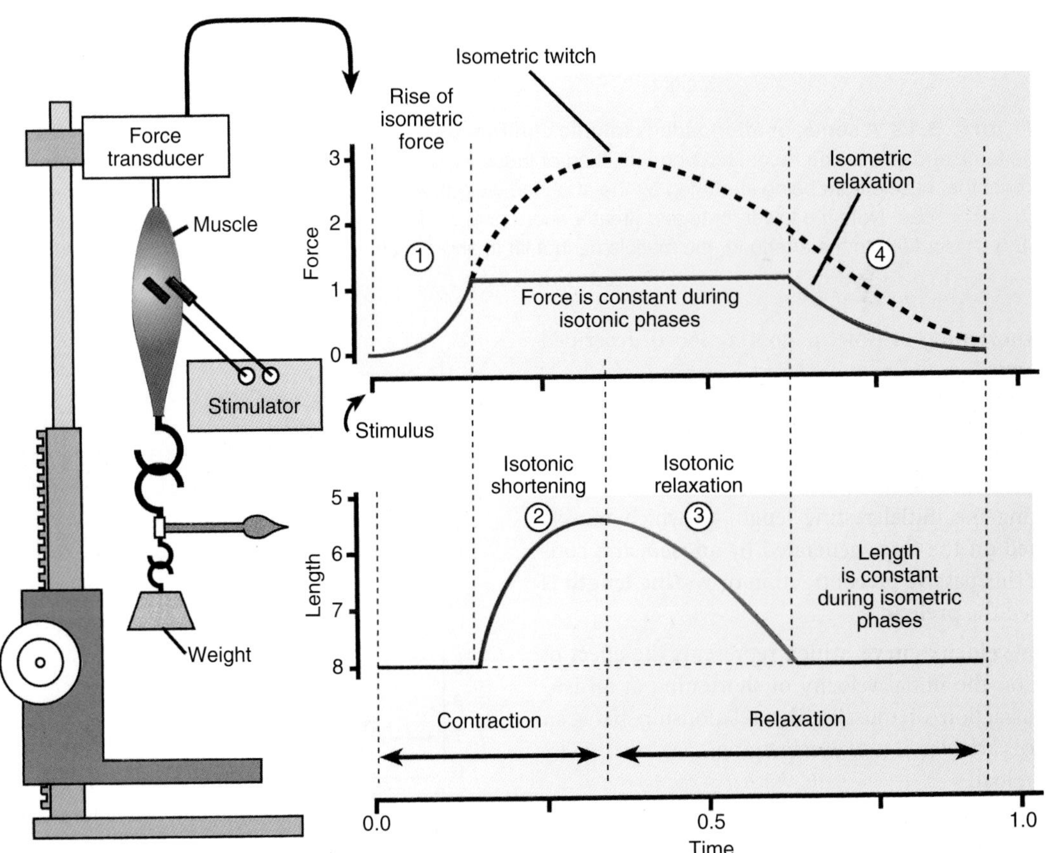

Figure 8.12 A modified apparatus showing the recording of a single isotonic twitch. The pen at the lower end of the muscle marks its length, and the weight attached to the muscle provides the afterload, whereas the platform beneath the weight prevents the muscle from being overstretched at rest. The first part of the contraction, until sufficient force has developed to lift the weight, is isometric. During shortening and isotonic relaxation, the force is constant (isotonic conditions), and during the final relaxation, conditions are again isometric because the muscle no longer lifts the weight. The *dotted lines* in the force and length traces show the isometric twitch that would have resulted if the force had been too large (greater than three units) for the muscle to lift.

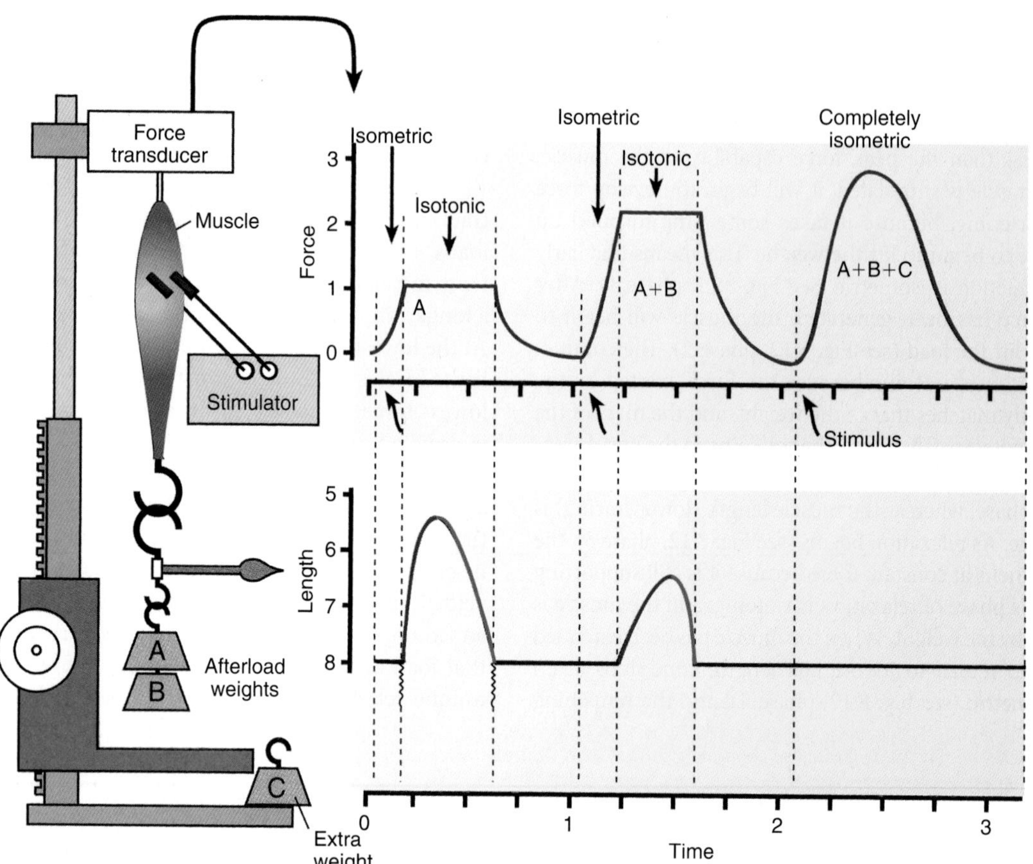

Figure 8.13 A series of afterloaded isotonic contractions. The curves labeled A and A + B correspond to the force and shortening records during the lifting of those weights. In each case, the adjustable platform prevents the muscle from being stretched by the attached weight, and all contractions start from the same muscle length, or preload. Note the lower force and greater shortening with the lower weight (A). If weight C (total weight = A + B + C) is added to the afterload, the muscle cannot lift it, and the entire contraction remains isometric.

velocity of shortening of isotonic contraction is described in detail below.

The mechanics of isometric and isotonic contraction of muscle can be described in terms of two important relationships:

- The **length–tension curve**, which represents the effect of changing the initial resting length to which muscle is stretched on the force generated by an *isometric* contraction. This passive precontraction or **resting length** is often called the **preload**.
- The **force–velocity curve**, which represents the effect of afterload on the initial velocity of shortening of an isotonic contraction of muscle. The relationship between these two variables is not unchanging; it is altered by muscle preload.

Preload affects the force of isometric contraction by changing the number of crossbridge sites available for contraction.

Resting, noncontracting skeletal muscle has nonlinear elastic properties when stretched to different lengths. When it is short, it is slack and will not resist passive extension. As it is made longer and longer, however, its resisting force increases more and more (Fig. 8.14, upper right inset). This resisting

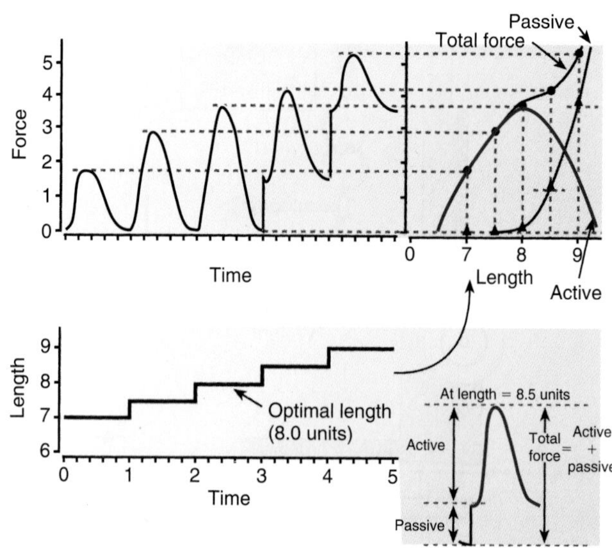

Figure 8.14 A length–tension curve for skeletal muscle. Contractions are made at several resting lengths, and the resting (passive) and peak (total) forces for each twitch are transferred to the graph at the right. Subtraction of the passive curve from the total curve yields the active force curve. These curves are further illustrated in the lower right corner of the figure.

force is called **passive force** or **resting force** and is related to the preload on the muscle.

In an isometric muscle contraction, the relationship between active force and passive length is much different. Figure 8.14 shows the force produced by a series of isometric twitches made over the range of muscle lengths. The resulting tension (passive tension plus active force) at each muscle length is plotted on the length–tension curve at the right with the total peak force from each twitch related to its initial passive length (connected by *dotted lines* in the figure). The muscle length is changed only when the muscle is not stimulated, and it is held constant (isometric) during contraction. The difference between the total force and the passive force is the active force. Note that the active length–tension curve is parabolic. Consequently, isometric force has a maximum in relation to its preload or initial resting length. The length corresponding to this maximum isometric force is called the **optimum length** (**L_o**) (represented by length "8" in the upper right inset of Fig. 8.14). If the muscle is set to a shorter length and then stimulated, it produces less force. At an extremely short length, it produces no force at all. If the muscle is made longer than its optimal length, it also produces less force when stimulated.

Preload alters isometric force in muscle by altering the overlap of actin and myosin prior to contraction.

The parabolic length–tension relationship for isometric contraction in the whole muscle results from changes in actin and myosin overlap within the sarcomere as the muscle is stretched prior to contraction. The effect of sarcomere length on force generation is summarized in Figure 8.15. At lengths near L_o, the amount of isometric force that the muscle can generate is a maximum. At this length, the maximum number of actin-binding sites sits opposite the maximum number of myosin heads available for contraction. This

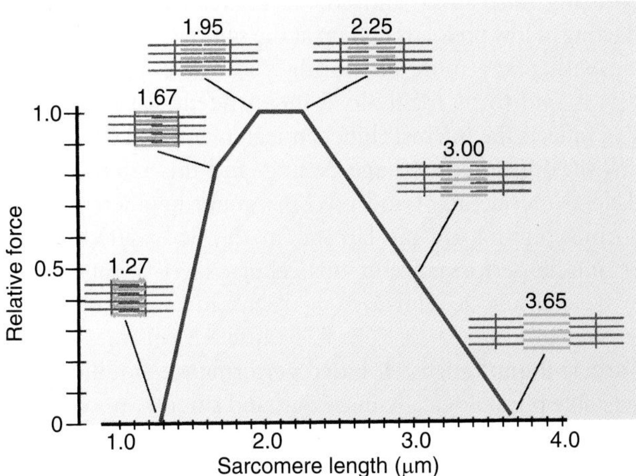

Figure 8.15 Effect of filament overlap on force generation. The force a muscle can produce depends on the amount of overlap between the thick and thin filaments, because this determines how many crossbridges can interact effectively. (See text for details.)

"maximum" isometric force is not a true peak corresponding to a minute location in the sarcomere. Note in Figure 8.15 that force does not vary with the degree of overlap between 1.95 and 2.25 μm. This occurs because of the bare zone along the thick filaments at the center of the A band (where no myosin heads are present). Over this small region, further overlap does not lead to an increase in the number of attached crossbridges, and, therefore, isometric force capability does not change.

When the muscle is stretched beyond its normal resting length, decreased filament overlap occurs with increasing stretch. This limits the amount of force that can be produced as precontraction length increases, because less length of the thin filaments interdigitates with A-band thick filaments, and fewer crossbridges are available for any subsequent contraction. Thus, over this region of lengths, force is negatively proportional to the degree of passive stretch on the muscle prior to contraction (i.e., the preload).

At passive resting lengths shorter than L_o, additional geometric and physical factors play a role in myofilament interactions. Because muscle is a "telescoping" system, there is a physical limit to the degree to which resting muscle length can decrease. As thin myofilaments penetrate the A band from opposite sides, they begin to meet in the middle and interfere with each other (1.67 μm). At the extreme, further reduction in resting lengths are limited by the thick filaments of the A band being forced against the structure of the Z lines (1.27 μm). Note, however, if one were to start at this extreme short length and then stretch the muscle to various passive lengths up to and including L_o, any resulting isometric contraction would increase as the passive initial length of the muscle increases. This region of the isometric length–tension curve is sometimes called the *ascending limb* of the length–tension curve. In this region, isometric force is positively proportional to initial resting length (preload).

It should be noted that, in the body, many skeletal muscles are confined by their attachments to a relatively short region of the isometric length–tension curve near L_o. Although this positions preload on the muscle to an area that can produce the best possible force generation in the muscle, it also means that skeletal muscle is essentially incapable of adjusting its force-generating capacity by repositioning itself at different locations on the length–tension curve prior to initiating contractions. (This modulation of contraction is possible, however, in hollow distensible organs like the heart, where it is exploited to alter the performance of that organ; see Chapter 13, "Cardiac Muscle Mechanics and the Cardiac Pump.")

Extent and velocity of shortening of skeletal muscle are negatively related to afterload but positively related to preload.

Everyday experience shows that light loads are lifted faster than heavy ones. A detailed analysis of this observation can be performed by arranging a muscle so that it can be presented with a series of afterloads (Fig. 8.16; also see Fig. 8.13). When all the motor units in the muscle are stimulated (i.e., no motor units left to recruit), the muscle's performance can

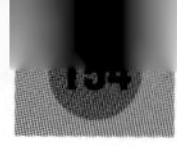

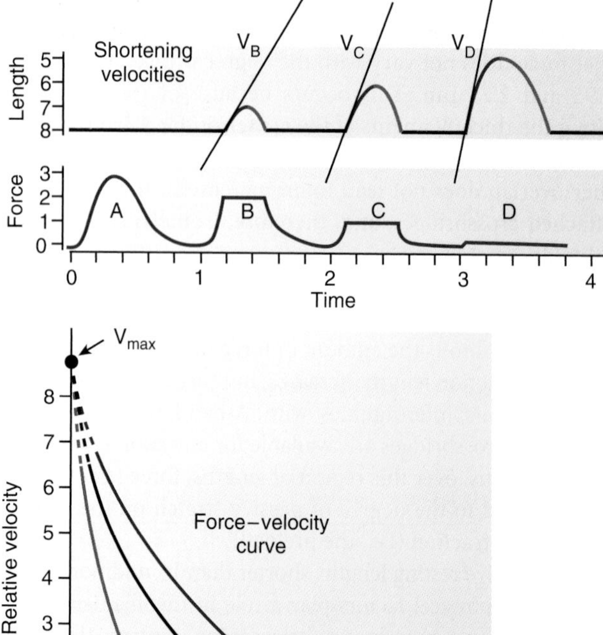

Figure 8.16 Force–velocity curves for skeletal muscle. Contractions at four different afterloads (decreasing left to right) with muscle starting at three different preloads are shown. The initial shortening velocity (slope) is measured (V_B, V_C, V_D) and the corresponding force and velocity points plotted on the axes in the *bottom* graph. Note that the initial velocity of shortening increases when afterload is decreased. The *black line* represents the force–velocity curve of a muscle starting at a less than maximum preload. The *red line* represents this relationship with the muscle starting at its optimum preload, and the *blue line* shows the relationship at the lowest preload in this example. Notice how preload affects the velocity of shortening of any isotonic contraction and F_{max}, but neither it nor afterload affects V_{max}, or the maximal velocity of shortening.

still be affected by loading conditions. Measuring the **initial velocity**—the velocity during the earliest part of the shortening—at various afterloads shows that lighter loads are lifted more quickly and heavier loads more slowly. (Initial velocity is measured because the muscle soon begins to slow down upon moving the load.) When all the initial velocity measurements are related to each corresponding afterload lifted, an inverse relationship known as the force–velocity curve is obtained (see Fig. 8.16). This relationship shows that as the afterload is decreased, the velocity of shortening increases in a hyperbolic manner; that is, velocity increases more and more as the load is reduced. Note that if the applied load is greater than the *maximal force* capability of the muscle, no shortening will result, and the contraction will be isometric. This point is known as F_{max}. Also note from the force–velocity relationship that there is a theoretical velocity that the muscle could obtain if it does not have to move any afterload. This represents the *maximum velocity* of shortening capability of that muscle and is, therefore, known as V_{max}. In practice, a completely unloaded contraction is very difficult to arrange. Instead, V_{max} is obtained by mathematical extrapolation of the force–velocity curve. The value of V_{max} represents the maximal rate of crossbridge cycling for that muscle; it is directly related to the biochemistry of the actin–myosin ATPase activity in a particular muscle type and can be used to compare the properties of different muscles.

Figure 8.16 shows a force–velocity curve for a fully activated muscle. When measurements are made on a fully activated muscle, the force–velocity curve defines the upper limits of the muscle's isotonic capability. Factors that modify muscle performance, such as fatigue or incomplete stimulation (e.g., fewer motor units activated), result in operation *below* the limits defined by this force–velocity curve. The force–velocity relationship is also affected by the preload on the muscle prior to contraction (see red and blue relationships in Fig. 8.16). The curve is shifted to the right with increasing preload and to the left with decreased preload, providing the muscle is not operating at lengths longer than L_o. When on the ascending limb of the length–tension relationship, the effect of preload on the force–velocity relationship reflects the number of crossbridges available at the start of the contraction. With increasing preload, a given afterload can be moved faster, or, at any given velocity, a larger afterload can be moved as compared to velocity and loads moved with a lesser preload. In short, muscle performance is enhanced by bringing more crossbridges into play at the start of a contraction. Although this feature of muscle behavior does not have direct physiologic application to skeletal muscle *in situ*, which has a fixed initial length through attachment to the skeleton, this relationship becomes important in muscles in hollow organs were preload may be altered prior to a contraction (e.g., the heart and blood vessels).

Figure 8.16 also reveals an important feature of the physiology of skeletal muscle with regards to crossbridge cycling. Note that although preload alters the force–velocity position for isotonic contractions, the V_{max} is altered neither by preload nor any afterloads moved; force–velocity relationships starting at low preloads end up at the same V_{max} as those starting with many more crossbridges available. Consequently, V_{max} is said to be totally load independent. This is because V_{max} reflects the intrinsic biochemical interaction of actin and myosin during crossbridge cycling, and this interaction for skeletal muscle actin and myosin cannot be altered by any normal physiologic mechanism in the body; modification of muscle performance at this cellular level is not possible in skeletal muscle. However, as discussed later in this chapter and in Chapter 13, "Cardiac Muscle Mechanics and the Cardiac Pump," such cell-based performance modification is available physiologically in cardiac and smooth muscle.

Extent of skeletal muscle shortening is increased by increasing preload but decreased by increasing afterload.

The length–tension curve represents the effect of initial resting length, or preload, on the isometric contraction of skeletal muscle. At initial lengths less than and up to L_o, such

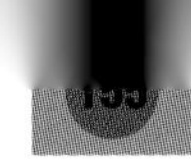

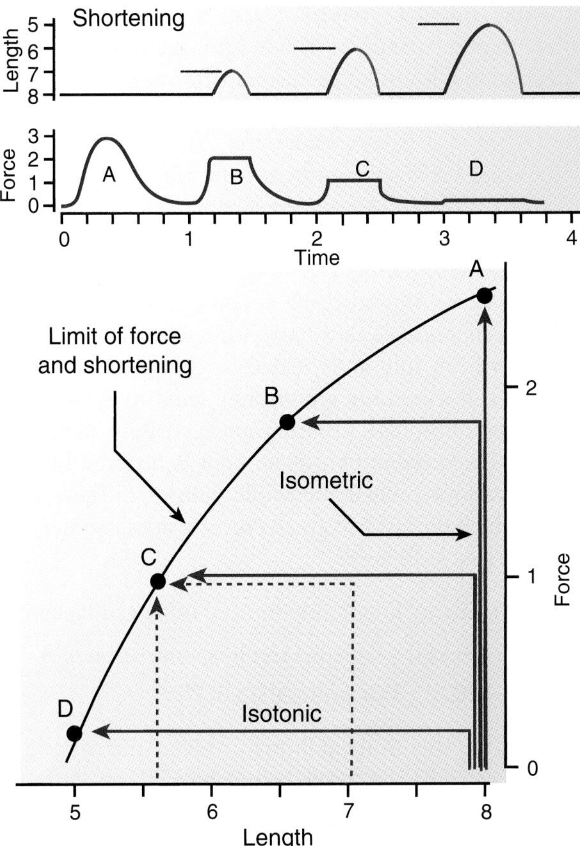

Figure 8.17 The relationship between isotonic and isometric contractions. The *top* graphs show the contractions from Figure 8.16, with different amounts of shortening. The *bottom* graph shows that, for contractions B, C, and D, the initial portion is isometric (the line moves upward at constant length) until the afterload force is reached. The muscle then shortens at the afterload force (the line moves to the left) until its length reaches a limit determined (at least approximately) by the isometric length–tension curve. The *dotted lines* show that the same final force/length point can be reached by several different approaches. Relaxation data, not shown on the graph, would trace out the same pathways in reverse.

a curve represents the maximum force that can be generated at any given preload. No shortening occurs in these situations, and the curve depicts muscle contraction as if all its energy was channeled into the development of force only. Points A, B, C, and D in Figure 8.17, for example, are the maximum force generation capability of a muscle at initial lengths (arbitrary units) of 8, 6.6, 5.6, and 5, respectively, with point A in this diagram being representative of the isometric force-generating capacity of the muscle at L_0.

Although the length–tension curve depicts isometric contraction of muscle, it also represents the limit of shortening of any isotonic contraction of that muscle. Isotonic contractions involve an initial isometric phase during which force is generated up to the point it equals the load the muscle is trying to move (i.e., the afterload). Once that point is reached, the muscle then shortens while moving that load (the isotonic phase). Such contractions are depicted in Figure 8.17 by the solid arrows that point to B, C, and D in the diagram. Note that during an isotonic contraction, the muscle will eventually reach a length where the maximum isometric force capable by the muscle matches the afterload the muscle is trying to move (see *dashed lines* in Fig. 8.17 that converge on point C). At this point, the muscle cannot shorten anymore; to do so would place the muscle at a length where its maximum isometric force-generating capability was less than the load it was trying to move—an impossibility. In other words, isotonic contractions of a muscle moving any afterload cannot move past a limit defined by the isometric length–tension curve for that muscle.

As Figure 8.17 shows, a lightly afterloaded muscle will shorten farther than one starting from the same initial length, or preload, but bearing a heavier load. Also note that if the muscle begins its shortening from a reduced initial length, its subsequent extent, or amount, of shortening will be reduced because the muscle starts closer to the limit imposed by the isometric-length tension curve but is still constrained from shortening beyond that curve. The following constraints apply to the extent of shortening of an isotonic contraction of skeletal muscle: (1) increasing preload up to L_0 increases the extent of shortening (the distance the afterload is moved); (2) at any given preload, increasing the afterload decreases the extent of shortening; (3) at any given preload, the final length of the muscle at the end of an isometric contraction is proportional to the afterload; and (4) the isometric length–tension curve sets the limit to the extent of shortening of any isotonic contraction of skeletal muscle.

The example above demonstrates that the distance a muscle can move an object is progressively reduced as the muscle tries to move increasingly heavy loads. One might postulate that skeletal muscle could move a heavy load as far as it could a light load if there was some way to change (enhance) the intrinsic contractile ability of the muscle cell itself. Such intrinsic enhancement would reveal itself as a shift of the ascending limb of the length–tension curve upward and to the left relative to its original position. Such a shift would move the limit to extent of shortening to the left, thereby letting the muscle move the heavy load as far as it could move the light load in its original state. Unfortunately, such intrinsic or cellular/molecular enhancement of the contractile mechanism is not possible under any normal physiologic conditions for skeletal muscle fibers. In essence, skeletal muscle shortening is constrained by its loading conditions; it *cannot* modify its cellular contractile capabilities independently of preload and afterload. Such is not the case with cardiac muscle, however (see Chapter 13, "Cardiac Muscle Mechanics and the Cardiac Pump").

Skeletal attachments create lever systems that modify muscle action.

Anatomic location also places restrictions on muscle function by limiting the amount of shortening or determining the kinds of loads encountered. Skeletal muscle is generally attached to bone, and bones are attached to each other. The bones and muscles together constitute a **lever system** (Fig. 8.18). In most cases, the system works at a mechanical disadvantage with respect to the force exerted. For example, curling the forearm requires muscles that are attached near

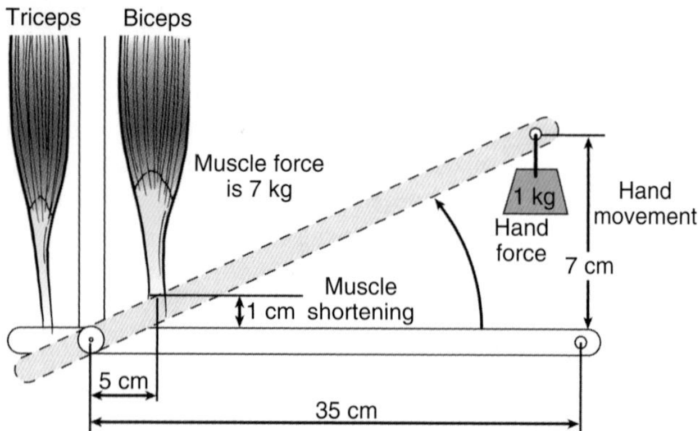

Figure 8.18 Antagonistic pairs and the lever system of skeletal muscle. Contraction of the biceps muscle lifts the lower arm (flexion) and elongates the triceps, whereas contraction of the triceps lowers the arm and hand (extension) and elongates the biceps. The bones of the lower arm are pivoted at the elbow joint (the fulcrum of the lever). The force of the biceps is applied through its tendon close to the fulcrum. The hand is seven times as far away from the elbow joint. Thus, the hand will move seven times as far (and fast) as the biceps shortens (lever ratio, 7:1), but the biceps will have to exert seven times as much force as the hand is supporting.

the fulcrum of the arm "lever," rather than at the more mechanically advantageous position near the hand. This means that in order to pull the forearm toward the upper arm while holding a weight in the hand, the muscle must exert a much greater force than the actual weight of the load being lifted (the muscle force is increased by the same ratio that the length change at the end of the extremity is increased). In the case of the human forearm, the biceps brachii, when moving a force applied to the hand, must exert a force at its insertion on the radius that is approximately seven times as great. However, the shortening capability of skeletal muscle by itself is rather limited. The benefit of the skeletal lever system is that it multiplies the distance over which an extremity can be moved. Consequently, it is possible to create large distances moved by the end of the limb with only small actual shortening of the muscle responsible for moving the limb. In this forearm example, the resulting movement of the hand is approximately seven times as far and seven times as rapid as the shortening of the muscle itself.

Finally, acting independently, a muscle can only shorten. Force must be used to relengthen the muscle, and this force must be provided externally. In some cases, gravity provides the external restoring force. However, in the body, muscles are often arranged in **antagonistic pairs** of **flexors** and **extensors**. In this manner, for example, a shortened biceps can be relengthened by the action of the triceps; the triceps, in turn, is relengthened by contraction of the biceps.

Adenosine triphosphate derived from glucose metabolism is the primary energy supplying mechanism for skeletal muscle contraction.

Cellular processes within skeletal muscle cells must supply biochemical energy to the contractile mechanism because contracting muscles perform work. Additional energy is required to pump the calcium ions involved in the control of contraction and for other cellular functions as well. In muscle cells, as in other cells, this energy ultimately comes from the universal high-energy compound, ATP (Fig. 8.19).

Although ATP is the immediate fuel for the contraction process, its concentration in the muscle cell is never high enough to sustain a long series of contractions. Most of the immediate energy supply is held in an "energy pool" of the compound creatine phosphate or phosphocreatine (PCr), which is in chemical equilibrium with ATP. After a molecule of ATP has been split and yielded its energy, the resulting ADP molecule is readily rephosphorylated to ATP by the high-energy phosphate group from a creatine phosphate molecule. The creatine phosphate pool is restored by ATP from the various cellular metabolic pathways. These reactions (of which the last two are the reverse of each other) can be summarized as follows:

ATP $\rightarrow$ P_i + Energy for contraction and other cell systems

ADP + PCr $\rightarrow$ ATP + Creatine (rephosphorylation of ATP)

ATP + Cr $\rightarrow$ ADP + PCr (restoration of PCr)

Because of the chemical equilibria involved, the concentration of PCr can fall to low levels before the ATP concentration shows a significant decline. It has been shown experimentally that when 90% of PCr has been used, the ATP concentration has fallen by only 10%. This situation results in a steady source of ATP for contraction that is maintained despite variations in energy supply and demand. Creatine phosphate is the most important storage form of high-energy phosphate; together with some other smaller sources, this energy reserve is sometimes called the **creatine phosphate pool**.

Two major metabolic pathways supply energy to the energy-requiring reactions in the cell and to the mechanisms that replenish the creatine phosphate pool. Their relative contributions depend on the muscle type and conditions of contraction. The first of the supply pathways is the **glycolytic pathway** or **glycolysis** (see Fig. 8.19). This is an **anaerobic** pathway; glucose is broken down without the use of oxygen to regenerate two molecules of ATP for every molecule of glucose consumed. Glucose for the glycolytic pathway may be derived from circulating blood glucose or from **glycogen**, which is the polymer storage form of glucose in skeletal muscle and liver cells. Glycolysis is rapid but extracts only a small fraction of the energy contained in the glucose molecule.

The end-product of anaerobic glycolysis is **lactic acid** or **lactate**. Under conditions of sufficient oxygen, this is converted to **pyruvic acid** or **pyruvate**, which enters another cellular (mitochondrial) pathway called the **Krebs cycle**. In Krebs cycle reactions, substrates are made available for **oxidative phosphorylation**. The Krebs cycle and oxidative phosphorylation are **aerobic** processes that require a continuous supply of oxygen. In this pathway, an additional 36 molecules of ATP are regenerated from the energy in the original glucose molecule; the final products are carbon dioxide and water. Although the oxidative phosphorylation

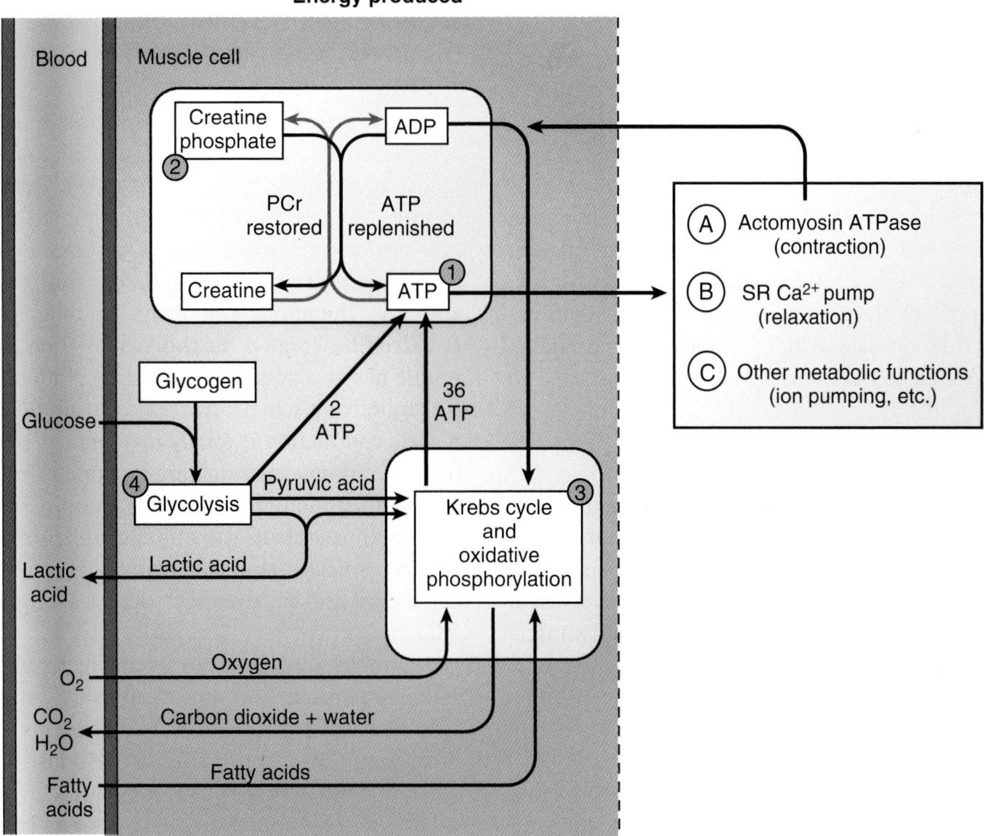

Figure 8.19 The major metabolic processes of skeletal muscle. These processes center on the supply of adenosine triphosphate (ATP) for the actomyosin ATPase of the crossbridges. Energy sources are numbered in order of their proximity to the actual reactions of the crossbridge cycle. Energy is used by the cell in the A, B, and C order shown. The scheme shown here is typical for all types of muscle, although there are specific quantitative and qualitative variations. ADP, adenosine diphosphate; PCr, phosphocreatine; SR, sarcoplasmic reticulum.

pathway provides the greatest amount of energy, it cannot be used if the oxygen supply is insufficient. In this case, glycolytic metabolism predominates.

Glucose is the preferred fuel for skeletal muscle contraction at higher levels of exercise. At maximal work levels, almost all the energy used is derived from glucose produced by glycogen breakdown in muscle tissue and from bloodborne glucose from dietary sources. Glycogen breakdown increases rapidly during the first tens of seconds of vigorous exercise. Sustained exercise can lead to substantial depletion of glycogen stores, which can restrict further muscle activity.

Muscle has performance limitations based on its structure and energy-conversion processes; as such, its efficiency is much <100%, and it produces relatively large quantities of heat, which must be dealt with by the organism that it is serving.

Metabolic and structural adaptations fit skeletal muscle for a variety of roles.

Although the basic structural features of the sarcomeres and the thick–thin filament interactions are essentially the same among skeletal muscles, the chemical reactions that supply the contractile system with energy vary. The enzymatic properties (i.e., the rate of ATP hydrolysis) of actomyosin ATPase also vary. A typical skeletal muscle usually contains a mixture of fiber types with different metabolic properties. However, in most muscles, a particular type predominates.

Red muscle fibers

Color differences exist among skeletal muscles and arise from differences in the amount of **myoglobin** the muscles contain. Similar to the related red blood cell protein hemoglobin, myoglobin is a protein that can bind, store, and release oxygen. It is abundant in muscle fibers that depend heavily on aerobic metabolism for their ATP supply, where it facilitates oxygen diffusion (and serves as a minor auxiliary oxygen source) in times of heavy demand. Such muscles have a red appearance and generally perform heavy sustained tasks. The quadriceps and gluteal muscles that continually maintain posture while standing are examples of this type of muscle. Red muscle fibers are divided into **slow-twitch fibers** and **fast-twitch fibers** on the basis of their contraction speed. The differences in rates of contraction (shortening velocity or force development) arise from differences in actomyosin ATPase activity (i.e., in the basic crossbridge cycling rate).

White muscle fibers

White muscle fibers, which contain little myoglobin, are fast-twitch fibers that rely primarily on glycolytic metabolism. They contain significant amounts of stored glycogen, which can be broken down rapidly to provide a quick source of energy. Although they contract rapidly and powerfully, their endurance is limited by their ability to sustain an oxygen deficit (i.e., to tolerate the buildup of lactic acid). They require a period of recovery (and a supply of oxygen) after heavy use.

Fast muscles, both white and red, not only contract rapidly but also relax rapidly. Rapid relaxation requires a high rate of calcium pumping by the SR, which is abundant in these muscles. In such muscles, the energy used for calcium pumping can be as much as 30% of the total consumed. Fast muscles are supplied by large motor axons with high conduction velocities; this correlates with their ability to make quick and rapidly repeated contractions.

SMOOTH MUSCLE

Smooth muscle participates in the organ processes that help regulate our internal environment. It is by far the most physiologically complicated and versatile of the muscle tissues in the body. This complexity is consistent with the diversity of tasks this contractile tissue must perform. Smooth muscle cells are small in comparison to those of skeletal muscle, which allows precise control of very small structures. Unlike skeletal muscles, most smooth muscles are not discrete organs but are, instead, intimate components of larger organs. For example, circular layers of smooth muscle are contained within the walls of all arteries and veins as well as those of the bronchi of the lungs. Unlike skeletal muscle, which is only activated intermittently, smooth muscle in blood vessels must keep these tube-like organs partially contracted at all times. Smooth muscle is able to do this task, which would seem to require huge amounts of energy expenditure, at a fraction of the estimated ATP cost because of a crossbridge cycling mechanism unique to it.

Smooth muscle outside the cardiovascular system is often labeled *visceral smooth muscle*. This muscle plays a key role in the functions of the digestive system, where it physically manipulates gut contents in complex ways during digestion. It is also responsible for the controlled movement of materials and fluids along the alimentary tract. Visceral smooth muscle in the gut, therefore, must be able to contract or relax phasically in response to the many different types of neural, chemical, and physical signals involved in digestion.

Some smooth muscles such as that in sphincters (circular bands of muscle that can stop flow in tubular organs) can remain contracted and closed for long periods but then relax and open transiently before contracting and closing again. Sphincters at the ends of the esophagus, the stomach, and rectum operate in this fashion (see Chapter 25, "Neurogastroenterology and Motility"). Smooth muscle in still other visceral organs remains relaxed most of the time but then can contract strongly in response to physiologic stimuli. Smooth muscles in the esophagus, urinary bladder, and gall bladder behave in this manner. As an ultimate case of this type of behavior, the smooth muscle of the uterus is basically inactive during most of a woman's life but contracts and relaxes rapidly and powerfully over several hours during parturition.

Contraction and relaxation of smooth muscle can be produced by neural, humoral, chemical, physical, and intrinsic stimuli.

Contractile activity in smooth muscle is initiated in markedly different ways compared to that for skeletal muscle. In the previous section, it was shown that skeletal muscle cannot contract without neural input to the muscle cells. Furthermore, the neural inputs to skeletal muscle fibers (via motor neurons) are often initiated from conscious thought and are, therefore, voluntary. Although smooth muscle is indeed innervated (in this case by branches of the autonomic nervous system [ANS]), innervation is not necessary for smooth muscle contraction, and modulation of smooth muscle contraction is totally involuntary. Neurotransmitters from autonomic neurons are but one of many factors that modulate smooth muscle contraction. Contractile activity in smooth muscle is also modulated by different hormones, local chemicals, physical forces, and numerous receptor-dependent and independent (direct acting) contractile and relaxing agents.

Smooth muscle is an excitable tissue that can form action potentials and contractions in response to signals from the ANS or other contractile agents. Smooth muscle can also summate contractions in response to a train or small burst of ANS-driven action potentials in a manner similar to temporal summation and tetanus in skeletal muscle. However, there are many other contractile agents that can mobilize calcium in smooth muscle and cause contraction without the formation of any action potentials. Furthermore, calcium levels and associated contraction of smooth muscle can be *graded* in response to changes in its resting membrane potential or in response to changes in the concentration of a vasoactive agent acting on the smooth muscle. Progressively larger depolarizations of the resting membrane potential in smooth muscle cause proportional increases in smooth muscle tone (contractile force) without creating action potentials. Similarly, graded contractions of smooth muscle can be elicited by progressively increasing concentrations of a contractile agonist to which the smooth muscle is exposed.

Finally, recall that skeletal muscle relaxation is a passive process or, essentially, the absence of contraction. In contrast, smooth muscle can be actively relaxed in a graded fashion either by hyperpolarizing the cell membrane or by exposure to agonists that directly activate relaxation of the muscle cells. Thus, whereas skeletal muscle is like a lamp with an on-and-off switch that will not work unless it is linked to a source of electricity, smooth muscle is like a light with a dimmer switch and its own source of electricity.

Altering smooth muscle contraction changes the internal dimensions of hollow organs based on the arrangements of muscle cells in the wall of the organ.

Just a few basic tissue types accomplish a wide variety of smooth muscle tasks. The simplest smooth muscle arrangement is found in the arteries and veins of the circulatory system. Smooth muscle cells are oriented in the circumference of a vessel so that their shortening results in reducing the vessel's diameter. This reduction may range from a slight narrowing to a complete obstruction of the vessel lumen, depending on the physiologic needs of the body or organ. Smooth muscle in arteries and veins is tonically (i.e., continually) partially contracted. In the larger muscular arteries, there may be many layers of cells, and the force of contraction may be quite high. Tonic contraction there helps the vessels counteract the expanding force on the arteries due to exposure to high blood pressures. The blood pressure provides a force that would otherwise relengthen the cells in the vessel walls. In small arterioles, the muscle layer may consist of single cells wrapped around the vessel, where it functions as an on/off spigot to control blood flow to very precise regions within an organ.

The circular arrangement of smooth muscle is also prominent in the airways of the lungs, where it regulates the resistance to airflow entering and exiting that organ. The circular smooth muscles making up sphincters, such as those in the gastrointestinal and urogenital tracts, have a special nerve supply and participate in complex reflex behavior. The muscle in sphincters is characterized by the ability to remain contracted for long periods with little metabolic cost.

The intestinal tracts (small and large) contain two regions of smooth muscle cells within their walls. The outermost muscle layer, which is relatively thin, runs along the length of the intestine. The inner muscle layer, thicker and more powerful, has a circular arrangement. Coordinated alternating contractions and relaxations of these two layers mix and propel the contents of the intestine, although most of the motive power is provided by circular muscle. Control of the motility of the intestine through these smooth muscle layers is very complex and discussed in more detail in Chapter 25, "Neurogastroenterology and Motility."

The most complex arrangement of smooth muscle is found in organs such as the urinary bladder and uterus. Numerous layers and orientations of muscle fibers are present, and the effect of their contraction is an overall reduction of the volume of the organ. The relengthening force, in the case of these hollow organs, is provided by the gradual accumulation of contents. In the urinary bladder, for example, the muscle is gradually stretched as the emptied organ fills again with urine from the kidneys.

Smooth muscle cells lack the ultrastructural organization of skeletal muscle fibers.

In contrast to skeletal muscle fibers, smooth muscle cells are small relative to the tissue they make up. Individual smooth muscle cells are 100- to 300-μm long and 5 to 10 μm in diameter. When isolated from the tissue, the cells are roughly cylindrical along most of their length and taper at the ends. Electron microscopy reveals that the cell margins contain many areas of small membrane invaginations, called **caveolae**, which are sites of specialized cell functions. There is no organized T tubular system in smooth muscle, and its SR is not as abundant as that of skeletal muscle. The bulk of the cell interior is occupied by three types of myofilaments: thick, thin, and intermediate (Fig. 8.20). The *thin filaments* are similar to those of skeletal muscle but lack the troponin protein complex. The *thick filaments* are composed of myosin molecules, as in skeletal muscle, but the arrangement of the individual molecules into filaments is different; the filaments are called *side-polar*, with one crossbridge orientation on one side of the filament and the opposite orientation on the other. There is no central bare region without crossbridges. The **intermediate filaments** are so named because their diameter of 10 nm is between that of the thick and thin filaments. Intermediate filaments appear to have a cytoskeletal, rather than a contractile, function. Prominent throughout the cytoplasm are small, dark-staining areas called **dense bodies**. They are associated with the thin and intermediate filaments and are considered analogous to the Z lines of skeletal muscle. Dense bodies associated with the cell margins are often called **membrane-associated dense bodies** (or **patches**) or **focal adhesions**. They appear to serve as anchors for thin filaments and transmit the force of contraction to adjacent cells.

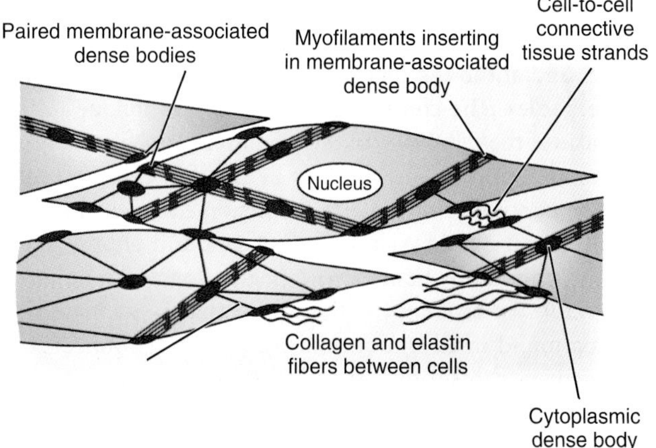

Figure 8.20 The contractile system and cell-to-cell connections in smooth muscle. Note regions of association between thick and thin filaments that are anchored by the cytoplasmic and membrane-associated dense bodies. A network of intermediate filaments provides some spatial organization (see, especially, the *left side*). Several types of cell-to-cell mechanical connections are shown, including direct connections and connections to the extracellular connective tissue matrix. Structures are not necessarily drawn to scale.

Smooth muscle lacks the regular sarcomere structure of skeletal muscle. There is some association among dense bodies down the length of a cell and a tendency of thick filaments to show a degree of lateral grouping. However, it appears that the lack of a strongly periodic arrangement of the contractile apparatus is an adaptation of smooth muscle associated with its ability to function over a wide range of lengths and to develop high forces.

Cell-to-cell transmission of force and electrical activity

A proposed arrangement of the smooth muscle contractile and force transmission system is also shown in Figure 8.20, which represents a current consensus view. Assemblies of myofilaments are anchored within the cell by the dense bodies and at the cell margins by the membrane-associated dense bodies. The contractile apparatus lies oblique to the long axis of the cell. Many of the membrane-associated dense bodies are opposite one another in adjacent cells and may provide continuity of force transmission between the contractile apparatus in each cell. Collagen and elastin fibers run throughout the tissue. These fibers, along with reticular connective tissue, comprise the **connective tissue matrix** or **stroma** found in all smooth muscle tissues. It connects the cells and gives integrity to the whole tissue.

Smooth muscle cells are also coupled electrically, although this functional connectivity is less prominent in vascular smooth muscle than it is in the muscle tissues of the gut and urogenital system. The structure most responsible for this coupling is the **gap junction**, which allows a low-resistance pathway for electrical currents to flow between cells. This facilitates the propagation of action potentials or other changes in membrane potential throughout tissue containing many smooth muscle cells.

Calcium required to contract smooth muscle is obtained from extracellular and intracellular sources.

An elevation of intracellular calcium concentration to about 1 μM is an absolute requirement for smooth muscle contraction, and a reduction in smooth muscle calcium concentration is considered a primary mechanism of smooth muscle relaxation. Figure 8.21 gives an overall picture of calcium regulation in smooth muscle. These processes may be grouped into those concerned with *calcium entry*, intracellular *calcium liberation*, and *calcium exit* from the cell. Calcium enters the cell through several pathways, including voltage-gated calcium channels (1,4 dihydropyridine), numerous different ligand-gated channels, stretch-activated calcium channels, and a relatively small number of unregulated **"leak" channels**. Calcium leaks permit the continual passive entry of small amounts of extracellular calcium. (Leak channels do not contribute significantly to smooth muscle contraction in normal physiologic conditions but do contribute to excessive smooth muscle contraction in diseases such as hypertension.)

Membrane depolarization opens voltage-gated calcium channels in smooth muscle that operate in a similar fashion to the voltage-gated sodium channels in nerves and skeletal muscle. These channels have an activation gate that opens during membrane depolarization starting at approximately −55 to −40 mV, along with an inactivation gate that closes with membrane depolarization several milliseconds after opening of the activation gate. The opening and closing kinetics of these types of calcium channels are much slower compared to that seen in voltage-gated sodium channels. The kinetics of voltage-gated calcium channels provides a short window of opportunity through which calcium can enter the cell from the extracellular fluid through the channel down

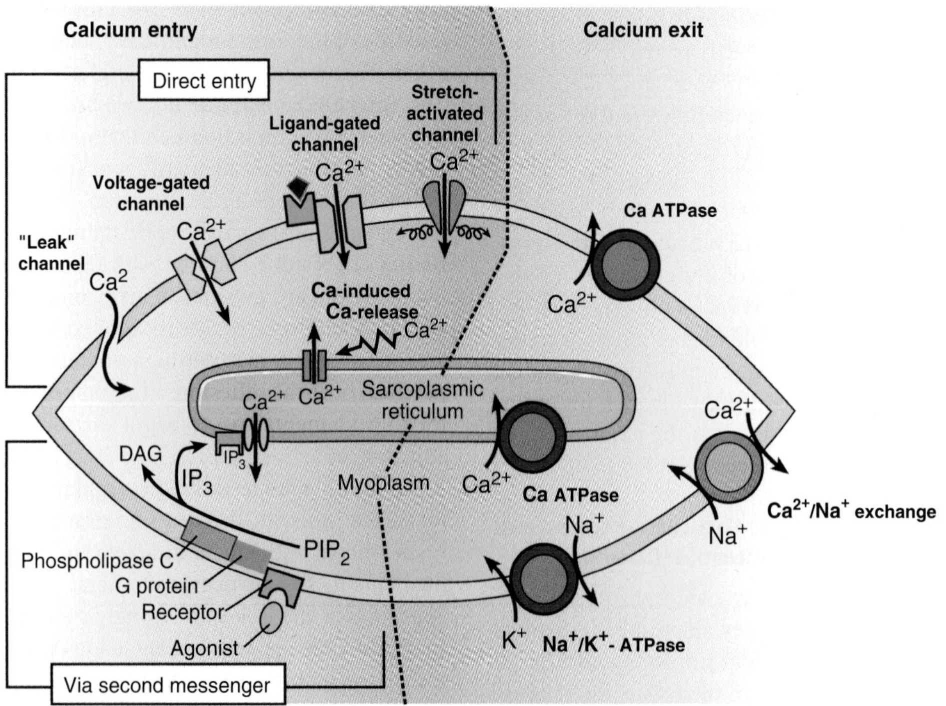

Figure 8.21 Major routes of calcium entry and exit from the cytoplasm of smooth muscle. Calcium enters the smooth muscle cell through voltage-gated, ligand-gated, stretch-activated, and "leak" calcium channels. Calcium can also be released into the cell from the sarcoplasmic reticulum (SR) by inositol 1,4,5-triphosphate (IP_3) and Ca^{2+}-stimulated calcium release. The processes on the *left side* increase cytoplasmic calcium and promote contraction; those on the *right side* decrease internal calcium and promote relaxation. DAG, diacylglycerol; IP_3, inositol 1,4,5-trisphosphate; PIP_2, phosphatidylinositol 4,5-bisphosphate.

an enormous electrochemical gradient. (The resting concentration gradient for calcium is more than a millionfold. The cell interior is negatively charged, and calcium ions contain a double positive charge.) This calcium influx is a significant source of activator calcium for smooth muscle contraction, and the rapid influx of positive charge from the calcium also creates an action potential in the smooth muscle cell. This action potential, however, is of smaller amplitude and has a slower upstroke than those caused by voltage-gated sodium channels in nerve and skeletal muscles.

Literally, dozens of different ligand-gated calcium channels exist among the smooth muscles in the body, with each type designed to be activated by a specific agonist, or ligand. Many of these channels open, allow calcium entry into the cell, and initiate contraction when a specific agonist binds to a specific receptor on the channel. Many hormones and neurotransmitters contract smooth muscle through this mechanism. Activation of ligand-gated calcium channels generally does not generate action potentials in smooth muscle except at very high agonist concentrations. Contraction of smooth muscle via ligand-gated calcium channels without the generation of an action potential is called **pharmacomechanical coupling**. Some ligand-gated calcium channels operate by decreasing the open time of the channel or decreasing the amount of time the channel is open. These ligand channels decrease intracellular calcium concentration and cause the muscle to relax.

Within the cell, the major storage site of calcium is the SR. In some types of smooth muscles, its capacity is quite small and these tissues are strongly dependent on extracellular calcium for their function. Nevertheless, calcium from the SR can contribute to smooth muscle contraction. Calcium is released from the SR by at least two mechanisms, including inositol 1,4,5-triphosphate (IP_3)-induced release and a **calcium-induced calcium release**. IP_3 formation is caused by activation of phospholipase C (PLC) by ligand-activated membrane receptors, which forms IP_3 and diacylglycerol (DAG) from phosphatidylinositol 4,5-bisphosphate (PIP_2) in the cell. In this latter mechanism, calcium that has entered the cell via a membrane channel causes additional calcium release from the SR, amplifying its activating effect.

Smooth muscle cells also have calcium channels that open whenever the smooth muscle cell is stretched. These channels are distinct entities and not simply voltage-gated or ligand-gated calcium channels that are opened by stretch (although there is some evidence to suggest that voltage-gated channels can be affected this way in some circumstances). Stretch-activated calcium channels play a key role in the regulation of vascular smooth muscle contraction because they are linked to the control of organ blood flow (see Chapter 15, "Microcirculation and Lymphatic System").

During a contraction, the cytoplasmic calcium level is significantly elevated. This sustained level is the result of a balance between mechanisms allowing calcium entry and those favoring its removal from the cytoplasm. Calcium leaves the myoplasm in two directions: A portion of it is returned to storage in the SR by an active transport system (a Ca^{2+}-ATPase), and the rest is ejected from the cell by two mechanisms. The more important of these is another ATP-dependent active transport system located in the cell membrane. The second mechanism, also located in the plasma membrane, is sodium–calcium exchange, a process in which the entry of three sodium ions is coupled to the extrusion of one calcium ion. This mechanism derives its energy from the large sodium gradient across the plasma membrane; thus, it depends critically on the operation of the cell membrane Na^+/K^+-ATPase.

Phosphorylation of myosin filament proteins is necessary for smooth muscle contraction.

Calcium ions are central to the control of contraction in smooth muscle, where they modulate the crossbridge cycle. However, the thin filaments of smooth muscle lack troponin, and the actin-linked regulation of skeletal muscle is not present in smooth muscle. Control of smooth muscle contraction relies instead on the thick filaments and is, therefore, called **myosin-linked regulation**. In actin-linked regulation, the contractile system is in a constant state of "inhibited readiness" and calcium ions remove the inhibition. In the case of smooth muscle myosin-linked regulation, the role of calcium is to cause *activation* of a resting state of the contractile system. The general outlines of this process are well understood and appear to apply to all types of smooth muscles, although a variety of secondary regulatory mechanisms are being found in different tissue types. This general scheme is shown in Figure 8.22. At rest, there is little cyclic interaction between the myosin and actin filaments because of a special feature of its myosin molecules. The S2 portion of each myosin molecule (the paired "head" portion) contains four protein *light chains*. Two of these have a molecular weight of 16,000 and are called essential light chains; their presence is necessary for actin–myosin interaction, but they do not appear to participate in the regulatory process. The other two light chains have a molecular weight of 20,000 and are called regulatory light chains; their role in smooth muscle contraction is critical. These chains contain specific locations (amino acid residues) to which the terminal phosphate group of an ATP molecule can be attached via the process of **phosphorylation**. The enzyme responsible for promoting this reaction is **myosin light-chain kinase (MLCK)**. When the regulatory light chains are phosphorylated, the myosin heads can interact in a cyclic fashion with actin, and the reactions of the crossbridge cycle (and its mechanical events) take place much as in skeletal muscle. It is important to note that the ATP molecule that phosphorylates a myosin light chain is separate and distinct from the one consumed as an energy source by the mechanochemical reactions of the crossbridge cycle.

For myosin phosphorylation to occur, the MLCK must be activated, and this step is also subject to control. Closely associated with the MLCK is **calmodulin (CaM)**, a smaller protein that binds calcium ions. When four calcium ions are bound, the CaM protein activates its associated MLCK and light-chain phosphorylation can proceed. It is this MLCK-activating step that is sensitive to the cytoplasmic calcium

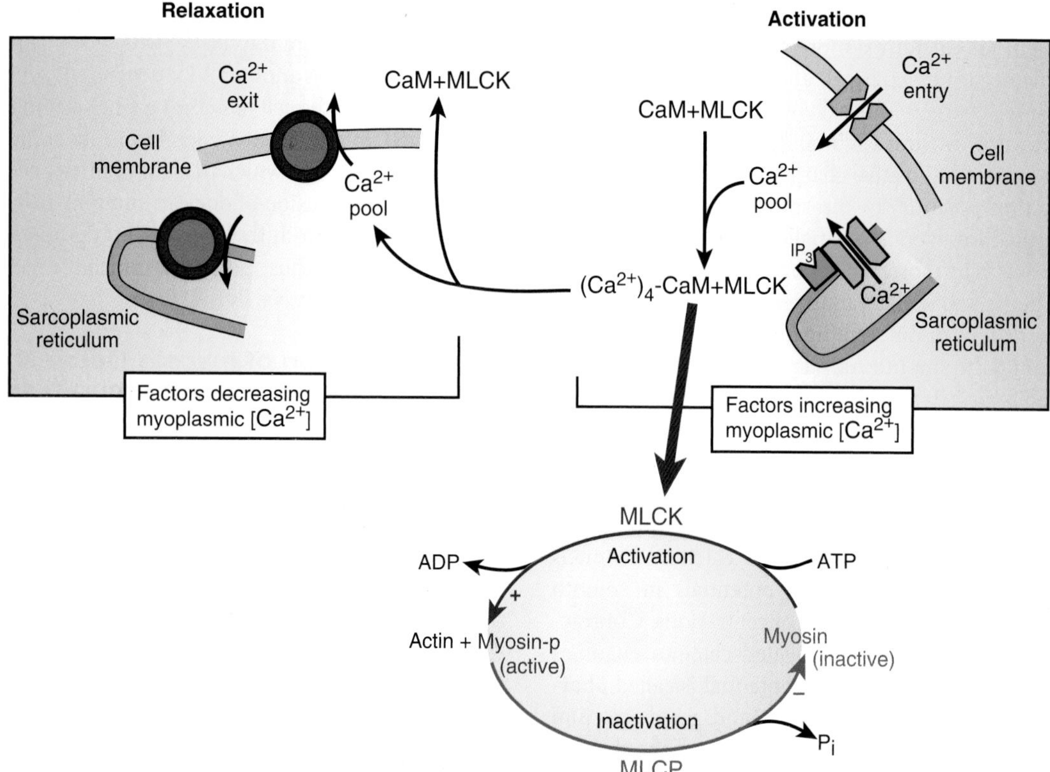

Figure 8.22 Reaction pathways involved in the regulation of the crossbridge cycle in smooth muscle. Activation begins (*upper right*) when cytoplasmic calcium levels are increased and calcium binds to calmodulin (CaM), activating the myosin light-chain kinase (MLCK). The kinase (*lower right*) catalyzes the phosphorylation of myosin, changing it to an active form (myosin-P or Mp). The phosphorylated myosin can then participate in a mechanical crossbridge cycle much like that in skeletal muscle, although much slower. When calcium levels are reduced, calcium leaves CaM, the kinase is inactivated, and the myosin light-chain phosphatase (MLCP) dephosphorylates the myosin, making it inactive. The crossbridge cycle stops, and the muscle relaxes. A, actin; ADP, adenosine diphosphate; ATP, adenosine triphosphate; P_i, inorganic phosphate ion.

concentration. At levels below 10^{-7} M Ca^{2+}, no calcium is bound to CaM and no contraction can take place. When cytoplasmic calcium concentration is $>10^{-4}$ M, the binding sites on CaM are fully occupied, light-chain phosphorylation proceeds at maximal rate, and contraction occurs. Between these extreme limits, variations in the internal calcium concentration can cause corresponding gradations in the contractile force. This, then, is the biochemical mechanism responsible for the fact that smooth muscle contraction can be graded between total relaxation and total contraction in contrast to the all-or-none type limitation to the activation of skeletal muscle contraction. Graded modulation of smooth muscle contraction is essential for its regulatory functions, especially in the vascular system.

The phosphorylation of myosin can be reversed by the enzyme **myosin light-chain phosphatase (MLCP)**. The activity of this phosphatase appears to be only partially regulated; that is, there is always some enzymatic activity, even while the muscle is contracting. During contraction, however, MLCK-catalyzed phosphorylation proceeds at a significantly higher rate, and phosphorylated myosin predominates. There is also some evidence that during ligand-gated contraction, a DAG-PKC (protein kinase C)–dependent pathway results in the formation of the protein CPI-17, which inhibits MLCP, thereby favoring the phosphorylated myosin state (Fig. 8.23). When the cytoplasmic calcium concentration falls, MLCK activity is reduced because the calcium dissociates from the CaM, and myosin dephosphorylation (catalyzed by the phosphatase) predominates. Because dephosphorylated myosin has a low affinity for actin, the reactions of the crossbridge cycle can no longer take place. Relaxation is, thus, brought about by mechanisms that lower cytoplasmic calcium concentrations, decrease MLCK activity, or increase MLCP activity. It also appears that actin itself can be made more or less available to interact with phosphorylated myosin and, thereby, control the extent of contraction of vascular smooth muscle. PKC, in particular, appears to activate a cascade of signal proteins within the smooth muscle cell to increase the availability of actin to then interact with phosphorylated myosin and promote vascular smooth muscle contraction.

Latch state is a mechanism that reduces the energy cost of continual smooth muscle contraction.

Smooth muscle contractile activity cannot be divided clearly into twitch and tetanus, as in skeletal muscle. In some cases, smooth muscle makes rapid phasic contractions, followed by complete relaxation. In other cases, smooth muscle can

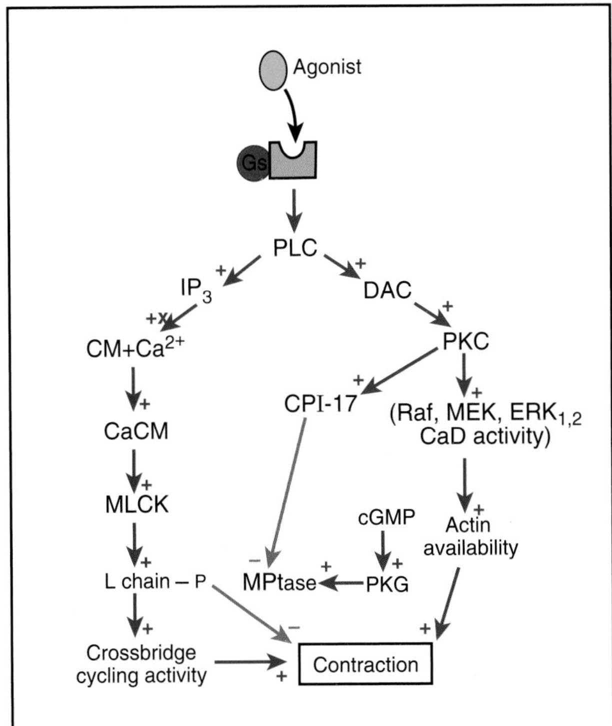

Figure 8.23 Regulation of contraction and relaxation in vascular smooth muscle. Myosin must be phosphorylated in vascular smooth muscle in order for crossbridge cycling to occur. The extent of myosin phosphorylation is believed to result from the net effect of phosphorylation by myosin light-chain kinase (MLCK) and dephosphorylation by myosin light chain phosphatase (MPtase). Contraction of vascular smooth muscle occurs through activation of MLCK and inactivation of MPtase. Other proteins also regulate contraction/relaxation by making actin more or less available to participate in cross-bridge cycling with phosphorylated myosin.

maintain a low level of active tension for long periods without cyclic contraction and relaxation; a long-maintained contraction is called *tonus* (rather than tetanus) or a *tonic contraction*. This is typical of smooth muscle activated by ligand-gated calcium channels (hormonal, pharmacologic, or metabolic factors), whereas phasic activity is more closely associated with activation of voltage-gated calcium channels (i.e., by membrane depolarization).

In vascular smooth muscle, for example, the force of contraction may be maintained for long periods. This extended maintenance of force capability is called the **latch state** (Fig. 8.24). The latch state is not a rigor state but, rather, appears to be a reduced rate of crossbridge cycling so that each remains attached for a longer portion of its total cycle. This condition is characterized by low levels of ATP utilization and has a lower calcium concentration requirement. Thus, the latch state can be thought of as a means of reducing the energy requirements of long, sustained contractions over what might be expected based on energy usage in short-duration contractions. Current understanding of the latch state suggests that it may occur when myosin is dephosphorylated while still attached to actin during the crossbridge cycle. In this sense, this dephosphorylation of myosin is

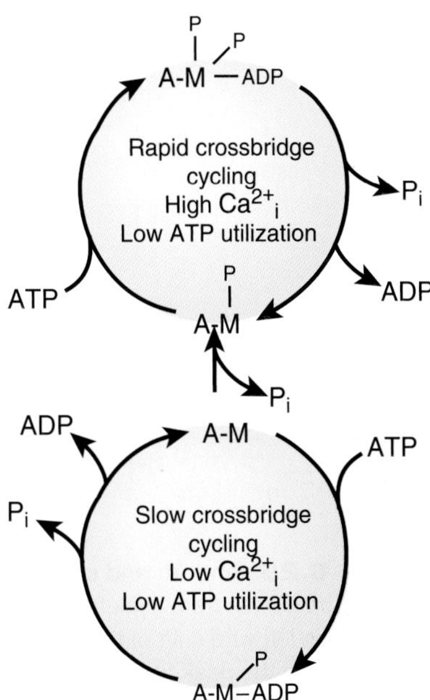

Figure 8.24 Representation of the latch state of the crossbridge cycle. When myosin is dephosphorylated while attached to actin, the crossbridge cycling slows because actin and myosin remain attached for longer durations than normal. This allows the muscle to maintain tonic contractions for long periods of time with a reduced energy requirement for adenosine triphosphate (ATP) and at a reduced intracellular calcium concentration. P, protein; P_i, inorganic phosphate; ADP, adenosine diphosphate; A-M, actin bound to myosin.

different than that which appears to be involved in smooth muscle relaxation.

Mechanical activity in smooth muscle is adapted for its specialized physiologic roles.

The contraction of smooth muscle is much slower than that of skeletal or cardiac muscle, whereas it can maintain contraction far longer and relax much more slowly than the striated muscles. The source of these differences lies largely in the chemistry of the interaction between the actin and myosin of smooth muscle. The inherent rate of the actomyosin ATPase correlates strongly with the velocity of shortening of the intact muscle. Most smooth muscles require several seconds (or even minutes) to develop maximal isometric force. A smooth muscle that contracts 100 times more slowly than a skeletal muscle will have an actomyosin ATPase that is 100 times as slow. The major source of this difference in rates is the myosin molecules; the actin found in smooth and skeletal muscles is rather similar. There is also a close association in smooth muscle between maximal shortening velocity and degree of myosin light-chain phosphorylation.

The high economy of tension maintenance, typically 300 to 500 times greater than that in skeletal muscle of similar size, is vital to the physiologic function of smooth muscle. Compared with that in skeletal muscle, the crossbridge cycle in smooth muscle is hundreds of times slower, and,

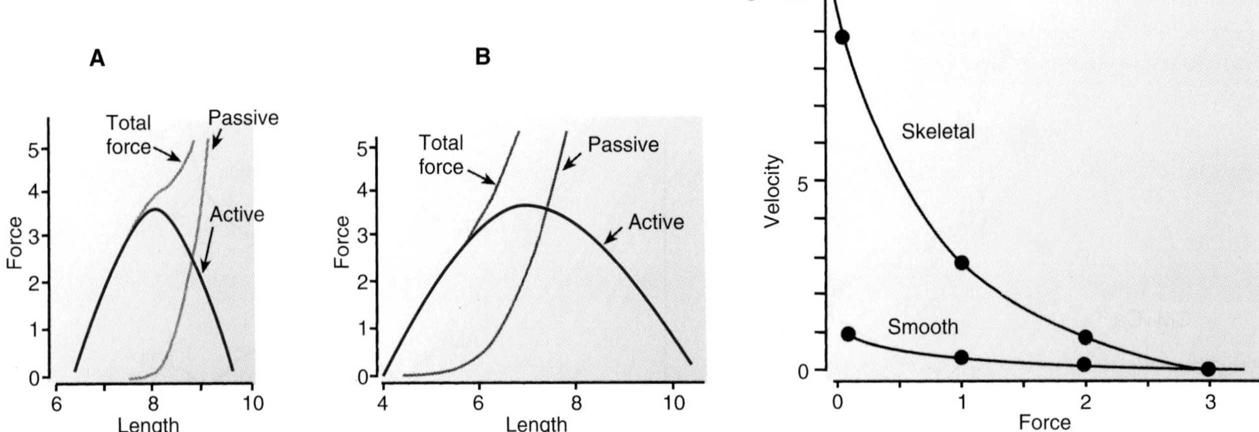

Figure 8.25 Smooth and skeletal muscle mechanical characteristics compared. (A and B) Typical length–tension curves from skeletal and smooth muscle. Note the greater range of operating lengths for smooth muscle and the leftward shift of the passive (resting) tension curve. **(C)** Skeletal and smooth muscle force–velocity curves. Although the peak forces may be similar, the maximum shortening velocity of smooth muscle is typically 100 times lower than that of skeletal muscle. (Force and length units are arbitrary.)

therefore, much more time is spent with the crossbridges in the attached phase of the cycle.

The force–velocity curve for smooth muscle (Fig. 8.25) reflects the differences in crossbridge functions described previously. Although smooth muscle contains one third to one fifth as much myosin as skeletal muscle, the longer smooth muscle myofilaments and the slower crossbridge cycling rate allow it to produce as much force per unit of cross-sectional area as does skeletal muscle. Thus, the maximum values for smooth muscle on the force axis would be similar, whereas the maximum (and intermediate) velocity values are different. Furthermore, smooth muscle can have a set of force–velocity curves, each corresponding to a different level of myosin light-chain phosphorylation. This is often revealed by the particular maximum contractions and contraction velocities produced by different receptor-mediated contractile agonists in smooth muscle. That is, three different contractile agonists of the same concentration on the same smooth muscle can produce three different levels of maximum force generation and contraction velocities in that smooth muscle.

Other mechanical properties of smooth muscle are also related to its physiologic roles. Although its underlying cellular basis is uncertain, smooth muscle has a length–tension curve somewhat similar to that of skeletal muscle, although there are some significant differences (see Fig. 8.25). At lengths at which the maximal isometric force is developed, many smooth muscles bear a substantial passive force. This is mostly a result of the network of connective tissue that supports the smooth muscle cells and resists overextension. In some cases, it may be partly a result of residual interaction between actin and attached but noncycling myosin crossbridges. Compared with skeletal and cardiac muscle, smooth muscle can function over a significantly greater range of lengths; its length–tension relationship is much broader with less of a defined L_0 compared to that of skeletal muscle. This is a beneficial property in that smooth muscle is not constrained by skeletal attachments, and it makes up several organs that vary greatly in volume (and therefore vary greatly in muscle stretch) during the course of their normal functioning.

Smooth muscle has several modes of relaxation.

The central cause of smooth muscle relaxation is a reduction in the internal (cytoplasmic) calcium concentration, a process that is itself the result of several mechanisms. Electrical repolarization of the plasma membrane leads to a decrease in the influx of calcium ions, whereas the plasma membrane calcium pump and the sodium–calcium exchange mechanism (to a lesser extent) actively promote calcium efflux. Most important quantitatively is the uptake of calcium back into the SR. The net result of lowering the calcium concentration is a reduction in MLCK activity so that dephosphorylation of myosin can predominate over phosphorylation.

The cyclic nucleotides cAMP and cGMP are second messenger links between smooth muscle relaxation and activation of specific smooth muscle membrane receptors by chemical ligands that mediate relaxation of smooth muscle. For example, binding of catecholamines to β-adrenergic receptors in vascular and bronchial smooth muscle activates **adenylyl cyclase**, which results in the formation of **cyclic adenosine monophosphate (cAMP)**. cAMP triggers a cascade of events within the smooth muscle that lead to relaxation. These include enhanced calcium uptake by the SR and decrease myosin light-chain phosphorylation.

Nitric oxide and nitrodilators such as nitroglycerin relax smooth muscle by activating guanyl cyclase and producing the second messenger cyclic guanosine monophosphate. This leads to the activation of **cGMP-dependent protein kinase (PKG)**, which promotes the opening of calcium-activated potassium ion channels in the cell membrane, leading to hyperpolarization and subsequent relaxation. PKG also activates MLCP and blocks the activity of agonist-evoked **phospholipase C (PLC)**. This action reduces the liberation of stored calcium ions by IP_3.

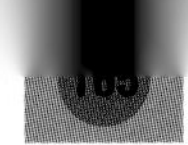

Chapter Summary

- The contractile apparatus of skeletal muscle is made up of thick (myosin-containing) and thin (actin-containing) myofilaments arranged in interdigitating arrays called *sarcomeres*.
- Contraction is brought about by a cyclic enzymatic interaction between the myofilaments. The biochemical reactions of the crossbridge cycle require adenosine triphosphate (ATP) as a metabolic fuel. The power stroke of the myosin head provides the force to cause relative myofilament sliding and sarcomere shortening.
- The neuromuscular junction is a specialized chemical synapse that links motor nerve impulses to skeletal muscle activation.
- In skeletal muscle, membrane excitation is coupled to contraction by the action of calcium on the troponin regulatory complex on the thin filaments.
- Calcium ions are stored in the sarcoplasmic reticulum (SR) and released to the myofilaments upon depolarization of the T-tubule membranes. Binding of calcium ions to troponin allows the interaction of actin and myosin filaments in the crossbridge cycle. Relaxation occurs when calcium is pumped back into the SR.
- Muscle contraction is fueled by ATP derived primarily from the metabolism of carbohydrates and also by metabolism of amino acids and fatty acids.
- Contractile force in whole muscle is adjusted by varying the number of motor units stimulated and by the temporal pattern of stimulation. Brief, single contractions are twitches; summed contractions are tetani.
- Isometric contraction occurs at constant muscle length, and isotonic contraction occurs at constant force. The length–tension curve relates initial, passive muscle length to isometric tension, and the force–velocity curve relates speed of shortening to the force exerted by the muscle while moving a load.
- Skeletal attachments act as a lever system, increasing the effective range of muscle shortening while reducing the amount of force a limb can exert. These attachments also confine the muscle operation to near the optimal length on its length–tension curve.
- Metabolic and structure adaptations fit skeletal muscle for a variety of roles. Red muscle fibers are associated with aerobic metabolism, whereas white muscles operate anaerobically.
- Smooth muscle is specialized for slow and maintained contraction. It is an important component of visceral organs, where it controls flow, propels contents, and expels the contents of reservoir organs. It has an actin–myosin contractile system, although it lacks the regular ultrastructure of skeletal muscle.
- Regulation of smooth muscle contraction is under the control of calcium ions, which promote, via the enzyme myosin light-chain kinase, phosphorylation of the light chains on the myosin molecule. This enables it to interact with actin and participate in the crossbridge cycle.
- Smooth muscle cells are electrically coupled, and activation spreads from cell to cell throughout the tissue. Nerves may initiate contraction or modify its duration and extent.
- The latch state of smooth muscle is a condition of slow crossbridge cycling that conserves energy during tonic contractions.
- Relaxation in smooth muscle is associated with a lowering of cytoplasmic calcium and dephosphorylation of the myosin through the actin of a phosphatase enzyme.

thePoint *Visit http://thePoint.lww.com for chapter review Q&A, case studies, animations, and more!*

9 Blood Components

ACTIVE LEARNING OBJECTIVES

Upon mastering the material in this chapter, you should be able to:

- Understand the four major functions of blood: transport, hemostasis, homeostasis, and immunity.
- Explain how the blood lipid profile can be applied to assess cardiovascular health.
- Discuss the clinical usefulness of both the basic metabolic panel and the complete metabolic panel.
- Understand the clinically relevant difference between blood density and viscosity.
- Describe the principles and general uses of serum protein electrophoresis, erythrocyte sedimentation rate, and hematocrit.
- Explain the value of the tests included in the complete blood cell count.
- Explain the rationale for having different blood collection tubes and know which tubes are used to obtain serum and plasma.
- Explain how pathologic changes in erythrocyte morphology can affect function.
- Explain how erythrocyte dysfunction causes different types of anemia.
- Explain how the iron profile can assist in diagnosing iron deficiency or overload.
- Explain the function and clinical significance of erythropoietin as well as its involvement in blood doping.
- Track the roles and components of the blood clotting phases, from immediate actions to wound healing.
- Explain the relevance of blood type in blood transfusions.

To understand the function of the body as a whole, it is crucial to know about the component that affects every single cell: the blood. Blood flows through the arteries, carrying oxygen and nutrients to every part of the body and transporting carbon dioxide and other waste products to the lungs, kidneys, and liver for disposal. Blood interacts with individual cells via an extensive capillary network, maintaining a constant surrounding environment that includes pH, salt concentration, temperature, and blood sugar. Blood tests, therefore, are common diagnostic tools to detect homeostatic imbalances. Blood also plays a role in fighting infection, which is discussed in Chapter 10, "Immunology, Organ Interaction, and Homeostasis."

BLOOD FUNCTIONS

Blood is a dynamic and complex living tissue carrying out four major roles: transport of substances, arrest of bleeding (hemostasis), maintenance of a stable internal environment (homeostasis), and aid in resisting infection or disease (immunity). Cellular and plasma components work in concert to perform these functions.

Transport. Blood is the primary means of long-distance transport in the body. It carries an abundance of important substances from one area of the body to another, including antibodies, acids and bases, ions, vitamins, cofactors, hormones, nutrients, lipids, pigments, metabolites, and minerals. These substances might be freely dissolved in solution or bound to carrier proteins and other molecules. For example, about half of Ca^{2+} is present in its free form, whereas the other half is bound to albumin and small diffusible anions. Some substances are preferentially transported within blood cells, such as O_2 and CO_2 bound to **hemoglobin (Hgb or Hb)** within **red blood cells**. In addition to transporting materials, blood also transports heat. The circulatory system is responsible for homeothermy by distributing heat throughout the body.

Hemostasis. Complex and efficient mechanisms have evolved to prevent blood loss from a damaged blood vessel. The arrest of bleeding is called **hemostasis**. The failure to stop bleeding after injury is called **hemorrhage** and can quickly lead to bleeding to death.

Homeostasis. **Homeostasis** is a steady state that provides an optimal internal environment for cell function. By maintaining pH, ion concentrations, osmolality, temperature, nutrient supply, and vascular integrity, the blood system plays a crucial role in preserving homeostasis. Homeostasis is the result of normal functioning of the blood's transport, immune, and hemostatic systems.

Immunity. Blood **leukocytes** are involved in the body's battle against infection by microorganisms (see Chapter 10, "Immunology, Organ Interaction, and Homeostasis"). Although the skin and mucous membranes physically limit the entry of infectious agents, microbes constantly overcome these barriers and threaten to cause infection. Blood leukocytes, working in conjunction with proteins, are continuously on patrol in the blood to detect microorganisms and other foreign substances. In most cases, the blood's defense system is efficient enough to eliminate the pathogens or to prevent their spreading before they can cause substantial bodily harm.

WHOLE BLOOD

The average adult has about 5 L (5.3 quarts) of whole blood, with 5.0 to 6.0 L in men and 4.5 to 5.5 L in women. About 2.75 L (55%) is the liquid portion of blood; the rest (45%) is the cellular portion of blood. Blood accounts for 6% to 8% of the body weight of a healthy adult.

Blood is a specialized connective tissue containing formed elements: red cells, white cells, and platelets.

Whole blood belongs histologically to connective tissue because it is, like other connective tissue, composed of formed elements and intercellular matter. The formed elements are red and **white blood cells** and **platelets**, and the intercellular matter is the liquid component of blood (**plasma**).

Whole blood is bright red in color, which derives from the oxygenated iron in hemoglobin (Hgb). Deoxygenated blood is darker red, which can be seen in venous blood samples. Veins, when viewed through the skin, typically appear blue in color as a result of the deflection of light when it penetrates the skin.

Medical terms related to whole blood often begin with **hem/o-** or **hemat/o-** from the Greek word *haima* for blood. For instance, ***hemo*lysis** is the premature destruction of **erythrocytes**, and a ***hemato*logist** is a blood specialist.

To obtain whole blood for laboratory analysis, an **anticoagulant** must be added to keep the blood from clotting. Clinically, there are not many assays in which whole blood is used, because it is so viscous and cloudy that it interferes with the chemical reactions in the test solutions. However, there is a continual effort to develop immunoassays that overcome these problems and allow the use of whole blood for fast diagnosis in emergency situations.

SOLUBLE COMPONENTS OF BLOOD AND THEIR TESTS

Blood is the transport medium for many organic and inorganic substances in the body. These are dissolved in the plasma and participate in many cellular processes.

The most abundant electrolytes in the plasma include Na^+ and Cl^- with smaller amounts of HCO_3^-, K^+, and Ca^{2+}. Some of the functions of these electrolytes include their role in membrane excitability, osmotic pressure, pH buffers, and fluid flux between interstitial fluid and cells. In addition, the liquid portion of the blood transports large amounts of internal heat to the skin.

The most abundant organic constituents of the plasma are the plasma proteins. Other constituents include nutrients (glucose, amino acids, lipids, and vitamins), which are being transported to tissues.

Plasma becomes serum after the removal of clotting factors.

Plasma is the liquid portion of the blood. It comprises mostly water (93%), with the remaining 7% being various dissolved solutes (6% organic substances and 1% inorganic substances). Plasma is obtained by collecting whole blood in a tube containing an anticoagulant, such as ethylenediaminetetra-acetic acid (EDTA), citrate, or heparin. Centrifugation at 3,000 × g for about 5 minutes will separate the blood into an upper plasma layer, a lower red blood cell layer, and a thin interface, called **buffy coat**, containing white blood cells and platelets.

Plasma is a cloudy, pale, or grayish-yellow liquid. When it appears milky, it might be high in lipids, such as that from a patient with hyperlipidemia or possibly even from someone who did not fast long enough before the blood draw. Reddish samples might be due to hemolysis, which indicates that red blood cells have lysed and Hgb was released.

When blood is allowed to clot or coagulate, usually at room temperature for 15 to 30 minutes, the remaining liquid after centrifugation is referred to as **serum**. Serum is a clear, watery fluid.

For many biochemical tests, plasma and serum can be used interchangeably. For some tests, only serum can be used because the clotting factors in plasma interfere with the assay. For coagulation tests, only plasma can be used because all clotting factors need to be present.

Plasma can be stored frozen below −20°C for future analysis, but it must be frozen within 6 to 8 hours after donation to preserve clotting factors. Such **fresh frozen plasma (FFP)** can be stored in a blood bank for 1 to 7 years and can be used for therapeutic plasma exchange called **plasmapheresis**. FFP frozen within 6 and 24 hours of whole blood collection has a lower level of labile coagulation factors, such as factors V and VIII.

Serum is also used as a supplement to cell culture media. The numerous proteins, nutrients, hormones, and attachment factors present in serum provide for the successful growth of cells *in vitro*. For the growth of human cells, fetal calf serum is used most often.

Analyzing blood samples reveals a patient's health status.

A blood draw is necessary for hematologic testing. A blood sample is typically drawn from a patient's arm vein (**venipuncture**). In newborns, blood is collected by a **heel stick**. Blood from an artery is collected mainly to measure arterial blood gases, such as O_2, CO_2, and HCO_3^-, to determine their levels related to lung function before tissue extraction changes the values. For most other natural body substances, values from the arterial and venous sides are equal. For many drugs, arterial concentrations are higher than venous concentrations due to their extraction by tissues. A needle stick from an artery is more painful than from a vein.

Blood is collected into evacuated collection tubes, designed to fill with a predetermined volume (typically 7 mL) of blood by vacuum. The rubber stoppers are color coded according to the additive that the tube contains. Lavender (purple) tubes contain EDTA to bind calcium, a critical component for blood clotting. They are for hematology tests such as **complete blood counts (CBCs)** and are used at the blood bank for blood typing. Light blue tubes used for coagulation tests contain sodium citrate, which is another way to bind calcium and thereby produce plasma. In green tubes, the anticoagulant heparin (sodium, lithium, or ammonium heparin) is used to obtain plasma for a variety of clinical chemistry tests. Red tubes (serum separator tubes) do not have any additives except for clot activators and are, therefore, used to produce serum for chemistry tests.

Before being sent to the clinical laboratory, each blood specimen must be labeled with the patient's name and identification number and accompanied by a requisition form that links the requested test with a specific symptom/sign of the patient. Each tube must also display the phlebotomist's initials, the date, and the time of draw.

Plasma contains key electrolytes and dissolved nutrients: glucose, amino acids, and fatty acids.

About 55% of whole blood is made up of plasma, the liquid portion of blood. Plasma contains electrolytes, which are negatively (**anions**) or positively (**cations**) charged substances that conduct electricity. Because of the principle of **electroneutrality** in blood plasma, the sum of positive charges equals the sum of negative charges. Na^+ is the most abundant cation in plasma, and Cl^- and HCO_3^- are the major anions. These three ions are the main osmotically active solutes in the extracellular fluid. A gain or loss of Na^+ will cause volume expansion or contraction, respectively. Some electrolytes are proteins, which bear negative charges at physiologic pH. Some of the components of plasma are listed in Table 9.1.

There are close to 1,400 different *plasma proteins* identified, which amounts to a total of about 6 to 8 g of proteins/dL. Proteins are needed as building blocks of cells and tissues; they function as enzymes, hormones, antibodies, and transporters; they contribute to plasma osmolality and acid-base balance; and they serve as an energy source under limiting conditions.

Albumins are plasma proteins that are small in size but account for about 60% of the total plasma protein concentration. A primary function of albumin is to maintain **oncotic pressure**, which is about 0.5% of the total osmotic pressure. Oncotic pressure plays an important role in the exchange of fluid across capillaries, because albumin cannot cross the capillary membrane easily. It is also called *colloidal osmotic blood pressure*, because when combined with water, albumin and other globular proteins form a colloid, or a solution that appears homogenous. A second function of albumin is to transport fatty acids,

TABLE 9.1 Basic Components of Plasma

Class	Substance	Reference Range
Cations	Sodium (Na^+)	136–145 mEq/L
	Potassium (K^+)	3.5–5.0 mEq/L
	Calcium (Ca^{2+})	4.2–5.2 mEq/L
	Magnesium (Mg^{2+})	1.5–2.0 mEq/L
	Iron (Fe^{3+})	50–170 μg/dL
	Copper (Cu^{2+})	70–155 μg/dL
	Hydrogen (H^+)	35–45 nmol/L
Anions	Chloride (Cl^-)	95–105 mEq/L
	Bicarbonate (HCO_3^-)	22–26 mEq/L
	Lactate$^-$	0.67–1.8 mEq/L
	Sulfate (SO_4^{2-})	0.9–1.1 mEq/L
	Phosphate (HPO_4^{2-}/$H_2PO_4^-$)	3.0–4.5 mg/dL
Proteins	Total	6–8 g/dL
	Albumin	3.5–5.5 g/dL
	Globulin	2.3–3.5 g/dL
Fats	Cholesterol	150–200 mg/dL
	Phospholipids	150–220 mg/dL
	Triglycerides	35–160 mg/dL
Carbohydrates	Glucose	70–110 mg/dL
Other examples	Alkaline phosphatase	20–140 U/L
	Alanine transaminase	8–20 U/L
	Aspartate transaminase	9–40 U/mL
	Bilirubin (total)	0.1–1.0 mg/dL
	Blood urea nitrogen	7–18 mg/dL
	Creatinine	0.6–1.2 mg/dL
	Ketones	0.2–2.0 mg/dL
	Vitamin A	0.15–0.6 μg/mL
	Vitamin B12	200–800 pg/mL
	Vitamin C	0.4–1.5 mg/dL

hormones, drugs, and other substances through blood. Albumins are produced by the liver, and low concentration in serum might indicate liver disease and/or malnutrition.

Globulins constitute about 36% of the total plasma proteins. α and β globulins are produced in the liver and perform diverse functions as transporters or substrates. For instance, the transporter for thyroxin is an α2 globulin, and plasminogen, the inactive form of the enzyme plasmin, is a β globulin. **Gamma globulins**, produced by lymphoid tissue, are antibodies necessary for immune defense (see Chapter 10, "Immunology, Organ Interaction, and Homeostasis"). Serum protein electrophoresis (described in greater detail below) separates and quantifies globulins.

Lastly, *fibrinogens* are large molecules synthesized in the liver, which account for about 4% of the total plasma protein concentration. When they are converted to the insoluble protein **fibrin**, they form the structure of a blood clot. *In vitro*, plasma freed of fibrinogen will not coagulate. Higher levels of plasma fibrinogen are correlated with an increased risk of stroke.

The main *plasma lipids* are cholesterol, phospholipids, and triglycerides. Cholesterol is an important component of cell membranes as well as a precursor for certain hormones. Phospholipids are formed in the liver and are mainly used for building plasma membranes because they can form lipid bilayers. Triglycerides are important for the transfer of food-derived energy into cells. Because of their hydrophobic nature, plasma lipids are transported together with proteins.

Dietary triglycerides and dietary cholesterol are assembled in the intestinal mucosa as **chylomicrons** before being released via lymph into the circulation. Triglycerides and cholesterol from the liver are packaged as **very low–density lipoproteins (VLDLs)** and released into the plasma for the delivery of lipids to other tissues. VLDLs can be converted to **intermediate-density** or **low-density (LDL) lipoprotein**. The liver and small intestine also form and release **high-density lipoprotein (HDL)**, which is initially nearly devoid of any cholesterol or cholesterol ester, so that the high proportion of protein makes them appear "dense." HDL is capable of picking up cholesterol from cells, including **macrophages** of atherosclerotic arteries, and delivering it to the liver, where it is passed from the body via bile. The blood lipid profile (described in greater detail below) tests individual serum lipids.

The major *plasma carbohydrate* is glucose. There are only traces of other sugars present in the blood. Glucose is the primary source of energy for the body's cells, including red blood cells. Serum glucose level is tightly regulated throughout the day within 70 to 110 mg/dL (4–8 mmol/L), except for the time immediately following a meal. Hence, blood glucose tests are standardized in regard to food intake (e.g., fasting test, 2-hour postprandial test). When quantifying blood glucose, the serum must be promptly separated from the red blood cells because they can use up glucose without the presence of insulin.

Blood lipid profile helps determine a patient's cardiovascular disease risk.

A standard **blood lipid profile** includes values for total cholesterol, LDL cholesterol, HDL cholesterol, and total triglycerides. Blood lipid profiles give important information about a patient's risk of developing heart-related conditions, with high LDL, low HDL, and/or high triglycerides being positive indicators of health risk. With high LDL levels, the risk of cholesterol plaque buildup in arterial walls (**atherosclerosis**) is increased because LDLs are the primary plasma transporters of cholesterol to tissues. Low HDL levels indicate decreased ability to remove excess cholesterol. Many people with obesity, heart disease, and/or diabetes also have high triglyceride levels. Several ratios, such as the LDL/HDL ratio and the total cholesterol/HDL ratio, are used as risk predictors.

Basic and complete metabolic panels comprise specific blood chemistry tests.

The **basic metabolic panel (BMP)** is a group of blood chemistry tests that includes the analysis of glucose, Ca^{2+}, Na^+, K^+, CO_2 (or HCO_3^-), Cl^-, blood urea nitrogen (commonly referred to as *BUN*), and creatinine. The latter two parameters are metabolites that are filtered out of plasma by the kidneys, so their increase in plasma might indicate kidney dysfunction. Depending on the clinical problem, electrolytes only (an *electrolyte panel*) or any individual parameters might be monitored.

The **complete metabolic panel (CMP)** includes the tests of the BMP and the following additional tests: albumin, total protein, alkaline phosphatase (ALP), alanine transaminase (ALT, also called *serum glutamic pyruvic transaminase*, or *SGPT*), aspartate transaminase (AST, also called *serum glutamic oxaloacetic transaminase*, or *SGOT*), and bilirubin. The latter four components aim to assess liver function. For instance, ALP is elevated in hepatitis, ALT is often increased with bile duct obstruction, and AST detects general liver damage. Bilirubin is a waste product of the liver, produced from disassembled red blood cells. Depending on its form (conjugated or not), bilirubin levels can be used to identify a problem that occurs before the liver (e.g., hemolytic anemia), a problem within the liver (e.g., metabolic problem), or a problem after the liver (e.g., bile duct blockage).

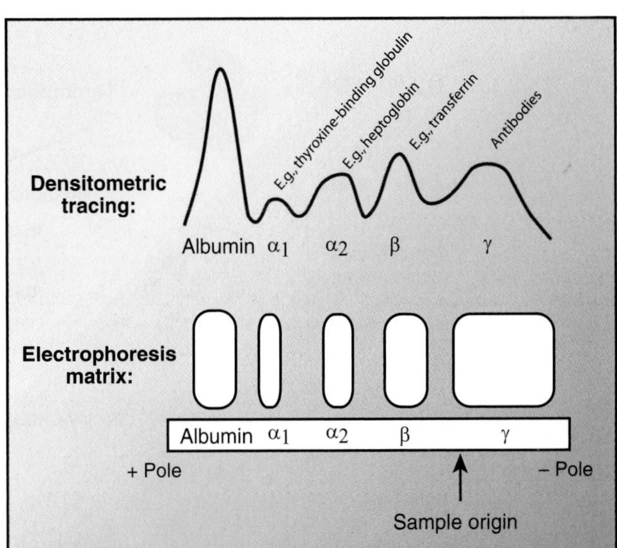

Figure 9.1 Serum protein electrophoresis. Globular serum proteins are electrophoretically separated on a matrix and stained. The intensity of the dye can be densitometrically quantified, and changes to the normal pattern are clinically relevant.

Abnormal serum protein patterns in electrophoresis reveal health problems.

Serum protein electrophoresis is a common method for separating blood proteins in a solid matrix (mostly cellular acetate) according to their size and charge (Fig. 9.1). Globular blood proteins (albumin, α, β, and γ globulins) normally form five main peaks, or zones, on the matrix: albumin (59%), α1-zone (4%), α2-zone (7.5%), β-zone (12%), and γ-zone (17.5%). These zones might differ in size and/or pattern from normal in a patient with certain types of anemias, during acute inflammation, or in the presence of an autoimmune disease. For instance, an increased β1 peak is typical of iron deficiency anemia due to an increased level of the iron-binding protein transferrin. Another example is a largely increased γ globulin peak in most patients with multiple myeloma due to the presence of Bence Jones protein produced by abnormal plasma cells.

Immunologic assays detect and measure serum antigens and antibodies.

A wide variety of serum immune assays are available in which a precipitation reaction occurs as a result of the combination of antibodies and antigen. The detection and quantification of either antibodies or antigens are diagnostically valuable. For example, elevation of smooth muscle cell antibodies and antinuclear antibodies point toward autoimmune hepatitis. Elevation of gluten or antigliadin antibodies can be used to diagnose celiac disease. On the other hand, detection of the protein antigen present on the surface of the hepatitis B virus can identify infected people before symptoms appear.

In some cases, large antigens produce large aggregates with antibodies visible to the naked eye. For instance, this is the case for the instantaneously readable ABO-Rh blood typing test kits, in which the antigens on the surface of a person's red blood cells agglutinate with added antibodies. In most cases, however, the antigen–antibody complexes must be labeled for detection, either with dyes and fluorescent reagents or with radioactive isotopes, respectively. The antibody titer is established by the highest dilution of serum that produces a visible reaction with antigen. The more the sample needs to be diluted, the greater is the antibody concentration.

There are many additional important blood tests available to support the clinician's work. Please see Chapter 10, "Immunology, Organ Interaction, and Homeostasis," for acute phase proteins; Part IV, "Cardiovascular Physiology," for important cardiac tests; Part VI, "Renal Physiology and Body Fluids," to see how the body deals with toxins; Part VII, "Gastrointestinal Physiology," for a discussion on vitamins; and Part IX, "Endocrine Physiology," for an analysis of hormones.

FORMED ELEMENTS OF BLOOD AND THEIR TESTS

The formed elements of blood are blood cells, which include erythrocytes (red blood cells), leukocytes (white blood cells), and thrombocytes (platelets). Thrombocytes are derived from special bone marrow cells, but in blood, they are present as cell fragments. Each microliter of blood contains 4 to 6 million erythrocytes, 4,500 to 10,000 leukocytes, and 150,000 to 400,000 thrombocytes (Fig. 9.2). Erythrocytes and the various subtypes of leukocytes, as defined by morphologic and functional characteristics and capabilities, are described in more detail in the sections *Red Blood Cells* and *White Blood Cells*, respectively, later in this chapter. More details on thrombocytes are in the section *Blood Clotting*, later in this chapter.

Blood density and viscosity convey the functional status of blood.

Density is defined as mass per unit volume, and relative density with respect to water is called **specific gravity**. The specific gravity for pure water is 1.000 g/mL and that for whole blood is approximately 1.050 g/mL. The exact value depends on the number of blood cells present and the composition of the plasma. In venous blood, it is slightly higher when a person is standing compared to sitting. The density of blood

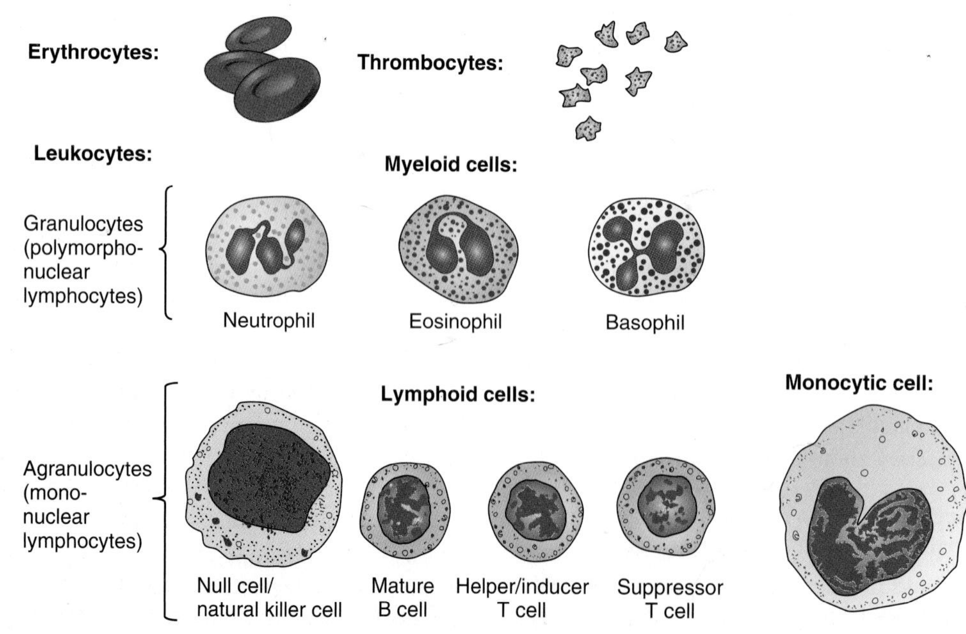

Figure 9.2 Formed elements of blood. This figure shows the size and appearance of cells and cell fragments (thrombocytes) found in the circulation.

plasma is approximately 1.025 g/mL. The density of individual blood cells varies according to cell type and ranges from 1.115 g/mL for erythrocytes to 1.070 g/mL for certain leukocytes.

Specific gravity is measured clinically in urine samples or blood by the extent of light bending in a refractometer. For fast dipstick technology (a test strip held in the tested liquid), specific gravity is estimated by the color change of an indicator dye in the presence of ions. A high specific gravity is most often indicative of dehydration, but other interpretations are possible.

Although blood is only slightly heavier than water, it is certainly much thicker. The **viscosity** of blood, a measure of resistance to flow, is 3.5 to 5.5 times that of water. The viscosity of plasma is about 1.5 to 1.8 times that of water.

Blood's viscosity increases as the total number of cells or platelets in a given volume increases and, to a much smaller extent, when the concentration of macromolecules in plasma increases. Increased viscosity impairs blood flow so that a higher blood pressure is required to achieve tissue perfusion. Furthermore, high viscosity blood tends to coagulate more easily. In healthy people, a slight increase in blood cell concentration and, hence, blood viscosity, for instance due to dehydration from the flu, can be easily tolerated. However, in those with more viscous blood, such as in patients with lung disease, the additional slight increase may lead to a stroke or myocardial infarction. Blood viscosity also increases with decreasing temperature by about 2% for each degree Celsius.

Decreased viscosity of blood can lead to turbulent flow, rather than the normal, more energy-efficient laminar flow.

Measurements of plasma viscosity (together with erythrocyte sedimentation rate, see below) are inexpensive and useful tests to monitor inflammation beyond the acute phase (after the first 24 hours). To analyze viscosity, a small sample of plasma is drawn through a narrow capillary using constant pressure, and the time that the sample takes to move a known distance is measured.

High red blood cell sedimentation rates indicate inflammation.

Erythrocytes have a slightly higher density than the suspending plasma, so that they slowly settle out of whole blood. This **erythrocyte sedimentation rate (ESR)** is an inexpensive, important diagnostic monitoring index, because values are often significantly elevated in patients with infection, autoimmune diseases, and inflammatory diseases. The relevance of the test for diagnoses of specific diseases is decreasing, however, because of the development of new methods for evaluating disease.

Erythrocytes sediment faster with increased plasma protein concentration. One explanation is **rouleaux phenomenon**, in which the red cells stack like coins, and therefore, sediment faster and elevate ESR. Any condition that elevates fibrinogen may also elevate the ESR, as do macrocytic red blood cells. In **sickle cell anemia** and hyperglycemia, the ESR is slower than normal.

To determine the ESR, blood anticoagulated with EDTA or sodium citrate is placed in a long, thin, graduated tube (Fig. 9.3). As the red cells sink, they leave behind the less dense leukocytes and platelets in the suspending plasma. Erythrocytes in the blood of healthy men sediment at a rate of up to 15 mm/h and those in the blood of healthy women often sediment slightly faster, at a rate of up to 20 mm/h. The normal range varies depending on the type of test tube used. Traditionally, the result is read after 1 hour. A newer method is available that measures the rate of infrared light blockage by settling erythrocytes for 15 minutes and predicts the 1-hour regular ESR value.

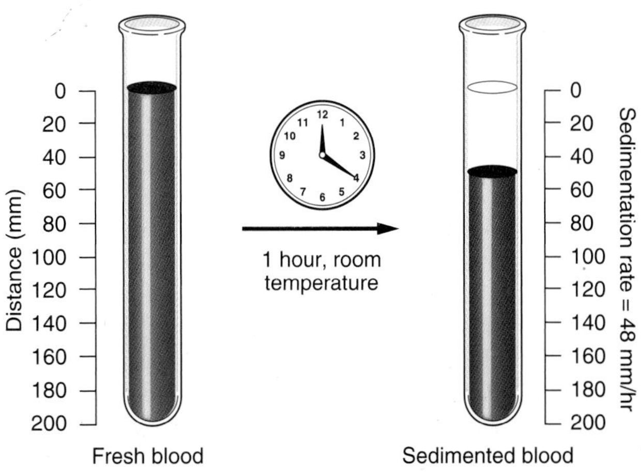

Figure 9.3 Determination of the erythrocyte sedimentation rate (ESR). Fresh, anticoagulated blood is allowed to settle at room temperature in a thin, graduated tube. After a fixed time interval (1 hour), the height (in millimeters) that the erythrocytes sediment is measured.

Altered hematocrit indicates a blood disorder.

The **hematocrit (Hct)** is the fraction of the total blood volume that is made up of red blood cells. Hct can be determined by centrifugation of anticoagulated blood within small capillary tubes to separate blood cells from the suspending fluid.

Hct and plasma volume can be used to calculate blood volume.

$$\text{Blood volume} = \frac{\text{Plasma volume}}{1 - \text{hematocrit}} \quad (1)$$

Determination of Hct values is a simple and important screening diagnostic procedure in the evaluation of hematologic disease. Hct values for men are normally ~48% and ~38% for women.

Decreased Hct indicates the presence of **anemia** and often reflects blood loss or deficiencies in blood cell production. Immediately after hemorrhage, the Hct does not adequately reflect the extent of red blood cell loss because cells and plasma are lost in equal proportions. However, several hours after hemorrhage, the Hct will have decreased because interstitial fluid shifts into the vasculature to replace the lost volume before new red blood cells are produced.

Increased Hct values indicate **polycythemia** and may result from an increased production of erythrocytes or decreased rate of their destruction. Dehydration, which decreases the water content and thus the volume of plasma, also results in an increase in Hct. When packed red blood cells are given to a patient to correct anemia, the Hct rises by about 3% to 4% per 250 to 300 mL unit given.

Complete blood count is a more definitive test for blood disorders.

The CBC, or complete blood count (also known as *full blood count*), is part of a routine medical examination. The CBC includes the following tests, which are summarized in Table 9.2: red blood cell (RBC) count, Hgb, Hct, and RBC indices; total white blood cell count and differential white blood cell count; and platelet count and mean platelet volume.

The CBC is usually an automated procedure, and several instruments are available that count and sort the cells according to their differences in electronic impedance, light absorption, and scattering of light. The cell counts can be given as absolute values and as percentages. Additionally, a histogram and/or a scattergram give information about the quantity and size of the individual cells in a graphic manner. In most cases, the automated CBC is accurate, and the results are available the same day that the blood is drawn. Patient values that are out of range when compared with reference values are automatically marked or flagged. In Table 9.2, the reference values of a healthy adult are given both in commonly used and in standardized international (SI) units. The values vary with age, sex, and physiologic condition (e.g., pregnancy, elevation above sea level) and also vary slightly from lab to lab.

In evaluating patients for hematologic diseases, it is important to determine the total number of circulating erythrocytes (the RBC) and the Hgb concentration in the blood. This information is used to determine whether the patient is anemic. Because there is no substantial storage site for erythrocytes, almost all functioning cells are present in the blood.

The Hct can only be used to determine anemia when fluid status is taken into account. With a given RBC, Hgb, and Hct, several other important **blood indices** can be derived.

Clinical Focus / 9.1

Blood Group Systems

The blood types A, B, AB, and O are based on the presence or absence of distinct antigens A or B on red blood cells. People with red blood cells that are covered with A or B molecules are said to have *type A* blood or *type B* blood, respectively. If both molecules, A and B, are present, the blood is *type AB*; if neither is present, it is called *type O*.

Knowing about a person's blood type is important for blood transfusions. Inappropriate combinations between the blood groups of the recipient and the donor can lead to potentially fatal agglutination, because the recipient's immune system will have antibodies against the donor's red blood cells. Agglutination of red blood cells may lead to improper blood circulation, release of hemoglobin (Hgb) that crystallizes, and eventually, kidney failure.

A person with type A blood can receive type A or O, not type B or AB blood, because his or her blood contains anti-B immunoglobulin (Ig) M antibodies. Correspondingly, a person with type B can receive type B or O blood. Those with AB blood can receive blood from anyone and are, therefore, universal recipients. Type O persons can receive blood from a type O person only but can donate to all groups and are, thus, called **universal donors**.

Blood type can further be useful for a person's identification, such as in forensic medicine and for the identification of family relationships in disputed parentage. The latter is possible because the blood type of the child is related to that of the parents. For instance, if the mother and the child have type O blood, the father's blood type must be A, B, or O. If the mother has type O blood, but the child presents with type B blood, the father's blood type must be B or AB.

In the United States, type O blood is most common, followed by types A, B, and AB, in this order. Because blood groups are genetically determined, the frequency of the ABO blood groups varies in populations throughout the world, even in different groups within a given country. Most likely, these differences will gradually disappear because of increased mobility and a greater societal acceptance of interracial marriages. In the United States, whites are currently characterized by a higher frequency of type A and a lower frequency of type B in comparison with African and Asian Americans, whereas Indians have a low frequency of both A and B types.

The blood type of a person is unrelated to his or her Hgb type because blood antigens are membrane factors, whereas Hgb is dissolved within the cell cytoplasm.

The other blood group system that is often grouped with the ABO system involves **Rhesus factors (Rh)**, which are unique cell surface proteins. The Rh system is complex, involving three genes producing Rh antigens C, D, and E. Rh D, the most important, is found in the blood of 85% of people, who are classified as Rh^+ (Rhesus positive). The remaining 15% are said to be Rh^- (Rhesus negative).

Knowing about the Rh factor is important in pregnant women because a baby's life can be endangered if it inherits Rh^+ blood from its father but the mother is Rh^-. Late in pregnancy, or at parturition, the baby's blood may cross to the mother's system, where the erythrocytes are recognized as foreign and antibodies are formed against them. These antibodies pose a serious threat to any future Rh^+ babies of hers in following pregnancies. To prevent this, the mother receives anti-Rhesus D immunoglobulin around the 28th week of pregnancy and right after delivery. The Ig attaches to Rh^+ cells from the baby in the mother's bloodstream and destroys them, preventing the triggers for the mother's immune system to produce its own anti-D antibody.

Similar problems of incompatibility between mother and child generally do not exist for the ABO blood groups because the antibodies formed belong to the IgM class and cannot cross the placenta. In rare cases, however, a type O mother with a fetus that is A, B, or AB can deliver a newborn with signs of immunologic defense reactions such as jaundice, anemia, and elevated bilirubin levels. Because of these complications, all pregnant women are advised to determine their ABO/Rh blood type during their first prenatal visit.

More than 600 additional red blood cell antigens have been described since the discovery of the ABO and Rh blood systems in the early 20th century, but they are considered minor antigens. They might be included in the screening of patients with frequent blood transfusions. Minor antigens, for example, are part of the Duffy, Kell, Kidd, MNS, and P antigen systems.

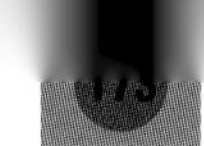

TABLE 9.2	Complete Blood Count			
Abbreviation	**Test**	**Reference Range**	**Standardized International Reference**	**Significance (Examples)**
RBC	Red blood cells			
	Men	4.3–5.9 × $10^6/\mu L$	4.3–5.9 × $10^{12}/L$	
	Women	3.5–5.5 × $10^6/\mu L$	3.5–5.5 × $10^{12}/L$	
Hgb or Hb	Hemoglobin			↓: anemia, severe bleeding
	Men	13.5–17.5 g/dL	2.09–2.71 mmol/L	↑: too many made, fluid loss, polycythemia
	Women	12–16 g/dL	1.86–2.48 mmol/L	
Hct or Ht	Hematocrit			
	Men	41%–53%	0.41–0.53	
	Women	36%–46%	0.36–0.46	
MCV	Mean cell volume	80–100 μm^3	80–100 fL	↓: iron deficiency, thalassemia ↑: B12 and folate deficiency (MCH variable)
MCH	Mean cell hemoglobin	25.4–34.6 pg/cell	0.39–0.54 fmol/cell	
MCHC	Mean cell hemoglobin concentration	31%–36% Hb/cell	4.81–5.58 mmol Hb/L	↓: deficient Hgb synthesis ↑: spherocytosis
RDW	RBC distribution width	11.7%–14.2%	0.12–0.14	↑: mixed population, immature cells
WBC	White blood cells	4,500–11,000/μL	4.5–11 × $10^{-3}/L$	↓: some medication, autoimmune diseases, bone marrow diseases, severe infections ↑: infection (abscess, meningitis, pneumonia, appendicitis, tonsillitis), inflammation, leukemia, stress, dead tissue (burns, heart attack, gangrene)
Neutrophil	Neutrophils (54%–62%)	4,000–7,000/μL	4–7 × $10^{-3}/L$	
Lymph	Lymphocytes (25%–33%)	2,500–5,000/μL	2.5–5 × $10^{-3}/L$	
Mono	Monocytes (3%–7%)	100–1,000/μL	0.1–1 × $10^{-3}/L$	
Eos	Eosinophils (1%–3%)	0–500/μL	0–0.5 × $10^{-3}/L$	
Baso	Basophils (0%–1%)	0–100/μL	0–0.1 × $10^{-3}/L$	
Plt	Platelet count	0.15–0.4 × $10^6/\mu L$	0.15–0.4 × $10^{12}/L$	↓: not enough made, bleeding, systemic lupus erythematosus, pernicious anemia, hypersplenism, leukemia, chemotherapy ↑: too many made, young cells
MPV	Mean platelet volume	7.5–11.5 μm^3	7.5–11.5 fL	

The **mean corpuscular** (or **cell**) **volume (MCV)** is the index most often used because it reflects the average volume of each red blood cell. It is calculated as follows:

$$\text{MCV} = \frac{\text{Hematocrit}}{\text{RBC (cells/L)}} \tag{2}$$

Example:

$$\text{MCV} = \frac{0.450}{5 \times 10^{12}\ \text{cells}} = 0.090 \times 10^{-12}\ \text{L/cell} = 90\,\text{fL}(1\,\text{fL} = 10^{-15}\ \text{L})$$

Cells of normal size are described as *normocytic*. Cells with a low MCV are *microcytic*, and the ones with a high MCV are *macrocytic*. These size categories are used to classify anemias. MCV is less useful when mixed cell populations are present.

The **mean corpuscular** (or **cell**) **hemoglobin (MCH)** value is an estimate of the average Hgb content of each red blood cell. It is derived as follows:

$$\text{MCH} = \frac{\text{Blood hemoglobin (g/L)}}{\text{RBC (cells/L)}} \tag{3}$$

Example:

$$\text{MCH} = \frac{150\,\text{g/L}}{5 \times 10^{12}\,\text{cells}} = 30 \times 10^{-12}\ \text{g/cell} = 30\ \text{pg/cell}$$

MCH values usually rise or fall as the MCV is increased or decreased. The MCH is also often related to the **mean corpuscular** (or **cell**) **hemoglobin concentration (MCHC)** because the RBC is usually related to the Hct. Exceptions to this rule yield important diagnostic clues.

The MCHC provides an index of the average Hgb content in the mass of circulating red blood cells. It is calculated as follows:

$$\text{MCHC} = \frac{\text{Blood hemoglobin (g/L)}}{\text{Hematocrit}} = \frac{\text{MCH}}{\text{MCV}} \tag{4}$$

Example:

$$\text{MCHC} = \frac{150\ \text{g/L}}{0.45} = 333\ \text{g/L}$$

Low MCHC indicates deficient Hgb synthesis, and the cells are described as *hypochromic*. High MCHC values are rare, because normally, the Hgb concentration is close to the saturation point in red cells. MCHC can only be elevated when sphere-shaped erythrocytes are produced rather than the normal biconcave disk-shaped cells.

The **red cell distribution width (RDW)** measures the degree of the average size dispersion of erythrocytes, thereby indicating the degree of **anisocytosis** (variation in erythrocyte size). RDW helps to classify anemias, particularly in the case of mixed macro- and microcytic anemias.

The **oxygen carrying capacity (OCC)** is not part of the CBC but is nevertheless an important equation in clinical practice because anomia can lead to severe hypoxemia. The OCC is the maximum amount of oxygen that can be carried in 1 dL (100 mL) of blood, including both oxygen bound to Hgb and oxygen dissolved in plasma. Each gram of Hgb can combine with and transport 1.34 mL of oxygen. For each 1 mm Hg of arterial oxygen pressure (PaO_2), there is 0.003 mL of oxygen dissolved in 1 dL of blood. Thus, OCC in 100 mL of blood can be calculated as follows:

$$\text{OCC} = (\text{Blood hemoglobin} \times 1.34) + (0.003 \times PaO_2) \tag{5}$$

Example:

$$\begin{aligned}\text{OCC} &= \left(15\ \text{g Hgb} \times 1.34 \frac{\text{mL } O_2}{\text{g Hgb}}\right) + \left(0.003 \frac{\text{mL } O_2}{\text{mm Hg}} \times 100\ \text{mm Hg}\right)\\ &= 20.1\ \text{mL } O_2 + 0.3\ \text{mL } O_2\\ &= 20.4\ \text{mL } O_2\ \text{(per 100 mL of whole blood)}\end{aligned}$$

The **white blood cell count (WBC)** measures the concentration of leukocytes in blood as an indicator of the body's immunologic stress state. A high WBC is called **leukocytosis** and occurs in infection, allergy, systemic illness, inflammation, tissue injury, and leukemia. A low WBC is called **leukopenia** and can be present in some viral infections, immunodeficiency states, and bone marrow failures. The WBC is often used to monitor a patient's recovery from illness.

For diagnostic purposes, a **differential white blood count** (abbreviated as *diff*) will determine which type of white blood cells is affected. Automated diffs are obtained as part of the CBC. They are typically based on the count of 10,000 cells and reveal the number of cells per 1 μL of blood. This information provides invaluable clues to the type of illness but also requires careful analysis. Unlike for red blood cells, there are various storage pools for white blood cells outside of blood, such as spleen, lymph nodes, and lymphoid tissues. Therefore, changes in blood concentrations indicate changes in the storage pool equilibrium. Second, the number and percentage of leukocytes vary somewhat with age. Newborns have a high WBC for the first weeks after birth with a high percentage of **neutrophils**. In childhood, **lymphocytes** are predominant. At senescence, WBC might be decreased. Third, many medications and supplements are known to affect the WBC results.

Last, the CBC provides a **platelet count (Plt)**, which is used as a starting point in the diagnosis of hemostasis. Decreased Plts may be the result of bone marrow failure or of peripheral platelet destruction. The **mean platelet volume (MPV)** can assist in evaluating the size of the platelets, similar to the way in which MCV provides information about erythrocytes. High MPV can indicate young platelets, whereas low MPV can be part of bone marrow failure.

Blood smears detect blood parasites and other hematologic disorders.

If there are significant abnormalities in one or more of the blood cell populations, visual analysis of the blood cells on a **blood smear** is required to validate the automated results. In this test, a drop of blood is placed on a slide, smeared into a thin layer with a second glass slide, allowed to dry, and then dyed with a special stain (Fig. 9.4). Microscopic analysis allows the differentiation of leukocytes according to their morphologic appearance and staining characteristics. Visual

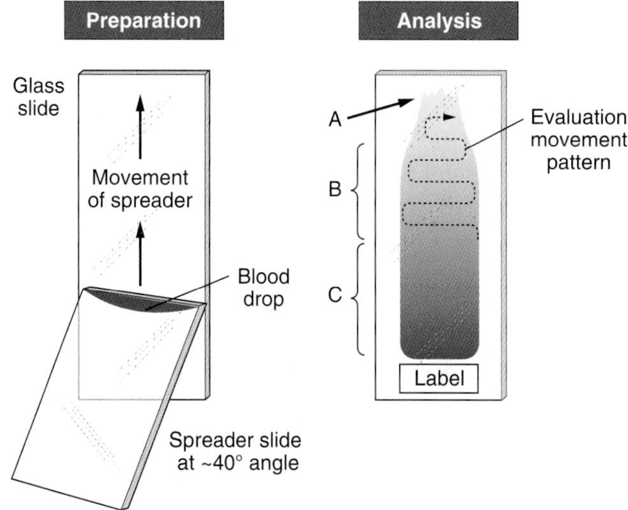

Figure 9.4 Preparation and analysis of a blood smear. This figure shows how to prepare a thin film of blood (*left*) with a feathered edge (*right*, A). Microscopic analysis occurs in the thin, single-cell layer section (B) but not in the thick section of the smear (C).

inspection also allows the identification of immature cells, so-called **band cells**. Regarding red blood cells, blood smear analysis is necessary to detect the characteristic "sickle" shapes of red blood cells, which occur in sickle cell anemia. Red blood cell–infecting parasites such as the malaria parasite can be seen in a blood smear, whereas they are not detectable using automated cell counters. Lastly, automated Plts might be incorrect because of platelet clumping or the misinterpretation of microcytic red blood cells as platelets. Visual inspection might clarify these errors.

Blood smears are commonly stained with a polychrome stain such as **Wright-Giemsa** after fixing the cells with methanol. The orange **eosin** dye stains the basic components of the cell, such as the granules of eosinophils and the Hgb of red blood cells. The **methylene blue** dye stains the acid components of the cell, such as the DNA and RNA. There are many other stains available that are specific to the type of blood analysis.

The visual count on a blood smear is based on the analysis of 100 cells, from which the percentage of each white blood cell type is calculated. These *relative counts* show which cell type is expanded relative to the other types of cells in blood. When multiplied by the WBC, *absolute counts* are obtained to show the total quantity of each cell type available to fight disease. It is possible for a person to have a relative lymphopenia but an absolute neutrophilia.

The **hemocytometer** method is used for manual cell counts. A manual blood cell count is rarely performed in the clinical laboratory unless the cell concentration is too low to obtain accurate results from the automated instruments, which might be the case in an immunocompromised person. The hemocytometer, also called *hemacytometer*, is an etched glass chamber with raised sides that will hold a quartz coverslip exactly 0.1 mm above the chamber floor. Calculation of the cell concentration is based on the known volume underneath the coverslip.

RED BLOOD CELLS

Red blood cells (also referred to as *erythrocytes*) are the most common cell type in the blood and are the principle means of delivering oxygen to cells via the circulatory system. The cell's cytoplasm is rich in Hgb, an iron-containing biomolecule that readily binds with oxygen and is responsible for the blood's red color. In humans, mature red blood cells are flexible biconcave disks. Mammalian RBCs are unique among those of vertebrates in that mature cells lack a nucleus and most organelles. Erythrocytes are produced in the bone marrow and circulate for about 100 to 120 days before they are recycled by macrophages.

Erythrocytes consist mainly of hemoglobin, a unique pigment containing heme groups where iron atoms bind to oxygen.

The largest population of blood cells is erythrocytes, or red blood cells. Erythrocytes have the shape of biconcave disks, with a diameter of about 7 μm and a maximum thickness of 2.5 μm (Fig. 9.5). This form optimizes the area of the cell surface for gas exchange. Erythrocytes are responsible for providing oxygen to tissue and partly for recovering carbon dioxide as waste. Erythrocytes also play a major role in the homeostasis of blood pH. When CO_2 is converted to H_2CO_3 within erythrocytes and is then ionized, the hydrogen ions (H^+) are buffered by Hgb. Hgb acts as a buffer by ionizing the imidazole ring of histidines in the protein.

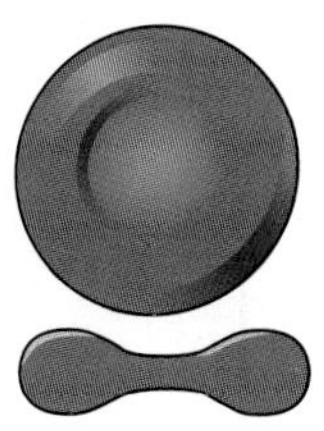

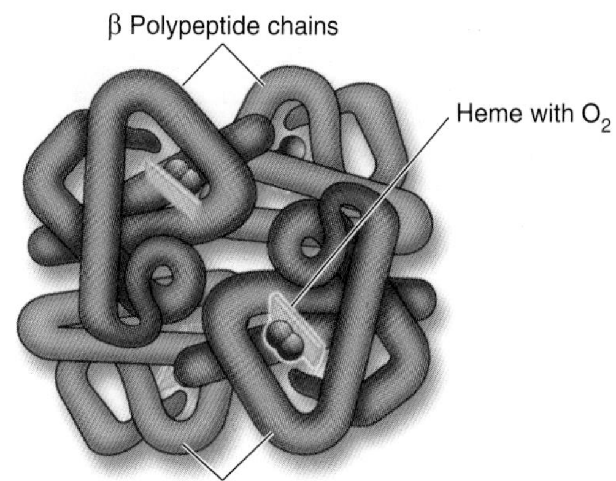

Figure 9.5 Structure of hemoglobin A. Erythrocytes have the shape of biconcave disks, with a diameter of about 7 μm and a maximum thickness of 2.5 μm. They contain several hundreds of hemoglobin molecules, each consisting of four polypeptide chains (two α and two β), with each chain containing iron bound to its heme group. (Modified from McArdle WD, Katch FI, Katch VL. *Essentials of Exercise Physiology*. Baltimore, MA: Lippincott Williams and Wilkins, 2005.)

Hgb, the red, oxygen-transporting protein of erythrocytes, consists of a globin portion and four heme groups, the iron-carrying portion. This complex protein of about 64,500 daltons possesses four polypeptide chains: two α-globin molecules and two molecules of another type of globin chain (β, γ, δ, or ε). Each red blood cell contains several hundred Hgb molecules.

Four types of Hgb molecules can be found in human erythrocytes, designated by their polypeptide composition. The most prevalent adult Hgb—HbA (Fig. 9.5)—consists of two α polypeptide chains and two β polypeptide chains ($\alpha_2\beta_2$). HbA_2, which makes up about 1.5% to 3% of total Hgb in an adult, has the subunit formula $\alpha_2\delta_2$. *Fetal hemoglobin* ($\alpha_2\gamma_2$) is the major Hgb component during intrauterine life. Its levels in circulating blood cells decrease rapidly during infancy and reach a concentration of 0.5% in adults. *Embryonic hemoglobin* is found earlier in development. It consists of two α chains and two ε chains ($\alpha_2\varepsilon_2$). The production of ε chains ceases at about the third month of fetal development.

The production of each type of globin chain is controlled by an individual structural gene with five different loci. The resultant four protein chains join in developing red blood cells and remain together during the lifetime of the cell. Mutations can occur anywhere in the five gene loci and in many different ways. This has resulted in the production of over 550 types of abnormal Hgb molecules leading to **hemoglobinopathies**. Most mutations have no known clinical significance; some of them, however, cause problems. The best-known one causes sickle cell anemia. The abnormal Hgb is called *sickle cell hemoglobin (HbS)*, which differs from normal adult HbA because of the substitution of a single amino acid in each of the two β chains.

Oxyhemoglobin (HbO_2), the oxygen-saturated form of Hgb, transports oxygen from the lungs to tissues, where the oxygen is released. When oxygen is released, HbO_2 becomes reduced hemoglobin. Carbon monoxide (CO) can rapidly replace oxygen in HbO_2, resulting in the formation of the almost irreversible compound **carboxyhemoglobin (HbCO)**. The formation of HbCO accounts for the asphyxiating properties of CO. Nitrates and certain other chemicals oxidize the iron in Hb from the ferrous to the ferric state, resulting in the formation of **methemoglobin (metHb)**. MetHb contains oxygen so tightly bound to ferric iron that it will not be released at the tissues and is therefore useless in respiration.

Cyanosis, the dark-blue coloration of skin associated with anoxia, becomes evident when the concentration of reduced hemoglobin exceeds 5 g/dL. It is reversible if the condition is caused only by a diminished oxygen supply, but is irreversible by administration of oxygen alone if it is caused by the accumulation of stabilized metHb.

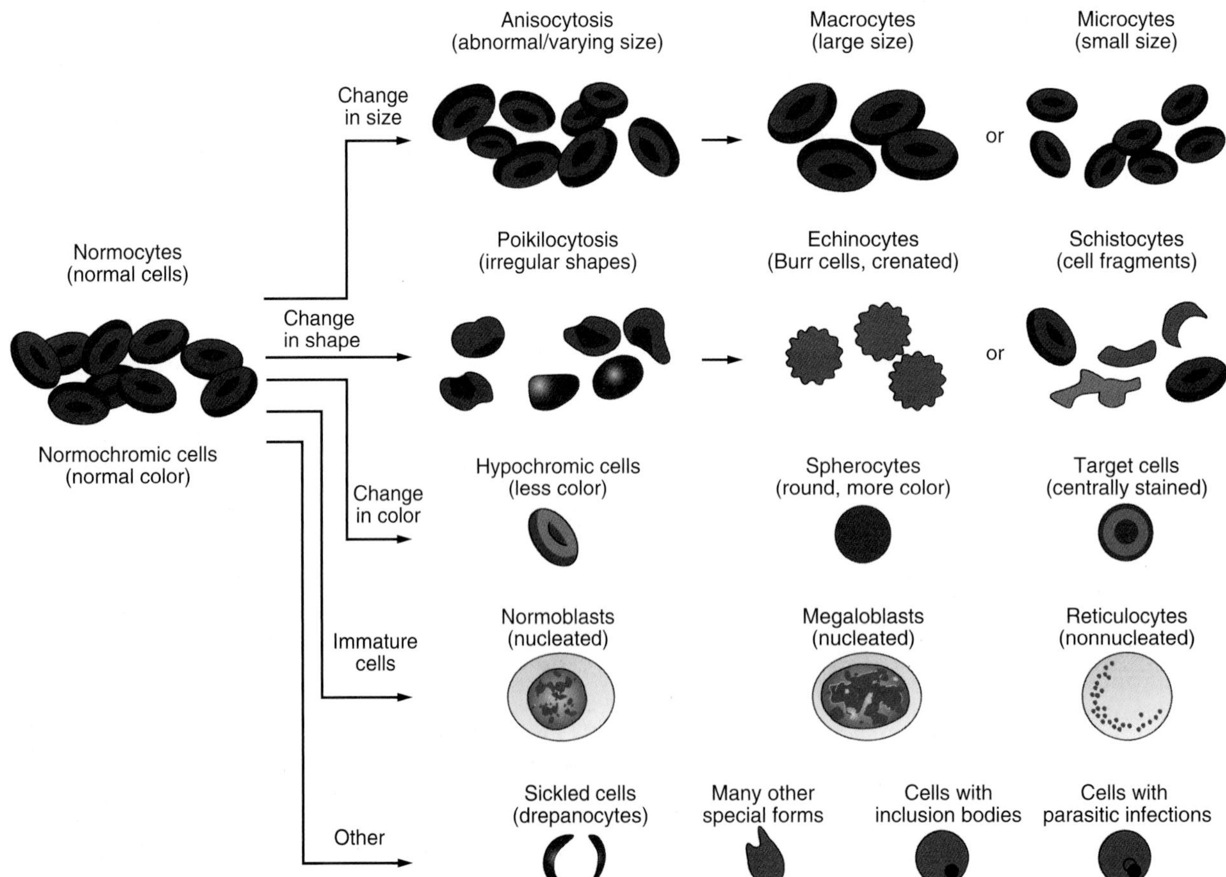

Figure 9.6 Pathological changes in erythrocyte morphology. This synopsis of erythrocyte abnormalities presents a variety of possible deviations that are helpful for the diagnosis of anemias and other diseases.

Changes in erythrocyte morphology provide insights into specific blood disorders.

Distortions of the red blood cell composition or the red blood cell membrane may lead to morphologic changes, which can provide valuable clinical information (Fig. 9.6). The red blood cell membrane is attached to a complex skeleton of fibrous intracellular proteins. This structural arrangement gives the cell stability as well as great flexibility as it twists and turns through small, curved vessels.

- *Changes in Size and Shape*. Large variation in the size of cells is referred to as anisocytosis. Larger-than-normal erythrocytes are termed **macrocytes**; smaller-than-normal erythrocytes are referred to as **microcytes**. **Poikilocytes** are irregularly shaped erythrocytes. **Echinocytes**, or *Burr cells,* are spiked erythrocytes generated by alterations in the plasma environment. **Schistocytes** are fragments of RBCs damaged during blood flow through abnormal blood vessels or cardiac prostheses.
- *Changes in Color*. Abnormal Hgb content of erythrocytes may lead to changes in the staining pattern of cells on dried films. Normal cells appear red–orange throughout, with a slight central pallor as a result of the cell shape (**normochromic**). **Hypochromic** cells appear pale with only a ring of deeply colored Hgb on the periphery. Other pathologic variations in red cell appearance include **spherocytes**, small, densely staining red cells with a loss of biconcavity as a result of membrane abnormalities, and **target cells** (also known as *codocytes* or *leptocytes*), cells with a densely staining central area and a pale surrounding area. Target cells are observed in liver disease and after splenectomy.
- *Immaturity*. Normal mature erythrocytes do not have cell nuclei, and therefore, the presence of nucleated red cells in peripheral blood is of diagnostic significance. One type of nucleated red cell, the **normoblast**, is seen in several types of anemias, especially when the marrow is actively responding to a demand for new erythrocytes. In seriously ill patients, the appearance of normoblasts in peripheral blood is a grave prognostic sign preceding death, often by several hours. Another nucleated erythrocyte, the **megaloblast**, is seen in peripheral blood in pernicious anemia and folic acid deficiency. **Reticulocytes** are immature cells that have shed their nucleus but still retain residual nuclear material that can be made visible with a specific vital stain. The percentage of reticulocytes in blood is an indicator for the level of erythropoiesis in bone marrow. For instance, if a patient is anemic and the bone marrow is functioning properly, the reticulocyte count will be elevated in response.
- *Other Changes*. There are an abundance of other erythrocyte abnormalities, which may result from hemoglobinopathies (e.g., *drepanocytes*, or sickled cells), liver diseases (e.g., *stomatocytes*, or mouth cells), and hematopoiesis outside the bone marrow cavity (e.g., *dacryocytes*, or teardrop cells). Cells may contain **protein inclusion bodies** as a result of abnormal Hgb precipitation, lead poisoning, or abnormal erythropoiesis. The parasitic protozoan of malaria (*Plasmodium*) and babesiosis (*Babesia*) infect red blood cells. *Trypanosomes* causing sleeping sickness, Chagas disease, and other illnesses live in blood and lymph but do not enter red blood cells.

Hemoglobin is recycled to create new products.

RBCs are transported by the circulatory system. They circulate for about 120 days in blood after they are released from the bone marrow before they become senescent and die. About 200 billion new red blood cells are made every day to replace the dead ones. Some of the old red cells break up in the bloodstream, causing **hemolysis**, but the majority are engulfed by macrophages of the **reticuloendothelial system (RES)**.

Either way, the Hgb of the red blood cells is recycled. Hgb released by red blood cells that lyse in the circulation either binds to **haptoglobin**, a protein in plasma, or is broken down to globin and heme. Heme binds a second plasma carrier protein, **hemopexin**, which, like haptoglobin, is cleared from the circulation by macrophages in the liver. Hgb from red blood cells that are engulfed by macrophages is broken into globin and heme within cells. The globin portion is catabolized by proteases into constituent amino acids that are reused for protein synthesis. Heme is broken down into free iron and **biliverdin**, a green substance that is further reduced to **bilirubin**. Bilirubin is then carried by albumin to the liver, where it is chemically conjugated to glucuronide before it is excreted in the bile. Unconjugated and conjugated bilirubin are measured clinically to detect and monitor liver or gallbladder dysfunction. Blood samples with high bilirubin are said to be *icteric*, which is evident as a yellow–green tint and increased stickiness. Those with icteric blood develop jaundice.

The premature destruction of erythrocytes can lead to **hemolytic anemia**, in which the production of new red blood cells cannot compensate for the cell loss. Hemolytic anemia represents about 5% of all anemias. Depending on the severity and onset of destruction, a patient's reaction ranges from asymptomatic to life-threatening symptoms. For example, red blood cell destruction in patients with sickle cell anemia leads to mild symptoms for some people and painful, life-threatening crises for others.

Body recycles most of the iron from senescent erythrocytes.

Most of the iron that is needed for new Hgb synthesis is obtained from the heme of senescent red blood cells that is metabolized by RES macrophages. Iron is released from macrophages into plasma by the integral cell membrane protein **ferroportin**. In blood, iron is transported in the *ferric state* (Fe^{3+}) bound to the protein **transferrin**. Cells that need iron possess membrane receptors to which transferrin binds and is then internalized. Inside the cell, the ferric iron is released from transferrin, reduced to the *ferrous state* (Fe^{2+}), and either incorporated into new heme or stored as **ferritin**, a complex of protein and Fe^{2+}. A portion of the ferritin is catabolized to **hemosiderin**, an insoluble compound consisting of crystalline aggregates of ferritin. The accumulation of large amounts of hemosiderin formed during periods of massive hemolysis can result in damage to vital organs.

Although iron recycling is efficient, small amounts are continuously lost and must be replenished by dietary intake. In women, iron loss is increased during menstrual bleeding. The majority of iron in the diet is derived from heme in meat (*organic iron*), but iron can also be provided by the absorption of *inorganic iron* (mainly Fe^{2+}, some Fe^{3+}) by intestinal epithelial cells. In intestinal epithelial cells, iron attached to heme is released, and Fe^{3+} is reduced to Fe^{2+} by intracellular flavoprotein. The reduced iron (both released from heme and absorbed as the inorganic ion) is transported through the cytoplasm bound to a protein. Iron is released into the plasma by ferroportin, where it is oxidized to the ferric state (Fe^{3+}) and bound to transferrin for use in new heme synthesis.

Hepcidin regulates iron homeostasis by inhibiting ferroportin.

Hepcidin, a hormone of the liver, regulates iron homeostasis by inhibiting the ferroportin of duodenal enterocytes, macrophages, and other iron-exporting cells. The current dogma is that iron homeostasis is mainly controlled in the gastrointestinal tract by the level of dietary iron uptake and hepcidin-mediated iron release into plasma. Recent evidence of hepcidin production in mammalian renal tubular cells suggests an iron excretory route, but its importance remains unconfirmed in humans. Hepcidin also provides a link between inflammation and anemia. In **anemia of inflammation**, also called *anemia of chronic disease*, hepcidin is overproduced such that the restricted iron supply for Hgb synthesis causes anemia.

Iron profile evaluates iron stores of the body.

To sufficiently differentiate between iron deficiency anemias and the anemia of inflammation as well as to diagnose iron overload, an **iron profile** is analyzed (Table 9.3).

Serum iron levels (“iron” for short) are measured with a simple colorimetric method. However, the body’s iron stores are better represented by *ferritin* levels, which are measured by radioimmunoassays or enzyme immunoassays. **Total iron-binding capacity (TIBC)** is the amount of iron needed to bind to all iron-binding proteins. Because *transferrin* represents the largest quantity of iron-binding proteins, TIBC is an indirect assessment of the amount of transferrin present. The **unsaturated iron-binding capacity (UIBC)** is the calculated amount of transferrin that is not occupied by iron. UIBC equals TIBC minus iron.

The molecule transferrin can nowadays also be independently measured by an enzyme immunoassay. In this case, the clinical transferrin value indicates the transferrin that is not bound to iron. Some laboratories measure TIBC, some measure UIBC, and some, transferrin. Last, the **transferrin saturation** is calculated as the ratio of serum iron and TIBC values and is expressed as a percentage. Thus, it indicates the percentage of transferrin with iron bound to it, which is 20% to 40% in healthy people. In addition to serum analyses, bone marrow aspiration and biopsies are helpful to determine total body iron stores.

In iron deficiency, the iron level is low, but the TIBC is increased so that transferrin saturation becomes low. TIBC is also increased in pregnancy. In iron overload, the iron level will be high and the TIBC will be low or normal, resulting in an increase of transferrin saturation. To evaluate a patient’s nutritional status or liver function, typically the protein transferrin, because it is produced in the liver, will be monitored.

Anemias often present with characteristic iron profiles. For instance, **iron-deficiency anemia** resulting from abnormal blood loss first causes a decrease in serum ferritin. Once ferritin stores are depleted, free iron is used for new red blood cells, and consequently, serum iron levels decrease. As a result, the percentage of transferrin that is not bound to iron, UIBC, is increased. For the test, more iron is needed to saturate all transferrin molecules as indicated in an elevated TIBC.

WHITE BLOOD CELLS

White blood cells, or leukocytes, are delivered by the blood to sites of infection or tissue disruption, where they defend

TABLE 9.3 Iron Profile

Name	Reference Range	Standardized International Reference	Description
Iron	50–170 μg/dL	9–30 μmol/L	Amount of iron bound to transferrin in blood
Ferritin	150–200 ng/mL	15–200 μg/L	Storage form of excess iron
Total iron-binding capacity (TIBC)	252–479 μg/dL	45–86 μmol/L	Amount of iron needed to bind to all transferrin
Unsaturated iron-binding capacity ([UIBC] TIBC – iron)	202–309 μg/dL	36–56 μmol/L	Transferrin not bound to iron
Transferrin (measured)	200–380 mg/dL	2–3.8 g/L	Transferrin not bound to iron
Transferrin saturation (iron/TIBC)	20%–50%	0.2–0.5	Percentage of transferrin with iron bound to it

the body against infecting organisms and foreign agents in conjunction with antibodies and protein cofactors in blood.

Leukocytes comprise five diverse cell types and constitute part of the immune system.

The five main types of white blood cells are neutrophils, **eosinophils, basophils**, lymphocytes, and **monocytes** (see Fig. 9.2). These cells are from three developmental cell lines, the **myeloid, lymphoid**, and **monocytic series**, which make up two main groups of mature leukocytes: **granulocytes** and agranulocytes.

Mature cells of the myeloid series (neutrophils, eosinophils, and basophils) are termed granulocytes based on their appearance after staining with polychromatic dyes such as the Wright stain. The nuclei of most mature granulocytes are divided into two to five oval lobes connected by thin strands of chromatin. This nuclear separation imparts a multinuclear appearance to granulocytes, which are therefore also known as **polymorphonuclear leukocytes**.

Lymphocytes and monocytes are often referred to as *agranular leukocytes*. Although monocytes and lymphocytes may also possess cytoplasmic granules, they are not as common or as distinct on a regularly stained microscopic slide. To distinguish them from polymorphonuclear leukocytes, they are also sometimes called **mononuclear leukocytes**.

Neutrophils defend against bacterial and fungal infection through phagocytosis.

Neutrophils are usually the most prevalent leukocyte in peripheral blood of a healthy person (40%–75% of all leukocytes). They are amoeba-like phagocytic cells. Neutrophils are the first defensive cell type to be recruited to a site of inflammation. Defects in neutrophil function quickly lead to massive infection and, quite often, death. The primary mission of neutrophils is to find bacteria or fungi and neutralize them by **phagocytosis**. This process can be described in four steps (Fig. 9.7).

Step 1: Recognition of foreign invader. When bacteria or their products bind to circulating antibodies, the bacteria release chemotactic factors that attract neutrophils. Neutrophils recognize the bacteria as foreign by binding to the antibodies via Fc receptors. Bacteria can also interact with tissue cells, lymphocytes, or platelets that then release factors that attract and activate neutrophils.

Step 2: Invagination of cell membrane. At the site of infection, neutrophils engulf the invading pathogen by phagocytosis. Phagocytosis is facilitated when the bacteria are coated with host defense proteins known as **opsonins** (for details see Chapter 10, “Immunology, Organ Interaction, and Homeostasis”).

Step 3: Phagosome formation. In the generated phagocytic vacuole, or **phagosome**, the bacterium is exposed to enzymes that were originally positioned on the cell surface. Therefore, phagocytosis involves invagination and then vacuolization of the segment of membrane to which a pathogen is bound.

Step 4: Killing of pathogens. Membrane-bound enzymes work in conjunction with enzymes secreted from intracellular granules into the phagocytic vacuole to destroy the pathogen.

An important step for the effective destruction of pathogens is the activation of the enzyme nicotinamide adenine dinucleotide phosphate (also known as *NADPH*) oxidase. The oxidase is dormant in resting cells but activated by its interaction with a G protein and cytosolic molecules that are generated during phagocytosis. Enzyme activation leads to the catalytic production of **superoxide ion**, a toxic free radical, within the phagosome. The generation of superoxide and other potent reactive agents is collectively termed the **respiratory burst** or **oxidative burst**. The reactive agents kill bacteria directly or participate in secondary free radical reactions to generate other potent antimicrobial agents such as **hydrogen peroxide**.

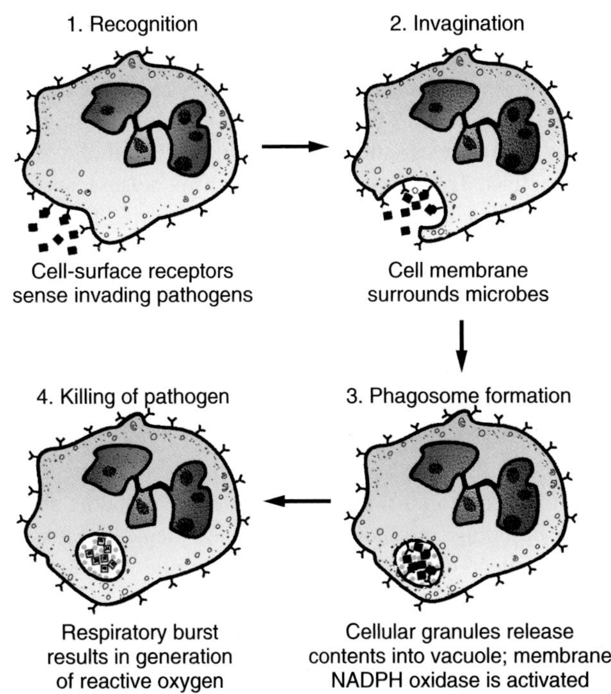

Figure 9.7 Steps in phagocytosis and intracellular killing by neutrophils. Neutrophils are the first defensive cells to be recruited to a site of inflammation. NADPH, nicotinamide adenine dinucleotide phosphate.

The important role of the NADPH oxidase for efficient host protection against invading pathogens becomes evident in **chronic granulomatous disease**, which is caused by the lack of phagocytic NADPH oxidase and is characterized by recurrent life-threatening bacterial and fungal infections.

In addition to the oxidative burst, other bactericidal agents and processes operate in neutrophils to ensure efficient bacterial killing. For instance, **defensins**, intracellular cationic proteins, can enter phagosomes and bind to bacteria to inhibit their replication. Agents stored in neutrophil granules include **lysozyme**, a bacteriolytic enzyme, and myeloperoxidase, which reacts with hydrogen peroxide to generate potent, bacteria-killing oxidants. One of these oxidants is hypochlorous acid (*HOCl*), a chemical typically found in household bleach. Granules also contain **collagenase** and other proteases.

Eosinophils are inflammatory cells that defend against parasitic infections.

A healthy person has >350 eosinophils/μL, which amounts to about 1 in a 100-cell differential count. Together with basophils, eosinophils represent 1% to 6% of total leukocytes.

Although eosinophils are rare, they are nevertheless easy to spot because of their characteristic appearance. As the name implies, *eosin*ophils take on a deep orange–red color during staining with eosin. Like neutrophils, eosinophils migrate to sites where they are needed and exhibit a metabolic burst when activated. Eosinophils participate in defense against certain multicellular parasites such as helminths. They also play a role in wound healing. On the negative side, eosinophils are involved in allergic reactions and are likely to contribute to chronic inflammation.

Basophils release histamine, causing the inflammation of allergic and antigen reactions.

With 0% to 2% in a healthy individual, basophils are even rarer than eosinophils, and many differential counts show none. Basophils are polymorphonuclear leukocytes with multiple pleomorphic, deep-staining granules throughout their cytoplasm. These granules contain **heparin** and **histamine**, which have anticoagulant and vasodilating properties, respectively. The release of these and other mediators by basophils increases regional blood flow and attracts other leukocytes, including eosinophils, to areas of infection. In addition to the release of preformed mediators, other mediators such as prostaglandins and leukotrienes are newly generated from arachidonic acid of the surrounding tissue and released at the site of infection.

Basophils are capable of synthesizing many of the same mediators as **mast cells**. However, they appear to be a distinct cell type despite the close similarities. Basophils and mast cells must be activated to release their inflammatory mediators in a process called **degranulation**. One means of cell activation occurs when antigens that are already tightly bound (at their constant end via FcεRI receptors) to basophils and mast cells bind to immunoglobulin E (*IgE*). Other direct or supportive stimuli for degranulation include mechanical injury, activated complement proteins, and drugs. When antigens are allergens such as pollen, the release of the chemical mediators can cause the characteristic symptoms of allergy.

Lymphocytes comprise three cell types participating in the immune system.

About 16% to 45% of leukocytes are lymphocytes composed of B cells and several subtypes of T cells. In a stained light microscopic slide, the B and T cells cannot be distinguished. Identification occurs with fluorescent monoclonal antibodies but is not routinely done in the clinical laboratory. Circulating lymphocytes possess a deeply stained nucleus that is large in relation to the remainder of the cell so that often only a small rim of cytoplasm appears around the nucleus (see Fig. 9.2). Some lymphocytes such as **natural killer cells** have a broader band of cytoplasm, closely resembling monocytes.

The majority of circulating lymphocytes are **T cells** or T lymphocytes (for *thymus-dependent lymphocytes*). They participate in cell-mediated immune defenses and are divided into several subtypes, such as helper T cells, cytotoxic T cells, and natural killer cells (see Chapter 10, "Immunology, Organ Interaction, and Homeostasis"). Some 20% to 30% of circulating lymphocytes are **B cells**, which are the cells ultimately responsible for producing antibodies. B cells are bone marrow–derived lymphocytes. When a B cell encounters its specific antigen, it becomes activated and proceeds to replicate itself. Many of these clonal B cells will mature into **plasma cells**. Plasma cells are highly specialized cells with the ability to produce large amounts of target-specific antibody. They are larger and stain darker than B cells because of the large amount of protein being synthesized within.

Lymphocytes not characteristic of either T cells or B cells are called **null cells**. The entire scope of the function of null cells, which comprise only 1% to 5% of circulating lymphocytes, is unknown, but it has been established that null cells are capable of destroying tumor cells and virus-infected cells.

Monocytes migrate from the blood stream and become macrophages.

Monocytes belong, together with lymphocytes, to the category of mononuclear leukocytes. Monocytes comprise about 4% to 10% of total leukocytes. Monocytes are phagocytic cells and are distinguished from lymphocytes based on their pale blue or blue–gray cytoplasm when stained with a Wright stain. The cytoplasm contains multiple fine, reddish–blue granules. Although the monocyte nucleus may have a characteristic kidney bean or horseshoe-shaped appearance, it can also be round or ovoid.

On activation, monocytes leave blood vessels, migrate into tissue, and transform into macrophages, which are large, active mononuclear phagocytes. Macrophages contain granules, which are packed with enzymes and chemicals that are used to ingest and destroy microbes, antigens, and other foreign substances. In addition to phagocytosis, macrophages also serve as antigen-presenting cells to T cells (see Chapter 10, "Immunology, Organ Interaction, and Homeostasis").

BLOOD CELL FORMATION

Blood cells must be continuously replenished. Erythrocytes survive in circulation for about 120 days. Platelets have, on average, a lifespan of 15 to 45 days in circulation, but many of them are immediately consumed as they participate in the day-to-day hemostasis. Leukocytes have a variable lifespan. Some lymphocytes circulate for 1 year or longer after production. Neutrophils, constantly guarding body fluids and tissues against infection, have a circulating half-life of only a few hours. According to the circulating lifespan, the interruption of hematopoiesis (e.g., by cancer treatment) results in a matter of hours in the depletion of neutrophils, called **neutropenia**. **Thrombocytopenia**, the depletion of platelets, occurs next, followed by the depletion of erythrocytes and lymphocytes.

Hematopoiesis takes place in bone marrow and lymphatic tissue.

Hematopoiesis, the process of blood cell generation, occurs in healthy adults in the bone marrow and lymphatic tissues, such as the spleen, thymus, and lymph nodes (Fig. 9.8). During fetal development, hematopoietic cells are present in high levels in the liver, spleen, and blood. Shortly before birth, blood cell production gradually begins to shift to the marrow. By age 20 years, the marrow in the cavities of many long bones becomes inactive, and blood cell production mainly occurs in the iliac crest, sternum, pelvis, and ribs. Within the bones, hematopoietic cells germinate in extravascular sinuses called **marrow stroma**. Active cellular marrow is called **red marrow**; inactive marrow that is infiltrated with fat is called **yellow marrow**.

In some disease states such as leukemia, liver and lymphatic tissue can resume their hematopoietic function in adulthood, which is called **extramedullary hematopoiesis**.

Mature blood cells originate from a multipotent stem cell.

Blood cell production begins with the proliferation of **multipotent stem cells**. Depending on the stimulating factors, the progeny of multipotent stem cells may be other uncommitted stem cells or stem cells committed to development along a certain lineage. The committed stem cells include **myeloblasts**, which form cells of the myeloid series (neutrophils, basophils, and eosinophils), **erythroblasts**, **lymphoblasts**, and **monoblasts**.

Promoted by **hematopoietins** and other cytokines, each of these blast cells differentiates further, a process that ultimately results in the formation of mature blood cells. This is a highly dynamic process, also influenced by factors from capillary endothelial cells, stromal fibroblasts, and mature blood cells.

Erythropoiesis is regulated by the renal hormone erythropoietin.

Erythropoiesis is the process by which red blood cells are produced. It is stimulated by decreased O_2 delivery to the kidneys, which then secrete the hormone **erythropoietin**. This, in turn, activates increased erythropoiesis in the hemopoietic tissues (Fig. 9.9). Erythropoietin regulates the differentiation of the uncommitted stem cell along the erythrocyte lineage, forming normoblasts (also referred to as *erythroblasts* or *burst-forming*

From Bench to Bedside / 9.1

Stem Cell Hematotherapy

Patients with leukemias that cannot be cured by chemotherapy may be cured by allogenic bone marrow transplantation. This procedure involves the destruction of the patient's entire hematopoietic system, including the most primitive multipotent stem cells, and its restoration with cells from a healthy donor. However, most patients in need of such a transplant do not have an appropriate immune type–matched sibling donor or are unable to identify a potential unrelated donor in a timely manner (for donor matching, see Chapter 10, "Immunology, Organ Interaction, and Homeostasis"). Additionally, the procedure can be hindered by significant complications including **graft-versus-host disease** (see Chapter 10, "Immunology, Organ Interaction, and Homeostasis") and infection.

Addressing these and other limitations has led to rapid advancement in stem cell hematotherapy. The general idea is to identify and isolate hematopoietic stem cells, to manipulate the cells *ex vivo*, and to introduce them into the patient for expansion to new blood cells.

The source of the stem cells can be from a human embryo, from umbilical cord blood, from an adult donor, or from the patient him- or herself. Embryonic stem cells are most desirable because they have the capacity to become any mature cell (called **totipotency**). The umbilical cords of newborns are rich in multipotent stem cells, both because the baby's blood cells have not yet developed their typical set of antigens and because umbilical cord blood lacks the well-developed immune cells that cause graft-versus-host disease. Hematopoietic stem cells from a matched donor (**allogenic**) or from the patient itself (**autologous**) can be collected from the bone marrow. However, multipotent peripheral blood stem cells that are present in the bloodstream are commonly collected by filtering the blood (apheresis). Last, some somatic cells (e.g., fibroblasts) can be reprogrammed by viral or other vector transfer of defined genes to create pluripotent stem cells.

Purified hematopoietic populations can be "engineered" before reinfusion into the patient to commit to differentiation along a specific lineage into the mature cells that are not present, that do not function in a patient, or that are helpful for the recovery of a patient with a particular disease. These tumor-free grafts ideally create a hematopoietic microenvironment that favors normal cells versus malignant or aberrant stem cells.

The techniques for isolating and regulating stem cells can be combined with any modern cell and molecular biology methods. For instance, patients with tuberculosis, malaria, and leprosy might be treated with stem cells that are genetically modified to produce cells that resist colonization by the infectious agent. Implanting the cell nucleus of a patient's somatic cell into embryonic stem cells will produce offspring cells that resemble the patient's own cells, thus avoiding any complications from immune system rejection. Scientists hope that stem cell manipulation will soon be achieved noninvasively, for instance, by administering a regulating gene or drug.

Beneficiaries of stem cell hematotherapy would include a very wide variety of patients. For instance, patients with AIDS could be supported by periodic infusions of T-cell precursors. Infusion of neutrophil progenitors may help patients with cancer to recover from aggressive cancer therapy. Patients with certain anemias will benefit from infusions of red cell progenitors, and platelet progenitors should be useful in patients with one of the many forms of inborn or acquired thrombocytopenia.

Although stem cell hematotherapy has enormous potential to revolutionize the field of regenerative medicine, the past decade of stem cell therapy has not advanced medically as fast as predicted. The complexity of human cell interactions, the limitations of biotechnology companies for risk investment, and the competition for research support and fame that has led some researchers to falsify data have to be taken into account when discussing the transfer of knowledge from bench to bedside.

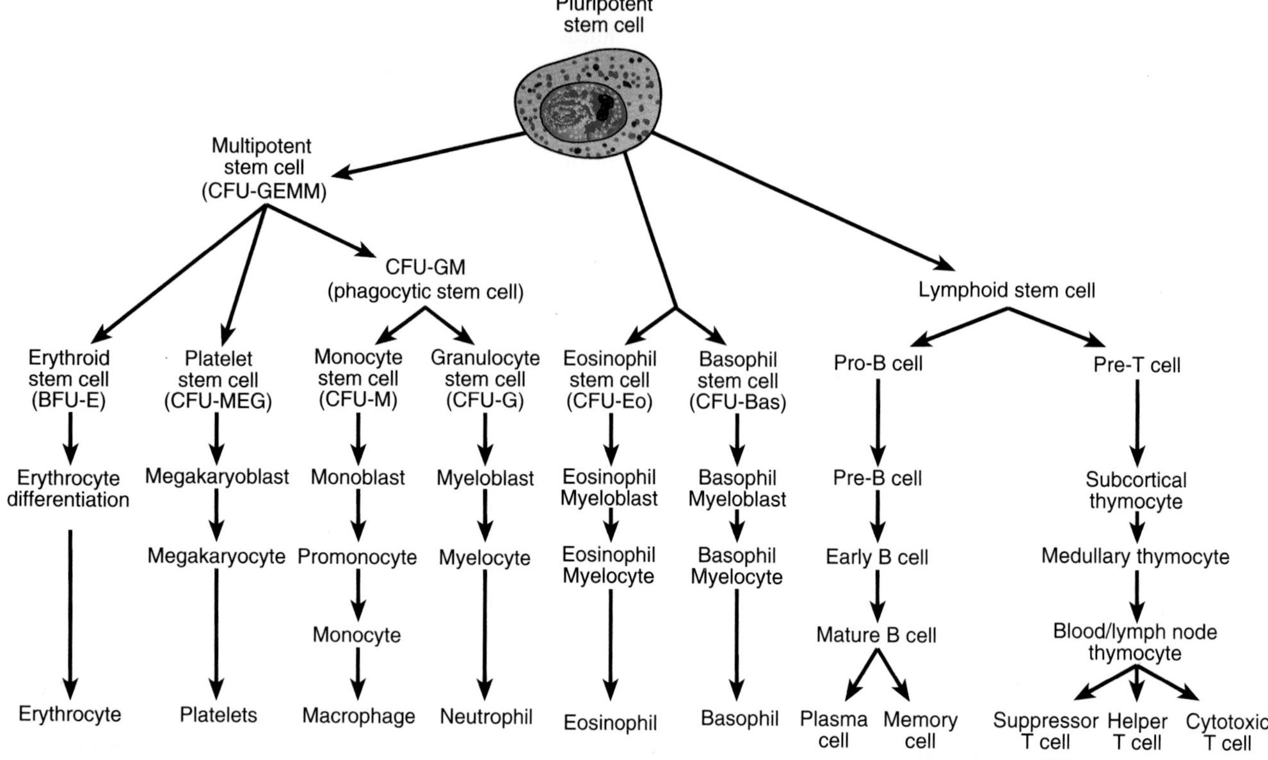

Figure 9.8 Hematopoiesis. All circulating blood cells are believed to be derived from a common, uncommitted bone marrow progenitor—the multipotent stem cell. Its differentiation along different lineages via increasingly more committed stem cells toward the final blood cells depends on the encountered conditions and hematopoietins. CFU, colony-forming unit; BFU, burst-forming unit; -G, granulocyte; -E, erythrocyte; -M, macrophage; -MEG, megakaryocyte; -Eo, eosinophil; -Bas, basophil.

cells), reticulocytes, and finally, mature erythrocytes, which enter the bloodstream. A common symptom of patients with chronic kidney disease is anemia from the lack of erythropoietin. Genetically engineered human erythropoietin and some longer lasting analogs are used for the treatment of anemia due to chemotherapy, chronic kidney failure, and treatment for human immunodeficiency virus. The risks for excessive erythropoietin are discussed in Clinical Box 9.2.

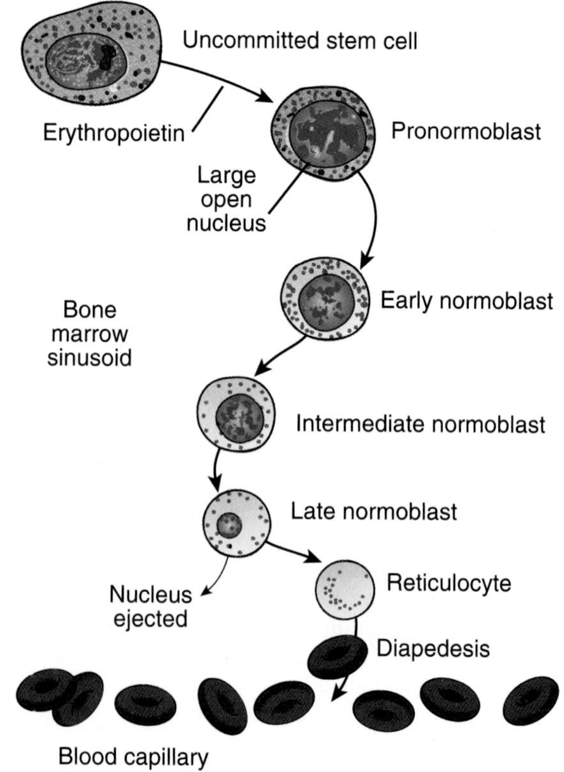

Figure 9.9 Erythropoiesis. Erythrocytes are the result of a process that begins with uncommitted stem cells and involves a series of differentiations in the bone marrow sinusoids until the final erythrocytes enter the bloodstream by diapedesis.

BLOOD CLOTTING

Damage to the vasculature quickly leads to massive bruising and, if unrepaired, to extreme blood loss and consequent organ failure.

Hemostasis (the cessation of blood loss from a damaged vessel) can be organized into four separate but interrelated events: compression and vasoconstriction; the formation of a temporary loose platelet plug (also called *primary hemostasis*); formation of the more stable fibrin clot (also called *secondary hemostasis*), and finally, clot retraction and dissolution (Fig. 9.10). The four steps are explained in more detail in the following sections.

Hemostasis step 1: Blood clots form immediately upon endothelial damage.

Immediately after tissue injury, blood flow through the disrupted vessel is slowed by the interplay of several important *physical factors*. These include back pressure exerted by

Clinical Focus / 9.2

Blood Doping and Erythropoietin

Charging an athlete with "blood doping" is unfortunately a common newspaper headline during world-class endurance sports events. It refers to methods of enhancing an athlete's red blood cell count in advance of a competition to increase oxygen delivery capacity and to reduce muscle fatigue.

One blood doping method is autologous blood transfusion, in which the athlete's own red blood cells are stored and then transfused back in, a week or less before competition. This technique is not only illegal but also dangerous. An increase in hematocrit (Hct) results in increased blood viscosity, which is a major contributor for developing heart disease.

Another way to increase Hct is by increasing the concentration, production, or activity of the hormone erythropoietin. Erythropoietin is a glycoprotein produced primarily by peritubular cells in adult kidneys. It promotes the proliferation and differentiation of erythrocyte precursors by rescuing them from apoptosis. It may also have effects on the brain resembling those of older performance-enhancing drugs such as amphetamines, cortisone, and anabolic steroids.

The human gene for erythropoietin was cloned in 1985, leading to the production of recombinant human erythropoietin (rhEPO) and genetically engineered darbepoetin, which has an increased lifetime compared to the recombinant hormone form. The large-scale production of synthetic erythropoietin made it available for the treatment of anemias. Currently, rhEPO and darbepoetin are approved in the United States to treat patients with severe anemias associated with chronic renal failures, AIDS, and chemotherapy. Usage in the treatment of severe anemias is limited in order to balance the benefits of increased red cell production with the health risks associated with an elevated Hct, especially for those who are ill.

Within the first 4 years after the synthetic versions of erythropoietin became available, more than 17 high-performance athletes died from the consequences of drug-induced blood clots. The fatal dangers of excess erythropoietin include sudden death, consequent to drastic reduction in heart rate and development of antierythropoietin antibodies, which ironically, cause the destruction of red blood cells.

During drug screening, increased Hct and elevated reticulocyte blood count indicate potential erythropoietin abuse. In this case, urine and blood tests will be done. Erythropoietin in urine can only be detected for a few days after injection. Erythropoietin in blood has a half-life of about 2 weeks. This creates a challenge for the correct testing time because the drug-induced increase in red cell mass can last up to a few months.

Because repeated pre-event testing for every athlete is not realistic, exposing illegal blood doping will remain a challenge. Additionally, drug advancements will require concomitant advancements in drug-detection procedures. Currently, exogenous hormone can be distinguished from the body's normal hormone by electrophoresis. Blood doping by erythropoietin gene transfer, however, similar to novel therapeutic drug–delivery systems, would only reveal an increase in the natural hormone and would necessitate a careful determination of the person's normal hormone level.

the tissue around the injured area and vasoconstriction. The degree of compression varies in different tissues; for example, bleeding below the eye is not readily deterred because the skin in this area is easily distensible. A "black eye" is the consequence. The back pressure increases as blood leaks out of the disrupted capillaries and accumulates in the surrounding tissue. Sometimes, contraction of underlying muscles further compresses the blood vessels. This is one of the physiologic actions to minimize blood loss from the uterus after childbirth.

In addition, damaged cells at the site of injured tissue release potent *chemical substances* that directly cause blood vessels to constrict. These include serotonin, thromboxane A_2, epinephrine, and fibrinopeptide B.

Hemostasis step 2: Circulating platelets are activated to form a temporary plug.

The immediate goal of hemostasis is the fast production of a physical barrier that covers the opening in the blood vessel. This initial plug is composed of platelets (thrombocytes) bound together by fibrinogen. The plug is still weak and somewhat fragile, which accounts for the clinical rule to never wipe blood away from tissue but instead lightly dab it, using the capillary action of the dry towel.

Platelets are irregularly shaped, disk-like fragments of their precursor cell, the megakaryocyte. They are one fourth to one third the size of erythrocytes (1.5–3.0 μm). As megakaryocytes develop, they undergo a process of fragmentation that results in the release of over 1,000 platelets per cell. Several factors stimulate megakaryocytes to release platelets within the bone marrow sinusoids. This includes the hormone **thrombopoietin**, which is mainly generated by the liver and the kidneys and released in response to low numbers of circulating platelets. Platelets have no defined nucleus but possess important proteins, which are stored in intracellular granules and secreted when platelets are activated during coagulation.

Platelet adherence can be initiated by a variety of substances. For instance, factors released by platelets cause the upregulation of adherence proteins (integrins) on endothelial cells. One critical substance released by endothelial cells and also megakaryocytes is called von Willebrand factor. It enhances platelet adhesion to the endothelium by forming a bridge between platelet surface receptors and collagen in the subendothelial matrix. The most common hereditary bleeding disorder is von Willebrand disease, caused by an inherited deficiency of the factor. Ruptured cells at the site of tissue injury release adenosine diphosphate (ADP), causing the aggregation of more platelets, which are, in turn, stabilized by fibrinogen. Clinically, penicillin in high doses can coat platelets and prevent aggregate formation.

Platelet count and bleeding time

Thrombocytopenia, a low number of platelets, can be due to decreased production of platelets (e.g., in patients with serious bone marrow disease), unavailability of platelets

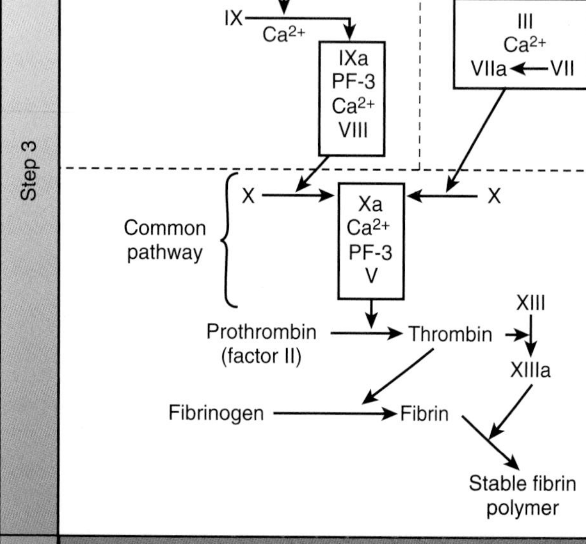

Figure 9.10 Hemostasis. The blood's response to blood vessel injury can be viewed as four interrelated steps. Roman numerals refer to coagulation factors. ADP, adenosine diphosphate; PF-3, platelet factor 3.

(e.g., platelets can be sequestered in an enlarged spleen), or accelerated destruction of platelets (e.g., coating of platelets with IgG and removal by phagocytes). **Immune thrombocytopenia purpura** is a common autoimmune disorder caused by autoantibodies to platelets formed in the spleen and by the platelet–antibody aggregates being destroyed by the spleen.

In addition to serum platelet counts and bone marrow biopsies, a standardized test is available to determine **bleeding time**. The time it takes to form a platelet plug to stop bleeding is recorded after a 1-mm-deep incision has been cut in the skin of the forearm and an inflated blood pressure cuff has provided 40 mm Hg pressure. Using this *Ivy method*, bleeding should stop within 1 to 9 minutes. A less invasive test, the *Duke method*, involves a standardized prick on the fingertip or earlobe, and bleeding should stop within 1 to 3 minutes.

Hemostasis step 3: Thrombin triggers the conversion of fibrinogen into fibrin to form a stable clot.

In the next step, the relatively weak and soluble platelet–fibrinogen aggregate formed in step 2 matures to a strong, insoluble fibrin polymer aggregate. The major actor is the enzyme **thrombin**. Thrombin cleaves four small peptides (fibrinopeptides) from a molecule of fibrinogen, which without the fibrinopeptides, is called *fibrin monomer*. Fibrin monomers spontaneously assemble into ordered fibrous arrays of fibrin. Fibrous strands of fibrin eventually form a highly organized and firm network, which traps more platelets, red blood cells, and leukocytes at the site of vascular damage, thereby forming a stable blood clot. A plasma enzyme, **fibrin stabilizing factor (factor XIII)**, catalyzes the formation of covalent bonds between strands of polymerized fibrin, stabilizing and tightening the blood clot.

Coagulation cascade

The formation of a stable fibrin clot, the ultimate goal of step 3, is preceded and regulated by the sequential activation of coagulation factors in events called the **coagulation cascade**. The coagulation cascade is mediated by the sequential activation of a series of **coagulation factors**, proteins synthesized in the liver that circulate in the plasma in an inactive state. They are referred to by Roman numerals in a sequence based on the order in which they were discovered. The plasma coagulation factors and their common names are listed in Table 9.4.

Intrinsic and extrinsic coagulation pathways

Two separate coagulation cascades, the *intrinsic coagulation pathway* and the *extrinsic coagulation pathway*, result in blood clotting in different circumstances (see Fig. 9.10). The final steps in fibrin formation are common to both pathways.

The intrinsic pathway is initiated when blood comes in contact with exposed negatively charged surfaces and by the activation of factor XII through contact with an exposed surface such as collagen in the subendothelial matrix. The intrinsic pathway is also called the *contact activation pathway*. The activation of factor XII requires several cofactors, including kallikrein and high-molecular-weight kininogen.

For the initiation of the extrinsic pathway, a factor extrinsic to blood but released from injured tissue, called **tissue thromboplastin (factor III)** or *tissue factor*, is required. Phospholipids are required for the activation of both coagulation pathways.

Any attempt to describe a distinct division of coagulation into two separate pathways is an oversimplification, and the cascade theory has been extensively modified. There are many points of interaction between the two pathways, and no one pathway alone will account for

TABLE 9.4 Factors of the Coagulation Cascade

Scientific Name	Common Names	Pathway
Factor I	Fibrinogen	Both
Factor II	Prothrombin	Both
Factor III	Tissue factor; tissue thromboplastin	Extrinsic
Factor IV	Calcium	Both
Factor V	Proaccelerin; labile factor; accelerator (Ac-) globulin	Both
Factor VI (Va)	Accelerin	
Factor VII	Proconvertin; serum prothrombin conversion accelerator (SPCA); cothromboplastin	Extrinsic
Factor VIII	Antihemophilic factor A; platelet cofactor 1; antihemophilic globulin (AHG)	Intrinsic
Factor IX	Christmas factor; platelet thromboplastin component (PTC); antihemophilic factor B	Intrinsic
Factor X	Prothrombinase; Stuart–Prower factor	Both
Factor XI	Plasma thromboplastin antecedent (PTA)	Intrinsic
Factor XII	Hageman factor; contact factor	Intrinsic
Factor XIII	Fibrin-stabilizing factor (FSF); protransglutaminase; fibrinoligase	Both

hemostasis. Although the concept of independently acting intrinsic versus extrinsic coagulation pathways is incorrect, the pathways are still important to understand because they are monitored individually in the clinical coagulation tests: the activated partial thromboplastin time monitors the intrinsic pathway, and prothrombin time monitors the extrinsic pathway, as outlined below.

Common coagulation pathway

The final events leading to fibrin formation by both pathways result from the activation of the common pathway. It is initiated by the conversion of inactive clotting **prothrombinase (factor X)** to its active form, factor Xa (see Fig. 9.10). In the extrinsic pathway, factor X is activated by a complex consisting of activated factor VII, Ca^{2+}, and factor III (tissue factor). Activation of this complex by factor III bypasses the requirement for coagulation factors VIII, IX, XI, and XII used in the intrinsic pathway. In the intrinsic pathway, factor X is activated by a complex consisting of factor VIII, factor IXa, platelet factor 3, and Ca^{2+}.

Calcium is a necessary component for the activation of both pathways. Hence, sequestering Ca^{2+} by calcium chelators such as EDTA is used as a means to inhibit blood coagulation independently of the coagulation pathway.

The activation of factor X results in the conversion of prothrombin to *thrombin,* thereby catalyzing the generation of *fibrin*. Thrombin also enhances the activity of clotting factors V and VIII, accelerating "upstream" events in the coagulation pathway. Finally, thrombin is a potent platelet and endothelial cell stimulus and enhances the participation of these cells in coagulation during hemostasis step 2, as explained above.

To a large extent, the interaction of coagulation factors occurs on the surfaces of platelets and endothelial cells. Although plasma can eventually clot in the absence of surface contact, localization and assembly of coagulation factors on cell surfaces amplifies reaction rates by several orders of magnitude.

Prothrombin time

The extrinsic system is evaluated by determination of the **prothrombin time (PT)**. The PT, reported as time in seconds, represents how long a plasma sample takes to clot after a mixture of thromboplastin (factor III) and calcium chloride is added. A time longer than 11 to 13 seconds indicates a deficiency in prothrombin or other clotting factors that affect prothrombin. To resolve the problem of the use of variable thromboplastins in different laboratories, patient values are normalized to a reference value and an international sensitivity index and reported as an **international normalized ratio** (commonly referred to as *INR*).

PT is commonly used to measure the effectiveness of **coumarin**-type anticoagulant drugs. If a patient with a prolonged PT must have surgery, it is important that the PT be brought within a normal range before surgery. This is often done with **vitamin K** injections. Vitamin K is a coenzyme for a carboxylase, which makes it possible for seven clotting factors to bind calcium ions.

Activated partial thromboplastin time

The test used to monitor the activity of the intrinsic system is the **activated partial thromboplastin time (aPTT)**. The aPTT measures the clotting time of plasma, from the activation of factor XII by a commercial biologic aPTT reagent through the formation of a fibrin clot. If a patient's value does not fall into the reference range of 25 to 38 seconds, additional tests are necessary to determine the exact cause of the coagulation problem.

The aPTT is used to monitor **heparin** therapy. Heparin is an injectable anticoagulant, which prevents the formation and growth of clots and may be advised for patients with severe heart disease. All anticlotting therapies carry the risk of bleeding, and so appropriate heparin dosing is critical. The heparin dose is set to ensure the aPTTs are 1.5 to 2.5 times the mean normal values of the patient. If the patient is resistant to heparin, or if low-molecular-weight heparin is given because of its smaller bleeding risk, the heparin anti-Xa test is used. This test quantifies heparin, insofar as heparin inhibits the activity of clotting factor Xa.

The deficiency or deletion of any one factor of the cascade can lead to **hemophilia**. For instance, individuals deficient in factor VIII (antihemophilic factor) have *hemophilia A,* a severe X chromosome–linked condition with prolonged bleeding time on tissue injury as a result of delayed clotting. *Hemophilia B* is the result of factor IX deficiency.

Hemostasis step 4: Plasmin mediates fibrinolysis.

Fibrinolysis is the process of breaking down the product of coagulation, a fibrin clot. The main enzyme in fibrinolysis is **plasmin**, which cuts the fibrin mesh at various places, leading to the formation of circulating fragments that are cleared by other proteases or by the kidney and liver. Plasmin is a serine protease that circulates as inactive proenzyme **plasminogen**. Plasminogen is converted to plasmin by a **tissue plasminogen activator (TPA)**, which is released by activated endothelial cells.

Clot retraction is a phenomenon that may occur within minutes or hours after clot formation. The clot draws together, which pulls the torn edges of the vessel closer together, reducing residual bleeding and stabilizing the injury site. The retraction requires the action of platelets, which contain actin and myosin. Clot retraction reduces the size of the injured area, making it easier for fibroblasts, smooth muscle cells, and endothelial cells to start wound healing.

There are several other important mechanisms that regulate and eventually reverse the final consequence of coagulation to allow healing to proceed. Platelet function is strongly inhibited, for example, by the endothelial cell metabolite prostacyclin, which is generated from arachidonic acid during cellular activation. Thrombin bound to thrombomodulin on the surface of endothelial cells converts protein C to an active protease. Activated protein C and its cofactor, protein S, restrain further coagulation by proteolysis of factors Va and VIIIa. Furthermore, activated protein C augments fibrinolysis by blocking an inhibitor of TPA. Finally, antithrombin III is a potent inhibitor of proteases involved in the coagulation cascade such as thrombin. The activity of antithrombin III is accelerated by small amounts of heparin, a mucopolysaccharide present in the cells of many tissues.

Abnormalities in the proteins that are involved in step four may result in hypercoagulation or thrombotic disorders, in which intravascular clot formation leads to severe problems, including embolism and stroke. The most common inherited disorder is **protein S deficiency**, in which about two thirds of patients will have had a thrombotic event by age 35 years. Other disorders include abnormalities in protein C, antithrombin III, and plasminogen.

Chemoattractants, mitogens, and growth factors

While the blood clot resolves, multiple factors participate in wound healing. Optimal wound healing requires the recruitment or generation of new tissue cells as well as new blood vessels to nourish the repairing tissue. **Chemoattractants** attract smooth muscle cells, inflammatory cells, and fibroblasts to the wound, and **mitogens** induce them to proliferate. **Growth factors**, a subclass of cytokines, induce primitive cells to differentiate. There are many different growth factors involved in this proliferative phase of wound repair, including **platelet-derived growth factor** (PDGF) and **epidermal growth factor** (EGF). They control the wound contraction and continuous remodeling of tissue and collagen over an extended period for ultimate healing.

An important event during wound healing is **angiogenesis**, the formation of new blood vessels. Platelets play an important role because they secrete factors that induce proliferation, migration, and differentiation of two of the major components of blood vessels: endothelial cells and smooth muscle cells. There are at least 20 angiogenic growth factors and 30 angiogenesis inhibitors found naturally in the body. These keep most blood vessels in a quiescent state but shift the balance toward formation of new vessels after injury. The new capillaries deliver nutrients to the wound and help maintain the granulation tissue bed as scaffolding on which connective tissue will form.

A detailed understanding of the events that regulate angiogenesis has profound therapeutic potential. For example, exogenously applied *angiogenesis-inducing agents* may prove useful in accelerating the repair of tissue damaged by thrombi in the pulmonary, cerebral, or cardiac circulation. In addition, angiogenic factors may assist in the repair of lesions that normally repair slowly—or not at all—such as skin ulcers in patients who are bedridden or diabetic. On the other hand, *inhibition of angiogenesis* may prove particularly useful in the treatment of patients with cancer, because growing tumors require the development of blood vessels to survive. New cancer therapies are being tested that interfere with the angiogenic response, by acting on either the factors involved or the cells that respond to them.

Chapter Summary

- Blood is a dynamic connective tissue with four major functions: to transport substances throughout the body; to protect the body from losing blood due to an injury; to maintain body homeostasis in regard to pH, water, heat, osmolality, and other factors; and to circulate cells and chemicals that defend the body against disease.
- Adult humans have approximately 5 L of whole blood, with about 45% of formed elements suspended in 55% plasma and solutes.
- Plasma is the liquid portion of blood. It contains 93% water and 7% of molecules such as proteins (e.g., albumin, globulins, fibrinogen, enzymes, and hormones), lipids (e.g., cholesterol and triglycerides), carbohydrates (glucose), electrolytes (e.g., Na^+, Cl^-, and HCO_3^-), and cellular waste (e.g., bilirubin, urea, and creatinine).
- Serum contains all components of plasma except for substances involved in blood clotting.
- Erythrocytes are anuclear, disc-shaped cells that deliver oxygen throughout the body via hemoglobin (Hgb). Adult Hgb is made of an α and a β part, which must be bound together for normal functioning. Although changes in Hgb are rare, several hundred abnormal variants exist, producing clinical disorders such as sickle cell disease.
- Erythrocytes are characterized by number, shape, size, color, and maturity, providing valuable indicators for the diagnosis of anemias, hemoglobinopathies, infections, and other illnesses.
- Anemias can be caused by a loss or destruction of blood (e.g., hemorrhagic anemia and hemolytic anemia) or by decreased or faulty erythrocyte production (e.g., sickle cell anemia, pernicious anemia, iron deficiency anemia, and anemia of inflammation).
- At the end of about a 4-month lifetime, old erythrocytes are engulfed by macrophages. Their hemoglobin, including iron, is recycled while generating the diagnostic waste product bilirubin. Hyperbilirubinemia produces characteristic icteric blood samples and jaundice.
- Leukocytes are classified morphologically as granulocytes (eosinophils, basophils, and neutrophils) and agranulocytes (monocytes and lymphocytes). Lymphocytes are functionally divided into T cells, which participate in cellular immune defenses, and B cells, which are part of humoral immune defenses.
- Leukocytes defend the body against infection using phagocytosis and various antimicrobial weapons, release mediators to control inflammation, and contribute to wound healing.
- Neutrophils are early inflammatory mediators, killing and engulfing microbes. Eosinophils support neutrophils and can also attack multicellular parasites. Basophils support eosinophils but may overshoot in their response and are, therefore, associated with asthma and allergies. T lymphocytes are the key players for cell-mediated immunity. B lymphocytes make antibodies that attack bacteria and toxins.
- Hematopoiesis is the development of circulating blood cells from the uncommitted multipotent stem cell of bone marrow. Immature cells differentiate along cell lineages into mature cells promoted by hematopoietins and other cytokines.
- Erythropoietin is a hormone produced by the kidney and promotes erythropoiesis in bone marrow. Patients on dialysis often require erythropoietin intake to maintain hematocrit (Hct). Some aerobic athletes abuse erythropoietin to illegally increase their Hct.
- Thrombocytes (platelets) are irregularly shaped, small, anuclear, cell-derived structures that, together with plasma proteins, control blood clotting and promote wound healing.
- Blood coagulation involves the fast formation of a weak platelet plug stabilized by fibrinogen, which is expanded and stabilized into a more robust plug made of cells, platelets, and insoluble fibrin molecules. Endothelial cells, blood coagulation factors, and Ca^{2+}, which are mediators released by platelets, organize the process. Thrombin is necessary for fibrin clot formation and plasmin for its dissolution.
- The basic metabolic panel reveals blood values for glucose, electrolytes, blood urea nitrogen, and creatinine; the complete metabolic panel additionally gives values for albumin, total protein, alkaline phosphatase, alanine transaminase, aspartate transaminase, and bilirubin.
- The complete blood count includes white blood cell counts, red blood cell counts, platelet counts, hemoglobin (Hgb), hematocrit, red blood cell indices (mean cell volume, mean cell Hgb, and mean cell Hgb concentration), and mean platelet volume.
- Other routinely performed blood tests include blood viscosity, specific gravity, serum protein electrophoresis, erythrocyte sedimentation rate, coagulation tests (e.g., bleeding time, platelet count, prothrombin time, and activated partial prothrombin time), and iron profile (serum iron, ferritin, total iron-binding capacity, unsaturated iron-binding capacity, transferrin, and transferrin saturation).
- For blood transfusions, donor and recipient blood must be compatible to avoid agglutination between erythrocyte-associated A, B, and Rh antigens and anti-A, anti-B, and anti-Rh antibodies.

thePoint *Visit http://thePoint.lww.com for chapter review Q&A, case studies, animations, and more!*

10 Immunology, Organ Interaction, and Homeostasis

ACTIVE LEARNING OBJECTIVES

Upon mastering the material in this chapter you should be able to:

- Understand immune system triggers and explain self-tolerance.
- Apply the roles of both noncellular and cellular components of innate immunity in maintaining body homeostasis.
- Explain how the adaptive immune system achieves its three main features: specificity, diversity, and memory.
- Understand the mechanisms for both exogenous and endogenous antigen presentation in cell-mediated immunity.
- Know the roles of T-cell subtypes in adaptive immunity.
- Apply the clinical relevance of both innate and adaptive immunity.
- Know three ways that the five classes of antibodies work to eliminate antigens.
- Recognize the signs and functions of both acute and chronic inflammation.
- Understand the clinical roles of pro- and anti-inflammatory cytokines.
- Explain the immune-related requirements for organ transplantation.
- Recognize immune system disorders by immune system reactions.
- Apply the principle of immune surveillance in hematologic malignancies.
- Understand the interaction among the immune, neuronal, and hormonal systems in maintaining homeostasis.

Immunology covers the study of the immune system and organ interaction. It deals with the physiologic function of the immune system in both health and disease. Immunologic disorders (autoimmune diseases, hypersensitivities, immune deficiency, and transplant rejection) result from malfunctions of the immune system. Immunology has made great advances in medicine following Louis Pasteur's development of vaccination and his theory that microorganisms are the cause of infection. The aim of this chapter is to present basic components of the immune system and its role in organ function and homeostasis.

IMMUNE SYSTEM COMPONENTS

The immune system is the body's defense mechanism against disease by identifying and killing pathogens and tumor cells. The immune system detects a wide variety of agents, from viruses to parasitic worms. In addition to identifying pathogens, it must also distinguish them from the body's own healthy cells and tissues to function properly. Detection is a complex, ongoing process because pathogens evolve rapidly, producing adaptations that avoid detection by the immune system and allow successful infection.

The immune system protects the body against infection with layered defenses of increasing specificity. The physical barrier (e.g., skin and membrane secretions) is the first line of defense to prevent pathogens such as bacteria and virus from entering the body. If pathogens breach these barriers, the **innate immune system** provides an immediate nonspecific response. The innate response is triggered when microbes are identified by pattern recognition receptors. If pathogens successfully evade the innate response, the body engages a third layer of protection, the **adaptive immune system** , which is activated by the innate response. That is, the system adapts its response during an infection to improve recognition of the pathogen through signature antigens.

The immune system comprises various organs, tissues, and special cells and molecules within the body that work together in the body's defense system (Fig. 10.1).

Lymphocytes are specialized leukocytes involved in adaptive immunity. The thymus and bone marrow are called **primary** (or **central**) **lymphoid organs**. These are the organs where immune cells mature to become *immunocompetent*. Most cells mature to immune cells within the bone marrow and, after release, begin a life of patrol in the blood. The exception is the pre-T cell, which first undergoes maturation in the thymus before circulating in blood.

Lymph nodes, tonsils, lymph follicles of mucous membranes, and the spleen belong to the list of **secondary** (or **peripheral**) **lymphoid organs**. These are the organs where mature immune cells participate in specific immune defense reactions.

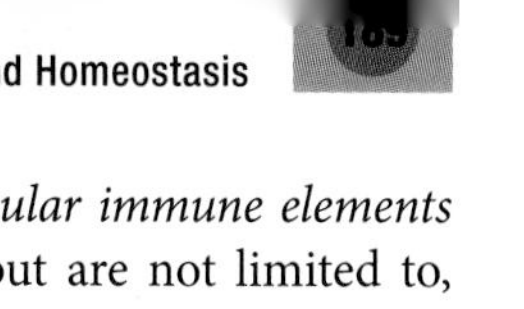

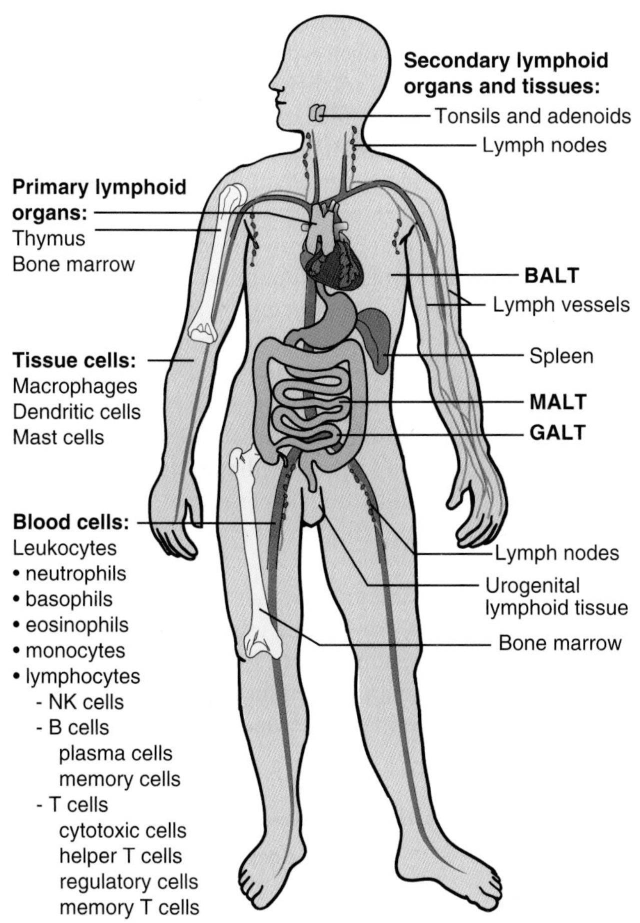

Figure 10.1 Organs and cells of the immune system. The thymus and bone marrow are primary lymphoid organs. Lymph nodes, tonsils, the spleen, and lymphoid tissues of the lung, gut, and urogenital tract are secondary lymphoid organs. Leukocytes are immune cells that circulate in blood. Macrophages, dendritic cells, and mast cells are primarily found in tissues. NK cells, natural killer cells; BALT, bronchus-associated lymphoid tissue; MALT, mucosa-associated lymphoid tissue; GALT, gut-associated lymphoid tissue.

Lymph nodes are the sites through which blood, lymph, and immune cells are filtered. These encapsulated organs are located throughout the body at junctions of **lymphatic vessels** and are optimized for interaction between **antigen-presenting cells** (**APCs**) and T and B lymphocytes. The movement of lymph through lymph vessels and lymph nodes is supported by skeletal muscle movement but is otherwise passive. There is no organ similar to the heart to pressurize the lymph flow.

During a bacterial infection, lymph nodes swell as a result of proliferation of immune cells. Palpation of lymph nodes is part of most orderly clinical examinations and gives an indication of the activity level of the immune system.

Figure 10.1 on the left side shows the immune cells present in tissue (macrophages, dendritic cells, mast cells) and present in blood (leukocytes). Their morphology and basic function are introduced in Chapter 9, "Blood Components." Their immune function is presented as part of the sections on innate and adaptive immunity below.

There is an abundance *of noncellular immune elements* in blood and lymph. They include, but are not limited to, antibodies, communication molecules, and complement and are presented later in the chapter. Additionally, the innate immune system encompasses a variety of fluids and other noncellular components with antimicrobial character.

IMMUNE SYSTEM ACTIVATION

Organisms with the potential to cause disease are called **pathogens**. The paramount function of the immune system is to recognize and destroy pathogens that enter the body. Infection activates the immune system. More specifically, infection activates B and T cells through receptor recognition of **antigens**. **Immunogens** are antigens that activate the immune system, whereas **haptens** are antigens that do not activate the immune system. However, haptens do become immunogenic when linked to a carrier molecule.

Pathogens may enter the body through a cut in the skin (as in the case of the hepatitis B virus), through membranes of the respiratory tract (as in the case of the measles virus), or through membranes of the digestive tract (as in the case of *Salmonella* bacteria). Pathogens may also be transmitted by insect bites (as in the case of malaria) or by sexual encounter (as in the case of the human immunodeficiency virus).

In addition to pathogens, other organic and nonorganic foreign substances can be antigenic. Pesticides, cosmetics, and exhaust particles are examples of such substances that can activate the immune system.

Large complex molecules and proteins best activate the immune system.

Proteins are by far the best immunogens. Their complex three-dimensional conformation, their electrical charges, and their ability to aggregate make them good candidates to be recognized by immune defenses. Polysaccharides, lipids, and nucleic acids by themselves induce either weak or no immune responses. But, similar to proteins, their immunogenicity increases with increasing complexity. For instance, polymers made of only one type of nucleotide are poor immunogens, whereas polymers made of more than one base are usually well recognized. Lastly, size matters. Typically, only molecules with a molecular weight of 4,000 daltons or above elicit an immune response.

Because proteins are immunologically best recognized, they are used for many vaccines to greatly improve their antigenicity. For instance, the vaccine targeting the carbohydrate capsule of *Streptococcus pneumoniae* is conjugated to protein carriers to enhance its effectiveness.

Severe injury and necrosis activate immune defenses, whereas mild injury and apoptosis do not.

Immune activation is tied in with other body defenses and reactions. When the body sustains injury, whether from physical force, cellular breakdown, or biologic infection, the decision for immune activation depends on the severity and type of the injury.

If the cells are not too severely injured, the injured and surrounding cells may adapt to the event by changing their size, number, and functions, without activating immune responses. For instance, if skeletal muscle cells are stressed within tolerable limits, they become larger and improve their functions. This is the principle for strength exercise. On the other hand, severe cellular stress can activate the immune system, for instance, in the case of airway cells that respond strongly to irritating smoke particles with inflammation.

If the damage to a cell is too great to cope with, the cell will die by **necrosis**. The immune system is activated when the cells collapse and their cellular contents, including lysosomal enzymes, leak into the surrounding tissue, causing further damage and starting an intense inflammatory response. As a result, chemokines (chemicals leaked from the damaged cells) attract various types of immune cells to the injury site, which further activate the immune response. An example is Duchenne muscular dystrophy. In this case, severe muscle fiber stress and necrosis as a result of genetically induced muscle weakness lead to immune responses that further promote the death of dystrophic muscle.

In contrast to necrotic cell death, cell death by **apoptosis** will not cause immune activation. Apoptotic cells die without bursting apart, and as a result, no damaging substances are released from the cells and an inflammatory response is avoided. Apoptosis is the body's nonpathologic process of removing cells. This is especially important for the role of the immune system in maintaining homeostasis and regulating tolerance. Every day, several million B and T cells are generated, and most die apoptotically by negative selection (or activation-induced cell death).

Activation of innate and adaptive immunity involves different, yet complementary, processes.

Exogenous antigens, such as extracellular bacteria, intracellular bacteria that are still identifiable from outside the cell, and foreign particles, are first attacked by the **innate immunity** response (Fig. 10.2). Phagocytic cells, such as polymorphonuclear cells, macrophages, and dendritic cells, might engulf the antigen or destroy it with enzymes and oxygen radicals. The pathogen might directly activate **complement** proteins, another important component of the innate immune system. This activation can result in **lysis** of the organism and promote activation of the *inflammatory response*, or **inflammation**.

If some antigenic parts of the microorganism are present on the plasma membrane of the host cell, antibodies can bind directly to the infected cell and activate complement (discussed later in further detail).

If the pathogen evades the innate immune system and no antibodies are present, then the **adaptive immune system** response is triggered. Adaptive immunity is also activated by viruses, protozoan parasites, and intracellular bacteria that cannot be recognized from outside the cell (endogenous antigens). In this case, the antigen is degraded intracellularly into smaller pieces. The pieces are then incorporated into the host cells' plasma membranes and, together with **major histocompatibility complex (MHC) proteins**, are presented to T cells (discussed later in further detail). These microorganisms are ultimately eliminated together with the infected host cell.

Although innate and adaptive immune systems are characterized by contrasting functions and timing, they work together in ways that obscure their differences. For instance, the initiation and adequate functioning of the innate system often depend on the presence of elements of the adaptive immune system such as small amounts of specific antibody in blood plasma. The reverse is true as well. Antibodies and other mediators of the adaptive immune system depend on elements that are typically associated with the innate immune system such as phagocytic neutrophils.

Only when working together can the innate and adaptive immune systems prevent the establishment and long-term survival of infectious agents. For this reason, Figure 10.2 is presented as an analogy of the Chinese yin and yang symbol, in which opposites intertwine and complement each other toward a greater whole.

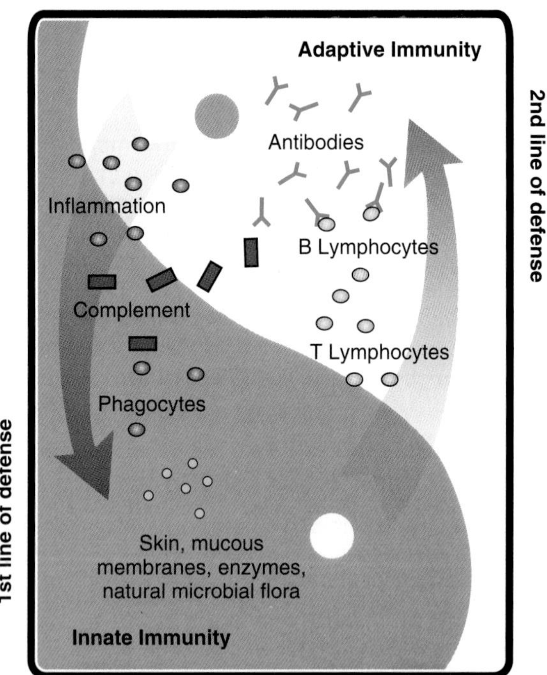

Figure 10.2 Overview of the immune system. As part of the innate immunity, the human body has natural measures to prevent entry of microbes. When these barriers are overcome, a foreign invader first encounters phagocytes. Complement may bind to the invader and facilitate its phagocytosis. Inflammation may develop. If the innate immune system cannot destroy the invader, the adaptive immune system is activated with T and B lymphocytes as its effectors. Although the innate and adaptive immune systems are characterized by contrasting functions and timing, they work closely together and rely on each other to succeed in removing invading pathogens.

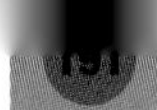

IMMUNE SYSTEM DETECTION

The task of the immune system is to screen millions of molecules and eliminate them when recognized as not being a normal part of the body. For that, it has to discriminate *self* from *nonself*. To destroy only pathogens while leaving the body's own cells intact requires a complex system because microorganisms and human tissue are made of closely related material.

Self-tolerance, the action of the immune system to forgo attacking the body's own cells and proteins, is a multistep process that begins in the thymus during fetal development. This process, called *central tolerance*, is discussed later in this chapter. Self-tolerance then continues with additional mechanisms outside of the thymus (called *peripheral tolerance*).

In addition to self-antigen, a few important foreign antigens also do not activate the immune system (called *acquired* or *induced tolerance*). We generally tolerate food-derived foreign particles (although some foods induce allergic reactions). A mother does not reject the fetal molecules that are derived from both maternal and paternal genes. Identical twins can accept skin grafts from each other without causing immune reactions.

Any mistake in the process of self/nonself discrimination can have devastating consequences to the organism. When a body erroneously mounts an immune response against its own tissues, it leads to *autoimmune diseases*.

IMMUNE SYSTEM DEFENSES

The immune system employs three mechanisms to mount its defense against invaders: physical barriers, innate immunity, and adaptive immunity.

Physical barriers are the immune system's first line of defense.

As mentioned earlier, our immune system protects against infection with layered defenses of increasing specificity. The physical barrier (e.g., skin and membrane secretions) is the first line of defense to prevent pathogens, such as bacteria and virus, from entering the body. Our skin is an effective anatomical and physiologic barrier against microorganisms. First, the cells of the epidermal layers are dry and densely packed, making them an inhospitable environment to many bacteria. Furthermore, salty secretions from sweat glands and oily secretions from the sebaceous glands associated with hair follicles create a hyperosmotic and slightly acidic skin environment, which dehydrates bacteria and discourages those that prefer a neutral pH for colonization. Additionally, the continuous desquamation of skin cells eliminates bacteria adhering to epithelial cells. Lastly, many noninvasive, nonpathogenic commensal (part of normal flora) microorganisms on the skin prevent growth of harmful microorganisms in a process called *competitive exclusion*.

Acidity is also used in other places of the body as an antimicrobial tool. For example, natural microflora alter the fluids of the vaginal and urinary tracts to an acidic pH below 4.5 so that yeast and other microorganisms cannot grow. Parietal cells of the stomach create highly acidic gastric juice below pH 3.

Other factors such as low oxygen tension and fever are said to contribute to the barrier defenses, but their physiologic impact is still under discussion. The role of fever is thought to have a direct negative effect on certain microorganisms and may also enhance the efficiency of **phagocytosis**.

Cellular secretions

Various types of cellular secretions destroy and eliminate germs. *Mucus* prevents microorganisms from adhering to epithelial cells and contains antibacterial components. The mucus of the gut blocks, inactivates, or destroys pathogens associated with food before they can enter the body. A thin layer of mucus covering the airway from the nose to the bronchioles traps inhaled viruses, bacteria, pollens, and other particles and facilitates their removal before they can damage the airway lining cells. The flow of *saliva* helps to wash away bacteria attached to food particles while attacking them with thiocyanates and lysozymes. Saliva can also contain antibodies that destroy oral bacteria in certain people. *Tears* and *nasal secretions* contain similar antibacterial components.

The list of chemical factors with antimicrobial characteristics is long. It includes **pepsin** in the stomach, **defensins** produced by immunologic cells, and **surfactant** of the lung, to name a few. **Interferons** are a group of proteins that are produced by cells following viral infection. Complement is a group of serum proteins that circulate in an inactive state and can be activated by a variety of specific and nonspecific immunologic mechanisms. Their actions, sometimes called the *humoral component* of the innate immune system, are discussed in more detail below. An important role of these chemical factors is to connect the three lines of immune system defenses.

Innate immunity is the immune system's second line of defense.

The ability of cells and tissues to respond to and to get rid of environmental challenges is an ancient evolutionary development that has persisted through vertebrate development as innate immunity. The zoologist Metchnikoff discovered that cells of a starfish could phagocytose invaders, which means that this 600-million-year-old invertebrate possesses an innate immune system. Innate immunity is also called *nonspecific* or *natural immunity*. Table 10.1 contrasts its basic characteristics in humans with the characteristics of the adaptive immune system (also called the *acquired immune system*). The innate immune system:

- is present at birth;
- persists throughout life;
- can be mobilized rapidly and acts quickly;
- attacks all antigens fairly equally; and
- maintains the quantity or quality of the response long term.

TABLE 10.1 Characteristics of the Innate and Adaptive Immune Systems

	Innate	Adaptive
System	Old, found in invertebrates	Evolved in early vertebrates
	Present at birth	Develops throughout life
Stimulation	Not required	Required
Specificity	Minimal	Highly specific
		Self vs. nonself discrimination
		Receptor mediated
Response	Within minutes	Develops over days
	No change in quality and quantity over time	Improved by previous exposure
Memory	No	Yes
Soluble factors	Lysozyme, complement, acute phase proteins, interferon, cytokines	Antibodies, cytokines (interleukins), interferon
Cells	Phagocytic leukocytes, natural killer cells	T cells, B cells

TABLE 10.2 Cellular Elements of the Innate Immune System

Cells	Main Function	Phagocytosis
Neutrophils	Kill microorganisms intracellularly and extracellularly	+
Macrophages	Kill microorganisms intracellularly and extracellularly	+
	Present antigen	
Dendritic cells	Phagocytose pathogens	+
	Present antigen	
NK and LAK cells	Destroy virus-infected and tumor cells	–
Eosinophils	Secrete factors that kill certain parasites and worms	–
Mast cells	Release factors that increase blood flow and vascular permeability	–

NK, natural killer; LAK, lymphokine-activated killer.

Phagocytic leukocytes

Innate leukocytes include phagocytes (neutrophils, macrophages, and dendritic cells) and nonphagocytic cells as outlined in Table 10.2. Phagocytic cells identify and eliminate pathogens, either by attacking larger pathogens through contact or by engulfing them. They can be activated by T-cell contact (e.g., neutrophils), by T-cell cytokines (e.g., macrophages), or by contact with pathogens. For the latter, phagocytes directly sense the pathogens through a group of transmembrane receptors, the so-called **toll-like receptors.**

On activation, phagocytes engulf the microorganism, particle, or cell debris. The engulfed matter is enclosed within vacuoles and enzymatically digested, after fusion with lysosomes. Some pathogens have been coated with **opsonins** in a process called opsonization to render them more attractive to phagocytosis. Examples of opsonins are immunoglobulin G (IgG) antibody and the C3b molecule of the complement system.

Neutrophils (or *polymorphonuclear leukocytes*) recognize chemicals produced by bacteria in a cut or scratch and migrate toward them. Once arrived, they ingest the bacteria and kill them. For killing, neutrophils use proteolytic enzymes and reactive oxygen species produced as part of the respiratory burst (see Chapter 9, "Blood Components").

Macrophages are derived from circulating monocytes. Macrophages kill, like neutrophils, by using the respiratory burst and proteolytic enzymes. Macrophages secrete various cytokines that attract other leukocytes to the infection site and initiate the acute phase inflammatory response. Finally, macrophages act as APCs to T cells and are hence an important bridge to the T-cell–mediated immunity of the adaptive immune response.

Macrophages can circulate in lymph vessels (*wandering, nonfixed macrophages*), or they can reside in connective tissue, in lymph nodules, along the digestive tract, in the lungs, in the spleen, and in other places (*mature, fixed macrophages*). Fixed macrophages are part of the **reticuloendothelial system (RES)**, which, in addition to removing pathogens, also removes old cells and cellular debris from the bloodstream. Some of these macrophages have their own names. For instance, the macrophages along certain blood vessels in the liver are called **Kupffer cells**, whereas the macrophages of the joints are called *synovial A cells. Microglial cells* are macrophages located in the brain.

Dendritic cells are, like macrophages, a critical link between the innate and adaptive immune systems. They exist in an immature form throughout the epithelium of the skin (e.g., **Langerhans cells**), the respiratory tract, and the gastrointestinal tract. After phagocytosis of pathogens, the cells mature and travel to regional lymph nodes, where they activate T cells, which then activate B cells to produce antibodies against the pathogen.

Nonphagocytic leukocytes

In addition to phagocytic cells, innate leukocytes also comprise nonphagocytic cells (NK and LAK cells, eosinophils, and mast cells) as shown in Table 10.2. NK cells attack aberrant body cells such as virus-infected cells and malignant cells. They release the cytolytic protein **perforin**, which forms a pore in the plasma membrane of the target cell. Proteolytic enzymes, such as **granzyme**, are also part of the NK's cytoplasmic granules. When released, it enters through the perforin pore of the target cell and induces apoptosis. On exposure to lymphocyte secretions, such as interleukin-2 and interferon-γ, NK cells become **lymphokine-activated killer (LAK) cells**, which are even more efficient in killing than NK cells.

Eosinophils are best known as participants in allergic reactions, where they might detoxify some of the inflammation-inducing substances. But they are also primarily evolved to secrete factors that punch small holes in worms and other parasites, causing them to die.

Mast cells are present in most tissues in the vicinity of blood vessels and contain many granules rich in **histamine** and **heparin**. They are especially prominent under coverings lining the body surfaces such as the skin, mouth, nose, lung mucosa, and digestive tract. Although best known for their roles in allergy and anaphylaxis of the adaptive immune system (see next section), mast cells play an important role in the innate system as well. Additionally, they are intimately involved in wound healing. They release factors that increase blood flow and vascular permeability, bringing components of immunity to the site of infection. In combination with IgE antibody from B cells, mast cells can also target parasites that are too large to be phagocytosed, such as intestinal worms.

Adaptive immunity is the immune system's third line of defense.

Adaptive immunity uses three important features in its method of attack: specificity, diversity, and memory. Microbes that escape the onslaught of cells and molecules of the innate immune system face attack by T cells and B-cell products of the adaptive immune system, also called the *acquired immune system*. Table 10.1 lists its basic characteristics in comparison to those of the innate immune system. The adaptive immune system:

- is a relatively recent evolutionary development and characteristic of jawed vertebrates;
- is activated by thousands of diverse antigens, which are presented as glycoproteins on the surface of bacteria, as coat proteins of viruses, as microbial toxins, or as membranes of infected cells;
- responds with the proliferation of cells and the generation of antibodies that specifically assault the invading pathogens;
- responds slowly, being fully activated about 4 days after the immunologic threat; and
- exhibits immunologic memory, so that repeated exposure to the same infectious agent results in improved resistance against it.

Specificity

The *specificity* of the adaptive immune system is created by **antigen-recognition molecules**, which are synthesized prior to the exposure to antigen and are then modified during the immune response to make them even more specific to the antigen.

Several different types of recognition molecules participate in this anticipatory defense system: (1) specific receptors on T and B lymphocytes; (2) molecules encoded in the MHC cells (recognition molecules on cell surfaces); and (3) antibodies, which are the secreted form of B-cell receptors (BCRs). Though the general structure for the types of molecules is very similar (e.g., every antibody has a typical Y-shaped form), a small region of each individual molecule is different (e.g., for antibodies, the hypervariable region) and allows only the binding of a specific antigen.

An antigen might be recognized by several recognition molecules. To stay with the example of antibodies, most protein antigens have several **epitopes** (the antigen part that binds the antibody), and hence, are recognized by different B cells, which release different antibodies to mount a **polyclonal antibody** response.

On the other hand, closely related antigens may share epitopes (**cross-reactivity**). For instance, antibodies that are induced by some microbial antigens cross-react with polysaccharide antigens found on red blood cells and are the basis for the ABO blood-type system (see Chapter 9, "Blood Components").

Diversity

The *diversity* of adaptive immune responses is based on a huge variety of antigen receptor configurations, essentially one receptor for each different antigen that might be encountered.

In a person, there are about 10^{18} different **T-cell receptors (TCRs)**, with each T cell expressing one type of receptor. The **B-cell receptor (BCR)** is a membrane-bound form of immunoglobulin. There are about 10^{14} BCRs, again one type of immunoglobulin per B cell. The molecule diversity is mainly achieved by variable recombination of gene segments prior to exposure to antigen and, in the case of immunoglobulins, additionally by mutation of the molecules after exposure to antigen.

The recognition of an antigen by the lymphocyte with the best-fitting receptor occurs mainly in the local lymph node and induces the activation, proliferation, and differentiation of the responsive cell, a process known as **clonal selection**. Figure 10.3 shows the process for four B cells with different BCRs and one recognizing the antigen. Not evident in the figure is that optimal activation and proliferation of B cells (and, equally, T cells) occur only when costimulatory signals, secondary to TCR and BCR bindings, are present.

Clonal selection amplifies the number of T or B lymphocytes that are programmed to specifically respond to the inciting stimulus. In Figure 10.3, this means all resulting cells have the same BCRs, ready to bind more of the specific antigen. However, not all cells possess the exact identical functional characteristics. For instance, antigen binding to the BCRs induces their development into memory cells (see below) and into **plasma cells.** Plasma cells are much

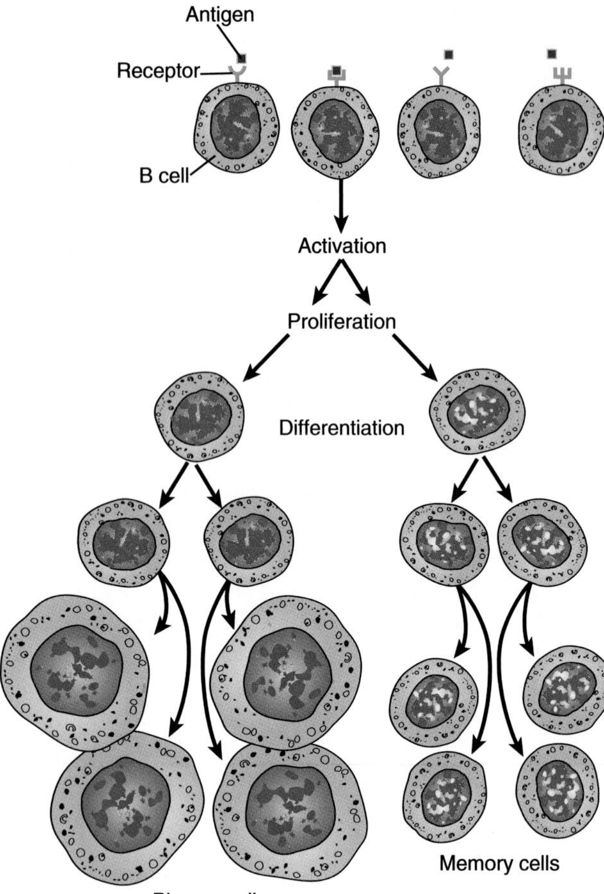

Figure 10.3 Clonal selection of lymphocytes. Only the clone of lymphocytes that has the unique ability to recognize the antigen of interest proliferates and generates progenitor cells. These cells are specific to the inducing antigen but may have different functions. In the case of B cells, plasma cell clones produce antibodies, and memory cell clones enhance subsequent immune responses to the specific antigen. Clonal selection occurs in secondary lymph organs such as the local lymph nodes.

larger and are capable of producing and secreting antibodies. Initially, the plasma cells produce IgM antibodies and later IgG, IgA, and/or IgE antibodies. This maturation process is called *Ig isotype switching.*

Similarly, clonal proliferation of T cells can lead to the generation of more *antigen-specific T cells* and/or to the production of **effector T cells,** such as T helper and cytotoxic T cells.

Memory

The *memory* of the adaptive immune system is based on the fact that some descendants in the expanded B-cell and T-cell clones function as **memory cells** (Fig. 10.3, *right*). These cells mimic the reactive specificity of the original lymphocytes that responded to the antigen and accelerate the responsiveness of the immune system when the antigen is encountered again (**anamnestic response**) and are the basis for immunization via vaccinations.

CELL-MEDIATED AND HUMORAL RESPONSES

The adaptive immune system has two major branches, cell-mediated and humoral immunity. As mentioned previously, though presented as distinct systems, no part of the immune system works separately. Rather, they all work in a cooperative fashion using cytokines and other means as a communication system.

B cells mediate humoral immune response, whereas T cells regulate cell-mediated immune response. The CD4 and CD8 ("CD" for cluster differentiation) T-cell subgroups of the cellular immunity branch of the adaptive immune system have different functions and are divided into additional subtypes according to their molecular appearance, cytokine sensitivity and production, and specific functions. These cells form, with each other and with other immune cells, a complex network of communication and immune response that is the basis for the efficiency, flexibility, and longevity of the adaptive immune system.

Figure 10.4 graphically summarizes important interactions between cells. The figure is organized according to the type of encountered antigen (exogenous or endogenous, left), to the sensing and responding processes (antigen recognition and presentation and immune response, top), and to the involvement of T cells and B cells (cellular and humoral immune responses, right).

Cell-mediated response involves activating T cells and releasing cytokines.

Cell-mediated responses in adaptive immunity do not involve antibodies or complement but, rather, activation of macrophages, NK cells, T lymphocytes, and the release of various cytokines in response to an antigen.

T cells continuously patrol the body and check the foreignness of antigen. T cells become activated when antigen binds to the specific TCR plus a costimulatory element. Antigen can only bind to the TCR when presented by APCs in combination with MHC proteins. The complex interaction between the T cell and the APC is called **immunologic synapse.** Figure 10.5 and Table 10.3 summarize the events at the synapse.

Major histocompatibility complex

The **major histocompatibility complex (MHC)** is a large genomic region or gene family found in most vertebrates. When activated, MHC genes produce MHC molecules that play an important role in the immune system, especially autoimmunity.

In the case that the antigen stems from extracellular proteins or phagocytosed bacteria, the antigen is digested intracellularly within phagolysosomes and associated with **MHC class II molecules,** which then present the antigen to the TCR of $CD4^+$ T helper cells (Fig. 10.4, upper part). Only macrophages, dendritic cells, and B cells can do so; hence, they are CD4 APCs.

Macrophages become APCs when they upregulate their MHC II molecules in response to infection and appropriate stimuli such as **lipopolysaccharide (LPS)** from

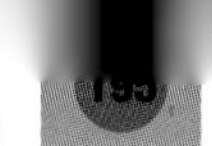

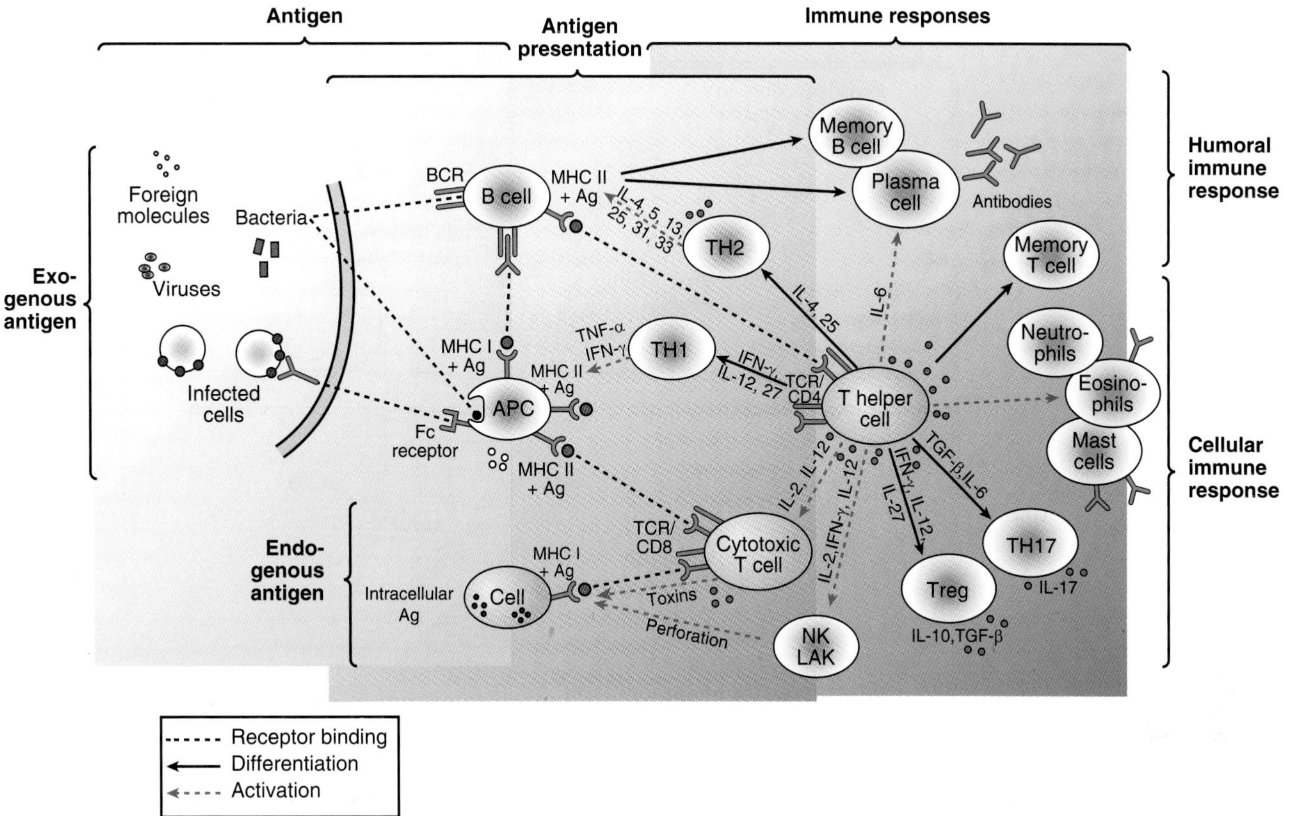

Figure 10.4 Cellular and humoral immune responses of the adaptive immune system. Activated T cells accomplish cellular immune responses, and B cells and antibodies mediate humoral immune responses. Exogenous antigen activates B cells by binding to the B-cell receptor (BCR) and T cells by binding to the T-cell receptor. However, TCR binding only occurs when antigen peptide is presented by antigen-presenting cells in association with major histocompatibility complex (MHC) proteins. The TCR is either associated with CD4 or CD8, depending on the T-cell type. Endogenous antigens are presented via MHC proteins to cytotoxic T cells, which destroy the host cell together with the intracellular pathogen. The immune response involves two paths, one using B cells (humoral immune response) and one using cytotoxic T cells (cellular immune response). Millions of different B and T cell types exist to recognize millions of different antigens. Ag, antigen; CD, cluster of differentiation; IFN, interferon; IL, interleukin; LAK, lymphokine-activated killer cell; NK, natural killer cell; TH, T helper cell; TNF, tumor necrosis factor; T reg, T regulatory cell.

gram-negative bacteria. *Dendritic cells* are mostly found in peripheral tissue, where they ingest, accumulate, and process antigens. *B cells* become activated by ligand binding to the BCR. Hence, they present the particular antigen to which the antibody that they express is directed. For other antigens, B cells are not very efficient APCs.

Almost every cell in the body can present antigen to the TCR of $CD8^+$ cytotoxic T cells. In this case, the antigen is an intracellular pathogen, which is degraded in the cytosol and associated with **MHC class I molecules** (Fig. 10.5, lower part).

T-cell differentiation

T cells, or *T lymphocytes*, play a central role in cell-mediated immunity. They are distinguished from other lymphocytes, such as B cells and NK cells, by the presence of a special receptor on their cell membrane called T-cell receptors. "T" stands for *thymus*, the principal organ responsible for the T cell's maturation.

There are more than 160 known clusters that coat the surface of leukocytes and many other cells. CD4 and CD8 are two that are used to identify functionally critical T-cell subsets. To give two other examples, CD14 is an LPS-binding protein receptor on monocytes and macrophages, and CD69 is used as a marker of cell activation for T, B, NK cells, and macrophages.

When the cells coming from the bone marrow reach the thymus, T cells have neither CD4 nor CD8 (*double-negative T cells*), but they then start expressing both markers (*double-positive T cells*). During development in the thymus, double-positive cells differentiate into cells with either CD4 or CD8 receptors. This is a critical step because during this time they develop their ability to distinguish self from nonself peptides.

First, cells that do not bind MHC/antigen complexes within 3 to 4 days will die. The rest of the cells undergo **positive selection**. This means that T cells proliferate that bind antigen (self or nonself) complexed with class I or II MHC proteins. This happens in the *cortex of the thymus*. At this time, cells that are specific to MHC II protein retain CD4, lose CD8, and become T-helper cells. T cells that are specific to MHC I protein retain CD8, lose CD4, and become cytotoxic T cells.

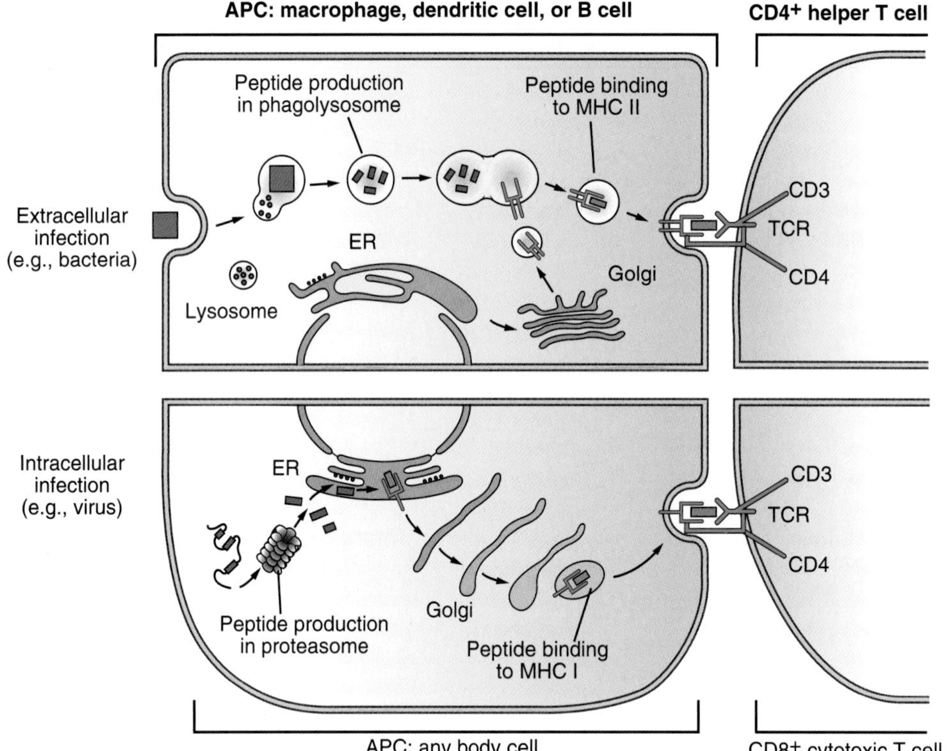

Figure 10.5 Antigen presentation. Extracellular antigen is phagocytosed, degraded within the phagolysosome, and associated with MHC class II molecules. The antigen is then presented on the surface of the cell, where it binds to the T-cell receptor (TCR) of $CD4^+$ T helper cells. Macrophages, dendritic cells, and B cells are such antigen-presenting cells. Intracellular antigen is digested within the proteasome, associated with MHC class I molecules, and presented on the surface of the cell to the TCR of $CD8^+$ cytotoxic T cells. Almost every cell in the body can present antigen in this latter way. CD, cluster of differentiation; ER, endoplasmic reticulum; MHC, major histocompatibility complex.

After positive selection, cells undergo **negative selection** in the *medulla of the thymus.* During this process, cells that bind with *high affinity* to MHC/self-antigen complexes die by apoptosis. This is important because these cells would have later reacted with self-peptides and caused autoimmune diseases. Cells that bind with low affinity to MHC/

TABLE 10.3 T Cell-Mediated Immunity Response According to the Source of Antigen

	Intracellular Infection (Viruses, Some Bacteria)	Extracellular Infection (Bacteria)	Extracellular Proteins (Vaccines, Toxins)
Location of antigen	Cytosol	Phagosomes	Endosomes
Antigen-presenting cell (APC)	Any cell	Macrophages, dendritic cells	B cells
Location of APC	Anywhere	Lymphoid and connective tissue, body cavities, epithelium	Lymphoid tissues, blood
Molecules of display	MHC class I	MHC class II	MHC class II
Antigen-recognition cells	$CD8^+$ CTL cells	$CD4^+$ T_H1 cells	$CD4^+$ T_H2 cells
Response	Release of cytotoxic effector molecules	Release of macrophage-activating effector molecules	B cell activation
Effect	Death of infected cell	Killing of bacteria and parasites	Secretion of antibody to eliminate bacteria and toxins

MHC, major histocompatibility complex; CD, cluster of differentiation; CTL, cytotoxic T cells; T_H1, T helper 1 cell; T_H2, T helper 2 cell.

self-antigen are allowed to leave the thymus. They will later only be activated by high-affinity binding to MHC/*foreign* antigen complex, the adequate signal for immune activation.

Similarly, exposure to foreign antigen from a tissue graft triggers an immune reaction in the body. This occurs when the tissues of the donor and the recipient are not *histocompatible*, which explains the origin for the name "major histocompatibility complex."

Killer, helper, and memory T cells

Killer T cells, helper T cells, and memory T cells can be distinguished based on their immunologic function. *Killer T cells*, or cytotoxic T cells ($CD8^+$), are lymphocytes that release lymphotoxins and are designed to kill cells that are infected with viruses or other pathogens.

Helper T cells ($CD4^+$) help determine which types of immune response the body will take to attack a particular pathogen. These cells have no cytotoxic activity and do not directly kill infected cells or clear pathogens. Instead, helper T cells control the immune response by directing other cells to perform these tasks. They function to regulate both the innate and adaptive immune responses. Helper T cells direct the immune response by secreting lymphokines. They stimulate proliferation of B cells and cytotoxic T cells, attract neutrophils, and activate macrophages. The differentiation of $CD4^+$ T-helper cells into best-known subtypes T helper 1 and helper 2 cells occurs after activation in the peripheral lymphoid system. Each subtype produces a distinct set of effector molecules.

For instance, *T helper 1 cells* release the macrophage-activating effector molecules interferon-γ and tumor necrosis factor-α (TNF-α). The resultant actions, called T helper 1 response, or *type 1 response*, support activities of macrophages and cytotoxic T cells of the cellular immune system.

On the other hand, *T helper 2 cells* produce effector molecules, such as interleukin 4, 5, 13, 25, 31, and 33, among numerous other cytokines. This T helper 2 response, or *type 2 response*, promotes the actions of B cells and hence the humoral immune system.

Research shows that the dual T helper cell model is far too simple. For instance, many helper T cells express

From Bench to Bedside / 10.1

Vaccination and DNA Vaccines Against AIDS

According to a 2008 estimate, about 1.4 million people in North America and over 33 million worldwide are infected with HIV, the human immunodeficiency virus causing acquired immunodeficiency syndrome (AIDS). Since its first diagnosis in 1981, more than 25 million people have died of AIDS. One preventative measure against the spreading of pandemic infectious diseases such as AIDS would be **vaccination**. For instance, vaccination has eradicated the poliovirus from the western hemisphere. Great effort is currently being undertaken to develop an AIDS vaccine, but so far has not been successful. Reasons include the fast mutation rate of the target antigen and the ability of the virus to hide from the antibody.

The principle of vaccination, also called immunization, is to stimulate the acquired immune system of the body to fight diseases. *Passive immunization* refers to the administration of antibodies and lymphocytes. These last only a few days and hence provide only transient protection. Nevertheless, it might be a life-saving measure in emergencies, as in the case of a person bitten by an animal with the rabies virus. ***Active immunization*** is necessary to provide long-term protection, because it generates immunologic memory.

To immunize against viral diseases, the vaccine generally contains an *attenuated* (weakened) or *killed* virus. The vaccine for immunization against bacterial diseases often contains only a portion of the bacteria. These vaccines might be generated by *recombinant technology,* in which the gene coding for the pathogenic epitope of the bacteria is isolated and expressed in an appropriate host cell.

In the fight against AIDS worldwide, several clinical trials are underway using another type of vaccine, *DNA vaccine*. In this case, genes that encode proteins of pathogens, rather than the pathogens themselves, are injected into a person. The gene of interest is grown in bacteria and called plasmid DNA. When injected in humans, it is taken up by antigen-presenting cells, which then produce the encoded protein. This antigenic protein elicits an immune response just as a conventional vaccine does.

DNA vaccines are preferable for protection against HIV, in which inoculation with a dead or attenuated virus is risky. It has been proven, at least in animal models, that DNA vaccines activate both cellular and humoral immune responses. This is advantageous because vaccines containing killed microorganisms, or subunits of them, induce the humoral response but only very poorly the cell-mediated immune response.

Other advantages of using a DNA vaccine include the abilities to favor a T helper-1 or a T helper-2 lymphocyte response and to induce a cytotoxic T-lymphocyte response. It is believed that these vaccines provide long-lasting immunity. DNA vaccines are stable at ambient temperatures, which make them useable in areas where constant refrigeration of the vaccine is not possible. Finally, DNA vaccines are cheaper and easier to produce than conventional vaccines and can be manipulated by the tools of recombinant technology.

DNA vaccines also have disadvantages. The potential that the vaccine plasmid DNA might become integrated into human chromosomal DNA raises concern. This could induce cancer as a result of the disruption of a cell division gene or the activation of an oncogene. Another problem of DNA vaccines includes the question of how to terminate their actions. Continuous antigenic stimulation may lead to immune tolerance or autoimmunity. Another safety consideration involves the potential induction of antibodies against the plasmid DNA, which again might lead to autoimmune diseases. These concerns are currently being addressed. Although the ultimate clinical effectiveness of DNA vaccines is unclear, they have shown promising results in preclinical trials against diseases such as AIDS.

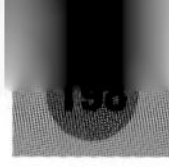

cytokines from both profiles. They are often named *T helper zero* cells. Additionally, other immune cells express many of the specific T-cell cytokines as well. Several new models have been proposed that include T helper 17 cells, but as of today there is no unanimously accepted approach.

About 10% of CD4⁺ T cells are *T regulatory cells* (previously known as *suppressor T cells*). Several subtypes have been named (e.g., Tr1 and Th3 cells). All have in common the ability to suppress the immune system and, hence, are important in maintaining immune homeostasis. For instance, they are known to inhibit the production of cytotoxic T cells when they are no longer needed. On the other hand, it has been shown that T helper cells are also capable of "regulating" their own responses, another indication that a general overhaul of the T-cell nomenclature might be forthcoming.

Lastly, *memory T cells* are lymophocytes that become "experienced" by having encountered an antigen during a prior infection or a previous vaccination, or a cancer cell. Memory T cells, when encountering an invader (pathogen) for the second time, mount a faster and stronger immune response than they did the first time. These cells are critical for the longevity of specific immunity.

Time delay in cell-mediated immunity

T cells and their products may exert their effects in concert with other effector cells, such as neutrophils, eosinophils, and mast cells. The secretion of T-cell factors that recruit and activate other cells takes time, and thus, the consequences of T-cell activation are not noticeable until 24 to 48 hours after antigen challenge.

An example illustrating this time delay is the *delayed-type hypersensitivity reaction* (see *Immunologic Disorders* below) to purified protein derivative (PPD), a response used to assess prior exposure to the bacteria that cause tuberculosis. Injected under the skin, PPD elicits the familiar inflammatory reaction characterized by local erythema and edema 1 to 2 days post injection.

Cell-mediated immune responses, although slow to develop, are potent and versatile. They provide the main defense against many pathogens. T cells are also responsible for the rejection of transplanted tissue grafts (see *Organ Transplantation and Immunology* below) and the containment of the growth of neoplastic cells (see *Immunologic Disorders* below). A deficiency in T-cell immunity, such as that associated with AIDS, predisposes the affected patient to a wide array of serious, life-threatening infections.

Humoral immunity is mediated by secreting antibodies produced in the cells of B lymphocytes.

B cells (with costimulation) transform into plasma cells that secrete antibodies. Binding of antigen to the BCR activates B lymphocytes, which then start proliferating and maturing in the presence of T helper cytokines (Fig. 10.4). In the classical, somewhat outdated model, B-cell–activating cytokines, such as IL-4, are part of the T helper 2 response. Most new cells become plasma cells, which produce antibodies for 4 to 5 days, resulting in a high level of antibodies in plasma and other body fluids. These antibodies can bind specifically to the antigenic determinant that induced their secretion. Other B-cell clones become long-lived *memory B cells* (see Figs. 10.3 and 10.4).

In contrast to the time-delayed response of cell-mediated immunity, antibodies are known to induce immediate responses to antigens and, thereby, provoke *immediate hypersensitivity reactions* (see Immunologic Disorders below).

Antibody structure

Antibodies are also called **immunoglobulins**. The primary structure of an antibody is illustrated in Figure 10.6. Each antibody molecule consists of four polypeptide chains (two *heavy chains* and two *light chains*) held together as a Y-shaped molecule by one or more disulfide bridges.

There are two isotypes of the light chain, κ and λ. Heavy chains have five different isotypes (α, δ, ε, γ, and μ), which constitute five different classes of antibody, each with different effector functions (see below). Each polypeptide chain possesses both a *constant region*, where the protein structure is highly conserved, and a *variable region*, where considerable amino acid sequence heterogeneity is found.

The amino terminal domains at each end of the forked portion of the "Y" of both the heavy and light chains are known as the *Fab regions* ("Fab" for fragment, antigen binding), which contain the antigen-binding regions. Antibodies are flexible in that the Fab arm can wave, bend, and rotate. This freedom of movement allows it to more easily conform to the shape of the antigen.

The carboxy terminal end of the heavy chain is termed the *Fc region* ("Fc" for fragment, crystallizable). Neutrophils, monocytes, and mast cells can recognize Fc regions via their Fc receptors.

Classes of antibodies

There are five classes of antibodies (IgA, IgD, IgE, IgG, and IgM), which have specific biologic properties that deal with particular antigens. Figure 10.6 and Table 10.4 summarize the shape and functions of the five major classes of antibodies. Figure 10.6 additionally shows the similarities of the TCR and the BCR to immunoglobulin.

IgG is the major antibody produced in response to secondary and higher-order antigen encounters. Secondary immune responses are faster and longer and produce more antibody with higher affinity to the antigen compared with primary responses. Hence, IgG is the most prevalent antibody in serum and is responsible for adaptive immunity to bacteria and other microorganisms. IgG exists in serum as a monomer. There are four IgG subclasses in humans (IgG1, 2, 3, and 4). IgG has a half-life of 23 days, the longest amongst the antibody classes. When bound to antigen, IgG can activate serum complement and cause opsonization. It can cross the placenta and is secreted into colostrum, protecting the fetus as well as the newborn from infection.

IgA usually exists as a polymer of the fundamental Y-shaped antibody unit. In most IgA molecules, a *joining*

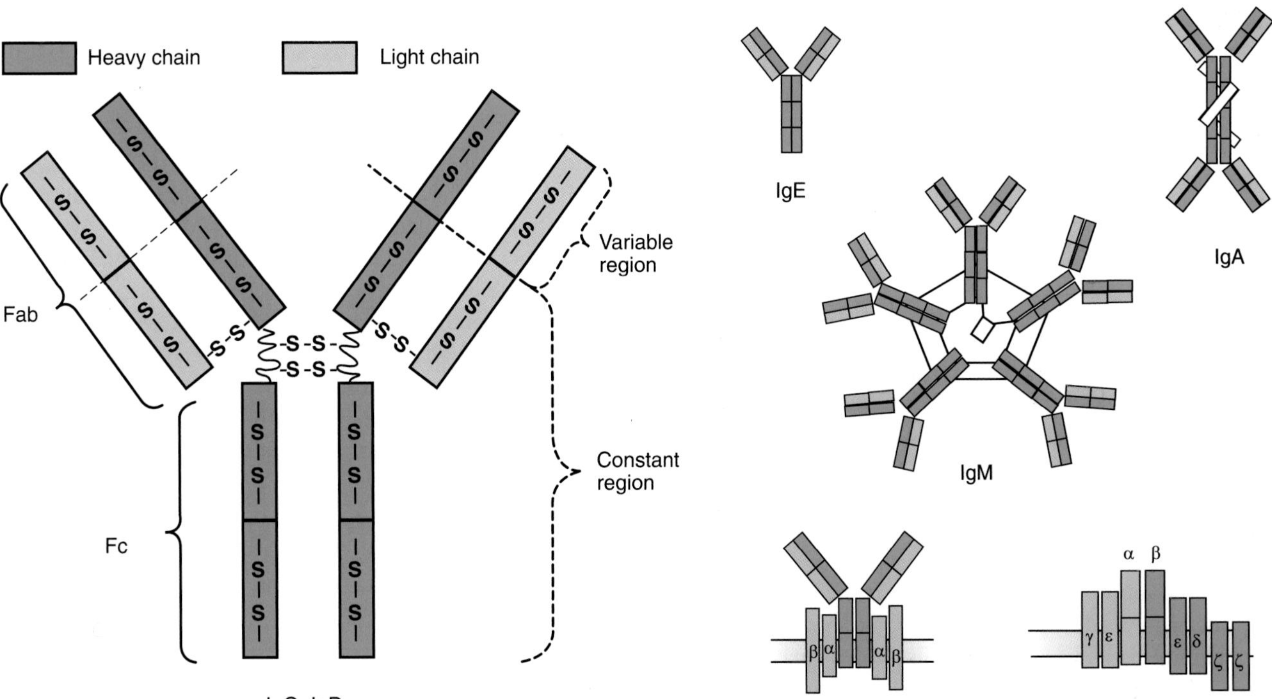

Figure 10.6 Antibodies and antigen receptors. Antibodies are immunoglobulins (Ig) with five isotypes (IgA, IgD, IgE, IgG, IgM). Not shown are the four IgG and two IgA subtypes and the secretory sIgA. The basic unit of each isotype is a monomer that consists of two heavy chains and two light chains held together in a Y-configuration by disulfide bonds. The heavy chains are different for the various isotypes. Each end portion of the "Y" is called *Fab* because it is the *fragment* that contains the *antigen-binding* site. A second, *crystallizable fragment* is called *Fc*. IgG, IgD, and IgE consist of one monomer, IgA of two, and IgM of five. The B-cell receptor resembles a membrane-associated Ig monomer. The T-cell receptor contains a Fab-like structure.

TABLE 10.4 Characteristics of Different Antibody Classes

Characteristic	IgG	IgA	IgM	IgD	IgE
Molecular weight (× 10^{-3} daltons)	150	150, 400	900	180	190
Y units/molecule	1	1–2	5	1	1
Serum concentration (mg/dL)	600–1,500	85–300	50–400	<15	0.01–0.03
Serum concentration (% of Ig)	~76	~15	8	1	0.002
Half-life (days)	~23	~6	~5	~3	~3
Crosses placenta	+ (not IgG4)	–	–	–	–
Enters secretions	+	+ +	–	–	–
Agglutinates particles	+	+	+ + +	–	–
Allergic reactions	+	–	–	–	+ + + +
Complement fixation	+ (not IgG4)	–	+ +	–	–
Fc receptor binding to monocytes and neutrophils	+ +	–	+	–	–
Dominant in secondary immune response	+				
Main antibody in external secretions		+			
Dominant in primary immune responses			+		
Responsible for hypersensitivity					+

Ig, immunoglobulin.

chain (*J chain*) holds together two antibody units (dimeric form). As the IgA passes through epithelial cells, an additional antigenic fragment, the *secretory piece,* is added. In this conformation, IgA is actively secreted into saliva, tears, colostrum, and mucus and hence is known as secretory immunoglobulin (sIgA). IgA is also found in serum, mainly as IgA1 isotype produced by bone marrow B cells, while IgA2 subtype is present in secretions. IgA has a half-life of about 6 days.

IgM consists of five "Y" units, also held together by J chains. The size of IgM and its many antigen-binding sites provide the molecule with an excellent capacity for agglutination of bacteria and blood cells. Although it has the potential to bind 10 antigens, in reality it only binds five. Fixed macrophages of the RES system efficiently and quickly remove such agglutinated antigens. IgM is the first antibody secreted after an initial immune challenge. It has a half-life of about 5 days.

IgE is a monomeric antibody that is slightly larger than IgG but has a relatively short half-life of about 3 days. IgE avidly binds via its Fc region to cells, such as mast cells and basophils, which are involved in allergic reactions.

IgD is found in plasma and on the surface of some immature B cells. It has a half-life of 3 days. An exact function is not known, but its ubiquitous presence in the animal kingdom suggests an important role. It is expressed early during B-cell differentiation and has been postulated to be involved in the induction of immune tolerance. IgD was also found to activate basophils and mast cells. IgD serum concentration does increase during chronic infection but is not associated with any particular disease.

Antibody action

Antibodies act against antigens in three ways: they neutralize the antigen, opsonize the antigen, or stimulate complement fixation to assist phagocytes.

1. **Neutralization.** Antibodies can bind to antigens, forming easily recognizable antibody–antigen complexes, which are removed by phagocytosis. Antibodies can also immobilize and agglutinate infectious agents so that a virus cannot penetrate the host cell or a microbe cannot colonize mucosal tissue.
2. **Opsonization.** IgG antibodies bind to bacteria or virus-infected cells at the Fab region. That way, the pathogen is "tagged" or "opsonized" for destruction by free radicals and enzymes or phagocytosis. For phagocytosis, the Fc portion of the antibody binds to Fc receptors on phagocytes. Some complement components (e.g., C3b and C4b) can also act as opsonins.
3. **Complement Fixation.** Complement is a group of at least nine distinct proteins that circulate in plasma, but involves about 25 proteins and protein fragments (Fig. 10.7). A cascade of events occurs when the first protein, C1, recognizes preformed IgM or IgG antibody–antigen complexes (*classical pathway*). A side entry to this pathway exists, in which complexes between bacterial mannose residues and the plasma protein mannose-binding lectin start the cascade (*lectin pathway*). The cascade leads to the formation of C3 convertase through the action of C4 and C2, and so the convertase has the form C4b2a.

A homologous variant of C3 convertase (C3bBb) is also formed in an *alternate pathway,* which is activated by small amounts of C3b that spontaneously hydrolyzed from C3. C3b becomes protected when it binds to microbial cell surfaces and forms C3 convertase through factors D and B. The complex is stabilized by properdin. Compared to the classical pathway, the alternate pathway is faster, but requires specific antigens for activation.

The C3 convertases of both pathways lead to the formation of complement C3b that activates C5 convertase,

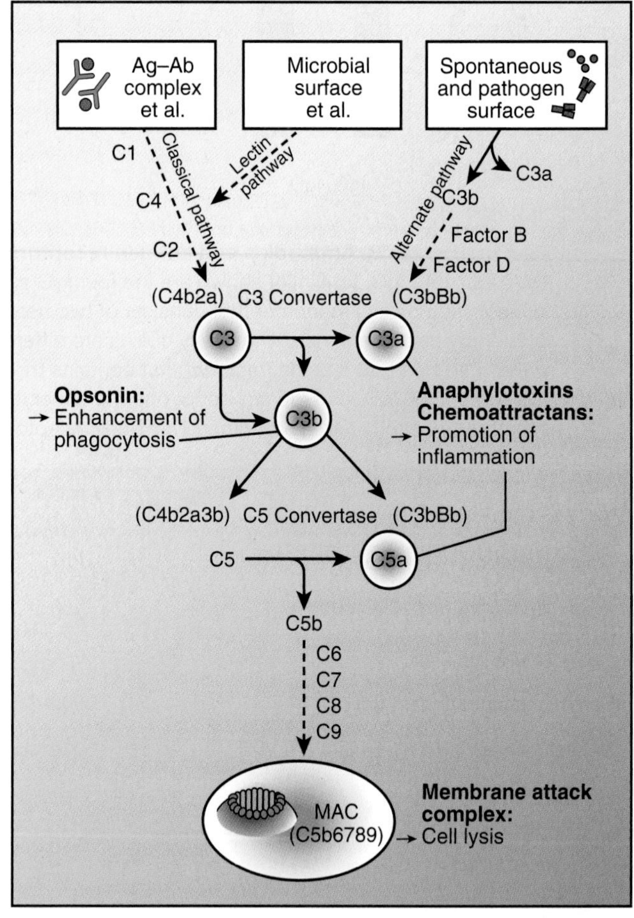

Figure 10.7 Complement. Complement (C) proteins supplement antibody activity in a cascade of events to eliminate pathogens. The goal is to produce opsonin C3b, anaphylotoxins and chemoattractants C3a and C5a, and the membrane attack complex (MAC). The MAC inserts into membranes of cells and causes their lysis. Antigen–antibody (Ag–Ab) complexes and other (e.g., C-reactive protein) activate the classical pathway. Spontaneously hydrolyzed C3 and pathogen surface molecules (e.g., lipopolysaccharides) activate the alternate pathway. The lectin pathway is initiated in response to microbial surfaces (mannose) and other (e.g., IgA). The classical and alternate pathways use structurally different, but functionally identical enzymes C3 convertase and C5 convertase.

another critical enzyme in the cascade that forms C5b. C5b and the four complement proteins, C6, C7, C8, and C9, form the **membrane attack complex (MAC).** The tubular MAC inserts into the membrane of the target cell because of its hydrophobic nature. This allows free passage of small molecules, ions, and water, resulting in the target cell's death.

Complement fixation is the basis for various diagnostic tests to detect a specific antibody or a specific antigen in a patient's serum. The test is based on the principle that the presence of antibody (or antigen) in the patient's serum reduces the lysis of added target cells (typically sheep red blood cells), since complement proteins are already used up.

ACUTE AND CHRONIC INFLAMMATION

Inflammation is the initial response of the body to infection or trauma (Fig. 10.8). Although *microbial infection* is probably the most common cause of inflammation, many types of *tissue injury* can also evoke inflammatory reactions. These include mechanical injury, radiation, burns, frostbites, chemical irritants, and tissue necrosis resulting from lack of oxygen or nutrients. The body instigates a nonspecific cascade of physiologic processes in vascularized tissues involving elements of the innate immune system, with the goal of repairing cellular damage and restoring the tissue to its normal function.

- In a normal response, inflammation is self-initiating, temporally self-propagating, and self-terminating.
- Inflammation is a necessary response to tissue injury, and human life without inflammation is unthinkable.
- On the other hand, inflammation often overshoots in its reactions, which leads to a vicious circle of repeated injury and persistent inflammation.
- Inflammation is closely connected with all kinds of illnesses, so that anti-inflammatory therapy is at the heart of many treatments.

Acute inflammation is a short-term process and is characterized by five cardinal signs.

Inflammation received its name from the clinical signs resembling an "internal fire."

The first sign is **rubor** (redness), which is generated by increased blood flow resulting from dilation of small blood vessels.

The second sign is **calor** (heat), which is generated by the metabolism of leukocytes and macrophages recruited to the site of damage and by increased blood flow. Additionally, systemic *fever* might develop as a result of the chemical mediators of inflammation, especially IL-1 acting at hypothalamic neurons.

The third sign is **tumor** (swelling) resulting from edema caused by leaky vessels.

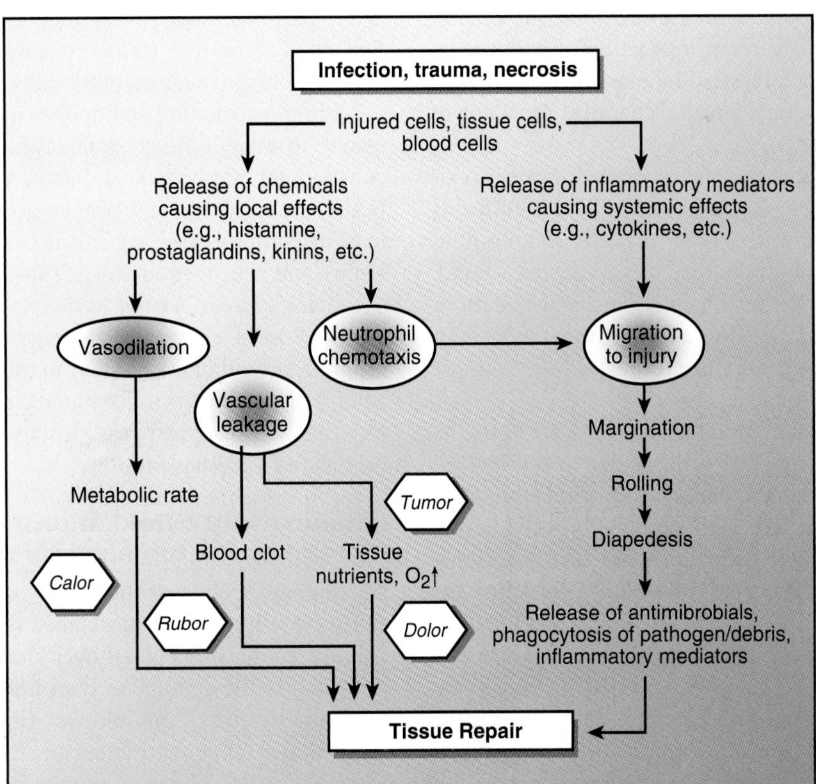

Figure 10.8 Acute inflammation. Inflammation can be caused by infection, physical agents, and tissue necrosis. Its goal is to destroy invaders and prepare for tissue healing and repair. The process is stereotypic, highly complex, and controlled by many soluble mediators. Principal signs are redness (rubor) and heat (calor) from vasodilation and increased blood flow, swelling (tumor) due to increased vascular leakage, and pain (dolor) from edema and nerve irritants. Neutrophils are among the first cells to arrive at an inflamed site.

The fourth sign is **dolor** (pain) caused by irritated nerves. Chemical mediators such as prostaglandin, histamine, bradykinin, and serotonin can directly affect nociceptors or they may sensitize them to touch or movement. In the chronic state, nerves from touch receptors that normally do not transmit pain become involved.

With increasing knowledge of the complexity of inflammation, many models have been put forward to expand on these four signs to categorize the great variety of inflammatory responses. However, they are not yet commonly applied in clinics beyond the addition of a fifth sign, **functio laesa** (the loss of function), which is a consequence of tissue operating at conditions out of homeostasis and spending energy on repair processes.

Acute inflammation is a three-step process involving vasodilation and cell migration from blood to tissue.

At the onset of infection, burn, or other injuries, acute inflammation is initiated by cells already present in the injured tissue—mainly, resident macrophages, dendritic cells, Kupffer cells, and mast cells. These cells release inflammatory mediators that bring about functional changes to tissues. One of the immediate changes seen in acute inflammation is vasodilation and, therefore, increased permeability. Vasodilation increases blood flow to the injured area, sometimes up to 10-fold. Additionally, the vessel walls become leaky, on the one hand, as a result of the injury-related necrosis of endothelial cells and, on the other hand, as a result of chemically directed retraction of endothelial cells. Histamine, which is released by mast cells, basophils, eosinophils, and platelets, is a major chemical mediator of this process.

Water, salts, and small proteins, such as fibrinogen, exit from the plasma into the damaged area. Fibrinogen forms fibrin networks to trap microorganisms. The leaking fluid is called **exudate,** or *pus* in the case of an infected wound. If an infected area is further liquefied and shielded from surrounding tissue, it is called an *abscess. Blisters,* pools of lymph fluid, may form in response to burns, infections, or irritating agents.

Blood leukocytes are attracted and migrate into the inflamed tissue in a regulated process that evolves three steps: In the first step, called **margination**, neutrophils are recruited and form bridges with endothelial cells at the inflamed sites. These bridges are made of *selectins,* P-selectins and E-selectins on endothelial cells and L-selectins on neutrophils.

In the second step, called **rolling**, the neutrophils move along the endothelium like tumbleweed since the bridges are initially loose and form and detach continuously. Eventually, a more firm connection is built between *integrins* on neutrophils and *intracellular* and *vascular adhesion molecules* on the endothelial side.

In the third step called **diapedesis**, or *transmigration*, neutrophils actively migrate through the blood vessel basement membrane into the tissue, with the aid of the firm connection but without damaging the endothelial cells. After neutrophils, *monocytes* are attracted and differentiate into *macrophages* when leaving the blood. Later, *lymphocytes* follow as well. The appearance and increase of the nongranular cells monocytes and lymphocytes mark the transition into *chronic inflammation*.

Inflammatory mediators develop and maintain the inflammatory response.

Inflammatory mediators are soluble molecules that act locally at the site of damage and coordinate the inflammatory responses.

Exogenous mediators (produced outside of the body) include **endotoxin**, which is a family of bacterial toxins associated with the LPS complex of gram-negative bacteria. Endotoxin is released from lysed bacteria and binds to receptors on monocytes and macrophages, thus activating them.

Endogenous mediators (produced within the body) are released from injured or activated cells at the inflammation site. Numerous substances are known to regulate inflammation, many of them with seemingly redundant functions. Table 10.5 lists a few of them. Some have previously been mentioned as part of the innate immune system, demonstrating the tight interplay between the two bodily responses.

Many mediators can interact with multiple receptors, often exerting diverse functions. For instance, histamine interacts with three receptors. The H_1 receptors mediate acute proinflammatory vascular effects, whereas activation of H_2 receptors results in anti-inflammatory actions. H_3 receptors are involved in the control of histamine release.

From a structural standpoint, inflammatory mediators belong to many different categories, such as proteins (e.g., complement, antibodies, and acute phase reactants), lipids (e.g., prostaglandins and platelet-activating factor), amines (e.g., histamine), gases (e.g., nitric oxide), kinins (e.g., bradykinin), and neuropeptides (e.g., substance P).

Many *plasma-derived mediators* are present as precursors and need enzymatic cleavage to become active. For example, thrombin is necessary to transform soluble fibrinogen into insoluble fibrin for blood clots. *Cell-derived mediators* can be preformed (e.g., histamine) or synthesized as needed (e.g., prostaglandin).

Inflammatory cytokine profiles provide a biomarker for the severity of inflammation.

An important class of mediators is **cytokines**, which are cell products synthesized *de novo* in response to immune stimuli. They generally act over short distances and short time spans. They comprise **lymphokines** (cytokines made by lymphocytes), **monokines** (made by monocytes), **chemokines** (chemoattractants made by various cells), and **interleukins** ([ILs] made by one leukocyte and acting on other leukocytes).

Certain inflammatory diseases present with characteristic cytokine profiles; therefore, inhibition of proinflammatory cytokines may aid in the treatment of the diseases. The classical proinflammatory cytokine triad

TABLE 10.5 Endogenous Inflammatory Mediators

Type	Name	Released by	Some Actions
Chemical mediators	Histamine	Mast cells, basophils, eosinophils, leukocytes, platelets	Vasodilation, vascular permeability
	Lysosomal compounds	Neutrophils, macrophages	Vascular permeability, complement activation
	Eicosanoids: Prostaglandins Thromboxanes Leukotrienes	Many cells, neutrophils	Vasoactive properties, platelet aggregation, prolongation of edema
	Platelet-activating factor	Neutrophils, monocytes, mast cells, eosinophils	Vascular permeability, neutrophil migration, bronchoconstriction
	Serotonin	Mast cells, platelets	Vasoconstriction
	Cytokines	Lymphocytes, monocytes	Vasoactive and chemotactic properties
	Chemokines	Tissue cells, endothelial cells, leukocytes	Chemotaxis of inflammatory effector cells
Gases	Nitric oxide	Endothelial cells, macrophages	Vascular smooth muscle relaxation and vasodilation, microbe killing, inhibitor of platelet actions
Neuropeptides	Tachykinins, kinins	Mainly sensory neurons	Vasodilation, vascular permeability, smooth muscle contraction, mucus secretion, pain
Plasma factors	Complement (C5a, C3a, and others)	Enzymes from dying cells; antigen-antibody complexes; endotoxins; products of kinin, coagulation, and fibrinolytic system	Chemotaxis, degranulation of phagocytes, mast cells and platelets, cytolytic activity, opsonization of bacteria
	Kinins	Coagulation factor XII	Vascular permeability, mediators of pain, nonvascular smooth muscle contraction
	Coagulation factors	Coagulation factor XII	Conversion of fibrinogen to fibrin
	Fibrinolytic system		Plasmin lyses fibrin

contains IL-1, IL-6, and TNF-α. The following presents the sources, major targets, and principal activities along with those of IL-8 and INF-γ, two additional clinically relevant factors.

Interleukin-1: IL-1 is produced by phagocytes, lymphocytes, endothelial cells, and others. It targets lymphocytes, macrophages, and endothelial cells. It activates cells, increases adhesion molecule expression, and induces fever and the release of acute phase proteins.

Interleukin-6: IL-6 is produced by lymphocytes, macrophages, fibroblasts, and endothelial cells. It targets B cells and hepatocytes. In addition to B-cell differentiation, it promotes the production of acute-phase proteins and fever.

Tumor necrosis factor-α: TNF-α derives from macrophages, mast cells, lymphocytes, and endothelial cells. Its targets and effects are similar to those of IL-1. It additionally stimulates angiogenesis and has other functions outside of inflammation.

Interleukin-8: IL-8 has been renamed *CXCL8*, because it is a chemokine produced by macrophages, endothelial cells, and fibroblasts. It targets neutrophils, basophils, and T cells. Its major actions are chemotactic and angiogenic.

Interferon-γ: IFN-γ is produced by T cells and NK cells. It targets leukocytes, tissue cells, and T helper cells. In inflammation, it activates phagocytes and enhances leukocyte–endothelial adherence.

In response to IL-1, IL-6, and others, the liver produces **acute-phase blood proteins**. An important member is *C-reactive protein*, which is used to monitor the severity and progression of some cardiovascular diseases. Another example is *acute-phase serum amyloid A proteins* (*A-SAAs*). One of their functions is to recruit immune cells to the inflammation site. Blood concentration of A-SAA might rise up to 1,000-fold during inflammation; therefore, A-SAAs are used as markers in some autoimmune diseases.

Acute inflammation is a stereotypic, highly complex process that self-terminates.

Modern molecular biology superimposes many additional layers of complexity onto the presented model of acute inflammation as a stereotypic process of the vascular system. For instance, it has been shown that factors independent of the vasculature can trigger and modulate aspects of both inflammation and repair. *Vibration* and *hypoxia* can lead to histamine release of mast cells (degranulation), hence initiating inflammatory events. *Mechanical load* on tendon fibroblasts can modulate their response during inflammation, promoting their destruction at high loads and protecting them from inflammatory apoptosis at low loads. Furthermore, proinflammatory molecules can be upregulated without any concomitant invasion and stimulation of inflammatory cells.

Under normal circumstances, acute inflammation terminates itself. Chemical mediators disappear either because of their short half-life or because they are enzymatically inactivated (e.g., kininases inactivate kinins). Substrates are consumed, and the lymph flow carries the mediators away faster than they can be produced. *Anti-inflammatory cytokines* such as IL-4, IL-10, and transforming growth factor-β (TGF-β) induce repair of damaged tissue. Following inflammation, tissue that is capable of regeneration will be almost completely restored; otherwise, *scar formation* occurs.

CHRONIC INFLAMMATION

Chronic inflammation is the failure to resolve inflammation and usually causes serious bodily harm. Chronic inflammation develops when neither agent nor host is strong enough to overwhelm the other (e.g., in the case of osteomyelitis, the infection of bone), when there is prolonged exposure to the toxic agent (e.g., silicosis from inhalation of crystalline silica dust), or as part of autoimmune disease processes (e.g., rheumatoid arthritis [RA]). It can also develop without being preceded by acute inflammation (e.g., in tuberculosis).

Chronic inflammation may last weeks, months, or years and leads to *chronic wounds,* which are composed of loosely arranged connective tissue (*granulation tissue*), infiltrated fibroblasts, and inflammatory cells. The vastly dominant inflammatory cell type is the *monocyte/macrophage.* It is the only cell type present in the case of chronic inflammation resulting from a nonantigenic agent such as a suture thread. In the case that the injurious agents are also antigenic, other cell types appear, such as lymphocytes, plasma cells, and eosinophils.

Chronic inflammation is characterized by ongoing tissue damage from reactive oxygen species and proteases that are secreted by inflammatory cells. Other products, such as arachidonic acid and proinflammatory cytokines, amplify as well as propagate the damage. This weakens the body and makes it even more susceptible to infection and further inflammation. It is a vicious cycle that leads to clinical symptoms typical of chronic inflammatory diseases, such as RA, atherosclerosis, or psoriasis.

Overwhelming inflammation combined with immune suppression can lead to *systemic inflammatory response syndrome*, which is called **sepsis**, or *septicemia,* when the inflammation is a result of infection. In severe cases, it might lead to organ failure and death.

Chronic inflammatory diseases can affect every body part and are usually indicated by adding the suffix "itis." For instance, *myocarditis* is inflammation of the heart, and *nephritis* is inflammation of the kidney. Some inflammatory conditions do not follow the conventional terminology. The most common one is *asthma.* Another example is *pneumonia,* which is more commonly used than "pneumonitis" to name the chronic inflammatory infection of the lung.

Chronic inflammation is both a symptom and a cause of disease.

The role of inflammation in diseases that were not traditionally categorized as inflammatory diseases has been viewed with considerable interest. In fact, inflammation is now recognized to be a critical pathologic component underlying many illnesses, ranging from Alzheimer and Parkinson diseases to diabetes and certain types of cancer (e.g., colon cancer).

Although the association between inflammation and disease has been recognized for a long time, inflammation was mainly considered a result, rather than a cause, of the illness. Studies clearly show that the destructive, self-promoting cycle between oxidative stress and resulting inflammation causes more oxidative stress and contributes to the development of chronic illnesses. This chain of events may initially occur at a low, asymptomatic level, which leads to the speculation that *subacute chronic inflammation* might be the cause of a number of illnesses.

ANTI-INFLAMMATORY DRUGS

Corticosteroids are the most powerful anti-inflammatory drugs currently available. They inhibit the accumulation of neutrophils at sites of inflammation (see Neuroendoimmunology below) but also have widespread effects on other inflammatory cells and processes. This is the reason for abundant side effects ranging from osteoporosis to disruption of the hypothalamic–hypophyseal axis.

Nonsteroidal anti-inflammatory drugs inhibit **cyclooxygenase (COX)**, which is involved in prostaglandin production. There are at least two isoforms of this enzyme. *COX-1* is found in platelets and catalyzes the production of thromboxane A2. COX-1 inhibition leads to diminished blood clotting, so COX-1 inhibitors are used as blood thinners. *COX-2* is expressed in blood vessels and macrophages in response to inflammation and leads to production of prostaglandin I2. Drugs that specifically block COX-2 can decrease inflammatory pain. However, both COX-1 and COX-2 are present at many more sites, which is a reason for the abundant pharmacologic side effects of COX inhibitors.

Alternative therapies include a wide variety of natural substances that have antioxidative and anti-inflammatory features but often have poorly understood mechanisms of action. These include nutrients such as vitamin E, herbs such as chamomile, and supplements such as glucosamines, to name a few.

Because the side effects or uncertainties associated with many existing anti-inflammatory drugs often outweigh their

Clinical Focus / 10.1

Allergies and Asthma

Epidemiologic data confirm what many parents suspect. Their children are more susceptible to allergies than they were when they were children. In the past 30 years, the prevalence of allergies has been greatly increasing in the United States and in the world, proportionally more so in wealthy countries compared with poorer countries and more so in urban areas compared with rural areas.

One hypothesis, often referred to as the hygiene model, proposes that the children who are not exposed to enough antigens or innocuous microorganisms from soil or water early in their childhood do not adequately develop their T helper 1 cell response. This leads to an imbalance toward T helper 2 cell responses, which are present from birth. Even though this hypothesis is disputed as being grossly oversimplified, it is undisputed that the various manifestations of allergic inflammation are the result of wrongly activated T helper 2 cells.

Asthma is one common allergic disorder. About 23 million people, including almost 7 million children, have asthma. It belongs to type I hypersensitivity disorders and is characterized by airways that narrow excessively in response to many provoking stimuli.

Typical allergens causing atopic (also called extrinsic or allergic) asthma are animal dander, dust mite excrement, grass pollen, and cockroach antigen. Strenuous exercise or worry might trigger nonatopic asthma.

The symptoms of both asthma types are similar and include wheezing, breathlessness, chest tightness, and coughing. They are caused by chronic inflammation of the airways, which leads to their swelling. Bronchoconstriction, excess production of mucus, and increased collagen deposition further exacerbate the restriction of the airflow to and from the lungs.

The development of asthmatic symptoms ultimately depends on the presence of T helper 2 cells, which secrete a characteristic cytokine repertoire. Of these, interleukin-4 activates B cells to produce IgE antibodies, which bind to mast cells, eosinophils, and other airway cells via receptors that are specific for the Fc portions of the antibodies. On subsequent encounters with the antigen, IgE-cross-linking on mast cells and eosinophils occurs and leads to the cells' activation and their release of histamine and leukotrienes. These and other inflammatory factors recruit inflammatory cells including additional T helper 2 cells to the lung, resulting in a damaging immunologic positive feedback cycle. IL-4, together with IL-9, is also necessary for mast cell maturation, and along with IL-5, for eosinophil recruitment.

Several treatments are available that interfere with the different allergic processes. *Antihistamine* drugs for treatment of atopic asthma target H_1 receptors or prevent the release of histamine from mast cells. *Epinephrine* and *β2-selective agonists* for treatment of systemic anaphylaxis modulate the sympathetic nervous system (see Neuroendoimmunology). *Corticosteroids* and *leukotriene receptor antagonists* aim at suppression of inflammation. *Immunotherapy* (also called *hyposensitization* or *desensitization*) involves subcutaneous injections that contain gradually increasing doses of an allergen extract. The immune system responds with gradually increasing levels of IgG antibodies, which seems to help a patient avoid dramatic reactions when the allergen is then encountered naturally.

benefits, a main focus of 21st century medical research is to develop novel anti-inflammatory strategies. Novel drugs generally aim to target specific pathways instead of general inflammatory processes. For instance, leukotriene modifiers, a relatively new class of anti-inflammatory medications for asthma patients, specifically block the products of arachidonic acid metabolism and can prevent an asthma episode before it even starts.

ORGAN TRANSPLANTATION AND IMMUNOLOGY

Both organs and tissues can be transplanted. Transplanted organs include heart, liver, lungs, pancreas, intestines, and thymus. Transplanted tissues include cornea, skin, heart valves, veins, bones, and tendons. The latter two are both referred to as *musculosketal grafts*. Worldwide, the kidney is the most common organ transplant, although musculoskeletal transplants (tissue) outnumber the kidney more than 10-fold.

Organ transplantation is indicated when irreversible damage has occurred and alternative treatments are not applicable. In the early 1950s, the first human kidney was transplanted to the donor's identical twin brother. Such transplantation is called an **isograft** because the donor and recipient are genetically identical. In 2008, about 14,000 kidneys were transplanted in the United States, most often as **allografts** between genetically different people.

Kidneys, lungs, and livers can come from living donors, because a person can live with one kidney or one lung and because liver tissue regenerates. Transplantations of heart, whole lung, pancreas, and cornea come from deceased donors. So-called nonvital transplantations refer to the replacement of body parts such as the hands, knees, uterus, trachea, and larynx.

Tissue that is transplanted in the same person from one place to another place is called an **autograft**. Skin and blood vessels are common tissues for such transplantation.

Although, technologically, transplantation is a viable therapeutic option, it is mainly limited by the immensely larger demand for transplantation organs compared with their availability. An attempt to overcome the donor organ shortage is **xenograft** transplantation, in which animal tissue is used to replace human organs.

Two other avenues for replacement of tissue are the development of *artificial organs* and the transplantation of *stem cells*, which may be able to regenerate full-functioning

Clinical Focus / 10.2 

Bone Marrow Transplantation

When a patient has a terminal bone marrow disease, such as leukemia or aplastic anemia, often the only possibility for a cure is *bone marrow transplantation*. In this procedure, healthy bone marrow cells are used to replace the patient's diseased hematopoietic system.

To identify a suitable donor, blood leukocytes of prospective donors, usually close relatives, are screened to determine whether their antigenic patterns match those of the patient. A transplant is successful only if the donor's human leukocyte antigens closely match those of the recipient. Because there are several antigenic determinants and each can be occupied by any one of several genes, there are thousands of possible combinations of leukocyte antigens. The odds that any person's cells will randomly match those of another are less than one in a million.

The antigenic composition of leukocytes in bone marrow and peripheral blood are identical, so analysis of blood leukocytes usually provides enough information to determine whether the transplanted cells will engraft successfully. Peripheral blood stem cells are currently being tested to see whether they can be used as effectively as bone marrow blood stem cells. This would minimize the pain and risks undertaken by donors.

If significantly different from the recipient's tissue type, transplanted leukocytes may be recognized as foreign by the patient's immune system and, therefore, be rejected, a phenomenon called *host-versus-graft response*. However, this often plays a minor role because the patient's immune system is immunocompromised and not able to react strongly.

The major problem of bone marrow transplantations is the fact that functional T cells in the graft recognize the host's tissue as foreign and mount an immune response. This is called **graft-versus-host disease (GVHD)**. The disease often begins with a skin rash, as transplanted lymphocytes invade the dermis, and may end in death as lymphocytes destroy every organ system in the marrow recipient.

There are several methods to decrease the morbidity of marrow transplantations. Immunosuppressive agents, including steroids, cyclosporine, and anti-T cell antiserum effectively decrease the immune function of the transplanted lymphocytes. Another useful approach involves the physical removal of T cells from bone marrow prior to transplantation. T cell-depleted bone marrow is much less capable than untreated marrow of causing acute GVHD. These techniques have substantially increased the potential pool of bone marrow donors for a given patient.

organs. The latter is the "holy grail" of organ replacement because the new tissue would not be recognized as foreign if the stem cells came from the recipient. However, as of today, these alternatives do not work nearly as well as allografted organs.

There are many issues to consider and challenges to overcome, involving ethical, social, and legal problems. To name just a few, xenotransplantation may introduce new diseases to humans and lead to disrespectful treatment of animals. The use of animal and artificial organs may raise problems of self-esteem as a result of the presence of nonhuman tissue in the body, and stem cell therapy may confer the risk of developing cancer.

Histocompatibility is the most important criterion for a match in organ donation.

The major reason for the donor organ shortage is that the blood and tissue of the donor and the blood and tissue of the recipient need to have similar immunogenic markers in order to avoid rejection. For instance, virtually all kidneys with unmatched *ABO blood group* between donor and recipient will be rejected within minutes.

For successful transplantation, it is critical that the recipient does not have, or develop, antibodies against the donor's **human leukocyte antigen (HLA)**. HLAs are the human antigens of the MHC complex. These antigens are present on the surface of most body cells but were first discovered on leukocytes. For minimum requirements, transplantation doctors look at only six of the many HLA antigens (two A, two B, and two DRB1 antigens) for matching donors to patients. However, research has shown that a match in more HLA antigens and additional factors such as race, age, and sex improves the patient's chances for successful transplantation.

Although tissue typing decreases the risk of transplant rejection, the match between donor and recipient is never perfect (except for identical twins), and the reality of clinical practice requires organ transfer between less well-matched pairs. So, clinical symptoms resulting from immune attacks on the transplant are fairly common. Table 10.6 summarizes the three different types of graft rejections according to time elapsed after transplantation.

Immunosuppressive drugs greatly decrease the risk of rejection. There are numerous drugs available that inhibit different immune processes. Many of them down-regulate the unwanted T-cell responses. Total destruction of T cells and other leukocytes by whole-body irradiation is used before bone marrow transplantation (see Chapter 9, "Blood Components").

IMMUNOLOGIC DISORDERS

Table 10.7 presents a list of immune disorders, which is not intended to be comprehensive but to provide a framework that can be expanded in pathology courses. It is organized into hypersensitivity, immunodeficiency, and autoimmune disorders.

Many diseases involve mechanisms that fit into several of the categories presented in the table. One such example is **rheumatoid arthritis (RA)**. An unknown antigen that stimulates antibody formation, characteristic of *type III hypersensitivity*, is thought to initiate RA. This leads to joint damage, and as a result, autoantigens

TABLE 10.6 Types of Graft Rejection

Type	Time	Reasons
Hyperacute	Minutes–hours	Preformed antidonor antibodies trigger type II hypersensitivity reactions
Acute	Days–weeks	HLA incompatibility or inappropriate connection to blood supply triggers T-cell activation and type IV hypersensitivity reactions; complicated by antibody-mediated rejection
Chronic	Months–years	Unclear; cellular and humoral mechanisms involved; recurrence of disease possible

HLA, human leukocyte antigen.

are released, which perpetuate the disease in the typical fashion of an *autoimmune disease*. T cells are recruited and activate macrophages, which leads to *chronic inflammation*. Immune responses dominated by T helper 1 cells, characteristic of *type IV hypersensitivity* reactions, maintain the disease.

Immune system dysfunction happens in three ways: overreaction, failure to respond, and inappropriate response.

As mentioned previously, under normal conditions, when infection resolves, the inflammatory response is actively terminated to prevent unnecessary damage to tissue. Failure to do so results in chronic inflammation and tissue damage. Accordingly, the immune system must be sensitive to pathogens and use sufficient power to defeat the pathogen, because even small numbers of residual microorganisms might soon endanger the body again. But this principle comes at the risk of overdoing it. Asthma, familial Mediterranean fever (recurrent episodes of peritonitis, pleuritis, and arthritis), and Crohn disease (an inflammatory bowel disease) are examples in which chronic inflammation leads to immunologic disorders and tissue damage.

Overreaction

The most common immunologic disorders due to immune system overreaction are known as **allergies**, or **anaphylaxis** in the case of severe forms. Allergies are categorized into four types, hypersensitivity disorder type I to type IV (Table 10.7). Depending on the sensitivity of the affected person, allergies can cause localized or systemic symptoms. *Localized reactions* are usually biphasic, with an immediate response mainly resulting from histamine (erythema, wheal, and flare) and a delayed response that may last several hours. *Systemic reactions* might be life-threatening because of the sudden loss of blood pressure as a result of general vasodilation and because of airway obstruction as a result of smooth muscle contraction.

Immunodeficiency

Immunodeficiency (or **immune deficiency**) is a condition in which the immune system's ability to a fight infectious is compromised or absent. Many disorders fall into this category, although the number of patients with congenital immunodeficiency diseases is low.

The study of these diseases reveals important insights into cellular and molecular immunology. For instance, *severe combined immunodeficiency (SCID)*, commonly known as "bubble-boy disease," represents a group of congenital disorders characterized by little or no immune response. Research showed that in the X-linked form of SCID, T cells are dysfunctional as a result of a mutation of the IL-2 receptor gamma chain. The immunologic consequences of missing IL-2 signaling are severe and may lead to the patient's death within the first year of life.

Autoimmunity

Autoimmunity is the failure of the body to recognize its own tissue as self, which allows an attack against its own cells and tissues. Diseases that result from such aberrant immune response are termed *autoimmune disease*. Examples of autoimmune diseases include type 1 diabetes mellitus, systemic lupus erythematosus (SLE), Graves disease, and RA. Genetic and environmental factors as well as gender contribute to the development of many of the autoimmune diseases. For example, approximately 75% of the 24 million Americans who suffer from autoimmune disease are women.

Although many contributing factors are involved, the trigger mechanism(s) for most autoimmune diseases are still poorly understood. For example, SLE is a chronic autoimmune disease that affects multiple organs. Its etiology is largely unknown, although it is clear that a genetic predisposition to its development exists, because an identical twin has 3 to 10 times increased risk for getting SLE. Hormones most likely play a role because 90% of patients with SLE are women. Lastly, it seems that environmental factors such as certain medications, stress, and other diseases can also lead to SLE.

Immune cells may become carcinogenic, leading to hematologic malignancies.

Not represented in Table 10.7 are *hematologic malignancies* because they typically form a separate category. Like any cell, cells of the immune system can grow out of control. **Multiple myeloma** is a typically incurable cancer of plasma cells. **Leukemia**, the abnormal proliferation of leukocytes, is the most common childhood cancer. **Lymphomas** encompass a diverse group of cancers that present as solid masses specific to the lymphatic system.

Immune **oncogenes**, genes that turn immune cells into tumors cells, can be induced spontaneously, but their persistence is increased in the presence of carcinogens or viruses. For example, the human T-cell leukemia virus type 1 can cause acute T-cell leukemia and lymphoma, and the Epstein-Barr virus of the herpes family is linked to Burkitt lymphoma and, possibly, Hodgkin lymphoma.

TABLE 10.7 Immune Disorders

Names	Mechanisms	Examples (Common Names)
Hypersensitivity disorders: Damaging immune responses elicited by allergens		
Immediate hypersensitivity		
Type I (Immediate hypersensitivity)	Allergens cause cross-linking of IgE bound to mast cells, which release vasoactive mediators (e.g., histamine); symptoms within minutes	• Atopic diseases: hives, asthma, hay fever • Food allergy • Systemic anaphylaxis
Type II (Ag–Ab cytotoxicity)	Cytotoxicity mediated by antibody (primarily IgG) directed against epitopes on surface membrane of host cells; symptoms within hours	• Newborn hemolytic anemia • Autoimmune hemolytic anemia • Blood transfusion reactions
Type III (Ag–Ab immune complex disease)	Circulating immune complexes (primarily IgG) escape phagocytosis and cause deposits in tissues or blood vessels; symptoms within hours	• Serum sickness • Glomerulonephritis • Farmer's lung
Delayed-type hypersensitivity		
Type IV (Cell-mediated hypersensitivity)	Memory TH_1 cells cause cell-mediated immunity resulting in tissue damage by macrophages; symptoms within day(s)	• Contact dermatitis • Photoallergic dermatitis • Celiac disease
Immunodeficiency disorders: Caused by deficiencies (primary) or by illness in previously healthy person (secondary)		
B cell	Deficiency in antibody-mediated immunity	• Bruton agammaglobulinemia • IgA deficiency
T cell	Deficiency in cell-mediated immunity	• DiGeorge syndrome
Combined	Combined deficiency of cellular and humoral immunity	• SCIDs
Nonspecific	Mediated by deficient phagocytic and/or natural killer cells	• Leukocyte adhesion deficiency • Chronic granulomatous disease
Complement	Defects of individual components or control proteins	• C1, C4, C2, C3, C5–C9 deficiencies • Hereditary angioedema with lack of C1 inhibitor
Acquired	Human immunodeficiency virus (HIV) infects and kills CD_4^+ T helper cells, which leads to massive immunosuppression and increased risk of cancer	• AIDS • Prolonged serious disorders such as cancers, kidney failure, and diabetes.
Autoimmune disorders: Immune responses against the body's own tissue		
Antibody mediated	A specific antibody targets a particular antigen, which leads to its destruction and the signs of the disease	• Autoimmune hemolytic anemia • Myasthenia gravis • Graves disease
Immune complex mediated	Antibodies complex with autoantigen into large molecules, which circulate around the body and cause destruction	• Systemic lupus erythematosus • Rheumatoid arthritis
T cell-mediated	T cells recognize autoantigen, which leads to tissue destruction without requiring the production of autoantibody	• Multiple sclerosis • Type 1 diabetes • Hashimoto thyroiditis
Mediated by complement deficiency	Deficiencies in complement components often lead to or predispose to the development of autoimmunity	• Systemic lupus erythematosus

Ag–Ab, antigen–antibody; CD, cluster of differentiation; Ig, immunoglobulin; SCIDs, severe combined immunodeficiencies; AIDS, acquired immunodeficiency syndrome; T_H1, T helper 1 cell.

Tumor antigens are useful in identifying tumor cells and serve as potential candidates for cancer immunotherapy.

Some leukemias and lymphomas are easy to differentiate histologically from healthy tissue, whereas others are not. For instance, lymphoblasts in acute lymphoblastic leukemia look different than normal lymphocytes. However, the cells in chronic lymphocytic leukemia are hard to distinguish from normal cells, and the disease might resemble a physiologically high lymphocyte count. In this case, identification of specific cell markers, which are not present in healthy cells, aids the diagnosis. Hence, tumor immunology identifies **tumor-specific antigens**.

Some tumor antigens, such as overexpressed normal antigen, mutated proteins, or viral antigens, can be recognized by the immune system, in a process called **immune surveillance**, to develop antitumor immunity. This is one reason why people immunocompromised as a result of previous infections, other diseases, or age, are more likely to develop cancers. For instance, although most people with Burkitt lymphoma remain healthy when infected, it is the most common childhood cancer in central Africa, where malaria and malnutrition contribute to immune weakness.

Nevertheless, most cancers occur in people with a normal immune system. The reason is that tumor cells develop mechanisms to evade or disturb appropriate immune responses. These include the loss or reduced expression of MHC proteins and the secretion of immunosuppressive cytokines such as TGF-β, to name two mechanisms.

The resulting combat of the immune system against cancer that has already developed has led to the rapidly advancing field of **cancer immunotherapy**, which aims to enhance the body's natural defense against malignant tumors. Active immunization with tumor antigens and cytokine or antitumor antibody therapies are examples of this type of treatment.

NEUROENDOIMMUNOLOGY

Neuroendoimmunology is the study of the interaction between the nervous, endocrine, and immune systems. These systems exchange an extensive amount of information between the three systems (Fig. 10.9). For example, they share a number of the same receptors and synthesize and secrete some of the same molecules, such as peptide hormones, neurotransmitters, and cytokines. Accordingly, it is not surprising that the strength of a person's immune response depends not only on the type of antigen or the person's age and genetic makeup but also on the person's lifestyle choices, emotional character, and ability to cope with stress.

Immune system is down-regulated by neuronal and hormonal stress signals.

It is well known that the physiologic stress response down-regulates the immune system. As a consequence, people who are stressed, for instance, from losing a relative or facing examinations, become sick more easily.

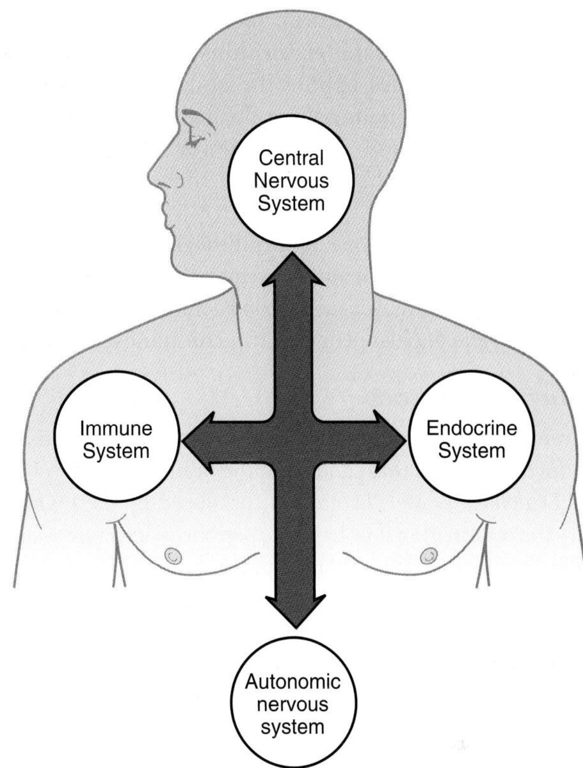

Figure 10.9 Integration of the immune system with the nervous and endocrine systems. Immune, endocrine, and neuronal systems synthesize some of the same hormones, neuropeptides, thymosins, and lymphokines, which act as communication molecules and regulate body homeostasis.

During an immune response, cytokines such as IL-1 stimulate the activity of paraventricular hypothalamic neurons to release corticotropin-releasing factor (CRF), which starts the physiologic *stress response* aimed at bringing a person back to physiologic homeostasis. In the pituitary gland, CRF stimulates the expression of pro-opiomelanocortin, which is converted to β-endorphin and adrenocorticotrophic hormone (ACTH). ACTH stimulates the release of corticosteroids from the adrenal gland. All these molecules can also bind directly to receptors on immune cells. The resulting actions are generally inhibitory to the immune system but can also be immunosupportive, especially during acute stress.

Activating the stress response via CRF also leads to activation of the *sympathetic nervous system* and the release of its neurotransmitter norepinephrine. Cytokines that are released by immune cells or by cells of the central nervous system (CNS), such as microglia and astrocytes, can also activate this autonomic path of the stress response.

A network of autonomic nerve fibers sends information from the CNS to the spleen and lymph nodes, where the fibers end in close proximity to T lymphocytes and macrophages. These immune cells, as well as B cells, express β-adrenergic receptors of the β-2 subtype, and their activation generally leads to inhibition of cell function. For instance, adrenergic activation of T cells decreases their expression of integrins, so that they cannot migrate out of the blood vessel into the tissue. This explains the high T-cell counts seen soon after a stressful event.

Furthermore, sympathetic nerve endings and the adrenal medulla release **endorphins** and **enkephalins**. These opioid peptides oppose the body's stress response and help maintain homeostasis during times of physical and psychologic stress. The immune responses to opioids vary, depending on the subtype and concentration of the neuropeptide.

Melatonin, a hormone of the pineal gland, seems to regulate the opioid network and consequently the immune system. Melatonin therapy is under consideration to support the immune system of older people to avoid **immunosenescence**.

Neural mediators

Immune cells that are present in neuronal and other tissue produce mediators that can act on the brain.

Lymphokines and *thymokines* (produced by the thymus) carry the information flow from the peripheral immune system to the CNS. For instance, IL-1 is a potent mitogen for astroglial cells. In addition, it acts at hypothalamic cells causing fever. IL-2 promotes the division and maturation of oligodendrocytes, and IL-3 supports the survival of cholinergic neurons.

$CD4^+$ T cells and macrophages manufacture *neuropeptides*, such as vasoactive intestinal peptide and pituitary adenylate cyclase-activating polypeptide, which were previously thought to be products solely of nonimmunologic cells.

Until recently, it was also believed that the brain was devoid of peripheral immune cells. It is now known that immune cells can pass the blood–brain barrier and enter the CNS. It seems that they are present to regulate physiologic neuronal events and potentially contribute to maintaining the brain's ability for cognitive functions. On the other hand, neuronal immune cells may trigger autoimmune disorders. For instance, autoimmune T cells that recognize myelin basic protein of myelin sheaths can be associated with the development of the demyelinating disease **multiple sclerosis**.

Hormonal mediators

The endocrine system and immune system work in concert to maintain homeostasis. Many hormones have been linked to immune functions. For instance, cells of the immune system also produce (e.g., monocytes and lymphocytes) and use (e.g., lymphocytes, monocytes, and NK cells) **thyroid-stimulating hormone**, best known as a regulator of metabolic functions. *Growth hormone, prolactin,* female and male *sex hormones,* and *leptin* are additional examples of hormones that unequivocally modulate the immune system.

In the typical fashion of hormones, they exert multiple actions in many target cells, depending on the hormone concentration, the target tissue, and the environment. This makes hormones optimal regulators of the body's homeostasis but usually causes unwanted side effects when used clinically.

Unpredictable side effects are one reason why use of *immune therapies* using hormones or neuropeptides is still limited. For example, the adrenal hormone epinephrine is given to reverse some effects of allergy. The field of psychoneuroimmunology builds on the fact that positive experiences and behavior may also boost the immune system. Thus, immunocompromised people are counseled to find ways for dealing with stress. However, many new applications are being developed, and targeted enhancement or suppression of immune cell functions will play a more important role in our approaches to help the body to maintain homeostasis.

Chapter Summary

- The immune system encompasses key immunologic organs such as the thymus, spleen, bone marrow, and lymph system. It additionally includes cells such as macrophages, dendritic cells, mast cells, and leukocytes as well as noncellular elements that are part of every tissue throughout the body.
- Antigens are substances that bind to T and B cells or antibodies. Haptens and immunogens are antigens, but only immunogens elicit an immune response. Proteins are by far the best immunogens.
- The immune system operates by distinguishing between self and foreign molecules inside our body, and then tolerating normal body substances and eliminating any invasion by foreign substances.
- Immune system defenses include physical, chemical, and mechanical barriers to entry of pathogens.
- The immune system has two additional systems: the innate system as the immediate response to common immunogens and the adaptive system, which changes throughout life and which has memory to respond to an encounter more strongly the second time compared with the first time.
- Innate and adaptive systems work together so that the immune response is adjusted to the type and severity of the threat.
- The innate leukocytes include phagocytes that encompass neutrophils, macrophages, and dendritic cells. They have receptors for pathogens and for complement-coated and antibody-coated antigens. They can engulf and destroy pathogens while influencing immune responses via immune mediators. Macrophages and dendritic cells can act as antigen-presenting cells to T cells.
- The innate leukocytes further include nonphagocytic cells such as killer cells that attack aberrant body cells, eosinophils that are involved in allergy and other immune responses, and mast cells that participate in the defense against pathogens and wound healing.
- Acute inflammation is the body's initial response to infectious microorganisms and injuries. It is characterized by the dilation and leaking of blood vessels and subsequent attraction of circulating leukocytes to the inflammatory site, where innate immune responses occur.
- Adaptive immunity is a delayed, but highly effective immune response against specific antigens. The adaptive immune system is divided into cell-mediated immunity, which targets infected cells and is managed by T cells, and humoral immunity, which deals with infectious agents in blood and tissues and is mediated by B cells and their antibodies.
- B cells and T cells have antigen-specific receptors, which bind pure antigen in the case of B cells or antigen presented on major histocompatibility complex (MHC) proteins in the case of T cells.
- Stimulation by antigen plus other signals causes T lymphocytes to proliferate into clones of effector cells and memory cells.
- Cytotoxic $CD8^+$ T cells recognize and kill cells that are infected with intracellular pathogen presented on class I MHC protein. $CD4^+$ T helper 1 cells activate macrophages that present antigen on class II major histocompatibility complex (MHC) protein. $CD4^+$ T helper 2 cells activate B cells that present antigen on class II MHC protein.
- Activated B cells become antibody-producing plasma cells and memory cells.
- Antibodies are antigen-specific binding proteins that neutralize and opsonize antigen and activate complement to promote various immune reactions. Antibodies occur in five isotypes with special functions.
- Graft rejection is the major complication of organ transplantation and depends mainly on the allogeneic differences in the human leukocyte antigens between organ donor and recipient.
- Every acute illness evokes an immunologic response, and nearly every prolonged serious illness interferes with the immune system.
- Chronic inflammation leads to tissue damage and clinical symptoms. It is associated with the predisposition to a multitude of diseases.
- Immune system diseases are caused by abnormal or absent immunologic mechanisms, whether humoral, cell-mediated, or both.
- The actions of the immune system are closely interrelated with the actions of the endocrine and the neuronal systems to keep body homeostasis. Forthcoming information on these interactions will lead to novel treatments.

thePoint *Visit http://thePoint.lww.com for chapter review Q&A, case studies, animations, and more!*

11 Overview of the Cardiovascular System and Hemodynamics

ACTIVE LEARNING OBJECTIVES

Upon mastering the material in this chapter, you should be able to:

- Explain how the series arrangement of the right and left ventricle links the performance of the ventricles to each other.
- Explain how the parallel arrangement of arteries in the systemic organ systems allows the blood flow of each organ to be controlled independently.
- Explain the effects of the primary physiologic vasodilator and vasoconstrictor agents on vascular resistance and vascular compliance.
- Explain how vascular compliance is altered by transmural pressure and vascular smooth muscle contraction; describe how aging affects these relationships.
- Explain why gain or loss of total blood volume in the cardiovascular system alters volume in the venous side of the circulation more than on the arterial side of the circulation.
- Explain why standing causes blood to drain from regions above the heart and pool in the lower extremities.
- Predict the consequences of a distended vessel on vascular wall stress and the ability of a blood vessel to contract.
- Explain how changes in arterial pressure and vascular resistance alter blood flow and how blood flow and vascular resistance affect arterial pressure.
- Predict how vasodilator and vasoconstrictor agents directly affect blood pressure or organ blood flow.
- Postulate the detrimental effects of high and low blood velocity on the ability of pressure to create flow through the cardiovascular system; explain the hemodynamic mechanism that increases workload on the heart following dehydration.
- Predict what changes in the output of the heart and blood composition will create heart murmurs.

The cardiovascular system is commonly described as a fluid transport system that delivers substances to the tissues of the body while removing the byproducts of metabolism. This definition is technically correct. However, lost in such a bland, generic, definition of the cardiovascular system is a sense of how absolutely essential its key function is for *human survival.* The primary energy source for cell functions in the body is adenosine triphosphate (ATP), which is generated predominantly by oxidative metabolism. Cells get the oxygen they need for oxidative metabolism and rid themselves of the CO_2 they then generate by simple passive diffusion across the cell membrane.

Transporting substances using passive diffusion is a benefit to the cell in that it does not require the cell to expend energy. However, there is a significant limitation involved with the transport of any substances into or out of cells by passive diffusion alone. For example, the consumption of oxygen by cells requires that oxygen must be transported across the cell membrane at a rate (flux) that is sufficient to match the rate at which it is consumed. Failure to do so will severely diminish or stop oxidative metabolism. The resulting deficit in ATP production would soon thereafter result in cell death. The key to this process is getting the oxygen to the cells "on time." Herein resides a problem with delivering oxygen into the cell by passive diffusion. The time required for any molecule to randomly diffuse from one point to another is severely affected by the distance to be traveled between the points. For example, it takes approximately 5 seconds for an O_2 molecule to diffuse 100 μm, a distance that is compatible with the size of our cells, their nearest source of O_2, and the rate of O_2 usage by cells during metabolism. However, the physics of random molecular motion are such that each 10-fold increase in the distance that must be traveled by passive random diffusion increases the average time required to traverse that distance by a factor of 100. Therefore, should the diffusion distance in the example just given increase from 100 to 1,000 μm (1 mm), the average time a molecule would take to traverse that distance would increase to 500 seconds. A further increase from 1 mm to 1 cm would increase the average time to 50,000 seconds or almost 14 hours!

The time factor limitation associated with passive diffusion of oxygen has important consequences for us as living organisms. First, we could not exist in a body as large as our own were it not for the existence of some sort of transport system that could bring oxygen very close to every cell of our body (<100 μm). Oxygen must be brought in close enough proximity to every cell body so that the time required for it to passively diffuse into the cell is compatible with the rate at which it is consumed. This is the ultimate function of any circulatory system; even the tiniest multicellular organisms must have one. Thus, it is not surprising that when our cardiovascular system malfunctions, by extension, we as a whole malfunction; and if this system fails, the whole body "fails." The physics of passive diffusion place another constraint on our body. If for any reason a cell becomes separated by more than 100 μm from its nearest source of oxygen, as arranged by the existing cardiovascular system, that cell will become ischemic (insufficient O_2 to meet its metabolic needs). In such a state, the cell will malfunction and likely die. As shall be seen later in this text, such conditions do occur in various human cardiovascular diseases.

In addition to its essential role in maintaining overall health and survival, the cardiovascular system is also exploited for other tasks in the body. For example, the body uses hormones for the control of important physiologic functions, and these hormones are transported from the site of their production to their target organs in the bloodstream by the cardiovascular system. In addition, blood serves as a reservoir for heat. The cardiovascular system plays an important role in the control of heat exchange between the body and external environment by controlling the amount of blood flowing through the skin, which is in contact with the environment surrounding the body as a whole.

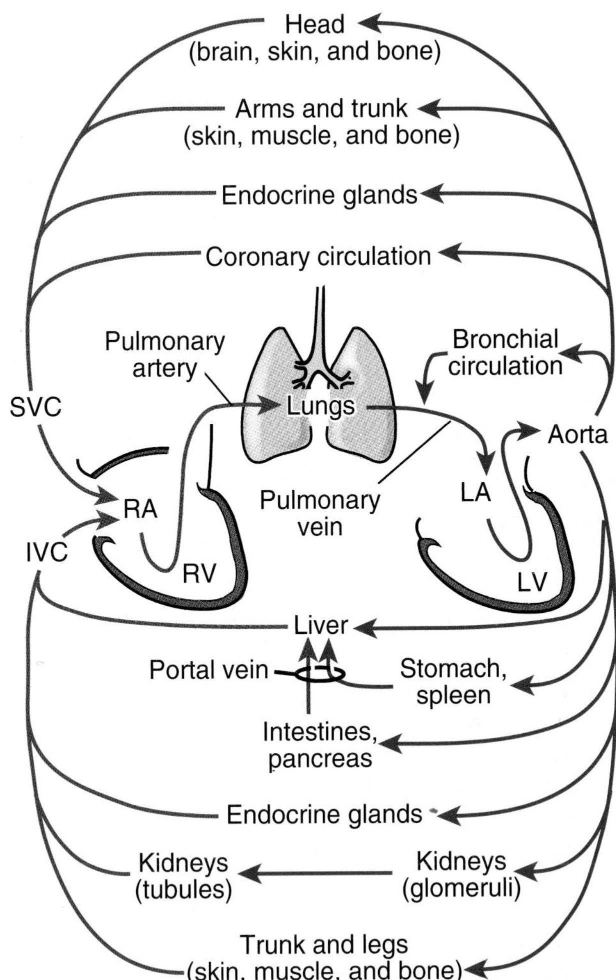

Figure 11.1 A model of the cardiovascular system. The right and left sides of the heart are aligned in series, as are the systemic circulation and the pulmonary circulation. In contrast, the circulations of the organs other than the lungs are in parallel. Each organ receives blood from the aorta and returns it to the vena cava. Exceptions to the series arrangements exist between the splanchnic veins and portal circulation of the liver as well as the glomerular and tubular capillary networks of the kidney. SVC, superior vena cava; IVC, inferior vena cava; RA, right atrium; RV, right ventricle; LA, left atrium; LV, left ventricle.

FUNCTIONAL ORGANIZATION

Students in medical physiology should be familiar with the anatomic organization of the heart and blood vessels from their courses in gross anatomy and histology. A brief overview is contained in this section. The cardiovascular system is a fluid transport system for the movement of blood throughout the body. This is represented in simplistic form in Figure 11.1. Blood is driven through the cardiovascular system through the actions of a hollow muscular pump called the **heart**. The heart is a four-chambered muscular organ that contracts and relaxes in a regular repeating cycle to pump blood. The period of time the heart spends in contraction is called **systole**, and the time it spends in relaxation is called **diastole**.

The heart is really two pumps connected in series. The left heart is composed of the **left atrium** and **left ventricle**, separated by the **mitral valve**. Contraction of the left ventricle is responsible for pumping blood to all systemic organs except the lungs. Blood exits the ventricle through the **aortic valve** into a single tubular conduit called the **aorta**. The aorta is classified as an **artery**, which, by definition, is any blood vessel that carries blood away from the heart and to the tissues of the body. The aorta branches into successively smaller arteries, many of which are given anatomic names. These arteries, in turn, branch into millions of smaller vessels of approximately 10 to 1,000 μm in external diameter, called **arterioles**. Arterioles, in turn, terminate into billions of **capillaries**, which are the main site of transport of water, gases, electrolytes, substrates, and waste products between the bloodstream and the extracellular fluid. Blood from the capillaries of all the systemic organs coalesces into thin-walled **venules**, which merge into **veins**. By definition, a vein is any blood vessel that returns blood from the tissues back into the heart. Small veins eventually merge to form two large single veins called the **superior vena cava (SVC)** and the **inferior vena cava (IVC)**. The SVC collects blood from the head and upper extremities above the level of the heart, whereas the IVC collects blood from all regions below the level of the heart. Both of these large veins empty into the **right atrium**. The right atrium is the upper chamber of the right heart. It is separated from the **right ventricle** by the **tricuspid valve.** The right ventricle pumps blood through the **pulmonic valve** into the **pulmonary artery** and thence into the lungs. Blood exiting the lungs is returned to the **left atrium** where it passes through the **mitral valve** and into the left ventricle, completing the circulatory loop.

Right and left heart are interdependent because they are connected in series.

Imagine a bucket of water interposed between two pumps. One pump removes water from the bucket and pumps it into the inflow of the second pump. The second pump takes this inflow and returns it into the bucket for the first pump to remove again. It is obvious that if the outputs of the two pumps are not matched identically, the level of water in the bucket will change; the bucket could be drained dry or could overflow. This is analogous to the situation in the pulmonary circulation that is interposed between the right and left heart. The right and left heart are said to be arranged *in series*, or in line, one after the other, right to left.

There are multiple consequences of this series arrangement of the two pumps in the heart. First, should the output of the left heart exceed that of the right by as little as 2%, the pulmonary circulation would be drained of blood in <10 minutes. Conversely, if the right heart output exceeded the left by a similar amount, the pulmonary circulation would overflow and the person would drown in his or her own body fluids. Clearly, neither of these situations arises in a healthy person. The implication of this arrangement is that some mechanism must function to closely match the outputs of the right and left heart. Such a mechanism exists at the level of the muscle cell and is based on the same principles of skeletal muscle cell mechanics you have read about in earlier chapters. Details of these principles as they apply to the heart are discussed in Chapter 13, "Cardiac Muscle Mechanics and the Cardiac Pump."

The series arrangement of the right and left heart also implies that malfunctions in the left heart will be transmitted back into the pulmonary circulation and the right heart, potentially causing the respiratory system to malfunction. Indeed, one of the first clinical signs of left heart failure is respiratory distress. Conversely, problems originating on the right side of the circulation affect the output of the left heart and imperil the blood supply to all systemic organs. Large blood clots can form in the major veins of the leg and abdomen following surgery. These clots can break away and slip through the tricuspid valve into the right ventricle and thence into the pulmonary artery, where they eventually lodge, to form what is called a **pulmonary embolus**. Pulmonary emboli compromising about 75% of the pulmonary circulation can block enough flow into the left heart to kill a person.

Parallel arrangement of organ circulations permits independent control of blood flow in individual organs.

The arterial system delivers blood to organ systems that are arranged in a *parallel*, or side-by-side, network. Therefore, in most cases, blood flow into one organ system is not dependent on blood flow through another organ upstream. The metabolic demands of our muscles, digestive system, brain, and so on, may be different relative to one another and relative to their own resting values, depending on the activity in the organ at a given time. The parallel arterial distribution system of organ blood supply allows the adjustment of blood flow to an individual organ to meet its own needs without creating major disturbances in the blood supply to other organs. A notable exception to this arrangement, however, is seen in the portal circulation. Venous outflow from the intestines and other splanchnic organs drains into the liver through the **portal vein** before being emptied into the IVC. The liver obtains blood from the portal vein as well as its own arterial supply (see Fig. 11.1) and can be considered to be arranged in series with much of the splanchnic circulation.

Vascular smooth muscle actively controls the diameter of arteries and veins.

All blood vessels, except capillaries, have a similar basic structure. All arteries and veins are lined with a single layer of epithelial cells called the **endothelium**. The media of vessels contain circular layers of smooth muscle cells, whereas the outermost layer, called the **adventitia**, is composed of collagen and elastin fibers that add flexible structural integrity to arteries and veins. Because the smooth muscle within blood vessels is arranged in circular layers, contraction or relaxation of these muscles will reduce or widen, respectively, the lumen diameter of arteries and veins. The changing of the diameter of blood vessels has a profound effect on the physical factors that determine blood flow and blood volume distribution in the cardiovascular system. These effects are discussed later in this chapter.

There are literally scores of normal physiologic, pathologic, and clinical pharmacologic agents that can alter the contraction and relaxation of arterial and venous smooth muscle. These take the form of direct physical forces and chemical agents, hormones, paracrine substances, and receptor-mediated hormonal and neurotransmitter agonists. For example, the contraction of arteries and veins is modified by transmitters released from sympathetic nerve endings that enter these vessels through their adventitial layer and act on specific receptors on the smooth muscle membrane. The endothelium, which stands at the interface between blood plasma and the rest of the vessel wall, is the source of important paracrine agents that have major, receptor-independent, direct effects on blood vessel contraction. A simplified, partial list of factors that contract or relax vascular smooth muscle is provided in Table 11.1.

TABLE 11.1 A Simplified List of Direct (Non-Receptor-Mediated) and Smooth Muscle Membrane Receptor-Mediated Factors that Contract or Relax Vascular Smooth Muscle

Vasoconstrictor/Vasodilator	Mechanism of Action
	Vasoconstrictors: Non–receptor-mediated
Ca^{2+}	Enters cells through membrane channels or is released from the SR by IP_3; binds to calmodulin, which activates myosin light chain kinase to initiate crossbridge attachment and cycling
Ba^{2+}	Substitutes for calcium in the contraction cascade
K^+	Membrane depolarization resulting in activation of L-type Ca^{2+} channels
O_2 (normoxia)	Maintenance of contraction
Membrane depolarization	Activation of L-type Ca^{2+} channels
Decreased transmembrane sodium gradient	Decreased activation of plasma membrane Na^+/Ca^{2+} exchange and increased intracellular Ca^{2+}
Increased transmural pressure	Activation of stretch-activated plasma membrane calcium channels
Action potentials	Membrane depolarization resulting in the activation of L-type calcium channels
	Vasoconstrictors: Receptor-mediated
Norepinephrine	Activation of α adrenoreceptors to open receptor-operated plasma membrane calcium channels, activation of phospholipase C, formation of IP_3 and DAG, and release of calcium from the SR
Epinephrine	Same as norepinephrine
Acetylcholine	Activation of muscarinic M_2 receptors to open receptor-operated plasma membrane calcium channels, activation of phospholipase C, formation of IP_3 and DAG, and release of calcium from the SR; inhibits cAMP formation
Serotonin	Activation of 5-HT_2 receptors to open receptor-operated plasma membrane calcium channels
Vasopressin	Activation of V receptors to open receptor-operated plasma membrane calcium channels
Angiotensin II	Activation of angiotensin II receptors to open receptor-operated plasma membrane channels and activate IP_3 /DAG pathways
Endothelin	Activation of ET_A receptors with stimulation of phospholipase C, formation of IP_3, and release of calcium from the SR
ATP	Activation of P_{2x} receptor to open receptor-operated plasma membrane channels
	Vasodilators: Direct
Nitric oxide	Activation of guanylate cyclase, formation of cGMP, and calcium removal
Hyperpolarization	Inhibition of voltage-gated Ca^{2+} channels
Decreased transmural pressure	Inhibition of stretch-operated Ca^{2+} channels
Hypoxia	Likely a result of decreased ATP supply and formation of adenosine
Hypercapnia	Likely from acidosis
Acidosis	Unknown
Hyperosmolarity	Unknown
cAMP	Activation of cAMP-dependent kinase resulting in a phosphorylation cascade that reduces intracellular calcium concentration
cGMP	Activation of cGMP-dependent kinase resulting in a phosphorylation cascade that reduces intracellular calcium concentration
Hyperkalemia (low levels)	Membrane hyperpolarization resulting from activation of the Na^+/K^+ pump

(Continued)

TABLE 11.1 A Simplified List of Direct (Non-Receptor-Mediated) and Smooth Muscle Membrane Receptor–Mediated Factors that Contract or Relax Vascular Smooth Muscle *(Continued)*

Vasoconstrictor/Vasodilator	Mechanism of Action
	Vasodilators: Receptor-mediated
Epinephrine	β_2 receptor-mediated activation of adenyl cyclase resulting in the formation of cAMP
Adenosine	A_1, A_{2A}, and A_{2B} receptor activation of ATP-dependent K^+ channels leading to membrane hyperpolarization and closure of voltage-gated calcium channels
Histamine	H_2 receptor activation of adenyl cyclase with formation of cAMP
PGI_2	Activation of adenyl cyclase with formation of cAMP

5-HT, 5-hydroxytryptamine; ATP, adenosine triphosphate; cAMP, cyclic adenosine monophosphate; cGMP, cyclic guanosine monophosphate; DAG, diacylglycerol; IP_3, inositol 1,4,5-triphosphate; PGI_2, prostacyclin; SR, sarcoplasmic reticulum.

PHYSICS OF BLOOD CONTAINMENT AND MOVEMENT

The study of the physical variables related to the containment and movement of blood in the cardiovascular system is called **hemodynamics**. From an engineering standpoint, an accurate description of all the hemodynamic phenomena in the cardiovascular system is complex. Fortunately, the human body deals with these phenomena and their control in a considerably simplified manner. The cardiovascular system behaves much as if the heart were producing an average steady flow through a series of solid pipes, similar to the flow of water through a city's water distribution system. Thus, basic principles of fluid dynamics can be applied to the understanding of cardiovascular phenomena.

Fluid cannot move through a system unless some energy is applied to it. In fluid dynamics, this energy is in the form of a difference in **pressure**, or pressure gradient, between two points in the system. Pressure is expressed as units of force, or weight, per unit area. A familiar example of this is in the pounds per square inch (psi) recommendation stamped on the side of tires. The psi indicates the pressure to which a tire should be inflated with air above atmospheric pressure. Inflating a tire to 32 psi signifies that 32 more pounds press against every square inch of the inner tire surface than against the outside of the tire.

The pressure exerted at any level within a column of fluid reflects the collective weight of all the fluid above that level as it is pulled down by the acceleration of gravity. It is defined as

$$P = \rho g h \qquad (1)$$

where P = pressure, ρ = the density of the fluid, g = the acceleration of gravity, and h = the height of the column of fluid above the layer where pressure is being measured. The force represented by pressure in a fluid system is often described as the force that is able to push a column of fluid in a tube straight up against gravity. In this way, the magnitude of the force resulting from fluid pressure can be measured by how high the column of fluid rises in the tube (Fig. 11.2A). In physiologic systems, this manner of expressing pressure is designated as centimeters H_2O, or the more convenient mm Hg, because mercury is much denser than water and therefore will not be pushed as far upward by typical pressures seen within the cardiovascular system.

Without going into mechanistic detail at this time, it should be noted that arterial pressure peaks shortly after the heart contracts and pumps blood into the aorta, and it falls to a lower value when the heart relaxes between beats and is therefore not pumping blood into the aorta. The peak pressure during contraction of the heart is called the **systolic pressure** and is typically about 120 mm Hg in humans, whereas the minimum arterial pressure value during relaxation of the heart is called the **diastolic pressure** and is about 80 mm Hg. Thus, if one end of a tube were to be inserted into the aorta with the other end connected to a column of mercury sitting perpendicular to the ground at the level of the heart, that column of mercury would rise 120 mm during systole and fall to 80 mm during diastole. In clinical practice, human arterial pressure is reported as systolic over diastolic pressure or, in this example, 120/80 mm Hg. (Our mean arterial pressure is not the arithmetic mean of systolic and diastolic pressure but is instead about 93 mm Hg, because the time the heart spends relaxing is longer than the time it spends contracting and ejecting blood into the aorta.)

The pressure outside the body or on the outside of any hollow structure within the body other than the intrapleural space is about the same as atmospheric pressure. Technically, atmospheric pressure at sea level is equivalent to 760 mm Hg. In reality, therefore, our average blood pressure is 93 + 760 mm Hg, or 863 mm Hg. However, pressures within our body are never expressed in this technically exact manner. Instead, the effect of atmospheric pressure is simply ignored and taken as a *zero reference point.* The pressure reported in our systems then is really the difference of pressure within that system relative to atmospheric pressure.

The effect of gravity on pressure within the veins is significant when we are standing, as shown in Figure 11.2B. In a recumbent position, pressure in the veins is between 2 and 10 mm Hg. However, when one stands, the influence of gravity subtracts approximately 40 mm Hg of pressure from

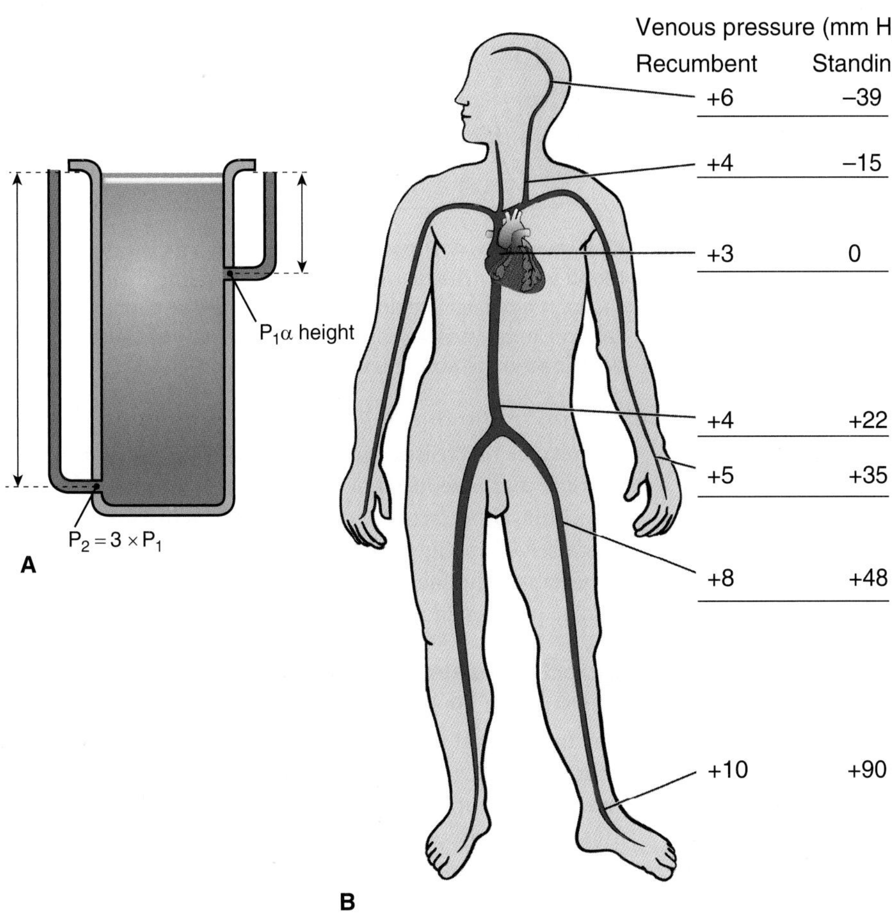

Figure 11.2 The effect of height on pressure at different levels within a column of fluid. **(A)** Fluid pressure is proportional to the height of the column of fluid above the point of measurement. Pressure is measured as the force capable of raising a column of mercury vertically against gravity; higher pressure raises the column of mercury higher. Pressure in the cardiovascular system is commonly reported in units of mm Hg. **(B)** The effect of gravity on pressure within veins. Blood is pulled down by the effect of gravity, subtracting pressure from veins in the head and adding pressure to veins in the lower extremities.

the arteries and veins in the head and adds about 90 mm Hg of pressure to those in the feet. If human veins were rigid tubes, this would not have a profound effect on the circulation and distribution of blood within sections of the cardiovascular system. However, veins are flexible structures. Therefore, when one stands, blood tends to pool in the veins of the lower extremities. This is responsible for the swelling and aching of the feet after standing for long periods of time and can even result in such extensive pooling that fainting will occur from an inability of enough blood to return to the heart to be pumped to the brain.

Blood vessel volume is a function of vessel flexibility and the pressure difference across the vessel wall.

The volume of fluid within a container made of inflexible walls, such as a glass bottle, is the same no matter what the pressure difference is between the inside and the outside of the bottle. In such a container, the walls cannot move, or flex, when any difference in pressure is applied across them. Arteries and veins are more like rubber balloons in that their walls are flexible. Consequently, the volume contained within them is a function of both the pressure difference across their wall, called the **transmural pressure**, and the degree of flexibility within the vascular wall. Transmural pressure, or P_T, is always defined as the difference in pressure *inside versus outside* a hollow structure. Large transmural pressures within flexible vessels create large intravascular volumes; small transmural pressures in stiff-walled vessels produce small intravascular volumes.

There are a couple of ways of depicting pressure/volume interrelationships in blood vessels. The volume of blood contained in the vessel for a given transmural pressure is called the **vascular capacitance** and is calculated as

$$\text{Capacitance} = \text{Volume}/P_T \quad (2)$$

where volume and transmural pressure are generally given the units of mL and mm Hg, respectively. However, physiologists are interested in how volume changes in a distensible blood vessel for a given change in pressure. The change in volume for a given change in transmural pressure is called the **compliance** and is given by the equation

$$C = \Delta V/\Delta P \quad (3)$$

where the Δ signifies the before/after change of volume or pressure. (Note that, for this equation, the pressure outside the vessel is taken to be atmospheric pressure, which is always given the baseline value of zero in normal physiologic systems. Nevertheless this pressure can become positive in certain pathologic conditions and thus reduce the distending effect of pressure within the blood vessel.) This equation can be rearranged to yield two useful relationships relating changes in either volume or pressure as a function of the other in a blood vessel. For example, one can write $\Delta V = C \times \Delta P_T$. This equation indicates that the change in volume contained within a vascular segment will be great in a vessel with high compliance and a large change in intravascular pressure. One can also write $\Delta P_T = \Delta V/C$, which indicates that adding volume into a vascular segment will produce a large increase in pressure within the vessel if the volume added is large and the vascular compliance is low.

Both capacitance and compliance can be used as measures of the distensibility, or flexibility, of a blood vessel; highly distensible vessels have a higher capacitance and compliance than vessels of the same dimensions with stiff walls. For this reason, veins have a higher capacitance and compliance than arteries of similar size. However, the use of these variables to measure vessel flexibility fails when vessels of significantly different sizes are compared. For example, a large stiff-walled vessel may have a higher capacitance value than a tiny flexible vessel. For this reason, one should use the *percentage* increase in volume for a given increase in pressure as a means of comparing distensibility between vessels and segments of the vasculature of different sizes. This value is sometimes called *vascular distensibility*. As explained in later chapters, vascular capacitance, compliance, and distensibility are critical determinants of the performance of the heart, the stress and workload placed on the heart, and the amount of oxygen the heart must receive to function properly. All of these factors are important parameters in understanding the consequences of heart diseases and their treatments.

Even within a given artery or vein, capacitance, distensibility, and compliance are not constants in the cardiovascular system. For example, as transmural pressure increases, compliance decreases (Fig. 11.3). In addition, arterial

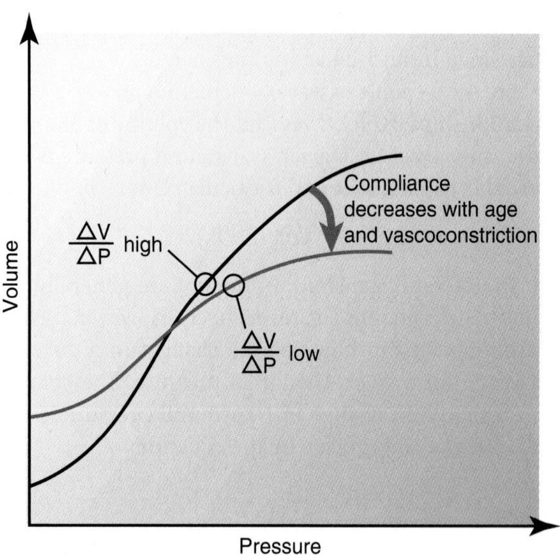

Figure 11.3 The relationship between volume and transmural pressure in arteries. Volume within an artery increases with an increase in transmural pressure because the arterial wall is flexible. In normal healthy people, the artery is stiffest when it is first filled and when it is near the limit of its volume capacity (i.e., at low and high transmural pressure, respectively). The slope of this arterial volume:pressure relationship at any point is a measure of arterial compliance. Arterial compliance tends to diminish when pressure increases in the high transmural pressure range. Arterial compliance decreases with aging and with the level of active smooth muscle contraction in the artery.

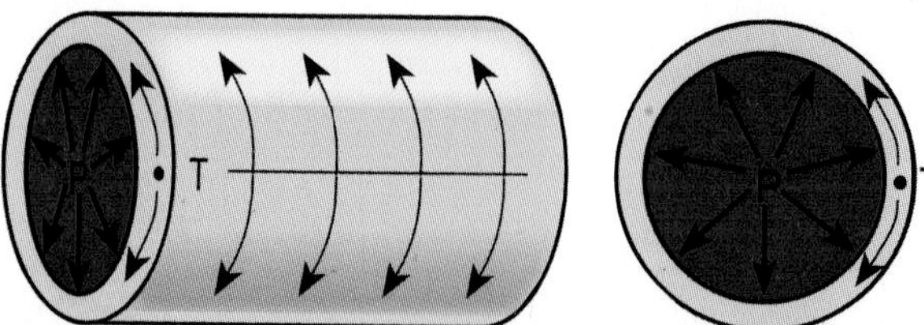

Figure 11.4 Pressure and tension in a cylindrical blood vessel. The tension is the force that would pull the vessel apart along an imaginary line along the length of the vessel. Tension in the vessel wall is related by the law of Laplace, as described in the text. P, pressure; T, tension.

compliance decreases with age at any given transmural pressure and is reduced by contraction of the smooth muscle within either arteries or veins. This active control over pressure/volume interrelationships in the vascular tree is an important component of the moment-to-moment control of cardiac performance in a person.

Blood vessels must overcome wall stress to be able to contract.

Any transmural pressure within an artery or a vein exerts a force on the vessel wall that would tend to rip the wall apart were it not for the opposing forces supplied by the muscle and connective tissue of the vessel wall, as shown in Figure 11.4. This force is called **tension** and is equal to the product of the transmural pressure and the vessel radius. This relationship is called the **Law of Laplace**. However, this relationship applies directly only to cylinders with thin walls. Blood vessel walls are sufficiently thick, so that in reality, this force is equal to a **wall stress** (the product of pressure [P] and the radius [r] divided by the wall thickness [w], or $S = P \times r/w$). There are many consequences of this relationship in distensible tubes. First, because tension and stress are related to vessel radius, small vessels are able to withstand higher pressures than vessels of larger diameters. For this reason capillaries (inner diameter ~10 μm) can withstand relatively high intravascular pressure even though they are composed only of a single cell layer of endothelial cells. In arteries and veins, vessels with thick walls relative to their radius are able to withstand higher pressure than vessels with small r/w ratios because wall stress is lower in the former. Finally, tension and stress, not simply pressure, are the true forces that must be overcome to contract any hollow organ, such as a blood vessel or the heart. As is discussed later, tension and stress are important determinants of energy requirements for the contraction of hollow organs.

PHYSICAL DYNAMICS OF BLOOD FLOW

An understanding of the physical factors and laws that govern the movement of blood in the cardiovascular system is critically important to the overall understanding of its function. The amount of blood that flows through any segment of the cardiovascular system is a function of blood pressure, vascular geometry, and the dynamic fluid characteristics of blood. In any tube of a given diameter, the amount of flow through the tube is proportional to the *difference* in pressure between one end of the tube and the other (Fig. 11.5); doubling the pressure difference doubles the flow, whereas halving the difference halves the flow. Flow through a cylindrical tube is related to the length of the tube in an inverse proportion. For example, keeping all other determinants of flow unchanged, doubling only the length of a cylindrical tube reduces flow through the tube by half. However, as shown in Figure 11.5, flow through a tube is profoundly affected by the tube radius in that flow is proportional to the radius^4. Consequently, doubling the tube radius results in a 16-fold increase in flow. Finally, flow through a tube is affected by the viscosity, or the "thickness and stickiness," of the fluid. Thick, sticky fluids will not flow as easily as thinner, watery fluid. The units of viscosity are given in poise, named after Louis Poiseuille. Water has a viscosity of approximately 0.01 P, or 1 cP, which is a handy reference point for comparing the viscosity of body fluids. Blood is a suspension of proteins and cells in a salt solution. As such, it is more viscous than water. Blood plasma has a viscosity of 1.7 cP; and whole blood, a viscosity of approximately 4 cP.

Relationships between pressure, fluid flow, and resistance are quantified by the Poiseuille law.

In the 1840s, Louis Poiseuille conducted experiments that resulted in a mathematical relationship to describe flow in a cylindrical tube. This has become known as the **Poiseuille law** (also known as *Poiseuille's law*). It states that:

$$Q = (P_1 - P_2)\pi r^4/8\eta l \qquad (4)$$

where Q = flow, $(P_1 - P_2)$ = the pressure difference between the beginning and the end of the tube, r = the tube radius, l = the tube length, η = viscosity, and π and 8 are constants of proportionality. This relationship can also be written as $(P_1 - P_2) = Q\ 8\eta l/\pi r^4$. In this form, the term $8\eta l/\pi r^4$ is the resistance to blood flow and is sometimes simply designated as R. Flow resistance is a measure of how easily the fluid can pass through a tube for any given pressure difference. In physiology, it is easier to calculate resistance as $(P_1 - P_2)/Q$ and to express it in units of mm Hg/mL/min with the name peripheral resistance unit (PRU).

The Poiseuille law gives two of the most fundamentally important relationships used to describe flow and pressure in the cardiovascular system, which are

$$Q = (P_1 - P_2)/R \quad \text{and} \quad (P_1 - P_2) = Q \times R \qquad (5)$$

These equations indicate that flow is proportional to the pressure difference between the entrance and exit points of the tube and inversely proportional to the resistance (i.e., as resistance increases, flow decreases). It also tells us that at any given flow, the pressure drop along any two points down the length of the tube is proportional to resistance. In the cardiovascular system, the final "end of the tube" is

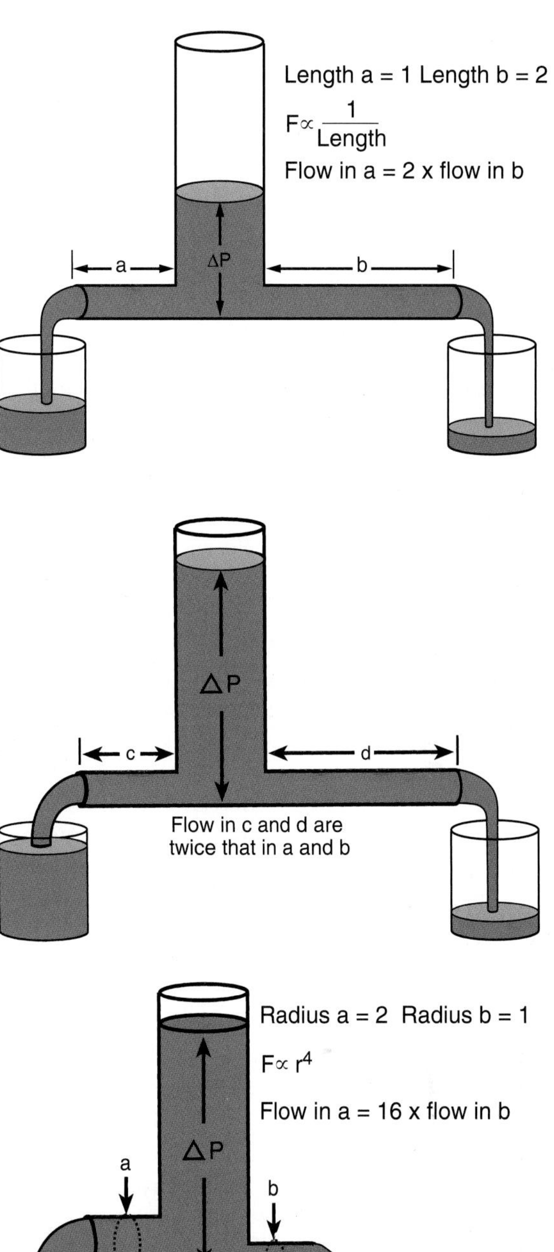

Figure 11.5 The influence of pressure difference, tube length, and tube radius on flow. The pressure difference (ΔP) driving flow is the result of the height of the column of fluid above the openings of tubes a and b. Increasing the pressure *difference* increases flow; if pressure was the same at the opening of the tubes as at the bottom of the column of fluid, there would be no flow. Flow is reduced in direct proportion to the length of the tube through which it flows; doubling the length reduces flow to half its original value, tripling the length reduces it to one third, and so on. Because flow is determined by the fourth power of the radius, small changes in radius have a much greater effect than small changes in length. F, flow; P, pressure.

considered the right atrium. Pressure in the right atrium is about 2 mm Hg. This is sufficiently close to zero and thus is ignored. Therefore, $(P_1 - P_2)$ becomes simply P. Furthermore, for the purpose of understanding basic hemodynamics, P is usually taken as the mean pressure within an artery or vein. Consequently, the Poiseuille law reduces to

$$Q + P/R \quad \text{or} \quad P = Q \times R \tag{6}$$

Applied to the whole cardiovascular system, this law indicates that arterial pressure is the product of the flow output of the heart, called the **cardiac output**, and the resistance to flow provided by all the blood vessels in the circulation. This resistance is termed the **total peripheral vascular resistance (TPR)**.

Strictly speaking, the Poiseuille law applies only to nonpulsatile flow of a homogenous fluid in uniform, rigid, nonbranching cylindrical tubes. Because none of these characteristics is met in the cardiovascular system, one might imagine that the Poiseuille law cannot be applied to blood flow. However, the Poiseuille law can be and is applied to the cardiovascular system. Note that this application is not used as a simplified expedient that ignores some of the complex realities of the physics of blood flow. Rather, the Poiseuille law can be used in the manner expressed in equation 6 because the cardiovascular system truly does behave as if the heart pumped blood at a steady flow, producing a mean arterial pressure as the result of a single TPR. Thus, equation 6 predicts that mean arterial pressure will increase if the cardiac output, TPR, or both increase, whereas it will decrease if TPR and/or cardiac output decrease. It also predicts that at a constant mean arterial pressure, blood flow through any portion of the vascular tree will increase if TPR decreases and will decrease if TPR increases. These simple cause-and-effect relationships are the critical determinants of arterial pressure and organ blood flow in the cardiovascular system and will be described more fully in the context of cardiovascular control mechanisms in Chapter 17, "Control Mechanisms in Circulatory Function."

The series and parallel arrangement of blood vessels within an organ affects vascular resistance in the organ.

The cardiovascular system is composed of millions of vessels of different sizes. There are two simple rules used to determine how many vessels of different sizes combine to give a single resistance to flow. In a system composed of different-sized tubes arranged in series (i.e., sequentially, or end to end), the total resistance of that system is simply the sum of the individual resistances. Or

$$R_{total} = \Sigma R_{individual} \tag{7}$$

For example, if one were to examine the resistance of a segment of a tapering artery where the proximal 1 cm of length had an R = 1 PRU, the next 1 cm had an R = 2 PRUs, and the 1 cm after that an R = 5 PRUs, the resistance across the entire 3-cm length of that artery would be 1 + 2 + 5 = 8 PRUs. Nevertheless, the arterial portion of the cardiovascular system is not a single long blood vessel; the aorta branches into thousands of arteries and capillaries, which are arranged in parallel. The total resistance in a system of parallel tubes is given by

$$1/R_{total} = \Sigma(1/R_{individual}) \tag{8}$$

That is, the reciprocal of the total resistance in vessels arranged in parallel is the sum of the reciprocals of the individual resistances. Using the arterial segments mentioned above but arranged in parallel, one would arrive at $1/R_{total}$ = 1/1 + 1/2 + 1/8 = 13/8, R_{total} = 8/13, or approximately 0.62 PRUs. In this case, the total resistance of the system of the three arteries arranged in parallel is not only less than it would have been if they were arranged in series but is actually less than the resistance of any one segment in the circuit. In most cases, this is a *general* rule of thumb; losing similar-sized vessels in parallel in the arterial system raises vascular resistance, whereas adding similar-sized vessels in parallel reduces resistance. This latter case is seen following long-term aerobic training such as distance running, in which the arteriolar and capillary network increase their numbers in parallel, thereby reducing resistance to blood flow.

The effect of summed series and parallel elements in the cardiovascular system is a bit more complicated than can be described solely by equations 7 and 8. In the cardiovascular system, vessels get smaller as they proceed from arteries down to arterioles and capillaries, which tends to increase resistance to flow. However, the number of vessels arranged in parallel also increases dramatically in this direction, which tends to decrease resistance. The effect of these two phenomena on the relative resistances of individual sections of the vasculature depends on whether the number of vessels added in parallel can compensate for the resistance effects of adding vessels that individually have a high resistance resulting from small radii. This interplay between series and parallel elements is a primary factor in creating the form of the pressure profile, or ΔP, along the arterial side of the circulation, as will be explained later in this chapter.

High blood flow velocity decreases lateral pressure and increases shear stress on the arterial wall.

In addition to the amount of blood flowing through a given vessel or organ system per minute, physiologists are often interested in how fast the bloodstream is flowing from one point to the next in the circulation. This variable is called **flow velocity** and is usually expressed in cm/s. Fluid flow velocity is given simply by the volume of flow per second (cm^3/s) divided by the cross-sectional area of the system through which the fluid is flowing (Fig. 11.6A). If a flow of 200 mL/s in a tube is forced through another narrower tube, the flow must go through that smaller opening faster to maintain a volume flow at 200 mL/s. Conversely, if this fluid is allowed to expand into a much larger cross-sectional area, it can move slower and still deliver 200 mL/s. This relationship holds whether applied to a single tube or a composite cross-sectional area of many tubes arranged in parallel such as a cross section of the vascular system. Thus, at a constant flow, a decrease in the cross-sectional area through which the flow is moving increases flow velocity, and an increase in area decreases flow velocity.

The blood pressure exerted against the walls of an artery and the energy required to move blood at a certain velocity are interrelated by the **Bernoulli principle**. The total energy of blood flow in a blood vessel is the sum of its potential energy (represented as pressure against the vascular wall) and its kinetic energy resulting from its velocity (kinetic energy = $\frac{1}{2}mv^2$, where m = mass and v = velocity). The total of potential and kinetic energy at any given point in a system is constant. Consequently, any increase in one form of the energy has to come at the expense of the other. For example, as flow velocity increases, lateral pressure must decrease to keep the total energy of the system constant (see Fig. 11.6A). This principle is seen with flow in the aorta, where high velocity reduces lateral pressure relative to that measured directly facing the flow stream, which equals the total energy in the system. This phenomenon is exploited clinically to evaluate the severity of the hemodynamic consequences of a stenotic (narrowed) aortic valve. A catheter with two pressure sensors is placed in the heart such that one lies within the ventricle and the other just across the narrowed aortic valve. A high-velocity jet of blood moving through the narrowed valve causes a significant drop in lateral pressure detected by the aortic sensor compared with the ventricular sensor; the hemodynamic severity of the stenosis is proportional to this pressure difference.

All flowing fluids exert a "rubbing" force against the inner wall of the cylinder in which they flow. This force is analogous to what one feels by rubbing the palms of the hands together and is called **shear stress**. Shear stress on the inner walls of arteries increases proportionally as the velocity of flow near the wall increases. This has several medical implications. The release of paracrine substances from the endothelium as well as other transport phenomena is stimulated by increased shear stress on the endothelial cells that line the arteries. These factors are instrumental in stimulating capillary growth and endothelial repair and may be the link between the growth of new blood vessels in skeletal muscle and periodic increases in muscle blood flow during aerobic exercise training. Conversely, the accumulation of atherosclerosis in arteries is enhanced in low shear areas of arteries, whereas it is diminished in high shear areas.

High blood flow velocity can create turbulent flow in arteries.

Flow velocity also affects the organization of fluid layers of blood in arteries. Normally, the cells in layers of fluid in arteries flow with a streamlined or bullet-shaped profile, as shown in Figure 11.6B. In three dimensions, this can be envisioned as a set of thin telescoping cylindrical layers projecting from the inner arterial wall out uniformly to the center of the vessel. Flow velocity is highest in the center of a

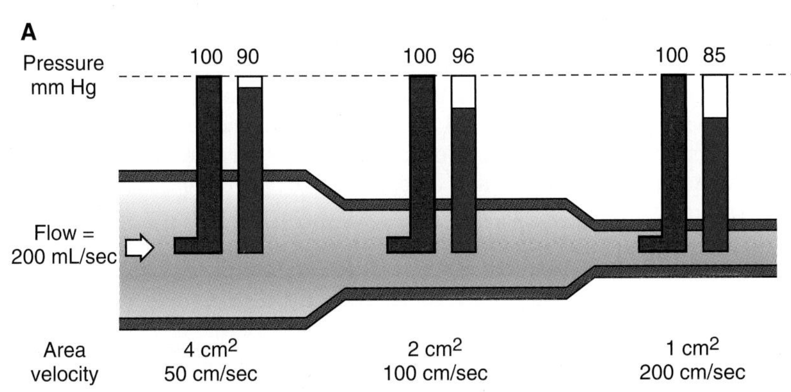

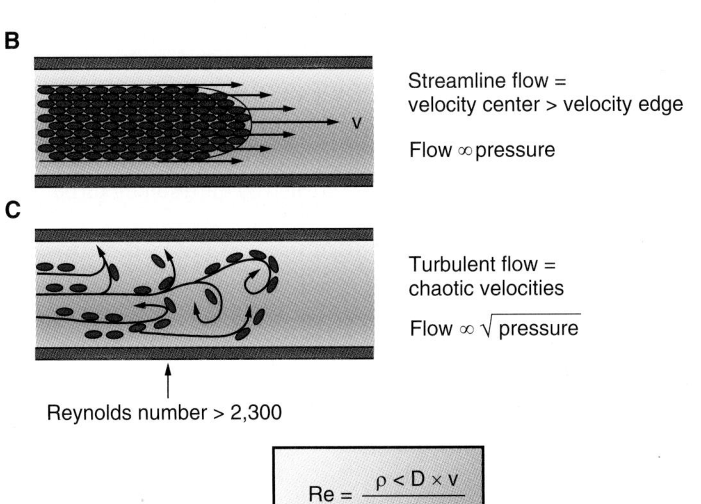

Figure 11.6 Effect of flow velocity on lateral intravascular pressure, axial streaming, and turbulent flow. (A) If any given flow of blood is forced through progressively smaller cross-sectional areas, the velocity of blood flow must increase. The Bernoulli principle states that increased flow velocity reduces the lateral pressure of the flow stream exerted against the wall of the vessel. **(B)** The distribution of red blood cells in a blood vessel depends on flow velocity. As flow velocity increases, red blood cells move toward the center of the blood vessel (axial streaming), where velocity is highest. Axial streaming of red blood cells creates a cell-free layer of plasma along the inner vessel wall. **(C)** At high flow velocity, the kinetic forces of flowing fluid overcome the viscous forces holding layers of fluid together, resulting in turbulent flow.

Clinical Focus / 11.1 

Vascular Abnormalities Associated with Chronic Primary Arterial Hypertension

Chronic arterial hypertension is a cardiovascular disease in which the main manifestation is a consistent elevation of arterial pressure ≥140/90 mm Hg. Diagnosis of chronic arterial hypertension in a patient is established following documentation of elevated pressure in a clinical setting when measured at several time points over a period of many weeks or months. Even with this criterion, it is estimated that more than 65,000,000 people in the United States alone suffer from chronic arterial hypertension. Hypertension is the leading risk factor in stroke and a major contributor to morbidity and mortality associated with heart failure, coronary artery disease, atherosclerosis, and renal insufficiency.

Arterial hypertension is largely asymptomatic until it results in end-stage organ disease, which is a major reason it goes unrecognized and untreated by the patient. However, people with untreated hypertension have significantly higher levels of cardiovascular morbidity with a reduced life expectancy. Left untreated, arterial hypertension becomes more severe over time, as the result of progressive arterial and renal abnormalities resulting from exposure to high arterial pressure. People with untreated arterial hypertension do not get better on their own and must stay on medication throughout their lives to keep their arterial pressure down. Keeping pressure down in hypertension has been clinically established to provide definitive benefits to the patient, whereas failure to control hypertension results in definitive harm.

Chronic arterial hypertension is classified broadly as either *primary* or *secondary hypertension*. Secondary hypertension indicates that the elevated blood pressure is a secondary result of some other primary disease. Hypertension is a secondary consequence of renal artery stenosis, renal parenchymal disease, primary aldosteronism, pheochromocytoma, aortic coarctation, and thyrotoxicosis. Although there are several mechanisms known to cause secondary hypertension, only 5% to 15% of all cases of chronic arterial hypertension fall into that category. The remaining individuals with hypertension have what is called *primary*, or *essential*, *hypertension*, which indicates that the hypertension is a primary condition in its own right of unknown cause. Currently, the etiology of primary hypertension is believed to be the result of multiple initiating factors and regulatory dysfunctions with strong genetic predispositions involving both. Current attention is being directed at such factors as they relate to how the kidney manages a salt and water balance.

Regardless of the mechanisms creating hypertension, chronic arterial hypertension is the result of elevated peripheral vascular resistance. For this reason, hypertension is considered a *vascular* disease involving the arterioles because these vessels are the major contributor to total peripheral vascular resistance. In hypertension, arteries have thickened vascular walls and narrowed lumens. This thickening increases the resistance of arteries, even in the absence of all vascular smooth muscle contraction, and acts as a structural, geometric amplifier of any vasoconstrictor stimuli. As a result, all arterial smooth muscle contraction, from maximum dilation to maximum contraction, results in exaggerated increases in arterial resistance in arteries from hypertensive, compared with normotensive, patients. In addition, the arterial smooth muscle cells in hypertensive patients are hypersensitive to vasoconstrictor stimuli and hyposensitive to vasodilators. This too, results in an exaggerated increase in vascular resistance with any vasoconstrictor stimuli and a reduced capacity of the arteries to counteract this exaggeration with any vasodilator stimuli. Lost production and bioavailability of arterial endothelial nitric oxide, which is a potent anticonstrictor agent, are also a consequence of hypertension and may be a contributing factor in the maintenance and inevitable progression of that disease.

streamline blood flow and lowest adjacent to the inner arterial wall. However, if total flow velocity becomes too high in an artery, the kinetic energy of the flow streams overcomes the tendency of the fluid layers to stick together from viscous forces. When this happens, the fluid layers break apart and become random and chaotic. This is a condition called **turbulence** (see Fig. 11.6C). Turbulence is a wasteful process that dissipates pressure energy in the cardiovascular system, which could otherwise be used to produce flow. The tendency to produce turbulent flow is expressed in a mathematical term called the **Reynolds number, Re**. The Reynolds number is a measure of the ratio of kinetic energy in the system (which will pull fluid layers apart) and the viscous component of the system (which holds the fluid layers together) as given by

$$Re = \rho Dv/\eta \qquad (9)$$

where ρ = fluid density, D = inner vessel diameter, v = flow velocity, and η = blood viscosity. Generally, a Re ≥ 2,300, indicates that turbulence will occur in the fluid stream. Clearly, large-diameter vessels, high flow velocity, and low blood viscosity favor turbulence in the cardiovascular system.

Whereas laminar flow in arteries is silent, turbulent flow creates sounds. These sounds are clinically known as **heart murmurs** or simply murmurs. Certain diseases, such as atherosclerosis and rheumatic fever, can scar the aortic or pulmonic valves, creating narrow openings and high flow velocities when blood is forced through them. A clinician can detect this problem by listening to the noise created by the resultant turbulence.

Blood viscosity is increased by red blood cells and low flow velocity.

The fluid dynamic properties of blood are more complicated than they are for a simple homogenous fluid such as water because blood is really a suspension of proteins and cells in an aqueous medium. The study of the fluid dynamic properties of blood is called **rheology**. The presence of proteins and cells in blood has two important hemodynamic consequences. First, blood viscosity increases exponentially as the blood hematocrit increases (Fig. 11.7A). Thus,

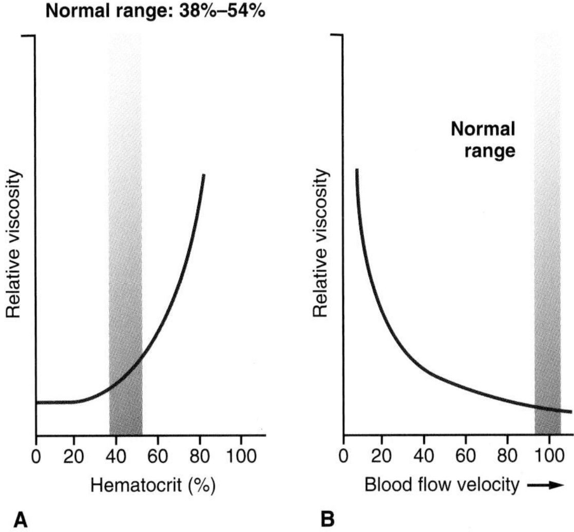

Figure 11.7 Effect of hematocrit and flow velocity on blood viscosity. (A) Increases in hematocrit above normal values produce a sharp increase in viscosity, causing marked increase in resistance to flow. **(B)** Blood viscosity in arteries increases dramatically whenever flow velocity slows to low levels. Thus, the resistance to flow is higher in a slow-moving arterial stream than in one that moves with high velocity.

as the hematocrit increases, the flow resistance against which the heart must pump increases, which substantially increases the cardiac workload. Although there are some pathologic instances in which a patient's hematocrit can increase because of overproduction of red blood cells (e.g., polycythemia vera), an abnormally increased hematocrit can come from the loss of blood plasma without a proportional loss of red blood cells. This can occur in severe dehydration, loss of plasma from severe burns, or inappropriate loss of water through the renal system as a consequence of kidney disease. Second, red blood cells tend to clump if blood flow velocity is sluggish (see Fig. 11.7B). This clumping raises blood viscosity, which can be a negative complicating factor in any condition that adversely diminishes the overall flow output of the heart, such as circulatory shock or heart failure.

Axial streaming of cells in flowing blood

Application of the Poiseuille law to fluid dynamics assumes that the fluid is homogeneous, that is, it is made of one element of uniform composition, such as water. Although all suspensions are nonhomogeneous fluids, the properties of blood cells flowing in arteries are such that blood behaves hemodynamically as if it were a homogeneous fluid. Blood cells tend to compact densely in the center of the flow stream, leaving a thin layer of cell-free plasma against the vascular wall. This is called **axial streaming**. As such, blood flows more as a compact bulk fluid than as a mixed conglomeration of particles in suspension. For this reason, the Poiseuille law can be applied to the cardiovascular system as it was written. However, axial streaming of blood does tend to separate this bulk flow into two components of different viscosities. The fluid near the vessel wall is essentially cell-free plasma and has a viscosity of only 1.7 cP, as opposed to 4 cP for whole blood. In large vessels such as

Clinical Focus / 11.2 

Hemodynamic Localization of Atherosclerosis

Atherosclerosis is a chronic inflammatory response of the walls of large arteries that is initiated by injury to the endothelium. The exact cause of this injury is unknown. However, exposure to high serum lipid levels in the form of low-density lipoprotein (LDL) cholesterol and triglycerides is considered a prime factor, with contributions from hypertension, the chemicals in cigarette smoke, viruses, toxins, and homocysteinemia. With injury, the normal endothelium barrier becomes compromised. Leukocytes, primarily monocytes, adhere and infiltrate the arterial intima. These cells and the intima itself accumulate large quantities of lipoproteins, mainly LDL, which, when oxidized, further damage the artery, stimulate the production of damaging reactive oxygen species, and set up a more aggressive local arterial inflammatory response. This results in mitotic activation of arterial smooth muscle cells, which migrate into the intima, and the activation of monocytes, which transform into macrophages that engulf lipid to become foam cells. These factors create an inwardly directed growth of the arterial wall, which encroaches on the arterial lumen and creates the appearance of fatty streaks on the inner arterial wall. Over time, this streak grows with the accumulation of extracellular matrix proteins, the development of a fibrous cap over the atheroma, and the development of a lipid-laden necrotic core containing debris, foam cells, crystallized cholesterol, and calcium deposits. This creates what is often called an *atherosclerotic plaque*.

Atherosclerotic plaques usually only develop in a portion of the circumference of the arterial wall. The wall opposite the plaque can actively contract in response to vasoconstrictor stimuli, whereas the wall beneath the plaque becomes weakened, creating an arterial aneurysm that can rupture. Atherosclerotic plaques stimulate platelet aggregation and blood clot formation that can totally occlude an artery. Furthermore, the plaque is friable and can rupture, spilling debris into the arterial lumen, which further stimulate clot formation. Plaque rupture in the arteries of the heart is a primary cause of death from heart attack.

Branch points and curvatures alter flow velocity and shear stress at the arterial wall; these are decreased along the inner curvature of a flow stream and at the upstream edge of branch points. It is well known that plaque formation in the vascular system does not occur either uniformly or totally randomly in the vascular tree. Instead, it develops at branch points and bifurcations and along the inner curvature of arteries. All these areas contain regions of low flow velocity and low shear stress. The anatomic characteristics of the coronary arteries, the bifurcation of the common carotid arteries, and the entry to the renal arteries make these regions especially susceptible to the accumulation of atherosclerotic plaques and thus place the blood supply to the heart, the brain, and the kidneys at risk.

the aorta, this low-viscosity layer is only a small percentage of the viscosity of the flow stream. Thus, for all practical purposes, the viscosity of the blood flowing through the entire aortic cross section can be considered to be 4 cP. However, in small arterioles (<300 μm interior diameter) and capillaries, this thin layer becomes a greater percentage of the total volume contained within the vessel and thus contributes a greater percentage to the total viscosity of the blood traveling through those vessels. When fluid flows through these smaller vessels, fluid viscosity, as a whole, decreases. This is called the *Fahraeus-Lindqvist effect*, and it is responsible for reducing blood viscosity, and therefore, flow resistance when blood flows through extremely small vessels such as capillaries. This makes it easier for blood to flow through vessels that otherwise have extremely high resistances.

DISTRIBUTION OF PRESSURE, FLOW, VELOCITY, AND BLOOD VOLUME

Meaningful insights into characteristics of the cardiovascular system can be obtained by examination of the distribution of flow, velocity, pressure, and volume within the system. For example, because veins are more compliant than arteries, one would expect that more of the total volume of blood in the cardiovascular system would reside in the venous rather than the arterial side of the circulation. This is precisely the case, as shown in Figure 11.8. Also, because cross-sectional area increases greatly from arteries to the arterioles and to the capillaries, the lowest blood flow velocity occurs through the capillary network (Fig. 11.9). This slow velocity through this exchange segment of the vascular system has the beneficial effect of allowing more time for the exchange of material between the cardiovascular system and the extracellular fluid.

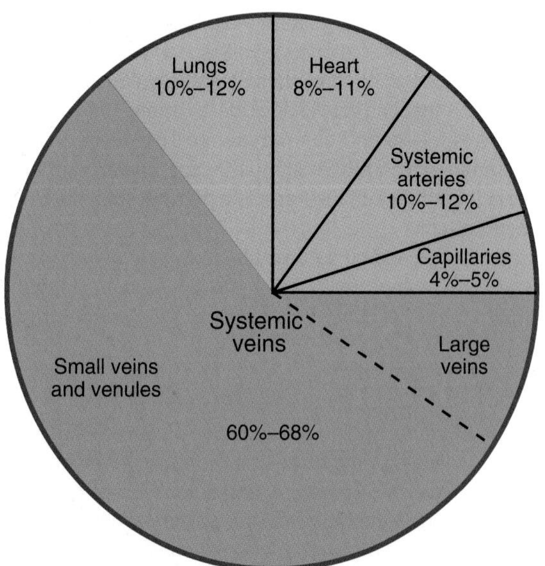

Figure 11.8 Blood volumes of various elements of the circulation in a person at rest. The majority of the blood volume is in systemic veins.

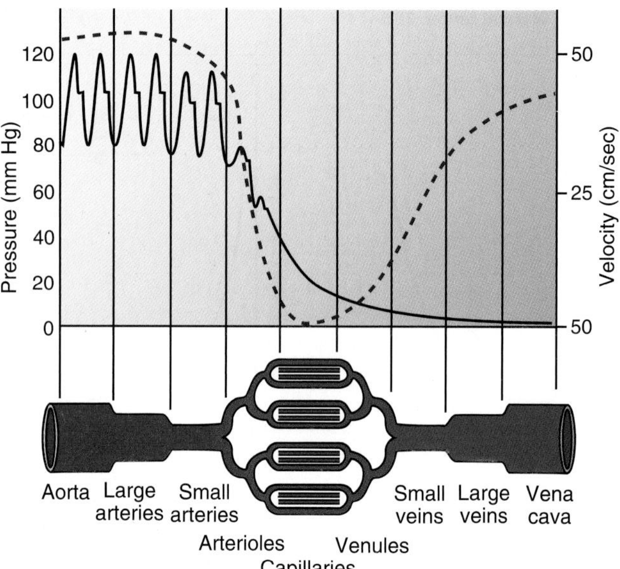

Figure 11.9 Pressure and flow velocity profile in the systemic circulation. The arterial portion of the circulation is characterized by high, pulsatile pressure and high flow velocity. This profile changes to one of low pressure and velocity without pulsatile character in the veins. The largest drop in mean arterial pressure occurs across the arteriolar segment of the circulation, indicating that this is the site of highest vascular resistance in the cardiovascular system.

The heart is an intermittent pump; it generates high pressure within the ventricles when it contracts during systole, which then drops to near zero during diastole. However, because arteries are compliant, some of the ejected blood into the arteries distends these vessels, like the expansion of a water-filled balloon. During diastole, recoil of the arteries pushes blood forward against the downstream vascular resistance, generating a significant diastolic pressure. For this reason, diastolic pressure drops to only about 80 mm Hg in the aorta as compared with near zero in the ventricles.

Examination of the pressure profile across the cardiovascular system (see Fig. 11.9) shows that the largest drop of pressure occurs across the arterioles, indicating that this is the site of the greatest vascular resistance in the cardiovascular system. Although there are many more arterioles than arteries in the cardiovascular system (resistances in parallel), this large pressure drop indicates that their reduction in individual size dominates over the addition of parallel vessels. Similarly, although individual capillaries are small, so many of these lie in parallel that resistance across the capillaries is actually lower than that across the arterioles; hence, the pressure drop across the capillary segment of the circulation is less than that across the arterioles.

Pressures within the arteries of the pulmonary circulation are not the same as those in the systemic circulation. Pulmonary arterial pressure is about 25 mm Hg during systole and 8 mm Hg during diastole. Because the outputs of the right and left heart are the same, the low pressure in the pulmonary circulation must indicate, according to the Poiseuille law, that vascular resistance is much lower in the pulmonary circulation than in all the organs combined that make up the systemic circulation.

From Bench to Bedside / 11.1

Emerging Noninvasive Imaging Techniques for the Evaluation of Atherogenic Pathology

Less than two decades ago, the primary means of detecting atherosclerosis in arteries was by dye contrast angiography. This technique, in which contrast dye was injected into the circulation and the coronary arteries imaged with x-ray imaging tools, was essentially nothing more than a *lumen*ography. What was produced by this technique was a simple two-dimensional image of the lumen of an artery and not the arterial wall that is the site of the atherosclerotic disease process. As such, coronary, carotid, or systemic arterial angiography could only show the stenosis caused by an atherosclerotic plaque and yielded no information about the diseased lesion. This simple imaging of the lumen in areas of atherosclerotic plaques appears inadequate in the context of modern understanding of the pathogenesis of atherosclerosis and the complexity of the atherosclerotic lesion.

Atherosclerosis is better understood as an inflammatory condition creating complex lesions characterized by necrotic tissue; abnormal smooth muscle cell growth; intimal thickening; variable lipid and calcium burdens within the lesion; and a fibrous cap structure that, upon rupture, can result in thrombosis-induced ischemia or infarction of the tissue perfused below the lesion. This latter condition, when it occurs in the coronary circulation, can result in sudden death. Rupture of unstable atherosclerotic plaques, for example, is known to be the underlying cause of sudden myocardial ischemia and heart attack. For these reasons, imaging techniques that could yield information about plaque composition, metabolism, and its evolving pathology could be of great utility in assessing lesions in younger, asymptomatic patients as well as help the physician manage risk factors and gauge the effectiveness of individual therapies used to treat atherosclerosis. More detailed analysis of plaque composition, plaque pathogenesis, and detection of inflammatory markers in regions of atherosclerotic lesions could further aid in assessing risk and designing treatments.

Currently, several noninvasive or moderately invasive imaging techniques are being used to measure intimal/medial thickening, lipid versus calcium composition, and inflammatory burden in atherosclerotic plaques. Magnetic resonance imaging (MRI) has proven to be a good tool in measuring intraplaque hemorrhage and thinning of the fibrous cap that overlays atherosclerotic lesion, which can be a prelude to plaque rupture. Experimental studies have also provided evidence that MRI may be especially useful in gauging the regression of plaques following treatment with lipid-lowering drugs. Multidetector and electron beam computed tomography (MDCT and EBCT) are proving to be excellent techniques for detecting high calcium burden in atherosclerotic lesions that are linked to increased risk of adverse cardiovascular events from atherosclerosis even if significant stenosis is not present. MDCT can also be used to differentiate between calcified and noncalcified plaque burden, the latter of which is sensitive to high cholesterol, complicating risk factors (i.e., diabetes mellitus), and lipid-lowering therapies. Whether MDCT can be used as a risk-screening test for young patients to help guide the level of subsequent therapies has yet to be evaluated fully.

Recently, MRI has been used in conjunction with targeted contrast agents to conduct "molecular" imaging of atherosclerotic lesions. Positron emission tomography (PET) scans have also been used in this manner to provide what can be considered *in vivo* immunohistochemistry of plaque cells and their components. PET scanning used with fludeoxyglucose is proving to be applicable to the analysis of inflammatory activity in the region of atherosclerotic lesions over time. This technique, more commonly used to detect organ metastases in cancer, has been shown to detect macrophage activity in the area of a lesion. Similarly, MRI used with inflammatory antigen–targeted contrast agents is being used to detect inflammatory activity in atherosclerotic plaques. Taken as a whole, new techniques designed to assess the severity of atherosclerosis as a disease, rather than just the geometric effect of atherosclerotic plaques, should prove to be of great benefit to clinicians, allowing them to match the aggressiveness of their treatment of atherosclerosis with the aggressiveness of the disease.

Chapter Summary

- The cardiovascular system is a fluid transport system that delivers substances to the tissues of the body while removing the byproducts of metabolism.
- The heart is composed of two pumps connected in series. The right heart pumps blood into the lungs. The left heart pumps blood through the rest of the body.
- Pressure is created within the atria and ventricles of the heart by contraction of the cardiac muscle. The directional opening of valves, which prevent back flow between chambers, ensures forward movement of blood through the heart.
- Arteries transport blood from the heart to the organs. Veins transport blood from the organs to the heart.
- Capillaries are the primary site of transport between blood and the extracellular fluid.
- Altering vascular smooth muscle contraction changes blood vessel radius.
- Vascular resistance is inversely proportional to the internal radius of the blood vessel.
- The volume contained within any vascular segment is a function of transmural pressure and the compliance of the vascular wall.
- Transmural pressure in blood vessels produces wall tension and stress that must be overcome for the vessel to contract.
- Blood flow and pressure throughout the vascular system is created in accordance with the principles of the Poiseuille law; flow is proportional to the pressure gradient between points in the circulation but inversely proportional to vascular resistance.
- The velocity of blood flow in an artery affects lateral pressure and inner wall shear stress in arteries as well as transport across the capillary wall. It is also a factor in the creation of turbulence in the arterial flow stream.
- The viscous nature of blood as a suspension affects flow resistance in arteries and is influenced by hematocrit and blood flow velocity.
- The hemodynamic profile in the cardiovascular system is the result of the combined effects of all the relationships and laws governing the containment and movement of blood in the cardiovascular system.

thePoint *Visit http://thePoint.lww.com for chapter review Q&A, case studies, animations, and more!*

12 Electrical Activity of the Heart

ACTIVE LEARNING OBJECTIVES

Upon mastering the material in this chapter, you should be able to:

- Contrast the mode of electrical activation of cardiac muscle with that of skeletal muscle and how those modes relate to different mechanical properties in the two muscles.
- Explain how changes in voltage-gated membrane channels for sodium, potassium, and calcium create the five phases of atrial and ventricular muscle action potentials.
- Explain the electrical characteristics of cardiac muscle that prevent the muscle from becoming tetanized.
- Predict how changes in ventricular conductances or plasma concentrations of sodium, calcium, and potassium will change the amplitude, duration, and refractory period of action potentials in ventricular muscle cells; predict how this will either enhance or inhibit the formation of ectopic foci in the myocardium.
- Explain how both hyperkalemia and hypokalemia can prevent the formation of action potentials in ventricular muscle cells.
- Relate the changes in membrane conductances for sodium, calcium, and potassium in cardiac nodal tissue to their automaticity.
- Predict how a change in either resting membrane potential or the rate of decay of K^+ conductance in the sinoatrial node affects its rhythmicity.
- Explain the mechanism for the effects of acetylcholine and norepinephrine on sinoatrial node rhythmicity and how that relates to the effect of these neurotransmitters on heart rate.
- Predict ways in which the conduction of action potentials can be slowed through the myocardium.
- Explain how electrocardiogram waveforms are determined by electrical events in the myocardium.
- Describe how the frontal and horizontal electrocardiogram lead systems can be used to determine the orientation of the heart, the direction of electrical activation of the heart, and changes in the muscle mass of atria and ventricles.
- Use electrocardiogram recordings to identify atrial and ventricular tachycardias, fibrillation, premature atrial contractions, premature ventricular contractions, and heart blocks.
- Use QRS vector analysis of electrocardiogram recordings to identify atrial and ventricular hypertrophy as well as abnormal myocardial conduction pathways.
- Identify myocardial ischemia, injury, infarction, and plasma electrolyte disturbances from electrocardiogram recordings.
- Predict changes in electrocardiograms caused by various cardiac drugs on the basis of their effect on ion conductances in the myocardium.

Although the heart is composed of striated muscle, it only shares a few functional similarities with skeletal muscle tissue. The basic mechanical molecular mechanism responsible for contraction (i.e., cyclic interaction of actin and myosin) is similar in cardiac and skeletal muscle. The links between action potential generation, calcium, and the initiation of contraction are similar on a general level, as are principles of muscle mechanics related to the effects of preload and afterload. However, beyond these few shared characteristics, the physiology of cardiac muscle is very different from that of skeletal muscle.

The activation of skeletal muscle is much like that of a table lamp with an electric cord that is plugged into an electric outlet. Skeletal muscle cannot contract without being connected to a motor neuron, which is analogous to a lamp that cannot be turned on if its cord is unplugged. Also, like the light bulb in the lamp, skeletal muscle fibers are either switched on (activated) or off (not activated). Unlike skeletal muscle, cardiac muscle does not require innervation to be activated. In addition, the intrinsic force generation capacity of each cardiac cell can be enhanced or depressed, much like a light on a dimmer switch rather than an on/off switch. In this and the following chapter, the unique electrical and mechanical properties of cardiac muscle are addressed. This chapter focuses on the electrophysiology of cardiac muscle, whereas the mechanical properties of this muscle tissue are examined in Chapter 13, "Cardiac Muscle Mechanics and the Cardiac Pump."

ELECTROPHYSIOLOGY OF CARDIAC MUSCLE

The heart is composed of muscle where the contraction is coupled to the generation of action potentials within its cells. However, cardiac muscle does not require action potentials from nerves to activate its own electrical activity (a fact that makes heart transplantation operations possible). Although both branches of the autonomic nervous system (ANS) innervate the heart, the ANS modulates cardiac function rather than initiates it. Furthermore, there are no anatomic or functional correlates of neuromuscular motor units in the heart. Therefore, the heart cannot recruit neuromuscular units to enhance its force-generating capacity.

Cardiac cells are electrically connected and can generate their own action potentials.

All myocardial cells are coupled electrically through gap junctions at points called *nexi* (Fig. 12.1). This allows the generation of an action potential in one myocardial cell to spread rapidly to all cells in the heart. This means that, electrically, the heart behaves as a **functional syncytium**, or as if it was one large cell. The advantage of this electrical connectivity between all cells is that it helps the heart contract as a large, coordinated mechanical unit for the purpose of pumping blood. The heart could not function as a pump if its millions of cells activated randomly. However, the syncytial character of the myocardium also means that the contractile force of the heart as a whole cannot be modulated by the recruitment of motor units, as occurs in skeletal muscle groups. During systole, all cardiac cells are activated; there are no cells left to recruit. In addition, the electrical connectivity of all myocardial cells means that the activation of any cell in the heart can inadvertently activate the heart as a whole, as is discussed later in this chapter.

Cardiac cells possess the unique property of **automaticity**; that is, they have the ability to generate their own action potentials without the need for chemical or electrical stimuli from other sources. Furthermore, cardiac cells display the property of **rhythmicity**, or the ability to generate these potentials in a regular, repetitive manner. In the normal heart, only specialized myocardial cells express automaticity and rhythmicity continuously. Unfortunately, however, in pathologic conditions, automaticity can occur in atrial and ventricular myocytes as well as in the **Purkinje fibers**.

Gated myocardial calcium and potassium channels create unique features of cardiac action potentials.

The appearance of cardiac action potential is significantly different from the rapid spike characteristics seen in neurons or skeletal muscle fibers. There are two broad types of cardiac action potentials: those characteristic of ventricular

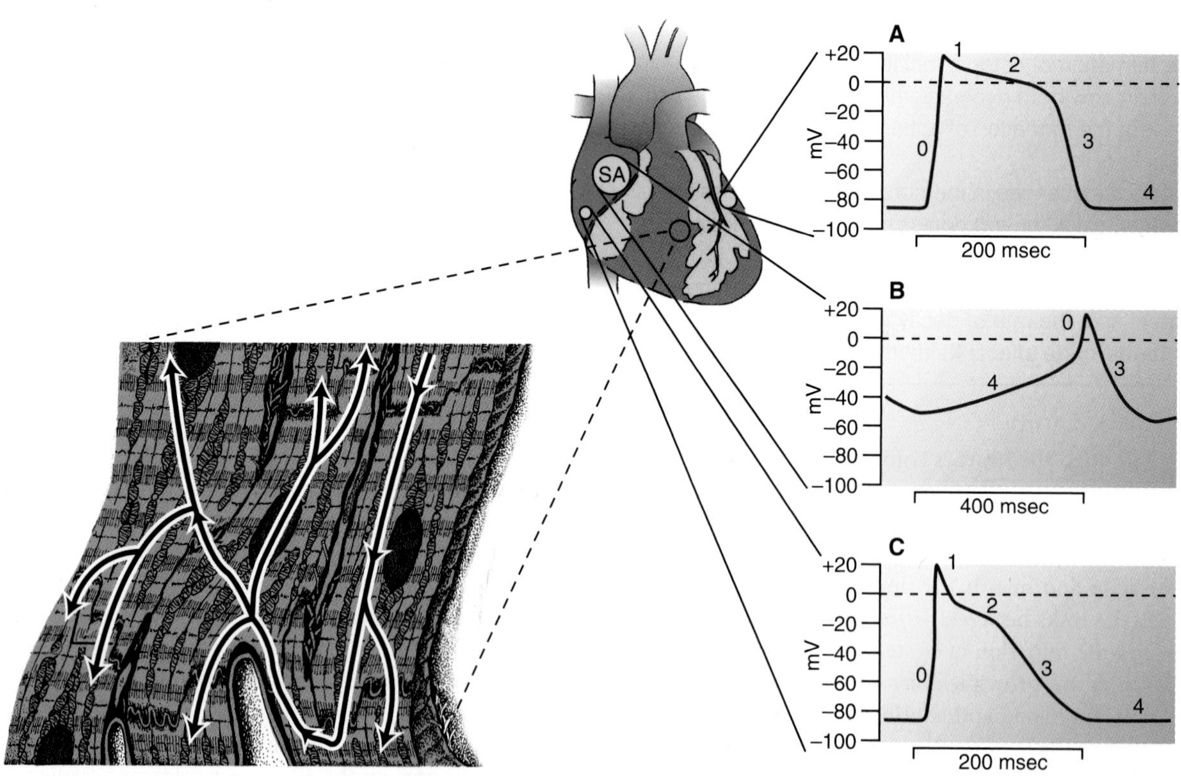

Figure 12.1 Cardiac action potentials (mV) recorded from ventricular (A), sinoatrial (B), and atrial cells (C). Note the difference in the time scale of the sinoatrial cell. Numbers 0 to 4 refer to the phases of the action potential (see text). *Inset*: Diagrammatic representation of the nature of the functional syncytium in the myocardium. The colored line represents the path of excitation through the myocardium as it passes from cell to cell through gap junctions at cardiac nexi. SA, sinoatrial.

and atrial muscle as well as of Purkinje fibers are called "fast response" action potentials, whereas those observed in the **sinoatrial (SA) node** and **atrioventricular (AV) node** are called "slow response" action potentials (see Fig. 12.1). The fast response is divided into five phases (Fig. 12.2). The initial rapid depolarization of the cell membrane is designated phase 0. Phase 1 represents the subsequent partial repolarization of the membrane, which is followed by phase 2. Phase 2 is unique to cardiac muscle and is often called the **plateau** region of the action potential. Phase 3 is the rapid repolarization phase of the action potential, and phase 4 is the resting membrane potential.

Changes in plasma potassium concentration markedly alter the myocardial resting membrane potential.

Many ion channels are involved in the overall makeup of the fast response (Table 12.1; also see Fig. 12.2). The resting membrane potential is primarily a K^+ diffusion potential, and thus, it is sensitive to changes in external K^+ concentration. High plasma K^+ concentrations (hyperkalemia) depolarize cardiac cells, whereas hypokalemia hyperpolarizes the tissue. Both of these conditions can adversely affect cardiac function, and for this reason, plasma K^+ levels are monitored carefully in a clinical setting. There is a small component of Na^+ influx to the cardiac resting membrane potential, making the resting membrane potential slightly more positive than the K^+ Nernst potential.

Cardiac cells have an intrinsic resting membrane potential buffer system that attenuates changes in membrane potential caused by changes in external K^+ concentration. Potassium conductance in myocardial cells is altered by the extracellular concentration of potassium surrounding the cells. Over the physiologic extremes of 2 to 7 mM, K^+ conductance increases when external K^+ concentration increases and decreases when external K^+ concentration decreases. Thus, the depolarizing effect of elevated external K^+ concentration is somewhat buffered by an increased K^+ efflux, allowed by an increase in K^+ conductance. However, this system cannot totally counteract the effect of external potassium on cardiac resting membrane potential. Hyperkalemia will always depolarize the cell at rest; it is simply that the magnitude of this depolarization is less than that predicted by the Nernst equation because of the effect of external K^+ on potassium conductance in cardiac myocytes.

Voltage-gated sodium and potassium channels initiate and terminate phase 0 of the fast response.

The rapid upstroke in phase 0 occurs by a mechanism similar to that seen in nerve or skeletal muscle. Membrane depolarization up to and more positive than approximately −55 mV opens voltage-sensitive Na^+ channels, allowing a rapid influx of Na^+ down its steep electrochemical gradient, which further depolarizes the membrane, opening more Na^+ channels, and so on. This causes a rapid, self-reinforcing depolarization of the cell. The cardiac Na^+ channel is self-limiting in that the same depolarization that causes conformational changes in

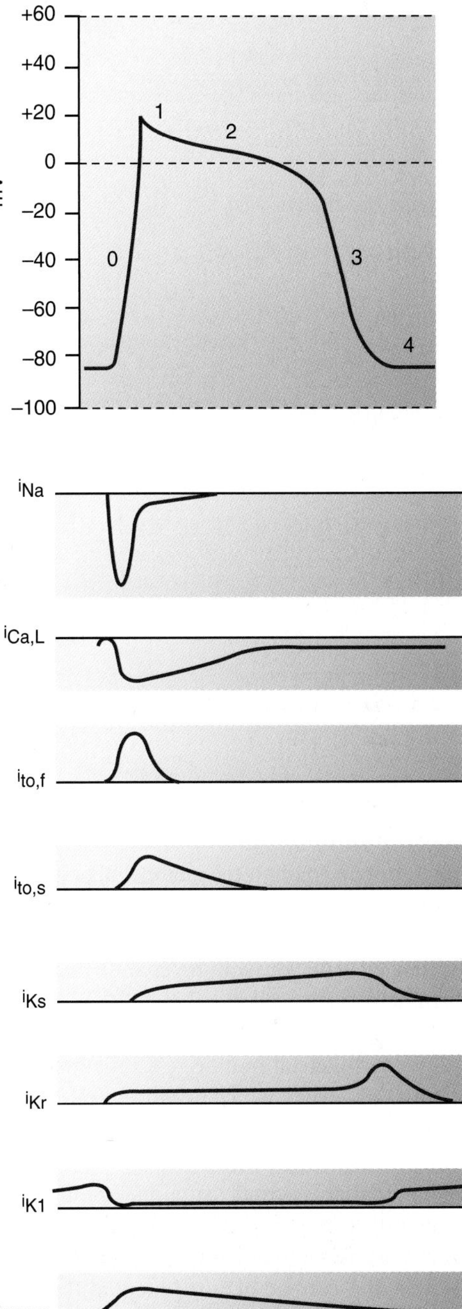

Figure 12.2 Changes in cationic currents responsible for the formation of the fast response action potential in ventricular muscle cells. I = current flow; downward deflections indicate current flow into the cell, whereas upward deflections indicate current flow out of the cell. The rise in action potential (phase 0) is caused by rapidly increasing Na^+ current carried by voltage-gated Na^+ channels. Na^+ current falls rapidly because voltage-gated Na^+ channels are inactivated by depolarization. K^+ current rises briefly, causing phase 1, because of the opening of transient outward K^+ channels, and then falls precipitously, because I_{K1} channels are closed by depolarization. Ca^{2+} channels are opened by depolarization and are responsible, along with closed I_{K1} channels, for phase 2. K^+ current begins to increase because of the delayed opening of I_{Kr} and I_{Ks} channels by depolarization. Once repolarization occurs, Na^+ and Ca^{2+} channels are returned to their resting state. Repolarization reopens I_{K1} channels and reestablishes phase 4.

TABLE 12.1 Major Ion Channels Involved in Purkinje and Ventricular Myocyte Membrane Potentials

Name	Voltage (V) Gated or Ligand (L) Gated	Functional Role
Voltage-gated Na^+ channel (fast, I_{Na})	V	Phase 0 of action potential (permits influx of Na^+)
Voltage-gated Ca^{2+} channel (slow, I_{Ca})	V	Contributes to phase 2 of action potential (permits influx of Ca^{2+} when membrane is depolarized); β-adrenergic agents increase the probability of channel opening and raise Ca^{2+} influx; acetylcholine (ACh) lowers the probability of channel opening
Inward rectifying K^+ channel (I_{K1})	V	Maintains resting membrane potential (phase 4) by permitting outflux of K^+ at highly negative membrane potentials
Outward (transient) rectifying K^+ channel (I_{to})	V	Contributes briefly to phase 1 by transiently permitting outflow of K^+ at positive membrane potentials
Outward (delayed) rectifying K^+ channels (i_{Kr}, i_{Ks})	V	Cause phase 3 of action potential by permitting outflow of K^+ after a delay when membrane depolarizes; I_{Kr} channel is also called HERG channel
G protein–activated K^+ channel ($i_{K.G}$, $i_{K.ACh}$, $i_{K.ado}$)	L	G protein–operated channel, opened by ACh and adenosine (ado); this channel hyperpolarizes membrane during phase 4 and shortens phase 2

the channel protein to open the channel also inactivates the channel several milliseconds later. At the peak of the cardiac action potential, more than 99% of the sodium channels are in the inactivated state. The sodium channels cannot be reset to open again unless the cell membrane becomes repolarized below −50 mV.

In nerve and skeletal muscle, a K^+ channel opens soon after the initial Na^+ event, which produces rapid cell repolarization. This does not occur in cardiac cells. In cardiac cells, the K^+ conductance (I_{K1}) largely responsible for the resting membrane potential is actually suppressed during depolarization, thus impairing any potential rapid repolarization of the tissue. There is some enhanced K^+ conductance at this time point in the cardiac action potential, but it is short lived, owing to the opening and then closing of two potassium channels that allow only K^+ to exit the cell. These channels are called **transient outward (TO) K^+ channels** and are designated $I_{to,fast}$ and $I_{to,slow}$, owing to slightly different channel time kinetics. $I_{to,fast}$ and $I_{to,slow}$ are responsible for phase 1 of the cardiac action potential. Coupled with the inactivation of the sodium channels in the membrane, the opening of these transient potassium channels causes the small, temporary repolarization of the cardiac cell that is designated as phase 1.

Cardiac cell refractory period is prolonged by opening of slow voltage-gated Ca^{2+} channels.

All cardiac cells contain **L-type Ca^{2+} channels**. These channels open and close with depolarization in a manner analogous to that seen for the Na^+ channel, except their opening and closing kinetics are much slower. They also open at slightly less negative potentials than do the fast sodium channels (approximately −40 to −50 mV). The electrochemical gradient for Ca^{2+} is enormous in cardiac cells, and the conductance of these channels for calcium is high. Once the calcium channels open, positive charge from Ca^{2+} rushes into the cell. During this phase of the action potential, hyperpolarizing effects of potassium efflux are diminished because of the inactivation of I_{K1} channels by membrane depolarization. In phase 2, the influx of positive current from calcium approximately matches the remaining positive efflux carried by K^+ exiting through a few open K^+ channels in the membrane. This balance causes the membrane potential to remain relatively constant at a positive value and helps create phase 2 of the action potential. Eventually, the Ca^+ channels close, preventing further depolarizing currents from entering the cell, which aids in the eventual repolarization of the cell membrane. The plateau phase of the ventricular action potential has a significant functional effect on cardiac muscle activation. Because of the plateau phase of the cardiac action potential, cardiac cells have a long refractory period, and a single contraction of cardiac muscle is completed before a second action potential can be generated. Thus, *cardiac cells cannot be tetanized.* This is a fortunate consequence in that the tetany of cardiac muscle would obviously not be compatible with the function of the heart as a blood pump.

Repolarization of cardiac muscle cells involves activation of K^+ channels.

A class of channels called **outwardly rectifying potassium channels** opens during phase 2 and beyond, with the cardiac muscle action potential. These currents, called outwardly rectifying currents because they counteract membrane depolarization, bring about a rapid repolarization of the cell membrane. I_{Ks} and I_{Kr}, which stand for slow and rapid outwardly rectifying potassium currents, are principally responsible for phase 3 of the cardiac action potential, although I_{K1} is responsible for the final reestablishment of the resting membrane potential. I_{Ks} and I_{Kr} channels are opened by depolarization and closed by membrane hyperpolarization starting at about −55 mV. The activation of I_{K1} late in phase 3 is responsible for finishing the repolarization of the cell membrane and the reestablishment of the resting membrane potential. It is believed that increased intracellular calcium concentration from calcium influx during phase 2 may activate I_{K1}.

Myocardial action potentials are significantly altered by changes in ion conductance and concentration.

A summary of the ionic events and channels responsible for cardiac muscle action potentials is shown in Figure 12.3. As can be surmised from this chart and the previous discussion, the size, shape, and refractory period of cardiac action potentials could be significantly influenced by factors that alter channel kinetics or extracellular and intracellular concentrations of Na^+, Ca^{2+}, or K^+. As is seen later in this chapter, these changes can significantly affect the excitability of cardiac cells as well as the propagation of action potentials through the myocardium. Some changes can cause unwanted activation of myocardial cells and abnormal heart rhythms.

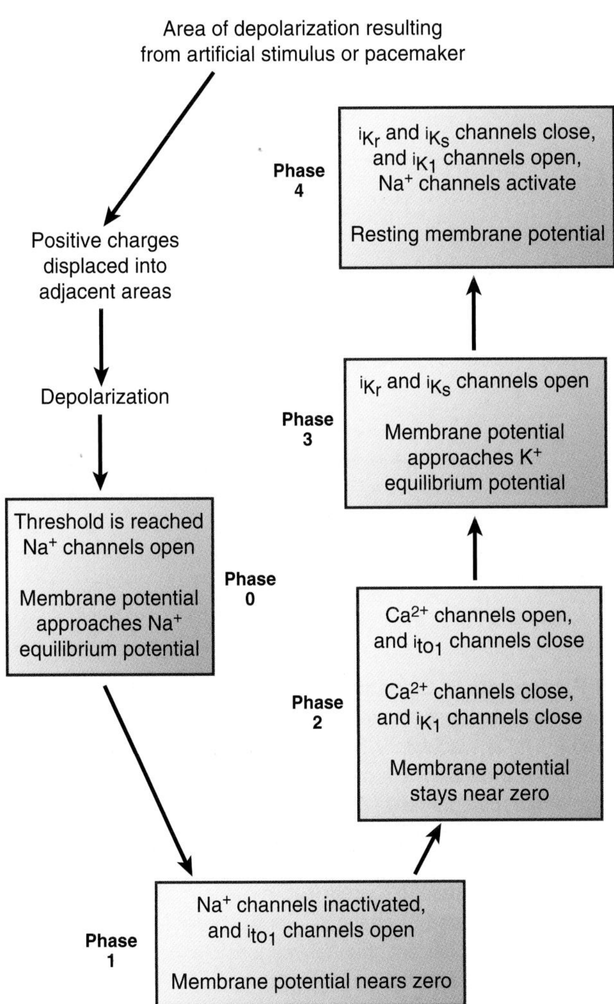

Figure 12.3 Events associated with the ventricular action potential. (See Table 12.1 and text for details.)

Catecholamines (epinephrine and norepinephrine) enhance Ca^{2+} movement and increase the size of the cardiac action potential. A class of drugs called calcium channel blockers, often used to treat abnormal cardiac rhythms and high blood pressure, depress Ca^{2+} movement and result in small action potentials. Norepinephrine increases phase 3 K^+ conductance, thereby shortening the cardiac refractory period. Hypokalemia and certain drugs used to treat abnormal cardiac rhythms have the opposite effect. Distention of the heart chambers, as in heart failure, or reducing the O_2 available to the heart (myocardial ischemia) partially depolarizes the resting membrane potential, bringing cardiac cells closer to their action potential threshold and making it easy for even weak nonphysiologic stimuli to activate the myocardium.

Heart muscle cells also contain a group of non–voltage-gated K^+ channels called **inwardly rectifying (IR) K^+ channels**, because they pass the inward current of K^+ more easily than the outward currents. However, this is simply an electrophysiologic classification in that there are never any inwardly directed potassium currents during the cardiac action potential (the cell membrane never reaches the potassium reversal potential of −90 to −100 mV). It is the outward current through these channels that is physiologically important. $I_{K,\,ATP}$ potassium channels are inactivated by intracellular adenosine triphosphate (ATP) and activated by adenosine diphosphate. They are thought to be one link between cardiac metabolism and membrane potential. These channels may be responsible for the reduced refractory period seen during **myocardial ischemia**, which is caused when oxygen demand in cardiac muscle exceeds its oxygen supply. The reduced refractory periods along with increased myocardial excitability increase the probability that the heart will experience abnormal heart rhythms or **arrhythmias**. Other IR channels, such as $I_{K,\,ACh}$ and $I_{K,\,ado}$ (ACh, acetylcholine; ado, adenosine), are ligand-gated potassium channels that hyperpolarize the ventricular muscle cell. These channels may mediate cholinergic-mediated atrial arrhythmias (abnormally slow heart rates) as well as certain antiarrhythmic effects of adenosine.

Sinoatrial node initiates action potentials in the heart.

For the heart to function as an efficient pump, action potentials and subsequent myocardial contraction must be generated and spread through the myocardium in a regular, repetitive, organized manner. This will not occur if cardiac cells express their automaticity in a random, unpredictable fashion. Normally, prior to each contraction of the heart, cardiac electrical activity is initiated by a modified set of muscle cells on the posterior aspect of the right atrium at the junction of the superior and inferior vena cava, called the sinoatrial

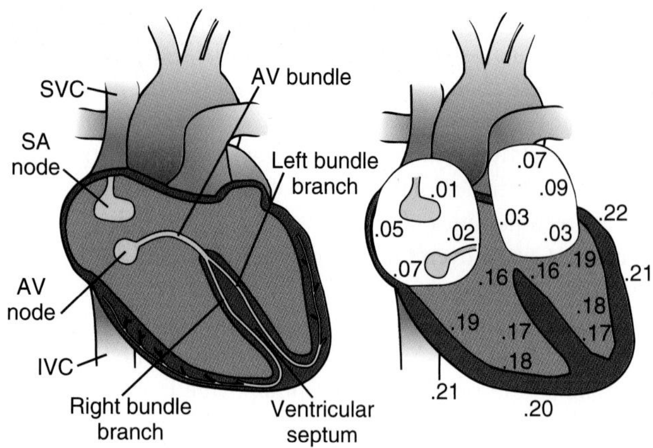

Figure 12.4 The timing of the excitation of various areas of the heart (in fractions of a second). AV, atrioventricular; SA, sinoatrial; SVC, superior vena cava; IVC, inferior vena cava.

or SA node (Fig. 12.4). Once the SA node initiates an action potential, it travels through both atria at a rate of 0.1 to 1.0 m/s and coalesces at a second area of specialized conduction tissue called the atrioventricular or AV node. This node lies at the junction between the atria and ventricles in the ventricular septum. Conduction through the AV node is slow (~0.05 m/s), which delays the movement of the cardiac action potential into the ventricles. This delay has the important effect of allowing more time for the ventricles to fill during diastole. The conduction of action potentials through the AV node shows directional preference; that is, action potentials travel more easily from the atria through the AV node toward the ventricles than in the opposite direction. The AV node is the only normal pathway by which action potentials from the atria travel into the ventricles because the connective tissue between the atria and ventricles act as an electrical insulator. Once the action potential emerges from the AV node, it enters the **bundle of His**, which splits into left and right bundle branches; these branches, in turn, give rise to Purkinje fibers, which line the entire endocardial surface of both ventricles. The ventricle itself is then activated in sequence from the septum/papillary muscle and endocardium to the epicardium and from the apex to the base of the heart. Action potentials travel so fast through Purkinje fibers (~4 m/s) that the underlying ventricular muscle is activated almost essentially simultaneously. Coupled with the syncytial nature of ventricular muscle cells, this ensures that the ventricular cells contract essentially en masse once activated, which allows for the effective pumping of blood out of the ventricles.

Unique recycling changes in ion conductances create automaticity and rhythmicity in cardiac nodal tissue.

Action potentials generated in the SA and AV nodes are smaller than those seen in cardiac muscle (Fig. 12.5). Nodal tissue does not contain fast voltage-gated Na^+ channels (Table 12.2). The action potential is carried entirely by slow L-type voltage-gated Ca^{2+} channels. Although these channels can carry small amounts of sodium current, the channel is conductive predominantly for calcium. The slow opening and closing of these channels with membrane depolarization creates a phase 0 that is slower compared with that in ventricular cells. In addition, nodal action potentials start from more positive resting membrane potential (approximately −65 mV), show no plateau, and exhibit a slow phase 3 as compared with that in cardiac muscle cells.

The most unique characteristic of action potentials in the SA and AV nodes is the spontaneous, progressive, and recycling depolarization that occurs in phase 4. This forms the basis for automaticity and rhythmicity in the SA node and cannot be considered a true "resting" potential. Shortly after phase 3, these cells experience a slight increase in Na^+ conductance (gNa^+_i), which results in a small inward depolarizing sodium current called $I_{Na,f}$, or "funny" sodium current. The depolarizing effects

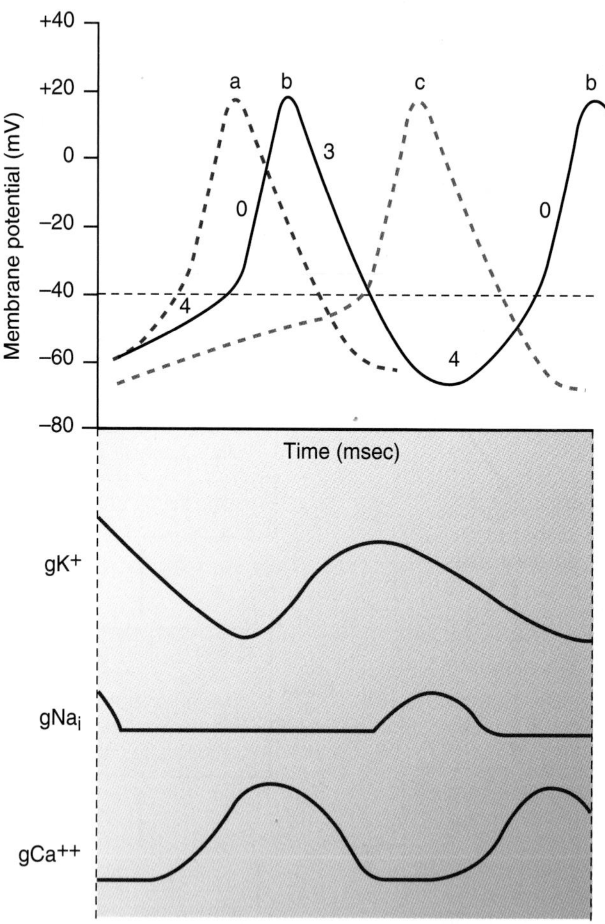

Figure 12.5 Sinoatrial (SA) nodal potential as a function of time. Normal pacemaker potential (b) is affected by norepinephrine (a) and acetylcholine (c). The dashed line indicates threshold potential. The more rapidly rising phase 4 in the presence of norepinephrine (a) results from enhanced Na^+ permeability. The hyperpolarization and slower rise in phase 4 in the presence of acetylcholine (ACh) results from decreased Na^+ permeability and increased K^+ permeability, as a result of the opening of ACh-activated K^+ channels. For the purpose of illustration only, the cationic current changes responsible for the formation of the SA nodal potential are aligned temporally with the normal pacemaker potential in the figure. The pacemaker potential reaches threshold primarily through a slow decay of potassium conductance (gK^+) in the SA node. g, ionic conductance.

TABLE 12.2 Major Channels Involved in Nodal Membrane Potentials

Name	Voltage (V) Gated or Ligand (L) Gated	Functional Role
Voltage-gated Ca^{2+} channel (slow, I_{Ca})	V	Phase 0 of sinoatrial and atrioventricular nodal action potential; carries influx of Ca^{2+} when membrane is depolarized; contributes to the early pacemaker potential of nodal cells; β-adrenergic agents increase the probability of channel opening and raise Ca^{2+} influx; ACh lowers the probability of channel opening
Voltage-gated Ca^{2+} channel (transient, I_{Ca})	V	Contributes to the pacemaker potential
Mixed cation channel (funny, I_f)	V	Carries Na^+ (mostly) and K^+ inward when activated by hyperpolarization; contributes to pacemaker potential
K^+ channel (delayed outward rectifier, I_K)	V	Contributes to phase 3 of action potential; closing early in phase 4 contributes to pacemaker potential
G protein–activated K^+ channel ($I_{K.G}$, $I_{K.ACh}$, $I_{K.ado}$)	L	G protein–operated channel, opened by ACh and adenosine (ado); this channel hyperpolarizes the membrane during phase 4, slowing pacemaker potential

of this current tend to halt and reverse further hyperpolarization of the cell membrane (see Fig. 12.5). The sodium channel mediating this effect belongs to a class of cyclic nucleotide–gated channels called *hyperpolarization-activated cyclic nucleotide–gated* (more commonly known as *HCN*) *channels*. Defects in this channel have been associated with abnormal hyperpolarization of nodal tissue, which results in an abnormally slow or irregular heart rate. The $I_{Na,f}$ current, however, is not responsible by itself for the automaticity of nodal tissue. Throughout phase 4 of nodal action potentials, there is a progressive reduction in K^+ conductance. This tends to progressively reduce K^+ efflux from the cell, allowing it to slowly depolarize over time. This depolarization eventually begins to activate slow Ca^+ channels, and the membrane becomes depolarized in a reinforcing manner that commonly results in an action potential. Depolarization later inactivates the Ca^{2+} channels, and the cell repolarizes as a result of K^+ efflux. The decay of gK^+ in phase 4 then occurs again, thus repeating the action potential cycle. The cycling decay of gK^+ during phase 4 of the nodal action potential is the primary source of the automaticity of the SA node and its resulting rhythm.

The SA node possesses the highest intrinsic rate of spontaneous action potential generation of the specialized conduction tissues. With a normal, intact ANS, its rate is about 80 impulses/min. This is faster than the intrinsic rates of the AV node (~50 impulses/min) or the Purkinje fibers (<20 impulses/min). Consequently, the SA node continually activates the heart before any other cardiac tissue can generate its own action potential. For this reason, the SA node is often called the heart's "pacemaker," because its rate of firing determines the **heart rate** or number of contractions per minute of the heart as a whole.

Both branches of the ANS affect phase 4 of the action potentials in the nodal tissue. ACh from the vagus nerve, which innervates the SA node, slows the heart rate. ACh hyperpolarizes the resting membrane potential and decreases the slope in phase 4; thus, it takes longer for the cells to spontaneously reach a threshold and fewer action potentials are generated in a given period of time. Conversely, norepinephrine or activation of the sympathetic nervous system increases the slope in phase 4 of the SA node and hence increases the heart rate (see Fig. 12.5).

Although the AV node can generate its own repetitive action potentials, it does not do so unless the intrinsic activity of the SA node is severely suppressed. This can happen with excess activation of the vagus nerve to the SA node, administration of drugs that increase ACh at the SA node, or in certain disease states. When this occurs, the heart does not stop but instead becomes paced by the AV node. In this situation, the AV node is considered to be a **secondary pacemaker**. Although the heart beats rhythmically when paced by the AV node, it does so at a substantially reduced rate compared with that seen with the SA node as the heart's pacemaker. In extreme conditions, when both the SA and AV nodes are not functional, the Purkinje fibers will pace the heart. However, the intrinsic Purkinje pacing is so slow that the flow output of the heart is insufficient for little more than keeping a person alive; normal activity is not possible.

Action potential conduction velocity through the myocardium is proportional to the amplitude and phase 0 upstroke of the cardiac action potential.

The conduction of action potentials through myocardial cells is affected by the characteristics of the action potentials themselves. Conduction velocity is increased when the amplitude of the action potential is increased. This increase in amplitude may occur as a result of factors that enhance Na^+ and Ca^+ influx into the cell. Alternately, it can also occur by hyperpolarization of the resting membrane potential, which results in

From Bench to Bedside / 12.1

Assessment of Sudden Cardiac Death Risk in Young Athletes

Athletes are commonly viewed as individuals with exceptional physical health. However, each year there are reports of young athletes, without any overt symptoms of any disease, who collapse and die unexpectedly on the field or court during practices or exercises associated with their sport. Although dehydration from heavy exercise in hot conditions can play a role in circulatory collapse and possible death, many of these athletes die instead from what is called *sudden cardiac death (SCD)*. SCD is characteristic of not only athletes but can also occur in any individual with cardiovascular disease. There are two major causes of SCD: a sudden thrombosis of a coronary artery or sudden cardiac arrhythmias. The latter often takes the form of ventricular tachyarrhythmias that lead to ventricular fibrillation and death.

SCD from a coronary thrombosis is most likely to occur in individuals with coronary artery disease and a high atherosclerotic plaque burden. This condition is not common in young fit individuals and, therefore, not a likely cause of SCD in young athletes. SCD in athletes appears to result more from sudden abnormal ventricular rhythms. In this regard, athletes do not exhibit the formation of supraventricular or atrioventricular ectopic beats at a higher frequency than do normal individuals. Athletes do exhibit sinus bradycardia from good aerobic conditioning, and this can sometimes even cause temporary second-degree heart block. However, this condition is often asymptomatic and resolves with exercise.

It is now recognized that undiagnosed hypertrophic cardiomyopathy (HCM), a type of congenital heart disease, is the leading cause of SCD in young athletes regardless of race. Almost 40% of SCD in athletes under age 30 years is associated with HCM. Myocardial hypertrophy in general renders the heart prone to ischemia and arrhythmias, especially tachyarrhythmias from triggered activity. In HCM, it appears that disordered myocardial tissue resulting from genetic mutations that generate the erratic alignment of muscle fibers make the heart prone to abnormal rhythms. Unfortunately for young athletes, extensive cardiovascular screening prior to participation in sports is rarely undertaken.

Recently, newly proposed guidelines have been recommended to aid the physician in identifying the risk for SCD in otherwise healthy young individuals planning on engaging in demanding athletic activities. Unexplained syncope, incidence of nonsustained ventricular tachycardias, and a family history of SCD are considered prime indicators of high risk for SCD in young athletes. A positive history warrants further cardiac evaluation. In this regard, the presence of left ventricular outflow obstruction, left ventricular wall thickening, scarring, and evidence of ischemia or heart murmur are considered evidence of HCM and risk of SCD. However, standard ECG screening of athletes as a means of preventing SCD is considered controversial at this time due to poor sensitivity of the technique and a high incidence of false positives.

a larger amplitude action potential once the cell is activated. Hypokalemia and increased K^+ conductance are two such mechanisms that can lead to hyperpolarization of myocardial muscle cells. Conduction of action potentials through the myocardium is also increased whenever the rate of depolarization in phase 0 is increased (increased dV/dt). This increase in depolarization will happen with any factor that increases the electrochemical gradient for Na^+ in phase 0, such as cell membrane hyperpolarization or increased extracellular to intracellular ratios of Na^+ concentration. Partial depolarization of the cell membrane at rest will reduce the size of subsequent cardiac muscle action potentials, which in turn will slow the conduction of these action potentials from cell to cell through the myocardium. The most common manner in which this partial depolarization occurs is following the inhibition of Na^+/K^+ pump activity in the myocytes during **myocardial ischemia** (insufficient oxygen to the cells) or in response to cardiac glycosides such as digitalis. If myocardial cells become depolarized to the extent that their fast sodium channels cannot reset from the inactive state, then these cells will generate action potentials from the activation of the slow Ca^{2+} channels in the membrane instead. However, these action potentials will exhibit a slow rate of depolarization in phase 0 and small amplitude. As such, they will be conducted slowly through the myocardial muscle tissue. Slow conduction of action potentials through the myocardium predisposes ventricular muscle to a specific kind of dangerous arrhythmia, called **reentry tachycardia** (see Clinical Focus box).

Conduction specifically through the AV node is affected by normal physiologic phenomena. ACh from the vagus nerve decreases, whereas norepinephrine from sympathetic nerves enhances conduction velocity through the AV node. This helps the activation of the ventricles to match the pace set by the rate in the SA node. In addition, conduction through the AV node is sensitive to repetitive stimulation. Continuous stimulation at high rates results in an increased refractoriness of the nodal tissue. This helps prevent the ventricles from being driven at abnormally high rates, which can impinge on ventricular diastolic filling and hence the output of the whole heart.

Abnormal pacemaker sites can appear anywhere in the heart in response to pathologic conditions.

Certain pathologic conditions, such as scar tissue formation after **myocardial infarction** or myocardial ischemia, can result in action potentials being generated in areas of the myocardium other than the SA node. These areas are called **ectopic foci**. Problems occur when such action potentials are generated in the myocardium in a random fashion or when the myocardium becomes paced by the ectopic foci at rates too high to allow proper filling of the ventricular chambers with blood (ventricular tachycardias). Ischemic conditions predispose the heart to the formation of ectopic foci. Ischemia inhibits Na^+/K^+ pump activity, resulting in intracellular accumulation of Na^+ and

Clinical Focus / 12.1

Mechanism of Ventricular Reentry Tachycardias

Abnormal rhythms of the heart are characteristic of many pathologic conditions in the heart and by themselves can pose serious health risks to a patient. However, cardiac arrhythmias are most problematic when they cause the ventricles to be paced at extremely high rates. Starting at heart rates of about 180 beats/min and above, the time available for the filling of the heart during diastole becomes so compromised that even though the heart is beating many times per minute, its output is reduced as a result of decreased filling time. The faster the heart rate is, the lower is the output of the heart. It is not uncommon for individuals with ventricular tachycardia to have heart rates in excess of 250 beats/min.

Myocardial ischemia and injury can induce severe ventricular arrhythmias, called **reentry tachycardias**. Following myocardial infarction, dead zones of myocardium exist within the heart. These zones cannot conduct electrical impulses, but this does not create a serious issue for the activation of the remaining myocardium. Because of the syncytial nature of the myocardium, electrical impulses simply flow around the dead zone. However, during conditions of ischemia and ischemic injury, portions of the myocardium will allow action potentials to proceed through them in one direction but not in the opposite direction. This is called unidirectional blockade. If action potentials are able to leak through this damaged area and emerge in the healthy myocardium after that myocardium is over its refractory period, this "reentry" action potential will reactivate the healthy myocardium, sending another action potential generation through the injured tissue path. This process will then repeat again and again in an endless circle. This condition is called a circus rhythm and, as impulses generated from this circus cycle radiate outward through the syncytium, they activate the heart as a whole, at the rate set by the circus rhythm. Such circuits thus become secondary pacemakers. They may involve a few myocardial cells or large portions of the myocardium and often end up pacing the heart at an abnormally high rate (>250 beats/min). The AV node, because it is naturally predisposed to unidirectional conduction, is a common site of such reentry problems.

Although the presence of a unidirectional conduction block in the myocardium is necessary for the creation of reentry tachycardias, it is not, by itself, sufficient to set up a circus rhythm. The area of myocardium that receives action potentials from the reentry circuit as they emerge from the injured tissue zone must be over their refractory period for the cycle to be initiated anew. Consequently, *any* condition that slows, but does not block, the conduction of action potentials through the myocardium and shortens the refractory period of healthy cardiac cells will increase the probability of creating ventricular reentry tachycardias. Furthermore, should these factors enhance automaticity in cardiac cells, there is a good chance that such tachycardias will be initiated. Unfortunately, myocardial ischemia, by causing partial membrane depolarization, brings cells closer to their activation threshold and creates action potentials of small amplitude, which are conducted slowly through the myocardium, thus enhancing the probability of creating reentry tachycardias.

Ca^{2+} and partial membrane depolarization. This places the resting membrane potential in the affected cells closer to their threshold for the activation of action potentials. In this state, small additional depolarizing stimuli, which would be insufficient to push normal cells to their threshold, will push the membrane potential of affected cells beyond their threshold, causing them to fire and activate the rest of the heart. In some cases, these ectopic foci fire only occasionally, but they can fire repetitively and at high intrinsic rhythms. The probability of the formation of ectopic foci in the heart is increased by drugs such as caffeine, nicotine, and norepinephrine. Fatigue and emotional and/or physical stress, all of which activate the sympathetic nervous system, can thus predispose a person to the formation of ectopic foci in the myocardium.

Triggered activity

Triggered activity is a type of ventricular arrhythmia in which the abnormal beats are not generated *de novo* from ectopic foci. Instead, such activity follows a previous action potential or action potentials, which "triggers" an abnormal depolarization of ventricular muscle before the prior action potential is complete. This type of arrhythmia is most likely to occur in ventricular muscle cells and can result in individual **premature ventricular contractions (PVCs)** or, when repeated, in more detrimental sustained tachycardias. Triggered activity is considered self-sustaining but not self-activating.

Triggered activity is classified by the timing of the appearance of the abnormal depolarization. **Early afterdepolarizations (EADs)** are depolarizations of the muscle cells that occur late in phase 2 or early in phase 3 of the ventricular action potential. They are more likely to be induced by factors that prolong the action potential duration, such that slow calcium channels have time to recover and, thus, fire an additional low-amplitude action potential. Consequently, bradycardias favor the formation of EADs as do factors that impair outward potassium current in phase 3 of the ventricular action potential. EADs are also more likely to occur with excessive stretch of the ventricular chambers, such as that which occurs with congestive heart failure.

Delayed afterdepolarizations (DADs) occur late in phase 3 or during phase 4 of the ventricular action potential. DADs are associated with increased intracellular calcium concentration within the myocytes, which can thus depolarize the cell membrane to the threshold for an action potential. Factors that increase intracellular calcium concentration in myocytes make the ventricles prone to DADs. Consequently, DAD formation is favored by tachycardias as well as by stimuli that either increase calcium entry into myocytes (e.g., norepinephrine) or reduce intracellular calcium removal (e.g., Na^+/K^+ ATPase inhibition by digitalis).

ELECTROCARDIOGRAM

When activated, the heart is a concentrated locus of time-varying electrical potentials in the body. When a portion of the myocardium becomes depolarized from an action potential, its polarity is temporarily reversed, becoming positive on the inside and negative on the outside relative to the neighboring inactivated tissue. When this reversal occurs, it temporarily creates two neighboring regions of opposite charge, or polarity, within the myocardium (Fig. 12.6). This difference in polarity between two locations is called a **dipole**. Electrical currents readily flow from one pole of a dipole to the other through any media between the poles that can conduct electrical current. The intracellular and extracellular fluids in the body are largely composed of electrolyte solution, which is a good conductor of electricity. Thus, any dipole formed at any time and in any direction within the myocardium between depolarized and nondepolarized regions is transmitted through the body as currents between the ends of the dipoles. These currents radiate outward through the body all the way to the surface of the skin.

An electrocardiogram (ECG) is an amplified, timed recording of the electrical activity of the heart, as detected on the surface of the body. The recording gives a plot of voltage as a function of time. It results from the composite effect of all the different types of action potentials generated in the myocardium during activation and the resulting magnitude and orientation of the dipoles created. Although it is correct to say that the electrical activity in the heart is responsible for creating the ECG, the physician looks at this process in reverse; that is, the physician examines the ECG to create a picture of the electrical activity in the heart.

The ECG is one of the most useful diagnostic tools available in medicine to the physician, but it is important to understand what information can and cannot be gained from the analysis of an ECG. The ECG can be used to detect abnormalities in heart rhythm and conduction, myocardial ischemia and infarction, plasma electrolyte imbalances, and effects of numerous drugs. One can also gain information from the ECG about the anatomic orientation of the heart, the size of the atria and ventricles, and the path taken by action potentials through the heart during normal or abnormal activation (e.g., the average direction of activation of the ventricles). The ECG, however, cannot give *direct* information about the contractile performance of the heart, which is equally important in the evaluation of myocardial status in a clinical setting. Other tools must be used for such an evaluation, and these will be discussed in later chapters.

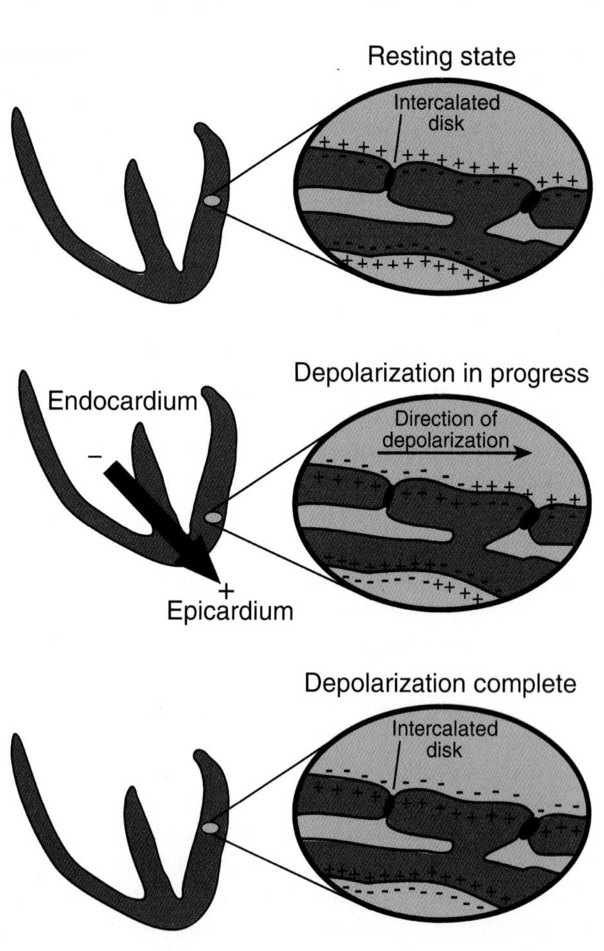

Figure 12.6 Example of a cardiac dipole. Partially depolarized myocardium creates a dipole. *Arrows* show the direction of depolarization. Dipoles are present only when myocardium is in the process of depolarization or repolarization. They are not formed when the entire myocardium is depolarized or repolarized.

Normal ECGs depict cyclic electrical activity and conduction in the atria and ventricles.

Although a detailed explanation of how all the myocardial action potentials create the common ECG is complex, it is useful to recognize that certain well-known elements of the ECG represent key events in the electrical activation of the heart. The basic normal ECG is represented in Figure 12.7. The first sign of electrical activity in the heart as revealed by the ECG is a small, rounded, upward (positive) deflection called a **P wave**. The P wave is caused by depolarization of the atria (not just the SA node). After a short interval, a complex, short-duration, high-amplitude, spike-like potential is observed. This potential is called the **QRS complex** and is caused by depolarization of the ventricles. By definition, within this complex, the first downward deflection after the P wave is called a **Q wave**, the next upward deflection is called an **R wave**, and the next subsequent downward deflection is called an **S wave**. (Depending on the location of the ECG recording on the body, the Q and S deflections may not be seen, and ventricular depolarization may appear only as an R wave on the ECG. Nevertheless, such a wave is still often called the QRS complex.) Following the QRS complex, the entire ventricular mass is depolarized and there are no potential differences between areas of the myocardium. For this reason, no deflections are registered on the ECG, and the ECG is said to be **isoelectric** or at zero potential. Once the ventricles begin to repolarize, a broad wave of modest amplitude called a **T wave** is observed. Repolarization of the atria is not seen in a typical ECG because it occurs during the same interval as the QRS complex and is lost in that signal.

Intervals between waves in the ECG are of physiologic and clinical importance. The **PR interval** is the

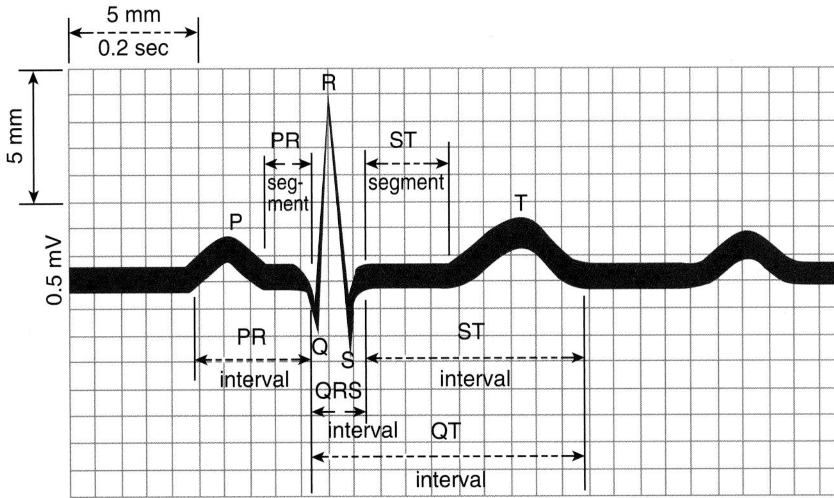

Figure 12.7 The major waveforms and intervals associated with the normal electrocardiogram (ECG). P, P wave; Q, Q wave. R, R wave; S, S wave; T, T wave.

time from the beginning of the P wave to the start of the QRS complex and represents the amount of time the action potential takes to travel from the SA node through the AV node. The PR interval typically lasts 0.12 to 0.20 seconds. Because most of this time represents the delay of the action potential conduction through the AV node, the inhibition of conduction through the AV node is often reflected as a lengthening of the PR interval. The **QRS interval** represents the interval of time that the action potential takes to travel from the end of the AV node through the ventricles (normally 0.06–0.1 seconds). The duration of the QRS complex is roughly equivalent to the duration of the P wave, despite the much greater muscle mass of the ventricles. The relatively brief duration of the QRS complex is the result of the rapid, synchronous excitation of the ventricles. The normal conduction pathway through the bundle branches, Purkinje fibers, and ventricular muscle is the most efficient and rapid mode of action potential travel. Therefore, any pathway other than this takes longer than normal and results in an abnormally long QRS interval.

The time between the initiation of the QRS complex and the end of the T wave is called the **QT interval**. This interval is inversely proportional to the heart rate and is often altered by drugs or conditions that alter the rate of myocardial repolarization in phase 3 (e.g., altered K^+ conductance). If the ventricular action potential and QT interval are compared, the QRS complex corresponds to depolarization, the ST segment corresponds to the plateau, and the T wave corresponds to repolarization. The relationship between a single ventricular action potential and the events of the QT interval are approximate because the events of the QT interval represent the combined influence of all of the ventricular action potentials. The QT interval measures the total duration of ventricular activation. If ventricular repolarization is delayed, the QT interval is prolonged. Delayed repolarization is associated with the genesis of ventricular arrhythmias. Therefore, the presence of a prolonged QT interval on the ECG is clinically significant.

Moment-to-moment orientation and magnitude of net dipoles in the heart determine the formation of the ECG.

The formation of the standard waveforms within the ECG can be explained as arising from the orientation and magnitude of the net, or collective average, dipoles that are created throughout the heart during the electrical activation of the myocardium. To explain further, consider the voltage changes produced in which the body serves as a volume conductor and the heart generates a collection of changing dipoles (Fig. 12.8). In this example, an electrocardiographic recorder is connected between points A and B such that when point A is positive relative to point B, the ECG is deflected upward, and when B is positive relative to A, a downward deflection results. The black arrows show (in two dimensions) the direction of the net dipole resulting from the many individual dipoles present at any one time. The lengths of the arrows are proportional to the magnitude (voltage) of the net dipole, which is related to the mass of myocardium generating the net dipole. The black arrows show the magnitude of the dipole component that is parallel to the line between points A and B (the recorder electrodes); this component determines the amplitude and polarity of voltage that will be recorded on the ECG. Atrial excitation results from a wave of depolarization that originates in the SA node and spreads over the atria, as indicated in panel 1 of Figure 12.8. The net dipole generated by this excitation has a magnitude proportional to the mass of the atrial muscle involved and a direction indicated by the black arrow. The head of the arrow points toward the positive end of the dipole, where the atrial muscle is not yet depolarized. The negative end of the dipole is located at the tail of the arrow, where depolarization has already occurred. Point A, therefore, is positive relative to point B, and there will be an upward deflection of the ECG. The magnitude of this upward deflection depends on two factors: (1) it is proportional to the amount of tissue generating the dipole (the magnitude of the net dipole), and (2) it depends on the orientation of the dipole relative to a parallel line connecting

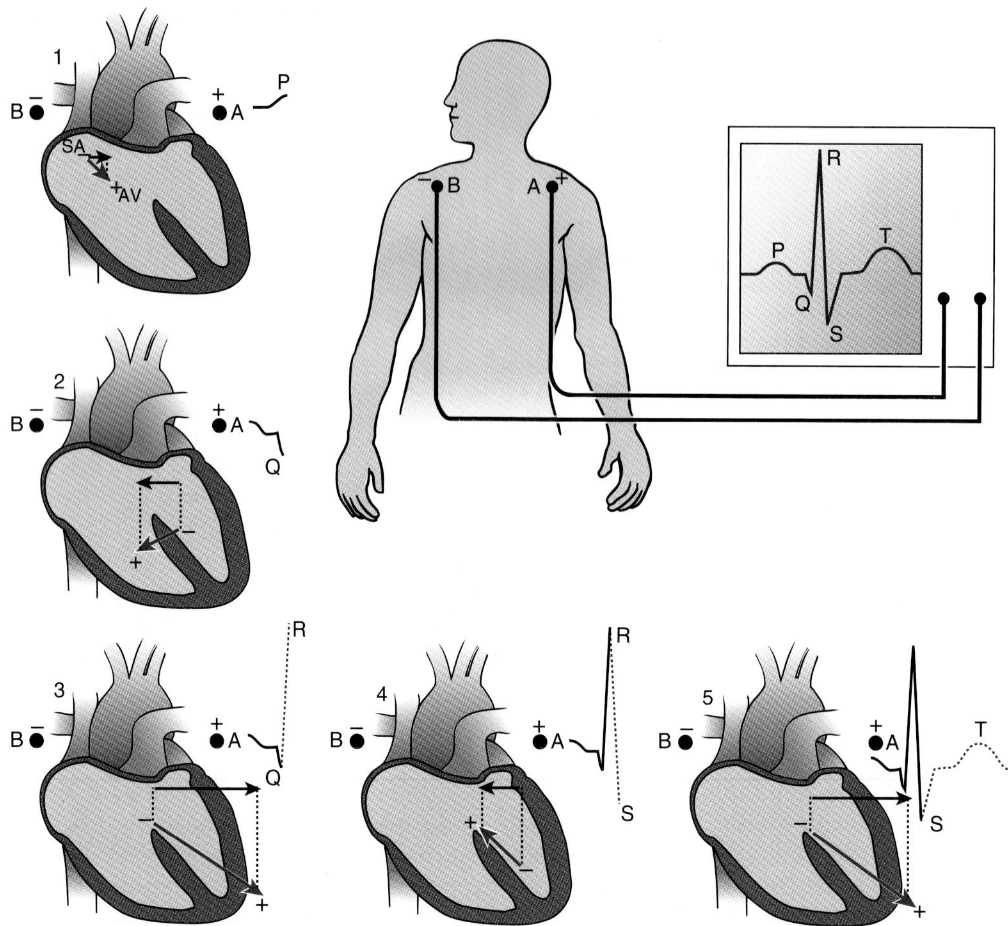

Figure 12.8 The sequence of major dipoles giving rise to electrocardiogram (ECG) waveforms. The *black arrows* are vectors that represent the magnitude and direction of a major dipole. The magnitude is proportional to the mass of myocardium involved. The direction is determined by the orientation of depolarized and polarized regions of the myocardium. The vertical dashed lines project the vector onto the A-B coordinate (analogous to lead I); it is this component of the vector (represented here by the *red arrow*) that is sensed and recorded on the ECG. In panel 5, the tail of the vector (*black arrow*) shows the yet-to-be-repolarized region of the myocardium (negative) and the head points to the repolarized region (positive). The last areas of the ventricles to depolarize are the first to repolarize (i.e., repolarization proceeds in a direction opposite to that of depolarization). The projection of the vector (*red arrow*) for repolarization points to the more positive electrode (A) as opposed to the less positive electrode (B) and so an upward deflection is recorded on this lead. AV, atrioventricular; SA, sinoatrial.

points A and B. This latter relationship is demonstrated in Figure 12.9. For example, imagine a wave of depolarization traveling through the atria muscle as a sagittal plane, perpendicular to the ground, proceeding directly along the line connecting point B to point A. This wave of depolarization is then aimed directly at the positive pole A and will create a positive deflection as described above. For the sake of example only, we shall assign this deflection an amplitude of +4 mm on the ECG recorder. Should this same wave of depolarization, however, proceed from point A toward point B, the wave would be aimed directly at the negative pole, resulting in a −4-mm or downward deflection of the wave. The amplitude of the deflection will thus vary in this example between −4 mm and +4 mm, depending on the angle of the wave of depolarization relative to the line connecting A and B. Should the wave proceed toward A at a 45° angle, the deflection would be a positive 2 mm; if it proceeds at 90° (perpendicular) to the line connecting A and B, it would not be pointing at either pole and no deflection would be recorded on the ECG. The deflection will also register zero once the atria are completely depolarized, because no voltage difference will exist between A and B (i.e., no dipole exists).

Although the preceding discussion is an oversimplification, it presents the basic principles of dipole magnitude and orientation relative to two recording points that create the pattern of the common ECG. For example, after the P wave, the ECG returns to its baseline or isoelectric level. During this time, the wave of depolarization moves through the AV node, the AV bundle, the bundle branches, and the Purkinje system. The dipoles created by the depolarization of these structures are too small to produce a deflection on the ECG. However, the depolarization of ventricular structures does create deflections on the ECG. The net dipole

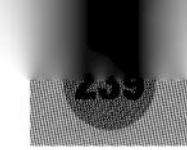

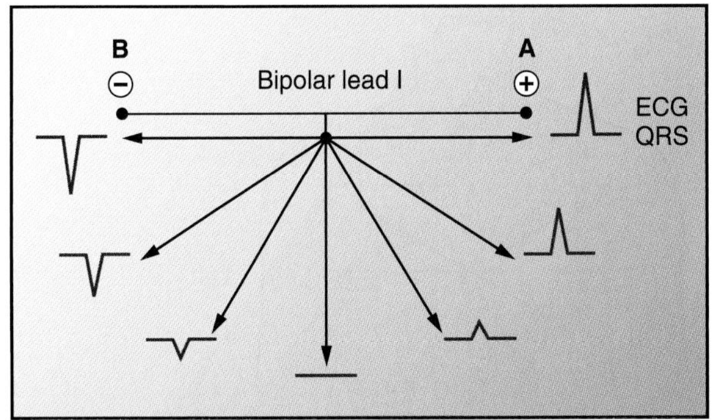

Figure 12.9 A representation of how the orientation of a dipole in the ventricle relative to a bipolar lead affects the magnitude and polarity of a hypothetical deflection on the electrocardiogram (ECG). Positive ends of the dipole pointed to the positive pole of the lead create positive deflections in the complex. Dipoles pointed directly to the positive pole, parallel to the lead, create the most positive deflection waves. Those pointed directly to the negative pole create the most negative deflection. Dipoles oriented perpendicularly to the lead have no component pointed to either pole of the lead and thus register no deflection of the ECG. QRS, QRS complex and interval.

that results from the initial depolarization of the septum is shown in panel 2 of Figure 12.8. This depolarization is pointed toward point B and away from point A because the left side of the septum depolarizes before the right side. This orientation creates a small downward deflection produced on the ECG, called the Q wave. The normal Q wave is often so small that it is not apparent. Next, the wave of depolarization spreads via the Purkinje system across the inside surface of the free walls of the ventricles. Depolarization of free-wall ventricular muscle proceeds from the innermost layers of muscle (subendocardium) to the outermost layers (subepicardium). Because the muscle mass of the left ventricle is much greater than that of the right ventricle, the net dipole during this phase has the direction indicated in panel 3. The deflection of the ECG is upward because the dipole is directed at point A and is large because of the great mass of tissue involved. This upward deflection is the R wave. The last portions of the ventricle to depolarize generate a net dipole with the direction shown in panel 4, and thus the deflection on the ECG is downward. This final deflection is the S wave. The ECG tracing returns to baseline when all of the ventricular muscle becomes depolarized and all dipoles associated with ventricular depolarization disappear. The ST segment, or the period between the end of the S wave and the beginning of the T wave, is generally isoelectric. This indicates that no dipoles large enough to influence the ECG exist because all ventricular muscle is depolarized (the action potentials of all ventricular cells are in phase 2).

Repolarization, like depolarization, generates a dipole because the voltage of the depolarized area is different from that of the repolarized areas. The dipole associated with atrial repolarization does not appear as a separate deflection on the ECG because it generates a low voltage and because it is masked by the much larger QRS complex, which is present at the same time. Ventricular repolarization is not as orderly as ventricular depolarization. The duration of ventricular action potentials is longer in subendocardial myocardium than in subepicardial myocardium. The longer duration of subendocardial action potentials means that even though subendocardial cells were the first to depolarize, they are the last to repolarize. Because subepicardial cells repolarize first, the subepicardium is positive relative to the subendocardium; that is, the polarity of the net dipole of repolarization is the same as the polarity of the dipole of depolarization. This results in an upward deflection because, as in depolarization, point A is positive with respect to point B. This deflection is the T wave (see panel 5, Fig. 12.8). The T wave has a longer duration than the QRS complex because repolarization does not proceed as a synchronized, propagated wave. Instead, the timing of repolarization is a function of the properties of individual cells, such as the number of particular K^+ channels.

ECG evaluation is standardized by the use of a designated 12-lead system.

There are two broad classes of evaluations performed using the ECG. One of these involves the evaluation of abnormalities in the basic ECG form. Arrhythmias, conduction abnormalities, electrolyte disturbances, drug effects, and myocardial metabolic disorders such as ischemia can be detected as abnormal patterns from a single ECG electrode placed at most any location on the surface of the body. However, as explained in the previous section, the amplitude—and in some cases, the polarity—of various waveforms within the ECG depend in part on the orientation of the net dipoles generated in the heart relative to the position of the electrode system measuring the ECG. Clearly then, the same electrical event in the heart will appear differently if it is viewed by two different electrode systems with different orientations relative to the event. For this reason, a standard system consisting of 12 specifically located "leads" or electrode connections is used in the clinical application of ECG recordings in patients. This system provides two benefits. First, it provides a standard framework for identifying patterns in the ECG, thus making the recognition of abnormalities easier and more consistent. Second, it enables an investigator to see the electrical activity in the heart from many different "views and angles" at any given point in time. Analysis of these different "views" is used to gain information about the orientation of the heart, the size of its chambers, and the general direction of activation in the myocardium during any interval.

Imaging cardiac electrical activity in the frontal plane uses six standardized frontal limb leads.

The first ECG lead system was developed by Willem Einthoven and was based on the idea that the heart sits at the center of a triangle in the frontal plane of the body, with vertices at the right shoulder, left shoulder, and pubic region (Fig. 12.10A). This triangle is called the **Einthoven triangle.** In this system, the arms and the legs are considered extensions of the vertices, and ECG electrodes are placed on the right arm, left arm, and left leg, with the right leg serving as

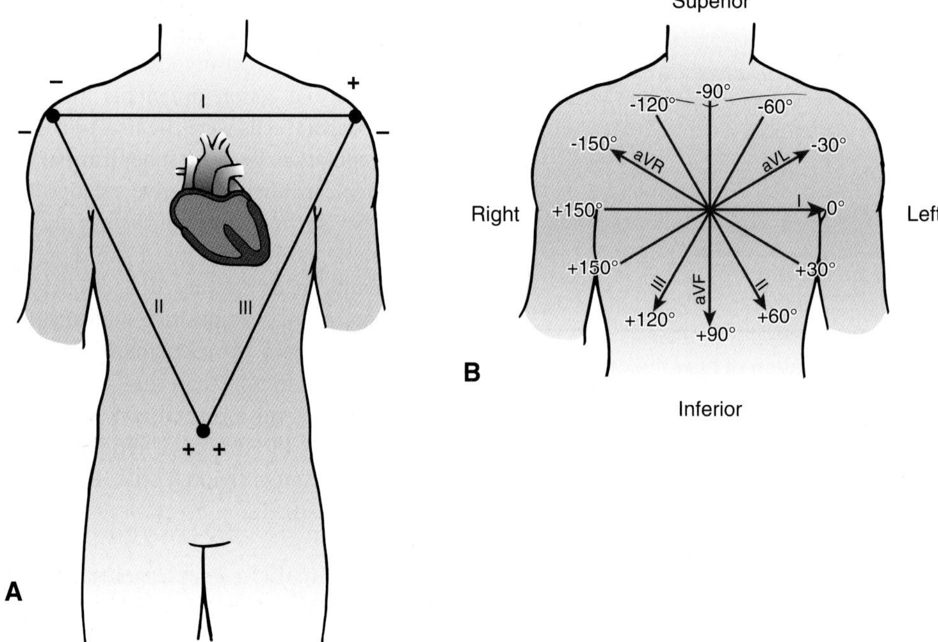

Figure 12.10 The Einthoven triangle. (A) Einthoven codified the analysis of electrical activity of the heart by proposing that certain conventions be followed. The heart is considered to be at the center of a triangle, each corner of which serves as the location for electrodes that send signals as a positive and negative pole pair to the electrocardiogram recorder. The three resulting bipolar leads are designated I, II, and III. By convention, positive dipoles generated within the myocardium that are pointed in the direction of the positive pole of any of the leads create a positive deflection that is recorded from that lead; when pointed at the negative pole, a negative deflection is recorded. **(B)** The hexaxial reference system for the standard bipolar and augmented unipolar chest leads. The limb leads give information on cardiac dipole vectors in the frontal plane and are referenced by the angle, in degrees, in which the vector points as if viewed in two dimensions in the frontal plane. aVF, augmented foot lead; aVL, augmented left lead; aVR, augmented right lead.

an electrical ground. This creates a series of connections, or leads, between each pair of electrodes. These leads are bipolar, meaning each has a positive and a negative pole. In lead I, the right arm is negative and the left arm is positive. In lead II, the right arm is negative and the left leg is positive. Finally, in lead III, the left arm is negative and the left leg positive. If one imagines the frontal plane of a person as being represented by a 360° circle, lead I has its positive pole placed at "3 o'clock" on the circle and is designated at 0°. The positive pole of lead II is at 60°, and that for lead III is at 120°, as shown in Figure 12.10B. This lead system is often called the **standard frontal lead system**.

Another lead system is arranged in the frontal plane of the body and consists of three unipolar leads created by special connections between the same electrodes placed on the arms and legs for the creation of the standard frontal system. These leads are single positive poles relative to the center of the chest, which is considered zero potential. These leads are often called **augmented leads** because the ECG recording device amplifies the signal from these leads relative to those obtained in the bipolar leads (see Fig. 12.10B). The pole of one lead sits at the right shoulder at 210° and is called the *augmented right*, or *aVR*, lead; another sits at the left shoulder at 330° and is called *augmented left* or *aVL*. The final augmented lead is oriented at 90° in the frontal plane and is called the *augmented foot* or *aVF lead*. The augmented and standard limb leads give a two-dimensional picture of the electrical activity of the heart as it would be viewed in the frontal plane of the body.

Imaging cardiac electrical activity in the horizontal plane uses six standardized chest leads.

In a typical 12-lead ECG system, a series of six unipolar leads is arranged in a horizontal plane around the chest as shown in Figure 12.11. These are sometimes called **precordial or chest leads** and are designated V_1 to V_6. These leads give a two-dimensional picture of the electrical activity in the heart as it would be viewed from above or below a horizontal plane bisecting the heart. In addition, because these leads lie so close to the surface of the heart, individual precordial leads give detailed information about the electrical activity in the small, specific portions of the heart that lie beneath each electrode.

Typical tracings from the six frontal and six horizontal lead systems are shown in Figure 12.12. In general clinical practice, all 12 leads are recorded in a patient at the same time. The chart speed is standardized at 25 mm/s, which allows the physician an easy means of converting intervals between two points on the ECG into seconds. This standard

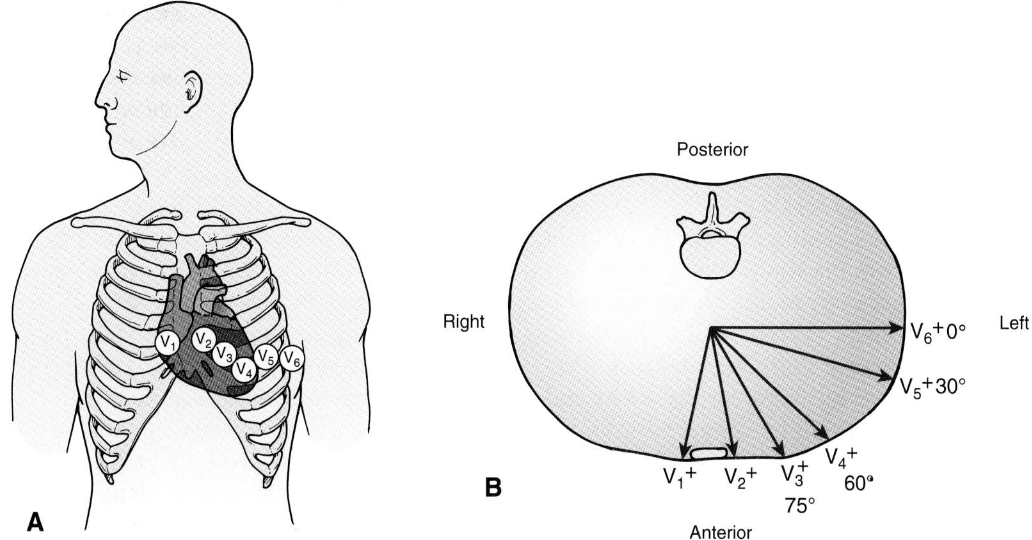

Figure 12.11 Unipolar chest leads. (A) V_1 is just to the right of the sternum in the 4th intercostal space. V_2 is just to the left of the sternum in the 4th intercostal space. V_4 is in the 5th interspace in the midclavicular line. V_3 is positioned midway between V_2 and V_4. V_5 is in the 5th intercostal space in the anterior axillary line. V_6 is in the 5th intercostal space in the midaxillary line. The reference, or zero, voltage for the unipolar chest leads is electronically combined out of the three limb leads. **(B)** The orientation of the unipolar chest leads in the horizontal plane. The chest leads give information on cardiac dipole vectors as if viewed in two dimensions in the horizontal plane.

method is often used to estimate heart rate in a patient. For example, R waves that are uniformly separated by 25 mm indicate that the heart is beating at 60 beats/min. Similar waves, separated by 20 mm, indicate a rate of 75 beats/min, and so on.

Figure 12.12 shows that a P wave is always followed by a QRS complex of uniform shape and size. The PR interval is 0.16 seconds (normal, 0.10–0.20 seconds). This measurement indicates that the conduction velocity of the action potential from the SA node to the ventricular muscle is normal. The average time between R waves (successive heart beats) is about 0.84 seconds, making the heart rate approximately 71 beats/min.

Information about the orientation of the heart, ventricular size, and conduction pathways is obtained through the frontal QRS vector.

As explained above, changes in the magnitude and direction of any momentary cardiac dipole will cause changes in a given ECG lead, as determined by the orientation of the dipole relative to the orientation of the specific lead. Theoretically, by comparing the magnitude of any portion of an ECG simultaneously in several leads, one could work backward and determine what the orientation of the net dipole was in the heart at that moment in time that gave rise to those magnitudes in the different leads. In other words, by comparing, for example, the magnitude of

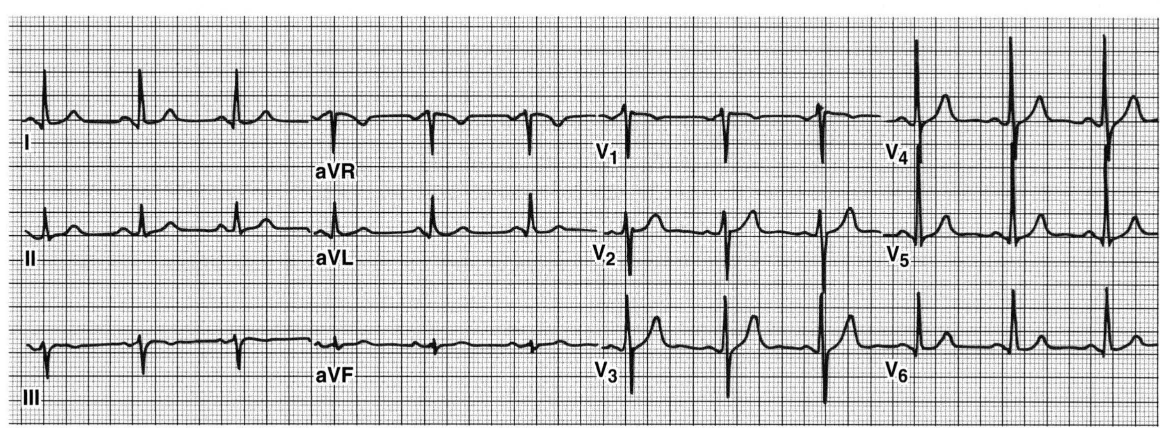

Figure 12.12 The clinical standard 12-lead electrocardiogram. Six limb leads and six chest leads are shown. Two dark horizontal lines (10 mm) are calibrated to be 1 mV. Dark vertical lines represent 0.2 seconds. aVF, augmented foot lead; aVL, augmented left lead; aVR, augmented right lead.

positive deflections at one point of the ECG in leads I, II, and III at the same time, one could "triangulate in" and get an estimate of the direction of the net dipole in the heart at that precise moment. That direction could then be represented by a vector on the hexaxial reference system shown in Figure 12.10B.

Hypothetically, one could construct a set of vectors resulting from the net dipoles produced in the myocardium during every millisecond of cardiac activation. However, this is impractical and is not necessary to glean important clinical information from the ECG. Instead, what is done is to focus on a major cardiac event that is represented within the ECG, construct a general average vector for that event, and then plot its vector in the frontal or horizontal plane as one wished. The most common event that is analyzed this way from the ECG is ventricular depolarization and its associated ECG QRS complex. In broad terms, the overall direction of depolarization through the myocardium is from the right atrium to the left atrium, down through the AV node, and on into the ventricles. The heart in most people sits in the chest at an angle with the apex pointing toward the lower left portions of the lungs. Thus, the long axis of the ventricles from base to apex is tilted downward from a line parallel to the horizontal axis of the body. Therefore, if one could "see" the *average* grand trek of depolarization through the ventricles projected in two dimensions onto a flat screen in the frontal plane of the body, it would appear as a vector proceeding from the upper right of the ventricles down toward the heart apex at an angle of approximately +60° on the hexaxial reference system. This average vector is called the **mean electrical QRS axis**, or mean QRS vector. Using the standard bipolar limb leads, the Einthoven triangle, and the hexaxial reference system for orientation, the *average* deflections of the QRS complexes giving rise to this mean vector would be most positive in lead II, least positive in lead III, and in between these two magnitudes in lead I. (In fact, the axiom that I + III = II for average QRS amplitudes always applies and can be used as a handy check for proper connection of the frontal ECG leads.)

Calculating the average deflection for the mean QRS vector is simplified for the purpose of clinical analysis. Basically, a line along the isoelectric line of the ECG from a given lead (e.g., lead I) is drawn through the QRS complex, and the amplitude of the peak of the R wave is measured. The peak of any negative deflections in the QRS complex below the isoelectric line (usually the negative peaks of the Q and S waves) is subtracted from the R wave peak to give the average QRS amplitude in that lead. This value is then plotted in the appropriate polarity direction along the line in the hexaxial system that corresponds to the lead from which the ECG was measured. A perpendicular line is then dropped from that point. This same procedure is repeated for at least one other lead in the plane in which the QRS vector is desired to be determined (e.g., either lead II or lead III in this example) until the perpendicular lines intersect. A line then drawn from the center of the hexaxial system to the point of intersection of the perpendiculars represents the mean QRS vector in the heart at that moment in time.

The mean QRS vector is influenced by the position of the heart in the chest, the properties of the cardiac conduction system, the excitation and repolarization properties of the ventricular myocardium, and the size of the ventricles. Because the last three of these influences are most significant, the mean QRS electrical axis can provide valuable information about a variety of cardiac diseases. In healthy people, the mean QRS vector is oriented within the range of −10° to +110° on the hexaxial system. Any mean QRS vector outside these limits indicates serious changes in either the path of depolarization in the ventricles or abnormal ventricular muscle mass, such as **left ventricular hypertrophy** or **right ventricular hypertrophy**, depending on which ventricle is involved. Ventricular hypertrophy is a serious condition resulting from either ventricle being exposed chronically to high arterial pressure or outflow obstruction. Its presence in an ECG recording indicates serious underlying cardiac diseases in the patient.

An example of the application of mean QRS vector determination is shown in Figure 12.13, which uses the ECG tracings in Figure 12.12. For this example, the Einthoven triangle is used to calculate the vector. The net magnitude of the QRS complex in leads I, II, and III is measured and plotted on the appropriate axis. A perpendicular line is dropped from each of the plotted points and a vector drawn between the center of the triangle and the intersection of the perpendicular lines. This gives a mean QRS axis of +3°. Another simple way of estimating the mean QRS vector in this example is to simply examine the QRS complex in the six frontal limb leads and select the lead in which the average QRS complex deflection is closest to zero. As discussed earlier, when the cardiac dipole is perpendicular to a particular lead, the net deflection is zero. Once the net QRS deflection closest to zero is identified, it follows that the mean electrical axis is perpendicular to that lead, and the hexaxial reference system can be consulted to determine the angle of that axis. In Figure 12.12, the lead in which the net QRS deflection is closest to zero is lead aVF (enlarged in Fig. 12.13). Lead I is perpendicular to the axis of lead aVF, and because the QRS complex is upward in lead I, the mean electrical axis points to the left arm. Thus, the mean QRS vector is estimated to be about 0°, which is not far from the +3° of the actual calculated vector from the frontal bipolar leads.

Any ECG can detect abnormalities in the electrical activation and conduction in the heart.

Several types of ECG recordings are listed in Figure 12.14. Unless stated otherwise, one may assume these ECGs are from lead II. The ECG in Figure 12.14A shows **respiratory sinus arrhythmia**, which is an increase in the heart rate with inspiration and a decrease with expiration. The presence of a P wave before each QRS complex indicates that these beats originate in the SA node. Intervals between successive R waves of 1.08, 0.88, 0.88, 0.80, 0.66, and 0.66 seconds correspond to heart rates of 56, 68, 68, 75, 91, and 91 beats/min. The interval between the beginning of the P wave and the end of the T wave is uniform, and the change in the interval between beats is primarily accounted for by the variation in time

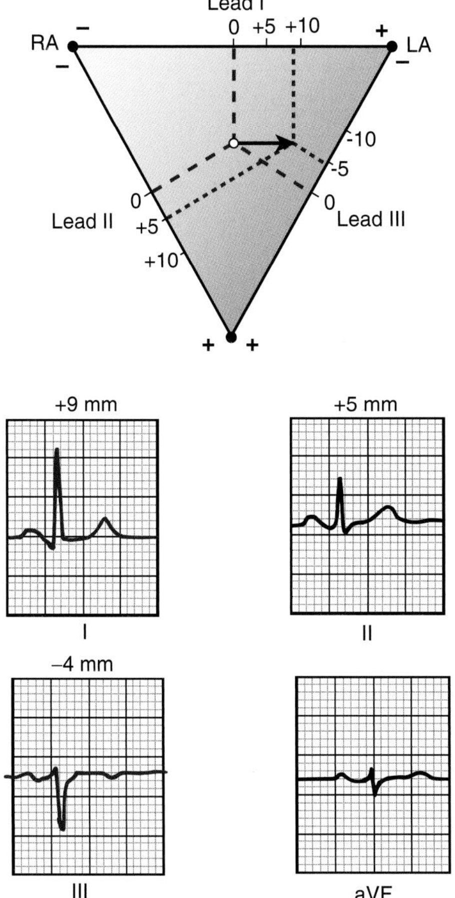

Figure 12.13 An example of the determination of the mean QRS axis. This axis can be estimated by using the Einthoven triangle and the net voltage of the QRS complex in any two of the bipolar limb leads. It can also be estimated by inspection of the six limb leads (see text for details). Electrocardiogram tracings are from Figure 12.12 and are enlarged to facilitate viewing. aVF, augmented foot; LA, left arm; LL, left leg; RA, right arm.

between the end of the T wave and the beginning of the P wave. Although the heart rate changes, the interval during which electrical activation of the atria and ventricles occurs does not change nearly as much as the interval between beats, indicating that it is variations in the activation of the heart at the SA node that is altering rhythm and not variation in conduction through the heart. Respiratory sinus arrhythmia is caused by cyclic changes in sympathetic and parasympathetic neural activity to the SA node that accompany respiration. It is observed in people with healthy hearts, although this particular example is somewhat dramatic. Respiratory sinus arrhythmia is usually accentuated in patients under anesthesia.

In the tracing in Figure 12.14B, the pattern of P, QRS, and T waves is normal but the interval between successive P, QRS, or T waves is much shorter than normal. This represents an SA node–driven increase in heart rate and is called a **sinus tachycardia**. A slow SA node–driven heart rate is called a **sinus bradycardia**. These patterns would be seen in a healthy person during acute exercise or in a well-conditioned athlete during rest, respectively.

Occasionally, the heart will be activated by the spontaneous generation of an action potential in one of the ventricular cells. Such an ectopic focus and its effect on the ECG are shown in Figure 12.14C. The first three QRS complexes in this panel are preceded by P waves; then, after the T wave of the third QRS complex, a QRS complex of increased voltage and longer duration occurs. This complex will result in PVCs. The term PVC is used to designate the appearance of the firing of the ventricular ectopic foci on the ECG even though the ECG is the electrical, not mechanical, representation of the event. This premature complex is not preceded by a P wave and is followed by a pause before the next normal P wave and QRS complex. In panel C, the ectopic focus is probably in the Purkinje system or ventricular muscle, where an aberrant pacemaker reached threshold before being depolarized by the normal wave of excitation. Once the ectopic focus triggers an action potential, the excitation is propagated over the ventricles. The abnormal pattern of excitation accounts for the greater voltage, change of mean electrical axis, and longer duration (inefficient conduction) of the QRS complex. Although the abnormal wave of excitation reached the AV node, retrograde conduction usually dies out in the AV node because the node preferentially conducts in the antegrade direction. The next normal atrial excitation (P wave) occurs but is hidden by the inverted T wave associated with the abnormal QRS complex. This normal wave of atrial excitation does not result in ventricular excitation because, when the impulse arrives at the AV node, a portion of the node is still refractory from excitation by the PVC. As a consequence, the next "scheduled" ventricular beat is missed. A prolonged interval following a premature ventricular beat is the **compensatory pause**.

Ectopic pacemakers can also occur in the atria, where they are called **premature atrial contractions (PACs)**. Occasional PVCs or PACs are not uncommon in healthy people. However, the probability of their occurrence is increased by cigarette smoking, nicotine in any form, physical or emotional stress, caffeine, and fatigue.

On some occasions, atrial ectopic foci fire in a random, or chaotic, manner. This is shown in Figure 12.14D, which is an ECG from a patient with **atrial fibrillation**. In this condition, atrial systole does not occur because the atria are excited by many small, random waves of depolarization. The AV node conducts action potentials whenever a wave of atrial excitation happens to reach it and it is not refractory to activation. Thus, the ventricles are activated in a "hit or miss" fashion, and the ventricular rate is highly irregular. This can be detected in the patient as a highly irregular arterial pressure pulse. Unless there are other abnormalities, conduction through the AV node and ventricles is normal, and the resulting QRS complex is normal. The ECG shows QRS complexes that are not preceded by P waves. Atrial fibrillation is associated with numerous disease states, such as cardiomyopathy, pericarditis, hypertension, and hyperthyroidism, but it sometimes occurs in otherwise healthy people.

Sometimes specific atrial ectopic foci can fire repetitively at a high rate, serving as an ectopic pacemaker that drives the ventricles. Such atrial ectopic foci can fire regularly at a rate of >200 times/min. This high rate of atrial activation

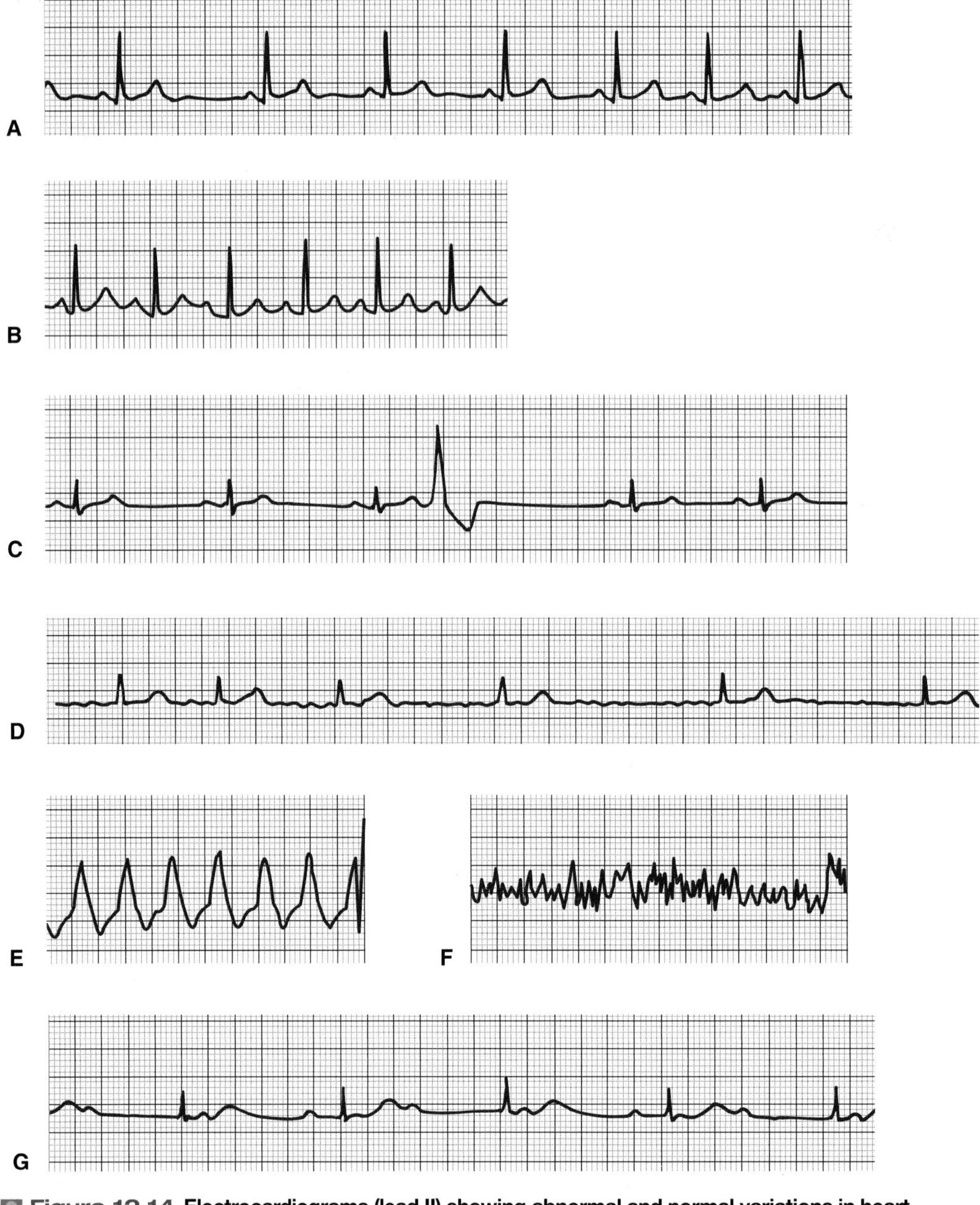

Figure 12.14 Electrocardiograms (lead II) showing abnormal and normal variations in heart rhythms. **(A)** Respiratory sinus arrhythmia. **(B)** Sinus tachycardia. **(C)** Premature ventricular complex. **(D)** Atrial fibrillation. **(E)** Ventricular tachycardia. **(F)** Ventricular fibrillation. **(G)** Complete atrioventricular block.

can get through the AV node and activate the ventricles at the same rate. This is called **supraventricular tachycardia**. Such a condition may pace the heart at a rate too fast to allow adequate ventricular filling during diastole. Because pacing of the heart in this condition results from an ectopic focus in the atria, the resulting ECG may look similar to that of a person with sinus tachycardia.

An abnormally high rate of ventricular activation generated from ventricular ectopic foci is called **ventricular tachycardia** (Fig. 12.14E). Ventricular tachycardia results in a bizarre inefficient activation of the ventricles and extremely high heart rates that limit the period of ventricular filling. It represents a serious threat to the person and is often a prelude to the life-threatening event of **ventricular fibrillation** (Fig. 12.14F). Ventricular fibrillation is characterized by the random, uncoordinated activation of millions of ventricular cells and results in no pumping of blood by the heart. Unless the heart can be converted through intervention to a regular rhythm, ventricular fibrillation will result in death.

In some situations, there is a disconnect between the electrical activation of the atria and the ventricles. This results from an impairment of the conduction of action potentials from the atria through the AV node and can result in a dissociation of P waves and QRS complexes in the ECG. In Figure 12.14G, both P waves and QRS complexes are present, but the timing of each is independent of the others. This situation is called **complete**

atrioventricular block (sometimes called *third-degree block* or *complete heart block*) because the AV node fails to conduct impulses from the atria to the ventricles. Because the AV node is the only electrical connection between these areas, the pacemaker activities of the two muscles become entirely independent. In addition, a secondary pacemaker must take over the regular pacing of the heart. Because this usually involves the AV node or Purkinje systems that have low intrinsic rhythmicity, the heart will be paced at a lower-than-normal rate. In this example, the distance between P waves is about 0.8 seconds, giving an atrial rate of 75 beats/min. However, the distance between R waves averages 1.2 seconds, giving a ventricular rate of 50 beats/min. It is this rate that is responsible for pumping blood from the heart. In this example, the atrial pacemaker is probably in the SA node, and the ventricular pacemaker is probably in a lower portion of the AV node or bundle of His. Patients with complete atrioventricular block cannot resume normal activity without having an electronic pacemaker device permanently implanted in their chest to pace the heart at a faster rate.

AV block is not always complete. Sometimes the PR interval is lengthened beyond normal limits (e.g., >0.2 seconds), but all atrial excitations are eventually conducted to the ventricles. This is **first-degree atrioventricular block**. When some, but not all, of the atrial excitations are conducted by the AV node, the block is called **second-degree atrioventricular block**. In some disease states, either the left or right branch of the bundle of His cannot transmit excitation. Depending on the branch affected, this type of heart block is called **right bundle branch block** or **left bundle branch block**. In these conditions, the portion of the heart with the blocked branch receives delayed activation from the unblocked side of the ventricles because the route of activation is less efficient. The resulting QRS complex becomes widened and will sometimes display a characteristic split in the R wave, creating a double-positive spiked appearance of the QRS complex. These twin spikes are designated as R and R[1].

Changes in electrically active ventricular mass are obtained by contrasting the ECGs from multiple leads.

In the healthy heart, the left ventricle is larger than the right and dominates the resulting QRS complex. Leads with positive poles most closely aligned on the left side of the body yield the most positive average QRS deflections (e.g., leads I, II, V_5, and V_6). Indeed, the only ECG lead that shows an average negative QRS deflection in a healthy person is aVR because the normal wave of activation of the ventricles proceeds away from this positive electrode. The ECGs from a healthy heart recorded from the precordial leads create a pattern of progressively greater average positive deflections from V_1 to V_6.

Right ventricular hypertrophy, which is usually caused by pulmonary hypertension or pulmonary emboli, creates a large muscle mass on the right side of the heart. This mass pulls the net dipoles during ventricular activation to the right and can create mirror image deflections of the QRS complex in all the leads compared with that seen in the healthy heart. For example, lead aVR becomes positively deflected; V_6, downwardly deflected; and so on. The recording in Figure 12.15 shows additional effects of right ventricular enlargement on the ECG. The increased mass of right ventricular muscle results in large R waves in lead V_1. The large S waves in lead I and the large R waves in lead aVF are also characteristic of a shift in the dipole of ventricular depolarization to the right. In patients with right ventricular hypertrophy, the mean frontal QRS vector is shifted to the right, into the lower left quadrant of the hexaxial reference system, beyond +110°.

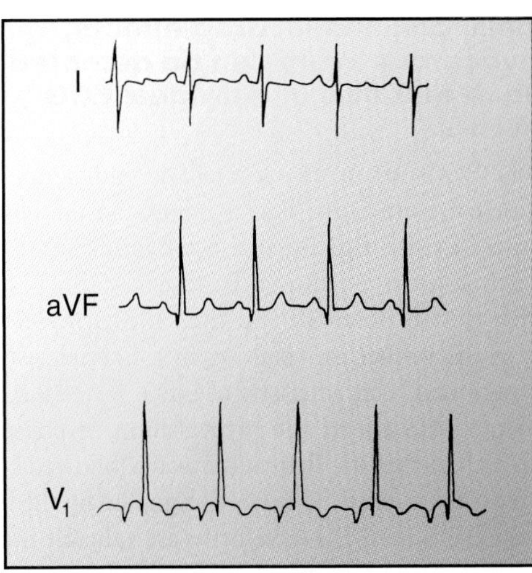

Figure 12.15 Effects of right ventricular hypertrophy on electrocardiogram recordings. Leads I, aVF, and V_1 of a patient are shown. aVF, augmented foot.

Other types of cardiac hypertrophy can be detected through the examination of simultaneous waveforms from different ECG leads. For example, Figure 12.16A shows the effects of atrial hypertrophy on the P waves of lead III, which have a higher amplitude than normal. Altered QRS complexes in leads V_1 and V_5 associated with left ventricular hypertrophy are shown in Figure 12.16B. Left ventricular hypertrophy rotates the direction of the major dipole associated with ventricular depolarization more to the left than usual. This causes large S waves in V_1 and large R waves in V_5. Left ventricular hypertrophy is usually indicated by an average frontal QRS vector that is shifted well into the upper left quadrant of the hexaxial reference system in the area of −30° and beyond.

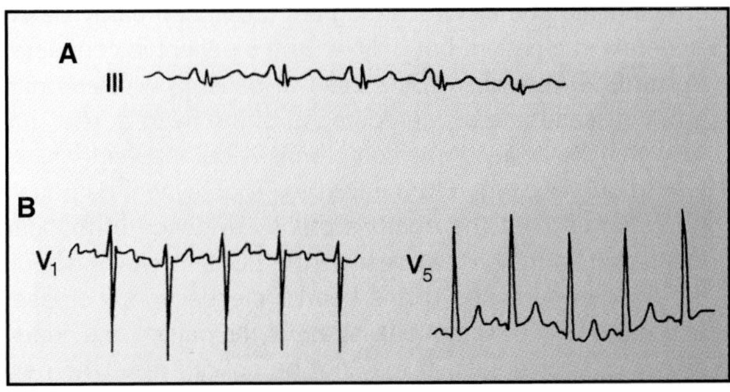

Figure 12.16 Effects of atrial and left ventricular hypertrophy on electrocardiogram recordings. **(A)** Large P waves (lead III) caused by atrial hypertrophy. **(B)** Altered QRS complex (leads V_1 and V_5) produced by left ventricular hypertrophy.

Metabolic conditions, drug effects, and myocardial injury can be detected by unique features of individual ECG waveforms.

Individual ECGs often contain telltale indicators of certain abnormal conditions. For example, plasma electrolyte disturbances create signature abnormalities in the ECG. Hyperkalemia alters potassium conductance in phase 3 of the cardiac action potential and thus affects repolarization patterns in the ventricles, resulting in a characteristic spiking or "mountain" characteristic of the T wave (Fig. 12.17). Hypokalemia also affects the myocardium in phase 3 and produces a characteristic flattened T wave followed by a secondary repolarization ECG wave, designated by the letter U. Both of these unique ECG waveforms are valuable indicators to clinicians because high plasma K^+ levels can induce fibrillation in the heart, whereas low K^+ can seriously suppress cardiac excitability and prevent cardiac activation. Calcium is one of the triggers for increasing potassium conductance during phase 3. Abnormally low serum Ca^{2+} levels can thus delay the repolarization of the ventricles, and this is revealed on the ECG as an abnormally long QT interval. Similarly, many antiarrhythmic drugs exert their effects by either increasing or decreasing the electrical refractory period in the ventricles. The toxic effects of such drugs can be detected by their effect on the QT interval of the ECG.

Finally, the clinically useful indication of myocardial ischemia, injury, and infarction (tissue death) can be detected by the ECG. One of the first signs of ischemia in the heart is an inversion of the T wave, as shown in Figure 12.18. Injury to the myocardium resulting from progressive ischemia results in an easily detected elevation of the ST segment. Infarction, or death of myocardial tissue, is recognized by the development of Q waves in front of these elevated ST segments. With myocardial ischemia, the cells in the ischemic region partially depolarize to a lower resting membrane potential because of a lowering of the potassium ion concentration gradient,

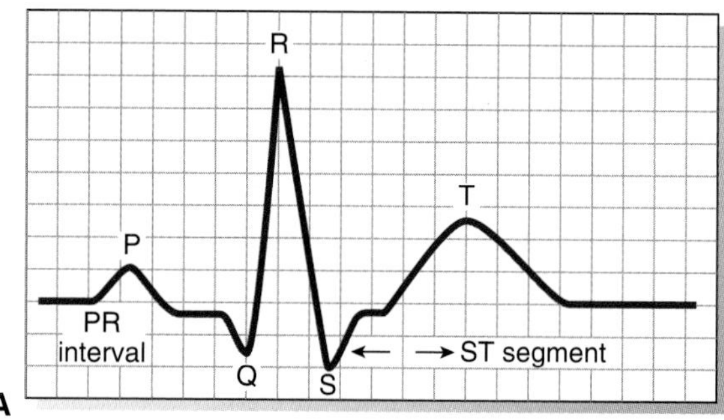

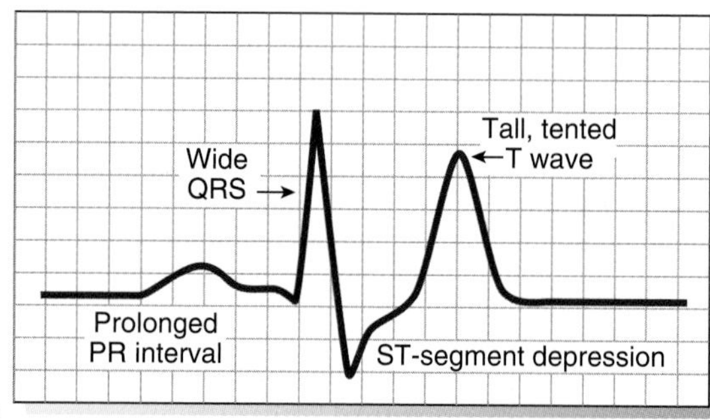

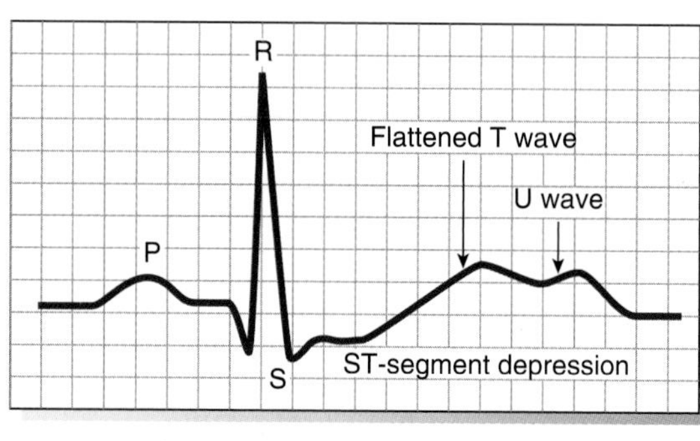

Figure 12.17 The effects of hyperkalemia and hypokalemia on the electrocardiogram (ECG). **(A)** A normal ECG tracing. **(B)** Effect of hyperkalemia. Notice the presence of the distinctive, tall, tented T wave. **(C)** Effect of hypokalemia with the appearance of a U wave.

Clinical Focus / 12.2

Electrocardiogram Exercise Testing

Myocardial ischemia is one of the most dangerous cardiac conditions existing in patients with common forms of coronary artery disease, hypertension, heart failure, and arrhythmias. The electrocardiogram (ECG) can easily detect ischemia in a patient but only while the patient is connected to the ECG recorder. In the case of patients with coronary artery disease, the oxygen demand of the heart at rest may be within the ability of the compromised coronary artery system to deliver oxygen to the tissues. Thus, when the patient is at rest (e.g., at the doctor's office), evidence of myocardial ischemia may not appear on the ECG. However, should the oxygen demands of the heart increase from increased physical activity or emotional stress in the patient, the heart's diseased arterial system may not be able to meet the new, higher oxygen demand and the heart will become ischemic. This condition can result in heart attack and death.

Cardiac stress testing uses the ECG coupled with exercise as a means to examine the limits of oxygen delivery in the heart of patients with coronary artery disease. This testing has many clinical modalities but all basically create increased oxygen demand in the heart through exercise while the patient's ECG and/or hemodynamic indicators are measured. Isometric stress (handgrip), dynamic (aerobic) stress (such as bicycle or treadmill ergometry), or combinations of the two are used to increase oxygen demand in the heart and in the body as a whole. The response of the patient as well as the response of the heart to this increased demand is monitored and analyzed. Extensive clinical guidelines for using and interpreting exercise stress testing have been developed and are continually updated by the American Heart Association and other professional cardiopulmonary or sports medicine groups.

Exercise testing has long been used for the diagnosis of obstructive coronary artery disease. Such obstructions do not often manifest themselves at rest but can be revealed by ECG and by other evidence of myocardial ischemia during exercise stress testing. The use of exercise stress testing to reveal this "silent coronary artery disease" is invaluable to the clinician.

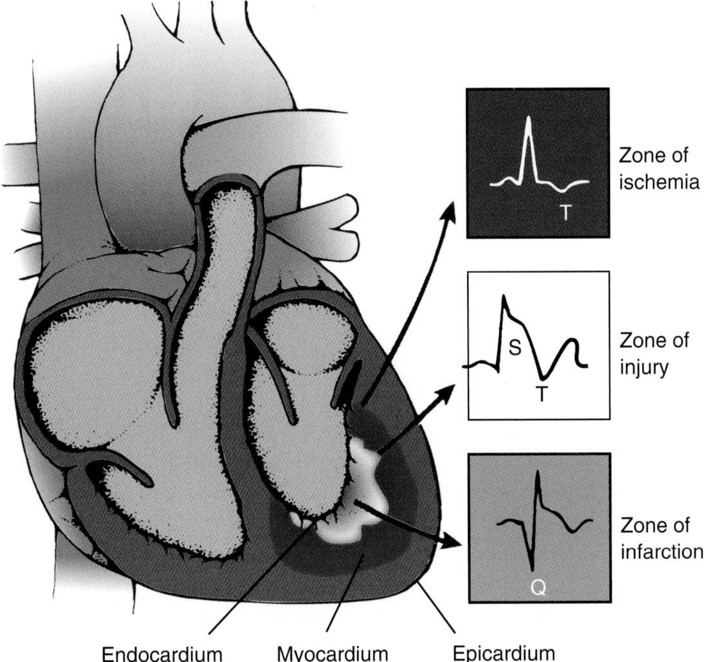

Figure 12.18 Electrocardiogram (ECG) changes associated with myocardial ischemia. The initial effects of ischemia create an inverted T wave in the ECG. The distinctive ST segment elevation in ischemic myocardium is associated with myocardial injury. The apparent zero baseline of the ECG before depolarization is below zero because of partial depolarization of the injured area. After depolarization (during the action potential plateau), all areas are depolarized and true zero is recorded. Because zero baseline is set arbitrarily (on the ECG recorder), a depressed diastolic baseline (TP segment) and an elevated ST segment cannot be distinguished. Regardless of the mechanism, this is referred to as an elevated ST segment. The ECG of a patient with acute myocardial infarction produces a Q wave in association with ST segment elevation.

although they are still capable of firing action potentials. As a consequence, a dipole is present during the TP interval in injured hearts because of the voltage difference between normal (polarized) and abnormal (partially polarized) tissue. However, no dipole is present during the ST interval because depolarization is uniform and complete in both injured and normal tissue (this is the plateau period of ventricular action potentials). Because the ECG is designed so that the TP interval is recorded as zero voltage, the true zero during the ST interval is recorded as a positive or negative deflection.

Chapter Summary

- Cardiac cells do not require signals from nerve fibers to generate action potentials.
- Cardiac cells exhibit the properties of automaticity and rhythmicity.
- Gap junctions at nexi between adjacent cells allow the heart to behave as a functional syncytium.
- Opening of voltage-gated sodium and calcium channels and the closing of voltage-gated potassium channels form action potentials in cardiac muscle cells.
- Action potentials in atrial and ventricular muscle cells have an extended depolarization plateau that creates an extended refractory period in the cardiac muscle cell.
- Cardiac muscle cells are repolarized following the depolarization phase by the closing of voltage-gated calcium channels and the opening of voltage-gated and ligand-gated potassium channels.
- A recycling decay and resetting of potassium conductance in the nodal cell membrane create recycling pacemaker potentials in the sinoatrial node.
- The sinoatrial node initiates electrical activity in the normal heart.
- Norepinephrine increases pacemaker activity and the speed of action potential conduction, whereas acetylcholine decreases pacemaker activity and the speed of action potential conduction.
- Electrical activity initiated at the sinoatrial node spreads preferentially in sequence across the atria, through the atrioventricular node, through the Purkinje system, and to ventricular muscle.
- The atrioventricular node delays the entry of action potentials into the ventricular system.
- Purkinje fibers transmit electrical activity rapidly into the inner layers of the ventricular myocardium.
- The conduction of electrical activity through the myocardium is a function of the amplitude of action potentials and the rate of rise of depolarization of action potentials in phase 0.
- An electrocardiogram is a recording of the time-varying voltage differences between repolarized and depolarized regions of the heart.
- The electrocardiogram provides clinically useful information about the rate, rhythm, pattern of depolarization, and mass of electrically active cardiac muscle.
- Changes in cardiac metabolism and plasma electrolytes as well as the effects of medicinal drugs on the electrical activity of the heart can be detected by an electrocardiogram.

13 Cardiac Muscle Mechanics and the Cardiac Pump

ACTIVE LEARNING OBJECTIVES

Upon mastering the material in this chapter, you should be able to:

- Explain the cellular mechanisms involving calcium that can alter the inotropic state of the heart.
- Identify the correct sequence of any set of hemodynamic variables associated with the cardiac cycle.
- Identify the mechanism responsible for the four primary heart sounds and relate abnormalities in these sounds to valve defects, changes in aortic pressure, or changes in ventricular filling.
- Explain how changes in myocardial contractility can counteract inhibitory effects of reduced preload or increased afterload on cardiac muscle contraction.
- Explain how changes in preload or afterload can counteract the inhibitory effects of reduced contractility on cardiac muscle contraction.
- Use left ventricular pressure volume loops to demonstrate how changes in left ventricular end-diastolic volume, aortic pressure, and contractility affect stroke volume.
- Explain how the Starling law of the heart is responsible for matching the output of the right and left ventricles.
- Explain how the mechanisms revealed by the Starling law of the heart help maintain cardiac output when either afterload or heart rate is increased.
- Predict how changes in heart rate, aortic pressure, stroke volume, and contractility, either alone or in combination, alter myocardial oxygen demand.
- Identify physiologic and pathologic factors that alter cardiac output through changes in ventricular filling.
- Identify pathologic factors that are associated with a negative inotropic state in the heart.
- Explain how the reciprocal relationship between heart rate and stroke volume helps maintain cardiac output at high heart rates and at low heart rates.
- Explain how indicator dilution techniques can be used to measure cardiac output.
- Explain the benefits and limitations associated with modern imaging techniques for determining ventricular volumes and cardiac output.

Clinical interest in cardiac electrophysiology stems from the fact that this activity is coupled to the contraction of the heart and movement of blood through the cardiovascular system. The metabolic demands of different organs of the body and the body as a whole vary widely in an individual, and the output of blood pumped by the heart must match those demands. Clearly, therefore, the mechanical force-generating capacity of the heart muscle must be able to be increased when tissue metabolic demands are high, such as during exercise or digestion.

The heart is made of striated muscle and shares several contractile properties with that tissue. However, two modes of enhancing contractile force in skeletal muscle, tetanic contraction and motor unit recruitment, are unavailable to cardiac muscle and therefore cannot be used to enhance contractile performance of the heart. The extended action potential refractory period of atrial and ventricular muscle fibers is too long to allow for the force of individual twitch contractions to be summated (Fig. 13.1). Consequently, cardiac muscle cannot be tetanized. This is actually an important property of the heart in that the heart could not function as a pump were it to simply generate single prolonged contractions.

Chapter 8, "Skeletal and Smooth Muscle," demonstrated that the force-generating capacity of a skeletal muscle group can be enhanced by the recruitment of motor units. Theoretically, such a mechanism could have been built into the heart such that it could enhance pumping force when needed. However, the heart behaves as a functional syncytium, and therefore, all cells that can contract in the heart during a single activation do so; there are no spare units to recruit.

In spite of its syncytial properties, cardiac muscle has the ability to enhance its force-generation capacity via two mechanisms that are not available to skeletal muscle. Recall that skeletal muscle exhibits an active isometric length–tension relationship whereby increases in passive length, up to an optimum length, generate more isometric force. However, because skeletal muscles are attached to fixed points on the skeleton, moving passively along the ascending

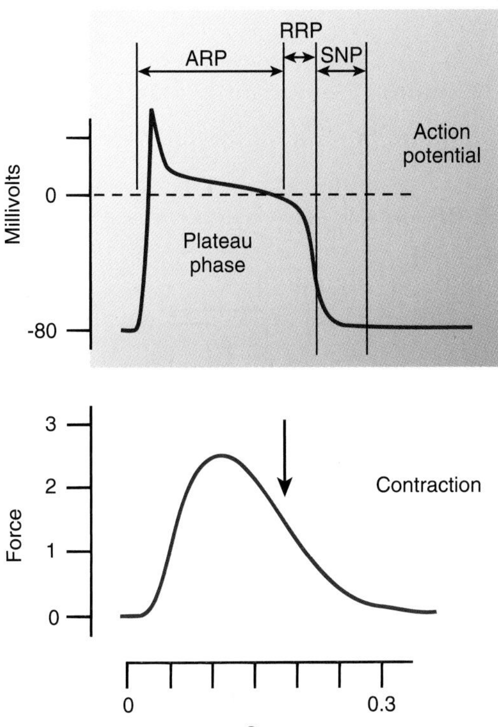

Figure 13.1 A cardiac muscle action potential and isometric twitch. Because of the duration of the action potential, cardiac muscle relaxes before it can be activated again. Therefore, an effective tetanic contraction cannot be produced. ARP, absolute refractory period; RRP, relative refractory period; SNP, period of supranormal excitability.

limb of the length–tension curve is not done by that tissue to adjust contractile force. The heart, in contrast, is a hollow, compliant, striated muscular organ. It can, and does, have its force generation modified by moving along the ascending limb of the length–tension relationship. Finally, skeletal muscle fibers do not contain mechanisms whereby the intrinsic force-generation capacity of individual cells can be enhanced. In contrast, the force-generation capacity of the heart muscle can be altered at the level of the individual cell.

CARDIAC EXCITATION–CONTRACTION COUPLING

Coupling of electrical activity in the heart to myocardial contraction uses mechanisms similar to those previously discussed for skeletal muscle. Cardiac cells contain an extensive network of T tubules that extend longitudinally along myocardial fibers (in contrast to the radial arrangement in skeletal muscle). As in skeletal muscle, these carry action potentials deep into the cell, where they open dihydropyridine (DHP) calcium channels that are coupled to ryanodine-sensitive calcium-release channels in the sarcoplasmic reticulum (SR). The actin myosin crossbridge cycling and its control through Ca^{2+}/troponin/tropomyosin interactions are essentially the same in both forms of striated muscle. Finally, both muscles use Ca^{2+} adenosine triphosphatase (ATPase) pumps in the SR to remove Ca^{2+} from the intracellular fluid while accumulating it into the SR against a concentration gradient. Ca^{2+} ATPase pumps and Na^+/Ca^{2+} in the cell membrane also reduce intracellular calcium levels and terminate contraction.

Force of contraction in cardiac muscle can be altered by changes in intracellular calcium concentration.

In contrast to skeletal muscle fibers, intracellular Ca^+ levels and force generation can be altered in individual myocardial cells by normal physiologic as well as abnormal conditions (Fig. 13.2). Physiologically, myocardial contraction is modified by altering the gating of the DHP Ca^{2+} channels in the cell membrane that are activated during phase 2 of the action potential. Calcium entering the cell through this channel, however, is not itself used to directly cause contraction. Only about 10% of this Ca^{2+} contributes to contraction of the muscle. Instead, increased intracellular Ca^{2+} stimulates the release of Ca^{2+} stored within the SR (**calcium-induced calcium release**). In cardiac muscle, the DHP channel is linked to the ryanodine-sensitive calcium-release channel in the SR membrane such that each quanta of calcium admitted through the DHP channel causes a small, localized release of calcium from one ryanodine receptor channel in the SR. These localized intracellular bursts of calcium are called **calcium sparks**. The localized high concentration caused by a single calcium spark does not diffuse within the intracellular fluid to neighboring receptors in the SR at a high enough concentration to cause them to open. Instead, multiple calcium sparks are summated temporally and spatially within the myocardial cell to produce overall increases in intracellular Ca^{2+} concentration. In addition, high intracellular calcium concentrations stimulate the pumping of Ca^{2+} into the SR by the Ca^{2+} ATPase. This creates a larger pool of calcium within the SR that can be released with the next opening of the SR calcium-release channels.

In general, anything that enhances calcium influx during the action potential in a cardiac myocyte results in a greater intracellular release of Ca^{2+} and thus a stronger subsequent contraction. Ca^{2+} influx through the voltage-dependent channels is modified by cellular regulatory mechanisms. For example, catecholamines, like norepinephrine released from sympathetic nerve endings or circulating epinephrine, activate β_1-adrenergic receptors on the myocardial cell membrane. These receptors are linked to adenylate cyclase within the myocardial cell by a stimulatory G protein (more commonly known as G_S). When catecholamine binds to the β_1 receptor, the adenylate cyclase is activated and converts adenosine triphosphate (ATP) to cyclic adenosine monophosphate (cAMP). In myocardial cells, cAMP binds to a cAMP-dependent protein kinase that phosphorylates the voltage-dependent Ca^{2+} channel. This phosphorylation increases the channel's probability of opening and increases the average time the channel remains open once activated. This enhances Ca^{2+} influx with every action potential, resulting in more intracellular calcium release and a stronger contraction in each myocardial cell. Extended to the whole heart,

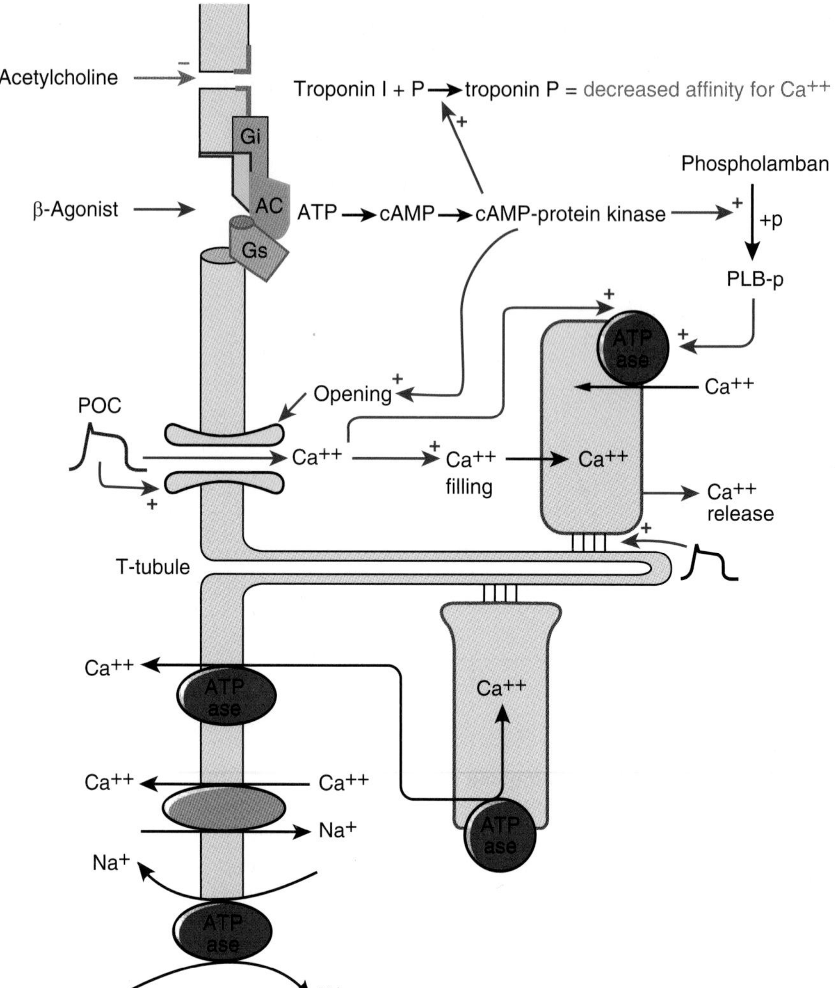

Figure 13.2 Handling of calcium in a cardiac muscle cell. Action potentials cause an influx of calcium through voltage-gated L-type dihydropyridine (DHP) Ca^{2+} channels. Influx through one channel is closely linked to the release of calcium from one calcium-release channel in the sarcoplasmic reticulum (SR) (calcium-induced calcium release), causing a "calcium spark." Modulation of the influx of calcium through DHP channels through G protein–linked receptor and non–receptor-mediated events gives rise to the capability of modulation of the inotropic state of a single myocardial cell. Cyclic adenosine monophosphate (cAmp) mediated mechanisms also speed relaxation during β-adrenergic receptor–mediated increases in cell contractility. Calcium pumps in the plasma and SR membranes as well as secondary active transport mechanisms for calcium in the plasma membrane return intracellular calcium concentration to low levels between action potential activations of the cell. AC, adenylate cyclase; ATPase, adenosine triphosphatase; Gi, inhibitory G protein; Gs, stimulatory G protein; P, phosphate; PLB-p, phosphorylated phospholamban; POC, potential-operated Ca^{2+} channel.

this increases the force-generating capacity of that organ and increases the output of blood with each contraction.

Modification of force generation in individual myocardial cells is independent of loading conditions on cardiac muscle.

The key feature of the ability of cardiac cells to alter their intrinsic contractile force through manipulation of their intracellular calcium is that the resulting changes in force are intrinsic to the cell itself. This modification of force capacity occurs *independently* of any changes in force caused by alterations in preload or afterload in the heart. In cardiac muscle, this additional intrinsic factor is superimposed on any effects of preload or afterload on the heart. Thus, at *any* given loading condition, the heart exposed to a catecholamine can generate more isometric force and move a greater load faster and farther compared with a heart not exposed to that agent. A modification of the contractile ability of muscle at the cellular level, independent of loading conditions, is said to be a modification of the **inotropic state**, or **contractility** of the muscle. The ability to modify muscle performance by changing the inotropic state of individual cells is a hallmark of the physiologic control of myocardial contraction. (Note that the term "contractility" is often applied loosely in clinical settings to any factor that alters the contraction of the heart. This is a misapplication of the term. It is important for both physiologists and clinicians to distinguish alterations in the performance of the heart resulting from changes in inotropic state from those changes caused by alterations in loading conditions. In this text, the term "contractility" will be used to refer to the inotropic state of the heart).

Factors that alter myocardial contractility

Many agents and conditions can alter the contractility, or inotropic state, of the heart (see Fig. 13.2 for reference). For example, β-receptor agonists are positive inotropic agents, whereas β-receptor antagonists are negative inotropic agents. Agents that block Ca^{2+} channels reduce myocardial contractility and thus exert a negative inotropic effect on the heart. Acetylcholine, released from parasympathetic nerve endings in the heart, is a negative inotropic agent. It inhibits adenyl cyclase through an inhibitory G protein and stimulates guanylate cyclase to produce cyclic guanosine monophosphate, which inhibits the opening of the Ca^{2+} channel.

The DHP channels require ATP for normal functioning. Therefore, myocardial ischemia, or lack of oxygen to the heart, results in an inhibition of the Ca^{2+} channel and is considered a negative inotropic condition. Also, acidosis (increased plasma H^+), which often occurs in ischemic conditions, inhibits myocardial contractility by interfering with the DHP channel.

Other non–receptor-mediated factors can alter intracellular calcium concentration and thus alter myocardial contractility. For example, Na^+/Ca^{2+} exchangers in the myocardial cell membrane remove Ca^{2+} from the interior of the cell through a secondary active transport mechanism linked to the influx of sodium moving down its electrochemical gradient. The reduction of the sodium gradient by the inhibition of Na^+/K^+ ATPase results in Ca^{2+} accumulation within the cell and a positive inotropic effect. This mechanism is responsible for the positive inotropic effects of cardiac glycosides such as digitalis and is the basis of their use to augment myocardial performance in heart failure. Methylxanthines, such as caffeine and theophylline (an asthma medication and component of tea), inhibit cAMP phosphodiesterase. These compounds therefore increase cAMP accumulation and induce a positive inotropic effect.

Activation of the sympathetic nervous system increases heart rate (HR) (a positive chronotropic effect) at the same time as it enhances myocardial contractility. Rapid HRs require the rapid removal of calcium and inactivation of actin/myosin interactions so that diastole can shorten to accommodate the high HR. Generation of cAMP and protein kinase A by norepinephrine phosphorylates a regulatory protein in the SR called **phospholamban**. This protein normally inhibits the Ca^{2+} ATPase pumps in the SR. However, when it is phosphorylated, its inhibitory effects are removed and the pumping activity of the Ca^{2+} ATPase increases, thus removing calcium from the intracellular fluid more rapidly. In addition, phosphorylation of troponin 1 by protein kinase A decreases its affinity for calcium, allowing the cellular actin–myosin crossbridges to break their attachments. The effects on calcium pumping in the SR and on calcium affinity accelerate the relaxation of myocardial cells.

CARDIAC CYCLE

The cyclic contraction and relaxation of the myocardium sets into motion a sequence of events resulting in time-dependent changes in ventricular and aortic pressure, ventricular volumes, and flow into and out of the heart. A graphical representation of the variations in hemodynamic variables associated with the cardiac cycle is depicted in Figure 13.3. Such representations usually depict hemodynamic variables in the left heart and systemic circulation along with a tracing of the electrocardiogram (ECG) and heart sounds.

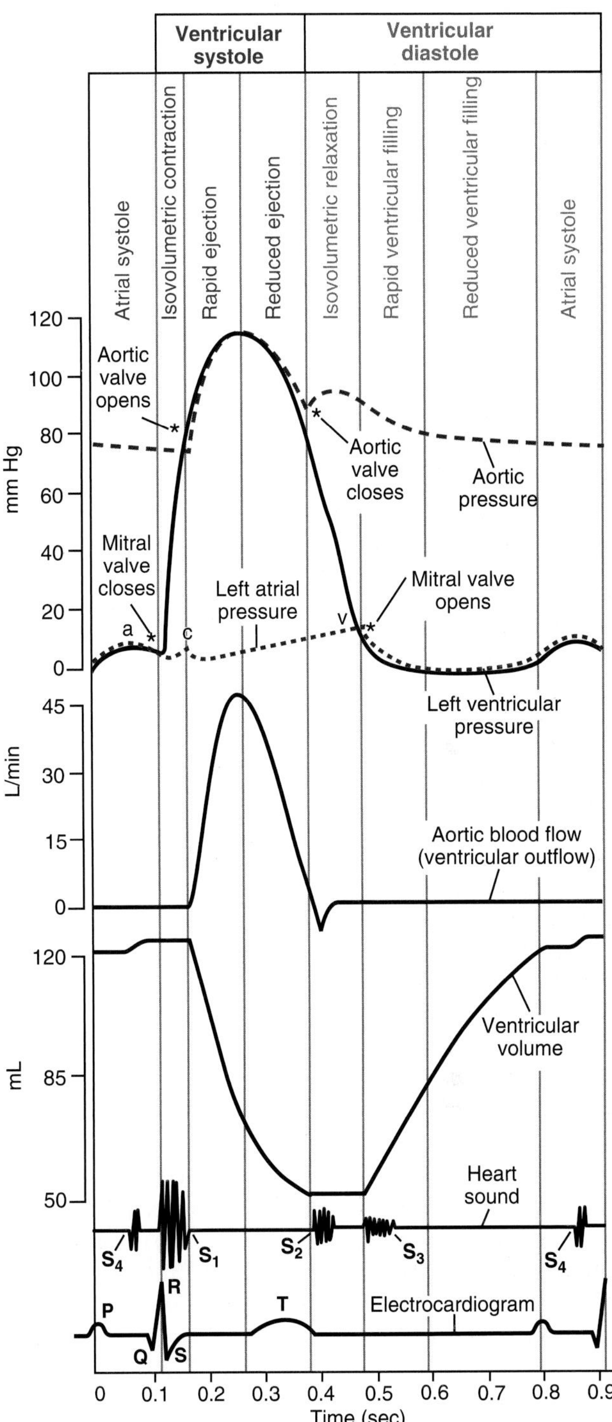

Figure 13.3 The timing of various hemodynamic and related events during the cardiac cycle. (See text for details.)

Ventricular systole consists of isovolumic contraction followed by a rapid, then reduced, blood ejection phase.

The peak of the R wave on the ECG is used to signify the start of systole, or the contraction phase of the cardiac cycle. This contraction increases intraventricular pressure above end-diastolic levels, which are about 0 to 5 mm Hg. When left ventricular pressure exceeds that in the left atrium, the mitral valve closes. This closure is associated with a series of broad low-pitched sounds that are a result of the vibrations of blood and the chordae tendineae in the ventricles. This is called the **first heart sound**. (Contrary to popular belief, these sounds are not made by valve leaflets "slapping against one another.") The intensity of the first heart sound is proportional to the strength of myocardial contraction, and its evaluation can be used in clinical diagnosis.

As the ventricle contracts, there is a period of time when intraventricular pressure exceeds that in the left atrium but is still less than that in the aorta. Consequently, the ventricle builds pressure without moving blood because the aortic valve remains closed. This phase of systole is called **isovolumic contraction**. Once the left ventricular pressure exceeds that in the aorta, the aortic valve opens and blood leaves the ventricle, initiating the ejection phase of the cardiac cycle. During systole, a healthy heart ejects only 50% to 60% of the volume of blood that was in the ventricles at the end of the previous diastole. Initially, with the opening of the aortic valve, ventricular volume decreases rapidly; about 70% of the volume to be ejected exits the ventricle during the first third of the ejection phase. This period is thus called the **rapid ejection phase**. Intraventricular and aortic pressures rise during this phase. As the outflow of blood distends the aortic wall, lateral aortic pressure may actually slightly exceed left ventricular pressure, but the combined potential and kinetic energy during ejection is still greater in the ventricle than in the aorta, and flow proceeds outward. However, as some of the kinetic energy of outflow is converted into the potential energy that distends the aorta walls, the rate at which blood exits the ventricle begins to decline. This is called the **reduced ejection phase**.

During the reduced ejection phase, the T wave of the ECG begins and the aortic and ventricular pressures start to decline. When ventricular pressure eventually declines below total aortic pressure (lateral and kinetic components), the aortic valve closes. The sudden cessation of ventricular outflow caused by valve closure creates a **second heart sound**, whose intensity is proportional to the intensity of the valve closure. Intensity is increased whenever aortic or pulmonary pressure is abnormally high and is taken as a clinical indicator of possible systemic or pulmonary hypertension, respectively. The closure of the aortic valve also causes a transient, sharp, small drop in aortic pressure, creating a notch in the aortic pressure profile, called the **incisura**. This event is used sometimes to demarcate the transition from systole to diastole in the cardiac cycle. However, systole is more correctly considered to be concluded when the T wave ends on the ECG.

The difference in the volume of blood in the ventricle at the end of diastole, or the **end-diastolic volume**, and the volume of blood in the ventricle at the end of systole, or the **end-systolic volume**, is called the **stroke volume**. Stroke volume represents the volume of blood ejected with each contraction of the ventricles (see Fig. 13.3). Note that the ventricles do not empty completely with each contraction. At the end of systole, a significant volume of blood remains in the ventricles and is called the **residual volume**. Residual volume is decreased by increases in myocardial contractility and HR and increased whenever the heart is weakened (e.g., heart failure) or faced with increased outflow resistance (e.g., aortic valve stenosis). Thus, examination of end-systolic volume is useful clinically as an indicator of conditions affecting the heart.

Ventricular diastole consists of isovolumic relaxation followed by a rapid, then reduced, filling phase.

During the relaxation, or diastolic, phase of the cardiac cycle, volume and pressure changes proceed in the reverse of that seen for systole. Early in left ventricular relaxation, the aortic and mitral valves are closed, and the ventricle relaxes isovolumetrically. Thus, this phase of the cardiac cycle is called **isovolumic relaxation**. Once ventricular pressure falls below that of the left atrium, the mitral valve opens and the filling of the ventricle begins. Just prior to this occurrence, abrupt cessation of ventricular distention and the deceleration of blood create a faint third heart sound not normally heard in healthy people. However, the third heart sound is amplified in abnormally distended ventricles, such as that associated with heart failure, and its presence is therefore considered a serious sign of underlying cardiac abnormalities.

During diastole, the ventricles fill in a rapid filling phase, followed by a reduced filling phase (**rapid ventricular filling** and **reduced ventricular filling** phases). During rapid ventricular filling, ventricular pressure actually continues to decline because ventricular relaxation occurs more rapidly than the filling of the ventricle. During the reduced filling phase, sometimes called diastasis, ventricular pressure starts to increase. The appearance of the P wave on the ECG and atrial contraction begin at the end of the reduced filling phase of the cardiac cycle. Atrial contraction produces a faint fourth heart sound. Diastole is considered to be concluded with the appearance of the R wave on the ECG.

Abnormal conditions in heart valves are revealed by changes in venous pressure waveforms.

Occasionally, pressure waveforms in the jugular vein are depicted on cardiac cycle diagrams. When the right atrium contracts, a retrograde pressure pulse wave is sent backward into the jugular vein. This is called the **A wave**. Factors that impede the flow of blood from the atria to the ventricles, such as tricuspid valve stenosis, increase the amplitude of the A wave. A second venous pulse wave, called the **C wave**, is seen as an increase followed by a decrease in venous pulse pressure during the early phase of systole. The upslope of this wave is created by the bulging of the tricuspid valve into the right atrium during ventricular contraction, which sends a wave into the jugular vein. This wave is combined with a lateral transmission of the carotid systolic arterial pulse to the adjacent jugular vein. The subsequent decrease in pressure in the C wave is caused by the descent of the base of the heart and atrial stretch. Failure of the tricuspid valve to completely close during ventricular systole results in the propulsion of blood back into the atrium and vena cava and results in a high-amplitude C wave. The **V wave** of the venous pulse is seen as a gradual pressure increase during reduced ejection and isovolumic relaxation followed by a pressure decrease during the rapid-filling phase of the cycle. This wave is created first by continual peripheral venous return of blood to the atrium against a closed tricuspid valve followed by the sudden decrease in atrial distention caused by rapid ventricular filling. Tricuspid valve stenosis increases resistance to the filling of the right ventricle, which is indicated by an attenuation of the descending phase of the V wave.

Analysis of venous waveforms was once used to provide clinical insights into cardiovascular disease involving cardiac valves. However, this type of indirect diagnosis has since been supplanted by several modern imaging techniques used to reveal valve condition and motion (as discussed later in this chapter).

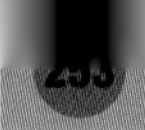

DETERMINANTS OF MYOCARDIAL PERFORMANCE

Evaluation of the mechanical properties of the heart as they relate to its function as a pump is important to physiologists and physicians. The heart must be able to enhance its pumping capacity to meet increased demands for blood flow during conditions such as exercise. In addition, myocardial performance is hindered in pathologic conditions such as ischemia and heart failure. In the previous section, it was shown that alteration of the inotropic state of individual myocardial cells is one means by which the contractile performance of the heart can be altered. However, as is the case with skeletal muscle, preload and afterload effects are also important determinants of myocardial performance. These factors interact collectively with inotropic conditions to modify the overall contractile performance of the heart.

Preload, afterload, and inotropic state interact to alter contractile performance of cardiac muscle.

Four factors influence the contractile strength of cardiac muscle: (1) preload, or the initial length to which the muscle is stretched prior to contraction, (2) afterload, or all the forces against which cardiac muscle must contract to generate pressure and shorten, (3) contractility, or inotropic state, and (4) the inotropic effect of increased HR (beats/min). Note that HR affects the inotropic state of the heart indirectly. High HRs bring Ca^{2+} into the cell with each action potential at a rate faster than it can be removed. Consequently, intracellular Ca^{2+} concentration increases with a resulting positive inotropic effect. This enhanced contractility helps the heart eject more blood with each contraction at high HRs and helps compensate for reduced filling time associated with tachycardia.

Cardiac muscle exhibits an active length–tension relationship similar to that of skeletal muscle (Fig. 13.4), which relates isometric force generation to the initial stretch, or preload, on the muscle. However, cardiac muscle is not anchored to a fixed object, as is the case with skeletal muscles. It is a distensible hollow organ and could theoretically, as a whole, have its preload altered by how much it was filled with blood prior to a contraction. In this regard, however, it is important to realize that cardiac muscle is intrinsically stiffer than skeletal muscle and exhibits significant passive resistance to stretch at a length corresponding to L_o. As a result, cardiac muscle is *constrained to contract from lengths* $\leq L_o$. Consequently, all other factors being equal, force generation in cardiac muscle is proportional to its preload, or its initial passive stretch, prior to the onset of contraction. This is shown as the ascending limb of active length–tension relationship for isometric contraction (see Fig. 13.4). In the whole heart, passive stretch, or preload, is proportional to the amount of filling of the heart during diastole. Thus, increased filling increases the potential isometric contractile force that can be generated in the heart.

Clinical Focus / 13.1

Cardiomyopathies: Abnormalities of Heart Muscle

Heart disease takes many forms. Although some of these are related to problems with the valves or the electrical conduction system, many are a result of malfunctions of the cardiac muscle itself. These conditions, called *cardiomyopathies*, result in impaired heart function that may range from asymptomatic conditions to malfunctions causing sudden death.

There are several types of cardiomyopathy, and they have several causes. In *hypertrophic cardiomyopathy*, an enlargement of the ventricular muscle fibers occurs because of a chronic overload, such as that caused by hypertension or a stenotic aortic valve. Such muscle may fail because its high metabolic demands cannot be met, or fatal electrical arrhythmias may develop. *Congestive* or *dilated cardiomyopathy* refers to cardiac muscle so weakened that it cannot pump strongly enough to empty the heart properly with each beat. In *restrictive cardiomyopathy*, the muscle becomes so stiffened and inextensible that the heart cannot fill properly between beats. Chronic poisoning with heavy metals, such as cobalt or lead, can produce *toxic cardiomyopathy*, and the skeletal muscle degeneration associated with muscular dystrophy is often accompanied by cardiomyopathy.

The cardiomyopathy arising from *viral myocarditis* is difficult to diagnose and may show no symptoms until death occurs. The action of some enteroviruses (e.g., coxsackievirus B) may cause an autoimmune response that does the actual damage to the muscle. This damage may occur at the subcellular level by interfering with energy metabolism while producing little apparent structural disruption. Such conditions, which can usually only be diagnosed by direct muscle biopsy, are difficult to treat effectively, although spontaneous recovery can occur. In tropical regions, infection with a trypanosome (Chagas disease) can produce chronic cardiomyopathy. The tick-borne spirochete infection called Lyme disease can cause heart muscle damage and lead to heart block. Excessive and chronic consumption of alcohol can also cause cardiomyopathy that is often reversible if total abstinence is maintained.

Another important kind of cardiomyopathy arises from ischemia. An acute ischemic episode may be followed by a *stunned myocardium*, with reduced mechanical performance, whereas chronic ischemia can further reduce mechanical performance. Ischemic tissue has impaired calcium handling, which can lead to destructively high levels of internal calcium. These conditions can be improved by reestablishing an adequate oxygen supply (e.g., following clot dissolution or coronary bypass surgery), but even this treatment is risky because rapidly restoring the blood flow to ischemic tissue can lead to the production of oxygen radicals that cause significant cellular damage.

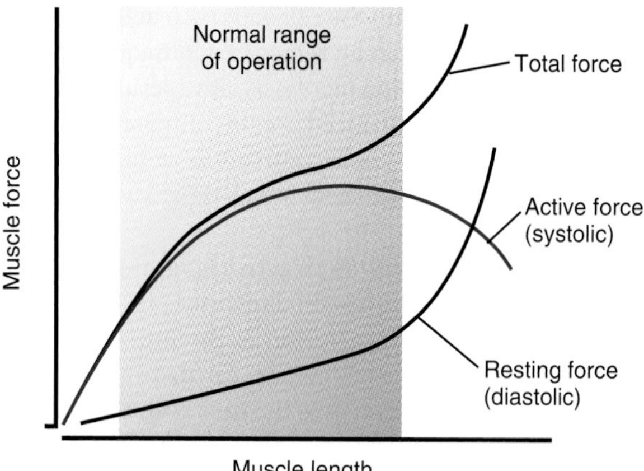

Figure 13.4 The isometric length–tension curve for isolated cardiac muscle. Cardiac muscle displays a parabolic active length–tension relationship similar to that of skeletal muscle but shows considerably more passive resistance to stretch at L_o.

Isotonic contraction of cardiac muscle proceeds along a loop depicting contraction and relaxation.

The heart muscle is constrained by its anatomy and functional arrangements to follow different pathways during contraction and relaxation. This pattern is seen clearly when the phases of the contraction–relaxation cycle are displayed on a length–tension curve. In physical terms, the area enclosed by the cardiac muscle pathway illustrated in Figure 13.5 represents work done by the muscle on the external load. If the afterload or the starting length (or both) is changed, then a different pathway will be traced. The area enclosed will differ with changes in the conditions of contraction, reflecting differing amounts of external work delivered to the load.

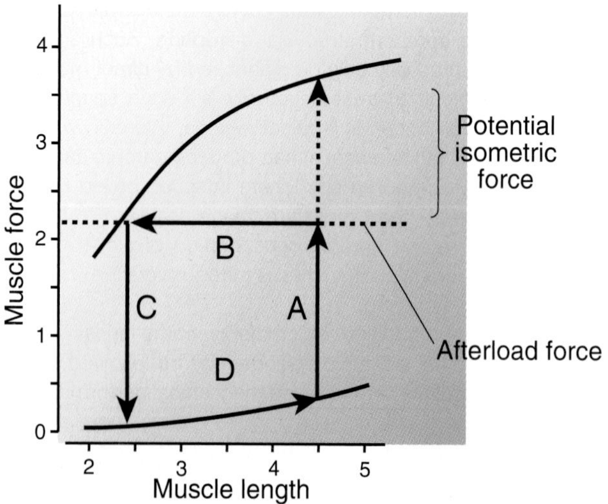

Figure 13.5 An afterloaded contraction of cardiac muscle, plotted in terms of the length–tension curve. The afterload provides the limit to force; the isometric length–tension curve provides the limit to shortening. **(A)** Isometric contraction phase. **(B)** Isotonic shortening phase. **(C)** Isometric relaxation phase. **(D)** Relengthening. (See text for details.)

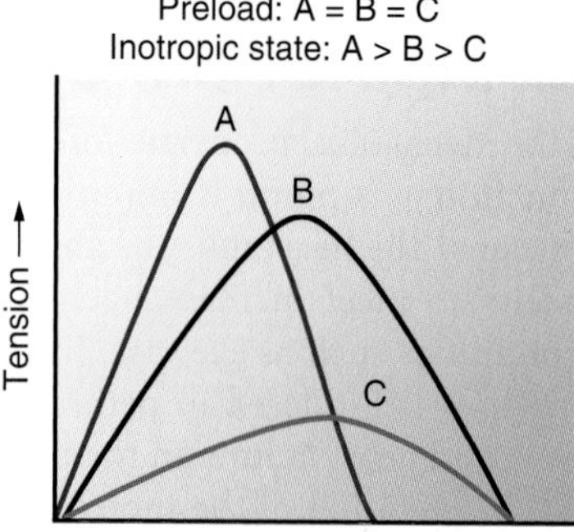

Figure 13.6 The effects of changes in inotropic state on the isometric length–tension curve of the heart. The slope of the ascending limb is proportional to the inotropic state so that positive inotropic influences enhance isometric force at any preload along the ascending limb.

Effects of preload and afterload on cardiac contraction are enhanced or depressed by changes in cardiac inotropic state.

A key to understanding the contraction of cardiac muscle is to understand that the effect of any given loading condition on the contractile force generated by the heart can be altered by the inotropic state of the heart. At any given preload, isometric force generation of cardiac muscle is increased by positive inotropic stimuli and decreased by negative inotropic stimuli (Fig. 13.6).

In cardiac muscle, changes in preload have effects on the isotonic force–velocity relationship that are similar to those seen in skeletal muscle (Fig. 13.7A). Decreasing preload reduces the velocity of shortening obtainable while contracting against any given afterload. It also reduces the magnitude of the afterload the cardiac muscle can move at any given velocity. Increasing preload, naturally, has the opposite effect on afterloaded contraction. However, the entire force–velocity relationship for any given preload can be shifted by the inotropic state of the heart. Positive inotropic influences shift the force–velocity curve at any given preload upward and to the right. In this manner, a positive inotropic agent allows cardiac muscle, at any given preload, to move a heavier load faster than that possible with normal myocardium at the same preload (Fig. 13.7B). Negative inotropic agents obviously have the opposite effect.

Unlike in skeletal muscle, the theoretical maximum rate of crossbridge cycling, or $\mathbf{V_{max}}$, in cardiac muscle is not constant. It is enhanced by positive inotropic stimuli and depressed by negative inotropic influences. Furthermore, V_{max} is not appreciably affected by loading conditions in cardiac muscle. Therefore, V_{max} can be thought of as a reflection of the inotropic state of the heart and indices that are related to V_{max}, can be used to estimate the state of myocardial contractility (see below).

Figure 13.8 demonstrates how preload, afterload, and inotropic state interact to define myocardial contraction. Positive inotropic stimuli shift the isometric length–tension relationship upward and to the left, enabling cardiac muscle to generate more isometric force at any preload. However, as is the case with skeletal muscle, the isometric length–tension relationship sets the boundary that determines the limits

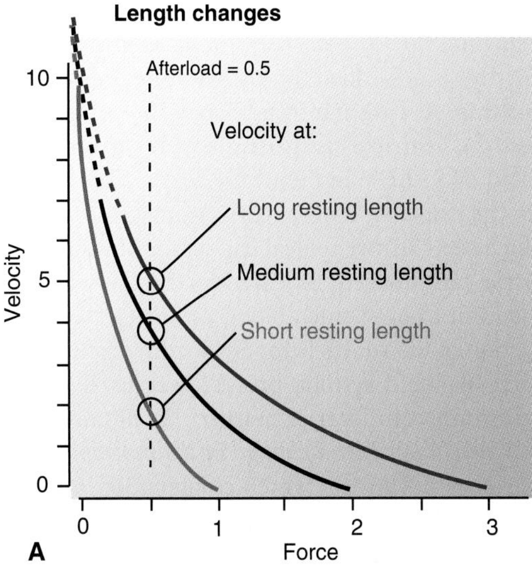

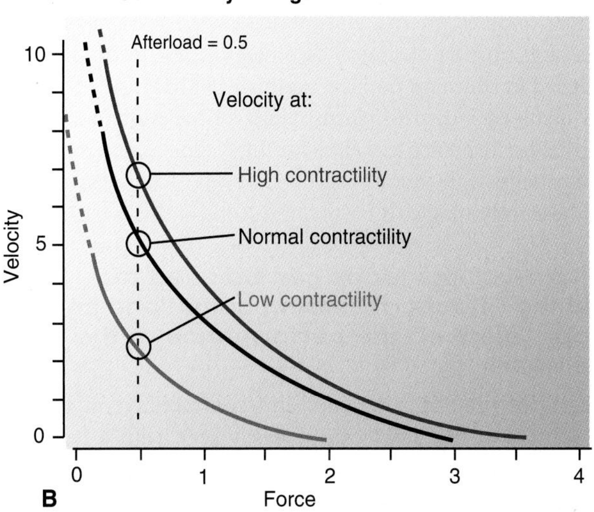

Figure 13.7 Effect of length and contractility changes on the force–velocity curves of cardiac muscle. (A) Decreased starting length (with constant contractility) produces lower velocities of shortening at a given afterload. Because of the presence of resting force (characteristic of heart muscle), it is impossible to make a direct measure of a zero-force contraction at each length. **(B)** Increased contractility produces increased velocity of shortening at a constant muscle length, but there is no tendency for the curves to converge at the low forces, and thus V_{max} is proportional to the inotropic state.

of how far muscle can shorten. Therefore, compared with normal muscle at the same preload and afterload, positive inotropic stimuli allow cardiac muscle to shorten farther or to move a greater load a given distance compared with that produced by the heart in the normal inotropic state.

Variables associated with contractile performance of the heart *in situ* are analogous to afterload, preload, and muscle shortening.

The relationships among preload, afterload, shortening, and contractility elucidated from studies of isolated cardiac muscle determine the output characteristics of the whole heart. To understand how loading conditions and contractility affect myocardial shortening and flow output, one needs to identify variables in the whole heart that are analogous to those in isolated muscle. For example, stroke volume, or the amount of blood ejected with each contraction of the heart, is analogous to muscle shortening. Similar analogs are used for preload and afterload. As the left ventricle fills with blood during diastole, it becomes stretched. The magnitude of this stretch is related to the blood volume in the heart at the end of diastole (**left ventricular end-diastolic volume [LVEDV]**). Therefore, LVEDV can be used as an indicator of the preload in the left ventricle in the heart *in situ*. Because changes in end-diastolic volume result in changes in end-diastolic pressure, left ventricular end-diastolic pressure (LVEDP) is sometimes used as an indicator of preload on the heart. However, caution must be used in applying LVEDP as a measure of preload. Myocardial compliance can be reduced by myocardial ischemia, infarction, or hypertrophy. All of these will result in an increased end-diastolic pressure at any given LVEDV. Consequently, a person recording an increased LVEDP in this situation could assume, erroneously, that preload in the heart was increased, when in fact it was not.

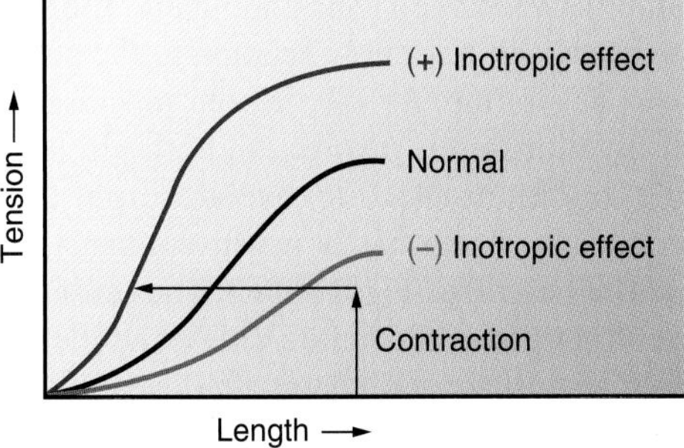

Figure 13.8 The effects of changes in inotropic state on isotonic contractions in the heart. Positive inotropic stimuli shift the isometric length–tension curve to the left, thereby increasing the extent of shortening of muscle at any given set of preload and afterload on the heart; negative inotropic stimuli have the opposite effect.

In the left ventricle as a whole, afterload is equal to all the forces the muscle must overcome to eject a given volume of blood. The major contributor to afterload in the heart is ventricular or aortic pressure. However, it is more difficult to contract a distended heart against a given pressure than a normal heart, because cardiac wall tension and stress increase at any pressure when the average radius of the ventricular chamber increases. Therefore, tension (T = pressure × chamber radius) and wall stress (S = pressure × chamber radius/chamber thickness) are considered more accurate estimators of cardiac afterload. Other forces, such as those needed to overcome blood inertia, accelerate blood, or overcome the resistance of the valves, also contribute to afterload, especially in pathologic conditions.

The use of technologies to give better determinations of ventricular volumes, preload, and afterload in the heart encompasses a large portion of clinical cardiology. These techniques are not used in common physical examination but instead are used by qualified specialists in cardiology for the evaluation of myocardial performance as well as for the diagnosis and treatment of patients with serious heart conditions. A full description of these technologies is beyond the scope of this book.

Pressure–volume loops can be used to reveal the effects of loading conditions and inotropic state on the performance of the whole heart.

The effects of preload, afterload, and contractility on the stroke volume of the heart can be easily seen with a tracing of the changes in intraventricular pressure and volume during the cardiac cycle, as shown in Figure 13.9. (The time for each portion of the cardiac cycle is not considered in such depictions.) The cardiac cycle starts at point A with the beginning of diastole. Filling of the ventricle proceeds along the line connecting point A to point B, where B represents LVEDV. Isovolumic contraction proceeds from this point to point C, where ventricular pressure equals that in the aorta. At this point, the aortic valve opens and blood is then ejected from the ventricle. The volume in the ventricle during ejection proceeds from point C to point D. This volume change is the stroke volume. Finally, the heart undergoes isovolumic relaxation (from point D back to point A).

Stroke volume is positively related to the level of the inotropic state of the heart.

The pressure and volume in the heart at the end of systole represent the point to which the heart has shortened such that the maximum force it can develop at that length just equals aortic afterload. The heart cannot shorten beyond this point because, at shorter circumferences, the number of cross bridges available would not enable the heart to generate enough force to move blood against existing ventricular stress. If one were to examine several cardiac cycles starting at various LVEDV (preloads) and proceeding against several different aortic pressures (afterloads), the points depicting

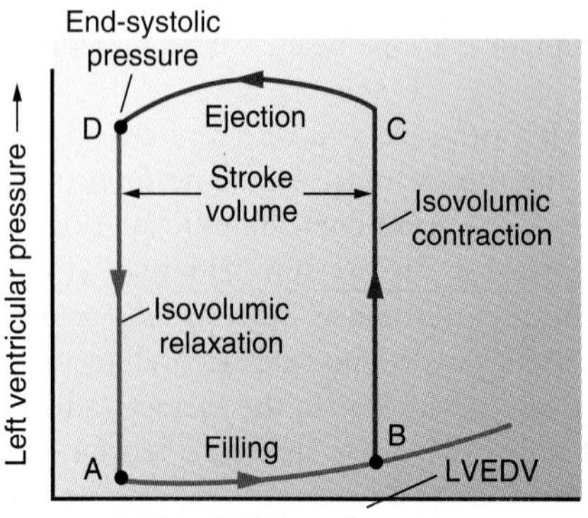

Figure 13.9 A pressure–volume loop representation of the cardiac cycle. (See text for details.) LVEDV, left ventricular end-diastolic volume.

pressure and volume at the end of systole would all fall on a straight line like the dashed line labeled no. 2 in Figure 13.10A. This line is called the **end-systolic pressure–volume relationship**. At a given inotropic state, it represents the most pressure the ventricle can generate with a given preload (as estimated by LVEDV in the whole heart). This line is analogous to the isometric length–tension curve for isolated muscle, which sets the mechanical barrier for how far shortening in muscle can proceed. By extension to the whole heart, it sets the limit to how much blood can be ejected by a single contraction of the ventricle for any loading condition. A plot of ventricular end-systolic pressure versus LVEDV represents this important barrier marker. *The position of this line can only be changed by a change in the inotropic state of the heart;* it does not matter which combination of afterloaded and preloaded contractions was used to create the line. As a result, the LVESP–LVEDV relationship is considered to be an ideal indicator of the inotropic state of the heart.

Relative to a normal heart, a positive inotropic influence will shift the LVESP–LVEDV relationship upward and to the left, whereas a negative inotropic influence will shift it downward and to the right. The consequence of such a shift for the whole heart is that, *at any given afterload and preload, stroke volume is increased by a positive inotropic influence and decreased by a negative inotropic influence* (see Fig. 13.10, nos. 3 and 1, respectively).

Stroke volume is increased by an increase in preload or a decrease in afterload.

By themselves, loading conditions also affect the output of the heart. In Figure 13.10B, cardiac cycles 1, 2, and 3 occur with the same afterload and inotropic state but at different LVEDVs and thus preloads. As LVEDV increases, the initial muscle length from which myocardial contraction proceeds increases as well. Once contraction is initiated, the heart ejects blood until the pressure and volume remaining in the ventricle meet the LVESP–LVEDV relationship. As can be seen in the figure, greater initial LVEDV results in a greater stroke volume. This relationship was elucidated by Starling in the early 1900s and is known now as the **Starling law of the heart** (also known as *Starling's law of the heart*), or *heterometric regulation of stroke volume*. Put simply, this law states, "*the more the heart fills, the more it pumps*." The relationship between ventricular filling and the stroke output of the heart is important in relation to the performance of the heart as a pump. As will be seen in subsequent sections, this relationship is involved in both physiologic and pathologic conditions affecting **cardiac output (CO)**.

Afterload, by itself, has a negative impact on stroke volume in the heart (see Fig. 13.10C). The cardiac cycles in this figure all start from the same LVEDV and inotropic state but proceed against three different afterloads. Blood cannot be ejected from the heart until left ventricular pressure meets aortic pressure. When the ventricle contracts at higher afterloads (pressure), it will not shorten much before encountering the LVESP–LVEDV relationship. This results in a lower stroke volume compared with contractions proceeding against lower afterloads. It also indicates that pumping against

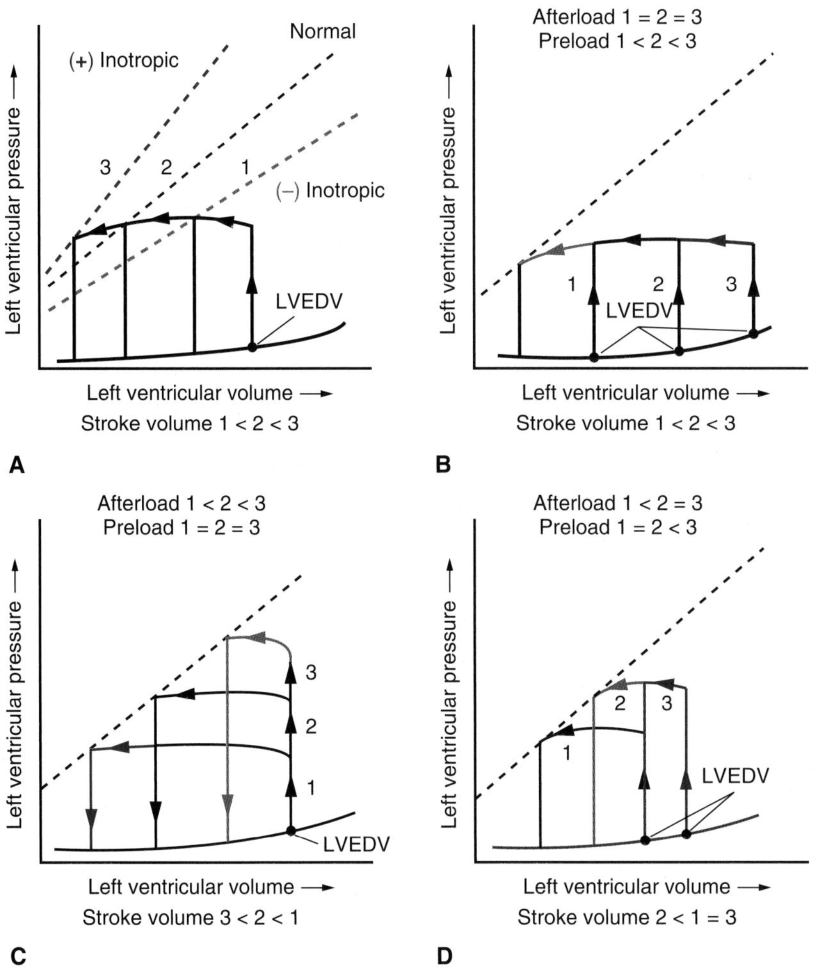

Normal
(+) Inotropic
3 2 1
(–) Inotropic
LVEDV
Left ventricular pressure →
Left ventricular volume →
Stroke volume 1 < 2 < 3
A

Afterload 1 = 2 = 3
Preload 1 < 2 < 3
1 2 3
LVEDV
Left ventricular pressure →
Left ventricular volume →
Stroke volume 1 < 2 < 3
B

Afterload 1 < 2 < 3
Preload 1 = 2 = 3
3 2 1
LVEDV
Left ventricular pressure →
Left ventricular volume →
Stroke volume 3 < 2 < 1
C

Afterload 1 < 2 = 3
Preload 1 = 2 < 3
2 3 1
LVEDV
Left ventricular pressure →
Left ventricular volume →
Stroke volume 2 < 1 = 3
D

Figure 13.10 Determinants of stroke volume as revealed by pressure–volume loop representations of the cardiac cycle. **(A)** Effect of inotropic state on stroke volume at a given left ventricular end-diastolic volume (LVEDV) and arterial pressure. **(B)** Effect of changes in LVEDV on stroke volume at constant arterial pressure and inotropic state (the Frank-Starling relationship). **(C)** Immediate effect of changes in afterload on ventricular stroke volume at constant LVEDV. **(D)** Changes in LVEDV following an initial increase in afterload and during the next cycle.

elevated arterial pressure will have a detrimental effect on the output of the heart unless certain compensatory mechanisms are brought into play. One such mechanism is shown in Figure 13.10D. Cardiac cycle 1 in this figure represents a contraction at normal LVEDV against a normal arterial pressure. Loop 2 represents the first contraction against an elevated arterial pressure from the same LVEDV and results in a reduced stroke volume. However, this reduced stroke volume means that the heart did not empty to its normal residual volume at the end of systole. In other words, the volume in the heart prior to its next filling is larger than normal. Thus, when the normal diastolic inflow in the next cardiac cycle is added to this increased residual volume, an increased LVEDV will be created for the next contraction (contraction 3 on the figure). By the Starling law, an augmented stroke volume will result. Heterometric regulation is thus a mechanism that enables the heart, within certain limits, to maintain a normal stroke volume against elevated arterial pressure.

The Starling law of the heart is responsible for the remarkable balancing of the output of the right and left ventricles. For example, if the output of the right heart should suddenly exceed that of the left, inflow into the left ventricle would also increase, increasing LVEDV and thus subsequent output of the left ventricle. In this manner, the Starling law automatically ensures that outputs of the right and left ventricle will always match.

In heart failure, the inability of individual cardiac cells to contract normally (a negative inotropic effect) can be somewhat compensated for by the fact that reduced ejection allows blood to pool in the left ventricle, increasing LVEDV and subsequently enhancing stroke volume. In heart failure, further compensatory mechanisms in the body increase water retention, which also results in increased filling of the ventricle and an augmentation of stroke volume. Similarly, when the heart is paced at a slow rate, such as from the atrioventricular node or Purkinje fibers, the detrimental effect of this low rate on the minute output of the heart is somewhat reduced by the fact that the ventricles have more time to fill and thus augment each contraction according to the Starling law.

Although the Starling law of the heart applies qualitatively to the heart in any condition, the quantitative effect of this law is attenuated whenever compliance of the myocardium is decreased. This is because when the heart is stiff, little stretch of the myocardium is produced with a normal amount of ventricular filling. Myocardial hypertrophy, from chronic ventricular overload, reduces ventricular compliance. Reduced ventricular compliance is especially problematic when it results from scar tissue formed in the heart following myocardial infarction, chronic inflammation, or infiltrative processes from infections. **Cardiac tamponade** is a condition in which bleeding occurs between the ventricles and the pericardium such as what may occur from the puncture of coronary arteries during cardiac catheterization procedures. This bleeding compresses the ventricles, making it difficult for them to expand during diastole, thus reducing diastolic filling. A similar situation can arise from scar tissue following **pericarditis**, in which inflamed pericardium eventually adheres to the epicardial tissue.

Increased afterload increases myocardial oxygen demand more than does increased preload, shortening, or inotropic state.

The heart is only 0.3% of adult body weight, yet at rest, it uses approximately 7% of the body's total O_2 consumption. The energy output of a system is classically defined as the sum of heat production and work. Work, in turn, is defined as a force acting through a distance, or work = force × distance. In the left ventricle, work is given as the product of stroke volume and arterial pressure, ventricular pressure, or wall stress during systole.

In relation to basic physics, it does not matter whether cardiac work results from a large stroke volume ejected against a low pressure or a small stroke volume ejected against a high pressure. In the heart, energy, and thus, oxygen must be used both to generate pressure and to pump blood. However, in the heart, *much more O_2 is required to meet increased work resulting from an increase in cardiac stress than from an increase in stroke volume*. Consequently, changes in cardiac work are poor indicators of myocardial oxygen demand.

Any index that could predict cardiac O_2 demand under a variety of hemodynamic conditions would be invaluable to clinical medicine. However, such an index that is both clinically practical and universally applicable to all cardiac O_2 demands has yet to be found. Nevertheless, much has been learned about the determinants of myocardial O_2 demand. Beyond the basal O_2 consumption needed to run membrane pumps and basic cellular activity, the major determinants of increased O_2 demand in the heart are (1) increased systolic pressure (or wall stress), (2) increased extent of muscle shortening (or stroke volume), (3) increased HR, and (4) positive inotropic stimuli.

Combinations of hemodynamic variables have been used as estimators of O_2 demand in the heart. The double product or systolic pressure × heart rate and the tension–time index (the area under the left ventricular systolic pressure curve) have been used as predictors of cardiac O_2 demand. Changes in these indices reflect changes in the oxygen needs of the heart caused by changes in HR and force generation, but they often fail in instances in which certain agents such as β-adrenergic agonists produce positive inotropic and chronotropic (HR) effects on the heart at the same time that their vascular effects cause a drop in blood pressure. Indices based on the end-systolic pressure–volume relationship can give accurate estimates of cardiac O_2 demand, but the determination of such relationships is currently impractical in a clinical setting.

Even without a good index available as a predictor of cardiac oxygen demand, it is useful to simply recognize factors proven to alter myocardial O_2 demand. For example, hypertension, or high blood pressure, greatly increases the stress load on the heart and thus increases O_2 demand. Similarly, exercise increases HR, myocardial contractility, and in unconditioned people, blood pressure as well, all of which synergize to create greatly increase myocardial O_2 demand. In patients who have compromised coronary circulations resulting from atherosclerosis, it is thus helpful to realize that decreasing blood pressure and HR will lessen the possibility that cardiac oxygen demand will exceed cardiac oxygen supply, a condition that will produce myocardial ischemia.

Due to the high O_2 requirements of the heart, this organ uses a wide variety of metabolic substrates. Two thirds of myocardial energy needs are derived from the metabolism of fatty acids, because on a molar basis, these compounds generate more ATP than sugars or amino acids. However, the metabolism of glucose supplies about 20% of the heart's energy needs, with the remaining needs supplied by the metabolism of lactate, pyruvate, amino acids, and certain ketones.

Ejection fraction and hemodynamic evaluation serve as simple clinical indices of myocardial performance.

Clinical evaluation of the condition of the heart in patients is technologically demanding. Although many technologies and devices are available to evaluate the heart, their implementation requires specialized skills in cardiology. Nevertheless, there are some simple, well-established means of estimating cardiac performance that do not require sophisticated technology or evaluation.

Clinicians can obtain a two-dimensional view of changes in the size of the ventricular chambers during the cardiac cycle through the use of ultrasound images employed in echocardiography. Ventricular chamber cross-sectional area during systole and diastole can be estimated from these real time images and used to estimate stroke volume in the heart. The change in areas between diastole and systole is expressed as a percentage of the area during diastole and is called the **ejection fraction**. In normal hearts, the ejection fraction is 55% to 67%; values ≤40% indicate impaired performance.

Variables that reflect changes in the inotropic state of the heart have proven difficult to develop and are often used on the basis of what works reasonably well in the clinic. For example, the peak rate of rise of pressure in the left ventricle during isovolumic contraction (peak dP/dt) has been shown to reflect alterations in myocardial contractility and to be affected little by changes in preload or afterload. A peak dP/dt <1,200 mm Hg/s indicates abnormally low myocardial contractility. Under certain conditions, inferences about the inotropic state of the heart can be made from the inspection of plots of stroke volume, CO, or stroke work versus LVEDV or LVEDP. For example, the only way a patient's stroke volume can decrease or remain unchanged in the face of an increased ventricular preload is if the patient's heart is experiencing a negative inotropic influence. Conversely, increased stroke volume in the face of an unchanged or decreased preload indicates that the heart is under a positive inotropic influence.

There are certain physiologic and pathologic conditions that are known to change loading conditions and the

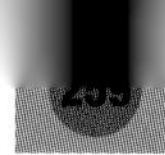

Clinical Focus / 13.2

Heart Failure and Muscle Mechanics

Heart failure is evident when the heart is unable to maintain sufficient output to meet the body's normal metabolic needs. It is usually a progressively worsening condition. The incidence of heart failure in patients in the United States has been increasing over the past 40 years, making it the only cardiovascular pathology that has not shown a decline over the same period of time. The condition is a result of either a deterioration of the heart muscle or worsening of the contributing factors external to the heart. The term *congestive heart failure* refers to fluid congestion in the lungs that often accompanies heart failure. However, a person can have impaired ventricular performance without pulmonary congestion or fluid congestion independent of a malfunctioning heart. Patients suffering from heart failure may be unable to perform simple, everyday tasks without fatigue or shortness of breath. In later stages, there may be significant distress even while resting. Although many intrinsic and extrinsic factors contribute to the condition, this discussion will focus on the muscle mechanics of heart failure.

Much of poor cardiac function can be understood in terms of the mechanics of the heart muscle as it interacts with several external factors that determine the resting muscle length or the load against which it must contract. Some heart failure is of the systolic type. If the heart has been damaged by a myocardial infarction (heart attack) or ischemia (impaired blood supply to the heart muscle) or by chronic overload (as with untreated high blood pressure), the muscle may become weakened and have reduced contractility. In this case, the load presented to the weakened heart by even normal blood pressure is too high for the muscle and, therefore, the muscle cannot shorten as far and with normal velocity. The stroke volume of the muscle will thus be reduced because the ability of the heart to eject blood declines. This increases ventricular volumes and therefore wall stress during contraction, which further increases afterload on the heart. Stroke volume further declines because of the combined effects of the weakened condition of the heart and the increased afterload, which tends to reduce shortening.

The pumping ability of the heart in heart failure can be improved with therapies that improve the contractility of the muscle and/or reduce the load on the heart. The cardiac glycoside digitalis is still used to enhance contractility in the failing heart. Vasodilator therapy can also reduce the increased volume and therefore wall stress in the heart while relieving congestion in the lungs. However, this type of therapy is more commonly used today to alleviate acute heart failure rather than chronic heart failure (CHF) because of the advent of new therapies to treat the chronic condition. In this latter case, the fact that the Starling curve and left ventricular end-systolic pressure–volume (LVESPV) relationship are both flattened in heart failure is exploited by vasodilator therapy. In heart failure, the relationship between stroke volume and preload is flat, and the slope of the LVESPV is decreased, which pushes the low volume end of the relationship to the left on a pressure volume loop diagram. The decreased slope of the LVESPV in the failing heart means that any reduction in afterload such as by arterial vasodilation produces a *greater* increase in stroke volume than that seen for the same reduction in afterload in a normal heart. The flattened Starling relationship in the failing heart also means that reduced preload on the heart caused by venodilation will result in less of a deficit in stroke volume than that seen in a normal heart that has a steep relationship between stroke volume and preload. Thus, it is possible to dilate both arteries and veins to the point of distributing volume out of the heart and lungs not only to relieve congestion but actually to improve the output of the heart.

Heart failure can also be of the diastolic type, which may or may not accompany systolic failure. Here, the relaxation is impaired, and the muscle is resistant to the stretch that must take place during its filling with blood. Some types of hypertrophy or connective tissue fibrosis can contribute to diastolic failure. Because the muscle cannot be sufficiently lengthened during its diastole, it begins its contraction at too short a length. As the length–tension curve would predict, the muscle is unable to shorten sufficiently to pump an adequate volume of blood with each beat.

The treatment of systolic heart failure involves approaches that affect several areas of muscle mechanics and systemic variables. (There are currently no accepted protocols for the treatment of diastolic failure.) Current therapies are also now directed at processes at the cellular level in heart failure, which seem to cause the heart cells themselves to malfunction. This includes counteracting apoptosis and ventricular remodeling caused by sustained exposure to catecholamine and angiotensin II. For this reason, treatment of CHF patients with β-adrenergic receptor antagonists and with agents that reduce the production of angiotensin II are now recommended throughout the duration of the disease, even before symptoms arise. These modern therapies and potential new therapies are discussed in From Bench to Bedside Box 13.1.

inotropic state of the heart. For example, drugs that relax venous smooth muscle lower central venous and ultimately right atrial pressure, thereby reducing the filling of the heart and ventricular preload. Standing causes pooling of blood in the lower extremities and also decreases pressure in the right atrium with a similar effect. Drugs that dilate veins such as nitroglycerin used to treat angina exacerbate this effect. People taking these medications are susceptible to fainting on standing as a result of excess pooling of blood in the veins. This pooling causes a precipitous drop in right atrial pressure, reduced output of the heart, and transient low blood pressure (postural hypotension). Conversely, lying down and raising the feet increases right atrial pressure and stroke output of the heart by the Starling mechanism. Contraction of skeletal muscle squeezes muscle veins, increasing venous pressure and forcing blood back to the heart. This factor augments ventricular output during exercise by increasing preload.

Certain hormones, neurotransmitters, and drugs are known to affect the contractility of the heart. For example, any condition that activates the sympathetic nervous system will enhance myocardial contractility. Conditions that activate the vagus nerve to the heart or inhibit sympathetic activity can have a negative inotropic effect. Such conditions are discussed in later chapters. Myocarditis and myocardial ischemia can be assumed to be negative inotropic influences. Certain anesthetics, such as barbiturates, antiarrhythmic drugs, and Ca^{2+} antagonists are known negative inotropic agents.

CARDIAC OUTPUT

Cardiac output (CO) is defined as the volume of blood ejected from the heart per unit time. The usual resting values for adults are 5 to 6 L/min, or approximately 8% of body weight per minute. CO divided by body surface area is called the **cardiac index**. When it is necessary to normalize CO to compare outputs among people of different sizes, either cardiac index or CO divided by body weight can be used. CO is the product of HR and stroke volume (SV):

$$CO = SV \times HR \quad (1)$$

Previously discussed factors that alter stroke volume as well as factors that alter HR combine to produce an overall effect on CO. These are outlined in Table 13.1. If HR remains constant, CO increases in proportion to stroke volume and, therefore, factors that increase stroke volume are also considered factors that can increase CO. If stroke volume remains constant, CO increases in proportion to HR of up to approximately 180 beats/min. Therefore, within this limit, factors that increase HR increase CO.

TABLE 13.1 Factors Influencing Cardiac Output
Positive Effects
Increased stroke volume
Increased force of contraction
Increased end-diastolic fiber length (Starling law, preload)
Increased end-diastolic pressure
Increased contractility
Sympathetic stimulation via norepinephrine acting on β_1 receptors
Circulating epinephrine acting on β_1 receptors (minor)
Intrinsic changes in contractility in response to changes in heart rate and afterload
Positive inotropic drugs, e.g., digitalis, β-receptor agonists, phosphodiesterase inhibitors
Hypertrophy (if filling is not impaired due to decreased compliance)
Negative Effects
Decreased ventricular compliance; hypertrophy, cardiac tamponade, scar tissue
Increased afterload
Increased ventricular radius
Increased ventricular systolic pressure
Negative inotropic drugs; calcium-channel blockers, general anesthetics, β blockers
Disease (coronary artery disease, myocarditis, cardiomyopathy, etc.)
Myocardial ischemia
Ventricular and supraventricular tachycardias
Irregular heart beat, atrial fibrillation

Cardiac output is maintained over a large range of heart rates through a reciprocal interaction between heart rate and stroke volume.

Heart rate can vary from <50 beats/min in a resting, physically fit person to >200 beats/min during maximal exercise. Changes in HR, however, do not necessarily cause proportional changes in CO because HR inversely affects stroke volume. As the HR increases, the duration of the cardiac cycle and diastole decrease. As the duration of diastole decreases, the time for the filling of the ventricles is diminished. Less filling of the ventricles leads to a reduced end-diastolic volume and decreased stroke volume. Events in the myocardium compensate to some degree for the decreased time available for filling. First, increases in HR reduce the duration of the action potential, and thus, the duration of systole, so the time available for diastolic filling decreases less than it would if the rate was changed through a change in the duration of diastole alone. Second, as discussed earlier, faster HRs are accompanied by an increase in the force of contraction, which tends to augment stroke volume. The increased contractility is sometimes called *treppe* or the *staircase phenomenon*. In reality, CO is not adversely affected by the decreased filling time during high HRs unless the rate exceeds approximately 180 beats/min. A person with abnormal tachycardia, however, (e.g., caused by an ectopic ventricular pacemaker) may have a reduction in CO despite an increased HR.

The reciprocal relationship between HR and the duration of diastole can also have a positive effect when HR is low. It follows from this reciprocal relationship that, within limits, decreasing the rate of a normal resting heart will not decrease CO. As the HR falls, the duration of ventricular diastole increases, and the longer duration of diastole results in greater ventricular filling. The resulting elevated end-diastolic fiber length increases stroke volume, which compensates for the decreased HR. This balance works until the HR is <20 beats/min. At this point, additional increases in end-diastolic fiber length cannot augment stroke volume further because the maximum of the ventricular function curve has been reached. At HRs <20 beats/min, CO falls in proportion to decreases in HR.

Changes in heart rate occur through the reciprocal activation of parasympathetic and sympathetic nerves to the heart.

Normal physiologic changes in HR are brought about by reciprocal changes in the activity of the parasympathetic (vagus) and sympathetic nerves to the sinoatrial (SA) node. Vagal effects predominate with regard to HR such that the HR is actually held down from what it would be if paced by the SA node without any autonomic influences present. Nevertheless, decreasing vagal activity and increasing sympathetic nerve activity to the heart bring about increases in HR. A decreased HR results from opposite changes in both branches of the autonomic nervous system.

The release of norepinephrine by sympathetic nerves not only increases the HR (see Chapter 12, "Electrical

Activity of the Heart") but also dramatically increases the force of contraction through its positive inotropic effect. Furthermore, norepinephrine increases conduction velocity in the heart, resulting in a more efficient and rapid ejection of blood from the ventricles. These effects, summarized in Figure 13.11, maintain the stroke volume as the HR increases. When the HR increases physiologically as a result of an increase in sympathetic nervous system activity (as during exercise), CO increases proportionately over a broad range.

Cardiac output is changed by factors that alter stroke volume and heart rate.

In summary, changing stroke volume and HR regulate CO. Stroke volume is influenced by the contractile force of the ventricular myocardium and by the force opposing ejection (the aortic pressure or afterload). Myocardial contractile force depends on ventricular end-diastolic fiber length (Starling law) and myocardial contractility. Contractility is influenced by four major factors:

1. Norepinephrine released from cardiac sympathetic nerves and, to a much lesser extent, circulating norepinephrine and epinephrine released from the adrenal medulla
2. Certain hormones and drugs, including glucagon, isoproterenol, and digitalis (which increase contractility), and anesthetics (which decrease contractility)
3. Disease states, such as coronary artery disease, myocarditis, bacterial toxemia, and alterations in plasma electrolytes and acid–base balance
4. Intrinsic changes in contractility with changes in HR

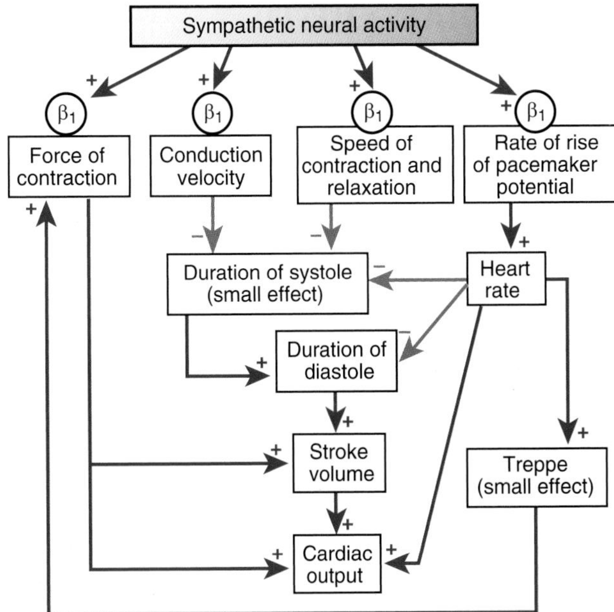

Figure 13.11 Effects of increased sympathetic neural activity on heart rate, stroke volume, and cardiac output (CO). Various effects of norepinephrine on the heart compensate for the decreased duration of diastole and hold stroke volume relatively constant, so that CO increases with increasing heart rate. *Red lines* indicate stimulatory influences. *Blue lines* indicate inhibitory influences.

Sympathetic and parasympathetic nerves to the heart (primarily) and circulating norepinephrine and epinephrine (to a lesser extent) influence HR. The effect of HR on CO depends on the extent of concomitant changes in ventricular filling and contractility.

From Bench to Bedside / 13.1

Inodilator Therapy for the Treatment of Acute Uncompensated and Chronic Heart Failure

Acute uncompensated or decompensated heart failure occurs when the body's natural mechanisms from stimulating cardiac output as the heart weakens are inadequate to prevent further worsening of the failure. In the acute setting, such a situation can be life threatening. The therapeutic goal for the treatment of acute heart failure is to stimulate myocardial contractility, while avoiding increasing myocardial afterload and improving peripheral perfusion. An ancillary therapeutic goal is to reduce pulmonary congestion if it is present.

The ability to increase myocardial contractility without increasing afterload, improve peripheral perfusion, and relieve congestion has been a long-sought goal for the treatment of chronic heart failure (CHF), especially in an acute crisis stage. Inodilators are drugs that can both increase myocardial contractility and induce arterial vasodilation. Therefore, such drugs seem well suited for treating acute uncompensated heart failure. Early drugs of this class, such as dobutamine and milrinone, were shown to achieve positive inotropy with reduced afterload but did so at the expense of increasing myocardial oxygen consumption and intracellular calcium concentration. Both of these side effects markedly increase the risk of arrhythmias developing in the heart during exposure to the drugs.

Levosimendan is a newer type of inodilator that dilates arteries and veins while increasing myocardial contractility. However, this drug produces a positive inotropic effect on the heart, without increasing myocardial calcium concentration. Levosimendan produces its effects by sensitizing the heart to calcium during systole through a stabilizing effect on troponin C. Positive inotropy is achieved without increasing myocardial adenosine triphosphate consumption or affecting the duration of diastole, which affects cardiac preload. Early clinical trials with levosimendan demonstrated superior mortality outcomes in patients with New York Heart Association Class 4 CHF compared to that associated with dobutamine, although more work needs to be done to demonstrate the useful applications of this agent. Regardless, the pharmacology of this inodilator agent has promising properties for numerous applications in cardiac medicine. At this writing, there are >25 clinical trials in various stages of completion operating worldwide to investigate the usefulness of this dual action agent in the clinical management of cardiac disease, including acute and advanced CHF.

CARDIAC OUTPUT MEASUREMENT

The ability to measure CO is important for performing physiologic studies involving the heart and managing clinical problems in patients with heart disease or heart failure. In current clinical practice, CO in an individual is most often estimated as the product of HR obtained by an ECG and stroke volume. However, older techniques that were used to measure CO did so accurately by applying the principles of mass balance with indicator dilution techniques. This technique, in a slightly different iteration, is still employed today in experimental settings as a means for measuring oxygen consumption in an organ. An overview of the basic principles involved in determining CO and organ oxygen consumption by mass balance and indicator dilution techniques is summarized below.

Indicator–dilution techniques are applications of the principle of mass balance to determine cardiac output.

The use of mass balance to measure CO is best understood by considering the measurement of an unknown volume of liquid in a beaker. Dispersing a known quantity of dye throughout the liquid and then measuring the concentration of dye in a sample of liquid can determine the volume. Because mass is conserved, the quantity of dye (A) in the liquid is equal to the concentration of dye in the liquid (C) times the volume of liquid (V), $A = C \times V$. Because A is known and C can be measured, V can be calculated as $V = A/C$.

When the principle of mass balance is applied to CO, the goal is to measure the volume of blood flowing through the heart per unit of time. A known amount of dye or other indicator is injected and the concentration of the dye or indicator is measured over time. This method for determining CO is called the **indicator-dilution technique**. In this application, a known amount of indicator (A) is injected into the circulation, and the blood downstream is serially sampled after the indicator has had a chance to mix (Fig. 13.12). The indicator is usually injected on the venous side of the circulation (often into the right ventricle or pulmonary artery but, occasionally, directly into the left ventricle), and sampling is performed from a distal artery. The resulting concentration of indicator in the distal arterial blood (C) changes with time. First, the concentration rises as the portion of the indicator carried by the fastest-moving blood reaches the arterial sampling point. Concentration rises to a peak as the majority of indicator arrives and falls off as the indicator carried by the slower moving blood arrives. Before the last of the indicator arrives, the indicator carried by the blood flowing through the shortest pathways comes around again (recirculation). To correct for this recirculation, the downslope of the curve is assumed to be semilogarithmic, and the arterial value is extrapolated to zero indicator concentration. The average concentration of indicator can be determined by measuring the indicator concentration continuously from its first appearance (t_1) until its disappearance (t_2). The average concentration during that period (C_{ave}) is determined, and CO is calculated as

$$CO = A/C_{ave}(t_1 - t_2) \quad (2)$$

Note the similarity between this equation and the one for calculating volume in a beaker. On the left is volume per minute (rather than volume, as in the beaker example). In the numerator on the right is the amount of indicator and in the denominator is the average concentration over time (rather than absolute concentration, as in the beaker example). Concentration, volume, and amount appear in both examples, but time is present in the denominator on both sides in equation 2.

A better means of determining CO by the indicator dilution method involves the process of **thermodilution**.

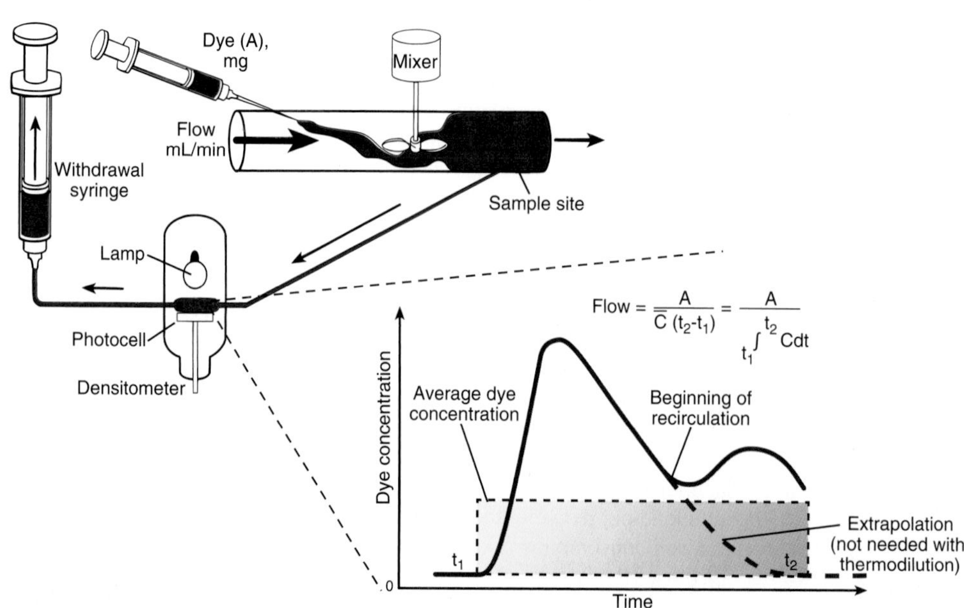

Figure 13.12 The indicator-dilution method for determining flow through a tube. The volume per minute flowing in the tube equals the quantity of indicator (in this example, an injected dye (A) divided by the average dye concentration (C) at the sample site, multiplied by the time between the appearance (t_1) and disappearance (t_2) of the dye. The downslope of the dye concentration curve shows the effects of recirculation of the dye (*solid line*) and the semilogarithmic extrapolation of the downslope (*dashed line*) used to correct for recirculation.

A Swan-Ganz catheter (a soft, flow-directed catheter with a balloon at the tip) is placed into a large vein and threaded through the right atrium and ventricle so that its tip lies in the pulmonary artery. The catheter is designed to allow a known amount of ice-cold saline solution to be injected into the right side of the heart via a side port in the catheter. This solution decreases the temperature of the surrounding blood. The magnitude of the decrease in temperature depends on the volume of blood that mixes with the solution, which depends on CO. A thermistor on the catheter tip (located downstream in the pulmonary artery) measures the fall in blood temperature. The CO can be determined using calculations similar to those described for the indicator-dilution method.

Another way in which the principle of mass balance is used to calculate CO takes advantage of the continuous entry of oxygen into the blood via the lungs (Fig. 13.13). This measurement is based on the **Fick principle** (also known as *Fick's principle*) and is the oldest means of accurately measuring CO. In a steady state, the oxygen leaving the lungs (per unit time) via the pulmonary veins must equal the oxygen entering the lungs via the mixed venous blood and respiration (in a steady state, the amount of oxygen entering the blood through respiration is equal to the amount consumed by body metabolism). Therefore, O_2 in blood leaving the lungs equals O_2 in blood and air entering the lungs, or O_2 output via pulmonary veins equals O_2 input via the pulmonary artery plus O_2 added by respiration.

The O_2 output via the pulmonary veins is equal to the pulmonary vein O_2 content multiplied by the CO. Because O_2 is neither added nor subtracted from the blood as it passes from the pulmonary veins through the left heart to the systemic arteries, the O_2 output via pulmonary veins is also equal to the arterial O_2 content (aO_2) multiplied by the CO. Similarly, O_2 input via the pulmonary artery is equal to mixed venous blood oxygen input to the right heart and is mixed venous blood O_2 content ($\dot{V}O_2$) multiplied by the CO. As indicated above, in the steady state, O_2 added by respiration is equal to oxygen consumption ($\dot{V}o_2$). Thus,

$$CO \times aO_2 = (CO \times vO_2) + Vo_2 \tag{3}$$

which rearranges to

$$CO = \dot{V}o_2 (aO_2 - vO_2) \tag{4}$$

Systemic arterial blood oxygen content, pulmonary arterial (mixed venous) blood oxygen content, and oxygen consumption can all be measured and, therefore, CO can be calculated. This method measures CO accurately. The theory behind this method is sounder than the theory behind the indicator-dilution method because it avoids the need for extrapolation. However, cardiac catheterization and measurement of whole body oxygen consumption is required with this method, which makes it impractical to use in the clinical setting. The relationship above, however, can be rearranged to yield cardiac O_2 consumption if total coronary blood flow and the arterial–venous oxygen difference across the ventricle (called *oxygen extraction*) are known. This application is a more common application of the Fick principle and is used to measure myocardial oxygen consumption in experimental settings.

IMAGING TECHNIQUES FOR MEASURING CARDIAC STRUCTURES, VOLUMES, BLOOD FLOW, AND CARDIAC OUTPUT

A variety of clinical techniques that employ imaging modalities can be used to measure cardiac structures and estimate ventricular volumes and CO. Those that use time-dependent

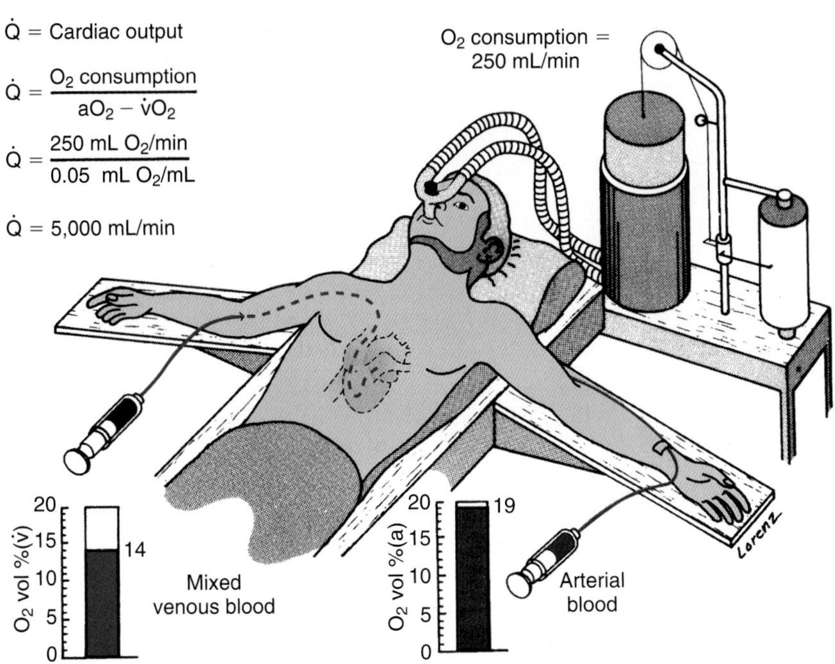

Figure 13.13 Calculating cardiac output using the Fick principle. Oxygen is the "indicator" that is "added" to the mixed venous blood. For oxygen, 1 vol % = 1 mL oxygen/100 mL blood.

images of the heart to estimate the difference between end-diastolic and end-systolic volumes, which gives stroke volume, can be combined with HR to estimate CO.

Technetium-99 scans use tagging of radioactive technetium-99 to red blood cells or albumin as a means of measuring blood distribution. The radiation (gamma rays) emitted by the large pool(s) of blood in the cardiac chambers is measured using a specially designed gamma camera. The emitted radiation is proportional to the amount of technetium bound to the blood (easily determined by sampling the tagged blood) and the volume of blood in the heart. Using computerized analysis, the amount of radiation emitted by the left (or right) ventricle during various portions of the cardiac cycle can be determined (Fig. 13.14). Stroke volume is determined by comparing the amount of radiation measured at the end of systole with that at the end of diastole; multiplying this number by the HR yields CO.

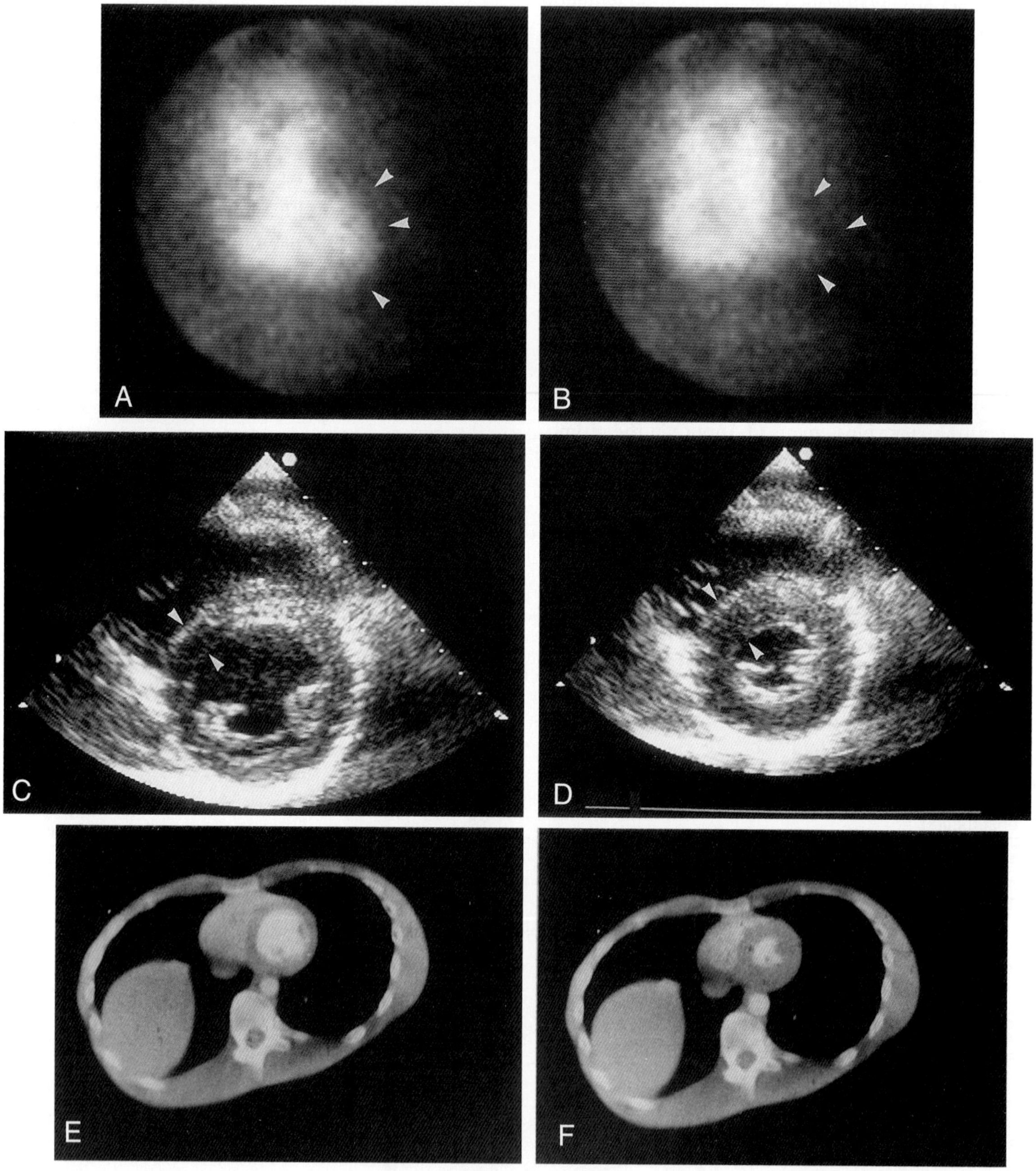

A B C D E F

Figure 13.14 Imaging techniques for measuring cardiac output (CO). (A and **B)** Radionuclide angiograms. The *white arrows* in **(A)** show the boot-shaped left ventricle during cardiac diastole when it is maximally filled with radionuclide-labeled blood. In B, much of the apex appears to be missing (*white arrows*) because cardiac systole has caused the blood to be ejected as the intraventricular volume decreases. **(C** and **D)** Two-dimensional echocardiograms. In this cross-sectional view, the left ventricle appears as a ring. *White arrows* indicate wall thickness. In diastole **(C)** the ventricle is large and the wall is thinned; during systole **(D)** the wall thickens and the ventricular size decreases. **(E** and **F)** Ultrafast (cine) computed tomography. The ventricular size and the wall thickness can be assessed during diastole and systole, and the change in ventricular size can be used to calculate CO.

Echocardiography is a noninvasive technique that uses ultrasound waves to produce one-dimensional and two-dimensional, real time images of the heart. This is the same technology that is used to image the fetus during pregnancy. The one-dimensional mode of this technique sends a single beam of ultrasound across a single position to reveal a reflection of the components of the heart. This can reveal the thickness of the ventricular walls and chambers as well as of the valves and the great vessels as the heart beats. Consequently, it detects the motion of these structures during the cardiac cycle and is therefore designated **M-mode echocardiography**, in which the M stands for "motion." Ventricular wall motion abnormalities are important clinical signs of underlying muscle disease. Ischemic myocardium does not contract and infarcted areas of the heart weaken and actually bulge outward, rather than contract inward, during systole. These abnormalities in wall motion can be detected by M-mode echocardiography but also through two-dimensional echocardiography (**2D-echocardiography**). In 2D-echocardiography, the ultrasound probe rapidly scans back and forth to give a two-dimensional cross-sectional image of the heart. Simply changing the position of the probe can change the plane of this cross section and the resulting image. This method is used as a noninvasive means of estimating stroke volume and ejection fraction in the heart of a patient.

Doppler ultrasound is also used to detect blood flow across the valves of the heart and in blood vessels. The velocity of blood flow can be determined by measuring the Doppler shift (change in sound frequency) that occurs when the ultrasound wave is reflected off moving blood, much the same way a radar device measures the speed of a moving object. In cardiac evaluation, the velocity of blood flow is color-coded by computer. Turbulence and unusually high flow velocities, such as what occurs when blood is ejected through a stenotic aortic valve, can be detected. The direction of flow can also be color-coded. This is useful in detecting regurgitation through valves such as in **mitral valve prolapse**.

In sum, echocardiography can be used to measure changes in ventricular chamber size (see Fig. 13.14C,D), aortic diameter, and aortic blood flow velocity occurring throughout the cardiac cycle. CO can also be estimated with this information in one of two ways. First, the change in ventricular volume occurring with each beat (stroke volume) can be determined and multiplied by the HR. Second, the average aortic blood flow velocity can be measured (just above or below the aortic valve) and multiplied by the measured aortic cross-sectional area to give aortic blood flow (which is nearly identical to CO).

Ultrafast (cine) computed tomography and magnetic resonance imaging provide cross-sectional views of the heart during different phases of the cardiac cycle (see Fig. 13.14E,F). Stroke volume (and CO) can be calculated using the same principles described for radionuclide techniques or echocardiography. When ventricular volume changes are estimated from cross-sectional data, assumptions are made about ventricular geometry. The heart is not a sphere, and therefore its volume cannot be estimated from a single dimension. Estimates using two or three axis approximations of an ellipse can be used with imaging techniques to give a better estimate of volume changes, but these assumptions can lead to errors in calculating CO. Nevertheless, such estimates of ventricular volumes and CO from imaging ventricular dimensions have proven useful in clinical applications.

Chapter Summary

- Tetanic contraction is prevented in cardiac muscle because the cardiac muscle action potential is longer than the duration of the contraction.
- Cardiac muscle operates at lengths along the ascending limb of the isometric length–tension curve.
- The velocity of shortening of cardiac muscle is inversely related to the force being exerted.
- Cardiac muscle contraction can be regulated by changes in contractility.
- The contractility of cardiac muscle is changed by inotropic interventions that include changes in the heart rate, the presence of circulating catecholamines, or sympathetic nerve stimulations.
- Calcium enters a cardiac muscle cell during the plateau of the action potential and promotes the release of internal calcium stores in the sarcoplasmic reticulum.
- Changes in cardiac muscle contractility are associated with changes in the amount of calcium released by calcium-induced calcium-release mechanisms.
- Ventricular ejection is divided into rapid and reduced ejection phases.
- Stroke volume is the amount of blood ejected from the ventricles during one systole; it is the difference between ventricular end-diastolic and end-systolic volumes.
- The ventricles do not empty completely during systole, leaving a residual volume in the ventricle for the next filling cycle.
- Ventricular filling is divided into rapid and reduced filling phases.
- Heart sounds during the cardiac cycle are related to the opening and closing of valves in the heart.
- Venous pressure waves can detect abnormalities in atrioventricular valves.
- Cardiac output is the product of stroke volume times heart rate.
- Stroke volume is determined by end-diastolic fiber length, afterload, and contractility. Heart rate influences ventricular filling time and stroke volume so that changes in cardiac output caused by changes in heart rate are attenuated.
- The influence of the heart rate on cardiac output depends on simultaneous effects that enhance ventricular contractility at high rates.
- Cardiac energy demands are determined by collective contributions from ventricular wall stress, heart rate, stroke volume, and contractility.
- The energy cost of work in the heart is greater for work done to generate pressure than for work done to eject blood.
- Cardiac output can be measured by methods that rely on mass balance or cardiac imaging.

thePoint *Visit http://thePoint.lww.com for chapter review Q&A, case studies, animations, and more!*

14 Systemic Circulation

ACTIVE LEARNING OBJECTIVES

Upon mastering the material in this chapter, you should be able to:

- Predict how changes in arterial compliance and stroke volume will change pulse pressure.
- Analyze the arterial pressure waveform to determine the changes in heart rate, stroke volume, and vascular resistance that created the waveform.
- Explain how shifts between central and systemic venous volumes affect cardiac output.
- Explain how an equilibrium point for cardiac output and central venous pressure arises from the interdependent relationship between cardiac output and central venous pressure.
- Explain how and why the equilibrium point between cardiac output and central venous pressure is shifted by changes in blood volume, venous tone, arteriolar resistance, and the inotropic state of the heart.

The aorta and other large-diameter arteries do not contribute significantly to total vascular resistance, and therefore, do not play a role in the regulation of organ blood flow or systemic arterial pressure. However, their elastic characteristics influence a variety of important cardiovascular variables, which include the arterial pressure waveform, cardiac work, and the interaction between cardiac output (CO) and vascular function.

DETERMINANTS OF ARTERIAL PRESSURES

Four principal pressures of physiologic interest are contained within the arterial pressure waveform. These are mean arterial pressure, systolic and diastolic arterial pressures, and pulse pressure. Clinically, arterial pressure is often reported as the ratio of systolic over diastolic pressure. Typical values for such pressures are, for example, reported as 120/80 or 110/75 mm Hg. However, these values really only list the peak systolic and minimum diastolic values for arterial pressure. In reality, arterial pressures vary around average values from heartbeat to heartbeat and from minute to minute.

Mean arterial pressure is determined by cardiac output and systemic vascular resistance.

Mean arterial pressure ($\bar{P}_a$) is determined mathematically as indicated in Figure 14.1 but is often approximated from the equation

$$\bar{P}_a = P_d + (P_s - P_d)/3 \qquad (1)$$

where P_d is the diastolic pressure, P_s is the systolic pressure, and $P_s - P_d$ is the **pulse pressure**. Mean arterial pressure is closer to P_d, instead of halfway between P_s and P_d, because the duration of diastole is about twice as long as that of systole. The difference between $\bar{P}_a$ and right atrial pressure (P_{ra}) is equal to the product of CO and systemic vascular resistance (SVR) in accordance with the Poiseuille law (also known as *Poiseuille's* law), as discussed in Chapter 11, "Overview of the Cardiovascular System and Hemodynamics." However, because right atrial pressure is small compared with mean arterial pressure, mean arterial pressure is considered to result from the product of CO and SVR.

Changes in stroke volume and arterial compliance alter pulse pressure.

Figure 14.2A shows the effect of a change in aortic volume on aortic pressures if aortic compliance, SVR, and heart rate are constant. During rapid ejection, blood enters the aorta

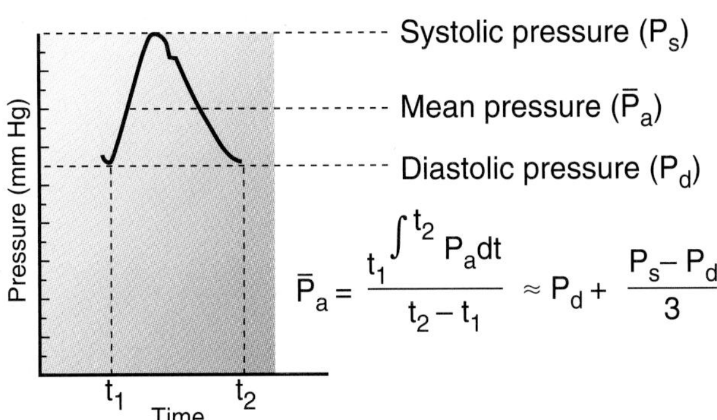

Figure 14.1 Definition of mean arterial pressure. Mean pressure is the area under the pressure curve divided by the time interval. This can be approximated as the diastolic pressure plus one third the pulse pressure.

faster than it can run off, thus increasing aortic volume. This increased volume increases pressure in the aorta to a value determined by the compliance of the arteries and the amount of volume change. During reduced ejection and diastole, blood exits the arteries faster than it enters. Volume, thus, declines, as does the pressure associated with it. If the amount of blood ejected into the aorta doubled, without a change in heart rate or SVR, this would double CO and, from the Poiseuille law, double mean arterial pressure. Changes in pressure resulting from increases and decreases in arterial blood volume would oscillate about this new mean. In the example in which ejection doubles, most of this volume would be ejected in the rapid ejection phase. Therefore, the amount ejected that exceeds runoff would also double, and this would be reflected by an increase in the systolic pressure. Runoff would be little affected by this increased ejection, so diastolic pressure would fall at about the same rate as before, but from this new, higher systolic pressure. Therefore, pressure would drop to a minimum diastolic pressure that is considerably higher than before. Thus, when stroke volume increases without any change in SVR, arterial compliance, and heart rate, an increase in mean arterial pressure, systolic arterial pressure, diastolic arterial pressure, and pulse pressure will result.

In the human body, however, arterial compliance is a nonlinear variable that depends on the pressure in the aorta and major arteries; aortic compliance decreases as transmural pressure is increased, as shown in Figure 14.2B. Because of this, a given change in aortic volume at a low initial volume and pressure causes a relatively small change in pressure with ejection, but the same change in volume at a high initial volume and pressure causes a much larger change in pressure.

The general effect of a change in arterial compliance on arterial pressures is shown in Figure 14.2C. When arteries are stiff, they flex less during systole and recoil less during diastole. Therefore, for a given change in volume, decreased arterial compliance causes an increase in systolic pressure and a *decrease* in diastolic pressure. If there is no change in arterial resistance or heart rate, mean arterial pressure will *not* change with a change in arterial compliance.

Interactions among stroke volume, heart rate, and systemic vascular resistance alter arterial pressures.

When CO changes in the face of a constant vascular resistance, mean arterial pressure is influenced according to the formula $\overline{P}_a = CO \times SVR$. The influence of a change in CO on mean arterial pressure is independent of the cause of the change, whether it comes from heart rate or stroke volume. In contrast, the effect of a change in CO on pulse pressure greatly depends on whether and how stroke volume or heart rate change.

If an increase in heart rate is balanced by a proportional and opposite change in stroke volume, *mean* arterial pressure does not change because CO remains constant. However, the decrease in stroke volume that occurs in this situation results in a diminished pulse pressure; the diastolic pressure increases, while the systolic pressure decreases around an unchanged mean arterial pressure. An increase in stroke volume accompanied by a decreased heart rate such that there is no change in CO likewise causes no change in mean arterial pressure. The increased stroke volume, however, produces a rise in pulse pressure. The increased volume ejected with each contraction causes a greater rise in systolic pressure. However, because heart rate is reduced, there is more time available for blood to run out of the arteries during diastole. Arterial volume, therefore, drops to a lower value than before, and diastolic pressure decreases as a result.

Another way to think about these events is depicted in Figure 14.3A. The first two pressure waves have a diastolic pressure of 80 mm Hg, systolic pressure of 120 mm Hg, and mean arterial pressure of 93 mm Hg. Heart rate is

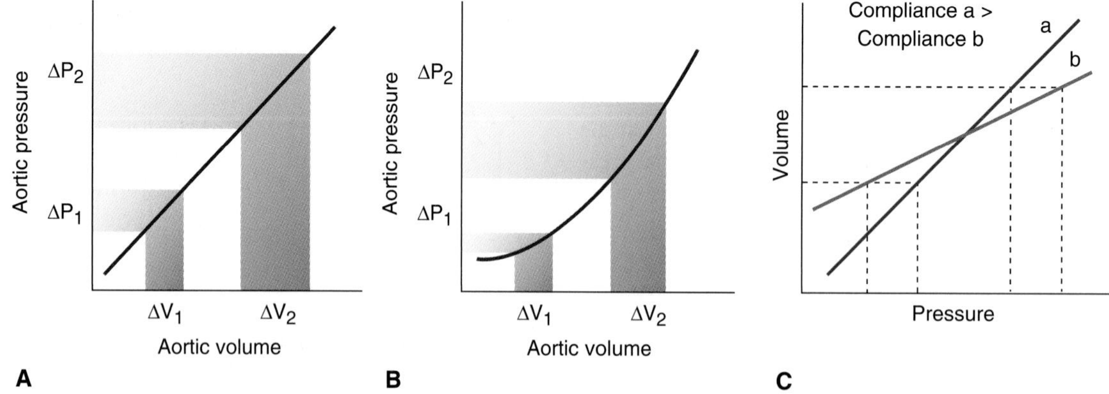

Figure 14.2 Relationship between aortic volume and pressure. (A) Aortic compliance is independent of aortic volume. The change in volume (ΔV_1) causes the change in pressure (ΔP_1). A larger volume increment (ΔV_2), without changing heart rate or arterial compliance, produces a larger mean pressure and larger change in pressure (ΔP_2). **(B)** Aortic compliance decreases as aortic volume and pressure increase. A change in volume (ΔV_1) is associated with a change in pressure (ΔP_1). The same change in volume at a higher initial volume (ΔV_2) causes a much larger change in pressure (ΔP_2). **(C)** At the same ΔV, ΔP increases with a decrease in compliance.

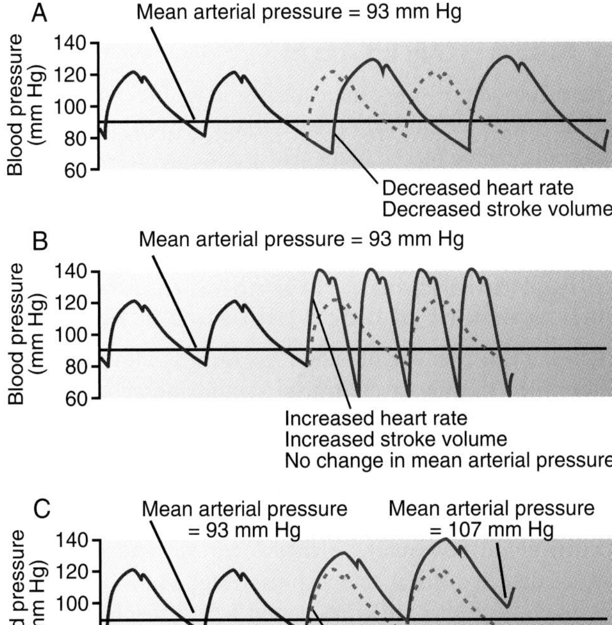

Figure 14.3 Effects of changes in heart rate (HR), stroke volume, and systemic vascular resistance (SVR) on arterial pressure. (A) Effect of increased stroke volume on arterial pressure with constant cardiac output (CO) and SVR. When cardiac output is held constant by lowering HR, there is no change in mean arterial pressure (93 mm Hg), but systolic pressure increases while diastolic pressure decreases. **(B)** Effect of increased HR *and* stroke volume with no change in mean arterial pressure because of decreased SVR. After the first two beats, stroke volume and HR are increased. Pulse pressure increases around an unchanged mean arterial pressure, and systolic pressure is higher and diastolic pressure is lower than the control. **(C)** Effect of increased stroke volume, with constant HR and SVR. CO, mean arterial pressure, systolic pressure, diastolic pressure, and pulse pressure are all increased.

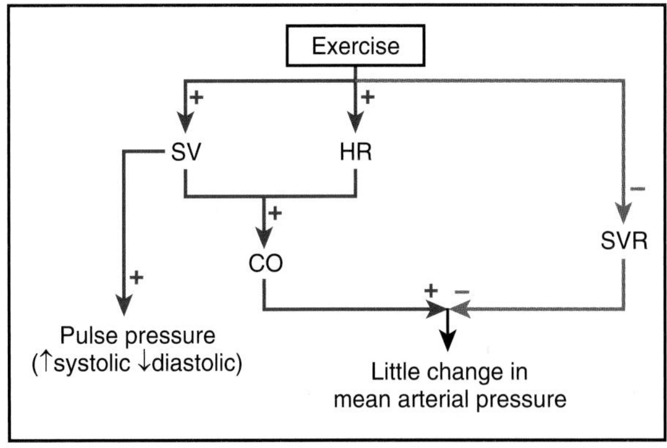

Figure 14.4 Effect of dynamic exercise on mean arterial pressure and pulse pressure. Heart rate (HR) and stroke volume (SV) increase, resulting in an increase in cardiac output (CO). However, the dilation of resistance vessels in skeletal muscle lowers the systemic vascular resistance (SVR), balancing the increase in CO and causing little change in mean arterial pressure. Lower SVR allows more runoff during diastole and lowers diastolic blood pressure. *Red (+) arrows* represent positive effects between the cause and effect variables. *Blue (–) lines* represent a negative effect between the cause and effect variables.

72 beats/min. After the second beat, the heart rate is slowed to 60 beats/min, but stroke volume is increased sufficiently to maintain the same CO. The longer time interval between beats allows the diastolic pressure to fall to a new (lower) value of 70 mm Hg. The next systole, however, produces an increase in pulse pressure because of the ejection of a greater stroke volume, so systolic pressure rises to 130 mm Hg. The pressure then falls to the new (lower) diastolic pressure, and the cycle is repeated. Mean arterial pressure does not change because CO and SVR are constant, and the increased pulse pressure is distributed around the same mean arterial pressure. If an increase in heart rate is balanced by a decrease in stroke volume so that there is no change in CO, the result is no change in mean arterial pressure but a decrease in pulse pressure; systolic pressure decreases and diastolic pressure increases.

In some instances, mean arterial pressure may remain constant despite a change in CO because of an alteration in SVR. A good example of this is **dynamic exercise** (e.g., running or swimming). Dynamic exercise often produces little change in mean arterial pressure because the increase in CO is balanced by a decrease in SVR. The increase in CO is caused by increases in both heart rate and stroke volume. The elevated stroke volume results in a higher pulse pressure; systolic pressure is higher because of the elevated stroke volume. Diastolic pressure is lower because the fall in SVR allows greater runoff from the aorta during diastole (Figs. 14.3B and 14.4).

Figure 14.3C shows what happens if CO is increased by increasing stroke volume with no change in heart rate or SVR. The increased stroke volume occurs at the time of the next expected beat, and the diastolic pressure is, as for previous beats, 80 mm Hg. After a transition beat, the increased stroke volume results in an elevation in systolic pressure to 140 mm Hg, after which the pressure falls to a new diastolic pressure of 90 mm Hg. In this new steady state, systolic, diastolic, and mean arterial pressures are all higher; the latter being consistent with the fact that CO is increased without a corresponding decrease in SVR. A further change in pulse pressure is created in this situation also, because the increase in mean arterial pressure (to 107 mm Hg) pushes oscillations in pressure to a portion of the compliance curve where compliance decreases (see Fig. 14.2). Consequently, the increase in pulse pressure results from both higher stroke volume and decreased arterial compliance.

When total peripheral arterial resistance increases, flow out of the larger arteries transiently decreases. If CO is unchanged, the volume in the aorta and large arteries must increase (Fig. 14.5). Mean arterial pressure also increases, until it is sufficient to drive the blood out of the larger vessels and into the smaller vessels at the same rate as it enters from the heart (i.e., CO). At a higher volume (and mean arterial pressure), arterial compliance is lower, and therefore pulse pressure is greater for a given stroke volume

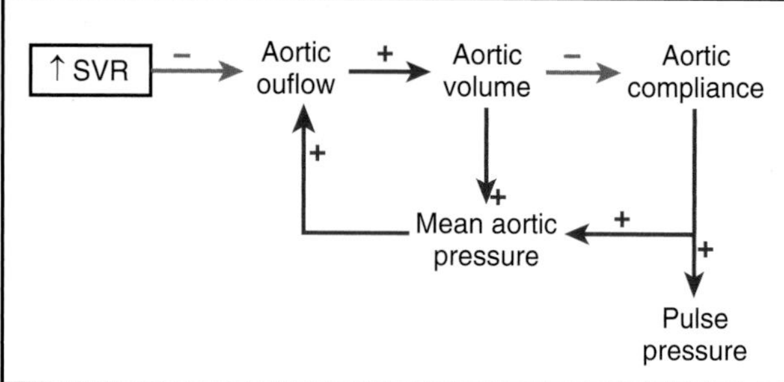

Figure 14.5 Effect of increased systemic vascular resistance (SVR) on mean arterial and pulse pressures. Increased SVR impedes outflow from the aorta and large arteries, increasing their volume and pressure. The increase in aortic pressure brings the outflow from the aorta back to its original value, but at a higher aortic volume. The larger volume lowers aortic compliance and, thereby, raises pulse pressure at a constant stroke volume. The word "increase" in smaller type indicates a secondary change. *Red (+) arrows* represent positive effects between the cause and effect variables. *Blue (–) lines* represent a negative effect between the cause and effect variables.

(see Fig. 14.2 for reference). The net result is an increase in mean arterial, systolic, and diastolic pressures. The extent of the increase in pulse pressure depends on how much arterial compliance decreases with the rise in mean arterial pressure and arterial volume.

ARTERIAL PRESSURE MEASUREMENT

Arterial blood pressure can be measured by direct or indirect (noninvasive) methods. In the laboratory or hospital setting, a cannula can be placed in an artery and the pressure measured directly using electronic transducers. In clinical practice, however, blood pressure is usually measured indirectly.

Sphygmomanometry is a noninvasive measurement of blood pressure in humans.

Clinical measurement of blood pressure is almost always determined noninvasively by sphygmomanometry. In this method, a **sphygmomanometer** that registers pressure is connected to an inflatable cuff that is wrapped around the patient's arm and inflated so that the external pressure on the arm exceeds systolic blood pressure. When this occurs, blood flow into the arm is cut off, and no pulses can be detected distal to the pulse (Fig. 14.6). The external pressure in the cuff is measured by the height of a column of mercury in the manometer connected to the cuff or by means of a mechanical or digital electronic manometer that has been calibrated to a column of mercury. To measure blood pressure, the air in the cuff is first slowly released until blood can leak past the occlusion at the peak of systole. At this point, blood spurts past the point of partial occlusion at high velocity, resulting in turbulence.

Clinical Focus / 14.1 

Arteriosclerosis and Systolic Hypertension in Older Adults

Arteriosclerosis is broadly defined as any disease leading to the general hardening of the arteries. This includes reduced arterial compliance associated with the aging process, restenosis phenomena, hardening following transplantations of vascular segments, and even atherosclerosis. However, these conditions all differ significantly with respect to key immune, mitogenic, and lipid components to their pathology. Typically, the term arteriosclerosis is used to describe the general progressive decrease in arterial compliance that accompanies the aging process (see Chapter 11, "Overview of the Cardiovascular System and Hemodynamics"). Even in the absence of true systemic arterial hypertension and atherosclerosis, people in their 70s have considerably stiffer arterial walls than those in their 20s. The process that leads to this arterial stiffening is thought to be related to the continual, day-by-day exposure of arteries to oxygen radicals produced in the arterial wall as the by-product of natural metabolism. However, it likely also results from accumulated insults to the arteries from lifestyle and dietary factors known to stimulate excess radical production.

Arteries in young people owe their suppleness to the orderly arrangement of elastin and collagen in the arterial wall. Oxygen radicals attack the collagen and elastin matrix within the arteries, causing strand breaks and rearrangement of elastic fibers into a more random configuration. This random configuration results in a stiffer artery. In the absence of any other pathologic factors, an older person with this type of arteriosclerosis exhibits an increased pulse pressure associated with an increased systolic pressure, lower diastolic pressure, and normal mean arterial pressure compared with a healthy younger person (healthy meaning with a blood pressure of 120/80 mm Hg). People with this condition have what is called **systolic hypertension** because only their systolic pressure is elevated.

Until somewhat recently, older patients with systolic hypertension were not treated with typical antihypertensive drug therapies for their condition. These people did not have the underlying vascular abnormalities associated with the increased arterial resistance of essential hypertension because their mean arterial pressure was not changed. Furthermore, their age led some to believe that there may be no long-term benefit to drug treatments, especially when such treatments were associated with serious side effects. However, systolic hypertension is no longer neglected in older adults and is instead treated with therapies designed to reduce arterial pressure. The reason for this is that systolic pressure is a key component of the wall stress that must be overcome for the heart to effectively contract. High systolic pressure leads to high stress that in turn increases oxygen demand by the heart. This places the heart at risk for ischemia and arrhythmias, which can be reduced if the systolic pressure is decreased. In addition, clinical trials have shown that the morbidity and mortality of older patients with systolic hypertension are improved with treatment to lower the systolic pressure, thus providing evidence-based support for antihypertensive therapies in older adults with systolic hypertension.

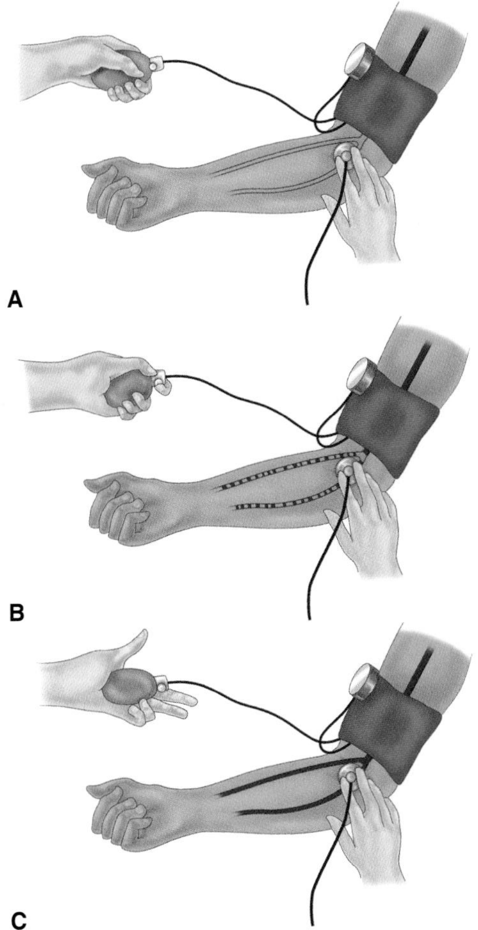

Figure 14.6 Indirect measurement of arterial blood pressure by sphygmomanometry and radial artery auscultation. **(A)** To measure arterial blood pressure, an inflatable cuff connected to a pressure gauge calibrated against a mercury manometer is placed around the upper arm above the brachial artery. The cuff is initially inflated to a pressure approximately 20 mm Hg above the expected systolic pressure for the patient, and a stethoscope is placed lightly over the brachial region. Initially, there is no audible sound from the region. **(B)** The cuff is deflated at a rate of 2 to 3 mm Hg/sec while the clinician listens for sounds created by pulsatile squirting of blood past the partially occluded artery. These sounds are called *Korotkoff sounds* and are characterized by five phases: Phase I: the level at which the first appearance of clear, repetitive, tapping sounds occur. This coincides approximately with the resumption of a palpable pulse. The pressure at which the first two tapping sounds occur is considered to be the peak systolic pressure. Phase II: interval in which a swishing murmur is added to the initial tapping as the cuff is deflated. Phase III: characterized by distinct, crisp, and louder sounds as the artery opens more with cuff deflation. Phase IV: Start of a distinct muffled sound with a soft, blowing quality. **(C)** Phase V: Final silence due to the resumption of full laminar flow in the brachial artery. The last sound heard before the onset of Phase V is considered to represent diastolic arterial pressure. If sounds persist from Phase IV to zero pressure, then the pressure recorded at the start of Phase IV is considered the diastolic blood pressure.

The vibrations associated with the turbulence are in the audible range and can be heard with a stethoscope placed over the brachial artery. These noises are known as **Korotkoff sounds**. The pressure corresponding to the first Korotkoff sound is the systolic pressure. As pressure in the cuff continues to fall, the brachial artery returns toward its normal shape and both the turbulence and Korotkoff sounds cease. The pressure at which the Korotkoff sounds cease is the diastolic pressure.

Although the cuff method has become a standard means of evaluating blood pressure in clinical settings, it is not without artifact errors. The width of the inflatable cuff is an important factor that can affect pressure measurements. A cuff that is too narrow will give a falsely high pressure because the pressure in the cuff is not fully transmitted to the underlying artery. Ideally, cuff width should be approximately 1.5 times the diameter of the limb at the measurement site. Obesity may contribute to an inaccurate assessment if the cuff used is too small. In older adults (or those who have "stiff" or hard-to-compress blood vessels from other causes, such as arteriosclerosis), additional external pressure may be required to compress the blood vessels and stop the flow. This extra pressure can give a falsely high estimate of blood pressure.

Age, race, gender, diet, and body weight affect the range of arterial pressure seen in humans.

Although the range of blood pressures in the population as a whole is rather broad, changes in a given patient are of diagnostic importance. Normal arterial blood pressure in adults is often stated to be 120/80 mm Hg. However, these values were originally derived from insurance actuary data in acculturated Western societies after World War II. A more extensive examination of arterial pressures in the human population suggests that true normal human systolic pressure should be 100 to 110 mm Hg, with a corresponding diastolic pressure of 60 to 70 mm Hg.

In Western societies, arterial pressure is dependent on age. Systolic blood pressure rises throughout life, whereas diastolic blood pressure rises until the sixth decade of life, after which it stays relatively constant. Blood pressure is higher among African Americans than among Caucasian Americans, but it is lower in aboriginal Negroid populations than in western Caucasian populations. Blood pressure is higher among men than among premenopausal women of similar age. Dietary fat and salt as well as obesity are associated with higher blood pressure. Other factors that affect blood pressure are excessive alcohol intake and psychosocial stress (which elevates pressure) as well as potassium and calcium intake and physical activity (which are associated with reductions in arterial pressure).

PERIPHERAL AND CENTRAL BLOOD VOLUME

As discussed in Chapter 11, "Overview of the Cardiovascular System and Hemodynamics," blood volume is distributed

Clinical Focus / 14.2

Effects of Age, Race, and Diet on Arterial Pressure

The common assignment of normal arterial pressure as 120/80 mm Hg is a convenient benchmark that does not reveal that arterial pressure varies within populations as a result of a variety of genetic and environmental factors. Human blood pressure values are a continuum. There is no clear dividing point within the human population that separates subpopulations with normal distributions around a "normal" mean arterial pressure from a different subpopulation distributed around a distinctly elevated arterial pressure. As stated in this chapter, the 120/80 mm Hg dividing point for normalcy has more to do with data sampling than with what is intrinsically supposed to be normal arterial pressure in healthy human beings.

In Western societies, mean arterial pressure and pulse pressure drift upward with age, even if the patient does not have clinically established classic arterial hypertension. However, members of primitive societies do not show this increase in pressure with age, have lower systolic and diastolic pressures as a group compared with members of western societies, do not develop standard arteriosclerosis, and actually exhibit a slow decline in blood pressure with aging. An interesting observation in members of these primitive groups is that they consume potash as a seasoning, which contains high amounts of potassium. There is laboratory evidence that potassium reduces wall stiffening.

African Americans have a higher incidence and severity of arterial hypertension than do Caucasians of the same gender and similar age. In particular, their arterial pressure seems to be sensitive to salt in the diet. However, Negroid groups in primitive societies do not show this propensity to hypertension at all. Human populations that evolve in hot, dry environments have restricted access to water and salt. Those living in such environments possess traits that better enable them to aggressively extract and conserve salt and water from their diet and have a survival advantage. However, placing such people in a western society, where salt is a ubiquitous component of canned and processed foods, restaurant meals, and so on, can create a situation whereby too much salt is added into the vascular system, resulting in hypertension. It is believed also that such salt loading leads to a stiffening of the arterial walls as well as hypervolemia in the circulation. Thus, the dietary intake of salt may affect arterial pressure. The influence of diet, rather than race, on hypertension is further supported by the observation that populations such as native Asians that consume a largely vegetarian diet have lower arterial pressures than the same populations raised and living in western cultures, whereas native Japanese, who consume large amounts of salted fish, have a high incidence of arterial hypertension and hypertension-related stroke.

among the various segments of the circulatory system according to their basic geometry and compliance. Total blood volume in a 70-kg adult is 5.0 to 5.6 L. Approximately 80% of the total blood volume is located in the **systemic circulation** (i.e., the total volume minus the volume in the heart and lungs), and about 60% of the total blood volume (or 75% of the systemic blood volume) is located on the venous side of the circulation. The blood present in the arteries and capillaries is only about 20% of the total blood volume. Because most of the systemic blood volume is in the veins, it is not surprising that changes in systemic blood volume primarily reflect changes in venous volume.

Large compliance in the veins allows them to accommodate high volumes with little change in pressure.

Systemic veins are approximately 20 times more compliant than systemic arteries; small changes in venous pressure are, therefore, associated with large changes in venous volume. If 500 mL of blood is infused into the circulation, about 80% (400 mL) locates in the systemic circulation. This increase in systemic blood volume raises **mean circulatory filling pressure (MCP)** by a few mm Hg. MCP is the pressure that equilibrates into all segments of the vascular system when the heart is stopped. It is normally about 7 mm Hg and is a measure of how "full" the vascular system is or how "tightly" blood is contained within the vasculature as a result of vascular tone. A small rise in filling pressure, distributed throughout the systemic circulation, has a much larger effect on the volume of systemic veins than systemic arteries. Because of the much higher compliance of veins than arteries, 95% of the 400 mL from the example above (or 380 mL) is found in veins, and only 5% (20 mL) is found in arteries.

Central blood volume is useful for evaluating the effects of changes in blood volume on cardiac output.

The filling of the heart is one of the key determinants of CO. In considering the role of distribution of blood volume in filling the heart, it is useful to divide the blood volume into central (or intrathoracic) and extrathoracic portions. The **central blood volume** includes the blood in the superior vena cava and intrathoracic portions of the inferior vena cava, right atrium and ventricle, pulmonary circulation, and left atrium; this constitutes approximately 25% of the total blood volume. The central blood volume can be decreased or increased by shifts in blood to and from the extrathoracic blood volume. From a functional standpoint, the most important components of the **extrathoracic blood volume** are the *veins of the extremities and abdominal cavity*. Blood shifts readily between these veins and the vessels containing the central blood volume. Blood in the central and extrathoracic arteries can be ignored in these shifts because the low compliance of these vessels means that little change in their volume occurs. Furthermore, the extrathoracic blood volume in the neck and head is of little importance because there is far less blood in these regions,

and the blood volume inside the cranium cannot change much because the skull is rigid. The volume of blood in the veins of the abdomen and extremities is about equal to the central blood volume; therefore, about half of the total blood volume is involved in shifts in distribution that affect the filling of the heart.

Central venous pressure is a qualitative measure of central blood volume.

Central venous pressure can be measured by placing the tip of a catheter in the right atrium. Changes in central venous pressures are a good reflection of central blood volume because the compliance of the intrathoracic vessels tends to be constant. In general, the use of central venous pressure to assess changes in central blood volume depends on the assumption that the right heart is capable of pumping normally. In certain situations, however, the physiologic meaning of central venous pressure is changed. For example, if the tricuspid valve is incompetent (i.e., cannot close completely), right ventricular pressure is transmitted to the right atrium during ventricular systole, creating an abnormally high central venous pressure that is not primarily a result of increased central venous volume. Also, central venous pressure does not necessarily reflect left atrial or left ventricular filling pressure. Abnormalities in right or left heart function or in pulmonary vascular resistance can make it difficult to predict left atrial pressure from central venous pressure.

Unfortunately, measurements of the peripheral venous pressure, such as the pressure in an arm or leg vein, are subject to too many influences (e.g., partial occlusion caused by positioning or venous valves) to be helpful in most clinical situations.

Cardiac output is sensitive to changes in central blood volume.

Consider what happens if blood is steadily infused into the inferior vena cava of a healthy person. As this occurs, the volume of blood returning to the chest—the venous return—is transiently greater than the volume leaving it, that is, the CO. This difference between the input and output of blood produces an increase in central blood volume. It will occur first in the right atrium, where the accompanying increase in pressure enhances right ventricular filling, end-diastolic fiber length (preload), and stroke volume. Increased flow into the lungs increases pulmonary blood volume and filling of the left atrium. Left CO will increase according to the Starling law, so that the output of the two ventricles exactly matches. CO will increase until it equals the sum of the previous venous return to the heart plus the infusion of new blood.

It can be surmised from the discussion above that central blood volume is altered by two events: changes in total blood volume and changes in the distribution of total blood volume between central and extrathoracic regions. An increase in total blood volume can occur as a result of an infusion of fluid, the retention of salt and water by the kidneys, or a shift in fluid from the interstitial space to plasma. A decrease in blood volume can occur as a result of hemorrhage, fluid losses through sweat, vomiting or diarrhea, or the transfer of fluid from plasma into the interstitial space. In the absence of compensatory events, changes in blood volume result in proportional changes in both central and extrathoracic blood volume. For example, a moderate hemorrhage (10% of blood volume) with no distribution shift would cause a 10% decrease in central blood volume. The reduced central blood volume would, in the absence of compensatory events, lead to decreased filling of the ventricles and diminished stroke volume and CO.

Central blood volume can be altered by a shift in blood volume to or away from the periphery. Shifts in the distribution of blood volume occur for two reasons: a change in transmural pressure or a change in venous compliance. Changes in the transmural pressure of vessels in the chest or periphery enlarge or diminish their size. Because there is a finite volume of blood, the volume shifts in response to changes in transmural pressure in one region affects the volume of the other regions. Imagine a long balloon filled with water. If it is slowly turned vertically on its long axis, the lower end of the balloon has the greatest transmural pressure because of the weight of the water pressing from above. Consequently, the lower end of the balloon will bulge and the upper end will shrink; volume increases in the lower end at the expense of a loss at the upper end.

The best physiologic example of a change in transmural pressure occurs when a person stands up. Standing increases the transmural pressure in the blood vessels of the legs because it creates a vertical column of blood between the heart and the blood vessels of the legs. The arterial and venous pressures at the ankles during standing can easily be increased almost 100 mm Hg higher than in the recumbent position. The increased transmural pressure (outside pressure is still atmospheric) results in little distention of arteries because of their low compliance but results in considerable distention of veins because of their high compliance. In fact, approximately 550 mL of blood is needed to fill the stretched veins of the legs and feet when an average person stands up. Filling of the veins of the buttocks and pelvis also increases but to a lesser extent than the lower extremities, because the increase in transmural pressure is less. When a person stands, blood continues to be pumped by the heart at the same rate and stroke volume for one or two beats. However, much of the blood reaching the legs remains in the veins as they become passively stretched to their new size by the increased venous (transmural) pressure. This decreases the return of blood to the chest. Since CO exceeds venous return for a few beats, the central blood volume falls (as does the end-diastolic fiber length, stroke volume, and CO). Once the veins of the legs reach their new steady-state volume, the venous return again equals CO. Thus, the equality between venous return and CO is reestablished even though the central blood volume is reduced by 550 mL. However, the new CO and venous return are decreased (relative to what they were before standing) because of the reduction in central blood volume. Without compensation, the decrease in systemic arterial pressure resulting from this decreased CO would cause a drop in brain blood flow and loss of consciousness. Compensatory mechanisms are then obviously required to maintain arterial pressure in the face of decreased CO. Such compensatory mechanisms are discussed in detail in Chapter 17, "Control Mechanisms in Circulatory Function."

When the smooth muscle of the systemic veins contracts, the compliance of the systemic veins decreases. This results in a redistribution of blood volume toward the central blood volume. The redistribution of blood toward the central blood volume helps to maintain ventricular filling and CO. It is not surprising, therefore, that venoconstriction is one important compensatory mechanism for supporting CO following a drop in this output due to hemorrhage.

COUPLING OF VASCULAR AND CARDIAC FUNCTION

Characteristics associated with arteries and veins, such as vascular compliance, blood volume, and vascular resistance, affect the function of the heart and variables associated with that function. However, because the cardiovascular system is a true "circulatory" system, it follows that the performance of the heart influences volumes and pressures within the vasculature. Therefore, an equilibrium must exist between cardiac and vascular function. This section will explain how interactions between the cardiac and vascular systems combine to affect CO. For this purpose, a graphical analysis of the hemodynamics associated with each system is used to describe new equilibriums in the cardiovascular system as a whole that result from a variety of perturbations to cardiac or vascular function.

Compliant arteries reduce cardiac work.

One of the more important consequences of the elastic nature of large arteries is that it reduces cardiac work. Consider a situation in which the heart pumps blood at a constant flow of 100 mL/s (6 L/min) into rigid arteries with a resistance of 16.6 peripheral resistance units (PRUs) for 4 seconds. This would then generate a constant pressure of 100 mm Hg, and cardiac work over the 4 seconds would be simply pressure (P) × volume (V) or 100 mm Hg × 400 mL = 40,000 mm Hg ml. If the heart pumped intermittently and ejected blood at 100 mL/s into noncompliant arteries during the first half-second of the cycle only (i.e., 200 mL/s for 0.5 seconds), pressure would rise to 200 mm Hg during each ejection and drop

From Bench to Bedside / 14.1

Supersonic Shear Imaging for Noninvasive Assessment of Arterial Compliance

As humans age, the normally linear, ordered, elastin–collagen fiber orientation in arteries breaks down, becoming more and more entangled and disorganized as time passes. This change in the adventitial layer of arteries is thought to relate at least in part to the cumulative effects of years of exposure of the fibers to reactive oxygen species (oxygen radicals) that are generated in the body continuously as by-products of oxidative metabolism, NADPH oxidase (more commonly known as *NOX*), and inflammatory events. This oxygen radical–mediated destruction of elastin and collagen in arteries is similar to the same process that destroys these fibers in the skin, which then results in the appearance of skin wrinkles as we age.

Arteries with age-related damage to elastin and collagen lose their compliance over time. This age-related stiffening of arteries is called **arteriosclerosis**, or as known in lay terms, "hardening of the arteries." The decreased compliance of arteries raises systolic arterial pressure, and thus, systolic stress and oxygen demand in the heart. For this reason, systolic hypertension in the elderly is now considered a condition that requires antihypertensive treatment even if mean arterial pressure is normal. Furthermore, as discussed in this chapter, the work of the heart and its oxygen demand increases as arterial compliance decreases.

Measurement of compliance or other measures related to elasticity in an artery is relatively easy if the vessel can be removed from the body. In this way pressure–volume and stress–strain relationships can be produced through simple techniques. However, such assessments in human arteries are relegated to postmortem determinations. For obvious reasons, they cannot be used to assess arterial compliance changes over the lifetime of an individual. Instead, such longitudinal evaluation requires some sort of noninvasive means of assessing arterial compliance. For decades, arterial compliance in a patient was determined by estimating the pulse wave velocity from near the aorta to the periphery. This was done by timing the difference in the appearance of the arterial pulse in the carotid or other suitable site versus the dorsal pedal artery. Pulse wave velocity increases as arterial compliance decreases and, thus, the interval between the pulses was used as a crude estimate of arteriosclerosis.

Currently however, a new, noninvasive technique called *supersonic shear imaging (SSI)* is being tested as a means of creating a quantitative elasticity map of all major arteries in the body over a patient's lifetime. This technique is a sophisticated application of the principles of ultrasound transmission and was originally designed for the detection of tissue density irregularities and calcium deposits in human breast tumors. This technique produces ultrasonic shear waves in tissues at supersonic speeds, such that they produce a shock wave analogous to the sonic boom created in air when planes exceed the speed of sound. These shock waves set up micropalpations of soft or distensible structures such as the arterial wall, which when combined with ultrafast imaging techniques (5,000–10,000 images/s), produce quantified values for arterial stiffness. The value of this technique for clinical applications is that it is noninvasive, does not require exceptionally complex equipment, and can produce elastic maps of the entire arterial tree. For this reason, clinical trials are underway to use SSI to map local arterial wall elasticity and correlate it with a similar determination of local pulse wave velocity. It is hoped that this trial will lay the groundwork for the clinical use of SSI to monitor age-related changes in arterial compliance in patients.

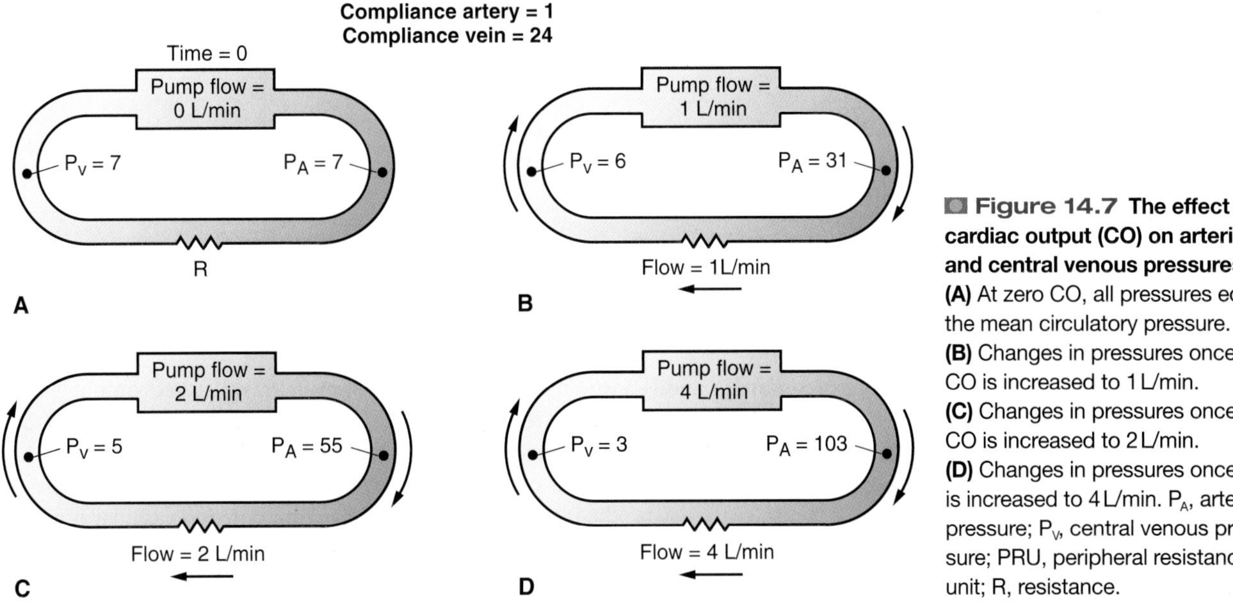

Figure 14.7 The effect of cardiac output (CO) on arterial and central venous pressures. **(A)** At zero CO, all pressures equal the mean circulatory pressure. **(B)** Changes in pressures once CO is increased to 1 L/min. **(C)** Changes in pressures once CO is increased to 2 L/min. **(D)** Changes in pressures once CO is increased to 4 L/min. P_A, arterial pressure; P_V, central venous pressure; PRU, peripheral resistance unit; R, resistance.

to 0 mm Hg during relaxation. Although no work would be done during relaxation, work done during the contraction would be 80,000 mm Hg ml. If this same intermittent flow was ejected into arteries with infinite compliance (flexibility), pressure would not rise during systole or fall during diastole and would remain at an average of 100 mm Hg. Work in this situation would then again equal 40,000 mm Hg ml. In reality, arteries are neither totally rigid nor infinitely compliant. Clearly, however, the example above indicates that increased arterial compliance reduces cardiac work, whereas decreased compliance increases cardiac work. Thus, myocardial oxygen demand will be increased by any factor that reduces arterial compliance, even if all other factors, such as arterial pressure, stroke volume, and heart rate, do not change. For this reason, the heart of an older person is confronted by increased oxygen demand from the simple fact that arterial compliance decreases with aging.

Increasing venous filling pressure increases cardiac output, but increasing cardiac output decreases venous pressure.

Central venous pressure is one of the key determinants of the filling of the right heart and, by extension of the Starling law of the heart, it is a key determinant of CO. However, the heart is a pump set in a *circulatory* system. Increased CO into the arterial segment of the circulation must come at the expense of volume and pressure in the venous side of the circulation. However, decreased venous pressure decreases CO. This poses an interesting set of questions: (1) how are values of CO above or below the resting level ever achieved or maintained, and (2) what determines the resting equilibrium between CO and central venous pressure?

A demonstration of how CO alters central venous and arterial pressure is shown in Figure 14.7A–D, where a theoretical circulation is depicted with a pump, a "venous" side of the circulation, and an "arterial" side of the circulation. For the sake of example, peripheral vascular resistance is set at 25 PRUs, and the veins are 24 times more compliant than the arteries. When the heart is stopped (CO equals zero), arterial and venous pressures are in equilibrium and dependent only on blood volume and the compliance of the vascular system. This value is called the *MCP* and is generally about 7 mm Hg. MCP is the value from which arterial pressure increases and venous pressure decreases once the heart starts to pump blood.

Using the example in Figure 14.7 with an MCP of 7 mm Hg, let us start the heart at an output of 1 L/min. The effect of this pumping by the heart will be to first translocate blood from the venous side into the arterial side of the circulation. This will increase pressure in the arterial component of the circulation while it reduces pressure on the venous side. These changes in pressure will continue until enough pressure difference between the arteries and veins has been created to move blood at a rate of 1 L/min across the resistance vessels in the circulation. At that point, the rate of blood moving across the resistance and into the "venous" side of the circulation will match the rate of blood exiting the "venous" side of the circulation through the pump and into the "arterial side." By the Poiseuille law, this will occur when the pressure difference is 25 mm Hg. What needs to be determined next is how much of this 25 mm Hg is added to the arterial side and how much is subtracted from the venous side of the circulation. In this example, the veins are 24 times more compliant than the arteries, or stated another way, the arteries are 24 times stiffer than the veins. This means that for every 25 mm Hg of pressure difference between the arterial and venous sides of the circulation, 24 mm Hg is added to the MCP on the arterial side (increasing its pressure to 31 mm Hg) and 1 mm Hg is subtracted from the MCP on the venous side (reducing it to 6 mm Hg). Thus, for every 1 L/min of increase in CO, 24 mm Hg will be added to arterial pressure and 1 mm Hg subtracted from the venous pressure once the system reaches its steady state. In our example, at

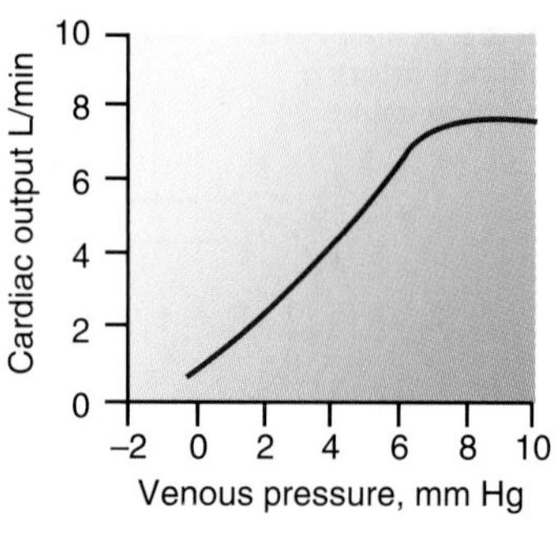

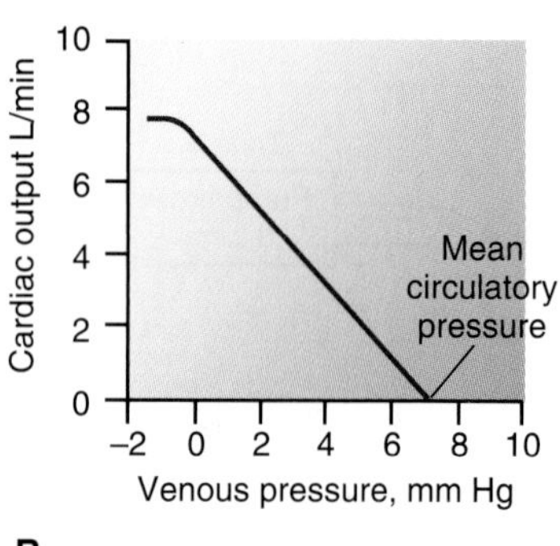

Figure 14.8 (A) The cardiac function curve. The curve is a representation of the Starling law of the heart and is a function of characteristics of the heart. **(B) The vascular function curve.** The relationship shows how venous pressure changes in response to a change in cardiac output (CO). The independent variable, CO, in this depiction is placed, by convention, on the *y* axis.

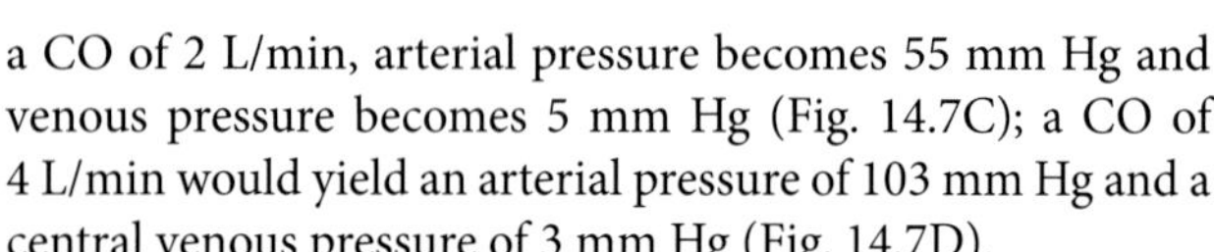

a CO of 2 L/min, arterial pressure becomes 55 mm Hg and venous pressure becomes 5 mm Hg (Fig. 14.7C); a CO of 4 L/min would yield an arterial pressure of 103 mm Hg and a central venous pressure of 3 mm Hg (Fig. 14.7D).

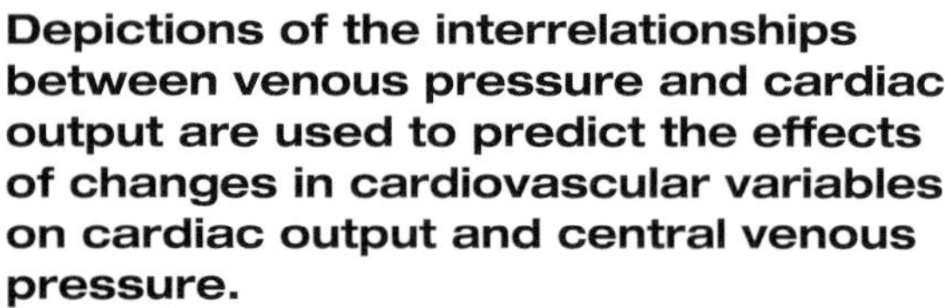

Depictions of the interrelationships between venous pressure and cardiac output are used to predict the effects of changes in cardiovascular variables on cardiac output and central venous pressure.

Two graphical means can be used to depict the interrelationship between CO and central venous pressure. One relationship, called the **cardiac function curve**, plots CO as a function of central venous pressure (Fig. 14.8A). This is simply an extension of the Starling law of the heart showing that factors augmenting ventricular filling result in increased CO. The cardiac function curve is characteristic of the heart itself in that only factors affecting the heart affect the position and shape of this curve. This curve can be produced even in a heart separated from the vasculature.

A second relationship, called the **vascular function curve**, shows how central venous pressure changes as a function of CO (Fig. 14.8B). Contrary to mathematical convention, the independent variable of the vascular function curve, CO, is placed on the *y* axis and the dependent variable in this relationship, the venous pressure, is placed on the *x* axis. (For reasons to be seen below, this facilitates combining the cardiac function curve with the vascular function curve on the same plot.) The position and slope of the vascular function curve are affected only by factors associated with blood vessels, such as vascular resistance, vascular compliance, and blood volume. The vascular function curve is independent of characteristics of the heart and can be observed even if an artificial pump replaces the heart. This curve shows that when CO is zero, venous pressure equals MCP and that venous pressure decreases as CO increases, until it results in venous collapse (at approximately −2 mm Hg), which limits further increases in CO.

Figure 14.9A to F depicts vascular and cardiac function curves plotted together and demonstrates how equilibrium between CO and venous pressure is obtained in a variety of altered states. Equilibrium for this system exists at the point of intersection between the two curves (see Fig. 14.9A). If venous pressure were to suddenly increase, an increased CO would initially result. However, this elevated output would tend to reduce venous pressure, which would then reduce CO, and so on. The end result, after an initial perturbation in either venous pressure or CO, will be to return these two variables to the intersection of the cardiac function and vascular function curves. This is therefore the equilibrium point for the system.

Several factors influence the vascular and cardiac function curves. For example, increases in blood or plasma volume (hypervolemia) "fill" the cardiovascular system more and thus raise MCP. This does not significantly alter venous compliance but does shift the entire vascular function curve in parallel fashion to the right of the normal relationship (Fig. 14.9B). This shift will also be seen with increased venous tone (venous smooth muscle contraction), which "squeezes" the blood contained in the veins, thus raising their internal pressure. Conversely, hypovolemia or decreased venous tone has the opposite effect.

A change in arteriolar tone has a different effect on the vascular function curves than does a change in venous tone (Fig. 14.9C). Because little blood volume is contained in the arterioles, changes in arteriolar tone do not significantly affect MCP. However, reduced arteriolar resistance makes it easier for the heart to eject a given stroke volume with any venous filling pressure, whereas increased resistance has the opposite effect. Therefore, at any central venous pressure, reducing arteriolar resistance augments CO and increasing such resistance impedes CO.

Changes in the inotropic state of the heart alter the cardiac but not the vascular function curve. By definition, a positive inotropic influence will allow the heart to produce a larger output at any given venous pressure, whereas a negative inotropic influence will have the opposite effect. This axiom is reflected by the cardiac function curves in Figure 14.9D.

The net effect of changes in cardiac and venous function curves on the equilibrium values for CO and venous pressure are shown in Figure 14.9E, where point "A" represents the normal equilibrium. Factors that increase blood volume such as an intravenous infusion of blood or fluid will shift the vascular function curve to the right, whereas loss of volume (i.e., hemorrhage) will shift that curve to the left. The new equilibrium points for these conditions, points "B" and "C," respectively, correctly show the interrelationship between CO and venous pressure that would be predicted by the Starling law of the heart; an increased CO is associated with an elevated venous (and hence filling) pressure, whereas a decreased CO is associated with a

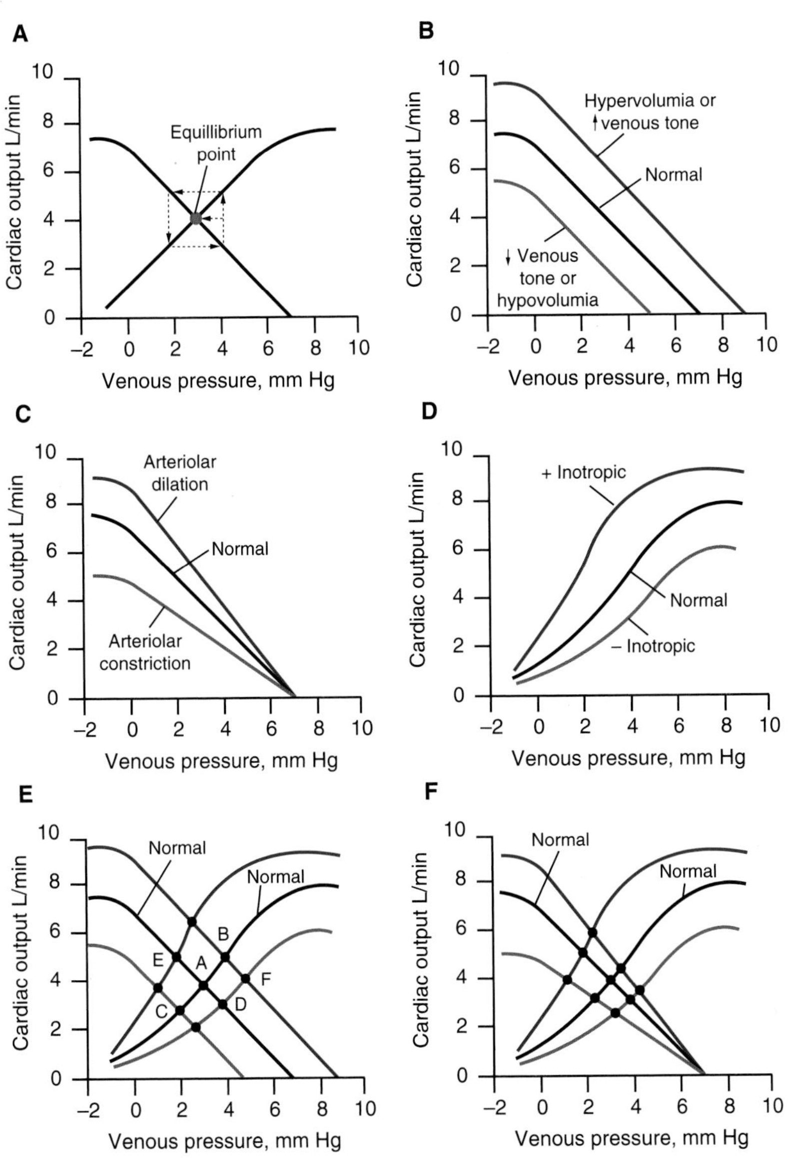

Figure 14.9 Various equilibriums in cardiac output (CO) and venous pressure resulting from changes in cardiac function or vascular variables. (A) Vascular function curves. **(B)** Effect of changing either blood volume or venous tone on the vascular function curve. **(C)** Effect of changes in arteriolar resistance on the vascular function curve. **(D)** Effect of changes in inotropic state on the cardiac function curve. **(E)** Possible equilibrium points obtained by the conjunction of altered myocardial contractility and vascular function curves. **(F)** Effects of alteration in vascular resistance on cardiac and vascular function curves and equilibrium points for CO.

decreased venous pressure. Similarly, the equilibrium points associated with changes in the inotropic state of the heart, "D" and "E," also correspond with our understanding of the definition of contractility. A positive inotropic influence allows for an elevated CO at a lower-than-normal venous pressure, whereas a negative inotropic influence has the opposite effect. Stated another way, by the Starling law, a decreased filling pressure must decrease CO, and therefore, the only way the heart can produce a larger-than-normal output in spite of reduced filling is to enhance its contraction by inotropic mechanisms. A positive inotrope, according to cardiac muscle mechanics, will allow the heart to empty more in spite of reduced preload. Figure 14.9E also serves to illustrate how the body can compensate for a failing heart. In acute heart failure, a patient's condition is likely best represented by point "D" on the graph, with reduced CO in spite of an elevated venous pressure because of the inability of the weakened heart to move blood from the veins to the arteries. However, with continued failure, the kidneys retain salt and water, resulting in a hypervolemic shift in the vascular function curve. This shift, through the Starling mechanism, helps augment the CO of the failing myocardium, as shown by point "F" on the diagram. Again, stated another way, one way in which a heart in a negative inotropic state (as in heart failure) can increase its output is to exploit the Starling law and increase its preload.

Figure 14.9F demonstrates the effects of alterations in vascular resistance on the cardiac and vascular function curves. Elevated resistance impedes CO and also requires an increased drop in venous pressure to establish the elevated gradient across the elevated resistance that will be needed to produce peripheral flow that matches a given CO. Decreased resistance has the opposite effect. The precise location of the new equilibrium depends on the relative degree to which the vascular and cardiac function curves are affected by changes in resistance.

Chapter Summary

- Cardiac output and systemic vascular resistance determine mean arterial pressure.
- Stroke volume and arterial compliance are the main determinants of pulse pressure.
- Arterial compliance decreases as arterial pressure increases.
- Changes in arterial pulse waveforms can be used to ascertain changes in hemodynamic variables that affect blood pressure.
- Shifts in blood volume between the periphery (extra-thoracic blood volume) and the chest (central blood volume) influence preload and cardiac output.
- Central venous pressure and cardiac output are interrelated.
- The cardiac function curve depicts how cardiac output varies as a function of central venous pressure.
- The vascular function curve depicts how central venous pressure varies as a function of cardiac output.
- Cardiac and vascular function curves can predict new equilibriums for cardiac output and central venous pressure whenever cardiac or vascular properties are changed.

thePoint *Visit http://thePoint.lww.com for chapter review Q&A, case studies, animations, and more!*

15 Microcirculation and Lymphatic System

ACTIVE LEARNING OBJECTIVES

Upon mastering the material in this chapter, you should be able to:

- Predict how changes in the blood concentration, tissue concentration, and capillary permeability of a substance alter its transport by diffusion across the capillaries.
- Explain how changes in perfused capillary density affect transport of substances across capillaries by diffusion.
- Predict whether a capillary will reabsorb or filter water based on changes in arteriolar pressure, plasma oncotic pressure, interstitial hydrostatic pressure, or interstitial oncotic pressure.
- Predict whether tissue edema will form based on changes in factors that affect capillary fluid filtration and lymph drainage of tissue as well as identify changes in these factors associated with common disease states and traumas.
- Explain how changes in precapillary and postcapillary resistances can increase or decrease fluid movement out of capillaries and how these changes may induce or attenuate edema formation.
- Explain myogenic and metabolic mechanisms. responsible for autoregulation of blood flow.
- Predict the effects of the loss of endothelial nitric oxide on arterial function.

The **microcirculation** is the portion of the vascular system comprising arterioles, capillaries, and venules. It is the part of the circulation where nutrients, water, gases, hormones, and waste products are exchanged between the blood and cells. The microcirculation minimizes diffusion distances, which facilitates material exchange between blood and tissue. This is its most important function. Virtually every cell in the body is in close contact with a microvessel. In fact, most cells are in direct contact with at least one microvessel. Regulation of vascular resistance in the microcirculation is an important aspect of total health. It regulates blood flow to individual organs, the distribution of blood flow within organs, diffusion distances between an organ's blood supply and tissues, and the exchange of fluid between the intravascular and extravascular compartments. In conjunction with cardiac output, it helps maintain arterial blood pressure by altering total peripheral vascular resistance (see Chapter 11, "Overview of the Cardiovascular System and Hemodynamics").

Normally, all microvessels, other than capillaries, are partially constricted by the contraction of their vascular smooth muscle (VSM) cells. If all microvessels were to dilate fully because of the relaxation of their smooth muscle cells, the arterial blood pressure would plummet. Cerebral blood flow in a standing person would then become inadequate, resulting in fainting, or syncope.

There is a constant conflict between the regulation of vascular resistance to maintain arterial pressure and the regulation required to allow each tissue to receive sufficient blood flow to sustain its metabolism. The compromise results in the preservation of mean arterial pressure, if needed, by increasing arterial resistance at the expense of reduced blood flow to most organs while preserving flow to the heart and brain. The organs survive this conflict by extracting more of the oxygen and nutrients from the blood flow they manage to get in the microvessels.

COMPONENTS OF THE MICROVASCULATURE

The microvasculature (Fig. 15.1) is considered to begin where the smallest arteries enter organs and to end where the smallest veins, the venules, exit organs. In between are microscopic arteries, the arterioles, and the capillaries. Depending on an animal's size, the largest arterioles have an inner diameter of 100 to 400 μm, and the largest venules have a diameter of 200 to 800 μm. The arterioles divide into progressively smaller vessels so that each section of the tissue has its own specific microvessels.

Control of total peripheral vascular resistance is effected at the level of the arterial microvasculature.

As discussed in Chapter 11, "Overview of the Cardiovascular System and Hemodynamics," large arteries have a low resistance to blood flow and function primarily

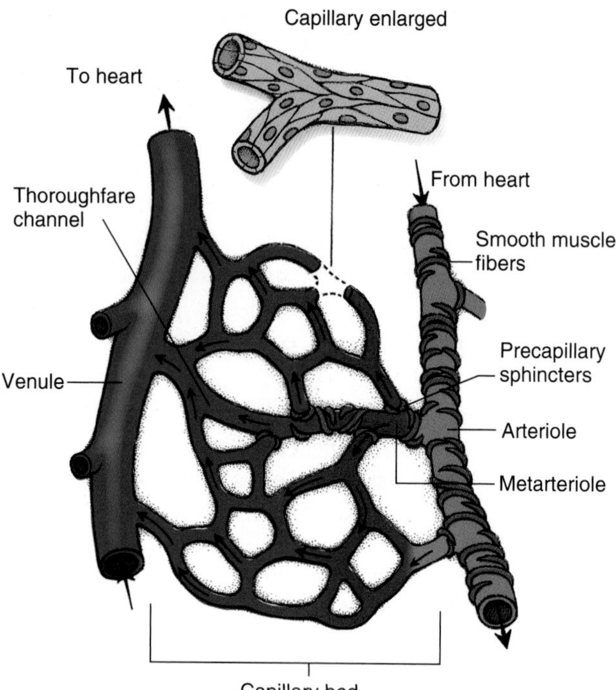

Figure 15.1 A diagrammatic representation of the components of the microvasculature. Arterioles control blood flow into a region of tissue and, along with precapillary sphincters, control the distribution of blood flow within the capillary network.

as conduits. As arteries approach the organ they supply, they divide into many small arteries both just outside and within the organ. In most organs, these small arteries, which are 500 to 1,000 μm in diameter, control about 30% to 40% of the total vascular resistance. These smallest arteries, combined with the arterioles of the microcirculation, constitute the resistance blood vessels; together they regulate about 70% to 80% of the total vascular resistance, with the remainder of the resistance about equally divided between the capillary beds and venules. Tonic constriction of small arteries and arterioles maintains the relatively high vascular resistance in organs. Constriction results from the release of norepinephrine by the sympathetic nervous system, from a myogenic mechanism intrinsic to the smooth muscle (to be discussed later), and from other chemical and physical factors that are active at the level of the smooth muscle cell.

Most arterioles, whether large or small, are tubes of endothelial cells surrounded by a connective tissue basement membrane, a single or double layer of VSM cells, and a thin outer layer composed of connective tissue cells, nerve axons, and mast cells. The arteriolar smooth muscle cells are 70 to 90 μm long when fully relaxed. The muscle cells are anchored to the basement membrane and to each other in a way such that any change in their length changes the diameter of the vessel. A single muscle cell will not completely encircle a larger vessel but may encircle a smaller vessel almost two times. VSM cells wrap around the arterioles at an approximately 90° angle to the long axis of the vessel. This arrangement is efficient because the tension developed by the VSM cell can then be almost totally directed toward maintaining or changing vessel diameter against the blood pressure within the vessel.

In most organs, arteriolar muscle cells operate at about half their maximal length. If the muscle cells fully relax, the diameter of the vessel can nearly double to increase blood flow dramatically, because flow increases as the fourth power of the vessel radius. When the muscle cells contract, the arterioles constrict, and with intense stimulation, the arterioles can literally close for brief periods of time.

The regulation of vascular resistance by the smallest arteries and all arterioles is primarily responsible for regulating organ blood flow. Vessel radius is determined by the transmural pressure gradient and wall tension, as expressed by the law of Laplace (see Chapter 11, "Overview of the Cardiovascular System and Hemodynamics"). Changes in wall tension developed by arteriolar smooth muscle cells directly alter vessel radius and, thus, vascular resistance. Most arterioles can dilate 60% to 100% from their resting diameter and can maintain a 40% to 50% constriction for long periods because of the latch state properties of VSM (see Chapter 8, "Skeletal and Smooth Muscle"). Therefore, large decreases and increases in organ vascular resistance and blood flow are well within the capability of the microscopic blood vessels. For example, a 20-fold increase in blood flow can occur in contracting skeletal muscle during exercise, and blood flow in the same vasculature can be reduced to 20% to 30% of normal during reflex increases in sympathetic nerve activity.

Exchange of water and materials between blood and tissues occurs across capillaries.

Capillaries provide for most of the exchange between blood and tissue cells. The capillaries are supplied by the smallest of arterioles, the **terminal arterioles**, and their outflow is collected by the smallest venules, **postcapillary venules**. A capillary is an endothelial tube surrounded by a basement membrane composed of dense connective tissue (Fig. 15.2). Capillaries in mammals do not have VSM cells and are unable to appreciably change their inner diameter. Pericytes (Rouget cells), wrapped around the outside of the basement membrane, may be a primitive form of VSM cell and may add structural integrity to the capillary.

Capillaries, with inner diameters of about 4 to 8 μm, are the smallest vessels of the vascular system. Although they are small in diameter and individually have a high vascular resistance, the parallel arrangement of many thousands of capillaries per cubic millimeter of tissue minimizes their collective resistance. However, the capillary lumen is so small that red blood cells must fold into a shape resembling a parachute as they pass through and virtually fill the entire lumen. The small diameter of the capillary and the thin endothelial wall minimize the diffusion path for molecules from the capillary core to the tissue just outside the vessel. In fact, the diffusion path is so short that most gases and inorganic ions can pass through the capillary wall in <2 ms.

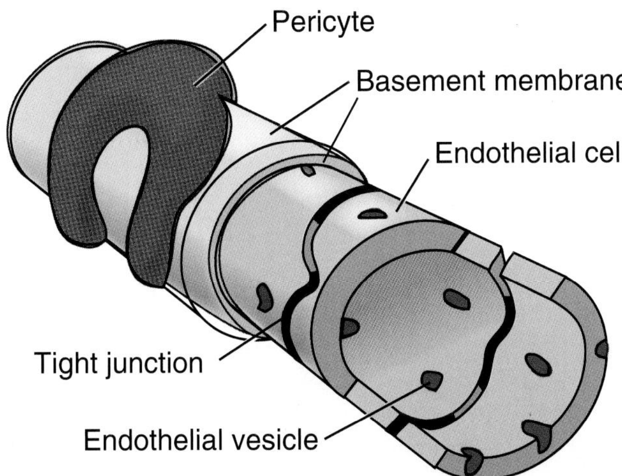

Figure 15.2 The various layers of a mammalian capillary. Adjacent endothelial cells are held together by tight junctions, which have occasional gaps. Water-soluble molecules pass through pores formed where tight junctions are imperfect. Vesicle formation and the diffusion of lipid-soluble molecules through endothelial cells provide other pathways for exchange.

Passage of molecules occurs between and through capillary endothelial cells.

The exchange function of the capillary is intimately linked to the structure of its endothelial cells and basement membrane. Lipid-soluble molecules, such as oxygen and carbon dioxide, readily pass through the lipid components of endothelial cell membranes. Water-soluble molecules, however, must diffuse through water-filled pathways formed in the capillary wall between adjacent endothelial cells. These pathways, known as **pores**, are not cylindrical holes but complex passageways formed by irregular tight junctions (see Fig. 15.2).

In capillaries, there are sufficient open areas in adjacent tight junctions to provide pores filled with water for the diffusion of small molecules. The pores are partially filled with a matrix of small fibers of submicron dimensions. The potential importance of this fiber matrix is that it acts partially to sieve the molecules approaching a water-filled pore. The combination of the fiber matrix and the small spaces in the basement membrane and between endothelial cells explains why the vessel wall behaves as if only about 1% of the total surface area were available for the exchange of water-soluble molecules. Most pores permit only molecules with a radius <3 to 6 nm to pass through the vessel wall. These small pores only allow water, inorganic ions, glucose, amino acids, and similar small, water-soluble solutes to pass; they exclude large molecules, such as serum albumin and globular proteins as well as blood cell components. A limited number of large pores, or possibly defects, allow virtually any large molecule in blood plasma to pass through the capillary wall. Even though few large pores exist, there is enough that nearly all the serum albumin molecules leak out of the cardiovascular system each day.

The porosity of capillaries is not the same in all organs. The capillaries of the brain and spinal cord have virtually continuous tight junctions between adjacent endothelial cells; consequently, only the smallest water-soluble molecules pass through their capillary walls. Capillaries in cardiac and skeletal muscle also have relatively low porosity to water and small water-soluble molecules. In contrast, capillaries in the intestines, the liver, and the glomerulus of the kidney have capacities for large water transport, whereas those in the spleen and bone marrow have capillary pores so large that they allow for the passage of cellular elements between the blood and those organs.

An alternative pathway for water-soluble molecules through the capillary wall is via **endothelial vesicles** (see Fig. 15.2). Membrane-bound vesicles form on either side of the capillary wall by pinocytosis, and exocytosis occurs when the vesicle reaches the opposite side of the endothelial cell. The vesicles appear to migrate randomly between the luminal and abluminal sides of the endothelial cell. Even the largest molecules may cross the capillary wall in this way. The importance of transport by vesicles to the overall process of transcapillary exchange remains unclear. Occasionally, continuous interconnecting vesicles have been found that bridge the endothelial cell. This open channel could be a random error or a purposeful structure, but in either case, it would function as a large pore to allow the diffusion of large molecules.

Venules collect blood from capillaries and act as a blood reservoir.

After the blood passes through the capillaries, it enters the venules—endothelial tubes usually surrounded by a monolayer of VSM cells. In general, the vascular muscle cells of venules are much smaller in diameter but longer than those of arterioles. The muscle size may reflect the fact that venules do not need a powerful muscular wall because they operate at intravascular pressures of 10 to 16 mm Hg, compared with 30 to 70 mm Hg in arterioles. The smallest venules are unique because they are more permeable than capillaries to large and small molecules. This increased permeability seems to exist because tight junctions between adjacent venular endothelial cells have more frequent and larger discontinuities or pores. It is probable that much of the exchange of large water-soluble molecules occurs as the blood passes through small venules. The permeability of this venular microvasculature can be affected by local agents such as histamine, which increase their permeability. This is part of the mechanism responsible for local tissue fluid accumulation in allergic reactions.

Venules are an important component of the blood reservoir system in the venous circulation. At rest, approximately two thirds of the total blood volume is within the venous system, and perhaps more than half of this volume is within venules. Although the blood moves within the venous reservoir, it moves slowly, much like water in a reservoir behind a river dam. If venule radius is increased or decreased, the volume of blood in tissue can change up to 20 mL/kg of tissue; therefore, the volume of blood readily available for circulation would increase by more than 1 L in a 70-kg person. Such a large change in the available blood volume can substantially improve the venous return of blood to the heart following the depletion of blood volume caused by hemorrhage or dehydration. For example, the volume of blood typically removed from blood donors is about 500 mL, or about 10% of the total blood volume. Usually no ill effects are experienced, in part because the venules and veins decrease their reservoir volume to restore the circulating blood volume.

LYMPHATIC SYSTEM

Lymphatic vessels are microvessels that form an interconnected system of simple endothelial tubes within tissues. They do not carry blood but instead transport fluid, serum proteins, lipids, and even foreign substances from the interstitial spaces back to the circulation. Collecting lymphatic fluid from the organs is important because a volume of fluid equal to the plasma volume is filtered from the blood to the tissues every day.

The gastrointestinal tract, the liver, and the skin have the most extensive lymphatic systems. The lymphatic network is also extensive in the pulmonary circulation, in which preventing water accumulation is essential for proper gas transport between the environment and blood in that organ. The central nervous system, in contrast, may not contain any lymph vessels.

The lymphatic system typically begins as blind-ended tubes, or **lymphatic bulbs**, which drain into the meshwork of interconnected lymphatic vessels (Fig. 15.3). The lymphatic vessels coalesce into increasingly more developed and larger collection vessels. These larger vessels in the tissue and the macroscopic lymphatic vessels outside the organs have contractile cells similar to VSM cells. In connective tissues of the mesentery and skin, even the simplest of lymphatic vessels and bulbs spontaneously contract, perhaps as a result of contractile endothelial cells. Even if the lymphatic bulb or vessel cannot contract, external compression of these lymphatic structures by movements of the organ (e.g., intestinal movements or skeletal muscle contractions) changes lymphatic vessel size.

Lymphatic vessels mechanically collect fluid from tissue fluid between cells.

In all organ systems, more fluid is filtered than absorbed by the capillaries, and plasma proteins diffuse into the interstitial spaces through the large pore system. By removing the fluid, the lymphatic vessels also collect the proteins. This function is essential because higher protein concentration in the plasma versus that in the interstitium helps retain water in the vascular system (see below). The ability of lymphatic vessels to change diameter is important for lymph formation and protein removal, whether initiated by the lymphatic vessel or by compressive forces generated within a contractile organ. In the smallest lymphatic vessels and, to some extent, in the larger lymphatic vessels in a tissue, the endothelial cells are overlapped rather than fused together, as in blood capillaries. The overlapped portions of the cells

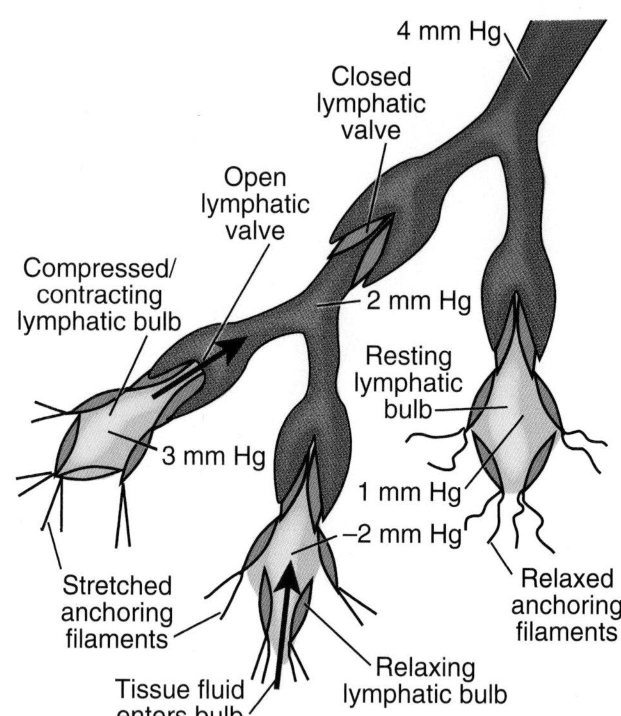

Figure 15.3 Lymphatic vessels: Basic structure and functions. The contraction–relaxation cycle of lymphatic bulbs (*bottom*) is the fundamental process that removes excess water and plasma proteins from the interstitial spaces. Pressures along the lymphatics are generated by lymphatic vessel contractions and by organ movements.

are attached to anchoring filaments, which extend into the tissue (see Fig. 15.3). When stretched, anchoring filaments pull apart the free edges of the endothelial cells when the lymphatic vessels relax after a compression or contraction. The openings created in this process allow tissue fluid and molecules carried in the fluid to easily enter the lymphatic vessels.

The movement of fluid from tissue to the lymphatic vessel lumen is passive. When lymphatic vessels are allowed to passively relax, the pressure in the lumen becomes slightly lower than in the interstitial space, and tissue fluid enters the lymphatic vessel. Once the interstitial fluid is in a lymphatic vessel, it is called **lymph**. When the lymphatic bulb or vessel next actively contracts or is compressed, the overlapped cells are mechanically sealed to hold the lymph. The pressure developed inside the lymphatic vessel forces the lymph into the next downstream segment of the lymphatic system. Because the anchoring filaments are stretched during this process, the overlapped cells can again be parted during the relaxation of the lymphatic vessel the way in which a stretched rubber band snaps back after the release of its stretch. This compression–relaxation cycle facilitates the uptake and flow of fluid from the interstitium into and down the lymphatic channels. The compression–relaxation cycle increases in frequency and vigor when excess water is in the lymph vessels, whether lymphatic smooth muscle cells or the contractile lymphatic endothelial cells control it. Conversely, less fluid in the lymphatic vessels allows the vessels to become quiet and pump less fluid. This simple regulatory system ensures that the fluid status of the organ's interstitial environment is appropriate.

The active and passive compression of lymphatic bulbs and vessels also provides the force needed to propel the lymph back to the venous side of the blood circulation. To maintain directional lymph flow, microscopic lymphatic bulbs and vessels as well as large lymphatic vessels have one-way valves (see Fig. 15.3). These valves allow lymph to flow only from the tissue toward the progressively larger lymphatic vessels and, finally, into large veins in the chest cavity.

Lymphatic pressures are only a few millimeters Hg in the bulbs and smallest lymphatic vessels and as high as 10 to 20 mm Hg during contractions of larger lymphatic vessels. This progression from lower to higher lymphatic pressures is possible because, as each lymphatic segment contracts, it develops a slightly higher pressure than in the next lymphatic vessel, and the lymphatic valve momentarily opens to allow lymph flow. When the activated lymphatic vessel relaxes, its pressure is again lower than that in the next vessel, and the lymphatic valve closes.

MATERIAL EXCHANGE BETWEEN THE VASCULATURE AND TISSUES

The increasing numbers of vessels through successive branches of the arterial tree and on into the capillaries and venules dramatically increase the surface area of the microvasculature. The length, diameter, and number of vessels determine the surface area. In the small intestine, for example, the total surface area of the capillaries and smallest venules is >10 cm^2 for 1 cm^3 of tissue. The large surface area of the capillaries and smallest venules is important because most exchange of nutrients, wastes, and fluid occurs across these tiny vessels. Exchange between the interstitial fluid and capillaries occur via three processes: (1) diffusion, (2) filtration, and (3) vesicular transport.

Transport across capillaries is enhanced by increasing their collective surface area and reducing diffusion distances from capillaries to cells.

The spacing of microvessels in the tissues determines the distance molecules must diffuse from the blood to the interior of the tissue cells. In the example shown in Figure 15.4A, nutrients are supplied to the cell by a single capillary. The density of dots at various locations represents the concentration of bloodborne molecules across the cell interior. Note that as molecules travel farther from the capillary, their concentration decreases substantially because the volume into which diffusion proceeds increases as the square of the distance. Furthermore, the time required for traversing a given distance by diffusion increases as an exponential function of distance (see Chapter 11, "Overview of the Cardiovascular System and Hemodynamics"). Thus, if molecules are consumed by cells, the concentration of substances surrounding the cells will decrease dramatically if the distance between the cell and its nearest capillary increases.

Increasing the number of microvessels reduces diffusion distances from a given point inside a cell to the nearest capillary. Doing so minimizes the dilution of molecules within the cells caused by large diffusion distances. At any given moment during resting conditions, only about 40% to 60% of the capillaries are perfused by red blood cells in most organs. The capillaries not in use do contain blood, but it is not moving. As shown in Figure 15.4B,C, increasing the number of perfused capillaries with moving blood elevates cell concentrations of molecules derived from the blood. For this reason, however, it is equally true that decreasing the number of capillaries perfused with blood lengthens diffusion distances and decreases exchange. This can occur by severely constricting arterioles or by destroying existing capillaries, as in diabetes mellitus.

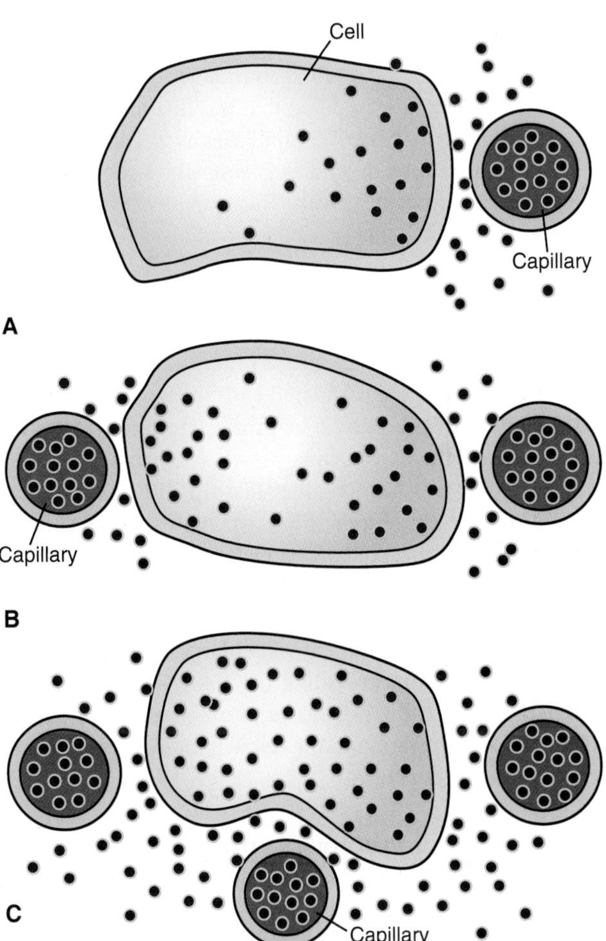

Figure 15.4 The effect of the number of perfused capillaries on cell concentration of blood borne molecules Molecules are represented by *dots* in the figure. **(A)** With one capillary, the left side of the cell has a low concentration. **(B)** The concentration can be substantially increased if a second capillary is perfused. **(C)** The perfusion of three capillaries around the cell increases concentrations of bloodborne molecules throughout the cell.

Transport of solutes across capillaries is enhanced by increasing the capillary permeability and concentration gradient for the solute across the capillaries.

Diffusion is by far the most important means for moving solutes across capillary walls. The Fick law (see Chapter 2, "Plasma Membrane, Membrane Transport, and Resting Membrane Potential") gives the rate of diffusion of a solute between blood and tissue as

$$J_s = PS(C_b - C_t) \quad (1)$$

J_s is the net movement of solute (often expressed in mol/min per 100 g of tissue), PS is the **permeability surface area coefficient**, and C_b and C_t are the blood and tissue concentrations of the solute, respectively. The PS coefficient is directly related to the diffusion coefficient of the solute in the capillary wall and the vascular surface area available for exchange and is inversely related to the diffusion distance. The surface area and diffusion distance are determined, in part, by the number of microvessels with active blood flow. The diffusion coefficient is relatively constant, unless the capillaries are damaged, because it depends on the anatomic properties of the vessel wall (e.g., the size and abundance of pores) and the chemical nature of the material that is diffusing.

The number of perfused capillaries as well as blood and tissue concentrations of solutes is constantly changing, and chronic changes occur as well. Therefore, the diffusion distance and surface area for exchange can be influenced by physiologic events. The same is true for concentrations in the tissue and blood. In this context, microvascular exchange can be dynamically altered by many physiologic events. For example, perfused capillary density increases significantly in metabolic active tissue compared with the tissue at rest. This is seen in exercising muscle or in the intestinal circulation during digestion.

It is important to remember that the diffusion rate also depends on the *difference* between the high and low concentrations, not the specific concentrations. For example, if the cell consumes a particular solute, the concentration in the cell will decrease, and for a constant concentration in blood plasma, the diffusion gradient will enlarge and thus increase the rate of diffusion, providing the tissue is supplied with sufficient blood. If the cell ceases to use as much of a given solute, the concentration in the cell will increase and the rate of diffusion will decrease.

Increases in vascular permeability, surface area, or blood flow enhance the diffusion of small molecules from the blood.

As a result of diffusional losses and gains of molecules as blood passes through the tissues, the concentrations of various molecules in venous blood can be very different from those in arterial blood. The **extraction** (E), or **extraction ratio**, of material from blood perfusing a tissue can be calculated from the arterial (C_a) and venous (C_v) blood concentration as

$$E = (C_a - C_v)/C_a \quad (2)$$

If the blood loses material to the tissue, the value of E is positive and has a maximum value of 1 if all material is removed from the arterial blood ($C_v = 0$). An E value of 0 indicates that no loss or gain occurred. A negative E value indicates that the tissue added material to the blood. The total mass of material lost or gained by the blood can be calculated as

$$\text{Amount lost or gained} = E \times \dot{Q} \times C_a \quad (3)$$

where E is extraction, $\dot{Q}$ is blood flow, and C_a is the arterial concentration. Although this equation is useful for calculating the total amount of material exchanged between tissue and blood, it does not allow a direct determination of how changes in vascular permeability and exchange surface area influence the extraction process. The extraction can be related to the permeability (P) and surface area (A) available for exchange as well as the blood flow ($\dot{Q}$):

$$E = 1 - e^{-PA/\dot{Q}} \quad (4)$$

The e is the base of the natural system of logarithms. This equation predicts that extraction increases when either permeability or exchange surface area increases or blood flow decreases. Extraction decreases when permeability and surface area decrease or blood flow increases. Consequently, physiologically induced changes in the number of perfused capillaries, which alter surface area, and changes in blood flow are important determinants of exchange processes. The inverse effect of blood flow on extraction occurs because, if flow increases, less time is available for exchange. Conversely, a slowing of flow allows more time for exchange.

Ordinarily, the blood flow and total perfused surface area change in the same direction, although by different relative amounts. For example, surface area can generally, at most, double or be reduced by about half; however, blood flow can increase threefold to fivefold or more in some tissues such as skeletal muscle during exercise or decrease by about half in many organs, yet maintain viable tissue. The net effect is that extraction is rarely more than doubled or decreased by half relative to the resting value in most organs. This is still an important range because changes in extraction can compensate for reduced blood flow or enhance exchange when blood flow is increased.

Diffusion-limited and flow-limited transport

Simple passive diffusion is the major process by which nutrients and metabolites cross the capillary barrier. Gases, small lipid-soluble molecules, water, simple sugars, and ions can diffuse so rapidly across capillaries that their transport from capillary to tissue or vice versa is not limited by their ability to diffuse across the tissue. Their transport is only limited by their rate of delivery into the capillaries from blood flow perfusing the capillary network. Thus, the transport of these substances across the capillary network is said to occur by **flow-limited transport**. Larger lipophobic molecules, such as sucrose, polysaccharides, and proteins, have difficulty diffusing across the capillary membrane or pores, and therefore, diffusion is rate limiting in their transport into or out of capillaries. Such transport is called **diffusion-limited transport**.

Diffusion of a substance across the capillary is proportional to its concentration difference within and outside the vessel and inversely proportional to the distance over which diffusion must occur. For diffusible, flow-limited transport substances, increased blood flow increases the effective concentration of a substance in the capillary and accelerates its outward diffusion; that is, the faster materials are brought into the capillary, relative to their diffusion out of the capillary, the higher will be the concentration of the substance inside the capillary.

In pathologic conditions, substances that otherwise exhibit flow-limited transport can become diffusion limited. This can occur when diffusion distances between capillaries and cells become too great to allow the rapid exchange of materials, as seen in the lungs when the transport of normally highly diffusible oxygen is impaired by infection or fluid accumulation in the pulmonary interstitium. This phenomenon can also occur in any organ when the number of perfused capillaries is radically reduced (by low flow, blood clots, etc.), thereby increasing the distance between tissue cells and the nearest capillaries with blood flow.

Interplay between net hydrostatic and oncotic forces determines the net direction of fluid exchange across the capillaries.

To force the blood through microvessels, the heart pumps blood into the elastic arterial system and provides the pressure needed to move the blood. This hemodynamic, or hydrostatic pressure, although absolutely necessary, favors the pressurized filtration of water through pores because the hydrostatic pressure on the blood side of the pore is greater than on the tissue side. In organs other than the kidney, capillary hydrostatic pressure is highest at the arteriolar end of the capillary and declines to its lowest value at the venule end of the capillary. Typically, capillary hydrostatic pressure is approximately 40 mm Hg at its arteriolar end and approximately 15 mm Hg at the venule end. The average capillary pressure is different in each organ, ranging from about 15 mm Hg in intestinal villus capillaries to 55 mm Hg in the kidney glomerulus. The interstitial hydrostatic pressure ranges from slightly negative to 8 to 10 mm Hg.

The primary defense against excessive fluid filtration from the capillaries is the **colloid osmotic pressure**, also called *plasma oncotic pressure*, which is generated by plasma proteins that generally cannot cross the capillary endothelium. Because interstitial protein concentration is lower than that in plasma, there is a net oncotic force pulling water into the capillaries. Colloid osmotic pressure is conceptually similar to osmotic pressures for small molecules in that both primarily depend on the number of molecules in the solution. The major plasma protein contributing to capillary oncotic pressure is serum albumin because it has the highest molar concentration of all plasma proteins. The plasma colloid osmotic pressure of plasma proteins is typically 18 to 25 mm Hg in mammals.

In the body as a whole, this net inward oncotic force does not balance the net outward hydrostatic pressure difference. Thus, there exists a slightly positive overall force for the filtration of fluid out of the capillaries. Most organs continuously form lymph, which supports the concept that capillary and venular filtration pressures generally are larger than absorption pressures. The balance of pressures is likely +1 to +2 mm Hg outward in most organs. However, this is not true for each capillary in the body all the time. Some capillaries can be net reabsorbers of water, others filter along their entire length, whereas still others can filter at their arteriolar end and reabsorb at their venule end, depending on existing hydrostatic pressure in both capillary locations. Based on directly measured capillary hydrostatic and plasma colloid osmotic pressures, the entire length of the capillaries in skeletal muscle filter slightly all of the time, whereas the lower capillary pressures in the intestinal mucosa and brain primarily favor absorption along the entire capillary length.

Effects of capillary oncotic and hydrostatic pressure on fluid flux are counteracted by these pressures in the interstitium.

A small amount of plasma protein enters the interstitial space; these proteins and, perhaps, native proteins of the space generate the tissue colloid osmotic pressure. This pressure of 2 to 5 mm Hg offsets part of the colloid osmotic pressure in the plasma. This is, in a sense, a filtration pressure that opposes the blood colloid osmotic pressure. The hydrostatic pressure on the tissue side of the endothelial pores is the tissue hydrostatic pressure. The water volume in the interstitial space and tissue distensibility determine this pressure. Tissue hydrostatic pressure can be increased by organ compression such as in a muscle during contraction. The tissue hydrostatic pressure in various tissues during resting conditions is a matter of debate. Tissue pressure is probably slightly below atmospheric pressure (negative) to slightly positive (<+3 mm Hg) during normal hydration of the interstitial space and becomes positive when excess water is in the interstitial space. Tissue hydrostatic pressure is a filtration force when negative and is an absorption force when positive.

If water is removed from the interstitial space, the hydrostatic pressure becomes negative (Fig. 15.5). If a substantial amount of water is added to the interstitial space, the tissue hydrostatic pressure is increased. Standing also causes high capillary hydrostatic pressures from gravitational effects on the blood in the arterial and venous vessels and results in excessive filtration.

However, a margin of safety exists over a wide range of tissue fluid volumes (see Fig. 15.5), and excessive tissue hydration is avoided. The ability of tissues to allow substantial changes in interstitial volume with only small changes in pressure indicates that the interstitial space is distensible. As a general rule, about 500 to 1,000 mL of fluid can be withdrawn from the interstitial space of the entire body to help replace water losses resulting from sweating, diarrhea, vomiting, or blood loss.

Starling-Landis equation quantifies fluid flow across the capillaries.

At the end of the 19th century, the English physiologist Ernest Starling first postulated the role of hydrostatic and colloid osmotic pressures in determining fluid movement

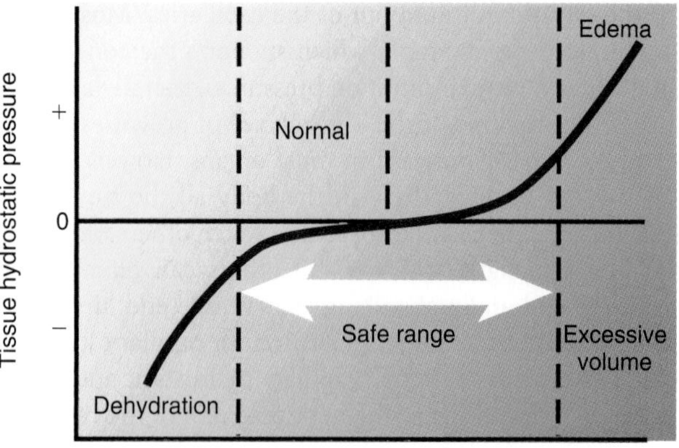

Figure 15.5 Variations in tissue hydrostatic pressure as interstitial fluid volume is altered. Under normal conditions, tissue pressure is slightly negative (subatmospheric), but an increase in volume can cause the pressure to become positive. If the interstitial fluid volume exceeds the "safe range," high tissue hydrostatic pressures and edema will be present. Tissue dehydration can cause negative tissue hydrostatic pressures.

across capillaries. In the 1920s, the American physiologist Eugene Landis obtained experimental proof for Starling's hypothesis. The relationship is defined for a single capillary by the **Starling-Landis equation**:

$$J_V = K_h A\,[(P_c - P_t) - \sigma(COP_p - COP_t)] \qquad (5)$$

J_V is the net volume of fluid moving across the capillary wall per unit of time (μm³/min). K_h is the **hydraulic conductivity for water**, which is the fluid permeability of the capillary wall. It can be thought of as the ease with which water crosses the capillary wall. K_h is expressed as μm³/min/mm Hg (μm² of capillary surface area per minute per mm Hg pressure difference). The value of K_h increases up to fourfold from the arterial to the venous end of a typical capillary. A, in equation 5, is the vascular surface area, P_c is the capillary hydrostatic pressure, and P_t is the tissue hydrostatic pressure. COP_p and COP_t represent the plasma and tissue colloid osmotic pressures, respectively, and σ is the **reflection coefficient** for plasma proteins. This coefficient is included because the microvascular wall is slightly permeable to plasma proteins, preventing the full expression of the two colloid osmotic pressures. The value of σ is 1 when molecules cannot cross the membrane (i.e., they are 100% "reflected") and 0 when molecules freely cross the membrane (i.e., they are not reflected at all). Typical σ values for plasma proteins in the microvasculature exceed 0.9 in most organs other than the liver and spleen, which have capillaries that are permeable to plasma proteins. The reflection coefficient is normally relatively constant but can be decreased dramatically by hypoxia, inflammatory processes, and tissue injury. This leads to increased fluid filtration because the effective fluid-retaining power of the colloid osmotic pressure is reduced when the vessel wall becomes more permeable to plasma proteins.

The extrapolation of fluid filtration or absorption for a single capillary to fluid exchange in a whole tissue is difficult. Within organs, there are regional variations in microvascular pressures, possible filtration and absorption of fluid in vessels other than capillaries, and physiologically and pathologically induced variations in the available surface area for capillary exchange. Therefore, for whole organs, a measurement of total fluid movement relative to the mass of the tissue is used. To take into account the various hydraulic conductivities and total surface areas of all vessels involved, the volume (mL) of fluid moved per minute for a change of 1 mm Hg in net capillary filtration pressure for each 100 g of tissue is determined. This value is called the **capillary filtration coefficient (CFC)**, even though it is likely that fluid exchange also occurs in venules. CFC values in tissues such as skeletal muscle and the small intestine are typically in the range of 0.025 to 0.16 mL/min/mm Hg per 100 g. The CFC replaces the combined hydraulic conductivity (K_h) and capillary surface area (A) variables in the Starling-Landis equation that apply for filtration across a single capillary.

The CFC can change if fluid permeability, the surface area (determined by the number of perfused microvessels), or both are altered. For example, during the intestinal absorption of foodstuff, particularly lipids, both capillary fluid permeability and perfused surface area increase. Therefore, CFC increases dramatically. In contrast, the skeletal muscle vasculature increases CFC primarily because of increased perfused capillary surface area during exercise whereas only small increases in fluid permeability occur.

The hydrostatic and colloid osmotic pressure differences across capillary walls, called the **Starling forces**, cause the movement of solutes along with the water into the interstitial spaces. These solute movements are, however, normally small and contribute minimally to tissue nutrition. Most solutes transferred to the tissues move across capillary walls by simple diffusion, not by bulk flow of fluid.

Capillary hydrostatic pressure is altered by changes in precapillary and postcapillary resistance and arteriolar and venule blood pressure.

Normal physiologic processes in the body do not control plasma colloid osmotic pressure, tissue hydrostatic pressure, or tissue colloid osmotic pressure. Thus, manipulation of these parameters cannot be used to regulate filtration and reabsorption of fluid at the capillary. Such regulation is accomplished, therefore, by the adjustment of capillary hydrostatic pressure.

Capillary pressure (P_c) is influenced by four major variables: precapillary (R_{pre}) and postcapillary (R_{post}) resistances and arterial (P_a) and venous (P_v) pressures. Precapillary and postcapillary resistances can be calculated from the pressure dissipated across the respective vascular regions divided by the total tissue blood flow ($\dot{Q}$), which is essentially equal for both regions:

$$R_{pre} = (P_a - P_c)/\dot{Q} \qquad (6)$$

$$R_{post} = (P_c - P_v)/\dot{Q} \qquad (7)$$

In most organ vasculatures, the precapillary resistance is three to six times higher than the postcapillary resistance. This has a substantial effect on capillary pressure. To demonstrate the effect of precapillary and postcapillary resistances on capillary pressure, we use the equations for the precapillary and postcapillary resistances to solve for blood flow:

$$\dot{Q} = (P_a - P_c)/R_{pre} = (P_c - P_v/R_{post}) \qquad (8)$$

The two equations to the right of the flow term can be solved for capillary pressure:

$$P_c = \frac{(R_{post}/R_{pre})P_a + P_v}{1 + (R_{post}/R_{pre})} \qquad (9)$$

Equation 9 indicates that the ratio of postcapillary to precapillary resistance, rather than the absolute magnitude of either resistance, determines the effect of arterial pressure (P_a) on capillary pressure. It also shows that venous pressure substantially influences capillary pressure and that the denominator influences both pressure effects. At a typical postcapillary to precapillary resistance ratio of 0.16:1, the denominator will be 1.16, which allows about 80% of a change in venous pressure to be reflected back to the capillaries.

Capillary hydrostatic pressure increases with the arteriolar vasodilation that accompanies increased tissue metabolism. With increased tissue metabolism, the postcapillary to precapillary resistance ratio increases because precapillary resistance decreases more than postcapillary resistance. Because the balance of hydrostatic and colloid osmotic pressures is usually −2 to +2 mm Hg, a 10- to 15-mm Hg increase in capillary pressure during maximum vasodilation can cause a profound increase in filtration. The increased filtration associated with microvascular dilation is usually associated with a large increase in lymph production, which removes excess tissue fluid.

When sympathetic nervous system stimulation causes a substantial increase in precapillary resistance and a proportionately smaller increase in postcapillary resistance, the capillary pressure can decrease up to 15 mm Hg and, thereby, greatly increase the absorption of tissue fluid. This process is an important compensatory mechanism the body uses to combat the early stages of circulatory shock (see Chapter 17, "Control Mechanisms in Circulatory Function").

Edema impairs diffusional transport across the capillaries.

Changes in capillary hydrostatic pressure and plasma protein concentration can have a profound effect on filtration at the capillary. The hydrostatic pressure involved in transcapillary fluid exchange depends on how the microvasculature dissipates the prevailing arterial and venous pressures. As seen above, certain interplay of these variables can lead to substantial filtration across the capillaries. Although plasma protein concentration does not vary moment to moment, its circulating value is determined largely by the rate of protein synthesis in the liver, where most of the plasma proteins are made. Disorders that impair albumin synthesis or promote the loss of albumin result in reduced plasma protein concentration, a lowered plasma colloid osmotic pressure, and excessive fluid filtration at the capillaries.

Edema is a condition in which there is excessive accumulation of fluid in tissue spaces. Edema formation in the abdominal cavity, known as **ascites**, can allow large quantities of fluid to collect in and grossly distend the abdominal cavity. Edema interferes with capillary transport by increasing diffusion distances between capillaries and tissue cells. Also, edema can cause circulatory collapse if edematous fluid is derived from a loss of plasma volume. Anything that causes excess fluid filtration at the capillaries or impairs fluid transport through the lymph channels will create edema (Fig. 15.6). For example, one of the common causes of edema formation is a loss of protein, especially albumin, from the plasma. Edema will occur when albumin concentrations drop below 2.5 g/100 mL. This is common in diseases in which the liver is unable to manufacture albumin or in kidney diseases in which albumin and other proteins are lost into the urine. Burns, by destroying capillary integrity, cause edema through increased capillary permeability, loss of albumin from damaged vessels, and inflammatory vasodilation. Hives, which are a form of localized edema associated with allergic reactions, result from an increase in capillary permeability, venule permeability, and arterial dilation, all caused by histamine release during the allergic response. Obstruction of veins, usually from blood clot formation after surgery, is another common cause of edema, as is the obstruction of lymph channels at lymph nodes by infections.

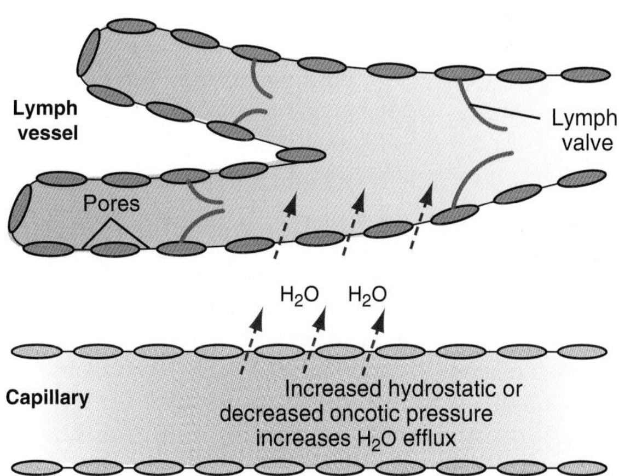

↑ **Lymph flow:**	↑ **Edema formation:**
↑ capillary hydrostatic pressure	↓ plasma protein concentration
↑ capillary surface area	arteriolar dilation
↑ capillary permeability	venous obstruction
↑ tissue metabolism	lymphatic obstruction
↑ muscle activity (massage)	histamine (↑ capillary permeability)
	all factors that ↑ lymph flow except muscle massage

Figure 15.6 Water efflux from the capillary and into a lymph vessel. Factors that increase lymph flow or favor the formation of edema are listed on the figure.

REGULATION OF MICROVASCULAR RESISTANCE

The VSM cells around arterioles and venules respond to a wide variety of physical and chemical stimuli, altering the diameter and resistance of the microvessels. Here, we consider the various physical and chemical conditions in tissues that affect the contractile state of VSM and, thus, affect resistance in the microvasculature.

Myogenic regulation causes arterioles to constrict in response to an elevated intravascular pressure.

VSM can contract rapidly when stretched and, conversely, can actively relax when passively shortened. This process, called *myogenic regulation*, is activated when microvascular pressure is increased or decreased. Myogenic mechanisms are extremely fast and appear to be able to adjust to the most rapid pressure changes. In fact, VSM may be able to contract or relax when the load on the muscle is increased or decreased, respectively, without any change in muscle length. These responses are known to persist for as long as the initial stimulus is present, unless vasoconstriction reduces blood flow to the extent that tissue becomes severely hypoxic.

The cellular mechanisms responsible for myogenic regulation are not entirely understood, but several possibilities are likely involved. The first mechanism is a calcium ion-selective channel that is opened in response to increased membrane stretch or tension. Stretch-activated calcium channels, distinct from receptor-operated or voltage-gated calcium channels, exist in VSM. VSM contraction is sensitive to external calcium concentration. Adding calcium to the cytoplasm from the extracellular fluid through membrane channels would activate the smooth muscle cell and result in contraction. Conversely, limiting calcium entry through channels would allow calcium pumps to remove calcium ions from the cytoplasm and favor relaxation.

The second mechanism postulated for the myogenic response is that a nonspecific cation channel is opened in proportion to cell membrane stretch or tension. The entry of sodium ions through these open channels would depolarize the cell and lead to the opening of voltage-activated calcium channels, followed by calcium entry into the cell and contraction. During reduced stretch or tension, the nonspecific channels would close and allow hyperpolarization to occur. Most recently, evidence suggests that the modulation of the calcium sensitivity of the contractile apparatus in VSM is a key factor in the myogenic response. The sensitivity of the apparatus to calcium is modified by the relative activities of myosin light chain kinase and myosin light chain phosphatase in the smooth muscle and appears to suggest a role for latch state mechanisms in the myogenic response. What is clear, however, is that regardless of the precise mechanism involved, arteries contract and depolarize as their intravascular pressure is increased and relax and hyperpolarize as their intravascular pressure is decreased.

The myogenic response is particularly helpful in preventing tissue edema when venous pressure is elevated by more than 5 to 10 mm Hg above the typical resting values. The elevation of venous pressure results in an increase in capillary pressure, which would normally favor filtration and edema formation. However, some of this pressure is transmitted back into the arteriolar segment of the microcirculation. Myogenic arteriolar constriction in response to the stretch caused by this increase in pressure lowers the transmission of arterial pressure into the capillaries and thus minimizes the risk of edema (although this occurs at the expense of decreased blood flow).

Myogenic regulation may also be one factor that helps control organ blood flow in the face of changes in arterial

Clinical Focus / 15.1 

Diabetes Mellitus and Microvascular Function

More than 95% of persons with diabetes experience periods of elevated blood glucose concentration, or hyperglycemia, as a result of inadequate insulin action and the resulting decreased glucose transport into the muscle and fat tissues and as a result of increased glucose release from the liver. The most common cause of diabetes mellitus is obesity, which increases the requirement for insulin to the extent that even the high insulin concentrations provided by the pancreatic beta cells are insufficient. This overall condition is called *insulin resistance*.

Obesity independent of periods of hyperglycemia does not injure the microvasculature. However, periods of hyperglycemia over time cause reduced nitric oxide (NO) production by endothelial cells, increased reactivity of vascular smooth muscle to norepinephrine, accelerated atherosclerosis, and a reduced ability of microvessels to participate in tissue repair. The consequences are as follows: cerebrovascular accidents (stroke) and coronary artery disease as a result of endothelial cell abnormalities; loss of toes or whole legs as a result of microvascular and atherosclerotic disease; and loss of retinal microvessels followed by a pathologic overgrowth of capillaries, leading to blindness. The kidney glomerular capillaries are also damaged, which may lead to renal failure.

The mechanism of many of these abnormalities appears to stem from the fact that hyperglycemia activates protein kinase C (PKC) in endothelial cells. PKC inhibits NO synthase, so NO formation is gradually suppressed. This leads to the loss of an important vasodilatory stimulus (NO) and results in vasoconstriction. PKC also activates phospholipase C, leading to increased diacylglycerol and arachidonic acid formation. The increased availability of arachidonic acid leads to increased prostaglandin synthesis and the generation of oxygen radicals that destroy part of the NO present. In addition, oxygen radicals damage cells of the microvasculature and produce long-term problems caused by DNA breakage.

pressure. This process is called **autoregulation** and is discussed in more detail below.

Organ blood flow is increased by increased tissue metabolism through local, nonneurogenic mechanisms.

In almost all organs, an increase in metabolic rate is associated with increased blood flow and the extraction of oxygen to meet the metabolic needs of the tissues. This phenomenon is called **active hyperemia**. In addition, a reduction in oxygen within the blood is associated with the dilation of the arterioles and increased blood flow, assuming neural reflexes to hypoxia are not activated. In both active hyperemia and local ischemia, the changes in arteriolar resistance are in the direction of increasing blood flow, thereby indirectly increasing oxygen delivery to the tissue. This helps tissues receive the oxygen necessary to meet their local demand.

The mechanisms behind the local regulation of the microvasculature in response to the metabolic needs of tissues involve many different types of cellular events. Decreased O_2, increased CO_2, increased H^+, and increased adenosine all accompany increased tissue metabolism, and all are vasodilators, but none of these alone seems to be able to explain active hyperemia or the local vascular response to ischemia.

Oxygen is not stored in appreciable amounts in tissues, and the oxygen concentration will fall to nearly zero in about 1 minute if blood flow is stopped in any organ. Therefore, an increase in metabolic rate would decrease the tissue oxygen concentration and possibly directly signal the vascular muscle to relax by limiting the production of adenosine triphosphate (ATP) needed for the contraction of smooth muscle cells. Alternatively, the depletion of ATP could release the inhibitory effect of ATP on K^+ channels, which would result in hyperpolarization of the cell membrane, reduced activation of voltage-gated calcium channels, and vasodilation. Oxygen, or the lack thereof, has been implicated to play a role at the molecular level in the operation of L-type calcium channels and ATP-gated K^+ channels. However, no known effect of oxygen on these channels appears to adequately explain the effects of oxygen on smooth muscle contraction. Furthermore, many arterioles have a normal to slightly increased periarteriolar oxygen tension during skeletal muscle contractions, because the increased delivery of oxygen through elevated blood flow offsets the increased use of oxygen by tissues immediately around the arteriole. Therefore, as long as blood flow is allowed to increase substantially, it is unlikely that oxygen availability at the arteriolar wall is a major factor in the sustained vasodilation that occurs during increased metabolism. In fact, it appears that the signal causing arteries to dilate when tissue metabolism increases may not exist in the arterial wall at all but, instead, may originate in the surrounding tissues.

Finally, recent studies indicate that VSM cells are not particularly responsive to a broad range of oxygen tensions. Only unusually low or high oxygen tensions seem to be associated with direct changes in VSM force. However, either oxygen depletion from an organ's cells or an increased metabolic rate does cause the formation and release of free adenosine, Krebs cycle intermediates and, in hypoxic conditions, lactic acid. It is possible then that these factors contribute to the phenomenon of active hyperemia.

High concentration of either CO_2 or H^+ causes relaxation of the VSM. However, the levels of CO_2 and H^+ produced at the tissue, even with high metabolism, cannot dilate arterioles to the level seen during the active hyperemia. Furthermore, usually only transient increases in venous blood and interstitial tissue acidity occur if blood flow through an organ with increased metabolism is allowed to increase appropriately.

Of all the metabolite concentrations that change during both tissue metabolism and hypoxic conditions, adenosine is the best situated to account for arteriolar dilation and increased blood flow. Adenosine increases during both hypoxia and increased tissue metabolism, readily diffuses out of tissue cells to the adjacent microcirculation, and is a potent vasodilator. It appears to be an important mediator of active hyperemia in the heart and brain but is not believed to be the sole mediator of the ability of tissues to match their blood flow to their moment-to-moment metabolic needs. A summary of the metabolic regulation of blood flow is shown in Figure 15.7.

Arteriolar contractile state is affected by vasoactive chemicals released by endothelial cells.

Substances released by endothelial cells are important contributors to local vascular regulation. The most important of these is **nitric oxide (NO)** (formerly known as **endothelium-derived relaxing factor** or **EDRF**), which is formed by the action of **NO synthase** on the amino acid L-arginine. NO is released from all arterial, microvascular, venular, and lymphatic endothelial cells and is a potent vasodilator (Fig. 15.8). NO causes the relaxation of VSM by activating muscle guanylate cyclase to increase the production of cyclic guanosine monophosphate (cGMP). cGMP activates various cGMP-dependent protein kinases that, in turn, activate calcium-lowering mechanisms in the smooth muscle cell, thus inhibiting contraction. Numerous compounds, such as acetylcholine, histamine, and adenine nucleotides (ATP and adenosine diphosphate) as well as hypertonic conditions and hypoxia, cause the release of NO. Adenosine causes NO

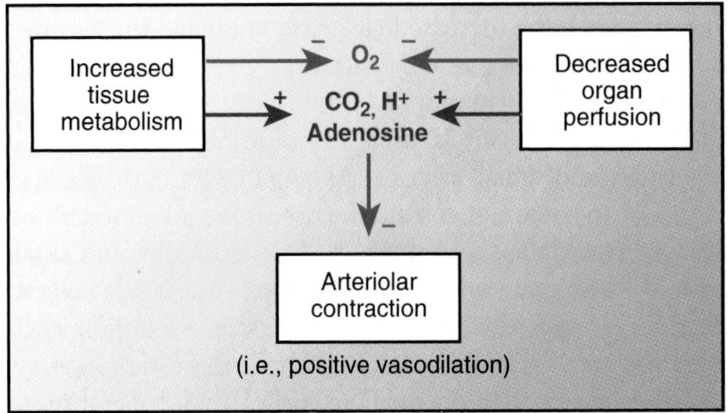

Figure 15.7 An overview of metabolic effects on arteriolar contraction. Decreased organ perfusion or increased tissue metabolism produce factors that dilate microcirculatory arteries and veins.

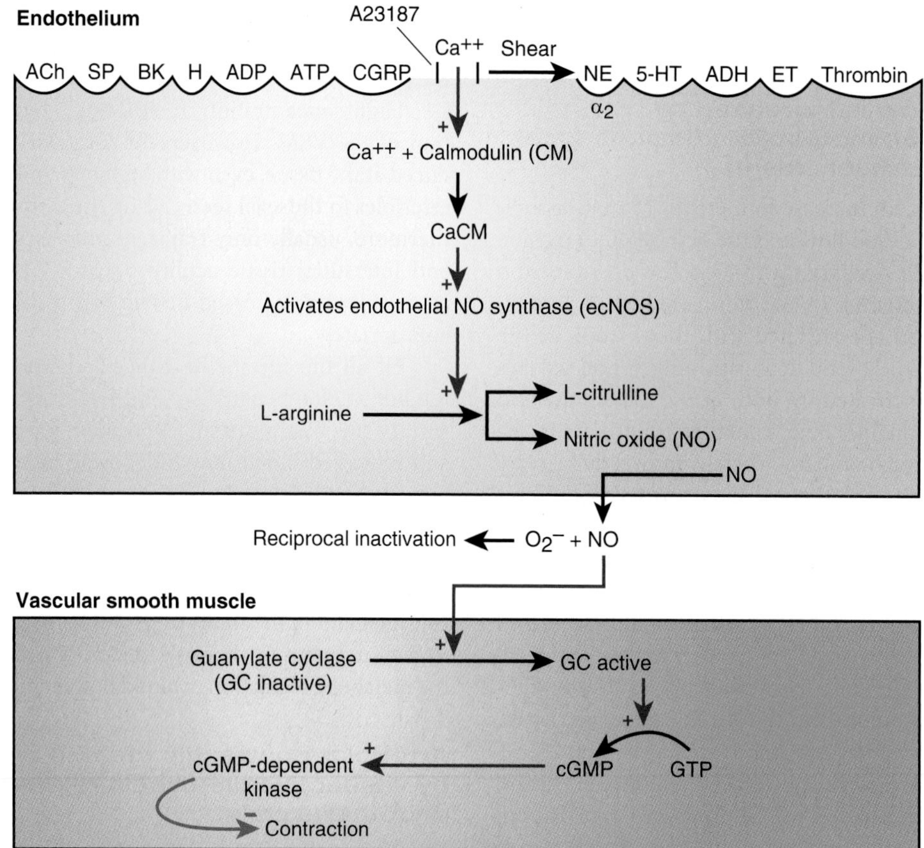

Figure 15.8 The mechanism of endothelium-dependent relaxation of vascular smooth muscle by nitric oxide (NO). Ach, acetylcholine; SP, substance P; BK, bradykinin; H, histamine; ADP, adenosine diphosphate; ATP, adenosine triphosphate; CGRP, calcitonin gene-related peptide; A23187, calcium ionophore; NE, norepinephrine; 5-HT, serotonin; ADH, antidiuretic hormone (vasopressin); ET, endothelin; +, stimulation; cGMP, cyclic guanosine monophosphate; GTP, guanosine triphosphate.

release from endothelial cells, although it also directly relaxes VSM cells through adenosine receptors.

Another important mechanism that releases NO from endothelial cells is the shear stress generated by blood moving past the cells. Frictional forces between moving blood and the stationary endothelial cells distort the endothelial cells, opening special potassium channels and causing endothelial cell hyperpolarization. This increases calcium ion entry into the cell down the increased electrical gradient. The elevated cytosolic calcium ion concentration activates endothelial NO synthase to form more NO, and the blood vessels dilate.

This mechanism is used to coordinate various sized arterioles and small arteries. As small arterioles dilate in response to some signal from the tissue, the increased blood flow increases the shear stress in larger arterioles and small arteries upstream, which prompts their endothelial cells to release NO and relax their smooth muscle. As larger arterioles and small arteries control much more of the total vascular resistance than do small arterioles, the cooperation of the larger resistance vessels is vital to adjusting blood flow to the needs of the tissue. Examples of this process, called **flow-mediated vasodilation**, have been observed in cerebral, skeletal muscle, and small intestinal vasculatures.

Flow-mediated dilation is also seen in conduit arteries. However, the dilation of these vessels has no significant effect on vascular resistance and likely represents a negative feedback mechanism to control shear stress on the arterial endothelium; small dilation of larger arteries has no effect on blood flow but does reduce flow velocity and wall shear stress. This effect may be important because many transport processes across the endothelium as well as the interaction of blood cell elements with this vascular barrier are altered by changes in endothelial shear stress. Endothelial cells of arterioles also release vasodilatory prostaglandins and an uncharacterized hyperpolarizing factor when blood flow and shear stress are increased. However, NO appears to be the dominant vasodilator molecule for flow-dependent regulation.

Basal release of NO exerts important bioregulatory effects in the cardiovascular system. Its vasodilatory contribution provides a major antihypertensive effect to the cardiovascular system as a whole. Blood pressure increases rapidly by 75% or more if a person is given an agent that blocks endothelial NO production. NO inhibits platelet aggregation and therefore serves as an anticoagulant. It also inhibits neutrophil–endothelial interactions, is antimitogenic to VSM, and promotes the recovery of the endothelium from injury. Thus, NO is considered an antiatherogenic agent.

Endothelial cells release a 21–amino acid peptide called **endothelin**, which is the most potent vasoconstrictor of blood vessels in the body. Although extremely small amounts are released under natural conditions, endothelin is thought to play a role in several vascular diseases such as hypertension and atherosclerosis. The vasoconstriction occurs because of a cascade of events beginning with phospholipase C activation and leading to activation of protein kinase C (PKC) (see Chapter 1, "Homeostasis and Cellular Signaling"). Two major types of endothelin receptors have been identified and others may exist. The constrictor function of endothelin is mediated by type B endothelin receptors. Type A endothelin receptors cause hyperplasia and hypertrophy of vascular muscle cells and the release of NO from endothelial cells. The precise function of endothelin in the normal vasculature is not clear; however, it is active during embryologic development. In knockout mice, the absence of the endothelin A receptor results in serious cardiac defects so that newborns are not viable. An absence of the type B receptor is associated with an enlarged colon, eventually leading to death.

In damaged heart tissue, such as in an infarct, cardiac endothelial cells increase endothelin production. The endothelin stimulates both VSM and cardiac muscle to contract more vigorously and induces the growth of surviving cardiac cells. However, excessive stimulation and hypertrophy of cells appear to contribute to heart failure, failure of contractility, and excessive enlargement of the heart. Part of the stimulation of endothelin production in the injured heart may be the cause of the damage per se. Also, increased formation of angiotensin II and norepinephrine during chronic heart disease stimulates endothelin production, probably at the gene expression level. The activation of PKC increases the expression of the *c-jun* proto-oncogene, which, in turn, activates the preproendothelin-1 gene.

Endothelin has also been implicated as a contributor to renal vascular failure, both the pulmonary hypertension and the systemic hypertension associated with insulin resistance, and the spasmodic contraction of cerebral blood vessels exposed to blood after a brain injury or stroke associated with blood loss to brain tissue.

Sympathetic nervous system regulates blood pressure and flow by constricting the microvessels.

Although the microvasculature uses local control mechanisms to adjust vascular resistance based on the physical and chemical environment of the tissue and vasculature, the dominant regulatory system of the vasculature is the sympathetic nervous system. As Chapter 17, "Control Mechanisms in Circulatory Function," explains, the arterial pressure is monitored moment to moment by neural reflex mechanisms that adjust the cardiac output and systemic vascular resistance as needed via the autonomic nervous system. This system is used to counteract deviations in blood pressure from normal. Sympathetic nerves communicate with the resistance vessels and venous system through the release of norepinephrine onto the surface of smooth muscle cells in the vessel walls. Because sympathetic nerves form an extensive meshwork of axons over the exterior of the microvessels, all VSM cells are likely to receive norepinephrine. Because the diffusion path is a few microns, norepinephrine rapidly reaches the vascular muscle and activates α_1-adrenergic receptors, which constrict arteries and veins within 2 to 5 seconds. Sympathetic nerve activation must occur quickly because rapid changes in body position or sudden exertion require immediate responses to maintain or increase arterial pressure. The sympathetic nervous system routinely overrides local regulatory mechanisms in most organs, with the exception of the brain, the heart, and the skeletal muscle during exercise.

Certain organs control their blood flow via local autoregulation mechanisms.

If the arterial blood pressure to an organ is decreased to the extent that blood flow is compromised, the arterioles of

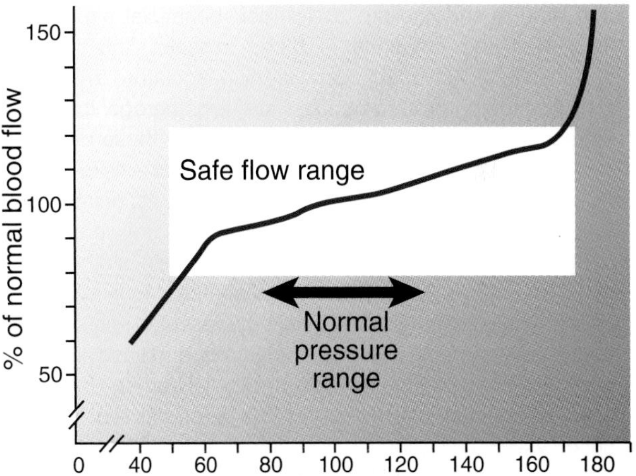

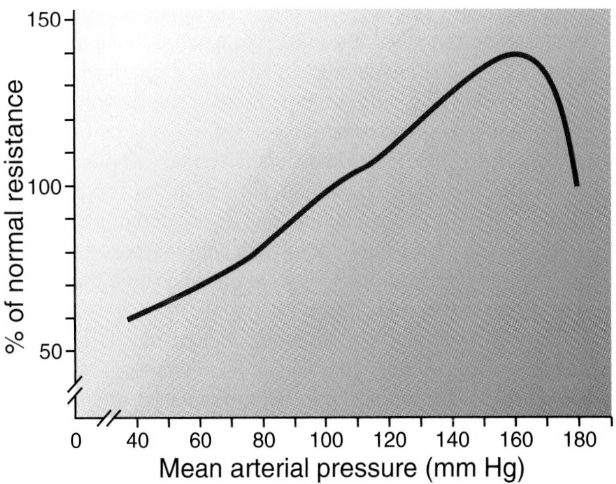

Figure 15.9 Autoregulation of blood flow and vascular resistance as mean arterial pressure is altered. The safe range for blood flow is about 80% to 125% of normal and usually occurs at arterial pressures of 60 to 160 mm Hg as a result of active adjustments of vascular resistance. At pressures above about 160 mm Hg, vascular resistance decreases because the pressure forces dilation to occur; at pressures below 60 mm Hg, the vessels are fully dilated, and resistance cannot be appreciably decreased further.

that organ respond by vasodilating so as to counteract the effects of reduced arterial pressure on organ blood flow. If arterial pressure is elevated, flow is initially increased, but the vascular resistance increases in response to the local increase in arterial pressure and restores the blood flow toward normal. The ability of an organ to maintain normal or near normal blood flow in the face of changes in the pressure driving flow is called *autoregulation of blood flow*, or simply, *autoregulation* (Fig. 15.9). Autoregulation is a *local* control phenomenon involving responses of the arterial smooth muscle to local stimuli. It occurs independently of any nerve activity on the organ vasculature and can be observed in circulations with impaired functions of arterial endothelium. Autoregulation appears to be primarily related to metabolic and myogenic control as well as to an increased release of NO if the tissue oxygen availability decreases.

The efficacy of autoregulation is not 100% in all organs, and there are upper and lower pressure limits to its effectiveness. When arterial pressure in vessels to an organ decreases, the vasodilation that ensues from autoregulation may help increase organ blood flow to a value higher than that which would occur following the same pressure drop in a rigid tube. However, the blood flow might not be brought all

From Bench to Bedside / 15.1

The Continuing Search for Methods of Assessing Endothelial Function in Cardiovascular Disease

Hypertension, atherosclerosis, and diabetes mellitus all damage arterial endothelium and impair beneficial nitric oxide (NO)–mediated functions in those vessels. This damage and loss of NO invariably causes the vasculature to become prospasmodic, prothrombotic, and proatherogenic, which worsens the cardiovascular outcomes with these diseases. Assessment of the patient's condition and future risks in these diseases often focuses on the key variable associated with the condition: blood pressure in hypertension, serum low-density lipoprotein and high-density lipoprotein in atherosclerosis, and blood glucose levels in diabetes. More recent use of biochemical markers of NO biology, vascular inflammation, endothelial damage, adhesion molecules, and vascular repair have also been used to assess endothelial damage in cardiovascular disease. However, from the perspective of the vascular system, the quantitative value of these risk factors and biochemical variables are not good markers of the function of the endothelium, owing to the large variability in individual production and response to these parameters. What is of greater importance for the individual is what these variables actually *do* to the vasculature in terms of its performance. A patient could have modest hypertension, borderline plasma lipoprotein levels, and minimal glucose intolerance but could be at extreme risk for cardiovascular complications if these conditions were causing serious damage to the arterial endothelium. A physician could intervene earlier and more aggressively than he or she might otherwise if there were indications that these modest risk factors were causing more than modest disease in the vasculature.

The determination of impaired function of arterial endothelium in cardiovascular disease is relatively easy when assessments are made invasively on isolated arteries or in experimental animals. Indeed, these types of invasive studies have been the primary means by which the impairment of the endothelium NO system has been shown to be associated so extensively with cardiovascular disease. This type of invasive analysis, however, cannot be employed as a clinical test for patients. To be able to detect impaired endothelial function in patients as a regular clinical test, any technique employed needs to be noninvasive.

Over the past 20 years, a noninvasive indicator for the assessment of vascular damage in cardiovascular disease has been examined based on the phenomenon of flow-mediated vasodilation (FMD). FMD is caused by shear stress–induced release of endothelial NO when arteries are exposed to high laminar flow. It is known that the endothelial NO system is impaired in all forms of cardiovascular disease and that it may be an early sign of the damaging effects of a disease on the vascular system. FMD can be induced in patients by placing an inflatable blood pressure cuff around a patient's arm, pressurizing the cuff to stop blood flow for 5 minutes, and then releasing the cuff pressure rapidly to allow a reactive hyperemia to occur. The hyperemia increases blood flow and shear in the brachial artery. This shear induces a FMD that can be measured as a diameter change of the brachial artery using Doppler ultrasound. Poor dilation is taken to be a measure of damage done to the artery by cardiovascular disease. Clinical studies continue to demonstrate enough correlations between impaired FMD and poor cardiovascular outcomes or cardiovascular disease to warrant further examination of this technique for its suitability as a regular part of standard clinical examinations in cardiovascular disease. Such an establishment as a regular clinical test, however, will still require additional validation of its ability to predict patient cardiovascular outcomes and disease risk. Nevertheless, the relative success of this technique has spurred the development and investigation of additional FDM applications and additional noninvasive measures designed to assess endothelial function in patients. Most recently, FDM has been investigated as a tool to assess the functional state of penile cavernous arteries in erectile dysfunction. Newer, non-FMD methods have been developed based on the assessment of vasodilation associated with reactive hyperemia or dilation produced by a local application of vasodilators on arteries at concentrations too low to have systemic effects. Digital pulse amplitude tonometry, where reactive hyperemia is measured by monitoring changes in the arterial pulse wave during the hyperemia, is one such hyperemia-based noninvasive method. Skin laser Doppler has been tried as a means of assessing endothelial function in the microcirculation by the same hyperemia principle. However, the future usefulness of hyperemia-based tools to assess endothelial health is questionable because there are simply too many variables that contribute to active hyperemic responses in organs for any change in such a response to solely represent any malfunction of the endothelium.

Clinical Focus / 15.2

Ischemia–Reperfusion Injury

Ischemia is defined as a condition of reduced oxygen availability to a tissue as a result of reduced blood flow, although it can arise from any condition in which the oxygen demand of the tissue is greater than its oxygen supply. In the heart, this most often occurs acutely by the development of an obstruction in one of the major coronary arteries, usually from a blood clot forming in the area of an advanced or ruptured atherosclerotic plaque. Ischemia cannot be tolerated for long in myocardial tissue. Immediately upon the start of ischemia, the myocardial tissue affected becomes acontractile. Within one hour, the oxygen-dependent cellular ionic and water balance mechanisms fail and the tissues swell, with a further loss of membrane integrity. Calcium extrusion mechanisms fail, and the resultant increase in intracellular calcium activates enzymes that destroy myocardial tissue. If acute ischemia involves enough myocardial tissue, the initial loss of contractile ability may be enough to cause acute failure of the heart and circulatory collapse (cardiogenic shock, see Chapter 17, "Control Mechanisms in Circulatory Function"). However, even with small ischemic episodes, the resulting tissue damage within hours after the initial insult may be enough to cause death from cardiac failure.

The best course of action to reduce ischemic damage in the heart would seem to be to restore myocardial blood flow as quickly as possible and/or change hemodynamic variables in such a way as to reduce myocardial oxygen demand. If the ischemia results from a blood clot in a major coronary artery, the physician can inject "clot-busting" drugs, such as streptokinase or tissue-type plasminogen activator, through a catheter into the major coronary artery affected by the clot. These drugs dissolve the clot and restore flow. However, in reality, restoration of blood flow is often associated with the induction of additional damage to the myocardium instead of the restoration of cell viability. This phenomenon is called *ischemia–reperfusion injury*. During the ischemic period, the arterioles downstream from the coronary arterial obstruction dilate through myogenic and metabolic mechanisms, lowering vascular resistance. On removal of the obstruction, this lowered resistance allows for a substantial reactive hyperemia, which floods the previously ischemic area with blood elements and oxygen. This hyperemia is associated with cell death from necrosis and apoptosis. The area also becomes infiltrated with neutrophils.

Tissue damage from ischemia–reperfusion injury results from the increased generation of oxygen radicals in the myocardium and endothelium as well as from infiltrating neutrophils. Myocardial sources may result from the ischemic conversion of xanthine dehydrogenase to xanthine oxidase, which produces superoxide from xanthine. Damaged mitochondria likely also act as a source of oxygen radicals. The radicals also impair mitochondrial recovery and adenosine triphosphate generation, which leads to cell death. The neutrophil infiltration also sets up a local inflammatory response, which produces damaging cytokines and additional reactive oxygen species in the tissues and complements activation. Recent studies suggest that nitric oxide attenuates ischemia–reperfusion injury, perhaps by quenching superoxide produced in the condition.

the way back to its original level before the pressure dropped. Autoregulation is still considered to have occurred; it is just that the efficacy of this autoregulation is <100%. The cerebral and cardiac vasculatures, followed closely by the renal vasculature, have the highest autoregulation efficacy. Skeletal muscle and intestinal vasculatures exhibit less well-developed autoregulation, whereas autoregulation is essentially absent in the circulation of the skin. Furthermore, in all organs, there are limits to the vasodilation of arteries in autoregulation; once vessels are maximally dilated, there is no capacity left to serve as an autoregulatory correction for further drops in pressure. Autoregulation of blood flow ceases at arterial pressures below 50 mm Hg in most organs. In an analogous manner, the ability of arterioles to constrict against progressively higher pressures has a maximum. Eventually, the pressure pushing the vessel open is too much for the contractile force generated by the arterial smooth muscle. Pressures beyond this point tend then to increase organ blood flow and capillary hydrostatic pressure. Thus, there is an upper pressure limit as well to the autoregulation of blood flow in organs.

An example of autoregulation, based on data from the cerebral vasculature, is shown in Figure 15.9. Note that the arterioles continue to dilate at arterial pressures below 60 mm Hg, when blood flow begins to decrease significantly as arterial pressure is further lowered. The vessels clearly cannot dilate sufficiently to maintain blood flow at low arterial pressures. At greater-than-normal arterial pressures, the arterioles constrict. If the mean arterial pressure is elevated appreciably above 150 to 160 mm Hg, the vessel walls cannot maintain sufficient tension to oppose passive distention by the high arterial pressure. The result is excessive blood flow and high microvascular pressures, eventually leading to the rupture of small vessels, excess fluid filtration into the tissue, and edema.

A phenomenon related to autoregulation is **reactive hyperemia**. When blood flow to any organ is stopped or reduced by vascular compression for more than a few seconds, vascular resistance dramatically decreases. The absence of blood flow allows vasodilatory chemicals to accumulate as hypoxia occurs; the vessels also dilate as a result of decreased myogenic stimulation (low intravascular pressure). As soon as the vascular compression is removed, blood flow is dramatically increased for a few minutes because the vessels dilated during the stoppage of flow. This phenomenon represents reactive hyperemia because it is a reaction to the previous period of ischemia. A good example of reactive hyperemia is the redness of skin seen after compression has been removed.

Chapter Summary

- The microcirculation controls the transport of water and substances between the tissues and blood.
- Transport of gases and lipid-soluble molecules occurs by diffusion across endothelial cells.
- Transport of water-soluble molecules occurs by diffusion through pores in adjacent endothelial cells.
- Transport of substances across a capillary by diffusion depends on the concentration gradient of the substance and the permeability of the capillary to the substance.
- Flow-limited transport of a substance is limited by the amount of blood flow to the tissues.
- Diffusion-limited transport of a substance is limited by the diffusion of the substance across the capillary membrane.
- Bulk filtration or absorption of water occurs across capillaries through pores in adjacent endothelial cells.
- Plasma hydrostatic and colloid osmotic pressures are the primary forces for fluid filtration and absorption across capillary walls.
- The ratio of postcapillary to precapillary resistance is a major determinant of capillary hydrostatic pressure.
- Lymphatic vessels collect excess water and protein molecules from the interstitial space between cells.
- Myogenic arteriolar regulation is a response to increased tension or stretch of the vessel wall muscle cells.
- By products of metabolism cause the dilation of arterioles.
- Nitric oxide from the endothelium is a major local vasodilator of arterioles.
- The axons of the sympathetic nervous system release norepinephrine, which constricts the arterioles and venules.
- Autoregulation of blood flow allows some organs to maintain nearly constant blood flow when arterial blood pressure is changed.

thePoint *Visit http://thePoint.lww.com for chapter review Q&A, case studies, animations, and more!*

16 Special Circulations

ACTIVE LEARNING OBJECTIVES

Upon mastering the material in this chapter, you should be able to:

- Explain why most coronary blood flow occurs during diastole.
- Explain why myocardial infarction is more severe in the endocardium than in the epicardium following an ischemic episode.
- Explain why the heart must depend solely on autoregulation of blood flow to maintain myocardial oxygen delivery in the face of decreased perfusion pressure.
- Predict how variables related to myocardial oxygen demand will affect blood flow to the heart.
- Explain why general sympathetic nervous system activation has no appreciable effect on blood flow to the brain.
- Explain how autoregulation of blood flow protects the blood–brain barrier.
- Explain the importance of cerebral hyperemia following exposure to CO_2 or H^+.
- Explain the mechanism responsible for adjusting the efficacy of autoregulation in the intestine and skeletal muscle on the basis of tissue metabolic demand.
- Explain the mechanisms responsible for active hyperemia in the intestine and skeletal muscle.
- Explain the benefit of the hepatic arterial buffer response.
- Explain the role of the sympathetic nervous system in the control of the cutaneous circulation.
- Explain the transition between a fetal and adult circulatory structure after birth.

The vascular system of every organ has special characteristics that are designed to meet the specific functions and specialized needs of that organ. In this chapter, the characteristics unique to the circulations of the heart, brain, small intestine, liver, skeletal muscle, and skin are described. Table 16.1 presents data on blood flow and oxygen use by these different organs and tissues. In addition, the anatomy and physiology of the fetal/placental circulation are presented along with the circulatory changes that occur at birth. The pulmonary and renal circulations are discussed in Chapter 19, "Gas Transfer and Transport," and Chapter 22, "Kidney Function."

CORONARY CIRCULATION

The coronary circulation provides blood flow to the heart. During resting conditions, the heart muscle consumes as much or more oxygen as does an equal mass of skeletal muscle during vigorous exercise (see Table 16.1). Coronary blood flow can normally increase about fourfold to fivefold during heavy exercise to provide for the increased oxygen needs of the heart. This increment in blood flow constitutes the **coronary reserve**. The ability to increase the blood flow to provide additional oxygen is imperative for the heart because it extracts almost the maximum amount of oxygen from blood during resting conditions. Coupled with the fact that the heart's ability to use anaerobic glycolysis is limited, the only practical way to provide for increased energy needs is to increase blood flow and oxygen delivery.

Cardiac blood flow occurs primarily during diastole because of inhibition of flow from cardiac contraction during systole.

An examination of the data in Table 16.1 reveals an interesting fact: On the basis of its oxygen demand (O_2 consumption), the heart is the most *under*perfused organ in the body! This may seem surprising considering the importance of the heart to the well-being of the body as a whole. However, it reflects the fact that the contraction of the heart gets in the way of its own blood supply. Blood flow through the left ventricle decreases to a minimum during systole because the small intramuscular blood vessels are compressed and actually "sheared" closed by compressive forces generated in the muscle. Blood flow in the left coronary artery during cardiac systole is only 10% to 30% of that during diastole. In diastole, the heart musculature is relaxed and blood flow to the heart is not impeded by cardiac muscle effects (Fig. 16.1). Compared to its effect in the left ventricle, the compression effect of systole on blood flow is minimal in the right ventricle as a result of the lower pressures developed by that chamber. The heart's dependence on receiving most of its blood flow during diastole has no obvious deleterious effects on total coronary blood flow even during maximal exercise. However, in people with compromised coronary arteries, an increased heart rate decreases the time spent in diastole to the extent that coronary blood flow increases needed to match myocardial demands are impaired.

TABLE 16.1 Blood Flow and Oxygen Consumption of the Major Systemic Organs Estimated for a 70-kg Adult Man

Organ	Mass (kg)	Flow (mL/100 g/min)	Total Flow (mL/min)	Oxygen Use (mL/100 g/min)	Total Oxygen Use (mL/min)
Heart	0.4–0.5				
Rest		60–80	250	7.0–9.0	25–40
Exercise		200–300	1,000–1,200	25.0–40.0	65–85
Brain	1.4	50–60	750	4.0–5.0	50–60
Small intestine	3				
Rest		30–40	1,500	1.5–2.0	50–60
Absorption		45–70	2,200–2,600	2.5–3.5	80–110
Liver	1.8–2.0				
Total		100–300	1,400–1,500	13.0–14.0	180–200
Portal		70–90	1,100	5.0–7.0	
Hepatic artery		30–40	350	5.0–7.0	
Muscle	28				
Rest		2–6	750–1,000	0.2–0.4	60
Exercise		40–100	15,000–20,000	8.0–15.0	2,400–?
Skin	2.0–2.5				
Rest		1–3	200–500	0.1–0.2	2–4
Exercise		5–15	1,000–2,500		

The heart musculature is perfused from the epicardial (outside) surface to the endocardial (inside) surface. The mechanical compression of systole has a more negative effect on the blood flow through the endocardial layers, where compressive forces are higher and microvascular pressures are lower than it does in the epicardium. In heart diseases of all types, therefore, the subendocardial layers of the heart suffer more severe impairment and ischemia than do the epicardial layers.

Coronary blood flow is closely linked to cardiac oxygen demand.

Mean coronary blood flow increases significantly whenever contraction of the heart or heart rate is increased. Most of the small arteries and arterioles of the heart are surrounded by cardiac muscle cells and are exposed to chemicals released by cardiac cells into the interstitial space. Many of these chemicals cause dilation of the coronary arterioles and therefore could be responsible for coronary active hyperemia. For example, adenosine, derived from the breakdown of adenosine triphosphate in cardiac cells, is a potent vasodilator. Its release increases whenever cardiac metabolism is increased or blood flow to the heart is experimentally or pathologically decreased. Blockade of the vasodilator actions of adenosine with theophylline, however, does not prevent coronary vasodilation when cardiac work is increased, blood flow is suppressed, or the arterial blood is depleted of oxygen. Therefore, although adenosine is likely an important contributor to cardiac vascular regulation, there are obviously other potent regulatory agents. Vasodilatory prostaglandins, H^+, CO_2, nitric oxide (NO), and decreased availability of oxygen as well as myogenic mechanisms, are capable of contributing to coronary vascular regulation. However, no single mechanism adequately explains the dilation of coronary arterioles and small arteries when the metabolic rate of the heart is increased or when pathologic or experimental means are used to restrict blood flow.

Direct and indirect actions of sympathetic nerves dilate coronary arteries.

Coronary arteries and arterioles are innervated by the sympathetic nervous system and contain, predominantly, β_2-adrenergic receptors. The large-surface epicardial arteries also contain some α_1-adrenergic receptors. Activation of the sympathetic nervous system to the heart, such as during exercise, results in coronary vasodilation and a marked increase in coronary blood flow. This increase is primarily an active hyperemia in response to increased cardiac metabolism brought about by sympathetic nerve stimulation of heart rate and myocardial contractility. However, direct β-adrenergic–mediated coronary vasodilation also contributes to the hyperemia following sympathetic nerve activation.

It appears that the α_1-adrenergic receptors of the epicardial arteries may play a role in supporting endocardial perfusion during exercise. Intramural pressure in the heart during contraction is greater in the endocardium than in the

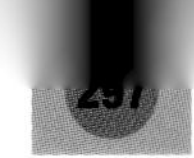

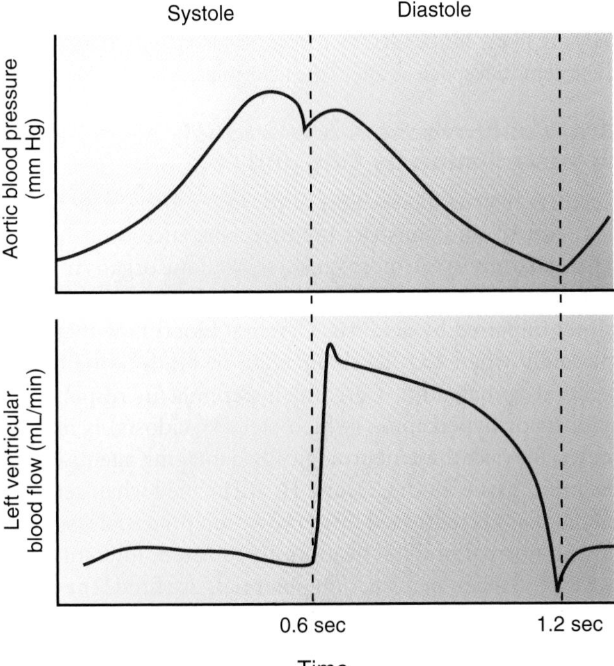

Figure 16.1 Aortic blood pressure and left coronary blood flow during the cardiac cycle. Note that left coronary artery blood flow decreases dramatically during the isovolumetric phase of systole, prior to opening of the aortic valve. Left coronary artery blood flow remains lower during systole than during diastole because of compression of the coronary blood vessels in the contracting myocardium. The left ventricle receives most of its arterial blood inflow during diastole.

epicardium. This creates a tendency for blood to be pushed backward away from the endocardium and toward the epicardium during systole. Adding some α-adrenergic constrictor influence to the epicardial vessels during heavy exercise helps minimize backflow and loss of blood from endocardial muscle.

Autoregulation of blood flow has different limits in the endocardium versus the epicardium.

All of the heart's capillaries receive blood flow, even at normal heart rates and cardiac outputs. Thus, at rest, extraction of oxygen from the coronary circulation by the heart is essentially maximized. The only way the heart can maintain oxygen delivery in the face of a reduction in **perfusion pressure** is by local autoregulation of blood flow. The heart is a strong autoregulator of blood flow and can maintain near normal flows over a wide range of perfusion pressures. However, compressive forces make blood flow more constrained during systole in the endocardium than in the epicardium. Consequently, the endocardial arterioles must dilate more during diastole to compensate for a more severe reduction of flow during systole. For this reason, the low pressure limit for autoregulation in the endocardial layer is greater than that in the epicardial layer. Endocardial arterial dilation reaches a maximum when arterial pressure drops to approximately 70 mm Hg, whereas maximum dilation in the epicardial arteries is not reached until pressure is approximately 40 mm Hg. For this reason, whenever blood flow to the heart is severely restricted, the endocardium is first to suffer damage and the resulting area of myocardial injury and infarction is larger in the endocardium than in the epicardium.

Collateral vessels in the human heart

Some mammalian species (e.g., the dog) have a network of microvascular connections between the major epicardial arteries. These are called **collateral vessels**. A total blockage of blood flow in one of the epicardial arteries does not necessarily create ischemia and infarction in the myocardium because blood enters the affected area from an unblocked epicardial artery through the collateral vessels. Human beings are not so fortunate. Humans have virtually no collateral connections between their major coronary arteries, and blockade of one of those arteries therefore results in a massive myocardial infarction. However, it appears that collateral vessel development is plastic and can be induced under certain conditions. One of the likely contributing factors to compensate for slowly developing coronary vascular disease is the enlargement of the few collateral blood vessels between the left and right coronary arterial systems or among parts of each system. The expansion of existing collateral vessels and the limited formation of new collaterals provide a partial bypass for blood flow to areas of muscle whose primary supply vessels are impaired. Subendocardial arteriolar collaterals usually enlarge more than epicardial collaterals in coronary artery disease because the ratio of flow between the endocardium and epicardium, which is normally 1:1, can become 1:2 or 1:3 during periods of ischemia.

The exact mechanism responsible for the development of collateral vessels is unknown. However, periods of inadequate blood flow to the heart muscle caused by experimental flow reduction do stimulate collateral enlargement in healthy animals. It is assumed that in humans with coronary vascular disease who develop functional collateral vessels, the mechanism is related to occasional or even sustained periods of hypoxia from inadequate blood flow.

CEREBRAL CIRCULATION

The ultimate organ of life is the brain. Even the determination of death often depends on whether the brain is viable. The most common cause of brain injury is some form of impaired brain blood flow. Such problems can develop as a result of accidents to arteries in the neck or brain, occlusion of vessels secondary to atherosclerotic processes, or **aneurysms** that occur as a result of vessel wall tearing.

Brain blood flow remains essentially constant over a large range of arterial blood pressures.

The cerebral circulation shares many of the physiologic characteristics of the coronary circulation. The heart and brain have a high metabolic rate (see Table 16.1), extract a large amount of oxygen from blood, and have a limited ability to use anaerobic glycolysis for metabolism. Their vessels have a limited ability to constrict, if at all, in response to sympathetic nerve stimulation. The brain, like the coronary vasculature, has an excellent ability to autoregulate blood flow at arterial pressures from about 50 to about 160 mm Hg. The vasculature of the brainstem exhibits the most precise autoregulation, with good but less precise regulation of blood flow in the cerebral cortex. This regional variation in autoregulatory ability has clinical implications because the region of the brain most likely to suffer at low arterial pressure is the cortex, where consciousness will be lost long before the automatic cardiovascular and ventilatory regulatory functions of the brainstem are compromised.

A variety of mechanisms appear to be responsible for cerebral vascular autoregulation. The identification of a specific chemical that causes cerebral autoregulation has not occurred. For example, when blood flow is normal, regardless of the arterial blood pressure, little extra adenosine, K^+, H^+, or other vasodilator metabolites are released, and brain tissue Po_2 remains relatively constant. However, increasing concentrations of any of these chemicals causes vasodilation and increased blood flow. The brain vasculature does exhibit myogenic vascular responses and may use this mechanism as a major contributor to autoregulation. Both the cerebral arteries and cerebral arterioles appear to be involved in cerebral vascular autoregulation and other types of vascular responses. In fact, the arteries can change their resistance almost proportionately to the arterioles during autoregulation. This may occur in part because cerebral arteries exhibit myogenic vascular responses and because they are partially to fully embedded in the brain tissues. As such, they are likely influenced by the same vasoactive chemicals in the interstitial space as affect the arterioles.

Brain microvessels are uniquely sensitive to vasodilation by CO_2 and H^+.

Cerebral arteries dilate significantly in response to increased CO_2 and H^+ and constrict if either substance is decreased. Its vasculature is the most sensitive of all the organs to these metabolites, perhaps because neural function is significantly impaired by acidosis. Cerebral blood flow increases markedly when CO_2 levels increase in brain tissue or the cerebral spinal fluid. Cerebral hyperemia in response to acidosis or hypercapnia (which creates acidosis) is likely a means to wash these neurologically damaging agents out of the brain tissue. Both CO_2 and H^+ are formed when cerebral metabolism is increased by nerve action potentials, such as during normal brain activation. In addition, interstitial K^+ is elevated when many action potentials are fired. The cause of dilation in response to both K^+ and CO_2 involves the formation of NO. However, the mechanism is not necessarily the typical endothelial formation of NO. The source of NO appears to be NO synthase in neurons as well as endothelial cells.

Reactions of cerebral blood flow to chemicals released by increased brain activity, such as CO_2, H^+, adenosine, and K^+, are part of the overall process of matching the brain's metabolic needs to the blood supply of nutrients and oxygen.

Clinical Focus / 16.1

Exertional Angina

Coronary artery disease can produce many focal atherosclerotic lesions in the main coronary arteries. In spite of what may be extensive vascular disease, patients with this disease often do not present with any symptoms when they are at rest. However, once they increase their activity, for example, by walking up stairs, doing yard work, etc., they suddenly exhibit shortness of breath and angina pectoris, or severe chest pain. Angina is a sign that the myocardium is ischemic and is a warning of potentially serious consequences. The angina induced by physical activity is often called "exertional angina" and is the most common form of angina in patients with coronary artery disease.

The reason that patients with this condition are asymptomatic at rest but show signs of myocardial ischemia on exertion is rooted in the flow autoregulation capacity of the myocardium and the strong link between coronary blood flow and tissue metabolism. When focal lesions within the large coronary arteries encroach on the lumen, they raise resistance to flow at that point. This increased resistance dissipates pressure downstream and would decrease flow in the artery as a whole if the downstream arterioles could not alter their resistance. However, the arterioles downstream are the site of autoregulation in the coronary circulation. When their internal pressure decreases, they respond by dilating in an effort to reduce resistance so that flow can be restored to its original value. Thus, any upstream increase in vascular resistance, such as that caused by a partial obstruction, is compensated by a decrease in resistance in the vessels downstream from the obstruction. In this manner, the total resistance of the arterial circuit is returned to normal and, therefore, so is blood flow.

It has been estimated that a single focal obstruction in a major coronary must reduce lumen diameter by more than 90% before arteriolar diameter is maximized at rest, thus allowing ischemia to ensue upon any further reduction of the lumen. However, with coronary obstructions that are not severe enough to cause ischemia at rest, problems occur once the metabolic demands of the heart increase. The same arterioles that are dilated to compensate for upstream obstructions in the main coronary arteries are also the vessels that need to dilate to increase blood flow to the myocardium whenever activity of the heart is increased. If a portion of their dilating capacity is used to simply maintain resting flow, there may not be enough dilating capacity (or reserve) left to augment blood flow to meet an increased oxygen demand by the heart. In that situation, the oxygen demand of the heart would exceed its oxygen supply, and ischemia, with the appearance of angina, would occur. This is why people with coronary artery disease feel fine if they are physically inactive but can suffer severe ischemic episodes and even heart attacks if they exert themselves by shoveling snow, playing touch football, or doing any other activity that requires increased output from the heart.

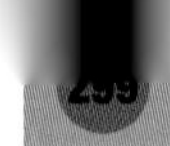

From Bench to Bedside / 16.1

Second-Generation Drug-Eluting Stents for the Treatment of Coronary Artery Disease—Addressing Thrombosis Versus Restenosis

Coronary artery bypass grafts (CABGs) were one of the first solutions for the prevention of angina and heart attack caused by blockage of coronary arteries from atherosclerotic plaques. However, CABG procedures involve significant invasive surgery and risk for the patient, and graphs are susceptible to restenosis over time. Balloon angioplasty, when first developed, was intended as a less invasive, more risk-adverse treatment for alleviating the hemodynamic consequences of atherosclerotic plaques in the coronary arteries. This technique used a coronary arterial catheter fitted with an inflatable balloon on its tip. Under fluoroscopic guidance, the physician would advance this catheter into the coronary artery to the site of the plaque and inflate/deflate the balloon repeatedly to "smash, flatten, and stretch" the plaque and artery, which was often sclerotic. This procedure would remove the narrowing at the site of the plaque and indirectly restore coronary dilator reserve in the downstream arterioles as they would no longer have to dilate to compensate for the added coronary resistance caused by the plaque. The problem with this procedure is that constant stretching and smashing of the atherosclerotic lesion during the procedure caused repeated trauma and injury to the artery as well. This would set up a local injury and inflammatory response that stimulated smooth muscle cell growth, resulting in intimal hyperplasia and significant restenosis of the artery.

An early solution to the restenosis problem was to outfit the balloon catheter with a cylindrical, flexible wire mesh that could be expanded to a wide diameter directly into the coronary arterial wall upon inflation of the balloon. This wire mesh tube is called a stent and provides a type of support scaffolding in the arterial wall at the site of plaques that holds the artery in an open position. Stenting has proven to be an improvement over balloon angioplasty, but is not without its complications as well. Early stents were composed of bare metal. As such, these stents proved to act as foreign material in the body and became a nidus for blood clot formation and focal inflammation. Laser polishing of the stent surface to an ultrasmooth finish was an early attempt at making the stent less procoagulant. However, implanting a bare metal mesh into the artery acted as an injury stimulus of its own that stimulated local tissue growth and restenosis.

For this reason, drug-eluting stents (DESs) were developed to counteract the inflammatory tissue growth associated with earlier bare metal stents. Stents were created to serve as a local drug-delivery system to help stop the process of restenosis, atherosclerosis, and blood clot formation at the site of coronary plaques. The first DESs were coated with polymers containing antimitogenic agents, or antirejection agents, or with reabsorbable polymer coatings that reduced the risk of blood clot formation.

The first U.S. Food and Drug Administration–approved DESs contained sirolimus (an antirejection drug) or paclitaxel (an antimitotic) as the eluent. However, an unexpected consequence resulted from the use of these stents in human coronary arteries. The effects of antimitogenic drugs have been shown to be different in healthy coronary arteries, such as those of the animals used to first test the stents, compared to the effects seen in arteries with atherosclerotic disease, such as those in human patients for which the stents were intended. In healthy arteries, antimitotic eluents prevent intimal hyperplasia without affecting the endothelial growth and integrity surrounding the metal stent. However, in atherosclerotic arteries, the function and health of the endothelium are severely impaired. In these vessels, antimitotic eluents suppress endothelial growth over the metal stents. The endothelial covering of the stent becomes thin and patchy, thus making the stented region a prime area for the stimulation of blood clot formation. Consequently, in humans with coronary artery disease, these stents exchange one side effect of stenting (restenosis) for the creation of another (thrombosis). In one study, it was shown that the incidence of thrombosis to be greater in patients with DESs compared to those with bare metal stents. For this reason, a second generation of DESs has been developed and is undergoing multiple clinical trials to test efficacy and the occurrence of thrombosis and artery hyperplasia. These stents improve drug delivery into the stented area, use thinner metal components, and use biocompatible polymers for drug delivery. Early results indicate that these new generation DESs have improved efficacy and excellent safety compared to antimitogenic first-generation DESs. Such improved DESs could, thus, provide a long-term solution to the hemodynamic problems associated with the use of arterial stents in for the treatment of atherosclerotic lesions in the coronary arteries.

The 10% to 30% increase in blood flow in brain areas excited by peripheral nerve stimulation, mental activity, or visual activity may be related to these three substances released from active nerve cells. The cerebral vasculature also dilates when the oxygen content of arterial blood is reduced, but the vasodilatory effect of elevated CO_2 is much more powerful.

Cerebral vessels are insensitive to hormones and sympathetic nerve activity.

Circulating vasoconstrictor and vasodilator hormones and the release of norepinephrine by sympathetic nerve terminals on cerebral blood vessels do not play much of a role in moment-to-moment regulation of cerebral blood flow. The blood–brain barrier effectively prevents constrictor and dilator agents in blood plasma from reaching the cerebral vascular smooth muscle. Sympathetic nerves sparsely innervate the cerebral arteries and arterioles, and these vessels are essentially devoid of α-adrenergic or β-adrenergic receptors. Most of the adrenergic innervations appear to be located on the pial vessels rather than on the brain parenchyma. Stimulation of these nerves produces only mild vasoconstriction in most cerebral vessels. If, however, sympathetic activity to the cerebral vasculature is permanently interrupted, the cerebral vasculature has a decreased ability to autoregulate blood flow at high arterial pressures, and the integrity of the blood–brain barrier is more easily disrupted. Therefore, some aspect of sympathetic nerve activity other than the routine regulation of vascular resistance is important for the maintenance of normal cerebral vascular function. This may relate to a neurogenic trophic factor that promotes the health of endothelial and smooth muscle cells in the cerebral microvessels.

Cerebral vasculature adapts to chronic high blood pressure.

In conditions of chronic hypertension, cerebral vascular resistance increases, thereby allowing cerebral blood flow and, presumably, capillary pressures to be normal. The adaptation of cerebral vessels to sustained hypertension lets them maintain vasoconstriction at arterial pressures that would overcome the contractile ability of a normal vasculature (Fig. 16.2).

The mechanisms that enable the cerebral vasculature to adjust the autoregulatory range upward appear to be hypertrophy of the vascular smooth muscle and a mechanical constraint to vasodilation, as a result of more muscle tissue, more connective tissue, or both. The drawback to such adaptation is partial loss of the ability to dilate and regulate blood flow at low arterial pressures. This loss occurs because the passive structural properties of the resistance vessels restrict increases in vessel diameter at subnormal pressures and, in doing so, increase resistance. In fact, the lower pressure limit of constant blood flow (autoregulation) can be almost as high as the normal mean arterial pressure (see Fig. 16.2). This can be problematic if the arterial blood pressure is rapidly lowered to normal in a person whose vasculature has adapted to hypertension. The person may faint from inadequate brain blood flow, even though the arterial pressure is in the normal range. Fortunately, a gradual reduction in arterial pressure over weeks or months returns autoregulation to a more normal pressure range.

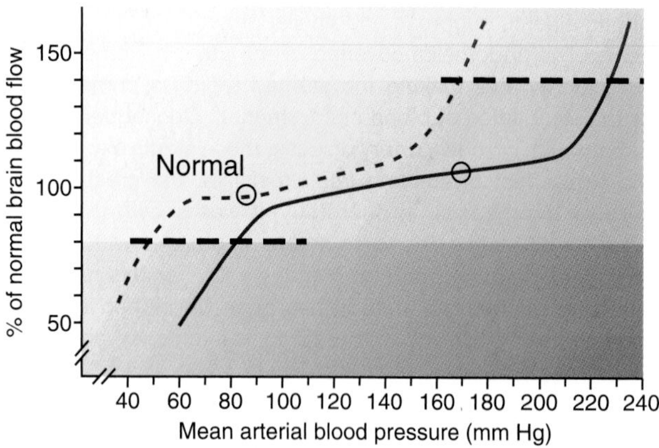

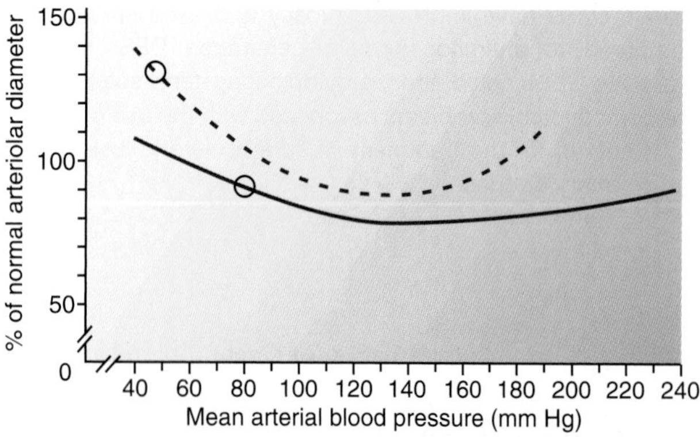

Figure 16.2 Effects of chronic hypertension on cerebral autoregulation. This condition is associated with a rightward shift in the arterial pressure range over which autoregulation of cerebral blood flow occurs (*upper panel*) because, for any given arterial pressure, resistance vessels of the brain have smaller-than-normal diameters (*lower panel*). As a consequence, people with hypertension can tolerate high arterial pressures that would cause vascular damage in healthy people. However, they risk reduced blood flow and brain hypoxia at low arterial pressures that are easily tolerated by healthy people.

Cerebral edema impairs blood flow to the brain.

The brain is encased in a rigid bony case, the cranium. As such, should the brain begin to swell, the intracranial pressure will dramatically increase. There are many causes of **cerebral edema**, or an excessive accumulation of fluid in the brain tissue, including infection, tumors, trauma to the head that causes massive arteriolar dilation, and bleeding into the brain tissue after a hemorrhagic stroke or trauma. In each case, as the intracranial pressure increases, the venules and veins are partially collapsed because their intravascular pressure is low. As these outflow vessels collapse, their resistance increases and capillary pressure rises (see Chapter 15, "Microcirculation and Lymphatic System"). The increased capillary pressure favors increased filtration of fluid into the brain to further raise the intracranial pressure. The end result is a positive feedback system in which intracranial pressure will become so high as to begin to compress small arterioles and decrease blood flow.

Excessive intracranial pressure is a major clinical problem. Hypertonic mannitol can be given to promote water extraction from swollen brain cells. Sometimes opening of the skull and drainage of cerebrospinal fluid or hemorrhaged blood, if any, may be necessary. Hemorrhaged blood is particularly a problem because clotted blood contains denatured hemoglobin that destroys NO. This, in turn, leads to inappropriate vasoconstriction of the arterioles in the area of the hemorrhage, which further compromises blood supply to the brain.

If blood flow to the pons and medulla of the brain is decreased, tissue hypoxia activates sympathetic nervous system control centers in the brain. This results in massive sympathetic outflow to the organs of the body, resulting in severe vasoconstriction. This response is called the **Cushing reflex**. During this reflex, flow to the kidney may be so compromised as to prevent the formation of urine. The skin pales from removal of blood from that circulation, and ischemia can be produced in the intestine. However, this massive systemic vasoconstriction creates a marked elevation of mean arterial pressure (up to ~270 mm Hg), which helps open brain arterioles in the face of high external compression. However, although blood flow may improve, the increase in arterial pressure elevates microvascular pressures, which worsens cerebral edema.

SMALL INTESTINE CIRCULATION

The small intestine completes the digestion of food and then absorbs the nutrients to sustain the remainder of the body. At rest, the intestine receives about 20% of the cardiac output and uses about 20% of the body's oxygen consumption. Both of these numbers nearly double after a large meal. Unless the blood flow can increase, food digestion and absorption simply do not occur. For example, if intense exercise is required in the midst of digesting a meal, blood flow through the small intestine can be reduced to half of normal by the sympathetic nervous system activation associated with the exercise. This delays food absorption but otherwise produces no ill effects. Once the stress imposed on the body is over, intestinal blood flow again increases and the process of digesting and absorbing food resumes.

The vasculature of the small intestine is elaborate. Small arteries and veins penetrate the muscular wall of the bowel and form a microvascular distribution system in the submucosa (Fig. 16.3). The muscle layers receive small arterioles from the submucosal vascular plexus; other small arterioles continue into individual vessels of the deep submucosa around glands and to the villi of the mucosa. Small arteries and larger arterioles preceding the separate muscle and submucosal–mucosal vasculatures control about 70% of the intestinal vascular resistance. The small arterioles of the muscle, submucosa, and mucosal layers can partially adjust blood flow to meet the needs of these small areas of tissue.

Autoregulation efficiency in the small intestine is modulated by intestinal oxygen consumption.

Compared with other major organs, the circulation of the small intestine has a poorly developed autoregulatory response, and, as a result, blood flow usually declines because resistance does not adequately decrease in response to locally decreased arterial pressure. However, this is not true for all metabolic states of the intestine. It appears that the intestine is a poor autoregulator of blood flow, when it is not occupied with digestion, and thus, its oxygen consumption is low, whereas it is a much better autoregulator during digestion, when its oxygen consumption is high. To explain this phenomenon, it is helpful to acknowledge that what the intestine needs during high metabolic demand is the oxygen in the blood more so than the blood itself. Thus, if perfusion pressure to the intestine were to drop, causing an initial drop in flow, the intestine could obtain the *oxygen* it needed to compensate for the drop in blood flow by simply taking the flow it gets and extracting more oxygen from it, rather than initiating an autoregulatory vasodilation to restore total flow. It could do this by simply perfusing more capillaries. Indeed, this is exactly what happens as the permeability surface area coefficient increases, following a drop

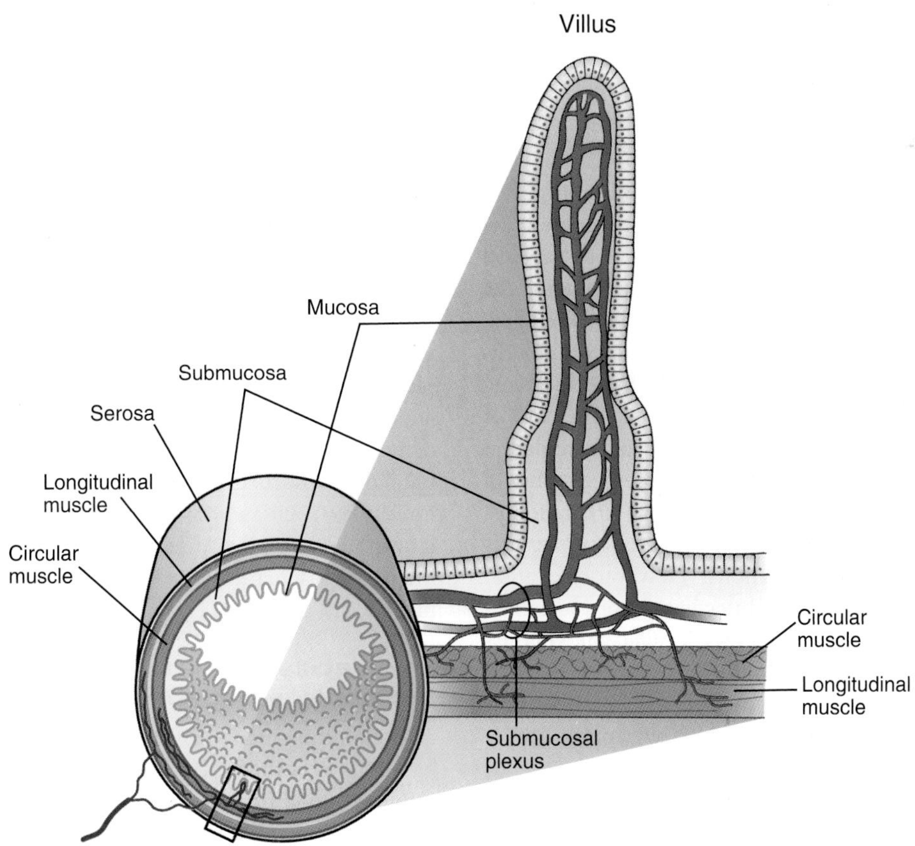

Figure 16.3 The vasculature of the small intestine. The intestinal vasculature is unusual because branches from a common vasculature located in the submucosa serve three different tissues, the muscle layers, submucosa, and mucosal layer. Small arteries and arterioles preceding the separate muscle, submucosal, and mucosal vasculatures regulate most of the intestinal vascular resistance.

in flow before frank vasodilation of the arterioles affecting total flow occurs. Thus, when intestinal metabolism is low, a local decrease in oxygen delivery, brought on by an initial drop in blood pressure and blood flow, is compensated for by first dilating precapillary sphincters to open more capillary surface area for diffusion of oxygen. Should demand increase or total flow be compromised further, more capillaries will open until perfused capillary surface area is maximized. At that point, any further compensation for the effects of a drop in perfusion pressure would have to be met by autoregulation of blood flow. Experimental evidence has validated this scenario repeatedly. In short, when oxygen delivery to the intestine is compromised, the intestine extracts more oxygen from its blood supply first and adjusts its total blood supply second. During digestion, all capillaries are open or nearly so in the intestine. Any reduction in oxygen supply to the tissues caused by a temporary local drop in perfusion pressure and, thus, flow is then compensated for by strong local autoregulation of blood flow.

The fact that the intestine seems to extract oxygen from blood before altering vascular resistance has a certain appeal for whole-body homeostatic mechanisms. Even though effects of changes in vascular resistance in the intestine on the body as a whole are lessened by the parallel arrangement of organ circulations in the body, the intestine and splanchnic circulations are still enormous. Manipulating extraction of oxygen before adjusting vascular resistance and flow might be a means of meeting the metabolic needs of the intestine while minimizing the effects of those needs on the blood supply to other organs.

Effects of elevated venous pressure on fluid filtration by intestinal capillaries

The intestine has one of the highest capillary filtration coefficients among the organ systems in the body. This could be potentially troublesome because the intestine is not a good autoregulator of blood flow for most of the time during the day (i.e., when it is not occupied with digestion). Consequently, its arteriolar constrictor response, which would otherwise keep capillary hydrostatic pressure in check during variation in systemic arterial pressure, is weak. Thus, loss of fluid from the capillaries of the huge intestinal circulation could be significant, especially if venous pressures should rise. The intestine is connected to the portal circulation of the liver, and such venous pressure rises can occur in liver disease or portal vein obstructions. However, elevation of venous pressure outside the intestine causes sustained myogenic arteriolar constriction. This is called the **venous-arteriolar response** and is a means by which major fluid loss from the capillaries is prevented in the intestine. This response occurs in other organ systems but appears to be strongest in the intestinal circulation.

High intestinal mucosal blood flow is required for absorption of nutrients by the intestine.

The intestinal mucosa receives about 60% to 70% of the total intestinal blood flow. Blood flows of 70 to 100 mL/min/100 g in this specialized tissue are probable and much higher than the average blood flow for the total intestinal wall (see Table 16.1). This blood flow can exceed the resting blood flow in the heart and brain. The intestinal mucosa is composed of individual projections of tissue called *villi.* The interstitial space of the villi is mildly hyperosmotic (~400 mOsm/kg H_2O) at rest as a result of NaCl. During food absorption, the interstitial osmolality increases to 600 to 800 mOsm/kg H_2O near the villus tip, compared with 400 mOsm/kg H_2O near the villus base. The primary cause of high osmolalities in the villi appears to be greater absorption than removal of NaCl and nutrient molecules. There is also a possible countercurrent exchange process in which materials absorbed into the capillary blood diffuse from the venules into the incoming blood in the arterioles.

Lipid absorption causes a greater increase in intestinal blood flow and oxygen consumption, a condition known as **absorptive hyperemia**, than either carbohydrate or amino acid absorption. During absorption of all three classes of nutrients, the mucosa releases adenosine and CO_2 and oxygen is depleted. The hyperosmotic lymph and venous blood that leave the villus to enter the submucosal tissues around the major resistance vessels are also major contributors to absorptive hyperemia. By an unknown mechanism, hyperosmolality resulting from NaCl induces endothelial cells to release NO and dilate the major resistance arterioles in the submucosa. Hyperosmolality resulting from large organic molecules that do not enter endothelial cells does not cause appreciable increases in NO formation, producing much less of an increase in blood flow than equivalent hyperosmolality resulting from NaCl. These observations suggest that NaCl entering the endothelial cells is essential to induce NO formation.

The active absorption of amino acids and carbohydrates and the metabolic processing of lipids into chylomicrons by mucosal epithelial cells place a major burden on the microvasculature of the small intestine. There is an extensive network of capillaries just below the villus epithelial cells. The villus capillaries are unusual in that portions of the cytoplasm are missing, so that the two opposing surfaces of the endothelial cell membranes appear to be fused. These areas of fusion, or closed fenestrae, are thought to facilitate the uptake of absorbed materials by capillaries. However, large molecules, such as plasma proteins, do not easily cross the fenestrated areas because the reflection coefficient for the intestinal vasculature is >0.9, about the same as in skeletal muscle and the heart.

Low capillary pressure in intestinal villi facilitates water absorption.

Although the mucosal layer of the small intestine has a high blood flow both at rest and during food absorption, the capillary blood pressure is usually 13 to 18 mm Hg and seldom higher than 20 mm Hg during food absorption. Therefore, plasma colloidal osmotic pressure is higher than capillary blood pressure, favoring the absorption of water brought into the villi. During lipid absorption, the plasma protein reflection coefficient for the overall intestinal vasculature is decreased from a normal value of more than 0.9 to about 0.7. It is assumed that most of the decrease in reflection coefficient occurs in the mucosal capillaries. This lowers the ability of plasma proteins to counteract capillary filtration, with the net result that fluid is added to the interstitial space. Eventually, this fluid must be removed. Not surprisingly, the highest rates of intestinal lymph formation normally occur during fat absorption.

Sympathetic nerve activity greatly decreases intestinal blood flow and volume.

The intestinal vasculature is richly innervated by sympathetic nerve fibers and contains predominantly α_1-adrenoceptors. Thus, major reductions in gastrointestinal blood flow and venous volume occur whenever sympathetic nerve activity is increased, such as during strenuous exercise or periods of pathologically low arterial blood pressure. Venoconstriction in the intestine during hemorrhage helps to mobilize blood to the central circulation and helps compensates for the blood loss. Gastrointestinal blood flow is about 25% of the cardiac output at rest; a reduction in this blood flow, by heightened sympathetic activity, allows more vital functions to be supported with the available cardiac output. However, gastrointestinal blood flow can be so drastically decreased by sympathetically mediated vasoconstriction in combination with a low arterial blood pressure (**hypotension**) that mucosal tissue damage can result. Nevertheless, this intestinal vasoconstriction can be seen as using the large size of the intestinal vasculature to the advantage of the body as a whole. In severe hypotension, perfusion to the heart and brain takes priority over that to other organs. Low blood pressure can be increased to help support blood flow to the brain and heart by systemic vasoconstriction in large organ systems such as the intestine.

HEPATIC CIRCULATION

The hepatic circulation perfuses one of the largest organs in the body, the liver. The liver is primarily an organ that maintains the organic chemical composition of the blood plasma. For example, all plasma proteins are produced by the liver, and the liver adds glucose from stored glycogen to the blood. The liver also removes damaged blood cells and bacteria and detoxifies many manmade or natural organic chemicals that have entered the body.

Hepatic circulation is perfused by gastrointestinal venous blood and hepatic arterial supply.

The human liver has a large blood flow, about 1.5 L/min or 25% of the resting cardiac output. It is perfused by both arterial blood through the hepatic artery and venous blood that has passed through the stomach, small intestine, pancreas, spleen, and portions of the large intestine. The venous blood arrives via the hepatic portal vein and accounts for about 67% to 80% of the total liver blood flow (see Table 16.1). The remaining 20% to 33% of the total flow is through the hepatic artery. The majority of blood flow to the liver is determined by the flow through the stomach and small intestine.

About half of the oxygen used by the liver is derived from venous blood, even though the splanchnic organs have removed one third to one half of the available oxygen. The hepatic arterial circulation provides additional oxygen. The liver tissue efficiently extracts oxygen from the blood. The liver has a high metabolic rate and is a large organ; consequently, it has the largest oxygen consumption of all organs in a resting person. The metabolic functions of the liver are discussed in Chapter 27, "Physiology of the Liver."

The liver vasculature is arranged into subunits that allow the arterial and portal blood to mix and provide nutrition for the liver cells (Fig. 16.4). Each subunit, called an *acinus*, is about 300 to 350 μm long and wide. In humans, usually three acini occur together. The core of each acinus is supplied by a single terminal portal venule; sinusoidal capillaries originate from this venule (see Fig. 16.4). The endothelial cells of the capillaries have fenestrated regions with discrete openings that facilitate exchange between the plasma and interstitial spaces. The capillaries do not have a basement membrane, which partially contributes to their high permeability. The terminal hepatic arteriole to each acinus is paired with the terminal portal venule at the acinus core, and blood from the arteriole and blood from the venule jointly perfuse the capillaries. The intermixing of the arterial and portal blood

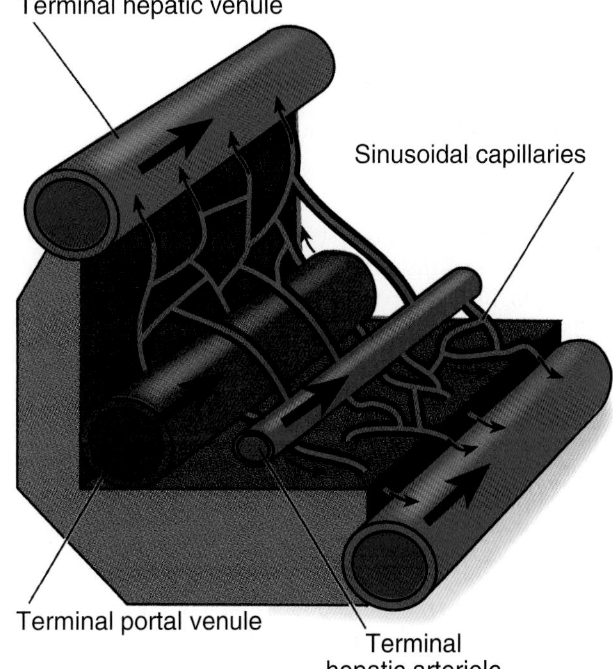

Figure 16.4 Liver acinus microvascular anatomy. A single liver acinus, the basic subunit of liver structure, is supplied by a terminal portal venule and a terminal hepatic arteriole. The mixture of portal venous and arterial blood occurs in the sinusoidal capillaries formed from the terminal portal venule. Usually two terminal hepatic venules drain the sinusoidal capillaries at the external margins of each acinus.

tends to be intermittent because the vascular smooth muscle of the small arteriole alternately constricts and relaxes. This prevents arteriolar pressure from causing a sustained reversed flow in the sinusoidal capillaries, where pressures are 7 to 10 mm Hg. The best evidence for this is that hepatic artery and portal venous blood first mix at the level of the capillaries in each acinus. The sinusoidal capillaries are drained by the terminal hepatic venules at the outer margins of each acinus; usually at least two hepatic venules drain each acinus.

Regulation of hepatic arterial and portal venous blood flow requires an interactive control system.

The regulation of portal venous and hepatic arterial blood flows is an interactive process: Hepatic arterial flow increases and decreases reciprocally with the portal venous blood flow. This mechanism, known as the **hepatic arterial buffer response**, can compensate or buffer about 25% of the decrease or increase in portal blood flow. Exactly how this is accomplished is still under investigation, but vasodilatory metabolite accumulation, possibly adenosine, during decreased portal flow, as well as increased metabolite removal during elevated portal flow, is thought to influence the resistance of the hepatic arterioles.

One might suspect that during digestion, when gastrointestinal blood flow, and therefore, portal venous blood flow is increased, the gastrointestinal hormones in portal venous blood would influence hepatic vascular resistance. However, at concentrations in portal venous blood equivalent to those during digestion, none of the major hormones appears to influence hepatic blood flow. Therefore, the increased hepatic blood flow during digestion would appear to be determined primarily by vascular responses of the gastrointestinal vasculatures.

The vascular resistances of the hepatic arterial and portal venous vasculatures are increased during sympathetic nerve activation, and the buffer mechanism is suppressed. When the sympathetic nervous system is activated, about half the blood volume of the liver can be expelled into the general circulation. Because up to 15% of the total blood volume is in the liver, constriction of the hepatic vasculature can significantly increase the circulating blood volume during times of cardiovascular stress.

SKELETAL MUSCLE CIRCULATION

The circulation of skeletal muscle involves the largest mass of tissue in the body: 30% to 40% of an adult's body weight. At rest, the skeletal muscle vasculature accounts for about 25% of systemic vascular resistance, even though individual muscles receive a low blood flow of about 2 to 6 mL/min/100 g. The dominant mechanism controlling skeletal muscle resistance at rest is the sympathetic nervous system. Resting skeletal muscle has remarkably low oxygen consumption per 100 g of tissue, but its large mass makes its metabolic rate a major contributor to the total oxygen consumption in a resting person.

Skeletal muscle blood flow varies over a large range of tissue metabolism.

Skeletal muscle blood flow can increase 10- to 20-fold or more during the maximal vasodilation associated with high-performance aerobic exercise. Comparable increases in metabolic rate occur. Under such circumstances, total muscle blood flow may be equal to three or more times the resting cardiac output. (Obviously, cardiac output must increase during exercise to maintain the normal-to-increased arterial pressure; see Chapter 29, "Exercise Physiology.")

Skeletal muscle arteries contain both α-adrenoceptors and β-adrenoceptors, but the former predominate. Thus, activation of the sympathetic nervous system causes vasoconstriction in skeletal muscle. Vascular resistance can easily double as a result of increased sympathetic nerve activity, with a resulting decrease in muscle blood flow. Skeletal muscle cells can survive long periods with minimal oxygen supply; consequently, low blood flow is not a problem. The increased vascular resistance helps preserve arterial blood pressure when cardiac output is compromised. In addition, contraction of the skeletal muscle venules and veins forces blood in these vessels to enter the central circulation. This action helps counteract the effects of any loss of blood volume such as that which occurs in hemorrhage and dehydration.

Neurogenic vasoconstriction in skeletal muscle can override local metabolic demands of the tissue to such an extent that the muscle can become modestly ischemic. In effect, the skeletal muscle vasculature can either place major demands on the cardiopulmonary system during exercise or perform as if expendable during a cardiovascular crisis. Like a similar response in the small intestine, this ability enables essential tissues to be perfused with the available cardiac output when such output is compromised.

Muscle blood flow is markedly affected by numerous local vasoactive agents.

As discussed in Chapter 15, "Microcirculation and Lymphatic System," many potential local regulatory mechanisms adjust blood flow to the metabolic needs of the tissues. In skeletal muscle, as in the small intestine, autoregulation efficiency is dependent on local metabolism with increased efficiency associated with high metabolism. The most important local vascular regulatory phenomenon in skeletal muscle, however, is the coupling of muscle blood flow to its activity (i.e., active hyperemia). In fast-twitch muscles, which primarily depend on anaerobic metabolism, the accumulation of hydrogen ions from lactic acid is potentially a major contributor to the vasodilation that occurs. In slow-twitch skeletal muscles, oxidative metabolic requirements can increase by more than 10 to 20 times during heavy exercise. It is not hard to imagine that whatever causes metabolically linked vasodilation is in ample supply at high metabolic rates. Increased CO_2, H^+, K^+, and adenosine as well as hypoxia are all vasodilatory to skeletal muscle arterioles, and such changes in these metabolites occur in exercising muscle. However, none of these metabolites wholly explains muscle active hyperemia. During rhythmic muscle contractions, the blood flow during the relaxation phase can be high, and it is unlikely that the muscle becomes significantly hypoxic during submaximal aerobic exercise. Studies in humans and animals indicate that lactic acid formation, an indication of hypoxia and anaerobic metabolism, is present only during the first several minutes of submaximal exercise. Once the vasodilation and increased blood flow associated with exercise are established, after 1 to 2 minutes, the microvasculature is probably capable of maintaining ample oxygen for most workloads, perhaps up to 75% to 80% of maximum performance because remarkably little additional lactic acid accumulates in the blood. Although the tissue oxygen content likely decreases as exercise intensity increases, the reduction does not compromise the high aerobic metabolic rate except with the most demanding forms of exercise.

To ensure the best possible supply of nutrients, particularly oxygen, even mild exercise causes sufficient vasodilation to perfuse virtually all of the capillaries, rather than just 25% to 50% of them as occurs at rest. However, near-maximum or maximum exercise exhausts the ability of the microvasculature to meet tissue oxygen needs and hypoxic conditions rapidly develop, limiting the performance of the muscles. The burning sensation and muscle fatigue during maximum exercise or at any time that muscle blood flow is inadequate to provide adequate oxygen are partially a consequence of hypoxia. This type of burning sensation is particularly evident when a muscle must hold a weight in a steady position. In this situation, the contraction of the muscle compresses the microvessels stopping the blood flow and, with it, the availability of oxygen.

The vasodilation associated with exercise is dependent upon NO. If endothelial production of NO is curtailed by the inhibition of endothelial NO synthase, the increased muscle blood flow during exercise is strongly suppressed. However, exactly which chemicals released or consumed by skeletal muscle to induce the increased release of NO from endothelial cells is unknown. In addition, skeletal muscle cells can make NO and, although not yet tested, may produce a substantial fraction of the NO that causes the dilation of the arterioles.

CUTANEOUS CIRCULATION

The structure of the skin vasculature differs according to location in the body. In all areas, an arcade of arterioles exists at the boundary of the dermis and the subcutaneous tissue over fatty tissues and skeletal muscles (Fig. 16.5). From this arteriolar arcade, arterioles ascend through the dermis into the superficial layers of the dermis, adjacent to the epidermal layers. These arterioles form a second network in the superficial dermal tissue and perfuse the extensive capillary loops that extend upward into the dermal papillae just beneath the epidermis. The dermal vasculature also provides the vessels that surround hair follicles, sebaceous glands, and sweat glands. All the capillaries from the superficial skin layers are drained by venules, which form a venous plexus in the superficial dermis and eventually drain into many large venules and small veins beneath the dermis.

The vascular pattern just described is modified in the tissues of the hand, feet, ears, nose, and some areas of the face, in that direct vascular connections between arterioles and venules, known as **arteriovenous anastomoses**, occur primarily in the superficial dermal tissues (see Fig. 16.5). By contrast, relatively few arteriovenous anastomoses exist in the major portion of human skin over the limbs and torso.

Control of the cutaneous circulation is dominated by the sympathetic nervous system. At rest in a cool room, cutaneous vessels are significantly constricted by norepinephrine from sympathetic nerves that innervate all areas of the circulation except the cutaneous capillaries. Innervation is especially dense on the arteriovenous anastomoses. Nerve blockade of the cutaneous circulation results in maximal vasodilation of its vessels because the vessels possess essentially no intrinsic active smooth muscle tone.

Adjustment of cutaneous blood flow is used for temperature regulation.

The skin is a large organ, representing 10% to 15% of total body mass, and is positioned at the interface of the body with the external environment. The primary functions of the skin

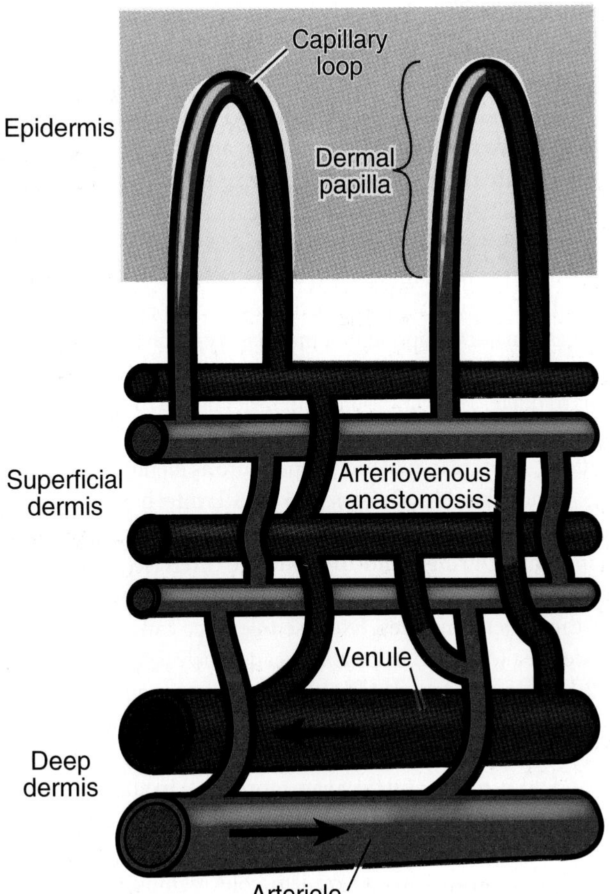

Figure 16.5 The vasculature of the skin. The skin vasculature is composed of a network of large arterioles and venules in the deep dermis, which send branches to the superficial network of smaller arterioles and venules. Arteriovenous anastomoses allow direct flow from arterioles to venules and greatly increase blood flow when dilated. The capillary loops into the dermal papillae beneath the epidermis are supplied and drained by microvessels of the superficial dermal vasculature.

are protection of the body from the external environment and dissipation or conservation of heat during body temperature regulation. If a great amount of heat must be conserved, the sympathetic nerves to the cutaneous circulation are activated and all vessels there, especially the anastomoses, constrict and restrict blood from entering skin near the external environment. If a great amount of heat must be dissipated, dilation of the arteriovenous anastomoses allows substantially increased skin blood flow to warm the skin, thereby increasing heat loss to the environment. This allows vasculatures of the hands and feet and, to a lesser extent, the face, neck, and ears to lose heat efficiently in a warm environment.

The skin has one of the lowest metabolic rates in the body and requires relatively little blood flow for purely nutritive functions. Consequently, despite its large mass, its resting metabolism does not place a major flow demand on the cardiovascular system. However, in warm climates, body temperature regulation requires that warm blood from the body core be carried to the external surface, where heat transfer to the environment can occur. Therefore, at typical indoor temperatures and during warm weather, skin blood flow is usually far in excess of the need for tissue nutrition. The reddish color of the skin during exercise in a warm environment reflects the large blood flow and dilation of skin arterioles and venules (see Table 16.1).

The increase in the skin's blood flow probably occurs through two main mechanisms. First, an increase in body core temperature causes a reflex increase in the activity of sympathetic cholinergic nerves, which release acetylcholine. Acetylcholine release near sweat glands leads to the breakdown of a plasma protein (kininogen) to form bradykinin, a potent dilator of skin blood vessels, which also increases the release of NO as a major component of the dilatory mechanism. Second, simply increasing skin temperature will cause the blood vessels to dilate. This can result from heat applied to the skin from the external environment, heat from underlying active skeletal muscle, or increased blood temperature as it enters the skin.

Total skin blood flows of 5 to 8 L/min have been estimated in humans during vigorous exercise in a hot environment. During mild-to-moderate exercise in a warm environment, skin blood flow can equal or exceed blood flow to the skeletal muscles. Exercise tolerance can, therefore, be lower in a warm environment because the vascular resistance of the skin and muscle is too low to maintain an appropriate arterial blood pressure, even at maximum cardiac output. One of the adaptations to exercise is an ability to increase blood flow in skin and dissipate more heat. In addition, aerobically trained humans are capable of higher sweat production rates; this increases vasodilation of the skin arterioles.

Most humans live in cool-to-cold regions, where body heat conservation is imperative. The sensation of cool or cold skin or a lowered body core temperature elicits a reflex increase in sympathetic nerve activity, which causes vasoconstriction of blood vessels in the skin. Heat loss is minimized because the skin becomes a poorly perfused insulator, rather than a heat dissipator. As long as the skin temperature is higher than about 10°C to 13°C (50°F–55°F), the neurally induced vasoconstriction is sustained. However, at lower tissue temperatures, the vascular smooth muscle cells progressively lose their contractile ability, and the vessels passively dilate to various extents. The reddish color of the hands, face, and ears on a cold day demonstrates increased blood flow and vasodilation as a result of low skin temperatures. To some extent, this cold-mediated vasodilation is useful because it lessens the chance of cold injury to exposed skin. However, if this process included most of the body surface, such as occurs when the body is submerged in cold water or inadequate clothing is worn in a cold environment, heat loss would be rapid and hypothermia would result. (Chapter 29, "Exercise Physiology," discusses skin blood flow and temperature regulation.)

FETAL AND PLACENTAL CIRCULATIONS

The development of a human fetus depends on nutrient, gas, water, and waste exchange in the maternal and fetal portions of the placenta. The human fetal placenta is supplied by two **umbilical arteries**, which branch from the internal iliac arteries, and is drained by a single **umbilical vein** (Fig. 16.6). The umbilical vein of the fetus returns oxygen and nutrients from the mother's body to the fetal cardiovascular system, and the umbilical arteries bring in blood laden with carbon dioxide and waste products to be transferred to the mother's blood. Although many liters of oxygen and carbon dioxide, together with hundreds of grams of nutrients and wastes, are exchanged between the mother and fetus each day, the exchange of red blood cells or white blood cells is a rare event. This large chemical exchange without cellular exchange is possible because the fetal and maternal blood are kept completely separate, or nearly so.

The fundamental anatomic and physiologic structure for fetal/maternal exchange is the placental villus. As the umbilical arteries enter the fetal placenta, they divide into many branches that penetrate the placenta toward the maternal system. These small arteries divide in a pattern similar to a fir tree, the placental villi being the small branches. The fetal capillaries bring in the fetal blood from the umbilical arteries, and then, blood leaves through sinusoidal capillaries to the umbilical venous system. Exchange occurs in the fetal capillaries and probably to some extent in the sinusoidal capillaries. The mother's vascular system forms a reservoir around the treelike structure such that her blood envelops the placental villi.

As shown in Figure 16.6, the outermost layer of the placental villus is the syncytiotrophoblast, where exchange by passive diffusion, facilitated diffusion, and active transport between fetus and mother occurs through fully differentiated epithelial cells. The underlying cytotrophoblast is composed of less differentiated cells, which can form additional syncytiotrophoblast cells as required. As cells of the syncytiotrophoblast die, they form syncytial knots, and eventually these break off into the mother's blood system surrounding the fetal placental villi.

The placental vasculatures of both the fetus and the mother adapt to the size of the fetus as well as to the oxygen

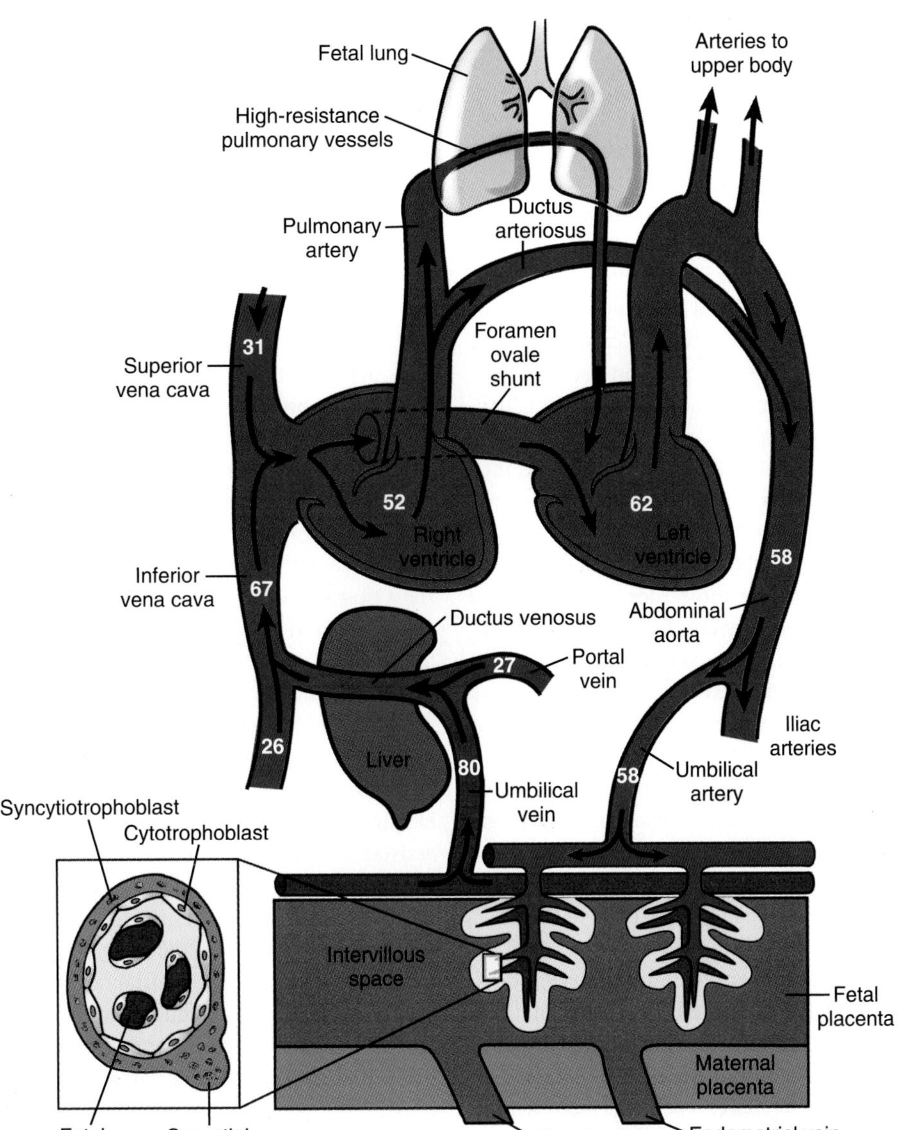

Figure 16.6 The fetal and placental circulations. The left and right sides of the fetal heart are separated for the purpose of illustration to emphasize the right-to-left shunt of blood through the open foramen ovale in the atrial septum and the right-to-left shunt through the ductus arteriosus. Arrows indicate the direction of blood flow. The numbers represent the percentage of saturation of blood hemoglobin with oxygen in the fetal circulation. Closure of the ductus venosus, foramen ovale, ductus arteriosus, and placental vessels at birth and the dilation of the pulmonary vasculature establish the adult circulation pattern. The insert is a cross-sectional view of a fetal placental villus, one of the branches of the treelike fetal vascular system in the placenta. The fetal capillaries provide incoming blood, and the sinusoidal capillaries act as the venous drainage. The villus is completely surrounded by the maternal blood, and the exchange of nutrients and wastes occurs across the fetal syncytiotrophoblast.

available within the maternal blood. For example, a minimal placental vascular anatomy will provide for a small fetus, but as the fetus develops and grows, a complex tree of placental vessels is essential to provide the surface area needed for the fetal–maternal exchange of gases, nutrients, and wastes. If the mother moves to a higher altitude, where less oxygen is available, the complexity of the placental vascular tree increases, compensating with additional areas for exchange. If this type of adaptation does not take place, the fetus may be underdeveloped or may die from a lack of oxygen.

During fetal development, the fetal tissues invade and cause partial degeneration of the maternal endometrial lining of the uterus. The result, after about 10 to 16 weeks of gestation, is an intervillous space between fetal placental villi that is filled with maternal blood. Instead of microvessels, there is a cavernous blood-filled space. The intervillous space is supplied by 100 to 200 **spiral arteries** of the maternal endometrium and is drained by the **endometrial veins**. During gestation, the spiral arteries enlarge in diameter and simultaneously lose their vascular smooth muscle layer. It is

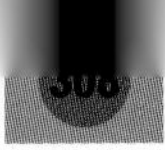

the arteries preceding them that actually regulate blood flow through the placenta. At the end of gestation, the total maternal blood flow to the intervillous space is approximately 600 to 1,000 mL/min, which represents about 15% to 25% of the resting cardiac output. In comparison, the fetal placenta has a blood flow of about 600 mL/min, which represents about 50% of the fetal cardiac output.

The exchange of materials across the syncytiotrophoblast layer follows the typical pattern for all cells. Gases, primarily oxygen and carbon dioxide, and nutrient lipids move by simple diffusion from the site of highest concentration to the site of lowest concentration. Small ions are moved predominantly by active transport processes. The GLUT 1 transport protein passively transfers glucose, and amino acids require primarily facilitated diffusion through specific carrier proteins in the cell membranes, such as the system A transporter protein.

Large molecular-weight peptides and proteins and many large, charged, water-soluble molecules used in pharmacologic treatments do not readily cross the placenta. Part of the transfer of large molecules probably occurs between the cells of the syncytiotrophoblast layer and by pinocytosis and exocytosis. Lipid-soluble molecules diffuse through the lipid bilayer of cell membranes. For example, lipid-soluble anesthetic agents in the mother's blood do enter and depress the fetus. As a consequence, anesthesia during pregnancy is somewhat risky for the fetus.

Placental exchange of oxygen and carbon dioxide is limited.

In spite of intimate contact between fetal and maternal circulations in the placenta the combined structures of this interface still create a significant diffusion barrier for oxygen. Special fetal adaptations are required for oxygen exchange, because of the limitations of passive exchange across the placenta. The Po_2 of maternal arterial blood is about 80 to 100 mm Hg, whereas that of the incoming blood in the umbilical artery is about 20 to 25 mm Hg. This difference in oxygen tension provides a large driving force for exchange, but results in an increase in the Po_2 of fetal blood in the umbilical vein to only 30 to 35 mm Hg. Fortunately, **fetal hemoglobin** carries more oxygen at a low Po_2 than adult hemoglobin carries at a Po_2 two to three times higher. In addition, the concentration of hemoglobin in fetal blood is about 20% higher than in adult blood. The net result is that the fetus has sufficient oxygen to support its metabolism and growth but does so at low oxygen tensions, using the unique properties of fetal hemoglobin. After birth, when much more efficient oxygen exchange occurs in the lung, the newborn gradually replaces the red cells containing fetal hemoglobin with red cells containing adult hemoglobin.

Absence of lung ventilation requires a unique circulatory arrangement in the fetus.

After the umbilical vein leaves the fetal placenta, it passes through the abdominal wall at the future site of the umbilicus (navel). The umbilical vein enters the liver's portal venous circulation, although the bulk of the oxygenated venous blood passes directly through the liver in the **ductus venosus** (see Fig. 16.6). The low oxygen-content venous blood from the lower body and the high oxygen-content placental venous blood mix in the inferior vena cava. The oxygen content of the blood returning from the lower body is about twice that of venous blood returning from the upper body in the superior vena cava. The two streams of blood from the superior and inferior vena cava do not completely mix as they enter the right atrium. In addition, in the fetus the lungs are collapsed and have a high vascular resistance. The net result is that oxygen-rich blood from the inferior vena cava bypasses the lungs through the open **foramen ovale** in the atrial septum to the left atrium. The upper-body blood generally enters the right ventricle as in the adult.

The preferential passage of oxygenated venous blood into the left atrium and the minimal amount of venous blood returning from the lungs to the left atrium allow blood in the left ventricle to have oxygen content about 20% higher than that in the right ventricle. This relatively high oxygen-content blood supplies the coronary vasculature, the head, and the brain.

The right ventricle actually pumps at least twice as much blood as the left ventricle during fetal life. In fact, the infant at birth has a relatively more muscular right ventricular wall than the adult. Perfusion of the collapsed lungs of the fetus is minimal because the pulmonary vasculature has a high resistance. The elevated pulmonary resistance occurs because the lungs are not inflated and probably because the pulmonary vasculature has the unusual characteristic of vasoconstriction at low oxygen tensions. The right ventricle pumps blood into the systemic arterial circulation via a shunt called the **ductus arteriosus**, which connects the pulmonary artery and aorta (see Fig. 16.6). For ductus arteriosus blood to enter the initial part of the descending aorta, the right ventricle must develop a higher pressure than the left ventricle, which is the exact opposite of circumstances in the adult. The blood in the descending aorta has less oxygen content than that in the left ventricle and ascending aorta because of the mixture of less well-oxygenated blood from the right ventricle. This difference is crucial because about two thirds of this blood must be used to perfuse the placenta and pick up additional oxygen. In this situation, a lack of oxygen content is useful.

Transition from fetal to neonatal life requires complex changes in the fetal circulatory structure after birth.

After the newborn is delivered and the initial ventilatory movements cause the lungs to expand with air, pulmonary vascular resistance decreases substantially, as does pulmonary arterial pressure. At this point, the right ventricle can perfuse the lungs, and the circulation pattern in the newborn switches to that of an adult. In time, the reduced workload on the right ventricle causes its hypertrophy to subside.

The highly perfused, ventilated lungs allow a large amount of oxygen-rich blood to enter the left atrium. The increased oxygen tension in the aortic blood may provide the signal for closure of the ductus arteriosus, although suppression of vasodilator prostaglandins cannot be discounted.

Clinical Focus / 16.2 

Atrial Septal Defects

The four-chamber configuration of the adult heart is necessary for the proper oxygenation of blood and the delivery of carbon dioxide to the lungs. The separation of the ventricles into the mammalian right and left chambers occurs early during normal fetal development of the ventricular septum. The atria in the fetus communicate through the foramen ovale. This is a normal configuration in the fetus, which provides a beneficial mechanism for diverting oxygenated blood from the inferior vena cava away from the lungs (which are nonfunctional in utero) and into the systemic circulation. In some people, the foramen ovale does not close after birth. This is called an *atrial septal defect*.

The consequences of retaining this defect through childhood and on into adulthood vary. People with atrial septal defects (ASDs) can live well into their seventh or eighth decade, although life expectancy is not normal; mortality increases about 6% per year after age 40. ASD is twice as prevalent in women as in men. The primary hemodynamic abnormality associated with ASD is shunting of blood from the left to the right atrium, which can cause right ventricular overload and mild pulmonary hypertension. Mixing of blood between the right and left heart is minimal and patients are largely asymptomatic with regard to signs of reduced oxygen in arterial blood. However, young patients often exhibit increased endurance after surgical correction of ASD. More serious effects from ASD can arise about the fourth decade of life, as the patient accumulates additional cardiovascular diseases, such as stiffened ventricles from chronic arterial hypertension and coronary artery disease. In those cases, the communication between the atria allows for more serious right-side overload, resulting in symptomatic pulmonary hypertension and distended atria. The latter can cause tricuspid regurgitation and atrial fibrillation. Surgical correction is required to remedy ASD. This often involves simply inserting a catheter tipped with a collapsed patch through the ASD, opening the patch, pulling it against the septal opening, and thereby sealing the ASD. Tissue then grows over the patch, making the repair permanent.

In any event, the ductus arteriosus constricts to virtual closure and, over time, becomes anatomically fused. Simultaneously, the increased oxygen to the peripheral tissues causes constriction in most body organs, and the sympathetic nervous system also stimulates the peripheral arterioles to constrict. The net result is that the left ventricle now pumps against a higher resistance. The combination of greater resistance and higher blood flow raises the arterial pressure and, in doing so, increases the mechanical load on the left ventricle. Over time, the left ventricle hypertrophies.

During all the processes just described, the open foramen ovale must be sealed to prevent blood flow from the left to the right atrium. After birth, left atrial pressure increases from the returning blood from the lungs and exceeds right atrial pressure. This pressure difference passively pushes the tissue flap on the left side of the foramen ovale against the open atrial septum. In time, the tissues of the atrial septum fuse; however, an anatomic passage that is probably only passively sealed can be documented in some adults. The ductus venosus in the liver is open for several days after birth but gradually closes and is obliterated within 2 to 3 months.

After the fetus begins breathing, the fetal placental and umbilical vessels undergo progressive vasoconstriction to force placental blood into the fetal body, minimizing the possibility of fetal hemorrhage through the placental vessels. Vasoconstriction is related to physical trauma, increased oxygen availability, sensitivity of these vessels to circulating catecholamines, and less of a signal for vasodilator chemicals and prostaglandins in the fetal tissue.

The final event of gestation is separation of the fetal and maternal placenta as a unit from the lining of the uterus. The separation process begins almost immediately after the fetus is expelled, but external delivery of the placenta can require up to 30 minutes. The separation occurs along the decidua spongiosa, a maternal structure, and requires that blood flow in the mother's spiral arteries be stopped. The cause of the placental separation may be mechanical, as the uterus surface area is greatly reduced by removal of the fetus and folds away from the uterine lining. Normally about 500 to 600 mL of maternal blood is lost in the process of placental separation. However, as maternal blood volume increases 1,000 to 1,500 mL during gestation, this blood loss is not a significant concern.

Chapter Summary

- Coronary blood flow is intimately linked to oxygen consumption in the heart.
- The heart is a strong autoregulator of blood flow.
- Activation of the sympathetic nerves to the heart results in direct and indirect coronary arterial dilation.
- The blood vessels of the brain are largely unaffected by circulating hormones and vasoactive compounds because of the blood–brain barrier.
- Cerebral arteries are more sensitive to dilating effects of CO_2 and H^+ than other systemic arteries.
- Control of cerebral blood flow is dominated by local metabolic factors.
- Brain blood flow increases when the neurons are active and require additional oxygen.
- The regulation of intestinal blood flow during nutrient absorption depends on the elevated sodium chloride concentration in the tissue and the release of nitric oxide.
- Autoregulation efficiency in the intestine and skeletal muscle is linked to oxygen consumption in the tissue.
- The liver receives the portal venous blood from the gastrointestinal organs as its main blood supply, supplemented by hepatic arterial blood.
- Skeletal muscle tissue receives minimal blood flow at rest because of its limited oxygen requirements, but flow and oxygen use can increase up to or beyond 20-fold during intense muscle activity.
- The skin has a low oxygen requirement, but the high blood flow during warm temperatures or exercise supplies a large amount of heat for dissipation to the external environment.
- The circulations of the intestines, skeletal muscles, and skin participate in defense of the whole body against severe hypotension.
- The fetus obtains nutrients and oxygen from the mother's blood supply using the combined maternal and fetal placental circulations.
- The heart chambers have radically different roles in pumping blood in the fetus and adult.

thePoint *Visit http://thePoint.lww.com for chapter review Q&A, case studies, animations, and more!*

17 Control Mechanisms in Circulatory Function

ACTIVE LEARNING OBJECTIVES

Upon mastering the material in this chapter, you should be able to:

- Correctly identify changes in cardiovascular variables that can cause hypotension and explain their mechanism of action.
- Explain how hypotension leads to the activation of neurogenic reflexes that correct the hypotension, and explain the cardiovascular mechanism of the correction.
- Explain the hormonal mechanisms involved in the defense against hypotension and how these synergize with neurogenic reflexes to control blood pressure.
- Explain the mechanism of pressure diuresis and how this sets the long-term level of mean arterial pressure.
- Explain how standing results in a decrease in cardiac output and blood pressure and what mechanisms are set into motion to restore normal blood pressure and cardiac output upon standing.
- Predict the effect of hypotensive events on the metabolism and viability based on the neurohumoral mechanisms activated by the hypotension.
- Explain the mechanisms responsible for progressive shock.
- Correctly identify the initiating causes and unique complications of shock caused by hemorrhage, severe vomiting, sweating, diarrhea, decreased fluid and electrolyte intake, kidney damage, adrenal cortical destruction, severe burns, intestinal obstructions, general or spinal anesthesia, fever, emotional stress, anaphylaxis, and sepsis.

The mechanisms controlling the circulation involve individual and cooperative effects among neural control mechanisms, hormonal control mechanisms, and local control mechanisms. Local vascular control mechanisms were discussed in Chapter 15, "Microcirculation and Lymphatic System." This chapter will focus on neural and hormonal mechanisms. Neural and hormonal mechanisms are primarily involved with the control of central blood volume and arterial pressure. Adequate central blood volume is necessary to ensure proper cardiac output. In addition, autoregulation of blood flow would not function properly without some mechanism to maintain a relatively constant arterial blood pressure.

Neural control of the cardiovascular system involves sympathetic and parasympathetic branches of the autonomic nervous system (ANS). Blood volume and arterial pressure are monitored by stretch receptors in the heart and arteries. Afferent nerve traffic from these receptors is integrated with other afferent information in the medulla oblongata, which leads to activity in sympathetic and parasympathetic nerves that adjust heart rate, myocardial contraction, arterial resistance, and venous tone. In this way, cardiac output and systemic vascular resistance (SVR) are adjusted to maintain arterial pressure. Sympathetic nerve activity and, more important, hormones, such as **arginine vasopressin (AVP)** (i.e., **antidiuretic hormone** or ADH), **angiotensin II**, **aldosterone**, and **atrial natriuretic peptide (ANP)**, serve as effectors for the regulation of blood volume by regulating salt and water balance.

Neural control of cardiac output and SVR plays a larger role in the moment-to-moment regulation of arterial pressure, whereas hormones play a larger role in the long-term regulation of arterial pressure. In some situations, factors other than blood volume and arterial pressure regulation strongly influence cardiovascular control mechanisms. These situations include the fight-or-flight response, diving, thermoregulation, standing, and exercise.

AUTONOMIC NEURAL CONTROL OF THE CIRCULATORY SYSTEM

Neural regulation of the cardiovascular system involves the firing of postganglionic parasympathetic and sympathetic neurons, triggered by preganglionic neurons in the brain (parasympathetic) and spinal cord (sympathetic and parasympathetic). Afferent inputs influencing these neurons come from specified locations in the cardiovascular system and operate as sensors for arterial pressure and blood volume as well as from other organs and the external environment.

Neurogenic control of the heart involves reciprocal activation of parasympathetic and sympathetic nerves.

Autonomic control of the heart and blood vessels was described in Chapter 6, "Autonomic Nervous System." Briefly, the heart is innervated by parasympathetic (vagus) and sympathetic (cardioaccelerator) nerve fibers (Fig. 17.1). Parasympathetic fibers to the heart are tonically active. That is, they exhibit a steady stream of action potential firing at rest. Acetylcholine (ACh) released from these fibers binds to muscarinic receptors of the sinoatrial (SA) and atrioventricular nodes as well as the specialized conducting tissues. Stimulation of parasympathetic fibers causes a slowing of the heart rate and conduction velocity. The ventricular muscle is only sparsely innervated by parasympathetic nerve fibers, and stimulation of these fibers has only a small negative inotropic effect on the heart. Some cardiac parasympathetic fibers end on sympathetic nerves and inhibit the release of norepinephrine (NE) from sympathetic nerve fibers. Therefore, in the presence of sympathetic nervous system activity, parasympathetic activation reduces cardiac contractility.

Sympathetic fibers to the heart are also tonically active and release NE, which binds to β_1-adrenergic receptors in the SA node, the atrioventricular node and specialized conducting tissues, and cardiac muscle. Stimulation of these fibers causes increased heart rate, conduction velocity, and contractility. Activity along the two divisions of the ANS changes in a reciprocal manner to create changes in heart rate. For example, an increase in heart rate is brought about by a simultaneous decrease in parasympathetic and an increase in sympathetic nerve activity to the heart. However, control of heart rate is dominated by parasympathetic effects. Activation of the parasympathetic system can slow the heart even when the sympathetic system is maximally activated. At submaximal sympathetic rates, activation of the vagus nerve can totally suppress the SA node and temporarily cause the heart to stop. In contrast to the relationships controlling heart rate, control of cardiac contractility is dominated by sympathetic over parasympathetic effects. Inotropic state is only minimally affected by vagal influence, and myocardial contractility is, therefore, primarily modulated by the level of the activity in the sympathetic nerves to ventricular muscle.

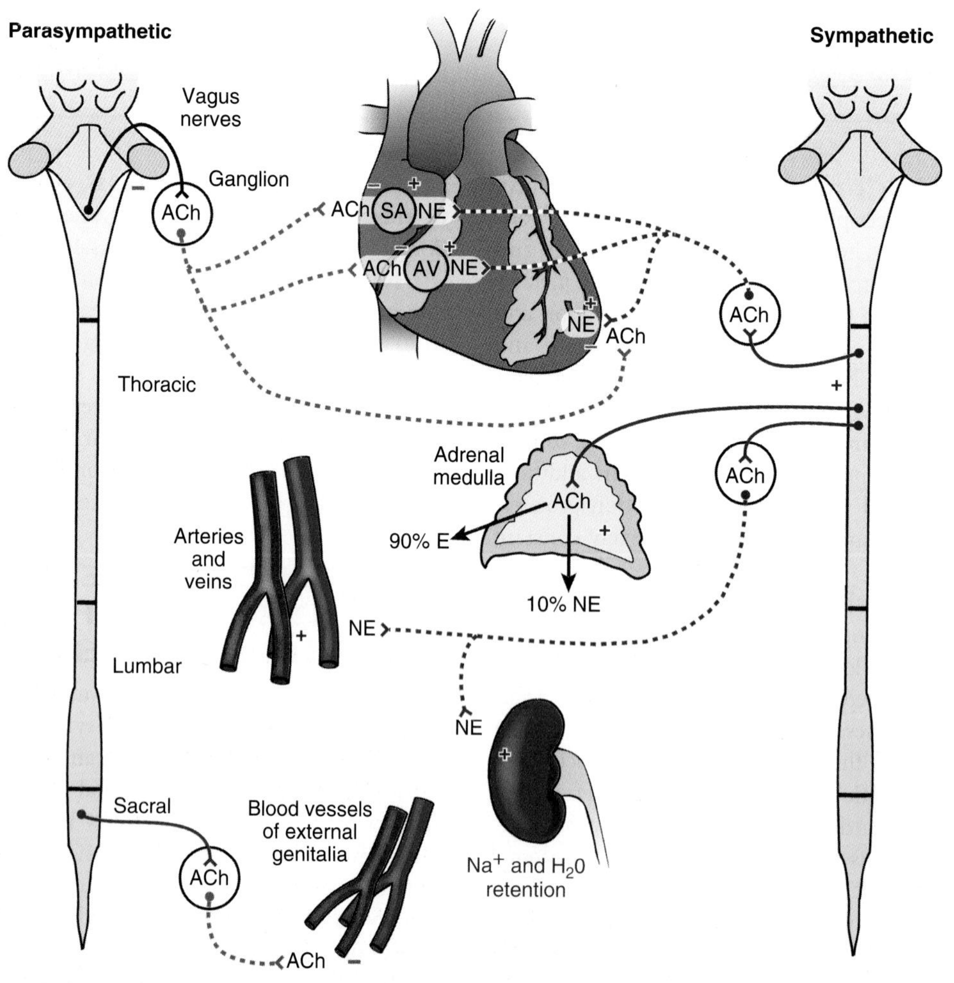

Figure 17.1 Autonomic innervation of the cardiovascular system. ACh, acetylcholine; NE, norepinephrine; E, epinephrine; SA, sinoatrial node; AV, atrioventricular node.

There is no known parasympathetic innervation of blood vessels in systemic organs with the exception of those of the external genitalia. Sympathetic fibers innervate arteries and veins of all the major systemic organs except the brain (see Fig. 17.1). These fibers tonically release NE, which binds to α_1-adrenergic and β_2-adrenergic receptors on blood vessels. However, because the arteries of all vascular beds except the heart and brain contain more α_1-adrenergic than β_2-adrenergic receptors, activation of the sympathetic nerves to the systemic circulations causes vasoconstriction and an increase in SVR. Circulating epinephrine, released from modified sympathetic nerve endings in the adrenal medulla, binds to α_1-adrenergic and β_2-adrenergic receptors of vascular and smooth muscle cells, as well. However, the affinity of both β_1 and β_2 receptors for epinephrine is greater than that for NE. Therefore, at low circulating concentrations, epinephrine essentially activates only β receptors, with the effect of increasing cardiac output (chronotropic and inotropic effects) but *decreasing* SVR.

Postganglionic parasympathetic fibers release ACh and nitric oxide (NO) to blood vessels in the external genitalia. ACh causes the further release of NO from endothelial cells, which results in vascular smooth muscle relaxation and vasodilation. These fibers mediate erection in males and engorgement of the female genitalia.

Arterial effects of spinal cord injury

The steady train of sympathetic nerve activity, or tone, to blood vessels, the heart, and the adrenal medulla produces a background level of sympathetic vasoconstriction, cardiac stimulation, and adrenal catecholamine secretion in the body. All of these factors contribute to the maintenance of normal blood pressure. This tonic activity is generated by excitatory signals from the medulla oblongata. When the spinal cord is acutely transected and these excitatory signals can no longer reach sympathetic preganglionic fibers, their tonic firing is reduced and blood pressure falls. This effect is known as **spinal shock**. Humans have spinal reflexes of cardiovascular significance. For example, the stimulation of pain fibers entering the spinal cord below the level of a chronic spinal cord transection can cause reflex vasoconstriction and increased blood pressure.

Cardiovascular reflex integration by the medulla oblongata

The medulla oblongata has three major cardiovascular functions: (1) generating tonic excitatory signals to spinal sympathetic preganglionic fibers, (2) integrating cardiovascular reflexes, and (3) integrating signals from supramedullary neural networks, circulating hormones and drugs. Specific pools of neurons are responsible for elements of these functions. Neurons in the rostral ventrolateral nucleus (RVL) are normally active and provide tonic excitatory activity to the spinal cord. Specific pools of neurons within the RVL have actions on the heart and blood vessels. RVL neurons are critical in mediating reflex inhibition or activating sympathetic firing to the heart and blood vessels. The cell bodies of cardiac preganglionic parasympathetic neurons are located in the nucleus ambiguus; the activity of these neurons is influenced by reflex input as well as input from respiratory neurons. Respiratory sinus arrhythmia is primarily the result of the influence of medullary respiratory neurons that inhibit firing of preganglionic parasympathetic neurons during inspiration and excite these neurons during expiration. Other inputs to the RVL and nucleus ambiguus will be described below.

Baroreceptor reflexes maintain the moment-to-moment level of arterial pressure.

The most important reflex control of the cardiovascular system originates in mechanoreceptors located in the aorta, carotid sinuses, atria, ventricles, and pulmonary vessels. These mechanoreceptors are sensitive to the stretch of the walls of these structures. The firing rate of nerves from these mechanoreceptors increases when the wall is stretched by increased transmural pressure. For this reason, mechanoreceptors in the aorta and carotid sinuses are called **baroreceptors**, *arterial baroreceptors*, or *high-pressure receptors*. Mechanoreceptors in the atria, ventricles, and pulmonary vessels primarily sense pressure changes brought about by changes in blood volume. Therefore, these receptors are referred to as *volume-receptors*, *low-pressure baroreceptors*, or **cardiopulmonary baroreceptors**.

Changes in the firing rate of the arterial baroreceptors and cardiopulmonary baroreceptors initiate reflex responses of the ANS that alter cardiac output and SVR. The central terminals for these receptors are located in the *nucleus tractus solitarii* (NTS) in the medulla oblongata. Neurons from the NTS project to the RVL and nucleus ambiguus, where they influence the firing of sympathetic and parasympathetic nerves.

Baroreceptor reflex modulates cardiac output and total peripheral resistance to control mean arterial pressure.

Increased pressure in the carotid sinus and aorta stretches **carotid sinus baroreceptors** and **aortic baroreceptors** and raises their firing rate. Nerve fibers from carotid sinus baroreceptors join the glossopharyngeal (cranial nerve IX) nerves and travel to the NTS. Nerve fibers from the aortic baroreceptors, located in the wall of the arch of the aorta, travel with the vagus (cranial nerve X) nerves to the NTS. The increased action potential traffic reaching the NTS leads to excitation of nucleus ambiguus neurons and inhibition of firing of RVL neurons. This results in increased parasympathetic neural activity to the heart and decreased sympathetic neural activity to the heart, resistance vessels (primarily arterioles), and veins (Fig. 17.2). Collectively, these effects cause decreased cardiac output and SVR. Because mean arterial pressure is the product of SVR and cardiac output, mean arterial pressure is returned toward the normal level. This completes a negative-feedback loop by which increase in mean arterial pressure can be attenuated.

Conversely, decreases in arterial pressure (and decreased stretch of the baroreceptors) increase sympathetic neural activity and decrease parasympathetic neural activity,

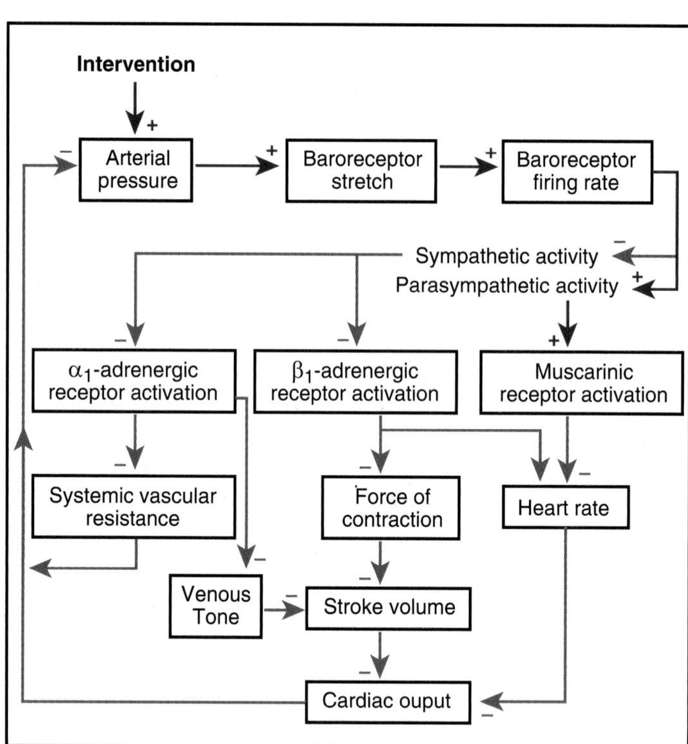

■ **Figure 17.2 Baroreceptor neural reflex responses to increased arterial pressure.** An intervention elevates arterial pressure (either mean arterial pressure or pulse pressure), stretches the baroreceptors, and initiates the reflex. The resulting reduced systemic vascular resistance and cardiac output return arterial pressure toward the level existing before the intervention. Hormonal responses (not shown in the figure) involving adrenal epinephrine and the renin–angiotensin system are also involved in the reflex but more so as a response to hypotension rather than to hypertension. *Red (+) arrows* signify positive effects. *Blue (–) arrows* signify negative effects.

resulting in increased heart rate, stroke volume, and SVR; this returns blood pressure toward the normal level. If the fall in mean arterial pressure is large, increased sympathetic neural activity to veins is added to the above responses, causing contraction of the venous smooth muscle and reducing venous compliance. Decreased venous compliance shifts blood toward the central blood volume, increasing right atrial pressure and, in turn, stroke volume. The reflex activation of the cardiovascular system in response to changes in mean arterial pressure in order to maintain that pressure within narrow limits is called the **baroreceptor reflex**. This reflex is extremely sensitive in that firing rate from nerves exiting the baroreceptors can sense changes in arterial pressure as little as 0.001 mm Hg. These receptors also respond rapidly to the rate of rise in arterial pressure and changes in pulse pressure; firing rate is greater in systole compared to diastole and greater early in systole than later. Increased pulse pressure will activate baroreceptor firing even in the absence of a change in mean arterial pressure.

Baroreceptor reflex activates hormonal systems affecting blood pressure.

The baroreceptor reflex influences hormone levels in addition to vascular and cardiac muscles. The most important influence is on the renin–angiotensin–aldosterone system (RAAS). A decreased baroreceptor firing from decreased systemic arterial pressure results in increased sympathetic nerve activity to the kidneys, which, through activation of renal β_2 receptors, causes the kidneys to release renin. Renin converts a precursor peptide called *angiotensinogen* into the peptide angiotensin I, which, in turn, is enzymatically cleaved by pulmonary endothelium to produce an active peptide called *angiotensin II*. A reduction in arterial pressure in the renal artery also stimulates renin release. Angiotensin II is a potent vasoconstrictor and stimulates the release of a steroid hormone called *aldosterone* from the adrenal gland, which causes the kidney to reabsorb salt and water. The activation of this system increases vascular resistance and blood volume, ultimately causing blood pressure to rise. The details of the RAAS are discussed later in this chapter and in Chapter 23, "Regulation of Fluid and Electrolyte Balance."

The information on the firing rate of the baroreceptors is also projected to the paraventricular nucleus of the hypothalamus, where the release of AVP by the posterior pituitary is controlled (see Chapter 31, "Hypothalamus and the Pituitary Gland"). AVP release is increased by a decrease in the firing rate of the baroreceptors. AVP is a vasoconstrictor that also activates receptors in the kidney, causing the kidneys to save water, which results in an increase in blood volume. An increase in arterial pressure causes decreased AVP release and increased excretion of water by the kidneys. Hormonal effects on salt and water balance and, ultimately, on cardiac output and blood pressure are powerful, but they occur more slowly (a timescale of many hours to days) than ANS effects (seconds to minutes).

Baroreceptor reflex preserves flow to the brain and heart.

The defense of arterial pressure by the baroreceptor reflex results in maintenance of blood flow to two vital organs: the heart and brain. If resistance vessels of the heart and brain participated in the sympathetically mediated vasoconstriction found in skeletal muscle, skin, and the splanchnic region during the reflex, it would lower blood flow to these organs. This does not happen.

The combination of a minimal or no vasoconstrictor effect of sympathetic nerves on cerebral blood vessels and a robust autoregulatory response keep brain blood flow nearly normal despite modest decreases in arterial pressure (see Chapter 15, "Microcirculation and Lymphatic System"). Activation of sympathetic nerves to the heart causes β_2-adrenergic receptor-mediated dilation of coronary arterioles and β_1-adrenergic receptor–mediated increases in cardiac muscle metabolism (see Chapter 16, "Special Circulations"). The net effect is a marked increase in coronary blood flow. In summary, when arterial pressure drops, the generalized vasoconstriction caused by the baroreceptor reflex restores blood pressure without vasoconstricting the brain and heart. This prevents blood flow from decreasing to the heart and brain whenever blood pressure falls.

Baroreceptor activation is site-, pressure-, and time-dependent.

The effective range of the carotid sinus baroreceptor mechanism is approximately 40 mm Hg (when the receptor stops firing) to 180 mm Hg (when the firing rate reaches a maximum) (Fig. 17.3). Aortic baroreceptors initiate activation at about 70 mm Hg and reach maximum firing at a higher pressure.

An important property of the baroreceptor reflex is that it adapts during a period of 1 to 2 days to the prevailing mean arterial pressure. When the mean arterial pressure is suddenly raised, baroreceptor firing increases. If arterial pressure is held at the higher level, baroreceptor firing declines during the next few seconds. Firing rate then continues to decline more slowly until it returns to the original firing rate over the next 1 to 2 days. Consequently, if the mean arterial pressure is maintained at an elevated level, the tendency for the baroreceptors to initiate a decrease in cardiac output and SVR quickly disappears. This occurs, in part, because of a reduction in the rate of baroreceptor firing for a given mean arterial pressure (see Fig. 17.3). This is an example of **receptor adaptation**. A "resetting" of the reflex in the central nervous system (CNS) occurs as well.

The baroreceptor mechanism can be viewed as the "first line of defense" in the maintenance of normal blood pressure; it makes possible the rapid control of blood pressure needed with changes in posture or blood loss. However, this control mechanism does not provide for the long-term control of blood pressure.

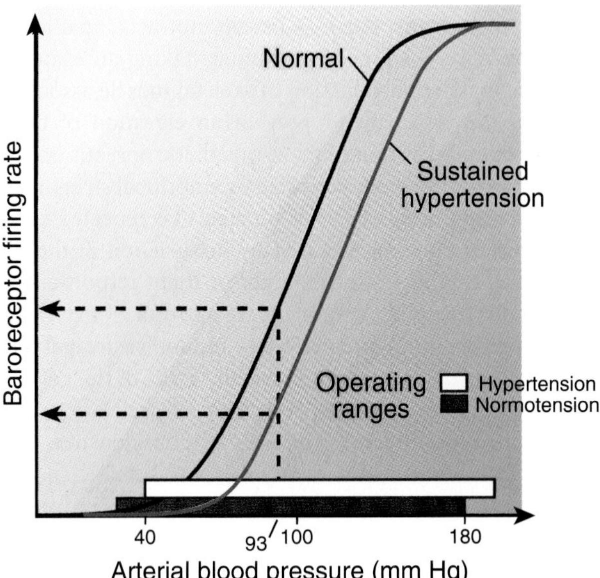

Figure 17.3 Carotid sinus baroreceptor nerve firing rate and mean arterial pressure. With normal conditions, a mean arterial pressure of 93 mm Hg is near the midrange of the firing rates for the nerves. Sustained hypertension causes the operating range to shift to the right, putting 93 mm Hg at the lower end of the firing range for the nerves. Aortic baroreceptors show similar relationships, except the point at which pressure activates the receptor and reaches maximum response is higher than that seen in carotid receptors.

Cardiopulmonary baroreceptors sense central blood volume.

Cardiopulmonary baroreceptors are located in the cardiac atria, at the junction of the great veins and atria, in the ventricular myocardium, and in pulmonary vessels. Their nerve fibers run in the vagus nerve to the NTS, with projections to supramedullary areas as well. Unloading (i.e., decreasing the stretch) of the cardiopulmonary receptors by reducing central blood volume results in increased sympathetic nerve activity to the heart and blood vessels and decreased parasympathetic nerve activity to the heart. In addition, the cardiopulmonary reflex interacts with the baroreceptor reflex. Unloading of the cardiopulmonary receptors enhances the baroreceptor reflex, and loading the cardiopulmonary receptors, by increasing central blood volume, inhibits the baroreceptor reflex. Like the arterial baroreceptors, decreased stretch of the cardiopulmonary baroreceptors activates the RAAS and increases the release of AVP.

Chemoreceptors for P_{CO_2}, pH, and P_{O_2} affect mean arterial pressure.

The **carotid** and **aortic bodies** are specialized structures located in the areas of the carotid sinus and aortic arch that sense changes in blood O_2, CO_2, and pH. These structures are sometimes referred to as **chemoreceptors**. The carotid and aortic chemoreceptors are primarily involved with control of ventilation (see Chapter 21, "Control of Ventilation"), but they also affect the cardiovascular system through neurogenic reflexes. Peripheral chemoreceptors send impulses to the NTS and exhibit increased firing rate when, either the P_{O_2} or pH of the arterial blood is low, the P_{CO_2} of arterial blood is increased, the flow through the bodies is low or stopped, or a chemical is given that blocks oxidative metabolism in the chemoreceptor cells. There are also central medullary chemoreceptors that increase their firing rate primarily in response to elevated arterial P_{CO_2}, which causes a decrease in brain pH.

The increased firing of both peripheral and central chemoreceptors (via the NTS and RVL) leads to profound peripheral vasoconstriction that significantly elevates arterial pressure. If respiratory movements are voluntarily stopped, the vasoconstriction is more intense and a striking bradycardia and decreased cardiac output occur. This response pattern is typical of the diving response (discussed later). As in the case of the baroreceptor reflex, the coronary and cerebral circulations are not subject to the sympathetic vasoconstrictor effects and instead exhibit vasodilation as a result of the combination of the direct effect of the abnormal blood gases and local metabolic effects.

In addition to its importance when arterial blood gases are abnormal, the chemoreceptor reflex is important in the cardiovascular response to severe hypotension. As blood pressure falls, blood flow through the carotid and aortic bodies decreases and chemoreceptor firing increases, probably because of changes in local P_{CO_2}, pH, and P_{O_2}. The chemoreceptor reflex, however, does not respond to a change in blood pressure itself until mean arterial pressure drops to about 80 mm Hg. Therefore, this reflex is not involved in maintenance of normal blood pressure on a moment-to-moment basis, but rather serves as a secondary emergency reflex if blood pressure continues to fall in spite of activation of the baroreceptor reflex.

Pain and myocardial ischemia initiate cardiovascular reflexes.

Two reflex cardiovascular responses to pain occur. In the most common reflex, pain causes increased sympathetic activity to the heart and blood vessels, coupled with decreased parasympathetic activity to the heart. These events lead to increases in cardiac output, SVR, and mean arterial pressure. An example of this reaction is the **cold pressor response**, which is the elevated blood pressure that normally occurs from pain associated with placing an extremity in ice water. The increase in blood pressure produced by this challenge is exaggerated in several forms of hypertension.

A second type of response is produced by deep pain. The stimulation of deep pain fibers associated with crushing injuries, disruption of joints, testicular trauma, or distention of the abdominal organs results in diminished sympathetic activity and enhanced parasympathetic activity with decreased cardiac output, SVR, and blood pressure. This hypotensive response contributes to cardiovascular shock from severe trauma (see below).

Myocardial ischemia in the posterior and inferior myocardium causes reflex bradycardia and hypotension. The bradycardia results from increased parasympathetic tone. Dilation of systemic arterioles and veins in this situation is caused by withdrawal of sympathetic tone. This response mimics that following an injection of bradykinin, 5-hydroxytryptamine (serotonin), certain prostaglandins, or various other compounds into the coronary arteries supplying the posterior and inferior regions of the ventricles. This reflex is responsible for the bradycardia and hypotension that can occur in response to acute infarction of the posterior or inferior myocardium.

Higher-order ANS responses alter blood pressure and cardiac output.

The highest levels of organization in the ANS are the supramedullary networks of neurons with weigh stations in the limbic cortex, amygdala, and hypothalamus. These supramedullary networks orchestrate cardiovascular responses to specific patterns of emotion and behavior by their projections to the ANS. Unlike the medulla, supramedullary networks do not contribute to the tonic maintenance of blood pressure, nor are they necessary for most cardiovascular reflexes. However, they do modulate reflex reactivity and can affect the behavior of the heart and systemic circulation.

Fear

On stimulation of certain areas in the hypothalamus, cats demonstrate a stereotypical rage response, with spitting, clawing, tail lashing, and back arching. This is accompanied by the autonomic **fight-or-flight response** described in Chapter 6, "Autonomic Nervous System." This reaction occurs naturally whenever the cat feels threatened and/or experiences fear. Cardiovascular responses include elevated heart rate and blood pressure. The initial behavioral pattern during the fight-or-flight response includes increased skeletal muscle tone and general alertness. There is increased sympathetic neural activity to blood vessels and the heart. There is some evidence that sympathetic cholinergic fibers to the muscle arteries elicit a neurogenic vasodilation in skeletal muscles. The result of this cardiovascular response is an increase in cardiac output (by increasing both heart rate and stroke volume), SVR, and arterial pressure. When the fight-or-flight response is consummated by fight or flight, arterioles in skeletal muscle dilate because of accumulation of local metabolites from the exercising muscles. This vasodilation may outweigh the sympathetic vasoconstriction in other organs, and SVR may actually fall. With a fall in SVR, mean arterial pressure returns toward normal despite the increase in cardiac output.

Emotional stress

Emotional situations often provoke the fight-or-flight response in humans, but it is usually not accompanied by muscle exercise (e.g., medical students taking an examination). The massive vasodilation in skeletal muscle associated with exercise, which helps prevent an elevation of blood pressure upon activation of the sympathetic nervous system, is lost when the system is activated by emotional stress alone. For this reason, it has been postulated that repeated elevations in arterial pressure caused by dissociation of the cardiovascular component of the fight-or-flight response from the muscular exercise component are harmful.

Certain emotional experiences induce **vasovagal syncope** (fainting). Stimulation of specific areas of the cerebral cortex can lead to a sudden relaxation of skeletal muscles, depression of respiration, and loss of consciousness. The cardiovascular events accompanying these somatic changes include profound parasympathetic-induced bradycardia and withdrawal of resting sympathetic vasoconstrictor tone. There is a dramatic drop in heart rate, cardiac output, and SVR. The resultant decrease in mean arterial pressure results in unconsciousness (fainting) because of lowered cerebral blood flow. Vasovagal syncope appears in lower animals as the "playing dead" response typical of the opossum.

Exercise

Exercise causes activation of supramedullary neural networks that inhibit the activity of the baroreceptor reflex. The inhibition of medullary regions involved in the baroreceptor reflex is called *central command.* Central command results in withdrawal of parasympathetic tone to the heart, with a resulting increase in heart rate and cardiac output. The increased cardiac output supplies the added requirement for blood flow to exercising muscle. As exercise intensity increases, central command adds sympathetic tone that further increases heart rate and contractility. It also recruits sympathetic vasoconstriction that redistributes blood flow away from splanchnic organs and resting skeletal muscle to exercising muscle. The local metabolic vasodilator influences in exercising muscle overwhelm any enhanced sympathetic activity to arteries in the muscle, thereby overriding the constrictor influences of the sympathetic nerves in that muscle. Finally, afferent impulses from exercising skeletal muscle terminate in the RVL, where they further augment sympathetic tone. During exercise, blood flow of the skin is largely influenced by temperature regulation, as described in Chapter 16, "Special Circulations."

Diving response

The **diving response** is best observed in seals and ducks, but it also occurs in humans. An experienced diver can exhibit intense slowing of the heart rate (parasympathetic) and peripheral vasoconstriction (sympathetic) of the extremities and splanchnic regions when his or her face is submerged in cold water. With breath holding during the dive, arterial Po_2 and pH fall and Pco_2 rises, and the chemoreceptor reflex reinforces the diving response. The arterioles of the brain and heart do not constrict and, therefore, cardiac output is distributed to these organs. This heart–brain circuit makes use of the oxygen stored in the blood that would normally be used by the other tissues, especially skeletal muscle. Once the diver surfaces, the heart rate and cardiac output increase substantially; vasodilation replaces peripheral vasoconstriction, restoring nutrient flow and washing out accumulated waste products.

Behavioral conditioning

Cardiovascular responses can be conditioned. Both classical and operant conditioning techniques have been used to raise and lower the blood pressure and heart rate of animals. Humans can also be taught to alter their heart rate and blood pressure using a variety of behavioral techniques, such as biofeedback. Behavioral conditioning of cardiovascular responses has significant clinical implications. Animal and human studies indicate that psychological stress can raise blood pressure, increase atherogenesis, and predispose the person to fatal cardiac arrhythmias. These effects are thought to result from activation of the fight-or-flight response. Other studies have shown beneficial effects of behavior patterns designed to introduce a sense of relaxation and well-being. Some clinical regimens for the treatment of cardiovascular disease take these factors into account.

Baroreceptor override

Supramedullary responses can override the baroreceptor reflex. For example, the fight-or-flight response causes the heart rate to rise above normal levels despite a simultaneous rise in arterial pressure. In such circumstances, the neurons connecting the hypothalamus to medullary areas inhibit the baroreceptor reflex and allow the corticohypothalamic response to predominate. Also, during exercise, input from supramedullary regions inhibits the baroreceptor reflex, promoting increased sympathetic tone and decreased parasympathetic tone despite an increase in arterial pressure. Moreover, the various cardiovascular response patterns do not necessarily occur in isolation, as previously described. Many response patterns interact, reflecting the extensive neural interconnections between all levels of the CNS and interaction with various elements of the local control systems. For example, the baroreceptor reflex interacts with thermoregulatory responses. Cutaneous sympathetic nerves participate in body temperature regulation but also serve the baroreceptor reflex. At moderate levels of heat stress, the baroreceptor reflex can cause cutaneous arteriolar constriction despite elevated core temperature. However, with severe heat stress, the baroreceptor reflex cannot overcome the cutaneous vasodilation; as a result, arterial pressure regulation may fail.

HORMONAL CONTROL OF THE CARDIOVASCULAR SYSTEM

Various hormones play a role in the control of the cardiovascular system. Important hormonal control mechanisms involved in cardiovascular homeostasis include epinephrine from the adrenal medulla, AVP (antidiuretic hormone) from the posterior pituitary gland, renin from the kidney, and ANP from the cardiac atrium.

Circulating epinephrine exerts different cardiovascular effects from those caused by sympathetic nerves.

When the sympathetic nervous system is activated, the adrenal medulla releases epinephrine (>90%) and NE (<10%) into the bloodstream. Changes in the circulating NE concentration are small relative to changes in NE resulting from the direct release from nerve endings close to vascular smooth muscle and cardiac cells. Increased circulating epinephrine, however, contributes to skeletal muscle vasodilation during the fight-or-flight response and exercise. In these cases, epinephrine binds to β_2-adrenergic receptors of skeletal muscle arteriolar smooth muscle cells and causes relaxation. In the heart, circulating epinephrine binds to cardiac cell β_1-adrenergic receptors and reinforces the effect of NE released from sympathetic nerve endings.

A comparison of the responses to infusions of low concentrations of epinephrine and NE illustrates not only the different effects of the two hormones but also the different reflex response each one elicits (Fig. 17.4). Epinephrine and NE have similar direct effects on the heart, but NE elicits a powerful baroreceptor reflex because it causes systemic vasoconstriction and increases mean arterial pressure. The reflex masks some of the direct cardiac effects of NE by significantly increasing cardiac parasympathetic tone.

Clinical Focus / 17.1 

Nitric Oxide in Cardiovascular Disease

The nitric oxide (NO) system found in all vascular endothelium possesses some of the most important and beneficial functions for the maintenance of the cardiovascular system. Its intense vasodilatory effect antagonizes arterial spasm that might otherwise restrict organ blood flow while it also prevents significant resting hypertension. It has antithrombotic and antiatherogenic properties, attenuates ischemia–reperfusion injury, and suppresses smooth muscle growth in vascular injury while stimulating endothelial wound healing. In addition, it chemically quenches superoxide (O_2^-) that causes extensive vascular damage.

In this section of the textbook, the student may realize that the body often seems to initiate processes that compensate for the effects of cardiovascular malfunctions and disease. These can be thought of as extensions of the body's homeostatic mechanisms and are revealed in such cardiovascular phenomena as autoregulation of blood flow, volume retention in heart failure, and neural/humoral responses used to combat circulatory shock. It is reasonable to postulate that the endothelial NO system might be recruited by the cardiovascular system to counteract the effects of various cardiovascular diseases. Unfortunately, nothing could be further from the truth. To date, the endothelial NO system has been shown to be impaired and malfunctional in all forms of cardiovascular disease. This system does not work properly, if at all, in acute and chronic hypertensions, hyperlipidemia and atherosclerosis, diabetes mellitus, ischemic injury, stroke, vascular transplantations, and heart failure.

As a result of an impaired NO system, people with cardiovascular disease are left with cardiovascular systems that are prone to vascular spasm and hypertension, thrombosis, atherosclerosis, and stenosis from abnormal vascular wall growth. In addition, NO from the endothelium cannot abate the presence of oxygen radicals, or reactive oxygen species, in the vasculature. Such radicals, which arise from extravascular and intravascular sources, are overproduced in all cardiovascular disease. The enzyme NADP oxidase appears to be a prime source of the radicals. Exposure of vessels to reactive oxygen species damages the arterial endothelium and converts NO synthase to a form that makes oxygen radicals from l-arginine instead of NO.

Several studies are investigating ways to restore the NO system in cardiovascular disease. NO itself seems to be able to attenuate ischemia–reperfusion injury, but widespread use of nitrates as substitutes for endogenous NO has not proven effective in treating many cardiovascular diseases. Instead, additional therapies are being directed to combat the effects of reactive oxygen species on the vasculature. Attenuation of these effects may cause the vascular endothelium to sustain less damage, thereby enhancing their ability to produce NO, while at the same time improving the bioavailability of NO to the arterial smooth muscle by removing chemical species that quench any NO produced.

In contrast, low concentrations of epinephrine cause vasodilation in skeletal muscle and splanchnic beds because it preferentially activates the β_2 over the α_1-adrenoceptors on the arterial system of those vascular beds. SVR may actually fall and mean arterial pressure does not rise. Consequently, the baroreceptor reflex is not elicited, parasympathetic tone to the heart is not increased, and the direct cardiac effects of epinephrine are evident. At high concentrations, epinephrine binds to α_1-adrenergic receptors and causes peripheral vasoconstriction; this level of epinephrine is probably never reached except when it is administered as a drug.

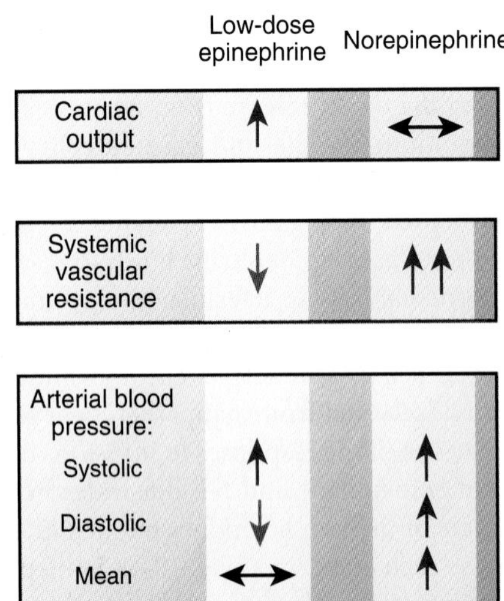

Figure 17.4 A comparison of the effects of intravenous infusions of low concentrations of epinephrine and norepinephrine. ↑ = increase; ↓ = decrease; ↔ = no change; ↑↑ = marked increase.

Denervated organs, such as transplanted hearts, are hyperresponsive to circulating levels of epinephrine and NE. This increased sensitivity to neurotransmitters is referred to as **denervation hypersensitivity**. Several factors contribute to denervation hypersensitivity, including the absence of sympathetic nerve endings to take up circulating NE and epinephrine actively, leaving more transmitter available for binding to receptors. In addition, denervation results in up-regulation of neurotransmitter receptors in target cells. During exercise, circulating levels of NE and epinephrine increase. Because of their enhanced response to circulating catecholamines, transplanted hearts can perform almost as well as normal hearts.

Renin-angiotensin-aldosterone system contributes to regulation of blood pressure and volume.

The control of total blood volume is extremely important in regulating arterial pressure. Because changes in total blood volume lead to changes in central blood volume, the long-term influence of blood volume on ventricular end-diastolic volume and cardiac output is paramount. Cardiac output, in turn, strongly influences arterial pressure. Hormonal control

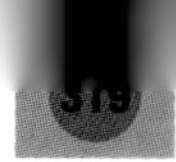

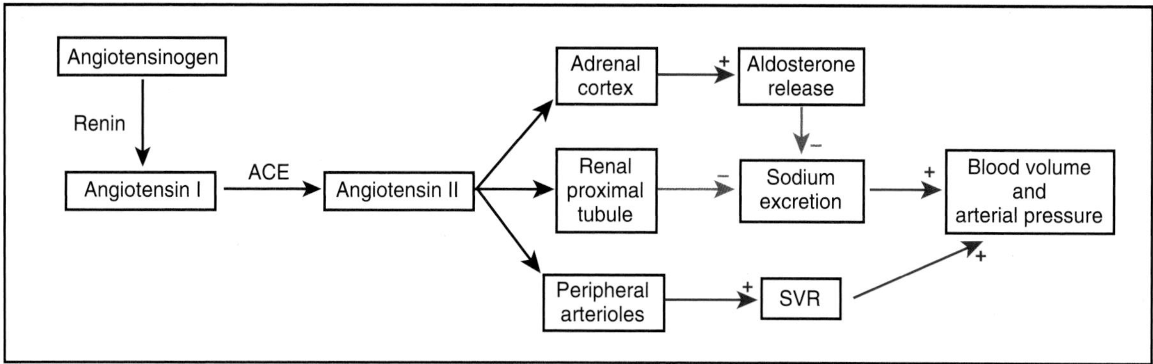

Figure 17.5 Renin–angiotensin–aldosterone system. This system plays an important role in the regulation of arterial blood pressure and blood volume. ACE, angiotensin-converting enzyme; SVR, systemic vascular resistance. *Red (+) arrows* signify positive effects. *Blue (–) arrows* signify negative effects.

of blood volume depends on hormones that regulate salt and water balance as well as red blood cell formation.

Reduced arterial pressure and blood volume cause the release of **renin** from the kidneys. Renin release is mediated by the sympathetic nervous system and by the direct effect of lowered arterial pressure within the afferent arterioles of the kidneys. Renin is a proteolytic enzyme that catalyzes the conversion of angiotensinogen, a plasma protein, to angiotensin I which, in turn, is converted to angiotensin II in the lung by **angiotensin-converting enzyme (ACE)** (Fig. 17.5). Angiotensin II counteracts any decrease in mean arterial pressure by the following actions: (1) it is a powerful direct arteriolar vasoconstrictor. In some circumstances, it is present in plasma in concentrations sufficient to increase SVR; (2) it reduces sodium excretion by increasing sodium reabsorption by proximal tubules of the kidney; (3) it causes the release of aldosterone from the adrenal cortex, which in turn promotes sodium reabsorption in the distal convoluted tubules of the kidney; (4) it causes the release of AVP from the posterior pituitary gland, which promotes water reabsorption from the collecting ducts of the kidney and directly causes systemic arterial vasoconstriction; (5) it potentiates the effects of the sympathetic nervous system to blood vessels by enhancing NE release from sympathetic nerve endings, reducing neuronal reuptake of NE, and sensitizing vascular smooth muscle to the actions of NE; and (6) it stimulates thirst to promote ingestion of water.

Renin is released during blood loss, even before blood pressure falls, and the resulting rise in plasma angiotensin II from renin release increases the SVR.

One of the effects of aldosterone is to reduce renal excretion of sodium, the major cation of the extracellular fluid. Retention of sodium paves the way for increasing blood volume. Renin, angiotensin, aldosterone, and the factors that control their release and formation are discussed in detail in Chapter 23, "Regulation of Fluid and Electrolyte Balance." Although the RAAS is important in the normal maintenance of blood volume and blood pressure, it is critical when salt and water intake is reduced. Angiotensin II plays an important role in increasing SVR, as well as blood volume, in people on a low-salt diet. If an ACE inhibitor is given to such people, blood pressure falls.

Rarely, renal artery stenosis, which reduces afferent arteriolar pressure in the kidney, causes hypertension that can be attributed solely to elevated renin and angiotensin II levels. In addition, the RAAS plays an important (but not unique) role in maintaining elevated pressure in more than 60% of patients with essential hypertension. In patients with congestive heart failure, renin and angiotensin II are increased and contribute to elevated SVR, sodium and water retention, and myocardial hypertrophy. These latter actions of angiotensin II contribute to pulmonary congestion and dilated ventricular hypertrophy seen in chronic congestive heart failure.

Arginine vasopressin primarily regulates blood volume.

AVP is released by the posterior pituitary gland controlled by the hypothalamus. Three primary classes of stimuli lead to AVP release: increased plasma osmolality; decreased baroreceptor and cardiopulmonary receptor firing; and various types of stress, such as physical injury or surgery. In addition, circulating angiotensin II stimulates AVP release. Although AVP is a vasoconstrictor, it is not ordinarily present in plasma in high enough concentrations to exert an effect on blood vessels. However, in special circumstances (e.g., severe hemorrhage), it probably contributes to increased SVR. AVP exerts its major effect on the cardiovascular system by causing the retention of water by the kidneys (see Chapter 23, "Regulation of Fluid and Electrolyte Balance"), which is an important part of the neural/humoral mechanisms that regulate blood volume.

Stretch-activated release of atrial natriuretic peptide counteracts volume overload.

ANP is a 28-amino acid polypeptide synthesized and stored in the atrial muscle cells and released into the bloodstream when the atria are stretched. When central blood volume and atrial stretch are increased, ANP secretion rises, leading to higher sodium excretion and a reduction in blood volume (see Chapter 23, "Regulation of Fluid and Electrolyte Balance"). It also inhibits renin release as well as aldosterone and AVP secretion. The role of ANP in controlling blood volume appears to be important in submammalian sea-dwelling animals (e.g.,

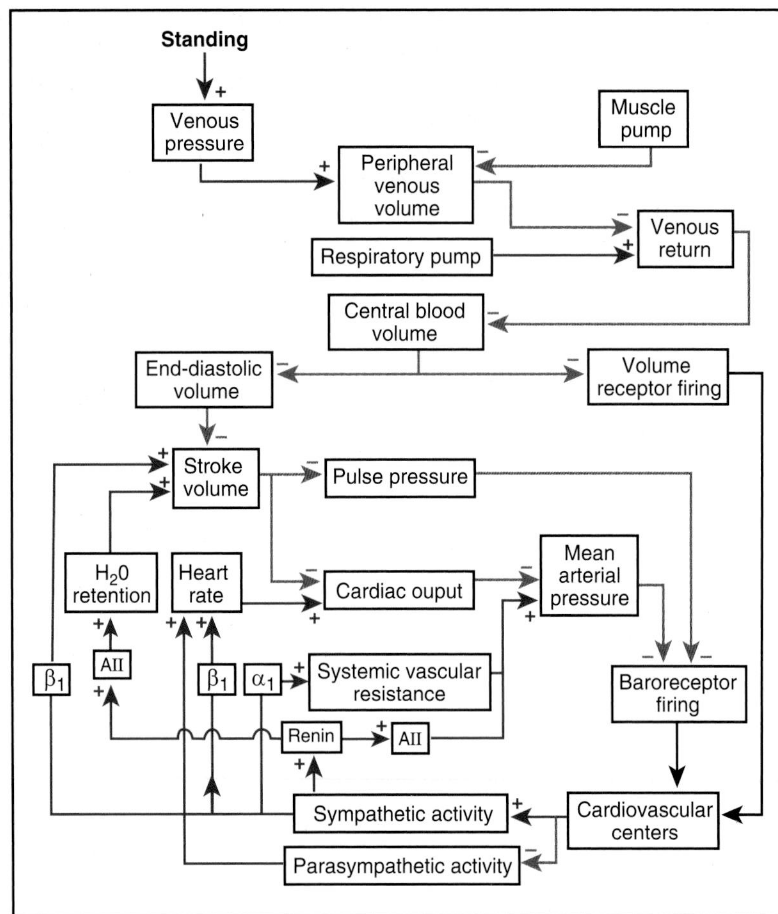

Figure 17.6 Cardiovascular events associated with standing. Small type represents compensatory changes that return variables toward the original values; α and β refer to adrenoceptors. *Red (+) arrows* signify positive effects. *Blue (–) arrows* signify negative effects. AII, angiotensin II.

fish), and its role in the day-to-day control of volume and pressure in the cardiovascular system is questionable. However, increased ANP (along with decreased aldosterone and AVP) may be partially responsible for the reduction in blood volume that occurs with prolonged bed rest.

Renal hypoxia stimulates red blood cell production.

The final step in blood volume regulation is production of erythrocytes. **Erythropoietin** is a hormone released by the kidneys in response to either hypoxia or reduced hematocrit that causes bone marrow to increase production of red blood cells, raising the total mass of circulating red cells. An increase in circulating AVP and aldosterone that enhances salt and water retention results in an elevated plasma volume, which lowers (dilutes) the hematocrit. The decrease in hematocrit stimulates erythropoietin release, which stimulates red blood cell synthesis and, therefore, balances the increase in plasma volume with a larger red blood cell mass. Erythropoietin is also responsible for stimulating erythrocyte production when a person ascends from low altitude to high altitude and resides there for several weeks.

Short- and long-term blood pressure control involve different cardiovascular mechanisms.

Different mechanisms are responsible for the short-term and long-term control of blood pressure. Short-term control depends on activation of neural reflexes and hormonal control mechanisms as described earlier. A good example of short-term control of blood pressure occurs in the body's response to standing (Fig. 17.6). The effect of gravity on pooling of blood in veins of the lower limbs was described in earlier chapters. This pooling results in an initial decrease in cardiac output and blood pressure that activates neural mechanisms, primarily the baroreceptor reflex, that rapidly restores cardiac output and mean arterial pressure. Standing also activates the low-pressure, volume-receptor control mechanisms involving AVP and ANP as well as the RAAS. These mechanisms are not critical in restoring pressure in the short term but can become important if the person is required to stand for long periods of time without moving. (Moving the lower limbs helps compress veins and move blood into the central circulation.) The activation of the RAAS on acute standing, however, can confound clinical measurements of plasma renin activity. Therefore, such measurements are obtained with the patient in the supine position.

None of the neural and humoral mechanisms that are used by the body to control blood pressure in the short term seems to be involved in the long-term setting of mean arterial pressure. It appears that the kidney is responsible for setting the absolute level of mean arterial pressure, about which the neural/humoral mechanisms described above try to control on a moment-to-moment basis. The setting of mean arterial pressure and its long-term control depend on salt and water excretion by the kidneys. Although the excretion of salt and water by the kidneys is regulated by some of the neural and hormonal mechanisms mentioned earlier in this chapter, it is also regulated by arterial pressure. Increased arterial pressure results in increased excretion of salt and water. This phenomenon is known as **pressure diuresis** (Fig. 17.7). As long as mean arterial pressure is elevated, salt and water excretion will exceed the normal rate because of pressure diuresis. Importantly, pressure diuresis persists until it lowers blood volume and cardiac output sufficiently to return mean arterial pressure to its original set level. A decrease in mean arterial pressure has the opposite effect on salt and water excretion; reduced pressure diuresis increases blood volume and cardiac output until mean arterial pressure is returned to its original set level.

Pressure diuresis is a slow but persistent mechanism for regulating arterial pressure. Because it persists in altering salt and water excretion and blood volume as long as arterial pressure is above or below a set level, it will eventually return pressure to that level. In hypertensive patients, salt and water excretion are normal at a higher arterial pressure. If this were not the case, pressure diuresis would inexorably bring arterial pressure back to normal.

CIRCULATORY SHOCK

Circulatory shock is a condition of generalized cardiovascular failure characterized by insufficient organ blood flow that is often accompanied by hypotension. This condition can result in deterioration of all tissues in the body and eventual death. The basic cause of shock is a loss of support of cardiac output. This may arise from direct cardiac dysfunction caused by myocardial ischemia, infarction, arrhythmias, and so on or from diminished venous filling pressure resulting from loss of whole blood, plasma volume, or venous tone.

Shock is divided into three stages of increasing severity.

Shock can be divided into three, progressively more serious stages, as diagrammed in Figure 17.8. The mildest form of shock is called *nonprogressive* or **compensated shock** because the normal cardiovascular regulatory mechanisms will compensate for the initial decrease in cardiac output and/or arterial pressure. These mechanisms will eventually lead to recovery of the person without the need for physician intervention. Donation of a unit of blood is a common form of compensated shock.

The compensatory mechanisms in nonprogressive shock are the same as those activated by an acute decrease in blood pressure and include activation of baroreceptor and other neurogenic pressor reflexes, stimulation of the renin–angiotensin system, and release of AVP. These reflexes and hormones tend to increase blood pressure and cardiac output by increasing heart rate, myocardial contractility, and vascular resistance (especially in skin, splanchnic organs, skeletal muscle, and the kidney), while also promoting renal Na^+ and H_2O retention to increase central venous pressure in support of stroke volume. In addition, low arterial pressure and increased

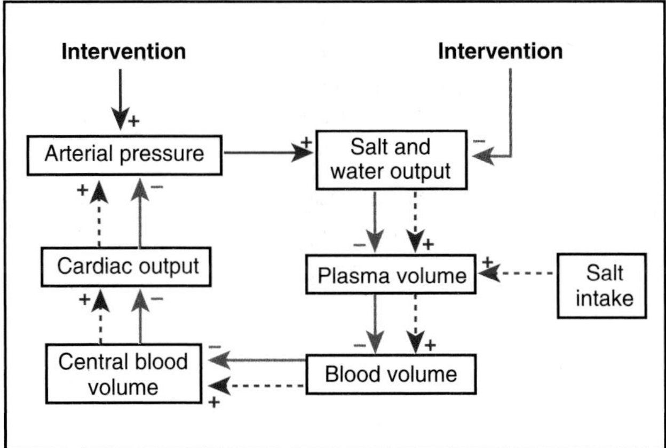

Figure 17.7 Regulation of arterial pressure by pressure diuresis. A higher output of salt and water in response to increased arterial pressure reduces blood volume. Blood volume is reduced until pressure returns to its normal level. The curve on the left shows the relationship in a person with normal blood pressure. A primary reduction in salt and water output (salt and water retention) or a primary increase in salt intake results in an elevation in arterial pressure. *Red (+) arrows* signify positive effects. *Blue (–) arrows* signify negative effects.

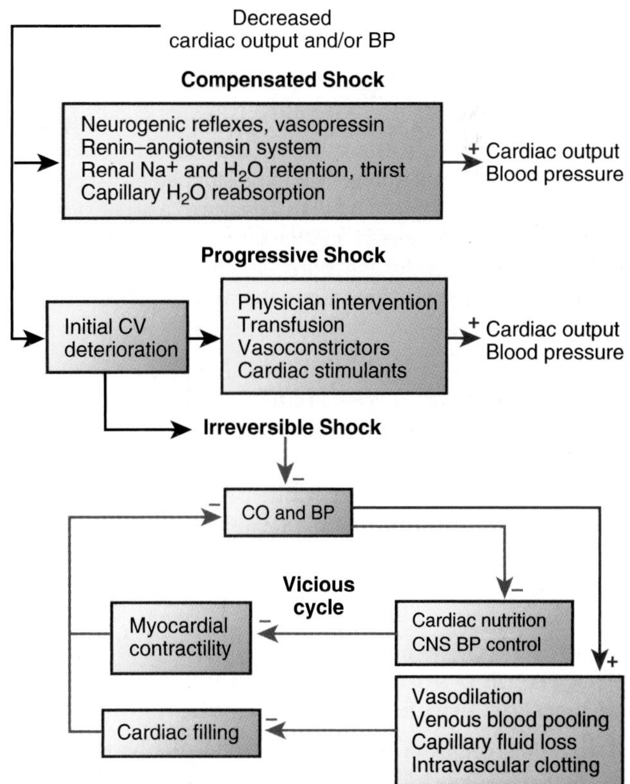

Figure 17.8 Mechanisms of circulatory shock. (See text for details.) *Red (+) arrows* signify positive effects. *Blue (–) arrows* signify negative effects. BP, blood pressure; CNS, central nervous system; CO, cardiac output; CV, cardiovascular.

Clinical Focus / 17.2 

The Evolving Understanding of the Potential Benefits of Estrogen in Cardiovascular Disease

The physiologic processes presented in most medical physiology textbooks are somewhat generic in what is represented. With the exception of reproduction, such presentations concern the basic physiology of the healthy adult male. It has become increasingly recognized that, even when reproductive differences are excluded, the physiology of adult women cannot be considered in all ways similar to men.

It is widely known that the incidence, morbidity, and mortality associated with cardiovascular disease are much lower in premenopausal women than in men of similar age. After menopause, however, whether surgically induced or from natural processes, women rapidly catch up to men such that by age 60 to 70 their incidence and mortality of cardiovascular disease is as great or greater than it is for men. This basic observation has led many to the assumption that estrogen is cardioprotective in women and has led to numerous studies and clinical trials to investigate this possibility. A persistent distressing issue in this regard, however, concerns an apparent disconnect between supportive evidence for beneficial effects of estrogen that have arisen from experimental laboratory experiments and the lack of such effects, or even exacerbation of disease phenomena, in clinical trials.

Estrogen is a coronary vasodilatory. Although much of this effect is a nonspecific response to the phenolic or steroid chemical structure of estrogen at high doses, there is some evidence for receptor-mediated vasorelaxation of smooth muscle in response to estrogen. More importantly, exposure of vascular tissue to physiologic concentrations of estrogen appears to induce production of NO and enhance cell mechanisms that inactivate oxygen radicals. Such actions of estrogen seem perfectly positioned to provide benefits to a person in combating the effects of virtually any cardiovascular disease. Nevertheless, clinical trials testing the effectiveness of hormone replacement therapy on cardiovascular outcomes in postmenopausal women had to be stopped early because of both a lack of demonstrated efficacy and a slight increase in thrombosis-associated myocardial infarctions.

An explanation for the extreme difference in suggested benefits of estrogen based on laboratory findings and the reality of clinical trials can be rooted in the concept that curatives and preventatives are not the same thing. It is likely that estrogen in premenopausal women protects their cardiovascular system for all the reasons suggested by laboratory studies. However, after menopause, this preventative aspect is lost. Cardiovascular disease then proceeds unabated in a manner similar to that seen in men. Recent laboratory studies support this postulate at least as it relates to the age of an individual receiving estrogen. For example, protective effects of estrogen on inflammatory responses and mitochondrial oxygen radical production are diminished in older versus younger animals. Women live decades past menopause, and during that time, many do not take any hormonal replacement therapy. Women in this state likely have accumulated cardiovascular disease that cannot be reversed by the simple addition of estrogen into the woman's system, even though such estrogen prevented the progression of the disease earlier in her life. Indeed, estrogen is slightly prothrombotic, and addition of this steroid to a woman with existing cardiovascular disease, especially coronary artery disease, might be enough to tip the balance into a thrombotic state that causes myocardial infarction. It is interesting that the average age of women was about 65 years in recent clinical trials that were stopped because of adverse cardiovascular events associated with estrogen and hormone replacement therapy and included women several years or decades past menopause. Thus, the lack of benefit from estrogen at this stage may have resulted from the buildup of cardiovascular insults prior to taking estrogen rather than from a direct lack of effect of estrogen on the cardiovascular system. Still, additional evidences involving a growing understanding of genomics and metabolomics suggest that not all women may have the genetic/metabolic infrastructure that supports beneficial effects of estradiol on the cardiovascular system. Such women may benefit from more targeted application of specific different types of estrogens, even while their systems are not able to respond favorably to others.

arterial resistance in the compensated stage of shock reduce capillary hydrostatic pressure. This augments fluid reabsorption from the interstitial fluid, especially in the intestine and kidney. Collectively, these compensatory mechanisms result in the initial clinical presentation of shock, which includes pale, cold skin, a rapid pulse, a sensation of thirst, hypotension, and reduced urine output.

Progressive shock causes a vicious cycle of cardiac and brain deterioration.

If the initial causes of shock are severe or the body is unable to compensate fully for the malfunctioning cardiovascular system, shock enters into a vicious positive-feedback cycle known as **progressive shock**. This occurs when organs maintaining the cardiovascular system, notably the heart and brain, deteriorate as a result of poor blood supply. This usually ensues when blood pressure decreases to a level below the limit of autoregulation of blood flow in the heart and brain, in spite of the efforts of short-term neural/hormonal control mechanisms. At this point, the body's own compensatory mechanisms cannot correct progressive shock and a person in this stage cannot recover without cardiovascular support intervention.

The most important factor involved in the progression of shock is deterioration of the heart itself. During progressive shock, the coronary circulation is compromised and myocardial contractility progressively decreases, which further causes a drop in arterial pressure and further reduction of coronary blood supply. In addition, progressive shock is exacerbated by deterioration of vasomotor centers in the brain that are needed to support the heart and increase vascular resistance during hypotension. Furthermore, acidotic conditions in peripheral organs brought on by poor oxygen delivery as well as release of toxins from deteriorating tissues, promote arterial blood clotting, thus exacerbating the already reduced tissue oxygen delivery. Acidosis and toxins also increase capillary permeability, which enhances fluid loss from the vascular compartment.

Without proper intervention, progressive shock will enter into an irreversible phase. In **irreversible shock**, cardiac and cerebral function are so compromised that no intervention is able to restore normal cardiovascular function. Death is inevitable. In general, severe shock is characterized by decreased body temperature resulting from the effects of poor tissue oxygen delivery on tissue metabolism, depressed mental function or unconsciousness, renal failure, and generalized muscle weakness.

Shock can arise from cardiac, vascular, volume, and brain malfunctions.

Several different conditions can lead to circulatory shock. In addition to blood volume losses resulting from hemorrhage, any loss of plasma volume, or hypovolemia, can reduce cardiac output and induce shock. This can occur from dehydration as a result of severe vomiting, sweating, diarrhea, decreased fluid and electrolyte intake, kidney damage, or adrenal cortical destruction, which results in the loss of the hormone aldosterone, which promotes Na^+ and H_2O reabsorption in the kidney. Severe burns result in capillary destruction with loss of albumin from the vascular space. This causes transcapillary loss of plasma. A similar capillary fluid loss is associated with intestinal obstructions that greatly increase intestinal venous pressure. Capillary damage from severe physical injury or trauma also causes transcapillary loss of plasma. An important complication of shock resulting from plasma loss is an increase in blood viscosity, which further impairs the heart's ability to move blood through the peripheral circulation.

Neurogenic shock is a form of circulatory collapse brought about by loss of neurogenic tone to veins and arteries, secondary to inhibition or dysfunction of the CNS. This can result from general or spinal anesthesia (i.e., spinal blocks used in childbirth), traumatic brain injury, or depressed vasomotor center function resulting from fever, stress, insomnia, or even severe emotional distress (the last resulting in emotional fainting).

Many people are severely allergic to certain antigens (bee venom, certain foods, etc.). In a serious antigen:antibody reaction, called **anaphylaxis**, tremendous amounts of histamine and other toxins are released into the tissue spaces. This results in destruction of surrounding tissue that increases interstitial osmolality, thus promoting fluid efflux from the capillaries. Histamine also causes arteriolar dilation while increasing the permeability of capillaries and venules to water and proteins. These factors further exacerbate fluid loss from the vascular compartment and result in a decrease in circulating blood volume. A characteristic of anaphylactic shock is that it is not caused by a loss of volume but rather from translocation of volume out of the vascular space into the interstitium,

Septic shock, formerly called "blood poisoning," is a form of shock caused by disseminated infection throughout the body. Next to cardiogenic shock, this is the leading cause of death from shock in the United States. It is often brought on by peritonitis secondary to female reproductive organ infection, rupture of the gut during appendicitis, or the introduction of skin bacteria into the bloodstream. Unlike other forms of shock, septic shock is characterized by high body temperature as a result of infection and high cardiac output as a result of intense vasodilation in the infected area. This excess cardiac output, however, represents useless blood flow to infected regions at the expense of perfusion of other organ systems. Bacterial products stimulate the production of NO by the vascular tissue itself, producing the intense vasodilation associated with septic shock. This NO arises from bacterial induction, from endotoxins and lipopolysaccharides, of an inducible form of NO synthase (**iNOS**). In addition, tissue destruction in this form of shock produces widespread microvascular blood clots, called *disseminated intervascular coagulation*. This condition greatly impairs O_2 delivery to tissues, thus exacerbating the primary shock condition.

From Bench to Bedside / 17.1

Novel Approaches to the Dietary Prevention of Cardiovascular Disease

Much of modern medicine is directed at detecting, diagnosing, and fixing problems after they occur. In the cardiovascular system, this includes angioplasty for coronary artery disease, antihypertensive drugs for hypertension, rehabilitation therapy following heart attacks, and all the diagnostic techniques and evaluations that are used by physicians to correctly diagnose and treat disease. However, more attention is now being directed at ways of preventing cardiovascular disease before it occurs.

It has been known for some time that people who consume diets rich in fruits and vegetables have less cardiovascular disease than those whose diet is lacking in these foods. Although there are numerous hypotheses to explain this observation, some of these postulate that some components of fruits and vegetables are beneficial for cardiovascular health. Unfortunately, much of the evidence for "food as a cure" claims is anecdotal, and these claims have flooded the public forum.

Some scientifically sound studies have been performed to examine food components that, theoretically, might have a benefit in cardiovascular disease. Most of these have centered on examination of antioxidant vitamins such as vitamins C and E as well as β-carotene. None of these studies has proven that such antioxidants provide any cardiovascular benefit, in spite of the well-established connection between diet and cardiovascular health. For this reason, it has been proposed that some other component of food, previously disregarded as nonnutritional, might be the source of cardiovascular protection in the diet. In this regard, recent studies have focused on several polyphenolic compounds belonging to the broad class of compounds called *flavonoids* as candidates. A special subclass of this group, called *anthocyanins*, seems to have particularly useful characteristics that would be supportive of cardiovascular health.

Anthocyanins are a group of glycosylated polyphenolic compounds that make up the intense red, blue, and purple pigments of fruits. They exist naturally in highest concentration in chokeberry, bilberry, elderberry, concord grapes, and blueberries, although the commercially available fruit sources are not as rich in anthocyanins as those now obtainable at farmer's markets. Anthocyanins are one of the few natural polyphenolics that can be absorbed intact from the digestive tract. These polyphenolics are the most potent scavengers of oxygen radicals yet found in dietary foodstuffs, with efficacy on a per gram basis far in excess of that seen for vitamin C, vitamin E, and β-carotene combined. More importantly, at concentrations that are seen in plasma after modest consumption, anthocyanins protect arteries and endothelial cells from oxidant injury and at slightly higher levels can directly stimulate the production of NO in arteries. Their protective effects are likely related to the combination of their powerful oxygen radical–scavenging capacity coupled with the fact that they incorporate into the membrane and cytosol of vascular endothelial cells. Should these effects of anthocyanins translate into long-term effects in the human cardiovascular system, their potential for alleviating the diseases associated with all forms of cardiovascular disease is enormous.

Although the anthocyanin class of polyphenols is known for strong antioxidant protective effects in vascular and other tissues, these polyphenolics have long been known to *appear* to be chemically unstable at physiologic pH. Anthocyanin monomers disappear quickly in solutions of neutral pH, leading many to wonder how these compounds are able to confer antioxidant protective effects to living tissues. Recent experiments, however, have shown that anthocyanins do not necessarily degrade at neutral pH but instead from polymers and addition products out of the single anthocyanin molecules. The resulting polymers are more stable than the native monopolyphenol and are antioxidants in their own right. Of significance to the potential health benefits of these compounds is recent evidence that shows that the rate of anthocyanin polymerization is enhanced by exposure of the monomers to biologic fluids such as plasma. This polymerization property may underlie protective properties of anthocyanins on the arterial NO system that appear to last beyond the half-life of the monomer in solution.

In spite of evidence indicating the potential value of polyphenolics such as anthocyanins in cardiovascular disease, actual clinical investigations designed to establish cause-and-effect relationships between polyphenol consumption, antioxidant effects, and cardiovascular benefits have been scarce. Recent controlled clinical studies have shown however that flow-mediated dilation (FMD) responses (an indicator of endothelial NO functional capacity) are enhanced following consumption of purple grape juice and black tea. Purple grapes are rich in anthocyanins, and black tea contains large amounts of tannins and epicatechins, which are antioxidant polyphenols. Additional studies comparing high and low flavonol consumption show a definitive link between high flavonol content and enhanced FMD. Still other studies have provided evidence that polyphenols may be more than just radical scavengers in that they appear affect transcription of factors involved with inflammation and cell growth.

Although evidence continues to accumulate showing the beneficial effects of antioxidant properties of polyphenolics in biologic systems, it should be noted that these systems are often assessed in the healthy state. Cardiovascular disease states, in contrast, are associated with an oxygen radical burden far above the levels seen in healthy tissue. Consequently, the need to test the efficacy of polyphenolics in disease states, in which the oxygen radical burden is high, must be carried out before true cardiovascular benefits can be ascribed to consumption of these compounds.

Chapter Summary

- The sympathetic nervous system acts on the heart primarily via β-adrenergic receptors.
- The parasympathetic nervous system acts on the heart via muscarinic cholinergic receptors.
- The sympathetic nervous system acts on blood vessels primarily via α-adrenergic receptors.
- Control of heart rate is predominantly by the parasympathetic nervous system, whereas control of contractility of the heart and vascular tone is predominantly by the sympathetic nervous system.
- Reflex control of the blood pressure involves neurogenic mechanisms that control heart rate, stroke volume, and systemic vascular resistance.
- Behaviors involved in emotion and stress affect cardiovascular responses.
- Baroreceptors and cardiopulmonary receptors are important in the moment-to-moment neural reflex regulation of arterial pressure.
- The renin–angiotensin–aldosterone system, arginine vasopressin, and atrial natriuretic peptide are important in the longer-term hormonal regulation of blood volume and arterial pressure.
- Pressure diuresis in the kidney is the mechanism that ultimately adjusts arterial pressure to a set level.
- The defense of arterial pressure during standing involves the integration of neural and humoral mechanisms.
- Circulatory shock occurs in the stages of compensation, progression, and irreversibility.
- Compromised function of the heart and brain as well as abnormalities in the function of blood vessels and blood volume control mechanisms, can cause shock.

18 Ventilation and the Mechanics of Breathing

ACTIVE LEARNING OBJECTIVES

Upon mastering the material in this chapter, you should be able to:

- Explain how a pleural pressure is generated.
- Describe how transairway pressure maintains airway patency.
- Explain how changes in alveolar pressure move air in and out of the lungs.
- Explain how spirometry measures lung volumes and airflow in patients.
- Explain why alveolar ventilation measures the amount of fresh air that enters the lung.
- Describe how expired carbon dioxide can be used to measure alveolar ventilation.
- Explain how alveolar ventilation influences blood arterial P_{CO_2}.
- Explain how lung elastic recoil affects lung compliance.
- Explain how regional lung compliance affects airflow in the lung.
- Explain how surfactant stabilizes alveoli at low lung volumes.
- Explain how airways become compressed during forced explanation.
- Predict how restrictive and obstructive lung disorders will affect the work of breathing.

Breathing is essential to life. The inrush of air at birth sets off a series of events that allows the newborn to progress from a dependent, placental life-support system to an independent air-breathing system. A breath in, a breath out, 12 to 15 times every minute may seem like a simple process to build the entire human respiratory system, which brings in, on an average, 7 L (~1.85 gal) of air per minute into the lungs. At rest, breathing 7 L of air per minute supplies enough oxygen to sustain the metabolism of trillions of cells in the body. However, this simplicity is deceptive because breathing is amazingly responsive to small changes in blood chemistry, mood, level of alertness, and body activity. The study of respiratory physiology is important to medicine, because many of the respiratory diseases (e.g., **cystic fibrosis**, **asthma**, **emphysema**, **pulmonary hypertension**, and **pneumonia**) impact many of the subspecialties, from pediatrics, to internal medicine, to surgery, and to geriatrics.

The process of taking up oxygen and removing carbon dioxide from cells in the body is known as **respiration**. Respiration takes place in two stages. The first stage is known as **gas exchange**, and the second as **cellular respiration**. Gas exchange occurs at two levels. The first level involves the transfer of oxygen and carbon dioxide between the atmosphere and the lungs. The second level involves the exchange of oxygen and carbon dioxide and occurs between the systemic blood and the metabolically active tissue. The movement of oxygen and carbon dioxide in and out of cells occurs by simple diffusion.

The human lungs are so efficiently designed that gas exchange can increase >20-fold to remove carbon dioxide and to supply oxygen to tissues in order to meet the body's energy demands. The gas exchange process rarely limits our activity. For example, a marathon runner who staggers across the 26-mile finish line in <3 hours or someone who swims the English Channel in record time is rarely limited by the amount of oxygen taken up by the lungs. These examples of human activity not only underscore the functional capacity of the lungs but also illustrate the important role respiration plays in our extraordinary adaptability to our environment. The second stage of respiration is known as cellular respiration, a series of complex metabolic reactions that break down molecules of food, releasing carbon dioxide and energy. Recall that oxygen is required in the final step of cellular respiration to serve as an electron acceptor in the process by which cells obtain energy.

The functions of the respiratory system can be divided into the following: ventilation and the mechanics of breathing, gas transfer and transport, pulmonary circulation, and the control of breathing. This chapter discusses ventilation, the mechanics of breathing, and the work of breathing. Chapter 19 discusses gas uptake and transport. Chapter 20 discusses the pulmonary circulation and the matching of airflow with blood flow in the lungs. Chapter 21 deals with basic breathing rhythms, breathing reflexes, integrated control of breathing, and the control of breathing in unusual environments.

LUNG STRUCTURAL AND FUNCTIONAL RELATIONSHIPS

Although the lungs are clearly an essential component to the process of breathing, they do not provide the entire picture. The lungs alone cannot bring air in and out of the lungs or exchange oxygen and carbon dioxide from the blood. For example, without muscles and an airtight chest wall to create a negative pressure within the chest, the lungs would be nonfunctional gas exchangers. Moreover, without blood flow going to the lungs and matching blood and airflow, there would be essentially no gas exchange.

Airway tree divides repeatedly to increase lung surface area for gas exchange.

The human gas exchange organ consists of two lungs, each divided into several lobes. The lungs comprise two tree-like structures, the vascular tree and the airway tree, which are embedded in highly elastic connective tissue. The vascular tree consists of arteries and veins connected by capillaries (see Chapter 20, "Pulmonary Circulation and Ventilation/Perfusion"). The airway tree consists of a series of hollow branching tubes that decrease in diameter at each branching (Fig. 18.1A). The main airway, the **trachea**, branches into two **bronchi**. Each bronchus enters a lung and branches many times into progressively smaller bronchi, which, in turn, form **bronchioles**.

A functional model of the airway tree is presented in Figure 18.1B. The trachea and the first 16 generations of airway branches make up the **conducting zone**. The trachea, bronchi, and bronchioles of the conducting zone have three important functions: (1) to warm and humidify inspired air; (2) to distribute air evenly to the deeper parts of the lungs; and (3) to serve as part of the body's defense system (removal of dust, bacteria, and noxious gases from the lungs). The first four generations of the conducting zone are subjected to changes in negative and positive pressures and contain a considerable amount of cartilage to prevent airway collapse. In the trachea and main bronchi, the cartilage consists of U-shaped rings. Further down, in the lobar and segmental bronchi, the cartilaginous rings give way to small plates of cartilage. In the bronchioles, the cartilage disappears altogether. The smallest airways in the conducting zone are the terminal bronchioles. Bronchioles are suspended by elastic tissue in the lung parenchyma, and the elasticity of the lung tissue helps keep these airways open. The conducting zone has its own separate circulation, the **bronchial circulation**, which originates from the descending aorta and drains into the pulmonary veins. No gas exchange occurs in the conducting zone.

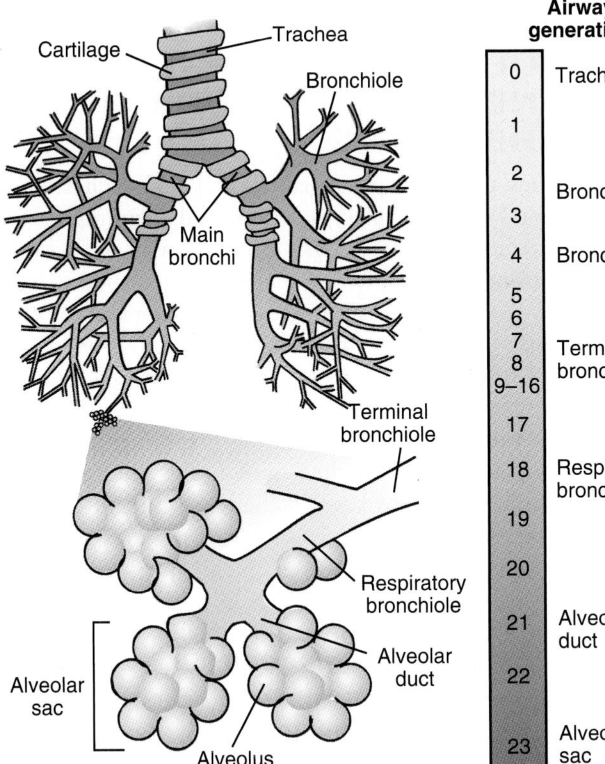

Figure 18.1 Airway tree divides repeatedly. The airway tree consists of a series of highly branched hollow tubes that become narrower, shorter, and more numerous as they penetrate the deeper parts of the lung. The airways are divided into two functional zones with the first 16 generations of branches comprising the conducting zone and functioning to conduct air to the deeper parts of the lungs. The last seven generations participate in gas exchange and comprise the respiratory zone.

The last seven generations of airways make up the respiratory zone, the site of gas exchange. The exchange of gases is accomplished in the mosaic of millions of specialized cells that form thin-walled air sacs called **alveoli.** Like the conducting zone, the respiratory zone has its own separate and distinct circulation, the pulmonary circulation. The lungs have the most extensive capillary network of any organ in the body. Pulmonary capillaries occupy 70% to 80% of the alveolar surface area. The pulmonary circulation receives all of the cardiac output and, therefore, blood flow is high. One pulmonary arterial branch accompanies each airway and branches with it. Red blood cells can pass through the pulmonary capillaries in <1 second.

The formation of outpockets from the small airways to form alveoli accomplishes an increase in internal surface area (see Fig. 18.1). As mentioned above, a network of capillaries surrounds each alveolus and brings blood into close proximity with air inside the alveolus. Oxygen and carbon dioxide move across the thin-walled alveolus by diffusion. Adult lungs contain 300 to 500 million alveoli, with a combined internal surface area of approximately 75 m^2, approximately the size of a tennis court. This represents one of the largest biologic membranes in the body. During growth, alveolar surface area increases in two ways: in alveolar number and in diameter, as seen in Table 18.1. However, after adolescence, alveoli only increase in size and, if damaged, have limited ability to repair themselves. Cigarette smoke, for example, can destroy alveoli and lead to concomitant decrease in alveolar surface area for gas exchange.

Vascular and airway trees merge to form a blood–gas interface for gas diffusion.

In the respiratory zone, a group of alveolar ducts and their alveoli merge with pulmonary capillaries to form a terminal respiratory unit; there are approximately 60,000 of these units in both lungs. The alveolar–capillary membrane of these units forms a blood–gas interface that separates the blood in the pulmonary capillaries from the gas in the alveoli (Fig. 18.2). The alveolar–capillary barrier serves several functions. It provides a large interface (surface area) for the diffusion of oxygen and carbon dioxide and also exists to prevent both air bubbles from forming in the blood and blood from entering the alveoli. It is exceedingly thin (in some places <0.5 μm) and is composed of alveolar epithelium, interstitial fluid layer, and capillary endothelium. Air is brought to one side of the interface by ventilation—the movement of air to and from the alveoli. Blood is brought to the other side of the interface by the pulmonary circulation. As the blood perfuses the alveolar capillaries, oxygen is taken up and carbon dioxide crosses the blood–gas interface by diffusion.

TABLE 18.1 Age-Related Changes in Alveolar Number and Surface Area in the Human Lung

Age	Number of Alveoli (10^6)	Alveolar Surface Area (m^2)	Skin Surface Area (m^2)
Birth	24	2.8	0.2
8 y	300	32.0	0.9
Adult	300	75.0	1.8

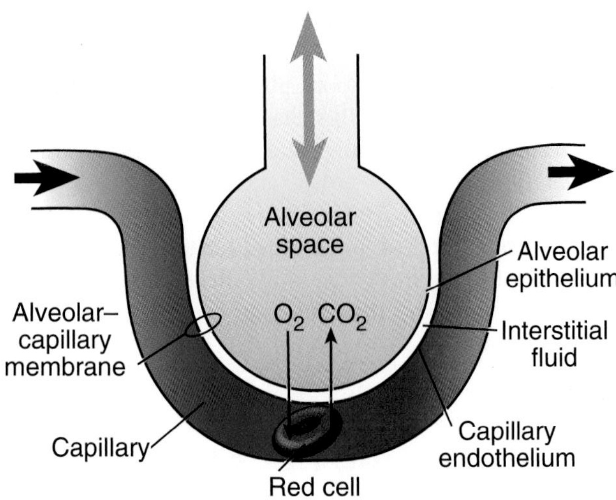

Figure 18.2 Blood–gas interface is the site of gas exchange. The pulmonary capillaries and alveoli form a blood–gas interface. *Thick arrows* indicate direction of blood flow and ventilation, and *thin arrows* indicate the diffusion paths for O_2 and CO_2. The alveolar–capillary membrane is thin (~0.5 μm), and therefore, the diffusion distance for gases is small.

PULMONARY PRESSURES AND AIRFLOW DURING BREATHING

The ventilatory pump consists of the lungs, chest wall, intercostal muscles, pleural lining, and diaphragm. The lungs are housed in an airtight thoracic cavity and are separated from the abdomen by a large dome-shaped skeletal muscle, the **diaphragm** (Fig. 18.3). The thoracic cavity is made up of 12 pairs of ribs, the sternum, and internal and external intercostal muscles, which lie between the ribs. The rib cage is hinged to the vertebral column, allowing it to be raised and lowered during breathing. The space between the lungs and chest wall is the **pleural space,** which contains a thin layer of fluid (~10 μm thick), that functions, in part, as a lubricant so the lungs can slide against the chest wall.

Diaphragm expands the thoracic cavity and is the main muscle of breathing.

Breathing is largely driven by the muscular diaphragm at the bottom of the thorax. Contraction of the diaphragm enlarges the cavity in which the lung is enclosed, increasing volume

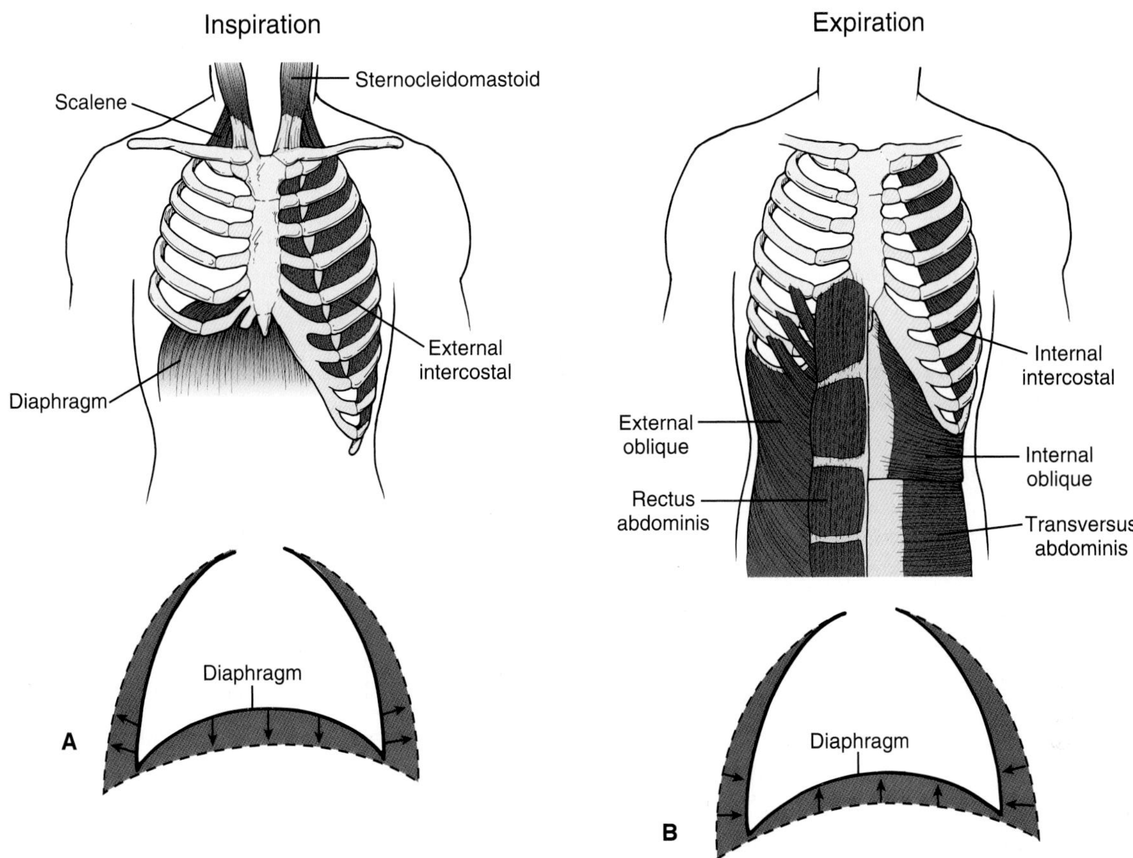

Figure 18.3 Changes in thoracic volume during the breathing cycle. Movements of the diaphragm and rib cage change thoracic volume, which allows the lungs to inflate during inspiration and deflate during expiration. **(A)** At rest, during inspiration, the diaphragm contracts and pushes the abdominal contents downward. The downward movement also pushes the rib cage outward. With deep and heavy breathing, the accessory muscles (the external intercostals and sternocleidomastoids) also contract and pull the rib cage upward and outward. **(B)** Expiration is passive during resting conditions. The diaphragm relaxes and returns to its dome shape, and the rib cage is lowered. During forced expiration, however, the internal intercostal muscles contract and pull the rib cage downward and inward. The abdominal muscles also contract and help pull the rib cage downward, compressing thoracic volume.

and, therefore, decreasing pressure, causing air to flow into the airways (see Fig. 18.3). This enlargement of the airtight thoracic cavity is accomplished in two ways. First, when the diaphragm (which is attached to the lower ribs and sternum) contracts, the abdominal contents are pushed downward, enlarging the thoracic cavity in the vertical plane. Second, when the diaphragm descends and pushes down on the abdominal contents, it also pushes the rib cage outward, further enlarging the thoracic cavity. The effectiveness of the diaphragm in changing thoracic volume is related to the strength of its contraction and its dome-shaped configuration when relaxed. With a normal breath, the diaphragm moves only about 1 to 2 cm, but with forced inspiration, a total excursion of 10 to 12 cm can occur. Obesity, pregnancy, and tight clothing around the abdominal wall can impede the effectiveness of the diaphragm in enlarging the thoracic cavity. Damage to the phrenic nerves (the diaphragm is innervated by two phrenic nerves, one to each lateral half) can lead to paralysis of the diaphragm. When a phrenic nerve is damaged, that portion of the diaphragm moves up rather than down during inspiration.

During forced inspiration, in which a large volume of air is taken in, the accessory muscles are also used. These include the external intercostal muscles, which raise the anterior end of the rib cage, causing the rib cage to be pulled upward and outward. To see how this works, stand sideways in front of a mirror and take a deep breath. The chest wall moves upward while the sternum moves outward, thereby enlarging the thoracic cavity. The **scalene muscles** of the neck and the **sternocleidomastoids**, which are inserted into the top of the sternum, are brought into play during deep and heavy breathing, such as that in exhaustive exercise, and are used to elevate the upper rib cage to further increase the thoracic volume.

During normal resting ventilation, breathing out is much simpler; expiration is passive, and no muscles are contracted. At the end of a normal inspiration, the diaphragm relaxes, the rib cage drops, the thoracic volume decreases, pressure around the lungs increases, and they deflate. However, with exercise or forced expiration, the expiratory muscles become active. These muscles include not only the diaphragm, but also those of the abdominal wall and the internal intercostal muscles (see Fig. 18.3B). Contraction of the abdominal wall pushes the diaphragm upward into the chest, and the internal intercostal muscles pull the rib cage down, reducing thoracic volume. These accessory respiratory muscles are necessary for such functions as coughing, straining, vomiting, and defecating. The expiratory muscles are extremely important in endurance running and are one of the reasons competitive long-distance runners, as part of their training program, often do exercises to strengthen their abdominal and chest muscles.

Partial pressure drives the diffusion of oxygen and carbon dioxide.

At this point, the ventilatory pump has been examined at the systems level. We have reviewed the components, some properties of lung and the chest wall, and the interaction between these structures during inspiration and expiration. A more complete understanding of how the lungs are inflated/deflated and air is inhaled/exhaled, however, requires knowledge of the pressure changes at the biophysical level. A brief review of the gas laws is in order before explaining how changes in pressures within the thoracic cavity and lungs cause changes in lung volumes and air movement.

The atmosphere, the air we breathe and live in, exerts a pressure (P) known as **barometric pressure** (P_B). At sea level, P_B is equal to 760 mm Hg. The relationship between the total pressure exerted by a mixture of gases and the pressure of individual gases is governed by the **Dalton law**, which states that the total barometric pressure (P_B) is equal to the sum of the **partial pressures** of the individual gases. Each gas has a partial pressure, which is the pressure the gas would exert if each gas were present alone in the volume occupied by the whole mixture at the same temperature. Dalton law can be written as follows:

$$P_B = P_{N_2} + P_{O_2} + P_{H_2O} + P_{CO_2} \tag{1}$$

where P_{N_2} equals partial pressure of nitrogen, P_{O_2} equals partial pressure of oxygen, P_{H_2O} equals partial pressure of water vapor, and P_{CO_2} equals partial pressure of carbon dioxide.

Therefore, the partial pressure of oxygen (P_{O_2}), according to the Dalton law, is determined as $P_{O_2} = P_B \times F_{O_2}$, where F_{O_2} is the fractional concentration of oxygen. Because 21% of air is made up of oxygen, the partial pressure (P_{O_2}) exerted by oxygen is 160 mm Hg (760 × 0.21) at sea level. If all of the other gases in a container of air were removed, the remaining oxygen would still exert a pressure of 160 mm Hg. Partial pressure of a gas is often referred to as **gas tension**, and partial pressure and gas tension are used synonymously.

When air is inspired, it is warmed and humidified. The inspired air becomes saturated with water vapor at 37°C. The water vapor exerts a partial pressure that is a function of body temperature, not barometric pressure. At 37°C, water vapor exerts a partial pressure (P_{H_2O}) of 47 mm Hg. Water vapor pressure does not change the percentage of oxygen or nitrogen in a dry gas mixture; however, water vapor does lower the partial pressure of oxygen inside the lungs. The partial pressures of gases in the lungs are calculated on the basis of a dry gas pressure; therefore, water vapor pressure is subtracted when the partial pressure of a gas is determined. The dry gas pressure in the trachea is 760 – 47 = 713 mm Hg, and the individual partial pressures of O_2 and N_2 are

$$P_{O_2} = 0.21 \times (760 - 47) = 150 \text{ mm Hg} \tag{2}$$

and

$$P_{N_2} = 0.79 \times (760 - 47) = 563 \text{ mm Hg} \tag{3}$$

Table 18.2 lists normal partial pressures of respiratory gases in different locations in the body. When calculating gas tensions in the lung, a good way to remember is as soon as the air hits the nose, always subtract water vapor pressure when converting gas fraction to partial pressure.

In respiratory physiology, pressures are measured both in mm Hg and centimeters of water (cm H_2O). Gas tensions are measured in mm Hg, but airflow and respiratory pressures are so small that they are measured in cm H_2O. A pressure of 1 cm H_2O is equal to 0.74 mm Hg (or 1 mm Hg = 1.36 cm H_2O). Changes in respiratory pressures during breathing are often expressed as *relative pressure*, a pressure relative to

TABLE 18.2 Partial Pressures and Percentages of Respiratory Gases at Sea Level (P_B = 760 mm Hg)

Gas	Ambient Dry Air		Moist Tracheal Air		Alveolar Air		Systemic Arterial Blood	Mixed Venous Blood
	mm Hg	%	mm Hg	%	mm Hg	%	mm Hg	mm Hg
O_2	160	21	150	20	102	14	95	40
CO_2	0	0	0	0	40	5	40	46
Water vapor	0	0	47	6	47	6	47	47
N_2	600	79	563	74	571	75[a]	571	571
Total	760	100	760	100	760	100	760	704[b]

[a]Alveolar P_{N_2} increased by 1% because R < 1.
[b]Total pressure in venous blood is reduced because P_{O_2} decreases more than P_{CO_2} increases.

atmospheric pressure. For example, the pressure inside the alveoli can be −2 cm H_2O during inspiration. The minus sign indicates that the pressure is subatmospheric, that is, −2 cm below PB. Conversely, during expiration, the pressure inside the alveoli can be +3 cm H_2O. This means that the pressure is 3 cm H_2O above PB. A positive or negative pressure indicates that the pressure is relative to atmospheric pressure and is, respectively, above or below PB. When relative pressures are used, it is important to remember that PB is set at zero. If airway pressure is zero, the pressure inside the airway equals atmospheric pressure. Unless otherwise specified, the pressures of breathing are relative pressures and the unit is cm H_2O. A list of symbols and abbreviations used in respiratory physiology is shown in Table 18.3.

Pleural pressure is critical for lung inflation and deflation.

Because the thoracic cavity is airtight, an increase in thoracic volume causes the **pleural pressure (P_{pl})**, the pressure in the pleural fluid between the lung and chest wall, to fall. A decrease in P_{pl} causes the lungs to expand and fill with air. This key pressure–volume relationship in breathing is based on two gas laws. The **Boyle law** states that, at a constant temperature, the pressure (P) of the gas varies inversely with the volume (V) of gas, or P = 1/V. If either pressure or volume changes and if temperature remains constant, the product of pressure and volume remains constant:

$$P_1V_1 = P_2V_2 \quad (4)$$

The **Charles law** states that if pressure is constant, the volume of a gas and its temperature vary proportionately, or V ≈ T. If either temperature or volume changes and pressure remains constant, then

$$V_1/T_1 = V_2/T_2 \quad (5)$$

These two gas laws can be combined into the **general gas law**:

$$P_1V_1/T_1 = P_2V_2/T_2 \quad (6)$$

From the general gas law, at constant temperature, an increase in thoracic volume leads to a decrease in pleural pressure.

Transpulmonary and transairway pressures prevent lung and airway collapse.

In addition to pleural pressure, several other pressures are associated with breathing and airflow (Fig. 18.4). **Alveolar pressure** (PA) is the pressure inside the alveoli. **Transmural pressure** (P_{tm}) is the pressure difference across a wall. In respiration, transmural pressure is the pressure across the airway or across the lung wall or the alveolar wall. Two major transmural pressures are involved in breathing. First, is the pressure difference across the lung wall termed **transpulmonary pressure** (PL). Transpulmonary pressure describes the difference between the alveolar pressure and the pleural pressure in the lungs and is measured by subtracting pleural

TABLE 18.3 Symbols and Terminology Used in Respiratory Physiology

Symbol	Term
Primary	
C	Compliance
D	Diffusion
F	Fractional concentration of a gas
f	Frequency
P	Pressure or partial pressure
$\dot{Q}$	Volume of blood per unit time (blood flow or perfusion)
R	Resistance
S	Saturation
T	Time
V	Gas volume
$\dot{V}$	Volume of gas per unit time (airflow)
Secondary	
A	Alveolar
a	Arterial
Aw	Airway
B	Barometric
D	Dead space
E	Expiratory
I	Inspiratory
L	Lung
c'	Pulmonary end-capillary
pl	Pleural
pw	Pulmonary wedge
s	Shunt
T	Tidal
Tp	Transpulmonary
v	Venous
Examples of combinations	
CL	Lung compliance
DLCO	Lung diffusing capacity for carbon monoxide

(Continued)

TABLE 18.3 Symbols and Terminology Used in Respiratory Physiology (*Continued*)

Symbol	Term
Examples of Combinations	
F_{IO_2}	Fractional concentration of inspired O_2
P_B	Barometric pressure
P_{CO_2}	Partial pressure of carbon dioxide
P_A	Alveolar pressure
P_{O_2}	Partial pressure of oxygen
Pa_{CO_2}	Partial pressure of carbon dioxide in arterial blood
P_{ACO_2}	Partial pressure of carbon dioxide in alveoli
P_{IO_2}	Partial pressure of inspired O_2
P_{ECO_2}	Partial pressure of CO_2 in expired gas
$P_{AO_2} - Pa_{O_2}$	Alveolar–arterial difference in partial pressure of O_2
P_{pl}	Pleural pressure
R_{aw}	Airway resistance
Sa_{O_2}	Saturation of hemoglobin with oxygen in O_2 in arterial blood
T_I	Inspiratory time
T_E	Expiratory time
$\dot{V}_A$	Alveolar ventilation
$\dot{V}_A\dot{Q}$	Alveolar ventilation/perfusion ratio
V_D	Dead space volume
$\dot{V}_D$	Dead space ventilation
$\dot{V}_E$	Expired minute ventilation
$\dot{V}_{O_2}$	Oxygen consumption per minute

Note: A dot above a primary symbol denotes flow per unit time.

pressure from alveolar pressure ($P_L = P_A - P_{pl}$). At rest, pleural pressure is –5 cm H_2O (P_{pl} = –5 cm H_2O) and alveolar pressure is zero (P_A = 0 cm H_2O). This means, for example, that transpulmonary pressure at rest is 5 cm H_2O [$P_L = 0 - (-5) = 5$ cm H_2O]. It is important to remember that transpulmonary pressure is the pressure that keeps the lungs inflated and prevents the lungs from collapsing. The more positive it becomes, the more the lungs are distended or inflated. In the example above, at rest, the P_L is 5 cm H_2O. An increase in P_L is responsible for inflating the lungs above the resting volume. The second transmural pressure is **transairway pressure** (P_{ta}), the pressure difference across the airways ($P_{ta} = P_{aw} - P_{pl}$), where P_{aw} is the pressure inside the airway. Transairway pressure is important in keeping the airways open during forced expiration. One way to remember how to calculate transairway or transpulmonary pressure is “in minus out,” where P_{pl} is always the pressure outside the lung or airway.

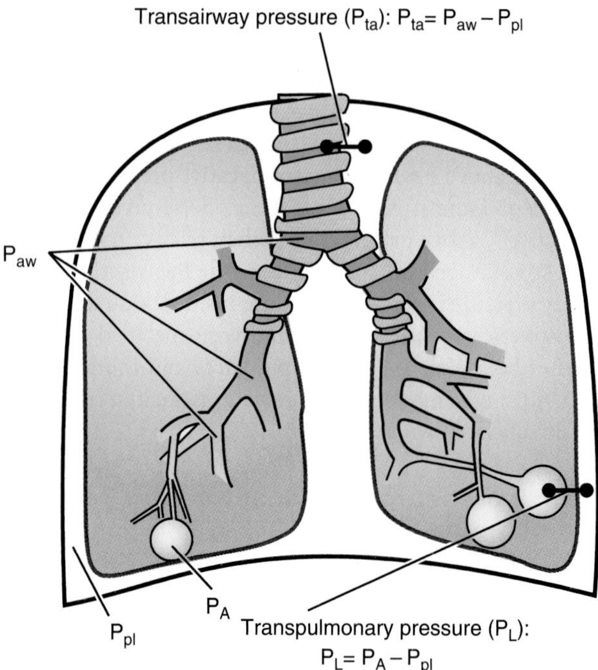

Figure 18.4 Pressures change during breathing. Several important pressures are involved in breathing: airway pressure (P_{aw}), alveolar pressure (P_A), pleural pressure (P_{pl}), transpulmonary pressure (P_L), and transairway pressure (P_{ta}). Both transpulmonary pressure and transairway pressure can be defined as the pressure inside minus the pressure outside. In both cases, the pressure outside is pleural pressure (P_{pl}).

Why is pleural pressure negative or subatmospheric? The reason is because of the elastic recoil of the lungs and chest wall—that is, their capability of being stretched and then recoiling to an unstretched configuration, like a spring. At the end of a normal expiration, the lungs and chest wall are stretched in equal but opposite directions (Fig. 18.5). The stretched lungs have the potential to recoil inwardly, and the stretched chest wall has the potential to recoil outwardly. These equal but opposing forces cause the pleural pressure to decrease below atmospheric pressure. Pleural pressure is negative or subatmospheric during quiet breathing and becomes more negative with deep inspiration. Only during forced expiration does pleural pressure become positive or rise above atmospheric pressure.

The importance of pleural pressure is seen when the chest wall is punctured (see Fig. 18.5B) and air enters into the pleural space. The stretched lung collapses immediately (recoils inwardly), and the rib cage simultaneously expands outwardly (recoils outwardly). Because the normal pleural pressure is subatmospheric, air will rush into the pleural

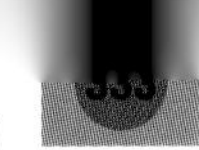

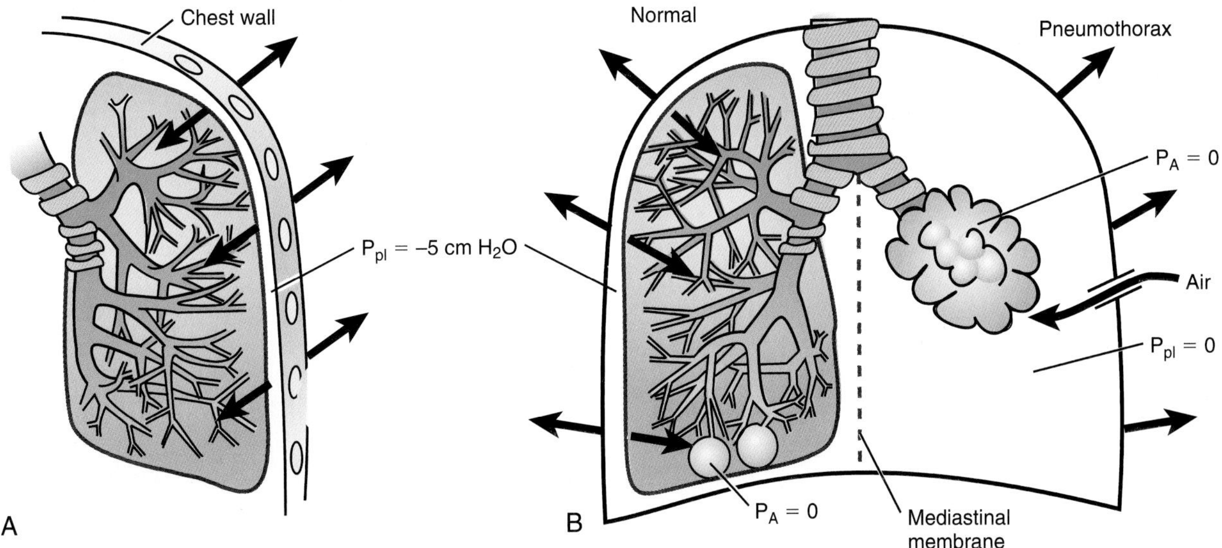

Figure 18.5 A negative pleural pressure results from the elastic recoil of the lungs and chest wall pulling in opposite directions. (A) The stretched lung (at the end of a normal expiration) tends to recoil inwardly, and the chest wall tends to recoil outwardly, but in equal and opposite directions. Consequently, pleural pressure (P_{pl}) becomes negative (i.e., less than atmospheric pressure). **(B)** Rupture or puncture of the lung or chest wall results in a pneumothorax, during which the transpulmonary pressure becomes zero and elastic recoil causes the lung to collapse. The mediastinal membrane prevents the other lung from collapsing. P_A, alveolar pressure.

space any time the chest wall or lung is punctured, and the pleural pressure will become equal to atmospheric pressure because air moves from regions of high to low pressure. In this situation, transpulmonary pressure is zero (P_L = 0) because the pressure difference across the lung is eliminated. This condition, in which air or gas accumulates in the pleural space and the lung collapses, is known as **pneumothorax** (see Fig. 18.5B, *right side*). A pneumothorax occurs with a knife or gunshot wound in which the chest wall is punctured or when the lung ruptures from an abscess or severe coughing. In the treatment of some lung disorders (e.g., tuberculosis), a pneumothorax is purposely created by inserting a sterile needle between the ribs and injecting nitrogen into the pleural fluid to rest the diseased lung. It is important to note that the mediastinal membrane keeps the other lung from collapsing.

Changes in alveolar pressure move air in and out of the lungs.

Pressure changes during a normal breathing cycle are illustrated in Figure 18.6. At the end of expiration, the respiratory muscles are relaxed and there is no airflow. At this point, alveolar pressure is zero (equal to atmospheric pressure or P_B). Pleural pressure is −5 cm H_2O, and transpulmonary pressure is, therefore, 5 cm H_2O [$P_{pl} = 0 - (-5 \text{ cm } H_2O) = 5 \text{ cm } H_2O$].

Inflation of the lungs is initiated by contraction of the diaphragm. If inspiration is started from the end of a maximal expiration, the chest wall can be felt to expand during inhalation. At no time, as our lungs fill, do we feel the need to close our epiglottis to keep the air in. This is because only a slight pressure difference holds air in our lungs. In the example shown in Figure 18.6A, pleural pressure goes from −5 to −8 cm H_2O. One of the basic characteristics of gases, such as air, is that the pressures between two regions tend to equilibrate. Therefore, when pleural pressure decreases, transpulmonary pressure increases, and the lungs inflate. Inflation of the lungs causes the alveolar diameter to increase and alveolar pressure to decrease below atmospheric pressure (see Fig. 18.6B). This produces a pressure difference between the mouth and alveoli, which causes air to rush into the alveoli. Airflow stops at the end of inspiration because alveolar pressure again equals atmospheric pressure (see Fig. 18.6C). The sequence of events is summarized in Figure 18.7.

During expiration, the inspiratory muscles relax, the rib cage drops, pleural pressure becomes less negative, transpulmonary pressure decreases, and the stretched lungs deflate. When alveolar diameter decreases during deflation, alveolar pressure becomes greater than atmospheric pressure and pushes air out of the lungs. Airflow out of the lungs occurs until alveolar pressure equals atmospheric pressure.

SPIROMETRY AND LUNG VOLUMES

Spirometry is the most common of the pulmonary function tests and specifically measures the amount (volume) and the speed (flow) of air that can be inhaled and exhaled from the lungs (Fig. 18.8). A spirometer is a volume recorder consisting of a double-walled cylinder in which an inverted

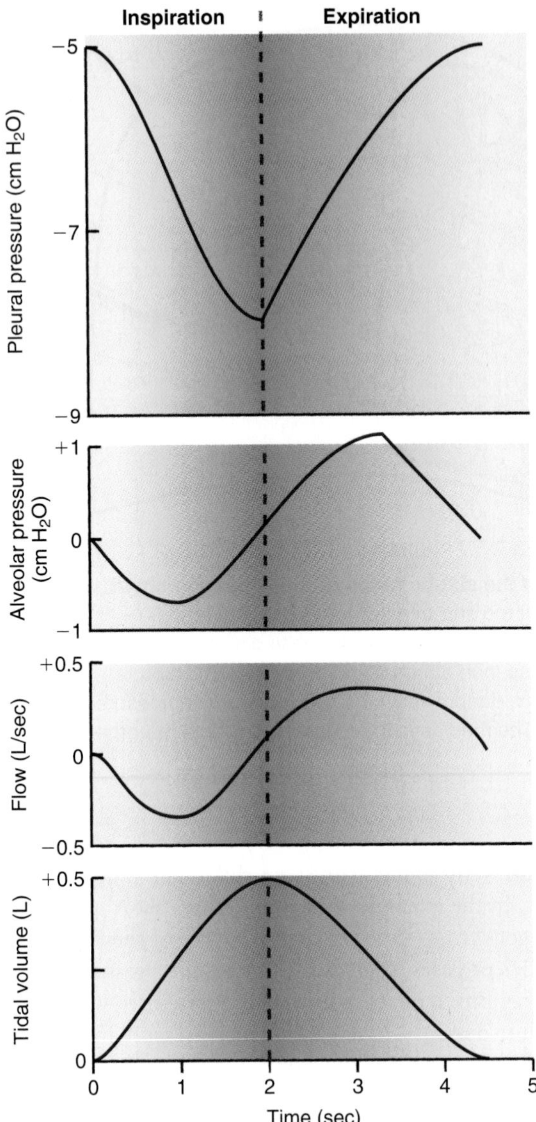

Figure 18.6 Airflow resistance affects expiratory time. The inspiratory time (T_I) is 2 seconds and is less than the expiratory time (T_E) of 3 seconds. This difference is a result of, in part, a higher airflow resistance during expiration, as is reflected by a higher alveolar pressure (P_A) change during expiration (1.2 cm H_2O) than during inspiration (0.8 cm H_2O). An increase in airway resistance will decrease the T_I/T_E ratio. Only a small pressure change between the mouth and alveoli is required for a normal tidal volume.

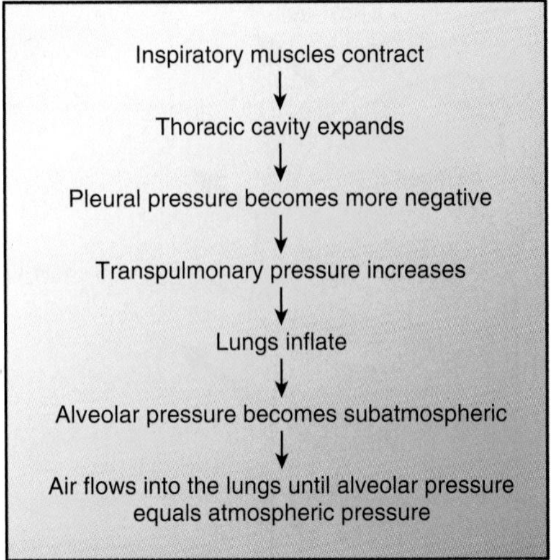

Figure 18.7 Airflow parallels alveolar pressure during breathing. The sequence during inspiration results in a fall in alveolar pressure causing air to flow into the lungs.

bell is immersed in water to form a seal. A pulley attaches the bell to a marker that writes on a rotating drum. When air enters the spirometer from the lungs, the bell rises. Because of the pulley arrangement, the marker is lowered. Therefore, a downward deflection represents expiration and an upward deflection represents inspiration. The recording is known as a **spirogram**. The slope of the spirogram on the moving drum measures rate of airflow, and the amplitude of the deflection measures the volume of air. Volume is plotted on the vertical axis (*y* axis), and time is plotted on the horizontal axis (*x* axis).

Spirometry measures patient lung volumes and airflow.

The volume of air leaving the lungs during a single breath is called **tidal volume** (V_T). Under resting conditions, V_T is approximately 500 mL and represents only a fraction of the air in the lungs. The maximum amount of air in the lungs at the end of a maximal inhalation is the **total lung capacity (TLC)** and is approximately 6 L in an adult man. Another important spirometry measurement is the **functional residual capacity (FRC)**, the volume of air remaining in the lungs at the end of a normal tidal volume (end of expiration). Note the use of "volume" in the first term and "capacity" in the next two. *Volume* is used when only one volume is involved, and *capacity* is used when a volume can be broken down into two or more smaller volumes; for example, FRC equals **expiratory reserve volume (ERV)** plus **residual volume (RV)**. The various lung volumes and capacities are summarized in Table 18.4.

Forced vital capacity assesses ventilatory function.

The maximum volume of air that can be exhaled after a maximum inspiration is **vital capacity (VC)**. When expiration is performed as rapidly and as forcibly as possible into a spirometer, this volume is called **forced vital capacity (FVC)** and is about 5 L in an adult man (see Fig. 18.8). VC and FVC are the same volume. VC is determined by the sum of ERV, tidal volume, and inspiratory reserve volume. FVC is a direct volume measurement from spirometry and is one of the most useful measurements to assess ventilatory function of the lungs. To measure FVC, the person inspires maximally and then exhales into the spirometer as forcefully, rapidly, and completely as possible.

Two additional measurements can be obtained from the FVC spirogram (Fig. 18.9). One is **forced expiratory**

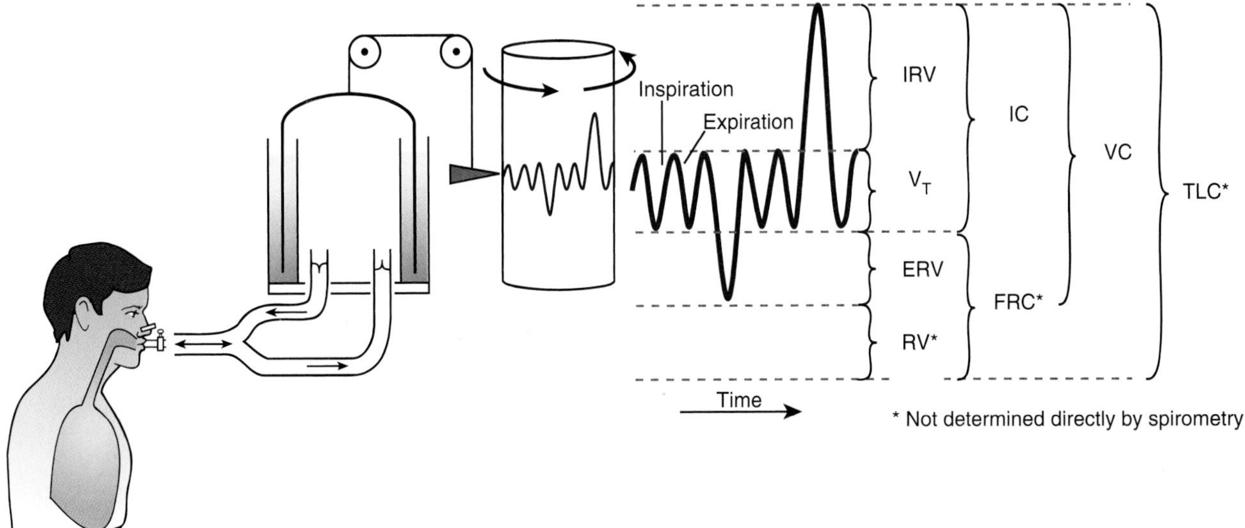

Figure 18.8 Spirometry measures lung volume. With expiration, the marker records a downward deflection. Note that residual volume (RV), functional residual capacity (FRC), and total lung capacity (TLC) cannot be measured directly by spirometry. IRV, inspiratory reserve volume; V_T, tidal volume; ERV, expiratory reserve volume; IC, inspiratory capacity; VC, vital capacity.

volume of air exhaled in 1 second (FEV_1). This volume has the least variability of the measurements obtained from a forced expiratory maneuver and is considered one of the most reliable spirometry measurements. Another useful way of expressing FEV_1 is as a percentage of FVC (i.e., $FEV_1/FVC \times 100$), which corrects for differences in lung size. Normally, the FEV_1/FVC ratio is 0.8, which means that 80% of a person's FVC can be exhaled in the first second of FVC. This is rather remarkable because small pressure changes are involved to move this volume of air. FVC and FEV_1 are important measurements in the diagnosis of certain types of lung diseases.

A second measurement obtained from the FVC spirogram is forced expiratory flow (FEF_{25-75}); it has the greatest sensitivity in terms of detecting early airflow obstruction (see Fig. 18.9). This measurement represents the expiratory flow rate over the middle half of the FVC (between 25% and 75%). FEF_{25-75} is obtained by identifying the 25% and 75% volume points of the FVC and then measuring the volume and time between these two points. The calculated flow rate is expressed in liters per second.

Residual lung volume cannot be measured directly by spirometry.

Because the lungs cannot be emptied completely following forced expiration, neither RV nor FRC can be measured directly by simple spirometry. Instead, they are measured indirectly using a dilution technique involving helium, an inert and relatively insoluble gas that is not readily taken up by blood in the lungs. The subject is connected to a spirometer filled with 10% helium in oxygen (Fig. 18.10). The lungs initially contain no helium. After the subject rebreathes the helium–oxygen mixture and equilibrates with the spirometer, the helium concentration in the lungs will become the same as in the spirometer. From the conservation of mass principle, we can write

$$C_1 \times V_1 = C_2(V_1 + V_2) \tag{7}$$

where C_1 equals the initial concentration of helium in the spirometer, V_1 equals the initial volume of helium–oxygen mixture in the spirometer, C_2 equals helium concentration after equilibration, and V_2 equals unknown volume in the lungs.

$$V_2 = \frac{V_1(C_1 - C_2)}{C_2} \tag{8}$$

Starting the test at precisely the right time is important. If the test begins at the end of a normal tidal volume (end of expiration), the volume of air remaining in the lungs represents FRC. If the test begins at the end of an FVC, then the test will measure RV. Similarly, if the test starts after a maximal inspiration, then V_2 would equal TLC. In practice, carbon dioxide is absorbed and oxygen is added to the spirometer to make up for the oxygen consumed by the person during the test. Although the helium-dilution technique is an excellent test for the measurement of FRC and RV in healthy people, it has a major limitation in patients whose lungs are poorly ventilated because of plugged airways or high airway resistance. In these diseased lungs, helium gives a falsely low FRC value.

An entirely different way to measure FRC that overcomes the problem of trapped gas in the lungs is the use of the **body plethysmograph** or *body box* (Fig. 18.11). The person is seated comfortably in an airtight box in which changes in pressure and volume can be measured accurately. The subject takes a breath against a closed mouthpiece starting at FRC. During inspiration against a closed mouthpiece, the pressure in the lungs decreases, the chest volume increases, and the surrounding volume increases. An expiratory effort against a closed mouthpiece produces just the opposite effect:

TABLE 18.4 Abbreviations and Definitions Used in Pulmonary Function

Abbreviation	Term	Definition	Normal Value[a]
EPP	Equal pressure point	The point at which the pressure inside the airway equals the pressure outside of the airway (i.e., P_{pl}).	
ERV	Expiratory reserve volume	The maximum volume of air exhaled at the end of the tidal volume.	1.2 L
FEF_{25-75}	Forced expiratory flow	The maximum midexpiratory flow rate, measured by drawing a line between points representing 25% and 75% of the forced vital capacity.	5 L/sec
FEV_1	Forced expiratory volume	The maximum volume of air forcibly exhaled in 1 s.	4.0 L
FEV_1/FVC%	Forced expired volume/ forced vital capacity ratio	The percentage of FVC forcibly exhaled in 1 s.	80 %
FRC	Functional residual capacity	The volume of air remaining in the lungs at the end of a normal tidal volume.	2.4 L
FVC	Forced vital capacity	The maximum volume of air forcibly exhaled after a maximum inhalation.	4.8 L
IC	Inspiratory capacity	The maximum volume of air inhaled after a normal expiration.	3.6 L
IRV	Inspiratory reserve volume	The maximum volume of air inhaled at the end of a normal inspiration.	3.1 L
PEF	Peak expiratory flow	The maximal expiratory flow during an FVC maneuver.	7.5 L/sec
RV	Residual volume	The volume of air remaining in the lungs after maximum expiration.	1.2 L
RV/TLC	Residual volume/total lung capacity ratio	The percentage of total lung capacity made up of residual volume.	20 %
TLC	Total lung capacity	The volume of air in the lungs at the end of a maximum inspiration.	6.0 L
VC	Vital capacity	The maximum volume of air that can be exhaled. (Note that the values for FVC and VC are the same.) VC is calculated from static lung volumes (VC = ERV + VT + IRV). FVC is determined from direct spirometry.	4.8 L
V_D/V_T	Dead space–tidal volume ratio	The fraction of tidal volume made up of dead space.	30 %
V_T	Tidal volume	The volume of air inhaled or exhaled with each breath.	0.5 %

[a]Values are for an average, healthy, young man.

the pressure inside the lung increases as the chest volume decreases. In Figure 18.11, the lung volume is determined by applying the Boyle law.

MINUTE VENTILATION

So far, ventilation has been described in terms of measurements of static lung volumes and forced lung volumes. Breathing is a dynamic process involving how much air is brought in and out of the lungs in a minute. If 500 mL of air is inspired with each breath (V_T) and the breathing rate (f) is 14 times a minute, then the total **minute ventilation** ($\dot{V}_A$) is the amount of air that enters the lungs each minute (500 × 14 = 7,000 mL/min or 7 L/min). Expired minute ventilation ($\dot{V}_E$) is calculated from the amount of expired air per minute and can be represented by the equation:

$$\dot{V}_E = V_T \times f \quad (9)$$

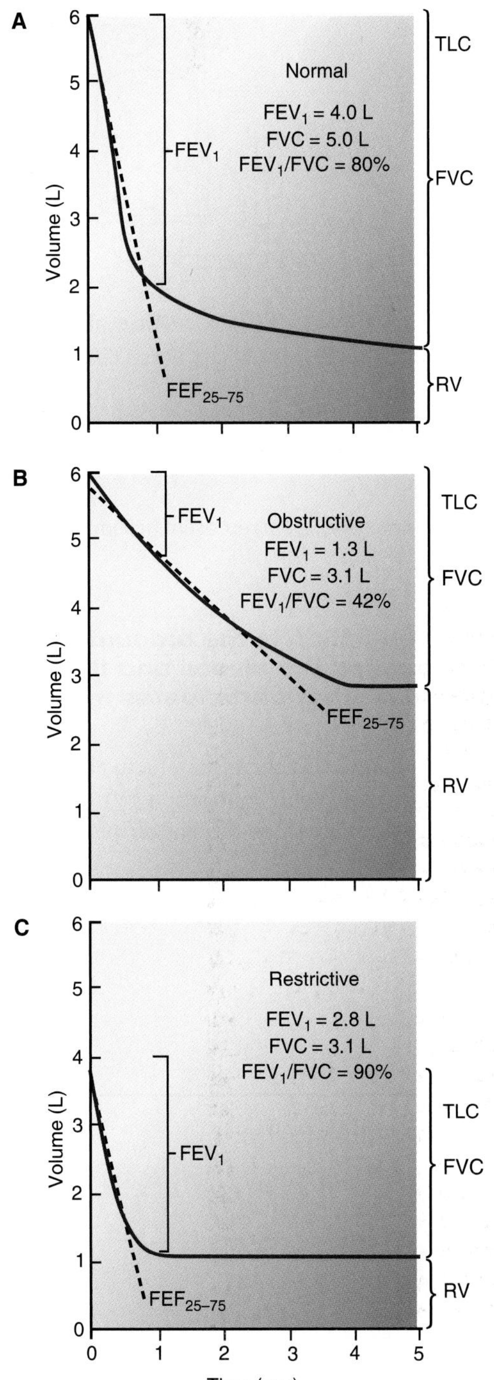

Figure 18.9 Forced vital capacity (FVC) is a useful measurement in assessing ventilatory function. (A) A healthy subject inspires maximally to total lung capacity and then exhales as forcefully and completely as possible into the spirometer. Two other measurements can be obtained from this maneuver: the forced expiratory volume in 1 second (FEV_1) and the flow rate over the middle half of the forced vital capacity (FEF_{25-75}). Measurements of FVC, FEV_1, FEV_1/FVC ratio, and FEF_{25-75} are used to detect obstructive and restrictive disorders. **(B)** In an obstructive disorder, expiratory flow rate is significantly decreased, and the FEV_1/FVC ratio is low. **(C)** In a restrictive disorder, lung inflation is decreased, resulting in reduced residual volume (RV) and total lung capacity (TLC). Although FVC is decreased, it is important to note that the FEV_1/FVC ratio is normal or increased in a restrictive disorder.

Minute ventilation and expired minute ventilation are the same, based on the assumption that the volume of air inhaled equals the volume exhaled. This is not quite true because more oxygen is consumed than carbon dioxide is produced. This difference, for all practical purposes, is ignored.

Not all inspired air reaches the alveoli and becomes wasted air.

The tidal volume is distributed between the conducting airways and alveoli. Because gas exchange occurs only in the alveoli and not in the conducting airways, a fraction of the minute ventilation is wasted air. For each 500 mL of air inhaled, approximately 150 mL remain in the conducting airways and are not involved in gas exchange (Fig. 18.12). This volume of wasted air is known as **dead space volume** (V_D). Because V_D is a result of the anatomy of the airways, the volume is often referred to as *anatomic* V_D.

Picture what occurs during a normal breathing cycle. A normal tidal volume of 500 mL is expired. During the next inspiration, another 500 mL is taken in, but the first 150 mL of air entering the alveoli is V_D (old alveolar gas left behind). Thus, only 350 mL of fresh air reaches the alveoli and 150 mL is left in the conducting airways. The normal ratio of dead space volume to tidal volume (V_D/V_T) is in the range of 0.25 to 0.35. In this example, the ratio (150:500) is 0.30, which means that 30% of the tidal volume or 30% of the minute ventilation does not participate in gas exchange and constitutes dead air.

Dead space air is not confined to the conducting airways alone. Any time gases in the alveoli do not participate in gas exchange; these gases also become part of the wasted air. For example, if inspired air is distributed to alveoli that have no blood flow, this constitutes dead space and is referred to as **alveolar dead space volume** (Fig. 18.13A). Alveolar dead space volume is not confined to alveoli without blood flow. Alveoli that have reduced blood flow exchange less inspired air than normal (see Fig. 18.13B); any portion of alveolar air in excess of that needed to maintain normal gas exchange constitutes alveolar dead space volume. Therefore, dead space volume (V_D) may be either anatomic or alveolar in nature. The sum of the two types of dead space is **physiologic dead space volume**. Therefore, physiologic V_D = anatomic V_D + alveolar V_D. Physiologic dead space can be measured using the following equation:

$$\frac{V_D}{V_T} = \frac{F_{ACO_2} - F_{ECO_2}}{F_{ACO_2}} \tag{10}$$

Normal values for the V_D/V_T ratio are 0.2 to 0.35. Once the ratio is determined, physiologic dead space can be computed. For example, if a patient has a tidal volume of 500 mL and the V_D/V_T ratio is 0.35, then the physiologic dead space is 175 mL (physiologic $V_D = 500 \times 0.35$). Basically, anatomic V_D, alveolar V_D, and physiologic V_D are terms that denote the volume of inspired gas that does not participate in gas exchange. In one case, there is a fraction

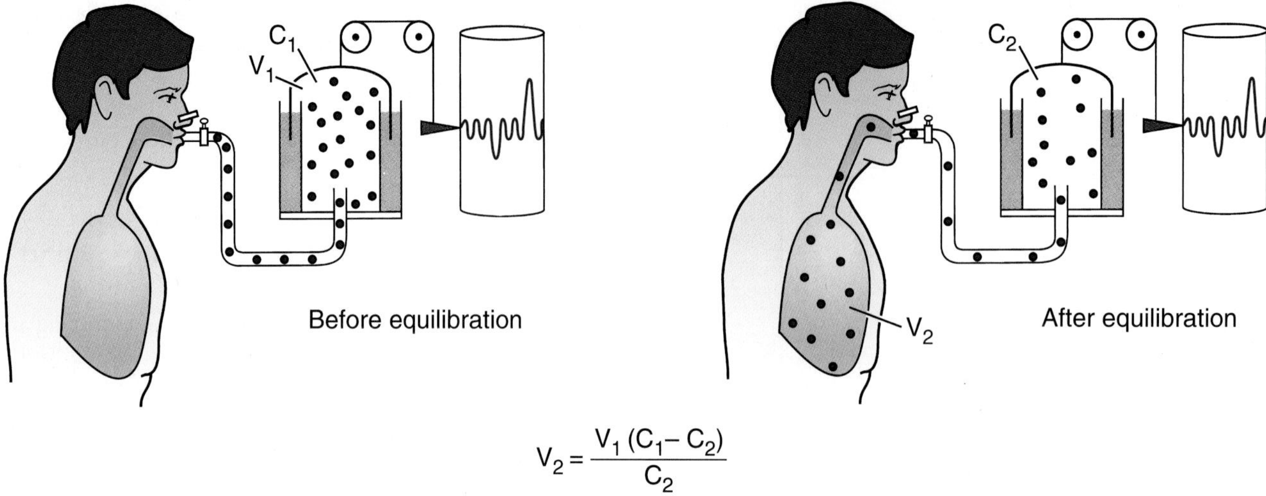

$$V_2 = \frac{V_1 (C_1 - C_2)}{C_2}$$

Figure 18.10 Helium dilution is used to measure residual volume (RV). *Dots* represent helium before and after equilibration. C, concentration; V, volume.

of VT that does not reach the alveoli (anatomic dead space volume). In another, a fraction of the VT reaches the alveoli, but there is reduced or no blood flow, leading to alveolar dead space volume. Physiologic dead space volume represents the sum of anatomic and alveolar dead space volumes. In healthy people, physiologic VD is approximately the same as anatomic dead space.

Alveolar ventilation is the amount of fresh air that reaches the alveoli and the fraction of inspired air that participates in gas exchange.

The volume of fresh air per minute actually reaching the alveoli is known as **alveolar ventilation** ($\dot{V}_A$). To determine how much fresh air reaches the alveoli per minute, dead

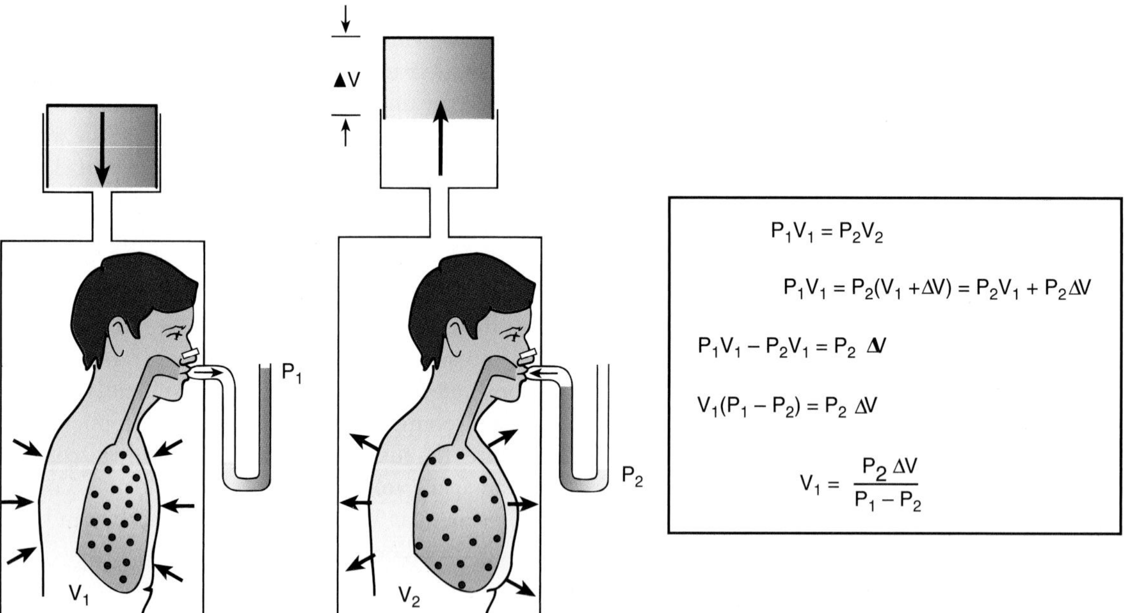

Figure 18.11 The body plethysmograph measures functional residual capacity (FRC). The subject sits in an airtight box and breathes air through a mouthpiece. The mouthpiece is closed at the end of a normal expiration, and the subject continues to try to breathe against a closed mouthpiece. P_1 is the lung pressure at the end of expiration and P_2 at the end of inspiration. ΔV is the volume change accompanying inspiration and expiration; it results from changes in thoracic volume against a closed mouthpiece. The body plethysmograph is used clinically for measuring FRC in patients with airway obstruction.

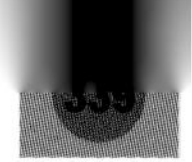

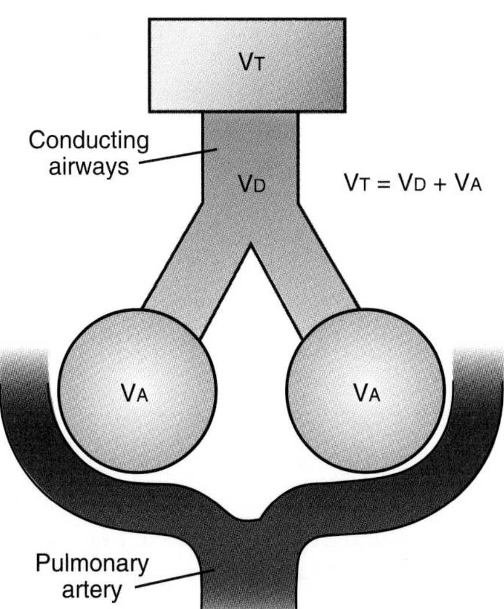

Figure 18.12 Some of the air in the tidal volume does not participate in gas exchange. Tidal volume (V_T) is represented by the *rectangle* and is the volume of air that will be drawn in during inspiration. V_T will be distributed between the conducting airways and alveoli. The volume of air in the conducting airways does not participate in gas exchange and constitutes dead space volume (V_D). V_A is the volume of fresh air added to the alveoli.

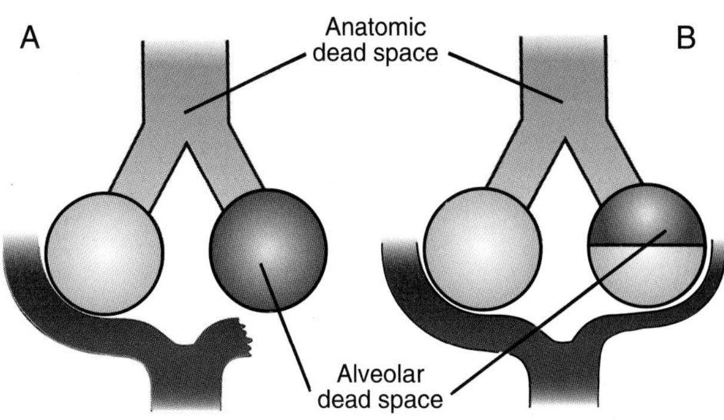

Figure 18.13 Total wasted air in the lungs is computed from the physiologic dead space volume. Dead space volume occurs in the conducting airways and in alveoli with poor capillary circulation. **(A)** There is no blood flow to an alveolar region. **(B)** There is reduced blood flow. In both cases, a portion of alveolar air does not participate in gas exchange and constitutes alveolar dead space volume. Note that physiologic dead space is the sum of alveolar dead space plus anatomic dead space.

space volume is subtracted from the tidal volume and the result is multiplied by breathing frequency, f, and can be represented by

$$\dot{V}_A = (V_T - V_D) \times f \qquad (11)$$

where V_T equals tidal volume, V_D equals dead space volume, and f equals breathing frequency. For example, if a person has a breathing rate of 14 breaths/min, a V_T of 500 mL, and a V_D of 150 mL, then the volume of air entering the alveoli is 4.9 L/min [(500 − 150 mL) × 14 = 4,900 mL/min]. Only alveolar ventilation represents the amount of fresh air reaching the alveoli, and it is the only air that participates in gas exchange. For instance, a swimmer using a snorkel breathes through a tube that increases dead space volume. Similarly, a patient connected to a mechanical ventilator also has increased dead

Clinical Focus / 18.1

Chronic Obstructive Pulmonary Disease

Chronic bronchitis, emphysema, and asthma are major pathophysiologic disorders of the lungs. Bronchitis is an inflammatory condition that affects one or more bronchi. Emphysema stems from loss of lung elastic recoil and overdistention of alveoli. Asthma is marked by spasmodic contractions of smooth muscle in the bronchi. Although the pathophysiology and etiology of chronic bronchitis, emphysema, and asthma are different, they all are classified collectively as **chronic obstructive pulmonary disease (COPD)**. The hallmark of COPD is decreased airflow during forced expiration, which leads to a decrease in **forced vital capacity (FVC)** and **forced expiratory volume in 1 second (FEV_1)**.

COPD is characterized by chronic obstruction of the small airways. Airflow can become obstructed in three ways: excessive mucus production (as in bronchitis), airway narrowing caused by bronchial spasms (as in asthma), and airway collapse during expiration (as in emphysema). In the last case, airway collapse stems from abnormally high compliance and concomitant loss of lung elastic recoil. In severe COPD, air is trapped in the lungs during forced expiration, leading to abnormally high residual volume.

Chronic bronchitis and emphysema often coexist. Bronchitis leads to excessive mucus production, which plugs the small airways. A cough is produced to clear the excess mucus from the airways; on repeated exposure to a bronchial irritant, such as tobacco smoke, a persistent cough develops. As a result, the alveoli become overdistended and can rupture from excessive pressure, especially in plugged airways.

Late in COPD, patients often experience hypoxemia (low blood oxygen) and hypercapnia (high blood carbon dioxide). These two conditions, especially hypoxemia, cause pulmonary arteries to constrict. Hypoxia-induced pulmonary vasoconstriction leads to pulmonary hypertension (high blood pressure in the pulmonary circulation), which causes right heart failure. Based on changes in blood oxygen levels, patients with COPD can be categorized into two types. Those exhibiting predominantly emphysema are referred to as “pink puffers” because their oxygen levels are usually satisfactory and their skin remains pink. They develop a puffing style of breathing. Those manifesting predominantly chronic bronchitis are called “blue bloaters” because low oxygen levels give their skin a blue cast and fluid retention from heart failure gives them a bloated appearance.

COPD is by far the most common chronic lung disease in the United States. More than 10 million Americans suffer from COPD, and it is the fifth leading cause of death nationwide. The single most common cause of chronic bronchitis and emphysema is tobacco smoke.

TABLE 18.5 Effect of Breathing Patterns on Alveolar Ventilation

Subject	Tidal Volume	×	Frequency	=	Minute Ventilation	–	Dead Space Ventilation[a]	=	Alveolar Ventilation
	(mL)		(breaths/min)		(mL/min)		(mL/min)		(mL/min)
A	150	×	40	=	6,000	–	150 × 40	=	0
B	500	×	12	=	6,000	–	150 × 12	=	4,200
C	1,000	×	6	=	6,000	–	150 × 6	=	5,100

[a]Dead space volume is 150 mL in all three subjects.

space volume. Indeed, if minute ventilation is held constant, then alveolar ventilation is decreased with snorkel breathing or with mechanical ventilation.

The significance of dead space volume, minute ventilation, and alveolar ventilation is shown in Table 18.5. Subject C's breathing is slow and deep, B's breathing is normal, and A's breathing is rapid and shallow. Note that each subject has the same minute ventilation (i.e., the total amount of expired air per minute), but each has marked differences in alveolar ventilation. Subject A has no alveolar ventilation and would die in a matter of minutes, whereas subject C has an alveolar ventilation greater than normal. The important lesson from the examples presented in Table 18.5 is that increasing the depth of breathing is far more effective in elevating alveolar ventilation than is increasing the frequency or rate of breathing. A good example is exercise because in most exercise situations, increased alveolar ventilation is accomplished by increases in the depth of breathing rather than in the rate. A well-trained athlete can often increase alveolar ventilation during moderate exercise with little or no increase in breathing frequency.

Alveolar ventilation in patients is derived by measuring the volume of expired carbon dioxide.

Alveolar ventilation is easy to calculate if dead space volume is known. However, dead space volume is not easily determined in a human subject or a patient. Often, dead space is approximated for a seated subject by assuming that dead space (in milliliters) is equal to the subject's weight in pounds (e.g., a subject who weighs 170 lb would have a dead space volume of 170 mL). This assumption is fairly reliable for healthy people but is not in patients with respiratory problems. Alveolar ventilation is calculated in the pulmonary function laboratory from the volume of expired carbon dioxide per minute and fractional concentration of carbon dioxide in the alveolar gas (Fig. 18.14). This pulmonary function test is based on the concept that (1) no gas exchange occurs in the conducting airways, (2) the inspired air contains essentially no carbon dioxide, and (3) all of the expired carbon dioxide originates from alveoli.

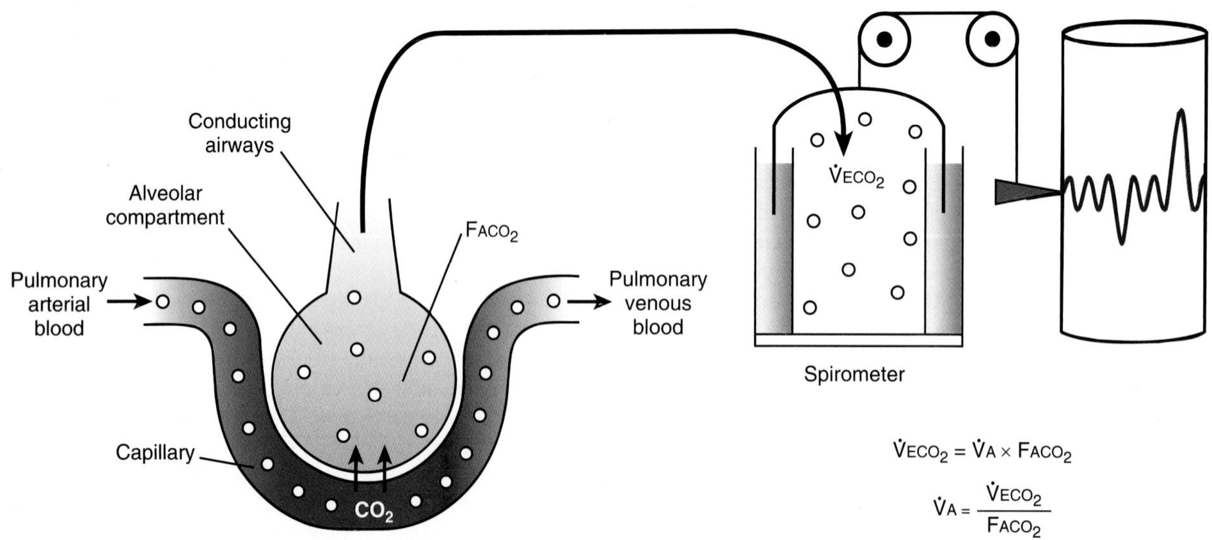

Figure 18.14 Expired carbon dioxide is used to calculate alveolar ventilation. Because all of the CO_2 (represented by *dots*) in the expired air originates from the alveoli, the fractional concentration of CO_2 in alveolar gas (F_{ACO_2}) can be obtained at the end of a tidal volume (often referred to as *end-tidal CO_2*). $\dot{V}_A$, alveolar ventilation; $\dot{V}_{ECO_2}$, volume of carbon dioxide expired per minute.

Therefore,

$$\dot{V}_{ECO_2} = \dot{V}_A \times F_{ACO_2} \quad (12)$$

where $\dot{V}_{ECO_2}$ equals the volume of carbon dioxide expired per minute and F_{ACO_2} equals the fractional concentration of carbon dioxide in alveolar gas. Rearranging yields the alveolar ventilation equation:

$$\dot{V}_A = \dot{V}_{ECO_2}/F_{ACO_2} \quad (13)$$

The carbon dioxide concentration in the alveoli can be obtained by sampling the last portion of the tidal volume during expiration (end-tidal volume), which contains alveolar gas. Alveolar ventilation can also be determined from the partial pressure of carbon dioxide in the alveoli (P_{ACO_2}) based on the fact that P_{ACO_2} is equal to F_{ACO_2} times total alveolar gas pressure. Equation 13 can be written as

$$\dot{V}_A(\text{L/min}) = \frac{\dot{V}_{ECO_2}(\text{mL/min}) \times 0.863}{P_{ACO_2}(\text{mm Hg})} \quad (14)$$

where 0.863 is a constant with dimensions mm Hg × L/mL. Carbon dioxide diffuses readily between the blood and alveoli, and as a result P_{ACO_2} is in equilibrium with the partial pressure of carbon dioxide in the arterial blood (Pa_{CO_2}). Recall that a capital "A" signifies alveolar gas and the lower case "a" denotes arterial gas. Because carbon dioxide is in equilibrium between the alveoli and the arterial blood, arterial carbon dioxide tension (P_{ACO_2}) can be used to calculate alveolar ventilation as follows:

$$\dot{V}_A = \frac{\dot{V}_{ECO_2} \times 0.863}{P_{ACO_2}} \quad (15)$$

From the above equation, it is important to recognize the inverse relationship between $\dot{V}_A$; and Pa_{CO_2} (Fig. 18.15). If alveolar ventilation is halved, alveolar P_{CO_2} will double (assuming a steady state and constant carbon dioxide production). This decrease in alveolar ventilation is called **hypoventilation**. Conversely, increased ventilation, referred to as **hyperventilation**, leads to a fall in blood Pa_{CO_2}. Clinically, the adequacy of alveolar ventilation is usually evaluated in terms of arterial P_{CO_2}. An increase in blood Pa_{CO_2} reflects hypoventilation and a decrease in blood Pa_{CO_2} reflects hyperventilation. It is important to make the distinction between *hyperventilation* and *hyperpnea*. Remember, hyperventilation is increased alveolar ventilation with a concomitant decrease in arterial P_{CO_2}. **Hyperpnea** is increased alveolar ventilation with *no change* in arterial P_{CO_2}. The reason for this physiologic response is that alveolar ventilation increases proportionally to carbon dioxide production. An example of hyperpnea is seen with exercise. With exercise, the production of CO_2 increases and offsets the increased CO_2 blown off with increased alveolar ventilation.

What happens to Pa_{O_2} when alveolar ventilation increases? When alveolar ventilation increases, Pa_{O_2} will also increase. However, doubling alveolar ventilation will not lead to a doubling of Pa_{O_2}. The quantitative relationship between alveolar ventilation and Pa_{O_2} is more complex than that for Pa_{CO_2}, for two reasons. First, the inspired P_{O_2} is obviously not zero.

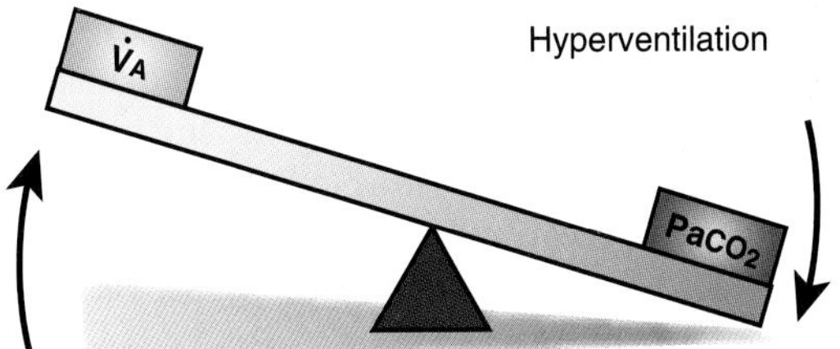

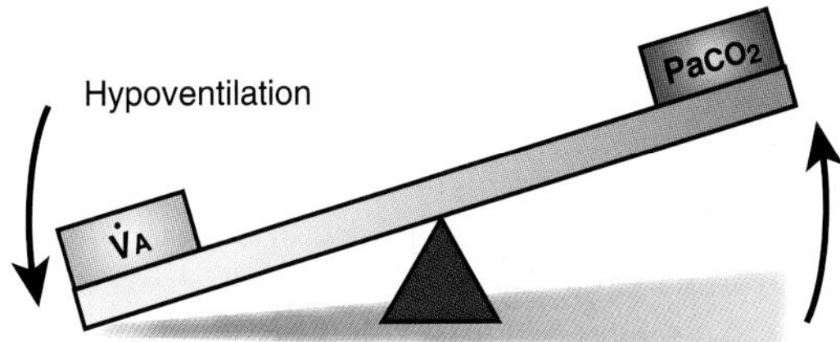

Figure 18.15 Alveolar ventilation and Pa_{CO_2} are inversely related. Alveolar ventilation ($\dot{V}_A$) is a key determinant of Pa_{CO_2}. In the above figure the inverse relationship is illustrated. For example, if alveolar ventilation is halved, Pa_{CO_2} will double, and conversely if alveolar ventilation is doubled, Pa_{CO_2} will be halved.

Second, the respiratory exchange ratio (R), defined as the ratio of the volume of carbon dioxide exhaled to the volume of oxygen taken up ($\dot{V}_{CO_2}/\dot{V}_{O_2}$), is usually <1, which means more oxygen is removed from the alveolar gas per unit time than carbon dioxide is added. The alveolar partial pressure of oxygen (P_{AO_2}) can be calculated by using the alveolar gas equation:

$$P_{AO_2} = P_{IO_2} - P_{ACO_2}[F_{IO_2} + (1 - F_{IO_2})/R] \quad (16)$$

where P_{IO_2} equals the partial pressure of inspired oxygen (moist tracheal air), Pa_{CO_2} equals the partial pressure of carbon dioxide in the alveoli, F_{IO_2} equals the fractional concentration of oxygen in the inspired air, and R equals the **respiratory exchange ratio**. When R is 1, the complex term in the brackets equals 1. Also, if a subject breathes 100% oxygen (F_{IO_2} = 1.0) for a brief period, the correction factor in the brackets reduces to 1.

In a normal resting person breathing air at sea level with an R of 0.82, a Pa_{CO_2} of 40 mm Hg, and a P_{IO_2} of 150 mm Hg, Pa_{O_2} is calculated as follows:

$$\begin{aligned} P_{AO_2} &= 150 \text{ mm Hg} - 40 \text{ mm Hg } [0.21 + (1 - 0.21)/0.82] \\ P_{AO_2} &= 150 \text{ mm Hg} - 40 \text{ mm Hg} \times [1.2] = 102 \text{ mm Hg} \end{aligned} \quad (17)$$

Because P_{IO_2} and F_{IO_2} stay fairly constant and P_{ACO_2} equals Pa_{CO_2}, the alveolar gas equation can be simplified to:

$$P_{AO_2} = 150 \times 1.2(P_{ACO_2}) \quad (18)$$

LUNG AND CHEST WALL MECHANICAL PROPERTIES

The lungs, airway tree, and vascular tree are embedded in elastic tissue. When the lungs are inflated, this elastic component is stretched. The degree of lung expansion at any given time in the breathing cycle is proportional to transpulmonary pressure. How well a lung inflates and deflates with a change in transpulmonary pressure depends on its elastic properties. An important feature of an elastic material is that, once stretched, it will recoil to its unstretched position. Therefore, the lung is like a spring; it recoils when stretched.

Lung inflation and deflation is directly affected by the elastic properties of the lungs.

When discussing the elastic properties of the lung, there are three basic components that are involved with respiration. These include elastic recoil, stiffness, and lung distensibility. **Distensibility** is the term applied to the ease with which the lungs can be stretched or inflated. Stiffness is defined as resistance to stretch or to inflation. **Elastic recoil** is defined as the ability of a stretched or inflated lung to return to its resting volume (FRC). Elastic recoil of the lung is directly related to lung stiffness, that is, the stiffer the lung, the greater the elastic recoil. An analogy of this relationship is a coiled spring—the more difficult it is to stretch the spring (greater stiffness), the greater the ability to snap back (greater elastic recoil); similarly, a lung that is stiff is more difficult to stretch (inflate), but the inflated lung has a greater ability to recoil back. Elastic recoil plays a key role during expiration in forcing air out of the lungs. Distensibility and elastic recoil, however, are inversely related to each other. A lung that is easily inflated has less elastic recoil. Overstretching causes the lungs to lose elastic recoil. Lung distensibility and elastic recoil arise from the elastin and collagen fibers enmeshed around the alveolar walls, adjacent bronchioles, and small blood vessels. Elastin fibers are highly distensible and can be stretched to almost double their resting length. Collagen fibers, however, resist stretch and limit lung expansion at high lung volumes. As the lungs expand during inflation, the fiber network around alveoli, small blood vessels, and small airways unfolds and rearranges—similar to stretching a nylon stocking. When the stocking is stretched, there is not much change in individual fiber length, but the unfolding and rearrangement of the nylon mesh allows the stocking to be easily stretched out to fit the contour of the legs. However, if the nylon stocking is overstretched, it loses its elastic recoil, no longer fits the contour of the legs, and sags or becomes "baggy." In the same way, lungs that lose their elastic recoil also become "baggy." In other words, "baggy lungs" are easy to inflate (stretched) but are difficult to deflate because of their inability to recoil inwardly and force air out of the lungs.

Lung compliance measures distensibility and is determined from a pressure–volume curve.

Lung distensibility and elastic recoil can be determined from a pressure–volume curve. A simple analogy is the inflation of a balloon with a syringe (Fig. 18.16). For each change in pressure, the balloon inflates to a new volume. The slope of the line of the pressure–volume curve is known as **lung compliance (C_L)**. Compliance is a measure of distensibility and is represented by

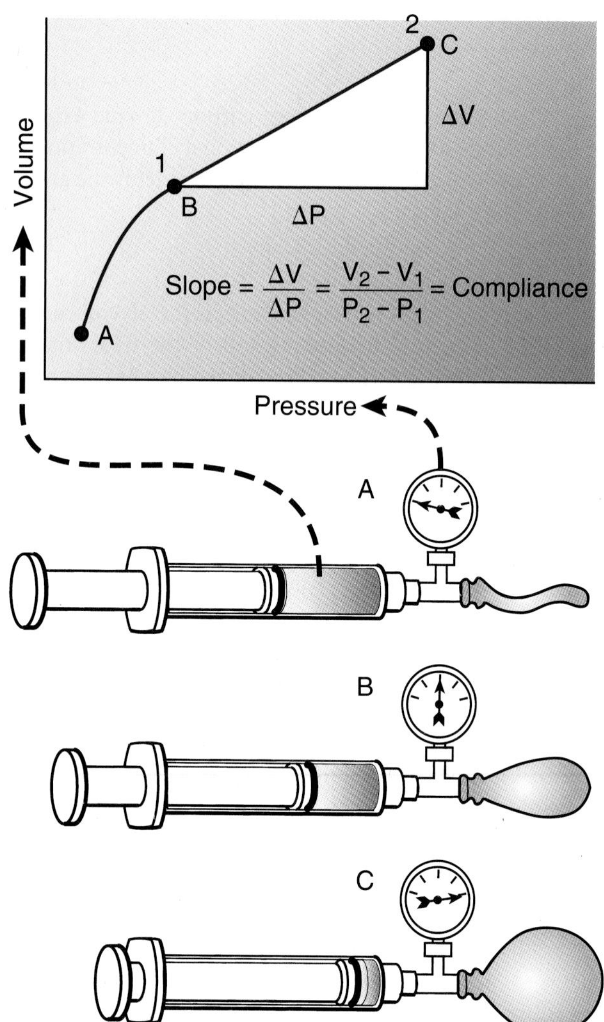

Figure 18.16 Pressure–volume measurements of a balloon illustrate its elastic properties. For each change in pressure (shown by movement of the *arrow* on the manometer dial), the balloon inflates to a new volume (plotted on the graph at points A, B, and C). The slope of the line determined by ΔV/ΔP between any two points on a pressure–volume curve is compliance. Compliance and elastic recoil are inversely related (C_L = 1/elastic recoil). For example, the lower the compliance the greater the elastic recoil. P, pressure; V, volume.

$$C_L = \Delta\text{volume}/\Delta\text{pressure} \quad (19)$$

where Δvolume equals change in volume and Δpressure equals change in pressure (see Chapter 11, "Overview of the Cardiovascular System and Hemodynamics").

A similar pressure–volume curve can be generated for the human lungs by simultaneously measuring changes in lung volume with a spirometer and changes in pleural pressure with a pressure gauge (Fig. 18.17). Because the esophagus passes through the thorax, changes in pleural pressure can be obtained indirectly from the pressure changes in the esophagus by using a balloon catheter. In practice, a pressure–volume curve is obtained by having the person first inspire maximally to TLC and then expire slowly. During the slow

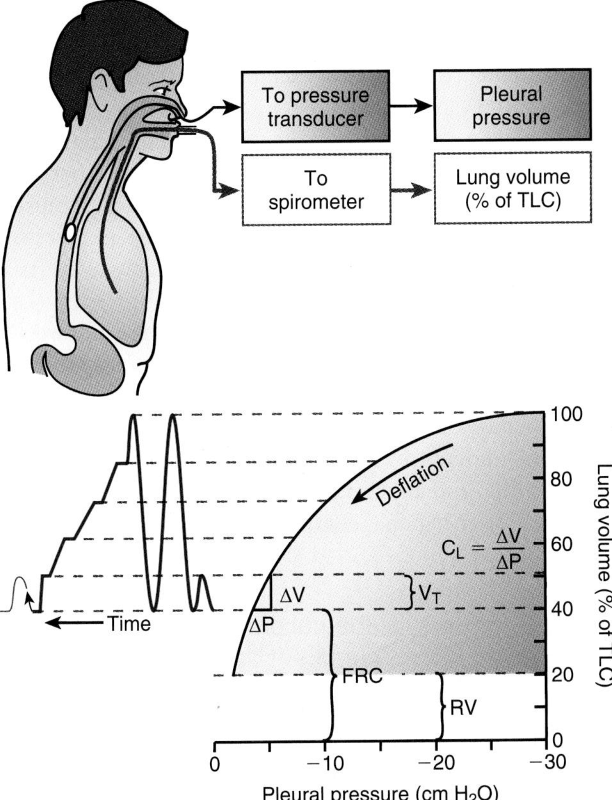

Figure 18.17 Lung compliance is measured from a pressure–volume curve. The subject first inspires maximally to total lung capacity (TLC) and then expires slowly, while airflow is periodically stopped to simultaneously measure pleural pressure and lung volume. Lung compliance (C_L) is measured in L/cm H_2O. Note that pleural pressure is determined by measuring esophageal pressure. FRC, functional residual capacity; P, pressure; RV, residual volume; V_T, tidal volume.

expiration, airflow is periodically interrupted (so that alveolar pressure is zero), and lung volume and pleural pressure are measured. Under these conditions, in which no airflow is occurring, the volume change per unit pressure change ($\Delta V/\Delta P$) is called *static compliance.* Lung compliance is affected by lung volume, lung size, and surface tension inside the alveoli. Because the pressure–volume curve of the lung is nonlinear, compliance is not the same at all lung volumes; it is high at low lung volumes and low at high lung volumes. In the midrange of the pressure–volume curve, lung compliance is about 0.2 L/cm H_2O in adult humans. Lung size affects lung compliance. A mouse lung, for example, has a different compliance than an elephant lung. To allow comparisons among lungs of different sizes, lung compliance is normalized by dividing it by FRC to give a specific compliance.

What is the significance of an abnormally low or abnormally high compliance? Remember lung compliance is inversely related to lung stiffness (C_L = 1/stiffness). For example, abnormally low lung compliance indicates a stiff lung, which means more work is required to inflate the lungs to bring in a normal tidal volume. An example of abnormally low compliance is seen in a restrictive disorder, a disease state that is characterized by restriction of lung inflation (stiff lungs). The compliance of these lungs is decreased compared with that of a normal lung (Fig. 18.18). Restrictive disorders can be caused by infection or by toxic environmental insults in which the elastic properties are altered. As a result, these lungs become stiffer and lung compliance is decreased. An abnormally high compliance is just as detrimental as a low compliance. Consider the disease **emphysema**, a disease in which lungs have been overstretched from chronic coughing and congested airways. Lungs of patients with emphysema (which is strongly linked to smoking) have a high compliance (high distensibility) and are extremely easy to inflate. However, getting air out again is another matter. Lungs with abnormally high compliance have low elastic recoil, and additional effort is required to force air out of the lungs. Emphysema is part a broader category termed **chronic obstructive pulmonary disease (COPD)**, which is a diseased state defined as an obstruction of airflow out of the lungs. It is important to remember that lungs with increased compliance have high static lung volumes (increased TLC, FRC, and RV) and low airflow. Therefore, patients with COPD have high lung volumes and retain an abnormally high RV of air.

Elastic recoil of the chest wall affects lung volumes.

Just as the lungs have elastic properties, so does the chest wall. The outward elastic recoil of the chest wall aids lung expansion, whereas the inward elastic recoil of the lungs pulls in the chest wall. The elastic recoil of the chest wall is such that if the chest were unopposed by the recoil of the lung, it would expand to about 70% of TLC. This volume represents the resting position of the chest wall unopposed by the lung. If the chest wall is mechanically expanded beyond its resting position, it recoils inward. At volumes <70% of TLC, the recoil of the chest wall is directed outward and is opposite to the elastic recoil of the lung. Therefore,

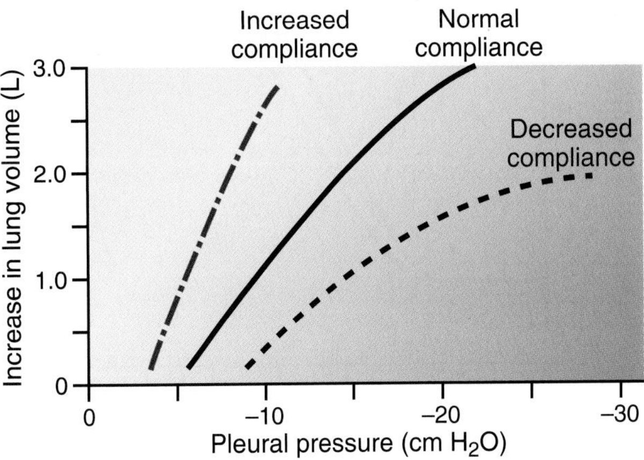

Figure 18.18 Obstructive and restrictive diseases alter lung compliance. Patients with a chronic obstructive lung disease such as emphysema, have abnormally high lung compliance. Patients with restrictive diseases, such as respiratory distress syndrome, have abnormally low lung compliance. See Clinical Focus Box 18.2.

the outward elastic recoil of the chest wall is greatest at RV, whereas the inward elastic recoil of the lung is greatest at TLC.

The lung volume at which lung and chest are at equilibrium (i.e., equal elastic recoil but in opposite directions) is represented by FRC. Because the lungs and chest walls are recoiling equally but in opposite directions at FRC, FRC is often referred to as the resting volume of the lungs. A change in the elastic properties of either the lungs or chest wall has a significant effect on FRC. For example, if the elastic recoil of the lungs is increased (i.e., lower C_L), a new equilibrium is established between the lungs and chest wall, resulting in a decreased FRC. Conversely, if the elastic recoil of the chest wall is increased, FRC is higher than normal. The elastic recoil of the chest wall at low lung volumes is a major determinant of RV in young people.

Uneven ventilation in the lungs is caused by differences in regional lung compliance.

Lung compliance also affects ventilation and causes inspired air to be unevenly distributed in the lungs. In a healthy upright person, compliance at the top part of the lungs is less than that at the base. This difference in compliance between the apex and base, known as **regional compliance**, is caused by gravity (Fig. 18.19). The gravitational effect occurs because lung tissue is approximately 80% water and gravity exerts a "downward pull," resulting in a lower pleural pressure (i.e., more negative) at the apex than at the base of the lungs. As a result, there is a greater transpulmonary pressure at the apex (see Fig. 18.19A). The higher transpulmonary pressure at the apex causes the alveoli to be more expanded and leads to regional differences in compliance. At any given volume from the FRC and above, the apex of the lung is less distensible (i.e., lower regional compliance) than at the base. This

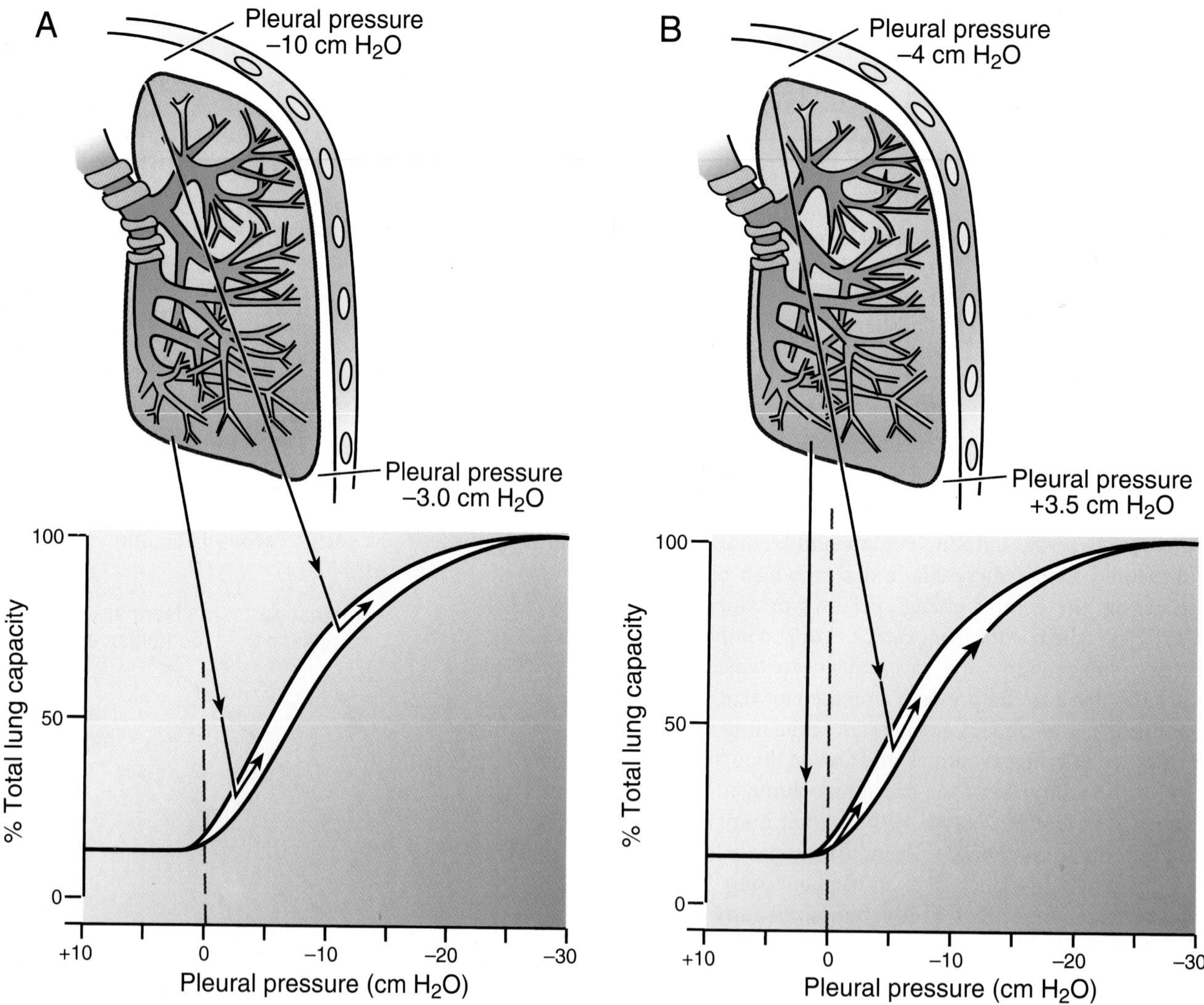

Figure 18.19 Regional differences in lung compliance affect airflow. (A) Because of gravity, pleural pressure in the upright person is more negative at the apex than at the base of the lungs. Consequently, the base of the lungs is more compliant at functional residual capacity. Because the base of the lung is more distensible, proportionally more of the tidal volume will go to the base with inspiration. **(B)** At low lung volumes (e.g., residual volume), pleural pressure at the base becomes positive. This results in more air going to the top part of the lung during a tidal volume because the apical region of the lung is more compliant.

means the base of the lung has both a larger change in volume for the same pressure change and a smaller resting volume than at the apex. In other words, as one takes a breath in from FRC, a greater portion of the tidal volume will go to the base of the lungs, resulting in greater ventilation at the base.

At low lung volumes, alveoli at the apex of the lung are more compliant than that at the base (see Fig. 18.19B). At lung volumes approaching RV, the pleural pressure at the base of the lungs actually exceeds the pressure inside the airways, leading to airway closure. At RV, the base is compressed, and ventilation in the base is impossible until pleural pressure falls below atmospheric pressure. By contrast, the apex of the lung is in a more favorable portion of the compliance curve. Consequently, the first portion of the breath taken in from RV enters alveoli in the apex, and the distribution of ventilation is inverted at low lung volumes (i.e., the apex is better ventilated than the base).

Alveolar surface tension affects lung compliance.

Another property that significantly affects lung compliance is surface tension at the air–liquid interface of the alveoli. The surface of the alveolar membrane is moist and is in contact with air, producing a large air–liquid interface. Surface tension (measured in dyne/cm) arises because water molecules are more strongly attracted to one another than to air molecules. In the alveoli, surface tension produces an inwardly directed force that tends to reduce alveolar diameter.

Surface tension of the lungs can be studied by examining a pressure–volume curve. Figure 18.20 shows results from lungs that have been removed from the chest (excised lungs) and inflated and deflated in a stepwise fashion, first with air and then with saline. With air-filled lungs, the gas–liquid interface creates surface tension. However, with saline-filled lungs, the air–liquid interface is eliminated and the surface tension is eliminated. In comparing these two pressure–volume curves, two important observations can be made. First, the slope of the deflation limb of the saline curve is much steeper than that of the air curve. This means that when surface tension is eliminated, the lung is far more compliant (more distensible). Second, the different areas to the left of the saline and air inflation curves show that surface tension significantly contributes to the work required to inflate the lungs. Because the area to the left of each curve is equal to work, which can be defined as force (change in pressure) times distance (change in volume), the elastic forces and surface tension can be separated. The area to the left of the saline inflation curve is the work required to overcome the elastic recoil of the lung

Clinical Focus / 18.2

Acute Respiratory Distress Syndrome

Scot was a healthy 20-year-old man who was starting his second year of college. He was in good condition and worked out regularly by lifting weights and swimming 1 mile each day. However, one fall afternoon he began to feel ill. He went to the student health center with complaints of a severe headache and pain in the neck and lower abdomen. He was told that he was coming down with the flu and was advised to go back to the dorm and rest. The next day his symptoms, which included fever, cold sweats, skin rash, and nausea became worse. His roommate took him back to the health center, and Scot was placed in one of the student beds for closer observation. When his condition started deteriorating, he was rushed to the emergency department at the university medical center. On arrival, Scot was in shock with a high fever and a breathing rate of 36 (normal is 12–15/min). Because of the danger of respiratory failure, the attending physician immediately had Scot intubated and placed on assisted ventilation. The decision was made to transfer him to the intensive care unit (ICU). Upon arrival at the ICU, Scot's chest radiograph showed patchy infiltrate and pulmonary edema. He was diagnosed as having **acute respiratory distress syndrome (ARDS)**, with a secondary widespread infection of unknown origin.

Scot's case is similar to those of the other 200,000 to 250,000 persons who develop ARDS each year. Despite Scot's youth and previous history, his anticipated mortality for ARDS was approximately 80%. Scot was lucky and was able to return to his classes after 12 weeks in the hospital and a bill of >$200,000.

ARDS is the name given to diffuse lung injury of various causes. Few patients have a history of previous lung disorders. The causes include trauma from chest injury (e.g., car accidents), long bone injury, and pelvic injury. During the Vietnam era, the trauma that caused ARDS was coined "Da Nang lung." Two other major causes include sepsis (presence of pathogen or toxin in the blood) and aspiration of gastric contents. The latter occurs with gastric reflux and usually occurs at night during sleep. Other causes include cardiopulmonary bypass, smoke inhalation, high altitude, and exposure to irritant gases. Exposure to irritant gases was the cause of the widespread ARDS cases that occurred at the disaster in Bhopal, India.

Although the etiology for ARDS is varied in different cases, the pathophysiology is nearly identical. ARDS is characterized by decreased lung compliance, pulmonary edema, focal atelectasis, and hypoxemia (low partial pressure of O_2 in the arterial blood), with an inflammatory reaction that leads to an infiltration of neutrophils into the lung. The increase in lung stiffness (i.e., decreased compliance) is a result of loss of surfactant, which is unrelated to edema.

Neutrophil aggregation is a key underlying mechanism of ARDS. Aggregation of neutrophils causes capillary endothelial damage by releasing a number of toxic products. These include oxygen free radicals, proteolytic enzymes, arachidonic acid metabolites (leukotrienes, thromboxane, prostaglandin), and platelet-activating factor.

New approaches to therapy involve ways to reduce neutrophil chemoattraction and aggregation in the pulmonary capillaries and ways to reduce the amount of toxic substances released by neutrophils.

tissue. The area to the left of the air inflation curve is the work required to overcome both elastic tissue recoil and surface tension. Subtracting the area to the left of the saline curve from the area to the left of the air curve shows that approximately two thirds of the work required to inflate the lungs is needed to overcome surface tension. Lung distensibility and the work of breathing are significantly affected by surface tension.

Surface tension has important ramifications for maintaining alveolar stability. In a sphere such as an alveolus, surface tension produces a force that pulls inwardly and creates an internal pressure. The relationship between surface tension and pressure inside a sphere is shown in Figure 18.21. The pressure developed in an alveolus is given by the **Laplace law** (see Chapter 11, "Overview of the Cardiovascular System and Hemodynamics"):

$$P = 2T/r \qquad (20)$$

where T equals surface tension and r equals radius. Because alveoli are interconnected and vary in diameter, the Laplace equation assumes functional importance in the lung. In the example shown in Figure 18.21A, surface tension is constant at 50 dyne/cm, and an unstable condition results because pressure is greater in the smaller alveolus than in the larger one. Consequently, smaller alveoli tend to collapse, especially at low lung volumes, a phenomenon known as **atelectasis**. Larger alveoli become overdistended when atelectasis occurs. Therefore, two questions arise: How do alveoli of different sizes coexist in the intact lung when interconnected? How does the normal lung prevent atelectasis at low lung volumes? The answers lie, in part, in the fact that the alveolar surface tension is not constant at 50 dyne/cm, as in other biologic fluids.

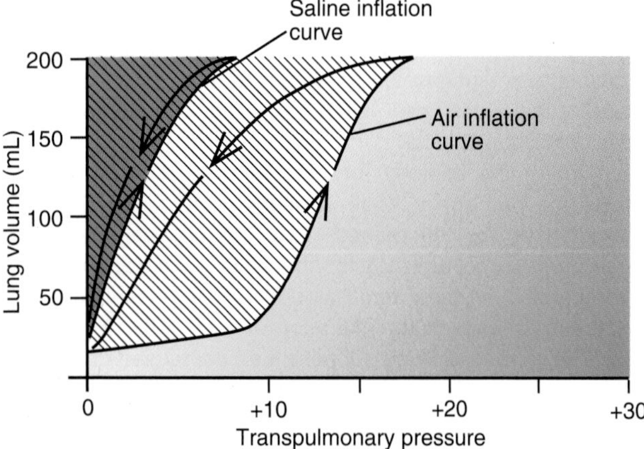

Figure 18.20 Alveolar surface tension affects lung compliance. Inflating lungs with saline instead of air eliminates surface tension. Two pressure–volume curves for lungs inflated to the same volume are shown, first with air and then with saline. The differences in the two curves occur because surface tension contributes significantly to lung compliance in the air-filled lungs. When lungs are inflated and deflated with saline, the lungs are more distensible, with a concomitant increase in lung compliance. The *shaded areas* are equal to the work done in inflating the lungs.

Surfactant lowers surface tension and stabilizes alveoli at low lung volumes.

The alveolar lining is coated with a special surface-active agent, **pulmonary surfactant**, which not only lowers surface tension at the gas–liquid interface but also changes surface tension with changes in alveolar diameter (see Fig. 18.21B). Pulmonary surfactant is a lipoprotein rich in phospholipid. The principal agent responsible for its surface

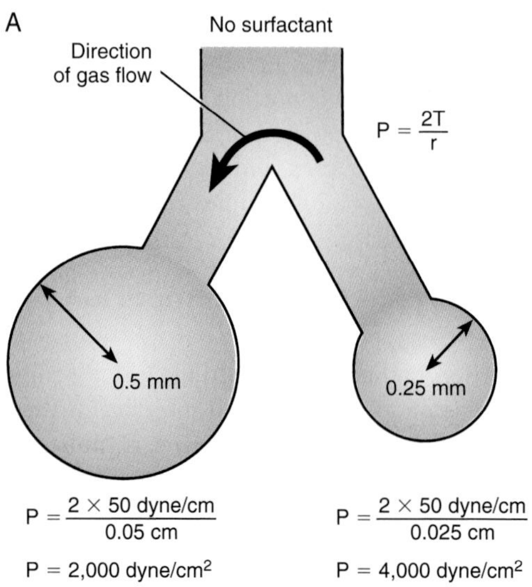

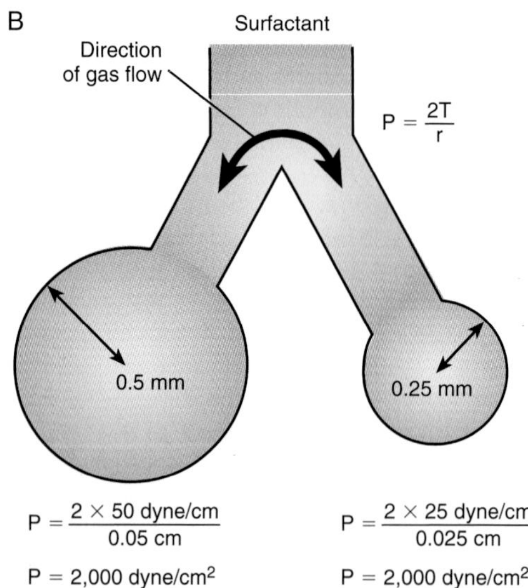

Figure 18.21 Surfactant promotes alveolar stability at low lung volumes. (A) If surface tension remains constant (50 dyne/cm), alveoli that are interconnected but differ in diameter become unstable and cannot coexist. Pressure in the smaller alveolus is greater than that in the larger alveolus, which causes air from the smaller alveolus to empty into the larger alveolus. At low lung volumes, the smaller alveoli tend to collapse, a phenomenon known as atelectasis. **(B)** Surfactant lowers surface tension proportionately more in the smaller alveolus. As a result, pressures in the two alveoli are equal, and alveoli of different diameters can coexist. P, pressure; r, radius; T, tension.

tension-reducing properties is **dipalmitoylphosphatidylcholine (DPPC)**.

The functional importance of surfactant can be demonstrated by using a surface tension balance (Fig. 18.22). Surface tension is measured by placing a platinum flag connected to a force transducer into a trough of liquid. Surface tension creates a meniscus on each side of the platinum flag, and the greater the contact angles of the meniscus, the greater the surface tension. The surface is repeatedly expanded and compressed by a movable barrier that skims the surface of the liquid (simulating lung inflation and deflation). Surface tension of pure water is 72 dyne/cm, a value that is independent of the surface area of water in the balance. Therefore, when the surface is expanded and compressed, surface tension does not change. When a detergent is added, surface tension is reduced but again is independent of surface area. However, when a **lung lavage** is added that contains pulmonary surfactant, surface tension not only is reduced but also changes in a nonlinear fashion with surface area. Therefore, pulmonary surfactant makes it possible for alveoli of different diameters that are connected in parallel to coexist and be stable at low lung volumes, by lowering surface tension proportionately more in the smaller alveoli (see Fig. 18.21B).

Surfactant works by reducing surface tension at the gas–liquid interface. When the molecules are compressed during lung deflation, surfactant causes a decrease in surface tension. At low lung volumes, when the molecules are tightly compressed, some surfactant is squeezed out of the surface and forms micelles (Fig. 18.23). On expansion (reinflation), new surfactant is required to form a new film that is spread on the alveolar surface lining. When surface area remains fairly constant during quiet or shallow breathing, the spreading of surfactant is often impaired. A deep sigh or yawn causes the lungs to inflate to a larger volume and new surfactant molecules spread onto the gas–liquid interface. Patients recovering from anesthesia are often encouraged to breathe deeply to enhance the spreading of surfactant. Patients who have undergone abdominal or thoracic surgery often find it too painful to breathe deeply; poor surfactant spreading results, causing part of their lungs to become atelectatic.

The alveolar epithelium consists basically of two cell types: alveolar type I and type II cells (Fig. 18.24). Alveolar type II cells are often referred to as **type II pneumocytes**. The ratio of these cells in the epithelial lining is about 1:1, but type I cells occupy approximately two thirds of the surface area. Type II cells seem to aggregate around the alveolar septa. Surfactant is synthesized in the alveolar type II cell. Compared with type I cells, they are rich in mitochondria and are metabolically active. Electron-dense **lamellar inclusion bodies** are a distinguishing feature of the type II cell. These lamellar inclusion bodies, rich in surfactant, are thought to be the storage sites for surfactant.

The process of surfactant synthesis is shown in Figure 18.25. Basic substrates, such as glucose, palmitate, and

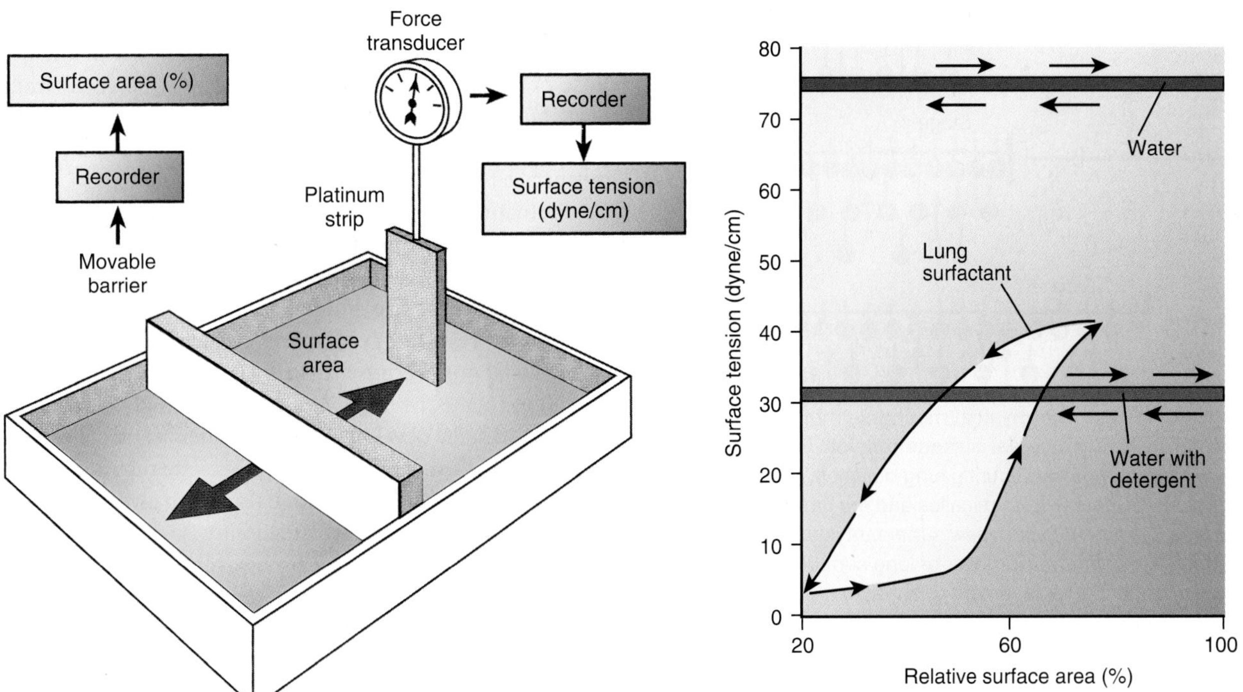

Figure 18.22 Lung surfactant lowers alveolar surface tension. A surface tension balance is used to measure surface tension at the air–liquid interface. When distilled water is placed in the balance, surface tension is independent of surface area (a constant 72 dynes/cm). The addition of a detergent reduces surface tension, but it is still independent of surface area. When lung surfactant (obtained from a lung lavage—a volume of saline flushed into the airways to rinse the alveoli) is placed in the balance, surface tension not only is decreased but also changes with surface area.

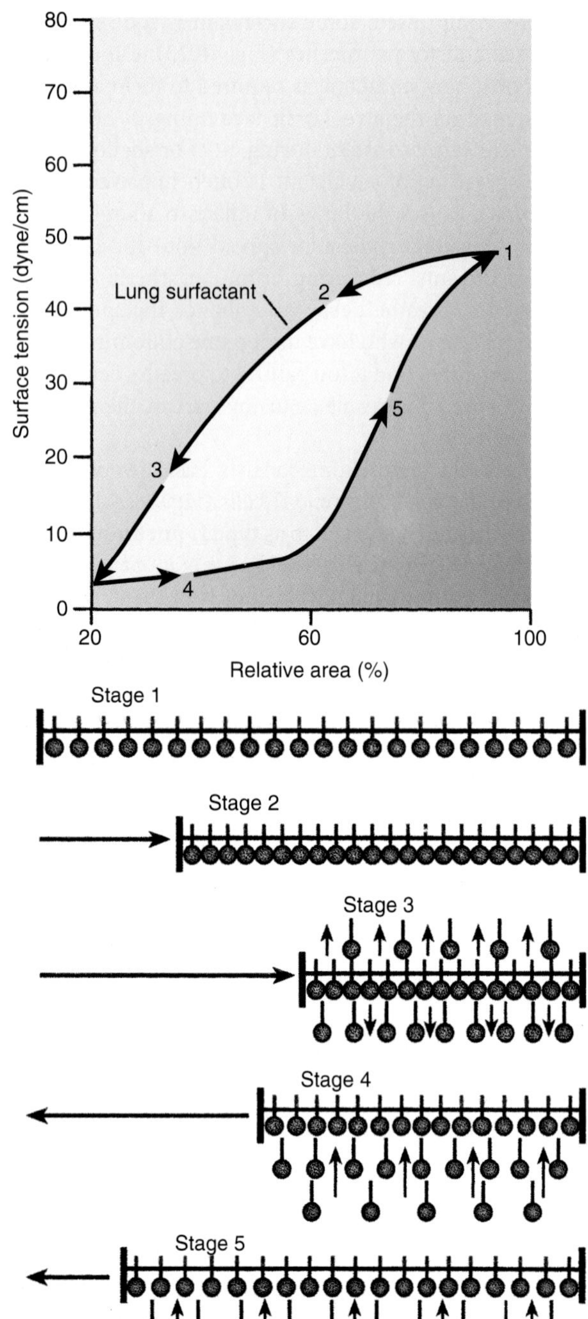

Figure 18.23 Biophysical mechanism of lung surfactant for lowering alveolar surface tension. Surfactant molecules are compressed during lung deflation. At stage 3, surfactant molecules form micelles and are removed from the surface. On lung inflation, new surfactant is spread onto the surface film (stage 4). Turnover of lung surfactant is high because of continual replacement of surfactant during lung expansion.

choline are taken up from the circulating blood and synthesized into DPPC by alveolar type II cells. Stored surfactant from the lamellar inclusion bodies is discharged onto the alveolar surface. The turnover of surfactant is high because of the continual renewal of surfactant at the alveolar surface during each expansion of the lung. The high rate of replacement of surfactant probably accounts for the active lipid synthesis

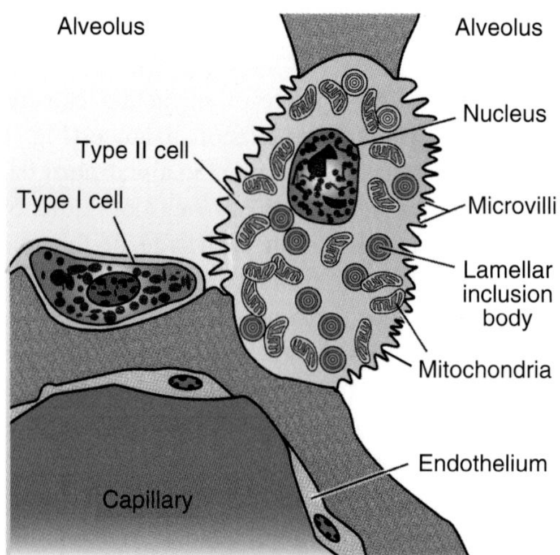

Figure 18.24 Alveolar type II cells synthesize lung surfactant. Alveolar type I cells occupy most of the alveolar surface, and alveolar type II cells are located in the corners between two adjacent alveoli. Alveolar type II cells produce lung surfactant. Also shown are endothelial cells that line the pulmonary capillaries.

that occurs in the lung. Because the lungs are among the last organs to develop, the synthesis of surfactant appears rather late in gestation. In humans, surfactant appears at about week 34 (a full-term pregnancy is 40 weeks). Regardless of the total duration of gestation in any mammalian species, the process of lung maturation seems to be "triggered" about the time gestation is 85% to 90% complete. Clearly, the fetal lung is endowed with a special regulatory mechanism to control the timing and appearance of surfactant. Failure of proper lung maturation during the perinatal period is still a major cause of death in newborns. The lung may be structurally intact but functionally immature if inadequate amounts of surfactant are available to reduce surface tension and stabilize surface forces during breathing.

Premature birth and certain hormonal disturbances (such as those seen in diabetic pregnancies) interfere with the normal control and timing of lung maturation. These infants have immature lungs at birth, which often leads to **neonatal respiratory distress syndrome**. Breathing is extremely labored because surface tension is high, making it difficult to inflate the lungs. Because of the high surface tension, these infants develop pulmonary edema and atelectasis. They are at high risk until the lungs become mature enough to secrete surfactant. In addition to lowering alveolar surface tension and promoting alveolar stability, surfactant helps to prevent edema in the lung. The inwardly contracting force that tends to collapse alveoli also tends to lower interstitial pressure, which "pulls" fluid from the capillaries. Pulmonary surfactant reduces this tendency by lowering surface forces. Some pulmonary physiologists think that keeping the lungs dry may be the major role of surfactant, especially in adults.

Alveolar interdependence promotes alveolar stability.

Another mechanism that plays a role in promoting alveolar stability is interdependence or mutual support among

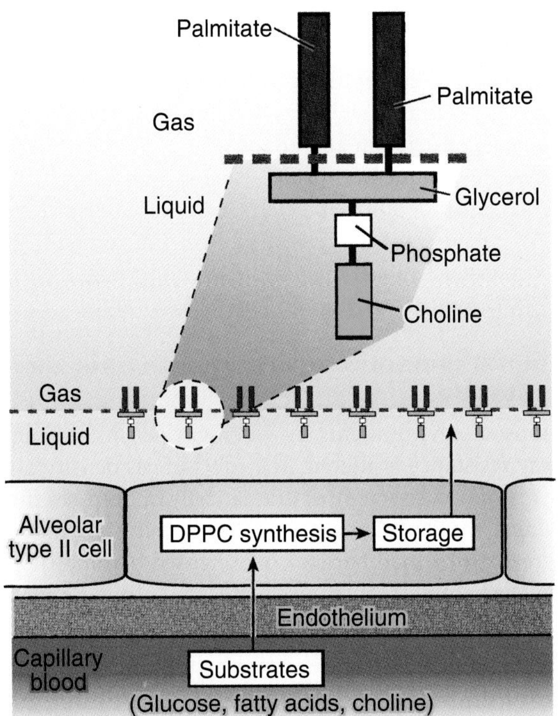

Figure 18.25 Metabolic synthesis of lung surfactant. Substrates for lung surfactant synthesis are taken up by alveolar type II cells from the pulmonary capillary blood. Surfactant (dipalmitoylphosphatidylcholine or DPPC) is stored in lamellar inclusion bodies and subsequently discharged onto the alveolar surface. Surfactant is oriented perpendicularly to the gas–liquid interface at the alveolar surface; the polar end is immersed in the liquid phase, whereas the nonpolar portion is in the gas phase.

adjacent alveoli. Because alveoli (except those next to the pleural surface) are interconnected with surrounding alveoli, they support each other. Studies have shown that this type of structural arrangement, with many connecting links, prevents the collapse of adjacent alveoli. For example, if alveoli tend to collapse, surrounding alveoli would develop large expanding forces. Therefore, interdependence can play a role in preventing atelectasis as well as in opening up lungs that have collapsed. Alveolar interdependence seems to be more important in adults than in newborns because newborns have fewer interconnecting links.

AIRFLOW AND THE WORK OF BREATHING

Two basic types of airflow occur in the lung. Turbulent flow occurs at high flow rates in the large airways (the trachea and large bronchi). Turbulent flow consists of completely disorganized patterns of airflow, resulting in a sound that can be heard with inspiration and expiration. The faster and deeper the breathing, the more noise produced from turbulence. In contrast, laminar flow is characterized by a streamlined flow that runs parallel to the sides of the airways. Laminar flow is silent because layers of air molecules slide over each other. This type of flow occurs in the small peripheral airways, where airflow is exceedingly slow.

Airway resistance decreases airflow in the lung.

Both turbulent and laminar flow cause resistance to air moving in the airways. Airway resistance is expressed in centimeters H_2O per liter per second and is defined as the ratio of driving pressure (ΔP) to airflow ($\dot{V}$). For total airway resistance (R_{aw}), the driving pressure is the pressure difference between the mouth (P_{mouth}) and the alveoli (P_A). The equation can be written as

$$R_{aw} = \frac{P_{mouth} - P_A}{\dot{V}} \quad (21)$$

The major site of airway resistance is the medium bronchi (lobar and segmental) and bronchi down to about the seventh generation (Fig. 18.26). One would expect the major site of resistance, based on the **Poiseuille law** (see Chapter 11, "Overview of the Cardiovascular System and Hemodynamics"), to be located in the narrow airways (the bronchioles), which have the smallest radius. However, measurements show that only 10% to 20% of total airway resistance can be attributed to the small airways (those <2 mm in diameter). This apparent paradox results because so many small airways are arranged in parallel and their resistances are added as reciprocals. Resistance of each individual bronchiole is relatively high, but the great number of them results in a large total cross-sectional area, causing their total combined resistance to be low.

Many airway diseases often begin in the small airways. Diagnosing a disease in the small airways is difficult,

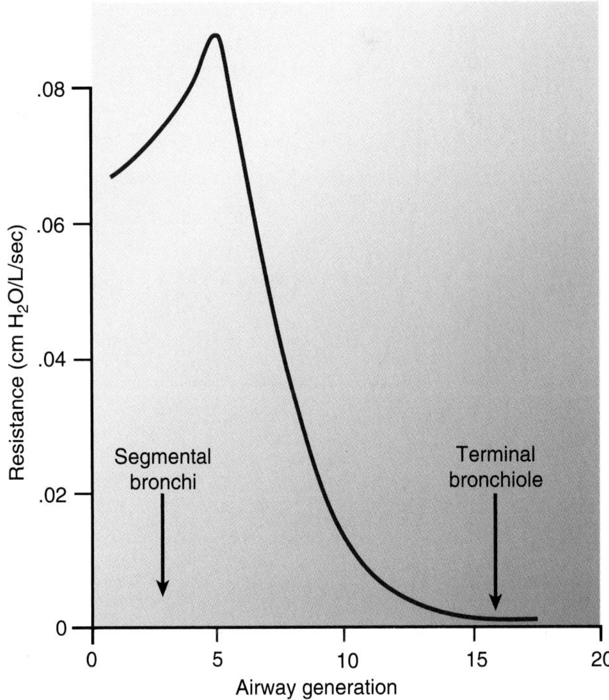

Figure 18.26 Airflow turbulence significantly increases airway resistance. The major sites of resistance are the lobar and segmental bronchi, where airway turbulence is the greatest, down to about the seventh generation of airway branches.

however, because the small airways account for such a low percentage of the total R_{aw}. Early detection is difficult because changes in airway resistance are not noticeable until the disease becomes severe.

Airways are compressed at low lung volumes, causing an increase in airway resistance.

One of the major factors affecting airway diameter, especially that of bronchioles, is lung volume. Like lung tissue, the smaller airways are capable of being distended or compressed. Bronchi and smaller airways are embedded in lung parenchyma and are connected by "guy wires" to surrounding tissue. As the lung enlarges, airway diameter increases, which results in a concomitant decrease in airway resistance during inspiration (Fig. 18.27). Conversely, at low lung volumes, airways are compressed, and airway resistance rises. Note that the inverse relationship between lung volume and airway resistance is nonlinear. At low lung volumes, airway resistance rises sharply.

Airway patency is affected by changes in smooth muscle tone.

Bronchial smooth muscle tone also affects airway diameter with a concomitant change in airway resistance. A change in smooth muscle tone will change airway diameter. The smooth muscles in the airway, from the trachea down to the terminal bronchioles, are under autonomic control. Stimulation of parasympathetic cholinergic postganglionic fibers causes bronchial constriction as well as increased mucus secretion. Stimulation of sympathetic adrenergic fibers causes dilation of bronchial and bronchiolar airways and inhibition of glandular secretion. Drugs, such as isoproterenol and epinephrine, which stimulate β_2-adrenergic receptors in the airways, cause dilation. These drugs alleviate bronchial constriction and are often used to treat asthmatic attacks. Environmental insults, such as breathing chemical irritants, dust, or smoke particles, can cause reflex constriction of the airways. Increased P_{CO_2} in the conducting airways can cause a local dilation. More important, a decrease in P_{CO_2} causes airway smooth muscle to contract.

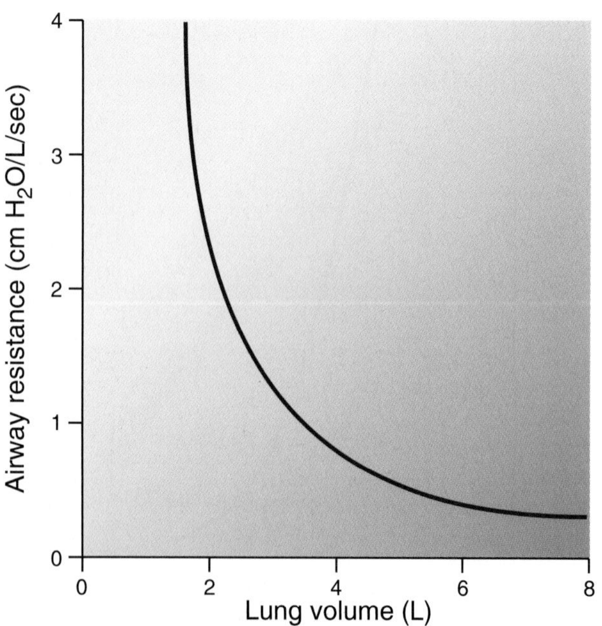

Figure 18.27 Low lung volumes increase airway resistance. Airways are connected to lung parenchyma. During inflation, airway diameter is increased. During deflation, airways are compressed resulting in a concomitant increase in airway resistance.

Unusual gaseous environments alter airway resistance.

In unusual environments in which gas density is changed, airway resistance is altered. The effect of gas density on airway resistance is seen most dramatically in deep-sea diving, in which the diver breathes air the density of which may be greatly increased because of increased barometric pressure. The barometric pressure increases 1 atm for each 10 m or 33 ft underwater (e.g., at 10 m below the surface, P_B = 2 atm or 1,520 mm Hg). As a result of increased resistance, a large pressure gradient is required just to move a normal tidal volume. Instead of breathing air during diving, the diver uses a helium–oxygen mixture because helium is less dense than air and, consequently, makes breathing easier. The fact that density has a marked effect on airway resistance again indicates that the medium airways are the main site of resistance and that airflow here is primarily turbulent (see Chapter 11, "Overview of the Cardiovascular System and Hemodynamics").

Forced expiration causes airway compression and increases airway resistance.

Airway resistance does not change much during normal quiet breathing. It is significantly increased, however, during forced expiration, such as in vigorous exercise. The marked change during forced expiration is a result of airway compression. This effect of airway compression on airway resistance can be demonstrated with a **flow–volume curve**. A flow–volume curve shows the relationship between airflow and lung volume during a forced inspiratory and expiratory effort (Fig. 18.28). The curve is generated by taking the spirometer tracings of forced inspiratory and expiratory flow and then closing them by connecting the end of maximum expiration back to the beginning of maximum inspiration. The small loop in Figure 18.28 is the flow–volume curve for a normal tidal volume. The large loop shows the maximal flow–volume curve. During forced inspiration, inspiratory flow is limited only by effort—that is, how hard the person tries. During forced expiration (FVC), flow rises rapidly to a maximum value, **peak expiratory flow (PEF)**, and then decreases linearly over most of expiration as lung volume decreases.

The first part of the FEF_{25-75}-volume curve is dependent on effort. Once PEF is achieved, flow is independent of effort in the last part of the flow–volume curve. Actually, over most of the expiratory flow–volume curve, flow is virtually independent of effort. The FEF_{25-75} limitation illustrates the importance of dynamic airway compression. Dynamic airway compression increases airway resistance, which effectively limits the FEF_{25-75}.

How does dynamic airway compression occur during forced expiration? The mechanism is related to changes in transairway pressure (i.e., pressure across the airway) and the compressibility of the airways. For example, in Figure 18.29, pleural pressure (P_{pl}) is approximately 5 cm H_2O and airway pressure (P_{aw}) is zero before inspiration (no airflow). Transairway pressure ($P_{aw} - P_{pl}$) is 5 cm H_2O [$0 - (-5) = +5$] and holds the airways open. At the start of maximum inspiration, pleural pressure decreases to approximately 7 cm H_2O and alveolar pressure falls to approximately 2 cm H_2O. The difference between alveolar pressure and pleural pressure is still 5 cm H_2O [$-2 - (-7) = +5$]. However, there is a pressure drop from the mouth to the alveoli because of resistance to airflow, and the transairway pressure will change along the airway. At the end of maximum inspiration, pleural pressure decreases further, to approximately 12 cm H_2O, and airway pressure is again zero because of no airflow. During maximum inspiration, airway resistance actually decreases because transairway pressure increases, which enlarges the diameter of the airways, especially the small airways.

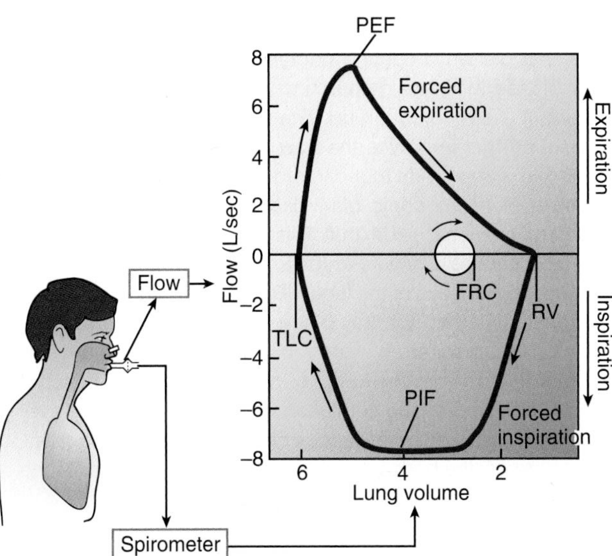

Figure 18.28 Flow–volume curve illustrates the functional importance of airflow. The relationship between lung volume and airflow can be seen from a flow–volume curve. During tidal breathing, airflow is small (small loop labeled FRC [functional residual capacity]). However, during forced expiration, airflow rises rapidly and reaches a maximum, termed *peak expiratory flow (PEF)*. PEF occurs before there is much change in lung volume. However, during the last part of the expiratory portion of the flow–volume curve (or a forced vital capacity), airflow is effort independent because of dynamic airway compression. Once PEF is achieved, airflow rate decreases proportionately as lung volume decreases to residual volume (RV). Maximum airflow during inspiration, termed *peak inspiratory flow (PIF)*, is maintained over a large change in lung volume because airways are distended and not compressed. TLC, total lung capacity.

On forced expiration, pleural pressure is no longer negative but rises above atmospheric pressure and can increase up to +30 cm H_2O. The added pressure in the alveoli is a result of the elastic recoil of the lungs, is termed *recoil pressure* (P_{recoil}), and can be written as

$$P_{recoil} = P_A - P_{pl} \quad (22)$$

where P_A equals alveolar pressure and P_{pl} equals pleural pressure.

The recoil pressure in Figure 18.29 is 12 cm H_2O, because at the beginning of expiration, lung volume has not appreciably decreased. Note that at the beginning of FVC, a pressure drop occurs along the airway because of airway resistance. Airway pressure falls progressively from the alveolar region to the airway opening (the mouth). The transairway pressure gradient along the airways reverses and tends to compress the airways. For example, at a point inside the airway where the pressure is 21 cm H_2O, the transairway pressure would be 9 cm H_2O, which would tend to close the airway.

At some point along the airway, the airway pressure equals pleural pressure and transairway pressure is zero (Fig. 18.30). This is the **equal pressure point (EPP)**. Theoretically, the EPP divides the airways into an upstream segment (from alveoli to EPP) and a downstream segment (EPP to the mouth). In the downstream segment, the airway pressure is below pleural pressure and the transairway pressure becomes negative. As a result, airways in the downstream segment are compressed or collapsed. The large airways (the trachea

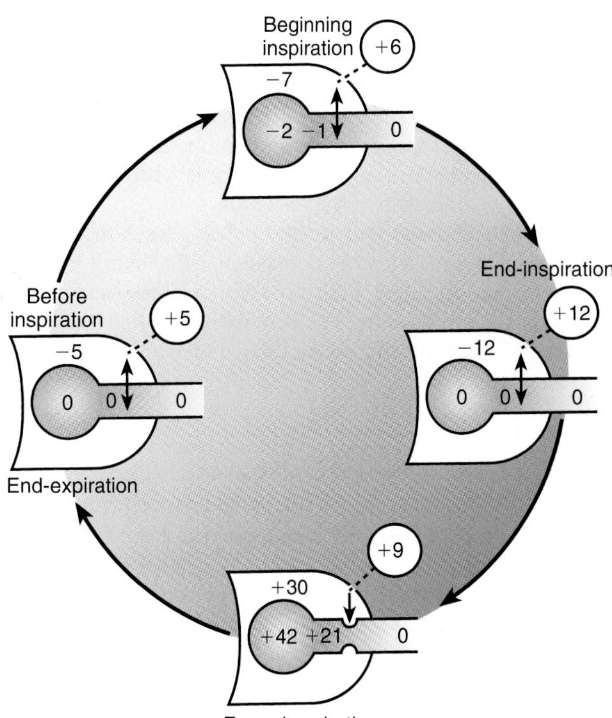

Figure 18.29 Airways are compressed during forced expiration. Transairway pressure is +5 cm H_2O before inspiration and reaches +12 cm H_2O at the end of inspiration. During forced expiration, transairway pressure becomes negative and the small airways are compressed.

From Bench to Bedside / 18.1 

Cystic Fibrosis: Potential New Therapies

Cystic fibrosis (CF) is a genetic disease that affects the exocrine glands of the body. These glands are responsible for secreting material (mucus, tears, and digestive juices) onto the body's internal surfaces (e.g., lungs and gut) and sweat on the external surface (skin). CF causes abnormal secretion of these materials. For example, normal mucus forms a gel-like barrier that protects the lining of cells. In the lung, mucus also functions as a mucociliary transport system that removes dust, pollen, and bacterial and other particulate matter out of the lungs to prevent infection. Patients with CF have excess mucus production as well as altered mucus proteins. The latter causes the mucus to become thick. The most characteristic feature of CF is the excessive production of thick, sticky mucus, which obstructs airflow in the small airways and alveolar ducts, not only making breathing difficult but also allowing bacterial infection and inflammation to occur. In response, white cells (neutrophils) are recruited to fight infection in the lungs. White cells migrate into the infected area, die, and release elastase and their DNA into the mucus, which further thickens the mucus and further exacerbates the situation. Thus, a viscous cycle is set up causing more airway obstruction/infection/inflammation as a result of the excessive sticky mucus production. Patients with CF cough frequently and require daily chest and back clapping to dislodge the mucus from the plugged airways. The pancreas is also damaged in these patients. The pancreas supplies digestive enzymes and bicarbonate to neutralize stomach acid so the enzymes can work properly in the intestines. Pancreatic function is lost as a result of thick secretion clogging the pancreatic ducts. Consequently, most patients with CF have an insufficient amount of digestive enzymes for normal digestion.

In addition to the pancreas, CF also affects the reproductive organs, causing infertility in almost all males and some females. Male infertility occurs because the vas deferens, the tubules that transport sperm from the testes, are either absent or underdeveloped. In the female, fertility is reduced because of irregularity in the menstrual cycle.

CF affects males and females equally, but is much more prevalent in the Caucasian population. CF affects 1 in 3,000 infants, making it one of the most common genetic diseases among Caucasians. The risk is much lower (1 in 13,000) in African Americans and is rare (one in 50,000) among Asian Americans. The genetic defect responsible for CF is the mutation in one gene called the **cystic fibrosis transmembrane conductance regulator (CFTR)**, which results in nonfunctional CFTR proteins in the exocrine glands. There are approximately 30,000 Americans with CF and an estimated 8 million people carrying one copy of the defective gene responsible for the disease. Fortunately, these people do not exhibit the symptoms of CF, because they must inherit two defective genes, one from each parent, to develop CF.

Pseudomonas bacteria are the leading cause of lung infection and death among patients with CF. New therapies are currently being developed to improve antibody therapy and clearance of lung mucus. Until recently, bacterial infections of the lung were treated intravenously with the antibody. Systemic treatment required high doses of antibodies so that the drug reached the lung. These high systemic doses are not only expensive but, more important, also cause irreversible damage to hearing and to kidney function. Recently, new nanomist devices have been developed to deliver the antibodies as an aerosol to the peripheral airways. This new procedure requires less antibody dosage, reduces the cost, and eliminates the systemic side effects.

A second aerosol device being developed is a DNAse atomizer. A naturally occurring enzyme called DNAse can cut long DNA molecules into shorter strands. Giving DNAse as an aerosol cuts the DNA strands that are released from white cells that die while fighting bacterial infection. The aerosolized DNAse thins the mucus and makes it less sticky, thus making it easier to cough up. The third new treatment being developed that improves the flow of mucus in patients with CF is two compounds found in red tide toxin. Florida red tide consists of microscopic plant-like organisms that produce a potent chemical toxin that kills fish, contaminates shellfish, and creates a severe irritation of the eyes, nose, throat, tongue, and airways of the lungs. Researchers have identified two antitoxins that not only block the effects of the red tide toxin on the respiratory system but also enhance the clearance of mucus from the lungs of patients with CF. The latter is accomplished by the combination of increased cilia movement and the thinning of mucus. The new therapeutic treatments are on the horizon and show much promise for those who suffer from CF.

and bronchi) are protected from collapse because they are supported by cartilage. However, small airways without this structural support are easily compressed and can collapse.

Lung compliance affects where the equal pressure point is established in airways.

The EPP is established after the PEF is achieved (see Fig. 18.28). As the forced expiratory effort continues, the EPP moves down the airways from larger to smaller airways because the recoil pressure decreases. A greater length of the downstream segment collapses. The driving pressure for airflow, once the EPP is established, is no longer the difference between alveolar pressure and mouth pressure but is alveolar pressure minus pleural pressure (see Fig. 18.30).

Two basic conclusions follow. First, regardless of the forcefulness of the expiratory effort, airflow cannot be increased because pleural pressure increases, causing more airway compression. This explains why the last part of the FVC is independent of effort.

Second, elastic recoil of the lung determines maximum flow rates because it is the elastic recoil pressure that generates the alveolar–pleural pressure difference. As lung volume decreases, so does elastic recoil force. The decrease in elastic recoil is the main reason maximum flow falls so rapidly at low lung volumes.

The effect of elastic recoil on expiratory airflow can be demonstrated by comparing a normal lung with an emphysematous lung, in which the compliance is abnormally high.

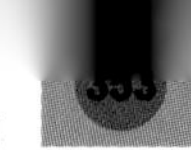

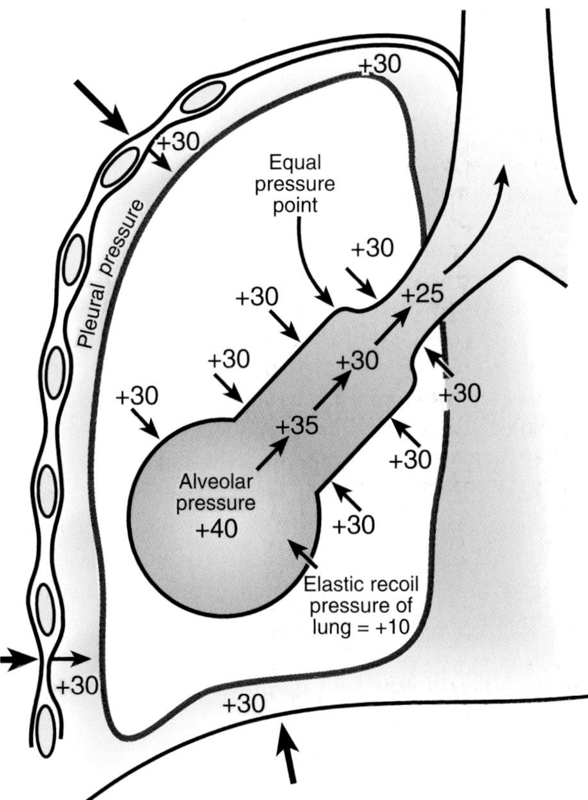

Figure 18.30 An equal pressure point establishes the driving pressure in the airways. An equal pressure point (EPP) divides the airways into downstream and upstream segments. The EPP is established at peak expiratory flow (PEF) and occurs when the pressure inside the airway equals the pressure outside the airway (pleural pressure). The upstream segment is represented from the alveoli to the EPP, and the downstream segment is represented from the EPP to the mouth. The driving pressure for airflow is now alveolar pressure minus the pleural pressure. Airways are subjected to compression during forced expiration from the EPP to the trachea.

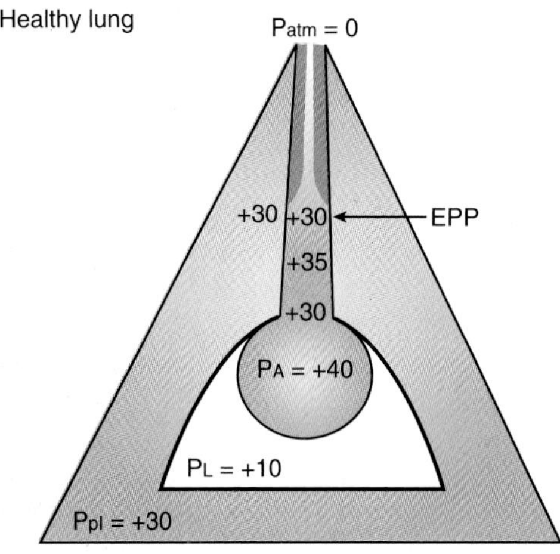

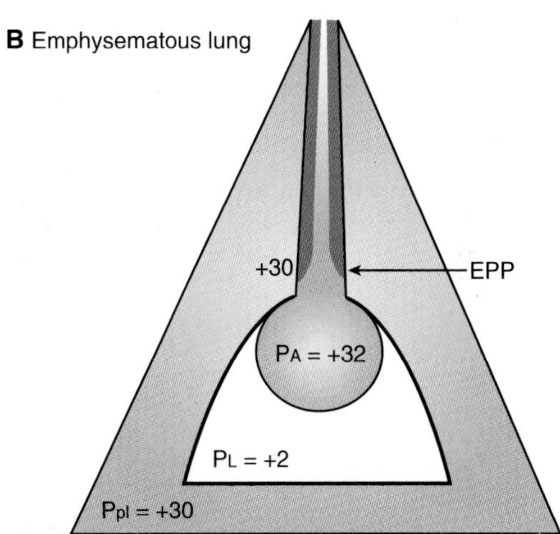

Figure 18.31 The equal pressure point is influenced by the lung elastic recoil. (A) In healthy lungs, elastic recoil adds 10 cm H_2O pressure, to produce an alveolar pressure of 40 cm H_2O at the beginning of a forced expiration. As a result, the equal pressure point (EPP) is established in the larger airways. Airway collapse is minimal in the large airways because cartilage supports the airways. **(B)** A loss of elastic recoil causes the EPP to shift downward and to be established in the small airways. In emphysematous lungs, the elastic recoil pressure is low, and little recoil pressure is added to the alveolus. As a result, the EPP is shifted downward and is established in the smaller airways. The small airways are more easily compressed because they are thin and lack cartilaginous support. P_A, alveolar pressure; P_{atm}, atmospheric pressure; P_L, transpulmonary pressure; P_{pl}, pleural pressure.

As seen in Figure 18.31, in both instances, pleural pressure rises to +30 cm H_2O with forced expiration. In the normal lung, the recoil pressure of 10 cm H_2O is added to produce an alveolar pressure of +40 cm H_2O. With forced expiration, there is a progressive fall in airway pressure. Because of the elastic recoil, the normal lung has "added" pressure to the small airways that keeps their pressures above pleural pressure, and less collapse occurs. With emphysema, however, the lungs have diminished elastic recoil, resulting in less elastic recoil pressure. In the example shown in Figure 18.31, elastic recoil adds only 2 cm H_2O to alveolar pressure, resulting in an alveolar pressure of +32 cm H_2O. With expiratory effort, flow proceeds along the pressure gradient. But in the emphysematous lungs, the pressure inside the small airways falls below pleural pressure well before the large airways, resulting in more airway compression and collapsed airways at low lung volumes. As a result, flow stops temporarily in these collapsed airways. Airway pressure upstream to the collapsed segment then rises to equal alveolar pressure, causing the airways to open again. This process repeats itself continually, leading to the "wheeze" that is often heard in emphysematous patients.

In lungs with abnormally high compliance, the position of the EPP with forced expiration is established further down in the small airways, where there is no cartilage to keep them distended. Therefore, these patients are much more vulnerable to compression and airway collapse. The greatest problem for patients with emphysema is not getting air into the lungs but getting it out. Consequently, they tend to breathe at higher lung volumes, thereby increasing elastic recoil, which reduces airway resistance and facilitates expiration.

Inspiratory muscles inflate the lungs and overcome airway resistance.

During inspiration, muscular work is involved in expanding the thoracic cavity, inflating the lungs, and overcoming airway resistance. Because work can be measured as force times distance, the amount of work involved in breathing can be expressed as a change in lung volume (distance) multiplied by the change in transpulmonary pressure (force). Thus, work (W) is equal to the product of pressure (P) and volume (V). With a volume change, the work involved in taking a breath is defined by this equation:

$$W = P \times \Delta V \quad (23)$$

where P equals transpulmonary pressure and ΔV equals change in lung volume. During work, energy is expended with muscular contraction to create a force (transpulmonary pressure) to inflate the lungs. When a greater transpulmonary pressure is required to bring more air into the lungs, more muscular work and, hence, greater energy are required.

Figure 18.32 shows how a pressure–volume curve can be used to determine the work required for breathing. The *shaded area* in blue represents the inspiratory work of breathing. In healthy people at rest, the energy needed for breathing represents approximately 5% of the body's total energy expenditure. During heavy exercise, about 20% of the total energy expenditure is involved in breathing. Breathing is efficient and is most economical when elastic and resistive forces yield the lowest work. Note that the total inspiratory work of breathing in a restrictive lung disorder, compared with the normal lung, is increased and is a result of a greater inspiratory effort required. It is important to remember that lungs with a marked decrease in lung compliance (i.e., restrictive lung disorder) require more work to overcome increase in elastic recoil and surface forces. Patients with a restrictive disorder economize their ventilation by taking rapid and shallow breaths. In contrast, patients with severe airway obstruction tend to do the opposite; they take deeper breaths and breathe more slowly, to reduce their work resulting from the increased airway resistance. Despite this tendency, patients with obstructive disease still expend a considerable portion of their basal energy for breathing. The reason for this is that the expiratory muscles must do additional work to overcome the increased airway resistance. These different breathing patterns help minimize the amount of work required for breathing.

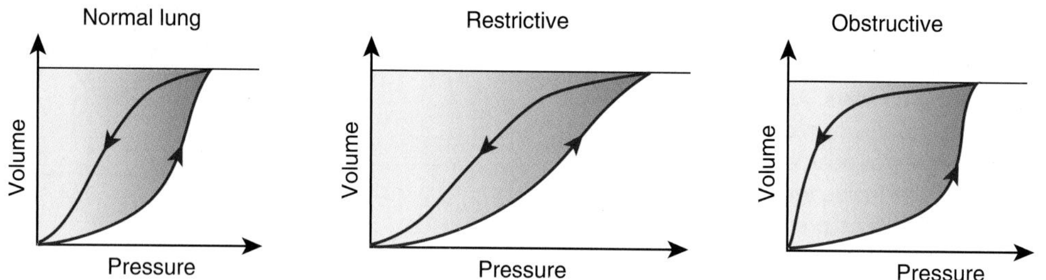

Figure 18.32 Work of breathing is measured from a pressure–volume curve. Pressure is plotted on the *x* axis, and volume is plotted on the *y* axis. Work equals force times distance and is represented by the *blue area* above the pressure–volume curve. This area represents the inspiratory work of breathing. Note that with a restrictive disorder, the inspiratory work of breathing is increased. Airway resistance is increased in an obstructive disorder, which requires more energy to force air out of the lungs during expiration.

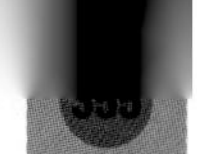

Chapter Summary

- The primary function of the lungs is gas exchange, which involves five steps (ventilation, gas uptake, blood flow, matching airflow and blood flow, and gas transport).
- The alveolar–capillary membrane forms a large blood–gas interface for diffusion of oxygen and carbon dioxide.
- Airflow parallels alveolar pressure.
- A negative alveolar pressure is created during inspiration for airflow into the lungs.
- A positive alveolar pressure is created during expiration for airflow out of the lungs.
- Forced vital capacity is one of the most useful spirometry tests to assess ventilatory function.
- Helium dilution is an indirect method for determining residual lung volume.
- Minute ventilation ($\dot{V}$) is the volume of air expired per minute and is equal to expired minute ventilation ($\dot{V}_E$).
- Alveolar ventilation ($\dot{V}_A$) is the amount of fresh air that reaches alveoli and regulates carbon dioxide levels in the blood.
- Physiologic dead space volume is the portion of tidal volume that is wasted air and does not participate in gas exchange. It is the sum of anatomic dead space volume minus the volume of air in the conducting zone alveolar dead space volume minus volume of air in alveoli that does not participate in gas uptake.
- Compliance is a measure of lung distensibility.
- Surfactant and alveolar interdependence maintain alveolar stability.
- Airway turbulence has a marked effect on airway resistance.
- Lung compliance affects airway compression during forced expiration.
- The work of breathing is required to expand the lungs and overcome airway resistance.
- Major lung diseases are categorized into two groups: obstructive and restrictive. Chronic obstructive pulmonary disease obstructs airflow out of the lungs. Restrictive disease restricts lung inflation.

thePoint *Visit http://thePoint.lww.com for chapter review Q&A, case studies, animations, and more!*

19 Gas Transfer and Transport

ACTIVE LEARNING OBJECTIVES

Upon mastering the material in this chapter, you should be able to:

- Describe how alveolar surface area and membrane thickness affect gas diffusion.
- Describe how pulmonary blood flow and blood hematocrit affect lung diffusion capacity.
- Describe how oxygen binds to hemoglobin.
- Describe the concept P_{50} and its effect on hemoglobin's affinity for oxygen.
- Predict how changes in blood pH, blood P_{CO_2}, body temperature, and carbon monoxide will affect the oxyhemoglobin–equilibrium curve.
- Predict the effect of hematocrit on arterial P_{O_2} and O_2 content.
- Describe the mechanisms by which CO_2 is transported by the blood.
- Describe the physiologic causes of hypoxemia.
- Describe the relationship between venous admixture and the alveolar–arterial oxygen gradient.
- Describe how a low ventilation/perfusion ratio ($\dot{V}_A/\dot{Q}$) differs from an anatomic shunt as a cause of hypoxemia.

GAS DIFFUSION AND UPTAKE

There are two types of gas movements in the lungs: **bulk flow** and **diffusion**. Gas moves in the airways, from the trachea down to the alveoli, by bulk flow, analogous to water coming out of a faucet, in which all molecules move as a unit. The driving pressure (P) for bulk flow in the airways is barometric pressure (PB) at the mouth minus alveolar pressure (PA).

Oxygen and carbon dioxide move across the alveolar-capillary membrane by diffusion.

The movement of gases in the alveoli and across the alveolar–capillary membrane is by diffusion, partly due to partial pressure gradients of individual gases (see Chapter 2, "Plasma Membrane, Membrane Transport, and Resting Membrane Potential"). Recall that partial pressure or gas tension can be determined by measuring barometric pressure and the fractional concentration (F) of the gas (the **Dalton law**; see Chapter 18, "Ventilation and the Mechanics of Breathing"). At sea level, P_{O_2} is 160 mm Hg (760 mm Hg × 0.21). F_{O_2} does not change with altitude, which means that the percentage of O_2 in the atmosphere is essentially the same at 30,000 ft (about 9,000 m) as it is at sea level. Therefore, the difficulty in breathing at high altitudes is due to a decrease in P_{O_2} rather than a decrease in F_{O_2} (Fig. 19.1).

Oxygen is taken up by blood in the lungs and is transported to the tissues. **Oxygen uptake** is the transfer of oxygen from the alveolar spaces, across the alveolar–capillary membrane to the blood. Gas uptake is determined by three factors: the diffusion properties of the alveolar–capillary membrane, the partial pressure gradient of oxygen, and pulmonary capillary blood flow.

The diffusion of gases is a function of the partial pressure difference of the individual gases. For example, oxygen diffuses across the alveolar–capillary membrane because of the difference in P_{O_2} between the alveoli and pulmonary capillaries (Fig. 19.2). The partial pressure difference for oxygen is referred to as the **oxygen diffusion gradient**. In the normal lung, the initial oxygen diffusion gradient, alveolar P_{O_2} (102 mm Hg) minus venous P_{O_2} (40 mm Hg), is 62 mm Hg. The initial diffusion gradient across the alveolar–capillary membrane for carbon dioxide (venous P_{CO_2} [$P_{\bar{v}CO_2}$] minus alveolar P_{CO_2} [P_{ACO_2}]) is about 6 mm Hg, which is much smaller than that of oxygen.

When gases are exposed to a liquid such as blood plasma, gas molecules diffuse into the liquid and exist in a dissolved state. The dissolved gases also exert a partial pressure. A gas will continue to dissolve in the liquid until the partial pressure of the dissolved gas equals the partial pressure above the liquid. The **Henry law** states that at equilibrium, the amount of gas dissolved in a liquid at a given temperature is directly proportional to the partial pressure and the solubility of the gas. The Henry law accounts only for the gas that is physically dissolved and not for chemically combined gases (e.g., oxygen bound to hemoglobin).

Gas diffusion in the lungs can be described by the **Fick law**, which states that the volume of gas diffusing per minute (V_{gas}) across a membrane is directly proportional to the membrane surface area (A_s), the diffusion coefficient of the gas (D), and

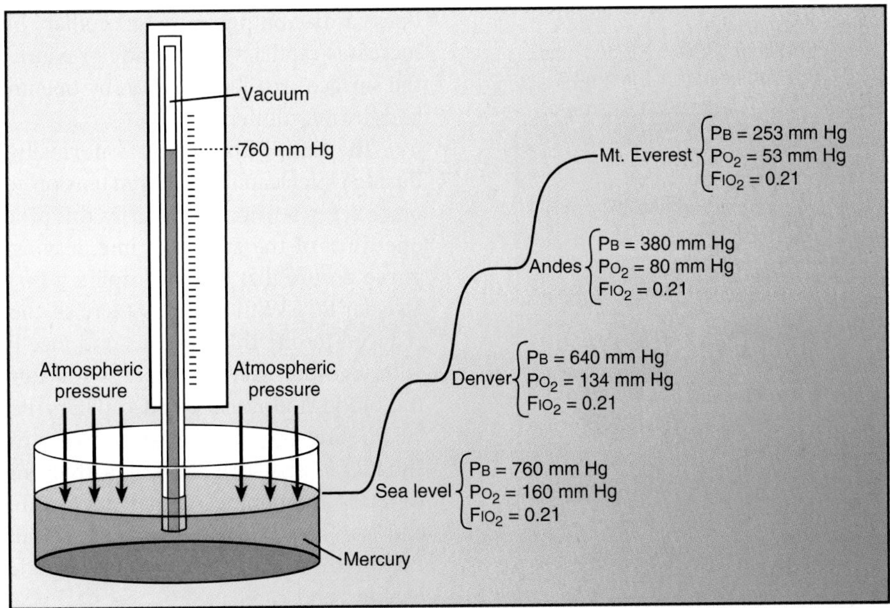

Figure 19.1 Oxygen tension changes with altitude. The height of the column of mercury that is supported by air pressure decreases with altitude, which is a result of a fall in barometric pressure (PB). Because the fractional concentration of inspired O_2 (F_{IO_2}) does not change with altitude, the decrease in P_{O_2} with altitude is caused entirely by a decrease in PB.

the partial pressure difference (ΔP) of the gas and is inversely proportional to the membrane thickness (T) (Fig. 19.3):

$$\dot{V}_{gas} = \frac{As \times D \times \Delta P}{T} \qquad (1)$$

The diffusion coefficient of a gas is directly proportional to its solubility and inversely related to the square root of its molecular weight (MW):

$$D = \frac{\text{solubility}}{(MW)^{1/2}} \qquad (2)$$

Therefore, a small molecule or one that is very soluble will diffuse at a fast rate; for example, the diffusion coefficient of carbon dioxide in aqueous solutions is about 20 times greater than that of oxygen because of its higher solubility, even though it is a larger molecule than O_2.

Therefore, the Fick law states that the rate of gas diffusion is inversely related to membrane thickness. This means that the diffusion of a gas will be halved if membrane thickness is doubled. The Fick law also states that the rate of diffusion is directly proportional to surface area (A_s). If two lungs have the same oxygen diffusion gradient and membrane thickness but one has twice the alveolar–capillary surface area, the rate of diffusion will differ by twofold.

Under steady-state conditions, approximately 250 mL of oxygen per minute are transferred to the pulmonary circulation ($\dot{V}_{O_2}$), whereas 200 mL of carbon dioxide per minute are removed ($\dot{V}_{CO_2}$). The ratio $\dot{V}_{CO_2}/\dot{V}_{O_2}$ is the **respiratory exchange ratio (R)**, and it is 0.8 in this case.

Pulmonary capillary blood flow is a major determinant in limiting oxygen uptake in the lungs.

Pulmonary capillary blood flow has a significant influence on oxygen uptake, and its effect is illustrated in Figure 19.4. The time required for the red cells to move through the capillary, referred to as **transit time**, is approximately 0.75 seconds. During transit time, the gas tension in the blood equilibrates with the alveolar gas tension. Transit time can change dramatically with cardiac output. For example, when cardiac output increases, blood flow through the pulmonary capillaries increases, but transit time decreases (i.e., the time blood is in capillaries is less). Figure 19.4 illustrates the effect of blood flow on the uptake of three test gases. In the first case, a trace amount of nitrous oxide (laughing gas), a common dental anesthetic, is breathed. Nitrous oxide (N_2O) is chosen because it diffuses across the alveolar–capillary membrane and dissolves in the blood but does not combine with hemoglobin. The partial pressure in the blood rises rapidly and virtually reaches equilibrium with the partial pressure of N_2O in the alveoli by the time the blood has spent one tenth of the time in

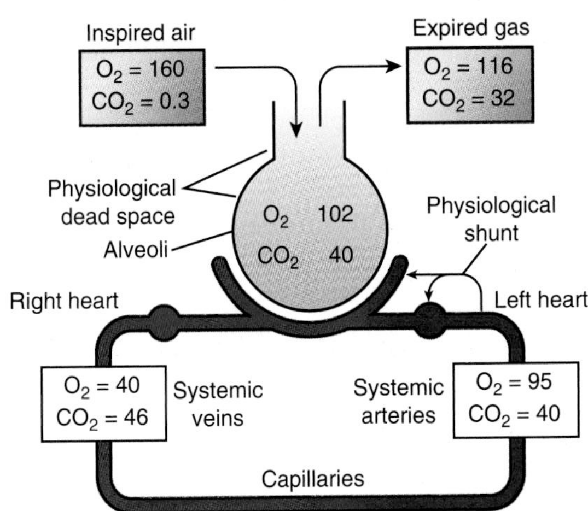

Figure 19.2 Carbon and oxygen tensions vary between systemic and pulmonary circulation. Partial pressure of oxygen (P_{O_2}) is highest when it leaves the lungs, and that of carbon dioxide (P_{CO_2}) is highest when it enters the lungs.

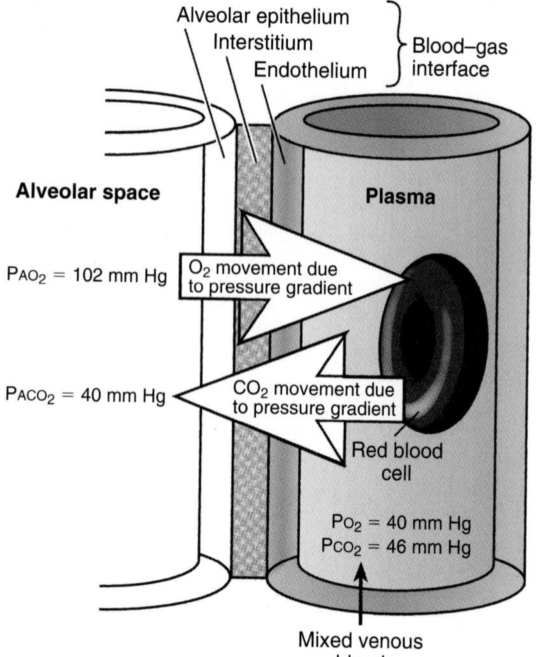

Figure 19.3 Movement of O_2 and CO_2 across the alveolar–capillary membrane is by diffusion. Gases move across the blood–gas interface (alveolar–capillary membrane) by diffusion, following the Fick law. P_{AO_2}, partial pressure of alveolar oxygen; P_{ACO_2}, partial pressure of alveolar carbon dioxide.

the capillary. At this point, the diffusion gradient for N_2O is zero. Once the pressure gradient becomes zero, no additional N_2O is transferred. The only way the transfer of N_2O can be increased is by increasing blood flow. The amount of N_2O that can be taken up is entirely limited by blood flow, not by diffusion of the gas. Therefore, the net transfer or uptake of N_2O is **perfusion limited.**

When a trace amount of carbon monoxide (CO) is breathed, the transfer shows a different pattern (see Fig. 19.4). CO readily diffuses across the alveolar–capillary membrane, but, unlike N_2O, CO has a strong affinity for hemoglobin. As the red cell moves through the pulmonary capillary, CO rapidly diffuses across the alveolar–capillary membrane into the blood and binds to hemoglobin. When a trace amount of CO is breathed, most is chemically bound in the blood, resulting in low partial pressure (P_{CO}). Consequently, equilibrium for CO across the alveolar–capillary membrane is never reached, and the transfer of CO to the blood is, therefore, **diffusion limited** and not limited by the blood flow.

Figure 19.4 shows that the equilibration curve for oxygen lies between the curves for N_2O and CO. Oxygen combines with hemoglobin but not as readily as CO because it has a lower binding affinity. As blood moves along the pulmonary capillary, the rise in P_{O_2} is much greater than the rise in P_{CO} because of differences in binding affinity. Under resting conditions, the capillary P_{O_2} equilibrates with alveolar P_{O_2} when the blood has spent about one third of its time in the capillary. Beyond this point, there is no additional transfer of oxygen. Under normal conditions, oxygen transfer is more like that of N_2O and is limited primarily by blood flow in the capillary (perfusion limited). Hence, an increase in cardiac output will increase oxygen uptake. Not only does cardiac output increase capillary blood flow, but it also increases capillary hydrostatic pressure. The latter increases the surface area for diffusion by opening up more capillary beds by recruitment.

The transit time at rest is normally about 0.75 seconds, during which capillary oxygen tension equilibrates with alveolar oxygen tension. Ordinarily, this process takes only about one third of the available time, leaving a wide safety margin to ensure that the end-capillary P_{O_2} is equilibrated with alveolar P_{O_2}. With vigorous exercise, the transit time may be reduced to one third of a second (see Fig. 19.4). Therefore, with vigorous exercise, there is still time to fully oxygenate the blood. Pulmonary end-capillary P_{O_2} still equals alveolar P_{O_2} and rarely falls with vigorous exercise. In abnormal situations, in which there is a thickening of the alveolar–capillary membrane so that oxygen diffusion is impaired, end-capillary P_{O_2} may not reach equilibrium with alveolar P_{O_2}. In this case, there is measurable difference between alveolar and end-capillary P_{O_2}.

DIFFUSING CAPACITY

In practice, direct measurements of A_s, T, and D in intact lungs are impossible to make. To circumvent this problem, the Fick law can be rewritten as shown in Figure 19.5, in which the three terms are combined as lung **diffusing capacity** (D_L), and is a measure of the lung's ability to transfer gases.

Diffusing capacity measures the rate of oxygen transfer across the alveolar–capillary membrane.

The diffusing capacity provides a measure of the rate of gas transfer in the lungs per partial pressure gradient. For example, if 250 mL of O_2 per minute are taken up and the average alveolar–capillary P_{O_2} difference during a normal transit time is 14 mm Hg, then the D_L for oxygen is 18 mL/min/mm Hg. Because the initial alveolar–capillary difference for oxygen cannot be measured and can only be estimated, CO is used to determine the lung diffusing capacity in patients. CO offers several advantages for measuring D_L:

- Its uptake is limited by diffusion and not by blood flow.
- There is essentially no CO in the venous blood.
- The affinity of CO for hemoglobin is 210 times greater than that of oxygen, which causes the partial pressure of CO to remain essentially zero in the pulmonary capillaries.

To measure the diffusing capacity in a patient with CO, the equation is

$$DL = \frac{\dot{V}_{CO}}{P_{ACO}} \quad (3)$$

where $\dot{V}_{CO}$ equals CO uptake in milliliters per minute and P_{ACO} equals alveolar partial pressure of CO.

The most common technique for making this measurement is called the **single-breath test**. The patient inhales a single breath of a dilute mixture of CO and holds his or her breath for about 10 seconds. By determining the percentage

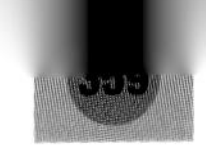

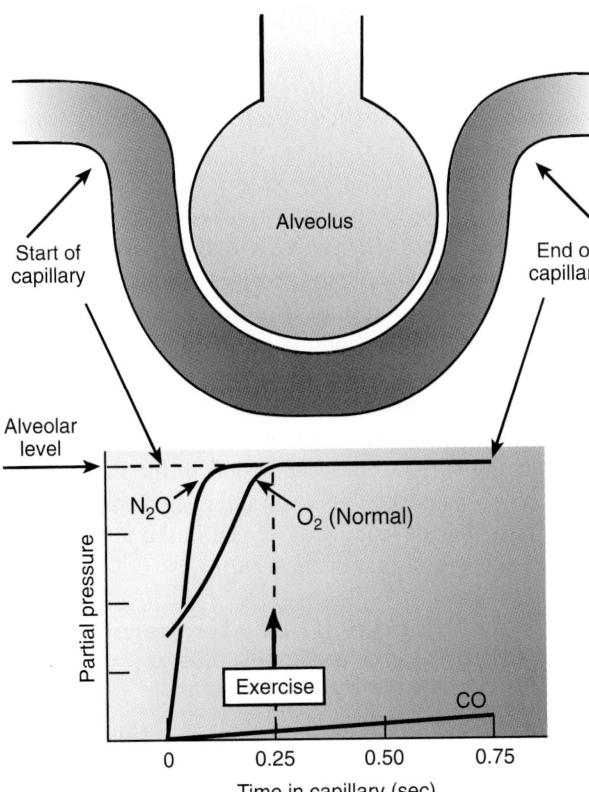

Figure 19.4 Blood flow limits oxygen uptake. Gas transfer across the alveolar–capillary membrane is affected by pulmonary capillary blood flow. The horizontal axis shows time in the capillary. The average transit time it takes blood to pass through the pulmonary capillaries is 0.75 seconds. The vertical axis indicates gas tension in the pulmonary capillary blood, and the top of the vertical axis indicates gas tension in the alveoli. Individual curves indicate the time it takes for the partial pressure of a specific gas in the pulmonary capillaries to equal the partial pressure in the alveoli. Nitrous oxide (N_2O) is used to illustrate how gas transfer is limited by blood flow; carbon monoxide (CO) illustrates how gas transfer is limited by diffusion. The profile for oxygen is more like that of N_2O, which means oxygen transfer is limited primarily by blood flow. Pulmonary capillary Po_2 equilibrates with the alveolar Po_2 in about 0.25 seconds (*arrow*).

of CO in the alveolar gas at the beginning and the end of 10 seconds and by measuring lung volume, one can calculate $\dot{V}_{CO}$. The single-breath test is reliable. The normal resting value for D_{LCO} depends on age, sex, and body size. D_{LCO} ranges from 20 to 30 mL/min/mm Hg and decreases with pulmonary edema or a loss of alveolar membrane (e.g., emphysema).

Blood hematocrit and pulmonary capillary blood volume affect lung diffusing capacity for oxygen.

Diffusing capacity does not depend solely on the diffusion properties of the lungs; it is also affected by blood hematocrit and pulmonary capillary blood volume. Both the hematocrit and capillary blood volume affect D_L in the same direction (i.e., a decrease in either the hematocrit or capillary blood volume will lower the diffusing capacity in otherwise normal lungs). For example, if two people have the same pulmonary diffusion properties but one is anemic (reduced hematocrit), the anemic person will have a decreased lung diffusing capacity. An abnormally low cardiac output lowers the pulmonary capillary blood volume, which decreases the alveolar capillary surface area and will, in turn, decrease the diffusing capacity in otherwise normal lungs.

GAS TRANSPORT BY THE BLOOD

The transport of O_2 and CO_2 by the blood, often referred to as *gas transport*, is an important step in the overall gas exchange process and is one of the important functions of the systemic circulation.

Most oxygen is transported by hemoglobin.

Oxygen is transported to the tissues in two forms: combined with **hemoglobin (Hb)** in the red cell or physically dissolved in the blood. Approximately 98% of the oxygen is carried by hemoglobin, and the remaining 2% is carried in the physically dissolved form. The amount of physically dissolved oxygen in the blood can be calculated from the following equation:

$$\text{Dissolved } O_2(\text{mL/dL}) = 0.003(\text{mL/dL/mm Hg}) \times Pao_2(\text{mm Hg}) \quad (4)$$

If arterial oxygen tension (Pao_2) equals 100 mm Hg, then dissolved O_2 equals 0.3 mL/dL.

The hemoglobin molecule consists of four oxygen-binding heme sites and a globular protein chain. When hemoglobin binds with oxygen, it is called **oxyhemoglobin (Hbo_2)**. The hemoglobin that does not bind with O_2 is called **deoxyhemoglobin (Hb)**. Each gram of hemoglobin can bind with 1.34 mL of oxygen. Oxygen binds rapidly and reversibly to hemoglobin: $O_2 + Hb \leftrightarrow Hbo_2$. The amount of oxyhemoglobin is a function of the partial pressure of oxygen in the blood. In the pulmonary capillaries, in which Po_2 is high, the reaction is shifted to the right to form oxyhemoglobin. In tissue capillaries, in which Po_2 is low, the reaction is shifted to the left; oxygen is unloaded from hemoglobin and becomes available to the cells. The maximum amount of oxygen that can be carried by hemoglobin is called the **oxygen-carrying capacity**—about 20 mL O_2/dL blood in a healthy young adult. This value is calculated assuming a normal hemoglobin concentration of 15 g Hb/dL of blood (1.34 mL O_2/g Hb × 15 g Hb/dL blood = 20.1 mL O_2/dL blood).

Oxygen content is the amount of oxygen actually bound to hemoglobin (whereas capacity is the amount that can potentially be bound). The percentage saturation of hemoglobin with oxygen (So_2) is calculated from the ratio of oxyhemoglobin content over capacity:

$$So_2 = \frac{Hbo_2\text{ content}}{Hbo_2\text{ capacity}} \times 100 \quad (5)$$

Therefore, the oxygen saturation is the ratio of the quantity of oxygen *actually bound* to the quantity that can be *potentially*

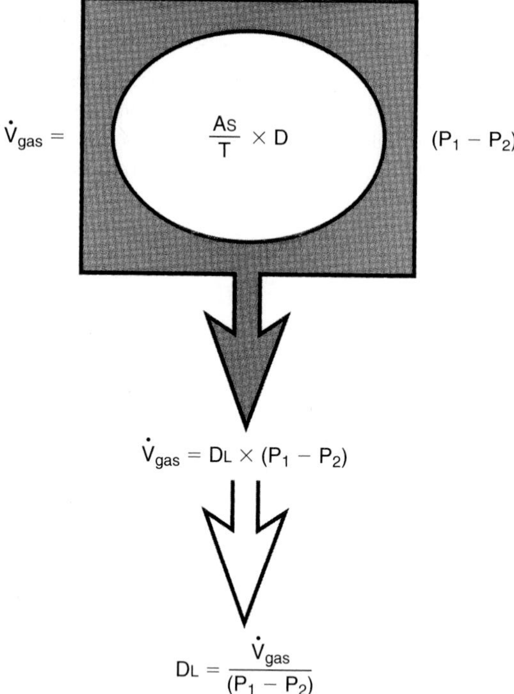

Figure 19.5 Lung diffusing capacity measures the volume of O_2 taken up per minute. Membrane surface area (As), the gas diffusion coefficient (D), and membrane thickness (T) affect gas diffusion in the lungs. These properties are combined into one term, lung diffusing capacity (DL), which can be measured in a human subject. DL is equal to the volume of gas transferred per minute ($\dot{V}_{gas}$) divided by the mean partial pressure gradient for the gas. P, pressure.

bound. For example, if oxygen content is 16 mL O_2/dL blood and oxygen capacity is 20 mL O_2/dL blood, then the blood is 80% saturated. Arterial blood saturation of hemoglobin with oxygen (Sao_2) is normally about 98%.

Oxyhemoglobin-equilibrium curve illustrates how plasma Po_2 affects the loading and unloading of oxygen from hemoglobin.

Blood Po_2, O_2 saturation, and oxygen content are three closely related indices of oxygen transport. The relationship between Po_2, oxygen saturation, and oxygen content is illustrated by the **oxyhemoglobin-equilibrium curve**, an S-shaped curve over a range of arterial oxygen tensions from 0 to 100 mm Hg (Fig. 19.6). The shape of the curve results because the hemoglobin affinity for oxygen increases progressively as blood Po_2 increases.

The shape of the oxyhemoglobin-equilibrium curve reflects several important physiologic advantages. The *plateau region* of the curve is the **loading phase**, in which oxygen is loaded onto hemoglobin to form oxyhemoglobin in the pulmonary capillaries. The plateau region illustrates how oxygen saturation and content remain fairly constant despite wide fluctuations in alveolar Po_2. For example, if Pao_2 were to rise from 100 to 120 mm Hg, hemoglobin would become only slightly more saturated (97%–98%). For this reason, oxygen content cannot be raised appreciably by hyperventilation. The *steep* **unloading phase** of the curve allows large quantities of oxygen to be released or unloaded from hemoglobin in the tissue capillaries, in which a lower capillary Po_2 prevails. The S-shaped oxyhemoglobin-equilibrium curve enables oxygen to saturate hemoglobin under high partial pressures in the lungs and to give up large amounts of oxygen with small changes in Po_2 at the tissue level.

A change in the binding affinity of hemoglobin for O_2 shifts the oxyhemoglobin-equilibrium curve to the right or left of normal (Fig. 19.7). The P_{50}—the Po_2 at which 50% of the hemoglobin is saturated—provides a functional way to assess the binding affinity of hemoglobin for oxygen. The normal P_{50} for arterial blood is 26 to 28 mm Hg. A high P_{50} signifies a decrease in hemoglobin's affinity for oxygen and results in a rightward shift in the oxyhemoglobin-equilibrium curve, whereas a low P_{50} signifies the opposite and shifts the curve to the left. A shift in the P_{50} in either direction has the greatest effect on the steep phase and only a small effect on the loading of oxygen in the normal lung, because loading occurs at the plateau.

Changes in blood pH, body temperature, and arterial Pco_2 drastically alters hemoglobin's affinity for oxygen.

Several factors affect the binding affinity of hemoglobin for O_2, including blood temperature, arterial carbon dioxide tension, and arterial pH. A rise in Pco_2, a fall in pH, and a rise in temperature all shift the curve to the right (see Fig. 19.7). The effect of carbon dioxide and hydrogen ions on hemoglobin's affinity for oxygen is known as the **Bohr effect**. A shift of the oxyhemoglobin-equilibrium curve to the right is physiologically advantageous at the tissue level because the affinity is lowered (increased P_{50}). A rightward shift enhances the unloading of oxygen for a given Po_2 in the tissue, and a leftward

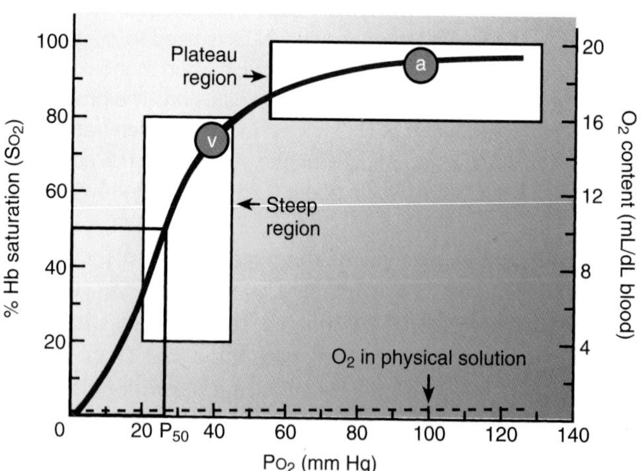

Figure 19.6 The oxyhemoglobin-equilibrium curve is nonlinear. The oxygen saturation (left vertical axis) or oxygen content (right vertical axis) is plotted against the partial pressure of oxygen (horizontal axis) to generate an oxyhemoglobin equilibrium curve. The curve is S shaped and can be divided into a *plateau region* and a *steep region*. The *dashed line* indicates the amount of oxygen dissolved in the plasma. a, arterial; v, venous; Hb, hemoglobin; So_2, oxygen saturation; P_{50}, partial pressure of O_2 required to saturate 50% of the hemoglobin with oxygen.

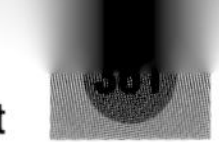

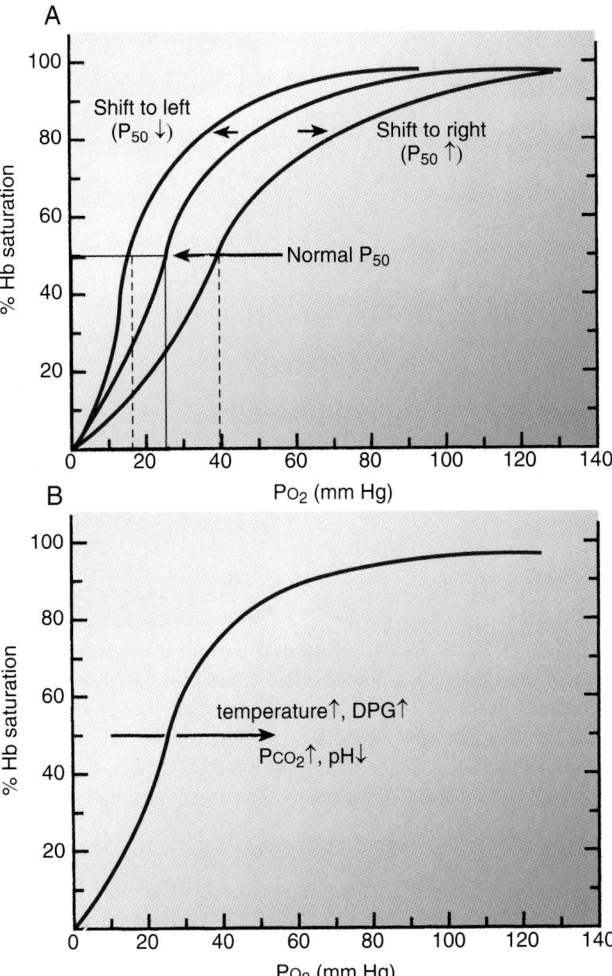

Figure 19.7 Oxygen has a strong binding affinity for hemoglobin (Hb). (A) P_{50} is a measure of Hb's affinity to bind with oxygen. **(B)** An increase in temperature, [H^+], or arterial Pco_2 causes a rightward shift of the oxyhemoglobin–equilibrium curve. An increase in P_{50} indicates a lower binding affinity for oxygen, which favors the unloading of O_2 from Hb at the tissue level. An increase in red cell levels of 2,3-diphosphoglycerate (DPG) will also shift the curve to the right. The increase in DPG occurs with hypoxemic conditions. P_{50}, partial pressure of O_2 required to saturate 50% of the hemoglobin with oxygen; temp., temperature.

shift increases the affinity of hemoglobin for oxygen, thereby lowering the ability to release oxygen to the tissues. A simple way to remember the functional importance of these shifts is that an exercising muscle is hot and acidic and produces large amounts of carbon dioxide (high Pco_2), all of which favor the unloading of more oxygen to metabolically active muscles.

Red blood cells contain 2,3-diphosphoglycerate (2,3-DPG), an organic phosphate compound that can also affect the affinity of hemoglobin for oxygen. In red cells, 2,3-DPG levels are much higher than in other cells because erythrocytes lack mitochondria. An increase in 2,3-DPG facilitates unloading of oxygen from the red cell at the tissue level (shifts the curve to the right). An increase in red cell 2,3-DPG occurs with exercise and with hypoxia (e.g., high altitude, chronic lung disease).

It is important to remember that oxygen content, rather than Po_2 or Sao_2, is what keeps us alive and serves as a better gauge for oxygenation. For example, a person can have a normal arterial Po_2 and Sao_2 but reduced oxygen content. This situation is seen in patients who have anemia (a decreased number of circulating red cells). A patient with anemia who has a hemoglobin concentration half of normal (7.5 g/dL instead of 15 g/dL) will have a normal arterial Po_2 and Sao_2, but oxygen content will be reduced to half of normal. A patient with anemia has a normal Sao_2 because that content and capacity are proportionally reduced. The usual oxyhemoglobin–equilibrium curve does not show changes in blood oxygen content, because the vertical axis is saturation. If the vertical axis is changed to oxygen content (mL O_2/dL blood), then changes in content are seen (Fig. 19.8). The shape of the oxyhemoglobin–equilibrium curve does not change, but the curve moves down to reflect the reduction in oxygen content. A good analogy for comparing an anemic patient with a healthy patient is comparing a bicycle tire with a truck tire: Both can have the same air pressure, but the amount of air each tire holds is different.

Binding affinity of carbon monoxide for hemoglobin is much stronger than that of oxygen.

CO interferes with oxygen transport by competing for the same binding sites on hemoglobin. CO binds to hemoglobin to form **carboxyhemoglobin (Hbco)**. The reaction (Hb + CO ↔ Hbco) is reversible and is a function of Pco. This means that breathing higher concentrations of CO will favor the reaction to the right and form more Hbco. Breathing fresh air will favor the reaction to the left, which will cause CO to be released from the hemoglobin. A striking feature of CO

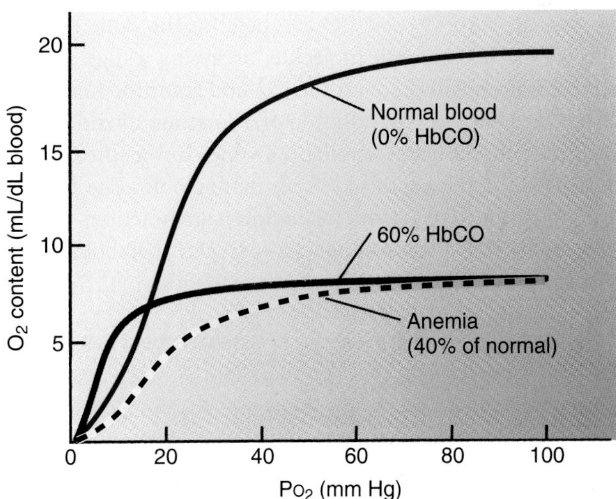

Figure 19.8 Hematocrit and CO poisoning affect the oxyhemoglobin–equilibrium curve. Severe anemia can lower the O_2 content to 40% of normal. The blood O_2 content of a person exposed to CO is shown for comparison. When the blood is 60% saturated with carbon monoxide (Hbco), O_2 content is reduced to about 8 mL/dL of blood. Note the leftward shift of the oxyhemoglobin equilibrium curve when CO binds with hemoglobin.

is a binding affinity about 210 times that of oxygen. Consequently, CO will bind with the same amount of hemoglobin as oxygen at a partial pressure 210 times lower than that of oxygen. For example, breathing normal air (21% O_2) contaminated with 0.1% CO would cause half of the hemoglobin to be saturated with CO and half with O_2. With the high affinity of hemoglobin for CO, breathing a small amount of CO can result in the formation of large amounts of Hbco. Arterial P_{O_2} in the plasma will still be normal because the oxygen diffusion gradient has not changed. However, oxygen content will be greatly reduced because oxygen cannot bind to hemoglobin. This is seen in Figure 19.8, which shows the effect of CO on the oxyhemoglobin equilibrium curve. When the blood is 60% saturated with CO (carboxyhemoglobin), the oxygen content is reduced to <10 mL/dL. The presence of CO also shifts the curve to the left, making it more difficult to unload or release oxygen to the tissues. CO is dangerous for several reasons:

- It has a strong binding affinity for hemoglobin.
- As an odorless, colorless, and nonirritating gas, it is virtually undetectable.
- Pa_{O_2} is normal, and there is no feedback mechanism to indicate that oxygen content is low.
- There are no physical signs of hypoxemia (i.e., **cyanosis** or bluish color around the lips and fingers) because the blood is bright cherry red when CO binds with hemoglobin.

Therefore, a person can be exposed to CO and have oxygen content reduced to a level that becomes lethal, by causing tissue anoxia, without the person being aware of the danger. The brain is one of the first organs affected by lack of oxygen. CO can alter reaction time and cause blurred vision and, if severe enough, unconsciousness.

The best treatment for CO poisoning is breathing 100% oxygen. Because O_2 and CO compete for the same binding site on the hemoglobin molecule, breathing a high oxygen concentration will drive off the CO and favor the formation of oxyhemoglobin. The addition of 5% carbon dioxide to the inspired gas stimulates ventilation, which lowers the CO and enhances the release of CO from hemoglobin. The loading and unloading of CO from hemoglobin are functions of Pco. Oxygen is not always beneficial—oxygen metabolism can produce harmful products that injure tissues.

Most carbon dioxide is transported as bicarbonate.

Figure 19.9 illustrates the processes involved in carbon dioxide transport. Carbon dioxide is carried in the blood in three forms:

- Physically dissolved in the plasma (10%)
- As bicarbonate ions in the plasma and in the red cells (60%)
- As carbamino proteins (30%)

The high P_{CO_2} in the interstitial fluid drives carbon dioxide from the tissue into the blood, but only a small amount stays as dissolved CO_2 in the plasma. The bulk of the carbon dioxide diffuses into the red cell, in which it forms either carbonic acid (H_2CO_3) or **carbaminohemoglobin**. In the red cell, carbonic acid is formed in the following reaction:

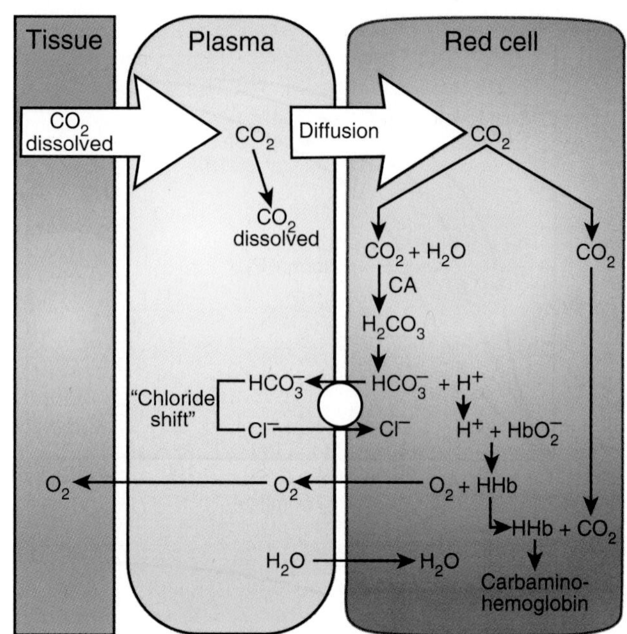

Figure 19.9 Bicarbonate is the main transporter of carbon dioxide. CO_2 is transported in the blood in three forms: physically dissolved, as HCO_3^-, and as carbaminohemoglobin in the red cell (see text for details). The uptake of CO_2 favors the release of O_2. A major portion of the CO_2 is carried in the form of bicarbonate. CA, carbonic anhydrase; Hb, hemoglobin.

$$CO_2 + H_2O \xleftrightarrow{CA} H_2CO_3 \leftrightarrow H^+ + HCO_3^- \quad (6)$$

The hydration of CO_2 would take place slowly if it were not accelerated about 1,000 times in red cells by the enzyme **carbonic anhydrase (CA)**. This enzyme is also found in renal tubular cells, gastrointestinal mucosa, muscle, and other tissues, but its activity is highest in red blood cells.

Carbonic acid readily dissociates in red blood cells to form bicarbonate (HCO_3^-) and H^+. HCO_3^- leaves the red blood cells, and chloride diffuses in from the plasma to maintain electrical neutrality (see Fig. 19.9). The chloride movement is known as the *chloride shift* and is facilitated by a chloride–bicarbonate exchanger (anion exchanger) in the red blood cell membrane. The H^+ cannot readily move out because of the low permeability of the membrane to H^+. Most of the H^+ is buffered by hemoglobin: $H^+ + HbO_2^- \leftrightarrow HHb + O_2$. As H^+ binds to hemoglobin, it decreases oxygen binding and shifts the oxyhemoglobin–equilibrium curve to the right. This promotes the unloading of oxygen from hemoglobin in the tissues and favors the carrying of carbon dioxide. In the pulmonary capillaries, the oxygenation of hemoglobin favors the unloading of carbon dioxide.

Carbaminohemoglobin is formed in red cells from the reaction of carbon dioxide with free amine groups (NH_2) on the hemoglobin molecule:

$$CO + HbNH2 \leftrightarrow HbNHCOOH \quad (7)$$

Deoxygenated hemoglobin can bind much more CO_2 in this way than oxygenated hemoglobin. Although major reactions related to CO_2 transport occur in the red cells, the bulk of the CO_2 is actually carried in the plasma in the form of bicarbonate.

A **carbon dioxide equilibrium curve** can be constructed in a fashion similar to that for oxygen (Fig. 19.10). The carbon dioxide equilibrium curve is nearly a straight-line function of P_{CO_2} in the normal arterial CO_2 range. Note that a higher P_{O_2} will shift the curve downward and to the right. This is known as the **Haldane effect**, and its advantage is that it allows the blood to load more CO_2 in the tissues and unload more CO_2 in the lungs.

Important differences are observed between the carbon dioxide and oxygen equilibrium curves (Fig. 19.11). First, 1 L of blood can hold much more carbon dioxide than oxygen. Second, the CO_2 equilibrium curve is steeper and more linear, and because of its shape, large amounts of CO_2 can be loaded and unloaded from the blood with a small change in P_{CO_2}. This is important not only in gas exchange and transport but also in the regulation of acid–base balance.

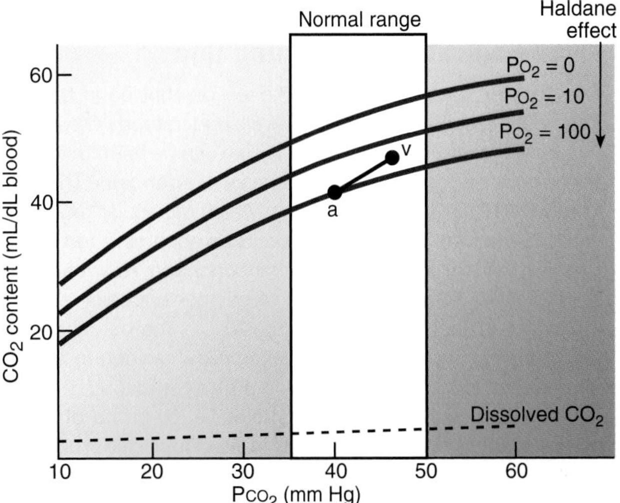

Figure 19.10 Increased O_2 tension shifts the carbon dioxide equilibrium curve. The carbon dioxide equilibrium curve is relatively linear. An increase in P_{O_2} tension causes a rightward and downward shift of the curve. The P_{O_2} effect on the CO_2 equilibrium curve is known as the *Haldane effect*. The *dashed line* indicates the amount dissolved in plasma. a, CO_2 content in arterial blood; v, CO_2 content in mixed venous blood.

RESPIRATORY CAUSES OF HYPOXEMIA

Under normal conditions, hemoglobin is 100% saturated with oxygen when the blood leaves the pulmonary capillaries, and the P_{O_2} at the end of the capillary (end-capillary P_{O_2}) equals alveolar P_{O_2}. However, the blood that leaves the lungs (via the pulmonary veins) and returns to the left side of the heart has a lower P_{O_2} than the pulmonary end-capillary blood. As a result, the systemic arterial blood has an average oxygen tension (Pa_{O_2}) of about 95 mm Hg and is only 98% saturated.

Alveolar-arterial oxygen gradient originates because bronchial circulation mixes with oxygenated blood.

The difference between alveolar oxygen tension (P_{AO_2}) and arterial oxygen tension (Pa_{O_2}) is the **alveolar–arterial oxygen gradient** or A–a O_2 gradient (Fig. 19.12). Because alveolar P_{O_2} is normally 100 to 102 mm Hg and arterial P_{O_2} is 85 to 95 mm Hg, a normal A–a O_2 gradient is 5 to 15 mm Hg. The A–a O_2 gradient is obtained from blood gas measurements and the alveolar gas equation to determine P_{AO_2}. Recall from Chapter 18, "Ventilation and the Mechanics of Breathing," that the simplified equation is $P_{AO_2} = F_{IO_2} \times (P_B - 47) - 1.2 \times Pa_{CO_2}$.

The A–a O_2 gradient arises in the healthy person because a fraction of venous blood mixes with oxygenated blood. This mixing of unoxygenated and oxygenated blood is known as **venous admixture**. Two physiologic causes of venous admixture are the result of a small anatomical shunt (e.g., bronchial circulation) and regional variations in the ventilation/perfusion ($\dot{V}_A/\dot{Q}$); ratio. The $\dot{V}_A/\dot{Q}$ ratio is simply the ratio of minute alveolar ventilation to pulmonary blood flow in any unit of lung. Proper oxygenation of blood in the pulmonary circulation leaving any region of lung (regional alveoli and their blood supply) occurs when ventilation and perfusion are quantitatively matched in that region. Oxygenation of blood

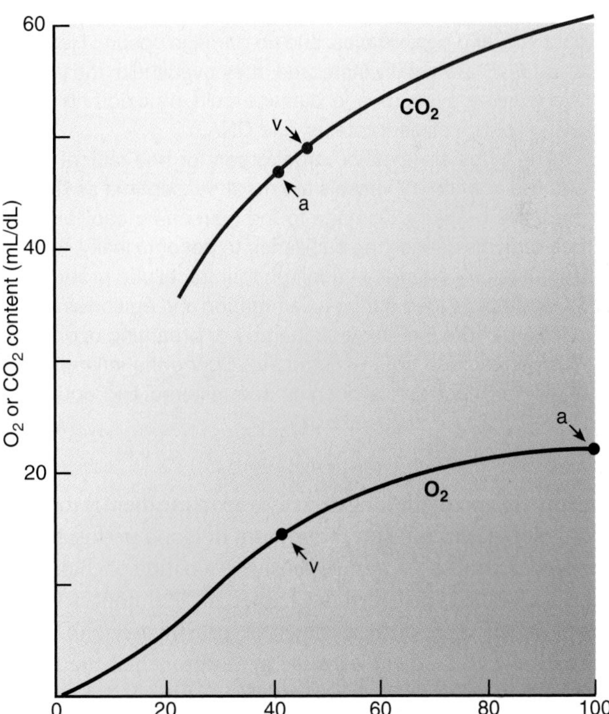

Figure 19.11 Blood has a greater carrying capacity for CO_2 than O_2. A comparison of the oxyhemoglobin and CO_2 equilibrium curves reveals that the carrying capacity for CO_2 is much greater than for O_2. The increased steepness and linearity of the CO_2 equilibrium curve allow the lungs to remove large quantities of CO_2 from the blood with a small change in CO_2 tension. a, gas content in arterial blood; v, gas content in mixed venous blood.

Clinical Focus / 19.1

Free Radical–Induced Lung Injury

An "oxygen paradox" has long been recognized in biology, but only recently has it been well understood: Oxygen is essential for life, but too much oxygen can be harmful to both cells and the organism. The synthesis of **adenosine triphosphate (ATP)** involves reactions in which molecular oxygen is reduced to form water. This reduction is accomplished by the addition of four electrons by the mitochondrial electron transport system. About 98% of the oxygen consumed is reduced to water in the mitochondria. "Leaks" in the mitochondrial electron transport system, however, allow oxygen to accept fewer than four electrons, forming a **free radical**.

A free radical is any atom, molecule, or group of molecules with an unpaired electron in its outermost orbit. Free radicals include the **superoxide ion (O_2•$^-$)** and the **hydroxyl radical (•OH)**. The single unpaired electron in the free radical is denoted by a dot. The •OH radical is the most reactive and most damaging to cells. **Hydrogen peroxide (H_2O_2)**, although not a free radical, is also reactive to tissues and has the potential to generate the hydroxyl radical (•OH). These three substances are collectively called **reactive oxygen species (ROS)**. In addition to free radicals produced by leaks in the mitochondrial transport system, ROS can also be formed by cytochrome P450, in the production of nicotinamide adenine dinucleotide phosphate and in arachidonic acid metabolism. A superoxide ion in the presence of nitric oxide will form peroxynitrite, another free radical that is also extremely toxic to cells. Under normal conditions, ROS are neutralized by the protective enzymes superoxide dismutase, catalase, and peroxidases, and no damage occurs. However, when ROS are greatly increased, they overwhelm the protective enzyme systems and damage cells by oxidizing membrane lipids, cellular proteins, and DNA.

The lungs are a major target organ for free radical injury, and the pulmonary vessels are most susceptible as the primary site of injury. Damage to the pulmonary capillaries by free radicals causes the capillaries to become leaky, leading to pulmonary edema. In addition to intracellular production, ROS are produced during inflammation and episodes of oxidant exposure (i.e., oxygen therapy or breathing ozone and nitrogen dioxide from polluted air). During the inflammatory response, neutrophils become sequestered and activated; they undergo a respiratory burst (which produces free radicals) and release catalytic enzymes. This release of free radicals and catalytic enzymes functions to kill bacteria, but endothelial cells can become damaged in the process.

Paraquat, an agricultural herbicide, is another source of free radical–induced injury to the lungs. Crop dusters and migrant workers are particularly at risk because of exposure to paraquat through the lungs and skin. Tobacco or marijuana that has been sprayed with paraquat and subsequently smoked can also produce lung injury from ROS.

Ischemia reperfusion is another cause of free radical–induced injury in organs. In the lungs, ischemia-reperfusion injury results from a blood clot that gets lodged in the pulmonary circulation. Tissues beyond the clot (or embolus) become ischemic, cellular ATP decreases, and hypoxanthine accumulates. When the clot dissolves, blood flow is reestablished. During the reperfusion phase, hypoxanthine, in the presence of oxygen, is converted to xanthine and then to urate. These reactions, catalyzed by the enzyme xanthine oxidase on the pulmonary endothelium, result in the production of superoxide ions. Neutrophils also become sequestered and activated in these vessels during the reperfusion phase. Therefore, the pulmonary vasculature and surrounding lung parenchyma become damaged from a double hit of free radicals—those produced from the oxidation of hypoxanthine and those from activated neutrophils.

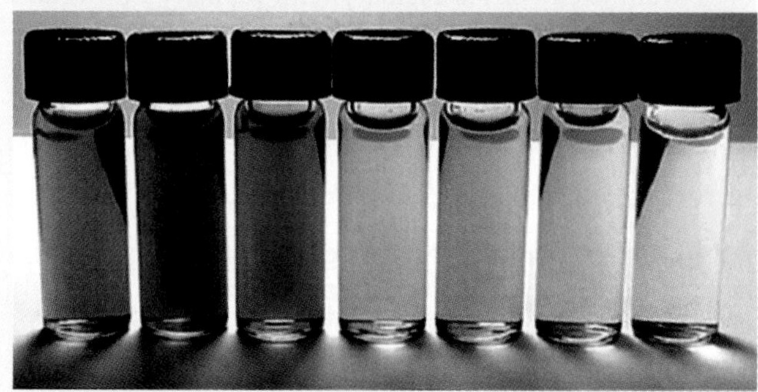

Microscopic gold particles and nanoshells (shown here in solution) can be tuned to absorb different wavelengths of light—a feature that is proving useful for cancer treatment.

leaving a region of the lung decreases any time there is too little ventilation per minute for the amount of blood perfusing that region per minute. (A more extensive evaluation of changes in the $\dot{V}_A/\dot{Q}$; ratio is provided in Chapter 20, "Pulmonary Circulation and Ventilation/Perfusion.") Approximately half of the normal A–a O_2 gradient is caused by the bronchial circulation and half is caused by regional variations of the $\dot{V}_A/\dot{Q}$ ratio. In some pathophysiologic disorders, the A–a O_2 gradient can be greatly increased. A value of >15 mm Hg is considered abnormal and usually leads to low oxygen in the blood or **hypoxemia**. The normal ranges of blood gases are shown in Table 19.1. Values for $Pa{O_2}$ below 85 mm Hg indicate hypoxemia. A $Pa{CO_2}$ <35 mm Hg is called **hypocapnia**, and a $Pa{CO_2}$ >48 mm Hg is called **hypercapnia**. A pH value for arterial blood of <7.35 or >7.45 is called **acidemia** or **alkalemia**, respectively.

Respiratory dysfunction is the major cause of hypoxemia.

The causes of hypoxemia are classified as respiratory or nonrespiratory (Table 19.2). Respiratory dysfunction is by far the most common cause of hypoxemia in adults. Nonrespiratory causes include anemia, CO poisoning, and a decreased inspired oxygen tension (as occurs at high altitude). The respiratory causes of hypoxemia are listed in order of frequency in Table 19.2. **Regional hypoventilation** is by far the most common cause of hypoxemia (about 90% of cases) and reflects a local $\dot{V}_A/\dot{Q}$ ratio imbalance. The matching of airflow and blood flow is best examined by considering the **ventilation/perfusion ratio**, which compares alveolar ventilation with blood flow in lung regions. Because resting healthy people have an alveolar ventilation ($\dot{V}_A$) of 4 L/min and a cardiac output ($\dot{Q}$) of 5 L/min, the

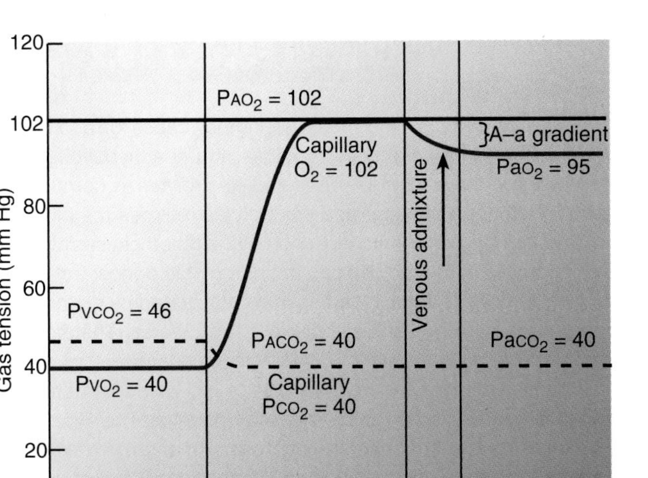

Figure 19.12 An oxygen gradient is established between the alveoli and arterial blood. The diagram shows O_2 and CO_2 tensions in blood in the pulmonary artery, pulmonary capillaries, and systemic arterial blood. The Po_2 leaving the pulmonary capillary has equilibrated with alveolar Po_2. However, systemic arterial Po_2 is below alveolar Po_2. Venous admixture results in the alveolar–arterial oxygen gradient (A–a O_2). P_{ACO_2}, partial pressure of carbon dioxide, alveolar; $Paco_2$, partial pressure of carbon dioxide, arterial; P_{AO_2}, partial pressure of oxygen, alveolar; Pao_2, partial pressure of oxygen, arterial; $Pvco_2$, partial pressure of carbon dioxide, venous; Pvo_2, partial pressure of oxygen, venous.

ideal alveolar ventilation/perfusion ratio ($\dot{V}_A/\dot{Q}$) should be 0.8 (there are no units; this is a ratio). When a partially obstructed airway occurs, a fraction of the blood that passes through the capillary bed of the obstructed airway does not get fully oxygenated, resulting in an increase in venous admixture. Only a small amount of venous admixture is required to lower systemic arterial Po_2 as a result of the nature of the oxyhemoglobin–equilibrium curve. This can be seen from Figure 19.13, which depicts oxygen content from three groups of alveoli with low, normal, and high ($\dot{V}_A/\dot{Q}$) ratios. The oxygen content of the blood leaving these alveoli is 16.0, 19.5, and 20.0 mL/dL of blood, respectively. As Figure 19.13 shows, a low ($\dot{V}_A/\dot{Q}$) ratio is far more serious because it has the greatest effect on lowering both the Po_2 and the O_2 content because of the nonlinear shape of the oxyhemoglobin equilibrium curve. Patients who have an abnormally low ($\dot{V}_A/\dot{Q}$) ratio have a high A–a O_2 gradient, low Po_2, and low O_2 content, but usually a normal or slightly elevated $Paco_2$. $Paco_2$ does not change much because the CO_2 equilibrium curve is nearly linear, which allows excess CO_2 to be removed from the blood by the lungs.

Another cause for a regionally low ($\dot{V}_A/\dot{Q}$) ratio is a large blood clot that occludes a major artery in the lungs. When a major pulmonary artery becomes occluded, a greater portion of the cardiac output is redirected to another part of the lungs, resulting in overperfusion with respect to alveolar ventilation. This causes a regionally low ($\dot{V}_A/\dot{Q}$) and leads to an increase in venous admixture.

TABLE 19.1 Arterial Blood Gases

Parameter	Normal Range[a]
Pao_2	85–95 mm Hg
$Paco_2$	35–48 mm Hg
Sao_2	94%–98%
pH	7.35–7.45
HCO_3^-	23–28 mEq/L

[a]Normal range at sea level.

Pao_2, partial pressure of oxygen, arterial; $Paco_2$, partial pressure of carbon dioxide, arterial; Sao_2, saturation of arterial oxygen.

The next most common cause of hypoxemia is a **shunt**, either a right-to-left anatomic shunt or an absolute intrapulmonary shunt. The latter occurs when an airway is totally obstructed by a foreign object (such as a peanut) or by a lung tumor. Patients with hypoxemia stemming from a shunt also have a high A–a O_2 gradient, low Po_2, low O_2 content, and a normal or slightly elevated $Paco_2$. A test that is often used to distinguish between an abnormally low ($\dot{V}_A/\dot{Q}$) ratio and a shunt is to have the patient breathe 100% O_2 for 15 minutes. If the Pao_2 is >150 mm Hg, the cause is a low ($\dot{V}_A/\dot{Q}$) ratio. If the patient's Pao_2 is <150 mm Hg, the cause of hypoxemia is a shunt. The principle for using 100% O_2 is illustrated in Figure 19.14. The patient with regional hypoventilation who breathes 100% O_2 compensates for the low ($\dot{V}_A/\dot{Q}$) ratio, and because all of the blood leaving the pulmonary capillaries is now fully saturated, the venous admixture is eliminated. However, the low arterial Po_2 does not get corrected by breathing 100% O_2 in a patient with a shunt because the

TABLE 19.2 Pathophysiologic Causes of Hypoxemia

Causes	Effect on A–a O_2 Gradient
Respiratory	
Regional low $\dot{V}_A/\dot{Q}$ ratio	Increased
Anatomic shunt	Increased
Generalized hypoventilation	Normal
Diffusion block	Increased
Nonrespiratory	
Intracardiac right-to-left shunt	Increased
Decreased P_{IO_2}, low P_B, low F_{IO_2}	Normal
Reduced oxygen content (anemia and CO poisoning)	Normal

A–a O_2, alveolar–arterial oxygen; F_{IO_2}, fractional concentration of inspired oxygen; P_B, barometric pressure; P_{IO_2}, inspired partial pressure of oxygen; $\dot{V}_A/Q$ ratio, ventilation/perfusion ratio.

Clinical Focus / 19.2

Anemia

Anemia, an abnormally low hematocrit or hemoglobin concentration, is by far the most common disorder affecting erythrocytes. The different causes of anemia can be grouped into three categories: decreased erythropoiesis by bone marrow, blood loss, and increased rate of red cell destruction (hemolytic anemia).

Several mechanisms lead to decreased production of red cells by the bone marrow, including aplastic anemia, malignant neoplasms, chronic renal disease, defective DNA synthesis, defective hemoglobin synthesis, and chronic liver disease. **Aplastic anemia** is the result of stem cell destruction in the bone marrow, which leads to decreased production of white cells, platelets, and erythrocytes. Malignant neoplasms (e.g., leukemia) cause an overproduction of immature red cells. Patients with chronic renal disease have a decreased production of erythropoietin, with a concomitant decrease in red cell production.

Patients with defective DNA synthesis have **megaloblastic anemia**, a condition in which red cell maturation in the bone marrow is abnormal; this may result from vitamin B12 or folic acid deficiency. These cofactors are essential for DNA synthesis. Vitamin B12 is present in high concentration in liver and, to some degree, in most meat, but it is absent in plants. Vitamin B12 deficiency is rare except in strict vegetarians. Folic acid is widely distributed in leafy vegetables; folic acid deficiency commonly occurs when malnutrition is prevalent. **Pernicious anemia** is a form of megaloblastic anemia resulting from vitamin B12 deficiency. Most common in adults over 60, it results not from deficient dietary intake but from a decreased vitamin B12 absorption by the small intestine. Pernicious anemia is linked to an autoimmune disease in which there is immunologic destruction of the intestinal mucosa, particularly the gastric mucosa.

Iron-deficiency anemia is the most common cause of anemia worldwide. Although it occurs in both developed and undeveloped countries, the causes are different. In developed countries, the cause is usually a result of pregnancy or chronic blood loss resulting from gastrointestinal ulcers or neoplasms. In undeveloped countries, hookworm infections account for most cases of iron-deficiency anemia.

Acute or chronic blood loss is another cause of anemia. With hemorrhage, red cells are lost and the hypovolemia causes the kidneys to retain water and electrolytes as compensation. Retention of water and electrolytes restores the blood volume, but the concomitant dilution of the blood causes a further decrease in the red cell count, hemoglobin concentration, and hematocrit. Chronic bleeding is compensated by erythroid hyperplasia, which eventually depletes iron stores. Therefore, chronic blood loss results in iron-deficiency anemia.

The last category, increased rate of red cell destruction, includes the Rh factor and sickle cell anemia. The Rhesus (Rh) blood group antigens are involved in maintaining erythrocyte structure. Patients who lack Rh antigens (Rh null) have severe deformation of the red cells.

Sickle cell anemia, associated with the abnormal hemoglobin HbS gene, is common in Africa, India, and among African Americans but is rare in the Caucasian and Asian populations. In the sickle cell trait, which occurs in about 9% of African Americans, one abnormal gene is present. A single point mutation occurs in the hemoglobin molecule, causing the normal glutamic acid at position 6 of the beta chain to be replaced with valine, resulting in HbS. The amino acid substitution is on the surface, resulting in a tendency for the hemoglobin molecule to crystallize with anoxia. However, heterozygous people have no symptoms, and oxygen transport by fetal (HbF) and adult hemoglobin (HbA) is normal. The sickle cell trait (i.e., heterozygous people) offers protection against malaria, and this selective advantage is thought to have favored the persistence of the HbS gene, especially in regions in which malaria is common. Sickle cell disease represents the homozygous condition (S/S) and occurs in about 0.2% of African Americans. The onset of sickle cell anemia occurs in infancy as HbS replaces HbF; death often occurs early in adult life. Patients with sickle cell anemia have >80% HbS in their blood with a decrease or an absence of normal HbA.

Whatever the cause of anemia, the pathophysiologic effect is the same—hypoxemia. Symptoms include pallor of the lips and skin, weakness, fatigue, lethargy, dizziness, and fainting. If the anemia is severe, myocardial hypoxia can lead to angina pain.

---

enriched oxygen mixture never comes into contact with the shunted blood.

Generalized hypoventilation, the third most common cause of hypoxemia, occurs when alveolar ventilation is abnormally decreased. This situation can arise from a chronic obstructive pulmonary disorder (such as emphysema) or depressed respiration (as a result of a head injury or a drug overdose, for example). Because alveolar ventilation is depressed, there is also a significant increase in arterial P_{CO_2}, with a concomitant decrease in arterial pH. In generalized hypoventilation, total ventilation is insufficient to maintain normal systemic arterial P_{O_2} and P_{CO_2}. A feature that distinguishes generalized hypoventilation from the other causes of hypoxemia is a normal A–a O_2 gradient, as a result of the alveolar and arterial P_{O_2} being lowered equally. If a patient has a low Pa_{O_2} and a normal A–a O_2 gradient, the cause of hypoxemia is entirely a result of generalized hypoventilation. The best corrective measure for generalized hypoventilation is to place the patient on a mechanical ventilator, breathing room air. This treatment will return both arterial P_{O_2} and P_{CO_2} to normal. Administering supplemental oxygen to a patient with generalized hypoventilation will correct hypoxemia but not hypercapnia because ventilation is still depressed.

The least common cause of hypoxemia is a **diffusion block**. This condition occurs when the diffusion distance across the alveolar–capillary membrane is increased or the permeability of the alveolar–capillary membrane is decreased. It is characterized by a low Pa_{O_2}, a high A–a O_2 gradient, and a high Pa_{CO_2}. Pulmonary edema is one of the major causes of diffusion block.

In summary, there are four basic respiratory disturbances that cause hypoxemia. Examining the A–a O_2 gradient or Pa_{CO_2} and/or breathing 100% oxygen distinguishes the four types. For example, if a patient has a low Pa_{O_2}, high Pa_{CO_2}, and normal A–a O_2 gradient, the cause of hypoxemia is generalized hypoventilation. If the Pa_{O_2} is low and the A–a O_2 gradient is high, then the cause can be a shunt, a regional low ($\dot{V}_A/\dot{Q}$) ratio, or a diffusion block. Breathing 100% O_2 will distinguish between a low ($\dot{V}_A/\dot{Q}$) ratio and a shunt. Diffusion impairment is the least likely cause and can be deduced if the other three causes have been eliminated.

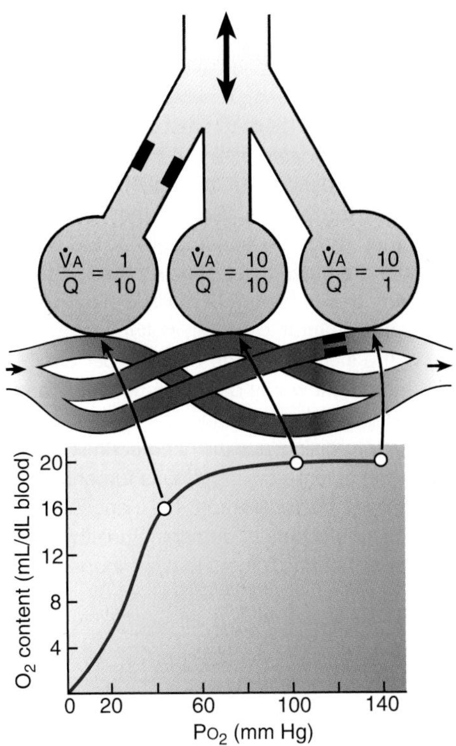

Figure 19.13 Venous admixture lowers O_2 content. Because of the S-shaped oxyhemoglobin–equilibrium curve, a high ventilation/perfusion ratio ($\dot{V}_A/\dot{Q}$ ratio) has little effect on arterial O_2 content. However, mixing with blood from a region with a low $\dot{V}_A/\dot{Q}$ ratio can dramatically lower P_{O_2} in blood leaving the lungs.

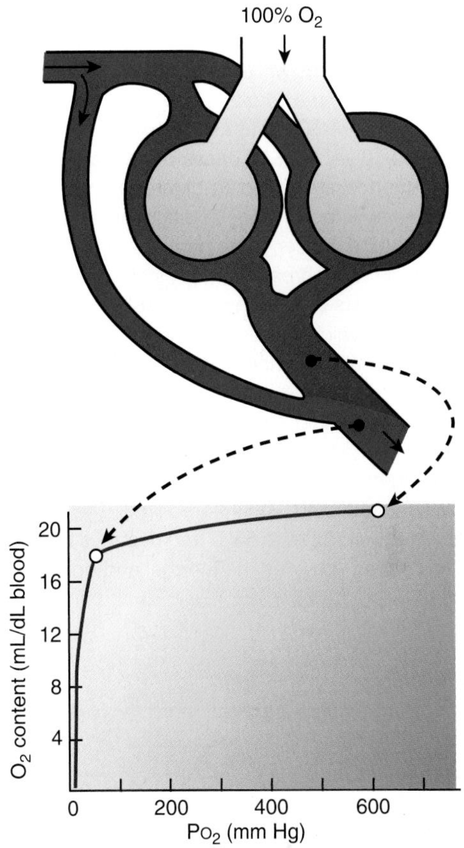

Figure 19.14 Breathing oxygen can be used to diagnose a shunt. A shunt can be diagnosed by having the subject breathe 100% O_2 for 15 minutes. P_{O_2} in systemic arterial blood in a patient with a shunt does not increase above 150 mm Hg during the 15-minute period. The shunted blood is not exposed to 100% O_2, and the venous admixture reduces arterial P_{O_2}.

From Bench to Bedside / 19.1 

New Remotely Activated Nanoparticles Join the Fight Against Lung Cancer

Each year more than 100,000 men and 50,000 women in the United States are diagnosed with primary carcinoma of the lung. Most of these patients die within 1 year of diagnosis, making lung cancer the leading cause of cancer death. Lung cancer interferes with gas uptake/transport and causes edema and hypoxemia. Most cases of lung cancer are directly linked to carcinogens and tumor promoters inhaled from cigarette smoke. Over 90% of cases of lung cancer diagnosed in North America are linked to cigarette smoking. Although great strides have been made in diagnosis and in pinpointing the cause of lung cancer, more people still die from lung cancer than from colon, breast, and prostate cancers combined. Most lung cancers are inoperable or will not respond to current treatment, which makes treatment difficult.

A new generation of nanotechnology-based cancer treatments that are in phase I clinical trials holds much promise for destroying tumors that are inoperable, such as those in lung cancer. The new approach places gold-coated, glass, spherical nanoparticles, called *nanoshells*, inside tumors, which become heat activated with infrared light. The heat activation raises cell temperatures to 55°C and destroys the tumor cells. The gold coating provides the thermal and optical properties for the heat activation process and prevents the body from generating antibodies against the nanoshells. They are also nontoxic and are eliminated from the body through the liver over several weeks.

In the clinical trials, nanoshells (or *nanospheres*) are injected into the bloodstream, in which targeting agents are applied to them to allow them to seek out and attach to surface receptors of malignant cells. The nanospheres are small enough (~100 nm wide) to slip through the gaps in blood vessels that feed the tumors. So, as they circulate in the bloodstream, they gradually accumulate at the tumor sites. Because tumors are highly vascularized, these nanospheres become concentrated in the tumor mass. Following the injection, the nanoshell-infiltrated tumor is illuminated with laser either through the skin or with an optical fiber down the airways. The infrared light works best because it penetrates the body the farthest.

Early trials are promising, and the nanotechnology approach has the advantage of targeting cancerous tumors without surgery. By eliminating chemotherapy, the nanoshell approach will eliminate the side effects of damaging healthy tissues and make it possible to destroy tumors that are inoperable.

Chapter Summary

- The uptake of oxygen is determined by the diffusion properties of the alveolar–capillary membrane, the O_2 partial pressure gradient, and pulmonary capillary blood flow.
- Oxygen in the lungs is limited primarily by blood flow.
- Approximately 250 mL of oxygen per minute is transferred to the pulmonary circulation. Lung diffusion capacity is a measure of the total amount of oxygen transferred.
- Oxygen is transported by the blood in two forms: oxyhemoglobin and dissolved.
- P_{50} is a measure of hemoglobin's (Hb) affinity to bind with oxygen. When Hb affinity increases, the P_{50} changes in the opposite direction.
- Changes in blood chemistry (pH, $Paco_2$, and temperature) alter the oxyhemoglobin–equilibrium curve.
- Oxygen content is a better determinant for adequate tissue oxygenation than arterial Po_2 or percentage O_2 saturation. Hematocrit, $Paco_2$, and CO poisoning can affect oxygen content.
- Carbon dioxide is transported in three forms: dissolved, bicarbonate, and hemoglobin.
- Venous admixture causes an alveolar–arterial oxygen gradient.
- A normal alveolar–arterial oxygen gradient is present because alveolar ventilation and capillary blood flow are not evenly matched in regions of the lung and because bronchial circulation mixes with oxygenated blood.
- Regional hypoventilation (low ventilation/perfusion ratio) is the major cause of hypoxemia.

thePoint *Visit http://thePoint.lww.com for chapter review Q&A, case studies, animations, and more!*

20 Pulmonary Circulation and Ventilation/Perfusion

ACTIVE LEARNING OBJECTIVES

Upon mastering the material in this chapter, you should be able to:

- Predict how changes in cardiac output will affect pulmonary vascular resistance.
- Describe the relationship between capillary recruitment and vascular resistance.
- Describe how low oxygen affects pulmonary vascular resistance.
- Explain the difference between regional and generalized hypoxia and their effect on pulmonary arterial pressure.
- Explain how changes in surface tension will affect interstitial fluid pressure in the lungs.
- Describe how gravity alters blood flow in the base and apex of the lungs.
- Describe how regional ventilation and regional blood flow are matched in the lungs.
- Describe how an anatomic shunt affects the regional ventilation/perfusion ratio.

FUNCTIONAL ORGANIZATION

The heart drives two separate and distinct circulatory systems in the body: the pulmonary circulation and the systemic circulation. The pulmonary circulation carries venous blood from the heart to the lungs and arterialized blood from the lungs back to the heart. Pulmonary circulation is analogous to the entire systemic circulation, because the pulmonary circulation receives all of the cardiac output. Therefore, the pulmonary circulation is not a regional circulation like the renal, hepatic, or coronary circulations. A change in pulmonary vascular resistance has the same implications for the right ventricle as a change in systemic vascular resistance has for the left ventricle.

The pulmonary arteries branch in the same tree-like manner as do the airways. Each time an airway branches, the arterial tree branches so that the two parallel each other (Fig. 20.1). Blood in the pulmonary blood vessels comprises >40% of lung weight. The total blood volume of the pulmonary circulation (main pulmonary artery to left atrium) is approximately 500 mL or 10% of the total circulating blood volume (5,000 mL). The pulmonary veins contain more blood (~270 mL) than do the arteries (~150 mL). The blood volume in the pulmonary capillaries is approximately equal to the stroke volume of the right ventricle (~80 mL), under most physiologic conditions.

Pulmonary circulation has many secondary functions that facilitate gas exchange.

The primary function of the pulmonary circulation is to bring venous blood from the superior and inferior vena cavae (i.e., mixed venous blood) into contact with alveoli for gas exchange. In addition to gas exchange, the pulmonary circulation has three secondary functions: It serves as a filter, a metabolic organ, and a blood reservoir.

Pulmonary vessels protect the body against **thrombi** (blood clots) and **emboli** (fat globules or air bubbles), preventing them from entering important vessels in other organs. Thrombi and emboli often occur after surgery or injury, and enter the systemic venous blood. Small pulmonary arterial vessels and capillaries trap the thrombi and emboli and prevent them from obstructing the vital coronary, cerebral, and renal vessels. Endothelial cells lining the pulmonary vessels release fibrinolytic substances that help dissolve thrombi. Emboli, especially air emboli, are absorbed through the pulmonary capillary walls. If a large thrombus occludes a large pulmonary vessel, gas exchange can be severely impaired and can cause death. A similar situation occurs if emboli are extremely numerous and lodge all over the pulmonary arterial tree.

Vasoactive hormones are metabolized in the pulmonary circulation. One such hormone is angiotensin I (AI), which is activated and converted to angiotensin II (AII) in the lungs by **angiotensin-converting enzyme**

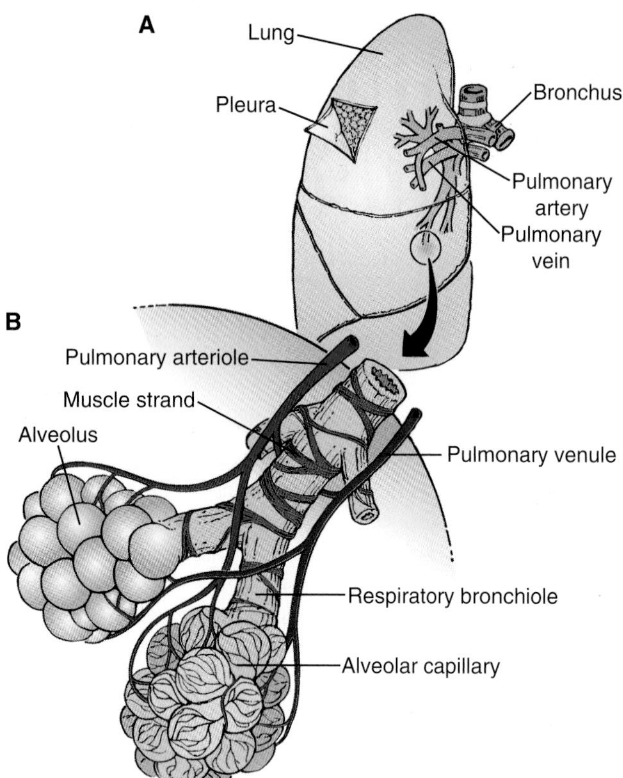

Figure 20.1 Pulmonary vessels branch in the same treelike fashion as the airways. (A) Systemic venous blood flows through the pulmonary arteries into the alveolar capillaries and back to the heart via the pulmonary veins, to be pumped into the systemic circulation. **(B)** A mesh of capillaries surrounds each alveolus. As the blood passes through the capillaries, it gives up carbon dioxide and takes up oxygen.

(ACE) located on the surface of the pulmonary capillary endothelial cells. Activation is rapid; 80% of AI can be converted to AII during a single passage through the pulmonary circulation. In addition to being a potent vasoconstrictor, AII has other important actions in the body (see Chapter 23, "Regulation of Fluid and Electrolyte Balance"). Metabolism of vasoactive hormones by the pulmonary circulation appears to be selective. Pulmonary endothelial cells inactivate bradykinin, serotonin, and the prostaglandins E_1, E_2, and $F_{2\alpha}$. Other prostaglandins, such as PGA_1 and PGA_2, pass through the lungs unaltered. Norepinephrine is inactivated, but epinephrine, histamine, and arginine vasopressin pass through the pulmonary circulation unchanged. With acute lung injury (e.g., oxygen toxicity, fat emboli), the lungs can release histamine, prostaglandins, and leukotrienes, which can cause vasoconstriction of pulmonary arteries and pulmonary endothelial damage.

The lungs serve as a blood reservoir. Approximately 500 mL or 10% of the total circulating blood volume is in the pulmonary circulation. During hemorrhagic shock, some of this blood can be mobilized to improve the cardiac output.

HEMODYNAMIC FEATURES

In contrast to the systemic circulation, the pulmonary circulation is a high-flow, low-pressure, low-resistance system. The pulmonary artery and its branches have much thinner walls than the aorta and are more compliant. The pulmonary artery is much shorter and contains less elastin and smooth muscle in its walls. The pulmonary arterioles are thin-walled and contain little smooth muscle and, consequently, have less ability to constrict than the thick-walled, highly muscular systemic arterioles. The pulmonary veins are also thin-walled, highly compliant, and contain little smooth muscle compared with their counterparts in the systemic circulation.

The pulmonary capillary bed is also different. Unlike the systemic capillaries, which are often arranged as a network of tubular vessels with some interconnections, the pulmonary capillaries mesh together in the alveolar wall so that blood flows as a thin sheet. It is, therefore, misleading to refer to pulmonary capillaries as a capillary network; they comprise a dense capillary bed. The walls of the capillary bed are exceedingly thin, and a whole capillary bed can collapse if local alveolar pressure exceeds capillary pressure.

The systemic and pulmonary circulations differ strikingly in their pressure profiles (Fig. 20.2). Mean pulmonary arterial pressure is 15 mm Hg, compared with 93 mm Hg in the aorta. The driving pressure (10 mm Hg) for pulmonary flow is the difference between the mean pressure in the pulmonary artery (15 mm Hg) and the pressure in the left atrium (5 mm Hg). These pulmonary pressures are measured using a Swan-Ganz catheter, a thin, flexible tube with an inflatable rubber balloon surrounding the distal end. The balloon is inflated by injecting a small amount of air through the proximal end. Although the Swan-Ganz catheter is used for several pressure measurements, the most useful is the **pulmonary wedge pressure** (Fig. 20.3). To measure wedge pressure, the catheter tip with the balloon inflated is "wedged" into a small branch of the pulmonary artery. When the inflated balloon interrupts blood flow, the tip of the catheter measures downstream pressure. The downstream pressure in the occluded arterial branch represents pulmonary venous pressure, which, in turn, reflects left atrial pressure. Changes in pulmonary venous and left atrial pressures have a profound effect on gas exchange, and pulmonary wedge pressure provides an indirect measure of these important pressures.

The right ventricle pumps mixed venous blood through the pulmonary arterial tree, the alveolar capillaries (where oxygen is taken up and carbon dioxide is removed), the pulmonary veins, and then on to the left atrium. All of the cardiac output is pumped through the pulmonary circulation at a much lower pressure than through the systemic circulation. As shown in Figure 20.2, the 10 mm Hg pressure gradient across the pulmonary circulation drives the same blood flow (5 L/min) as in the systemic circulation, where the pressure gradient is almost 100 mm Hg. Remember that vascular resistance (R) is equal to the pressure gradient (ΔP)

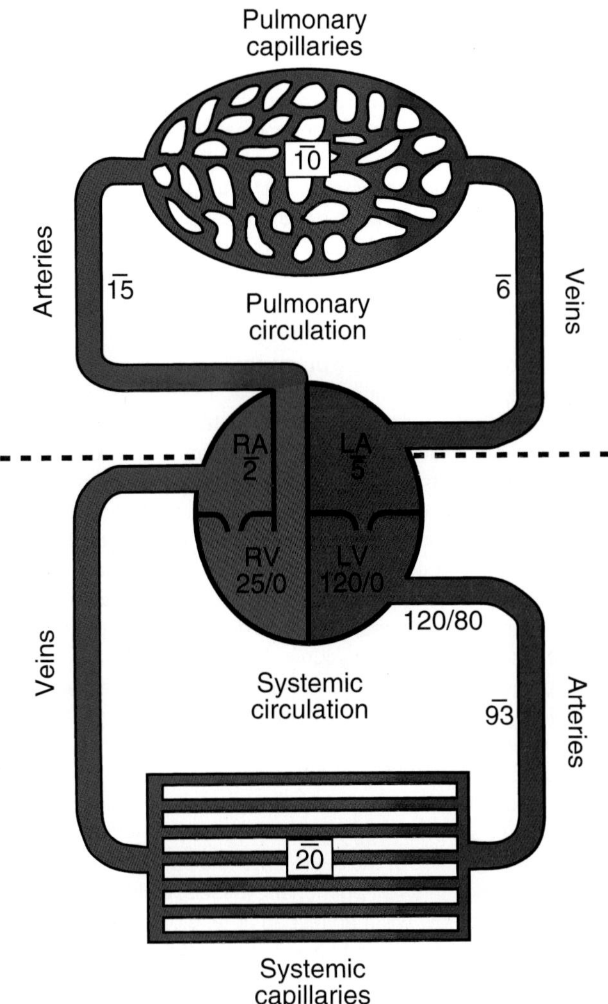

Figure 20.2 Pulmonary circulation has unique hemodynamic features. Unlike the systemic circulation, the pulmonary circulation is a low-pressure and low-resistance system. Pulmonary circulation is characterized as normally dilated, whereas the systemic circulation is characterized as normally constricted. Pressures are given in mm Hg; a bar over the number indicates mean pressure. LA, left atrium; LV, left ventricle; RA, right atrium; RV, right ventricle.

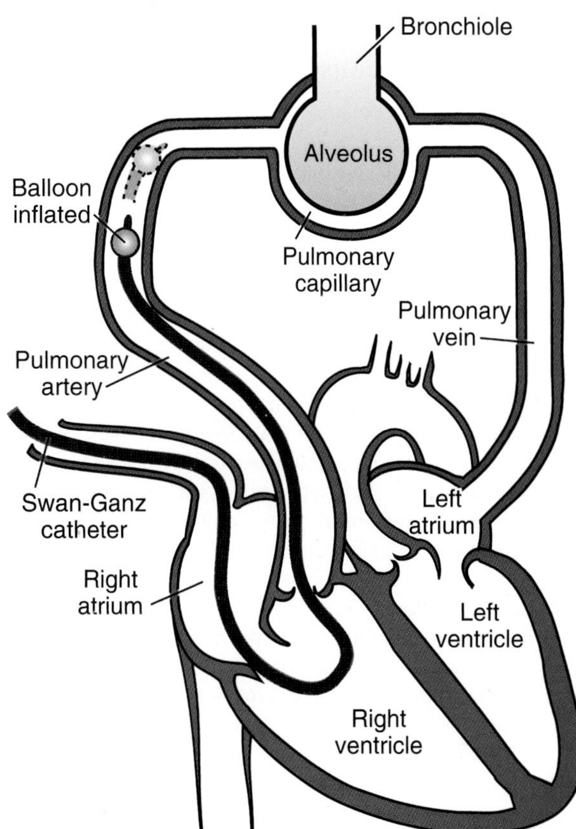

Figure 20.3 Pulmonary wedge pressure measures capillary pressure. A catheter is threaded through a peripheral vein in the systemic circulation, through the right heart, and into the pulmonary artery. The wedged catheter temporarily occludes blood flow in a part of the vascular bed. The wedge pressure is a measure of downstream pressure, which is pulmonary venous pressure. Pulmonary venous pressure reflects left atrial pressure.

divided by blood flow (see Chapter 12, "Electrical Activity of the Heart"):

$$R = \Delta P/\dot{Q}$$

Pulmonary vascular resistance is extremely low, about 1/10 that of systemic vascular resistance. The difference in resistances is a result, in part, of the enormous number of small pulmonary resistance vessels that are dilated. By contrast, systemic arterioles and precapillary sphincters are partially constricted.

Pulmonary vascular resistance falls with increased cardiac output.

Another unique feature of the pulmonary circulation is its ability to decrease resistance when pulmonary arterial pressure rises. When cardiac output increases, pulmonary pressure rises, resulting in a marked decrease in pulmonary vascular resistance (Fig. 20.4). Similarly, increasing pulmonary venous pressure causes pulmonary vascular resistance to fall. These responses are different from those of the systemic circulation, where an increase in perfusion pressure increases vascular resistance. Two local mechanisms in the pulmonary circulation are responsible (Fig. 20.5). The first mechanism is known as **capillary recruitment**. Under normal conditions, some capillaries are partially or completely closed, particularly in the top part of the lungs, because of the low perfusion pressure. As blood flow increases, the pressure rises, and the closed vessels are opened, lowering the overall resistance. This process of opening capillaries is the primary mechanism for the fall in pulmonary vascular resistance when cardiac output increases. The second mechanism is **capillary distention** or widening of capillary segments, which occurs because the pulmonary arterioles and capillaries are exceedingly thin and highly compliant.

The fall in pulmonary vascular resistance with increased cardiac output has two beneficial effects. It opposes the tendency of blood velocity to speed up with increased flow rate, maintaining adequate time for pulmonary capillary blood to

Clinical Focus / 20.1 

Pulmonary Embolism

Pulmonary embolism is clearly one of the more important disorders affecting the pulmonary circulation. The incidence of pulmonary embolism exceeds 500,000 per year, with a mortality rate of approximately 10%. Pulmonary embolism is often misdiagnosed and, if improperly diagnosed, the mortality rate can exceed 30%.

The term *pulmonary embolism* refers to the movement of a blood clot or other plug from the systemic veins through the right heart and into the pulmonary circulation, where it lodges in one or more branches of the pulmonary artery. Although most pulmonary emboli originate from thrombosis in the leg veins, they can originate from the upper extremities as well. A thrombus is the major source of pulmonary emboli; however, air bubbles introduced during intravenous injections, hemodialysis, or the placement of central catheters can also cause emboli. Other sources of pulmonary emboli include fat emboli (a result of multiple long-bone fractures), tumor cells, amniotic fluid (secondary to strong uterine contractions), parasites, and various foreign materials in intravenous drug abusers.

The etiology of pulmonary emboli focuses on three factors that potentially contribute to the genesis of venous thrombosis: (1) hypercoagulability (e.g., a deficiency of antithrombin III, malignancies, the use of oral contraceptives, the presence of lupus anticoagulant); (2) endothelial damage (e.g., caused by atherosclerosis); and (3) stagnant blood flow (e.g., varicose veins). Risk factors for thrombi include immobilization (e.g., prolonged bed rest, prolonged sitting during travel, or immobilization of an extremity after a fracture), congestive heart failure, obesity, underlying carcinoma, and chronic venous insufficiency.

When a thrombus migrates into the pulmonary circulation and lodges in pulmonary vessels, several pathophysiologic consequences ensue. When a vessel is occluded, blood flow stops, perfusion to pulmonary capillaries ceases, and the ventilation/perfusion ratio in that lung unit becomes high because ventilation is wasted. As a result, there is a significant increase in physiologic dead space. Besides the direct mechanical effects of vessel occlusion, thrombi release vasoactive mediators that cause bronchoconstriction of small airways, which leads to hypoxemia. These vasoactive mediators also cause endothelial damage that leads to edema and atelectasis. If the pulmonary embolus is large and occludes a major pulmonary vessel, an additional complication occurs in the lung parenchyma distal to the site of the occlusion. The distal lung tissue becomes anoxic because it does not receive oxygen (either from airways or from the bronchial circulation). Oxygen deprivation leads to necrosis of lung parenchyma (pulmonary infarction). The parenchyma will subsequently contract and form a permanent scar.

Pulmonary emboli are difficult to diagnose because they do not manifest any specific symptoms. The most common clinical features include dyspnea and sometimes pleuritic chest pains. If the embolism is severe enough, a decreased arterial P_{O_2}, decreased P_{CO_2}, and increased pH result. The major screening test for pulmonary embolism is the perfusion scan, which involves the injection of aggregates of human serum albumin labeled with a radionuclide into a peripheral vein. These albumin aggregates (~10–50 µm wide) travel through the right side of the heart, enter the pulmonary vasculature, and lodge in small pulmonary vessels. Only lung areas receiving blood flow will manifest an uptake of the tracer; the nonperfused region will not show any uptake of the tagged albumin. The aggregates fragment and are removed from the lungs in about a day.

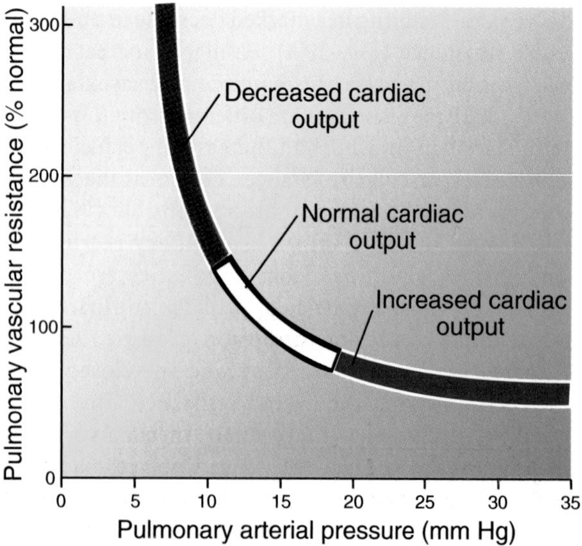

Figure 20.4 Pulmonary vascular resistance falls with a rise in cardiac output. Unlike in the systemic circulation, vascular resistance decreases when perfusion pressure rises (pulmonary arterial pressure). Note that if cardiac output increases, there is a rise in pulmonary arterial pressure and a concomitant fall in pulmonary vascular resistance.

take up oxygen and dispose of carbon dioxide. It also results in an increase in capillary surface area, which enhances the diffusion of oxygen into and carbon dioxide out of the pulmonary capillary blood.

Capillary recruitment and distention also have a protective function. High capillary pressure is a major threat to the lungs and can cause **pulmonary edema**, an abnormal accumulation of fluid, which can flood the alveoli and impair gas exchange. When cardiac output increases from a resting level of 5 L/min to 25 L/min with vigorous exercise, the decrease in pulmonary vascular resistance not only minimizes the load on the right heart but also keeps the capillary pressure low and prevents excess fluid from leaking out of the pulmonary capillaries.

Pulmonary vascular resistance increases at high and low lung volumes.

Pulmonary vascular resistance is also significantly affected by lung volume. Because pulmonary capillaries have little structural support, they can be easily distended or collapsed, depending on the pressure surrounding them. It is the change in transmural pressure (pressure across the capillaries) that influences vessel diameter. From a functional point

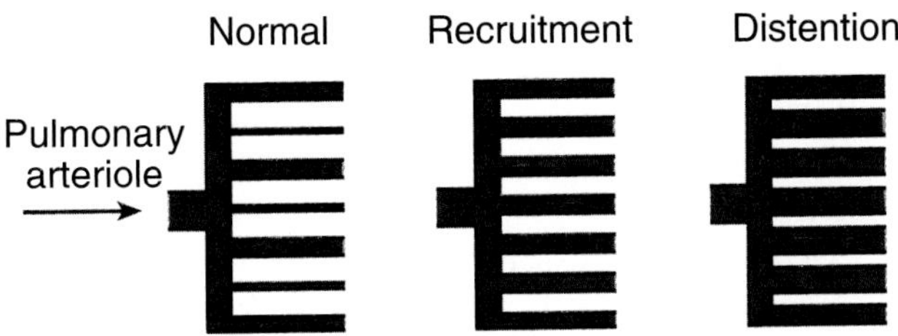

Figure 20.5 Capillary recruitment and capillary distention are the mechanisms responsible for decreasing pulmonary vascular resistance. In the normal condition, not all capillaries are perfused. Capillary recruitment (the opening up of previously closed vessels) results in the perfusion of an increased number of vessels with a concomitant decrease in resistance. Capillary distention (an increase in the caliber of vessels) resulting from high vessel compliance also results in a lower resistance and higher blood flow.

of view, pulmonary vessels can be classified into two types: **extra-alveolar vessels** (pulmonary arteries and veins) and alveolar vessels (arterioles, capillaries, and venules). The *extra-alveolar vessels* are subjected to pleural pressure—any change in pleural pressure affects pulmonary vascular resistance in these vessels by changing the transmural pressure. *Alveolar vessels*, however, are subjected primarily to alveolar pressure.

At high lung volumes, the pleural pressure is more negative. The transmural pressure in the extra-alveolar vessels increases, and they become distended (Fig. 20.6A). However, alveolar diameter increases at high lung volumes, causing the transmural pressure in alveolar vessels to decrease. As the alveolar vessels become compressed, pulmonary vascular resistance increases. At low lung volumes, pulmonary vascular resistance also increases, as a result of more positive pleural pressure, which compresses the extra-alveolar vessels. Because alveolar and extra-alveolar vessels can be viewed as two groups of resistance vessels connected in series, their resistances are additive at any lung volume. Pulmonary vascular resistance is lowest at **functional residual capacity (FRC)** and increases at both higher and lower lung volumes (Fig. 20.6B).

Because smooth muscle plays a key role in determining the caliber of extra-alveolar vessels, drugs can also cause a change in resistance. Serotonin, norepinephrine, histamine, thromboxane A_2, and leukotrienes are potent vasoconstrictors, particularly at low lung volumes when the vessel walls are already compressed. Drugs that relax smooth muscle in the pulmonary circulation include adenosine, acetylcholine, prostacyclin (prostaglandin I_2), and isoproterenol. The pulmonary circulation is richly innervated with sympathetic nerves but, surprisingly, pulmonary vascular resistance is virtually unaffected by autonomic nerves under normal conditions.

Low oxygen tension in alveoli and/or pulmonary blood induces pulmonary vasoconstriction.

Although changes in pulmonary vascular resistance are accomplished mainly by passive mechanisms, resistance can be increased by low oxygen in the alveoli, alveolar

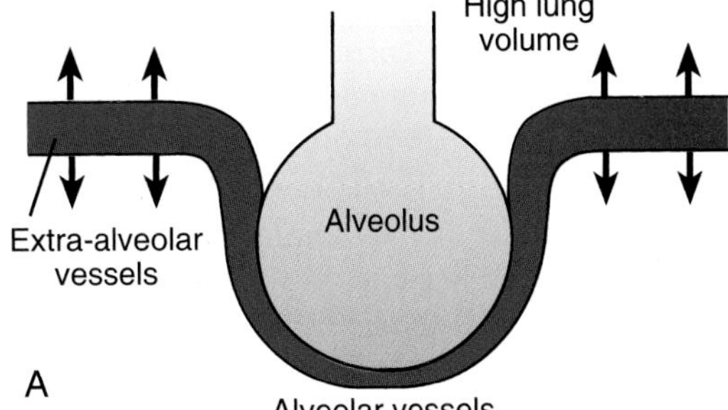

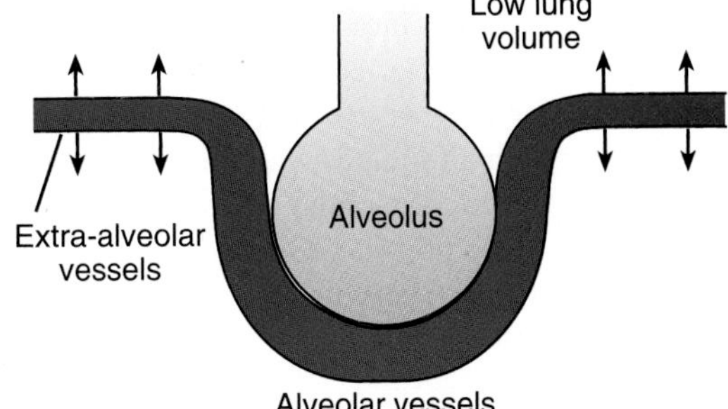

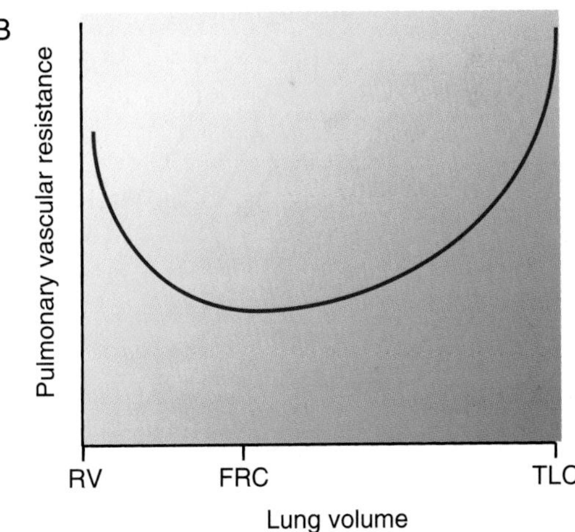

Figure 20.6 Changes in lung volumes affect pulmonary vascular resistance. (A) At high lung volumes, extra-alveolar vessels are actually distended because of the lower pleural pressure. However, alveolar vessels are compressed, causing a rise in pulmonary vascular resistance. At low lung volumes, alveolar vessels are distended, but the extra-alveolar vessels are compressed from the rise in pleural pressure, which results in a rise in pulmonary vascular resistance. **(B)** Total pulmonary vascular resistance as a function of lung volumes follows a U-shaped curve, with resistance lowest at functional residual capacity (FRC). RV, residual volume; TLC, total lung capacity.

hypoxia, and low oxygen in the blood, **hypoxemia**. Hypoxemia causes vasodilation in systemic vessels but, in pulmonary vessels, hypoxemia or alveolar hypoxia causes vasoconstriction of small pulmonary arteries. This unique phenomenon of **hypoxia-induced pulmonary vasoconstriction** is accentuated by high carbon dioxide and low blood pH. The exact mechanism is not known, but hypoxia can directly act on pulmonary vascular smooth muscle cells, independent of any agonist or neurotransmitter released by hypoxia.

Two types of alveolar hypoxia (**regional hypoxia** and **generalized hypoxia**) are encountered in the lungs, with different implications for pulmonary vascular resistance. In regional hypoxia, pulmonary vasoconstriction is localized to a specific region of the lungs and diverts blood away from a poorly ventilated region, which minimizes the effect on gas exchange (Fig. 20.7A). Regional hypoxia is often caused by bronchial obstruction. Regional hypoxia has little effect on pulmonary arterial pressure, and when alveolar hypoxia no longer exists, the vessels dilate and blood flow is restored. Generalized hypoxia causes vasoconstriction throughout both lungs, leading to a significant rise in resistance and pulmonary artery pressure (Fig. 20.7B). Generalized hypoxia occurs when the partial pressure of alveolar oxygen (P_{AO_2}) is decreased with high altitude or with the chronic hypoxia seen in certain types of respiratory diseases (e.g., asthma, emphysema, and cystic fibrosis). Generalized hypoxia can lead to pulmonary hypertension (high pulmonary arterial pressure), which leads to pathophysiologic changes (hypertrophy and proliferation of smooth muscle cells, narrowing of arterial lumens, and a change in contractile function). Pulmonary hypertension causes a substantial increase in workload of the right heart, often leading to right heart hypertrophy.

Generalized hypoxia plays an important nonpathophysiologic role before birth. In the fetus, pulmonary vascular resistance is extremely high as a result of generalized hypoxia: <15% of the cardiac output goes to the lungs, and the remainder is diverted to the left side of the heart via the foramen ovale and to the aorta via the ductus arteriosus. When alveoli are oxygenated on the newborn's first breath, pulmonary vascular smooth muscle relaxes, the vessels dilate, and vascular resistance falls dramatically. The foramen ovale and the ductus arteriosus close and pulmonary blood flow increases enormously.

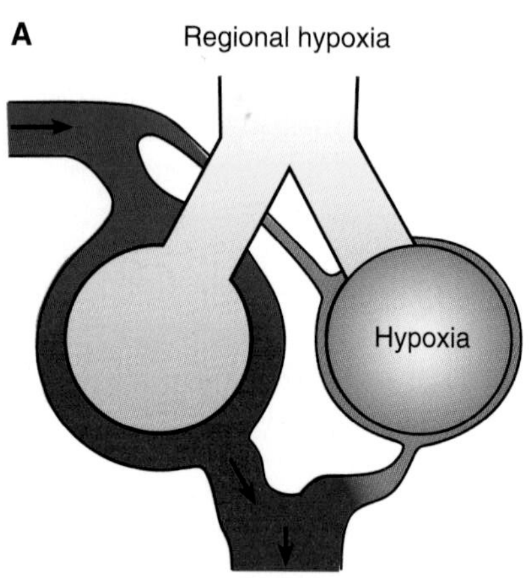

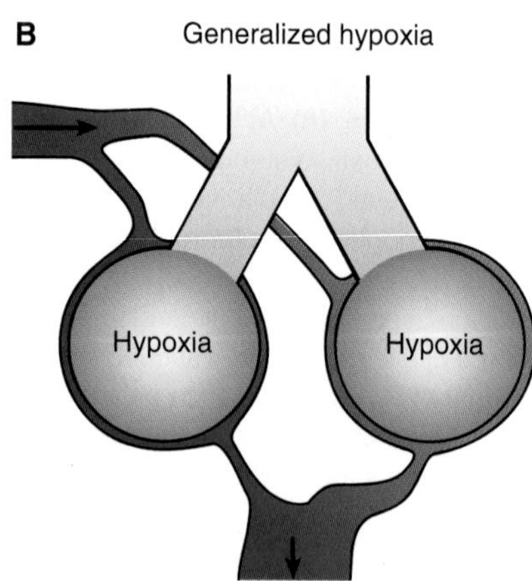

Figure 20.7 Hypoxia-induced vasoconstriction is a feature unique to the pulmonary circulation. Low oxygen tension in the alveoli (alveolar hypoxia) is the major mechanism regulating blood flow within normal lungs. **(A)** With regional hypoxia, precapillary constriction diverts blood flow from poorly ventilated regions, with little change in pulmonary arterial pressure. **(B)** In generalized hypoxia, which can occur with high altitude or with certain lung diseases, precapillary constriction occurs throughout the lungs and there is a marked increase in pulmonary arterial pressure.

FLUID EXCHANGE IN PULMONARY CAPILLARIES

Starling forces, which govern the exchange of fluid across capillary walls in the systemic circulation (see Chapter 16, "Special Circulations"), also operate in the pulmonary capillaries. Net fluid transfer across the pulmonary capillaries depends on the difference between hydrostatic and colloid osmotic pressures inside and outside the capillaries. In the pulmonary circulation, two additional forces play a role in fluid transfer—surface tension and alveolar pressure. The force of alveolar surface tension (see Chapter 19, "Gas Transfer and Transport") pulls inwardly, which tends to lower the interstitial pressure and draw fluid into the interstitial space. By contrast, the alveolar pressure tends to compress the interstitial space and the interstitial pressure is increased (Fig. 20.8).

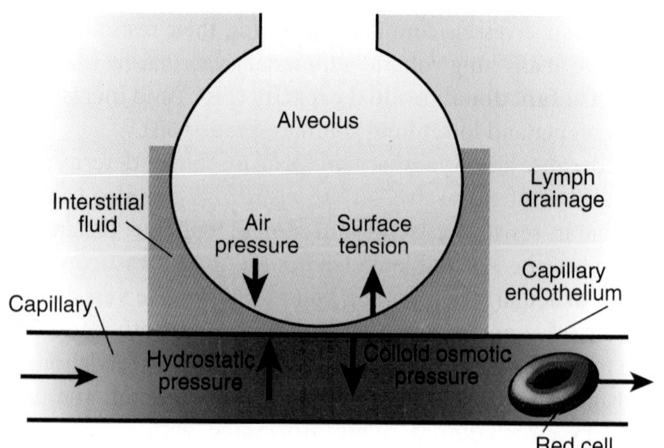

Figure 20.8 Alveolar surface tension and alveolar pressure affect fluid exchange in pulmonary capillaries. Fluid movement in and out of capillaries depends on the net difference between hydrostatic and colloidal osmotic pressures. In the lung, two additional factors (alveolar surface tension and pressure) are involved in fluid exchange. Alveolar surface tension enhances filtration, whereas alveolar pressure opposes filtration. The relatively low pulmonary capillary hydrostatic pressure helps keep the alveoli "dry" and prevents pulmonary edema.

Pulmonary edema is caused by an abnormal increase in capillary pressure, capillary permeability, or alveolar surface tension or a decrease in plasma colloidal osmotic pressure.

Mean pulmonary capillary hydrostatic pressure is normally 8 to 10 mm Hg, which is lower than the plasma colloid osmotic pressure (25 mm Hg). This is functionally important because the low hydrostatic pressure in the pulmonary capillaries favors the net absorption of fluid. Alveolar surface tension tends to offset this advantage and results in a net force that still favors a small continuous flux of fluid out of the capillaries and into the interstitial space. This excess fluid travels through the interstitium to the perivascular and peribronchial spaces in the lungs, where it then passes into the lymphatic channels (see Fig. 20.8). The lungs have a more extensive lymphatic system than most organs. The lymphatic vessels are not found in the alveolar–capillary area but are strategically located near the terminal bronchioles to drain off excess fluid. Lymphatic channels, like small pulmonary blood vessels, are held open by tethers from surrounding connective tissue. Total lung lymph flow is about 0.5 mL/min, and the lymph is propelled by smooth muscle in the lymphatic walls and by ventilatory movements of the lungs.

Pulmonary edema occurs when excess fluid accumulates in the lung interstitial spaces and alveoli and usually results when capillary filtration exceeds fluid removal. Pulmonary edema can be classified as *cardiogenic* pulmonary edema (due to heart dysfunction) or *noncardiogenic* pulmonary edema (due to lung injury). Cardiogenic pulmonary edema is caused by an increase in capillary hydrostatic pressure, capillary permeability, or alveolar surface tension or by a decrease in plasma colloidal osmotic pressure. Increased capillary hydrostatic pressure is the most frequent cause of pulmonary edema and is often the result of an abnormally high pulmonary venous pressure (e.g., with mitral stenosis, left heart failure, or heart attack).

The second major cause of pulmonary edema is increased permeability of the alveolar–capillary membrane, which results in excess fluid and plasma proteins flooding the interstitial spaces and alveoli. Protein leakage makes pulmonary edema more severe because additional water is pulled from the capillaries to the alveoli when plasma proteins enter the interstitial spaces and alveoli. Increased capillary permeability occurs with pulmonary vascular injury, usually from oxidant damage (e.g., oxygen therapy, ozone toxicity), an inflammatory reaction (endotoxins), or neurogenic shock (e.g., head injury). High surface tension is the third major cause of pulmonary edema. Loss of surfactant causes high surface tension, lowering interstitial hydrostatic pressure and resulting in an increase in capillary fluid entering the interstitial space. A decrease in plasma colloidal osmotic pressure occurs when plasma protein concentration is reduced (e.g., starvation).

Pulmonary edema is a hallmark of **acute respiratory distress syndrome (ARDS)** and is often associated with abnormally high surface tension. Pulmonary edema is a serious problem because excess fluid enters alveoli and hinders gas exchange, causing arterial Po_2 to fall below normal (i.e., $Pao_2 < 85$ mm Hg) and arterial Pco_2 to rise above normal ($Paco_2 > 45$ mm Hg). As mentioned earlier, abnormally low arterial Po_2 produces hypoxemia and abnormally high arterial Pco_2 produces hypercapnia. Pulmonary edema can also flood small airways, thereby obstructing airflow and increasing airway resistance. Lung compliance is decreased with pulmonary edema because of interstitial swelling and

Clinical Focus / 20.2

Hypoxia-Induced Pulmonary Hypertension

Hypoxia has opposite effects on the pulmonary and systemic circulations. Hypoxia relaxes vascular smooth muscle in systemic vessels and elicits vasoconstriction in the pulmonary vasculature. Hypoxic pulmonary vasoconstriction is the major mechanism regulating the matching of regional blood flow to regional ventilation in the lungs. With regional hypoxia, the matching mechanism automatically adjusts regional pulmonary capillary blood flow in response to alveolar hypoxia and prevents blood from perfusing poorly ventilated regions in the lungs. Regional hypoxic vasoconstriction occurs without any change in pulmonary arterial pressure. However, when hypoxia affects all parts of the lung (generalized hypoxia), it causes pulmonary hypertension because all of the pulmonary vessels constrict. Hypoxia-induced pulmonary hypertension affects people who live at a high altitude (8,000–12,000 ft) and those with **chronic obstructive pulmonary disease (COPD),** especially patients with emphysema.

With chronic *hypoxia-induced pulmonary hypertension,* the pulmonary artery undergoes major remodeling during several days. An increase in wall thickness results from hypertrophy and hyperplasia of vascular smooth muscle and an increase in connective tissue. These structural changes occur in both large and small arteries. Also, there is abnormal extension of smooth muscle into peripheral pulmonary vessels, where muscularization is not normally present; this is especially pronounced in precapillary segments. These changes lead to a marked increase in pulmonary vascular resistance. With severe, chronic hypoxia-induced pulmonary hypertension, the obliteration of small pulmonary arteries and arterioles as well as pulmonary edema, eventually occurs. The latter is caused, in part, by the hypoxia-induced vasoconstriction of pulmonary veins, which results in a significant increase in pulmonary capillary hydrostatic pressure.

A striking feature of the vascular remodeling is that both the pulmonary artery and the pulmonary vein constrict with hypoxia; however, only the arterial side undergoes major remodeling. The postcapillary segments and veins are spared the structural changes seen with hypoxia. Because of the hypoxia-induced vasoconstriction and vascular remodeling, pulmonary arterial pressure increases. Pulmonary hypertension eventually causes right heart hypertrophy and failure, the major cause of death in patients with COPD.

the increase in alveolar surface tension. Decreased lung compliance, together with airway obstruction, greatly increases the work of breathing. The treatment of pulmonary edema is directed toward reducing pulmonary capillary hydrostatic pressure. This is accomplished by decreasing blood volume with a diuretic drug, increasing left ventricular function with digitalis, and administering a drug that causes vasodilation in systemic blood vessels.

Although freshwater drowning is often associated with aspiration of water into the lungs, the cause of death is not pulmonary edema but ventricular fibrillation. The low capillary pressure that normally keeps the alveolar–capillary membrane free of excess fluid becomes a severe disadvantage when freshwater accidentally enters the lungs. The aspirated water is rapidly pulled into the pulmonary capillary circulation via the alveoli because of the low capillary hydrostatic pressure and high colloidal osmotic pressure. Consequently, the plasma is diluted and the hypotonic environment causes red cells to burst (hemolysis). The resulting elevation of plasma K^+ level and depression of Na^+ level alter the electrical activity of the heart. Ventricular fibrillation often occurs as a result of the combined effects of these electrolyte changes and hypoxemia. In saltwater drowning, the aspirated seawater is hypertonic, which leads to increased plasma Na^+ and pulmonary edema. The cause of death in this case is asphyxia.

BLOOD FLOW DISTRIBUTION IN THE LUNGS

As previously mentioned, blood accounts for approximately half the weight of the lungs. The effects of gravity on pulmonary blood flow are dramatic and result in an uneven distribution of blood in the lungs. In an upright person, the gravitational pull on the blood is downward. Because the vessels are highly compliant, gravity causes the blood volume and flow to be greater at the bottom of the lung (the base) than at the top (the apex). Pulmonary vessels can be compared with a continuous column of fluid. The difference in arterial pressure between the apex and the base of the lungs is about 30 cm H_2O. Because the heart is situated midway between the top and the bottom of the lungs, the arterial pressure is about 11 mm Hg less (15 cm H_2O ÷ 1.36 cm H_2O per mm Hg = 11 mm Hg) at the lungs' apex (15 cm above the heart) and about 11 mm Hg more than the mean pressure in the middle of the lungs at the lungs' base (15 cm below the heart). As a result, the low pulmonary arterial pressure results in reduced blood flow in the capillaries at the lungs' apex, whereas capillaries at the base are distended because of increased pressure and blood flow is augmented.

Gravity causes lungs to be underperfused at the apex and overperfused at the base.

In an upright person, pulmonary blood flow decreases almost linearly from the base to the apex (Fig. 20.9). Blood flow distribution is affected by gravity and can be altered by changes in body positions. For example, when a person

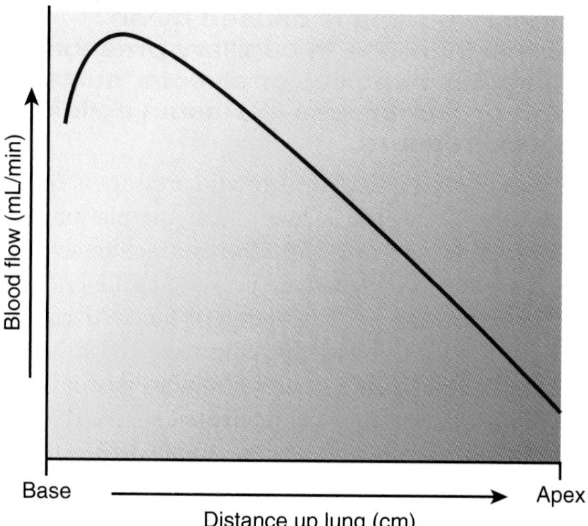

Figure 20.9 Gravity causes uneven pulmonary blood flow in the upright person. The downward pull of gravity causes a lower blood pressure at the apex of the lungs. Consequently, pulmonary blood flow is low at the apex. Toward the base of the lungs, gravity has an added effect on blood pressure, causing an increase in blood flow.

is lying down, blood flow is distributed relatively evenly from the base to the apex. The measurement of blood flow in a subject suspended upside-down would reveal an apical blood flow exceeding basal flow in the lungs. Exercise tends to offset the gravitational effects in an upright person. As cardiac output increases with exercise, the increased pulmonary arterial pressure leads to capillary recruitment and distention in the lungs' apex, resulting in increased blood flow and minimizing regional differences in blood flow in the lungs.

Because gravity causes capillary beds to be underperfused in the apex and overperfused in the base, the lungs are often divided into zones to describe the effect of gravity on pulmonary capillary blood flow (Fig. 20.10). Zone 1 occurs when alveolar pressure is greater than pulmonary arterial pressure; pulmonary capillaries collapse and there is no blood flow. Pulmonary arterial pressure (Pa) is still greater than pulmonary venous pressure (Pv), hence the pressure gradient in zone 1 is represented as $P_A > Pa > Pv$. Zone 1 is usually small or nonexistent in healthy people because the pulsatile pulmonary arterial pressure is sufficient to keep the capillaries partially open at the apex. However, when zone 1 does occur, alveolar dead space is increased in the lungs. This occurs because, in zone 1, that region is still being ventilated but not perfused (no blood flows through the pulmonary capillaries). Zone 1 may easily be created by conditions that elevate alveolar pressure or decrease pulmonary arterial pressure. For example, a zone 1 condition can be created when a patient is placed on a mechanical ventilator, which results in an increase in alveolar pressure with positive ventilation pressures. Hemorrhage or low blood pressure can create a zone 1 condition by lowering pulmonary arterial pressure. A zone 1 condition can also be created in the lungs of astronauts during a spacecraft launching. The rocket acceleration makes

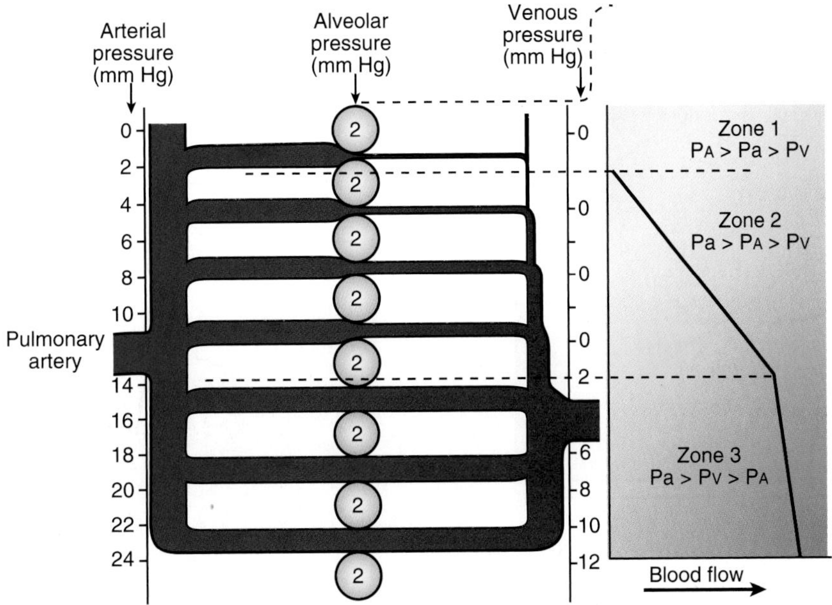

Figure 20.10 Zones are established as a result of gravitational effects. The three zones are established in an upright person and are dependent on the relationship between pulmonary arterial pressure (Pa), pulmonary venous pressure (Pv), and alveolar pressure (PA). A zone 1 is established when alveolar pressure exceeds arterial pressure and there is no blood flow. Zone 1 occurs toward the apex of the lung and occurs only in abnormal conditions in which alveolar pressure is increased (e.g., positive pressure ventilation) or when arterial pressure is decreased below normal (e.g., the gravitational pull while standing at attention or during the launching of a spacecraft). A zone 2 is established when arterial pressure exceeds alveolar pressure, and blood flow depends on the difference between arterial and alveolar pressures. Blood flow is greater at the bottom than at the top of this zone. In zone 3, both arterial and venous pressures exceed alveolar pressure, and blood flow depends on the normal arterial–venous pressure difference. Note that arterial pressure increases down each zone, vessel transmural pressure also becomes greater, capillaries become more distended, and pulmonary vascular resistance falls.

the gravitational pull even greater, causing arterial pressure in the top part of the lung to fall. To prevent or minimize a zone 1 condition from occurring, astronauts are placed in a supine position during blastoff.

A zone 2 condition occurs in the middle of the lungs, where pulmonary arterial pressure, caused by the increased hydrostatic effect, is greater than alveolar pressure (see Fig. 20.10). Venous pressure is less than alveolar pressure. As a result, blood flow in a zone 2 condition is determined not by the arterial–venous pressure difference but by the difference between arterial pressure and alveolar pressure. The pressure gradient in zone 2 is represented as Pa > PA > Pv. The functional importance of this is that venous pressure in zone 2 has no effect on flow (i.e., lowering venous pressure will not increase capillary blood flow in this zone). In zone 3, venous pressure exceeds alveolar pressure and blood flow is determined by the usual arterial–venous pressure difference. The increase in blood flow down this region is primarily a result of capillary distention.

Gravity causes a mismatch of regional ventilation and blood flow in the lungs.

Thus far, we have assumed that if ventilation and cardiac output are normal, gas exchange will also be normal because ventilation and blood flow are matched. Unfortunately, this is not the case. Even though total ventilation and total blood flow (i.e., cardiac output) may be normal, there are regions in the lung where ventilation and blood flow are not matched and so a certain fraction of the cardiac output is not fully oxygenated.

The matching of airflow and blood flow is best examined by considering the **ventilation/perfusion ratio** ($\dot{V}A/\dot{Q}$ ratio), which compares alveolar ventilation with blood flow in lung regions. Because resting healthy people have an alveolar ventilation ($\dot{V}A$) of 4 L/min and a cardiac output ($\dot{Q}$) of 5 L/min, the ideal alveolar ventilation/perfusion ratio ($\dot{V}A/\dot{Q}$) should be 0.8 (there are no units, because this is a ratio). We have already seen that gravity can cause regional differences in blood flow and alveolar ventilation. In an upright person, the base of the lungs is better ventilated and better perfused than the apex. Regional alveolar ventilation and blood flow are illustrated in Figure 20.11. Three points are apparent from this figure:

- Ventilation and blood flow are both gravity dependent; airflow and blood flow increase down the lung.
- Blood flow shows about a fivefold difference between the top and the bottom of the lung, whereas ventilation shows about a twofold difference. This causes gravity-dependent regional variations in the $\dot{V}A/\dot{Q}$ ratio, which range from 0.6 at the base to 3 or higher at the apex.
- Blood flow is proportionately greater than ventilation at the base, and ventilation is proportionately greater than blood flow at the apex.

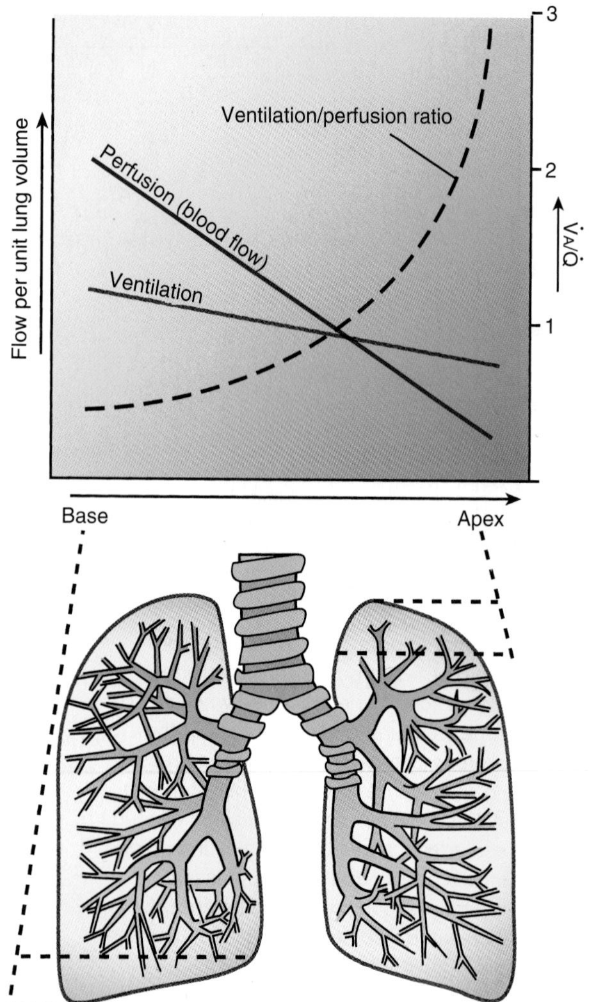

Figure 20.11 Gravity causes a mismatch of blood flow and ventilation in the alveoli. Both ventilation and perfusion are gravity dependent. At the base of the lungs, alveolar capillary blood flow exceeds alveolar ventilation, resulting in a low ventilation/perfusion ($\dot{V}_A/\dot{Q}$) ratio. At the apex, the opposite occurs; alveolar ventilation is greater than capillary blood flow, resulting in a high ventilation/perfusion ($\dot{V}_A/\dot{Q}$) ratio.

	$\dot{V}_A$ (L/min)	$\dot{Q}$ (L/min)	$\dot{V}_A/\dot{Q}$	Pa_{O_2} (mm Hg)	Pa_{CO_2} (mm Hg)
Apex	0.25	0.07	3.6	130	28
Base	0.8	1.3	0.6	88	42

Figure 20.12 Ventilation/perfusion ratios ($\dot{V}_A/\dot{Q}$) affect capillary blood gas tension. A high ($\dot{V}_A/\dot{Q}$) ratio leads to a high Pa_{O_2} and a low Pa_{CO_2} in the blood leaving the apex of the lung as a result of overventilation with respect to blood flow. Blood leaving the base of the lung has a low Pa_{O_2} and a high Pa_{CO_2} as a result of a low ($\dot{V}_A/\dot{Q}$) ratio. Pa_{CO_2}, partial pressure of carbon dioxide, arterial; Pa_{O_2}, partial pressure of oxygen, arterial; $\dot{Q}$, perfusion; $\dot{V}_A$, alveolar ventilation.

The functional importance of lung ventilation/perfusion ratios is that the crucial factor in gas exchange is the matching of regional ventilation and blood flow, as opposed to total alveolar ventilation and total pulmonary blood flow. At the apical region, where the $\dot{V}_A/\dot{Q}$ ratio is >0.8, there is overventilation relative to blood flow. At the base, where the ratio is <0.08, the opposite occurs (i.e., overperfusion relative to ventilation). In the latter case, a fraction of the blood passes through the pulmonary capillaries at the base of the lungs without becoming fully oxygenated.

The effect of regional $\dot{V}_A/\dot{Q}$ ratio on blood gases is shown in Figure 20.12. At the apex, where the lungs are overventilated relative to blood flow, the Pa_{O_2} is high, but the Pa_{CO_2} is low in this region of the lungs. Oxygen tension (P_{O_2}) in the blood leaving pulmonary capillaries at the base of the lungs is low because the blood is not fully oxygenated as a result of underventilation relative to blood flow. Regional differences in $\dot{V}_A/\dot{Q}$ ratios tend to localize some diseases to the top or bottom parts of the lungs. For example, tuberculosis tends to be localized in the apex because of a more favorable environment (i.e., higher oxygen levels for *Mycobacterium tuberculosis*).

SHUNTS AND VENOUS ADMIXTURE

Matching of the lungs' airflow and blood flow is not perfect. On one side of the alveolar–capillary membrane there is "wasted air" (i.e., physiologic dead space), and on the other side there is "wasted blood" (Fig. 20.13). Wasted blood refers to any fraction of the venous blood that does not get fully oxygenated. The mixing of unoxygenated blood with oxygenated blood is known as **venous admixture**. There are two causes for venous admixture: a **shunt** and a low $\dot{V}_A/\dot{Q}$ ratio.

An *anatomic shunt* has a structural basis and occurs when blood bypasses alveoli through a channel, such as from the right to left heart through an atrial or ventricular septal defect or from a branch of the pulmonary artery connecting directly to the pulmonary vein. An anatomic shunt is often called a *right-to-left shunt*. The bronchial circulation also constitutes shunted blood because bronchial venous blood (deoxygenated blood) drains directly into the pulmonary veins, which are carrying oxygenated blood.

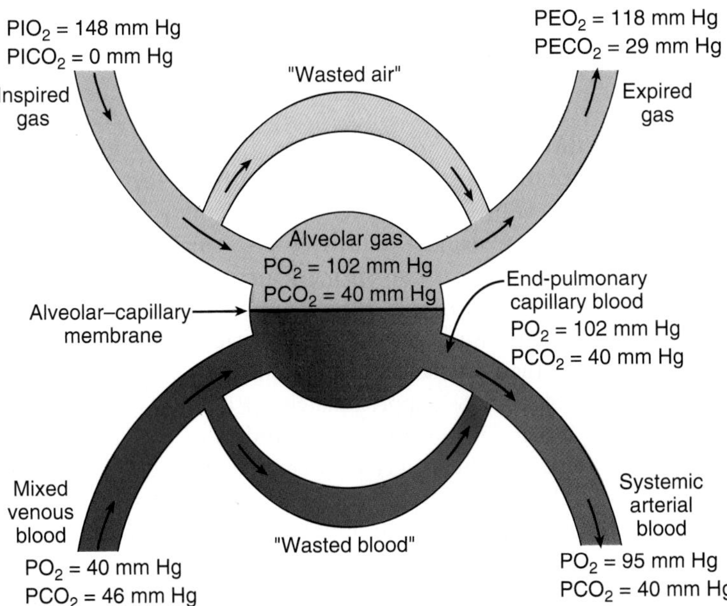

Figure 20.13 On both sides of the alveolar–capillary membrane there is some "wasted air" and "wasted blood." The plumbing on both sides of the alveolar–capillary membrane is imperfect. All of the inspired air does not participate in gas exchange, resulting in some "wasted air." All of the blood entering the lung is not fully oxygenated, leading to some "wasted blood." The total amount of wasted air constitutes physiologic dead space, and the total amount of wasted blood (venous admixture) constitutes physiologic shunt. P_{ECO_2}, partial pressure of expired carbon dioxide; P_{EO_2}, partial pressure of expired oxygen; P_{ICO_2}, partial pressure of inspired carbon dioxide; P_{IO_2}, partial pressure of inspired oxygen.

The second cause for venous admixture is a low regional $\dot{V}_A/\dot{Q}$ ratio. This occurs when a portion of the cardiac output goes through the regular pulmonary capillaries but there is insufficient alveolar ventilation to fully oxygenate all of the blood. With a low regional $\dot{V}_A/\dot{Q}$ ratio, there is no abnormal anatomic connection and the blood does not bypass the alveoli. Rather, blood that passes through the alveolar capillaries is not completely oxygenated. In a healthy person, a low $\dot{V}_A/\dot{Q}$ ratio occurs at the base of the lung (i.e., gravity dependent). A low regional $\dot{V}_A/\dot{Q}$ ratio can also occur with a partially obstructed airway (Fig. 20.14), in which lung regions are underventilated with respect to blood flow, resulting in regional hypoventilation. A fraction of the blood passing through a hypoventilated region is not fully oxygenated, resulting in an increase in venous admixture.

The total amount of venous admixture as a result of anatomic shunt and a low $\dot{V}_A/\dot{Q}$ ratio equals physiologic shunt and represents the total amount of wasted blood that does not get fully oxygenated. *Physiological shunt* is analogous to physiologic dead space; the two are compared in Table 20.1, in which one represents wasted blood flow and the other represents wasted air. It is important to remember that in healthy people, there is some degree of physiologic dead space as well as physiologic shunt in the lungs.

In summary, venous admixture results from anatomic shunt and a low regional $\dot{V}_A/\dot{Q}$ ratio. In healthy people, approximately 50% of the venous admixture comes from an anatomic shunt (e.g., bronchial circulation) and 50% from a low $\dot{V}_A/\dot{Q}$ ratio at the base of the lungs as a result of gravity. Physiologic shunt (i.e., total venous admixture) represents about 1% to 2% of cardiac output in healthy people. This amount can increase up to 15% of cardiac output with some bronchial diseases, and, in certain congenital disorders, a right-to-left anatomic shunt can account for up to 50% of cardiac output. It is important to remember that any deviation of $\dot{V}_A/\dot{Q}$ ratio from the ideal condition (0.8) impairs gas exchange and lowers oxygen tension in the arterial blood. A good way to remember the importance of a shunt is that it always leads to venous admixture and reduces the amount of oxygen carried in the systemic blood.

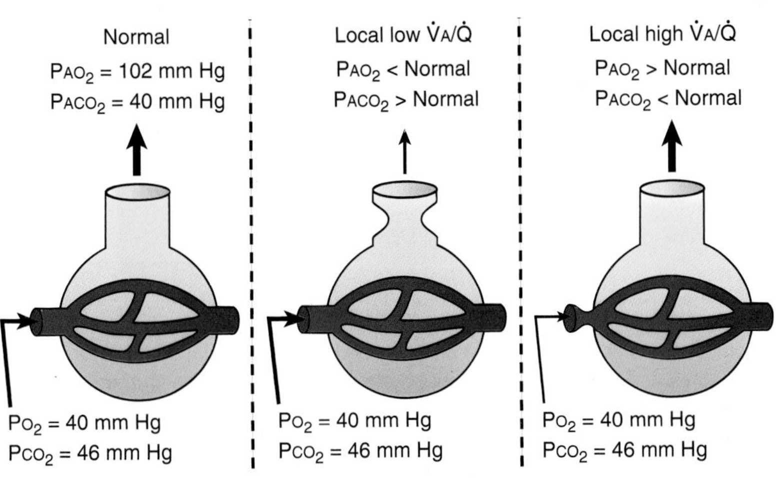

Figure 20.14 Abnormal ventilation/perfusion ($\dot{V}_A/\dot{Q}$) ratios alter gas tensions. Airway obstruction (*middle panel*) causes a low regional ventilation/perfusion ($\dot{V}_A/\dot{Q}$) ratio. A partially blocked airway causes this region to be underventilated relative to blood flow. Note the alveolar gas composition. A low regional $\dot{V}_A/\dot{Q}$ ratio causes venous admixture and will increase the physiologic shunt. A partially obstructed pulmonary arteriole (*right panel*) will cause an abnormally high $\dot{V}_A/\dot{Q}$ ratio in a lung region. Restricted blood flow causes this region to be overventilated relative to blood flow, which leads to an increase in physiologic dead space. P_{ACO_2}, partial pressure of carbon dioxide, alveolar; P_{AO_2}, partial pressure of oxygen, alveolar.

TABLE 20.1 Shunts and Dead Spaces Compared

Shunt	Dead Space
Anatomic	Anatomic
+	+
Low $(\dot{V}_A/\dot{Q})$ ratio	Alveolar
=	=
Physiological shunt (calculated total "wasted blood")	Physiologic dead space (calculated total "wasted air")

BRONCHIAL CIRCULATION

The conducting airways have a separate circulation known as the bronchial circulation, which is distinct from the pulmonary circulation. The primary function of the bronchial circulation is to nourish the walls of the conducting airways and surrounding tissues by distributing blood to the supporting structures of the lungs. Under normal conditions, the bronchial circulation does not supply blood to the terminal respiratory units (respiratory bronchioles, alveolar ducts, and alveoli); they receive their blood from the pulmonary circulation. Venous return from the bronchial circulation is by two routes: bronchial veins and pulmonary veins. About half of the bronchial blood flow returns to the right atrium by way of the bronchial veins, which empty into the azygos vein. The remainder returns through small bronchopulmonary anastomoses into the pulmonary veins.

Bronchial arterial pressure is approximately the same as aortic pressure, and bronchial vascular resistance is much higher than resistance in the pulmonary circulation. Bronchial blood flow is approximately 1% to 2% of cardiac output but in certain inflammatory disorders of the airways (e.g., chronic bronchitis), it can be as high as 10% of cardiac output.

The bronchial circulation is the only portion of the circulation in the adult lung that is capable of undergoing angiogenesis, the formation of new vessels. This is extremely important in providing collateral circulation to the lung parenchyma, especially when the pulmonary circulation is compromised. When a clot or embolus obstructs pulmonary blood flow, the adjacent parenchyma is kept alive by the development of new blood vessels.

From Bench to Bedside / 20.1

Superthin Three-Dimensional Endoscope to Scan Airways

Most ventilation/perfusion problems arise from airway diseases that lead to regional hypoventilation and a concomitant low ventilation/perfusion ratio. Endoscopes are used to examine the patency of airways. Endoscopes are like eyes that can see far into the body and allow physicians to use a minimally invasive procedure to examine the lung. However, today's endoscopes that produce the clearest three-dimensional (3-D) images use cameras several millimeters in diameter and are limited to the upper airways. Most airway diseases occur in the smaller airways, and the current endoscopes are too big to image small airways.

A new endoscope has been developed that is as thin as a human hair and nearly as flexible, which will allow 3-D images of the small airways. The new endoscope, which is undergoing safety tests and clinical trials, is just 350 μm wide and can send back 3-D images that are as clear as those produced by conventional endoscopes. The breakthrough of the new device is better use of light. The conventional endoscope works like a periscope and uses white light that shines down a bundle of glass fibers. The light bounces off tissues and returns to the doctor's eye, creating an image that looks like a photograph. With the new scope, white light moves down a single glass fiber and is broken into a rainbow of colors by an optical device called *diffraction grating*. Each color hits a different part of the tissue being imaged. The colors are reflected back through the glass fiber and fed into a spectrophotometer outside the patient's body, which measures two qualities of the reflected rays. First is intensity, which in the final image translates as shadows. Second, it measures how the returning colors compare with rays bounced off a flat reference object, creating topography. Each color provides a separate pixel of information about the makeup of the tissue. The computer then compares the reflections with a reference beam to create a 3-D image that looks like the topography of the small airway. The new scope may allow patients to have their airways examined without administration of anesthesia.

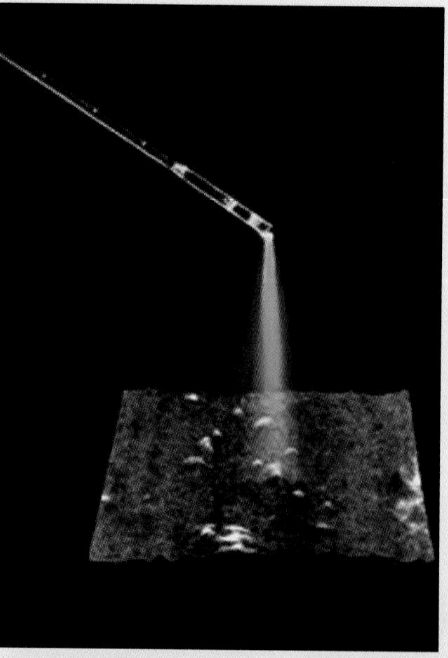

A new, hair-thin endoscope beams a fanned-out rainbow at tissues inside the body to make a clear three-dimensional image that doctors can see. An image of tumors (*white bumps*) detected in a mouse's abdomen is shown in gray.

Chapter Summary

- Pulmonary circulation is a high-flow, low-resistance, and low-pressure system.
- Capillary recruitment is the primary cause for pulmonary vascular resistance to decrease with increased cardiac output.
- Alveolar hypoxia is the major stimulus for pulmonary vasoconstriction.
- Hypoxia-induced pulmonary vasoconstriction shunts blood away from poorly ventilated regions of the lung.
- Gravity causes regional differences in ventilation/perfusion ($\dot{V}_A/\dot{Q}$) ratios in the lungs.
- Venous admixture occurs when venous blood mixes with oxygenated arterial blood.
- An abnormally low ventilation/perfusion ratio exists when a lung region is overperfused relative to airflow, resulting in an increase in venous admixture.
- Physiologic shunt (wasted blood) is analogous to physiologic dead space (wasted air).

21 Control of Ventilation

ACTIVE LEARNING OBJECTIVES

Upon mastering the material in this chapter, you should be able to:

- Describe the major cell groups in the medulla controlling breathing.
- Describe how depth and rate of breathing are controlled.
- Describe how alveolar ventilation is matched to the metabolic needs of the body.
- Predict how changes in blood pH and arterial P_{CO_2} (Pa_{CO_2}) will affect the control of breathing.
- Describe how the respiratory system responds to changes in arterial oxygen levels.
- Describe how breathing patterns change during sleep.
- Describe the ventilatory response to high-altitude acclimatization.
- Describe the sequence of events causing underwater blackout.

GENERATION OF THE BREATHING PATTERN

The most important function of breathing is the exchange of O_2 and CO_2, and the control of breathing is centered primarily on maintaining adequate gas exchange to meet metabolic demands. Ventilation is normally controlled by the autonomic nervous system and occurs without any conscious effort while we are awake, asleep, or under anesthesia. We can, however, override automatic control and exert some conscious effort over our own ventilation by voluntarily changing the rate and depth of breathing. We can also voluntarily stop breathing for a short period of time until carbon dioxide builds up in the blood, which then will stimulate breathing regardless of how hard we try to hold our breath. The basic pattern of breathing generated in the medulla is extensively modified by several control mechanisms. Figure 21.1 illustrates the overall control of breathing, with signals from the blood, higher cortical regions, and stretch receptors from within and outside the lung modulating our breathing. This chapter examines our basic rhythm and how our breathing pattern is modified to meet changes in body functions.

The control of breathing is critical for understanding respiratory responses to activity, changes in the environment, and lung diseases. Our breathing depends on the cyclic excitation of many muscles that can influence the volume of the thorax. Control of that excitation is the result of multiple neuronal interactions involving all levels of the nervous system. Furthermore, the muscles used for breathing must often be used for other purposes, as well. For example, talking while walking requires that some muscles simultaneously attend to the tasks of posturing, walking, phonation, and breathing. Because it is impossible to study extensively the subtleties of such a complex system in humans, much of what is known about the control of breathing has been obtained from the study of other species. Much, however, remains unexplained.

The control of upper and lower airway muscles that affect airway tone is integrated with control of the muscles that start tidal air movements. During quiet breathing, inspiration is brought about by a progressive increase in activation of inspiratory muscles, most importantly the diaphragm (Fig. 21.2). This nearly linear increase in activity with time causes the lungs to fill at a nearly constant rate until tidal volume has been reached. The end of inspiration is associated with a rapid decrease in excitation of inspiratory muscles, after which expiration occurs passively by elastic recoil of the lungs and chest wall. Some excitation of inspiratory muscles resumes during the first part of expiration, slowing the initial rate of expiration. As more ventilation is required—for example, during exercise—other inspiratory muscles (external intercostals, cervical muscles) are recruited. In addition, when ventilation is elevated, expiration becomes an active process through the use, most notably, of the muscles of the abdominal wall. The neural basis of these breathing patterns depends on the generation and subsequent tailoring of cyclic changes in the activity of cells primarily located in the medulla oblongata in the brain.

Basic breathing rhythm is controlled by two major cell groups in the medulla oblongata.

The central pattern for the basic breathing rhythm has been localized to fairly discrete areas in the medulla oblongata that discharge action potentials in a phasic pattern with respiration. Cells in the medulla oblongata associated with breathing have been identified by noting the correlation between their activity and mechanical events of the breathing

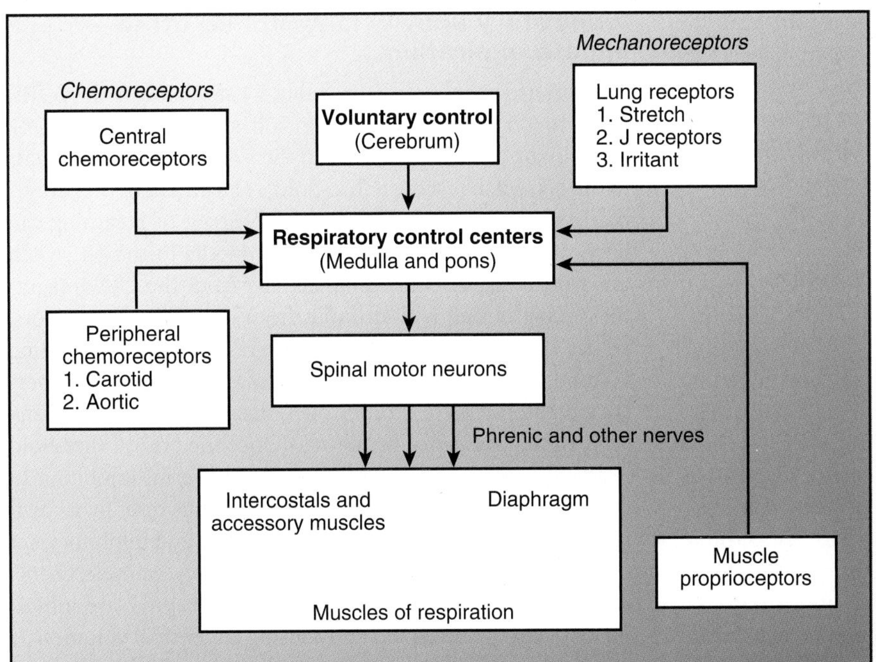

Figure 21.1 Breathing is regulated by various input signals. The schematic illustrates the overall control of breathing with various mechanoreceptors, proprioceptors, and chemoreceptors that are involved in adjusting breathing to meet various metabolic demands.

cycle. Two different groups of cells have been found, and their anatomic locations are shown in Figure 21.3. The **dorsal respiratory group (DRG)**, named for its dorsal location in the region of the nucleus tractus solitarii, predominantly contains cells that are active during inspiration. The **ventral respiratory group (VRG)** is a column of cells in the general region of the nucleus ambiguus that extends caudally nearly to the bulbospinal border and cranially nearly to the bulbopontine junction. The VRG contains both inspiration- and expiration-related neurons. Both groups contain cells projecting ultimately to the bulbospinal motor neuron pools. The DRG and VRG are bilaterally paired, but cross-communication enables them to respond in synchrony; as a consequence, respiratory movements are symmetric.

The neural networks forming the central pattern generator for breathing are contained within the DRG/VRG complex, but the exact anatomic and functional description remains uncertain. Central pattern generation probably does not arise from a single pacemaker or by reciprocal inhibition of two pools of cells, one having inspiratory-related

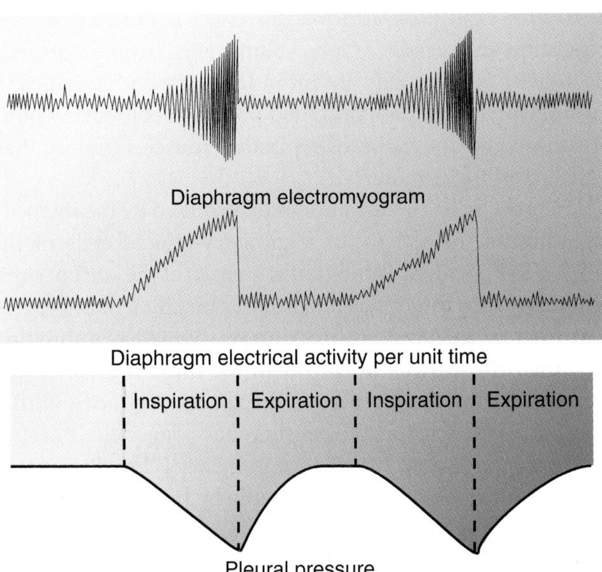

Figure 21.2 Changes in pleural pressure correspond to the electrical activity during breathing at rest. During inspiration, the number of active muscle fibers and the frequency at which each fires increase progressively, leading to a mirror image fall in pleural pressure as the diaphragm descends.

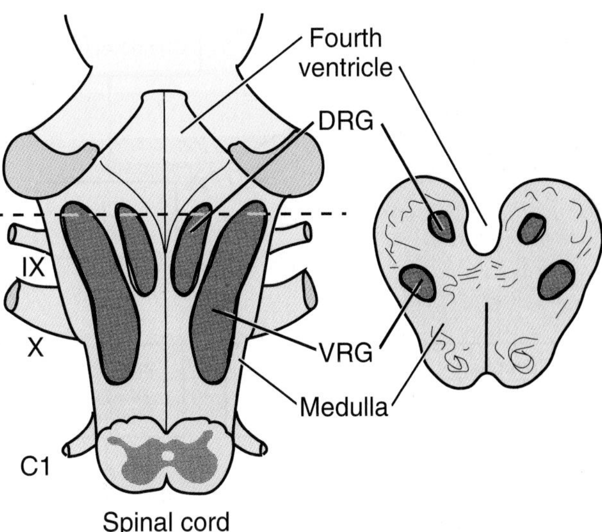

Figure 21.3 The dorsal respiratory group (DRG) and the ventral respiratory group (VRG) are located in the medulla. These drawings show the dorsal aspect of the medulla oblongata and the cross section of the region of the fourth ventricle. C1, first cervical nerve; IX, glossopharyngeal nerve; X, vagus nerve.

and the other having expiratory-related activity. Instead, the progressive rise and abrupt fall of inspiratory motor activity associated with each breath can be modeled by the starting, stopping, and resetting of an integrator of background ventilatory drive. An integrator-based theoretical model, as described below, is suitable for a first understanding of respiratory pattern generation.

Activation of integrator neurons helps synchronize the onset of inspiration.

Many different signals (e.g., volition, anxiety, musculoskeletal movements, pain, chemosensor activity, and hypothalamic temperature) provide a background ventilatory drive to the medulla. Inspiration begins by the abrupt release from inhibition of a group of cells, **central inspiratory activity (CIA) integrator** neurons, located within the medullary reticular formation that integrate this background drive (Fig. 21.4). Integration results in a progressive rise in the output of the integrator neurons, which, in turn, excites a similar rise in activity of inspiratory premotor neurons of the DRG/VRG complex. The rate of rising activity of inspiratory neurons and, therefore, the rate of inspiration itself can be influenced by changing the characteristics of the CIA integrator. Inspiration is ended by abruptly switching off the rising excitation of inspiratory neurons. The CIA integrator is reset before the beginning of each inspiration, so that activity of the inspiratory neurons begins each breath from a low level.

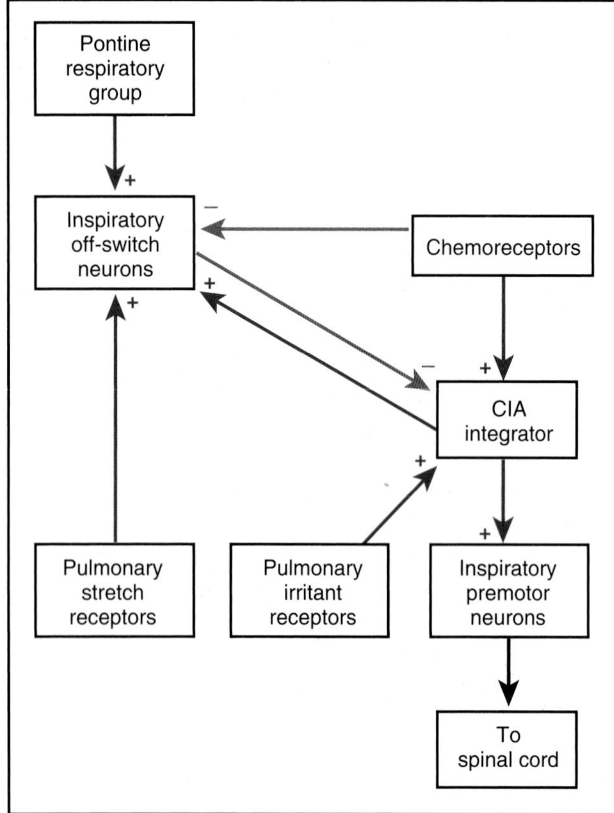

Figure 21.4 Inspiration is initiated by the medullary inspiratory generator. Inspiration begins by the abrupt release from inhibition of a group of cells called the central inspiratory activity (CIA) integrator.

Inspiratory activity is switched off to initiate expiration.

Two groups of neurons, probably located within the VRG, seem to serve as an inspiratory off switch (see Fig. 21.4). Switching occurs abruptly when the sum of excitatory inputs to the off switch reaches a threshold. Adjustment of the threshold level is one of the ways in which depth of breathing can be varied. Two important excitatory inputs to the off switch are a progressively increasing activity from the CIA integrator's rising output and an input from lung stretch receptors, whose afferent activity increases progressively with rising lung volume. (The first of these is what allows the medulla to generate a breathing pattern on its own; the second is one of many reflexes that influence breathing.) Once the critical threshold is reached, off-switch neurons apply a powerful inhibition to the CIA integrator. The CIA integrator is thus reset by its own rising activity. Other inputs, both excitatory and inhibitory, act on the off switch and change its threshold. For example, chemical stimuli, such as hypoxemia and hypercapnia, are inhibitory, raising the threshold and causing larger tidal volumes.

An important excitatory input to the off switch comes from a group of spatially dispersed neurons in the rostral pons called the **pontine respiratory group**. Electrical stimulation in this region causes variable effects on breathing, dependent not only on the site of stimulation but also on the phase of the respiratory cycle in which the stimulus is applied. It is believed that the pontine respiratory group may serve to integrate many different autonomic functions in addition to breathing.

Expiration comprises two phases.

Shortly after the abrupt termination of inspiration, some activity of inspiratory muscles resumes. This activity serves to control expiratory airflow. This effect is greatest early in expiration and recedes as lung volume falls. Inspiratory muscle activity is essentially absent in the second phase of expiration, which includes continued passive recoil during quiet breathing and activation of expiratory muscles if more than quiet breathing is required.

The duration of expiration is determined by the intensity of inhibition of activity of inspiratory-related cells of the DRG/VRG complex. Inhibition is greatest at the start of expiration and falls progressively until it is insufficient to prevent the onset of inspiration. The progressive fall of inhibition amounts to a decline of threshold for initiating the switch from expiration to inspiration. The rate of decline of inhibition and the occurrence of events that trigger the onset of inspiration are subject to several influences. The duration of expiration can be controlled not only by neural information arriving during expiration but also in response to the pattern of the preceding inspiration. How the details of the preceding inspiration are stored and later recovered is unresolved.

Increasing metabolic demands require various control mechanisms to match alveolar ventilation.

Multiple controls provide a greater capability for regulating breathing under a larger number of conditions. Their interactions modify each other and provide for backup in case of failure. The set of strategies for controlling a given variable, such as minute ventilation, typically includes individual schemes that differ in several respects, including choices of sensors and effectors, magnitudes of effects, speeds of action, and optimum operating points. The use of multiple control mechanisms in breathing can be illustrated by considering some of the ways breathing changes in response to exercise. Perhaps the simplest strategies are feedforward mechanisms, in which breathing responds to some component of exercise but without recognition of how well the response meets the demand. One such mechanism would be for the central nervous system (CNS) to vary the activity of the medullary pattern generator in parallel with, and in proportion to, the excitation of the muscles used during exercise. Another prospective feedforward scheme involves sensing the magnitude of the carbon dioxide load delivered to the lungs by systemic venous return and then driving ventilation in response to the magnitude of that load. Experimental evidence supports this mechanism, but the identity of the required intrapulmonary sensor remains uncertain. Still another recognized feedforward mechanism is the enhancement of breathing in response to increased receptor activity in skeletal joints as joint motion increases with exercise.

Although feedforward methods bring about changes in the appropriate direction, they do not provide control in response to the difference between desired and prevailing conditions, as can be done with feedback control. For example, if arterial P_{CO_2} (Pa_{CO_2}) deviates from a reference point, say 40 mm Hg, ventilation could be adjusted by feedback control to reduce the discrepancy. This well-known control system, diagrammed according to the principles given in Chapter 1, "Homeostasis and Cellular Signaling," is shown in Figure 21.5. Unlike feedforward control, feedback control requires a sensor, a reference (set point), and a comparator that together generate an error signal, which drives the effector. Negative-feedback systems provide good control in the presence of considerable variations of other properties of the system, such as lung stiffness or respiratory muscle strength. They can, if sufficiently sensitive, act quickly to reduce discrepancies from reference points to low levels. Too much sensitivity, however, may lead to instability and undesirable excursions of the regulated variable.

Other mechanisms involve minimization or optimization. For example, evidence indicates that rate and depth of breathing are adjusted to minimize the work expenditure for ventilation of a given magnitude. In other words, the controller decides whether to use a large breath with its attendant large elastic load or more frequent smaller breaths with their associated higher resistive load. This strategy requires afferent neural information about lung volume, rate of volume change, and transpulmonary pressures, which can be provided by lung and chest wall mechanoreceptors. During exercise, such a controller would act in concert with, among other things, the feedback control of carbon dioxide described earlier. As a final example, an optimization model using two pieces of information is illustrated in Figure 21.6. Breathing is adjusted to minimize the sum of the muscle effort and the sensory "cost" of tolerating a raised Pa_{CO_2}.

Muscles of the upper airways are also under phasic control.

The same rhythm generator that controls the chest wall muscles also controls muscles of the nose, pharynx, and larynx. However, unlike the inspiratory ramp-like rise of the stimulation of chest wall muscles, the excitation of upper airway muscles quickly reaches a plateau and is sustained until inspiration is ended. Flattening of the expected ramp excitation waveform probably results from progressive inhibition by the rising afferent activity of airway stretch reflexes as lung volume increases. Excitation during inspiration causes contractions of upper airway muscles, airway widening, and reduced resistance from the nostrils to the larynx.

During the first phase of quiet expiration, when expiration is slowed by renewed inspiratory muscle activity, there is also expiratory braking caused by active adduction of the

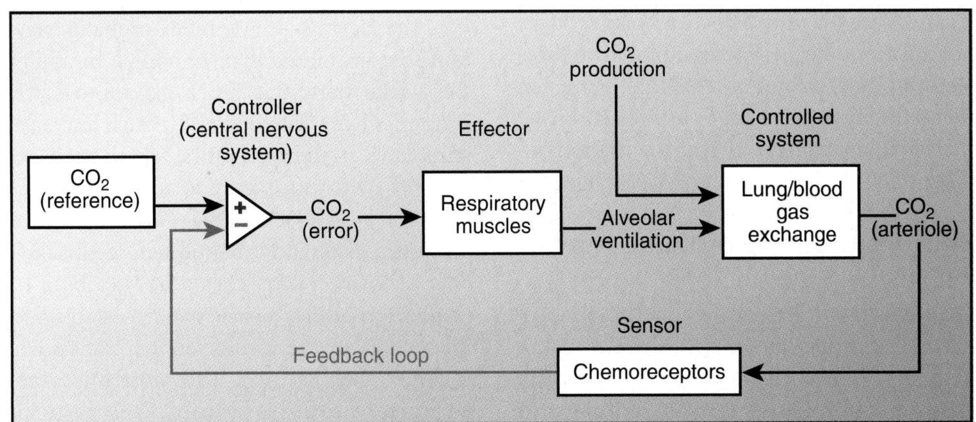

Figure 21.5 Arterial P_{CO_2} is controlled by a negative-feedback system. Variations in CO_2 production lead to changes in arterial P_{CO_2} that are sensed by chemoreceptors. The chemoreceptor signal is subtracted from a reference value. The absolute value of the difference is taken as an input by the central nervous system and passed on to respiratory muscles as new minute ventilation. The loop is completed as the new ventilation alters blood gas composition through the mechanism of lung/blood gas exchange.

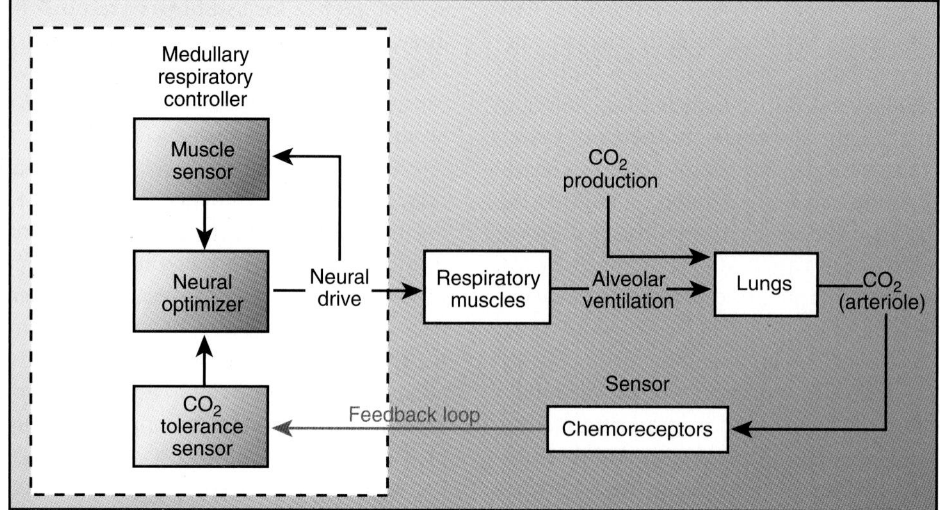

Figure 21.6 Rate and depth of breathing are optimized by a medullary controller to minimize the work of breathing. A proposed model in which the components inside the *dashed box* constitute the medullary controller is represented. In this strategy for breathing, the conflicting needs to maintain chemical homeostasis and to minimize respiratory effort are resolved by selecting an optimal ventilation. The sensors for muscle use and CO_2 tolerance convert neural drive and the output of the chemoreceptors to a form an integrated signal to the neural optimizer, which, in turn, selects optimal ventilation to minimize the work of breathing.

vocal cords. However, during exercise-induced **hyperpnea** (increased depth and rate of breathing), the cords are separated during expiration and expiratory resistance is reduced.

LUNG AND CHEST WALL REFLEXES

Reflexes arising from the periphery provide feedback for fine tuning, which adjusts frequency and tidal volume to minimize the work of breathing. Reflexes from the upper airways and lungs also act as defensive reflexes, protecting the lungs from injury and environmental insults. This section considers reflexes that arise from the lungs and chest wall. Among reflexes influencing breathing, the lung and chest wall mechanoreceptors and the chemoreflexes responding to blood pH and gas tension changes are the most widely recognized (see Fig. 21.1). Although many other less well-explored reflexes also influence breathing, most are not covered in this chapter. Examples are reflexes induced by changes in arterial blood pressure, cardiac stretch, epicardial irritation, sensations in the airway above the trachea, skin injury, and visceral pain.

Mechanoreceptors mediate reflexes that protect the lung.

Pulmonary mechanoreceptors are located in the airways and lung parenchyma and play an important role in the regulation of breathing. These mechanoreceptors can be divided into three groups: **pulmonary stretch**, **irritant**, and **J receptors** (see Fig. 21.1). Afferent fibers of all three types lie predominantly in the vagus nerves, although some pass with the sympathetic nerves to the spinal cord. The role of the sympathetic afferents is uncertain and is not considered further.

Pulmonary stretch receptors

Pulmonary stretch receptors are sensory terminals of myelinated afferent fibers that lie within the smooth muscle layer of conducting airways. Stretch receptors fire in proportion to applied airway transmural pressure, and their usual role is to sense lung volume. When stimulated, an increased firing rate is sustained as long as stretch is imposed; that is, they adapt slowly and are also called *slowly adapting receptors*. Stimulation of these receptors causes an excitation of the inspiratory off switch and a prolongation of expiration. Because of these two effects, inflating the lungs with a sustained pressure at the mouth terminates an inspiration in progress and prolongs the time before a subsequent inspiration occurs. This sequence is known as the **Hering–Breuer reflex** or **lung-inflation reflex**.

The Hering–Breuer reflex probably plays a more important role in infants than in adults. In adults, particularly in the awake state, this reflex may be overwhelmed by more prominent central control. Because increasing lung volume stimulates stretch receptors, which then excite the inspiratory off switch, it is easy to see how they could be responsible for a feedback signal that results in cyclic breathing. However, as already mentioned, feedback from vagal afferents is not necessary for cyclic breathing to occur. Instead, feedback modifies a basic pattern established in the medulla. The effect may be to shorten inspiration when tidal volume is larger than normal. The most important role of slowly adapting receptors is probably their participation in regulating expiratory time, expiratory muscle activation, and functional residual capacity. Stimulation of the stretch receptors also relaxes airway smooth muscle, reduces systemic vasomotor tone, increases heart rate, and, as previously noted, influences laryngeal muscle activity.

Irritant receptors

The irritant receptors are sensory terminals of myelinated afferent fibers that are found in the larger conducting airways. Also called **rapid adapting receptors**, they have nerve endings, which lie in the airway epithelium, and respond to irritation of the airways by touch or by noxious substances, such as smoke and dust. Irritant receptors are stimulated by histamine, serotonin, and prostaglandins released locally in response to allergy and inflammation. They are also stimulated by lung inflation and deflation, but their firing rate rapidly declines when a volume change is sustained. Because of this rapid adaptation, bursts of activity occur that are in proportion to the change of volume and the rate at which that change occurs. Acute congestion of the pulmonary vascular bed also stimulates these receptors, but, unlike the effect of inflation, their activity may be sustained when congestion is maintained.

Background activity of the irritant receptors is inversely related to lung compliance, and they are thought to serve as sensors of compliance change. These receptors are probably nearly inactive in normal quiet breathing. Based on what stimulates them, their role would seem to be to sense the onset of pathologic events. In spite of considerable information about what stimulates them, the effect of their stimulation remains controversial. As a general rule, stimulation causes excitatory responses such as coughing, gasping, and prolonged inspiration time.

J Receptors

J receptors are also called **juxtapulmonary capillary receptors**, or **C-fiber endings**, and belong to unmyelinated nerves. These nerve endings are classified as two populations in the lungs. One group, *pulmonary C fibers*, is located adjacent to the alveoli and is accessible from the pulmonary circulation, hence the name *juxtapulmonary receptors*. A second group, *bronchial C fibers*, is accessible from the bronchial circulation, and consequently, is located in airways. Like rapidly adapting receptors, both groups play a protective role. They are both stimulated by lung injury, large inflation, acute pulmonary vascular congestion, and certain chemical agents.

J receptors are sensitive to mechanical events (e.g., edema, congestion, and pulmonary embolism) but are not as sensitive to products of inflammation, whereas the opposite is true of bronchial C fibers. Their activity excites breathing, and they probably provide a background excitation to the medulla. When stimulated, they cause rapid shallow breathing, bronchoconstriction, increased airway secretion, and cardiovascular depression (bradycardia, hypotension). **Apnea** (cessation of breathing) and a marked fall in systemic vascular resistance occur when they are stimulated acutely and severely. An abrupt reduction of skeletal muscle tone is an intriguing effect that follows intense stimulation of pulmonary C fibers, the homeostatic significance of which remains unexplained.

Chest wall proprioceptors provide information about body movement and muscle tension.

Joint, tendon, and muscle spindle receptors—collectively called **proprioceptors**—may play a role in breathing, particularly when more than quiet breathing is called for or when breathing efforts are opposed by increased airway resistance or reduced lung compliance. Muscle spindles are present in considerable numbers in the intercostal muscles but are rare in the diaphragm. It has been proposed, but not fully verified, that muscle spindles may adjust breathing effort by sensing the discrepancy between tensions of the intrafusal and extrafusal fibers of the intercostal muscles. If a discrepancy exists, information from the spindle receptor alters the contraction of the extrafusal fiber, thereby minimizing the discrepancy. This mechanism provides increased motor excitation when movement is opposed. Evidence also shows that chest wall proprioceptors play a major role in the perception of breathing effort, but other sensory mechanisms may also be involved (see Fig. 21.1).

FEEDBACK CONTROL OF BREATHING

Another set of receptors that profoundly affect breathing is the **chemoreceptors** (see Fig. 21.4). These receptors are stimulated by the hydrogen ion concentration and respiratory gas composition of the arterial blood. The general rule is that breathing activity is inversely related to arterial blood P_{O_2} but directly related to P_{CO_2} and $[H^+]$. Figures 21.7 and 21.8 show the ventilatory responses of a typical person when alveolar P_{CO_2} (P_{ACO_2}) and P_{O_2} are individually varied by controlling the composition of inspired gas. Responses to carbon dioxide and, to a lesser extent, blood pH depend on sensors in the brainstem, carotid arteries, and aorta. In contrast, responses to hypoxia are brought about only by the stimulation of arterial receptors.

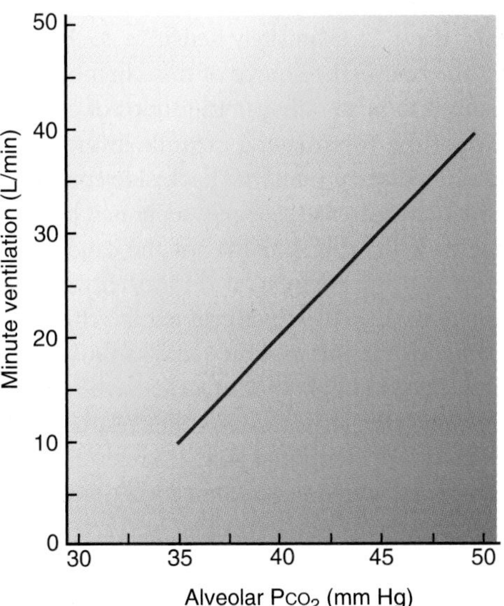

Figure 21.7 CO_2 is a powerful stimulus to ventilation. Ventilatory responses to increasing alveolar CO_2 tension are shown with the line, representing the response when alveolar P_{O_2} (P_{AO_2}) is held at ≥100 mm Hg to essentially eliminate O_2-dependent activity of the chemoreceptors.

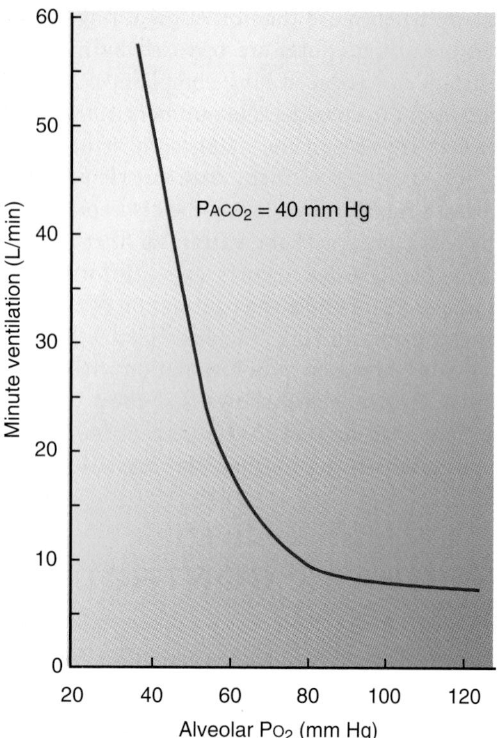

Figure 21.8 Ventilatory responses to hypoxia are mediated via the peripheral chemoreceptors. Inspired oxygen is lowered while alveolar Pco_2 ($PACO_2$) is held at 40 mm Hg by adding CO_2 to the inspired air in order to eliminate the chemoresponse to CO_2-dependent activity. Note that when alveolar carbon dioxide is held constant at 43 mm Hg, minute ventilation does not significantly change until alveolar Po_2 is reduced to 60 mm Hg.

Cerebrospinal fluid pH stimulates neuronal cells in the medulla.

Ventilatory drive is exquisitely sensitive to Pco_2 of blood perfusing the brain. The source of this chemosensitivity has been localized to bilaterally paired groups of cells just below the surface of the ventrolateral medulla immediately caudal to the pontomedullary junction. Each side contains a rostral and a caudal chemosensitive zone, separated by an intermediate zone in which the activities of the caudal and rostral groups converge and are integrated together with other autonomic functions. Exactly which cells exhibit chemosensitivity is unknown, but they are not the same as those of the DRG/VRG complex. Although specific cells have not been identified, the chemosensitive neurons that respond to the [H^+] of the surrounding interstitial fluid are referred to as **central chemoreceptors**. The H^+ concentration in the interstitial fluid is a function of Pco_2 in the cerebral arterial blood and the bicarbonate concentration of cerebrospinal fluid (CSF).

Cerebrospinal fluid has a weak buffering system and is sensitive to changes in carbon dioxide partial pressure pH.

CSF is an ultrafiltrate of plasma and is formed mainly by the **choroid plexuses** of the ventricular cavities of the brain. The epithelium of the choroid plexus provides a barrier between blood and CSF that severely limits the passive movement of large molecules, charged molecules, and inorganic ions. However, choroidal epithelium actively transports several substances, including ions, and this active transport participates in determining the composition of CSF. This selective barrier is termed the **blood–brain barrier** and is illustrated in Figure 21.9. CSF formed by the choroid plexuses is exposed to brain interstitial fluid across the surface of the brain and spinal cord, with the result that the composition of CSF away from the choroid plexuses is closer to that of interstitial fluid than it is to CSF as first formed. Brain interstitial fluid is also separated from blood by the blood–brain barrier (capillary endothelium), which has its own transport capability.

Because of the properties of the limiting membranes, CSF is essentially free of protein, but it is not just a simple ultrafiltrate of plasma. CSF differs most notably from an ultrafiltrate by its lower bicarbonate and higher sodium and chloride ion concentrations. Potassium, magnesium, and calcium ion concentrations also differ somewhat from plasma; moreover, they change little in response to marked changes in plasma concentrations of these cations. Bicarbonate serves as the only significant buffer in CSF, but the mechanism that controls bicarbonate concentration is controversial.

Most proposed regulatory mechanisms invoke the active transport of one or more ionic species by the epithelial and endothelial membranes. Because of the relative

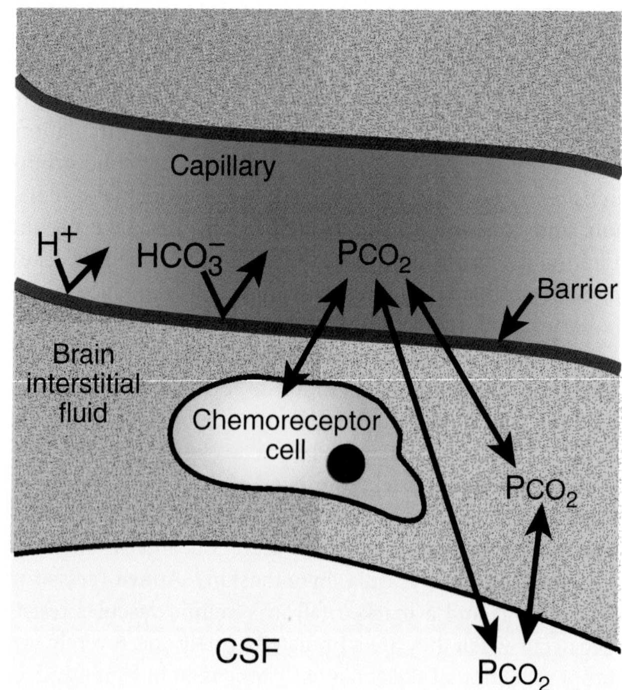

Figure 21.9 The blood–brain barrier is impermeable to blood H^+ and HCO_3^-. Movement of H^+, HCO_3^-, and molecular CO_2 is illustrated between capillary blood, brain interstitial fluid, and cerebrospinal fluid (CSF). Because the blood–brain barrier is permeable to CO_2 and not to H^+ and HCO_3^-, the acid–base status of the chemoreceptors can be quickly changed only by changing $Paco_2$.

impermeabilities of the choroidal epithelium and capillary endothelium to H^+, changes in H^+ concentration of blood are poorly reflected in CSF. By contrast, molecular carbon dioxide diffuses readily; therefore, blood P_{CO_2} can influence the pH of CSF. The pH of CSF is primarily determined by its bicarbonate concentration and P_{CO_2}. The relative ease of movement of molecular carbon dioxide in contrast to hydrogen ions and bicarbonate is depicted in Figure 21.9.

In healthy people, the P_{CO_2} of CSF is about 6 mm Hg higher than that of arterial blood, approximating that of brain tissue. The pH of CSF, normally slightly below that of blood, is held within narrow limits. CSF pH changes little in states of metabolic acid–base disturbances (see Chapter 24, "Acid–Base Homeostasis")—about 10% of that in plasma. In respiratory acid–base disturbances, however, the change in pH of the CSF may exceed that of blood. During chronic acid–base disturbances, the bicarbonate concentration of CSF changes in the same direction as in blood, but the changes may be unequal. In metabolic disturbances, the CSF bicarbonate changes are about 40% of those in blood but, with respiratory disturbances, CSF and blood bicarbonate changes are essentially the same. When acute acid–base disturbances are imposed, CSF bicarbonate changes more slowly than does blood bicarbonate, and may not reach a new steady state for hours or days. As already noted, the mechanism of bicarbonate regulation is unsettled. Irrespective of how it occurs, the bicarbonate regulation that occurs with acid–base disturbances is important because, by changing buffering, it influences the response to a given P_{CO_2}.

Peripheral chemoreceptors respond to plasma oxygen and carbon dioxide partial pressures and pH.

Peripheral chemoreceptors are located in the carotid and aortic bodies and detect changes in arterial blood plasma P_{O_2}, P_{CO_2}, and pH (see Fig. 21.1). **Carotid bodies** are small (~2 mm wide) sensory organs located bilaterally near the bifurcations of the common carotid arteries near the base of the skull. Afferent nerves travel to the CNS from the carotid bodies in the glossopharyngeal nerves. **Aortic bodies** are located along the ascending aorta and are innervated by vagal afferents.

As with the medullary chemoreceptors, increasing Pa_{CO_2} stimulates peripheral receptors. H^+ formed from H_2CO_3 within the peripheral chemoreceptors (glomus cells) is the stimulus and not molecular CO_2. About 40% of the effect of Pa_{CO_2} on ventilation is brought about by peripheral chemoreceptors, whereas central chemoreceptors bring about the rest. Unlike the central sensor, peripheral chemoreceptors are sensitive to rising arterial blood H^+ and falling P_{O_2}. They alone cause the stimulation of breathing by hypoxia; hypoxia in the brain has little effect on breathing unless it is severe, at which point breathing is depressed.

Carotid chemoreceptors play a more prominent role than do aortic chemoreceptors; because of this and their greater accessibility, they have been studied in greater detail. The discharge rate of carotid chemoreceptors (and the resulting minute ventilation) is approximately linearly related to Pa_{CO_2}. The linear behavior of the receptor is reflected in the linear ventilatory response to carbon dioxide illustrated in Figure 21.7. When expressed using pH, the response curve is no longer linear but shows a progressively increasing effect as pH falls below normal. This occurs because pH is a logarithmic function of $[H^+]$, so the absolute change in $[H^+]$ per unit change in pH is greater when brought about at a lower pH.

The response of peripheral chemoreceptors to oxygen depends on arterial P_{O_2} (Pa_{O_2}) and not oxygen content. Therefore, anemia or carbon monoxide poisoning, two conditions that exhibit reduced oxygen content but have normal Pa_{O_2}, have little effect on the response curve. The shape of the response curve is not linear; instead, hypoxia is of increasing effectiveness as P_{O_2} falls below about 90 mm Hg. The behavior of the receptors is reflected in the ventilatory response to hypoxia illustrated in Figure 21.8. The shape of the curve relating ventilatory response to P_{O_2} resembles that of the oxyhemoglobin equilibrium curve when plotted upside down (see Chapter 20, "Pulmonary Circulation and Ventilation/Perfusion"). As a result, the ventilatory response is inversely related in an approximately linear fashion to arterial blood oxygen saturation.

The nonlinearities of the ventilatory responses to P_{O_2} and pH and the relatively low sensitivity across the normal ranges of these variables cause ventilatory changes to be apparent only when P_{O_2} and pH deviate significantly from the normal range, especially toward hypoxemia or acidemia. By contrast, ventilation is sensitive to P_{CO_2} within the normal range, and carbon dioxide is normally the dominant chemical regulator of breathing through the use of both central and peripheral chemoreceptors (compare Figs. 21.7 and 21.8).

There is a strong interaction among stimuli, which causes the slope of the carbon dioxide response curve to increase if determined under hypoxic conditions (see Fig. 21.7), causing the response to hypoxia to be directly related to the prevailing P_{CO_2} and pH (see Fig. 21.8). As discussed in the next section, these interactions and interaction with the effects of the central carbon dioxide sensor profoundly influence the integrated chemoresponses to a primary change in arterial blood composition.

Carotid and aortic bodies can also be strongly stimulated by certain chemicals, particularly cyanide ion and other poisons of the metabolic respiratory chain. Changes in blood pressure have only a small effect on chemoreceptor activity, but responses can be stimulated if arterial pressure falls below about 60 mm Hg. This effect is more prominent in aortic bodies than in carotid bodies. Afferent activity of peripheral chemoreceptors is under some degree of efferent control capable of influencing responses by mechanisms that are not clear. Afferent activity from the chemoreceptors is also centrally modified in its effects by interactions with other reflexes, such as the lung stretch reflex and the systemic arterial baroreflex (see Chapter 17, "Control Mechanisms in Circulatory Function"). Although the breathing interactions are not well understood in humans, they serve as examples of the complex interactions of cardiorespiratory regulation. Interactions among chemoreflexes, however, are easily demonstrated.

CHEMORESPONSES TO ALTERED OXYGEN AND CARBON DIOXIDE

Before the interactions of oxygen and carbon dioxide chemoresponses are discussed, a brief review of terminology is order. Hypoxia leads to **hypoxemia** (a condition in which the Pa_{O_2} is below normal). Hypoxemia stimulates respiration, causing **hyperventilation** (increased alveolar ventilation greater than required to meet metabolic needs). Hyperventilation results in excess carbon dioxide being blown off, which, in turn, leads to **hypocapnia** (a condition in which Pa_{CO_2} is below normal). A low Pa_{CO_2} decreases blood [H^+], leading to **alkalosis** (a condition in which blood pH is above normal). Because alkalosis in this case is caused by hyperventilation, the condition is termed **respiratory alkalosis.** When the lungs are underventilated, alveolar ventilation is less than required to meet metabolic needs (**hypoventilation**) leading to a concomitant decrease in Pa_{O_2} and an elevated in Pa_{CO_2} (**hypercapnia**). The latter leads to an increase in blood [H^+], causing **acidosis** (a condition in which blood pH is below normal). Again, because the acidotic condition is caused by a respiratory dysfunction, the condition is referred to as **respiratory acidosis. Metabolic acidosis** (a condition caused by accumulation of nonvolatile acids such as lactic acid) and respiratory acidosis (accumulation of carbon dioxide) both show low blood pH; the only difference between them is the cause of the acidotic condition.

A distinction needs to be made between hyperventilation and **hyperpnea** (increased alveolar ventilation in proportion to elevated metabolism, e.g., exercise). Remember that *hyperventilation* is characterized by an increase in alveolar ventilation with a concomitant decrease in arterial P_{CO_2}, whereas *hyperpnea* is an increase in alveolar ventilation relative to metabolic carbon dioxide production. The distinguishing feature of hyperpnea is increased ventilation with no change in arterial blood P_{CO_2}, because alveolar ventilation increases in proportion to carbon dioxide production.

Chemoresponses to altered blood oxygen and carbon dioxide are interdependent.

The effect of P_{O_2} on the response to carbon dioxide and the effect of carbon dioxide on the response to P_{O_2} have already been noted. By virtue of this interdependence, the subsequent increased ventilation is blunted in a hyperventilatory response, unless P_{ACO_2} is somehow maintained, because P_{ACO_2} ordinarily falls as ventilation is stimulated (see Fig. 21.8). The central chemoreceptors, which respond more potently than the peripheral receptors to low P_{ACO_2}, are mainly responsible for blunting the stimulating effect of hyperventilation. The sequence of events in the response to hypoxia (e.g., ascent to high altitude) exemplifies interactions among chemoresponses. For example, if 100% oxygen is given to a person just arriving at high altitude, ventilation is quickly restored to its sea-level value. During the next few days, ventilation in the absence of supplemental oxygen progressively rises further, but it is no longer restored to sea-level value by breathing oxygen. Rising ventilation while acclimatizing to altitude could be explained by a reduction of blood and CSF bicarbonate concentrations. This would reduce the initial increase in pH created by the increased ventilation and allow the hypoxic stimulation to be less strongly opposed. However, this mechanism is not the full explanation of altitude acclimatization. CSF pH is not fully restored to normal, and the increasing ventilation raises Pa_{O_2} while further lowering Pa_{CO_2}, changes that should inhibit the stimulus to breathe. In spite of much inquiry, the reason for persistent hyperventilation in altitude-acclimatized subjects, the full explanation for altitude acclimatization, and the explanation for the failure of increased ventilation in acclimatized subjects to be relieved promptly by restoring a normal Pa_{O_2} are still unknown.

In *metabolic acidosis*, the increase in blood [H^+] initiates and sustains hyperventilation by stimulating the peripheral chemoreceptors (Fig. 21.10). Because of the restricted movement of H^+ into CSF, the fall in blood pH cannot directly stimulate the central chemoreceptors. The central effect of the hyperventilation, brought about by decreased pH via the peripheral chemoreceptors, results in a paradoxical rise of CSF pH (i.e., an alkalosis as a result of reduced Pa_{CO_2}) that actually restrains the hyperventilation. With time, CSF bicarbonate concentration is adjusted downward, although it changes less than that of blood, and the pH of CSF remains somewhat higher than pH of blood. Ultimately, ventilation increases more than it did initially as the paradoxical CSF alkalosis is removed.

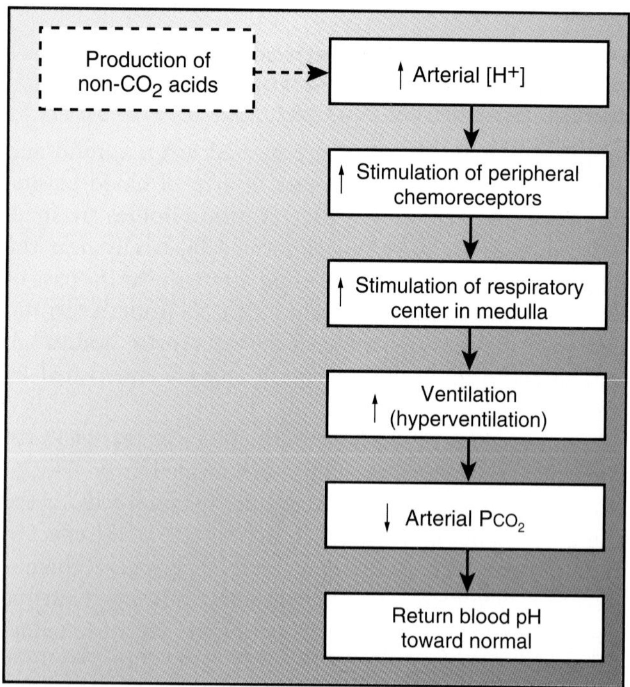

Figure 21.10 Metabolic acidosis stimulates hyperventilation via the peripheral chemoreceptors. The accumulation of non-CO_2 acids increases the blood [H^+] and stimulates breathing. The sustained hyperventilation occurs by stimulating the respiratory center in the medulla via the peripheral chemoreceptors. The sustained hyperventilation blows off alveolar CO_2, which lowers Pa_{CO_2} and compensates by returning blood pH back toward normal.

Respiratory acidosis is rarely a result of elevated environmental CO_2, although this can occur in submarine mishaps, in wet limestone caves, and in physiology laboratories where responses to carbon dioxide are measured. Under these conditions, the response is a vigorous increase in minute ventilation proportional to the level of $Pa{CO_2}$; $Pa{O_2}$ actually rises slightly and arterial pH falls slightly, but these have relatively little effect. If mild hypercapnia can be sustained for a few days, the intense hyperventilation subsides, probably as CSF bicarbonate is raised. More commonly, respiratory acidosis occurs from generalized hypoventilation and results from failure of the controller to respond to carbon dioxide (e.g., during anesthesia, following brain injury, and in some patients with chronic obstructive lung disease). The hypoventilation response is illustrated in Figure 21.11. Another cause of respiratory acidosis is a failure of the breathing apparatus to provide adequate ventilation at an acceptable effort, as may be the case in some patients with obstructive lung disease. When these subjects breathe room air, hypercapnia caused by reduced alveolar ventilation is accompanied by significant hypoxia and acidosis. If the hypoxic component alone is corrected—for example, by breathing oxygen-enriched air—a significant reduction in the ventilatory stimulus may promote underventilation, thereby increasing hypercapnia and acidosis. A more appropriate treatment is providing mechanical assistance for restoring adequate ventilation.

Exercise-induced hyperpnea does not depend on chemoresponses to changes in blood oxygen and carbon dioxide partial pressures and pH.

With light-to-moderate exercise, $Pa{O_2}$, $P{CO_2}$, and pH are essentially unchanged from normal values. Yet ventilation can be substantially increased in these exercise regimes. Even with heavy exercise, a person with a normal cardiovascular system does not become hypoxemic or acutely hypercapnic. What controls ventilation during exercise if chemoresponses to changes in $Pa{O_2}$, $P{CO_2}$, and pH are not involved? The answer to this question involves various control systems, many of which are not well defined. The control of breathing with exercise occurs in three phases and includes the *neurological phase (phase I)*, the *metabolic phase (phase II)*, and the *compensatory phase (phase III)*. In phase I, ventilation increases almost instantly with the onset of exercise and involves a neurologic response insofar as the increase occurs in the absence of any changes in blood gases and metabolism. The neurologic phase involves the medullary generator (see Fig. 21.4) and a feedforward mechanism. During this phase, depth and rate of breathing change in response to the activity of the medullary generator, which increases in parallel with the excitation of the muscles used during exercise. Another recognized feedforward mechanism is stimulation of ventilation in response to increased receptor activity to skeletal joints as joint motion increases with exercise.

In phase II, as exercise continues, ventilation increases linearly with the increase in carbon dioxide production. Exercise-induced hyperpnea occurs in this phase, because the increase in alveolar ventilation is proportional to carbon dioxide production, and, as a result, there is no change in $Pa{CO_2}$. Remember, not only does $Pa{CO_2}$ not change with exercise-induced hyperpnea, but there is no change in $Pa{O_2}$ and arterial pH as well. Although the increased ventilation is tightly coupled to carbon dioxide production, it is still not known how the medullary controller monitors carbon dioxide production. One hypothesis is that a feedforward scheme senses the magnitude of the carbon dioxide load delivered to

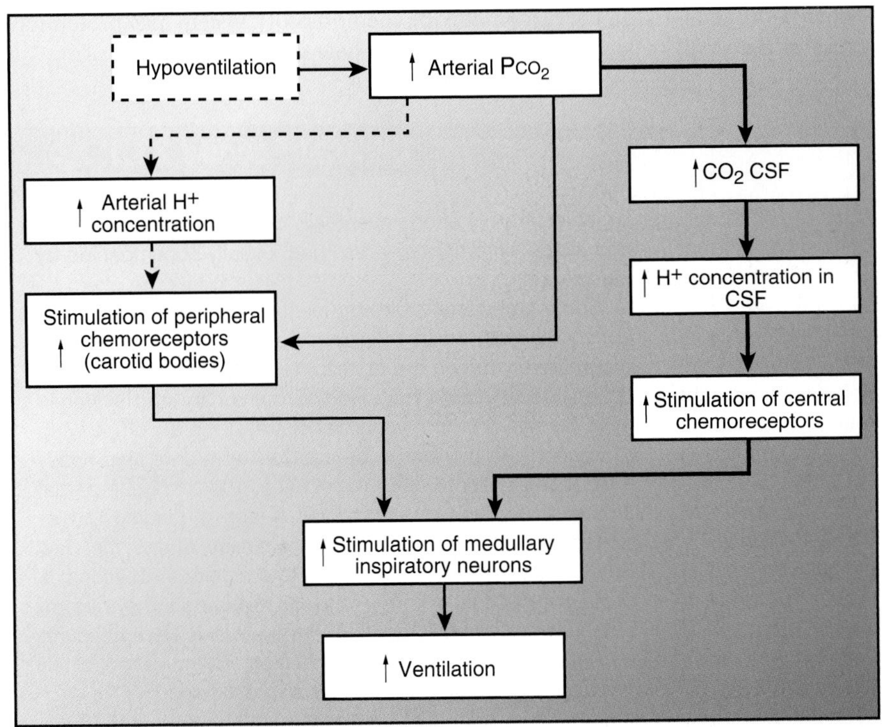

Figure 21.11 Generalized hypoventilation is the primary cause of respiratory acidosis. When the lungs are underventilated, arterial blood CO_2 rises (hypercapnia), leading to respiratory acidosis. The common cause for respiratory acidosis is failure of the controller to respond to carbon dioxide (e.g., anesthesia, head injury, and patients with severe chronic obstructive pulmonary disease). CSF, cerebrospinal fluid.

the lungs by systemic venous blood, which in return drives ventilation. Experimental evidence supports this mechanism, but specific intrapulmonary sensors have not been identified.

In phase III, the intensity of exercise increases (i.e., exhaustive exercise), and the energy needs outstrip the ability of the cardiovascular system to supply sufficient oxygen for aerobic metabolism. Muscle cells shift to anaerobic metabolism with increased blood lactic acid as a by-product. Sustained ventilation is compensation for the development to metabolic acidosis, hence the term *compensatory phase.* During this phase, the sustained ventilation is driven by the acidotic condition.

CONTROL OF BREATHING DURING SLEEP

We spend about one third of our lives asleep. Sleep disorders and disordered breathing during sleep are common and often have physiologic consequences. Chapter 7, "Integrative Functions of the Nervous System," describes the two different neurophysiologic sleep states: **rapid eye movement (REM) sleep** and **slow-wave sleep**. Sleep is a condition that results from withdrawal of the wakefulness stimulus that arises from the brainstem reticular formation. This wakefulness stimulus is one component of the tonic excitation of brainstem respiratory neurons, and one would predict correctly that sleep results in a general depression of breathing. There are, however, other changes, and the effects of REM and slow-wave sleep on breathing differ.

Sleep changes the breathing frequency and inspiratory flow rate.

During *slow-wave sleep*, breathing frequency and inspiratory flow rate are reduced, and minute ventilation falls. These responses partially reflect the reduced physical activity that accompanies sleep. However, because of the small rise in $Paco_2$ (about 3 mm Hg), there must also be a change in either the sensitivity or the set point of the carbon dioxide controller. In the deepest stage of slow-wave sleep (stage 4), breathing is slow, deep, and regular. But in stages 1 and 2, the depth of breathing sometimes varies periodically. The explanation is that during light sleep, withdrawal of the wakefulness stimulus varies over time in a periodic fashion. When the stimulus is removed, sleep is deepened and breathing is depressed; when returned, breathing is excited not only by the wakefulness stimulus but also by the carbon dioxide retained during the interval of sleep. This periodic pattern of breathing is known as **Cheyne–Stokes breathing** (Fig. 21.12).

In *REM sleep*, breathing frequency varies erratically, whereas tidal volume varies little. The net effect on alveolar ventilation is probably a slight reduction, but this is achieved by averaging intervals of frank **tachypnea** (excessively rapid breathing) with intervals of apnea. Unlike those in slow-wave sleep, the variations during REM sleep do not reflect a changing wakefulness stimulus but instead represent responses to increased CNS activity of behavioral, rather than autonomic or metabolic, control systems.

Sleep blunts the responses to respiratory stimuli of carbon dioxide.

Responsiveness to carbon dioxide is reduced during sleep. In slow-wave sleep, the reduction in sensitivity seems to be secondary to a reduction in the wakefulness stimulus and its tonic excitation of the brainstem rather than to a suppression of the chemosensory mechanisms. It is important to note that breathing remains responsive to carbon dioxide during slow-wave sleep, although at a less sensitive level, and that carbon dioxide stimulus may provide the major background brainstem excitation in the absence of the wakefulness stimulus or behavioral excitation. Hence, pathologic alterations in the carbon dioxide chemosensory system may profoundly depress breathing during slow-wave sleep.

Clinical Focus / 21.1 

Sleep Apnea Syndrome

The analysis of multiple physiologic variables recorded during sleep, known as polysomnography, is an important method for research into the control of breathing and has been increasingly used in clinical evaluations of sleep disturbances. In normal sleep, reduced dilatory upper-airway muscle tone may be accompanied by brief intervals of no breathing movements. Some people, typically overweight men, exhibit more severe disruption of breathing, referred to as **sleep apnea syndrome**. Sleep apnea is classified in two broad groups: obstructive and central.

In *central sleep apnea*, breathing movements cease for a longer-than-normal interval. In obstructive sleep apnea, the fault seems to lie in a failure of the pharyngeal muscles to open the airway during inspiration. This may be the result of decreased muscle activity, but the obstruction is worsened by an excessive amount of neck fat with which the muscles must contend. With *obstructive sleep apnea*, progressively larger inspiratory efforts eventually overcome the obstruction and airflow is temporarily resumed, usually accompanied by loud snoring.

Some patients exhibit both central and obstructive sleep apneas. In both types, hypoxemia and hypercapnia develop progressively during the apnea intervals. Frequent episodes of repeated hypoxia may lead to pulmonary and systemic hypertension and to myocardial distress; the accompanying hypercapnia is thought to be a cause of the morning headache these patients often experience. There may be partial arousal at the end of the periods of apnea, leading to disrupted sleep and resulting in drowsiness during the day. Daytime sleepiness, often leading to dangerous situations, is probably the most common and most debilitating symptom. The cause of this disorder is multivariate and often obscure, but mechanically assisted ventilation during sleep often results in significant symptomatic improvement.

From Bench to Bedside / 21.1

Narcolepsy and the Secret of Sleep

Sleep is a vital part of our existence. We spend a third of our lives asleep, and if we become sleep deprived, many of our body functions are impaired. In particular, our immune system, memory, and motor skills are compromised with the lack of sleep. Although several diseases are linked to sleep disorders (e.g., cardiovascular, kidney, and Alzheimer diseases), no one knows why we need sleep.

Sleep disorders constitute a major health problem but still remain a scientific mystery. One sleep disorder that is not well understood is narcolepsy. Narcolepsy is a neurologic disorder characterized by excessive daytime sleepiness. The main characteristic of narcolepsy, the overwhelming excessive daytime sleepiness, occurs even after adequate nighttime sleep. A person with narcolepsy is likely to become drowsy or to fall asleep, often at inappropriate times and places. Daytime naps may occur with or without warning and may be irresistible. Narcoleptic patients are at risk for accidents and often cannot drive or operate machinery because of the danger of falling asleep. These naps can occur several times a day. They are typically refreshing, but only for a few hours. Drowsiness may persist for prolonged periods of time. In addition, nighttime sleep may be fragmented with frequent awakenings.

Narcolepsy is not well understood and is often difficult to diagnose. In the United States, it is estimated that narcolepsy afflicts as many as 200,000 Americans, but fewer than 50,000 are diagnosed. Narcolepsy is as widespread as Parkinson disease or multiple sclerosis and is more prevalent than cystic fibrosis, but is less well known. Narcolepsy is often mistakenly diagnosed as depression, epilepsy, or a side effect of medications. Narcolepsy can occur in both men and women at any age, although its symptoms are usually first noticed in teenagers or young adults. There is strong evidence that narcolepsy may be genetic: 8% to 12% of people with narcolepsy have a close relative with this neurologic disorder. There is an average 15-year delay between onset and correct diagnosis, which contributes to the disabling features of the disorder that often lead to accidents or poor work performance.

Although the cause of narcolepsy has not yet been determined, clinical investigators believe that the disease is rooted in subtle wiring problems. Normally, when a person is awake, brain waves show a regular rhythm. When a person falls asleep, the brain waves become slower and less regular. This sleep status is called *non–rapid eye movement (NREM) sleep*. After about an hour and a half of NREM sleep, the brain waves begin to show a more active pattern again. This sleep state, called rapid eye movement (REM) sleep, is when most remembered dreaming occurs. In narcolepsy, the order and length of NREM and REM sleep periods are disturbed, with REM sleep occurring at the onset of sleep instead of after a period of NREM sleep. Thus, narcolepsy is a disorder in which REM sleep appears at an abnormal time.

Two new approaches are being explored to understand the secrets of sleep and abnormal sleep such as narcolepsy. One approach incorporates a new device called diffusion tensor imaging (DTI). The DTI is a newly developed modification of magnetic resonance imaging (MRI) that allows clinical investigators to study the connections between different areas of the brain for the first time. Conventional MRI images reveal major anatomic features of the brain (e.g., gray matter), which is made up of nerve cell bodies. However, many sleep disorders are rooted in subtle wiring problems involving axons, the long, thin tails of the neuron that carry the electrical signals. The DTI uses specific radiofrequency and magnetic field–gradient pulses to track the movement of water in the brain tissue. In most neuronal tissue, water molecules diffuse in different directions but do diffuse along the length of axons. The white fatty myelin coating holds the water in and allows investigators to create pictures of axons by analyzing the direction of the net movement of water.

Once an axonal picture is established, a new algorithm method is explored to analyze the electrical activity. The new algorithm can detect subtle but statistically significant changes in neuronal activity. By incorporating the axonal picture using DTI with the new algorithm function, a unique temporal resolution can be used to correlate neuronal water movement with electrical activity. The success of these new techniques will enable neuroscientists and neurologists to pinpoint specific wiring defects that lead to sleep disorders such as narcolepsy.

During intervals of REM sleep in which there is little sign of increased activity, the breathing response to carbon dioxide is slightly reduced, as in slow-wave sleep. However, during intervals of increased activity, responses to carbon dioxide during REM sleep are significantly reduced, and breathing seems to be regulated by the brain's behavioral control system. It is interesting that regulation of breathing during REM sleep by the behavioral control system, rather than by carbon dioxide, is similar to the way breathing is controlled during speech.

Ventilatory responses to hypoxia are probably reduced during both slow-wave and REM sleep, especially in people who have high sensitivity to hypoxia while awake. There does not seem to be a difference between the effects of slow-wave and REM sleep on hypoxic responsiveness, and the irregular breathing of REM sleep is unaffected by hypoxia.

Both slow-wave and REM sleep cause an important change in responses to airway irritation. Specifically, a stimulus that causes cough, tachypnea, and airway constriction during wakefulness will cause apnea and airway dilation during sleep unless the stimulus is sufficiently intense to cause arousal. The lung stretch reflex appears to be unchanged or somewhat enhanced during arousal from sleep, but the effect of stretch receptors on upper airways during sleep may be important.

Arousal mechanisms protect the sleeper.

Several stimuli cause arousal from sleep; less intense stimuli cause a shift to a lighter sleep stage without frank arousal. In general, arousal from REM sleep is more difficult than from slow-wave sleep. In humans, hypercapnia is a more potent arousal stimulus than does hypoxia, the former requiring a

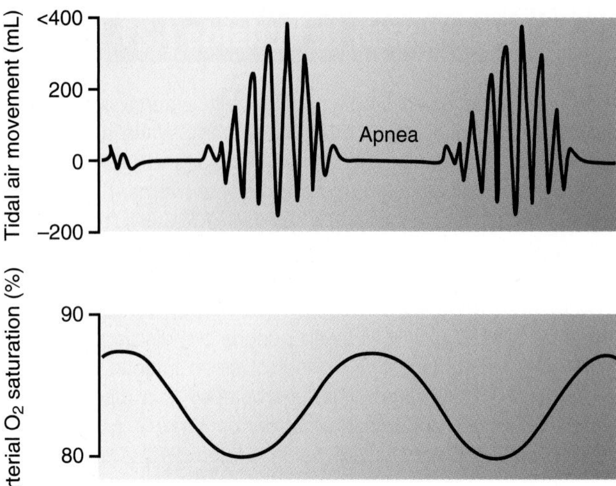

Figure 21.12 Cheyne–Stokes breathing lowers arterial O_2 saturation. Cheyne–Stokes breathing occurs frequently during sleep, especially in subjects at high altitude, as in this example. In the presence of preexisting hypoxemia secondary to high altitude or other causes, the periods of apnea may result in further decreases of O_2 saturation to dangerous levels. Falling P_{O_2} and rising P_{CO_2} during the apnea intervals ultimately induce a response, and breathing returns, reducing the stimuli and leading to a new period of apnea.

Pa_{CO_2} of about 55 mm Hg and the latter requiring a Pa_{O_2} <40 mm Hg. Airway irritation and airway occlusion induce arousal readily in slow-wave sleep but much less readily during REM sleep.

All of these arousal mechanisms probably operate through the activation of a reticular arousal mechanism similar to the wakefulness stimulus. They play an important role in protecting the sleeper from airway obstruction, alveolar hypoventilation of any cause, and the entrance into the airways of irritating substances. Recall that coughing depends on the aroused state and without arousal airway irritation leads to apnea. Obviously, wakefulness altered by other-than-natural sleep—such as during drug-induced sleep, brain injury, or anesthesia—leaves the person exposed to risk because arousal from those states is impaired or blocked. From a teleologic point of view, the most important role of sensors of the respiratory system may be to cause arousal from sleep.

Upper airway tone may be compromised during REM sleep.

A prominent feature during REM sleep is a general reduction in skeletal muscle tone. Muscles of the larynx, pharynx, and tongue share in this relaxation, which can lead to obstruction of the upper airways. Airway muscle relaxation may be enhanced by the increased effectiveness of the lung inflation reflex.

A common consequence of airway narrowing during sleep is snoring. In many people, usually men, the degree of obstruction may at times be sufficient to cause essentially complete occlusion. In these people, an intact arousal mechanism prevents suffocation, and this sequence is not in itself unusual or abnormal. In some people, obstruction is more complete and more frequent, and the arousal threshold may be raised. Repeated obstruction leads to significant hypercapnia and hypoxemia, and repeated arousals cause sleep deprivation that leads to excessive daytime sleepiness, often interfering with normal daily activities.

CONTROL OF BREATHING IN UNUSUAL ENVIRONMENTS

Changes in activity and the environment initiate integrated ventilatory responses that involve changes in the cardiopulmonary system. Examples include the response to exercise (see Chapter 29, "Exercise Physiology") and the response to altered gas tensions. The importance of understanding integrated ventilatory responses is that similar interactions occur under pathophysiologic conditions in patients with respiratory illnesses.

Breathing with low inspired oxygen tension (**hypoxia**) leads to low Pa_{O_2} and is frequently encountered during ascent to high altitudes. How the body responds to high altitude has fascinated physiologists for centuries. The French physiologist Paul Bert first recognized that the harmful effects of high altitude are caused by low oxygen tension. Recall from Chapter 20, "Pulmonary Circulation and Ventilation/Perfusion," that the percentage of oxygen does not change at high altitude but P_{O_2} decreases due to a drop in barometric pressure (see Fig. 20.1). Thus, the hypoxic response at high altitude is caused by a decrease in P_{IO_2} resulting from a decrease in P_B and not from a change in fraction of inspired oxygen (F_{IO_2}). At high altitude, when the P_{IO_2} decreases and oxygen supply in the body is threatened, several compensations are made in an effort to deliver normal amounts of oxygen to the tissues. Somewhat surprising is the fact that hypoxia-induced hyperventilation is not significantly increased until the alveolar P_{O_2} decreases to <60 mm Hg (see Fig. 21.8). In a healthy adult, a drop in alveolar P_{O_2} to 60 mm Hg occurs at an altitude of approximately 4,500 m (14,000 ft). As altitude increases, the body makes noticeable efforts to deliver more oxygen to the tissues. Chief among these responses to altitude is induced hyperventilation—that is, deeper breaths taken more rapidly in which alveolar ventilation is increased. The sequence of events for hypoxia-induced hyperventilation is as follows: (1) a fall in P_{IO_2}, (2) decreased alveolar tension (P_{AO_2}), (3) decreased arterial oxygen tension (Pa_{O_2}), (4) increased firing of peripheral chemoreceptors (the carotid bodies), (5) hyperventilation, and (6) increased alveolar and arterial oxygen tension. As mentioned previously, the stimulus for ventilation during hypoxia is a decrease in Pa_{O_2} rather than O_2 content or percent O_2 saturation. It is important to realize that there is also an immediate rise in cardiac output to match the increase in ventilation. The increase in cardiac output leads to an increase in pulmonary circulation.

Remember, an increase in pulmonary blood flow reduces both the capillary transit time and pulmonary vascular resistance. The latter is due to capillary recruitment, which occurs primarily in the top part of the lungs, resulting in more even blood flow throughout the lungs. As a result, regional blood flow and airflow in the lungs are better matched (i.e., improved ventilation/perfusion ratio). Thus, the increased cardiac output indirectly increases gas exchange by decreasing transit time and improving the overall ventilation/perfusion ratio in the lungs. Figure 21.13 shows how ventilation and P_{ACO_2} change with hypoxia. The hypoxia-induced hyperventilation appears in two stages. First, there is an immediate increase in ventilation, which is primarily a result of hypoxia-induced stimulation via the carotid bodies. However, the increase in ventilation seen in the first stage is small compared with that of the second stage, in which ventilation continues to rise slowly over the next 8 hours. After 8 hours of hypoxia, minute ventilation is sustained. The reason for the small rise in ventilation seen in the first stage is that the hypoxic stimulation is strongly opposed by the decrease in P_{aCO_2} as a result of excess carbon dioxide blown off with altitude-induced hyperventilation. Remember that hyperventilation elevates P_{aO_2} but significantly lowers P_{aCO_2}, resulting in a concomitant increase in arterial pH. The hypocapnia (due to the decrease in arterial P_{CO_2}) and the rise in blood pH work in concert to strongly blunt the hypoxic drive.

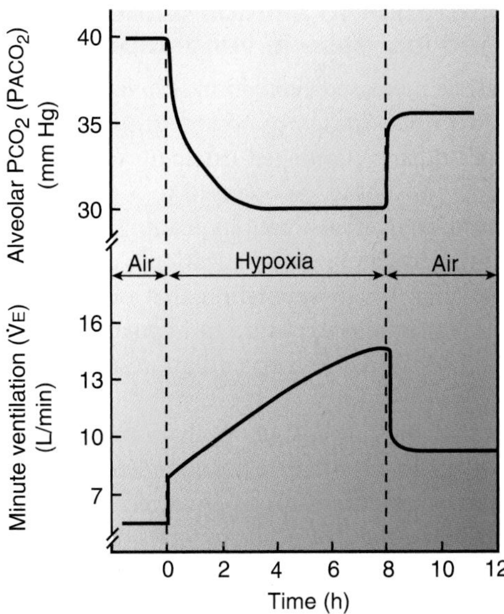

Figure 21.13 Hypoxia stimulates minute ventilation and causes hypocapnia. Hypoxia was induced by having a healthy subject breathe 12% O_2 for 8 hours. With hypoxia-induced hyperventilation, excess CO_2 is blown off, resulting in a decrease in alveolar P_{CO_2} and arterial P_{CO_2} with a concomitant rise in blood pH. Minute ventilation remains elevated for awhile after the subject returns to room air.

Clinical Focus / 21.2

Acute Mountain Sickness

Acute mountain sickness (AMS) is a pathophysiologic condition that is caused by acute exposure to high altitude. AMS is also known as *altitude sickness* and is caused by hypoxia resulting from a decrease in oxygen tension (P_{O_2}) and not from a decrease in the percentage of oxygen in the inspired air.

AMS commonly occurs above 2,400 m (~8,000 ft) and dates back to 30 BCE, as indicated by the mention of the "Great Headache Mountain" in Chinese documents. Symptoms are often manifested 6 to 10 hours after ascent and generally subside in 1 to 2 days. AMS symptoms include headache, dizziness, nausea, and sleep disturbance. In most cases, the symptoms are only temporary and usually abate with time as altitude acclimation occurs. Exertion aggravates the situation, and in extreme cases pulmonary edema and cerebral edema can occur, which can be fatal. The rate of ascent, the altitude, the degree of physical activity, and individual susceptibility are contributing factors to the incidence and severity of AMS.

Many AMS symptoms are caused by hypoxia-induced cerebral blood vessel dilation, which puts pressure on adjacent neuronal tissue. The hypoxia-induced cerebral dilation accounts for the headache, dizziness, nausea, and sleep disturbance. Dehydration can exacerbate these symptoms. The most serious symptoms of AMS are a result of edema (abnormal fluid accumulation) in the lungs and brain. Severe hypoxia causes endothelial cells to become leaky. The edema that is the most life threatening is pulmonary edema. The problem arises because of the oxygen paradox (i.e., hypoxia causes blood vessels to dilate in systemic circulation but causes vasoconstriction in the pulmonary circulation). The hypoxia-induced pulmonary vasoconstriction increases pulmonary vascular resistance. The heart trying to pump blood through "water-soaked lungs" further complicates the situation. As the lungs fill up with water, victims become increasingly short of breath and often spit up foamy blood. Severe pulmonary edema causes death as a result of heart failure.

AMS can be avoided with care. One precaution is to avoid caffeine and alcohol, which tend to exacerbate AMS. A second precaution is to ascend the mountain slowly. Most AMS occurs following a rapid ascent. The third precaution of high-altitude climbing is to follow the climbers' "golden rule": climb high and sleep low, meaning that a climber should stay a few days at base camp to acclimatize, then climb slowly to a higher camp, and return to base camp for the night. This procedure is repeated a few times, each time extending the time spent at higher altitudes to let the body acclimatize to the lower oxygen tension. The general rule is not to ascend >300 m (1,000 ft) per day to sleep. This means a climber can climb from 3,000 m (10,000 ft) to 4,500 m (15,000 ft) in 1 day, but then descend back to 3,300 m (11,000 ft) to sleep. This explains why climbers need to spend days and/or weeks at a time to acclimatize before they can reach a high peak.

Acetazolamide may help speed up the acclimatization process and can treat mild cases of altitude sickness. It reduces fluid formation and helps reduce the severity of cerebral edema. The drug also forces the kidneys to excrete bicarbonate, the base form of carbon dioxide, thus counteracting the effects of hyperventilation that occurs at altitude.

Acclimatization to altitude leads to a sustained increase in ventilation.

The sustained increased ventilation seen in the second stage is referred to as *ventilatory acclimatization*. Acclimatization occurs during prolonged exposure to hypoxia and is a physiologic response, as opposed to a genetic or evolutionary change over generations leading to a permanent **adaptation**. Ventilatory acclimatization is defined as a time-dependent increase in ventilation that occurs over hours to days of continuous exposure to hypoxia. After 2 weeks, the hypoxia-induced hyperventilation reaches a stable plateau.

Although the physiologic mechanisms responsible for ventilatory acclimatization are not completely understood, it is clear that two mechanisms are involved. One involves the chemoreceptors, and the second involves the kidneys. When ventilation is stimulated by hypoxia, CSF pH becomes more alkaline. The elevated CSF pH is brought closer to normal by the movement of bicarbonate out of the CSF. Also, during prolonged hypoxia, the carotid bodies increase their sensitivity to Pa_{O_2}. These changes result in a further increase in ventilation.

The second mechanism responsible for ventilatory acclimatization involves the kidneys. The alkaline blood pH resulting from the hypoxia-induced hyperventilation is antagonistic to the hypoxic drive. Blood pH is regulated by both the lungs and the kidneys (see Chapter 24, "Acid–Base Homeostasis"), which is illustrated in the following equation:

$$\frac{pH = pK_a + \log[HCO_3^-]}{[H_2CO_3]} \quad (1)$$

Because pK_a stays constant and H_2CO_3 dissociates into CO_2 and H_2O, the equation can be represented by

$$pH = \frac{[HCO_3^-]}{Pa_{CO_2}} \frac{20}{1} \frac{\text{Kidney}}{\text{Lung}} \quad (2)$$

Equation 2 also illustrates that the blood $[HCO_3^-]$ is regulated primarily by the kidney and that the Pa_{CO_2} is regulated primarily by the lungs. A normal pH of 7.4 is maintained by a ratio of 20:1. The functional importance of the second equation illustrates that it is not the absolute amount of $[HCO_3^-]$ and Pa_{CO_2} that is important in maintaining a normal pH but the 20:1 ratio. During hypoxic-induced hyperventilation, excess CO_2 is blown off, lowering Pa_{CO_2}, which, in turn, increases the ratio and makes the blood pH more alkaline. The kidneys compensate by excreting more bicarbonate, which brings the ratio back to a normal 20:1 and a blood pH toward 7.4. This process occurs over a 2- to 3-day period, and the antagonistic effect resulting from the hypoxia-induced alkaline pH is minimized, allowing the hypoxic drive to increase minute ventilation further. Thus, it is important to remember that the regulation of pH by ventilatory Pa_{CO_2} is fast (seconds to minutes), whereas the regulation of pH by the kidney to adjust the sodium bicarbonate concentration is slow (hours to days).

Cardiovascular acclimatization improves the delivery of oxygen to the tissues.

In addition to ventilatory acclimatization, the body undergoes other physiologic changes to acclimatize to low oxygen levels. These include increased pulmonary blood flow, increased red cell production, and improved oxygen and carbon dioxide transport. There is an increase in cardiac output at high altitude, resulting in increased blood flow to the lungs and other organs of the body. As mentioned previously, the increase in pulmonary blood flow reduces capillary transit time and improves the overall ventilation/perfusion ratio in the lungs, both of which result in an increase in oxygen uptake by the lungs. Low P_{O_2} causes vasodilation in the systemic circulation. The increase in blood flow resulting from the combined increased vasodilation and increased cardiac output sustains oxygen delivery to the tissues at high altitude.

Red cell production is also increased at high altitude, which improves oxygen delivery to the tissues. Hypoxia stimulates the kidneys to produce and release **erythropoietin**, a hormone that stimulates the bone marrow to produce erythrocytes, which are released into the circulation. The increased hematocrit resulting from the hypoxia-induced **polycythemia** enables the blood to carry more oxygen to the tissues. However, the increased viscosity, as a result of the elevated hematocrit, tends to increase the workload on the heart. In some cases, the polycythemia becomes so severe (hematocrit > 70%) at high altitude that blood has to be withdrawn periodically to permit the heart to pump effectively. Oxygen delivery to the cells is also favored by an increased concentration of 2,3-diphosphoglycerate in the red cells, which shifts the oxyhemoglobin equilibrium curve to the right and favors the unloading of oxygen in the tissues (see Chapter 19, "Gas Transfer and Transport"). Although the body undergoes many beneficial changes that allow acclimatization to high altitude, there are some undesirable effects. One of these is pulmonary hypertension (abnormally high pulmonary arterial blood pressure). Alveolar hypoxia causes pulmonary vasoconstriction. Remember that regional hypoxia redirects blood away from poorly ventilated regions in the lung without any change in pulmonary pressure (Chapter 20, "Pulmonary Circulation and Ventilation/Perfusion"). However, with generalized hypoxia, pulmonary pressure rises because all of the prealveolar vessels constrict. In addition, prolonged hypoxia causes vascular remodeling in which pulmonary arterial smooth muscle cells undergo hypertrophy and hyperplasia. The vascular remodeling results in narrowing of the small pulmonary arteries and increases pulmonary vascular resistance, leading to a further significant increase in pulmonary vascular hypertension. With severe hypoxia, the pulmonary veins are also constricted. The increase in venous pressure elevates the filtration pressure in the alveolar capillary beds and under severe conditions causes pulmonary edema. Pulmonary hypertension also increases the workload of the right heart, causing right heart hypertrophy, and under severe conditions often leads to death.

Diving reflex involves a number of cardiopulmonary responses.

Another respiratory movement is breath holding, in which normal breathing patterns are voluntarily suspended. Breath holding can occur until the $Paco_2$ rises and overrides the conscious, voluntary effort. Breath holding is a frequent maneuver with diving and is usually harmless. However, breath holding becomes dangerous if hyperventilation precedes the event. For example, if a diver hyperventilates before going under water, the diver can hold his or her breath longer, because of the strong inhibitory effect of blowing off excess CO_2. The problem with hyperventilation is that $Paco_2$ is drastically decreased, with no appreciable change in oxygen content. As a result of the low arterial CO_2 tension, the respiratory centers become depressed. At the same time, the exercising muscles rapidly take up oxygen, which lowers both the arterial oxygen content and Po_2. The low Pao_2 stimulates the carotid chemoreceptors. However, the hypoxia-induced stimulation is overridden by the strong inhibitory effect of the low Pco_2. As a result, the brain becomes hypoxic, causing the swimmer to black out under water and drown, a condition commonly referred to as **shallow-water blackout**, which is illustrated in Figure 21.14.

Another breathing pattern associated with underwater breathing is the **diving reflex**. The diving reflex is most pronounced in newborns and young children and can be elicited when the face is submerged in water. The diving reflex initiates a number of cardiopulmonary responses. The reflex causes the person to gasp; breathing stops (apnea), heart rate decreases (bradycardia), and blood flow to the peripheral tissue also decreases (peripheral vasoconstriction). Diverting blood away from the peripheral tissues allows for a heart–brain perfusion. The diving reflex that causes these cardiopulmonary changes is a protective reflex. The gasp component of the diving reflex stops breathing and prevents water from entering the lung. The reflex component that induces bradycardia reduces energy requirements for cardiac tissue, and the peripheral vasoconstriction conserves oxygen to the brain. A number of cases have been reported in which young children have broken through ice on a frozen pond or have fallen into a lake and survived under water for 20 to 30 minutes. Often, the diving reflex by itself is not enough to save victims who remain under water for longer periods of time than 10 to 15 minutes. Lowered body temperature caused by the cold water is equally important. A lower body temperature means a lower metabolic rate, which means less oxygen is required. The reason children survive better than adults is twofold. First, children have a more pronounced diving reflex. Second, their body temperature drops at a faster rate because of their high surface area to body weight ratio. For further details about the relationship between body temperature and surface area, see Chapter 28, "Regulation of Body Temperature."

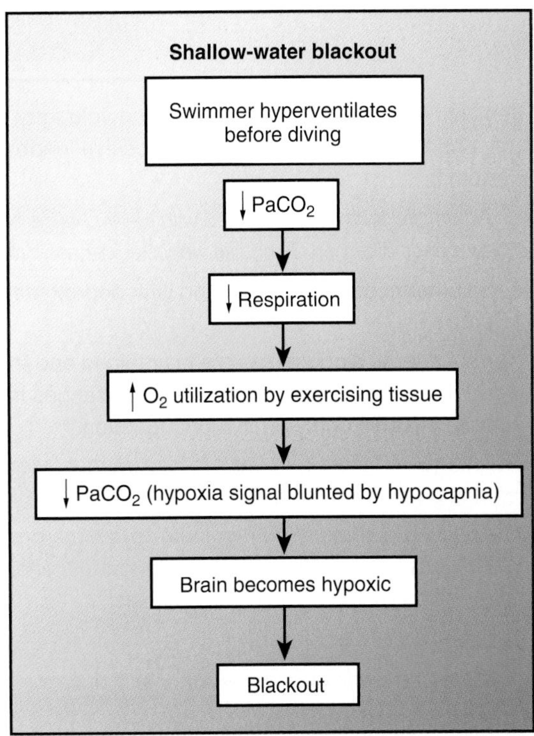

Figure 21.14 Shallow-water blackout is a result of failure of the controller to respond to low oxygen. When hyperventilation precedes underwater diving, hypocapnia occurs. The hypocapnia blunts the hypoxia-induced ventilation, causing severe hypoxia in the brain.

Chapter Summary

- Ventilation involves both automatic and voluntary control, and both require negative and positive feedback systems.
  - Automatic control is concerned with the exchange of oxygen and carbon dioxide as well as acid–base balance.
  - Voluntary control is concerned with coordinated activities.
- Normal arterial blood gases are maintained and the work of breathing is minimized despite changes in activity, the environment, and lung function.
- The neural structure responsible for automatic control resides primarily in the medulla.
- The basic breathing rhythm (minute-to-minute breathing) is generated by neurons in the brainstem and can be modified by three ventilatory reflexes:
  - center controller (driver)
  - effector (muscles of breathing)
  - sensors (feedback)
- The rate and depth of breathing are finely regulated by vagal nerve endings that are sensitive to lung stretch.
- The autonomic nerves and vagal sensory nerves maintain local control of airway function.
- Mechanical or chemical irritation of the airways and lungs induces coughing, bronchoconstriction, shallow breathing, and excess mucus production.
- Arterial P_{CO_2} is the most important factor in determining the ventilatory drive in healthy individuals at rest.
- Peripheral chemoreceptors detect changes in arterial P_{O_2}, P_{CO_2}, and pH, whereas central chemoreceptors detect changes only in arterial P_{CO_2}.
- Control of breathing during exercise involves three phases (neurogenic, metabolic, and compensatory).
- Sleep is induced by the withdrawal of a wakefulness stimulus arising from the brainstem reticular formation and results in a general depression of breathing.
- The hypoxia-induced stimulation of ventilation is not significantly increased until the arterial P_{O_2} drops to <60 mm Hg.
- Chronic hypoxemia causes ventilatory acclimatization that results in increased sustained breathing.
- The diving reflex invokes several cardiopulmonary responses.
- Shallow-water blackout occurs when the diving reflex is overridden by hypoxia.

thePoint *Visit http://thePoint.lww.com for chapter review Q&A, case studies, animations, and more!*

Part VI • Renal Physiology and Body Fluids

22 Kidney Function

ACTIVE LEARNING OBJECTIVES

Upon mastering the material in this chapter, you should be able to:

- Define the renal clearance of a substance.
- Explain how glomerular filtration rate is measured, and the factors that affect filtration rate.
- Describe how renal blood flow can be determined from the clearance of *p*-aminohippurate and the hematocrit and discuss the factors that influence renal blood flow.
- Calculate rates of net tubular reabsorption or secretion of a substance, given the filtered and excreted amounts of the substance.
- Explain what is meant by tubular transport maximum, threshold, and splay for glucose.
- Discuss the magnitude and mechanisms of solute and water reabsorption in the proximal convoluted tubule, loop of Henle, and distal nephron. Explain why sodium reabsorption is a key operation in the kidneys.
- Describe the active tubular secretion of organic anions and organic cations in the proximal tubule and passive transport via nonionic diffusion.
- Explain how arginine vasopressin increases collecting duct water permeability.
- Discuss the countercurrent mechanisms responsible for production of osmotically concentrated urine. Explain how osmotically dilute urine is formed.

The kidneys are paired organs that perform several functions. They are an essential part of the urinary system and participate in many homeostatic functions. The kidneys serve as the body's main filtering system for the blood and extracellular fluid (ECF) by removing wastes, such as urea and uric acid. They are also responsible for the reabsorption of water, glucose, and amino acids. Finally, the kidneys produce hormones, including calcitriol (the activated form of vitamin D), renin, and erythropoietin. This chapter considers the basic renal processes that determine the excretion of various substances.

OVERVIEW OF STRUCTURE AND FUNCTION

Each kidney in an adult weighs about 150 g and is roughly the size of one's fist. Despite their relatively small size, they receive approximately 20% of the cardiac output. If the kidney is sectioned (Fig. 22.1), two regions are seen: an outer part, called the **cortex**, and an inner part, called the **medulla**. The cortex is typically reddish brown and has a granulated appearance. All of the glomeruli, convoluted tubules, and cortical collecting ducts are located in the cortex. The medulla is lighter in color and has a striated appearance that results from the parallel arrangement of loops of Henle, medullary collecting ducts, and blood vessels of the medulla. The medulla can be further subdivided into an *outer medulla*, which is closer to the cortex, and an *inner medulla*, which is farther from the cortex.

The human kidney is organized into a series of lobes, usually 8 to 10 in number. Each lobe consists of a pyramid of medullary tissue, plus the cortical tissue overlying its base and covering its sides. The tip of the medullary pyramid forms a renal papilla. Each renal papilla drains its urine into a *minor calyx*. The minor calices unite to form a *major calyx*, and the urine then flows into the renal pelvis. Peristaltic movements propel the urine down the **ureters** to the urinary bladder, which stores the urine until the bladder is emptied. The medial aspect of each kidney is indented in a region called the **hilum**, where the ureter, blood vessels, nerves, and lymphatic vessels enter or leave the kidney.

Kidneys are highly innervated and have a rich blood supply.

Each kidney is typically supplied by a single renal artery, which branches into anterior and posterior divisions, which, in turn, give rise to a total of five segmental arteries. The segmental arteries branch into interlobar arteries, which pass toward the cortex between the kidney lobes (see Fig. 22.1). At the junction of the cortex and medulla, the interlobar arteries branch to form arcuate arteries. These, in turn, give rise to smaller cortical radial arteries, which pass through the cortex toward the surface of the kidney. Several short, wide, muscular *afferent arterioles* arise from the cortical radial arteries. Each afferent arteriole gives rise to a **glomerulus**. The glomerular capillaries are followed by an *efferent arteriole*. The efferent arteriole then divides into a second capillary network, the peritubular capillaries, which surround the kidney tubules. Venous vessels, in general, lie parallel to the arterial vessels and have similar names.

The blood supply to the medulla is derived from the efferent arterioles of juxtamedullary glomeruli. These vessels give rise to two patterns of capillaries: peritubular capillaries, which are similar to those in the cortex, and vasa recta, which are long, straight capillaries (Fig. 22.2). Some vasa recta reach deep into the inner medulla. In the outer medulla, many of the descending and ascending vasa recta are grouped in vascular bundles and are in close contact with each other. This arrangement greatly facilitates the exchange of substances between blood flowing into and out of the medulla.

The kidneys are richly innervated by *sympathetic nerve fibers*, which travel to the kidneys mainly in thoracic spinal nerves X, XI, and XII and lumbar spinal nerve I. Stimulation of sympathetic fibers causes constriction of renal blood vessels and a decrease in renal blood flow (RBF). Sympathetic nerve fibers also innervate tubular cells and may cause an increase in Na^+ reabsorption by a direct action on these cells. In addition, stimulation of sympathetic nerves increases the release of renin by the kidneys. Afferent (sensory) renal nerves are stimulated by mechanical stretch or by various chemicals in the renal parenchyma.

Renal lymphatic vessels drain the kidneys, but little is known about their functions.

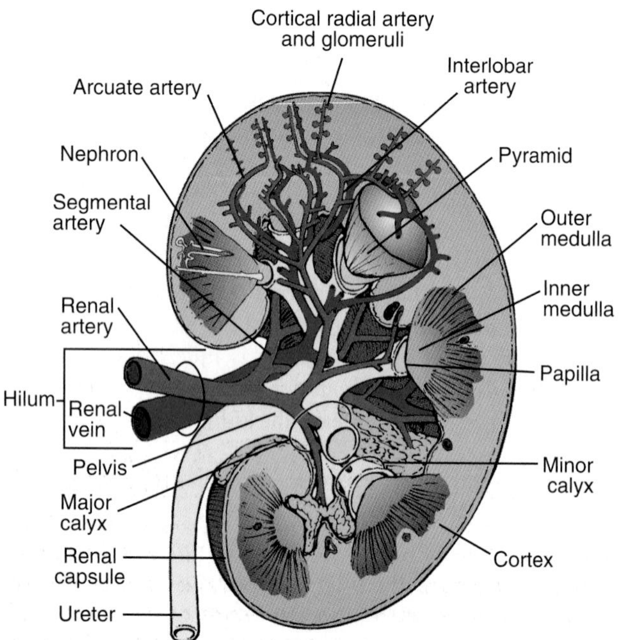

Figure 22.1 The human kidney, sectioned vertically.

Kidney is designed to maintain several homeostatic functions.

The kidneys are the organs responsible for many of the homeostatic functions of the body. These important functions include the following:

- They regulate the osmotic pressure (osmolality) of the body fluids by excreting osmotically dilute or concentrated urine.

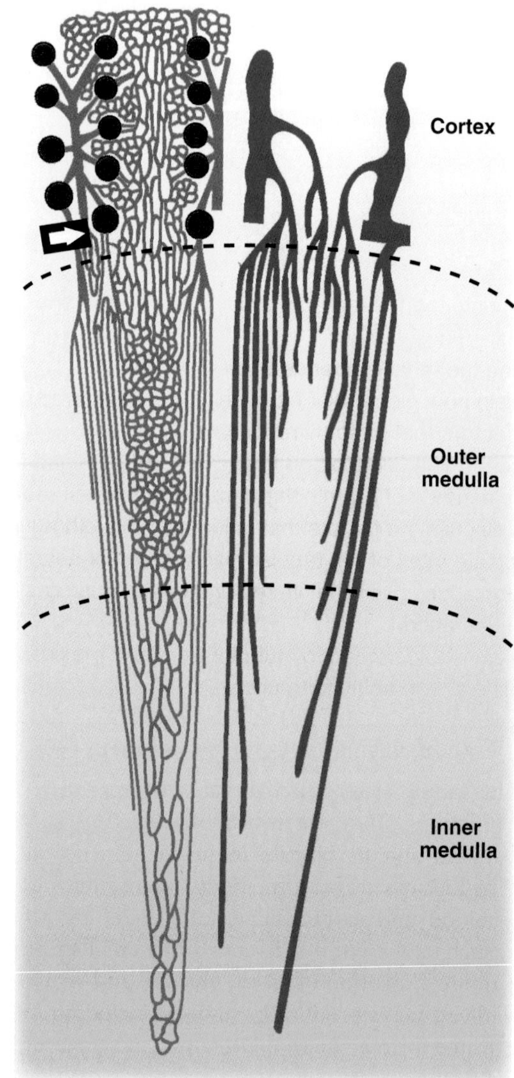

Figure 22.2 The kidney microvasculature. The left side (*red*) shows the arterial vessels, glomeruli, and capillaries. A cortical radial artery arises from an arcuate artery (*white arrow*) and gives rise to afferent arterioles, which supply the glomeruli (*red balls*). Efferent arterioles leave the glomeruli and give rise to the extensive peritubular capillary network that surrounds tubules in the cortex. The efferent arterioles of juxtamedullary glomeruli give rise to peritubular capillaries directly and indirectly via the vasa recta that descend into the medulla. The right side of the figure (*blue*), which may be superimposed on the left side, depicts the venous vessels. Ascending vasa recta drain into interlobular veins or arcuate veins. In the outer medulla, many of the ascending and descending vasa recta are grouped together in vascular bundles; this facilitates the exchange of substances between blood flowing into and out of the medulla.

Clinical Focus / 22.1 

Dialysis and Transplantation

Chronic kidney disease is usually progressive and may lead to renal failure. Common causes include diabetes mellitus, hypertension, inflammation of the glomeruli (glomerulonephritis), urinary reflux and infections (pyelonephritis), and polycystic kidney disease. Renal damage may occur over many years and may be undetected until a considerable loss of functioning nephrons has occurred. When glomerular filtration rate (GFR) has declined to 5% of normal or less, the internal environment becomes so disturbed that patients usually die within weeks or months if they are not dialyzed or provided with a functioning kidney transplant.

Most of the signs and symptoms of renal failure can be relieved by **dialysis**, the separation of smaller molecules from larger molecules in solution by diffusion of the small molecules through a selectively permeable membrane. Two methods of dialysis are commonly used to treat patients with severe, irreversible ("end-stage") renal failure.

In **continuous ambulatory peritoneal dialysis**, the peritoneal membrane, which lines the abdominal cavity, acts as a dialyzing membrane. About 1 to 2 L of a sterile glucose/salt solution are introduced into the abdominal cavity, and small molecules (e.g., K^+ and urea) diffuse into the introduced solution, which is then drained and discarded. The procedure is usually done several times every day.

Hemodialysis is more efficient in terms of rapidly removing wastes. The patient's blood is pumped through an artificial kidney machine. The blood is separated from a balanced salt solution by a cellophanelike membrane, and small molecules can diffuse across this membrane. Excess fluid can be removed by applying pressure to the blood and filtering it. Hemodialysis is usually done three times a week (4–6 hours per session) in a medical facility or at home.

Dialysis can enable patients with otherwise fatal renal disease to live useful and productive lives. However, many physiologic and psychologic problems persist, including bone disease, disorders of nerve function, hypertension, atherosclerotic vascular disease, and disturbances of sexual function. There is a constant risk of infection and, with hemodialysis, clotting and hemorrhage. Dialysis does not maintain normal growth and development in children. Anemia (primarily resulting from deficient erythropoietin production by damaged kidneys) was once a problem but can now be treated with recombinant human erythropoietin.

Renal transplantation is the only real cure for patients with end-stage renal failure. Patients who have had a kidney transplant have more energy, can enjoy a less restricted diet, have fewer complications, and live much longer than if they had stayed on dialysis. In 2009, about 17,000 kidney transplantation operations were performed in the United States. At present, about 95% of kidneys grafted from a living donor related to the recipient function for 1 year; about 90% of kidneys from cadaver donors function for 1 year.

Several problems complicate kidney transplantation. The immunologic rejection of the kidney graft is a major challenge. The powerful drugs used to inhibit graft rejection compromise immune defensive mechanisms so that unusual and difficult-to-treat infections often develop. The limited supply of donor organs is also a major, unsolved problem; there are many more patients who would benefit from kidney transplantation than there are donors. The median waiting time for a kidney transplant is currently about 1,100 to 1,200 days. Finally, the cost of transplantation (or dialysis) is high. Fortunately for people in the United States, Medicare covers the cost of dialysis and transplantation, but these life-saving therapies are beyond the reach of most people in developing countries.

- They regulate the concentrations of numerous ions in blood plasma, including Na^+, K^+, Ca^{2+}, Mg^{2+}, Cl^-, bicarbonate (HCO_3^-), phosphate, and sulfate.
- They play an essential role in acid–base balance by excreting H^+ when there is excess acid or HCO_3^- when there is excess base.
- They regulate the volume of the extracellular fluid by controlling Na^+ and water excretion.
- They help regulate arterial blood pressure by adjusting Na^+ excretion and producing various substances (e.g., renin) that can affect blood pressure.
- They eliminate the waste products of metabolism, including urea (the main nitrogen-containing end-product of protein metabolism in humans), uric acid (an end-product of purine metabolism), and creatinine (an end-product of muscle metabolism).
- They remove many drugs (e.g., penicillin) and foreign or toxic compounds.
- They are the major sites of production of certain hormones, including erythropoietin (see Chapter 9, "Blood Components") and 1,2,5-dihydroxy vitamin D3 (see Chapter 35, "Endocrine Regulation of Calcium, Phosphate, and Bone Homeostasis").
- They degrade several polypeptide hormones, including insulin, glucagon, and parathyroid hormone.
- They synthesize ammonia, which plays a role in acid–base homeostasis (see Chapter 24, "Acid–Base Homeostasis").
- They synthesize substances that affect RBF and Na^+ excretion, including arachidonic acid derivatives (prostaglandins and thromboxane A_2) and kallikrein (a proteolytic enzyme that results in the production of kinins).

When the kidneys fail, a host of problems ensue. Dialysis and kidney transplantation are commonly used treatments for advanced (end-stage) renal failure.

Nephron is the structural and functional unit of the kidney.

Each human kidney contains about 1 million **nephrons** (Fig. 22.3), each of which consists of a *renal corpuscle* and a *renal tubule.* The renal corpuscle consists of a tuft of capillaries, the **glomerulus**, surrounded by the Bowman capsule. The renal tubule is divided into several segments. The part of the tubule nearest the glomerulus is the proximal tubule. This is subdivided into a *proximal convoluted tubule* and *proximal straight tubule.* The straight portion heads toward the medulla, away from the surface of the kidney. The **loop of Henle** includes the proximal straight tubule, *thin limb*, and *thick ascending limb*. Connecting tubules connect the next segment, the short distal convoluted tubule, to the collecting duct system. Several nephrons drain into a cortical collecting duct, which passes into an *outer medullary collecting duct*. In the inner medulla, *inner medullary collecting ducts* unite to form large papillary ducts.

The collecting ducts perform the same types of functions as the renal tubules, so they are often considered to be part of the nephron. The collecting ducts and nephrons differ in embryologic origin, and because the collecting ducts form a branching system, there are many more nephrons than collecting ducts. The entire renal tubule and collecting duct system consists of a single layer of epithelial cells surrounding fluid (urine) in the tubule or duct lumen. Cells in each segment have a characteristic histologic appearance. Each segment has unique transport properties (discussed later).

Three types of nephrons are classified by location and architectural design.

Three groups of nephrons are distinguished, based on the location of their glomeruli in the cortex: *superficial*, *midcortical*, and *juxtamedullary nephrons*. The juxtamedullary nephrons, whose glomeruli lie in the cortex next to the medulla, comprise about one eighth of the total nephron population. They differ in several ways from the other nephron types: they have a longer loop of Henle, longer thin limb (both descending and ascending portions), lower renin content, different tubular permeability and transport properties, and different type of postglomerular blood supply. Figure 22.3 shows superficial and juxtamedullary nephrons; note the long loop of the juxtamedullary nephron.

Juxtaglomerular apparatus is the site of renin production.

Each nephron forms a loop, and the thick ascending limb touches the vascular pole of the glomerulus (see Fig. 22.3). At this site is the **juxtaglomerular apparatus**, a region comprising the macula densa, extraglomerular mesangial cells, and granular cells (Fig. 22.4). The **macula densa** (dense spot) consists of densely crowded tubular epithelial cells on the side of the thick ascending limb that faces the glomerular tuft; these cells monitor the composition of the fluid in the tubule lumen at this point. The *extraglomerular mesangial cells* are continuous with mesangial cells of the glomerulus; they may transmit information from macula densa cells to the granular cells. The **granular cells** (also known as *juxtaglomerular cells*) are modified vascular smooth muscle cells with an epithelioid appearance, located mainly in the afferent arterioles close to the glomerulus. These cells synthesize and release **renin**, a proteolytic enzyme that results in angiotensin formation (see Chapter 23, "Regulation of Fluid and Electrolyte Balance").

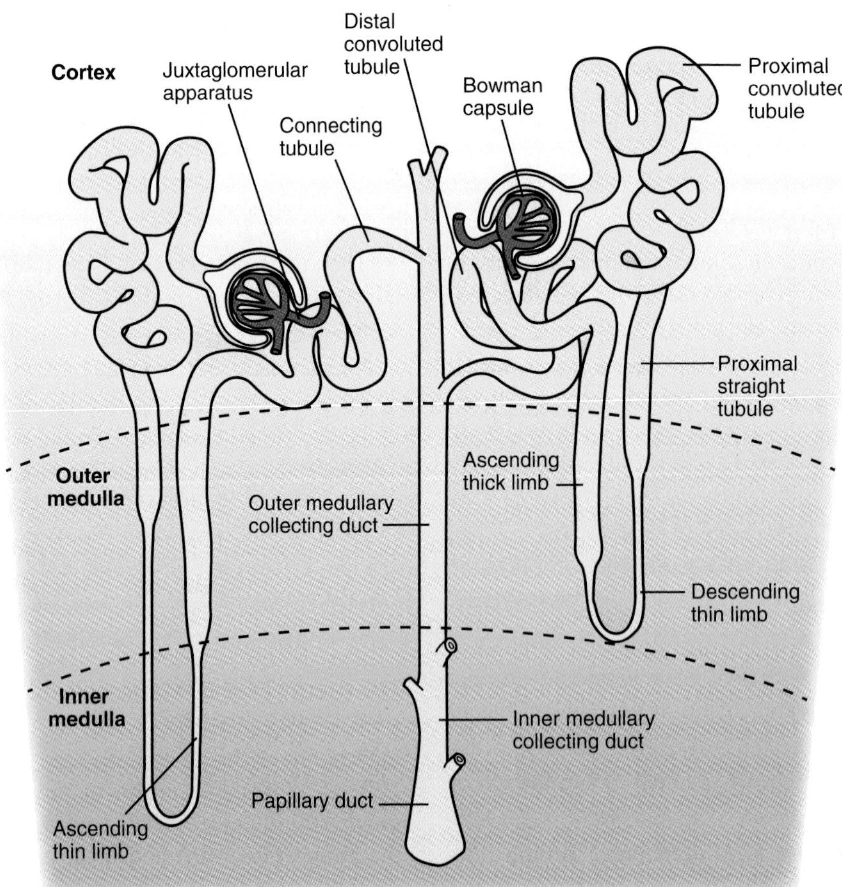

Figure 22.3 Components of the nephron and the collecting duct system. On the left is a long-looped juxtamedullary nephron; on the right is a superficial cortical nephron.

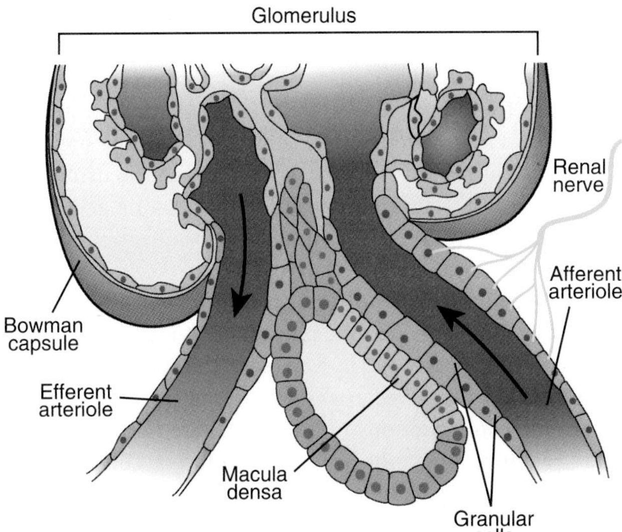

Figure 22.4 Histologic appearance of the juxtaglomerular apparatus. A cross section through a thick ascending limb is below, and part of a glomerulus is on top.

URINE FORMATION

Three processes are involved in forming urine: glomerular filtration, tubular reabsorption, and tubular secretion (Fig. 22.5). **Glomerular filtration** involves the ultrafiltration of plasma in the glomerulus. The filtrate enters the urinary space of the Bowman capsule and then flows downstream through the tubule lumen. The tubule epithelial cells decrease the volume and change the composition of the fluid in the tubule lumen. **Tubular reabsorption** involves the transport of substances out of tubular urine. These substances are then returned to the capillary blood, which surrounds the kidney tubules. Reabsorbed substances include many important ions (e.g., Na^+, K^+, Ca^{2+}, Mg^{2+}, Cl^-, HCO_3^-, and phosphate), water, important metabolites (e.g., glucose and amino acids), and even some waste products (e.g., urea and uric acid). **Tubular secretion** involves the transport of substances into the tubular urine. For example, many organic anions and cations are taken up by the tubular epithelium from the blood surrounding the tubules and added to the tubular urine. Some substances (e.g., H^+ and ammonia) are produced in the tubular cells and secreted into the tubular urine. The terms *reabsorption* and *secretion* indicate movement out of and into tubular urine, respectively. Tubular transport (reabsorption and secretion) may be either active or passive, depending on the particular substance and other conditions.

Excretion refers to elimination via the urine. In general, the amount excreted is expressed by the following equation:

$$\text{Excreted} = \text{Filtered} - \text{Reabsorbed} + \text{Secreted} \qquad (1)$$

The functional state of the kidneys can be evaluated using several tests based on the renal clearance concept (see below). These tests measure the rates of glomerular filtration, RBF, and tubular reabsorption or secretion of various substances. Some of these tests, such as the measurement of glomerular filtration rate (GFR), are routinely used to evaluate kidney function.

Clearance is the measurement of the kidney's excretion ability.

A useful way of looking at kidney function is to think of the kidneys as clearing substances from the blood plasma. When a substance is excreted in the urine, a certain volume of plasma is, in effect, freed (or cleared) of that substance. The renal **clearance** of a substance can be defined as the volume of plasma from which that substance is completely removed (cleared) per unit time. The clearance formula is

$$C_X = \frac{U_X \times \dot{V}}{P_X} \qquad (2)$$

where X is the substance of interest, C_X is the clearance of substance X, U_X is the urine concentration of substance X, P_X is the plasma concentration of substance X, and $\dot{V}$ is the urine flow rate. The product of U_X times $\dot{V}$ equals the excretion rate and has dimensions of amount per unit time (e.g., mg/min or mEq/d). Thus, clearance equals the urinary excretion rate divided by plasma concentration. The clearance of a substance can easily be determined by measuring the concentrations of a substance in urine and plasma and the urine flow rate (urine volume/time of collection) and substituting these values into the clearance formula.

Inulin clearance equals the glomerular filtration rate.

An important measurement in the evaluation of kidney function is the **glomerular filtration rate (GFR)**, the rate at which plasma is filtered by the kidney glomeruli. GFR is equal to the sum of the filtration rates of all functional nephrons and is used as index of renal function. A decrease in GFR generally indicates that kidney function is impaired. Thus, GFR is an important diagnostic tool to evaluate kidney disease.

If we had a substance that was cleared from the plasma only by glomerular filtration, it could be used to measure GFR. The ideal substance to measure GFR is **inulin**, a fructose polymer with a molecular weight of about 5,000.

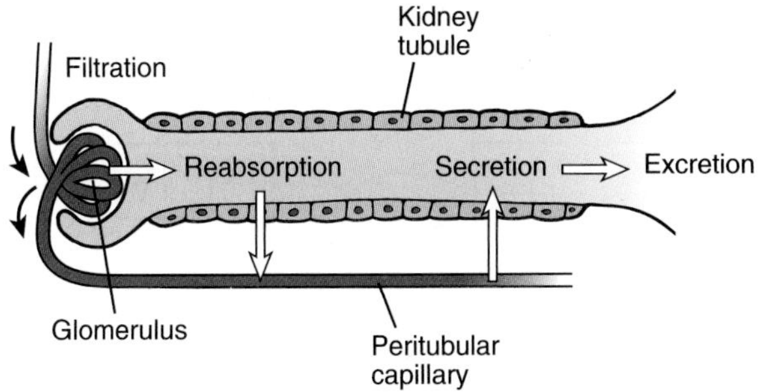

Figure 22.5 Processes involved in urine formation. This highly simplified drawing shows a nephron and its associated blood vessels.

Inulin (IN) is suitable for measuring GFR for the following reasons: (1) IN is freely filtered by the glomeruli; (2) IN is not reabsorbed or secreted by the kidney tubules; (3) IN is not synthesized, destroyed, or stored in the kidneys; (4) IN is not toxic; and (5) IN's concentration in plasma and urine can be determined by simple analysis.

The principle behind the use of IN is illustrated in Figure 22.6. The amount of IN filtered per unit time, the **filtered load**, is equal to the product of the plasma concentration of IN (P_{IN}) times GFR. The rate of IN excretion is equal to the urine IN concentration (U_{IN}), times $\dot{V}$, the urine flow rate. Because IN is not reabsorbed, secreted, synthesized, destroyed, or stored by the kidney tubules, the filtered IN load equals the rate of IN excretion. The equation can be rearranged by dividing by the plasma IN concentration.

The expression $U_{IN}\dot{V}/P_{IN}$ is defined as the **inulin clearance**. Therefore, IN clearance equals GFR.

Normal values for IN clearance or GFR (corrected to a body surface area of 1.73 m^2) are 110 ± 15 (SD) mL/min for young adult women and 125 ± 15 mL/min for young adult men. In newborns, even when corrected for body surface area, GFR is low, about 20 mL/min per 1.73 m^2 body surface area. Adult values (when corrected for body surface area) are attained by the end of the first year of life. After the age of 45 to 50, GFR declines, and it is typically reduced by 30% to 40% by age 80.

If GFR is 125 mL plasma/min, then the volume of plasma filtered in a day is 180 L (125 mL/min × 1,440 min/d). Plasma volume in a 70-kg young adult man is only about 3 L, and so the kidneys filter the plasma some 60 times in a day. The glomerular filtrate contains essential constituents (salts, water, and metabolites), most of which are reabsorbed by the kidney tubules.

Plasma creatinine clearance is used clinically to estimate GFR.

IN clearance is the gold standard for measuring GFR and is used whenever highly accurate measurements of GFR are desired. The clearance of iothalamate, an iodinated organic compound, also provides a reliable measure of GFR. It is not common, however, to use these substances in the clinic. They must be infused intravenously and the bladder is usually catheterized, because short urine collection periods are used; these procedures are inconvenient. It would be simpler to use an endogenous substance (i.e., one native to the body) that is only filtered, is excreted in the urine, and normally has a stable plasma value that can be accurately measured. There is no such known substance, but creatinine comes close.

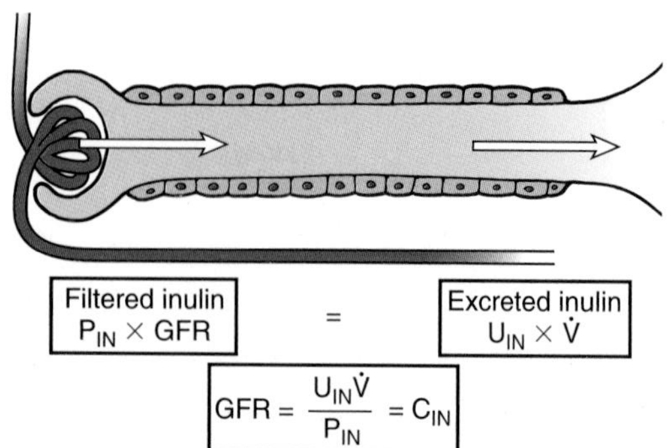

Figure 22.6 The principle behind the measurement of glomerular filtration rate (GFR). P_{IN}, plasma inulin concentration; U_{IN}, urine inulin concentration; $\dot{V}$, urine flow rate; C_{IN}, inulin clearance.

Creatinine is an end-product of muscle metabolism, a derivative of muscle creatine phosphate. It is produced continuously in the body and is excreted in the urine. Long urine collection periods (e.g., a few hours) can be used, because creatinine concentrations in the plasma are normally stable and creatinine does not have to be infused; consequently, there is no need to catheterize the bladder. Plasma and urine concentrations can be measured using a simple colorimetric method. The **endogenous creatinine clearance** is calculated from the formula

$$C_{CREATININE} = \frac{U_{CREATININE} \times \dot{V}}{P_{CREATININE}} \quad (3)$$

or, creatinine clearance equals the ratio of urine to plasma creatinine concentration times the urine flow rate. There are two potential drawbacks of using creatinine to measure GFR. First, creatinine is not only filtered but also secreted by the human kidney. This elevates urinary excretion of creatinine, normally causing a 20% increase in the numerator of the clearance formula. The second drawback is related to errors of measuring creatinine concentration in the plasma. The colorimetric method usually used also measures other plasma substances, such as glucose, leading to a 20% increase in the denominator of the clearance formula. Because both numerator and denominator are 20% too high, the two errors cancel, and so the endogenous creatinine clearance fortuitously affords a good approximation of GFR when it is about normal. However, when GFR in an adult has been reduced to about 20 mL/min because of renal disease, the endogenous creatinine clearance may overestimate the GFR by as much as 50%. This results from higher plasma creatinine levels and increased tubular secretion of creatinine. Drugs that inhibit tubular secretion of creatinine or elevated plasma concentrations of chromogenic (color-producing) substances other than creatinine may cause the endogenous creatinine clearance to underestimate GFR.

Plasma creatinine concentration can be used to estimate GFR.

Because the kidneys continuously clear creatinine from the plasma by excreting it in the urine, the GFR and plasma creatinine concentration are inversely related. Figure 22.7 shows the steady-state relationship between these variables—that is, when creatinine production and excretion are equal. Halving the GFR from a normal value of 180 to 90 L/d results in a doubling of plasma creatinine concentration from a normal

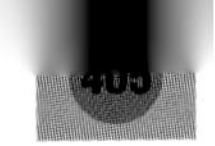

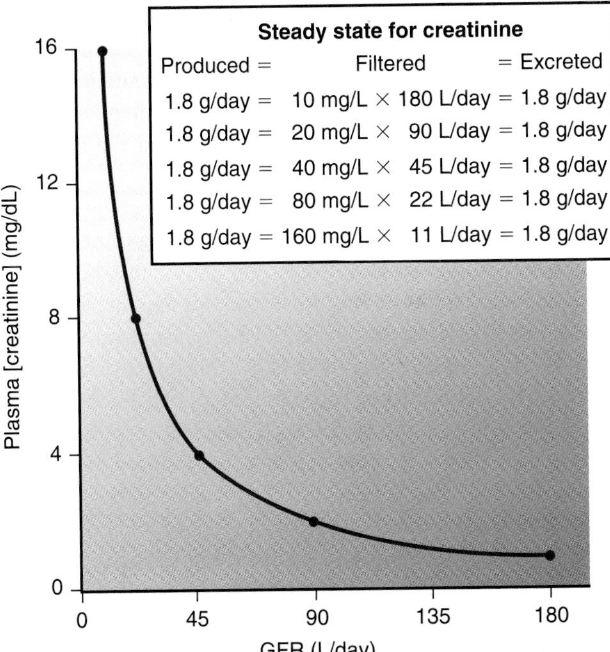

Figure 22.7 The inverse relationship between plasma creatinine concentration and glomerular filtration rate (GFR). If GFR is decreased by half, plasma creatinine concentration is doubled when the production and excretion of creatinine are in balance in a new steady state.

value of 1 to 2 mg/dL after a few days. A reduction in GFR from 90 to 45 L/d results in a greater increase in plasma creatinine, from 2 to 4 mg/dL. Figure 22.7 shows that with low GFR values, small absolute changes in GFR lead to much greater changes in plasma creatinine concentration than at high GFR values.

The inverse relationship between GFR and plasma creatinine concentration allows the use of plasma or serum creatinine to estimate GFR, provided that these three cautions are kept in mind:

1. It takes a certain amount of time for changes in GFR to produce detectable changes in plasma creatinine concentration.
2. Plasma creatinine concentration is also influenced by muscle mass. A young, muscular man will have a higher plasma creatinine concentration than an older woman with reduced muscle mass.
3. Some drugs inhibit tubular secretion of creatinine, leading to a raised plasma creatinine, even though GFR may be unchanged.

The relationship between plasma creatinine and GFR is one example of how a substance's plasma concentration can depend on GFR. The same relationship is observed for several other substances whose excretion depends on GFR. For example, the plasma urea concentration or blood urea nitrogen rises when GFR falls. The plasma level of a 13-kDa protein molecule called *cystatin C* also rises when GFR falls, and it has been suggested that serum cystatin C levels can be used to estimate GFR.

Several empirical equations have been developed that allow physicians to estimate GFR from serum creatinine concentration. These equations often take into consideration such factors as age, gender, race, and body size. The equation for adults, GFR (in mL/min per 1.73 m^2) = 186 × (serum creatinine in mg/dL)$^{-1.154}$ × (age in years)$^{-0.203}$ × 0.742 (if the subject is female) or × 1.212 (if the subject is black), is recommended by the National Kidney Disease Education Program, which provides GFR calculators on its Web site: www.nih.nkdep.gov. Staging of chronic kidney disease is usually based on estimated GFR measurements; a value <60 mL/min may indicate renal disease.

Para-aminohippurate clearance nearly equals renal plasma flow.

Renal blood flow (RBF) can be determined from measurements of **renal plasma flow (RPF)** and blood hematocrit, using the following equation:

$$RBF = RPF/(1 - \text{Hematocrit}) \tag{4}$$

The hematocrit is easily determined by centrifuging a blood sample. RPF is estimated by measuring the clearance of the organic anion *p*-aminohippurate (PAH), infused intravenously. PAH is filtered and so vigorously secreted that it is nearly completely cleared from all of the plasma flowing through the kidneys. The renal clearance of PAH, at low plasma PAH levels, approximates the RPF.

The equation for calculating the true value of the RPF is

$$RPF = C_{PAH}/E_{PAH} \tag{5}$$

where C_{PAH} is the PAH clearance and E_{PAH} is the extraction ratio (see Chapter 15, "Microcirculation and Lymphatic System") for PAH—the difference between the arterial and renal venous plasma PAH concentrations ($P^a{}_{PAH} - P^{rv}{}_{PAH}$) divided by the arterial plasma PAH concentration ($P^a{}_{PAH}$). The equation is derived as follows. In the steady state, the amounts of PAH per unit time entering and leaving the kidneys are equal. The PAH is supplied to the kidneys in the arterial plasma and leaves the kidneys in urine and renal venous plasma, or:

$$\text{PAH entering kidneys} = \text{PAH leaving kidneys}$$

$$RPF \times P^a{}_{PAH} = U_{PAH} \times \dot{V} + RPF \times P^{rv}{}_{PAH} \tag{6}$$

Rearranging, we get:

$$RPF = U_{PAH} \times \dot{V}/(P^a{}_{PAH} - P^{rv}{}_{PAH}) \tag{7}$$

If we divide the numerator and denominator of the right side of the equation by $P^a{}_{PAH}$, the numerator becomes C_{PAH} and the denominator becomes E_{PAH}.

If we assume extraction of PAH is 100% (E_{PAH} = 1.00), then the RPF equals the PAH clearance. When this assumption is made, the RPF is usually called the *effective RPF*, and the blood flow calculated is called the *effective RBF*. The extraction of PAH by healthy kidneys at low plasma PAH concentrations, however, is not 100% but averages about 91%, so the assumption of 100% extraction results in about a 10% underestimation of the true RPF. To calculate the true RPF or blood flow, it is necessary to sample renal venous blood to measure its plasma PAH concentration, a procedure not often done.

Net tubular reabsorption or secretion of a substance can be calculated from filtered and excreted amounts.

The rate at which the kidney tubules reabsorb a substance can be calculated if we know how much is filtered and how much is excreted per unit time. If the filtered load of a substance exceeds the rate of excretion, the kidney tubules must have reabsorbed the substance. The equation is

$$T_{reabsorbed} = P_X \times GFR - U_X \times \dot{V} \quad (8)$$

where T is the tubular transport rate.

The rate at which the kidney tubules secrete a substance is calculated from this equation:

$$T_{secreted} = U_X \times \dot{V} - P_X \times GFR \quad (9)$$

Note that the quantity excreted exceeds the filtered load, because the tubules secrete X.

In equations 8 and 9, we assume that substance X is freely filterable. If, however, substance X is bound to the plasma proteins, which are not filtered, then it is necessary to correct the filtered load for this binding. For example, about 40% of plasma Ca^{2+} is bound to plasma proteins, and so 60% of plasma Ca^{2+} is freely filterable.

Equations 8 and 9, which quantify tubular transport rates, yield the *net* rate of reabsorption or secretion of a substance. It is possible for a single substance to be both reabsorbed and secreted; the equations do not give unidirectional reabsorptive and secretory movements, only the net transport.

Renal glucose reabsorption retrieves filtered glucose and prevents glucose loss in the urine.

Insights into the nature of glucose handling by the kidneys can be derived from a glucose titration study (Fig. 22.8). The plasma glucose concentration is elevated to increasingly higher levels by the infusion of glucose-containing solutions. IN is infused to permit measurement of GFR and calculation of the filtered glucose load (plasma glucose concentration × GFR). The rate of glucose reabsorption is determined from the difference between the filtered load and the rate of excretion. At normal plasma glucose levels (about 100 mg/dL), all of the filtered glucose is reabsorbed and none is excreted. When the plasma glucose concentration exceeds a certain value (about 200 mg/dL in Fig. 22.8), significant quantities of glucose appear in the urine; this plasma concentration is called the **glucose threshold**. Further elevations in plasma glucose lead to progressively more excreted glucose. Glucose appears in the urine because the filtered amount of glucose exceeds the capacity of the tubules to reabsorb it. At high filtered glucose loads, the rate of glucose reabsorption reaches a constant maximal value, called the **tubular transport maximum (Tm)** for glucose (G). At Tm_G, the tubule glucose carriers are all saturated and transport glucose at the maximal rate.

The glucose threshold is not a fixed plasma concentration but depends on three factors: GFR, Tm_G, and amount of splay. A low GFR leads to an elevated threshold, because the filtered glucose load is reduced and the kidney tubules can reabsorb all the filtered glucose despite an elevated plasma glucose concentration. A reduced Tm_G lowers the threshold, because the tubules have a diminished capacity to reabsorb glucose.

Splay is the rounding of the glucose reabsorption curve. Figure 22.8 shows that tubular glucose reabsorption does not abruptly attain Tm_G when plasma glucose is progressively elevated. One reason for splay is that not all nephrons have the same filtering and reabsorbing capacities. Thus, nephrons with relatively high filtration rates and low glucose reabsorptive rates excrete glucose at a lower plasma concentration than nephrons with relatively low filtration rates and high reabsorptive rates. A second reason for splay is the fact that the glucose carrier does not have an infinitely high affinity for glucose, so glucose escapes in the urine even before the carrier is fully saturated. An increase in splay causes a decrease in glucose threshold.

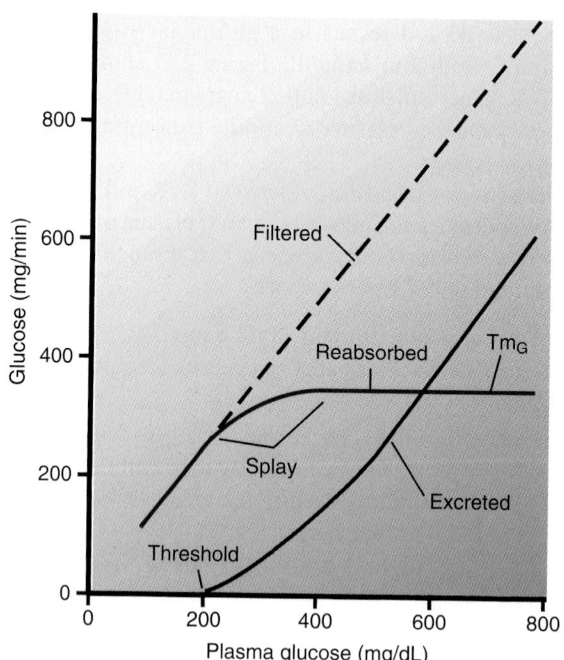

Figure 22.8 Glucose titration study in a healthy man. The plasma glucose concentration was elevated by infusing glucose-containing solutions. The amount of glucose filtered per unit time (*top dashed line*) is determined from the product of the plasma glucose concentration and glomerular filtration rate (measured with inulin). Excreted glucose (*bottom line*) is determined by measuring concentration of glucose in the urine and urine flow rate. Reabsorbed glucose is calculated from the difference between filtered and excreted glucose. Tm_G, tubular transport maximum for glucose.

In uncontrolled **diabetes mellitus**, plasma glucose levels are abnormally elevated, so more glucose is filtered than can be reabsorbed. Urinary excretion of glucose, **glucosuria**, produces an osmotic diuresis. A diuresis is an increase in urine output. In osmotic diuresis, the increased urine flow results from the excretion of osmotically active solute. *Diabetes* (from the Greek for "*syphon*") gets its name from this increased urine output and secondary increased water intake.

Tm for PAH provides a measure of functional proximal secretory tissue.

PAH is secreted only by proximal tubules in the kidneys. At low plasma PAH concentrations, the rate of secretion increases linearly with the plasma concentration of PAH. At high plasma PAH concentrations, the secretory carriers are saturated and the rate of PAH secretion stabilizes at a constant maximal value, called the *tubular transport maximum for PAH* (Tm_{PAH}). The Tm_{PAH} is directly related to the number of functioning proximal tubules and therefore provides a measure of the mass of proximal secretory tissue. Figure 22.9 illustrates the pattern of filtration, secretion, and excretion of PAH observed when the plasma PAH concentration is progressively elevated by intravenous infusion.

RENAL BLOOD FLOW

The kidneys have a high blood flow. This allows them to filter the blood plasma at a high rate. Many factors, both intrinsic (autoregulation and local hormones) and extrinsic (nerves and bloodborne hormones), affect the rate of RBF.

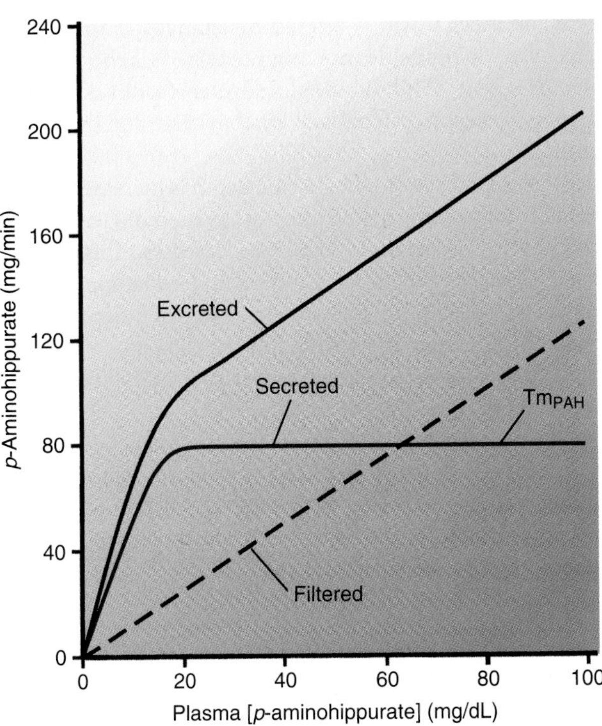

Figure 22.9 Rates of excretion, filtration, and secretion of *p*-aminohippurate (PAH) as a function of plasma PAH. More PAH is excreted than is filtered; the difference represents secreted PAH. Tm_{PAH}, tubular transport maximum for PAH.

High renal blood flow is necessary for an effective GFR.

In resting, healthy, young adult men, RBF averages about 1.2 L/min. This is about 20% of the cardiac output (5–6 L/min). Both kidneys together weigh about 300 g, so blood flow per gram of tissue averages about 4 mL/min. The high blood flow to the kidneys is necessary for a high GFR and is not a result of excessive metabolic demands.

The kidneys use about 8% of total resting oxygen consumption, but they receive much more oxygen than they need. Consequently, renal extraction of oxygen is low, and renal venous blood has a bright red color (resulting from its high oxyhemoglobin content). The anatomic arrangement of the vessels in the kidney permits a large fraction of the arterial oxygen to be shunted to the veins before the blood enters the capillaries. Therefore, the oxygen tension in the tissue is not as high as one might think, and the kidneys are sensitive to ischemic damage.

Renal blood flow is higher in the renal cortex than in the renal medulla.

Blood flow rates differ in different parts of the kidney (Fig. 22.10). Blood flow is highest in the cortex, averaging about 4 to 5 mL/min/g of tissue. The high cortical blood flow permits a high rate of filtration in the glomeruli. Blood flow (per gram of tissue) is about 0.7 to 1 mL/min in the outer medulla and 0.20 to 0.25 mL/min in the inner medulla. The relatively low blood flow in the medulla helps maintain a hyperosmolar environment in this region of the kidney.

Kidneys maintain optimal blood flow by autoregulation.

Despite changes in mean arterial blood pressure (from 80–180 mm Hg), RBF is kept at a relatively constant level, a process known as **autoregulation** (see Chapter 15, "Microcirculation

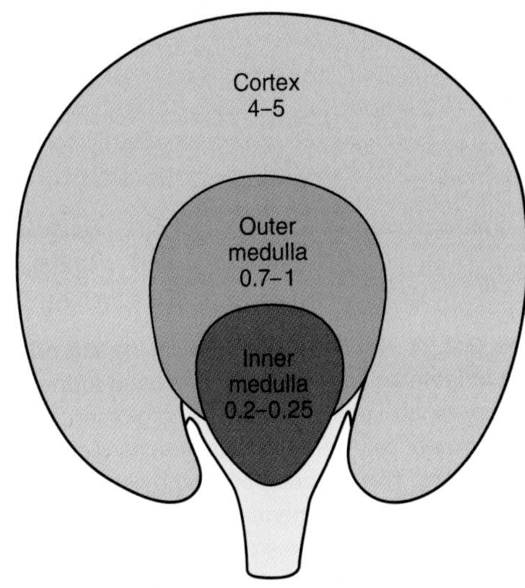

Figure 22.10 Blood flow rates (in mL/min/g tissue) in different parts of the kidney.

and Lymphatic System"). Autoregulation is an intrinsic property of the kidneys and is observed even in an isolated, denervated, perfused kidney. GFR is also autoregulated (Fig. 22.11). When the blood pressure is raised or lowered, vessels upstream of the glomerulus (cortical radial arteries and afferent arterioles) constrict or dilate, respectively, thereby maintaining relatively constant glomerular blood flow and capillary pressure. Below or above the autoregulatory range of pressures, blood flow and GFR change appreciably with arterial blood pressure.

Two mechanisms account for renal autoregulation: the myogenic mechanism and the tubuloglomerular feedback mechanism. In the **myogenic mechanism**, an increase in pressure stretches blood vessel walls and opens stretch-activated cation channels in smooth muscle cells. The ensuing membrane depolarization opens voltage-dependent Ca^{2+} channels and intracellular Ca^{2+} rises, causing smooth muscle contraction. Vessel lumen diameter decreases and vascular resistance increases. Decreased blood pressure causes the opposite changes.

In the **tubuloglomerular feedback mechanism**, the transient increase in GFR resulting from an increase in blood pressure leads to increased NaCl delivery to the macula densa (Fig. 22.12). This increases NaCl reabsorption and adenosine triphosphate (ATP) release from macula densa cells. ATP is metabolized to adenosine diphosphate (ADP), adenosine monophosphate (AMP), and adenosine in the juxtaglomerular interstitium. Adenosine combines with receptors in the afferent arteriole and causes vasoconstriction, and blood flow and GFR are lowered to a more normal value. Sensitivity of the tubuloglomerular feedback mechanism is altered by changes in local renin activity, but adenosine, not angiotensin II, is the vasoconstrictor agent. The tubuloglomerular feedback mechanism is a negative-feedback system that stabilizes RBF and GFR.

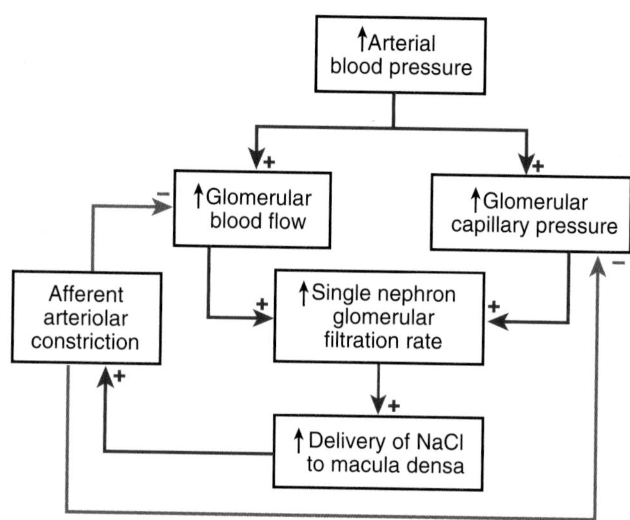

Figure 22.12 The tubuloglomerular feedback mechanism. When single-nephron glomerular filtration rate (GFR) is increased—for example, because of an increase in arterial blood pressure—more NaCl is delivered to and reabsorbed by the macula densa, leading to constriction of the nearby afferent arteriole. This negative-feedback system plays a role in autoregulation of renal blood flow and GFR.

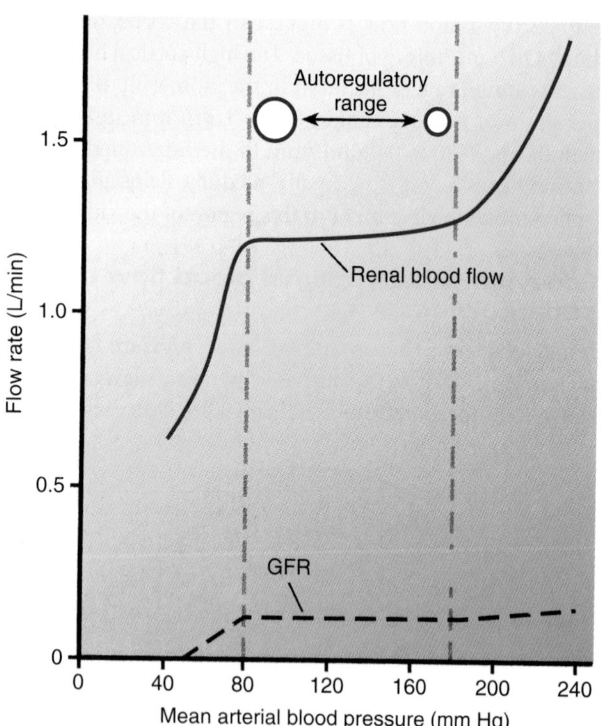

Figure 22.11 Renal autoregulation, based on measurements in isolated, denervated, perfused kidneys. In the autoregulatory range, renal blood flow and glomerular filtration rate (GFR) stay relatively constant despite changes in arterial blood pressure. This is accomplished by changes in the resistance (caliber) of preglomerular blood vessels. The circles indicate that vessel radius (r) is smaller when blood pressure is high and larger when blood pressure is low. Because resistance to blood flow varies as r^4, changes in vessel caliber are greatly exaggerated in this figure.

If NaCl delivery to the macula densa is increased experimentally by perfusing the lumen of the loop of Henle, filtration rate in the perfused nephron decreases. This suggests that the purpose of tubuloglomerular feedback may be to control the amount of Na^+ presented to distal nephron segments. These segments have a limited capacity to reabsorb Na^+; if they are overwhelmed, excessive urinary excretion of Na^+ might ensue.

Renal autoregulation minimizes the impact of changes in arterial blood pressure on Na^+ excretion. Without renal autoregulation, increases in arterial blood pressure would lead to dramatic increases in GFR and potentially serious losses of NaCl and water from the ECF.

Renal blood flow is altered by sympathetic stimulation and hormones.

The stimulation of renal sympathetic nerves or the release of various hormones may change RBF. Sympathetic nerve stimulation causes renal vasoconstriction and a consequent decrease in RBF. Renal sympathetic nerves are activated under stressful conditions, including cold temperatures, deep anesthesia, fearful situations, hemorrhage, pain, and strenuous exercise. In these conditions, renal vasoconstriction may be viewed as an emergency mechanism that increases total peripheral vascular resistance, raises arterial blood pressure, and allows more of the cardiac output to perfuse other vital organs, such as the brain and heart, which are more important for short-term survival.

Many substances cause vasoconstriction in the kidneys, including adenosine, angiotensin II, endothelin, epinephrine, norepinephrine, thromboxane A_2, and vasopressin. Other substances cause vasodilation in the kidneys, including atrial natriuretic peptide, dopamine, histamine, kinins, nitric oxide, and prostaglandins E_2 and I_2. Some of these substances (e.g., prostaglandins E_2 and I_2) are produced locally in the kidneys. An increase in sympathetic nerve activity or plasma angiotensin II concentration stimulates the production of renal vasodilator prostaglandins. These prostaglandins then oppose the pure constrictor effect of sympathetic nerve stimulation or angiotensin II, thereby reducing the fall in RBF and preventing renal damage.

GLOMERULAR FILTRATION

Glomerular filtration involves the **ultrafiltration** of plasma. The term *ultrafiltration* indicates that the glomerular filtration barrier functions as a molecular sieve that allows filtration of small molecules but restricts the passage of macromolecules (e.g., most of the plasma proteins).

The glomerular filtration barrier is made up of three layers.

An ultrafiltrate of plasma passes from glomerular capillary blood into the space of the Bowman capsule through the **glomerular filtration barrier** (Fig. 22.13). This barrier consists of three layers. The first, the capillary **endothelium**, is called the *lamina fenestra*, because it contains pores or windows (fenestrae). At about 50 to 100 nm in diameter, these pores are too large to restrict the passage of plasma proteins. The second layer, the **basement membrane**, consists of a meshwork of fine fibrils embedded in a gel-like matrix. The third layer is composed of **podocytes**, which constitute the visceral layer of the Bowman capsule. Podocytes ("foot cells") are epithelial cells with extensions that terminate in foot processes that rest on the outer layer of the basement membrane (see Fig. 22.13). The space between adjacent foot processes, called a **filtration slit**, is about 40 nm wide and is bridged by a diaphragm. A key component of the diaphragm is a molecule called **nephrin**, which forms a zipperlike structure; between the prongs of the zipper are rectangular pores. Nephrin is mutated in **congenital nephrotic syndrome**, a rare, inherited condition characterized by excessive filtration of plasma proteins. The glomerular filtrate normally takes an extracellular route, through holes in the endothelial cell layer, the basement membrane, and the pores between adjacent nephrin molecules.

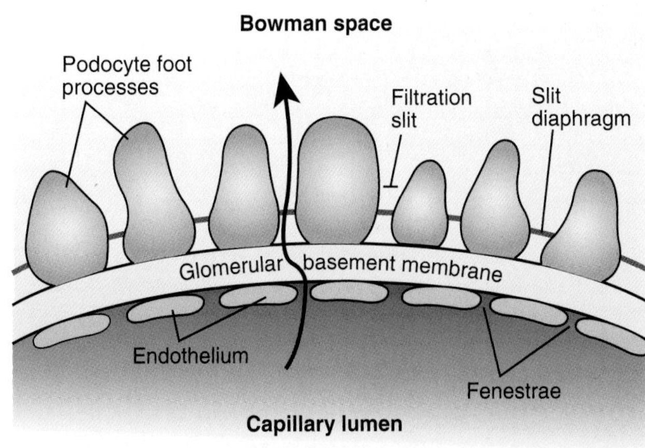

Figure 22.13 Schematic of the three layers of the glomerular filtration barrier: endothelium, basement membrane, and podocytes. The pathway for filtration is indicated by the *arrow*.

Physical properties determine the ease with which molecules are filtered.

The permeability properties of the glomerular filtration barrier have been studied by determining how well molecules of different sizes pass through it. Table 22.1 lists several molecules that have been tested. Molecular radii were calculated from diffusion coefficients. The concentration of the molecule in the glomerular filtrate (fluid collected from the Bowman capsule) is compared with its concentration in plasma water. A ratio of 1 indicates complete filterability, and a ratio of 0 indicates complete exclusion by the glomerular filtration barrier. Figure 22.14 shows an optical section through a glomerulus *in vivo* and demonstrates the extremely low filterability of serum albumin.

Molecular size is an important factor affecting filterability. All molecules with weights <10 kDa are freely filterable, provided they are not bound to plasma proteins. Molecules with weights >10 kDa experience more and

TABLE 22.1 Restrictions to the Glomerular Filtration of Molecules

Substance	Molecular Weight	Molecular Radius (nm)	Filtrate/ Plasma Water Concentration
Water	18	0.10	1.00
Glucose	180	0.36	1.00
Inulin	5,000	1.4	1.00
Myoglobin	17,000	2.0	0.75
Hemoglobin	68,000	3.3	0.01
Serum albumin	66,000	3.6	0.001

Figure 22.14 *In vivo* two-photon microscope image of a rat glomerulus. Serum albumin was tagged with a red fluorescent dye and is seen in the capillary plasma. The urinary space of the Bowman capsule, which surrounds the glomerulus, appears black, demonstrating that very little albumin is filtered by a normal glomerulus.

more restriction to passage through the glomerular filtration barrier. Large molecules (e.g., molecular weight of 100 kDa) cannot get through at all. Most plasma proteins are large molecules, so they are not appreciably filtered. From studies with molecules of different sizes, it has been calculated that the glomerular filtration barrier behaves as though cylindrical pores of about 7.5 to 10 nm in diameter penetrated it. No one, however, has ever seen such pores in electron micrographs of the glomerular filtration barrier.

Molecular shape influences the filterability of macromolecules. For a given molecular weight, a long, slender molecule will pass through the glomerular filtration barrier more easily than a spherical molecule. Also, passage of a macromolecule through the barrier is favored by greater deformability.

Electrical charge is thought, by most investigators, to influence the passage of macromolecules through the glomerular filtration barrier. The barrier bears fixed negative charges. Glomerular endothelial cells and podocytes have a negatively charged surface coat (glycocalyx), and the glomerular basement membrane contains negatively charged sialic acid, sialoproteins, and heparan sulfate. These negative charges could impede the passage of negatively charged macromolecules by electrostatic repulsion. The net negative charge on serum albumin at physiologic pH may be a factor that reduces its filterability, but the main reason for its very low filterability is its large size.

Filtered serum albumin is reabsorbed in the proximal tubule by endocytosis, but when excessive amounts are filtered, some will escape in the urine, a condition called **albuminuria**, a type of **proteinuria. Microalbuminuria**, defined as excretion of 30 to 300 mg serum albumin/d, may be an early sign of kidney damage in patients with diabetes mellitus or hypertension or an indication of cardiovascular disease. A normal albumin excretion rate is about 5 to 20 mg/d.

Proteinuria is a hallmark of glomerular disease. Proteinuria not only is a sign of kidney disease but also results in tubular and interstitial damage and contributes to the progression of chronic renal disease. When too much protein (along with bound materials such as lipids) is filtered, proximal tubule cells endocytose abnormally large amounts. This may have a toxic effect on the cells, and they then express a number of vasoactive and chemotactic factors. These factors lead to ischemia, inflammation, and interstitial fibrosis.

The layer of the glomerular filtration barrier primarily responsible for limiting the filtration of macromolecules such as serum albumin is a matter of debate. Some investigators hold that the basement membrane is the principal size-selective barrier, whereas others believe that it is the filtration slit diaphragms. It is likely that both together contribute to the size-selective properties of the barrier. Why the glomerular filtration barrier does not clog is not fully understood.

Glomerular capillary pressure is the major force that determines glomerular filtration.

GFR depends on the balance of hydrostatic and colloid osmotic pressures (COPs) acting across the glomerular filtration barrier, the Starling forces (see Chapter 15, "Microcirculation and Lymphatic System"), and therefore, it is determined by the same factors that affect fluid movement across capillaries in general. In the glomerulus, the driving force for fluid filtration is the glomerular capillary hydrostatic pressure (P_{GC}). This pressure ultimately depends on the pumping of blood by the heart, an action that raises the blood pressure on the arterial side of the circulation. Filtration is opposed by the hydrostatic pressure in the space of the Bowman capsule (P_{BS}) and by the COP exerted by plasma proteins in glomerular capillary blood. Because the glomerular filtrate is virtually protein free, we neglect the COP of fluid in the Bowman capsule. The *net ultrafiltration pressure gradient* (UP) is equal to the difference between the pressures favoring and opposing filtration:

$$\text{GFR} = K_f \times \text{UP} = K_f \times (P_{GC} - P_{BS} - \text{COP}) \quad (10)$$

where K_f is the glomerular ultrafiltration coefficient. Estimates of average, normal values for pressures in the human kidney are as follows: P_{GC}, 55 mm Hg; P_{BS}, 15 mm Hg; and COP, 30 mm Hg. From these values, we calculate a net UP of +10 mm Hg.

Glomerular hemodynamics are characterized by high capillary pressure and low vascular resistance.

Figure 22.15 shows how pressures change along the length of a glomerular capillary, in contrast to those seen in a

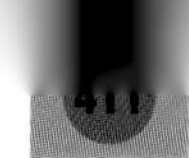

Clinical Focus / 22.2

Glomerular Disease

The kidney glomeruli may be injured by many immunologic, toxic, hemodynamic, and metabolic disorders. Glomerular injury impairs filtration barrier function and consequently increases the filtration and excretion of plasma proteins (proteinuria). Red cells may appear in the urine, and sometimes the glomerular filtration rate (GFR) is reduced. Three general syndromes are encountered: nephritic diseases, nephrotic diseases (nephrotic syndrome), and chronic glomerulonephritis.

In the nephritic diseases, the urine contains red blood cells, red cell casts, and mild-to-modest amounts of protein. A red cell cast is a mold of the tubule lumen formed when red cells and proteins clump together; the presence of such casts in the final urine indicates that bleeding had occurred in the kidneys (usually in the glomeruli), not in the lower urinary tract. Nephritic diseases are usually associated with a fall in GFR, accumulation of nitrogenous wastes (urea, creatinine) in the blood, and hypervolemia (hypertension and edema). Most nephritic diseases are a result of immunologic damage. The glomerular capillaries may be injured by antibodies directed against the glomerular basement membrane, by deposition of circulating immune complexes along the endothelium or in the mesangium, or by cell-mediated injury (infiltration with lymphocytes and macrophages). A renal biopsy and tissue examination by light and electron microscopy and immunostaining are often helpful in determining the nature and severity of the disease and in predicting its most likely course.

Poststreptococcal glomerulonephritis is an example of a nephritic condition that may follow a sore throat caused by certain strains of streptococci. Immune complexes of antibody and bacterial antigen are deposited in the glomeruli, complement is activated, and polymorphonuclear leukocytes and macrophages infiltrate the glomeruli. Endothelial cell damage, accumulation of leukocytes, and the release of vasoconstrictor substances reduce the glomerular surface area and fluid permeability and lower glomerular blood flow, causing a fall in GFR.

Nephrotic syndrome is a clinical state that can develop as a consequence of many different diseases causing glomerular injury. It is characterized by heavy proteinuria (>3.5 g/d per 1.73 m^2 body surface area), hypoalbuminemia (<3 g/dL), generalized edema, and hyperlipidemia. Abnormal glomerular leakiness to plasma proteins leads to increased proximal tubular reabsorption and catabolism of filtered proteins and increased protein excretion in the urine. The resulting loss of protein (mainly serum albumin) leads to a fall in plasma protein concentration (and colloid osmotic pressure). The edema results from the hypoalbuminemia and renal Na^+ retention. Also, a generalized increase in capillary permeability to proteins (not just in the glomeruli) may lead to a decrease in the effective colloid osmotic pressure of the plasma proteins and may contribute to the edema. The hyperlipidemia (elevated serum cholesterol and elevated triglycerides in severe cases) is probably a result of increased hepatic synthesis of lipoproteins and decreased lipoprotein catabolism. Most often, nephrotic syndrome in young children cannot be ascribed to a specific cause; this is called idiopathic nephrotic syndrome. Nephrotic syndrome in children or adults can be caused by infectious diseases, neoplasia, certain drugs, various autoimmune disorders (such as lupus), allergic reactions, metabolic disease (such as diabetes mellitus), or congenital disorders.

The distinctions between nephritic and nephrotic diseases are sometimes blurred, and both may result in chronic **glomerulonephritis**. This disease is characterized by proteinuria and/or hematuria (blood in the urine), hypertension, and renal insufficiency that progresses over years. Renal biopsy shows glomerular scarring and increased numbers of cells in the glomeruli and scarring and inflammation in the interstitial space. The disease is accompanied by a progressive loss of functioning nephrons and proceeds relentlessly even though the initiating insult may no longer be present. The exact reasons for disease progression are not known, but an important factor may be that surviving nephrons hypertrophy when nephrons are lost. This leads to an increase in blood flow and pressure in the remaining nephrons, a situation that further injures the glomeruli. Also, increased filtration of proteins causes increased tubular reabsorption of proteins, and the latter results in production of vasoactive and inflammatory substances that cause ischemia, interstitial inflammation, and renal scarring. Dietary manipulations (such as a reduced protein intake) or antihypertensive drugs (such as angiotensin-converting enzyme inhibitors) may slow the progression of chronic glomerulonephritis. Glomerulonephritis in its various forms is the major cause of renal failure in people.

capillary in other vascular beds (in this case, skeletal muscle). Note that average capillary hydrostatic pressure in the glomerulus is much higher (55 vs. 25 mm Hg) than in a skeletal muscle capillary. Also, capillary hydrostatic pressure declines little (perhaps 1–2 mm Hg) along the length of the glomerular capillary, because the glomerulus contains many (30–50) capillary loops in parallel, thereby making the resistance to blood flow in the glomerulus low. In the skeletal muscle capillary, there is a much higher resistance to blood flow, resulting in an appreciable fall in capillary hydrostatic pressure with distance. Finally, note that in the glomerulus, the COP increases substantially along the length of the capillary, because a large volume of filtrate (about 20% of the entering plasma flow) is pushed out of the capillary, and the proteins remain in the circulation. The increase in COP opposes the outward movement of fluid.

In the skeletal muscle capillary, the COP hardly changes with distance, because little fluid moves across the capillary wall. In the "average" skeletal muscle capillary, outward filtration occurs at the arterial end and absorption occurs at the venous end. At some point along the skeletal muscle capillary, there is no net fluid movement; this is the point of so-called **filtration pressure equilibrium**. Filtration pressure equilibrium is probably not attained in the normal human glomerulus; in other words, outward filtration of fluid probably occurs all along the glomerular capillaries.

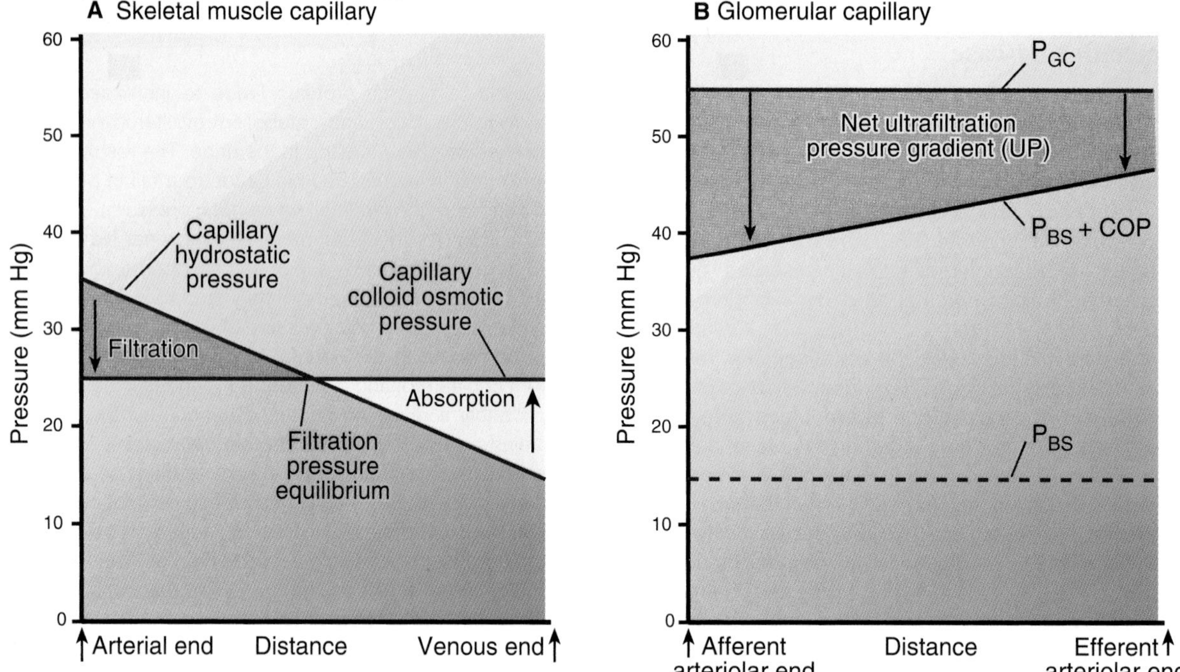

Figure 22.15 Pressure profiles along a skeletal muscle capillary and a glomerular capillary. **(A)** In the "typical" skeletal muscle capillary, filtration occurs at the arterial end and absorption at the venous end of the capillary. Interstitial fluid hydrostatic and colloid osmotic pressures (COP) are neglected here because they are roughly equal and counterbalance each other. **(B)** In the glomerular capillary, glomerular hydrostatic pressure (P_{GC}) (*top line*) is high and declines only slightly with distance. The bottom (*dashed*) line represents the hydrostatic pressure in the Bowman capsule (P_{BS}). The middle line is the sum of P_{BS} and the glomerular capillary COP. The difference between P_{GC} and P_{BS} + COP is equal to the net ultrafiltration pressure gradient (UP). In the normal human glomerulus, filtration probably occurs along the entire capillary. Assuming that K_f is uniform along the length of the capillary, filtration rate would be highest at the afferent arteriolar end and lowest at the efferent arteriolar end of the glomerulus.

The GFR depends on the magnitudes of the different terms in equation 10. Thus, GFR varies with changes in K_f, hydrostatic pressures in the glomerular capillaries and the Bowman capsule, and the glomerular capillary COP. These factors are discussed next.

Glomerular ultrafiltration coefficient depends on the properties of the glomerular filtration barrier.

The **glomerular ultrafiltration coefficient** (**K_f**) is the glomerular equivalent of the capillary filtration coefficient encountered in Chapter 15, "Microcirculation and Lymphatic System." It depends on both the hydraulic conductivity (fluid permeability) and surface area of the glomerular filtration barrier. In chronic renal disease, functioning glomeruli are lost, leading to a reduction in surface area available for filtration and a fall in GFR. Acutely, a variety of drugs and hormones appear to change glomerular K_f and thus alter GFR, but the mechanisms are not completely understood.

Changes in glomerular capillary hydrostatic pressure profoundly affect the GFR.

Glomerular capillary hydrostatic pressure (P_{GC}) is the driving force for filtration; it depends on the arterial blood pressure and the resistances of upstream and downstream blood vessels. Because of autoregulation, P_{GC} and GFR are maintained at relatively constant values when arterial blood pressure is varied from 80 to 180 mm Hg. Below a pressure of 80 mm Hg, however, P_{GC} and GFR decrease, and GFR ceases at a blood pressure of about 40 to 50 mm Hg. One of the classic signs of hemorrhagic or cardiogenic shock is an absence of urine output, which is a result of an inadequate arterial blood pressure, P_{GC}, and GFR.

The caliber of afferent and efferent arterioles can be altered by a variety of hormones and by sympathetic nerve stimulation, leading to changes in P_{GC}, glomerular blood flow, and GFR. Some hormones act preferentially on afferent or efferent arterioles. Afferent arteriolar dilation increases glomerular blood flow and P_{GC} and therefore produces an increase in GFR. Afferent arteriolar constriction produces the exact opposite effects. Efferent arteriolar dilation increases glomerular blood flow but leads to a fall in GFR because P_{GC} is decreased. Constriction of efferent arterioles increases P_{GC} and decreases glomerular blood flow. With modest efferent arteriolar constriction, GFR increases because of the increased P_{GC}. With extreme efferent arteriolar constriction, however, GFR decreases because of the marked decrease in glomerular blood flow.

Normally, about 20% of the plasma flowing through the kidneys is filtered in the glomeruli. This percentage (or fraction) is called the **filtration fraction**. It can be estimated from the IN clearance/PAH clearance. Changes in filtration fraction will result from constriction or dilation of afferent or efferent arterioles. For example, afferent arteriolar dilation or efferent arteriolar constriction leads to an increase in filtration fraction. An increase in filtration fraction will increase the protein concentration of the blood exiting the glomerulus and hence will increase the COP in the peritubular capillaries. Such a change will increase sodium reabsorption in the kidneys (see Chapter 23, "Regulation of Fluid and Electrolyte Balance").

Capillary osmotic pressure and hydrostatic pressure in the Bowman capsule oppose glomerular filtration.

Hydrostatic pressure in the Bowman capsule (P_{BS}) depends on the input of glomerular filtrate and the rate of removal of this fluid by the tubule. This pressure opposes filtration. It also provides the driving force for fluid movement down the length of the tubule. If there is obstruction anywhere along the urinary tract—for example, because of stones, ureteral obstruction, or prostate enlargement—then pressure upstream to the block is increased and GFR consequently falls. If tubular reabsorption of water is inhibited, pressure in the tubular system is increased because an increased pressure head is needed to force a large volume flow through the loops of Henle and collecting ducts. Consequently, a large increase in urine output caused by a diuretic drug may be associated with a tendency for GFR to fall.

The COP opposes glomerular filtration. Dilution of the plasma proteins (e.g., by intravenous infusion of a large volume of isotonic saline) lowers the plasma COP and leads to an increase in GFR. Part of the reason glomerular blood flow has important effects on GFR is that it changes the COP profile along the length of a glomerular capillary. Consider, for example, what would happen if glomerular blood flow were low. Filtering a small volume of fluid out of the glomerular capillary would lead to a sharp rise in COP early along the length of the glomerulus. As a consequence, filtration would soon cease and GFR would be low. On the other hand, a high blood flow would allow a high rate of filtrate formation with a minimal rise in COP. In general, then, RBF and GFR change hand in hand, but the exact relationship between GFR and RBF depends on the magnitude of the other factors that affect GFR.

High GFR in the human kidney is the result of several factors.

The rate of plasma ultrafiltration in the kidney glomeruli (180 L/d) far exceeds that in all other capillary beds, for several reasons:

- The filtration coefficient is unusually high in the glomeruli. Compared with most other capillaries, the glomerular capillaries behave as though they have more pores per unit surface area; consequently, they have an unusually high hydraulic conductivity. The total glomerular filtration barrier area, about 2 m^2, is large.
- Capillary hydrostatic pressure is higher in the glomeruli than in any other capillaries.
- The high rate of RBF helps sustain a high GFR by limiting the rise in COP, thereby favoring filtration along the entire length of the glomerular capillaries.

In summary, glomerular filtration is high because the glomerular capillary blood is exposed to a large, porous surface and there is a high transmural pressure gradient favoring filtration.

TRANSPORT IN THE PROXIMAL TUBULE

Glomerular filtration is a rather nonselective process, because both useful and waste substances are filtered. By contrast, tubular transport is selective; different substances are transported by different mechanisms. Some substances are reabsorbed, others are secreted, and still others are both reabsorbed and secreted. For most, the amount excreted in the urine depends in large measure on the magnitude of tubular transport. Transport of various solutes and water differs in the various nephron segments. Here we describe transport along the nephron and collecting duct system, starting with the proximal convoluted tubule. Remember that the proximal convoluted tubule is the first part of the renal tubule and comes right after the Bowman capsule.

The proximal convoluted tubule comprises the first 60% of the length of the proximal tubule. Because the proximal straight tubule is inaccessible to study *in vivo*, most quantitative information about function in the living animal is confined to the convoluted portion. Studies on isolated tubules *in vitro* indicate that the two segments of the proximal tubule are functionally similar. The proximal tubule is responsible for reabsorbing all of the filtered glucose and amino acids, reabsorbing the largest fraction of the filtered Na^+, K^+, Ca^{2+}, Cl^-, HCO_3^-, and water, and secreting various organic anions and organic cations.

Most of the filtered water is reabsorbed by the proximal convoluted tubule.

The percentage of filtered water reabsorbed along the nephron has been determined by measuring the degree to which IN is concentrated in tubular fluid using the kidney micropuncture technique in laboratory animals. Samples of tubular fluid from surface nephrons are collected and analyzed, and the site of collection is identified by nephron microdissection. Because IN is filtered but not reabsorbed by the kidney tubules, as water is reabsorbed the IN becomes increasingly concentrated. For example, if 50% of the filtered water is reabsorbed by a certain point along the tubule, the concentration of IN in tubular fluid (TF_{IN}) will be twice its concentration in the plasma (P_{IN}). If two third (67%) of the filtered water is reabsorbed, the ratio TF_{IN}/P_{IN} will be 3; if three quarters (75%), the ratio TF_{IN}/P_{IN} will be 4, and so forth. At the end of the proximal convoluted tubule the ratio TF_{IN}/P_{IN} is usually slightly higher than 3, leading to the conclusion that about 70% of filtered water is reabsorbed by this nephron segment. In the final urine, the IN concentration may be 100 times that in the plasma, indicating that 99% of the filtered water was reabsorbed by the kidney tubules.

Proximal tubular fluid is essentially isosmotic to plasma.

Samples of fluid collected from the proximal convoluted tubule are always essentially isosmotic to plasma (Fig. 22.16), a consequence of the high water permeability of this segment. Overall, 70% of filtered solutes and water are reabsorbed along the proximal convoluted tubule.

Na^+ salts are the major osmotically active solutes in the plasma and glomerular filtrate. Because osmolality does not change appreciably with proximal tubule length, it is not surprising that Na^+ concentration also does not change under ordinary conditions.

If an appreciable quantity of nonreabsorbed solute is present (e.g., the sugar alcohol mannitol), proximal tubular fluid Na^+ concentration falls to values below the plasma concentration. This is evidence that Na^+ can be reabsorbed against a concentration gradient in an active process. The fall in the concentration of Na^+ in the proximal tubular fluid increases diffusion of Na^+ into the tubule lumen and results in reduced net Na^+ and water reabsorption, leading to increased excretion of Na^+ and water, an osmotic diuresis. This situation occurs, for example, in uncontrolled diabetes mellitus.

Two major anions, Cl^- and HCO_3^-, accompany Na^+ in plasma and glomerular filtrate. HCO_3^- is preferentially reabsorbed along the proximal convoluted tubule, leading to a fall in tubular fluid HCO_3^- concentration, mainly because of H^+ secretion (see Chapter 24, "Acid–Base Homeostasis"). The Cl^- lags behind; as water is reabsorbed, and, therefore, Cl^- concentration rises (see Fig. 22.16). The result is a tubular fluid-to-plasma concentration gradient that favors Cl^- diffusion out of the tubule lumen. Outward movement of Cl^- in the late proximal convoluted tubule creates a small (1–2 mV), lumen-positive transepithelial potential difference that favors the passive reabsorption of Na^+.

Figure 22.16 shows that the K^+ concentration hardly changes along the proximal convoluted tubule. If K^+ were not reabsorbed, its concentration would increase as much as that of IN. The fact that the concentration ratio for K^+ remains about 1 in this nephron segment indicates that 70% of filtered K^+ is reabsorbed along with 70% of the filtered water.

The concentrations of glucose and amino acids fall steeply in the proximal convoluted tubule. This nephron segment and the proximal straight tubule are responsible for complete reabsorption of these substances. Separate, specific mechanisms reabsorb glucose and various amino acids. The concentration ratio for urea rises along the proximal tubule, but not as much as the IN concentration ratio, because about 50% of the filtered urea is reabsorbed. The concentration ratio for PAH in proximal tubular fluid increases more steeply than the IN concentration ratio because of PAH secretion.

In summary, though the osmolality (total solute concentration) does not change detectably along the proximal convoluted tubule, it is clear that the concentrations of individual solutes vary widely. The concentrations of some substances fall (glucose, amino acids, and HCO_3^-), others rise (IN, urea, Cl^-, and PAH), and still others do not change (Na^+ and K^+). By the end of the proximal convoluted tubule, only about one third of the filtered Na^+, water, and K^+ are left; almost all of the filtered glucose, amino acids, and HCO_3^- have been reabsorbed; and many solutes destined for excretion (PAH, IN, and urea) have been concentrated in the tubular fluid.

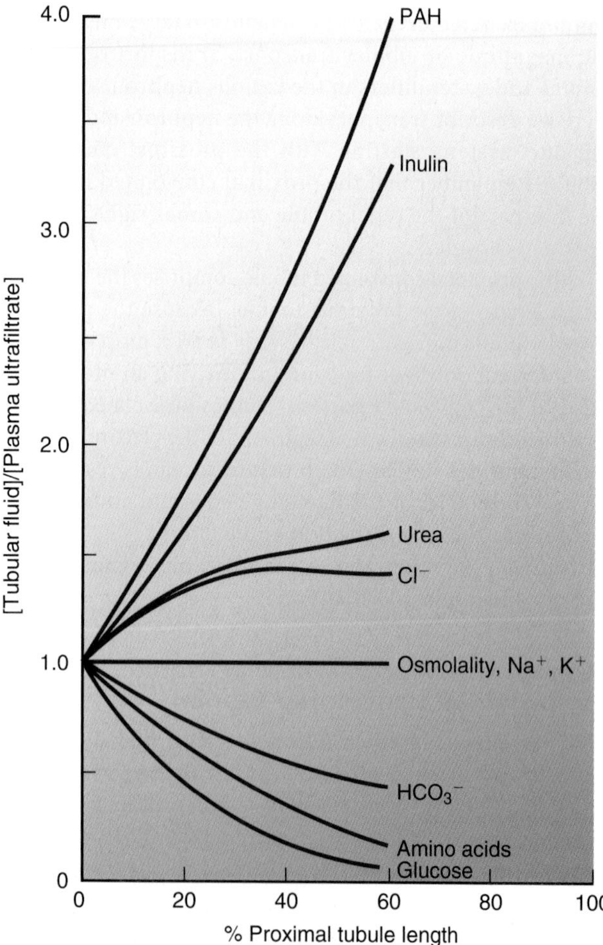

Figure 22.16 Tubular fluid/plasma ultrafiltrate concentration ratios for various solutes as a function of proximal tubule length. All values start at a ratio of 1, because the fluid in the Bowman capsule (0% proximal tubule length) is a plasma ultrafiltrate. PAH, *p*-aminohippurate.

Active Na^+/K^+-ATPase pump is essential for sodium reabsorption, which serves as the major driving force for reabsorption of solutes and water in the proximal tubule.

Figure 22.17 is a model of a proximal tubule cell. Na^+ enters the cell from the lumen across the apical cell membrane

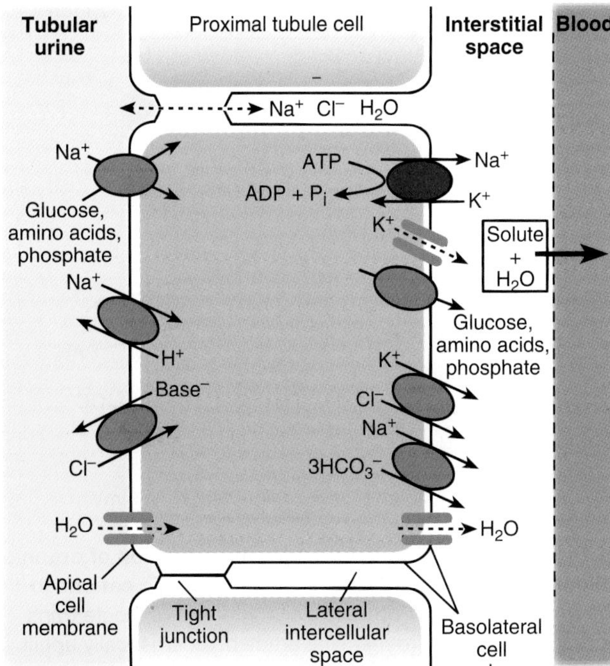

Figure 22.17 Cell model for transport in the proximal tubule. The luminal (apical) cell membrane in this nephron segment has a large surface area for transport because of the numerous microvilli that form the brush border (*not shown*). Glucose, amino acids, phosphate, and numerous other substances are transported by separate carriers. ADP, adenosine diphosphate; ATP, adenosine triphosphate.

and is pumped out across the basolateral cell membrane by the Na^+/K^+-ATPase. The blood surrounding the tubules then takes up the Na^+, accompanying anions, and water. Filtered Na^+ salts and water are thus returned to the circulation.

At the luminal cell membrane (brush border) of the proximal tubule cell, Na^+ enters the cell down combined electrical and chemical potential gradients. The inside of the cell is about −70 mV compared with tubular fluid, and intracellular Na^+ is about 30 to 40 mEq/L compared with a tubular fluid concentration of about 140 mEq/L. Na^+ entry into the cell occurs via a number of cotransport and antiport mechanisms. Na^+ is reabsorbed together with glucose, amino acids, phosphate, and other solutes by way of separate, specific cotransporters. The downhill (energetically speaking) movement of Na^+ into the cell drives the uphill transport of these solutes. In other words, glucose, amino acids, phosphate, and so on are reabsorbed by secondary active transport. Na^+ is also reabsorbed across the luminal cell membrane in exchange for H^+. The Na^+/H^+ exchanger, an antiporter, is also a secondary active transport mechanism; the downhill movement of Na^+ into the cell energizes the uphill secretion of H^+ into the lumen. This mechanism is important in the acidification of urine (see Chapter 24, "Acid–Base Homeostasis"). Cl^- may enter the cells by way of a luminal cell membrane Cl^-/base (formate or oxalate) exchanger.

Once inside the cell, Na^+ is pumped out the basolateral side by a vigorous Na^+/K^+-ATPase that keeps intracellular Na^+ low. This membrane ATPase pumps three Na^+ out of the cell and two K^+ into the cell and splits one ATP molecule for each cycle of the pump. K^+ pumped into the cell diffuses out the basolateral cell membrane mostly through a K^+ channel. Glucose, amino acids, and phosphate, accumulated in the cell because of secondary active transport across the luminal cell membrane, exit across the basolateral cell membrane by way of separate, Na^+-independent facilitated diffusion mechanisms. HCO_3^- exits together with Na^+ by an electrogenic mechanism; the carrier transports three HCO_3^- for each Na^+. Cl^- may leave the cell by way of an electrically neutral K–Cl cotransporter.

The reabsorption of Na^+ and accompanying solutes establishes an osmotic gradient across the proximal tubule epithelium that is the driving force for water reabsorption. Because the water permeability of the proximal tubule epithelium is extremely high, only a small gradient (a few mOsm/kg H_2O) is needed to account for the observed rate of water reabsorption. Some experimental evidence indicates that proximal tubular fluid is slightly hypo-osmotic to plasma. Because the osmolality difference is so small, it is still proper to consider the fluid as essentially isosmotic to plasma. Water crosses the proximal tubule epithelium through the cells (via water channel proteins—aquaporins—in the cell membranes) and between the cells (tight junctions and lateral intercellular spaces).

The final step in the overall reabsorption of solutes and water is uptake by the peritubular capillaries. This mechanism involves the usual Starling forces that operate across capillary walls. Recall that blood in the peritubular capillaries was previously filtered in the glomeruli. Because a protein-free filtrate was filtered out of the glomeruli, the protein concentration (hence, COP) of blood in the peritubular capillaries is high, thereby providing an important driving force for the uptake of reabsorbed fluid. The hydrostatic pressure in the peritubular capillaries (a pressure that opposes the capillary uptake of fluid) is low, because the blood has passed through upstream resistance vessels. The balance of pressures acting across peritubular capillaries favors the uptake of reabsorbed fluid from the interstitial spaces surrounding the tubules.

Proximal tubule secretion eliminates many toxins and drugs from the blood.

The proximal tubule, both convoluted and straight portions, secretes a large variety of organic anions and organic cations (Table 22.2). Many of these substances are endogenous compounds, drugs, or toxins. The organic anions are mainly carboxylates and sulfonates (carboxylic and sulfonic acids in their protonated forms). A negative charge on the molecule appears to be important for secretion of these compounds. Examples of organic anions actively secreted in the proximal tubule include penicillin and PAH. Organic anion transport becomes saturated at high plasma organic anion concentrations (see Fig. 22.9), and the organic anions compete with each other for secretion.

Figure 22.18 shows a cell model for active secretion. Proximal tubule cells actively take up PAH from the blood

TABLE 22.2 Some Organic Compounds Secreted by Proximal Tubules[a]

Compound	Use
Organic anions	
Phenol red (phenolsulfonphthalein)	pH indicator dye
p-Aminohippurate	Measurement of renal plasma flow and proximal tubule secretory mass
Penicillin	Antibiotic
Probenecid (Benemid)	Inhibitor of penicillin secretion and uric acid reabsorption
Furosemide (Lasix)	Loop diuretic drug
Acetazolamide (Diamox)	Carbonic anhydrase inhibitor
Creatinine[b]	Normal end-product of muscle metabolism
Organic cations	
Histamine	Vasodilator, stimulator of gastric acid secretion
Cimetidine (Tagamet)	Drug for treatment of gastric and duodenal ulcers
Cisplatin	Cancer chemotherapeutic agent
Norepinephrine	Neurotransmitter
Quinine	Antimalarial drug
Tetraethylammonium	Ganglion-blocking drug
Creatinine[b]	Normal end-product of muscle metabolism

[a]This list includes only a few of the large variety of organic anions and cations secreted by kidney proximal tubules.
[b]The creatinine molecule may be present in both anionic and cationic forms at physiological pH, and this property may explain its secretion by both organic anion and organic cation pathways.

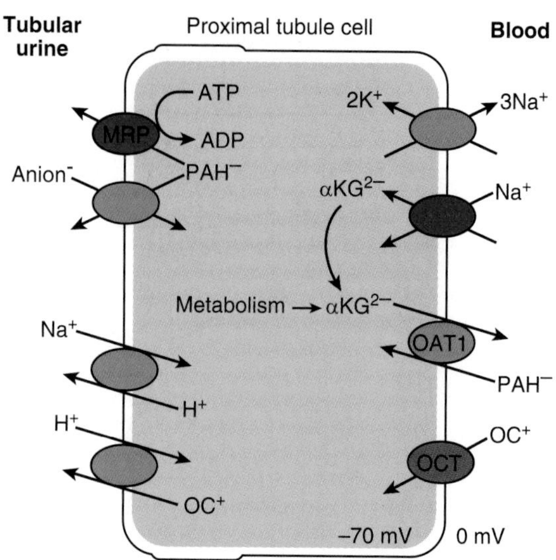

Figure 22.18 Cell model for the secretion of organic anions (*p*-aminohippurate [PAH]) and organic cations in the proximal tubule. Upward slanting arrows indicate transport against an electrochemical gradient (energetically uphill transport) and downward slanting arrows indicate downhill transport. PAH accumulation in the cell is mediated by a basolateral membrane organic anion transporter (OAT) that exchanges PAH for α-ketoglutarate (α-KG^{2-}). The α-KG^{2-} level in the cell is higher than in the blood because of metabolic production and Na^+-dependent uptake of α-KG^{2-}. PAH exits the cell passively via a luminal membrane PAH/anion exchanger or can be actively pumped into the lumen by a multidrug resistance-associated protein (MRP) that consumes adenosine triphosphate (ATP). Organic cations (OC^+) enter the cell down the electrical gradient, a process mediated by a basolateral membrane organic cation transporter (OCT), and are transported uphill into the lumen by an organic cation/H^+ exchanger. ADP, adenosine diphosphate; ATP, adenosine triphosphate.

side by exchange for cell α-ketoglutarate. This exchange is mediated by an organic anion transporter (OAT). The cells accumulate α-ketoglutarate from metabolism and because of the cell-membrane Na^+-dependent dicarboxylate transporters. PAH accumulates in the cells at a high concentration and then moves downhill into the tubular urine in an electrically neutral fashion, by exchanging for an inorganic anion (e.g., Cl^-) or an organic anion via another OAT. Organic anions may also be actively pumped into the tubular urine via a multidrug resistance-associated protein, which is an ATPase.

The organic cations are mainly amine and ammonium compounds and are secreted by other transporters. Entry into the cell across the basolateral membrane is favored by the inside negative membrane potential and occurs via facilitated diffusion, mediated by an organic cation transporter. The exit of organic cations across the luminal membrane is accomplished by an organic cation/H^+ antiporter (exchanger) and is driven by the lumen-to-cell H^+ concentration gradient established by Na^+/H^+ exchange. The transporters for organic anions and organic cations show broad substrate specificity and accomplish the secretion of a large variety of chemically diverse compounds.

In addition to being actively secreted, some organic compounds passively diffuse across the tubular epithelium (Fig. 22.19). Organic anions can accept H^+ and organic cations can release H^+, so their charge is influenced by pH. The nonionized (uncharged) form, if it is lipid-soluble, can diffuse through the lipid bilayer of cell membranes down concentration gradients. The ionized (charged) form passively penetrates cell membranes with difficulty.

Consider, for example, the carboxylic acid probenecid (pK_a = 3.4). This compound is filtered by the glomeruli and secreted by the proximal tubule. When H^+ is secreted into the tubular urine (see Chapter 24, "Acid–Base Homeostasis"), the anionic form (A^-) is converted to the nonionized acid (HA, in Figure 22.19). The concentration of nonionized

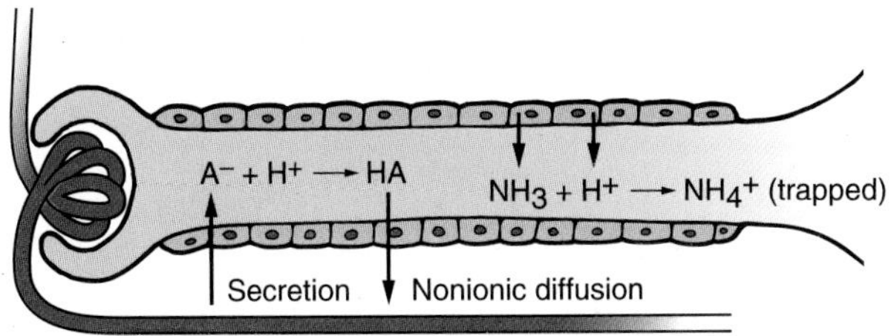

Figure 22.19 Nonionic diffusion of lipid-soluble weak organic acids and bases. Acidification of the urine converts the organic anion A^- to the undissociated (nonionized) acid HA, which is reabsorbed by diffusion. NH_3, a lipid-soluble base, diffuses into the tubular urine, where it is converted to NH_4^+, thereby trapping the ammonia in the acidic urine.

acid is also increased because of water reabsorption. A concentration gradient for passive reabsorption across the tubule wall is created, and appreciable quantities of probenecid are passively reabsorbed. This occurs in most parts of the nephron, but particularly in those where pH gradients are largest and where water reabsorption has resulted in the greatest concentration (i.e., the collecting ducts). The excretion of probenecid is enhanced by making the urine more alkaline (by administering $NaHCO_3$) and by increasing urine flow rate (e.g., by drinking water). In the case of a lipid-soluble base, such as ammonia (NH_3), excretion is favored by making the urine more acidic and enhancing the urine flow.

Finally, a few organic anions and cations are also actively reabsorbed. For example, uric acid is both secreted and reabsorbed in the proximal tubule. Normally, the amount of uric acid excreted is equal to about 10% of the filtered uric acid, so reabsorption predominates. In **gout**, plasma levels of uric acid are increased. One treatment for gout is to promote urinary excretion of uric acid by administering drugs that inhibit its tubular reabsorption.

TUBULAR TRANSPORT IN THE LOOP OF HENLE

The loop of Henle includes several distinct segments with different structural and functional properties. As noted earlier, the proximal straight tubule has transport properties similar to those of the proximal convoluted tubule. The thin descending, thin ascending, and thick ascending limbs of the loop of Henle differ in permeability and transport properties.

Ascending limb of the loop of Henle is water impermeable.

Tubular fluid entering the loop of Henle is isosmotic to plasma, but fluid leaving the loop is distinctly hypo-osmotic. Fluid collected from the earliest part of the distal convoluted tubule has an osmolality of about 100 mOsm/kg H_2O, compared with 285 mOsm/kg H_2O in plasma, because more solute than water is reabsorbed by the loop of Henle. The loop of Henle reabsorbs about 20% of filtered Na^+, 25% of filtered K^+, 30% of filtered Ca^{2+}, 65% of filtered Mg^{2+}, and 10% of filtered water. The descending limb of the loop of Henle (except for its terminal portion) is highly water permeable. The ascending limb is impermeable to water. The reason the ascending limb is impervious to water is not due to "tight junctions," but because of the lack of **aquaporins**, which are water channel proteins. These water channels exist in all cells except the cells of the ascending loop of Henle. Because solutes are reabsorbed along the ascending limb and water cannot follow, fluid along the ascending limb becomes more and more dilute. Deposition of reabsorbed solutes (mainly Na^+ salts) in the interstitial space of the kidney medulla is critical in the operation of the urinary concentrating mechanism (discussed below).

Sodium enters the luminal cell of the thick ascending limb by a Na-K-2Cl cotransporter.

Figure 22.20 is a model of a thick ascending limb cell. Na^+ enters the cell across the luminal cell membrane by an electrically neutral Na-K-2Cl cotransporter that is specifically inhibited by the "loop" diuretic drugs, bumetanide and furosemide. The downhill movement of Na^+ into the cell results in secondary active transport of one K^+ and two Cl^-. Na^+ is pumped out of the basolateral cell membrane by a vigorous Na^+/K^+-ATPase. K^+ recycles back into the lumen via a luminal cell membrane K^+ channel. Cl^- leaves through the basolateral side by a K-Cl cotransporter or Cl^- channel. The luminal cell membrane is predominantly permeable to K^+, and the basolateral cell membrane is predominantly permeable to Cl^-. Diffusion of these ions out of the cell produces a transepithelial potential difference, with the lumen about +6 mV compared with the interstitial space around the tubules. This potential difference drives small cations (Na^+, K^+, Ca^{2+}, Mg^{2+}, and NH_4^+) out of the lumen, between the cells. The tubular epithelium is extremely impermeable to water; there is no measurable water reabsorption along the ascending limb despite a large transepithelial gradient of osmotic pressure.

TUBULAR TRANSPORT IN THE DISTAL NEPHRON

The so-called **distal nephron** includes several distinct segments: the distal convoluted tubule; the connecting tubule; and the cortical, outer medullary, and inner medullary collecting ducts (see Fig. 22.3). Note that the distal nephron includes the collecting duct system, which, strictly speaking, is not part of the nephron but, from a functional perspective, this is justified. Transport in the distal nephron differs from that in the proximal tubule in several ways:

- The distal nephron reabsorbs much smaller amounts of salt and water. Typically, the distal nephron reabsorbs 9% of the filtered Na^+ and 19% of the filtered water, compared with 70% for both substances in the proximal convoluted tubule.

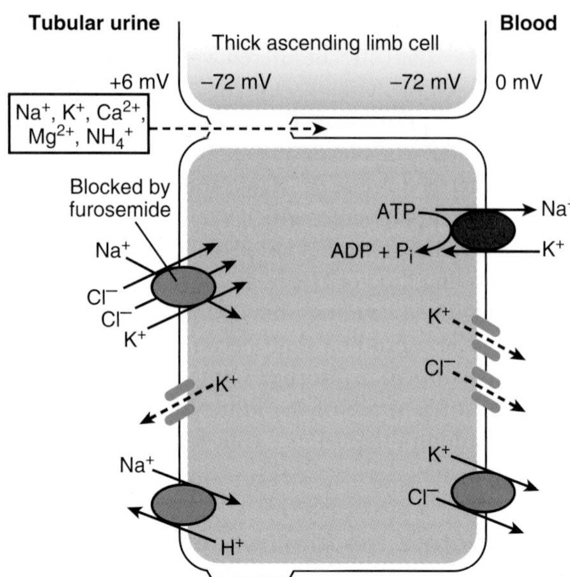

Figure 22.20 Cell model for ion transport in the thick ascending limb. ADP, adenosine diphosphate; ATP, adenosine triphosphate.

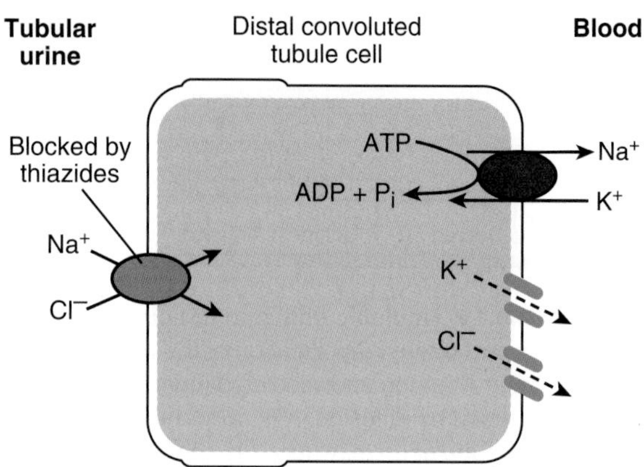

Figure 22.21 Cell model for ion transport in the distal convoluted tubule. ADP, adenosine diphosphate; ATP, adenosine triphosphate.

- The distal nephron can establish steep gradients for salt and water. For example, the concentration of Na^+ in the final urine may be as low as 1 mEq/L (vs. 140 mEq/L in plasma) and the urine osmolality can be almost one tenth that of plasma. By contrast, the proximal tubule reabsorbs Na^+ and water along small gradients, and the Na^+ concentration and osmolality of its tubule fluid are normally close to those of plasma.
- The distal nephron has a "tight" epithelium, whereas the proximal tubule has a "leaky" epithelium (see Chapter 2, "Plasma Membrane, Membrane Transport, and Resting Membrane Potential"). This explains why the distal nephron can establish steep gradients for small ions and water, whereas the proximal tubule cannot.
- Na^+ and water reabsorption in the proximal tubule are normally closely coupled, because epithelial water permeability is always high. By contrast, Na^+ and water reabsorption can be uncoupled in the distal nephron, because water permeability may be low and variable.

Proximal reabsorption overall can be characterized as a coarse operation that reabsorbs large quantities of salt and water along small gradients. By contrast, distal reabsorption is a finer process.

The collecting ducts are at the end of the nephron system, and what happens there determines the final excretion of Na^+, K^+, H^+, and water. Transport in the collecting ducts is finely tuned by hormones. Specifically, aldosterone increases Na^+ reabsorption and K^+ and H^+ secretion, and arginine vasopressin increases water reabsorption at this site.

Sodium and chloride are transported through the luminal cell membrane of the distal convoluted tubule by a Na-Cl cotransporter.

Figure 22.21 is a model of a distal convoluted tubule cell. In this nephron segment, Na^+ and Cl^- are transported from the lumen into the cell by a Na–Cl cotransporter that is inhibited by thiazide diuretics. Na^+ is pumped out the basolateral side by the Na^+/K^+-ATPase. Water permeability of the distal convoluted tubule is low and is not changed by arginine vasopressin.

Cortical collecting duct is the primary site for potassium secretion.

Under normal circumstances, the cortical collecting ducts secrete most of the excreted K^+. With great K^+ excess (e.g., a high-K^+ diet), the cortical collecting ducts may secrete so much K^+ that more K^+ is excreted than was filtered. With severe K^+ depletion, the cortical collecting ducts reabsorb K^+.

K^+ secretion appears to be a function primarily of the collecting duct principal cell (Fig. 22.22). K^+ secretion involves active uptake by a Na^+/K^+-ATPase in the basolateral cell membrane, followed by diffusion of K^+ through luminal membrane K^+ channels. Outward diffusion of K^+ from the cell is favored by concentration gradients and opposed by electrical gradients. Note that the electrical gradient opposing exit from the cell is smaller across the luminal cell membrane than across the basolateral cell membrane, thereby favoring movement of K^+ into the lumen rather than back into the blood. The luminal cell membrane potential difference is low (e.g., 20 mV, cell inside negative) because this membrane has a high Na^+ permeability and is depolarized by Na^+ diffusing into the cell. Recall that the entry of Na^+ into a cell causes membrane depolarization (see Chapter 3, "Action Potential, Synaptic Transmission, and Maintenance of Nerve Function").

The magnitude of K^+ secretion is affected by several factors (see Fig. 22.22):

- The activity of the basolateral membrane Na^+/K^+-ATPase is a key factor affecting secretion: the greater the pump activity, the higher the rate of secretion. A high plasma K^+ concentration promotes K^+ secretion. Increased amounts of Na^+ in the collecting duct lumen (e.g., as a result of inhibition of Na^+ reabsorption by a loop diuretic

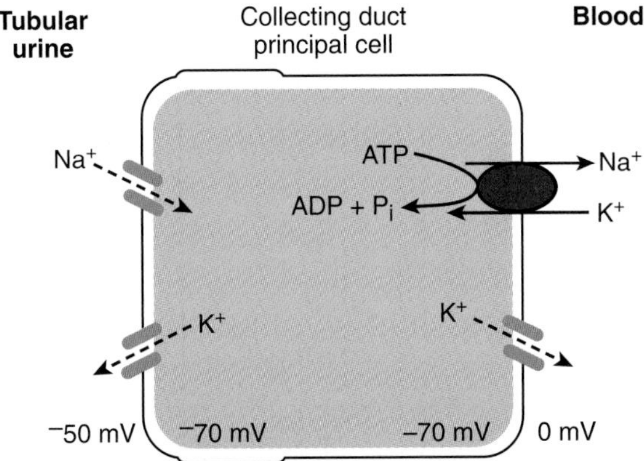

Figure 22.22 Model for Na^+ and K^+ transport by a collecting duct principal cell. ADP, adenosine diphosphate; ATP, adenosine triphosphate.

drug) result in increased entry of Na^+ into principal cells, increased activity of the Na^+/K^+-ATPase, and increased K^+ secretion.

- The lumen-negative transepithelial electrical potential promotes K^+ secretion.
- An increase in permeability of the luminal cell membrane to K^+ favors secretion.
- A high fluid flow rate through the collecting duct lumen maintains the cell-to-lumen concentration gradient, which favors K^+ secretion.

The hormone aldosterone promotes K^+ secretion by several actions (see Chapter 23, "Regulation of Fluid and Electrolyte Balance"). Na^+ entry into the collecting duct cell is by diffusion through a Na^+ channel (see Fig. 22.22). This channel has been cloned and sequenced and is known as **ENaC**, for **epithelial sodium (Na) channel**. The entry of Na^+ through this channel is rate-limiting for overall Na^+ reabsorption and is increased by aldosterone.

Intercalated cells are scattered among collecting duct principal cells; they are important in acid–base transport (see Chapter 24, "Acid–Base Homeostasis"). A H^+/K^+-ATPase is present in the luminal cell membrane of α-intercalated cells and contributes to renal K^+ conservation when dietary intake of K^+ is deficient.

URINARY CONCENTRATION AND DILUTION

The human kidney can form urine with a total solute concentration greater or lower than that of plasma. Maximum and minimum urine osmolalities in humans are about 1,200 to 1,400 mOsm/kg H_2O and 30 to 40 mOsm/kg H_2O, respectively. We next consider the mechanisms involved in producing osmotically concentrated or dilute urine.

Ability to concentrate urine osmotically is an important adaptation to life on land.

When the kidneys form osmotically concentrated urine, they save water for the body. The kidneys have the task of getting rid of excess solutes (e.g., urea and various salts), which requires the excretion of solvent (water). Suppose, for example, we excrete 600 mOsm of solutes per day. If we were only capable of excreting urine that is isosmotic to plasma (~300 mOsm/kg H_2O), then we would need to excrete 2.0 L H_2O/d. If we can excrete the solutes in urine, which is four times more concentrated than plasma (1,200 mOsm/kg H_2O), then only 0.5 L H_2O/d would be required. By excreting solutes in osmotically concentrated urine, the kidneys in effect save 1.5 L H_2O (2.0–0.5 L H_2O) for the body. The ability to concentrate the urine decreases the amount of water we are obliged to find and drink each day.

Urine concentrating ability can be looked at in two ways. We can determine what the urine osmolality (or specific gravity) is compared to the plasma, or the U_{osm}/P_{osm} ratio. In people, a maximal value is about 4 to 5, a value that might be observed in a dehydrated, otherwise healthy, individual. Or, we can calculate how much solute-free water per unit time the kidneys save or eliminate in the urine; this quantity is called the *free water clearance* (or *free water production*), abbreviated C_{H_2O}. C_{H_2O} is calculated from the following equation:

$$C_{H_2O} = \dot{V} - C_{osm} \quad (11)$$

where $\dot{V}$ is the urine flow rate and C_{osm} (the osmolal clearance) is defined as $U_{osm} \times \dot{V}/P_{osm}$. If we factor out $\dot{V}$ in equation 11, we get:

$$C_{H_2O} = \dot{V}(1 - U_{osm}/P_{osm}) \quad (12)$$

From equation 12, we see that if the U_{osm}/P_{osm} ratio is >1 (osmotically concentrated urine), C_{H_2O} is negative; if $U_{osm}/P_{osm} = 1$ (urine isosmotic to plasma), then C_{H_2O} is zero; and if U_{osm}/P_{osm} is <1 (osmotically dilute urine), then C_{H_2O} is positive.

Arginine vasopressin promotes the excretion of osmotically concentrated urine.

Changes in urine osmolality are normally brought about largely by changes in plasma levels of **arginine vasopressin (AVP)**, also known as **antidiuretic hormone** (see Chapter 31, "Hypothalamus and the Pituitary Gland"). In the absence of AVP, the kidney collecting ducts are relatively water impermeable. Reabsorption of solute across a water-impermeable epithelium leads to osmotically dilute urine. In the presence of AVP, collecting duct water permeability is increased. Because the medullary interstitial fluid is hyperosmotic, water reabsorption in the medullary collecting ducts can lead to the production of osmotically concentrated urine.

A model for the action of AVP on cells of the collecting duct is shown in Figure 22.23. When plasma osmolality is increased, plasma AVP levels increase. The hormone binds to a specific vasopressin (V_2) receptor in the basolateral cell membrane of principal cells. By way of a guanine nucleotide stimulatory protein (G_s), the membrane-bound

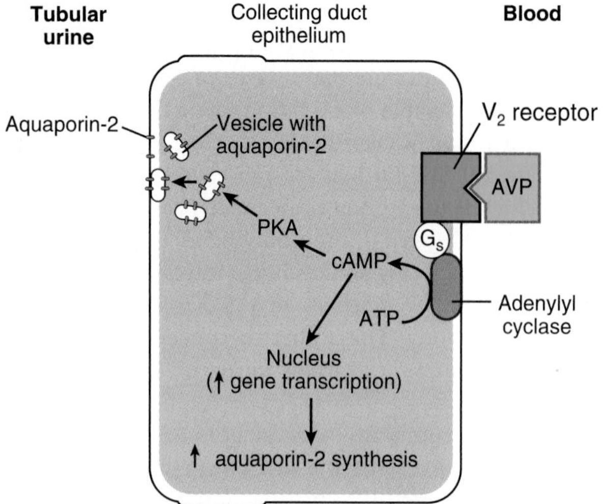

Figure 22.23 Model for the action of arginine vasopressin (AVP) on the epithelium of the collecting duct. The second messenger for AVP is cyclic adenosine monophosphate (cAMP). AVP has both prompt effects on luminal membrane water permeability (the movement of aquaporin-2–containing vesicles to the luminal cell membrane) and delayed effects (increased aquaporin-2 synthesis). ATP, adenosine triphosphate; G_s, guanine nucleotide stimulatory protein; PKA, protein kinase A; V_2, type 2 vasopressin receptor.

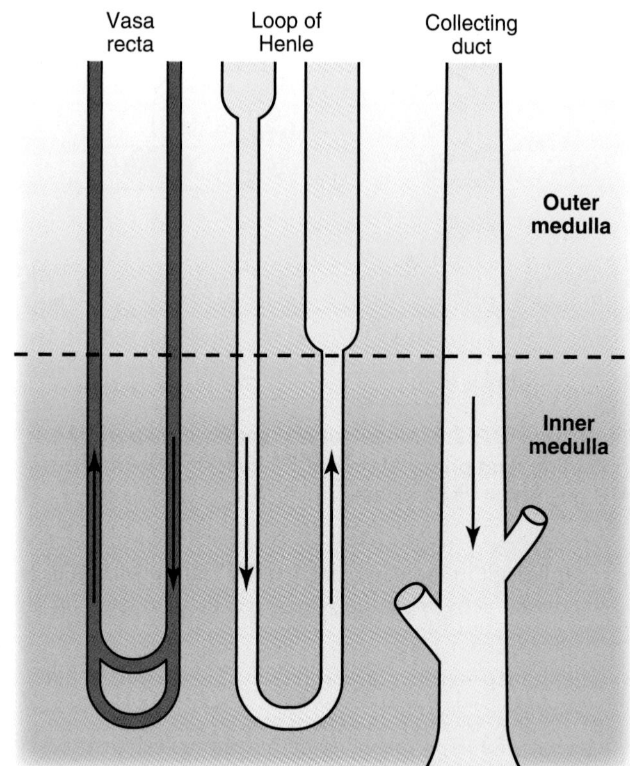

Figure 22.24 Elements of the urinary concentrating mechanism. The vasa recta are countercurrent exchangers, the loops of Henle are countercurrent multipliers, and the collecting ducts are osmotic equilibrating devices. Most loops of Henle and vasa recta do not reach the tip of the papilla but turn at higher levels in the outer and inner medulla. There are no thick ascending limbs in the inner medulla.

enzyme adenylyl cyclase is activated. This enzyme catalyzes the formation of cyclic AMP (cAMP) from ATP. cAMP then activates a cAMP-dependent protein kinase (protein kinase A, or *PKA*) that phosphorylates other proteins. This leads to the insertion, by exocytosis, of intracellular vesicles that contain water channels (aquaporin-2) into the luminal cell membrane. The resulting increase in number of luminal membrane water channels leads to an increase in water permeability. Water leaves the lumen and then exits the cells via aquaporin-3 and aquaporin-4 in the basolateral cell membrane. The solutes in the collecting duct lumen become concentrated as water leaves. This response to AVP occurs in minutes. AVP also has delayed effects on collecting ducts; it increases the transcription of aquaporin-2 genes and increases the total number of aquaporin-2 molecules per cell.

Countercurrent multiplication in the loops of Henle is the underlying mechanism for urine concentration.

A gradient of osmolality exists in the kidney medulla, with the highest osmolality present at the tips of the renal papillae. This gradient is explained by the *countercurrent mechanism.* Two countercurrent processes occur in the kidney medulla—**countercurrent multiplication** and **countercurrent exchange**. The term *countercurrent* indicates a flow of fluid in opposite directions in adjacent structures (Fig. 22.24). The loops of Henle are *countercurrent multipliers.* Fluid flows toward the tip of the papilla along the descending limb of the loop and toward the cortex along the ascending limb of the loop. The loops of Henle set up the osmotic gradient in the medulla (see below). The vasa recta are *countercurrent exchangers.* Blood flows in opposite directions along juxtaposed descending (arterial) and ascending (venous) vasa recta, and solutes and water are exchanged passively between these capillary blood vessels. The vasa recta help maintain the gradient in the medulla. The collecting ducts act as *osmotic equilibrating devices*; depending on the plasma level of AVP, the collecting duct urine is allowed to equilibrate more or less with the hyperosmotic medullary interstitial fluid.

Countercurrent multiplication

Countercurrent multiplication is the process whereby a modest gradient established at any level of the loop of Henle is increased (multiplied) into a much larger gradient along the axis of the loop. A simplified model for countercurrent multiplication (Fig. 22.25) shows how this works. Initially the loop is filled with fluid isosmotic to plasma (~300 mOsm/kg H_2O) (Fig. 22.25A). Next, we assume that at any level of the loop, the loop can establish an osmotic gradient of 200 mOsm/kg H_2O (Fig. 22.25B). This so-called single effect occurs by active transport of solute (salt) out of the water-impermeable ascending limb and deposition of the salt in the tiny interstitial space; water leaves the water-permeable descending limb and equilibrates osmotically with the interstitial space. Next, we add new fluid to the loop and push the fluid in the loop around the bend (Fig. 22.25C). We repeat the single effect (Fig. 22.25D) and continue the process (Fig. 22.25E–H). The final result (see Fig. 22.25H) is a much larger gradient (~400 mOsm/kg H_2O) along the axis of the loop. The interstitial space shares this axial gradient and in the kidney the highest osmolalities are reached at the bends of the longest loops of Henle, i.e., deep within the inner medulla or, in other words, at the tips of the renal papillae.

The extent to which countercurrent multiplication can establish a large axial gradient in a model of the kidney depends on several factors, including the magnitude of the single effect, the rate of fluid flow, and the length of the loop. The larger the single effect, the larger the axial gradient. If flow rate through the loop is too high, not enough time is allowed for establishing a significant single effect, and consequently, the axial gradient is reduced. Finally, if the loops are long, there is more opportunity for multiplication, and a larger axial gradient can be established.

Countercurrent multiplication is an energy-demanding process. To build an osmotic gradient requires work. In the kidney outer medulla, the energy is derived from ATP, which powers active pumping of Na^+ by the Na^+/K^+-ATPase in thick ascending limbs. The energy source for building the gradient in the inner medulla is less well understood.

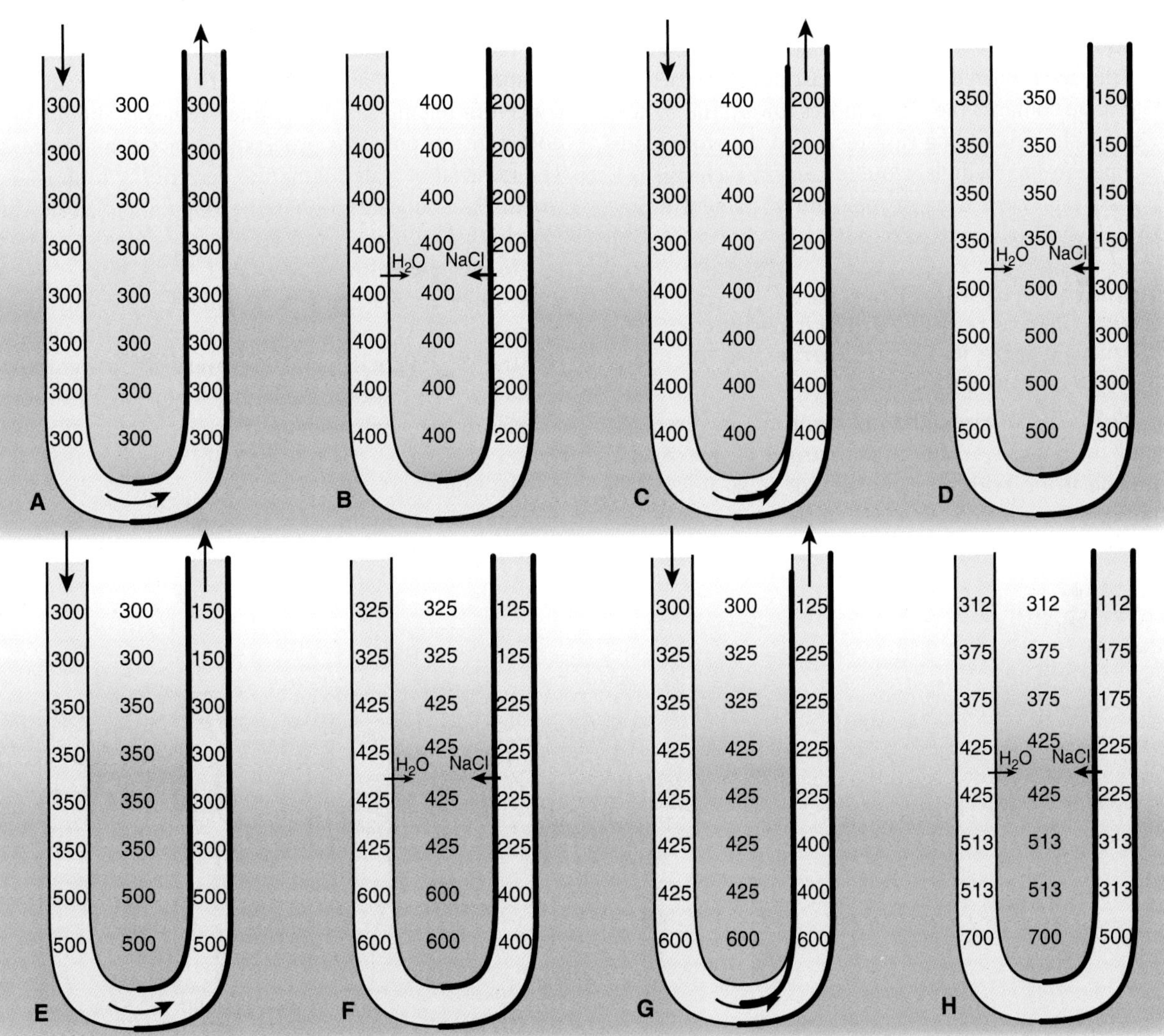

Figure 22.25 Model for countercurrent multiplication. The numbers represent osmolality (mOsm/kg H_2O) of tubule fluid and interstitium. The establishment of an osmotic gradient along the vertical axis of the loop (or with increasing depth in the medulla) is viewed as the resultant of two successive processes: (1) a shift of fluid within the loop (**A, C, E,** and **G**) and (2) development of an osmotic gradient of 200 mOsm/kg H_2O at any horizontal level of the loop, the so-called "single effect" (**B, D, F,** and **H**). The single effect involves active transport of solute (mostly NaCl) out of the ascending limb across a water-impermeable barrier (the latter is indicated by heavy outlining along the ascending limb of the loop of Henle) into a tiny interstitial space and osmotic withdrawal of water from the water-permeable descending limb. The single effect of 200 mOsm/kg H_2O is multiplied (magnified) into a larger (~400 mOsm/kg H_2O) gradient along the length of the loop by a stepwise shift of fluid, countercurrent flow, and repetition of the single effect. The interstitium of the medulla shares in the increased osmolality.

Countercurrent exchange

Countercurrent exchange is a common process in the vascular system. In many vascular beds, arterial and venous vessels lie close to each other, and exchanges of heat or materials can occur between these vessels. For example, because of the countercurrent exchange of heat between blood flowing toward and away from its feet, a penguin can stand on ice and yet maintain a warm body (core) temperature. Countercurrent exchange between descending and ascending vasa recta in the kidney reduces dissipation of the solute gradient in the medulla. The descending vasa recta tend to give up water to the more concentrated interstitial fluid. The ascending vasa recta, which come from more concentrated regions of the medulla, take up this water. In effect, then, much of the water in the blood short-circuits across the tops of the vasa recta and does not flow deep into the medulla, where it would tend to dilute the accumulated solute. The ascending vasa recta tend to give up solute as the blood moves toward the cortex. Solute enters the descending vasa recta and therefore tends to be trapped in the medulla. Countercurrent exchange is a purely passive process; it helps maintain a gradient established by some other means.

Mechanism for generating urine hyperosmolarity requires the integrated functioning of the loops of Henle, vasa recta, and collecting ducts.

Figure 22.26 summarizes the mechanisms involved in producing osmotically concentrated urine. A maximally concentrated urine, with an osmolality of 1,200 mOsm/kg H_2O and a low urine volume (0.5% of the original filtered water), is being excreted.

About 70% of filtered water is reabsorbed along the proximal convoluted tubule, and so 30% of the original filtered volume enters the loop of Henle. As discussed earlier, proximal reabsorption of water is essentially an isosmotic process, so fluid entering the loop is isosmotic to plasma. As the fluid moves along the descending limb of the loop of Henle in the medulla, it becomes increasingly concentrated. This rise in osmolality could in principle be a result of two processes:

1. The movement of water out of the descending limb because of the hyperosmolality of the medullary interstitial fluid.
2. The entry of solute from the concentrated medullary interstitial fluid.

The relative importance of these processes may depend on the species of animal. For the most efficient operation of the concentrating mechanism, water removal should be predominant, so only this process is depicted in Figure 22.26. The removal of water along the descending limb leads to a rise in NaCl concentration in the loop fluid to a value higher than in the interstitial fluid.

In the ascending limb, NaCl is reabsorbed across a water-impermeable epithelium, and the NaCl is deposited in the medullary interstitial fluid. In the thick ascending limb, Na^+ transport is active and is powered by a vigorous Na^+/K^+-ATPase. In the thin ascending limb, NaCl reabsorption appears to be mainly passive. It occurs because the concentration of NaCl in the tubular fluid is higher than that in the interstitial fluid and because the passive permeability of the thin ascending limb to Na^+ is high. There is also some evidence for a weak active Na^+ pump in the thin ascending

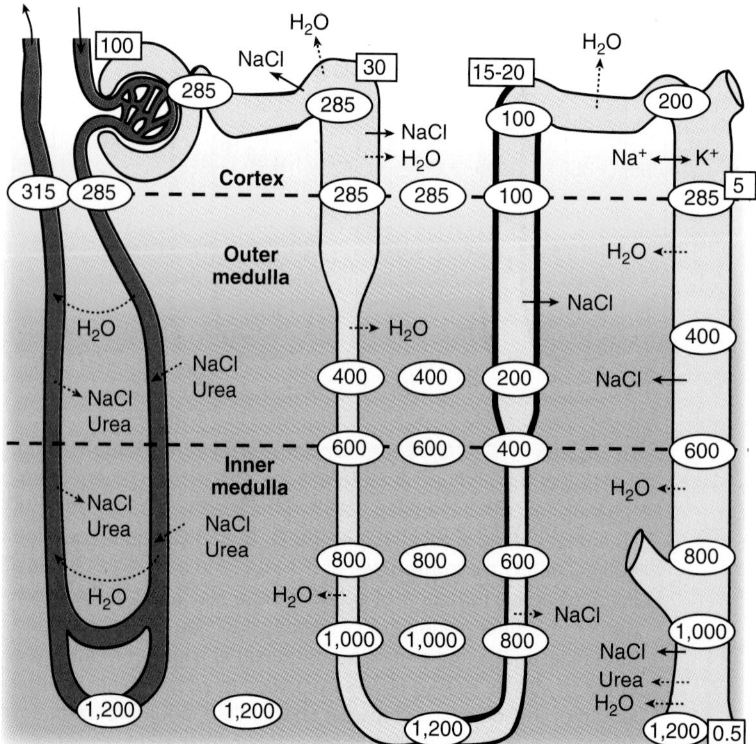

Figure 22.26 Formation of osmotically concentrated urine. This diagram summarizes movements of ions, urea, and water in the kidney during production of maximally concentrated urine (1,200 mOsm/kg H_2O). Numbers in *ovals* represent osmolality in mOsm/kg H_2O. Numbers in *boxes* represent relative amounts of water present at each level of the nephron. *Solid arrows* indicate active transport; *dashed arrows* indicate passive transport. The *heavy outlining* along the ascending limb of the loop of Henle indicates relative water-impermeability.

limb. The net addition of solute to the medulla by the loops is essential for the subsequent osmotic concentration of urine in the collecting ducts.

Fluid entering the distal convoluted tubule is hypo-osmotic compared with plasma (see Fig. 22.26) because of the removal of solute without water along the ascending limb. In the presence of AVP, the cortical collecting ducts become water permeable and water is passively reabsorbed into the cortical interstitial fluid. The high blood flow to the cortex rapidly carries away this water, and so there is no detectable lowering of cortical tissue osmolality. Before the tubular fluid re-enters the medulla, it is isosmotic, and its volume is only about 5% of the original filtered volume. The reabsorption of water in the cortical collecting ducts is important for the overall operation of the urinary concentrating mechanism. If this water were not reabsorbed in the cortex, an excessive amount would enter the medulla. It would tend to wash out the gradient in the medulla, leading to an impaired ability to concentrate the urine maximally.

All nephrons drain into collecting ducts that pass through the medulla. In the presence of AVP, the medullary collecting ducts are permeable to water. Water moves out of the collecting ducts into the more concentrated interstitial fluid. At high levels of AVP, the fluid equilibrates with the interstitial fluid, and the final urine becomes as concentrated as the tissue fluid at the tips of the papillae.

Many different models for the countercurrent mechanism have been proposed; each must take into account the principle of conservation of matter (mass balance). In the steady state, the inputs of water and every nonmetabolized solute must equal their respective outputs. This principle must be obeyed at every level of the medulla. Figure 22.27 presents a simplified scheme that applies the mass balance principle to the medulla as a whole. It provides some additional insights into the countercurrent mechanism. Notice that fluids entering the medulla (from the proximal tubule, descending vasa recta, and cortical collecting ducts) are isosmotic; they all have an osmolality of about 285 mOsm/kg H_2O. Fluid leaving the medulla in the urine is hyperosmotic. It follows from mass balance considerations that somewhere a hypo-osmotic fluid has to leave the medulla; this occurs in the ascending limb of the loop of Henle.

The input of water into the medulla must equal its output. Because water is added to the medulla along the descending limbs of the loops of Henle and the collecting ducts, this water must be removed at an equal rate. The ascending limbs of the loops of Henle cannot remove the added water because they are water-impermeable. The water is removed by the vasa recta; this is why blood flow in the ascending vasa recta exceeds blood flow in the descending vasa recta (see Fig. 22.27). Blood leaving the medulla is hyperosmotic because it drains a region of high osmolality.

Urea plays a special role in the concentrating mechanism.

It has been known for many years that animals or humans on a low-protein diet have an impaired ability to concentrate

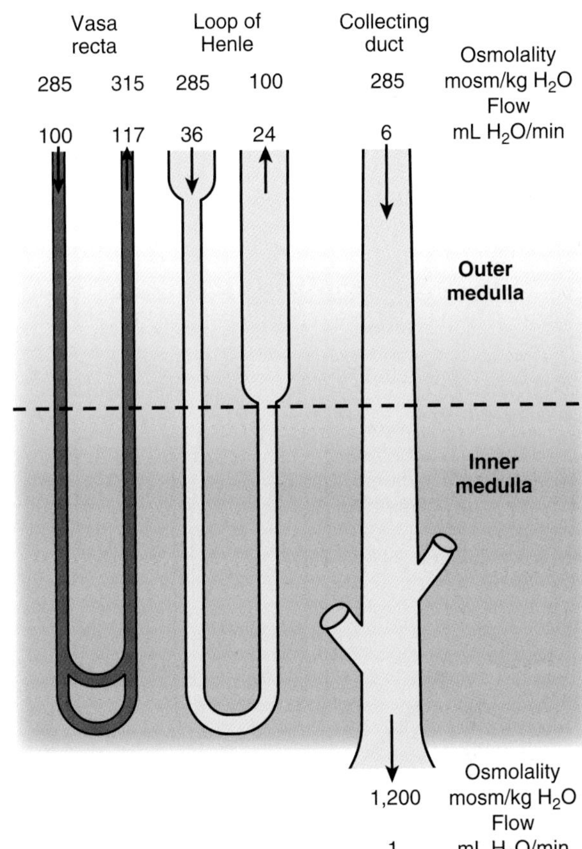

Figure 22.27 Mass balance considerations for the medulla as a whole. In the steady state, the inputs of water and solutes must equal their respective outputs. Water input into the medulla from the cortex (100 + 36 + 6 = 142 mL/min) equals water output from the medulla (117 + 24 + 1 = 142 mL/ min). Solute input (28.5 + 10.3 + 1.7 = 40.5 mOsm/min) is likewise equal to solute output (36.9 + 2.4 + 1.2 = 40.5 mOsm/min).

the urine maximally. A low-protein diet is associated with a decreased urea concentration in the kidney medulla.

Figure 22.28 shows how urea is handled along the nephron. The proximal convoluted tubule is fairly permeable to urea and reabsorbs about 50% of the filtered urea. Fluid collected from the distal convoluted tubule, however, has as much urea as the amount filtered. Therefore, urea is secreted in the loop of Henle.

The thick ascending limb, distal convoluted tubule, connecting tubule, cortical collecting duct, and outer medullary collecting duct are relatively impermeable to urea. The urea concentration rises as water is reabsorbed along cortical and outer medullary collecting ducts. The result is the delivery to the inner medulla of a concentrated urea solution.

The inner medullary collecting duct has a facilitated urea transporter, which is activated by AVP and favors urea diffusion into the interstitial fluid of the inner medulla. Urea may reenter the loop of Henle and be recycled (see Fig. 22.28), thereby building up its concentration in the inner medulla. Urea is also added to the inner medulla by diffusion from the urine surrounding the papillae (calyceal urine). Urea accounts for about half of

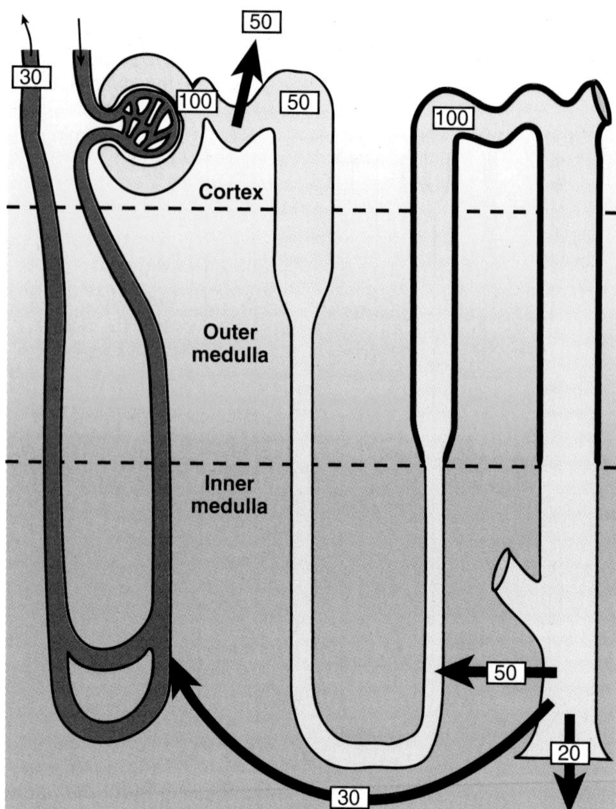

Figure 22.28 Movements of urea along the nephron. The numbers indicate relative amounts (100 = filtered urea), not concentrations. The heavy outline from the thick ascending limb to the outer medullary collecting duct indicates relatively urea-impermeable segments. Urea is added to the inner medulla by its collecting ducts; most of this urea reenters the loop of Henle, and the vasa recta remove some.

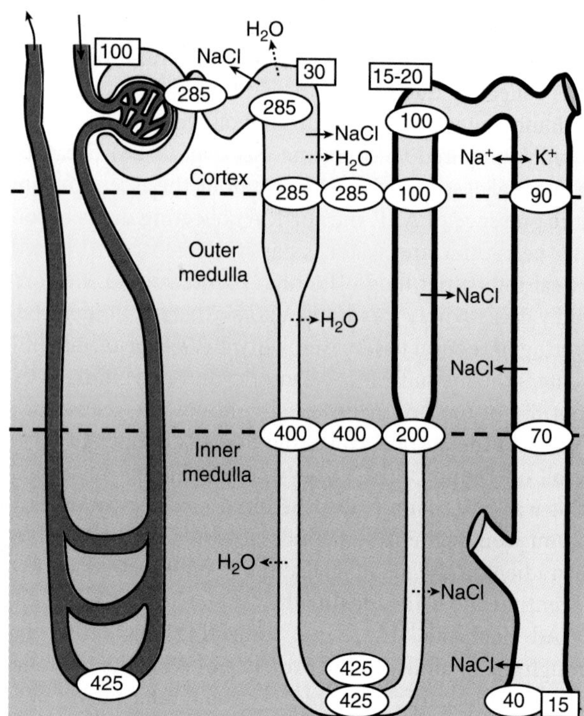

Figure 22.29 Osmotic gradients during excretion of osmotically dilute urine. The collecting ducts are relatively water-impermeable (heavy outlining) because arginine vasopressin (AVP) is absent. The medulla is still hyperosmotic, but less so than in a kidney producing osmotically concentrated urine.

the osmolality in the inner medulla. The urea in the interstitial fluid of the inner medulla counterbalances urea in the collecting duct urine, allowing the other solutes (e.g., NaCl) in the interstitial fluid to counterbalance osmotically the other solutes (e.g., creatinine and various salts) that need to be concentrated in the urine. This enhances the urinary concentrating ability and allows urea to be excreted with less water.

Dilute urine is excreted when plasma AVP levels are low.

Figure 22.29 depicts osmolalities during excretion of dilute urine, as occurs when plasma AVP levels are low. Tubular fluid is diluted along the ascending limb and becomes even more dilute as solute is reabsorbed across the relatively water-impermeable distal portions of the nephron and collecting ducts. Because as much as 15% of filtered water is not reabsorbed, a high urine flow rate results. In these circumstances, the osmotic gradient in the medulla is reduced but not abolished. The decreased gradient results from several factors. First, medullary blood flow is increased, which tends to wash out the osmotic gradient. Second, less urea is added to the inner medulla interstitium because the urea in the collecting duct urine is less concentrated than usual and there is less of a concentration gradient for passive reabsorption of urea. Furthermore, the inner medulla collecting ducts are less permeable to urea when AVP levels are low. Finally, because of diminished water reabsorption in the cortical collecting ducts, too much water may enter and be reabsorbed in the medulla, lowering its osmotic gradient.

INHERITED DEFECTS IN KIDNEY TUBULE EPITHELIAL CELLS

Recent studies have elucidated the molecular basis of several inherited kidney disorders. In many cases, the normal and mutated molecules have been cloned and sequenced. It appears that inherited defects in kidney tubule carriers, ion channels, receptors, or other molecules may explain the disturbed physiology of these conditions.

Table 22.3 lists a few of these inherited disorders. Specific molecular defects have been identified in the proximal tubule (renal glucosuria and cystinuria), thick ascending limb (**Bartter syndrome**), distal convoluted tubule (**Gitelman syndrome**), and collecting duct (**Liddle syndrome**, pseudohypoaldosteronism, distal renal tubular acidosis, nephrogenic diabetes insipidus, nephrogenic syndrome of inappropriate antidiuresis). Although these disorders are rare, they shed light on the pathophysiology of disease in general. For example, the finding that increased epithelial Na^+ channel activity in Liddle syndrome

TABLE 22.3 Inherited Defects in Kidney Tubule Epithelial Cells

Condition	Molecular Defect	Clinical Features
Renal glucosuria	Na^+-dependent glucose cotransporter	Glucosuria, polyuria, polydipsia, polyphagia
Cystinuria	Amino acid transporter	Kidney stone disease
Bartter syndrome	Na-K-2Cl cotransporter, K channel, Cl channel, or barttin (recruits Cl channel to basolateral membrane) in thick ascending limb	Salt wasting, hypokalemic metabolic alkalosis
Gitelman syndrome	Thiazide-sensitive Na–Cl cotransporter in distal convoluted tubule	Salt wasting, hypokalemic metabolic alkalosis, hypocalciuria
Liddle syndrome	Increased open time and number of principal cell epithelial sodium channels	Hypertension, hypokalemic metabolic alkalosis
Pseudohypoaldosteronism	Decreased activity of epithelial sodium channels or defective mineralocorticoid receptor	Salt wasting, hyperkalemic metabolic acidosis
Distal renal tubular acidosis	α-Intercalated cell Cl^-/HCO_3^- exchanger, H^+-ATPase	Metabolic acidosis, osteomalacia
Nephrogenic diabetes insipidus	Vasopressin-2 (V_2) receptor or aquaporin-2 deficiency	Polyuria, polydipsia
Nephrogenic syndrome of inappropriate antidiuresis	Increased vasopressin-2 (V_2) receptor activity	Hyponatremia, inappropriately elevated urine osmolality

is associated with hypertension strengthens the view that excessive salt retention leads to high blood pressure. Recent studies in patients have identified defects in voltage-gated potassium channels in kidney tubule cells that can lead to abnormal excretion of sodium, chloride, and magnesium ions.

From Bench to Bedside / 22.1

Polycystic Kidney Disease

Polycystic kidney disease (PKD) is a disorder, usually inherited, in which numerous cysts develop in both kidneys. The cysts are fluid-filled epithelial sacs that arise from nephrons or collecting ducts. The growth of cysts can produce massive enlargement of the kidneys and ultimately complete renal failure. PKD is the most common of all life-threatening genetic diseases and affects 600,000 Americans and millions more worldwide.

PKD in people may be produced by several genes. The most common (about 85% of patients) and more severe form is autosomal dominant PKD1 (ADPKD1) and is due to a defective gene on chromosome 16. The PKD1 gene encodes a large transmembrane protein called *polycystin-1*. Its extracellular domain contains a variety of protein motifs, which may serve as targets (receptors) for mostly unknown ligands. The PKD2 gene, which accounts for about 10% to 15% of patients, encodes a transmembrane protein called *polycystin-2*, which behaves as a Ca^{2+}-permeable nonselective cation channel. Polycystins-1 and -2 interact with each other and form a cell-signaling complex that transfers information from the cell's environment to intracellular signals that affect cell behavior. Both polycystins are associated with the primary cilium that is present in most renal tubule cells. The primary cilium is a nonmotile sensory process located at the apical side of the cell that senses changes in tubule fluid flow. In many cystic kidney diseases, the primary cilium is poorly developed, suggesting a link between abnormalities in this structure and the cystic phenotype. Genetic testing for ADPKD1 and ADPKD2 is now available.

Autosomal recessive PKD (ARPKD), which occurs in about 1:10,000 to 1:40,000 live births, results in a high infant mortality. The defective gene (*PKHD1*) is on chromosome 6 and produces a protein called *fibrocystin* or *polyductin*, which is found in cilia, plasma membranes, and the cytoplasm. There are, additionally, other inherited and acquired forms of renal cystic disease.

The phenotypic expression of ADPKD is quite variable. Some people show symptoms in childhood, whereas others may lead a long and healthy life with cystic kidneys detected only upon autopsy. The usual pattern is for patients to develop symptoms (such as hypertension or pain in the back and sides) in their 30s and 40s. Extrarenal manifestations, such as hepatic cysts, are common. End-stage renal failure (requiring dialysis or a kidney transplant) occurs in about 50% of patients by the age of 60 years. The specific gene affected, the nature of the mutation in a particular gene, the genetic background of an individual, and environmental factors may all play a role in determining development of the disease.

(*Continued*)

Polycystic Kidney Disease (*Continued*)

It is generally believed that only a minority (about 1%) of nephrons produce cysts in ADPKD, even though every kidney cell has a mutant copy of a dominant gene. The "two-hit hypothesis" explains why the disease is so variable and why only relatively few nephrons produce cysts. According to this idea, the production of a cyst requires a second (somatic) mutation, so that the gene (allele) accompanying the inherited defective gene is also abnormal. Only when this happens does a cyst develop. Researchers have demonstrated that in many cysts in human ADPKD, two abnormal genes are indeed present.

An elevated blood pressure is a common finding in patients with ADPKD and is associated with a faster progression of renal disease in general and increased cardiovascular mortality. Angiotensin-converting enzyme inhibitors are currently recommended to control blood pressure and reduce urinary protein excretion in PKD patients. Whether this treatment is really effective in slowing the progression of ADPKD is currently being investigated in a large-scale clinical trial.

It is possible that dietary manipulations might affect the course of the disease. Studies on rats with PKD had demonstrated that a low protein intake results in improved kidney function. In a large-scale, multicenter, randomized clinical trial (the Modification of Diet in Renal Disease Study), patients with ADPKD were provided with a reduced protein intake, but no beneficial effect could be demonstrated. The relatively short duration of this study (patients were studied for an average of 2.2 years) and the fact that treatment was started at a relatively late stage of the disease may have resulted in these disappointing results. In rats with PKD, a diet containing soy protein (instead of animal protein), flaxseed oil, or citrate (an alkalinizing diet) leads to greatly improved kidney function; whether these treatments would benefit people with PKD is not known.

Researchers have succeeded in ameliorating PKD in animals with the use of epidermal growth hormone receptor tyrosine kinase inhibitors, c-myc antisense RNA, vasopressin-2 receptor antagonists, and mTOR (mammalian target of rapamycin) inhibitors. Vasopressin stimulates the production of cyclic adenosine monophosphate (cAMP), a second messenger that stimulates cyst fluid secretion and cyst cell proliferation. mTOR is a serine/threonine kinase that regulates cell growth, proliferation, motility, and survival. Current active clinical trials include studies on the effects of vasopressin-2 receptor antagonists, long-lasting somatostatin analogues (drugs that reduce cAMP levels in kidney and liver), and mTOR inhibitors. One of the difficulties in studying treatments for ADPKD in patients is the long time-course of the disease (decades). Clinical investigators have recently shown that changes in kidney volumes determined by magnetic resonance imaging are a good short-term marker for the progression of the disease; this should enable more clinical treatment trials in the future.

Further information on PKD can be obtained from the PKD Foundation and its Web site: http://www.pkdcure.org.

Chapter Summary

- The formation of urine involves glomerular filtration, tubular reabsorption, and tubular secretion.
- The renal clearance of a substance is equal to its rate of excretion divided by its plasma concentration.
- Inulin clearance provides the most accurate measure of glomerular filtration rate.
- The clearance of *p*-aminohippurate is equal to the effective renal plasma flow.
- The rate of net tubular reabsorption of a substance is equal to its filtered load minus its excretion rate. The rate of net tubular secretion of a substance is equal to its excretion rate minus its filtered load.
- The kidneys, especially the cortex, have a high blood flow.
- Kidney blood flow is autoregulated; it is also profoundly influenced by nerves and hormones.
- The glomerular filtrate is an ultrafiltrate of plasma.
- Glomerular filtration rate is determined by the glomerular ultrafiltration coefficient, glomerular capillary hydrostatic pressure, hydrostatic pressure in the space of the Bowman capsule, and glomerular capillary colloid osmotic pressure.
- The proximal convoluted tubule reabsorbs about 70% of filtered Na^+, K^+, and water and nearly all of the filtered glucose and amino acids. It also secretes many organic anions and organic cations.
- The transport of water and most solutes across tubular epithelia is dependent on active reabsorption of Na^+.
- The thick ascending limb is a water-impermeable segment that reabsorbs Na^+ via a Na-K-2Cl cotransporter in the apical cell membrane and a vigorous Na^+/K^+-ATPase in the basolateral cell membrane.
- The distal convoluted tubule epithelium is water impermeable and reabsorbs Na^+ via a thiazide-sensitive apical membrane Na-Cl cotransporter.
- Cortical collecting duct principal cells reabsorb Na^+ and secrete K^+.
- The kidneys save water for the body by producing urine with a total solute concentration (i.e., osmolality) greater than that of plasma.
- The loops of Henle are countercurrent multipliers; they set up an osmotic gradient in the kidney medulla. Vasa recta are countercurrent exchangers; they passively help maintain the medullary gradient. Collecting ducts are osmotic equilibrating devices; they have a low water permeability, which is increased by arginine vasopressin.
- Genetic defects in kidney epithelial cells account for several disorders.

thePoint *Visit http://thePoint.lww.com for chapter review Q&A, case studies, animations, and more!*

23 Regulation of Fluid and Electrolyte Balance

ACTIVE LEARNING OBJECTIVES

Upon mastering the material in this chapter, you should be able to:

- Discuss the major fluid compartments of the body and their relative volumes for an average young adult man and woman. Explain the indicator–dilution principle and its application to the measurement of body fluid volumes.
- Describe the ionic composition of extracellular and intracellular fluids.
- Describe how arginine vasopressin and thirst regulate water balance.
- Compare the amounts of sodium filtered and excreted in a day. State the percentage of filtered sodium reabsorbed by the proximal convoluted tubule, loop of Henle, distal convoluted tubule, and collecting ducts.
- Explain how the following affect renal sodium excretion: glomerular filtration rate, the renin–angiotensin–aldosterone system, intrarenal pressure, natriuretic hormones and factors, renal sympathetic nerves, estrogens, glucocorticoids, osmotic diuretics, poorly reabsorbed anions, and diuretic drugs.
- Explain the relation between sodium chloride, extracellular fluid volume (or effective arterial blood volume), and blood pressure.
- Discuss the importance, amount, and distribution of potassium in the body.
- Discuss the factors that affect renal excretion of potassium.
- Explain how the kidneys keep us in calcium, magnesium, and phosphate balance. State the effects of the parathyroid hormone on renal tubular reabsorption of calcium and phosphate and on renal synthesis of $1,25(OH)_2$ vitamin D_3.
- Describe the genesis and events of the micturition reflex.

A major function of the kidneys is a homeostatic mechanism involved in regulating electrolyte concentration, extracellular fluid (ECF) volume, osmolality, and acid–base balance. The kidney accomplishes these homeostatic functions both independently and in concert with other organs, especially the endocrine system. Endocrine hormones involved in coordinating these functions include renin, angiotensin II, aldosterone, antidiuretic hormone (ADH), and atrial natriuretic peptide (ANP).

The kidney is a remarkable filtration organ. The kidneys generate approximately 180 L of filtrate/d. Although a large percentage is retained, about 2 L of urine is excreted. The kidney accomplishes this by relatively simple mechanisms of filtration, reabsorption, and secretion, which take place in the nephron. This chapter discusses fluid compartments of the body (i.e., their location, size, and composition), water and electrolyte balance (specifically, sodium, potassium, calcium, magnesium, and phosphate balance), and finally the role of the ureters, urinary bladder, and urethra in the transport, storage, and elimination of urine.

FLUID COMPARTMENTS OF THE BODY

There are two main fluid compartments in the human body: **intracellular fluid (ICF)** and **extracellular fluid (ECF)**. The ICF compartment makes up approximately 60% to 67% of body water, and the ECF compartment makes up the other 33% to 40%. Water is the major solvent of all body fluid compartments. **Total body water** averages about 60% of body weight in young adult men and about 50% of body weight in young adult women (Table 23.1). The percentage of body weight that water occupies depends on the amount of adipose tissue (fat) a person has. A lean person has a high percentage of body weight that is water, and an obese person a low one, because adipose tissue contains a low percentage of water (about 10%), whereas most other tissues contain a much higher percentage of water. For example, muscle is about 75% water. Newborns have a high percentage of body weight as water because of a relatively large ECF volume and little fat (see Table 23.1). Adult women have relatively less water than men, because, on average, they have more

TABLE 23.1 Average Total Body Water as a Percentage of Body Weight

Age	Men (%)	Both Sexes (%)	Women (%)
0–1 mo		76	
1–12 mo		65	
1–10 y		62	
10–16 y	59		57
17–39 y	61		50
40–59 y	55		47
60 y and over	52		46

subcutaneous fat and less muscle mass. As people age, they tend to lose muscle and add adipose tissue; hence, water content declines with age.

Body water is distributed between intracellular fluid and extracellular fluid compartments.

Total body water can be divided into ICF and ECF. ICF is found inside the bilayered cell plasma membrane and is the medium in which cellular organelles are suspended and where many chemical reactions take place. Under normal circumstances, ICF is in osmotic equilibrium with the ECF.

In a young adult man, two thirds of the body water is in the ICF and one third is in the ECF (Fig. 23.1). These two fluids differ strikingly in terms of their electrolyte composition. Their total solute concentrations (osmolalities), however, are normally equal, because of the high water permeability of most cell membranes, and so an osmotic difference between cells and the ECF rapidly disappears.

The ECF is composed of fluid outside of the cells and consists of three subdivisions: the *interstitial compartment*, the *plasma compartment*, and the *third space*. The *interstitial compartment* is the fluid space that surrounds the cells of a given tissue. It is filled with interstitial fluid and comprises about three fourths of the ECF. Interstitial fluid provides the microenvironment that allows for movement of ions, proteins, and nutrients across the cell membrane. The fluid is not static, but is continuously being turned over and recollected by the lymphatic channels. When excess fluid accumulates in the interstitial compartment, **edema** develops. The *intravascular plasma* found within the vascular system and comprises approximately one fourth of the ECF, or about 3.5 L for the average 70-kg male (see Fig. 23.1). The *third space* is space where fluids do not normally collect in large amounts. The third space is part of the ECF compartment and is often referred to as **transcellular fluid**. Major examples of third spaces include the peritoneal fluid, pleural fluid, cerebrospinal fluid, fluid within the digestive tract, aqueous humor of the eye, synovial fluid, and renal tubular fluid. Although the volume is small and normally ignored in calculations (not shown in Fig. 23.1), it is physiologically important (i.e., it functions as a lubricant in the pleural space and the fluid surrounding joints). There is a constant turnover of transcellular fluids; they are continuously formed and absorbed or removed. Impaired formation, abnormal loss from the body, or blockage of fluid removal can have serious consequences (e.g., hydrocephalus).

Indicator–dilution method measures fluid compartment size.

The **indicator–dilution method** can be used to determine the size of body fluid compartments. A known amount of a substance (the indicator), which should be confined to the compartment of interest, is administered. After allowing sufficient time for uniform distribution of the indicator throughout the compartment (total body water, ECF, or plasma), a plasma sample is collected. At equilibrium, the concentration of the indicator will be the same in the entire compartment, including the plasma. The plasma concentration is measured, and the distribution volume is calculated from this formula:

$$\text{Volume} = \frac{\text{Amount of indicator}}{\text{Concentration of indicator}} \quad (1)$$

If indicator was lost from the fluid compartment, the amount lost is subtracted from the amount administered.

To measure total body water, heavy water (deuterium oxide [D_2O]), tritiated water (HTO), or antipyrine (a drug that distributes throughout all of the body water) is used as an indicator. For example, suppose we want to measure total body water in a 60-kg woman. We inject 30 mL of D_2O as an isotonic saline solution into an arm vein. After a 2-hour equilibration period,

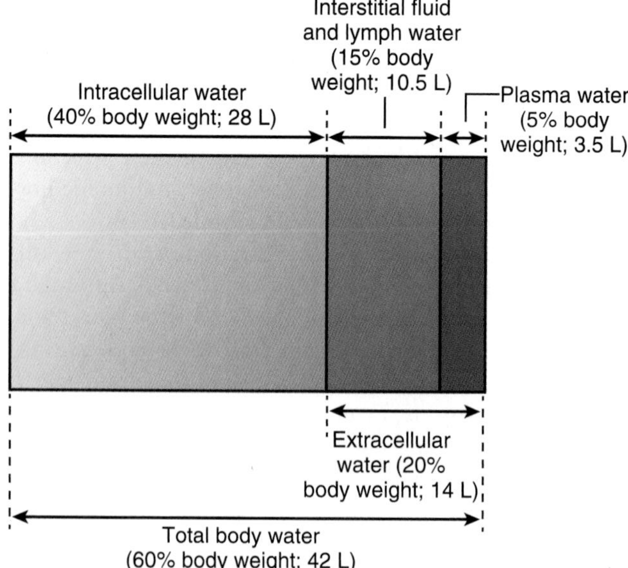

Figure 23.1 Distribution of water in the body. This diagram is for an average young adult man weighing 70 kg. In an average young adult woman, total body water is 50% of body weight, intracellular water is 30% of body weight, and extracellular water is 20% of body weight.

a blood sample is withdrawn, and the plasma is separated and analyzed for D_2O. A concentration of 0.001 mL D_2O/mL plasma water is found. Suppose that, during the equilibration period, urinary, respiratory, and cutaneous losses of D_2O are 0.12 mL. Substituting these values into the indicator–dilution equation, we get: Total body water = (30 – 0.12 mL D_2O) ÷ 0.001 mL D_2O/mL water = 29,880 mL or 30 L. Therefore, total body water as a percentage of body weight equals 50% in this woman.

To measure extracellular water volume, the ideal indicator should distribute rapidly and uniformly outside the cells and should not enter the cell compartment. Unfortunately, there is no such ideal indicator, and so the exact volume of the ECF cannot be measured. A reasonable estimate, however, can be obtained using two different classes of substances: impermeant ions and inert sugars. ECF volume has been determined from the volume of distribution of these ions: radioactive Na^+, radioactive Cl^-, radioactive sulfate, thiocyanate (SCN^-), and thiosulfate ($S_2O_3^{2-}$)—of which radioactive sulfate ($^{35}SO_4^{2-}$) is probably the most accurate. However, ions are not completely impermeant; they slowly enter the cell compartment, and so measurements tend to lead to an overestimate of ECF volume. Measurements with inert sugars (such as mannitol, sucrose, and inulin) tend to lead to an underestimate of ECF volume, because they are excluded from some of the extracellular water—for example, the water in dense connective tissue and cartilage. Special techniques are required when using these sugars, because the kidneys rapidly filter and excrete them after their intravenous injection.

Cellular water cannot be determined directly with any indicator. It can, however, be calculated from the difference between measurements of total body water and extracellular water. Plasma water is determined by using Evans blue dye, which avidly binds serum albumin, or radioiodinated serum albumin, and by collecting and analyzing a blood plasma sample. In effect, the plasma volume is measured from the distribution volume of serum albumin. The assumption is that serum albumin is completely confined to the vascular compartment, but this is not entirely true. Indeed, serum albumin is slowly (3%–4% per hour) lost from the blood by diffusive and convective transport through capillary walls. To correct for this loss, repeated blood samples can be collected at timed intervals, and the concentration of albumin at time zero (the time at which no loss would have occurred) can be determined by extrapolation. Alternatively, the plasma concentration of indicator 10 minutes after injection can be used; this value is usually close to the extrapolated value. If plasma volume and hematocrit are known, total circulating blood volume can be calculated (see Chapter 11, "Overview of the Cardiovascular System and Hemodynamics").

Interstitial fluid and lymph volume cannot be determined directly but can be calculated from the difference between ECF and plasma volumes.

Interstitial fluid and plasma are similar in composition, but the intra- and extracellular fluids are markedly different.

Body fluids contain many uncharged molecules (e.g., glucose and urea), but quantitatively speaking, **electrolytes** (ionized substances) contribute most to the total solute concentration (or osmolality) of body fluids. Osmolality is of prime importance in determining the distribution of water between ICF and ECF compartments.

The importance of ions (particularly Na^+) in determining the plasma osmolality (P_{osm}) is exemplified by an equation that is of value in the clinic:

$$P_{osm} = 2 \times \left[Na^+\right] + \frac{\left[\text{glucose}\right]\text{in mg/dL}}{18} + \frac{\left[\text{blood urea nitrogen}\right]\text{in mg/dL}}{2.8} \quad (2)$$

If the plasma Na^+ level is 140 mmol/L, blood glucose level is 100 mg/dL (5.6 mmol/L), and blood urea nitrogen level is 10 mg/dL (3.6 mmol/L), then the calculated osmolality is 289 mOsm/kg H_2O. The equation indicates that Na^+ and its accompanying anions (mainly Cl^- and HCO_3^-) normally account for more than 95% of the plasma osmolality. In some special circumstances (e.g., alcohol intoxication), plasma osmolality calculated from the above equation may be much lower than the measured osmolality because of the presence of solutes other than sodium salts, glucose, and urea (e.g., ethanol).

The concentrations of various electrolytes in plasma, interstitial fluid, and ICF are summarized in Table 23.2. The ICF values are based on determinations made in skeletal muscle cells. These cells account for about two thirds of the cell mass in the human body. Concentrations are expressed in terms of milliequivalents (mEq) per liter or per kilogram H_2O.

An **equivalent** contains one mole of univalent ions, and a **milliequivalent (mEq)** is 1/1,000th of an equivalent. Equivalents are calculated as the product of moles times valence and represent the concentration of charged species. For singly charged (univalent) ions, such as Na^+, K^+, Cl^-, or HCO_3^-, 1 mmol is equal to 1 mEq. For doubly charged (divalent) ions, such as Ca^{2+}, Mg^{2+}, or SO_4^{2-}, 1 mmol is equal to 2 mEq. Some electrolytes, such as proteins, are polyvalent, and so there are several mEq per mmol. The usefulness of expressing concentrations in terms of mEq/L arises from the fact that in solutions we have electrical neutrality; that is:

$$\Sigma \text{ cations} = \Sigma \text{ anions} \quad (3)$$

If we know the total concentration (mEq/L) of all cations in a solution and know the concentration of only some of the anions, we can easily calculate the concentration of the remaining anions. This was done in Table 23.2 for the anions labeled "Others."

Plasma concentrations are listed in the first column of numbers in Table 23.2. Na^+ is the major cation in plasma, and Cl^- and HCO_3^- are the major anions. The plasma proteins (mainly serum albumin) bear net negative charges at physiologic pH. The electrolytes are actually dissolved in the plasma water, and so the next column in Table 23.2 expresses concentrations per kilogram H_2O. The water content of plasma is usually about 93%; about 7% of plasma volume is occupied by solutes, mainly the plasma proteins. To convert

TABLE 23.2 Electrolyte Composition of the Body Fluids

Cation or Anion	Plasma Concentration (mEq/L)	Plasma Water Concentration (mEq/kg H_2O)	Interstitial Fluid (mEq/kg H_2O)	Intracellular Fluid[a] (mEq/kg H_2O)
Cations				
Na^+	142	153	145	10
K^+	4	4.3	4	159
Ca^{2+}	5	5.4	3	1
Mg^{2+}	2	2.2	2	40
Total	153	65	154	210
Anions				
Cl^-	103	111	117	3
HCO_3^-	25	27	28	7
Proteins	17	18	—	45
Others	8	9	9	155
Total	153	165	154	210

[a]Skeletal muscle cells.

concentration in plasma to concentration in plasma water, we divide the plasma concentration by the plasma water content (0.93 L H_2O/L plasma). Thus, 142 mEq Na^+/L plasma becomes 153 mEq/L H_2O or 153 mEq/kg H_2O (because 1 L of water weighs 1 kg).

Interstitial fluid (next column of Table 23.2) is an ultrafiltrate of plasma. It contains all of the small electrolytes in essentially the same concentration as in plasma, but little protein. The proteins are largely confined to the plasma because of their large molecular size. Differences in small ion concentrations between plasma and interstitial fluid arise because of the different protein concentrations in these two compartments. Two factors are involved. The first is an electrostatic effect: because the plasma proteins are negatively charged, they cause a redistribution of small ions, so that the concentrations of diffusible cations (such as Na^+) are lower in interstitial fluid than in plasma and the concentrations of diffusible anions (such as Cl^-) are higher in interstitial fluid than in plasma. Second, plasma proteins bind Ca^{2+} and Mg^{2+} (about 40% and 20%, respectively), and only the free (unbound) ions can diffuse through capillary walls. Hence, the total (bound plus free) Ca^{2+} and Mg^{2+} concentrations are higher in plasma than in interstitial fluid.

ICF composition (Table 23.2, last column) is different from ECF composition. The cells have higher K^+, Mg^{2+}, and protein concentrations than in the surrounding interstitial fluid. The intracellular Na^+, Ca^{2+}, Cl^-, and HCO_3^- levels are lower than outside the cell. The anions in skeletal muscle cells labeled "Others" are mainly organic phosphate compounds important in cell energy metabolism, such as creatine phosphate, adenosine triphosphate (ATP), and adenosine diphosphate (ADP). As described in Chapter 2, "Plasma Membrane, Membrane Transport, and Resting Membrane Potential," the high intracellular K^+ level and low intracellular Na^+ level are a consequence of plasma membrane Na^+/K^+-ATPase activity; this enzyme extrudes Na^+ from the cell and takes up K^+. The low intracellular Cl^- and HCO_3^- in skeletal muscle cells is primarily a consequence of the inside negative membrane potential (−90 mV), which favors the outward movement of these small negatively charged ions. The intracellular Mg^{2+} is high; most is not free but is bound to cell proteins. Intracellular Ca^{2+} is low; as discussed in Chapter 1, "Homeostasis and Cellular Signaling," the cytosolic Ca^{2+} in resting cells is about 10^{-7} M (0.0002 mEq/L). Most of the cell Ca^{2+} is sequestered in organelles, such as the sarcoplasmic reticulum in skeletal muscle.

Extracellular fluid osmolality changes cause changes in intracellular volume.

Despite the different compositions of ICF and ECF, the total solute concentration (**osmolality**) of these two fluid compartments is normally the same. ICF and ECF are in osmotic equilibrium because of the high water permeability of cell membranes, which does not permit an osmolality difference to be sustained. If the osmolality changes in one compartment, water moves so as to restore a new osmotic equilibrium (see Chapter 2, "Plasma Membrane, Membrane Transport, and Resting Membrane Potential").

The volumes of ICF and ECF depend primarily on the volume of water present in these compartments. But the latter depends on the amount of solute present and the osmolality. This fact follows from the definition of the term *concentration:* concentration = amount/volume; hence, volume = amount/concentration. The main osmotically active solute in cells is K^+; therefore, a loss of cell K^+ will cause cells to lose water and shrink (see Chapter 2, "Plasma Membrane, Membrane Transport, and Resting Membrane Potential"). The main osmotically active solute in the ECF is Na^+; therefore, a gain or loss of Na^+ from the body will cause the ECF volume to swell or shrink, respectively.

The distribution of water between intracellular and extracellular compartments changes in various circumstances. Figure 23.2 provides some examples. Total body water is divided into the two major compartments, ICF and ECF. The *y* axis represents total solute concentration, and the *x* axis represents the volume; the area of a box (concentration times volume) gives the amount of solute present in a compartment. Note that the height of the boxes is always equal, because osmotic equilibrium (equal osmolalities) is achieved.

In the normal situation (shown in Fig. 23.2A), two thirds (28 L for a 70-kg man) of total body water is in the ICF and one third (14 L) is in the ECF. The osmolality of both fluids is 285 mOsm/kg H_2O. Hence, the cell compartment contains 7,980 mOsm and the ECF contains 3,990 mOsm.

In Figure 23.2B, 2 L of pure water was added to the ECF (e.g., by drinking water). Plasma osmolality is thereby lowered, and water moves into the cell compartment along the osmotic gradient. The entry of water into the cells causes them to swell, and intracellular osmolality falls until a new equilibrium (*solid lines*) is achieved. Because 2 L of water was added to an original total body water volume of 42 L, the new total body water volume is 44 L. No solute was added, and so the new osmolality at equilibrium is (7,980 + 3,990 mOsm)/44 kg = 272 mOsm/kg H_2O. The volume of the ICF at equilibrium is calculated by solving the following equation: 272 mOsm/kg $H_2O \times$ volume = 7,980 mOsm, or 29.3 L. The volume of the ECF at equilibrium is 14.7 L. From these calculations, we conclude that two thirds of the added water ends up in the cell compartment and one third stays in the ECF. This description of events is artificial, because in reality the kidneys would excrete the added water over the course of a few hours, thereby minimizing the fall in plasma osmolality and cell swelling.

In Figure 23.2C, 2 L of isotonic saline (0.9% NaCl solution) was added to the ECF. Isotonic saline is isosmotic to plasma or ECF and, by definition, causes no change in cell volume. Therefore, all of the isotonic saline is retained in the ECF and there is no change in osmolality.

Figure 23.2D shows the effect of intravenously infusing 1 L of a 5% NaCl solution (osmolality about 1,580 mOsm/kg H_2O). All the salt stays in the ECF. The cells are exposed to a hypertonic environment, and water leaves the cells. Solutes left behind in the cells become more concentrated as water leaves. A new equilibrium will be established, with the final osmolality higher than normal but equal inside and outside of the cells. The final osmolality can be calculated from the amount of solute present (7,980 + 3,990 + 1,580 mOsm) divided by the final volume (28 + 14 + 1 L), which equals 315 mOsm/kg H_2O.

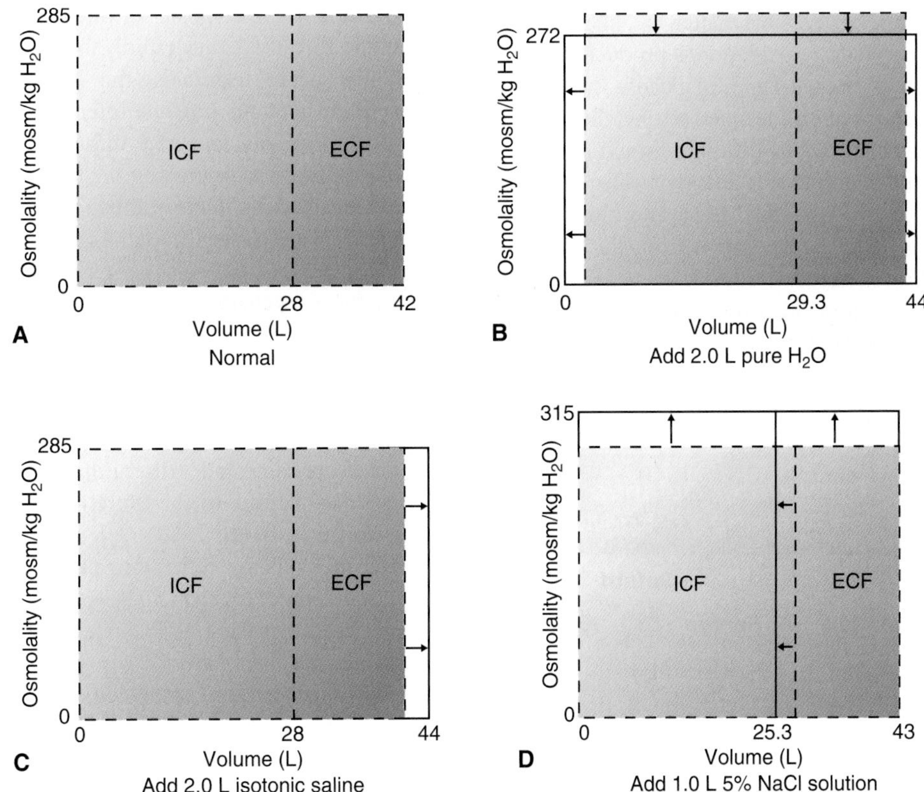

Figure 23.2 The effects of various disturbances on the osmolalities and volumes of intracellular fluid (ICF) and extracellular fluid (ECF). The *dashed lines* indicate the normal condition, and the *solid lines,* the situation after a new osmotic equilibrium has been attained. (See text for details.)

The final volume of the ICF equals 7,980 mOsm divided by 315 mOsm/kg H_2O, or 25.3 L, which is 2.7 L less than the initial volume. The final volume of the ECF is 17.7 L, which is 3.7 L more than its initial value. Thus, the addition of hypertonic saline to the ECF led to its considerable expansion in large part because of loss of water from the cell compartment.

FLUID BALANCE

Homeostatic fluid balance is dependent on maintaining a balance between input and output. Regulation of fluid balance involves the control of ECF volume and control of ECF **osmolality**. *Osmolality* is the measure of solute concentration and is defined as the number of osmoles (Osm) of solute per kilogram of water (Osm/kg), whereas *osmolarity* is a measure of the osmoles of solute per liter of solution (Osm/L). The kidneys control ECF volume by regulating salt balance, and control ECF osmolality by regulating water balance. As a result of these homeostatic mechanisms, body fluid volumes and plasma osmolality are kept remarkably constant.

To maintain a stable extracellular fluid osmolality, water input must equal its output.

A balance chart for water for an average 70-kg man is presented in Table 23.3. The person is in a stable balance (or steady state) because the total input and total output of water from the body are equal (2,500 mL/d). On the input side, water is found in the beverages we drink and in the foods we eat. Solid foods, which consist of animal or vegetable matter, are, like our own bodies, mostly water. Water of oxidation is produced during metabolism; for example, when 1 mol of glucose is oxidized, 6 mol of water are produced. In a hospital setting, the input of water resulting from intravenous infusions would also need to be considered. On the output side, losses of water occur via the skin, lungs, gastrointestinal (GI) tract, and kidneys. We always lose water by simple evaporation from the skin and lungs; this is called **insensible water loss**.

Appreciable water loss from the skin, in the form of sweat, occurs at high temperatures or with heavy exercise. As much as 4 L of water per hour can be lost in sweat.

TABLE 23.3 Daily Water Balance in an Average 70-kg Man

Input		*Output*	
Source	**Amount**	**Source**	**Amount**
Water in beverages	1,000 mL	Skin and lungs	900 mL
Water in food	1,200 mL	Gastrointestinal tract (feces)	100 mL
Water of oxidation	300 mL	Kidneys (urine)	1,500 mL
Total	2,500 mL	Total	2,500 mL

Sweat, which is a hypo-osmotic fluid, contains NaCl; excessive sweating can lead to significant losses of salt. GI losses of water are normally small (see Table 23.3), but with diarrhea, vomiting, or drainage of GI secretions, massive quantities of water and electrolytes may be lost from the body.

The kidneys are the sites of adjustment of water output from the body. Renal water excretion changes to maintain balance. If there is a water deficiency, the kidneys diminish the excretion of water and urine output falls. If there is water excess, the kidneys increase water excretion and urine flow, to remove the extra water. The renal excretion of water is controlled by **arginine vasopressin (AVP)**.

The water needs of an infant or young child, per kilogram body weight, are several times higher than those of an adult. Children have, for their body weight, a larger body surface area and higher metabolic rate. They are much more susceptible to volume depletion.

Arginine vasopressin control of water balance is critical in regulating extracellular fluid osmolality.

AVP, also known as **antidiuretic hormone (ADH)**, is a nonapeptide synthesized in the body of nerve cells located in the supraoptic and paraventricular nuclei of the anterior hypothalamus (Fig. 23.3) (see Chapter 31, "Hypothalamus and the Pituitary Gland"). The hormone travels by axoplasmic flow down the hypothalamic–neurohypophyseal tract and is stored in vesicles in nerve terminals in the median eminence and, mostly, the posterior pituitary. When the cells are brought to threshold, they rapidly fire action potentials, Ca^{2+} enters the nerve terminals, the AVP-containing vesicles release their contents into the interstitial fluid surrounding the nerve terminals, and AVP diffuses into nearby capillaries. The hormone is carried by the bloodstream to its target tissue, the collecting ducts of the kidneys, where it increases water reabsorption (see Chapter 22, "Kidney Function").

Arginine vasopressin release

Many factors increase the release of AVP, including pain, trauma, emotional stress, nausea, fainting, most anesthetics, nicotine, morphine, and angiotensin II. These conditions or agents therefore produce a decline in urine output and more concentrated urine. Ethanol and ANP decrease AVP release, leading to the excretion of a large volume of dilute urine.

The main factor controlling AVP release under ordinary circumstances is a change in plasma osmolality. Figure 23.4 shows how plasma AVP concentrations vary as a function of plasma osmolality. When plasma osmolality rises, neurons called **osmoreceptor cells**, located in the anterior hypothalamus, shrink. This stimulates the nearby neurons in the paraventricular and supraoptic nuclei to release AVP, and plasma AVP concentration rises. The result is the formation of osmotically concentrated urine. Not all solutes are equally effective in stimulating the osmoreceptor cells; for example, urea, which can enter these cells and therefore does not cause the osmotic withdrawal of water, is ineffective.

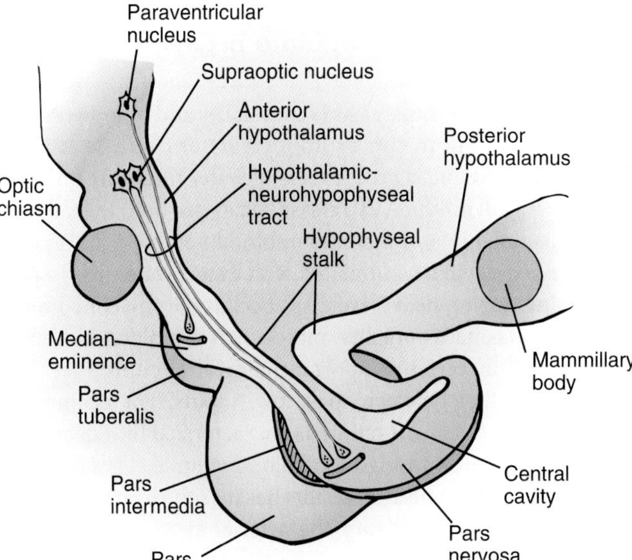

Figure 23.3 The pituitary and hypothalamus. Arginine vasopressin (AVP) is synthesized primarily in the supraoptic nucleus and to a lesser extent in the paraventricular nuclei in the anterior hypothalamus. It is then transported down the hypothalamic–neurohypophyseal tract and stored in vesicles in the median eminence and posterior pituitary (pars nervosa), where it can be released into the blood.

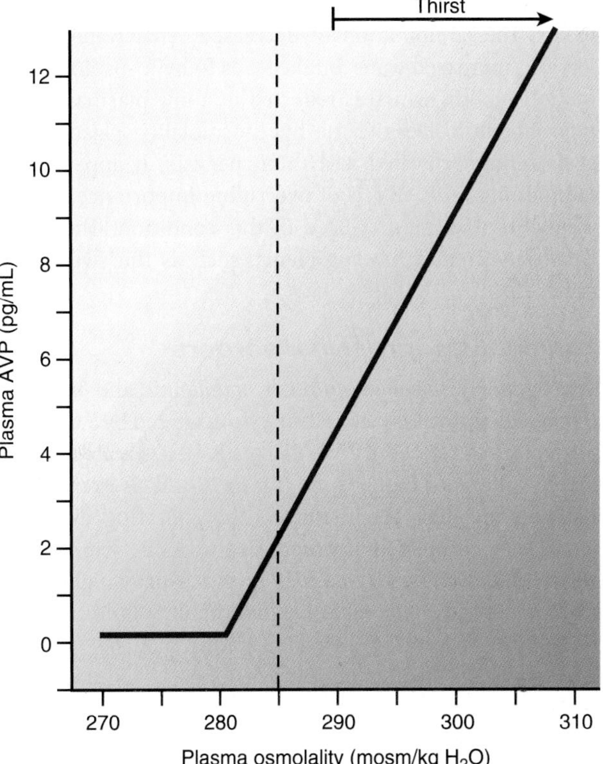

Figure 23.4 The relationship between plasma arginine vasopressin (AVP) level and plasma osmolality in healthy people. Decreases in plasma osmolality were produced by drinking water and increases by fluid restriction. Plasma AVP levels were measured by radioimmunoassay. The plasma AVP level is essentially zero at plasma osmolalities <280 mOsm/kg H_2O. Above this plasma osmolality (threshold), plasma AVP increases linearly with plasma osmolality. Normal plasma osmolality averages about 285 mOsm/kg H_2O (*dashed vertical line*), and so we live above the threshold for AVP release. The thirst threshold is attained at a plasma osmolality of 290 mOsm/kg H_2O; the thirst mechanism "kicks in" only when there is an appreciable water deficit. Changes in plasma AVP and consequent changes in renal water excretion are normally capable of maintaining a normal plasma osmolality below the thirst threshold.

Extracellular NaCl, on the other hand, is an effective stimulus for AVP release. When plasma osmolality falls in response to the addition of excess water, the osmoreceptor cells swell, AVP release is inhibited, and plasma AVP levels fall. In this situation, the collecting ducts express their intrinsically low water permeability, less water is reabsorbed, dilute urine is excreted, and plasma osmolality can be restored to normal by elimination of the excess water. Figure 23.5 shows that the entire range of urine osmolalities, from dilute to concentrated urine, is a linear function of plasma AVP in healthy people.

A second important factor controlling AVP release is the blood volume—more precisely, the **effective arterial blood volume (EABV)**. An increased blood volume inhibits AVP release, whereas a decreased blood volume (**hypovolemia**) stimulates AVP release. Intuitively, this makes sense, because with excess volume, a low plasma AVP level would promote the excretion of water by the kidneys. With hypovolemia, a high plasma AVP level would promote conservation of water by the kidneys.

The receptors for blood volume include stretch receptors in the left atrium of the heart and in the pulmonary veins within the pericardium. More stretch results in more impulses transmitted to the brain via vagal afferents and inhibition of AVP release. The common experiences of producing a large volume of dilute urine, a **water diuresis**—when lying down in bed at night, when exposed to cold weather, or when immersed in a pool during the summer—may be related to the activation of this pathway. In all these situations, an increased central blood volume stretches the atria. Arterial baroreceptors in the carotid sinuses and aortic arch also reflexly change AVP release; a fall in pressure at these sites stimulates AVP release. Finally, a decrease in renal blood flow stimulates renin release, which leads to increased angiotensin II production. Angiotensin II stimulates AVP release by acting on the brain.

Relatively large blood losses (more than 10% of blood volume) are required to increase AVP release significantly (Fig. 23.6). With a loss of 15% to 20% of total blood volume,

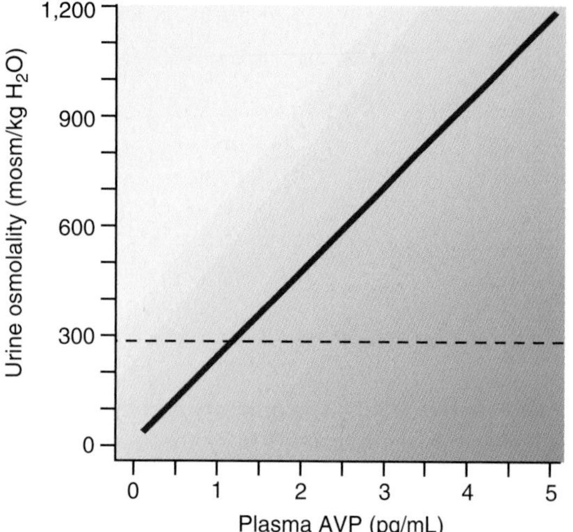

Figure 23.5 The relationship between urine osmolality and plasma arginine vasopressin (AVP) levels. The *dashed horizontal line* reports the normal plasma osmolality (285 mOsm/kg H_2O). With low plasma AVP levels, urine hypo-osmotic to plasma is excreted, and with high plasma AVP levels, hyperosmotic urine is excreted. Maximally concentrated urine (1,200 mOsm/kg H_2O) is produced when the plasma AVP level is about 5 pg/mL.

however, very large increases in plasma AVP are observed. Plasma levels of AVP may rise to levels much higher (e.g., 50 pg/mL) than are needed to concentrate the urine maximally (e.g., 5 pg/mL) (compare Figs. 23.5 and 23.6.) With severe hemorrhage, high circulating levels of AVP exert a significant vasoconstrictor effect, which helps compensate by raising the blood pressure.

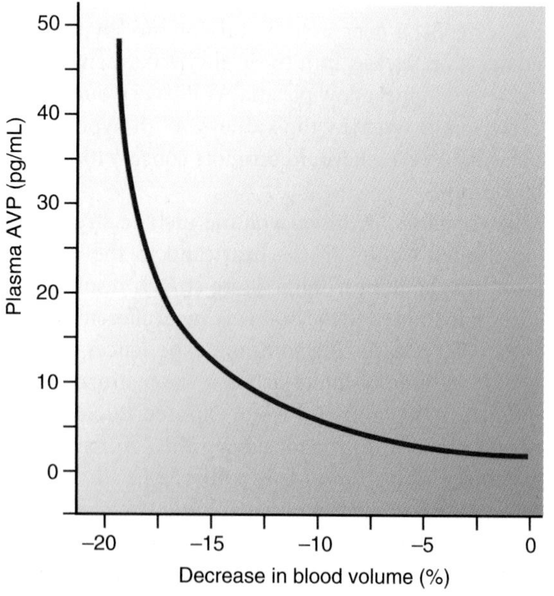

Figure 23.6 The relationship between plasma arginine vasopressin (AVP) and blood volume depletion in the rat. Severe hemorrhage (loss of 20% of total blood volume) causes a striking increase in plasma AVP. In this situation, the vasoconstrictor effect of AVP becomes significant and helps counteract the low blood pressure.

Plasma osmolality and blood volume work in concert to regulate arginine vasopressin release.

The two stimuli, plasma osmolality and blood volume, most often work in concert to increase or decrease AVP release. For example, a great excess of water intake in a healthy person will inhibit AVP release because of both the fall in plasma osmolality and increase in blood volume. Conversely, AVP release will be stimulated with excess water loss (e.g., sweating during heavy exercise) because of the concomitant rise in plasma osmolality and decrease in blood volume. However, in certain important clinical circumstances, there is a conflict between these two inputs. For example, severe congestive heart failure is characterized by a decrease in the EABV, even though total blood volume is greater than normal. This condition results because the heart does not pump sufficient blood into the arterial system to maintain adequate tissue perfusion. The arterial baroreceptors signal less volume, and AVP release is stimulated. The patient will produce osmotically concentrated urine and will also be thirsty from the decreased EABV, with consequent increased water intake. The combination of decreased renal water excretion and increased water intake leads to hypo-osmolality of the body fluids, which is reflected in a low plasma Na^+, or **hyponatremia**. Despite the hypo-osmolality, plasma AVP levels remain elevated and thirst persists. It appears that maintaining an EABV is of overriding importance, and so osmolality may be sacrificed in this condition. The hypo-osmolality creates new problems, such as the swelling of brain cells.

Arginine vasopressin disorders

Neurogenic (also called *central, hypothalamic,* and *pituitary*) **diabetes insipidus** is a condition characterized by a deficient production or release of AVP. Plasma AVP levels are low, and a large volume of dilute urine (up to 20 L/d) is excreted. In **nephrogenic diabetes insipidus**, the collecting ducts are partially or completely unresponsive to AVP. Urine output is increased, but the plasma AVP level is usually higher than normal (secondary to excessive loss of dilute fluid from the body). Nephrogenic diabetes insipidus may be acquired (e.g., via drugs such as lithium) or inherited. Mutations in the collecting duct AVP receptor gene or in the water channel (aquaporin-2) gene have now been identified in some families. In the **syndrome of inappropriate ADH (SIADH)**, plasma AVP levels are inappropriately high for the existing osmolality. Plasma osmolality is low because the kidneys form concentrated urine and save water. This condition is sometimes caused by a bronchogenic tumor that produces AVP in an uncontrolled fashion.

Habit and thirst govern water intake.

People drink water largely from habit, and this water intake normally covers a person's water needs. Most of the time, we operate below the threshold for thirst. **Thirst**, a conscious desire to drink water, is mainly an emergency mechanism that comes into play when there is a perceived water deficit. Its function is to encourage water intake to repair the water

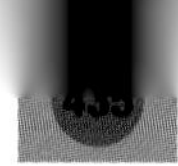

Clinical Focus / 23.1

Hyponatremia

Hyponatremia, defined as a plasma Na^+ level of <135 mEq/L, is the most common disorder of body fluid and electrolyte balance in hospitalized patients. Most often it reflects a problem of too much water, not too little Na^+, in the plasma. Because Na^+ is the major solute in the plasma, it is not surprising that hyponatremia is usually associated with hypo-osmolality. Hyponatremia, however, may also occur with a normal or even elevated plasma osmolality.

Drinking large quantities of water (20 L/d) rarely causes frank hyponatremia because of the large capacity of the kidneys to excrete dilute urine. If, however, plasma arginine vasopressin (AVP) is not decreased when plasma osmolality is decreased or if the ability of the kidneys to dilute the urine is impaired, hyponatremia may develop even with a normal water intake.

Hyponatremia with hypo-osmolality can occur in the presence of a decreased, normal, or even increased total body Na^+. Hyponatremia and decreased body Na^+ content may be seen with increased Na^+ loss, such as with vomiting, diarrhea, and diuretic therapy. In these instances, the decrease in extracellular fluid (ECF) volume stimulates thirst and AVP release. More water is ingested, but the kidneys form osmotically concentrated urine, and so plasma hypo-osmolality and hyponatremia result. Hyponatremia and normal body Na^+ content are seen in hypothyroidism, cortisol deficiency, and the syndrome of inappropriate secretion of antidiuretic hormone (SIADH). SIADH occurs with neurologic disease, severe pain, certain drugs (such as hypoglycemic agents), and some tumors. For example, a bronchogenic tumor may secrete AVP without control by plasma osmolality. The result is renal conservation of water. Hyponatremia and increased total body Na^+ content are seen in edematous states, such as congestive heart failure, hepatic cirrhosis, and nephrotic syndrome. The decrease in effective arterial blood volume stimulates thirst and AVP release. Excretion of dilute urine may also be impaired because of decreased delivery of fluid to diluting sites along the nephron and collecting ducts. Although both Na^+ and water are retained by the kidneys in the edematous states, relatively more water is conserved, leading to a dilutional hyponatremia.

Hyponatremia and hypo-osmolality can cause a variety of symptoms, including muscle cramps, lethargy, fatigue, disorientation, headache, anorexia, nausea, agitation, hypothermia, seizures, and coma. These symptoms, mainly neurologic, are a consequence of the swelling of brain cells as plasma osmolality falls. Excessive brain swelling may be fatal or may cause permanent damage. Treatment requires identifying and then treating the underlying cause. If Na^+ loss is responsible for the hyponatremia, isotonic or hypertonic saline or NaCl by mouth is usually given. If the blood volume is normal or the patient is edematous, water restriction is recommended. Hyponatremia should be corrected slowly and with constant monitoring, because too rapid correction can be harmful.

Hyponatremia in the presence of increased plasma osmolality is seen in hyperglycemic patients with uncontrolled diabetes mellitus. In this condition, the high plasma glucose causes the osmotic withdrawal of water from cells, and the extra water in the ECF space leads to hyponatremia. Plasma Na^+ level falls by 1.6 mEq/L for each 100-mg/dL rise in plasma glucose.

Hyponatremia and a normal plasma osmolality are seen with so-called **pseudohyponatremia.** This occurs when plasma lipids or proteins are greatly elevated. These molecules do not significantly elevate plasma osmolality. They do, however, occupy a significant volume of the plasma, and because the Na^+ is dissolved only in the plasma water, the Na^+ measured in the entire plasma is low.

deficit. The **thirst center** is located in the anterior hypothalamus, close to the neurons that produce and control AVP release. This center relays impulses to the cerebral cortex, so that thirst becomes a conscious sensation.

Several factors affect the thirst sensation (Fig. 23.7). The major stimulus is an increase in osmolality of the blood, which is detected by osmoreceptor cells in the hypothalamus. These cells are distinct from those that affect AVP release. Ethanol and urea are not effective stimuli for the osmoreceptors because they readily penetrate these cells and therefore do not cause them to shrink. NaCl is an effective stimulus. An increase of 1% to 2% (i.e., about 3–6 mOsm/kg H_2O) in plasma osmolality is needed to reach the thirst threshold.

Hypovolemia or a decrease in the EABV stimulates thirst. Blood volume loss must be considerable for the thirst threshold to be reached; most blood donors do not become thirsty after donating 500 mL of blood (10% of blood volume). A larger blood loss (15%–20% of blood volume), however, evokes *intense* thirst. A decrease in EABV resulting from severe diarrhea or vomiting or congestive heart failure may also provoke thirst.

The receptors for blood volume that stimulate thirst include the arterial baroreceptors in the carotid sinuses and aortic arch and stretch receptors in the cardiac atria and great veins in the thorax. The kidneys may also act as volume receptors. When blood volume is decreased, the kidneys release renin into the circulation. This results in production of angiotensin II, which acts on neurons near the third ventricle of the brain to stimulate thirst.

The thirst sensation is reinforced by dryness of the mouth and throat, which is caused by a reflex decrease in secretion by salivary and buccal glands in a water-deprived person. The GI tract also monitors water intake. Moistening of the mouth or distention of the stomach, for example, inhibit thirst, thereby preventing excessive water intake. For example, if a dog is deprived of water for some time and is then presented with water, it will commence drinking but will stop before all of the ingested water has been absorbed by the small intestine. Monitoring of water intake by the mouth and stomach in this situation limits water intake, thereby preventing a dip in plasma osmolality below normal.

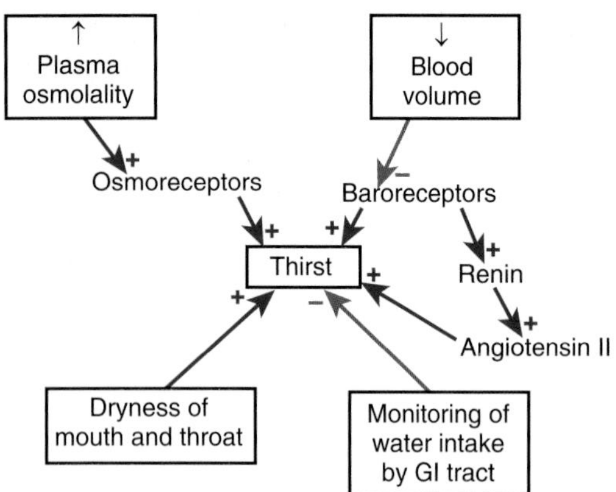

Figure 23.7 Factors affecting the thirst sensation. A plus sign indicates stimulation of thirst, whereas a minus sign indicates an inhibitory influence. GI, gastrointestinal.

SODIUM BALANCE

Na^+ is the most abundant cation in the ECF and, along with its accompanying anions Cl^- and HCO_3^-, largely determines the osmolality of the ECF. Because AVP, the kidneys, and thirst closely regulate the osmolality of the ECF, it follows that the Na^+ content of the ECF compartment determines the amount of water in it (and hence its volume). The kidneys are primarily involved in the regulation of Na^+ balance. We consider first the renal mechanisms involved in Na^+ excretion and then overall Na^+ balance.

Kidneys excrete only a small percentage of the filtered sodium load.

Table 23.4 shows the magnitude of filtration, reabsorption, and excretion of ions and water for a healthy adult man on an average American diet. The amount of Na^+ filtered was calculated from the product of the plasma Na^+ and the **glomerular filtration rate (GFR)**. The quantity of Na^+ reabsorbed was calculated from the difference between filtered and excreted amounts. Note that 99.6% (25,100 ÷ 25,200 mEq/d) of the filtered Na^+ was reabsorbed or, in other words, percentage excretion of Na^+ was only 0.4% of the filtered load. In terms of overall Na^+ balance for the body, the quantity of Na^+ excreted by the kidneys is of key importance, because ordinarily about 95% of the Na^+ that we consume is excreted by way of the kidneys. Tubular reabsorption of Na^+ must be finely regulated to keep us in Na^+ balance.

Figure 23.8 shows the percentage of filtered Na^+ reabsorbed in different parts of the nephron. Seventy percent of filtered Na^+, together with the same percentage of filtered water, is reabsorbed in the proximal convoluted tubule. The loop of Henle reabsorbs about 20% of filtered Na^+, but only 10% of filtered water. The distal convoluted tubule reabsorbs about 6% of filtered Na^+ (and no water), and the collecting ducts reabsorb about 3% of the filtered Na^+ (and 19% of the filtered water). Only about 1% of the filtered Na^+ (and water) is usually excreted. The distal nephron (distal convoluted tubule, connecting tubule, and collecting duct) has a lower capacity for Na^+ transport than more proximal segments and can be overwhelmed if too much Na^+ fails to be reabsorbed in proximal segments. The distal nephron is of critical importance in determining the final excretion of Na^+.

Glomerular filtration rate, hydrostatic pressure, and hormonal and neural factors influence renal sodium excretion.

Multiple factors affect renal Na^+ excretion; these are discussed in the subsections that follow. A factor may promote Na^+ excretion by increasing the amount of Na^+ filtered by the glomeruli, by decreasing the amount of Na^+ reabsorbed by the kidney tubules or, in some cases, by affecting both processes.

Effect of glomerular filtration rate

Na^+ excretion tends to change in the same direction as GFR. If GFR rises—for example, from an expanded ECF volume—then the tubules reabsorb the increased filtered load less completely and Na^+ excretion increases. If GFR falls—for example, as a result of blood loss—then the tubules can reabsorb the reduced filtered Na^+ load more completely and Na^+ excretion falls. These changes are of benefit in restoring a normal ECF volume.

Small changes in GFR could potentially lead to massive changes in Na^+ excretion, if it were not for a phenomenon

TABLE 23.4 Magnitude of Daily Filtration, Reabsorption, and Excretion of Ions and Water in a Healthy Young Man on a Typical American Diet

Substance	Plasma (mEq/L)	GFR (L/d)	Filtered (mEq/d)	Excreted (mEq/d)	Reabsorbed (mEq/d)	% Reabsorbed
Sodium	140	180	25,200	100	25,100	99.6
Chloride	105	180	18,900	100	18,800	99.5
Bicarbonate	24	180	4,320	2	4,318	99.9
Potassium	4	180	720	100	620	86.1
Water	0.93[a]	180	167 L/d	1.5 L/d	165.5 L/d	99.1

[a]Plasma contains about 0.93 L H_2O/L.

GFR, glomerular filtration rate.

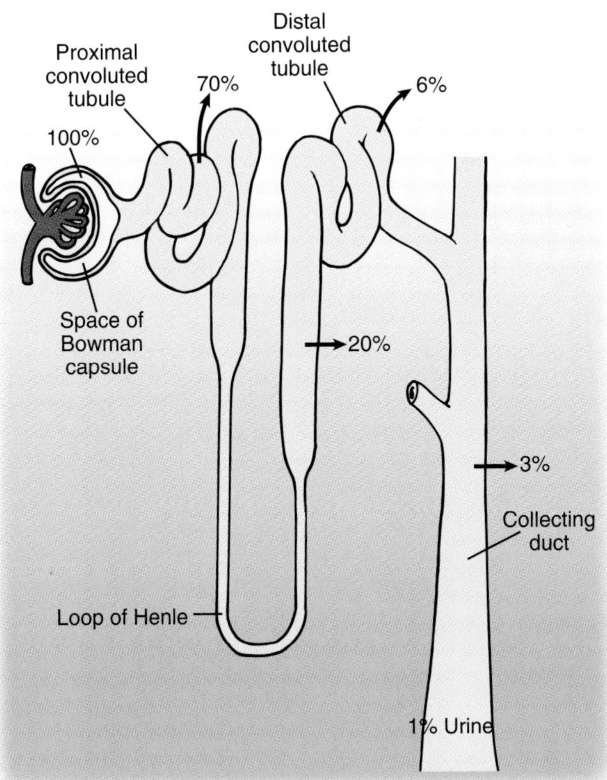

Figure 23.8 The percentage of the filtered load of sodium reabsorbed along the nephron. About 1% of the filtered Na^+ is usually excreted.

TABLE 23.5 Glomerulotubular Balance[a]

Period	Filtered (mEq/min)	– Reabsorbed (mEq/min)	= Excreted (mEq/min)
1	6.00	5.95	0.05
Increase GFR by one third			
2	8.00	7.90	0.10

[a]Results from an experiment performed on a 10-kg dog. In response to an increase in glomerular filtration rate (GFR) (produced by infusing a drug that dilated afferent arterioles), tubular reabsorption of Na^+ also increased, and so only a modest increase in Na^+ excretion occurred. If there had been no glomerulotubular balance and if tubular Na^+ reabsorption had stayed at 5.95 mEq/min, the kidneys would have excreted 2.05 mEq/min in period 2. If we assume that the extracellular fluid (ECF) volume in the dog is 2 L (20% of body weight) and if plasma Na^+ is 140 mEq/L, an excretion rate of 2.05 mEq/min would result in excretion of the entire ECF Na^+ (280 mEq) in a little over 2 h. The dog would have been dead long before this could happen, which underscores the importance of glomerulotubular balance.

called **glomerulotubular balance** (Table 23.5). There is a balance between the amount of Na^+ filtered and the amount of Na^+ reabsorbed by the tubules. Proximal convoluted tubules and loops of Henle reabsorb an essentially constant fraction of the filtered sodium or, in other words, the tubules increase the rate of Na^+ reabsorption when GFR is increased and decrease the rate of Na^+ reabsorption when GFR is decreased. Glomerulotubular balance reduces the impact of changes in GFR on Na^+ excretion.

Effect of the renin–angiotensin–aldosterone system

The **renin–angiotensin–aldosterone system** is a hormone system that regulates blood pressure and water (fluid) balance. **Renin** is a proteolytic enzyme produced by granular cells, which are located in afferent arterioles in the kidneys (see Chapter 22, "Kidney Function," Fig. 22.4). There are three main stimuli for renin release:

1. A decrease in pressure in the afferent arteriole, with the granular cells sensitive to stretch and functioning as an intrarenal **baroreceptor**
2. Stimulation of sympathetic nerve fibers to the kidneys, via β_2-adrenergic receptors on the granular cells
3. A decrease in luminal sodium chloride concentration at the macula densa region of the nephron, resulting, for example, from a decrease in GFR

All three of these pathways are activated and reinforce each other when there is a decrease in the EABV—for example, following hemorrhage, transudation of fluid out of the vascular system, diarrhea, severe sweating, or a low salt intake. Conversely, an increase in the EABV inhibits renin release. Long-term stimulation causes vascular smooth muscle cells in the afferent arteriole to differentiate into granular cells and leads to further increases in renin supply.

Renin in the blood plasma acts on a plasma α_2-globulin produced by the liver, called **angiotensinogen** (or renin substrate), and splits off the decapeptide **angiotensin I** (Fig. 23.9). Angiotensin I is converted to the octapeptide **angiotensin II** as the blood courses through the lungs. The **angiotensin-converting enzyme (ACE)**, which is present on the surface of endothelial cells, catalyzes this reaction. All the components of this system (renin, angiotensinogen, ACE) are present in some organs (e.g., the kidneys and brain), so that angiotensin II may also be formed and act locally.

The renin–angiotensin–aldosterone system is a salt-conserving system. Angiotensin II has several actions related to Na^+ and water balance:

- It stimulates the production and secretion of aldosterone from the zona glomerulosa of the adrenal cortex (see Chapter 33, "Adrenal Gland"). This mineralocorticoid hormone then acts on the distal nephron to increase Na^+ reabsorption.
- Angiotensin II directly stimulates tubular Na^+ reabsorption by proximal tubules.
- Angiotensin II stimulates thirst and the release of AVP by the posterior pituitary.

Angiotensin II is also a potent vasoconstrictor of both resistance and capacitance vessels; increased plasma levels following hemorrhage, for example, help sustain blood pressure. Inhibiting angiotensin II production, by giving

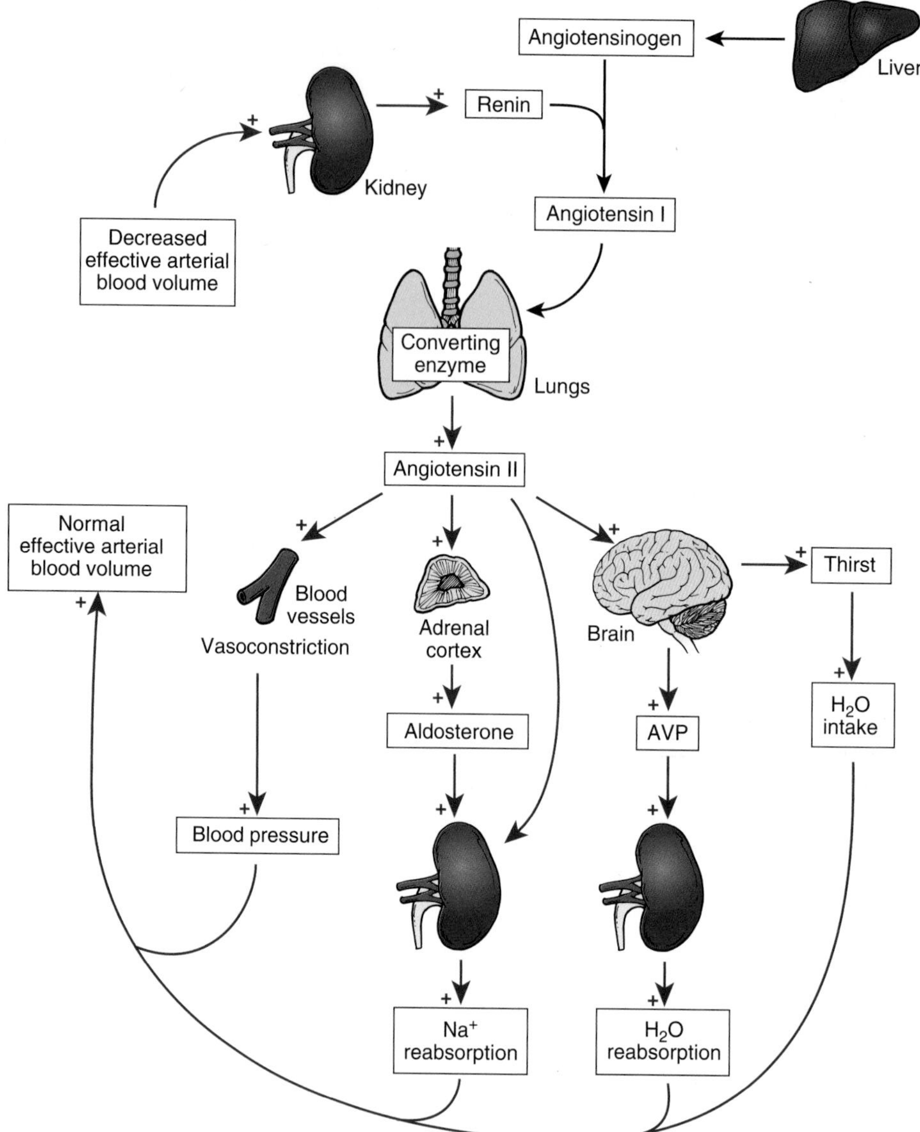

Figure 23.9 Components of the renin–angiotensin–aldosterone system. This system is activated by a decrease in the effective arterial blood volume (e.g., following hemorrhage) and results in compensatory changes that help restore arterial blood pressure and blood volume to normal. AVP, arginine vasopressin.

an ACE inhibitor or inhibiting the binding of angiotensin II to its receptor by using an angiotensin-receptor blocker, lowers blood pressure and is used in the treatment of hypertension. Recently, an orally active nonpeptide renin inhibitor called *aliskiren* has been developed; this drug is an effective antihypertensive that may also slow the progression of chronic renal disease by inhibiting intrarenal renin activity.

The renin–angiotensin–aldosterone system plays an important role in the day-to-day control of Na^+ excretion. It favors Na^+ conservation by the kidneys when there is a Na^+ or volume deficit in the body. When there is an excess of Na^+ or volume, diminished activity of the renin–angiotensin–aldosterone system permits enhanced Na^+ excretion. In the absence of aldosterone (e.g., in an adrenalectomized person) or in a person with adrenal cortical insufficiency, excessive amounts of Na^+ are lost in the urine. Percentage reabsorption of Na^+ may decrease from a normal value of about 99.6% to a value of 98%. This change (1.6% of the filtered Na^+ load) may not seem like much, but if the kidneys filter 25,200 mEq/d (see Table 23.4) and excrete an extra $0.016 \times 25{,}200 =$ 403 mEq/d, this is the amount of Na^+ in almost 3 L of ECF (assuming a Na^+ of 140 mEq/L). Such a loss of Na^+ would lead to a decrease in plasma and blood volume, circulatory collapse, and even death.

When there is an extra need for Na^+, people and many animals display a **sodium appetite**, an urge for salt intake, which can be viewed as a brain mechanism, much like thirst, that helps compensate for a deficit. Patients with primary adrenal cortical insufficiency (**Addison disease**) often show a well-developed sodium appetite, which helps keep them alive.

In the case of mineralocorticoid excess, it has been observed that large doses of a potent mineralocorticoid will cause a person to retain about 200 to 300 mEq Na^+ (equivalent to about 1.4–2 L of ECF), and then the person will no longer continue to retain Na^+. Obvious edema is not seen after retaining this much fluid. The escape from the salt- retaining action of the mineralocorticoid is called **mineralocorticoid escape.** The fact that the person will not continue to accumulate Na^+ and water is a result of numerous factors that promote renal Na^+ excretion when ECF volume is expanded. These factors overpower the salt-retaining action of mineralocorticoid hormones and include increases in GFR, changes in intrarenal pressures, and release of natriuretic factors (see below).

Effect of hydrostatic pressure

An increase in the hydrostatic pressure or a decrease in the colloid osmotic pressure (the so-called **Starling forces**) in peritubular capillaries results in reduced fluid uptake by the capillaries. Such changes would occur, for example, after intravenous infusion of a large volume of isotonic saline. The resulting accumulation of the reabsorbed fluid in the kidney interstitial spaces widens the tight junctions between proximal tubule cells, and the epithelium becomes even more leaky than normal. The result is increased back-leak of salt and water into the tubule lumen and an overall reduction in net reabsorption. The increased sodium and water excretion produced by an increase in pressure in the kidneys is called a *pressure* **natriuresis** or *pressure diuresis*. The term "natriuresis" means an increase in Na^+ excretion.

An increase in blood pressure rapidly causes Na^+/H^+ exchangers to be removed from the apical cell membrane of proximal tubule cells and internalized. Basolateral cell membrane Na^+/K^+-ATPase activity is also decreased. These changes result in diminished tubular sodium reabsorption and enhanced sodium excretion. This transcellular pathway is another way the kidneys can dispose of excess sodium when arterial or intrarenal pressure is elevated.

Effect of natriuretic hormones

Atrial natriuretic peptide (ANP) is a 28–amino acid polypeptide synthesized and stored in myocytes of the cardiac atria (Fig. 23.10). It is released on stretch of the atria—for example, following volume expansion. This hormone has several actions that increase Na^+ excretion. ANP acts on the kidneys to increase glomerular blood flow and filtration rate and inhibits Na^+ reabsorption by the inner medullary collecting ducts. The second messenger for ANP in the collecting duct is cyclic guanosine monophosphate (cGMP). ANP directly inhibits aldosterone secretion by the adrenal cortex; it also indirectly inhibits aldosterone secretion by diminishing renal renin release. ANP is a vasodilator and therefore lowers blood pressure. Some evidence suggests that ANP inhibits AVP secretion. The actions of ANP are in many respects just the opposite of those of the renin–angiotensin–aldosterone system; ANP promotes salt and water loss by the kidneys and lowers blood pressure, whereas activation of the renin–angiotensin–aldosterone system results in salt and water conservation and a higher blood pressure.

Several other natriuretic hormones and factors have been described. **Urodilatin** (kidney natriuretic peptide) is a 32–amino acid polypeptide derived from the same prohormone

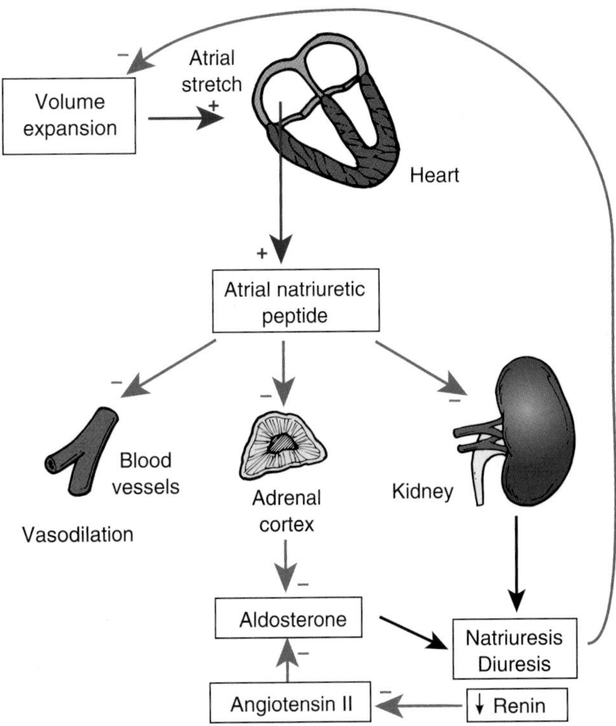

Figure 23.10 Atrial natriuretic peptide (ANP) and its actions. ANP release from the cardiac atria is stimulated by blood volume expansion, which stretches the atria. ANP produces effects that bring blood volume back toward normal, such as increased Na^+ excretion.

as ANP. It is synthesized primarily by intercalated cells in the cortical collecting duct, is secreted into the tubule lumen, and inhibits Na^+ reabsorption by inner medullary collecting ducts via cGMP. **Brain natriuretic peptide (BNP)** was first isolated from the brain but is also produced by myocytes in the cardiac ventricles. Increased plasma levels serve as a marker of cardiac injury. Recombinant BNP is used to promote renal sodium excretion in patients with decompensated congestive heart failure. **Guanylin** and **uroguanylin** are polypeptide hormones produced by the small intestine in response to salt ingestion. They activate guanylyl cyclase and produce cGMP as a second messenger, as their names suggest, and induce a natriuresis. **Bradykinin** is produced locally in the kidneys and inhibits Na^+ reabsorption.

Prostaglandins E_2 and I_2 (prostacyclin) increase Na^+ excretion by the kidneys. These locally produced hormones are formed from arachidonic acid, which is liberated from phospholipids in cell membranes by the enzyme phospholipase A_2. A **cyclooxygenase (COX)** enzyme that has two isoforms, COX-1 and COX-2, mediates further processing. In most tissues, COX-1 is constitutively expressed, whereas COX-2 is generally induced by inflammation. In the kidney, COX-1 and COX-2 are both constitutively expressed in the cortex and medulla. In the cortex, COX-2 may be involved in macula densa–mediated renin release. COX-1 and COX-2 are present in high amounts in the renal medulla, where the main role of the prostaglandins is to inhibit Na^+ reabsorption. The inhibition of Na^+ reabsorption occurs via direct effects on the tubules and collecting ducts and via

From Bench to Bedside / 23.1

Mesenchymal Stem Cells: A New Treatment for Acute Kidney Injury?

Acute kidney injury, or **AKI** (also known as *acute renal failure*), is a serious condition characterized by a rapid decline in glomerular filtration rate, retention of nitrogenous waste products, and disturbances of extracellular fluid volume and electrolyte and acid–base balance. It is most commonly caused by renal ischemia (inadequate renal blood flow) or nephrotoxins (drugs or chemicals that damage the kidneys). AKI affects about 5% of hospitalized patients. Often patients with AKI have abnormal function in several organ systems, and consequently, overall mortality rate is exceedingly high, ranging from 30% to 50%. Even so, the kidneys have a remarkable capacity to regenerate after AKI. Numerous pharmacologic treatments have been shown to improve kidney function in animal models of AKI, but few have proved effective in the clinic.

There is widespread agreement that AKI needs to be recognized and treated early. Plasma creatinine and blood urea nitrogen rise after the onset of AKI, but this rise is delayed, and it would be best to have sensitive and specific markers that can identify renal damage early so that therapy can be initiated right away, thereby preventing further damage. Various markers for AKI have been suggested, including neutrophil gelatinase-associated lipocalin (NGAL), kidney injury molecule-1 (KIM-1), and interleukin-18 (IL-18), but the search continues.

A novel proposed treatment for AKI is the administration of mesenchymal stem cells. These cells can be harvested from adult bone marrow and have the potential to form bone, cartilage, muscle, fat, nerve, and other cell types. In studies on rats with ischemia-induced AKI, intra-arterial injections of mesenchymal stem cells suspended in tissue culture medium were given immediately after renal blood flow had been stopped for 40 minutes or 24 hours after blood reflow was allowed. The animals receiving the mesenchymal stem cells showed smaller rises in serum creatinine concentration and less histologic damage than when vehicle alone was injected. Likewise, when mesenchymal stem cells were administered to mice with AKI induced by a nephrotoxin (the antitumor drug, cisplatin), renal function and structure were much better preserved. It appears that the beneficial effects of mesenchymal stem cells are not primarily due to transdifferentiation into kidney tubule epithelial cells or endothelial cells but, rather, are due to several other effects of the stem cells, namely, stem cells produce various growth factors (such as vascular endothelial growth factor and cytokines) and they have anti-inflammatory and immune-modulating effects that promote reparative processes in damaged kidneys.

The possible beneficial effects of mesenchymal stem cell administration are currently being investigated in clinical trials for a number of disorders, including myocardial infarction, systemic lupus erythematosus, transplant rejection, vascular diseases, cancer, and many other conditions (see www.clinicaltrials.gov). A trial is currently underway to see whether administration of mesenchymal stem cells is of benefit in preventing the AKI that frequently occurs in patients subjected to open-heart surgery. Preliminary results indicate that mesenchymal stem cell administration is a safe procedure. Also, mesenchymal stem cell-treated patients do considerably better; for example, 20% of patients who were not treated developed AKI, whereas all treated individuals remained at baseline kidney function levels and none required hemodialysis. Hopefully, this new treatment modality will also prove effective in the many of the other conditions that can cause AKI.

hemodynamic effects, because the prostaglandins (PGE_2, PGI_2) are vasodilators (see Chapter 22, "Kidney Function"). Inhibition of the formation of prostaglandins with common nonsteroidal anti-inflammatory drugs (NSAIDs), such as aspirin, may lead to a fall in renal blood flow and to Na^+ retention.

Effect of other hormones

Estrogens decrease Na^+ excretion, probably by the direct stimulation of tubular Na^+ reabsorption. Most women tend to retain salt and water during pregnancy, which may be related, in part, to the high plasma estrogen levels during this time.

Glucocorticoids, such as cortisol (see Chapter 33, "Adrenal Gland"), increase tubular Na^+ reabsorption but also cause an increase in GFR, which may mask the tubular effect. Usually, a decrease in Na^+ excretion is seen. Glucocorticoids circulate in the blood at much higher free concentrations than does aldosterone, and they can bind to and activate mineralocorticoid receptors in the kidney. Their binding and actions in distal nephron cells, however, are minimized by conversion, catalyzed by the enzyme 11β-hydroxysteroid dehydrogenase, to metabolites that do not bind the mineralocorticoid receptor.

Effect of sympathetic stimulation

The stimulation of renal sympathetic nerves reduces renal Na^+ excretion in at least three ways:

1. It produces a decline in GFR and renal blood flow, leading to a decreased filtered Na^+ load and peritubular capillary hydrostatic pressure, both of which favor diminished Na^+ excretion.
2. It has a direct stimulatory effect on Na^+ reabsorption by the renal tubules.
3. It causes renin release, which results in increased plasma angiotensin II and aldosterone levels, both of which increase tubular Na^+ reabsorption.

Activation of the sympathetic nervous system occurs in a number of stressful circumstances (such as hemorrhage) in which the conservation of salt and water by the kidneys is of clear benefit.

Effect of osmotic diuretics

Osmotic diuretics are solutes that are excreted in the urine and increase urinary excretion of Na^+ and K^+ salts and water. Examples are urea, glucose (when the reabsorptive capacity of the tubules for glucose has been exceeded), and mannitol (a six-carbon sugar alcohol used in the clinic to promote Na^+ excretion or cell shrinkage). Osmotic diuretics decrease the reabsorption of Na^+ in the proximal tubule. This response results from the development of a Na^+ concentration gradient (lumen Na^+ < plasma Na^+) across the proximal tubular epithelium when there is a high concentration of unreabsorbed solute in the tubule lumen. This gradient leads to significant back-leak of Na^+ into the tubule lumen and consequently decreased net Na^+ reabsorption. Because the proximal tubule is the place where most of the filtered Na^+ is normally reabsorbed, osmotic diuretics, by interfering with this process, can potentially cause the excretion of large amounts of Na^+. Osmotic diuretics may also increase Na^+ excretion by inhibiting distal Na^+ reabsorption (similar to the proximal inhibition) and by increasing medullary blood flow.

Effect of reabsorbed anions

Poorly reabsorbed anions result in increased Na^+ excretion. Solutions are electrically neutral, and so whenever there are more anions in the urine, more cations must also be present. If there is increased excretion of phosphate, ketone body acids (as occurs in uncontrolled diabetes mellitus), HCO_3^-, or SO_4^{2-}, then more Na^+ is also excreted. To some extent, the Na^+ in the urine can be replaced by other cations, such as K^+, NH_4^+, and H^+.

Effect of diuretic drugs

Most of the diuretic drugs used today are specific Na^+ transport inhibitors. For example, the loop diuretic drugs (furosemide, bumetanide) inhibit the Na-K-2Cl cotransporter in the thick ascending limb, the thiazide diuretics inhibit the Na-Cl cotransporter in the distal convoluted tubule, and amiloride blocks the epithelial Na^+ channel in the collecting ducts (see Chapter 22, "Kidney Function"). Spironolactone promotes Na^+ excretion by competitively inhibiting the binding of aldosterone to the mineralocorticoid receptor. The diuretic drugs are really natriuretic drugs; they produce an increased urine output (diuresis) because water reabsorption is diminished whenever Na^+ reabsorption is decreased. The loop diuretic drugs produce an especially large increase in Na^+ excretion, because normally 20% of filtered Na^+ is reabsorbed in the loop of Henle. Furthermore, the diminished osmotic gradient in the kidney medulla, produced by inhibition of thick ascending limb Na^+ reabsorption, may result in a striking increase in urine output. Diuretics commonly are prescribed for treating hypertension and edema.

Kidneys play a dominant role in regulating sodium balance.

Figure 23.11 summarizes Na^+ balance in a healthy adult man. Dietary intake of Na^+ varies and, in a typical American diet, amounts to about 100 to 300 mEq/d, mostly in the form of NaCl. Ingested Na^+ is mainly absorbed in the small intestine and is added to the ECF, where it is the major determinant of the osmolality and the amount of water in (or volume of) this fluid compartment. About 50% of the body's Na^+ is in the ECF, about 40% in bone, and 10% within cells.

Losses of Na^+ occur via the skin, GI tract, and kidneys. Skin losses are usually small but can be considerable with sweating, burns, or hemorrhage. Likewise, GI losses are also usually small, but they can be large and serious with vomiting, diarrhea, or iatrogenic suction or drainage of GI secretions. The kidneys are ordinarily the major route of Na^+ loss from the body, excreting about 95% of the ingested Na^+ in a healthy person. Thus, the kidneys play a dominant role in the control of Na^+ balance. The kidneys can adjust Na^+ excretion over a wide range, reducing it to low levels when there is a Na^+ deficit and excreting more Na^+ when there is Na^+ excess in the body. Adjustments in Na^+ excretion occur by engaging many of the factors discussed above.

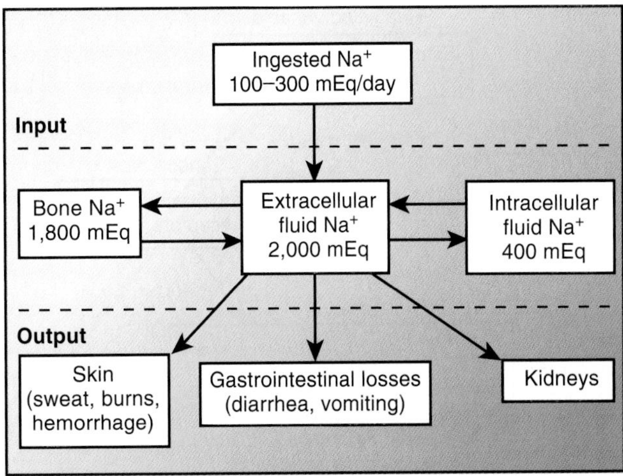

Figure 23.11 Sodium balance. Most of the Na^+ that we consume in our diet is excreted by the kidneys.

In a healthy person, one can think of the ECF volume as the regulated variable in a negative feedback control system (Fig. 23.12). The kidneys are the effectors, and they change Na^+ excretion in an appropriate manner. An increase in ECF volume promotes renal Na^+ loss, which restores a normal volume. A decrease in ECF volume leads to decreased renal Na^+ excretion, and this Na^+ retention (with continued dietary Na^+ intake) leads to the restoration of a normal ECF volume. Closer examination of this idea, particularly when considering pathophysiologic states, however, suggests that it is of limited usefulness. A more considered view suggests that the EABV is actually the regulated variable. In a healthy person, ECF volume and EABV usually change together in the same direction. In an abnormal condition, such as congestive heart failure, however, EABV is low when the ECF volume is abnormally increased. In this condition, there is a potent stimulus for renal Na^+ retention that clearly cannot be the ECF volume.

When EABV is diminished, the degree of fullness of the arterial system is less than normal and tissue blood flow is inadequate. Arterial baroreceptors in the carotid sinuses and aortic arch sense the decreased arterial stretch. This will produce reflex activation of sympathetic nerve fibers to the kidneys, with consequent decreased GFR and renal blood flow and increased renin release. These changes favor renal Na^+ retention. Reduced EABV is also "sensed" in the kidneys in three ways:

1. A low pressure at the level of the afferent arteriole stimulates renin release via the intrarenal baroreceptor mechanism.

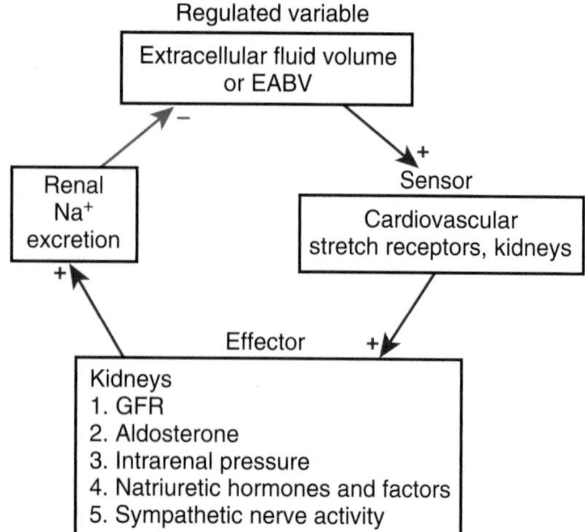

Figure 23.12 The regulation of extracellular fluid (ECF) volume or effective arterial blood volume (EABV) by a negative feedback control system. Arterial baroreceptors and the kidneys sense the degree of fullness of the arterial system. The kidneys are the effectors, and they change Na^+ excretion to restore EABV to normal. GFR, glomerular filtration rate.

2. Decreases in renal perfusion pressure lead to a reduced GFR and hence diminished Na^+ excretion.
3. Decreases in renal perfusion pressure also reduce peritubular capillary hydrostatic pressure, thereby increasing the uptake of reabsorbed fluid and diminishing Na^+ excretion.

When kidney perfusion is threatened, the kidneys retain salt and water, a response that tends to improve their perfusion.

In a number of important diseases, including heart, liver, and some kidney diseases, abnormal renal retention of Na^+ contributes to the development of **generalized edema**, a widespread accumulation of salt and water in the interstitial spaces of the body. The condition is often not clinically evident until a person has accumulated more than 2.5 to 3 L of ECF in the interstitial space. Expansion of the interstitial space has two components: (1) an altered balance of Starling forces exerted across capillaries and (2) the retention of extra salt and water by the kidneys. Total plasma volume is only about 3.5 L; if edema fluid were derived solely from the plasma, hemoconcentration and circulatory shock would ensue. Conservation of salt and water by the kidneys is clearly an important part of the development of generalized edema.

Patients with congestive heart failure may accumulate many liters of edema fluid, which is easily detected as weight gain (because 1 L of fluid weighs 1 kg). Because of the effect of gravity, the ankles become swollen and pitting edema develops. As result of heart failure, venous pressure is elevated, causing fluid to leak out of the capillaries because of their elevated hydrostatic pressure. Inadequate pumping of blood by the heart leads to a decrease in EABV, and so the kidneys retain salt and water. Alterations in many of the factors discussed above—decreased GFR, increased renin–angiotensin–aldosterone activity, changes in intrarenal pressures, and increased sympathetic nervous system activity—contribute to the renal salt and water retention. To minimize the accumulation of edema fluid, patients are often placed on a reduced Na^+ intake and given diuretic drugs.

POTASSIUM BALANCE

Potassium (K^+) is the most abundant ion in the ICF compartment. It has many important effects in the body, and its plasma concentration is closely regulated. The kidneys play a dominant role in regulating K^+ balance.

Potassium influences cell volume, excitability, acid–base balance, and metabolism.

As K^+ is the major osmotically active solute in cells, the amount of cellular K^+ is the major determinant of the amount of water in (and therefore the volume of) the ICF compartment, in the same way that extracellular Na^+ is a major determinant of ECF volume. When cells lose K^+ (and accompanying anions), they also lose water and shrink; the converse is also true.

The distribution of K^+ across cell membranes—that is, the ratio of intracellular to extracellular K^+ concentrations—is the major determinant of the resting membrane potential of cells and hence their excitability (see Chapter 3, "Action Potential, Synaptic Transmission, and Maintenance of Nerve Function"). Disturbances of K^+ balance often produce altered excitability of nerves and muscles. A low plasma K^+ level leads to membrane hyperpolarization and reduced excitability; muscle weakness is a common symptom. Excessive plasma K^+ levels lead to membrane depolarization and increased excitability. High plasma K^+ levels cause cardiac arrhythmias and eventually ventricular fibrillation, usually a lethal event.

K^+ balance is linked to acid–base balance in complex ways (see Chapter 24, "Acid–Base Homeostasis"). K^+ depletion, for example, can lead to metabolic alkalosis, and K^+ excess can lead to metabolic acidosis. A primary disturbance in acid–base balance can also lead to abnormal K^+ balance.

K^+ affects the activity of enzymes involved in carbohydrate metabolism and electron transport. K^+ is needed for tissue growth and repair. Tissue breakdown or increased protein catabolism results in a loss of K^+ from cells.

Most of the body's potassium is in cells.

Total body content of K^+ in a healthy, young adult, 70-kg man is about 3,700 mEq. About 2% of this, roughly 60 mEq, is in the functional ECF (blood plasma, interstitial fluid, and lymph); this number was calculated by multiplying the plasma K^+ level of 4 mEq/L times the ECF volume (20% of body weight or 14 L). About 8% of the body's K^+ is in bone, dense connective tissue, and cartilage, and another 1% is in transcellular fluids. Ninety percent of the body's K^+ is in the cell compartment.

A normal plasma K^+ level ranges from 3.5 to 5.0 mEq/L. By definition, a plasma K^+ level below 3.5 mEq/L is **hypokalemia** and a plasma K^+ level above 5.0 mEq/L is **hyperkalemia**. The K^+ level in skeletal muscle cells is about 150 mEq/L cell water. Skeletal muscle cells constitute the largest fraction of the cell mass in the human body and contain about two thirds of the body's K^+. One can easily appreciate that abnormal leakage of K^+ from muscle cells, as a result of trauma for example, may lead to dangerous hyperkalemia.

A variety of factors influence the distribution of K^+ between cells and ECF (Fig. 23.13):

- A key factor is the Na^+/K^+-ATPase, which pumps K^+ into cells. If this enzyme is inhibited—as a result of an inadequate tissue oxygen supply or digitalis overdose, for example—then hyperkalemia may result.
- A decrease in ECF pH (an increase in ECF H^+) tends to produce a rise in ECF K^+. This results from an exchange of extracellular H^+ for intracellular K^+. When a mineral acid such as HCl is added to the ECF, a fall of 0.1 unit in blood pH leads to roughly a 0.6-mEq/L rise in plasma K^+ level. When an organic acid (which can penetrate cell membranes) is added, the rise in plasma K^+ level for a given fall in blood pH is considerably less. The fact that blood pH influences plasma K^+ level is sometimes used in the emergency treatment of hyperkalemia; intravenous infusion of a $NaHCO_3$ solution (which makes the blood more alkaline) causes H^+ to move out of cells and K^+, in exchange, to move into cells.
- Insulin promotes the uptake of K^+ by skeletal muscle and liver cells. This effect appears to be a result of stimulation of cell membrane Na^+/K^+-ATPase pumps. Insulin (administered with glucose) is also used in the emergency treatment of hyperkalemia.
- Epinephrine increases K^+ uptake by cells, an effect mediated by β_2 receptors.
- Hyperosmolality (e.g., resulting from hyperglycemia) tends to raise plasma K^+ level; hyperosmolality causes cells to shrink and raises intracellular K^+ level, which then favors outward diffusion of K^+ into the ECF.
- Tissue trauma, infection, ischemia, hemolysis, and severe exercise release K^+ from cells and can cause significant hyperkalemia. An artifactual increase in plasma K^+ level, **pseudohyperkalemia**, results if blood has been mishandled and red cells have been injured and allowed to leak K^+.

The plasma K^+ level is sometimes taken as a rough guide to total body K^+ stores. For example, if a condition is known to produce an excessive loss of K^+ (such as taking a diuretic drug), a decrease of 1 mEq/L in plasma K^+ level may correspond to a loss of 200 to 300 mEq K^+. Clearly, however, many factors affect the distribution of K^+ between cells and ECF, and so in many circumstances the plasma K^+ is not a good index of the amount of K^+ in the body.

Kidneys normally maintain potassium balance.

Figure 23.14 depicts K^+ balance for a healthy adult man. Most of the foods that we eat contain K^+. K^+ intake (50–150 mEq/d) and absorption by the small intestine are unregulated. On the output side, GI losses are normally small, but they can be large, especially with diarrhea. Diarrheal fluid may contain as much as 80 mEq K^+/L. K^+ loss in sweat is clinically unimportant. Normally, the kidneys excrete 90% of the ingested K^+. The kidneys are the major site of control of K^+ balance; they increase K^+ excretion when there is too much K^+ in the body and conserve K^+ when there is too little.

Abnormal renal potassium excretion can result in hyperkalemia or hypokalemia.

The major cause of K^+ imbalances is abnormal renal K^+ excretion. The kidneys may excrete too little K^+; if the dietary intake of K^+ continues, hyperkalemia can result. For example, in Addison disease, a low plasma aldosterone level leads to deficient K^+ excretion. Inadequate renal K^+ excretion also occurs with acute renal failure; the hyperkalemia resulting from inadequate renal excretion is often compounded by tissue trauma, infection, and acidosis, all of which raise plasma K^+ level. In chronic renal failure, hyperkalemia usually does not develop until the GFR falls below 15 to 20 mL/min, because of the remarkable ability of the kidney collecting ducts to adapt and increase K^+ secretion.

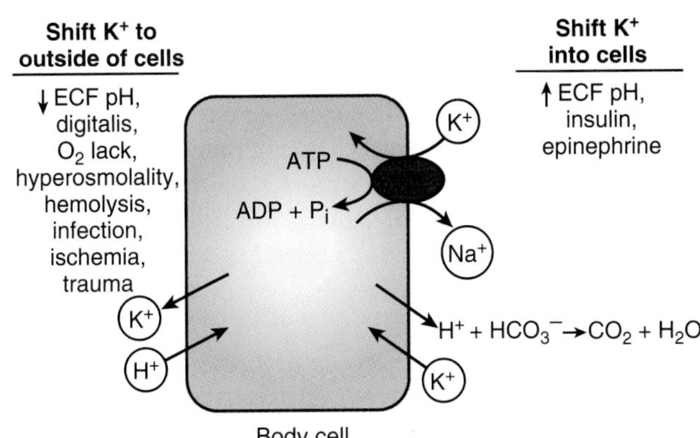

Figure 23.13 Factors influencing the distribution of potassium between intracellular and extracellular fluids. ADP, adenosine diphosphate; ATP, adenosine triphosphate; ECF, extracellular fluid; P_i, inorganic phosphate.

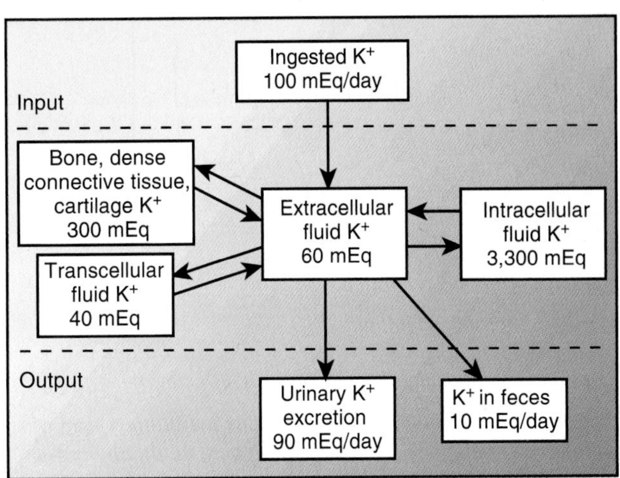

Figure 23.14 Potassium balance for a healthy adult. Most of the body's K^+ is in the cell compartment. Renal K^+ excretion is normally adjusted to keep a person in balance.

Excessive loss of K^+ by the kidneys leads to hypokalemia. The major cause of renal K^+ wasting is iatrogenic, an unwanted side effect of diuretic drug therapy. Hyperaldosteronism causes excessive K^+ excretion. In uncontrolled diabetes mellitus, K^+ loss is increased because of the osmotic diuresis resulting from the glucosuria and an elevated rate of fluid flow in the cortical collecting ducts. Several rare inherited defects in tubular transport, including Bartter, Gitelman, and Liddle syndromes, also lead to excessive renal K^+ excretion and hypokalemia (see Table 22.3).

Changes in dietary potassium intake change renal potassium excretion in an appropriate direction.

As was discussed in Chapter 22, "Kidney Function," K^+ is filtered, reabsorbed, and secreted in the kidneys. Most of the filtered K^+ is reabsorbed in the proximal convoluted tubule (70%) and loop of Henle (25%), and most K^+ excreted in the urine is usually the result of secretion by cortical collecting duct principal cells. The percentage of filtered K^+ excreted in the urine is typically about 15% (Fig. 23.15). With prolonged K^+ depletion, the kidneys may excrete only 1% of the filtered load. On the other hand, excessive K^+ intake may result in the excretion of an amount of K^+ that exceeds the amount filtered; in this case, there is greatly increased K^+ secretion by cortical collecting ducts.

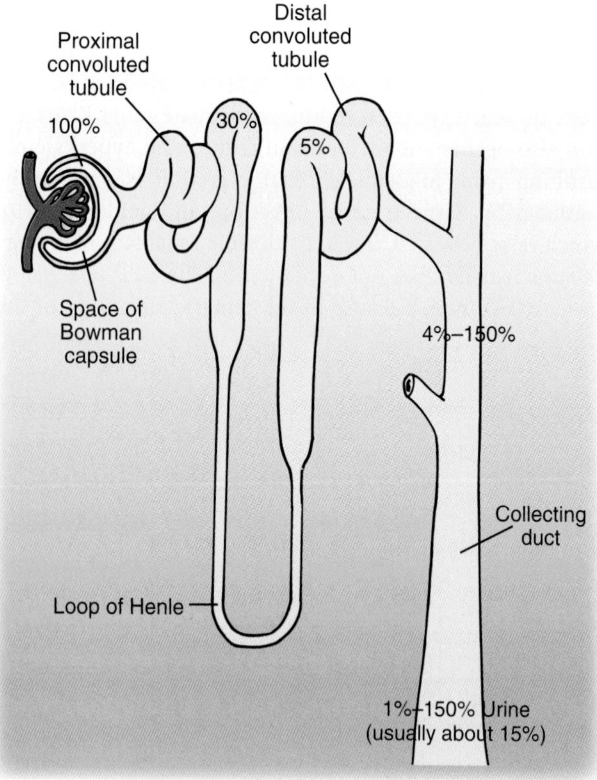

Figure 23.15 The percentage of the filtered load of potassium remaining in tubular fluid as it flows down the nephron. K^+ is usually secreted in the cortical collecting duct. With K^+ loading, this secretion is so vigorous that the amount of K^+ excreted may actually exceed the filtered load. With K^+ depletion, K^+ is reabsorbed by the collecting ducts.

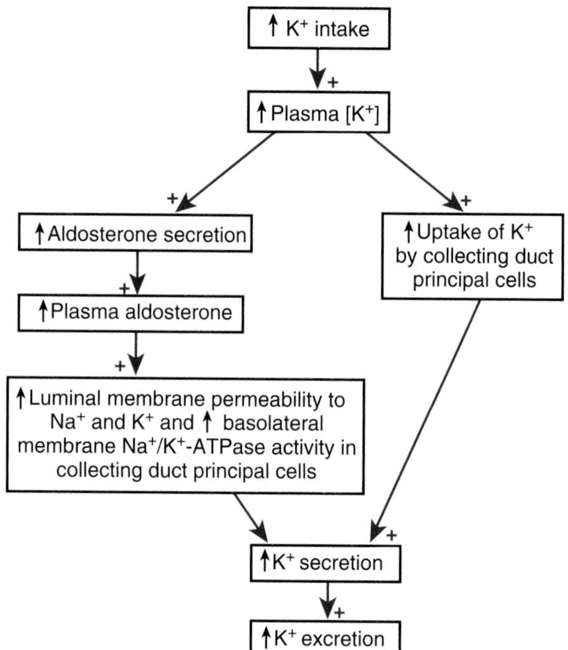

Figure 23.16 The effect of increased dietary potassium intake on potassium excretion. K^+ directly stimulates aldosterone secretion and leads to an increase in cell K^+ in collecting duct principal cells. Both effects lead to enhanced secretion and, hence, excretion of K^+.

When the dietary intake of K^+ is changed, renal excretion changes in the same direction. An important site for this adaptive change is the cortical collecting duct. Figure 23.16 shows the response to an increase in dietary K^+ intake. Two pathways are involved. First, an elevated plasma K^+ level leads to increased K^+ uptake by the basolateral cell membrane Na^+/K^+-ATPase in collecting duct principal cells, resulting in increased intracellular K^+, K^+ secretion, and K^+ excretion. Second, an elevated plasma K^+ level has a direct effect (i.e., not mediated by renin and angiotensin) on the adrenal cortex to stimulate the synthesis and release of aldosterone. Aldosterone acts on collecting duct principal cells to (1) increase the Na^+ permeability of the luminal cell membrane, (2) increase the number and activity of basolateral cell membrane Na^+/K^+-ATPase pumps, (3) increase the luminal cell membrane K^+ permeability, and (4) increase cell metabolism. All these changes result in increased K^+ secretion.

In cases of decreased dietary K^+ intake or K^+ depletion, the activity of the luminal cell membrane H^+/K^+-ATPase found in α-intercalated cells is increased. This promotes K^+ reabsorption by the collecting ducts. The collecting ducts can greatly diminish K^+ excretion, but it takes a couple of weeks for K^+ loss to reach minimal levels.

Net renal potassium excretion may be determined by counterbalancing effects.

Considering the fact that aldosterone stimulates both Na^+ reabsorption and K^+ secretion, why does Na^+ deprivation, a stimulus that raises plasma aldosterone levels, not lead to enhanced K^+ excretion? The explanation is related to the fact that Na^+ deprivation tends to lower the GFR and increase

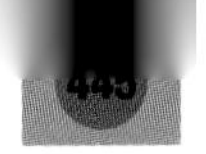

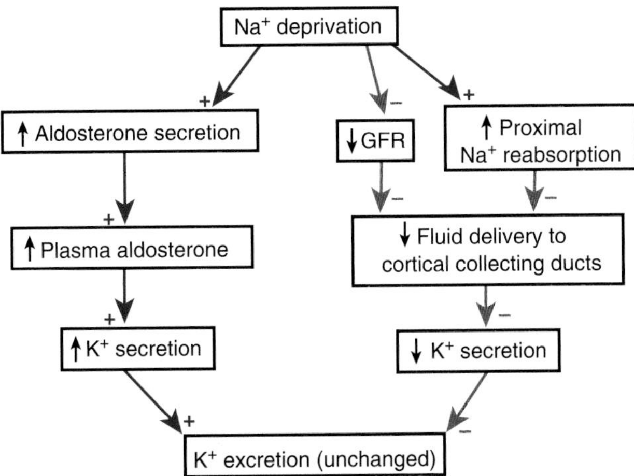

Figure 23.17 Why sodium depletion does not lead to enhanced potassium excretion. GFR, glomerular filtration rate.

proximal Na^+ reabsorption (Fig. 23.17). This response leads to a fall in Na^+ delivery and decreased fluid flow rate in the cortical collecting ducts, which diminishes K^+ secretion (see Chapter 22, "Kidney Function"), counterbalancing the stimulatory effect of aldosterone. Consequently, K^+ excretion is unaltered.

Another puzzling question is, why doesn't K^+ excretion increase during a water diuresis? In Chapter 22, we mentioned that an increase in fluid flow through the cortical collecting ducts increases K^+ secretion. AVP, in addition to its effects on water permeability, stimulates K^+ secretion by increasing the activity of luminal membrane K^+ channels in cortical collecting duct principal cells. Because plasma AVP levels are low during a water diuresis, this will reduce K^+ secretion, opposing the effects of increased flow, with the result that K^+ excretion hardly changes.

CALCIUM BALANCE

The kidneys play an important role in the maintenance of Ca^{2+} balance. Ca^{2+} intake is about 1,000 mg/d and mainly comes from dairy products in the diet. About 300 mg/d is absorbed by the small intestine, a process controlled by 1,25$(OH)_2$ vitamin D_3. About 150 mg/d of Ca^{2+} is secreted into the GI tract (via saliva, gastric juice, pancreatic juice, bile, and intestinal secretions), so that net absorption is only about 150 mg/d. Fecal Ca^{2+} excretion is about 850 mg/d, and urinary excretion about 150 mg/d.

A normal plasma Ca^{2+} level is about 10 mg/dL, which is equal to 2.5 mmol/L (because the atomic weight of calcium is 40) or 5 mEq/L. About 40% of plasma Ca^{2+} is bound to plasma proteins (mainly serum albumin), 10% is bound to small diffusible anions (such as citrate, bicarbonate, phosphate, and sulfate), and 50% is free or ionized. It is the ionized Ca^{2+} in the blood that is physiologically important and closely regulated (see Chapter 35, "Endocrine Regulation of Calcium, Phosphate, and Bone Homeostasis"). Most of the body's Ca^{2+} is in bone (99%), which constantly turns over. In a healthy adult, the rate of release of Ca^{2+} from old bone exactly matches the rate of deposition of Ca^{2+} in newly formed bone (500 mg/d).

Ca^{2+} that is not bound to plasma proteins (i.e., 60% of the plasma Ca^{2+}) is freely filterable in the glomeruli. About 60% of the filtered Ca^{2+} is reabsorbed in the proximal convoluted tubule (Fig. 23.18). Two thirds is reabsorbed via a paracellular route in response to solvent drag and the small lumen positive potential found in the late proximal convoluted tubule. One third is reabsorbed via a transcellular route that includes Ca^{2+} channels in the apical cell membrane and a primary Ca^{2+}-ATPase or 3 Na^+/1 Ca^{2+} exchanger in the basolateral cell membrane. About 30% of filtered Ca^{2+} is reabsorbed along the loop of Henle. Most of the Ca^{2+} reabsorbed in the thick ascending limb is reabsorbed by passive transport through the tight junctions, propelled by the lumen positive potential.

Reabsorption continues along the distal convoluted tubule. Reabsorption here is increased by thiazide diuretics, which may be prescribed in cases of excess Ca^{2+} in the urine (**hypercalciuria**) and **nephrolithiasis** (kidney stone disease). Thiazides inhibit the luminal membrane Na-Cl cotransporter in distal convoluted tubule cells, which leads to a fall in intracellular Na^+ level. In turn, this promotes Na^+/Ca^{2+} exchange and increased basolateral extrusion of Ca^{2+} and, hence, increased Ca^{2+} reabsorption. Ca^{2+} reabsorption continues in the connecting tubules and collecting ducts. The parathyroid hormone (PTH) is the primary hormonal regulator of Ca^{2+} excretion and increases Ca^{2+}

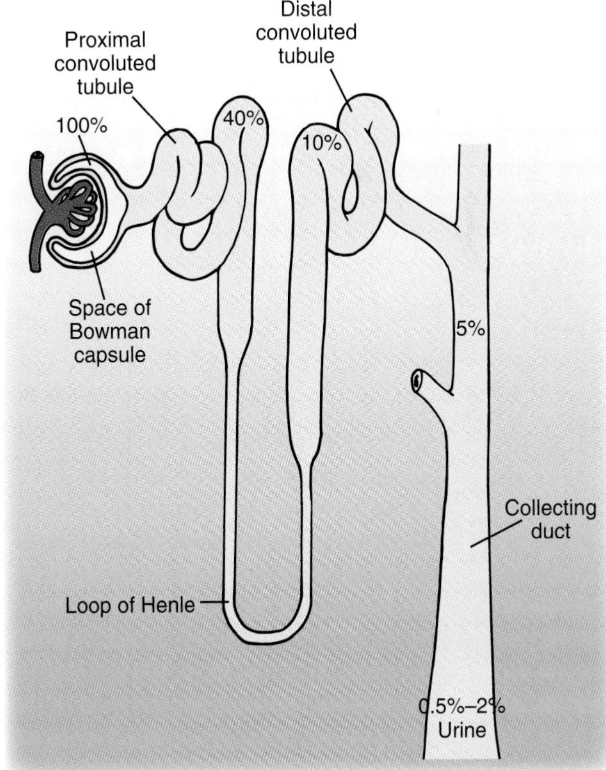

Figure 23.18 The percentage of the filtered load of calcium remaining in tubular fluid as it flows down the nephron. The kidneys filter about 10,800 mg/d (0.6 × 100 mg/L × 180 L/d) and reabsorb about 60% of the filtered load in the proximal convoluted tubule and another 30% in the loop of Henle. About 0.5% to 2% of the filtered load (about 50–200 mg Ca^{2+}/d) is excreted.

reabsorption in the thick ascending limb, distal convoluted tubule, and connecting tubule. Only about 0.5% to 2% of the filtered Ca^{2+} is usually excreted. (Chapter 35, "Endocrine Regulation of Calcium, Phosphate, and Bone Homeostasis," discusses Ca^{2+} balance and its control by several hormones in more detail.)

MAGNESIUM BALANCE

An adult body contains about 2,000 mEq of Mg^{2+}, of which about 60% is present in bone, 39% is in cells, and 1% is in the ECF. Mg^{2+} is the second most abundant cation in cells, after K^+ (see Table 23.2). The bulk of intracellular Mg^{2+} is not free but is bound to a variety of organic compounds such as ATP. Mg^{2+} is present in the plasma at a concentration of about 1 mmol/L (2 mEq/L). About 20% of plasma Mg^{2+} is bound to plasma proteins, 20% is complexed with various anions, and 60% is free or ionized.

About 25% of the Mg^{2+} filtered by the glomeruli is reabsorbed in the proximal convoluted tubule (Fig. 23.19); this is a lower percentage than for Na^+, K^+, Ca^{2+}, or water. The proximal tubule epithelium is rather impermeable to Mg^{2+} under normal conditions, and so there is little passive Mg^{2+} reabsorption. The major site of Mg^{2+} reabsorption is the loop of Henle (mainly the thick ascending limb), which reabsorbs about 65% of filtered Mg^{2+}. Reabsorption here is mainly passive and occurs through the tight junctions, driven by the lumen positive potential. Recent studies have identified a tight junction protein that is a channel that facilitates Mg^{2+} movement. Changes in Mg^{2+} excretion result mainly from changes in loop transport. More distal portions of the nephron reabsorb only a small fraction of filtered Mg^{2+} and, under normal circumstances, appear to play a minor role in controlling Mg^{2+} excretion.

An abnormally low plasma Mg^{2+} level is characterized by neuromuscular and central nervous system hyperirritability and cardiac arrhythmias. Abnormally high plasma Mg^{2+} levels have a sedative effect and may cause cardiac arrest. Dietary intake of Mg^{2+} is usually 20 to 50 mEq/d; two thirds is excreted in the feces and one third in the urine. The kidneys are mainly responsible for regulating the plasma Mg^{2+} level. The kidneys rapidly excrete excess amounts of Mg^{2+}. In Mg^{2+}-deficient states, Mg^{2+} virtually disappears from the urine.

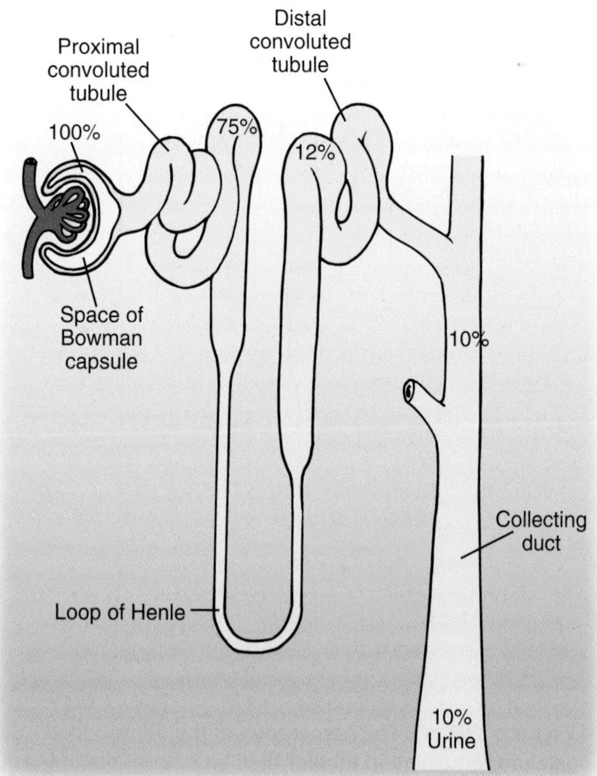

Figure 23.19 The percentage of the filtered load of magnesium remaining in tubular fluid as it flows down the nephron. The loop of Henle, specifically the thick ascending limb, is the major site of reabsorption of filtered Mg^{2+}.

PHOSPHATE BALANCE

A normal plasma concentration of inorganic phosphate is about 1 mmol/L. At a normal blood pH of 7.4, 80% of the phosphate is present as HPO_4^{2-} and 20% is present as $H_2PO_4^-$. Phosphate plays a variety of roles in the body. It is an important constituent of bone; it plays a critical role in cell metabolism, structure, and regulation (as organic phosphates); and it is a pH buffer.

Phosphate is mainly unbound in the plasma and is therefore freely filtered by the glomeruli. About 60% to 70% of filtered phosphate is actively reabsorbed in the proximal convoluted tubule, and another 15% is reabsorbed by the proximal straight tubule via a Na^+-phosphate cotransporter in the luminal cell membrane (Fig. 23.20). The remaining portions of the nephron and collecting ducts reabsorb little, if any, phosphate. The proximal tubule is the major site of phosphate reabsorption. Only about 5% to 20% of filtered phosphate is usually excreted. Phosphate in the urine is an important pH buffer and contributes to titratable acid excretion (see Chapter 24, "Acid–Base Homeostasis").

Phosphate reabsorption is tubular transport maximum (Tm)–limited (see Chapter 22, "Kidney Function"), and the amounts of phosphate filtered usually exceed the maximum reabsorptive capacity of the tubules for phosphate. This is different from the situation for glucose, in which normally less glucose is filtered than can be reabsorbed. If more phosphate is ingested and absorbed by the intestine, then plasma phosphate rises, more phosphate is filtered, and the filtered load exceeds the Tm more than usual, so the extra phosphate is excreted. Thus, the kidneys participate in regulating the plasma phosphate by an "overflow"-type of mechanism. When there is an excess of phosphate in the body, they automatically increase phosphate excretion. In cases of phosphate depletion, the kidneys filter less phosphate and the tubules reabsorb a larger percentage of the filtered phosphate.

Phosphate reabsorption in the proximal tubule is controlled by a variety of factors. PTH is of particular importance; it decreases the phosphate Tm, thereby increasing phosphate excretion.

The response to PTH is rapid (minutes) and involves endocytosis of Na^+-phosphate cotransporters from the apical cell membrane and subsequent degradation of the cotransporters in lysosomes. **Fibroblast growth factor-23 (FGF-23)** is another protein hormone that inhibits tubular phosphate

Clinical Focus / 23.2 

Kidney Stone Disease (Nephrolithiasis)

A kidney stone is a hard mass that forms in the urinary tract. About 5% of American women and 12% of men develop a kidney stone some time in their life. A stone lodged in the ureter causes bleeding and intense pain. **Nephrolithiasis** causes considerable suffering and loss of time from work, and it may lead to kidney damage. Once a stone forms in a person, stone formation often recurs.

Stones form when poorly soluble substances in the urine precipitate out of solution, causing crystals to form, aggregate, and grow. About 80% of kidney stones are composed of insoluble Ca^{2+} salts of oxalate and phosphate. There may be excessive amounts of Ca^{2+} or oxalate in the urine as a result of diet, a genetic defect, or unknown causes. Stones may also form from precipitated ammonium magnesium phosphate (struvite), uric acid, and cystine. Struvite stones (about 10% of all stones) are the result of infection with bacteria, usually *Proteus* species. Uric acid stones (9% of all stones) may form in patients with excessive uric acid production and excretion, as occurs in some patients with gout. Defective tubular reabsorption of cystine (in patients with **cystinuria**) leads to cystine stones (1% of stones). The rather insoluble amino acid cystine was first isolated from a urinary bladder stone by Wollaston in 1810, hence its name. Since a low urine flow rate raises the concentration of all poorly soluble substances in the urine, thereby favoring precipitation, a key to prevention of kidney stones is to drink plenty of water and maintain a high urine output day and night.

Fortunately, most stones are small enough to be passed down the urinary tract and spontaneously eliminated. Microscopic and chemical examination of the eliminated stones is used to determine the nature of the stone and helps guide treatment. Sometimes a change in diet is recommended to reduce the amount of potential stone-forming material (e.g., Ca^{2+}, oxalate, or uric acid). Thiazide diuretics are useful in reducing Ca^{2+} excretion if excessive urinary Ca^{2+} excretion (hypercalciuria) is the problem. Potassium citrate is useful in treating most stone disease, because citrate complexes Ca^{2+} in the urine and inhibits the crystallization of Ca^{2+} salts. It also makes the urine more alkaline (since citrate is oxidized to HCO_3^- in the body). This treatment is also helpful in reducing the risk of uric acid stones, because urates (favored in alkaline urine) are more soluble than uric acid (the form favored in acidic urine). Administering an inhibitor of uric acid synthesis, such as allopurinol, can help reduce the amount of uric acid in the urine.

If the stone is not passed spontaneously, extracorporeal shock wave lithotripsy, using a lithotripter, is widely used. The patient is placed in a tub of water, and the stone is localized by *x* ray imaging. Sound waves are generated in the water by high-voltage electric discharges and are focused on the stone through the body wall. The shock waves break the stone into small pieces that can pass down the urinary tract and be eliminated. Some renal injury is produced by this procedure, and so it may not be entirely innocuous. Other procedures include passing a tube with an ultrasound transducer through the skin into the renal pelvis; stone fragments can be removed directly. A ureteroscope with a laser can also be used to break up stones.

reabsorption; elevated plasma levels cause hypophosphatemia and, consequently, rickets or osteomalacia.

Patients with chronic renal disease often have an elevated plasma phosphate (**hyperphosphatemia**) level, depending on the severity of the disease. When the GFR falls, the filtered phosphate load is diminished and the tubules reabsorb phosphate more completely. Phosphate excretion is inadequate in the face of continued intake of phosphate in the diet. Hyperphosphatemia is dangerous because of the precipitation of calcium phosphate in soft tissues. For example, when calcium phosphate precipitates in the walls of blood vessels, blood flow is impaired. Hyperphosphatemia can lead to myocardial failure and pulmonary insufficiency.

When plasma phosphate level rises, the plasma ionized Ca^{2+} level tends to fall, for two reasons. First, phosphate forms a complex with Ca^{2+}. Second, hyperphosphatemia decreases production of 1,25(OH)$_2$ vitamin D$_3$ in the kidneys by inhibiting the 1α-hydroxylase enzyme that forms this hormone. With decreased plasma levels of 1,25(OH)$_2$ vitamin D$_3$, there is less Ca^{2+} absorption by the small intestine and hence a tendency for hypocalcemia.

A low plasma ionized Ca^{2+} level stimulates hyperplasia of the parathyroid glands and increased secretion of PTH. A high plasma phosphate level also stimulates PTH secretion directly. PTH then inhibits phosphate reabsorption by the proximal tubules, promotes phosphate excretion, and helps return plasma phosphate levels back to normal. Elevated PTH levels, however, also cause mobilization of Ca^{2+} and phosphate from bone. Increased bone resorption results, and the bone minerals are replaced with fibrous tissue that renders the bone more susceptible to fracture.

Patients with advanced chronic renal failure are often advised to restrict phosphate intake and consume substances (such as Ca^{2+} salts) that bind phosphate in the intestines, so as to avoid the many problems caused by hyperphosphatemia. Administration of synthetic 1,25(OH)$_2$ vitamin D$_3$ may compensate for deficient renal production of this hormone. This hormone opposes hypocalcemia and inhibits PTH synthesis and secretion. Parathyroidectomy is sometimes necessary in patients with advanced chronic renal failure.

URINARY TRACT

Recall that the urinary system is the system that produces, stores, and eliminates urine. In humans, the system includes two kidneys, two ureters, the bladder, the urethra, and two sphincter muscles.

Urinary tract provides the pathway for transporting, storing, and eliminating urine.

The kidneys form urine continuously. The urine is transported by the ureters to the urinary bladder. The bladder is

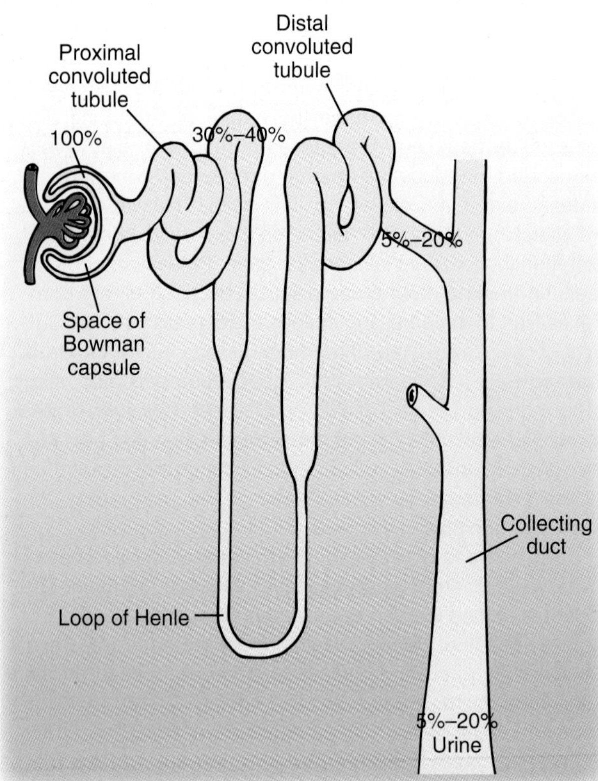

Figure 23.20 The percentage of the filtered load of phosphate remaining in tubular fluid as it flows down the nephron. The proximal tubule is the major site of phosphate reabsorption, and downstream nephron segments reabsorb little, if any, phosphate.

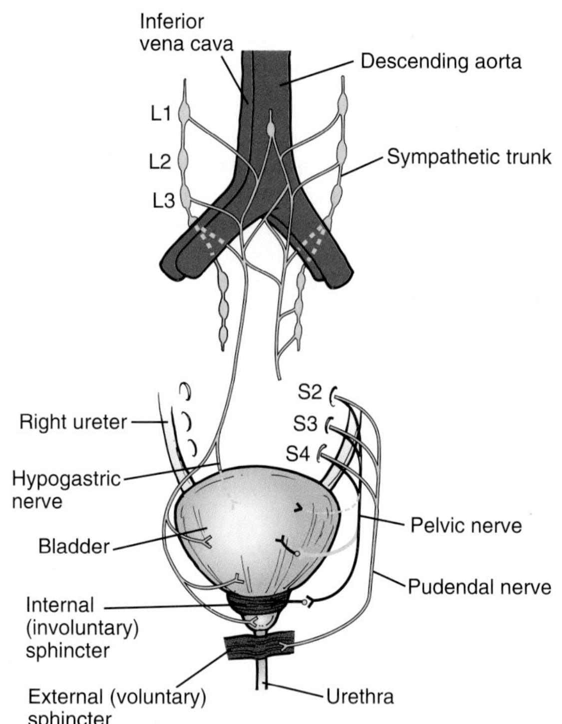

Figure 23.21 The innervation of the urinary bladder. The parasympathetic pelvic nerves arise from the S2 to S4 segments of the spinal cord and supply motor fibers to the bladder musculature and internal (involuntary) sphincter. Sympathetic motor fibers supply the bladder via the hypogastric nerves, which arise from lumbar segments of the spinal cord. The pudendal nerves supply somatic motor innervation to the external (voluntary) sphincter. Sensory afferents (*yellow dashed lines*) from the bladder travel mainly in the pelvic nerves but also to some extent in the hypogastric nerves.

specialized to fill with urine at a low pressure but then to empty its contents when appropriate. Contractions of the bladder and its sphincters are controlled by the nervous system.

Ureters

The **ureters** are muscular tubes that propel the urine from the pelvis of each kidney to the urinary bladder. Peristaltic movements originate in the region of the calices, which contain specialized smooth muscle cells that generate spontaneous pacemaker potentials. These pacemaker potentials trigger action potentials and contractions in the muscular regions of the renal pelvis that propagate distally to the ureter. Peristaltic waves sweep down the ureters at a frequency of one every 10 seconds to one every 2 to 3 minutes. The ureters enter the base of the bladder obliquely, thus forming a valvular flap that passively prevents the reflux of urine during contractions of the bladder. The ureters are innervated by sympathetic and parasympathetic nerve fibers. Sensory fibers mediate the intense pain that is felt when a stone distends or blocks a ureter.

Bladder

The urinary bladder is a distensible hollow vessel containing smooth muscle in its wall (Fig. 23.21). The muscle is called the *detrusor,* from Latin for "that which pushes down." The neck of the bladder, the involuntary internal sphincter, also contains smooth muscle. The parasympathetic pelvic nerves and sympathetic hypogastric nerves innervate the body of the bladder and bladder neck. The external sphincter is composed of skeletal muscle and is innervated by somatic nerve fibers that travel in the pudendal nerves. Pelvic, hypogastric, and pudendal nerves contain both motor and sensory fibers.

The bladder has two functions: to serve as a distensible reservoir for urine and to empty its contents at appropriate intervals. When the bladder fills, it adjusts its tone to its content, so that minimal increases in bladder pressure occur. The external sphincter is kept closed by discharges along the pudendal nerves. The first sensation of bladder filling is experienced at a volume of 100 to 150 mL in an adult, and the first desire to void is elicited when the bladder contains about 150 to 250 mL of urine. A person becomes uncomfortably aware of a full bladder when the volume is 350 to 400 mL; at this volume, hydrostatic pressure in the bladder is about 10 cm H_2O. With further volume increases, bladder pressure rises steeply, in part because of reflex contractions of the detrusor. An increase in volume to 700 mL creates pain and, often, loss of control. The sensations of bladder filling, of conscious desire to void, and of painful distention are mediated by afferents in the pelvic nerves.

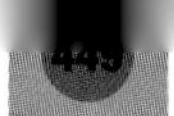

Urethra

The **urethra** is a tube that connects the urinary bladder to the exterior of the body. In males, the urethra travels through the penis and carries both urine and semen. In females, the urethra is shorter and emerges in front of the vaginal opening.

Micturition

Micturition (urination), the periodic emptying of the bladder, is a complex act involving both autonomic and somatic nerve pathways and several reflexes that can be either inhibited or facilitated by higher centers in the brain. The basic reflexes occur at the level of the sacral spinal cord and are modified by centers in the midbrain and cerebral cortex. Distention of the bladder is sensed by stretch receptors in the bladder wall; these induce reflex contraction of the detrusor and relaxation of the internal and external sphincters. This reflex is released by removing inhibitory influences from the cerebral cortex. Fluid flow through the urethra reflexly causes further contraction of the detrusor and relaxation of the external sphincter. Increased parasympathetic nerve activity stimulates contraction of the detrusor and relaxation of the internal sphincter. Sympathetic innervation is not essential for micturition. During micturition, the perineal and levator ani muscles relax, thereby shortening the urethra and decreasing urethral resistance. Descent of the diaphragm and contraction of abdominal muscles raise intra-abdominal pressure and aid in the expulsion of urine from the bladder.

Micturition is, fortunately, under voluntary control in healthy adults. In the young child, however, it is purely reflex and occurs whenever the bladder is sufficiently distended. At about age 2.5 years, it begins to come under cortical control, and, in most children, complete control is achieved by age 3 years. Damage to the nerves that supply the bladder and its sphincters can produce abnormalities of micturition and incontinence. An increased resistance of the upper urethra commonly occurs in older men and is a result of enlargement of the surrounding prostate gland. This condition is called *benign prostatic hyperplasia* (also called *BPH*), and it results in decreased urine stream, overdistention of the bladder as a result of incomplete emptying, and increased urgency and frequency of urination.

Chapter Summary

- Total body water is distributed in two major compartments: intracellular water and extracellular water. In an average young adult man, total body water, intracellular water, and extracellular water amount to 60%, 40%, and 20% of body weight, respectively. The corresponding figures for an average young adult woman are 50%, 30%, and 20% of body weight.
- The volumes of body fluid compartments are determined by using the indicator–dilution method and the following equation: Volume = Amount of indicator/ Concentration of indicator at equilibrium.
- Electrical neutrality is present in solutions of electrolytes; that is, the sum of the cations is equal to the sum of the anions (both expressed in milliequivalents).
- Sodium (Na^+) is the major osmotically active solute in the extracellular fluid (ECF) compartment, and potassium (K^+) has the same role in the intracellular fluid compartment. Cells are typically in osmotic equilibrium with their external environment. The amount of water in (and hence volume of) cells depends on the amount of K^+ they contain and, similarly, the amount of water in (and hence volume of) the ECF is determined by its Na^+ content.
- Plasma osmolality is closely regulated by arginine vasopressin, which governs renal excretion of water, and by habit and thirst, which govern water intake.
- Arginine vasopressin (AVP) is synthesized in the hypothalamus, is released from the posterior pituitary gland, and acts on the collecting ducts of the kidney to increase their water permeability. The major stimuli for the release of AVP are an increase in effective plasma osmolality (detected by osmoreceptors in the anterior hypothalamus) and a decrease in blood volume (detected by stretch receptors in the left atrium, carotid sinuses, and aortic arch).
- The kidneys are the primary site of control of Na^+ excretion. Only a small percentage of the filtered Na^+ is usually excreted in the urine, but this amount is of critical importance in overall Na^+ balance.
- Multiple factors affect Na^+ excretion, including glomerular filtration rate, angiotensin II and aldosterone, intrarenal pressures, natriuretic hormones, such as atrial natriuretic peptide, and renal sympathetic nerves. Changes in these factors may account for altered Na^+ excretion in response to excess Na^+ or Na^+ depletion. Estrogens, glucocorticoids, osmotic diuretics, poorly reabsorbed anions in the urine, and diuretic drugs also affect renal Na^+ excretion.
- The effective arterial blood volume (EABV) depends on the degree of filling of the arterial system and determines the perfusion of the body's tissues. A decrease in EABV leads to Na^+ retention by the kidneys and contributes to the development of generalized edema in pathophysiologic conditions such as congestive heart failure. The kidneys play a major role in the control of K^+ balance. K^+ is reabsorbed by the proximal convoluted tubule and loop of Henle and is secreted by cortical collecting duct principal cells. Inadequate renal K^+ excretion produces hyperkalemia, and excessive K^+ excretion produces hypokalemia.
- Calcium balance is regulated on both input and output sides. The absorption of Ca^{2+} from the small intestine is controlled by $1,25(OH)_2$ vitamin D_3, and the excretion of Ca^{2+} by the kidneys is controlled by parathyroid hormone.
- Magnesium in the body is mostly in bone, but it is also an important intracellular ion. The kidneys regulate the plasma Mg^{2+}.
- Filtered phosphate usually exceeds the maximal reabsorptive capacity of the kidney tubules for phosphate (Tm_{PO4}), and about 5% to 20% of filtered phosphate is usually excreted. Phosphate reabsorption occurs mainly in the proximal tubules and is inhibited by parathyroid hormone and fibroblast growth factor-23. Phosphate is an important pH buffer in the urine. Hyperphosphatemia is a significant problem in chronic renal failure.
- The urinary bladder stores urine until it can be conveniently emptied. Micturition is a complex act involving both autonomic and somatic nerves.

thePoint *Visit http://thePoint.lww.com for chapter review Q&A, case studies, animations, and more!*

24

Acid–Base Homeostasis

ACTIVE LEARNING OBJECTIVES

Upon mastering the material in this chapter you should be able to:

- Define acid, base, acid dissociation constant, weak and strong acids, pK_a, pH, and buffer.
- Describe the metabolic processes that produce and consume hydrogen ions, and explain why the body is threatened by net acid gain on a mixed diet.
- Describe the three mechanisms that defend blood pH.
- List the main chemical buffers present in extracellular fluid, intracellular fluid, and bone.
- Use the Henderson–Hasselbalch equation for the HCO_3^-/CO_2 buffer pair to calculate the blood pH, given values for plasma HCO_3^- and partial pressure of carbon dioxide (P_{CO_2}).
- Explain why the bicarbonate/CO_2 buffer pair is so effective in pH buffering in the body.
- State the isohydric principle and its significance.
- Describe the respiratory responses to an increase or decrease of arterial blood pH or P_{CO_2}.
- Calculate renal net acid excretion from excretion of titratable acid, NH_4^+, and bicarbonate.
- Describe the three processes involved in urinary acidification: reabsorption of filtered bicarbonate, formation of titratable acid, and excretion of NH_4^+. Indicate where along the nephron these processes take place and which processes generate new bicarbonate.
- Discuss the factors that influence renal secretion and excretion of hydrogen ions.
- Describe the mechanisms that maintain stability of intracellular pH.
- List the four simple acid–base disturbances and describe for each the primary defect, changes in arterial blood chemistry (pH, P_{CO_2}, and plasma HCO_3^-), common causes, chemical buffering processes, and respiratory and renal compensations.
- Identify the type of acid–base disturbance present from blood acid–base data and a patient history.
- Calculate the anion gap from plasma electrolyte concentrations, and interpret its meaning.

Acid–base homeostasis is concerned with maintaining the proper balance between acids and bases in the body (i.e., pH). The body is extremely sensitive to changes in pH, and, as a result, powerful mechanisms exist to tightly regulate the body's acid–base balance to maintain it in a very narrow range. Outside the acceptable range of pH, proteins are denatured, and enzyme activity and nerve and cardiac function are altered, which ultimately results in death.

Most of this chapter discusses the regulation of H^+ in extracellular fluid (ECF), because ECF is easier to analyze than intracellular fluid (ICF) and is the fluid used in the clinical evaluation of acid–base balance. In practice, systemic *arterial* blood is used as the reference for this purpose. Measurements on whole blood with a pH meter give values for the H^+ of plasma and, therefore, provide an ECF pH measurement.

BASIC PRINCIPLES OF ACID–BASE CHEMISTRY

In this section, we briefly review some principles of acid–base chemistry. We define acid, base, acid dissociation constant, weak and strong acids, pK_a, pH, and the Henderson–Hasselbalch equation and explain buffering.

Acids dissociate to release hydrogen ions in solution.

An **acid** is a substance that can release, or donate, H^+; a **base** is a substance that can combine with, or accept, H^+. When an acid (generically written as HA) is added to water, it dissociates reversibly according to the reaction: $HA \rightleftharpoons H^+ + A^-$. The species A^- is a base, because it can combine with an H^+ to form HA. In other words, when an acid dissociates, it yields a free H^+ and its conjugate base. (*Conjugate* means "joined in a pair.")

Acid dissociation constant shows the strength of an acid.

At equilibrium, the rate of dissociation of an acid to form $H^+ + A^-$ and the rate of association of H^+ and base A^- to form HA are equal. The equilibrium constant (K_a), which is also called the *ionization constant* or **acid dissociation constant**, is given by the expression

$$K_a = \frac{[H^+] \times [A^-]}{[HA]} \quad (1)$$

The higher the acid dissociation constant, the more an acid is ionized and the greater is its strength. Hydrochloric acid (HCl) is an example of a **strong acid**. It has a high K_a and is almost completely ionized in aqueous solutions. Other strong acids include sulfuric acid (H_2SO_4), phosphoric acid (H_3PO_4), and nitric acid (HNO_3).

An acid with a low K_a is a **weak acid**. For example, in a 0.1 M solution of acetic acid ($K_a = 1.8 \times 10^{-5}$) in water, most (99%) of the acid is nonionized, so that little (1%) is present as acetate$^-$ and H^+. The acidity (concentration of free H^+) of this solution is low. Other weak acids are lactic acid, carbonic acid (H_2CO_3), ammonium ion (NH_4^+), and dihydrogen phosphate ($H_2PO_4^-$).

pK_a is a logarithmic expression of K_a.

Acid dissociation constants vary widely and often are small numbers. It is convenient to convert K_a to a logarithmic form, defining pK_a as

$$pK_a = \log_{10}(1/K_a) = -\log_{10}K_a \qquad (2)$$

In aqueous solution, each acid has a characteristic pK_a, which varies slightly with temperature and the ionic strength of the solution. Note that pK_a is *inversely* proportional to acid strength. A strong acid has a high K_a and a low pK_a. A weak acid has a low K_a and a high pK_a.

pH is inversely related to hydrogen ion concentration.

H^+ is often expressed in pH units. The following equation defines **pH**:

$$pH = \log_{10}(1/[H^+]) = -\log_{10}[H^+] \qquad (3)$$

where H^+ concentration is in mol/L. Note that pH is *inversely* related to H^+ concentration. Each whole number on the pH scale represents a 10-fold (logarithmic) change in acidity. A solution with a pH of 5 has 10 times the H^+ of a solution with a pH of 6.

Henderson–Hasselbalch equation relates pH to the ratio of the concentrations of conjugate base and acid.

For a solution containing an acid and its conjugate base, we can rearrange the equilibrium expression (equation 1) as

$$[H^+] = \frac{K_a \times [HA]}{[A^-]} \qquad (4)$$

If we take the negative logarithms of both sides,

$$-\log[H^+] = -\log K_a + \log\frac{[A^-]}{[HA]} \qquad (5)$$

Substituting pH for $-\log [H^+]$ and pK_a for $-\log K_a$, we get

$$pH = pK_a + \log\frac{[A^-]}{[HA]} \qquad (6)$$

This equation is known as the **Henderson–Hasselbalch equation**. It shows that the pH of a solution is determined by the pK_a of the acid and the ratio of the concentrations of conjugate base to acid.

Buffers promote the stability of pH.

The stability of pH is protected by the action of buffers. A **pH buffer** is defined as an agent that *minimizes* the change in pH produced when an acid or base is added. Note that a buffer *does not prevent* a pH change. A **chemical pH buffer** is a mixture of a weak acid and its conjugate base (or a weak base and its conjugate acid). Following are examples of buffers:

Weak Acid		Conjugate Base	
H_2CO_3 (carbonic acid)	$\rightleftharpoons$	$HCO_3^- + H^+$ (bicarbonate)	(7)
$H_2PO_4^-$ (dihydrogen phosphate)	$\rightleftharpoons$	$HPO_4^{2-} + H^+$ (monohydrogen phosphate)	(8)
NH_4^+ (ammonium ion)	$\rightleftharpoons$	$NH_3 + H^+$ (ammonia)	(9)

Generally speaking, the equilibrium expression for a buffer pair can be written in terms of the Henderson–Hasselbalch equation:

$$pH = pK_a + \log\frac{[\text{conjugate base}]}{[\text{acid}]} \qquad (10)$$

For example, for $H_2PO_4^-/HPO_4^{2-}$

$$pH = 6.8 + \log\frac{[HPO_4^{2-}]}{[H_2PO_4^-]} \qquad (11)$$

The effectiveness of a buffer—how well it minimizes pH changes when an acid or base is added—depends on its concentration and its pK_a. A good buffer is present in high concentrations and has a pK_a close to the desired pH.

Figure 24.1 shows a titration curve for the phosphate buffer system. As a strong acid or strong base is progressively added to the solution (shown on the *x* axis), the resulting pH is recorded (shown on the *y* axis). Going from right to left, as strong acid is added, H^+ combines with the basic form of phosphate: $H^+ + HPO_4^{2-} \rightleftharpoons H_2PO_4^-$. Going from left to right, as strong base is added, OH^- combines with H^+ released from the acid form of the phosphate buffer: $OH^- + H_2PO_4^- \rightleftharpoons HPO_4^{2-} + H_2O$. These reactions lessen the fall or rise in pH.

At the pK_a of the phosphate buffer, the ratio $HPO_4^{2-}/H_2PO_4^-$ is 1, and the titration curve is flattest (the change in pH for a given amount of an added acid or base is at a minimum). In most cases, pH buffering is effective when the solution pH is within ±1 pH unit of the buffer pK_a. Beyond that range, the pH shift that a given amount of acid or base produces may be quite large, so the buffer becomes relatively ineffective.

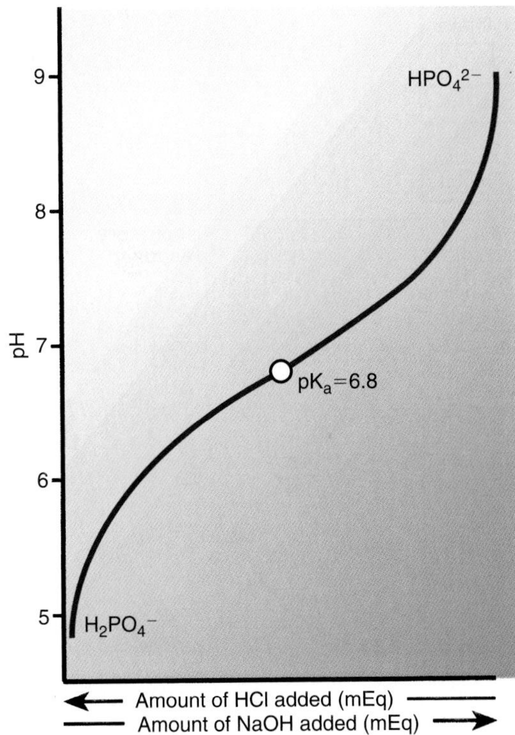

Figure 24.1 Titration curve for a phosphate buffer. The pK_a for $H_2PO_4^-$ is 6.8. A strong acid (HCl) (*right to left*) or strong base (NaOH) (*left to right*) was added and the resulting solution pH recorded (*y* axis). Notice that buffering is best (i.e., the change in pH on the addition of a given amount of acid or base is least) when the solution pH is equal to the pK_a of the buffer.

ACID PRODUCTION

Acids are continuously produced in the body and threaten the normal pH of the ECF and ICF. Physiologically, acids fall into two groups: (1) H_2CO_3 (carbonic acid) and (2) all other acids (noncarbonic; also called "nonvolatile" or "fixed" acids). The distinction between these groups arises because H_2CO_3 is in equilibrium with the volatile gas CO_2, which can leave the body via the lungs. The concentration of H_2CO_3 in arterial blood is, therefore, set by respiratory activity. By contrast, noncarbonic acids in the body are not directly affected by breathing. Noncarbonic acids are buffered in the body and are then excreted by the kidneys.

Metabolism is a constant source of carbon dioxide.

A normal adult produces about 300 L of CO_2 daily from metabolism. CO_2 from the tissues enters the capillary blood, where it reacts with water to form H_2CO_3, which dissociates instantly to yield H^+ and HCO_3^-: $CO_2 + H_2O \rightleftharpoons H_2CO_3 \rightleftharpoons H^+ + HCO_3^-$. Blood pH would rapidly fall to lethal levels if the H_2CO_3 formed from CO_2 were allowed to accumulate in the body.

Fortunately, H_2CO_3 is converted to CO_2 and water in the pulmonary capillaries and the CO_2 is expired (see Chapter 19, "Gas Transfer and Transport"). In the lungs, the reactions reverse:

$$H^+ + HCO_3^- \rightleftharpoons H_2CO_3 \rightleftharpoons H_2O + CO_2 \qquad (12)$$

As long as CO_2 is expired as fast as it is produced, then arterial blood CO_2 tension, H_2CO_3 concentration, and pH do not change.

Incomplete carbohydrate and fat metabolism produces nonvolatile acids.

Normally, carbohydrates and fats are completely oxidized to CO_2 and water. If carbohydrates and fats are *incompletely* oxidized, nonvolatile acids are produced. Incomplete oxidation of carbohydrates occurs when the tissues do not receive enough oxygen, as during strenuous exercise or hemorrhagic or cardiogenic shock. In such states, glucose metabolism yields lactic acid (pK_a = 3.9), which dissociates into lactate$^-$ and H^+, lowering the blood pH. Incomplete fatty acid oxidation occurs in uncontrolled diabetes mellitus, starvation, and alcoholism, and produces ketone body acids (acetoacetic and β-hydroxybutyric acids). These acids have pK_a values around 4 to 5. At blood pH, they mostly dissociate into their anions and H^+, making the blood more acidic.

Protein metabolism generates strong acids.

The metabolism of dietary proteins is a major source of H^+. The oxidation of proteins and amino acids produces strong acids, such as H_2SO_4, HCl, and H_3PO_4. The oxidation of sulfur-containing amino acids (methionine, cysteine, and cystine) produces H_2SO_4, and the oxidation of cationic amino acids (arginine, lysine, and some histidine residues) produces HCl. The oxidation of phosphorus-containing proteins and nucleic acids produces H_3PO_4.

Mixed diet of meat and vegetables produces a net acid gain that threatens pH.

A diet containing both meat and vegetables results in a net production of acids, largely from protein oxidation. To some extent, acid-consuming metabolic reactions balance H^+ production. Food also contains basic anions, such as citrate, lactate, and acetate. When these are oxidized to CO_2 and water, hydrogen ions are consumed (or what amounts to the same thing, HCO_3^- is produced). The balance of acid-forming and acid-consuming metabolic reactions results in net production of about 1 mEq H^+/kg body weight per day in an adult person who eats a mixed diet. Vegetarians generally have less of a dietary acid burden and a more alkaline urine pH than nonvegetarians, because most fruits and vegetables contain large amounts of organic anions that are metabolized to HCO_3^-. The body generally has to dispose of more or less nonvolatile acid, a function performed by the kidneys.

Whether a particular food has an acidifying or an alkalinizing effect depends on whether and how its constituents are metabolized. Cranberry juice has an acidifying effect because of its content of benzoic acid, an acid that cannot be broken down in the body. Orange juice has an alkalinizing effect, despite its acidic pH of about 3.7, because it contains citrate, which is metabolized to HCO_3^-. The citric acid in orange juice is converted to CO_2 and water, has only a transient effect on blood pH, and has no effect on urine pH.

BLOOD pH REGULATION

The body contains several different buffers that reversibly bind H^+ and blunt any change in pH. These buffers include bicarbonate, protein, phosphate, and others.

The bicarbonate buffering system is especially important, as carbon dioxide (CO_2) can be shifted through carbonic acid (H_2CO_3) to hydrogen ions and bicarbonate (HCO_3^-), as shown previously in equation 12. Figure 24.2 shows key buffering mechanisms that tightly regulate pH despite the daily net acid gain. Buffering is accomplished by chemical buffers, the lungs, and the kidneys.

- *Chemical buffering.* Chemical buffers in ECF and ICF and in bone are the first line of defense of blood pH. Chemical buffering minimizes a change in pH but does not remove acid or base from the body.
- *Respiratory response.* The respiratory system is the second line of defense of blood pH. Normally, breathing removes CO_2 as fast as it is produced. Large loads of acid stimulate breathing (respiratory compensation), which removes CO_2 from the body and thus lowers the H_2CO_3 in arterial blood, reducing the acidic shift in blood pH.
- *Renal response.* The kidneys are the third line of defense of blood pH. Although chemical buffers in the body can bind H^+ and the lungs can change the H_2CO_3 of blood, the burden of removing excess H^+ falls directly on the kidneys. Hydrogen ions are excreted in combination with urinary buffers. At the same time, the kidneys add new HCO_3^- to the ECF to replace HCO_3^- used to buffer strong acids. The kidneys also excrete the anions (phosphate, chloride, and sulfate) that are liberated from strong acids. The kidneys affect blood pH more slowly than other buffering mechanisms in the body; full renal compensation may take 1 to 3 days.

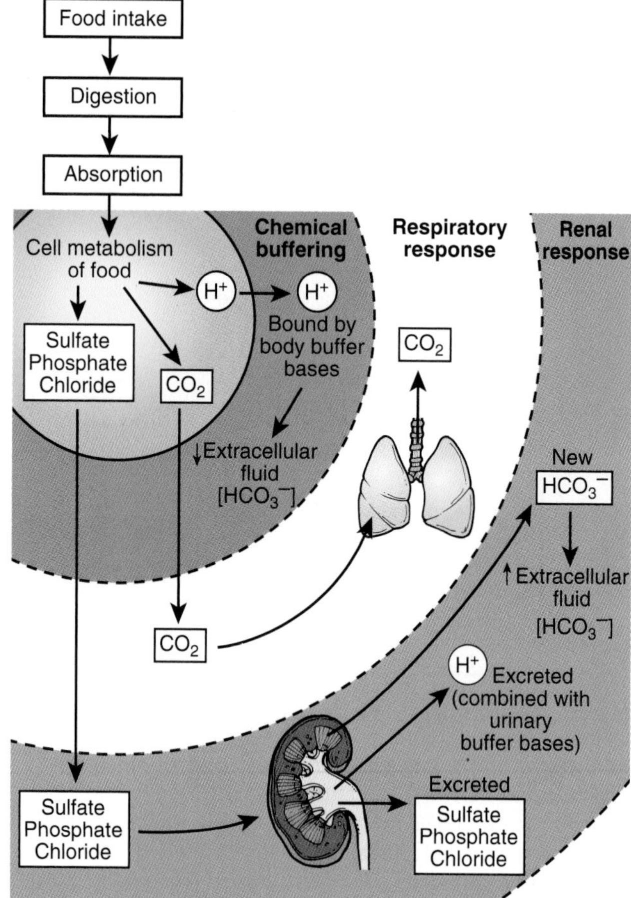

Figure 24.2 The maintenance of normal blood pH by chemical buffers, the respiratory system, and the kidneys. On a mixed diet, pH is threatened by the production of strong acids (sulfuric, hydrochloric, and phosphoric) mainly as a result of protein metabolism. These strong acids are buffered in the body by chemical buffer bases, such as extracellular fluid (ECF) HCO_3^-. The kidneys eliminate hydrogen ions (combined with urinary buffers) and anions in the urine. At the same time, they add new HCO_3^- to the ECF to replace the HCO_3^- consumed in buffering strong acids. The respiratory system disposes of CO_2.

Chemical buffers are the first line of defense against changes in pH.

The body contains many conjugate acid–base pairs that act as chemical buffers (Table 24.1). In the ECF, the main chemical buffer pair is HCO_3^-/CO_2. Plasma proteins and inorganic phosphate are also ECF buffers. Cells have large buffer stores, particularly proteins and organic phosphate compounds. HCO_3^- is present in cells, although at a lower concentration than in ECF. Bone contains large buffer stores, specifically salts of phosphate and carbonate.

When an acid or base is added to the body, the buffers just mentioned bind or release H^+, thereby minimizing the change in pH. Buffering in ECF occurs rapidly, in minutes. Acids or bases also enter cells and bone, but this generally occurs more slowly, over hours, allowing cell buffers and bone to share in buffering.

Phosphate buffer

The pK_a for phosphate, $H_2PO_4^- \rightleftharpoons H^+ + HPO_4^{2-}$, is 6.8, close to the desired blood pH of 7.4, so phosphate is a good buffer. In the ECF, phosphate is present as inorganic phosphate. Its concentration, however, is low (about 1 mmol/L), so it plays a minor role in extracellular buffering.

Phosphate is an important intracellular buffer, however, for two reasons. First, cells contain large amounts of phosphate in such organic compounds as adenosine triphosphate (ATP), adenosine diphosphate (ADP), and creatine phosphate. Although these compounds primarily function in energy metabolism, they also act as pH buffers. Second, intracellular pH is generally lower than the pH

TABLE 24.1 Major Chemical pH Buffers in the Body

Buffer	Reaction
Extracellular fluid	
Bicarbonate/CO_2	$CO_2 + H_2O \rightleftharpoons H_2CO_3 \rightleftharpoons H^+ + HCO_3^-$
Inorganic phosphate	$H_2PO_4^- \rightleftharpoons H^+ + HPO_4^{2-}$
Plasma proteins (Pr)	$HPr \rightleftharpoons H^+ + Pr^-$
Intracellular fluid	
Cell proteins (e.g., hemoglobin [Hb])	$HHb \rightleftharpoons H^+ + Hb^-$
Organic phosphates	organic-$HPO_4^- \rightleftharpoons H^+$ + organic-PO_4^{2-}
Bicarbonate/CO_2	$CO_2 + H_2O \rightleftharpoons H_2CO_3 \rightleftharpoons H^+ + HCO_3^-$
Bone	
Mineral phosphates	$H_2PO_4^- \rightleftharpoons H^+ + HPO_4^{2-}$
Mineral carbonates	$HCO_3^- \rightleftharpoons H^+ + CO_3^{2-}$

of ECF and is closer to the pK_a of phosphate. (The cytosol of skeletal muscle, e.g., has a pH of 6.9.) Phosphate is thus more effective in this environment than in one with a pH of 7.4. Bone has large phosphate salt stores, which also help in buffering.

Protein buffers

Proteins are the largest buffer pool in the body and are excellent buffers. Proteins can function as both acids and bases, so they are **amphoteric**. They contain many ionizable groups, which can release or bind H^+. Serum albumin and plasma globulins are the major extracellular protein buffers, present mainly in the blood plasma. Cells also have large protein stores. Recall that the buffering properties of hemoglobin play an important role in the transport of CO_2 and O_2 by the blood (see Chapter 19, "Gas Transfer and Transport").

Bicarbonate/carbon dioxide buffer

For several reasons, the HCO_3^-/CO_2 buffer pair is especially important in acid–base physiology:

- Its components are abundant; the concentration of HCO_3^- in plasma or ECF normally averages 24 mmol/L. Although the concentration of dissolved CO_2 is lower (1.2 mmol/L), metabolism provides a nearly limitless supply.
- Despite a pK of 6.10, a little far from the desired plasma pH of 7.40, it is effective because the system is "open" (i.e., its components can be added to or removed from the body at controlled rates).
- It is controlled by the lungs and kidneys.

CO_2 exists in the body in several different forms: as gaseous CO_2 in the lung alveoli and as dissolved CO_2, H_2CO_3, HCO_3^-, carbonate (CO_3^{2-}), and carbamino compounds in the body fluids. CO_3^{2-} is present at appreciable concentrations only in alkaline solutions, and so we will ignore it. We will also ignore any CO_2 that is bound to proteins in the carbamino form. The most important forms are gaseous CO_2, dissolved CO_2, H_2CO_3, and HCO_3^-.

Dissolved CO_2 in pulmonary capillary blood equilibrates with gaseous CO_2 in the lung alveoli. Consequently, the partial pressures of CO_2 (P_{CO_2}) in alveolar air and systemic arterial blood are normally identical. The concentration of dissolved CO_2 ($CO_{2(d)}$) is related to the P_{CO_2} by the Henry law (see Chapter 19, "Gas Transfer and Transport"). The solubility coefficient for CO_2 in plasma at 37°C is 0.03 mmol CO_2/L per mm Hg P_{CO_2}. Therefore, $CO_{2(d)} = 0.03 \times P_{CO_2}$. If P_{CO_2} is 40 mm Hg, then $CO_{2(d)}$ is 1.2 mmol/L.

In aqueous solutions, $CO_{2(d)}$ reacts with water to form H_2CO_3: $CO_{2(d)} + H_2O \rightleftharpoons H_2CO_3$. The reaction to the right is called the **hydration reaction** and the reaction to the left is called the **dehydration reaction.** These reactions are slow if uncatalyzed. In many cells and tissues, such as the kidneys, pancreas, stomach, and red blood cells, the reactions are catalyzed by **carbonic anhydrase**, a zinc-containing enzyme. At equilibrium, $CO_{2(d)}$ is greatly favored; at body temperature, the ratio of $CO_{2(d)}$ to H_2CO_3 is about 400:1. If $CO_{2(d)}$ is 1.2 mmol/L, then H_2CO_3 equals 3 μmol/L. H_2CO_3 dissociates instantaneously into H^+ and HCO_3^-: $H_2CO_3 \rightleftharpoons H^+ + HCO_3^-$. The Henderson–Hasselbalch expression for this reaction is as follows:

$$pH = 3.5 + \log\frac{[HCO_3^-]}{[H_2CO_3]} \tag{13}$$

Note that H_2CO_3 is a fairly strong acid ($pK_a = 3.5$). Its low concentration in body fluids lessens its impact on acidity.

Henderson–Hasselbalch equation.

Because plasma H_2CO_3 is so low and hard to measure and because $H_2CO_3 = CO_{2(d)}/400$, we can use $CO_{2(d)}$ to represent the acid in the Henderson–Hasselbalch equation:

$$pH = 3.5 + \log\frac{[HCO_3^-]}{[CO_{2(d)}]/400} = 3.5 + \log 400 + \log\frac{[HCO_3^-]}{[CO_{2(d)}]} \tag{14}$$

$$= 6.1 + \log\frac{[HCO_3^-]}{[CO_{2(d)}]}$$

We can also use $0.03 \times P_{CO_2}$ in place of $CO_{2(d)}$:

$$pH = 6.1 + \log\frac{[HCO_3^-]}{0.03P_{CO_2}} \tag{15}$$

This form of the Henderson–Hasselbalch equation is useful in understanding acid–base problems. Note that the "acid" in this equation appears to be $CO_{2(d)}$ but is really H_2CO_3 "represented" by CO_2. Therefore, this equation is valid only if $CO_{2(d)}$ and H_2CO_3 are in equilibrium with each other, which is usually (but not always) the case.

Many clinicians prefer to work with H^+ rather than pH. The following expression results if we take antilogarithms of the Henderson–Hasselbalch equation:

$$[H^+] = 24 \times P_{CO_2} / [HCO_3^-] \qquad (16)$$

In this equation, H^+ is expressed in nmol/L, HCO_3^- in mmol/L or mEq/L, and P_{CO_2} in mm Hg. If the P_{CO_2} is 40 mm Hg and plasma HCO_3^- is 24 mmol/L, H^+ is 40 nmol/L.

Bicarbonate/carbon dioxide open system.

As noted previously, the pK of the HCO_3^-/CO_2 system (6.10) is far from 7.40, the normal pH of arterial blood. From this, one might view this buffer pair as rather poor. On the contrary, it is remarkably effective because it operates in an "open" system; that is, the two buffer components can be added to or removed from the body at controlled rates.

The HCO_3^-/CO_2 system is open in several ways:

- Metabolism provides an endless source of CO_2, which can replace any H_2CO_3 consumed by a base added to the body.
- The respiratory system can change the amount of CO_2 in body fluids by hyperventilation or hypoventilation.
- The kidneys can change the amount of HCO_3^- in the ECF by forming new HCO_3^- when excess acid has been added to the body or by excreting HCO_3^- when excess base has been added.

How the kidneys and respiratory system influence ECF pH by operating on the HCO_3^-/CO_2 system is described below. For now, the advantages of an open buffer system are best explained by an example (Fig. 24.3). Suppose we have 1 L of ECF containing 24 mmol of HCO_3^- and 1.2 mmol of dissolved $CO_{2(d)}$ (P_{CO_2} = 40 mm Hg). Using the special form of the Henderson–Hasselbalch equation described above, we find that the ECF pH is 7.40:

$$pH = 6.10 + \log \frac{[HCO_3^-]}{0.03 P_{CO_2}} = 6.10 + \log \frac{[24]}{[1.2]} = 7.40 \qquad (17)$$

Suppose we now add 10 mmol of HCl, a strong acid. HCO_3^- is the major buffer base in the ECF (we will neglect the contributions of other buffers). We predict that the HCO_3^- level will fall by nearly 10 mmol and, from the reaction $H^+ + HCO_3^- \rightleftharpoons H_2CO_3 \rightleftharpoons H_2O + CO_2$, we predict that nearly 10 mmol of $CO_{2(d)}$ will form. If the system were closed and no CO_2 could escape, the new pH would be:

$$pH = 6.10 + \log \frac{[24 - 10]}{[1.2 + 10]} = 6.20 \qquad (18)$$

This is an intolerably low—indeed a fatal—pH.

If the system is open, then CO_2 can escape into the air. If all of the extra CO_2 is expired by the lungs and the $CO_{2(d)}$ is kept at 1.2 mmol/L, then the pH would be

$$pH = 6.10 + \log \frac{[24 - 10]}{[1.2]} = 7.17 \qquad (19)$$

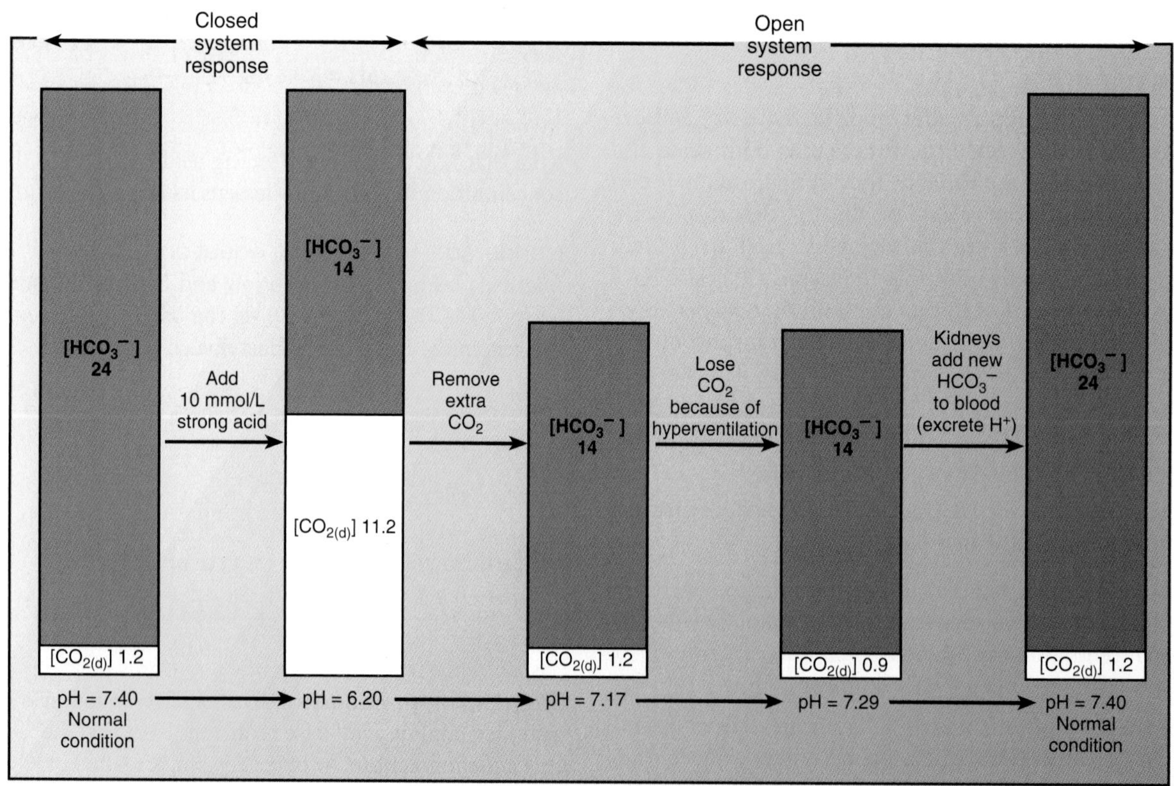

Figure 24.3 The HCO_3^-/CO_2 system. This system is remarkably effective in buffering added strong acid in the body, because it is open. HCO_3^- and $CO_{2(d)}$ are in mmol/L. (See text for details.)

Although this pH is low, it is compatible with life.

Still another mechanism promotes the escape of CO_2 from the body. An acidic blood pH stimulates breathing, which can make the P_{CO_2} lower than 40 mm Hg. If P_{CO_2} falls to 30 mm Hg ($CO_{2(d)}$ = 0.9 mmol/L), the pH would be

$$pH = 6.10 + \log\frac{[24-10]}{[0.9]} = 7.29 \quad (20)$$

This pH is much closer to normal, demonstrating the effectiveness of an open system in buffering acid. The system is also open at the kidneys, and new HCO_3^- can be added to the plasma to correct the HCO_3^-. Once the pH of the blood is normal, the stimulus for hyperventilation disappears.

Changes in acid production

Another way in which blood pH may be protected is by changes in endogenous acid production (Fig. 24.4). An increase in blood pH caused by the addition of base to the body stimulates production of lactic acid and ketone body acids, which then reduces the alkaline shift in pH. A decrease in blood pH inhibits production of lactic acid and ketone body acids, which diminishes the acidic shift in pH.

This scenario is especially important when the endogenous production of these acids is high, as occurs during strenuous exercise or other conditions of circulatory inadequacy (lactic acidosis) or during ketosis resulting from uncontrolled diabetes, starvation, or alcoholism. These effects of pH on endogenous acid production result from changes in enzyme activities brought about by the pH changes, and they are part of a negative feedback mechanism regulating blood pH.

Isohydric principle

We have discussed the various buffers separately, but in the body they all work together. In a solution containing multiple buffers, all are in equilibrium at the same H^+. This idea is known as the **isohydric principle** (*isohydric* means "same H^+"). For plasma, for example, we can write:

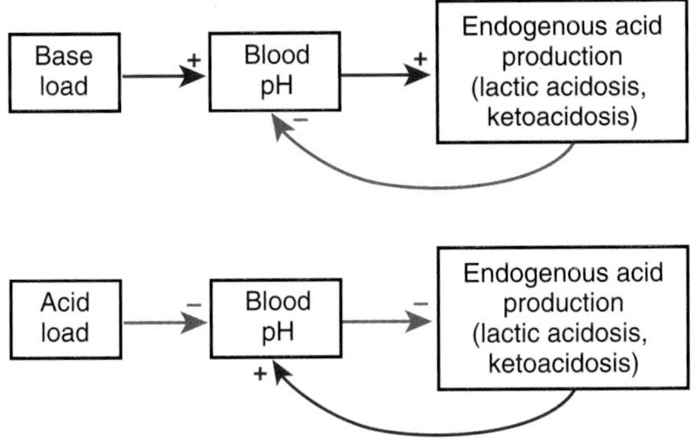

Figure 24.4 Negative feedback control of endogenous acid production. A base load, by raising pH, stimulates the endogenous production of acids. The addition of an exogenous acid load or increased endogenous acid production results in a fall in pH, which in turn inhibits the production of ketone body acids and lactic acid. These negative feedback effects attenuate changes in blood pH.

$$pH = 6.80 + \log\frac{[HPO_4^{2-}]}{[H_2PO_4^-]} = 6.10 + \log\frac{[HCO_3^-]}{0.03\ P_{CO_2}} = pK_{protein} + \log\frac{[proteinate^-]}{[H-protein]} \quad (21)$$

If an acid or a base is added to such a complex mixture of buffers, all buffers take part in buffering and shift from one form (base or acid) to the other. The relative importance of each buffer depends on its amount, pK, and availability.

The isohydric principle underscores the fact that it is the concentration ratio for any buffer pair, along with its pK, that sets the pH. We can focus on the concentration ratio for one buffer pair, and all other buffers will automatically adjust their ratios according to the pH and their pK values.

The rest of this chapter emphasizes the role of the HCO_3^-/CO_2 buffer pair in setting the blood pH. Other buffers, however, are present and participate in buffering. The HCO_3^-/CO_2 system is emphasized because physiologic mechanisms (lungs and kidneys) regulate pH by acting on components of this buffer system.

Lungs are the second line of defense against changes in pH.

Reflex changes in ventilation help to defend blood pH. By changing the P_{CO_2}, and hence H_2CO_3 of the blood, the respiratory system can rapidly and profoundly affect blood pH. As discussed in Chapter 21, "Control of Ventilation," a fall in blood pH stimulates ventilation, primarily by acting on peripheral chemoreceptors. An elevated arterial blood P_{CO_2} is a powerful stimulus to increase ventilation; it acts on both peripheral and central chemoreceptors but primarily on the latter. CO_2 diffuses into brain interstitial and cerebrospinal fluids, where it causes a fall in pH, which stimulates chemoreceptors in the medulla oblongata. When ventilation is stimulated, the lungs blow off more CO_2, thereby making the blood less acidic. Conversely, a rise in blood pH inhibits ventilation; the consequent rise in blood H_2CO_3 reduces the alkaline shift in blood pH. Respiratory responses to disturbed blood pH begin within minutes and are maximal in about 12 to 24 hours.

Kidneys are the third line of defense against changes in pH.

The kidneys play a critical role in maintaining acid–base balance. The kidneys remove H^+ if there is excess acid in the body or HCO_3^- if there is excess base. The usual challenge is to remove excess acid. As we have learned, body buffer bases, particularly HCO_3^-, first buffer the strong acids produced by metabolism. The kidneys then must eliminate H^+ in the urine and restore the depleted HCO_3^-.

Little of the H^+ excreted in the urine is present as free H^+. For example, if the urine has its lowest pH value (pH = 4.5), the H^+ is only 0.03 mEq/L. With a typical daily urine output of 1 to 2 L, the amount of acid the body must dispose of daily (roughly 70 mEq) obviously is not excreted in the free form. Most of the H^+ combines with urinary buffers to be excreted as titratable acid and as NH_4^+.

Titratable acid is measured in the clinical laboratory by titrating the urine with a strong base (NaOH) and measuring the number of milliequivalents of OH^- needed to bring the urine pH back to the pH of the blood (usually 7.40). It represents the amount of hydrogen ions that are excreted combined with urinary buffers, such as phosphate, creatinine, and other bases. The largest component of titratable acid is normally phosphate, that is, $H_2PO_4^-$.

Hydrogen ions secreted by the renal tubules also combine with the free base NH_3 and are excreted as NH_4^+. Ammonia (a term that collectively includes both NH_3 and NH_4^+) is produced by the kidney tubule cells and is secreted into the urine. Because the pK_a for NH_4^+ is high (9.0), most of the ammonia in the urine is present as NH_4^+. For this reason, too, NH_4^+ is not appreciably titrated when titratable acid is measured. Urinary ammonia is measured by a separate, colorimetric or enzymatic, method.

Net renal acid excretion

In stable acid–base balance, net acid excretion by the kidneys equals the net rate of H^+ addition to the body by metabolism or other processes, assuming that other routes of loss of acid or base (e.g., gastrointestinal losses) are small and can be neglected, which normally is the case. The net loss of H^+ in the urine can be calculated from the following equation, which shows typical values in the parentheses:

$$\begin{aligned}&\text{Renal net acid excretion (70 mEq/d)}\\&= \text{urinary titratable acid (24 mEq/d)}\\&\quad + \text{urinary ammonia (48 mEq/d)}\\&\quad - \text{urinary } HCO_3^- \text{ (2 mEq/d)}\end{aligned} \tag{22}$$

Urinary ammonia (as NH_4^+) ordinarily accounts for about two thirds of the excreted H^+ and titratable acid for about one third. Excretion of HCO_3^- in the urine represents a loss of base from the body. Therefore, it must be subtracted in the calculation of net acid excretion. If the urine contains significant amounts of organic anions such as citrate, that potentially could have yielded HCO_3^- in the body, these should also be subtracted. Because the amount of free H^+ excreted is negligible, this is omitted from the equation.

Urinary acidification

As the urine flows along the tubule, from the Bowman capsule on through the collecting ducts, three processes occur: filtered HCO_3^- is reabsorbed, titratable acid is formed, and ammonia is added to the tubular urine. All three processes involve H^+ secretion (urinary acidification) by the tubular epithelium. The nature and magnitude of these processes vary in different nephron segments. Figure 24.5 summarizes measurements of tubular fluid pH along the nephron and shows ammonia movements in various nephron segments.

H^+ secretion in the proximal convoluted tubule

The pH of the glomerular ultrafiltrate, at the beginning of the proximal tubule, is identical to that of the plasma from which it is derived (7.4). Hydrogen ions are secreted by the proximal tubule epithelium into the tubule lumen; about two thirds of this is accomplished by a Na^+/H^+ exchanger and about one third by a H^+-ATPase in the brush border membrane. Tubular fluid pH falls to a value of about 6.7 by the end of the proximal convoluted tubule (see Fig. 24.5).

The drop in pH is modest for two reasons: buffering of secreted H^+ and the high permeability of the proximal tubule epithelium to H^+. The glomerular filtrate and tubule fluid contain abundant buffer bases, especially HCO_3^-, which soak up secreted H^+, thereby minimizing a fall in pH. The proximal tubule epithelium is also rather "leaky" to H^+, so that any gradient from urine to blood, established by H^+ secretion, is soon limited by the diffusion of H^+ out of the tubule lumen into the blood surrounding the tubules

Most of the hydrogen ions secreted by the nephron are secreted in the proximal convoluted tubule and are used to bring about the reabsorption of filtered HCO_3^-. Secreted

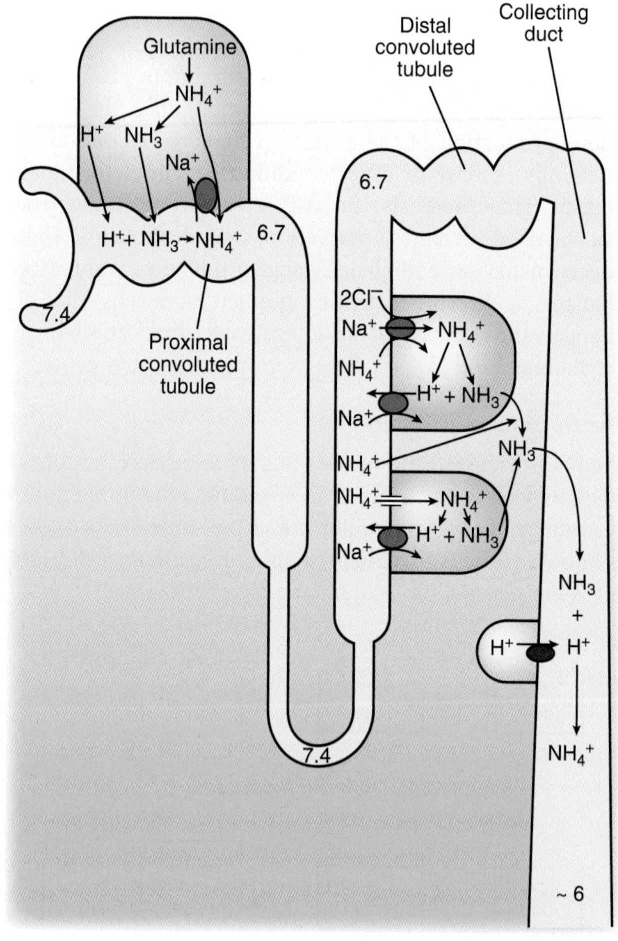

Figure 24.5 Acidification along the nephron. The pH of tubular urine decreases along the proximal convoluted tubule, rises along the descending limb of the Henle loop, falls along the ascending limb, and reaches its lowest values in the collecting ducts. Ammonia (NH_3 + NH_4^+) is chiefly produced in proximal tubule cells and is secreted into the tubular urine. NH_4^+ is reabsorbed in the thick ascending limb and accumulates in the kidney medulla. NH_3 diffuses into acidic collecting duct urine, where it is trapped as NH_4^+.

hydrogen ions are also buffered by filtered phosphate to form titratable acid. Proximal tubule cells produce ammonia, mainly from the amino acid glutamine. Ammonia is secreted into the tubular urine by the diffusion of NH_3, which then combines with a secreted H^+ to form NH_4^+, or via the brush border membrane Na^+/H^+ exchanger, which can operate in a Na^+/NH_4^+ exchange mode.

H^+ secretion in the loop of Henle

Along the descending limb of the loop of Henle, the pH of tubular fluid rises (from 6.7 to 7.4). This rise is explained by an increase in intraluminal HCO_3^- caused by water reabsorption. Ammonia is secreted along the descending limb.

The tubular fluid is acidified by the secretion of H^+ along the ascending limb via a Na^+/H^+ exchanger. Along the thin ascending limb, ammonia is passively reabsorbed. Along the thick ascending limb, the Na–K–2Cl cotransporter in the luminal cell membrane (NH_4^+ substitutes for K^+) actively reabsorbs NH_4^+. Some NH_4^+ can be reabsorbed via a luminal cell membrane K^+ channel. Also, some NH_4^+ can be passively reabsorbed between cells in this segment; the driving force is the lumen positive transepithelial electrical potential difference (see Chapter 22, "Kidney Function," Fig. 22.20). Ammonia may undergo countercurrent multiplication in the loop of Henle, leading to an ammonia concentration gradient in the kidney medulla. The highest concentrations are at the tip of the papilla.

H^+ secretion in the distal nephron

The distal nephron (distal convoluted tubule, connecting tubule, and collecting duct) differs from the proximal portion of the nephron in its H^+ transport properties. It secretes far fewer hydrogen ions, and they are secreted primarily via an electrogenic H^+-ATPase or an electroneutral H^+/K^+-ATPase. The distal nephron is also lined by "tight" epithelia, so little secreted H^+ diffuses out of the tubule lumen, making steep urine-to-blood pH gradients possible (see Fig. 24.5). Final urine pH is typically around 6 but may be as low as 4.5.

The distal nephron usually almost completely reabsorbs the small quantities of HCO_3^- that were not reabsorbed by more proximal nephron segments. Considerable amounts of titratable acid form as the urine is acidified. Ammonia that was reabsorbed by the ascending limb of the loop of Henle and accumulated in the medullary interstitial space diffuses as lipid-soluble NH_3 into collecting duct urine and combines with secreted H^+ to form NH_4^+. The collecting duct epithelium is impermeable to the lipid-insoluble NH_4^+, so ammonia is trapped in an acidic urine and is excreted as NH_4^+ (see Fig. 24.5).

The **intercalated cells** of the collecting duct are involved in acid–base transport and are of two major types: an acid-secreting *α-intercalated cell* and a bicarbonate-secreting *β-intercalated cell.* The α-intercalated cell has a vacuolar type of H^+-ATPase (the same kind as is found in lysosomes, endosomes, and secretory vesicles) and a H^+/K^+-ATPase (similar to that found in stomach and colon epithelial cells) in the luminal cell membrane and a Cl^-/HCO_3^- exchanger in the basolateral cell membrane (Fig. 24.6). The β-intercalated cell has the opposite polarity.

A more acidic blood pH results in the insertion of cytoplasmic H^+ pumps into the luminal cell membrane of α-intercalated cells and enhanced H^+ secretion. If the blood is made alkaline, HCO_3^- secretion by β-intercalated cells is increased. Because the amounts of HCO_3^- secreted are ordinarily small compared with the amounts filtered and reabsorbed, HCO_3^- secretion will be omitted from the remaining discussion.

Reabsorption of filtered bicarbonate

The kidney glomeruli filter about 4,320 mEq of HCO_3^- per day (180 L/d × 24 mEq/L). Urinary loss of even a small portion of this HCO_3^- would lead to an acidic blood and impair the body's ability to buffer its daily load of metabolically produced H^+. The kidney tubules have the important task of recovering the filtered HCO_3^- and returning it to the blood.

Figure 24.7 shows how HCO_3^- filtration, reabsorption, and excretion normally vary with plasma HCO_3^-. This type of graph should already be familiar (see Chapter 22, "Kidney Function," Fig. 22.8). The *y* axis of the graph is unusual, however, because amounts of HCO_3^- per minute are factored by the glomerular filtration rate (GFR). The data are expressed in this way because the maximal rate of tubular reabsorption of HCO_3^- varies with GFR. The amount of HCO_3^- excreted in the urine per unit time is calculated as the difference between filtered and reabsorbed amounts. At low plasma concentrations

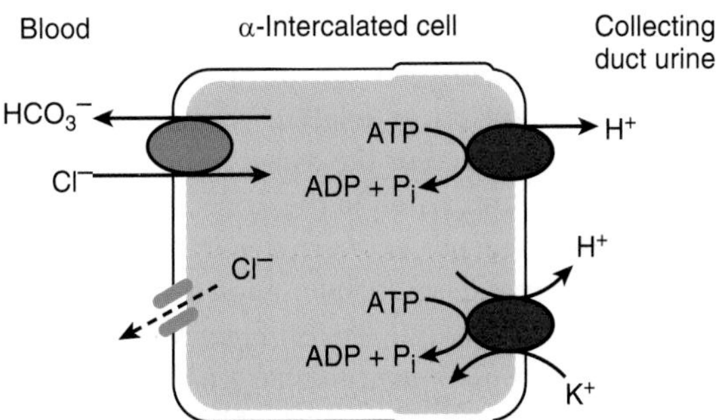

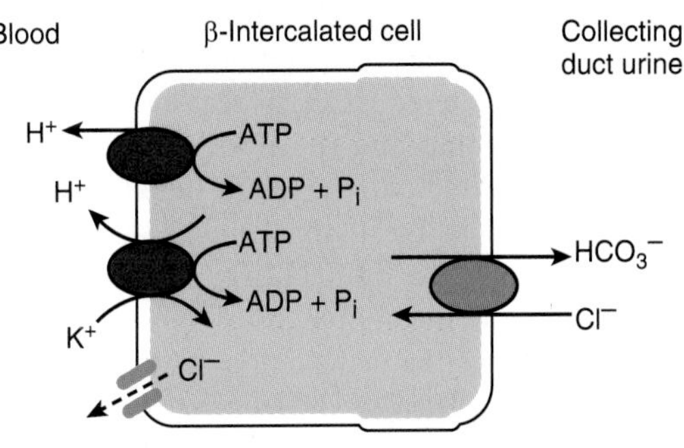

Figure 24.6 Collecting duct intercalated cells. The α-intercalated cell secretes H^+ via an electrogenic, vacuolar H^+-ATPase, and electroneutral H^+/K^+-ATPase and adds HCO_3^- to the blood via a basolateral cell membrane Cl^-/HCO_3^- exchanger. The β-intercalated cell, which is located in cortical collecting ducts, has the opposite polarity and secretes HCO_3^-. ADP, adenosine diphosphate; ATP, adenosine triphosphate; P_i, inorganic phosphate.

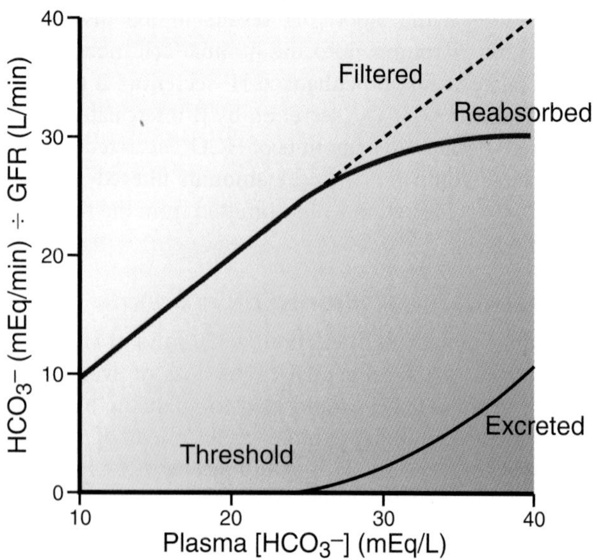

Figure 24.7 The filtration, reabsorption, and excretion of HCO_3^-. The plasma HCO_3^- was decreased by ingestion of an acidifying salt (NH_4Cl) and increased by intravenous infusion of an $NaHCO_3$ solution. All of the filtered HCO_3^- was reabsorbed below a plasma concentration of about 26 mEq/L. Above this value ("threshold"), appreciable quantities of filtered HCO_3^- were excreted in the urine. GFR, glomerular filtration rate.

of HCO_3^- (below about 26 mEq/L), all of the filtered HCO_3^- is reabsorbed; this is an appropriate compensatory response to metabolic acidosis (see *Metabolic acidosis results from a gain of noncarbonic acid or a loss of bicarbonate*, below).

If the plasma HCO_3^- is raised to high levels, for example, by intravenous infusion of solutions containing $NaHCO_3$, then filtered HCO_3^- exceeds the reabsorptive capacity of the tubules and HCO_3^- excretion in the urine increases (see Fig. 24.7). This response tends to return the pH of the blood back to its normal value and is a compensation for metabolic alkalosis (see *Metabolic alkalosis results from a gain of strong base or bicarbonate or a loss of noncarbonic acid*, below).

At the cellular level (Fig. 24.8), filtered HCO_3^- is not reabsorbed directly across the tubule's luminal cell membrane as is, for example, glucose. Instead, filtered HCO_3^- is reabsorbed indirectly via H^+ secretion, in the following way. About 90% of the filtered HCO_3^- is reabsorbed in the proximal convoluted tubule, and so we will emphasize events at this site. H^+ is secreted into the tubule lumen mainly via the Na^+/H^+ exchanger in the luminal membrane. It combines with filtered HCO_3^- to form H_2CO_3. Carbonic anhydrase (CA) in the luminal membrane (brush border) of the proximal tubule catalyzes the dehydration of H_2CO_3 to CO_2 and water in the lumen. The CO_2 diffuses back into the cell.

Inside the cell, the hydration of CO_2 (catalyzed by intracellular CA) yields H_2CO_3, which instantaneously forms H^+ and HCO_3^-. The H^+ is secreted into the lumen, and the HCO_3^- moves into the blood surrounding the tubules. In proximal tubule cells, this movement is favored by the inside negative membrane potential of the cell and by an electrogenic cotransporter in the basolateral membrane that simultaneously transports three HCO_3^- and one Na^+. In addition,

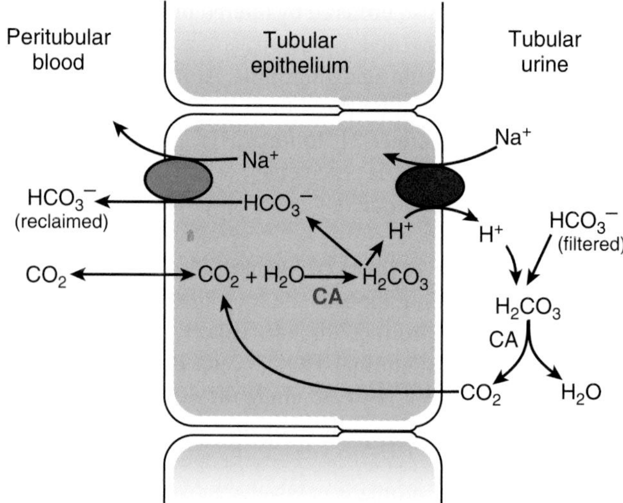

Figure 24.8 A cell model for HCO_3^- reabsorption. Filtered HCO_3^- combines with secreted H^+ and is reabsorbed indirectly. Carbonic anhydrase (CA) is present in the cells and, in the proximal tubule, on the brush border.

HCO_3^- can leave the cell via a Cl^-/HCO_3^- exchanger in the basolateral cell membrane.

The reabsorption of filtered HCO_3^- does not result in H^+ excretion or the formation of any "new" HCO_3^-. The secreted H^+ is not excreted because it combines with filtered HCO_3^- that is, indirectly, reabsorbed. There is no net addition of HCO_3^- to the body in this operation. It is simply a recovery or reclamation process.

New bicarbonate generation

When H^+ is excreted as titratable acid and ammonia, new HCO_3^- is formed and is added to the blood. New HCO_3^- replaces the HCO_3^- used to buffer the strong acids produced by metabolism (see Fig. 24.2).

The formation of new HCO_3^- and the excretion of H^+ are like two sides of the same coin. This fact is apparent if we assume that H_2CO_3 is the source of H^+:

$$CO_2 + H_2O \rightleftharpoons H_2CO_3 \begin{array}{l} \nearrow H^+ \text{(urine)} \\ \searrow HCO_3^- \text{(blood)} \end{array} \quad (23)$$

A loss of H^+ in the urine is equivalent to adding new HCO_3^- to the blood. The same is true if H^+ is lost from the body via another route such as by vomiting of acidic gastric juice. This process leads to a rise in plasma HCO_3^-. *Conversely, a loss of HCO_3^- from the body is equivalent to adding H^+ to the blood.*

Figure 24.9 shows a cell model for the formation of titratable acid. In this figure, $H_2PO_4^-$ is the titratable acid formed. H^+ and HCO_3^- are produced in the cell from H_2CO_3. The secreted H^+ combines with the basic form of the phosphate (HPO_4^{2-}) to form the acid phosphate ($H_2PO_4^-$). The secreted H^+ replaces one of the Na^+ ions accompanying the basic phosphate. The new HCO_3^- generated in the cell moves into the blood, together with Na^+. For each milliequivalent of H^+ excreted in the urine as titratable acid, a milliequivalent of new HCO_3^- is added to the blood. This process gets rid of

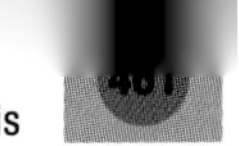

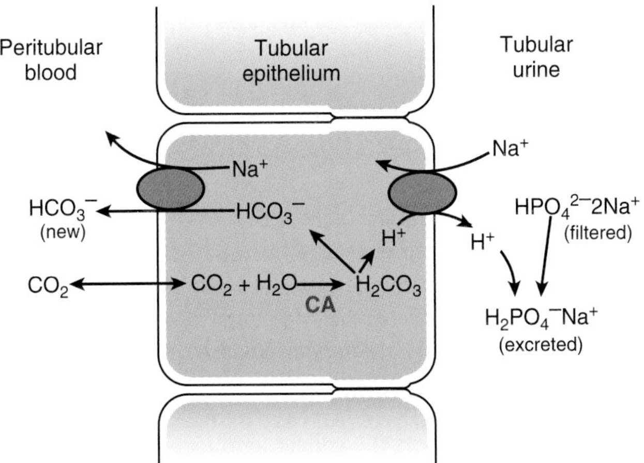

Figure 24.9 A cell model for the formation of titratable acid. Titratable acid (e.g., $H_2PO_4^-$) is formed when secreted H^+ is bound to a buffer base (e.g., HPO_4^{2-}) in the tubular urine. For each milliequivalent of titratable acid excreted, a milliequivalent of new HCO_3^- is added to the peritubular capillary blood. CA, carbonic anhydrase.

H^+ in the urine, replaces ECF HCO_3^-, and helps to restore a normal blood pH.

The amount of titratable acid excreted depends on two factors: the pH of the urine and the availability of buffer. If the urine pH is lowered, more titratable acid can form. The supply of phosphate and other buffers is usually limited. To excrete large amounts of acid, the kidneys must rely on increased ammonia excretion.

Figure 24.10 shows a cell model for the excretion of ammonia. The majority of ammonia is synthesized in proximal tubule cells by deamidation and deamination of glutamine:

$$\text{Glutamine} \xrightarrow[\text{Glutaminase}]{\nearrow NH_4^+} \text{Glutamate}^- \xrightarrow[\text{Glutamate dehydrogenase}]{\nearrow NH_4^+} \alpha\text{-Ketoglutarate}^{2-} \quad (24)$$

As discussed earlier, ammonia is secreted into the urine by two mechanisms. As NH_3, it diffuses into the tubular urine; as NH_4^+, it substitutes for H^+ on the Na^+/H^+ exchanger. In the lumen, NH_3 combines with secreted H^+ to form NH_4^+, which is excreted.

For each milliequivalent of H^+ excreted as NH_4^+, one milliequivalent of new HCO_3^- is added to the blood. The hydration of CO_2 in the tubule cell produces H^+ and HCO_3^-, as described earlier. Two H^+s are consumed when the anion α-ketoglutarate^{2-} is converted into CO_2 and water or into glucose in the cell. The new HCO_3^- returns to the blood along with Na^+.

If excess acid is added to the body, urinary ammonia excretion increases for two reasons. First, a more acidic urine traps more ammonia (as NH_4^+) in the urine. Second, renal ammonia synthesis from glutamine increases over a period of several days. Enhanced renal ammonia synthesis and excretion is a life-saving adaptation, because it allows the kidneys to remove large H^+ excesses and add more new HCO_3^- to the blood. Also, the excreted NH_4^+ can substitute in the urine for Na^+ and K^+, thereby diminishing the loss of these cations. With severe metabolic acidosis, ammonia excretion may increase almost 10-fold.

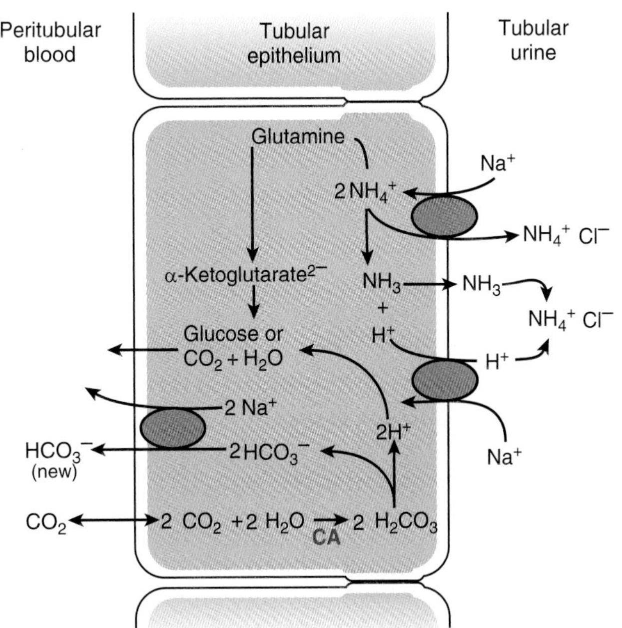

Figure 24.10 A cell model for renal synthesis and excretion of ammonia. Ammonium ions are formed from glutamine in the cell and are secreted into the tubular urine (*top*). H^+ from H_2CO_3 (*bottom*) is consumed when α-ketoglutarate is converted into glucose or CO_2 and H_2O. New HCO_3^- is added to the peritubular capillary blood—1 milliequivalent for each milliequivalent of NH_4^+ excreted in the urine. CA, carbonic anhydrase.

Factors influencing renal excretion of hydrogen ions

Many factors influence the renal excretion of H^+, including intracellular pH, arterial blood P_{CO_2}, carbonic anhydrase activity, Na^+ reabsorption, plasma K^+, and aldosterone (Fig. 24.11).

Decrease in intracellular pH

The pH in kidney tubule cells is a key factor influencing the secretion and, therefore, the excretion of H^+. A fall in pH (increased H^+) enhances H^+ secretion. A rise in pH (decreased H^+) lowers H^+ secretion. A decreased pH increases the activity of Na^+/H^+ exchangers in the luminal cell membrane. The activity of these exchangers is increased by the increased supply of H^+ and also by changes in exchanger conformation induced by binding of H^+. A low pH increases the recruitment of H^+-ATPases from intracellular vesicles and their insertion into the luminal membrane. A fall in intracellular pH also stimulates renal ammonia synthesis, which allows the kidneys to excrete more H^+ as NH_4^+.

Changes in arterial partial pressure of carbon dioxide

An increase in P_{CO_2} increases the formation of H^+ from H_2CO_3, leading to enhanced renal H^+ secretion and excretion—a useful compensation for any condition in which the blood contains too much H_2CO_3. (We will discuss this later when we consider respiratory acidosis.) A decrease in P_{CO_2} results in lowered H^+ secretion and, consequently, less complete reabsorption of filtered HCO_3^- and a loss of base in the urine (a useful compensation for respiratory alkalosis, also discussed later).

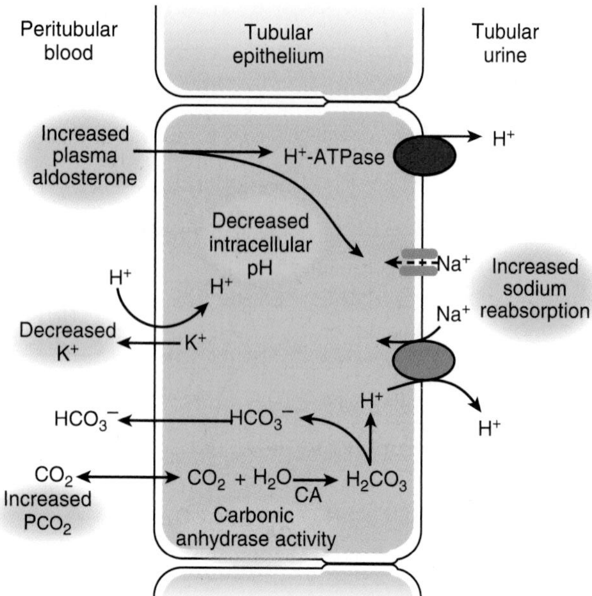

Figure 24.11 Factors leading to increased H^+ secretion by the kidney tubule epithelium. CA, carbonic anhydrase. (See text for details.)

Carbonic anhydrase activity

The enzyme carbonic anhydrase catalyzes two key reactions in urinary acidification:

1. The hydration of CO_2 in the cells, forming H_2CO_3 and yielding H^+ for secretion.
2. The dehydration of H_2CO_3 to H_2O and CO_2 in the proximal tubule lumen, an important step in the reabsorption of filtered HCO_3^-.

Large amounts of filtered HCO_3^- may escape reabsorption if carbonic anhydrase is inhibited (usually by a drug). This situation leads to a fall in blood pH.

Increased sodium reabsorption

Na^+ reabsorption is closely linked to H^+ secretion. In the proximal tubule, the two ions are directly linked, both being transported by the Na^+/H^+ exchanger in the luminal cell membrane. The relation is less direct in the collecting ducts. Enhanced Na^+ reabsorption in the ducts leads to a more negative intraluminal electrical potential, which favors H^+ secretion by its electrogenic H^+-ATPase. The avid renal reabsorption of Na^+ seen in states of volume depletion is accompanied by a parallel rise in urinary H^+ excretion.

Changes in plasma potassium concentration

Changes in plasma K^+ influence the renal excretion of H^+. A fall in plasma K^+ favors the movement of K^+ from body cells into interstitial fluid (or blood plasma) and a reciprocal movement of H^+ into cells. In the kidney tubule cells, these movements lower intracellular pH and increase H^+ secretion. K^+ depletion also stimulates ammonia synthesis by the kidneys (probably by lowering the intracellular pH); this increases NH_4^+ excretion. Finally, low plasma K^+ results in increased expression and activity of the H^+/K^+-ATPase in α-intercalated cells; although this allows the kidneys to conserve K^+, it also enhances H^+ secretion. The result is the complete reabsorption of filtered HCO_3^- and the enhanced generation of new HCO_3^- as more titratable acid and ammonia are excreted. Consequently, hypokalemia (or a decrease in body K^+ stores) leads to increased plasma HCO_3^- (*metabolic alkalosis*).

Hyperkalemia (or excess K^+ in the body) results in the opposite changes: an increase in intracellular pH, decreased H^+ secretion, incomplete reabsorption of filtered HCO_3^-, and a fall in plasma HCO_3^- (*metabolic acidosis*).

Aldosterone

Aldosterone stimulates the collecting ducts to secrete H^+ by three actions:

1. It directly stimulates the H^+-ATPase in collecting duct α-intercalated cells.
2. It enhances collecting duct Na^+ reabsorption, which leads to a more negative intraluminal potential and consequently promotes H^+ secretion by the electrogenic H^+-ATPase.
3. It promotes K^+ secretion. This response leads to hypokalemia, which increases renal H^+ secretion.

Hyperaldosteronism results in enhanced renal H^+ excretion and an alkaline blood pH (metabolic alkalosis); the opposite occurs with **hypoaldosteronism**.

High transepithelial pH gradient

The secretion of H^+ by the kidney tubules and collecting ducts is gradient limited. The collecting ducts cannot lower the urine pH below 4.5, corresponding to a urine/plasma H^+ gradient of $10^{-4.5}/10^{-7.4}$ or 800/1 when the plasma pH is 7.4. If more buffer base (NH_3, HPO_4^{2-}) is available in the urine, more H^+ can be secreted before the limiting gradient is reached. In some kidney tubule disorders, the secretion of H^+ is gradient limited.

INTRACELLULAR pH REGULATION

The intracellular and ECFs are linked by exchanges across cell membranes of H^+, HCO_3^-, various acids and bases, and CO_2. By stabilizing ECF pH, the body helps to protect intracellular pH.

If hydrogen ions were passively distributed across cell membranes, intracellular pH would be lower than what is seen in most body cells. In skeletal muscle cells, for example, we can calculate from the Nernst equation (see Chapter 2, "Plasma Membrane, Membrane Transport, and Resting Membrane Potential") and a membrane potential of −90 mV that cytosolic pH should be 5.9 if ECF pH is 7.4; actual measurements, however, indicate a pH of 6.9. From this discrepancy, two conclusions are clear: hydrogen ions are not at equilibrium across the cell membrane, and the cell must use active mechanisms to extrude H^+.

Cells are typically threatened by acidic metabolic end products and by the tendency for H^+ to diffuse into the cell down the electrical gradient (Fig. 24.12). H^+ is extruded by Na^+/H^+ exchangers, which are present in nearly all body cells.

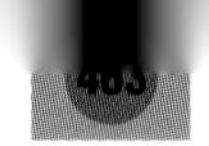

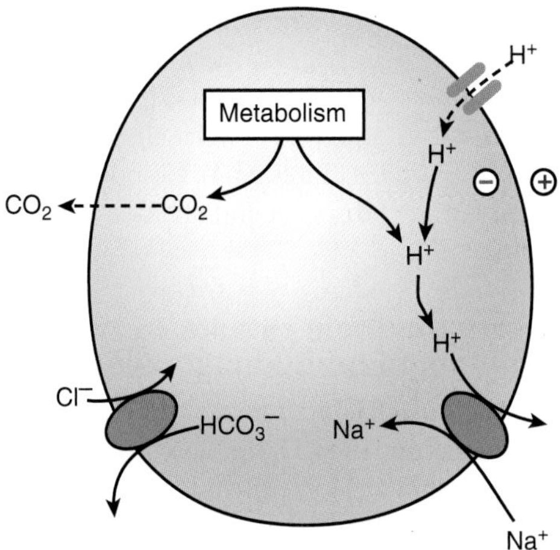

Figure 24.12 Cell acid–base balance. Body cells usually maintain a constant intracellular pH. The cell is acidified by the production of H^+ from metabolism and the influx of H^+ from the extracellular fluid (favored by the inside negative cell membrane potential). To maintain a stable intracellular pH, the cell must extrude hydrogen ions at a rate matching their input. Many cells also possess various HCO_3^- transporters (one example is shown) that defend against excess acid or base.

Eight different isoforms of these exchangers (designated NHE1, NHE2, etc.), with different tissue distributions, have been identified. These transporters exchange one H^+ for one Na^+ and therefore function in an electrically neutral fashion. Active extrusion of H^+ keeps the internal pH within narrow limits.

The activity of the Na^+/H^+ exchanger is regulated by intracellular pH and a variety of hormones and growth factors. Not surprisingly, an increase in intracellular H^+ stimulates the exchanger, but not only because of more substrate (H^+) for the exchanger. H^+ also stimulates the exchanger by protonating an activator site on the cytoplasmic side of the exchanger, thereby making the exchanger more effective in dealing with the threat of intracellular acidosis. Many hormones and growth factors, via intracellular second messengers, activate various protein kinases that stimulate or inhibit the Na^+/H^+ exchanger. In this way, they produce changes in intracellular pH, which may lead to changes in cell activity.

Besides extruding H^+, the cell can deal with acids and bases in other ways. First, in some cells, various HCO_3^- transporters (e.g., Na^+-dependent and Na^+-independent Cl^-/HCO_3^- exchangers, electrogenic Na^+/HCO_3^- cotransporters) may be present in plasma membranes. These exchangers may be activated by changes in intracellular pH. Second, cells have large stores of protein and organic phosphate buffers, which can bind or release H^+. Third, various chemical reactions in cells can also use up or release H^+. For example, the conversion of lactic acid to CO_2 and water or to glucose effectively disposes of acid. Fourth, various cell organelles may sequester H^+. For example, H^+-ATPases in endosomes and lysosomes pump H^+ out of the cytosol into these organelles. In summary, ion transport, buffering mechanisms, and metabolic reactions all ensure a relatively stable intracellular pH.

ACID–BASE BALANCE DISTURBANCES

Table 24.2 lists the normal values for the pH (or H^+), P_{CO_2}, and HCO_3^- of arterial blood plasma. A blood pH below 7.35 (H^+ concentration > 45 nmol/L) indicates **acidemia.** A blood pH above 7.45 (H^+ concentration < 35 nmol/L) indicates **alkalemia.** The range of pH values compatible with life is approximately 6.8 to 7.8 (H^+ concentration = 160–16 nmol/L).

From Bench to Bedside / 24.1

Alkali Therapy for Chronic Kidney Disease

Chronic kidney disease (CKD) is typically a slowly progressive disorder that results in an abnormal internal environment. When glomerular filtration rate falls to 20–40 mL/min, **metabolic acidosis** is usually present. Metabolic acidosis develops because of impaired renal ability to excrete the acids produced in the body as a result of food metabolism and is largely due to decreased renal synthesis and excretion of ammonia, a key buffer in the urine. The metabolic acidosis adversely affects bone, metabolism, nutrition, and other functions. Might it also contribute to the relentless progression of CKD?

Most, but not all, studies of base (alkali) administration in laboratory models of CKD have demonstrated a beneficial effect in preserving renal function. The base has often been given in the diet (food or drinking water) to rats and mice as sodium bicarbonate or as sodium or potassium citrate. One molecule of citrate, when completely oxidized in the body, is converted to three molecules of bicarbonate. One might think that giving extra sodium to animals or people with CKD is unwise because of the threat of hypertension. The sodium salts of citrate or bicarbonate, however, do not produce high blood pressure; the chloride salt (NaCl) does. Potassium citrate administration may not be advisable with advanced CKD (glomerular filtration rate [GFR] <20% of normal) because of the risk of hyperkalemia.

In the early part of the last century, there were several case reports of beneficial effects of bicarbonate administration in patients with CKD. This work, however, was not adequately controlled and has been largely forgotten. Recent controlled clinical studies have demonstrated a beneficial effect of sodium bicarbonate or sodium citrate administration to CKD patients. In one study, for example, the decline in GFR over an 18- to 30-month period was significantly less in the alkali-treated CKD patients than in untreated CKD patients. Alkali supplementation may be an inexpensive and safe treatment for CKD.

Will these recent results stand up to further tests on patients with a variety of causes of CKD? Could the beneficial effects of a low-protein diet in CKD patients be partially due to a reduced acid burden? Would vegetarian or vegan diets (which tend to be more alkalinizing than a meat plus vegetables diet) benefit patients with CKD? Further controlled clinical trials, with more subjects, are needed to verify the limited studies done so far and answer these questions.

TABLE 24.2 Normal Arterial Blood Plasma Acid-Base Values

Measure	Mean	Range[a]
pH	7.40	7.35–7.45
$[H^+]$ (nmol/L)	40	45–35
P_{CO_2} (mm Hg)	40	35–45
$[HCO_3^-]$ (mEq/L)	24	22–26

[a]The range extends from two standard deviations below to two standard deviations above the mean and encompasses 95% of the population of healthy people.

P_{CO_2}, partial pressure of carbon dioxide. Brackets denote concentration.

Four **simple acid–base disturbances** may lead to an abnormal blood pH: respiratory acidosis, respiratory alkalosis, metabolic acidosis, and metabolic alkalosis. The word "simple" indicates a single primary cause for the disturbance. **Acidosis** is an abnormal process that tends to produce acidemia. **Alkalosis** is an abnormal process that tends to produce alkalemia. If there is too much or too little CO_2, then a *respiratory disturbance* is present. If the problem is too much or too little HCO_3^-, then a *metabolic* (or *nonrespiratory*) *disturbance* of acid–base balance is present. Table 24.3 summarizes the changes in blood pH (or H^+ concentration), plasma HCO_3^- concentration, and P_{CO_2} that occur in each of the four simple acid–base disturbances.

In considering acid–base disturbances, it is helpful to recall the Henderson–Hasselbalch equation for HCO_3^-/CO_2:

$$pH = 6.10 + \log \frac{[HCO_3^-]}{0.03P_{CO_2}} \quad (25)$$

If the primary problem is a change in HCO_3^- concentration or P_{CO_2}, the pH can be brought closer to normal by changing the other member of the buffer pair *in the same direction*. For example, if P_{CO_2} is primarily decreased, a decrease in plasma HCO_3^- concentration will minimize the change in pH. In various acid–base disturbances, the lungs adjust the blood P_{CO_2} and the kidneys adjust the plasma HCO_3^- concentration to reduce departures of pH from normal; these adjustments are called *compensations* (Table 24.4). Compensations do not correct the underlying disorder and often do not bring about a normal blood pH.

Respiratory acidosis results from accumulation of carbon dioxide in the body.

Respiratory acidosis is an abnormal process characterized by CO_2 accumulation. The CO_2 buildup pushes the following reactions to the right:

$$CO_2 + H_2O \rightleftharpoons H_2CO_3 \rightleftharpoons H^+ + HCO_3^- \quad (26)$$

Blood H_2CO_3 concentration increases, leading to an increase in H^+ concentration or a fall in pH. Respiratory acidosis is usually caused by a failure to expire metabolically produced CO_2 at an adequate rate. This disturbance may be a result of a decrease in overall alveolar ventilation (hypoventilation) or, as occurs commonly in lung disease, a mismatch between ventilation and perfusion. Respiratory acidosis also occurs if a person breathes CO_2-enriched air.

Cell chemical buffering during respiratory acidosis

In respiratory acidosis, more than 95% of the chemical buffering occurs within cells. The cells contain many proteins and organic phosphates that can bind H^+. For example, hemoglobin (Hb) in red blood cells combines with H^+ from H_2CO_3, thereby minimizing the increase in free H^+. Recall from Chapter 19, "Gas Transfer and Transport," the buffering reaction:

$$H_2CO_3 + HbO_2^- \rightleftharpoons HHb + O_2 + HCO_3^- \quad (27)$$

This reaction raises the plasma HCO_3^-. In *acute* respiratory acidosis, such chemical buffering processes in the body

TABLE 24.3 Directional Changes in Arterial Blood Plasma Values in the Four Simple Acid-Base Disturbances

Disturbance	Arterial Plasma				Compensatory Response
	pH	$[H^+]$	$[HCO_3^-]$	P_{CO_2}	
Respiratory acidosis	↓	↑	↑	⇑	Kidneys increase H^+ excretion
Respiratory alkalosis	↑	↓	↓	⇓	Kidneys increase HCO_3^- excretion
Metabolic acidosis	↓	↑	⇓	↓	Alveolar hyperventilation; kidneys increase H^+ excretion
Metabolic alkalosis	↑	↓	⇑	↑	Alveolar hypoventilation; kidneys increase HCO_3^- excretion

Double arrows indicate the main effect. Brackets denote concentration.
P_{CO_2}, partial pressure of carbon dioxide.

TABLE 24.4 Compensatory Responses in Acid-Base Disturbances[a]

Type	Response
Respiratory acidosis	
Acute	1 mEq/L increase in plasma $[HCO_3^-]$ for each 10-mm Hg increase in P_{CO_2}[b]
Chronic	4 mEq/L increase in plasma $[HCO_3^-]$ for each 10-mm Hg increase in P_{CO_2}[c]
Respiratory alkalosis	
Acute	2 mEq/L decrease in plasma $[HCO_3^-]$ for each 10-mm Hg decrease in P_{CO_2}[b]
Chronic	4 mEq/L decrease in plasma $[HCO_3^-]$ for each 10-mm Hg decrease in P_{CO_2}[c]
Metabolic acidosis	1.3-mm Hg decrease in P_{CO_2} for each 1-mEq/L decrease in plasma $[HCO_3^-$[d]$]$
Metabolic alkalosis	0.7-mm Hg increase in P_{CO_2} for each 1-mEq/L increase in plasma $[HCO_3^-$[d]$]$

[a]Empirically determined average changes measured in people with simple acid–base disorders.

[b]This change is primarily a result of chemical buffering.

[c]This change is primarily a result of renal compensation.

[d]This change is a result of respiratory compensation.

P_{CO_2}, partial pressure of carbon dioxide. Brackets denote concentration.

lead to an increase in plasma HCO_3^- concentration of about 1 mEq/L for each 10-mm Hg increase in P_{CO_2} (see Table 24.4). Bicarbonate is not a buffer for H_2CO_3 because the reaction

$$H_2CO_3 + HCO_3^- \rightleftharpoons HCO_3^- + H_2CO_3 \qquad (28)$$

is simply an exchange reaction and does not affect the pH.

An example illustrates how chemical buffering reduces a fall in pH during respiratory acidosis. Suppose P_{CO_2} increased from a normal value of 40 mm Hg to 70 mm Hg ($CO_{2(d)}$ = 2.1 mmol/L). If there were no body buffer bases that could accept H^+ from H_2CO_3 (i.e., if there were no measurable increase in HCO_3^- concentration), the resulting pH would be 7.16:

$$pH = 6.10 + \log\frac{[24]}{[2.1]} = 7.16 \qquad (29)$$

In acute respiratory acidosis, a 3 mEq/L increase in plasma HCO_3^- concentration occurs with a 30-mm Hg rise in P_{CO_2} (see Table 24.4). Therefore, the pH is 7.21:

$$pH = 6.10 + \log\frac{[24+3]}{[2.1]} = 7.21 \qquad (30)$$

The pH of 7.21 is closer to a normal pH because body buffer bases (mainly intracellular buffers) such as proteins and phosphates combine with H^+ liberated from H_2CO_3.

Ventilation stimulation

Respiratory acidosis produces a rise in P_{CO_2} and a fall in pH and is often associated with hypoxia. These changes stimulate breathing (see Chapter 21, "Control of Ventilation") and, thereby, diminish the severity of the acidosis. In other words, a person would be worse off if the respiratory system did not reflexly respond to the abnormalities in blood P_{CO_2}, pH, and P_{O_2}.

Renal compensation for respiratory acidosis

The kidneys compensate for respiratory acidosis by adding more H^+ to the urine and adding new HCO_3^- to the blood. The increased P_{CO_2} stimulates renal H^+ secretion, which allows the reabsorption of all filtered HCO_3^-. Excess H^+ is excreted as titratable acid and NH_4^+; these processes add new HCO_3^- to the blood, causing plasma HCO_3^- concentration to rise. This compensation takes several days to develop fully.

With *chronic* respiratory acidosis, plasma HCO_3^- increases, on the average, by 4 mEq/L for each 10-mm Hg rise in P_{CO_2} (see Table 24.4). This rise exceeds that seen with acute respiratory acidosis because of the renal addition of HCO_3^- to the blood. One would expect a person with chronic respiratory acidosis and a P_{CO_2} of 70 mm Hg to have an increase in plasma HCO_3^- of 12 mEq/L. The blood pH would be 7.33:

$$pH = 6.10 + \log\frac{[24+12]}{[2.1]} = 7.33 \qquad (31)$$

With chronic respiratory acidosis, time for renal compensation is allowed, so blood pH (in this example, 7.33) is much closer to normal than that is observed during acute respiratory acidosis (pH 7.21).

Excessive loss of carbon dioxide from hyperventilation causes respiratory alkalosis.

Respiratory alkalosis is most easily understood as the opposite of respiratory acidosis; it is an abnormal process causing the loss of too much CO_2. This loss causes blood H_2CO_3 concentration and, thus, H^+ concentration to fall (pH rises). Alveolar hyperventilation causes respiratory alkalosis by flushing out CO_2 from the lung alveoli at an excessive rate, thereby lowering alveolar and arterial blood P_{CO_2}. Hyperventilation and respiratory alkalosis can be caused by voluntary effort, anxiety, direct stimulation of the medullary respiratory center by some abnormality (e.g., meningitis, fever, and aspirin intoxication), or hypoxia resulting from severe anemia or high altitude.

Although hyperventilation causes respiratory alkalosis, it also causes changes (a fall in P_{CO_2} and a rise in blood pH) that inhibit ventilation, and therefore, limit the extent of hyperventilation.

Cell chemical buffering during respiratory alkalosis

As with respiratory acidosis, during respiratory alkalosis, more than 95% of chemical buffering occurs within cells.

Cell proteins and organic phosphates liberate hydrogen ions, which are added to the ECF and lower the plasma HCO_3^- concentration, thereby reducing the alkaline shift in pH.

With *acute* respiratory alkalosis, plasma HCO_3^- concentration falls by about 2 mEq/L for each 10 mm Hg drop in P_{CO_2} (see Table 24.4). For example, if P_{CO_2} drops from 40 to 20 mm Hg ($CO_{2(d)}$ = 0.6 mmol/L), plasma HCO_3^- concentration falls by 4 mEq/L and the pH will be 7.62:

$$pH = 6.10 + \log\frac{[24-4]}{[0.6]} = 7.62 \qquad (32)$$

If plasma HCO_3^- had not changed, the pH would have been 7.70:

$$pH = 6.10 + \log\frac{[24]}{[0.6]} = 7.70 \qquad (33)$$

Kidney compensation for respiratory alkalosis

The kidneys compensate for respiratory alkalosis by excreting HCO_3^- in the urine, thereby getting rid of base. A reduced P_{CO_2} reduces H^+ secretion by the kidney tubule epithelium. As a result, some of the filtered HCO_3^- is not reabsorbed. When the urine becomes more alkaline, titratable acid excretion vanishes, and little ammonia is excreted. The enhanced output of HCO_3^- causes plasma HCO_3^- concentration to fall.

Chronic respiratory alkalosis is accompanied by a 4 mEq/L fall in plasma HCO_3^- concentration for each 10-mm Hg drop in P_{CO_2} (see Table 24.4). For example, in a person with chronic hyperventilation and a P_{CO_2} of 20 mm Hg, the blood pH is

$$pH = 6.10 + \log\frac{[24-8]}{[0.6]} = 7.53 \qquad (34)$$

This pH is closer to normal than the pH of 7.62 of acute respiratory alkalosis. The difference between the two situations is largely a result of renal compensation.

Metabolic acidosis results from a gain of noncarbonic acid or a loss of bicarbonate.

Metabolic acidosis is an abnormal process characterized by a gain of acid (other than H_2CO_3) or a loss of HCO_3^-. Either causes plasma HCO_3^- concentration and pH to fall. If a strong acid is added to the body, the reactions

$$H^+ + HCO_3^- \rightleftharpoons H_2CO_3 \rightleftharpoons H_2O + CO_2 \qquad (35)$$

are pushed to the right. The added H^+ consumes HCO_3^-. If much acid is infused rapidly, P_{CO_2} rises, as the equation predicts. This increase occurs only transiently, however, because the body is an open system, and the lungs expire CO_2 as it is generated. P_{CO_2} actually falls below normal because an acidic blood pH stimulates ventilation (see Fig. 24.3).

Many conditions can produce metabolic acidosis, including renal failure, uncontrolled diabetes mellitus, lactic acidosis, the ingestion of acidifying agents such as NH_4Cl, excessive renal excretion of HCO_3^-, and diarrhea. In renal failure, the kidneys cannot excrete H^+ fast enough to keep up with metabolic acid production, and in uncontrolled diabetes mellitus, the production of ketone body acids increases. Lactic acidosis results from tissue hypoxia. Ingested NH_4Cl is converted into urea and a strong acid, HCl, in the liver. Diarrhea causes a loss of alkaline intestinal fluids.

ECF, cell, and bone chemical buffering during metabolic acidosis

Excess acid is chemically buffered in ECF and ICF and bone. In metabolic acidosis, roughly half the buffering occurs in cells and bone. HCO_3^- is the principal buffer in the ECF.

Respiratory compensation for metabolic acidosis

The acidic blood pH stimulates the respiratory system to lower blood P_{CO_2} by hyperventilation. This action lowers blood H_2CO_3 concentration and, thereby, tends to alkalinize the blood, opposing the acidic shift in pH. Metabolic acidosis is accompanied on average by a 1.3-mm Hg fall in P_{CO_2} for each 1-mEq/L drop in plasma HCO_3^- concentration (see Table 24.4). Suppose, for example, the infusion of a strong acid causes the plasma HCO_3^- concentration to drop from 24 to 12 mEq/L. If there were no respiratory compensation and the P_{CO_2} did not change from its normal value of 40 mm Hg, the pH would be 7.10:

$$pH = 6.10 + \log\frac{[12]}{[1.2]} = 7.10 \qquad (36)$$

With respiratory compensation, the P_{CO_2} falls by 16 mm Hg (12 × 1.3) to 24 mm Hg ($CO_{2(d)}$ = 0.72 mmol/L) and pH is 7.32:

$$pH = 6.10 + \log\frac{[12]}{[0.72]} = 7.32 \qquad (37)$$

This value is closer to normal than a pH of 7.10. The respiratory response develops promptly (within minutes) and is maximal after 12 to 24 hours.

Renal compensation for metabolic acidosis

The kidneys respond to metabolic acidosis by adding more H^+ to the urine. Because the plasma HCO_3^- concentration is primarily lowered, the filtered load of HCO_3^- drops, and the kidneys can accomplish the complete reabsorption of filtered HCO_3^- (see Fig. 24.7). More H^+ is excreted as titratable acid and NH_4^+. With chronic metabolic acidosis, the kidneys make more ammonia. The kidneys can therefore add more new HCO_3^- to the blood, to replace lost HCO_3^-. If the underlying cause of metabolic acidosis is corrected, then healthy kidneys can correct the blood pH in a few days.

Plasma anion gap is calculated from sodium, chloride, and bicarbonate concentrations.

The **anion gap** is a useful concept, especially when trying to determine the possible etiology of a metabolic acidosis. In any body fluid, the sum of the cations and the sum of the anions are equal because solutions are electrically neutral. For blood plasma, we can write:

$$\sum \text{cations} = \sum \text{anions} \quad (38)$$

or

$$[Na^+] + [\text{unmeasured cations}] = [Cl^-] + [HCO_3^-] + [\text{unmeasured anions}] \quad (39)$$

The unmeasured cations include K^+, Ca^{2+}, and Mg^{2+} ions and, because these are present at relatively low concentrations (compared with Na^+) and are usually fairly constant, we choose to neglect them. The unmeasured anions include plasma proteins, sulfate, phosphate, citrate, lactate, and other organic anions. If we rearrange the above equation, we get

$$[\text{unmeasured anions}] \text{ or "anion gap"} = [Na^+] - [Cl^-] - [HCO_3^-] \quad (40)$$

In a healthy person, the anion gap falls in the range of 8 to 14 mEq/L. For example, if plasma Na^+ concentration is 140 mEq/L, Cl^- concentration is 105 mEq/L, and HCO_3^- concentration is 24 mEq/L, the anion gap is 11 mEq/L. If an acid such as lactic acid is added to plasma, the reaction of lactic acid + $HCO_3^- \rightleftharpoons$ lactate$^-$ + H_2O + CO_2 will be pushed to the right. Consequently, the plasma HCO_3^- will be decreased and, because the Cl^- is not changed, the anion gap will be increased. The unmeasured anion in this case is lactate$^-$.

In several types of metabolic acidosis, the low blood pH is accompanied by a high anion gap (Table 24.5). (These can be remembered from the mnemonic MULEPAKS, formed from the first letters of this list.) In other types of metabolic acidosis, the low blood pH is accompanied by a normal anion gap (see Table 24.5). For example, with diarrhea and a loss of alkaline intestinal fluid, plasma HCO_3^- concentration falls but plasma Cl^- concentration rises, and the two changes counterbalance each other so the anion gap is unchanged. Again, the chief value of the anion gap concept is that it allows a clinician to narrow down possible explanations for metabolic acidosis in a patient.

Metabolic alkalosis results from a gain of strong base or bicarbonate or a loss of noncarbonic acid.

Metabolic alkalosis is an abnormal process characterized by a gain of a strong base or HCO_3^- or a loss of an acid (other than carbonic acid). Plasma HCO_3^- concentration and pH rise; P_{CO_2} rises because of respiratory compensation. These changes are opposite of those seen in metabolic acidosis (see Table 24.3). A variety of situations can produce metabolic alkalosis, including the ingestion of antacids, vomiting of acidic gastric juice, and enhanced renal H^+ loss (e.g., resulting from hyperaldosteronism or hypokalemia).

Clinical Focus / 24.1

Metabolic Acidosis in Diabetes Mellitus

Diabetes mellitus is a common disorder characterized by an insufficient secretion of insulin, or insulin resistance by the major target tissues (skeletal muscle, liver, and adipocytes). A severe **metabolic acidosis** may develop in uncontrolled diabetes mellitus.

Acidosis occurs because insulin deficiency leads to decreased glucose use, a diversion of metabolism toward the use of fatty acids, and an overproduction of ketone body acids (acetoacetic acid and β-hydroxybutyric acids). Ketone body acids are fairly strong acids (pK_a 4–5); they are neutralized in the body by HCO_3^- and other buffers. Increased production of these acids leads to a fall in plasma HCO_3^- concentration, an increase in plasma anion gap, and a fall in blood pH (acidemia).

Severe acidemia, whatever its cause, has many adverse effects on the body. It impairs myocardial contractility, resulting in a decrease in cardiac output. It causes arteriolar dilation, which leads to a fall in arterial blood pressure. Hepatic and renal blood flows are decreased. Re-entrant arrhythmias and a decreased threshold for ventricular fibrillation can occur. The respiratory muscles show decreased strength and fatigue easily. Metabolic demands are increased, as a result in part of activation of the sympathetic nervous system, but at the same time, anaerobic glycolysis and adenosine triphosphate (ATP) synthesis are reduced by acidemia. Hyperkalemia is favored and protein catabolism is enhanced. Severe acidemia causes impaired brain metabolism and cell volume regulation, leading to progressive obtundation and coma.

An increased acidity of the blood stimulates pulmonary ventilation, resulting in a compensatory lowering of alveolar and arterial blood partial pressure of carbon dioxide (P_{CO_2}). The consequent reduction in blood H_2CO_3 concentration acts to move the blood pH back toward normal. The labored, deep breathing that accompanies severe uncontrolled diabetes is called **Kussmaul respiration.**

The kidneys compensate for metabolic acidosis by reabsorbing all the filtered HCO_3^-. They also increase the excretion of titratable acid, part of which is composed of ketone body acids. These acids can only be partially titrated to their acid form in the urine, because the urine pH cannot go below 4.5. Thus, ketone body acids are excreted mostly in their anionic form; because of the requirement of electroneutrality in solutions, increased urinary excretion of Na^+ and K^+ results.

An important compensation for the acidosis is increased renal synthesis and excretion of ammonia. This adaptive response takes several days to develop fully, but it allows the kidneys to dispose of large amounts of H^+ in the form of NH_4^+. The NH_4^+ in the urine can replace Na^+ and K^+ ions, thereby resulting in conservation of these valuable cations.

The severe acidemia, electrolyte disturbances, and volume depletion that accompany uncontrolled diabetes mellitus may be fatal. Correction of the acid–base disturbance is best achieved by addressing the underlying cause, rather than just treating the symptoms. Therefore, the administration of a suitable dose of insulin is usually the key element of therapy. In some patients with marked acidemia (pH < 7.10), $NaHCO_3$ solutions may be infused intravenously to speed recovery, but this does not correct the underlying metabolic problem. Losses of Na^+, K^+, and water should be replaced.

TABLE 24.5 High and Normal Anion Gap Metabolic Acidosis

Condition	Explanation
High anion gap metabolic acidosis	
Methanol intoxication	Methanol metabolized to formic acid
Uremia	Sulfuric, phosphoric, uric, and hippuric acids retained as a result of renal failure
Lactic acid	Lactic acid buffered by HCO_3^- and accumulated as lactate
Ethylene glycol intoxication	Ethylene glycol metabolized to glyoxylic, glycolic, and oxalic acids
p-Aldehyde intoxication	*p*-Aldehyde metabolized to acetic and chloroacetic acids
Ketoacidosis	Production of β-hydroxybutyric and acetoacetic acids
Salicylate intoxication	Impaired metabolism leading to production of lactic acid and ketone body acids; accumulation of salicylate
Normal anion gap metabolic acidosis	
Diarrhea	Loss of HCO_3^- in stool; kidneys conserve Cl^-
Renal tubular acidosis	Loss of HCO_3^- in urine or inadequate excretion of H^+; kidneys conserve Cl^-
Ammonium chloride ingestion	NH_4^+ is converted to urea in liver, a process that consumes HCO_3^-; excess Cl^- is ingested

Chemical buffering during metabolic alkalosis

Chemical buffers in the body limit the alkaline shift in blood pH by releasing H^+ as they are titrated in the alkaline direction. About one third of the buffering occurs in cells.

Respiratory system compensation for metabolic alkalosis

The respiratory compensation for metabolic alkalosis is hypoventilation. An alkaline blood pH inhibits ventilation. Hypoventilation raises the blood P_{CO_2} and H_2CO_3 concentration, thereby reducing the alkaline shift in pH. A 1 mEq/L rise in plasma HCO_3^- concentration caused by metabolic alkalosis is accompanied by a 0.7 mm Hg rise in P_{CO_2} (see Table 24.4). If, for example, the plasma HCO_3^- concentration rose to 40 mEq/L, what would the plasma pH be with and without respiratory compensation? With respiratory compensation, the P_{CO_2} should rise by 11.2 mm Hg (0.7 × 16) to 51.2 mm Hg ($CO_{2(d)}$ = 1.54 mmol/L). The pH is 7.51:

$$pH = 6.10 + \log\frac{[40]}{[1.54]} = 7.51 \qquad (41)$$

Without respiratory compensation, the pH would be 7.62:

$$pH = 6.10 + \log\frac{[40]}{[1.2]} = 7.62 \qquad (42)$$

Respiratory compensation for metabolic alkalosis is limited because hypoventilation leads to hypoxia and CO_2 retention, and both increase breathing.

Renal compensation for metabolic alkalosis

The kidneys respond to metabolic alkalosis by lowering the plasma HCO_3^- concentration. The plasma HCO_3^- concentration is primarily raised, so more HCO_3^- is filtered than can be reabsorbed (see Fig. 24.7); in addition, HCO_3^- is secreted in the collecting ducts. Both of these changes lead to increased urinary HCO_3^- excretion. If the cause of the metabolic alkalosis is corrected, the kidneys can often restore the plasma HCO_3^- concentration and pH to normal in a day or two.

Clinical evaluation of acid–base disturbances requires a comprehensive study.

Acid–base data should always be interpreted in the context of other information about a patient. To identify an acid–base disturbance from laboratory values, it is best to look first at the pH. A low blood pH indicates acidosis; a high blood pH indicates alkalosis. If acidosis is present, for example, it could be either respiratory or metabolic. A low blood pH and elevated P_{CO_2} point to respiratory acidosis; a low pH and low plasma HCO_3^- indicate metabolic acidosis. If alkalosis is present, it could be either respiratory or metabolic. A high blood pH and low plasma P_{CO_2} indicate respiratory alkalosis; a high blood pH and high plasma HCO_3^- indicate metabolic alkalosis.

Whether the body is making an appropriate response for a simple acid–base disorder can be judged from the values in Table 24.4. Inappropriate values suggest that more than one acid–base disturbance may be present. Patients sometimes have two or more of the four simple acid–base disturbances at the same time, in which case

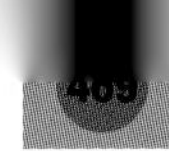

Clinical Focus / 24.2

Vomiting and Metabolic Alkalosis

Vomiting of acidic gastric juice results in **metabolic alkalosis** and fluid and electrolyte disturbances. Gastric juice contains about 0.1 M HCl. The acid is secreted by stomach parietal cells; these cells have an H^+/K^+-ATPase in their luminal cell membrane and a Cl^-/HCO_3^- exchanger in their basolateral cell membrane. When HCl is secreted into the stomach lumen and lost to the outside, there is a net gain of HCO_3^- in the blood plasma and no change in the anion gap. The HCO_3^-, in effect, replaces lost plasma Cl^-.

Ventilation is inhibited by the alkaline blood pH, resulting in a rise in P_{CO_2}. This respiratory compensation for the metabolic alkalosis, however, is limited, because hypoventilation leads to a rise in partial pressure of carbon dioxide (P_{CO_2}) and a fall in partial pressure of oxygen (P_{O_2}), both of which stimulate breathing.

The logical renal compensation for metabolic alkalosis is enhanced excretion of HCO_3^-. In people with persistent vomiting, however, the urine is sometimes acidic and renal HCO_3^- reabsorption is enhanced, thereby maintaining the elevated plasma HCO_3^-. This situation arises because vomiting is accompanied by losses of extracellular fluid (ECF) and K^+. Fluid loss leads to a decrease in effective arterial blood volume and engagement of mechanisms that reduce Na^+ excretion, such as decreased glomerular filtration rate and increased plasma renin, angiotensin, and aldosterone levels (see Chapter 23, "Regulation of Fluid and Electrolyte Balance"). Aldosterone stimulates H^+ secretion by collecting duct α-intercalated cells. Renal tubular Na^+/H^+ exchange is stimulated by volume depletion because the tubules reabsorb Na^+ more avidly than usual. With more H^+ secretion, more new HCO_3^- is added to the blood. The kidneys reabsorb filtered HCO_3^- completely, even though the plasma HCO_3^- level is elevated and maintain the metabolic alkalosis.

Vomiting results in K^+ depletion because of a loss of K^+ in the vomitus, decreased food intake, and, most important quantitatively, enhanced renal K^+ excretion. Extracellular alkalosis results in a shift of K^+ into cells (including renal cells) and thereby promotes K^+ secretion and excretion. Elevated plasma aldosterone levels also favor K^+ loss in the urine.

Treatment for the metabolic alkalosis primarily depends on eliminating the cause of vomiting. Correction of the alkalosis by administering an organic acid, such as lactic acid, does not make sense, because this acid would simply be converted to CO_2 and H_2O; this approach also does not address the Cl^- deficit. The ECF volume depletion and the Cl^- and K^+ deficits can be corrected by administering isotonic saline and appropriate amounts of KCl. Because replacement of Cl^- is a key component of therapy, this type of metabolic alkalosis is said to be "chloride-responsive." After Na^+, Cl^-, water, and K^+ deficits have been replaced, excess HCO_3^- (accompanied by surplus Na^+) will be excreted in the urine, and the kidneys will return blood pH to normal.

they have a **mixed acid–base disturbance**. The pH-bicarbonate diagram (Fig. 24.13) is a useful way to look at arterial blood data and determine what type of acid–base disturbance may be present in a patient. The ellipse in the center of this diagram shows the normal range of values for arterial blood pH, P_{CO_2}, and plasma bicarbonate concentration. Values on the left side of the diagram (acidemia, or a low blood pH) are caused by respiratory acidosis or metabolic acidosis. Values on the right side of the diagram (alkalemia, or an elevated blood pH) are caused by respiratory alkalosis or metabolic alkalosis. The curved lines that slope upward and to the right are P_{CO_2}

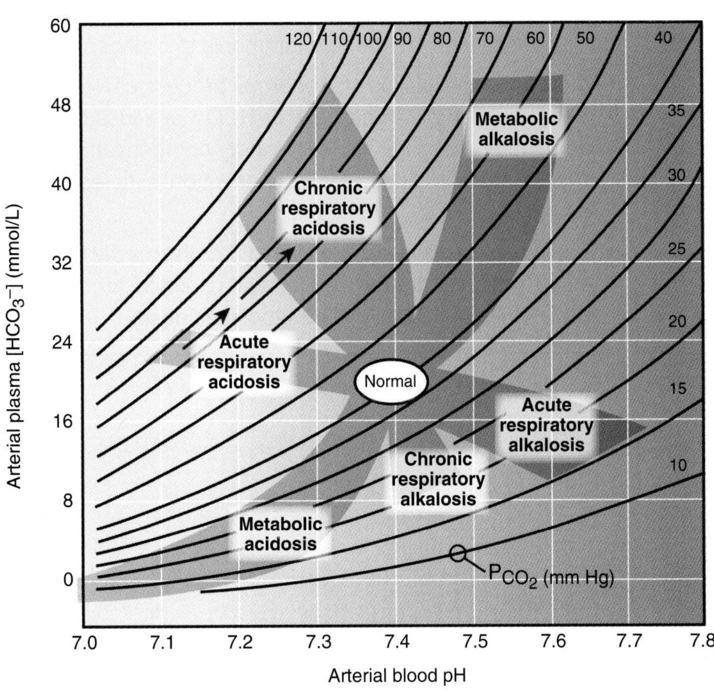

Figure 24.13 pH-bicarbonate diagram. The diagram shows the relationship between arterial pH, bicarbonate, and P_{CO_2} and provides a useful way of determining a patient's acid–base disturbance. Normal values are shown in the *elliptical area* in the center. Values left of normal represent acidemia, which may be caused by respiratory acidosis or metabolic acidosis. Values to the right of normal represent alkalemia, which may be caused by respiratory alkalosis or metabolic alkalosis. P_{CO_2}, partial pressure of carbon dioxide (mm Hg).

isobars; all along each line, the P_{CO_2} is the same (*iso* = "same," *bar* = "pressure"). Each point on the diagram must satisfy the Henderson–Hasselbalch equation, so that, for example, if pH and HCO_3^- concentration are known, the P_{CO_2} is automatically defined. The shaded areas include 95% of people with the designated simple acid–base disturbance. Note that a distinction is made between acute and chronic *respiratory* disturbances of acid–base balance but not between acute and chronic *metabolic* disturbances of acid–base balance. This is because the renal compensation for a respiratory disturbance may take days, whereas the respiratory compensation for a metabolic disturbance is prompt (minutes to hours). Note that compensations tend to return the blood pH closer to normal. For example, if the P_{CO_2} is acutely raised (move to a higher P_{CO_2} isobar), then the pH will fall, but with chronic respiratory acidosis, the pH is closer to normal at the same P_{CO_2} (because of renal addition of bicarbonate to the blood). Mixed acid–base disturbances often, but do not always, fall outside of the shaded areas. For example, values for a patient with chronic respiratory acidosis (e.g., as a result of pulmonary disease) and metabolic acidosis (e.g., resulting from shock) might just happen to fall in the area that indicates acute respiratory acidosis. Values for a patient with a simple disturbance could fall outside the shaded area if insufficient time has elapsed, especially for renal compensation. For example, as illustrated in Figure 24.13 (*arrows on left*), data from a person might have been collected during the period of transition from acute to chronic respiratory acidosis. A complete history and physical examination provide important clues in deciding what acid–base disturbances may be present in a patient.

Chapter Summary

- The body is constantly threatened by acid resulting from diet and metabolism. The stability of blood pH is maintained by the concerted action of chemical buffers, the lungs, and the kidneys.
- Numerous chemical buffers (e.g., HCO_3^-/CO_2, phosphates, and proteins) work together to minimize pH changes in the body. The concentration ratio (base/acid) of any buffer pair, together with the pK of the acid, automatically defines the pH.
- The bicarbonate/CO_2 buffer pair is effective in buffering in the body because its components are present in large amounts and the system is open.
- The respiratory system influences plasma pH by regulating the P_{CO_2} by changing the level of alveolar ventilation. The kidneys influence plasma pH by getting rid of acid or base in the urine.
- Renal acidification involves three processes: reabsorption of filtered HCO_3^-, excretion of titratable acid, and excretion of ammonia. New HCO_3^- is added to the plasma and replenishes depleted HCO_3^- when titratable acid (normally mainly $H_2PO_4^-$) and ammonia (as NH_4^+) are excreted.
- The stability of intracellular pH is ensured by membrane transport of H^+ and HCO_3^-, by intracellular buffers (mainly proteins and organic phosphates), and by metabolic reactions.
- Respiratory acidosis is an abnormal process characterized by an accumulation of CO_2 and a fall in arterial blood pH. The kidneys compensate by increasing the excretion of H^+ in the urine and adding new HCO_3^- to the blood, thereby diminishing the severity of the acidemia.
- Respiratory alkalosis is an abnormal process characterized by an excessive loss of CO_2 and a rise in pH. The kidneys compensate by increasing the excretion of filtered HCO_3^-, thereby diminishing the alkalemia.
- Metabolic acidosis is an abnormal process characterized by a gain of acid (other than H_2CO_3) or a loss of HCO_3^-. Respiratory compensation is hyperventilation, and renal compensation is an increased excretion of H^+ bound to urinary buffers (ammonia and phosphate).
- Metabolic alkalosis is an abnormal process characterized by a gain of strong base or HCO_3^- or a loss of acid (other than H_2CO_3). Respiratory compensation is hypoventilation, and renal compensation is increased excretion of HCO_3^-.
- The plasma anion gap is equal to the plasma $Na^+ - Cl^- - HCO_3^-$ concentration and is most useful in narrowing down possible causes of metabolic acidosis.

25 Neurogastroenterology and Motility

ACTIVE LEARNING OBJECTIVES

Upon mastering the material in this chapter, you should be able to:

- Distinguish the differences between electromechanical and pharmacomechanical coupling in initiation of contractions in gastrointestinal muscles.
- Explain the concept of "functional electrical syncytium," as applied to the gastrointestinal smooth musculature.
- Understand the forms of electrical activity in the gastrointestinal musculature.
- Understand how the sympathetic, parasympathetic, and enteric divisions of the autonomic nervous system innervate the gut and explain why the enteric nervous system is called a "brain in the gut."
- Apply the locations of postsynaptic receptors, presynaptic receptors, and presynaptic autoreceptors on enteric neurons to their physiologic significance.
- Distinguish the role of presynaptic inhibition from that of presynaptic facilitation in the function of enteric neural microcircuits.
- Explain why inhibitory musculomotor neurons are necessary for control of contractile behavior of the intestinal musculature by the enteric nervous system.
- Relate the mechanisms involved in the production of physiologic as opposed to pathophysiological ileus in the intestine.
- Explain how inhibitory neuronal motor control of gastrointestinal sphincters differs from inhibitory control of the longitudinal and circular muscle coats of the small and large intestine.
- Explain how function of the gastric reservoir differs from that of the antral pump in the determination of rate of gastric emptying and the role of the vagovagal reflex.
- Diagram and explain the reflex pathways and reflex mechanisms for three kinds of relaxation of the musculature of the gastric reservoir.
- Explain why symptoms of early satiety, bloating, and vomiting are experienced by patients after disruption of transmission in vagal nerves.
- Explain how the motor behavior of the stomach and small intestine in the interdigestive state differs from that in the digestive state.
- Describe how the mechanisms of neural control of the internal anal sphincter differ from those of the external anal sphincter and puborectalis muscle.
- Explain how neural deficits lead to gastrointestinal problems.

Gastrointestinal (GI) physiology is a branch of physiology addressing the structure and function of the GI system. The major functional processes occurring in the GI system are motility, secretion, digestion, circulation, and regulation. The coordination and control of these processes are vital in maintaining GI health. This chapter presents concepts and principles of neurogastroenterology in relation to motor functions of the specialized organs and muscle groups of the digestive tract. **Neurogastroenterology** is a subspecialty of clinical gastroenterology and digestive science. As such, it encompasses the investigative sciences dealing with functions, malfunctions, and malformations in the brain and spinal cord and the sympathetic, parasympathetic, and enteric divisions of the autonomic innervation of the digestive tract. Somatomotor systems (i.e., spinal cord and skeletal muscles) are included insofar as pharyngeal phases of swallowing and pelvic floor involvement in defecation and continence are concerned. Basic physiology of smooth muscles, as it relates to enteric neural control of motor movements, is a part of neurogastroenterology. Psychological and psychiatric relations to GI disorders are significant components of the neurogastroenterologic domain, especially in brain–gut interactions that underlie symptoms of abdominal pain, fecal urgency, diarrhea, and constipation.

Gastrointestinal tract is divided into upper and lower tracts.

The *upper GI tract* consists of the esophagus, stomach, and duodenum. The *lower GI tract* consists of the small and large intestine. Digestive motility refers to wall movement or lack thereof in the digestive tract. Integrated function of multiple muscles, nerves, and sometimes endocrine cells is necessary for the generation of various motility patterns that emerge at the organ level along the digestive tract. Digestive motor movements involve the application of forces of muscle contraction to contents present in the mouth, pharynx, esophagus, stomach, gallbladder, or small and large intestine. The musculature is striated muscle in the mouth, pharynx, upper esophagus, and pelvic floor and visceral-type smooth muscle elsewhere. Specialized pacemaker cells, called **interstitial cells of Cajal (ICCs)**, are associated with the smooth musculature. The nervous system, with its different kinds of neurons and glial cells, organizes muscular activity into functional patterns of wall behavior needed for propulsion, mixing, and storage. Functions of the nervous system are influenced by chemical signals released from **enterochromaffin cells**, **enteroendocrine cells**, and cells associated with the enteric immune system (e.g., **mast cells** and polymorphonuclear leukocytes).

Motility in the various organs of the digestive tract is organized to fulfill the specialized function of each individual organ. Esophageal motility, for example, differs from gastric motility, and gastric motility differs from small intestinal motility. The motility in the different organs reflects coordinated contractions and relaxations of the smooth musculature. Contractions are organized to produce the propulsive forces that move digesta along the tract, triturate large particles to smaller particles, mix ingested foodstuffs with digestive enzymes, and bring nutrients into contact with the mucosa for efficient absorption. Relaxation of spontaneous tone in the smooth muscle allows sphincters to open and ingested material to be accommodated in reservoirs of the stomach and large intestine. The enteric nervous system (ENS), together with its input from the central nervous system (CNS), organizes motility into patterns of efficient behavior suited to differing digestive states (e.g., fasting and processing of a meal) as well as abnormal patterns (e.g., during emesis).

MUSCULATURE OF THE DIGESTIVE TRACT

The smooth muscles of the digestive tract are organized in distinct layers. Two important muscle layers for motility in the small and large intestine are the longitudinal and circular layers (Fig. 25.1). The two layers form the intestinal **muscularis externa.** Unlike that of the esophagus and small and large intestines, the muscularis externa of the stomach consists of three layers with the long axis of the muscle fibers oriented in the longitudinal, circular, or oblique directions.

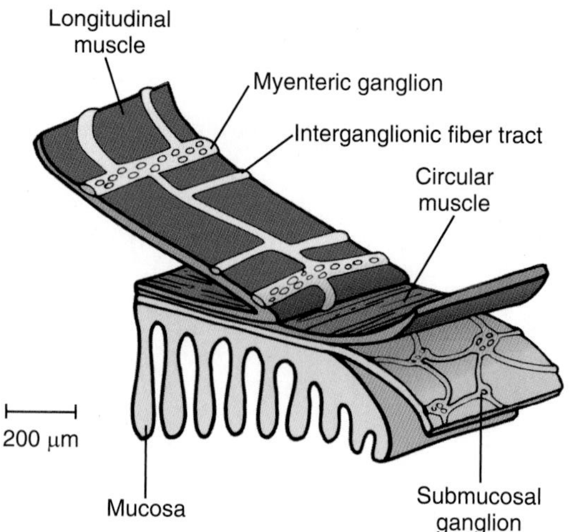

Figure 25.1 Structural relationship of the intestinal musculature and the enteric nervous system. Ganglia and interganglionic fiber tracts form the myenteric plexus between longitudinal and circular muscle layer and the submucosal plexus between the mucosa and circular muscle layer.

Circular and longitudinal muscle layers of the intestine differ in both structure and function.

The circular muscle is a thicker layer than the longitudinal and more powerful in exerting contractile forces on the contents of the lumen. The long axis of the fibers of the circular muscle is oriented in the circumferential direction. Consequently, contraction reduces the diameter of the lumen of an intestinal segment and increases its length. Because the long axis of fibers in the longitudinal muscle layer is oriented in the longitudinal direction, contraction shortens the segment of intestine in which it occurs and expands the lumen. Interactive changes in length and diameter occur because the intestine is a cylinder with constant surface area for which a change in diameter is accompanied by an opposite change in length.

Both longitudinal and circular muscle layers are innervated by motor neurons of the ENS. The longitudinal muscle layer is innervated mainly by excitatory musculomotor neurons and the circular muscle layer by both excitatory and inhibitory musculomotor neurons. Nonneural pacemaker cells and excitatory musculomotor neurons activate contraction of the circular muscle, and excitatory musculomotor neurons are the main triggers for contraction of the longitudinal muscle. More **gap junctions** (i.e., electrical connections between muscle fibers; see Chapter 8, "Skeletal and Smooth Muscle") are found in the circular than in the longitudinal muscle layer. Calcium influx from outside the muscle cells is important for excitation–contraction coupling in longitudinal muscle fibers. Intracellular release from internal stores is more important for excitation–contraction coupling in the muscle fibers of the circular layer.

Smooth muscles of the stomach and intestine contract spontaneously in the absence of neural or endocrine influence.

Smooth muscles are classified based on their behavioral properties and associations with nerves (see Chapter 8, "Skeletal and Smooth Muscle"). Muscles of the stomach and intestine behave like **unitary-type smooth muscle**. These muscles contract spontaneously in the absence of neural or endocrine influence and contract in response to stretch. There are no structured neuromuscular junctions, and neurotransmitters travel over extended diffusion distances to influence relatively large numbers of muscle fibers simultaneously. The smooth muscle of the esophagus and gallbladder is more like **multiunit-type smooth muscle**. These muscles do not contract spontaneously in the absence of nervous input and do not contract in response to stretch. Activation to contract is by nervous input at structured junctions to relatively small groups of muscle fibers.

Electromechanical and pharmacomechanical coupling trigger contractions in gastrointestinal muscles.

GI smooth muscle differs from skeletal muscle in having two mechanisms that initiate the processes leading to contractile shortening and development of tension. In both skeletal muscle and GI smooth muscle, depolarization of the membrane electrical potential leads to the opening of voltage-gated calcium channels, followed by the elevation of cytosolic calcium, which in turn activates the contractile proteins. This mechanism is called **electromechanical coupling**. Smooth muscles have an additional mechanism, in which the binding of ligands to G-protein-coupled receptors on the muscle membrane leads to the opening of calcium channels and the elevation of cytosolic calcium without any change in the membrane electrical potential. This mechanism is called **pharmacomechanical coupling**. The ligands may be chemical substances released as signals from nerves (neurocrine), from nonneural cells in close proximity to the muscle (paracrine), or from endocrine cells as hormones delivered to the muscle by the blood.

Gastrointestinal and esophageal smooth muscles have properties of a functional electrical syncytium.

Smooth muscle fibers are connected to their neighbors by **gap junctions**, which are permeable to ions and thereby transmit electrical current from muscle fiber to muscle fiber. Ionic connectivity, without cytoplasmic continuity from fiber to fiber, accounts for the **electrical syncytial** properties of smooth muscle, which confers electrical behavior analogous to that of cardiac muscle (see Chapter 12, "Electrical Activity of the Heart"). Electrical activity and associated contractions spread from a point of initiation (e.g., a pacemaker region) in three dimensions throughout the bulk of the muscle. The distance and direction of spread of electrical activity in the electrical syncytium are controlled by the ENS. A failure of nervous control can lead to disordered motility that includes spasm and associated cramping abdominal pain.

Electrical activity in gastrointestinal muscles consists of slow waves and action potentials.

Electrical slow waves are always present and responsible for triggering action potentials in some regions, whereas in other regions (e.g., the gastric antrum and large intestinal circular muscle), they represent the only form of electrical activity (Fig. 25.2). They are always present in the small intestine, where they decrease in frequency along a gradient from the duodenum to the ileum. In the gastric antrum, the terms *slow wave* and *action potential* are used interchangeably for the same electrical event. When action potentials are associated with electrical slow waves, they occur during the plateau phase of the slow wave.

Action potentials

Action potentials in GI smooth muscle are mediated by changes in conductance in calcium and potassium ionic channels. The depolarization phase of the action potential is produced by an all-or-nothing increase in calcium conductance, with the inward calcium current carried by L-type calcium channels. The opening of potassium channels as the calcium channels are closing at or near the peak of the action potential accounts for the repolarization phase. The L-type calcium channels in GI smooth muscle are essentially the same as those found in cardiac and vascular smooth muscle. Therefore, disordered GI motility may be a side effect of treating cardiovascular disease with drugs that block L-type calcium channels.

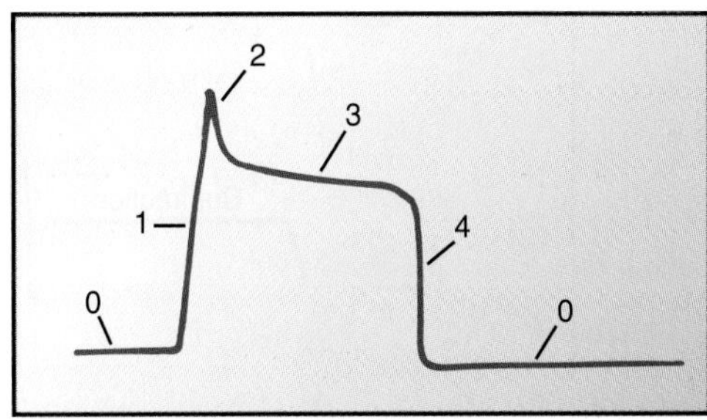

Figure 25.2 Electrical slow waves in gastrointestinal muscles occur in four phases determined by specific ionic mechanisms. Phase 0: Resting membrane potential; outward potassium current. Phase 1: Rising phase (upstroke depolarization); activation of voltage-gated calcium channels and activation of voltage-gated potassium channels. Phase 3: Plaeau phase; balance of inward calcium current and outward potassium current. Phase 4: Falling phase (repolarization); inactivation of voltage-gated calcium channels and activation of calcium-gated potassium channels.

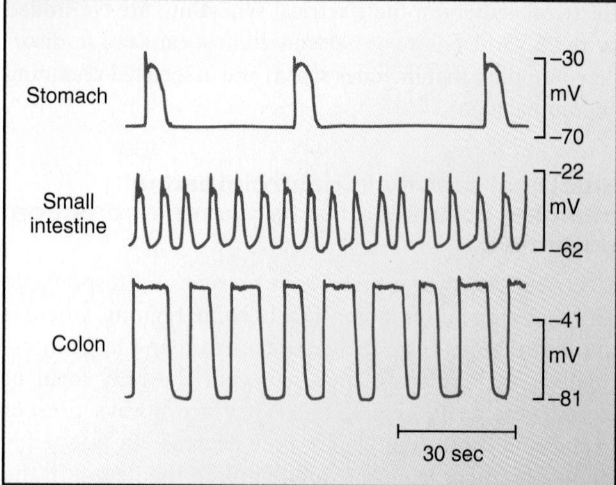

Figure 25.3 Electrical slow waves with similar waveforms occur at different frequencies in the stomach, small intestine, and colon.

Electrical slow-wave frequencies differ in the stomach, small intestine, and colon.

Electrical slow waves with similar waveforms occur at different frequencies in the gastric antrum and small and large intestinal circular muscle when recorded with intracellular electrodes (Fig. 25.3). Slow waves occur at 3 waves/min in the human antrum, 11 to 12 waves/min in the duodenum, and 2 to 13 waves/min in the colon. The maximum contractile frequency of the muscle does not exceed the frequency of the slow waves, but it may occur at a lower frequency because all slow waves may not trigger contractions. The ENS determines the nature of the contractile response during each slow wave in the integrated functional state of the whole organ.

Electrical slow waves may occur with or without action potentials in the small intestine.

As a general rule, slow waves in the small intestinal circular muscle trigger action potentials and action potentials trigger contractions. Slow waves are omnipresent in virtually all mammalian species and may or may not be accompanied by action potentials. Contractions do not occur in the absence of action potentials. The electrical slow waves in Figure 25.4 were recorded with an extracellular electrode attached to the serosal surface of the small intestine. This method records from many circular muscle fibers. Shallow contractions appearing in the absence of action potentials on the slow waves reflect the responses of a few of the total population of muscle fibers under the electrode (see Fig. 25.4A). In this case, the action potential currents from the small number of fibers are too small to be detected by the surface electrode. With this method of recording, the size of an action potential appears larger when greater numbers

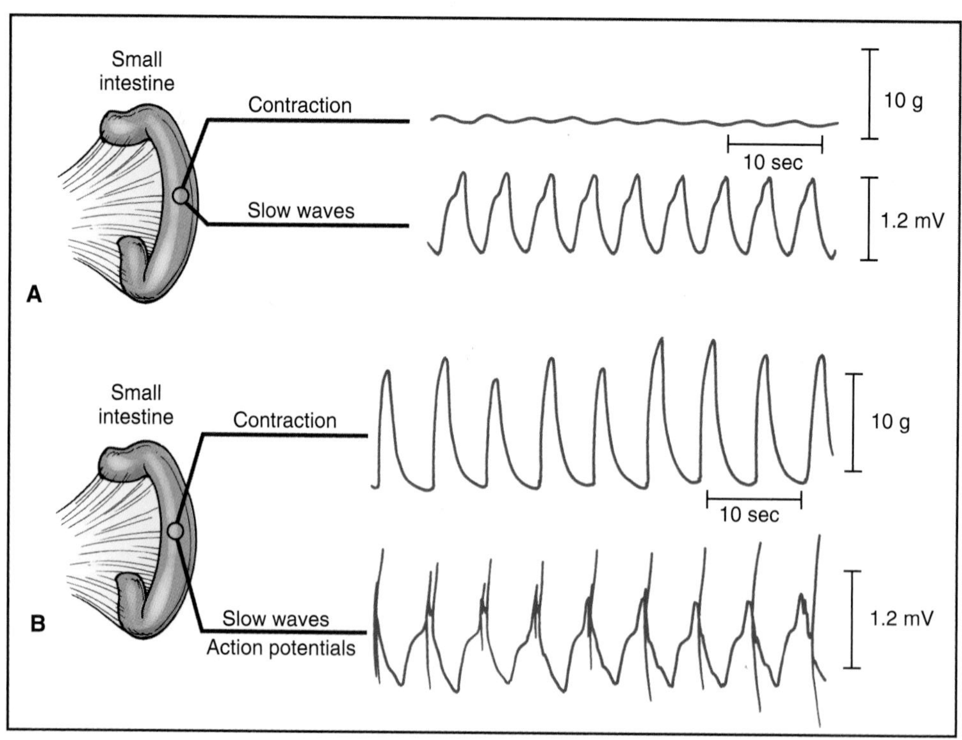

Figure 25.4 Electrical slow waves trigger action potentials and action potentials trigger contractions. Electrical slow waves in the small intestine are always present. Action potentials are not always associated with a slow wave. **(A)** No action potentials appear at the crests of the slow waves and the muscle contractions associated with each slow wave are small. **(B)** Muscle action potentials appear as sharp upward–downward deflections at the crests of the slow waves. Large-amplitude muscle contractions are associated with each slow wave when action potentials are present. (See text for details.)

of the total population of muscle fibers are depolarized to action potential threshold by each slow wave. The amplitude of phasic contractions associated with each electrical slow wave increases in direct relation to the number of muscle fibers recruited to firing threshold by each slow-wave cycle (see Fig. 25.4B).

Interstitial cells of Cajal generate electrical slow waves.

ICCs generate the electrical slow waves in the stomach and small and large intestine (Fig. 25.5). Gap junctions that impart the properties of a functional electrical syncytium interconnect the ICCs into networks. Gap junctions likewise electrically connect the ICCs to the circular muscle. Electrical current flows from the ICC network across the gap junctions to depolarize the membrane potential of the circular muscle fibers to the threshold for action potential discharge, which in turn initiates a contraction.

Pacemaker networks of ICCs are located surrounding the small intestinal circular muscle at the border with the longitudinal muscle layer (myenteric border) and at its border with the submucosa. Slow waves generated by the ICC network at the submucosal border spread passively across gap junctions into the bulk of circular muscle, and those at the myenteric border spread passively into both longitudinal and circular muscle. Slow wave electrical current, after crossing into muscle layer, spreads from muscle fiber to muscle fiber via gap junctions in the musculature.

NEURAL CONTROL OF DIGESTIVE FUNCTIONS

The innervation of the digestive tract controls muscle contraction, secretion, and absorption across the mucosal lining and blood flow inside the walls of the esophagus, stomach, intestines, and gallbladder. Depending on the kind of neurotransmitter released, the motor components of the innervation can activate or inhibit muscle contraction. Secretion of water, electrolytes, and mucus into the lumen and absorption from the lumen are determined by motor components of the innervation called **secretomotor neurons**. Amounts of blood flow within the wall and the distribution of flow between the muscle layers and mucosa are controlled also by nervous activity inside the gut.

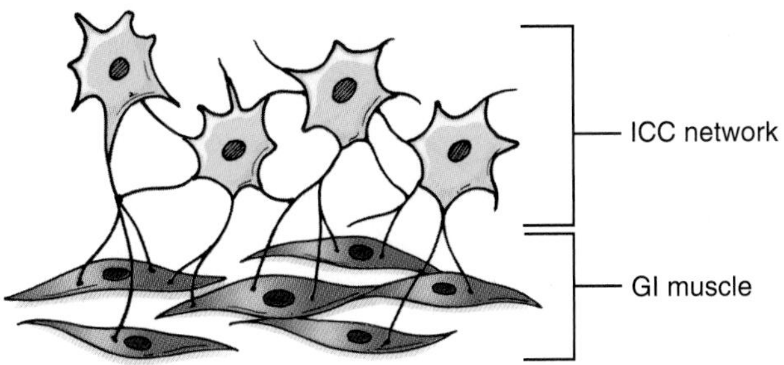

Figure 25.5 Interstitial cells of Cajal (ICCs) generate electrical slow waves. ICCs are interconnected into networks, which contact the gastrointestinal musculature. Electrical slow waves originate in the networks of ICCs. ICCs are the generators (pacemaker sites) for the slow waves. Gap junctions connect the ICCs to the circular muscle. Ionic current flows across the gap junctions to depolarize the membrane potential of the circular muscle fibers to the threshold for the discharge of action potentials.

Sensory nerves transmit information on the state of the gut to the brain for processing. Sensory transmission and central processing account for sensations that are localized to the digestive tract. These include sensations of discomfort (such as upper abdominal fullness), abdominal pain, and chest pain (heartburn). Control by the CNS includes the sensory inflow of information from the gut to the brain and spinal cord and outflow from the brain and spinal cord to the gut. Outflow may originate in higher processing centers of the brain (the frontal cortex) and accounts for the projection of an individual's emotional state (psychogenic stress) to the gut. This kind of brain–gut interaction underlies the symptoms of diarrhea and cramping lower abdominal pain sometimes reported by students during anticipation of stressful life events (a difficult exam or oral interview).

Network of neural integrative centers controls the moment-to-moment motor activity of the gastrointestinal tract.

The sympathetic, parasympathetic, and ENS divisions of the autonomic nervous system (ANS) innervate the digestive tract. Neural control of the gut is hierarchical, with five basic levels of integrative organization (Fig. 25.6). Level 1 is the ENS, which behaves like an independent integrative nervous system (minibrain) inside the walls of the gut. Level 2 consists of the prevertebral ganglia of the sympathetic nervous system. Levels 3, 4, and 5 are within the CNS. Sympathetic and parasympathetic signals to the digestive tract originate at levels 3 and 4 (central sympathetic and parasympathetic centers) in the medulla oblongata and represent the final

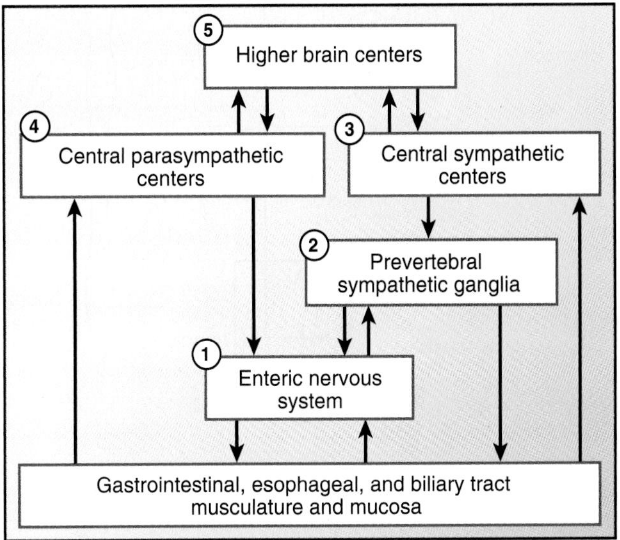

Figure 25.6 A hierarchy of five levels of neural organization determines the moment-to-moment motor behavior of the digestive tract. (See text for details.)

common pathways for the outflow of information from the brain to the gut. Level 5 includes higher brain centers that provide input for integrative functions at levels 3 and 4.

Autonomic signals to the gut are carried from the brain and spinal cord by sympathetic and parasympathetic nervous pathways, which represent the extrinsic component of innervation. Neurons of the ENS form the local intramural control networks that make up the intrinsic component of the autonomic innervation. The parasympathetic and sympathetic subdivisions are identified by the positions of the ganglia that contain the cell bodies of the postganglionic neurons and by the point of outflow from the CNS. Comprehensive autonomic innervation of the digestive tract consists of interactive interconnections between the brain, the spinal cord, and the ENS.

Parasympathetic neurons innervate the gut from the medulla oblongata and sacral spinal cord.

The cell bodies of parasympathetic neurons projecting to the gut are in both the brainstem and the sacral region of the spinal cord (Fig. 25.7). The nerve fibers projecting from these regions of the CNS are *motor* **efferents**. Neuronal cell bodies for motor efferents in the parasympathetic cranial division reside in the medulla oblongata and transmit information to the gut via vagus nerves. Motor neuronal cell bodies of the sacral division are located in the sacral region of the spinal cord and transmit information via pelvic nerves to the large intestine. Efferent fibers in the pelvic nerves make synaptic contact with neurons in ganglia located on the serosal surface of the colon and in ganglia of the ENS deeper within the wall of the large intestine. Efferent vagal fibers make synaptic connections with neurons of the ENS in the esophagus, stomach, small intestine, and colon as well as the gallbladder and pancreas.

Efferent vagal nerves transmit signals to the ENS innervation of the GI musculature to control digestive processes both in anticipation of food intake and following a meal. This involves the stimulation and inhibition of contractile behavior in the stomach, which results from activation of the ENS circuits that control excitatory or inhibitory musculomotor neurons, respectively. Parasympathetic efferents to the small and large intestinal musculature are predominantly stimulatory due to their input to the ENS microcircuits that control the activity of excitatory musculomotor neurons.

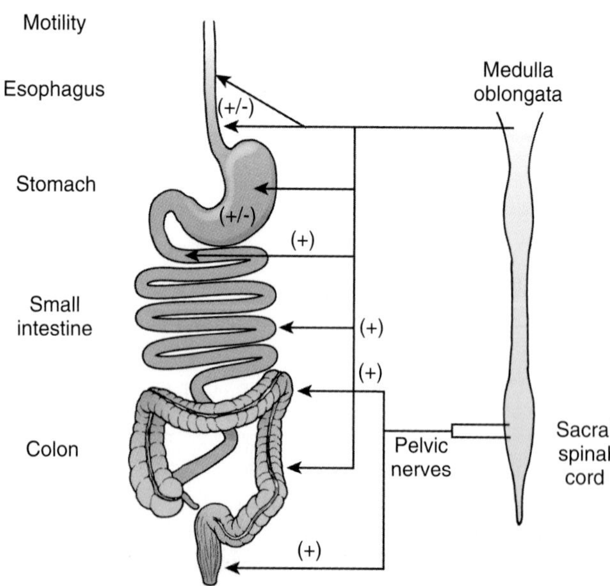

Figure 25.7 The digestive tract is innervated by the parasympathetic division of the autonomic nervous system. Signals from parasympathetic centers in the central nervous system are transmitted to the enteric nervous system by the vagus and pelvic nerves. These signals result in either contraction (+) or relaxation (–) of the digestive musculature.

Dorsal vagal complex in the medulla controls the upper gastrointestinal tract.

The **dorsal vagal complex** consists of the **dorsal motor nucleus** of the vagus, **nucleus tractus solitarius**, **area postrema**, and **nucleus ambiguus**; it is the vagal integrative center in the CNS (Fig. 25.8). This center is more directly involved in the control of the specialized digestive functions of the esophagus and stomach and the functional cluster of duodenum, gallbladder, and pancreas than those of the distal small intestine and large intestine. The neural networks in the dorsal vagal complex and their interactions with higher centers are responsible for the rapid and precise control necessary for adjustments to rapidly changing conditions in the upper digestive tract during anticipation, ingestion, and digestion of meals of varied composition.

Vagovagal reflex controls contractions of gastrointestinal muscle layers in response to food stimuli.

Vagovagal reflex refers to the GI reflex circuits in which afferent and efferent fibers of the vagus nerve coordinate response to gut stimuli. One of the functions of the vagovagal

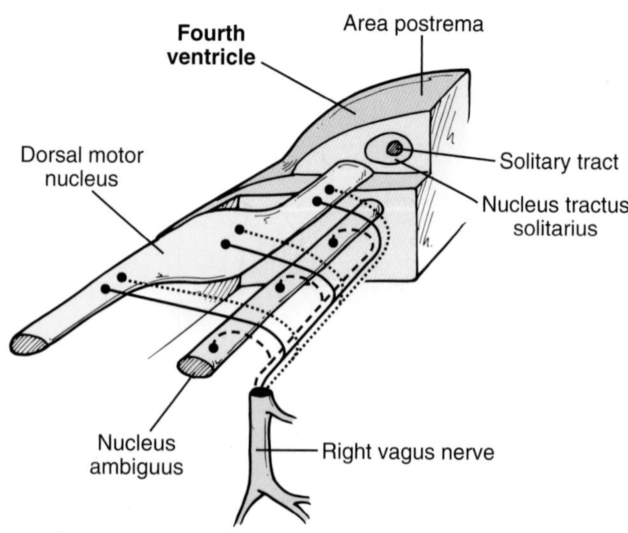

Figure 25.8 The dorsal vagal complex in the medulla oblongata is the central vagal integrative center for nervous control of the upper gastrointestinal tract. The dorsal motor nucleus of the vagus and nucleus tractus solitarius form the dorsal vagal complex.

reflex is to control the contraction of GI smooth muscle in response to distension of the tract from food. For example, the vagovagal reflex is activated to cause the stomach muscles to relax in response to swallowing food. This reflex also allows for the accommodation of large amounts of food in the GI tract.

The sensory side of the reflex arc consists of vagal afferent neurons connected with a variety of sensory receptors specialized for the detection and signaling of mechanical parameters, such as muscle tension and mucosal brushing, or luminal chemical parameters, including pH, osmolality, and glucose concentration. Cell bodies of the vagal afferents are in **nodose ganglia**. The afferent neurons are synaptically connected with neurons in the dorsal motor nucleus of the vagus and in the nucleus tractus solitarius. Neurons in the nucleus tractus solitarius, which lies directly above the dorsal motor nucleus of the vagus (see Fig. 25.8), project to form chemical synapses with the neuronal pool in the vagal motor nucleus. Synaptic networks, formed by processes from neurons in both nuclei, tightly link the two into an integrative center, which, together with the area postrema and nucleus ambiguus, forms the dorsal vagal complex. The dorsal vagal neurons are second-order or third-order neurons representing the efferent arms of the reflex circuits, which project to the gut in right or left vagal nerves. They are the final common command pathways out of the brain to the integrative neural networks in the ENS that innervate and control behavior of the musculature and secretory glands.

Efferent vagal fibers form synapses with neurons in the ENS to activate circuits that ultimately drive the outflow of signals in motor neurons to the effector systems. When the effector system is the musculature, its innervation consists of both inhibitory and excitatory musculomotor neurons that participate in reciprocal control. If the effector systems are gastric glands or digestive glands, the secretomotor neurons are excitatory and stimulate secretory behavior.

The circuits for CNS control of the upper GI tract are organized much like those dedicated to the control of skeletal muscle movements (see Chapters 5, "Motor System" and 7, "Integrative Functions of the Nervous System"), in which fundamental reflex circuits are located in the spinal cord. Inputs to spinal reflex circuits from higher-order integrative centers in the brain (motor cortex and basal ganglia) organize skeletal muscle contractile activity into functional motor behavior. Memory, the processing of incoming information from outside the body, and the integration of proprioceptive information are ongoing functions of the higher brain centers responsible for the logical organization of outflow to the skeletal muscles by way of basic spinal reflex circuits. The basic connections of the vagovagal reflex circuit are like somatic motor reflexes, in being "fine-tuned" from moment to moment by input from higher integrative centers in the brain.

Sympathetic nerves innervate blood vessels, mucosa, and muscularis of the gut and suppress gut activity when stimulated.

Sympathetic innervations to the gut exit from the thoracic and lumbar regions of the spinal cord (Fig. 25.9). Efferent sympathetic fibers leave the spinal cord in the ventral roots to make their first synaptic connections with neurons in **prevertebral sympathetic ganglia** located in the abdomen. The prevertebral ganglia are the **celiac, superior mesenteric**, and **inferior mesenteric ganglia**. Cell bodies in the prevertebral ganglia project to the digestive tract, where they synapse with neurons of the ENS in addition to innervating the blood vessels, mucosa, and specialized regions of the musculature. The sympathetic neurons leaving the spinal cord are called **preganglionic sympathetic neurons**; the neurons leaving prevertebral ganglia to the gut are **postganglionic sympathetic neurons**.

CNS activation of sympathetic input to the GI tract shunts blood from the splanchnic to the systemic circulation during exercise and stressful environmental change. Sympathetic suppression of digestive functions, including motility and secretion, occurs as a coincident adaptation for reduced

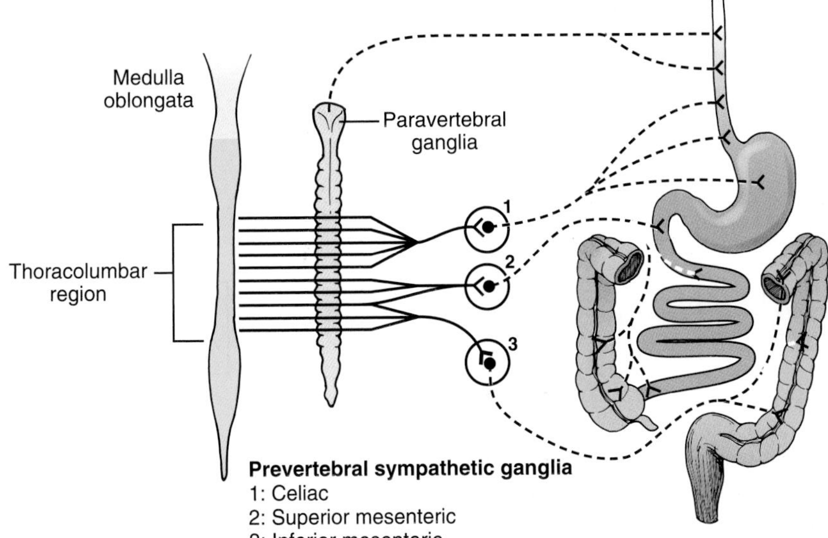

Figure 25.9 The digestive tract is innervated by the sympathetic division of the autonomic nervous system. Preganglionic neurons of the sympathetic division of the autonomic nervous system project to the gut from thoracic and upper lumbar segments of the spinal cord. Efferent sympathetic fibers leave the spinal cord in the ventral roots to make their first synaptic connections with neurons in the prevertebral ganglia in the abdomen. Cell bodies in the prevertebral ganglia are postganglionic neurons, which project to the digestive tract where they synapse with neurons of the enteric nervous system in addition to innervating the blood vessels, mucosa, and specialized regions of the musculature.

blood flow. The release of **norepinephrine (NE)** from sympathetic postganglionic neurons is the main mediator of these effects. NE acts directly on sphincteric muscles (**lower esophageal sphincter** [**LES**] and internal anal sphincter) to increase tension and keep the sphincter closed. Presynaptic inhibitory action of NE at synapses in the control circuitry of the ENS suppresses gastric and intestinal motility. Postsynaptic inhibitory action of NE at secretomotor neurons, which innervate the intestinal secretory glands, prevents secretion when blood flow is reduced (Fig. 25.10). Aside from smooth muscle sphincters, most of the sympathetic innervation goes to the ENS, not the musculature.

Suppression of synaptic transmission by NE release from sympathetic nerves occurs at both fast and slow excitatory synapses in the neural networks of the ENS. This inactivates the neural circuits that normally generate intestinal motor behavior. Activation of the sympathetic inputs allows only continuous discharge of inhibitory musculomotor neurons to the nonsphincteric smooth muscles. The overall effect is a state of paralysis of intestinal motility in conjunction with reduced intestinal blood flow. This state is called **physiological ileus** when it occurs transiently and **paralytic ileus** when it persists abnormally (e.g., after abdominal surgery).

Splanchnic nerves are paired nerves that innervate the gut and carry sensory information from and efferent sympathetic signals to the gastrointestinal tract.

Splanchnic nerves are mixed nerves in the mesentery that have both sympathetic efferent and sensory afferent fibers. Sensory nerves, on their way to the spinal cord, course side by side with the sympathetic fibers; nevertheless, they are not part of the sympathetic nervous system. The term, *sympathetic afferent,* which is sometimes used, is incorrect. The cell bodies of the sympathetic innervation of the GI tract are positioned in the intermediolateral cell columns of the spinal cord descending from the first thoracic to the third lumbar segment. Efferent sympathetic fibers leave the spinal cord in the ventral roots to make synaptic connections with neurons in prevertebral sympathetic ganglia located in the abdomen (see Fig. 25.9).

Sensory afferent fibers in the splanchnic nerves have their cell bodies in dorsal root spinal ganglia (see Chapter 4, "Sensory Physiology"). They transmit information from the GI tract and gallbladder to the CNS for processing. These fibers bifurcate within the gut wall to transmit a steady stream of information to the local processing circuits in the ENS. The afferent fibers branch and send collaterals to form synapses with neurons in prevertebral sympathetic ganglia before projecting on to the spinal cord.

Mechanoreceptors, chemoreceptors, and thermoreceptors are sensory receptors in the gut. Mechanoreceptors sense mechanical events in the mucosa, musculature, serosal surface, and mesentery. They supply both the ENS and the CNS with information on stretch-related tension and muscle length in the wall and on the movement of luminal contents as they brush the mucosal surface. Mesenteric receptors code for gross movements of the organ. Chemoreceptors generate information on the concentration of nutrients, kind of nutrient (e.g., lipids), osmolality, and pH in the luminal contents. Recordings of sensory information exiting the gut in splanchnic afferent fibers and in afferents in the vagal nerves show that most receptors are multimodal in that they respond to both mechanical and chemical stimuli. The presence in the GI tract of pain receptors (nociceptors) equivalent to C-fibers and A-delta fibers elsewhere in the body is likely but not unequivocally confirmed, except in the gallbladder. The sensitivity of splanchnic afferents, including nociceptors, may be elevated when inflammation is present in the intestine or gallbladder.

Enteric nervous system serves as a "minibrain" controlling the gut.

The ENS is a subdivision of the ANS embedded in the walls of the GI tract. It serves as a "minibrain" located near the effector systems it controls. Effector systems of the digestive tract

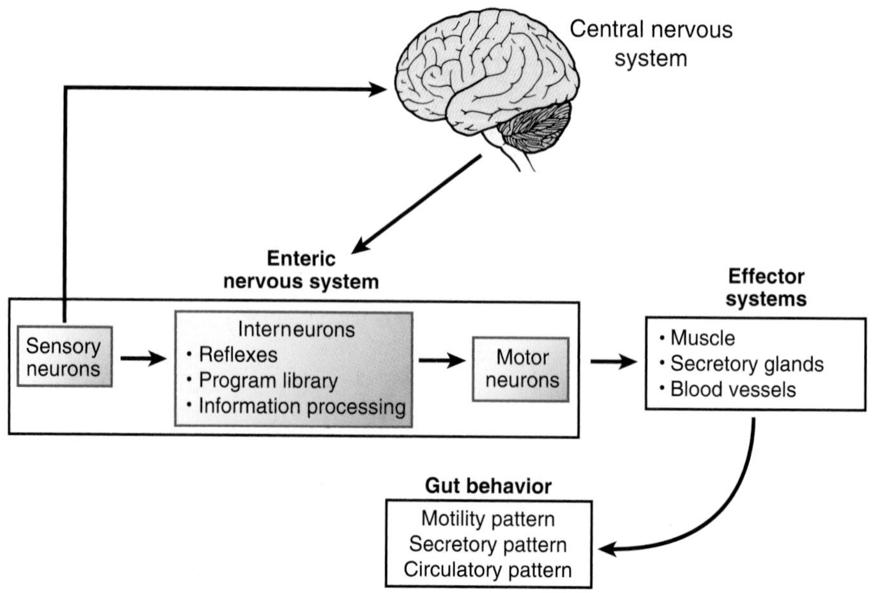

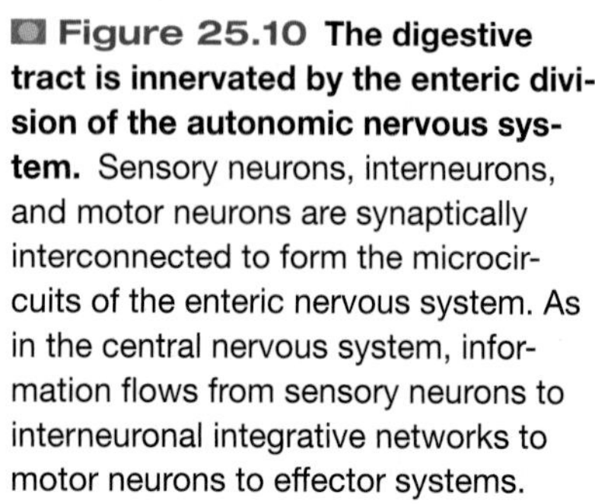
Figure 25.10 The digestive tract is innervated by the enteric division of the autonomic nervous system. Sensory neurons, interneurons, and motor neurons are synaptically interconnected to form the microcircuits of the enteric nervous system. As in the central nervous system, information flows from sensory neurons to interneuronal integrative networks to motor neurons to effector systems.

are the musculature, secretory glands, and blood vessels (see Fig. 25.10). The ENS has as many neurons as the spinal cord. Rather than crowding the large number of neurons required for controlling digestive functions into the cranium as part of the cephalic brain and relying on signal transmission over long and unreliable pathways, the integrative microcircuits are distributed along the 7 m or so of human small intestine and 1.5 m of large intestine near the borders of the effector systems they control and regulate. The circuits connected with each effector have evolved as an organized array of different kinds of neurons interconnected by chemical synapses. Function in the circuits is determined by the generation of action potentials within single neurons and chemical transmission of information at the synapses in the same manner as in the CNS.

The enteric microcircuits in the various specialized regions of the digestive tract are "wired" with large numbers of neurons and synaptic sites, in which information processing occurs. Multisite computation generates output behavior from the integrated circuits that could not be predicted from properties of their individual neurons and synapses. As in the brain and spinal cord, emergence of complex behaviors is a fundamental property of the neural networks of the ENS.

The processing of sensory signals is one of the major functions of the neural networks of the ENS. Sensory signals are generated by sensory nerve endings and are coded in the form of action potentials. The code could represent the status of an effector system (e.g., change of tension in a muscle) or could signal a change in an environmental parameter such as luminal pH. Sensory signals are computed by the neural networks to generate output signals that initiate homeostatic adjustments in the behavior of the effector system.

The cell bodies of the neurons that make up the neural networks of the ENS are clustered in ganglia that are interconnected by fiber tracts to form a plexus (see Fig. 25.1). Structure, function, and neurochemistry of the ganglia differ from those of other ANS ganglia. Unlike autonomic ganglia elsewhere in the body, which function mainly as relay-distribution centers for signals transmitted from the brain and spinal cord, enteric ganglia are interconnected to form a nervous system with mechanisms for the integration and processing of information like those found in the CNS. This is why the ENS is sometimes referred to as the "minibrain in the gut."

Myenteric and submucosal plexuses control gut motility and secretion.

As subdivisions of the ENS, the myenteric and submucosal plexuses are the major nerve supply controlling gut motility and secretion. The ENS consists of ganglia, primary interganglionic fiber tracts, and secondary and tertiary fiber projections to the effector systems (i.e., musculature, glands, and blood vessels). These structural components of the ENS are interlaced to form a plexus. Two ganglionated plexuses are the most obvious constituents of the ENS (see Fig. 25.1). The **myenteric plexus**, also known as the **Auerbach plexus**, is positioned between the longitudinal and circular muscle layers along most of the digestive tract. The **submucosal plexus**, also known as the **Meissner plexus**, is situated in the submucosal region between the circular muscle and mucosa. The submucosal plexus is most prominent as a ganglionated network in the small and large intestine. It does not exist as a ganglionated plexus in the esophagus and is sparse in the submucosal space of the stomach.

Cell bodies of motor neurons (secretomotor neurons) to the intestinal secretory glands (i.e., Brunner glands in the duodenum and the crypts of Lieberkühn in the small and large intestine) originate in the submucosal plexus. Neurons in submucosal ganglia send fibers to the myenteric plexus and also receive synaptic input from axons projecting from the myenteric plexus. The interconnections link the two networks into a functionally integrated nervous system.

Sensory neurons, interneurons, and motor neurons form the microcircuits that integrate the enteric nervous system.

The heuristic model for the ENS is the same as for the brain and spinal cord (see Fig. 25.10). As in the CNS, sensory neurons, interneurons, and motor neurons in the ENS are connected synaptically for the flow of information from sensory neurons to interneuronal integrative networks to motor neurons to effector systems. The ENS organizes and coordinates the activity of each effector system into the meaningful behavior seen in each of the integrated organs. Bidirectional communication occurs between the CNS and ENS. Musculomotor neurons control behavior of the musculature. Secretomotor neurons control the mucosal secretory glands, and interneurons are synaptically interconnected to form information-processing networks.

Afterhyperpolarization-type and synaptic-type neurons in the enteric nervous system are distinguished by their electrophysiologic and synaptic behavior.

Two primary types of ENS neurons are distinguished by their electrophysiologic behavior and morphology in studies in which electrical activity is recorded with intracellular microelectrodes containing injectable markers. The two types are designated as afterhyperpolarization (AH)-type or synaptic (S)-type. AH-type neurons have multiple long processes, any of which might be an axon or dendrite. S-type neurons have a single long axon with multiple short dendrites (Fig. 25.11).

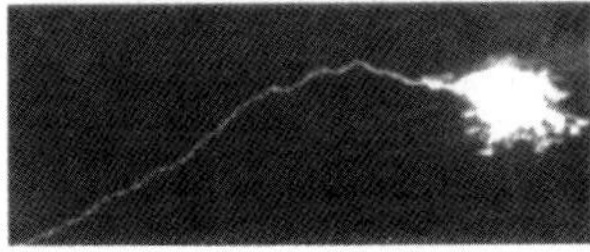

S-Type Neuron

AH-Type Neuron

Figure 25.11 Neurons with uniaxonal morphology and neurons with multipolar morphology are the two primary morphologic categories of enteric nervous system neurons. Examples of the two kinds of neurons were obtained by injecting a fluorescent marker from the microelectrode during intracellular electrophysiologic studies. Synaptic-type neurons have a single long axon. Afterhyperpolarization-type neurons have multipolar morphology.

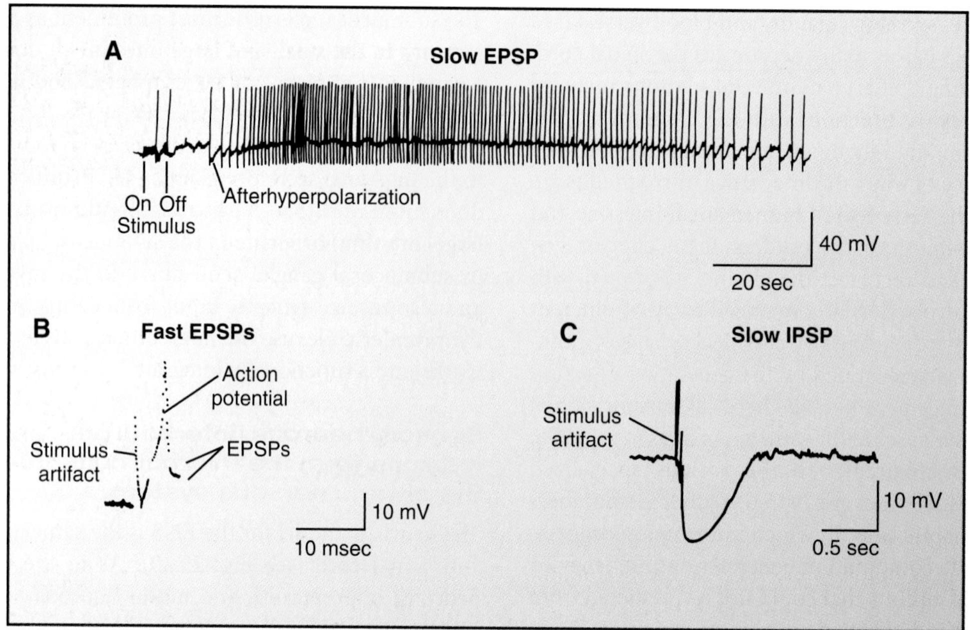

Figure 25.12 Fast excitatory postsynaptic potentials (EPSPs), slow EPSPs, and slow inhibitory postsynaptic potentials (IPSPs) are synaptic events in enteric neurons. (A) The slow EPSP was evoked by repetitive electrical stimulation of the synaptic input to the neuron. Slowly activating membrane depolarization of the membrane potential continues for almost 2 minutes after the termination of the stimulus. During the slow EPSP, repetitive discharge of action potentials reflects enhanced neuronal excitability. **(B)** The fast EPSPs were also evoked by single electrical shocks applied to the axon that synapsed with the recorded neuron. Two fast EPSPs were evoked by successive stimuli and are shown as superimposed records. Only one of the EPSPs reached the threshold for the discharge of an action potential. The time course of the EPSPs is in the millisecond range. **(C)** The slow IPSP was also evoked by the stimulation of an inhibitory input to the neuron. This hyperpolarizing synaptic potential will suppress excitability (i.e., decrease the probability of action potential discharge), compared with enhanced excitability during the slow EPSP.

Afterhyperpolarization-type neurons

AH-type neurons are so named because a characteristic long-lasting hyperpolarizing potential (i.e., an AH) occurs after discharge of an action potential (Fig. 25.12). These were suggested earlier to be *intrinsic primary afferent neurons* (a misleading term); however, these are not sensory neurons with properties like those of vagal and spinal sensory afferents. AH-type neurons fulfill the role of interneurons in the ENS microcircuits. Neurons with AH-type properties comprise the largest proportion of neurons in the myenteric plexus and the smallest proportion of submucosal plexus neurons.

Synaptic-type neurons

Although AH-type neurons were named for their characteristic postspike afterhyperpolarizing potentials, S-type neurons were so named because *nicotinic* fast excitatory postsynaptic potentials (EPSPs) were found in virtually all S-type enteric neurons, whereas only restricted subpopulations of AH-type neurons express nicotinic receptors and fast EPSP-like depolarizing responses to the application of acetylcholine (ACh). S-type enteric neurons can be distinguished from AH-type neurons by absence of an AH, lower resting membrane conductance, greater excitability, and, unlike AH-type neurons, application of the sodium channel blocker tetrodotoxin, which abolishes the action potentials in most S-type neurons.

All motor neurons in the myenteric and submucosal plexuses (i.e., **musculomotor neurons**, secretomotor neurons, and vasculomotor neurons) are S-type neurons. Subpopulations of interneurons can also exhibit the properties of S-type neurons. Unlike the low representation of AH-type neurons in the submucosal plexus, most of the neurons in the submucosal plexus are S-type neurons.

SYNAPTIC TRANSMISSION IN THE ENTERIC NERVOUS SYSTEM

Multiple forms of synaptic transmission occur in the microcircuits of the ENS. Both "fast" synaptic potentials with durations <50 ms and slow synaptic potentials lasting several seconds are found in cell bodies of ENS neurons. These synaptic events may be EPSPs or inhibitory postsynaptic potentials (IPSPs). In experimental protocols, they can be evoked by electrical stimulation of presynaptic axons or may occur spontaneously as a reflection of ongoing discharge of spikes by other synaptically connected neurons in the same microcircuit.

Figure 25.12 shows three forms of synaptic transmission that can occur in enteric neurons. The synaptic potentials in this illustration were evoked by placing fine stimulating electrodes on interganglionic fiber tracts in the myenteric or submucosal plexus and applying electrical shocks to fire the presynaptic axons and release the neurotransmitter at their synapses with neuronal cell bodies in one or the other of the plexuses.

Metabotropic receptors mediate enteric slow excitatory postsynaptic potentials.

The *slow* EPSP in Figure 25.12 was evoked by repetitive shocks (5 Hz) applied to the fiber tract for 5 seconds. Slowly activating depolarization of the membrane potential with a time course lasting more than 2 minutes after termination of the stimulus is apparent. Repetitive discharge of action potentials reflects enhanced neuronal excitability during the EPSP. The record shows hyperpolarizing after-potentials associated with the first four spikes of the train. As stated earlier, hyperpolarizing after-potentials are a property of enteric neurons called AH-type neurons. As the slow EPSP develops, the hyperpolarizing after-potentials are suppressed and can be seen to recover at the end of the spike train as the slow EPSP subsides. Suppression of the after-potentials is part of the mechanism of slow synaptic excitation that permits the AH-type neurons to convert from a low to a high state of excitability. In the resting state, AH-type neurons will either not fire at all or fire only one or two action potentials because development of the AH prevents further firing. Suppression of the AH during a slow EPSP permits repetitive firing at increased frequencies.

Slow EPSPs are mediated by multiple chemical messengers acting at many different G-protein-coupled **metabotropic receptors**. Different kinds of receptors, each of which mediates slow EPSP-like responses, are found in varied combinations on each neuron. The common mode of signal transduction for slow EPSPs in AH-type neurons involves receptor activation of adenylyl cyclase and second messenger function of cyclic adenosine monophosphate (cAMP), which links several different chemical messages to the behavior of a common set of ionic channels responsible for the generation of the slow EPSP responses. Postreceptor signal transduction for slow EPSPs in S-type neurons involves activation of phospholipase C, synthesis of inositol trisphosphate, and elevation of free intraneuronal calcium.

Serotonin, substance P, and ACh are examples of enteric neurotransmitters that evoke slow EPSPs. In **paracrine signaling**, mediators released from nonneural cells also evoke slow EPSP-like responses when released in the vicinity of the ENS. Histamine, for example, is released from mast cells during hypersensitivity reactions (e.g., food allergies and parasitic infections) and acts at the histamine H_2 receptor subtype to evoke slow EPSP-like responses in ENS neurons in rodents. Subpopulations of ENS neurons in specialized regions of the gut (e.g., the upper duodenum) have receptors for hormones such as gastrin and cholecystokinin, which also evoke slow EPSP-like responses.

Slow excitatory postsynaptic potentials prolong neural excitation or inhibition of gastrointestinal effector systems.

The long-lasting discharge of action potentials during the slow EPSP drives the release of neurotransmitter from the neuron's axon for the duration of the spike discharge. This could result in either prolonged excitation or inhibition at neuronal synapses and neuroeffector junctions.

Contractile responses within the musculature and glandular secretory responses are slow events that span time courses of several seconds from start to completion. The train-like discharge of spikes during slow EPSPs is the neural correlate of long-lasting responses of the gut effectors during physiologic stimuli. Figure 25.13 illustrates how the occurrence of slow EPSPs in excitatory musculomotor neurons to the intestinal musculature or mucosal secretory glands might evoke prolonged contraction of the muscle or prolonged secretion. The occurrence of slow EPSPs in inhibitory musculomotor neurons to the musculature results in prolonged inhibition of contractile force.

Ionotropic receptors mediate enteric fast excitatory postsynaptic potentials.

The fast EPSPs in Figure 25.12 were evoked by single shocks applied to an interganglionic fiber tract in the myenteric plexus. Two EPSPs were evoked by successive stimuli and are shown superimposed. One of the two EPSPs reached the threshold for the discharge of an action potential.

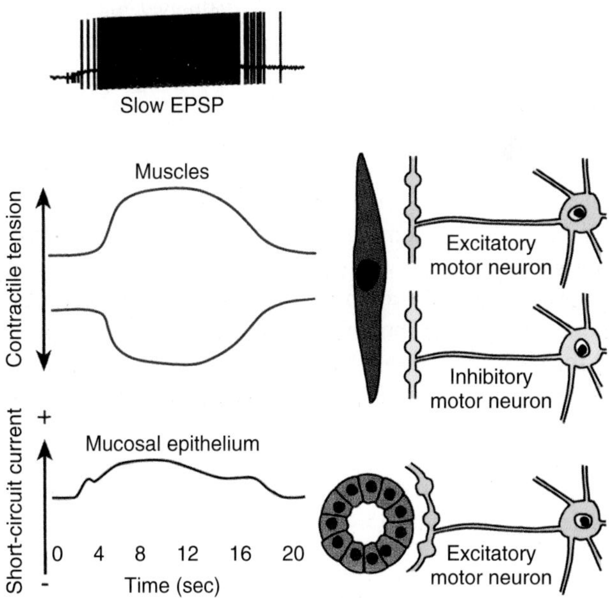

Figure 25.13 Slow excitatory postsynaptic potentials (EPSPs) in excitatory motor neurons to the muscles or mucosal epithelium result in prolonged muscle contraction or mucosal crypt secretion. Stimulation of secretion in experiments is seen as an increase in ion movement (short-circuit current). Slow inhibitory postsynaptic potentials in inhibitory motor neurons to the muscles result in prolonged inhibition of contractile activity, which is observed as decreased contractile tension.

Fast EPSPs are transient depolarizations of membrane potential that have durations <50 ms. They occur in the ENS throughout the digestive tract. Most fast EPSPs are mediated by ACh acting at **ionotropic receptors**. Ionotropic receptors are coupled directly to ion channels. Fast EPSPs function in the rapid transfer and transformation of neurally coded information between the elements of the enteric microcircuits. They are "bytes" of information in the information-processing operations of the logic circuits.

Multiple kinds of receptors mediate enteric slow inhibitory postsynaptic potentials.

The slow IPSP in Figure 25.12 was evoked by stimulation of an interganglionic fiber tract in the submucosal plexus. This kind of hyperpolarizing synaptic potential suppresses excitability (decreases the probability of spike discharge), compared with enhanced excitability during the slow EPSP.

Experimental application of several different chemical messenger substances, which may be peptidergic, purinergic, noradrenergic, somatostatinergic, or cholinergic, evokes slow IPSP-like responses. Opioid peptides and opiate analgesics (morphine) are slow IPSP mimetics. This action is limited to subpopulations of neurons. The effects of opiates and opioid peptides are blocked by the antagonist naloxone. Addiction to morphine may be seen in enteric neurons, and withdrawal is observed as high-frequency spike discharge on the addition of naloxone during chronic morphine exposure.

NE released from sympathetic postganglionic neurons acts at α_2-adrenoreceptors to mimic slow IPSPs in S-type neurons in the submucosal plexus. This occurs primarily in secretomotor neurons of the submucosal plexus that, when active, stimulate secretion. Slow IPSPs in submucosal neurons reflect a mechanism by which the sympathetic innervation suppresses intestinal secretion during physical exercise, when the brain directs that blood be shunted from the splanchnic to systemic circulation. Somatostatin released from ENS neurons in the submucosal plexus at synapses on secretomotor neurons mimics the inhibitory action of NE.

Galanin is a 29-amino acid polypeptide that simulates slow synaptic inhibition when applied to any of the neurons of the myenteric plexus. The application of adenosine, adenosine triphosphate (ATP) or other purinergic analogues also mimics slow IPSPs. The inhibitory action of adenosine is at adenosine A_1 receptors. Inhibitory actions of adenosine A_1 agonists result from the suppression of the enzyme adenylyl cyclase and the reduction in intraneuronal cAMP.

Presynaptic inhibition occurs at synapses and neuromuscular junctions.

Presynaptic inhibition (Fig. 25.14) is a significant function at fast nicotinic synapses, at slow excitatory synapses, and at sympathetic inhibitory synapses in the neural networks of the ENS and at excitatory neuromuscular junctions. It is a specialized form of neurocrine transmission whereby neurotransmitter released from one axon acts at receptors on a second axon (i.e., axoaxonal transmission) to prevent the release of neurotransmitter from the second axon. Presynaptic inhibition resulting from actions of paracrine or endocrine mediators on receptors at presynaptic release sites is an alternative to axoaxonal synapses as a mechanism for enhancing or suppressing transmission at a synapse or neuroeffector junction.

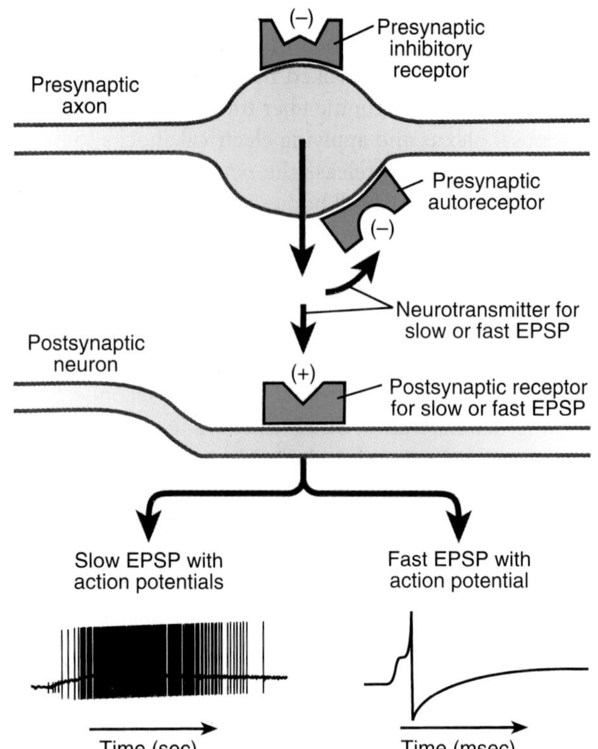

Figure 25.14 Presynaptic inhibitory receptors are found on axons at neurotransmitter release sites for both slow and fast excitatory postsynaptic potentials (EPSPs). Different neurotransmitters act at presynaptic inhibitory receptors to suppress axonal release of the transmitters for slow and fast EPSPs. Presynaptic autoreceptors are involved in a special form of presynaptic inhibition whereby the transmitter for slow or fast EPSPs accumulates at the synapse and acts on the autoreceptor to suppress further release of the neurotransmitter. (+), excitatory receptor; (–), inhibitory receptor.

Presynaptic inhibition in the ENS is mediated by multiple chemical mediators and their receptors, with variable combinations of the receptors involved at each release site. Among the known mediators are peptides (cholecystokinin), purinergic substances (ATP), amines (histamine), and ACh. NE, after its release from sympathetic nerve terminals, acts at α_2-presynaptic adrenoreceptors to suppress fast EPSPs at nicotinic synapses and slow EPSPs and cholinergic transmission at neuromuscular junctions. Serotonin suppresses both fast and slow EPSPs in the myenteric plexus. Opiates or opioid peptides suppress some fast EPSPs in the intestinal myenteric plexus.

ACh acts at muscarinic presynaptic receptors to suppress fast EPSPs in the myenteric plexus. This is a form of autoinhibition in which ACh released at synapses with nicotinic postsynaptic receptors feeds back onto presynaptic muscarinic receptors to suppress ACh release in negative-feedback fashion (see Fig. 25.14). Histamine, after its release from enteric mast cells, acts at histamine H_3 presynaptic receptors to suppress fast EPSPs in the ENS microcircuits of rodents. Presynaptic inhibition mediated by paracrine or endocrine release of mediators is significant in pathophysiologic states (e.g., inflammation). The release of histamine from enteric mast cells in response to food allergens and infectious organisms is an important example of paracrine-mediated presynaptic suppression in the enteric neural networks.

Presynaptic inhibition operates normally as a mechanism for selective shutdown or de-energizing of an entire microcircuit. Preventing transmission among the neural elements of a circuit inactivates the circuit. For example, a major component of shutdown of gut function by the sympathetic nervous system involves the presynaptic inhibitory action of NE at the fast nicotinic synapses.

Presynaptic facilitation enhances the amounts of neurotransmitter released by the arrival of action potentials at axonal release sites.

Presynaptic facilitation refers to an enhancement of synaptic transmission, which results from the actions of chemical mediators at neurotransmitter release sites on ENS axons (Fig. 25.15). The phenomenon enhances the amplitude of fast EPSPs in the myenteric plexus of the small intestine and gastric antrum and at noradrenergic inhibitory synapses in the submucosal plexus. It is also an action of the gut hormone, cholecystokinin, in the ENS of the gallbladder. Presynaptic facilitation is evident experimentally as an increase in amplitude of fast EPSPs at nicotinic synapses and reflects enhanced release of ACh from axonal release sites. At noradrenergic inhibitory synapses in the submucosal plexus, presynaptic facilitation involves the elevation of cAMP in the postganglionic sympathetic fiber and appears as an enhancement of the slow IPSPs evoked when action potentials propagate into the transmitter release site.

Presynaptic facilitation is the mechanism of action of some prokinetic drugs (e.g., tegaserod and cisapride). **Prokinetic drugs** are therapeutic agents that enhance propulsive motility in the GI tract. These agents act to enhance the amplitude of fast nicotinic EPSPs in the ENS networks that program propulsive motor functions. In both the stomach and the intestine, increases in EPSP amplitudes and rates of rise decrease the probability of transmission failure at the synapses, thereby increasing the speed of information transfer. This "energizes" the network circuits and thereby enhances propulsive motility (i.e., gastric emptying and intestinal transit).

ENTERIC MOTOR NEURONS

Musculomotor neurons innervate the muscles of the digestive tract and, like spinal motor neurons, are the final pathways for signal transmission from the integrative microcircuits of the minibrain in the gut to the musculature (Fig. 25.16; also see Fig. 25.10). The ENS motor neuronal pool has both excitatory and inhibitory neurons.

The neuromuscular junction is the site where neurotransmitters released from axons of motor neurons act on smooth muscle fibers and ICCs. Neuromuscular junctions in the unitary-type smooth muscle of the digestive tract are simpler structures than the motor endplates of skeletal muscle (see Chapter 5, "Motor System"). Most motor axons in the unitary-type smooth muscle do not release neurotransmitters from terminals as such; instead, release is from multiple varicosities, which are spaced out along the axon. The neurotransmitter is released when a passing action potential depolarizes the membrane potential of the varicosity. Once released, the neurotransmitter diffuses over relatively long distances before reaching the muscle and/or ICCs. This structural organization is an adaptation for the simultaneous application of a chemical neurotransmitter to multiple muscle fibers from a small number of motor axons. Because the muscle fibers are electrically coupled one to another by gap junctions, the membrane potential in expanded regions of the musculature can be depolarized to action potential threshold or hyperpolarized below spike threshold by release from a reduced number of motor axons.

Enteric nervous system excitatory musculomotor neurons evoke contractions in gastrointestinal musculature.

Excitatory musculomotor neurons release neurotransmitters that evoke contraction and increased contractile force in GI smooth muscles. ACh and **substance P** are the main excitatory neurotransmitters released from ENS musculomotor neurons.

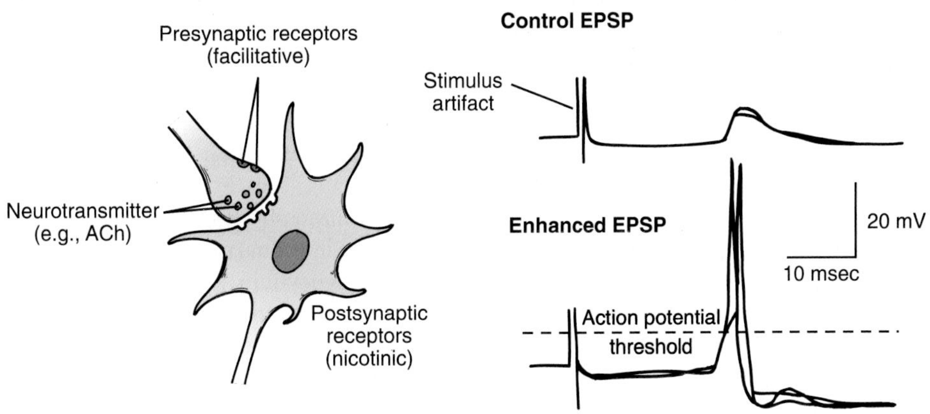

Figure 25.15 Presynaptic facilitation enhances release of acetylcholine (Ach) and increases the amplitude of fast excitatory postsynaptic potentials (EPSPs) at a nicotinic synapse.

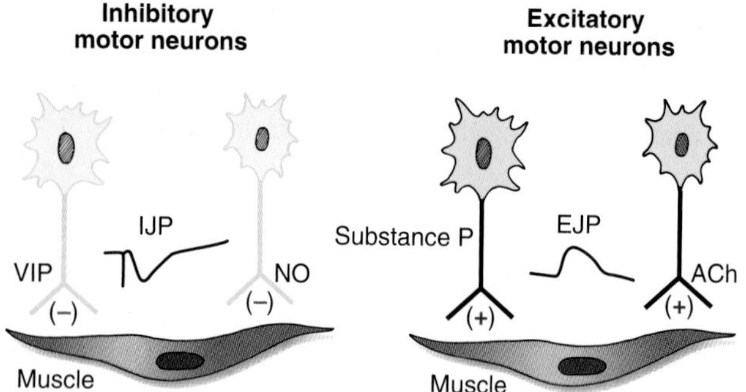

Figure 25.16 Enteric musculomotor neurons are final pathways from the enteric nervous system (ENS) to the gastrointestinal musculature. The motor neuron pool of the ENS consists of both excitatory and inhibitory neurons. Release of vasoactive intestinal peptide (VIP) or nitric oxide (NO) from inhibitory motor neurons evokes inhibitory junction potentials (IJPs). Release of acetylcholine (ACh) or substance P from excitatory motor neurons evokes excitatory junction potentials (EJPs).

Two mechanisms of excitation–contraction coupling are involved in the neural initiation of muscle contraction in the GI tract. Transmitters released from excitatory musculomotor axons can trigger muscle contraction by depolarizing the muscle fiber membrane to the threshold for the discharge of action potentials or by the activation of G-protein-coupled receptors linked to the release of calcium from stores inside the muscle fiber (i.e., pharmacomechanical coupling). Neurally evoked depolarizations of the muscle membrane potential, called **excitatory junction potentials (EJPs)**, can be recorded with microelectrodes, which distinguishes them from EPSPs and IPSPs (see Fig. 25.16).

Direct release of calcium by the neurotransmitter fits the definition of pharmacomechanical coupling. In this case, occupation of receptors on the muscle cell membrane by the neurotransmitter leads to the release of intracellular calcium. Once released, calcium triggers contractions independent of any changes in membrane electrical activity (i.e., EJPs or action potentials).

The cell bodies of excitatory musculomotor neurons are generally found in the myenteric plexus. In the small and large intestine, they project their axons over relatively short distances to innervate the longitudinal muscle coat and for longer distances away from the cell body in the oral direction to innervate the circular muscle.

Excitatory motor neurons to mucosal secretory glands (i.e., the crypts of Lieberkühn and Brunner glands) are called secretomotor neurons. Secretomotor neurons release neurotransmitters that evoke glandular secretion of NaCl, HCO_3, and H_2O and secretion of mucus from goblet cells. ACh and vasoactive intestinal peptide (VIP) are the principal excitatory neurotransmitters released by secretomotor neurons. The cell bodies of secretomotor neurons are mainly in the submucosal plexus. Hyperactivity of these neurons, for example, evoked by histamine release from enteric mast cells during allergic responses, can lead to **neurogenic secretory diarrhea**. Suppression of secretomotor neuronal excitability and its stimulatory action on secretion (e.g., by morphine or other opioid analgesics) underlies the constipating side effects of these drugs.

Enteric nervous system inhibitory musculomotor neurons relax contractile tone in sphincters and prevent electrical slow waves from triggering action potentials and contractions.

Inhibitory neurotransmitters released from inhibitory musculomotor neurons activate receptors on the muscle cell membranes to generate **inhibitory junction potentials (IJPs)** (see Fig. 25.16). IJPs are hyperpolarizing potentials that move the membrane potential away from the threshold for the discharge of action potentials and thereby reduce the excitability of the muscle fibers in which they occur. Hyperpolarization during IJPs prevents depolarization to the action potential threshold by the ever-present electrical slow waves and suppresses propagation of action potentials from muscle fiber to muscle fiber at the level of the electrically coupled syncytium.

Early evidence supported a purine nucleotide, possibly ATP, as the inhibitory transmitter released by ENS inhibitory musculomotor neurons. Consequently, the term **purinergic neuron** temporarily became synonymous with ENS inhibitory musculomotor neuron. The evidence for ATP as the inhibitory transmitter is now integrated with evidence for VIP, pituitary adenylate cyclase-activating peptide, and nitric oxide (NO) as inhibitory transmitters released by ENS musculomotor neurons. Inhibitory musculomotor neurons that express VIP and/or NO synthase (NOS) innervate the circular muscle of the stomach, intestine, gallbladder, and various sphincters. Cell bodies of the inhibitory musculomotor neurons are primarily in the myenteric plexus. In the stomach and small and large intestine, they project their single axon in the aboral direction away from the cell body to innervate the circular muscle.

The longitudinal muscle layer of the small intestine appears not to have a significant inhibitory musculomotor innervation. On the other hand, there is significant inhibitory musculomotor innervation of the teniae coli of the large intestine in humans and in animal species that have this specialized arrangement of the longitudinal muscle. In contrast to the innervation of the longitudinal muscle, inhibitory innervation of the circular muscle coat of the small and large intestine is essential for ENS programming of the various forms of intestinal motility.

An absolute requirement for inhibitory neural control of the circular muscle layer is a demand that emerges from the specialized physiology of the musculature. As mentioned earlier, the intestinal musculature behaves like a self-excitable electrical syncytium, resulting from cell-to-cell communication across gap junctions and the presence of the pacemaker system of electrical slow waves generated by networks of ICCs. Action potentials triggered anywhere in the muscle can spread from muscle fiber to muscle fiber in three dimensions throughout the syncytium, which might be the entire length and circumference of the bowel. Action potentials trigger contractions as they spread through the muscular syncytium. A nonneural pacemaker system of electrical slow waves (i.e., ICCs) accounts for the self-excitable characteristic of the electrical syncytium and the spontaneous contractile activity that characterizes unitary-type smooth muscle. In the integrated system, the electrical slow waves are a separate extrinsic factor to which the circular muscle responds in addition to its motor innervation.

Why does the circular muscle not always respond with action potentials and a contraction during every slow-wave cycle? Why do action potentials and contractions not spread in the syncytium throughout the entire length and circumference of the intestine each time they occur? The short answer is that inhibitory neuronal motor activity determines when a slow wave can evoke a contraction and determines the distance and direction over which a contraction spreads within the muscular syncytium.

Subset of inhibitory musculomotor neurons continuously inhibits the intestinal circular muscle layer.

Figure 25.17A shows the spontaneous discharge of action potentials occurring in bursts, as recorded extracellularly from a neuron in the myenteric plexus of the small intestine of a cat. This kind of continuous discharge of action potentials by subsets of intestinal inhibitory musculomotor neurons occurs in all mammals. The outcome is continuous inhibition of myogenic activity, because in intestinal segments in which neuronal discharge in the myenteric plexus is prevalent, slow wave-associated muscle action potentials and associated contractile activity are absent or the contractions occur with reduced force with each electrical slow wave. Continuous release of the inhibitory neuromuscular transmitters VIP and NO from intestinal preparations can be detected experimentally in these kinds of intestinal states. When the inhibitory neuronal discharge is blocked experimentally with tetrodotoxin, every electrical slow-wave cycle triggers an intense discharge of muscle action potentials associated with a forceful phasic contraction superimposed on a tonic contracture (i.e., a sustained increase in baseline tension). Figure 25.17B shows how phasic contractions, occurring at slow-wave frequency, progressively increase to maximal amplitude during progressive blockade of inhibitory neural activity after the application of tetrodotoxin in the small intestine. This response coincides with a progressive increase in baseline tension.

Tetrodotoxin has been a valuable pharmacologic tool for demonstrating ongoing inhibition of contractile activity in the intestine, because it selectively blocks neural activity without affecting the muscle. This action of tetrodotoxin results from a selective blockade of sodium channels in ENS neurons, including inhibitory musculomotor neurons. The rising phase of circular muscle action potentials reflects opening of calcium channels and inward current that is unaffected by tetrodotoxin (see Chapter 3, “Action Potential, Synaptic Transmission, and Maintenance of Nerve Function” and Chapter 8, “Skeletal and Smooth Muscle”).

As a general rule, any treatment or condition that removes or inactivates inhibitory musculomotor neurons results in tonic contracture and continuous, uncoordinated contractile activity of the circular musculature. Situations in which the activity of inhibitory musculoneurons is absent are associated with conversion from a hypoirritable condition of the circular muscle to a hyperirritable state. Situations in which this takes place are the application of local anesthetics, hypoxia from restricted blood flow to an intestinal segment, an autoimmune attack that destroys enteric neurons, congenital absence in Hirschsprung disease, treatment with opiate drugs, and inhibition of the synthesis of NO (see From Bench to Bedside Box 25.1).

Inhibitory musculomotor activity determines the force of contraction associated with each electrical slow-wave cycle.

The force of a circular muscle contraction evoked by a slow-wave cycle is a function of the number of inhibitory musculomotor neurons in an active state (i.e., firing action potentials and releasing inhibitory neurotransmitter at neuromuscular

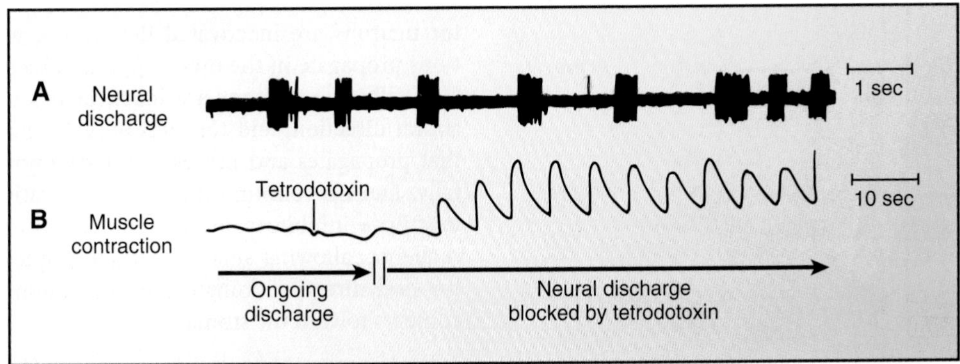

Figure 25.17 Ongoing firing of a subpopulation of inhibitory musculomotor neurons to the intestinal circular muscle prevents electrical slow waves from triggering the action potentials that trigger contractions. When the inhibitory neural discharge is blocked with tetrodotoxin, every cycle of the electrical slow wave triggers discharge of action potentials and large-amplitude contractions. **(A)** Electrical record of ongoing burst-like firing of an inhibitory motor neuron. **(B)** Record of muscle contractile activity before and after application of tetrodotoxin. Application of tetrodotoxin blocked neuronal firing.

junctions). The circular muscle in an intestinal segment can respond to the electrical slow waves only when the inhibitory musculomotor neurons are inactivated (turned off) by inhibitory synaptic input from other neurons in the control circuits. This means that inhibitory musculomotor neurons determine when the continuously running slow waves initiate a contraction as well as the force of the contraction that is initiated by each slow wave cycle. The force of each contraction is determined by the proportion of muscle fibers in the total population that can respond during a given slow wave cycle, which in turn is determined by the proportion exposed to inhibitory transmitters released by musculomotor neurons. With maximum inhibition, no contractions are permitted during a slow-wave cycle (see Fig. 25.4A); contractions of maximum strength occur after all inhibition is removed and all of the muscle fibers in a segment can be activated by each slow-wave cycle (see Fig. 25.4B). Contractions between the two extremes are graded in force according to the number of inhibitory musculomotor neurons that are inactivated by the ENS minibrain during each slow-wave cycle.

Inhibitory musculomotor activity determines the distance and direction a contraction travels within the functional electrical syncytium.

The activity state of inhibitory musculomotor neurons determines the length of a contracting segment by controlling the distance of spread of action potentials within the three-dimensional electrical geometry of the muscular syncytium (Fig. 25.18). This occurs coincidently with control of contractile force. Contractions can only occur in the circular muscle layer of segments where ongoing inhibition has been

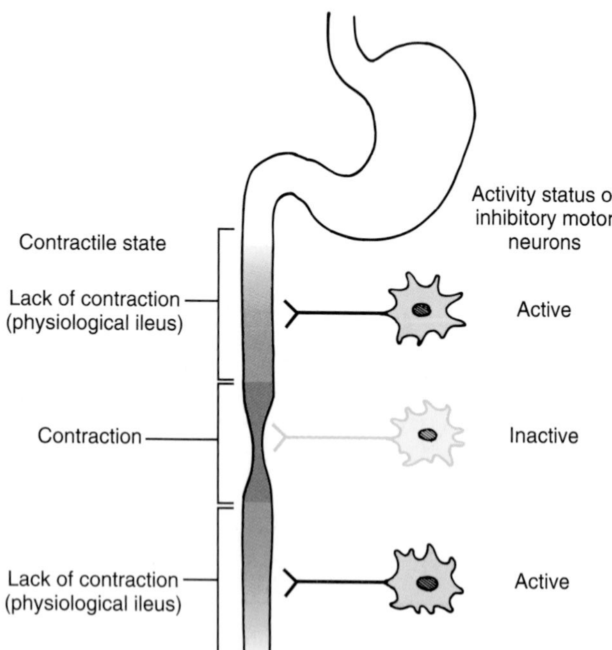

Figure 25.18 Myogenic contractions can occur in segments of intestine where inhibitory musculomotor neurons are inactive. Physiological ileus occurs in segments of intestine where the inhibitory neurons are actively firing.

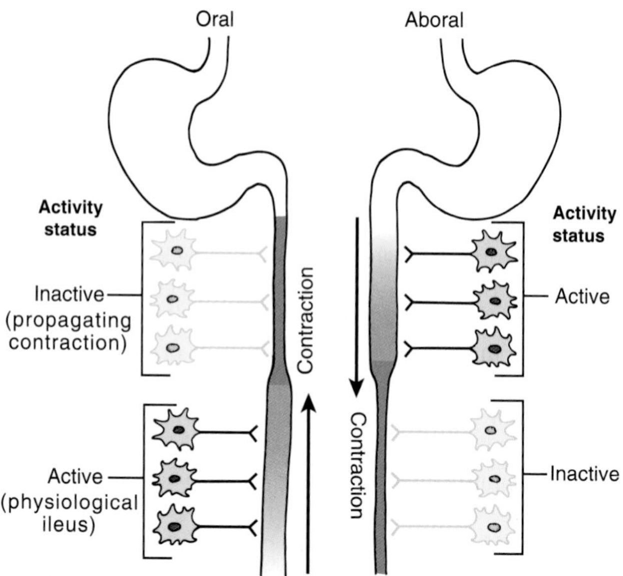

Figure 25.19 Contractions propagate into intestinal segments where inhibitory musculomotor neurons are inactivated. Sequential inactivation in the oral direction permits oral propagation of contractions. Sequential inactivation in the aboral direction permits aboral propagation.

inactivated, whereas they are prevented in adjacent segments where the inhibitory innervation is active. The oral and aboral circumferential boundaries of a contracted segment reflect the transition zone from inactive to active inhibitory musculomotor neurons. This is the way in which the ENS generates short contractile segments during the digestive (mixing) pattern of small intestinal motility and longer contractile segments during propulsive motor patterns, such as "power propulsion" that travels over extended distances along the intestine.

The functional electrical syncytial properties of the musculature underlie the mandate for inhibitory musculomotor neuronal control in the determination of the direction in which contractions travel along the intestine. The directional sequence in which inhibitory musculomotor neurons are inactivated determines whether contractions propagate in the oral or aboral direction (Fig. 25.19). Normally, the neurons are inactivated sequentially in the aboral direction, and this accounts for contractile activity that propagates and moves the intraluminal contents distally. During vomiting, the ENS integrative microcircuits inactivate inhibitory musculomotor neurons in a reverse sequence, allowing small intestinal propulsion to travel in the oral direction from the mid-jejunum and propel the contents toward the stomach.

Inhibitory motor innervation of smooth muscle sphincters is silent when the sphincter is closed.

The circular muscle of sphincters remains tonically contracted to occlude the lumen and prevent the passage of contents between adjacent compartments (e.g., between

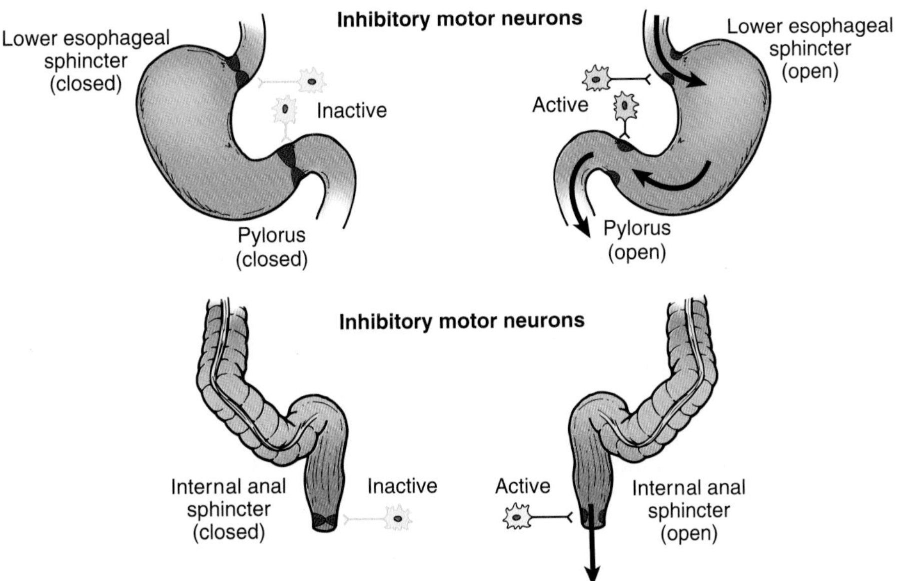

Figure 25.20 Smooth muscle sphincters are closed when their inhibitory innervation is inactive. The sphincters are opened by active firing of the inhibitory musculomotor neurons.

stomach and esophagus). Inhibitory motor neurons are normally inactive in the sphincters and are switched on with timing appropriate to coordinate the opening of the sphincter with physiologic events in adjacent regions (Fig. 25.20). When this occurs, the released inhibitory neurotransmitter relaxes the ongoing muscle contraction in the sphincteric muscle and prevents excitation and contraction in the adjacent muscle from spreading into and closing the sphincter.

GASTROINTESTINAL MOTILITY PATTERNS

ICCs cause spontaneous cycles of slow-wave potentials that can cause action potentials in smooth muscle cells. GI motility can be divided into two distinct patterns, **peristalsis** and **segmentation**. *Peristalsis* is responsible for propeling food down the GI tract, and *segmentation* is involved in mixing ingested food for orderly processing and the elimination of waste. **Propulsion** is the controlled movement of ingested foods, liquids, GI secretions, and sloughed cells from the mucosa through the digestive tract. It moves the food from the stomach into the small intestine and along the small intestine with appropriate timing for efficient digestion and absorption. Propulsive forces move undigested/unabsorbed material into the large intestine and eliminate waste through defecation. **Trituration** decreases particle size, thereby increasing the surface area for action by digestive enzymes in the small intestine. The **mixing motility pattern** blends pancreatic, biliary, and intestinal secretions with nutrients in the stomach and bring products of digestion into contact with the absorptive surfaces of the mucosa. *Reservoir functions* are performed by the stomach and colon. The body of the stomach stores ingested food and exerts steady mechanical compressive forces, which are important determinants of gastric emptying. The colon holds material during the time required for the absorption of excess water and stores the residual material until defecation is convenient. Each of the specialized organs along the digestive tract has motility patterns that are specialized for the normal functions of the organ. These patterns differ depending on factors such as time after a meal, awake or sleeping state, and the presence of disease. Motor patterns that accomplish propulsion in the esophagus and small and large intestine are derived from a basic peristaltic reflex circuit in the ENS.

Peristalsis is a polysynaptic neural reflex.

The gut motility responsible for transport of luminal contents in the esophagus and small and large intestines is **peristalsis**. Contractions occur in waves directly behind a bolus of food, forcing it into the next relaxed section of the GI tract. The muscle layers of the intestine behave in a stereotypic pattern during peristaltic propulsion (Fig. 25.21). This pattern is determined by the neural networks of the ENS. During peristaltic propulsion, the longitudinal muscle coat in the segment ahead of the advancing intraluminal contents contracts, whereas contractile force in the circular muscle layer relaxes simultaneously. The intestinal tube behaves like a cylinder with constant surface area. The shortening of the longitudinal axis of the cylinder is accompanied by a widening of the cross-sectional diameter. The simultaneous shortening of the longitudinal axis and relaxation of the circular muscle results in expansion of the lumen, which prepares a *receiving segment* for the forward-moving intraluminal contents during peristaltic propulsion.

The second component of stereotypic peristaltic behavior is contraction of the circular muscle in the segment behind the advancing intraluminal contents. The longitudinal muscle layer in this segment lengthens simultaneously

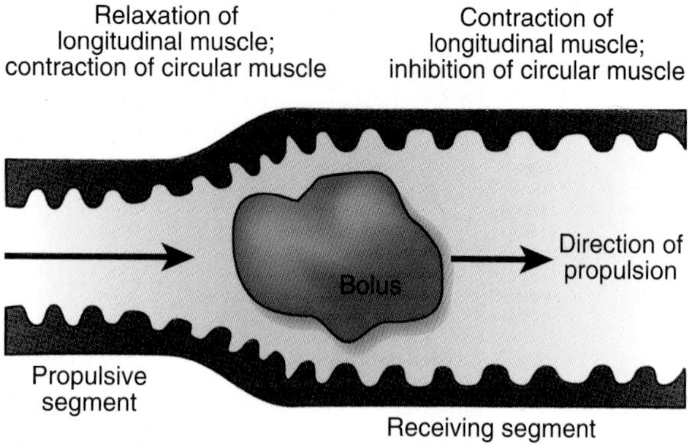

Relaxation of longitudinal muscle; contraction of circular muscle

Contraction of longitudinal muscle; inhibition of circular muscle

Bolus

Direction of propulsion

Propulsive segment

Receiving segment

Figure 25.21 Peristaltic propulsion involves the formation of a propulsive and a receiving segment, mediated by reflex control of the intestinal musculature.

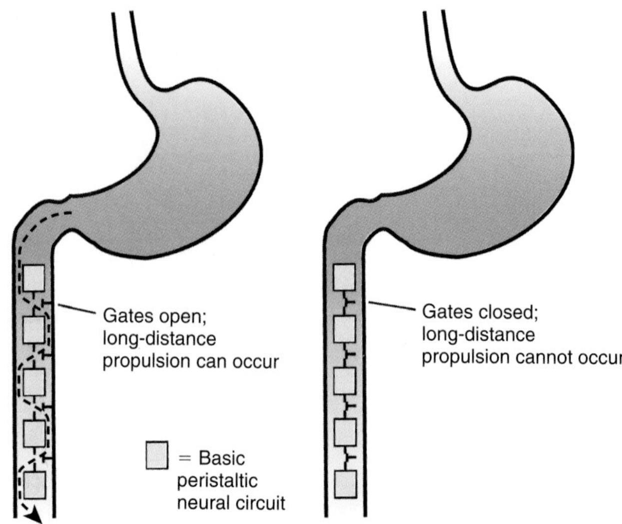

Figure 25.22 Excitatory synapses connect successive blocks of the basic peristaltic reflex circuit and behave like gates between blocks of circuitry. Transmission at the synapses opens the gates between successive blocks of the basic peristaltic reflex circuit and accounts for propagation of peristaltic propulsion over long distances. Long-distance propulsion is prevented when all gates are closed.

with contraction of the circular muscle, resulting in the conversion of this region to a *propulsive segment*, which propels the luminal contents forward into the receiving segment. Understanding of the lengthening in the longitudinal axis in the propulsive segment is unclear. It might be taking place as the longitudinal muscle continues to contract and reflect contraction of the more powerful circular layer, reduction in diameter, and passive geometric elongation of the cylinder. Circular and longitudinal layers, if this were the case, are antagonistic muscles when they contract simultaneously. A second possibility for the elongation in the propulsive segment is "switching off" of excitatory musculomotor input to the long muscle and a third is that excitatory musculomotor input is "switched off" while inhibitory musculomotor input to the longitudinal muscle is "turned on." Intestinal segments ahead of the advancing front become receiving segments and then propulsive segments in succession, as the peristaltic complex of propulsive and receiving segments travels along the intestine.

Peristaltic reflex circuit

The peristaltic reflex (i.e., the formation of propulsive and receiving segments) can be triggered experimentally by distending the intestinal wall or by "brushing" the mucosa. Involvement of the reflex in the neural organization of peristaltic propulsion is like the reflex behavior mediated by the CNS for somatic movements of skeletal muscles. Reflex circuits with fixed connections in the spinal cord automatically reproduce a stereotypic pattern of behavior each time the circuit is activated (e.g., Achilles and patellar tendon reflexes; see Chapter 5, "Motor System"). Connections for these reflexes remain, irrespective of the destruction of adjacent regions of the spinal cord. The peristaltic reflex circuit is similar, with the exception that the basic circuit is repeated continuously within the ENS along and around the entire length of the intestine. In the same way that the monosynaptic reflex circuit of the spinal cord is the terminal circuit for the production of almost all skeletal muscle movements (see Chapter 5, "Motor System"), the basic peristaltic reflex circuit underlies all patterns of propulsive motility. Blocks of the same basic circuit are connected in series up and down the intestine (Fig. 25.22). The basic peristaltic circuit consists of synaptic connections between sensory neurons, interneurons, and musculomotor neurons. Distances over which peristaltic propulsion travels are determined by the number of blocks recruited in sequence along the bowel. Synaptic gates between blocks of the basic circuit determine whether recruitment occurs for the next circuit in the sequence.

When the gates are opened, neural signals pass between successive blocks of the basic circuit, resulting in propagation of the peristaltic event over extended distances. Long-distance propulsion is prevented when all gates are closed.

Presynaptic inhibition and/or facilitation is involved in gating the transfer of signals between sequentially positioned blocks of peristaltic reflex circuitry. Synapses between the neurons that carry excitatory signals to the next block of circuitry are gating points for control of the distance over which peristaltic propulsion travels (Fig. 25.23). Messenger substances, which act at presynaptic receptors to inhibit the release of transmitter at excitatory synapses, close the gates to the transfer of information, thereby determining the distance of propagation. Prokinetic drugs, which facilitate the release of neurotransmitters at the excitatory synapses (e.g., cisapride and tegaserod), have therapeutic application by increasing the probability of information transfer at the synaptic gates, thereby enhancing propulsive motility.

Emesis

Vomiting, or **emesis**, is the forceful expulsion of the stomach contents back into the environment. Vomiting

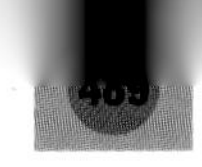

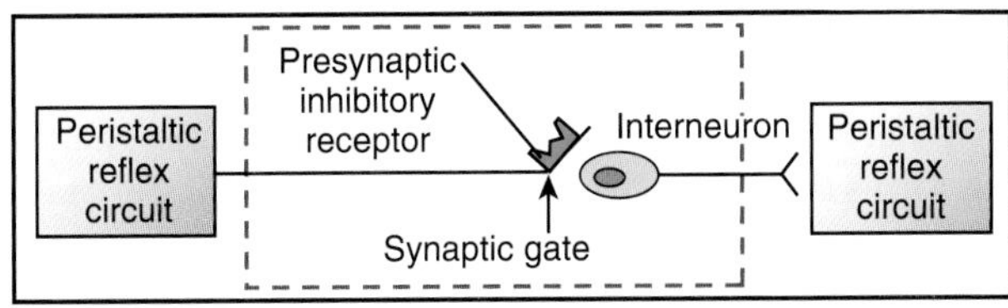

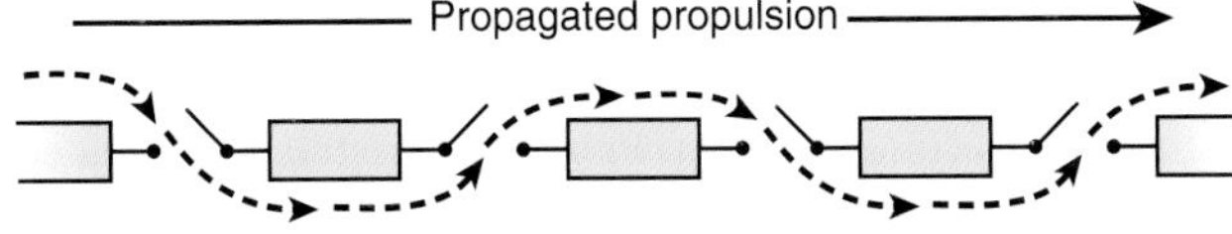

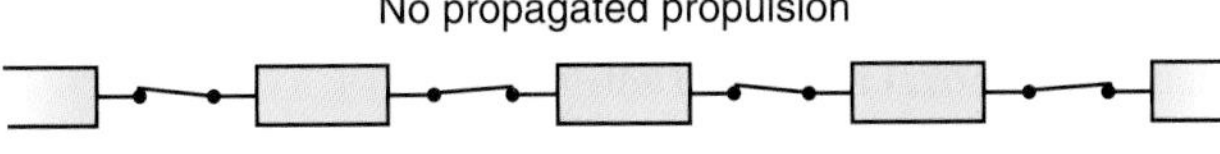

Figure 25.23 Presynaptic inhibitory receptors determine the open and closed states of synaptic gates between blocks of the peristaltic reflex circuit. When the gating synapses are uninhibited (i.e., no presynaptic inhibition), propagation can proceed in the direction in which the gates are open. Activation of presynaptic inhibitory receptors closes the gates and no propagated propulsion can occur.

may result from several causes, ranging from gastritis, food poisoning, binge drinking, brain tumors, and brain trauma. Although the terms "vomiting" and "regurgitation" are often used interchangeably, regurgitation is different. *Regurgitation*, most commonly occurring in the neonate, is the return of undigested food back up the esophagus, without the forced expulsion or displeasure associated with *vomiting*.

The enteric neural circuits are programmed to evoke peristaltic propulsion in either direction along the intestine. If forward passage of the intraluminal contents is impeded in the large intestine, reverse propulsion propels the bolus over a variable distance away from the obstructed segment. **Retropulsion** then stops and forward propulsion moves the bolus again in the direction of the obstruction. During the act of vomiting, retropulsion occurs in the small intestine. In this case, as well as in the case of the obstructed intestine, the ENS programs the behavior of the musculature for propulsion in the reverse direction, with the receiving segment leading the propulsive segment as the propulsive complex travels in the direction of the stomach.

Ileus

Physiological ileus is the absence of motility in the small and large intestine and reflects the output of a specific motor program stored in the program library of the ENS. It is a fundamental behavioral state of the intestine in which quiescence of motor function is neurally programmed. The state of physiological ileus disappears after ablation of the ENS. When enteric neural functions are destroyed by pathologic processes, disorganized and nonpropulsive contractile behavior occurs continuously because of the myogenic properties of the musculature in the absence of ENS control. Quiescence of the intestinal circular muscle (i.e., physiological ileus) reflects the operation of a neural program in which all the gates within and between basic peristaltic circuits are held shut (see Fig. 25.23). In this state, a subset of inhibitory musculomotor neurons is continuously active and responsiveness of the circular muscle coat to the ever-present electrical slow waves is suppressed. Physiological ileus is a normal state that remains in effect for varying periods of time in different intestinal regions, which depends on factors such as the time after a meal.

The normal state of motor quiescence becomes pathologic when the gates for the particular motor patterns are rendered inoperative for abnormally long periods. In this state of **pathological ileus** (sometimes called **paralytic ileus**), the basic circuits are locked in an inoperable state, while unremitting activity of the inhibitory motor neurons continuously suppresses myogenic activity.

Sphincters prevent reflux between specialized compartments.

Smooth muscle sphincters are found at the gastroesophageal junction, gastroduodenal junction, opening of the bile duct, ileocolonic junction, and termination of the large intestine in the anus. They consist of rings of smooth muscle, which remain in a continuous state of contraction. The effect of the tonic contractile state is to occlude the lumen in a region that separates two specialized compartments. With the exception of the internal anal sphincter, sphincters function to prevent the backward reflux of intraluminal contents.

Lower esophageal sphincter

The **LES** prevents the reflux of gastric contents, including acid, into the esophagus. An incompetent sphincter permits chronic exposure of the esophageal mucosa to acid, which can lead to heartburn, also known as **pyrosis** or **acid indigestion**. Heartburn is a burning sensation in the chest caused by regurgitation of gastric acid into the esophagus (gastric reflux). If chronic gastric reflux persists, it can lead to inflammation and damage to the esophageal mucosal lining and becomes the major symptom of **gastroesophageal reflux disease (GERD)**. GERD patients are at risk for **Barrett esophagus**, a condition in which damaged esophageal cells are transformed into metaplastic cells, which are a precursor to carcinoma. Although the risk of cancer is present, the progression from precancerous metaplasia to dysplasia is less than 20% in patients diagnosed with Barrett disease.

Pyloric sphincter

The sphincter at the gastroduodenal junction (i.e., **pyloric sphincter**) prevents reflux of duodenal contents into the stomach. Incompetence of this sphincter can result in the reflux of bile acids and digestive enzymes from the duodenum. Bile acids are damaging to the protective barrier in the gastric mucosa; prolonged exposure can lead to gastritis, formation of ulcers, and risk of perforation.

Sphincter of Oddi

The **sphincter of Oddi** surrounds the opening of the bile duct as it enters the duodenum. It acts to prevent the reflux of intestinal contents into the ducts leading from the liver, gallbladder, and pancreas. Failure of this sphincter to open leads to distention in the biliary tree and the associated biliary tract pain that can be felt in the epigastric region and right upper abdominal quadrant. An incompetent sphincter of Oddi is associated with a risk of reflux-induced pancreatitis.

Ileocolonic sphincter

The **ileocolonic sphincter** prevents the back flux of colonic contents into the ileum. Incompetence of this sphincter can allow the entry of bacteria into the ileum from the colon, which can result in small intestinal bacterial overgrowth. Bacterial counts are normally low in the small intestine. Bacterial overgrowth in the small intestine has been associated with symptoms of bloating and abdominal pain in a subgroup of patients diagnosed with the irritable bowel syndrome.

Internal anal sphincter

The **internal anal sphincter** helps to control the movement of flatus and feces out of the anus. Failure of this sphincter to relax underlies the fecal retention that occurs in children with Hirschsprung disease and adults with Ogilvie syndrome. Fecal incontinence is associated with an incompetent internal anal sphincter.

Ongoing contractile tone in each of the smooth muscle sphincters is generated by **myogenic** mechanisms. The tonic contractile state is an inherent property of the muscle and independent of the nervous system. Transient relaxation of the sphincter to permit the forward passage of material is accomplished by activation of inhibitory motor neurons (see Fig. 25.20). **Achalasia** is a pathologic state in which smooth muscle sphincters fail to relax. Loss of the ENS and its complement of inhibitory musculomotor innervation of the sphincters is most often the underlying cause of sphincteric achalasia.

ESOPHAGEAL MOTILITY

The esophagus is divided into three functionally distinct regions: the upper esophageal sphincter, the esophageal body, and the LES. Esophageal motor behavior reflects contraction of striated muscle in the upper esophagus and smooth muscle in the lower one third of the esophagus in humans. The sole motor function of the esophagus is the transport of ingested solids and liquids from the pharynx to the stomach during a swallow and as a conduit for expulsion of gastric contents during emesis. Transport during a swallow is accomplished by peristaltic propulsion similar to that in the intestinal tract. At the onset of the swallow, the entire esophageal body behaves like a peristaltic receiving segment extending down the entire length of esophagus to the gastroesophageal junction, where relaxation of the sphincteric circular muscle becomes a counterpart of circular muscle relaxation in a peristaltic receiving segment (see Fig. 25.21). The circular muscle then starts to contract at the junction with the pharynx and forms the propulsive segment of stereotypic peristaltic reflex behavior. As the propulsive segment is propagated toward the stomach, the receiving segment becomes progressively shorter until it ends in the opened sphincter at the gastric junction.

Peristaltic propulsion in the esophagus can occur as a primary or secondary event. **Primary peristalsis** is initiated by the act of swallowing, irrespective of the presence of food in the mouth. **Secondary peristalsis** is triggered by distension caused by failed transport of a large food bolus or by the presence of acid due to gastric reflux. Secondary peristalsis in concert with stimulation of salivary secretion is a mechanism for removal of acid irritation.

When not involved in the act of swallowing, the muscles of the esophageal body are relaxed and the LES is tonically contracted. In contrast to the intestine, the relaxed state of the esophageal body is not produced by the ongoing activity of inhibitory motor neurons. Excitability of the muscle is low, and there are no electrical slow waves to trigger contractions. Activation of excitatory motor neurons, rather than myogenic mechanisms, accounts for the coordinated contractions of the circular and longitudinal muscles during peristaltic propulsion.

Esophageal motility disorders are diagnosed with manometric catheters and electrical impedance probes.

Two approaches are available for clinical diagnosis of esophageal motor disorders. One approach uses **manometric catheters**, which consist of multiple small catheters fused into a single assembly with pressure sensors positioned at various levels (see Clinical Focus Box 25.1). The second approach is with a catheter with pairs of electrodes spaced equally along its length. Electrical current passed between each pair of electrodes measures the impedance across the esophageal lumen. Passage of a swallowed bolus past an electrode pair is detected and recorded as a change in impedance. Both kinds of catheters are placed into the esophagus *via* the nasal cavity. Manometric catheters record a distinctive pattern of motor behavior following a swallow (Fig. 25.24). At the onset of the swallow, contractile tone in the LES relaxes. This is recorded as a fall in pressure in the sphincter that lasts throughout the swallow and until the esophagus empties its contents into the stomach. The pressure-sensing ports along the catheter assembly show transient increases in pressure as the segment with the sensing port becomes the peristaltic propulsive segment on its way to the stomach.

GASTRIC MOTILITY

The function of the stomach is threefold: (1) storage of large quantities of food, (2) mixing, and (3) emptying. *Mixing* involves food blending with digestive juices to

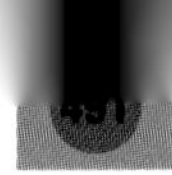

Clinical Focus / 25.1

Esophageal Motor Disorders

Dysphagia is defined as difficulty in swallowing. It can result from the failure of mechanisms involving the skeletal muscles of the pharynx, failure of peristalsis in the esophageal body, or failure of the lower esophageal sphincter (LES) to relax. A combination of liquid and solid food dysphagia is a reliable sign of disordered motility in the esophageal body and/or failure of the LES to relax. Dysphagia limited to solid food is most often a symptom of a mechanical obstruction (e.g., malignancies or strictures).

Some dysphagic patients have pressure waves of abnormally high amplitude as peristalsis propagates past the recording ports during diagnostic assessment with manometric catheters. This condition, which is called nutcracker esophagus, is often associated with angina-like chest pain. A manometric diagnosis of nutcracker esophagus is usually made subsequent to confirmation that the patient's chest pain does not reflect blockages in the coronary arteries.

In the absence of chest pain, liquid and solid food dysphagia is likely to be a reflection of LES achalasia. A diagnosis of achalasia is made when manometric recording of esophageal motility fails to show the typical relaxation and reduction of intraluminal pressure in the sphincter coincident with the onset of a swallow (see Fig. 25.24). Successful therapy involves mechanical dilation of the sphincter either by pneumatic inflation of a balloon placed in the sphincter or by passage of a solid bougie dilator. A class of smooth muscle relaxant drugs, known as phosphodiesterase-5 inhibitors (e.g., sildenafil), have been used recently as an alternative treatment for achalasia.

Achalasia, in the majority of cases, reflects the loss of the enteric inhibitory motor innervation of the sphincter. Loss or malfunction of inhibitory motor neurons, which in most cases is an inflammatory enteric nervous system (ENS) neuropathy, is the pathophysiologic starting point for achalasia. Achalasia associated with paraneoplastic syndrome, Chagas disease, and idiopathic ENS degenerative disease are recognizable forms of inflammatory ENS neuropathy. In paraneoplastic syndrome, antigens expressed by a carcinoma (usually small-cell carcinoma) in the lung have sufficient similarity to antigenic epitopes expressed by enteric neurons that the immune system simultaneously attacks the tumor and the patient's enteric neurons. Most patients with a diagnosis of achalasia, in combination with small-cell lung carcinoma, have circulating immunoglobulin-G autoantibodies that react with and destroy their enteric neurons. The detection of anti-enteric neuronal antibodies is a supplemental means to a specific diagnosis of achalasia. The association of enteric neuronal loss and symptoms of achalasia in Chagas disease also reflects autoimmune attack on ENS neurons. *Trypanosoma cruzi*, the blood-borne parasite that causes Chagas disease, has antigenic epitopes similar to enteric neuronal antigens. This antigenic commonality activates the immune system to assault the ENS coincident with its attack on the parasite.

When chest pain is associated with liquid and solid food dysphagia in the absence of nutcracker-like contractions, the problem is most likely a form of failure of peristaltic propulsion, called diffuse esophageal spasm. A diagnosis of diffuse spasm is made when manometric recording of esophageal motility demonstrates that the act of swallowing results in simultaneous contractions all along the length of the smooth muscle region of the esophageal body. A barium study in patients with diffuse spasm shows the morphometric correlate of diffuse spasm to be a contorted esophageal body, which has been described as a "corkscrew esophagus." Diffuse spasm and LES achalasia are often seen as concurrent pathophysiologies, both of which probably reflect a disorder of enteric inhibitory motor neurons. There are no fully satisfactory treatments for diffuse spasm other than attempts to treat with smooth muscle relaxant drugs.

form chyme. The stomach structure is divided into four sections, each of which has different cells and functions. They include: (1) **cardia**, (2) **fundus**, (3) **corpus** (body), and (4) **pylorus** (Fig. 25.25) The *cardia* is where the esophageal contents are emptied. The *fundus* forms the upper curvature of the organ, and the *corpus* forms the main or central region of the stomach. The *pylorus* is the lower section of the stomach that facilitates gastric emptying into the small intestine.

The functional regions of the stomach do not correspond to the anatomic regions. Functionally, the stomach is divided into a proximal **gastric reservoir** and distal **antral pump** on the basis of distinct differences in motility between the two compartments (see Fig. 25.25). The *reservoir* consists of the fundus and approximately one third of the corpus; the *antral pump* includes the caudal two thirds of the corpus, the antrum, and the pylorus.

Differences in motility between the reservoir and antral pump reflect adaptations for different functions. The muscles of the proximal stomach are adapted for maintaining continuous contractile tone (tonic contraction) and do not contract phasically. In contrast, the muscles of the antral pump contract phasically. The spread of phasic contractions in the region of the antral pump propels the gastric contents toward the gastroduodenal junction. Strong propulsive waves of this nature do not occur in the proximal stomach.

Motor behavior of the antral pump consists of leading and trailing contractile components triggered by gastric action potentials.

Gastric action potentials determine the duration and strength of the phasic contractions of the antral pump. They are initiated by a dominant pacemaker located in the corpus distal to the midregion. Gastric ICCs are believed to be the pacemaker. Once started at the pacemaker site, the action potentials propagate rapidly around the gastric circumference and trigger a ringlike contraction. The action potentials

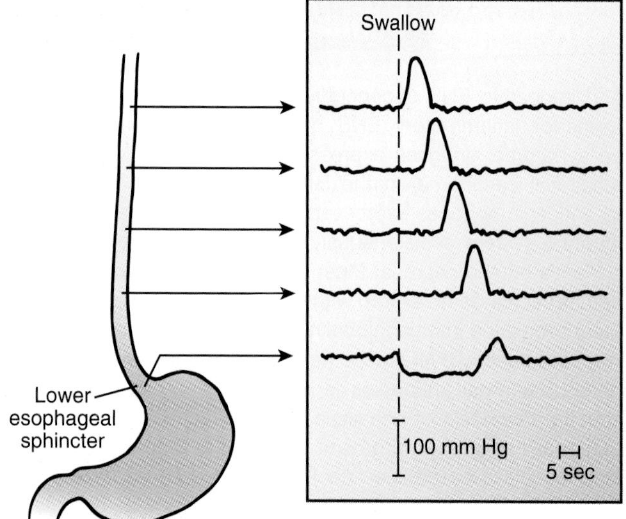

Figure 25.24 Manometric recording of pressure events in the esophageal body and lower esophageal sphincter are used to investigate esophageal motility following a swallow. The propulsive segment of the peristaltic behavioral complex produces a positive pressure wave at each recording site in succession as it travels down the esophagus. Pressure falls in the lower esophageal sphincter shortly after the onset of the swallow, and the sphincter remains relaxed until the propulsive complex has transported the swallowed material into the stomach.

and associated ringlike contraction then travel more slowly toward the gastroduodenal junction.

Electrical syncytial properties of the gastric musculature account for the propagation of the action potentials from the pacemaker site to the gastroduodenal junction. The pacemaker region in humans generates action potentials and associated antral contractions at a frequency of three per minute. The gastric action potential lasts about 5 seconds and has a rising phase (depolarization), a plateau phase, and a falling phase (repolarization) (see Fig. 25.2).

The gastric action potential is responsible for two components of the propulsive contractile behavior in the antral pump. An antral *leading contraction*, which has relatively constant amplitude, is associated with the rising phase of the action potential, and an antral *trailing contraction*, of variable amplitude, is associated with the plateau phase (Fig. 25.26). Gastric action potentials are generated continuously by the pacemaker; nevertheless, they do not trigger a trailing contraction when the plateau phase is reduced below the threshold voltage. Trailing contractions appear when the plateau phase is above

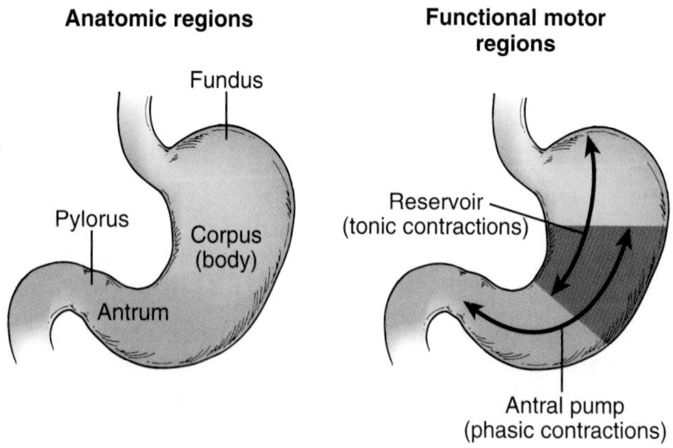

Figure 25.25 The stomach is divided into three anatomic and two functional regions. The gastric reservoir is specialized for receiving and storing a meal. The musculature in the region of the antral pump exhibits phasic contractions that function in the mixing and trituration of the gastric contents. No distinctly identifiable boundary exists between the reservoir and the antral pump.

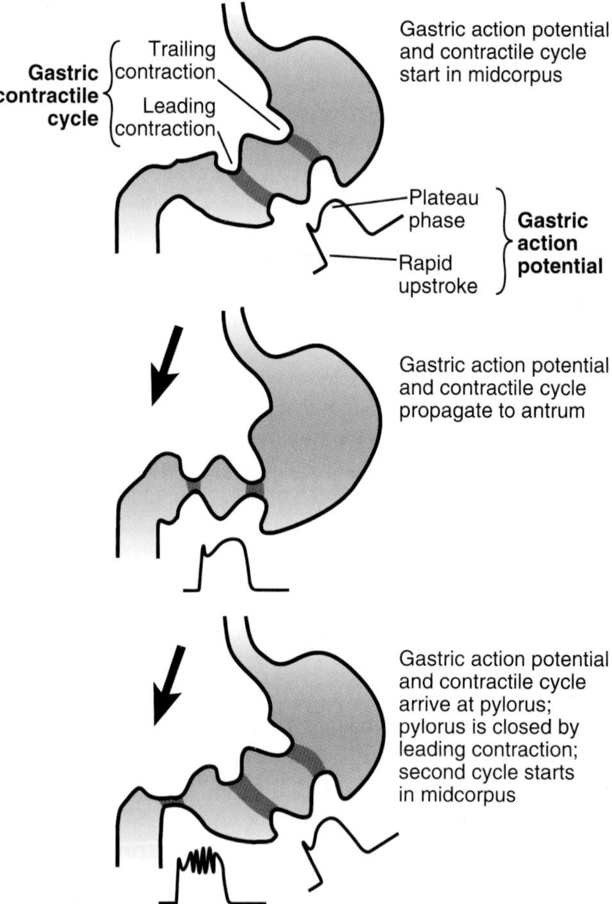

Figure 25.26 A pacemaker in the antral pump generates gastric action potentials that evoke ring-like contractions as they propagate to the gastroduodenal junction. Gastric action potentials are characterized by an initial rapidly rising upstroke followed by a plateau phase and then a falling phase back to the baseline membrane potential (see Fig. 25.2). The rising phase of the gastric action potential accounts for a leading contraction that propagates toward the pylorus during one propulsive cycle. The plateau phase accounts for the trailing contraction of the cycle. The strength of the leading contraction is relatively constant; the strength of the trailing contraction is variable and increases in direct relation to neurally evoked increases in amplitude of the plateau phase of the action potential.

threshold. They increase in strength in direct relation to increases in the amplitude of the plateau potential above threshold.

The leading contractions produced by the rising phase of the gastric action potential have negligible amplitude as they propagate to the pylorus. As the rising phase reaches the terminal antrum and spreads into the pylorus, contraction of the pyloric muscle closes the orifice between the stomach and duodenum. The trailing contraction follows the leading contraction by a few seconds. As the trailing contraction approaches the closed pylorus, the gastric contents are forced into an antral compartment of ever-decreasing volume and progressively increasing pressure. This results in jetlike retropulsion through the orifice formed by the trailing contraction (Fig. 25.27). Trituration and reduction in particle size occur as the material is forcibly retropulsed through the advancing orifice and back into the gastric reservoir to wait for the next propulsive cycle. Repetition, at 3 cycles/min, reduces particle size to the 1-mm to 7-mm range that is necessary before a particle can be emptied into the duodenum during the digestive phase of gastric motility.

Excitatory musculomotor neurons in the enteric nervous system determine the strength of trailing antral contractions.

Action potentials in the antral pump are myogenic (i.e., an inherent property of the muscle) and occur in the absence of any neurotransmitters or other chemical messengers. The myogenic characteristics of the action potential are modulated by musculomotor neurons in the gastric ENS. Neurotransmitters released by musculomotor neurons mainly change the amplitude of the plateau phase of the action potential and thereby control the strength of the contractile event triggered by the plateau phase. Neurotransmitters, including ACh, released from excitatory musculomotor neurons increase the amplitude of the plateau phase and of the contraction initiated by the plateau. Inhibitory neurotransmitters at the neuromuscular junctions, including NE and VIP, act to decrease the amplitude of the plateau and the strength of the associated contraction.

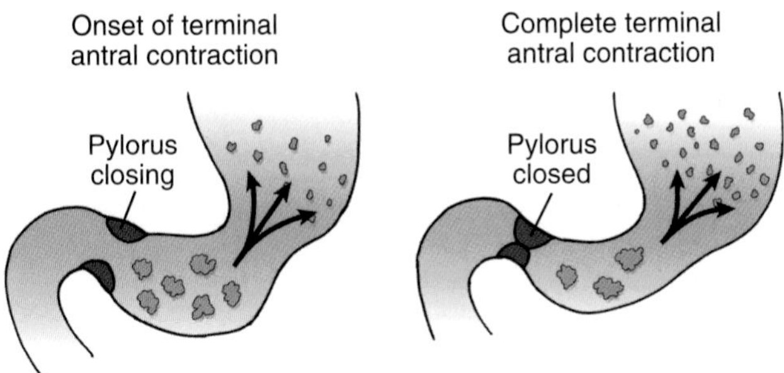

Figure 25.27 Jetlike retropulsion through the orifice of the antral contraction triturates solid particles in the stomach. The force for retropulsion is increased pressure in the terminal antrum as the trailing antral contraction approaches the closed pylorus.

The magnitude of the effects of transmitter release at the neuromuscular junctions increases with increasing concentration of the transmitter substance in the junction. Progressive increase in the frequency of action potential discharge by musculomotor neurons releases progressively increasing amounts of neurotransmitter. In this way, musculomotor neurons determine, through the actions of their neurotransmitters on the plateau phase, whether or not a trailing antral contraction occurs. With sufficient release of transmitter, the plateau amplitude grows and starts a contraction as it crosses the depolarization threshold. Beyond threshold, the strength of contraction is determined by the amount of neurotransmitter released and present at receptors on the muscle, which in itself determines the extent of membrane depolarization beyond threshold.

The action potentials in the terminal antrum and pylorus differ somewhat in configuration from those in the more proximal regions. The principal difference is the occurrence of spike potentials on the plateau phase (see Fig. 25.26), which trigger short-duration phasic contractions superimposed on the phasic contraction associated with the plateau. These may contribute to the sphincteric function of the pylorus in preventing the reflux of duodenal contents back into the stomach.

Motor behavior of the gastric reservoir differs from the gastric antrum.

The gastric reservoir has two primary functions. One is to accommodate the arrival of a meal, without a significant increase in intragastric pressure and distention inside the reservoir. Failure of this mechanism leads to the uncomfortable sensations of bloating, epigastric pain, and nausea (sometimes called *dyspepsia*). The second function is to maintain constant compressive forces on the contents of the reservoir, which act to push the contents into the 3-cycles/min motor activity of the antral pump. Agents that relax the musculature of the gastric reservoir (e.g., insulin) neutralize this function and thereby suppress gastric emptying.

The musculature of the gastric reservoir is innervated by both excitatory and inhibitory musculomotor neurons in the ENS (see Figs. 25.10 and 25.16). Efferent vagal nerves and the intramural microcircuits of the ENS control the activity of the musculomotor neurons. Their level of activity adjusts the volume and pressure of the reservoir to the amount of solid and/or liquid present, while sustaining constant compressive forces on the contents. Continuous adjustments of the volume and pressure within the reservoir are required during both the ingestion and the emptying of a meal.

An increased firing frequency of excitatory musculomotor neurons, in concert with decreased firing of inhibitory musculomotor neurons, results in increased contractile tone in the reservoir, a decrease in its volume, and an increase in intraluminal pressure (Fig. 25.28). An increase in activity of inhibitory musculomotor neurons in concert with decreased activity of excitatory musculomotor neurons results in

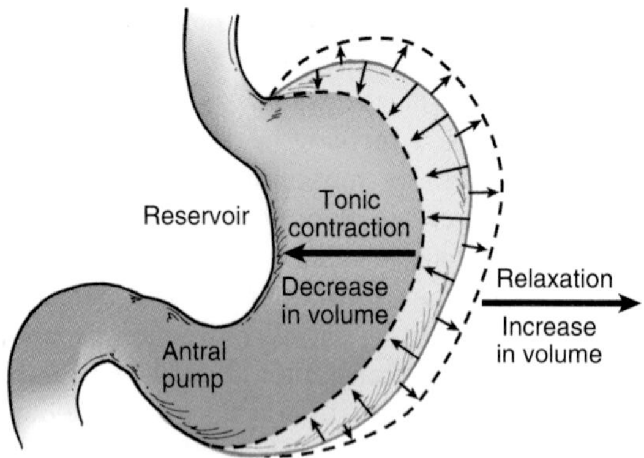

Figure 25.28 Tonic contraction of the musculature of the gastric reservoir decreases the volume and exerts compressive force on the contents. Tonic relaxation of the musculature expands the volume of the gastric reservoir. Neural mechanisms of feedback control determine intramural contractile tone in the reservoir (see Fig. 25.29).

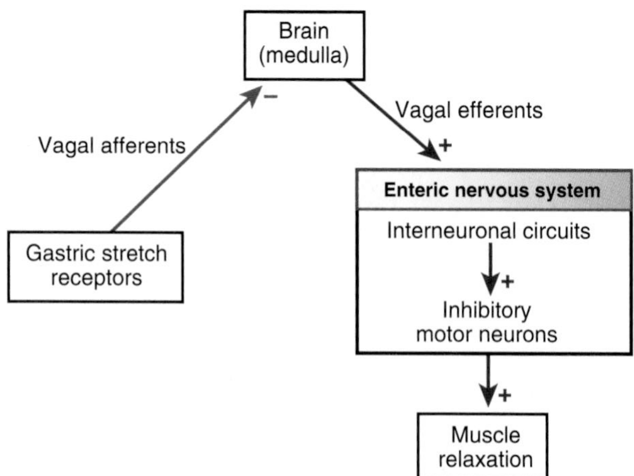

Figure 25.29 Adaptive relaxation in the gastric reservoir is a vagovagal reflex in which information from gastric stretch receptors is the afferent arm of the reflex and outflow from the medullary region of the brain is the efferent component. Vagal efferents transmit to the enteric nervous system, which controls the activity of inhibitory motor neurons that relaxes contractile tone in the reservoir.

decreased contractile tone in the reservoir, expansion of its volume, and a decrease in intraluminal pressure.

Three kinds of relaxation occur in the gastric reservoir.

Neurally mediated decreases in tonic contracture of the musculature are responsible for relaxation in the gastric reservoir (i.e., increased volume). Three kinds of relaxation are recognized. The act of swallowing initiates **receptive relaxation**. It is a reflex triggered by stimulation of mechanoreceptors in the pharynx followed by transmission over afferents to the dorsal vagal complex and activation of efferent vagal fibers to inhibitory musculomotor neurons in the gastric ENS. Distention of the gastric reservoir triggers **adaptive relaxation**. It is a vagovagal reflex triggered by stretch receptors in the gastric wall, transmission over vagal afferents to the dorsal vagal complex, and efferent vagal fibers back to inhibitory musculomotor neurons in the gastric ENS (Fig. 25.29). The presence of nutrients in the small intestine triggers **feedback relaxation**. It can involve both local reflex connections between receptors in the small intestine and the gastric ENS or hormones that are released from endocrine cells in the small intestinal mucosa and transported by the blood to signal the gastric ENS and stimulate firing in vagal afferent terminals in the stomach.

Adaptive relaxation is lost in patients who have undergone vagotomy, which was an earlier treatment for gastric acid disease (e.g., peptic ulcer) or can occur as an iatrogenic result of fundoplication surgery for the treatment of an incompetent LES and acid reflux disease. Following a vagotomy, increased tone in the musculature of the reservoir decreases the wall compliance, which in turn changes responses of gastric stretch receptors to distention of the reservoir. Pressure–volume curves obtained before and after vagotomy reflect the decrease in compliance of the gastric wall (Fig. 25.30). The loss of adaptive relaxation after a vagotomy is associated with a lowered threshold for sensations of epigastric fullness and pain. This response is explained by increased stimulation of the gastric mechanoreceptors as they sense distention of a reservoir with elevated wall stiffness. These effects of vagotomy may explain disordered gastric sensations in diseases with a component of vagal nerve pathology (e.g., autonomic neuropathy of diabetes mellitus).

Type of meal and conditions in the duodenum determine the rate of gastric emptying.

Together with storage in the reservoir and mixing and grinding by the antral pump, another important function of gastric motility is the orderly delivery of the gastric chyme to the duodenum at a rate that does not overload the digestive and absorptive functions of the small intestine (see Clinical Focus Box 25.2). Neural control mechanisms adjust the rate of gastric emptying to compensate for variations in

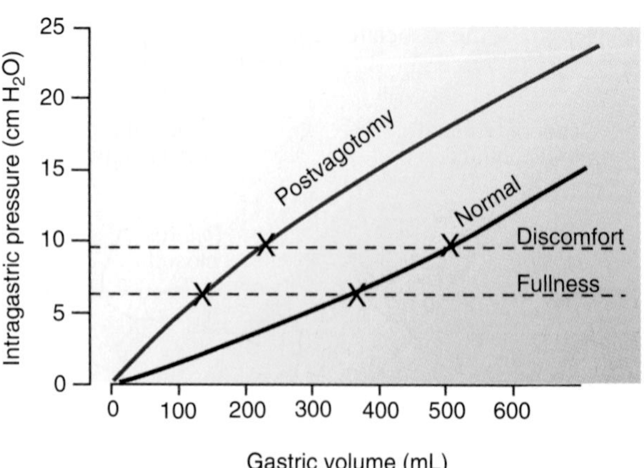

Figure 25.30 Adaptive relaxation of the gastric reservoir is lost following a vagotomy. Loss of adaptive relaxation in the gastric reservoir is associated with a lowered threshold for sensations of fullness and epigastric pain.

the volume, composition, and physical state of the gastric contents.

The volume of liquid in the stomach is one of the major determinants of gastric emptying. The rate of emptying of isotonic, noncaloric liquids (e.g., 0.9% NaCl) is proportional to the initial volume in the reservoir: the larger the initial volume, the more rapid the emptying.

With a mixed meal in the stomach, liquids empty faster than solids. If an experimental meal consisting of solid particles of various sizes suspended in H_2O is instilled in the stomach, emptying of the particles lags behind emptying of the liquid (Fig. 25.31). With digestible particles (e.g., studies with isotopically labeled chunks of liver), the **lag phase** is the time required for the grinding action of the antral pump to reduce the particle size. If the particles are plastic spheres of various sizes, the smallest spheres are emptied first; however, spheres up to about 7-mm diameter empty at a slow but steady rate when digestible food is in the stomach. The selective emptying of smaller particles first is referred to as the **sieving action** of the distal stomach. Inert spheres greater than 7 mm are not emptied while food is in the stomach; they empty at the start of the first migrating motor complex (MMC) as the digestive tract converts from the digestive motility state to the motor program for the interdigestive state.

Osmolality, acidity, and caloric content of the gastric chyme are major determinants of the rate of gastric emptying. Hypotonic and hypertonic liquids empty more slowly than isotonic liquids. The rate of gastric emptying decreases as the acidity of the gastric contents increases. Meals with a high caloric content empty from the stomach at a slower rate than meals with a low caloric content. The neurophysiologic control mechanisms for gastric emptying keep the rate of delivery of calories to the small intestine within a narrow range, regardless of whether the calories are presented as carbohydrate, protein, fat, or a mixture.

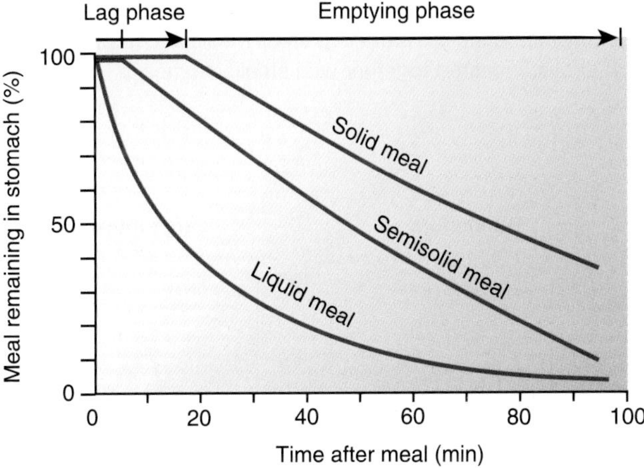

Figure 25.31 The rate of gastric emptying varies with the composition of the meal. Solid meals empty more slowly than semisolid or liquid meals. The emptying of a solid meal is preceded by a lag phase, which is the time required for particles to be reduced to a sufficiently small size for emptying.

Of the food groups, fat is emptied at the slowest rate; fat is the most potent inhibitor of gastric emptying. Part of the inhibition of gastric emptying by fats involves the enteroendocrine release of cholecystokinin in the upper small intestine. Cholecystokinin is a potent inhibitor of gastric emptying.

The intraluminal milieu of the small intestine is different from that of the stomach (see Chapter 26, "Gastrointestinal Secretion, Absorption, and Digestion"). Undiluted stomach contents have a composition that is poorly tolerated by the duodenum. Mechanisms of control of gastric emptying automatically adjust the delivery of gastric chyme to an optimal rate that is acceptable for the small intestine. This prevents overloading of the small intestinal mechanisms for the neutralization of acid, dilution to isosmolality, and enzymatic digestion of the foodstuff (see Clinical Focus Box 25.2).

SMALL INTESTINAL MOTILITY

Although the small intestine is three times longer than the large intestine, its name is derived from its comparatively smaller diameter. The small intestine is divided into three functional components (duodenum, jejunum, and ileum) and is the site where most food is digested and absorbed. Contractions of the small intestine are divided into segmental (mixing) and propulsion contractions.

Time required for transit of experimentally labeled meals from the stomach to the small intestine and on to the large intestine is measured in hours (Fig. 25.32). Transit time through the stomach is the most rapid of the three compartments; transit in the large intestine is the slowest. Three fundamental patterns of motility, which influence the transit of material through the small intestine, are the interdigestive pattern, the digestive pattern, and power propulsion. Each pattern is configured by a separate program in the library of programs in the small intestinal ENS (See Fig. 25.10).

Digestive and interdigestive motor patterns in the stomach and small intestine reflect the presence and absence of intraluminal nutrients, respectively.

The small intestinal ENS "runs" the **digestive state** motor program when nutrients are present and digestive processes are ongoing. Conversion to the **interdigestive state** motor program starts when digestion and absorption of nutrients are complete, 2 to 3 hours after a meal. The motility pattern of the interdigestive state in the small intestine is called the **MMC**. The MMC is detected by placing pressure sensors in the lumen of the intestine or by attaching electrodes to the intestinal serosal surface (Fig. 25.33). Sensors in the stomach show the MMC starting as large-forceful contractions at 3 per minute in the antral pump. Elevated contractile force in the LES coincides with the onset of the MMC in the stomach. The MMC activity, starting in the stomach, migrates into the duodenum and on down the small intestine to the ileum.

Clinical Focus / 25.2 

Disorders of Gastric Motility

Patients with disordered gastric motility fall either into the category of delayed gastric emptying or into a category in which emptying is too rapid. The symptoms associated with accelerated gastric emptying and delayed emptying overlap.

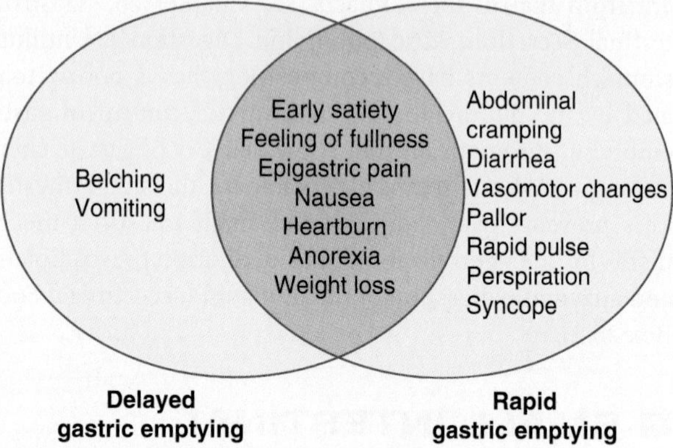

Figure 25.2A Some of the symptoms associated with delayed and rapid gastric emptying are overlapping.

Delayed Gastric Emptying (Gastric Retention)

Delayed gastric emptying occurs in 20% to 30% of patients with diabetes mellitus and is related to vagal neuropathy as part of a spectrum of diabetic autonomic neuropathy. Conduction in vagal afferent and efferent nerve fibers is impaired in diabetic neuropathy. This compromises the vagovagal reflexes, which underlie receptive, adaptive, and feedback relaxation of the gastric reservoir, during intake of a meal (see Figs. 25.28 and 25.29). Vagal nerves are sometimes damaged during laparoscopic surgery for repair of a hiatal hernia or fundoplication as a treatment for gastroesophageal reflux disease. Iatrogenic vagotomy results in a rapid emptying of liquids and a delayed emptying of solids. Truncal vagotomy impairs adaptive relaxation and results in increased contractile tone in the gastric reservoir (see Fig. 25.29). Elevated contractile tone increases pressure in the gastric reservoir, which ejects liquids more forcefully through the antral pump into the duodenum.

Paralysis with a loss of propulsive motility and, consequently, the trituration of solids by the antral pump occur in diabetic autonomic neuropathy and after a vagotomy. The result is **gastroparesis**, which can account for the delayed emptying of solids after a vagotomy.

Delayed gastric emptying with no demonstrable underlying condition is common. Up to 80% of patients with anorexia nervosa have delayed gastric emptying of solids. Another such condition is idiopathic gastric stasis, in which no evidence of an underlying condition can be found. These kinds of disorders for which no physiologic or biochemical explanation for the patient's symptoms can be found are called **functional gastrointestinal motility disorders.** Motility-stimulating drugs (e.g., cisapride, erythromycin, or domperidone) are used successfully in treating patients for both diabetic gastroparesis and the idiopathic form of gastroparesis.

In children, hypertrophic pyloric stenosis impedes gastric emptying. This is a thickening of the muscles of the pyloric canal, which is associated with a loss of the enteric nervous system (ENS) including inhibitory motor neurons to the musculature. The absence of inhibitory motor neurons and achalasia of the circular muscle in the pyloric canal are factors that account for the obstructive stenosis.

Rapid Gastric Emptying

Resection of the distal stomach might be done as a treatment for cancer or peptic ulcer disease. Surgical pyloroplasty used to be done together with vagotomy for the treatment of peptic ulcer disease and might still be done at times in patients with idiopathic gastroparesis. Both resection and pyloroplasty compromise mixing and renders the stomach incontinent for solids. Premature and rapid gastric emptying ("dumping") of solids and liquids into the duodenum in these cases causes hyperglycemia. Vagotomy, done at the same time as pyloroplasty, impairs receptive relaxation and accommodation in the gastric reservoir and exacerbates the dumping symptoms of anxiety, sweating, strong hunger, dizziness, weakness, and palpitations. The **dumping syndrome** is managed by restricting the patient to small meals of complex carbohydrates ingested together with small volumes of liquids.

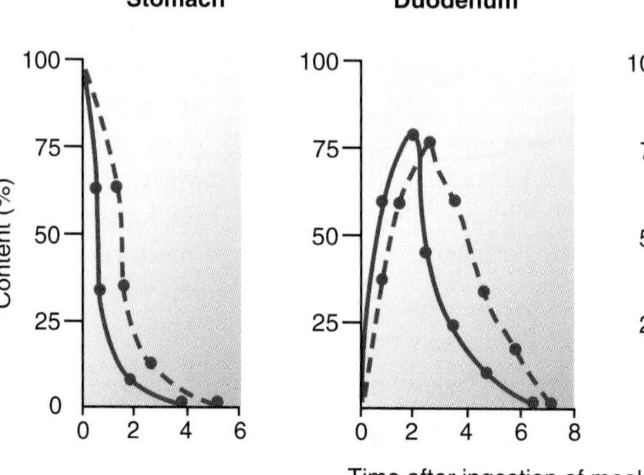

Figure 25.32 The time between entry of solid and liquid meals and final emptying of the meal for the stomach, duodenum, and large intestine is measured in hours.

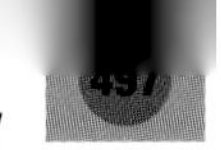

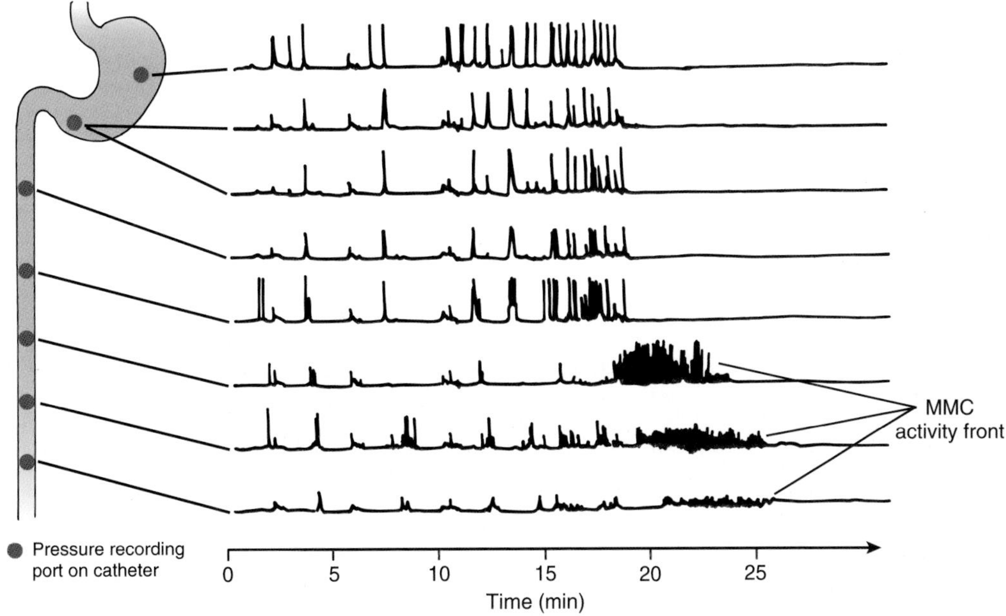

Figure 25.33 The small intestinal migrating motor complex (MMC) consists of an activity front that starts in the gastric antrum and slowly migrates through the small intestine to the ileum. Repetitive contractions, which reflect peristaltic propulsion, occur within the activity front.

At a single recording site in the small intestine, the MMC consists of three consecutive phases:

- Phase I: motor silence without contractile activity; corresponds to physiologic ileus
- Phase II: irregularly occurring contractions
- Phase III: regularly occurring contractions

Phase I returns after phase III, and the cycle repeats at the single recording site after 80 to 120 minutes (see Fig. 25.33). With multiple sensors positioned along the intestine, slow propagation of the phase II and phase III activity down the intestine becomes evident.

At a given time, the MMC occupies a limited length of intestine called the **activity front**, which has an oral and aboral boundary (Fig 25.34; also see Fig. 25.33). The activity front slowly advances (migrates) along the intestine at a rate that progressively slows as it approaches the ileum. Peristaltic propulsion of luminal contents occurs in the aboral direction between the oral and aboral boundaries of the activity front. The frequency of the peristaltic propulsive waves within the activity front is the same as the frequency of electrical slow waves in that intestinal segment. Each peristaltic wave consists of propulsive and receiving segments, as described earlier (see Fig. 25.21). Successive peristaltic waves start on average slightly farther in the aboral direction and propagate on average slightly beyond the boundary where the previous one stopped. Therefore, the entire activity front slowly migrates down the intestine, sweeping the lumen clean as it goes.

"Phase II" and "phase III" are commonly used descriptive terms that have minimal value for understanding of the MMC. Contractile activity described as phase II or phase III occurs because of the irregularity of the arrival of peristaltic waves at the aboral boundary of the activity front. On average, each consecutive peristaltic wave within the activity front propagates farther in the aboral direction than the previous wave. Nevertheless, at the lower boundary of the activity front, some waves terminate early and others travel farther (see Fig. 25.34). Therefore, as the lower boundary of the front passes the recording point, only the waves that reach the sensor are recorded, giving the appearance of irregular contractions. As propagation continues and the midpoint of the activity front reaches the recording point, the propulsive segment of every peristaltic wave is detected. Because the peristaltic waves occur with the same rhythmicity as the electrical slow waves, the contractions are described as being "regular." The regular contractions, which are seen when the central region of the front passes a single recording site, last for 8 to 15 minutes. This time is shortest in the duodenum and progressively increases as the MMC migrates toward the ileum. The MMC occurs in most mammals, including humans, in conscious states and during sleep. It starts in the antrum of the stomach as an increase in the strength of the regularly occurring antral contractile complexes and accomplishes the emptying of indigestible particles (e.g., pills and capsules) larger than 7 mm. In humans, 80 to 120 minutes is required for the activity front of the MMC to travel from the antrum to the ileum. As one activity front terminates in the ileum, another begins in the antrum. In humans, the time between cycles is longer during the day than that at night. The activity front travels at about 3 to 6 cm/min in the duodenum and progressively slows to about 1 to 2 cm/min in the ileum. It is important not to confuse the speed of travel of the activity front of the MMC with that of the electrical slow waves, action potentials, and peristaltic waves within the activity

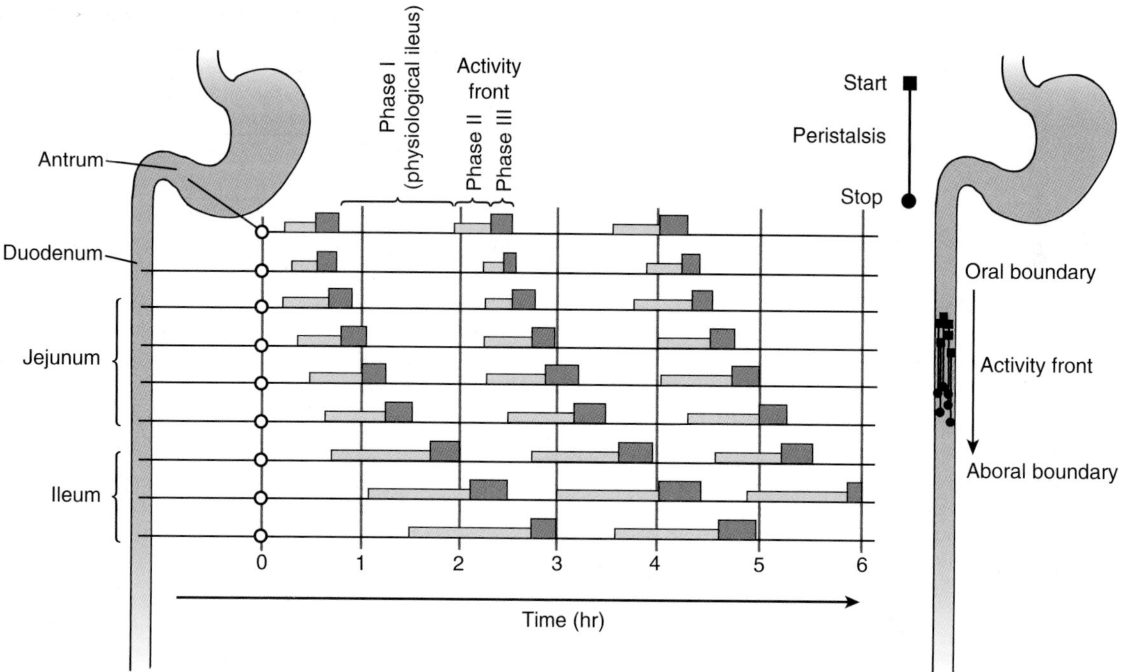

Figure 25.34 The migrating motor complex (MMC) consists of three phases. Phase I of physiological ileus refers to the intestinal region with no activity. Phases II and III reflect the migrating activity front, which has an aboral boundary where propulsive contractions stop and an oral boundary where propulsive contractions start. (See text for details.)

front. Slow waves with associated action potentials and associated contractions of circular muscle travel about 10 times faster.

Cycling of the MMC continues until it is ended by the ingestion of food. A sufficient nutrient load terminates the MMC wherever it is located in its travel down the intestine. Termination requires the physical presence of a meal in the upper digestive tract; intravenous feeding does not stop the MMC. The speed, with which the MMC is terminated, wherever it is along the intestine suggests a neural or hormonal mechanism. Gastrin and cholecystokinin, both of which are released during a meal, terminate the MMC in the stomach and upper small intestine, but not in the ileum, when injected intravenously.

The MMC is organized by a program stored in the ENS neural networks (See Fig. 25.10). It continues in the small intestine after a vagotomy or sympathectomy but stops when it encounters a region of the intestine where the ENS has been interrupted. Presumably, command signals to the enteric neural circuits are necessary for initiating the MMC, but whether the commands are neural, hormonal, or both remains unknown. Although levels of the hormone **motilin** increase in the blood at the onset of the MMC, it is unclear whether motilin is the trigger or is released as a consequence of MMC occurrence.

Migrating motor complex acts as "housekeeper" for the small intestine.

Gallbladder contraction and delivery of bile to the duodenum are coordinated with the movement of the MMC into the antroduodenal region. The activity front of the MMC propels the bile from the duodenum through the jejunum to the terminal ileum, in which it is reabsorbed into the hepatic portal circulation. This mechanism minimizes the accumulation of concentrated bile in the gallbladder and increases the movement of bile acids in the enterohepatic circulation during the interdigestive state (see Chapter 26, "Gastrointestinal Secretion, Absorption, and Digestion").

Adaptive significance of the MMC appears also to be a mechanism for clearing indigestible debris from the intestinal lumen during the fasting state. Large indigestible particles are emptied from the stomach only during the MMC.

Bacterial overgrowth in the small intestine is associated with an absence of the MMC. This condition suggests that the MMC might play a "housekeeper" role in preventing the overgrowth of microorganisms that would occur in the small intestine if the contents were allowed to stagnate in the lumen.

Enteric nervous system "calls-up" digestive motor program from its program library when nutrients are present in the upper gastrointestinal tract.

The mixing pattern of motility replaces the MMC when small intestinal motility is "switched" to the digestive motility program following ingestion of a meal. The mixing movements are sometimes called segmenting movements or segmentation because of their appearance on x-ray and magnetic resonance images of the working small intestine. Peristaltic contractions, which propagate for only short distances and occur simultaneously at

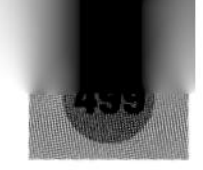

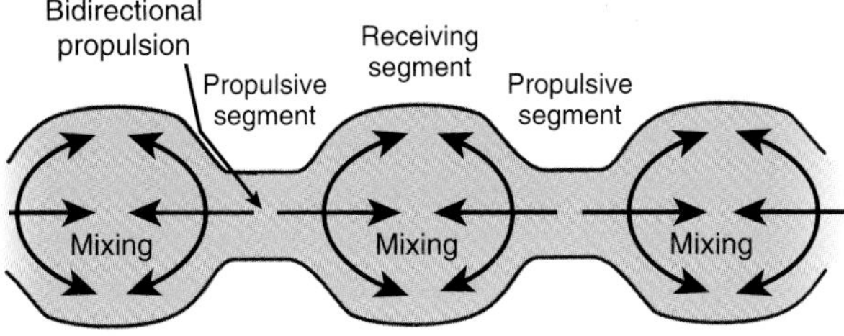

Same length of intestine later in time

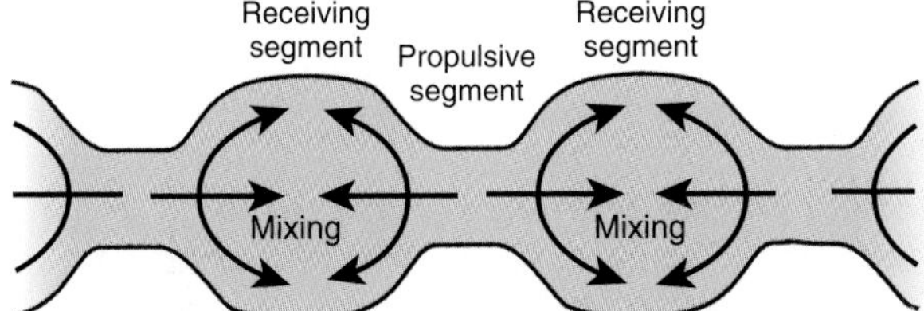

Figure 25.35 The mixing pattern of small intestinal motility is characteristic of the digestive state. Propulsive segments separated by receiving segments occur randomly at multiple sites along the small intestine. Mixing of the luminal contents occurs in the receiving segments. Receiving segments convert to propulsive segments, whereas propulsive segments become receiving segments.

multiple sites along the bowel, account for the segmentation appearance. Receiving segments with an expanded lumen separate circular muscle contractions that form short propulsive segments on either end of the receiving segment (Fig. 25.35). Each propulsive segment jets the chyme in both directions into the opened receiving segments, where stirring and mixing occur. This happens continuously at closely spaced sites along the entire length of the small intestine. The intervals of time between mixing contractions are the same as for electrical slow waves or are multiples of the shortest slow-wave interval in the particular region of intestine. A higher frequency of electrical slow waves and associated contractions in more proximal regions and the peristaltic nature of the mixing movements result in a net aboral propulsion of the luminal contents over time.

The ENS programs the digestive mixing pattern of small intestinal motility. Signals transmitted by vagal efferent nerves to the ENS interrupt the MMC and initiate mixing motility during ingestion of a meal. After the vagus nerves are cut, a larger quantity of ingested food is necessary to interrupt the interdigestive motor pattern, and termination of the MMCs is often incomplete. Evidence of vagal commands for the mixing pattern has been obtained in animals with cooling cuffs placed surgically around each vagus nerve. During the digestive state, cooling and consequent blockade of impulse transmission in the nerves result in the interruption of the pattern of mixing movements. When the vagus nerves are blocked during the digestive state, MMCs reappear in the intestine but not in the stomach. With warming of the nerves and release of the neural blockade, the mixing motility pattern returns.

Power propulsion is a specialized motor program that moves large volumes of luminal content rapidly over extended distances along the bowel.

Power propulsion involves strong, long-lasting contractions of the circular muscle of peristaltic propulsive segments that propagate for extended distances in both the small and large intestine. The *giant migrating contractions* are much stronger than the circular muscle contractions in the propulsive segments during the MMC or mixing pattern. Giant migrating contractions last 18 to 20 seconds and span several cycles of the electrical slow waves. They are a component of a highly efficient propulsive mechanism that rapidly strips the lumen clean as it travels at about 1 cm/s over long lengths of intestine.

Intestinal power propulsion differs from peristaltic propulsion during the MMC, and mixing movements in that circular contractions in the propulsive segment are stronger and more open gates permit propagation over longer reaches of intestine (see Figs. 25.22 and 25.23). The powerful circular muscle contractions are not time locked to the electrical slow waves and reflect strong activation of the muscle by excitatory musculomotor neurons.

Power propulsion occurs in the retrograde direction, from mid-jejunum to stomach, during emesis in the small intestine and in the orthograde direction in response to noxious stimulation in both the small and large intestine. Abdominal cramping sensations and sometimes diarrhea can be associated with the running of this ENS motor program. Application of irritants to the mucosa, the introduction of luminal parasites, enterotoxins from pathogenic bacteria, allergic reactions, and exposure to ionizing radiation each can call-up the power propulsion motor program. This suggests that power propulsion is a defensive adaptation for rapid clearance of undesirable contents from the intestinal lumen. It also accomplishes mass movement of intraluminal material in normal states, especially during defecation in the large intestine.

LARGE INTESTINAL MOTILITY

The large intestine consists of the cecum and the colon. Food is no longer broken down at this stage of digestion and the large intestine simply absorbs water, minerals, and vitamins. It also functions in compacting feces. Contractile activity occurs continuously in the normally functioning large intestine. Whereas the contents of the small intestine move through sequentially with no mixing of individual meals, the large bowel contains a mixture of the remnants of several meals ingested over 3 to 4 days. The arrival of undigested residue from the ileum does not predict the time of its elimination in the stool. The large intestine is subdivided into functionally distinct compartments, which correspond approximately to the ascending colon, transverse colon, descending colon, rectosigmoid region, and internal anal sphincter (Fig. 25.36). γ-Scintigraphy or small solid pellets that can be followed fluoroscopically are methods that

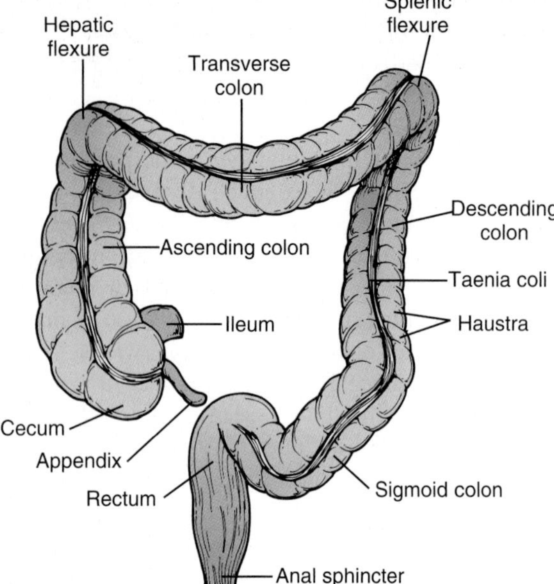

Figure 25.36 The main anatomical regions of the large intestine are the ascending, transverse, descending, and sigmoid colon and the rectum. The hepatic flexure is the boundary between ascending and transverse colon; the splenic flexure between transverse and descending colon. The sigmoid colon is well defined by its shape. The rectum is the most distal region. The cecum is the blind ending of the colon at the ileocecal junction. The appendix is an evolutionary vestige. Internal and external anal sphincters close the terminus of the large intestine. The longitudinal muscle layer in humans is restricted to bundles of fibers called *taeniae coli*.

are used to evaluate rates of transit in diagnostic workup of patients with chronic constipation. The time required for normal transit of small pellets through the large intestine is 36 to 48 hours, on average.

Ascending colon receives large volumes of chyme from the terminal ileum.

The ascending colon is specialized for processing chyme delivery from the terminal ileum. Power propulsion in the terminal length of ileum rapidly delivers relatively large volumes of liquid chyme into the ascending colon, especially during the digestive state. Neuromuscular mechanisms, analogous to adaptive relaxation in the gastric reservoir, permit filling of the ascending colon to occur without large increases in intraluminal pressure. Chemoreceptors and mechanoreceptors in the cecum and ascending colon provide feedback information for controlled delivery from the ileum, analogous to the feedback control of gastric emptying from the small intestine.

Dwell time of material in the ascending colon is short when studied with γ-scintigraphic imaging of the movements of radiolabeled markers. When radiolabeled chyme is instilled into the cecum in humans, half of the instilled volume empties in 87 minutes on average. This time period is long compared with that in an equivalent length of small intestine but short compared with that in the transverse colon. It suggests that the ascending colon is not the primary site for the large intestinal functions of storage, mixing, and removal of water from the feces.

The motor patterns programmed by the ENS of the ascending colon consist of peristaltic propulsion that can be seen to travel sometimes in the orthograde direction and at times in the retrograde direction. The significance of backward propulsion in this region is uncertain; it may be a mechanism for temporary retention of the chyme in the ascending colon. Forward propulsion in this region is probably controlled by feedback signals on the fullness of the transverse colon. The programming capacity of the ENS for propulsion in forward and reverse directions illustrates the neural plasticity of the brainlike ENS.

Motility of the transverse colon is specialized for storage and removal of water from the feces.

The transverse colon is specialized for storage and dehydration of feces. Radioscintigraphy shows that labeled marker moves relatively quickly into the transverse colon (Fig. 25.37), where it is retained for about 24 hours. This suggests that the transverse colon is the primary location for the removal of water and electrolytes and the storage of solid feces.

A segmental pattern of motility, which is programmed by the ENS, accounts for the ultraslow forward movement and compacting of feces in the transverse colon. Ring-like contractions of the circular muscle divide the colon into pockets called *haustra* (Fig. 25.38). The **haustration** motility pattern is reminiscent of mixing (segmentation) movements in the small intestine and the "wiring" of the synaptic connections in the neural networks is no doubt the same or very similar (see Fig. 25.35). Nevertheless, haustral formation differs from small intestinal segmentation in that the contracting and the receiving segments on either side remain in their respective states for extended periods of time.

Haustrations are dynamic in that they form and reform at different sites. The most common pattern in the fasting individual is for the contracting segment to propel the contents in both directions into receiving segments (i.e., haustra). This mechanism mixes and compresses the semiliquid feces in the haustral pockets and probably facilitates the absorption of water without any net forward propulsion.

Net forward propulsion occurs when sequential migration of the haustra occurs along the length of the bowel. The contents of one haustral pocket are propelled into the next region, where a second pocket is formed, and from there to the next segment, where the same events occur. This pattern results in slow forward progression and is believed to be a mechanism for compaction of the feces during storage.

Power propulsion underlies the mass movement of feces in the colon.

Power propulsion, which was discussed earlier in the chapter, occurs as one of the ENS-programmed motor events in the transverse and descending colon. This motor behavior fits the general pattern of neurally coordinated peristaltic propulsion and accomplishes the mass movement of feces over long distances. Increased delivery of ileal chyme into the ascending

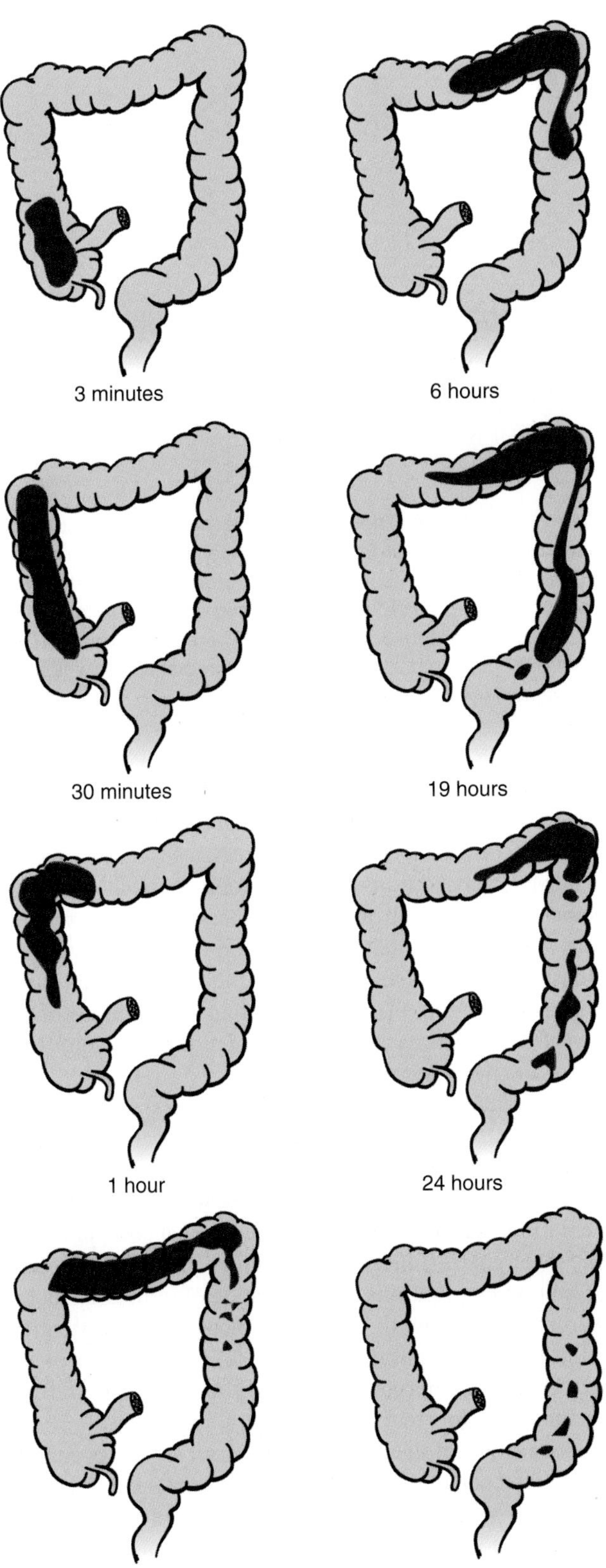

Figure 25.37 Radiosintigraphy is used to study transit times for compartments of the large intestine. Successive scintigrams reveal that the longest dwell time for intraluminal isotope injected initially into the cecum is in the transverse colon. The image is faint after 48 hours, indicating that most of the marker has been excreted with the feces.

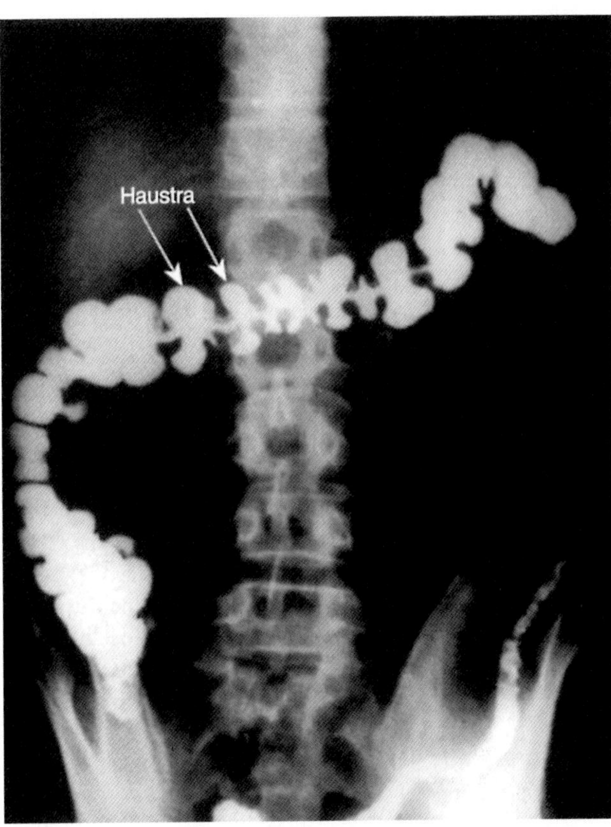

Figure 25.38 X-ray shows haustral contractions in the ascending and transverse colon. Between the haustral pockets are segments of contracted circular muscle. Ongoing activity of inhibitory motor neurons maintains the relaxed state of the circular muscle in the pockets. Inactivity of inhibitory motor neurons permits the contractions between the pockets.

colon, following a meal, often triggers mass movements in the aboral direction in the colon. The increased incidence of mass movements and generalized increase in segmental movements following a meal is called the **gastrocolic reflex**. Irritant laxatives (e.g., castor oil and senna) act to initiate the motor program for power propulsion in the normally functioning colon. Threatening agents, which might find their way into the colonic lumen (e.g., parasites, enterotoxins, and food allergens), also initiate power propulsion.

Mass movement of feces (power propulsion) in the healthy bowel usually starts in the middle of the transverse colon and is preceded by relaxation of the circular muscle and the downstream disappearance of haustral contractions. An extended length (e.g., 300 cm) of the colon might be emptied as the contents are propelled at rates up to 5 cm/min as far as the rectosigmoid region. Haustration returns after the passage of the power contractions.

Power propulsion in the descending colon is responsible for mass movements of feces into the sigmoid colon and rectum.

The descending colon is a conduit from transverse to sigmoid colon. Radioscintigraphic studies in humans show that feces do not have long dwell times in the descending colon (see

Fig. 25.37). Labeled feces begin to accumulate in the sigmoid colon and rectum about 24 hours after the label is instilled in the cecum. This suggests that the transverse colon is the main fecal storage reservoir, whereas the descending colon serves as a conduit without long-term retention of the feces. The neural program for power propulsion in the descending colon is responsible for mass movements of feces into the sigmoid colon and rectum.

Rectosigmoid, anal canal, and pelvic floor musculature preserve fecal continence.

The sigmoid colon and rectum are reservoirs with a capacity of up to 500 mL in humans. Distensibility in this region is an adaptation for temporarily accommodating the mass movements of feces from upstream. The rectum begins at the level of the third sacral vertebra and follows the curvature of the sacrum and coccyx for its entire length. It connects to the anal canal, which is surrounded by the internal and external anal sphincters. Overlapping sheets of striated muscle called the **levator ani** form the pelvic floor. This muscle group, which includes the **puborectalis** and the striated **external anal sphincter**, is a functional unit that maintains fecal continence. These skeletal muscles behave in many respects like the somatic muscles that maintain posture elsewhere in the body (see Chapter 5, "Motor System").

The pelvic floor musculature can be imagined as an inverted funnel consisting of the levator ani and external sphincter muscles in a continuous sheet from the bottom margins of the pelvis to the anal verge, which is the transition zone between mucosal epithelium and stratified squamous epithelium of the skin. The pelvic floor descends during defecation. When defecation is complete, the levator ani contracts to restore the perineum to its normal position. Fibers of the puborectalis muscle join behind the anorectum and pass around it on both sides to insert on the pubis. This forms a U-shaped sling that pulls the anorectal tube anteriorly such that the long axis of the anal canal lies at nearly a right angle to that of the rectum (Fig. 25.39). Tonic contractile pull of the puborectalis narrows the anorectal tube from side to side at the bend of the angle, resulting in a physiologic valve that is important in the mechanisms that prevent leakage of feces and flatus.

The puborectalis sling and the upper margins of the internal and external sphincters form the anorectal ring, which marks the boundary of the anal canal and rectum. Surrounding the anal canal over a length of about 2 cm are the internal and external anal sphincters. The external anal sphincter is skeletal muscle attached to the coccyx posteriorly and the perineum anteriorly. When contracted, it compresses the anus into a slit, thereby closing the orifice. The internal anal sphincter is a modified extension of the circular muscle coat of the rectum. It is composed of smooth muscle that, like other sphincteric muscles in the digestive tract, contracts tonically to sustain closure of the anal canal.

Anal canal is innervated by somatosensory nerves.

Mechanoreceptors in the rectum detect distention and supply the ENS with the sensory information needed for feedback control of this region. Unlike the rectum, the anal canal in the region of skin at the anal verge is innervated by somatosensory nerves that transmit signals to the spinal cord and on to higher processing centers in the brain. This region has sensory receptors that detect and transmit information on touch, pain, and temperature with high sensitivity. Processing of information from these receptors allows the person to discriminate consciously between the presence of gas, liquid, and solids in the anal canal. In addition, stretch receptors in the muscle of the pelvic floor detect changes in the orientation of the anorectum as feces are propelled into the region.

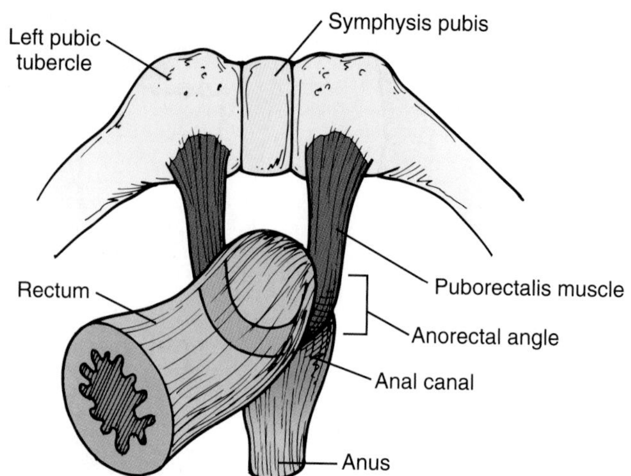

Figure 25.39 The puborectalis muscle contributes to the maintenance of fecal continence. One end of the puborectalis muscle inserts on the left pubic tubercle and the other on the right pubic tubercle, forming a loop around the junction of the rectum and anal canal. Thereby, contraction of the puborectalis muscle forms the anorectal angle, which blocks passage of feces.

Contraction of the internal anal sphincter and the puborectalis muscle blocks the passage of feces and maintains continence with small volumes in the rectum. When the rectum is distended, the **rectoanal reflex** pathway is activated to relax contractile tone in the internal anal sphincter. Like other enteric reflexes, this one involves a stretch receptor, enteric interneurons, and excitation of inhibitory motor neurons to the smooth muscle sphincter. Rectal distention results also in the sensation of rectal fullness, which is mediated by central nervous processing of information from mechanoreceptors in the pelvic floor musculature.

Relaxation of the internal anal sphincter allows contact of the rectal contents with the sensory receptors in the lining of the anal canal. Signals from the receptors in the anal canal reach conscious perception and provide the person with an early warning of the possibility of a breakdown of continence. When this occurs, voluntary contraction of the external anal sphincter and puborectalis muscle maintains continence. The external sphincter closes the anal canal and the puborectalis sharpens the anorectal angle (see Fig. 25.39). An increase in the anorectal angle works in concert with increases in intra-abdominal pressure to create a "**flap valve**." This valve is formed by the collapse of the anterior rectal wall onto the upper end of the anal canal, thereby occluding the lumen.

From Bench to Bedside / 25.1

Neural Stem Cells for the Treatment of Disorders of the Enteric Nervous System

The physiology in this chapter is basic to understanding a variety of esophageal and gastrointestinal (GI) motility disorders. Among the esophageal motility disorders are diffuse spasm and lower esophageal sphincter achalasia. The GI motility disorders are infantile hypertrophic pyloric stenosis and small intestinal pseudoobstruction in association with paraneoplastic syndrome, Chagas disease, or autoimmune enteric neuropathy. Hirschsprung disease, which is characterized by a tonically constricted segment of variable length at the anal terminal of the large intestine and a grossly distended colon proximal to the constricted segment, is well known among the large intestinal motility disorders. Each of these disorders is characterized by loss of the enteric nervous system from the affected region of the gut. Inhibitory motor neurons are included among the neurons that are lost when the enteric nervous system is destroyed by autoimmune attack or when it fails to develop in the fetus as in the case of Hirschsprung disease. Absence of inhibitory braking action accounts for hyperactive and uncoordinated contractile activity of the denervated autogenic musculature.

Neural stem cells (NSC) are multipotential cells that have the ability to divide and differentiate into mature neurons and glial cells. Research on NSCs, especially on transplantation of NSCs, has advanced rapidly over the last two decades. NSC transplantation is now considered to be a plausible approach for the treatment of several of the neural degenerative disorders of the central nervous system (CNS) and peripheral nervous system. Transplantation of CNS-derived NSCs, as a promising approach for the restoration of the ENS in the motility disorders described in this chapter, is moving forward with animal research. NSC transplantation has not yet been done successfully in the human digestive tract; nevertheless, successes in animal studies suggest that it is on the horizon for humans.

NSCs have now been successfully isolated, proliferated in cell culture, and transplanted into the stomachs of mice. The NSCs used in these studies are isolated from embryonic mice, which have been transgenically engineered to express green fluorescent protein, and transplanted into the gastric pylorus. Expression of green fluorescent protein enables the investigators to visualize the grafted NSCs with fluorescence microscopy and thereby track their migration, survival, and differentiation into neurons or glia after placement in the gut wall. Early on, posttransplant survival was a limiting factor for successful functional repopulation of the host gut. Most of the grafted NSCs in animal and human studies die with cell death occurring as programmed cell death (i.e., apoptosis). There is interest in the fact that the degeneration of neurons in the ENS, which take place in the autoimmune inflammatory neuropathies in human gut, reflects apoptotic neuronal death. An example of an advance for NSC transplantation in the gut was the discovery that blockade of apoptosis, by drugs that inhibit the apoptotic caspase-1 enzyme, enhances survival and proliferation of transplanted CNS–NSCs in the mouse stomach. A second example is a recent discovery that stimulation of the 5-HT4 hydroxytryptamine receptor subtype stimulates proliferation of enteric neuronal stem cells in situ.

Mice with a gene mutation (nNOS–/–) became useful for the investigation of restoration of disordered function by NSC transplantation in the gut. The mutation prevents the development of neuronal nitric oxide (NO) synthase (NOS) and therefore the synthesis and release of the inhibitory neurotransmitter, NO, by inhibitory motor neurons in the ENS. Consequently, the autogenic pyloric musculature in the mutant mice remains in a tonic state of contraction in the absence of functional inhibitory innervation. A reflection of the loss of enteric inhibitory function in the nNOS–/– mice is a grossly dilated stomach that is associated with delayed emptying of both liquids and solids. The condition of the stomach in these mice is reminiscent of hypertrophic pyloric stenosis in human infants. NSCs, labeled with green fluorescent protein, have now been isolated from the brains of mice, allowed to proliferate in cell culture and then injected into the pyloric region of nNOS–/– mutants. One week after transplantation, the NSCs are found to differentiate into neurons and express neuronal NOS. Evidence that the NSCs become incorporated into functional ENS networks is the findings that gastric emptying improves significantly, relative to controls, in the mice with the NSC transplants.

The results from animal research are promising and support feasibility for manipulation of NSCs for neuronal replacement in the diseased ENS in the future. Earlier social issues related to stem cell procurement and experimentation are being resolved and the techniques and methodologies are under continuous refinement, which offers hope that long-term neuronal replacement therapy in the gut can become a reality for disorders of the ENS.

Whereas the ENS mediates the rectoanal reflex, synaptic circuits for the neural reflexes of the external anal sphincter and other pelvic floor muscles reside in the sacral portion of the spinal cord. The mechanosensory receptors are believed to be muscle spindles and Golgi tendon organs like those found in skeletal muscles elsewhere in the body. Sensory input from the anorectum and pelvic floor is transmitted over dorsal spinal roots into the sacral cord, and motor outflow to these areas is in ventral sacral root motor nerve fibers. The spinal circuits mediate reflex increases in contraction of the external sphincter and pelvic floor muscles, which can occur during behaviors that raise intra-abdominal pressure (e.g., coughing, sneezing, and lifting weights) and threaten a breakdown in continence.

Neural control of defecation involves the central and enteric nervous systems.

Distention of the rectum by the mass movement of feces or gas evokes an urge to defecate or release flatus. Processing in the spinal cord and brain of mechanosensory information from the rectum is the underlying mechanism for these sensations. Local processing of the mechanosensory information in the ENS activates the motor program for relaxation of the internal anal sphincter. At this stage of rectal distention, voluntary contraction of the external anal sphincter and the puborectalis muscle prevents leakage. The decision to defecate at this stage is voluntary. When the decision is made, commands from the brain to the sacral spinal cord shut off the excitatory input to the external sphincter and levator ani muscles. Additional skeletal motor commands contract the abdominal muscles and diaphragm to increase intra-abdominal pressure. Coordination of the skeletal muscle components of defecation results in a straightening of the anorectal angle, descent of the pelvic floor, and opening of the anus.

Programmed behavior of large intestinal smooth muscle during defecation includes shortening of the longitudinal muscle layer in the sigmoid colon and rectum, followed by strong contraction of the circular muscle layer. This behavior corresponds to the basic stereotypic pattern of peristaltic propulsion (see Fig. 25.21). It represents terminal intestinal peristalsis in that the circular muscle of the distal colon and rectum becomes the final propulsive segment, whereas the outside environment receives the forwardly propelled luminal contents (i.e., the equivalent of a receiving segment).

A voluntary decision to resist the urge to defecate is eventually accompanied by relaxation of the circular muscle of the rectum. This form of adaptive relaxation accommodates the increased volume in the rectum. As wall tension relaxes, the stimulus for the rectal mechanoreceptors is removed, and the urge to defecate subsides. Receptive relaxation of the rectum is accompanied by a return of contractile tension in the internal anal sphincter, relaxation of tone in the external sphincter, and increased pull by the puborectalis muscle sling. When this occurs, the feces remain in the rectum until the next mass movement further increases the rectal volume and stimulation of mechanoreceptors again signals the neural mechanisms for defecation.

Chapter Summary

- The musculature of the digestive tract is mainly smooth muscle.
- Electrical slow waves and action potentials are the primary forms of electrical activity in the gastrointestinal (GI) musculature.
- GI smooth muscles have properties of a functional electrical syncytium.
- A hierarchy of neural integrative centers in the brain, spinal cord, and periphery determines moment-to-moment behavior of the digestive tract.
- The digestive tract is innervated by the sympathetic, parasympathetic, and enteric divisions of the autonomic nervous system.
- Vagus nerves transmit afferent sensory information to the brain and parasympathetic autonomic efferent signals to the digestive tract.
- Splanchnic nerves transmit sensory information to the spinal cord and sympathetic autonomic efferent signals to the digestive tract.
- The enteric nervous system functions as an independent minibrain in the gut.
- Fast and slow excitatory postsynaptic potentials, slow inhibitory postsynaptic potentials, presynaptic inhibition, and presynaptic facilitation are key synaptic events in the enteric nervous systems.
- Enteric musculomotor neurons may be excitatory or inhibitory.
- Enteric inhibitory musculomotor neurons to the intestinal circular muscle are continuously active and transiently inactivated to permit muscle contraction.
- Enteric inhibitory musculomotor neurons to the musculature of sphincters are inactive and transiently activated for the timed opening and passage of luminal contents.
- A polysynaptic reflex circuit determines the behavior of the intestinal musculature during peristaltic propulsion.
- Physiological ileus is the normal absence of contractile activity in the intestinal musculature.
- Peristaltic propulsion and relaxation of the lower esophageal sphincter are the main motility events in the esophagus.
- The gastric reservoir and antral pump have different kinds of functional motor behavior.
- Vagovagal reflexes are important in the control of gastric motor functions.
- Feedback signals from the duodenum determine the rate of gastric emptying.
- The migrating motor complex is the small intestinal motility pattern of the interdigestive state.
- Mixing movements are the small intestinal motility pattern of the digestive state.
- Intestinal power propulsion is a protective response to harmful agents.
- Motor functions of the large intestine are specialized for storage and dehydration of feces.
- Physiologic functions of the rectosigmoid region, anal canal, and pelvic floor musculature are responsible preserving fecal continence.

thePoint *Visit http://thePoint.lww.com for chapter review Q&A, case studies, animations, and more!*

26 Gastrointestinal Secretion, Digestion, and Absorption

ACTIVE LEARNING OBJECTIVES

Upon mastering the material in this chapter, you should be able to:

- Describe how salivary secretion is regulated.
- Explain the mechanism by which the stomach secretes hydrochloric acid.
- Describe the phases of acid secretion associated with digestion.
- Explain the phasic secretion of the pancreatic enzymes.
- Explain the role that bile salts play in the absorption of intestinal lipids.
- Describe the difference between primary and secondary bile acids.
- Explain why gallstones are usually formed in the gallbladder.
- Describe how enterocytes transport the digested products of carbohydrate.
- Explain how the enterocytes digest and transport triglycerides.
- Explain the function of the fat-soluble vitamins.
- Explain the mechanism of ion absorption by the small intestine.

The gastrointestinal (GI) tract refers to the esophagus, stomach, and intestines. The GI tract is divided into the upper GI (esophagus, stomach, and duodenum) and the lower GI tract (small and large intestines). The major processes occurring in the GI tract are that of motility, secretion, regulation, and digestion. The GI hormones (gastrin, **cholecystokinin** [**CCK**], secretin, and glucose-dependent insulin tropic polypeptide) play an important role in regulating GI function, especially secretion of digestive enzymes.

Some absorption occurs in the stomach, including that of medium-chain fatty acids and some drugs, but most digestion and absorption of nutrients take place in the small intestine. The digestion of nutrients, such as proteins and fat, occurs as soon as food is emptied into the stomach. Digestion in the stomach prepares the small intestine to complete the digestion process and to uptake nutrients. The small intestine is equipped with specialized epithelial cells called **enterocytes** that take up the digested products of nutrients for transport into either the portal circulation or the lymphatic system. Villi and the brush border membrane of the enterocytes in the small intestine increase the surface area for uptake from the lumen of the small intestine. The digestion and absorption in the GI tract is significantly facilitated by secretions from the salivary glands, stomach, pancreas, and liver. Secretory glands serve two primary functions. First, digestive enzymes are secreted from the mouth to the small intestine to break down food. Second, glands secrete mucus for lubrication and protection. Because the surface of the GI tract is literally exposed to the external environment, the secretion of mucus also plays a prominent role in the immune system and is actively involved in preventing pathogens from entering the blood stream and lymph (see Chapter 10, "Immunology, Organ Interaction, and Homeostasis," for review). This chapter discusses the basic structure, function, and regulation of GI secretion and the role that the GI tract plays in the absorption of carbohydrate, fat, protein, fat-soluble and water-soluble vitamins, electrolytes, bile salts, and water.

SALIVARY SECRETION

In adults, approximately 7 to 10 L of fluid are added to the GI tract daily as various GI secretions for the digestion and absorption of nutrients occur. GI secretions include saliva, gastric juice, pancreatic juice, and bile.

Salivary glands are a heterogeneous group of **exocrine glands** that produce two types of protein secretion. One secretion is made up of **serous cells** and contains **amylase**, an enzyme that breaks down starch. The other secretion is made up of **mucous cells** and contains **mucin** for lubrication and protection. Three major pairs of salivary glands are situated at the beginning of the GI tract to initiate the digestion process. These include the (1) **parotid**, (2) **submandibular**, and (3) **sublingual glands**, all of which empty saliva into the mouth via secretory ducts (Fig. 26.1). Like other GI secretions, the secretion of saliva is stimulated primarily by contact with food and is unique in that

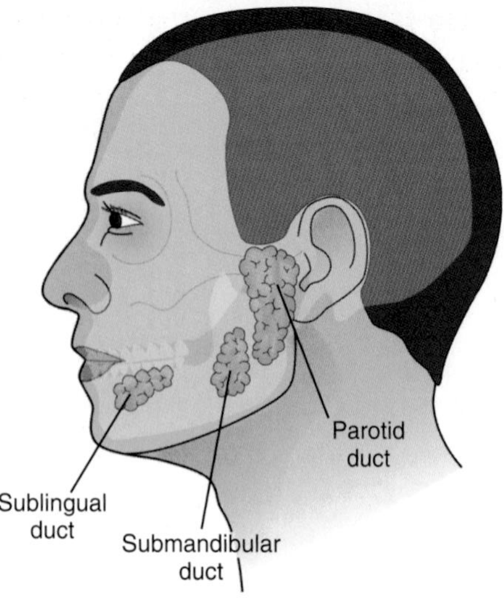

Figure 26.1 Salivary glands.

it is regulated almost exclusively by parasympathetic and sympathetic nerves. Saliva performs several functions. It facilitates chewing and swallowing by lubricating food, carries immunoglobulins that combat pathogens, and assists in carbohydrate digestion.

Salivon is the functional unit of the salivary gland.

The human salivary glands contain different proportions of **serous cells** and **mucous cells**. The cells are organized as a single layer surrounding a sac-like structure called an **acinus.** The functional unit, the **salivon** (Fig. 26.2), consists of the acinus, the **intercalated duct**, the **striated duct**, and the **excretory (collecting) duct**. Serous cells secrete digestive enzymes, and mucous cells secrete mucin. Serous cells contain an abundance of rough endoplasmic reticulum (RER), which reflects active protein synthesis, and numerous **zymogen granules**. Salivary amylase is an important digestive enzyme synthesized and stored in the zymogen granules and secreted by the serous acinar cells. The mucous acinar cells store numerous mucin droplets.

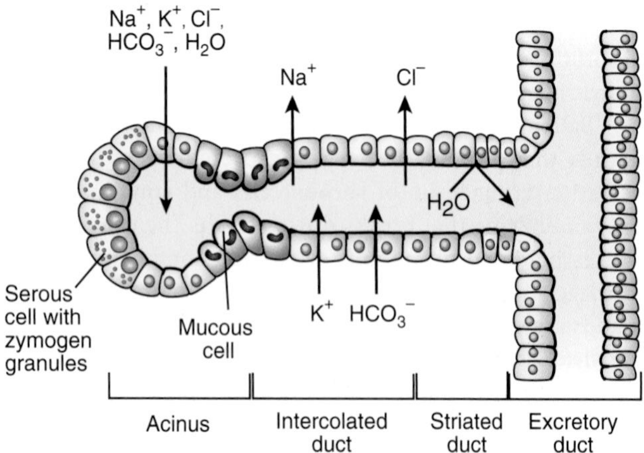

Figure 26.2 The salivon is the basic unit of the salivary gland. The acinus and associated ductal system form the salivon.

Mucin is composed of glycoproteins of various molecular weights.

The secretions of the major glands differ significantly. The *parotid gland* secretes primarily serous secretion that is rich in water and electrolytes, whereas the *submandibular* and *sublingual glands* secrete both serous and mucous secretions. The salivary glands are endowed with a rich blood supply, and both the parasympathetic and sympathetic divisions of the autonomic nervous system (ANS) innervate them. Although hormones may modify the composition of saliva, their physiologic role is questionable, and it is generally believed that salivary secretion is mainly under autonomic control.

Salivary ducts

The *intercalated ducts* contain secretory granules and are involved in the secretions of proteins. The intercalated ducts are connected to the striated duct, which eventually empties into the excretory duct. The *striated duct* is lined with columnar cells. Its major function is to modify the ionic composition of the saliva. The large *excretory ducts*, lined with columnar cells, also play a role in modifying the ionic composition of saliva. Although the acinar cells synthesize and secrete most proteins, the duct cells also synthesize several proteins, such as epidermal growth factor, ribonuclease, α-amylase, and proteases.

Electrolyte composition of the saliva is modified in the striated ducts by selectively reabsorbing sodium and selectively secreting potassium and bicarbonate.

Saliva lubricates the mucosal surface, reducing the frictional damage caused by the rough surfaces of food. It helps small food particles stick together to form a bolus, which makes them easier to swallow. Moistening of the oral cavity with saliva facilitates speech. Saliva can dissolve flavorful substances, stimulating the different taste buds located on the tongue. Finally, saliva plays an important role in water intake; the sensation of dryness of the mouth as a result of low salivary secretion urges a person to drink.

The electrolyte composition of the primary secretion produced by the acinar cells and intercalated ducts resembles that of plasma. However, samples from the striated and excretory (collecting) ducts are hypotonic relative to plasma, indicating modification of the primary secretion. The modification of the primary secretion takes place in the striated ducts. There is less sodium (Na^+) and chloride (Cl^-) and more potassium (K^+) and bicarbonate (HCO_3^-) in saliva than in plasma. This is because the cells in the striated duct actively absorb Na^+ and actively secrete K^+ and HCO_3^- into the lumen (see Fig. 26.2). Cl^- ions leave the lumen either in exchange for HCO_3^- ions or by passive diffusion along the electrochemical gradient created by Na^+ absorption.

The electrolyte composition of saliva depends on the rate of secretion. As the secretion rate increases, the electrolyte composition of saliva approaches the ionic composition of plasma, but at low flow rates it differs significantly. At low secretion rates, the ductal epithelium has more time to modify and, thus, reduce the osmolality of the primary secretion, so the saliva has a much lower osmolality than plasma. The opposite is true at high secretion rates.

Although the absorption and secretion of ions may explain changes in the electrolyte composition of saliva, these processes do not explain why the osmolality of saliva is lower than that of the primary secretion of the acinar cells. Saliva is hypotonic to plasma because of a net absorption of ions by the ductal epithelium, a result of the action of a **Na^+/K^+-ATPase** in the basolateral cell membrane. The Na^+/K^+-ATPase transports three Na^+ ions out of the cell in exchange of two K^+ ions taken up by the cell. The epithelial lining of the duct is not permeable to water, so water does not follow the absorbed salt, thus resulting in a net absorption of ions.

Salivary proteins

The two major proteins present in saliva are **amylase** and **mucin**. Salivary α-amylase (ptyalin) is produced predominantly by the parotid glands, and mucin is produced mainly by the sublingual and submandibular salivary glands. Amylase catalyzes the hydrolysis of polysaccharides with α-1,4- glycosidic linkages. It is a hydrolytic enzyme involved in the digestion of starch. It is synthesized by the RER and transferred to the Golgi apparatus, where it is packaged into zymogen granules. The zymogen granules are stored at the apical region of the acinar cells and released with appropriate stimuli. Because some time usually passes before acids in the stomach can inactivate the amylase, a substantial amount of the ingested carbohydrate can be digested before reaching the duodenum. (The action of amylase is described later in this chapter.)

Mucin is the most abundant protein in saliva. The term describes a family of glycoproteins, each associated with different amounts of different sugars. Mucin is responsible for most of saliva's viscosity.

Mucus performs immunologic functions by lysing bacteria and killing HIV-infected leukocytes.

Saliva plays an important role in the general hygiene of the oral cavity. Saliva's pH is almost neutral (pH = 7.0), containing HCO_3^-, which neutralizes any acidic substance entering the oral cavity, including regurgitated gastric acid. Also present in saliva are small amounts of **muramidase**, a lysozyme that can lyse the muramic acid of certain bacteria (e.g., *Staphylococcus*); **lactoferrin**, a protein that binds iron, depriving microorganisms of a source of iron for growth; **epidermal growth factor**, which stimulates gastric mucosal growth; immunoglobulins (mainly IgA); and ABO blood group substances.

Recent research has indicated that the hypotonic property of saliva protects against certain infections. Hypotonic saliva has been shown to kill the human immunodeficiency virus (HIV)-infected mononuclear leukocytes to prevent further transmission of the virus.

Acidic food is a potent stimulus to salivary secretion.

As mentioned earlier, salivary secretion is predominantly under the control of the ANS. In the resting state, salivary secretion is low, amounting to about 30 mL/h. The submandibular glands contribute about two thirds of resting salivary secretion, the parotid glands about one fourth, and the sublingual glands the remainder. Stimulation increases the rate of salivary secretion, most notably in the parotid glands, up to 400 mL/h. The most potent stimuli for salivary secretion are acidic-tasting substances such as citric acid. Other types of stimuli that induce salivary secretion include the smell of food and chewing. Anxiety, fear, dehydration, and medications (e.g., antihistamines) inhibit secretion.

Autonomic control

Parasympathetic stimulation of the salivary glands results in increased activity of the acinar and ductal cells and increased salivary secretion. The parasympathetic nervous system plays an important role in controlling the secretion of saliva. The centers involved are located in the medulla oblongata. Preganglionic fibers from the inferior salivatory nucleus are contained in cranial nerve IX and synapse in the otic ganglion. They send postganglionic fibers to the parotid glands. Preganglionic fibers from the superior salivatory nucleus course with cranial nerve VII and synapse in the submandibular ganglion. They send postganglionic fibers to the submandibular and sublingual glands.

Blood flow is low in resting salivary glands and can increase as much as 10-fold when salivary secretion is stimulated. This increase in blood flow is also under parasympathetic control. Parasympathetic stimulation induces the acinar cells to release the serine protease **kallikrein**, which acts on a plasma globulin, **kininogen** (endogenous peptides present in body fluid), to release **lysyl-bradykinin (kallidin)**, which causes dilation of the blood vessels supplying the salivary glands (Fig. 26.3). Atropine, an anticholinergic agent, is a potent inhibitor of salivary secretion. Agents that inhibit

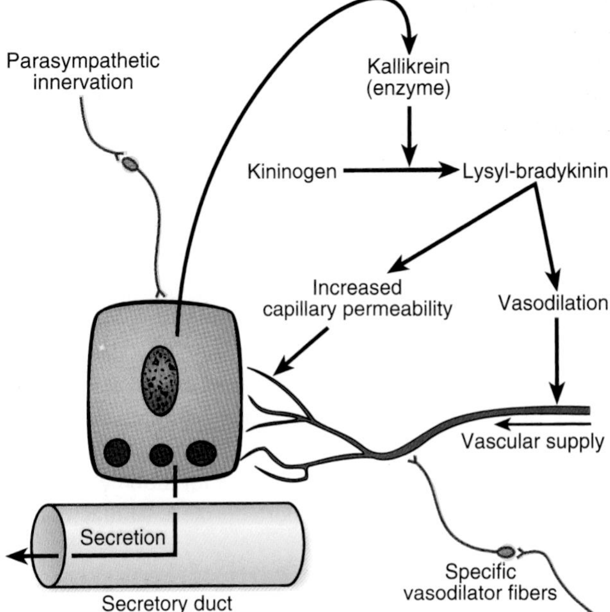

Figure 26.3 Blood flow in the salivary glands is controlled by the parasympathetic nervous system. The parasympathetic nervous system stimulates bradykinin release, which causes vasodilation and augments blood flow.

acetylcholinesterase (e.g., pilocarpine) enhance salivary secretion. Some parasympathetic stimulation also increases blood flow to the salivary glands directly, apparently via the release of the neurotransmitter **vasoactive intestinal peptide (VIP)**.

The sympathetic nervous system also innervates the salivary glands. Sympathetic fibers arise in the upper thoracic segments of the spinal cord and synapse in the **superior cervical ganglion**. Postganglionic fibers leave the superior cervical ganglion and innervate the acini, ducts, and blood vessels. Sympathetic stimulation tends to result in a short-lived and much smaller increase in salivary secretion than parasympathetic stimulation. The increase in salivary secretion observed during sympathetic stimulation is mainly via β-adrenergic receptors, which are more involved in stimulating the contraction of myoepithelial cells to increase salivary flow. Although both sympathetic and parasympathetic stimulation increase salivary secretion, the responses produced are different. Parasympathetic stimulation increases saliva secretion, which is rich in electrolytes and salivary amylase. In contrast, sympathetic stimulation, which also increases saliva secretion, although to a lesser extent, increases mucus secretion, making the saliva much more viscous.

Mineralocorticoid administration reduces the Na^+ concentration of saliva with a corresponding rise in K^+ concentration. Mineralocorticoids act mainly on the striated and excretory ducts. Arginine vasopressin, also known as vasopressin, reduces the Na^+ concentration in saliva by increasing Na^+ reabsorption by the ducts. Some GI hormones (e.g., VIP and substance P) have been experimentally demonstrated to evoke salivary secretory responses.

GASTRIC SECRETION

The stomach is a muscular, hollow, dilated part of the GI tract that functions as the primary storage organ. In addition to storage, the stomach secretes proteolytic enzymes and strong acids that initiate the digestion of foods. The stomach also absorbs water-soluble and lipid-soluble substances (e.g., alcohol and some drugs). An important function of the stomach is to prepare the **chyme** for digestion in the small intestine. Chyme is the semifluid material produced by the gastric digestion of food. Chyme results partly from the conversion of large solid particles into smaller particles via the combined peristaltic movements of the stomach and contraction of the pyloric sphincter. The propulsive, grinding, and retropulsive movements associated with antral peristalsis are discussed in Chapter 25, "Neurogastroenterology and Motility." A combination of the squirting of antral content into the duodenum, the grinding action of the antrum, and retropulsion provides much of the mechanical action necessary for the emulsification of dietary fat, which plays an important role in fat digestion.

Mucus and bicarbonate are secreted by epithelial cells to protect the stomach from the acidic condition in the lumen.

The stomach wall consists of four distinct layers (from inside to outside): (1) **gastric mucosa**, (2) **submucosa**, (3) **muscularis externa**, and (4) **serosa**. The mucosa layer consists of epithelium that houses the secretory glands. The submucosa consists of fibrous connective tissue that separates the mucosa from the muscularis externa. The smooth muscle layer of the muscularis externa of the stomach differs from other portions of the GI tract in that it contains three layers of smooth muscle instead of two. The serosa is the outer layer of the stomach consisting of connective tissue and is continuous with the peritoneum.

Epithelial mucosa

The epithelium of the stomach's mucosa form deep pits and contains two main types of glands: **pyloric** and **oxyntic** (Fig. 26.4). The **pyloric glands** secrete gastrin and mucus for protection, and are located in the distal 20% of the stomach, the antral portion. They contain cells similar to mucous neck cells but differ from oxyntic glands in having many gastrin-producing cells called **G cells**. The **oxyntic glands**, sometimes referred to as *gastric glands*, the most abundant glands in the stomach, secrete mainly hydrochloric acid and pepsinogen and are found in the fundus and the corpus, which occupies 80% of the stomach.

Parietal cells of the oxyntic glands are the only cells that secrete hydrochloric acid.

The oxyntic glands contain **parietal (oxyntic) cells, chief cells, mucous neck cells**, and some endocrine cells (see Fig. 26.4).

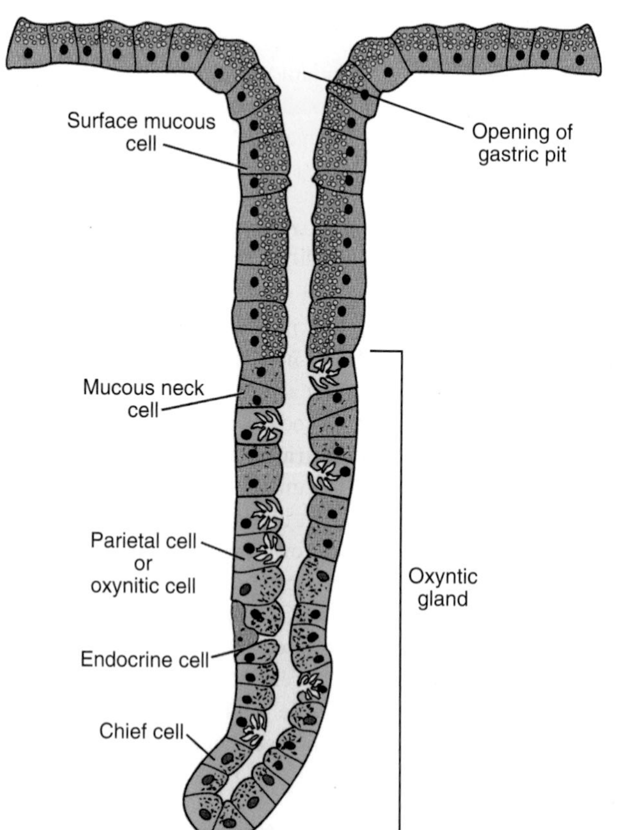

Figure 26.4 Different cell types contribute to gastric secretion. The oxyntic gland in the corpus of a mammalian stomach is illustrated in this diagram. The glands contain parietal (oxyntic) cells that open into a common pit.

Parietal cells are the most distinctive cells in the stomach. Most of the mucous cells are located in the neck region, and the base of the oxyntic gland contains mostly chief cells, along with some parietal and endocrine cells. Mucous neck cells secrete mucus, parietal cells principally secrete **hydrochloric acid (HCl)** and **intrinsic factor**, and chief cells secrete **pepsinogen**. (Intrinsic factor and pepsinogen are discussed later in this chapter.)

HCl is secreted across the parietal cell microvillar membrane and flows out of the intracellular canaliculi into the oxyntic gland lumen (Fig. 26.5). The capacity of the stomach to secrete HCl is proportionate to parietal cell number. As mentioned, surface mucous cells line the entire surface of the gastric mucosa and the openings of the cardiac, pyloric, and oxyntic glands. These cells secrete mucus and HCO_3^- to protect the gastric surface from the acidic environment of the stomach. The distinguishing characteristic of a surface mucous cell is the presence of numerous mucus granules at its apex. The number of mucus granules in storage varies depending on synthesis and secretion. The mucous neck cells of the oxyntic glands are similar in appearance to surface mucous cells.

Chief cells are morphologically distinguished primarily by the presence of zymogen granules in the apical region and an extensive RER. The zymogen granules contain pepsinogen, a precursor of the enzyme pepsin.

Also present in the stomach are various neuroendocrine cells, such as G cells, located predominantly in the antrum. These cells produce the hormone **gastrin**, which stimulates acid secretion by the stomach. An overabundance of gastrin secretion, a condition known as *Zollinger-Ellison syndrome*, results in gastric hypersecretion and peptic ulceration. In most cases, the tumors arise within the pancreas and/or the duodenum. **D cells**, also present in the antrum, produce **somatostatin**, another important GI hormone.

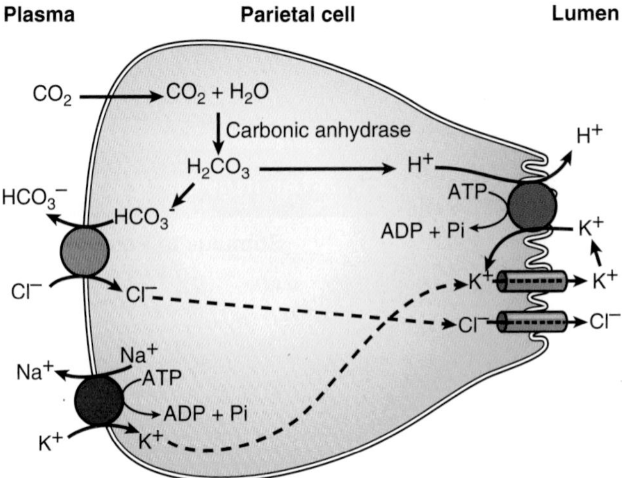

Figure 26.5 Parietal cells secrete gastric acid. The mechanism of HCl secretion is depicted in the above schematic. ATPase in the apical cell membrane pumps H^+ out of the cell in exchange for K^+. ADP, adenosine diphosphate; ATP, adenosine triphosphate; Pi, inorganic phosphate.

Mucous gel layer

The important constituents of human gastric juice are HCl, electrolytes, pepsinogen, and intrinsic factor. The pH is low, about 0.7 to 3.8. This raises a question: How does the gastric mucosa protect itself from acidity? As mentioned earlier, the surface mucous cells secrete a fluid containing mucus and HCO_3^- ions. The mucus forms a **mucous gel layer** covering the surface of the gastric mucosa. Bicarbonate trapped in the mucous gel layer neutralizes acid, preventing damage to the mucosal cell surface.

H^+/K^+-ATPase pumps hydrogen ions out of the parietal cell in exchange for potassium to form hydrochloric acid.

The parietal cells of the corpus and fundus secrete the HCl present in the gastric lumen. The mechanism of HCl production is depicted in Figure 26.5. An **H^+/K^+-ATPase** in the apical (luminal) cell membrane of the parietal cell actively pumps H^+ out of the cell in exchange for K^+ entering the cell. Omeprazole inhibits the H^+/K^+-ATPase. Omeprazole, an acid-activated prodrug that is converted in the stomach to the active drug, binds to two cysteines on the ATPase, resulting in an irreversible inactivation. Omeprazole is used to treat ulcers, gastroesophageal reflux disease (GERD) (often associated with heartburn), and other conditions resulting from acid production by the stomach. Although the secreted H^+ is often depicted as being derived from carbonic acid (see Fig. 26.5), the source of H^+ is probably mostly the dissociation of H_2O. Carbon dioxide (CO_2) and H_2O form carbonic acid (H_2CO_3) in a reaction catalyzed by carbonic anhydrase. Acetazolamide inhibits carbonic anhydrase. Metabolic sources inside the cell and from the blood provide the CO_2.

For the H^+/K^+-ATPase to work, an adequate supply of K^+ ions must exist outside the cell. Although the mechanism is still unclear, there is an increase in K^+ conductance (through K^+ channels) in the apical membrane of the parietal cells simultaneous with active acid secretion. This surge of K^+ conductance ensures plenty of K^+ in the lumen. The H^+/K^+-ATPase recycles K^+ ions back into the cell in exchange for H^+ ions. As shown in Figure 26.6, the basolateral cell membrane has an electroneutral Cl^-/HCO_3^- exchanger that balances the entry of Cl^- into the cell with an equal amount of HCO_3^- entering the bloodstream. The Cl^- inside the cell then leaks into the lumen through Cl^- channels, down an electrochemical gradient. Consequently, HCl is secreted into the lumen.

The parietal cells can secrete a large amount of HCl. This is balanced by an equal amount of HCO_3^- added to the bloodstream. The blood coming from the stomach during active acid secretion contains much HCO_3^-, a phenomenon called the **alkaline tide**. The osmotic gradient created by the HCl concentration in the gland lumen drives water passively into the lumen, thereby maintaining the iso-osmolality of the gastric secretion.

Gastric secretion occurs in three phases.

Gastric secretion resulting from the ingestion of food can be divided into three phases: the cephalic phase, the gastric phase, and the intestinal phase (Table 26.1). These phases

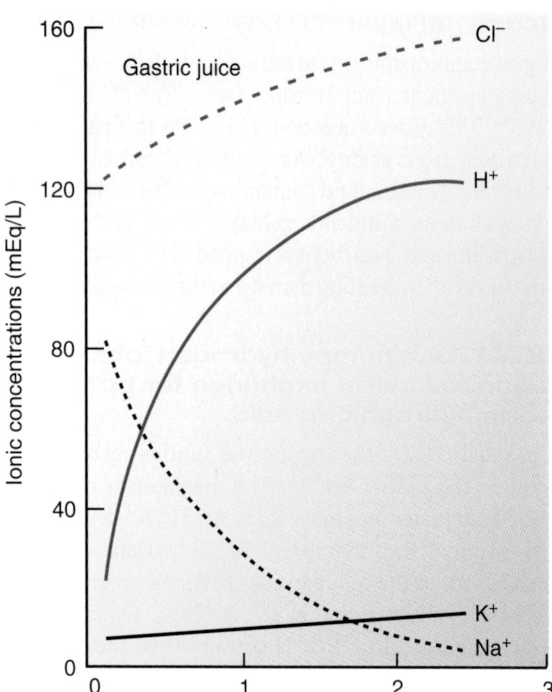

Figure 26.6 Rate of secretion affects the composition of the gastric juice. The concentration of electrolytes in the gastric juice is depicted in a healthy young adult male. The K^+ and H^+ ion concentration parallels the secretion rates.

are named accordingly to where the stimuli originates. The **cephalic phase** involves the central nervous system. Smelling, chewing, and swallowing food (or merely the thought of food) sends impulses via the vagus nerves to the parietal and G cells in the stomach. The nerve endings release **acetylcholine (ACh)**, which directly stimulates acid secretion from parietal cells. The nerves also release **gastrin-releasing peptide**, which stimulates G cells to release gastrin, indirectly stimulating parietal cell acid secretion. The cephalic phase probably accounts for about 40% of total acid secretion.

The **gastric phase** is mainly a result of gastric distention and chemical agents such as digested proteins. Distention of the stomach stimulates mechanoreceptors, which stimulate the parietal cells directly through short local (enteric) reflexes and by long **vagovagal reflexes**. Afferent and efferent impulses traveling in the vagus nerves mediate vagovagal reflexes. Digested proteins in the stomach are also potent stimulators of gastric acid secretion, an effect mediated through gastrin release. Several other chemicals, such as alcohol and caffeine, stimulate gastric acid secretion through mechanisms that are not well understood. The stimulation of acid secretion by alcohol appears to be related to its alcohol content—beverages lower in alcohol content, such as beer, tend to stimulate acid secretion more effectively than beverages with higher alcohol content. The gastric phase accounts for about 50% of total gastric acid secretion.

During the **intestinal phase**, protein digestion products in the duodenum stimulate gastric acid secretion through the action of the circulating amino acids on the parietal cells. Distention of the small intestine, probably via the release of the hormone **entero-oxyntin** from intestinal endocrine cells, stimulates acid secretion. The intestinal phase accounts for only about 10% of the total gastric acid secretion.

Acid production in the stomach parallels rates of gastric secretion.

Figure 26.6 depicts the changes in the electrolyte composition of gastric juice at different secretion rates. At a low secretion rate, gastric juice contains high concentrations of Na^+ and Cl^- and low concentrations of K^+ and H^+. When the rate of secretion increases, the concentration of Na^+ decreases whereas that of H^+ increases significantly. Also coupled with this increase in gastric secretion is an increase in Cl^- concentration. To understand the changes in electrolyte composition of gastric juice at different secretion rates, remember that gastric juice is derived from the secretions of two major sources: parietal cells and nonparietal cells. Secretion from nonparietal cells is probably constant; therefore, it is parietal secretion (HCl secretion) that contributes mainly to the changes in electrolyte composition with higher secretion rates.

TABLE 26.1 The Three Phases of Stimulation of Acid Secretion after Ingesting a Meal

Phase	Stimulus	Pathway	Stimulus to Parietal Cell
Cephalic	Thought of food, smell, taste, chewing, and swallowing	Vagus nerve to: Parietal cells G cells	ACh Gastrin
Gastric	Stomach distention	Local (enteric) reflexes and vagovagal reflexes to: Parietal cells G cells	ACh Gastrin
Intestinal	Protein digestion products in duodenum Distention	Amino acids in blood Intestinal endocrine cell	Amino acids Entero-oxyntin

ACh, acetylcholine.

Pepsin is the main gastric enzyme in protein digestion.

The enzymes that get secreted in the stomach are referred to as *gastric enzymes* and include **pepsinogens, pepsins, gastric amylase**, **gastric lipase**, and **intrinsic factor**. Pepsin, an **endopeptidase**, is the main gastric enzyme that cleaves proteins into smaller peptides. The optimal pH for pepsin activity is 1.8 to 3.5; therefore, it is extremely active in the highly acidic medium of gastric juice. The chief cells of the oxyntic glands release inactive pepsinogen. Acid in the gastric lumen activates pepsinogen to form the active enzyme **pepsin**. Pepsin also catalyzes its own formation from pepsinogen.

Gastric amylase degrades starch but otherwise appears to be of minor significance. Gastric lipase acts almost exclusively on butterfat. The intrinsic factor, produced by stomach parietal cells, is necessary for the absorption of vitamin B_{12} in the terminal ileum.

Gastric secretion is under neural and hormonal control.

Gastric secretion is mediated through neural and hormonal pathways. Vagus nerve stimulation is the neural effector; histamine and gastrin are the hormonal effectors (Fig. 26.7). Parietal cells possess special histamine receptors, **H_2 receptors**, whose stimulation results in increased acid secretion. Special neuroendocrine cells of the stomach, known as **enterochromaffin-like cells**, are believed to be the source of this histamine. They are located mostly in the acid-secreting regions of the stomach. The mechanisms that stimulate the enterochromaffin-like cells to release histamine are poorly understood. The effectiveness of cimetidine, an H_2 blocker, in reducing acid secretion has indirectly demonstrated the importance of histamine as an effector of gastric acid secretion. H_2 blockers are commonly used for the treatment of peptic ulcer disease or GERD. Ranitidine, a longer-acting H_2-receptor antagonist with fewer side effects, has largely replaced cimetidine.

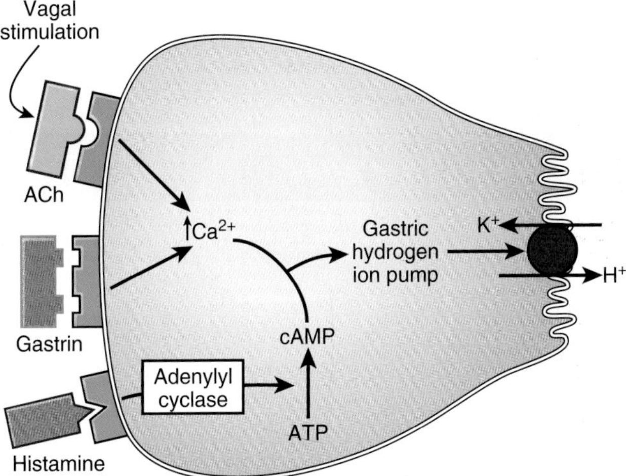

Figure 26.7 The parietal cell is under neural and hormonal control. Histamine, gastrin, and acetylcholine (ACh) stimulate acid secretion. cAMP, cyclic adenosine monophosphate; ATP, adenosine triphosphate.

The effects of each of these three stimulants (ACh, gastrin, and histamine) augment those of the others, a phenomenon known as **potentiation**. Potentiation is said to occur when the effect of two stimulants is greater than the effect of either stimulant alone. For example, the interaction of gastrin and ACh molecules with their respective receptors results in an increase in intracellular Ca^{2+} concentration, and the interaction of histamine with its receptor results in an increase in cellular **cyclic adenosine monophosphate (cAMP)** production. The increased intracellular Ca^{2+} and cAMP interact in numerous ways to stimulate the gastric H^+/K^+-ATPase, which increases acid secretion (see Fig. 26.7). Exactly how the increase in intracellular Ca^{2+} and cAMP greatly enhances the effect of the other in stimulating gastric acid secretion is not well understood.

Acid secretion is inhibited by gastric hormones.

The inhibition of gastric acid secretion is physiologically important for two reasons. First, the secretion of acid is important only during the digestion of food. Second, excess acid can damage the gastric and the duodenal mucosal surfaces, causing ulcerative conditions. The body has an elaborate system for regulating the amount of acid secreted by the stomach. Gastric luminal pH is a sensitive regulator of acid secretion. Proteins in food provide buffering in the lumen; consequently, the gastric luminal pH is usually above 3 after a meal. However, if the buffering capacity of protein is exceeded or if the stomach is empty, the pH of the gastric lumen will fall below 3. When this happens, the endocrine cells (D cells) in the antrum secrete somatostatin, which inhibits the release of gastrin and, thus, gastric acid secretion.

Another mechanism for inhibiting gastric acid secretion is acidification of the duodenal lumen. Acidification stimulates the release of **secretin**, which inhibits the release of gastrin and several peptides, collectively known as **enterogastrones**, which are released by intestinal endocrine cells. Acid, fatty acids, and hyperosmolar solutions in the duodenum stimulate the release of enterogastrones, which inhibit gastric acid secretion. **Gastric inhibitory peptide**, an enterogastrone produced by the small intestinal endocrine cells, inhibits parietal cell acid secretion. There are also several currently unidentified enterogastrones.

PANCREATIC SECRETION

The pancreas is located adjacent to the duodenum and functions as both an endocrine and exocrine gland (Fig. 26.8). The *endocrine* function produces several important hormones including insulin, glucagon, and somatostatin. (The endocrine function is further discussed in Chapter 34, "Endocrine Pancreas.") The *exocrine* gland secretes pancreatic digestive enzymes that pass to the small intestine and break down carbohydrates, protein, and fat in the chyme. The pancreatic enzymes responsible for the digestion of

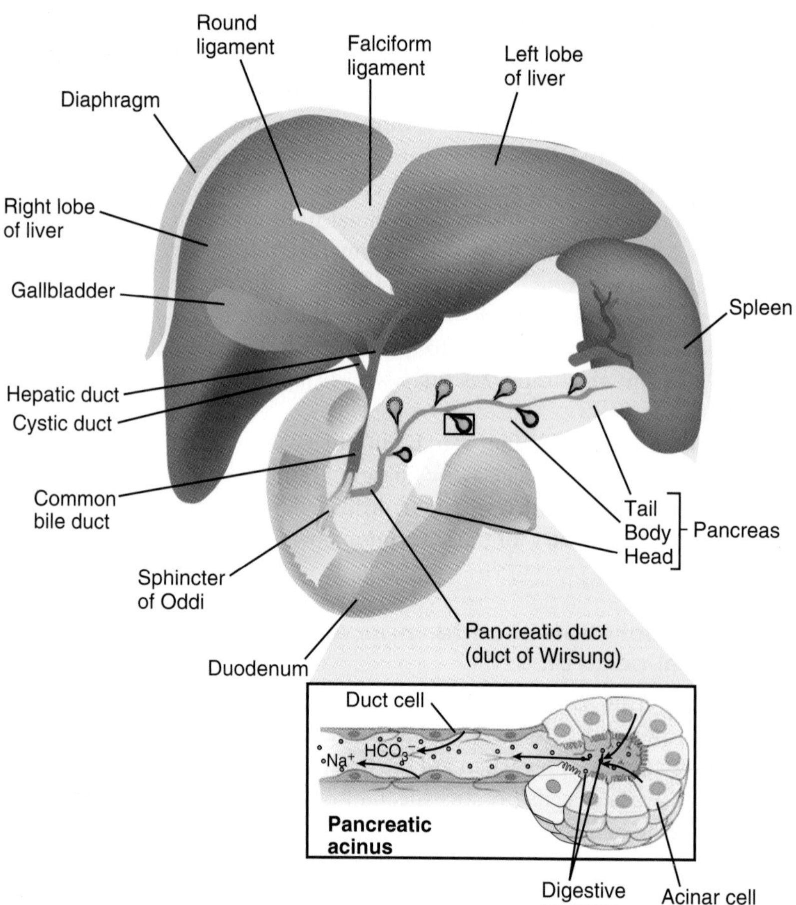

Figure 26.8 The pancreas performs both endocrine and exocrine functions. The exocrine cells of the pancreas secrete digestive juices into the duodenal lumen. Digestive enzymes are secreted by acinar cells (*inset*), and an aqueous $NaHCO_3$ solution is secreted by duct cells.

proteins are **trypsin** (by far the most abundant), **chymotrypsin**, and **carboxypolpeptidase**. Trypsin and chymotrypsin split proteins into polypeptides, and carboxypolpeptidase splits the polypeptides into amino acids. The pancreatic enzyme responsible for digesting carbohydrates is **pancreatic amylase**, and the main enzyme for fat digestion is **pancreatic lipase**. Table 26.2 summarizes the various enzymes present in pancreatic juice. Some are secreted as proenzymes, which are activated in the duodenal lumen to form the active enzymes. (The digestion of nutrients by these enzymes is discussed later in this chapter.)

Pancreatic enzymes are produced in the acinar cells.

One of the major functions of pancreatic secretion is to neutralize the acids in the chyme when it enters the duodenum from the stomach. This mechanism is important because pancreatic enzymes operate optimally near the neutral pH. Another important function is the production of enzymes involved in the digestion of dietary carbohydrate, fat, and protein. The exocrine pancreas is composed of numerous small, sac-like dilatations called *acini*, which are composed of a single layer of pyramidal **acinar cells** (see Fig. 26.8). These

TABLE 26.2 The Pancreas is the Main Digestive Gland of the Body

Enzyme Secreted	Hydrolytic Action
Trypsin	Protease that breaks down proteins at the basic amino acids
Chymotrypsin	Protease that breaks down proteins at the aromatic amino acids
Lipase	Degrades triglycerides into fatty acids and glycerol
Carboxypeptidase	Protease that takes off the terminal acid group from a protein
Elastases	Degrade the protein elastin and some other proteins
Nucleases	Degrade nucleic acids, like DNAase and RNAase
Pancreatic amylase	Besides starch and glycogen, degrades most other carbohydrates; humans lack the enzyme to digest the carbohydrate cellulose

cells are actively involved in the production of enzymes. Their cytoplasm is filled with an elaborate system of RERs and a Golgi apparatus. Zymogen granules are observed in the apical region of acinar cells. A few **centroacinar cells** line the lumen of the acinus. In contrast to acinar cells, these cells lack elaborate RER and Golgi apparatus systems. Their major function seems to be modification of the electrolyte composition of the pancreatic secretion. Because the processes involved in the secretion or uptake of ions are active, centroacinar cells have numerous mitochondria in their cytoplasm.

The acini empty their secretions into intercalated ducts, which join to form intralobular and then interlobular ducts. The interlobular ducts empty into two pancreatic ducts: a major duct, the duct of Wirsung, and a minor duct, the duct of Santorini. The duct of Santorini enters the duodenum more proximally than the duct of Wirsung, which enters the duodenum usually together with the common bile duct. A ring of smooth muscle, the **sphincter of Oddi**, surrounds the opening of these ducts in the duodenum. The sphincter of Oddi not only regulates the flow of bile and pancreatic juice into the duodenum but also prevents the reflux of intestinal contents into the pancreatic ducts.

Pancreatic secretions are rich in bicarbonate ions.

Although pancreatic enzymes are secreted for the digestion of food, two other important components of pancreatic juices, bicarbonate and water, are secreted to neutralize the HCl that is emptied into the duodenum from the stomach. The pancreas secretes about 1 L a day of HCO_3^--rich fluid. The osmolality of pancreatic fluid, unlike that of saliva, is equal to that of plasma at all secretion rates. The Na^+ and K^+ concentrations of pancreatic juice are the same as those in plasma, but unlike plasma, pancreatic juice is enriched with HCO_3^- and has a relatively low Cl^- concentration (Fig. 26.9). When the pancreas is stimulated, bicarbonate ion concentration can increase about five times that of plasma, reaching a maximal concentration of about 140 mEq/L, with a pH of 8.2. A reciprocal relationship exists between the Cl^- and HCO_3^- concentrations in pancreatic juice. As the concentration of HCO_3^- increases with secretion rate, the Cl^- concentration falls accordingly, resulting in a combined total anion concentration that remains relatively constant (150 mEq/L) regardless of the pancreatic secretion rate.

Two separate mechanisms have been proposed to explain the secretion of a HCO_3^--rich juice by the pancreas and the HCO_3^- concentration changes. The first mechanism proposes that some cells, probably the acinar cells, secrete a plasma-like fluid containing predominantly Na^+ and Cl^-, whereas other cells, probably the centroacinar and duct cells, secrete a HCO_3^--rich solution when stimulated. Depending on the different rates of secretion from these three different cell types, the pancreatic juice can be rich in either HCO_3^- or Cl^-. The second mechanism depicts the primary secretion as rich in HCO_3^-. As the HCO_3^- solution moves down the ductal system, HCO_3^- ions are exchanged for Cl^- ions. When

Clinical Focus / 26.1

Peptic Ulcers: When Bacteria Break the Barrier

Peptic ulcers, also known as **peptic ulcer disease**, are erosions of the mucosal lining of the stomach. Ulcerative lesions occur when the mucus barrier is disrupted, and pepsin and hydrochloric acid attack the stomach lining instead of digesting the food. Peptic ulcers affect approximately 4 million patients in the United States, where duodenal ulcers predominate. In Japan, however, gastric ulcers are more prevalent.

Symptoms include bloating, fullness, and abdominal pain. The symptoms are exacerbated with the ingestion of food (intensifies 3 hours after eating). In contrast, duodenal ulcers are classically relieved by food. If gastric ulcers go untreated, symptoms can escalate to include nausea and vomiting of blood that is accompanied by loss of appetite and weight loss. Many patients confuse heartburn with gastric ulcers. Heartburn, a burning sensation in the chest, is not associated with gastric ulcers. Heartburn is associated with regurgitation of gastric acid (gastric reflux), which is a major symptom of gastroesophageal reflux disease. Ulcers can also be caused or worsened by drugs such as aspirin or other nonsteroidal anti-inflammatory drugs (NSAIDs) that inhibit cyclooxygenase and most glucocorticoids (e.g., dexamethasone and prednisolone) for the treatment of arthritis. NSAIDs interfere with prostaglandin-mediated defense mechanisms against gastric acidity, mucus production, and bicarbonate secretion. Gastrointestinal bleeding is the most common complication. Sudden large bleeding can be life threatening and occurs when an ulcer erodes a blood vessel.

In the 1990s, a surprising discovery was made in the field of peptic ulcer disease linking *Helicobacter pylori* infection to the cause of gastric and duodenal ulcers. As many as 70% to 90% of peptic ulcers are linked to *H. pylori*, a spiral-shaped bacterium that lives in the acidic environment of the stomach. *H. pylori* appears to be protected by producing large quantities of urease, which hydrolyzes urea to produce ammonia. The ammonia neutralizes acid in the gastric lumen, thus protecting the bacteria from the injurious effects of HCl. The underling mechanism that causes ulcerations appears to be the release of bacterial toxins that leads to chronic inflammation of the gastric mucosal lining. The immune system appears to be unable to clear the infection, despite the appearance of antibodies.

The investigator, Dr. Barry Marshall, who discovered this link, was awarded the Nobel Prize for Medicine and Physiology in 2005 for his pioneering discovery.

Recently, it has been demonstrated that H_2 receptor-antagonists (i.e., cimetidine and ranitidine) have no effect on *H. pylori* infection. In contrast, omeprazole (an inhibitor of the H^+/K^+-ATPase) appears to be bacteriostatic. A combined therapy using omeprazole and the antibiotic amoxicillin appears to be quite effective in the eradication of *H. pylori* in 50% to 80% of patients with peptic ulcer disease, resulting in a significant reduction of duodenal ulcer recurrence.

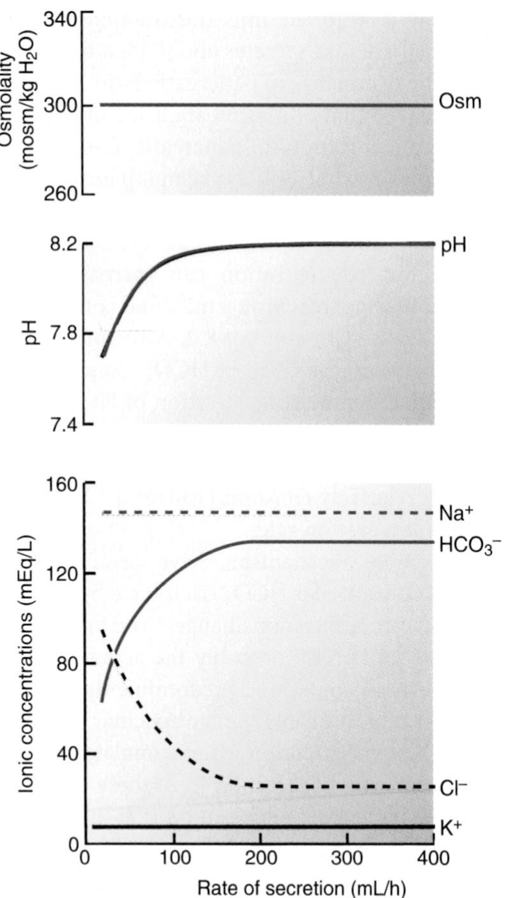

Figure 26.9 Pancreatic pH and electrolytes are altered with flow. Pancreatic fluid is rich in bicarbonate ions. The HCO_3^- concentration and, hence, the pH parallel the increased rates of secretion. *Note*: Na^+, K^+, and pancreatic osmolality are independent of flow.

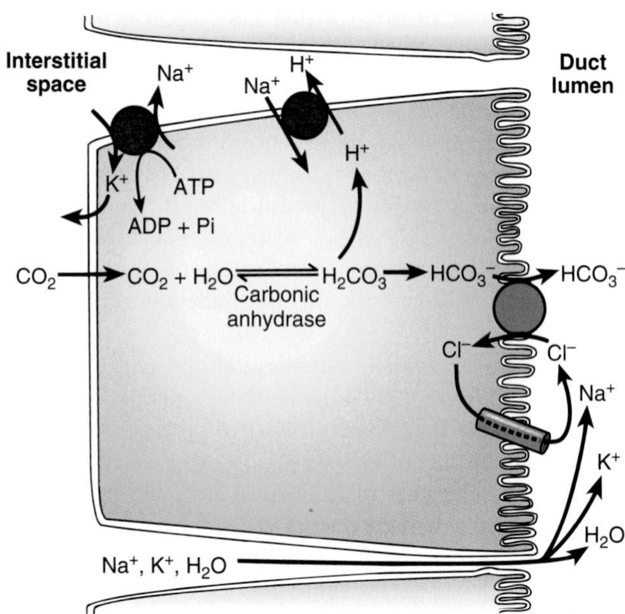

Figure 26.10 The pancreatic duct supplies the pancreatic juices for digestion. The above model depicts the mechanism for electrolyte secretion by pancreatic duct cells. The luminal membrane Cl^- channel is CFTR (cystic fibrosis transmembrane conductance regulator). ADP, adenosine diphosphate; ATP, adenosine triphosphate; Pi, inorganic phosphate.

the flow is fast, there is little time for this exchange, so the concentration of HCO_3^- is high. The opposite is true when the flow is slow.

The secretion of electrolytes by pancreatic duct cells is depicted in Figure 26.10. A Na^+/H^+ exchanger is located in the basolateral cell membrane. The energy required to drive the exchanger is provided by the Na^+/K^+-ATPase-generated Na^+ gradient. Carbon dioxide diffuses into the cell and combines with H_2O to form H_2CO_3, a reaction catalyzed by carbonic anhydrase, which dissociates to H^+ and HCO_3^-. The Na^+/H^+ exchanger extrudes the H^+, and HCO_3^- is exchanged for luminal Cl^- via a Cl^-/HCO_3^- exchanger. Also located in the luminal cell membrane is a protein called **cystic fibrosis transmembrane conductance regulator**. It is an ion channel belonging to the ABC (**adenosine triphosphate [ATP]**-binding cassette) family of proteins. Regulated by ATP, its major function is to secrete Cl^- ions out of the cells, providing Cl^- in the lumen for the Cl^-/HCO_3^- exchanger to work. The Na^+/K^+-ATPase removes cell Na^+ that enters through the Na^+/H^+ antiporter. Sodium from the interstitial space follows secreted HCO_3^- by diffusing through a paracellular path (between the cells). Movement of H_2O into the duct lumen is passive, driven by the osmotic gradient. The net result of pancreatic HCO_3^- secretion is the release of H^+ into the plasma; thus, pancreatic secretion is associated with an **acid tide** in the plasma.

Pancreatic enzyme secretion is under neural and hormonal control.

Pancreatic secretion, like gastric secretion, occurs in three phases: (1) cephalic, (2) gastric, and (3) the intestinal phase. Three basic stimuli are involved in the phases of pancreatic secretion and include (1) **ACh**, (2) **CCK**, and (3) **secretin**. Parasympathetic fibers in the vagus nerve that release ACh simulate pancreatic secretion. Stimulation of the vagus nerve results predominantly in an increase in enzyme secretion—fluid and HCO_3^- secretion are marginally stimulated or unchanged. Sympathetic nerve fibers mainly innervate the blood vessels supplying the pancreas, causing vasoconstriction. Stimulation of the sympathetic nerves neither stimulates nor inhibits pancreatic secretion, probably because of the reduction in blood flow.

Circulating GI hormones, particularly secretin and **CCK**, greatly influence the secretion of electrolytes and enzymes by the pancreas. Secretin tends to stimulate an HCO_3^--rich secretion. CCK stimulates a marked increase in enzyme secretion. The small intestine produces both hormones, and the pancreas has receptors for them.

Structurally similar hormones have effects similar to those of secretin and CCK. For example, VIP, structurally similar to secretin, stimulates the secretion of HCO_3^- and H_2O. However, because VIP is much weaker than secretin, it produces a weaker pancreatic response when given together

TABLE 26.3 Factors Regulating Pancreatic Secretion after a Meal

Phase	Stimulus	Mediators	Response
Cephalic	Thought of food, smell, taste chewing, and swallowing	Release of ACh and gastrin by vagal stimulation	Increased secretion, with greater effect on enzyme output
Gastric	Protein in food	Gastrin	Increased secretion, with greater effect on enzyme output
	Gastric distention	ACh release by vagal stimulation	Increased secretion, with a greater effect on enzyme output
Intestinal	Acid in chyme	Secretin	Increased H_2O and HCO_3^- secretion
	Long-chain fatty acids	CCK and vagovagal reflex	Increased secretion, with greater effect on enzyme output
	Amino acids and peptides	CCK and vagovagal reflex	Increased secretion, with greater effect on enzyme output

ACh, acetylcholine; CCK, cholecystokinin.

with secretin than when secretin is given alone. Similarly, gastrin can stimulate pancreatic enzyme secretion because of its structural similarity to CCK, but unlike CCK, it is a weak agonist for pancreatic enzyme secretion.

Table 26.3 summarizes the regulation of pancreatic secretion by various hormonal and neural factors. Seeing, smelling, tasting, chewing, swallowing, or thinking about food results in the secretion of a pancreatic juice rich in enzymes. In this cephalic phase, stimulation of pancreatic secretion is mainly mediated by direct efferent impulses sent by vagal centers in the brain to the pancreas and, to a minor extent, by the indirect effect of parasympathetic stimulation of gastrin release. The gastric phase is initiated when food enters the stomach and distends it. The vagovagal reflex then stimulates pancreatic secretion. Gastrin may also be involved in this phase.

During the most important phase, the intestinal phase, the entry of acidic chyme from the stomach into the small intestine stimulates the release of secretin by the **S cells** (a type of endocrine cell) in the intestinal mucosa. The secretin concentration in the plasma increases when the pH of the lumen in the duodenum decreases. An increase in HCO_3^- output by the pancreas follows this response. Circulating CCK and parasympathetic stimulation increase the secretion of pancreatic enzymes through a vagovagal reflex. Exposure of the intestinal mucosa to long-chain fatty acids (lipid digestion products) and free amino acids stimulates the release of CCK by the **I cells** (a type of endocrine cell) in the intestinal mucosa.

Potentiation, as previously described for gastric secretion, also exists in the pancreas. Its effect in pancreatic secretion is a result of the different receptors used for ACh, CCK, and secretin. Secretin binding triggers an increase in adenylyl cyclase activity, which, in turn, stimulates the formation of cAMP (Fig. 26.11). ACh, CCK, and the neuropeptides gastrin-releasing peptide and substance P bind to their respective receptors and trigger the release of Ca^{2+} from intracellular stores. The increase in intracellular Ca^{2+} release and cAMP formation results in an increase in pancreatic enzyme secretion. The mechanism by which this takes place is not well understood.

BILIARY SECRETION

Bile is a dark green to yellowish brown fluid produced by the **hepatocytes** in the liver, draining through many bile ducts that penetrate the liver (see Fig. 26.8). Bile facilitates the digestion of lipids in the small intestine by emulsifying fat and forming aggregates around fat droplets called **micelles**. The dispersion of fat droplets into micelles greatly

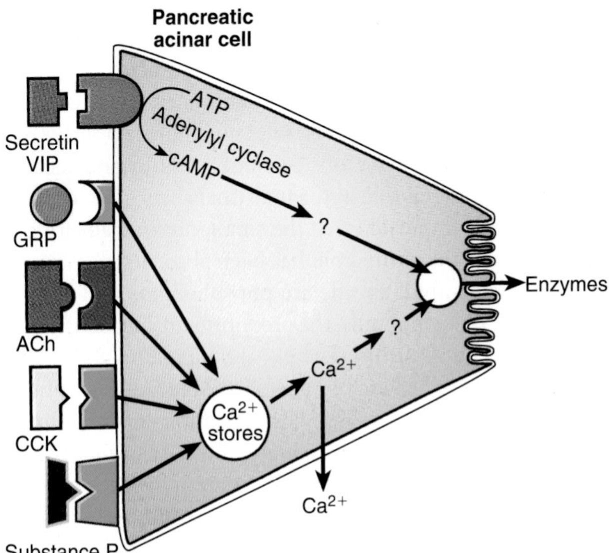

Figure 26.11 Calcium release from intracellular stores stimulates pancreatic enzyme secretion. This diagram illustrates the mechanisms involved in pancreatic secretion by hormones and neurotransmitters. ACh, acetylcholine; ATP, adenosine triphosphate; cAMP, cyclic adenosine monophosphate; CCK, cholecystokinin; GRP, gastrin-releasing peptide; VIP, vasoactive intestinal peptide.

increases the surface area, which acts to enhance the action of pancreatic lipase. Bile is stored in the gallbladder and, with eating, is discharged into the duodenum. The human liver can produce close to 1 L of bile a day (depending on body size). Bile contains bile salts, bile pigments (e.g., bilirubin), cholesterol, phospholipids, and proteins and performs several important functions. For example, bile salts play an important role in the intestinal absorption of lipid. Bile salts are derived from cholesterol and, therefore, constitute a path for its excretion. Biliary secretion is an important route for the excretion of bilirubin from the body (see Fig. 26.8).

Major component of cholesterol metabolism in the liver involves the formation of bile acids.

The electrolyte composition of human bile collected from the hepatic ducts is similar to that of blood plasma, except that the HCO_3^- concentration may be higher, resulting in an alkaline pH. **Bile acids** are formed in the liver by the cytochrome P450 oxidation of cholesterol.

During the conversion, hydroxyl groups and a carboxyl group are added to the steroid nucleus. Bile acids are stored in the gall bladder and are concentrated by removing water. Bile acids are classified as primary or secondary. The hepatocytes synthesize the *primary bile acids*, which include **cholic acid** and **chenodeoxycholic acid**. When bile enters the GI tract, bacteria present in the lumen act on the primary bile acids and convert them to *secondary bile acids* by dehydroxylation. Cholic acid is converted to **deoxycholic acid** and chenodeoxycholic acid to **lithocholic acid**.

At a neutral pH, the bile acids are mostly ionized and are referred to as **bile salts**. Conjugated bile acids ionize more readily than the unconjugated bile acids and, thus, usually exist as salts of various cations (e.g., sodium glycocholate). Bile salts are much more polar than bile acids and have greater difficulty penetrating cell membranes. Consequently, the small intestine absorbs bile salts much more poorly than bile acids. This property of bile salts is important because they play an integral role in the intestinal absorption of lipid. Therefore, it is important that the small intestine absorb bile salts only after the entire lipid has been absorbed.

The major lipids in bile are phospholipids and cholesterol. Of the phospholipids, the predominant species is phosphatidylcholine (lecithin). The phospholipid and cholesterol concentrations of hepatic bile are 0.3 to 11 mmol/L and 1.6 to 8.3 mmol/L, respectively. The concentrations of these lipids in the gallbladder bile are even higher because of the absorption of water by the gallbladder. Cholesterol in bile is responsible for the formation of cholesterol gallstones.

The total bile flow is composed of the ductular secretion and the canalicular bile flow. The ductular secretion is from the cells lining the bile ducts. These cells actively secrete HCO_3^- into the lumen, resulting in the movement of water into the lumen of the duct. Another mechanism that may contribute to ductular secretion of fluid is the presence of a cAMP-dependent Cl^- channel that secretes Cl^- into the ductule lumen.

Cholecystokinin released from the duodenum stimulates bile secretion from the gallbladder.

Bile **canaliculi** are fine tubular canals running between the hepatocytes. Bile then flows through the canaliculi to the bile ducts, which drain into the gallbladder. Between meals, the sphincter of Oddi is closed, thereby preventing bile from draining into the intestine to instead flow into the gallbladder, where it is stored and concentrated up to five times its original concentration. When food is released by the stomach into the duodenum in the form of chyme, the duodenum releases CCK. The release of CCK into the blood causes the gallbladder to contract and release the concentrated bile into the duodenum to complete the digestion of fat. Canalicular bile flow can be conceptually divided into two components: *bile acid–dependent secretion* and *bile acid–independent secretion*.

Canalicular bile acid–dependent flow

Hepatocyte uptake of free and conjugated bile salts is Na^+-dependent and mediated by **bile salt–sodium symport** (Fig. 26.12). The transmembrane Na^+ gradient generated by the Na^+/K^+-ATPase provides the energy required. This mechanism is a type of secondary active transport because the energy required for the active uptake of bile acid, or its conjugate, is not directly provided by ATP but by an ionic gradient. The free bile acids are reconjugated with taurine or glycine before secretion. Hepatocytes also make new bile

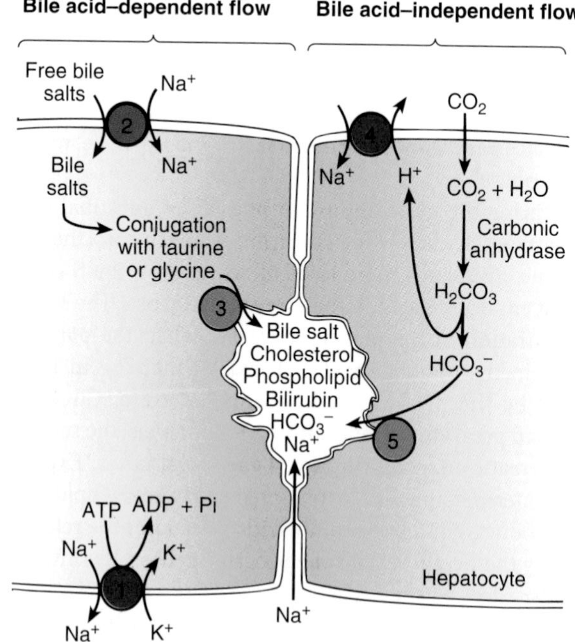

Figure 26.12 Bile salt secretions and bile flow are linked to Na^+/K^+-ATPase activity. This schematic illustrates the mechanism of bile acid secretion and flow. A marked reduction in flow and secretion occurs when Na^+/K^+-ATPase activity is inhibited. (1) Na^+/K^+-ATPase. (2) Bile salt-sodium symport. (3) Canalicular bile acid carrier. (4) Na^+/H^+ exchanger. (5) HCO_3^- transport system. ADP, adenosine diphosphate; ATP, adenosine triphosphate; Pi, inorganic phosphate.

acids from cholesterol. Hepatocytes secrete bile salts by a carrier located at the canalicular membrane. This secretion is not Na^+-dependent; instead, it is driven by the electrical potential difference between the hepatocyte and the canaliculus lumen.

Other major components of bile, such as phospholipid and cholesterol, are secreted in concert with bile salts. Hepatocytes secrete bilirubin via an active process involving a transporter. Although the secretion of cholesterol and phospholipid is not well understood, it is closely coupled to bile salt secretion. The osmotic pressure generated as a result of the secretion of bile salts draws water into the canaliculus lumen through the paracellular pathway.

Canalicular bile acid–independent flow

As the name implies, this component of canalicular flow is not dependent on the secretion of bile acids (see Fig. 26.12). The Na^+/K^+-ATPase plays an important role in bile acid–independent bile flow, a role that is clearly demonstrated by the marked reduction in bile flow when an inhibitor of this enzyme is applied. Another mechanism responsible for bile acid–independent flow is canalicular HCO_3^- secretion.

Bile acids are potentially toxic to cells, and their concentrations are tightly regulated.

Because bile acids act like a detergent, their concentration is tightly regulated in order to prevent damage the lining of the gut. The major determinant of bile acid synthesis and secretion by hepatocytes is the bile acid concentration in hepatic portal blood, which exerts a negative-feedback effect on the synthesis of bile acids from cholesterol. The concentration of bile acids in portal blood also determines bile acid-dependent secretion. Between meals, the portal blood concentration of bile salts is usually extremely low, resulting in increased bile acid synthesis but reduced bile acid-dependent flow. After a meal, there is increased delivery of bile salts in the portal blood, which not only inhibits bile acid synthesis but also stimulates bile acid-dependent secretion.

The intestinal mucosa secretes CCK when fatty acids or amino acids are present in the lumen. CCK causes contraction of the gallbladder, which, in turn, causes increased pressure in the bile ducts. As the bile duct pressure rises, the sphincter of Oddi relaxes (another effect of CCK), and bile is delivered into the lumen.

When the mucosa of the small intestine is exposed to acid in the chyme, it releases secretin into the blood. Secretin stimulates HCO_3^- secretion by the cells lining the bile ducts. As a result, bile contributes to the neutralization of acid in the duodenum.

Gastrin stimulates bile secretion directly by affecting the liver and indirectly by stimulating increased acid production that results in increased secretin release. Steroid hormones (e.g., estrogen and some androgens) are inhibitors of bile secretion, and reduced bile secretion is a side effect associated with the therapeutic use of these hormones (e.g., administration of oral contraceptives or postmenopausal hormone replacement therapy). During pregnancy, the high circulating level of estrogen can reduce bile acid secretion.

Parasympathetic and sympathetic nerves supply the biliary system. Parasympathetic (vagal) stimulation results in contraction of the gallbladder and relaxation of the sphincter of Oddi as well as increased bile formation. Bilateral vagotomy results in reduced bile secretion after a meal, suggesting that the parasympathetic nervous system plays a role in mediating bile secretion. By contrast, stimulation of the sympathetic nervous system results in reduced bile secretion and relaxation of the gallbladder.

Gallbladder bile differs from hepatic bile.

Gallbladder bile has a different composition from hepatic bile. The principal difference is that gallbladder bile is more highly concentrated. Water absorption is the major mechanism involved in concentrating hepatic bile by the gallbladder. Water absorption by the gallbladder epithelium is passive and is secondary to active Na^+ transport via a Na^+/K^+-ATPase in the basolateral membrane of the epithelial cells lining the gallbladder. As a result of isotonic fluid absorption from the gallbladder bile, the concentration of the various unabsorbed components of hepatic bile increases dramatically—as much as 20-fold.

Bile salts are recycled between the small intestine and the liver.

The **enterohepatic circulation** of bile salts is the recycling of bile salts between the small intestine and the liver. The total amount of bile acids in the body, primary or secondary, conjugated or free, at any time is defined as the **total bile acid pool**. In healthy people, the bile acid pool ranges from 2 to 4 g. The enterohepatic circulation of bile acids in this pool is physiologically extremely important. By cycling several times during a meal, a relatively small bile acid pool can provide the body with sufficient amounts of bile salts to promote lipid absorption. In a light eater, the bile acid pool may circulate 3 to 5 times a day, whereas in a heavy eater, it may circulate 14 to 16 times a day. The intestine is normally extremely efficient in absorbing the bile salts by carriers located in the distal ileum. Inflammation of the ileum can lead to their malabsorption and result in the loss of large quantities of bile salts in the feces. Depending on the severity of illness, malabsorption of fat may result.

Bile salts in the intestinal lumen are absorbed via four pathways (Fig. 26.13). First, they are absorbed throughout the entire small intestine by passive diffusion, but only a small fraction of the total amount of bile salts is absorbed in this manner. Second, and most important, bile salts are absorbed in the terminal ileum by an active carrier-mediated process, an extremely efficient process in which usually less than 5% of the bile salts escapes into the colon. Third, bacteria in the terminal ileum and colon deconjugate the bile salts to form bile acids, which are much more lipophilic than bile salts and, thus, can be absorbed passively. Fourth, these same bacteria are responsible for transforming the primary bile acids to secondary bile acids (deoxycholic and lithocholic acids)

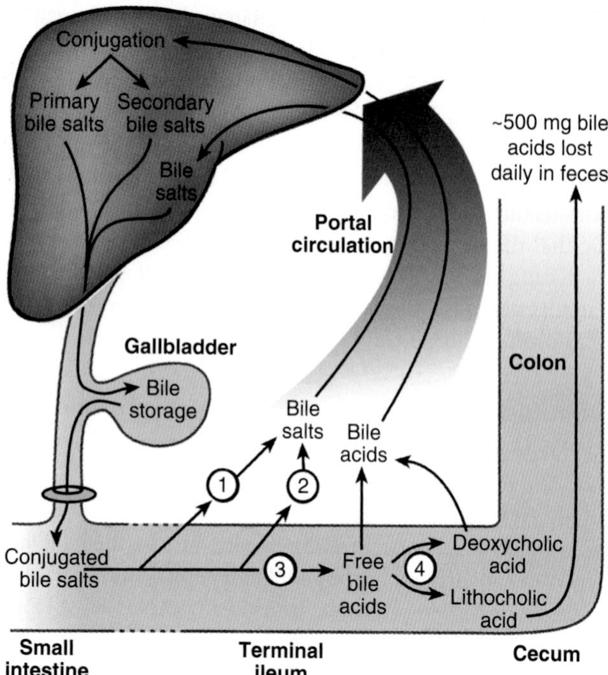

Figure 26.13 Recycling mechanisms for bile salt. The enterohepatic circulation of bile salts is illustrated in the schematic. Approximately 95% of the bile salts secreted are reabsorbed in the terminal ileum and recycled by four mechanisms: **(1)** passive diffusion along the small intestine (playing a relatively minor role), **(2)** carrier-mediated active absorption in the terminal ileum (the most important absorption route), **(3)** deconjugation to primary bile acids before being absorbed either passively or actively, and **(4)** conversion of primary bile acids to secondary bile acids with subsequent absorption of deoxycholic acid.

by dehydroxylation. Deoxycholic acid may be absorbed, but lithocholic acid is poorly absorbed.

Although bile salt and bile acid absorption is extremely efficient, some salts and acids are nonetheless lost with every cycle of the enterohepatic circulation. About 500 mg of bile acids is lost daily. The synthesis of new bile acids from cholesterol replenishes them. The loss of bile acid in feces is, therefore, an efficient way to excrete cholesterol.

Absorbed bile salts are transported in the portal blood bound to albumin or **high-density lipoproteins (HDLs)**. The uptake of bile salts by hepatocytes is extremely efficient. In just one pass through the liver, more than 80% of the bile salts in the portal blood is removed. Once taken up by hepatocytes, bile salts are secreted into bile. The uptake of bile salts is a primary determinant of bile salt secretion by the liver.

Bilirubin is the major component of bile pigments.

The major pigment present in bile is the orange compound **bilirubin**, an end product of hemoglobin degradation in the monocyte–macrophage system in the spleen, bone marrow, and liver (Fig. 26.14). Hemoglobin is first converted to biliverdin with the release of iron and globin. Biliverdin is

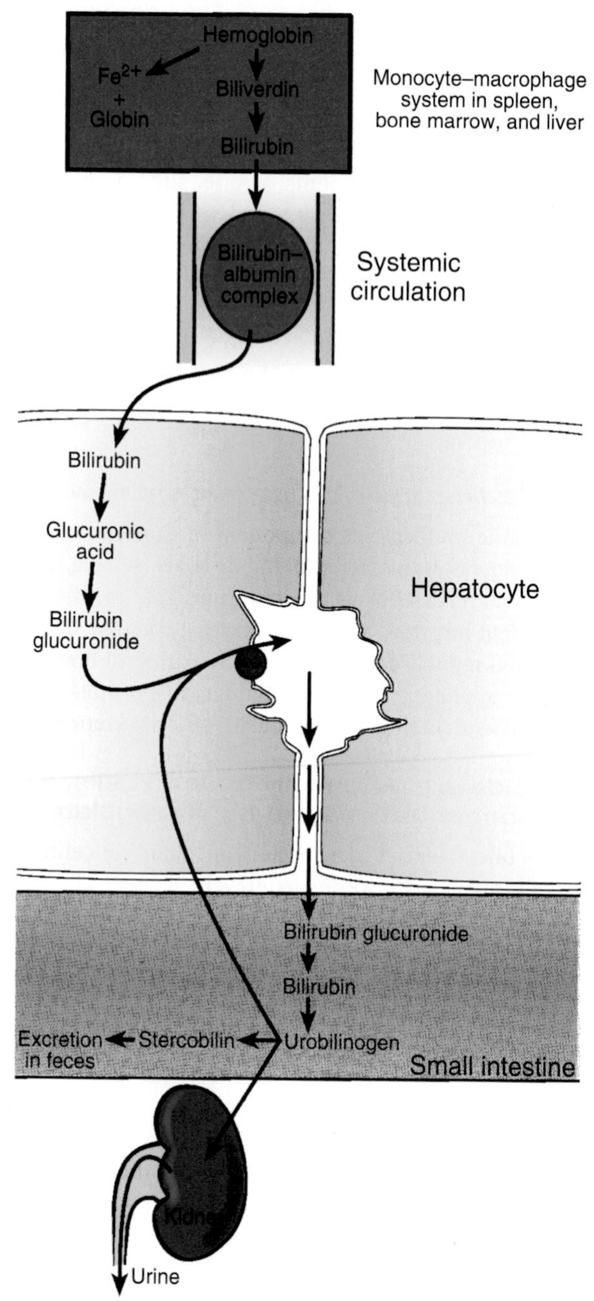

Figure 26.14 Bilirubin is the end product of hemoglobin degradation. The liver removes the bile pigment, bilirubin, and is conjugated with glucuronic acid.

then converted into bilirubin, which is transported in blood bound to albumin. The liver removes bilirubin from the circulation rapidly and conjugates it with glucuronic acid. The glucuronide is secreted into the bile canaliculi through an active carrier-mediated process.

In the small intestine, bilirubin glucuronide is poorly absorbed. In the colon, however, bacteria deconjugate it, and part of the bilirubin released is converted to the highly soluble, colorless compound called urobilinogen. Urobilinogen can be oxidized in the intestine to stercobilin or absorbed by the small intestine. It is excreted in either urine or bile. Stercobilin is responsible for the brown color of the stool.

Gallstones form when cholesterol becomes supersaturated.

Bile salts and lecithin in the bile help solubilize cholesterol. However, when bile contains too much cholesterol and not enough bile salts, it starts to crystallize and forms **gallstones**. Gallstones can occur anywhere within the biliary tree, including the gallbladder and the common bile duct. Obstruction of the biliary tree can cause jaundice, and obstruction of the outlet of the pancreatic exocrine system can cause pancreatitis. Gallstones can easily be detected on an x-ray image because calcium deposits are formed in the stones that increase their opacity. Gallstones are more prevalent in women than in men.

INTESTINAL SECRETION

The small intestine is where the vast majority of digestion and absorption of food takes place, and it secretes 2 to 3 L/d of isotonic alkaline fluid to aid in the digestive process. The secretion of watery fluid is derived mainly from cells in the **crypts of Lieberkühn**. These are tubular glands that dip down into the mucosal surface between the villi (Fig. 26.15). Unlike the gastric pits, the intestinal crypts do not secrete digestive enzymes, but do secrete mucus, electrolytes, and water. In addition, the crypts of Lieberkühn function to replace epithelial cells. Epithelial cells lining the small intestine are continually being sloughed off and are replaced at a rapid rate as a result of the high mitotic index of epithelial stem cells in the crypts. The new epithelial cells that are continually being produced migrate up the villi and replace older epithelial cells at the tips of the villi. Epithelial cells are sloughed off into the lumen, and the epithelial lining of the small intestine is replaced approximately every 3 days. The intestinal crypts, due to a high rate of cell division, are very sensitive to damage by radiation and to anticancer drugs used in chemotherapy. This is particularly problematic with patients who have been treated for colorectal cancer because their treatment is specially designed to target and kill tumors in the epithelial lining.

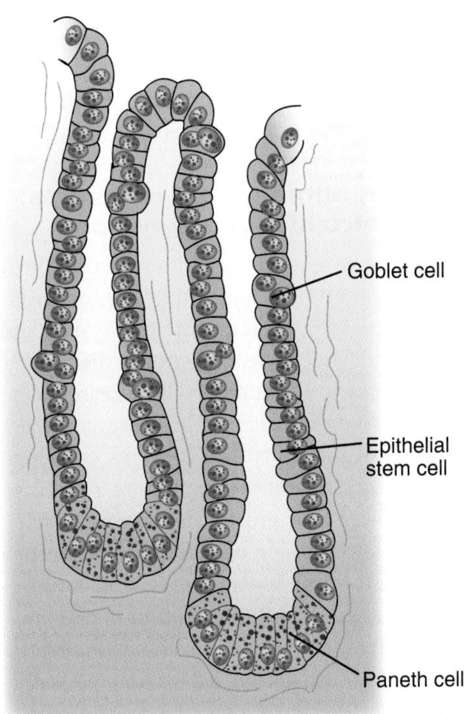

Figure 26.15 Crypt of Lieberkühn. Intestinal crypts secrete intestinal fluid and are found in all regions of the small intestine sandwiched in between villi.

The **Paneth cells** are involved in the host defense system in the small intestine. They function like neutrophils and produce antimicrobial substances that provide a protective barrier. The Paneth cells are particularly important in protecting the stem cells from damage. Also present in intestinal secretions are various mucins (mucoproteins) secreted by **goblet cells**. Mucins are glycoproteins high in carbohydrate, which form gels in solution. They are extremely diverse in structure and are usually large molecules. The mucus lubricates the mucosal surface and protects it from mechanical damage by solid food particles. It may also provide a physical barrier in the small intestine against the entry of microorganisms into the mucosa.

Intestinal secretions contain mucoproteins for lubrication and protective functions.

Intestinal secretion probably helps maintain the fluidity of the chyme and may also play a role in diluting noxious agents and washing away infectious microorganisms. The HCO_3^- in intestinal secretions protects the intestinal mucosa by neutralizing any H^+ present in the lumen. This is important in the duodenum and also in the ileum, where bacteria degrade certain foods to produce acids (e.g., dietary fibers to short-chain fatty acids).

The small intestine and colon usually absorb the fluid and electrolytes from intestinal secretions, but if secretion surpasses absorption (e.g., in cholera), watery diarrhea may result. If uncontrolled, this can lead to the loss of large quantities of fluid and electrolytes, which can result in dehydration and electrolyte imbalances and, ultimately, death. Several noxious agents, such as bacterial toxins (e.g., cholera toxin), can induce intestinal hypersecretion. Cholera toxin binds to the brush border membrane of crypt cells and increases intracellular adenylyl cyclase activity. Adenylyl cyclase is the enzyme that synthesizes cAMP from ATP. The result is a dramatic increase in intracellular cAMP that stimulates active Cl^- and HCO_3^- secretion into the lumen.

It is well documented that tactile stimulation, or an increase in intraluminal pressure, stimulates intestinal secretion. Other potent stimuli are certain noxious agents and the toxins produced by microorganisms. With the exception of toxin-induced secretion, our understanding of the normal control of intestinal secretion is meager. VIP is known to be a potent stimulator of intestinal secretion. This is demonstrated by a form of endocrine tumor of the pancreas that results in the secretion of large amounts of VIP into the circulation. In this condition, intestinal secretion rates are high.

CARBOHYDRATE DIGESTION AND ABSORPTION

Sucrose and lactose are two major sources of carbohydrate in the human diet. Sucrose (present in cane sugar and honey) is a disaccharide, and is composed of glucose and fructose. Lactose (also a disaccharide) is the main sugar in milk and is made up of galactose and glucose.

Starches, large polysaccharides, are the third major source of carbohydrate in our diet. This polysaccharide is produced by all green plants and is the most important carbohydrate in the human diet. Sources include such stable foods as potatoes and most types of grain (e.g. wheat, corn, and rice). All three major dietary sources (sucrose, lactose, and starch) must be hydrolyzed to monosaccharides before they can be absorbed in the small intestine. Although some carbohydrates are digested in the mouth and stomach, most are digested in the small intestine by powerful pancreatic enzymes.

To ensure the optimal absorption of carbohydrates and other nutrients, the GI tract has several unique features. For instance, after a meal, the small intestine undergoes rhythmic contractions called *segmentations* (see Chapter 25, "Neurogastroenterology and Motility"), which ensure proper mixing of the small intestinal contents, exposure of the contents to digestive enzymes, and maximum exposure of digestion products to the small intestinal mucosa. The rhythmic segmentation has a gradient along the small intestine, with the highest frequency in the duodenum and the lowest in the ileum. This gradient ensures slow but forward movement of intestinal contents toward the colon.

Another unique feature of the small intestine is its architecture. Spiral or circular concentric folds increase the surface area of the intestine by about three times (Fig. 26.16). Finger-like projections of the mucosal surface called **villi** further increase the surface area of the small intestine, by about 30 times. To amplify the absorptive surface further, numerous closely packed **microvilli** cover each epithelial cell, or **enterocyte**. The total surface area is increased to 600 times. The GI tract absorbs the various nutrients, vitamins, and bile salts and water by passive, facilitated, or active transport. (The site and mechanism of absorption are discussed below.) The GI tract has a large reserve for the digestion and absorption of various nutrients and vitamins. Malabsorption of nutrients is usually not detected unless a large portion of the small intestine has been lost or damaged because of disease.

The duodenum and jejunum absorb most nutrients and vitamins, but because bile salts are involved in the intestinal absorption of lipids, it is important that they not be absorbed prematurely. For effective fat absorption, the small intestine has adapted to absorb the bile salts in the terminal ileum through a bile salt transporter. The enterocytes along the villus that are involved in the absorption of nutrients are replaced every 2 to 3 days.

The digestion and absorption of dietary carbohydrates take place in the small intestine. These are extremely efficient processes, in that essentially all of the carbohydrates consumed are absorbed. Carbohydrates are an extremely important component of food intake, because they constitute about 45% to 50% of the typical Western diet and provide the greatest and least expensive source of energy. Carbohydrates must be digested to monosaccharides before absorption, which are then taken up by the enterocytes.

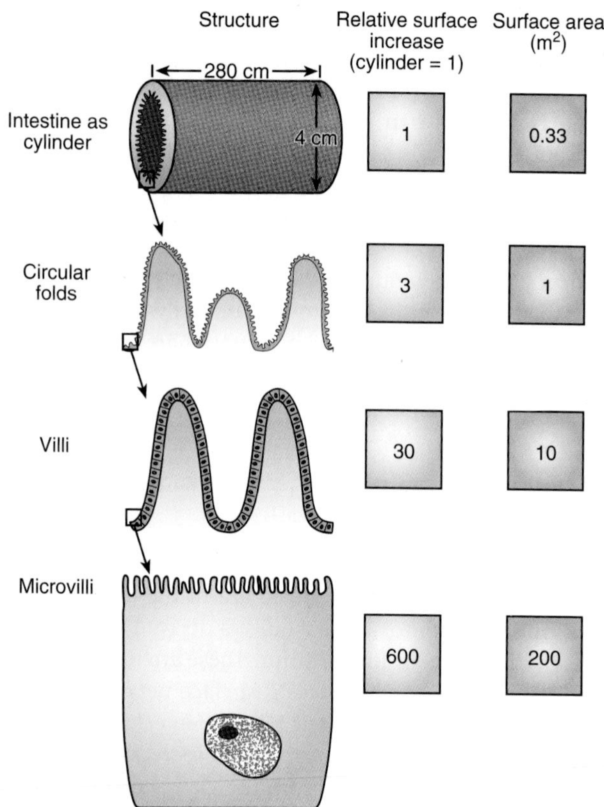

Figure 26.16 The unique architecture of the small intestine greatly enhances the surface area. Circular folds, villi, and microvilli are three functional features that greatly increase the surface area of the small intestine for the digestion of food.

Dietary fiber, although not digestible, benefits gastrointestinal function.

Humans can digest most carbohydrates except dietary fiber. Dietary fiber, sometimes called *roughage*, consists of non-starch polysaccharides, such as cellulose, waxes, and pectin. Although it is not digestible, consuming dietary fiber offers several advantages to GI function. It increases bulk, softens stools, and shortens transit time in the intestinal tract, all of which facilitates motility and prevents constipation. For example, many fruits and vegetables are rich in fiber, and their frequent ingestion greatly decreases intestinal transit time. Dietary fiber has also been shown to reduce the incidence of colon cancer by shortening GI transit time. A shortened transit time inhibits the formation of carcinogenic bile acids, such as lithocholic acid, as well as reducing the contact time of ingested carcinogens to act on tissue. Another benefit of dietary fiber is the lowering of blood cholesterol. This is accomplished by dietary fiber binding with bile acids, which are formed from cholesterol. Thus, fiber consumption lowers blood cholesterol by promoting excretion.

Clinical Focus / 26.2 

Celiac Sprue (Gluten-Sensitive Enteropathy)

Celiac sprue, also called **gluten-sensitive enteropathy**, is an autoimmune disorder of the small intestine that occurs in genetically predisposed individuals of all ages. The disease is caused by a reaction to a protein (gluten) found in grain (wheat, barely, oats, and rye). Upon exposure to gluten, the immune system cross reacts, causing chronic inflammation and primary lesions of the intestinal mucosa. This disorder can result in the malabsorption of all nutrients as a result of the shortening or total loss of intestinal villi, which reduces the mucosal enzymes for nutrient digestion and the mucosal surface for absorption. For example, as the small intestine becomes more damaged, a degree of lactose intolerance may develop. Symptoms include diarrhea, weight loss, failure to gain weight (in young children), and anemia. The latter is due to iron malabsorption. Celiac sprue occurs in about 1 to 6 of 10,000 people in the Western world. The highest incidence is in western Ireland, where the prevalence is as high as 3 of 1,000 people. People of African, Chinese, and Japanese descent are rarely diagnosed. Although the disease may occur at any age, it is more common during the first few years and the third to fifth decades of life. Celiac disease has been present since humans started harvesting grains in an organized fashion (about 5,000 BCE); however, the link to wheat was not discovered until the 1940s, by Dutch pediatrician, Dr. Wilhelm-Karel Dicke. He observed during World War II that when flour and bread were scarce, the death rate among children due to celiac sprue dropped to zero. The specific link to gluten was made later in 1952 by a team of physicians in England. The elimination of dietary gluten is a standard treatment for patients with celiac sprue. Occasionally, intestinal absorptive function and intestinal mucosal morphology of patients with celiac sprue are improved with glucocorticoid therapy. Presumably, such treatment is beneficial because of the immunosuppressive and anti-inflammatory actions of these hormones.

Carbohydrate is present in food as monosaccharides, disaccharides, oligosaccharides, and polysaccharides. The monosaccharides are mainly hexoses (six-carbon sugars), and glucose is by far the most abundant of these. Glucose is obtained directly from the diet or from the digestion of disaccharides, oligosaccharides, or polysaccharides. The next most common monosaccharides are galactose, fructose, and sorbitol. Galactose is present in the diet only as milk lactose, a disaccharide composed of galactose and glucose. Fructose is present in abundance in fruit and honey and is usually present as disaccharides or polysaccharides. Sorbitol is derived from glucose and is almost as sweet as glucose, but is absorbed much more slowly and, thus, maintains a high blood sugar level for a longer period when similar amounts are ingested. It has been used as a weight reduction aid to delay the onset of hunger sensations and is often used in diet foods and diet drinks.

The digestible polysaccharides are starch, dextrins, and glycogen. Starch, by far the most abundant carbohydrate in the human diet, is made of amylose and amylopectin. Amylose is composed of a straight chain of glucose units; amylopectin is composed of branched glucose units. Dextrins, formed from heating (e.g., toasting bread) or the action of the enzyme amylase, are intermediate products of starch digestion. Glycogen is a highly branched polysaccharide that stores carbohydrates in the body. Normally, about 300 to 400 g of glycogen is stored in the liver and muscle, with more stored in muscle than in the liver. Muscle glycogen is used exclusively by muscle, and liver glycogen is used to provide blood glucose during fasting.

The digestion of carbohydrates starts when food is mixed with saliva during chewing (Fig. 26.17). The enzyme salivary amylase acts on starches to release the disaccharide maltose. Because salivary amylase works best at neutral pH, its digestive action terminates rapidly after the bolus mixes with acid in the stomach. However, if the food is thoroughly mixed with amylase during chewing, a substantial amount of complex carbohydrates is digested before this point. **Pancreatic amylase** continues the digestion of the remaining carbohydrates. However, pancreatic secretions must first neutralize the chyme because pancreatic amylase works best at neutral pH. The products of pancreatic amylase digestion of polysaccharides are also maltose, and maltose to glucose. Enzymes located at the brush border membrane further digest the digestion products of starch and glycogen, together with disaccharides (sucrose and lactose). Figure 26.17 shows the enzymes involved in the digestion of carbohydrates. The final products are glucose, fructose, and galactose.

Enterocytes hydrolyze disaccharides and absorb monosaccharides.

Enterocytes are simple columnar cells that line the villi of the small intestine. They contain enyzymes that hydrolyze disaccharides into monosaccharides, which are then

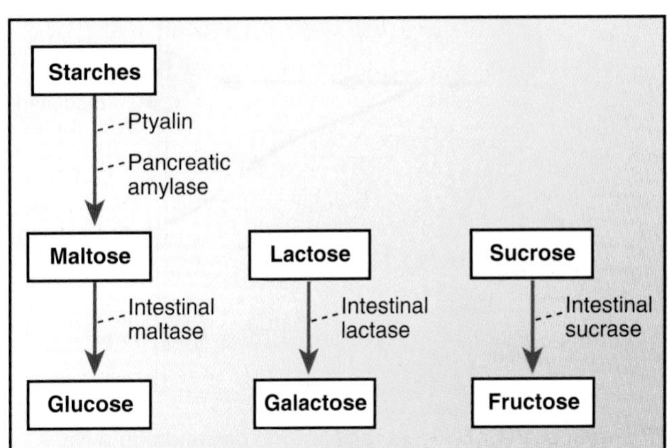

Figure 26.17 Salivary and pancreatic enzymes break down carbohydrates. Starch, sucrose, and lactose are the three main dietary sources of carbohydrates. All three are hydrolyzed to monosaccharides for absorption.

absorbed. The enterocytes are often referred to as the *absorptive cells* of the intestine. Absorption of monosaccharides by enterocytes involves both active and facilitated transport. A symporter (see Chapter 2, "Plasma Membrane, Membrane Transport, and Resting Membrane Potential") that transports two Na^+ for every molecule of monosaccharide absorbs glucose and galactose via secondary active transport (Fig. 26.18). The movement of Na^+ into the cell, down concentration and electrical gradients, affects the uphill movement of glucose into the cell. The basolateral membrane Na^+/K^+-ATPase maintains the low intracellular Na^+ concentration. The osmotic effects of sugars increase the Na^+/K^+-ATPase activity and the K^+ conductance of the basolateral membrane. Sugars accumulate in the cell at a higher concentration than in plasma and leave the cell by Na^+-independent facilitated transport or passive diffusion through the basolateral cell membrane. Glucose and galactose share a common transporter at the brush border membrane of enterocytes and, thus, compete with each other during absorption.

Fructose is taken up by facilitated transport. Although facilitated transport is carrier-mediated, it is not an active process (see Chapter 2, "Plasma Membrane, Membrane Transport, and Resting Membrane Potential"). Fructose absorption is much slower than glucose and galactose absorption and is not Na^+-dependent. Although in some animal species both galactose and fructose can be converted to glucose in enterocytes, this mechanism is probably not important in humans.

The portal blood transports the sugars absorbed by enterocytes to the liver, where they are converted to glycogen or remain in the blood. After a meal, the level of blood glucose rises rapidly, usually peaking at 30 to 60 minutes. The concentration of glucose can be as high as 150 mg/dL, even higher in diabetic patients. Although enterocytes can use glucose for fuel, glutamine is preferred. Both galactose and glucose are used in the glycosylation of proteins in the Golgi apparatus of the enterocytes.

Salivary or pancreatic dysfunction impairs carbohydrate absorption.

Impaired carbohydrate absorption caused by the absence of salivary or pancreatic amylase almost never occurs because these enzymes are usually present in great excess. However, impaired absorption as a result of a deficiency in membrane disaccharidases is common. Such deficiencies can be either genetic or acquired. Among congenital deficiencies, lactase deficiency is, by far, the most common. Affected people suffer from **lactose intolerance.** Lactose intolerance stems from the inability to metabolize lactose because of lactase deficiencies. Because disaccharides cannot be absorbed by the small intestine, the absence of lactase allows ingested dairy products to remain uncleaved and pass into the colon. Bacteria in the colon quickly switch over to lactose metabolism, resulting in *in vivo* fermentation that produces copious amounts of gas. This, in turn, causes a number of abdominal symptoms including stomach cramps, bloating, and flatulence. Fermentation products also raise the osmotic pressure of the colon contents, resulting in excess fluid accumulation and diarrhea (Fig. 26.19).

Congenital sucrase deficiency results in symptoms similar to those of lactase deficiency. Sucrase deficiency can be inherited or acquired through disorders of the small intestine, such as tropical sprue or Crohn disease.

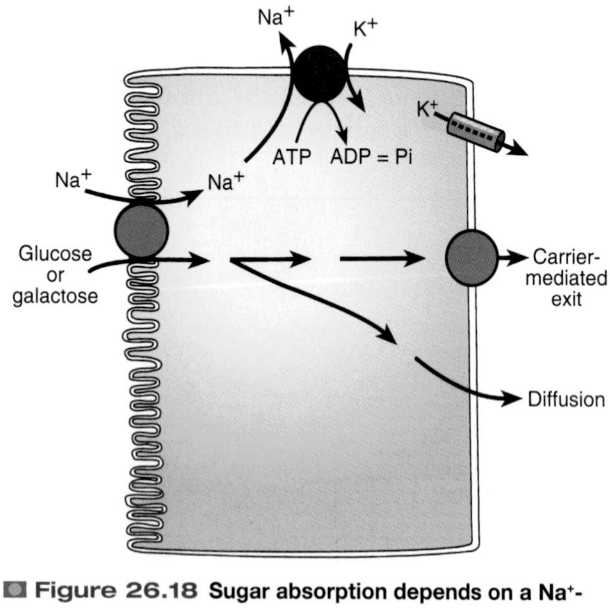

Figure 26.18 Sugar absorption depends on a Na^+-dependent carrier system. Sodium moves into the cell down a concentration and electrical gradient, which favors the uphill movement of glucose and galactose into the enterocytes. ADP, adenosine diphosphate; ATP, adenosine triphosphate; Pi, inorganic phosphate.

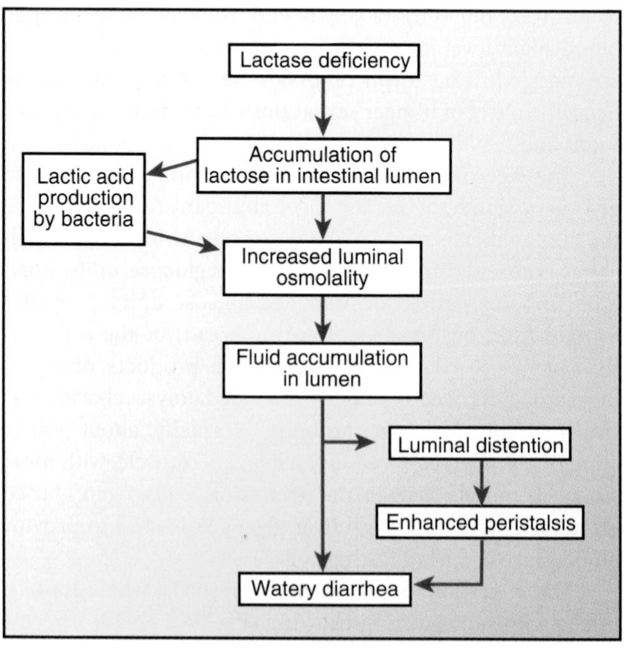

Figure 26.19 Lactase deficiency causes diarrhea. Undigested lactose increases the osmolality of the gut, which results in an increase in osmolality. The high osmolality leads to a concomitant increase in a net water secretion, causing diarrhea.

LIPID DIGESTION AND ABSORPTION

Lipids are a group of naturally occurring molecules in the human diet that includes fats, sterols, monoglycerides, diglycerides, phospholipids, and fat-soluble vitamins (A, D, E, and vitamin K). The physiologic function of lipids includes energy storage, structural components for cell membranes, and cell signaling molecules. Although the term "lipid" is sometimes used synonymously with fats, fats are a subgroup of lipids called **triglycerides**. Triglycerides in adipose tissue are the major form of concentrated energy storage in animals. Triglycerides provide 30% to 40% of the daily caloric intake in the Western diet. The human body is capable of synthesizing most of the lipids it requires, with the exception of the **essential fatty acids**. Essential fatty acids are fatty acids that humans and other animals must ingest for normal cell function, because they cannot be made from other foods. There are only three essential fatty acids that cannot be made and need to be ingested in the adult human diet. These include linolenic acid, linoleic acid, and arachidonic acid. Linolenic acid belongs to the omega-3 fatty acids, and linoleic and arachidonic both belong to omega-6 fatty acids. These essential fatty acids are abundant in seafood, flaxseed (linseed), soya oil, canola oil, sunflower seeds, leafy vegetables, and walnuts. Researchers have recently provided convincing evidence that **eicosapentaenoic acid** (C 20:5) and **docosahexaenoic acid** (C 22:6) are also essential for the normal development of vision in newborns.

Lipids comprise several classes of compounds that are insoluble in water but soluble in organic solvents. By far the most abundant dietary lipids are triglycerides, also known as *triacylglycerols*. More than 90% of the daily dietary lipid intake is in the form of triglycerides.

The other lipids in the human diet are cholesterol and phospholipids. **Cholesterol** is a sterol derived exclusively from animal fat. Humans also ingest a small amount of plant sterols, notably β-sitosterol and campesterol. The phospholipid molecule is similar to a triglyceride, with fatty acids occupying the first and second positions of the glycerol backbone (Fig. 26.20). However, the third position of the glycerol backbone is occupied by a phosphate group coupled to a nitrogenous base (e.g., choline or ethanolamine), for which each type of phospholipid molecule is named. For example, phosphatidylcholine (lecithin) is an important phospholipid in cell membranes and lung surfactant.

Bile serves as an endogenous source of cholesterol and phospholipids. Bile contributes about 12 g/d of phospholipid to the intestinal lumen, mostly in the form of phosphatidylcholine, whereas dietary sources contribute 2 to 3 g/d. Another important endogenous source of lipid is desquamated intestinal villus epithelial cells.

Lipases are water-soluble enzymes that break down fats into monoglycerides and free fatty acids.

Lipid digestion occurs mainly in the intestinal lumen but starts in the stomach. The stomach secretes a lipase called acid lipase because it works best in the acidic medium. Acid lipase prefers triglycerides containing medium-chain fatty acids (8–10 carbons in length), which are abundant in milk. Acid lipase plays an important role in the digestion of

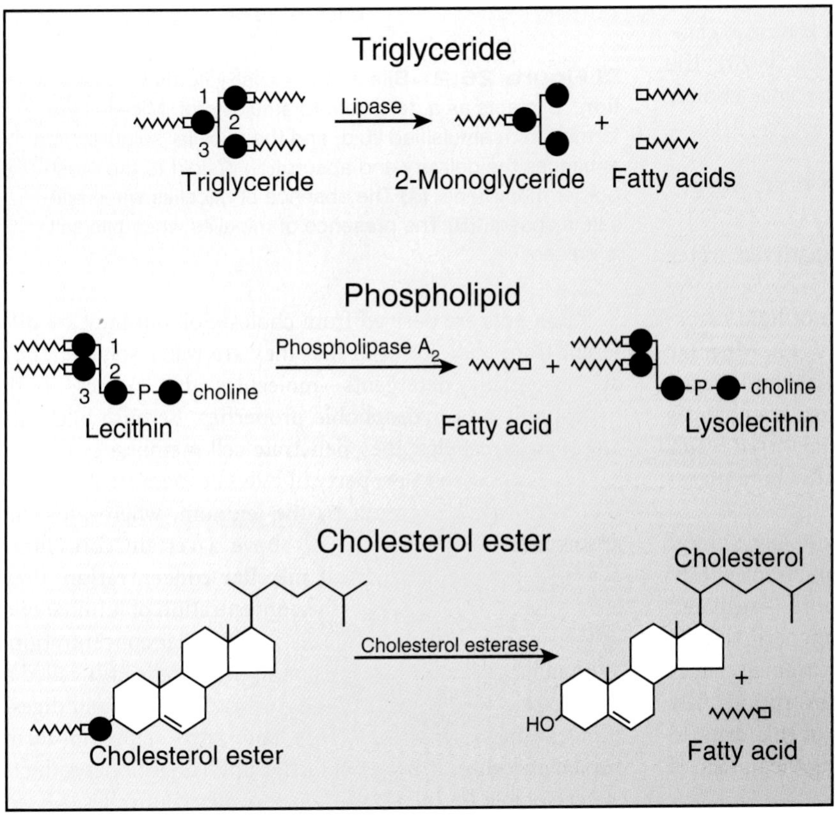

Figure 26.20 Pancreatic lipases digest dietary lipids. Different pancreatic lipases carry out lipid hydrolysis in the gut. Solid circles represent oxygen atoms.

fat in infants because they do not have sufficient pancreatic lipase to digest the milk fat, which is their major source of fat intake. In adults, acid lipase continues to play a role in the digestion of fat, and the partial digestion of products of fat aid in its emulsification. The emulsified fat greatly facilitates the hydrolysis of fat by the pancreatic lipase. Humans secrete an overabundance of **pancreatic lipase**. Depending on the substrate being digested, pancreatic lipase has an optimal pH of 7 to 8, allowing it to work well in the intestinal lumen after the acidic contents from the stomach have been neutralized by pancreatic HCO_3^- secretion. Pancreatic lipase hydrolyzes the triglyceride molecule to a 2-monoglyceride and two fatty acids (see Fig. 26.20). It works on the triglyceride molecule at the oil–water interface; thus, the rate of lipolysis depends on the surface area of the interface. The products from the partial digestion of dietary triglyceride by gastric lipase and the churning action of the stomach produce a suspension of oil droplets (an emulsion) that helps increase the area of the oil–water interface. Pancreatic juice also contains the peptide **colipase**, which is necessary for the normal digestion of fat by pancreatic lipase. Colipase binds lipase at a molar ratio of 1:1, thereby allowing the lipase to bind to the oil–water interface where lipolysis takes place. Colipase also counteracts the inhibition of lipolysis by bile salt, which, despite its importance in intestinal fat absorption, prevents the attachment of pancreatic lipase to the oil–water interface.

Phospholipase A_2 is the major pancreatic enzyme for digesting phospholipids and forming lysophospholipids and fatty acids. For instance, phosphatidylcholine (lecithin) is hydrolyzed to form lysophosphatidylcholine (lysolecithin) and fatty acid (see Fig. 26.20).

Dietary cholesterol is presented as a free sterol or as a sterol ester (cholesterol ester). The pancreatic enzyme carboxylester hydrolase, also called **cholesterol esterase**, catalyzes the hydrolysis of cholesterol ester (see Fig. 26.20). The digestion of cholesterol ester is important because cholesterol can be absorbed only as the free sterol.

Emulsification by bile acids is the first step in fat digestion.

A layer of poorly stirred fluid called the **unstirred water layer** coats the surface of the intestinal villi (Fig. 26.21A). The unstirred water layer reduces the absorption of lipid digestion products because they are insoluble in water. **Micellar emulsification** by bile salts in the small intestinal lumen renders them water soluble. This mechanism greatly enhances the concentration of these products in the unstirred water layer (Fig. 26.21B). Enterocytes then absorb the lipid digestion products, mainly by passive diffusion. Fatty acid and monoglyceride molecules are taken up individually. Similar mechanisms seem to operate for cholesterol and lysolecithin. However, recent evidence supports a transporter-mediated uptake of cholesterol. Ezetimibe, an antihyperlipidemic medication to lower blood cholesterol, is extremely effective in decreasing intestinal cholesterol absorption. The mechanism appears to involve the binding of this drug to the cholesterol transporter, thereby preventing the uptake of cholesterol.

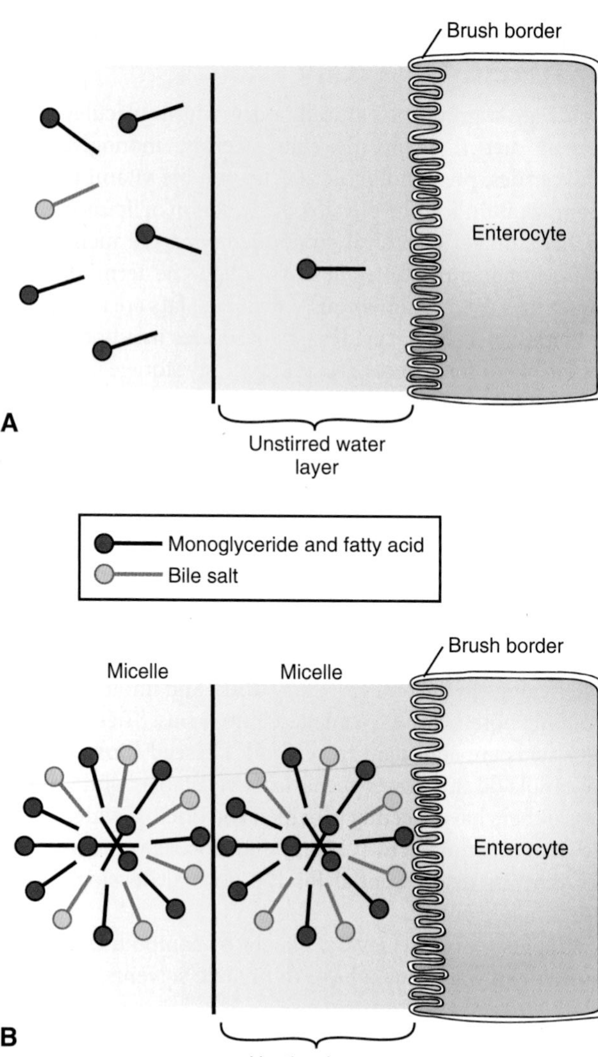

Figure 26.21 Bile salts emulsify lipids for absorption. Bile acts as a detergent to emulsify fat. Micelles are formed from emulsified lipid, and the micellar solubilization enhances the delivery and absorption of lipid to the brush border membrane. **(A)** The absence of micelles when bile salt is absent. **(B)** The presence of micelles when bile salt is present.

Bile salts are derived from cholesterol, but they are different from cholesterol in that they are water-soluble. They are essentially detergents—molecules that possess both hydrophilic and hydrophobic properties. Because bile salts are polar molecules, they penetrate cell membranes poorly. This is an important property of bile salts because it ensures their minimal absorption by the jejunum, where most fat absorption takes place. At or above a certain concentration of bile salts, the **critical micellar concentration**, they aggregate to form micelles; the concentration of luminal bile salts is usually well above the critical micellar concentration. When bile salts alone are present in the micelle, it is called a **simple micelle**. Simple micelles incorporate the lipid digestion products—monoglyceride and fatty acids—to form **mixed micelles**. This renders the lipid digestion products water-soluble by incorporation into mixed micelles. Mixed micelles diffuse across the unstirred water layer and deliver lipid digestion products to the enterocytes for absorption.

Enterocytes secrete chylomicrons and very-low-density lipoproteins.

After entering the enterocytes, the fatty acids and monoglycerides migrate to the smooth endoplasmic reticulum (SER). A fatty acid-binding protein may be involved in the intracellular transport of fatty acids, but whether a protein carrier is involved in the intracellular transport of monoglycerides is still unclear. In the SER, monoglycerides and fatty acids are rapidly reconstituted to form triglycerides (Fig. 26.22). Fatty acids are first activated to form acyl-coenzyme A (acyl-CoA), which is then used to esterify monoglyceride to form diglyceride, which is transformed into triglyceride. The lysolecithin absorbed by the enterocytes can be reesterified in the SER to form lecithin.

Cholesterol can be transported out of the enterocytes as free cholesterol or as esterified cholesterol. The enzyme responsible for the esterification of cholesterol to form cholesterol ester is **acyl-CoA cholesterol acyltransferase**.

The reassembled triglycerides, lecithin, cholesterol, and cholesterol esters are then packaged into **lipoproteins** and exported from the enterocytes. The intestine produces two major classes of lipoproteins: **chylomicrons** and **very low-density lipoproteins (VLDLs)**. Both are triglyceride-rich lipoproteins with densities <1.006 g/mL. The small intestine exclusively makes chylomicrons, and their primary function is to transport the large amount of dietary fat absorbed by the small intestine from the enterocytes to the lymph. Chylomicrons are large, spherical lipoproteins with diameters of 80 to 500 nm. They contain less protein and phospholipid than VLDLs and are, therefore, less dense than VLDLs. The small intestine continuously makes VLDLs during both fasting and feeding, although the liver contributes significantly more VLDLs to the circulation. VLDL is converted in the blood stream to **low-density lipoprotein (LDL)**. Since cholesterol is carried by LDL, cholesterol content can be measured on LDL particles. High levels of LDL particles can promote medical problems and they are often referred to as *bad cholesterol* particles (as opposed to HDL, which is often referred to as *good cholesterol*). One of the functions of LDL particles is the transport of cholesterol into the arterial wall where it is retained by arterial proteoglycans. **Macrophages** are attracted to these sites and function to engulf the LDL particles. The process of engulfing LDL particles by macrophages initiates plaque formation in the arterial wall. Increased levels of plaque formation are linked to atherosclerosis. Over time, plaques become vulnerable to rupturing, which, in turn, activates blood clotting and leads to arterial stenosis.

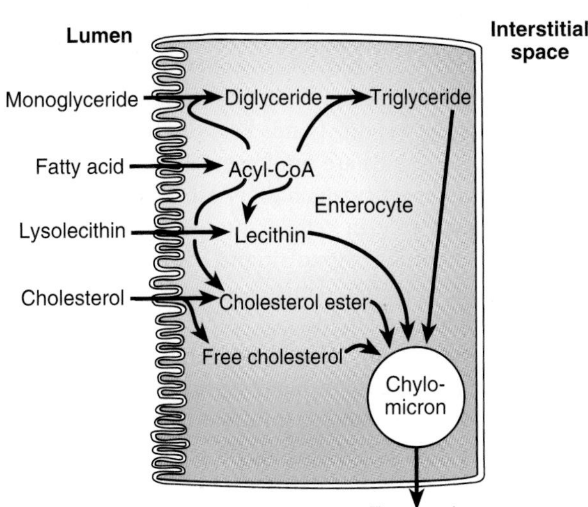

Figure 26.22 Endocytes form chylomicrons for lipid transport. Chylomicrons are large lipoproteins and transport exogenous lipids to the liver and to adipose, cardiac, and skeletal tissue. Acyl-CoA, acyl-coenzyme A.

Apoproteins—apo A-I, apo A-IV, and apo B—are among the major proteins associated with the production of chylomicrons and VLDLs. Apo B is the only protein that seems to be necessary for the normal formation of intestinal chylomicrons and VLDLs. This protein is made in the small intestine. It has a molecular weight of 250,000 and is extremely hydrophobic. Apo A-I is involved in a reaction catalyzed by the plasma enzyme **lecithin cholesterol acyltransferase (LCAT)**. Plasma LCAT is responsible for the esterification of cholesterol in the plasma to form cholesterol ester with the fatty acid derived from the 2-position of lecithin. After the chylomicrons and VLDLs enter the plasma, apo A-I is rapidly transferred from chylomicrons and VLDLs to HDLs. Apo A-I is the major protein present in plasma HDLs. The small intestine and the liver make Apo A-IV. Recently, it was shown that apo A-IV, secreted by the small intestine, may be an important factor contributing to anorexia after fat feeding.

Newly synthesized lipoproteins in the SER are transferred to the Golgi apparatus, where they are packaged in vesicles. Chylomicrons and VLDLs are released into the intercellular space by exocytosis. From there, they are transferred to the central lacteals (the beginnings of lymphatic vessels) by a process that is not well understood. Experimental evidence seems to indicate that the transfer occurs mostly by diffusion. Intestinal lipid absorption is associated with a marked increase in lymph flow called the **lymphagogic effect** of fat feeding. This increase in lymph flow plays an important role in the transfer of lipoproteins from the intercellular spaces to the central lacteal.

Fatty acids can also travel in the blood bound to albumin. Whereas most of the long-chain fatty acids are transported from the small intestine as triglycerides packaged in chylomicrons and VLDLs, some are transported in the portal blood bound to serum albumin. Most of the medium-chain (8–12 carbons) and all of the short-chain fatty acids are transported via the hepatic portal route.

Altered lipase and bile secretion significantly impair lipid digestion and absorption.

In several clinical conditions, lipid digestion and absorption are impaired, resulting in the malabsorption of lipids and other nutrients and fatty stools. Abnormal lipid absorption

From Bench to Bedside / 26.1

Cholesterol Absorption and Transporters

As discussed in this chapter, the absorption of cholesterol is believed to be passive. This concept, however, is challenged by the following three intriguing clinical observations. First, humans normally absorb cholesterol derived from animals efficiently but absorb plant sterols inefficiently. Consequently, the level of plant sterols, for example, β-sitosterol, in the blood is usually low. However, several families are documented whose members have high circulating levels of β-sitosterol. This interesting observation remained a mystery to clinicians as well as biologists for years. The second observation is of a case study of a farmer who consumed 25 eggs a day and yet had a normal blood cholesterol level. When his ability to absorb cholesterol was tested, his physicians found that the farmer was atypical compared with most humans in that he did not absorb cholesterol efficiently. Additionally, the farmer had an increased capacity to convert cholesterol into bile acids, the primary mechanism by which the body rids itself of cholesterol. Third, ezetimibe, a drug that reduces cholesterol absorption by the gut, has recently been approved by the FDA and is currently being prescribed by clinicians to lower cholesterol concentration. The interesting aspect of this drug is that it is efficient and that its efficacy is reached with only a small dose. These three observations are difficult to reconcile based on the fact that the uptake of cholesterol by the small intestine is passive.

Several cholesterol transporters have been proposed such as the scavenger receptor class BI and the Niemann-Pick C1 like 1 (NPC1L1). The presence of such transporters for cholesterol may help to explain the observations indicated above. Even so, the presence of a cholesterol transporter for the absorption of such abundant molecules as cholesterol is not yet understood. Perhaps these transporters help the enterocyte sort out the amount of cholesterol that should be retained for its own use versus the amount that should be transported. We are at an interesting juncture, and in the next few years, we will probably witness many exciting findings about transporters and signal molecules in the gut and their role in the absorption of lipid-soluble molecules such as fatty acids and cholesterol. A thorough understanding of this process will require the close interaction and collaboration of both bench scientists and clinicians.

can result in numerous problems because the body requires certain fatty acids (e.g., linoleic and arachidonic acid, precursors of prostaglandins) to function normally. These are called essential fatty acids because the human body cannot synthesize them and is, therefore, totally dependent on the diet to supply them. Recent studies suggest that the human body may also require omega-3 fatty acids in the diet during development. These include linolenic, docosahexaenoic, and eicosapentaenoic acids. Linolenic acid is abundant in plants, and docosahexaenoic and eicosapentaenoic acids are abundant in fish. Docosahexaenoic acid is an important fatty acid present in the retina and parts of the brain.

Pancreatic deficiency significantly reduces the ability of the exocrine pancreas to produce digestive enzymes. Because the pancreas normally produces an excess of digestive enzymes, enzyme production has to be reduced to about 10% of normal before symptoms of malabsorption develop. One characteristic of pancreatic deficiency is **steatorrhea** (fatty stool), resulting from the poor digestion of fat by the pancreatic lipase. Normally about 5 g/d of fat is excreted in human stool. With steatorrhea, as much as 50 g/d of fat can be excreted.

Fat absorption subsequent to the action of pancreatic lipase requires solubilization by bile salt micelles. Acute or chronic liver disease can cause defective biliary secretion, resulting in bile salt concentrations lower than that necessary for micelle formation. The normal absorption of fat is thereby inhibited.

Abetalipoproteinemia, an autosomal recessive disorder, is characterized by a complete lack of apo B in the circulation, which is required for the formation and secretion of chylomicrons and VLDLs. Apo B-containing lipoproteins in the circulation, including chylomicrons, VLDLs, and LDLs, are absent. Plasma LDLs are absent because they are derived mainly from the metabolism of VLDLs. Because people with abetalipoproteinemia do not produce any chylomicrons or VLDLs in the small intestine, they are unable to transport absorbed fat, resulting in an accumulation of lipid droplets in the cytoplasm of enterocytes. They also suffer from a deficiency of fat-soluble vitamins (A, D, E, and K). Using molecular biologic techniques, it has been determined that apo B deficiency is caused by a mutation of a transfer protein called microsomal triglyceride transfer protein. This protein facilitates the transfer of triglyceride formed in the cytoplasm to the ER.

PROTEIN DIGESTION AND ABSORPTION

Dietary proteins are basically long-chain amino acids that are joined together by peptide bonds between the carboxyl and amino groups. Proteins are fundamental structural components of cells and participate in every aspect of cell function. Aside from their role in cellular protein synthesis, amino acids are also an important nutritional source of nitrogen.

Most proteins are found in muscle, with the remainder in other cells, blood, body fluids, and body secretions. Enzymes and many hormones are proteins. Proteins are composed of amino acids and have molecular weights of a few thousand to a few hundred thousand. More than 20 common amino acids form the building blocks for proteins (Table 26.4). Of these, nine are considered essential and must be provided by the diet. Although the **nonessential amino acids** are also required for normal protein synthesis, the body can synthesize them from other amino acids.

Complete proteins are those that can supply all of the essential amino acids in amounts sufficient to support

TABLE 26.4 Amino Acids Found in Proteins

Essential	Nonessential
Histidine	Alanine
Isoleucine	Arginine
Leucine	Aspartic acid
Lysine	Citrulline
Methionine	Glutamic acid
Phenylalanine	Glycine
Threonine	Hydroxyglutamic acid
Tryptophan	Hydroxyproline
Valine	Norleucine
	Proline
	Serine
	Tyrosine

normal growth and body maintenance. Examples are eggs, poultry, and fish. The proteins in most vegetables and grains are called **incomplete proteins** because they do not provide all of the essential amino acids in amounts sufficient to sustain normal growth and body maintenance. Vegetarians need to eat a variety of vegetables and soy protein to avoid amino acid deficiencies.

Some dietary proteins are often the cause of allergies and/or allergic reactions. These reactions occur because the structure of each form of protein is slightly different. Some may trigger an immune response, whereas other structures are perfectly safe. For example, many individuals are allergic to gluten (the protein in wheat and other grains), to casein (the protein in milk), and to particular proteins found in peanuts. Some individuals are also allergic to shellfish and to specific proteins found in other types of seafood.

Not all proteins come from the diet.

The average American adult takes in 70 to 110 g/d of protein. The minimum daily protein requirement for adults is about 0.8 g/kg body weight (e.g., 56 g for a 70-kg person). Pregnant and lactating women require 20 to 30 g of protein above the recommended daily allowance to meet the extra demand. A lactating woman can lose as much as 12 to 15 g of protein per day as milk protein. Children need more protein for body growth; the recommended daily allowance for infants is about 2 g/kg body weight.

Although most of the protein entering the GI tract is dietary protein, there are also proteins derived from endogenous sources such as pancreatic, biliary, and intestinal secretions and the cells shed from the intestinal villi. About 20 to 30 g/d of protein enters the intestinal lumen in pancreatic juice and about 10 g/d in bile. Enterocytes of the intestinal villi are continuously shed into the intestinal lumen, and as much as 50 g/d of enterocyte proteins enter the intestinal lumen. An average of 150 to 180 g/d of total protein is presented to the small intestine, of which more than 90% is absorbed.

Digestion of proteins starts in the stomach and continues in the small intestine.

Most of the protein in the intestinal lumen is completely digested into amino acids, dipeptides, or tripeptides before it is taken up by the enterocytes (Fig. 26.23). Protein digestion typically begins in the stomach when **pepsinogen** is converted to **pepsin** by the action of HCl. Pepsin hydrolyzes protein to form smaller polypeptides. It is classified as an endopeptidase because it attacks specific peptide bonds inside the protein molecule. This phase of protein digestion is normally not important other than in people suffering from pancreatic exocrine deficiency.

Most of the digestion of proteins and polypeptides takes place in the small intestine by **trypsin** and **chymotrypsin.** Most proteases are secreted in the pancreatic juice as inactive proenzymes. When the pancreatic juice enters the duodenum, **enteropeptidase** (also known as *enterokinase*), an enzyme found on the luminal surface of enterocytes, converts trypsinogen to trypsin. The active **trypsin** then converts the other proenzymes to active enzymes.

The pancreatic proteases are classified as endopeptidases or exopeptidases (see Table 26.2). **Endopeptidases** hydrolyze certain internal peptide bonds of proteins or polypeptides to release the smaller peptides. The three endopeptidases present in pancreatic juice are trypsin, chymotrypsin, and elastase.

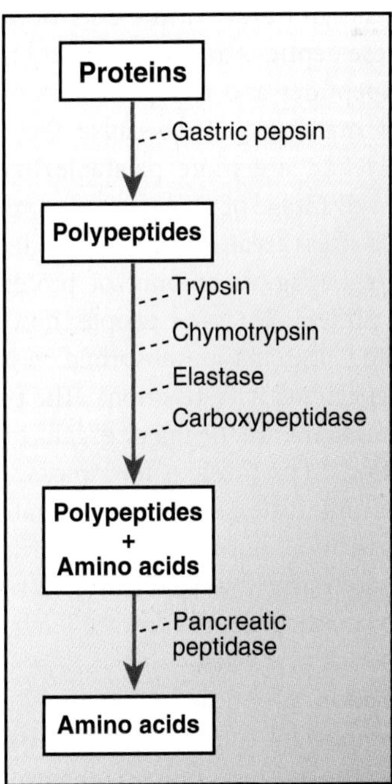

Figure 26.23 Protein absorption begins in the stomach. The first step involves the hydrolysis of proteins by gastric pepsin to form smaller polypeptides. Polypeptides are further acted upon by intestinal enzymes and proteases from both the small intestine and pancreas.

Trypsin splits off basic amino acids from the C-terminus of a protein, **chymotrypsin** attacks peptide bonds with an aromatic carboxyl terminal, and **elastase** attacks peptide bonds with a neutral aliphatic C-terminus. The **exopeptidases** in pancreatic juice are carboxypeptidase A and carboxypeptidase B. Like the endopeptidases, the exopeptidases are specific in their action. **Carboxypeptidase A** attacks polypeptides with a neutral aliphatic or aromatic C-terminus. **Carboxypeptidase B** attacks polypeptides with a basic C-terminus. The final products of protein digestion are amino acids and small peptides.

Intestinal enterocytes contain specific transporters to take up amino acids and peptides.

Enterocytes take up amino acids in the small intestine via secondary active transport. Six major amino acid carriers in the small intestine have been identified; they transport related groups of amino acids. The amino acid transporters favor the L-form over the D-form. As in the uptake of glucose, the uptake of amino acids depends on a Na^+ concentration gradient across the enterocyte brush border membrane.

The absorption of peptides by enterocytes was once thought to be less efficient than amino acid absorption. However, subsequent studies in humans clearly demonstrated that dipeptide and tripeptide uptake is significantly more efficient than the uptake of amino acids. Dipeptides and tripeptides use different transporters than those used by amino acids. The peptide transporter prefers dipeptides and tripeptides with either glycine or lysine residues. Furthermore, the peptide transporter only poorly transports tetrapeptides and more complex peptides. The **peptidases** (exopeptidases) located on the brush border of the enterocytes can further break down these peptides to dipeptides and tripeptides (see Fig. 26.23). Dipeptides and tripeptides are given to people suffering from malabsorption because they are absorbed more efficiently and are more palatable than free amino acids. Another advantage of peptides over amino acids is the smaller osmotic stress created as a result of delivering them.

In adults, a negligible amount of protein is absorbed as undigested protein. In some people, however, intact or partially digested proteins are absorbed, resulting in anaphylactic or hypersensitivity reactions. The pulmonary and cardiovascular systems are the major organs involved in anaphylactic reactions. For the first few weeks after birth, the newborn's small intestine absorbs considerable amounts of intact proteins. This is possible because of low proteolytic activity in the stomach, low pancreatic secretion of peptidases, and poor development of intracellular protein degradation by lysosomal proteases.

The absorption of immunoglobulins (predominantly IgG) plays an important role in the transmission of **passive immunity** from the mother's milk to the newborn in several animal species (e.g., ruminants and rodents). In humans, the absorption of intact immunoglobulins does not appear to be an important mode of transmission of antibodies for two reasons. First, passive immunity in humans is derived almost entirely from the intrauterine transport of maternal antibodies. Second, human **colostrum**, the thin, yellowish, milky fluid secreted by the mammary glands a few days before or after parturition, contains mainly IgA, which is poorly absorbed by the small intestine. The ability to absorb intact proteins is rapidly lost as the gut matures—a process called **closure**. Colostrum contains a factor that promotes the closure of the small intestine.

After the enterocytes take up dipeptides and tripeptides, peptidases in the cytoplasm break them down further to amino acids. The amino acids are transported in the portal blood. The small amount of protein that is taken up by the adult intestine is largely degraded by lysosomal proteases, although some proteins escape degradation.

Genetic disorders lead to impaired protein absorption.

Although pancreatic deficiency has the potential to affect protein digestion, it only does so in severe cases. Pancreatic deficiency seems to affect lipid digestion more than protein digestion. There are several extremely rare genetic disorders of amino acid carriers. In **Hartnup disease**, the membrane carrier for neutral amino acids (e.g., tryptophan) is defective. **Cystinuria** involves the carrier for basic amino acids (e.g., lysine and arginine) and the sulfur-containing amino acids (e.g., cystine). Cystinuria was once thought to involve only the kidneys because of the excretion of amino acids such as cystine in urine, but the small intestine is involved as well.

Because the peptide transport system remains unaffected, disorders of some amino acid transporters can be treated with supplemental dipeptides containing these amino acids. However, this treatment alone is not effective if the kidney transporter is also involved, as in the case of cystinuria.

VITAMIN ABSORPTION

Vitamins are organic compounds required as a nutrient in small quantities by an animal. Thus, an organic compound is called a *vitamin* when it cannot be synthesized by the body and must be obtained from the diet. By convention, the term "vitamin" excludes dietary minerals, which are inorganic. Currently, 13 vitamins are universally recognized (Table 26.5).

Vitamins are classified in many ways, but, in terms of absorption, they are classified as either water-soluble or fat-soluble. In humans, there are 13 vitamins: 4 fat-soluble vitamins (A, D, E, and K) and 9 water-soluble vitamins (8 B vitamins and vitamin C). Because water-soluble vitamins are easily dissolved in water, they are readily be excreted from the body. Thus, urinary output is a strong predictor of the B-vitamin and C-vitamin consumption.

Vitamins have many diverse cell functions, and many are essential for metabolic reactions.

Vitamins act as both catalysts and substrates in chemical reactions of the body. When acting as a catalyst, vitamins are bound to enzymes and are called cofactors (e.g., vitamin K forms part of the proteases involved in blood clotting).

TABLE 26.5 Vitamins: Their Solubility and Dietary Sources

Vitamin	Solubility	Food Source
Vitamin A (retinol)	Fat	Cod liver oil, carrots
Vitamin B1 (thiamine)	Water	Rice bran
Vitamin C (ascorbic acid)	Water	Citrus, most fresh foods
Vitamin D (calciferol)	Fat	Cod liver oil
Vitamin B2 (riboflavin)	Water	Meat, eggs
Vitamin E (tocopherol)	Fat	Wheat germ oil, unrefined vegetable oils
Vitamin B12 (cobalamins)	Water	Liver, eggs, animal products
Vitamin K (phylloquinone/phytol naphthoquinone)	Fat	Leafy green vegetables
Vitamin B5 (pantothenic acid)	Water	Meats, whole grains, and many foods
Vitamin B7 (biotin)	Water	Meats, dairy products, eggs
Vitamin B6 (pyridoxine)	Water	Meat, dairy products
Vitamin B3 (niacin)	Water	Meat, eggs, grains
Vitamin B9 (folic acid)	Water	Leafy green vegetables

Vitamins can also act as coenzymes to carry chemical groups between enzymes (e.g., folic acid carries a carbon group methylene in the cell).

Fruits and vegetables are good sources of vitamins. Until the early 1900s, vitamins were obtained solely through food intake. Therefore, changes in diet often led to vitamin deficiencies and deadly diseases (e.g., lack of citrus fruit led to scurvy, a particularly deadly disease in which collagen is not properly formed, and causes poor wound healing, bleeding gums, and severe pain). However, today, vitamins can be produced commercially and made widely available as inexpensive supplementation of dietary intake.

Fat-soluble vitamins: A, D, E, and K

The principal form of **vitamin A** is retinol; the aldehyde (retinal) and the acid (retinoic acid) are also active forms of vitamin A. Retinol can be derived directly from animal sources or through conversion from β-carotene (found abundantly in carrots) in the small intestine. Micellar solubilization renders vitamin A water-soluble, and the small intestine absorbs it passively. The unesterified retinol is then complexed with retinol-binding protein type 2, and the complex serves as a substrate for the re-esterification of the retinol by the enzyme lecithin:retinol acyltransferase. The retinyl ester is incorporated in chylomicrons and taken up by the liver. Vitamin A is stored in the liver and released to the circulation bound to retinol-binding protein only when needed. Vitamin A is important in the production and regeneration of rhodopsin of the retina and in the normal growth of the skin. People with vitamin A deficiency develop night blindness and skin lesions.

Vitamin D is a group of fat-soluble compounds collectively known as the *calciferols*. Vitamin D3 (also called *cholecalciferol* or *activated dehydrocholesterol*) in the human body is derived from two main sources: the skin, which contains a rich source of 7-dehydrocholesterol, which is rapidly converted to cholecalciferol when exposed to ultraviolet light, and dietary vitamin D3. Like vitamin A, vitamin D3 is absorbed by the small intestine passively and is incorporated into chylomicrons. During the metabolism of chylomicrons, vitamin D3 is transferred to a binding protein in plasma called the **vitamin D–binding protein.**

Unlike vitamin A, vitamin D is not stored in the liver but is distributed among the various organs depending on their lipid content. In the liver, vitamin D3 is converted to 25-hydroxycholecalciferol, which is subsequently converted to the active hormone **1,25-dihydroxycholecalciferol** in the kidneys. The latter enhances Ca^{2+} and phosphate absorption by the small intestine and mobilizes Ca^{2+} and phosphate from bones.

Vitamin D is essential for normal development and growth and the formation of bones and teeth. Vitamin D deficiency can result in **rickets**, a disorder of normal bone ossification manifested by distorted bone movements during muscular action.

The major dietary **vitamin E** is α-tocopherol. Vegetable oils are rich in vitamin E. It is absorbed by the small intestine by passive diffusion and incorporated into chylomicrons. Unlike vitamins A and D, vitamin E is transported in the circulation associated with lipoproteins and erythrocytes.

Vitamin E is a potent antioxidant and therefore prevents lipid peroxidation. Tocopherol deficiency is associated with increased red cell susceptibility to lipid peroxidation, which may explain why the red cells are more fragile in people with vitamin E deficiency than in healthy people.

Vitamin K can be derived from green vegetables in the diet or the gut flora. The vitamin K derived from green vegetables is in the form of **phylloquinones**. Vitamin K derived from bacteria in the small intestine is in **menaquinones**. The small intestine takes up phylloquinones via an energy-dependent process from the proximal small intestine. In contrast, menaquinones are absorbed from the small intestine passively, depending only on the micellar solubilization of these compounds by bile salts. Vitamin K is incorporated into chylomicrons. It is rapidly taken up by the liver and secreted together with VLDLs. No carrier protein for vitamin K has been identified.

Vitamin K is essential for the synthesis of various clotting factors by the liver. Vitamin K deficiency is associated with bleeding disorders.

Water-soluble vitamins: C, B1, B2, B6, B12, niacin, biotin, and folic acid

Most of the water-soluble vitamins are absorbed by the small intestine by both passive and active processes. The water-soluble vitamins are summarized in Table 26.5.

The major source of **vitamin C (ascorbic acid)** is green vegetables and fruits. It plays an important role in many oxidative processes by acting as a coenzyme or cofactor. It is absorbed mainly by active transport through the transporters in the ileum. The uptake process is sodium-dependent. Vitamin C deficiency is associated with **scurvy**, a disorder characterized by weakness, fatigue, anemia, and bleeding gums.

Vitamin B1 (thiamine) plays an important role in carbohydrate metabolism. The jejunum absorbs thiamine passively as well as by an active, carrier-mediated process. Thiamine deficiency results in **beriberi**, characterized by anorexia and disorders of the nervous system and heart.

Vitamin B2 (riboflavin) is a component of the two groups of flavoproteins—flavin adenine dinucleotide and flavin mononucleotide. Riboflavin plays an important role in metabolism. It is absorbed by a specific, saturable, active transport system located in the proximal small intestine. Its deficiency is associated with anorexia, impaired growth, impaired use of food, and nervous disorders.

Niacin plays an important role as a component of the coenzymes NAD(H) and NADP(H), which participate in a wide variety of oxidation–reduction reactions involving H^+ transfer.

At low concentrations, the small intestine absorbs niacin by Na^+-dependent, carrier-mediated facilitated transport. At high concentrations, it is absorbed by passive diffusion. Niacin has been used to treat hypercholesterolemia, for the prevention of coronary artery disease. It decreases plasma total cholesterol and LDL cholesterol, yet increases plasma HDL cholesterol.

Niacin deficiency is characterized by many clinical symptoms including anorexia, indigestion, muscle weakness, and skin eruptions. Severe deficiency leads to **pellagra**, a disease characterized by dermatitis, dementia, and diarrhea.

Vitamin B6 (pyridoxine) is involved in amino acid and carbohydrate metabolism. Vitamin B6 is absorbed throughout the small intestine by simple diffusion. A deficiency of this vitamin is often associated with anemia and central nervous system disorders.

Biotin acts as a coenzyme for carboxylase, transcarboxylase, and decarboxylase enzymes, which play an important role in the metabolism of lipids, glucose, and amino acids. At low luminal concentrations, biotin is absorbed by the small intestine by Na^+-dependent active transport. At high concentrations, biotin is absorbed by simple diffusion. Biotin is so common in food that deficiency is rarely observed. However, biotin deficiency occurs frequently when parenteral nutrition (nutrition administered intravenously) is administered for a long period.

Folic acid is usually found in the diet as polyglutamyl conjugates (pteroylpolyglutamates). It is required for the formation of nucleic acids, the maturation of red blood cells, and growth. An enzyme on the brush border degrades pteroylpolyglutamates to yield a monoglutamylfolate, which is taken up by enterocytes by facilitated transport. Inside enterocytes, the monoglutamylfolate is released directly into the bloodstream or converted to 5-methyltetrahydrofolate before exiting the cell. A folate-binding protein binds the free and methylated forms of folic acid in plasma. Folic acid deficiency causes a fall in plasma and red cell folic acid content and, in its most severe form, the development of megaloblastic anemia, dermatologic lesions, and poor growth. A link between folic acid deficiency and spina bifida (opening in the spine) has been demonstrated. Consequently, the U.S. Public Health Service recommends that women of childbearing age should take 0.4 mg of folic acid daily during the first 3 months of pregnancy.

The discovery of **vitamin B12 (cobalamin)** followed from the observation that patients with **pernicious anemia** who ate large quantities of raw liver recovered from the disease. Subsequent analysis of liver components isolated the cobalt-containing vitamin, which plays an important role in the production of red blood cells. A glycoprotein secreted by the parietal cells in the stomach called the **intrinsic factor** binds strongly with vitamin B12 to form a complex that is then absorbed in the terminal ileum through a receptor-mediated process (Fig. 26.24). Vitamin B12 is transported in the portal blood bound to the protein **transcobalamin**. People who lack the intrinsic factor fail to absorb vitamin B12 and develop pernicious anemia.

ELECTROLYTE AND MINERAL ABSORPTION

Nearly all of the dietary nutrients and approximately 95% to 98% of the water and electrolytes that enter the upper small intestine are absorbed. The absorption of electrolytes and minerals involves both passive and active processes, resulting in the movement of electrolytes, water, and metabolic substrates into the blood for distribution and use throughout the body.

Dietary minerals are chemical elements required for physiologic function.

Dietary minerals are the chemical elements required by the body for normal function. They can be either bulk minerals

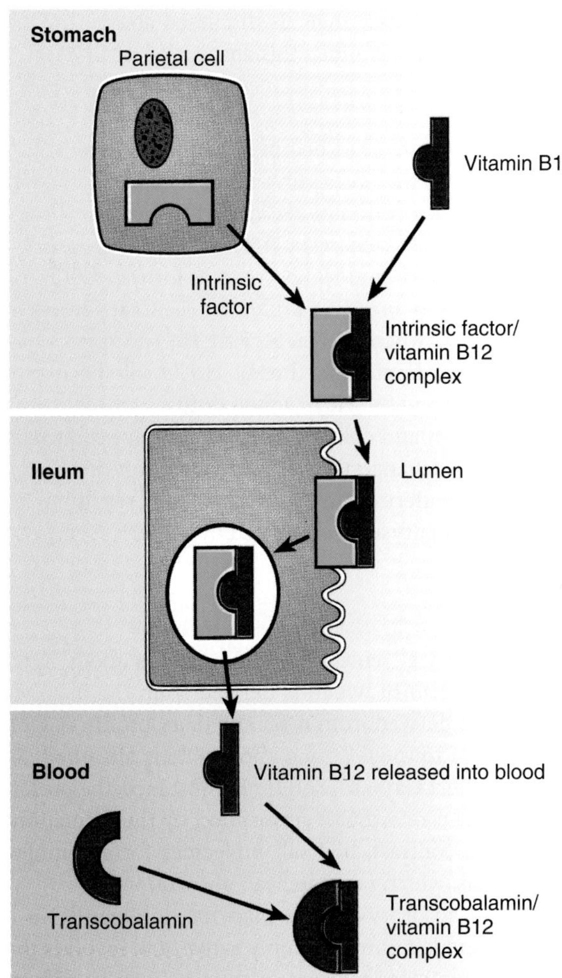

Figure 26.24 Vitamin B12 absorption depends on a gastric intrinsic factor. Vitamin B12 is one of the eight water-soluble B vitamins that play a vital role in cell metabolism. The intestinal absorption of B12 depends on an intrinsic factor secreted from the stomach parietal cell.

(required in relatively large amounts) or trace amounts (required in minute amounts). These can be acquired naturally in dietary food or added to the diet in mineral form (e.g., calcium carbonate or sodium chloride). Bulk minerals include calcium, magnesium, phosphorus, potassium, sodium, and sulfur. Trace minerals include chromium, cobalt, copper, fluorine, iodine, iron, manganese, molybdenum, selenium, and zinc.

Sodium

The GI system is well equipped to handle the large amount of Na^+ entering the GI lumen daily—on average, about 25 to 35 g of Na^+ every day. Around 5 to 8 g is derived from the diet and the rest from salivary, gastric, biliary, pancreatic, and small intestinal secretions. The GI tract is extremely efficient in conserving Na^+: Only 0.5% of intestinal Na^+ is lost in the feces. The jejunum absorbs more than half of the total Na^+, and the ileum and colon absorb the remainder. The small intestine absorbs the bulk of the Na^+ presented to it, but the colon is most efficient in conserving Na^+.

Several different mechanisms operating at varying degrees in different parts of the GI tract absorb sodium. When a meal that is hypotonic to plasma is ingested, considerable absorption of water from the lumen to the blood takes place, predominantly through tight junctions and intercellular spaces between the enterocytes, resulting in the absorption of small solutes such as Na^+ and Cl^- ions. This mode of absorption, called **solvent drag**, is responsible for a significant amount of the Na^+ absorption by the duodenum and jejunum, but it probably plays a minor role in Na^+ absorption by the ileum and colon because more distal regions of the intestine are lined by a "tight" epithelium (see Chapter 2, "Plasma Membrane, Membrane Transport, and Resting Membrane Potential").

In the jejunum, a Na^+/K^+-ATPase actively pumps Na^+ out of the basolateral surface of enterocytes (Fig. 26.25A). The result is low intracellular Na^+ concentration, and the luminal Na^+ enters enterocytes down the electrochemical

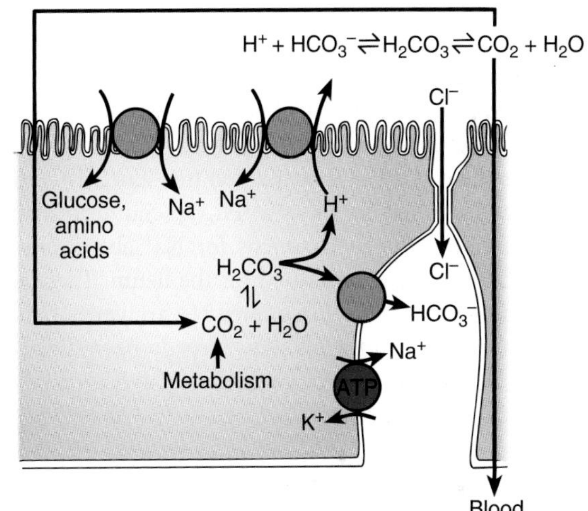

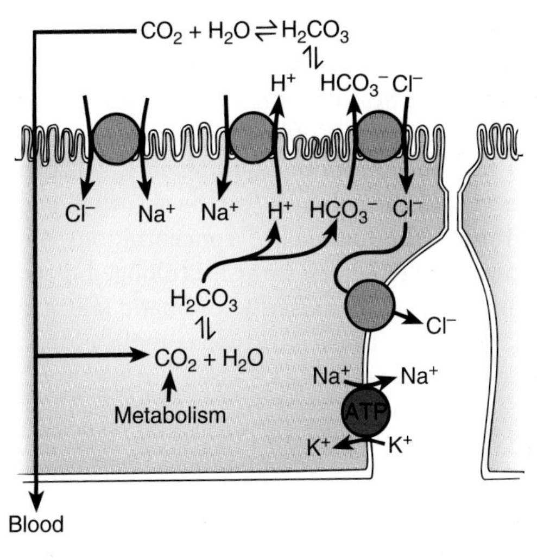

Figure 26.25 Na^+ absorption in the gut depends on an Na^+/K^+-ATPase pump. Approximately 98% of the electrolytes in the gut are reabsorbed. An Na^+/K^+-ATPase actively pumps Na^+ into the basal lateral surface. **(A)** Na^+ absorption by the jejunum. **(B)** Na^+ absorption by the ileum. ATP, adenosine triphosphate.

gradient, providing energy for the extrusion of H^+ into the lumen (via a Na^+/H^+ exchanger). The H^+ then reacts with HCO_3^- in bile and pancreatic secretions in the intestinal lumen to form H_2CO_3. Carbonic acid dissociates to form CO_2 and H_2O. The CO_2 readily diffuses across the small intestine into the blood. Another mode of Na^+ uptake is via a carrier located in the enterocyte brush border membrane, which transports Na^+ together with a monosaccharide (e.g., glucose) or an amino acid molecule (a symport type of transport).

In the ileum, the presence of an Na^+/K^+-ATPase at the basolateral membrane also creates a low intracellular Na^+ concentration, and luminal Na^+ enters enterocytes down the electrochemical gradient. Sodium absorption by Na^+-coupled symporters is not as great as in the jejunum because the small intestine has already absorbed most of the monosaccharides and amino acids (see Fig. 26.25B). Sodium chloride is transported via two exchangers located at the brush border membrane. One is a Cl^-/HCO_3^- exchanger, and the other is an Na^+/H^+ exchanger. The downhill movement of Na^+ into the cell provides the energy required for the uphill movement of the H^+ from the cell to the lumen. Similarly, the downhill movement of HCO_3^- out of the cell provides the energy for the uphill entry of Cl^- into the enterocytes. The Cl^- then leaves the cell through facilitated transport. This mode of Na^+ uptake is called Na^+/H^+–Cl^-/HCO_3^- countertransport.

In the colon, the mechanisms for Na^+ absorption are mostly similar to those described for the ileum. There is no sugar-coupled or amino acid-coupled Na^+ transport because most sugars and amino acids have already been absorbed. Sodium is also absorbed here via Na^+-selective ion channels in the apical cell membrane (electrogenic Na^+ absorption).

Potassium

The average daily intake of K^+ is about 4 g. Absorption takes place throughout the intestine by passive diffusion through the tight junctions and lateral intercellular spaces of the enterocytes. The driving force for K^+ absorption is the difference between luminal and blood K^+ concentration. The absorption of water results in an increase in luminal K^+ concentration, resulting in K^+ absorption by the intestine. In the colon, K^+ can be absorbed or secreted depending on the luminal K^+ concentration. With diarrhea, considerable K^+ can be lost. Prolonged diarrhea can be life-threatening, because the dramatic fall in extracellular K^+ concentration can cause complications such as cardiac arrhythmias.

Chloride

Most of the Cl^- ions added to the GI tract from the diet and from the various secretions of the GI system are absorbed. Intestinal chloride absorption involves both passive and active processes. In the jejunum, active Na^+ absorption generates a potential difference across the small intestinal mucosa, with the serosal side more positive than the lumen. Chloride ions follow this potential difference and enter the bloodstream via the tight junctions and lateral intercellular spaces. In the ileum and colon, enterocytes actively take up Cl^- via Cl^-/HCO_3^- exchange, as discussed above. The presence of other halides inhibits this absorption of Cl^-.

Bicarbonate

Bicarbonate ions are absorbed in the jejunum together with Na^+. In humans, the absorption of HCO_3^- by the jejunum stimulates the absorption of Na^+ and H_2O (see Fig. 26.25A). Through an Na^+/H^+ exchanger, H^+ is secreted into the intestinal lumen, where H^+ and HCO_3^- react to form H_2CO_3, which then dissociates to form CO_2 and H_2O. The CO_2 diffuses into the enterocytes, where it reacts with H_2O to form H_2CO_3 (catalyzed by carbonic anhydrase). H_2CO_3 dissociates into HCO_3^- and H^+, and the HCO_3^- then diffuses into the blood.

In the ileum and colon, HCO_3^- is actively secreted into the lumen in exchange for Cl^-. This secretion of HCO_3^- is important in buffering the decrease in pH resulting from the short-chain fatty acids produced by bacteria in the distal ileum and colon.

Calcium

The amount of Ca^{2+} entering the GI tract is about 1 g/day, approximately half of which is derived from the diet. Most dietary Ca^{2+} is derived from meat and dairy products. Of the Ca^{2+} presented to the GI tract, about 40% is absorbed. Several factors affect Ca^{2+} absorption. For instance, the presence of fatty acid can retard Ca^{2+} absorption by the formation of Ca^{2+} soap. In contrast, bile salt molecules form complexes with Ca^{2+} ions, which facilitate Ca^{2+} absorption.

Calcium absorption takes place predominantly in the duodenum and jejunum, is mainly active, and involves three steps: (1) Enterocytes take up calcium by passive diffusion through a Ca^{2+} channel, because there is a large Ca^{2+} concentration gradient; the luminal Ca^{2+} is about 5 to 10 mM, whereas free intracellular Ca^{2+} is about 100 nM. (2) Once inside the cell, Ca^{2+} is complexed with **Ca^{2+}-binding protein, calbindin D (CaBP)**. (3) At the basolateral membrane, Ca^{2+} is extruded from the enterocytes via the Ca^{2+}-ATPase pump. Calcium uptake by enterocytes, the level of CaBP in the cells, and transport by Ca^{2+}-ATPase pumps are increased by 1,25-dihydroxyvitamin D3. Once inside the cell, the Ca^{2+} ions are sequestered in the ER and Golgi membranes by binding to the CaBP in these organelles.

Calcium absorption by the small intestine is regulated by the circulating plasma Ca^{2+} concentration. Lowering of the Ca^{2+} concentration stimulates the release of parathyroid hormone, which stimulates the conversion of vitamin D to its active metabolite, 1,25-dihydroxyvitamin D3, in the kidney. This, in turn, stimulates the synthesis of CaBP and the Ca^{2+}-ATPase by the enterocytes (Fig. 26.26). Because protein synthesis is involved in the stimulation of Ca^{2+} uptake by parathyroid hormone, a lapse of a few hours usually occurs between the release of parathyroid hormone and the increase in Ca^{2+} absorption by the enterocytes.

Magnesium

Humans ingest about 0.4 to 0.5 g/d of Mg^{2+}. The absorption of Mg^{2+} seems to take place along the entire small intestine, and the mechanism involved seems to be passive.

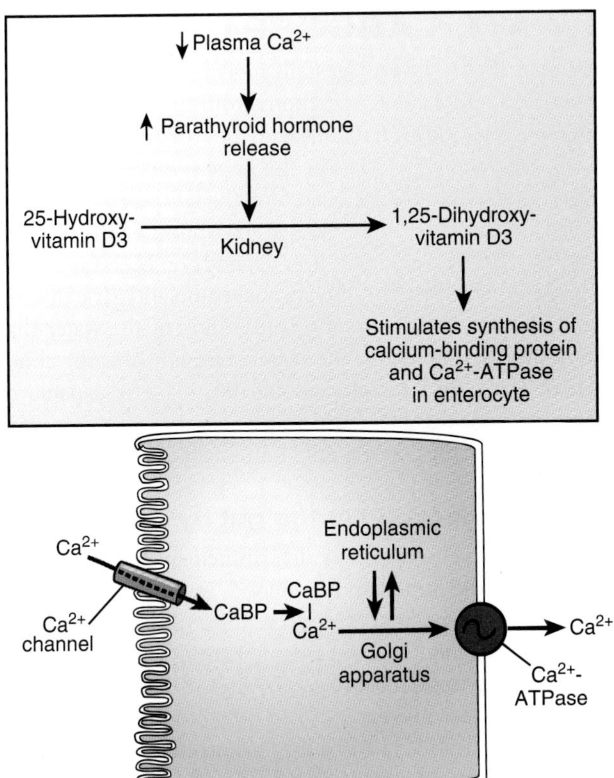

Figure 26.26 Parathyroid hormone and vitamin D are required for calcium absorption by enterocytes. Parathyroid hormone stimulates the conversion of vitamin D3 in the kidney to its active metabolite 1,25-dihydroxycholecalciferol (1,25-diOH-vitamin D3), which stimulates Ca^{2+} uptake via the Ca^{2+} channels. Additionally, it stimulates the synthesis of both Ca^{2+}-binding protein (CaBP) and the Ca^{2+}-ATPase.

Zinc

The average daily zinc intake is 10 to 15 mg, about half of which is absorbed, primarily in the ileum. A carrier located in the brush border membrane actively transports zinc from the lumen into the cell, where it can be stored or transferred into the bloodstream. Zinc plays an important role in several metabolic activities. For example, a group of metalloenzymes (e.g., alkaline phosphatase, carbonic anhydrase, and lactic dehydrogenase) require zinc to function.

Iron

Iron plays an important role not only as a component of heme but also as a participant in many enzymatic reactions. About 12 to 15 mg/day of iron enters the GI tract, where mainly the duodenum and upper jejunum absorb it (Fig. 26.27). There are two forms of dietary iron: heme and nonheme. Enterocytes absorb the heme iron intact. Nonheme iron absorption depends on both pH and concentration. Ferric (Fe^{3+}) salts are not soluble at pH 7, whereas ferrous (Fe^{2+}) salts are. Consequently, in the duodenum and upper jejunum, unless Fe^{3+} ion is chelated, it forms a precipitate. Several compounds such as tannic acid in tea and phytates in vegetables form insoluble complexes with iron, preventing absorption. Iron is absorbed by an active process via a carrier(s) located in the

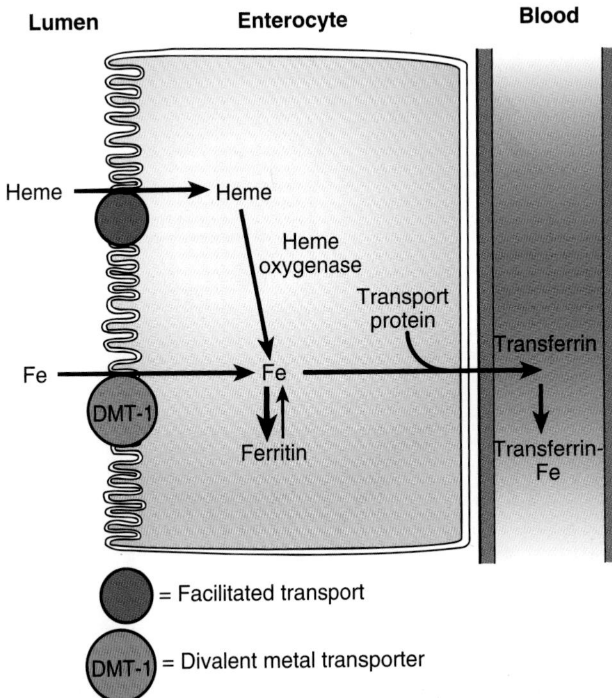

Figure 26.27 Enterocytes absorb iron in the duodenum. Iron must be in ferrous form (Fe^{2+}) to be absorbed. A ferric reductase enzyme on the enterocyte brush border reduces ferric iron Fe^{3+} to Fe^{2+}. A protein metal transporter (DMT-1) transports the iron across the enterocyte's cell membrane and into the cell. DMT-1, divalent metal transporter.

brush border membrane. One such transporter, the divalent metal transporter, is expressed abundantly in the duodenum.

Once inside the cell, heme iron is released by the action of heme oxygenase and mixed with the intracellular free iron pool. Iron is either stored in the enterocyte cytoplasm bound to the storage protein **apoferritin** to form **ferritin** or transported across the cell bound to transport proteins, which carry the iron across the cytoplasm and release it into the intercellular space. **Transferrin,** a β-globulin synthesized by the liver, binds and transports iron in the blood.

Iron absorption is closely regulated by iron storage in enterocytes and iron concentration in the plasma. Enterocytes are continuously shed into the lumen, and the ferritin contained within is also lost. Normally, iron in enterocytes is derived from the lumen and the blood (Fig. 26.28). The amount of iron stored in enterocytes regulates the amount of iron absorbed. In iron deficiency, the circulating plasma iron concentration is low, which stimulates the absorption of iron from the lumen and the transport of iron into the blood. Moreover, in a deficient state, less iron is stored as ferritin in the enterocytes, so the loss of iron through this means is significantly reduced. In iron-loaded patients, there is less absorption of iron because of the large amount of mucosal iron storage, which increases iron loss as a result of enterocyte shedding. Furthermore, because of the high level of circulating plasma iron, the transfer of iron from enterocytes to the blood is reduced. Through a combination of various mechanisms, body iron homeostasis is maintained.

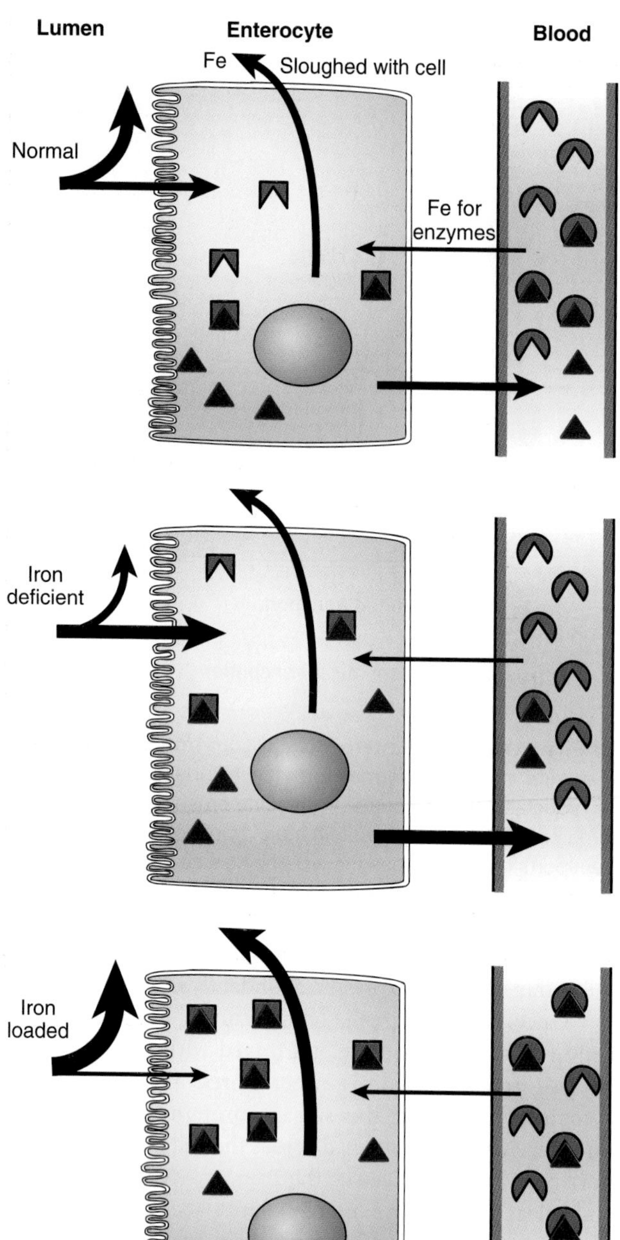

Figure 26.28 Endocyte pathways regulate the body's iron levels. In healthy subjects, the intestinal lining cells can store iron as ferritin (in which case the iron leaves the body when the cell dies and is sloughed off into feces), or the cell can move it into the blood using a protein transporter and carry it as transferrin to the rest of the body. The body regulates iron levels by regulating these pathways. For example, in patients with iron deficiency (middle figure), cells produce more ferrous reductase, DMT-1, and the protein transporter, which results in a net increase in the amount of iron going to transferrin that can be taken up by red cells. In the bottom figure, iron overload results in more iron converted to ferritin and less converted to transferrin.

WATER ABSORPTION

In human adults, the average daily intake of water is about 2 L. As shown in Table 26.6, secretions from the salivary glands, pancreas, liver, and GI tract make up most of the fluid entering the GI tract (about 7 L). Despite this large volume of fluid, only 100 mL is lost in the feces. Therefore, the GI tract is extremely efficient in absorbing water. Water absorption by the GI tract is passive. The rate of absorption depends on both the region of the intestinal tract and the luminal osmolality. The duodenum, jejunum, and ileum absorb the bulk of the water that enters the GI tract daily. The colon normally absorbs about 1.4 L of water and excretes about 100 mL. It is capable of absorbing considerably more water (about 4.5 L), however, and watery diarrhea occurs only if this capacity is exceeded.

Water is absorbed in the gut by osmosis.

Because water absorption is determined by the osmolality difference of the lumen and the blood, water can move both ways in the intestinal tract (i.e., secretion and absorption). The osmolality of blood is about 300 mOsm/kg H_2O. The ingestion of a hypertonic meal (e.g., 600 mOsm/kg H_2O) initially leads to net water movement from blood to lumen; however, as the small intestine absorbs the various nutrients and electrolytes, the luminal osmolality falls, resulting in the net water movement from lumen to blood. The water of a hypertonic meal is therefore absorbed mainly in the ileum and colon. In contrast, if a hypotonic meal is ingested (e.g., 200 mOsm/kg H_2O), net water movement is immediately from the lumen to the blood, resulting in the absorption of most of the water in the duodenum and jejunum.

TABLE 26.6 Water Intake, Absorption, and Excretion by the Gastrointestinal (GI) Tract

Source	Amount (mL)
Water added to GI tract	
Food and beverages	2,000
Salivary secretion	1,000
Biliary secretion	1,000
Gastric secretion	2,000
Pancreatic secretion	1,000
Intestinal mucosal secretion	2,000
Water absorbed in feces	
Duodenum and jejunum	4,000
Ileum	3,500
Colon	1,400
Water lost in feces	100

Chapter Summary

- Saliva assists in the swallowing of food, carbohydrate digestion, and the transport of immunoglobulins that combat pathogens.
- The major function of the gastrointestinal tract is storage, mixing, digestion, and absorption of nutrients.
- The stomach prepares chyme to aid in the digestion of food in the small intestine.
- Parietal cells secrete hydrochloric acid and intrinsic factor, and chief cells secrete pepsinogen.
- Gastrin plays an important role in stimulating gastric acid secretion.
- The acidity of gastric secretion provides a barrier to microbial invasion of the gastrointestinal tract.
- Gastric secretion is under neural and hormonal control and consists of three phases: cephalic, gastric, and intestinal.
- Gastric inhibitory peptide, secreted by K cells, is a potent inhibitor of gastric acid secretion and enhances insulin release in elevated concentrations of plasma glucose.
- Pancreatic secretion neutralizes the acids in chyme and contains enzymes involved in the digestion of carbohydrates, fat, and protein.
- Secretin stimulates the pancreas to secrete a bicarbonate-rich fluid, thereby neutralizing acidic chyme.
- Cholecystokinin stimulates the pancreas to secrete an enzyme-rich secretion.
- Pancreatic secretion is under neural and hormonal control and consists of three phases: cephalic, gastric, and intestinal.
- Bile salts play an important role in the intestinal absorption of lipids.
- Proteins are digested to form amino acids, dipeptides, and tripeptides before being taken up by enterocytes and transported in the blood.
- The gastrointestinal tract absorbs water-soluble vitamins and ions by different mechanisms.
- Most of the salt and water entering the intestinal tract, whether in the diet or in gastrointestinal secretions, is absorbed in the small intestine.
- Lipids absorbed by enterocytes are packaged and secreted as chylomicrons into the lymph.
- Carbohydrates, when digested, form maltose, maltotriose, and α-limit dextrins, which are cleaved to brush border enzymes to monosaccharides and taken up by enterocytes.
- Calcium-binding protein is involved in calcium absorption.
- Heme and nonheme iron is absorbed in the small intestine by different mechanisms.
- Minerals are classified as bulk minerals and trace minerals.

thePoint *Visit http://thePoint.lww.com for chapter review Q&A, case studies, animations, and more!*

27 Liver Physiology

ACTIVE LEARNING OBJECTIVES

Upon mastering the material in this chapter you should be able to:

- Describe how the liver lobule is functionally organized.
- Describe how the architecture of the hepatocytes allows for the rapid exchange of molecules.
- Describe how the liver is oxygenated.
- Explain the difference between phase I and phase II reactions in drug metabolism of the liver.
- Explain the mechanism by which the liver regulates blood glucose concentration.
- Explain how the liver is involved in the uptake and metabolism of lipids.
- Explain how vitamin A is stored in the liver and transported to other parts of the body when needed.
- Describe the importance and risks of iron storage in the liver.

The liver is a vital organ and is the largest internal organ in the body, constituting about 2.5% of an adult's body weight. It has a wide range of functions, including glycogen storage, protein synthesis, detoxification, red cell destruction, hormone production, and bile production for lipid emulsification. The liver also plays an important role in maintaining blood glucose levels and regulating blood lipids by the amount of very low–density lipoproteins (VLDLs) it secretes.

The liver has two unique features. It has a dual blood supply, and it is the only internal human organ capable of natural regeneration of lost tissue. With as little as 25% of the liver remaining, it can regenerate into a whole liver again. This is due to liver cells (**hepatocytes**) reentering the cell cycle. That is, hepatocytes go from a quiescent G_0 phase to the G_1 phase and undergo mitosis.

In terms of blood supply, the hepatic portal vein and hepatic arteries provide the dual blood supply to the liver. During rest, it receives 25% of the cardiac output via the hepatic portal vein and hepatic artery. Supplying approximately 75% of the liver's blood supply, the hepatic portal vein carries venous blood drained from the spleen and gastrointestinal (GI) tract and its associated organs. About half of the liver's oxygen demand is met by the hepatic portal vein and the other half is met by hepatic arteries.

This chapter summarizes the liver's many functions including (1) detoxification of hormones, drugs, and waste products; (2) metabolism of carbohydrates, proteins, and fats; (3) storage of iron and vitamins; and (4) hormone production.

LIVER STRUCTURE AND FUNCTION

The liver is a vital organ and is essential for maintaining optimal function of other organs in the body. In addition, the liver interacts with the cardiovascular, renal, and immune systems to maintain homeostasis. Given the wide range of functions, there is surprisingly little specialization among cells within the liver. The basic functional unit of the liver is the **liver lobule**, which is hexagonal in shape and built around a central vein (Fig. 27.1). The lobule itself is made up of many cellular plates that radiate from the central vein like spokes in a bicycle wheel. Each plate consists of specialized liver cells, hepatocytes. These hepatocytes perform the same wide variety of metabolic functions and secretory tasks. Their specialization comes from discrete organelles within each hepatocyte.

Architectural arrangement of hepatocytes in the liver lobule enhances the rapid exchange of material.

Hepatocytes are highly specialized cells. The bile canaliculus is usually lined by two hepatocytes and is separated from the pericellular space by tight junctions, which are impermeable and, thus, prevent the mixing of contents between the bile canaliculus and the pericellular space (see Fig. 27.1). The bile from the bile canaliculus drains into a series of ducts and eventually joins the pancreatic duct near which it enters the duodenum. A sphincter located at the junction between the bile duct and the duodenum, the sphincter of Oddi, partly regulates drainage of bile into the duodenum. The pericellular space, the space between two hepatocytes, is continuous with the perisinusoidal space (see Fig. 27.1). A layer of **sinusoidal endothelial cells** separates the **perisinusoidal space** from the sinusoid. Hepatocytes possess numerous, fingerlike projections that extend into the perisinusoidal space, greatly increasing the surface area over which hepatocytes contact the perisinusoidal fluid.

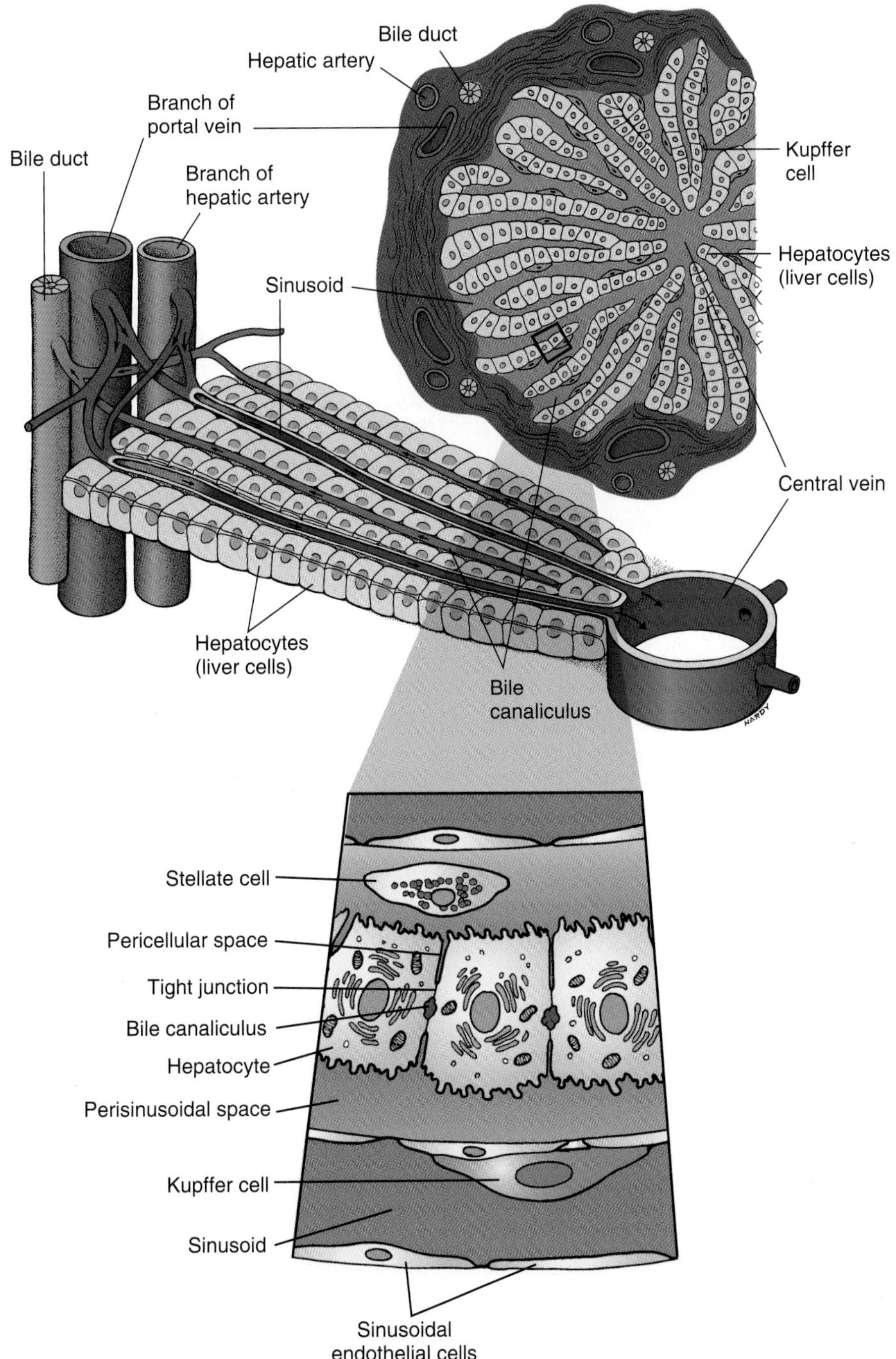

Figure 27.1 Functional anatomy of the liver. The basic unit of the liver is the lobule, which is arranged in a hexagonal fashion. The lobule is delineated by vascular and bile channels. The lobule contains specialized cells, such as hepatocytes, sinusoidal cells, and Kupffer cells.

Endothelial cells of the liver, unlike those in other parts of the cardiovascular system, lack a basement membrane. Furthermore, they have sieve-like plates that permit the ready exchange of materials between the perisinusoidal space and the sinusoid. Electron microscopy has demonstrated that even particles, as big as chylomicrons (80–500 nm wide), can penetrate these porous plates. Although the barrier between the perisinusoidal space and the sinusoid is permeable, it does have some sieving properties. For example, the protein concentration of hepatic lymph, assumed to derive from the perisinusoidal space, is lower than that of plasma by about 10%.

The only liver function not carried out by the hepatocytes is phagocytic activity. This function is accomplished by resident macrophages termed **Kupffer cells** that line the hepatic sinusoids and are part of the reticuloendothelial system. These are resident macrophages that play an important role in removing unwanted material (e.g., bacteria, virus particles, fibrin–fibrinogen complexes, damaged erythrocytes, and immune complexes) from the circulation via **endocytosis**.

Some perisinusoidal cells contain distinct lipid droplets in the cytoplasm. These fat-storage cells are called **stellate cells** or **Ito cells**. The lipid droplets contain vitamin A. Through complex and typically inflammatory processes, stellate cells become transformed to myofibroblasts, which then become capable of secreting collagen and extracellular matrix into the perisinusoidal space.

Portal vein provides most of the blood supply to the liver.

The liver is highly vascularized and is characterized as a high-flow, low-pressure system. The hepatic portal vein provides about 70% to 80% of the liver's blood supply; the hepatic artery provides the rest. Blood in the hepatic portal vein is poorly oxygenated, unlike that from the hepatic artery. However, oxygen is provided from both sources; approximately half of the liver's oxygen demand is met by the hepatic portal vein and half is met by the hepatic arteries. What accounts for the high oxygen uptake from the portal vein in which the oxygen tension is so low? Recall from Chapter 19, "Gas Transfer and Transport," that oxygen uptake is highly dependent on blood flow. The large flow in the hepatic portal vein offsets the low oxygen tension and accounts for half of the oxygen taken up by the liver. The portal vein branches repeatedly, forming smaller venules that eventually empty into the sinusoids. The hepatic artery branches to form arterioles and then capillaries, which also drain into the sinusoids. Liver sinusoids can be considered specialized capillaries. As previously mentioned, the hepatic sinusoid is porous and allows the rapid exchange of materials between the perisinusoidal space and the sinusoid. The sinusoids empty into the central veins, which subsequently join to form the hepatic vein, which then joins the inferior vena cava.

Liver as blood reservoir

The vasculature in the liver is characterized as a high-flow, high-compliance, and low-resistance system. Because of the high compliance and low resistance, large quantities of blood can be stored in its vascular tree. Under normal conditions, the liver contains about 10% of the body's total blood volume.

Hepatic blood volume and flow vary inversely with GI activity. Blood flow increases after eating, whereas blood volume decreases. Conversely, during sleep, flow decreases but blood volume slightly increases. The splanchnic arterioles predominantly regulate blood flow to the intestines and spleen and, in turn, in the portal vein. In this way, eating results in increased blood flow to the intestines followed by increased liver blood flow. Portal vein pressure is normally low. Increased resistance to portal blood flow results in **portal hypertension**. Portal hypertension is the most common complication of chronic liver disease and accounts for a large percentage of the morbidity and mortality associated with chronic liver diseases.

The blood storage function is especially important in pathophysiologic situations. For example, congestive heart failure causes the pressure to rise in the liver. The liver vasculature can expand and accommodate an extra liter of blood. Conversely, during hemorrhaging, the liver is capable of supplying extra blood to help maintain central circulating blood volume.

Clinical Focus / 27.1

Viral Hepatitis

Viral hepatitis refers to the inflammation of the liver caused by the invasion of a number of specific viruses. Many hepatitis viruses have been identified; they are named types A, B, C, D, E, F (not confirmed), and G. Undoubtedly, the known number of viruses that attack the liver will grow as our knowledge of the subject expands. Because of the limitation of space, discussion will focus on the three most common types of hepatitis, A, B, and C.

Viral hepatitis A is a common type of acute liver inflammation caused by the infection of the hepatitis A virus. It is acute but never becomes chronic. It is highly contagious and can spread from person to person like other viral infections. The spread of viral hepatitis A can be rapid and widespread because it is usually caused by unsanitary conditions that allow water or food to become contaminated by human waste containing the hepatitis A virus. Hepatitis A can also spread from person to person through close contact, such as through the passage of oral secretions via intimate kissing and other sexual practices or contact with feces resulting from improper or poor handwashing. Interestingly, some people can harbor the virus with no apparent symptoms; others, however, exhibit symptoms that mimic severe flu. Patients who are infected with hepatitis A usually recover on their own and do not develop chronic hepatitis or cirrhosis.

Hepatitis B, unlike hepatitis A, is a chronic inflammatory liver disease. It is prevalent in the United States, with 200,000 to 300,000 new cases of viral hepatitis B virus infection reported each year. Hepatitis B is spread through sexual contact and the transfer of blood or serum through shared needles in drug abusers, accidental needle sticks with needles contaminated with infected blood, blood transfusions, and hemodialysis and by infected mothers to their newborns. The hepatitis B virus can remain in a person for years without manifesting any symptoms. About 6% to 10% of patients with hepatitis B viruses develop chronic liver inflammation, lasting between 6 months and years or decades. Hepatitis B infections can potentially develop into more advanced stages of liver disease associated with cirrhosis of the liver. Cirrhosis is characterized by abnormal structure and function of the liver. The diseases that lead to cirrhosis do so because they injure and kill liver cells and because the inflammation and repair that are associated with the dying liver cells cause scar tissue to form.

The other common virus that attacks the liver is *hepatitis C*. The spread of hepatic C virus is similar to that of hepatitis B. However, the transmission of the hepatitis C virus through sexual contact is not as prevalent as hepatitis B. Patients with chronic hepatitis C infection can continue to infect others. Patients with chronic hepatitis C infection are at risk for developing cirrhosis, liver failure, and liver cancer.

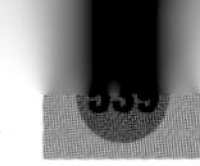

Liver permeability results in large quantities of lymph formation with high flow rates.

The hepatic sinusoids are very porous and allow a large flux of fluid and protein to move into the perisinusoidal space. Because of the high permeability, sinusoidal epithelia form large quantities of lymph. Consequently, under resting conditions, about half of the formed lymph in the body comes from the liver. To accommodate the large quantities of lymph and high lymph flow, the hepatic lymphatic system is extensive and is organized around three main areas: adjacent to the central veins, adjacent to the portal veins, and coursing along the hepatic artery. As in other organs, these channels drain fluid and proteins, but the protein concentration is highest in lymph from the liver.

In the liver, the largest space drained by the lymphatic system is the perisinusoidal space. Disturbances in the balance of filtration and drainage are the primary causes of **ascites**, the accumulation of serous fluid in the peritoneal cavity. Ascites is another common cause of morbidity in patients with chronic liver disease.

Tissue regeneration in the adult liver is a unique feature to maintain optimal metabolic function.

As mentioned previously, of all of the solid organs in the body, the adult liver is the only one that can regenerate. There appears to be a critical ratio between functioning liver mass and body mass. Deviations in this ratio trigger a modulation of either hepatocyte proliferation or apoptosis to maintain the liver's optimal size for metabolic function. Peptide growth factors—transforming growth factor-α, hepatocyte growth factor, and epidermal growth factor—have been the best-studied stimuli of hepatocyte DNA synthesis. After these peptides bind to their receptors on the remaining hepatocytes and work their way through a myriad of transcription factors, gene transcription is accelerated, resulting in increased cell number and increased liver mass.

Alternatively, enhanced hepatocyte apoptosis rates achieve a decrease in liver volume. Apoptosis is a carefully programmed process by which cells kill themselves while maintaining the integrity of their cellular membranes. In contrast, cell death that results from necroinflammatory processes is characterized by a loss of cell membrane integrity and the activation of inflammatory reactions. Proapoptotic signals, such as tumor necrosis factor, mediate liver cell suicide. The liver's ability to regenerate is a "double-edged sword." Although regeneration is most beneficial under normal circumstances, its ability to rapidly regenerate tissue becomes a liability with certain diseases. For example, patients diagnosed with primary hepatic carcinoma often undergo a partial hepatectomy to remove malignant tumors. The hepatectomy not only stimulates tissue regeneration but also activates dormant cancer cells, causing the malignancy to further spread in the liver.

DRUG METABOLISM IN THE LIVER

The liver is well suited for its ability to metabolize and detoxify many drugs, including penicillin, ampicillin, erythromycin, and sulfonamides. In a similar manner, many hormones secreted by endocrine glands are either metabolized and/or excreted by the liver. These include thyroxin and the steroid hormones (aldosterone, estrogen, and cortisol). The liver also regulates calcium and is the major route of calcium secretion via the bile acids, which pass into the gut and are excreted in feces.

Hepatocytes also play an important role in the metabolism of **xenobiotics**—compounds that are foreign to the body, some of which are toxic. Most xenobiotics are introduced into the body with food, absorbed by the skin, or taken in with breathing. The kidneys ultimately dispose of these substances, but for effective elimination, the drug or its metabolites must be first rendered hydrophilic (polar,

Clinical Focus / 27.2

Esophageal Varices, a Common Manifestation of Portal Hypertension

Chronic liver injury can lead to a sequence of changes that terminates with fatal bleeding from esophageal blood vessels. In most forms of chronic liver injury, stellate cells are transformed into collagen-secreting myofibroblasts. These cells are responsible for the deposition of collagen into the sinusoids, thereby interfering with the exchange of compounds between the blood and hepatocytes and increasing resistance to portal venous flow. The resistance appears to be further increased when stellate cells contract. The combination of increased resistance and reduced clearance of vasoconstrictors that have affected arteriolar beds results in increased portal pressure and decreased liver blood flow. This disorder is seen in approximately 80% of patients with **cirrhosis**. In a compensatory effort, new channels are formed or dormant venous tributaries are expanded, resulting in the formation of varicose veins in the abdomen. Although varicose veins develop in many areas, portal pressure increases are least opposed in the esophagus because of the limited connective tissue support at the base of the esophagus. This structured condition, along with the negative intrathoracic pressure, favors the formation and rupture of **esophageal varices**. Approximately 30% of those who develop an esophageal variceal hemorrhage die during the episode of bleeding, making it one of the most lethal medical illnesses.

Currently, there are no well-recognized treatments to reverse cirrhosis, but numerous strategies are used to reduce portal hypertension and bleeding. Chief among these is the use of nonselective β blockers, which have a vasodilator effect. Additionally, bleeding varices are frequently treated by endoscopic ligation of the varices and radiologically or surgically placed portal venous shunts to reduce the portal pressure.

water-soluble). This is because reabsorption of a substance by the renal tubules depends on its hydrophobicity. The more hydrophobic (nonpolar, lipid-soluble) a substance is, the more likely it will be reabsorbed. Many drugs and metabolites are hydrophobic, and the liver converts them into hydrophilic compounds.

Liver metabolism of drugs and xenobiotics occurs in two phases.

Two reactions (phases I and II), catalyzed by different enzyme systems, are involved in the conversion of xenobiotics and drugs into hydrophilic compounds. In **phase I reactions**, the introduction of one or more polar groups biotransforms the parent compound into more polar compounds. The common polar groups are hydroxyl (OH) and carboxyl (COOH). Most phase I reactions involve oxidation of the parent compound. The enzymes involved are mostly located in the smooth endoplasmic reticulum (SER); some are located in the cytoplasm. For example, alcohol dehydrogenase is located in the cytoplasm of hepatocytes and catalyzes the rapid conversion of alcohol to acetaldehyde. It may also play a role in the dehydrogenation of steroids. Often the metabolites from phase I reactions are inactive, although in some instances the activity is only modified.

The enzymes involved in phase I reactions of drug biotransformation are present as an enzyme complex composed of the **NADPH–cytochrome P450 reductase** and a series of hemoproteins called **cytochrome P450** (Fig. 27.2). The drug combines with the oxidized cytochrome $P450^{3+}$ to form the cytochrome $P450^{3+}$–drug complex. This complex is then reduced to the cytochrome $P450^{2+}$–drug complex, catalyzed by the enzyme NADPH–cytochrome P450 reductase. The reduced complex combines with molecular oxygen to form an oxygenated intermediate. One atom of the molecular oxygen then combines with two H^+ atoms and an electron to form water. The other oxygen atom remains bound to the cytochrome $P450^{3+}$–drug complex and is transferred from the cytochrome $P450^{3+}$ to the drug molecule. The drug product with an oxygen atom incorporated is released from the complex. The cytochrome $P450^{3+}$ released can then be recycled for the oxidation of other drug molecules.

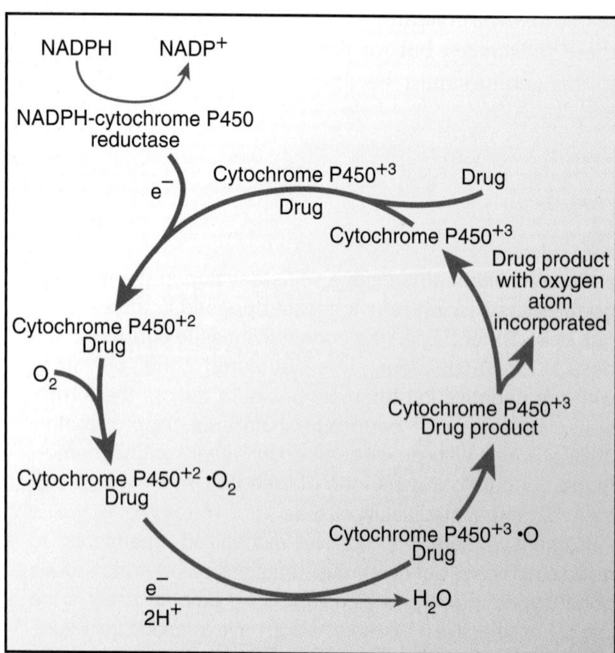

Figure 27.2 Two different enzyme systems catalyze drug metabolism by the liver. The phase I reaction is illustrated here.

In **phase II reactions**, the phase I reaction products undergo conjugation with several compounds to render them more hydrophilic. Glucuronic acid is the substance most commonly used for conjugation, and the enzymes involved are the glucuronosyltransferases. Other molecules used in conjugation are glycine, taurine, and sulfates. Drug conjugations were once believed to represent the terminal inactivation events, thus rendering the drug nontoxic. However, this is not always the case, because we know that certain conjugation reactions, for example, *N*-acetylation of isoniazid, may lead to the formation of reactive species responsible for the hepatotoxicity of the drug. Isoniazid is used alone or in conjunction with other drugs to treat tuberculosis.

Genetics, nutrition, and age significantly affect the liver's ability to metabolize drugs and detoxify foreign material.

The enzyme systems in phase I and II reactions are age dependent. These systems are poorly developed in human newborns because newborns' ability to metabolize any given drug is lower than that of adults. Older adults also have a lower capacity than young adults to metabolize drugs.

Nutritional factors can also affect the enzymes involved in phase I and II reactions. Insufficient protein in the diet to sustain normal growth results in the production of fewer of the enzymes involved in drug metabolism.

It is well known that certain factors, such as polycyclic aromatic hydrocarbons, can induce drug-metabolizing enzymes. Cigarette smokers inhale polycyclic aromatic hydrocarbons, thereby increasing the metabolism of certain drugs such as caffeine.

The role of genetics in the regulation of drug metabolism by the liver is less well understood. Briefly, a single gene or several genes (polygenic control) can control drug metabolism by the liver. Careful study of the metabolism of a certain drug in a controlled population can provide important clues as to whether its metabolism is under single gene or polygenic control. Genetic variability combined with the induction or inhibition of P450 enzymes by other drugs or compounds can have a profound effect on what is a safe and effective dose.

ENERGY METABOLISM IN THE LIVER

The liver is a pivotal organ in regulating the metabolism of carbohydrates, lipids, and proteins. The liver is especially important in maintaining blood glucose homeostasis by storing glycogen, which removes excess glucose, or by increasing blood glucose through the conversion of other substances, such as amino acids, into glucose (**gluconeogenesis**).

Hepatic portal vein transports nutrients, water-soluble vitamins, and minerals to the liver.

Most water-soluble nutrients and water-soluble vitamins and minerals absorbed from the small intestine are transported via the portal blood to the liver. The nutrients transported in portal blood include amino acids, monosaccharides, and fatty acids (predominantly short-chain and medium-chain forms). Short-chain fatty acids are largely derived from the fermentation of dietary fibers by bacteria in the colon. Some dietary fibers, such as pectin, are almost completely digested to form short-chain fatty acids (or volatile fatty acids), whereas cellulose is not well digested by the bacteria. The portal blood transports only a small amount of long-chain fatty acids, bound to albumin; most is transported in intestinal lymph as triglyceride-rich lipoproteins (chylomicrons).

Liver plays an important role in blood glucose buffering.

The liver is especially important in maintaining normal blood glucose levels. The ability to store glycogen allows the liver to remove excess glucose from the blood and then return it when blood glucose levels begin to fall. This process is called **glucose buffering** (Fig. 27.3). Glucose buffering by the liver occurs in the following manner. After the ingestion of a meal (usually within 1–2 hours), blood glucose increases to a concentration of 120 to 150 mg/dL (6.6–8.3 mmol/L). The hepatocytes take up glucose by a facilitated carrier-mediated process, and glucose is converted to glucose 6-phosphate and subsequently uridine diphosphate-glucose (UDP-glucose). UDP-glucose can be used for glycogen synthesis, or **glycogenesis**. It is generally believed that blood glucose is the major precursor of glycogen. However, recent evidence seems to indicate that the lactate in blood (from the peripheral metabolism of glucose) is also a major precursor of glycogen. Amino acids (e.g., alanine) can supply pyruvate to synthesize glycogen.

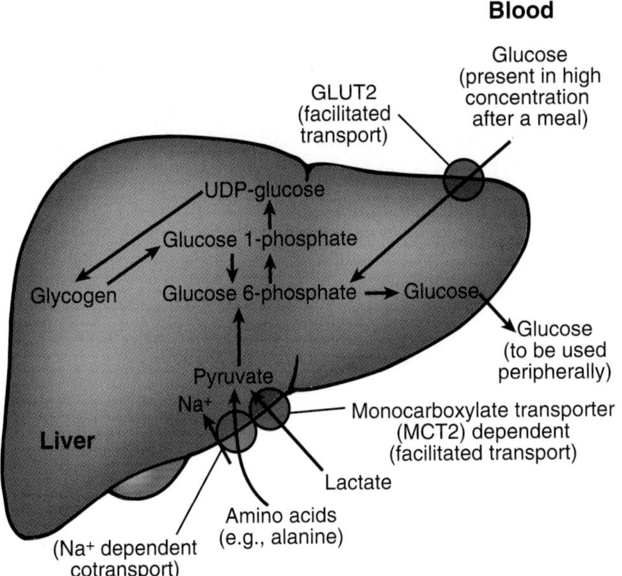

Figure 27.3 The liver is a key organ in regulating carbohydrate metabolism. Glycogen is the main carbohydrate stored by the liver. UDP-glucose (uridine diphosphate-glucose) is a precursor to glycogen.

Glycogen is the main carbohydrate stored in the liver, amounting to as much as 7% to 10% of the weight of a normal, healthy liver. During fasting, **glycogenolysis** breaks down glycogen. The enzyme **glycogen phosphorylase** catalyzes the cleavage of glycogen into glucose 1-phosphate. Glycogen phosphorylase acts only on the α-1,4-glycosidic bond, and the enzyme α-1,6-glucosidase is used to break the α-1,6-glycosidic bonds.

Glucose 1-phosphate is converted to glucose 6-phosphate by the enzyme phosphoglucomutase. The enzyme **glucose-6-phosphatase,** which is present in the liver but not in muscle or the brain, converts glucose 6-phosphate to glucose. This last reaction enables the liver to release glucose into the circulation. Glucose 6-phosphate is an important intermediate in carbohydrate metabolism because it can be channeled either to provide blood glucose or to form glycogen.

Both glycogenolysis and glycogenesis are hormonally regulated. The pancreas secretes insulin into the portal blood. Therefore, the liver is the first organ to respond to changes in plasma insulin levels, to which it is sensitive. For instance, a doubling of portal insulin concentration completely shuts down hepatic glucose production. About half the insulin in portal blood is removed as it passes through the liver. Insulin tends to lower blood glucose by stimulating glycogenesis and suppressing glycogenolysis and gluconeogenesis. Glucagon, in contrast, stimulates glycogenolysis and gluconeogenesis, raising blood sugar levels. Epinephrine stimulates glycogenolysis.

The liver regulates the blood glucose concentrations within a narrow limit, 70 to 100 mg/dL. Although patients with liver disease might be expected to have difficulty regulating blood glucose, this is usually not the case because of the relatively large reserve of hepatic function. However, those with chronic liver disease occasionally have reduced glycogen synthesis and reduced gluconeogenesis. Some patients with advanced liver disease develop portal hypertension, which induces the formation of portosystemic shunting, resulting in elevated arterial blood levels of insulin and glucagon as a result of suppressed liver removal.

Glucokinase is a liver enzyme specific for monosaccharide metabolism.

A reaction catalyzed by the enzyme hexokinase first phosphorylates monosaccharides. In the liver (but not in the muscle), there is a specific enzyme (**glucokinase**) for the phosphorylation of glucose to form glucose 6-phosphate. Depending on the energy requirement, the glucose 6-phosphate is channeled to glycogen synthesis or used for energy production by the glycolytic pathway.

The liver takes up fructose, and fructokinase phosphorylates fructose to form fructose 1-phosphate. This molecule is either isomerized to form glucose 6-phosphate or metabolized by the glycolytic pathway. The glycolytic pathway uses fructose 1-phosphate more efficiently than glucose 6-phosphate.

Galactose is an important sugar used not only to provide energy but also in the biosynthesis of glycoproteins and glycolipids. When the liver takes up galactose, it is phosphorylated to form galactose 1-phosphate, which then reacts with uridine diphosphate-glucose, or **UDP-glucose**, to form UDP-galactose and glucose 1-phosphate. The UDP-galactose can be used for glycoprotein and glycolipid biosynthesis or converted to UDP-glucose, which can then be recycled.

Gluconeogenesis in the liver is an important mechanism by which blood glucose is maintained during fasting.

Gluconeogenesis is the production of glucose from noncarbohydrate sources, such as fat, amino acids, and lactate. The process is energy dependent and starts with pyruvate as the substrate. The energy required is derived predominantly from the β-oxidation of fatty acids. Pyruvate can be derived from lactate and the metabolism of glucogenic amino acids—those that can contribute to the formation of glucose. The two major organs involved in the production of glucose from noncarbohydrate sources are the liver and the kidneys. However, because of its size, the liver plays a far more important role than the kidney in the production of sugar from noncarbohydrate sources.

Gluconeogenesis is important in maintaining blood glucose concentrations, especially during fasting. The red blood cells and renal medulla totally depend on blood glucose for energy, and glucose is the preferred substrate for the brain. Most amino acids can contribute to the carbon atoms of the glucose molecule, and alanine from muscle is the most important. The rate-limiting factor in gluconeogenesis is not the liver enzymes but the availability of substrates. Epinephrine and glucagon stimulate gluconeogenesis, but insulin greatly suppresses it. Thus, in type I diabetic patients, gluconeogenesis is greatly stimulated, contributing to the hyperglycemia observed in these patients (see Chapter 15, "Microcirculation and Lymphatic System").

Liver plays a vital role in lipid metabolism.

Most cells in the body can metabolize fatty acids. However, certain aspects of lipid metabolism must occur first in the liver before cells can metabolize fatty acids (Fig. 27.4). The liver takes up free fatty acids and lipoproteins (complexes of lipid and protein) from the plasma. Lipid circulates in the plasma as lipoproteins because lipid and water are not miscible; the lipid droplets coalesce in an aqueous medium. The protein and phospholipid on the surface of the lipoprotein particles stabilize the hydrophobic triglyceride core of the particle.

During fasting, fatty acids mobilize from adipose tissue, and the liver takes them up. The hepatocytes use them to provide energy via β-oxidation to generate ketone bodies and to synthesize the triglyceride necessary for **very-low-density lipoprotein (VLDL)** formation. After feeding, chylomicrons from the small intestine are metabolized peripherally, and the liver rapidly takes up the chylomicron remnants that are formed. The fatty acids derived from the triglycerides of the chylomicron remnants are used for the formation of VLDLs or for energy production via β-oxidation.

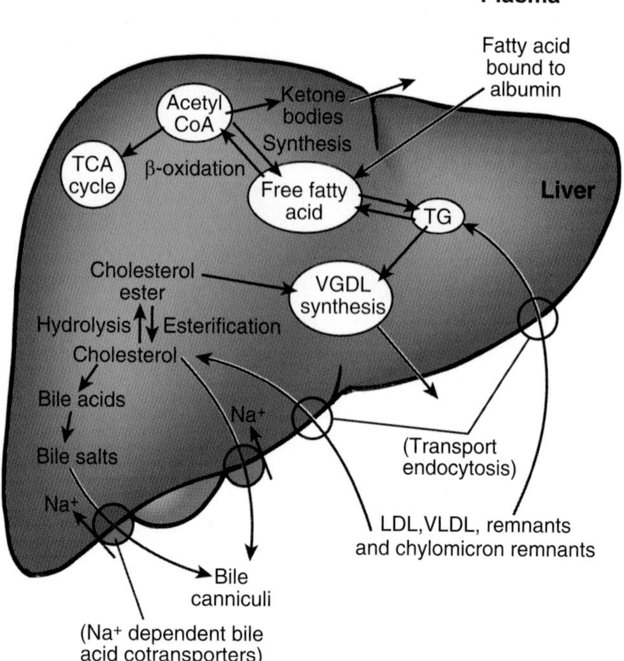

Figure 27.4 The liver plays an important role in lipid metabolism. The liver is involved in the oxidation of fatty acids, the synthesis of very low–density lipoproteins (VLDLs), and the regulation of blood triglycerides, low-density lipoproteins (LDLs), and circulating high-density lipoprotein (HDLs). Acetyl-CoA, acetyl coenzyme A; TCA cycle, tricarboxylic acid cycle; TG, triglyceride.

Fatty acid metabolism

Fatty acids derived from plasma can be metabolized in the mitochondria of hepatocytes by β-oxidation to provide energy. Fatty acids are broken down to form acetyl coenzyme A (acetyl-CoA). Acetyl-CoA can be used in the tricarboxylic acid cycle to produce adenosine triphosphate, to synthesize fatty acids, and to form ketone bodies. Because fatty acids are synthesized from acetyl-CoA, any substances that contribute to acetyl-CoA, such as carbohydrate and protein sources, enhance fatty acid synthesis.

The liver is one of the main organs involved in fatty acid synthesis. Palmitic acid is synthesized in the hepatocellular cytosol; the other fatty acids synthesized in the body are derived by shortening, elongating, or desaturating the palmitic acid molecule.

Lipoprotein metabolism

One of the major functions of the liver in lipid metabolism is synthesis of lipoproteins. The four major classes of circulating plasma lipoproteins are chylomicrons, VLDLs, low-density lipoproteins (LDLs), and high-density lipoproteins (HDLs) (Table 27.1). These lipoproteins, which differ in chemical composition, are usually isolated from plasma according to their flotation properties.

TABLE 27.1 Plasma Lipoproteins

Lipoprotein	Source	Function
Chylomicron	Intestine	Composition: 1% protein and 99% lipid; chylomicron is responsible for transporting lipids from the gut to liver, heart, and skeletal muscle
VLDL	Liver	Composition: 7%–10% protein, 90%–93% lipid; VLDL is converted to LDL in liver and functions as the body's internal transport system for lipids (triglycerides, phospholipids, and cholesterol)
LDL	Chylomicron/VLDL	Composition: 20% protein, 80% lipid; transports cholesterol into cells
HDL	Chylomicron/VLDL	Composition: 35%–60% protein, 40%–65% lipid; smallest in size of all the lipoproteins and functions in the transport of cholesterol and triglycerides; HDL carries about 30% of blood cholesterol and is able to remove cholesterol from atheroma from blood vessels and carry it back to the liver; sometimes called *good cholesterol*

VLDL, very-low-density lipoprotein; LDL, low-density lipoprotein; HDL, high-density lipoprotein.

Chylomicrons are the lightest of the four lipoprotein classes, with a density of <0.95 g/mL. They are made only by the small intestine and are produced in large quantities during fat ingestion. Their major function is to transport the large amount of absorbed fat to the bloodstream via the lymphatic system.

VLDLs are denser and smaller than chylomicrons. The liver synthesizes about 10 times more circulating VLDLs than the small intestine. Like chylomicrons, VLDLs are triglyceride-rich and carry most of the triglyceride from the liver to the other organs. **Lipoprotein lipase** breaks down the triglyceride of VLDLs to yield fatty acids, which can be metabolized to provide energy. The human liver normally has a considerable capacity to produce VLDLs, but in acute or chronic liver disorders, this ability is significantly compromised. Liver VLDLs are associated with an important class of proteins, the **apo B proteins**. The two forms of circulating apo B are B48 and B100. The human liver makes only apo B100, which has a molecular weight of about 500,000. Apo B100 is important for the hepatic secretion of VLDL. In **abetalipoproteinemia**, apo B and the secretion of VLDLs are blocked. Large lipid droplets can be seen in the cytoplasm of the hepatocytes of patients with abetalipoproteinemia.

Although considerable amounts of circulating plasma LDLs and HDLs are produced in the plasma, the liver also produces a small amount of these two cholesterol-rich lipoproteins. LDLs are denser than VLDLs, and HDLs are denser than LDLs. The function of LDLs is to transport cholesterol ester from the liver to the other organs. HDLs are believed to remove cholesterol from the peripheral tissue and transport it to the liver.

Precursors and hormones, such as estrogen and thyroid hormones, regulate the formation and secretion of lipoproteins. For instance, during fasting, the fatty acids in VLDLs are derived mainly from fatty acids mobilized from adipose tissue. In contrast, during fat feeding, fatty acids in VLDLs produced by the liver are largely derived from chylomicrons.

As noted earlier, the fatty acids taken up by the liver can be used for β-oxidation and ketone body formation. The relative amounts of fatty acid channeled for these various purposes largely depend on the person's nutritional and hormonal status. More fatty acid is channeled to ketogenesis or β-oxidation when the supply of carbohydrate is short (during fasting) or under conditions of high circulating glucagon or low circulating insulin (diabetes mellitus). In contrast, more of the fatty acid is used for synthesis of triglyceride for lipoprotein export when the supply of carbohydrate is abundant (during feeding) or under conditions of low circulating glucagon or high circulating insulin.

Familial hypercholesterolemia, a disorder in which the liver fails to produce the LDL receptor, exemplifies the importance of the liver in lipoprotein metabolism. When LDL binds its receptor, it is internalized and catabolized in the hepatocyte. Consequently, the LDL receptor is crucial for the removal of LDL from the plasma. People suffering from familial hypercholesterolemia usually have high plasma LDLs, which predisposes them to early coronary heart disease. Often the only effective treatment is a liver transplantation.

The liver also plays an important role in the uptake of chylomicrons after their metabolism. After the chylomicrons produced by the small intestine enter the circulation, lipoprotein lipase on the endothelial cells of blood vessels acts on them to liberate fatty acids and glycerol from the triglycerides. As metabolism progresses, the chylomicrons shrink, resulting in the detachment of free cholesterol, phospholipid, and proteins and the formation of HDL. Chylomicrons are converted to **chylomicron remnants** during metabolism, and the liver rapidly takes up chylomicron remnants via chylomicron remnant receptors.

Ketone body production

Most organs, not including the liver, can use ketone bodies as fuel. For example, during prolonged fasting, the brain shifts to use ketone bodies (acetoacetate and β-hydroxybutyrate) for energy, although glucose is the preferred fuel for the brain. Their formation by the liver is not only normal but also physiologically important. For instance, during fasting, a rapid depletion of the glycogen stores in the liver occurs, resulting in a shortage of substrates (e.g., oxaloacetate) for the citric acid cycle. There is also a rapid mobilization of fatty acids from adipose tissues to the liver. Under these circumstances, the acetyl-CoA formed from β-oxidation is channeled to ketone bodies.

The liver is efficient in producing ketone bodies. In humans, it can produce half of its equivalent weight of ketone bodies per day. However, it lacks the ability to metabolize the ketone bodies formed because it lacks the necessary enzyme ketoacid-CoA transferase.

The level of ketone bodies circulating in the blood is usually low, but during prolonged starvation and in diabetes mellitus it is highly elevated, a condition known as **ketosis.** In patients with diabetes, large amounts of β-hydroxybutyric acid can make the blood pH acidic, a state called **ketoacidosis.**

Blood cholesterol homeostasis

Cholesterol is essential for life, and the liver plays an important role in cholesterol homeostasis. Liver cholesterol is derived from both *de novo* synthesis and the lipoproteins taken up by the liver. Hepatic cholesterol can be used in the formation of bile acids, biliary cholesterol secretion, the synthesis of VLDLs, and the synthesis of liver membranes. Because the absorption of biliary cholesterol and bile acids by the GI tract is incomplete, this method of eliminating cholesterol from the body is essential and efficient. However, patients with high plasma cholesterol levels might be given additional drugs, such as statins, to lower their plasma cholesterol levels. Statins act by inhibiting enzymes that play an essential role in cholesterol synthesis. VLDLs secreted by the liver provide cholesterol to organs that need it for the synthesis of steroid hormones (e.g., the adrenal glands, ovaries, and testes).

PROTEIN AND AMINO ACID METABOLISM IN THE LIVER

The liver is one of the major organs involved in synthesizing nonessential amino acids from the essential amino acids. The body can synthesize all but nine of the amino acids necessary for protein synthesis.

Circulating plasma proteins originate from the liver.

The liver synthesizes many of the circulating plasma proteins, albumin being the most important (Fig. 27.5). It synthesizes about 3 g of albumin a day. Albumin plays an important role in preserving plasma volume and tissue fluid balance by maintaining the colloid osmotic pressure of plasma. This important function of plasma proteins is illustrated by the fact that both liver disease and long-term starvation result in generalized edema and ascites. Plasma albumin plays a pivotal role in the transport of many substances in blood, such as free fatty acids and certain drugs, including penicillin and salicylate.

The other major plasma proteins synthesized by the liver are components of the complement system, components of the **coagulation cascade** (fibrinogen and prothrombin), and proteins involved in iron transport (transferrin, haptoglobin, and hemopexin).

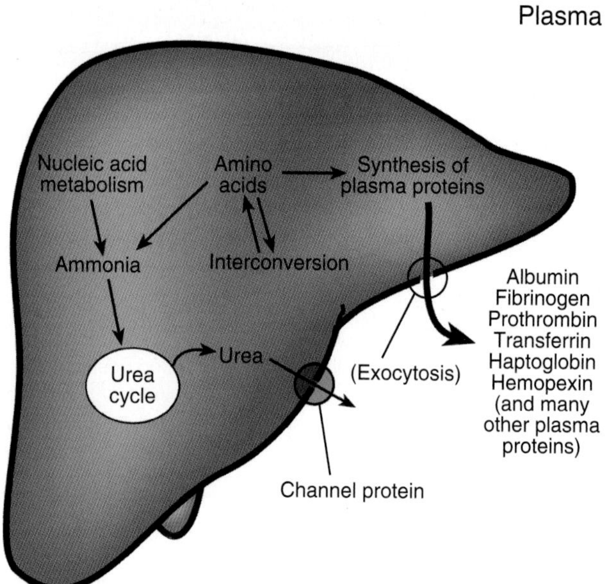

Figure 27.5 The liver synthesizes nonessential amino acids. The liver produces most of the circulating plasma proteins and is the body's only source of nonessential amino acids.

Urea production

Ammonia, derived from protein and nucleic acid catabolism, plays a pivotal role in nitrogen metabolism and is needed in the biosynthesis of nonessential amino acids and nucleic acids. Ammonia metabolism is a major function of the liver. The liver has an ammonia level 10 times higher than the plasma ammonia level. High circulating ammonia levels are highly neurotoxic, and hyperammonemia causes irritability, vomiting, cerebral edema, and finally coma, which leads to death.

The liver synthesizes most of the urea in the body. Protein intake regulates the enzymes involved in the urea cycle. In humans, starvation stimulates these enzymes causing increased urea production and, therefore, an increase in the circulating urea level.

Nonessential amino acids

The **essential amino acids** (see Chapter 26, "Gastrointestinal Secretion, Digestion, and Absorption") must be supplied in the diet. The liver can form nonessential amino acids from the essential amino acids. For instance, tyrosine can be synthesized from phenylalanine, and cysteine can be synthesized from methionine.

Glutamic acid and glutamine play an important role in the biosynthesis of certain amino acids in the liver. Glutamic acid is derived from the amination of α-ketoglutarate by ammonia. This reaction is important because ammonia is used directly in the formation of the α-amino group and constitutes a mechanism for shunting nitrogen from wasteful urea-forming products. Glutamic acid can be used in the amination of other α-keto acids to form the corresponding amino acids. It can also be converted to glutamine by coupling with ammonia, a reaction catalyzed by glutamine synthetase. Following urea, glutamine is the second most important metabolite of ammonia in the liver. It plays an important role in the storage and transport of ammonia in the blood. Through the action of various transaminases, glutamine can be used to aminate various keto acids to their corresponding amino acids. It also acts as an important oxidative substrate and, in the small intestine, it is the primary substrate for providing energy.

LIVER AS STORAGE ORGAN

Another important role of the liver is the storage and metabolism of fat-soluble vitamins and iron. Some water-soluble vitamins, particularly vitamin B12, are also stored in the liver. These stored vitamins are released into the circulation when a need for them arises.

Liver is the site for uptake, storage, and release of fat-soluble vitamins.

Vitamin A comprises a family of compounds related to retinol. Vitamin A is important in vision, growth, the maintenance of epithelia, and reproduction. The liver plays a pivotal role in the uptake, storage, and maintenance of circulating plasma vitamin A levels by mobilizing its vitamin A store (Fig. 27.6). Retinol (an alcohol) is transported in chylomicrons mainly as an ester of long-chain fatty acids. When chylomicrons enter the circulation, lipoprotein lipase rapidly acts on the triglyceride; the triglyceride content of the particles is significantly reduced, whereas the retinyl ester content remains unchanged. Receptors in the liver mediate the rapid uptake of chylomicron remnants, which are degraded, and the retinyl ester is stored.

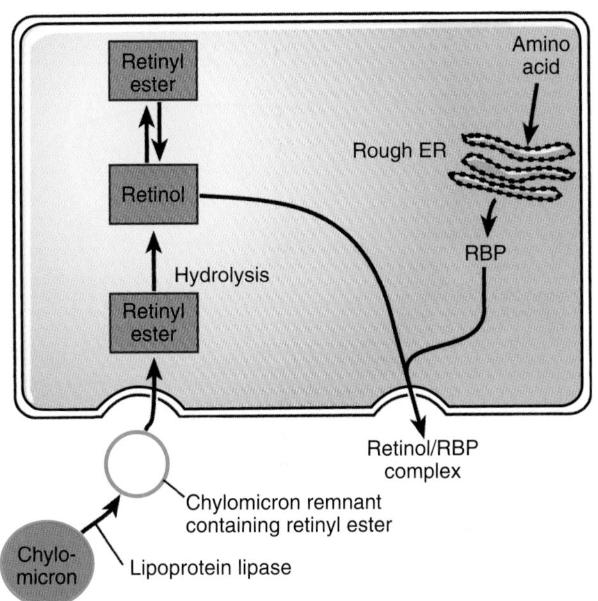

Figure 27.6 The liver stores fat-soluble vitamins. The metabolism of vitamin A (retinol) by the hepatocyte is illustrated here. When circulating vitamin A levels fall, the liver hydrolyzes the stored retinyl ester. ER, endoplasmic reticulum; RBP, retinol-binding protein.

When the vitamin A level in blood falls, the liver mobilizes the vitamin A store by hydrolyzing the retinyl ester (see Fig. 27.6). The retinol formed is bound with **retinol-binding protein (RBP)**, which the liver synthesizes before it is secreted into the blood. The amount of RBP secreted into the blood depends on vitamin A status. Vitamin A deficiency significantly inhibits the release of RBP, whereas vitamin A loading stimulates its release. RBP binds retinol in a 1:1 stoichiometry. The binding results in the solubilization of retinol and protects it from oxidation.

Hypervitaminosis A develops when massive quantities of vitamin A are consumed. Because the liver is the storage organ for vitamin A, hepatotoxicity is often associated with hypervitaminosis A. The continued ingestion of excessive amounts of vitamin A eventually leads to portal hypertension and cirrhosis. Most cases of vitamin A toxicity and reported overdose were found in arctic explorers who consumed the livers of polar bears and seals, both of which contain high vitamin A content. β-Carotene (the name of which is derived from the Latin name for carrot) belongs to a family of natural compounds called *carotenes* or *carotenoids*. It is abundant in plants and gives fruits and vegetables their rich orange and yellow colors. The body converts β-carotene to vitamin A (retinol). Although excessive amounts of vitamin A can be toxic, the body converts only as much vitamin A from β-carotene as needed, thereby making β-carotene a safe source of vitamin A.

Vitamin D is thought to be stored mainly in skeletal muscle and adipose tissue. However, the liver is responsible for the initial activation of vitamin D by converting vitamin D3 to 25-hydroxyvitamin D3. Additionally, vitamin D synthesizes vitamin D–binding protein.

Vitamin K is a fat-soluble vitamin important in the hepatic synthesis of prothrombin. Prothrombin is synthesized as a precursor that is converted to the mature prothrombin, a reaction that requires the presence of vitamin K (Fig. 27.7). Vitamin K deficiency, therefore, leads to impaired blood clotting.

The largest vitamin K store is in skeletal muscle, but the physiologic significance of this and other body stores is unknown. The dietary vitamin K requirement is small and is adequately supplied by the average Western diet. Bacteria in the GI tract also provide vitamin K. This appears to be an important source of vitamin K because prolonged administration of wide-spectrum antibiotics sometimes results in hypoprothrombinemia. Because vitamin K absorption depends on normal fat absorption, any prolonged malabsorption of lipid can result in its deficiency. The vitamin K store in the liver is relatively limited and, therefore, hypoprothrombinemia can develop within a few weeks. Parenteral administration of vitamin K usually provides a cure.

Liver plays a key role in the transport, storage, and metabolism of iron.

Iron is essential for life, and the regulation of iron metabolism is important in many aspects of human health and disease. Most of the body's iron is contained in red blood cells and, as a result, both hemoglobin deficiency anemia and

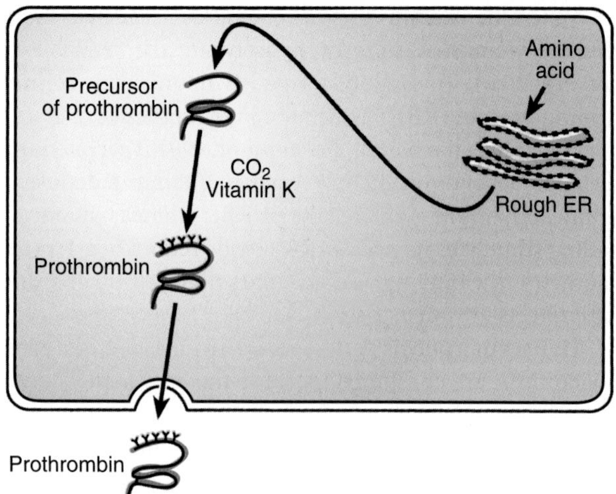

Figure 27.7 The liver synthesizes blood-clotting proteins. The figure illustrates how the hepatocyte synthesizes and releases prothrombin, a key blood-clotting protein. ER, endoplasmic reticulum.

iron deficiency anemia are the two most common types of diseases associated with iron deficiency.

Iron acts as a carrier protein that plays an important role in many metabolic processes. The most important group of iron-binding proteins contains the heme molecules, all of which contain iron at their center. These heme-carrier proteins carry out the redox reactions and electron transport required for oxidative phosphorylation, the process that is the principal source of energy for cells. The iron–sulfur proteins are another important group of iron-containing proteins, some of which are an essential component in oxidative phosphorylation.

Heme-binding protein is also found in hemoglobin for carrying and transporting oxygen and carbon dioxide. Iron is also an essential component of myoglobin for the transport and unloading of oxygen in skeletal and cardiac muscle cells.

The liver is the major site for the synthesis of several proteins involved in iron transport and metabolism. The protein **transferrin** plays a critical role in the transport and homeostasis of iron in the blood. The circulating plasma transferrin level is inversely proportional to the iron load of the body—the higher the concentration of ferritin in the hepatocyte, the lower the rate of transferrin synthesis. During iron deficiency, liver synthesis of transferrin is significantly stimulated, enhancing the intestinal absorption of iron. **Haptoglobin**, a large glycoprotein with a molecular weight of 100,000, binds free hemoglobin in the blood. The liver rapidly removes the hemoglobin–haptoglobin complex, conserving iron in the body. **Hemopexin** is another protein synthesized by the liver that is involved in the transport of free heme in the blood. It forms a complex with free heme, and the liver rapidly removes the complex.

The spleen is the organ that removes red blood cells that are slightly altered. Kupffer cells of the liver also have the capacity to remove damaged red blood cells, especially those that are moderately damaged (Fig. 27.8). Secondary lysosomes rapidly digest the red cells taken up by Kupffer cells to release heme. Microsomal **heme oxygenase** releases iron from the heme, which then enters the free iron pool and is stored as ferritin or released into the bloodstream (bound to apotransferrin). Some of the ferritin iron may be

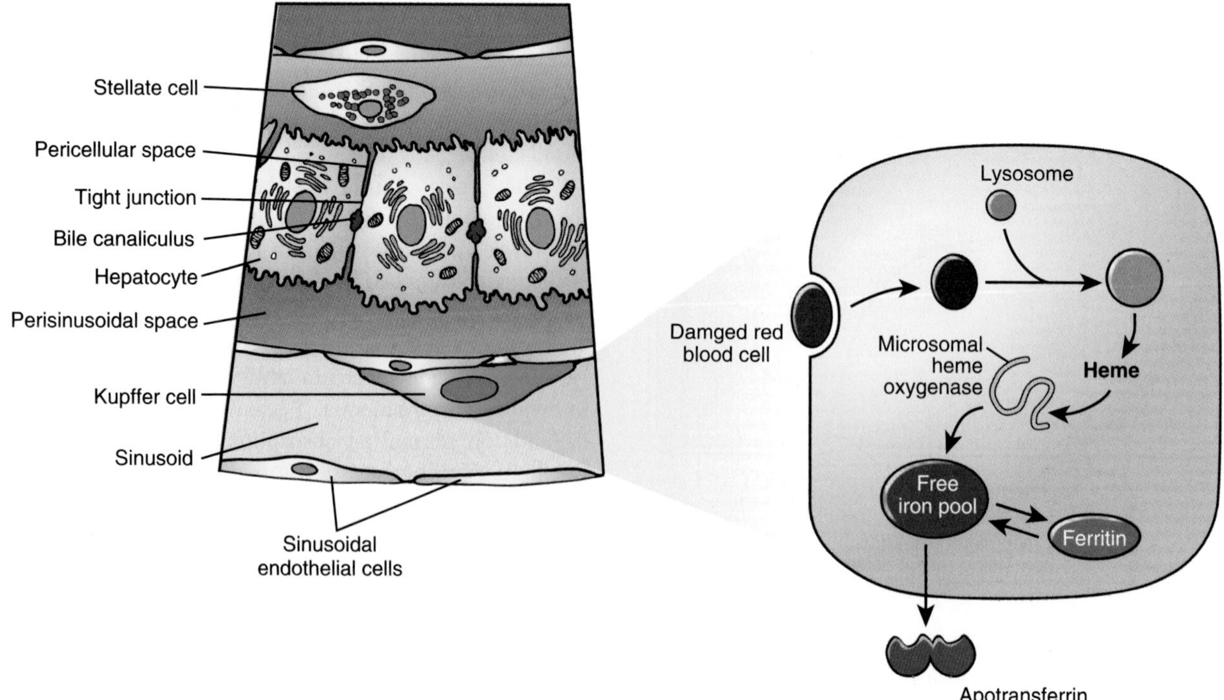

Figure 27.8 Kupffer cells remove damaged red cells from the circulation. Kupffer cells rapidly phagocytize damaged red cells. The iron released from the heme molecule during phagocytosis becomes part of the free iron pool.

converted to **hemosiderin granules**. It is unclear whether the iron from the hemosiderin granules is exchangeable with the free iron pool.

It was long believed that Kupffer cells were the only cells involved in iron storage, but recent studies suggest that hepatocytes are the major sites of long-term iron storage. Transferrin binds to receptors on the surface of hepatocytes, and the entire transferrin–receptor complex is internalized and processed (Fig. 27.9). The apotransferrin (not containing iron) is recycled back to the plasma, and the released iron enters a labile iron pool. The iron from transferrin is probably the major source of iron for the hepatocytes, but they also derive iron from haptoglobin–hemoglobin and hemopexin–heme complexes. When hemoglobin is released inside the hepatocytes, it is degraded in the secondary lysosomes and heme is released. Heme is processed in the SER, and free iron is released and enters the labile iron pool. A significant portion of the free iron in the cytosol probably combines rapidly with apoferritin to form ferritin. Like Kupffer cells, hepatocytes may transfer some of the iron in ferritin to hemosiderin.

The human body tightly regulates iron absorption and recycling. However, an "iron paradox" exists in physiology; that is, iron is an essential element for life, but the human body has no physiologic regulatory mechanism for excreting iron. Iron overload is prevented solely by the regulation of iron absorption; therefore, those who cannot regulate absorption get disorders of iron overload. With iron overload, iron becomes toxic to cells. Iron toxicity occurs when the amount of circulating iron exceeds the amount of transferrin available to bind it. Toxicity arises with iron that is not bound, existing as free iron within the cell, where it can catalyze the conversion of hydrogen peroxide into free radicals. These free radicals cause extensive cell damage to the heart, liver, and kidney as well as other metabolic organs. Iron toxicity is usually the result of iron supplement overdose (rare), repeated transfusions, and with genetic diseases. A classic example of genetic iron overload is **hereditary hemochromatosis (HH)**. The defect in HH causes excessive iron absorption from food, in turn causing the hepatocytes to become defective and fail to perform normal functions. The cellular dysfunction is due to free radical damage.

Some iron is stored in the body. Most stored iron is bound by ferritin, and the largest amount of ferritin-bound iron is found in the liver hepatocytes, bone marrow, and the spleen. However, the liver's stores of ferritin are the primary storage site for iron.

Coagulation proteins are metabolized by the liver.

Liver cells are important in both the production and the clearance of coagulation proteins. Hepatocytes secrete most of the known clotting factors and inhibitors, some of them exclusively. Additionally, several coagulation and anticoagulation proteins require a vitamin K–dependent modification following synthesis, specifically factors II, VII, IX, and X and proteins C and S, to make them effective.

The monocyte–macrophage system of the liver, predominantly Kupffer cells, is an important system for clearing

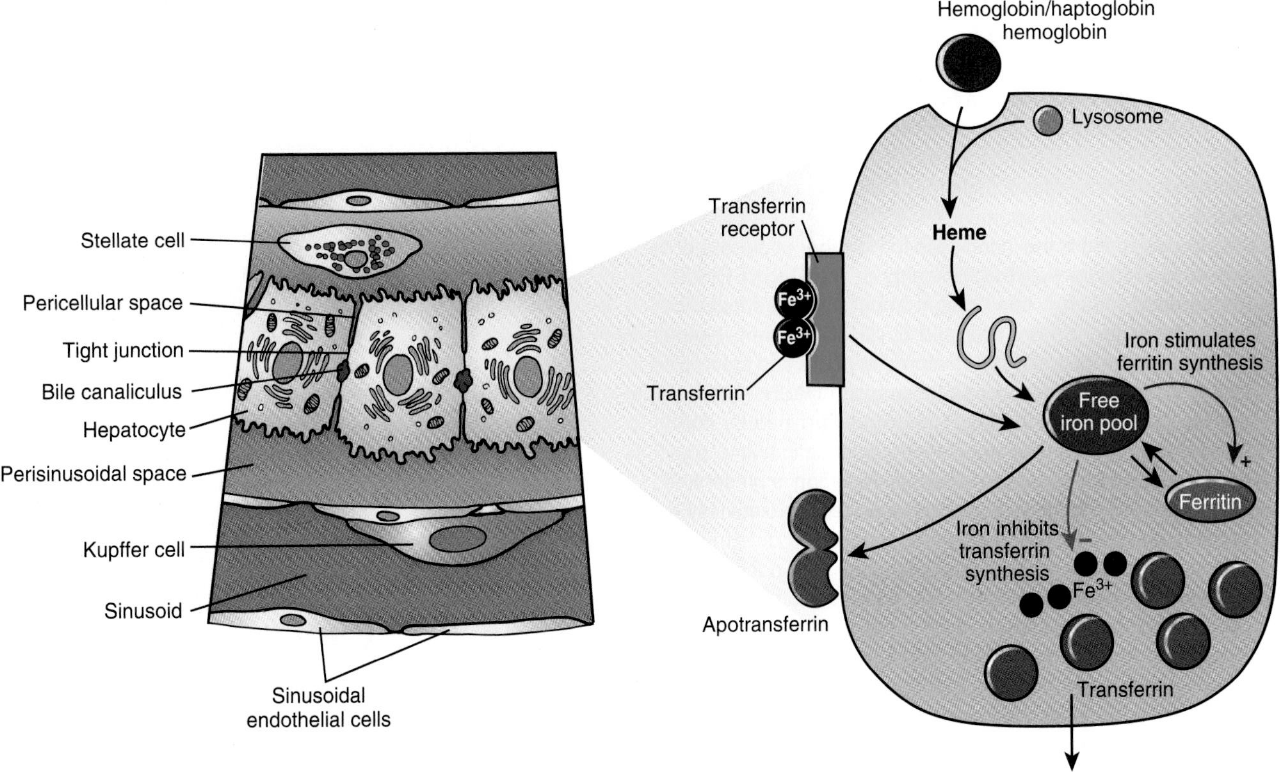

Figure 27.9 The hepatocyte is the major storage site for iron. Transferrin binds to receptors on hepatocytes and, as seen in this figure, the entire transferrin–receptor complex is internalized and processed.

clotting factors and factor–inhibitor complexes. Disturbances in liver perfusion or a compromise in liver function can result in the ineffective clearance of activated coagulation proteins, so patients with advanced liver failure may be predisposed to developing **disseminated intravascular coagulation (DIC)**. DIC is the pathologic activation of blood-clotting mechanisms, leading to the formation of small blood clots inside the blood vessels throughout the body, thereby disrupting normal coagulation and causing abnormal bleeding and possible organ malfunction.

ENDOCRINE FUNCTIONS OF THE LIVER

The liver is important in regulating the endocrine functions of hormones. It can amplify the action of some hormones. It is also the major organ for the removal of peptide hormones.

Liver activates and degrades many hormones.

As previously discussed, the liver converts vitamin D3 to 25-hydroxyvitamin D3, an essential step before conversion to the active hormone 1,25-dihydroxyvitamin D3 in the kidneys. The liver is also a major site of conversion of the thyroid hormone thyroxine (T_4) to the biologically more potent hormone triiodothyronine (T_3). The regulation of the hepatic T_4 to T_3 conversion occurs at both uptake and conversion. Because of the liver's capacity to convert T_4 to T_3, hypothyroidism is uncommon in patients with liver disease. However, in advanced chronic liver disease, signs of hypothyroidism can be more evident.

The liver modifies the function of growth hormone secreted by the pituitary gland. **Insulin-like growth factors** made by the liver mediate some growth hormone actions (see Chapter 31, "Hypothalamus and the Pituitary Gland").

The liver also helps to remove and degrade many circulating hormones. Insulin is degraded in many organs, but the liver and kidneys are by far most important. The presence of insulin receptors on the surface of hepatocytes suggests that the binding of insulin to these receptors results in degradation of some insulin molecules. There is also degradation of insulin by proteases of hepatocytes that does not involve the insulin receptor.

Glucagon and growth hormone are degraded mainly by the liver and the kidneys. Consequently, patients with cirrhotic livers often have high circulating levels of hormones such as glucagon. The liver is also involved in the degradation of various GI hormones (e.g., gastrin), but the kidneys and other organs probably contribute more significantly to inactivating these hormones.

From Bench to Bedside / 27.1

Obesity and Nonalcoholic Steatohepatitis

Nonalcoholic steatohepatitis, also called *NASH*, is a silent but potentially serious liver disease. As the term implies, patients suffering from NASH usually consume little or no alcohol. NASH presents as fat in the liver, along with inflammation and fibrosis. Patients with NASH often feel healthy and are unaware they have a liver problem. However, NASH can be severe and can lead to cirrhosis and consequent failure of the liver. It is estimated that 10% to 20% of people living in the United States have fatty livers, yet only a fraction of this population has hepatitis.

More alarming is the increase in the number of overweight and obese children and adolescents in the United States. The obesity epidemic has brought about a marked increase in the incidence of fatty liver and NASH in children. Figure 27.A shows the infiltration of inflammatory cells and steatosis in a liver biopsy sample from a patient with NASH. Although the presence of fat in the liver is easily determined by computerized tomography or magnetic resonance imaging, physicians cannot be certain whether inflammation is present as well. Currently, a biopsy of the liver analyzed histologically is the only method by which inflammation can be determined. Although this procedure is relatively safe, it is nonetheless invasive and has associated risks, such as damage to the liver. The biopsy procedure itself is not particularly uncomfortable; lying still for several hours afterward to prevent bleeding at the site presents, perhaps, the most discomfort.

Currently, the finding of liver enzymes in the circulation serves as the primary indicator of liver disease. However, the many advances in molecular biology and the discovery of numerous biomarkers in the circulation may well lead to great advances in the management of NASH if the disease can be identified by elevated liver marker(s) in the circulation. Observed elevation of such marker(s) would indicate liver inflammation, thereby possibly eliminating the need to perform a liver biopsy. For example, the level of circulating bile acids in the liver may serve as a more sensitive biomarker indicating the early stages of liver disease.

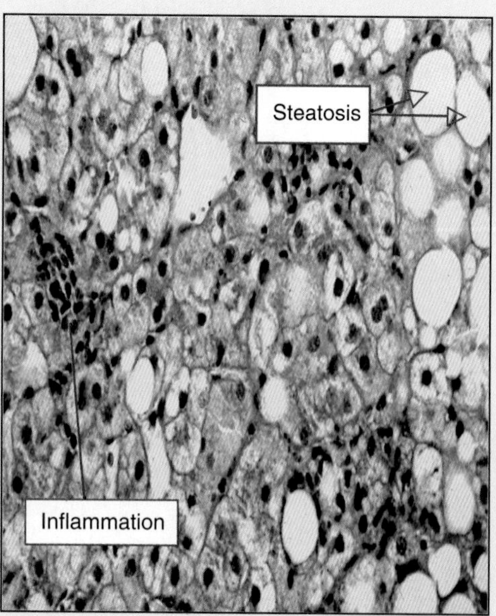

Figure 27.A A liver biopsy sample from a patient with nonalcoholic steatohepatitis illustrates the infiltration and steatosis of inflammatory cells.

Chapter Summary

- The basic functional unit of the liver is the liver lobule that contains specialized cells, the hepatocytes.
- The liver plays an important role in maintaining blood glucose levels and in metabolizing drugs and toxic substances.
- The liver is the only organ in the adult that can be regenerated.
- Gluconeogenesis regulates blood glucose during fasting.
- The liver is the first organ to be affected by and respond to changes in plasma insulin levels.
- The liver is one of the main organs involved in fatty acid synthesis.
- The liver aids in the elimination of cholesterol from the body.
- The liver plays an important role in iron metabolism.
- The liver modifies the action of hormones released by other organs.

thePoint *Visit http://thePoint.lww.com for chapter review Q&A, case studies, animations, and more!*

Part VIII • Temperature Regulation and Exercise Physiology

28 Regulation of Body Temperature

ACTIVE LEARNING OBJECTIVES

Upon mastering the material in this chapter, you should be able to:

- Explain how the body produces heat and exchanges energy with the environment.
- Explain how sweating, skin blood flow, and metabolic heat production help regulate body temperature.
- Explain how the thermoregulatory set point varies cyclically and is elevated during fever.
- Explain the reflex control of physiologic thermoregulatory responses.
- Explain temperature alterations during exercise in various environments.
- Explain the process of acclimatization to hot and cold environments.
- Explain the adverse effects of excessive heat stress, heat injury, and excessive cold stress.

Humans are warm-blooded animals or **homeotherms**. Like other mammals, humans are able to regulate their internal body temperatures within a narrow range near 37°C, despite wide variations in environmental temperature (Fig. 28.1). In contrast, internal body temperatures of **poikilotherms**, or cold-blooded animals, are governed by environmental temperature. The range of temperatures that living cells and tissues can tolerate without harm extends from just above freezing to nearly 45°C—far wider than the limits within which homeotherms regulate body temperature. What biologic advantage do homeotherms gain by maintaining a stable body temperature? As we shall see, tissue temperature is important for two reasons.

First, temperature extremes injure tissue directly. High temperatures alter the three-dimensional structure of protein molecules, even though the sequence of amino acids remains unchanged. Such alteration of protein structure is called **denaturation**. A familiar example of denaturation by heat is the coagulation of albumin in the white of a cooked egg. Because the biological activity of a protein molecule depends on its configuration and charge distribution, denaturation inactivates a cell's proteins and injures or kills the cell. Injury occurs at tissue temperatures higher than about 45°C, which is also the point at which heating the skin becomes painful. The severity of injury depends on the temperature to which the tissue is heated and its duration.

Excessive cold can also injure tissues. As a water-based solution freezes, ice crystals consisting of pure water form and all dissolved substances in the solution are left in the unfrozen liquid. Therefore, as more ice forms, the remaining liquid becomes more and more concentrated. Freezing damages cells through two mechanisms. First, ice crystals mechanically injure the cell. The increase in solute concentration of the cytoplasm as ice forms denatures the proteins by removing their water of hydration, increasing the ionic strength of the cytoplasm and causing other changes in the physicochemical environment in the cytoplasm.

Second, temperature changes profoundly alter biologic function through specific effects on such specialized functions as electrochemical properties and fluidity of cell membranes and through a general effect on most chemical reaction rates. In the physiologic temperature range, most reaction rates vary approximately as an exponential function of temperature (T); increasing T by 10°C increases the reaction rate by a factor of two to three. For any particular reaction, the ratio of the rates at two temperatures 10°C apart is called the $\mathbf{Q_{10}}$ for that reaction, and the effect of temperature on reaction rate is called the Q_{10} effect. The notion of Q_{10} may be generalized to apply to a group of reactions that have some measurable overall effect (such as O_2 consumption) in common and, thus, are thought of as comprising a physiologic process. The Q_{10} effect is clinically important in managing patients who have high fevers and are receiving fluid and nutrition intravenously. A commonly used rule is that a patient's fluid and caloric needs are increased 10% to 13% above normal for each 1°C of fever.

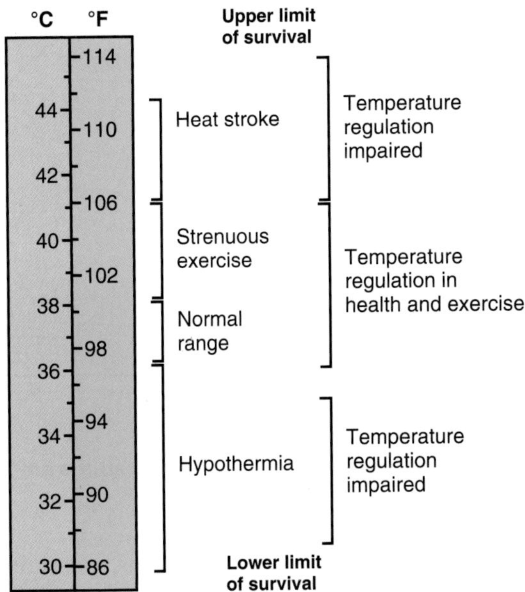

Figure 28.1 Survival ranges of body temperatures in humans.

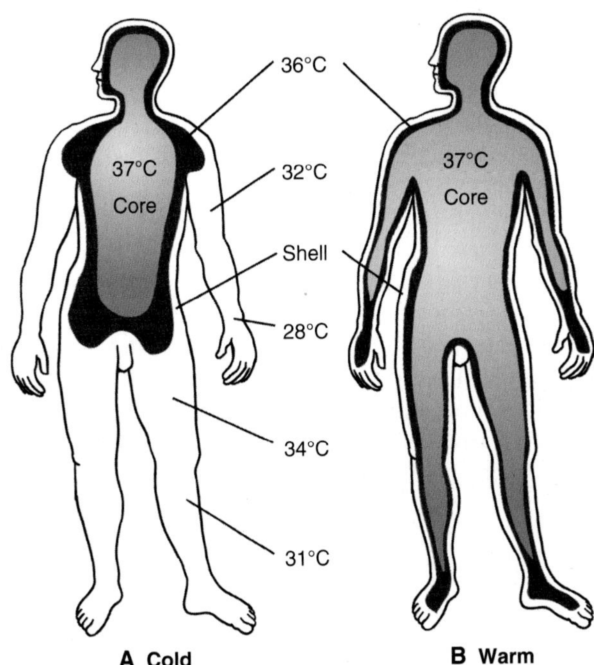

Figure 28.2 Distribution of temperatures in the body's core and shell. (A) During exposure to cold. **(B)** In a warm environment. Because the temperatures of the surface and the thickness of the shell depend on environmental temperature, the shell is thicker in the cold and thinner in the heat.

The sluggishness of a reptile that comes out of its burrow in the morning chill and becomes active only after being warmed by the sun illustrates the profound effect of temperature on biochemical reaction rates. Homeotherms avoid such a dependence of metabolic rate on environmental temperature by regulating their internal body temperatures within a narrow range. A disadvantage of homeothermy is that, in most homeotherms, certain vital processes cannot function at low levels of body temperature that poikilotherms tolerate easily. For example, shipwreck victims immersed in cold water die of respiratory or circulatory failure (through disruption of the electrochemical activity of the brainstem or heart) at body temperatures of about 25°C, even though such a temperature produces no direct tissue injury and fish thrive in the same water.

BODY TEMPERATURE AND HEAT TRANSFER

The body is divided into a warm internal core and a cooler outer shell (Fig. 28.2). Because the environment greatly influences the temperature of the shell, its temperature is not regulated within narrow limits as is the internal body temperature. This is true despite the thermoregulatory responses that strongly affect the temperature of the shell, especially its outermost layer, the skin. The thickness of the shell depends on the thermal environment and the body's need to conserve heat. In a warm environment, the shell may be <1-cm thick, but in a subject conserving heat in a cold environment, it may extend several centimeters below the skin. The regulated internal body temperature is the temperature of the vital organs inside the head and trunk, which, together with a variable amount of other tissue, comprise the warm internal core.

Heat is produced in all tissues of the body but is lost to the environment only from tissues in contact with the environment—predominantly from the skin and, to a lesser degree, from the respiratory tract. Therefore, we need to consider heat transfer within the body, especially heat transfer (1) from major sites of heat production to the rest of the body and (2) from the core to the skin. Heat is transported within the body by two means: **conduction** through the tissues and **convection** by the blood, whereby flowing blood carries heat from warmer tissues to cooler tissues.

Heat flow by conduction varies directly with the thermal conductivity of the tissues, the change in temperature over the distance the heat travels, and the area (perpendicular to the direction of heat flow) through which the heat flows. It varies inversely with the distance the heat must travel. As Table 28.1 shows, the tissues are rather poor heat conductors.

TABLE 28.1 Thermal Conductivities and Rates of Heat Flow

Material	Conductivity (kcal/s·m·°C)	Rate of Heat Flow[a] kcal/h	W
Copper	0.092	33,120	38,474
Epidermis	0.00005	18	21
Dermis	0.00009	32	38
Fat	0.00004	14	17
Muscle	0.00011	40	46
Oak (across grain)	0.00004	14	17
Fiberglass insulation	0.00001	3.6	4.2

[a]Values are calculated for slabs 1 m^2 in area and 1-cm thick, with a 1°C-temperature difference between the two faces of the slab.

Heat flow by convection depends on the rate of blood flow and the temperature difference between the tissue and the blood supply. Because the vessels of the microvasculature have thin walls and, collectively, a large total surface area, the blood temperature equilibrates with that of the surrounding tissue before it reaches the capillaries. Changes in skin blood flow in a cool environment change the thickness of the shell. When skin blood flow is reduced in a cold environment, the affected skin becomes cooler and the underlying tissues—which in the cold may include most of the limbs and the more superficial muscles of the neck and trunk—become cooler as they lose heat by conduction to the cool overlying skin and, ultimately, to the environment. In this way, these underlying tissues, which in a hot environment were part of the body core, now become part of the shell. In addition to organs in the trunk and head, the core includes a greater or lesser amount of more superficial tissues—mostly skeletal muscle—depending on the body's thermal state.

Because the shell lies between the core and the environment, all heat leaving the body core, except that which is lost through the respiratory tract, must pass through the shell before being lost to the environment. Thus, the shell insulates the core from the environment. In a cool subject, the skin blood flow is low, and so conduction dominates core-to-skin heat transfer; the shell is also thicker, providing more insulation to the core, because heat flow by conduction varies inversely with the distance the heat must travel. Changes in skin blood flow, which directly affect core-to-skin heat transfer by convection, also indirectly affect conductive core-to-skin heat transfer by changing the thickness of the shell. In a cool subject, the subcutaneous fat layer contributes to the insulation value of the shell because the fat layer increases the thickness of the shell and because fat has conductivity about 0.4 times that of dermis or muscle (see Table 28.1). Thus, fat is a correspondingly better insulator. In a warm subject, however, the shell is relatively thin and provides little insulation. Furthermore, a warm subject's skin blood flow is high, and so convection dominates heat flow from the core to the skin. In these circumstances, the subcutaneous fat layer, which affects conduction but not convection, has little effect on heat flow from the core to the skin.

Core temperature approximates central blood temperature.

Core temperature varies slightly from one site to another, depending on such local factors as metabolic rate, blood supply, and the temperatures of neighboring tissues. However, temperatures at different places in the core are all similar to the temperature of the central blood and tend to change together. The notion of a single, uniform core temperature, although not strictly correct, is a useful approximation. The value of 98.6°F, often given as the "normal" body temperature, may give the misleading impression that body temperature is regulated so precisely that it deviates less than a few tenths of a degree. In fact, 98.6°F is simply the Fahrenheit equivalent of 37°C, and body temperature does vary somewhat (see Fig. 28.1). The effects of heavy exercise and fever are familiar; variation among people and such factors as time of day and method of measurement (Fig. 28.3), phase of the menstrual cycle, and acclimatization to heat can also cause differences of up to about 1°C in resting core temperature.

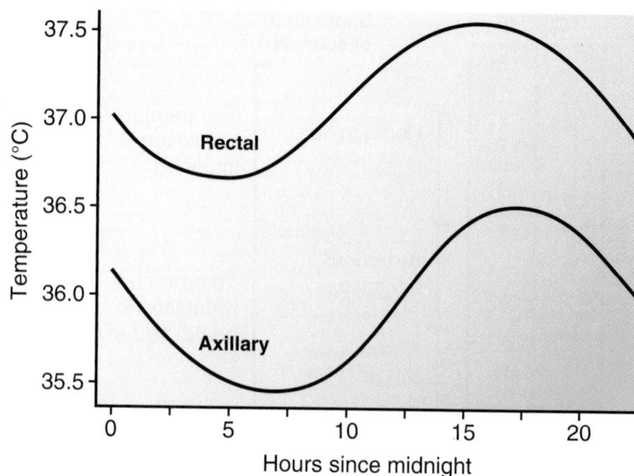

Figure 28.3 Core and axillary temperatures are influenced by circadian rhythm.

To maintain core temperature within a narrow range (36.5°C–37.5°C), the thermoregulatory system needs continuous information about the level of core temperature. Temperature-sensitive neurons and nerve endings in the abdominal viscera, great veins, spinal cord, and especially, the brain provide this information. We consider how the thermoregulatory system processes and responds to this information later in the chapter.

Core temperature should be determined at a site where the measurement is not biased by environmental temperature. Clinically used sites include the rectum, the mouth, and occasionally, the axilla. The rectum is well insulated from the environment; its temperature is independent of environmental temperature and is a few tenths of 1°C warmer than arterial blood and other core sites. The tongue is richly supplied with blood; oral temperature under the tongue is usually close to blood temperature (and 0.4°C–0.5°C below rectal temperature), but cooling the face, neck, or mouth can make oral temperature misleadingly low. If a patient holds his or her upper arm firmly against the chest to close the axilla, axillary temperature will eventually come reasonably close to core temperature. However, because this may take 30 minutes or more, axillary temperature is used infrequently. Infrared ear (aural) thermometers are convenient and widely used in the clinic, but temperatures of the tympanum and external auditory meatus are loosely related to more accepted indices of core temperature, and ear temperature in collapsed hyperthermic runners may be 3°C to 6°C below rectal temperature.

Skin is an important organ in heat exchange and thermoregulatory control.

Most heat is exchanged between the body and the environment at the skin surface. Consequently, skin temperature is much more variable than core temperature; thermoregulatory responses such as skin blood flow and sweat secretion, the temperatures of underlying tissues, and environmental factors, such as air temperature, air movement, and thermal radiation, affect it. Skin temperature is one of the major factors in heat exchange with the environment. For these reasons, it provides the thermoregulatory system with important information about the need to conserve or dissipate heat.

Because skin temperature usually is not uniform over the body surface, mean skin temperature ($\overline{T}_{sk}$) is frequently calculated from temperatures at several skin sites, usually weighting each temperature according to the fraction of body surface area it represents. $\overline{T}_{sk}$ is used to summarize the input to the central nervous system (CNS) from temperature-sensitive nerve endings in the skin. $\overline{T}_{sk}$ is also commonly used, along with core temperature, to calculate a mean body temperature and to estimate the quantity of heat stored in the body because the direct measurement of shell temperature would be difficult and invasive.

Peripheral thermoreceptors in the skin code absolute and relative temperatures.

Cutaneous thermosensation is mediated by various primary afferent nerve fibers that transduce, encode, and transmit thermal information. These peripheral **thermoreceptors** are located in the skin and in the oral and urogenital mucosa and include warm-sensitive and cold-sensitive populations. The most numerous and most superficially situated thermosensors are cold-sensitive. Skin cold sensors are located in or immediately beneath the epidermis, and their afferent input is conveyed by thin myelinated Aδ fibers. The less common warm-sensitive receptors (10-fold fewer than cold sensors) are located slightly deeper in the dermis, and their input travels via unmyelinated C fibers. In contrast to central thermosensitive neurons, peripheral thermosensors are not pacemakers. The mechanisms of peripheral thermosensitivity involve changes in the resting membrane potential. Importantly, the response of most peripheral thermosensors shows a powerful dynamic (phasic) component. These neurons are very active when the temperature changes but quickly adapt when the temperature stabilizes, enabling rapid reaction to thermal environmental changes.

The mechanism of thermal transduction is believed to involve the action of **transient receptor potential (TRP) ion channels** whose activity depends on the temperature of their environment. Of these, members of three families, the vanilloid TRP channels, the melastatin (or long) TRP channels, and the ankyrin transmembrane protein channels are of particular interest as thermoreceptors. Each of these receptors operates over a specific temperature range, thereby providing a potential molecular basis for peripheral thermosensation. These specialized thermal receptors are embedded in the membranes of afferent fiber terminals as free nerve endings in the skin. Activation of all thermo TRP channels results in an inward, nonselective cationic current and a consequent depolarization of the resting membrane potential. Although individual thermo TRP channels are activated within a relatively narrow temperature range, collectively, their range is quite broad, from noxious cold to noxious heat (see Fig. 28.10). Thermotransduction in these nerve endings by TRP-independent means is also likely.

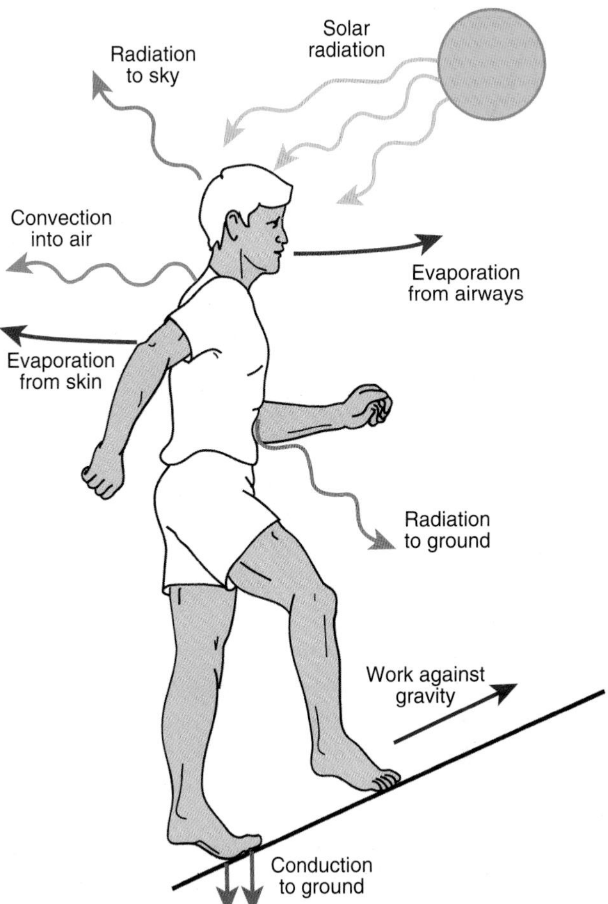

Figure 28.4 Exchange of energy with the environment. This hiker gains heat from the sun by radiation and loses heat by conduction to the ground through the soles of his feet, convection into the air, radiation to the ground and sky, and evaporation of water from his skin and respiratory passages. In addition, some of the energy released by his metabolic processes is converted into mechanical work, rather than heat, because he is walking uphill.

Experimental evidence indicates cold sensing through cold-induced inhibition of K^+ conductances that are normally active at resting membrane potential, leading to membrane depolarization and action potential generation, whereas warm sensing may involve the otherwise mechanosensitive K_2P channels and other nociceptive C fibers.

BALANCE BETWEEN HEAT PRODUCTION AND HEAT LOSS

All animals exchange energy with the environment. Some energy is exchanged as mechanical work, but most is exchanged as heat (Fig. 28.4). Heat is exchanged by conduction, convection, and radiation and as latent heat through evaporation or (rarely) condensation of water. If the sum of energy production and energy gain from the environment does not equal energy loss, the extra heat is "stored" in, or lost from, the body. This relationship is summarized in the following heat balance equation:

$$M = E + R + C + K + W + S \quad (1)$$

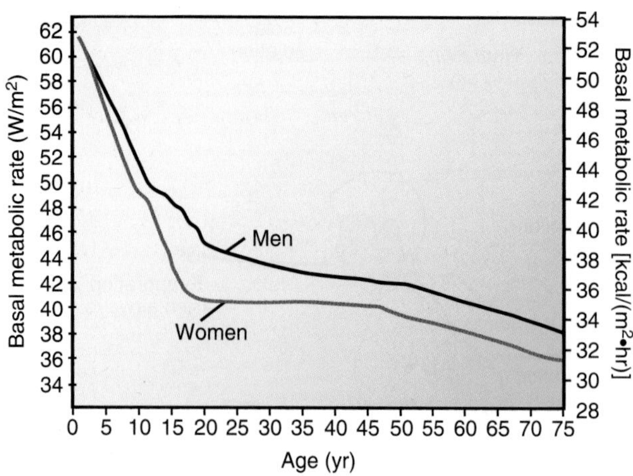

Figure 28.5 Effects of age and sex on the basal metabolic rate of healthy subjects. Metabolic rate here is expressed as the ratio of energy consumption to body surface area.

where M is metabolic rate; E is the rate of heat loss by evaporation; R and C are rates of heat loss by radiation and convection, respectively; K is the rate of heat loss by conduction; W is the rate of energy loss as mechanical work; and S is the rate of heat storage in the body, manifested as changes in tissue temperatures.

M is always positive, but the terms on the right side of equation 1 represent energy exchange with the environment and storage and may be either positive or negative. E, R, C, K, and W are positive if they represent energy losses from the body and negative if they represent energy gains. When S = 0, the body is in heat balance and body temperature neither rises nor falls. When the body is not in heat balance, its mean tissue temperature increases if S is positive and decreases if S is negative. This situation commonly lasts only until the body's responses to the temperature changes are sufficient to restore balance. However, if the thermal stress is too great for the thermoregulatory system to restore balance, the body will continue to gain or lose heat until either the stress diminishes sufficiently or the animal dies.

The traditional units for measuring heat are a potential source of confusion because the word calorie refers to two units differing by a 1,000-fold. The *calorie* used in chemistry and physics is the quantity of heat that will raise the temperature of 1 g of pure water by 1°C; it is also called the *small calorie* or *gram calorie*. The *Calorie* (capital C) used in physiology and nutrition is the quantity of heat that will raise the temperature of 1 kg of pure water by 1°C; it is also called the *large calorie, kilogram calorie*, or (the usual practice in thermal physiology) the **kilocalorie** (kcal). Because heat is a form of energy, it is now often measured in joules, the unit of work (1 kcal = 4,186 J), and rate of heat production or heat flow in watts, the unit of power (1 W = 1 J/s). This practice avoids confusing calories and Calories. However, kilocalories are still used widely enough that it is necessary to be familiar with them, and there is a certain advantage to a unit based on water because the body itself is mostly water.

METABOLIC RATE AND HEAT PRODUCTION AT REST

Metabolic energy is used for active transport via membrane pumps, for energy-requiring chemical reactions, such as the formation of glycogen from glucose and proteins from amino acids, and for muscular work. Most of the metabolic energy used in these processes is converted into heat within the body. This conversion may occur almost immediately as with energy used for active transport or heat produced as a by-product of muscular activity. Other energy is converted to heat only after a delay, as when the energy used in forming glycogen or protein is released as heat when the glycogen is converted back into glucose or the protein is converted back into amino acids.

Among subjects of different body sizes, metabolic rate at rest varies approximately in proportion to body surface area. In a resting and fasting young adult man, it is about 45 W/m^2 (81 W or 70 kcal/h for 1.8 m^2 body surface area), corresponding to an O_2 consumption of about 240 mL/min. About 70% of energy production at rest occurs in the body core—trunk viscera and the brain—even though it comprises only about 36% of the body mass (Table 28.2). As a by-product of their metabolic processes, these organs produce most of the heat needed to maintain heat balance at comfortable environmental temperatures; only in the cold must heat produced expressly for thermoregulation supplement such by-product heat.

Factors other than body size that affect metabolism at rest include age, sex (Fig. 28.5), hormones, and digestion. The ratio of metabolic rate to surface area is highest in infancy and declines with age, most rapidly in childhood and adolescence and more slowly thereafter. Children have high metabolic rates in relation to surface area because of the energy used to synthesize the fats, proteins, and other tissue components needed to sustain growth. Similarly, a woman's metabolic rate increases during pregnancy to supply the energy needed for the growth of the fetus. However, a nonpregnant woman's metabolic rate is 5% to 10% lower than that of a man of the same age and surface area probably because a higher proportion of the female body is composed of fat, a tissue with low metabolism.

TABLE 28.2 Relative Masses and Metabolic Heat Production Rates During Rest and Heavy Exercise

		Percentage of Heat Production	
Region	**Percentage of Body Mass**	**Rest**	**Exercise**
Brain	2	16	1
Trunk viscera	34	56	8
Muscle and skin	56	18	90
Other	8	10	1

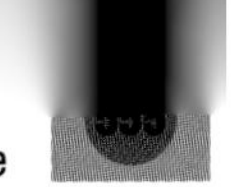

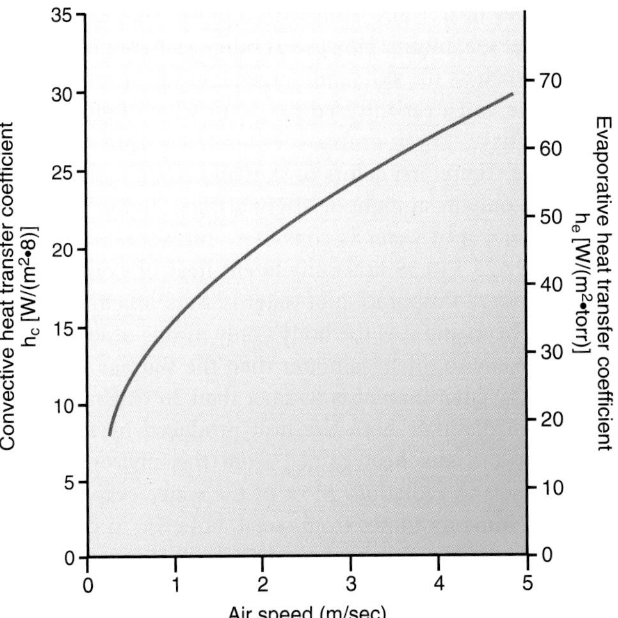

Figure 28.6 Dependence of convection and evaporation on air movement. This figure shows the convective heat transfer coefficient, h_c (left), and the evaporative heat transfer coefficient, h_e (right), for a standing human as a function of air speed. The convective and evaporative heat transfer coefficients are related by the equation $h_e = h_c \times 2.2°C/torr$. The horizontal axis can be converted into English units by using the relation 5 m/sec = 16.4 ft/sec = 11.2 miles/hr.

The catecholamines and thyroxine are the hormones that have the greatest effect on metabolic rate. Catecholamines cause glycogen to break down into glucose and stimulate many enzyme systems, increasing cellular metabolism. Hypermetabolism is a clinical feature of some cases of pheochromocytoma, a catecholamine-secreting tumor of the adrenal medulla. Thyroxine magnifies the metabolic response to catecholamines, increases protein synthesis, and stimulates oxidation by the mitochondria. The metabolic rate is typically 45% above normal in hyperthyroidism (but up to 100% above normal in severe cases) and 25% below normal in hypothyroidism (but 45% below normal with complete lack of thyroid hormone). Other hormones have relatively minor effects on metabolic rate.

A resting person's metabolic rate increases 10% to 20% after a meal. This effect of food, called the **thermic effect of food** (formerly known as *specific dynamic action*), lasts several hours. The effect is greatest after consuming protein and less after consuming carbohydrate and fat; it appears to be associated with processing the products of digestion in the liver.

Metabolic rate is the amount of energy expended under standard conditions.

Because so many factors affect metabolism at rest, metabolic rate is often measured under a set of standard conditions to compare it with established norms. Metabolic rate measured under these conditions is called **basal metabolic rate (BMR)**. The commonly accepted conditions for measuring BMR are that the person must have fasted for 12 hours; the measurement must be made in the morning after a good night's sleep, beginning after the person has rested quietly for at least 30 minutes; and the air temperature must be comfortable, about 25°C (77°F). BMR is "basal" only during wakefulness because the metabolic rate during sleep is somewhat lower than the BMR.

Heat exchange with the environment can be measured directly by using a human calorimeter. In this insulated chamber, heat can exit either in the air ventilating the chamber or in water flowing through a heat exchanger in the chamber. By measuring the flow of air and water and their temperatures as they enter and leave the chamber, one can determine the subject's heat loss by conduction, convection, and radiation, and, by measuring the moisture content of air entering and leaving the chamber, one can determine heat loss by evaporation. This technique is called **direct calorimetry**, and though conceptually simple, it is cumbersome and costly.

Metabolic rate is often estimated by **indirect calorimetry**, which is based on measuring a person's rate of O_2 consumption because virtually all energy available to the body depends ultimately on reactions that consume O_2. Consuming 1 L of O_2 is associated with releasing 21.1 kJ (5.05 kcal) if the fuel is carbohydrate, 19.8 kJ (4.74 kcal) if the fuel is fat, and 18.6 kJ (4.46 kcal) if the fuel is protein. An average value often used for the metabolism of a mixed diet is 20.2 kJ (4.83 kcal) per liter of O_2. The ratio of CO_2 produced to O_2 consumed in the tissues is called the **respiratory quotient (RQ)**. The RQ is 1.0 for the oxidation of carbohydrate, 0.71 for the oxidation of fat, and 0.80 for the oxidation of protein. In a steady state in which CO_2 is exhaled from the lungs at the same rate it is produced in the tissues, RQ is equal to the respiratory exchange ratio, R (see Chapter 19, "Gas Transfer and Transport"). One can improve the accuracy of indirect calorimetry by also determining R and either estimating the amount of protein oxidized—which usually is small compared with fat and carbohydrate—or calculating it from urinary nitrogen excretion.

Skeletal muscle is the main source of heat during external work.

Even during mild exercise, the muscles are the principal source of metabolic heat, and during intense exercise, they may account for up to 90%. Moderately intense exercise by a healthy, but sedentary, young man may require a metabolic rate of 600 W (in contrast to about 80 W at rest) and intense activity by a trained athlete 1,400 W or more. Because of their high metabolic rate, exercising muscles may be almost 1°C warmer than the core. The blood perfusing these muscles is warmed and, in turn, warms the rest of the body, raising the core temperature.

Muscles convert most of the energy in the fuels they consume into heat rather than mechanical work. During phosphorylation of **adenosine diphosphate (ADP)** to form **adenosine triphosphate (ATP)**, 58% of the energy released from the fuel is converted into heat, and only about 42% is captured in the ATP that is formed in the process. When a muscle contracts, some of the energy in the ATP that was hydrolyzed is converted into heat rather than mechanical work. The efficiency at this stage varies enormously; it is zero in isometric muscle contraction, in which a muscle's length does not change while it develops tension so that no work is done even though metabolic energy is required. Finally, friction converts some of the mechanical work produced into heat within the body. (This is, e.g., the fate of all the mechanical work done by the heart in pumping blood.) At best, no more than 25% of the metabolic energy released during exercise is converted into mechanical work outside the body, and the other 75% or more is converted into heat within the body.

Sweating is the primary means of heat loss in humans.

There are four avenues of heat loss: *convection, conduction, radiation*, and *evaporation*. Convection is the transfer of heat resulting from the movement of a fluid, either liquid or gas. In thermal physiology, the fluid is usually air or water in the environment; in the case of heat transfer inside the body, the fluid is blood. To illustrate, consider an object immersed in a fluid that is cooler than the object. Heat passes from the object to the immediately adjacent fluid by conduction. If the fluid is stationary, conduction is the only means by which heat can pass through the fluid, and over time, the rate of heat flow from the body to the fluid will diminish as the fluid nearest the object approaches the temperature of the object. In practice, however, fluids are rarely stationary. If the fluid is moving, heat will still be carried from the object into the fluid by conduction, but once the heat has entered the fluid, the movement of the fluid will carry it—by convection. The same fluid movement that carries heat away from the surface of the object constantly brings fresh cool fluid to the surface, and so the object gives up heat to the fluid much more rapidly than if the fluid were stationary. Although conduction plays a role in this process, convection so dominates the overall heat transfer that we refer to the heat transfer as if it were entirely convection. Therefore, the conduction term (K) in the heat balance equation is restricted to heat flow between the body and other solid objects, and it usually represents only a small part of the total heat exchange with the environment.

Every surface emits energy as an electromagnetic radiation, with a power output proportional to the area of the surface, the fourth power of its absolute temperature (i.e., measured from absolute zero), and the **emissivity** (e) of the surface, a number between 0 and 1 that depends on the nature of the surface and the wavelength of the radiation. (In this discussion, the term *surface* is broadly defined, so that a flame and the sky, e.g., are surfaces.) Such radiation, called *thermal radiation*, has a characteristic distribution of power as a function of wavelength, which depends on the temperature of the surface. The emissivity of any surface is equal to the **absorptivity**—the fraction of incident radiant energy that the surface absorbs. If two bodies exchange heat by thermal radiation, radiation travels in both directions, but because each body emits radiation with an intensity that depends on its temperature, the net heat flow is from the warmer to the cooler body.

At ordinary tissue and environmental temperatures, virtually all thermal radiation is in a region of the infrared range where most surfaces, other than polished metals, have emissivities near 1 and emit with a power output near the theoretical maximum. However, bodies that are hot enough to glow, such as the sun, emit large amounts of radiation in the visible and near-infrared range, in which light-colored surfaces have lower emissivities and absorptivities than dark ones. Therefore, colors of skin and clothing affect heat exchange only in sunlight or bright artificial light.

When 1 g of water is converted into vapor at 30°C, it absorbs 2,425 J (0.58 kcal), the **latent heat of evaporation**, in the process. Evaporation of water is, thus, an efficient way of losing heat, and it is the body's only means of losing heat when the environment is hotter than the skin, as it usually is when the environment is warmer than 36°C. Evaporation must then dissipate both the heat produced by metabolic processes and any heat gained from the environment by convection and radiation. Most of the water evaporated in hot environments comes from sweat, but even in cold environments, the skin loses some water by the evaporation of **insensible perspiration**, water that diffuses through the skin rather than being secreted. In equation 1, E is nearly always positive, representing heat loss from the body. However, E is negative in the rare circumstances in which water vapor gives up heat to the body by condensing on the skin (as in a wet sauna).

Heat exchange is proportional to surface area and obeys thermodynamic principles.

Heat exchange always occurs from a higher to a cooler temperature and, thus, follows the second law of thermodynamics. Animals exchange heat with their environment through both the skin and the respiratory passages, but only the skin exchanges heat by radiation. In panting animals, respiratory heat loss may be large and may be an important means of achieving heat balance. In humans, however, respiratory heat exchange is usually relatively small and (though hyperthermic subjects may hyperventilate) is not predominantly under thermoregulatory control. Therefore, we do not consider it further here.

Convective heat exchange between the skin and the environment is proportional to the difference between skin and ambient air temperatures, as expressed by the following equation:

$$C = h_c \times A \times (\overline{T}_{sk} - T_a) \tag{2}$$

where A is the body surface area, $\overline{T}_{sk}$ and T_a are mean skin and ambient temperatures, respectively, and h_c is the convective heat transfer coefficient.

The value of h_c includes the effects of the factors other than temperature and surface area that influence convective heat exchange. For the whole body, air movement is the most important of these factors and convective heat exchange (and, thus, h_c) varies approximately as the square root of the air speed, except when air movement is slight (Fig. 28.6). Other factors that affect h_c include the direction of air movement and the curvature of the skin surface. As the radius of curvature decreases, h_c increases, and so the hands and fingers are effective in convective heat exchange disproportionately to their surface area.

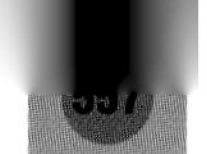

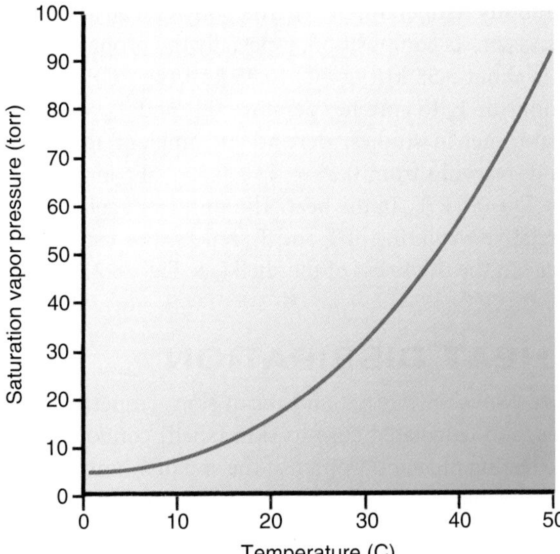

Figure 28.7 Saturation vapor pressure of water as a function of temperature. For any given temperature, the water vapor pressure is at its saturation value when the air is "saturated" with water vapor (i.e., holds the maximum amount possible at that temperature). At 37°C, Ph_2o equals 47 torr.

Radiative heat exchange is proportional to the difference between the fourth powers of the absolute temperatures of the skin and of the radiant environment (T_r) and to the emissivity of the skin (e_{sk}): $R \propto e_{sk} \times (\overline{T}_{sk}^4 - T_r^4)$. However, if T_r is close enough to $\overline{T}_{sk}$ that $\overline{T}_k - T_r$ is much smaller than the absolute temperature of the skin, R is nearly proportional to $e_{sk} \times (\overline{T}_{sk} - T_r)$. Some parts of the body surface (e.g., the inner surfaces of the thighs and arms) exchange heat by radiation with other parts of the body surface, and so the body exchanges heat with the environment as if it had an area smaller than its actual surface area. This smaller area, called the effective radiating surface area (A_r), depends on the body's posture and is closest to the actual surface area in a spread-eagle position and least in a curled-up position. Radiative heat exchange can be represented by the following equation:

$$R = h_r \times e_{sk} \times A_r \times (\overline{T}_{sk} - T_r) \quad (3)$$

where h_r is the radiant heat transfer coefficient, 6.43 W/(m^2·°C) at 28°C.

Evaporative heat loss from the skin to the environment is proportional to the difference between the water vapor pressure at the skin surface and the water vapor pressure in the ambient air. These relationships are summarized as follows:

$$E = h_e \times A \times (P_{sk} - P_a) \quad (4)$$

where P_{sk} is the water vapor pressure at the skin surface, P_a is the ambient water vapor pressure, and h_e is the evaporative heat transfer coefficient.

Moving air carries away water vapor, like heat, and so geometric factors and air movement affect E and h_e in the same way that they affect C and h_c. If the skin is completely wet, the water vapor pressure at the skin surface is the saturation water vapor pressure at the temperature of

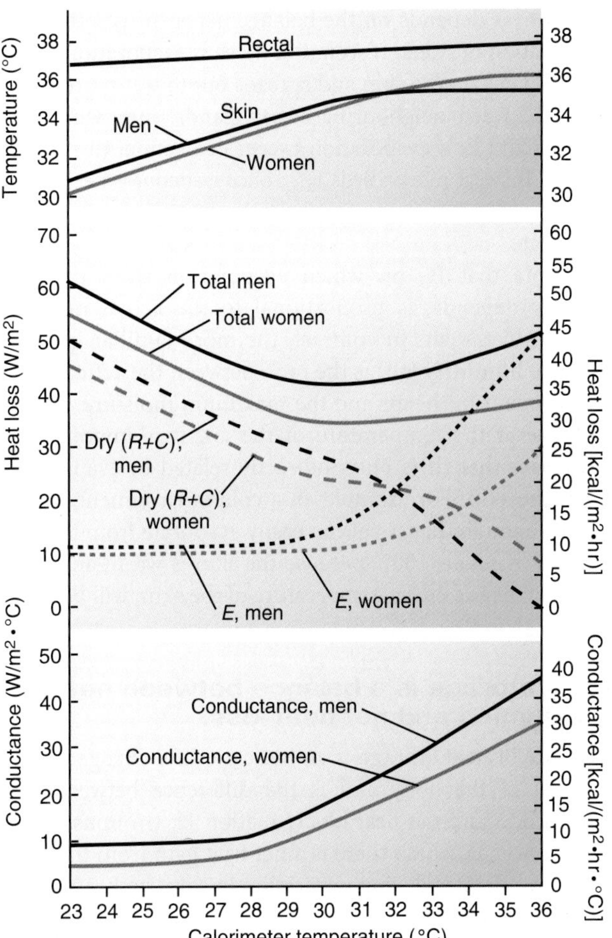

Figure 28.8 Heat dissipation. These graphs show the average values of rectal and mean skin temperatures, heat loss, and core-to-skin thermal conductance for nude resting men and women near steady state after 2 hours at different environmental temperatures in a calorimeter. (All energy exchange quantities in this figure have been divided by body surface area to remove the effect of individual body size.) Total heat loss is the sum of dry heat loss, by radiation (R) and convection (C), and evaporative heat loss (E). Dry heat loss is proportional to the difference between skin temperature and calorimeter temperature and decreases with increasing calorimeter temperature.

the skin (Fig. 28.8) and evaporative heat loss is E_{max}, the maximum possible for the prevailing skin temperature and environmental conditions. This condition is described as follows:

$$E_{max} = h_e \times A \times (P_{sk,sat} - P_a) \quad (5)$$

where $P_{sk,sat}$ is the saturation water vapor pressure at skin temperature. When the skin is not completely wet, it is impractical to measure P_{sk}, the actual average water vapor pressure at the skin surface. Therefore, a coefficient called skin *wettedness* (w) is defined as the ratio E/E_{max}, with $0 < w < 1$. Skin wettedness depends on the hydration of the epidermis and the fraction of the skin surface that is wet. We can now rewrite equation 4 as follows:

$$E = h_e \times A \times w \times (P_{sk,sat} - P_a) \quad (6)$$

Wettedness depends on the balance between secretion and evaporation of sweat. If secretion exceeds evaporation, sweat accumulates on the skin and spreads out to wet more of the space between neighboring sweat glands, increasing wettedness and E; if evaporation exceeds secretion, the reverse occurs. If sweat rate exceeds E_{max}, once wettedness becomes 1, the excess sweat drips from the body because it cannot evaporate.

Note that P_a, on which evaporation from the skin directly depends, is proportional to the actual moisture content in the air. In contrast, the more familiar quantity **relative humidity** (rh) is the ratio between the actual moisture content in the air and the maximum moisture content possible at the temperature of the air. It is important to recognize that rh is only indirectly related to evaporation from the skin. For example, in a cold environment, P_a will be low enough that sweat can easily evaporate from the skin even if rh equals 100% because the skin is warm and $P_{sk,sat}$, which depends on the temperature of the skin, will be much greater than P_a.

Heat storage is a balance between net heat production and net heat loss.

The rate of **heat storage** is an indicator of a change in heat content of the body and is the difference between heat production and net heat loss (equation 1). (In unusual circumstances in which there is a net heat gain from the environment, such as during immersion in a hot bath, storage is the sum of heat production and net heat gain.) It can be determined experimentally from simultaneous measurements of metabolism by indirect calorimetry and heat gain or loss by direct calorimetry. Storage of heat in the tissues changes their temperature, and the amount of heat stored is the product of body mass, the body's mean specific heat, and a suitable mean body temperature (T_b). The body's mean specific heat depends on its composition, especially the proportion of fat, and is about 3.55 kJ/(kg·°C) [0.85 kcal/(kg·°C)]. Empirical relations of T_b to core temperature (T_c) and T_{sk}, determined in calorimetric studies, depend on ambient temperature, with T_b varying from $0.65 \times T_c + 0.35 \times T_{sk}$ in the cold to $0.9 \times T_c + 0.1 \times T_{sk}$ in the heat. The shift from cold to heat in the relative weighting of T_c and T_{sk} reflects the accompanying change in the thickness of the shell (see Fig. 28.2).

HEAT DISSIPATION

Figure 28.9 shows rectal and mean skin temperatures, heat losses, and calculated core-to-skin (shell) conductances for nude resting men and women at the end of 2-hour exposures in a calorimeter to ambient temperatures of 23°C to 36°C. Shell conductance represents the sum of heat transfer by two parallel modes: conduction through the tissues of the shell and convection by the blood. It is calculated by dividing heat flow through the skin (HF_{sk}) (i.e., total heat loss from the body minus heat loss through the respiratory tract) by the difference between core and mean skin temperatures:

$$C = HF_{sk}/(T_c - \bar{T}_{sk}) \quad (7)$$

where C is shell conductance and T_c and $\bar{T}_{sk}$ are core and mean skin temperatures.

From 23°C to 28°C, conductance is minimal because the skin is vasoconstricted and its blood flow is low. The minimal level of conductance attainable depends largely on the thickness of the subcutaneous fat layer, and women's thicker layer allows them to attain a lower conductance than men. At about 28°C, conductance begins to increase, and above 30°C, conductance continues to increase and sweating begins.

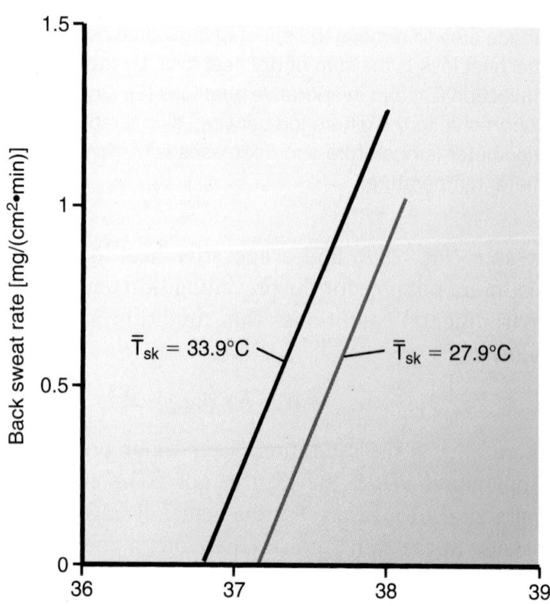

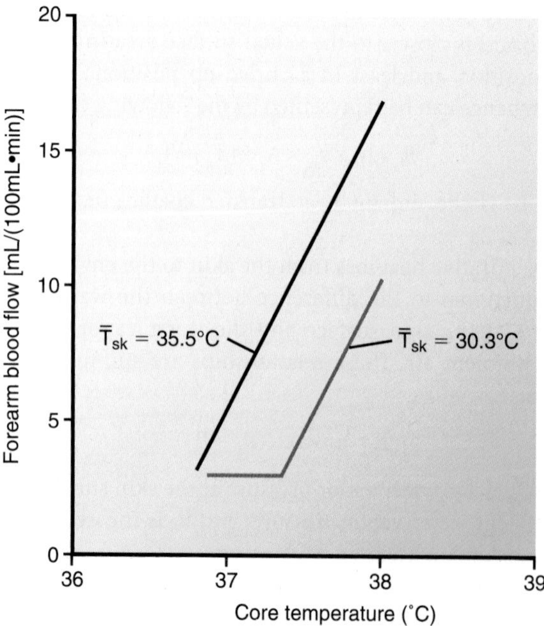

Figure 28.9 Control of heat-dissipating responses. These graphs show the relations of back (scapular) sweat rate (*left*) and forearm blood flow (*right*) to core temperature and mean skin temperatures ($\bar{T}_{sk}$). In these experiments, exercise increased core temperature.

For these subjects, 28°C to 30°C is the zone of **thermoneutrality**, or thermoneutral zone, the range of comfortable environmental temperatures in which thermal balance is maintained without either shivering or sweating. In this zone, controlling conductance and $\overline{T}_{sk}$ and, thus, R and C entirely maintain heat balance. As equations 2 to 4 show, C, R, and E all depend on skin temperature, which, in turn, depends partly on skin blood flow. E depends also, through skin wettedness, on sweat secretion. Therefore, all these modes of heat exchange are partly under physiologic control.

Sweat evaporation can dissipate large amounts of heat.

In Figure 28.9, evaporative heat loss is nearly independent of ambient temperature below 30°C and is 9 to 10 W/m^2, corresponding to evaporation of about 13 to 15 g/(m^2·h), of which about half is moisture lost in breathing and half is insensible perspiration. This evaporation occurs independent of thermoregulatory control. As the ambient temperature increases, the body depends more and more on the evaporation of sweat to achieve heat balance.

The two histologic types of sweat glands are **eccrine** and **apocrine**. Eccrine sweat glands, the dominant type in all human populations, are more important in human thermoregulation and number about 2,500,000. Eccrine secretion is controlled through postganglionic sympathetic C fibers that release acetylcholine (ACh) rather than norepinephrine. Cholinergic stimulation of the sweat gland elicits the secretion of the so-called *precursor fluid*, the composition of which resembles that of plasma, except that it does not contain the plasma proteins. Therefore, sodium and chloride are the primary electrolytes in sweat, with potassium, calcium, and magnesium present in smaller amounts. As the fluid moves through the duct portion of the sweat gland, its composition is modified by active reabsorption of sodium and chloride. As a result, sweat is hyposmotic compared to plasma, and sodium concentration typically ranges from 10 to 70 mmol/L, whereas chloride ranges from 5 to 60 mmol/L, depending on diet, sweat rate, and degree of heat acclimatization. Eccrine sweat also includes lactate, urea, ammonia, serine, ornithine, citrulline, aspartic acid, heavy metals, organic compounds, and proteolytic enzymes. In contrast, apocrine sweat consists mainly of sialomucin. Although initially odorless, as apocrine sweat comes in contact with normal bacterial flora on the surface of the skin, an odor develops. Apocrine sweat is more viscous and produced in much smaller amounts than eccrine sweat. The exact function of apocrine glands is unclear, though they are thought to represent scent glands. During puberty, **apoeccrine** glands appear. These are hybrid sweat glands that are found in the axilla and may play a role in axillary hyperhidrosis. Their secretory structures have both a small-diameter portion similar to those found in eccrine glands and a large-diameter portion that resembles an apocrine gland. Functionally similar to eccrine glands, they respond mainly to cholinergic stimuli, and their ducts are long and open directly onto the skin surface. Interestingly, apoeccrine glands secrete nearly 10 times as much sweat as eccrine glands do.

A healthy man unacclimatized to heat can secrete up to 1.5 L/h of sweat. Although the number of functional sweat glands is fixed before the age of 3, the secretion capacity of the individual glands can change, especially with endurance exercise training and heat acclimatization. Men well acclimatized to heat can attain peak sweat rates of 1.0 to 2.5 L/h, with sweat rates of more than 2.5 L/h possible when the ambient temperature is high. However, such rates cannot be maintained as the maximum daily sweat output is probably about 15 L.

Increased vasodilation of the skin augments heat loss.

Heat produced in the body must be delivered to the skin surface to be eliminated. When skin blood flow is minimal, shell conductance is typically 5 to 9 W/°C/m^2 of body surface. For a lean resting subject with a surface area of 1.8 m^2, minimal whole body conductance of 16 W/°C (i.e., 8.9 W/[°C·m^2] × 1.8 m^2), and a metabolic heat production of 80 W, the temperature difference between the core and the skin must be 5°C (i.e., 80 W ÷ 16 W/°C) for the heat produced to be conducted to the surface. In a cool environment, $\overline{T}_{sk}$ may easily be low enough for this to occur. However, in an ambient temperature of 33°C, $\overline{T}_{sk}$ is typically about 35°C, and without an increase in conductance, core temperature would have to rise to 40°C—a high, although not yet dangerous, level—for the heat to be conducted to the skin. If the rate of heat production was increased to 480 W by moderate exercise, the temperature difference between core and skin would have to rise to 30°C—and core temperature to well beyond lethal levels—to allow all the heat produced to be conducted to the skin. In the latter circumstances, the conductance of the shell must increase greatly for the body to reestablish thermal balance and continue to regulate its temperature. This is accomplished by increasing the skin blood flow.

Effectiveness of skin blood flow in heat transfer

Assuming that blood on its way to the skin remains at core temperature until it reaches the skin, reaches skin temperature as it passes through the skin, and then stays at skin temperature until it returns to the core, we can compute the rate of heat flow (HF_b) as a result of convection by the blood as follows:

$$HF_b = SkBF \times (T_c - \overline{T}_{sk}) \times 3.85\ kJ/(L\cdot°C) \qquad (8)$$

where SkBF is the rate of skin blood flow, expressed in liter per second rather than the usual liter per minute, to simplify computing HF in W (i.e., J/s), and 3.85 kJ/(L·°C) (0.92 kcal/[L·°C]) is the volume-specific heat of blood. Conductance as a result of convection by the blood (C_b) is calculated as follows:

$$C_b = HF_b / (T_c - \overline{T}_{sk}) = SkBF \times 3.85\ kJ/(L\cdot°C) \qquad (9)$$

Of course, heat continues to flow by conduction through the tissues of the shell, and so total conductance is the sum of conductance as a result of convection by the blood and that

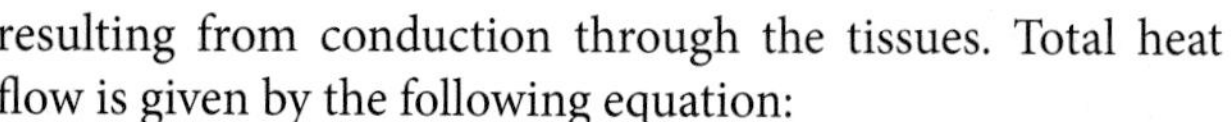

resulting from conduction through the tissues. Total heat flow is given by the following equation:

$$HF = (C_b + C_0) \times (T_c - \overline{T}_{sk}) \qquad (10)$$

where C_0 is thermal conductance of the tissues when skin blood flow is minimal and, thus, is predominantly a result of conduction through the tissues.

The assumptions made in deriving equation 8 are somewhat artificial and represent the conditions for maximum efficiency of heat transfer by the blood. In practice, blood exchanges heat also with the tissues through which it passes on its way to and from the skin. Heat exchange with these other tissues is greatest when skin blood flow is low; in such cases, heat flow to the skin may be much less than predicted by equation 8, as discussed further below. However, equation 8 is a reasonable approximation in a warm subject with moderate-to-high skin blood flow. Although measuring whole-body SkBF directly is not possible, it is believed to reach several liters per minute during heavy exercise in the heat. The maximum obtainable is estimated to be nearly 8 L/min. If SkBF = 1.89 L/min (0.0315 L/s), according to equation 9, skin blood flow contributes about 121 W/°C to the conductance of the shell. If conduction through the tissues contributes 16 W/°C, total shell conductance is 137 W/°C, and if T_c = 38.5°C and T_{sk} = 35°C, this will produce a core-to-skin heat transfer of 480 W, the heat production in our earlier example of moderate exercise. Therefore, even a moderate rate of skin blood flow can have a dramatic effect on heat transfer.

When a person is not sweating, raising skin blood flow brings skin temperature nearer to blood temperature, and lowering skin blood flow brings skin temperature nearer to ambient temperature. Under such conditions, the body can control dry (convective and radiative) heat loss by varying skin blood flow and, thus, skin temperature. Once sweating begins, skin blood flow continues to increase as the person becomes warmer. In these conditions, however, the tendency of an increase in sweating to cool the skin approximately balances the tendency of an increase in skin blood flow to warm the skin. Therefore, after sweating has begun, further increases in skin blood flow usually cause little change in skin temperature or dry heat exchange and serve primarily to deliver to the skin the heat that is being removed by the evaporation of sweat. Skin blood flow and sweating work in tandem to dissipate heat under such conditions.

Sympathetic control of skin circulation

Because the nutritional needs of the skin are low, the skin blood flow is controlled mainly by reflexes or changes in temperature. In most of the skin, the vasodilation that occurs during heat exposure depends on sympathetic nerve signals that cause the blood vessels to dilate, and regional nerve block can prevent or reverse this vasodilation. Because it depends on the action of neural signals, such vasodilation is sometimes referred to as *active vasodilation*. Active vasodilation occurs in almost all the skin, except in so-called acral regions—hands, feet, lips, ears, and nose. In skin areas where active vasodilation occurs, vasoconstrictor activity is minimal at thermoneutral temperatures, and active vasodilation during heat exposure does not begin until close to the onset of sweating. Therefore, small temperature changes within the thermoneutral range do not much affect skin blood flow in these areas.

Reflex vasodilation is complex, with potentially many layers of redundancy. With whole-body heating, sympathetic adrenergic tone is withdrawn and a sympathetic active vasodilator system is activated, most likely mediated by a cholinergic neurotransmitter system. Neuronal **nitric oxide (NO)** synthase (nNOS)-mediated NO is directly required for 30% to 40% of this response, may also contribute by inhibiting adrenergic function (see below), and appears to have no specific role in response to local skin warming. Other mediators of this response include **vasoactive intestinal peptide (VIP)**, histamine, prostanoids, endothelial-derived hyperpolarizing factor, and **substance P**.

Increased local skin temperature (T_{loc}) also results in cutaneous vasodilation, although through an independent mechanism. Local vasodilatory mechanisms involved in local skin warming can increase SkBF to maximal levels, a biphasic response involving sensory nerves that mediate an initial transient vasodilatory "peak" followed by a prolonged vasodilatory "plateau" that is mediated primarily by endothelial NOS generation of NO.

Active vasodilation operates in tandem with sweating in the heat and is impaired or absent in **anhidrotic ectodermal dysplasia**, a congenital disorder in which sweat glands are sparse or absent. For these reasons, the existence of a mechanism linking active vasodilation to the sweat glands has long been suspected but never established. Recent studies have demonstrated an attenuated sweat rate after NOS inhibition, suggesting that NO augments sweat secretion. Additionally, NOS inhibition attenuates the sweat response to exercise in a warm environment, suggesting that NO has a role in modifying thermoregulatory sweating, specifically by changing sweat gland output. While the overall importance of NO's role remains unknown, the magnitude of reduction in local sweat rate by NOS inhibition is as large as that induced by dehydration. Accordingly, NO appears to play an important role in thermoregulatory control of sweating during exercise in warm environments.

Reflex vasoconstriction, occurring in response to cold and as part of certain nonthermal reflexes such as baroreflexes, is mediated primarily through adrenergic sympathetic fibers distributed widely over most of the skin. The process is initiated when the mean whole-body skin temperature decreases below thermoneutral levels of around 33°C and is mediated by **norepinephrine (NE)** and neuropeptide Y released from vasoconstrictor nerves. Although evidence suggests that endogenous NO can inhibit or modulate this response, these two neurotransmitters appear to be all that is necessary. Accordingly, reducing the flow of impulses in these nerves allows the blood vessels to dilate. In the acral regions and superficial veins (whose role in heat transfer is discussed below), vasoconstrictor fibers are the predominant vasomotor innervation, and the vasodilation that occurs during heat exposure is largely a result of the withdrawal of vasoconstrictor activity. Blood flow in these skin regions is sensitive to small temperature changes even in the thermoneutral range and may be responsible for "fine-tuning" heat loss to maintain heat balance in this range.

There also exists a local vasoconstrictive response to skin cooling that has two components. An adrenergic element includes reduced secretion of norepinephrine that is more than overcome by an increase in α_{2C}-receptor sensitivity via Rho-kinase-mediated signaling. The second element involves a withdrawal of tonic NO production through inhibition of NOS and subsequent steps in the NO system.

THERMOREGULATORY CONTROL

In discussions of control systems, the words "regulation" and "regulate" have meanings distinct from those of the word "control" (see Chapter 1, "Homeostasis and Cellular Signaling"). The variable that a control system acts to maintain within narrow limits (e.g., temperature) is called the *regulated* variable, and the quantities it controls to accomplish this (e.g., sweating rate, skin blood flow, metabolic rate, and thermoregulatory behavior) are called *controlled* variables.

Humans have two distinct subsystems for regulating body temperature: behavioral thermoregulation and physiologic thermoregulation. Behavioral thermoregulation—through the use of shelter, space heating, air conditioning, and clothing—enables humans to live in the most extreme climates in the world, but it does not provide fine control of body heat balance. In contrast, physiologic thermoregulation is capable of fairly precise adjustments of heat balance but is effective only within a relatively narrow range of environmental temperatures.

Temperature is perceived through thermal sensation.

Sensory information about body temperatures is an essential part of both behavioral and physiologic thermoregulation. The distinguishing feature of behavioral thermoregulation is the involvement of consciously directed efforts to regulate body temperature. Thermal discomfort provides the necessary motivation for thermoregulatory behavior, and behavioral thermoregulation acts to reduce both the discomfort and the physiologic strain imposed by a stressful thermal environment. For this reason, both thermal comfort and the absence of shivering and sweating characterize the zone of thermoneutrality.

Warmth and cold on the skin are felt as either comfortable or uncomfortable, depending on whether they decrease or increase the physiologic strain—a shower temperature that feels pleasant after strenuous exercise may be uncomfortably chilly on a cold winter morning. The processing of thermal information in behavioral thermoregulation is not as well understood as it is in physiologic thermoregulation. However, perceptions of thermal sensation and comfort respond much more quickly than core temperature or physiologic thermoregulatory responses to changes in environmental temperature and, thus, appear to anticipate changes in the body's thermal state. Such an anticipatory feature would be advantageous because it would reduce the need for frequent small behavioral adjustments.

Thermal defense mechanisms operate through changes in heat production and heat loss.

Familiar inanimate control systems, such as most refrigerators and heating and air-conditioning systems, operate at only two levels: on and off. In a typical home-heating system, for example, when the indoor temperature falls below the desired level, the thermostat detects the change and turns on the burner and blower in the furnace; when the temperature is restored to the desired level, the thermostat turns the furnace off. Rather than operating at only two levels, most physiologic control systems produce a graded response according to the size of the disturbance in the regulated variable. In many instances, changes in the controlled variables are proportional to displacements of the regulated variable from some threshold value.

The control of heat-dissipating responses is an example of a proportional control system. Figure 28.10 shows how reflex control of two heat-dissipating responses, sweating and skin blood flow, depends on body core temperature and mean skin temperature. Each response has a core temperature threshold—a temperature at which the response starts to increase—and this threshold depends on mean skin temperature. At any given skin temperature, the change in each response is proportional to the change in core temperature, and increasing the skin temperature lowers the threshold level of core temperature and increases the response at any given core temperature. In humans, a change of 1°C in core temperature elicits about nine times as great a thermoregulatory response as a 1°C change in mean skin temperature. (Besides its effect on the reflex signals, skin temperature has a local effect that modifies the response of the blood vessels and sweat glands to the reflex signal, discussed later.)

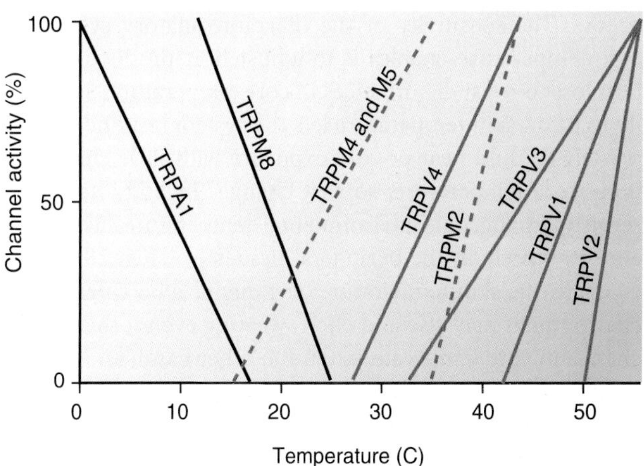

Figure 28.10 Dependence of the activity of cold-activated (*black*) and heat-activated (*green*) thermoTRP channels on temperature. TRP, transient receptor potential.

Cold stress elicits increases in metabolic heat production through shivering and nonshivering thermogenesis. **Shivering** is a rhythmic oscillating tremor of skeletal muscles. On a local level, movement is uncoordinated as reflected by rapid (i.e., 250 Hz) electromyographic activity. However, there is an overall 4 to 8 cycle/min "waxing-and-waning" activity that is clinically apparent during intense shivering. The CNS network generating the cold-evoked bursts of α-motor neuron activity is not well understood, but it includes transmission of cutaneous cold afferent signals through the lateral parabrachial nucleus, integration of thermoregulatory signals in the preoptic area (POA), lifting of its inhibition, and activation of fusimotor neurons (β- and γ-motor neurons), which is dependent on descending neurons in the rostral ventromedial medulla. Although these efferent impulses are not rhythmic, their overall effect is to increase muscle tone, thereby increasing metabolic rate. Once the tone exceeds a critical level, the contraction of one group of muscle fibers stretches the muscle spindles in other fiber groups in series with it, eliciting contractions from those groups of fibers via the stretch reflex, and so on; thus, the rhythmic oscillations that characterize frank shivering begin.

Shivering occurs in bursts, and signals from the cerebral cortex inhibit the "shivering pathway," so that voluntary muscular activity and attention can suppress shivering. Because the limbs are part of the shell in the cold, trunk and neck muscles are preferentially recruited for shivering—the centralization of shivering—to help retain the heat produced during shivering within the body core; and the familiar experience of teeth chattering is one of the earliest signs of shivering. As with heat-dissipating responses, the control of shivering depends on both core and skin temperatures, but the details of its control are not precisely understood.

Hypothalamus integrates thermal information from the core and the skin.

Temperature receptors in the body core and skin transmit information about their temperatures through afferent nerves to the brainstem and, especially, the hypothalamus, where much of the integration of temperature information occurs. The sensitivity of the thermoregulatory system to core temperature enables it to adjust heat production and heat loss to resist disturbances in core temperature. Sensitivity to mean skin temperature lets the system respond appropriately to mild heat or cold exposure with little change in body core temperature, so that changes in body heat as a result of changes in environmental temperature take place almost entirely in the peripheral tissues (see Fig. 28.2). For example, the skin temperature of someone who enters a hot environment may rise and elicit sweating even if there is no change in core temperature. On the other hand, an increase in heat production within the body, as during exercise, elicits the appropriate heat-dissipating responses through a rise in core temperature.

Core temperature receptors involved in controlling thermoregulatory responses are unevenly distributed and are concentrated in the POA of the anterior hypothalamus. Most of these are warm sensitive, meaning they increase their activity with increased brain temperature. In experimental mammals, temperature changes of only a few tenths of 1°C in the hypothalamus elicit changes in the thermoregulatory effector responses. Although much less common, cold-sensitive neurons (i.e., those that increase their activity with a decrease in brain temperature) also exist. However, the cold sensitivity of most of them seems to be due to inhibitory synaptic input from the nearby warm-sensitive neurons. The roles of cold- and warm-sensitive POA neurons are not reciprocal. Studies involving thermal and chemical stimulation of POA cells have shown that both cold-defense and heat-defense autonomic responses are initiated by the corresponding changes in the activity of warm-sensitive neurons; increased activity of warm-sensitive POA neurons triggers heat-defense responses, whereas decreased activity triggers cold-defense responses. Because warm-sensitive POA neurons display spontaneous membrane depolarization, these cells are considered pacemakers; their thermosensitivity is due to currents that determine the rate of spontaneous depolarization between successive action potentials. Furthermore, morphologic identification of thermosensitive neurons in the POA has shown their orientation to be ideally suited to receiving afferent projections from peripheral thermoreceptors via the medial forebrain bundle. In addition to the POA cold- and warm-sensitive neurons, there are peripheral deep-body sensors that respond to changes in core body temperature. They are located in the esophagus, stomach, large intra-abdominal veins, and other organs.

Consider what happens when some disturbance—say, an increase in metabolic heat production resulting from exercise—upsets the thermal balance. Additional heat is stored in the body, and core temperature rises. The central thermoregulatory controller receives information about these changes from the thermal receptors and elicits appropriate heat-dissipating responses. Core temperature continues to rise, and these responses continue to increase until they are sufficient to dissipate heat as fast as it is being produced, restoring heat balance and preventing further increases in body temperatures. In the language of control theory, the rise in core temperature that elicits heat-dissipating responses sufficient to reestablish thermal balance during exercise is an example of a load error. A load error is characteristic of any proportional control system that is resisting the effect of some imposed disturbance or "load." Although the disturbance in this example is exercise, the same principle applies if the disturbance is a decrease in metabolic rate or a change in the environment. However, if the disturbance is in the environment, most of the temperature change will be in the skin and shell rather than in the core. If the disturbance produces a net loss of heat, the body will restore heat balance by decreasing heat loss and increasing heat production.

Set point is the body's "thermostat."

The primary autonomic defenses against heat are sweating and active cutaneous vasodilation, whereas the primary autonomic defenses against cold are cutaneous vasoconstriction and shivering. The core temperature, or reference temperature, triggering each response defines its activation threshold. Body temperature is normally maintained between the sweating and vasoconstriction thresholds, temperatures that define the *interthreshold range* (also referred to as the "null zone"). Onset of sweating, thus, usually defines the upper boundary of normal body temperature, and cutaneous vasoconstriction defines the lower boundary. Typically, the interthreshold range is only a few tenths of a degree, and it is thus likely that temperatures within this range are accurately sensed but do not generate an error signal and trigger thermoregulatory responses. Under normal circumstances, thermoregulatory control can thus be roughly modeled as a discrete yet adjustable "**set point**" (T_{set}) (Fig. 28.11). That thermoregulatory responses are likely to have unique thresholds within the thermal integrative network is exemplified by the fact that the threshold for shivering is typically a degree less than the vasoconstriction threshold. Shivering is thus a "last resort" response and appropriately so because it is metabolically inefficient.

Effect of nonthermal inputs on thermoregulatory responses

Each thermoregulatory response may be affected by inputs other than body temperatures and factors that influence the set point. We have already noted that voluntary activity affects shivering and certain hormones affect metabolic heat production. In addition, nonthermal factors may produce a burst of sweating at the beginning of exercise, and emotional effects on sweating and skin blood flow are matters of common experience. Skin blood flow is the thermoregulatory response most influenced by nonthermal factors because of its potential involvement in reflexes that function to maintain cardiac output, blood pressure, and tissue O_2 delivery under a variety of disturbances, including heat stress, postural changes, hemorrhage, exercise, hydration state, sleep, and narcosis.

Biologic rhythms and altered physiologic states can change the thermoregulatory set point.

Fever elevates core temperature at rest, heat acclimatization decreases it, and time of day and (in women) the phase of the menstrual cycle change it in a cyclic fashion. Core temperature at rest varies in an approximately sinusoidal fashion with time of day. The minimum temperature occurs at night, several hours before awaking, and the maximum, which is 0.5°C to 1°C higher, occurs in the late afternoon or evening (see Fig. 28.3). This pattern coincides with patterns of activity and eating but does not depend on them, and it occurs even during bed rest in fasting subjects. This pattern is an example of a **circadian rhythm**, a rhythmic pattern in a physiologic function with a period of about 1 day. During the menstrual cycle, core temperature is at its lowest point just before ovulation. During the next few days, it rises from 0.5°C to 1°C to a plateau that persists through most of the luteal phase. Each of these factors—fever, heat acclimatization, the circadian rhythm, and the menstrual cycle—changes the core temperature at rest by changing the thermoregulatory set point, producing corresponding changes in the thresholds for all of the thermoregulatory responses.

Skin temperature alters cutaneous blood flow and sweat gland responses.

The skin is the organ most directly affected by environmental temperature. Skin temperature influences heat-loss responses not only through reflex actions (see Fig. 28.10) but also through direct effects on the skin blood vessels and sweat glands.

Local temperature changes act on skin blood vessels in at least two ways. First, local cooling potentiates (and heating weakens) the constriction of blood vessels in response to nerve signals and vasoconstrictor substances. (At low temperatures, however, cold-induced vasodilation (CIVD) increases skin blood flow, as discussed later.) Second, in

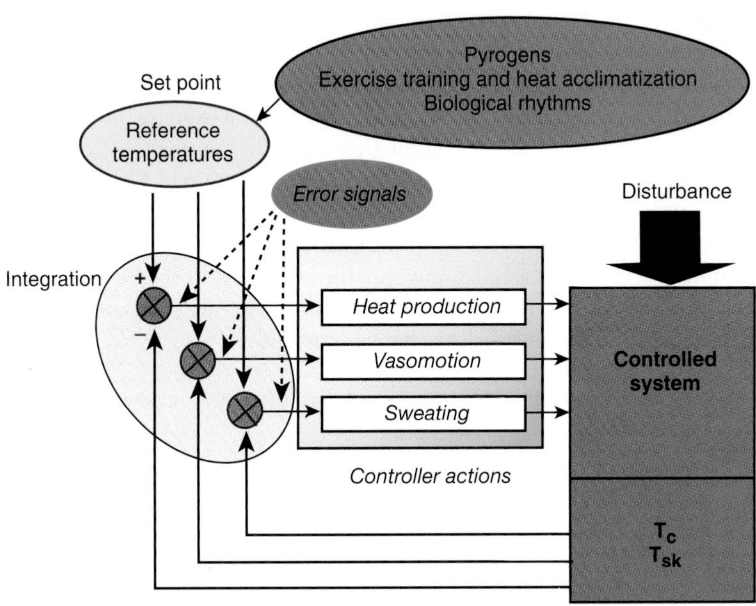

Figure 28.11 A control system model for temperature regulation. Multiple reference temperatures (one for each thermoeffector) comprise the adjustable set point, and active (controlling system) and passive (controlled system) components are part of the system. A disturbance to the passive system is detected by the sensors, and the information from these compare with reference temperatures. The resultant error signal drives the thermoeffectors comprising the controlling system. The aim of such a negative feedback control system is the minimization of the error signal. T_c is core temperature, and T_{sk} is skin temperature.

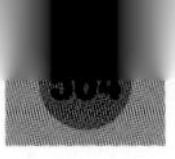

skin regions where active vasodilation occurs, local heating causes vasodilation (and local cooling causes vasoconstriction) through a direct action on the vessels, independent of nerve signals. The local vasodilator effect of skin temperature is especially strong above 35°C; and, when the skin is warmer than the blood, increased blood flow helps cool the skin and protect it from heat injury, unless this response is impaired by vascular disease. Local thermal effects on sweat glands parallel those on blood vessels, and so local heating potentiates (and local cooling diminishes) the local sweat gland response to reflex stimulation or ACh, and intense local heating elicits sweating directly, even in skin whose sympathetic innervation has been interrupted surgically.

Sweat gland "fatigue" alters thermoregulation.

During prolonged heat exposure (lasting several hours) with high sweat output, sweating rates gradually decline and the response of sweat glands to local cholinergic drugs is reduced. This reduction of sweat gland responsiveness is sometimes called sweat gland "fatigue." Wetting the skin makes the stratum corneum swell, mechanically obstructing the sweat gland ducts and causing a reduction in sweat secretion, an effect called **hidromeiosis**. The glands' responsiveness can be at least partly restored if air movement increases or humidity is reduced, allowing some of the sweat on the skin to evaporate. Sweat gland fatigue may involve processes besides hidromeiosis because prolonged sweating also causes histologic changes, including the depletion of glycogen, in the sweat glands.

THERMOREGULATORY RESPONSES DURING EXERCISE

During intense physical exercise, metabolic heat production can increase 10- to 20-fold, with <30% of the heat generated converted into mechanical energy. Consequently, more than 70% of metabolic heat generated must be transported to the skin to be dissipated to the environment. Heat starts to accumulate in the body when the heat-dissipating mechanisms are unable to cope with metabolic heat production, leading to an increase in body temperature. Although hot environments also elicit heat-dissipating responses, exercise ordinarily is responsible for the greatest demands on the thermoregulatory system for heat dissipation. Exercise provides an important example of how the thermoregulatory system responds to a disturbance in heat balance. In addition, exercise and thermoregulation impose competing demands on the circulatory system because exercise requires large increases in blood flow to exercising muscle, whereas the thermoregulatory responses to exercise require increases in skin blood flow. Muscle blood flow during exercise is several times as great as skin blood flow, but the increase in skin blood flow is responsible for disproportionately large demands on the cardiovascular system, as discussed below. Finally, if the water and electrolytes lost through sweating are not replaced, the resulting reduction in plasma volume will eventually create a further challenge to cardiovascular homeostasis.

Core temperature rises during exercise, triggering heat-loss responses.

As previously mentioned, the increased metabolic heat production during exercise causes an increase in core temperature, which in turn elicits heat-loss responses. Core temperature continues to rise until heat loss has increased enough to match heat production, and core temperature and the heat-loss responses reach new steady-state levels. Because the heat-loss responses are proportional to the increase in core temperature, the increase in core temperature at steady state is proportional to the rate of heat production and, thus, to the metabolic rate.

A change in ambient temperature causes changes in the levels of sweating and skin blood flow necessary to maintain any given level of heat dissipation. However, the change in ambient temperature also elicits, via direct and reflex effects of the accompanying skin temperature changes, altered responses in the right direction. For any given rate of heat production, there is a certain range of environmental conditions within which an ambient temperature change elicits the necessary changes in heat-dissipating responses almost entirely through the effects of skin temperature changes, with virtually no effect on core temperature. (The limits of this range of environmental conditions depend on the rate of heat production and such individual factors as skin surface area and state of heat acclimatization.) Within this range, the core temperature reached during exercise is nearly independent of ambient temperature. For this reason, it was once believed that the increase in core temperature during exercise is caused by an increase in the thermoregulatory set point, as during fever. As noted, however, the increase in core temperature with exercise is an example of a load error rather than an increase in set point.

This difference between fever and exercise is shown in Figure 28.12. Note that, although heat production may increase substantially (through shivering), when core temperature is rising early during fever, it need not stay high to maintain the fever. In fact, it returns nearly to prefebrile levels once the fever is established. During exercise, however, an increase in heat production not only causes the elevation in core temperature but is necessary to sustain it. Also, although the core temperature is rising during fever, the rate of heat loss is, if anything, lower than it was before the fever began. During exercise, however, the heat-dissipating responses and the rate of heat loss start to increase early and continue to increase as the core temperature rises.

Exercise in the heat can impair cardiac filling.

The rise in core temperature during exercise increases the temperature difference between the core and the skin somewhat but not nearly enough to match the increase in metabolic heat production. Therefore, as we saw earlier, skin blood flow must increase to carry all of the heat that is produced to the skin. In a

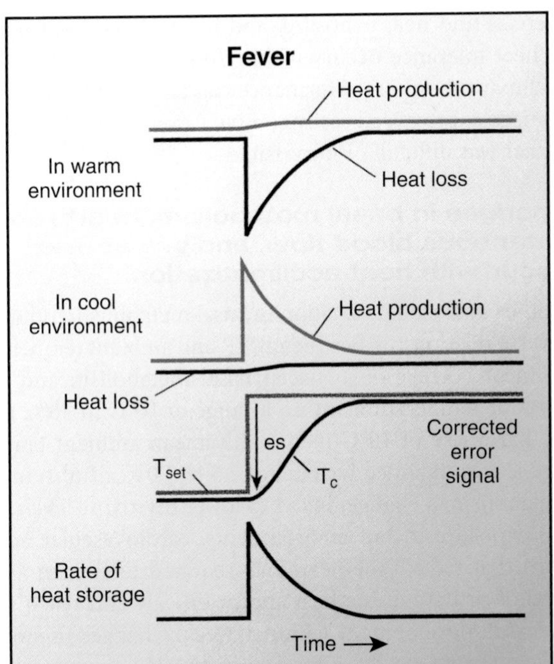

A

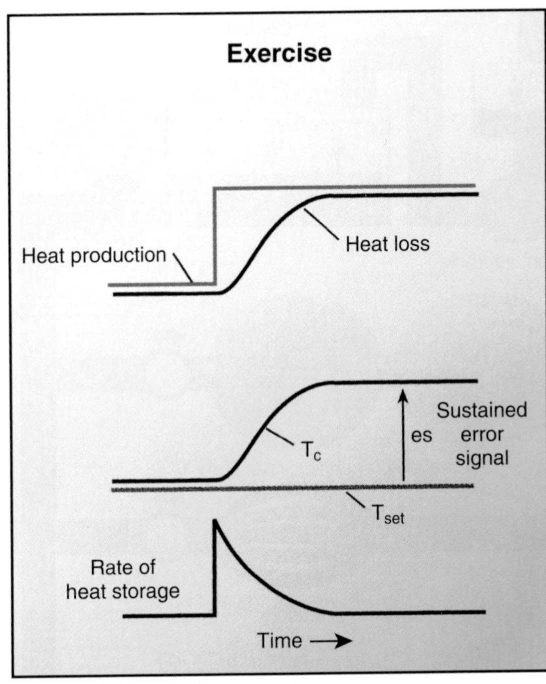

B

Figure 28.12 Thermal events during fever and exercise. (A) The development of fever. **(B)** The increase in T_c during exercise. The es is the difference between the T_c and the T_{set}. At the start of a fever, T_{set} has risen, so that T_{set} is higher than T_c and es is negative. At steady state, T_c has risen to equal the new level of T_{set} and es is corrected (i.e., it returns to zero.) At the start of exercise, $T_c = T_{set}$ so that es = 0. At steady state, T_{set} has not changed but T_c has increased and is greater than T_{set}, producing a sustained error signal, which is equal to the load error. (The error signal, or load error, is here represented with an *arrow* pointing downward for $T_c < T_{set}$ and with an *arrow* pointing upward for $T_c > T_{set}$.)

warm environment, where the temperature difference between core and skin is relatively small, the necessary increase in skin blood flow may be several liters per minute.

The work of providing the skin blood flow required for thermoregulation in the heat may impose a heavy burden on a diseased heart, but in healthy people, the major cardiovascular burden of heat stress results from impaired venous return. As skin blood flow increases, the dilated vascular bed of the skin becomes engorged with large volumes of blood, reducing central blood volume and cardiac filling (Fig. 28.13). Stroke volume is decreased, and a higher heart rate is required to maintain cardiac output. A decrease in plasma volume, if the large amounts of salt and water lost in the sweat are not replaced, aggravates these effects. Because the main cation in sweat is sodium, a disproportionately large amount of the body water lost in sweat is at the expense of extracellular fluid, including plasma, although this effect is mitigated if the sweat is dilute.

Compensatory responses during exercise in the heat

Several reflex adjustments help maintain cardiac filling, cardiac output, and arterial pressure during exercise and heat stress. The most important of these is constriction of the renal and splanchnic vascular beds. A reduction in blood flow through these beds allows a corresponding diversion of cardiac output to the skin and the exercising muscles. In addition, because the splanchnic vascular beds are compliant, a decrease in their blood flow reduces the amount of blood pooled in them (see Fig. 28.13), helping compensate for decreases in central blood volume caused by reduced plasma volume and blood pooling in the skin.

The degree of vasoconstriction is graded according to the levels of heat stress and exercise intensity. During strenuous exercise in the heat, renal and splanchnic blood flows may fall to 20% of their values in a cool resting subject. Such intense splanchnic vasoconstriction may produce mild ischemic injury to the gut, helping explain the intestinal symptoms that some athletes experience after endurance events. The cutaneous veins constrict during exercise; because most of the vascular volume is in the veins, constriction makes the cutaneous vascular bed less easily distensible and reduces peripheral pooling. Because of the essential role of skin blood flow in thermoregulation during exercise and heat stress, the body preferentially compromises splanchnic and renal flow for the sake of cardiovascular homeostasis. Above a certain level of cardiovascular strain, however, skin blood flow, too, is compromised.

HEAT ACCLIMATIZATION

As in other mammals, thermoregulation is an important aspect of human homeostasis. Humans have been able to adapt to a great diversity of climates, including hot humid and hot

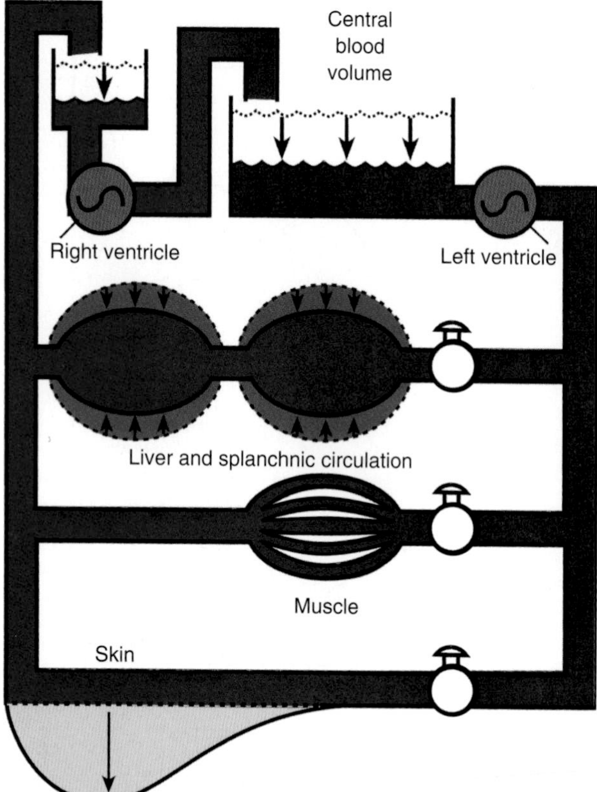

Figure 28.13 Cardiovascular strain and compensatory responses during heat stress. This figure first shows the effects of skin vasodilation on peripheral pooling of blood and the thoracic reservoirs from which the ventricles are filled and, second, the effects of compensatory vasomotor adjustments in the splanchnic circulation. The valves on the *right* represent the resistance vessels that control blood flow through the liver/splanchnic, muscle, and skin vascular beds. *Arrows* show the direction of the changes during heat stress.

arid. High temperatures pose serious stresses for the human body, placing it in great danger of injury and even death. For humans, adjusting to varying climatic conditions includes both physiologic mechanisms as a by-product of evolution and the conscious development of cultural adaptations.

Prolonged or repeated exposure to stressful environmental conditions elicits significant physiologic changes, called **acclimatization**, the process of an individual organism adjusting to a change in its environment. Acclimatization occurs in a short time (days or weeks) and can be reversed once removed from the stress, unlike **adaptation**, which is permanent and passed on to the next generation. Some degree of heat *acclimatization* occurs either by heat exposure alone or by regular strenuous exercise, which raises core temperature and provokes heat-loss responses. Indeed, the first summer heat wave produces enough heat acclimatization that most people notice an improvement in their level of energy and general feeling of well-being after a few days. However, the acclimatization response is greater if heat exposure and exercise are combined, causing a greater rise of internal temperature and more profuse sweating. Evidence of acclimatization appears in the first few days of combined exercise and heat exposure, and most of the improvement in heat tolerance occurs within 10 days. The effect of heat acclimatization on performance can be dramatic, and acclimatized subjects can easily complete exercise in the heat that earlier was difficult or impossible.

Changes in basal metabolism, heart rate, cutaneous blood flow, and sweat rate occur with heat acclimatization.

Studies of seasonal variation in basal metabolism indicate an inverse relationship between BMR and ambient temperature. A linear correlation between basal metabolism and mean outdoor temperature within a range of 10°C to 25°C exists, so that a rise of 10°C in monthly mean ambient temperature is accompanied by a fall of 2.5 to 3.0 kcal/m^2/h in basal metabolism, a change linked to lower thyroxine levels. During exposure to hot environments, cardiovascular adaptations that reduce the heart rate required to sustain a given level of activity in the heat appear quickly and reach nearly their full development within 1 week. Changes in sweating develop more slowly. After acclimatization, sweating begins earlier and at a lower core temperature (i.e., the core temperature threshold for sweating is reduced). The sweat glands become more sensitive to cholinergic stimulation, and a given elevation in core temperature elicits a higher sweat rate. In addition, the glands become resistant to hidromeiosis and fatigue, and so higher sweat rates can be sustained. These changes reduce the levels of core and skin temperatures reached during a period of exercise in the heat, increase the sweat rate, and enable one to exercise longer. The threshold for cutaneous vasodilation is reduced along with the threshold for sweating, and so the heat transfer from the core to the skin is maintained. The lower heart rate, lower core temperature, and higher sweat rate are the three classic signs of heat acclimatization (Fig. 28.14).

Heat acclimatization modifies fluid and electrolyte balance.

During the first week, total body water and, especially, plasma volume increase. These changes likely contribute to the cardiovascular adaptations. Later, the fluid changes seem to diminish or disappear, although the cardiovascular adaptations persist. In an unacclimatized person, sweating occurs mostly on the chest and back, but during acclimatization, especially in humid heat, the fraction of sweat secreted on the limbs increases to make better use of the skin surface for evaporation. An unacclimatized person who is sweating profusely can lose large amounts of sodium. With acclimatization, the sweat glands become able to conserve sodium by secreting sweat with a sodium concentration as low as 5 mmol/L. This effect is mediated through aldosterone, which is secreted in response to sodium depletion and to exercise and heat exposure. The sweat glands respond to aldosterone more slowly than the kidneys, requiring several days. Unlike the kidneys, the sweat glands do not escape the influence of aldosterone when sodium balance has been restored but continue to conserve sodium for as long as acclimatization persists. The cell membranes are freely permeable to water

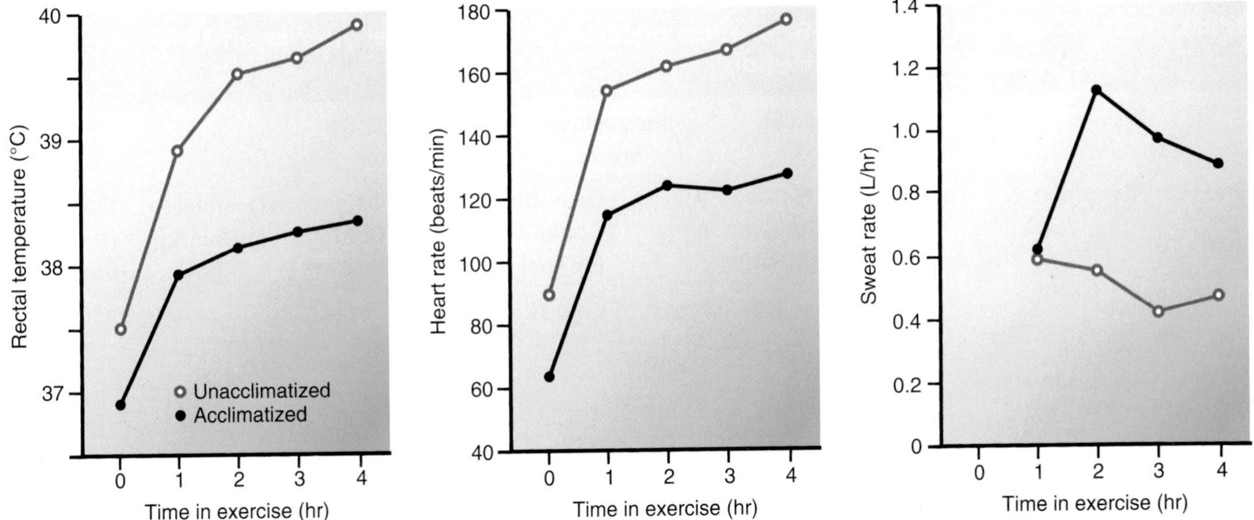

Figure 28.14 Heat acclimatization. These graphs show rectal temperatures, heart rates, and sweat rates during 4-hour exercise (bench stepping, 35 W mechanical power) in humid heat (33.9°C dry bulb, 89% relative humidity, 35 torr ambient vapor pressure) on the first and last days of a 2-week program of acclimatization to humid heat.

so that the movement of water across the cell membranes rapidly corrects any osmotic imbalance between the intracellular and extracellular compartments (see Chapter 2, "Plasma Membrane, Membrane Transport, and Resting Membrane Potential"). One important consequence of the salt-conserving response of the sweat glands is that the loss of a given volume of sweat causes a smaller decrease in the volume of the extracellular space than if the sodium concentration of the sweat is high (Table 28.3).

Heat acclimatization is transient, disappearing in a few weeks if not maintained by repeated heat exposure. The components of heat acclimatization are lost in the order in which they were acquired. The cardiovascular changes decay more quickly than the reduction in exercise core temperature and sweating changes.

RESPONSES TO COLD

When the body is exposed to cold temperatures, heat loss is minimized in several ways: (1) sweating stops; (2) minute muscles under the skin contract (**piloerection**), causing "gooseflesh," or "goose bumps"; (3) cutaneous arterioles constrict, thereby rerouting the blood away from the skin; (4) shivering, which increases heat production in skeletal muscles; and (5) fat is converted into energy by mitochondria. Brown fat is specialized for this purpose and is most abundant in newborns.

Arteriole vasoconstriction reroutes blood away from the skin to conserve heat.

Reducing **conductance** is the chief physiologic means of heat conservation during cold exposure. The constriction of cutaneous arterioles reduces skin blood flow and shell conductance. However, in extremely cold conditions, excessive vasoconstriction leads to numbness and pale skin. Frostbite only occurs when water within the cells begins to freeze; this destroys the cell, causing the characteristic damage.

Constriction of the superficial limb veins further improves heat conservation by diverting venous blood to the deep limb veins, which lie close to the major arteries of the limbs and do not constrict in the cold. (Many penetrating veins connect the superficial veins to the deep veins so that venous blood from anywhere in the limb potentially can return to the heart via either superficial or deep veins.) In the deep veins, cool venous blood returning to the core can take up heat from the warm blood in the adjacent deep limb arteries. Therefore, some of the heat contained in the arterial blood as it enters the limbs takes a "short circuit" back to the core. When the arterial blood reaches the skin, it is already cooler than the core, and so it loses less heat to the skin than it otherwise would. (When the superficial veins dilate in the heat, most venous blood returns via superficial veins so as to maximize core-to-skin heat flow.) The transfer of heat from arteries to veins by this short circuit is called **countercurrent heat exchange**. This mechanism can cool the blood in the radial artery of a cool but comfortable subject to as low as 30°C by the time it reaches the wrist.

As we saw earlier, the shell's insulating properties increase in the cold as its blood vessels constrict and its thickness increases. Furthermore, the shell includes a fair amount of skeletal muscle in the cold, and although muscle blood flow is believed not to be affected by thermoregulatory reflexes, direct cooling reduces it. In a cool subject, the resulting reduction in muscle blood flow adds to the shell's insulating properties. As the blood vessels in the shell constrict, blood is shifted to the central blood reservoir in the thorax. This shift produces many of the same effects as an increase in blood volume, including the so-called **cold diuresis** as the kidneys respond to the increased central blood volume.

TABLE 28.3 Effect of Sweat Secretion on Body Fluid Compartments and Plasma Sodium Concentration[a]

		Extracellular Space		Intracellular Space		Total Body Water			
Subject	Condition	Volume (L)	Osmotic Content (mOsm)	Volume (L)	Osmotic Content (mOsm)	Volume (L)	Osmotic Content (mOsm)	Osmolality (mOsm/kg)	Plasma [Na^+] (mmol/L)
A	Initial	15	4,350	25	7,250	40	11,600	290	140
	Loss of 5 L of sweat, 120 mOsm/L, 60 mmol Na^+/L	11.9	3,750	23.1	7,250	35	11,000	314	151
	Above condition accompanied by intake of 5 L water	13.6	3,750	26.4	7,250	40	11,000	275	132
B	Loss of 5 L of sweat, 20 mOsm/L, 10 mmol Na^+/L; accompanied by intake of 5 L water	12.9	4,250	22.1	7,250	35	11,500	329	159
	Above condition accompanied by intake of 5 L water	14.8	4,250	25.2	7,250	40	11,500	288	139

[a]Each subject has total body water of 40 L. The sweat of subject A has a relatively high [Na^+] of 60 mmol/L, whereas that of subject B has a relatively low [Na^+] of 10 mmol/L. Volumes of the extracellular and intracellular spaces are calculated assuming that water moves between the two spaces as needed to maintain osmotic balance.

Shivering is a heat-producing mechanism in response to hypothermia.

Shivering is a physiologic response to early hypothermia in mammals. When core body temperature drops, the shivering reflex, mediated through the hypothalamus, is triggered to maintain homeostasis. Muscle groups around vital organs begin to contract in an attempt to generate heat by expending energy. There are two types of shivering: low intensity and high intensity. During *low intensity*, shivering occurs constantly at low levels over long periods during cold conditions. During *high intensity*, shivering occurs violently over a short period. Both processes consume energy. Low intensity burns fat, whereas glucose is burned with high-intensity shivering.

Once skin blood flow is near minimal, metabolic heat production increases—almost entirely through shivering in human adults. Shivering may increase metabolism at rest more than fourfold—that is, to 350 to 400 W. Although it is often stated that shivering diminishes substantially after several hours and is impaired by exhaustive exercise, such effects are not well understood. In most laboratory mammals, chronic cold exposure also causes **nonshivering thermogenesis**, an increase in metabolic rate that is not a result of muscle activity. Nonshivering thermogenesis appears to be elicited through sympathetic stimulation and circulating catecholamines. It occurs in many tissues, especially in the liver and in **brown adipose tissue**, also called *brown fat*, specialized for nonshivering thermogenesis whose color is imparted by high concentrations of iron-containing respiratory enzymes. Brown adipose tissue is found in human infants, and nonshivering thermogenesis is important for their thermoregulation. The existence of brown adipose tissue and nonshivering thermogenesis in human adults is controversial, but recent evidence strongly suggests the presence of functioning brown adipose deposits in a substantial fraction of adult humans. These are located symmetrically in the supraclavicular and the neck regions with some additional paravertebral, mediastinal, para-aortic, and suprarenal (but no interscapular) localizations and respond with increased activity to sympathetic stimulation and exposure to cold.

Cold acclimatization is an important characteristic in thermal regulation and maintaining homeostasis.

The pattern of human cold acclimatization depends on the nature of the cold exposure. It is partly for this reason that the occurrence of cold acclimatization in humans was controversial for a long time. Our knowledge of human cold acclimatization comes from both laboratory studies and studies of populations whose occupation or way of life exposes them repeatedly to cold temperatures.

Metabolic changes in cold acclimatization

At one time, it was believed that humans must acclimatize to cold as laboratory mammals do—by increasing their metabolic rate. There are a few reports of increased BMR and, sometimes, thyroid activity in the winter. More often,

Clinical Focus / 28.1

Water and Salt Depletion as a Result of Sweating

Changes in fluid and electrolyte balance are probably the most frequent physiologic disturbances associated with sustained exercise and heat stress. Water loss via the sweat glands can exceed 1 L/h for many hours. Salt loss in the sweat is variable; however, because sweat is more dilute than plasma, sweating always results in an increase in the osmolality of the fluid remaining in the body and increased plasma [Na^+] and [Cl^-], as long as the lost water is not replaced. Because people who secrete large volumes of sweat usually replace at least some of their losses by drinking water or electrolyte solutions, the final effect on body fluids may vary. In Table 28.3, the second and third conditions (subject A) represent the effects on body fluids of sweat losses alone and combined with replacement by an equal volume of plain water, respectively, for someone producing sweat, with a [Na^+] and [Cl^-] in the upper part of the normal range. By contrast, the fourth and fifth conditions (subject B) represent the corresponding effects for a heat-acclimatized person secreting dilute sweat. Comparing the effects on these two people, we note: (1) The more dilute the sweat that is secreted, the greater the increase in osmolality and plasma [Na^+] if no fluid is replaced; (2) extracellular fluid volume, a major determinant of plasma volume (see Chapter 18, "Ventilation and the Mechanics of Breathing"), is greater in subject B (secreting dilute sweat) than in subject A (secreting saltier sweat), whether water is replaced; and (3) drinking plain water allowed subject B to maintain plasma sodium and extracellular fluid volume almost unchanged while secreting 5 L of sweat. In subject A, however, drinking the same amount of water reduced plasma [Na^+] by 8 mmol/L and failed to prevent a decrease of almost 10% in extracellular fluid volume. In 5 L of sweat, subject A lost 17.5 g of salt, somewhat more than the daily salt intake in a normal Western diet, and is becoming salt depleted.

Increased osmolality of the extracellular fluid and decreased plasma volume via a reduction in the activity of the cardiovascular stretch receptors stimulates thirst (see Chapter 18, "Ventilation and the Mechanics of Breathing"). When sweating is profuse, however, thirst usually does not elicit enough drinking to replace fluid as rapidly as it is lost so that people exercising in the heat tend to become progressively dehydrated—in some cases losing as much as 7% to 8% of body weight—and restore normal fluid balance only during long periods of rest or at meals. Depending on how much of his fluid losses he replaces, subject B may either be hypernatremic and dehydrated or be in essentially normal fluid and electrolyte balance. (If he drinks fluid well in excess of his losses, he may become overhydrated and hyponatremic, but this is an unlikely occurrence.) However, subject A, who is somewhat salt depleted, may be dehydrated and hypernatremic, normally hydrated but hyponatremic, or somewhat dehydrated with plasma [Na+] anywhere in between these two extremes. Once subject A replaces all the water lost as sweat, his extracellular fluid volume will be about 10% below its initial value. If he responds to the accompanying reduction in plasma volume by continuing to drink water, he will become even more hyponatremic than shown in Table 28.3.

The disturbances shown in Table 28.3, although physiologically significant and useful for illustration, are not likely to require clinical attention. Greater disturbances, with correspondingly more severe clinical effects, may occur. The consequences of the various possible disturbances of salt and water balance can be grouped as effects of decreased plasma volume secondary to decreased extracellular fluid volume, effects of hypernatremia, and effects of hyponatremia. The circulatory effects of decreased volume are nearly identical to the effects of peripheral pooling of blood (see Fig. 28.12), and the combined effects of peripheral pooling and decreased volume will be greater than the effects of either alone. These effects include impairment of cardiac filling and cardiac output and compensatory reflex reductions in renal, splanchnic, and skin blood flow. Impaired cardiac output leads to fatigue during exertion and decreased exercise tolerance. If skin blood flow is reduced, heat dissipation will be impaired. Exertional rhabdomyolysis, the injury of skeletal muscle fibers, is a frequent result of unaccustomed intense exercise. Myoglobin released from injured skeletal muscle cells appears in the plasma, rapidly enters the glomerular filtrate, and is excreted in the urine, producing myoglobinuria and staining the urine brown if enough myoglobin is present. This process may be harmless to the kidneys if urine flow is adequate. However, a reduction in renal blood flow reduces urine flow, increasing the likelihood that the myoglobin will cause renal tubular injury.

Hypernatremic dehydration is believed to predispose to heatstroke. Both hypernatremia and reduced plasma volume often accompany dehydration. Hypernatremia impairs the heat-loss responses (sweating and increased skin blood flow) independently of any accompanying reduction in plasma volume and elevates the thermoregulatory set point. Hypernatremic dehydration promotes the development of high core temperature in multiple ways through the combination of hypernatremia and reduced plasma volume.

Even in the absence of sodium loss, overdrinking that exceeds the kidneys' ability to compensate dilutes all the body's fluid compartments, producing dilutional hyponatremia, which is also called water intoxication if it causes symptoms. The development of water intoxication requires either massive overdrinking or a condition, such as the inappropriate secretion of arginine vasopressin, that impairs the excretion of free water by the kidneys. Overdrinking sufficient to cause hyponatremia may occur in patients with psychiatric disorders or disturbance of the thirst mechanism or may be done with a mistaken intention of preventing or treating dehydration. However, people who secrete copious amounts of sweat with a high sodium concentration, such as subject A or people with cystic fibrosis, may easily lose enough salt to become hyponatremic because of sodium loss. Some healthy young adults who come to medical attention for salt depletion after profuse sweating are found to have genetic variants of cystic fibrosis, which cause these people to have salty sweat without producing the characteristic digestive and pulmonary manifestations of cystic fibrosis.

(Continued)

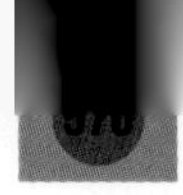

Clinical Focus / 28.1

Water and Salt Depletion as a Result of Sweating *(Continued)*

As sodium concentration and osmolality of the extracellular space decrease, water moves from the extracellular space into the cells to maintain osmotic balance across the cell membranes. The resulting swelling of the brain cells causes most of the manifestations of hyponatremia. Mild hyponatremia is characterized by nonspecific symptoms such as fatigue, confusion, nausea, and headache and may be mistaken for heat exhaustion. Severe hyponatremia can be a life-threatening medical emergency and may include seizures, coma, herniation of the brainstem (which occurs if the brain swells enough to exceed the capacity of the cranium), and death. In the setting of prolonged exertion in the heat, symptomatic hyponatremia is far less common than heat exhaustion but potentially far more dangerous. Therefore, it is important not to treat a presumed case of heat exhaustion with large amounts of low-sodium fluids without first ruling out hyponatremia.

however, increased metabolic rate has not been observed in studies of human cold acclimatization. In fact, several reports indicate the opposite response, consisting of a lower core temperature threshold for shivering, with a greater fall in core temperature and a smaller metabolic response during cold exposure. Such a response would spare metabolic energy and might be advantageous in an environment that is not so cold that a blunted metabolic response would allow core temperature to fall to dangerous levels.

Increased tissue insulation in cold acclimatization

A lower core-to-skin conductance (i.e., increased insulation by the shell) has often been reported in studies of cold acclimatization in which a reduction in the metabolic response to cold occurred. This increased insulation is not a result of subcutaneous fat (in fact, it has been observed in lean subjects) but apparently results from lower blood flow in the limbs or improved countercurrent heat exchange in the acclimatized subjects. In general, the cold stresses that elicit a lower shell conductance after acclimatization involve either cold water immersion or exposure to air that is chilly but not so cold as to risk freezing the vasoconstricted extremities.

Cold-induced vasodilation and the Lewis "hunting" response

As the skin is cooled below about 15°C, its blood flow begins to increase somewhat, a response called **cold-induced vasodilation (CIVD)**. This response is elicited most easily in comfortably warm subjects and in skin rich in arteriovenous anastomoses (in the hands and feet). The mechanism has not been established but may involve a direct inhibitory effect of cold on the contraction of vascular smooth muscle or on neuromuscular transmission. The CIVD response varies greatly among people and is usually rudimentary in hands and feet unaccustomed to cold exposure. After repeated cold exposure, CIVD begins earlier during cold exposure, produces higher levels of blood flow, and takes on a rhythmic pattern of alternating vasodilation and vasoconstriction. This is called the **Lewis hunting response** because the rhythmic pattern of blood flow suggests that it is "hunting" for its proper level. This response is often well developed in workers whose hands are exposed to cold such as fishermen and surfers exposed to cold water. The hunting response is observed more frequently in populations native to cold environments, such as Eskimos, in whom the response is far stronger. Because the Lewis hunting response may actually increase heat loss from the body, whether it is truly an example of acclimatization to cold is debatable. However, the response is advantageous because it keeps the extremities warmer, more comfortable, and functional and probably protects them from cold injury.

CLINICAL ASPECTS OF THERMOREGULATION

Temperature is important clinically because of the presence of fever in many diseases, the effects of many factors on tolerance to heat or cold stress, and the effects of heat or cold stress in causing or aggravating certain disorders.

Fever is one of the body's immunologic responses to a bacterial or viral infection.

Fever is due to an increase in the body temperature regulatory set point. This increase in set point increases muscle tone and shivering. Infection or noninfectious conditions (e.g., inflammatory processes such as collagen vascular diseases, trauma, neoplasms, acute hemolysis, and immunologically mediated disorders) may cause fever. **Pyrogens** are substances that cause fever and may be either exogenous or endogenous. *Exogenous pyrogens* are derived from outside the body; most are microbial products, microbial toxins, or whole microorganisms. The best studied of these is the lipopolysaccharide endotoxin of gram-negative bacteria. Exogenous pyrogens stimulate a variety of cells, especially monocytes and macrophages, to release *endogenous pyrogens*, polypeptides that cause the thermoreceptors in the hypothalamus (and perhaps elsewhere in the brain) to alter their firing rate and input to the central thermoregulatory controller, raising the thermoregulatory set point. The local synthesis and release of **prostaglandin E_2** mediate this effect of endogenous pyrogens. Aspirin and other drugs that inhibit the synthesis of prostaglandins also reduce fever.

Fever accompanies disease so frequently and is such a reliable indicator of the presence of disease that body temperature is probably the most commonly measured clinical index. A group of polypeptides called **cytokines** elicits many of the body's defenses against infection and cancer;

the endogenous pyrogen is usually a member of this group, **interleukin-1**. However, other cytokines, particularly **tumor necrosis factor**, **interleukin-6**, and the **interferons**, are also pyrogenic in certain circumstances. Elevated body temperature enhances the development of these defenses. How cytokines interact with the hypothalamus remains unclear. However, they are believed to either cross the blood–brain barrier by facilitated transport mechanisms, diffuse into areas of the brain where there is no blood–brain barrier, or interact with peripheral neural components of the immune system to signal the hypothalamus to increase the thermal set point. There is evidence to support each mechanism, and it seems likely that each contributes to some extent in various circumstances. If laboratory animals are prevented from developing a fever during experimentally induced infection, survival rates may be dramatically reduced. (Although, in this chapter, fever specifically means an elevation in core temperature resulting from pyrogens, some authors use the term more generally to mean any significant elevation of core temperature.)

Tolerance to heat and cold is dependent on factors affecting thermoregulatory responses.

Regular physical exercise and heat acclimatization increase heat tolerance and the sensitivity of the sweating response. Aging has the opposite effect; in healthy 65-year-old men, the sensitivity of the sweating response is half of that in 25-year-old men. Many drugs inhibit sweating, most obviously those used for their anticholinergic effects, such as atropine and scopolamine. In addition, some drugs used for other purposes, such as glutethimide (a sleep-inducing drug), tricyclic antidepressants, phenothiazines (tranquilizers and antipsychotic drugs), and antihistamines, have some anticholinergic action. All of these and several others have been associated with heatstroke. Congestive heart failure and certain skin diseases (e.g., ichthyosis and anhidrotic ectodermal dysplasia) impair sweating, and in patients with these diseases, heat exposure and especially exercise in the heat may raise body temperature to dangerous levels. Lesions that affect the thermoregulatory structures in the brainstem can also alter thermoregulation. Such lesions can produce **hypothermia** (abnormally low core temperature) if they impair heat-conserving responses. However, **hyperthermia** (abnormally high core temperature) is a more usual result of brainstem lesions and is typically characterized by a loss of both sweating and the circadian rhythm of core temperature.

Certain drugs, such as barbiturates, alcohol, and phenothiazines, and certain diseases, such as hypothyroidism, hypopituitarism, congestive heart failure, and septicemia, may impair the defense against cold. (Hypothermia, instead of the usual febrile response to infection, may accompany septicemia, especially in debilitated patients.) Furthermore, newborns and many healthy older adults are less able than older children and younger adults to maintain adequate body temperature in the cold. This failing appears to be a result of an impaired ability to conserve body heat by reducing heat loss and to increase metabolic heat production in the cold.

Heat stress can lead to pathophysiologic disorders.

The harmful effects of heat stress are exerted through cardiovascular strain, fluid and electrolyte loss, and, especially in heatstroke, tissue injury whose mechanism is uncertain. In a patient suspected of having hyperthermia secondary to heat stress, temperature should be measured in the rectum because hyperventilation may render oral temperature spuriously low.

Heat syncope

Heat syncope is circulatory failure resulting from a pooling of blood in the peripheral veins, with a consequent decrease in venous return and diastolic filling of the heart, resulting in decreased cardiac output and a fall of arterial pressure. Symptoms range from lightheadedness and giddiness to loss of consciousness. Thermoregulatory responses are intact, and so core temperature typically is not substantially elevated, and the skin is wet and cool. The large thermoregulatory increase in skin blood flow in the heat is probably the primary cause of the peripheral pooling. Heat syncope affects mostly those who are not acclimatized to heat presumably because the plasma–volume expansion that accompanies acclimatization compensates for the peripheral pooling of blood. Treatment consists in laying the patient down out of the heat to reduce the peripheral pooling of blood and improve the diastolic filling of the heart.

Heat exhaustion

Heat exhaustion, also called *heat collapse*, is probably the most common heat disorder and represents a failure of cardiovascular homeostasis in a hot environment. Collapse may occur either at rest or during exercise and may be preceded by weakness or faintness, confusion, anxiety, ataxia, vertigo, headache, and nausea or vomiting. The patient has dilated pupils and usually sweats profusely. As in heat syncope, reduced diastolic filling of the heart appears to have a primary role in the pathogenesis of heat exhaustion. Although blood pressure may be low during the acute phase of heat exhaustion, the baroreflex responses are usually sufficient to maintain consciousness and may be manifested in nausea, vomiting, pallor, cool or even clammy skin, and rapid pulse. Patients with heat exhaustion usually respond well to rest in a cool environment and oral fluid replacement. In more severe cases, however, intravenous replacement of fluid and salt may be required. Core temperature may be normal or only mildly elevated in heat exhaustion. However, heat exhaustion accompanied by hyperthermia and dehydration may lead to heatstroke. Therefore, patients should be actively cooled if rectal temperature is 40.6°C (105°F) or higher.

The reasons underlying the reduced diastolic filling in heat exhaustion are not fully understood. Hypovolemia contributes if the patient is dehydrated, but heat exhaustion often occurs without significant dehydration. In rats heated to the point of collapse, compensatory splanchnic vasoconstriction develops during the early part of heating but is reversed shortly before the maintenance of blood pressure fails. A similar process may occur in heat exhaustion.

From Bench to Bedside / 28.1

Therapeutic Hypothermia and Cardioprotection

Despite cardiopulmonary resuscitation (CPR), many cardiac arrest patients die within a few hours after return of spontaneous circulation (ROSC). This is due to the onset of tissue injury related to an increase in tissue oxidative stress caused by ischemia/reperfusion (I/R) and is observed within minutes of CPR and ROSC. Recently, studies have shown that significant levels of oxidative stress in the heart following cardiac arrest stem from the mitochondria within minutes of reperfusion. In isolated cardiomyocytes, simulated I/R induces mitochondrial reactive oxygen species (ROS) generation; contractile dysfunction; mitochondrial release of cytochrome c; caspase activation; opening of the mitochondrial permeability transition pore; and, ultimately, cardiomyocyte death. **Therapeutic hypothermia (TH)** (also called *controlled hypothermia*) is the only treatment currently known to improve survival in clinical post–cardiac arrest patients. This type of cooling is now being induced by emergency personnel during CPR, prior to ROSC, in an attempt to improve cardiovascular function and outcome. In some cardiac arrest survivors, cooling is induced after ROSC, while they are transported to a cardiac catheterization laboratory for possible percutaneous coronary intervention of acute myocardial infarction (AMI). Although intra-ischemic cooling prior to reperfusion may be highly cardioprotective, the general use of cooling for AMI (noncardiac arrest) patients remains unproven. Recent studies in mouse cardiomyocytes suggest that cardioprotective hypothermia (5°C drop in temperature) does more than simply slow reperfusion ROS generation. Cardiomyocyte cooling increases Akt phosphorylation resulting from relative inhibition of regulatory phosphatases. Akt is a survival kinase that mediates mitochondrial protection by modulating ROS and nitric oxide (NO) generation. In addition, mild cooling can alter the strength of weak bonds that determine the conformation and association of individual proteins/enzymes that increase protein function—temperature-mediated alterations in mRNA translational efficiency have also been observed. Mildly decreased temperature, such as that used in TH, seems to result in significant alterations in the cellular activity of proteins such as Akt and HSP27, a regulator of Akt. Cardioprotection is also associated with several kinase-signaling pathways including the G-coupled receptors, the reperfusion injury salvage kinase (RISK), and the JAK STAT pathway. The enhanced Akt-NO signaling with reduction of ROS independent of cGMP suggests that TH is associated with RISK pathway signaling. This is of particular clinical relevance because it suggests that the beneficial effects of TH soon may be replaced or enhanced by suitable pharmacologic agents that target these protein pathways.

Heatstroke

High core temperature and the development of serious neurologic disturbances with a loss of consciousness, and frequently, convulsions characterize the most severe and dangerous heat disorder. **Heatstroke** occurs in two forms, classical and exertional. In the classical form, the primary factor is environmental heat stress that overwhelms an impaired thermoregulatory system, and most patients have preexisting chronic disease. In exertional heatstroke, the primary factor is high metabolic heat production. Patients with exertional heatstroke tend to be younger and more physically fit (typically, soldiers and athletes) than patients with the classical form. Rhabdomyolysis, hepatic and renal injury, and disturbances of blood clotting are frequent accompaniments of exertional heatstroke. The traditional diagnostic criteria of heatstroke—coma, hot dry skin, and rectal temperature above 41.3°C (106°F)—are characteristic of the classical form. However, patients with exertional heatstroke may have somewhat lower rectal temperatures and often sweat profusely. Heatstroke is a medical emergency, and prompt appropriate treatment is critically important to reducing morbidity and mortality. The rapid lowering of core temperature is the cornerstone of treatment, and immersion in cold water most effectively accomplishes it. With prompt cooling, vigorous hydration, maintenance of a proper airway, avoidance of aspiration, and appropriate treatment of complications, most patients will survive, especially if they were previously healthy.

The pathogenesis of heatstroke is not well understood, but it seems clear that factors other than hyperthermia are involved, even if the action of these other factors partly depends on the hyperthermia. Exercise may contribute more to the pathogenesis than simply metabolic heat production. Elevated plasma levels of several inflammatory cytokines have been reported in patients presenting with heatstroke, suggesting a systemic inflammatory component. No trigger for such an inflammatory process has been established, although several possible candidates exist. One possible trigger is some product(s) of the bacterial flora in the gut, perhaps including lipopolysaccharide endotoxins. Several lines of evidence suggest that sustained splanchnic vasoconstriction may produce a degree of intestinal ischemia sufficient to allow these products to "leak" into the circulation and activate inflammatory responses. Other causes relate to reticuloendothelial system deactivation and heat-induced changes in proinflammatory and anti-inflammatory cytokines.

The preceding diagnostic categories are traditional. However, they are not entirely satisfactory for heat illness associated with exercise because many patients have laboratory evidence of tissue and cellular injury, but they are classified as having heat exhaustion because they do not have the serious neurologic disturbances that characterize heatstroke. Some more recent literature uses the term **exertional heat injury** for such cases. The boundaries of exertional heat injury, with heat exhaustion on one hand and heatstroke on the other, are not clearly and consistently defined, and these categories probably represent parts of a continuum.

Malignant hyperthermia, a rare process triggered by depolarizing neuromuscular blocking agents or certain inhalational anesthetics, was once thought to be a form of heatstroke but is now known to be a distinct disorder that occurs in people with a mutation of the ryanodine receptor (type 1) on the sarcoplasmic reticulum. In 90% of susceptible people, biopsied skeletal muscle tissue contracts on exposure to caffeine or halothane in concentrations having little effect on normal muscle. Susceptibility may be associated with any of several myopathies, but most susceptible people have no other clinical manifestations. The control of free (unbound) calcium ion concentration in skeletal muscle cytoplasm is severely impaired in susceptible people, and when an attack is triggered, calcium concentration rises abnormally, activating myosin ATPase and leading to an uncontrolled hypermetabolic process that rapidly increases core temperature. Treatment with dantrolene sodium, which appears to act by reducing the release of calcium ions from the sarcoplasmic reticulum, has dramatically reduced the mortality rate of this disorder.

Aggravation of disease states by heat exposure

Other than producing specific disorders, heat exposure aggravates several other diseases. Epidemiologic studies show that during unusually hot weather, mortality may be two to three times that normally expected for the months in which heat waves occur. Deaths ascribed to specific heat disorders account for only a small fraction of the excess mortality (i.e., the increase above the expected mortality). Deaths from diabetes, various diseases of the cardiovascular system, and diseases of the blood-forming organs account for most of the excess mortality.

Drop in core temperature results in hypothermia.

Core temperature drops when the body is exposed to cold and the internal thermoregulatory mechanisms are unable to replenish the heat that is being lost. Hypothermia exhibits characteristic symptoms as body temperature decreases, such as shivering and mental confusion.

Much of our knowledge about the physiologic effects of hypothermia comes from observations of surgical patients.

During the initial phases of cooling, stimulation of shivering through thermoregulatory reflexes overwhelms the Q_{10} effect. Metabolic rate, therefore, increases, reaching a peak at a core temperature of 30°C to 33°C. At lower core temperatures, however, the Q_{10} effect dominates metabolic rate, and thermoregulation is lost. A vicious circle develops, wherein a fall in core temperature depresses metabolism and allows core temperature to fall further, so that at 17°C, the O_2 consumption is about 15% and cardiac output 10% of precooling values.

Hypothermia that is not induced for therapeutic purposes is called **accidental hypothermia**. It occurs in people whose defenses are impaired by drugs (especially ethanol, in the United States), disease, or other physical conditions and in healthy people who are immersed in cold water or become exhausted working or playing in the cold.

Hypothermia is also often induced during surgical procedures, based on the idea that hypothermia reduces metabolic rate via the Q_{10} effect and prolongs the time that tissues can safely tolerate a loss of blood flow. Because the brain is damaged by ischemia soon after circulatory arrest, *controlled hypothermia* is often used to protect the brain during surgical procedures in which its circulation is occluded or the heart is stopped.

Clinical Focus / 28.2

Hypothermia

Hypothermia is classified according to the patient's core temperature as mild (32°C–35°C), moderate (28°C–32°C), or severe (below 28°C). Shivering is usually prominent in mild hypothermia but diminishes in moderate hypothermia and is absent in severe hypothermia. The pathophysiology is characterized chiefly by the depressant effect of cold (via the Q_{10} effect) on multiple physiologic processes and differences in the degree of depression of each process. Other than shivering, the most prominent features of mild and moderate hypothermia are a result of depression of the central nervous system. Beginning with mood changes (commonly, apathy, withdrawal, and irritability), they progress to confusion and lethargy, followed by ataxia and speech and gait disturbances, which may mimic a cerebrovascular accident (stroke). In severe hypothermia, voluntary movement, reflexes, and consciousness are lost and muscular rigidity appears. Cardiac output and respiration decrease as core temperature falls. Myocardial irritability increases in severe hypothermia, causing a substantial danger of ventricular fibrillation, with the risk increasing as cardiac temperature falls. The primary mechanism presumably is that cold depresses conduction velocity in Purkinje fibers more than in ventricular muscle, favoring the development of circus-movement propagation of action potentials. Myocardial hypoxia also contributes. In more profound hypothermia, cardiac sounds become inaudible and pulse and blood pressure are unobtainable because of circulatory depression; the electrical activity of the heart and brain becomes unmeasurable; and extensive muscular rigidity may mimic rigor mortis. The patient may appear clinically dead, but patients have been revived from core temperatures as low as 17°C, and so "no one is dead until warm and dead." The usual causes of death during hypothermia are respiratory cessation and the failure of cardiac pumping because of either ventricular fibrillation or direct depression of cardiac contraction.

Depression of renal tubular metabolism by cold impairs the reabsorption of sodium, causing a diuresis and leading to dehydration and hypovolemia. Acid–base disturbances in hypothermia are complex. Respiration and cardiac output typically are depressed more than metabolic rate, and a mixed respiratory and metabolic acidosis results because of CO_2 retention and lactic acid accumulation and the cold-induced shift of the hemoglobin-O_2 dissociation curve to the left. Acidosis aggravates the susceptibility to ventricular fibrillation.

(Continued)

Clinical Focus / 28.2 

Hypothermia *(Continued)*

Treatment consists of preventing further cooling and restoring fluid, acid–base, and electrolyte balance. Patients in mild-to-moderate hypothermia may be warmed solely by providing abundant insulation to promote the retention of metabolically produced heat; those who are more severely affected require active rewarming. The most serious complication associated with treating hypothermia is the development of ventricular fibrillation. Vigorous handling of the patient may trigger this process, but an increase in the patient's circulation (e.g., associated with warming or skeletal muscle activity) may itself increase the susceptibility to such an occurrence, as follows. Peripheral tissues of a hypothermic patient are, in general, even cooler than the core, including the heart, and acid products of anaerobic metabolism will have accumulated in underperfused tissues while the circulation was most depressed. As the circulation increases, a large increase in blood flow through cold, acidotic peripheral tissue may return enough cold, acidic blood to the heart to cause a transient drop in the temperature and pH of the heart, increasing its susceptibility to ventricular fibrillation.

The diagnosis of hypothermia is usually straightforward in a patient rescued from the cold but may be far less clear in a patient in whom hypothermia is the result of a serious impairment of physiologic and behavioral defenses against cold. A typical example is the older person, living alone, who is discovered at home, cool and obtunded or unconscious. The setting may not particularly suggest hypothermia, and when the patient comes to medical attention, the diagnosis may easily be missed because standard clinical thermometers are not graduated low enough (usually only to 34.4°C) to detect hypothermia and, in any case, do not register temperatures below the level to which the mercury has been shaken. Because of the depressant effect of hypothermia on the brain, the patient's condition may be misdiagnosed as cerebrovascular accident or other primary neurologic disease. Recognition of this condition depends on the physician's considering it when examining a cool patient whose mental status is impaired and obtaining a true core temperature with a low-reading glass thermometer or other device.

Chapter Summary

- Humans are homeotherms and regulate temperature within a narrow range.
- Physiologic and physical processes operate continuously to balance heat production and heat loss and to maintain constant body temperature.
- The body can exchange heat energy with the environment via conduction, convection, radiation, and evaporation.
- Internal conduction and circulatory convection transfer heat from the body's core to the shell.
- Most of the metabolic energy used in cell processes is converted into heat within the body, and metabolic rate is the quantity of food energy converted to heat per unit of time.
- Basal metabolic rate is a measure of the metabolic cost of living measured under standard conditions.
- Metabolic rate is estimated by indirect calorimetry, based on measuring a person's rate of oxygen consumption.
- Heat exchange is proportional to surface area and obeys biophysical principles.
- Sweating is essential for heat loss and involves the glandular secretion of a dilute electrolyte solution onto the skin surface that rapidly evaporates in hot environments, cooling the body.
- Raising skin blood flow brings skin temperature nearer to blood temperature, and lowering skin blood flow brings skin temperature nearer to ambient temperature, enabling convective and radiative heat loss.
- Humans have two distinct subsystems for regulating body temperature: behavioral and physiologic thermoregulation; the latter operates through graded control of heat-production and heat-loss responses.
- Temperature receptors in the body core and skin transmit information about their temperatures through afferent nerves to the brainstem and, especially, the hypothalamus, where much of the integration of temperature information takes place.
- Signals for both heat-dissipating and heat-generating mechanism recruitment are based on the information about core and skin temperatures as well as nonthermal information received by the central nervous system and on the thermoregulatory set point.
- Fever elevates core temperature at rest, heat acclimatization decreases it, and time of day changes it in a cyclic fashion by changing the thermoregulatory set point.
- Intense exercise may increase heat production within the body 10-fold or more, requiring large increases in skin blood flow and sweating to reestablish the body's heat balance.
- Prolonged or repeated exposure to stressful environmental conditions elicits significant physiologic changes that reduce the resulting strain.
- The body maintains core temperature in the cold by minimizing heat loss and, when this response is insufficient, by increasing heat production.
- Fever enhances defense mechanisms, whereas heat and cold stress can cause significant harm.

29 Exercise Physiology

ACTIVE LEARNING OBJECTIVES

Upon mastering the material in this chapter you should be able to:

- Explain the concept of exercise to predict acute or chronic physiologic responses.
- Explain how maximal oxygen uptake predicts work performance.
- Explain the mechanism whereby substantial regional blood flow shifts occur during dynamic and isometric exercise.
- Explain training effects on both myocardial muscle and the coronary circulation.
- Explain how the respiratory system responds predictably to increased O_2 consumption and CO_2 production with exercise.
- Explain how, in healthy people, muscle fatigue during exercise is mediated.
- Explain how chronic physical activity affects insulin sensitivity and glucose entry into cells.
- Explain how exercise affects age-related functional changes.

Exercise is bodily activity that enhances or maintains physical fitness and overall health or wellness. It is performed for various reasons, including strengthening muscles and the cardiovascular system, honing athletic skills, weight loss and maintenance, and for enjoyment. Frequent and regular physical exercise boosts the immune system and helps prevent the "diseases of affluence," such as heart disease, cardiovascular disease, type 2 diabetes, and obesity. It also improves mental health, helps prevent depression, helps to promote or maintain positive self-esteem, and can even augment an individual's sex appeal or body image. Exercise, in its many forms, is so common that true physiologic "rest" is rarely observed. During exercise, defined ultimately in terms of skeletal muscle contraction, the body must initiate rapidly integrated adjustments in every organ system in a coordinated response to increased muscular energy demands. So too must the body adjust integratively by adaptive responses in these same systems to days, weeks, and months of repeated bouts of activity. Exercise physiology is thus the branch of physiology that studies the physical and biochemical events that enable physical movement and underlie cellular and systemic adaptations to repeated or chronic activity.

OXYGEN UPTAKE AND EXERCISE

Exercise physiology is the study of acute and chronic changes in response to a wide range of exercise conditions. In addition, many experiments are carried out to study the effect of exercise on pathophysiology and the mechanisms by which exercise can reduce or reverse disease progression. Oxygen uptake is used to quantitate energy expenditure during these experiments.

Exercise is as varied as it is ubiquitous. A single episode of exercise, or "acute" exercise, may provoke responses different from the adaptations seen when activity is chronic—that is, during *training*. The forms of exercise vary as well. The amount of muscle mass at work (one finger? one arm? both legs?), the intensity of the effort, its duration, and the type of muscle contraction (isometric, rhythmic) all influence the body's responses and adaptations.

These many aspects of exercise imply that its interaction with disease is multifaceted. There is no simple answer as to whether or how exercise promotes health. In fact, physical activity can be healthful, harmful, or irrelevant, depending on the patient, the disease, and the specific exercise in question.

Maximal oxygen uptake is used to quantitate energy expenditure during dynamic exercise.

Dynamic exercise is defined as skeletal muscle contractions at changing lengths and with rhythmic episodes of relaxation. Fundamental to any discussion of dynamic exercise is a description of its intensity. Because dynamically exercising muscle primarily generates energy from oxidative metabolism, a traditional standard is to measure, by mouth, the oxygen uptake ($\dot{V}o_2$) of an exercising subject. This measurement is limited to dynamic exercise and usually to the steady state, when exercise intensity and oxygen consumption are stable and no net energy is provided from nonoxidative sources. Three implications of the original oxygen consumption measurements deserve mention. First, the centrality of oxygen usage to work output gave rise to the term "aerobic" exercise. Second, the apparent excess in oxygen consumption during the first minutes of recovery has been termed the **oxygen debt** (Fig. 29.1). The "excess"

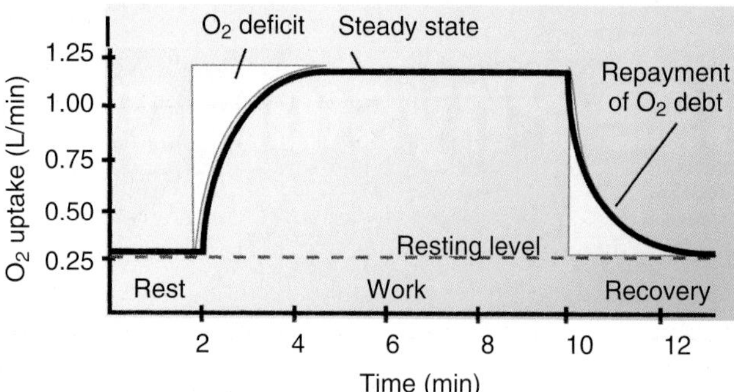

Figure 29.1 Oxygen uptake during rest, exercise, and recovery.

oxygen consumption of recovery results from a multitude of physiologic processes, and little usable information is obtained from its measurement. Third, and more useful, during dynamic exercise that uses a large muscle mass, each person has a **maximal oxygen uptakeV** ($\dot{V}O_{2max}$), a ceiling up to 20 times basal consumption that cannot be exceeded, although it can be increased by appropriate training. ($\dot{V}O_{2max}$) is defined as the highest rate at which oxygen can be taken up and used by the body during severe exercise. It is one of the main variables in the field of exercise physiology and is commonly used to indicate cardiorespiratory fitness. Maximal oxygen uptake is decreased, all else being equal, by age, bed rest, or increased body fat.

Maximal oxygen uptake is also used to express relative work capacity. Obviously, a world-class marathon runner has a greater capacity to consume oxygen than a novice. However, when both are exercising at intensities requiring two third of their respective maximal oxygen uptakes (the world-class athlete is moving much faster in doing this, as a result of higher capacity), both become exhausted at roughly the same time and for the same physiologic reasons

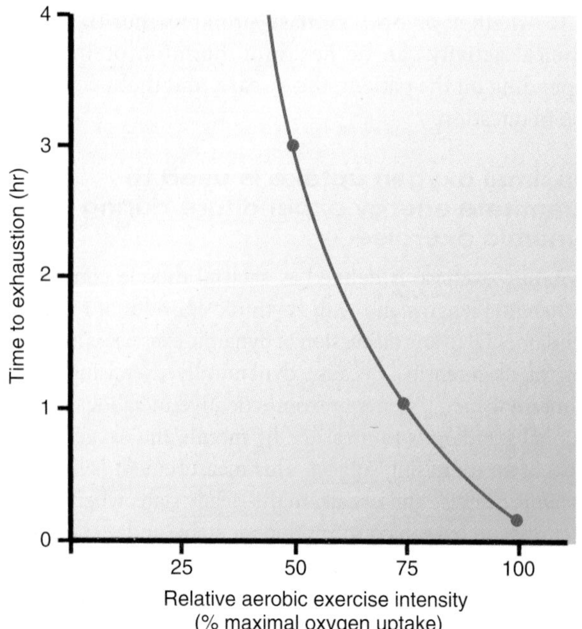

Figure 29.2 Time to exhaustion during dynamic exercise. Exhaustion is predictable on the basis of relative demand on the maximal oxygen uptake.

TABLE 29.1 Absolute and Relative Costs of Daily Activities

		% Maximal Oxygen Uptake	
Activity	**Energy Cost (kcal/min)**	**Sedentary 22-Year-Old**	**Sedentary 70-Year-Old**
Sleeping	1	6	8
Sitting	2	12	17
Standing	3	19	25
Dressing, undressing	3	19	25
Walking (3 mi./h)	4	25	33
Making a bed	5	31	42
Dancing	7	44	58
Gardening/ shoveling	8	50	67
Climbing stairs	11	69	92
Crawl swimming (50 m/min)	16	100	
Running (8 mi./h)	16	100	

(Fig. 29.2). In the discussion that follows, relative as well as absolute (expressed as L/min of oxygen uptake) work levels are used to explain physiologic responses. The energy costs and relative demands of some familiar activities are listed in Table 29.1.

What is it that causes oxygen uptake to reach an upper limit? Historically, many arguments claim primacy for either cardiac output (oxygen delivery) or muscle metabolic capacity (oxygen use) limitations. However, it may be that every link in the chain taking oxygen from the atmosphere to the mitochondrion reaches its capacity at about the same time. In practical terms, this means that any lung, heart, vascular, or musculoskeletal illness that reduces oxygen flow capacity will diminish a patient's functional capacity. It has been shown that in untrained subjects, the oxygen consumption dominates ($\dot{V}O_{2max}$), whereas in endurance-trained athletes, the oxygen supply is the main limiting factor.

In **isometric exercise**, force is generated at constant muscle length and without rhythmic episodes of relaxation. Isometric work intensity is usually described as a percentage of the **maximal voluntary contraction (MVC)**, the peak isometric force that can be briefly generated for that specific exercise. Analogous to work levels relative to maximal oxygen uptake, the ability to endure isometric effort and many physiologic responses to that effort are predictable when the percentage of MVC among people is held constant.

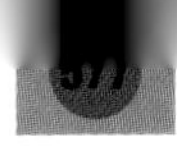

CARDIOVASCULAR RESPONSES TO EXERCISE

Increased energy expenditure with exercise demands more energy production. For prolonged work, the oxidation of foodstuff supplies this energy, with the cardiovascular system carrying the oxygen to working muscles.

Blood flow is preferentially directed to working skeletal muscle during exercise.

Local control of blood flow ensures that only working muscles with increased metabolic demands receive increased blood and oxygen delivery. If the legs alone are active, leg muscle blood flow should increase, whereas arm muscle blood flow remains unchanged or is reduced. At rest, skeletal muscle receives only a small fraction of the cardiac output. In dynamic exercise, both total cardiac output and relative and absolute output directed to working skeletal muscle increase dramatically (Table 29.2).

Cardiovascular control during exercise involves systemic regulation (cardiovascular centers in the brain, with their autonomic nervous output to the heart and systemic resistance vessels) in tandem with local control. For millennia, our ancestors successfully used exercise both to escape being eaten and to catch food; therefore, it is no surprise that cardiovascular control in exercise is complex and unique. It is as if a brain software program titled "Exercise" were inserted into the brain as work begins. Initially, the motor cortex is activated; the total neural activity is roughly proportional to the muscle mass and its work intensity. This neural activity communicates with the cardiovascular control centers, reducing vagal tone on the heart (which raises heart rate) and resetting the arterial baroreceptors to a higher level. As work rate is increased further, lactic acid is formed in actively contracting muscles, which stimulates muscle afferent nerves to send information to the cardiovascular center that increases sympathetic outflow to the heart and systemic resistance vessels. However, despite this **muscle chemoreflex** activity, within these same working muscles, low Po_2, increased nitric oxide, vasodilator prostanoids, and associated local vasoactive factors dilate arterioles despite rising sympathetic vasoconstrictor tone. Increased sympathetic drive does elevate heart rate and cardiac contractility, resulting in increased cardiac output; local factors in the coronary vessels mediate coronary vasodilation. Increased sympathetic vasoconstrictor tone in the renal and splanchnic vascular beds and in inactive muscle reduces blood flow to these tissues. Blood flow to these inactive regions can fall 75% if exercise is strenuous. Increased vascular resistance and decreased blood volume in these tissues help maintain blood pressure during dynamic exercise. In contrast to blood flow reductions in the viscera and in inactive muscle, the brain autoregulates blood flow at constant levels independent of exercise. The skin remains vasoconstricted only if thermoregulatory demands are absent. Table 29.3 shows how a profound fall in systemic vascular resistance matches the enormous rise in cardiac output during dynamic exercise.

Dynamic exercise, at its most intense level, forces the body to choose between maximum muscle vascular dilation and defense of blood pressure. Blood pressure is, in fact, maintained. During strenuous exercise, sympathetic drive can begin to limit vasodilation in active muscle. When exercise is prolonged in the heat, increased skin blood flow and sweating-induced reduction in plasma volume both contribute to the risk of hyperthermia and hypotension (heat exhaustion). Although chronic exercise provides some heat acclimatization, even highly trained athletes are at risk of hyperthermia and hypotension if work is prolonged and water is withheld in demanding environmental conditions.

Isometric exercise causes a somewhat different cardiovascular response. Muscle blood flow increases relative to the resting condition, as does cardiac output, but the higher mean intramuscular pressure limits these flow increases much more than when exercise is rhythmic. Because the blood flow increase is blunted inside a statically contracting muscle, the consequences of hard work with too little oxygen appear quickly: a shift to anaerobic metabolism, the production of lactic acid, a rise in the **adenosine diphosphate (ADP)/adenosine triphosphate (ATP)** ratio, and fatigue. Maintaining just 50% of the MVC is agonizing after about 1 minute and usually cannot be

TABLE 29.2 Blood Flow Distribution During Rest and Heavy Exercise in an Athlete

	Rest		Heavy Exercise	
Area	**mL/min**	**%**	**mL/min**	**%**
Splanchnic	1,400	24	300	1
Renal	1,100	19	900	4
Brain	750	13	750	3
Coronary	250	4	1,000	4
Skeletal muscle	1,200	21	22,000	86
Skin	500	9	600	2
Other	600	10	100	0.5
Total cardiac output	5,800	100	25,650	100

TABLE 29.3 Cardiac Output, Mean Arterial Pressure, and Systemic Vascular Resistance Changes with Exercise

Measure	Rest	Strenuous Dynamic Exercise
Cardiac output (L/min)	6	21
Mean arterial pressure (mm Hg)	90	105
Systemic vascular resistance [(mm Hg • min)/L]	15	5

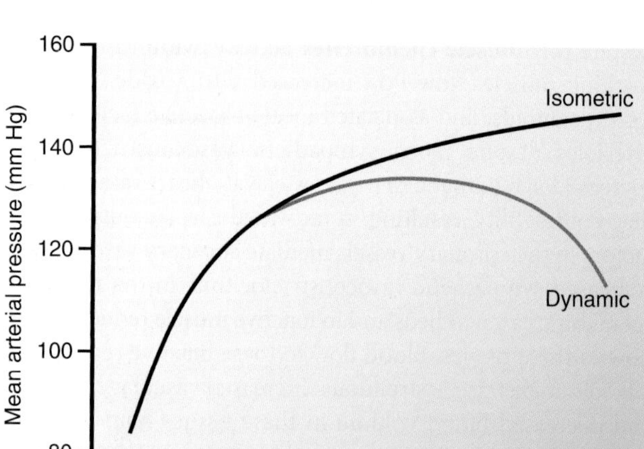

Figure 29.3 Effect of active muscle mass on mean arterial pressure during exercise. The highest pressures during dynamic exercise occur when an intermediate muscle mass is involved; pressure continues to rise in isometric exercise as more muscle is added.

TABLE 29.4 Acute Cardiac Response to Graded Exercise in a 30-Year-Old Untrained Woman

Exercise Intensity	Oxygen Uptake (L/min)	Heart Rate (beats/min)	Stroke Volume (mL/beat)	Cardiac Output (L/min)
Rest	0.25	72	70	5
Walking	1.0	110	90	10
Jogging	1.8	150	100	15
Running fast	2.5	190	100	19

continued after 2 minutes. A long-term sustainable level is only about 20% of maximum. These percentages are much less than the equivalent for dynamic work, as defined in terms of maximal oxygen uptake. Rhythmic exercise requiring 70% of the maximal oxygen uptake can be maintained in healthy people for about an hour, whereas work at 50% of the maximal oxygen uptake may be prolonged for several hours (see Fig. 29.2).

The reliance on anaerobic metabolism in isometric exercise triggers muscle ischemic chemoreflex responses that raise blood pressure more and cardiac output and heart rate less than that in dynamic work (Fig. 29.3). Oddly, for dynamic exercise, the elevation of blood pressure is most pronounced when a medium muscle mass is working. This response results from the combination of a small, dilated active muscle mass with powerful central sympathetic vasoconstrictor drive. Typically, the arms exemplify a medium muscle mass; shoveling snow is a good example of primarily arm and heavily isometric exercise. Shoveling snow can be risky for people in danger of stroke or heart attack because it substantially raises systemic arterial pressure. The elevated pressure places already-compromised cerebral arteries at risk and presents an ischemic or failing heart with a greatly increased afterload.

Cardiovascular responses differ with acute and chronic exercise.

In acute dynamic exercise, vagal withdrawal and increases in sympathetic outflow elevate heart rate and contractility in proportion to exercise intensity (Table 29.4). Factors enhancing venous return also aid cardiac output in dynamic exercise. These include the "muscle pump," which compresses veins as muscles rhythmically contract, and the "respiratory pump," which increases breath-by-breath oscillations in intrathoracic pressure (see Chapter 18, "Ventilation and the Mechanics of Breathing"). The importance of these factors is clear in patients with heart transplants who lack extrinsic cardiac innervation. Stroke volume rises in cardiac transplant patients with increasing exercise intensity as a result of increased venous return that enhances cardiac preload. In addition, circulating epinephrine and norepinephrine from the adrenal medulla and norepinephrine from sympathetic nerve spillover augment heart rate and contractility.

Maximal dynamic exercise yields a maximal heart rate: Further vagal blockade (e.g., via pharmacologic means) cannot elevate heart rate further. Stroke volume, in contrast, reaches a plateau in moderate work and is unchanged as exercise reaches its maximum intensity (see Table 29.4). This plateau occurs in the face of ever-shortening ventricular filling time, testimony to the increasing effectiveness of the mechanisms that enhance venous return and those that promote cardiac contractility. Sympathetic stimulation decreases left ventricular volume and pressure at the onset of cardiac relaxation (as a result of increased ejection fraction), leading to more rapid ventricular filling early in diastole. This helps maintain stroke volume as diastole shortens. Even in untrained people, the ejection fraction (stroke volume as a percentage of end-diastolic volume) reaches 80% in strenuous exercise.

The increased blood pressure, heart rate, stroke volume, and cardiac contractility seen in exercise all increase myocardial oxygen demands. A linear increase in coronary blood flow during exercise that can reach a value five times the basal level meets these demands. Local, metabolically linked factors (nitric oxide, adenosine, and the activation of ATP-sensitive K^+ channels) acting on coronary resistance vessels in defiance of sympathetic vasoconstrictor tone drive this increase in flow. Coronary oxygen extraction, high at rest, increases further with exercise (up to 80% of delivered oxygen). In healthy people, there is no evidence of myocardial ischemia under any exercise condition and there may be a coronary vasodilator reserve in even the most intense exercise.

The heart adapts to chronic exercise overload much as it does to high-demand pathologic states: by increasing left ventricular volume when exercise requires high blood flow and by left ventricular hypertrophy when exercise creates high systemic arterial pressure (high afterload). Consequently, the hearts of people adapted to prolonged, rhythmic exercise that involves relatively low arterial pressure exhibit large left ventricular volumes with normal wall thickness, whereas wall thickness is increased at normal volume in those adapted to activities involving isometric contraction and greatly elevated arterial pressure such as lifting weights.

The larger left ventricular volume in people chronically active in dynamic exercise leads directly to larger resting and exercise stroke volume. A simultaneous increase in vagal tone and decrease in β-adrenergic sensitivity enhance the resting and exercise bradycardia seen after training, so that in effect the trained heart operates further up the ascending limb of its length–tension relationship. Nevertheless, resting bradycardia is a poor indicator of endurance fitness because genetic factors explain a much larger proportion of the individual variation in resting heart rate than does training.

The effects of endurance training on coronary blood flow are partly mediated through changes in myocardial oxygen uptake. Because myocardial oxygen consumption is roughly proportional to the rate–pressure product (heart rate × mean arterial pressure) and because heart rate falls after training at any absolute exercise intensity, coronary flow at a fixed submaximal workload is reduced in parallel. Training, however, increases the peak coronary blood flow, just as it increases cardiac muscle capillary density and peak capillary exchange capacity. Training also improves endothelium-mediated regulation, responsiveness to adenosine, and control of intracellular free calcium ions within coronary vessels. Preserving endothelial vasodilator function may be the primary benefit of chronic physical activity on the coronary circulation.

Exercise increases levels of "good" cholesterol.

Chronic, dynamic exercise is associated with increased circulating levels of **high-density lipoproteins (HDLs)** and reduced **low-density lipoproteins (LDLs)**, such that the ratio of HDL to total cholesterol is increased. These changes in cholesterol fractions occur at any age if exercise is regular. Weight loss and increased insulin sensitivity, which typically accompany increased chronic physical activity in sedentary people, undoubtedly contribute to these changes in plasma lipoproteins. Nonetheless, in people with lipoprotein levels that place them at high risk for coronary heart disease, exercise appears to be an essential adjunct to dietary restriction and weight loss for lowering LDL cholesterol levels. Because exercise acutely and chronically enhances fat metabolism and cellular metabolic capacities for β-oxidation of free fatty acids, it is not surprising that regular activity increases both muscle and adipose tissue lipoprotein lipase activity. Changes in lipoprotein lipase activity, in concert with increased lecithin–cholesterol acyltransferase activity and apo A-I synthesis, enhance the levels of circulating HDLs. Evidence from interventional studies repeatedly demonstrates the effectiveness of proper diet and exercise in the prevention of metabolic syndrome and type 2 diabetes in a broad range of subjects.

Clinical Focus / 29.1

Stress Testing

To detect coronary artery disease, physicians often record an electrocardiogram (ECG), but at rest, many disease sufferers have a normal ECG. To increase demands on the heart and coronary circulation, an ECG is performed while the patient walks on a treadmill or rides a stationary bicycle. It is sometimes called a stress test. Exercise increases the heart rate and the systemic arterial blood pressure. These changes increase cardiac work and the demand for coronary blood flow. In many patients, coronary blood flow is adequate at rest but, because of coronary arterial blockage, cannot rise sufficiently to meet the increased demands of exercise. During a stress test, specific ECG changes can indicate that cardiac muscle is not receiving sufficient blood flow and oxygen delivery.

As heart rate increases during exercise, the length of any portion of the ECG (e.g., the R wave) becomes shorter (Fig. 29.A, A and B). In patients suffering from ischemic heart disease, however, other changes occur. Most common is an abnormal depression between the S and T waves, known as *ST segment depression* (see Fig. 29.A, B). Depression of the ST segment arises from changes in cardiac muscle electrical activity secondary to lack of blood flow and oxygen delivery.

During the stress test, the ECG is continuously analyzed for changes while blood pressure and arterial blood oxygen saturation are monitored. At the start of the test, the exercise load is mild. The load is increased at regular intervals, and the test ends when the patient becomes exhausted, the heart rate safely reaches a maximum, significant pain occurs, or abnormal ECG changes are noted. With proper supervision, the stress test is a safe method for detecting coronary artery disease. Because the exercise load is gradually increased, the test can be stopped at the first sign of problems.

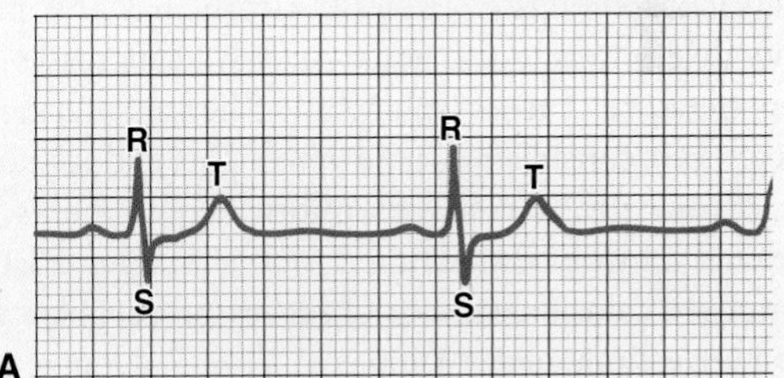

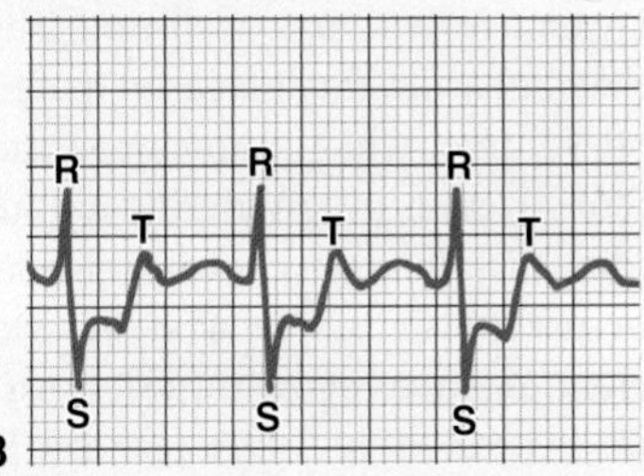

Figure 29.A Effect of exercise on the electrocardiogram (ECG) in a patient with ischemic heart disease. (A) The ECG is normal at rest. **(B)** During exercise, the interval between R waves is reduced, and the ECG segment between the S and T waves is depressed.

Exercise prevents cardiovascular diseases and reduces mortality.

There is a direct relationship between physical inactivity and cardiovascular mortality, and physical inactivity is an independent risk factor for the development of coronary artery disease. Changes in the ratio of HDL to total cholesterol that take place with regular physical activity reduce the risk of atherogenesis and coronary artery disease in active people, as compared with those who are sedentary. A lack of exercise is established as a risk factor for coronary heart disease similar in magnitude to hypercholesterolemia, hypertension, and smoking. A reduced risk grows out of the changes in lipid profiles noted above, reduced insulin requirements and increased insulin sensitivity, and reduced cardiac β-adrenergic responsiveness and increased vagal tone. When coronary ischemia does occur, increased vagal tone may reduce the risk of fibrillation.

Regular exercise often, but not always, reduces resting blood pressure. Why some people respond to chronic activity with a resting blood pressure decline and others do not remains unknown. Responders typically show diminished resting sympathetic tone, so that systemic vascular resistance falls. In obesity-linked hypertension, declining insulin secretion and increasing insulin sensitivity with exercise may explain the salutary effects of combining training with weight loss. Nonetheless, because some obese people who exercise and lose weight show no blood pressure changes, exercise remains adjunctive therapy for hypertension.

Cardiovascular response to pregnancy is similar to chronic exercise.

The physiologic demands and adaptations of pregnancy in some ways are similar to those of chronic exercise. Both increase blood volume, cardiac output, skin blood flow, and caloric expenditure. Exercise clearly has the potential to be deleterious to the fetus. Acutely, it increases body core temperature, causes splanchnic (hence, uterine and umbilical) vasoconstriction, and alters the endocrinologic milieu; chronically, it increases caloric requirements. This last demand may be devastating if food shortages exist: The superimposed caloric demands of successful pregnancy and lactation are estimated at 80,000 kcal. Given adequate nutritional resources, however, there is little evidence of other damaging effects of maternal exercise on fetal development. The failure of exercise to harm well-nourished pregnant women may relate in part to the increased maternal and fetal mass and blood volume, which reduce specific heat loads, moderate vasoconstriction in the uterine and umbilical circulations, and diminish the maternal exercise capacity.

At least in previously active women, even the most intense concurrent exercise regimen (unless associated with excessive weight loss) does not alter fertility, implantation, or embryogenesis, although the combined effects of exercise on insulin sensitivity and central obesity can restore ovulation in anovulatory obese women suffering from polycystic ovary disease. Regular exercise may reduce the risk of spontaneous abortion of a chromosomally normal fetus. Continued exercise throughout pregnancy characteristically results in normal-term infants after relatively brief labor. These infants are usually normal in length and lean body mass but reduced in fat. Maternal exercise through improved glucose tolerance reduces the risk of large infant size for gestational age, which is increased in diabetic mothers. The incidences of umbilical cord entanglement, abnormal fetal heart rate during labor, stained amniotic fluid, and low fetal responsiveness scores may all be reduced in women who are active throughout pregnancy. Further, when examined 5 days after birth, newborns of exercising women perform better in their ability to orient to environmental stimuli and their ability to quiet themselves after sound and light stimuli than weight-matched children of nonexercising mothers.

RESPIRATORY RESPONSES TO EXERCISE

Increased breathing is perhaps the single most obvious physiologic response to acute dynamic exercise. Figure 29.4 shows that minute ventilation (the product of breathing frequency and tidal volume) initially increases linearly with work intensity and then supralinearly beyond that point. Ventilation of the lungs in exercise is linked to the twin goals of oxygen uptake and carbon dioxide removal.

Ventilation matches metabolic demands during a wide range of exercise conditions, but the control mechanisms are poorly understood.

Exercise increases oxygen consumption and carbon dioxide production by working muscles, and the pulmonary response is precisely calibrated to maintain homeostasis of these gases in arterial blood. In mild or moderate work, arterial P_{O_2} (and, hence, oxygen content), P_{CO_2}, and pH all remain unchanged from resting levels (Table 29.5). The respiratory muscles accomplish this severalfold increase in ventilation primarily by increasing tidal volume, without provoking sensations of dyspnea.

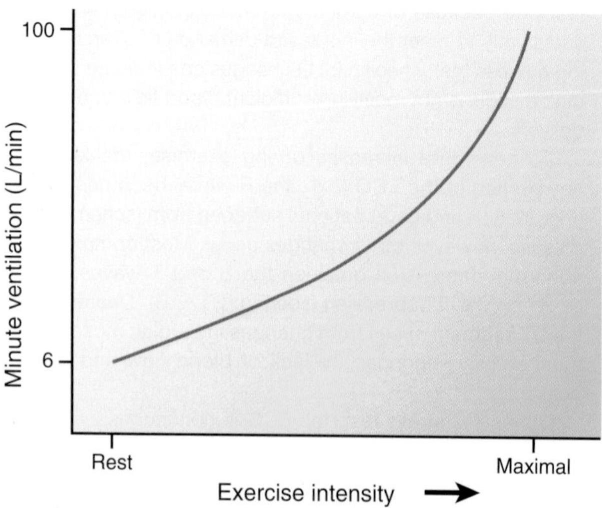

Figure 29.4 Minute ventilation is dependent on the intensity of dynamic exercise. Ventilation rises linearly with intensity until exercise nears maximal levels.

TABLE 29.5 Acute Respiratory Response to Graded Dynamic Exercise in a 30-Year-Old Untrained Woman

Exercise Intensity	Ventilation (L/min)	Ventilation/Perfusion Ratio	Alveolar P_{O_2} (mm Hg)	Arterial P_{O_2} (mm Hg)	Arterial P_{CO_2} (mm Hg)	Arterial pH
Rest	5	1	103	100	40	7.40
Walking	20	2	103	100	40	7.40
Jogging	45	3	106	100	36	7.40
Running fast	75	4	110	100	25	7.32

More intense exercise presents the lungs with tougher challenges. Near the halfway point from rest to maximal dynamic work, lactic acid formed in working muscles begins to appear in the circulation. This occurs when it is produced faster than it can be removed (metabolized). This point, which depends on the type of work involved and the person's training status, is called the **anaerobic threshold**, or lactate threshold. Lactate concentration gradually rises with work intensity, as more and more muscle fibers must rely on anaerobic metabolism. Almost fully dissociated, lactic acid causes metabolic acidosis. During exercise, healthy lungs respond to lactic acidosis by further increasing ventilation, lowering the arterial P_{CO_2}, and maintaining arterial blood pH at normal levels. It is the response to acidosis that spurs the supralinear ventilation rise seen in strenuous exercise (see Fig. 29.4). Through a range of exercise levels, the respiratory system fully compensates the pH effects of lactic acid. However, eventually in the hardest work—near exhaustion—ventilatory compensation becomes only partial, and both pH and arterial P_{CO_2} may fall well below resting values (see Table 29.5). Tidal volume continues to increase until pulmonary stretch receptors limit it, typically at or near half of the vital capacity. Frequency increases at high tidal volume produce the remainder of the ventilatory volume increases.

Hypercapnia relative to carbon dioxide production in heavy exercise helps maintain arterial oxygenation. The blood returned to the lungs during exercise is more thoroughly depleted of oxygen because active muscles with high oxygen extraction receive most of the cardiac output. Because the pulmonary arterial P_{O_2} is reduced in exercise, blood shunted past ventilated areas can profoundly depress systemic arterial oxygen content. Other than having a diminished oxygen content, pulmonary arterial blood flow (cardiac output) rises during exercise. In compensation, ventilation rises faster than cardiac output: The ventilation/perfusion ratio of the lung rises from near 1 at rest to >4 with strenuous exercise (see Table 29.5). Healthy people maintain nearly constant arterial P_{O_2} with acute exercise, although the alveolar-to-arterial P_{O_2} gradient does rise. This increase shows that despite the increase in the ventilation/perfusion ratio, areas of relative pulmonary underventilation and, possibly, some mild diffusion limitation exist even in highly trained, healthy people.

The ventilatory control mechanisms in exercise remain undefined. Where there are stimuli—such as in mixed venous blood, which is hypercapnic and hypoxic in proportion to exercise intensity—there are seemingly no receptors. Conversely, where there are receptors—the carotid bodies, the lung parenchyma or airways, the brainstem bathed by cerebrospinal fluid—no stimulus proportional to the exercise demand exists. Paradoxically, the central chemoreceptor is immersed in increasing alkalinity as exercise intensifies a consequence of blood–brain barrier permeability to CO_2 but not hydrogen ions. Perhaps exercise respiratory control parallels cardiovascular control, with a central command proportional to muscle activity directly stimulating the respiratory center and feedback modulation from the lung, respiratory muscles, chest wall mechanoreceptors, and carotid body chemoreceptors.

Training does not significantly alter the respiratory system.

The effects of training on the pulmonary system are minimal. Lung diffusing capacity, lung mechanics, and even lung volumes change little, if at all, with training. The widespread assumption that training improves vital capacity is false; even exercise designed specifically to increase inspiratory muscle strength elevates vital capacity by only 3%. The demands placed on respiratory muscles increase their endurance, an adaptation that may reduce the sensation of dyspnea in exercise. Nevertheless, the primary respiratory changes with training are secondary to lower lactate production that reduces ventilatory demands at previously heavy absolute work levels.

Exercise is used to detect lung dysfunction.

Any compromise of lung or chest wall function is much more apparent during exercise than at rest. One hallmark of lung disease is **dyspnea** (difficult or labored breathing) during exertion, when this exertion previously was unproblematic. Restrictive lung diseases limit tidal volume, reducing the ventilatory reserve volumes and exercise capacity. Obstructive lung diseases increase the work of breathing, exaggerating dyspnea and limiting work output. Lung diseases that compromise oxygen diffusion from alveolus to blood exaggerate exercise-induced widening of the alveolar-to-arterial P_{O_2} gradient. This effect contributes to potentially dangerous systemic arterial hypoxia during exercise. The signs and symptoms of a respiratory limitation to exercise include exercise cessation with low maximal heart rate, oxygen desaturation of

Clinical Focus / 29.2 

Exercise in Patients with Emphysema

Normally, the respiratory system does not limit exercise tolerance. In healthy people, arterial blood saturation with oxygen, which averages 98% at rest, is maintained at or near 98% in even the most strenuous dynamic or isometric exercise. The healthy response includes the ability to augment ventilation more than cardiac output; the resulting rise in the ventilation/perfusion ratio counterbalances the falling oxygen content of mixed venous blood.

In patients with emphysema, ventilatory limitations to exercise occur long before either skeletal muscle oxidative capacity or the ability of the cardiovascular system to deliver oxygen to exercising muscle imposes ceilings. These limitations are manifest during a stress test on the basis of three primary measurements. First, patients with ventilatory limitations typically cease exercise at a relatively low heart rate, indicating that exhaustion is a result of factors unrelated to cardiovascular limitations. Second, their primary complaint is usually shortness of breath, or dyspnea. In fact, patients with chronic obstructive pulmonary disease often first seek medical evaluation because of dyspnea experienced during such routine activities as climbing a flight of stairs. In healthy people, exhaustion is rarely associated solely with dyspnea. In emphysematous patients, exercise-induced dyspnea results, in part, from respiratory muscle fatigue exacerbated by diaphragmatic flattening brought on by loss of lung elastic recoil. Third, in emphysematous patients, arterial oxygen saturation characteristically falls steeply and progressively with increasing exercise, sometimes reaching dangerously low levels. In emphysema, the inability to fully oxygenate blood at rest is compounded during exercise by increased pulmonary blood flow, and by increased exercise oxygen extraction that more fully desaturates blood returning to the lungs.

arterial blood, and severe shortness of breath. The prospects of training-based rehabilitation are modest, although locomotor muscle-based adaptations can reduce lactate production and ventilatory demands in exercise. Specific training of respiratory muscles to increase their strength and endurance is of minimal benefit to patients with compromised lung function.

Exercise causes bronchoconstriction in nearly every asthmatic patient and is the sole provocative agent for asthma in many people. In healthy people, catecholamine release from the adrenal medulla and sympathetic nerves dilates the airways during exercise. Sympathetic bronchodilation in people with asthma is outweighed by constrictor influences, among them heat loss from airways (cold, dry air is a potent bronchoconstrictor), release of inflammatory mediators, and increases in airway tissue osmolality. Leukotriene-receptor antagonists block exercise-induced symptoms in most people. The effects of exercise on airways are a result of increased ventilation per se; the exercise is incidental. People with exercise-induced bronchoconstriction are simply the most sensitive people along a continuum; for example, breathing high volumes of cold, dry air provokes at least mild bronchospasm in everyone.

SKELETAL MUSCLE AND BONE RESPONSES TO EXERCISE

Events within exercising skeletal muscle are a primary factor in fatigue. These same events, when repeated during training, lead to adaptations that increase exercise capacity and retard fatigue during similar work. Skeletal muscle contraction also increases stresses placed on bone, leading to specific bone adaptations.

Muscle fatigue is independent of lactic acid.

Muscle fatigue is defined as a loss of muscle power that results from a decline in both force and velocity. Muscle fatigue differs from *muscle weakness* or *damage* in that the loss of power with fatigue is reversible by rest. Although the exact causes of muscle fatigue and the relative importance of particular factors remains controversial, it is clear that an individual's state of fitness, dietary status, fiber-type composition, and intensity and duration of the exercise all affect the process. Historically, an increase in intracellular H^+ (decrease in cell pH) was thought to contribute to muscle fatigue by direct inhibition of the actin–myosin cross bridge, leading to a reduction in contractile force. Although strenuous exercise can reduce intramuscular pH to values as low as 6.8 (arterial blood pH may fall to 7.2), evidence suggests that elevation in $[H^+]$, although a substantial factor in declining force, is not the sole cause of fatigue. The best correlate of fatigue in healthy people is ADP accumulation in the face of normal or slightly reduced ATP, such that the ADP/ATP ratio is high. ADP is implicated in slowing crossbridge cycling. During periods of high energy demand, the ATP concentration initially remains almost constant, whereas creatine phosphate (CrP) breaks down to Cr and inorganic phosphate (Pi). Although Cr has little effect on contractile function, Pi may cause a marked decrease of myofibrillar force production and Ca^{2+} sensitivity as well as sarcoplasmic reticular Ca^{2+} release. Accordingly, increased Pi is considered to be a major cause of fatigue. There is widespread use of extra Cr intake among athletes, not only in elite athletes but also among people exercising on a recreational level. Cr enters muscle cells via a Na^+-dependent transporter in the sarcolemma. Inside the cell, Cr is phosphorylated by creatine kinase, and the [PCr]/[Cr] ratio basically depends on the energy state of the cell. Extra Cr intake increases the total muscle Cr content by up to approximately 20%. Cr supplementation has a modest positive effect on muscle performance during bouts of short-term (~10 s), high-power exercise, whereas it does not improve function during more long-lasting types of muscle activity. This corresponds to the fact that PCr breakdown contributes to a relatively large fraction of the ATP supply during the first seconds of high-intensity muscle activity, whereas the PCr contribution is minimal during prolonged exercise. Because the complete oxidation of glucose, glycogen, or free fatty acids to carbon dioxide and water is the major source of energy in prolonged work, people with defects in glycolysis or electron transport exhibit a reduced ability to sustain exercise.

These metabolic defects are distinct from another group of disorders exemplified by the various muscular dystrophies. In these illnesses, the loss of active muscle mass as a result of fat infiltration, cellular necrosis, or atrophy reduces exercise tolerance despite normal capacities (in healthy fibers) for ATP production. For this reason, Cr supplementation is also used in an attempt to improve muscle function in these diseases. Other potential factors in the etiology of fatigue may occur centrally (pain from fatigued muscle may feed back to the brain to lower motivation and, possibly, to reduce motor cortical output) or at the level of the motor neuron or the neuromuscular junction. Also, skeletal muscles produce ammonia during exercise. This ammonia is taken up by the brain in proportion to its cerebral arterial concentration. Ammonia accumulation is currently thought to be a factor in the sensation of fatigue.

Endurance training enhances muscle oxidative capacity.

Within skeletal muscle, adaptations to training are specific to the form of muscle contraction. Increased activity with low loads results in increased oxidative metabolic capacity without hypertrophy; increased activity with high loads produces muscle hypertrophy. Increased activity without overload increases capillary and mitochondrial density, myoglobin concentration, and virtually the entire enzymatic machinery for energy production from oxygen (Table 29.6). Coordination of energy-producing and energy-using systems in muscle ensures that even after atrophy the remaining contractile proteins are adequately supported metabolically. In fact, the easy fatigability of atrophied muscle is a result of the requirement that more motor units be recruited for identical external force; the fatigability per unit cross-sectional area is normal. Factors outside the muscle limit the magnitude of the skeletal muscle endurance training response, because cross-innervation or chronic stimulation of muscles in animals can produce adaptations five times larger than those created by the most intense and prolonged exercise.

Local adaptations of skeletal muscle to endurance activity reduce reliance on carbohydrate as a fuel and allow more metabolism of fat, prolonging endurance and decreasing lactic acid accumulation. Decreased circulating lactate, in turn, reduces the ventilatory demands of heavier work. Because metabolites accumulate less rapidly inside trained muscle, there is reduced chemosensory feedback to the central nervous system at any absolute workload. This reduces sympathetic outflow to the heart and blood vessels, reducing cardiac oxygen demands at a fixed exercise level.

Isometric contraction stimulates muscle hypertrophy.

It is common knowledge that walking downhill is easier than walking uphill, but the mechanisms underlying this phenomenon are complex. Forces generated by muscles are identical in the two situations. However, moving the body uphill against gravity involves muscle shortening, or **concentric contractions**. In contrast, walking downhill primarily involves muscle tension development that resists muscle lengthening, or **eccentric contractions**. All routine forms of physical activity, in fact, involve combinations of concentric, eccentric, and isometric contractions. Because less ATP is required for force development during a contraction when external forces lengthen the muscle, the number of active motor units is reduced and energy demands are less for eccentric work. However, perhaps because the force per active motor unit is greater in eccentric exercise, eccentric contractions can readily cause muscle damage. This damage includes weakness (apparent the first day), soreness and edema (delayed 1–3 days in peak magnitude), and elevated plasma levels of intramuscular enzymes (delayed 2–6 days). Histologic evidence of damage may persist for 2 weeks. Damage is accompanied by an acute-phase reaction that includes complement activation, increases in circulating cytokines, neutrophil mobilization, and increased monocyte cell adhesion capacity. Training adaptation to the eccentric components of exercise is efficient; soreness after a second episode is minimal if it occurs within 2 weeks of a first episode.

TABLE 29.6 Effects of Training and Immobilization on the Human Biceps Brachii Muscle in a 22-Year-Old Woman

Measurement	After Strength Training	After 4 Months' Immobilization	Sedentary	After Endurance Training
Total number of cells	300,000	300,000	300,000	300,000
Total cross-sectional area (cm^2)	10	10	13	6
Isometric strength (% control)	100	100	200	60
Fast-twitch fibers (% by number)	50	50	50	50
Fast-twitch fibers, average area ($\mu m^2 \times 10^2$)	67	67	87	40
Capillaries/fiber	0.8	1.3	0.8	0.6
Succinate dehydrogenase activity/unit area (% control)	100	150	77	100

Eccentric contraction-induced muscle damage and its subsequent response may be the essential stimulus for muscle hypertrophy. Although standard resistance exercise involves a mixture of contraction types, careful studies show that when one limb works purely concentrically and the other purely eccentrically at equivalent force, only the eccentric limb hypertrophies. The immediate changes in actin and myosin production that lead to hypertrophy are mediated at the post-translational level; after a week of loading, mRNA for these proteins is altered.

Despite our understanding that feeding and resistance exercise affect muscle protein synthesis and growth, researchers have now begun to study the signaling pathways that are activated with either stimulus in humans. It is known that proteins in the Akt/protein kinase B–rapamycin (mTOR)–S6 kinase pathway are involved in stimulating ribosomal assembly, biogenesis, and global protein synthesis, but the extent of signal to response is far from clear. Upstream of mTOR is the tuberous sclerosis complex (TSC), which is a heterodimeric complex of the TSC1 and TSC2 gene products, hamartin and tuberin, respectively. Studies have shown that the TSC1/TSC2 heterodimer regulates cell growth and cell proliferation as a downstream component of the PI3K (phosphoinositide 3-kinase)-Akt signaling, which modulates signal transduction through mTOR. The cellular mechanisms for hypertrophy are also known to include the induction of insulin-like growth factor-1 and upregulation of several members of the fibroblast growth factor family, although skeletal muscle hypertrophy can occur in the absence of circulating pituitary hormones, insulin, or local growth factors in response to increased mechanical loading. Although the actual biochemical events that link the mechanical loading of a muscle or muscle fiber to mTOR signaling remain unclear, it is known that mechanical loading or stretch acts independently of amino acids and growth factors to activate mTOR signaling in skeletal muscle. Passive mechanical loading is sufficient to activate mTOR signaling in skeletal muscle, and studies have shown that there is a cytoskeletal requirement for communication of the signaling pathway from mechanical stretch to mTOR. It is likely that this mechanical pathway acts synergistically with changes in amino acid uptake and growth factor availability to contribute to the prolonged activation of mTOR signaling following high-resistance contractions or mechanical overload.

Exercise plays an important role in calcium homeostasis.

Skeletal muscle contraction acts across joints and thus applies force to bone. Because the architecture of bone remodeling involves osteoblast and osteoclast activation in response to loading and unloading, physical activity is a major site-specific influence on bone mineral density and geometry. Repetitive physical activity can create excessive strain, leading to inefficiency in bone remodeling and stress fracture; however, extreme inactivity allows osteoclast dominance and bone loss.

The forces applied to bone during exercise are related both to the weight borne by the bone during activity and to the strength of the involved muscles. Consequently, bone strength and density appear to be closely related to applied gravitational forces and to muscle strength. This suggests that exercise programs to prevent or treat **osteoporosis** should emphasize weight-bearing activities and strength as well as endurance training. Adequate dietary calcium is essential for any exercise effect: weight-bearing activity enhances spinal bone mineral density in postmenopausal women only when calcium intake exceeds 1 g/d. Because exercise may also improve gait, balance, coordination, proprioception, and reaction time, even in older and frail persons, chronic activity reduces the risks of falls and osteoporosis. In fact, the incidence of hip fracture is reduced nearly 50% when older adults are involved in regular physical activity. However, even when activity is optimal, genetic contributions to bone mass are greater than those of exercise. Perhaps 75% of the population variance is genetic, and 25% is a result of different levels of activity. In addition, the predominant contribution of estrogen to homeostasis of bone in young women is apparent when amenorrhea occurs secondary to chronic heavy exercise. These exceptionally active women are typically thin and exhibit low levels of circulating estrogens, low trabecular bone mass, and a high fracture risk (Fig. 29.5).

Exercise also plays a role in the treatment of **osteoarthritis**. Controlled clinical trials find that appropriate, regular exercise decreases joint pain and degree of disability, although it fails to influence the requirement for anti-inflammatory drug treatment. In **rheumatoid arthritis (RA)**, exercise also increases muscle strength and functional capacity without increasing pain or medication requirements. Whether exercise alters disease progression in either RA or osteoarthritis is not known.

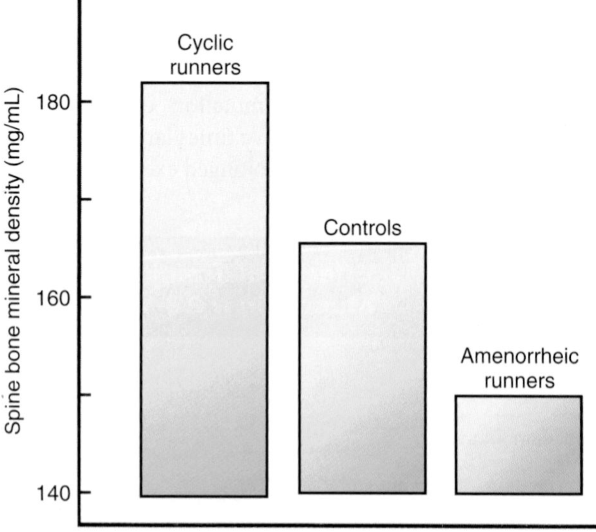

Figure 29.5 Exercise stimulates bone density. This graph shows spine bone density in young adult women who are nonathletes (controls), distance runners with regular menstrual cycles (cyclic runners), and distance runners with amenorrhea (amenorrheic runners). Differences from controls indicate the roles that exercise and estrogen play in determination of bone mineral density.

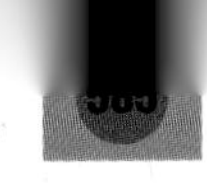

GASTROINTESTINAL, METABOLIC, AND ENDOCRINE RESPONSES TO EXERCISE

The effects of exercise on gastrointestinal (GI) function remain unclear. Physical exercise can be both beneficial and harmful to the GI tract in a dose–effect relationship between its intensity and health. However, chronic physical activity plays a major role in the control of obesity and type 2 diabetes.

Exercise can modify the rate of gastric emptying and intestinal absorption.

Dynamic exercise must be strenuous (demanding more than 70% of the maximal oxygen uptake) to slow gastric emptying of liquids. Little is known of the neural, hormonal, or intrinsic smooth muscle basis for this effect. Although acute exercise of any intensity does not change gastric acid secretion, nothing is known about the effects of exercise on other factors relevant to the development or healing of peptic ulcers. There is some evidence that strenuous postprandial dynamic exercise provokes gastroesophageal reflux by altering esophageal motility.

Chronic physical activity accelerates gastric emptying rates and small intestinal transit. These adaptive responses to chronically increased energy expenditure lead to more rapid processing of food and increased appetite. Animal models of **hyperphagia** show specific adaptations in the small bowel (increased mucosal surface area, height of microvilli, and content of brush border enzymes and transporters) that lead to more rapid digestion and absorption. These same effects likely take place in humans who are rendered hyperphagic by regular physical activity.

Blood flow to the gut decreases in proportion to exercise intensity as sympathetic vasoconstrictor tone rises. Water, electrolyte, and glucose absorption may be slowed in parallel, and acute diarrhea is common in endurance athletes during competition. However, these effects are transient, and malabsorption as a consequence of acute or chronic exercise does not occur in healthy people. Although exercise may not improve symptoms or disease progression in inflammatory bowel disease, there is some evidence that repetitive dynamic exercise may reduce the risk for this illness.

Although exercise is often recommended as treatment for postsurgical ileus, uncomplicated constipation, or irritable bowel syndrome, little is known in these areas. However, chronic dynamic exercise does substantially decrease the risk for colon cancer, possibly via increases in food and fiber intake, with consequent acceleration of colonic transit.

In obese patients, chronic exercise preferentially increases caloric expenditures over increased appetite.

Obesity is common in sedentary societies. Obesity increases the risk for hypertension, heart disease, and diabetes and is characterized, at a descriptive level, as an excess of caloric intake over energy expenditure. Because exercise enhances energy expenditure, increasing physical activity is a mainstay of treatment for obesity.

The metabolic cost of exercise averages 100 kcal/mi. walked. For exceptionally active people, exercise expenditure can exceed 3,000 kcal/d, added to the basal energy expenditure, which for a 55-kg woman averages about 1,400 kcal/d. At high levels of activity, appetite and food intake match caloric expenditure (Fig. 29.6). The biologic factors that allow this precise balance have never been defined. In obese people, modest increases in physical activity increase energy expenditure more than food intake, so progressive weight loss can be instituted if exercise can be regularized. This method of weight control is superior to dieting alone, because substantial caloric restriction (>500 kcal/d) results in both a lowered basal metabolic rate and a substantial loss of fat-free body mass.

Exercise has other, subtler, positive effects on the energy balance equation as well. A single exercise episode may increase basal energy expenditure for several hours and may increase the thermal effect of feeding. The greatest practical problem remains compliance with even the most precise exercise "prescription"; patient dropout rates from even short-term programs typically exceed 50%.

Exercise is a potent tool in regulating blood glucose in diabetic patients.

Although skeletal muscle is omnivorous, its work intensity and duration, training status, inherent metabolic capacities, and substrate availability determine its energy sources. For short-term exercise, stored phosphagens (ATP and CrP) are sufficient for cross-bridge interaction between actin and myosin. Even maximal efforts lasting 5 to 10 seconds

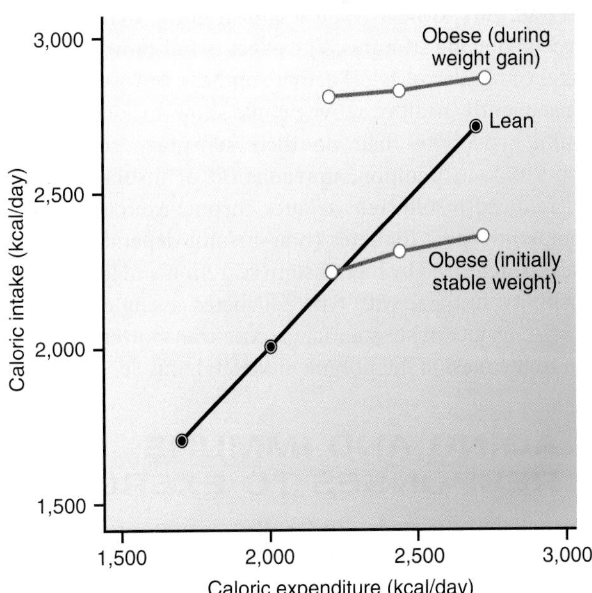

Figure 29.6 Caloric intake as a function of exercise-induced increases in daily caloric expenditure. For lean people, intake matches expenditure over a wide range. For obese people during periods of weight gain or periods of stable weight, increases in expenditure are not matched by increases in caloric intake.

require little or no glycolytic or oxidative energy production. When work to exhaustion is paced to be somewhat longer in duration, high intramuscular ADP concentrations drive glycolysis (particularly in fast glycolytic fibers), and this form of anaerobic metabolism, with its by-product lactic acid, is the major energy source. The carbohydrate provided to glycolysis comes from stored, intramuscular glycogen or blood-borne glucose. Exhaustion from work in this intensity range (50%–90% of the maximal oxygen uptake) is associated with carbohydrate depletion. Accordingly, factors that increase carbohydrate availability improve fatigue resistance. These include prior high dietary carbohydrate, cellular training adaptations that increase the enzymatic potential for fatty acid oxidation (thereby sparing carbohydrate stores), and oral carbohydrate intake during exercise. Frank hypoglycemia rarely occurs in healthy people during even the most prolonged or intense physical activity. When it does, it is usually in association with the depletion of muscle and hepatic stores and a failure to supplement carbohydrate orally.

Exercise suppresses insulin secretion by increasing sympathetic tone at the pancreatic islets. Despite acutely falling levels of circulating insulin, both non–insulin-dependent and insulin-dependent muscle glucose uptake, increase during exercise. Exercise recruits glucose transporters from their intracellular storage sites to the plasma membrane of active skeletal muscle cells. Because exercise increases insulin sensitivity, patients with **type 1 diabetes** (insulin-dependent) require less insulin when activity increases. However, this positive result can be treacherous because exercise can accelerate hypoglycemia and increase the risk of insulin coma in these people. Chronic exercise, through its reduction of insulin requirements, upregulates insulin receptors. This effect appears to be a result less of training than simply of a repeated acute stimulus. The effect is full-blown after 2 to 3 days of regular physical activity and can be lost as quickly. Consequently, healthy active people show strikingly greater insulin sensitivity than do their sedentary counterparts (Fig. 29.7). In addition, upregulation of insulin receptors and reduced insulin release after chronic exercise are ideal therapy in **type 2 diabetes** (non–insulin-dependent), a disease characterized by high insulin secretion and low receptor sensitivity. In those with type 2 diabetes, a single episode of exercise results in substantial glucose transporter translocation to the plasma membrane in skeletal muscle.

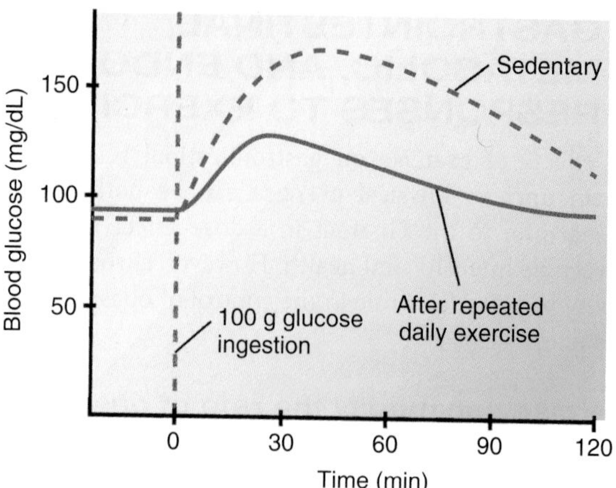

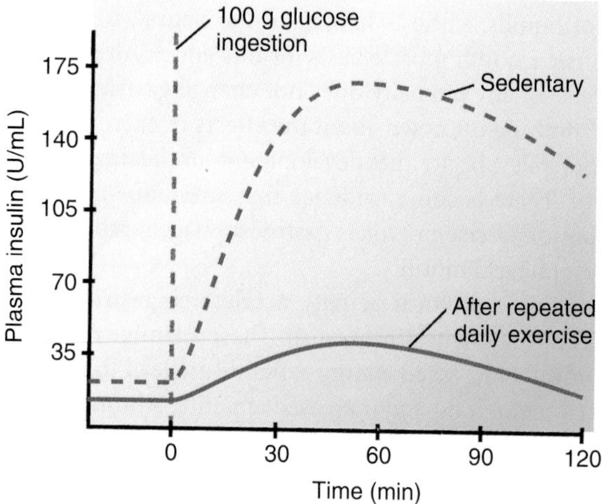

Figure 29.7 Daily exercise blunts blood glucose and insulin responses to glucose ingestion. Repeated exercise significantly decreases both responses, demonstrating increased insulin sensitivity.

AGING AND IMMUNE RESPONSES TO EXERCISE

Maximal dynamic and isometric exercise capacities are lower at age 70 years than at age 20 years. There is overwhelming evidence, however, that training can substantially mitigate declines in strength and endurance with advancing age. Changes in functional capacity, as well as protection against heart disease and diabetes, do increase longevity in active persons. However, it remains controversial whether chronic exercise enhances lifespan or whether exercise boosts the immune system, prevents insomnia, or enhances mood.

Exercise affects aging more profoundly than longevity.

The influence of exercise on strength and endurance at any age is dramatic. Although the ceiling for oxygen uptake during work gradually falls with age, the ability to train toward an age-appropriate ceiling is as intact at age 70 years as it is at age 20 years (Fig. 29.8). In fact, a highly active 70-year-old person, otherwise healthy, will typically display an absolute exercise capacity greater than that of a sedentary 20-year-old person. Aging affects all the links in the chain of oxygen transport and use, so aging-induced declines in lung elasticity, lung diffusing capacity, cardiac output, and muscle metabolic potential take place in concert. Consequently, the physiologic mechanisms underlying fatigue are similar at all ages.

Regular dynamic exercise, compared with inactivity, increases longevity in rats and humans. In descriptive terms, the effects of exercise are modest; all-cause mortality is reduced but only in amounts sufficient to increase longevity by 1 to 2 years. These facts leave open the possibility that exercise might alter biologic aging. Although physical activity increases cellular oxidative stress, it simultaneously increases antioxidant capacity. Food-restricted rats experience increased lifespan and exhibit elevated spontaneous

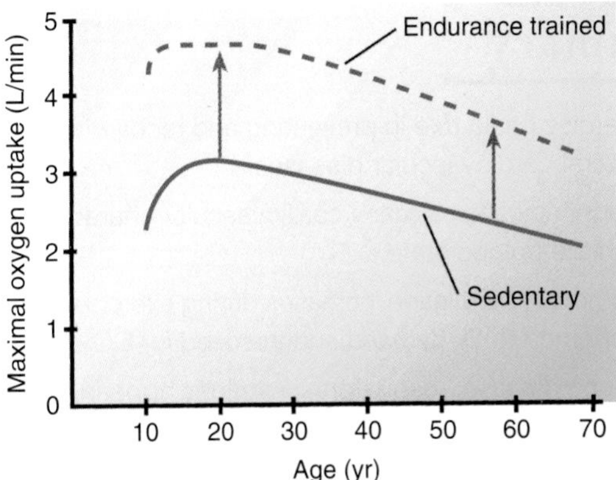

Figure 29.8 Maximal oxygen uptake is enhanced with endurance training. Endurance-trained subjects possess greater maximal oxygen uptake than sedentary subjects, regardless of age.

activity levels, but the role exercise may play in the apparent delay of aging in these animals remains unclear.

Exercise has modest effect on immune function.

In protein–calorie malnutrition, the catabolism of protein for energy lowers immunoglobulin levels and compromises the body's resistance to infection. Clearly, in this circumstance, exercise merely speeds the starvation process by increasing daily caloric expenditure and would be expected to diminish the immune response further. Nazi labor camps of the early 1940s became death camps, partly by severe food restrictions and incessant demands for physical work—a combination guaranteed to cause starvation.

If nutrition is adequate, it is less clear whether adopting an active versus a sedentary lifestyle alters immune responsivity. In healthy people, an acute episode of exercise briefly increases blood leukocyte concentration and transiently enhances neutrophil production of microbicidal reactive oxygen species and natural killer cell activity. However, it remains unproven that regular exercise over time can lower the frequency or reduce the intensity of, for example, upper respiratory tract infections (URTIs). In human immunodeficiency virus (HIV)–positive men and in men with AIDS and advanced muscle wasting, strength and endurance training yield normal gains. There is also incomplete evidence that training may slow progression to AIDS in HIV-positive men, with a corresponding increase in CD4 lymphocytes. In contrast to moderate or intermittent physical activity, prolonged and very intense exertions are associated with numerous changes in the immune system that reflect physiologic stress and suppression. Athletes subjected to regular strenuous exercise may be at increased risk of URTI during periods of heavy exercise and for a several weeks following race events. It has been suggested that the relationship between exercise and URTIs follows a "J curve," with moderate and regular exercise improving the ability to resist infections and heavy acute or chronic exercise decreasing it. Exercise-induced immunodeppression has a multifactorial origin, depending on mechanisms related to neuro–immune–endocrine systems. Prolonged periods of intense training may lead to high numbers of neutrophil and low numbers of lymphocytes; impaired phagocytic and neutrophilic function; and decreases in oxidative burst activity, natural killer cell cytolytic activity, and mucosal immunoglobulin levels. Highly intense training may also enhance release of injury-related proinflammatory cytokines, such as tumor necrosis factor-α, interleukin (IL)-1β, and IL-6. Whether these observations have relevance to clinically immune deficiency remains to be determined.

From Bench to Bedside / 29.1

Mobilizing Patients in the Intensive Care Unit—Benefits of Exercise

Due to a decline in mortality resulting from critical illness in recent years, a new focus on patient outcomes after hospital discharge has emerged. Prolonged stays in the intensive care unit (ICU) are commonly associated with the development of persistent and often severe neuromuscular weakness. Prominent in the development of ICU-acquired weakness is immobility related to prolonged bed rest. Studies in unrelated patient populations have demonstrated that moderate exercise is beneficial in counteracting the inflammatory conditions and atrophy that are associated with immobility and in improving muscle strength and physical function. Prolonged bed rest associated with critical illness typically leads to decreased muscle protein synthesis, increased urinary nitrogen excretion (indicating muscle catabolism), and decreased muscle mass, especially in the lower extremities. Such immobility leads to a proinflammatory state caused by increased systemic inflammation due to proinflammatory cytokines such as interleukin (IL)-1β, IL-2, and interferon-γ proinflammatory cytokines. This potentiates the production of reactive oxygen species with a decrease in antioxidative defenses and activation of both nuclear factor-κB and Akt/forkhead transcription factor, class O (FoxO) signaling pathways, resulting in protein loss. Recent studies in the ICU have demonstrated improved clinical outcomes with early mobility that included ambulation and physical therapy. Early mobilization and decreased sedation were found to be synergistically beneficial in mechanically ventilated patients. Their application was associated with decreased lengths of stay in the ICU and hospital, with a trend toward decreased duration of mechanical ventilation, resulting in overall improvements in physical function and quality of life for ICU survivors.

Chapter Summary

- Exercise is the performance of some activity that involves muscle contraction and joint flexion or extension and places unique stresses on the body's organ systems.
- Measuring maximal oxygen uptake is the most common method of quantifying dynamic exercise.
- Excess in oxygen consumption during the first minutes of recovery is called the oxygen debt.
- Blood flow is preferentially directed to working skeletal muscle during exercise.
- During exercise, blood pressure, heart rate, stroke volume, and cardiac contractility are all increased.
- Hearts of people adapted to prolonged, rhythmic exercise involving low arterial pressure exhibit large left ventricular volumes with normal wall thickness; in those adapted to activities involving isometric contraction and elevated arterial pressure, wall thickness is increased at normal volume.
- Chronic, dynamic exercise is associated with increased circulating levels of high-density lipoproteins (HDLs) and reduced low-density lipoproteins, such that the ratio of HDL to total cholesterol is increased.
- Exercise has a role in preventing and recovering from several cardiovascular diseases.
- Pregnancy shares many cardiovascular characteristics with the trained state.
- Pulmonary ventilation increases during exercise in proportion to O_2 demand and the need for CO_2 removal.
- In lung disease, respiratory limitations appear as shortness of breath or decreased oxygen content of arterial blood and become more apparent during exercise than at rest.
- Muscle fatigue is defined as an exercise-induced reduction in the maximal force capacity of the muscle and is independent of lactic acid.
- Endurance activity at low loads enhances muscle oxidative capacity without hypertrophy, whereas increased activity at high loads produces muscle hypertrophy.
- Although exercise effects on gastrointestinal function remain poorly understood, chronic physical activity plays a major role in the control of obesity and type 2 diabetes.
- As people age, the effects of exercise on functional capacity are more profound than its effect on longevity.

thePoint *Visit http://thePoint.lww.com for chapter review Q&A, case studies, animations, and more!*

30 Endocrine Control Mechanisms

ACTIVE LEARNING OBJECTIVES

Upon mastering the material in this chapter you should be able to:

- Explain how, in addition to specific endocrine disease states, other prominent diseases, including atherosclerosis, cancer, and even psychiatric disorders, likely have an underlying endocrine component.
- Explain how hormone receptors, restricted distribution of hormones, and hormone activation processes determine target tissues for a specific hormone.
- Explain how normal feedback relationships permit clinical diagnoses of functional and dysfunctional endocrine systems.
- Explain how secretagogues are used clinically to provide meaningful diagnostic information.
- Explain how chemical differences in hormones influence their endocrine functionality.
- Explain how altered target tissue hormonal responses reflected by dose–response curves can provide useful clinical information regarding the underlying cause of a particular disease.

Endocrinology is the branch of physiology concerned with the description and characterization of processes involved in the regulation and integration of cells and organ systems by a group of specialized chemical substances called **hormones**. The diagnosis and treatment of many endocrine disorders are important aspects of any general medical practice. Certain endocrine disease states, such as diabetes mellitus, thyroid disorders, and reproductive disorders, are fairly common in the general population; therefore, it is likely that they will be encountered repeatedly in the practice of medicine.

In addition, because hormones either directly or indirectly affect virtually every cell or tissue in the body, some other prominent diseases not primarily classified as endocrine diseases may have an important endocrine component. Atherosclerosis, certain forms of cancer, and even certain psychiatric disorders are examples of conditions in which an endocrine disturbance may contribute to the progression or severity of disease.

GENERAL CONCEPTS OF ENDOCRINE CONTROL

The word "hormone" is derived from the Greek *hormaein*, which means to "excite" or to "stir up." Glands that comprise the endocrine system produce hormones. The major morphologic feature of endocrine glands is that they are ductless; that is, they release their secretory products directly into the bloodstream and not into a duct system. Figure 30.1 illustrates classical endocrine glands. Many other "nonclassical" hormone-producing glands also exist. These include the central nervous system, kidneys, stomach, small intestine, skin, heart, lung, and placenta. Recent advances in cellular and molecular biology continually broaden our views of the endocrine system. For example, obesity-related research discovered leptin, a hormone formed in adipocytes that signals to the central nervous system to regulate appetite and energy expenditure. Here, general themes and principles that underpin the functionality of the endocrine system as a whole will be presented.

Hormones are bloodborne chemicals released by a cell that regulates and coordinates many functions.

Hormones function as homeostatic blood-borne chemicals to regulate and coordinate various biologic functions. They are highly potent, specialized, organic molecules produced by endocrine cells in response to specific stimuli and exert their actions on specific target cells. These target cells are equipped with receptors that bind hormones with high affinity and specificity; when bound, they initiate characteristic biologic responses by the target cells.

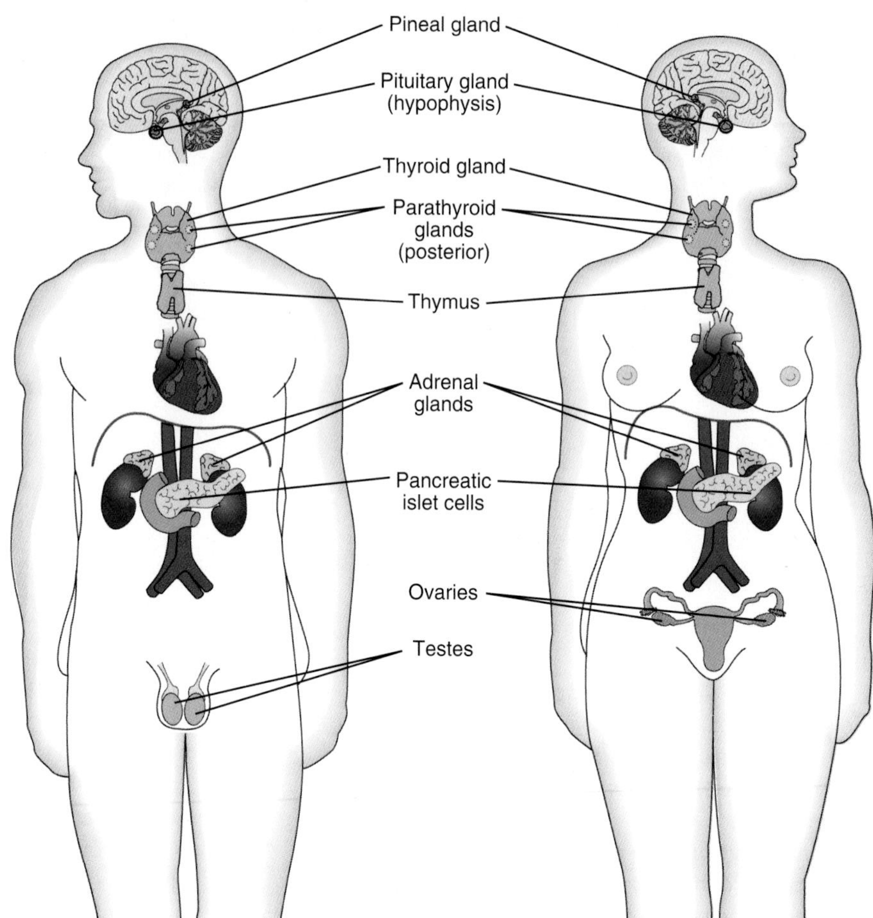

Figure 30.1 Location of many endocrine glands, organs containing endocrine tissue, and associated structures.

Although the effects of hormones are many and varied, their actions are involved in (1) regulating ion and water balance; (2) responding to adverse conditions, such as infection, trauma, and emotional stress; (3) sequentially integrating features of growth and development; (4) contributing to basic processes of reproduction, including gamete production, fertilization, nourishment of the embryo and fetus, delivery, and nourishment of the newborn; and (5) digesting, using, and storing nutrients.

In the past, definitions or descriptions of hormones usually included a phrase indicating that these substances were secreted into the bloodstream and carried by the blood to a distant target tissue. Although many hormones travel by this mechanism, we now realize that there are many hormones or hormone-like substances that play important roles in cell-to-cell communication that are not secreted directly into the bloodstream. Instead, these substances reach their target cells by diffusion through the interstitial fluid. Recall the discussion of autocrine and paracrine mechanisms in Chapter 1, "Homeostasis and Cellular Signaling."

Hormones initiate a cellular response by binding to specific receptors in a target organ.

In the endocrine system, a hormone molecule secreted into the blood is free to circulate and contact almost any cell in the body. However, only **target cells**, those cells that possess specific receptors for the hormone, will respond to that hormone. A **hormone receptor** is the molecular entity (usually a protein or glycoprotein) either outside or within a cell that recognizes and binds a particular hormone. When a hormone binds to its receptor, biologic effects characteristic of that hormone are initiated. Therefore, in the endocrine system, the basis for specificity in cell-to-cell communication rests at the level of the receptor. Similar concepts apply to autocrine and paracrine mechanisms of communication.

The restricted distribution of some hormones ensures a certain degree of specificity. For example, several hormones produced by the hypothalamus regulate hormone secretion by the anterior pituitary. These hormones are carried via small blood vessels directly from the hypothalamus to the anterior pituitary, prior to entering the general systemic circulation. The anterior pituitary is, therefore, exposed to considerably higher concentrations of these hypothalamic hormones than the rest of the body; as a result, the actions of these hormones focus on cells of the anterior pituitary.

Another mechanism that restricts the distribution of active hormone is the local transformation of a hormone within its target tissue from a less active to a more active form. An example is the formation of dihydrotestosterone from testosterone, occurring in such androgen target tissues as the prostate gland. Dihydrotestosterone is a much more potent androgen than testosterone. Because the enzyme that catalyzes this conversion is found only in certain locations, its cell or tissue distribution partly localizes the actions of the androgens to these sites. Therefore, although receptor distribution is the primary factor in determining the target tissues for a specific hormone, other factors may also focus the actions of a hormone on a particular tissue.

Hormone synthesis, storage, and secretion are regulated by homeostatic feedback regulation.

Feedback mechanisms regulate the endocrine system, just as in many other physiologic systems. The mechanism is usually negative feedback, although a few positive feedback mechanisms are known. Both types of feedback control occur because the endocrine cell, in addition to synthesizing and secreting its own hormone product, has the ability to sense the biologic consequences of secretion of that hormone. This enables the endocrine cell to adjust its rate of hormone secretion to produce the desired level of effect, ensuring the maintenance of homeostasis.

In the simplest form, negative feedback is a closed loop in which hormone A stimulates the production of hormone B, which in turn acts on the cells producing A to decrease its rate of secretion (Fig. 30.2A). In the less common positive feedback system, hormone B further stimulates the production of A instead of diminishing it (Fig. 30.2B). Many examples of negative feedback regulation exist and are discussed throughout the endocrine chapters. A typical example of a positive feedback loop is that which exists between luteinizing hormone (LH) and estradial. As detailed in Chapter 37, "Female Reproductive System," positive feedback by estradial is required to generate an abrupt increase in LH secretion during the menstrual cycle. Endocrine control systems also have feedforward loops, which can be negative or positive and which direct the flow of hormonal information. Unlike negative feedback, which promotes stability, feedforward control anticipates change. Recall discussion of the physiologic regulation in Chapter 1, "Homeostasis and Cellular Signaling."

More commonly, feedback regulation in the endocrine system is complex, involving second-order or third-order feedback loops. For example, multiple levels of feedback regulation may be involved in regulating hormone production by various endocrine glands under the control of the anterior pituitary (Fig. 30.2C). The regulation of the secretion of target gland hormones, such as adrenal steroids or thyroid hormones, begins with the production of a **releasing hormone** by the hypothalamus. The releasing hormone stimulates the production of a **trophic hormone** by the anterior pituitary, which in turn, stimulates the production of the target endocrine cell hormone by the target gland. As indicated by the blue lines in Figure 30.2C, the target endocrine cell hormone may have negative-feedback effects to inhibit secretion of both the trophic hormone from the anterior pituitary and the releasing hormone from the hypothalamus. In addition, the trophic hormone may inhibit releasing-hormone secretion from the hypothalamus, and in some cases, the releasing hormone may inhibit its own secretion by the hypothalamus.

The more complex multilevel form of regulation appears to provide certain advantages compared with the simpler system. Theoretically, it permits a greater degree of fine-tuning of hormone secretion, and the multiplicity of regulatory steps minimizes changes in hormone secretion in the event that one component of the system is not functioning normally.

Normal feedback relationships that control the secretion of each individual hormone are discussed in the chapters that follow. Clinical diagnoses are often made based on the evaluation of hormone–effector pairs relative to normal feedback relationships. For example, in the case of anterior pituitary hormones, measuring both the trophic hormone and the target gland hormone concentration provides important information to help determine whether a defect in hormone production exists at the level of the pituitary or at the level of the target gland. Furthermore, most dynamic tests of endocrine function performed clinically are based on our knowledge of these feedback relationships. Dynamic tests involve a prescribed perturbation of the feedback relationship(s).

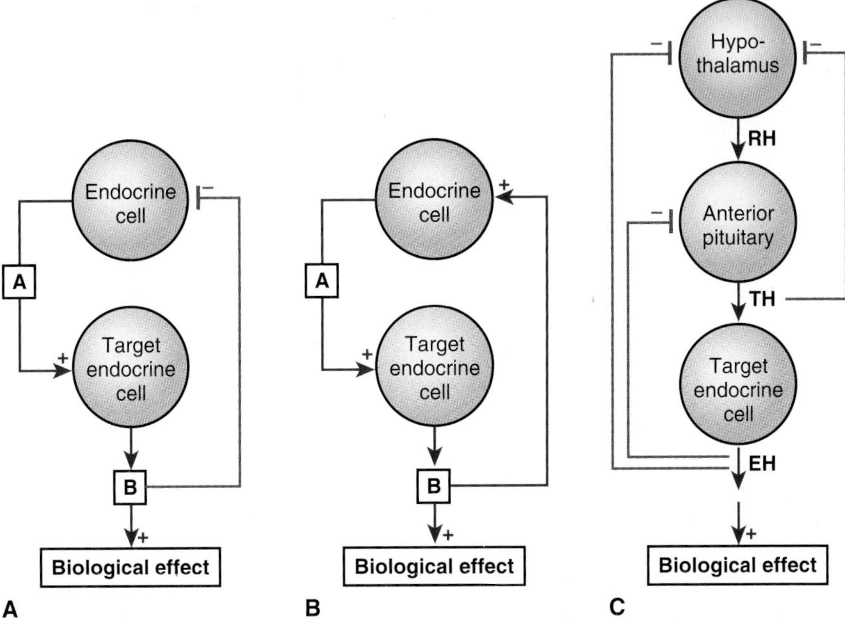

Figure 30.2 Negative, positive, and complex feedback mechanisms in the endocrine system. (A) Negative feedback loop. **(B)** Positive feedback loop. **(C)** A complex, multilevel feedback loop: the hypothalamic–pituitary–target gland axis. Red lines indicate stimulatory effects; blue lines indicate inhibitory, negative-feedback effects. EH, endocrine cell hormone; RH, releasing hormone; TH, trophic hormone.

The range of response in a healthy person is well established, whereas a response outside the normal range is indicative of abnormal function at some level and greatly enhances information gained from static measurements of hormone concentrations.

Signal amplification is part of the overall mechanism of hormone action.

Another important feature of the endocrine system is *signal amplification*, a mechanism that increases the amplitude of the blood-borne signal. Blood concentrations of hormones are exceedingly low, generally 10^{-9} to 10^{-12} mol/L. Even at the higher concentration of 10^{-9} mol/L, only one hormone molecule is present for roughly every 50 billion water molecules. Therefore, for hormones to be effective regulators of biologic processes, amplification must be part of the overall mechanism of hormone action.

Amplification generally results from the activation of a series of enzymatic steps involved in hormone action. At each step, many times more signal molecules are generated than were present at the prior step, leading to a cascade of ever-increasing numbers of signal molecules. The self-multiplying nature of the hormone action pathways provides the molecular basis for amplification in the endocrine system.

Multiplicity characterizes the endocrine system in terms of its action on target tissue and its regulation.

Most hormones have multiple actions in their target tissues and are, therefore, said to have **pleiotropic** effects. This phenomenon occurs when a single hormone regulates several functions in a target tissue. For example, in skeletal muscle, insulin stimulates glucose uptake, stimulates glycolysis, stimulates glycogenesis, inhibits glycogenolysis, stimulates amino acid uptake, stimulates protein synthesis, and inhibits protein degradation.

In addition, some hormones are known to have different effects in several different target tissues. For example, testosterone, the male sex steroid, promotes normal sperm formation in the testes, stimulates growth of the accessory sex glands, such as the prostate and seminal vesicles, and promotes the development of several secondary sex characteristics, such as beard growth and deepening of the voice.

Multiplicity of regulation is also common in the endocrine system. The input of information from several sources allows a highly integrated response to many stimuli, which is of ultimate benefit to the whole animal. For example, several different hormones, including insulin, glucagon, epinephrine, thyroid hormones, and adrenal glucocorticoids, may regulate liver glycogen metabolism.

Hormones are often secreted in rhythmic patterns.

A defined set of chemical substances in the blood or environmental factors either stimulate or inhibit the secretion of any particular hormone. In addition to these specific **secretagogues**, many hormones are secreted in a defined, rhythmic pattern. These rhythms can take several forms. For example, they may be pulsatile, episodic spikes in secretion lasting just a few minutes, or they may follow a daily, monthly, or seasonal change in overall pattern. Pulsatile secretion may occur in addition to other longer secretory patterns.

For these reasons, a single randomly drawn blood sample for determining a certain hormone concentration may be of little or no diagnostic value. A dynamic test of endocrine function in which a known agent specifically stimulates hormone secretion often provides much more meaningful information.

Clinical Focus / 30.1 

Growth Hormone and Pulsatile Hormone Secretion

Growth hormone (GH) is a 191–amino acid protein hormone that is synthesized and secreted by somatotrophs of the anterior lobe of the pituitary gland. As described in Chapter 31, "Hypothalamus and the Pituitary Gland," GH plays a role in regulating bone growth and energy metabolism in skeletal muscle and adipose tissue. A deficiency in GH production during adolescence results in dwarfism, and overproduction results in gigantism. Measurements of circulating growth hormone levels are, therefore, desirable in children whose growth rate is not appropriate for their age. However, as with many other hormones, the pulsatile nature of GH secretion needs to be considered.

Most healthy children experience episodes or "bursts" of GH secretion throughout the day, most prominently within the first several hours of sleep. To obtain reliable information about growth hormone secretion, insertion of a continuous withdrawal pump or patent indwelling catheter with unrestricted food intake and physical activity is required. Increasing sampling frequencies from every 20 minutes to 5-minute or 30-second sampling intervals enhances the threshold for detecting more pulses per hour. Although cumbersome and expensive, this method eliminates the error of isolated peak or trough measurements that might otherwise be obtained by single or multiple random GH samplings.

A less arduous approach with similar discriminating power in diagnosing GH deficiency entails measuring GH levels in the blood shortly after appropriate pituitary stimulation. Pharmacologic agents that stimulate GH secretion include L-dopa, clonidine, propanolol, glucagon, insulin, and arginine. By perturbing the system in a well-prescribed fashion, the endocrinologist is able to gain important information about growth hormone secretion that would not be possible if a random blood sample were used. As further presented in Clinical Focus Box 31.2, an alternative means of testing for GH deficiency is to measure the levels of insulin-like growth factor-1 and IGF-binding protein-3 in the blood because both predict the presence of GH.

HORMONE CLASSES

Hormones can be categorized by several criteria. Grouping them by chemical structure is convenient, because in many cases hormones with similar structures also use similar mechanisms to produce their biologic effects. In addition, tissues with similar embryonic origins usually produce hormones with similar chemical structures. The three chemical classes include (1) amine-derived hormone, (2) peptide-derived hormone, and (3) lipid/phospholipid-derived hormone.

Structurally, the simplest hormones consist of one or two modified amino acids.

Hormones derived from one or two amino acids are small in size and often hydrophilic. These hormones are formed by conversion from a commonly occurring amino acid; **norepinephrine** and **thyroxine**, for example, are derived from tyrosine (Fig. 30.3A). Each of these hormones is synthesized by a particular sequence of enzymes that are primarily localized in the endocrine gland involved in its production. Many environmental or pharmacologic agents can influence the synthesis of amino acid–derived hormones in a relatively specific fashion. The steps involved in the synthesis of these hormones are discussed in detail in later chapters.

Polypeptide hormones are diverse in size and complexity.

Hormones in the polypeptide group may be as small as the tripeptide **thyrotropin-releasing hormone (TRH)** or as large as **human chorionic gonadotropin**, which is composed of separate α and β subunits, has a molecular weight of approximately 34 kDa, and is a glycoprotein consisting of 16% carbohydrate by weight. Figure 30.3B provides examples of this second category of polypeptide and protein hormones.

Within the polypeptide class of hormones are many families of hormones, some of which are listed in Table 30.1. Hormones can be grouped into these families as a result of considerable homology with regard to amino acid sequence and structure. Presumably, the similarity of structure in these families resulted from the evolution of a single ancestral hormone into each of the separate and distinct hormones. In many cases, there is also considerable homology among receptors for the hormones within a family.

Amines
(Norepinephrine)
Thyroid Hormones
(3,5,3*-triiodothyronine, T_3)
A

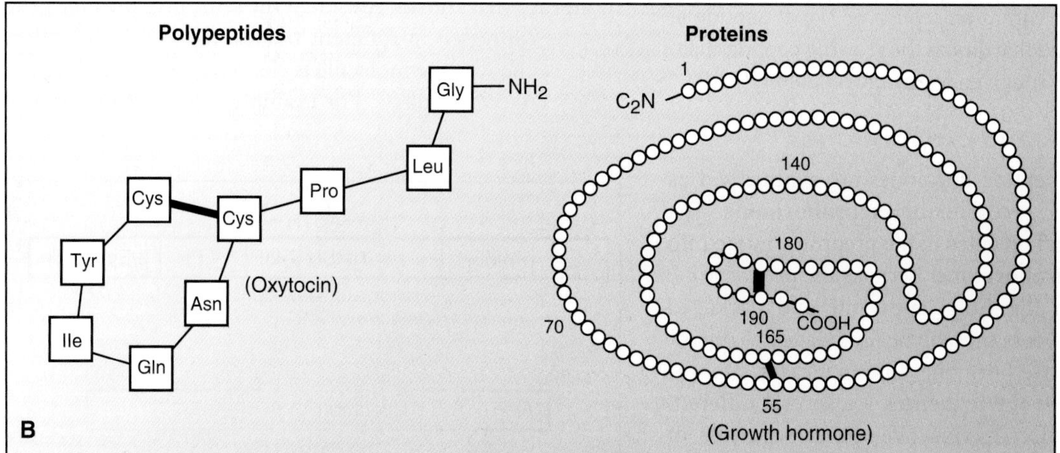

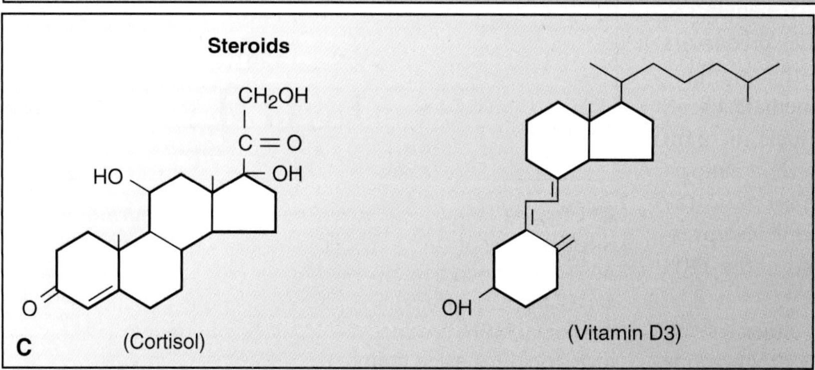

Figure 30.3 Examples of different hormone chemical structures. (A) Amine hormones. **(B)** Peptide hormones. **(C)** Steroid hormones.

TABLE 30.1 Examples of Peptide Hormone Families

Insulin Family	Glycoprotein Family	Growth Hormone Family	Secretin Family
Insulin	Luteinizing hormone	Growth hormone	Secretin
Insulin-like growth factor-1	Follicle-stimulating hormone	Prolactin	Vasoactive intestinal peptide
Insulin-like growth factor-2	Thyroid-stimulating hormone	Human placental lactogen	Glucagon
Relaxin	Human chorionic gonadotropin		Gastric inhibitory peptide

Steroid hormones are derived from cholesterol.

Steroids are lipid-soluble, hydrophobic molecules synthesized from cholesterol. Examples include **aldosterone, cortisol**, and **androgen**, which are secreted by the cortex (outer zone) of the adrenal glands; **testosterone**, secreted by the testes; and **estrogen** and **progesterone**, secreted by the ovaries. These adrenal and gonadal steroid hormones contain an intact steroid nucleus, as illustrated for cortisol in Figure 30.3C. In contrast, other hormone derivatives of cholesterol, such as vitamin D and its metabolites, have a broken steroid nucleus (the B ring) (see Fig. 30.3C). Steroid hormones are synthesized and secreted on demand (specific mechanisms are discussed in later chapters). The next section and its subsections detail the synthesis and secretion of polypeptide and protein hormones.

Peptide-derived hormones are synthesized and stored in advance of need.

Like other proteins destined for secretion, polypeptide hormones are synthesized with a *prepeptide* or *signal* peptide at their amino terminal end that directs the growing peptide chain into the cisternae of the rough endoplasmic reticulum (RER).

Preprohormones and prohormones

Most, if not all, polypeptide hormones are synthesized as part of an even larger precursor or **preprohormone**. The prepeptide is cleaved off on entry of the preprohormone into the RER to form the **prohormone**. Because the prohormone is processed through the Golgi apparatus and packaged into secretory vesicles, it is proteolytically cleaved at one or more sites to yield active hormone. In many cases, preprohormones may contain the sequences for several different biologically active molecules. Inactive spacer segments of peptide, in some cases, may separate these active elements.

Examples of prohormones that are the precursors for polypeptide hormones, which illustrate the multipotent nature of these precursors, are shown schematically in Figure 30.4. Note, for example, that **proopiomelanocortin (POMC)** actually contains the sequences for several biologically active signal molecules. Propressophysin serves as the precursor for the nonapeptide hormone **arginine vasopressin**. The precursor for TRH contains five repeats of the TRH tripeptide in one single precursor molecule.

In general, two basic amino acid residues, either lys–arg or arg–arg, demarcate the point(s) at which the prohormone will be cleaved into its biologically active components. Presumably, these two basic amino acids serve as specific recognition sites for the trypsinlike endopeptidases thought to be responsible for cleavage of the prohormones. Although somewhat rare, there are documented cases of inherited diseases in which a point mutation involving an amino acid residue at the cleavage site results in an inability to convert the prohormone into active hormone, resulting in a state of hormone deficiency. Partially cleaved precursor molecules having limited biologic activity may be found circulating in the blood in some of these cases.

In some disease states, large amounts of intact precursor molecules are found in the circulation. This situation may be the result of endocrine cell hyperactivity or even uncontrolled production of hormone precursor by nonendocrine tumor cells. Although precursors usually have relatively low biologic activity, if they are secreted in sufficiently high amounts, they may still produce biologic effects. In some cases, these effects may be the first recognized sign of neoplasia.

Tissue-specific differences in the processing of prohormones are well known. Although the same prohormone gene may be expressed in different tissues, tissue-specific differences in the way the molecule is cleaved give rise to different final secretory products. For example, within **alpha cells** of

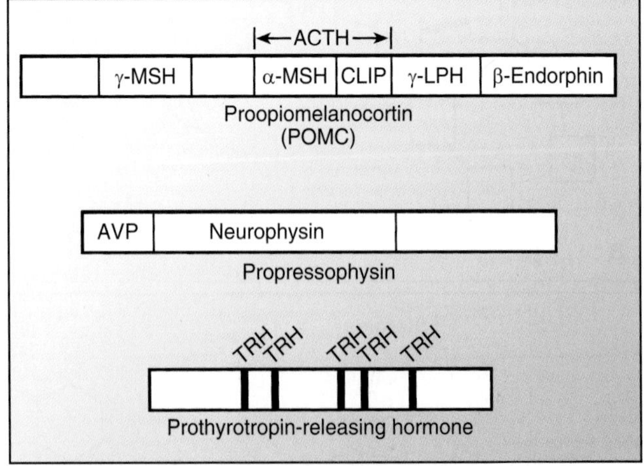

Figure 30.4 The structure of three prohormones. Relative sizes of individual peptides are only approximations. ACTH, adrenocorticotropic hormone; AVP, arginine vasopressin; CLIP, corticotropin-like intermediate lobe peptide; LPH, lipotropin; MSH, melanocyte-stimulating hormone; TRH, thyrotropin-releasing hormone.

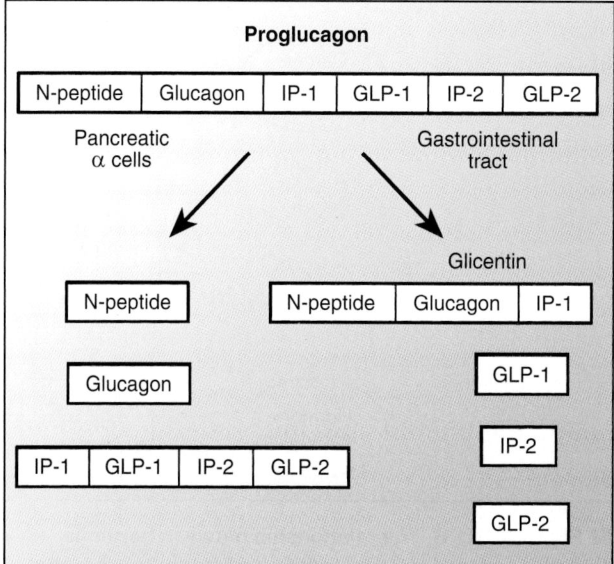

Figure 30.5 The differential processing of prohormones. In alpha cells of the pancreas (*left*), the major bioactive product formed from proglucagon is glucagon itself. It is not currently known whether the other peptides are processed to produce biologically active molecules. In intestinal cells (*right*), proglucagon is cleaved to produce the four peptides shown. Glicentin is the major glucagon-containing peptide in the intestine. GLP-1, glucagon-like peptide-1; GLP-2, glucagon-like peptide-2; IP-1, intervening peptide-1; IP-2, intervening peptide-2.

the pancreas, proglucagon is cleaved at two positions to yield three peptides, illustrated in Figure 30.5 (left). **Glucagon,** an important hormone in the regulation of carbohydrate metabolism, is the best characterized of the three peptides. In contrast, in other cells of the gastrointestinal tract in which proglucagon is also produced, the molecule is cleaved at three different positions such that **glicentin, glucagon-like peptide-1**, and **glucagon-like peptide-2** are produced (see Fig. 30.5, right).

Secretory vesicles and exocytosis

On insertion of the preprohormone into the cisternae of the RER, the prepeptide or signal peptide is rapidly cleaved from the amino terminal end of the molecule. The resulting prohormone is translocated to the Golgi apparatus in which it is processed and packaged for export. After processing in the Golgi apparatus, peptide hormones are stored in membrane-enclosed secretory vesicles. Secretion of the peptide hormone occurs by **exocytosis**; the secretory vesicle is translocated to the cell surface, its membrane fuses with the plasma membrane, and its contents are released into the extracellular fluid. An increase in cytosolic calcium stemming from an influx of calcium into the cytoplasm from internal organelles or the extracellular fluid triggers movement of the secretory vesicle and membrane fusion. In some cells, an increase in **cyclic adenosine monophosphate** and the subsequent activation of protein kinases are also involved in the stimulus-secretion coupling process. Elements of the microtubule–microfilament system play a role in the movement of secretory vesicles from their intracellular storage sites toward the cell membrane.

The cleavage of prohormone into active hormone molecules typically takes place during transit through the Golgi apparatus or, perhaps, soon after entry into secretory vesicles. Secretory vesicles, therefore, contain not only active hormone but also the excised biologically inactive fragments. When active hormone is released into the blood, a quantitatively similar amount of inactive fragment is also released. In some instances, this forms the basis for an indirect assessment of hormone secretory activity. Other types of processing of peptide hormones that may occur during transit through the Golgi apparatus include glycosylation and coupling of subunits.

Many hormones are secreted directly into the blood stream to reach their target cells.

According to the classical definition, the bloodstream carries hormones from their site of synthesis to their target tissues. However, the manner in which different hormones are carried in the blood varies.

Amino acid and polypeptide hormone transport

Most amino acid–derived and polypeptide hormones dissolve readily in the plasma, and thus no special mechanisms are required for their transport. Steroid and thyroid hormones are relatively insoluble in plasma. Mechanisms are present to promote their solubility in the aqueous phase of the blood and their ultimate delivery to a target cell.

Clinical Focus / 30.2

Pancreatic β-Cell Function and C-Peptide

β cells of the human pancreas produce and secrete **insulin.** The product of the insulin gene is a peptide known as *preproinsulin*. As with other secretory peptides, the prepeptide or signal peptide is cleaved off early in the biosynthetic process, yielding proinsulin. Proinsulin is an 86–amino acid protein that is subsequently cleaved at two sites to yield insulin and a 31–amino acid peptide known as **C-peptide.** Insulin and C-peptide are, therefore, localized within the same secretory vesicle and are cosecreted into the bloodstream.

For these reasons, measurements of circulating C-peptide levels can provide a valuable indirect assessment of β-cell insulin secretory capacity. In diabetic patients who are receiving exogenous insulin injections, the measurement of circulating insulin levels would not provide any useful information about their own pancreatic function because it would primarily be the injected insulin that would be measured. However, an evaluation of C-peptide levels in such patients would provide an indirect measure of how well the β cells were functioning with regard to insulin production and secretion.

Steroid and thyroid hormone transport

In most cases, 90% or more of steroid and thyroid hormones in the blood are bound to plasma proteins. Some of the plasma proteins that bind hormones are specialized, in that they have a considerably higher affinity for one hormone over another, whereas others, such as serum albumin, bind many hydrophobic hormones. The extent to which a hormone is protein-bound and the extent to which it binds to specific versus nonspecific transport proteins vary from one hormone to another. The principal binding proteins involved in specific and nonspecific transport of steroid and thyroid hormones are listed in Table 30.2. The liver synthesizes and secretes these proteins, and changes in various nutritional and endocrine factors influence their production.

Typically, for hormones that bind to carrier proteins, only 1% to 10% of the total hormone present in the plasma exists free in solution. However, only this free hormone is biologically active. Bound hormone cannot directly interact with its receptor and, thus, is part of a temporarily inactive pool. However, free hormone and carrier-bound hormone are in a dynamic equilibrium with each other (Fig. 30.6). The size of the free hormone pool and, therefore, the amount available to receptors are influenced not only by changes in the rate of secretion of the hormone but also by the amount of carrier protein available for hormone binding and the rate of degradation or removal of the hormone from the plasma.

In addition to increasing the total amount of hormone that can be carried in plasma, transport proteins also provide a relatively large reservoir of hormone that buffers rapid changes in free hormone concentrations. As unbound hormone leaves the circulation and enters cells, additional hormone dissociates from transport proteins and replaces free hormone that is lost from the free pool. Similarly, following a rapid increase in hormone secretion or the therapeutic administration of a large dose of hormone, most newly appearing hormone is bound to transport proteins because

TABLE 30.2	Circulating Transport Proteins
Transport Protein	**Principal Hormone(s) Transported**
Specific	
Corticosteroid-binding globulin (transcortin)	Cortisol, aldosterone
Thyroxine-binding globulin	Thyroxine, triiodothyronine
Sex hormone–binding globulin	Testosterone, estrogen
Nonspecific	
Serum albumin	Most steroids, thyroxine, triiodothyronine
Transthyretin (prealbumin)	Thyroxine, some steroids

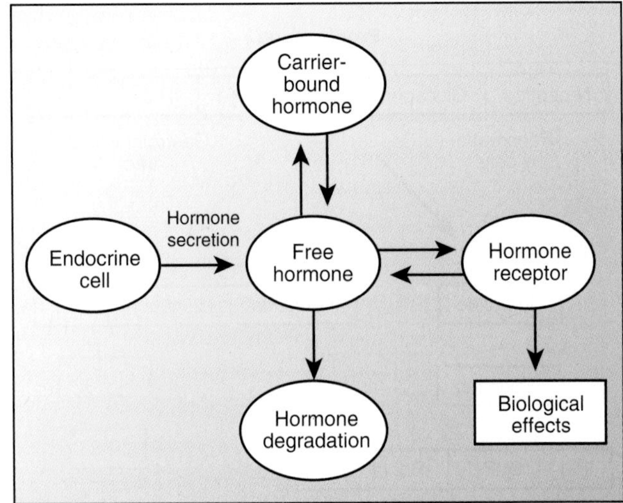

Figure 30.6 The relationship between hormone secretion, carrier protein binding, and hormone degradation. This relationship determines the amount of free hormone available for receptor binding and the production of biologic effects.

under most conditions these are present in considerable excess.

Protein binding greatly slows the rate of clearance of hormones from plasma. It not only slows the entry of hormones into cells, slowing the rate of hormone degradation, but also prevents loss by filtration in the kidneys.

From a diagnostic standpoint, it is important to recognize that most hormone assays are reported in terms of total concentration (i.e., the sum of free and bound hormone), not just free hormone concentration. The amount of transport protein and the total plasma hormone content are known to change under certain physiologic or pathologic conditions, whereas the free hormone concentration may remain relatively normal. For example, increased concentrations of binding proteins are seen during pregnancy, and decreased concentrations are seen with certain forms of liver or kidney disease. Assays of total hormone concentration might be misleading, because free hormone concentrations may be in the normal range. In such cases, it is helpful to determine the extent of protein binding, so that free hormone concentrations can be estimated.

The proportions of a hormone that are free, bound to a specific transport protein, and bound to albumin vary depending on the hormone's solubility, its relative affinity for the two classes of transport proteins, and the relative abundance of the transport proteins. For example, the affinity of cortisol for **corticosteroid-binding globulin (CBG)** is more than 1,000 times greater than its affinity for albumin, but albumin is present in much higher concentrations than CBG. Therefore, about 70% of plasma cortisol is bound to CBG, 20% is bound to albumin, and the remaining 10% is free in solution. Aldosterone also binds to CBG, but with a much lower affinity, such that only 17% is bound to CBG, 47% associates with albumin, and 36% is free in solution.

As this example indicates, more than one hormone may be capable of binding to a specific transport protein. When

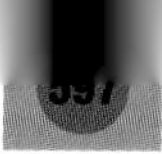

several such hormones are present simultaneously, they compete for a limited number of binding sites on these transport proteins. For example, cortisol and aldosterone compete for CBG binding sites. Increases in plasma cortisol result in displacement of aldosterone from CBG, raising the unbound (active) concentration of aldosterone in the plasma. Similarly, prednisone, a widely used synthetic corticosteroid, can displace about 35% of the cortisol normally bound to CBG. As a result, with prednisone treatment, the free cortisol concentration is higher than might be predicted from measured concentrations of total cortisol and CBG.

Hormonal activity is determined, in part, by its transformation, degradation, and excretion.

As a general rule, the gland or tissue of origin of a hormone produces it in an active form. However, for a few notable exceptions, the peripheral transformation of a hormone plays an important role in its action.

Peripheral transformation

Well-known transformations are the conversion of testosterone to dihydrotestosterone, as mentioned above and detailed in Chapter 36, "Male Reproductive System," and the conversion of thyroxine to triiodothyronine (see Chapter 32, "Thyroid Gland"). Other examples are the formation of the octapeptide angiotensin II from its precursor, angiotensinogen (see Chapter 33, "Adrenal Gland"), and the formation of 1,25-dihydroxycholecalciferol from cholecalciferol (see Chapter 35, "Endocrine Regulation of Calcium, Phosphate, and Bone Homeostasis"). Specific hormone transformations may be impaired because of a congenital enzyme deficiency or drug-induced inhibition of enzyme activity, resulting in endocrine abnormalities.

Hormonal degradation and excretion

As in any regulatory control system, the hormonal signal must dissipate or disappear once appropriate information has been transferred and the need for further stimulus has ceased. As described earlier, not only the rate of secretion but also the rate of degradation determine steady-state plasma concentrations of hormone. Thus, any factor that significantly alters the degradation of a hormone can potentially alter its circulating concentration. Commonly, however, secretory mechanisms can compensate for altered degradation such that plasma hormone concentrations remain within the normal range. Processes of hormone degradation show little, if any, regulation; alterations in the rates of hormone synthesis or secretion in most cases provide the primary mechanism for altering circulating hormone concentrations.

For most hormones, the liver is quantitatively the most important site of degradation; for a few others, the kidneys play a significant role as well. Diseases of the liver and kidneys may, therefore, indirectly influence endocrine status as a result of altering the rates at which hormones are removed from the circulation. Various drugs also alter normal rates of hormone degradation; thus, the possibility of indirect drug-induced endocrine abnormalities also exists. In addition to the liver and kidneys, target tissues may take up and degrade quantitatively smaller amounts of hormone. In the case of peptide and protein hormones, this occurs via receptor-mediated endocytosis.

The nature of specific structural modification(s) involved in hormone inactivation and degradation differs for each hormone class. As a general rule, however, specific enzyme-catalyzed reactions are involved. Inactivation and degradation may involve complete metabolism of the hormone to entirely different products, or it may be limited to a simpler process involving one or two steps, such as a covalent modification to inactivate the hormone. Urine is the primary route of excretion of hormone degradation products, but small amounts of intact hormone may also appear in the urine. In some cases, measuring the urinary content of a hormone or hormone metabolite provides a useful, indirect, noninvasive means of assessing endocrine function.

The degradation of peptide and protein hormones has been studied only in a few cases. However, it appears that proteolytic attack inactivates peptide and protein hormones in many tissues. The first step appears to involve attack by specific peptidases, resulting in the formation of several distinct hormone fragments. Several nonspecific peptidases then metabolize these fragments to yield the constituent amino acids, which can be reused.

The metabolism and degradation of steroid hormones have been studied in much more detail. The primary organ involved is the liver, although some metabolism also takes place in the kidneys. Complete steroid metabolism generally involves a combination of one or more of five general classes of reactions: reduction, hydroxylation, side-chain cleavage, oxidation, and esterification. Reduction reactions are the principal reactions involved in the conversion of biologically active steroids to forms that possess little or no activity. Esterification (or conjugation) reactions are also particularly important. Groups added in esterification reactions are primarily glucuronate and sulfate. The addition of such charged moieties enhances the water solubility of the metabolites, facilitating their excretion. Steroid metabolites are eliminated from the body primarily via the urine, although smaller amounts also enter the bile and leave the body in the feces.

At times, quantitative information concerning the rate of hormone metabolism is clinically useful. One index of the rate at which a hormone is removed from the blood is the **metabolic clearance rate (MCR)**. The metabolic clearance of a hormone is analogous to that of renal clearance (see Chapter 22, "Kidney Function"). The MCR is the volume of plasma cleared of the hormone in question per unit time. It is calculated from the equation:

$$\text{MCR} = \frac{\text{Hormone removed per unit time (mg/min)}}{\text{Plasma concentration (mg/mL)}} \quad (1)$$

and is expressed in mL plasma/min.

One approach to measuring MCR involves injecting a small amount of radioactive hormone into the subject and then collecting a series of timed blood samples to determine the amount of radioactive hormone remaining. Based on the rate of disappearance of hormone from the blood, its half-life and MCR can be calculated. The MCR and half-life are inversely related—the shorter the half-life, the greater the MCR. The half-lives of different hormones vary considerably, from 5 minutes or less for some to several hours for others. The circulating concentration of hormones with short half-lives can vary dramatically over a short period of time. This is typical of hormones that regulate processes on an acute minute-to-minute basis, such as many of those involved in regulating blood glucose. Hormones for which rapid changes in concentration are not required, such as those with seasonal variations and those that regulate the menstrual cycle, typically have longer half-lives.

Hormonal concentration is a measure of its biologic response.

The concentration of hormone present in a biologic fluid is often measured to make a clinical diagnosis of a suspected endocrine disease or to study basic endocrine physiology. Most hormones are measured in blood or urine, but alternate testing sources, such as saliva and transdermal membrane monitors, are emerging as new tools in endocrinology. Possibly the greatest contribution to recent advances in the field of endocrinology has derived from immunochemical methods.

Bioassay measurements

Even before hormones were chemically characterized, they were quantitated in terms of biologic responses they produced. Thus, early assays for measuring hormones were bioassays that depended on a hormone's ability to produce a characteristic biologic response. As a result, hormones came to be quantitated in terms of units, defined as an amount sufficient to produce a response of specified magnitude under a defined set of conditions. A unit of hormone is, thus, arbitrarily determined. Although bioassays are rarely used today for diagnostic purposes, many hormones are still standardized in terms of **biologic activity units.** For example, commercial insulin is still sold and dispensed based on the number of units in a particular preparation, rather than by the weight or the number of moles of insulin.

Bioassays in general suffer from a number of shortcomings, including a relative lack of specificity and a lack of sensitivity. In many cases, they are slow and cumbersome to perform, and often they are expensive, because biologic variability often requires the inclusion of many animals in the assay.

Radioimmunoassay

Development of the **radioimmunoassay (RIA)** in the late 1950s and early 1960s was a major step forward in clinical and research endocrinology. Much of our current knowledge of endocrinology is based on this method. An RIA or closely related assay is now available for virtually every known hormone. In addition, RIAs have been developed to measure circulating concentrations of many other biologically relevant proteins, drugs, and vitamins.

The RIA is a prototype for a larger group of assays termed **competitive binding assays.** These are modifications and adaptations of the original RIA, relying to a large degree on the principle of competitive binding on which the RIA is based. It is beyond the scope of this text to describe in detail the competitive binding assays currently used to measure hormone concentrations, but the principles are the same as those for the RIA.

The two key components of an RIA are a specific antibody (Ab) that has been raised against the hormone in question and a radioactively labeled hormone (H*). If the hormone being measured is a peptide or protein, the molecule is commonly labeled with a radioactive iodine atom (^{125}I or ^{131}I) that can be readily attached to tyrosine residues of the

From Bench to Bedside / 30.1

Artificial Pancreas

Of great importance to the practicing physician is the increased number of patients who have **diabetes mellitus.** As described in Chapter 34, "Endocrine Pancreas," diabetes mellitus is an endocrine disorder of metabolic dysregulation, most notably a dysregulation of glucose metabolism. In patients with type 1 diabetes, control of blood glucose may prevent long-term complications. Unfortunately, achieving near-normal glucose control is difficult, and the incidence of hypoglycemia increases when glucose control approaches normal glucose levels. Over the years, scientists have been trying to develop an "artificial pancreas." Now, with recent advances in *insulin pump* and *continuous glucose monitoring* technologies, the road has been paved for the development of an artificial pancreas system.

This mechanical closed-loop system would determine glucose levels on a continuous basis, via subcutaneous sensors, and transmit data to an insulin pump. Future artificial pancreas systems may have physiologic sensors that determine the onset of a meal, determine the level and duration of physical activity, and measure the concentration of insulin in the blood. These additional measures will advance the computer algorithm contained in the pump to provide the appropriate rate of insulin delivery at the right time.

Moreover, new artificial pancreas systems under development include multiple pancreatic peptide hormones, because the healthy endocrine pancreas controls blood glucose through the balanced and coordinated release of insulin, glucagon, amylin, somatostatin, and other pancreatic hormones (see Chapter 34, "Endocrine Pancreas"). Along these lines, development of a closed-loop system that delivers both insulin and glucagon as directed by continuous blood glucose readings has been shown to provide a feasible way to control blood glucose and avoid treatment-requiring hypoglycemia. Because amylin is normally cosecreted with insulin and functionally enhances insulin action and continuous delivery of somatostatin lowers the insulin requirement for diabetic subjects, coadministration of amylin and/or somatostatin with reduced insulin dosing might also be used to decrease hypoglycemic risk.

peptide chain. For substances lacking tyrosine residues, such as steroids, labeling may be accomplished by incorporating radioactive carbon (^{14}C) or hydrogen (^{3}H). In either case, the use of the radioactive hormone permits detection and quantification of small amounts of the substance.

The RIA is performed *in vitro* using a series of test tubes. Fixed amounts of Ab and of H* are added to all tubes (Fig. 30.7A). Samples (plasma, urine, cerebrospinal fluid, etc.) to be measured are added to individual tubes. Varying known concentrations of unlabeled hormone (the standards) are added to a series of identical tubes. The principle of the RIA, as indicated in Figure 30.7B, is that labeled and unlabeled hormones compete for a few antibody-binding

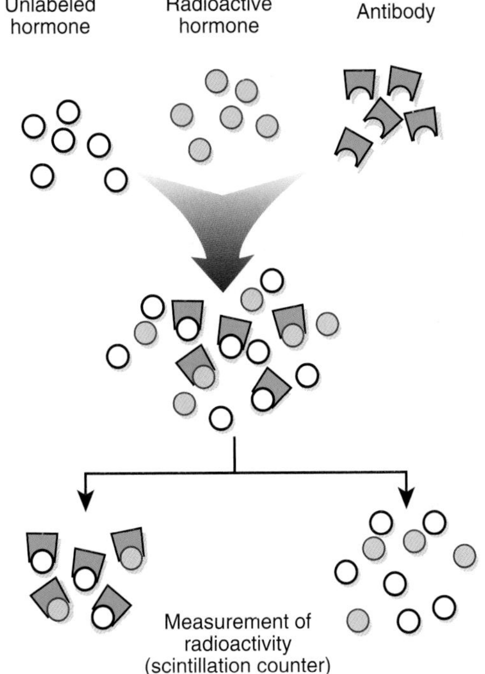

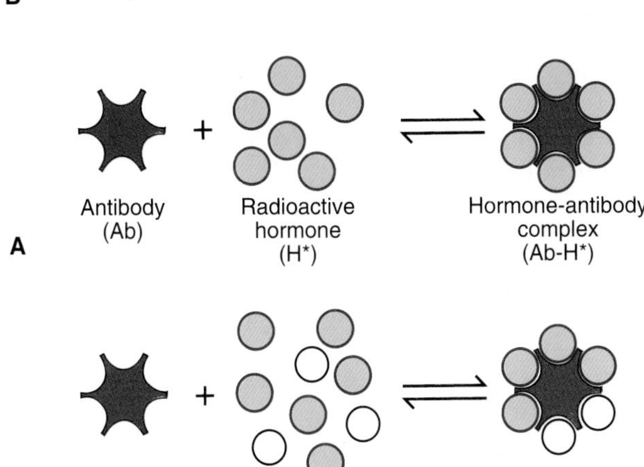

Figure 30.7 The principles of radioimmunoassay (RIA). (A) Specific antibodies (Ab) bind with radioactive hormone (H*) to form hormone–antibody complexes (Ab-H*). **(B)** When unlabeled hormone (*open circles*) is also introduced into the system, less radioactive hormone binds to the antibody.

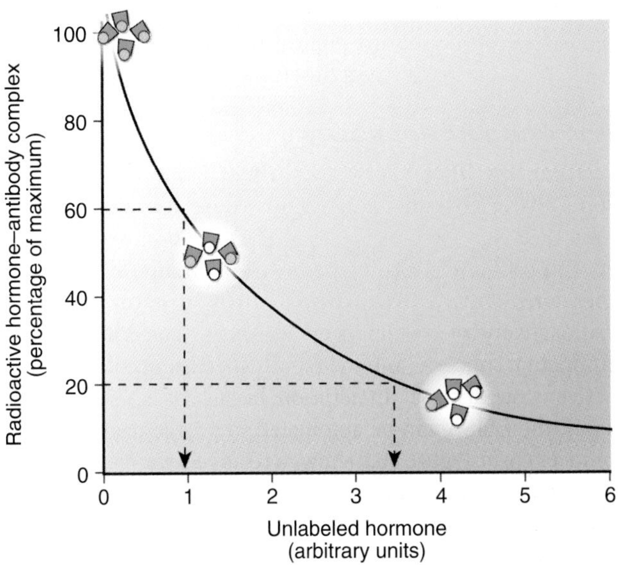

Figure 30.8 A typical radioimmunoassay standard curve. As indicated by the dashed lines, the hormone concentration in unknown samples can be deduced from the standard curve.

sites. The amount of each hormone that is bound to antibody is a proportion of that present in solution. In a sample containing a high concentration of hormone, less radioactive hormone is able to bind to the antibody, and in a sample containing a low concentration of hormone, more radioactive hormone is able to bind to the antibody. In each case, the amount of radioactivity present as antibody-bound H* is determined. The response produced by the standards is used to generate a **standard curve** (Fig. 30.8). Responses produced by the unknown samples are then compared with the standard curve to determine the amount of hormone present in the unknowns (see *dashed lines* in Fig. 30.8).

One major limitation of RIAs is that they measure immunoreactivity rather than biologic activity. The presence of an immunologically related but different hormone or the presence of heterogeneous forms of the same hormone can complicate the interpretation of the results. For example, POMC, the precursor of **adrenocorticotropic hormone (ACTH)**, is often present in high concentrations in the plasma of patients with bronchogenic carcinoma. Antibodies for ACTH may cross-react with POMC. The results of an RIA for ACTH in which such an antibody is used may suggest high concentrations of ACTH when actually POMC is being detected. Because POMC has of the biologic potency of ACTH, there may be little clinical evidence of significantly elevated ACTH. If appropriate measures are taken, however, such possible pitfalls can be overcome in most cases and reliable results from the RIA can be obtained.

One important modification of the RIA is the **radioreceptor assay**, which uses specific hormone receptors rather than antibodies as the hormone-binding reagent. In theory, this method measures biologically active hormone because receptor binding rather than antibody recognition is assessed. However, the need to purify hormone receptors and the somewhat more complex nature of this assay limit its usefulness for routine clinical measurements. It is more likely to be used in a research setting.

Immunosorbent assay

The **enzyme-linked immunosorbent assay (ELISA)** is a solid-phase, enzyme-based assay whose use and application have increased considerably over the past two decades. A typical ELISA is a colorimetric or fluorometric assay, and therefore, the ELISA, unlike the RIA, does not produce radioactive waste, which is an advantage, considering environmental concerns and the rapidly increasing cost of radioactive waste disposal. In addition, because it is a solid-phase assay, the ELISA can be automated to a large degree, which reduces costs. Figure 30.9 shows a relatively simple version of an ELISA. More complex assays using similar principles have been developed to overcome a variety of technical problems, but the basic principle remains the same. In recent years the RIA has been the primary assay used clinically; its use has expanded considerably, and it will likely be the predominant assay in the future because of the advantages listed above.

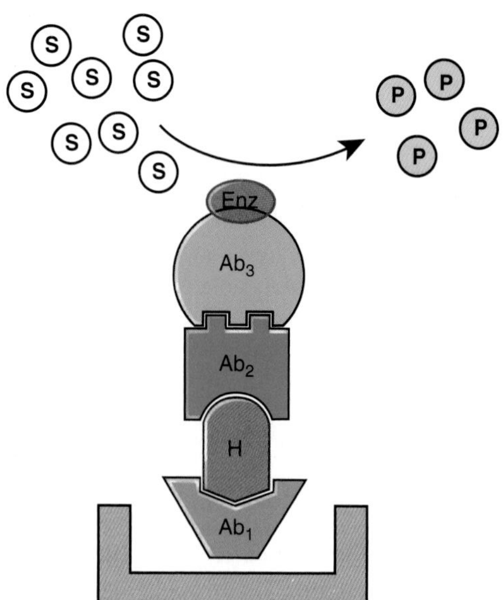

Figure 30.9 The basic components of an enzyme-linked immunosorbent assay (ELISA). A typical ELISA is performed in a 3 × 5-in. plastic plate containing 96 small wells. Each well is precoated with an antibody (Ab_1) that is specific for the hormone (H) being measured. Unknown samples or standards are introduced into the wells, followed by a second hormone-specific antibody (Ab_2). A third antibody (Ab_3), which recognizes Ab_2, is then added. Ab_3 is coupled to an enzyme (Enz) that will convert an appropriate substrate (S) into a colored or fluorescent product (P). The amount of product formed can be determined using optical methods. After the addition of each antibody or sample to the wells, the plates are incubated for an appropriate period of time to allow antibodies and hormones to bind. Any unbound material is washed out of the well before the addition of the next reagent. The amount of colored product formed is directly proportional to the amount of hormone present in the standard or unknown sample. Concentrations are determined using a standard curve. For simplicity, only one Ab_1 molecule is shown in the bottom of the well when, in fact, there is an excess of Ab_1 relative to the amount of hormone to be measured.

MECHANISMS OF HORMONE ACTION

As indicated earlier, hormones are one mechanism by which cells communicate with one another. Fidelity of communication in the endocrine system depends on each hormone's ability to interact with a specific receptor in its target tissues. This interaction results in the activation (or inhibition) of a series of specific events in cells that results in precise biologic responses characteristic of that hormone.

The binding of a hormone to its receptor with subsequent activation of the receptor is the first step in hormone action and also the point at which specificity is determined within the endocrine system. The second step in hormone action entails the receptor-mediated transduction of the extracellular message into an intracellular signal or second messenger. Many intracellular signal cascades exist (recall the discussion of cellular signaling in Chapter 1, "Homeostasis and Cellular Signaling"). Abnormal interactions of hormones with their receptors and/or disturbances in cell signaling are involved in the pathogenesis of many endocrine disease states, and therefore, considerable attention has been paid to these aspects of hormone action. Here discussion is confined to key principles of hormone–receptor interaction.

Biologic response is partly determined by hormone–receptor binding kinetics.

The probability that a hormone–receptor interaction will occur is related to both the abundance of cellular receptors and the receptor's affinity for the hormone relative to the ambient hormone concentration. The more receptors available to interact with a given amount of hormone, the greater the likelihood there is of a response. Similarly, the higher the affinity of a receptor for the hormone, the greater the likelihood that an interaction will occur. The circulating hormone concentration is, of course, a function of the rate of hormone secretion relative to hormone degradation.

The association of a hormone with its receptor generally behaves as if it were a simple, reversible chemical reaction that can be described by the following kinetic equation:

$$[H]+[R] \rightleftharpoons [HR] \quad (2)$$

where [H] is the free hormone concentration, [R] is the unoccupied receptor concentration, and [HR] is the *hormone–receptor complex* (also referred to as *bound hormone or occupied receptor*). Assuming a simple chemical equilibrium, it follows that

$$K_a = [HR]/[H]\times[R] \quad (3)$$

where K_a is the association constant. If R_0 is defined as the total receptor number (i.e., [R] + [HR]), then after substituting and rearranging, we obtain the following relationship:

$$[HR]/[H] = -K_a[HR] + K_aR_0 \quad (4)$$

Literally translated, this equation states

$$\frac{\text{Bound hormone}}{\text{Free hormone}} = -K_a \times \text{Bound hormone} + K_a \times \text{Total receptor number} \quad (5)$$

Notice that equations 4 and 5 have the general form of an equation for a straight line: y = mx + b.

To obtain information regarding a particular hormone–receptor system, a fixed number of cells (and, therefore, a fixed number of receptors) are incubated *in vitro* in a series of test tubes with increasing amounts of hormone. At each higher hormone concentration, the amount of receptor-bound hormone is increased until hormone occupies all receptors. Receptor number and affinity can be obtained by using the relationships given in equation 5 above and plotting the results as the ratio of receptor-bound hormone to free hormone ([HR]/[H]) as a function of the amount of bound hormone ([HR]). This type of analysis is known as a **Scatchard plot** (Fig. 30.10). In theory, a Scatchard plot of simple, reversible equilibrium binding is a straight line (see Fig. 30.10A), with the slope of the line being equal to the negative of the association constant ($-K_a$) and the x-intercept being equal to the total receptor number (R_0). Other equally valid mathematical and graphic methods can be used to analyze hormone–receptor interactions, but the Scatchard plot is probably the most widely used.

In practice, Scatchard plots are not always straight lines but instead can be curvilinear (see Fig. 30.10B). Insulin is a classic example of a hormone that gives curved Scatchard plots. One interpretation of this result is that cells contain two separate and distinct classes of receptors, each with a different binding affinity. Typically, one receptor population has a higher affinity but is fewer in number compared with the second population. Therefore, as indicated in Figure 30.10B, $K_{a_1} > K_{a_2}$ but $R_{0_2} > R_{0_1}$. Computer analysis is often required to fit curvilinear Scatchard plots accurately to a two-site model.

Another explanation for curvilinear Scatchard plots is that occupied receptors influence the affinity of adjacent, unoccupied receptors by **negative cooperativity**. According to this theory, when one hormone molecule binds to its receptor, it causes a decrease in the affinity of nearby unoccupied receptors, making it more difficult for additional hormone molecules to bind. The greater the amount of hormone bound, the lower the affinity of unoccupied receptors. Therefore, as shown in Figure 30.10B, as bound hormone increases, the affinity (slope) steadily decreases. Whether curvilinear Scatchard plots in fact result from two-site receptor systems or from negative cooperativity between receptors is unknown.

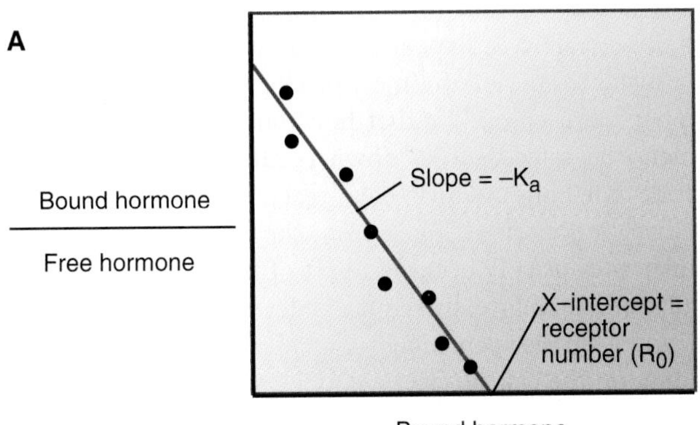

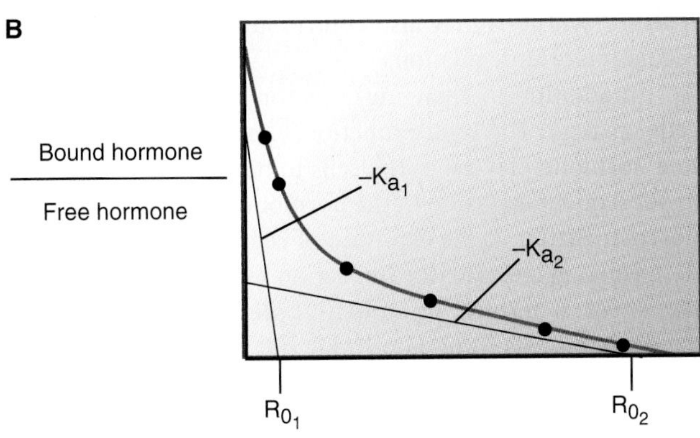

Figure 30.10 Scatchard plots of hormone-receptor binding data. (A) A straight-line plot typical of hormone binding to a single class of receptors. **(B)** A curvilinear Scatchard plot typical of some hormones. Several models have been proposed to account for nonlinearity of Scatchard plots. K_a, association constant; R_0, total receptor number. (See text for details.)

Dose–response curves determine changes in responsiveness and sensitivity.

Hormone effects are generally not all-or-none phenomena—that is, they generally do not switch from totally off to totally on and then back again. Instead, target cells exhibit graded responses proportional to the concentration of free hormone present.

The dose–response relationship for a hormone generally exhibits a sigmoid shape when plotted as the biologic response on the *y* axis versus the log of the hormone concentration on the *x* axis (Fig. 30.11). Regardless of the biologic pathway or process being considered, cells typically exhibit an intrinsic *basal level* of activity in the absence of added hormone, even well after any previous exposure to hormone. As the hormone concentration surrounding the cells increases, a minimal **threshold concentration** must be present before any measurable increase in the cellular response can be produced. At higher hormone concentrations, a **maximal response** by the target cell is produced, and increasing the hormone concentration cannot elicit greater response. The concentration of hormone required to produce a response half-way between the maximal and basal responses, the **median effective dose (ED_{50})**, is a useful index of the *sensitivity* of the target cell for that particular hormone (see Fig. 30.11).

For some peptide hormones, the maximal response may occur when only a small percentage (5%–10%) of the total receptor population is occupied by hormone. The remaining 90% to 95% of the receptors are called *spare receptors*, because on initial inspection they do not appear necessary to produce a maximal response. This term is unfortunate, because the receptors are not "spare" in the sense of being unused. Although at any one point in time only 5% to 10% of the

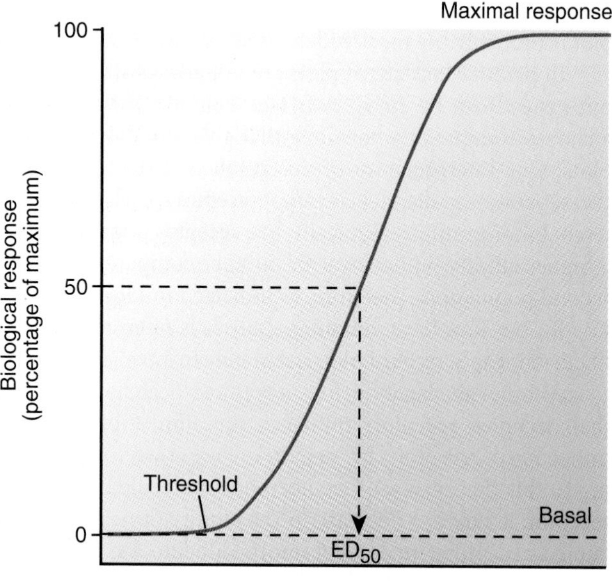

Figure 30.11 A normal dose–response curve of hormone activity. ED_{50}, median effective dose.

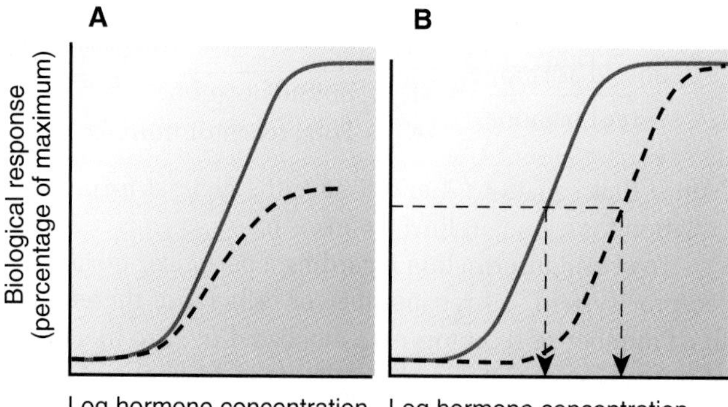

Figure 30.12 Altered target tissue responses reflected by dose–response curves. (A) Decreased target tissue responsiveness. **(B)** Decreased target tissue sensitivity.

receptors may be occupied, hormone–receptor interactions are an equilibrium process and hormones continually dissociate and reassociate with their receptors. Therefore, from one point in time to the next, different subsets of the total population of receptors may be occupied, but presumably all receptors participate equally in producing the biologic response.

Physiologic or pathophysiologic alterations in target tissue responses to hormones can take one of two general forms, as indicated by changes in the hormones' dose–response curves (Fig. 30.12). Although changes in dose–response curves are not routinely assessed in the clinical setting, they can serve to distinguish between a *receptor defect* and a *postreceptor defect* in hormone action, providing useful information regarding the underlying cause of a particular disease state. Changes in responsiveness are indicated by an increase or decrease in the maximal response of the target tissue and may be the result of one or more factors (see Fig. 30.12A). Altered responsiveness can be caused by a change in the number of functional target cells in a tissue, by a change in the number of receptors per cell for the hormone in question or, if receptor function itself is not rate-limiting for hormone action, by a change in the specific rate-limiting postreceptor step in the hormone action pathway.

A change in sensitivity is reflected as a right or left shift in the dose–response curve and, thus, a change in the ED_{50}; a right shift indicates decreased sensitivity and a left shift indicates increased sensitivity for that hormone (see Fig. 30.12B). Changes in sensitivity reflect (1) an alteration in receptor affinity or, if submaximal concentrations of hormone are present, (2) a change in receptor number. Dose–response curves may also reflect combinations of changes in responsiveness and sensitivity in which there is both a right or left shift of the curve (a sensitivity change) and a change in maximal biologic response to a lower or higher level (a change in responsiveness).

Cells can regulate their receptor number and/or function in several ways. Exposing cells to an excess of hormone for a sustained period of time typically results in a decreased number of receptors for that hormone per cell. This phenomenon is referred to as **down-regulation**. In the case of peptide hormones, which have receptors on cell surfaces, a redistribution of receptors from the cell surface to intracellular sites usually occurs as part of the process of down-regulation. Therefore, there may be fewer total receptors per cell, and a smaller percentage may be available for hormone binding on the cell surface. Although somewhat less prevalent than down-regulation, **upregulation** may occur when certain conditions or treatments cause an increase in receptor number compared with normal. Changes in rates of receptor synthesis may also contribute to long-term down-regulation or upregulation.

In addition to changing receptor number, many target cells can regulate receptor function. Chronic exposure of cells to a hormone may cause the cells to become less responsive to subsequent exposure to the hormone by a process termed **desensitization**. If the exposure of cells to a hormone has a desensitizing effect on further action by that same hormone, the effect is termed **homologous desensitization**. If the exposure of cells to one hormone has a desensitizing effect with regard to the action of a different hormone, the effect is termed **heterologous desensitization**.

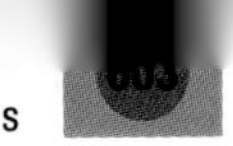

Chapter Summary

- The endocrine system integrates organ function via hormones that are secreted from classical endocrine glands and also from organs whose primary function is not endocrine.
- Hormones may signal to the cells that produced them (autocrine) or neighboring cells (paracrine) but are classically defined as blood-borne chemicals that signal to distant target tissues.
- Target cell recognition of hormones depends on specific, high-affinity receptors, which may be located on the cell surface, within the cytoplasm, or in the target cell's nucleus.
- Hormonal signals are organized in a hierarchy of feedback systems, millionfold amplification cascades, and often definable secretion patterns.
- Most hormones display pleiotropic effects and share their ability to control vital physiologic parameters with other hormones.
- Chemically, hormones may be metabolites of single amino acids, peptides, or metabolites of cholesterol and, depending on their solubility, are either transported in the bloodstream free (amine and peptide hormones) or bound to transport proteins (steroid and thyroid hormones).
- Radioimmunoassay and enzyme-linked immunosorbent assay have provided major advancements in the field of endocrinology, but each type of assay has limitations.
- Disturbances in cell signaling and abnormal interactions of hormones with their receptors may lead to endocrine abnormalities.

thePoint *Visit http://thePoint.lww.com for chapter review Q&A, case studies, animations, and more!*

31 Hypothalamus and the Pituitary Gland

ACTIVE LEARNING OBJECTIVES

Upon mastering the material in this chapter, you should be able to:

- Explain how the functional anatomy of the hypothalamic-pituitary axis results in different mechanisms of synthesis for posterior pituitary hormones (arginine vasopressin, oxytocin) versus anterior pituitary hormones (adrenocorticotropic hormone, thyroid-stimulating hormone, growth hormone, follicle-stimulating hormone, luteinizing hormone, and prolactin).
- Explain how arginine vasopressin (AVP) acts on water reabsorption in the kidneys and predict the AVP secretory response to a rise and fall in blood osmolality.
- Explain the smooth muscle response to oxytocin in the breast and uterus, and predict the effect of suckling or cervical dilation on oxytocin secretion.
- Explain the response of pituitary corticotrophs to corticotropin-releasing hormone, and predict the effect of adrenocorticotropic hormone to regulate glucocorticoid release from the adrenal cortex in response to physical or emotional stress, arginine vasopressin, the sleep–wake cycle, and an increase in glucocorticoid in the blood.
- Explain the response of pituitary thyrotrophs to thyrotropin-releasing hormone, and predict the effect of thyroid-stimulating hormone to regulate triiodothyronine and thyroxine release from the thyroid follicles in response to cold temperatures, the sleep–wake cycle, and an increase in thyroid hormone in the blood.
- Describe the hypothalamic-pituitary-growth hormone axis, explain the response of pituitary somatotrophs to growth hormone–releasing hormone and somatotropin release–inhibiting factor, and predict the effect of growth hormone (GH), insulin-like growth factor, aging, deep sleep, stress, exercise, and hypoglycemia on GH secretion.
- Describe the hypothalamic-pituitary-gonadal axis, and explain the response of pituitary gonadotrophs to luteinizing hormone–releasing hormone.
- Explain the effect of estrogen and dopamine on pituitary lactotrophs and the effect of prolactin on alveolar cells in the mammary gland.

The **pituitary gland** or **hypophysis** is an endocrine gland that protrudes off of the bottom of the hypothalamus at the base of the brain. The pituitary gland is composed of two lobes: The anterior pituitary and the posterior pituitary, and both lobes are functionally linked to the hypothalamus by the pituitary stalk. The hypothalamus releases hypothalamic-releasing factors that descend down the pituitary stalk and into the pituitary gland where they stimulate the release of pituitary hormones. Although the pituitary gland is sometimes referred to as the "*master gland*," both of its lobes are under hypothalamic control.

The pituitary gland secretes an array of peptide hormones that have important actions on almost every aspect of body function. Some pituitary hormones influence key cellular processes involved in preserving the volume and composition of body fluids. Others bring about changes in body function, which enable the person to grow, reproduce, and respond appropriately to stress and trauma. The pituitary hormones produce these physiologic effects by either acting directly on their target cells or stimulating other endocrine glands to secrete hormones, which, in turn, bring about changes in body function.

Stimuli that affect the secretion of pituitary hormones may originate within or outside the body. The brain, which signals the pituitary gland to increase or decrease the rate of secretion of a particular hormone, perceives and processes these stimuli. Thus, the brain links the pituitary gland to events occurring within or outside the body, which call for changes in pituitary hormone secretion. This important functional connection between the brain and the pituitary, in which the hypothalamus plays a central role, is called the **hypothalamic–pituitary axis**.

HYPOTHALAMIC-PITUITARY AXIS

The human pituitary is composed of two morphologically and functionally distinct glands connected to the hypothalamus. The pituitary gland is located at the base of the brain and is connected to the hypothalamus by a stalk. It sits in a depression in the sphenoid bone of the skull called the **sella turcica**. The two morphologically and functionally distinct glands comprising the human pituitary are the adenohypophysis and the neurohypophysis as shown in Figure 31.1. The

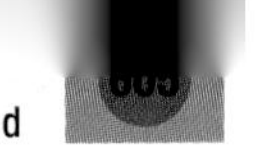

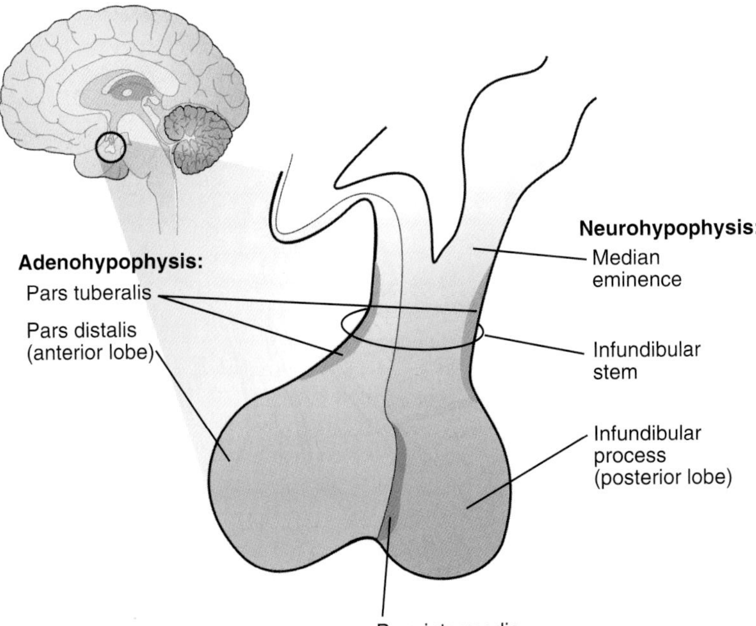

Figure 31.1 A midsagittal section of the human pituitary gland. The neurohypophysis is composed of the median eminence, the infundibular stem, and the posterior lobe. The adenohypophysis is composed of the pars tuberalis and the anterior lobe.

adenohypophysis, or *anterior pituitary*, comprises the **pars tuberalis**, which forms the outer covering of the pituitary stalk, and the **pars distalis** or **anterior lobe**. The **neurohypophysis**, or *posterior pituitary*, comprises the **infundibular stem**, which forms the inner part of the stalk, the **infundibular process** or **posterior lobe**, and the **median eminence** of the hypothalamus (though some do not consider this a part of the pituitary gland). In most vertebrates, the pituitary contains a third anatomically distinct lobe, the **pars intermedia** or **intermediate lobe**. In adult humans, only a vestige of the intermediate lobe is found, consisting of a thin, diffuse region of cells between the anterior and posterior lobes.

The adenohypophysis and neurohypophysis have different embryologic origins. The adenohypophysis is formed from an evagination of the oral ectoderm called the **Rathke pouch**. The neurohypophysis forms as an extension of the developing hypothalamus, which fuses with the Rathke pouch as development proceeds. The posterior lobe is, therefore, composed of neural tissue and is a functional part of the hypothalamus.

Hypothalamic neurons whose axons terminate in the posterior lobe synthesize posterior pituitary hormones.

The infundibular stem of the pituitary gland contains bundles of nonmyelinated nerve fibers, which terminate on the capillary bed in the posterior lobe. These fibers are the axons of neurons that originate in the **supraoptic nuclei** and **paraventricular nuclei** of the hypothalamus. The cell bodies of these neurons are large compared with those of other hypothalamic neurons; hence, they are called **magnocellular neurons**. The hormones **arginine vasopressin (AVP)** and **oxytocin** are synthesized as parts of larger precursor proteins (prohormones) in the cell bodies of these neurons. Prohormones are then packaged into granules and enzymatically processed to produce AVP and oxytocin. Axoplasmic flow transports the granules down the axons; they accumulate at the axon terminals in the posterior lobe.

Events occurring within or outside the body may generate stimuli for the secretion of posterior lobe hormones. The central nervous system (CNS) processes these stimuli, and the signal for the secretion of AVP or oxytocin is then transmitted to neurosecretory neurons in the hypothalamus. Secretory granules containing the hormone are then released into the nearby capillary circulation, from which the hormone is carried into the systemic circulation.

Anterior pituitary hormones are synthesized and secreted in response to hypothalamic-releasing hormones carried in the hypophyseal portal circulation.

The anterior lobe contains clusters of histologically distinct types of cells closely associated with blood sinusoids that drain into the venous circulation. These cells produce anterior pituitary hormones and secrete them into the blood sinusoids. Distinctly different types of cells produce the six well-known anterior pituitary hormones. **Adrenocorticotropic hormone (ACTH)**, also known as **corticotropin**, is secreted by corticotrophs, **thyroid-stimulating hormone (TSH)** by **thyrotrophs, growth hormone (GH)** by **somatotrophs, prolactin (PRL)** by **lactotrophs**, and **follicle-stimulating hormone (FSH)** and **luteinizing hormone (LH)** by **gonadotrophs**.

The cells that produce anterior pituitary hormones are not innervated and, therefore, are not under direct neural control. Rather, **releasing hormones** called **hypophysiotropic hormones**, synthesized by neural cell bodies in the hypothalamus, regulate their secretory activity. Granules containing releasing hormones are stored in the axon terminals of these neurons, located in capillary networks in the median eminence of the hypothalamus and lower infundibular stem. These capillary networks give rise to the principal blood supply to the anterior lobe of the pituitary.

The blood supply to the anterior pituitary is shown in Figure 31.2. The superior and inferior hypophyseal arteries bring blood to the hypothalamic–pituitary region. The **superior hypophyseal arteries** give rise to a rich capillary network in the median eminence. The capillaries converge into long veins that run down the pituitary stalk and empty into the blood sinusoids in the anterior lobe. They are considered to be portal veins because they deliver blood to the anterior pituitary rather than joining the venous circulation that carries blood back to the heart; therefore, they are called **long hypophyseal portal vessels.** The **inferior hypophyseal arteries** provide arterial blood to the posterior lobe. They also penetrate into the lower infundibular stem, where they form another important capillary network. The capillaries of this network converge into **short hypophyseal portal vessels,** which also deliver blood into the sinusoids of the anterior pituitary. The special blood supply to the anterior lobe of the pituitary gland is known as the **hypophyseal portal circulation.**

When a neurosecretory neuron is stimulated to secrete, the releasing hormone is discharged into the hypophyseal portal circulation (see Fig. 31.2). Releasing hormones travel only a short distance before they come in contact with their target cells in the anterior lobe. Neurosecretory neurons only deliver the amount of releasing hormone to hypophyseal portal circulation that is needed to control anterior pituitary hormone secretion. Consequently, releasing hormones are almost undetectable in systemic blood.

A releasing hormone either stimulates or inhibits the synthesis and secretion of a particular anterior pituitary hormone. **Corticotropin-releasing hormone (CRH), thyrotropin-releasing hormone (TRH),** and **GH-releasing hormone (GHRH)** stimulate the synthesis and secretion of ACTH, TSH, and GH, respectively (Table 31.1). **LH-releasing hormone (LHRH),** also known as **gonadotropin-releasing hormone (GnRH),** stimulates the synthesis and release of FSH and LH. In contrast, **somatostatin,** also called **somatotropin release–inhibiting factor (SRIF),** inhibits GH secretion. All of the releasing hormones are peptides, with the exception of **dopamine,** which is a catecholamine that inhibits the synthesis and secretion of PRL. Releasing hormones can be produced synthetically and used in the diagnosis and treatment of diseases of the endocrine system. For example, synthetic GnRH is now used for treating infertility in women.

Releasing hormones are secreted in response to neural inputs from other areas of the CNS. External events that affect the body or changes occurring within the body itself generate these signals. For example, sensory nerve excitation, emotional or physical stress, biologic rhythms, changes in sleep patterns or in the sleep–wake cycle, and changes in circulating levels of certain hormones or metabolites all affect the secretion of particular anterior pituitary hormones. Signals generated in the CNS by such events are transmitted to the neurosecretory neurons in the hypothalamus. Depending on the nature of the event and the signal generated, the secretion of a particular releasing hormone may be either stimulated or inhibited. In turn, this response affects the rate of secretion of the appropriate anterior pituitary hormone.

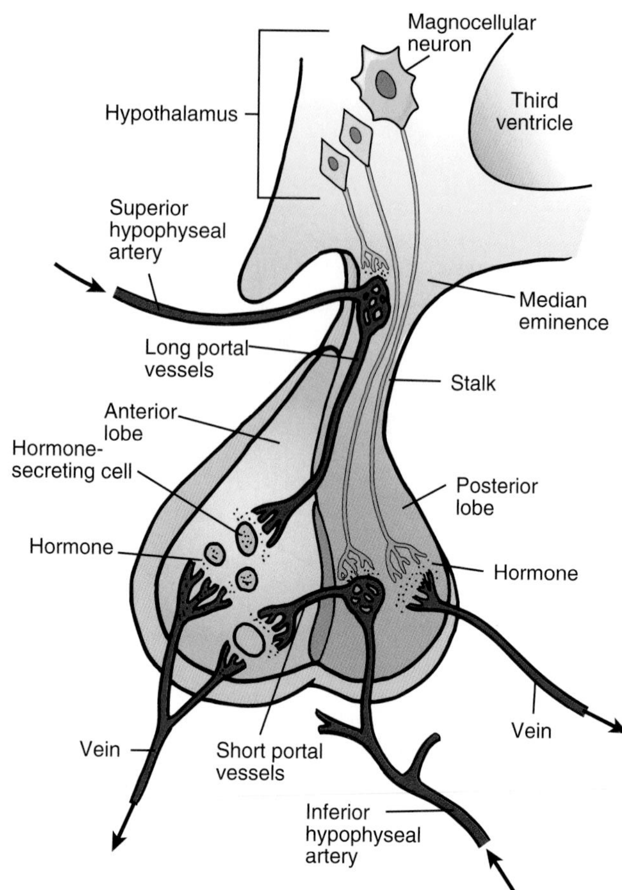

Figure 31.2 The blood supply to the anterior pituitary. This illustration shows the relationship of the pituitary blood supply to hypothalamic magnocellular neurons and to hypothalamic neurosecretory cells that produce releasing hormones. The magnocellular neuron (large yellow cell body) releases arginine vasopressin or oxytocin at its axon terminals into capillaries that give rise to the venous drainage of the posterior lobe. The neurons with smaller cell bodies are secreting releasing factors into capillary networks that give rise to the long and short hypophyseal portal vessels, respectively. Releasing hormones are shown reaching the hormone-secreting cells of the anterior lobe via the portal vessels.

The neural pathways involved in transmitting these signals to the neurosecretory neurons in the hypothalamus are not well defined.

POSTERIOR PITUITARY HORMONES

Magnocellular neurons in the supraoptic and paraventricular nuclei of the hypothalamus produce AVP (also known historically as *antidiuretic hormone*) and oxytocin. Individual neurons make either AVP or oxytocin, but not both. The axons of these neurons form the infundibular stem and terminate on the capillary network in the posterior lobe, where they discharge AVP and oxytocin into the systemic circulation.

AVP and oxytocin are closely related small peptides, each consisting of nine amino acid residues. Different mammals make two forms of vasopressin, one containing arginine

TABLE 31.1 Hypothalamic-Releasing Hormones

Hormone	Chemistry	Actions on Anterior Pituitary
Corticotropin-releasing hormone (CRH)	Single chain of 41 amino acids	Stimulates ACTH secretion by corticotrophs; stimulates expression of POMC gene in corticotrophs
Thyrotropin-releasing hormone (TRH)	Peptide of three amino acids	Stimulates TSH secretion by thyrotrophs; stimulates expression of genes for α and β subunits of TSH in thyrotrophs; stimulates PRL synthesis by lactotrophs
Growth hormone–releasing hormone (GHRH)	Two forms in human: single chain of 44 amino acids, single chain of 40 amino acids	Stimulates GH secretion by somatotrophs; stimulates expression of GH gene in somatotrophs
Luteinizing hormone–releasing hormone (LHRH)	Single chain of 10 amino acids	Stimulates FSH and LH secretion by gonadotropin-releasing hormone (GnRH) gonadotrophs
Somatostatin, somatotropin release–inhibiting factor (SRIF)	Single chain of 14 amino acids	Inhibits GH secretion by somatotrophs; inhibits TSH secretion by thyrotrophs
Dopamine	Catecholamine	Inhibits PRL synthesis and secretion by lactotrophs

ACTH, adrenocorticotropic hormone; FSH, follicle-stimulating hormone; GH, growth hormone; LH, luteinizing hormone; POMC, proopiomelanocortin; PRL, prolactin; TSH, thyroid-stimulating hormone.

and the other containing lysine. AVP is made in humans. Although AVP and oxytocin differ by only two amino acid residues, the structural differences are sufficient to give these two molecules different hormonal activities. They are similar enough, however, for AVP to have slight oxytocic activity and for oxytocin to have slight antidiuretic activity.

The genes for AVP and oxytocin are located near one another on chromosome 20. They code for much larger prohormones that contain the amino acid sequences for AVP or oxytocin and for a 93-amino acid peptide called **neurophysin** (Fig. 31.3). The neurophysin coded by the AVP gene has a slightly different structure than that coded by the oxytocin gene. Neurophysin is important in the processing and secretion of AVP, and mutations in the neurophysin portion of the AVP gene are associated with **hypothalamic** (or **central**) **diabetes insipidus**, a condition in which AVP secretion is impaired. Prohormones for AVP and oxytocin are synthesized in the cell bodies of magnocellular neurons and transported in secretory granules to axon terminals in the posterior lobe, as described earlier. During the passage of the granules from the Golgi apparatus to axon terminals, proteolytic enzymes cleave prohormones to produce AVP or oxytocin and their associated neurophysins.

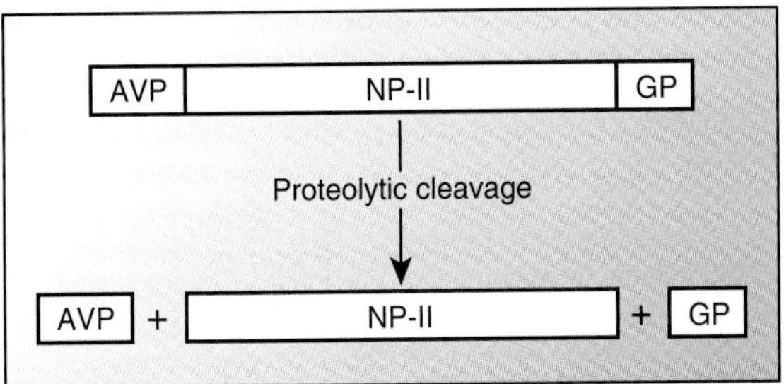

Figure 31.3 The structural organization and proteolytic processing of arginine vasopressin (AVP) from its prohormone. Mutations in neurophysin, which is important in processing AVP, result in diabetes insipidus. NP, neurophysin; GP, glycoprotein.

The hypothalamus and posterior pituitary act as a functional unit in the secretion of AVP and oxytocin. When magnocellular neurons receive neural signals for AVP or oxytocin secretion, action potentials are generated in these cells, triggering the release of AVP or oxytocin and neurophysin from the axon terminals. These substances diffuse into nearby capillaries and then enter the systemic circulation.

Arginine vasopressin increases the reabsorption of water by the kidneys.

Two physiologic signals, a rise in the osmolality of the blood and a decrease in blood volume, generate the CNS stimulus for AVP secretion. Chemical mediators of AVP release include catecholamines, angiotensin II, and **atrial natriuretic peptide (ANP)**. The main physiologic action of AVP is to increase water reabsorption by the collecting ducts of the kidneys. The result is decreased water excretion and the formation of osmotically concentrated urine (see Chapter 22, "Kidney Function"). This action of AVP works to counteract the conditions that stimulate its secretion. For example, reducing water loss in the urine limits a further rise in the osmolality of the blood and conserves blood volume. Low blood AVP levels lead to diabetes insipidus and the excessive production of dilute urine (see Chapter 23, "Regulation of Fluid and Electrolyte Balance").

Oxytocin stimulates the contraction of smooth muscle in the mammary glands and uterus.

Two physiologic signals stimulate the secretion of oxytocin by hypothalamic magnocellular neurons. Breast-feeding stimulates sensory nerves in the nipple. Afferent nerve impulses enter the CNS and eventually stimulate oxytocin-secreting magnocellular neurons. These neurons fire in synchrony and release a bolus of oxytocin into the bloodstream. Oxytocin stimulates the contraction of **myoepithelial cells**, which surround the milk-laden alveoli in the lactating mammary gland, aiding in milk ejection.

Neural input from the female reproductive tract during childbirth also stimulates oxytocin secretion. Cervical dilation before the beginning of labor stimulates stretch receptors in the cervix. Afferent nerve impulses pass through the CNS to oxytocin-secreting neurons. Oxytocin release stimulates the contraction of smooth muscle cells in the uterus during labor, aiding in the delivery of the newborn and placenta. The actions of oxytocin on the mammary glands and the female reproductive tract are discussed further in Chapter 38, "Fertilization, Pregnancy, and Fetal Development."

ANTERIOR PITUITARY HORMONES

The anterior pituitary secretes six protein hormones, all of which are small, ranging in molecular size from 4.5 to 29 kDa. Their chemical and physiologic features are given in Table 31.2.

Four of the anterior pituitary hormones have effects on the morphology and secretory activity of other endocrine glands; they are called *tropic* (Greek, meaning "to turn to") or *trophic* ("to nourish") hormones. For example, ACTH maintains the size of certain cells in the adrenal cortex and stimulates these cells to synthesize and secrete the **glucocorticoid hormones, cortisol**, and **corticosterone**. Similarly, TSH maintains the size of the cells of the thyroid follicles and stimulates these cells to produce and secrete the thyroid hormones **thyroxine (T_4)** and **triiodothyronine (T_3)**. The two other tropic hormones, FSH and LH, are called **gonadotropins** because both act on the ovaries and testes. FSH stimulates the development of follicles in the ovaries and regulates the process of **spermatogenesis** in the testes. LH causes **ovulation** and **luteinization** of the ovulated follicle in the ovary of the human female and stimulates the production of the female sex hormones **estrogen** and **progesterone** by the ovary. In the male, LH stimulates the **Leydig cells** of the testis to produce and secrete the male sex hormone, testosterone.

The two remaining anterior pituitary hormones, GH and PRL, are not usually thought of as tropic hormones because their main target organs are not other endocrine glands. As discussed later, however, these two hormones have certain effects that can be regarded as "tropic." The main physiologic action of GH is its stimulatory effect on the growth of the body during childhood. In humans, PRL is essential for the synthesis of milk by the mammary glands during **lactation.**

The following discussion focuses on ACTH, TSH, and GH. Regulation of the secretion of the gonadotropins and PRL and descriptions of their actions are covered in greater detail in Part X, "Reproductive Physiology" (Chapters 36–38).

Adrenocorticotropic hormone regulates the function of the adrenal cortex.

The adrenal glands sit on top of the kidneys, but, although connected to the kidneys, these glands do not play a role in renal function. Each adrenal gland is separated into two distinct functional units, the adrenal cortex and medulla. The *adrenal cortex* produces the glucocorticoid hormones, cortisol and corticosterone, in the cells of its two inner zones, the **zona fasciculata** and the **zona reticularis**. These cells also synthesize **androgens**, or male sex hormones. The main androgen synthesized is **dehydroepiandrosterone**.

Glucocorticoids alter gene transcription on many different target cells. Glucocorticoids permit metabolic

TABLE 31.2 Hormones of the Anterior Pituitary

Hormone	Chemistry	Physiologic Actions
Adrenocorticotropic hormone (ACTH, corticotropin)	Single chain of 39 amino acids; 4.5 kDa	Stimulates production of glucocorticoids and androgens by adrenal cortex; maintains size of zona fasciculata and zona reticularis of cortex
Thyroid-stimulating hormone (TSH, thyrotropin)	Glycoprotein having two subunits, α and β; 28 kDa	Stimulates production of thyroid hormones, T_4 and T_3, by thyroid follicular cells; maintains size of follicular cells
Growth hormone (GH)	Single chain of 191 amino acids; 22 kDa	Stimulates postnatal body growth; stimulates triglyceride lipolysis; inhibits insulin action on carbohydrate and lipid metabolism
Follicle-stimulating hormone (FSH)	Glycoprotein having two subunits, α and β; 28–29 kDa	Stimulates development of ovarian follicles; regulates spermatogenesis in testes
Luteinizing hormone (LH)	Glycoprotein having two subunits, α and β; 28–29 kDa	Causes ovulation and formation of corpus luteum in ovaries; stimulates production of estrogen and progesterone by ovaries; stimulates testosterone production by testes
Prolactin (PRL)	Single chain of 199 amino acids	Essential for milk production by lactating mammary glands

T_3, triiodothyronine; T_4, thyroxine.

adaptations during fasting, which prevent the development of **hypoglycemia** or low blood glucose level. They also play an essential role in the response of the body to physical and emotional stress. Other actions of glucocorticoids include inhibition of inflammation, suppression of the immune system, and regulation of vascular responsiveness to norepinephrine.

The cells of the outer zone of the cortex, the **zona glomerulosa**, produce **aldosterone**, the other physiologically important hormone made by the adrenal cortex. It acts to stimulate sodium reabsorption by the kidneys.

ACTH is the physiologic regulator of the synthesis and secretion of glucocorticoids by the zona fasciculata and zona reticularis. The smallest of the six anterior pituitary hormones, consisting of a single chain of 39 amino acids with a molecular size of 4.5 kDa, ACTH stimulates the synthesis of these steroid hormones and promotes the expression of the genes for various enzymes involved in steroidogenesis. It also maintains the size and functional integrity of the cells of the zona fasciculata and zona reticularis. ACTH is not an important regulator of aldosterone synthesis and secretion.

The actions of ACTH on glucocorticoid synthesis and secretion and details about the physiologic effects of glucocorticoids are described in Chapter 33, "Adrenal Gland."

Adrenocorticotropic hormone secretion

Glucocorticoids, physical and emotional stress, AVP, and the sleep–wake cycle regulate ACTH secretion.

Proopiomelanocortin

ACTH is synthesized in corticotrophs as part of a larger 30-kDa prohormone, **proopiomelanocortin (POMC)**. Enzymatic cleavage of POMC by **prohormone convertase 1** in the anterior pituitary results in ACTH, an amino terminal protein, and **β-lipotropin** (Fig. 31.4). β-Lipotropin has effects on lipid metabolism, but its physiologic function in humans has not been established. Although POMC can be cleaved into other peptides, such as β-endorphin, only ACTH and β-lipotropin are produced from POMC in the human corticotroph. Proteolytic processing of POMC occurs after it is packaged into secretory granules. Therefore, when the corticotroph receives a signal to secrete, ACTH and β-lipotropin are released into the bloodstream in a 1:1 molar ratio.

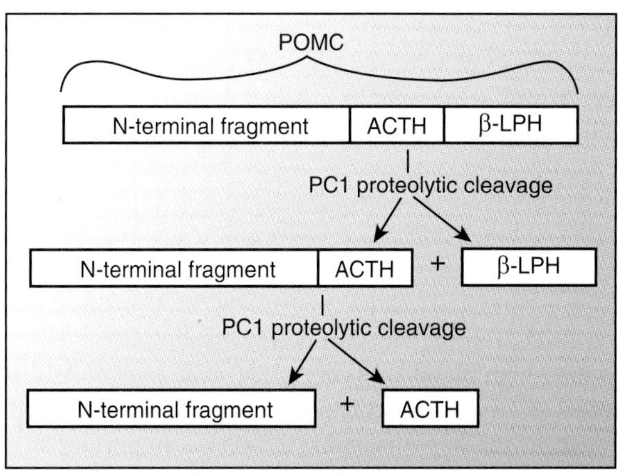

Figure 31.4 The proteolytic processing of adrenocorticotropic hormone (ACTH) from its prohormone proopiomelanocortin (POMC) in the human corticotroph. PC1, prohormone convertase 1; β-LPH, β-lipotropin.

Cells of the intermediate lobe of the pituitary gland and neurons in the hypothalamus also synthesize POMC. In the intermediate lobe, the ACTH sequence of POMC is cleaved by **prohormone convertase 2** to release a small peptide,

Clinical Focus / 31.1

Pituitary Tumors

Pituitary tumors arise from hormone-secreting adenohypophyseal cells, and the secretory product of the tumor depends on the cell of origin. Generally, tumors of the corticotroph, somatotroph, lactotroph, and thyrotroph hypersecrete their respective hormones. In contrast, gonadotrophic tumors usually do not secrete hormones efficiently. Pituitary adenomas with no clinical symptoms have been reported in about 11% of autopsies. Use of sensitive imaging techniques such as magnetic resonance imaging (MRI) for nonpituitary indications has increased the detection of pituitary abnormalities such that microadenoma is detected in 10% of the normal adult population undergoing MRI of the head. Suspected pituitary tumors are usually examined directly with MRI to assess the effect of tumor mass on surrounding soft tissue structures such as the optic chiasm, which can become compressed by larger tumors, resulting in partial loss of vision.

Testing for aberrant hormone levels in the blood can be used to diagnose the type of pituitary adenoma that is present. For Cushing disease, an elevated 24-hour urinary free cortisol level can be an effective screen for disease. An elevation in insulin-like growth factor I (IGF-I) level, compared with age-matched and gender-matched controls, would indicate a growth hormone (GH)–secreting adenoma. Following diagnosis, three modes of therapy are available: surgical, radiotherapeutic, and medical. The goals of therapy are to alleviate compressive mass effects and suppress hormone hypersecretion while maintaining intact function of the unaffected portions of the pituitary.

Pituitary surgery is generally recommended to alleviate pressure effects resulting from tumor mass and to reduce hormone hypersecretion. The transsphenoidal surgical approach is associated with the lowest mortality. Radiotherapy, the delivery of high-energy ionizing radiation, is generally indicated for persistent hormone hypersecretion or residual mass effects after surgery or when surgery is contraindicated. Medical therapy is useful in several cases because pituitary tumors often express receptors for hypothalamic control of hormone secretion. Pharmaceutical ligands that activate the somatotropin release–inhibitory factor (SRIF) receptor on tumors arising from somatotrophs can suppress prolactin (PRL) or GH hypersecretion, block tumor growth, and shrink tumor size. Thus, GH-secreting or PRL-secreting tumors usually respond well to medical therapy.

α-melanocyte-stimulating hormone (**α-MSH**), and, therefore, little ACTH is produced. α-MSH acts in lower vertebrates to produce temporary changes in skin color by causing the dispersion of melanin granules in pigment cells. As noted earlier, the adult human has only a vestigial intermediate lobe and does not produce and secrete significant amounts of α-MSH or other hormones derived from POMC. However, because ACTH contains the α-MSH amino acid sequence at its N-terminal end, it has melanocyte-stimulating activity when present in the blood at high concentrations. Humans who have high blood levels of ACTH, as a result of **Addison disease** or an ACTH-secreting tumor, are often hyperpigmented. In the hypothalamus, α-MSH is important in the regulation of feeding behavior.

Corticotropin-releasing hormone

CRH is the main physiologic regulator of ACTH secretion and synthesis. In humans, CRH consists of 41 amino acid residues in a single peptide chain.

A group of neurons with small cell bodies, called **parvicellular neurons**, synthesize CRH in the paraventricular nuclei of the hypothalamus. The axons of parvicellular neurons terminate on capillary networks that give rise to hypophyseal portal vessels. Secretory granules containing CRH are stored in the axon terminals of these cells. On receiving the appropriate stimulus, these cells secrete CRH into the capillary network; CRH enters the hypophyseal portal circulation and is delivered to the anterior pituitary gland.

CRH binds to receptors on the plasma membranes of corticotrophs. Stimulatory G proteins couple these receptors to adenylyl cyclase. The binding of CRH to its receptor increases the activity of adenylyl cyclase, which catalyzes the formation of **cyclic adenosine monophosphate (cAMP)** from **adenosine triphosphate (ATP)** (Fig. 31.5). The rise in cAMP concentration in the corticotroph activates **protein kinase A (PKA)**, which then phosphorylates cellular proteins. CRH stimulates the secretion of ACTH and β-lipotropin, most likely through PKA-mediated phosphorylation of voltage-dependent calcium channels, although the exact mechanism is still not completely understood.

Increased cAMP production in the corticotroph by CRH also stimulates expression of the gene for POMC, increasing the level of POMC messenger RNA (mRNA) in these cells (see Fig. 31.5). Thus, CRH not only stimulates ACTH secretion but also maintains the capacity of the corticotroph to synthesize the precursor for ACTH.

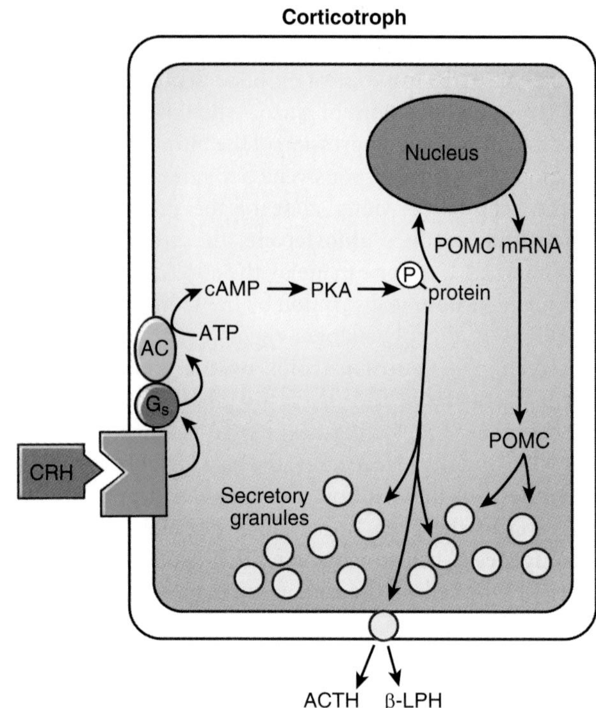

Figure 31.5 The main actions of corticotropin-releasing hormone (CRH) on a corticotroph. CRH binds to membrane receptors that are coupled to adenylyl cyclase (AC) by stimulatory G proteins (G_s). AC is stimulated, and cyclic adenosine monophosphate (cAMP) rises in the cell. cAMP activates protein kinase A (PKA), which then phosphorylates proteins (P-proteins) that stimulate adrenocorticotropic hormone (ACTH) secretion and the expression of the proopiomelanocortin (POMC) gene. β-LPH, β-lipotropin

Glucocorticoid hormones

A rise in glucocorticoid concentration in the blood resulting from the action of ACTH on the adrenal cortex inhibits the secretion of ACTH. Thus, glucocorticoids have a negative-feedback effect on ACTH secretion, which, in turn, reduces the rate of secretion of glucocorticoids by the adrenal cortex. If the blood glucocorticoid level begins to fall for some reason, this negative-feedback effect is reduced, stimulating ACTH secretion and restoring the blood glucocorticoid concentration. This interactive relationship is called the **hypothalamic-pituitary-adrenal axis** (Fig. 31.6). This control loop ensures that the level of glucocorticoids in the blood remains relatively stable in the resting state, although there is a diurnal variation in glucocorticoid secretion. As discussed later, physical and emotional stress can alter the mechanism regulating glucocorticoid secretion.

The negative-feedback effect of glucocorticoids on ACTH secretion results from actions on both the hypothalamus and the corticotroph (see Fig. 31.6). When the concentration of glucocorticoids rises in the blood, CRH secretion from the hypothalamus is inhibited. As a result, the stimulatory effect of CRH on the corticotroph is reduced and the rate of ACTH secretion falls. Glucocorticoids act directly on parvicellular neurons to inhibit CRH release and indirectly through neurons in the hippocampus that project to the hypothalamus to affect the activity of parvicellular neurons. At the corticotroph, glucocorticoids inhibit the actions of CRH to stimulate ACTH secretion.

If the blood concentration of glucocorticoids remains high for a long period of time, expression of the gene for POMC is inhibited. As a result, the amount of POMC mRNA falls in the corticotroph, and gradually the production of ACTH and the other POMC peptides declines as well. Because CRH stimulates POMC gene expression and glucocorticoids inhibit CRH secretion, glucocorticoids inhibit POMC gene expression, in part, by suppressing CRH

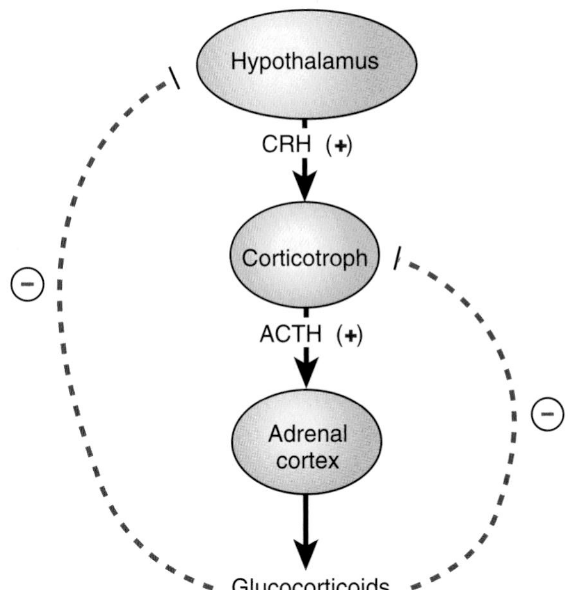

Figure 31.6 The hypothalamic-pituitary-adrenal axis. *Blue dashed lines* indicate the negative-feedback actions of glucocorticoids on the corticotroph and the hypothalamus. ACTH, adrenocorticotropic hormone; CRH, corticotropin-releasing hormone.

secretion. Glucocorticoids also act directly in the corticotroph itself to suppress POMC gene expression.

The negative-feedback actions of glucocorticoids are essential for the normal operation of the hypothalamic-pituitary-adrenal axis. The disturbances that occur when disease or glucocorticoid administration drastically changes blood glucocorticoid levels vividly illustrate this relationship. For example, if a person's adrenal glands have been surgically removed or damaged by disease (e.g., Addison disease), the resulting lack of glucocorticoids allows corticotrophs to secrete large amounts of ACTH. As noted earlier, this response may result in hyperpigmentation as a result of the melanocyte-stimulating activity of ACTH. People with glucocorticoid deficiency caused by inherited genetic defects affecting enzymes involved in steroid hormone synthesis by the adrenal cortex have high blood ACTH levels resulting from the absence of the negative-feedback effects of glucocorticoids on ACTH secretion. Because a high blood concentration of ACTH causes hypertrophy of the adrenal glands, these genetic diseases are collectively called **congenital adrenal hyperplasia** (see Chapter 33, "Adrenal Gland"). By contrast, in people treated chronically with large doses of glucocorticoids, the adrenal cortex atrophies because the high level of glucocorticoids in the blood inhibits ACTH secretion, resulting in the loss of its trophic influence on the adrenal cortex.

Stress-induced secretion

Stress greatly influences the hypothalamic-pituitary-adrenal axis. When a person experiences physical or emotional stress, ACTH secretion is increased. As a result, the blood level of glucocorticoids rises rapidly. Regardless of the blood glucocorticoid concentration, stress stimulates the hypothalamic-pituitary-adrenal axis because stress-induced neural activity generated at higher CNS levels stimulates parvicellular neurons in the paraventricular nuclei to secrete CRH at a greater rate. Thus, stress can override the normal operation of the hypothalamic-pituitary-adrenal axis. If the stress persists, the blood glucocorticoid level remains high because the glucocorticoid negative-feedback mechanism functions at a higher set point.

Arginine vasopressin

Glucocorticoid deficiency and certain types of stress also increase the concentration of AVP in hypophyseal portal blood. The physiologic significance is that AVP, like CRH, can stimulate corticotrophs to secrete ACTH. Acting along with CRH, AVP amplifies the stimulatory effect of CRH on ACTH secretion.

AVP interacts with a specific receptor on the plasma membrane of the corticotroph. G proteins couple these receptors to the enzyme **phospholipase C (PLC)**. The interaction of AVP with its receptor activates PLC, which, in turn, hydrolyzes **phosphatidylinositol 4,5-bisphosphate (PIP_2)** present in the plasma membrane. This generates the intracellular second messengers **inositol trisphosphate (IP_3)** and **diacylglycerol (DAG)**. IP_3 mobilizes intracellular calcium stores, and DAG activates the phospholipid-dependent and calcium-dependent **protein kinase C (PKC)** to mediate the stimulatory effect of AVP on ACTH secretion.

As noted earlier, magnocellular neurons of the supraoptic and paraventricular nuclei of the hypothalamus produce AVP and oxytocin. These neurons terminate in the posterior lobe, where they secrete AVP and oxytocin into capillaries that feed into the systemic circulation. However, parvicellular neurons in the paraventricular nuclei also produce AVP, which they secrete into hypophyseal portal blood. It appears that much of the AVP secreted by parvicellular neurons is made in the same cells that produce CRH. It is assumed that the AVP in hypophyseal portal blood comes from these cells and from a small number of AVP-producing magnocellular neurons whose axons pass through the median eminence of the hypothalamus on their way to the posterior lobe.

Diurnal variation

Under normal circumstances, the hypothalamic-pituitary-adrenal axis in humans functions in a pulsatile manner, resulting in several bursts of secretory activity over a 24-hour period. This pattern appears to be a result of rhythmic activity in the CNS, which causes bursts of CRH secretion and, in turn, bursts of ACTH and glucocorticoid secretion (Fig. 31.7). The diurnal oscillation in secretory activity of the axis is thought to result from changes in the sensitivity of CRH-producing neurons to the negative-feedback action of glucocorticoids, altering their rate of CRH secretion. As a result, there is a diurnal oscillation in the rate of ACTH and glucocorticoid secretion. This **circadian rhythm** is reflected in the daily pattern of glucocorticoid secretion. In people who are awake during the day and sleep at night, the blood glucocorticoid level begins to rise during the early morning hours, reaches a peak sometime before noon, and then falls gradually to a low

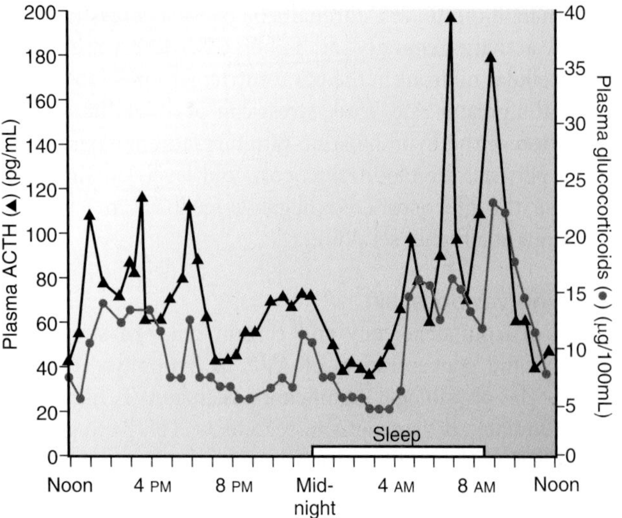

Figure 31.7 Adrenocorticotropic hormone (ACTH) secretion and the sleep–wake cycle. Pulsatile changes in the concentrations of ACTH and glucocorticoids in the blood of a young individual over a 24-hour period. Note that the amplitude of the pulses in ACTH and glucocorticoids is lower during the evening hours and increases greatly during the early morning hours. This pattern is a result of the diurnal oscillation of the hypothalamic-pituitary-adrenal axis.

level around midnight (see Fig. 31.7). This pattern is reversed in people who sleep during the day and are awake at night. This inherent biologic rhythm is superimposed on the normal operation of the hypothalamic-pituitary-adrenal axis.

Thyroid-stimulating hormone regulates thyroid gland function.

TSH is the physiologic regulator of T_4 and T_3 synthesis and secretion by the thyroid gland. The thyroid gland consists of two lobes and is composed of aggregates of follicle cells, which are formed from a single layer of cells. The follicular cells produce and secrete T_4 and T_3, thyroid hormones that are iodinated derivatives of the amino acid tyrosine.

The thyroid hormones control the body's basal metabolic rate, heat production, protein synthesis, growth, and CNS development. They act on many cells, changing gene expression and the production of particular proteins in the cell.

In addition to regulating thyroid hormones, TSH also promotes nucleic acid and protein synthesis in the cells of the thyroid follicles, maintaining their size and functional integrity. The actions of TSH on thyroid hormone synthesis and secretion and the physiologic effects of the thyroid hormones are described in detail in Chapter 32, "Thyroid Gland."

Thyroid-stimulating hormone synthesis

The α subunit of TSH is a single peptide chain of 92 amino acid residues with two carbohydrate chains linked to its structure. The β subunit is a single peptide chain of 112 amino acid residues, to which a single carbohydrate chain is linked. Noncovalent bonds hold the α and β subunits together. The two subunits combined give the TSH molecule a molecular weight of about 28,000. Neither subunit has significant TSH activity by itself. The two subunits must be combined in a 1:1 ratio to form an active hormone. The gonadotropins FSH and LH are also composed of two noncovalently combined subunits. The α subunits of TSH, FSH, and LH are derived from the same gene and are identical, but the β subunit gives each hormone its particular set of physiologic activities.

Thyrotrophs synthesize the peptide chains of the α and β subunits of TSH from separate mRNA molecules, which are transcribed from two different genes. The peptide chains of the α and β subunits are combined and undergo glycosylation in the rough endoplasmic reticulum (RER). These processes are completed as TSH molecules pass through the Golgi apparatus and are packaged into secretory granules. Normally, thyrotrophs make more α subunits than β subunits. As a result, secretory granules contain excess α subunits. When a thyrotroph is stimulated to secrete TSH, it releases both TSH and free α subunits into the bloodstream. In contrast, little free TSH β subunit is in the blood.

Thyrotropin-releasing hormone regulation of thyroid-stimulating hormone

TRH is the main physiologic stimulator of TSH secretion and synthesis by thyrotrophs. TRH is a small peptide consisting of three amino acid residues produced by neurons in the hypothalamus. These neurons terminate on the capillary networks that give rise to the hypophyseal portal vessels. Normally, these neurons secrete TRH into the hypophyseal portal circulation at a constant or tonic rate. It is assumed that the TRH concentration in the blood that perfuses the thyrotrophs does not change greatly; therefore, the thyrotrophs are continuously exposed to TRH.

TRH binds to receptors on the plasma membranes of thyrotrophs. G proteins couple these receptors to PLC (Fig. 31.8). The interaction of TRH with its receptor activates PLC, causing the hydrolysis of PIP_2 in the membrane. This action releases the intracellular messengers IP_3 and DAG. IP_3 causes the concentration of Ca^{2+} in the cytosol to rise, which stimulates the secretion of TSH into the blood. The rise in cytosolic Ca^{2+} and the increase in DAG activate PKC, which phosphorylates proteins to promote TSH secretion.

TRH also stimulates the expression of the genes for the α and β subunits of TSH (see Fig. 31.8). As a result, the amount of mRNA for the α and β subunits is maintained in the thyrotroph and the production of TSH is fairly constant.

Thyroid hormone regulation of thyroid-stimulating hormone

The thyroid hormones exert a direct negative-feedback effect on TSH secretion. For example, when the blood concentration of thyroid hormones is high, the rate of TSH secretion falls. In turn, the stimulatory effect of TSH on the follicular cells of the thyroid is reduced, resulting in a decrease in T_4 and T_3 secretion. However, when the circulating levels of T_4 and T_3 are low, their negative-feedback effect on TSH release is reduced and more TSH is secreted from thyrotrophs, increasing the rate of thyroid hormone secretion. This control system is part of the hypothalamic-pituitary-thyroid axis (Fig. 31.9).

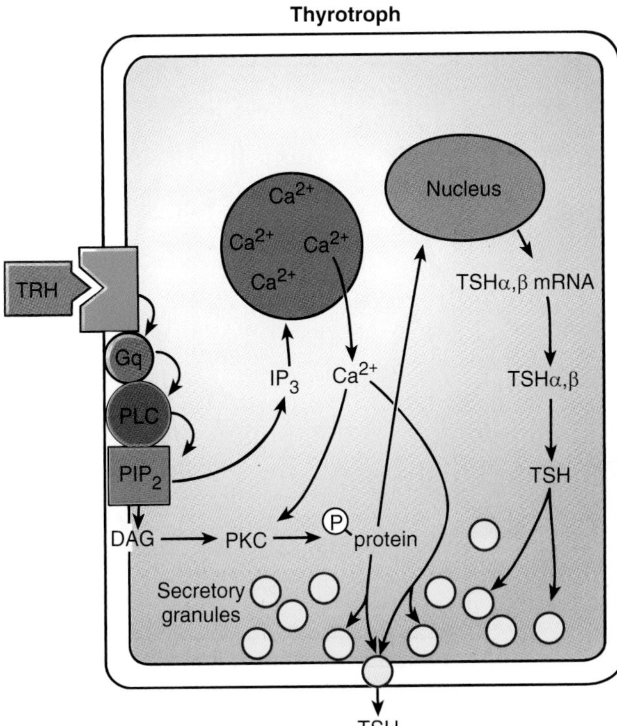

Figure 31.8 The actions of thyrotropin-releasing hormone (TRH) on a thyrotroph. TRH binds to membrane receptors, which are coupled to phospholipase C (PLC) by G proteins (G_q). PLC hydrolyzes phosphatidylinositol 4,5-bisphosphate (PIP_2) in the plasma membrane, generating inositol trisphosphate (IP_3) and diacylglycerol (DAG). IP_3 mobilizes intracellular stores of Ca^{2+}. The rise in Ca^{2+} stimulates thyroid-stimulating hormone (TSH) secretion. Ca^{2+} and DAG activate protein kinase C (PKC), which phosphorylates proteins (P-proteins) that stimulate TSH secretion and the expression of the genes for the α and β subunits of TSH.

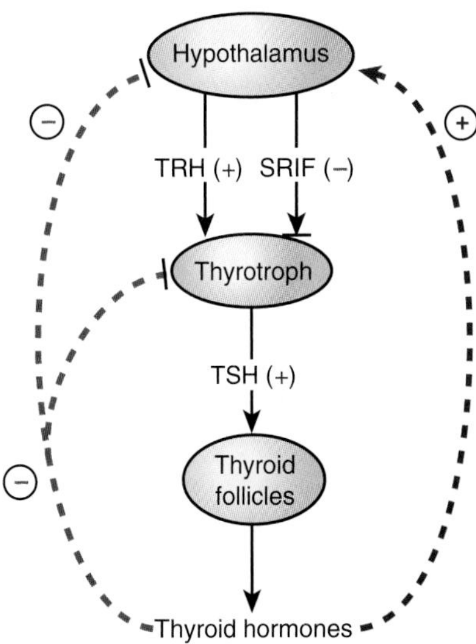

Figure 31.9 The hypothalamic-pituitary-thyroid axis. Thyrotropin-releasing hormone (TRH) stimulates and somatostatin (SRIF) inhibits thyroid-stimulating hormone (TSH) release by acting directly on the thyrotroph. The negative-feedback loops shown in *blue* inhibit TRH secretion and action on the thyrotroph, causing a decrease in TSH secretion. The feedback loops shown in *red*, stimulate somatostatin secretion, causing a decrease in TRH secretion. SRIF, somatostatin or somatotropin release–inhibiting factor.

The thyroid hormones exert negative-feedback effects on both the hypothalamus and the pituitary. In the hypothalamic TRH-secreting neurons, thyroid hormones reduce TRH mRNA and TRH prohormone to decrease TRH release. The thyroid hormones also increase the release of somatostatin from the hypothalamus. Somatostatin (SRIF) inhibits the release of TSH from the thyrotroph (see Fig. 31.9). In the pituitary, thyroid hormones reduce the sensitivity of the thyrotroph to TRH and inhibit TSH synthesis.

The negative-feedback effects of the thyroid hormones on thyrotrophs are produced primarily through the actions of T_3. Both T_4 and T_3 circulate in the blood bound to plasma proteins, with only a small percentage (<1%) unbound or free (see Chapter 32, "Thyroid Gland"). Thyrotrophs take up the free T_4 and T_3 molecules, and the enzymatic removal of one iodine atom converts T_4 to T_3. The newly formed T_3 molecules and those taken up directly from the blood enter the nucleus, where they bind to thyroid hormone receptors in the chromatin. The interaction of T_3 with its receptors changes gene expression in the thyrotroph, decreasing the synthesis and secretion of TSH. For example, T_3 inhibits the expression of the genes for the α and β subunits of TSH, directly decreasing the synthesis of TSH. T_3 also decreases thyrotroph sensitivity to TRH, in part through a reduction in the number of TRH receptors in thyrotroph plasma membrane.

Diurnal variation in thyroid-stimulating hormone and effect of cold exposure

The hypothalamic-pituitary-thyroid axis, like the hypothalamic-pituitary-adrenal axis, follows a diurnal circadian rhythm in humans. Peak TSH secretion occurs in the early morning, and a low point is reached in the evening. Physical and emotional stress can alter TSH secretion, but the effects of stress on the hypothalamic-pituitary-thyroid axis are not as pronounced as on the hypothalamic-pituitary-adrenal axis.

The exposure of certain animals to a cold environment stimulates TSH secretion. This makes sense from a physiologic perspective because the thyroid hormones are important in regulating body heat production (see Chapter 32, "Thyroid Gland"). Brief exposure of experimental animals to a cold environment stimulates the secretion of TSH, presumably a result of enhanced TRH secretion. Newborn humans behave much the same way, in that they respond to brief cold exposure with an increase in TSH secretion. This response to cold does not occur in adult humans.

Growth hormone regulates growth during childhood and remains important throughout life.

GH is a peptide hormone secreted by the anterior pituitary. As its name implies, it promotes the growth of the human body. It does not appear to stimulate fetal growth, and it is not an important growth factor during the first few months after birth. Thereafter, it is essential for the normal rate of body growth during childhood and adolescence.

The anterior pituitary secretes GH throughout life, and GH remains physiologically important even after growth has stopped. In addition to its growth-promoting action, GH has effects on many aspects of carbohydrate, lipid, and protein metabolism. For example, GH is thought to be one of the physiologic factors that counteract and, thus, modulate some of the actions of insulin on the liver and peripheral tissues.

Growth hormone synthesis

GH is a globular 22-kDa protein consisting of a single chain of 191 amino acid residues with two intrachain disulfide bridges, produced in somatotrophs of the anterior pituitary. It is synthesized in the RER as a larger preprohormone that is processed to a prohormone consisting of an N-terminal signal peptide and the 191–amino acid hormone. The signal peptide is then cleaved from the prohormone as it traverses the Golgi apparatus, and the GH is packaged in secretory granules. Human GH has considerable structural similarity to human PRL and placental lactogen.

GHRH regulates the production of GH by stimulating the expression of the GH gene in somatotrophs. Thyroid hormones also stimulate expression of the GH gene. As a result, the normal rate of GH production depends on these hormones. For example, a thyroid hormone–deficient person is also GH deficient. This important action of thyroid hormones is discussed further in Chapter 32, "Thyroid Gland."

Growth hormone secretion

Two opposing hypothalamic-releasing hormones regulate the secretion of GH. GHRH stimulates GH secretion, and somatostatin inhibits GH secretion by inhibiting the action of GHRH. The net effect of these counteracting hormones on somatotrophs determines the rate of GH secretion. When GHRH predominates, GH secretion is stimulated. When somatostatin predominates, GH secretion is inhibited.

Human GHRH is a peptide composed of a single chain of 44 amino acid residues. A slightly smaller version of GHRH consisting of 40 amino acid residues is also present in humans. The two forms of GH result from posttranslational modification of a larger prohormone. GHRH is synthesized in the cell bodies of neurons in the **arcuate nuclei** and **ventromedial nuclei** of the hypothalamus. The axons of these cells project to the capillary networks giving rise to the portal vessels. When these neurons receive a stimulus for GHRH secretion, they discharge GHRH from their axon terminals into the hypophyseal portal circulation.

GHRH binds to receptors in the plasma membranes of somatotrophs (Fig. 31.10). A stimulatory G protein, G_s, couples these receptors to adenylyl cyclase. The interaction of GHRH with its receptors activates adenylyl cyclase, increasing the concentration of cAMP in the somatotroph. The rise in cAMP activates PKA, which, in turn, phosphorylates proteins that stimulate GH secretion and GH gene expression. GHRH binding to its receptor also increases intracellular Ca^{2+}, which stimulates GH secretion. GHRH also stimulates PLC, causing the hydrolysis of membrane

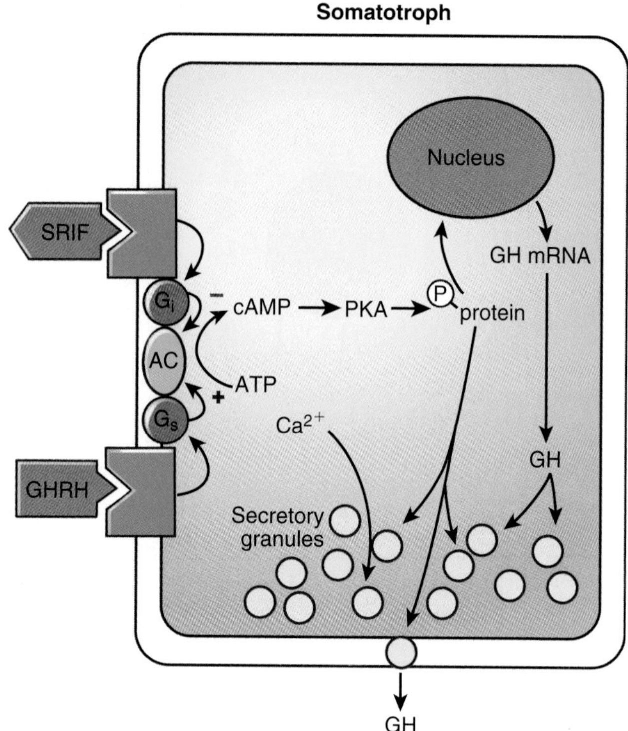

Figure 31.10 The actions of growth hormone–releasing hormone (GHRH) and somatostatin (SRIF) on a somatotroph. GHRH binds to membrane receptors that are coupled to adenylyl cyclase (AC) by stimulatory G proteins (G_s). Cyclic adenosine monophosphate (cAMP) rises in the cell and activates protein kinase A (PKA), which then phosphorylates proteins (P-proteins) that stimulate growth hormone (GH) secretion and the expression of the gene for GH. Ca^{2+} also facilitates GH secretion. The possible involvement of the phosphatidylinositol pathway in GHRH action is not shown. SRIF binds to membrane receptors that are coupled to adenylyl cyclase by inhibitory G proteins (G_i). This action inhibits the ability of GHRH to stimulate adenylyl cyclase, blocking its action on GH secretion.

PIP_2 in the somatotroph. The importance of this phospholipid pathway for the stimulation of GH secretion by GHRH is not established.

Somatostatin is a small peptide consisting of 14 amino acid residues. Although made by neurosecretory neurons in various parts of the hypothalamus, somatostatin neurons are especially abundant in the **anterior periventricular region** (close to the third ventricle). The axons of these cells terminate on the capillary networks giving rise to the hypophyseal portal circulation, where they release somatostatin into the blood.

Somatostatin binds to receptors in the plasma membranes of somatotrophs. These receptors, like those for GHRH, are also coupled to adenylyl cyclase but are coupled by an inhibitory G protein (see Fig. 31.10). The binding of somatostatin to its receptor decreases adenylyl cyclase activity, reducing intracellular cAMP. Somatostatin binding to its receptor also lowers intracellular Ca^{2+}, reducing GH secretion. When the somatotroph is exposed to both somatostatin and GHRH, the effects of somatostatin are dominant and intracellular cAMP and Ca^{2+} are reduced. Thus, somatostatin has a negative modulating influence on the action of GHRH.

Insulin-like growth factor I

GH is not a traditional trophic hormone that directly stimulates the growth of its target tissue. Rather, GH stimulates the production of a trophic hormone called **insulin-like growth factor I (IGF-I)**. IGF-I is a potent **mitogenic** agent that mediates the growth-promoting action of GH. IGF-I was originally called *somatomedin C* or *somatotropin-mediating hormone* because of its role in promoting growth. Somatomedin C was renamed IGF-I because of its structural similarity to proinsulin. Although most effects of GH are mediated by IGF-I, it is important to note that some effects of GH, such as stimulation of lipolysis in adipose tissue (see below) and stimulation of amino acid transport in muscle, occur independently of IGF-I action.

Insulin-like growth factor II (IGF-II) is structurally similar to IGF-I and has many of the same metabolic and mitogenic actions. However, the synthesis and release of IGF-II depends less on GH. Thus, IGF-I is the more important mediator of GH action.

IGF-I is a 7.5-kDa protein consisting of a single chain of 70 amino acids. Because of its structural similarity to proinsulin, IGF-I can produce some of the effects of insulin. Many cells of the body produce IGF-I; however, the liver is the main source of IGF-I in the blood. Most IGF-I in the blood is bound to specific IGF-I–binding proteins; only a small amount circulates in the free form. The bound form of circulating IGF-I has little insulin-like activity, so it does not play a physiologic role in the regulation of blood glucose level.

GH increases the expression of the genes for IGF-I in various tissues and organs, such as the liver, and stimulates the production and release of IGF-I. Excessive secretion of GH results in a greater-than-normal amount of IGF-I in the blood. People with GH deficiency have lower-than-normal levels of IGF-I, but there is still some present, because many hormones and factors, in addition to GH, regulate the production of IGF-I by cells.

Inhibition of growth hormone secretion

An increase in the blood concentration of GH has direct feedback effects on its own secretion, independent of the production of IGF-I. These effects of GH are a result of the inhibition of GHRH secretion and the stimulation of somatostatin secretion by hypothalamic neurons (Fig. 31.11). GH circulating in the blood can enter the interstitial spaces of the median eminence of the hypothalamus because there is no blood–brain barrier in this area.

IGF-I also has a negative-feedback effect on the secretion of GH (see Fig. 31.11). It acts directly on the somatotrophs to inhibit the stimulatory action of GHRH on GH secretion. IGF-I also inhibits GHRH secretion and stimulates the secretion of somatostatin by neurons in the hypothalamus. By stimulating IGF-I production, GH inhibits its own secretion. This mechanism is analogous to the way ACTH and TSH regulate their own secretion through the respective negative-feedback effects of the glucocorticoid and thyroid

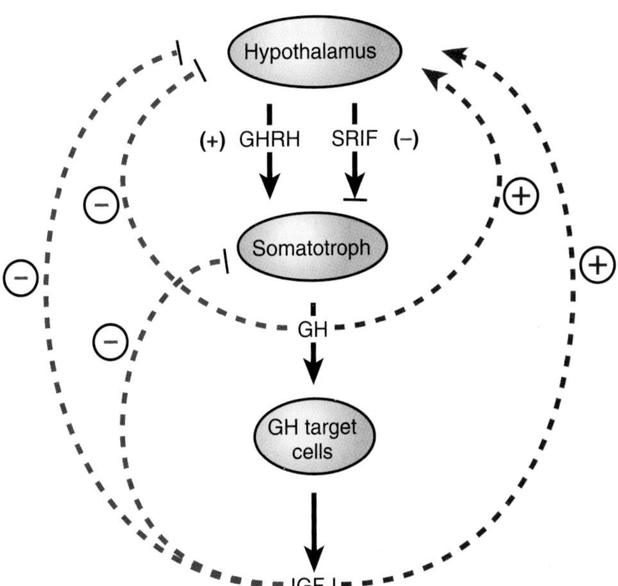

Figure 31.11 The hypothalamic-pituitary-GH axis. Growth hormone–releasing hormone (GHRH) stimulates and somatostatin (SRIF) inhibits growth hormone (GH) secretion by acting directly on the somatotroph. The negative-feedback loops shown in *blue* inhibit GHRH secretion and action on the somatotroph, causing a decrease in GH secretion. The positive feedback loops shown in *red* stimulate somatostatin secretion, causing a decrease in GH secretion. IGF-I, insulin-like growth factor I.

hormones. This interactive relationship involving GHRH, somatostatin, GH, and IGF-I comprises the hypothalamic-pituitary-GH axis.

Pulsatile secretory pattern

In humans, GH is secreted in periodic bursts, which produce large but short-lived peaks in GH concentration in the blood. Between these episodes of high GH secretion, somatotrophs release little GH; as a result, the blood concentration of GH falls to low levels. It is believed that an increase in the rate of GHRH secretion and a fall in the rate of somatostatin secretion cause these periodic bursts of GH secretion. Increased somatostatin secretion is thought to cause the intervals between bursts, when GH secretion is suppressed. These changes in GHRH and somatostatin secretion result from neural activity generated in higher levels of the CNS, which affects the secretory activity of GHRH and somatostatin-producing neurons in the hypothalamus.

Bursts of GH secretion occur during both awake and sleep periods of the day; however, GH secretion is maximal at night. The bursts of GH secretion during sleep usually occur within the first hour after the onset of deep sleep (stages 3 and 4 of slow-wave sleep). Mean GH levels in the blood are highest during adolescence (peaking in late puberty) and decline in adults. The reduction in blood GH with aging is mainly a result of a decrease in the size of the GH secretory burst but not the number of pulses (Fig. 31.12).

Many factors affect the rate of GH secretion in humans. These factors are thought to work by changing the secretion of

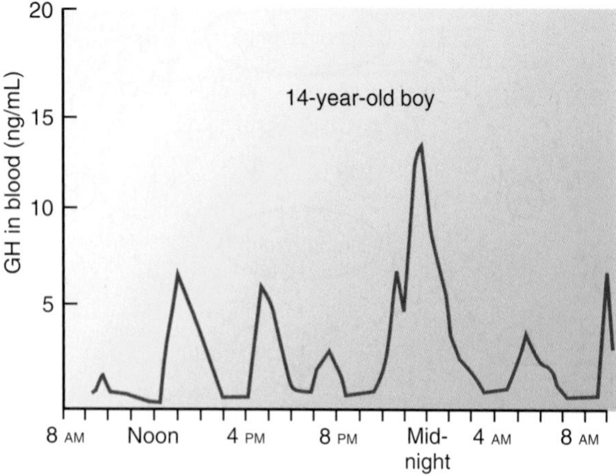

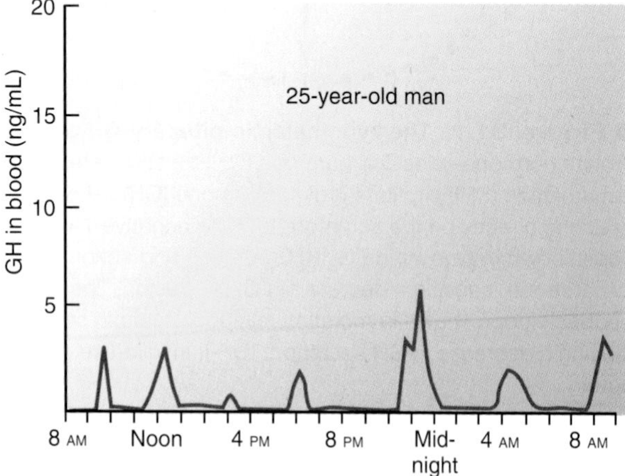

Figure 31.12 Pulsatile growth hormone (GH) secretion in an adolescent boy and in an adult. In the adult, GH levels are reduced as a result of smaller pulse width and amplitude rather than a decrease in the number of pulses.

GHRH and somatostatin by neurons in the hypothalamus. For example, emotional or physical stress causes a great increase in the rate of GH secretion. Vigorous exercise also stimulates GH secretion. Obesity results in reduced GH secretion.

Changes in the circulating levels of metabolites also affect GH secretion. A decrease in blood glucose concentration stimulates GH secretion, whereas hyperglycemia inhibits it. An increase in the blood concentration of the amino acids arginine and leucine also stimulates GH secretion.

Regulation of growth and metabolism

The cells of many tissues and organs of the body have receptors for GH in their plasma membranes. The interaction of GH with these receptors produces its growth-promoting and other metabolic effects, but the mechanisms that produce these effects are not fully understood. The binding of GH to its receptor activates an associated tyrosine kinase (JAK2), which is not an integral part of the GH receptor. Activation of JAK2 initiates changes in the phosphorylation pattern of cytoplasmic and nuclear proteins, ultimately stimulating the transcription of specific genes, including that for IGF-I.

IGF-I mediates many of the mitogenic effects of GH; however, evidence indicates that GH has direct growth-promoting actions on progenitor cells or stem cells, such as **prechondrocytes** in the growth plates of bone and **satellite cells** of skeletal muscle. GH stimulates such progenitor cells to differentiate into cells with the capacity to undergo cell division. An important action of GH on the differentiation of progenitor cells is stimulation of the expression of the IGF-I gene; these cells produce and release IGF-I. IGF-I exerts an autocrine mitogenic action on the cells that produced it or a paracrine action on neighboring cells. In response to IGF-I, these cells undergo division, causing the tissue to grow mainly through cell replication.

As mentioned earlier, GH deficiency in childhood causes a decrease in the rate of body growth. If left untreated, the deficiency results in **pituitary dwarfism**. People with this condition may be deficient in GH only, or they may have multiple anterior pituitary hormone deficiencies. A defect in the mechanisms that control GH secretion or the production of GH by somatotrophs can cause GH deficiency. In some people, the target cells for GH fail to respond normally to the hormone because of several different mutations in the GH receptor.

The excessive secretion of GH during childhood, caused by a defect in the mechanisms regulating GH secretion or a GH-secreting tumor, results in **gigantism**. Affected people may grow to a height of 7 to 8 ft (2.1–2.4 m). When excessive GH secretion occurs in an adult, further linear growth does not occur because the growth plates of the long bones have calcified. Instead, it causes the bones of the face, hands, and feet to become thicker and certain organs, such as the liver, to undergo hypertrophy. This condition, known as **acromegaly**, can also result from the chronic administration of excessive amounts of GH to adults.

Although the main physiologic action of GH is on body growth, it also has important effects on certain aspects of fat and carbohydrate metabolism. Its main action on fat metabolism is to stimulate the mobilization of triglycerides from the fat depots of the body. This process, known as **lipolysis**, involves the hydrolysis of triglycerides to fatty acids and glycerol by a number of lipases such as **hormone-sensitive lipase**. The fatty acids and glycerol are released from adipocytes and enter the bloodstream. GH has been shown to directly stimulate lipolysis in some experiments, but most evidence suggests that it causes adipocytes to be more responsive to other lipolytic stimuli, such as fasting and catecholamines.

GH is also thought to function as one of the counterregulatory hormones that limit the actions of insulin on muscle, adipose tissue, and the liver. For example, GH inhibits glucose use by muscle and adipose tissue and increases glucose production by the liver. These effects are opposite those of insulin. Also, GH makes muscle and fat cells resistant to the action of insulin itself. Thus, GH normally has a tonic inhibitory effect on the actions of insulin, much like the glucocorticoid hormones (see Chapter 33, "Adrenal Gland").

The insulin-opposing actions of GH can produce serious metabolic disturbances in people who secrete excessive amounts of GH (people with acromegaly) or who are given large amounts of GH for an extended time. They may develop insulin resistance and an elevated insulin level in the blood.

They may also have hyperglycemia caused by the underuse and overproduction of glucose. These disturbances are much like those in people with non–insulin-dependent (type 2) diabetes mellitus. For this reason, this metabolic response to excess GH is called its **diabetogenic action.**

In GH-deficient people, GH has a transitory insulin-like action. For example, intravenous injection of GH in a person who is GH deficient produces hypoglycemia. The hypoglycemia is caused by the ability of GH to stimulate the uptake and use of glucose by muscle and adipose tissue and to inhibit glucose production by the liver. After about 1 hour, the blood glucose level returns to normal. If this person is given a second injection of GH, hypoglycemia does not occur because the person has become insensitive or refractory to the insulin-like action of GH and remains so for some hours. Healthy people do not respond to the insulin-like action of GH, presumably because they are always refractory from being exposed to their own endogenous GH. The actions of GH in humans are summarized in Table 31.3.

TABLE 31.3 Actions of Growth Hormone

Name	Action
Growth promoting	Stimulates IGF-I gene expression by target cells; IGF-I produced by these cells has autocrine or paracrine stimulatory effect on cell division, resulting in growth
Lipolytic	Stimulates mobilization of triglycerides from fat deposits
Diabetogenic	Inhibits glucose use by muscle and adipose tissue and increases glucose production by the liver
	Inhibits the action of insulin on glucose and lipid metabolism by muscle and adipose tissue
Insulin-like	Transitory stimulatory effect on uptake and use of glucose by muscle and adipose tissue in GH-deficient people
	Transitory inhibitory effect on glucose production by liver of GH-deficient people

GH, growth hormone; IGF-I, insulin-like growth factor I.

Gonadotropins are anterior pituitary hormones that regulate sexual development and reproductive function.

The testes and ovaries have two essential functions in human reproduction. The first is to produce sperm cells and ova (egg cells), respectively. The second is to produce an array of steroid and peptide hormones, which influence virtually every aspect of the reproductive process. The gonadotropic hormones FSH and LH regulate both of these functions. The hypothalamic-releasing hormone LHRH and the hormones produced by the testes and ovaries in response to gonadotropic stimulation, in turn, regulate the production and secretion of the gonadotropins by the anterior pituitary. The regulation of human reproduction by this hypothalamic-pituitary-gonad axis is discussed in Chapter 36, "Male Reproductive System," and Chapter 37, "Female Reproductive System." Here, we describe the chemistry and formation of the gonadotropins.

Clinical Focus / 31.2

Growth Hormone Deficiency and Recombinant Human Growth Hormone

Growth hormone (GH) is species specific, and humans do not respond to GH derived from animals. In the past, the only human GH available for treating children who were GH deficient was a limited amount made from human pituitaries obtained at autopsy, but there was never enough to meet the need. Use of pituitary-derived GH was then halted because of concern of a causal relationship between therapy and development of **Creutzfeldt–Jakob disease (CJD)**, a rare and fatal spongiform encephalopathy. The problems of short supply and potential disease transmission were solved when the gene for human GH was cloned and the production of large amounts of recombinant human GH, with all the activities of the natural substance, became possible. During the 1980s, careful clinical trials established that recombinant human GH was safe to use in GH-deficient children to promote growth. The hormone was approved for clinical use and is now produced and sold worldwide.

Despite the availability of recombinant GH, the diagnosis of GH deficiency has remained controversial. As presented in Figure 31.12, GH release from the pituitary occurs episodically; thus, a random measure of GH in the blood is not useful for diagnosing GH deficiency. However, a random blood sample may be used to detect GH resistance, a syndrome in which the patient exhibits symptoms of GH deficiency but presents with high GH levels in the blood.

An alternative means of testing for GH deficiency is to measure the levels of insulin-like growth factor I (IGF-I) and the IGF-binding protein 3 (IGFBP3) in the blood. IGF-I mediates many of the mitogenic effects of GH on tissues in the body, and IGF-I binds to IGFBP3 in the blood. IGFBP3 extends the half-life of the IGF-I, transports it to target cells, and facilitates its interaction with the IGF receptor. GH stimulates the production of IGF-I and IGFBP3, which are present in the blood at fairly constant, readily detectable levels in healthy people. In children with GH deficiency, the concentrations of IGF-I and IGFBP3 are low. Treatment with recombinant GH will increase IGF-I and IGFBP3 in the blood, which will result in increased long bone growth. The epiphyseal growth plate in the bone becomes less responsive to GH and IGF-I several years after puberty, and long bone growth stops in adulthood (see Chapter 35, "Endocrine Regulation of Calcium, Phosphate, and Bone Homeostasis"). In general, GH therapy may be continued after cessation of long bone growth because GH has important metabolic effects, including support of normal gonadal function and attainment of normal adult bone density.

Like TSH, human FSH and LH are composed of two structurally different glycoprotein subunits, called α and β, which are held together by noncovalent bonds. The β subunit of human FSH consists of a peptide chain of 111 amino acid residues, to which two chains of carbohydrate are attached. The β subunit of human LH is a peptide of 121 amino acid residues. It is also glycosylated with two carbohydrate chains. The combined α and β subunits of FSH and LH give these hormones a molecular size of about 28 to 29 kDa.

As with TSH, the individual subunits of the gonadotropins have no hormonal activity. They must be combined with each other in a 1:1 ratio to have activity. Again, it is the β subunit that gives the gonadotropin molecule either FSH or LH activity because the α subunits are identical.

The same gonadotrophs in the anterior pituitary produce FSH and LH. There are separate genes for the α and β subunits in the gonadotroph; hence, the peptide chains of these subunits are translated from separate mRNA molecules. Glycosylation of these chains begins as they are synthesized and before they are released from the ribosome. The folding of the subunit peptides into their final three-dimensional structure, the combination of an α subunit and a β subunit, and the

From Bench to Bedside / 31.1

Hypothalamic-Pituitary Regulation of Energy Intake and Energy Expenditure

Body weight in adults is fairly stable over many years. This stability in body weight occurs despite large fluctuations in caloric intake, demonstrating that energy intake and energy expenditure are precisely matched. The central nervous system (CNS), primarily the hypothalamus, is responsible for matching energy intake and energy expenditure, receiving information relevant to energy balance through metabolic, neural, and hormonal signals. Some signals regulate energy intake over short time periods, for example, acting to terminate a feeding episode, whereas others are active in the long-term regulation of energy intake, ensuring the maintenance of adequate energy stores. Leptin is a hormone that provides information to the hypothalamus on the amount of energy stored in the body as adipose tissue.

Leptin is a 16-kDa protein product of the *LEP* gene (originally termed *ob* gene), which is most highly expressed in adipose tissue but which is also detectable in other tissues including muscle and placenta. Serum leptin increases as adipose tissue mass increases; thus, leptin is significantly greater in obese subjects than in lean subjects.

The receptor for leptin is detectable in several areas of the CNS but is highly expressed in the arcuate nucleus of the hypothalamus (Fig. 31.1A). Activation of leptin receptors reduces the expression of neuropeptides that stimulate food intake (neuropeptide Y and agouti-related protein) and increases expression of neuropeptides that reduce feeding (α-melanocyte–stimulating hormone). Administration of exogenous leptin to increase serum levels has been tested as a therapy for weight loss in humans, but leptin treatment had only modest effects on appetite and body weight.

In addition to regulating food intake, leptin signaling in the hypothalamus also alters anterior pituitary hormone secretion to influence energy expenditure. Reduced caloric intake and starvation initiate a complex series of biochemical and behavioral adaptations to promote survival, one of which is to reduce whole-body energy expenditure. Thyroid hormones stimulate metabolism and increase energy use (discussed in detail in Chapter 32, "Thyroid Gland"); thus, it is adaptive to reduce thyroid hormone levels during periods of insufficient food intake. Growth and the ability to reproduce are both energy-intensive processes that are also curtailed during starvation. Although leptin circulates in the blood in proportion to the amount of body fat, serum leptin falls rapidly with restriction of food intake, providing a signal to the hypothalamus to conserve body energy stores. Evidence that leptin coordinates the hypothalamic-pituitary response to starvation was originally derived through replacement experiments in rodents. Preventing the starvation-induced fall in leptin by infusion of recombinant protein blunted the reduction in gonadal, adrenal, and thyroid hormones that would normally occur in starved mice. Similar types of replacement experiments have demonstrated that leptin can regulate anterior pituitary hormone secretion in humans. Thus, leptin may prove useful in treatment of disease resulting from reduced hypothalamic-pituitary function secondary to reductions in adipose tissue mass caused by lipodystrophy, or excess energy expenditure such as that in highly trained, female athletes.

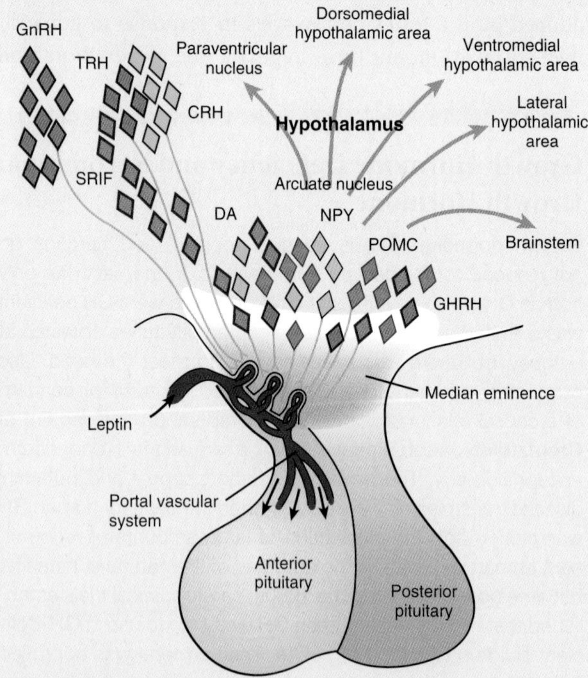

Figure 31.1A Leptin regulation of hypothalamic-pituitary axis function. TRH, thyrotropin-releasing hormone; GnRH, gonadotropin-releasing hormone; CRH, corticotropin-releasing hormone; SRIF, somatotropin release–inhibitory factor; DA, dopamine; NPY, neuropeptide Y; POMC, proopiomelanocortin; GHRH, growth hormone–releasing hormone.

completion of glycosylation all occur as these molecules pass through the Golgi apparatus and are packaged into secretory granules. As with the thyrotroph, the gonadotroph produces an excess of α subunits over FSH and LH β subunits. Therefore, the rate of β subunit production is considered to be the rate-limiting step in gonadotropin synthesis.

The hormones of the hypothalamic-pituitary-gonad axis regulate the synthesis of FSH and LH. For example, LHRH stimulates gonadotropin production. The steroid and peptide hormones produced by the gonads in response to stimulation by the gonadotropins also affect gonadotropin production. Such hormonally regulated changes in gonadotropin production are caused mainly by changes in the expression of the genes for the gonadotropin subunits. More information about the regulation of gonadotropin synthesis and secretion is found in Chapter 36, "Male Reproductive System," and Chapter 37, "Female Reproductive System."

Prolactin is a peptide hormone that regulates milk secretion from the mammary gland.

Lactation is the final phase of the process of human reproduction. During pregnancy, alveolar cells of the mammary glands develop the capacity to synthesize milk in response to stimulation by many steroid and peptide hormones. Milk synthesis by these cells begins shortly after childbirth. To continue to synthesize milk, PRL must periodically stimulate these cells, and this is thought to be the main physiologic function of PRL in the human female. What role, if any, PRL has in the human male is unclear. It is known to have some supportive effect on the action of androgenic hormones on the male reproductive tract, but whether this is an important physiologic function of PRL is not established.

Human PRL is a globular protein consisting of a single peptide chain of 199 amino acid residues with three intrachain disulfide bridges. Its molecular size is about 23 kDa. Human PRL has considerable structural similarity to human GH and to a PRL-like hormone produced by the human placenta called *placental lactogen (hPL)*. It is thought that these hormones are structurally related because their genes evolved from a common ancestral gene during the course of vertebrate evolution. Because of its structural similarity to human PRL, human GH has substantial PRL-like or lactogenic activity. However, PRL and hPL have little GH-like activity. Human placental lactogen is discussed further in Chapter 38, "Fertilization, Pregnancy, and Fetal Development."

Lactotrophs in the anterior pituitary synthesize and secrete PRL. PRL is synthesized in the RER as a larger peptide. Its N-terminal signal peptide sequence is then removed and the 199–amino acid protein passes through the Golgi apparatus and is packaged into secretory granules.

Estrogens and other hormones, such as TRH, that increase the expression of the PRL gene stimulate the synthesis and secretion of PRL. However, dopamine inhibits the synthesis of PRL. Dopamine produced by hypothalamic neurons plays a major role in the regulation of PRL synthesis and secretion by the hypothalamic-pituitary axis. The regulation of the synthesis and secretion of PRL and its physiologic actions are discussed in Chapter 38, "Fertilization, Pregnancy, and Fetal Development."

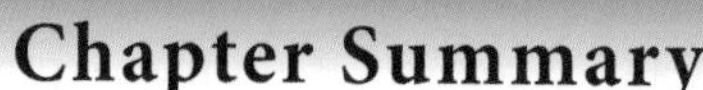

Chapter Summary

- The hypothalamic-pituitary axis is composed of the hypothalamus, infundibular stalk, posterior pituitary, and anterior pituitary.
- Arginine vasopressin and oxytocin are synthesized in hypothalamic neurons whose axons terminate in the posterior pituitary.
- Arginine vasopressin increases water reabsorption by the kidneys in response to a rise in blood osmolality or a fall in blood volume.
- Oxytocin stimulates milk letdown in the breast in response to suckling and muscle contraction in the uterus in response to cervical dilation during labor.
- Adrenocorticotropic hormone, thyroid-stimulating hormone, growth hormone, follicle-stimulating hormone, luteinizing hormone, and prolactin are synthesized in the anterior pituitary and secreted in response to hypothalamic-releasing hormones carried in the hypophyseal portal circulation.
- Hypothalamic corticotropin-releasing hormone stimulates adrenocorticotropic hormone release from corticotrophs, which, in turn, stimulates glucocorticoid release from the adrenal cortex, to comprise the hypothalamic-pituitary-adrenal axis.
- Glucocorticoids, physical and emotional stress, arginine vasopressin, and the sleep–wake cycle regulate adrenocorticotropic hormone secretion.
- Hypothalamic thyrotropin-releasing hormone stimulates thyroid-stimulating hormone release from thyrotrophs, which in turn, stimulates triiodothyronine and thyroxine release from the thyroid follicles, to comprise the hypothalamic-pituitary-thyroid axis.
- The thyroid hormones, cold temperatures, and the sleep–wake cycle regulate thyroid-stimulating hormone secretion.
- Hypothalamic growth hormone–releasing hormone increases and hypothalamic somatostatin decreases growth hormone secretion from somatotrophs, which, in turn, stimulates the release of insulin-like growth factor I from liver and other target cells, to comprise the hypothalamic-pituitary-growth hormone axis.
- Growth hormone (GH) itself, insulin-like growth factor I, aging, deep sleep, stress, exercise, and hypoglycemia regulate GH secretion.
- Luteinizing hormone–releasing hormone stimulates the secretion of follicle-stimulating hormone and luteinizing hormone from the anterior pituitary, which, in turn, affects functions of the ovaries and testes, to comprise the hypothalamic-pituitary-reproductive axis.
- Hypothalamic dopamine inhibits prolactin release from lactotrophs in the anterior pituitary.
- Prolactin regulates milk production in the breast.

thePoint *Visit http://thePoint.lww.com for chapter review Q&A, case studies, animations, and more!*

32 Thyroid Gland

ACTIVE LEARNING OBJECTIVES

Upon mastering the material in this chapter you should be able to:

- Explain the function of the thyroid follicles and the role of the iodination and coupling reactions catalyzed by the enzyme thyroid peroxidase, in synthesis of the thyroid hormones thyroxine and triiodothyronine.
- Explain how thyroid hormones are released from the thyroid gland.
- Explain the function of thyroid-stimulating hormone in regulating the synthesis and release of thyroid hormones.
- Predict the effect of changes in the concentration of thyroid hormones in the circulation on thyroid-stimulating hormone release from the anterior pituitary.
- Explain the function of the peripheral tissue deiodinases in synthesizing the physiologically active thyroid hormone triiodothyronine.
- Explain how triiodothyronine interacts with its receptor and activates transcription of target genes.
- Describe the effects of thyroid hormones on central nervous system development, growth hormone release, and target tissues such as bone.
- Explain how thyroid hormone regulates basal metabolic rate and intermediary metabolism.
- Predict the effects of excess thyroid hormone and thyroid hormone deficiency on metabolic rate, mental status, and body weight.

The development of the human body, from embryo to adult, is an orderly, programmed process. The timing of developmental events is remarkably consistent from one person to the next, with developmental milestones reached at about the same time in all of us. For example, the early development of motor skills, body growth, the start of puberty, and final sexual and physical maturation occur within narrow timeframes during the human lifespan.

At the level of the individual cell, the timing or rate of metabolic processes is also tightly regulated. For example, energy metabolism occurs at a rate needed to make the amount of **adenosine triphosphate (ATP)** required for activities such as excitability, secretion, maintaining osmotic integrity, and countless biosynthetic processes. The cell not only meets its basic metabolic "housekeeping" needs but also remains poised to do its own special work in the body, such as conducting nerve impulses and contracting, absorbing, and secreting. During its lifespan, the cell continues to make the enzymatic and structural proteins that ensure the maintenance of an appropriate rate of metabolism.

The thyroid hormones, **thyroxine (T_4)** and **triiodothyronine (T_3)**, play key roles in the regulation of body development and govern the rate at which metabolism occurs in individual cells. Although these hormones are not essential for life, without them, life would lose its orderly nature. Without adequate levels of thyroid hormones, the body fails to develop on time. Cellular housekeeping moves at a slower pace, eventually influencing the ability of individual cells to carry out their physiologic functions. The thyroid hormones exert their regulatory functions by influencing gene expression and affecting the developmental program and the amount of cellular constituents needed for the normal rate of metabolism.

FUNCTIONAL ANATOMY

The human thyroid gland consists of two lobes attached to either side of the trachea by connective tissue. A band of thyroid tissue or isthmus, which lies just below the cricoid cartilage, connects the two lobes and gives the gland a bowtie shape in appearance. The thyroid gland is one of the largest endocrine glands in the body and, in a healthy adult, it weighs about 20 g.

Each lobe of the thyroid receives its arterial blood supply from a superior and an inferior thyroid artery, which arise from the external carotid and subclavian arteries, respectively. Blood leaves the lobes of the thyroid by a series of thyroid veins that drain into the external jugular and innominate veins. This circulation provides a rich blood supply to the thyroid gland, giving it a higher rate of blood flow per gram than even that of the kidneys.

The thyroid gland receives adrenergic innervation from the cervical ganglia and cholinergic innervation from the vagus nerves. This innervation regulates vasomotor function to increase the delivery of **thyroid-stimulating hormone (TSH)**, iodide, and metabolic substrates to the thyroid gland. The adrenergic system can also affect thyroid function by direct effects on the cells.

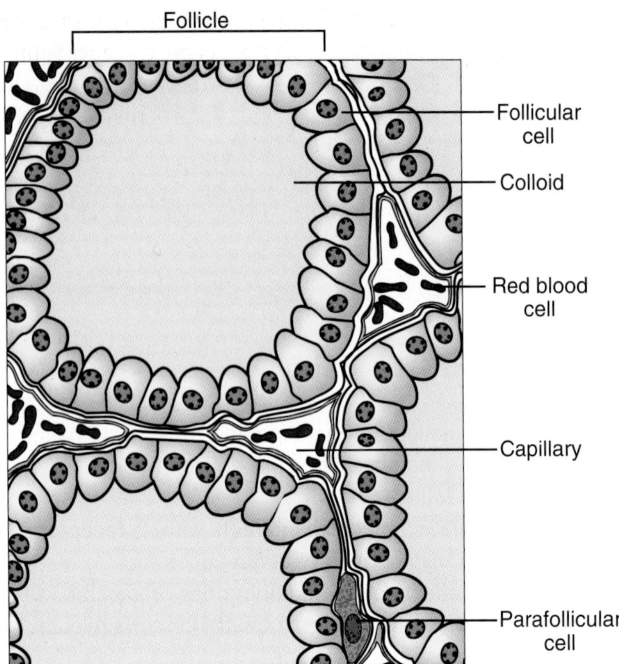

Figure 32.1 A cross-sectional view through a portion of the human thyroid gland. (See text for details.)

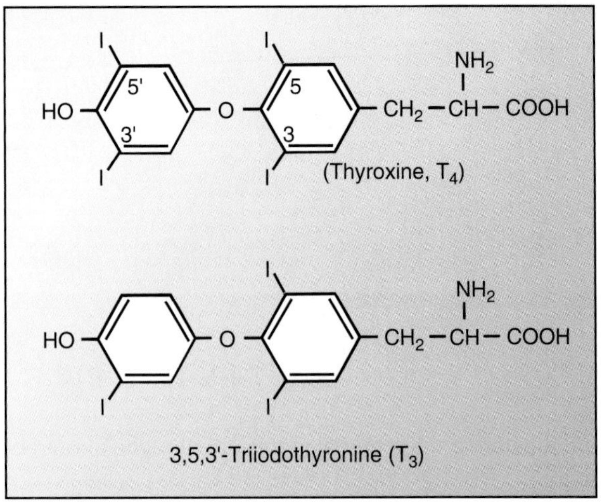

HO–(3′,5′)–O–(3,5)–CH_2–CH(NH_2)–COOH (Thyroxine, T_4)

HO–(3′)–O–(3,5)–CH_2–CH(NH_2)–COOH 3,5,3′-Triiodothyronine (T_3)

Figure 32.2 The molecular structure of the thyroid hormones. The numbering of the iodine atoms on the iodothyronine ring structure is shown.

Thyroid follicles contain the major cells that synthesize and secrete thyroxine and triiodothyronine.

The lobes of the thyroid gland consist of aggregates of many spherical **follicles**, lined by a single layer of epithelial cells (Fig. 32.1). Microvilli cover the apical membranes of the follicular cells, which face the lumen. Pseudopods formed from the apical membrane extend into the lumen. Tight junctions, which provide a seal for the contents of the lumen, connect the lateral membranes of the follicular cells. The basal membranes of the follicular cells are close to the rich capillary network that penetrates the stroma between the follicles.

The lumen of the follicle contains a thick, proteinaceous gel-like substance called **colloid** (see Fig. 32.1). The colloid is a solution composed primarily of **thyroglobulin (Tg)**, a large protein that is a storage form of the thyroid hormones. The high viscosity of the colloid is a result of the high concentration (10%–25%) of thyroglobulin.

The thyroid follicle produces and secretes two thyroid hormones, T_4 and T_3. The molecular structures of these hormones are shown in Figure 32.2. T_4 and T_3 are iodinated derivatives of the amino acid tyrosine; the coupling of the phenyl rings of two iodinated tyrosine molecules in an ether linkage forms them. The resulting structure is called an **iodothyronine**. The mechanism of this process is discussed in detail later.

T_4 contains four iodine atoms on the 3, 5, 3′, and 5′ positions of the thyronine ring structure, whereas T_3 has only three iodine atoms, at ring positions 3, 5, and 3′ (see Fig. 32.2). This explains why thyroxine is usually abbreviated as T_4 and triiodothyronine as T_3. Because T_4 and T_3 contain the element iodine, their synthesis by the thyroid follicle depends on an adequate supply of iodine in the diet.

Parafollicular cells synthesize calcitonin.

In addition to the epithelial cells that secrete T_4 and T_3, the wall of the thyroid follicle contains a few **parafollicular cells** (see Fig. 32.1). The parafollicular cell is usually embedded in the wall of the follicle, inside the basal lamina surrounding the follicle, and its plasma membrane does not form part of the wall of the lumen. Parafollicular cells produce and secrete the hormone calcitonin. Calcitonin and its effects on calcium metabolism are discussed in Chapter 35, "Endocrine Regulation of Calcium, Phosphate, and Bone Homeostasis."

THYROID HORMONE SYNTHESIS, SECRETION, AND METABOLISM

The thyroid follicle does not directly synthesize T_4 and T_3 in their final form. Instead, they are formed by the chemical modification of tyrosine residues in the peptide structure of Tg as it is secreted by the follicular cells into the lumen of the follicle. Therefore, the T_4 and T_3 formed by this chemical modification are actually part of the amino acid sequence of Tg.

The high concentration of Tg in the colloid provides a large reservoir of stored thyroid hormones for later processing and secretion by the follicle. The synthesis of T_4 and T_3 is completed when Tg is retrieved through **pinocytosis** of the colloid by the follicular cells. Lysosomal enzymes then hydrolyze Tg to its constituent amino acids, releasing T_4 and T_3 molecules from their peptide linkage. T_4 and T_3 are then secreted into the blood.

Thyroid follicular cells synthesize and secrete iodinated thyroglobulin precursor.

The steps involved in the synthesis of iodinated Tg are shown in Figure 32.3. This process involves the synthesis of a thyroglobulin precursor, the uptake of iodide, and the formation of iodothyronine residues.

Thyroglobulin precursor protein

The synthesis of the protein precursor for Tg is the first step in the formation of T_4 and T_3. This substance is a 660-kDa glycoprotein composed of two similar 330-kDa subunits held together by disulfide bridges. Ribosomes synthesize the subunits on the rough endoplasmic reticulum (ER), and then the subunits undergo dimerization and glycosylation in the smooth ER. The Golgi apparatus packages the completed glycoprotein into vesicles. These vesicles migrate to the apical membrane of the follicular cell and fuse with it. The thyroglobulin precursor protein is then extruded onto the apical surface of the cell, where iodination takes place.

Iodide uptake and release

The iodide used for iodination of the thyroglobulin precursor protein comes from the blood perfusing the thyroid gland. The **sodium-iodide symporter (NIS)** is located on the basal plasma membrane of the follicular cell, near the capillaries that supply the follicle (see Fig. 32.3). The NIS uses a sodium gradient across the basal membrane to transport one iodide atom along with two Na^+ ions into the cytosol of the follicular cell. Although NIS does not require ATP to transport iodide, the Na^+K^+ ATPase is required to pump Na^+ back out of the cell and to maintain the Na^+ gradient. The NIS also transports other anions, such as bromide, thiocyanate, and perchlorate, across the basal membrane, and these can be used to reduce iodide uptake by the gland. Through the activity of NIS, the follicular cell concentrates iodide many times over the concentration of iodide present in the blood; therefore, follicular cells are efficient extractors of the small amount of iodide circulating in the blood. Once inside follicular cells, the iodide ions diffuse rapidly to the apical membrane, where **pendrin** transports it into the follicle to be used for iodination of the thyroglobulin precursor.

Formation of iodothyronine residues

The next step in the formation of Tg is the addition of one or two iodine atoms to certain tyrosine residues in the thyroglobulin precursor protein, a process termed **organification**. The precursor protein contains 134 tyrosine residues but only a small fraction of these become iodinated. A typical Tg molecule contains only 20 to 30 atoms of iodine.

The enzyme **thyroid peroxidase (TPO)**, which is bound to the apical membrane of the follicular cells, catalyzes the iodination of Tg. TPO binds an iodide ion and a tyrosine residue in the thyroglobulin precursor, bringing them in

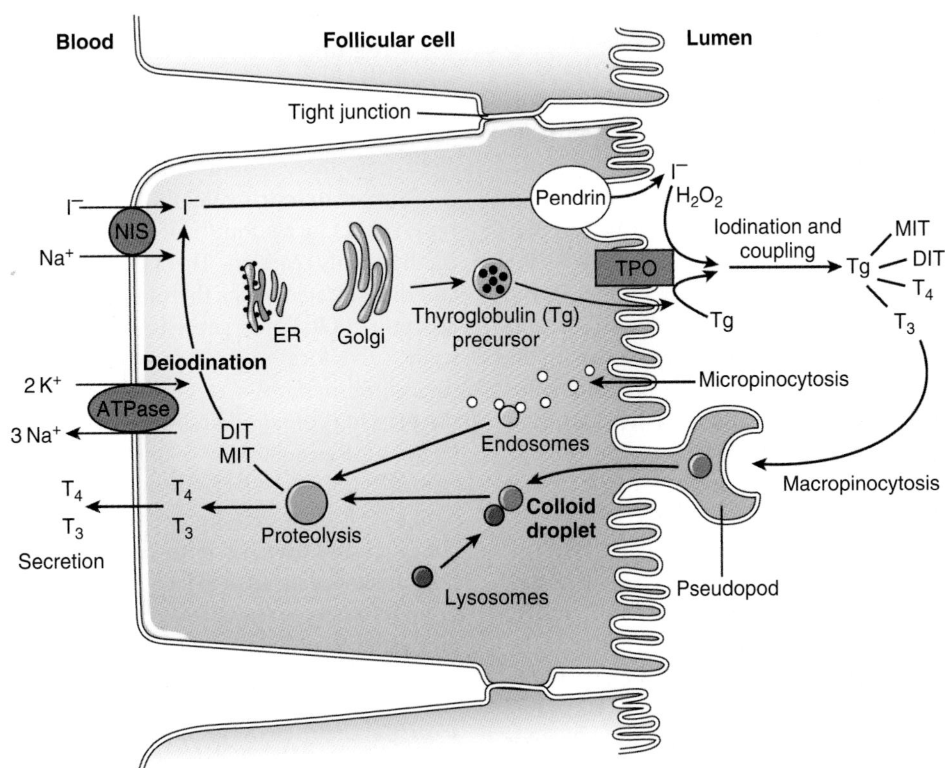

Figure 32.3 Thyroid hormone synthesis and secretion. (See text for details.) ATPase, adenosine triphosphatase; DIT, diiodotyrosine; MIT, monoiodotyrosine; T_3, triiodothyronine; T_4, thyroxine; NIS, sodium iodide symporter; TPO, thyroid peroxidase.

close proximity. The enzyme oxidizes the iodide ion and the tyrosine residue to short-lived free radicals, using hydrogen peroxide that has been generated within the mitochondria of follicular cells. The free radicals then undergo addition. The product formed is a **monoiodotyrosine (MIT)** residue, which remains in peptide linkage in the thyroglobulin structure. This same enzymatic process may add a second iodine atom to an MIT residue, forming a **diiodotyrosine (DIT)** residue.

Iodinated tyrosine residues that are close together in the thyroglobulin precursor molecule undergo a **coupling reaction**, which forms the iodothyronine structure. TPO, the same enzyme that initially oxidizes iodine, is believed to catalyze the coupling reaction through the oxidation of neighboring iodinated tyrosine residues to short-lived free radicals. These free radicals undergo addition to produce an iodothyronine residue and a **dehydroalanine residue**, both of which remain in peptide linkage in the Tg structure. For example, when two neighboring DIT residues couple by this mechanism, T_4 is formed. After being iodinated, the Tg molecule is stored as part of the colloid in the lumen of the follicle (see Fig. 32.3).

Only about 20% to 25% of the DIT and MIT residues in Tg become coupled to form iodothyronines. For example, a typical Tg molecule contains five to six uncoupled residues of DIT and two to three residues of T_4. However, T_3 is formed in only about one of five Tg molecules. As a result, the thyroid secretes substantially more T_4 than T_3.

Secretion of thyroid hormones involves phagocytosis of thyroglobulin by the follicular cells.

When the thyroid gland is stimulated to secrete thyroid hormones, vigorous pinocytosis occurs at the apical membranes of follicular cells. Pseudopods from the apical membrane reach into the lumen of the follicle, engulfing bits of the colloid (see Fig. 32.3). Endocytotic vesicles formed by this pinocytotic activity migrate toward the basal region of the follicular cell. Lysosomes, which are mainly located in the basal region of resting follicular cells, migrate toward the apical region of the stimulated cells. The lysosomes fuse with the Tg-containing droplet and hydrolyze the Tg to its constituent amino acids. As a result, T_4, T_3, and the other iodinated amino acids are released into the cytosol.

Thyroxine and triiodothyronine secretion

T_4 and T_3 formed from the hydrolysis of Tg are released from the follicular cell and enter the nearby capillary circulation; however, the mechanism of transport of T_4 and T_3 across the basal plasma membrane has not been defined. The DIT and MIT generated by the hydrolysis of Tg are rapidly deiodinated in the follicular cell. The follicular cell reuses the released iodide for the continued iodination of Tg (see Fig. 32.3).

Most of the T_4 and T_3 molecules that enter the bloodstream become bound to plasma proteins. About 70% of the T_4 and 80% of the T_3 are noncovalently bound to **thyroxine-binding globulin (TBG)**, a 54-kDa glycoprotein that is synthesized and secreted by the liver. Each molecule of TBG has a single binding site for a thyroid hormone molecule. The remaining T_4 and T_3 in the blood are bound to **transthyretin** or to albumin. Less than 1% of the T_4 and T_3 in blood is in the free form, and it is in equilibrium with the large protein-bound fraction. It is this small amount of free thyroid hormone that interacts with target cells.

The protein-bound form of T_4 and T_3 represents a large reservoir of preformed hormone that can replenish the small amount of circulating free hormone as it is cleared from the blood. This reservoir provides the body with a buffer against drastic changes in circulating thyroid hormone levels as a result of sudden changes in the rate of T_4 and T_3 secretion. The protein-bound T_4 and T_3 molecules are also protected from metabolic inactivation and excretion in the urine. As a result of these factors, the thyroid hormones have long half-lives in the bloodstream. The half-life of T_4 is about 7 days; the half-life of T_3 is about 1 day.

Peripheral tissues metabolize thyroid hormones.

Deiodination reactions in the peripheral tissues both activate and inactivate thyroid hormones. The enzymes that catalyze the various deiodination reactions are regulated, resulting in different thyroid hormone concentrations in various tissues under different physiologic and pathophysiologic conditions.

Thyroxine conversion to triiodothyronine

As noted earlier, T_4 is the major secretory product of the thyroid gland and is the predominant thyroid hormone in the blood. However, about 40% of the T_4 secreted by the thyroid gland is converted to T_3 by enzymatic removal of the iodine atom at position 5′ of the thyronine ring structure in a reaction termed **outer ring deiodination** (Fig. 32.4). **Deiodinase type 1 (D1)** located in the liver, kidneys, and thyroid gland catalyzes this reaction. The T_3 formed by this deiodination and that secreted by the thyroid react with **thyroid hormone receptors (TRs)** in target cells; therefore, T_3 is the physiologically active form of the thyroid hormones. Skeletal muscle, the central nervous system (CNS), the pituitary gland, and the placenta contain **deiodinase type 2 (D2)**. D2 is believed to function primarily to maintain intracellular T_3 in target tissues, but it may also contribute to the generation of circulating T_3. A third **deiodinase (type 3 or D3)** catalyzes **inner ring deiodination** reactions during degradation of thyroid hormones as discussed below. All of the deiodinases contain **selenocysteine** in the active center. This rare amino acid has properties that make it ideal for catalysis of oxidoreductive reactions.

Inactivation of thyroxine and triiodothyronine

Whereas the 5′-deiodination of T_4 to produce T_3 can be viewed as a metabolic activation process, both T_4 and T_3 undergo enzymatic deiodinations, particularly in the liver and kidneys, which inactivate them (see Fig. 32.4). For example, about 40% of the T_4 secreted by the human

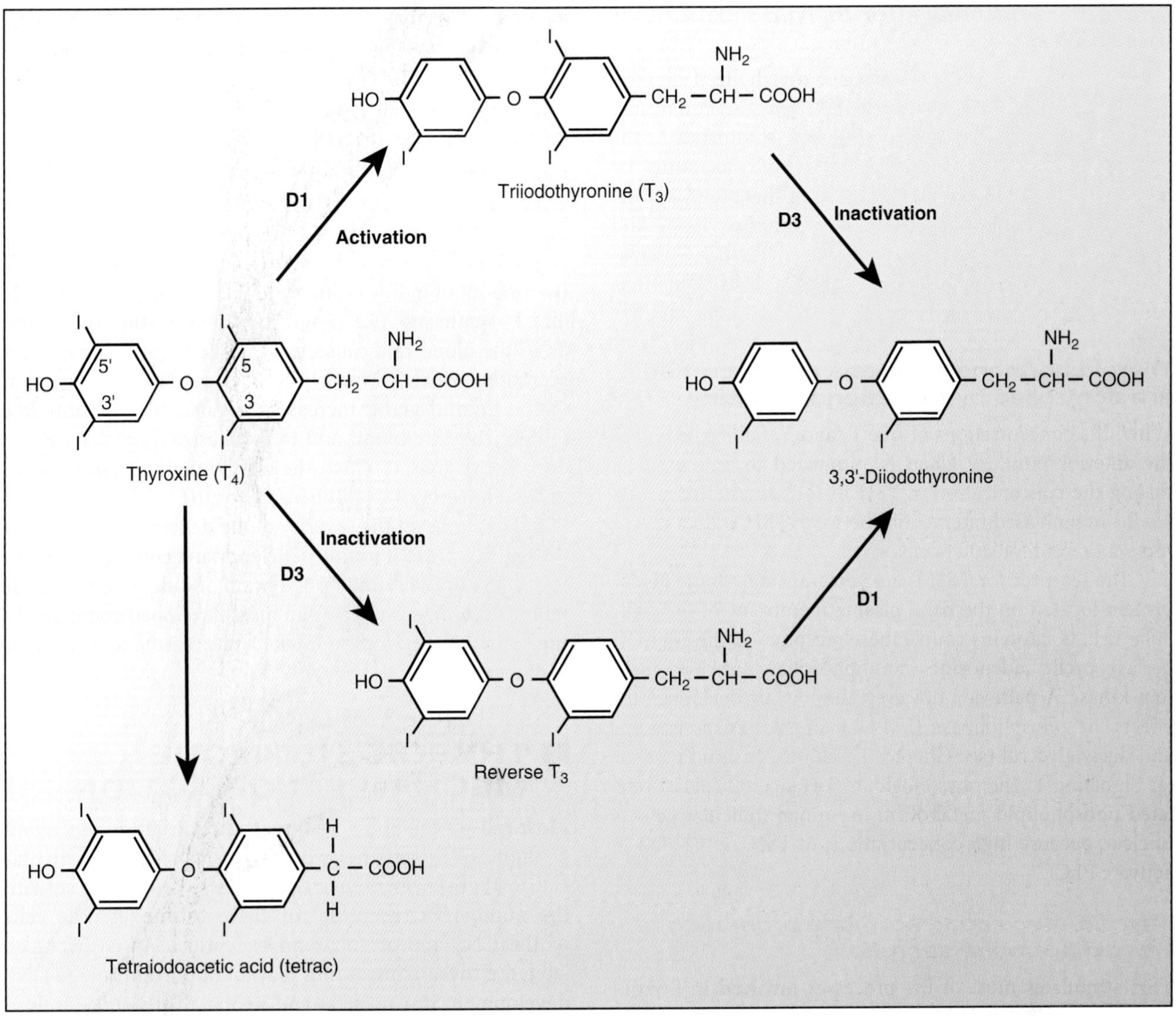

Figure 32.4 The metabolism of thyroxine. Deiodinase type 1 (D1) deiodinates thyroxine (T_4) at the 5′ position to form triiodothyronine (T_3), the physiologically active thyroid hormone. Deiodinase type 3 (D3) also enzymatically deiodinates some T_4 at the 5′ position to form the inactive metabolite, reverse T_3. T_3 and reverse T_3 undergo additional deiodinations to 3,3′-diiodothyronine before being excreted. A small amount of T_4 is also decarboxylated and deaminated to form the metabolite, tetraiodoacetic acid (tetrac). Tetrac may then be deiodinated before being excreted.

thyroid gland is deiodinated at the 5 position on the thyronine ring structure, primarily by D3 to produce **reverse T_3 (rT_3)**. Because rT_3 has little or no thyroid hormone activity, this deiodination reaction is a major pathway for the metabolic inactivation or disposal of T_4. T_3 and rT_3 also undergo deiodination to yield 3,3′-diiodothyronine. This inactivated metabolite may be further deiodinated before being excreted.

5′-Deiodination

Certain physiologic and pathologic factors influence the 5′-deiodination reaction. The result is a change in the relative amounts of T_3 and rT_3 produced from T_4. For example, a human fetus produces less T_3 from T_4 than does a child or adult because the 5′-deiodination reaction is less active in the fetus. Also, 5′-deiodination is inhibited during fasting, particularly in response to carbohydrate restriction, but it can be restored to normal when the person is fed again. Trauma as well as most acute and chronic illnesses also suppresses the 5′-deiodination reaction. Under all of these circumstances, the amount of T_3 produced from T_4 is reduced and its blood concentration falls. The amount of rT_3 rises in the circulation, not because its conversion from T_4 is increased, as originally believed, but rather because its clearance from the blood is reduced.

Note that during fasting or in the disease states mentioned above, the secretion of T_4 is usually not increased, despite the decrease of T_3 in the circulation. This response is likely a result of the fact that D2 in the CNS and pituitary is not affected under these conditions and continues to produce T_3 in quantities sufficient to maintain relatively normal hypothalamic-pituitary-thyroid axis function.

Alternative pathways for thyroid hormone metabolism

T_4 and, to a lesser extent, T_3 are also metabolized by conjugation with glucuronic acid in the liver. The conjugated hormones are secreted into the bile and eliminated in the feces. Many tissues also metabolize thyroid hormones by modifying the three-carbon side chain of the iodothyronine structure. These modifications include decarboxylation and deamination. The derivatives formed from T_4, such as **tetraiodoacetic acid (tetrac)**, may also undergo deiodinations before being excreted (see Fig. 32.4).

Thyroid hormone synthesis and secretion are controlled by the anterior pituitary.

When the concentrations of free T_4 and T_3 fall in the blood, the anterior pituitary gland is stimulated to secrete TSH, raising the concentration of TSH in the blood. This action results in increased interactions between TSH and its receptors on thyroid follicular cells.

The receptor for TSH is a seven-transmembrane glycoprotein located on the basal plasma membrane of the follicular cell. G_s proteins couple these receptors to the adenylyl cyclase–cyclic adenosine monophosphate (cAMP)–protein kinase A pathway; however, there is also evidence for effects via phospholipase C (PLC), inositol trisphosphate, and diacylglycerol (see Chapter 1, "Homeostasis and Cellular Signaling"). The physiologic importance of TSH-stimulated phospholipid metabolism in human follicular cells is unclear, because high concentrations of TSH are needed to activate PLC.

Thyroid-stimulating hormone promotion of thyroid hormone secretion

TSH stimulates most of the processes involved in thyroid hormone synthesis and secretion by increasing cAMP in the follicular cells. TSH stimulates synthesis of NIS and uptake of iodide by follicular cells, the iodination of tyrosine residues in the thyroglobulin precursor, and the coupling of iodinated tyrosines to form iodothyronines. Moreover, TSH stimulates the pinocytosis of colloid by the apical membranes, resulting in a great increase in endocytosis of thyroglobulin and its hydrolysis. The overall result of these effects of TSH is an increased release of T_4 and T_3 into the blood. In addition to its effects on thyroid hormone synthesis and secretion, TSH rapidly increases energy metabolism in the thyroid follicular cell.

Thyroid-stimulating hormone regulation of thyroid gland growth

Over the long term, TSH promotes protein synthesis in thyroid follicular cells, maintaining their size and structural integrity. Evidence of this trophic effect of TSH is seen in a hypophysectomized patient, whose thyroid gland atrophies, largely as a result of a reduction in the height of follicular cells. However, the chronic exposure of a person to excessive amounts of TSH causes the thyroid gland to increase in size. This enlargement results from an increase in follicular cell height and number. Such an enlarged thyroid gland is called a **goiter**. cAMP primarily mediates these trophic and proliferative effects of TSH on the thyroid.

Dietary iodide is essential for the synthesis of thyroid hormones.

Iodine atoms are constituent parts of the T_4 and T_3 molecules; therefore, a continual supply of iodide is required for the synthesis of these hormones. If a person's diet is severely deficient in iodide, as in some parts of the world, the amount of iodide available to the thyroid gland limits T_4 and T_3 synthesis. As a result, the concentrations of T_4 and T_3 in the blood fall, causing a chronic stimulation of TSH secretion, which in turn, produces a goiter. Enlargement of the thyroid gland increases its capacity to accumulate iodide from the blood and to synthesize T_4 and T_3. However, the degree to which the enlarged gland can produce thyroid hormones to compensate for their deficiency in the blood depends on the severity of the deficiency of iodide in the diet. To prevent iodide deficiency and consequent goiter formation in the human population, iodide is added to the table salt (iodized salt) sold in most developed countries. In the United States, iodine intake is decreasing as salt intake is reduced.

THYROID HORMONE MECHANISM OF ACTION

Most cells of the body are targets for the action of thyroid hormones. The sensitivity or responsiveness of a particular cell to thyroid hormones correlates to some degree with the number of receptors for these hormones. The cells of the CNS appear to be an exception. As is discussed later, the thyroid hormones play an important role in CNS development during fetal and neonatal life, and developing nerve cells in the brain are important targets for thyroid hormones. In the adult, however, brain cells show little responsiveness to the metabolic regulatory action of thyroid hormones, although they have numerous receptors for these hormones. The reason for this discrepancy is unclear.

Thyroid hormone receptor is a nuclear receptor that is activated by binding of triiodothyronine.

TRs are located in the nuclei of target cells bound to **thyroid hormone response elements (TREs)** in the DNA. TRs are protein molecules of about 50 kDa that are structurally similar to the nuclear receptors for steroid hormones and vitamin D. In humans, two TR genes (a and b) are found on two separate chromosomes and encode several different receptor proteins (TRa1, TRb1, TRb2, and TRb3). The TR receptors are expressed in a tissue-specific manner to impart different effects of thyroid hormone on various tissues. Thyroid receptors bound to the TRE in the absence of T_3 generally act to repress gene expression.

Target cells take up the free forms of T_3 and T_4 from the blood through a carrier-mediated process that requires ATP.

Once inside the cell, T_4 is deiodinated to T_3, which enters the nucleus of the cell and binds to its receptor in the chromatin. The TR with bound T_3 forms a complex (called a *heterodimer*) with another nuclear receptor called the **retinoid X receptor (RXR)**. TR can also complex with other TRs to form homodimers and activate transcription. Other transcription coactivators and corepressors also complex with the TR heterodimer or homodimer. As a result, the production of mRNA for certain proteins is either increased or decreased, changing the cell's capacity to make these proteins (Fig. 32.5). T_3 can influence differentiation by regulating the kinds of proteins produced by its target cells and can influence growth and metabolism by changing the amounts of structural and enzymatic proteins present in the cells. The mechanisms by which T_3 alters gene expression continue to be investigated.

The transcriptional response to T_3 is slow to appear. When T_3 is given to an animal or human, several hours elapse before its physiologic effects can be detected. This delayed action undoubtedly reflects the time required for transcriptional changes and consequent changes in the synthesis of key proteins to occur. When T_4 is administered, its course of action is usually slower than that of T_3 because of the additional time required for the body to convert T_4 to T_3.

Thyroid hormones also have effects on cells that occur much faster and do not appear to be mediated by nuclear TR receptors, including effects on signal transduction pathways that alter cellular respiration, cell morphology, vascular tone, and ion homeostasis. The physiologic relevance of these effects is currently being investigated.

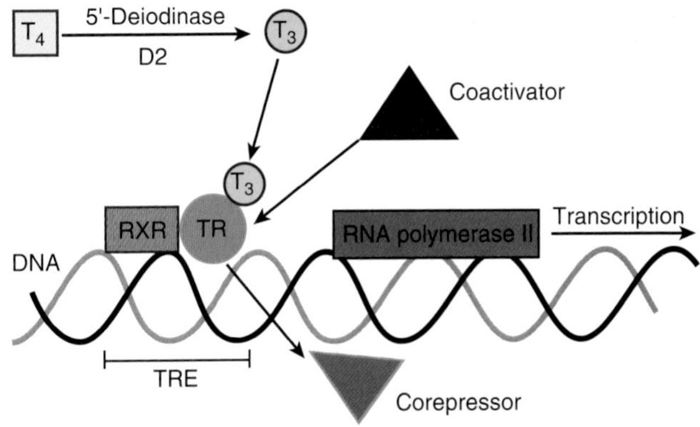

Figure 32.5 The activation of transcription by thyroid hormone. Thyroxine (T_4) is taken up by the cell and deiodinated to triiodothyronine (T_3), which then binds to the thyroid hormone receptor (TR). The activated TR heterodimerizes with a second transcription factor, 9-*cis* retinoic acid receptor (RXR) and binds to the thyroid hormone response element (TRE). The binding of TR/RXR to the TRE displaces repressors of transcription and recruits additional coactivators. The final result is the activation of RNA polymerase II and the transcription of the target gene.

THYROID HORMONE FUNCTION

Thyroid hormones play a critical role in the development of the CNS. They are also essential for normal body growth during childhood and in basal energy metabolism. Thyroid hormones also influence intermediary metabolism and regulate their own synthesis.

Clinical Focus / 32.1

Resistance to Thyroid Hormone

Resistance to thyroid hormone (RTH) results from mutations in one allele of the thyroid receptor β (TRb) gene. These mutations tend to cluster in areas of the thyroid hormone receptor that have important interactions with the hydrophobic ligand-binding domain; thus, the mutant TRb does not bind T_3 with high affinity. The mutations do not interfere with the DNA binding domain, the corepressor binding domain, or the region of the receptor that heterodimerizes with 9-*cis* retinoic acid receptor (RXR).

It is thought that RTH is produced when a mutant TRb heterodimerizes with RXR, a normal TRb, or TRa. Receptor dimers containing mutant TRb compete with wild-type (normal) TRb-containing dimers for binding to thyroid hormone response elements (TREs) of thyroid hormone–regulated genes (see Fig. 32.5). The mutant TRb heterodimers also bind corepressor molecules that cannot be released because of the defect in triiodothyronine (T_3) binding; therefore, genes become more repressed than they normally would be.

Over 400 families with RTH have been identified, and it is expected that many more cases likely exist. The frequency of mutation has been estimated at 1:50,000. Because TRb interferes with the function of the three normal TR-expressing genes, the effect of the mutation is characterized as a dominant negative inhibition. Patients with RTH may have features of hypothyroidism if the resistance is severe and affects all tissues. Such people are referred to as having *generalized resistance to thyroid hormone*. Alternatively, patients may present with hyperthyroidism if the hypothalamic-pituitary axis is more severely affected—a condition termed *pituitary resistance to thyroid hormone*. The hypothalamic-pituitary axis is primarily regulated by TRb and thus is more sensitive to mutations in the receptor than tissues such as the heart, in which TRa is the dominant thyroid hormone receptor. Although patients may present with symptoms of hyperthyroidism, thyroid-stimulating hormone (TSH) level is normal or only slightly elevated, a finding important in differentiating between a diagnosis of RTH and TSH-secreting pituitary tumor, which results in high TSH levels.

Many patients may present with a mixture of hypothyroid and hyperthyroid conditions. Treatment of patients with RTH is therefore difficult because thyroid hormone analogs designed to suppress the hypothalamic-pituitary axis and reduce the release of thyroxine (T_4) can worsen effects on the heart, which has functional TRa receptors. β-blockers may be of use in such situations. Development of thyroid hormone analogs that preferentially bind TRb or TRa may prove useful for treatment of RTH.

Thyroid hormones are crucial for brain maturation during fetal development.

The human brain undergoes its most active phase of growth during the last 6 months of fetal life and the first 6 months of postnatal life. During the second trimester of pregnancy, the multiplication of neuroblasts in the fetal brain reaches a peak and then declines. As pregnancy progresses and the rate of neuroblast division drops, neuroblasts differentiate into neurons and begin the process of synapse formation that extends into postnatal life.

Thyroid hormones first appear in the fetal blood during the second trimester of pregnancy, and levels continue to rise during the remaining months of fetal life. TRs increase about 10-fold in the fetal brain at about the time the concentrations of T_4 and T_3 begin to rise in the blood. These events are critical for normal brain development because thyroid hormones are essential for timing the decline in nerve cell division and the initiation of differentiation and maturation of these cells.

If thyroid hormones are deficient during these prenatal and postnatal periods of differentiation and maturation of the brain, mental retardation occurs. The cause is thought to be inadequate development of the neuronal circuitry of the CNS. Thyroid hormone therapy must be given to a thyroid hormone–deficient child during the first few months of postnatal life to prevent mental retardation. Starting thyroid hormone therapy after behavioral deficits have occurred cannot reverse the mental retardation (i.e., thyroid hormone must be present when differentiation normally occurs). Thyroid hormone deficiency during infancy causes both mental retardation and growth impairment, as discussed below. Fortunately, this occurs rarely today because thyroid hormone deficiency is usually detected in newborn infants and hormone therapy is given at the proper time.

The exact mechanism by which thyroid hormones influence differentiation of the CNS is unknown. Animal studies have demonstrated that thyroid hormones inhibit nerve cell replication in the brain and stimulate the growth of nerve cell bodies, the branching of dendrites, and the rate of myelinization of axons. These effects of thyroid hormones are presumably a result of their ability to regulate the expression of genes involved in nerve cell replication and differentiation. However, the details, particularly in humans, are unclear.

Thyroid hormones are required for normal cell differentiation, development, and body growth.

Thyroid hormones are important factors regulating the growth of the entire body. For example, a person who is deficient in thyroid hormones and who does not receive thyroid hormone therapy during childhood will not grow to a normal adult height.

Growth hormone expression

A major way thyroid hormones promote normal body growth is by stimulating the expression of the gene for growth hormone (GH) in the somatotrophs of the anterior pituitary gland. In a thyroid hormone–deficient person, GH synthesis by the somatotrophs is greatly reduced, and consequently, GH secretion is impaired; therefore, a thyroid hormone–deficient person will also be GH-deficient. If this condition occurs in a child, it will cause growth retardation, largely a result of the lack of the growth-promoting action of GH (see Chapter 31, "Hypothalamus and the Pituitary Gland").

Tissue and bone growth

Thyroid hormones have additional effects on growth. In tissues such as skeletal muscle, the heart, and the liver, thyroid hormones have direct effects on the synthesis of many structural and enzymatic proteins. For example, they stimulate the synthesis of structural proteins of mitochondria, as well as the formation of many enzymes involved in intermediary metabolism and oxidative phosphorylation.

Thyroid hormones also promote the general growth and calcification of the bones of the skeleton. Deficiency in the thyroid hormone early in life leads to delayed, abnormal development of bone and can lead to dwarfism. The effect of thyroid hormones to regulate bone growth is, in part, mediated through its effects on GH and insulin-like growth factor I (IGF-1) synthesis.

Thyroid hormones are the major determinant of basal metabolic rate.

When the body is at rest, about half of the ATP produced by its cells is used to drive energy-requiring membrane transport processes. The remainder is used in involuntary muscular activity, such as respiratory movements, peristalsis, contraction of the heart, and in many metabolic reactions requiring ATP such as protein synthesis. The energy required to do this work is eventually released as body heat.

The major site of ATP production is the mitochondria, where the oxidative phosphorylation of **adenosine diphosphate (ADP)** to ATP takes place. The rate of oxidative phosphorylation depends on the supply of ADP for electron transport. The ADP supply is, in turn, a function of the amount of ATP used to do work. For example, when more work is done per unit time, more ATP is used and more ADP is generated, increasing the rate of oxidative phosphorylation. The rate at which oxidative phosphorylation occurs is reflected in the amount of oxygen consumed by the body because oxygen is the final electron acceptor at the end of the electron transport chain.

Activities that occur when the body is not at rest such as voluntary movements use additional ATP for the work involved. The total amount of oxygen consumed and body heat produced over a 24-hour period is the combination of that needed at rest and that needed during activity.

Thyroid hormone action on thermogenesis

Thyroid hormones regulate the basal rate at which oxidative phosphorylation takes place in cells. As a result, they set the basal rates of body heat production and of oxygen consumed by the body. This is called the **thermogenic action** of thyroid hormones.

Thyroid hormone levels in the blood must be within normal limits for basal metabolism to proceed at the rate needed for a balanced energy economy of the body. For example, if thyroid hormones are present in excess, oxidative phosphorylation is accelerated and body heat production and oxygen consumption are abnormally high. The converse occurs when the blood concentrations of T_4 and T_3 are lower than normal. The fact that thyroid hormones affect the amount of oxygen consumed by the body has been used clinically to assess the status of thyroid function. The amount of oxygen consumed by the body under resting conditions is called the **basal metabolic rate (BMR)** and assumes normal thyroid function. A BMR higher than that expected for a person with normal thyroid function would indicate that concentrations of T_4 and T_3 are higher than normal.

Not all tissues are sensitive to the thermogenic action of thyroid hormones. Tissues and organs that give this response include skeletal muscle, the heart, the liver, and the kidneys. TRs are also abundant in these tissues. The adult brain, skin, lymphoid organs, and gonads show little thermogenic response to thyroid hormones. With the exception of the adult brain, these tissues contain few TRs, which may explain their poor response.

Oxidative phosphorylation

The thermogenic action of the thyroid hormones is poorly understood at the molecular level. The thermogenic effect takes many hours to appear after the administration of thyroid hormones to a human or animal, probably because of the time required for changes in the expression of genes involved. T_3 is known to stimulate the synthesis of cytochromes, cytochrome oxidase, and Na^+/K^+-ATPase in certain cells. This action suggests that T_3 may regulate the number of respiratory units in these cells, affecting their capacity to carry out oxidative phosphorylation. A greater rate of oxidative phosphorylation would result in greater heat production.

Thyroid hormone also stimulates the synthesis of **uncoupling protein-1 (UCP-1)** in brown adipose tissue. ATP synthase synthesizes ATP in the mitochondria when protons flow down their electrochemical gradient. UCP-1 acts as a channel in the mitochondrial membrane to dissipate the ion gradient without making ATP. As the protons move down their electrochemical gradient *uncoupled* from ATP synthesis, energy is released as heat. Adult humans do not appear to have enough brown adipose tissue for UCP-1 to make a significant contribution to nutrient oxidation or body heat production. However, several uncoupling proteins (UCP-2 and UCP-3) have recently been discovered in many tissues and thyroid hormones regulate their expression. These novel uncoupling proteins may have a role in the thermogenic action of thyroid hormones.

Thyroid hormones regulate carbohydrate, fat, and protein metabolism.

In addition to their ability to regulate the rate of basal energy metabolism, thyroid hormones influence the rate at which most of the pathways of intermediary metabolism operate in their target cells. When thyroid hormones are deficient, pathways of carbohydrate, lipid, and protein metabolism are slowed, and their responsiveness to other regulatory factors such as other hormones is decreased. However, these same metabolic pathways run at an abnormally high rate when thyroid hormones are present in excess. Thyroid hormones, therefore, can be viewed as amplifiers of cellular metabolic activity. The amplifying effect of thyroid hormones on intermediary metabolism is mediated through the activation of genes encoding enzymes involved in these metabolic pathways.

Thyroid hormones provide a negative feedback in the control of the hypothalamus-pituitary axis.

An important action of the thyroid hormones is the regulation of their own secretion. As discussed in Chapter 31, "Hypothalamus and the Pituitary Gland," T_3 exerts an inhibitory effect on TSH secretion by thyrotrophs in the anterior pituitary gland by decreasing thyrotroph sensitivity to **thyrotropin-releasing hormone (TRH)**. Consequently, when the circulating concentration of free thyroid hormone is high, thyrotrophs are relatively insensitive to TRH and the rate of TSH secretion decreases. The resulting fall in TSH levels in the blood reduces the rate of thyroid hormone release from the follicular cells in the thyroid. When the free thyroid hormone level falls in the blood, however, the negative-feedback effect of T_3 on thyrotrophs is reduced and the rate of TSH secretion increases. The rise in TSH in the blood stimulates the thyroid gland to secrete thyroid hormones at a greater rate. This action of T_3 on thyrotrophs is thought to be a result of changes in gene expression in these cells.

The physiologic actions of the thyroid hormones described above are summarized in Table 32.1.

TABLE 32.1 Physiologic Actions of Thyroid Hormones

Category	Specific Action
Development of central nervous system	Inhibit nerve cell replication Stimulate growth of nerve cell bodies Stimulate branching of dendrites Stimulate rate of axon myelinization
Body growth	Stimulate expression of gene for growth hormone in somatotrophs Stimulate synthesis of many structural and enzymatic proteins Promote calcification of bones
Basal energy economy of the body	Regulate basal rates of oxidative phosphorylation, body heat production, and oxygen consumption (thermogenic effect)
Intermediary metabolism	Stimulate synthetic and degradative pathways of carbohydrate, lipid, and protein metabolism
Thyroid-stimulating hormone (TSH) secretion	Inhibit TSH secretion by decreasing sensitivity of thyrotrophs to thyrotropin-releasing hormone

THYROID FUNCTION ABNORMALITIES IN ADULTS

A deficiency or excess of thyroid hormone produces characteristic changes in the body. These changes result from dysregulation of nervous system function and altered metabolism.

Hyperthyroidism increases energy expenditure and causes weight loss.

The most common cause of excessive thyroid hormone production in humans is **Graves disease**, an autoimmune disorder caused by antibodies directed against the TSH receptor in the plasma membrane of thyroid follicular cells. These antibodies bind to the TSH receptor, resulting in an increase in the activity of adenylyl cyclase. The consequent rise in cAMP in follicular cells produces effects similar to those caused by the action of TSH. The thyroid gland enlarges to form a **diffuse toxic goiter**, which synthesizes and secretes thyroid hormones at an accelerated rate, causing thyroid hormones to be chronically elevated in the blood. Feedback inhibition of thyroid hormone production by the thyroid hormones is also lost.

Less common conditions that cause chronic elevations in circulating thyroid hormones include adenomas of the thyroid gland that secrete thyroid hormones and excessive TSH secretion caused by malfunctions of the hypothalamic-pituitary-thyroid axis. Many changes in the functioning of the body characterize the disease state that develops in response to excessive thyroid hormone secretion, called **hyperthyroidism** or **thyrotoxicosis**.

People with hyperthyroidism are nervous and emotionally irritable, with a compulsion to be constantly moving around. However, they also experience physical weakness and fatigue. BMR is increased and, as a result, body heat production is increased. Vasodilation in the skin and sweating occur as compensatory mechanisms to dissipate excessive body heat. Heart rate and cardiac output are increased. Energy metabolism increases, as does appetite. However, despite the increase in food intake, a net degradation of protein and lipid stores occurs, resulting in weight loss. Reducing the rate of thyroid hormone secretion with drugs or by removal of the thyroid gland by radioactive ablation or surgery can reverse all of these changes.

Hypothyroidism decreases metabolism and causes weight gain.

Thyroid hormone deficiency in humans has many causes. For example, iodide deficiency may result in a reduction in thyroid hormone production. Autoimmune diseases such as **Hashimoto disease**, also known as *Hashimoto thyroiditis*, impair thyroid hormone synthesis. Other causes of thyroid hormone deficiency include heritable diseases that affect certain steps in the biosynthesis of thyroid hormones and hypothalamic or pituitary diseases that interfere with TRH or TSH secretion. Obviously, radioiodine ablation or surgical removal of the thyroid gland also causes thyroid hormone deficiency. **Hypothyroidism** is the disease state that results from thyroid hormone deficiency, and the effects on the body are the opposite of those caused by thyroid hormone excess.

Clinical Focus / 32.2

Autoimmune Thyroid Disease: Postpartum Thyroiditis

Certain diseases affecting the function of the thyroid gland occur when a person's immune system fails to recognize particular thyroid proteins as "self" and reacts to the proteins as if they were foreign. This usually triggers both humoral and cellular immune responses. As a result, antibodies to these proteins are generated, which then alter thyroid function. Two common autoimmune diseases with opposite effects on thyroid function are Hashimoto disease and Graves disease. In Hashimoto disease, the thyroid gland is infiltrated by lymphocytes, and elevated levels of antibodies against several components of thyroid tissue (e.g., antithyroid peroxidase and antithyroglobulin antibodies) are found in the serum. The thyroid gland is destroyed, resulting in hypothyroidism. In Graves disease, stimulatory antibodies to the thyroid-stimulating hormone (TSH) receptor activate thyroid hormone synthesis, resulting in hyperthyroidism (see text for details).

A third, fairly common autoimmune disease is **postpartum thyroiditis**, which usually occurs within 3 to 12 months after delivery. The disease is characterized by a transient destruction-induced thyrotoxicosis (hyperthyroidism), often followed by a period of hypothyroidism lasting several months. Many patients eventually return to the euthyroid state. Often, only the hypothyroid phase of the disease may be observed, occurring in more than 30% of women with antibodies to thyroid peroxidase that are detectable before conception. The disease is also observed in patients known to have Graves disease. The postpartum occurrence of the disorder is likely a result of increased immune system function following the suppression of its activity during pregnancy.

It has been estimated that 5% to 10% of women develop postpartum thyroiditis. The prevalence of the disease has prompted the clinical recommendation that thyroid function (serum thyroxin, triiodothyronine, and TSH levels) be surveyed postpartum at 2, 4, 6, and 12 months in all women with thyroid peroxidase antibodies or symptoms suggestive of thyroid dysfunction. Patients who have experienced one episode of postpartum thyroiditis should also be considered at risk for recurrence after pregnancy.

Treatment for postpartum thyroiditis depends on the time of diagnosis. If symptoms during the period of transient thyrotoxicosis are severe, a β-blocker such as propranolol can be given. Thioamides—drugs that inhibit the oxidation and organic binding of thyroid iodide—are generally not effective for this transient thyrotoxicosis because the release of excess thyroid hormone is not a result of hormone synthesis but rather of destruction of the gland. Thyroid hormone replacement is required to treat the hypothyroidism that occurs later in the progression of the disease.

Thyroid hormone deficiency impairs the functioning of most tissues in the body. As described earlier, a deficiency of thyroid hormones at birth that is not treated during the first few months of postnatal life causes irreversible mental retardation. Thyroid hormone deficiency later in life also influences the function of the nervous system. For example, all cognitive functions, including speech and memory, are slowed and body movements may be clumsy. These changes can usually be reversed with thyroid hormone therapy.

Metabolism is also reduced in thyroid hormone–deficient people. BMR is reduced, resulting in impaired body heat production. Vasoconstriction occurs in the skin as a compensatory mechanism to conserve body heat. Heart rate and cardiac output are reduced. The synthetic and degradative processes of intermediary metabolism are slowed and weight gain occurs, despite the fact that food intake is reduced. In severe hypothyroidism, a substance consisting of hyaluronic acid and chondroitin sulfate complexed with protein is deposited in the extracellular spaces of the skin, causing water to accumulate osmotically. This effect gives a puffy appearance to the face, hands, and feet, called **myxedema**. All of the above disorders in adults can be normalized with thyroid hormone therapy. See From Bench to Bedside Box 32.1 for a discussion of euthyroid sick syndrome, in which T_3 levels are reduced to nearly undetectable levels in critically ill patients.

From Bench to Bedside / 32.1

Euthyroid Sick Syndrome

Severe caloric restriction and starvation result in reductions in triiodothyronine (T_3) levels, with normal or slightly reduced thyroxine (T_4) levels and normal TSH levels. This is a result of the decreased release of thyroid hormone and possibly reduced D1 and D2 deiodinase activity in peripheral tissues. Reductions in T_3 result in lower basal oxygen consumption, slower heart rate, and attenuation of nitrogen loss—changes all considered beneficial in adapting to the reduced caloric intake. In a patient who is severely ill, T_3 levels can be reduced up to 90% and reverse T_3 (rT_3) increased several-fold, with only slight reductions in T_4. In the most severe cases, the release of TSH and other anterior pituitary hormones is reduced because of the loss of hypothalamic input resulting from endogenous causes and, in some instances, exacerbated by therapy such as glucocorticoids that is given to ill patients. This condition is called **euthyroid sick syndrome** or *low T_3 syndrome.*

The mechanisms resulting in almost undetectable serum T_3 in patients with euthyroid sick syndrome are not entirely clear. It appears that the reduction in serum T_3 is a result of the decreased release of T_3 from the thyroid gland and reduced T_4 outer ring deiodination. Lower transport of T_4 into tissues such as the liver and kidney, which express D1 deiodinase, and more rapid turnover of D2 deiodinase protein both contribute to reduced T_4 deiodination and lower T_3 levels. The increase in rT_3 in euthyroid sick patients was originally thought to indicate that the specificity of D1 deiodinase was changed from 5'-T_4 to 5-T_4 deiodination, but it has since been shown that increased rT_3 is a result of reduced clearance, not increased synthesis.

Therapy for euthyroid sick syndrome remains controversial. Despite the extremely low T_3 levels, it is still not clear that T_4 or T_3 supplementation should be done because most controlled studies have not shown a beneficial effect of increasing thyroid hormones alone. A promising alternative strategy that is currently under evaluation is the infusion of a combination of growth hormone (GH)–releasing peptides and thyrotropin-releasing hormone. The administration of these factors corrects the impaired pulsatile secretion of GH, thyrotropin, and gonadotropin, resulting in increased circulating concentrations of insulin-like growth factor I (IGF-I), GH-dependent IGF-binding proteins, T_4, T_3, and testosterone. Although hypothalamic pituitary function is clearly improved and catabolism in the severely ill patient reversed, a positive effect of this treatment on survival has still not been definitively established. Normalization of glycemia with exogenous insulin has recently been shown to have a positive impact on survival. Thus, studies to understand the neuroendocrine response to critical illness and the interaction of the hypothalamic-pituitary-adrenal axis with peripheral hormonal and metabolic alterations are ongoing in an attempt to improve survival of the critically ill patient.

Chapter Summary

- The thyroid gland consists of two lobes attached to either side of the trachea; its primary function is the synthesis and secretion of thyroxine, triiodothyronine, and calcitonin.
- Thyroid hormones are synthesized by iodination and the coupling of tyrosines in reactions catalyzed by the enzyme thyroid peroxidase.
- Most of the secreted thyroxine is converted to triiodothyronine outside the thyroid.
- The thyroid gland also produces calcitonin, which plays a key role in calcium homeostasis.
- The degradation of thyroglobulin within the follicular cells releases thyroid hormones from the thyroid gland.
- Thyroid-stimulating hormone regulates the synthesis and release of thyroid hormones through activation of adenylyl cyclase and generation of cyclic adenosine monophosphate. Most of the thyroid hormones circulating in the blood are bound to transport proteins. Only a small fraction of the circulating hormones are free (unbound) and are biologically active.
- The concentration of thyroid hormones in the circulation regulates thyroid-stimulating hormone release from the anterior pituitary.
- In peripheral tissues, 5′-deiodinase deiodinates thyroxine to the physiologically active hormone triiodothyronine.
- In target tissues, triiodothyronine binds to the thyroid hormone receptor (TR), which then associates with retinoic acid receptor or a second TR to regulate transcription.
- Thyroid hormone receptor regulates transcription by binding to specific thyroid hormone response elements in target genes and interacting with specific coactivators and corepressors of transcription.
- Brain cells are a major target for triiodothyronine and thyroxine and play a crucial role in brain maturation during fetal development.
- Thyroid hormones stimulate growth by regulating growth hormone release from the pituitary and by direct actions on target tissues such as bone.
- Thyroid hormones regulate the basal metabolic rate and intermediary metabolism through effects on mitochondrial adenosine triphosphate synthesis and the expression of genes encoding metabolic enzymes.
- Nervousness and increased metabolic rate, resulting in weight loss, characterize an excess of thyroid hormone (hyperthyroidism).
- Decreased metabolic rate, resulting in weight gain, characterizes a deficiency of thyroid hormone (hypothyroidism).

33 Adrenal Gland

ACTIVE LEARNING OBJECTIVES

Upon mastering the material in this chapter, you should be able to:

- Describe the anatomy of the adrenal gland and explain which histologically distinct zone synthesizes and secretes glucocorticoids, aldosterone, and adrenal androgens.
- Explain how cells obtain cholesterol to synthesize steroids and which step is common to steroid synthesis in all three zones of the adrenal cortex.
- Explain the role of the liver in metabolism of adrenal steroids.
- Explain the mechanism through which adrenocorticotropic hormone increases glucocorticoid and androgen synthesis in adrenal cortical cells.
- Describe the mechanism through which angiotensin II and angiotensin III stimulate aldosterone synthesis in the cells of the zona glomerulosa.
- Explain how glucocorticoids alter the transcription of specific genes in target cells and describe the role of glucocorticoids in the adaptation of the body to fasting, injury, and stress.
- Describe the synthesis of epinephrine and norepinephrine and explain how the cellular effects of catecholamines are achieved.
- Explain how stimuli such as injury, anger, pain, cold, strenuous exercise, and hypoglycemia cause secretion of catecholamines.
- Explain the effect of catecholamines on liver, muscle, and adipose tissue to regulate blood sugar.

To remain alive, the organs and tissues of the human body must have a finely regulated extracellular environment. This environment must contain the correct concentrations of ions to maintain body fluid volume and to enable excitable cells to function. The extracellular environment must also have an adequate supply of metabolic substrates for cells to generate adenosine triphosphate (ATP). Salts, water, and other organic substances are continually lost from the body as a result of perspiration, respiration, and excretion. Cells constantly use metabolic substrates. Under normal conditions, the intake of food and liquids replenishes these critical constituents of the body's extracellular environment. However, a person can survive for weeks on little else but water because the body has a remarkable capacity for adjusting the functions of its organs and tissues to preserve body fluid volume and composition.

The adrenal glands play a key role in making these adjustments. This is readily apparent from the fact that an adrenalectomized animal, unlike its normal counterpart, cannot survive prolonged fasting. Its blood glucose supply diminishes, ATP generation by the cells becomes inadequate to support life, and the animal eventually dies. Even when fed a normal diet, an adrenalectomized animal typically loses body sodium and water over time and eventually dies of circulatory collapse. Its death is caused by a lack of certain steroid hormones that the cortex of the adrenal gland produces and secretes.

FUNCTIONAL ANATOMY

The human adrenal glands are paired, pyramid-shaped organs located on the upper poles of each kidney. The adrenal gland is actually a composite of two separate endocrine organs, one inside the other, each secreting separate hormones and each regulated by different mechanisms. The outer portion or **cortex** of the adrenal gland completely surrounds the inner portion or **medulla** and makes up most of the gland. During embryonic development, the cortex forms from mesoderm; the medulla arises from neural ectoderm.

Adrenal cortex secretes steroids, and the medulla secretes catecholamines.

The glucocorticoid hormones, **cortisol** and **corticosterone**, play essential roles in adjusting the metabolism of carbohydrates, lipids, and proteins in liver, muscle, and adipose tissues during fasting, which ensures an adequate supply of glucose and fatty acids for energy metabolism despite the absence of food. The mineralocorticoid hormone **aldosterone**, another steroid hormone produced by the adrenal cortex, stimulates the kidneys to conserve sodium and, hence, body fluid volume.

The glucocorticoids also enable the body to cope with physical and emotional trauma or stress. The fact that adrenalectomized animals lose their ability to cope with physical or emotional stress emphasizes the physiologic importance of this action of the glucocorticoids. Even when given an appropriate diet to prevent blood glucose and body sodium depletion, an adrenalectomized animal may die when exposed to trauma that is not fatal to normal animals.

Hormones produced by the other endocrine component of the adrenal gland, the medulla, are also involved in compensatory reactions of the body to trauma or life-threatening situations. These hormones are the catecholamines **epinephrine** and **norepinephrine (NE)**, which have widespread effects on the cardiovascular system and muscular system and on carbohydrate and lipid metabolism in liver, muscle, and adipose tissues.

Adrenal cortex comprises three zones that produce and secrete distinct hormones.

In the adult human, the adrenal cortex consists of three histologically distinct zones or layers (Fig. 33.1). The outer zone, which lies immediately under the capsule of the gland, is called the **zona glomerulosa** and consists of small clumps of cells that produce the mineralocorticoid aldosterone. The **zona fasciculata** is the middle and thickest layer of the cortex and consists of cords of cells oriented radially to the center of the gland. The inner layer comprises interlaced strands of cells called the **zona reticularis**. The zona fasciculata and zona reticularis both produce the physiologically important glucocorticoids cortisol and corticosterone. These layers of the cortex also produce the androgen **dehydroepiandrosterone (DHEA)**, which is related chemically to the male sex hormone **testosterone**. The molecular structures of these hormones are shown in Figure 33.2.

Like all endocrine organs, the adrenal cortex is highly vascularized. Many small arteries branch from the aorta and

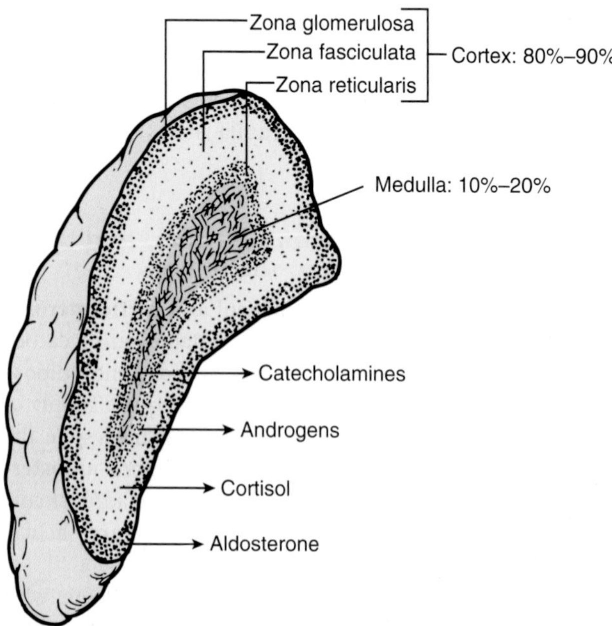

Figure 33.1 The three zones of the adrenal cortex and corresponding hormones secreted. (See text for details.)

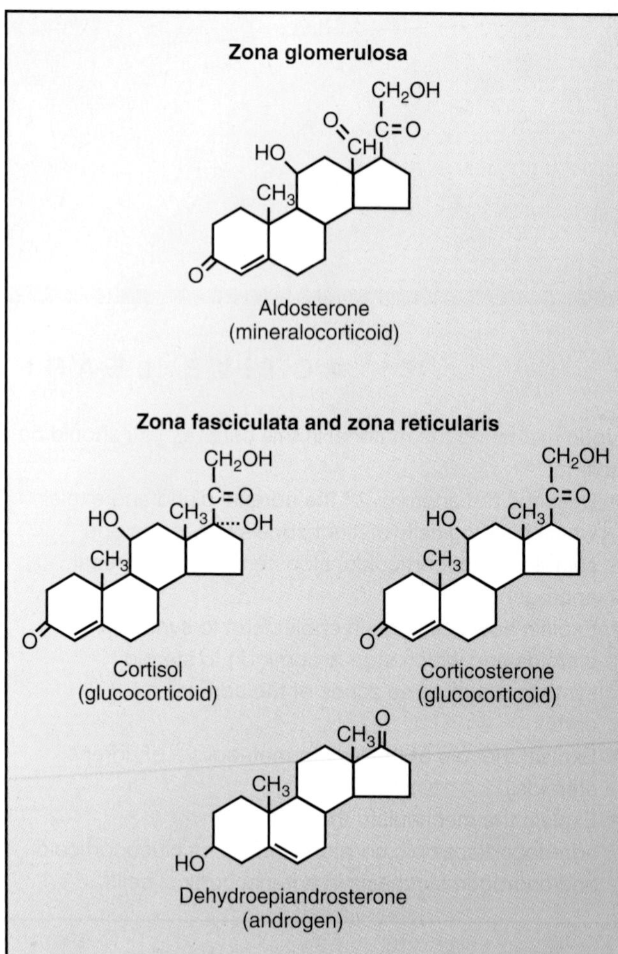

Figure 33.2 Molecular structures of the important hormones secreted by each zone of the adrenal cortex.

renal arteries and enter the cortex. These vessels give rise to capillaries that course radially through the cortex and terminate in venous sinuses in the zona reticularis and adrenal medulla; therefore, the hormones produced by the cells of the cortex have ready access to the circulation.

The cells of the adrenal cortex contain abundant lipid droplets. This stored lipid is functionally significant because **cholesterol esters** present in the droplets are an important source of the cholesterol used as a precursor for the synthesis of steroid hormones.

Adrenal medulla is a modified sympathetic ganglion.

The adrenal medulla can be considered a modified sympathetic ganglion. The medulla consists of clumps and strands of **chromaffin cells** interspersed with venous sinuses. Chromaffin cells, like the modified postganglionic neurons that receive sympathetic preganglionic cholinergic innervation from the splanchnic nerves, produce catecholamine hormones, principally epinephrine and NE. Epinephrine and NE are stored in granules in chromaffin cells and discharged into venous sinuses of the adrenal medulla when the adrenal branches of splanchnic nerves are stimulated (see Fig. 6.5.)

METABOLISM OF ADRENAL CORTEX HORMONES

Only small amounts of glucocorticoid, aldosterone, and adrenal androgens are found in adrenal cortical cells at a given time because the cells produce and secrete these hormones on demand, rather than storing them. Table 33.1 shows the daily production of adrenal cortex hormones in a healthy adult under resting (unstimulated) conditions. Because the molecular weights of these substances do not vary greatly, comparing the amounts secreted indicates the relative number of molecules of each hormone produced daily. Humans secrete about 10 times more cortisol than corticosterone during an average day, and corticosterone has only one fifth of the glucocorticoid activity of cortisol (Table 33.2). Cortisol is considered the physiologically important glucocorticoid in humans. Compared with the glucocorticoids, a much smaller amount of aldosterone is secreted each day.

Because of similarities in their structures, the glucocorticoids and aldosterone have overlapping actions. For example, cortisol and corticosterone have some mineralocorticoid activity; conversely, aldosterone has some glucocorticoid activity. However, given the amounts of these hormones secreted under normal circumstances and their relative activities, glucocorticoids are not physiologically important mineralocorticoids, nor does aldosterone function physiologically as a glucocorticoid.

As discussed in detail later, the amounts of glucocorticoids and aldosterone secreted by an individual can vary greatly from those given in Table 33.1. The amount secreted depends on the person's physiologic state. For example, in a person subjected to severe physical or emotional trauma, the rate of cortisol secretion may be 10 times greater than the resting rate shown in Table 33.1. Certain diseases of the adrenal cortex that involve steroid hormone biosynthesis can significantly increase or decrease the amount of hormones produced.

The adrenal cortex also synthesizes and secretes substantial amounts of androgenic steroids. DHEA, in both the free form and the sulfated form (DHEAS), is the main androgen secreted by the adrenal cortex of both men and women (see Table 33.1). Lesser amounts of other androgens are also produced. The adrenal cortex is the main source of androgens in the blood in human females. In the human male, however, androgens produced by the testes and those produced by the adrenal cortex both contribute to the male sex hormones circulating in the blood. Adrenal androgens normally have little physiologic effect other than a role in development before the start of puberty in both girls and boys. This is because the male sex hormone activity of the adrenal androgens is weak. Exceptions occur in people who produce inappropriately large amounts of certain adrenal androgens as a result of diseases affecting the pathways of steroid biosynthesis in the adrenal cortex.

TABLE 33.1 Average Daily Production of Hormones by the Adrenal Cortex

Hormone Amount	Produced (mg/d)
Cortisol	20
Corticosterone	2
Aldosterone	0.1
Dehydroepiandrosterone	30

TABLE 33.2 Comparison of Shared Activities of Adrenal Cortical Hormones

Hormone	Glucocorticoid Activity[a]	Mineralocorticoid Activity[b]
Cortisol	100	0.25
Corticosterone	20	0.5
Aldosterone	10	100

[a]Percentage activity, with cortisol being 100%.
[b]Percentage activity, with aldosterone being 100%.

Adrenal steroid hormones are synthesized from cholesterol.

Cholesterol is the starting material for the synthesis of steroid hormones. A cholesterol molecule consists of four interconnected rings of carbon atoms and a side chain of eight carbon atoms extending from one ring (Fig. 33.3). In all, there are 27 carbon atoms in cholesterol, numbered as shown in Figure 33.3.

The immediate source of cholesterol used in the biosynthesis of steroid hormones is the abundant lipid droplets in adrenal cortical cells. The cholesterol present in these lipid droplets is mainly in the form of cholesterol esters, single molecules of cholesterol esterified to single fatty acid molecules. The action of **cholesterol esterase (cholesterol ester hydrolase [CEH])**, which hydrolyzes the ester bond, generates the free cholesterol used in steroid biosynthesis from these cholesterol esters. The free cholesterol generated by that cleavage enters mitochondria near the lipid droplet. The process of remodeling the cholesterol molecule into steroid hormones is then initiated.

The cholesterol that has been removed from the lipid droplets for steroid hormone biosynthesis is replenished in two ways (Fig. 33.4). Most of the cholesterol converted to steroid hormones by the human adrenal gland comes from cholesterol esters contained in **low-density lipoprotein (LDL)** particles circulating in the blood. The LDL particles consist of a core of cholesterol esters surrounded by a coat of cholesterol and phospholipids. A 400-kDa protein molecule called **apoprotein B100** is also present on the surface of the LDL particle; LDL receptors localized to coated pits on the plasma membrane of adrenal cortical cells recognize it (see Fig. 33.4). The apoprotein binds to the LDL receptor, and the cell takes up both the LDL particle and the receptor through endocytosis. The endocytic vesicle containing the LDL

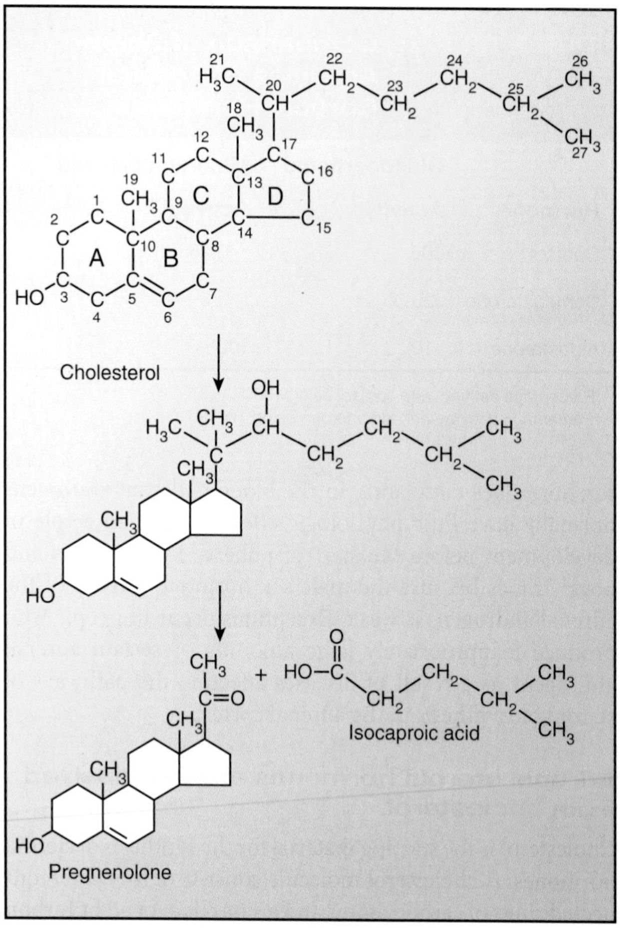

Figure 33.3 The formation of pregnenolone from cholesterol by the action of cholesterol side-chain cleavage enzyme (CYP11A1). Note the chemical structure of cholesterol, how the four rings are lettered (A–D), and how the carbons are numbered. The hydrogen atoms on the carbons composing the rings are omitted from the figure.

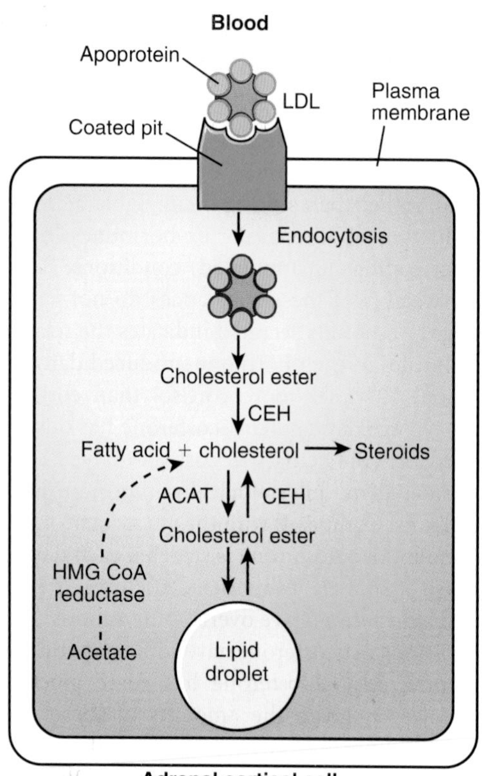

Figure 33.4 Sources of cholesterol for steroid biosynthesis by the adrenal cortex. Most cholesterol comes from low-density lipoprotein (LDL) particles in the blood, which bind to receptors in the plasma membrane and are taken up by endocytosis. The cholesterol in the LDL particle is used directly for steroidogenesis or stored in lipid droplets for later use. Some cholesterol is synthesized directly from acetate. ACAT, acyl-CoA:cholesterol acyltransferase; CEH, cholesterol ester hydrolase; CoA, coenzyme A; HMG, 3-hydroxy-3-methylglutaryl.

particles fuses with a lysosome, and the particle is degraded. The action of CEH hydrolyzes the cholesterol esters in the core of the particle to free cholesterol and fatty acid.

The action of the enzyme **acyl-CoA: cholesterol acyltransferase (ACAT)** converts any cholesterol not immediately used by the cell again to cholesterol ester. The esters are then stored in the lipid droplets of the cell to be used later. When steroid biosynthesis is proceeding at a high rate, cholesterol delivered to the adrenal cell may be diverted directly to mitochondria for steroid production rather than being reesterified and stored. There is also evidence that **high-density lipoprotein (HDL)** cholesterol may also be used as a substrate for adrenal steroidogenesis.

In humans, cholesterol that has been synthesized *de novo* from acetate by the adrenal glands is a significant but minor source of cholesterol for steroid hormone formation. The enzyme **3-hydroxy-3-methylglutaryl CoA reductase (HMG-CoA reductase)** catalyzes the rate-limiting step in this process. The newly synthesized cholesterol is then incorporated into cellular structures such as membranes or converted to cholesterol esters through the action of ACAT and stored in lipid droplets (see Fig. 33.4).

Oxidative cytochrome P450 enzymes

Cytochrome P450 enzymes (CYPs) are a large family of oxidative enzymes with a 450 nm absorbance maximum when complexed with carbon monoxide; hence, these molecules were once referred to as CYPs. The adrenal CYPs are often referred to by their trivial names, which denote their function in steroid biosynthesis (Table 33.3).

The conversion of cholesterol into steroid hormones begins with the formation of free cholesterol from the cholesterol esters stored in intracellular lipid droplets. The **steroidogenic acute regulatory protein (StAR)** facilitates the flow of free cholesterol into mitochondria.

Once inside a mitochondrion, single cholesterol molecules bind to the **cholesterol side-chain cleavage enzyme (CYP11A1)** embedded in the inner mitochondrial membrane. This enzyme catalyzes the first and rate-limiting reaction in steroidogenesis, which remodels the cholesterol molecule into a 21-carbon steroid intermediate called **pregnenolone**. The reaction occurs in three steps, as shown in Figure 33.3. The first two steps consist of the hydroxylation of carbons 20 and 22 by CYP11A1. Then the enzyme cleaves the side chain of cholesterol between carbons 20 and 22, yielding pregnenolone and **isocaproic acid**.

TABLE 33.3 Nomenclature for the Steroidogenic Enzymes

Common Name	Previous Form	Current Form	Gene
Cholesterol side-chain cleavage enzyme	$P450_{SCC}$	CYP11A1	$CYP_{11}A_1$
3β-Hydroxysteroid dehydrogenase	3β-HSD	3β-HSD II	HSD_3B_2
17α-Hydroxylase	$P450_{C17}$	CYP17	CYP_{17}
21-Hydroxylase	$P450_{C21}$	CYP21A2	$CYP_{21}A_2$
11β-Hydroxylase	$P450_{C11}$	CYP11B1	$CYP_{11}B_1$
Aldosterone synthase	$P450_{C11AS}$	CYP11B2	$CYP_{11}B_2$

Once formed, pregnenolone molecules dissociate from CYP11A1, leave the mitochondrion, and enter the smooth endoplasmic reticulum (ER) nearby. At this point, the further remodeling of pregnenolone into steroid hormones can vary, depending on whether the process occurs in the zona fasciculata and zona reticularis or the zona glomerulosa. We first consider what occurs in the zona fasciculata and zona reticularis. These biosynthetic events are summarized in Figure 33.5.

In cells of the zona fasciculata and zona reticularis, most of the pregnenolone is converted to cortisol and the main adrenal androgen, DHEA. Pregnenolone molecules bind to the enzyme **17α-hydroxylase (CYP17)**, embedded in the ER membrane, which hydroxylates pregnenolone at carbon 17. The product formed by this reaction is **17α-hydroxypregnenolone** (see Fig. 33.5).

The CYP17 has an additional enzymatic action that becomes important at this step in the steroidogenic process. Once the enzyme has hydroxylated carbon 17 of pregnenolone to form 17α-hydroxypregnenolone, it can lyse or cleave the carbon 20–21 side chain from the steroid structure. Some molecules of 17α-hydroxypregnenolone undergo this reaction and are converted to the 19-carbon steroid DHEA. This action of CYP17 is essential for the formation of androgens (19-carbon steroids) and estrogens (18-carbon steroids), which lack the carbon 20–21 side chain. The lyase activity of CYP17 is important in the gonads, where androgens and estrogens are primarily made. CYP17 does not exert significant lyase activity in children before the age of 7 or 8 years. As a result, young boys and girls do not secrete significant amounts of adrenal androgens. The appearance of significant adrenal androgen secretion in children of both sexes is termed **adrenarche**. It is not related to the onset of puberty, because it normally occurs before the activation of the hypothalamic-pituitary-gonad axis, which initiates puberty. The adrenal androgens produced as a result of adrenarche stimulate the growth of pubic and axillary hair.

Those molecules of 17α-hydroxypregnenolone that dissociate as such from CYP17 bind next to another ER enzyme, **3β-hydroxysteroid dehydrogenase (3β-HSD II)**. This enzyme acts on 17α-hydroxypregnenolone to isomerize the double bond in ring B to ring A and to dehydrogenate the 3β-hydroxy group, forming a 3-keto group. The product formed is **17α-hydroxyprogesterone** (see Fig. 33.5). This intermediate then binds to another enzyme, **21-hydroxylase (CYP21A2)**, which hydroxylates it at carbon 21. The mechanism of this hydroxylation is similar to that performed by the CYP17. The product formed is **11-deoxycortisol**, which is the immediate precursor for cortisol.

To be converted to cortisol, 11-deoxycortisol molecules must be transferred back into the mitochondrion to be acted on by **11β-hydroxylase (CYP11B1)** embedded in the inner mitochondrial membrane. This enzyme hydroxylates 11-deoxycortisol on carbon 11, converting it into cortisol. The 11β-hydroxyl group is the molecular feature that confers glucocorticoid activity on the steroid. Cortisol is then secreted into the bloodstream.

Some of the pregnenolone molecules generated in cells of the zona fasciculata and zona reticularis first bind to 3β-HSD II when they enter the ER. As a result, they are converted to **progesterone**. CYP21A2 hydroxylates some of these progesterone molecules to form the mineralocorticoid **11-deoxycorticosterone (DOC)** (see Fig. 33.5). The DOC formed may be either secreted or transferred back into the mitochondrion, where CYP11B1 acts on it to form corticosterone, which is then secreted into the circulation.

Progesterone may also undergo 17α-hydroxylation in the zona fasciculata and zona reticularis. It is then converted to either cortisol or the adrenal androgen **androstenedione.**

The CYP17 is not present in cells of the zona glomerulosa; therefore, pregnenolone does not undergo 17α-hydroxylation in these cells, and these cells do not form cortisol and adrenal androgens. Instead, the enzymatic pathway leading to the formation of aldosterone is followed (see Fig. 33.5). Enzymes in the ER convert pregnenolone to progesterone and DOC. The latter compound then moves into the mitochondrion, where it is converted to aldosterone. This conversion involves three steps: the hydroxylation of carbon 11 to form corticosterone, the hydroxylation of carbon 18 to form 18-hydroxycorticosterone, and the oxidation of the 18-hydroxymethyl group to form aldosterone. In humans, these three reactions are catalyzed by a single enzyme, **aldosterone synthase (CYP11B2),** an isozyme of CYP11B1, expressed only in glomerulosa cells. The CYP11B1 enzyme, which is expressed in the zona fasciculata and zona reticularis, although closely related to CYP11B2, cannot catalyze all three reactions involved in

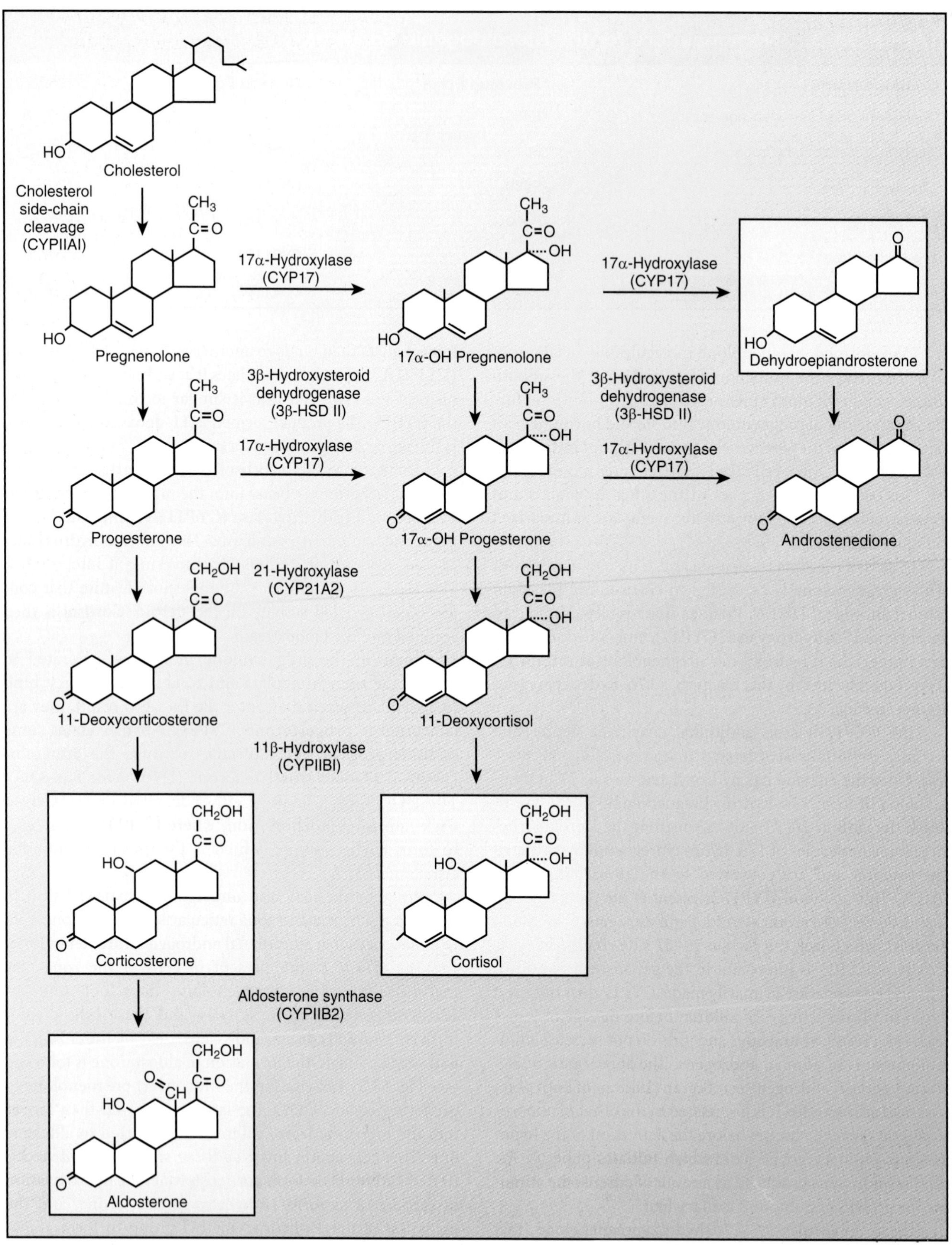

Figure 33.5 The synthesis of steroids in the adrenal cortex. The enzymes catalyzing each step are shown and the major steroidogenic products are boxed.

From Bench to Bedside / 33.1

Obesity and 11β Hydroxysteroid Dehydrogenase

Glucocorticoids regulate multiple metabolic processes, including glucose homeostasis, insulin sensitivity, and lipid metabolism (see text). In addition, glucocorticoids are important in the differentiation of preadipocytes to adipocytes, a process termed adipogenesis. In patients with Cushing syndrome, elevated blood glucocorticoid levels cause excess fat deposition in the abdomen surrounding the organs (visceral obesity), insulin resistance, dyslipidemia, and hypertension (see Clinical Focus Box 33.2). A similar collection of metabolic abnormalities is present in obese people, which has prompted investigation into the possibility that general obesity could be attributed to glucocorticoid excess. Although subtle alterations in the hypothalamic–pituitary–adrenal axis have been observed in obesity, no clear role for increased circulating glucocorticoid has been established.

Until recently, tissue glucocorticoid concentrations were thought to be determined by the level of glucocorticoid in the blood and tissue responses controlled by the availability of glucocorticoid receptors. It is now recognized that an enzyme, 11β hydroxysteroid dehydrogenase (11βHSD), has a central role in regulating the concentration of glucocorticoid in specific tissues. There are two 11βHSD isozymes. The type 2 enzyme is an NAD-dependent dehydrogenase, converting active cortisol to inactive cortisone (Fig. 33.A). Its main role is in aldosterone-sensitive target tissues such as the kidney, the colon, the salivary glands, and the placenta. Cortisol circulates in the blood at concentrations several-fold higher than those of aldosterone, but both hormones have an affinity for the nonselective mineralocorticoid receptor. By inactivating cortisol, 11βHSD2 prevents the flooding of mineralocorticoid receptors by cortisol, leaving free access for aldosterone. The second isozyme, 11βHSD1, is an NADPH-dependent reductase that converts inactive cortisone to active cortisol. 11βHSD1 is expressed primarily in glucocorticoid target tissues, such as the liver, adipose tissue, and the CNS, where it amplifies local glucocorticoid action.

Several lines of evidence point to an important role for 11βHSD1 in the pathogenesis of obesity. The expression and activity of 11βHSD1 are increased in subcutaneous fat of obese humans and in the adipose tissue of obese rodent models. Adipose tissue–specific overexpression of 11βHSD1 in transgenic mice results in a phenotype of central obesity, insulin resistance, and dyslipidemia. Mice with whole-body knockout of the 11βHSD1 allele are protected from the adverse metabolic consequences of obesity such as hyperglycemia that result from high-fat feeding. Furthermore, in mice prone to developing obesity, 11βHSD1 knockout attenuates the effect of a high-fat diet to induce weight gain. Taken together, these observations suggest that 11βHSD1 has a central role in the pathophysiology of obesity and that there could be significant benefit to 11βHSD1 inhibition.

Carbenoxolone, a liquorice-based compound previously used as an antiulcer medication, inhibits 11βHSD1 in the liver and has been shown to lower blood glucose in subjects with type 2 diabetes. However, carbenoxolone is poorly selective for the two 11βHSD isoforms, and an effect in adipose tissue has not been demonstrated. A class of selective 11βHSD1 inhibitors called *arylsulfonamidothiazoles* shows promise and is currently in development. Careful evaluation of potential side effects will be necessary during the development of 11βHSD1 inhibitors because the enzyme is found in many tissues. Despite this, 11βHSD1 remains a useful therapeutic target for the treatment of obesity and its associated comorbidities.

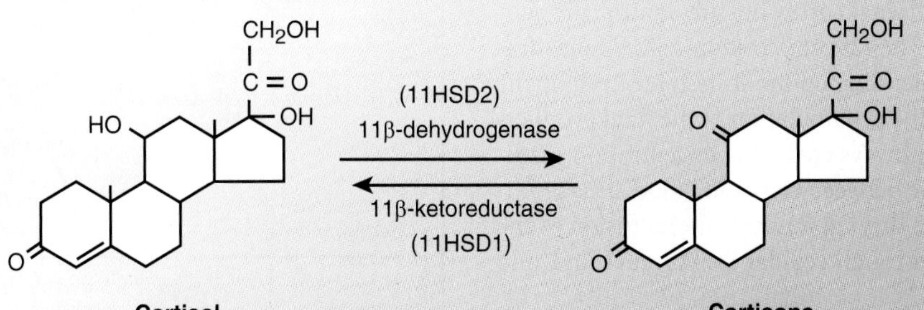

Cortisol ⇄ Cortisone: (11HSD2) 11β-dehydrogenase →; ← 11β-ketoreductase (11HSD1)

Figure 33.A Target tissue cortisol metabolism.

the conversion of DOC to aldosterone; therefore, aldosterone is not synthesized in the zona fasciculata and zona reticularis of the adrenal cortex.

Genetic defects

Inherited genetic defects can cause relative or absolute deficiencies in the enzymes involved in steroid hormone biosynthetic pathways. The immediate consequences of these defects are changes in the type and amount of steroid hormones secreted by the adrenal cortex. The end result is disease.

Most of the genetic defects affecting the steroidogenic enzymes impair the formation of cortisol. As discussed in Chapter 31, "Hypothalamus and the Pituitary Gland," a drop in cortisol concentration in the blood stimulates the secretion of **adrenocorticotropic hormone (ACTH)** by the anterior pituitary. The consequent rise in ACTH in the blood exerts a trophic (growth-promoting) effect on the adrenal cortex, resulting in adrenal hypertrophy. Because of this mechanism, people with genetic defects affecting adrenal steroidogenesis usually have hypertrophied adrenal glands.

These diseases are collectively called **congenital adrenal hyperplasia.**

In humans, inherited genetic defects occur that affect CYP11A1, CYP17, 3β-HSD II, CYP21A2, CYP11B1, and CYP11B2. The most common defect involves mutations in the gene for CYP21A2 and occurs in 1 of 7,000 people. The gene for CYP21A2 may be deleted entirely, or mutant genes may code for forms of CYP21A2 with impaired enzyme activity. The consequent reduction in the amount of active CYP21A2 in the adrenal cortex interferes with the formation of cortisol, corticosterone, and aldosterone, all of which are hydroxylated at carbon 21. Because of the reduction of cortisol (and corticosterone) secretion in these people, ACTH secretion is stimulated. This, in turn, causes hypertrophy of the adrenal glands and stimulates the glands to produce steroids.

Because 21-hydroxylation is impaired, the ACTH stimulus causes pregnenolone to be converted to adrenal androgens in inappropriately high amounts. Thus, women afflicted with CYP21A2 deficiency exhibit **virilization** from the masculinizing effects of excessive adrenal androgen secretion. In severe cases, the deficiency in aldosterone production can lead to sodium depletion, dehydration, vascular collapse, and death, if appropriate hormone therapy is not given.

Glucocorticoid and aldosterone deficiency also occur as a result of pathologic destruction of the adrenal glands by microorganisms or autoimmune disease. This disorder is called **Addison disease.** If sufficient adrenal cortical tissue is lost, the resulting decrease in aldosterone production can lead to vascular collapse and death, unless hormone therapy is given.

Adrenal steroid transport

As noted earlier, steroid hormones are not stored to any extent by cells of the adrenal cortex but are continually synthesized and secreted. The rate of secretion may change dramatically, however, depending on the stimuli received by the adrenal cortical cells. The accumulation of the final products of the steroidogenic pathways creates a concentration gradient for steroid hormone between cells and blood. This gradient is thought to be the driving force for the diffusion of the lipid-soluble steroids through cellular membranes and into the circulation.

A large fraction of the adrenal steroids that enter the bloodstream become bound noncovalently to certain plasma proteins. One of these is **corticosteroid-binding globulin (CBG),** a glycoprotein produced by the liver. CBG binds glucocorticoids and aldosterone but has a greater affinity for the glucocorticoids. Serum albumin also binds steroid molecules. Albumin has a high capacity for binding steroids but its interaction with steroids is weak. The binding of a steroid hormone to a circulating protein molecule prevents it from being taken up by cells or being excreted in the urine.

Circulating steroid hormone molecules not bound to plasma proteins are free to interact with receptors within cells and, therefore, are cleared from the blood. As this occurs, bound hormone dissociates from its binding protein and replenishes the circulating pool of free hormone. Because of this process, adrenal steroid hormones have long half-lives in the body, ranging from many minutes to hours.

Adrenal steroid catabolism

Adrenal steroid hormones are eliminated from the body primarily by excretion in the urine after they have been structurally modified to destroy their hormone activity and increase their water solubility. Although many cells are capable of carrying out these modifications, the modifications primarily occur in the liver.

The most common structural modifications made in adrenal steroids involve reduction of the double bond in ring A and conjugation of the resultant hydroxyl group formed on carbon 3 with glucuronic acid. Figure 33.6 shows how cortisol is modified in this manner to produce a major excretable metabolite, **tetrahydrocortisol glucuronide.**

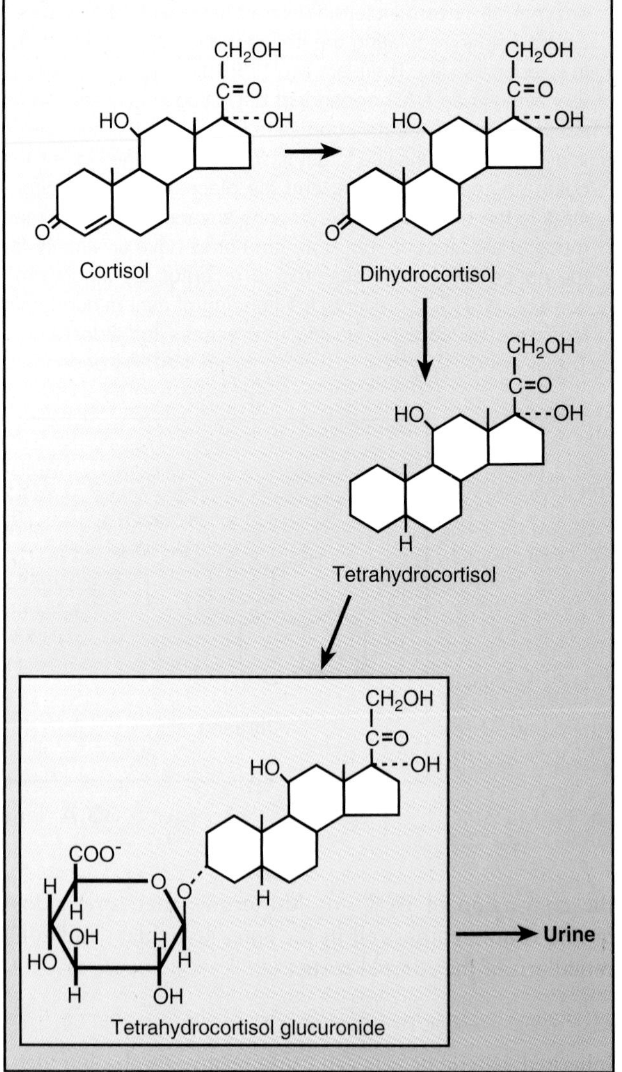

Figure 33.6 The metabolism of cortisol to tetrahydrocortisol glucuronide in the liver. The reduced and conjugated steroid is inactive. Because it is more water-soluble than cortisol, it is easily excreted in the urine.

Cortisol and other 21-carbon steroids with a 17α-hydroxyl group and a 20-keto group may undergo lysis of the carbon 20–21 side chain as well. The resultant metabolite, with a keto group on carbon 17, appears as one of the **17-ketosteroids** in the urine. Adrenal androgens are also 17-ketosteroids. They are usually conjugated with sulfuric acid or glucuronic acid before being excreted and normally comprise the bulk of the 17-ketosteroids in the urine. Before the development of specific methods to measure androgens and 17α-hydroxycorticosteroids in body fluids, the amount of 17-ketosteroids in urine was used clinically as a crude indicator of the production of these substances by the adrenal gland.

Adrenocorticotropic hormone regulates the synthesis of adrenal steroids.

ACTH is the physiologic regulator of the synthesis and secretion of glucocorticoids and androgens by the zona fasciculata and zona reticularis. It has a rapid stimulatory effect on steroidogenesis in these cells, which can result in a great rise in blood glucocorticoids within seconds. It also exerts several long-term trophic effects on these cells, all directed toward maintaining the cellular machinery necessary to carry out steroidogenesis at a high, sustained rate. These actions of ACTH are illustrated in Figure 33.7.

ACTH binds to the melanocortin-2 receptor on the adrenocortical cell. Stimulatory guanine nucleotide-binding proteins (G_s proteins) couple the melanocortin-2 receptor to the enzyme adenylyl cyclase. ACTH stimulates production of cyclic adenosine monophosphate (cAMP) from ATP, resulting in the activation of protein kinase A (PKA), which phosphorylates proteins that regulate steroidogenesis.

The rapid rise in cAMP produced by ACTH results in the phosphorylation of preexisting StAR protein and the rapid synthesis of new StAR protein to facilitate the flow of cholesterol into mitochondria. This action provides abundant cholesterol for side-chain cleavage enzyme, which carries out the rate-limiting step in steroidogenesis. As a result, the rates of steroid hormone formation and secretion rise greatly.

Steroidogenic enzyme expression

ACTH-induced transcription of genes for many of the enzymes involved in steroidogenesis maintains the capacity of the cells of the zona fasciculata and zona reticularis to produce steroid hormones. For example, transcription of the genes for side-chain cleavage enzyme, CYP17, CYP21A2, and CYP11B1 is increased several hours after ACTH stimulation of adrenal cortical cells. Because healthy people are

Clinical Focus / 33.1

Primary Adrenal Insufficiency: Addison Disease

Adrenal insufficiency may be caused by the destruction of the adrenal cortex (primary adrenal insufficiency), low pituitary adrenocorticotropic hormone secretion (secondary adrenal insufficiency), or deficient hypothalamic release of corticotropin-releasing hormone (tertiary adrenal insufficiency). Addison disease (primary adrenal insufficiency) results from the destruction of the adrenal gland by microorganisms or autoimmune disease. When Addison first described primary adrenal insufficiency in the mid-1800s, bilateral adrenal destruction by tuberculosis was the most common cause of the disease. Today, autoimmune destruction accounts for 70% to 90% of all cases in developed countries, with the remainder resulting from infection, cancer, or adrenal hemorrhage. The prevalence of primary adrenal insufficiency in the developed world is about 40 to 110 cases per 1 million adults.

In primary adrenal insufficiency, all three zones of the adrenal cortex are usually involved. The result is inadequate secretion of glucocorticoids, mineralocorticoids, and androgens. Major symptoms are not usually detected until 90% of the gland has been destroyed. The initial symptoms generally have a gradual onset, with only a partial glucocorticoid deficiency resulting in inadequate cortisol increase in response to stress. Mineralocorticoid deficiency may only appear as a mild postural hypotension. Progression to complete glucocorticoid deficiency results in a decreased sense of well-being and abnormal glucose metabolism. Lack of mineralocorticoid leads to decreased renal potassium secretion and reduced sodium retention, the loss of which results in hypotension and dehydration. The combined lack of glucocorticoid and mineralocorticoid can lead to vascular collapse, shock, and death. Adrenal androgen deficiency is observed in women only (men derive most of their androgen from the testes) as decreased pubic and axillary hair and decreased libido.

Antibodies that react with all three zones of the adrenal cortex have been identified in autoimmune adrenalitis and are more common in women than in men. The presence of antibodies appears to precede the development of adrenal insufficiency by several years. Antiadrenal antibodies are mainly directed to the steroidogenic enzymes: cholesterol side-chain cleavage enzyme (CYP11A1), 17α-hydroxylase (CYP17), and 21-hydroxylase (CYP21A2), although antibodies to other steroidogenic enzymes may also be present. In the initial stages of the disease, the adrenal glands may be enlarged with extensive lymphocyte infiltration. Genetic susceptibility to autoimmune adrenal insufficiency is strongly linked with the HLA-DR3 and HLA-DR4 alleles of human leukocyte antigen.

The onset of adrenal insufficiency is often insidious, and diagnosis is not made until an acute crisis occurs during another illness. Crisis is a medical emergency consisting of hypotension and acute circulatory failure. Treatment for acute adrenal insufficiency should be directed at the reversal of the hypotension and electrolyte abnormalities. Large volumes of 0.9% saline or 5% dextrose in saline should be infused as quickly as possible. Dexamethasone or a soluble form of injectable cortisol should also be given. Daily glucocorticoid and mineralocorticoid replacement allows the patient to lead a normal, active life.

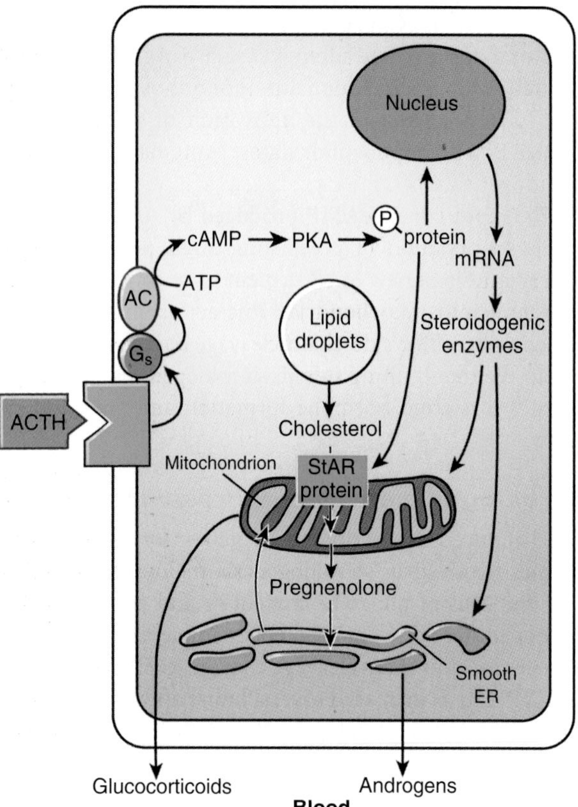

Figure 33.7 The main actions of adrenocorticotropic hormone (ACTH) on steroidogenesis. ACTH binds to the melanocortin-2 receptor, which is coupled to adenylyl cyclase (AC) by stimulatory G proteins (G_s). The intracellular rise in cyclic adenosine monophosphate (cAMP) activates protein kinase A (PKA), which then phosphorylates proteins (P-proteins) that stimulate the expression of genes for steroidogenic enzymes. PKA also phosphorylates steroidogenic acute regulatory (StAR) protein, which mediates the transfer of cholesterol into the mitochondria for steroidogenesis.

continually exposed to episodes of ACTH secretion (see Fig. 31.7), the messenger RNA (mRNA) for these enzymes is well maintained in the cells. Again, this long-term or maintenance effect of ACTH is a result of its ability to increase cAMP in the cells (see Fig. 33.7).

The importance of ACTH in gene transcription becomes evident in hypophysectomized animals or humans with ACTH deficiency. An example of the latter is a human treated chronically with large doses of cortisol or related steroids, which causes prolonged suppression of ACTH secretion by the anterior pituitary. The chronic lack of ACTH decreases the transcription of the genes for steroidogenic enzymes, causing a deficiency in these enzymes in the adrenals. As a result, the administration of ACTH to such a person does not cause a marked increase in glucocorticoid secretion. Chronic exposure to ACTH is required to restore mRNA levels for the steroidogenic enzymes and, hence, the enzymes themselves to obtain normal steroidogenic responses to ACTH. A patient receiving long-term treatment with glucocorticoid may suffer serious glucocorticoid deficiency if hormone therapy is halted abruptly; withdrawing glucocorticoid therapy gradually allows time for endogenous ACTH to restore steroidogenic enzyme levels to normal.

Cholesterol metabolism

There are several long-term effects of ACTH on cholesterol metabolism that support steroidogenesis in the zona fasciculata and zona reticularis. ACTH increases the abundance of LDL receptors and the activity of the enzyme HMG-CoA reductase in these cells, increasing the availability of cholesterol for steroidogenesis. It is not clear whether ACTH exerts these effects directly. The abundance of LDL receptors in the plasma membrane and the activity of HMG-CoA reductase in most cells are inversely related to the amount of cellular cholesterol. By stimulating steroidogenesis, ACTH reduces the amount of cholesterol in adrenal cells; therefore, the increased abundance of LDL receptors and high HMG-CoA reductase activity in ACTH-stimulated cells may merely result from the normal compensatory mechanisms that function to maintain cell cholesterol levels.

ACTH also stimulates the activity of cholesterol esterase in adrenal cells, which promotes the hydrolysis of the cholesterol esters stored in the lipid droplets of these cells, making free cholesterol available for steroidogenesis. The cholesterol esterase in the adrenal cortex appears to be identical to HSL, which is activated when a cAMP-dependent protein kinase phosphorylates it. The rise in cAMP concentration produced by ACTH might account for its effect on the enzyme.

Maintenance of adrenocortical cell size

ACTH maintains the two inner zones of the adrenal cortex, presumably by the stimulation of the synthesis of structural elements of the cells; however, ACTH does not affect the size of the cells of the zona glomerulosa. The trophic effect of ACTH is clearly evident in states of ACTH deficiency or excess. In hypophysectomized or ACTH-deficient people, the cells of the two inner zones atrophy. Chronic stimulation of these cells with ACTH causes them to hypertrophy. The exact mechanisms involved in this trophic action of ACTH are unclear.

Aldosterone regulates sodium and potassium to maintain fluid homeostasis.

Aldosterone is synthesized in the cells of the zona glomerulosa. These cells have melanocortin-2 receptors, which bind ACTH. The resulting rise in cAMP can increase aldosterone secretion to some extent. However, angiotensin II is the important physiologic regulator of aldosterone secretion, not ACTH. Other factors, such as an increase in serum potassium, can also stimulate aldosterone secretion, but normally, they play only a secondary role.

The principal site of aldosterone action on the kidney is in the distal and collecting tubules, where the hormone promotes Na^+ retention and enhances K^+ elimination during urine formation. A secondary effect of Na^+ retention is the osmotic retention of water in the extracellular fluid compartment. The secondary osmotic effect of aldosterone is important in regulating blood pressure.

Angiotensin II formation

Angiotensin II is a short peptide consisting of eight amino acid residues. It is formed in the bloodstream by the proteolysis of the a_2-globulin **angiotensinogen**, which is secreted by the liver. The formation of angiotensin II occurs in two stages (Fig. 33.8). First, the circulating protease **renin** cleaves angiotensinogen at its N-terminal end, releasing the inactive decapeptide **angiotensin I**. Granular (juxtaglomerular) cells in the kidneys produce and secrete renin (see Chapter 22, "Kidney Function"). The protease **angiotensin-converting enzyme**, present on the endothelial cells lining the vasculature, then removes a dipeptide from the C-terminal end of angiotensin I, producing angiotensin II. This step usually occurs as angiotensin I molecules traverse the pulmonary circulation. The rate-limiting factor for the formation of angiotensin II is the renin concentration of the blood.

Cleavage of the N-terminal aspartate from angiotensin II results in the formation of angiotensin III, which circulates at a concentration of 20% that of angiotensin II. Angiotensin III is as potent a stimulator of aldosterone secretion as angiotensin II.

Angiotensin II action

Angiotensin II stimulates the rate-limiting step in steroidogenesis (the movement of cholesterol into the inner mitochondrial membrane and its conversion to pregnenolone) to increase aldosterone synthesis. The primary mechanism is shown in Figure 33.9.

The stimulation of aldosterone synthesis is initiated when angiotensin II binds to its receptors on the plasma membranes of zona glomerulosa cells. A G protein transmits the signal generated by the interaction of angiotensin II with its receptors to phospholipase C (PLC), and the enzyme becomes activated. The PLC then hydrolyzes phosphatidylinositol 4,5 bisphosphate (PIP_2) in the plasma membrane, producing the intracellular second messengers inositol trisphosphate (IP_3) and diacylglycerol (DAG). The IP_3 mobilizes calcium, which is bound to intracellular structures, increasing the calcium concentration in the cytosol. This increase in intracellular calcium and DAG activates protein kinase C (PKC). The rise in intracellular calcium also activates calmodulin-dependent protein kinase (CMK). These enzymes phosphorylate proteins, which initiate steroidogenesis.

Although angiotensin II is the final mediator in the physiologic regulation of aldosterone secretion, its formation from angiotensinogen depends on the secretion of renin by the kidneys. The rate of renin secretion ultimately determines the rate of aldosterone secretion. The granular cells in the walls of the afferent arterioles of renal glomeruli secrete renin. These cells are stimulated to secrete renin by three signals that indicate a possible loss of body fluid: a fall in blood pressure in the afferent arterioles of the glomeruli, a drop in sodium chloride concentration in renal tubular fluid at the macula densa, and an increase in renal sympathetic nerve activity (see Chapter 22, "Kidney Function," and Chapter 23, "Regulation of Fluid and Electrolyte Balance"). Increased renin secretion results in an increase in angiotensin II formation in the blood, thereby stimulating aldosterone secretion by the zona glomerulosa. This series of events tends to conserve body fluid volume because aldosterone stimulates sodium reabsorption by the kidneys.

Aldosterone synthesis

An increase in the potassium concentration in extracellular fluid, caused by a direct effect of potassium on zona glomerulosa cells, also stimulates aldosterone secretion. Glomerulosa cells are sensitive to this effect of extracellular potassium and, therefore, increase their rate of aldosterone secretion in response to small increases in blood and interstitial fluid potassium concentration. This signal for aldosterone secretion is appropriate from a physiologic point of view because aldosterone promotes the renal excretion of potassium (see Chapter 23, "Regulation of Fluid and Electrolyte Balance").

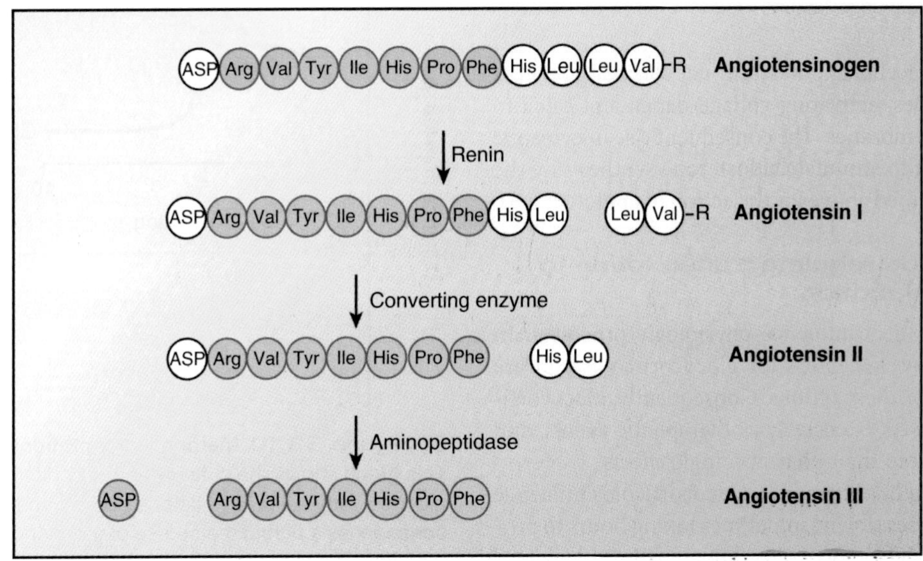

Figure 33.8 The formation of angiotensins I, II, and III from angiotensinogen.

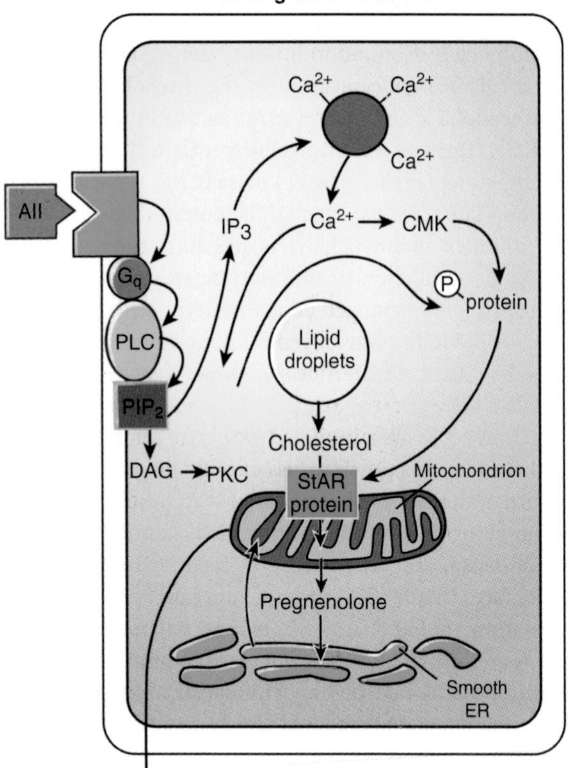

Figure 33.9 The action of angiotensin II on aldosterone synthesis. Angiotensin II binds to receptors on the plasma membrane of zona glomerulosa cells. This activates phospholipase C (PLC), which is coupled to the angiotensin II (AII) receptor by G proteins (G_q). PLC hydrolyzes phosphatidylinositol 4,5 bisphosphate (PIP_2) in the plasma membrane, producing inositol trisphosphate (IP_3) and diacylglycerol (DAG). IP_3 mobilizes intracellularly bound Ca^{2+}. The rise in Ca^{2+} and DAG activates protein kinase C (PKC) and calmodulin-dependent protein kinase (CMK). These enzymes phosphorylate proteins (P-proteins) that increase the transcription of steroidogenic enzymes and mediate the transfer of cholesterol into the mitochondria (StAR protein). StAR, steroidogenic acute regulatory protein; ER, endoplasmic reticulum.

A rise in extracellular potassium depolarizes glomerulosa cell membranes, activating voltage-dependent calcium channels in the membranes. The consequent rise in cytosolic calcium is thought to stimulate aldosterone synthesis by the mechanisms described above for the action of angiotensin II.

Glucocorticoids regulate transcription to influence cell function.

Glucocorticoids widely influence physiologic processes. In fact, most cells have receptors for glucocorticoids and are potential targets for their actions. Consequently, glucocorticoids have been used extensively as therapeutic agents, and much is known about their pharmacologic effects.

Unlike many other hormones, glucocorticoids influence physiologic processes slowly, sometimes taking hours to produce their effects. Glucocorticoids that are free in the blood diffuse through the plasma membranes of target cells; once inside, they bind tightly but noncovalently to receptor proteins present in the cytoplasm. The interaction between the glucocorticoid molecule and its receptor molecule produces an activated glucocorticoid–receptor complex, which translocates into the nucleus.

These complexes then bind to specific regions of DNA called **glucocorticoid response elements (GREs)**, located on glucocorticoid-sensitive target genes. The binding triggers events that either stimulate or inhibit the transcription of the target gene, resulting in changes in the amount of mRNA. This, in turn, affects the abundance of these proteins in the cell, which produces the physiologic effects of the glucocorticoids. The apparent slowness of glucocorticoid action is a result of the time required by the mechanism to change the protein composition of a target cell.

Glucocorticoids and fasting

During the fasting periods between food intakes in humans, metabolic adaptations prevent hypoglycemia. The maintenance of sufficient blood glucose is necessary because the brain depends on glucose for its energy needs. Many of the adaptations that prevent hypoglycemia are not fully expressed in the course of daily life because the person eats before they fully develop. Complete expression of these changes is seen only after many days to weeks of fasting. Glucocorticoids are necessary for the metabolic adaptation to fasting.

At the onset of a prolonged fast, there is a gradual decline in the concentration of glucose in the blood. Within 1 to 2 days, the blood glucose level stabilizes at a concentration of 60 to 70 mg/dL, where it remains even if the fast is prolonged for many days (Fig. 33.10). The production of glucose by the body and the restriction of its use by tissues other than the brain stabilize the blood glucose level. Although a limited

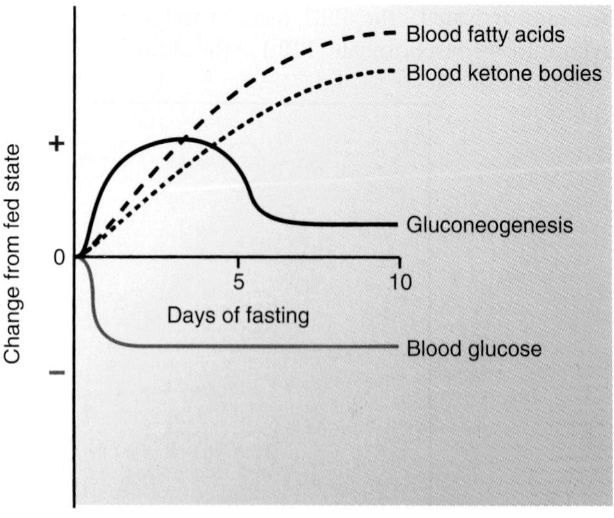

Figure 33.10 Metabolic adaptations during fasting. This graph shows the changes in the concentrations of blood glucose, fatty acids, and ketone bodies, and the rate of gluconeogenesis during the course of a prolonged fast. Only the *direction* of change over time, not the magnitude of change is indicated: increase (+) or decrease (–).

supply of glucose is available from glycogen stored in the liver, the more important source of blood glucose during the first days of a fast is gluconeogenesis in the liver and, to some extent, in the kidneys.

Gluconeogenesis begins several hours after the start of a fast. Amino acids derived from tissue protein are the main substrates. Fasting results in protein breakdown in the skeletal muscle and accelerated release of amino acids into the bloodstream. Insulin and glucocorticoids regulate protein breakdown and protein accretion in adult humans in an opposing fashion. During fasting, insulin secretion is suppressed and the inhibitory effect of insulin on protein breakdown is lost. As proteins are broken down, glucocorticoids inhibit the reuse of amino acids derived from tissue proteins for new protein synthesis, promoting the release of these amino acids from the muscle. The liver and kidneys extract from the blood at an accelerated rate amino acids released into the blood by the skeletal muscle. The amino acids then undergo metabolic transformations in these tissues, leading to the synthesis of glucose. The newly synthesized glucose is then delivered to the bloodstream.

The glucocorticoids are essential for the acceleration of gluconeogenesis during fasting. They play a permissive role in this process by maintaining gene expression and, therefore, the intracellular concentrations of many of the enzymes needed to carry out gluconeogenesis in the liver and the kidneys. For example, glucocorticoids maintain the amounts of transaminases, pyruvate carboxylase, phosphoenolpyruvate carboxykinase, fructose-1, 6-diphosphatase, fructose-6-phosphatase, and glucose-6-phosphatase needed to carry out gluconeogenesis at an accelerated rate. In an untreated, glucocorticoid-deficient person, the amounts of these enzymes in the liver are greatly reduced. As a consequence, the person cannot respond to fasting with accelerated gluconeogenesis and will die from hypoglycemia. In essence, the glucocorticoids maintain the liver and the kidney in a state that enables them to carry out accelerated gluconeogenesis should the need arise.

The other important metabolic adaptation that occurs during fasting involves the mobilization and use of stored fat. Within the first few hours of the start of a fast, the concentration of free fatty acids rises in the blood (see Fig. 33.10). This is a result of the acceleration of lipolysis in the fat depots, as a result of the activation of **adipose triglyceride lipase (ATGL)** and **hormone-sensitive lipase (HSL)**. ATGL and HSL hydrolyze the stored triglyceride to free fatty acids and glycerol, which are released into the blood. ATGL catalyzes the removal of the first fatty acid from the triglyceride, and HSL acts on the resulting DAG to cleave the second fatty acid. Monoacylglycerol lipase removes the remaining fatty acid from the glycerol backbone.

ATGL, mRNA, and protein expression is increased by glucocorticoids during fasting. HSL is activated when it is phosphorylated by cAMP-dependent PKA. As the level of insulin falls in the blood during fasting, the inhibitory effect of insulin on cAMP accumulation in the fat cell diminishes. There is a rise in the cellular level of cAMP, and HSL is activated. The glucocorticoids are essential for maintaining fat cells in an enzymatic state that permits lipolysis to occur during a fast. This is evident from the fact that accelerated lipolysis does not occur when a glucocorticoid-deficient person fasts.

Many tissues take up the abundant fatty acids produced by lipolysis. The fatty acids enter mitochondria, undergo β-oxidation to acetyl-CoA, and become the substrate for ATP synthesis. The enhanced use of fatty acids for energy metabolism spares the blood glucose supply. There is also significant gluconeogenesis in the liver from the glycerol released from triglycerides by lipolysis. In prolonged fasting, when the rate of glucose production from body protein has declined, a significant fraction of blood glucose is derived from triglyceride glycerol.

Within a few hours of the start of a fast, the increased delivery to and oxidation of fatty acids in the liver result in the production of the ketone bodies. As a result of these events in the liver, a gradual rise in ketone bodies occurs in the blood as a fast continues over many days (see Fig. 33.10). Ketone bodies become the principal energy source used by the central nervous system (CNS) during the later stages of fasting.

The increased use of fatty acids for energy metabolism by skeletal muscle results in less use of glucose in this tissue, sparing blood glucose for use by the CNS. Acetyl-CoA and citrate, two products resulting from the breakdown of fatty acids, inhibit glycolysis. As a result, the uptake and use of glucose from the blood are reduced.

In summary, the strategy behind the metabolic adaptation to fasting is to provide the body with glucose produced primarily from protein until the ketone bodies become abundant enough in the blood to be a principal source of energy for the brain. From that point on, the body uses mainly fat for energy metabolism, and it can survive until the fat depots are exhausted. Glucocorticoids do not trigger the metabolic adaptations to fasting but only provide the metabolic machinery necessary for the adaptations to occur. However, when present in excessive amounts, glucocorticoids can trigger many of the metabolic adaptations to the fasting state, resulting in disease.

Glucocorticoids, inflammation, and immunity

Tissue injury triggers a complex mechanism called **inflammation**, which precedes the actual repair of damaged tissue. Neighboring cells, adjacent vasculature, and phagocytic cells that migrate to the damaged site release a host of chemical mediators into the damaged area. Mediators released under these circumstances include **prostaglandins, leukotrienes, kinins, histamine, serotonin**, and **lymphokines**. These substances exert a multitude of actions at the site of injury and directly or indirectly promote the local vasodilation, increased capillary permeability, and edema formation that characterize the inflammatory response (see Chapter 9, "Blood Components," and Chapter 10, "Immunology, Organ Interaction, and Homeostasis").

Because glucocorticoids inhibit the inflammatory response to injury, they are used extensively as therapeutic anti-inflammatory agents; however, the mechanisms are not clear. Their regulation of the production of prostaglandins and leukotrienes is the best understood. These substances

Clinical Focus / 33.2 

Glucocorticoid Excess: Cushing Disease

Cushing syndrome is the name given to the signs and symptoms associated with prolonged exposure to inappropriately elevated glucocorticoids. The use of glucocorticoid in the definition covers both endogenous elevations in cortisol and increased glucocorticoid resulting from endogenous sources (e.g., prednisolone or dexamethasone treatment). Endogenous Cushing syndrome may be adrenocorticotropic hormone (ACTH)–dependent or ACTH-independent. The term *Cushing disease* is reserved for ACTH-dependent conditions, usually caused by a corticotroph adenoma that secretes excessive ACTH and stimulates the adrenal cortex to produce large amounts of cortisol. ACTH-independent Cushing syndrome is generally the result of an adrenocortical adenoma that secretes large amounts of cortisol.

Weight gain and obesity are the most common clinical signs of Cushing syndrome. In adults, weight gain occurs centrally with excess deposition in the visceral adipose depot (deep fat surrounding the organs), and this can be used to initially distinguish Cushing syndrome from generalized obesity. Patients also develop fat depots over the thoracocervical spine (buffalo hump) and over the cheeks, resulting in a moonlike face. Bright red–purple **striae** >1 cm wide are frequently found on the abdomen. In children, glucocorticoid excess often results in a phenotype lacking a central fat deposition but resulting in an overall generalized obesity. Thus, it is necessary to rule out Cushing syndrome in obese children before beginning treatment for general obesity.

Prolonged exposure of the body to large amounts of glucocorticoids causes the breakdown of skeletal muscle protein, increased glucose production by the liver, and mobilization of lipid from the fat depots. The increased mobilization of lipid provides abundant fatty acids for metabolism, and the increased oxidation of fatty acids by tissues reduces their ability to use glucose. The underuse of glucose by skeletal muscle, coupled with increased glucose production by the liver, results in hyperglycemia, which, in turn, stimulates the pancreas to secrete insulin. However, the rise in insulin is not effective in reducing the blood glucose concentration because glucose uptake and use are decreased in the skeletal muscle and adipose tissue. The net result is that the person becomes insensitive or resistant to the action of insulin and little glucose is removed from the blood, despite the high level of circulating insulin. The persisting hyperglycemia continually stimulates the pancreas to secrete insulin, resulting in a form of "diabetes" similar to type 2 diabetes mellitus (see Chapter 34, "Endocrine Pancreas").

The dexamethasone suppression test is useful for the diagnosis of Cushing syndrome of any cause. In normal subjects, the administration of supraphysiologic doses of glucocorticoid results in the suppression of ACTH and cortisol secretion. There is no suppression of ACTH and cortisol in patients with Cushing syndrome.

Treatment of Cushing syndrome depends on the cause. The removal of adrenal adenomas has a 100% cure rate. Transsphenoidal surgery for pituitary adenoma is also reasonably successful, depending on the type (micro or macro) of adenoma. Features of Cushing syndrome disappear after 2 to 12 months.

play a major role in mediating the inflammatory reaction. They are synthesized from the unsaturated fatty acid arachidonic acid, which is released from plasma membrane phospholipids by the hydrolytic action of **phospholipase A_2**. Glucocorticoids stimulate the synthesis of a family of proteins called **lipocortins** in their target cells. Lipocortins inhibit the activity of phospholipase A_2, reducing the amount of arachidonic acid available for conversion to prostaglandins and leukotrienes.

Glucocorticoids have little influence on the human immune system under normal physiologic conditions. When administered in large doses over a prolonged period, however, they can suppress antibody formation and interfere with cell-mediated immunity. Glucocorticoid therapy is therefore used to suppress the rejection of surgically transplanted organs and tissues.

Exposure to high concentrations of glucocorticoids can kill immature T cells in the thymus and immature B cells and T cells in lymph nodes, decreasing the number of circulating lymphocytes. The destruction of immature T and B cells by glucocorticoids also causes some reduction in the size of the thymus and lymph nodes.

Glucocorticoids and stress

Glucocorticoids are required for the normal responses of vascular smooth muscle to the vasoconstrictor action of NE and angiotensin II. In the absence of glucocorticoids, pressor agents are much less active on vascular smooth muscle. Glucocorticoids also reduce endothelial cell nitric oxide synthesis to limit the vasodilatory effects of this system.

Perhaps the most interesting, but least understood, of all glucocorticoid action is the ability to protect the body against stress. All that is really known is that the body cannot cope successfully with even mild stresses in the absence of glucocorticoids. One must presume that the processes that enable the body to defend itself against physical or emotional trauma require glucocorticoids. This emphasizes the permissive role glucocorticoids play in physiologic processes.

Both physical and emotional stresses stimulate the secretion of ACTH, which increases the secretion of glucocorticoids by the adrenal cortex (see Chapter 31, "Hypothalamus and the Pituitary Gland"). Cortisol levels rise in response to fever, surgery, burn injury, hypoglycemia, hypotension, and exercise. The increase in glucocorticoid secretion during stress is considered a normal counterregulatory response to the insult and is important for the appropriate defense mechanisms to be put into place. It is well known, for example, that glucocorticoid-deficient people receiving replacement therapy require larger doses of glucocorticoid to maintain their well-being during periods of stress.

Negative-feedback control

An important physiologic action of glucocorticoids is the ability to regulate their own secretion. This effect is achieved by a negative-feedback mechanism of glucocorticoids on the secretion of **corticotropin-releasing hormone (CRH)** and ACTH and on proopiomelanocortin (POMC) gene expression (see Chapter 31, "Hypothalamus and the Pituitary Gland").

ADRENAL MEDULLA HORMONES

The catecholamines, epinephrine and NE, are the two hormones synthesized by the chromaffin cells of the adrenal medulla. The human adrenal medulla produces and secretes about four times more epinephrine than NE. Postganglionic sympathetic neurons also produce and release NE from their nerve terminals but do not produce epinephrine.

Epinephrine and NE are formed in the chromaffin cells from the amino acid tyrosine. The pathway for the synthesis of catecholamines is illustrated in Figure 3.15.

Adrenal medulla consists of chromaffin cells, which are modified sympathetic postganglionic neurons.

The fusion of secretory granules with the plasma membrane releases epinephrine and some NE from chromaffin cells. The contents of the granules are extruded into the interstitial fluid. The catecholamines diffuse into capillaries and are transported in the bloodstream.

Neural stimulation of the cholinergic preganglionic fibers that innervate chromaffin cells triggers the secretion of catecholamines. Stimuli such as injury, anger, anxiety, pain, cold, strenuous exercise, and hypoglycemia generate impulses in these fibers, causing a rapid discharge of the catecholamines into the bloodstream.

Most cells of the body have receptors for catecholamines and therefore are their target cells. There are four structurally related forms of catecholamine receptors, all of which are transmembrane proteins: α_1, α_2, β_1, and β_2. All can bind epinephrine or NE to varying extents (see Chapter 3, "Action Potential, Synaptic Transmission, and Maintenance of Nerve Function").

Catecholamines are released during hypoglycemia, exercise, and trauma.

Epinephrine and NE produce widespread effects on the cardiovascular system, muscular system, and carbohydrate and lipid metabolism in the liver, muscle, and adipose tissues. In response to a sudden rise of catecholamines in the blood, heart rate accelerates, coronary blood vessels dilate, and blood flow to the skeletal muscles is increased as a result of vasodilation (but vasoconstriction occurs in the skin). Smooth muscles in the airways of the lungs, the gastrointestinal tract, and the urinary bladder relax. Muscles in the hair follicles contract, causing piloerection. Blood glucose level also rises. This overall reaction to the sudden release of catecholamines is known as the "fight-or-flight" response (see Chapter 6, "Autonomic Nervous System").

Hypoglycemia-induced catecholamine release

Catecholamines secreted by the adrenal medulla and NE released from sympathetic postganglionic nerve terminals are key agents in the body's defense against hypoglycemia. Catecholamine release usually starts when the blood glucose concentration falls to the low end of the physiologic range (60–70 mg/dL). A further decline in blood glucose concentration into the hypoglycemic range produces marked catecholamine release. Hypoglycemia can result from many situations, such as insulin overdosing, catecholamine antagonists, or drugs that block fatty acid oxidation. Hypoglycemia is always a dangerous condition because the CNS will die of ATP deprivation in extended cases. The length of time profound hypoglycemia can be tolerated depends on its severity and the person's sensitivity.

When the blood glucose concentration drops toward the hypoglycemic range, CNS receptors monitoring blood glucose are activated, stimulating the neural pathway leading to the fibers innervating the chromaffin cells. As a result, the adrenal medulla discharges catecholamines. Sympathetic postganglionic nerve terminals also release NE.

Catecholamines act on the liver to stimulate glucose production. They activate glycogen phosphorylase, resulting in the hydrolysis of stored glycogen, and stimulate gluconeogenesis from lactate and amino acids. Catecholamines also activate glycogen phosphorylase in skeletal muscle and adipose cells by binding to β receptors, activating adenylyl cyclase, and increasing cAMP in the cells. The glucose 6-phosphate generated in these cells is metabolized, although glucose is not released into the blood, because the cells lack glucose-6-phosphatase. Glycolysis converts the glucose 6-phosphate in muscle to lactate, much of which is released into the blood. The lactate taken up by the liver is converted to glucose via gluconeogenesis and returned to the blood.

In adipocytes, the rise in cAMP produced by catecholamines activates HSL, causing the hydrolysis of triglycerides and the release of fatty acids and glycerol into the bloodstream. These fatty acids provide an alternative substrate for energy metabolism in other tissues, primarily skeletal muscle, and block the phosphorylation and metabolism of glucose.

TABLE 33.4 Catecholamine-Mediated Responses to Hypoglycemia

Target Tissue	Action
Liver	Stimulation of glycogenolysis
	Stimulation of gluconeogenesis
Skeletal muscle	Stimulation of glycogenolysis
Adipose tissue	Simulation of glycogenolysis
	Stimulation of triglyceride lipolysis
Pancreatic islets	Inhibition of insulin secretion by β cells
	Stimulation of glucagon secretion by α cells

During profound hypoglycemia, the rapid rise in blood catecholamine levels triggers some of the same metabolic adjustments that occur more slowly during fasting. During fasting, these adjustments are triggered mainly in response to the gradual rise in the ratio of glucagon to insulin in the blood. The ratio also rises during profound hypoglycemia, reinforcing the actions of the catecholamines on glycogenolysis, gluconeogenesis, and lipolysis. The catecholamines released during hypoglycemia are thought to be partly responsible for the rise in the glucagon-to-insulin ratio by directly influencing the secretion of these hormones by the pancreas. Catecholamines stimulate the secretion of glucagon by the alpha cells and inhibit the secretion of insulin by β cells (see Chapter 34, "Endocrine Pancreas"). These catecholamine-mediated responses to hypoglycemia are summarized in Table 33.4.

Chapter Summary

- The adrenal gland comprises an outer cortex surrounding an inner medulla. The cortex contains three histologically distinct zones (from outside to inside): the zona glomerulosa, zona fasciculata, and zona reticularis.
- Hormones secreted by the adrenal cortex include glucocorticoids, aldosterone, and adrenal androgens.
- The glucocorticoids cortisol and corticosterone are synthesized in the zona fasciculata and zona reticularis of the adrenal cortex.
- The mineralocorticoid aldosterone is synthesized in the zona glomerulosa of the adrenal cortex.
- Cholesterol, used in the synthesis of the adrenal cortical hormones, comes from cholesterol esters stored in the cells. Stored cholesterol is derived mainly from low-density lipoprotein particles circulating in the blood, but it can also be synthesized *de novo* from acetate within the adrenal gland.
- The conversion of cholesterol to pregnenolone in mitochondria is the common first step in the synthesis of all adrenal steroids and occurs in all three zones of the cortex.
- The liver is the main site for the metabolism of adrenal steroids, which are conjugated to glucuronic acid and excreted in the urine.
- Adrenocorticotropic hormone (ACTH) increases glucocorticoid and androgen synthesis in adrenal cortical cells in the zona fasciculata and zona reticularis by increasing intracellular cyclic adenosine monophosphate. ACTH also has a trophic effect on these cells.
- Angiotensin II and angiotensin III stimulate aldosterone synthesis in the cells of the zona glomerulosa by increasing cytosolic calcium and activating protein kinase C.
- Glucocorticoids bind to glucocorticoid receptors in the cytosol of target cells. The glucocorticoid-bound receptor translocates to the nucleus and then binds to glucocorticoid response elements in the DNA to increase or decrease the transcription of specific genes.
- Glucocorticoids are essential to the adaptation of the body to fasting, injury, and stress.
- The chromaffin cells of the adrenal medulla synthesize and secrete the catecholamines epinephrine and norepinephrine.
- Catecholamines interact with four adrenergic receptors (α_1, α_2, β_1, and β_2) that mediate the cellular effects of the hormones.
- Stimuli such as injury, anger, pain, cold, strenuous exercise, and hypoglycemia generate impulses in the cholinergic preganglionic fibers innervating the chromaffin cells, resulting in the secretion of catecholamines.
- To counteract hypoglycemia, catecholamines stimulate glucose production in the liver, lactate release from muscle, and lipolysis in adipose tissue.

thePoint *Visit http://thePoint.lww.com for chapter review Q&A, case studies, animations, and more!*

34 Endocrine Pancreas

ACTIVE LEARNING OBJECTIVES

Upon mastering the material in this chapter, you should be able to:

- Explain how cells of the exocrine and endocrine pancreas coordinate to direct many processes related to the digestion, uptake, and use of metabolic fuels.
- Explain how islet cell junctions, vascular supply, and innervation influence islet cell functionality.
- Explain how a key mechanism of insulin secretion led to the development of sulfonylureas.
- Explain how amylin and pancreatic polypeptide work in conjunction with insulin to control fuel metabolism during periods of feeding and why amylin treatment may be important in patients with types 1 and 2 diabetes mellitus.
- Explain the molecular aspects of proinsulin biosynthesis that are of clinical significance.
- Explain how mechanistic aspects of the incretin effect have led to a new class of antidiabetic drugs.
- Explain the mechanism(s) underlying type 1 diabetes, type 2 diabetes, prediabetes, and gestational diabetes.
- Explain how both genetic and nongenetic factors can impair insulin responsiveness.
- Explain how, in addition to promoting diabetic complications, a series of related variables in metabolic syndrome apparently worsen insulin sensitivity.

The development of mechanisms for the storage of large amounts of metabolic fuel was an important adaptation in the evolution of complex organisms. The processes involved in the digestion, storage, and use of fuels require a high degree of regulation and coordination. The pancreas, which plays a vital role in these processes, consists of two functionally different groups of cells. The pancreas is both an endocrine gland producing several hormones, including **insulin, glucagon, amylin**, and **somatostatin**, and an exocrine gland secreting pancreatic enzymes. Cells of the **exocrine pancreas** produce and secrete digestive enzymes and fluids into the upper part of the small intestine. The **endocrine pancreas**, an anatomically small portion of the pancreas (1%–2% of the total mass), produces hormones involved in regulating fuel storage and use.

For convenience, functions of the exocrine and endocrine portions of the pancreas are usually discussed separately. Although this chapter focuses primarily on hormones of the endocrine pancreas, the overall function of the pancreas is to coordinate and direct many processes related to the digestion, uptake, and use of metabolic fuels.

ISLETS OF LANGERHANS

The endocrine pancreas consists of numerous discrete clusters of cells, known as the **islets of Langerhans**, which are located throughout the pancreatic mass. The islets contain specific types of cells responsible for the synthesis and secretion of the hormones amylin, ghrelin, glucagon, insulin, and somatostatin. Many circulating nutrients regulate the secretion of these hormones.

Islets of Langerhans are the functional units of the endocrine pancreas.

The islets of Langerhans contain from a few hundred to several thousand hormone-secreting endocrine cells. The islets are found throughout the pancreas but are most abundant in the tail region of the gland. The human pancreas contains, on average, about 1 million islets, which vary in size from 50 to 300 μm wide. A connective tissue sheath separates each islet from the surrounding acinar tissue.

Islets are composed of five (formerly thought to be only four) major hormone-producing cell types: insulin-secreting and amylin-secreting **β cells**, glucagon-secreting **α cells**, somatostatin-secreting **δ cells**, and **pancreatic polypeptide (PP)**-secreting **F cells**. The recently discovered fifth mammalian islet cell type is the **ghrelin**-producing **ε cells**. Studies performed mainly in rodents suggest that this is a developmentally regulated islet cell with the cells appearing early in development and then dissipating as β cells appear. Interestingly, the number of ε cells is increased in mouse models of β-cell deficiency. In human and rodent pancreases of neonates and adults, a few ε cells remain visible in marginal areas of the islets. Although studies continue to investigate the exact role that ghrelin-secreting ε cells may have in the pancreas, blockade of ghrelin or ghrelin action in pancreatic islets all markedly enhance glucose-induced insulin release.

Immunofluorescent staining techniques have shown that cell types are arranged in each islet in a pattern suggesting a highly organized cellular community, in which paracrine influences may play an important role in determining hormone secretion rates (Fig. 34.1). Further evidence that cell-to-cell communication within the islet may play a role in regulating hormone secretion comes from the finding that islet cells have both gap junctions and tight junctions. Gap junctions link different cell types in the islet cells and potentially provide a means for the transfer of ions, nucleotides, or electrical current between cells. The presence of tight junctions between outer membrane leaflets of contiguous cells could result in the formation of microdomains in the interstitial space, which may also be important for paracrine communication. Although the existence of gap junctions and tight junctions in pancreatic islets is well documented, their exact function has not been fully defined.

The arrangement of the vascular supply to islets is also consistent with paracrine involvement in regulating islet secretion. Afferent blood vessels penetrate nearly to the center of the islet before branching out and returning to the surface of the islet. The innermost cells of the islet, therefore, receive arterial blood, whereas those cells nearer the surface receive blood-containing secretions from inner cells. Because there is a definite anatomic arrangement of cells in the islet (see Fig. 34.1), one cell type could affect the secretion of others. In general, the effluent from smaller islets passes through neighboring pancreatic acinar tissue before entering into the hepatic portal venous system. By contrast, the effluent from larger islets passes directly into the venous system without first perfusing adjacent acinar tissue. Therefore, islet hormones arrive in high concentrations in some areas of the exocrine pancreas before reaching peripheral tissues. However, the exact physiologic significance of these arrangements is unknown.

Neural inputs also influence islet cell hormone secretion. Islet cells receive sympathetic and parasympathetic innervation. Responses to neural input occur as a result of activation of various adrenergic and cholinergic receptors (described below). Neuropeptides released together with the neurotransmitters may also be involved in regulating hormone secretion.

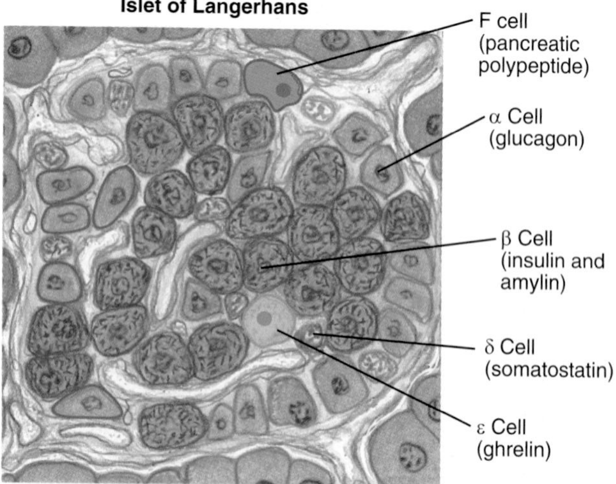

Figure 34.1 Major cell types in a typical islet of Langerhans. Note the distinct anatomic arrangement of the various cell types.

β Cells

In the early 1900s, M. A. Lane established a histochemical method by which two kinds of islet cells could be distinguished. He found that alcohol-based fixatives dissolved the secretory granules in most of the islet cells but preserved them in a few cells. Water-based fixatives had the opposite effect. He named cells containing alcohol-insoluble granules α cells and those containing alcohol-soluble granules β cells. Many years later, other investigators used immunofluorescence techniques to demonstrate that β cells produce insulin and α cells produce glucagon.

β Cells are the most numerous cell type of the islet, comprising 70% to 90% of the endocrine cells. β Cells typically occupy the most central space of the islets (see Fig. 34.1). They are generally 10 to 15 μm wide and contain secretory granules that measure 0.25 μm. β Cells secrete insulin, proinsulin, C-peptide, and amylin. Amylin is a newly described protein, also known as *islet amyloid-associated peptide*. Insulin, as detailed in the following sections, orchestrates body fuel metabolism both during periods of fasting and during feeding.

α Cells

α Cells comprise most of the remaining cells of the islets. They are generally located near the periphery, where they form a cortex of cells surrounding the more centrally located β cells. Blood vessels pass through the outer zone of the islet before extensive branching occurs. Inward extensions of the cortex may be present along the axes of blood vessels toward the center of the islet, giving the appearance that the islet is subdivided into small lobules. α Cells express the proglucagon gene, processed to yield glucagon. Recall that the proglucagon gene is also expressed in the intestinal L cell and processed to yield **glucagon-like peptide 1 (GLP-1), GLP-2**, and several spacer or intervening peptides in the gut (see Chapter 30, "Endocrine Control Mechanisms," Fig. 30.5). Pancreatic glucagon is a 29-amino acid peptide that regulates plasma glucose levels via effects on gluconeogenesis and glycogenolysis.

δ Cells

δ Cells are the sites of production of somatostatin in the pancreas. These cells are typically located in the periphery of the islet, often between β cells and the surrounding mantle of α cells. In both pancreatic δ cells and the hypothalamus, somatostatin exists as 14-amino acid and 28-amino acid peptides. In δ cells, the 28-amino acid peptide predominates; in the hypothalamus, the 14-amino acid peptide form is predominant. The 14-amino acid form is the C-terminal portion of the 28-amino acid form, and it is in the C-terminal 14 amino acids where the biologic activity of somatostatin resides. This biologic activity inhibits the secretion of several hormones including growth hormone (see Chapter 31, "Hypothalamus and the Pituitary Gland"), insulin, glucagon, PP, **gastrin, vasoactive intestinal peptide (VIP)**, and **thyroid-stimulating hormone**. The concentrations of somatostatin found in pancreatic venous drainage are sufficiently high to inhibit insulin

Clinical Focus / 34.1

Amylin

Amylin is a 37-amino acid peptide that is almost exclusively expressed within pancreatic β cells, where it is copackaged with insulin in secretory granules. Consequently, amylin is normally cosecreted with insulin, and the plasma concentrations of the two hormones display a similar diurnal pattern of low fasting levels and rapid and robust increases in response to meals.

As might be expected, because of the colocalization of both hormones within β cells, patients with type 1 diabetes have an absolute deficiency of both insulin and amylin, whereas patients with type 2 diabetes have a relative deficiency of both hormones, including a significantly impaired amylin and insulin response to meals. These findings have led to questions of whether amylin deficiency contributes to the metabolic derangements in patients with type 1 or type 2 diabetes and, if so, whether amylin replacement might convey clinical benefit when used in conjunction with insulin replacement.

Extensive studies with amylin and amylin antagonists in rodents and with pramlintide (a synthetic analogue of the human amylin hormone) in patients with type 1 or type 2 diabetes have provided a solid understanding of the role of amylin in glucose homeostasis. Preclinical data indicate that amylin acts as a neuroendocrine hormone that complements the actions of insulin in postprandial glucose homeostasis via several mechanisms. These include a suppression of postprandial glucagon secretion and a slowing of the rate at which nutrients are delivered from the stomach to the small intestine for absorption. The net effect of these actions is to mitigate the influx of endogenous (liver-derived) and exogenous (meal-derived) glucose into the circulation and thus to better match the rate of insulin-mediated glucose clearance from the circulation.

secretion. However, as blood flows from the centrally located β cells to the peripherally residing α cells, an endocrine effect of somatostatin on insulin secretion is minimized. Whether somatostatin has important paracrine actions on some α and β cells is unclear.

F cells

F cells are the least abundant of the hormone-secreting cells of islets, representing only about 1% of the total cell population. The distribution of F cells is generally similar to that of δ cells. F cells secrete PP. Recent study reveals that nutrients, hormones, neurotransmitters, gastric distention, insulin-induced hypoglycemia, and direct vagal nerve stimulation regulate PP secretion, whereas hyperglycemia, bombesin, and somatostatin inhibit PP secretion. Furthermore, we now know that a receptor termed *Y4*, a G-protein–coupled receptor that inhibits **cyclic adenosine monophosphate (cAMP)** accumulation, mediates the actions of PP. The Y4 receptor is expressed in the stomach, small intestine, colon, pancreas, prostate, and enteric nervous system and in select central nervous system neurons. Accumulating evidence suggests that PP reduces gastric acid secretion and increases intestinal transit times by reducing gastric emptying and upper intestinal motility. PP also appears to inhibit postprandial exocrine pancreas secretion via a vagal-dependent pathway. Transgenic mice that overexpress PP exhibit reduced weight gain and rate of gastric emptying and decreased fat mass. These new findings certainly merit further PP-related basic and clinical research.

Glucose is the most important physiologic factor in the regulation and synthesis of insulin.

Insulin secretion and biosynthesis are stimulated when islets are exposed to glucose. Apparently the metabolism of glucose by the β cell triggers both of these events. Whereas defined molecular steps by which glucose metabolites regulate insulin secretion have emerged, the molecular basis of glucose-regulated insulin biosynthesis is not entirely clear. However, studies document that within minutes after raising plasma glucose levels, the rate of proinsulin biosynthesis increases 5-fold to 10-fold. At the molecular level, an enhanced efficiency of initiation of translation of proinsulin RNA is observed in the β cell following glucose exposure.

Proinsulin biosynthesis

The gene for insulin is located on the short arm of chromosome 11 in humans. Like other hormones and secretory proteins, insulin is first synthesized by ribosomes of the rough endoplasmic reticulum (RER) as a larger precursor peptide that is then converted to the mature hormone prior to secretion (see Chapter 30, "Endocrine Control Mechanisms").

The insulin gene product is a 110-amino acid peptide, preproinsulin. Proinsulin consists of 86 amino acids (Fig. 34.2); residues 1 to 30 constitute what will form the B chain of insulin, residues 31 to 65 form the connecting peptide, and residues 66 to 86 constitute the A chain. (Note that "connecting peptide" should not be confused with "C-peptide.") In the process of converting proinsulin to insulin, two pairs of basic amino acid residues are clipped out of the proinsulin molecule, resulting in the formation of insulin and **C-peptide**, which are ultimately secreted from the β cell in equimolar amounts.

In addition, small amounts of intact proinsulin and proinsulin conversion intermediates are released. Proinsulin and its related conversion intermediates can be detected in the circulation, where they constitute 20% of total circulating insulin-like immunoreactivity. *In vivo*, proinsulin has a biologic potency that is only about 10% of that of insulin, and the potency of split proinsulin intermediates is between those of proinsulin and insulin.

It is of clinical significance that insulin and C-peptide are cosecreted in equal amounts. The measurement of peripheral insulin concentrations by radioimmunoassay is limited by the fact that 50% to 60% of the insulin produced by the pancreas is extracted by the liver without ever

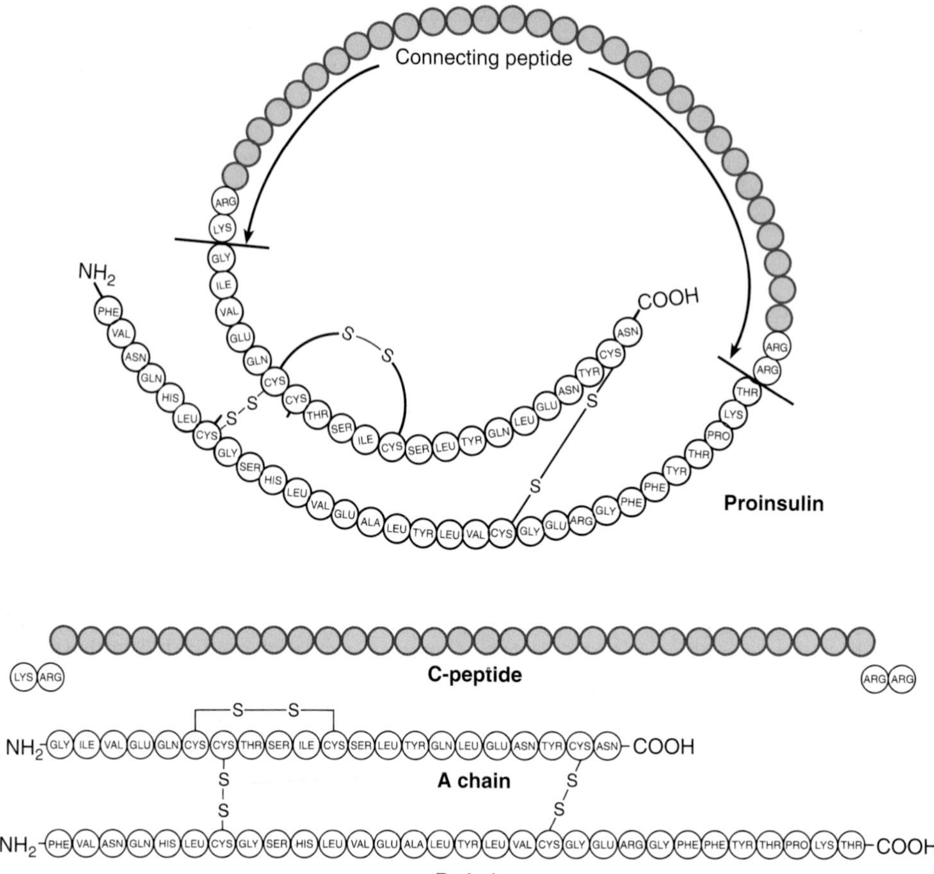

Figure 34.2 The structure of proinsulin, C-peptide, and insulin. Note that with the removal of two pairs of basic amino acids, proinsulin is converted into insulin and C-peptide.

reaching the systemic circulation. In contrast, the liver does not extract C-peptide. Because C-peptide is secreted in equimolar concentrations with insulin and is not extracted by the liver, β-cell insulin secretion rates can be calculated. In addition, as discussed earlier in Chapter 30, "Endocrine Control Mechanisms," another advantage of measuring C-peptide is that the standard insulin radioimmunoassay does not distinguish between endogenous and exogenous insulin, making it an ineffective measure of endogenous β-cell function in an insulin-treated diabetic patient (see Clinical Focus Box 30.2).

Insulin-secretion regulation

Glucose-stimulated insulin secretion is dose-related. Dose-dependent increases in insulin release have been observed after oral and intravenous glucose loads. The insulin secretory response is greater after oral than after intravenous glucose administration. Known as the **incretin effect**, this enhanced response to oral glucose has been interpreted as an indication that absorption of glucose by way of the gastrointestinal tract stimulates the release of hormones and other mechanisms that ultimately enhance the sensitivity of the β cell to glucose (see following discussion).

Insulin secretion does not respond as a linear function of glucose concentration. The relationship of glucose concentration to the rate of insulin secretion follows a sigmoidal curve, with a threshold corresponding to the glucose levels normally seen under fasting conditions and with the steep portion of the dose–response curve corresponding to a normal postprandial range of glucose levels (Fig. 34.3A).

When glucose is infused intravenously at a constant rate, a biphasic secretory response is observed that consists of a rapid, early insulin peak followed by a second, more slowly rising peak (Fig. 34.3B). In contrast to the slow rise in plasma glucose concentration following an oral glucose challenge, glucose given intravenously promotes a rapid rise in plasma glucose concentration (compare Fig. 34.3A and B). Sensing a rapid rise in plasma glucose concentration, the β cells first secrete their stores of presynthesized insulin. Following this *acute phase*, the cells begin to secrete newly synthesized insulin in the *chronic phase*, which lasts as long as the glucose challenge. *In vitro* studies of isolated islet cells and the perfused pancreas have identified a *third phase* of insulin secretion, commencing 1.5 to 3.0 hours after exposure to glucose and characterized by a spontaneous decline in secretion to 15% to 25% of the amount released during peak secretion: a level maintained for more than 48 hours.

β-Cell signaling

Recall from earlier discussion of membrane transport (see Chapter 2, "Plasma Membrane, Membrane Transport, and Resting Membrane Potential") that movement of glucose

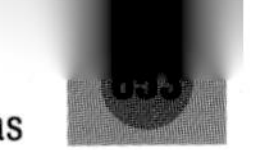

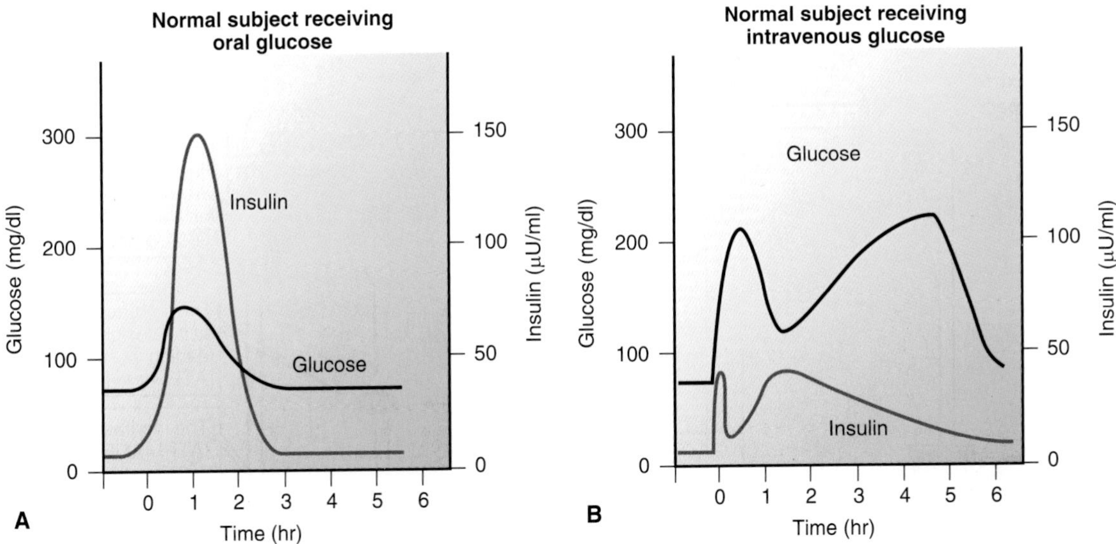

Figure 34.3 Plasma glucose and insulin following an oral (A) or intravenous (B) glucose challenge.

into a cell requires a transport system. This step and others underlying glucose-stimulated insulin secretion are illustrated in Figure 34.4. The glucose transporter **GLUT2** mediates movement of glucose into the β cell. In the cytosol, glycolysis converts glucose to pyruvate in a series of enzymatic steps; the first rate-limiting step in this process is the phosphorylation of glucose to glucose-6-phosphate by the enzyme **glucokinase**. There is considerable evidence that glucokinase, by determining the rate of **glycolysis**, functions as the glucose sensor of the β cell and that this is the primary mechanism whereby the rate of insulin secretion adapts to changes in blood glucose. **Adenosine triphosphate (ATP)** is synthesized during this process, and even more is produced during the subsequent oxidation of pyruvate within the mitochondrion. The increase in cytosolic ATP is a key signal that initiates insulin secretion by causing blockade of the **ATP-dependent K^+ channel (K_{ATP})** on the β-cell membrane. Blockade of this channel induces membrane depolarization, which activates voltage-gated Ca^{2+} channels. The associated Ca^{2+} influx increases cytosolic Ca^{2+} and additionally boosts cytosolic Ca^{2+} levels via triggering Ca^{2+}-induced Ca^{2+} release. The increase in cytosolic Ca^{2+} is the main spark for exocytosis, the process by which insulin-containing secretory granules fuse with the plasma membrane, leading to release of insulin into the circulation. Interestingly, in addition to the initial engagement of exocytosis by ATP (i.e., ATP-mediated K_{ATP} closure), ATP also serves as a major permissive factor for movement of insulin granules and for the priming of exocytosis.

The essential role of K_{ATP} channels in insulin secretion is the basis for **sulfonylureas**, a class of drugs used orally in the treatment of **type 2 diabetes (T2D)**. K_{ATP} channels comprise sulfonylurea receptors, and sulfonylureas are a class of drugs used orally as an insulin **secretagogue** in the treatment of T2D.

In addition to glucose, several other factors such as amino acids, hormones, and neural communication serve as important regulators of insulin secretion (Table 34.1). Among the amino acids, leucine, arginine, and lysine are potent secretagogues. The effects of arginine and lysine on the β cell appear to be more potent than that of leucine. Mechanistically, the amino acids appear to trigger insulin secretion via ATP production generated via their metabolism. Various lipids and their metabolites also impact insulin secretion. Acutely, lipids and their metabolites enhance glucose-stimulated insulin secretion. However, long-term exposure of islets to lipid impairs glucose-stimulated insulin secretion and biosynthesis, potentially contributing to β-cell failure.

Many gastrointestinal peptide hormones, including **glucose-dependent insulinotropic peptide, cholecystokinin**, and GLP-1, facilitate the release of insulin from the β cell after a meal. These hormones are released from small intestinal cells following a meal and travel in the blood to reach β cells, where they act through second messengers to increase the sensitivity of these islet cells to glucose. In general, these hormones are not themselves secretagogues, and their effects are evident only in the presence of hyperglycemia. The release of these peptides may explain the incretin effect: that is, the greater β-cell response observed after oral as opposed to intravenous glucose administration. Whether impaired postprandial secretion of incretin hormones plays a role in the inadequate insulin secretory response to oral glucose and to meals in prediabetes and T2D is controversial, yet pharmacologic doses of these peptides may have a future therapeutic benefit.

Other intestinal peptide hormones, including VIP, secretin, and gastrin, may also influence the postprandial insulin secretory response, but the exact role of these hormones is currently unclear. The hormones of the neighboring α and δ pancreatic islet cells also modulate β-cell insulin release. Whereas glucagon has a stimulatory effect via a $G_{\alpha s}$-stimulation of adenylyl cyclase, somatostatin suppresses this effect by engaging $G_{\alpha i}$ in the β cell (see Fig. 34.4). Other hormones that have been reported to exert a stimulatory effect on insulin secretion include growth hormone, glucocorticoids, prolactin, placental lactogen, and the sex steroids.

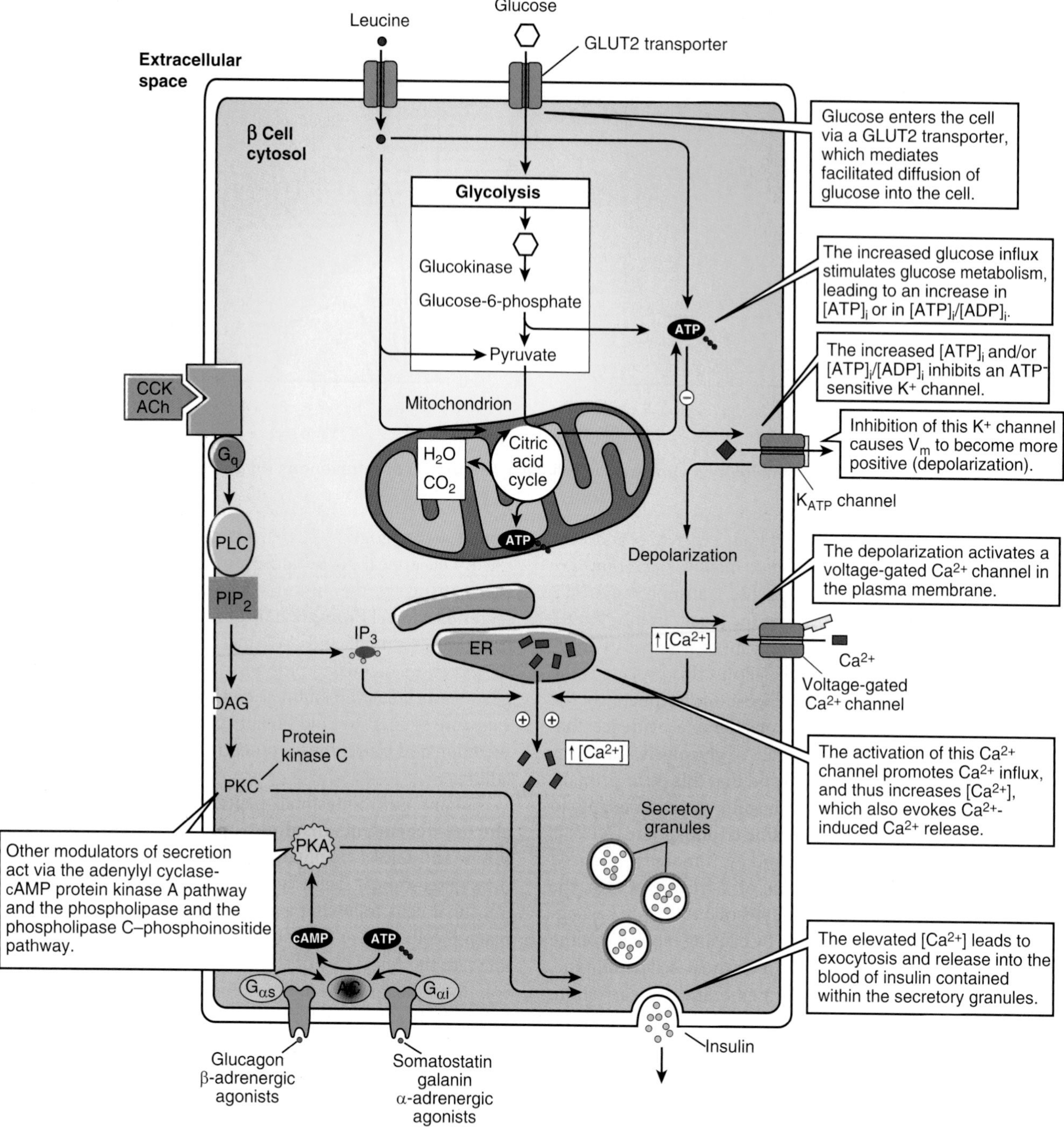

Figure 34.4 Stimuli and signals mediating insulin secretion by the pancreatic β cell. AC, adenylyl cyclase; $[ADP]_i$, intracellular concentration of adenosine diphosphate; ATP, adenosine triphosphate; $[ATP]_i$, intracellular concentration of adenosine triphosphate; cAMP, cyclic adenosine monophosphate; CCK, cholecystokinin; DAG, diacylglycerol; ER, endoplasmic reticulum; $G_{\alpha i}$, guanine nucleotide-binding protein α-inhibitory subclass; $G_{\alpha s}$, guanine nucleotide-binding protein α-stimulatory subclass; G_q, G proteins; GLUT2, glucose transporter; IP_3, inositol trisphosphate; K_{ATP}, ATP-dependent K^+ channel; PIP_2, phosphatidylinositol 4,5-bisphosphate; PKA, protein kinase A; PKC, protein kinase C; PLC, phospholipase C; V_m, electrical potential difference across membrane.

Sympathetic and parasympathetic divisions of the autonomic nervous system richly innervate islets. Activation of α-adrenoceptors inhibits insulin secretion, whereas β-adrenoceptor stimulation augments insulin secretion. Norepinephrine released by the sympathetic neurons of the pancreas stimulates α-adrenoceptors more than β-adrenoceptors. This response would appear appropriate because during periods of stress and high catecholamine secretion, the desired response is mobilization of glucose and other nutrient stores. Direct infusion of acetylcholine into the pancreatic circulation stimulates insulin secretion, reflecting the role of parasympathetic innervation in regulating insulin secretion.

TABLE 34.1 Factors Regulating Insulin Secretion from the Pancreas

Type	Name
Stimulatory agents or conditions	Hyperglycemia
	Amino acids
	Fatty acids, especially long-chain
	Gastrointestinal hormones, especially gastric inhibitory peptide, gastrin, and secretin
	Acetylcholine
	Sulfonylureas
Inhibitory agents or conditions	Somatostatin
	Norepinephrine
	Epinephrine

Hypoglycemia stimulates glucagon secretion.

Similar to insulin, glucagon is first synthesized as part of a larger precursor protein. Many of the factors that regulate insulin secretion also regulate glucagon secretion. In most cases, however, these factors have the opposite effect on glucagon secretion.

Proglucagon synthesis

Glucagon is a simple 29-amino acid peptide. The initial gene product for glucagon, *preproglucagon*, is a much larger peptide. As with other peptide hormones, the "pre" piece is removed in the RER, and the prohormone is converted into a mature hormone as it is packaged and processed in secretory granules. Recall that, within α cells of the pancreas, proglucagon is cleaved at two positions to yield three peptides (see Chapter 30, "Endocrine Control Mechanisms"). Glucagon, an important hormone in the regulation of carbohydrate metabolism, is best characterized of the three peptides. In contrast, in other cells of the gastrointestinal tract in which *proglucagon* is also produced, the molecule is cleaved at three different positions such that **glicentin**, GLP-1, and GLP-2 are produced.

Glucagon secretion

The principal factors that influence glucagon secretion are listed in Table 34.2. With a few exceptions, this table is nearly a mirror image of Table 34.1, which lists the factors that regulate insulin secretion. The primary regulator of glucagon secretion is blood glucose; specifically, when blood glucose falls below about 72 mg/dL, the α cells become active and deliver glucagon to the blood. Amino acids are also major secretagogues; however, the concentrations of amino acids required to provoke secretion of glucagon in vitro are higher than those generated in vivo. This observation suggests that other neural or humoral factors amplify the response in vivo, analogous to the effects of incretins on insulin secretion. Somatostatin inhibits glucagon secretion, as it does insulin secretion.

Hyperglycemia and glucagon stimulate somatostatin secretion.

As described earlier, somatostatin is first synthesized as a larger peptide precursor, *preprosomatostatin*. The hypothalamus also produces this protein, but the regulation of somatostatin secretion from the hypothalamus is independent of that from the pancreatic δ cells. On insertion of preprosomatostatin into the RER, it is initially cleaved and converted to prosomatostatin. The prohormone is converted into active hormone during packaging and processing in the Golgi apparatus.

Clinical Focus / 34.2 

Dipeptidyl Peptidase IV Inhibitors

Since 1955, sulfonylurea therapy has been effectively used to bolster insulin secretion in type 2 diabetics. As these drugs stimulate insulin secretion from the pancreatic β cells via glucose-independent closure of adenosine triphosphate-sensitive potassium channels, they are subject to inducing severe hypoglycemia. As a result, the pharmaceutical industry is constantly striving to develop safer and more efficacious antidiabetic drugs. Two newly emerging antidiabetic drug concepts are long-acting analogues of incretin hormones and enzyme inhibitors of dipeptidyl peptidase IV (DPP IV).

The incretin hormones glucagon-like peptide 1 (GLP-1) and glucose-dependent insulinotropic peptide are gastrointestinal insulin-releasing peptides involved in the regulation of postprandial nutrient homeostasis. These peptides augment nutrient-induced insulin release in a glucose-dependent fashion and thereby minimize risk of hypoglycemia. GLP-1 is known to inhibit glucagon secretion, stimulate insulin biosynthesis, and enlarge pancreatic β cell mass. Antidiabetic actions of GLP-1 are limited as a result of degradation and inactivation by DPP IV. Although DPP IV is the primary culprit for GLP-1 inactivation, it is a ubiquitous enzyme responsible for the inactivation of numerous regulatory peptides, including GIP and pituitary adenylate cyclase-activating polypeptide. The U.S. Food and Drug Administration has recently approved (April 2005) the first DPP IV-resistant GLP-1 analogue, exenatide (Byetta), and several others are undergoing clinical trials. Another promising area of investigation concerns the inhibition of DPP IV to prolong the actions of incretin hormones. Administration of DPP IV inhibitors to animal models of type 2 diabetes and diabetic patients lowers DPP IV activity, increases levels of intact GLP-1, and improves glucose tolerance.

TABLE 34.2 Factors Regulating Glucagon Secretion from the Pancreas

Type	Name
Stimulatory agents or conditions	Hypoglycemia Amino acids Acetylcholine Norepinephrine Epinephrine
Inhibitory agents or conditions	Fatty acids Somatostatin Insulin

Factors that stimulate pancreatic somatostatin secretion include hyperglycemia, glucagon, and amino acids. Glucose and glucagon are generally considered the most important regulators of somatostatin secretion.

The exact role of somatostatin in regulating islet hormone secretion has not been fully established. Somatostatin clearly inhibits both glucagon and insulin secretion from the α and β cells of the pancreas, respectively, when it is given exogenously. The anatomic and vascular relationships of δ cells to α and β cells further suggest that somatostatin may play a role in regulating both glucagon and insulin secretion. Although many of the data are circumstantial, it is generally accepted that somatostatin plays a paracrine role in regulating insulin and glucagon secretion from the pancreas.

INSULIN AND GLUCAGON INFLUENCE ON METABOLIC FUELS

The endocrine pancreas secretes hormones that direct the storage and use of fuels during times of nutrient abundance (fed state) and nutrient deficiency (fasting). The two main hormones are insulin and glucagon. Insulin is secreted in the fed state and is called the "hormone of nutrient abundance." By contrast, glucagon is secreted in response to an overall deficit in nutrient supply. It should be noted, as discussed earlier, that accumulating evidence highlights that the other islet hormones play a very important ancillary role.

Insulin regulates carbohydrate and fat metabolism.

Insulin is the primary regulator of carbohydrate and fat metabolism and controls the storage of these nutrients as well. Besides metabolism of glucose and lipid, signaling by insulin affects other processes such as amino acid transport and metabolism, protein synthesis, cell growth, differentiation, and apoptosis. The three principal targets of insulin action are the liver, skeletal muscle, and adipose tissue.

Pleiotropic action of insulin

The insulin receptor (IR) is a *heterotetramer*, consisting of a pair of α/β subunit complexes held together by disulfide bonds (Fig. 34.5). The α subunit is an extracellular protein containing the insulin-binding component of the receptor. The β subunit is a transmembrane protein that couples the extracellular event of insulin binding to its intracellular actions.

Activation of the β subunit of the IR results in autophosphorylation, involving the phosphorylation of a few selected tyrosine residues in the intracellular portion of the receptor. This event further activates the tyrosine kinase portion of the β subunit, leading to tyrosine phosphorylation of a host of proteins, including members of the insulin-receptor substrate (IRS) family (IRS-1, -2, -3, -4, -5, and -6), Gab-1, Shc, $p62^{dok}$, c-Cbl, SIRP (signal regulatory protein) family members, and APS (adapter protein containing a pleckstrin homology and Src-homology 2 domain). A cascade of phosphorylation and dephosphorylation events, second messenger generation, and protein–protein interactions result, leading to several cellular activities, including gene expression and growth, protein synthesis, antiapoptosis, antilipolysis, glycogen synthesis, and glucose transport, to name a few.

Internalization and dephosphorylation by protein tyrosine phosphatases (PTPs) terminate the activity of the IR. Protein tyrosine phosphatase 1B (PTP1B) and leukocyte antigen-related (LAR) phosphatases are two PTPs shown to attenuate IR activity. A later section on diabetes mellitus highlights that T2D is a complicated, multifactorial disease in which interactions between multiple genes and multiple environmental factors ultimately result in resistance to the action of insulin. In that regard, both PTP1B and LAR have been reported to be elevated in patients displaying **insulin resistance**. In addition to dampening IR function by tyrosine phosphatase activity, serine–threonine phosphorylation of the β subunit of the IR also has been suggested to decrease the ability of the receptor to autophosphorylate. Many protein kinase C isoforms that catalyze the serine or threonine phosphorylation of the IR are elevated in insulin-resistant tissue from animals and humans.

GLUT4

An essential aspect of insulin action is its impact on regulating plasma glucose concentration. The ability of insulin to efficiently stimulate the storage and metabolism of glucose achieves glucose homeostasis, helping to return elevated plasma glucose to normal levels. In muscle and adipose tissues, insulin promotes the removal of excess circulatory glucose by regulating the subcellular trafficking of the glucose transporter **GLUT4**. In the basal state, GLUT4 cycles continuously between the plasma membrane and one or more intracellular compartments, with most of the transporter residing in the cell interior. Activation of the IR triggers a large increase in the rate of GLUT4 vesicle exocytosis and a smaller but important decrease in the rate of internalization by endocytosis. The overall insulin-dependent shift in the cellular dynamics of GLUT4 vesicle trafficking results in a

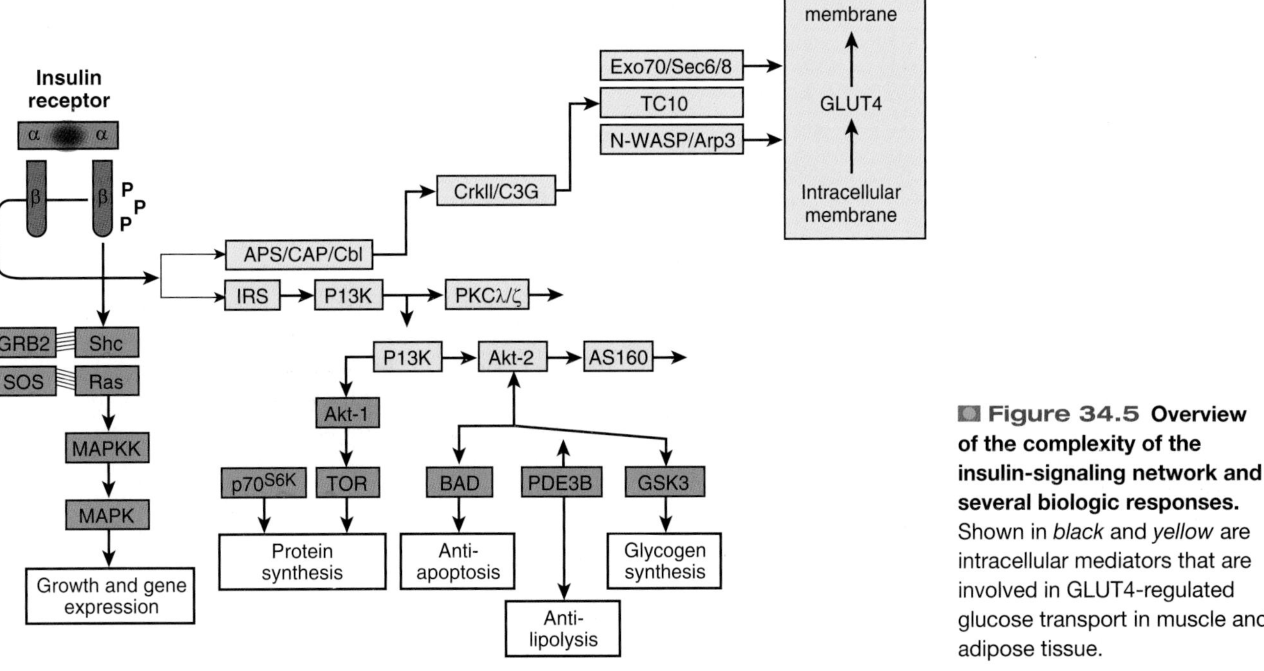

Figure 34.5 Overview of the complexity of the insulin-signaling network and several biologic responses. Shown in *black* and *yellow* are intracellular mediators that are involved in GLUT4-regulated glucose transport in muscle and adipose tissue.

net increase of GLUT4 protein levels at the cell surface. These plasma membrane-localized transporters subsequently facilitate the influx of plasma glucose into the cell.

On a molecular level, insulin resistance is associated with insufficient recruitment of GLUT4 to the plasma membrane despite normal GLUT4 protein expression. This finding emphasizes the importance of understanding the molecular mechanisms of insulin-regulated GLUT4 translocation and glucose uptake so that cellular perturbations associated with insulin resistance can be elucidated and effective treatment and preventative strategies can be designed to bypass or ameliorate insulin resistance. Although details linking insulin to GLUT4 translocation have yet to be fully resolved, substantial progress has been made within the last two decades to delineate the molecular events. For example, activation of the IR triggers a cascade of phosphorylation events that ultimately promote GLUT4 vesicle exocytosis (see Fig. 34.5). The classical insulin signaling pathway involves the following: docking of IRS-1/2 to the IR; activation of phosphatidylinositol 3-kinase (PI3K), which leads to formation of plasma membrane phosphatidylinositol 3,4,5-trisphosphate (PI 3,4,5-P_3); subsequent PI 3,4,5-P_3-mediated activation of Akt-2 and atypical protein kinase C (aPKC); and phosphorylation of AS160 (Akt substrate of 160 kDa) by Akt-2. Additionally, a PI3K-independent pathway involving c-Cbl, c-Cbl-associated protein, and the GTPase TC10 may also regulate GLUT4 translocation, although this pathway appears to be exclusive to adipocytes.

Glycogen synthesis

Besides promoting glucose uptake into target cells, insulin promotes its storage. Glucose carbon is stored in the body in two primary forms: as **glycogen** and (by metabolic conversion) as triglycerides. Glycogen is a short-term storage form that plays an important role in maintaining normal blood glucose levels. The primary glycogen storage sites are the liver and skeletal muscle; other tissues, such as adipose tissue, also store glycogen but in quantitatively small amounts. Insulin promotes glycogen storage primarily through two enzymes (Fig. 34.6). It activates **glycogen synthase** by promoting its dephosphorylation and concomitantly inactivates

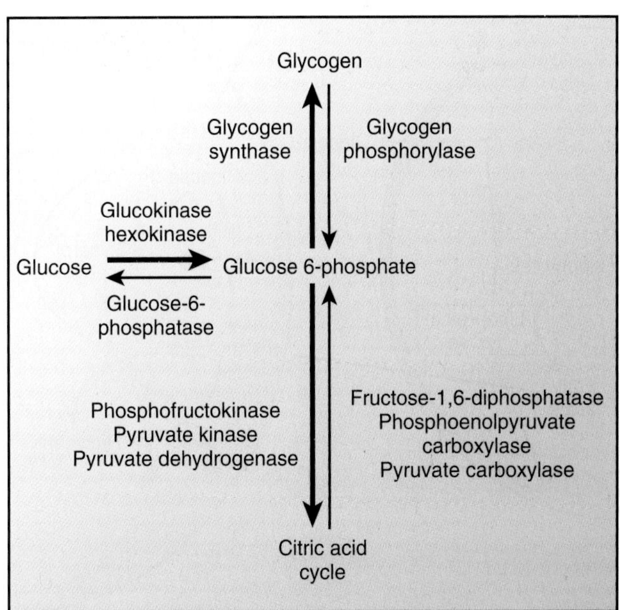

Figure 34.6 Insulin stimulation of glycogen synthesis and glucose metabolism. Insulin promotes glucose uptake into target tissues, stimulates glycogen synthesis, and inhibits glycogenolysis. In addition, it promotes glycolysis in its target tissues. *Bold arrows* indicate processes stimulated by insulin; *light arrows* indicate processes inhibited by insulin.

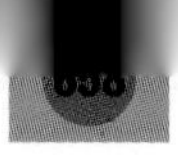

glycogen phosphorylase, also by promoting its dephosphorylation. The result is that glycogen synthesis is promoted and glycogen breakdown is inhibited.

Glycolysis

Insulin also enhances glycolysis. In addition to increasing glucose uptake and providing a mass action stimulus for glycolysis, insulin activates the enzymes glucokinase (liver), **hexokinase** (muscle), **phosphofructokinase, pyruvate kinase**, and **pyruvate dehydrogenase** of the glycolytic pathway (see Fig. 34.6). Also shown is the inhibition of gluconeogenesis by insulin inhibition of **fructose-1,6-diphosphatase, phosphoenolpyruvate carboxylase**, and **pyruvate carboxylase**.

Lipogenic and antilipolytic effects of insulin

In adipose tissue and liver tissue, insulin promotes **lipogenesis** and inhibits **lipolysis** (Fig. 34.7). Insulin has similar actions in muscle, but because muscle is not a major site of lipid storage, the discussion here focuses on actions in adipose tissue and the liver. By promoting the flow of intermediates through glycolysis, insulin promotes the formation of α-glycerol phosphate and fatty acids necessary for triglyceride formation. In addition, it stimulates **fatty acid synthase**, leading directly to increased fatty acid synthesis. Insulin inhibits the breakdown of triglycerides by inhibiting **hormone-sensitive lipase**, which is activated by many counterregulatory hormones, such as epinephrine and adrenal glucocorticoids. By inhibiting this enzyme, insulin promotes the accumulation of triglycerides in adipose tissue.

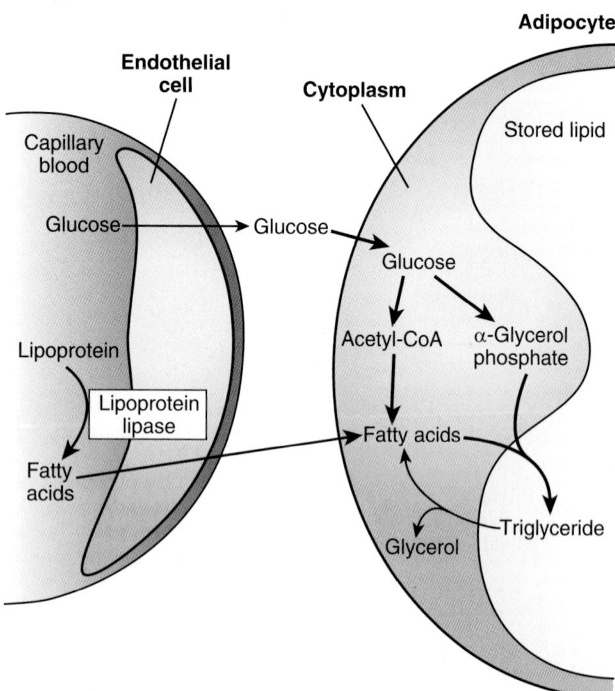

Figure 34.7 Effects of insulin on lipid metabolism in adipocytes. Insulin promotes the accumulation of lipid (triglycerides) in adipocytes by stimulating the processes shown by the *bold arrows* and inhibiting the processes shown by the *light arrows*. Similar stimulatory and inhibitory effects occur in liver cells.

In addition to promoting *de novo* fatty acid synthesis in adipose tissue, insulin increases the activity of **lipoprotein lipase**, which plays a role in the uptake of fatty acids from the blood into adipose tissue. As a result, adipose tissue takes up lipoproteins synthesized in the liver, and fatty acids are ultimately stored as triglycerides.

Protein synthesis and degradation

Insulin promotes protein accumulation in its primary target tissues—liver, adipose tissue, and muscle—in three specific ways (Fig. 34.8). First, it stimulates amino acid uptake. Second, it increases the activity of several factors involved in protein synthesis. For example, it increases the activity of protein synthesis initiation factors, promoting the start of translation and increasing the efficiency of protein synthesis. Insulin also increases the amount of protein synthesis machinery in cells by promoting ribosome synthesis. Third, insulin inhibits protein degradation by reducing lysosome activity and possibly other mechanisms as well.

Glucagon influences the metabolism of carbohydrates, lipids, and proteins primarily in the liver.

The primary physiologic actions of glucagon are exerted in the liver. Numerous effects of glucagon have been documented in other tissues, primarily adipose tissue, when the hormone has been added at high, nonphysiologic concentrations in experimental situations. Although these effects may play a role in certain abnormal situations, the normal daily effects of glucagon occur primarily in the liver.

cAMP-mediated glucagon action

Glucagon initiates its biologic effects by interacting with one or more types of cell membrane receptors. Glucagon receptors are coupled to G proteins and promote increased intracellular cAMP, via the activation of adenylyl cyclase or elevated cytosolic calcium as a result of phospholipid breakdown to form inositol trisphosphate.

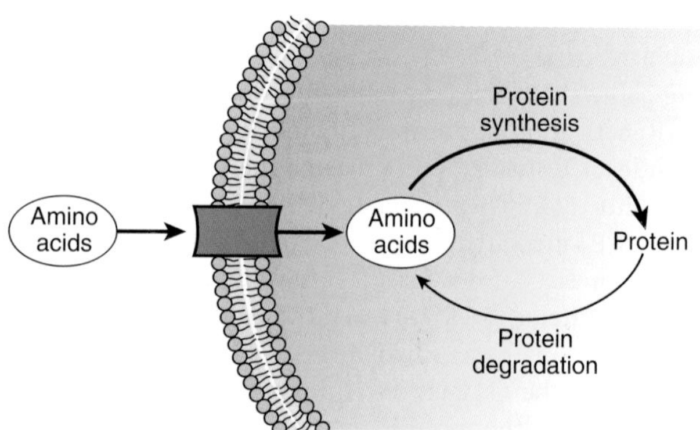

Figure 34.8 Effects of insulin on protein synthesis and protein degradation. Insulin promotes the accumulation of protein by stimulating (*bold arrows*) amino acid uptake and protein synthesis and by inhibiting (*light arrow*) protein degradation in liver, skeletal muscle, and adipose tissue.

Glycogenolysis

Glucagon is an important regulator of hepatic glycogen metabolism. It produces a net effect of glycogen breakdown (**glycogenolysis**) by increasing intracellular cAMP levels, initiating a cascade of phosphorylation events that ultimately results in the phosphorylation of phosphorylase *b* and its activation by conversion into phosphorylase *a*. Similarly, glucagon promotes the net breakdown of glycogen by promoting the inactivation of glycogen synthase (Fig. 34.9).

Gluconeogenesis

In addition to promoting hepatic glucose production by stimulating glycogenolysis, glucagon stimulates hepatic **gluconeogenesis** (Fig. 34.10). It does this principally by increasing the transcription of messenger RNA coding for the enzyme **phosphoenolpyruvate carboxykinase (PEPCK)**, a key rate-limiting enzyme in gluconeogenesis. Glucagon also stimulates amino acid transport into liver cells and the degradation of hepatic proteins, helping provide substrates for gluconeogenesis.

Ureagenesis

The glucagon-enhanced conversion of amino acids into glucose leads to increased formation of ammonia. Glucagon stimulates **ureagenesis**, a process that assists in the disposal of ammonia by increasing the activity of the urea cycle enzymes in liver cells (see Fig. 34.10).

Lipolysis

Glucagon promotes lipolysis in liver cells (Fig. 34.11), although the quantity of lipids stored in liver is small compared with that in adipose tissue.

Ketogenesis

Glucagon promotes **ketogenesis**, the production of **ketones**, by lowering the levels of **malonyl CoA**, relieving an inhibition of **palmitoyl transferase**, and allowing fatty acids to enter the mitochondria for oxidation to ketones (see Fig. 34.11). Ketones are an important source of fuel for muscle cells and heart cells during times of starvation, sparing blood glucose for other tissues that are obligated glucose users, such as the central nervous system. During prolonged starvation, the brain adapts its metabolism to use ketones as a fuel source, lessening the overall need for hepatic glucose production (see Chapter 27, "Liver Physiology").

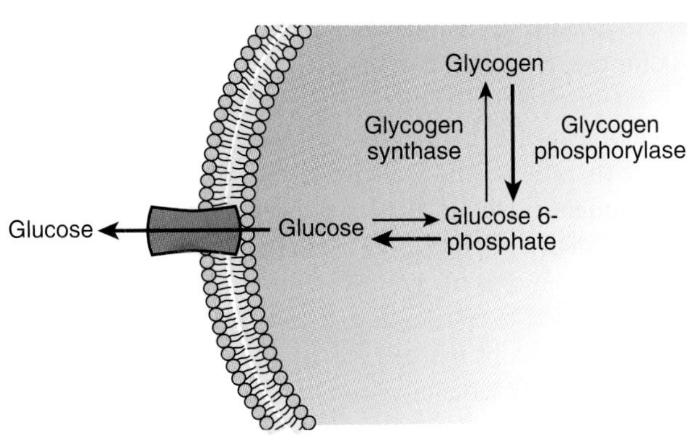

Figure 34.9 The role of glucagon in glycogenolysis and glucose production in liver cells. *Bold arrows* indicate processes stimulated by glucagon; *light arrows* indicate processes inhibited by glucagon.

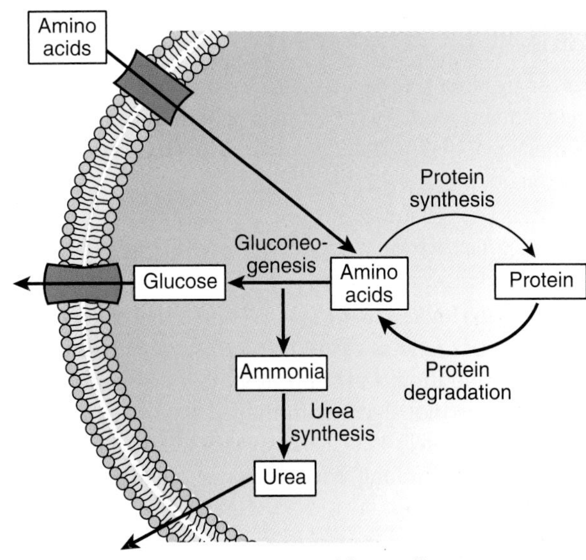

Figure 34.10 The role of glucagon in gluconeogenesis and ureagenesis in liver cells. *Bold arrows* indicate processes stimulated by glucagon; the *light arrow* indicates a process inhibited by glucagon.

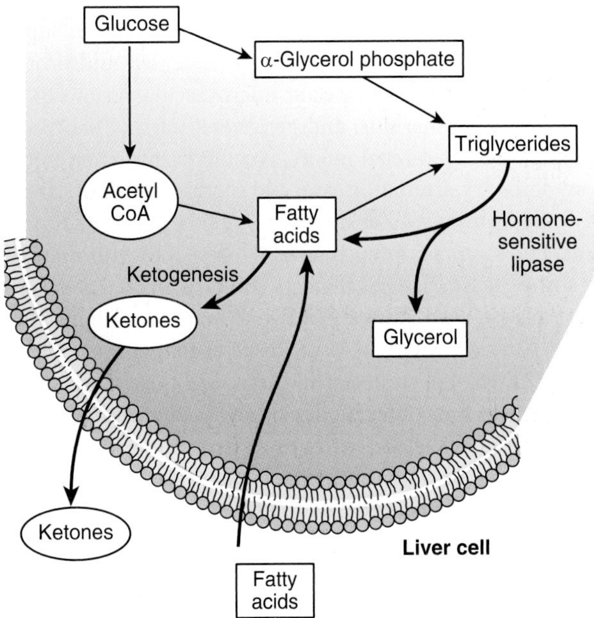

Figure 34.11 The role of glucagon in lipolysis and ketogenesis in liver cells. *Bold arrows* indicate processes stimulated by glucagon; *light arrows* indicate processes inhibited by glucagon.

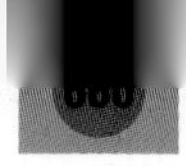

Insulin/glucagon ratio determines metabolic status.

In most instances, insulin and glucagon produce opposing effects. Therefore, the relative levels of both hormones in the blood plasma, the **insulin/glucagon ratio (I/G ratio)**, determine the net physiologic response.

Insulin/glucagon ratio during eating and fasting

The I/G ratio is highest in the fed state and is the lowest during fasting. The I/G ratio can vary 100-fold or more because the plasma concentration of each hormone can vary considerably in different nutritional states. In the fed state, the molar I/G ratio is approximately 30. After an overnight fast, it may fall to about 2, and with prolonged fasting, it may fall to as low as 0.5.

Insulin/glucagon ratio in type 1 diabetes

A good example of the profound influence of the I/G ratio on metabolic status is in **type 1 diabetes (T1D)** (see below), in which the lack of insulin alters the I/G ratio. Insulin levels are low, so pathways that insulin stimulates operate at a reduced level. However, insulin is also necessary for α cells to sense blood glucose appropriately; in the absence of insulin, the secretion of glucagon is inappropriately elevated. The result is an imbalance in the I/G ratio and an accentuation of glucagon effects well above what would be seen in normal states of low insulin, such as in fasting.

DIABETES MELLITUS

Diabetes mellitus is a systemic disease in which the body does not produce and/or use insulin and is characterized most notably by **hyperglycemia**. In addition to abnormal glucose metabolism, disorders in the metabolism of lipid and protein intensify the seriousness of the disease. Superimposed on the disorders of carbohydrate, fat, and protein metabolism are diabetic-specific microvascular lesions in the retina, renal glomerulus, and peripheral nerve. Diabetes, if untreated, leads to renal failure, erectile dysfunction, blindness, coronary arterial disease, and increased risk of cancer. Of these, cardiovascular disease (CVD) is the most prominent. For example, more than 65% of people with diabetes die from heart disease. In fact, adults with diabetes have heart disease death rates about two to four times higher than adults without diabetes. Also, stroke accounts for approximately 20% of diabetes-related deaths, and the risk for stroke is also two to four times higher among people with diabetes. The cause of diabetes continues to be a mystery, although both genetics and environmental factors such as obesity and lack of exercise appear to play roles.

There are 23.6 million children and adults in the United States, or 7.8% of the population, who have diabetes. Although an estimated 17.9 million have been diagnosed with diabetes, unfortunately 5.7 million people (or nearly one quarter) are unaware that they have the disease. Diabetes has several clinical forms, each of which has a distinct etiology, clinical presentation, and course.

Diagnosing diabetes mellitus is not difficult to do. Symptoms usually include frequent urination, increased thirst, increased food consumption, and weight loss. Diabetes mellitus is a heterogeneous disorder. The causes, symptoms, and general medical outcomes are variable. There are three main type of diabetes: (1) T1D results from the body failure to produce insulin, (2) T2D results from insulin resistance and β cell failure, and (3) **gestational diabetes** results from hyperglycemia during pregnancy.

Autoimmune disorder underlies type 1 diabetes.

T1D is one of the most intensively studied autoimmune disorders. Although most cases of T1D are a consequence of an inappropriate autoimmune destruction of the β cell, an autoimmune-independent subtype of T1D has recently been described, and recommendations have been put forth to divide T1D into type 1A (immune-mediated) and type 1B (other forms of diabetes with severe insulin deficiency). **Type 1B diabetes** appears to be a rare form of T1D in which histologic examination of pancreatic sections demonstrates inflammation but no anti-islet autoantibodies.

Whereas studies to gain deeper insight into type 1B diabetes are currently underway, it is known that **type 1A diabetes** results from a selective destruction of the β cells within the islets. A subtype of type 1A diabetes that has a latent onset in adults also exists, and recent findings suggest that its correct diagnosis and tailored insulin therapy may delay the progression and worsening of the disease (see From Bench to Bedside Box 34.1). The development of type 1A diabetes is usually divided into a series of stages, beginning with genetic susceptibility and ending with essentially complete β-cell destruction. Central to the pathology is **insulitis** β-cell injury, involving a lymphocytic attack on β cells.

Studies of identical twins have provided important information regarding the genetic basis of T1D. If one twin develops T1D, the odds that the second will develop the disease are much higher than for any random person in the population, even when the twins are raised apart under different socioeconomic conditions. In addition, people with certain cell-surface human leukocyte antigens bear a higher risk for the disease than others.

Environmental factors are involved as well because the development of T1D in one twin predicts only a 50% or less chance that the second will develop the disease. The specific environmental factors have not been identified, although much evidence implicates viruses. Therefore, it appears that a combination of genetics and environment is a strong contributing factor to the development of T1D.

Because the primary defect in T1D is the inability of β cells to secrete adequate amounts of insulin, these patients must be treated with injections of insulin. In an attempt to match insulin concentrations in the blood with the metabolic requirements of the person, various formulations of insulin with different durations of action have been developed. Patients inject an appropriate amount of these different insulin forms to match their dietary and lifestyle requirements.

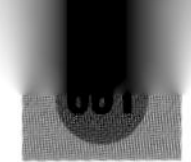

From Bench to Bedside / 34.1

Beyond Type 1 and Type 2 Diabetes

It is becoming increasingly apparent that the main forms of diabetes, type 1 diabetes (T1D) and type 2 diabetes (T2D), have subtypes that require the correct diagnosis. A subtype of new concern is termed *latent autoimmune diabetes in adults* ([*LADA*] though some prefer *type 1.5*). This subtype was first revealed in the 1970s while testing a method to identify autoantibodies in the blood of people with T1D. Scientists found that T1D-related autoantibodies were virtually absent in the general population, but they showed up in about 10% of people diagnosed with T2D, which is not an autoimmune disease. Although much debate exists as to whether LADA is distinct from T1D, researchers today are working on a set of criteria for its diagnosis: (1) the presence of autoantibodies in the blood, (2) adult age at onset, and (3) lack of need for insulin treatment in the first 6 months after diagnosis. The first criterion distinguishes LADA from T2D because of autoantibodies in the blood, and the third criterion differs from T1D because T1D patients need to start insulin immediately. We now know that those with T1D have higher levels and more types of autoantibodies than do those with LADA, which may be the reason β cells are destroyed faster in T1D than in LADA.

Importantly, a recent Japanese study found that early insulin treatment in adults with LADA may help them avoid total dependence on insulin. This study also compared the use of insulin to that of sulfonylurea treatment (a common treatment for people with T2D that enhances β-cell insulin production) and found that, unlike those treated with insulin, patients treated with sulfonylurea lost β-cell function faster and progressed rapidly to insulin dependency. Thus, it is coming to light that a better treatment for this subtype of diabetes may be insulin, not sulfonylureas.

New technologies such as insulin pumps and continuous glucose monitors have not only aided treatment but also are paving the road to an artificial pancreas (see From Bench to Bedside Box 30.1).

The long-term control of T1D depends on maintaining a balance between several factors including insulin, diet, and exercise. To strictly control their blood glucose, patients are advised to monitor their diet and level of physical activity, as well as their insulin dosage. Exercise per se, much like insulin, increases glucose uptake by muscle. Diabetic patients must take this into account and make appropriate adjustments in diet and/or insulin whenever general exercise levels change dramatically.

Insulin resistance underlies prediabetes, type 2 diabetes, and gestational diabetes.

T2D is the predominant form of diabetes worldwide. An epidemic of T2D is underway in both developed and developing countries. Globally, the number of people with diabetes is expected to rise from the current estimate of 220 million to 366 million in 2030. Even more alarmingly, it is estimated that twice as many people have prediabetes, a condition that occurs when a person's blood glucose levels are higher than normal but not high enough for the diagnosis of T2D. Recent research has shown that some long-term damage to the body, especially the heart and circulatory system, may already be occurring during prediabetes.

Central to prediabetes and T2D is insulin resistance, a term that indicates the presence of an impaired biologic response to either exogenously administered or endogenously secreted insulin. Insulin resistance is manifested by decreased insulin-stimulated glucose transport and metabolism in adipocytes and skeletal muscle and by impaired suppression of hepatic glucose output.

It is clear that the molecular etiology of insulin resistance involves multiple genetic and nongenetic mechanisms contributing to the final phenotype. Furthermore, the resultant hyperglycemia of the syndrome and the compensatory rise in insulin in turn exacerbate insulin resistance, contributing significantly to the pathogenesis of the disease. Superimposed on that backdrop is the negative contribution of the obese state. Although pinpointing the site of abnormality induced by these conditions is still an intensive area of current research, evidence strongly suggests that the defects induced by each occur at different loci in the insulin-signaling network.

Mutations in the IR are associated with rare forms of insulin resistance. These mutations affect IR number, splicing, trafficking, binding, and phosphorylation. The affected patients demonstrate severe insulin resistance, including type A syndrome, leprechaunism, Rabson-Mendenhall syndrome, and lipoatrophic diabetes.

In some cases, insulin resistance results from the secretion of counterregulatory hormones in a normal (e.g., pregnant) or pathophysiologic (e.g., Cushing syndrome) state. Gestational diabetes occurs in about 6% of all pregnancies but usually disappears after delivery. Women who have had gestational diabetes have an increased risk of developing T2D later in life.

Obesity and insulin resistance

Obesity is closely linked to insulin resistance. The marked increase in the prevalence of obesity has played an important role in the prevalence of diabetes. According to data from the National Health and Nutrition Examination Survey, two thirds of the adult men and women in the United States with a diagnosis of T2D had a **body mass index (BMI)** of 27 kg/m^2 or greater. This index is calculated by dividing weight (in kilograms) by height (in meters squared). There is a strong curvilinear relation between BMI and relative body fat mass. Men and women with a BMI of 25.0 to 29.9 kg/m^2 are considered overweight, and those with a BMI of 30 kg/m^2 or greater are considered obese.

Type 2 diabetes treatment

Early on, or in mild cases, diet, weight loss, and exercise can be extremely effective in diabetes therapy and may be the only

treatment necessary. Commonly, however, lifestyle intervention is supplemented by treatment with one or more oral agents. Multiple classes of drugs independently address different pathophysiologic features that contribute to the development of T2D. The available oral antidiabetic agents can be divided by mechanism of action into insulin sensitizers with primary action in the liver (e.g., **biguanides**), insulin sensitizers with primary action in peripheral tissues (e.g., **glitazones**), insulin secretagogues (e.g., sulfonylureas), and agents that slow the absorption of carbohydrates (e.g., **α-glucosidase inhibitors**). In some cases, persons with T2D may be treated with insulin, just like the patient with T1D. Although insulin therapy apparently provides some benefit to type 2 diabetics, this approach is limited in controlling elevated glucose levels or in controlling obesity that predisposes to this disease.

Diabetes mellitus leads to organ dysfunction and tissue damage.

If left untreated or if glycemic control is poor, diabetes leads to acute complications that may prove fatal. However, even with reasonably good control of blood glucose, over a period of years, most diabetics develop secondary complications of the disease that result in tissue damage, primarily involving the cardiovascular and nervous systems.

Metabolic and renal complications

The nature of acute complications that develop in type 1 and type 2 diabetics differs. Persons with poorly controlled T1D often exhibit hyperglycemia, **glucosuria**, dehydration, and **diabetic ketoacidosis**. As blood glucose becomes elevated above the renal plasma threshold, glucose appears in the urine. As a result of osmotic effects, water follows glucose, leading to **polyuria**, excessive loss of fluid from the body, and dehydration. With fluid loss, the circulating blood volume is reduced, compromising cardiovascular function, which may lead to circulatory failure.

Excessive ketone formation leads to acidosis and electrolyte imbalances in persons with T1D. If uncontrolled, ketones may be elevated in the blood to such an extent that the odor of **acetone** (one of the ketones) is noticeable on the breath. Production of the primary ketones, **β-hydroxybutyric acid** and **acetoacetic acid**, results in the generation of excess hydrogen ions and a metabolic acidosis. Ketones may accumulate in the blood to such a degree that they exceed renal transport capacities and appear in the urine. As a result of osmotic effects, water is also lost in the urine. In addition, the pK of ketones is such that even with the most acidic urine a normal kidney can produce, about half of the excreted ketones are in the salt (or base) form. To ensure electrical neutrality, these must be accompanied by a cation, usually either sodium or potassium. The loss of ketones in the urine, therefore, also results in a loss of important electrolytes. Excessive ketone production in T1D results in acidosis, a loss of cations, and a loss of fluids. Emergency department procedures are directed toward immediate correction of these acute problems and usually involve the administration of base, fluids, and insulin.

The complex sequence of events that can result from uncontrolled T1D is shown in Figure 34.12. If left unchecked, many of these complications can have an additive effect to further the severity of the disease state.

Persons with T2D are generally not ketotic and do not develop acidosis or the electrolyte imbalances characteristic of T1D. Hyperglycemia leads to fluid loss and dehydration. Severe cases may result in hyperosmolar coma as a result of excessive fluid loss. The initial objective of treatment in these people is the administration of fluids to restore fluid volumes to normal and eliminate the hyperosmolar state.

Microcirculatory complications

With good control of their disease, most persons with diabetes can avoid the acute complications described above; however, it is rare that they will not suffer from some of the chronic secondary complications of the disease. In most instances, such complications will ultimately lead to reduced life expectancy.

Most lesions occur in the circulatory system, although the nervous system is also often affected. Both small and large blood vessels systemic circulation are damaged, producing what are known, respectively, as microvascular and macrovascular complications in both T1D and T2D. The common finding in affected vessels is a thickening of the basement membrane. This condition leads to impaired delivery of nutrients and hormones to the tissues and inadequate removal of waste products, resulting in irreparable tissue damage. Interestingly, micro- and macrovascular complications do not

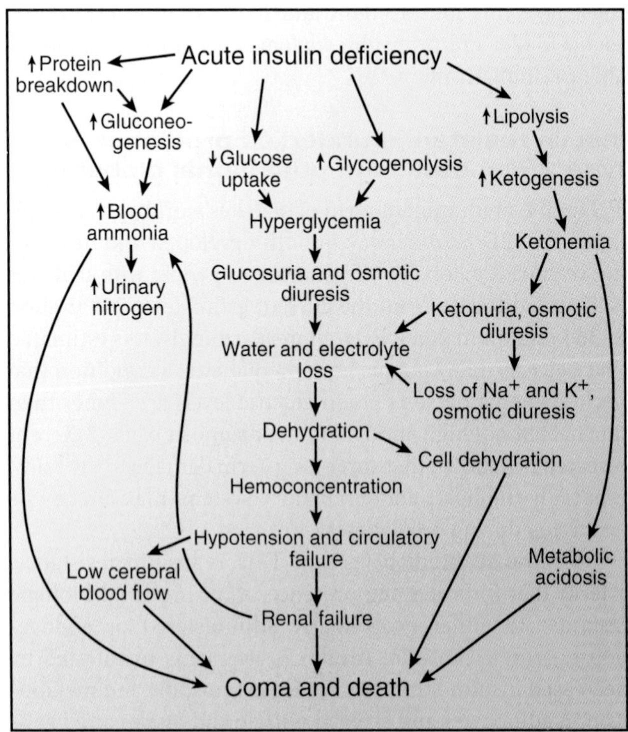

Figure 34.12 Events resulting from acute insulin deficiency in type 1 diabetes mellitus. If left untreated, insulin deficiency may lead to several complications, which may have additive or confounding effects that may ultimately result in death.

occur in the pulmonary circulation. Although the pulmonary circulation receives all of the cardiac output as well as the same concentration of insulin and glucose as the systemic circulation, it is refractory to diabetes-induced vascular damage.

"Microvascular complications" is a term defining the effects of diabetes on small blood vessels, which include the following: deterioration of blood flow to the retina of the eye, causing **retinopathy** and blindness; deterioration of blood flow to the extremities, causing, in some cases, the need for foot or leg amputation; and deterioration of glomerular filtration in the kidneys, leading to renal failure.

"Macrovascular complications" is a term defining the effect of diabetes on large blood vessels. Diabetes is a major cause of myocardial infarction, stroke, and CVD. Recognized risk factors for CVD (i.e., hypertension, high triglycerides, low high-density lipoprotein cholesterol) are recognized as common in people with diabetes, especially T2D. Patients displaying a constellation of these derangements have been designated as having **metabolic syndrome**.

Diabetic neuropathy

Another common complication of long-standing diabetes is **diabetic neuropathy**, a disorder affecting the peripheral sensory nerves also known as *peripheral neuropathy*. Diabetic neuropathy is thought to be a microvascular injury involving small blood vessels that supply the nerves. Diabetic neuropathy affects all peripheral nerves (pain fibers and motor neurons) and autonomic nerves.

Many patients with diabetes experience diminished sensation in the extremities, especially in the feet and legs, which compounds the problem of diminished blood flow to these areas. Often, impaired sensory nerve function results in lack of awareness of severe ulcerations of the feet caused by reduced blood flow. Men may develop erectile dysfunction leading to impotence, and both men and women may have impaired bladder and bowel function. Unfortunately, cancer patients (especially colon cancer) suffer from peripheral neuropathy as a side effect of radiation/chemotherapy and manifest the same symptoms.

Chapter Summary

- The relative distribution of α, β, δ, F, and ε cells within each islet of Langerhans shows a distinctive pattern and suggests that there may be some paracrine regulation of secretion.
- Plasma glucose is the primary physiologic regulator of insulin and glucagon secretion, but amino acids, fatty acids, and some gastrointestinal hormones also play a role.
- Insulin has anabolic effects on carbohydrate, lipid, and protein metabolism in its target tissues, where it promotes the storage of nutrients.
- Effects of glucagon on carbohydrate, lipid, and protein metabolism occur primarily in the liver and are catabolic in nature.
- Type 1 diabetes mellitus results from the destruction of β cells, whereas prediabetes, type 2 diabetes, and gestational diabetes often result from a lack of responsiveness to circulating insulin.
- Diabetes mellitus may produce both acute complications, such as ketoacidosis, and chronic secondary complications, such as microvascular disease, macrovascular disease, neuropathy, and nephropathy.

thePoint *Visit http://thePoint.lww.com for chapter review Q&A, case studies, animations, and more!*

35 Endocrine Regulation of Calcium, Phosphate, and Bone Homeostasis

ACTIVE LEARNING OBJECTIVES

Upon mastering the material in this chapter, you should be able to:

- Explain the mechanisms that control typical daily changes of both calcium and phosphorus between different tissue compartments in a healthy adult.
- Explain how chondrocytes, osteoblasts, osteocytes, and osteoclasts regulate bone formation and bone reabsorption.
- Explain how bones become brittle in osteogenesis imperfecta.
- Explain how parathyroid hormone, calcitonin, and 1,25-dihydroxycholecalciferol affect calcium and phosphate metabolism.
- Explain the underlying mechanisms of osteoporosis, osteomalacia, rickets, and Paget disease.

The plasma calcium concentration is among the most closely regulated of all physiologic parameters in the body. Typically, it varies by only 1% to 2% daily or even weekly. Such stringent regulation in a biologic system usually implies that the parameter plays an important role in one or more critical processes.

Phosphate also plays many important roles in the body, although its concentration is not as tightly regulated as that of calcium. Many of the factors involved in regulating calcium also affect phosphate.

OVERVIEW OF CALCIUM AND PHOSPHATE IN THE BODY

Calcium is the most abundant mineral in the body and plays a key role in many physiologically important processes. The average adult contains approximately 1 kg of calcium in the body, with 99% contained in bone in the form of calcium phosphate. A significant decrease in plasma calcium can rapidly lead to death. A chronic increase in plasma calcium can lead to soft tissue calcification and formation of stones. Phosphate also plays important roles in the body and, because it is a principal component of bone with calcium, phosphate homeostasis is tightly coupled to that of calcium. In addition, the same hormones regulate both calcium and phosphate balance.

Calcium plays a key role in nerve and muscle function, enzyme function, and mineral balance in bone.

Calcium affects nerve and muscle excitability, neurotransmitter release from axon terminals, and excitation–contraction coupling in muscle cells. It serves as a second or third messenger in several intracellular signal transduction pathways. Some enzymes use calcium as a cofactor, including some in the blood-clotting cascade. Finally, calcium is a major constituent of bone. Of all these roles, the one that demands the most careful regulation of plasma calcium is the effect of calcium on nerve excitability. Calcium affects the sodium permeability of nerve membranes, which influences the ease with which action potentials are triggered. Low plasma calcium (**hypocalcemia**) can lead to the generation of spontaneous action potentials in nerves. When motor neurons are affected, tetany of the muscles of the motor unit may occur; this condition is called **hypocalcemic tetany**. Certain diagnostically important signs may reveal latent tetany. The **Trousseau sign** is a characteristic spasm of the muscles of the forearm that causes flexion of the wrist and thumb and extension of the fingers. It may occur spontaneously or be elicited by inflation of a blood pressure cuff placed on the upper arm. The **Chvostek sign** is a unilateral spasm of the facial muscles that can be elicited by tapping the facial nerve at the point at which it crosses the angle of the jaw.

Although rare, hypocalcemia can also influence myocardial function by changing QT intervals or QRS and ST segments. These changes may mimic acute myocardial infarction or conduction abnormalities. A considerable reduction in serum calcium may lead to **grand mal seizures** or **laryngospasm**.

Phosphate participates in pH buffering and is a major constituent of macromolecules and bones.

Phosphorus (usually as phosphate) is a key constituent of bone. In addition, phosphate serves as an important component of intracellular pH buffering and cellular energy metabolism as part of the adenosine triphosphate molecule. Phosphate is a constituent of many other macromolecules, such as nucleic acid, phospholipids, and phosphoproteins.

Distributions of calcium and phosphorus differ in bone and cells.

Table 35.1 shows the relative distributions of calcium and phosphate in a healthy adult. The average adult body contains approximately 1 to 2 kg of calcium, roughly 99% of it in bones. Despite its critical role in excitation–contraction coupling, only about 0.3% of total body calcium is located in muscles. About 0.1% of total calcium is in extracellular fluid.

Of the roughly 600 g of phosphorus in the body, most is in bones (86%). Compared with calcium, a much larger percentage of phosphorus is located in cells (14%). The amount of phosphorus in extracellular fluid is low (0.08% of body content).

Bones also contain a relatively high percentage of the total body content of several other inorganic substances (Table 35.2). About 80% of the total carbonate in the body is located in bones. This carbonate can be mobilized into the blood to combat acidosis; thus, bone participates in pH buffering in the body. Long-standing uncorrected acidosis can result in considerable loss of bone mineral. Significant percentages of the body's magnesium and sodium and nearly 10% of its total water content are in bones.

Calcium and phosphorus exist in the plasma in several forms.

In humans, the normal plasma calcium concentration is 9.0 to 10.5 mg/dL. Plasma calcium exists in three forms: ionized or free calcium (50% of the total), protein-bound calcium (40%), and calcium bound to small diffusible anions, such as citrate, phosphate, and bicarbonate (10%). The association of calcium with plasma proteins is pH-dependent. At an alkaline pH, more calcium is bound; the opposite is true at an acidic pH.

Plasma phosphorus concentrations may fluctuate significantly during the course of a day (±50% of the average value) for any particular person. In adults, the normal range of plasma concentrations is 3.0 to 4.5 mg/dL (expressed in terms of milligrams of phosphorus).

Phosphorus circulates in the plasma primarily as inorganic **orthophosphate (PO_4)**. At a normal blood pH of 7.4, 80% of the phosphate is in the HPO_4^{2-} form and 20% is in the $H_2PO_4^-$ form. Nearly all plasma inorganic phosphate is ultrafilterable. In addition to free orthophosphate, phosphate is present in small amounts in the plasma in organic form, such as in hexose or in lipid phosphates.

TABLE 35.1 Body Content and Tissue Distribution of Calcium and Phosphorus in a Healthy Adult

Region	Calcium	Phosphorus
Total body content (g)	1,300	600
Relative tissue distribution (% of total body content)		
Bones and teeth	99	86
Extracellular fluid	0.1	0.08
Intracellular fluid	1.0	14

TABLE 35.2 Inorganic Constituents of Bone

Constituent	Percentage of Total Body Content Present in Bone
Calcium	99
Phosphate	86
Carbonate	80
Magnesium	50
Sodium	35
Water	9

Homeostatic pathways for calcium and phosphorus involve the gastrointestinal tract, kidneys, and bones.

Both calcium and phosphate are obtained from the diet. The gastrointestinal (GI) tract, the kidneys, and the bones primarily determine the ultimate fate of each substance.

Calcium homeostasis

The approximate tissue distribution and average daily flux of calcium among tissues in a healthy adult are shown in Figure 35.1. Dietary intakes may vary widely, but an "average" diet contains approximately 1,000 mg/d of calcium. Intakes up to twice that amount are usually well tolerated, but excessive calcium intake can result in soft tissue calcification or kidney stones. Only about one third of ingested calcium is actually absorbed from the GI tract; the remainder is excreted in the feces. The efficiency of calcium uptake from the GI tract varies with the person's physiologic status. The percentage uptake of calcium may be increased in young growing children and pregnant or nursing women; often it is reduced in older adults.

Figure 35.1 also indicates that approximately 150 mg/d of calcium actually enters the GI tract from the body. This component of the calcium flux partly results from sloughing of mucosal cells that line the GI tract and also

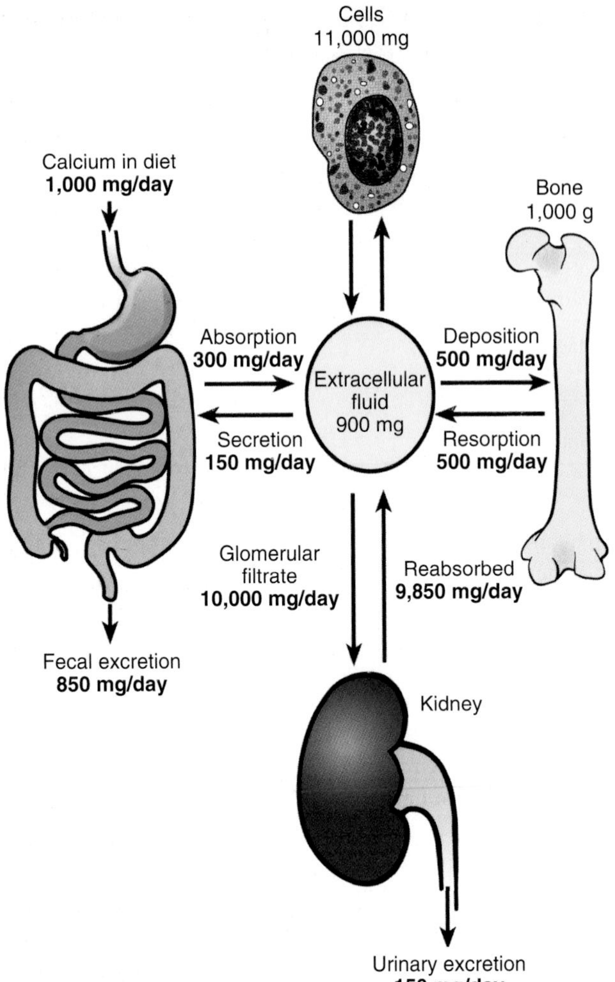

Figure 35.1 Typical daily exchanges of calcium between different tissue compartments in a healthy adult. Fluxes of calcium (mg/d) are shown in **bold font**. Total calcium content in each compartment is shown in *regular font*. Note that most ingested calcium is eliminated from the body via the feces.

from calcium that accompanies various secretions into the GI tract. This component of calcium metabolism is relatively constant, and so the primary determinant of *net* calcium uptake from the GI tract is calcium absorption. Intestinal absorption is important in regulating calcium homeostasis.

Bone in an average person contains approximately 1,000 g of calcium. Bone mineral is constantly reabsorbed and deposited in the remodeling process. As much as 500 mg/d of calcium may flow in and out of the bones (see Fig. 35.1). Because bone calcium serves as a reservoir, both bone resorption and bone formation are important in regulating plasma calcium concentration.

In overall calcium balance, the net uptake of calcium from the GI tract presents a daily load of calcium that will eventually require elimination. The primary route of elimination is via the urine and, therefore, the kidneys play an important role in regulating calcium homeostasis. The 150 mg/d of calcium excreted in the urine represents only about 1% of the calcium initially filtered by the kidneys; the remaining 99% is reabsorbed and returned to the blood. Therefore, small changes in the amount of calcium reabsorbed by the kidneys can have a dramatic impact on calcium homeostasis.

Phosphorus homeostasis

Figure 35.2 shows the overall daily flux of phosphate in the body. A typical adult ingests approximately 1,400 mg/d of phosphorus. In marked contrast to calcium, most (1,300 mg/d) of this phosphorus is absorbed from the GI tract, typically as inorganic phosphate. There is an obligatory contribution of phosphorus from the body to the contents of the GI tract (about 200 mg/d), much like that for calcium, resulting in a net uptake of phosphorus of 1,100 mg/d and excretion of 300 mg/d via the feces. Thus, most ingested phosphate is absorbed from the GI tract and little passes through to the feces.

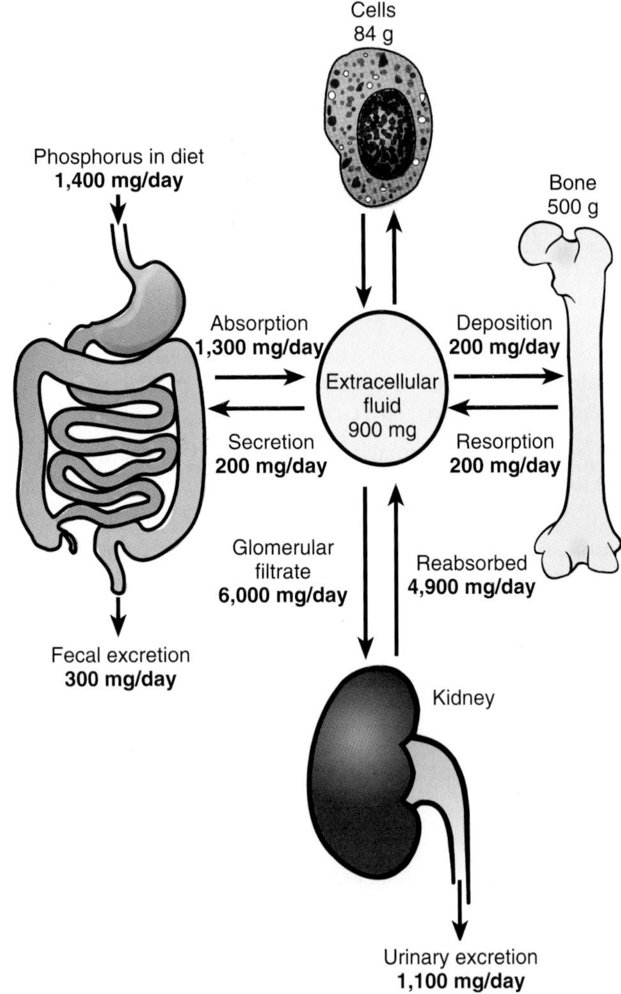

Figure 35.2 Typical daily exchanges of phosphorus between different tissue compartments in a healthy adult. Fluxes of phosphorus (mg/d) are shown in **bold font**. Total phosphorus content in each compartment is shown in *regular font*. Note that most ingested phosphorus is absorbed and eventually eliminated from the body via the urine.

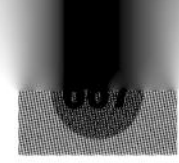

Because most ingested phosphate is absorbed, renal excretory mechanisms greatly influence phosphate homeostasis. Because most circulating phosphate is readily filtered in the kidneys, tubular phosphate reabsorption is a major process regulating phosphate homeostasis.

CALCIUM AND PHOSPHATE METABOLISM

Homeostasis is responsible for bringing calcium and phosphate into the body, maintaining normal blood levels of these minerals, transporting them into the cell, and transporting them out again when needed.

Calcium and phosphate are absorbed primarily by the small intestine.

Calcium absorption in the small intestine occurs by both active transport and diffusion. The relative contribution of each process varies with the region and with total calcium intake. Uptake of calcium by active transport predominates in the duodenum and jejunum; in the ileum, passive diffusion predominates. The relative importance of active transport in the duodenum and jejunum versus passive diffusion in the ileum depends on several factors. At high levels of calcium intake, active transport processes are saturated and most of the uptake occurs in the ileum, partly because of its greater length, compared with other intestinal segments. With moderate or low calcium intake, however, active transport predominates because the gradient for diffusion is low.

Active transport is the regulated variable in controlling calcium uptake from the small intestine. Metabolites of vitamin D provide a regulatory signal to increase intestinal calcium absorption. Under the influence of 1,25-dihydroxycholecalciferol, calcium-binding proteins in intestinal mucosal cells increase in number, enhancing the capacity of these cells to transport calcium actively (see Chapter 26, "Gastrointestinal Secretion, Digestion, and Absorption").

The small intestine is also a primary site for phosphate absorption. Uptake occurs by active transport and passive diffusion, but active transport is the primary mechanism. As indicated in Figure 35.2, phosphate is efficiently absorbed from the small intestine; typically, 80% or more of ingested phosphate is absorbed. However, phosphate absorption from the small intestine is regulated to a minor extent. Active transport of phosphate is coupled to calcium transport, and thus, when active transport of calcium is low, as with vitamin D deficiency, phosphate absorption is also low.

Kidneys play an important role in regulating plasma concentrations of calcium and phosphate.

As a result of regulating the urinary excretion of calcium and phosphate, the kidneys are in a key position to regulate the total body balance of these two ions. Hormones are an important signal to the kidneys to direct the excretion or retention of calcium and phosphate.

Renal handling of calcium

As discussed in Chapter 24, "Acid–Base Homeostasis," filterable calcium comprises about 60% of the total calcium in the plasma. It consists of free calcium ions and calcium bound to small diffusible anions. The remaining 40% of the total calcium is bound to plasma proteins and is not filterable by the glomeruli. Ordinarily, 99% of the filtered calcium is reabsorbed by the kidney tubules and returned to the plasma. Reabsorption occurs both in the proximal and distal tubules and in the loop of Henle. Approximately 60% of filtered calcium is reabsorbed in the proximal tubule, 30% in the loop of Henle, and 9% in the distal tubule; the remaining 1% is excreted in the urine. Renal calcium excretion is controlled primarily in the late distal tubule; parathyroid hormone (PTH) stimulates calcium reabsorption here, promoting calcium retention and lowering urinary calcium. As detailed later, PTH is an important regulator of plasma calcium concentration.

Renal handling of phosphate

Most ingested phosphate is absorbed from the GI tract, and the primary route of excretion of this phosphate is via the urine. Therefore, the kidneys play a key role in regulating phosphate homeostasis. Ordinarily about 85% of filtered phosphate is reabsorbed and 15% is excreted in the urine. Phosphate reabsorption occurs via active transport, mainly in the proximal tubule, where 65% to 80% of filtered phosphate is reabsorbed. PTH inhibits phosphate reabsorption in the proximal tubule and has a major regulatory effect on phosphate homeostasis. It increases urinary phosphate excretion, leading to the condition of **phosphaturia**, with an accompanying decrease in the plasma phosphate concentration.

Calcium and phosphate in bone are in continuous flux.

Although bone may be considered as being a relatively inert material, it is active metabolically. Considerable amounts of calcium and phosphate both enter and exit bone each day, and these processes are hormonally controlled.

Bone composition

Mature bone can be simply described as inorganic mineral deposited on an organic framework. The mineral portion of bone is composed largely of calcium phosphate in the form of **hydroxyapatite crystals**, which have the general chemical formula $Ca_{10}(PO_4)_6(OH)_2$. The mineral portion of bone typically comprises about 25% of its volume, but because of its high density, the mineral fraction is responsible for approximately half the weight of bone. Bone contains considerable amounts of the body's content of carbonate, magnesium, and sodium, in addition to calcium and phosphate (see Table 35.2).

The organic matrix of bone on which the bone mineral is deposited is called **osteoid. Type I collagen** is the primary constituent of osteoid, comprising 95% or more. Collagen in bone is similar to that of skin and tendons, but bone collagen exhibits some biochemical differences that impart increased mechanical strength. The remaining noncollagen portion (5%) of organic matter is referred to as **ground substance**. Ground substance consists of a mixture of various **proteoglycans**, high-molecular-weight compounds consisting of different types of polysaccharides linked to a polypeptide backbone. Typically, they are 95% or more carbohydrate.

Electron microscopic study of bone reveals needlelike hydroxyapatite crystals lying alongside collagen fibers. This orderly association of hydroxyapatite crystals with the collagen fibers is responsible for the strength and hardness characteristic of bone. A loss of either bone mineral or organic matrix greatly affects the mechanical properties of bone. Complete demineralization of bone leaves a flexible collagen framework, and the complete removal of organic matrix leaves a bone with its original shape, but extremely brittle. **Osteogenesis imperfecta**, or *brittle bone disease*, is a genetic disorder characterized by bones that break easily as a result of defects in the production of the fibrillar collagen matrix.

Osteoblasts, osteocytes, and osteoclasts constitute the major cell types in bone.

Osteoblasts are located on the bone surface and are responsible for osteoid synthesis (Fig. 35.3). Like many cells that actively synthesize proteins for export, osteoblasts have an abundant rough endoplasmic reticulum and Golgi apparatus. Cells actively engaged in osteoid synthesis are cuboidal, whereas those less active are more flattened. Numerous cytoplasmic processes connect adjacent osteoblasts on the bone surface and connect osteoblasts with osteocytes deeper in the bone. Osteoid produced by osteoblasts is secreted into the space adjacent to the bone. Eventually, new osteoid becomes mineralized, and in the process, mineralized bone surrounds osteoblasts.

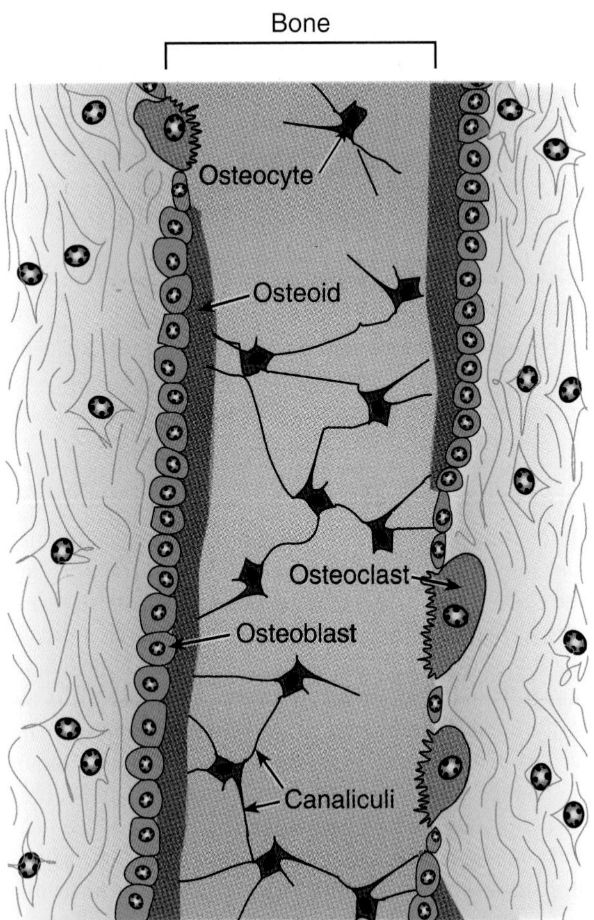

Figure 35.3 The location and relationship of the three primary cell types involved in bone metabolism.

As osteoblasts are progressively incorporated into mineralized bone, they lose much of their bone-forming ability and become quiescent. At this point, they are called **osteocytes**. Many of the cytoplasmic connections in the osteoblast stage are maintained into the osteocyte stage. These connections become visible channels or **canaliculi** that provide direct contact for osteocytes deep in bone with other osteocytes and with the bone surface. It is generally believed that these canaliculi provide a mechanism for the transfer of nutrients, hormones, and waste products between the bone surface and its interior.

Osteoclasts are cells responsible for bone resorption. They are large, multinucleated cells located on bone surfaces. Osteoclasts promote bone resorption by secreting acid and proteolytic enzymes into the space adjacent to the bone surface. Surfaces of osteoclasts facing bone are ruffled to increase their surface area and promote bone resorption. Bone resorption is a two-step process. First, osteoclasts create a local acidic environment that increases the solubility of surface bone mineral. Second, proteolytic enzymes secreted by osteoclasts degrade the organic matrix of bone.

Bone formation starts with neonatal development and continues throughout life.

Early in fetal development, the skeleton consists of little more than a cartilaginous model of what will later form the bony skeleton. The process of replacing this cartilaginous model with mature, mineralized bone begins in the center of the cartilage and progresses toward the two ends of what will later form the bone. As mineralization progresses, the bone increases in thickness and in length.

The **epiphyseal plate** is a region of growing bone of particular interest because it is here that the elongation and growth of bones occur after birth. Histologically, the epiphyseal plate shows considerable differences between its leading and trailing edges. The leading edge consists primarily of **chondrocytes**, which are actively engaged in the synthesis of cartilage of the epiphyseal plate. These cells gradually become engulfed in their own cartilage and are replaced by new cells on the cartilage surface, allowing the process to continue. The cartilage gradually becomes calcified, and the embedded chondrocytes die. The calcified cartilage begins to erode, and osteoblasts migrate into the area. Osteoblasts secrete osteoid, which eventually becomes mineralized, and a new mature bone is formed. In the epiphyseal plate, therefore, the continuing processes of cartilage synthesis, calcification, erosion, and osteoblast invasion result in a zone of active bone formation that moves away from the middle or center of the bone toward its end.

Hormones control chondrocytes of epiphyseal plates. Insulin-like growth factor I, primarily produced by the liver in

Clinical Focus / 35.1

Osteogenesis Imperfecta

Osteogenesis imperfecta is a genetic disorder characterized by bones that break easily, often from little or no apparent cause. Most patients with osteogenesis imperfecta have defects in the genes for type I collagen. Bones, ligaments, skin, sclera, and teeth are affected. Although the number of people affected with osteogenesis imperfecta in the United States is unknown, the best estimate suggests a minimum of 20,000 and possibly as many as 50,000. There are at least four recognized forms of the disorder (types I–IV), representing a range of severities.

Type I osteogenesis imperfecta is the most common and mildest form, resulting from decreased collagen production but normal molecular structure. The more severe forms, *types II* and *III*, involve mutations in the collagen molecule that prevent normal assembly. *Type II* osteogenesis imperfecta is frequently lethal at or shortly after birth, often as a result of respiratory problems. These patients have numerous fractures, severe bone deformity, and small stature. *Type III* patients feature progressively deforming bones, usually with moderate deformity at birth. *Type IV* osteogenesis imperfecta falls between types I and III in severity. Microscopic analyses revealed that some people who are clinically within the type IV group had a distinct pattern to their bones yet did not have evidence of mutations in the type I collagen genes. These people have been designated as *types V* and *VI* osteogenesis imperfecta.

Treatment is directed toward preventing or controlling the symptoms, maximizing independent mobility, and developing optimal bone mass and muscle strength. Care of fractures, extensive surgical and dental procedures, and physical therapy are often recommended for people with osteogenesis imperfecta. Use of wheelchairs, braces, and other mobility aids is common, particularly (although not exclusively) among people with more severe types of osteogenesis imperfecta. A surgical procedure called "rodding" is frequently considered. This treatment involves inserting metal rods through the length of the long bones to strengthen them and prevent and/or correct deformities. Several medications and other treatments are being explored for their potential use to treat osteogenesis imperfecta. For example, antiresorptive therapy with intravenous pamidronate has been shown to decrease fractures in children with severe osteogenesis imperfecta, even before age 3 years.

response to growth hormone, serves as a primary stimulator of chondrocyte activity and, ultimately, of bone growth. Insulin and thyroid hormones provide an additional stimulus for chondrocyte activity.

Beginning a few years after puberty, the epiphyseal plates in long bones (as in the legs and arms) gradually become less responsive to hormonal stimuli and, eventually, are totally unresponsive. This phenomenon is referred to as *closure of the epiphyses.* In most people, epiphyseal closure is complete by about age 20; adult height is reached at this point, because further linear growth is impossible. Not all bones undergo closure. For example, those in the fingers, feet, skull, and jaw remain responsive, which accounts for the skeletal changes seen in acromegaly, the condition of growth hormone overproduction (see Chapter 31, "Hypothalamus and the Pituitary Gland").

The flux of calcium and phosphate into and out of bone each day reflects a turnover of bone mineral and changes in bone structure generally referred to as **bone remodeling**. Bone remodeling occurs along most of the outer surface of the bone, making it either thinner or thicker, as required. In long bones, remodeling can also occur along the inner surface of the bone shaft, next to the marrow cavity. Remodeling is an adaptive process that allows bone to be reshaped to meet changing mechanical demands placed on the skeleton. It also allows the body to store or mobilize calcium rapidly.

PLASMA CALCIUM AND PHOSPHATE REGULATION

Regulatory mechanisms for calcium include rapid nonhormonal mechanisms with limited capacity and somewhat slower hormonally regulated mechanisms with much greater capacity. Similar mechanisms are also involved in regulating plasma phosphate concentrations.

Small changes in plasma free calcium are rapidly buffered by nonhormonal mechanisms.

The calcium bound to plasma proteins and a small fraction of that in bone mineral can help prevent a rapid decrease in the plasma calcium concentration.

Protein-bound calcium

The association of calcium with proteins is a simple, reversible, chemical equilibrium process. Protein-bound calcium, therefore, has the capacity to serve as a buffer of free plasma calcium concentrations. This effect is rapid and does not require complex signaling pathways; however, the capacity is limited, and the mechanism cannot serve a long-term role in calcium homeostasis.

Exchangeable bone calcium pool

Recall that approximately 99% of total body calcium is present in bones, and a healthy adult body has about 1 to 2 kg of calcium. Most of the calcium in bones exists as mature, hardened bone mineral that is not readily exchangeable but can be moved into the plasma via hormonal mechanisms (described below). However, approximately 1% (or 10 g) of the calcium in bones is in a simple chemical equilibrium with plasma calcium. This readily exchangeable calcium source is primarily located on the surface of newly formed bones. Any change in free calcium in the plasma or extracellular fluid results in a shift of calcium either into or out of the bone mineral until a new equilibrium is reached. Although this mechanism, like that described above, provides for a rapid defense against changes in free calcium concentrations, it is limited in capacity and can provide for only short-term adjustments in calcium homeostasis.

Long-term regulation of plasma calcium and phosphate is under hormonal control.

The hormonal mechanisms described here have a large capacity and the ability to make long-term adjustments in calcium and phosphate fluxes, but they do not respond instantaneously. It may take several minutes or hours for the response to occur and adjustments to be made. Nevertheless, these are the principal mechanisms that regulate plasma calcium and phosphate concentrations.

Parathyroid hormone

One of the primary regulators of plasma calcium concentrations is **parathyroid hormone (PTH)**. PTH is an 84-amino acid polypeptide produced by the parathyroid glands. Synthetic peptides containing the first 34 amino terminal residues appear to be as active as the native hormone.

There are two pairs of parathyroid glands, located on the dorsal surface of the left and right lobes of the thyroid gland. Because of this close proximity, a common cause of chronic hypocalcemia is postsurgical removal of all parathyroid tissue, either during a thyroidectomy and removal of neck-associated malignancies or after inadvertent interruption of blood supply to the parathyroid glands during head and neck surgery. *Hypoparathyroidism* is a rare complication of radioactive iodine ablation of the thyroid gland in **Graves disease**.

The primary physiologic stimulus for PTH secretion is a *decrease* in plasma calcium. Figure 35.4 shows the relationship between serum PTH concentration and total plasma calcium concentration. A decrease in the ionized calcium concentration actually triggers an increase in PTH secretion. The net effect of PTH is to increase the flow of calcium into plasma and return the plasma calcium concentration toward normal.

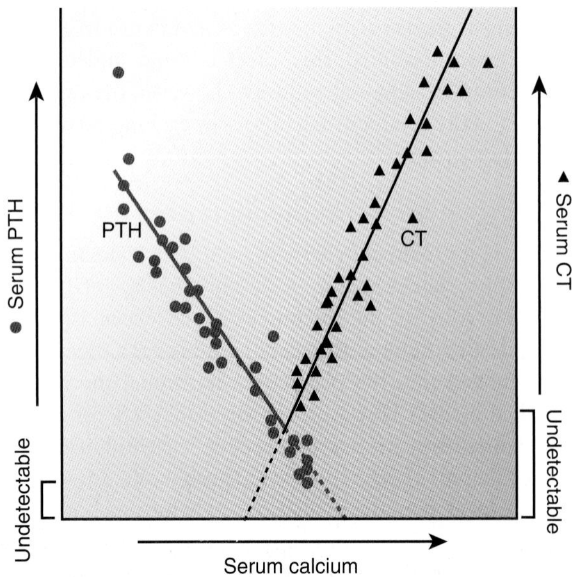

Figure 35.4 Effect of changes in plasma calcium on parathyroid hormone (PTH) and calcitonin (CT) secretion.

Calcitonin

Parafollicular cells of the thyroid gland produce a 32–amino acid polypeptide termed **calcitonin (CT)**, also known as *thyrocalcitonin* (see Fig. 32.1). Unlike PTH activity, for which only the initial amino terminal segment is required, CT activity requires the full polypeptide. Salmon CT differs from human CT in 9 of the 32 amino acid residues and is 10 times more potent than human CT in its hypocalcemic effect. The higher potency may be a result of a greater affinity for receptors and slower degradation by peripheral tissues. CT is often used clinically as a synthetic peptide matching the sequence of salmon CT.

In contrast to PTH, CT secretion is stimulated by an increase in plasma calcium (see Fig. 35.4). Hormones of the GI tract, especially gastrin, also promote CT secretion. Because the net effect of CT is to promote calcium deposition in bone, the stimulation of CT secretion by GI hormones provides an additional mechanism for facilitating the uptake of calcium into bone after the ingestion of a meal.

Vitamin D3 and 1,25-dihydroxycholecalciferol

The third key hormone involved in regulating plasma calcium is **vitamin D3** (cholecalciferol). More precisely, a metabolite of vitamin D3 serves as a hormone in calcium homeostasis. The D vitamins, a group of lipid-soluble compounds derived from cholesterol, have long been known to be effective in the prevention of rickets. Research during the past 30 years indicates that vitamin D exerts its effects through a hormonal mechanism.

Figure 35.5 shows the structure of vitamin D3 and the related compound **vitamin D2** (also known as *ergocalciferol*). Ergocalciferol is the form principally found in plants and yeasts and is commonly used to supplement human foods because of its relative availability and low cost. Although it is less potent on a mole-per-mole basis, vitamin D2 undergoes the same metabolic conversion steps and, ultimately, produces the same biologic effects as vitamin D3. The physiologic actions of vitamin D3 also apply to vitamin D2.

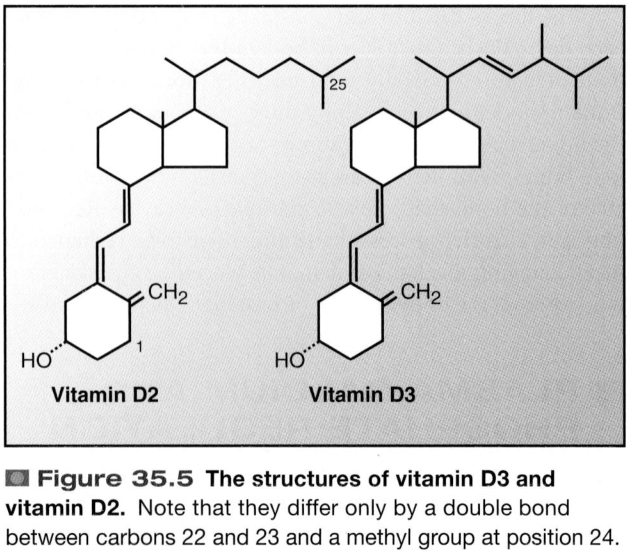

Figure 35.5 The structures of vitamin D3 and vitamin D2. Note that they differ only by a double bond between carbons 22 and 23 and a methyl group at position 24.

Vitamin D3 can be provided by the diet or formed in the skin by the action of ultraviolet light on a precursor, **7-dehydrocholesterol**, derived from cholesterol (Fig. 35.6). In many countries where food is not systematically supplemented with vitamin D, this pathway provides the major source of vitamin D. Because of the number of variables involved, it is difficult to specify a minimum exposure time. However, exposure to moderately bright sunlight for 30 to 120 min/d usually provides enough vitamin D to satisfy the body's needs without any dietary supplementation.

Vitamins D3 and D2 are by themselves relatively inactive. However, they undergo a series of transformations in the liver and kidneys that convert them into powerful calcium-regulatory hormones (see Fig. 35.6). The first step occurs in the liver and involves the addition of a hydroxyl group to carbon 25 to form 25-hydroxycholecalciferol (25-OH D_3). This reaction is largely unregulated, although certain drugs and liver diseases may affect this step. Next, 25-OH D_3 is released into the blood and undergoes a second hydroxylation reaction on carbon 1 in the kidney. The product is **1,25-dihydroxycholecalciferol**, also known as *1,25-dihydroxyvitamin D3* or *calcitriol*, the principal hormonally active form of the vitamin. The biologic activity of 1,25-dihydroxycholecalciferol is approximately 100 to 500 times greater than that of 25-OH D_3. The enzyme 1α-hydroxylase, which is located in tubule cells, catalyzes the reaction in the kidney.

The final step in 1,25-dihydroxycholecalciferol formation is highly regulated. The activity of 1α-hydroxylase is regulated primarily by PTH, which stimulates its activity. Therefore, if plasma calcium levels fall, PTH secretion increases; in turn, PTH promotes the formation of 1,25-dihydroxycholecalciferol. In addition, enzyme activity increases in response to a decrease in plasma phosphate. This does not appear to involve any intermediate hormonal signals but apparently involves direct activation of either the enzyme or cells in which the enzyme is located. Both a decrease in plasma calcium, which triggers PTH secretion, and a

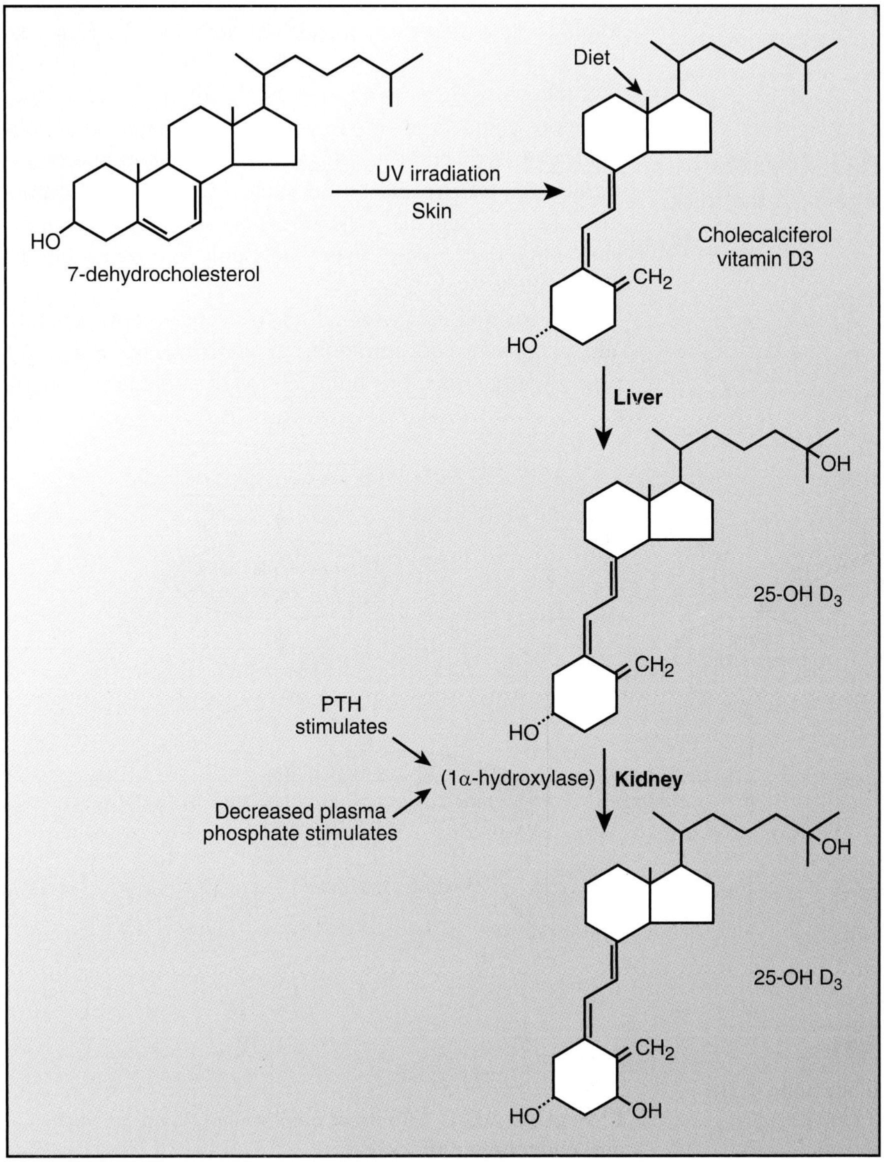

Figure 35.6 The conversion pathway of vitamin D3 into 1,25-dihydroxycholecalciferol [1,25-$(OH)_2$ D_3]. 25-OH D_3, 25-hydroxycholecalciferol; PTH, parathyroid hormone.

decrease in circulating phosphate result in the activation of 1α-hydroxylase and an increase in 1,25-dihydroxycholecalciferol synthesis.

Several target tissues are involved in hormonal regulation of plasma calcium and phosphate.

Most hormones generally improve the quality of life and the chance for survival when an animal is placed in a physiologically challenging situation. However, PTH is essential for life. The complete absence of PTH causes death from hypocalcemic tetany within just a few days. The condition can be avoided with hormone replacement therapy.

The net effects of PTH on plasma calcium and phosphate and its sites of action are shown in Figure 35.7. PTH causes an increase in plasma calcium concentration while decreasing plasma phosphate. This decrease in phosphate concentration is important with regard to calcium homeostasis. At normal plasma concentrations, calcium and phosphate are at or near chemical saturation levels. If PTH were to increase both calcium and phosphate levels, these substances would simply crystallize in bone or soft tissues as calcium phosphate, and the necessary increase in plasma calcium concentration would not occur. Thus, the effect of PTH to lower plasma phosphate is an important aspect of its role in regulating plasma calcium.

PTH has several important actions in the kidneys (see Fig. 35.7). It stimulates calcium reabsorption in the thick ascending limb and late distal tubule, decreasing calcium loss in the urine and increasing plasma concentrations. It also inhibits phosphate reabsorption in the proximal tubule, leading to increased urinary phosphate excretion and a decrease in plasma phosphate. Another important effect of PTH is to increase the activity of 1α-hydroxylase in the kidney, which is involved in forming active vitamin D.

In bone, PTH activates osteoclasts to increase bone resorption and the delivery of calcium from bone into plasma (see Fig. 35.7). In addition to stimulating active osteoclasts, PTH stimulates the maturation of immature osteoclasts into mature, active osteoclasts. PTH also inhibits collagen synthesis by osteoblasts, resulting in decreased bone matrix formation and decreased flow of calcium from plasma into bone mineral. 1,25-Dihydroxycholecalciferol augments the actions of PTH to promote bone resorption.

PTH does not appear to have any major direct effects on the GI tract. However, because it increases active vitamin D formation, it ultimately increases the absorption of both calcium and phosphate from the GI tract (see Fig. 35.7).

CT is important in several lower vertebrates, but despite its many demonstrated biologic effects in humans, it appears to play only a minor role in calcium homeostasis. This conclusion mostly stems from two lines of evidence. First, CT loss following surgical removal of the thyroid gland (and, therefore, removal of CT-secreting parafollicular cells) does not lead to overt clinical abnormalities of calcium homeostasis. Second, CT hypersecretion, such as from thyroid tumors involving parafollicular cells, does not cause any overt problems. On a daily basis, CT probably only fine-tunes the calcium regulatory system.

The overall action of CT is to decrease both calcium and phosphate concentrations in plasma (Fig. 35.8). The primary target of CT is bone, although some lesser effects

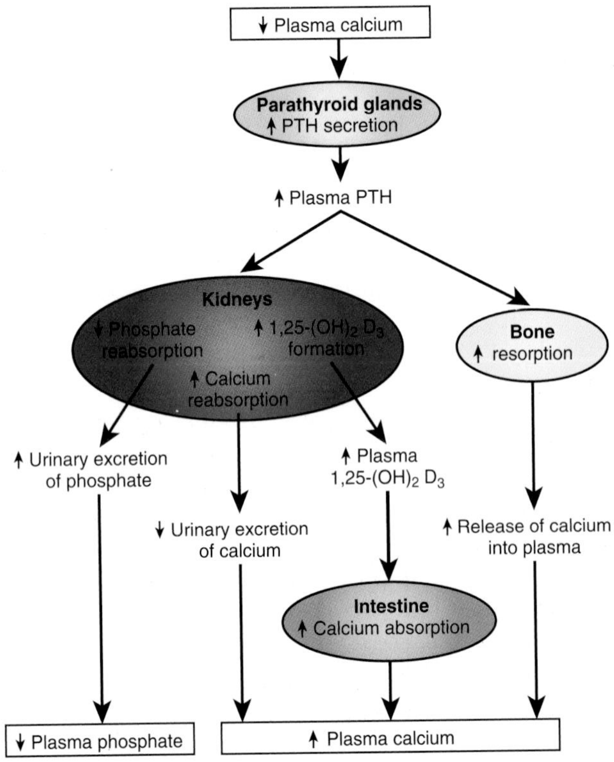

↓ Plasma calcium
Parathyroid glands ↑ PTH secretion
↑ Plasma PTH
Kidneys: ↓ Phosphate reabsorption; ↑ 1,25-$(OH)_2$ D_3 formation; ↑ Calcium reabsorption
Bone ↑ resorption
↑ Urinary excretion of phosphate
↑ Plasma 1,25-$(OH)_2$ D_3
↓ Urinary excretion of calcium
↑ Release of calcium into plasma
Intestine ↑ Calcium absorption
↓ Plasma phosphate
↑ Plasma calcium

Figure 35.7 Effects of parathyroid hormone (PTH) on calcium and phosphate metabolism. 1,25-$(OH)_2$ D_3, 1,25-dihydroxycholecalciferol.

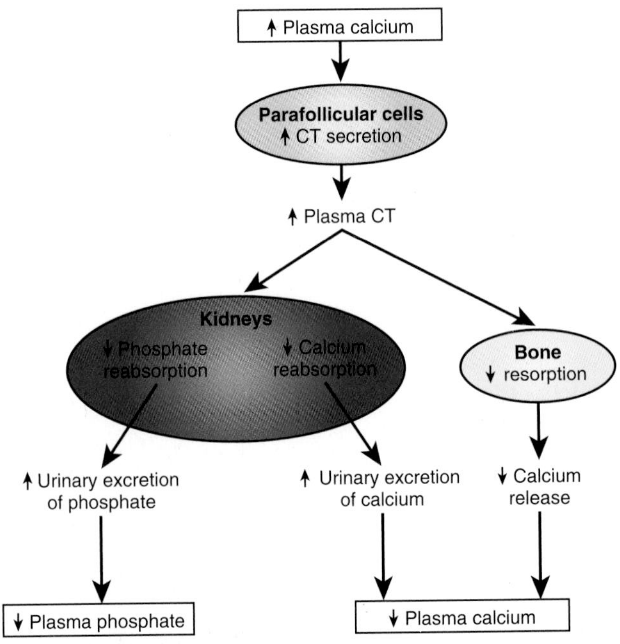

Figure 35.8 Effects of calcitonin (CT) on calcium and phosphate metabolism.

also occur in the kidneys. In the kidneys, CT decreases the tubular reabsorption of calcium and phosphate. This leads to an increase in urinary excretion of both calcium and phosphate and, ultimately, to decreased levels of both ions in the plasma. In bones, CT opposes the action of PTH on osteoclasts by inhibiting their activity. This leads to decreased bone resorption and an overall net transfer of calcium from plasma into bone. CT has little or no direct effect on the GI tract.

The net effect of 1,25-dihydroxycholecalciferol is to increase both calcium and phosphate concentrations in plasma (Fig. 35.9). The activated form of vitamin D primarily influences the GI tract, although it has actions in the kidneys and bones as well.

In the kidneys, 1,25-dihydroxycholecalciferol increases the tubular reabsorption of calcium and phosphate, promoting the retention of both ions in the body. However, this is a weak and probably only minor effect of the hormone. In bones, the hormone promotes actions of PTH on osteoclasts, increasing bone resorption (see Fig. 35.9).

In the GI tract, 1,25-dihydroxycholecalciferol stimulates calcium and phosphate absorption by the small intestine, increasing plasma concentrations of both ions. This effect is mediated by increased production of calcium transport proteins resulting from gene transcription events and usually requires several hours to appear.

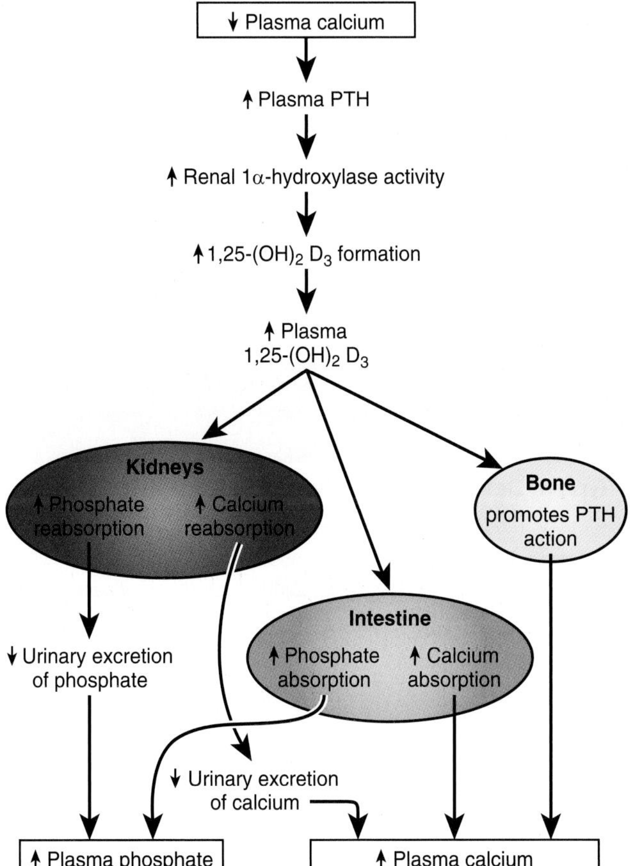

Figure 35.9 Effects of 1,25-dihydroxycholecalciferol [1,25-$(OH)_2$ D_3] on calcium and phosphate metabolism. PTH, parathyroid hormone.

BONE DYSFUNCTION

There are several **metabolic bone diseases**, all typified by ongoing disruption of the normal processes of either bone formation or bone resorption. The conditions most frequently encountered clinically are osteoporosis, osteomalacia, and Paget disease.

Osteoporosis leads to decreased bone density and increased risk of fracture.

Osteoporosis is a major health problem, particularly because older adults are more prone to this disorder and the average age of the population is increasing.

Osteoporosis involves a reduction in total bone mass, with an equal loss of both bone mineral and organic matrix. Several factors are known to contribute directly to osteoporosis. Long-term dietary calcium deficiency can lead to osteoporosis because bone mineral is mobilized to maintain plasma calcium levels. Vitamin C deficiency can also result in a net loss of bone because vitamin C is required for normal collagen synthesis to occur. A defect in matrix production and the inability to produce new bone eventually result in a net loss of bones. For reasons that are not entirely understood, a reduction in the mechanical stress placed on bone can lead to bone loss. Immobilization or disuse of a limb, such as with a cast or paralysis, can result in localized osteoporosis of the affected limb. Space flight can produce a type of disuse osteoporosis resulting from the condition of weightlessness.

Most commonly, osteoporosis is associated with advancing age in both men and women, and it cannot be assigned to any specific definable cause. For several reasons, postmenopausal women are more prone to develop the disease than are men. Figure 35.10 shows the average bone mineral content (as grams of calcium) for men and women versus age. Until about the time of puberty, males and females have similar bone mineral content. However, at puberty, males begin to acquire bone mineral at a greater rate; peak bone mass in men may be approximately 20% greater than that of women. Maximum bone mass is attained between 30 and 40 years of age and then tends to decrease in both sexes. Initially this occurs at an approximately equivalent rate, but women begin to experience a more rapid bone mineral loss at the time of menopause (about age 45–50 years). This loss appears to result from the decline in estrogen secretion that occurs at menopause. Low-dose estrogen supplementation of postmenopausal women is usually effective in retarding bone loss without causing adverse effects. This condition of increased bone loss in women after menopause is called **postmenopausal osteoporosis**. A new mechanistic perspective links immune cells to the etiology of postmenopausal osteoporosis as well as to bone loss caused by many other endocrine conditions.

Rickets is a softening of bones in children, potentially leading to fractures and deformity.

Rickets is characterized by the inadequate mineralization of new bone matrix, such that the ratio of bone mineral to matrix is reduced. As a result, bones may have reduced strength and

Clinical Focus / 35.2

The Toll of Osteoporosis

Osteoporosis is often called the "silent disease" because bone loss initially occurs without symptoms. People may not know that they have significant bone loss until their bones become so weak that a sudden strain, bump, or fall causes a fracture. Osteoporosis is a major public health threat for an estimated 44 million Americans, including 55% of people 50 years of age and older. In the United States, 10 million people are estimated to have the disease and almost 34 million more are estimated to have low bone mass, placing them at increased risk for osteoporosis. Approximately 80% of those affected by osteoporosis are women. Although osteoporosis is often thought of as an older person's disease, it can strike at any age. The seeds of osteoporosis are often sown in childhood, and it takes a lifetime of effort to prevent the disease. A frighteningly large number of children do not get sufficient exercise, vitamin D, and calcium to ensure protection from developing the disease later in life.

Osteoporosis is responsible for more than 1.5 million fractures annually, including 300,000 hip fractures, 700,000 vertebral fractures, 250,000 wrist fractures, and 300,000 fractures at other sites. The estimated national direct care expenditures (including hospitals, nursing homes, and outpatient services) for osteoporotic fractures are >18 billion dollars per year. Additional losses in wages and productivity greatly increase this number. Nearly one third of people who have hip fractures end up in nursing homes within a year; nearly 20% die within a year.

What causes osteoporosis, and what can be done to prevent or treat the disease? Although it is known that a diet low in calcium or vitamin D, certain medications, such as glucocorticoids and anticonvulsants, and excessive ingestion of aluminum-containing antacids can cause osteoporosis, in most cases, the exact cause is unknown. However, several identified risk factors associated with the disease include the following: being a woman (especially a postmenopausal woman); being Caucasian or Asian; being of advanced age; having a family history of the disease; having low testosterone levels (in men); having an inactive lifestyle; cigarette smoking; and an excessive use of alcohol.

A comprehensive program to help prevent osteoporosis includes a balanced diet rich in calcium and vitamin D, regular weight-bearing exercise, a healthy lifestyle with no smoking or excessive alcohol use, and bone density testing and medication when appropriate. Although at present there is no cure for osteoporosis, there are several U.S. Food and Drug Administration (FDA)–approved medications to either prevent or treat the disease in postmenopausal women: estrogens; estrogens and progestins; parathyroid hormone; alendronate, ibandronate, and risedronate (bisphosphonates); calcitonin; and raloxifene, a selective estrogen receptor modulator. Although 20% of all osteoporosis cases occur in men, only alendronate and risedronate are currently FDA approved for use in men and only for cases of corticosteroid-induced osteoporosis. Also, parathyroid hormone is approved for the treatment of osteoporosis in men who are at high risk of fracture. Testosterone replacement therapy is often helpful in a man with a low testosterone level.

are subject to distortion in response to mechanical loads. When the disease occurs in adults, it is called **osteomalacia**; when it occurs in children, it is called *rickets*. In children, the condition often produces a bowing of the long bones in the legs. In adults, it is often associated with severe bone pain.

The primary cause of osteomalacia and rickets is a deficiency in vitamin D *activity*. Vitamin D may be deficient in the diet; the small intestine may not adequately absorb it; it may not be converted into its hormonally active form; or target tissues may not adequately respond to the active hormone (Table 35.3). Dietary deficiency is generally not a problem in the United States, where vitamin D is added to many foods; however, it is a major health problem in other parts of the world. Because the liver and kidneys are involved in converting vitamin D3 into its hormonally active form, primary disease of either of these organs may result in vitamin D deficiency. Impaired vitamin D actions are somewhat rare but can be produced by certain drugs. In particular,

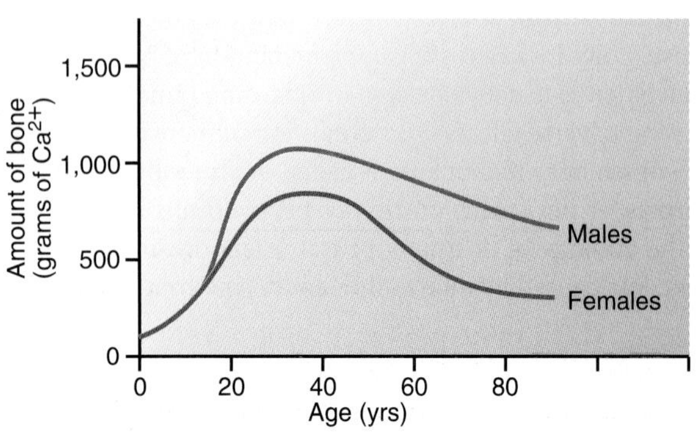

Figure 35.10 Changes in bone calcium content as a function of age in males and females. These changes can be roughly extrapolated into changes in bone mass and bone strength.

TABLE 35.3 Causes of Osteomalacia and Rickets

Type of Defect	Cause
Inadequate availability of vitamin D	Dietary deficiency or lack of exposure to sunlight Fat-soluble vitamin malabsorption
Defects in metabolic activation of vitamin D	25-Hydroxylation (liver) Liver disease Certain anticonvulsants, such as phenobarbital 1-Hydroxylation (kidney) Renal failure Hypoparathyroidism
Impaired action of 1,25-dihydroxycholecalciferol on target tissues	Certain anticonvulsants 1,25-Dihydroxycholecalciferol receptor defects Uremia

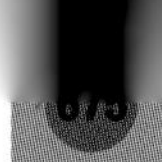

From Bench to Bedside / 35.1

New Anabolic Therapies for Osteoporosis

Osteoporosis is a major health problem with a significant consequence being fragility fractures. Mainstay therapies for this disease are agents that reduce *bone resorption*. These antiresorptive agents stabilize bone architecture and the incidence of fractures in osteoporosis. Although antiresorptive therapy has been useful in the management of osteoporosis, this therapy cannot restore the bone structure that has been lost because of increased remodeling. New insight into the mechanisms of bone formation may provide new *anabolic therapies* for osteoporosis. For example, **bone morphogenetic proteins (BMPs)** and Wnt have been identified and found to induce the differentiation of mesenchymal cells toward osteoblasts. Also insulin-like growth factor-1 (IGF-1) has been shown to enhance the function of mature osteoblasts. In addition to regulated synthesis and receptor binding, studies have found that the activities of BMPs, Wnt, and IGF-1 are tightly controlled by specific extracellular and intracellular regulatory proteins. Therefore, enhancing the synthesis and/or activity of BMPs, Wnt, or IGF-1, or targeting specific extracellular and intracellular proteins that regulate BMP, Wnt, or IGF-1 may provide a new therapeutic strategy to enhance bone formation in osteoporosis.

One interesting example is a protein named *sclerostin* that has BMP and Wnt antagonistic properties. It is expressed by osteoblasts and osteocytes. Interestingly sclerostin synthesis is suppressed by parathyroid hormone (PTH), and the study suggests that it is the enhanced Wnt signaling that contributes to the anabolic effect of PTH in bone. Along with these observations, clinical findings in *high-bone–mass syndrome* indicate that inactivation or neutralization of sclerostin could be used as an approach to enhance Wnt signaling and obtain an anabolic response in bone. Also, humanized monoclonal antibodies to sclerostin cause enhanced Wnt signaling and an increase in bone mass in animal studies. Of potential therapeutic significance, clinical trials have demonstrated that antisclerostin antibodies can increase bone mass density and biochemical markers of bone formation in humans. With regards to IGF-1, the decline in its secretion with aging may play a role in the pathogenesis of osteoporosis. New studies suggest that at low doses, IGF-1 increases bone formation without an effect on bone resorption. As the later effect is seen with higher doses of IGF-1, low-dose IGF-1 therapy may provide a means to increase osteoblast function without a concomitant effect on bone resorption.

some anticonvulsants used in the treatment of epilepsy may produce osteomalacia or rickets with prolonged treatment.

Paget disease is a chronic disorder leading to enlarged, deformed bones.

Paget disease is rarely diagnosed in patients younger than 40 years. Men are more commonly affected than women, and prevalence of Paget disease in people older than age 40 years is about 3%. It is typified by disordered bone formation and resorption (remodeling) and may occur at a single local site or at multiple sites in the body. Radiographs of the affected bone often exhibit increased density, but the abnormal structure makes the bone weaker than normal. Often, those with Paget disease experience considerable pain and, in severe cases, crippling deformities may lead to serious neurologic complications.

The cause of the disease is not well understood. Both genetic and environmental factors (probably viral) appear to be important. Several therapies are available for treating the disease, including treatment with CT, but these typically offer only temporary relief from pain and complications.

Chapter Summary

- When plasma calcium levels fall below normal, spontaneous action potentials can be generated in nerves, leading to tetany of muscles, which, if severe, can result in death.
- About half of the circulating calcium is in the free or ionized form, about 10% is bound to small anions, and about 40% is bound to plasma proteins. Most of the phosphorus circulates free as orthophosphate.
- Most ingested calcium is not absorbed by the gastrointestinal (GI) tract and leaves the body via the feces; by contrast, phosphate is almost completely absorbed by the GI tract and leaves the body mostly via the urine.
- A decrease in plasma ionized calcium stimulates the secretion of parathyroid hormone (PTH), a polypeptide hormone produced by the parathyroid glands. PTH plays a vital role in calcium and phosphate homeostasis and acts on bones, kidneys, and intestine to raise the plasma calcium concentration and lower the plasma phosphate concentration.
- Sequential hydroxylation reactions in the liver and kidneys convert vitamin D to the active hormone 1,25-dihydroxycholecalciferol. This hormone stimulates intestinal calcium absorption and, thereby, raises the plasma calcium concentration.
- Calcitonin, a polypeptide hormone produced by the thyroid glands, tends to lower the plasma calcium concentration, but its physiologic importance in humans has been questioned.
- Osteoporosis, osteomalacia and rickets, and Paget disease are the most common forms of metabolic bone disease.

thePoint *Visit http://thePoint.lww.com for chapter review Q&A, case studies, animations, and more!*

36 Male Reproductive System

ACTIVE LEARNING OBJECTIVES

Upon mastering the material in this chapter you should be able to:

- Explain how mechanisms of brain centers, the hypothalamus, the anterior pituitary, and the testes interactively regulate testicular function.
- Explain how castration causes a large increase in basal luteinizing hormone levels and why testosterone replacement therapy does not completely correct follicle-stimulating hormone levels.
- Explain how the pampiniform plexus and the cremasteric muscle maintain testes at a cooler temperature.
- Explain the mechanistic bases, causes, and consequences of sperm antibody formation.
- Explain how mitosis, meiosis, and spermiogenesis mechanistically coordinate the continuous production of fresh spermatozoa.
- Explain how the spermatozoon's head, middle piece, and tail mechanistically function to reach, recognize, and fertilize an egg.
- Explain the mechanism of testosterone synthesis in Leydig cells and function in extratesticular sites.
- Explain how normal spermatogenesis almost never occurs with defective steroidogenesis, but normal steroidogenesis can be present with defective spermatogenesis.

A central theme in physiology is that organ systems and most physiologic processes are aimed at homeostasis and to ensure survival of the individual. However, there is an exception in which one system is aimed neither at homeostasis nor at ensuring survival of the individual. That system is the reproductive system, which is designed solely for the purpose of survival of the *species*.

Although the reproductive system does not contribute to homeostasis and individual survival, it still plays an important role in physiology and behavior. The reproductive system includes the gonads, reproductive tract, and accessory glands. The overall function of male reproduction is the production of sperm and delivery of the sperm to the female.

The testes produce mature male gametes, known as **spermatozoa**, which function to fertilize the female counterpart, the **oocyte**, during conception. This fertilization, in turn, produces a single-celled individual known as a **zygote**. These processes are the cornerstone of sexual reproduction. This chapter focuses on functional aspects of the male reproductive system.

The testes have two primary functions: (1) **spermatogenesis**, the process of producing mature sperm, and (2) **steroidogenesis**, the synthesis of testosterone. Both functions are under hormonal control via the pituitary gonadotropins **luteinizing hormone (LH)** and **follicle-stimulating hormone (FSH)**. Testosterone is the primary sex hormone in the male and is responsible for primary and secondary sex characteristics. The primary sex characteristics include those structures responsible for promoting the development, preservation, and delivery of sperm. The secondary sex characteristics are those structures and behavioral features that make men externally different from women and include the typical male hair pattern, deep voice, and large muscle and bone masses.

ENDOCRINE GLANDS OF THE MALE REPRODUCTIVE SYSTEM

A diagram of reproduction regulation in the male is presented in Figure 36.1. The system is divided into factors affecting male function: brain centers, which control pituitary release of hormones and sexual behavior; gonadal structures, which produce sperm and hormones; a ductal system, which stores and transports sperm; and accessory glands, which support sperm viability. The endocrine glands of the male

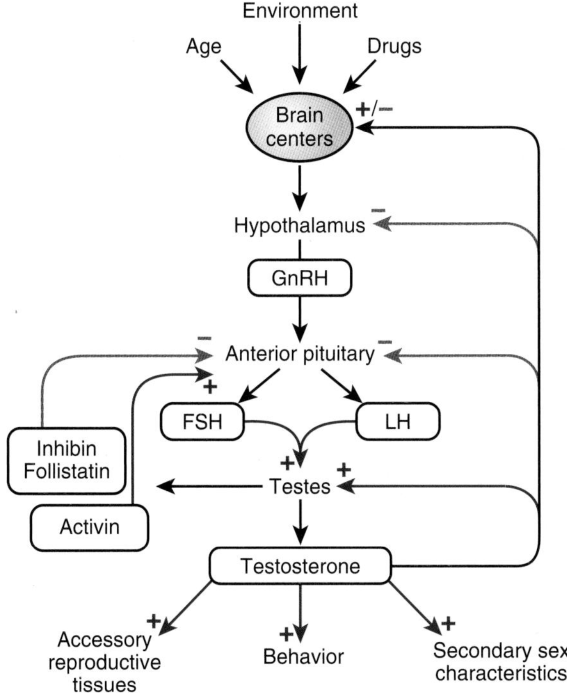

Figure 36.1 Regulation of reproduction in the male. The main reproductive hormones are shown in boxes. Plus and minus signs depict positive and negative regulations, respectively. FSH, follicle-stimulating hormone; GnRH, gonadotropin-releasing hormone; LH, luteinizing hormone.

reproductive system include the (1) hypothalamus, (2) anterior, (3) pituitary, and (4) testes. The hypothalamus processes information obtained from the external and internal environment, using neurotransmitters that regulate the secretion of **gonadotropin-releasing hormone (GnRH)**. GnRH moves down the hypothalamic-pituitary portal system and stimulates the secretion of LH and FSH by the gonadotrophs of the anterior pituitary. LH binds to receptors on the **Leydig cells**, and FSH binds to receptors on the **Sertoli cells**. Leydig cells reside in the interstitium of the testes, between seminiferous tubules, and produce **testosterone**. Sertoli cells are located within the seminiferous tubules, support spermatogenesis, contain FSH and testosterone receptors, and produce estradiol, albeit at low levels.

Testosterone belongs to a class of steroid hormones, the *androgens*, that promote "maleness." It carries out multiple functions, including the following: feedback on the hypothalamus and anterior pituitary; the support of spermatogenesis; the regulation of behavior, including sexual behavior; and the development and maintenance of secondary sex characteristics. Sertoli cells also produce glycoprotein hormones—inhibin, activin, and follistatin—that regulate the secretion of FSH.

The duct system that transports sperm from the testis to the outside through the penis includes the epididymis, vas deferens, and urethra. The sperm acquire motility and the capability to fertilize in the epididymis; they are stored in the epididymis and in the vas deferens. They are transported via the urethra through the penis and are ultimately expelled by ejaculation. The accessory structures of the male reproductive tract include the prostate gland, seminal vesicles, and bulbourethral glands. These glands contribute several constituents to the seminal fluid that are necessary for maintaining functional sperm.

TESTICULAR FUNCTION AND REGULATION

LH and FSH regulate testicular function. LH regulates the secretion of testosterone by the Leydig cells, and FSH, in synergy with testosterone, regulates the production of spermatozoa.

Hypothalamic gonadotropin-releasing hormone regulates both luteinizing and follicle-stimulating hormone secretion.

Hypothalamic neurons produce **gonadotropin-releasing hormone** ([**GnRH**] also called *luteinizing hormone–releasing hormone [LHRH]*), a decapeptide, which regulates the secretion of luteinizing hormone (LH) and follicle-stimulating hormone (FSH). Although neurons that produce GnRH can be located in various areas of the brain, their highest concentration is in the medial basal hypothalamus, in the region of the infundibulum and arcuate nucleus. GnRH enters the hypothalamic-pituitary portal system and binds to receptors on the plasma membranes of pituitary cells, resulting in the synthesis and release of LH and FSH.

Many external cues and internal signals influence the secretion of GnRH, LH, and FSH. For example, the amounts of GnRH, FSH, and LH secreted change with age, stress levels, and hormonal state. In addition, various disease states lead to hyposecretion of GnRH. Little, if any, secretion of hypothalamic GnRH occurs in patients with prepubertal **hypopituitarism**, resulting in a failure of the development of the testes, primarily a result of a lack of LH, FSH, and testosterone.

Male patients with **Kallmann syndrome** are hypogonadal from a deficiency in LH and FSH secretion because of a failure of GnRH neurons to migrate from the olfactory bulbs, their embryologic site of origin. These patients do not have a sufficient hypothalamic source of GnRH to maintain secretion of LH and FSH, and the testes fail to undergo significant development.

GnRH originates from a large precursor molecule called *preproGnRH* (Fig. 36.2). PreproGnRH consists of a signal peptide, native GnRH, and a GnRH-associated peptide (GAP). The signal peptide (or leader sequence) allows the protein to cross the membrane of the rough endoplasmic reticulum (RER). However, both the signal peptide and GAP are enzymatically cleaved at the RER prior to GnRH secretion.

Luteinizing and follicle-stimulating hormones from the anterior pituitary regulate testosterone secretion and sperm production.

Three distinct pituitary LH-secreting and FSH-secreting cells have been identified: (1) **gonadotrophs**, (2) Leydig cells, and (3) Sertoli cells. *Gonadotrophs* contain either LH or FSH, and

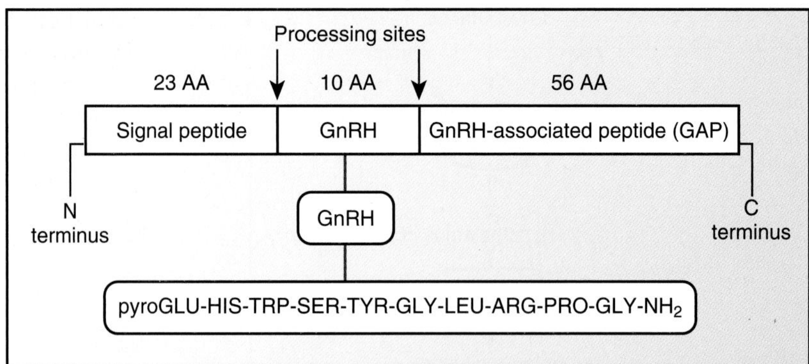

Figure 36.2 The precursor molecule, preproGnRH, that contains gonadotropin-releasing hormone (GnRH). The amino acid sequence of GnRH, a decapeptide, is indicated at the bottom.

some cells contain both LH and FSH. GnRH can induce the secretion of both hormones simultaneously because GnRH receptors are present on all of these cell types.

LH and FSH each contain two polypeptide subunits, referred to as α and β *chains*, that are about 15 kDa in size. These hormones contain the same α subunit but different β subunits. Each hormone is glycosylated prior to release into the general circulation. Glycosylation regulates the half-life, protein folding for receptor recognition, and biologic activity of the hormone.

LH and FSH regulate testicular function. They bind membrane receptors on *Leydig* and *Sertoli cells*, respectively. The activation of LH and FSH receptors on these cells increases the intracellular second messenger **cyclic adenosine monophosphate (cAMP)**. The two gonadotropin receptors are linked to G proteins and adenylyl cyclase for the production of cAMP from **adenosine triphosphate (ATP)**. For the most part, cAMP can account for all of the actions of LH and FSH on testicular cells. cAMP binds to protein kinase A (PKA), which activates transcription factors such as *steroidogenic factor-1 (SF-1)* and *cAMP response element–binding protein (CREB)*. These factors activate the promoter region of the genes of steroidogenic enzymes that control testosterone production by Leydig cells. Similar signal-transduction events occur in Sertoli cells that regulate the production of estradiol. The testis converts testosterone and some other androgens to estradiol by the process of aromatization, although estradiol production is low in males.

Another major function of the testis is the production of mature sperm and of several regulatory proteins (androgen-binding protein, inhibin, activin, and follistatin), characteristics of which are detailed in later sections.

Low-frequency GnRH pulses lead to follicle-stimulating hormone release, whereas high-frequency GnRH pulses stimulate luteinizing hormone release.

GnRH in the hypothalamus is secreted in a pulsatile manner into the hypothalamic-hypophyseal portal blood. GnRH pulsatility is ultimately necessary for proper functioning of the testes because it regulates the secretion of FSH and LH, which are also released in a pulsatile fashion (Fig. 36.3). Continuous exposure of gonadotrophs to GnRH results in desensitization of GnRH receptors, leading to a decrease in LH and FSH release. Therefore, the pulsatile pattern of GnRH release serves an important physiologic function. The administration of GnRH at an improper frequency results in a decrease in circulating concentrations of LH and FSH.

Most evidence for GnRH pulses has come from animal studies because GnRH must be measured in hypothalamic-hypophyseal portal blood, an extremely difficult area from which to obtain blood samples in humans. Because distinct pulses of LH follow discrete pulses of GnRH, measurements of LH in serum indirectly indicate that GnRH pulses have occurred. Unlike LH secretion, FSH secretion is not always pulsatile, and even when it is pulsatile, there is only partial concordance between LH and FSH pulses. Numerous human studies measuring LH and FSH secretion in peripheral blood at various times have provided much of the information regarding the role of LH and FSH in regulating testicular development and function. However, the exact relationship between endogenous GnRH pulses and LH and FSH secretion in humans is unknown.

Hypogonadal **eunuchoid** men exhibit low levels of LH in serum and do not exhibit pulsatile secretion of LH. Pulsatile injections of GnRH restore LH and FSH secretion and increase sperm counts. FSH pulses tend to be smaller in

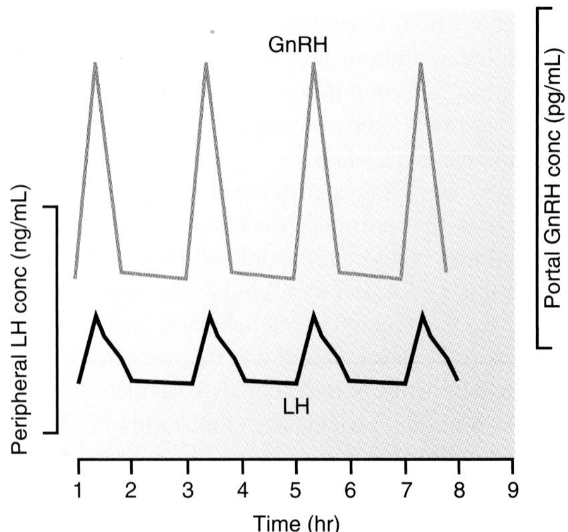

Figure 36.3 A diagram of the pulsatile release of gonadotropin-releasing hormone (GnRH) in portal blood and luteinizing hormone in peripheral blood. conc, concentration.

amplitude than LH pulses, mostly because FSH has a longer half-life than LH in the circulation.

The development of mice in which genes for green fluorescent protein have been put under the regulation of GnRH promoter has revealed that many, but not all, GnRH neurons show a bursting pattern of electrical activity. An existing question in the field of reproductive neuroendocrinology is what generates GnRH pulsatility. Studies using a line of clonal GnRH neurons have shown that these neurons grown in culture can release GnRH in a pulsatile pattern, suggesting that the pulse-generating capacity of GnRH neurons may be intrinsic. The presence of a *GnRH pulse generator* in the hypothalamus is postulated to be the basis of pulsatile GnRH.

The activity of the pulse generator is under negative-feedback control by gonadal steroids at the level of the hypothalamus and the pituitary. For example, castration causes a large increase in basal LH levels in serum, as evidenced by an increase in frequency and amplitude of LH pulses, indicating that negative feedback of gonadal steroids controls the GnRH pulse generator. Both testosterone and estradiol provide negative feedback. Studies indicate that the negative feedback of gonadal steroids controls LH at the hypothalamic level. Evidence also supports the concept that testosterone has negative-feedback control on LH secretion directly at the pituitary level. In addition to estradiol, dihydrotestosterone, another active metabolite of testosterone, exerts negative-feedback control on LH secretion, suggesting that testosterone does not need to be converted to estradiol.

Steroids and polypeptides from the testis inhibit both luteinizing and follicle-stimulating hormone secretion.

Testosterone, estradiol, inhibin, activin, and follistatin are major testicular hormones that regulate the release of the gonadotropins LH and FSH. Generally, testosterone, estradiol, and inhibin reduce the secretion of LH and FSH in the male. Activin stimulates the secretion of FSH, whereas follistatin inhibits FSH secretion.

Testosterone inhibits LH release by decreasing the secretion of GnRH and, to a lesser extent, by reducing gonadotroph sensitivity to GnRH. Estradiol formed from testosterone by aromatase also has an inhibitory effect on GnRH secretion. Acute testosterone treatment does not alter pituitary responsiveness to GnRH, but prolonged exposure significantly reduces the secretory response to GnRH.

Removal of the testes results in increased circulating levels of LH and FSH. Replacement therapy with physiologic doses of testosterone restores LH to precastration levels but does not completely correct FSH levels. This observation led to a search for a gonadal factor that specifically inhibits FSH release. The polypeptide hormone **inhibin** was eventually isolated from seminal fluid. Inhibin is produced by Sertoli cells and has a molecular weight of 32 to 120 kDa, the 32-kDa form being the most prominent. Inhibin is composed of two dissimilar subunits, α and β, which are held together by disulfide bonds. There are two β-subunit forms, called A and B. Inhibin B consists of the α subunit bound by a disulfide bridge to the βB subunit and is the physiologically important form of inhibin in the human male. Inhibin acts directly on the anterior pituitary and inhibits the secretion of FSH, but not LH.

Activin is produced by Sertoli cells, stimulates the secretion of FSH, has an approximate molecular weight of 30 kDa, and has multiple forms based on the βA and βB subunits of inhibin. The multiple forms of activin are called *activin A* (two βA subunits linked by a disulfide bridge), *activin B* (two βB subunits), and *activin AB* (one βA subunit and one βB subunit). The major form of activin in the male is currently unknown, although both Sertoli and Leydig cells have been implicated in its secretion.

Follistatin is a 31- to 45-kDa single-chain protein hormone, with several isoforms, that binds and deactivates activin. Thus, the deactivation of activin by binding to follistatin reduces FSH secretion. Follistatin is produced by Sertoli cells and acts as a paracrine factor on the developing spermatogenic cells.

MALE REPRODUCTIVE ORGANS

The male reproductive organs include the *penis* and *testes*. The testes produce spermatozoa and transport them through a series of ducts in preparation for fertilization. The testes also produce testosterone, which regulates development of the male gametes, male sex characteristics, and male behavior.

Testis is the site of sperm and seminal fluid formation.

During embryonic stages of development, the testes lie attached to the posterior abdominal wall. As the embryo elongates, the testes move to the inguinal ring. Between the seventh month of pregnancy and birth, the testes descend through the inguinal canal into the scrotum. The location of the testes in the scrotum is important for sperm production, which is optimal at 2°C to 3°C lower than the core body temperature. Two systems help maintain the testes at a cooler temperature. One is the **pampiniform plexus** of blood vessels, which serves as a countercurrent heat exchanger between warm arterial blood reaching the testes and cooler venous blood leaving the testes. The second is the **cremasteric muscle**, which responds to changes in temperature by moving the testes closer or farther away from the body. Prolonged exposure of the testes to elevated temperature, fever, or thermoregulatory dysfunction can lead to temporary or permanent sterility as a result of a failure of spermatogenesis, whereas steroidogenesis is unaltered.

A thick, fibrous connective tissue layer, the tunica albuginea, encapsulates the testes. Each human testis contains hundreds of tightly packed **seminiferous tubules**, ranging from 150 to 250 μm wide and from 30 to 70 cm long. The tubules are arranged in lobules, separated by extensions of the tunica albuginea, and open on both ends into the rete testis. Examination of a cross-section of a testis reveals distinct morphologic compartmentalization. Sperm production

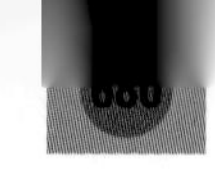

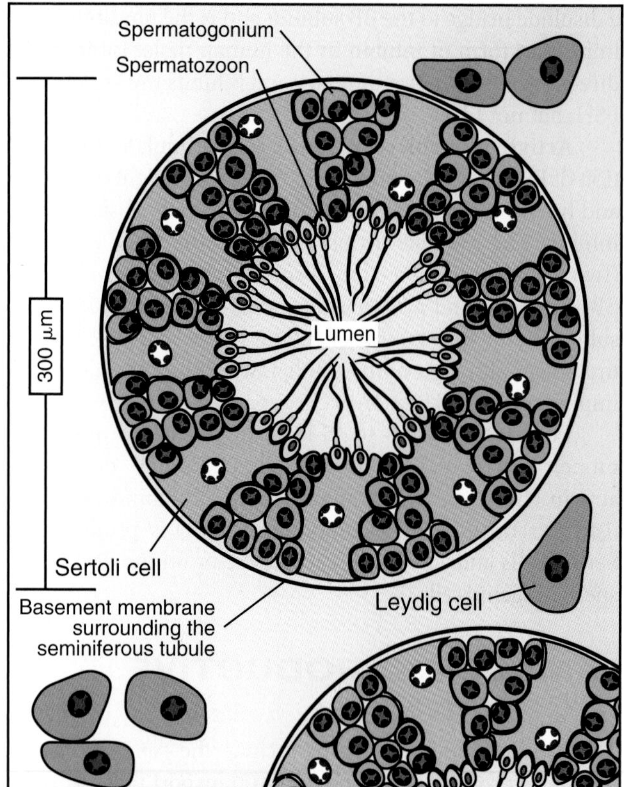

Figure 36.4 The testis. This cross-sectional view shows the anatomic relationship of the Leydig cells, basement membrane, seminiferous tubules, Sertoli cells, spermatogonia, and spermatozoa.

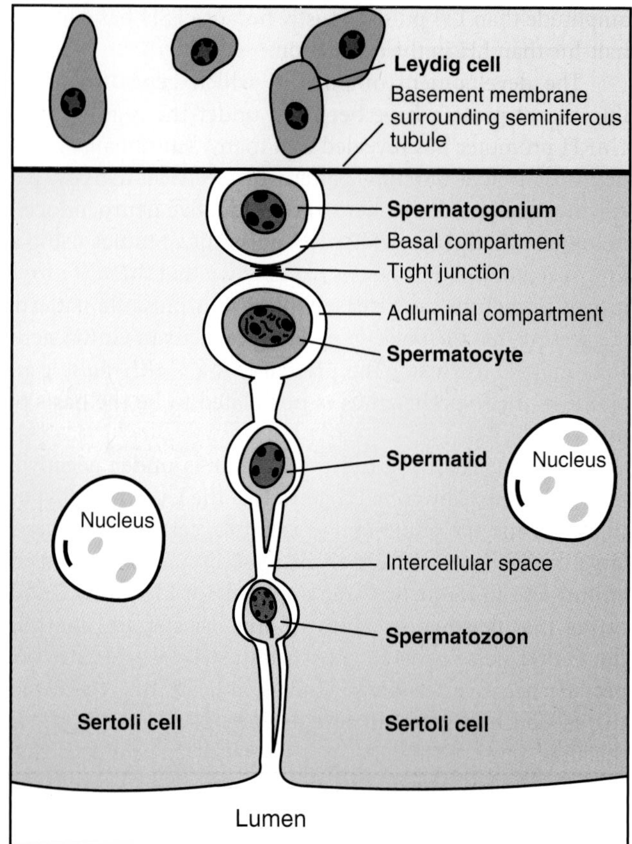

Figure 36.5 Sertoli cells. Sertoli cells are connected by tight junctions, which divide the intercellular space into a basal compartment and an adluminal compartment. Spermatogonia are located in the basal compartment and maturing sperm in the adluminal compartment. Spermatocytes are formed from the spermatogonia and cross the tight junctions into the adluminal compartment, they mature into spermatozoa.

is carried out in the avascular seminiferous tubules, whereas the Leydig cells, which are scattered in a vascular, loose connective tissue between the seminiferous tubules in the interstitial compartment, produce testosterone.

Each seminiferous tubule is composed of two somatic cell types (myoid cells and Sertoli cells) and germ cells. A basement membrane (basal lamina) with myoid cells on its perimeter, which define its outer limit, surrounds the seminiferous tubule. On the inside of the basement membrane are large, irregularly shaped Sertoli cells, which extend from the basement membrane to the lumen (Fig. 36.4). Tight junctions attach Sertoli cells to one another near their base (Fig. 36.5). The tight junctions divide each tubule into a basal compartment, whose constituents are exposed to circulating agents, and an adluminal compartment, which is isolated from blood-borne elements. The tight junctions limit the transport of fluid and macromolecules from the interstitial space into the tubular lumen, forming the *blood–testis barrier*.

Located between the nonproliferating Sertoli cells are germ cells at various stages of division and differentiation. Mitosis of the **spermatogonia** (diploid progenitors of spermatozoa) occurs in the basal compartment of the seminiferous tubule (see Fig. 36.5). The early meiotic cells (primary spermatocytes) move across the junctional complexes into the adluminal compartment, in which they mature into spermatozoa or **gametes** after meiosis. The adluminal compartment is an immunologically privileged site. The immune system does not recognize spermatozoa that develop in the adluminal compartment as "self." Consequently, males can develop antibodies against their own sperm, resulting in infertility. Sperm antibodies neutralize the ability of sperm to function. Sperm antibodies are often present after vasectomy or testicular injury and in some autoimmune diseases in which the adluminal compartment is ruptured, allowing sperm to mingle with immune cells from the circulation.

Main function of Sertoli cells is to nurture the development of sperm cells through spermatogenesis.

Sertoli cells are critical to germ cell development, as indicated by their close contact. As many as 6 to 12 spermatids may be attached to a Sertoli cell. Sertoli cells phagocytose residual bodies (excess cytoplasm resulting from the transformation of spermatids to spermatozoa) and damaged germ cells, provide structural support and nutrition for germ cells, secrete fluids, and assist in **spermiation**—the final detachment of mature spermatozoa from the Sertoli cell into the lumen. Spermiation may involve **plasminogen activator**, which converts plasminogen to plasmin, a proteolytic enzyme that assists in the release of the mature sperm into the lumen. Sertoli cells also synthesize large amounts of transferrin, an iron-transport protein important for sperm development.

During the fetal period, Sertoli cells and gonocytes form the seminiferous tubules as Sertoli cells undergo numerous rounds of cell divisions. Shortly after birth, Sertoli cells cease proliferating, and throughout life, the number of sperm produced is directly related to the number of Sertoli cells. At puberty, the capacity of Sertoli cells to bind FSH and testosterone increases. Receptors for FSH, present only on the plasma membranes of Sertoli cells, are glycoproteins linked to adenylyl cyclase via G proteins. FSH exerts multiple effects on the Sertoli cell, most of which are mediated by cAMP and PKA (Fig. 36.6). FSH stimulates the production of androgen-binding protein and plasminogen activator, increases secretion of inhibin, and induces aromatase activity for the conversion of androgens to estrogens. The testosterone receptor is within the nucleus of the Sertoli cell.

Androgen-binding protein (ABP) is a 90-kDa protein, made of a heavy and a light chain, that has a high binding affinity for dihydrotestosterone and testosterone. It is similar in function, with some homology in structure, to another binding protein, **sex hormone–binding globulin (SHBG)**, synthesized in the liver. ABP is found at high concentrations in the human testis and epididymis. It serves as a carrier of testosterone in Sertoli cells, as a storage protein for androgens in the seminiferous tubules, and as a carrier of testosterone from the testis to the epididymis. Recall that other products of the Sertoli cell include inhibin, follistatin, and activin. Whereas activin stimulates the secretion of FSH, inhibin and follistatin decrease the levels of FSH by suppressing FSH release from the pituitary gonadotrophs and reducing FSH secretion induced by activin, respectively.

Luteinizing hormone induces Leydig cells to produce testosterone.

Leydig cells are large polyhedral cells that are often found in clusters near blood vessels in the interstitium between seminiferous tubules. They are equipped to produce steroids because they have numerous mitochondria, a prominent smooth endoplasmic reticulum (SER), and conspicuous lipid droplets.

Leydig cells undergo significant changes in quantity and activity throughout life. This mechanism may depend on a nuclear transcription factor, SF-1, that recognizes a sequence in the promoter of all genes encoding CYP enzymes. In the human fetus, the period from week 8 to week 18 is marked by active steroidogenesis, which is obligatory for differentiation of the male genital ducts. Leydig cells at this time are prominent and active, reaching their maximal steroidogenic activity at about 14 weeks, when they constitute more than 50% of the testicular volume. Because the fetal hypothalamic-pituitary axis is still underdeveloped, **human chorionic gonadotropin (hCG)** from the placenta, rather than LH from the fetal pituitary, controls steroidogenesis (see Chapter 38, "Fertilization, Pregnancy, and Fetal Development"); LH and hCG bind the same receptor. After this period, Leydig cells slowly regress. At about 2 to 3 months of postnatal life, male infants have a significant rise in LH and testosterone production (infantile testosterone surge). The regulation and function of this short-lived surge remain uncertain, yet it is clearly independent of sex steroids. Leydig cells remain quiescent throughout childhood but increase in number and activity at the onset of puberty.

Leydig cells do not have FSH receptors, but FSH can increase the number of developing Leydig cells by stimulating the production of growth stimulators from Sertoli cells, which subsequently enhance the growth of Leydig cells. In addition, androgens stimulate the proliferation of developing Leydig cells. Estrogen receptors are present on Leydig cells and reduce the proliferation and activity of these cells.

Leydig cells have LH receptors, and the major effect of LH is to stimulate androgen secretion via a cAMP-dependent mechanism (see Fig. 36.6). The main product of Leydig cells is testosterone, but two other androgens of less biologic activity, dehydroepiandrosterone (DHEA) and androstenedione, are also produced.

There are bidirectional interactions between Sertoli and Leydig cells (see Fig. 36.6). The Sertoli cell is incapable of producing testosterone but contains testosterone receptors as well as FSH-dependent aromatase. The Leydig cell does not produce estradiol but contains receptors for it, and estradiol can suppress the response of the Leydig cell to LH.

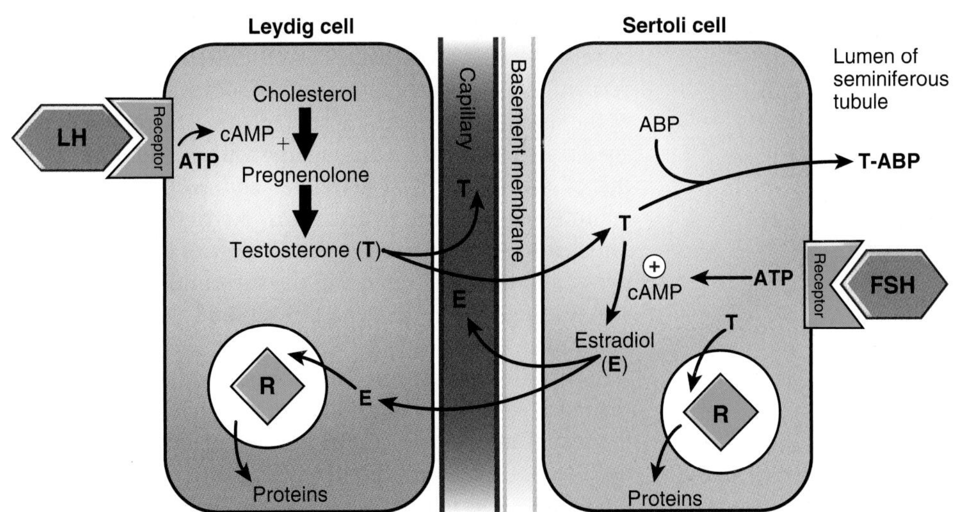

Figure 36.6 Regulation, hormonal products, of and interactions between Leydig and Sertoli cells. ABP, androgen-binding protein; ATP, adenosine triphosphate; cAMP, cyclic adenosine monophosphate; E, estradiol; FSH, follicle-stimulating hormone; LH, luteinizing hormone; R, receptor; T, testosterone.

Testosterone diffuses from the Leydig cells, crosses the basement membrane, enters the Sertoli cell, and binds to ABP. As a result, androgen levels can reach high local concentrations in the seminiferous tubules. Testosterone is obligatory for spermatogenesis and the proper functioning of Sertoli cells. In Sertoli cells, testosterone also serves as a precursor for estradiol production. The daily role of estradiol in the functioning of Leydig cells is unclear, but it may modulate responses to LH. For example, exposure of Leydig cells to LH or hCG leads to an LH or hCG desensitization. Studies show that estradiol is elevated within 30 minutes after hCG administration and that estrogen antagonists can block the hCG-induced down-regulation phenomena.

Duct system functions in sperm maturation, storage, and transport sites.

After formation in the seminiferous tubules, spermatozoa are transported to the rete testes and, from there, through the efferent ductules to the **epididymis**. Ciliary movement in the efferent ductules, muscle contraction, and the flow of fluid accomplish this movement of sperm.

Figure 36.7A shows a sagittal section of the male pelvis. The two major elements of the male sexual anatomy are the gonads (i.e., testes) and the sex accessories (i.e., the epididymis, vas deferens, seminal vesicles, ejaculatory duct, prostate, bulbourethral or Cowper glands, urethra, and penis). The epididymis is a single, tightly coiled, 4- to 5-m–long duct. It is composed of a head (caput), a body (corpus), and a tail (cauda) (Fig. 36.7B). The functions of the epididymis are storage, protection, transport, and maturation of sperm cells. Maturation at this point includes a change in functional capacity as sperm make their way through the epididymis. The sperm become capable of forward mobility during migration through the body of the epididymis. A significant portion of sperm maturation is carried out in the caput, whereas the sperm are stored in the cauda.

Frequent ejaculations result in reduced sperm numbers and increased numbers of immotile sperm in the ejaculate. The cauda connects to the vas deferens, which forms a dilated tube, the ampulla, prior to entering the prostate. The ampulla also serves as a storage site for sperm. Vasectomy—the cutting and ligation of the vas deferens—is an effective method of male contraception. Because the sperm are stored in the ampulla, men remain fertile for 4 to 5 weeks after vasectomy.

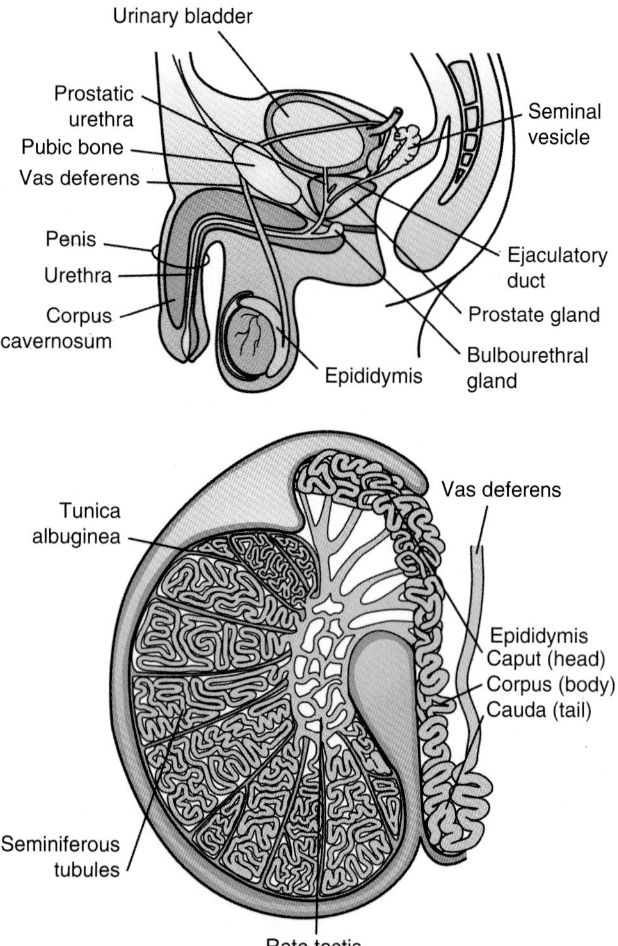

Figure 36.7 The male reproductive organs. The *top drawing* is a general side view. The *bottom enlargement* shows a sagittal section of the testis, epididymis, and vas deferens.

Male sex acts of erection and ejaculation are under neural control.

Erection is associated with sexual arousal emanating from sexually related psychic and/or physical stimuli. During sexual arousal, impulses from the genitalia, together with nerve signals originating in the limbic system, elicit motor impulses in the spinal cord. The parasympathetic nerves in the sacral region of the spinal cord carry these neuronal impulses via the cavernous nerve branches of the prostatic plexus that enter the penis. Those signals cause smooth muscle vasodilation of the arterioles and corpora cavernosa. An important biochemical regulator of arterial and cavernosal relaxations is **nitric oxide (NO)**, which is derived from the nerve terminals innervating the corpora cavernosa, the endothelial lining of penile arteries, and cavernosal sinuses. The actions of NO on the cavernosal smooth muscle and arterial blood flow are mediated through the activation of guanyl cyclase and production of **cyclic guanosine monophosphate (cGMP)**. As an intracellular second messenger, cGMP causes smooth muscle relaxation by lowering intracellular calcium. The relaxation of the smooth muscles in those structures dilates blood vessels, which then begin to engorge with blood. The swelling of the blood-filled arterioles and corpora cavernosa compresses the thin-walled veins, restricting blood flow. The result is a reduction in the outflow of blood from the penis, and blood is trapped in the surrounding erectile tissue, leading to engorgement, rigidity, and elongation of the penis in an erect position.

Erectile dysfunction (ED), formerly called *impotence*, is the repeated inability to get or keep an erection firm enough for sexual intercourse. The word "impotence" may also be used to describe other problems that interfere with sexual intercourse and reproduction, such as lack of sexual desire and problems with ejaculation or orgasm. Using the term *erectile dysfunction* makes it clear that those other problems are not involved.

Semen, consisting of sperm and associated fluids, is expelled by a neuromuscular reflex that is divided into two sequential phases: (1) emission and (2) ejaculation. Seminal emission moves the sperm and associated fluids from the cauda epididymis and vas deferens into the urethra. The latter process involves efferent stimuli originating in the lumbar areas (L1 and L2) of the spinal cord and is mediated by adrenergic sympathetic (hypogastric) nerves that induce contraction of smooth muscles of the epididymis and vas deferens. This action propels sperm through the ejaculatory ducts and into the urethra. Sympathetic discharge also closes the internal urethral sphincter, which prevents retrograde ejaculation into the urinary bladder. **Ejaculation** is the expulsion of the semen from the penile urethra; it is initiated after emission. The filling of the urethra with sperm initiates sensory signals via the pudendal nerves that travel to the sacrospinal region of the cord. A spinal reflex mechanism that induces rhythmic contractions of the striated bulbospongiosus muscles surrounding the penile urethra results in propelling the semen out of the tip of the penis.

The secretions of the accessory glands promote sperm survival and fertility. The accessory glands that contribute to the secretions are the seminal vesicles, prostate gland, and bulbourethral glands. The semen contains only 10% sperm by volume, with the remainder consisting of the combined secretions of the accessory glands. The normal volume of semen is 3 mL, with 20 to 50 million sperm per milliliter; normal is considered more than 20 million sperm per milliliter. The seminal vesicles contribute about 75% of the semen volume. Their secretion contains fructose (the principal substrate for glycolysis of ejaculated sperm), ascorbic acid, and prostaglandins. In fact, prostaglandin concentrations are high and were first discovered in semen but were mistakenly considered to be the product of the prostate. Seminal vesicle secretions are also responsible for coagulation of the semen seconds after ejaculation. Prostate gland secretions (~0.5 mL) include fibrinolysin, which is responsible for liquefaction of the coagulated semen 15 to 30 minutes after ejaculation, releasing sperm.

SPERMATOGENESIS

Spermatogenesis is the process of producing mature male gametes (sperm) and functions to fertilize the female counterpart, the oocyte. In humans, it occurs in the male testes and epididymis in a stepwise fashion in approximately 64 days. It is a continual process involving mitosis of the male germ cells, which undergo extensive morphologic changes in cell shape and, ultimately, meiosis to produce the haploid spermatozoa.

Spermatogenesis produces an abundance of highly specialized, mobile sperm.

Although sperm production continues throughout life, beginning at puberty, it declines with age. Spermatogenesis is the process of transformation of male germ cells into spermatozoa. This process can be divided into three distinct phases. The phases include cellular proliferation by **mitosis**, two reduction divisions by **meiosis** to produce haploid spermatids, and cell differentiation by a process called **spermiogenesis**, in which the spermatids differentiate into

From Bench to Bedside / 36.1

Male Infertility and a Potential Ring of Concern

An active area of bench research is aimed at better understanding the cause of male infertility with an underlying goal of improving semen quality and chances of pregnancy. It is estimated that infertility affects one in every six couples who are trying to conceive, with about half of the cases resulting from the inability of the male to initiate pregnancy. Causes of male infertility include abnormal sperm production or function, impaired delivery of sperm, general health and lifestyle issues, and exposure to certain environmental factors. Of new concern are findings that a modern-day environmental factor, the cell phone, commonly carried close to the testicles (i.e., in a trouser pocket or clipped to a waist belt) can reduce a male's fertilizing potential.

Recent studies have highlighted that spermatozoa of infertile men have higher levels of DNA damage, which adversely affects the ability of spermatozoa to reach, recognize, and fertilize an egg. Mechanistically, a key contributing factor of DNA damage in spermatozoa appears to be oxidative stress. Oxidative stress results from an imbalance of **reactive oxygen species (ROS)** and antioxidants. Immature spermatozoa with residual cytoplasm produce ROS by having excess nicotinamide adenine diphosphate hydrogen generated via glucose-6-phosphate dehydrogenase. Countering the harmful effect of excess ROS production are antioxidants that are present in normal, healthy seminal ejaculates.

Findings suggesting that electromagnetic waves (EMWs) emitted from cell phones may negatively impact the ROS/antioxidant balance in semen represent a new, modern-day concern. In a large prospective study, conducted by researchers at the Cleveland Clinic, the effects of EMWs emitted by cell phones (900–1,900 MHz) on various markers of sperm quality, such as count, motility, morphology, and viability, were assessed. These studies revealed that exposure of ejaculated semen samples to commercially available cell phones for 1 hour caused a significant decrease in sperm motility and viability, increased ROS levels, and decreased ROS–total antioxidant capacity score when compared with neat semen from a nonexposed group.

As importantly noted by this research team, many men carry their cell phones in a trouser pocket (or clipped to a waist belt) exposing testes to high-power cell phone density. As the effects of frequency, distance of the phone from the source, and the duration of talk time on spermatozoa are not known, current investigations by these researchers are now employing a two-dimensional anatomical model of the tissue to extrapolate the effects seen in *in vitro* condition to real-life conditions.

spermatozoa (Fig. 36.8). Spermatogenesis begins at puberty, so the seminiferous tubules are quiescent throughout childhood. Spermatogenesis is initiated shortly before puberty, under the influence of the rising levels of gonadotropins and testosterone, and continues throughout life, with a slight decline during old age.

The time required to produce mature spermatozoa from the earliest stage of spermatogonia is 65 to 70 days. Because several developmental stages of spermatogenic cells occur during this time frame, the stages are collectively known as the *spermatogenic cycle*. There is synchronized development of spermatozoa within the seminiferous tubules, and each stage is morphologically distinct. A spermatogonium becomes a mature spermatozoon after going through several rounds of mitotic divisions, a couple of meiotic divisions, and a few weeks of differentiation. Hormones can alter the number of spermatozoa, but they generally do not affect the duration of the cycle. Spermatogenesis occurs along the length of each seminiferous tubule in successive cycles. New cycles are initiated at regular time intervals (every 2 to 3 weeks) before the previous ones are completed. Consequently, cells at different stages of development are spaced along each tubule in a "spermatogenic wave." Such a succession ensures the continuous production of fresh spermatozoa. Approximately 200 million spermatozoa are produced daily in the adult human testes, which is about the same number of sperm present in a normal ejaculate.

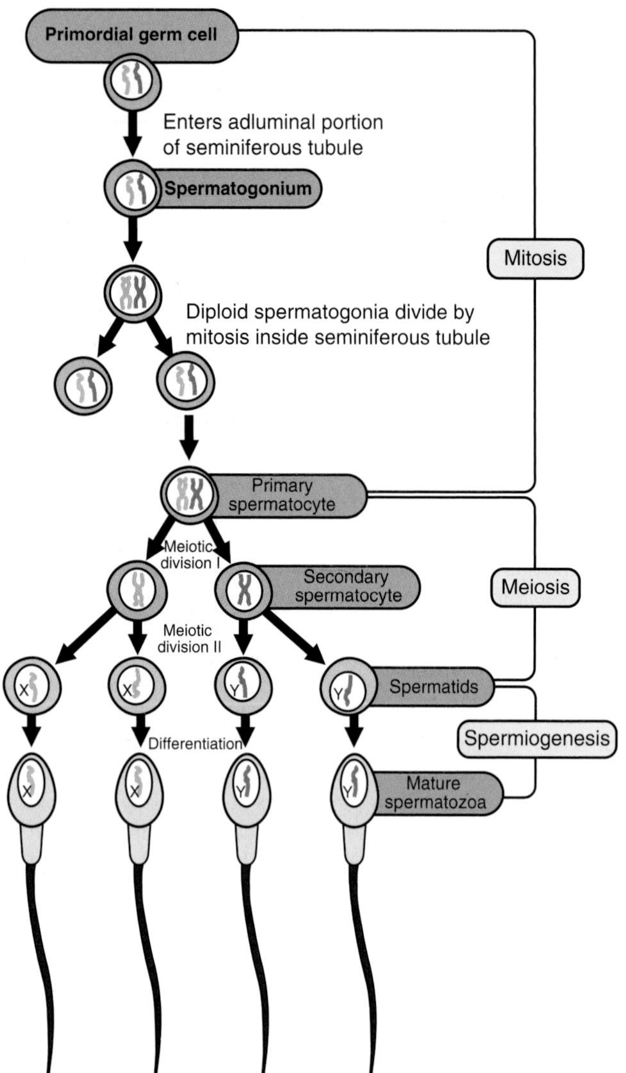

Figure 36.8 The process of spermatogenesis, showing successive cell divisions and remodeling leading to the formation of haploid spermatozoa.

Because sperm cells are rapidly dividing and undergoing meiosis, they are sensitive to external agents that alter cell division. Chemical carcinogens, chemotherapeutic agents, certain drugs, environmental toxins, irradiation, and extreme temperatures are factors that can reduce the number of replicating germ cells or cause chromosomal abnormalities in individual cells. Although the immune system normally detects and destroys defective somatic cells, the blood–testis barrier isolates advanced germ cells from immune surveillance.

If physical injury or infection ruptures the blood–testis barrier and sperm cells within the barrier are exposed to circulating immune cells, antibodies can develop to the sperm cells. In the past, it was thought that the development of antisperm antibodies could lead to male infertility. It appears that men with high levels of antisperm antibodies may exhibit some infertility problems. However, men who have developed low or moderate levels of antisperm antibodies after vasectomy and who have had their vasa deferens reconnected have normal fertility if the vasectomy was for a relatively short time, according to studies. Vasectomy does not appear to change hormone or sperm production by the testes. Nevertheless, in some cases, a high level of antisperm antibodies in men and women leads to infertility.

Spermatogonia divide mitotically to produce primary spermatocytes, which in turn, undergo meiosis to become spermatids.

Spermatogonia undergo several rounds of mitotic division prior to entering the meiotic phase (see Fig. 36.8). The spermatogonia remain in contact with the Sertoli cells, migrate away from the basal compartment near the walls of the seminiferous tubules, and cross into the adluminal compartment of the tubule (see Fig. 36.5). After crossing into the adluminal compartment, the cells differentiate into spermatocytes prior to undergoing meiosis I. The first meiotic division of **primary spermatocytes** gives rise to diploid ($2n$ chromosomes) **secondary spermatocytes**.

The second meiotic division produces haploid (one set of chromosomes) cells called **spermatids**. Of every four spermatids emanating from a primary spermatocyte, two contain X chromosomes and two have Y chromosomes (see Fig. 36.8). Because of the numerous mitotic divisions and two rounds of meiosis, each spermatogonium committed to meiosis should yield 256 spermatids if all cells survive.

There are numerous developmental disorders of spermatogenesis. The most frequent is **Klinefelter syndrome**, which causes hypogonadism and infertility in men. Patients with this disorder have an accessory X chromosome caused by meiotic nondisjunction. The typical karyotype is 47 XXY, but there are other chromosomal mosaics. Testicular volume is reduced >75%, and ejaculates contain few, if any, spermatozoa. Spermatogenic cell differentiation beyond the primary spermatocyte stage is rare.

Formation of a mature spermatozoon requires extensive cell remodeling.

Spermatids are small, round, and nondistinctive cells. During the second half of the spermatogenic cycle, they undergo considerable restructuring to form mature spermatozoa. Notable changes include alterations in the nucleus, the formation of a tail, and a massive loss of cytoplasm. The nucleus becomes eccentric and decreases in size, and the chromatin becomes condensed. The **acrosome**, a lysosome-like structure unique to spermatozoa, buds from the Golgi apparatus, flattens, and covers most of the nucleus. The centrioles, located near the Golgi apparatus, migrate to the caudal pole and form a long axial filament made of nine peripheral doublet microtubules surrounding a central pair (called the "9 + 2 arrangement"). This becomes the **axoneme** or major portion of the tail. Throughout this reshaping process, the cytoplasmic content is redistributed and discarded. During spermiation, most of the remaining cytoplasm is shed in the form of residual bodies.

The reasons for this lengthy and metabolically costly process become apparent when the unique functions of this cell are considered. Unlike other cells, the spermatozoon serves no apparent purpose in the organism. Its only function is to reach, recognize, and fertilize an egg; hence, it must fulfill several prerequisites: It should possess an energy supply and means of locomotion, it should be able to withstand a foreign and even hostile environment, it should be able to recognize and penetrate an egg, and it must carry all the genetic information necessary to create a new person.

The mature spermatozoon exhibits a remarkable degree of structural and functional specialization well adapted to carry out these functions. The cell is small, compact, and streamlined; it is about 1 to 2 μm wide and can exceed 50 μm in length in humans. It is packed with specialized organelles and long axial fibers but contains only a few of the normal cytoplasmic constituents, such as ribosomes, ER, and Golgi apparati. It has a prominent nucleus, a flexible tail, numerous mitochondria, and an assortment of proteolytic enzymes.

The spermatozoon consists of three main parts: a head, a middle piece, and a tail. The two major components in the head are the condensed chromatin and the acrosome. The haploid chromatin is transcriptionally inactive throughout the life of the sperm until fertilization, when the nucleus decondenses and becomes a pronucleus. The acrosome contains proteolytic enzymes, such as hyaluronidase, acrosin, neuraminidase, phospholipase A, and esterases. They are inactive until the **acrosome reaction** occurs on contact of the sperm head with the egg (see Chapter 38, "Fertilization, Pregnancy, and Fetal Development"). Their proteolytic action enables the sperm to penetrate through the egg membranes. The middle piece contains spiral sheaths of mitochondria that supply energy for sperm metabolism and locomotion. The tail is composed of a 9 + 2 arrangement of microtubules, which is typical of cilia and flagella, and is surrounded by a fibrous sheath that provides some rigidity. The tail propels the sperm by a twisting motion, involving interactions between tubulin fibers and dynein side arms and requiring ATP and magnesium.

Testosterone is essential for sperm production and maturation.

Spermatogenesis requires high intratesticular levels of testosterone, secreted from the LH-stimulated Leydig cells. The testosterone diffuses across the basement membrane of the seminiferous tubule, crosses the blood–testis barrier, and complexes with ABP. Sertoli cells, but not spermatogenic, cells contain receptors for testosterone. Sertoli cells also contain FSH receptors. FSH function in male subjects is not readily apparent but probably mediates spermatozoa development from spermatids in concert with testosterone, especially as failed spermatogenesis leads to elevated FSH levels.

The actions of FSH and testosterone at each point of sperm cell production are unknown. On entering meiosis, spermatogenesis appears to depend on the availability of FSH and testosterone. In human males, FSH is thought to be required for the initiation of spermatogenesis before puberty. When adequate sperm production has been achieved, LH alone (through stimulation of testosterone production) or testosterone alone is sufficient to maintain spermatogenesis.

ENDOCRINE FUNCTION OF THE TESTIS

A secondary function of the testis is steroidogenesis, the production of the steroid hormones. Testosterone is the main hormone produced by steroidogenesis and is primarily secreted by the testes of males and the ovaries of females. In males, testosterone production is about 10 times greater than that in females and plays a key role in the development of the testes and prostate as well as promoting secondary sexual characteristics, such as increased hair growth, muscle, and bone mass. It is also essential for health and well-being as well as the prevention of osteoporosis. Testosterone is converted to **dihydrotestosterone (DHT)**, the most biologically active androgen, and to **estradiol**, the most biologically active estrogen.

Testosterone is primarily synthesized in Leydig cells and is derived from cholesterol, involving many enzymatic steps.

The adrenal cortex, ovaries, testes, and placenta produce steroid hormones from cholesterol. Cholesterol, a 27-carbon (C27) steroid, can be obtained from the diet or synthesized within the body from acetate. Each organ uses a similar steroid biosynthetic pathway, but the relative amount of the final products depends on the particular subset of enzymes expressed in that tissue and the trophic hormones (LH, FSH, and adrenocorticotropic hormone) stimulating specific cells within the organ. The major steroid produced by the testis is testosterone, but other androgens, such as androstenediol, androstenedione, and DHEA, as well as a small amount of estradiol, are also produced.

Cholesterol from low-density lipoprotein (LDL) and high-density lipoprotein (HDL) is released in Leydig cells and transported from the outer mitochondrial membrane to the inner mitochondrial membrane, a process regulated by **steroidogenic acute regulatory protein**. Under the influence of LH, with cAMP as a second messenger, cholesterol side-chain cleavage enzyme (CYP11A1), which removes six carbons attached to the 21 position, converts cholesterol to **pregnenolone** (C21). Pregnenolone is a key intermediate for all steroid hormones in various steroidogenic organs (Fig. 36.9; see also Chapter 33, "Adrenal Gland," Fig. 33.5). Specific transport proteins transport pregnenolone out of mitochondria. The pregnenolone then moves by diffusion to the SER, where the remainder of sex hormone biosynthesis takes place.

Pregnenolone can be converted to testosterone via two pathways, the *δ5 pathway* and the *δ4 pathway.* In the δ5 pathway, the double bond is in ring B; in the δ4 pathway, the double bond is in ring A (see Fig. 36.9). The δ5 intermediates include 17α-hydroxypregnenolone, DHEA, and androstenediol, whereas the δ4 intermediates are progesterone, 17α-hydroxyprogesterone, and androstenedione.

The conversion of C21 steroids (the progestins) to androgens (C19 steroids) proceeds in two steps: first, 17α-hydroxylation of pregnenolone (to form 17α-hydroxypregnenolone) and second, C17,20 cleavage; therefore, two carbons are removed to form DHEA. A single enzyme, 17α-hydroxylase or 17,20-lyase

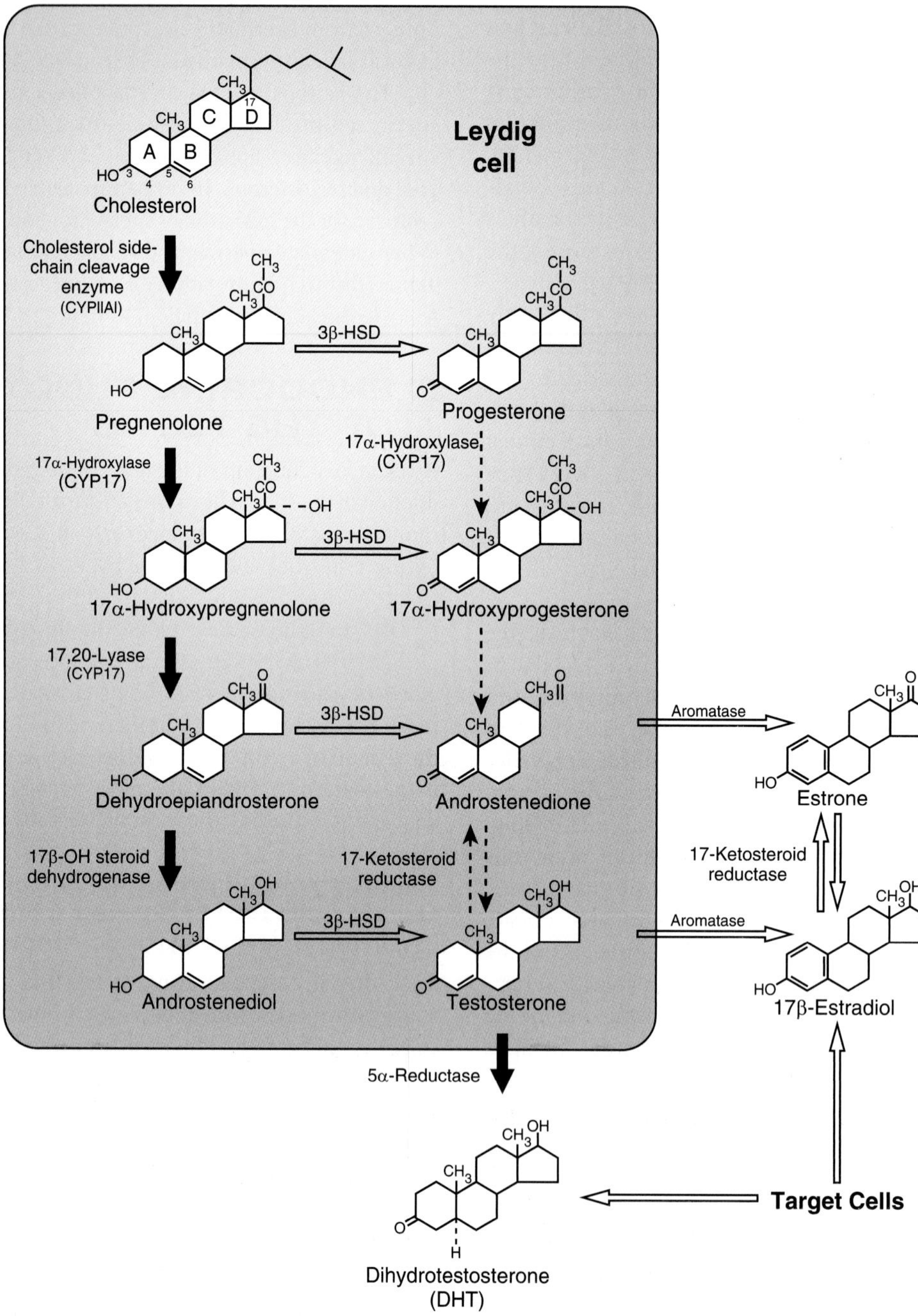

Figure 36.9 Steroidogenesis in Leydig cells and further modifications of androgens in target cells. *Solid arrows* represent the δ5 pathway. *Dashed arrows* represent the δ4 pathway. 3β-HSD, 3β-hydroxysteroid dehydrogenase.

(CYP17), accomplishes this hydroxylation and cleavage. Another two-step enzymatic reaction converts DHEA to androstenedione: dehydrogenation in position 3 (catalyzed by 3β-hydroxysteroid dehydrogenase [3β-HSD]) and shifting of the double bond from ring B to ring A (catalyzed by δ4,5-ketosteroid isomerase); these two may be the same enzyme. 17-Ketosteroid reductase (17β-hydroxysteroid dehydrogenase), which substitutes the keto group in position 17 with a hydroxyl group, carries out the final reaction yielding testosterone. Unlike all the preceding enzymatic reactions, this is a reversible step but tends to favor testosterone.

Although estrogens are only minor products of testicular steroidogenesis, they are normally found in low concentrations in men. The action of the enzyme complex aromatase (CYP19) converts androgens (C19) to estrogens (C18). Aromatization involves the removal of the methyl group in position 19 and the rearrangement of ring A into an unsaturated aromatic ring. The products of aromatization of testosterone and androstenedione are estradiol and estrone, respectively (see Fig. 36.9). In the testis, the Sertoli cell is the main site of aromatization, which is stimulated by FSH; however, aromatization may also occur in peripheral tissues that lack FSH receptors (e.g., adipose tissue).

cAMP regulates luteinizing hormone, which, in turn, regulates the number of Leydig cells.

The action of LH on Leydig cells is mediated through specific LH receptors on the plasma membrane. A Leydig cell has about 15,000 LH receptors, and occupancy of <5% of these is sufficient for maximal steroidogenesis. This is an example of "spare receptors" (see Chapter 30, "Endocrine Control Mechanisms"). Excess receptors increase target cell sensitivity to low circulating hormones by increasing the probability that sufficient receptors will be occupied to induce a response. After exposure to a high LH concentration, the number of LH receptors and testosterone decrease. However, in response to the initial high LH testosterone will increase and then decrease. Thereafter, subsequent challenges with LH lead to no response or decreased responses. This so-called *desensitization* involves a loss of surface LH receptors as a result of internalization and receptor modification by phosphorylation.

The LH receptor is a single 93-kDa glycoprotein composed of three functional domains: (1) a glycosylated extracellular hormone-binding domain, (2) a transmembrane spanning domain that contains seven noncontiguous segments, and (3) an intracellular domain. The receptor is coupled to a stimulatory G protein (G_s) via a loop of one of the LH receptor transmembrane segments. The activation of G_s results in increased adenylyl cyclase activity, the production of cAMP, and the activation of PKA (Fig. 36.10).

Low doses of LH can stimulate testosterone production without detectable changes in total cell cAMP concentration. However, the amount of cAMP bound to the regulatory subunit of PKA increases in response to such low doses of LH. This response emphasizes the importance of compartmentalization for both enzymes and substrates in mediating hormonal action. Other intracellular mediators, such as the phosphatidylinositol system or calcium, have roles in regulating Leydig cell steroidogenesis, but it appears that the PKA pathway may predominate.

The proteins phosphorylated by PKA are specific for each cell type. Some of these, such as CREB, which functions as a

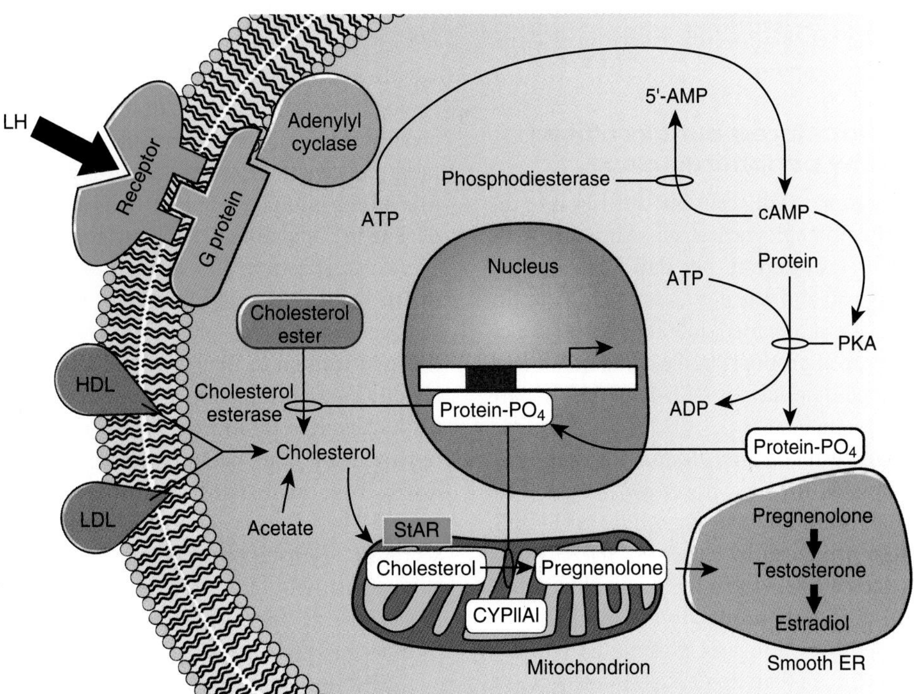

Figure 36.10 A proposed intracellular mechanism by which luteinizing hormone stimulates testosterone synthesis. ADP, adenosine diphosphate; AMP, adenosine monophosphate; ATP, adenosine triphosphate; cAMP, cyclic adenosine monophosphate; ER, endoplasmic reticulum; HDL, high-density lipoprotein; LDL, low-density lipoprotein; LH, luteinizing hormone; PKA, protein kinase A; StAR, steroidogenic acute regulatory protein.

DNA-binding protein, regulate the transcription of cholesterol side-chain cleavage enzyme (CYP11A1), the rate-limiting enzyme in the conversion of cholesterol to pregnenolone. **Phosphodiesterase** inactivates cAMP to AMP. This enzyme plays a major role in regulating LH (and, possibly, FSH) responses because gonadotropin stimulation activates phosphodiesterase. The increase in phosphodiesterase reduces the response to LH (and FSH). Certain drugs can inhibit phosphodiesterase; gonadotropin hormone responses will increase dramatically in the presence of those drugs. Numerous isoforms of phosphodiesterase and adenylyl cyclase exist; specific types of each in the testis have not yet been revealed.

LH stimulates steroidogenesis by two principal activations. One is the phosphorylation of cholesterol esterase, which releases cholesterol from its intracellular stores. The other is the activation of CYP11A1.

Leydig cells also contain receptors for **prolactin (PRL)**. *Hyperprolactinemia* in men with pituitary tumors, usually microadenomas, is associated with decreased testosterone levels. This condition is a result of a direct effect of elevated circulating levels of PRL on Leydig cells, reducing the number of LH receptors or inhibiting downstream signaling events. In addition, hyperprolactinemia may decrease LH secretion by reducing the pulsatile nature of its release. Under nonpathologic conditions, however, PRL may synergize with LH to stimulate testosterone production by increasing the number of LH receptors.

ANDROGEN ACTION AND MALE DEVELOPMENT

DHT enhances development of the male reproductive tract, accompanying accessory ducts and glands, and male sex characteristics, including behavior. A lack of androgen secretion or action causes feminization.

Testosterone is not stored but circulated and metabolized by peripheral tissue.

Testosterone is not stored in Leydig cells but diffuses into the blood immediately after being synthesized. An adult male produces 6 to 7 mg of testosterone per day. This amount slowly declines after age 50 and reaches about 4 mg/d in the seventh decade of life. Therefore, men do not undergo a sudden cessation of sex steroid production on aging as women do during their postmenopausal period when the ova are completely depleted.

Testosterone circulates bound to plasma proteins, with only 2% to 3% present as the free hormone. About 30% to 40% is bound to albumin and the remainder to SHBG, a 94-kDa glycoprotein produced by the liver. SHBG binds both estradiol and testosterone, with a higher binding affinity for testosterone. Because its production is increased by estrogens and decreased by androgens, plasma SHBG concentration is higher in women than in men. SHBG serves as a reservoir for testosterone and, therefore, a sudden decline in newly formed testosterone may not be evident because of the large pool bound to proteins. SHBG, in effect, deactivates testosterone because only the unbound hormone can enter the cell. It also prolongs the half-life of circulating testosterone because testosterone is cleared from the circulation much more slowly if bound to a protein. Any type of liver damage or disease will generally reduce SHBG production. The latter can upset the hormonal balance between LH and testosterone. For example, if SHBG declines acutely, then free testosterone may increase while the total amount of circulating testosterone would decrease. In response to the increase in free testosterone, LH levels would decline in a homeostatic attempt to reduce testosterone production.

Once testosterone is released into the circulation, its fate is variable. In most target tissues, testosterone functions as a prohormone and is converted to the biologically active derivatives DHT by **5α-reductase** or estradiol by aromatase (Fig. 36.11). Skin, hair follicles, and most of the male reproductive tract contain an active 5α-reductase. The enzyme irreversibly catalyzes the reduction of the double bond in ring A and generates DHT (see Fig. 36.9). DHT has a high binding affinity for the androgen receptor and is two to three times more potent than testosterone.

Congenital deficiency of 5α-reductase in males results in ambiguous genitalia containing female and male characteristics, because DHT is critical for directing the normal development of male external genitalia during embryonic life (see Chapter 38, "Fertilization, Pregnancy, and Fetal Development"). Without DHT, the female pathway may predominate, even though the genetic sex is male and small, undescended testes are present in the inguinal region. DHT is nonaromatizable and cannot be converted to estrogens.

Drugs that inhibit 5α-reductase are currently used to reduce prostatic hypertrophy because DHT induces hyperplasia of prostatic epithelial cells. In addition, analogs of GnRH, as either agonists or antagonists, can be given to patients to reduce the secretion of androgen in androgen-dependent neoplasia or cancer. In the case of the GnRH antagonist, this analog blocks the secretion of LH. In contrast, GnRH agonists given in large quantities initially induce the secretion of LH (and androgen). However, down-regulation of GnRH receptors on the pituitary gonadotrophs and, ultimately, a dramatic decline in circulating LH and androgen follow this response.

Aromatization of some androgens to estrogens occurs in fat, liver, skin, and brain cells. Circulating levels of total estrogens (estradiol plus estrone) in men can approach those in women in their early follicular phase. Men are protected from feminization as long as production of and tissue responsiveness to androgens are normal. The treatment of hypogonadal male patients with high doses of aromatizable testosterone analogs (or testosterone), the use of anabolic steroids by athletes, abnormal reductions in testosterone secretion, estrogen-producing testicular tumors, and tissue insensitivity to androgens can lead to **gynecomastia** or breast enlargement. All of these conditions are characterized by a decrease in the testosterone/estradiol ratio.

Androgens are metabolized in the liver to biologically inactive water-soluble derivatives suitable for excretion by the kidneys. The major products of testosterone metabolism are two 17-ketosteroids, androsterone and etiocholanolone. These, as well as native testosterone, are conjugated in position

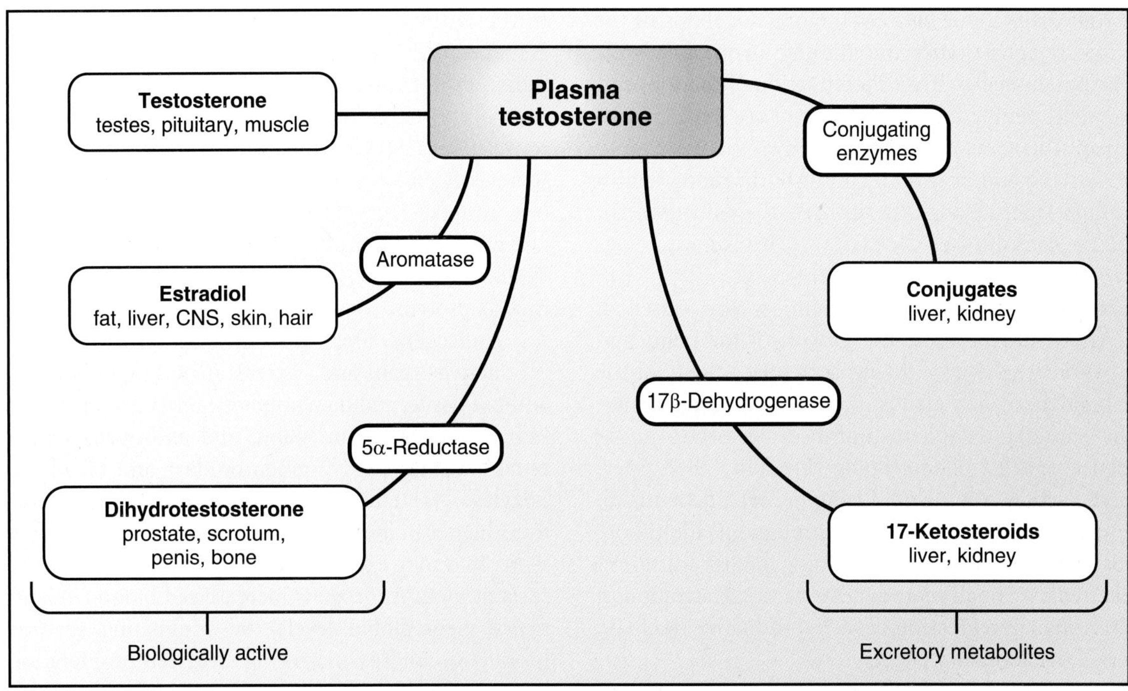

Figure 36.11 Conversion of testosterone to different products in extratesticular sites. CNS, central nervous system.

3 to form sulfates and glucuronides, which are water soluble, and excreted into the urine (see Fig. 36.11).

Androgens target both reproductive and nonreproductive tissues.

An **androgen** is a substance that stimulates the growth of the male reproductive tract and the development of secondary sex characteristics. Androgens have effects on almost every tissue, including alteration of the primary sex structures (i.e., the testes and genital tract), stimulation of the secondary sex structures (i.e., accessory glands), and development of secondary sex characteristics responsible for masculine phenotypic expression. Androgens also affect both sexual and nonsexual behavior. The relative potency ranking of androgens is as follows: DHT > testosterone > androstenedione > DHEA. The action of sex steroid hormones on somatic tissue, such as muscle, is referred to as "anabolic" because the end result is increased muscle size. The same molecular mechanisms that result in virilization mediate this action.

Between 8 and 18 weeks of fetal life, androgens mediate differentiation of the male genitalia. Testosterone, which reaches these target tissues by diffusion rather than by a

Clinical Focus / 36.1

Prostate Cancer

Some prostate cancers highly depend on androgens for cellular proliferation; therefore, physicians attempt to totally ablate the secretion of androgens by the testes. Generally, two options for those patients are surgical castration and chemical castration. Surgical castration is irreversible and requires the removal of the testes, whereas chemical castration is reversible.

One option for chemical treatment of these patients is the use of analogs of gonadotropin-releasing hormone (GnRH), the hormone that regulates the secretion of luteinizing hormone (LH) and follicle-stimulating hormone (FSH). Long-acting GnRH agonists or antagonists reduce LH and FSH secretion by different mechanisms. GnRH agonists reduce gonadotropin secretion by desensitization of the pituitary gonadotrophs to GnRH, leading to a reduction of LH and FSH secretion. GnRH agonists initially stimulate GnRH receptors on pituitary cells but ultimately reduce their numbers. GnRH antagonists bind to GnRH receptors on the pituitary cells, prevent endogenous GnRH from binding to those receptors, and subsequently reduce LH and FSH secretion. Shortly after treatment, testicular concentrations of androgens decline because of the low levels of circulating LH and FSH. The expectation is that androgen-dependent cancer cells will cease or slow proliferation and, ultimately, die.

GnRH agonists (leuprolide acetate [trade name, Lupron]) are usually used in combination with other drugs to block most effectively androgenic activity. For example, one of the androgen-blocking drugs includes 5a-reductase inhibitors that prevent the conversion of testosterone to the highly active androgen dihydrotestosterone. In addition, antiandrogens, such as flutamide, bind to the androgen receptor and prevent binding of endogenous androgen. Some prostate cancers are androgen-independent, and the treatment requires nonhormonal therapies, including chemotherapy and radiation.

systemic route, directly influences the organogenesis of the Wolffian (mesonephric) ducts into the epididymis, vas deferens, and seminal vesicles. The differentiation of the urogenital sinus and the genital tubercle into the penis, scrotum, and prostate gland depends on testosterone being converted to DHT. Toward the end of fetal life, testosterone and insulin-like hormones from Leydig cells promote the descent of the testes into the scrotum (see Chapter 38, "Fertilization, Pregnancy, and Fetal Development").

Enhanced androgenic activity marks the onset of puberty. Androgens promote the growth of the penis and scrotum, stimulate the growth and secretory activity of the epididymis and accessory glands, and increase the pigmentation of the genitalia. Enlargement of the testes occurs under the influence of the gonadotropins (LH and FSH). Spermatogenesis, which is initiated during puberty, depends on adequate amounts of testosterone. Throughout adulthood, androgens are responsible for maintaining the structural and functional integrity of all reproductive tissues. Castration of adult men results in regression of the reproductive tract and involution of the accessory glands.

Androgens are responsible for secondary sex characteristics and the masculine phenotype.

Androgens effect changes in hair distribution, skin texture, pitch of the voice, bone growth, and muscle development. Hair is classified by its sensitivity to androgens into the following: nonsexual (eyebrows and extremities); ambisexual (axilla), which is responsive to low levels of androgens; and sexual (face, chest, and upper pubic triangle), which is responsive only to high androgen levels. Hair follicles metabolize testosterone to DHT or androstenedione. Androgens stimulate the growth of facial, chest, and axillary hair; however, along with genetic factors, they also promote temporal hair recession and loss. Normal axillary and pubic hair growth in women is also under androgenic control, whereas excess androgen production in women causes the excessive growth of sexual hair (**hirsutism**).

The growth and secretory activity of the sebaceous glands on the face, upper back, and chest are stimulated by androgens, primarily DHT, and inhibited by estrogens. Increased sensitivity of target cells to androgenic action, especially during puberty, is the cause of **acne vulgaris** in both males and females. Skin derived from the urogenital ridge (e.g., the prepuce, scrotum, clitoris, and labia majora) remains sensitive to androgens throughout life and contains an active 5α-reductase. Growth of the larynx and thickening of the vocal cords are also androgen-dependent. Eunuchs maintain the high-pitched voice typical of prepubertal boys because they were castrated prior to puberty.

A complex interplay between androgens, growth hormone (GH), nutrition, and genetic factors influences the growth spurt of adolescent males. The growth spurt includes growth of the vertebrae, long bones, and shoulders. The mechanism by which androgens (likely DHT) alter bone metabolism is unclear. Androgens accelerate closure of the epiphyses in the long bones, eventually limiting further growth. Because of the latter, precocious puberty is associated with a final short adult stature, whereas delayed puberty or eunuchoidism usually results in tall stature. Androgens have multiple effects on skeletal and cardiac muscle. Because 5α-reductase activity in muscle cells is low, the androgenic action is a result of testosterone. Testosterone stimulates muscle hypertrophy, increasing muscle mass; however, it has minimal or no effect on muscle hyperplasia. Testosterone, in synergy with GH, causes a net increase in muscle protein.

Androgens affect, directly or indirectly, other nonreproductive organs and systems, including the liver, kidneys, adipose tissue, and hematopoietic and immune systems. The kidneys are larger in males, and androgens induce some renal enzymes (e.g., β-glucuronidase and ornithine decarboxylase). HDL levels are lower and triglyceride concentrations higher in men, compared with premenopausal women, a fact that may explain the higher prevalence of atherosclerosis in men. Androgens increase red blood cell mass (and, hence, hemoglobin levels) by stimulating erythropoietin production and by increasing stem cell proliferation in the bone marrow.

Androgen is involved in sexual differentiation of the brain.

Many sites in the brain contain androgen receptors, with the highest density in the hypothalamus, preoptic area, septum, and amygdala. Most of those areas also contain aromatase, and many of the androgenic actions in the brain result from the aromatization of androgens to estrogens. The pituitary also has abundant androgen receptors, but no aromatase. The enzyme 5α-reductase is widely distributed in the brain, but its activity is generally higher during the prenatal period than in adults. Sexual dimorphism in the size, number, and arborization of neurons in the preoptic area, amygdala, and superior cervical ganglia has been recently recognized in humans.

Unlike most species, which mate only to produce offspring, in humans, sexual activity and procreation are not tightly linked. Superimposed on the basic reproductive mechanisms dictated by hormones are numerous psychologic and societal factors. In normal men, no correlation is found between circulating testosterone levels and sexual drive, frequency of intercourse, or sexual fantasies. Similarly, there is no correlation between testosterone levels and impotence or homosexuality. Castration of adult men results in a slow decline in, but not a complete elimination of, sexual interest and activity.

MALE REPRODUCTIVE DISORDERS

Male reproductive dysfunctions may be caused by endocrine disruption, morphologic alterations in the reproductive tract, neuropathology, and genetic mutations. Several medical tests, including serum hormone levels, physical examination of the reproductive organs, and sperm count, are important in ascertaining causes of reproductive dysfunctions.

Clinical Focus / 36.2

Effects of Testosterone Administration

Although testosterone has a role in stimulating spermatogenesis, infertile men with a low sperm count do not benefit from testosterone treatment. Unless given at supraphysiologic doses, exogenous testosterone cannot achieve the required local high concentration in the testis. One function of androgen-binding protein in the testis is to sequester testosterone, which significantly increases its local concentration.

Exogenous testosterone given to men would normally inhibit endogenous luteinizing hormone (LH) release through a negative-feedback effect on the hypothalamic-pituitary axis and lead to a suppression of testosterone production by Leydig cells and a further decrease in testicular testosterone concentrations. Ultimately, because LH levels decrease when exogenous testosterone is administered, testicular size decreases, as has been reported for men who abuse androgens.

High androgens have an anabolic effect on muscle tissue, leading to increased muscle mass, strength, and performance, a desired result for bodybuilders and athletes. Androgen abuse has been associated with abnormally aggressive behavior and the potential for increased incidence of liver and brain tumors.

Hypogonadism leads to decreased functional activity of the gonads and can stem from several defects.

Male **hypogonadism** may result from defects in spermatogenesis, steroidogenesis, or both. It may be a primary defect in the testes or secondary to hypothalamic-pituitary dysfunction, and determining whether the onset of gonadal failure occurred before or after puberty is important in establishing the cause. However, several factors must be considered. First, normal spermatogenesis almost never occurs with defective steroidogenesis, but normal steroidogenesis can be present with defective spermatogenesis. Second, primary testicular failure removes feedback inhibition from the hypothalamic-pituitary axis, resulting in elevated plasma gonadotropins. In contrast, decreased gonadotropin and steroid levels and reduced testicular size almost always accompany hypothalamic and/or pituitary failure. Third, gonadal failure before puberty results in the absence of secondary sex characteristics, creating a distinctive clinical presentation called **eunuchoidism**. In contrast, men with a postpubertal testicular failure retain masculine features but exhibit low sperm counts or a reduced ability to produce functional sperm.

To establish the cause(s) of reproductive dysfunction, physical examination and medical history, semen analysis, hormone determinations, hormone stimulation tests, and genetic analysis are performed. Physical examination should establish whether eunuchoidal features (i.e., infantile appearance of external genitalia and poor or absent development of secondary sex characteristics) are present. In men with adult-onset reproductive dysfunction, physical examination can uncover problems such as *cryptorchidism* (nondescendent testes), testicular injury, *varicocele* (an abnormality of the spermatic vasculature), testicular tumors, prostatic inflammation, or gynecomastia. Medical and family history help determine delayed puberty, anosmia (an inability to smell, often associated with GnRH dysfunction), previous fertility, changes in sexual performance, ejaculatory disturbances, or impotence (an inability to achieve or maintain erection).

One step in the evaluation of fertility is semen analysis. Semen is analyzed on specimens collected after 3 to 5 days of sexual abstinence, as the number of sperm ejaculated remains low for a couple of days after ejaculation. Initial examination includes determination of viscosity, liquefaction, and semen volume. The sperm are then counted and the percentage of sperm showing forward motility is scored. The spermatozoa are evaluated morphologically, with attention to abnormal head configuration and defective tails. Chemical analysis can provide information on the secretory activity of the accessory glands, which is considered abnormal if semen volume is too low or sperm motility is impaired. Fructose and prostaglandin levels are determined to assess the function of the seminal vesicles and levels of zinc, magnesium, and acid phosphatase to evaluate the prostate. Terms used in evaluating fertility include aspermia (no semen), hypospermia and hyperspermia (too small or too large semen volume), azoospermia (no spermatozoa), and oligozoospermia (reduced number of spermatozoa).

Serum testosterone, estradiol, LH, and FSH analyses are performed using radioimmunoassays. Free and total testosterone levels should be measured; because of the pulsatile nature of LH release, several consecutive blood samples are needed. Dynamic hormone stimulation tests are most valuable for establishing the site of abnormality. A failure to increase LH release on treatment with **clomiphene**, an antiestrogen, likely indicates a hypothalamic abnormality. Clomiphene blocks the inhibitory effects of estrogen and testosterone on endogenous GnRH release. An absence of or blunted rise in testosterone after hCG injection suggests a primary testicular defect. Genetic analysis is used when congenital defects are suspected. Karyotyping of cultured peripheral lymphocytes or direct detection of specific Y antigens on cell surfaces can reveal the presence of the Y chromosome.

Most male reproductive disorders are linked to hypogonadotropic or hypergonadotropic states.

Endocrine factors are responsible for approximately 50% of hypogonadal or infertility cases. Those remaining are of unknown etiology or the result of injury, deformities, and environmental factors. Endocrine-related hypogonadism can be classified as hypothalamic-pituitary defects (hypogonadotropic because of the lack of LH and/or FSH), primary gonadal defects (hypergonadotropic because gonadotropins are high as a result of a lack of negative feedback from the testes), and defective androgen action (usually the result of absence of androgen receptors or 5α-reductase). Each of these is further subdivided into several categories, but only a few examples are discussed here.

Hypogonadotropic hypogonadism can be congenital, idiopathic, or acquired. The most common congenital form is Kallmann syndrome, which results from decreased or absent GnRH secretion, as mentioned earlier. It is often associated with anosmia or hyposmia and is transmitted as an autosomal-dominant trait. Patients do not undergo pubertal development and have eunuchoidal features. Plasma LH, FSH, and testosterone levels are low, and the testes are immature and have no sperm. There is no response to clomiphene, but intermittent treatment with GnRH can produce sexual maturation and full spermatogenesis.

Another category of hypogonadotropic hypogonadism, **panhypopituitarism** or **pituitary failure**, can occur before or after puberty and is usually accompanied by a deficiency of other pituitary hormones. **Hyperprolactinemia**, whether from hypothalamic disturbance or pituitary adenoma, often results in decreased GnRH production, hypogonadotropic state, impotence, and decreased libido. It can be treated with dopaminergic agonists (e.g., bromocriptine), which suppress PRL release (see Chapter 37, "Female Reproductive System"). Excess androgens can also result in suppression of the hypothalamic-pituitary axis, resulting in lower LH levels and impaired testicular function. This condition often results from **congenital adrenal hyperplasia** and increased adrenal androgen production from 21-hydroxylase (CYP21A2) deficiency (see Chapter 33, "Adrenal Gland").

Hypergonadotropic hypogonadism usually results from impaired testosterone production, which can be congenital or acquired. The most common disorder is Klinefelter syndrome, as discussed earlier.

Pseudohermaphroditism is a reproductive paradox that results from insensitivity to androgens.

Pseudohermaphroditism produces gonads of one sex and the genitalia of the other in the same person. The term *male pseudohermaphrodite* is used when a testis is present, and the term *female pseudohermaphrodite* is used when an ovary is present. The term *hermaphrodite* is reserved for the very rare cases in which both ovarian and testicular tissues are present. One of the most interesting causes of male reproductive abnormalities is the insensitivity of the end organ to androgens. The best characterized syndrome is **testicular feminization**, an X-linked recessive disorder caused by a defect in the testosterone receptor. In the classical form, patients are male pseudohermaphrodites with a female phenotype and an XY male genotype. They have abdominal testes that secrete testosterone but no other internal genitalia of either sex (see Chapter 38, "Fertilization, Pregnancy, and Fetal Development"). They commonly have female external genitalia, but with a short vagina ending in a blind pouch. Breast development is typical of a female (as a result of peripheral aromatization of testosterone), but axillary and pubic hair, which are androgen-dependent, are scarce or absent. Testosterone levels are normal or elevated, estradiol levels are above the normal male range, and circulating gonadotropin levels are high. The inguinally located testes usually have to be removed because of an increased risk of cancer. After orchiectomy, patients are treated with estradiol to maintain a female phenotype.

Chapter Summary

- In the testes, luteinizing hormone controls the synthesis of testosterone by Leydig cells, and follicle-stimulating hormone increases the production of androgen-binding protein, inhibin, and estrogen by Sertoli cells.
- Spermatozoa are produced within the seminiferous tubules of both testes. Sperm develop from spermatogonia through a series of developmental stages that include spermatocytes and spermatids.
- The sperm mature and are stored in the epididymis. At the time of ejaculation, muscular contractions of the epididymis and vas deferens move sperm through the ejaculatory ducts into the prostatic urethra. The sperm are finally moved out of the body through the urethra in the penis.
- Gonadotropin-releasing hormone controls luteinizing hormone and follicle-stimulating hormone secretion by the anterior pituitary.
- Testosterone mainly reduces luteinizing hormone secretion, whereas inhibin reduces the secretion of follicle-stimulating hormone. The testicular hormones complete a negative-feedback loop with the hypothalamic-pituitary axis.
- Androgens have several target organs and have roles in regulating the development of secondary sex characteristics, the libido, and sexual behavior.
- The most potent natural androgen is dihydrotestosterone, which the action of the enzyme 5a-reductase produces from testosterone.
- Male reproductive dysfunction is often a result of a lack of luteinizing hormone and follicle-stimulating hormone secretion or abnormal testicular morphology.

thePoint *Visit http://thePoint.lww.com for chapter review Q&A, case studies, animations, and more!*

37 Female Reproductive System

ACTIVE LEARNING OBJECTIVES

Upon mastering the material in this chapter you should be able to:

- Explain the mechanism through which pulses of hypothalamic gonadotropin-releasing hormone regulate secretion of luteinizing hormone and follicle-stimulating hormone and the effect of these two gonadotropins on follicular development, steroidogenesis, ovulation, and formation of the corpus luteum.
- Explain how luteinizing hormone and follicle-stimulating hormone, in coordination with ovarian theca and granulosa cells, regulate the secretion of follicular estradiol.
- Explain how positive feedback of follicular estradiol on the hypothalamic-pituitary axis induces luteinizing hormone and follicle-stimulating hormone surges and causes ovulation.
- Describe the sequence of distinct steps in follicular development and explain how appropriately timed luteinizing hormone and follicle-stimulating hormone surges, which induce inflammatory reactions in the graafian follicle, lead to follicular rupture (ovulation).
- Explain how withdrawal of gonadotropin support results in follicular atresia.
- Describe what is necessary for the formation of a functional corpus luteum.
- Describe the roles of estradiol and progesterone in the uterine cycle and the effect of these hormones on the uterine endometrium.
- Explain how changes in hypothalamic secretory function at the onset of puberty increase luteinizing hormone and follicle-stimulating hormone secretion, enhance ovarian function, and lead to the first ovulation.
- Explain the mechanism that results in menopause and describe the medical complications of this state for older women.

The fertility of the mature human female is cyclic. The release from the ovary of a mature female germ cell or ovum occurs at a distinct phase of the menstrual cycle. Cyclic changes in **luteinizing hormone (LH)** and **follicle-stimulating hormone (FSH)** from the pituitary gland and estradiol and progesterone from the ovaries control the secretion of ovarian steroid hormones, estradiol and progesterone, and the subsequent release of an ovum during the menstrual cycle. The cyclic changes in steroid hormone secretion cause significant changes in the structure and function of the uterus in preparing it for the reception of a fertilized ovum. At different stages of the menstrual cycle, progesterone and estradiol exert negative- and positive-feedback effects on the hypothalamus and pituitary gonadotrophs, generating the cyclic pattern of LH and FSH release characteristic of the female reproductive system. The hormonal events during the menstrual cycle are delicately synchronized; thus, stress and environmental, psychologic, and social factors can readily affect the menstrual cycle.

The female cycle is characterized by monthly bleeding, resulting from the withdrawal of ovarian steroid hormone support of the uterus, which causes shedding of the superficial layers of the uterine lining at the end of each cycle. The first menstrual cycle occurs during puberty. Menstrual cycles are interrupted during pregnancy and lactation and cease at menopause. Menstruation signifies a failure to conceive and results from regression of the corpus luteum and subsequent withdrawal of luteal steroid support of the superficial endometrial layer of the uterus.

HORMONAL REGULATION OF THE FEMALE REPRODUCTIVE SYSTEM

Unlike the male, in whom there is a continuous production of sperm and a constant secretion of testosterone, the female experiences cyclic swings in the menstrual cycle, ovulation, and secretion of sex hormones. Under normal conditions, these changes follow a 28-day cycle, but under stressful conditions, wide cyclic swings may result. These cyclic changes are brought about by the interactions of hormones on the female reproductive system. An overview of the female hormonal system is shown in Figure 37.1, and consists of the brain, pituitary, ovaries, and reproductive tract (oviduct, uterus, cervix, and vagina). In the brain, the hypothalamus produces **gonadotropin-releasing hormone** (**GnRH**; also called **luteinizing hormone–releasing hormone** [**LHRH**]), which controls the secretion of LH and FSH.

The mature ovary has two major functions: the maturation of germ cells and steroidogenesis. Each germ cell is ultimately enclosed within a follicle, a major source of steroid hormones during the menstrual cycle. At ovulation, the

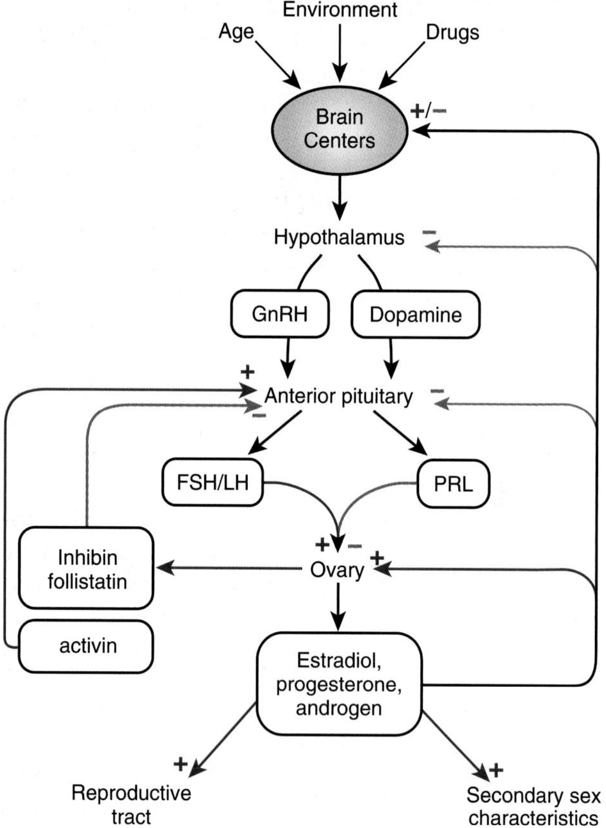

Figure 37.1 Regulation of the reproductive tract in the female. The main reproductive hormones are shown in boxes. Positive regulation is depicted by a plus sign. Negative regulation is depicted by a minus sign. FSH, follicle-stimulating hormone; GnRH, gonadotropin-releasing hormone; LH, luteinizing hormone; PRL, prolactin.

ovum or egg is released and the ruptured follicle is transformed into a corpus luteum, which secretes progesterone as its main product. FSH is primarily involved in stimulating the growth of ovarian follicles, whereas LH is involved in inducing ovulation. Both LH and FSH regulate follicular steroidogenesis and androgen and estradiol secretion, and LH regulates progesterone secretion from the corpus luteum. Ovarian steroids inhibit LH and FSH secretion with one exception: Just prior to ovulation (at midcycle), estradiol has a positive-feedback effect on the hypothalamic-pituitary axis and induces significant increases in the secretion of GnRH, LH, and FSH. The ovary also produces three polypeptide hormones: **Inhibin** suppresses the secretion of FSH. **Activin** (an inhibin-binding protein) increases the secretion of FSH, and **follistatin** (an activin-binding protein) reduces the secretion of FSH.

Shortly after fertilization, the embryo begins to develop placenta cells, which attach to the uterine lining and unite with the maternal placental cells. The **placenta** produces several pituitary-like and ovarian steroid-like hormones. These hormones support placental and fetal development throughout pregnancy and have a role in parturition. The mammary glands are also under the control of pituitary hormones and ovarian steroids and provide the baby with immunologic protection and nutritional support through lactation. **Prolactin (PRL)** from the anterior pituitary, which regulates milk production, and **oxytocin** from the posterior pituitary, which induces milk ejection from the breasts, hormonally control lactation.

Hypothalamic-pituitary axis regulates the menstrual cycle.

GnRH, a decapeptide produced in the hypothalamus and released in a pulsatile manner, controls the secretion of LH and FSH through a portal vascular system (see Chapter 31, "Hypothalamus and the Pituitary Gland"). Blockade of the portal system reduces the secretion of LH and FSH and leads to ovarian atrophy and a reduction in ovarian hormone secretion. Neurons from other brain regions regulate the secretion of GnRH by the hypothalamus. Neurotransmitters, such as epinephrine and norepinephrine, stimulate the secretion of GnRH, whereas dopamine and serotonin inhibit the secretion of GnRH. In addition, ovarian steroids and peptides and hypothalamic neuropeptides can regulate the secretion of GnRH. GnRH stimulates the pituitary gonadotrophs to secrete LH and FSH. It binds to high-affinity receptors on the gonadotrophs and stimulates the secretion of LH and FSH through a phosphoinositide–protein kinase C–mediated pathway.

LH release throughout the female life span is depicted in Figure 37.2. During the neonatal period, LH is released at

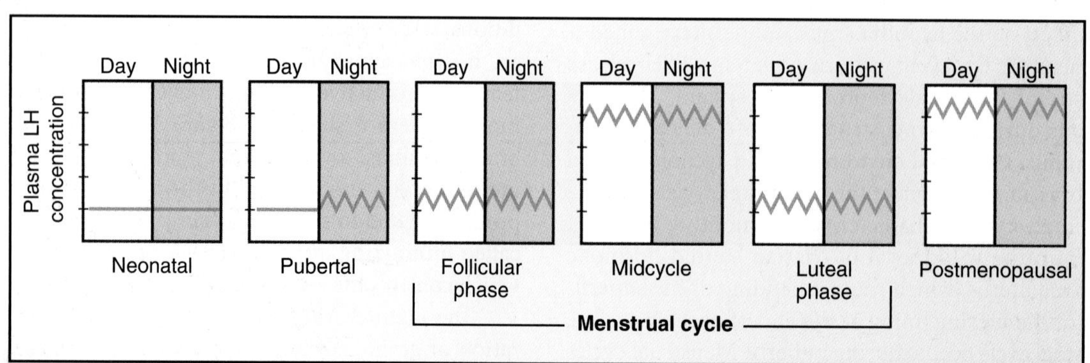

Figure 37.2 Relative levels of luteinizing hormone (LH) release in human females throughout life. Note that pulsatile LH release begins at puberty.

low and steady rates without pulsatility; this period coincides with lack of development of mature ovarian follicles and low-to-no ovarian estradiol secretion. Pulsatile release begins with the onset of puberty and, for several years, is expressed only during sleep; this period coincides with increased, but asynchronous, follicular development and with increased secretion of ovarian estradiol. On the establishment of regular functional menstrual cycles associated with regular ovulation, LH pulsatility prevails throughout the 24-hour period, changing in a monthly cyclic manner. In postmenopausal women whose ovaries lack sustained follicular development and exhibit low ovarian estradiol secretion, mean circulating LH levels are high and pulses occur at a high frequency.

FEMALE REPRODUCTIVE ORGANS

The female reproductive tract has two major components: the ovaries, which produce the mature ovum and secrete progestins, androgens, and estrogens; and the ductal system, which transports the ovum, is the place of the union of the sperm and egg, and maintains the developing conceptus until delivery. The morphology and function of these structures change in a cyclic manner under the influence of the reproductive hormones.

Female internal reproductive organs include the vagina, uterus, fallopian tubes, cervix, and ovaries.

The ovaries are in the pelvic portion of the abdominal cavity on both sides of the uterus and are anchored by ligaments (Fig. 37.3). An adult ovary weighs 8 to 12 g and consists of an outer cortex and an inner medulla, without a sharp demarcation. The cortex contains oocytes enclosed in **follicles** of various sizes, **corpora lutea, corpora albicantia**, and stromal cells. The medulla contains connective and interstitial tissues. Blood vessels, lymphatics, and nerves enter the medulla of the ovary through the hilus.

On the side that ovulates, the **oviduct** (fallopian tube) receives the ovum immediately after ovulation. The oviducts are the site of fertilization and provide an environment for development of the early embryo. The oviducts are 10 to 15 cm long and composed of sequential regions called the **infundibulum, ampulla**, and **isthmus**. The infundibulum is adjacent to the ovary and opens into the peritoneal cavity. It is trumpet-shaped with fingerlike projections, called **fimbria**, along its outer border that grasps the ovum at the time of follicular rupture. Densely ciliated projections, which facilitate ovum uptake and movement through this region, cover its thin walls. The ampulla is the site of fertilization. It has a thin musculature and well-developed mucosal surface. The isthmus is located at the uterotubal junction and has a

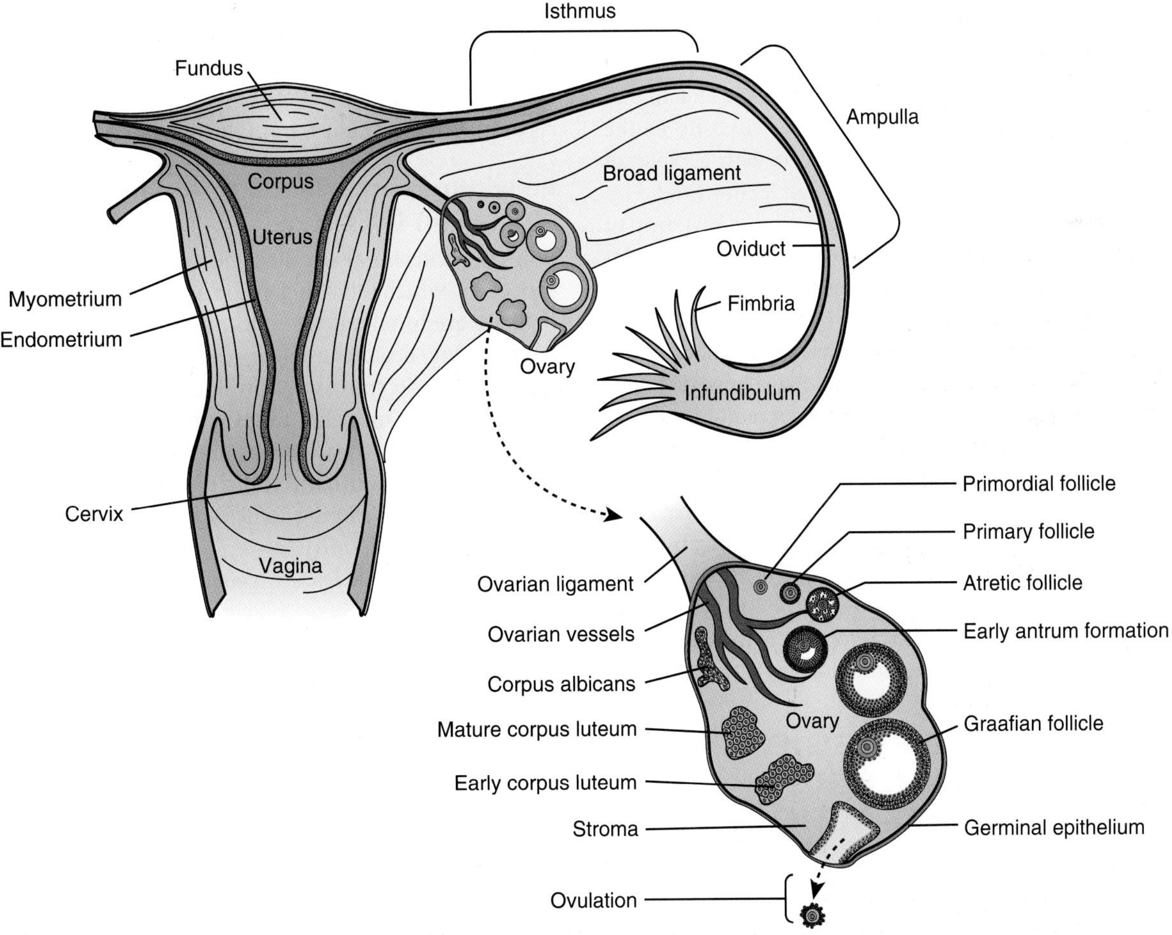

Figure 37.3 The female reproductive organs. (See text for details.)

narrow lumen surrounded by smooth muscle. It has sphincterlike properties and can serve as a barrier to the passage of germ cells. The oviducts transport the germ cells in two directions: Sperm ascend toward the ampulla and the zygote descends toward the uterus. This requires coordination between smooth muscle contraction, ciliary movement, and fluid secretion, all of which are under hormonal and neuronal control.

The **uterus** is situated between the urinary bladder and rectum. On each upper side, an oviduct opens into the uterine lumen, and on the lower side, the uterus connects to the vagina. The uterus is composed of two types of tissue: The outer part is the **myometrium**, composed of multiple layers of smooth muscle. The inner part, lining the lumen of the uterus, is the **endometrium**, which contains a deep **stromal layer** next to the myometrium and a superficial epithelial layer. The stroma is permeated by spiral arteries and contains much connective tissue. Uterine glands, which also penetrate the stromal layer and are lined by columnar secretory cells, interrupt the epithelial layer. The uterus provides an environment for the developing fetus, and eventually, the myometrium will generate rhythmic contractions that assist in expelling the fetus at delivery.

The **cervix** (neck) is a narrow muscular canal that connects the vagina and the body (corpus) of the uterus. It must dilate in response to hormones to allow the expulsion of the fetus. The cervix has numerous glands, with a columnar epithelium that produces mucus under the control of estradiol. As greater amounts of estradiol are produced during the follicular phase of the cycle, the cervical mucus changes from a scanty viscous material to a profuse, watery, and highly elastic substance called **spinnbarkeit**. The viscosity of the spinnbarkeit can be tested by touching it with a piece of paper and lifting vertically. The mucus can form a thread up to 6 cm under the influence of elevated estradiol. If a drop of the cervical mucus is placed on a slide and allowed to dry, it will form a typical **ferning** pattern when under the influence of estradiol.

The **vagina** is well innervated and has a rich blood supply. It is lined by several layers of epithelium that change histologically during the menstrual cycle. When estradiol levels are low, as during the prepubertal and postmenopausal periods, the vaginal epithelium is thin and the secretions are scanty, resulting in a dry and infection-susceptible area. Estradiol induces proliferation and **cornification (keratinization)** of the vaginal epithelium, whereas progesterone opposes those actions and induces the influx of polymorphonuclear leukocytes into the vaginal fluids. Estradiol also activates vaginal glands that produce lubricating fluid during coitus.

FOLLICULOGENESIS, STEROIDOGENESIS, ATRESIA, AND MEIOSIS

Each month, one of the ovaries releases a mature egg, known as an **oocyte**. A follicle is the functional structure in which the primary oocyte develops. They contain a single oocyte that is initiated to grow and develop into a single mature oocyte. These mature oocytes develop only once every menstrual cycle. Most follicles in the ovary undergo atresia. However, some develop into mature follicles, produce steroids, and ovulate. As follicles mature, oocytes also mature by entering meiosis, which produces the proper number of chromosomes in preparation for fertilization. After rupturing, the follicle becomes a **corpus luteum**.

Oogonia produce an oocyte, which is arrested in meiosis.

Female germ cells develop in the embryonic yolk sac and migrate to the genital ridge, where they participate in the development of the ovary (Table 37.1). Without germ cells, the ovary does not develop. The germs cells, called **oogonia**, actively divide by mitosis. Oogonia undergo mitosis only during the prenatal period. By birth, the ovaries contain a finite number of oocytes, estimated to be about 1 million. Most of them die by a process called **atresia**. By puberty, only 200,000 oocytes remain; by age 30, only 26,000 remain; and by the time of menopause, the ovaries are essentially devoid of oocytes.

When oogonia cease the process of mitosis, they are called oocytes. At that time, they enter the meiotic cycle to prepare for the production of a haploid ovum, become arrested in the prophase of the first meiotic division, and remain arrested in that phase until they either die or grow into mature oocytes at the time of ovulation. The **primordial follicle** (Fig. 37.4) is 20 μm wide and contains an oocyte, which may or may not be surrounded by a single layer of flattened (squamous) **pregranulosa cells**. When pregranulosa cells surround the oocyte, a basement membrane develops, separating the granulosa from the ovarian stroma.

Folliculogenesis leads to the maturation of a graafian follicle.

Folliculogenesis (also called *follicular development*) is the process by which follicles develop and mature (see Fig. 38.3). Follicles are in one of the following physiologic states: resting, growing, degenerating, or ready to ovulate. During each menstrual cycle, the ovaries produce a group of growing follicles, of which most fail to grow to maturity and undergo follicular atresia (death) at some stage of development. However, one **dominant follicle** generally emerges from the cohort of developing follicles and ovulates, releasing a mature haploid ovum.

Primordial follicles are generally considered the nongrowing resting pool of follicles, which gets progressively depleted throughout life. By the time of menopause, the ovaries are essentially devoid of all follicles. Primordial follicles are located in the ovarian cortex (peripheral regions of the ovary).

Progression from the primordial to the next stage of follicular development, the primary stage, occurs at a relatively constant rate throughout fetal, juvenile, prepubertal, and adult life. Once primary follicles leave the resting pool, they are committed to further development or atresia. Most become atretic, and typically only one fully developed follicle ovulates. Recent studies in genetically modified mouse models support the concept that the conversion

TABLE 37.1 Different Stages in the Development of an Ovum and Follicle

Stage	Process	Ovum	Follicle
Fetal life	Migration	Primordial germ cells	
	Mitosis	Oogonia	Primordial follicle
	First meiotic division begins	Primary oocyte	Primary follicle
Birth	Arrest in prophase	↓	↓
	Growth of oocyte and follicle	↓	↓
Puberty	Follicular maturation	↓	Secondary follicle
		↓	↓
Cycle		↓	Antral follicle
		↓	↓
Ovulation	Resumption of meiosis	Secondary oocyte	Graafian follicle
	Emission of first polar body	↓	↓
	Arrest in metaphase	↓	↓
		↓	Corpus luteum
Fertilization	Second meiotic division complete	Zygote	↓
	Emission of second polar body	↓	↓
Implantation	Mitotic divisions	Embryo	↓
	Blastocyst	↓	↓
Parturition	Body patterning	Fetus	Corpus albicans

from primordial to primary follicles is independent of pituitary gonadotropins. It is currently thought that signaling between the oocyte and surrounding granulosa cells, resulting in dramatic oocyte growth, initiates primordial follicle development. Furthermore, it appears that primordial follicles are maintained under a constant inhibitory influence that must be removed for follicle development to occur.

The first sign that a primordial follicle is entering the growth phase is a morphologic change of the flattened pregranulosa cells into cuboidal granulosa cells. The cuboidal granulosa cells proliferate to form a single continuous layer of cells surrounding the oocyte, which has enlarged from 20 μm in the primordial stage to 140 μm wide. At this stage, a glassy membrane, the **zona pellucida**, surrounds the oocyte and serves as the means of attachment through which the granulosa cells communicate with the oocyte. This is the **primary follicular stage** of development, consisting of one layer of cuboidal granulosa cells and a basement membrane.

Cells of the graafian follicle include the oocyte, granulosa cells, and the cells of the internal and external theca layers.

The follicle continues to grow, mainly through proliferation of its granulosa cells, so that several layers of granulosa cells exist in the **secondary follicular stage** of development (see Fig. 37.4). As the secondary follicle grows deeper into the cortex, stromal cells—near the basement membrane—begin to differentiate into cell layers called **theca interna** and **theca externa**, and a blood supply with lymphatics and nerves forms within the thecal component. The granulosa layer remains avascular.

The cells in the theca interna eventually become flattened, epithelioid, and steroidogenic. The granulosa cells of secondary follicles acquire receptors for FSH and start producing small amounts of estrogen. The theca externa remains fibroblastic and provides structural support to the developing follicle.

Development beyond the primary follicle is gonadotropin-dependent, begins at puberty, and continues in a cyclic manner throughout the reproductive years. As the follicle continues to grow, theca layers expand and fluid-filled spaces (or **antra**) begin to develop around the granulosa cells. This early antral stage of follicle development is referred to as the **tertiary follicular stage** (see Fig. 37.4). The critical hormone responsible for progression from the preantral to the antral stage is FSH. FSH stimulates mitosis of the granulosa cells. As the number of granulosa cells increases, the production of estrogens, the binding capacity for FSH, the size of the follicle, and the volume of the follicular fluid all increase significantly.

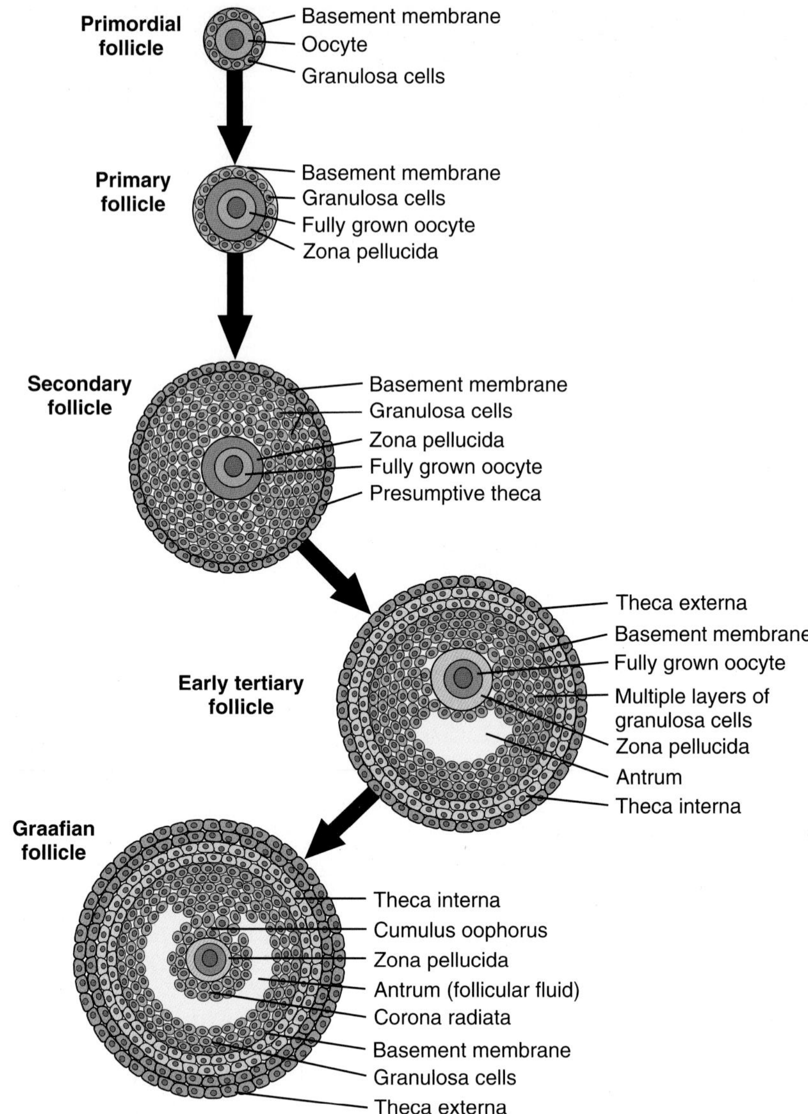

Figure 37.4 The developing follicle, from primordial through graafian. Growth from primordial to secondary follicle is independent of gonadotropins. Continued growth of the early tertiary to graafian follicle is dependent on gonadotropins.

As the antra increase in size, a single, large, coalesced antrum develops, pushing the oocyte to the periphery of the follicle and forming a large 2- to 2.5-cm–wide **graafian follicle** (preovulatory follicle; see Fig. 37.4). Three distinct granulosa cell compartments are evident in the graafian follicle. Granulosa cells surrounding the oocyte are **cumulus granulosa cells** (collectively called *cumulus oophorus*). Those cells lining the antral cavity are called **antral granulosa cells**, and those attached to the basement membrane are called **mural granulosa cells**. Mural and antral granulosa cells are more steroidogenically active than cumulus cells.

In addition to blood-borne hormones, antral follicles have a unique microenvironment in which the follicular fluid contains different concentrations of pituitary hormones, steroids, peptides, and growth factors. Some are present in the follicular fluid at a concentration 100 to 1,000 times higher than that in the circulation. Table 37.2 lists some parameters of human follicles at successive stages of development in the follicular phase. There is a 5-fold increase in follicular diameter and a 25-fold rise in the number of granulosa cells. As the follicle matures, the intrafollicular concentration of FSH does not change much, whereas that of LH increases and that of PRL declines. The concentrations of estradiol and progesterone increase 20-fold, whereas androgen levels remain unchanged.

The follicular fluid contains other substances, including inhibin, activin, GnRH-like peptide, growth factors, opioid peptides, oxytocin, and plasminogen activator. Inhibin and activin inhibit and stimulate, respectively, the release of FSH from the anterior pituitary. Granulosa cells secrete inhibin. In addition to its effect on FSH secretion, inhibin has a local effect on ovarian cells.

Both granulosa and theca cells participate in steroidogenesis.

The main physiologically active steroid produced by the follicle is **estradiol**, a steroid with 18 carbons. **Steroidogenesis**, the process of steroid hormone production, depends on the availability of cholesterol, which originates from several sources and serves as the main precursor to all of steroidogenesis. Ovarian cholesterol can come from plasma lipoproteins, *de novo* synthesis in ovarian cells, and cholesterol

TABLE 37.2 Different Parameters of Follicles During the First Half of the Menstrual Cycle

Cycle (d)	Diameter (mm)	Volume (mL)	Granulosa Cells (×10^6)	FSH	LH	PRL	A	E_2	P_4
1	4	0.05	2	2.5	—	60	800	100	—
4	7	0.15	5	2.5	—	40	800	500	100
7	12	0.50	15	3.6	2.8	20	800	1,000	300
12	20	0.50	50	3.6	2.8	5	800	2,000	2,000

Hormone concentrations in mIU or ng/ml.

FSH, follicle-stimulating hormone; LH, luteinizing hormone; PRL, prolactin; A, androstenedione; E_2, estradiol; P_4, progesterone.

esters within lipid droplets in ovarian cells. For ovarian steroidogenesis, the primary source of cholesterol is low-density lipoprotein (LDL).

The conversion of cholesterol to **pregnenolone** by **cholesterol side-chain cleavage enzyme** is a rate-limiting step, regulated by LH using the second messenger cyclic adenosine monophosphate (cAMP; Fig. 37.5). LH binds to specific membrane receptors on theca cells, activates adenylyl cyclase through a G protein, and increases the production of cAMP. cAMP increases LDL receptor mRNA, LDL cholesterol uptake, and cholesterol ester synthesis. cAMP also increases the transport of cholesterol from the outer to inner mitochondrial membrane, the site of pregnenolone synthesis, using a unique protein called **steroidogenic acute regulatory protein (StAR)**. Pregnenolone, a 21-carbon steroid of the progestin family, diffuses out of the mitochondria and enters the endoplasmic reticulum (ER), the site of subsequent steroidogenesis.

Two steroidogenic pathways may be used for subsequent steroidogenesis (see Fig. 36.9). In theca cells, the **delta 5 pathway** is predominant; in granulosa cells and the corpus luteum, the **delta 4 pathway** is predominant. Pregnenolone is converted either to **progesterone** by 3β-hydroxysteroid dehydrogenase in the δ4 pathway or to **17α-hydroxypregnenolone** by 17α-hydroxylase in the δ5 pathway. In the δ4 pathway, progesterone is converted to **17α-hydroxyprogesterone** (by 17α-hydroxylase), which is subsequently converted to **androstenedione** and **testosterone** by 17,20-lyase and 17β-hydroxysteroid dehydrogenase (17-ketosteroid reductase), respectively. In the δ5 pathway, 17α-hydroxypregnenolone is converted to **dehydroepiandrosterone** (**DHEA**; by 17,20-lyase), which is subsequently converted to androstenedione by 3β-hydroxysteroid dehydrogenase. The androgens contain 19 carbons. Testosterone and androstenedione diffuse from the thecal compartment, cross the basement membrane, and enter the granulosa cells.

In the granulosa cell, under the influence of FSH, with cAMP as a second messenger, testosterone and androstenedione are then converted to estradiol and **estrone**, respectively, by the enzyme aromatase, which aromatizes the A ring of the steroid and removes one carbon (see Fig. 37.5; see also Fig. 36.9). Estrogens typically have 18 carbons. Estrone can then be converted to estradiol by 17β-hydroxysteroid dehydrogenase in granulosa cells.

In summary, estradiol secretion by the follicle requires cooperation between granulosa and theca cells and coordination between FSH and LH. An understanding of this **two-cell, two-gonadotropin hypothesis** requires recognition that the actions of FSH are restricted to granulosa cells because all other ovarian cell types lack FSH receptors. The

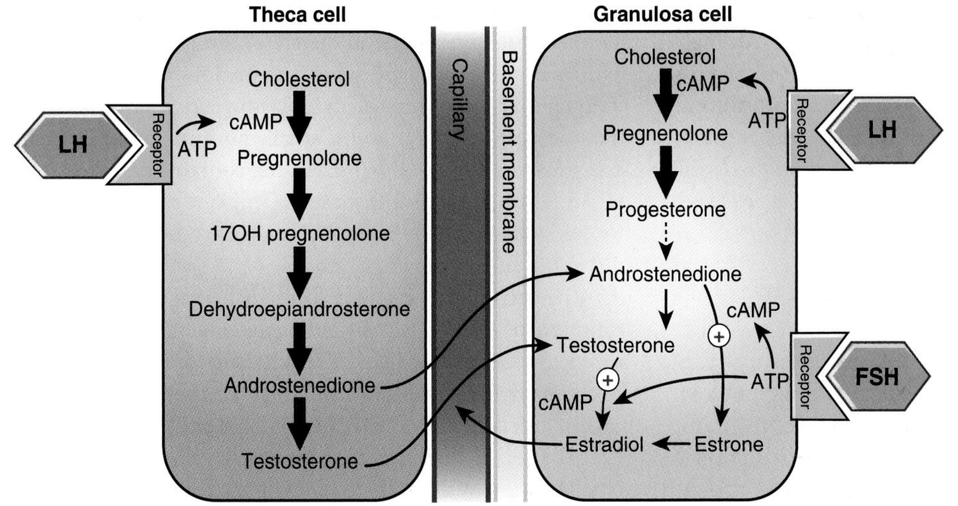

Figure 37.5 The two-cell, two-gonadotropin hypothesis. The follicular theca cells, under the control of luteinizing hormone (LH), produce androgens that diffuse to the follicular granulosa cells, where they are converted to estrogens via a follicle-stimulating hormone (FSH)–supported aromatization reaction. The *dashed arrow* indicates that granulosa cells cannot convert progesterone to androstenedione because of the lack of the enzyme 17α-hydroxylase.

actions of LH, on the other hand, are exerted on theca, granulosa, and stromal (interstitial) cells and the corpus luteum. The expression of LH receptors is time-dependent because theca cells acquire LH receptors at a relatively early stage, whereas FSH induces LH receptors on granulosa cells in the later stages of the maturing follicle.

The biosynthetic enzymes are differentially expressed in the two cells. Aromatase is expressed only in granulosa cells, and FSH regulates its activation and induction. Granulosa cells are deficient in 17α-hydroxylase and cannot proceed beyond the C-21 progestins to generate C-19 androgenic compounds (see Fig. 37.5). Consequently, estrogen production by granulosa cells depends on an adequate supply of exogenous aromatizable androgens, provided by theca cells. Under LH regulation, theca cells produce androgenic substrates, primarily androstenedione and testosterone, which reach the granulosa cells by diffusion. Aromatization then converts the androgens to estrogens.

In follicles, theca and granulosa cells are exposed to different microenvironments. Vascularization is restricted to the theca layer because blood vessels do not penetrate the basement membrane. Theca cells, therefore, have better access to circulating cholesterol, which enters the cells via LDL receptors. Granulosa cells, on the other hand, primarily produce cholesterol from acetate, a less efficient process than uptake. In addition, granulosa cells are bathed in follicular fluid and exposed to autocrine, paracrine, and juxtacrine control by locally produced peptides and growth factors. "Juxtacrine" describes the interaction of a membrane-bound growth factor on one cell with its membrane-bound receptor on an adjacent cell.

FSH acts on granulosa cells by a cAMP-dependent mechanism and produces a broad range of activities, including increased mitosis and cell proliferation, the stimulation of progesterone synthesis, the induction of aromatase, and the increased inhibin synthesis. As the follicle matures, the number of receptors for both gonadotropins increases. FSH stimulates the formation of its own receptors and induces the appearance of LH receptors. The combined activity of the two gonadotropins greatly amplifies estrogen production.

Theca and stromal cells produce androgens, which serve as precursors for estrogen synthesis and also have a distinct local action. At low concentrations, androgens enhance aromatase activity, promoting estrogen production. At high concentrations, androgens are converted by 5α-reductase to a more potent androgen, such as dihydrotestosterone (DHT). When androgens overwhelm follicles, the intrafollicular androgenic environment antagonizes granulosa cell proliferation and leads to apoptosis of the granulosa cells and subsequent follicular atresia.

Lack of gonadotropin support leads to follicular atresia.

Follicular atresia, the degeneration of follicles in the ovary, is characterized by the destruction of the oocyte and granulosa cells. Atresia is a continuous process and can occur at any stage of follicular development. During a woman's lifetime, approximately 400 to 500 follicles ovulate; those are the only follicles that escape atresia, and they represent a small percentage of the 1 to 2 million follicles present at birth. The cause of follicular atresia is likely a lack of gonadotropin support of the growing follicle. For example, at the beginning of the menstrual cycle, several follicles are selected for growth but only one follicle—the dominant follicle—goes on to ovulate. Because the dominant follicle has a preferential blood supply, it gets the most FSH (and LH). Other reasons for the lack of gonadotropin support of nondominant follicles could be a lack of FSH and LH receptors or the inability of granulosa cells to transduce the gonadotropin signals.

During atresia, granulosa cell nuclei become apoptotic and/or the oocyte undergoes **pseudomaturation**, characteristic of meiosis. During the early stages of oocyte death, the nuclear membrane disintegrates, the chromatin condenses, and the chromosomes form a metaphase plate with a spindle. The term *pseudomaturation* is appropriate because these oocytes are not capable of successful fertilization. During atresia of follicles containing theca cells, the theca layer may undergo hyperplasia and hypertrophy and may remain in the ovary for extended periods of time.

Meiosis resumes during the periovulatory period.

All healthy oocytes in the ovary remain arrested in prophase of the first meiosis. When a graafian follicle is subjected to a surge of gonadotropins (LH and FSH), the oocyte within undergoes the final stages of meiosis, resulting in the production of a mature gamete. This maturation is accomplished by two successive cell divisions in which the number of chromosomes is reduced, producing haploid gametes. At fertilization, the diploid state is restored.

Primary oocytes arrested in meiotic prophase 1 (of the first meiosis) duplicate their centrioles and DNA ($4n$ DNA) so that each chromosome has two identical **chromatids**. Crossing over and chromatid exchange occur during this phase, producing genetic diversity. The resumption of meiosis, ending meiotic prophase 1 and beginning meiotic metaphase 1, is characterized by disappearance of the nuclear membrane, condensation of the chromosomes, nuclear dissolution (germinal vesicle breakdown), and alignment of the chromosomes on the equator of the spindle. At meiotic anaphase 1, the homologous chromosomes move in opposite directions under the influence of the retracting meiotic spindle at the cellular periphery. At meiotic telophase 1, an unequal division of the cell cytoplasm yields a large **secondary oocyte** ($2n$ DNA) and a small, nonfunctional cell, the **first polar body** ($2n$ DNA). Each cell contains half the original $4n$ number of chromosomes. (Only one member of each homologous pair is present, but each chromosome consists of two unique chromatids.)

The secondary oocyte is formed several hours after the initiation of the LH surge, but before ovulation. It rapidly begins the second meiotic division and proceeds through a short prophase to become arrested in metaphase. At this stage, the secondary oocyte is expelled from the graafian follicle. The second arrest period is relatively short. In response to penetration by a spermatozoon during fertilization, second meiotic division resumes and is rapidly completed. A second unequal cell division soon follows, producing a small **second polar body** (1*n* DNA) and a large fertilized egg, the **zygote** (2*n* DNA, 1*n* from the mother and 1*n* from the father). The first and second polar bodies either degenerate or divide, yielding small nonfunctional cells. If fertilization does not occur, the secondary oocyte begins to degenerate within 24 to 48 hours.

FOLLICLE SELECTION AND OVULATION

The number of ovulating eggs is species-specific and is influenced by genetic, nutritional, and environmental factors. Usually in humans, only one follicle ovulates, but the timed administration of gonadotropins or antiestrogens can induce multiple ovulations in a single cycle (superovulation). The mechanism by which one follicle is selected from a cohort of growing follicles is poorly understood. It occurs during the first few days of the cycle, immediately after the onset of menstruation. Once selected, the follicle begins to grow and differentiate at an exponential rate and becomes the dominant follicle.

Dominant follicle is protected from atresia during follicle-stimulating hormone suppression.

In parallel with the growth of the dominant follicle, the rest of the preantral follicles undergo atresia. Two main factors contribute to atresia in the nonselected follicles. One is the suppression of plasma FSH in response to increased estradiol secretion by the dominant follicle. The decline in FSH support decreases aromatase activity and estradiol production and interrupts granulosa cell proliferation in those nondominant follicles. The dominant follicle is protected from a fall in circulating FSH levels because it has a healthy blood supply, FSH accumulated in the follicular fluid, and an increased density of FSH receptors on its granulosa cells. Another factor in selection is the accumulation of atretogenic androgens, such as DHT, in the nonselected follicles. The increase in DHT changes the intrafollicular ratio of estrogen to androgen and antagonizes the actions of FSH.

As the dominant follicle grows, vascularization of the theca layer increases. On day 9 or 10 of the cycle, the vascularity of the dominant follicle is twice that of the other antral follicles, permitting a more efficient delivery of cholesterol to theca cells and better exposure to circulating gonadotropins. At this time, the main source of circulating estradiol is the dominant follicle. Because estradiol is the primary regulator of LH and FSH secretion by positive and negative feedback, the dominant follicle ultimately determines its own fate.

Luteinizing hormone surge leads to maturation of dominant follicle.

The midcycle LH surge occurs as a result of rising levels of circulating estradiol and causes multiple changes in the dominant follicle, which occur within a relatively short time. These include the following: the resumption of meiosis in the oocyte (as already discussed); granulosa cell differentiation and transformation into luteal cells; the activation of proteolytic enzymes that degrade the follicle wall and surrounding tissues; increased production of prostaglandins, histamine, and other local factors that cause localized hyperemia; and an increase in progesterone secretion. Within 30 to 36 hours after the onset of the LH surge, this coordinated series of biochemical and morphologic events culminates in follicular rupture and ovulation. The midcycle FSH surge is not essential for ovulation because an injection of either LH or **human chorionic gonadotropin (hCG)** before the endogenous gonadotropin surge can induce normal ovulation. However, only follicles that have been adequately primed with FSH ovulate because they contain sufficient numbers of LH receptors for ovulation and subsequent luteinization. Four ovarian proteins are essential for ovulation: the progesterone receptor, the cyclooxygenase enzyme (which converts arachidonic acid to prostaglandins), cyclin D2 (a cell-cycle regulator), and a transcription factor called *C/EBPβ* (CCAAT/enhancer-binding protein). The mechanisms by which these proteins interact to regulate follicular rupture are largely unknown. However, mice with specific disruption of genes for any of these proteins fail to ovulate, and these proteins are likely to have a functional role in human ovulation.

Luteinizing hormone surge causes ovulation.

The earliest responses of the ovary to the midcycle LH surge are the release of vasodilatory substances, such as histamine, bradykinin, and prostaglandins, which mediate increased ovarian and follicular blood flow. The highly vascularized dominant follicle becomes hyperemic and edematous and swells to a size of at least 20 to 25 mm wide. There is also an increased production of follicular fluid, disaggregation of granulosa cells, and detachment of the oocyte–cumulus complex from the follicular wall, moving it to the central portion of the follicle. The basement membrane separating theca cells from granulosa cells begins to disintegrate, granulosa cells begin to undergo luteinization, and blood vessels begin to penetrate the granulosa cell compartment.

Just prior to follicular rupture, the follicular wall thins by cellular deterioration and bulges at a specific site called the **stigma**, the point on the follicle that actually ruptures. As ovulation approaches, the follicle enlarges and protrudes from the surface of the ovary at the stigma. In response to the LH surge, **plasminogen activator** is produced by theca and granulosa cells of the dominant follicle and converts plasminogen to **plasmin**. Plasmin is a proteolytic enzyme that acts directly on the follicular wall and stimulates the production of **collagenase**, an enzyme that digests the connective tissue matrix. The thinning and increased distensibility of the wall facilitate the rupture of the follicle. Smooth muscle contraction aids the extrusion of the oocyte–cumulus complex. At the time of rupture, the oocyte–cumulus complex and follicular fluid are ejected from the follicle.

The LH surge triggers the resumption of the first meiosis. Up to this point, unknown factors within the follicle have protected the primary oocyte from premature cell division. The LH surge also causes transient changes in plasma estradiol and a prolonged increase in plasma progesterone concentrations. Within a couple of hours after the initiation of the LH surge, the production of progesterone, androgens, and estrogens begins to increase. Progesterone, acting through the progesterone receptor on granulosa cells, promotes ovulation by releasing mediators that increase the distensibility of the follicular wall and enhance the activity of proteolytic enzymes. As LH levels reach their peak, plasma estradiol levels plunge because of down-regulation by LH of FSH receptors on granulosa cells and the inhibition of granulosa cell aromatase. Eventually, LH receptors on luteinizing granulosa cells escape the down-regulation, and progesterone production increases.

Corpus luteum forms from the postovulatory follicle.

In response to the LH and FSH surges and after ovulation, the wall of the graafian follicle collapses and becomes convoluted, blood vessels course through the luteinizing granulosa and theca cell layers, and the antral cavity fills with blood. The granulosa cells begin to cease their proliferation, begin to undergo hypertrophy, and produce progesterone as their main secretory product. The ruptured follicle develops a rich blood supply and forms a solid structure called the corpus luteum (Latin for "yellow body"). The mature corpus luteum develops as the result of numerous biochemical and morphologic changes, collectively referred to as **luteinization**. The granulosa cells and theca cells in the corpus luteum are called **granulosa lutein cells** and **theca lutein cells**, respectively.

Continued stimulation by LH is needed to ensure morphologic integrity (healthy luteal cells) and functionality (progesterone secretion). If pregnancy does not occur, the corpus luteum regresses, a process called **luteolysis** or **luteal regression**. Luteolysis occurs as a result of apoptosis and necrosis of the luteal cells. After degeneration, fibrous tissue replaces the luteinized cells, creating a nonfunctional structure, the **corpus albicans**. Therefore, the corpus luteum is a transient endocrine structure formed from the postovulatory follicle. It serves as the main source of circulating steroids during the luteal (postovulatory) phase of the cycle and is essential for maintaining pregnancy during the first trimester as well as maintaining menstrual cycles of normal length.

The process of luteinization begins before ovulation. After acquiring a high concentration of LH receptors, granulosa cells respond to the LH surge by undergoing morphologic and biochemical transformation. This change involves cell enlargement (hypertrophy) and the development of smooth ER and lipid inclusions, typical of steroid-secreting cells. Unlike the nonvascular granulosa cells in the follicle, luteal cells have a rich blood supply. Invasion by capillaries starts immediately after the LH surge and is facilitated by the dissolution of the basement membrane between theca and granulosa cells. Peak vascularization is reached 7 to 8 days after ovulation.

Differentiated theca and stroma cells as well as granulosa cells are incorporated into the corpus luteum, and all three classes of steroids—androgens, estrogens, and progestins—are synthesized. Although some progesterone is secreted before ovulation, peak progesterone production is reached 6 to 8 days after the LH surge. The life span of the corpus luteum is limited. Unless pregnancy occurs, it degenerates within about 13 days after ovulation. During the menstrual cycle, LH maintains the function of the corpus luteum. The lack of LH can lead to luteal insufficiency (see Clinical Focus Box 37.1).

Clinical Focus / 37.1

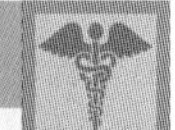

Luteal Insufficiency

Occasionally, the corpus luteum does not produce sufficient progesterone to maintain pregnancy during its early stages. Initial signs of early spontaneous pregnancy termination include pelvic cramping and the detection of blood, similar to indications of menstruation. If the corpus luteum is truly deficient then fertilization may occur around the idealized day 14 (ovulation), pregnancy will terminate during the deficient luteal phase, and menses will start on schedule. Without measuring levels of human chorionic gonadotropin (hCG), the pregnancy detection hormone, the woman would not know that she is pregnant because of the continuation of regular menstrual cycles. **Luteal insufficiency** is a common cause of infertility. Women are advised to see their physician if pregnancy does not result after 6 months of unprotected intercourse.

Analysis of the regulation of progesterone secretion by the corpus luteum provides insights into this clinical problem. There are several reasons for luteal insufficiency. First, the number of luteinized granulosa cells in the corpus luteum may be insufficient because of the ovulation of a small follicle or the premature ovulation of a follicle that was not fully developed. Second, the number of luteinizing hormone (LH) receptors on the luteinized granulosa cells in the graafian follicle and developing corpus luteum may be insufficient. LH receptors mediate the action of LH, which stimulates progesterone secretion. An insufficient number of LH receptors could be a result of insufficient priming of the developing follicle with follicle-stimulating hormone (FSH). It is well known that FSH increases the number of LH receptors in the follicle. Third, the LH surge could have been inadequate in inducing full luteinization of the corpus luteum, yet there was sufficient LH to induce ovulation. It has been estimated that only 10% of the LH surge is required for ovulation, but the amount required for full luteinization and adequate progesterone secretion to maintain pregnancy is not known.

If progesterone values are low in consecutive cycles at the midluteal phase and do not match endometrial biopsies, exogenous progesterone may be administered to prevent early pregnancy termination during a fertile cycle. Other options include the induction of follicular development and ovulation with clomiphene and hCG. This treatment would likely produce a large, healthy, estrogen-secreting graafian follicle with sufficient LH receptors for luteinization. The exogenous hCG is given to supplement the endogenous LH surge and to ensure full stimulation of the graafian follicle, ovulation, adequate progesterone, and luteinization of the developing corpus luteum.

Regression of the corpus luteum at the end of the cycle is not understood. Luteal regression is thought to be induced by locally produced luteolytic agents that inhibit LH action. Several ovarian hormones, such as estrogen, oxytocin, prostaglandins, and GnRH, have been proposed, but the exact function of each is controversial. The action of hCG, an LH-like hormone that is produced by the embryonic trophoblast during the implantation phase (see Chapter 38, "Fertilization, Pregnancy, and Fetal Development"), rescues the corpus luteum from degeneration in the late luteal phase. This hormone binds the LH receptor and increases cAMP and progesterone secretion.

MENSTRUAL CYCLE

The **menstrual cycle** describes the coordinated monthly changes in the ovary and endometrium (lining of the uterus) in women of reproductive age. The menstrual cycle is physically noted by the onset of menstruation, the flow of blood from the uterus through the vagina, when the lining of the uterus is shed. A woman's first menstruation, termed **menarche**, occurs around age 12 years. The end of a woman's reproductive phase is called **menopause**, which commonly occurs between ages 45 and 55 years.

The average menstrual cycle length in adult women is 28 days, with a range of 25 to 35 days. The interval from ovulation to the onset of menstruation is relatively constant, averaging 14 days in most women, and is dictated by the fixed lifespan of the corpus luteum. In contrast, the interval from the onset of menses to ovulation (the follicular phase) is more variable and accounts for differences in cycle lengths among ovulating women. Sexual intercourse may occur at any time during the cycle, but fertilization occurs only during the postovulatory period. Once pregnancy occurs, ovulation ceases, and after parturition, lactation also inhibits ovulation.

Menstrual cycles become irregular as menopause approaches at around age 50 years, and cycles cease thereafter. During the reproductive years, menstrual cycling is subjected to modulation by physiological, psychological, and social factors.

Puberty marks the start of cyclic reproductive function.

During the prepubertal period, the hypothalamic-pituitary-ovarian axis becomes activated—an event known as **gonadarche**—and gonadotropins increase in the circulation and stimulate ovarian estrogen secretion. The increase in gonadotropins results from increased secretion of GnRH. Factors that stimulate the secretion of GnRH include glutamate, norepinephrine, and neuropeptide Y emanating from synaptic inputs to GnRH-producing neurons. In addition, a decrease in γ-aminobutyric acid (GABA), an inhibitor of GnRH secretion, also occurs at this time. Further, the response of the pituitary to GnRH increases at the time of **puberty**. Collectively, numerous factors control the rise in ovarian estradiol secretion that triggers the development of physical characteristics of sexual maturation.

Estradiol induces the development of **secondary sex characteristics**, including the breasts and reproductive tract and increased fat on the hips and buttocks. Estrogens also regulate the growth spurt at puberty, induce closure of the epiphyses, have a positive effect in maintaining bone formation, and can antagonize the degrading actions of parathyroid hormone on bone. Therefore, estrogens have a positive effect on bone maintenance, and later in life, exogenous estrogens oppose the osteoporosis often associated with menopause.

As mentioned earlier, the first menstruation is called menarche and occurs around age 11–12 years. The first few menstrual cycles are usually irregular and anovulatory, as the result of delayed maturation of the positive feedback by estradiol on a hypothalamus that fails to secrete significant GnRH. During puberty, LH secretion occurs more during periods of sleep than during periods of wakefulness, resulting in a diurnal cycle.

During the pubertal period, the development of breasts, under the influence of estrogen, is known as **thelarche**. At this time, the appearance of axillary and pubic hair occurs, a development known as **pubarche**, controlled by adrenal androgens. The adrenals begin to produce significant amounts of androgens (dehydroepiandrosterone and androstenedione) 4 to 5 years prior to menarche, and this event is called **adrenarche**. The adrenal androgens are responsible in part for pubarche. Adrenarche is independent of gonadarche.

Menstrual cycle comprises four phases and requires synchrony among the ovary, brain, and pituitary.

The four phases of the menstrual cycle are illustrated in Figure 37.6. The **menstrual phase**, also called **menses** or **menstruation**, is the bleeding phase and lasts about 5 days. The ovarian **follicular phase** lasts about 10 to 16 days; follicle development occurs, estradiol secretion increases, and the uterine endometrium undergoes proliferation in response to rising estrogen levels. The **ovulatory phase** lasts 24 to 48 hours, and the **luteal phase** lasts 14 days. In the luteal phase, progesterone is produced, and the endometrium secretes numerous proteins in preparation for implantation of an embryo.

The menstrual cycle requires several coordinated elements: hypothalamic control of pituitary function, follicular and luteal changes in the ovary, and positive and negative feedback of ovarian hormones at the hypothalamic-pituitary axis. We have discussed separately the mechanisms that regulate the synthesis and release of the reproductive hormones; now we put them together in terms of sequence and interaction. For this purpose, we use a hypothetical cycle of 28 days (see Fig. 37.6), divided into four phases as follows: menstrual (days 0–5), follicular (days 0–13), ovulatory (days 13–14), and luteal (days 14–28).

Menstrual phase

During menstruation, estrogen, progesterone, and inhibin levels are low as a result of the luteal regression that has just occurred and the low estrogen synthesis by immature follicles. The plasma FSH levels are high whereas LH levels

Clinical Focus / 37.2

Precocious Puberty

The appearance of any sign of secondary sexual maturation at an age more than 2 standard deviations below the mean is referred to as *precocious puberty*. Because the average age of pubertal onset has gradually declined as a result in part to better nutrition and general health, present data indicate that the lower limit of puberty in normal boys is age 9 years, but that the lower limit for white girls is age 7 years and for black girls is age 6 years. Thus, signs of secondary sexual maturation earlier than these lower limits should be evaluated.

Two types of sexual precocity are recognized. If there is evidence of premature reactivation of the hypothalamic luteinizing hormone–releasing hormone (LHRH) pulse generator, the condition is referred to as *complete isosexual precocity*, *true precocious puberty*, or *central precocious puberty*. In this condition, there is pulsatile release of luteinizing hormone (LH) that mimics that seen in puberty, and LH increases in response to administration of LHRH. If extrapituitary gonadotropin secretion or gonadal steroid secretion occurs independent of pulsatile LHRH generation, the condition is termed *incomplete isosexual precocity*, *pseudoprecocious puberty*, or *LHRH-independent precocious puberty*. In this condition, there is no pulsatile release of LH as seen in puberty, and LH does not increase in response to administration of LHRH. In both forms of sexual precocity, increased gonadal steroid secretion results in an increased rate of linear growth, somatic development, and rate of skeletal maturation.

Hamartomas, congenital malformations composed of a heterotropic mass of nerve tissue usually located on the floor of the third ventricle or attached to the tuber cinereum, appear to be a major cause of central precocious puberty. Hamartomas are not neoplastic and do not progress or enlarge. LHRH-releasing cells have been found in hamartomas and appear to release LHRH in a pulsatile manner unrestrained by the central nervous system (CNS) mechanisms that inhibit the normal LHRH pulse generator. It has also been suggested that hamartomas release factors such as transforming growth factor α (TGF-α) that drive LHRH secretion from normal neurons in the hypothalamus. Central precocious puberty can also result from gliomas, CNS infection, CNS radiation, and head trauma.

LHRH agonists are the treatment of choice for central precocious puberty after the possibility of an expanding intracranial lesion has been eliminated. Chronic administration of LHRH analogs results in suppression of pulsatile LH and FSH release, gonadal steroidogenesis, and gametogenesis. This occurs because the constant presence of LHRH results in down-regulation of its receptors and uncoupling of the receptor from the intracellular signal transduction mechanism.

LHRH-independent precocious puberty results from inappropriate gonadal or adrenal steroid secretion. Congenital adrenal hyperplasia (see Chapter 33, "Adrenal Gland") leads to elevated androgens and masculinization of boys and girls. McCune–Albright syndrome, resulting from gain of function mutations in activating G proteins coupled to gonadotropin receptors, results in increased ovarian estrogen synthesis in girls and increased testicular testosterone synthesis in boys. In boys, human chorionic gonadotropin (hCG)–secreting tumors can result in precocious puberty because hCG has LH-like activity. (Serum LH and FSH are not elevated.) hCG-secreting tumors do not result in signs of precocity in girls, unless the tumor also secretes estrogen. Estrogen-secreting follicular cysts are the most common ovarian cause of precocity in girls.

Treatment of LHRH-independent precocious puberty centers on inhibition of steroidogenesis. Medroxyprogesterone acetate, which has a direct suppressive effect on gonadal steroidogenesis, is commonly prescribed for both boys and girls.

are low in response to the removal of negative feedback by estrogen, progesterone, and inhibin. A few days later, however, LH levels slowly begin to rise. FSH acts on a cohort of follicles recruited 20 to 25 days earlier from a resting pool of smaller follicles. The follicles on days 3 to 5 average 4 to 6 mm wide and are stimulated by FSH to grow into the preantral stages. In response to FSH, the granulosa cells proliferate, aromatase activity increases, and plasma estradiol levels rise slightly between days 3 and 7. The designated dominant follicle is selected between days 5 and 7 and increases in size and steroidogenic activity. Between days 8 and 10, plasma estradiol levels rise sharply, reaching peak levels above 200 pg/mL on day 12, the day before the LH surge.

Follicular phase

During the early follicular phase, LH pulsatility is of low amplitude and high frequency (about every hour). Coinciding pulses of GnRH are released about every hour. As estradiol levels rise, the pulse frequency in GnRH further increases, without a change in amplitude. The mean plasma LH level increases and further supports follicular steroidogenesis, especially because FSH has increased the number of LH receptors on growing follicles. During the midfollicular to late follicular phase, rising estradiol and inhibin from the dominant follicle suppress FSH release. The decline in FSH, together with an accumulation of nonaromatizable androgens, induces atresia in the nonselected follicles. The dominant follicle is saved by virtue of its high density of FSH receptors, the accumulation of FSH in its follicular fluid (see Table 37.2), and the acquisition of LH receptors by the granulosa cells.

Ovulatory phase

The midcycle surge of LH is short (24–36 hours) and is an example of positive feedback. For the LH surge to occur, estradiol must be maintained at a critical concentration (about 200 pg/mL) for a sufficient duration (36–48 hours) prior to the surge. Any reduction of the estradiol rise or a rise that is too small or too short eliminates or reduces the LH surge. In addition, in the presence of elevated progesterone, high concentrations of estradiol do not induce an LH surge. Paradoxically, although it exerts

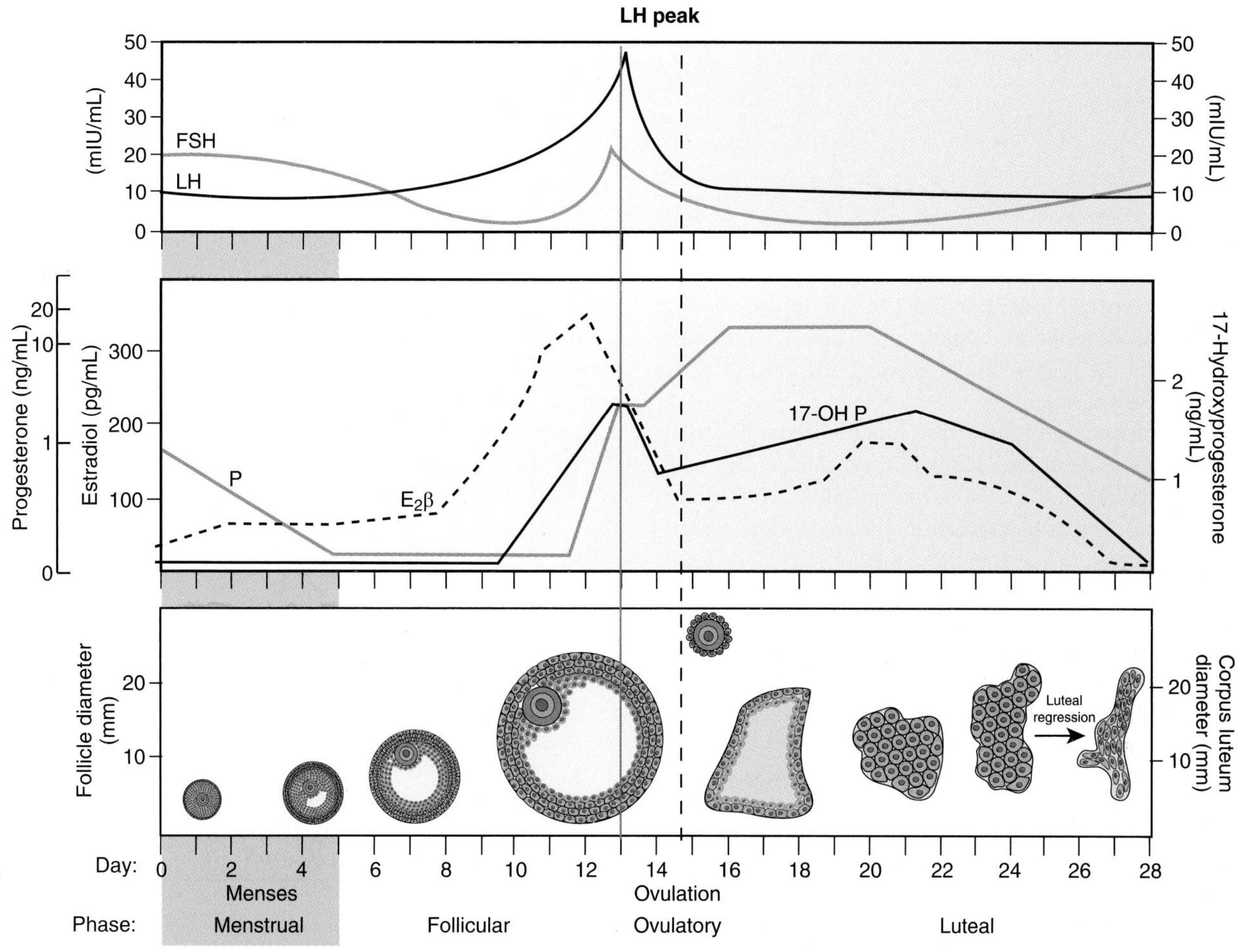

Figure 37.6 Hormonal and ovarian events during the menstrual cycle. 17-OH P, 17-hydroxyprogesterone; $E_2\beta$, estradiol; LH, luteinizing hormone; P, progesterone.

negative feedback on LH release most of the time, positive feedback by estradiol is required to generate the midcycle surge.

Estrogen exerts its effects directly on the anterior pituitary, with GnRH playing a permissive, albeit mandatory, role. This concept is derived from experiments in monkeys, whose medial basal hypothalamus, including the GnRH-producing neurons, was destroyed by lesioning, resulting in a marked decrease in plasma LH levels. The administration of exogenous GnRH at a fixed frequency restored LH release. When estradiol was given at an optimal concentration for an appropriate time, an LH surge was generated, in spite of maintaining steady and unchanging pulses of GnRH.

The mechanism that transforms estradiol from a negative to a positive regulator of LH release is unknown. One factor involves an increase in the number of GnRH receptors on the gonadotrophs, increasing pituitary responsiveness to GnRH. Another factor is the conversion of a storage pool of LH (perhaps within a subpopulation of gonadotrophs) to a readily releasable pool. Estrogen may also increase GnRH release, serving as a fine-tuning or fail–safe mechanism. A small but distinct rise in progesterone occurs before the LH surge. This rise is important for augmenting the LH surge and, together with estradiol, promotes a concomitant surge in FSH. There are indications that the midcycle FSH surge is important for inducing enough LH receptors on granulosa cells for luteinization, stimulating plasminogen activator for follicular rupture and oocyte release, and activating a cohort of follicles destined to develop in the next cycle.

The LH surge reduces the concentration of 17α-hydroxylase and subsequently decreases androstenedione production by the dominant follicle. Estradiol levels decline, 17-hydroxyprogesterone increases, and progesterone levels plateau. The prolonged exposure to high LH levels during the surge down-regulates the ovarian LH receptors, accounting for the immediate postovulatory suppression of estradiol. As the corpus luteum matures, it increases progesterone production and reinitiates estradiol secretion. Both reach high plasma concentrations on days 20 to 23, about 1 week after ovulation.

Some women experience pain around the time of ovulation. This condition is referred to as **mittelschmerz** or midcycle pain. Mittelschmerz is characterized by lower abdominal, pelvic, and/or lower back pain that can appear suddenly and usually subsides within hours, although in some cases the pain can last up to 2 to 3 days. It has been suggested that spillage of fluid from the ruptured follicle may irritate the peritoneum.

Luteal phase

During the luteal phase, the elevated steroids suppress circulating FSH levels. The LH pulse frequency is reduced during the early luteal phase, but the amplitude is higher than that during the follicular phase. LH is important at this time for maintaining corpus luteum function and sustaining steroid production. In the late luteal phase, a progesterone-dependent, opioid-mediated suppression of the GnRH pulse generator reduces both LH pulse frequency and amplitude.

After the demise of the corpus luteum on days 24 to 26, estradiol and progesterone levels plunge, causing the withdrawal of support of the uterine endometrium, culminating within 2 to 3 days in menstruation. The reduction in ovarian steroids acts centrally to remove feedback inhibition. The FSH level begins to rise and a new cycle is initiated.

Estradiol and progesterone influence cyclic changes comprising four phases in the reproductive tract.

The female reproductive tract undergoes cyclic alterations in response to the changing levels of ovarian steroids. The most notable changes occur in the function and histology of the oviduct and uterine endometrium, the composition of cervical mucus, and the cytology of the vagina (Fig. 37.7). At the time of ovulation, there is also a small but detectable rise in basal body temperature, caused by progesterone. All of the above parameters are clinically useful for diagnosing menstrual dysfunction and infertility.

The oviduct is a muscular tube lined internally with a ciliated, secretory, columnar epithelium with a deeper stromal tissue. Fertilization occurs in the oviduct, after which the zygote enters the uterus; therefore, the oviduct is involved in transport of the gametes and provides a site for fertilization and early embryonic development. Estrogens maintain the ciliated nature of the epithelium, and ovariectomy causes a loss of the cilia. Estrogens also increase the motility of the oviducts. Exogenous estrogen given around the time of fertilization can cause premature expulsion of the fertilized egg, whereas extremely high doses of estrogen can cause "tube locking"—the entrapment of the fertilized egg and an ectopic pregnancy. Progesterone opposes these actions of estrogen.

The endometrium (also called *uterine mucosa*) is composed of a superficial layer of epithelial cells and an underlying stromal layer. The epithelial layer contains glands that penetrate the stromal layer. A secretory columnar epithelium lines the glands.

The **endometrial cycle** consists of four phases, described in detail below.

Proliferative phase

The **proliferative phase** coincides with the midfollicular to late follicular phase of the menstrual cycle. Under the influence of the rising plasma estradiol concentration, the stromal and epithelial layers of the uterine endometrium undergo hyperplasia and hypertrophy and increase in size and thickness. The endometrial glands elongate and are lined with columnar epithelium. The endometrium becomes vascularized, and more spiral arteries—a rich blood supply to this region—develop. Estradiol also induces the formation of progesterone receptors and increases myometrial excitability and contractility.

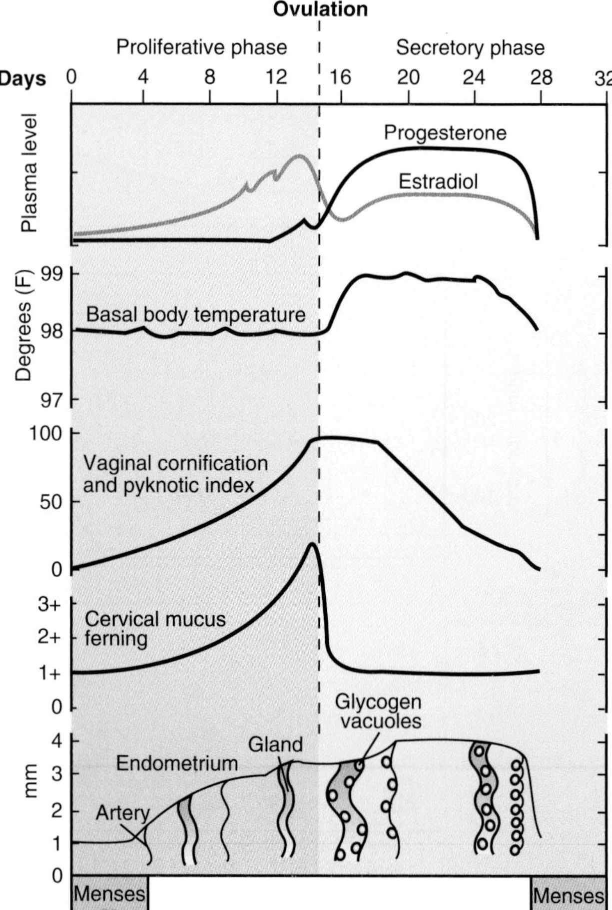

Figure 37.7 Cyclic changes in the uterus, cervix, vagina, and body temperature in relationship to estradiol, progesterone, and ovulation during the menstrual cycle.

Secretory phase

The **secretory phase** begins on the day of ovulation and coincides with the early to midluteal phase of the menstrual cycle. The endometrium contains numerous progesterone receptors. Under the combined action of progesterone and estrogen, the endometrial glands become coiled, store glycogen, and secrete large amounts of carbohydrate-rich mucus. The stroma increases in vascularity and becomes edematous, and the spiral arteries become tortuous (see Fig. 37.7). Peak secretory activity, edema formation, and overall thickness of the endometrium are reached on days 6 to 8 after ovulation in preparation for implantation of the blastocyst. Progesterone antagonizes the effect of estrogen on the myometrium and reduces spontaneous myometrial contractions.

Ischemic phase

The **ischemic phase**, generally not depicted graphically, occurs immediately before the menses and is initiated by the declining levels of progesterone and estradiol caused by regression of the corpus luteum. Necrotic changes and abundant apoptosis occur in the secretory epithelium as it collapses. The arteries constrict, reducing the blood supply to the superficial endometrium. Leukocytes and macrophages invade the stroma and begin to phagocytose the ischemic tissue. Leukocytes persist in large numbers throughout menstruation, providing resistance against infection to the denuded endometrial surface.

Menstrual phase

Desquamation and sloughing of the entire functional layer of the endometrium occur during the menstrual phase (*menses*). The mechanism leading to necrosis is only partly understood. The reduction in steroids destabilizes lysosomal membranes in endometrial cells, resulting in the liberation of proteolytic enzymes and increased production of vasoconstrictor prostaglandins such as prostaglandin F2. The prostaglandins induce vasospasm of the spiral arteries, and the proteolytic enzymes digest the tissue. Eventually, the blood vessels rupture and blood is released, together with cellular debris. The endometrial tissue is expelled through the cervix and vagina, with blood from the ruptured arteries. The menstrual flow lasts 4 to 5 days and averages 30 to 50 mL in volume. It does not clot because of the presence of fibrinolysin, but the spiral arteries constrict, resulting in a reduction in bleeding.

Cervical mucus changes

Changes in the properties of the cervical mucus promote the survival and transport of sperm and can thus be important for normal fertility. The cervical mucus undergoes cyclic changes in composition and volume. During the follicular phase, estrogen increases the quantity, alkalinity, viscosity, and elasticity of the mucus. The cervical muscles relax, and the epithelium becomes secretory in response to estrogen. By the time of ovulation, elasticity of the mucus or spinnbarkeit is greatest. Sperm can readily pass through the estrogen-dominated mucus. With progesterone rising either after ovulation, during pregnancy, or with low-dose administration of progestogen during the cycle, the quantity and elasticity of the mucus decline; it becomes thicker (low spinnbarkeit) and does not form a ferning pattern when dried on a microscope slide. With these conditions, the mucus provides better protection against infections and sperm do not easily pass through.

The vaginal epithelium proliferates under the influence of estrogen. Basophilic cells predominate early in the follicular phase. The columnar epithelium becomes cornified (keratinized) under the influence of estrogen and reaches its peak in the periovulatory period. During the postovulatory period, progesterone induces the formation of thick mucus, the epithelium becomes infiltrated with leukocytes, and cornification decreases (see Fig. 37.7).

Menopause is the cessation of ovarian function and reproductive cycles.

The time after which the final menses occurs is termed menopause. Generally, menstrual cycles and bleeding become irregular, and the cycles may become shorter from the lack of follicular development (shortened follicular phases). The ovaries atrophy and are characterized by the presence of few, if any, healthy follicles.

The decline in ovarian function is associated with a decrease in estrogen secretion and a concomitant increase in LH and FSH, which is characteristic of menopausal women (Table 37.3). It is used as a diagnostic tool. The elevated LH stimulates ovarian stroma cells to continue producing androstenedione. Estrone, derived almost entirely from the peripheral conversion of adrenal and ovarian androstenedione, becomes the dominant estrogen (see Fig. 36.9). Because the ratio of estrogens to androgens decreases, some women exhibit hirsutism, which results from androgen excess. The lack of estrogen causes atrophic changes in the breasts and reproductive tract, accompanied by vaginal dryness, which often causes pain and irritation. Similar changes in the urinary tract may give rise to urinary disturbances. The epidermal layer of the skin becomes thinner and less elastic.

Hot flashes, resulting from the loss of vasomotor tone, osteoporosis, and an increased risk of cardiovascular disease are not uncommon. Hot flashes are associated with episodic increases in upper body and skin temperatures, peripheral vasodilation, and sweating. They occur concurrently with LH pulses, but are not caused by the gonadotropins because they are evident in hypophysectomized women. Hot flashes, consisting of episodes of sudden warmth and sweating, reflect temporary disturbances in the hypothalamic

TABLE 37.3 Serum Gonadotropin and Steroid Levels in Premenopausal and Postmenopausal Women

		Menstrual Cycle			
Hormone	**Units**	**Follicular**	**Preovulatory**	**Luteal**	**Postmenopausal**
LH	mIU/mL	2.5–15	15–100	2.5–15	20–100
FSH	mIU/mL	2–10	10–30	2–6	20–140
Estradiol	pg/mL	70–200	200–500	75–300	10–60
Progesterone	ng/mL	≤0.5	≤1.5	4–20	≤0.5

LH, luteinizing hormone; FSH, follicle-stimulating hormone.

thermoregulatory centers, which are somehow linked to the GnRH pulse generator.

Osteoporosis increases the risk of hip fractures, and estrogen replacement therapy reduces the risk. Estrogen antagonizes the effects of PTH on bone but enhances its effect on kidney; that is, it stimulates retention of calcium. Estrogen also promotes the intestinal absorption of calcium through 1,25-dihydroxyvitamin D_3.

Menopausal symptoms are often treated with hormone replacement therapy, which includes estrogens and progestins (see Bench to Bedside Box 37.1).

ESTROGEN, PROGESTIN, AND ANDROGEN METABOLISM

The principal sex steroids in the female are estrogen, progestin, and androgen. Three **estrogens** are present in significant quantities: estradiol, estrone, and estriol. Estradiol is the most abundant and is 12 and 80 times more potent than estrone and estriol, respectively. Much of estrone is derived from peripheral conversion of either androstenedione or estradiol (see Fig. 36.9). During pregnancy, large quantities of estriol

From Bench to Bedside / 37.1

Hormone Replacement Therapy and the Role of Selective Estrogen Receptor Modulator in Menopause

Complications of menopause include hot flashes, vaginal dryness and irritation, bone loss, and increased risk of cardiovascular disease. Hormone replacement therapy (HRT), consisting of estrogen or estrogen with progestin, has been shown in large clinical trials to reduce hot flashes and vaginal discomfort in most women. The addition of progestin to the treatment does not increase therapeutic effectiveness, but it is given to protect the endometrium from estrogen-induced hyperplasia or carcinoma. In addition, HRT reduces spine and hip fractures resulting from osteoporosis, especially if given during the first 3 years of menopause. Recognized risks of HRT include a small increased incidence of breast cancer, an established greater incidence of endometrial cancer, and a modest increased incidence of deep vein thrombosis.

It has long been thought that estrogen protects against atherosclerosis and cardiovascular disease in women. In premenopausal women, total and low-density lipoprotein (LDL) cholesterol are lower, high-density lipoprotein (HDL) cholesterol is higher, and myocardial infarction is six times less common than in men of the same age. During menopause, total and LDL cholesterol levels rise, contributing to the increased risk of cardiovascular disease in postmenopausal women. Estrogen therapy in postmenopausal women has a favorable impact on lipid profile, lowering total and LDL cholesterol. This finding, in combination with observational studies suggesting improved cardiovascular risk with estrogen therapy in menopause, resulted in the conclusion that postmenopausal use of estrogen protected against cardiovascular disease. However, the Women's Health Initiative, a large randomized study of estrogen therapy, was halted early because of a small, but significant, increased incidence of cardiovascular disease and breast cancer. More recent analyses suggest that HRT may increase cardiovascular risk in older women starting therapy many years after menopause, but not in women starting therapy to relieve symptoms at the onset of menopause. Thus, use of HRT in menopausal women for primary or secondary prevention of cardiovascular disease remains a controversial topic.

Selective estrogen receptor modulators (or SERMs) are a new category of therapeutic agents available for treatment of the complications of menopause. They display an unusual tissue-selective pharmacology: They are activators of estrogen receptor (ER) in some tissues (bone, liver, and the cardiovascular system), inhibitors in other tissues (brain and breast), and mixed activator/inhibitors in the uterus. The goal of therapy with SERMs is to mimic the beneficial effects of estrogen on the bones and heart, but to inhibit estrogen action in the breast and uterus, minimizing harmful effects in these tissues. Two well-studied SERMs are tamoxifen (Nolvadex) and raloxifene (Evista), and other compounds are in development. Estrogen action is mediated through two distinct receptors, ERα and ERβ, that differ in the N-terminal region, which is important in initiation of transcription. The receptors also exhibit different expression patterns, with ERα found predominantly in breast, uterus, and vagina, and ERβ expressed in tissues of the central nervous system, cardiovascular system, immune system, gastrointestinal system, kidney, lungs, and bone. The differences in tissue distribution between the two receptors may explain some of the selective actions of estrogens in different tissues.

Several mechanisms have been proposed to explain the ability of SERMs to be estrogenic in some tissues and antiestrogenic in others. They can block estrogen binding to its receptor to inhibit transcription. In contrast, SERMs may promote interaction of the N-terminal domain of the ER with other transcription factors, resulting in activation of transcription. They may recruit different coactivators and coinhibitors in different tissues than does estrogen. Finally, the actions of SERMs can differ depending on which ER isoform they bind. SERMs function as pure antagonists when acting through ERβ on genes containing ER-binding elements, but can function as a partial agonist when acting on them through ERα.

A number of trials have been designed to test whether SERMs can provide cardiovascular protection in menopause. Raloxifene significantly decreases the risk of breast cancer and vertebral fracture, but it does not alter the risk for coronary events. Thus, at this time, SERMs appear to be an alternative to HRT for management of postmenopausal osteoporosis, but they do not provide additional cardiovascular benefits. Continuing research should lead to a better understanding of the mechanism(s) of action of SERMs and development of new molecules designed for specific and beneficial actions on estrogen-responsive tissues.

are produced from DHEA sulfate after 16α-hydroxylation by the fetoplacental unit (see Chapter 38, "Fertilization, Pregnancy, and Fetal Development"). Most estrogens are bound either to **albumin** (~60%) with a low affinity or to **sex hormone–binding globulin (SHBG)** (~40%) with high affinity. Estrogens are metabolized in the liver through oxidation or conversion to glucuronides or sulfates. The metabolites are then excreted in the urine.

The most important **progestin** is progesterone. It is secreted in significant amounts during the luteal phase of the menstrual cycle. During pregnancy, the corpus luteum secretes progesterone throughout the first trimester and the placenta continues progesterone production until parturition. Small amounts of 17-hydroxyprogesterone are secreted along with progesterone. Progesterone binds equally to albumin and to a plasma protein called **corticosteroid-binding protein (transcortin)**. Progesterone is metabolized in the liver to pregnanediol and, subsequently, excreted in the urine as a glucuronide conjugate.

Circulating **androgens** in the female originate from the ovaries and adrenals and from peripheral conversion. Androstenedione and DHEA originate from the adrenal cortex (see Chapter 33, "Adrenal Gland") and ovarian theca and stroma cells. Peripheral conversion from androstenedione provides an additional source of testosterone. Testosterone can also be converted in peripheral tissues to **dihydrotestosterone (DHT)** by **5α-reductase**. However, the primary biologically active androgen in women is testosterone. Androgens bind primarily to SHBG and bind to albumin by about half as much. Androgens are also metabolized to water-soluble forms by oxidation, sulfation, or glucuronidation and excreted in the urine.

INFERTILITY

Infertility affects one of five women in the United States. A thorough understanding of female endocrinology, anatomy, and physiology is critical to gaining insights into solving this major health problem. Several factors can cause infertility. Environmental factors, disorders of the central nervous system, hypothalamic disease, pituitary disorders, and ovarian abnormalities can interfere with follicular development and/or ovulation. If a normal ovulation occurs, structural, pathologic, and/or endocrine problems associated with the oviduct and/or uterus can prevent fertilization, impede the transport or implantation of the embryo, and ultimately, interfere with the establishment or maintenance of pregnancy.

Amenorrhea can result from anatomic or hormonal dysfunction.

As a general guideline, **amenorrhea** can be defined as the absence of menarche by age 16 years, or no menses for more than three cycles in a woman who previously had cyclic menses. Amenorrhea had been classically defined as primary (no prior menses) or secondary (cessation of menses in a normally cycling individual), but these distinctions are no longer recommended as they may lead to diagnostic error. A number of classification systems for diagnosis of amenorrhea have been developed based on whether the cause is anatomic or hormonal, inherited or acquired. Examples of amenorrhea with different etiologies are provided in the following paragraphs.

Turner syndrome

Also called *gonadal dysgenesis*, Turner syndrome is a congenital abnormality caused by a nondisjunction of one of the X chromosomes, resulting in a 45 X0 chromosomal karyotype. Because the two X chromosomes are necessary for normal ovarian development, women with this condition have rudimentary gonads and do not have a normal puberty. Because of ovarian steroid deficiency (lack of estrogen), secondary sex characteristics remain prepubertal, and plasma LH and FSH are elevated. Other abnormalities include short stature, a webbed neck, coarctation of the aorta, and renal disorders.

Hypogonadotropism

Hypogonadotropic hypogonadism with anosmia is similar to Kallmann syndrome in males (see Chapter 36, "Male Reproductive System"), but the specific mutations in the Kal-1 gene in females have not been identified, suggesting that there are other genetic mutations. Patients do not progress through normal puberty and have low and nonpulsatile LH and FSH levels. However, they can have normal stature, female karyotype, and anosmia. A failure of olfactory lobe development and GnRH deficiency cause the disorder.

Premature ovarian failure

Loss of oocytes and the surrounding support cells prior to age 40 years defines premature ovarian failure. Clinical findings include amenorrhea, low estrogen levels, and high gonadotropin (LH and FSH). Symptoms are similar to those of menopause, including hot flashes and an increased risk of osteoporosis. The etiology is variable, including chromosomal abnormalities; lesions resulting from irradiation, chemotherapy, or viral infections; and autoimmune conditions.

Polycystic ovarian syndrome

This syndrome has a range of presentations but is generally characterized by annovulation, amenorrhea and high androgen levels. Elevated LH/FSH ratio, polycystic ovaries, hirsutism, and obesity are often present. Although the etiology is unknown, excessive adrenal androgen production may initiate the syndrome during puberty or following stress that deranges the hypothalamic-pituitary axis secretion of LH. Elevated insulin due to insulin resistance in obese subjects can also cause hyperandrogenism. Androgens are converted peripherally to estrogens and stimulate LH release. Excess LH, in turn, increases ovarian stromal and thecal androgen production, resulting in impaired follicular maturation. The LH-stimulated ovaries are enlarged and contain many small follicles and hyperplastic and luteinized theca cells (the site of LH receptors). The elevated plasma androgen levels cause hirsutism, increased activity of sebaceous glands, and clitoral hypertrophy, which are signs of virilization in females.

Hyperprolactinemia

Elevated prolactin (PRL) secretion from the pituitary is a common cause of amenorrhea and is often associated with **galactorrhea**, a persistent milk-like discharge from the nipple in nonlactating women (also occurring in men). Pituitary prolactinomas account for about 50% of cases. Other causes are hypothalamic disorders, trauma to the pituitary stalk, and psychotropic medications, all of which are associated with a reduction in dopamine release, resulting in an increased PRL secretion. Hypothyroidism, chronic renal failure, and hepatic cirrhosis are additional causes of hyperprolactinemia. In some forms of hypothyroidism, increased hypothalamic thyrotropin-releasing hormone (TRH) is thought to contribute to excess PRL secretion, as experimental studies reveal that exogenous TRH increases the secretion of PRL. The mechanism by which elevated PRL levels suppress ovulation is not entirely clear but it appears suppression of GnRH release is a primary mechanism.

Female athletic triad and anorexia nervosa

Excessive exercise and nutritional, psychological, and social factors can cause amenorrhea. The *female athletic triad*, characterized by disordered eating, amenorrhea and osteoporosis, occurs in competitive athletes and dancers. Two main factors are thought to be responsible: a low level of body fat and the effect of stress itself through endorphins that are known to inhibit the secretion of LH. Other types of stress, such as from relocation, college examinations, general illness, and job-related pressures, have been known to induce amenorrhea.

Anorexia nervosa is a severe behavioral disorder associated with constant dieting to achieve a thin body habitus. The condition is characterized by extreme malnutrition and endocrine changes secondary to psychological and nutritional disturbances. Amenorrhea may not be immediately alleviated by weight gain.

Treatment of female infertility depends on the correct identification of the problem.

The diagnosis and treatment of amenorrhea present a challenging problem. After ruling out pregnancy, lactation and menopause as the cause, the next step is to determine whether the disorder originates in the hypothalamus and central nervous system, the anterior pituitary, the ovary, or the reproductive tract.

Several treatments can alleviate infertility problems. For example, some success has been achieved in hypothalamic disease with pulsatile administration of GnRH. When hypogonadotropism is the cause of infertility, sequential administration of FSH and hCG is a common treatment for inducing ovulation, although the risk of ovarian hyperstimulation and multiple ovulations is increased. Hyperprolactinemia can be treated surgically by removing the pituitary adenoma containing numerous lactotrophs (prolactin-secreting cells). It can also be treated pharmacologically with bromocriptine, a dopaminergic agonist that reduces the size and number of the lactotrophs and PRL secretion. Treatment with clomiphene, an antiestrogen that binds to and blocks estrogen receptors, can induce ovulation in women with endogenous estrogens in the normal range. Clomiphene reduces the negative-feedback effects of estrogen and thus increases endogenous FSH and LH secretion. When reproductive tract lesions are the cause of infertility, corrective surgery or *in vitro* fertilization is the treatment of choice.

Chapter Summary

- Pulses of hypothalamic gonadotropin–releasing hormone regulate the secretion of luteinizing and follicle-stimulating hormones, which enhance follicular development, steroidogenesis, ovulation, and formation of the corpus luteum.
- Luteinizing and follicle-stimulating hormones, in coordination with ovarian theca and granulosa cells, regulate the secretion of follicular estradiol.
- Ovulation occurs as the result of a positive feedback of follicular estradiol on the hypothalamic-pituitary axis that induces luteinizing and follicle-stimulating hormone surges.
- Follicular development occurs in distinct steps: primordial, primary, secondary, tertiary, and graafian follicle stages.
- Follicular rupture (ovulation) requires the coordination of appropriately timed luteinizing and follicle-stimulating hormone surges that induce inflammatory reactions in the graafian follicle, leading to dissolution at midcycle of the follicular wall by several ovarian enzymes.
- Follicular atresia results from the withdrawal of gonadotropin support.
- The formation of a functional corpus luteum requires the presence of a luteinizing hormone (LH) surge, adequate numbers of LH receptors, sufficient granulosa cells, and significant progesterone secretion.
- Estradiol and progesterone regulate the uterine cycle, such that estradiol induces proliferation of the uterine endometrium, whereas progesterone induces differentiation of the uterine endometrium and the secretion of distinct products.
- During puberty, the hypothalamus begins to secrete increasing quantities of gonadotropin-releasing hormone, which increases luteinizing and follicle-stimulating hormone secretion, enhances ovarian function, and leads to the first ovulation.
- Menopause ensues from the loss of numerous oocytes in the ovary and the subsequent failure of follicular development and estradiol secretion. Luteinizing and follicle-stimulating hormone levels rise from the lack of negative feedback by estradiol.

thePoint *Visit http://thePoint.lww.com for chapter review Q&A, case studies, animations, and more!*

38 Fertilization, Pregnancy, and Fetal Development

ACTIVE LEARNING OBJECTIVES

Upon mastering the material in this chapter you should be able to:

- Describe where fertilization of the ovum occurs and explain the roles of progesterone and estrogen in the preparation of the oviduct and uterus for receiving the developing embryo.
- Explain the process and timing of blastocyst implantation into the uterine wall.
- Describe the source of human chorionic gonadotropin and its role in activating the function of the corpus luteum early during pregnancy.
- Describe the development of the placenta and explain the importance of this organ for steroid production during pregnancy.
- Describe the major hormones produced by the fetoplacental unit and explain the role of each during the course of pregnancy.
- Explain the role of maternal insulin resistance during the latter half of pregnancy.
- Explain the function of oxytocin, estrogens, relaxin, and prostaglandins in the termination of pregnancy and birth.
- Explain the processes of lactogenesis and galactopoiesis and the roles prolactin, insulin, glucocorticoids, and oxytocin perform in each.
- Explain the mechanism initiated by lactation that results in anovulation and suppression of menstrual cycles.
- Explain the activation of the hypothalamic-pituitary axis during the late prepubertal period and the effect of this activation on steroid output by the gonads.
- Describe the effects of chromosomal and hormonal alterations on sexual development.

A mother is considered pregnant at the moment of fertilization—the successful union of a sperm and an egg. The life span of the sperm and an ovum is <2 days, so their rapid transport to the oviduct is required for fertilization to occur. Immediately after fertilization, the zygote or fertilized egg begins to divide, and a new life begins. The cell division produces a **morula**, a solid ball of cells, which then forms a **blastocyst**. Because the early embryo contains a limited energy supply, it enters the uterus within a short time and attaches to the uterine endometrium, a process that initiates the implantation phase. **Implantation** occurs only in a uterus that has been primed by gonadal steroids and is, therefore, receptive to accepting the blastocyst. At the time of implantation, the trophoblast cells of the early embryonic placenta begin to produce a hormone, **human chorionic gonadotropin (hCG)**, which signals the ovary to continue to produce progesterone–the major hormone required for the maintenance of pregnancy. As a signal from the embryo to the mother to extend the life of the corpus luteum (and progesterone production), hCG prevents the onset of the next menstruation and ovulatory cycle. The placenta, an organ produced by the mother and fetus, exists only during pregnancy; it regulates the supply of oxygen and the removal of wastes and serves as an energy supply for the fetus. It also produces protein and steroid hormones, which duplicate, in part, the functions of the pituitary gland and gonads. Some of the fetal endocrine glands have important functions before birth, including sexual differentiation.

Parturition, the expulsion of the fully formed fetus from the uterus, is the final stage of gestation. The onset of parturition is triggered by signals from both the fetus and the mother, and involves biochemical and mechanical changes in the uterine myometrium and cervix. After delivery, the mother's mammary glands must be fully developed and secrete milk to provide nutrition to the newborn baby. Milk is produced and secreted in response to suckling. The act of suckling, through neurohormonal signals, typically prevents new ovulatory cycles. Thus, suckling may act as a natural contraceptive. Thereafter, the mother regains metabolic balance, which has been reduced by the nutritional demands of pregnancy and lactation, and ovulatory cycles return. Sexual maturity of the offspring is attained during puberty, at approximately 11 to 12 years of age. The onset of puberty requires changes in the sensitivity, activity, and function of several endocrine organs, including those of the hypothalamic-pituitary-gonadal axis.

OVUM AND SPERM TRANSPORT

Fertilization (also known as *conception*) is the fusion of the male and female gametes (sperm with an ovum), which leads to the formation of an embryo and the development of a new organism. Sperm deposited in the female reproductive tract swim up the uterus and enter the oviduct, where fertilization of the ovum occurs. The developing embryo transits the oviduct, enters the uterus, and implants into the endometrium.

Egg and sperm enter the oviduct, the site of fertilization.

A meiotically active egg is released from the ovary in a cyclical manner in response to the luteinizing hormone (LH) surge. For a successful fertilization, fresh sperm must be present at the time the ovum enters the oviduct. To increase the probability that the sperm and egg will meet at an optimal time, the female reproductive tract facilitates sperm transport during the follicular phase of the menstrual cycle, prior to ovulation (see Chapter 37, "Female Reproductive System"). However, during the luteal phase, after ovulation, sperm survival and access to the oviduct are decreased. If fertilization does not occur, the egg and sperm begin to exhibit signs of degeneration within 24 hours after release.

Sperm transport

The volume of **semen** (ejaculatory fluids and sperm) in fertile men is 2 to 6 mL and contains some 20 to 30 million sperm per milliliter, which are deposited in the vagina. The liquid component of the semen, called **seminal plasma**, coagulates after ejaculation but liquefies within 20 to 30 minutes from the action of proteolytic enzymes secreted by the prostate gland. The coagulum forms a temporary reservoir of sperm, minimizing the expulsion of semen from the vagina. During intercourse, some sperm cells are immediately propelled into the cervical canal. Those remaining in the vagina do not survive long because of the acidic environment (pH 5.7), although the alkalinity of the seminal plasma provides some protection. The cervical canal constitutes a more favorable environment, enabling sperm survival for several hours. Under estrogen dominance, mucin molecules in the **cervical mucus** become oriented in parallel and facilitate sperm migration. Sperm stored in the cervical crypts constitute a pool for slow release into the uterus.

Sperm survival in the uterine lumen is short because of phagocytosis by leukocytes. The uterotubal junction also presents an anatomic barrier that limits the passage of sperm into the oviducts. Abnormal or dead spermatozoa may be prevented from entry to the oviduct. Of the millions of sperm deposited in the vagina, only 50 to 100, usually spaced in time, reach the oviduct. Major losses of sperm occur in the vagina, in the uterus, and at the uterotubal junction. Spermatozoa that survive can reach the ampulla within 5 to 10 minutes after coitus. The motility of sperm largely accounts for this rapid transit. However, muscular contractions of the vagina, cervix, and uterus; ciliary movement; peristaltic activity; and fluid flow in the oviducts assist transport. Semen samples with low sperm motility can be associated with male infertility.

There is no evidence for chemotactic interactions between the egg and sperm, although evidence exists for specific ligand–receptor binding between the egg and sperm. Sperm arrive in the vicinity of the egg at random, and some exit into the abdominal cavity. Although sperm remain motile for up to 4 days, their fertilizing capacity is limited to 1 to 2 days in the female reproductive tract. Sperm can be cryopreserved for years if agents such as glycerol are used to prevent ice crystal formation during freezing.

Freshly ejaculated sperm cannot immediately penetrate an egg. During maturation in the epididymis, the sperm acquire surface glycoproteins that act as stabilizing factors but also prevent sperm–egg interactions. To bind to and penetrate the zona pellucida, the sperm must undergo **capacitation**, an irreversible process that involves an increase in sperm motility, the removal of surface proteins, a loss of lipids, and merging of the acrosomal and plasma membranes of the sperm head. The uniting of these sperm membranes and change in acrosomal structure are called the **acrosome reaction**. The reaction occurs when the sperm cell binds to the zona pellucida of the egg. It involves a redistribution of membrane constituents, increased membrane fluidity, and a rise in calcium permeability. Capacitation takes place along the female genital tract and lasts 1 hour to several hours. Sperm can be capacitated in a chemically defined medium, a fact that has enabled *in vitro* fertilization.

Ovum transport

Because the oviduct does not entirely engulf the ovary, an active "pickup" of the released ovum is required. Ciliated fingerlike projections of the oviducts, called **fimbria**, grasp the ovum. Ciliary movement and muscle contractions facilitate the grasping of the egg, under the influence of estrogen secreted during the periovulatory period. Because the oviduct opens into the peritoneal cavity, eggs that the oviducts do not pick up can enter the abdominal cavity. An **ectopic pregnancy** may result if an abdominal ovum is fertilized. Egg transport from the fimbria to the **ampulla**, the swollen end of the oviduct, is accomplished by coordinated ciliary activity and depends on the presence of granulosa cells surrounding the egg.

The fertilizable life of the human ovum is about 24 hours, and fertilization occurs usually by 2 days after ovulation. The fertilized ovum remains in the oviduct for 2 to 3 days, develops into a solid ball of cells called a *morula*, and enters the uterus by day 3 or 4. While in the uterus, the morula further develops into a blastocyst, the **zona pellucida** is shed, and the blastocyst implants into the wall of the uterus on day 7. Progesterone and estrogen largely regulate the movement of the developing embryo from the oviduct to the uterus.

Clinical Focus / 38.1

In Vitro Fertilization

Candidates for *in vitro* fertilization (IVF) are women with disease of the oviducts, unexplained infertility, or endometriosis (occurrence of endometrial tissue outside the endometrial cavity, a condition that reduces fertility) and those whose male partners are infertile (e.g., low sperm count). Follicular development is induced with one or a combination of gonadotropin-releasing hormone (GnRH) analogs, clomiphene, recombinant follicle-stimulating hormone (FSH), and menopausal gonadotropins (a combination of luteinizing hormone [LH] and FSH). Follicular growth is monitored by measuring serum estradiol concentration and by ultrasound imaging of the developing follicles. When the leading follicle is 16 to 17 mm wide and/or the estradiol level is >300 pg/mL, hCG is injected to mimic an LH surge and induce final follicular maturation, including maturation of the oocyte. Approximately 34 to 36 hours later, oocytes are retrieved from the larger follicles by aspiration using laparoscopy or a transvaginal approach. Oocyte maturity is judged from the morphology of the cumulus (granulosa) cells and the presence of the germinal vesicle and first polar body. The mature oocytes are then placed in culture media.

The donor's sperm are prepared by washing, centrifuging, and collecting those that are most motile. About 100,000 spermatozoa are added for each oocyte. After 24 hours, the eggs are examined for the presence of two pronuclei (male and female). Embryos are grown to the four-cell to eight-cell stage, about 60 to 70 hours after their retrieval from the follicles. Approximately three embryos are often deposited in the uterine lumen to increase the chance for a successful pregnancy. To ensure a receptive endometrium, daily progesterone administrations begin on the day of retrieval. The rate of success for IVF techniques depends on several factors, including the age of the female partner and normal semen analysis.

FERTILIZATION AND IMPLANTATION

The initial stage of fertilization is the attachment of the sperm head to the zona pellucida of the egg. A successful fertilization restores the full complement of 46 chromosomes and subsequently initiates the development of an embryo. Fertilization involves several steps. Recognition of the egg by the sperm occurs first. The next step is the regulation of sperm entry into the egg. A series of key molecular events prevents **polyspermy**, or *multiple sperm*, from entering the egg. Coupled with fertilization is the completion of the second meiotic division of the egg, which extrudes the second polar body. At this point, the male and female pronuclei unite, followed by initiation of the first mitotic division (Fig. 38.1).

Binding of the spermatozoon to the sperm receptor in the zona pellucida initiates the acrosome reaction.

The zona pellucida contains specific glycoproteins that serve as sperm receptors. They selectively prevent the fusion of inappropriate sperm cells (e.g., from a different species) with the egg. Contact between the sperm and egg triggers the acrosome reaction, which is required for sperm penetration. Sperm proteolytic enzymes are released that dissolve the matrices of the cumulus (granulosa) cells surrounding the egg, enabling the sperm to move through this densely packed group of cells. The sperm penetrates the zona pellucida, aided by proteolytic enzymes and the propulsive force of the tail; this process may take up to 30 minutes. After entering the **perivitelline space**, the sperm head becomes anchored to the membrane surface of the egg, and microvilli protruding from the **oolemma** (plasma membrane of the egg) extend and clasp the sperm. The oolemma engulfs the sperm, and eventually, the whole head and then the tail are incorporated into the **ooplasm**.

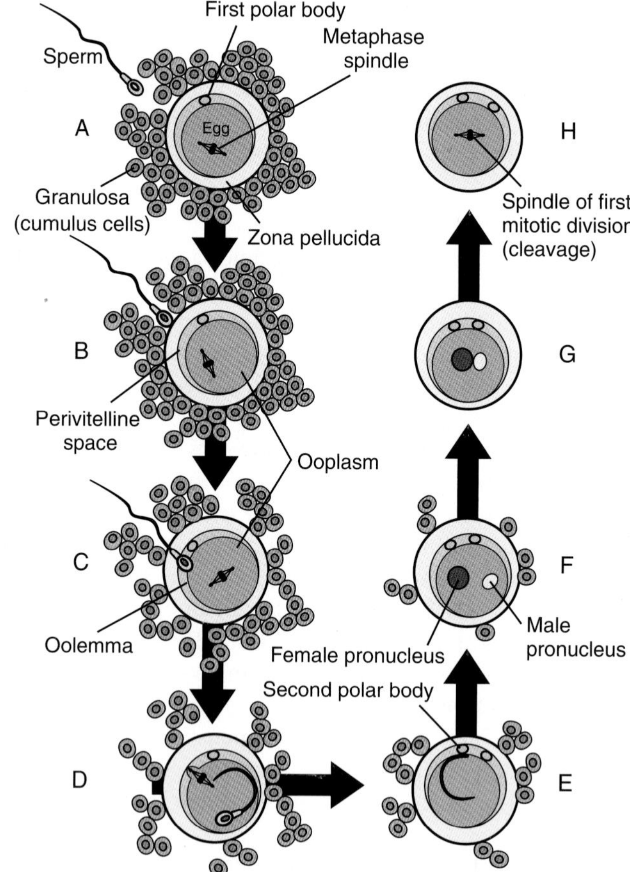

Figure 38.1 The process of fertilization. (A) A sperm cell approaches an egg. **(B)** Contact between the sperm and the zona pellucida. **(C)** The entry of the sperm and contact with the oolemma. **(D)** The resumption of the second meiotic division. **(E)** The completion of meiosis. **(F)** The formation of male and female pronuclei. **(G)** The migration of the pronuclei to the center of the cell. **(H)** The zygote is ready for the first mitotic division.

Shortly after the sperm enters the egg, **cortical granules**, which are lysosomelike organelles located underneath the oolemma, are released. The cortical granules fuse with the oolemma. Fusion starts at the point of sperm attachment and propagates over the entire egg surface. The content of the granules is released into the perivitelline space and diffuses into the zona pellucida, inducing the **zona reaction**, which is characterized by sperm receptor inactivation and a hardening of the zona. Consequently, once the first spermatozoon triggers the zona reaction, other sperm cannot penetrate the zona, and therefore, polyspermia is prevented.

An increase in intracellular calcium initiated by sperm incorporation into the egg triggers the next event, which is the activation of the egg for completion of the second meiotic division. The chromosomes of the egg separate, and half of the chromatin is extruded with the small second polar body. The remaining haploid nucleus with its 23 chromosomes is transformed into a **female pronucleus**. Soon after being incorporated into the ooplasm, the nuclear envelope of the sperm disintegrates; the **male pronucleus** is formed and increases four to five times in size. Contractions of microtubules and microfilaments move the two pronuclei, which are visible 2 to 3 hours after the entry of the sperm into the egg, to the center of the cell. Replication of the haploid chromosomes begins in both pronuclei. Pores are formed in their nuclear membranes, and the pronuclei fuse. The **zygote** (fertilized egg) then enters the first mitotic division (cleavage), producing two unequally sized cells, called **blastomeres**, within 24 to 36 hours after fertilization. Development proceeds with four- and eight-cell embryos and a morula, still in the oviduct, forming at approximately 48, 72, and 96 hours, respectively. The morula enters the uterine cavity at around 4 days after fertilization, and subsequently, a blastocyst develops at approximately 6 days after fertilization. The blastocyst implants into the uterine wall on approximately day 7 after fertilization.

Tropoblastic enzymes facilitate blastocyte implantation in the endometrium.

Cell division of the fertilized egg occurs without growth. The cells of the early embryo become progressively smaller, reaching the dimension of somatic cells after several cell divisions. The embryonic cells continue to cleave as the embryo moves from the ampulla toward the uterus (Fig. 38.2).

Until implantation, the embryo is enclosed in the zona pellucida. Retention of an intact zona is necessary for embryo transport, protection against mechanical damage or adhesion to the oviduct wall, and prevention of immunologic rejection by the mother.

At the 20- to 30-cell stage, a fluid-filled cavity (blastocoele) appears and enlarges until the embryo becomes a hollow sphere–the blastocyst. The cells of the blastocyst have undergone significant differentiation. A single outer layer

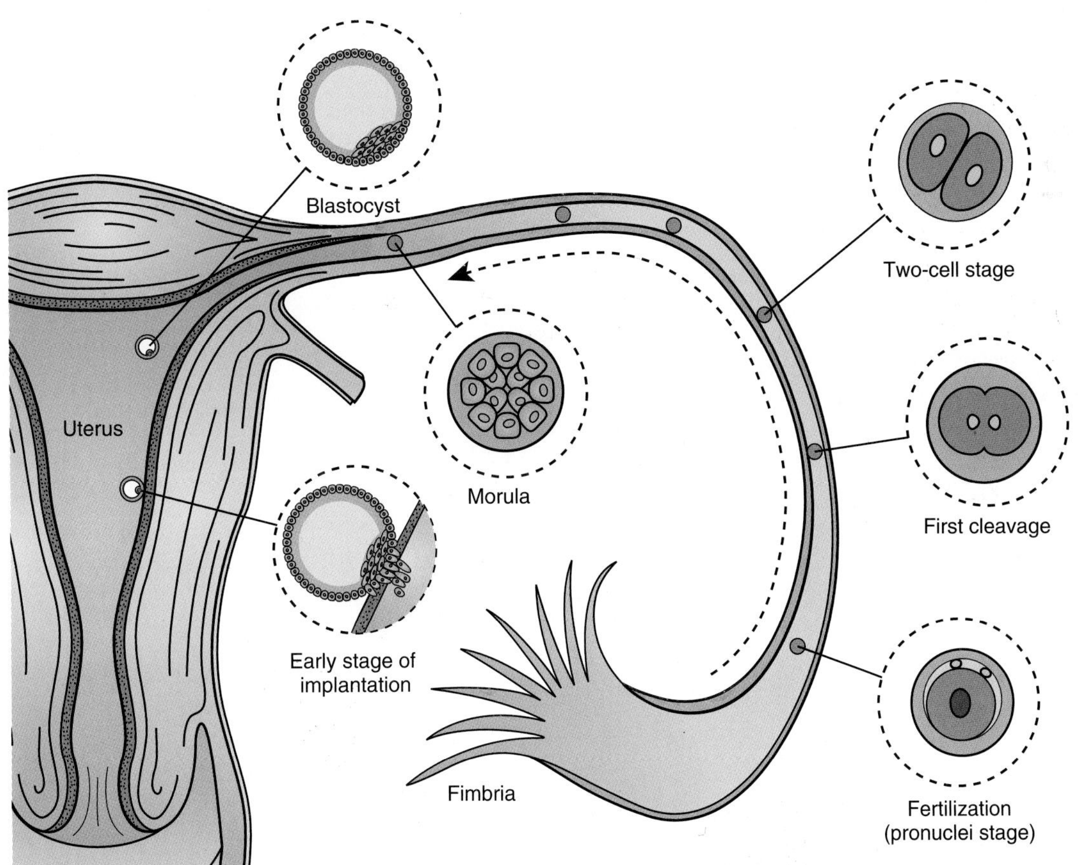

Figure 38.2 Transport of the developing embryo from the oviduct, the site of fertilization, to the uterus, the site of implantation.

of the blastocyst consists of extraembryonic ectodermal cells called the **trophoblast**, which participates in implantation, forms the embryonic contribution to the placenta and embryonic membranes, produces hCG, and provides nutrition to the embryo. A cluster of smaller, centrally located cells comprises the **embryoblast** or inner cell mass and gives rise to the fetus.

Implantation and decidualization

The morula reaches the uterus about 4 days after fertilization. It remains suspended in the uterine cavity for 2 to 3 days while developing into a blastocyst and is nourished by constituents of the uterine fluid during that time. Implantation of the blastocyst, which is attachment to the surface endometrial cells of the uterine wall, begins on days 7 to 8 after fertilization and requires proper priming of the uterus by estrogen and progesterone. In preparing for implantation, the blastocyst escapes from the zona pellucida. The zona is ruptured by expansion of the blastocyst and lysed by enzymes. The denuded trophoblast cells become negatively charged and adhere to the endometrium via surface glycoproteins. Microvilli from the trophoblast cells interdigitate with, and form, junctional complexes with the uterine endometrial cells.

In the presence of progesterone emanating from the corpus luteum, the endometrium undergoes **decidualization**, which involves the hypertrophy of endometrial cells that contain large amounts of glycogen and lipid. In some cases, the cells are multinucleated. This group of decidualized cells is called the **decidua**, which is the site of implantation and the maternal contribution to the placenta. In the absence of progesterone, decidualization does not occur and implantation fails. As the blastocyst implants into the decidualizing uterus, a **decidual reaction** occurs, involving the dilation of blood vessels, increased capillary permeability, edema formation, and increased proliferation of endometrial glandular and epithelial cells (Fig. 38.3). Many of the signals that mediate the interaction of the blastocyst with the endometrium have been identified and include steroids, prostaglandins, leukemia inhibitory factor, epidermal growth factor, transforming growth factor β, platelet-derived growth factor, and placental growth factor.

Embryoblast

The release of proteases produced by trophoblast cells adjacent to the uterine epithelium mediates invasion of the endometrium. By 8 to 12 days after ovulation, the human conceptus has penetrated the uterine epithelium and is embedded in the uterine stroma (see Fig. 38.3). The trophoblast cells have differentiated into large polyhedral **cytotrophoblasts**, surrounded by peripheral **syncytiotrophoblasts** lacking distinct cell boundaries. Maternal blood vessels in the endometrium dilate and spaces appear and fuse, forming blood-filled **lacunae**. Between weeks 2 and 3, villi originating from the embryo are formed, which protrude into the lacunae, establishing a functional communication between the developing embryonic vascular system and the maternal blood (see Fig. 16.6). At this time, the embryoblast has differentiated into three layers:

- *Ectoderm*, destined to form the epidermis, its appendages (nails and hair), and the entire nervous system
- *Endoderm*, which gives rise to the epithelial lining of the digestive tract and associated structures
- *Mesoderm*, which forms the bulk of the body, including connective tissue, muscle, bone, blood, and lymph

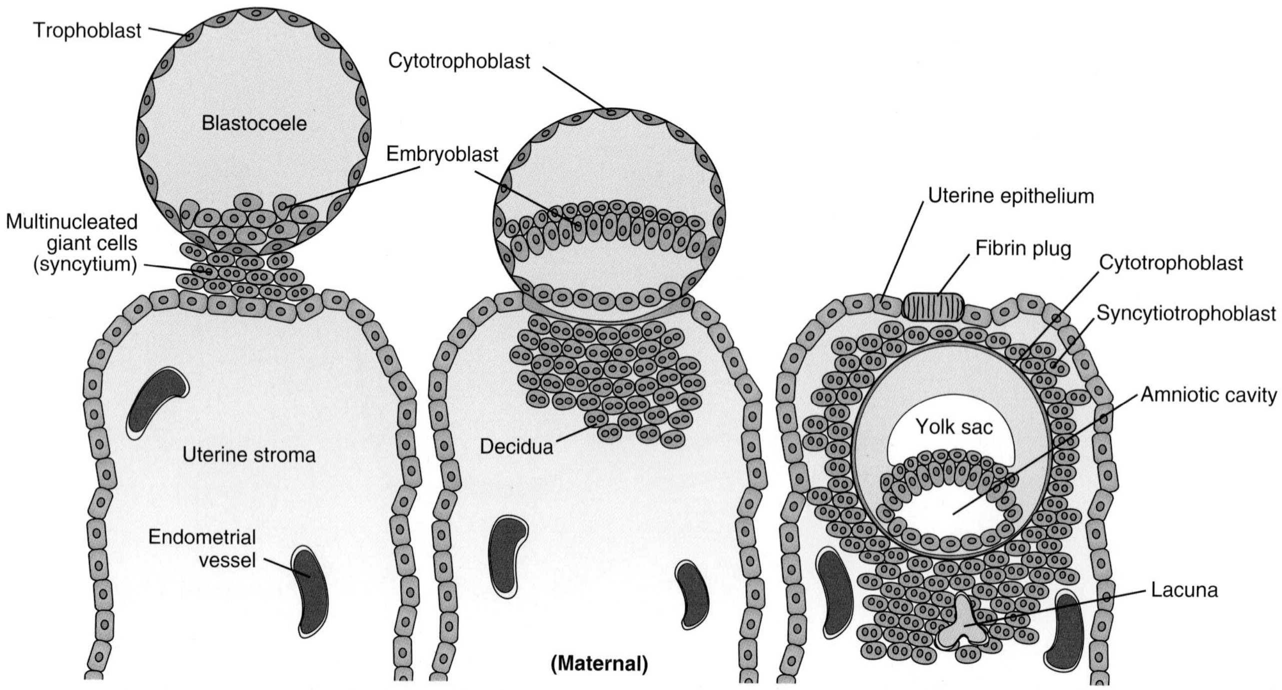

Figure 38.3 The process of embryo implantation and the decidual reaction.

PREGNANCY

Protein and steroid hormones from the mother's ovary and the placenta maintain pregnancy. The maternal endocrine system adapts to allow optimum growth of the fetus.

Placenta is the exchange organ between the mother and fetus.

The **placenta** is an organ that connects the developing fetus to the uterine wall to allow nutrient uptake, waste elimination, and gas exchange via the mother's blood supply. In the human placenta, the maternal and fetal components are interdigitated. The functional units of the placenta, the **chorionic villi** (see Fig. 16.6), form on days 11 to 12 and extend tissue projections into the maternal lacunae that form from endometrial blood vessels immediately after implantation. By week 4, the villi are spread over the entire surface of the chorionic sac. As the placenta matures, it becomes discoid in shape. During the third month, the chorionic villi are confined to the area of the **decidua basalis**, which is the maternal part of the placenta. The decidua basalis and **chorionic plate** together form the placenta proper (Fig. 38.4).

The **decidua capsularis** (which grows around the fetus) and the **decidua parietalis** (the endometrium away from the fetus) on the uterine wall fuse and occlude the uterine cavity. The **yolk sac** becomes vestigial, and the **amniotic sac** expands, pushing the chorion against the uterine wall. From the fourth month onward, the fetus is enclosed within the **amnion** and **chorion** and is connected to the placenta by the **umbilical cord**. Fetal blood flows through two umbilical arteries to capillaries in the villi, is brought into juxtaposition with maternal blood in the sinuses, and returns to the fetus through a single umbilical vein. The fetal and maternal circulations do not mix. The human placenta is a hemochorial type, in which maternal blood surrounds the fetal endothelium and fetal connective tissues. The chorionic villi aggregate into groups known as **cotyledons** and are surrounded by blood from the maternal spiral arteries that course through the decidua.

Major functions of the placenta are the delivery of nutrients to the fetus and the removal of its waste products. Oxygen diffuses from maternal blood to the fetal blood down an initial gradient of 60 to 70 mm Hg. **Fetal hemoglobin**, which has a high affinity for oxygen, enhances the oxygen-transporting capacity of fetal blood. The P_{CO_2} of fetal arterial blood is 2 to 3 mm Hg higher than that of maternal blood, allowing the diffusion of carbon dioxide toward the maternal compartment. Other compounds, such as glucose, amino acids, free fatty acids, electrolytes, vitamins, and some hormones, are transported by diffusion, facilitated diffusion, or pinocytosis. Waste products, such as urea and creatinine, diffuse away from the fetus down their concentration gradients. Large proteins, including most polypeptide hormones, do not readily cross the placenta, whereas the lipid-soluble steroids pass through easily. The **blood–placental barrier** allows the transfer of some immunoglobulins, viruses, and drugs from the mother to the fetus (Fig. 38.5).

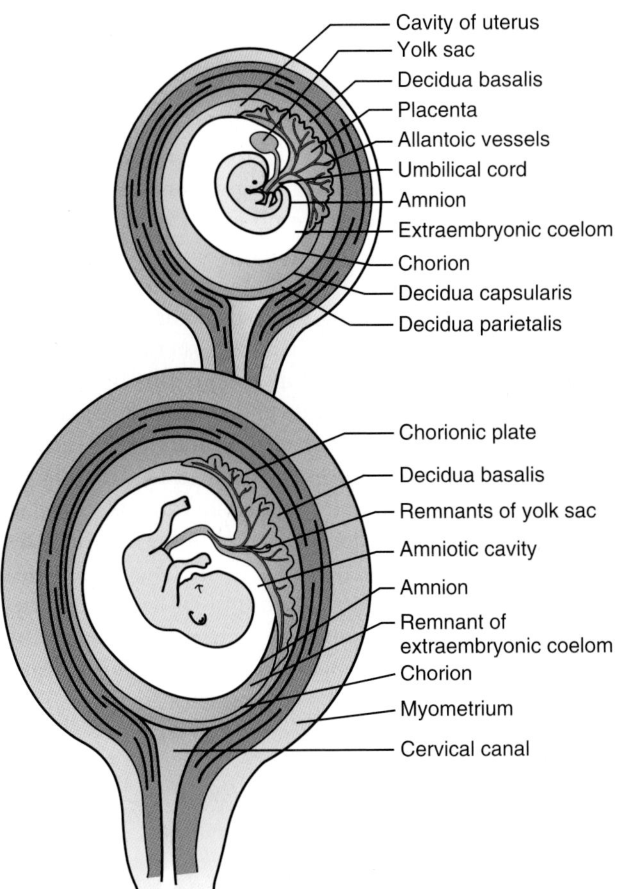

Figure 38.4 Two stages in the development of the placenta, showing the origin of the membranes around the fetus.

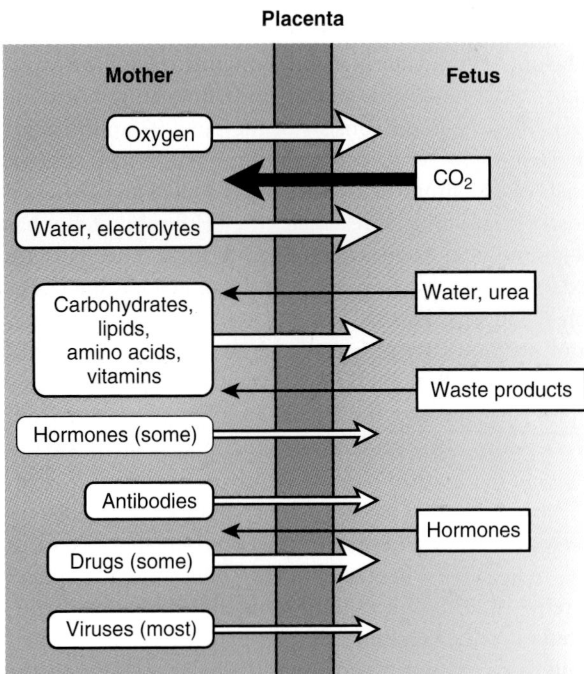

Figure 38.5 Role of the placenta in exchanges between the fetal and maternal compartments. The size of the *arrows* indicates the amount of exchange between the compartments.

Hormones secreted both the embryo and placenta are key to the maintenance of pregnancy.

The placenta is an endocrine organ that produces **progesterone** and **estrogens**, hormones essential for the continuance of pregnancy. The placenta also produces protein hormones unique to pregnancy, such as **human placental lactogen (hPL)** and human chorionic gonadotropin (hCG). Several polypeptides, including corticotropin-releasing hormone (CRH), gonadotropin-releasing hormone (GnRH), and insulin-like growth factors, are also synthesized by the placenta and function as paracrine factors.

During the menstrual cycle, the corpus luteum forms shortly after ovulation and produces significant amounts of progesterone and estrogen to prepare the uterus for receiving a fertilized ovum. If the egg is not fertilized, the corpus luteum regresses at the end of the luteal phase, as indicated by declining levels of progesterone and estrogen in the circulation. After losing ovarian steroidal support, the superficial endometrial layer of the uterus is expelled, resulting in menstruation. If the egg is fertilized, the developing embryo signals its presence by producing hCG, which extends the life of the corpus luteum. This signaling process is called the **maternal recognition of pregnancy**. Syncytiotrophoblast cells produce hCG 6 to 8 days after ovulation (fertilization), and hCG enters the maternal and fetal circulations. Similar to LH, hCG has a molecular weight of approximately 38 kDa, binds LH receptors on the corpus luteum, stimulates luteal progesterone production, and prevents menses at the end of the anticipated cycle. It can be detected in the pregnant woman's urine using commercial colorimetric kits available for home use.

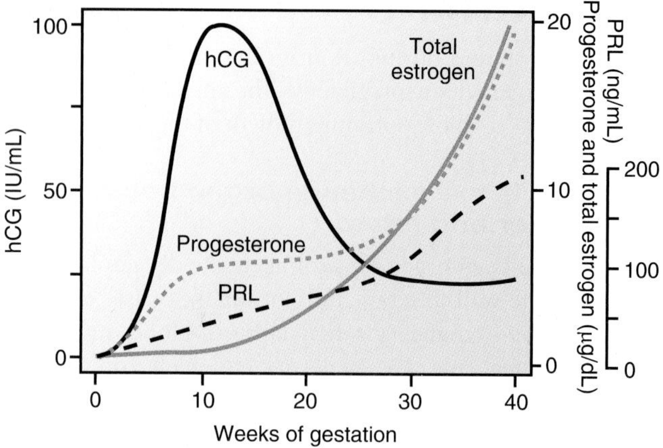

Figure 38.6 Profiles of human chorionic gonadotropin (hCG), progesterone, total estrogens, and prolactin (PRL) in the maternal blood throughout gestation.

Human chorionic gonadotropin

hCG is a glycoprotein made of two dissimilar subunits, α and β. It belongs to the same hormone family as LH, follicle-stimulating hormone (FSH), and thyroid-stimulating hormone (TSH). The α subunit consists of the same 92 amino acids as those in the other glycoprotein hormones. The β subunit consists of 145 amino acids, with six N-linked and O-linked oligosaccharide units. It resembles the LH β subunit, but has a 24–amino acid extension at the C-terminal end. Because of extensive glycosylation, the half-life of hCG in the circulation is longer than that of LH. Like LH, the major function of hCG in early pregnancy is the stimulation of luteal steroidogenesis. Both bind to the same or similar membrane receptors and increase the formation of pregnenolone from cholesterol by a cyclic adenosine monophosphate (cAMP)–dependent mechanism.

The hCG level in plasma doubles about every 2 to 3 days in early pregnancy and reaches peak levels at about 10 to 15 weeks of gestation. It is reduced by about 75% at week 25 and remains at that level until term (Fig. 38.6). Fetal concentrations of hCG follow a similar pattern. The hCG levels are higher in pregnancies with multiple fetuses. During the first trimester, GnRH locally produced by cytotrophoblasts appears to regulate hCG production by a paracrine mechanism. The suppression of hCG release during the second half of pregnancy is attributed to negative feedback by placental progesterone or other steroids produced by the fetus. Progesterone secretion by the corpus luteum is maximal 4 to 5 weeks after conception and declines, although hCG levels are still rising. Corpus luteum refractoriness to hCG results from receptor desensitization and the rising levels of placental estrogens. From weeks 7 to 10 of gestation, steroid production by the placenta gradually replaces steroid production by the corpus luteum. Removal of the corpus luteum after week 10 does not terminate the pregnancy. Other placenta-derived growth regulators affecting hCG production are activin, inhibin, and transforming growth factors α and β.

hCG has been shown to increase progesterone production by the trophoblast. Therefore, hCG may have a critical role in maintaining placental steroidogenesis throughout pregnancy and replacing luteal progesterone secretion after week 10, when the ovaries are no longer needed to maintain pregnancy. Other important targets of hCG are the fetal adrenal gland and testes. hCG is important in the sexual differentiation of the male fetus, which depends on testosterone production by the fetal testes. Peak testosterone production occurs 11 to 17 weeks after conception. This timing coincides with peak hCG production and predates the functional maturity of the fetal hypothalamic-pituitary axis. (Fetal LH levels are low.) hCG appears to regulate fetal Leydig cell proliferation as well as testosterone biosynthesis, especially because LH/hCG receptors are present in the early fetal testes. The role of hCG in fetal ovarian development is less clear because LH/hCG receptors are not present on fetal ovaries. There are some indications that increased hCG and thyroxine levels accompany maternal morning sickness, but a cause-and-effect relationship is not established.

Human placental lactogen

hPL has lactogenic and growth hormone (GH)–like actions. As a result, it is also called *human chorionic somatomammotropin* and *chorionic growth hormone*. This hormone is synthesized by syncytiotrophoblasts and secreted into the maternal circulation, where its levels gradually rise from the third week of pregnancy until term. Although the same cells produce hPL that produce hCG, the pattern of hPL secretion is different, indicating the possibility of control by different regulatory mechanisms. The hormone is composed of a single chain of 191 amino acids with two disulfide bridges and has a molecular weight of about 22 kDa. Its structure and function resemble those of prolactin (PRL) and growth hormone (GH).

hPL promotes cell specialization in the mammary gland, but is less potent than PRL in stimulating milk production and much less potent than GH in stimulating growth. Its main function is to alter fuel availability by antagonizing maternal glucose consumption and enhancing fat mobilization. This ensures adequate fuel supplies for the fetus. Its effects on carbohydrate, protein, and fat metabolism are similar to those of GH. The amniotic fluid also contains large amounts of PRL, produced mainly by the decidual compartments. Decidual PRL is indistinguishable from pituitary PRL, but its function and regulation are unclear.

Ovary and fetoplacental unit both produce steroids during pregnancy.

Progesterone is required to maintain normal human pregnancy. During the early stages of pregnancy (approximately the first 8 weeks), the ovaries produce most of the sex steroids; the corpus luteum produces primarily progesterone and estrogen. As the placenta develops, trophoblast cells gradually take over a major role in the production of progesterone and estrogen. Although the corpus luteum continues to secrete progesterone, the placenta secretes most of the progesterone. Levels of progesterone gradually rise during early pregnancy and plateau during the transition period from corpus luteal to placental production (see Fig. 38.6). Thereafter, plasma progesterone levels continue to rise and reach about 150 ng/mL near the end of pregnancy. Two major estrogens, estradiol and estriol, gradually rise during the first half of pregnancy and steeply increase in the latter half of pregnancy to more than 25 ng/mL near term.

Progesterone and estrogen have numerous functions throughout gestation. Estrogens increase the size of the uterus and uterine blood flow, are critical in the timing of implantation of the embryo into the uterine wall, induce the formation of uterine receptors for progesterone and oxytocin, enhance fetal organ development, stimulate maternal hepatic protein production, and increase the mass of breast and adipose tissues. Progesterone is essential for maintaining the uterus and early embryo, inhibits myometrial contractions, and suppresses maternal immunologic responses to fetal antigens. Progesterone also serves as a precursor for steroid production by the fetal adrenal glands and plays a role in the onset of parturition.

Beginning at approximately week 8 of gestation, the placenta carries out progesterone production, but its synthesis requires cholesterol, which is contributed from the mother. The placenta cannot make significant amounts of cholesterol from acetate and obtains it from the maternal blood via low-density lipoprotein (LDL) cholesterol. Trophoblast cells have **LDL receptors**, which bind the LDL cholesterol and internalize it. Cholesterol side-chain cleavage enzyme releases and uses free cholesterol to synthesize pregnenolone. 3β-Hydroxysteroid dehydrogenase converts pregnenolone to progesterone.

Dehydroepiandrosterone sulfate

The placenta lacks the 17α-hydroxylase for converting pregnenolone or progesterone to androgens (the precursors of the estrogens). Maternal 17α-hydroxyprogesterone can be measured during the first trimester and serves as a marker of corpus luteum function, because the placenta cannot make this steroid. The production of estrogens (estradiol, estrone, and estriol) during gestation requires cooperation between

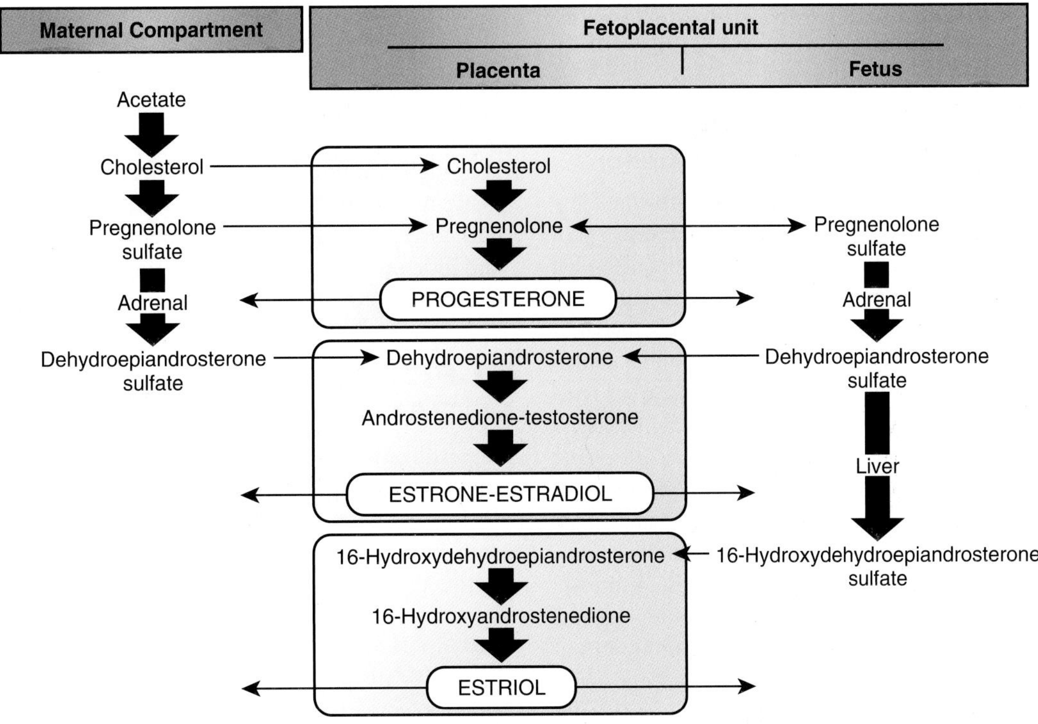

Figure 38.7 The fetoplacental unit and steroidogenesis. Note that estriol is the product of reactions occurring in the fetal adrenal, fetal liver, and placenta.

the maternal compartment and the placental and fetal compartments, referred to as the *fetoplacental unit* (Fig. 38.7). To produce estrogens, the placenta uses androgenic substrates derived from both the fetus and the mother. The primary androgenic precursor is **dehydroepiandrosterone sulfate (DHEAS)**, which is produced by the fetal zone of the fetal adrenal gland. The fetal adrenal gland is extremely active in the production of steroid hormones, but because it lacks 3β-hydroxysteroid dehydrogenase, it cannot make progesterone. Therefore, the fetal adrenals use progesterone from the placenta to produce androgens, which are ultimately sulfated in the adrenal glands. The conjugation of androgenic precursors to sulfates ensures greater water solubility, aids in their transport, and reduces their biologic activity while in the fetal circulation. DHEAS diffuses into the placenta and is cleaved by a sulfatase to yield a nonconjugated androgenic precursor. The placenta has an active aromatase that converts androgenic precursors to estradiol and estrone.

Estriol

The major estrogen produced during human pregnancy is **estriol**, which has relatively weak estrogenic activity. A unique biosynthetic pathway produces estriol (see Fig. 38.7). 16-Hydroxylation converts DHEAS from the fetal adrenal to 16-hydroxydehydroepiandrosterone sulfate (16-OH-DHEAS) in the fetal liver and, to a lesser extent, the fetal adrenal gland. This step is followed by desulfation using a placental sulfatase and conversion by 3β-hydroxysteroid dehydrogenase to 16-hydroxyandrostenedione, which is subsequently aromatized in the placenta to estriol. Although 16-OH-DHEAS can be made in the maternal adrenal from maternal DHEAS, the levels are low. It has been estimated that 90% of the estriol is derived from the fetal 16-OH-DHEAS. Therefore, the levels of estriol in plasma, amniotic fluid, or urine are used as an index of fetal well-being. Low levels of estriol would indicate potential fetal distress, although rare inherited sulfatase deficiencies can also lead to low estriol.

Maternal metabolism and endocrine function change in response to gestation.

The pregnant woman provides nutrients for her growing fetus, is the sole source of fetal oxygen, and removes fetal waste products. These functions necessitate significant adjustments in her pulmonary, cardiovascular, renal, metabolic, and endocrine systems. Among the most notable changes during pregnancy are hyperventilation, reduced arterial blood P_{CO_2} and osmolality, increased blood volume and cardiac output, increased renal blood flow and glomerular filtration rate, and substantial weight gain. These are brought about by the rising levels of estrogens, progesterone, hPL, and other placental hormones and by mechanical factors such as the expanding size of the uterus and the development of uterine and placental circulations.

The maternal endocrine system undergoes significant adaptations. The high levels of sex steroids suppress the hypothalamic-pituitary-ovarian axis. Consequently, circulating gonadotropins are low, and ovulation does not occur during pregnancy. In contrast, the rising levels of estrogens stimulate PRL release. PRL levels begin to rise during the first trimester, increasing gradually to reach a level 10 times higher near term (see Fig. 38.6). Pituitary lactotrophs undergo hyperplasia and hypertrophy and mostly account for the enlargement of the pregnant woman's pituitary gland. However, somatotrophs that produce GH are reduced, and GH levels are low throughout pregnancy.

The thyroid gland enlarges, but TSH levels are in the normal nonpregnant range. Triiodothyronine (T_3) and thyroxine (T_4) increase, but thyroxine-binding globulin (TBG) also increases in response to the rising levels of estrogen, which are known to stimulate TBG synthesis. Therefore, the pregnant woman remains in the euthyroid state. The parathyroid glands and their hormone, parathyroid hormone (PTH), increase mostly during the third trimester. PTH enhances calcium mobilization from maternal bone stores in response to the growing demands of the fetus for calcium. The rate of adrenal secretion of mineralocorticoids and glucocorticoids increases, and plasma-free cortisol is higher because of its displacement from transcortin, the cortisol-binding globulin, by progesterone, but hypercortisolism is not apparent during pregnancy.

Maternal adrenocorticotropic hormone (ACTH) levels rise throughout pregnancy such that, at 37 weeks, levels are fivefold higher than that prior to pregnancy. ACTH falls 50% just before parturition and then increases 15-fold during the stress of delivery. Both the maternal pituitary and the placenta contribute to the increase in ACTH.

Maternal metabolism responds in several ways to the increasing nutritional demands of the fetus. The major net weight gain of the mother occurs during the first half of gestation, mostly resulting from fat deposition. This response is attributed to progesterone, which increases appetite and diverts glucose into fat synthesis. The extra fat stores are used as an energy source later in pregnancy, when the metabolic requirements of the fetus are at their peak, and also during periods of starvation. Several maternal and placental hormones act together to provide a constant supply of metabolic fuels to the fetus. Toward the second half of gestation, the mother develops a resistance to insulin. This is brought about by combined effects of hormones antagonistic to insulin action, such as GH, PRL, hPL, glucagon, and cortisol. As a result, maternal use of glucose declines and gluconeogenesis increases, maximizing the availability of glucose to the fetus.

FETAL DEVELOPMENT AND PARTURITION

At fertilization, gonadal hormones determine genetic sex and control the subsequent sexual differentiation of the fetus. The fetal endocrine system participates in growth and development of the fetus, and interactions of fetal and maternal factors regulate parturition.

Fetal endocrine system develops early to regulate fetal homeostasis.

The protective intrauterine environment postpones the initiation of some physiologic functions that are essential for life after birth. For example, the fetal lungs and kidneys do not act as organs of gas exchange and excretion because the placenta carries out their functions. Constant isothermal surroundings alleviate the need to expend calories to maintain body temperature. The gastrointestinal (GI) tract does not carry out digestive activities, and fetal bones and muscles do not support weight or locomotion. Being exposed to low levels of external stimuli and environmental insults, the fetal nervous and immune systems develop slowly. However, the fetal endocrine system plays a vital role in fetal growth and development and homeostasis.

Given that the blood–placental barrier excludes most protein and polypeptide hormones from the fetus, the maternal endocrine system has little direct influence on the fetus. Instead, the fetus is almost self-sufficient in its hormonal requirements. Notable exceptions are some of the steroid hormones that are produced by the fetoplacental unit and cross easily between the different compartments to carry out integrated functions in both the fetus and the mother. By and large, fetal hormones perform the same functions as in the adult, but they also subserve unique processes such as sexual differentiation and the initiation of labor.

Fetal hormones

The fetal hypothalamic nuclei, including their releasing hormones, such as TRH, GnRH, and several of the neurotransmitters, are well developed by 12 weeks of gestation. At about week 4, the anterior pituitary begins its development from the Rathke pouch, an ectodermal evagination from the roof of the fetal mouth, and by week 8, most anterior pituitary hormones can be identified. The posterior pituitary or neurohypophysis is an evagination from the floor of the primitive hypothalamus, and its nuclei, supraoptic and paraventricular, containing arginine vasopressin (AVP) and oxytocin, can be detected around week 14. The hypothalamic-pituitary axis is well developed by midgestation, and well-differentiated hormone-producing cells in the anterior pituitary are also apparent at this time. Whether the fetal pituitary is tightly regulated by hypothalamic hormones or possesses some autonomy is unclear. However, the release of pituitary hormones can occur prior to the establishment of the portal system, indicating that the hypothalamic-releasing hormones may diffuse down to the pituitary from the hypothalamic sites.

Experiments with long-term catheterization of monkey fetuses indicate that by the last trimester, both LH and testosterone increase in response to GnRH administration. In the adult, GH largely regulates the secretion of the insulin-like growth factors (IGF-1 and IGF-2) from the liver. In the fetus, this may not be the case, because newborns with low GH have normal birth size; therefore, other mechanisms may control the secretion of IGFs in the fetus. GH levels increase in the fetus until midgestation and decline thereafter when fetal weight is increasing significantly, representing another dichotomy in GH and IGF in the fetus versus postnatal life. PRL levels increase in the fetus throughout gestation and can be inhibited by an exogenous dopamine agonist. Although the role of PRL in fetal growth is unclear, it has been implicated in adrenal and lung function, as well as in the regulation of amniotic fluid volume.

The fetal adrenal glands are unique in both structure and function. At month 4 of gestation, they are larger than the kidneys, as a result of the development of a fetal zone that constitutes 75% to 80% of the whole gland. The outer definitive zone forms the adult adrenal cortex, whereas the deeper fetal zone involutes after birth; the reason for the involution is unknown, but it is not caused by the withdrawal of ACTH support. The fetal zone produces large amounts of DHEAS and provides androgenic precursors for estrogen synthesis by the placenta (see Fig. 38.7). The definitive zone produces cortisol, which has multiple functions during fetal life, including the promotion of pancreas and lung maturation, the induction of liver enzymes, the promotion of intestinal tract cytodifferentiation, and possibly, the initiation of labor. ACTH is the main regulator of fetal adrenal steroidogenesis, partly evidenced by the observation that anencephalic fetuses have low ACTH and the fetal zone is small. The adrenal medulla develops by about week 10 and is capable of producing epinephrine and norepinephrine.

Fetal growth

The rate of fetal growth increases significantly during the last trimester. Surprisingly, GH of maternal, placental, or fetal origin has little effect on fetal growth, as judged by the normal weight of hypopituitary dwarfs or anencephalic fetuses. Fetal insulin is the most important hormone in regulating fetal growth. Glucose is the main metabolic fuel for the fetus. Fetal insulin, produced by the pancreas by week 12 of gestation, regulates tissue glucose use, controls liver glycogen storage, and facilitates fat deposition. It does not control the supply of glucose, however; this is determined by maternal gluconeogenesis and placental glucose transport. The release of insulin in the fetus is relatively constant, increasing only slightly in response to a rapid rise in blood glucose levels. When blood glucose levels are chronically elevated, as in women with **gestational diabetes**, the fetal pancreas becomes enlarged and circulating insulin levels increase. Consequently, fetal growth is accelerated, and infants of women with uncontrolled diabetes are overweight.

Calcium is in large demand because of the rapid growth and large amount of bone formation of the fetus during pregnancy. Maternal calcium is highly important for meeting this fetal requirement. During pregnancy, maternal calcium intake increases, and 1,25-dihydroxyvitamin D_3 and PTH increase to meet the increased calcium demands of the fetus. In the mother, total plasma calcium and phosphate decline without affecting free calcium. The placenta has a specialized calcium pump that transfers calcium to the fetus, resulting in sustained increases in calcium and phosphate throughout pregnancy. Although PTH and calcitonin are evident in the fetus near week 12 of gestation, their role in regulating fetal calcium is unclear. In addition, the placenta has 1α-hydroxylase and can convert 25-hydroxyvitamin D_3 to 1,25-dihydroxyvitamin D_3. At the end of gestation, calcium and phosphate levels in the fetus are higher than those in the mother. However, after delivery, neonatal calcium levels decrease and PTH levels rise to raise the levels of serum calcium.

Sex chromosomes dictate the development of the fetal gonads.

Sexual differentiation begins at the time of fertilization by a random unification of an X-bearing egg with either an X-bearing or Y-bearing spermatozoon and continues during early embryonic life with the development of male or female gonads. Therefore, at the time of fertilization, **chromosomal sex** or **genetic sex** is determined. Gonadal hormones that act at critical times during organogenesis control sexual differentiation. Testicular hormones induce masculinization, whereas feminization does not require (female) hormonal intervention. The process of sexual development is incomplete at birth; the secondary sex characteristics and a functional reproductive system are not fully developed until puberty.

Human somatic cells have 44 autosomes and 2 sex chromosomes. The female is **homogametic** (having two X chromosomes) and produces similar X-bearing ova. The male is **heterogametic** (having one X and one Y chromosomes) and generates two populations of spermatozoa, one with X chromosomes and the other with Y chromosomes. The X chromosome is large, containing 80 to 90 genes responsible for many vital functions. The Y chromosome is much smaller, carrying only a few genes responsible for testicular development and normal spermatogenesis. Mutation of genes on an X chromosome results in the transmission of X-linked traits, such as hemophilia and color-blindness, to male offspring, who, unlike females, cannot compensate with an unaffected allele.

Theoretically, by having two X chromosomes, the female has an advantage over the male, who has only one. However, because one of the X chromosomes is inactivated at the morula stage, the advantage is lost. Each cell randomly inactivates either the paternally or the maternally derived X chromosome, and this continues throughout the cell's progeny. The inactivated X chromosome is recognized cytologically as the sex chromatin or **Barr body**. In males with more than one X chromosome or in females with more than two, extra X chromosomes are inactivated and only one remains functional. This does not apply to the germ cells. The single active X chromosome of the spermatogonium becomes inactivated during meiosis, and a functional X chromosome is not necessary for the formation of fertile sperm. The oogonium, however, reactivates its second X chromosome, and both are functional in oocytes and important for normal oocyte development.

Testicular differentiation requires a Y chromosome and occurs even in the presence of two or more X chromosomes. A testis-determining gene designated ***SRY*** (sex-determining region, Y chromosome) regulates **gonadal sex** determination. Located on the short arm of the Y chromosome, *SRY* encodes an HMG-box transcription factor. The presence or absence of *SRY* in the genome determines whether male or female gonadal differentiation takes place. Thus, in normal XX (female) fetuses that lack a Y chromosome, ovaries, rather than testes, develop.

Whether possessing the XX or XY karyotype, every embryo initially goes through an **ambisexual stage** and has the potential to acquire either masculine or feminine characteristics. A 4- to 6-week-old human embryo possesses indifferent gonads and undifferentiated pituitary, hypothalamus, and higher brain centers.

The indifferent gonad consists of a **genital ridge**, derived from coelomic epithelium and underlying mesenchyme, and primordial germ cells, which migrate from the yolk sac to the genital ridges. Depending on genetic programming, the inner **medullary tissue** becomes the testicular component and the outer **cortical tissue** develops into an ovary. The primordial germ cells become oogonia or spermatogonia. In an XY fetus, the testes differentiate first. Between weeks 6 and 8 of gestation, the cortex regresses, the medulla enlarges, and the seminiferous tubules become distinguishable. Sertoli cells line the basement membrane of the tubules, and Leydig cells undergo rapid proliferation. Development of the ovary begins at weeks 9 to 10. Primordial follicles, composed of oocytes surrounded by a single layer of granulosa cells, are discernible in the cortex between weeks 11 and 12 and reach maximal development by weeks 20 to 25.

Hormones determine the differentiation of the genital ducts.

During the indifferent stage, the primordial genital ducts are the paired **mesonephric (wolffian) ducts** and the paired **paramesonephric (müllerian) ducts**. In the normal male fetus, the wolffian ducts give rise to the epididymis, vas deferens, seminal vesicles, and ejaculatory ducts, whereas the müllerian ducts become vestigial. In the normal female fetus, the müllerian ducts fuse at the midline and develop into the oviducts, uterus, cervix, and upper portion of the vagina, whereas the wolffian ducts regress (Fig. 38.8). The mesonephros is the embryonic kidney.

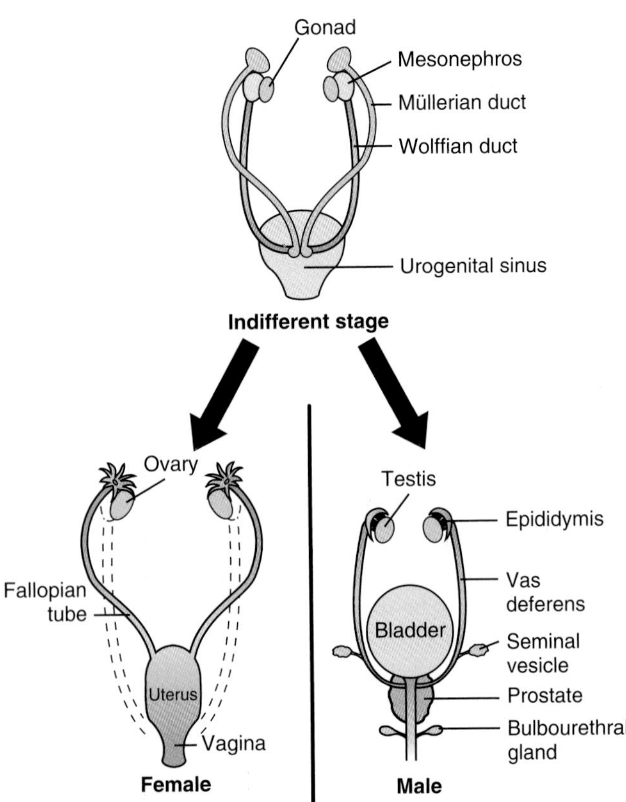

Figure 38.8 Differentiation of the internal genitalia and the primordial ducts.

The fetal testes differentiate between weeks 6 and 8 of gestation. Leydig cells, either autonomously or under regulation by hCG, start producing testosterone. Sertoli cells produce two nonsteroidal compounds. One is **müllerian-inhibiting substance** (**MIS**; also called *antimüllerian hormone* or *AMH*), a large glycoprotein with a sequence homologous to inhibin and transforming growth factor β, which inhibits cell division of the müllerian ducts. The second is **androgen-binding protein (ABP)**, which binds testosterone. Peak production of these compounds occurs between weeks 9 and 12, coinciding with the time of differentiation of the internal genitalia along the male line. The ovary, which differentiates later, does not produce hormones and has a passive role.

The primordial external genitalia include the genital tubercle, genital swellings, urethral folds, and urogenital sinus. Differentiation of the external genitalia also occurs between weeks 8 and 12, and is determined by the presence or absence of male sex hormones. Differentiation along the male line requires active **5α-reductase**, the enzyme that converts testosterone to dihydrotestosterone (DHT). Without DHT, regardless of the genetic, gonadal, or hormonal sex, the external genitalia develop along the female pattern. The structures that develop from the primordial structures are illustrated in Figure 38.9, and a summary of sexual differentiation during fetal life is shown in Figure 38.10. Androgen-dependent differentiation occurs only during fetal life and is thereafter irreversible. However, the exposure of females to high androgens either before or after birth can cause clitoral hypertrophy. Testicular descent into the scrotum, which occurs during the third trimester, is also controlled by androgens.

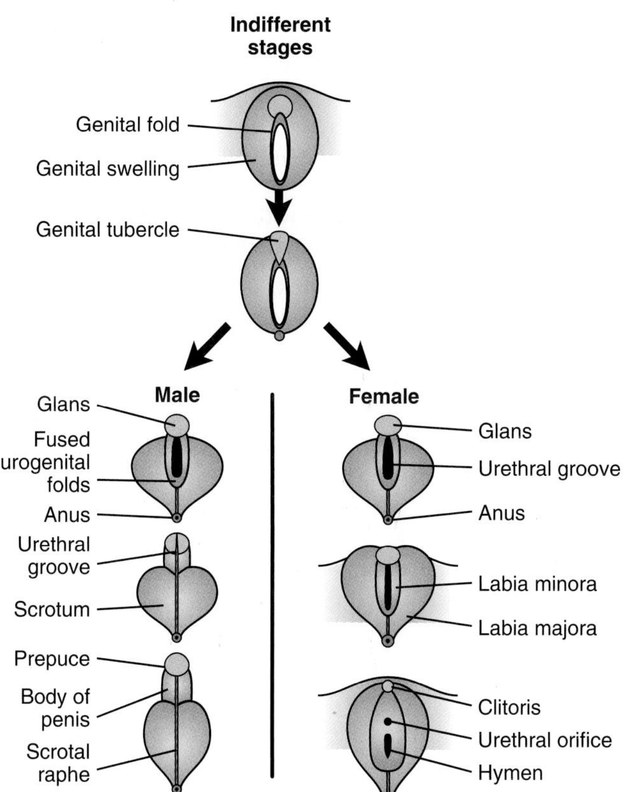

Figure 38.9 Differentiation of the external genitalia from bipotential primordial structures.

Maternal and fetal factors interact to induce parturition.

Parturition is the culmination of a pregnancy. The duration of pregnancy in women averages 270 ± 14 days from the time of fertilization. The interactions of fetal and maternal factors regulate parturition or the onset of birth. Uncoordinated uterine contractions start about 1 month before the end of gestation. Strong rhythmic contractions that may last several hours and eventually generate enough force to expel the fetus initiate the termination of pregnancy. Hormones and mechanical factors regulate the contraction of the uterine muscle. The hormones include progesterone, estrogen, prostaglandins, oxytocin, and relaxin. The mechanical factors include distention of the uterine muscle and stretching or irritation of the cervix.

Progesterone hyperpolarizes myometrial cells, lowers their excitability, and suppresses uterine contractions. It also prevents the release of phospholipase A_2, the rate-limiting enzyme in prostaglandin synthesis. Estrogen, in general, has the opposite effects. The maintenance of uterine quiescence throughout gestation, preventing premature delivery, is called the **progesterone block**. In many species, a sharp decline in the circulating levels of progesterone and a concomitant rise in estrogen precede birth. In humans, progesterone does not fall significantly before delivery. However, a rise in placental progesterone-binding protein or a decline in the number of myometrial progesterone receptors may alter its effective concentration.

Prostaglandins F_{2a} and E_2 are potent stimulators of uterine contractions and also cause significant ripening of the cervix and its dilation. They increase intracellular calcium concentrations of myometrial cells and activate the actin–myosin contractile apparatus. Shortly before the onset of parturition, the concentration of prostaglandins in amniotic fluid rises abruptly. The myometrium, decidua, and chorion produce prostaglandins. Aspirin and indomethacin, inhibitors of prostaglandin synthesis, delay or prolong parturition.

Oxytocin is also a potent stimulator of uterine contractions, and its release from both maternal and fetal pituitaries increases during labor. Oxytocin is used clinically to induce labor. The functional significance of oxytocin is that it helps expel the fetus from the uterus and, by contracting uterine muscles, reduces uterine bleeding when bleeding may be significant after delivery. Interestingly, oxytocin levels do not rise at the time of parturition.

Relaxin, a large polypeptide hormone produced by the corpus luteum and the decidua, assists parturition by softening the cervix, permitting the eventual passage of the fetus, and by increasing oxytocin receptors. However, the relative role of relaxin in parturition in humans is unclear, because its levels do not rise toward the end of gestation. Relaxin reaches its peak during the first trimester, declines to about half, and remains unchanged throughout the remainder of pregnancy.

The fetus may play a role in initiating labor. In sheep, the concentrations of ACTH and cortisol in the fetal plasma rise during the last 2 to 3 days of gestation. Ablation of the fetal lamb pituitary or removal of the adrenals prolongs gestation,

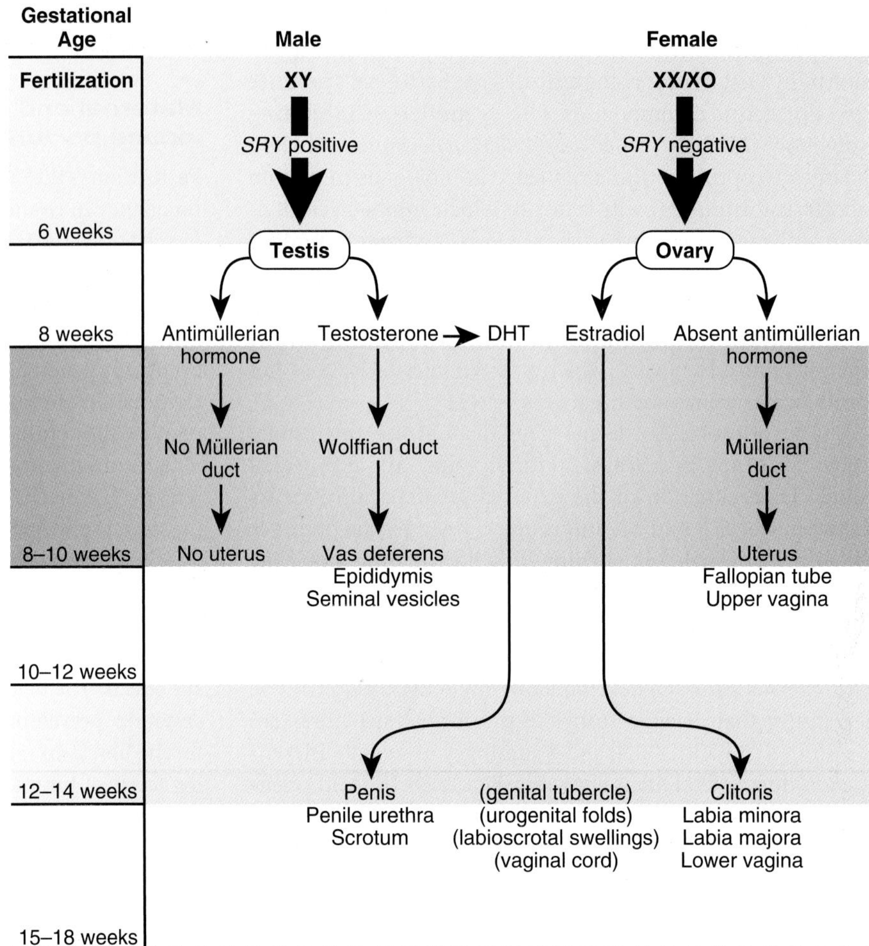

Figure 38.10 The process of sexual differentiation and its time course. DHT, dihydrotestosterone; *SRY*, sex-determining region, Y chromosome.

whereas administration of ACTH or cortisol leads to premature delivery. Cortisol enhances the conversion of progesterone to estradiol, changing the progesterone/estrogen ratio, and increases the production of prostaglandins. The role of cortisol and ACTH, however, has not been established in humans. Anencephalic or adrenal-deficient fetuses, which lack a pituitary and have atrophied adrenal glands, have an unpredictable length of gestation. These pregnancies also exhibit low estrogen levels because of the lack of adrenal androgen precursors. Injections of ACTH and cortisol in late pregnancy do not induce labor. Interestingly, the administration of estrogens to the cervix causes ripening, probably by increasing the secretion of prostaglandins.

POSTPARTUM AND PREPUBERTAL PERIODS

Postpartum is the period beginning immediately after childbirth and extending about 6 weeks. It is the time after birth in which the mother's body, including hormonal levels and uterine size, returns to nonpregnant conditions. Lactation (production and secretion of milk) is controlled by pituitary and ovarian hormones, requires suckling for continued milk production, and is the major source of nutrition for the newborn. As the child grows, puberty will occur around age 10 to 12 years because the hypothalamus activates secretion of pituitary hormones that cause secretion of estrogens and androgens from the gonads and adrenals during that time. Alterations in hormone secretion lead to abnormal onset of puberty and gonadal development.

Multiple hormones stimulate and sustain mammogenesis and lactogenesis.

Lactation (the secretion of milk) occurs at the final phase of the reproductive process. Several hormones participate in **mammogenesis**, the differentiation and growth of the mammary glands, and in the production and delivery of milk. **Lactogenesis** is milk production by alveolar cells. PRL regulates **galactopoiesis**, the maintenance of lactation. **Milk ejection** is the process by which stored milk is released from the mammary glands by the action of oxytocin.

Mammogenesis occurs at three distinct periods: embryonic, pubertal, and gestational. The **mammary glands** begin to differentiate in the pectoral region as an ectodermal thickening on the epidermal ridge during weeks 7 to 8 of fetal life. The prospective mammary glands lie along bilateral mammary ridges or milk lines extending from the axilla to groin on the ventral side of the fetus. Most of the ridge disintegrates except in the axillary region. However, in mammals with serially repeated nipples, a distinct milk line with several nipples persists, accounting for the accessory nipples that can occur in both sexes, although rarely. Mammary buds are derived from surface epithelium, which invades the underlying mesenchyme. During the fifth month, the buds elongate, branch, and sprout, eventually forming the **lactiferous ducts**—the primary milk ducts. They continue to branch and grow throughout life. The ducts unite, grow, and extend to the site of the future nipple. The primary buds give rise to secondary buds, which are separated into lobules by connective tissue and become surrounded by **myoepithelial cells** derived from epithelial progenitors. In response to oxytocin, myoepithelial cells contract and expel milk from the duct. The **nipple** and **areola**, which are first recognized as circular areas, are formed during the eighth month of gestation. The development of the mammary glands *in utero* appears to be independent of hormones but is influenced by paracrine interactions between the mesenchyme and epithelium.

The mammary glands of male and female infants are identical. Although underdeveloped, they have the capacity to respond to hormones, revealed by the secretion of small amounts of milk (witch's milk) in many newborns. Witch's milk results from the responsiveness of the fetal mammary tissue to lactogenic hormones of pregnancy and the withdrawal of placental steroids at birth. Sexual dimorphism in breast development begins at the onset of puberty. The male breast is fully developed at about age 20 years and is similar to the female breast at an early stage of puberty.

In females, estrogen exerts a major influence on breast growth at puberty. The first response to estrogen is an increase in size and pigmentation of the areola and accelerated deposition of adipose and connective tissues. In association with menstrual cycles, estrogen stimulates the growth and branching of the ducts, whereas progesterone acts primarily on the alveolar components. The action of both hormones, however, requires synergism with PRL, GH, insulin, cortisol, and thyroxine.

The mammary glands undergo significant changes during pregnancy. The ducts become elaborate during the first trimester, and new lobules and alveoli are formed in the second trimester. The terminal alveolar cells differentiate into secretory cells, replacing most of the connective tissue. The development of the secretory capability requires estrogen, progesterone, PRL, and placental lactogen. Insulin, cortisol, and several growth factors support their action. Lactogenesis begins during the fifth month of gestation, but only **colostrum** (initial milk) is produced. Elevated progesterone levels, which antagonize the action of PRL, prevent full lactation during pregnancy. The ovarian steroids synergize with PRL in stimulating mammary growth but antagonize its actions in promoting milk secretion.

Lactogenesis is fully expressed only after parturition, on the withdrawal of placental steroids. Lactating women produce up to 600 mL of milk each day, increasing to 800 to 1,100 mL/d by the sixth postpartum month. Milk is isosmotic with plasma, and its main constituents include proteins, such as casein and lactalbumin, lipids, and lactose. The composition of milk changes with the stage of lactation. Colostrum, produced in small quantities during the first postpartum days, is higher in protein, sodium, and chloride content and lower in lactose and potassium than normal milk. It also contains immunoglobulin A, macrophages, and lymphocytes, which provide passive immunity to the infant by acting on the GI tract. During the first 2 to 3 weeks, the protein content of milk decreases, whereas that of lipids, lactose, and water-soluble vitamins increases.

The milk-secreting **alveolar cells** form a single layer of epithelial cells, joined by junctional complexes (Fig. 38.11). The bases of the cells abut on the contractile myoepithelial cells, and their luminal surface is enriched with microvilli. They have a well-developed endoplasmic reticulum and Golgi apparatus and numerous mitochondria and lipid droplets. Alveolar cells contain plasma membrane receptors for PRL, which can be internalized after binding to the hormone. In synergism with insulin and glucocorticoids, PRL is critical for lactogenesis, promotes mammary cell division and differentiation, and increases the synthesis of milk constituents. This hormone also stimulates the synthesis of casein by increasing its transcription rate and stabilizing its messenger RNA (mRNA) and stimulates enzymes that regulate the production of lactose.

Suckling reflex maintains lactation and inhibits ovulation.

The suckling reflex is central to the maintenance of lactation in that it coordinates the release of PRL and oxytocin and delays the onset of ovulation. Lactation involves two components, milk secretion (synthesis and release) and milk removal, which are regulated independently. Milk secretion is a continuous process, whereas milk removal is intermittent. Milk secretion involves the synthesis of milk constituents by the alveolar cells, their intracellular transport, and the subsequent release of formed milk into the alveolar lumen (see Fig. 38.11). PRL is the major regulator of milk secretion in

Clinical Focus / 38.2

Pharmacologic Induction and Augmentation of Labor

Several drugs are currently used to assist in the therapeutic induction and augmentation of labor. Therapeutic induction implies that the use of a drug initiates labor. Augmentation indicates that labor has started and that a therapeutic agent further stimulates the process.

Oxytocin, the natural hormone produced from the posterior pituitary, is widely used to induce and augment labor. Several synthetic forms of oxytocin can be used by intravenous routes. Prostaglandins (F_{2a} and E_2) have also been used to induce and augment labor and cervical ripening. They promote dilatation and effacement of the cervix and can be used intravaginally, intravenously, or intra-amniotically. Another therapeutic agent being tested for efficacy in labor induction and augmentation is mifepristone (RU-486), a progesterone receptor blocker. It is used to induce labor and increase the sensitivity of the uterus to oxytocin and prostaglandins. Although animal studies suggest that RU-486 may be useful for cervical ripening, there have been mixed results in studies with humans, showing positive effects in small studies but no difference from placebo in larger studies.

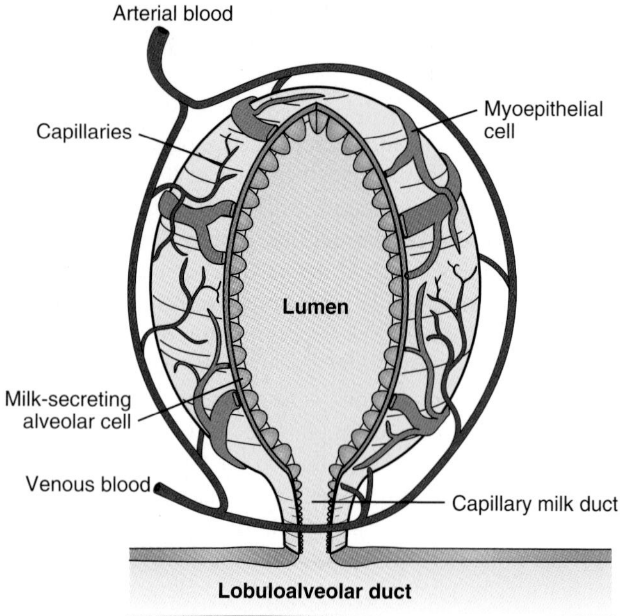

Figure 38.11 The structure of a mammary alveolus. A meshwork of contractile myoepithelial cells surrounds milk-producing cells.

women and most other mammals. Oxytocin is responsible for milk removal by activating **milk ejection** or **milk letdown**.

The stimulation of sensory nerves in the breast by the infant initiates the **suckling reflex**. Unlike ordinary reflexes with only neural components, the afferent arc of the suckling reflex is neural and the efferent arc is hormonal. The suckling reflex increases the release of PRL, oxytocin, and ACTH and inhibits the secretion of gonadotropins (Fig. 38.12). The neuronal component is composed of sensory receptors in the nipple that initiate nerve impulses in response to breast stimulation. These impulses reach the hypothalamus via ascending fibers in the spinal cord and then via the mesencephalon. Eventually, fibers terminating in the supraoptic and paraventricular nuclei trigger the release of oxytocin from the posterior pituitary into the general circulation (see Chapter 31, "Hypothalamus and the Pituitary Gland"). On reaching the mammary glands, oxytocin induces the contraction of myoepithelial cells, increasing intramammary pressure and forcing the milk into the main collecting ducts. The milk ejection reflex can be conditioned; milk ejection can occur because of anticipation or in response to a baby's cry.

PRL levels, which are elevated by the end of gestation, decline by 50% within the first postpartum week and decrease to near pregestation levels by 6 months. Suckling elicits a rapid and significant rise in plasma PRL. The intensity and duration of nipple stimulation determine the amount released. The exact mechanism by which suckling triggers PRL release is unclear, but the suppression of dopamine, the major inhibitor of PRL release, and the stimulation of PRL-releasing factor(s) have been considered. Dopaminergic agonists that reduce PRL or the discontinuation of suckling can terminate lactation. Swollen alveoli can depress milk production by exerting local pressure, resulting in vascular stasis and alveolar regression.

Lactation is associated with the suppression of cyclicity and **anovulation**. The contraceptive effect of lactation is moderate in humans. In non-breastfeeding women, the menstrual cycle may return within 1 month after delivery, whereas *fully* lactating women have a period of several months of **lactational amenorrhea**, with the first few menstrual cycles being anovulatory. The cessation of cyclicity results from the combined effects of the act of suckling and elevated PRL levels. PRL suppresses ovulation by inhibiting pulsatile GnRH release, suppressing pituitary responsiveness to GnRH, reducing LH and FSH, and decreasing ovarian activity. It is also possible that PRL may inhibit the action of the low circulating levels of gonadotropins on ovarian cells. Thus, a direct inhibitory action of PRL on the ovary would suppress follicular development. Although lactation reduces fertility, numerous other methods of contraception are more reliable.

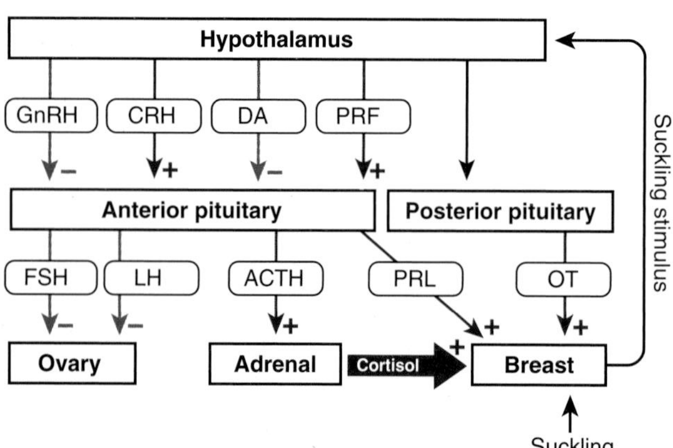

Figure 38.12 Effect of suckling on hypothalamic, pituitary, and adrenal hormones. GnRH, gonadotropin-releasing hormone; CRH, corticotropin-releasing hormone; DA, dopamine; PRF, prolactin-releasing factor; FSH, follicle-stimulating hormone; LH, luteinizing hormone; ACTH, adrenocorticotropic hormone; PRL, prolactin; OT, oxytocin.

Maturation of the hypothalamic GnRH pulse generator initiates the onset of puberty.

The onset of puberty depends on a sequence of maturational processes that begin during fetal life. The hypothalamic-pituitary-gonadal axis undergoes a prolonged and multiphasic activation–inactivation process. By midgestation, LH and FSH levels in fetal blood are elevated, reaching near-adult values. Experimental evidence suggests that the hypothalamic GnRH pulse generator is operative at this time, and gonadotropins are released in a pulsatile manner. The levels of FSH are lower in males than in females, probably because of suppression by fetal testosterone at midgestation. As the levels of placental steroids increase, they exert negative feedback on GnRH release, lowering LH and FSH levels toward the end of gestation.

Childhood

After birth, the newborn is deprived of maternal and placental steroids. The reduction in steroidal negative feedback stimulates gonadotropin secretion, which stimulates the

From Bench to Bedside / 38.1

Contraceptive Methods

Interfering with the association between the sperm and ovum, preventing ovulation or implantation, or terminating an early pregnancy can regulate fertility. Most current methods regulate fertility in women, with only a few contraceptives available for men.

Prevention of Egg and Sperm Contact

Methods to prevent contact between the germ cells include coitus interruptus (withdrawal before ejaculation), the rhythm method (no intercourse at times of the menstrual cycle, especially when an ovum is present in the oviduct), and barriers. Rhythm methods require meticulous monitoring of menstrual cycles and body temperature or cervical mucus characteristics, but can still fail as a result of the unpredictable duration of the follicular phase of the menstrual cycle. Both coitus interruptus and the rhythm method require cooperation and discipline in both partners. Barrier methods include condoms, diaphragms, and cervical caps. When combined with spermicidal agents, the barrier methods approach the high success rate of oral contraceptives. Condoms are the most widely used reversible contraceptives for men. Because they also provide protection against the transmission of venereal diseases and AIDS, their use has increased in recent years. However, if not used properly, condoms are associated with a high failure rate to prevent pregnancy. Diaphragms and cervical caps seal off the opening of the cervix. Postcoital douching with spermicides is not an effective contraceptive because some sperm enter the uterus and oviduct rapidly.

Surgical methods of contraception are generally considered nonreversible. Vasectomy is cutting of the two vasa deferentia, which prevents sperm from passing into the ejaculate. Tubal ligation is the closure or ligation of the oviducts. Restorative surgery for the reversal of a tubal ligation and a vasectomy can be performed; its success is limited.

Oral Contraceptives

Oral contraceptive steroids (birth control pills) prevent ovulation by reducing luteinizing hormone (LH) and follicle-stimulating hormone (FSH) secretion through negative feedback. The reduction in LH and FSH retards follicular development and prevents ovulation. Birth control pills also adversely affect the environment within the reproductive tract, making it unlikely for pregnancy to result even if fertilization were to occur. Exogenous estrogen and progesterone alter normal endometrial development, and progesterone thickens cervical mucus and reduces oviductal peristalsis, impeding gamete transport.

Noncontraceptive benefits of oral contraception include a reduction in excessive menstrual bleeding, alleviation of premenstrual syndrome, and some protection against pelvic inflammatory disease. In addition, oral contraception results in a significant decrease in incidence of endometrial and ovarian carcinoma. The most commonly reported adverse effects of oral contraception include nausea, headache, breast tenderness, water retention, and weight gain, some of which disappear after prolonged use. Breakthrough bleeding can also be expected in up to 30% of women during the first 3 months of use. Birth control pills are contraindicated in women with a history of thromboembolic disease, atherosclerosis, or breast carcinoma because of the effects of estrogen on these systems.

Emergency Contraception

Women can use emergency contraception to prevent pregnancy after known or suspected failure of birth control or after unprotected intercourse. Two prescription formulations are specifically available for emergency contraception, but the Food and Drug Administration has cleared for safety and efficacy the off-label use of oral contraceptives. The mechanism of action of emergency contraception is similar to that for standard oral contraceptive use: prevention of ovulation, insufficient corpus luteum function, and induction of an adverse environment in the female reproductive tract. Emergency contraception is most effective when started within the first 72 hours after unprotected intercourse. Thus, it has been recommended that physicians provide an advance prescription in states where the treatment is not available over the counter.

Prevention of Implantation

Long-acting progesterone preparations, high doses of estrogen, and progesterone receptor antagonists such as RU-486 (also called *mifepristone*), specifically interfere with zygote transport or implantation and can cause early termination of pregnancy. RU-486 blocks the action of the progesterone required for early pregnancy. Prostaglandins are given in combination with RU-486 to assist in the expulsion of the products of conception. The intrauterine device (IUD) also prevents implantation by provoking sterile inflammation of the endometrium and prostaglandin production. The contraceptive efficacy of IUDs, especially those impregnated with progestins, copper, or zinc, is high. The drawbacks to IUD use include a 5% to 6% rate of expulsion, cramps, and changes in uterine bleeding, depending on the type of IUD.

A considerable amount of basic research is focused on improving methods of contraception. **Immunocontraception**, or the use of antibodies against the egg, sperm or zygote, or against hormones such as hCG or FSH, is being studied in the laboratory. Another possibility might be to interfere with sperm capacitation or maturation in males.

gonads, resulting in transient increases in serum testosterone in male infants and estradiol in females. FSH levels in females are usually higher than those in males. At approximately 3 months of age, the levels of both gonadotropins and gonadal steroids are in the low-to-normal adult range. Circulating gonadotropins decline to low levels by 6 to 7 months in males and 1 to 2 years in females and remain suppressed until the onset of puberty.

Throughout childhood, the gonads are quiescent and plasma steroid levels are low. Gonadotropin release is also suppressed. Two mechanisms, both of which affect the hypothalamic GnRH pulse generator, explain the prepubertal restraint of gonadotropin secretion: One is a sex steroid–dependent mechanism that renders the pulse generator extremely sensitive to negative feedback by steroids. The other is an intrinsic central nervous system (CNS) inhibition of the GnRH pulse generator. γ-Aminobutyric acid (GABA) and endogenous opioid peptides in the CNS are the major inhibitors of the GnRH pulse generator. The combination of extreme sensitivity to negative feedback by steroids and CNS inhibition suppresses the amplitude and frequency of GnRH pulses, resulting in diminished secretion of LH, FSH, and gonadal steroids. Throughout this period of quiescence, the pituitary and the gonads can respond to exogenous GnRH and gonadotropins, but at a relatively low sensitivity.

Puberty

The hypothalamic-pituitary axis becomes reactivated during the late prepubertal period. This response involves a decrease in hypothalamic sensitivity to sex steroids and a reduction in the effectiveness of intrinsic CNS inhibition over the GnRH pulse generator. The exact mechanisms underlying the reduction in central inhibition are not understood, but they do involve stimulation of luteinizing hormone–releasing hormone (LHRH)–containing neurons by excitatory amino acids, neuropeptide Y, prostaglandins, and other signals. As a result of disinhibition, the frequency and amplitude of GnRH pulses increase. Initially, pulsatility is most prominent at night, entrained by deep sleep; later it becomes established throughout the 24-hour period. GnRH acts on the gonadotrophs of the anterior pituitary as a self-primer. It increases the number of GnRH receptors (upregulation) and augments the synthesis, storage, and secretion of the gonadotropins. The increased responsiveness of FSH to GnRH in females occurs earlier than that of LH, accounting for a higher FSH/LH ratio at the onset of puberty than during late puberty and adulthood. A reversal of the ratio is seen again after menopause.

The increased pulsatile GnRH release initiates a cascade of events. The sensitivity of gonadotrophs to GnRH is increased, the secretion of LH and FSH is augmented, the gonads become more responsive to the gonadotropins, and the secretion of gonadal hormones is stimulated. The rising circulating levels of gonadal steroids induce progressive development of the secondary sex characteristics and establish an adult pattern of negative feedback on the hypothalamic-pituitary axis. Activation of the positive-feedback mechanism in females and the capacity to exhibit an estrogen-induced LH surge is a late event, expressed in mid-puberty to late puberty.

The onset of puberty in humans begins on average at age 10 to 12 years (see Clinical Focus Box 37.2). Lasting 3 to 5 years, the process involves the development of secondary sex characteristics, a growth spurt, and the acquisition of fertility. Genetic, nutritional, climatic, and geographic factors determine the timing of puberty. Over the last 150 years, the age of puberty has declined by 2 to 3 months per decade; this pattern appears to correlate with improvements in nutrition and general health in Americans. Leptin is a hormone produced by adipose tissue that interacts with receptors in the hypothalamus to regulate satiety and energy metabolism. Leptin appears to have a permissive effect on the initiation of puberty, providing a signal to the CNS that there are sufficient energy stores to support reproduction. People with leptin deficiency resulting from mutations in the gene encoding leptin do not progress through puberty, but will do so if treated with exogenous hormone. See From Bench to Bedside Box 31.1 for an in-depth discussion of the role of leptin in regulation of the hypothalamic-pituitary axes.

Physical signs of puberty

The first physical signs of puberty in girls are breast budding, **thelarche**, and the appearance of pubic hair. Axillary hair growth and peak height spurt occur within 1 to 2 years. **Menarche**, the beginning of menstrual cycles, occurs at a median age of 12.8 years in Caucasian girls in the United States and at 12.2 years in African–American girls. The first few cycles are usually anovulatory. The first sign of puberty in boys is enlargement of the testes, followed by the appearance of pubic hair and enlargement of the penis. The peak growth spurt and appearance of axillary hair in boys usually occur 2 years later than in girls. The growth of facial hair, deepening of the voice, and broadening of the shoulders are late events in male pubertal maturation (Fig. 38.13).

Hormones other than gonadal steroids also regulate puberty. The adrenal androgens DHEA and DHEAS are primarily responsible for the development of pubic and axillary hair. Adrenal maturation (or **adrenarche**) precedes gonadal maturation (or **gonadarche**) by 2 years. The pubertal growth spurt requires a concerted action of sex steroids and GH. The principal mediator of GH is IGF-1. Plasma concentration of IGF-1 increases significantly during puberty, with peak levels observed earlier in girls than in boys. IGF-1 is essential for accelerated growth. The gonadal steroids appear to act primarily by augmenting pituitary GH release, which stimulates the production of IGF-1 in the liver and other tissues.

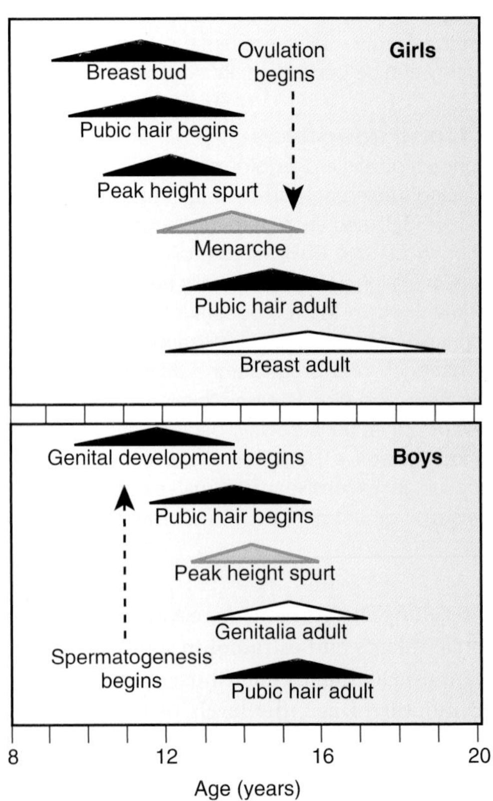

Figure 38.13 Peripubertal maturation of secondary sex characteristics in girls and boys.

SEXUAL DEVELOPMENT DISORDERS

Normal sexual development depends on a complex, orderly sequence of events that begins during early fetal life and is completed at puberty. Any deviation can result in infertility, sexual dysfunction, or various degrees of intersexuality or **hermaphroditism**. A true hermaphrodite possesses both ovarian and testicular tissues, either separate or combined as ovotestes. A **pseudohermaphrodite** has one type of gonads but a different degree of sexuality of the opposite sex. Sex is normally assigned according to the type of gonads.

Disorders of sexual differentiation can be classified as gonadal dysgenesis, female pseudohermaphroditism, male pseudohermaphroditism, or true hermaphroditism. Selected cases and their manifestations are briefly discussed here, as are disorders of pubertal development (see also Chapter 36, "Male Reproductive System," and Chapter 37, "Female Reproductive System").

Disorders of sexual differentiation occur when chromosomes, gonads, or development of reproductive anatomy is atypical.

Gonadal dysgenesis refers to incomplete differentiation of the gonads and is usually associated with sex chromosome abnormalities. These result from errors in the first or second meiotic division and occur by chromosomal nondisjunction, translocation, rearrangement, or deletion. The two most common disorders are Klinefelter syndrome (47,XXY) and Turner syndrome (45,XO). Because of a Y chromosome, a person with a 47,XXY karyotype has normal testicular function *in utero* in terms of testosterone and MIS production and no ambiguity of the genitalia at birth. The extra X chromosome, however, interferes with the development of the seminiferous tubules, which show extensive hyalinization and fibrosis, whereas the Leydig cells are hyperplastic. Such males have small testes, are azoospermic, and often exhibit some eunuchoidal features. Because of having only one X chromosome, a person with a 45,XO karyotype has no gonadal development during fetal life and is presented at birth as a phenotypic female. Given the absence of ovarian follicles, such patients have low levels of estrogens and primary amenorrhea and do not undergo normal pubertal development.

Female pseudohermaphrodites are 46,XX females with normal ovaries and internal genitalia, but with a different degree of virilization of the external genitalia, resulting from exposure to excessive androgens *in utero*. The most common cause is **congenital adrenal hyperplasia**, an inherited abnormality in adrenal steroid biosynthesis, with most cases of virilization resulting from 21-hydroxylase or 11β-hydroxylase deficiency (see Chapter 33, "Adrenal Gland"). In such cases, cortisol production is low, causing increased production of ACTH by activating the hypothalamic-pituitary axis (Fig. 38.14). The elevated ACTH

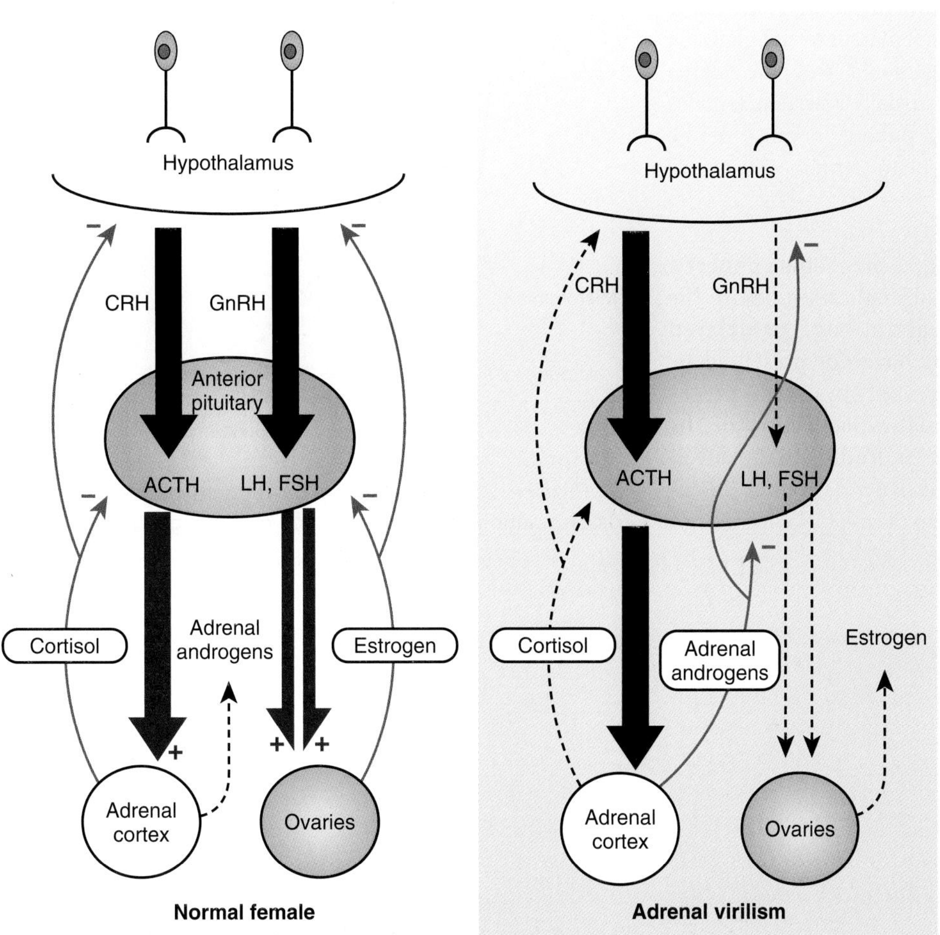

Figure 38.14 Hormonal interactions along the ovarian and adrenal axes during normal female development, compared with adrenal virilism. *Dashed arrows* indicate low production of the hormone. *Bold arrows* indicate increased hormone production. Plus and minus signs indicate positive and negative effects. GnRH, gonadotropin-releasing hormone; CRH, corticotropin-releasing hormone; FSH, follicle-stimulating hormone; LH, luteinizing hormone; ACTH, adrenocorticotropic hormone.

levels induce adrenal hyperplasia and an abnormal production of androgens and corticosteroid precursors. These infants are born with ambiguous external genitalia (i.e., clitoromegaly, labioscrotal fusion, or phallic urethra). The degree of virilization depends on the time of onset of excess fetal androgen production. When aldosterone levels are also affected, a life-threatening salt-wasting disease results. Untreated patients with congenital adrenal virilism develop progressive masculinization, amenorrhea, and infertility.

Male pseudohermaphrodites are 46,XY individuals with differentiated testes but underdeveloped and/or absent wolffian-derived structures and inadequate virilization of the external genitalia. These effects result from defects in testosterone biosynthesis, metabolism, or action. The 5α-reductase deficiency is an autosomal-recessive disorder caused by the inability to convert testosterone to DHT. Infants with 5α-reductase deficiency have ambiguous or female external genitalia and normal male internal genitalia (Fig. 38.15). They are often raised as females but undergo a complete or partial testosterone-dependent puberty, including enlargement of the penis, testicular descent, and the development of male psychosexual behavior. Azoospermia is common.

The **testicular feminization syndrome** is an X-linked recessive disorder caused by end-organ insensitivity to androgens, usually because of absent or defective androgen receptors. These 46,XY males have abdominal testes that secrete normal testosterone levels. Because of androgen insensitivity, the wolffian ducts regress and the external genitalia develop along the female line. The presence of MIS *in utero* causes regression of the müllerian ducts. These individuals have neither male nor female internal genitalia and phenotypic female external genitalia, with the vagina ending in a blind pouch. They are usually reared as females and undergo feminization during puberty because of the peripheral conversion of testosterone to estradiol.

Pubertal disorders

Disorders of puberty are classified as **precocious puberty**, defined as physical signs of sexual maturation before the age of 7 years in girls and 9 years in boys, or **delayed puberty**, in which physical signs have not presented by age 13 years in girls and 14 years in boys. True precocious puberty results from premature activation of the hypothalamic-pituitary-gonadal axis, leading to the development of secondary sex characteristics as well as gametogenesis. The most frequent causes are CNS lesions or infections, hypothalamic disease, and hypothyroidism. **Pseudoprecocious puberty** is the early development of secondary sex characteristics without gametogenesis. It can result from the abnormal exposure of immature boys to androgens and of immature girls to estrogens. Augmented steroid production can be of gonadal or adrenal origin. See Clinical Focus Box 37.2 for a detailed discussion of precocious puberty.

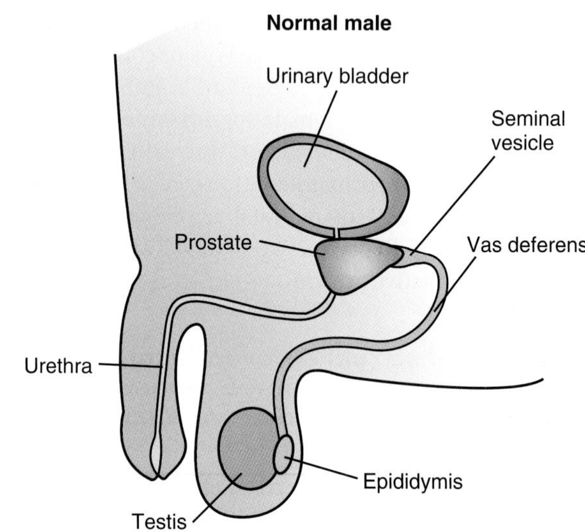

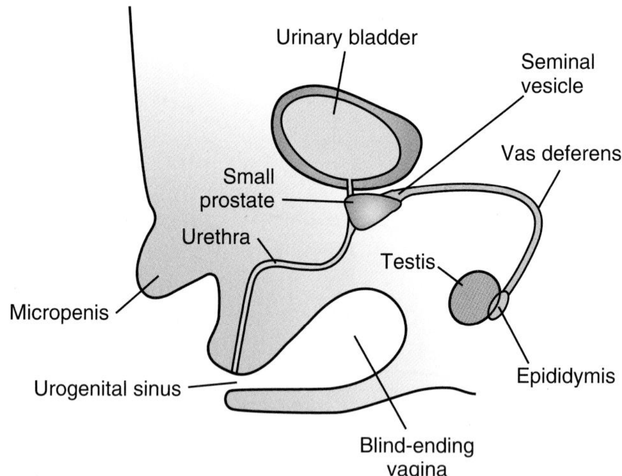

Figure 38.15 Effects of 5α-reductase deficiency on differentiation of the internal and external genitalia.

Chapter Summary

- Fertilization of the ovum occurs in the oviduct. Progesterone and estrogen released from the ovary prepare the oviduct and uterus for receiving the developing embryo.
- The blastocyst enters the uterus, leaves the surrounding zona pellucida, and implants into the uterine wall on day 7 of gestation.
- Human chorionic gonadotropin, produced by trophoblast cells of the developing embryo, activates the corpus luteum to continue producing progesterone and estradiol beyond its normal life span to maintain pregnancy.
- Shortly after the embryo implants into the uterine wall, a placenta develops from embryonic and maternal cells and becomes the major steroid-secreting organ during pregnancy.
- Major hormones produced by the fetoplacental unit are progesterone, estradiol, estriol, human chorionic gonadotropin, and human placental lactogen. Elevated estriol levels indicate fetal well-being, whereas low levels might indicate fetal stress. Human placental lactogen has a role in preparing the breasts for milk production.
- The pregnant woman becomes insulin resistant during the latter half of pregnancy to conserve maternal glucose consumption and make glucose available for the developing fetus.
- Strong uterine contractions induced by oxytocin initiate the termination of pregnancy. Estrogens, relaxin, and prostaglandins are involved in softening and dilating the uterine cervix so that the fetus may exit.
- Lactogenesis is milk production, which requires prolactin (PRL), insulin, and glucocorticoids. Galactopoiesis is the maintenance of an established lactation and requires PRL and numerous other hormones.
- Milk ejection is the process by which stored milk is released; oxytocin, which contracts the myoepithelial cells surrounding the alveoli and ejects milk into the ducts, regulates "milk letdown."
- Lactation is associated with the suppression of menstrual cycles and anovulation as a result of the inhibitory actions of prolactin on gonadotropin-releasing hormone release and the hypothalamic-pituitary-ovarian axis.
- The hypothalamic-pituitary axis becomes activated during the late prepubertal period, resulting in increased frequency and amplitude of gonadotropin-releasing hormone pulses, increased luteinizing and follicle-stimulating hormone secretion, and increased steroid output by the gonads.
- Chromosomal or hormonal alterations, which may result in infertility, sexual dysfunction, or various degrees of intersexuality (hermaphroditism), cause most disorders of sexual development.

thePoint *Visit http://thePoint.lww.com for chapter review Q&A, case studies, animations, and more!*

Appendix

COMMON ABBREVIATIONS IN PHYSIOLOGY

A	amount; area
A	alveolar
a	arterial; ambient
ABP	androgen-binding protein
AC	adenylyl cyclase
ACE	angiotensin-converting enzyme
ACh	acetylcholine
AChE	acetylcholinesterase
ACTH	adrenocorticotropic hormone (corticotropin)
ADH	antidiuretic hormone
ADP	adenosine diphosphate
AE	anion exchanger
AMH	antimüllerian hormone (müllerian-inhibiting substance)
AMP	adenosine monophosphate
ANP	atrial natriuretic peptide
ANS	autonomic nervous system
AQP	aquaporin
A_r	effective radiating surface area
ARDS	adult respiratory distress syndrome
ATP	adenosine triphosphate
AV	atrioventricular
A-V	arteriovenous
AVP	arginine vasopressin (antidiuretic hormone or ADH)
aw	airway
B	barometric
b	body; blood
BF	blood flow
BMR	basal metabolic rate
BS	urinary space of Bowman capsule
BSC	bumetanide-sensitive (Na^+-K^+-$2Cl^-$) cotransporter
BUN	blood urea nitrogen
C	concentration; compliance; capacitance; capacity; clearance; kilocalorie; conductance
C	heat loss by convection
c	core
CA	carbonic anhydrase
CaBP	calcium-binding protein (calbindin)
CaM	calmodulin
CAM	cell adhesion molecule
cAMP	cyclic AMP (cyclic adenosine 3′,5′-monophosphate)
CBG	corticosteroid-binding globulin
CCK	cholecystokinin
CFC	capillary filtration coefficient
CFTR	cystic fibrosis transmembrane conductance regulator
cGMP	cyclic GMP (cyclic guanosine monophosphate [guanosine 3′,5′-monophosphate])
CGRP	calcitonin-gene-related peptide
CIA	central inspiratory activity
CMK	calmodulin-dependent protein kinase
CNS	central nervous system
CO	cardiac output; carbon monoxide
COP	colloid osmotic pressure (oncotic pressure)
COPD	chronic obstructive pulmonary disease
COX	cyclooxygenase
CRH	corticotropin-releasing hormone
CSF	cerebrospinal fluid
CT	computed tomography; calcitonin (thyrocalcitonin)
CYP	cytochrome P450 enzyme
D	diffusion coefficient; diffusing capacity
D	dead space
DAG	diacylglycerol
DHEAS	dehydroepiandrosterone sulfate
DHT	dihydrotestosterone
DIT	diiodotyrosine
D_L CO	lung diffusion capacity
DMT	divalent metal transporter
DNA	deoxyribonucleic acid
DPG	diphosphoglycerate
DPPC	dipalmitoylphosphatidylcholine
E	extraction (extraction ratio)
E	evaporative heat loss
E	expiratory
e	emissivity
EABV	effective arterial blood volume
ECaC	epithelial calcium channel
ECF	extracellular fluid
ECG	electrocardiogram
ECL cell	enterochromaffin-like cell
ED_{50}	median effective dose
EDRF	endothelium-derived relaxing factor (NO)
EDV	end-diastolic volume
EEG	electroencephalogram
EF	ejection fraction
EGF	epidermal growth factor
E_{ion}	equilibrium potential for an ionic species
EJP	excitatory junction potential
ELISA	enzyme-linked immunosorbent assay
E_m	membrane potential
ENaC	epithelial sodium channel
ENS	enteric nervous system
EPP	equal pressure point
EPSP	excitatory postsynaptic potential
ER	endoplasmic reticulum
ERV	expiratory reserve volume
ESR	erythrocyte sedimentation rate
ESV	end-systolic volume
F	fractional concentration of gas; farad; Faraday constant
f	frequency
F_ECO_2	fractional concentration of expired carbon dioxide
FEF	forced expiratory flow
FEV	forced expiratory volume
FEV_1	forced expiratory volume in 1 second
FGF	fibroblast growth factor
F_I O_2	fractional concentration of inspired oxygen
FL	focal length
FM	frequency modulation
F_{max}	maximal force
FRC	functional residual capacity
FSH	follicle-stimulating hormone
FVC	forced vital capacity
g	ionic conductance
G protein	guanine nucleotide-binding protein
GABA	γ-aminobutyric acid

GAP	GnRH-associated peptide
GC	glomerular capillary
GDP	guanosine diphosphate
GFR	glomerular filtration rate
GH	growth hormone (somatotropin)
GHRH	growth hormone–releasing hormone
GI	gastrointestinal
GIP	gastric inhibitory peptide (glucose-dependent insulinotropic peptide)
GLP	glucagon-like peptide
GLUT	glucose transporter
GnRH	gonadotropin-releasing hormone (LHRH)
GPCR	G protein-coupled receptor
GRE	glucocorticoid response element
GRP	gastrin-releasing peptide
GTO	Golgi tendon organ
GTP	guanosine triphosphate
Hb	hemoglobin
h_c	convective heat transfer coefficient
hCG	human chorionic gonadotropin
HDL	high-density lipoprotein
h_e	evaporative heat transfer coefficient
HF	heat flow
HGF	hepatocyte growth factor
hPL	human placental lactogen (human chorionic somatomammotropin)
HR	heart rate
HRE	hormone response element
HSL	hormone-sensitive lipase
5-HT	5-hydroxytryptamine (serotonin)
5-HTP	5-hydroxytryptophan
I	inspiratory
IC	inspiratory capacity
ICC	interstitial cell of Cajal
ICF	intracellular fluid
IGF-I	insulin-like growth factor I (somatomedin C)
I/G ratio	insulin/glucagon ratio
I_{ion}	ionic current
IJP	inhibitory junction potential
IL	interleukin
IP_3	inositol 1,4,5-trisphosphate
IPSP	inhibitory postsynaptic potential
IRDS	infant respiratory distress syndrome
IRV	inspiratory reserve volume
J	flow (or flux) of a solute or water; joule
K	heat loss by conduction
K_f	ultrafiltration coefficient
K_h	hydraulic conductivity
L	lung
LDL	low-density lipoprotein
L-DOPA	L-3,4-dihydroxyphenylalanine
LH	luteinizing hormone
LHRH	luteinizing hormone–releasing hormone
L_o	optimum length
M	metabolic rate
MAP	microtubule-associated protein
MAP kinase	mitogen-activated protein kinase
MCH	mean cell (corpuscular) hemoglobin
MCHC	mean cell (corpuscular) hemoglobin concentration
MCR	metabolic clearance rate
M-CSF	macrophage-colony stimulating factor
MCV	mean cell (corpuscular) volume
MIT	monoiodotyrosine
MLCK	myosin light-chain kinase
MLCP	myosin light-chain phosphatase
MMC	migrating motor complex
MSH	melanocyte-stimulating hormone
MVC	maximal voluntary contraction
NANC	nonadrenergic noncholinergic
NaPi	sodium-coupled phosphate transporter
NE	norepinephrine
NHE	Na^+/H^+ exchanger
NK	natural killer
NMDA	*N*-methyl-D-aspartate
NO	nitric oxide
NOS	nitric oxide synthase
N_R	Reynolds number
OAT	organic anion transporter
OCT	organic cation transporter
P	pressure; partial pressure; permeability; permeability coefficient; plasma; plasma concentration
P_{50}	P_{O_2} at which 50% of hemoglobin is saturated
P_{ACO_2}	partial pressure of carbon dioxide in alveoli
PAH	*p*-aminohippurate
P_{aO_2}	partial pressure of oxygen in arterial blood
P_{AO_2}	partial pressure of oxygen in alveoli
P_{AO_2}–P_{aO_2}	alveolar arterial difference in partial pressure of oxygen
Pc′	pulmonary end-capillary
P_{CO_2}	partial pressure of carbon dioxide
PCr	phosphocreatine (creatine phosphate)
PDE	phosphodiesterase
PEF	peak expiratory flow
PG	prostaglandin
PI	phosphatidylinositol
P_i	inorganic phosphate
PIF	peak inspiratory flow
PIP_2	phosphatidylinositol 4,5-bisphosphate
PKA	protein kinase A
PKC	protein kinase C
PKG	cGMP-dependent protein kinase
pl	pleural
PLC	phospholipase C
PNS	peripheral nervous system
P_{O_2}	partial pressure of oxygen
POMC	proopiomelanocortin
PRL	prolactin
PRU	peripheral resistance unit
PT	prothrombin time
PTH	parathyroid hormone (parathormone)
PTT	partial thromboplastin time
PVC	premature ventricular complex
P_{VCO_2}	partial pressure of carbon dioxide in venous blood
pw	pulmonary wedge
$\dot{Q}$	blood flow
R	respiratory exchange ratio; resistance; universal gas constant
R	heat loss by radiation
r	radius; radiant environment
RAAS	renin-angiotensin-aldosterone system
RBF	renal blood flow
RBP	retinol-binding protein
RDA	recommended daily allowance
REM	rapid eye movement
rh	relative humidity
RIA	radioimmunoassay
RISA	radioiodinated serum albumin
ROS	reactive oxygen species
RPF	renal plasma flow
RQ	respiratory quotient
RV	residual volume
RVD	regulatory volume decrease
RVI	regulatory volume increase
RV/TLC	residual volume to tidal volume ration

S saturation; siemens
S rate of heat storage in the body
s shunt
SA sinoatrial
Sao_2 saturation of hemoglobin with oxygen in arterial blood
SF-1 steroidogenic factor-1
SGLT Na^+-glucose cotransporter
SH2 src homology domain
SHBG sex hormone–binding globulin
SIADH syndrome of inappropriate ADH
sk skin
SN single nephron
SPCA serum prothrombin conversion accelerator
SRIF somatotropin release inhibiting factor (somatostatin)
SRY sex-determining region, Y chromosome
SSRI selective serotonin reuptake inhibitor
StAR steroidogenic acute regulatory protein
SV stroke volume
SVR systemic vascular resistance (total peripheral resistance)
SW stroke work
T tension; temperature; time
T tidal
t time
T_3 triiodothyronine
T_4 thyroxine
ta transairway
TBG thyroxine-binding globulin
TBW total body water
TEA tetraethylammonium
TF tubular fluid
TGF transforming growth factor
TLC total lung capacity
tm transmural
Tm tubular transport maximum
Tn-C calcium-binding troponin
TNF tumor necrosis factor
Tn-I troponin that inhibits actin-myosin interactions
Tn-T tropomyosin-binding troponin
tp transpulmonary
TPA tissue plasminogen activator
TR thyroid hormone receptor
TRE thyroid hormone response element
TRH thyrotropin-releasing hormone
TSC thiazide-sensitive (Na^+-Cl^-) cotransporter
TSH thyroid-stimulating hormone
T tubule transverse tubule
U urine concentration
UCP uncoupling protein
UDP uridine diphosphate
UP ultrafiltration pressure gradient
V volume; volt; vasopressin; vacuolar
v velocity; venous
$\dot{V}$ gas volume per unit time (airflow); minute ventilation; urine flow rate
$\dot{V}_A$ alveolar ventilation
$\dot{V}_A/\dot{Q}$ alveolar ventilation/perfusion ratio
VC vital capacity
V_D dead space volume in the lung (i.e., volume of air that does not participate in gas exchange)
V_D/V_T ratio of dead space volume to tidal volume
$\dot{V}E$ expired minute ventilation (note: minute ventilation and expired minute ventilation are the same)
VEGF vascular endothelial growth factor
VIP vasoactive intestinal peptide
VLDL very-low-density lipoprotein
V_{max} maximum velocity
$\dot{V}o_2$ oxygen consumption per minute
W heat loss as mechanical work
w wettedness
z valence of an ion
φ osmotic coefficient
η viscosity
λ length (space) constant; wavelength
μ electrochemical potential
π osmotic pressure; 3.14 (pi)
ρ density
σ reflection coefficient
τ time constant

Glossary

A band band of muscle myofilament that appears dark under a microscope and contributes to muscles' striated appearance; contains myosin and overlapping projections of actin; *anisotropic*

A wave venous pulse wave created by backflow into the superior vena cava following contraction of the right atrium; amplified with stenosis of the right atrioventricular valve

abdominal wall muscles along the abdomen that are involved in forced expiration

abduction movement away from the midline

abetalipoproteinemia condition in which apo B synthesis cannot occur, thereby blocking production and secretion of very low–density lipoproteins by the liver; condition is marked by large accumulated lipid droplets in the cytoplasm of hepatocytes

absence seizure characterized by the lack of jerking or twitching movements of the eyes; the person seems to stare off into space; formerly known as *petit mal seizures*

absolute refractory period window of time following one action potential during which an axon cannot generate an action potential no matter how strong the stimulus

absolute threshold of hearing intensity level below which no sound can be heard; for the human ear, it has a value of 0.0002 dyne/cm^2

absorptive hyperemia increase in intestinal blood flow that accompanies digestion; local response to elevated tissue metabolism

absorptivity property of the body surface that depends on the temperature of the body and the wavelength of the incident radiation; dimensionless value measured as the fraction of incident radiation that is absorbed by the body

accessory muscles muscles that are brought into play during forced inspiration and expiration

accessory structures structures that either enhance the specific sensitivity of a sensory receptor or exclude unwanted stimuli such as the lens of the eye

accidental hypothermia drop of central temperature below 35°C not induced for therapeutic purposes; occurs in people impaired by drugs or ethanol, disease, and other physical conditions and in healthy people who are immersed in cold water or become exhausted in the cold

acclimation functional change in response to an environmental stress; changes are not permanent and can be reversed when the environmental stress is removed

acclimatization process of an organism adjusting to changes in its environment, often involving temperature or climate; usually occurs in a short time and within one organism's lifetime

accommodation 1. process of focusing the eye to adjust to varying object distances 2. decrease in sensory action potential frequency with a constant generator potential

acetoacetic acid unstable acid found in abnormal amounts in the blood and urine in some cases of impaired metabolism (such as diabetes mellitus or starvation)

acetone chemical formed in the blood when the body uses fat instead of glucose (sugar) for energy

acetylcholine (ACh) one of the two major neurotransmitters used by the autonomic nervous system; the transmitter molecule present in cholinergic synapses and the myoneural junction; the neurotransmitter present in cholinergic synapses in the brain; major neurotransmitter in cognitive function, learning, and memory

acetylcholinesterase (AChE) enzyme that hydrolyzes acetylcholine and terminates its action as a transmitter substance

achalasia failure of any sphincter to relax; in clinical practice, *achalasia* most often refers to a specific condition in which the lower esophageal sphincter fails to relax

acid substance that can release or donate a hydrogen ion

acid dissociation constant (K_a) ionization constant for an acid

acid indigestion indigestion resulting from hypersecretion of acid in the stomach, a synonym for pyrosis

acid tide condition in which the blood from the pancreas is acidic as a result of the bicarbonate secretion in pancreatic juice

acidemia condition in which arterial blood pH < 7.35

acidosis condition in which blood pH is below 7.35; also known as *acidemia*

acinar cells major cells found in the exocrine pancreas actively involved in the production of enzymes

acinus blind sac containing mainly pyramidal cells found in the salivary glands of humans and animals that aids in the secretion of saliva

acne vulgaris disease characterized by comedones, pustules, papules, nodules, and cysts; most common in teenagers and young adults

acoustic reflex automatic contractions of the stapedius and tensor tympani muscles that protect the ear from excessively loud noises

acromegaly thickening of the bones of the face, hands, feet, and liver hypertrophy resulting from excess growth hormone in an adult

acrosome caplike structure that surrounds the sperm head and contains enzymes that help penetration of the egg

acrosome reaction fusion of the outer acrosomal membrane of the sperm head with the sperm plasma membrane, forming a composite membrane with openings through which acrosomal enzymes necessary for sperm penetration of the egg pass

actin globular protein that, in its polymeric filamentous form, makes up the bulk of thin myofilaments in muscle and other contractile systems

α-actinin filamentous muscle protein that anchors thin filaments to the structures of Z lines

actin-linked regulation muscle control system based on controlled alteration of the constituents of the actin filaments

action potential short-lasting, all-or-none change in membrane potential of a nerve or muscle cell that arises when a graded membrane depolarization passes a threshold for opening voltage-gated sodium channels; usually becomes a propagated nerve or muscle cell action potential; also known as *spike potential*

action tremor tremor present during movement

activated partial thromboplastin time (aPTT) test to monitor the intrinsic and common coagulation pathways

activation gate molecular gate in voltage-gated ion channels that opens in response to a polarization change

active hyperemia blood flow increase in response to increased tissue metabolism or reduced oxygen content in the blood

active immunity immunity resulting from activation of lymphocytes (natural) or vaccination (artificial)

active tension component of muscle force that arises from cross-bridge interactions

active transport membrane transport of solute or ions against the prevailing chemical and/or electrical gradient and driven by input of metabolic energy usually in the form of adenosine triphosphate (ATP); when solute movement is directly coupled to ATP hydrolysis, the process is termed *primary active transport*, whereas *secondary active transport* of solute is driven by an ion gradient (usually Na^+) that is maintained by a separate primary active transport process (usually Na^+/K^+-ATPase)

active zones specialized regions of presynaptic terminals where the synaptic vesicles dock, fuse, and release their transmitter

activin nonsteroidal regulator, composed of two covalently linked β subunits, that is synthesized in the pituitary gland and gonads and stimulates the secretion of follicle-stimulating hormone; the actions of activins are the opposite of those of inhibins

activity front length of intestine with an oral and aboral boundary occupied by the migrating motor complex that advances (migrates) along the intestine at a rate that progressively slows as it approaches the ileum

actomyosin ATPase enzymic entity, formed by the interaction of actin and myosin, that hydrolyzes adenosine triphosphate and converts its chemical energy into a mechanical change

acute kidney injury (AKI) serious condition characterized by a rapid decline in glomerular filtration rate, retention of nitrogenous waste products, and disturbances of extracellular fluid volume and electrolyte and acid–base balance; also known as *acute renal failure*

acute-phase blood proteins serum proteins elevated during acute inflammation

acute mountain sickness (AMS) pathophysiologic condition caused by acute exposure to high altitude (from hypoxia resulting from a decrease in oxygen tension, not from a decrease in the percentage of oxygen in the inspired air); also known as *altitude sickness*

acute respiratory distress syndrome (ARDS) diffuse lung injury that is characterized by abnormally low compliance, pulmonary edema, and focal atelectasis

acyl-CoA: cholesterol acyltransferase (ACAT) enzyme responsible for the esterification of cholesterol to form cholesterol ester

adaptation genetic change that enhances the ability of the body to adjust to the environment; changes (e.g., sweat glands) are passed on to the next generation

adaptive immune system antigen-specific interacting components; also called *acquired immune system*

adaptive relaxation vagovagal reflex triggered by stretch receptors in the gastric wall, transmission over vagal afferents to the dorsal vagal complex, and efferent vagal fibers back to inhibitory musculomotor neurons in the gastric enteric nervous system

Addison disease disorder characterized by hypotension, weight loss, anorexia, weakness, and sometimes a bronzelike melanotic hyperpigmentation of the skin; results from primary deficiency of adrenal cortex function and inadequate secretion of aldosterone and cortisol

adduction movement toward the midline

adenohypophysis anterior pituitary

adenosine inhibitory neurotransmitter believed to play a role in promoting sleep and suppressing arousal, with levels increasing with each hour an organism is awake

adenosine diphosphate (ADP) nucleotide released upon platelet activation, causing platelet aggregation; the product of adenosine triphosphate (ATP) dephosphorylation by ATPases; ADP is converted back to ATP by ATP synthases

adenosine triphosphate (ATP) nucleotide important in transport of energy within cells for metabolism

adenylyl cyclase (AC) enzyme that catalyzes the synthesis of cyclic adenosine monophosphate from adenosine triphosphate

adequate stimulus biologically correct stimulus for a given sensory receptor

adipose triglyceride lipase (ATGL) triglyceride lipase that catalyses the removal of the first fatty acid from the glycerol backbone during triglyceride hydrolysis

adrenal cortex outer portion of the adrenal gland

adrenal medulla inner portion of the adrenal gland

adrenarche adrenal maturation during pubertal development in boys and girls

adrenergic characterized by nerves releasing or tissues responding to the catecholamines norepinephrine or epinephrine

adrenocorticotropic hormone (ACTH) hormone product of corticotrophs located in the anterior pituitary; also known as *corticotropin*

adventitia outer elastic layer of arteries and veins that gives some structural integrity to blood vessels but more so in arteries than in veins

aerobic pertaining to a biochemical or physiologic process that requires O_2 for its function

affect psychologic impression of the degree of "pleasantness" of a stimulus

affective disorders illnesses marked by abnormal mood regulation and associated signs and symptoms; also known as *mood disorders*

afferent carrying toward

afferent nerve (afferents) nerve fiber (usually sensory) carrying impulses from an organ or tissue toward the central nervous system or the information processing centers of the enteric nervous system; examples are vagal and spinal afferents

afterhyperpolarization transient period of increased membrane potential that follows the repolarization phase of the action potential

afterload force a muscle experiences (exerts) after it has begun to shorten

aganglionosis congenital absence of ganglion cells in the enteric nervous system in a segment of bowel; peristaltic propulsion is impossible in the absence of the enteric nervous system and is the pathophysiologic basis for Hirschsprung disease

age-related macular degeneration (AMD) decreased sharpness of central vision, especially in bright light; progressive condition whose causes are obscure and for which no cure is available

agonist 1. synthetic drug molecule that selectively binds to a specific receptor and triggers a signaling response in the cells; agonists mimic the action of endogenous biochemical molecules, such as hormones or neurotransmitters that bind to the same receptor 2. muscle producing the desired movement

agranulocyte mature cell, such as a monocyte or lymphocyte; also known as *monomorphonuclear leukocyte*

AH-type enteric neuron multipolar neuron with multiple long processes so named because a characteristic long-lasting hyperpolarizing potential (i.e., an *afterhyperpolarization* or *AH*) occurs after discharge of an action potential; fulfills the role of interneurons in the enteric neural networks; previously thought to be intrinsic primary afferent neurons

AIDS (acquired immunodeficiency syndrome) disease of the human immunodeficiency virus

airway resistance resistance of air movement in airways resulting from turbulent air flow

airway tree airway branching, which occurs in a treelike fashion in the lung

albumin main plasma protein and key regulator of oncotic pressure of blood

albuminuria presence of serum albumin in the urine

aldosterone mineralocorticoid hormone produced by the zona glomerulosa of the adrenal cortex that stimulates tubular reabsorption of sodium and secretion of potassium and hydrogen ions in the kidneys

aldosterone synthase (CYP11B2) substance that synthesizes aldosterone from pregnenolone

alkalemia condition in which arterial blood pH is greater than 7.45

alkaline tide phenomenon that occurs when blood coming from the stomach during active acid secretion contains an overabundance of bicarbonate

alkalosis condition in which blood pH is above 7.45; also known as *alkalemia*

allergy type I immediate hypersensitivity disorder

allo Greek for "other"

allograft transplantation between genetically different people

alpha cells see *α cell*

alpha motor neuron see *α motor neuron*

alpha waves electroencephalogram wave pattern with a rhythm ranging from 8 to 13 Hz, observed when the person is awake but relaxed with the eyes closed

alveolar arterial oxygen gradient pressure difference between the partial pressure of alveolar oxygen and arterial oxygen

alveolar cells milk-secreting cells in the breast

alveolar dead space volume fraction of air in the alveoli that does not participate in gas exchange; occurs because regional blood flow and airflow are not evenly matched (ventilation/perfusion inequalities)

alveolar gas equation formula used to calculate partial pressure of gases in the alveoli

alveolar pressure pressure inside the alveoli

alveolar ventilation amount of fresh air (L/min) brought into the alveoli

alveoli small air sacs in which oxygen and carbon dioxide are exchanged between the atmosphere and the pulmonary capillary blood

Alzheimer disease (AD) most common form of dementia characterized by plaques and neurofibrillary tangles; symptoms include confusion, long-term memory loss, irritability and aggression, mood swings, language breakdown, and sensory decline; disease is incurable, degenerative, and terminal

amacrine cells interneurons in the retina that are responsible for 70% of input to retinal ganglion cells and that regulate bipolar cells, which are responsible for the other 30% of input to retinal ganglia

ambisexual stage time in development when the embryo has the potential to acquire either masculine or feminine characteristics

amenorrhea absence of menstruation

γ-aminobutyric acid (GABA) inhibitory neurotransmitter

amnion extraembryonic membrane that forms a closed fluid-filled sac around the embryo and fetus

amniotic sac structure that forms the amnion around the developing fetus

amphipathic condition in which both hydrophilic and hydrophobic regions are present in the same molecule, for example, phospholipids

amphoteric capable of acting either as an acid or a base

ampulla woolen end of the oviduct, which receives the egg after it is passed into the duct by the fimbria

amygdala region of the cerebrum involved in visceral components of emotional responses

amylase important digestive enzyme secreted by both the salivary glands and the pancreas that is involved in the digestion of carbohydrates

amylin hormone formed by β cells in the pancreas that regulates the timing of glucose release into the bloodstream after eating by slowing the emptying of the stomach; also known as *islet amyloid-associated peptide*

anaerobic pertaining to a biochemical or physiologic process that does not require O_2 for its function

anaerobic threshold exercise intensity at which lactate (lactic acid) starts to accumulate in the blood, which happens when lactate is produced faster than it can be removed (metabolized)

anamnestic response increased immune response to repeated antigen exposure

anaphylaxis condition resulting from severe allergic reaction and characterized by intense local and sometimes systemic vasodilation that results in shock; histamine released during the reaction dilates arterioles and increases capillary permeability; fluid loss from transcapillary filtration contributes to plasma volume loss that exacerbates the shock caused by arteriolar dilation

anatomic dead space volume amount of air found in the lungs' airways that does not participle in gas exchange

anatomic shunt blood that bypasses alveoli and does not get oxygenated as a result of a structural defect in the circulation

androgen type of hormone that promotes the development and maintenance of male sex characteristics

androgen-binding protein (ABP) protein produced by Sertoli cells that binds testosterone

androstenedione adrenal androgen made by 17α-hydroxylation of progesterone; serves as a precursor for estrogen synthesis in the ovaries

anemia reduction in erythrocytes, which leads to a decrease in the oxygen-carrying capacity of the blood

anemia of inflammation long-term condition characterized by a lower-than-normal number of healthy red blood cells; also called *anemia of chronic disease*

aneurysm weakening of the arterial wall that causes it to bulge outward, leak blood, and eventually rupture

angiogenesis formation of new blood vessels

angiotensin I inactive decapeptide formed during the cleavage of angiotensinogen

angiotensin II peptide hormone formed from the proteolysis of angiotensin I by converting enzyme in the lungs; potent vasoconstrictor, enhances the effectiveness of sympathetic nerve stimulation to blood vessels, and promotes salt and water retention in the body primarily by stimulating the release of aldosterone from the adrenal cortex

angiotensin-converting enzyme (ACE) enzyme located in the capillary endothelial cells of the lung and kidney that converts angiotensin I to angiotensin II

angiotensinogen α_2-globulin produced by the liver that is the substrate for renin

anhidrotic ectodermal dysplasia hereditary condition characterized by abnormal development of the skin, hair, nails, teeth, and sweat glands and in which sweat glands are sparse or absent; affected adults are unable to tolerate a warm environment and require special measures to maintain normal body temperature

anion negatively charged ion, especially the ion that migrates to an anode in electrolysis

anion exchange protein (AE1) transport protein responsible for mediating the electroneutral exchange of chloride (Cl^-) for bicarbonate (HCO_3^-) across a plasma membrane

anion gap difference (in mEq/L) between plasma [Na^+] and plasma ([Cl^-] + [HCO_3^-])

anisocytosis large variation in the size of erythrocytes

anorectal angle (approximately) 90° angle between the rectum and the anal canal; becomes more obtuse during defecation and more acute when holding back stool; formed by the contraction of the puborectalis muscle

anorexia nervosa severe behavioral disorder associated with the lack of food intake, extreme malnutrition, and endocrine changes secondary to psychologic and nutritional disturbances

anorexigenic causing a loss of appetite

anovulation period in which no ovulation occurs

antagonist muscle opposing the desired movement

antagonistic pair pair of muscles arranged so that the shortening of one lengthens the other, and vice versa

anterior cavity forward section of the eyeball that contains the iris and the lens

anterior periventricular region portion of the hypothalamus close to the third ventricle

anterograde direction of axonal transport from the cell body toward the axon terminal

anterograde amnesia loss of memory going forward from a trauma

antibody secreted protein that can bind antigen

anticoagulant substance that prevents coagulation

antidiuretic hormone (ADH) peptide hormone synthesized in the anterior hypothalamus and released from the posterior pituitary, whose primary action is to increase the water permeability of the kidney collecting ducts; now known as *arginine vasopressin (AVP)*

antidromically conduction of nerve action potentials in the direction opposite to that which is usual

antigen substance that binds to antigen receptors and antibody

antigen-presenting cell (APC) cell displaying antigen on its surface together with major histocompatibility complex class II

antigen-recognition molecules T-cell receptor, B-cell receptor, and major histocompatibility complex proteins

antioxidant molecule capable of inhibiting the oxidation of another molecule, for example, the enzyme catalase reduces superoxide to oxygen and water

antiport membrane transport process in which the driver ion (usually Na^+) and the solute move in opposite directions

antra fluid-filled cavities in a tertiary ovarian follicle

antral dysrhythmia abnormality of electrical rhythm in the gastric antrum (bradygastria/tachygastria) analogous to cardiac arrhythmia

antral granulosa cells cells that line the antral cavity of a graafian follicle

antral pump caudal two thirds of the gastric corpus, the antrum, and the pylorus

aorta largest artery in the body; sole arterial outlet from the left ventricle

aortic bodies specialized innervated nodules located in the aorta that sense low P_{O_2} and pH as well as high P_{CO_2}; send impulses to the medullary area of the brain to elicit reflex activation of the sympathetic nervous system to increase blood pressure and do not respond to low arterial pressure unless the pressure is low enough to affect oxygen delivery to the bodies (<80 mm Hg)

aortic baroreceptors spray nerve endings in the wall of the aortic arch that sense stretch and, therefore, serve as sensory receptors for changes in arterial pressure

aortic valve thin, flexible, trileaflet structure that separates the aorta from the left ventricle; opens into the aorta and closes off the aorta when the left ventricle relaxes

aphasia loss of the ability to use or understand language, usually resulting from injury or stroke

apical that side of a cell or plasma membrane facing the lumen

aplastic anemia stem cell destruction in the bone marrow that leads to the decreased production of white blood cells, platelets, and red blood cells

apnea temporary cessation of breathing

Apo B proteins class of apolipoproteins that are a component of liver-synthesized very low–density lipoprotein particles; Apo B48 produced by the small intestine and apo B100 produced exclusively by the liver are the circulating forms of the apo B proteins

apocrine type of exocrine gland from which the apical portion of the secreting cell is released along with the secretory products, for example, sweat glands

apoeccrine hybrid sweat glands found in the axilla with secretory structures similar to those found in both apocrine and eccrine glands but are functionally similar to eccrine glands; they secrete nearly 10 times as much sweat as eccrine glands

apoferritin storage protein found in the enterocyte cytoplasm that binds and stores iron

apoprotein B_{100} 400-kDa protein present on the surface of the low-density lipoprotein (LDL) particle and recognized by LDL receptors

apoproteins major proteins associated with chylomicrons and very low–density lipoproteins

apoptosis deliberate type of programmed cell death that is mediated by an intracellular signaling pathway; a type of *programmed cell death*; contrast with *necrosis*

aquaporin (AQP) small (30-kDa) integral membrane protein that functions as a water channel and explains the rapid movement of polar water molecules across the lipid bilayer of plasma membranes

aqueous humor thin, clear liquid that fills the anterior chamber of the eyeball

arachidonic acid fatty acid released from phospholipids in the membrane when ligand-gated receptors activate phospholipase A2

arcuate nuclei (ARC) hypothalamic nuclei that synthesize growth hormone–releasing hormone and regulate feeding

area postrema area in the lateral wall of the fourth ventricle, which is one of the few locations in the brain where the blood–brain barrier is open; it contains neural networks that trigger vomiting

areflexia absence of muscle stretch reflexes

areola pigmented region of skin surrounding the nipple

arginine vasopressin (AVP) substance produced by magnocellular neurons in the supraoptic and paraventricular nuclei of the hypothalamus that regulates water resorption by the kidneys

arrhythmia abnormality in heart rate, irregularity in the rhythm of the heart or pattern

arterioles small arteries of approximately 10 to 1,000 μm in diameter that join large arteries to organ capillary networks; they have the same wall structure as larger arteries but generally have thicker walls and smaller lumen diameters and are the primary site of changes in vascular resistance and blood flow distribution within the cardiovascular system; they are also involved with the control of intracapillary hydrostatic pressure

arteriovenous anastomoses generally, a direct connection between the lumen of an artery and vein that bypasses the capillary bed of an organ; in the skin, arteriovenous anastomoses that are highly innervated by sympathetic nerves link arterioles to subcutaneous venous plexi that serve to conduct heat to or restrict loss of heat from the skin surface; these anastomoses play a key role, therefore, in temperature regulation in the body

artery blood vessel that carries blood away from the heart and composed of three layers: an outermost stiffly elastic adventitia, a middle zone composed of circular layers of smooth muscle, and a single-cell lining of epithelial cells called the *endothelium*; has structural rigidity in that it retains circular shape even without any positive transmural pressure

ascending reticular activating system system of neurons in the brainstem that project to the thalamus and are involved in arousal and motivation

ascites sites of edema in the organs of the abdominal cavity

aspartate amino acid used as an excitatory neurotransmitter

association cortex cortical region involved in higher order processing of information

asthma spasmodic contraction of smooth muscle in the bronchi leading to airway obstruction; symptoms include labored breathing accompanied especially by wheezing and coughing and by a sense of constriction in the chest; it is triggered by hyperreactivity to various stimuli (as allergens or rapid change in air temperature)

astigmatism vision defect caused by an eyeball with differing radii of curvature at differing orientations; causes vertical and horizontal lines to have different focal distances

ataxia inability to coordinate muscle activity during voluntary movement; most often due to disorders of the cerebellum or the posterior columns of the spinal cord; may involve the limbs, head, or trunk

atelectasis lack of gas exchange in alveoli due to alveolar collapse or fluid consolidation

atherosclerosis thickening of arterial walls from cholesterol plaque buildup, possibly leading to blockage

atopy genetic predisposition to allergens

ATP synthase see *F-type ATPase*

ATPase enzyme or membrane transport system that hydrolyses adenosine triphosphate (ATP) to adenosine diphosphate and free phosphate; *P-type ATPases* are ATPases that are phosphorylated during the hydrolysis of ATP and are found on the plasma membrane; *V-type ATPases* are found on intracellular vacuolar structures; *F-type ATPases* are located on the inner mitochondrial membrane and synthesize ATP rather than hydrolyze it

ATP-binding cassette (ABC) transporters superfamily of adenosine triphosphate–binding cassette transporters composed of two transmembrane domains and two cytosolic nucleotide binding domains; the transmembrane domains recognize specific solutes and transport them across the membrane by using a number of different mechanisms, including conformational change

ATP-dependent K^+ channel (K_{ATP}) class of K channels inhibited by the binding of intracellular adenosine triphosphate

atresia degeneration of follicles in the ovary

atrial fibrillation random, chaotic electrical activation of atrial muscle cells, often resulting in irregular pacing of the ventricles

atrial natriuretic peptide (ANP) peptide hormone that is released from the right atrium when it is stretched, usually from an increased blood volume; promotes the loss of sodium and, indirectly, water from the kidney, thereby helping to restore normal blood volume

atrioventricular (AV) node specialized conduction tissue in the septum of the heart between the atria and ventricles and the only normal path by which electrical activation of the atria spreads into the ventricles; also delays activation of the ventricles during normal cardiac activation, thereby increasing the time the ventricles have to fill with blood during diastole

attention ability to concentrate on a single stimulus

atypical antipsychotics drugs that treat both the positive and negative symptoms of schizophrenia

auditory cortex region of the brain responsible for the higher-level processing of auditory signals

Auerbach plexus ganglionated plexus of the enteric nervous system situated between the longitudinal and circular muscle coats of the muscularis externa of the digestive tract; also known as the *myenteric plexus*

augmented leads system of three unipolar frontal plane leads in the standard clinical recording of an electrocardiogram (ECG); leads are oriented toward the right shoulder, left shoulder, and feet and are designated aVR, aVL, and aVF, respectively; ECG recordings in these leads are amplified or "augmented" relative to the standard bipolar frontal plane ECG leads

auto Greek for "self"

autocrine signaling form of signaling in which a cell secretes a chemical messenger that signals the same cell

autograft transplanted tissue grafted into a new position on the same person

autoimmunity failure of the body to recognize its own tissue as self, allowing attacks against its own cells and tissues

automaticity property of all cardiac cells in which they have the ability to generate their own action potentials

autonomic nervous system (ANS) part of the nervous system that regulates involuntary functions

autoreceptor receptor found on the presynaptic membrane that responds to the transmitter released by the neuron; usually involved in feedback regulation of transmitter synthesis or release

autoregulation in the cardiovascular system, the ability of an organ to maintain near-normal blood flow in the face of changes in perfusion pressure

axial streaming tendency of blood cells and elements to compact into the center of a blood vessel, where flow velocity is highest, when blood flows through the vessel; results in the phenomenon of plasma skimming, whereby the cellular elements of blood are pushed away from the arterial wall, leaving the wall exposed to essentially cell-free plasma

axon hillock initial segment of an axon adjacent to the neuronal cell body; except in pseudounipolar cells, the region of the axon that initiates the action potential; also known as *initial segment* or *trigger zone*

axon terminal region of an axon most distal from the cell body and containing synaptic boutons that signal to the target cell

axoneme main core of a flagellum, consisting of two central microtubules surrounded by nine double microtubules

axons fibers that carry impulses away from the perikaryon of a nerve cell

axoplasmic transport molecular motor-driven mechanism to transport proteins, lipids, synaptic vesicles, and other organelles from the cell body to the axon and its terminals; also used to transport lysosomes and trophic factors from the axon terminal to the cell body

B cell type of lymphocyte with immunoglobulin on the cell membrane surface

band cell cell undergoing granulopoiesis to become a mature granulocyte

barometric pressure atmospheric pressure that surrounds us; at sea level, the barometric pressure is 760 mm Hg

baroreceptor stretch receptor sensitive to blood pressure and to the rate of change of blood pressure

baroreceptor reflex broad term used to describe the marshalling of the autonomic nervous system to the heart, arteries, veins, adrenal medulla, and kidney (renin release) in order to counteract changes in arterial pressure; also sometimes called *buffer reflex* in that it prevents large deviations in arterial blood pressure brought about by moment to moment changes in the internal or external environment of the body

Barr body condensed mass of chromatin in female cells resulting from inactivation of one X chromosome

Barrett esophagus precancerous metaplasia associated with abnormal exposure of the esophageal mucosa to reflux of gastric contents

Bartter syndrome rare, inherited condition of the thick ascending limb of the kidney thought to be caused by a defect in the kidney's ability to reabsorb sodium, causing a rise in aldosterone, and consequently, potassium wasting and hypokalemic alkalosis

basal forebrain nuclei subcortical groups of cholinergic neurons located at the base of the forebrain that are involved in cognition

basal ganglia subcortical groups of neurons involved in motor control

basal metabolic rate (BMR) oxygen consumption of the body measured under resting conditions

base substance that can bind or accept a hydrogen ion

basement membrane second layer of the glomerular membrane of the kidney, consisting of a meshwork of fine fibrils embedded in a gel-like matrix

basic metabolic panel (BMP) group of blood chemistry tests that includes the analysis of glucose, Ca^{2+}, Na^+, K^+, CO_2 (or HCO_3^-), Cl^-, blood urea nitrogen, and creatinine

basilar membrane vibrating membrane in the cochlea along which different frequencies of sound are detected

basket cell type of interneuron in the cerebellum

basolateral that side of a cell or plasma membrane facing the blood supply

basophil granulocyte of the myeloid series that plays a role in inflammation and immediate allergic reactions

B-cell receptor (BCR) antigen-recognition molecule of B cells

Becker muscular dystrophy (BMD) inherited disorder that involves slowly worsening muscle weakness of the legs and pelvis; less common than Duchenne muscular dystrophy, affecting 3 to 6 out of every 100,000 males

benign paroxysmal positional vertigo (BPPV) severe vertigo, with incidence increasing with age; thought to be a result of the presence of canaliths, debris in the lumen of one of the semicircular canals

benzodiazepines drugs used to treat anxiety that act at γ-aminobutyric acid receptors

beriberi disorder of the nervous system and heart characterized by anorexia caused by a deficiency in thiamine

Bernoulli principle principle derived from the fact that total energy in a system cannot be created or destroyed but only transferred between different forms. In the cardiovascular system, it is the principle that describes the transfer of energy between kinetic (from flow velocity) and potential (from lateral pressure) energy. The principle predicts that, as the velocity of flow in a fluid in a tube increases, the lateral pressure on the wall of the tube decreases.

beta cells see *β cell*

beta waves electroencephalogram wave pattern with a rhythm ranging from 13 to 30 Hz, observed when the person is awake but relaxed with the eyes open

biguanides class of drugs used in the treatment of type 2 diabetes that lower glucose via effects on peripheral tissues, primarily the liver

bile acids one of the major components of bile, they are formed in the liver from cholesterol. There are two types of bile acid: primary and secondary.

bile salts cholesterol derivatives that play an important role in the intestinal absorption of lipids. They are water-soluble and are essentially detergents. They are polar molecules that penetrate cell membranes poorly, an important property because it ensures their minimal absorption of fat.

bile salt–sodium symport mechanism of bile acid secretion and bile flow that involves hepatocyte uptake of free and conjugated bile salts and is calcium-dependent and calcium-mediated. Bile salts serve the important function of absorbing fat.

bilirubin major pigment present in bile; an orange compound that is an end product of hemoglobin degradation in the monocyte-macrophage system in the spleen, bone marrow, and liver

biliverdin green metabolite product of hemoglobin catabolism

biological activity unit quantitated amount of a hormone defined as an amount sufficient to produce a response of specified magnitude under a defined set of conditions

biological clock circadian pacemaker located in the hypothalamus that regulates sleep and wakefulness and metabolic pathways

biotin coenzyme for carboxylase, transcarboxylase, and decarboxylase enzymes, which play an important role in the metabolism of lipids, glucose, and amino acids

bipolar cell as part of the retina, these cells exist between photoreceptors (rod cells and cone cells) and ganglion cells and act, directly or indirectly, to transmit signals from the photoreceptors to the ganglion cells

bipolar disorder psychiatric mood disorder in which periods of profound depression are followed by periods of mania, in a cyclic pattern; formerly known as *manic-depressive disorder*

blastocyst stage in embryonic development after the morula stage that consists of an inner cell mass, an outer trophoblast layer, and a blastocoele

blastomeres cells produced by mitotic division of a zygote

bleeding time medical test of hemostasis

bloating sensation of abdominal distention that may or may not be visible

blood indices indices including mean cell (corpuscular) hemoglobin concentration, mean cell (corpuscular) hemoglobin, mean cell (corpuscular) volume, and blood oxygen-carrying capacity

blood lipid profile blood tests including total cholesterol, low-density lipoprotein (LDL), high-density lipoprotein (HDL), LDL/HDL ratio, and total triglycerides

blood smear spread of a blood drop for microscopic analysis

blood–brain barrier system of anatomical and transport barriers that controls the entry and rates of transport of substances into the brain extracellular space from the capillary blood

blood–gas interface interface between the atmosphere and blood, comprising the alveolar–capillary membrane; the site for gas exchange in the lung

blood–placental barrier barrier that regulates the transmission of substances from the maternal circulation to the fetus

body mass index (BMI) number, derived by using height and weight measurements, that gives a general indication of whether weight falls within a healthy range

body plethysmograph airtight box in which patient sits to measure the functional residual capacity and total lung capacity of the lungs; also called *pulmonary plethysmograph* and *body box*

Bohr effect rightward shift of the oxygen equilibrium curve resulting from a rise in blood carbon dioxide and hydrogen ions

bone conduction means of hearing in which sound vibrations are carried to the inner ear by the bones of the skull

bone morphogenetic proteins (BMPs) growth factors found to induce the differentiation of mesenchymal cells toward osteoblasts as well as orchestrating tissue architecture throughout the body

bone remodeling turnover of bone mineral and changes in bone structure from daily flux of calcium and phosphate into and out of bone; generally occurs along most of the outer surface of the bone, making it either thinner or thicker, as required

botulinum toxin toxin produced by the bacterium *Clostridium botulinum*, which, in small concentrations, prevents presynaptic transmitter release at the myoneural junction

Boyle law gas law that states that, at a constant temperature, the pressure of a gas varies inversely with gas volume

bradykinin nonapeptide hormone that causes a slowly developing contraction of intestinal smooth muscle, vasodilation, and increased renal sodium excretion

brain natriuretic peptide (BNP) peptide hormone produced in the brain and cardiac ventricles that increases renal sodium excretion

brain stem region of the central nervous system between the spinal cord and diencephalon comprising the midbrain, pons, and medulla

brain–gut axis bidirectional nervous connections between the brain (central nervous system [CNS]) and gut that serve various physiologic functions; visceral afferent fibers project to somatotypic, emotional, and cognitive centers of the CNS, producing many interpretations to the stimuli based on prior learning and one's cognitive and emotional state; in turn, the CNS can inhibit or facilitate afferent nociceptive signals, motility, secretory function, or inflammation

breast cancer–resistance protein (BRCP) member of the ABC transporter family of multidrug resistance proteins

Broca area area of the frontal lobe of the brain involved in motor control of speech, usually on the left hemisphere

bronchi bifurcation of the airway, or the two primary divisions of the trachea that lead respectively into the right and the left lung

bronchial circulation circulation that supplies oxygenated blood to the conducting zone (i.e., the first 17 generation of the lung airways), which originates from the descending aorta and drains into the pulmonary vein

bronchioles small, thin airways that lack cartilage and are subjected to lung pressures; they are branches of the bronchi and terminate by entering the circular sacs called *alveoli*; also called *bronchioli*

bronchioli see *bronchioles*

bronchitis inflammation of the bronchi

brown adipose tissue one of the two types of adipose tissue (the other being white adipose tissue) that are present in many newborn or hibernating mammals; in contrast to white adipocytes (fat cells), which contain a single, large fat vacuole, brown adipocytes contain several smaller vacuoles, a much higher number of mitochondria, and more capillaries because it has a greater need for oxygen than most tissues; also called *brown fat*

buffer agent that minimizes the change in pH produced when an acid or base is added

buffy coat fraction of an anticoagulated blood sample after centrifugation, containing most of the white blood cells and platelets

bundle of His specialized conduction tissue that receives impulses from the atrioventricular node. It branches into right and left structures, which in turn give rise to the Purkinje fiber system.

C wave venous pulse wave resulting from bulging of the right atrioventricular valve into the superior vena cava during systole followed by descent of the heart. It is amplified when the valve is incompetent (will not close).

Ca^{2+}-binding protein, calbindin D (CaBP) protein that binds with calcium in the duodenum and jejunum that aids in the absorption of calcium

calcitonin (CT) hormone secreted by the thyroid that lowers blood calcium; also known as *thyrocalcitonin*

calcium pump enzyme that uses the energy generated by the hydrolysis of adenosine triphosphate to move calcium ions out of the cytoplasm or, in the case of muscle cells, from the cytoplasm to the sarcoplasmic reticulum; also known as *Ca^{2+} ATPase*

calcium release channel membrane channel in the sarcoplasmic reticulum through which calcium ions exit into the myofilament space or cytoplasm

calcium sparks small localized releases of calcium intracellularly following the activation of one coupled voltage-gated calcium channel:sarcoplasmic reticular–calcium release channel pair

calcium-induced calcium release (CICR) release of calcium from the sarcoplasmic reticulum calcium release channel in response to a calcium entry signal through the plasma membrane voltage-gated calcium channel

calmodulin (CaM) small calcium-binding protein that confers calcium sensitivity to an enzyme by binding to it and makes its activity subject to calcium control; for example, calmodulin complexed with calcium binds to myosin light-chain kinase to promote its activation

calor heat associated with acute inflammation

cAMP-dependent protein kinase enzyme whose activity (such as regulation of glycogen, sugar, and lipid metabolism) is dependent on cyclic adenosine monophosphate; also known as *protein kinase A*

canaliculi tiny cytoplasmic connections between osteocytes, where much of calcium homeostasis takes place

canaliths debris in the lumen of one of the semicircular canals often causing vertigo

cancer immunotherapy therapeutic use of the body's immune system to fight cancer

capacitance two conductors separated by an insulator with the ability to store an electrical charge. The biologic capacitor is the lipid bilayer of the plasma membrane, which separates two conductive regions, the extracellular and intracellular fluids

capacitation maturation of sperm in the female genital tract, which involves a loss of surface glycoprotein, a loss of lipid, and an increase in motility

capillaries smallest elements of the vascular system, composed of a single layer of endothelial cells. They are the primary site of transport between the bloodstream and the extracellular fluid surrounding all tissues. Capillaries can be joined tightly or exist with intermittent openings in their walls, depending on the organ from which they come

capillary distention dilation of a bed of highly compliant small pulmonary arterioles with increased perfusion pressure

capillary filtration coefficient (CFC) constant relating Starling forces to movement of water across the capillaries. It is a measure of how much water can flow per unit of time per net positive pressure moving fluid out of the capillaries

capillary recruitment opening of a bed of pulmonary capillaries with increased perfusion pressure

carbamino proteins proteins that bind with carbon dioxide

carbaminohemoglobin compound of hemoglobin and carbon dioxide, which is one of the forms in which carbon dioxide is transported by the blood

carbon dioxide equilibrium curve curve that reflects the relationship between carbon dioxide content and partial pressure in the blood

carbonic anhydrase (CA) enzyme that catalyzes the reversible reaction $CO_2 + H_2O \Rightarrow H_2CO_3$, or, more correctly, $CO_2 + OH^- \Rightarrow HCO_3^-$

carboxyhemoglobin (Hbco) carbon monoxide bound to hemoglobin

carboxypeptidase A exopeptidase pancreatic protease found in pancreatic juice that is specific in its action to attack polypeptides with a neutral aliphatic or aromatic carboxyl terminal to produce the final products of protein digestion, amino acids, and small peptides

carboxypeptidase B exopeptidase pancreatic protease found in pancreatic juice that is specific in its action to attack polypeptides with a basic carboxyl terminal to produce the final products of protein digestion, amino acids, and small peptides

carboxypolypeptidase pancreatic enzyme that converts peptides into amino acid

cardia one of four stomach sections, where the esophageal contents are emptied

cardiac function curve relationship depicting how cardiac output varies as a function of central venous pressure

cardiac gland one of three main types of glands found in the mucosal lining of the stomach, specifically in a small area adjacent to the esophagus and lined by mucus-producing columnar cells, that secrete mucous and bicarbonate ions that protect the stomach from the acid in the lumen

cardiac glycoside drug used in the treatment of congestive heart failure and cardiac arrhythmia; see also *digoxin*

cardiac index cardiac output normalized for body surface area

cardiac output (CO) total flow output of the heart, often expressed as L/min or mL/min. It is typically between 4 and 5 L/min in healthy adults. It can also be thought of as the sum of all the individual organ blood flows in the body.

cardiac tamponade pericardial constriction of the ventricles resulting from bleeding beneath the pericardium. It reduces passive stretch of the ventricle during filling.

cardiopulmonary baroreceptors receptors located on the right side of the circulation that sense changes in circulating blood volume or central blood volume through the effects of volume on the stretch of the receptors. They elicit mechanisms to stimulate salt and water loss whenever blood or central volume is too high; also known as "*low pressure*" or "*volume*" receptors

carotid bodies specialized innervated nodules located in the bifurcation of the internal and external carotid arteries that sense low Po_2 and pH as well as high Pco_2. They send impulses to the medullary area of the brain to elicit reflex activation of the sympathetic nervous system to increase blood pressure. They do not respond to low arterial pressure unless the pressure is low enough to affect oxygen delivery to the bodies (<80 mm Hg)

carotid sinus slight dilation of the common carotid artery at its bifurcation into external and internal carotids; it contains baroreceptors that, when stimulated, cause slowing of the heart, vasodilation, and a fall in blood pressure and is innervated primarily by the glossopharyngeal nerve

carotid sinus baroreceptors spray nerve endings in the wall of the carotid sinus that sense stretch, and therefore, serve as sensory receptors for changes in arterial pressure

carrier-mediated diffusion see *passive transport*

caspases family of cysteine-*asp*artic prote*ases* involved in apoptosis, cell necrosis, and cell death

catalase enzyme that protects cells by catalyzing hydrogen peroxide

cataract increase in the opacity of the lens of the eye

catechol-O-methyltransferase (COMT) enzyme that degrades the catecholamines by methylating the 3-OH group on the catechol ring

cation ion or group of ions having a positive charge and characteristically moving toward the negative electrode in electrolysis

caudate nucleus one of the basal ganglia; receives input from the cerebral cortex

caveolae microdomain or lipid raft that contains the protein caveolin and forms an invagination of the plasma membrane

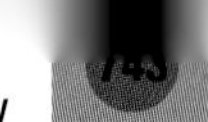

CCK cholecystokinin; satiety signal produced by endocrine cells in the gut wall mediated via the vagus

CD cluster of differentiation; distinct plasma membrane molecules used as cell markers

celiac ganglion prevertebral sympathetic ganglion involved in gastrointestinal control

celiac sprue (gluten-sensitive enteropathy) common disease involving a primary lesion of the intestinal mucosa caused by the sensitivity of the small intestine to gluten that results in the malabsorption of all nutrients as a result of the shortening or a total loss of intestinal villi, which reduces the mucosal enzymes for nutrient digestion and the mucosal surface for absorption

α cell one of five types of cells found in the islets of Langerhans of the endocrine pancreas; this one produces the hormone glucagon

β cell one of five types of cells found in the islets of Langerhans of the endocrine pancreas; this one produces the hormones insulin and amylin

δ cell one of five types of cells found in the islets of Langerhans of the endocrine pancreas; this one produces the hormone somatostatin

ε cell one of five types of cells found in the islets of Langerhans of the endocrine pancreas; this one produces ghrelin

cell adhesion molecules plasma membrane glycoproteins that promote cell adhesion

cellular immunity immunity mediated by activated T cells

cellular respiration final stage of respiration in which food molecules are broken down in the cell, releasing carbon dioxide and oxygen

central autonomic network neurons and axon pathways in the central nervous system that control autonomic function

central blood volume theoretical volume of blood contained in the veins of the thoracic cavity. It is usually measured as pressure in the right atrium.

central chemoreceptors chemosensitive neurons located adjacent to the dorsal and ventral respiratory group complex that respond to H^+ of the surrounding interstitial fluid

central inspiratory activity (CIA) integrator group of cells in the medulla that activates inspiration

central integrative system brain area involved in the coordinated activities of human behavior

central nervous system (CNS) brain and the spinal cord

central pattern generator (CPG) group of nerve cells producing rhythmic action potentials important in movements

central sulcus brain and the spinal cord

central venous pressure venous pressure in the right atrium. It is representative of the effect of changes in central blood volume.

centroacinar cells cells lining the lumen of the acinus that modify the electrolyte composition of pancreatic secretion

cephalic phase one of three phases that stimulate acid secretion during the ingestion of a meal; involves the central nervous system and is induced by the simple process of smelling, chewing, swallowing, and the thought of food. The cephalic phase accounts for about 40% of total acid secretion.

cerebellum brain stem structure important in movement coordination

cerebral cortex 1. sheet of neural tissue outermost to the cerebrum of the mammalian brain that plays a key role in memory, attention, perceptual awareness, thought, language, and consciousness 2. the outer shell of the brain, containing the cell bodies of neurons

cerebral edema edema formation in the brain. It causes an increase in intracranial pressure that results in restriction of cerebral blood flow

cerebral peduncles pathway in the upper brain stem for corticospinal tract axons

cerebrocerebellum region of the cerebellar cortex that receives input from cerebral cortex, active in control of limb muscles

cervical mucus viscous substance on the surface of the cervix, made up of mucopolysaccharide and water

cervix narrow muscular canal that connects the vagina and the body (corpus) of the uterus

C-fiber endings unmyelinated nerve endings that are located adjacent to alveoli and small bronchioles, that are sensitive to stretch, and that play a protective role in the lung

cGMP-dependent protein kinase enzyme that requires cyclic guanosine monophosphate to phosphorylate potential target substrates, such as calcium pumps in the sarcoplasmic reticulum or sarcolemma, leading to reduced cytoplasmic levels of calcium and, in turn, muscle relaxation; also known as *protein kinase G*

Chagas disease neuropathic degeneration of autonomic neurons resulting from autoimmune attack in patients infected with the bloodborne parasite *Trypanosoma cruzi*

channel gating regulation of the opening or closing of ion channels by changes in transmembrane voltage, mechanosensitivity, or ligand binding

channelopathies diseases caused by dysfunction of ion channels

Charcot-Marie-Tooth disease (CMT) group of inherited disorders that affect the peripheral nerves, specifically, involving damage to the covering (myelin sheath) around nerve fibers (either destruction of the myelin sheath or erosion of the central (axon) portion of the nerve cell); nerves that stimulate movement (the motor nerves) are most severely affected, and the nerves in the legs are affected first and most severely

Charles law gas law that states that, at a constant pressure, gas volume varies proportionately with temperature

chemical pH buffer mixture of a weak acid and its conjugate base (or a weak base and its conjugate acid)

chemical potential gradient created by different concentrations of an uncharged solute on opposite sides of a membrane

chemical synapse site of communication between a neuron and another neuron or other target cell mediated by the release of a chemical neurotransmitter

chemoattractants cytokines that attract white blood cells

chemokine cytokine that attracts and activates leukocytes

chemoreceptors neurosensitive endings that respond to changes in blood gases (oxygen, carbon dioxide, and pH)

chemosensitive nerve endings efferent arm of the chemoreflex that increases cardiac output and improved blood flow to the ischemic skeletal muscle

chenodeoxycholic acid primary bile acid

Cheyne–Stokes breathing abnormal breathing pattern characterized by periods of breathing with gradually increasing and decreasing tidal volume interspersed with periods of apnea

chief cells cells that reside in the oxyntic glands in the stomach and secrete pepsinogen

chloride bicarbonate exchanger exchange mechanism that allows the red blood cell to maintain electrical neutrality by exchanging Cl^- for HCO_3^- as it diffuses out of the cell

chloride shift movement of chloride ions into the red blood cell coupled with movement of HCO_3^- out to maintain electrical neutrality of the cell

cholecystokinin (CCK) hormone produced by the small intestine that is responsible for gallbladder contraction as well as the secretion of enzyme-rich pancreatic juice

cholesterol sterol derived exclusively from animal fat

cholesterol ester hydrolase (CEH) substance that hydrolyzes the ester bond of esterified cholesterol to generate free cholesterol

cholesterol esterase pancreatic enzyme that hydrolyzes cholesterol ester

cholesterol esters single molecules of cholesterol esterified to single fatty acid molecules

cholesterol side-chain cleavage enzyme (CYP11A1) rate-limiting enzyme in steroidogenesis

cholic acid primary bile acid

choline acetyltransferase enzyme that catalyzes synthesis of acetylcholine from acetyl-CoA and choline

cholinergic neurons releasing acetylcholine as a neurotransmitter

chondrocyte nondividing cartilage cell; occupies a lacuna within the cartilage matrix

chorion extraembryonic membrane that forms the fetal portion of the placenta

chorionic plate membrane forming the fetal portion of the placenta

chorionic villi fingerlike projections of the chorion that contact maternal blood

choroid plexuses region of the ventricular cavities of the brain that forms the cerebrospinal fluid

choroidal neovascularization (CNV) creation of new blood vessels in the choroid layer of the eye; common symptom of age-related macular degeneration

chromaffin cells cells in the adrenal medulla that synthesize and release epinephrine

chromatid one strand of a chromosome

chromophore part of a molecule responsible for its color

chronic granulomatous disease caused by the lack of phagocytic NADPH oxidase and characterized by recurrent life-threatening bacterial and fungal infections

chronic kidney disease (CKD) slowly progressive disorder that results in an abnormal internal environment and eventual loss of kidney function is typically a slowly progressive disorder that results in an abnormal internal environment

chronic myeloid leukemia (CML) cancer of the white blood cells characterized by increased and unregulated clonal proliferation of myeloid cells in the bone marrow; results from an inherited chromosomal abnormality that involves a reciprocal translocation or exchange of genetic material between chromosomes 9 and 22 and was the first malignancy to be linked to a genetic abnormality

chronic obstructive pulmonary disease (COPD) classification of lung diseases (e.g., asthma, bronchitis, and emphysema) that obstruct airflow out of the lung

chronotropic related to effects on heart rate

Chvostek sign spasm of the facial muscles elicited by tapping the facial nerve in the region of the parotid gland; seen in tetany

chylomicron remnants remains of chylomicron particles after being broken down during metabolism. These remnants can be rapidly cleared from circulation by the liver via chylomicron remnant receptors

chylomicrons least dense (<0.95 g/mL) of the four lipoprotein classes; produced by the small intestine; their production increases during ingestion of a fat-containing meal. Chylomicrons function in the transport of lipids to the bloodstream

chyme semifluid material produced by the gastric digestion of food that results partly from the conversion of large solid particles into smaller particles via the combined peristaltic movements of the stomach and contraction of the pyloric sphincter

chymotrypsin one of three endopeptidases present in pancreatic juice that attack peptide bonds with an aromatic carboxyl terminal

ciliary ganglion location of synapses between parasympathetic preganglionic and postganglionic neurons of the eye

cingulate gyrus area of the cerebral cortex involved in the limbic system

circadian rhythm roughly 24-hour cycle in organismal physiologic processes. Circadian rhythms are endogenously generated and modulated by external cues such as sunlight and temperature

circumventricular organs regions of the central nervous system that do not have a blood–brain barrier and thus can sense signals in the blood

cirrhosis chronic liver disease characterized by the replacement of normal liver tissue with fibrous tissue as a result of the transformation of stellate cells into collagen-secreting myofibroblasts. Stellate cells deposit collagen into the sinusoids, thereby leading to loss of functional hepatocytes

11-*cis*-retinal aldehyde form of vitamin A_1 (retinal); chromophore that starts the process of photoisomerization of retinal for vision

clearance volume of plasma per unit time from which a substance is completely removed (cleared)

climbing fibers afferent axons from the inferior olive of the medulla to the cerebellum that synapse on Purkinje cells

clomiphene fertility drug that stimulates the production of one or more follicles, and therefore, increases the chances of pregnancy

clonal selection collection of specific lymphocytes grouped by stimulation or inhibition of growth (positive or negative selection)

clonus repetitive, self-sustaining muscle stretch reflexes

closure inability to absorb intact proteins as a result of the maturation of the gut

clot retraction phenomenon that may occur within minutes or hours after clot formation in which the clot draws together, pulling the torn edges of the vessel closer together and reducing residual bleeding and stabilizing the injury site; requires the action of platelets, which contain actin and myosin; reduces the size of the injured area, making it easier for fibroblasts, smooth muscle cells, and endothelial cells to start wound healing

coagulation cascade pair of pathways (*intrinsic* and *extrinsic coagulation pathways*) in which a series of reactions ultimately result in a stable fibrin clot

coagulation factors series of 13 proteins (I–XIII) synthesized in the liver that circulate in the plasma in an inactive state

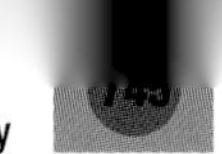

coated pit specialized regions of the cell membrane composed of pits made of the protein clathrin

cochlea bony structure of the inner ear, containing the organ of Corti, which is the organized transducer of sound vibrations into nerve impulses

cochlear microphonic generator potential of the inner ear

cognitive science interdisciplinary study of how information (e.g., concerning perception, language, speech, reasoning, and emotion) is represented and transformed in the brain

cold diuresis increased urine production on exposure to cold as a result of peripheral vasoconstriction leading to a transient increase in central blood volume, and consequently, an increase in glomerular filtration rate

cold pressor response reflex increase in heart rate, cardiac output, vascular resistance, and mean arterial pressure following pain associated with sudden exposure, usually of a hand or extremity, to extreme cold. It results from simultaneous activation of the sympathetic nervous system and reduced activity of the parasympathetics to the heart

cold-induced vasodilation (CIVD) acute increase in peripheral blood flow observed during cold exposures

colipase peptide in pancreatic juice necessary for the normal digestion of fat by pancreatic lipase

collagenase enzyme that digests the connective tissue matrix

collateral ganglia collective name for the three prevertebral sympathetic ganglia of the abdominal cavity

collateral vessels vascular connections between adjacent arteries such that when one artery is obstructed, the area it normally supplies with blood is supplied through the collaterals by the neighboring artery

colloid thick, proteinaceous gel-like substance in the thyroid follicle

colloid osmotic pressure osmotic pressure resulting from proteins in water

colostrum thin, yellowish, milky fluid secreted by the mammary glands a few days before or after parturition

commissures bundles of axons that connect neurons in the two sides of the brain

compensated shock stage of shock in which normal mechanisms within the body designed to elevate blood pressure and restore plasma volume correct the shock without the need for external intervention

compensatory pause pause following an extrasystole, when the pause is long enough to compensate for the prematurity of the extrasystole; the short cycle ending with the extrasystole plus the pause following the extrasystole together equal two of the regular cycles

competitive binding assay assay in which a biologically specific binding agent competes for radioactively labeled or unlabeled compounds, used especially to measure the concentration of hormone receptors in a sample by introducing a radioactively labeled hormone

complement serum proteins that can be activated and may lead to cell lysis, formation of opsonins, and regulation of inflammation

complete atrioventricular block any condition in which no action potentials from the atria reach the ventricles. It usually arises from blockage at the level of the atrioventricular node and results in atria and ventricles that are paced at two different rates.

complete blood count (CBC) blood tests including white blood cell count, red blood cell count, platelet count, hemoglobin, hematocrit, mean cell (corpuscular) volume, mean cell (corpuscular) hemoglobin, mean cell (corpuscular) hemoglobin concentration, and platelet indices; also known as *full blood count (FBC)*

complete metabolic panel (CMP) includes the tests of the basic metabolic panel (see also *basic metabolic panel*) and the following additional tests: albumin, total protein, alkaline phosphatase, alanine transaminase, aspartate transaminase, and bilirubin

complete proteins proteins that supply all of the essential amino acids in amounts sufficient to support normal growth and body maintenance, such as eggs, poultry, and fish

compliance 1. change in volume in a segment of blood vessel or vessels per unit of change in transmural pressure, or $\Delta V/\Delta P$; 2. capability of a region of the gut to adapt to an increased intraluminal volume

compression first step in process of encoding electrical signals to be sent to the central nervous system, involving increasing the density of a medium; opposite of *rarefaction*

concave lens lens that causes light rays to diverge

concentric contractions muscular contractions that permit the muscle to shorten, because the load on the muscle is less than the maximum tetanic tension the muscle is capable of generating

conductance 1. ease with which ions flow across the membrane through their channels; 2. property of a material that indicates its ability to conduct heat

conducting zone first 16 generations of airways (trachea down to the terminal bronchioles), with its own separate circulation (bronchial circulation). It has a major function of conducting air to the deeper parts of the lung.

conduction transfer of energy through matter from particle to particle; the transfer and distribution of heat energy from atom to atom within a substance

conduction velocity rate of impulse conduction in a peripheral nerve or its various component fibers, generally expressed in meters per second

conductive deafness hearing impairment due to sound vibrations being interrupted in the outer or middle ear and not reaching the inner ear and its nerve endings

cone cell cell in the retina responsible for color (photopic, chromatic) vision. There are separate cone cells for red, blue, and green wavelengths of light.

congenital adrenal hyperplasia inherited condition that affects the adrenal glands; these glands are located on top of the kidneys and produce three types of hormones, called *cortisol*, *aldosterone*, and *androgens*; females with classical congenital adrenal hyperplasia are born with masculine-appearing external genitals but with female internal sex organs; males with classical congenital adrenal hyperplasia appear normal at birth; males and females with classical congenital adrenal hyperplasia are likely to have trouble retaining salt, a condition that can be life-threatening

congenital nephrotic syndrome rare, inherited condition characterized by excessive filtration of plasma proteins due to a mutation of nephrin

connective tissue matrix complex ground substance between and among cells that provides mechanical integrity to a tissue; also called *stroma*

connexin four-pass transmembrane proteins that in a group of six form a hemichannel called a *connexon*

connexon assembly of six proteins called *connexins* that forms a bridge called a *gap junction* between the cytoplasm of two adjacent cells

constitutive exocytosis form of exocytosis in which the vesicles are continuously filled and released into the extracellular fluid at a constant rate

continuous ambulatory peritoneal dialysis (CAPD) procedure in which peritoneal membrane, which lines the abdominal cavity, acts as a dialyzing membrane: about 1 to 2 L of a sterile glucose/salt solution are introduced into the abdominal cavity, and small molecules (e.g., K^+ and urea) diffuse into the introduced solution, which is then drained and discarded; the procedure is usually done several times every day

contractile activity muscular activity of the gut wall, either of short duration (phasic contractions) or more sustained activity (tonic contractions)

contractility synonym for the inotropic state of the heart

convection transfer of heat by the actual movement of warmed (or cooled) matter. Convection is the transfer of heat energy in a gas or liquid by movement of currents.

convex lens converging lens that brings light rays to a focus; it can form a real image

cornea outer, transparent, covering of the front of the eyeball

corneal reflex involuntary blinking of the eye elicited by stimulation of the cornea, bright light, or very loud sounds; also known as the *blink reflex*

cornification (keratinization) formation of a group of intermediate filaments in epithelial cells

coronary reserve amount of blood flow possible in the coronary circulation above resting values; sometimes used to refer to the dilating capacity of the coronary circulation beyond its resting value

corpora albicantia or **corpus albicans** connective tissue–filled structure; remnant of the regressed corpus luteum

corpora lutea or **corpus luteum** yellowish endocrine gland formed from the wall of an ovulated follicle

corpus callosum major axonal fiber bundle in the brain that connects the two hemispheres

corpus of the stomach main or central region of the stomach (body); one of four stomach sections

cortex outer portion of an organ, such as the kidney, as distinguished from the inner, or medullary, portion

cortical granules granules in the cortical vesicles at the edge of the ovum cytoplasm; penetration of sperm causes these granules to release proteases and inhibitors to prevent more sperm from penetrating the egg

cortical motor area area of the cerebral cortex in which neurons active in movement initiation are located

cortical tissue part of the indifferent gonad of the embryo; will develop into the ovary

corticospinal tract axons arising from the cortical motor area

corticosteroid-binding globulin (CBG) glycoprotein produced by the liver; binds glucocorticoids and aldosterone in the blood

corticosteroid-binding protein (transcortin) binding protein for progesterone in the blood

corticosterone product of the adrenal cortex; regulates the body's response to fasting, injury, and stress

corticotroph cells located in the anterior pituitary that secrete adrenocorticotropic hormone

corticotropin-releasing hormone (CRH) polypeptide from the hypothalamus; stimulates secretion of adrenocorticotropic hormone

cortisol product of the adrenal cortex; regulates the body's response to fasting, injury, and stress

cotyledons aggregate groups of chorionic villi surrounded by maternal blood

coumarin common oral anticoagulant

countercurrent flowing in an opposite direction

countercurrent exchange passive exchange of solutes, water, or heat between two adjacent streams flowing in opposite directions

countercurrent heat exchange exchange of heat between two separated streams flowing in opposing directions, commonly used to minimize heat loss from the circulatory system by transferring heat from the arteries to the veins

countercurrent multiplication energy-demanding process that sets up a solute (osmotic) gradient along the length of two streams flowing in opposite directions

coupling reaction reaction that forms the iodothyronine structure from iodinated tyrosines

C-peptide peptide made when proinsulin is split into insulin and C-peptide on release from the pancreatic β cell into the blood

craniosacral regional locations of the parasympathetic preganglionic neurons

creatine phosphate pool cellular reservoir of high-energy phosphate that can rephosphorylate the adenosine triphosphate consumed in muscle contraction

creatinine anhydride of creatine, being the end product of creatine metabolism, found in muscle and blood and excreted in the urine

creep slow and time-dependent elongation of a viscoelastic substance subjected to an external force

cremasteric muscle muscle with origin from the internal oblique and inguinal ligament, with insertion into the cremasteric fascia and pubic tubercle, with nerve supply from the genitofemoral nerve, and whose action raises the testicle; in the male, the muscle envelops the spermatic cord and the testis; in the female, it envelopes the round ligament of the uterus

Creutzfeldt–Jakob disease (CJD) degenerative brain disorder caused by death of nerve cells leading to progressive dementia, speech impairment, ataxia, and seizures; most common form of transmissible spongiform encephalopathy

critical micellar concentration certain point of bile salt concentration in the intestinal lumen at which bile salts aggregate to form micelles

crossbridge transient structure in muscle formed by the myosin heads projecting from thick filaments and attached to binding sites on the thin filament actin

crossbridge cycle series of adenosine triphosphate–consuming and work-producing reactions between actin and myosin that produces muscle contraction

cross-reactivity antibody that reacts with antigen that is similar to the one that induced its formation

crypts of Lieberkühn tubular glands located at the base of intestinal villi in the small intestine. Cells in the crypts of Lieberkühn secrete isotonic alkaline fluid

cumulus granulosa cells granulosa cells surrounding the oocyte in a graafian follicle

curare drug that is a competitive inhibitor of acetylcholine at the myoneural junction

Cushing reflex sympathetic nerve reflex initiated by an increase in intracranial pressure. It significantly elevates arterial pressure by causing severe, sometimes occlusive, constriction of arterioles in systemic organs. It is a defense mechanism that helps maintain cerebral blood flow in the face of elevated intracranial pressure by forcing blood vessels open

Cushing syndrome name given to the signs and symptoms associated with prolonged exposure to inappropriately elevated glucocorticoids; the disease can result from excess adrenocorticotropic hormone secretion from the pituitary (specifically called *Cushing disease*) stimulating cortisol production from the adrenal gland, or it can result from an adrenocortical adenoma secreting large amounts of cortisol

cyanosis blue coloration of the skin and mucous membranes due to the presence of excess deoxygenated hemoglobin in blood vessels near the skin surface; it is caused by low hemoglobin concentration (>5 g/dL) or from breathing low oxygen (hypoxia)

cyclic adenosine monophosphate (cAMP) second messenger important in many biologic processes; derived from adenosine triphosphate and used for intracellular signal transduction in many different organisms

cyclic guanosine monophosphate (cGMP) intracellular second messenger derived from guanine triphosphate that causes smooth muscle relaxation by lowering intracellular calcium

cyclooxygenase (COX) enzyme that catalyzes the conversion of arachidonic acid to prostaglandins

cystic fibrosis hereditary disease prevalent especially in Caucasian populations that appears usually in early childhood. It is inherited as an autosomal recessive monogenic trait; involves functional disorder of the exocrine glands; and is marked by faulty digestion due to a deficiency of pancreatic enzymes, by difficulty in breathing due to excess mucus secretion and accumulation in airways and by excessive loss of salt in the sweat

cystic fibrosis transmembrane conductance regulator (CFTCR) ion channel belonging to the adenosine triphosphate–binding cassette family of proteins, whose major function is to secrete chloride ions out of the cells, providing chloride in the lumen for the chloride–bicarbonate exchanger to work

cystinuria autosomal recessive disorder associated with a defect in reabsorptive transport of cystine and the dibasic amino acids ornithine, arginine, and lysine by the kidney. Cystinuria also affects the transporter in the small intestine

cytochrome P450 complex of iron-containing proteins responsible for the oxidation–reduction reactions that transform a nonpolar xenobiotic compound to one that is more polar by introducing one or more polar groups onto the molecule. This particular complex contains the enzyme NADPH–cytochrome P450 reductase, which is responsible for the reduction of the cytochrome P450–drug complex being metabolized

cytokines group of protein-signaling compounds that, similar to hormones and neurotransmitters, are used extensively for intercellular communication

cytoskeleton internal meshwork of proteins adjacent and attached to the plasma membrane of the cell that maintains cell shape and other functions

cytosol liquid or matrix found within a living cell

cytotrophoblast cellular component of the trophoblast; gives rise to the chorion

D cells endocrine cells in the antrum that produce somatostatin, a gastrointestinal hormone that inhibits the release of gastrin and, thus, gastric acid secretion

D_1-like receptor dopamine receptor coupled to stimulatory G proteins, which activate adenylyl cyclase

D_2-like receptor dopamine receptor coupled to inhibitory G proteins, which inhibit adenylyl cyclase

Dalton law gas law that states that total barometric pressure is equal to the sum of the partial pressure of the individual gases

dead space volume air that is inhaled by the lungs but does not take part in gas exchange, because not all the air in each breath is able to be used for the gas exchange and represents the air in the conducting zone; about a third of every resting breath is exhaled exactly as it came into the body (in adults, it is usually in the range of 150 mL)

decerebrate rigidity describes the involuntary extension of the upper extremities in response to external stimuli: the head is arched back, the arms and elbows are extended by the sides, the legs are extended and rotated internally, and the patient is rigid, with the teeth clenched; these signs can be on just one or the other side of the body or on both sides, and it may be just in the arms and may be intermittent

decibel (dB) logarithmic unit that indicates the ration of a physical quantity relative to a specified or implied reference level; widely known as a measure of sound pressure level

decidua site of implantation and the maternal contribution to the placenta

decidua basalis maternal part of the placenta

decidua capsularis part of the endometrium that grows over and covers the fetus

decidua parietalis pregnant endometrium not contained in the placenta

decidual reaction reaction that occurs in the endometrium and consists of dilation of blood vessels, increasing capillary permeability, edema formation, and increased proliferation of endometrial glandular and epithelial cells

decidualization hypertrophy of endometrial cells that contain large amounts of glycogen and lipid to ready the endometrium for implantation

declarative memory memory of events (episodic) or facts (semantic); also known as *explicit memory*

defensins antibacterial proteins that can enter phagosomes

degranulation release of granules from cells, such as mast cells and basophils

dehydration removal of water

dehydration reaction chemical reaction in which a hydrogen atom is removed from the end of one molecule that was originally a hydroxyl; the remaining O, from the original OH that only the H was taken from, is then bonded with the other monosaccharide; the H and OH are then released as a molecule of water (H_2O)

dehydroalanine residue product of free radical reaction to produce an iodothyronine residue

7-dehydrocholesterol provitamin D; the presence of this compound in human skin enables humans to manufacture vitamin D_3 from ultraviolet rays in the sun light, via an intermediate isomer previtamin D_3

dehydroepiandrosterone (DHEA) androgenic precursor for estrogen and testosterone synthesis produced by ovaries and testes

dehydroepiandrosterone sulfate (DHEAS) sulfate salt of dehydroepiandrosterone

deiodinase type 1 (D1) substance located in the liver, kidneys, and thyroid gland; catalyzes outer-ring deiodination

deiodinase type 2 (D2) substance believed to function primarily to maintain intracellular T_3 in target tissues

deiodinase type 3 (D3) substance that catalyzes inner-ring deiodination reactions during degradation of thyroid hormones

delayed puberty condition in which physical signs of puberty have not presented by age 13 years in girls and 14 years in boys

delayed afterpolarizations (DADS) occur late in phase 3 or during phase 4 of the ventricular action potential and are associated with increased intracellular calcium concentration within myocytes, which can thus depolarize the cell membrane to the threshold for an action potential

δ (delta) 4 pathway dominant steroidogenic pathway in granulosa cells and the corpus luteum

δ (delta) 5 pathway dominant steroidogenic pathway in theca cells

delta cells see *δ cell*

delta waves electroencephalogram wave pattern with a rhythm ranging from 0.5 to 4 Hz, observed when the person is in deepest sleep

delusions fixed false beliefs

dementia progressive loss of brain function, including memory, behavior, and cognition

denaturation structural change in macromolecules caused by extreme conditions, for example, protein denaturation by extreme heat

dendrites fibers that receive synaptic inputs and transmit electrical signals toward the perikaryon of a nerve cell

dendritic cell phagocyte; antigen-presenting cell of epithelia and blood

dendritic spines small membrane protrusions from dendrites that receive synaptic input

denervation hypersensitivity heightened reactivity of an organ to a neurotransmitter following denervation of the organ

dense bodies intracellular protein complexes at the margins of smooth muscle cells that serve as attachment points between intracellular actin and extracellular connective tissue

density mass per unit volume

dentate nucleus largest of the four deep cerebellar nuclei, responsible for the planning, initiation, and control of volitional movements; receives its afferents from the premotor cortex and the supplementary motor cortex, and its efferents project via the superior cerebellar peduncle through the red nucleus to the ventrolateral thalamus

deoxycholic acid secondary bile acid

11-deoxycorticosterone (DOC) mineralocorticoid made from progesterone by 21-hydroxylase

11-deoxycortisol immediate precursor for cortisol synthesis

deoxyhemoglobin hemoglobin that does not bind with oxygen

depolarization change in electrical potential in the direction of a more positive potential inside cells; in excitable cells, depolarization to a threshold potential triggers an action potential

depolarizing blockers agents that interfere with neuromuscular transmission by causing the postsynaptic membrane to remain depolarized

desensitization process by which effector cells are rendered less reactive or nonreactive to a given stimulus

diabetes insipidus disorder characterized by excretion of a large volume of osmotically dilute urine, resulting from either deficient production or release of arginine vasopressin

diabetes mellitus disease in which plasma glucose control is defective because of insulin deficiency or decreased target cell response to insulin

diabetic ketoacidosis condition in which extremely high blood glucose levels, along with a severe lack of insulin, result in the breakdown of body fat for energy and an accumulation of ketones in the blood and urine

diabetic neuropathy family of nerve disorders caused by diabetes; also known as *peripheral neuropathy*

diabetogenic action effect of growth hormone to oppose the actions of insulin

diacylglycerol (DAG) intracellular second messenger generated by hydrolysis of phosphatidylinositol 4,5-bisphosphate

dialysis separation of smaller molecules from larger molecules in solution by diffusion of the small molecules through a selectively permeable membrane; two methods of dialysis are commonly used to treat patients with severe, irreversible ("end-stage") renal failure; see also *continuous ambulatory peritoneal dialysis (CAPD)* and *hemodialysis*

diapedesis movement of cells from blood to tissue during acute inflammation

diaphragm dome-shaped sheet of internal skeletal muscle that extends across the bottom of the rib cage and separates the thoracic cavity (heart, lungs, and ribs) from the abdominal cavity; it is the main muscle of breathing and is innervated by the phrenic nerves

diastole period of time in which the heart is in its relaxation phase

diastolic pressure blood pressure in the ventricle or arterial system during the relaxation of the heart, or diastole

diencephalon region of the central nervous system that includes the thalamus and hypothalamus

differential white blood cell count proportions of the different types of circulating white blood cells

diffuse esophageal spasm diagnosis of diffuse spasm is made when manometric recording of esophageal motility demonstrates that the act of swallowing results in simultaneous contractions all along the length of the smooth muscle region of the esophageal body

diffuse toxic goiter enlarged thyroid gland; secretes thyroid hormones at an accelerated rate

diffusion random movement of particles, such as ions and solutes in solution, that leads to mixing and elimination of concentration gradients

diffusion block abnormal increase in the diffusion distance across the alveolar–capillary membrane or to a decrease in the alveolar–capillary permeability leading to decrease in oxygen uptake by the lungs (hypoxemia); pulmonary edema is one of the major causes of a diffusion block and is characterized by a low PaO_2, elevated $PaCO_2$, and a high A-a O_2 gradient

diffusion capacity see *lung diffusion capacity*

diffusion limited condition in which the uptake of an alveolar gas is limited by the diffusion properties of the alveolar capillary membrane; see also *Fick law*

diffusion-limited transport transport of substances across the capillary in which the rate of diffusion is what limits the rate of transport

diffusive membrane transport see *passive transport*

digestive state gastrointestinal motor behavior (mixing/segmentation pattern) in effect when nutrients are present in the upper gut

digoxin binds to the extracellular aspect of the α-subunit of the Na^+/K^+ ATPase and decreases the transport function of the pump; often used to treat atrial fibrillation or atrial flutter; also known as *digitalis*

dihydropyridine receptors (DHRPs) voltage-sensitive protein molecules that respond to changes in the t-tubule membrane of skeletal muscle as an important step in the excitation–contraction coupling process

dihydrotestosterone (DHT) potent androgen derived from testosterone

1,25-dihydroxycholecalciferol substance that enhances calcium and phosphate absorption by the small intestine and mobilizes calcium and phosphate from bones; the active form of vitamin D; also known as *1,25-dihydroxycholecalciferol* or *calcitriol*

diiodotyrosine (DIT) intermediate in the biosynthesis of thyroid hormone

diopter (D) measure of lens converging power; the inverse of the focal length in meters

dimer chemical or biologic entity consisting of two structurally similar subunits that are joined by bonds, which can be strong or weak

dimerization 1. aggregation of two monomers of a protein (e.g., receptor), usually mediated by binding of a ligand 2. chemical reaction in which two monomers combine to form a dimer

dipalmitoylphosphatidylcholine (DPPC) surface-reducing material lining the alveoli that is responsible for reducing surface tension and increasing the distensibility of the lung

diplopia double-vision, caused by partial failure of the convergence mechanism driven by the extraocular muscles

dipole difference in the electrical polarity between two physical points

direct calorimetry measurement of the metabolic rate of an organism by measuring the amount of heat energy released as heat over a given period

direct pathway basal ganglia pathway for axons passing from the caudate directly to the globus pallidus internus

discomfort (upper abdomen) subjective, unpleasant sensation or feeling that is not interpreted as pain according to the patient and that, if fully assessed, can include nausea, fullness, bloating, and early satiety

disinhibitory motor disease disordered gastrointestinal motility resulting from neuropathic degeneration of enteric inhibitory motor neurons

disseminated intravascular coagulation (DIC) pathologic activation of blood-clotting mechanisms, leading to the formation of small blood clots inside the blood vessels throughout the body, thereby disrupting normal coagulation and causing abnormal bleeding and possible organ malfunction

distal nephron distal convoluted tubule, connecting tubule, and collecting duct, considered as one functional entity

distending pressure pressure required to stretch material (e.g., the pressure required to inflate the lung)

distensibility measure of the force required to stretch material

diuresis increased urine flow rate

diurnal rhythm 24-hour cycle of activity entrained to the day–night cycle

divergence tendency for spread of sympathetic responses from a lesser number of preganglionic neurons to a greater number of postganglionic ones

diving reflex reflex that occurs when the face is submerged under water, which is found in all mammals (including humans, although less pronounced). The reflex puts the body into an oxygen-saving mode to maximize the time that can be spent underwater. The reflex includes bradycardia, peripheral vasoconstriction, and a decreased metabolic rate

diving response cardiovascular reflex to immersion of the body or even the face underwater. It is characterized by intense slowing of heart rate and sympathetic peripheral vasoconstriction. It is a means of preserving oxygen to the brain and heart when submerged underwater

DNA vaccine gene that expresses, when incorporated in human cells, immunogenic protein

docosahexaenoic acid omega-3 fatty acid found abundantly in seafood; is present in parts of the brain and the retina and is essential for the normal development of vision in newborns

dolor pain associated with acute inflammation

dominant follicle one follicle from the cohort of developing follicles that will ovulate

dopamine (DA) catecholamine used as a neurotransmitter; important in motor function and limbic pathways; also inhibits the synthesis and secretion of prolactin

dopaminergic neurons relating to nerve cells or fibers that employ dopamine as their neurotransmitter

Doppler ultrasound ultrasound imaging technique that works similar to radar speed detection devices; gives an ultrasound image of the heart but is used primarily for detecting flow velocity abnormalities through the heart valves and in the aorta

dorsal motor nuclei region of the medulla where parasympathetic neurons involved in cardiac, respiratory, and gastrointestinal functions are located

dorsal motor nucleus of vagus location in the medulla oblongata (brainstem) containing cell bodies of vagal efferent (motor) fibers to the digestive tract, excluding parts of the esophagus

dorsal respiratory group (DRG) cluster of neuronal cells located in the dorsal portion of the medulla that is active during inspiration

dorsal vagal complex combined structures of the dorsal vagal motor nucleus, nucleus tractus solitarius, and area postrema in the medulla oblongata (brainstem)

down-regulation process by which a cell decreases the quantity of a cellular component, such as a protein, in response to an external variable

drusen localized deposits of cellular debris whose accumulation distorts the shape of the retina in age-related macular degeneration

Duchenne muscular dystrophy (DMD) rapidly worsening muscle weakness caused by a defective gene for dystrophin (a protein in the muscles); although inherited, it also occurs in those without a family history of the disease

ductus arteriosus fetal blood vessel connection between the pulmonary artery and aorta that shunts blood ejected from the pulmonary artery to the aorta

ductus venosus fetal blood vessel connection between the umbilical veins and the inferior vena cava that shunts oxygenated blood from the veins into the vena cava while bypassing the liver

dumping syndrome syndrome that occurs after a meal due to rapid delivery of gastric chyme into the duodenum, characterized by flushing, sweating, dizziness, weakness, and vasomotor collapse, resulting from rapid passage of large amounts of food into the small intestine, with an osmotic effect removing fluid from plasma and causing hypovolemia

dynamic exercise performance of some activity that involves muscle contraction and joint flexion or extension; in contrast to static exercise, in which joint movement is absent

dynein microtubule-associated protein involved in retrograde axonal transport of organelles and vesicles (from the plus to the minus ends of the microtubules) via the hydrolysis of adenosine triphosphate

dysphagia difficulty swallowing; sensation of abnormal bolus transit through the esophageal body

dyspnea shortness of breath; a perceived difficulty breathing or pain on breathing

dystrophin filamentous muscle protein that lies just inside the sarcolemma and participates in the transfer of force from the contractile system to the outside of the cells via integrins

early afterpolarizations (EADS) depolarizations of the muscle cells that occur late in phase 2 or early in phase 3 of the ventricular action potential and are more likely to be induced by factors that prolong the action potential duration, such that slow calcium channels have time to recover, and thus, fire an additional low-amplitude action potential

early satiety feeling that the stomach is overfilled soon after starting to eat; the sensation is out of proportion to the size of the meal being eaten, so that the meal cannot be finished

eccentric contraction contraction characterized by involving the lengthening of muscle fibers while they are maintaining contractile tension; related movements frequently involve deceleration and high passive tension

eccrine ordinary, or simple, sweat gland of the merocrine type. These unbranched, coiled, tubular glands are distributed over almost all of the body surface and promote cooling by evaporation of their secretion.

echinocyte spiked erythrocyte; also known as *Burr cell*

echocardiography general classification of noninvasive techniques by using ultrasound echoes to image the heart wall, its chambers, and movement of the valves during the cardiac cycle

2D-echocardiography 1. two-dimensional echocardiography; used to examine ventricular chambers and movement of valves in the beating heart *in situ*; 2. type of ultrasound cross-sectional image of the heart used to measure ventricular wall thickness, valve motion and abnormalities, and wall motion during the cardiac cycle; it is often used to estimate the ejection fraction of the left ventricle

ectopic foci portion of the myocardial muscle that fires its own action potentials, causing activation of the heart from an area other than the sinoatrial node

ectopic pregnancy implantation of an embryo outside of the uterus

edema clinically apparent increase in interstitial fluid volume

Edinger-Westphal nucleus nucleus region of cranial nerve III where parasympathetic neurons arise

effective arterial blood volume (EABV) degree of fullness of the arterial system, which determines the perfusion of the body's tissues

effector small molecule used to relay signals within a cell. Effectors bind to a protein and alter the activity of that protein

effector cell executing cell of immune responses; in adaptive immunity: plasma cells, T helper cells, and cytotoxic T cells

effector systems musculature, secretory epithelium, and blood–lymphatic vasculature in the digestive tract

efferent carrying away from

efferent nerve (efferents) nerve fibers carrying impulses away from the central nervous system that cause muscles to contract and glands to secrete (inhibitory efferent nerves)

efflux transfer of material from the interior to the exterior of a cell

eicosapentaenoic acid omega-3 fatty acid found in fish, essential for the normal development of vision in newborns

Einthoven triangle theoretical triangle in the frontal plane of the body created from the interconnected electrodes making up the standard bipolar electrocardiogram lead system

ejaculation abrupt discharge of fluid; the expulsion of seminal fluid from the urethra of the penis during orgasm

ejection fraction amount of blood ejected in one systole by the left ventricle expressed as a percentage of the left ventricular residual volume; it is used clinically to detect impaired performance of the left ventricle

elastase one of three endopeptidases present in pancreatic juice that attacks peptide bonds with a neutral aliphatic carboxyl terminal

elastic recoil degree to which stretched material returns to its unstretched position

elastic response proportionate and time-independent force response to the extension of a deformable body such as muscle

elasticity ability of a material when stretched to return to its unstretched position (e.g., a balloon). This property differs from plasticity (the capability of being stretched, but not returning to its unstretched position, e.g., putty)

electrical slow waves omnipresent form of electrical activity (rhythmic depolarization and repolarization) in gastrointestinal muscle cells

electrical synapse site of communication between neurons mediated by movement of ions from one cell to the other via gap junctions

electrical syncytium tissue (e.g., gastrointestinal and cardiac muscle) in which the cells are electrically coupled one to another at cell-to-cell appositions that do not include cytoplasmic continuity; accounts for three-dimensional spread of excitation in excitable tissues

electrochemical potential (gradient) combined effect of electrical and chemical gradients that determines the direction of movement of ions and charged solutes

electroencephalogram (EEG) recording of the electrical activity of the brain

electroencephalography measurement of the electrical activity of the brain

electrogastrography recording of gastric electrical activity from surface electrodes positioned on the abdominal wall

electrogenic transport membrane transport system that produces net movement of charge across the membrane, for example, H^+-ATPase, Na^+/glucose cotransport

electrolyte substance that is ionized in solution and thus becomes capable of conducting electricity

electromechanical coupling in gastrointestinal smooth muscle, depolarization of the membrane electrical potential that leads to the opening of voltage-gated calcium channels, followed by

the elevation of cytosolic calcium, which, in turn, activates the contractile proteins

electroneutral transport membrane transport system that produces no net movement of charge across the membrane, for example, Na^+/H^+ exchange

electroneutrality principle that, in an electrolytic solution, the concentrations of all the ionic species are such that the solution as a whole is neutral

electroretinogram generator potential of the retina, recorded extracellularly and at some distance from its source

electrotonic conduction local spread of electrical activity within a group of cells by ionic current flows that do not involve action potential mechanisms; also known as *passive conduction*

electrotonic potential nonpropagated potential in a nerve or muscle membrane generated by the flow of ions across the membrane

electrotonically way in which electrical activity spreads within a group of cells by ionic current flows that do not involve action potential mechanisms

emboli blood clots that form in the bloodstream

embryoblast cluster of small centrally located cells within the blastocyst that will give rise to the fetus

emesis vomiting

emissivity ratio of energy radiated to energy radiated by a black body at the same temperature. It is a measure of a material's ability to absorb and radiate energy. A true black body would have $\varepsilon = 1$, whereas any real object would have $\varepsilon < 1$

emmetropia normal vision, free from refractive errors

emphysema chronic lung disorder that leads to abnormally high compliance as a result of distension and eventual rupture of the alveoli with progressive loss of pulmonary elasticity; symptoms include shortness of breath with or without cough, which may lead to impaired heart action

end tidal volume volume of air measured at the end of a tidal volume

end-diastolic volume volume within the ventricle at the end of diastole

endocannabinoids endogenous ligands for the receptors that mediate the effects of the psychoactive component of marijuana, Δ^9-tetrahydrocannabinol

endocrine glands glands that secrete their products (hormones) directly into the bloodstream (ductless glands) or release hormones (paracrines) that affect only target cells nearby the release site

endocrine pancreas portion of the pancreas that secretes hormones such as insulin, glucagon, and somatostatin into the bloodstream

endocrinology branch of physiology concerned with the description and characterization of processes involved in the regulation and integration of cells and organ systems by specialized chemical substances called *hormones*

endocytosis invagination of the plasma membrane to pinch off and internalize substances that are otherwise unable to cross the plasma membrane. Some endocytic mechanisms involve binding of extracellular solute to a specific membrane receptor protein (receptor-mediated endocytosis)

endogenous originating or produced within the body

endogenous creatinine clearance renal clearance of endogenous (i.e., not administered) creatinine; used to estimate the glomerular filtration rate

endolymph fluid that fills the scala media of the cochlea and the semicircular canals of the vestibular apparatus

endometrial cycle life cycle of the endometrial lining of the uterus; consists of four parts

endometrial veins veins that drain the maternal blood-filled sinus of the placenta

endometrium inner membrane lining the lumen of the uterus

endopeptidases one of two classifications of pancreatic proteases; include trypsin, chymotrypsin, and elastase; are present in pancreatic juice; and hydrolyze certain internal peptide bonds of proteins or polypeptides to release the smaller peptides

endorphins endogenous opioid neurotransmitters that can produce a feeling of well-being

endothelial growth factor (EGF) cytokine involved in wound healing

endothelial vesicles pinocytotic vesicles used to transport substances that are too large to cross the capillary membrane or pores

endothelin extremely potent vasoconstrictor produced by the endothelium, especially in cardiovascular disease states; the most potent vasoconstrictor yet found

endothelium thin epithelial lining of all blood vessels and the inner surface of the chambers of the heart. It contains anticoagulant factors such as heparin and forms a nonthrombotic surface in arteries. It is also a barrier between the bloodstream and the inner vascular tissue. Endothelium is the prime source of nitric oxide and PGI_2, which have vasodilator, antithrombotic, and antiatherogenic properties as well as antimitogenic properties on the underlying vascular smooth muscle. The endothelium also produces growth factors that are thought to be responsible for angiogenesis and intimal repair in blood vessels. The endothelium and its ability to produce nitric oxide are impaired in all known forms of cardiovascular disease, especially those associated with hypertension, atherosclerosis, and diabetes

endothelium-derived relaxing factor (EDRF) original name for *endothelial nitric oxide*

endotoxin toxins of gram-negative bacteria associated with lipopolysaccharide

endplate current postsynaptic ionic current at the myoneural junction whose local flow excites adjacent areas of muscle membrane to produce action potentials

endplate membrane specialized postsynaptic membrane at the myoneural junction, containing the acetylcholine receptors

endplate potential proportionate depolarization of the endplate membrane caused by the permeability to Na^+ and K^+ ions that results from the binding of acetylcholine

end-product inhibition end product of a biosynthetic pathway, which inhibits the enzyme(s) that catalyze the first step in the pathway

end-systolic pressure–volume relationship considered the ideal measure of the inotropic state of the heart, this straight line sets the limit of contraction for the whole heart as it generates pressure and reduces volume during ejection that is formed from all the points of volume and pressure at the end of systole in the whole heart generated from multiple different preloaded and afterloaded contractions of the whole heart. The position, or slope, of this line is independent of the loading conditions on the myocardium but is shifted by positive or negative inotropic influences on the heart

end-systolic volume volume of blood in the ventricle (usually as concerns the left ventricle) and the end of systole; although it is sometimes called the *residual volume*, it actually best represents the volume in the ventricle at the end of ventricular ejection rather than the start of diastole

enkephalins endogenous ligand that binds to opioid receptors to regulate nociception

enteric minibrain term used in reference to the brainlike functions of the enteric nervous system

enteric nervous system division of the autonomic nervous system situated within the walls of the digestives tract and involved in independent integrative neural control of digestive functions

enteroceptors sensory receptors classified by "vantage point" in the body; detect stimuli from inside the body

enterochromaffin cells enteroendocrine cells located in the mucosal epithelium of the digestive and respiratory tracts that secrete most of the body's serotonin; cells are stimulated by "brushing" of the mucosa to release 5-hydroxytryptamine as a paracrine signal to neurons in the enteric nervous system

enterochromaffin-like (ECL) cells special neuroendocrine cells located mostly in the acid-secreting regions of the stomach that are believed to be the source of histamine

enterocytes specialized epithelial cells in the small intestine that take up the digested products of nutrients for transport into either the portal circulation or the lymphatic system

enteroendocrine cells cells found throughout the digestive tract that produce hormones and paracrine signals to enteric neurons and sensory neurons

enterogastrone hormone released by the small intestine that inhibits gastric motility and secretion

enterohepatic circulation recycling of bile salts between the small intestine and the liver

enterooxyntin hormone present in intestinal endocrine cells whose release is stimulated by the distention of the intestine; it stimulates acid secretion

enteropeptidase enzyme found on the luminal surface of enterocytes that converts trypsinogen to trypsin when pancreatic juice enters the duodenum; also known as *enterokinase*

enzyme-linked immunosorbent assay (ELISA) biochemical technique used mainly in immunology to detect the presence of an antibody or an antigen in a sample. It uses two antibodies, one of which is specific to the antigen and the other of which is coupled to an enzyme. This second antibody gives the assay its "enzyme-linked" name and will cause a chromogenic or fluorogenic substrate to produce a signal

eosin orange dye used in polychrome stains

eosinophil granulocyte of the myeloid series involved in allergic reactions and defense against parasites

epidermal growth factor (EGF) substance in saliva that stimulates gastric mucosal growth and the growth of epithelial cells in the skin and other organs

epididymis tiny tube where sperm collect after leaving the testis

epilepsy neurologic disorder of the brain characterized by spontaneous discharges of electrical activity

epinephrine (EPI) catecholamine neurotransmitter released by the adrenal medulla

epiphyseal plate cartilage (hyaline) region (between the epiphysis and diaphysis) in bone that allows lengthwise growth of a bone. In puberty, rapid growth in this region leads to height increase, which then ossifies in the adult

episodic memory memory of events

epithelial sodium channel (ENaC) channel through which Na^+ enters into the collecting duct cell during K^+ regulation

epitope part of the molecule that is recognized by the immune system

epsilon cells see *ε cell*

equal pressure point (EPP) point in the airway at which the pressure inside equals the pressure outside of the airway

equilibrioception physiologic sense of balance

equilibrium stable situation that occurs when the internal processes of a system are in balance and no overall change in the system occurs

equivalent in chemistry, a unit of measure that contains the Avogadro number (or a mole) of positive or negative charges, each the quantity of electricity possessed by a proton or electron

erectile dysfunction (ED) inability to attain and/or maintain an erection suitable for penetration

erythroblast stem cell committed to differentiate into erythrocytes

erythrocyte see *red blood cell*

erythrocyte sedimentation rate (ESR) test to measure the settling of erythrocytes per time

erythromelalgia rare neurovascular peripheral pain disorder in which blood vessels, usually in the lower extremities, are episodically blocked, then become hyperemic and inflamed; there is severe burning pain (in the small fiber sensory nerves) and skin redness

erythropoiesis process by which red blood cells are produced in hematopoietic tissues

erythropoietin hormone produced by the kidney that stimulates the bone marrow to produce red blood cells (such as in hypoxia)

esophageal varices varicose veins occurring at the base of the esophagus that form as a compensatory mechanism to the increased resistance to (and therefore, dramatically decreased) portal venous flow. Conditions at the base of the esophagus (including minimal connective tissue support and the associated negative intrathoracic pressure of the area) together with the increased demand on these vessels to support an increasing load of blood flow lead to a tendency for these varices to rupture, a condition leading to morbidity in 30% of occurrences

essential amino acids amino acids that cannot be synthesized by the liver, and therefore, supplied through dietary consumption. The essential amino acids include histidine, methionine, threonine, tryptophan, isoleucine, leucine, valine, phenylalanine, and lysine

essential fatty acids fatty acids that humans and other animals must ingest for normal cell function, because they cannot be made from other foods; the three essential fatty acids include linolenic acid, linoleic acid, and arachidonic acid

essential light chains small protein constituents of the myosin molecule that are necessary for its function (but not its regulation)

estradiol steroid with 18 carbons

estriol major estrogen produced by the placenta

estrogen steroid that promotes maturation of the female reproductive organs and secondary sexual characteristics

estrone estrogen secreted by the ovaries, placenta, and fat

eunuchoid sexually deficient person; especially one lacking in sexual differentiation and tending toward the intersexual state

eunuchoidism male hypogonadism; the state of being a eunuch (either because of lacking testicles or because they failed to develop)

euthyroid sick syndrome state of adaptation/dysregulation of feedback control with triiodothyronine and/or thyroxine at low levels but normal thyroid-stimulating hormone levels

excitable cells cells such as neurons and muscle with membranes capable of giving rise to action potentials and propagated impulses

excitation–contraction coupling series of steps that begins with the depolarization of a muscle cell membrane and the mechanical events of contraction

excitatory junction potentials (EJPs) neurally evoked depolarizations of the muscle membrane potential

excitatory postsynaptic potentials (EPSPs) depolarization of the neuronal membrane at the postsynaptic side of a synapse as a result of ions flowing through channels that open when neurotransmitter released from the presynaptic terminal binds to its receptor

excretion elimination of a substance from the body in the urine or feces

excretory ducts part of the network of ducts in the salivary glands that are lined with columnar cells and play a role in modifying the ionic composition of saliva

execution orchestration of complex sensorimotor sequences in a seamless path toward a goal

exercise performance of some activity that involves muscle contraction and joint flexion or extension; often to develop or maintain physical fitness and overall health

exertional heat injury syndrome occurring in association with strenuous exercise characterized by hyperthermia with exhaustion or collapse, which is accompanied by evidence of tissue and organ damage

exocrine glands glands that secrete their products (excluding hormones and other chemical messengers) into ducts leading directly into the external environment; contrast with *endocrine glands*

exocrine pancreas portion of the pancreas that secretes digestive enzymes and HCO_3^- into the intestinal lumen

exocytosis fusion of a secretory vesicle with the plasma membrane and expulsion of the vesicle contents into extracellular fluid

exopeptidases substance present in pancreatic secretion that could be a carboxypeptidase or an aminopeptidase, depending on whether it releases amino acid from the carboxyl end or the amino end of the peptides

expiratory reserve volume (ERV) maximum volume of air exhaled at the end of a tidal volume

expired minute ventilation (E) volume of air expired from the lungs in 1 minute

extension movement that increases the angle between two moving parts of the body

extensors skeletal muscles whose action extends a limb or body part

external anal sphincter ring of skeletal muscle surrounding the anal canal that can be voluntarily contracted to postpone defecation

external intercostal muscles muscles located between the ribs that are involved in forced inspiration

exteroceptors sensory receptors classified by "vantage point" in the body; detect stimuli from outside the body

extraalveolar vessels small pulmonary vessels that are not part of the blood vessels surrounding the alveoli

extracellular fluid (ECF) fluid outside of cells

extraction (E) or extraction ratio amount of a substance removed from the bloodstream from arterioles to venules by the capillaries

extrafusal muscle fibers population of muscle fibers that produce force during contraction

extramedullary hematopoiesis blood cell formation occurring outside bone medulla

extraocular muscles muscles within the orbit but outside of the eyeball, including the four rectus muscles (superior, inferior, medial, and lateral); two oblique muscles (superior and inferior), and the levator of the superior eyelid (levator palpebrae superioris)

extrathoracic blood volume blood volume in the veins of the systemic circulation

extrinsic coagulation pathway pathway in which blood clots in response to tissue injury

exudate tissue fluid that seeps out of blood vessels into inflammatory tissue

F cells one of five types of cells found in the islets of Langerhans of the endocrine pancreas; this one produces pancreatic polypeptide

F-type ATPase proton pump located in the inner mitochondrial membrane that synthesizes adensoine triphosphate using the energy stored in a gradient of protons that crosses the inner mitochondrial membrane from outside to inside the mitochondria down its electrochemical gradient; also called *ATP synthase*

facilitated diffusion see *passive transport*

factor III see *tissue thromboplastin*

factor X see *prothrombinase*

factor XIII see *fibrin stabilizing factor*

familial hypercholesterolemia disorder in which the liver fails to produce the low-density lipoprotein (LDL) receptor, which is essential for LDL clearance from plasma. The condition is characterized by elevated levels of plasma LDLs, which typically leads to a predisposition for early coronary heart disease.

farad (F) unit of measure of membrane capacitance

fast axoplasmic transport mechanism for transporting proteins, organelles, and other cellular materials needed for the maintenance of the cell along the length of axons

fastigial nucleus deals with antigravity muscle groups and other synergies involved with standing and walking; receives afferent input from the vermis, and most efferent connections travel via the inferior cerebellar peduncle to the vestibular nuclei

fast-twitch fibers skeletal muscle fibers specialized for rapid contractions

fatty acid synthase multienzyme complex required to convert carbohydrates to fatty acids

feedback relaxation involves both local reflex connections between receptors in the small intestine and the gastric enteric nervous system (ENS) or hormones that are released from endocrine cells in the small intestinal mucosa and transported by the blood to signal the gastric ENS and stimulate firing in vagal afferent terminals in the stomach

feedforward control kind of system that reacts to changes in the environment, usually to maintain some desired state of the system

female athletic triad condition of disorded eating, amenorrhea, and osteoporosis occurring in female athletes and dancers

female pseudohermaphrodite 46XX female with normal ovary and internal genitalia but virilization of the external genitalia

ferning crystal pattern formed by dried cervical mucus; indicates high estrogen levels

ferritin intracellular protein that binds and stores iron

ferroportin membrane protein for transport of iron from cell cytoplasm into the interstices

fertilization fusion of the male and female gametes (sperm with an ovum), which leads to the formation of an embryo and the development of a new organism; also known as *conception*

fetal hemoglobin major hemoglobin of the fetus, which exhibits much higher affinity for O_2 than does adult hemoglobin

fibrin fibers that form stable blood clots

fibrin stabilizing factor (factor XIII) clotting factor that catalyzes the formation of covalent bonds between strands of polymerized fibrin, stabilizing and tightening the blood clot

fibrinolysis process by which a fibrin clot is broken down

fibroblast growth factor-23 (FGF-23) protein hormone produced mainly by bone cells that inhibits tubular reabsorption of phosphate and 1α-hydroxylase activity in the kidneys and secretion of PTH; the result of these actions is a lower plasma phosphate concentration

Fick law gas law that states that the amount of gas that diffuses across a biologic membrane per minute is directly proportional to the membrane surface area (A_s), the diffusion coefficient of the gas (D), and the partial pressure gradient (ΔP) of the gas and inversely proportional to membrane thickness (T); also known as *Fick's law*

Fick principle principle based on mass conservation that uses oxygen content of the blood and whole body oxygen consumption to measure cardiac output

fight-or-flight response cardiovascular response to fear or stress that is similar to the baroreceptor reflex except that vasodilation occurs in the skeletal muscle circulation instead of vasoconstriction on initiation of the reflex

filopodia long, thin, transient actin-containing protrusions from growth cones

filtered load quantity of a substance, per unit time, that is filtered by the glomeruli and is therefore presented to the tubules; for a freely filterable substance, this is equal to the product of the plasma concentration and the glomerular filtration rate

filtration fraction fraction of the plasma flowing through the kidneys that is filtered; the ratio of glomerular filtration rate to renal plasma flow

filtration pressure equilibrium condition in which the pressures favoring fluid movement out of a capillary are exactly balanced by pressures favoring fluid movement into the capillary, so that there is no net fluid movement across the capillary wall

filtration slit space between adjacent glomerular podocytes about 40 nm wide and bridged by a diaphragm composed of nephrin

fimbria fingerlike projections on the edge of the infundibulum of each oviduct

final common path lower motor neurons and their axons that control skeletal muscles

first-degree atrioventricular block slower-than-normal transmission of action potentials from the atria to the ventricles. It is revealed on an electrocardiogram recording as a PR interval >0.2 seconds but with each P wave being followed by a QRS complex.

first heart sound series of low-pitched sounds following the start of ventricular contraction created by vibrations in the cordae tendinae and blood in the ventricles

first messengers the messenger that binds to a receptor to initiate a signaling cascade. Examples are hormones, peptides, gases such as nitric oxide or CO, ions such as Ca^{2+}, and adenosine triphosphate. Note that some first messengers, such as Ca^{2+}, can also be second messengers, depending on whether they initiate or amplify the signaling cascade.

first polar body small haploid cell produced by unequal first meiotic division of the oocyte

fixed monocyte-macrophage system cells that remain at a strategic fixed location and when encountering pathogenic material, thus being able to differentiate from monocytes to macrophages to ingest and thereby remove the material from the circulation

flap valve term used to describe the effects of contraction of the puborectalis muscle to interfere with the movement of feces or flatus in the direction of the anus; a component of the mechanisms that sustain fecal continence

flexion movement that decreases the angle between two moving parts of the body

flexor withdrawal reflex complex spinal reflex triggered by cutaneous stimulus resulting in withdrawal of the body part from the stimulus

flexors skeletal muscles whose action retracts a limb or body part

flow-limited transport transport of materials across the capillaries that is limited not by diffusion but instead by the rate at which the materials can be delivered by blood flow into the capillary network of an organ

flow velocity measure of how fast blood moves from one point to the next downstream in the cardiovascular system. It has units of cm/s and is an important determinant of lateral pressure, intimal shear stress, and flow turbulence when blood flows within blood vessels.

flow-mediated vasodilation dilation of arteries and arterioles following an increase in blood flow. It is caused by increases in shear stress stimulation of nitric oxide release from the arterial endothelium.

flow-volume curve curve that reflects airflow during forced expiration and forced inspiration

fluid mosaic model model conceived by S.J. Singer and G. Nicolson in 1972 to describe the structural features of biologic membranes

fluid phase endocytosis see *exocytosis*

flux net transport of ions or solutes in free solution or via membrane transport

focal adhesions dense bodies involved in cell-to-substrate attachment in smooth muscle

focal length (FL) distance behind a positive (converging) lens at which parallel rays from a distant object are brought to a focus

folia folds of the cerebellar cortex

folic acid water-soluble vitamin essential for the formation of nucleic acids, the maturation of red blood cells, and growth

follicles structures containing the oocyte in the ovary

follicle-stimulating hormone (FSH) in women, the pituitary hormone responsible for stimulating follicular cells in the ovary to grow, triggering egg development and production of the female hormone estrogen; in the male, the pituitary hormone that travels through the bloodstream to the testes and helps stimulate them to manufacture sperm

follicular phase part of the menstrual cycle during which follicle development occurs

folliculogenesis follicular development

follistatin single-chain protein hormone, with several isoforms, that binds and deactivates activin

foramen ovale opening between the right and left atria of the fetal heart. It shunts oxygenated blood from the right atrium of the fetus into the left atrium.

forced expiratory flow (FEF_{25-75}) maximum amount of air (L/s) forced out of the lungs between two given lung volumes (25% and 75%)

forced expiratory volume (FEV_1) maximum amount of air (L) forced out of the lungs in 1 second after maximal inhalation

forced vital capacity (FVC) maximum volume of air forcibly exhaled after maximal inhalation

force–velocity curve expression of the inverse relationship between muscle force and shortening

fornix fiber bundle in the brain containing axons from hippocampal neurons that project to the hypothalamus and basal forebrain and axons from neurons in those regions that project to the hippocampus

free radical atom and/or molecule with unpaired electrons that is damaging to cell structure and function

fresh frozen plasma (FFP) blood product frozen within 6 hours of collection

frontal lobe region of the brain located most anterior, which controls motor and cognitive functions

fructose 1,6-diphosphatase focal enzyme in gluconeogenesis via its conversion of fructose 1,6-diphosphate (FDP) to fructose 6-phosphate (F-6-P), which permits endogenous glucose production from gluconeogenic amino acids (e.g., alanine and glycine), glycerol, or lactate

functio laesa loss of function associated with acute inflammation

functional gastrointestinal motility disorder disorder for which the physical or biochemical cause of the patient's symptoms cannot be determined

functional residual capacity (FRC) volume of air remaining in the lungs at the end of a normal tidal volume

functional syncytium electrical property of the heart muscle created by gap junction connections between cells that allows all heart muscle to be activated following the generation of an action potential in any myocardial cell

fundus of the stomach upper curvature of the organ; one of four stomach sections

fused tetanus series of skeletal muscle twitches occurring in such rapid succession that there is no intervening relaxation

fusimotor system γ motor neurons and the intrafusal muscle fibers within muscle spindles

G cells neuroendocrine cells present in the stomach and located predominantly in the antrum that produce the hormone gastrin, which stimulates acid secretion by the stomach

G protein–coupled receptors (GPCRs) large protein family of transmembrane receptors that sense molecules outside the cell and activate signal transduction pathways and, ultimately, cellular responses, internally; these cell-surface receptors are implicated in many diseases and are the target of about 30% of all drug therapies

GABAergic neurons neurons that use γ-aminobutyric acid (GABA) as their neurotransmitter; in most cases inhibitory neurons

galactopoiesis maintenance of lactation; regulated by prolactin

galactorrhea persistent milklike discharge from the nipple in nonlactating people

gallstone crystalline concretion formed within the gallbladder by accretion of bile components; these calculi are formed in the gallbladder but may pass distally into other parts of the biliary tract such as the cystic duct, common bile duct, pancreatic duct, or the ampulla of Vater

gamete specialized reproductive cell through which sexually reproducing parents pass chromosomes to their offspring; a sperm or an egg

gamma globulin see *γ globulin*

gamma motor neuron see *γ motor neuron*

ganglia grouping of nerve cell bodies situated outside the central nervous system

ganglion cell type of neuron located near the inner surface (the ganglion cell layer) of the retina of the eye that receives visual information from photoreceptors via bipolar and amacrine cells

ganglionated plexus array of ganglia and interganglionic fiber tracts forming parts of the enteric nervous system

gap junction 1) specialized protein channels in the plasma membrane composed of *connexins* which form a *connexon*; allows the flow of ions (electrical current) from cell to cell 2) membrane-to-membrane apposition in gastrointestinal smooth muscle, which mediates electrotonic coupling and allows electrical (ionic) current to pass from one muscle fiber to another

gas diffusion process in which gas moves across the alveolar–capillary membrane

gas exchange transfer of oxygen and carbon dioxide between the atmosphere and blood via the alveoli

gas tension partial pressure of an individual gas; used interchangeably with partial pressure of oxygen and carbon dioxide (e.g., oxygen tension = Po_2; carbon dioxide tension = Pco_2)

gastric adaptive relaxation relaxation triggered by distention of the gastric reservoir; a vagovagal reflex triggered by stretch receptors in the gastric wall, transmission over vagal afferents to the dorsal vagal complex, and efferent vagal fibers to inhibitory motor neurons in the gastric enteric nervous system

gastric amylase digests starch that was not digested in the mouth; it is of minor significance in the stomach

gastric feedback relaxation relaxation triggered in the gastric reservoir by the presence of nutrients in the small intestine; it can involve both local reflex connections between receptors in the small intestine and the gastric enteric nervous system, or hormones that are released from endocrine cells in the small intestinal mucosa and transported by the blood to signal the gastric enteric nervous system and stimulate firing in vagal afferent terminals in the stomach

gastric glands see *oxyntic glands*

gastric inhibitory peptide (GIP) enterogastrone produced by the small intestinal endocrine cells that inhibits parietal cell acid secretion by the stomach. Recently, it has been named the glucose-dependent insulinotropic peptide

gastric lipase acidic lipase secreted by the gastric chief cells in the fundic mucosa in the stomach with a pH optimum of 3 to 6; acidic lipases do not require bile acid or colipase for optimal enzymatic activity and make up 30% of lipid hydrolysis occurring during digestion in the human adult, with gastric lipase contributing the most of the two acidic lipases

gastric mucosa mucous membrane layer of the stomach wall (one of four layers) containing the glands and the gastric pits; it is approximately 1 mm thick and its surface is smooth, soft, and velvety; it consists of epithelium, lamina propria, and the muscularis mucosae

gastric phase one of three phases that stimulates acid secretion resulting from the ingestion of food. Acid secretion during the gastric phase is mainly a result of gastric distention and the digested peptides in the gastric lumen. The gastric phase accounts for approximately 50% of total gastric acid secretion.

gastric receptive relaxation relaxation initiated in the gastric reservoir by the act of swallowing; a reflex triggered by stimulation of mechanoreceptors in the pharynx followed by transmission over afferents to the dorsal vagal complex and activation of efferent vagal fibers to inhibitory motor neurons in the gastric enteric nervous system

gastric reservoir reservoir consisting of the fundus and approximately one third of the corpus; the muscles of the gastric reservoir are adapted for maintaining continuous contractile tone (tonic contraction) and do not contract phasically

gastrin hormone that stimulates acid secretion by the stomach

gastrin-releasing peptide (GRP) peptide released by the nerves that stimulates G cells to release gastrin, which stimulates parietal cell acid secretion by the stomach. GRP is atropine-resistant, indicating that it works through a noncholinergic pathway

gastrocolic reflex mass movement of feces in the colon that is generally preceded by a similar movement in the small intestine, which often occurs immediately following the intake of a meal

gastroesophageal reflux retrograde flow of gastric contents into the esophagus

gastroesophageal reflux disease (GERD) chronic symptoms or mucosal damage produced by the abnormal reflux of stomach acid to the esophagus, typically accompanied by pyloris (heartburn)

gastrointestinal motility organized application of forces of muscle contraction that results in physiologically significant movement or nonmovement of intraluminal contents

gastroparesis weakness of phasic propulsive contractions in the antral pump, which results in delayed gastric emptying

general gas law gas equation that combines the Charles and Boyle laws

generalized edema widespread accumulation of salt and water in the interstitial spaces of the body

generalized hypoventilation condition in which the entire lung is underventilated, leading to low alveolar ventilation

generalized hypoxia condition in which the entire lung becomes hypoxic, as opposed to only a local region

generator potential local electrical activity of a sensory receptor that is proportional to the intensity of a stimulus. The magnitude of the generator potential controls the frequency of action potentials along the sensory nerve

genetic sex gender determined by the chromosomes; also known as *chromosomal sex*

genital ridge embryonic ridge on the dorsal wall of the abdominal cavity; will form the gonads

gestational diabetes form of diabetes that may develop during pregnancy in women who do not otherwise have diabetes

ghrelin signal produced by the epithelia of the stomach and small intestine that stimulates food intake by activating the NPY neurons in the arcuate nucleus of the hypothalamus

gigantism excess growth in height as a result of inappropriately high levels of growth hormone

Gitelman syndrome inherited defect in the distal convoluted tubule of the kidneys, causing the kidneys to pass sodium, magnesium, chloride, and potassium into the urine, rather than allowing them to be reabsorbed into the bloodstream

glandular gastric mucosa stomach's mucosal lining, which contains three main types of glands: cardiac, pyloric, and oxyntic

glaucoma condition that results when drainage of the ocular aqueous humor is impaired, pressure builds up in the anterior chamber, and internal structures are compressed, leading to optical nerve damage that can ultimately cause blindness

glial cells nonneuronal cells of the nervous system that mediate metabolic functions and myelination

glicentin peptide fragment cleaved from glucagon by prohormone convertase

glitazones antidiabetic sulfonylurea drug that stimulates insulin secretion

γ globulin electrophoretically similar serum proteins, including antibodies

globulins plasma proteins, including antibodies, enzymes, carriers, and transporters

globus pallidus (GP) group of nerve cells in the basal ganglia whose output influences the thalamus and cerebral cortex, divided into and external and internal segment

glomerular filtration process in the kidney glomeruli in which an essentially protein-free filtrate is pushed out of the glomerular capillaries into the urinary space of the Bowman capsule

glomerular filtration barrier structures that separate the blood plasma and the urinary space of the Bowman capsule, namely, capillary endothelium plus glomerular basement membrane plus podocytes

glomerular filtration rate (GFR) rate at which the kidneys filter the plasma

glomerular ultrafiltration coefficient (K_f) product of both the hydraulic conductivity (fluid permeability) and capillary surface area of the glomerular filtration barrier

glomerulonephritis disease characterized by proteinuria and/or hematuria, hypertension, and renal insufficiency that progresses over years

glomerulotubular balance proportionate change in sodium reabsorption by the proximal convoluted tubule and loop of Henle when glomerular filtration rate is changed

glomerulus in the kidney, the tuft of capillaries, surrounded by the Bowman capsule, where the blood is filtered

glucagon hormone produced by the α cells of the pancreas that increases the level of glucose in the blood

glucagon-like peptide-1 (GLP-1) peptide generated from proglucagon that is one of the most potent incretins

glucagon-like peptide-2 (GLP-2) peptide generated from proglucagon that is not an incretin and whose biologic actions are not known

glucocorticoid hormones products of the adrenal cortex

glucocorticoid response elements (GREs) specific regions of DNA that bind glucocorticoid receptor

glucokinase enzyme that facilitates phosphorylation of glucose to glucose 6-phosphate in cells in the liver, pancreas, gut, and brain of humans and most other vertebrates

gluconeogenesis occurring mainly in the liver and kidneys, the energy-dependent production of glucose from noncarbohydrate sources, such as fat, amino acids, and lactate; energy for this process is derived from B-oxidation of fatty acids

glucose simple sugar (monosaccharide) and an important carbohydrate used by cells as a source of energy and a metabolic intermediate; one of the main products of photosynthesis that starts cellular respiration

glucose buffering process by which the liver maintains normal blood glucose levels; its ability to store glycogen allows the liver to remove excess glucose from the blood and then return it when blood glucose levels begin to fall

glucose threshold plasma level at which glucose first appears in urine

glucose 6-phosphatase enzyme present exclusively in the liver that catalyzes the breakdown of glycogen to glucose molecules for release into the circulation

glucose 6-phosphate intermediate product in the conversion of glucose to glycogen and in the breakdown of glycogen to glucose

glucose-dependent insulinotropic peptide (GIP) gut hormone secreted by the duodenum that stimulates insulin secretion

glucose–galactose malabsorption (**GGM**) condition in which the cells lining the intestine cannot transport glucose and galactose, preventing proper absorption of these nutrients

α-glucosidase inhibitors oral medications used to treat type 2 diabetes that decrease the absorption of carbohydrates in the small intestine

glucosuria presence of glucose in the urine

GLUT2 facilitative glucose transporter expressed in the pancreas, liver, intestine, and kidney

GLUT4 insulin-responsive facilitative glucose transporter expressed in muscles and adipose tissues

glutamate (GLU) amino acid used as an excitatory neurotransmitter

glutamatergic neurons neurons that use glutamate as their neurotransmitter

glutamine precursor for glutamate

glycine (GLY) amino acid used as an inhibitory neurotransmitter

glycinergic neurons neurons that use glycine as their neurotransmitter

glycogen polysaccharide that is the main form of carbohydrate storage in animals and occurs mainly in liver and muscle tissue; readily converted to glucose

glycogen phosphorylase first acting enzyme in the breakdown of glycogen to glucose, responsible for the production of glucose 1-phosphate from the glycogen substrate, which is subsequently converted to glucose. The enzyme acts on the α-1,4-glycosidic linkage yielding glucose polymers.

glycogen synthase enzyme that catalyzes the transfer of glucose from uridine diphosphate–glucose to glycogen

glycogenesis occurring mainly in the liver, the process by which glycogen is synthesized from circulating glucose, lactate, and pyruvate after ingestion of a meal

glycogenolysis process by which glycogen is broken down to glucose 6-phosphate and then to glucose for subsequent release into circulation. The reaction is specific to the liver

glycolipid lipid molecule that contains one or more monosaccharide units

glycolysis adenosine triphosphate–generating metabolic process of most cells in which carbohydrates are converted to pyruvic acid

glycolytic pathway (or glycolysis) series of cytoplasmic chemical reactions that break down glucose molecules and produce adenosine triphosphate for energy-requiring cellular processes

glycoprotein protein that contains one or more monosaccharide units

goblet cells cells that serve to secrete various mucins (mucoproteins) found in intestinal secretions

goiter enlarged thyroid gland

Goldman equation value of the membrane potential when all the permeable ions are accounted for. In its usual form, it shows the relationship between intracellular and extracellular concentrations of Na^+, K^+, and Cl^-, the plasma membrane permeability to these ions, and the membrane potential

Golgi apparatus eukaryotic organelle that processes and packages proteins, lipids, and other molecules after their synthesis

Golgi cells type of cerebellar interneuron

Golgi tendon organs (GTOs) sensory receptors located in tendons that sense the force of muscle contraction

gonadal dysgenesis incomplete differentiation of the gonads

gonadal sex sex of the structures that produce the gametes

gonadarche gonadal maturation during pubertal development in boys and girls

gonadotrophs cells located in the anterior pituitary that secrete luteinizing hormone and follicle-stimulating hormone

gonadotropin-releasing hormone (GnRH) hormone made by the hypothalamus (part of the brain); causes the pituitary gland to make luteinizing hormone and follicle-stimulating hormone, which are involved in reproduction; also known as *luteinizing hormone–releasing hormone (LHRH)*

gonadotropins products of the gonadotrophs in the anterior pituitary; luteinizing hormone and follicle-stimulating hormone

gout disorder in which plasma uric acid is elevated and urate crystals precipitate in the joints, causing pain and inflammation and swelling of the affected joints (often in the big toe)

graafian follicle mature preovulatory follicle

graded response sensory response in approximate proportion to the intensity of the stimulus

graft tissue or organ transplanted from a donor to a recipient

graft-versus-host disease (GVHD) disease in which T cells of the graft attack tissue of the recipient as foreign

grand mal seizure see *tonic–clonic seizure*

granular cells modified pericytes of glomerular arterioles; also known as *juxtaglomerular cells*

granular layer nuclear layer in the cerebellar cortex rich in interneurons

granule cells small nerve cell bodies in the external and internal granular layers of the cerebral cortex

granulocyte mature cell of the myeloid series, including neutrophils, eosinophils, and basophils; also known as *polymorphonuclear leukocyte*

granulosa lutein cells cells that are found in the corpus luteum and formed from the granulose cells

granzyme enzyme family present in the granules of cytotoxic T cells and natural killer cells

Graves disease autoimmune disorder caused by antibodies directed against the thyroid-stimulating hormone receptor

gray ramus nerve branch that carries postganglionic sympathetic axons back to the spinal nerve

ground substance intercellular material in which the cells and fibers of connective tissue are embedded

growth cone leading edge of a growing axon that samples the environment and makes decisions about the direction of growth

growth factors naturally occurring proteins that are capable of stimulating cellular proliferation, differentiation, or other cellular responses

growth hormone (GH) hormone product of somatotrophs located in the anterior pituitary

growth hormone–releasing hormone (GHRH) polypeptide from hypothalamus; stimulates secretion of growth hormone

guanosine diphosphate (GDP) product of guanosine triphosphate phosphorylation

guanosine triphosphate (GTP) nucleotide that provides source of energy for protein synthesis; essential for G protein signal transduction by converting to guanosine diphosphate via GTPases

guanylin polypeptide hormone produced by the small intestine that increases renal salt excretion

guanylyl cyclase (GC) enzyme that catalyzes the conversion of guanosine triphosphate to 3′,5′-cyclic guanosine monophosphate and pyrophosphate

gustducin G protein associated with tasting bitterness in the gustatory system

gynecomastia excessive development of the male breasts

gyrus convolution on the surface of the brain

H^+-ATPases (proton pump) located in the membranes of lysosomes and the Golgi apparatus and pumps protons from the cytosol into the organelles to keep the inside of the organelle more acidic than the cytoplasm; classified as *V-type ATPases* because they were first discovered in intracellular vacuolar structures; now known to also exist in the plasma membranes

H^+/K^+-ATPase mechanism present in the apical (luminal) cell membrane of the parietal cell that is involved in the secretion of hydrochloric acid by the stomach

H zone pale region of A band that contributes to striated appearance in muscle

H_2 receptors special histamine receptors found in the parietal cells whose stimulation results in increased acid secretion by the stomach

habenula diencephalic component of the limbic system

hair cells mechanosensitive cells within the organ of Corti and the vestibular apparatus

haldane effect effect of oxygen on the carbon dioxide equilibrium curve. Oxygen causes a downward shift of the carbon dioxide equilibrium curve.

hallucination perception in the absence of a stimulus

hamartoma nonneoplastic congenital malformation composed of a heterotropic mass of nerve tissue usually located on the floor of the third ventricle or attached to the tuber cinereum

hapten nonimmunogenic antibody-binding molecule, unless linked to a carrier molecule

haptoglobin large glycoprotein (molecular weight of 100,000 Da) that binds free hemoglobin in blood, forming a hemoglobin-haptoglobin complex, which is rapidly removed from circulation by the liver, thereby conserving iron in the body

Hartnup disease extremely rare genetic disorder of amino acid carriers. The membrane carrier for neutral amino acids (e.g., tryptophan) is defective.

Hashimoto disease autoimmune disease resulting in impaired thyroid hormone synthesis; also known as *Hashimoto thyroiditis*

haustration process of formation of a haustrum (haustral sac) in the colon

heart four-chambered, hollow organ composed primarily of striated muscle that is responsible for pumping blood through the lungs and the systemic circulation. It is composed of two upper chambers, a right and left atrium, and two larger, lower chambers called the right and left ventricles. The right atrium receives blood from the peripheral veins and delivers it through the tricuspid valve to the right ventricle. The right ventricle pumps blood through the pulmonic valve into the pulmonary circulation. Blood return from the pulmonary circulation arrives in the left atrium, where it is pumped through the mitral valve into the left ventricle. This chamber pumps blood through the aortic valve into the aorta and thus into the systemic circulation. In this arrangement, the right and left sides of the heart are actually two pumps arranged in series

heart murmur sound heard over the chest with a stethoscope that results from turbulent flow in the heart, aorta, or pulmonary circulation, usually resulting from stenotic or incompetent heart valves

heart rate number of times the normal heart contracts in 1 minute; in a healthy individual, this rate can range between 50 and 180 contractions per minute

heartburn episodic retrosternal burning

heat exhaustion most common heat-related illness, involving mild to moderate dysfunction of temperature control associated with elevated ambient temperatures and/or strenuous exercise resulting in dehydration and salt depletion. It may rapidly progress to heatstroke when the body's thermoregulatory mechanisms become overwhelmed and fail

heat storage difference between heat production and net heat loss

heat syncope body temperature above 40°C (104°F) with fainting or weakness but without mental confusion; results from circulatory failure as a result of peripheral venous pooling, with a consequent decrease in venous return. Heat syncope is caused by mild overheating with inadequate water or salt

heatstroke extreme hyperthermia, typically above 100.4°F (40°C), associated with a systemic inflammatory response, which leads to end-organ damage with universal involvement of the central nervous system. Heatstroke traditionally is divided into exertional and classic varieties, which are defined by the underlying etiology but are clinically indistinguishable

heel stick procedure in which a newborn baby's heel is pricked and then a small amount of the blood is collected, usually with a narrow-gauge ("capillary") glass tube or a filter paper

hem(at)/o referring to blood

hematocrit (Hct or **Ht)** percentage of whole blood that is composed of erythrocytes

hematologist blood specialist

hematopoiesis generation of blood cells

hematopoietins subclass of cytokines regulating hematopoiesis

heme oxygenase microsomal enzyme responsible for the release of iron from heme. Iron then enters the cellular free iron pool and can be stored as ferritin or released into the bloodstream

hemochromatosis condition caused by iron overload that is characterized by excessive amounts of hemosiderin in the hepatocytes, rendering them defective and unable to perform many normal, essential functions. The excess iron is stored in the organs, especially the liver, heart, and pancreas, causing damage to these organs and leading to life-threatening illnesses such as cancer, heart complications, and liver disease

hemocytometer chamber for manual blood cell counts

hemodialysis procedure usually done three times a week (4 to 6 hours per session) in a medical facility or at home in which kidney-failure patient's blood is pumped through an artificial kidney machine; the blood is separated from a balanced salt solution by a cellophanelike membrane, and small molecules can diffuse across this membrane; excess fluid can be removed by applying pressure to the blood and filtering it

hemodynamics study of the physics of the containment and movement of blood in the cardiovascular system

hemoglobin (Hb) oxygen-transporting protein of erythrocytes

hemoglobinopathies diseases caused by mutated hemoglobin

hemolysis bursting (*lysis*) of red blood cells leading to the release of hemoglobin

hemolytic anemia anemia caused by destruction of erythrocytes that exceeds their generation

hemopexin protein synthesized by the liver that is involved in the transport of free heme in the blood. It forms a complex with free heme, and the liver rapidly removes the complex

hemophilia hereditary bleeding disorders

hemorrhage severe bleeding

hemosiderin iron-storing protein related to ferritin

hemosiderin granules iron stored as ferritin that is sometimes converted to these bodies

hemostasis cessation of bleeding occurring in four steps: (1) compression and vasoconstriction, (2) formation of a temporary loose platelet plug (also called *primary hemostasis*), (3) formation of stable fibrin clot (also called *secondary hemostasis*), and (4) clot retraction and dissolution

Henderson–Hasselbalch equation logarithmic form of the acid dissociation equation

Henry law gas law that states that the amount of gas dissolved in a liquid at a given temperature is directly proportional to the solubility and the partial pressure of the dissolved gas

heparin injectable anticoagulant

hepatic arterial buffer response phenomenon in which reduction in portal blood supply to the liver is compensated by an increase in supply from the hepatic artery and vice versa

hepatocytes highly specialized cells arranged along the liver sinusoids that facilitate the rapid exchange of molecules, as they possess numerous fingerlike projections that extend into the perisinusoidal space, thereby increasing the surface area over which they are in contact with the perisinusoidal fluid

hepcidin hormone produced by the liver that regulates iron homeostasis by inhibiting ferroportin

hereditary hemochromatosis (HH) inherited disorder that increases the amount of iron that the body absorbs from the gut, causing excess iron to be deposited in multiple organs of the body; excess iron in the liver causes cirrhosis (which may develop into liver cancer); iron deposits in the pancreas can result in diabetes; and excess iron stores can also cause cardiomyopathy, pigmentation of the skin, and arthritis

hereditary spherocytosis genetically transmitted (autosomal dominant) form of spherocytosis characterized by the production of red blood cells that are sphere-shaped and, therefore, more prone to hemolysis

Hering–Breuer reflex reflex that involves the pulmonary stretch receptors and that prevents overinflation of the lungs; also known as *lung-inflation reflex*

hermaphroditism condition in which one possesses both ovarian and testicular tissues

heterogametic having one X and one Y chromosome; a male

heterologous desensitization process by which effector cells are rendered less reactive or nonreactive to a given stimulus by a different stimulus

hexokinase enzyme that facilitates phosphorylation of glucose to glucose 6-phosphate in cells in yeast and muscle

hidromeiosis condition caused by excessive wetting of the skin and subsequent interference with the free flow of eccrine sweat during exercise in humid environments; may be compounded by fatigue of sweat glands as a result of a lack of neurotransmitters

high-density lipoprotein (HDL) particles circulating in the blood that contain cholesterol esters; removes cholesterol from cells and transports it back to the liver for excretion or reutilization; also known as *good cholesterol*

hilum depression or pit at that part of an organ where the blood vessels and nerves enter

hippocampus three-layered cortical structure in the temporal lobe involved in learning and memory

Hirschsprung disease congenital dilation and hypertrophy of the colon resulting from the absence of the enteric nervous system (aganglionosis) of the rectum and a varying but continuous length of gut above the rectum

hirsutism excessive increase of hair growth, sometimes leading to male-pattern hair growth in a female

histamine inflammatory mediator, released by mast cells and basophils

HMG-CoA reductase see *3-hydroxy-3-methylglutaryl CoA reductase*

homeostasis property of a living organism to regulate its internal environment to maintain a stable, constant condition, by means of multiple dynamic equilibrium adjustments, controlled by interrelated regulation mechanisms

homeotherms warm-blooded animals who keep their core body temperature at a nearly constant level regardless of the temperature of the surrounding environment

homogametic having two X chromosomes; a female

homologous desensitization loss of sensitivity only to the class of agonist used to desensitize the tissue

hopping reaction limb reflex that promotes automatic forward movement of a weight-bearing limb

horizontal cells laterally interconnecting neurons in the outer plexiform layer of the retina of mammalian eyes that help integrate and regulate the input from multiple photoreceptor cells; responsible for allowing eyes to adjust to see well under both bright and dim light conditions as well as other functions

hormone chemical or protein messenger made by one cell that serves as a signal to a target cell

hormone receptor molecular entity (usually a protein or glycoprotein) either outside or within a cell that recognizes and binds a particular hormone

hormone response element (HRE) short DNA sequence that is usually within the promoter or repressor regions of genes and binds a specific hormone/receptor complex to regulate transcription of the gene

hormone-sensitive lipase enzyme that plays an important role in controlling the balance of energy by breaking down fats

Horner syndrome rare condition resulting from damage to the sympathetic nervous system that affects the nerves to the eye and face and causes decreased sweating on the affected side of the face, ptosis, sinking of the eyeball into the face, and a mall (constricted) pupil

host-versus-graft response condition in which the recipient's immune system attacks transplanted leukocytes

human chorionic gonadotropin (hCG) protein secreted by the blastocyst and placenta; similar in biologic action to luteinizing hormone

human leukocyte antigen (HLA) human mean cell (corpuscular) hemoglobin protein

human placental lactogen (hPL) substance synthesized by the placenta; its structure and function resemble those of prolactin and growth hormone

humoral immunity part of adaptive immunity mediated by soluble factors, especially antibodies

Huntington disease disease of the basal ganglia producing uncontrollable spontaneous movements

hydration combination with water

hydration reaction chemical reaction, usually happening in a strong acidic solution, in which a hydroxyl group and a hydrogen cation (an acidic proton) are added to two carbon atoms

hydraulic conductivity for water measure of hydraulic resistance across a capillary network; high conductivity means that water passes easily across the network. It is a measure of how much water can be moved in bulk across the capillaries per minute for each 1 mm Hg of driving pressure.

hydrochloric acid (HCl) important constituent of gastric acid that aids in the digestion of food; secreted by the parietal cells and responsible for the change in the electrolyte composition of gastric juice

hydrogen peroxide (H_2O_2) reactive oxygen species that is that is not a free radical and is damaging to cell structure and function

hydrophilic highly soluble in water, for example, ions and polar molecules

hydrophobic having low solubility in water, for example, nonpolar molecules

3-hydroxy-3-methylglutaryl CoA reductase (HMG CoA reductase) substance that catalyzes the rate-limiting step in *de novo* cholesterol synthesis

hydroxyapatite crystals form of calcium phosphate in the mineral portion of bone

β-hydroxybutyric acid β derivative of hydroxybutyric acid that is found in the blood and urine in some cases of impaired metabolism

hydroxyl radical (•OH) reactive oxygen species that is damaging to cell structure and function

11β-hydroxylase (CYP11B1) substance that hydroxylates 11-deoxycortisol to form cortisol

17α-hydroxylase substance that hydroxylates pregnenolone to form 17α-hydroxypregnenolone

21-hydroxylase (CYP21A2) substance that converts 17α-hydroxyprogesterone to 11-deoxycortisol

17α-hydroxypregnenolone intermediate in steroid synthesis

17α-hydroxyprogesterone intermediate in steroid synthesis

3β-hydroxysteroid dehydrogenase (3β-HSD II) substance that acts on 17α-hydroxypregnenolone to produce 17α-hydroxyprogesterone

5-hydroxytryptamine (5-HT) monoamine neurotransmitter; also known as *serotonin*

hyperaldosteronism medical condition characterized by too much aldosterone being produced by the adrenal glands, which can lead to lowered levels of potassium in the blood

hypercalciuria excess calcium in the urine

hypercapnia high partial pressure of blood carbon dioxide (>45 mm Hg) resulting from underventilation of the lung

hyperglycemia presence of an abnormally high concentration of glucose in the blood

hypergonadotropic hypogonadism defective development of ovaries or testes; associated with excess pituitary gonadotropin secretion; results in delayed sexual development and growth delay

hyperkalemia condition in which the plasma K^+ level is >5.0 mEq/L

hyperkinesis excessive movement

hyperopia (farsightedness) condition caused by the eyeball being physically too short to focus on distant objects

hyperosmotic see *osmotic pressure*

hyperphagia abnormally increased appetite and consumption of food

hyperphosphatemia condition in which plasma phosphate concentration is elevated

hyperpnea increased minute ventilation via increased depth and frequency of breathing (unlike tachypnea, which is rapid, shallow breaths) to meet increased metabolic demand (such as during or following exercise); ventilation increases proportionally to carbon dioxide production, but Pa_{CO_2} remains the same, which is the hallmark feature of hyperpnea (unlike in hyperventilation, in which Pa_{CO_2} is significantly lowered, leading to alkalosis)

hyperpolarization change in electrical potential in the direction of a more negative potential inside cells; makes excitable cells less excitable by moving the membrane potential away from the threshold for discharge of an action potential

hyperprolactinemia elevated prolactin secretion from the pituitary causing amenorrhea

hypersensitivity damaging immune responses to allergens, resulting in inflammation and organ dysfunction

hyperthermia acute condition that occurs when the body produces or absorbs more heat than it can dissipate

hyperthyroidism or **thyrotoxicosis** excessive thyroid hormone secretion

hypertonic see *tonicity*

hyperventilation state of breathing faster and/or deeper than necessary that causes excess carbon dioxide to be blown off; it can result from a psychologic state such as a panic attack, can result from a physiologic condition such as metabolic acidosis, or can be brought about by voluntarily forced ventilation and is characterized by increased minute ventilation with a concomitant decrease in Pa_{CO_2}; also known as *overbreathing*

hypervitaminosis A condition resulting from consumption of massive quantities of vitamin A that is often associated with hepatotoxicity and that eventually leads to portal hypertension and cirrhosis

hypoaldosteronism condition characterized by decreased levels of aldosterone, which may result in hyperkalemia and urinary sodium wasting, leading to volume depletion and hypotension.

hypocalcemia deficiency of calcium in the blood

hypocalcemic tetany condition characterized by low blood calcium levels, resulting in convulsions

hypocapnia condition characterized by low partial pressure of blood carbon dioxide (<35 mm Hg) as a result of overventilation of the lung

hypochromic condition in which there is less hemoglobin per cell than normal

hypocretin/orexin neuropeptide found in neurons in the posterolateral hypothalamus that are sensitive to nutritional status, simulate appetite, and project to arousal areas of the brain; a functional deficit in hypocretin causes narcolepsy

hypoglycemia condition characterized by low blood sugar level

hypogonadism condition in which sex organs such as the testes or ovaries are underactive

hypogonadotropic hypogonadism absent or decreased function of the gonads, the male testes or the female ovaries, resulting from the absence of the gonadal stimulating pituitary hormones, follicle-stimulating hormone, and luteinizing hormone

hypokalemia condition in which the plasma K^+ level is <3.5 mEq/L

hypokinesis condition characterized by less than a normal amount of movement

hyponatremia condition in which the plasma Na^+ level is <135 mEq/L

hypophyseal portal circulation special blood supply to the anterior lobe of the pituitary gland

hypophysiotropic hormones releasing hormones synthesized by neural cell bodies in the hypothalamus and that regulate secretion of posterior pituitary hormones

hypopituitarism deficiency of one or more hormones of the pituitary gland

hyposmotic see *osmotic pressure*

hypotension condition characterized by lower-than-normal arterial pressure

hypothalamic diabetes insipidus condition in which arginine vasopressin secretion is impaired

hypothalamic hormones hormones released by the hypothalamus to reach the anterior pituitary lobe by a portal system of capillaries; designated as releasing factors because they regulate the secretion of most hormones of the endocrine system

hypothalamic–pituitary axis functional connection between the brain and the pituitary, in which the hypothalamus plays a central role: the brain links the pituitary gland to events occurring within or outside the body, which call for changes in pituitary hormone secretion

hypothalamic–pituitary–adrenal axis interactive relationship regulating glucocorticoid levels in the blood

hypothalamic–pituitary–gonad axis interactive relationship regulating gonadotropin levels in the blood

hypothalamic–pituitary–growth hormone axis interactive relationship regulating growth hormone levels in the blood

hypothalamic–pituitary–thyroid axis interactive relationship regulating thyroid hormone levels in the blood

hypothalamo-hypophyseal portal blood vessels system of blood vessels in the pituitary stalk into which neurons within specific nuclei of the hypothalamus secrete releasing factors that are then transported to the anterior pituitary, where they stimulate secretion of hormones

hypothalamo-hypophyseal tract axons within the pituitary stalk derived from magnocellular neurons whose cell bodies are located in the supraoptic and paraventricular hypothalamic nuclei; represents the efferent limbs of neuroendocrine reflexes that lead to the secretion of the hormones vasopressin and oxytocin into the blood within the posterior pituitary

hypothalamus region of the brain that integrates autonomic responses to temperature and hunger

hypothermia drop of central temperature below 35°C

hypothyroidism condition characterized by inadequate production of the thyroid hormone thyroxine and by low circulating levels of thyroxine and other thyroid hormones

hypotonia reduction in the amount of normal resistance to a muscle undergoing passive stretch

hypotonic see *tonicity*

hypoventilation decreased alveolar ventilation

hypovolemia condition characterized by an abnormally low circulating blood volume

hypoxemia condition characterized by an abnormally low oxygen content or low P_{O_2} in the arterial blood

hypoxia condition in which the inspired oxygen is below normal leading to abnormal low oxygen tension in the tissues

hypoxia-induced pulmonary vasoconstriction phenomenon in which small pulmonary vessels constrict with low oxygen, which is just the opposite of that which occurs in the systemic circulation

I band band of actin myofilament in muscle that shows low density under a microscope and contributes to muscles' striated appearance; *isotropic*

I cell type of endocrine cell in the small intestine that releases cholecystokinin

Ig immunoglobulin; a membrane-bound or soluble antigen-binding protein

ileocolonic sphincter prevents reflux of colonic contents into the ileum; ncompetence can allow entry of bacteria into the ileum from the colon that may result in bacterial overgrowth

immediate early genes genes that are activated transiently and rapidly in response to a cellular stimuli; the first round of response to stimuli, before any new proteins are synthesized

immune deficiency disease resulting from absent or malfunctioning immune components

immune surveillance theory that the immune system recognizes and destroys tumor cells that are constantly arising during the life of the individual.

immune thrombocytopenia purpura (ITP) common autoimmune disorder caused by autoantibodies to platelets formed in the spleen and by the platelet–antibody aggregates being destroyed by the spleen

immune tolerance process by which the immune system does not attack antigens and includes self-tolerance (in which the body does not mount an immune response to self-antigens) and induced tolerance (in which tolerance to external antigens can be created by manipulating the immune system); the three forms are central, peripheral, and acquired tolerance

immunocontraception use of antibodies against the egg, sperm or zygote, or against hormones such as human chorionic gonadotropin or follicle-stimulating hormone to prevent pregnancy (This method of contraception is still being developed for use in humans)

immunogen substance that provokes the immune system

immunologic synapse complex junction between a T cell and an antigen-presenting cell

immunosenescence gradual deterioration of the immune system because of advancing age

implantation attachment of the blastocyst to the surface endometrial cells of the uterine wall, beginning on days 7 to 8 after fertilization

impulse initiation region region of the receptor membrane that is the location of the formation of action potential in the neuron and constitutes an important link in the sensory process

inactivation gate molecular gate in voltage-gated sodium channels that closes after the channel opens and remains closed for a finite period of time and until the membrane is repolarized

incisura small indentation that appears in the aortic pressure waveform when the aortic valve closes

incomplete proteins proteins that do not provide all of the essential amino acids in amounts sufficient to sustain normal growth and body maintenance, such as those found in most vegetables and grains

incretin effect principle stating that a glycemic stimulus derived from enteral nutrients exerts a greater insulinotropic response than a comparable isoglycemic challenge achieved through parenteral glucose administration

indicator–dilution method technique used to determine cardiac output based on the disappearance of concentration of an indicator over time

indirect calorimetry measure of an organism's metabolic rate by calculating oxygen uptake, assuming that burning 1 cal (kilocalorie) requires 208.06 mL of oxygen

indirect pathway basal ganglia pathway in which axons from the caudate synapse in the globus pallidus externus before reaching the globus pallidus externus

inferior cervical ganglia lower-most of the sympathetic ganglia located in the neck; also known as *stellate ganglion*

inferior hypophyseal arteries arteries that provide arterial blood to the posterior lobe

inferior mesenteric ganglion most caudal of the three prevertebral sympathetic ganglia of the abdomen

inferior olive largest nucleus in the olivary body; closely associated with the cerebellum, meaning that it is involved in control and coordination of movements and likely also sensory processing and cognitive tasks, especially the timing of sensory input

inferior salivatory nucleus region of the medulla where neurons involved in the control of submandibular, sublingual, and parotid glands are located

inferior vena cava (IVC) single major vein that collects blood from the lower extremities and abdominal organs and that empties into the right atrium

inflammation pathologic process including cytologic changes, cellular infiltration, and mediator release; also known as *inflammatory response*

influx transfer of material from the exterior to the interior of a cell

infundibular process major portion of the neurohypophysis; also known as *posterior lobe*

infundibular stem inner part of the pituitary stalk

infundibulum end of the oviduct that receives the ovulated egg from the ovary; has fimbria

inhibin substance secreted by the testes and ovaries that regulates follicle-stimulating hormone levels

inhibitors of apoptosis (IAPs) family of functionally and structurally related proteins that serve as endogenous inhibitors of programmed cell death; there are eight human IAPs

inhibitory junction potential (IJP) membrane hyperpolarization leading to decreased excitability in gastrointestinal muscles; evoked by the release and action of inhibitory neurotransmitters from enteric motor neurons

inhibitory motor neuron neuron that releases neurotransmitter, which suppresses contractile or secretory behavior in the gastrointestinal tract

inhibitory postsynaptic potentials hyperpolarization of neuronal membrane at the postsynaptic side of a synapse resulting from ions flowing through channels that open when neurotransmitter released from the presynaptic terminal binds to its receptor

initial velocity earliest steady velocity of a contracting muscle

innate immune system nonspecific defense mechanisms; also called *nonspecific* or *natural immunity*

inner ear portion of the ear containing the cochlea, where the actual sound transduction takes place

inner ring deiodination enzymatic removal of an iodine atom at the 5 position on the thyronine ring structure

innervated supplied with intact nerves

iNOS inducible nitric oxide synthase; an enzyme that converts L-arginine to L-citrulline and nitric oxide that is not constitutively expressed in tissues. Its appearance in tissues such as vascular smooth muscle and leukocytes occurs after exposure of the tissues to an inducing factor such as endotoxins or lipopolysaccharides

inositol trisphosphate (IP_3) intracellular second messenger generated by hydrolysis of phosphatidylinositol 4,5-bisphosphate

inotropic state contractile state of cardiac muscle independent of effects of loading condition; a function of factors that modify the contractile force of cardiac muscle at the level of the cell

insensible perspiration those evaporative losses of water from the moist surfaces of the body (such as the skin and respiratory tree) not resulting from the secretory activity of glands

insensible water loss water loss from the skin and lungs that is not ordinary

insular area region of the temporal cerebral cortex involved in coordination of olfactory and gustatory stimuli for autonomic functions

insulin hormone produced by the β cells of the pancreas that decreases the level of glucose in the blood

insulin resistance inability of a known quantity of exogenous or endogenous insulin to increase glucose uptake and use in an individual as much as it does in a normal population

insulin/glucagon ratio (I/G ratio) net physiologic response determined by the relative levels of insulin and glucagon in the blood plasma

insulin-like growth factor-1 (IGF-1) potent trophic hormone whose secretion is stimulated by growth hormone; also know as *somatomedin C*

insulin-like growth factor-2 (IGF-2) trophic hormone

insulin-like growth factors (IGFs) class of small peptides responsible for mediation of some growth hormone actions that increase the rate of cartilage synthesis by promoting the uptake of sulfate and the synthesis of collagen; also known as *somatomedins*

insulitis penetration by lymphocytes into pancreatic islets of Langerhans that produces an inflammatory or autoimmune response and results in destruction of the β cells of the pancreas

integral (intrinsic) membrane protein protein that is embedded in or completely spans the phospholipid bilayer region of a cell membrane, for example, membrane transport proteins and ion channels

integration process of assembling parts together to make a whole; a neural network processing resulting in output being a function of input other than one; the organization of behavior of individual digestive effector systems into harmonious function of the whole organ

integrins protein important for platelet adhesion in hemostasis

intention shaping of behavior in accordance with internal motivation and external context

intercalated cells cells in the connecting tubules and collecting ducts of the kidneys that are involved in acid–base transport

intercalated duct one of three ducts that comprise the salivon of the human submandibular gland. The intercalated ducts are connected to the striated duct, which eventually empties into the excretory duct

intercostal muscles muscles between the ribs that are involved in forced expiration

interdigestive state fasting state in the upper gastrointestinal that begins with the absorption of all nutrients, which is characterized by the migrating motor complex behavioral pattern

interferons class of natural proteins produced by the cells of the immune system in response to challenges by foreign agents such as viruses, bacteria, parasites, and tumor cells. Interferons belong to the large class of glycoproteins known as cytokines

interganglionic fiber tracts bundles of nerve fibers connecting adjacent ganglia of the enteric nervous system

interleukin cytokine made by one leukocyte and acting on another

interleukin-1 cytokine secreted by macrophages, monocytes, and dendritic cells as part of the inflammatory response to infection. It enables transmigration of leukocytes to sites of infection and resets the hypothalamic thermoregulatory center, leading to an increased body temperature (fever), thereby acting as an endogenous pyrogen

interleukin-6 proinflammatory cytokine secreted by T cells and macrophages to stimulate immune response to trauma, especially burns or other tissue damage leading to inflammation. IL-6 is one of the most important mediators of fever.

intermediate filaments cellular constituents of smooth muscle that form part of the cytoskeleton

intermediate zones collection of spinal interneurons that receive input from brainstem pathways and influence motor neuron function

intermediate-density lipoprotein (IDL) particles circulating in the blood that contain cholesterol esters; falls between low-density and very low–density lipoprotein in density

internal anal sphincter smooth muscle sphincter surrounding the anal canal, which provides a passive barrier to the leakage of liquid and gas from the rectum

internal capsule pathway for corticospinal tract neurons within the deep cerebral white matter

international normalized ratio (INR) system established to report the results of blood coagulation (clotting) tests; results are standardized using the international sensitivity index for the particular thromboplastin reagent and instrument combination used to perform the test

interneuron internuncial neuron that is neither sensory nor motor and that connects neurons with neurons

interposed (interpositus) nucleus deep nucleus of the cerebellum composed of the globose nuclei and the emboliform nuclei; receives its afferent supply from the anterior lobe of the cerebellum and sends output via the superior cerebellar peduncle to the red nucleus; modulates muscle stretch reflexes of distal muscle groups

interstitial cells of Cajal (ICCs) specialized cells of mesodermal origin believed to be the pacemaker cells for intestinal electrical slow waves

interstitial fluid fluid outside blood vessels that directly bathes most body cells

intestinal phase one of three phases that stimulate acid secretion resulting from the ingestion of food. During the intestinal phase, protein digestion products in the duodenum stimulate gastric acid secretion through the action of the circulating amino acids on the parietal cells. Distention of the small intestine stimulates acid secretion. The intestinal phase accounts for about 10% of total gastric acid secretion

intracellular fluid (ICF) fluid within cells

intrafusal muscle fibers muscle cells within a muscle spindle whose contraction changes tension on spindle sensory endings

intralaminar nuclei of thalamus relay nuclei within the thalamus that innervate wide areas of the cerebral cortex and limbic system and receive inputs from the brainstem reticular formation; a component of the ascending reticular activating system

intrinsic coagulation pathway blood-clotting pathway in response to an abnormal vessel wall

intrinsic factor glycoprotein secreted by the parietal cells in the stomach that binds strongly with vitamin B_{12} to form a complex that is then absorbed in the terminal ileum through a receptor-mediated process

inulin fructose polymer whose renal clearance is used to measure glomerular filtration rate

inulin clearance procedure by which the filtering capacity of the kidney glomeruli is determined by measuring the rate at which inulin, the test substance, is cleared from blood plasma

inverse myotatic reflex type of muscle stretch reflex that helps regulate the tension of contracting muscle

inwardly rectifying K^+ channels channels responsible for the hyperpolarizing potassium currents that are the main determinant of the resting membrane potential in myocardial cells. The term “inwardly” is an electrophysiologic convention that refers to the fact that the channels pass potassium currents more readily in the inward than in the outward direction, even though such inward potassium currents never occur in myocardial cells

iodothyronine molecule formed by the coupling of the phenyl rings of two iodinated tyrosine molecules in an ether linkage

ion channel opening through the plasma membrane composed of several polypeptide subunits that span the membrane and contain a gate that determines whether the channel is open or closed; regulates the flow of ions across the membrane in all cells

ion pump integral membrane protein that uses energy derived from adenosine triphosphate hydrolysis to drive ion movement across the membrane against the electrochemical gradient, for example, Na^+/K^+-ATPase

ionotropic type of membrane receptor that is a channel that opens (or closes) when the receptor is bound by ligand

ionotropic receptor receptors that are coupled directly to ion channels; fast excitatory postsynaptic potentials in the enteric nervous system are mediated by ionotropic receptors

iron profile blood tests including serum iron, ferritin, total iron-binding capacity, transferrin, and transferrin saturation

iron-deficiency anemia condition characterized by decreased healthy red blood cell count because of a dietary iron deficiency, and hence, functioning hemoglobin

irreversible shock stage of shock in which the function of organs needed to maintain cardiac output and blood pressure (heart and brain) is so compromised that the person cannot survive even with prompt and aggressive medical intervention

irritable bowel syndrome (IBS) group of functional bowel disorders in which abdominal discomfort or pain is associated with defecation or a change in bowel habit, and with features of disordered defecation

irritant receptors rapidly adapting receptor that is found in the large conducting airways and that is sensitive to irritation by noxious substances

ischemia reperfusion condition in which blood flow is restricted, followed by normal blood flow being reestablished. This condition leads to the formation of reactive oxygen species, which often leads to tissue damage

ischemic phase (of the endometrial cycle) phase in which necrotic changes and abundant apoptosis occur in the secretory epithelium as it collapses

islets of Langerhans cell clusters in the pancreas that form the endocrine part of that organ and that secrete insulin and other hormones

isocaproic acid product of side-chain cleavage of cholesterol

isoelectric in cardiovascular physiology, the portion of the normal electrocardiogram where the tracing is at zero or near-zero potential, which occurs in the PR and ST intervals

isograft transplantation between genetically identical people

isohydric principle idea that all buffer systems in a solution are in equilibrium with the same hydrogen ion concentration

isometric contraction contraction at constant length that produces force but no muscle shortening

isometric exercise form of exercise involving the contraction of a muscle without the shortening of the angle of the joint

isosmotic see *osmotic pressure*

isotonic contraction muscle contraction at constant force that produces a change in length

isotonic solution see *tonicity*

isovolumic contraction portion of systole in which all valves are closed, and therefore, blood does not enter or exit the heart

isovolumic relaxation portion of diastole in which all valves are closed, and therefore, blood does not enter or exit the heart

isthmus portion of the oviduct connecting the ampulla to the uterus

Ito cells perisinusoidal cells that contain lipid droplets in the cytoplasm and that function in the storage of fat. During the inflammatory process, these cells have the ability to transform into myofibroblasts, which secrete collagen into the space of Disse and regulate sinusoidal portal pressure by contracting or relaxing. Ito cells may be involved in the pathologic fibrosis of the liver; also known as *stellate cells*

J receptors unmyelinated C-fiber nerves that are located adjacent to alveoli that are stimulated by stretch; also referred to as *juxtapulmonary capillary receptors* or *C-fiber endings*

junctional complex functional unit in muscle comprising both dihydropyridine and ryanodine receptors

juxtaglomerular apparatus structure in the kidney consisting of macula densa, extraglomerular mesangial cells, and granular cells, located where the end of the thick ascending limb touches afferent and efferent arterioles of its parent glomerulus

juxtamedullary nephron nephron with its glomerulus deep in the cortex, next to the medulla, and having long loops of Henle

juxtapulmonary capillary receptors see *J receptors*

kallidin see *lysyl-bradykinin*

kallikrein serine protease released by the acinar cells that acts on endogenous peptides present in body fluid to release lysyl-bradykinin (kallidin), which causes dilation of the blood vessels supplying the salivary glands

Kallmann syndrome syndrome characterized by a congenital absence of gonadotropin-releasing hormone in the hypothalamus (causing, in women, primary amenorrhea and anovulation and, in men, failure of puberty) in combination with a congenitally absent sense of smell

ketoacidosis condition found in diabetics that is characterized by large amounts of β-hydroxybutyric acid in the blood

ketogenesis process by which ketone bodies are produced as a result of fatty acid breakdown

ketones acidic substances produced when the body uses fat, instead of sugar, for energy

ketosis condition observed in diabetic patients and during periods of prolonged starvation that is marked by highly elevated levels of circulating ketone bodies

17-ketosteroids excretable metabolite formed by lysis of the carbon 20–21 side chain of 21-carbon steroids

kilocalorie unit of measurement for energy that approximates the heat energy needed to increase the temperature of 1 kg of water by 1°C, equivalent to the dietary "calorie"

kinesin microtubule-associated protein involved in anterograde axonal transport of organelles and vesicles (from the minus to the plus ends of the microtubules) via the hydrolysis of adenosine triphosphate

kininogen endogenous peptide present in body fluid that releases lysyl-bradykinin (kallidin)

kinins mediator of the inflammatory response

kinocilium nonmotile cilium-like structure on each hair cell of the human ear involved in converting mechanical movement of endolymph into electrical signals

Klinefelter syndrome congenital abnormality of men in which there is one X chromosome too many; men with this condition are usually sterile

Korotkoff sounds sounds made by the first spurts of blood escaping from a compressed artery as the compression is slowly released. They are detected by stethoscope in the measurement of arterial pressure by sphygmomanometer and used as the indicator of peak systolic pressure

Krebs cycle series of mitochondrial chemical reactions that produces adenosine triphosphate in the presence of O_2, a process called *oxidative phosphorylation*

Kupffer cells one of the types of cells lining hepatic sinusoids; resident macrophages of the "fixed monocyte–macrophage system" that act in the removal of unwanted material (bacteria, viral particles, etc.) from the circulation through endocytosis

Kussmaul respiration labored, deep breathing that accompanies severe uncontrolled diabetes

L-type Ca^{2+} channels voltage-gated channels that conduct primarily depolarizing Ca^{2+} currents into myocardial cells and nodal tissue during phase 2 of the action potential. The gates are opened and closed by membrane depolarization in a manner analogous to that seen for voltage-gated sodium channels except that their opening and closing kinetics are slower

labeled line concept that the central nervous system identifies a particular sensation on the basis of the pathway by which it arrives at the brain

lactation secretion of milk

lactational amenorrhea period of time with no menstrual cycles, resulting from lactation

lactic acid end product of glycolysis that accumulates under anaerobic conditions and is oxidized in the Krebs cycle in muscle

lactiferous ducts primary milk ducts

lactoferrin protein that binds iron

lactogenesis milk production by alveolar cells; also known as *lactogenic activity*

lactose intolerance condition in which the ingestion of milk products results in severe osmotic diarrhea

lactotrophs cells located in the anterior pituitary that secrete prolactin

lacunae blood-filled spaces formed in the endometrium

lag phase of gastric emptying time required for the grinding action of the antral pump to reduce the size of larger particles to particles of sufficiently small size for emptying (i.e., $< \approx 7$ mm)

Lambert–Eaton myasthenic syndrome (LEMS) rare presynaptic disorder of neuromuscular transmission in which quantal release of acetylcholine is impaired, causing a unique set of clinical characteristics, which include proximal muscle weakness, depressed tendon reflexes, posttetanic potentiation, and autonomic changes

lamellar inclusion bodies electron-dense bodies found in type II epithelial cells of the alveoli that are the storage sites of lung surfactant

laminar flow relative motion of elements of a fluid along smooth parallel paths, which occurs at lower values of Reynolds number

laminin external filamentous muscle protein that forms a link between integrins and the extracellular matrix

Langerhans cells dendritic epidermal cell present in lymph nodes and other organs

Laplace law mathematical relationship relating tension pulling on the wall of a thin-walled vessel to the transmural pressure across the wall. It is given by $T = P_r$, where T is wall tension and r is the internal vessel radius

laryngospasm severe constriction of the larynx in response to the introduction of water, allergies, or noxious stimuli

latch state modification of the crossbridge cycle in smooth muscle that results in long contractions with little expenditure of energy

latency period time period between stimulus onset and response occurrence

latent heat of evaporation energy required to overcome the molecular forces of attraction between the particles of a liquid and to bring them to the vapor state, in which such attractions are minimal

lateral inhibition capacity of an excited neuron to reduce the activity of its neighbors

LDL receptors low-density lipoprotein receptors; receptors present on the surface of steroid-producing and other cells; bind cholesterol containing LDL particles in the blood

leak channels channels that are always open, allowing the passage of sodium ions (Na^+) and potassium ions (K^+) across the membrane to maintain the resting membrane potential of −70 mV; also called *passive channels*

lecithin cholesterol acyltransferase (LCAT) plasma enzyme that is responsible for the esterification of cholesterol in the plasma to form cholesterol ester with the fatty acid derived from the 2 position of lecithin

left atrium left upper muscular chamber of the heart, which receives blood from the pulmonary veins and delivers it through the mitral valve into the left ventricle

left bundle branch block condition in which transmission of action potentials is blocked in the left branch of the bundle of His

left ventricle left lower muscular chamber of the heart, which receives blood from the left atrium and delivers it through the aortic valve into the aorta

left ventricular end-diastolic pressure (LVEDP) blood pressure within the left ventricle at the end of diastole; function of both the volume contained within the ventricle and ventricular compliance

left ventricular end-diastolic volume (LVEDV) volume of blood in the ventricle at the end of diastole and just before the initiation of ventricular contraction that determines the passive stretch on myocardium before contraction and, therefore, is considered the equivalent of muscle preload in the heart *in situ*

left ventricular hypertrophy abnormally increased muscle mass in the left ventricle, occurring in response to any factor that chronically increases wall stress on the ventricle, such as systemic arterial hypertension or a stenotic aortic valve

length–tension curve expression of the isometric force a muscle can produce at different fixed lengths

leptin anorexigenic hormone released by white fat cells (adipocytes); levels increase as fat stores increase

leukemia bone marrow cancer resulting in uncontrolled proliferation of leukocytes

leukocyte see *white blood cell*

leukocytosis condition characterized by a higher-than-normal number of circulating white blood cells

leukopenia condition characterized by a lower-than-normal number of circulating white blood cells

leukotrienes mediator of the inflammatory response

levator ani muscle compound muscle of the pelvic floor formed by pubococcygeus and iliococcygeus muscles that functions to resist prolapsing forces and draws the anus upward following defecation; supports the pelvic viscera

lever system system of bones and joints that confers a mechanical advantage (or disadvantage) on a skeletal muscle

Lewis hunting response response characterized by an intermittent supply of warm blood to cold exposed extremities via vasodilation that recurs in 5-minute to 10-minute cycles to provide some protection from the cold

Leydig cells cells within the seminiferous tubules of the testicles that produce testosterone

Liddle syndrome autosomal dominant disorder characterized by hypertension associated with low plasma renin activity, metabolic alkalosis due to hypokalemia, and hypoaldosteronism; one of several conditions with this unusual set of characteristics known collectively as *pseudohyperaldosteronism*

ligand molecule that binds to another, that is, a hormone or neurotransmitter that binds to a receptor

ligand-gated ion channels channels in membrane receptors that open (or close) when a ligand binds the receptor

limbic system interconnected structures in the central nervous system associated with olfaction, emotion, motivation, and behavior; includes amygdala, cingulate gyrus, hippocampus, hypothalamus, olfactory cortex, and thalamus

lipid bilayer structure formed by spontaneous assembly of amphipathic lipid molecules in an aqueous solution; hydrophobic regions or fatty acids are buried in the interior, and hydrophilic head groups are on the exterior, in contact with the solution

lipid rafts highly organized, free-floating "microdomains" within the plasma membrane that are aggregates of sphingolipids and cholesterol; serve to compartmentalize cellular processes by serving as organizing centers for the assembly of signaling molecules, influencing membrane fluidity and membrane protein trafficking, and regulating neurotransmission and receptor trafficking

lipocortin substance that inhibits the activity of phospholipase A_2

lipogenesis conversion of carbohydrates and organic acids to fat

lipolysis mobilization of fatty acids from triglycerides stored in adipose tissue

lipopolysaccharide (LPS) major component of gram-negative bacterial membrane

lipoprotein lipase enzyme responsible for hydrolysis of triacylglycerol molecules packaged and transported in very low–density lipoprotein particles to fatty acids, which can then in turn be metabolized for energy

lipoproteins complexes or compounds containing lipid and protein. The intestine produces two major classes of lipoproteins: chylomicrons and very low–density lipoproteins

β-lipotropin substance derived from proopiomelanocortin that may have effects on lipid metabolism

lithium-induced polyuria excretion of large volumes of dilute urine occurring in patients receiving chronic lithium therapy for psychiatric disorders

lithocholic acid carcinogenic secondary bile acid derived from chenodeoxycholic acid

liver lobule basic functional unit of the liver, which is hexagonal in shape and built around a central vein

loading phase flat portion of the oxygen-equilibrium curve that reflects the loading of oxygen onto blood hemoglobin

local axon reflex pathway for propagation of action potentials within a portion of a neuron; does not pass through the cell body

local excitatory current current through an axon membrane created by the generator potential that provides the link between the formation of the generator potential and the excitation of the nerve fiber membrane

locus ceruleus collection of noradrenergic neurons in the rostral pons that innervate widespread areas of the cerebrum, brainstem, and spinal cord

long hypophyseal portal vessels vessels that deliver blood to the anterior pituitary

long-loop reflexes reflexes that require processing by several spinal or higher levels

long-term memory memory that can last as little as a few days or as long as decades; includes *declarative* and *nondeclarative* subtypes

long-term potentiation increased synaptic excitability and altered chemical state on repeated synaptic stimulation with a persistence beyond the cessation of electrical stimulation; thought to underlie learning and memory and be dependent on Ca^{2+} entry through activation of *N*-methyl-D-aspartate receptors

loop of Henle portion of the nephron leading from the proximal straight tubule to the distal convoluted tubule; main function is to create a concentration gradient in the medulla of the kidney: a countercurrent multiplier system using sodium pumps creates an area of high concentration near the collecting duct; water present in the filtrate in the collecting duct flows through aquaporin channels out of the collecting duct, moving passively down its concentration gradient; this process reabsorbs water and creates a concentrated urine for excretion

low-density lipoprotein (LDL) complex containing a single apolipoprotein B-100 molecule, fatty acids, and cholesterol that transports cholesterol and fatty acids to cells; also known as *bad cholesterol*

lower esophageal sphincter smooth muscle sphincter separating the esophagus from the stomach

lung compliance (C_L) measure of lung distensibility; represented by Δvolume/Δpressure

lung diffusion capacity amount of oxygen taken up by the lung per minute per mm Hg of oxygen partial pressure

lung-inflation reflex see *Hering–Breuer reflex*

lung lavage medical procedure in which a bronchoscope is passed through the mouth or nose into the lungs and fluid is squirted into a small part of the lung and then recollected for examination; often referred to as *bronchoalveolar lavage*

luteal insufficiency condition in which the corpus luteum does not produce sufficient progesterone to maintain pregnancy in the early stages

luteal phase part of the menstrual cycle during which progesterone is produced; lasts 14 days

luteinization maturation of the corpus luteum

luteinizing hormone (LH) hormone secreted by the pituitary gland in the brain that stimulates the growth and maturation of eggs in females and sperm in males

luteinizing hormone–releasing hormone (LHRH) polypeptide from the hypothalamus that stimulates secretion of follicle-stimulating hormone and luteinizing hormone; also known as *gonadotropin-releasing hormone (GnRH)*

luteolysis process of regression of the corpus luteum; also known as *luteal regression*

lymph transparent, slightly yellow liquid, found in the lymphatic vessels and derived from tissue fluid

lymph node lymphoid organ filled with leukocytes

lymphagogic effect marked increase in lymph flow that plays an important role in the transfer of lipoproteins from the intercellular spaces to the central lacteal vessel

lymphatic bulbs blind lymphatic end tubes that are the origin of lymphatic vessels in the microcirculation

lymphatic vessels flexible, predominantly endothelial vessels that drain the interstitium of plasma filtrate from the capillaries and send the fluid back into the venous side of the circulation at the level of the right atrium. Larger vessels contain one-way valves directed away from the tissues and toward the heart. Muscle cells in the bulb ends of the vessels help with forward flow of fluid into larger lymphatic vessels.

lymphoblast stem cell committed to differentiate into lymphocytes

lymphocyte agranular leukocyte: B cell, T cell, or null cell; involved in body defenses against disease

lymphoid organs organs that include bone marrow and thymus (*primary*) and lymph follicles, lymph nodes, spleen, thymus, and tonsils (*secondary*)

lymphoid series series of morphologically distinguishable cells once thought to represent stages in lymphocyte development; now known to represent various forms of mature lymphocytes

lymphokine cytokine made by lymphocytes

lymphokine-activated killer cell (LAK) cell derived from natural killer cells

lymphoma cancer of the lymphatic system

lysis decomposition or destruction of cells by antibodies called *lysins*

lysozyme bacteriolytic enzyme

lysyl-bradykinin potent vasodilator that causes dilation of the blood vessels that supply the salivary glands; also known as *kallidin*

M line thin line in middle of H zone of muscle

macrocyte larger-than-normal erythrocyte

macrophage phagocytic tissue cell, derived from monocytes

macula sensory structure in each semicircular canal of the vestibular apparatus, used to sense rotation

macula densa closely packed group of densely staining cells in the distal tubular epithelium of a nephron, in direct apposition to the juxtaglomerular cells; they may function as either chemoreceptors or as baroreceptors feeding information to the juxtaglomerular cells

macula lutea area of the retina about 1 mm^2 in area, specialized for very sharp color vision

magnocellular neurons hypothalamic neurons with large cell bodies that synthesize arginine vasopressin and oxytocin

major depression one of a class of illnesses marked by abnormal mood regulation and associated signs and symptoms called *affective disorders* or *mood disorders*. Major depression can be so profound as to provoke suicide

major histocompatibility complex (MHC) genes encoding proteins involved in antigen presentation

male pseudohermaphrodite 46XY individuals with differentiated testes but underdeveloped external genitalia

malignant hyperthermia life-threatening condition resulting from a genetic sensitivity of skeletal muscles to volatile anesthetics and depolarizing neuromuscular blocking drugs that occurs during or after anesthesia

malonyl CoA condensation product of malonic acid and coenzyme A and an intermediate in fatty acid synthesis

mammary glands paired structures on the thoracic wall of both sexes; in females, they secrete milk

mammillothalamic tract neuroanatomical tract in the limbic system connecting the neurons in the mammillary bodies of the hypothalamus with the anterior nucleus of the thalamus

mammogenesis differentiation and growth of the mammary glands

manometric related to pressure measurements (in this context, within the gut lumen)

manometric catheter gut manometry is the measurement of intraluminal pressures in a segment of the gut to study its motor function: contractions of muscles within or around the gut wall produce intraluminal pressure waves that are detected by manometry

margination loose adhesion of leukocytes to endothelial cells

marrow stroma bone marrow tissue that is not directly involved in hematopoiesis but provides the necessary microenvironment and factors for it

mast cell connective tissue cell in the intestine that releases paracrine signals, such as histamine, to act at receptors on enteric neurons; functionally resembles basophil

maternal recognition of pregnancy the secretion of human chorionic gonadotropin by a developing embryo to sustain the corpus luteum

maximal oxygen uptake (Vo_2 max) highest rate at which oxygen can be taken up and used during exercise. The derivation is V = volume per time, O_2 = oxygen, max = maximum; expressed either as an absolute rate (L/min) or as a relative rate (mL/kg/min), the latter expression is often used to compare the performance of endurance sports athletes.

maximal response greatest, most complete, or best response

maximal voluntary contraction (MVC) the peak isometric force that can be briefly generated for a specific exercise; used as an index of muscle function and, more specifically, isometric work intensity

McCune–Albright syndrome (MAS) rare multisystem disorder characterized by replacement of normal bone tissue with areas of abnormal fibrous growth (fibrous dysplasia); patches of abnormal skin pigmentation (i.e., areas of light-brown skin [café au lait spots] with jagged borders); and abnormalities in the glands that regulate the body's rate of growth, its sexual development, and certain other metabolic functions (multiple endocrine dysfunction)

mean circulatory filling pressure equilibrium pressure throughout all components of the vascular system when cardiac output is zero

mean corpuscular hemoglobin (MCH) index of average hemoglobin per erythrocyte

mean corpuscular hemoglobin concentration (MCHC) index of average hemoglobin in circulating erythrocytes

mean corpuscular volume (MCV) index of average volume of each erythrocyte

mean electrical QRS axis average vector representing the wave of depolarization through the ventricles

mean platelet volume (MPV) evaluates size of platelets

mechanical obstruction any physical obstacle to the caudal movement of intestinal contents, including tumors, scar tissue, or volvulus (twisting); distinguished from a functional obstruction, which is a result of absent or abnormal contractions of the intestine

mechanoreceptors sensory receptors that respond to physical deformations both externally and internally

medial forebrain bundle fiber bundle interconnecting the brainstem (especially the midbrain tegmentum) with the hypothalamus, nucleus accumbens, and basal forebrain

medial prefrontal area region of frontal lobe cerebral cortex involved in coordination of autonomic responses to olfactory and emotional stimuli

median effective dose (ED_{50}) concentration of a substance (e.g., hormone) required to produce a response halfway between the maximal and basal responses

median eminence region in the floor of the hypothalamus that allows release of neurohormones and access to the hypothalamus of bloodborne substances

medium spiny neuron predominant type of neuron in the caudate and putamen of the basal ganglia

medulla innermost portion of an organ or structure

medullary reticulospinal tract descending inhibitory brainstem pathway to spinal interneurons that originates in the medulla

medullary tissue part of the indifferent gonad of the embryo, which will become the testicular components

megacolon colon filled with impacted feces above the terminal aganglionic region of large intestine in Hirschsprung disease

megakaryocyte bone marrow precursor cell for platelets

megaloblast nucleated, large, abnormal red blood cell

megaloblastic anemia condition in which red blood cell maturation in the bone marrow is abnormal, possibly resulting from vitamin B_{12} or folic acid deficiency

meiosis process of two consecutive cell divisions in the diploid progenitors of sex cells, resulting in four rather than two daughter cells, each with a haploid set of chromosomes

Meissner plexus named for Georg Meissner (1829–1905), a German histologist; see also *submucosal plexus*

melanin pigment that in humans is the primary determinant of skin color; in the pigment epithelium of the eye, melanin helps to sharpen an image by preventing the scattering of stray light

α-melanocyte-stimulating hormone (α-MSH) hormone derived from proopiomelanocortin that regulates feeding in humans

melatonin hormone released by the pineal gland in darkness; the mechanism by which the suprachiasmatic nucleus of the hypothalamus regulates diverse functions in a diurnal fashion

membrane lipid bilayer that acts as a boundary to various cellular structures or the environment

membrane attack complex (MAC) complement proteins that form a channel in the target cell membrane

membrane potential difference between voltage measured inside a cell compared with the value (0) detected by a reference electrode located in the extracelluar fluid

membrane-associated dense bodies dense bodies associated with cell margins that serve as anchors for thin filaments and transmit the force of contraction to adjacent cells; also called *focal adhesions*

memory cell T or B cell that recognizes previously encountered antigens

menaquinones vitamin K derived from bacteria in the small intestine

menarche beginning of menstrual cycles during pubertal development in girls

Ménière disease disorder of the inner ear that can affect hearing and balance to a varying degree and is characterized by episodes of both vertigo and tinnitus and by progressive hearing loss, usually in one ear

menopause final menses at the end of the reproductive cycle in the female

menstrual cycle monthly reproductive cycle in which periodic uterine bleeding occurs

menstrual phase or menses or menstruation bleeding phase of the menstrual cycle or endometrial cycle, which lasts about 5 days

mesangial cell specialized cell found between capillary loops in the glomerulus, which has supportive, phagocytic, and contractile functions

mesolimbic/mesocortical system projection of dopaminergic neurons from the midbrain ventral tegmental area to the limbic system, especially the prefrontal cortex, nucleus accumbens, and hypothalamus

mesonephric (wolffian) ducts ducts that give rise to the epididymis, vas deferens, seminal vesicles, and ejaculatory ducts in the male

messenger low–molecular weight diffusible molecule used in signal transduction to relay signals within a cell

metabolic acidosis abnormal condition in which the blood becomes too acidic, leading to a decrease in pH. The acidosis is a result of any cause other than elevated carbon dioxide level.

metabolic alkalosis abnormal process characterized by a gain of strong base or bicarbonate or a loss of acid (other than carbonic acid), which results in a rise in arterial blood pH

metabolic bone diseases diseases characterized by ongoing disruption of the normal processes of either bone formation or bone resorption

metabolic clearance rate (MCR) volume of plasma cleared of a substance, such as a hormone, in question per unit time

metabolic syndrome cluster of conditions that often occur together, including obesity, high blood glucose, hypertension, and high triglycerides, which can lead to cardiovascular disease

metabotropic characteristic of receptors that do not form an ion channel, instead, causing indirect alterations in current via signal transduction mechanisms or causing only metabolic changes

metabotropic receptor (mGLUR) type of receptor that is linked through G proteins to intracellular production of 1,2-diacylglycerol and inositol 1,4,5-trisphosphate

methemoglobin (metHb) abnormal hemoglobin with oxidized ferric iron

methylene blue component of blood dyes such as Wright stain

MHC class I molecule molecules found in every nucleated cell of the body that bind peptides mainly generated from degradation of cytosolic proteins; see also *major histocompatibility complex*

MHC class II molecule molecules found only on antigen-presenting cells (macrophages, dendritic cells, and B cells); see also *major histocompatibility complex*

micellar emulsification process that enhances the delivery of lipid to the brush border membrane by making it water soluble

micelle aggregate of surfactant molecules dispersed in a liquid colloid

microalbuminuria condition characterized by the excretion of 30 to 300 mg of serum albumin in the urine per day, usually a sign of kidney disease

microcirculation region of the circulation involved with the control of blood pressure, blood flow, and transcapillary transport of substances, composed of terminal arterioles, capillaries, and venules

microspherocyte erythrocyte that is smaller than normal

microfilaments elements of the neuronal cytoskeleton that are composed of actin, are 4 to 5 nm wide and about a micrometer in length (except in growth cones, where they can be much longer), and are concentrated in dendritic spines, growth cones, and presynaptic terminals

microtubules elements of the neuronal cytoskeleton responsible for the rapid movement of material in axons and dendrites; composed of a core polymer of 50-kDa tubulin subunits that align to form protofilaments, several of which join laterally to form a hollow tube whose diameter is about 23 nm

microtubule-associated proteins (MAPs) accessory proteins associated with microtubules that bind tubulin and, in neurons, are thought to be required for specific sorting of materials to axons or dendrites

microvilli numerous closely packed villi that cover the enterocytes to amplify further the villi that increase the surface area of the small intestine

micturition emptying of the urinary bladder; also known as *urination*

midbrain extrapyramidal area group of neurons in the upper brainstem that receive output from the globus pallidus and provide output to spinal interneurons

middle cervical ganglia middle of the three sympathetic ganglia located in the neck

middle ear portion of the ear, containing the ossicles, that couples eardrum movements to the sensory structures of the inner ear

migrating motor complex (MMC) organized, distinct, cyclically recurring, and distally propagating sequence of contractile activity occurring in the small intestine during the fasting state; may be associated with similar activities in the stomach and proximal colon

milk ejection release of stored milk from the mammary glands by the action of oxytocin; also known as *milk letdown*

milliequivalent (mEq) one-thousandth of an equivalent; calculated from the product of millimoles times valence

mineralocorticoid steroid hormone produced by the adrenal cortex and involved in the regulation of electrolyte and water balance

mineralocorticoid escape limitation of the salt-retaining effects of a mineralocorticoid hormone (such as aldosterone), expressed by increased renal sodium excretion despite the continued presence of high levels of mineralocorticoid

mitochondria membrane-bound organelle that supplies energy in the form of adenosine triphosphate to cells

mitochondrial permeability transition pore (MPTP) protein pore that is formed in the membranes of mitochondria under certain pathologic conditions; induction of MPTP can lead to mitochondrial swelling and cell death

mitochondrial voltage-dependent anion channel (VDAC) channel for metabolite diffusion across the mitochondrial outer membrane

mitogen agent that induces proliferation; many growth factors and hormones are mitogens

mitogenic agent agent that stimulates the growth of target tissues

mitosis replication of a cell to form two daughter cells with identical sets of chromosomes

mitral valve thin, flexible, bi-leaflet structure that separates the left atrium from the left ventricle. It opens into the left ventricle during diastole but closes off the left atrium when the left ventricle contracts

mitral valve prolapse condition whereby either the mitral valve bulges or the leaflets open backward into the pulmonary vein during systole. In the latter case, blood flows back into the pulmonary circulation from the left ventricle during systole

mittelschmerz (German for "middle pain") lower abdominal and pelvic pain that occurs roughly midway through a woman's menstrual cycle, associated with ovulation; pain can appear suddenly and usually subsides within hours

mixed acid–base disturbance simultaneous presence of two or more primary acid–base disturbances

mixed contractions skeletal muscle contractions that begin isometrically and become isotonic as force becomes sufficient to lift the afterload

mixed micelles substance that aids in rendering lipid digestion products water-soluble. Mixed micelles diffuse across the unstirred water layer and deliver lipid digestion products to the enterocytes for absorption

mixing motility pattern motility pattern of the digestive state

M-mode echocardiography ultrasound image of the heart used to examine motion of the ventricular walls, changes in chamber size and valve motion during the cardiac cycle

modulator peptide released at a synapse that modifies the response of a coreleased transmitter interacting with its own receptor in the synapse

molecular switch intracellular protein that can transition between "on" and "off" states in response to a cellular stimuli (e.g., G protein)

monoamine oxidase (MAO) a mitochondrial enzyme that degrades catecholamines by removing the amine

monoamine oxidase inhibitors (MAOIs) powerful class of antidepressant drugs

monoaminergic type of neuron that uses a catecholamine or indolamine as a neurotransmitter

monoblast stem cell committed to differentiate into monocytes

monocytic series succession of developing cells that ultimately culminates in the monocyte

monocyte agranular phagocytic white blood cell that is a precursor of macrophages

monoiodotyrosine (MIT) iodinated tyrosine in peptide linkage in the thyroglobulin structure

monokine cytokine made by monocytes

mononuclear leukocyte agranulocyte: either lymphocyte or monocyte

mood sustained emotional state

mood disorders see *affective disorders*

morula solid ball of cells that develops from a fertilized egg and enters the uterus at day 3 to 4 following fertilization

mossy fibers afferent axons to the cerebellum from pontine nuclei

motilin signal peptide released from enteroendocrine cells and associated with the onset of the migrating motor complex

motility movements within the intestinal tract, encompassing the phenomena of contractile activity, myoelectrical activity, tone, compliance, wall tension, and transit within the gastrointestinal tract

motor endplate see *myoneural junction*

motor neuron nerve cell that sends an axon to a muscle or, by extension, any effector

α motor neuron innervates the extrafusal muscle fibers, which are responsible for force generation

γ motor neuron controls contraction of the intrafusal muscle cells within a muscle spindle

motor unit alpha motor neuron, its axon, and all the muscle fibers controlled by that motor neuron

motor unit summation see *spatial summation*

mucin one of two major proteins (a glycoprotein high in carbohydrate) present in saliva that form gels in solution; the most abundant and produced mainly by the sublingual and submandibular salivary glands. Mucins lubricate the mucosal surface and protect it from mechanical damage by solid food particles. Mucins may also provide a physical barrier in the small intestine against the entry of microorganisms into the mucosa

mucous cells cells that line the entire surface of the gastric mucosa and the openings of the cardiac, pyloric, and oxyntic glands; secrete mucus and bicarbonate to protect the gastric surface from the acidic environment of the stomach

mucous gel layer layer covering the surface of the gastric mucosa that traps bicarbonate and neutralizes acid, thus preventing damage to the mucosal cells

mucous neck cells one of several types of cells in the oxyntic glands. The neck region in the gland is where mucous cells are most numerous. Mucous neck cells secrete mucus

müllerian-inhibiting substance (MIS) glycoprotein that inhibits cell division of the müllerian ducts; also known as *antimüllerian hormone*

multidrug resistance insensitivity of various tumors to a variety of chemically related anticancer drugs; mediated by a process of inactivating the drug or removing it from the target tumor cells

multidrug resistance–associated proteins (MRP) transporters class of ABC transporters that interfere with antibiotic therapy and chemotherapy

multiple myeloma typically incurable cancer of plasma cells

multiple sclerosis demyelinating autoimmune disease

multipotent stem cell cell, such as hematopoietic stem cell, that can give rise to different specialized cells

multiunit-type smooth muscle type of smooth muscle that predominates in the pupil of the eye and urinary bladder; it does not contract spontaneously, is not activated by stretch, and expresses structured neuromuscular junctions; contrast with *unitary-type smooth muscle*

mural granulosa cells cells attached to the basement membrane in a graafian follicle

muramidase lysozyme that can lyse the muramic acid of certain bacteria (e.g., *Staphylococcus*); present in saliva in small amounts

muscarine compound found in poisonous mushrooms that activates receptors for acetylcholine known as *muscarinic receptors*

muscarinic receptor metabotropic-type receptor for acetylcholine

muscle chemoreflex reflex initiated by stimulation of the chemosensitive nerve endings as a result of a mismatch between skeletal muscle blood flow and an increase in metabolites that stimulate

muscle fibers contractile cells of a muscle

muscle stretch reflex contraction of a skeletal muscle provoked by a sudden stretch of its tendon

muscularis externa longitudinal and circular muscle coats of the stomach wall (one of four layers); lies over the submucosa and differs from that of other GI organs in that it has three layers of smooth muscle instead of two

musculomotor neuron neuron that innervates the gastrointestinal musculature

myasthenia gravis autoimmune disease that produces weakness by reducing the number of postsynaptic acetylcholine receptors in skeletal muscle

myelin lipid structure formed by oligodendrocytes in the central nervous system or Schwann cells in the peripheral nervous system that wraps around neuronal axons and serves to increase the speed with which they conduct action potentials

myeloblast stem cell committed to differentiate into cells of the myeloid series

myeloid series various stages of cell development such as those seen in the granulocytic and erythrocytic series

myenteric plexus ganglionated plexus of the enteric nervous system situated between the longitudinal and circular muscle coats of the muscularis externa of the digestive tract; also known as *Auerbach plexus*

myocardial infarction pathologic condition is which myocardial tissue has died. This usually results from extended exposure of the tissue to ischemic conditions

myocardial ischemia any condition in which the oxygen supply to myocardial tissue cannot meet the oxygen demand of that tissue

myoclonus brief, involuntary muscle twitches that commonly occur during light sleep

myoepithelial cells cells that surround milk-secreting alveolar cells and that contract and expel milk from the lactiferous duct in response to oxytocin

myofibrils longitudinal columns of series-connected sarcomeres in a skeletal muscle

myogenic originating in muscle tissue

myogenic mechanism regulatory mechanism whereby arteries or veins contract in response to an increase in transmural pressure. It is related to activation of stress-operated calcium channels in the smooth muscle itself that open with stretch, thereby increasing calcium entry into the cells and causing contraction of vascular smooth muscle. It is one mechanism responsible for the ability of organs to autoregulate their blood flow in the face of changes in perfusion pressure

myoglobin oxygen-binding protein in skeletal muscle that facilitates oxygen diffusion

myometrium outer part of the uterus, composed of multiple layers of smooth muscle

myoneural junction see *neuromuscular junction*

myopathic pseudoobstruction pathologic failure of propulsion in the gastrointestinal tract related to muscular degeneration and weakened contractility

myopia (nearsightedness) condition caused by the eyeball being physically too long to focus on near objects

myosin fibrous protein dimer that makes up both the structural and chemically active portions of the thick filaments in muscle

myosin light-chain kinase (MLCK) calcium/calmodulin-dependent enzyme that phosphorylates the light chains of smooth muscle myosin and initiates contraction. In skeletal muscle, phosphorylation modulates tension during contraction

myosin light-chain phosphatase (MLCP) enzyme that catalyzes the dephosphorylation of the regulatory light chains of muscle, particularly those of smooth muscle myosin

myosin-linked regulation muscle control system that is based on controlled alteration of the constituents of the myosin filaments

myotatic another name for the muscle stretch reflex

myxedema puffy appearance to the face, hands, and feet as a result of hypothyroidism

Na^+/K^+-ATPase plasma membrane protein that uses the energy of adenosine triphosphate hydrolysis to transport three Na^+ ions out of the cell while transporting two K^+ ions into the cell; it is a *P-type ATPase* because the protein is phosphorylated during the transport cycle

NADPH oxidase enzyme producing superoxide as part of a respiratory burst

NADPH-cytochrome P450 reductase (NADPH-cytochrome C reductase) the enzyme in the cytochrome P450 complex that is responsible for the phase I reactions of the drug biotransformation. Specifically, it reduces the drug complexed with cytochrome P450 so as to make it a more polar compound

narcolepsy neurologic disorder characterized by excessive daytime sleepiness, cataplexy (sudden loss of muscle tone), and vivid hallucinations at sleep onset or on awakening

natriuresis increase in renal sodium excretion

natural killer (NK) cell cytotoxic lymphocyte and major component of the immune system; plays major role in the rejection of tumor and cells infected by virus; kills cells by releasing perforin and granzyme, which cause the target cell to die by apoptosis

nebulin filamentous muscle protein that may play a role in stabilizing thin filament length during muscle development

necrosis unprogrammed cell death, less orderly than apoptosis, results from acute tissue injury; initiates an immune response to remove cell debris (*immunogenic*)

negative cooperativity cooperativity in which the dissociation constant for each successive ligand bound is higher than for the one preceding it, so that the binding affinity is successively decreased

negative feedback type of feedback in which the system responds in an opposite direction to the perturbation

negative selection process by which T cells that recognize self-MHC (major histocompatibility complex) molecules with high affinity are deleted from the repertoire of cells, producing central tolerance and preventing autoimmunity; occurs after positive selection

neonatal respiratory distress syndrome syndrome of premature infants in which developmental insufficiency of surfactant production and structural immaturity of the lungs cause pulmonary edema and atelectasis; also known as *infant respiratory distress syndrome* or *IRDS*

nephrogenic diabetes insipidus condition characterized by lack of response of collecting ducts in the kidney to arginine vasopressin

nephrolithiasis kidney stone disease

nephrin protein necessary for the proper functioning of the renal filtration barrier, which consists of fenestrated endothelial cells, the glomerular basement membrane, and the podocytes of epithelial cells; nephrin is a transmembrane protein involved with the filtration slit diaphragm

nephron basic unit of kidney structure and function, consisting of a renal corpuscle and its attached tubule

nephrotic syndrome kidney disorder characterized by edema, proteinuria, hypoproteinemia, and, usually, elevated blood lipids

Nernst equation chemical equation that allows calculation of the equilibrium potential for any ion

Nernst equilibrium potential value of the electrical potential difference across the plasma membrane that is necessary for a specific ion to be at equilibrium (no net transport)

nerve growth factor protein molecule that enhances nerve cell development and stimulates the growth of axons

neural networks aggregates of interconnected neurons that produce a range of behaviors associated with the central and enteric nervous systems

neural presbycusis progressive loss of the ability to hear high frequency with increasing age as the hair cells "wear out" over time

neuregulin axonally derived epidermal growth factor–like substance that causes glial cells to myelinate the axon

neuroendoimmunology study of the interaction between the nervous, endocrine, and immune systems

neurofilaments cytoskeletal intermediate filaments in neurons that are found in both axons and dendrites and are thought to provide structural rigidity. Neurofilaments are composed of a 70-kDa core protein, are about 10 nm wide, and form ropelike filaments that are several micrometers in length

neurogastroenterology subdiscipline of gastroenterology that encompasses all basic and clinical aspects of nervous system involvement in normal and disordered digestive functions and sensations

neurogenic diabetes insipidus condition characterized by deficient production or release of arginine vasopressin; also called *central*, *hypothalamic*, and *pituitary diabetes insipidus*

neurogenic secretory diarrhea watery diarrhea that reflects stimulated activity of secretomotor neurons in the submucosal plexus

neurohypophysis neural portion of the hypophysis; also known as *posterior pituitary*

neuroleptic class of drugs used to treat schizophrenia and other psychotic disorders; also known as *antipsychotics*

neuromuscular junction (NMJ) excitatory nicotinic synapse between a motor nerve terminal and a skeletal muscle fiber; also known as *motor endplate* and *myoneural junction*

neuropathic pseudoobstruction pathological failure of propulsion in the gastrointestinal tract related to enteric neuropathy and loss of nervous control mechanisms

neuropeptide Y (NPY) neurotransmitter secreted by the hypothalamus associated with a number of physiologic processes in the brain, including the regulation of energy balance, memory and learning, and epilepsy, but its main effect is increased food intake and decreased physical activity

neurophysin part of the arginine vasopressin gene; important in the processing and secretion of arginine vasopressin

neuropil tangle of fine nerve fibers (dendrites and arborizations of axons) and their endings

neurosecretory (neuroendocrine) cells specialized nerve cells that directly convert a neural signal into a hormonal signal

neurotransmitter chemical substance released from nerve cells at synapses that amounts to a signal from one neuron to another or from motor neurons to effectors

neutropenia hematological disorder characterized by an abnormally low number of neutrophils

neutrophil most common type of granulocyte, which destroys pathogens by phagocytosis and respiratory burst

niacin water-soluble vitamin that is absorbed by the small intestine by calcium-dependent, carrier-mediated facilitated transport at low concentrations and by passive diffusion at high concentrations. Niacin plays an important role in coenzymes that participate in many oxidation–reduction reactions involving hydrogen transfer. It has been used to treat hypercholesterolemia for the prevention of coronary artery disease by decreasing plasma total cholesterol and low-density lipoprotein cholesterol and increasing plasma high-density lipoprotein cholesterol

nicotine compound found in tobacco that activates receptors for acetylcholine known as *nicotinic receptors*

nicotinic acetylcholine (ACh) receptor postsynaptic membrane-based protein assembly in the myoneural junction and at synapses in the central and autonomic nervous systems; its associated channel becomes permeable to Na^+ and K^+ ions when ACh is bound; stimulated selectively by nicotine and blocked by a specific class of receptor blocking drugs

nicotinic receptor ionotropic-type receptor for acetylcholine

nigrostriatal system projection of dopaminergic neurons from the midbrain substantia nigra to the striatum (caudate and putamen) in the cerebrum and involved in control of movement; loss of these projections results in parkinsonism

nipple conical structure on the breast that is the external opening to the lactiferous ducts

nitrergic neurons neurons that use nitric oxide as a neurotransmitter

nitric oxide (NO) endothelial autocoid (biological factor that acts like a local hormone) that is a potent relaxer of vascular smooth muscle; it is released from the endothelium in response to calcium activation of the enzyme NO synthase (calcium activation can occur in response to activation of various receptors on the endothelial cell by circulating agonists and by shear forces on the endothelial cells; putative inhibitory neurotransmitter released by enteric motor neurons to gastrointestinal muscles; previously called *endothelial derived relaxing factor (EDRF)*

nitric oxide synthase (NOS) enzyme that converts L-arginine to L-citrulline and nitric oxide

NMDA receptor glutamate receptor named for the synthetic analog that specifically activates it, *N*-methyl-D-aspartate; an ionotropic receptor that is a nonselective cation channel permeable to Na^+, K^+, and Ca^{2+}

nociception experience of a stimulus as harmful in contrast to a pleasant sensation

nociceptors receptors that sense unpleasant or harmful stimuli, external and internal

nodes of Ranvier apposition between two internodes of myelin along an axon; a high concentration of voltage-gated Na^+ channels are localized here

nodose ganglion location of cell bodies of vagal afferent neurons

nonadrenergic noncholinergic (NANC) autonomic neurons that release neither norepinephrine nor acetylcholine

nonalcoholic steatohepatitis (NASH) presents as fat in the liver, along with inflammation and fibrosis; patients with NASH often feel healthy and are unaware they have a liver problem, but NASH can be severe and can lead to cirrhosis and consequent failure of the liver

nondeclarative or procedural memory memory for new skills and procedures; sometimes referred to as *implicit memory*

nonessential amino acids amino acids that are required for normal protein synthesis and that the body synthesizes from other amino acids; comprise nine of the 20 total common amino acids that form the building blocks for proteins

nongated ion channels pores in the membrane that allow ions to flow across the membrane according to electrochemical parameters

nonrapid eye movement (NREM) sleep see *REM sleep*

nonshivering thermogenesis mechanism involving the generation of heat in brown adipose tissue as a result of uncoupled mitochondria located there; also known as *diet-induced thermogenesis*

noradrenergic neuron neuron that uses noradrenaline (norepinephrine) as a neurotransmitter

norepinephrine (NE) catecholamine produced by the adrenal medulla, postganglionic sympathetic axons (in the gut), and central neurons, especially those in the locus ceruleus

normoblast nucleated precursor of mature erythrocytes

normochromic erythrocyte with a normal concentration of hemoglobin

nuclear bag fibers type of intrafusal muscle fiber in muscle spindles

nuclear chain fibers type of intrafusal muscle fiber in muscle spindles

nuclear pore complex opening in membrane surrounding the nucleus, formed by a large assembly of proteins, through which molecules from the cytoplasm travel into the nucleus

nucleus accumbens cholinergic cell group in the base of the forebrain involved in reward pathways; part of the "brain's pleasure center"

nucleus ambiguus column of efferent neurons in the ventrolateral medulla oblongata from which its fibers leave with the vagus and glossopharyngeal nerves and innervate the striated muscle fibers of the pharynx, parts of the esophagus, and the vocal cord muscles of the larynx

nucleus reticularis gigantocellularis group of neurons in the medulla that are the origin of the medullary reticulospinal tract

nucleus reticularis pontis caudalis and oralis group of neurons in the pons that are the origin of the pontine reticulospinal tract

nucleus tractus solitarius located in the medulla oblongata, containing cell bodies of second-order neurons in vagal afferent sensory pathways

null cell lymphocyte without T-cell or B-cell markers

nutcracker esophagus condition in which propulsive esophageal contractions are found to have abnormally high amplitudes during manometric testing; can be a source of angina-like chest pain

nystagmus reflexive sideways movement of the eyeballs that allows visual fixation during rotation of the head. It occurs in response to information from the vestibular apparatus.

obesity excess of caloric intake over energy expenditure

occipital lobe posterior region of the cerebrum; contains the visual cortex

olfaction process of sensing odors

olfactory bulb complex multilayer neural architecture located in the forebrain that serves as a relay station to transmits information from the nose to the brain

oligomenorrhea infrequent or irregular menstruation

oncogenes genes that turn immune cells into tumors cells

oncotic pressure osmotic pressure exerted by colloids in a solution; also called *colloidal osmotic blood pressure*

oocyte immature ovum

oogonia germ cells

oolemma plasma membrane of the egg

ooplasm cytoplasm of an egg

opioids endogenous peptides that bind to opiate receptors

opsonin molecule that coats pathogens to facilitate their phagocytosis

optic chiasma part of the brain where the optic nerves (cranial nerve II) partially cross, located immediately below the hypothalamus

optic disc region of the retina where the optic nerve leaves the eye; the lack of photoreceptors in this region produces a blind spot

optic nerve transmits visual information from the retina to the brain; also called *cranial nerve II*

optic radiations collections of axons from relay neurons in the lateral geniculate nucleus of the thalamus carrying visual information to the visual cortex along the calcarine fissure; there is one such tract on each side of the brain; also known as *geniculostriate pathways*

optic tract continuation of the optic nerve that runs from the optic chiasm (where half of the information from each eye crosses sides, and half stays on the same side) to the lateral geniculate nucleus

optimum length (L_o) length corresponding to the maximum isometric force of muscle contraction

orbit bony socket for the eyeball

orexigenic causing stimulation of the appetite

organ of Corti complex organized structure of the inner ear that transforms sound vibrations into nerve impulses

organic anion transporting polypeptides (OATPs) highly expressed in liver, kidney, and brain, OATPs generally transport anionic and cationic chemicals, steroid, and peptide backbones into cells

organification addition of one or two iodine atoms to tyrosine residues in the thyroglobulin precursor protein

orthodromically conduction of nerve action potential in the conventional direction, efferently for motor and afferently for sensory

orthophosphate (Po_4) chemistry-based term that refers to an organic phosphate in which the phosphate is attached on the ortho position in a benzene ring

osmolality total solute concentration in a solution expressed as osmoles per kilogram of H_2O

osmolarity total solute concentration, expressed as osmoles per liter of solution

osmole Avogadro number (or a mole) of solute particles

osmoreceptor cells neurons in the anterior hypothalamus stimulated by increased extracellular fluid osmolality; they increase the release of arginine vasopressin and thirst

osmoregulation regulation of water potential in an organism by promoting either water retention or secretion, depending on the environmental situation

osmosis passive transport of water across a membrane in response to a gradient of osmotic pressure. Water always moves from the region of low osmotic pressure (low solute concentration) to the region of high osmotic pressure (high solute concentration). Water crosses cell membranes via water channel proteins (aquaporin).

osmotic diuretic solute excreted in the urine that increases urinary excretion of Na^+, K^+, and water

osmotic pressure property of a solution defined as the pressure necessary to stop the net movement of water across a selectively permeable membrane that separates the solution from pure water: A hyperosmotic solution has a greater solute concentration and exerts a greater pressure than another solution, whereas a hyposmotic solution has a lower solute concentration and exerts a lower pressure than another solution, and an isosmotic solution has the same solute concentration and exerts the same pressure as another solution

osteoarthritis arthritis characterized by erosion of articular cartilage, either primary or secondary to trauma or other conditions, which becomes soft, frayed, and thinned with eburnation of subchondral bone and outgrowths of marginal osteophytes; pain and loss of function result; mainly affects weight-bearing joints; more common in older persons

osteoblast cell that makes bone. It does so by producing a matrix that then becomes mineralized; bone mass is maintained by a balance between the activity of osteoblasts that form bone and other cells called *osteoclasts* that break it down.

osteoclast large multinucleate cell that breaks down bone and is responsible for bone resorption

osteocyte branched cell embedded in the matrix of bone tissue

osteogenesis imperfecta hereditary disease characterized by abnormally brittle, easily fractured bones

osteoid uncalcified bone matrix, the product of osteoblasts

osteomalacia softening of the bones as a result of a deficiency of vitamin D; also known as *adult rickets*

osteoporosis disease of bone in which the bone mineral density is reduced, bone microarchitecture is disrupted, and the amount and variety of noncollagenous proteins in bone is changed; osteoporotic bones are more susceptible to fracture

otic ganglion location of pre–postganglionic synapse of parasympathetic neurons from the inferior salivatory nucleus through cranial nerve IX, which are involved in control of the parotid gland

otolithic organs utricle and saccule in the vestibular apparatus responsible for sensing body position and acceleration

otoliths small particles, composed of a combination of a gelatinous matrix and calcium carbonate, in the viscous fluid of the saccule and utricle of the ear; inertia of these small particles causes them to stimulate hair cells when the head moves

outer-ring deiodination enzymatic removal of the iodine atom at position 5′ of the thyronine ring structure

outwardly rectifying potassium channels myocardial membrane channels that are responsible for carrying hyperpolarizing K^+ currents out of myocardial cells, resulting in cell repolarization during phase 3 of the myocardial cell action potential

overshoot part of the action potential in which the inside of the cell becomes positive with respect to the extracellular compartment

oviduct fallopian tube, which receives the ovum immediately after ovulation

ovulation release of an egg from a follicle in the ovary

ovulatory phase part of the menstrual cycle during which an egg is released; lasts 24 hours

oxidative burst cellular generation and release of antimicrobial reactive oxygen species; also known as *respiratory burst*

oxidative phosphorylation generation of adenosine triphosphate by the reactions of the Krebs cycle in the presence of O_2

oxidative stress imbalance between production and degradation of reactive oxygen species

oxidoreductase enzyme that catalyzes the transfer of electrons from one molecule (the reductant, also called the *hydrogen* or *electron donor*) to another (the oxidant, also called the *hydrogen* or *electron acceptor*)

oxygen-carrying capacity (OCC) total maximum amount of oxygen (mL/dL) carried by 100 mL of blood

oxygen content amount of oxygen (mL/dL) bound to hemoglobin in 100 mL of blood

oxygen debt measurably increased rate of oxygen consumption following strenuous activity. The extra oxygen is used in the processes that restore the body to a resting state and adapt it to the exercise just performed (hormone balancing, replenishment of fuel stores, cellular repair, innervation, and anabolism); also known as *excess postexercise oxygen consumption*.

oxygen diffusion gradient partial pressure difference for oxygen across the alveolar–capillary membrane

oxygen saturation ratio of the amount of oxygen bound to hemoglobin to the carrying capacity of oxygen (O_2 content/ O_2 capacity)

oxygen uptake transfer of oxygen from the alveoli to the pulmonary capillary blood

oxyhemoglobin (Hbo_2) oxygen-saturated form of hemoglobin

oxyhemoglobin equilibrium curve curve that reflects the relationship between oxygen saturation and the partial pressure of oxygen

oxyntic glands one of three types of glands found in the fundus and corpus of the stomach that contribute to gastric secretions; they are the most abundant and contain parietal (oxyntic) cells, chief cells, mucous neck cells, and some endocrine cells; also known as *gastric glands*

oxytocin substance that is produced by magnocellular neurons in the supraoptic and paraventricular nuclei of the hypothalamus and that stimulates smooth muscle contraction in the mammary glands and uterus

P wave initial, small positive deflection on the electrocardiogram resulting from depolarization of the atria

P-type ATPase see *ATPase*

Pacinian corpuscle very highly adaptive receptor that is responsible for sensitivity to pressure as well as pain

Paget disease condition in which the existing bone is resorbed, with the simultaneous overgrowth of new, poorly calcified, irregular bone

palmitoyl transferase dimeric complex of the endoplasmic reticulum that catalyzes S-palmitoylation, the addition of palmitate (C16:0), or other long-chain fatty acids to proteins at a cysteine residue

pampiniform plexus venous plexus accompanying the testicular artery up to the superficial inguinal ring and thought thereby to be involved in maintenance of the testicles at some 3°C below body temperature. It forms from veins draining the testis and epididymis at the mediastinum testis and unites within the inguinal canal to form the left and right testicular veins

pancreatic amylase substance that continues in the digestion of the remaining carbohydrates at the point at which salivary amylase stops. Pancreatic secretions must first neutralize chyme for pancreatic amylase to work best, at a neutral pH level. The products of pancreatic amylase digestion of polysaccharides are maltose, maltotriose, and α-limit dextrins

pancreatic lipase substance that hydrolyzes lipids, specifically the triglyceride molecule to a 2-monoglyceride and two fatty acids, in the intestinal lumen

pancreatic polypeptide (PP) 36-amino acid peptide secreted by islet cells of the pancreas in response to a meal; of uncertain physiologic function

Paneth cells found in the intestinal tract, these cells contain zinc and lysozyme as well as large eosinophilic refractile granules within their apical cytoplasm; their exact function is unknown but due to the presence of lysozyme, it is likely that Paneth cells contribute to host defense and protection of intestinal stem cells; when exposed to bacteria or bacterial antigens, Paneth cells secrete a number of lysozymes into the lumen of the crypt, thereby contributing to maintenance of the gastrointestinal barrier

panhypopituitarism generalized or particularly severe hypopituitarism, which in its complete form leads to absence of gonadal function and insufficiency of thyroid and adrenal cortical function. Dwarfism, regression of secondary sex characteristics, loss of libido, weight loss, fatigability, bradycardia, hypotension, pallor, depression, and many other manifestations may occur; also known as *pituitary failure*

parabrachial nucleus location of neurons in the upper pons that coordinate cardiac and respiratory functions

paracrine signaling signaling in which the signal substance is released into the local environment and moves to receptors on target cells by diffusion in the extracellular milieu

paradoxical sleep stage of sleep, also known as REM (rapid eye movement), in which the electroencephalogram exhibits unsynchronized, high-frequency, low-amplitude waves similar to the awake state, yet the subject is as difficult to arouse as when in stage 4 slow-wave (deep) sleep

parafollicular cells cells that are embedded in the wall of the thyroid follicle and that produce and secrete calcitonin

parahippocampal gyrus cortical region encasing the hippocampus

parallel fibers axons from cerebellar granule cells to the dendrites of the Purkinje cells

paralytic ileus obstruction of the bowel resulting from paralysis of the bowel wall, usually as a result of open abdominal surgery, peritonitis, or shock; sometimes called *adynamic ileus*

paramesonephric (müllerian) ducts ducts that fuse at the midline and develop into the oviducts, uterus, cervix, and upper portion of the vagina in females

paraneoplastic syndrome neuropathic degeneration of autonomic neurons resulting from autoimmune attack in patients with small-cell carcinoma of the lung

paraquat compound found in herbicides that is toxic to lung tissue and skin

parasympathetic division division of the autonomic nervous system, with its outflow from the central nervous system in certain cranial nerves and sacral nerves, respectively, and having its ganglia in or near the innervated viscera

parathyroid hormone (PTH) hormone made by the parathyroid gland that helps the body store and use calcium; also known as *parathormone* or *parathyrin*

paraventricular nuclei portion of the hypothalamus in which magnocellular neuron cell bodies are found

paravertebral sympathetic ganglia collections of sympathetic neurons located parallel to the spinal cord; site of most synapses between preganglionic and postganglionic sympathetic neurons

parietal lobe posterolateral region of the cerebrum

parietal (oxyntic) cells one of two primary sources of gastric secretion; secrete hydrochloric acid and intrinsic factor, a peptide that is crucial for the absorption of vitamin B_{12}

parietal–temporal–occipital association cortex interface between the parietal, temporal, and occipital lobes of the cerebrum, which integrates visual, somatosensory, auditory, and cognitive stimuli

Parkinson disease (PD) neurological disease characterized by hand tremor and slowed body movements as a result of loss of neurons in the substantia nigra

parotid gland largest salivary gland, found wrapped around the mandibular ramus, that secretes saliva into the oral cavity to facilitate mastication and swallowing

pars distalis major part of the adenohypophysis of the pituitary gland; also known as *anterior lobe*

pars intermedia thin diffuse region of cells between the anterior and posterior lobes of the pituitary in humans; also known as *intermediate lobe*

pars reticulata (SNr) group of neurons in the substantial nigra producing inhibitory output to eye movement control regions of the midbrain

pars tuberalis upward extension of the anterior lobe that wraps around the infundibular stalk; its cells, mostly gonadotropic, are arranged in cords and clusters; it is supplied by the superior hypophyseal arteries and contains the first capillary bed and the venules of a portal system that carries neurosecretory factors from the hypothalamus to a second capillary bed in the adenohypophysis where they regulate the release of hormones

partial pressure independent pressure exerted by an individual gas within a mixture of gases

partial seizures seizure preceded by an isolated disturbance of a cerebral function, such as twitching of a limb

partition coefficient measure of a molecule's solubility in oil compared with its solubility in water. The value will be high for a lipid-soluble molecule and low for a water-soluble molecule.

parturition birth

parvicellular neurons neurons located in the paraventricular nuclei of the hypothalamus that synthesize corticotropin-releasing hormone

passive force length-dependent component of muscle force that is produced by stretching an inactivated muscle

passive immunity transmission of antibodies from the mother to the fetus via intrauterine transport and to the infant via the mother's colostrum, the thin, yellowish, milky fluid secreted by the mammary glands a few days before or after parturition. The absorption of immunoglobulins (predominantly IgG) plays an important role in the transmission of passive immunity from the mother's milk to the newborn in several animal species

passive transport membrane transport of a solute or ions down the prevailing electrochemical gradient. Solute movement dissipates the gradient and an equilibrium state is eventually reached at which net movement ceases. Passive transport of lipid-soluble solutes can occur without a carrier protein (diffusive membrane transport). Passive transport of ions occurs through ion channel proteins, and polar hydrophilic solutes must use a carrier protein. These are examples of facilitated diffusion

patch clamp sophisticated laboratory technique for studying the behavior of individual ion channels, based on detecting the small current generated when ions flow through an open channel

pathogen agent, primarily an organism, that causes disease

peak expiratory flow (PEF) maximum flow at the outset of forced expiration, which is reduced in proportion to the severity of airway obstruction, as in asthma

pellagra severe niacin deficiency characterized by dermatitis, dementia, and diarrhea

pendrin substance that transports iodide into the follicle to be used for iodination of the thyroglobulin precursor

pepsin endopeptidase in gastric juice that cleaves protein molecules from the inside, resulting in the formation of smaller peptides. Protein digestion starts in the stomach with the action of pepsin, which is secreted as a proenzyme and activated by acid in the stomach

pepsinogen substance secreted by chief cells that is a precursor of the enzyme pepsin and is found in gastric juice

peptic ulcer disease ulcerative lesions of the gastroduodenal area caused by reduced bicarbonate secretion and, in many circumstances, increased acid secretion resulting from an increased mean number of gastric parietal cells that appear to have increased sensitivity to gastrin. The stomach-emptying rate is greatly increased in patients with duodenal ulcers

peptidase any enzyme capable of hydrolyzing one of the peptide links of a peptide

peptide hormones peptides, such as prolactin, secreted in the bloodstream that serve endocrine functions

perforin protein of natural killer and cytotoxic T cells that forms a pore in target cells

perfusion limited characterized by the limitation of the uptake of an alveolar gas, primarily by pulmonary capillary blood flow

perfusion pressure pressure force driving blood through the circulation of an organ; usually synonymous with the blood pressure in the artery supplying an organ or tissue

periaqueductal gray matter region of the midbrain where integration of afferent and nerve impulses from higher cerebral centers occurs

pericarditis inflammation of the pericardium. It often restricts the ability of the ventricle to fill during diastole

perikaryon cell body of a neuron containing the nucleus; also called *soma*

perilymph fluid that fills the scala vestibuli and scala tympani and is high in sodium and low in potassium

peripheral chemoreceptors neurosensitive endings that respond to changes in blood chemistry and that are located in the carotid artery and the aorta

peripheral protein protein that is attached to but does not penetrate the phospholipid bilayer region of a cell membrane, for example, spectrin in the red blood cell

perisinusoidal space (space of Disse) space that is continuous with the pericellular space and separated from the sinusoid by a layer of sinusoidal endothelial cells

peristalsis coordinated contraction of inner circular muscle layer and longitudinal layer of the gastrointestinal tract that propel a bolus of food through the tract

perivitelline space region between the zona pellucida and the vitelline membrane of an oocyte

permeability coefficient number expressing the ease with which a solute can cross a plasma membrane. It accounts for factors such as the thickness of the membrane and the lipid solubility of the solute

permeability surface area coefficient coefficient that characterizes how much material per unit of time can pass across the capillaries by diffusion for a given concentration gradient across the capillary membrane. It is a function of the capillary surface area over which diffusion occurs as well as the permeability of the capillary membrane to the diffusing substance in question

pernicious anemia chronic, progressive anemia found in older adults as a result of failure of vitamin B_{12} absorption, usually resulting from the lack of secretion of intrinsic factor. Pernicious anemia is characterized by numbness and tingling, weakness, diarrhea, loss of weight, and fever

peroxidases enzymes that neutralize peroxynitrites

peroxynitrite free radical formed in the presence of a superoxide ion and nitrous oxide

pH measure of the acidity or alkalinity of a solution. It is equal to the negative logarithm (to the base 10) of the hydrogen ion concentration (the latter is measured in mol/L)

phagocyte cell capable of engulfing particulate material

phagocytosis engulfment of large extracellular particles and microorganisms by expansion of the plasma membrane around the particle to eventually surround it. A typical example is the destruction of invading bacteria by macrophage cells

phagosome phagocytic vacuole for the killing of enclosed pathogens

pharmacomechanical coupling mechanism in gastrointestinal smooth muscles in which the binding of a ligand to its receptor on the muscle membrane leads to the opening of calcium channels and the elevation of cytosolic calcium without any change in the membrane electrical potential; the ligands may be chemical substances released as signals from nerves (neurocrine), from nonneural cells in close proximity to the muscle (paracrine), or from endocrine cells as hormones delivered to the muscle by the blood

phase I reactions reactions in which a parent xenobiotic compound is biotransformed into a more polar compound by the introduction of one or more polar groups, usually hydroxyl (OH) or carboxyl (COOH). Usually this is accomplished by the oxidation of the parent compound

phase II reactions reactions in which phase I products conjugate to various possible products (i.e., glucuronic acid, catalyzed by glucuronyltransferases), yielding a more hydrophilic end compound

phasic contraction contraction not long maintained (transitory), in contrast to a tonic contraction

pheromone chemical or set of chemicals produced by a living organism that transmits a message to other members of the same species and affects behavior or physiology

phlebotomist technician trained to draw blood

phorbol ester parent alcohol of the cocarcinogens, which are 12,13(9,9a) diesters of phorbol found in croton oil; the hydrocarbon skeleton is a cyclopropabenzazulene; phorbol esters mimic 1,2-diacylglycerol as activators of protein kinase C

phosphatidylinositol 4,5-bisphosphate (PIP_2) membrane lipid found in all eukaryotic cells, which regulates many important cellular processes, including ion channel activity

phosphaturia excess of phosphates in the urine

phosphodiesterase (PDE) enzyme that cleaves phosphodiester bonds to yield a phosphomonoester and a free hydroxyl group; for example, cyclic adenosine monophosphate (cAMP) is inactivated by phosphodiesterase action to 5′ AMP; this cleavage is mediated by a cyclic nucleotide phosphodiesterase enzyme

phosphoenolpyruvate carboxykinase (PEPCK) nzyme in the gluconeogenic pathway catalyzing oxaloacetate conversion to phosphoenolpyruvate

phosphoenolpyruvate carboxylase enzyme in the family of carboxy-lyases that catalyzes the addition of bicarbonate to phosphoenolpyruvate to form the four-carbon compound oxaloacetate

phosphofructokinase enzyme that catalyzes the conversion of fructose 6-phosphate to fructose 1,6-diphosphate

phospholamban regulatory protein that inhibits sarcoplasmic reticulum Ca^{2+} pumps. Its inhibitory effects are removed when it is phosphorylated

phospholipase A_2 major pancreatic enzyme for digesting phospholipids, thereby forming lysophospholipids and fatty acids

phospholipase C (PLC) lipase that hydrolyzes phosphatidylinositol 4,5-bisphosphate

phospholipid subclass of lipid molecules based on glycerol 3-phosphate to which are attached two long-chain fatty acids (hydrophobic), for example, palmitic acid, and a specific polar or charged head group (hydrophilic), for example, choline or inositol

phosphorylation transfer of the terminal phosphate of adenosine triphosphate to a protein, under the control of a kinase-type enzyme

photon quantum of the electromagnetic interaction and the basic unit of light and all other forms of electromagnetic radiation

photopic (daytime) vision color vision that is served by the cone cells of the retina

photoreceptors sensory receptors that respond to visual stimuli

phylloquinones vitamin K that is derived from green vegetables

physiological dead space volume total volume of wasted air that does not participate in gas exchange and is the sum of anatomical dead space and alveolar dead space volume

physiological ileus normal absence of motility in a region of the gastrointestinal tract; requires an intact enteric nervous system in the intestine

physiological shunt total amount of venous admixture from anatomical shunts and regions of low ventilation/perfusion ratios

physostigmine (eserine) blocker of the neuromuscular junction that inhibits the action of acetylcholinesterase

piloerection mechanism whereby furred or hairy animals can increase the thickness of their coat and insulating properties by making the hairs stand on end. This response makes a negligible contribution to heat conservation in humans, but manifests itself as gooseflesh

pineal gland gland located superficial to the dorsal midbrain and posterior dorsal thalamus that produces melatonin in response to signals from the suprachiasmatic nucleus of the hypothalamus relayed through the sympathetic nervous system

pinocytosis process by which a cell ingests extracellular fluid and its contents by forming narrow channels through its membrane that pinch off into vesicles and fuse with lysosomes that hydrolyze contents

pituitary dwarfism decrease in the rate of body growth as a result of growth hormone deficiency

pituitary gland gland lying in the sphenoid bone and connected to the hypothalamus; also known as *hypophysis*

place theory theory of hearing that states that our perception of sound depends on where each component frequency produces vibrations along the basilar membrane, such that the pitch of a musical tone is determined by the places where the membrane vibrates, based on frequencies corresponding to the tonotopic organization of the primary auditory neurons

placenta structure composed of maternal and fetal tissues attached to the inner surface of the uterus

placing reaction cortical reflex producing limb movement, provoked in a dependent limb touching a horizontal surface

plasma liquid portion of the blood; an extracellular fluid

plasma cell B cell–derived cell that secretes antibody

plasma membrane outer surface of the cell, which separates the contents of the cell from the extracellular fluid

plasma membrane calcium ATPase (PMCA) transport protein in the plasma membrane of cells that serves to remove calcium from the cells

plasmapheresis therapeutic plasma exchange

plasmin proteolytic enzyme that acts on the follicular wall during rupture

plasminogen inactive proenzyme that is converted to plasmin by tissue plasminogen activator, which is released by activated endothelial cells

plasminogen activator see *tissue plasminogen activator*

plateau in cardiovascular physiology, the extended depolarized plateau in phase 2 of the action potential in myocardial cells. It is the basis of the long refractory period seen in ventricular and atria muscle cells and is the reason tetanic contractions cannot be produced in those cells.

platelet anuclear, irregular disk-like fragment of megakaryocytes that is important for hemostasis

platelet count (Plt) used as a starting point in the diagnosis of hemostasis

platelet-derived growth factor (PDGF) cytokine involved in wound healing; stimulates the proliferation of vascular smooth muscle and endothelial cells

plegia complete paralysis of a muscle

pleiotropic producing more than one effect

plethysmograph airtight box designed to measure a patient's residual lung volume

pleural pressure (P_{pl}) pressure between the chest wall and the lungs (i.e., the pleural space)

pleural space fluid-filled space between the chest wall and the lungs

pneumonia disease of the lungs characterized by inflammation and consolidation of lung tissue and accompanied by fever, chills, cough, and difficulty in breathing; it is caused chiefly by infection and is categorized as *bronchopneumonia*, *lobar pneumonia*, or *primary atypical pneumonia*

pneumothorax collapsed lung

podocytes epithelial cells with extensions that terminate in foot processes that rest on the outer layer of the basement membrane of the glomerular membrane of the kidney

poikilocyte irregularly shaped erythrocyte

poikilotherms poikilothermic animal; denoting the so-called cold-blooded animals, such as the reptiles and amphibians

Poiseuille law law that quantifies the relationship between pressure, flow, and resistance in the cardiovascular system. It states that blood flow is proportional to the pressure difference across any two points along the vascular system and inversely proportional to the resistance to flow. It can also be used to show that pressure anywhere in the cardiovascular system is the direct product of blood flow and vascular resistance. The work of Poiseuille showed that vascular resistance is proportional to vessel length and blood viscosity but that it is inversely proportional to the fourth power of the internal vessel radius.

polyclonal antibody antibodies obtained from different B-cell resources; they are a combination of immunoglobulin molecules secreted against a specific antigen, each identifying a different epitope

polycystic kidney disease (PKD) most common life-threatening inherited disorder in which numerous cysts develop in both kidneys, producing massive enlargement of the kidneys and ultimately complete renal failure

polycystic ovarian syndrome heterogeneous group of disorders characterized by amenorrhea or anovulatory bleeding, an elevated luteinizing hormone/follicle-stimulating hormone ratio, high androgen levels, hirsutism, and obesity

polycythemia proportion of blood volume that is occupied by red blood cells increases abnormally due either to an increase in the mass of red blood cells (*absolute polycythemia*) or to a decrease in the volume of plasma (*relative polycythemia*); blood volume proportions can be measured as hematocrit level

polymorphonuclear leukocyte granulocyte, with a segmented or lobated nucleus

polyspermy fertilization of an egg by more than one sperm

polyuria condition characterized by excessive excretion of dilute urine, for example, diabetes insipidus

pontine respiratory group group of cells in the medulla responsible for integrating signals and acting as an "off" switch for inspiration and expiration

pontine reticulospinal tract descending brainstem pathway carrying excitatory impulses to spinal interneurons from the pons

pool group of lower motor neurons that control muscles of similar function

pores in the cardiovascular system, the gaps between endothelial cells, which allow the passage of water and solutes between the tissues and capillary blood

portal hypertension high blood pressure in the portal vein and its tributaries; it is often defined as a portal pressure gradient (the difference in pressure between the portal vein and the hepatic veins) of 10 mm Hg or greater

portal vein primary large vein that collects blood exiting the splanchnic organs, such as the stomach, pancreas, small intestine, and portions of the large intestine. It supplies blood into the hepatic circulation, where it is mixed with blood from the hepatic artery

positive feedback feedback that enhances the input and responds in the same direction as the perturbation and that involves the return of some of the output of the system as input, which helps control the process

positive selection process by which cells are recruited with a T-cell receptor able to bind major histocompatibility (MHC) class I and II molecules with at least a weak affinity, thereby eliminating (in "death by neglect") those T cells that would be nonfunctional due to an inability to bind MHC

postprandial after meals

postcapillary venules first veins in which blood enters from the capillaries

postganglionic axon axon of the postganglionic autonomic neuron

postganglionic sympathetic neurons neurons leaving prevertebral ganglia to the gut

postjunctional folds invaginations in the endplate membrane that increase its surface area

postmenopausal osteoporosis bone loss resulting from the deficiency of estrogen associated with menopause; also known as *type 1 osteoporosis*

postpartum thyroiditis usually occurring within 3 to 12 months after delivery, this disease is characterized by a transient destruction-induced thyrotoxicosis (hyperthyroidism), often followed by a period of hypothyroidism lasting several months; many patients eventually return to the euthyroid state

postprandial after meals

postrotatory nystagmus eyeball movements that follow an episode of body or head rotation, resulting from the continued motion of the endolymph in the semicircular canals

poststreptococcal glomerulonephritis nephritic condition that may follow a sore throat caused by certain strains of streptococci; endothelial cell damage, accumulation of leukocytes, and the release of vasoconstrictor substances reduce the glomerular surface area and fluid permeability and lower glomerular blood flow, causing a fall in the glomerular filtration rate

potentiation interaction between two or more drugs or agents, resulting in a pharmacologic response greater than the sum of individual responses to each drug or agent

power output rate at which muscle does work on the environment

power propulsion intestinal motility pattern associated with defense and pathologic conditions; rapid propulsion of luminal contents over extended lengths of intestine

power stroke conformational change in the myosin molecule (cross-bridge) that provides the driving force in muscle contraction

PR interval interval of time on the electrocardiogram between the end of the P wave and the start of the QRS complex. It is an estimate of the time it takes for action potentials to proceed from the atria into the ventricles and is a primary reflection of the delay of conduction through the atrioventricular node

prechondrocyte precursor to chondrocyte in the growth plates of bone

precocious puberty physical signs of sexual maturation before the age of 7 years in girls and 9 years in boys

precordial leads set of six unipolar electrocardiogram leads that are oriented in the horizontal plane of the body at the level of the heart. They are designated V_1 to V_6 and are placed in a semicircle around the left side of the chest, from just right of the sternum to the left midaxial point

prediabetes condition that occurs when a person's blood glucose levels are higher than normal but not high enough for a diagnosis of type 2 diabetes

prefrontal association cortex region of the cerebrum responsible for executive function (making decisions) and, by virtue of its connections with the limbic system, coordinating emotionally motivated behaviors

prefrontal cortex most anterior region of the cerebrum

preganglionic axon axon of the preganglionic autonomic neuron

preganglionic sympathetic neurons neurons in the intermediolateral cell column of the spinal cord that project axons to make synapse with one of the three prevertebral sympathetic ganglia (celiac, superior mesenteric, and inferior mesenteric ganglia)

pregnenolone precursor for steroid hormone synthesis derived from cholesterol

pregranulosa cells cells that surround the oocyte in a primordial follicle

preload passive stretch placed on a muscle prior to an active contraction; initial passive length to which muscle is stretched prior to the start of a contraction

premature atrial contraction (PAC) premature electrical activation of any portion of the atria outside the sinoatrial node, or atrial ectopic foci

premature ovarian failure condition characterized by amenorrhea, low estrogen levels, and high gonadotropin (luteinizing hormone and follicle-stimulating hormone) levels before age 40 years

premature ventricular atrial contraction (PVC) premature electrical activation of any portion of the ventricles or ventricular ectopic foci

premotor area region of the cerebrum just anterior to the primary motor cortex; serves as the association cortex for the motor system responsible for coordinating movements around a joint

preprohormone precursor form of a protein hormone extended at the amino termini by a hydrophobic amino acid sequence called *leader* or *signal peptide* and in addition containing internal cleavage sites that on enzymatic action yield different bioactive peptides

presbyopia condition characterized by reduced converging power (accommodation) of the eye, caused by increasing stiffness of the crystalline lens resulting from age

pressure technically, the force on a specific surface area and expressed as force per unit area (i.e., pounds per square inch); in physiological systems, it is measured as the force required to raise a column of water or mercury a given distance vertically against gravity and is expressed as mm Hg or cm H_2O

pressure diuresis response of the kidney to increase salt and water excretion whenever arterial pressure is elevated. It has the capacity to continue to increase excretion until blood pressure is returned to a level set by the body for mean arterial pressure. It is believed to be the mechanism whereby the kidney sets the long-term mean arterial pressure set point in the body, around which all cardiovascular reflexes operate

pressure–volume curve curve generated by measuring the amount of pleural pressure required to inflate the lungs to a given volume. The slope of the pressure–volume curve provides a measure of lung compliance

presynaptic facilitation effects of chemical mediators or drugs to increase the release of neurotransmitter at a synapse

presynaptic inhibition receptors on axons suppress the release of neurotransmitters at neural synapses and neuroeffector junctions

presynaptic inhibitory receptors receptors on axons that suppress the release of neurotransmitters at neural synapses and neuroeffector junctions

prevertebral sympathetic ganglia sympathetic ganglia (celiac, aorticorenal, and superior and inferior mesenteric) lying in front of the vertebral column, as distinguished from the ganglia of the sympathetic trunk (paravertebral ganglia); these ganglia occur mostly around the origin of the major branches of the

abdominal aorta, and all are in the abdominopelvic cavity; the neurons comprising the ganglia send postsynaptic sympathetic fibers to the gut

primary active transport mechanism see *active transport*

primary axons (type Ia) largest myelinated afferent axons from the muscle spindle

primary peristaltic wave peristaltic propulsion in the esophagus that is initiated by a swallow

primary follicular stage ovarian follicle consisting of one layer of cuboidal granulosa cells and a basement membrane surrounding the oocyte

primary lymphoid organs see *lymphoid organs*

primary motor cortex (M1) region of the cerebral cortex that fits all four parameters of a cortical motor area and is located just anterior to the central sulcus

primary oocyte oocyte in first meiotic arrest

primary peristaltic wave peristaltic propulsion in the esophagus initiated by swallowing

primary somatosensory cortex (S1) region of the cerebral cortex that receives tactile input and is located just posterior to the central sulcus

primary spermatocyte spermatocyte that divides into two secondary spermatocytes

prime mover contracting muscle that contributes most to a movement

primordial follicle ovarian follicle consisting of a primary oocyte surrounded by a layer of pregranulosa cells

progesterone major natural progestogen, promoting the secretory function of the uterus

progesterone block maintenance of uterine quiescence throughout gestation by progesterone to prevent premature delivery

progestin or progestogen substance that promotes secretory function of the uterus

programmed cell death death of a cell mediated by signaling pathways invoked in the cell in response to unhealthy growth conditions (e.g., loss of growth factors, infection by a virus); a deliberate cell suicide; this death is altruistic because the whole organism benefits from the death of a cell, thus it does not initiate an immune response (*nonimmunogenic*); *apoptosis* is one type of programmed cell death; contrast with *necrosis*

progressive shock stage of shock in which the function of organs needed to maintain cardiac output and blood pressure (heart and brain) is compromised but in which the person can survive with prompt and aggressive medical intervention

prohormone precursor form of a protein hormone extended at the amino termini by a hydrophobic amino acid sequence called *leader* or *signal peptide*

prohormone convertase 1 (PC1) enzyme encoded by the *PCSK1* gene that differentially cleaves proopiomelanocortin to adrenocorticotropic hormone and β-lipotropin

prohormone convertase 2 (PC2) enzyme responsible for the first step in the maturation of many neuroendocrine peptides from their precursors, such as the conversion of proinsulin to insulin intermediates

prokinetic drug agent that acts on enteric nerve endings to enhance propulsion of contents through the gut and that may act by direct muscle stimulation, release of motor neurotransmitters, or blockade of inhibitory neurotransmitters

prolactin (PRL) hormone product of lactotrophs located in the anterior pituitary

proliferative phase phase of the endometrial cycle in which stromal and epithelial layers of the uterine endometrium undergo hyperplasia

pronucleus haploid nucleus with 23 chromosomes. One female pronucleus and one male pronucleus present in an egg 2 to 3 hours after sperm penetration and fuse to form the nucleus of a fertilized egg

proopiomelanocortin (POMC) prohormone from which adrenocorticotropic hormone, β-lipotropin, and α-melanocyte-stimulating hormone are derived

proprioceptors sensory receptors that provide information as to body and limb position, muscle force, and similar properties from within the body

propriospinal cells interneurons that convey action potentials between adjacent spinal segments

propulsion controlled movement of ingested foods, liquids, gastrointestinal secretions, and sloughed cells from the mucosa through the digestive tract

propulsive segment intestinal segment identified by contraction of the circular muscle layer behind a receiving segment in which the circular muscle is inhibited and the longitudinal muscle layer contracted to produce a segment with expanded lumen

prostaglandin derivative of arachidonic acid that usually acts as a local hormone to affect many functions, such as blood flow and inflammation

prostaglandin E_2 substance released by blood vessel walls in response to infection or inflammation and acting on the brain to induce fever. The enzyme mPGES-1 is involved in the production of prostaglandin E_2 and is an important "switch" for activating the fever response

protein inclusion bodies protein aggregates that typically represent sites of viral multiplication in bacteria; can also signal genetic diseases such as Parkinson disease

protein kinase A (PKA) see *cAMP-dependent protein kinase*

protein kinase C (PKC) phospholipid-dependent and calcium-dependent kinase that, when activated, phosphorylates serine or threonine residues in target proteins to regulate their function

protein kinase G (PKG) see *cGMP-dependent protein kinase*

protein S deficiency most common inherited thrombotic disorder

proteinuria condition characterized by the presence of proteins in the urine

proteoglycans mortarlike substances made from protein and sugar that are the building blocks of cartilage

prothrombin time (PT) test to monitor the extrinsic coagulation pathway

prothrombinase (factor X) clotting factor that initiates the common coagulation pathway

proto-oncogenes normal gene that, when altered by mutation, becomes an oncogene

proton pump see *H^+-ATPases*

pseudohermaphroditism disorder in which the patient has external genitalia of one sex and internal sex organs of the other sex; this is not true hermaphroditism because there is no ambiguity in the sex of the external genitalia and hence no question about gender at birth

pseudohyperkalemia spurious elevation of the serum concentration of potassium occurring when potassium is released *in vitro* from cells in a blood sample collected for a potassium measurement

pseudohyponatremia artifactual decrease in plasma Na^+ resulting from greatly elevated levels of lipids or proteins in plasma

pseudomaturation process in which the oocyte enters meiosis during atresia of the follicle

pseudoobstruction pathological failure of propulsion in the gastrointestinal tract in the absence of mechanical obstruction

pseudoprecocious puberty early development of secondary sex characteristics in the absence of gametogenesis. Results from exposure of immature boys to androgens or immature girls to estrogen

pterygopalatine ganglion location of pre–postganglionic synapse of parasympathetic neurons from the superior salivatory nucleus through cranial nerve VII, which are involved in control of the lacrimal and nasal glands

ptosis drooping of the eyelid as a result of loss of sympathetic activation of the Müller muscle of the eyelid

pubarche development of pubic hair during pubertal development in boys and girls

puberty development of physical characteristics of sexual maturation

puborectalis muscle sling muscle that anchors to the symphysis pubis anteriorly and loops around the rectum to form the anorectal angle; this muscle is important to continence for stool

pulmonary artery single artery that transmits all blood pumped from the right ventricle into the lungs. This blood is deoxygenated venous blood from all the peripheral organs.

pulmonary circulation lung circulation that perfuses the respiratory zone and participates in alveolar gas exchange

pulmonary edema abnormal collection of fluid in the lung, which can flood the alveoli and impair gas exchange

pulmonary embolism movement of a blood clot or other plug from the systemic veins through the right heart and into the pulmonary circulation, where it lodges in one or more branches of the pulmonary artery; most pulmonary emboli originate from thrombosis in the leg veins but can also originate from the upper extremities; other major sources of pulmonary emboli include air bubbles introduced during intravenous injections, hemodialysis, placement of central catheters, fat emboli (a result of multiple long-bone fractures), tumor cells, amniotic fluid (secondary to strong uterine contractions), parasites, and various foreign materials in intravenous drug abusers

pulmonary embolus blood clot that has lodged in the microcirculation or farther upstream in the pulmonary arterial circulation; large pulmonary emboli can impede filling of the left heart to the extent that death results

pulmonary hypertension potentially severe disease characterized by an increase in blood pressure in the lung vasculature including the pulmonary artery, pulmonary vein, or pulmonary capillaries; symptoms include shortness of breath, dizziness, fainting, and others, all of which are exacerbated by exertion; markedly decreased exercise tolerance and heart failure may result

pulmonary stretch receptors slowly adapting receptors located in the smooth muscle of the conducting airways

pulmonary wedge pressure pressure measured by wedging a Swan–Ganz catheter-tipped balloon up against small pulmonary vessels

pulmonic valve thin, flexible, trileaflet structure that separates the right ventricle from the pulmonary artery. It opens into the pulmonary artery during systole but closes during diastole

pulse pressure difference in systolic and diastolic arterial pressure. It is the change in pressure associated with each cardiac cycle and is a function of arterial compliance and stroke volume

purinergic neuron enteric neurons that use a purine nucleotide (e.g., adenosine triphosphate or adenosine) as a neurotransmitter

Purkinje cells largest neurons of the cerebellum and the source of efferent cerebellar axons

Purkinje fibers specialized set of modified myocardial fibers that line the endocardium and provide rapid spread of electrical excitation to the ventricles

papillary light reflex in controlling the diameter of the pupil, this reflex regulates the intensity of light entering the eye

putamen one of the basal ganglia, receiving input from the cerebral cortex

pyloric glands one of three main types of glands in the stomach's mucosal lining. These glands contain mucous cells that secrete mucus and bicarbonate ions, which protect the stomach from the acid in the stomach lumen. The pyloric glands are located in a larger area adjacent to the duodenum and contain cells similar to mucous neck cells. The pyloric glands consist of many gastrin-producing cells, called *G cells*

pyloric sphincter sphincter at the boundary of the gastric pylorus and duodenum

pylorus lower section of four stomach sections that facilitates gastric emptying into the small intestine

pyramidal tract major motor control pathway for axons from the cerebral cortex to the spinal cord

pyrogens exogenous and endogenous substances that induce fever, such as the bacterial substance LPS and interleukin-1, which both increase the thermoregulatory set point in the hypothalamus. The endogenous pyrogens may also come directly from tissue necrosis

pyrosis burning sensation in the chest usually associated with gastroesophageal reflux; also called *heartburn* or *acid indigestion*

pyruvate carboxylase enzyme of the ligase class that catalyzes the irreversible carboxylation of pyruvate to form oxaloacetate

pyruvate dehydrogenase allosteric enzyme that transforms pyruvate into acetyl-CoA, which is then used in the citric acid cycle

pyruvate kinase enzyme that catalyzes the transfer of a phosphoryl group from phosphoenolpyruvate to adenosine diphosphate, yielding a pyruvate molecule

pyruvic acid end product of the glycolytic pathway

Q wave first downward (negative) deflection of the electrocardiogram following the isoelectric interval after completion of the P wave

Q_{10} phenomenon in which the rate of a biological process increases when the temperature is increased by 10°C; this value is obtained using the following formula: Q_{10} = rate at (T + 10°C)/rate at T, where T = starting temperature; for biological systems, the Q_{10} value is generally between 1.5 and 2.5

QRS complex complex, spiked, largely positive deflection on the electrocardiogram resulting from depolarization of the ventricles

QRS interval time interval associated with ventricular depolarization and represented by the time interval of the QRS complex on the electrocardiogram

QT interval interval of time associated with depolarization and repolarization of the ventricles. It measures the interval of time between the initiation of the Q wave and the end of the T wave on the electrocardiogram and is sometimes referred to as *electrical systole*

quantum number of transmitter molecules released by any one exocytosed synaptic vesicle

quantum content total number of quanta (vesicles filled with a certain number of transmitter molecules) released when the synapse is activated

R wave in the normal heart, it is the highest amplitude, sharp, positive deflection of the QRS complex of the electrocardiogram (ECG); it is associated with a positive deflection in all frontal ECG leads except aV_R and is caused by ventricular depolarization during electrical activation of the heart

radioimmunoassay (RIA) scientific method used to test antigens (e.g., hormone levels in the blood) without the need to use a bioassay. It involves mixing known quantities of radioactive antigen (frequently labeled with γ-radioactive isotopes of iodine attached to tyrosine) with antibody to that antigen, then adding unlabeled or "cold" antigen and measuring the amount of labeled antigen displaced

radioreceptor assay important modification of the radioimmunoassay that uses specific hormone receptors rather than antibodies as the hormone-binding reagent. In theory, this method measures biologically active hormone, because receptor binding rather than antibody recognition is assessed

rapid adaptation process in phasic sensory receptors in which the output to the central nervous system decreases quickly while the applied stimulus is maintained in intensity

rapid adapting receptors sensory terminals of myelinated afferent fibers found in large conducting airways; also referred to as *irritant receptors*

rapid ejection phase portion of the cardiac cycle starting immediately after the opening of the aortic valve, in which blood is ejected into the aorta from left ventricular contraction and continuing up to the peak ejection flow rate and peak aortic systolic pressure; it comprises the first third of the ejection phase of ventricular contraction, during which 70% of the stroke volume is ejected

rapid eye movement (REM) sleep see *REM sleep*

rapid ventricular filling first third of filling of the ventricle during diastole

rarefaction reduction of a medium's density; the opposite of *compression*

ras (RAt sarcoma) protein subfamily of small GTPases that are generally responsible for signal transduction leading

Rathke pouch evagination of the oral ectoderm from which the adenohypophysis is formed

reactive hyperemia response of a tissue to a period of ischemia or no blood flow in which blood flow exceeds the value seen before the ischemia or flow stoppage once the ischemia or stoppage is removed

reactive oxygen species (ROS) collective term to include free radicals (hydroxyl radical, superoxide ion, and hydrogen peroxide)

receptive relaxation reflex triggered by stimulation of mechanoreceptors in the pharynx followed by transmission over afferents to the dorsal vagal complex and activation of efferent vagal fibers to inhibitory musculomotor neurons in the gastric enteric nervous system

receptor protein molecule on the cell surface or located within the cell that receives and responds to a neurotransmitter, hormone, or other substance

receptors signaling systems that reside either in the plasma membrane or within cells and are activated by a variety of extracellular signals or first messengers, including peptides, protein hormones and growth factors, steroids, ions, metabolic products, gases, and various chemical or physical agents (e.g., light); divided into two categories: *cell-surface receptors* (G protein–coupled receptors, ion channel-linked receptors, and enzyme-linked receptors) and *intracellular receptors* (steroid and thyroid hormone receptors)

receptor (membrane) membrane surface proteins with unique conformations for specific binding of chemical signal substances and triggering of changes in the cell's behavior

receptor adaptation general loss in the efficacy of a receptor-mediated response after sustained stimulation of the receptor

receptor potential characteristic electrical record of a sensory receptor; a generator potential specific to a given organ, such as the electroretinogram or the cochlear microphonic

receptor-mediated endocytosis see *exocytosis*

reciprocal inhibition synaptic pattern that causes inhibition of antagonist muscles during contraction of an agonist

rectoanal reflex relaxation of the internal anal sphincter in response to mass movement of feces distension of the rectosigmoid region of the large intestine

red blood cells most common cell in the blood and the principal means of delivering oxygen to cells via the circulatory system; also referred to as *erythrocytes*

red cell distribution width (RDW) measures the degree of the average size dispersion of erythrocytes and thereby indicates the degree of anisocytosis

red marrow part of bone marrow that forms blood cells

red nucleus group of neurons in the midbrain that receive input from the cerebral cortex and cerebellum and produce output to the spinal cord via the rubrospinal tract

reduced ejection phase ejection phase of the cardiac cycle in which the rate of ventricular outflow starts to diminish. It is the last phase of systole

reduced ventricular filling filling phase of the cardiac cycle in which the rate of ventricular filling begins to diminish. It is the last phase of diastole

5α-reductase steroidogenic enzyme that converts testosterone to dihydrotestosterone

reentry tachycardia type of ventricular tachycardia caused by continual recycling of conducted action potentials through an abnormal pathway in the heart

referred pain pain from deep structures perceived as arising from a surface area remote from its actual origin; the area where the pain is appreciated is innervated by the same spinal segment(s) as the deep structure; also known as *reflective pain*

reflection coefficient measure of the relative permeability of a particular membrane to a particular solute; calculated as the ratio of observed osmotic pressure to that calculated from van't Hoff law; also equal to 1 minus the ratio of the effective pore areas available to solute and to solvent

reflex neuronal event occurring beyond volition; in neurophysiology, it is a relatively simple behavioral response of an effector produced by influx of sensory afferent impulses to a neural center and its reflection as efferent impulses back to the periphery to the effector (e.g., muscle); the neural center may consist of interneurons; the simplest reflex circuit consists only of sensory and motor neurons

reflex sympathetic dystrophy (RSD) chronic, painful, and progressive neurological condition that affects the skin, muscles, joints, and bones; usually develops in an injured limb, such as a broken leg, or following surgery; however, many cases of RSD involve only a minor injury, such as a sprain, and in some cases, no precipitating event can be identified; also called *complex regional pain syndrome*

regional compliance pressure–volume relationship at a specific region of the lung (i.e., the apex and base); *compliance* is the ability of the lungs to stretch during a change in volume relative to an applied change in pressure

regional hypoventilation condition in which a region of the lung is being underventilated with respect to an adjacent region of the lung

regional hypoxia hypoxic condition in a regional part of the lung, as contrasted with generalized hypoxia; see *generalized hypoxia*

regulated exocytosis form of exocytosis in which vesicles are filled, stored in the cytoplasm, and their contents released only when a specific extracellular stimulus arrives at the cell membrane

regulatory light chains small protein constituents of the myosin molecule that regulate its interaction with actin

regulatory volume decrease mechanism by which solute and water exit the cell to correct a previous increase in volume resulting from the osmotic entry of water

regulatory volume increase (RVI) mechanism by which solute and water enter the cell to correct a previous decrease in volume resulting from the osmotic exit of water

relative humidity quantity of water vapor that exists in a gaseous mixture of air and water

relative pressure pressure relative to atmospheric pressure

relative refractory period window of time following one action potential during which a greater-than-usual stimulus is required to elicit an action potential

relaxin a polypeptide hormone produced by the corpus luteum and the decidua that assists parturition by softening the cervix and increasing oxytocin receptors

releasing factors see *releasing hormones*

releasing hormones hypothalamic hormones that reach the anterior pituitary lobe by a portal system of capillaries and regulate the secretion of most hormones of the endocrine system; also known as *releasing factors*

REM sleep rapid eye movement sleep; a phase of sleep in which the eye movement becomes rapid and the electroencephalograph becomes desynchronized, with a loss of skeletal muscle tone

renal blood flow (RBF) volume of blood delivered to the kidneys per unit time

renal plasma flow (RPF) volume of blood plasma per unit time flowing through the kidneys

renin proteolytic enzyme produced by granular cells in the kidney that catalyzes the formation of angiotensin I from angiotensinogen

renin–angiotensin–aldosterone system salt-conserving system

renin–angiotensin–converting enzyme enzyme in lung endothelium that converts angiotensin I to angiotensin II

repolarization change in membrane potential that returns the membrane potential to a negative value after the depolarization phase of an action potential has just previously changed the membrane potential to a positive value

residual volume (RV) volume remaining in the ventricle at the end of systole

resistance capacity of a membrane that inhibits flow of ions across the membrane

resistance to thyroid hormone (RTH) rare syndrome in which thyroid hormone levels are elevated but thyroid-stimulating hormone level is not suppressed

respiration process of exchanging oxygen and carbon dioxide from the cells

respiratory acidosis abnormal condition caused by an accumulation of carbon dioxide causing the blood to become too acidic, with a concomitant decrease in pH

respiratory alkalosis abnormal process characterized by the loss of too much CO_2, which leads to a rise in arterial blood pH

respiratory burst rapid cellular release of reactive oxygen species; also known as *oxidative burst*

respiratory exchange ratio (R) ratio of the volume of carbon dioxide exhaled to the volume of oxygen taken up

respiratory quotient (RQ) steady-state ratio of carbon dioxide produced by tissue metabolism to oxygen consumed in the same metabolism; for the whole body, normally about 0.82 under basal conditions; in the steady state, the respiratory quotient is equal to the respiratory exchange ratio

respiratory sinus arrhythmia repetitive increase and decrease in heart rate seen on inspiration and expiration of the lungs

resting force force present in a muscle prior to stimulation

resting length length of an unstimulated muscle; its natural length in the body

resting membrane potential membrane potential of an excitable cell in the unstimulated state. It is negative relative to the extracellular fluid and is a steady-state potential defined by the electrochemical gradients for ions that can pass through leakage channels in the membrane and the contribution of an electrogenic pump

reticular formation region in the central portion of the pons and medulla where collections of nerve cells regulate integrative autonomic and spinal reflex functions and coordinate levels of consciousness and attention

reticuloendothelial system (RES) phagocytic cells in connective tissue responsible for removing old cells, debris, and pathogens from the bloodstream

reticulocyte young erythrocyte with residual nuclear material

retinohypothalamic tract projection of axons of retinal ganglion cells in the eye to the suprachiasmatic nucleus of the hypothalamus

retinoid X receptor (RXR) nuclear hormone receptor that binds to other nuclear receptors to form heterodimers that activate transcription

retinol principal form of vitamin A; plays a primary role in vision as well as growth and differentiation. It is derived directly from animal sources or through conversion from beta-carotene (found abundantly in carrots) in the small intestine

retinol-binding protein (RBP) protein synthesized by the liver that binds with retinol in response to falling levels of vitamin A. RBP secretion depends on vitamin A availability

retinopathy deterioration of the blood vessels of the retina, leading to blindness

retinyl ester reesterification of retinol by the enzyme lecithin: retinol acyltransferase. Retinyl ester is incorporated in chylomicrons and taken up by the liver

retrograde direction of movement of material from the tip of an axon toward the cell body of the neuron

retrograde amnesia loss of memory laid down prior to a certain point in time

retropulsion movement of luminal contents in the oral direction

reversal potential voltage across the endplate membrane determined by the Na^+ and K^+ concentrations inside and out

reverse T_3 (rT_3) reverse triiodothyronine; a substance generated by inner ring deiodination of thyroxine

reward system circuit in the brain involved in pleasurable sensation comprising projections from neurons in the midbrain ventral tegmental area to the nucleus accumbens in the cerebrum and other limbic structures

Reynolds number (Re) calculated unit used to predict when turbulence will occur in a fluid flow stream. It takes into account the contribution of fluid flow velocity, fluid viscosity, and the geometry of the tube through which the flow is occurring

rheology study of the fluid properties of blood, which is a suspension, and therefore, a nonhomogeneous fluid

rhesus factor (Rh) protein on the erythrocyte surface of some people

rheumatoid arthritis (RA) traditionally considered a chronic, inflammatory autoimmune disorder that causes the immune system to attack the joints. It is a disabling and painful inflammatory condition, which can lead to substantial loss of mobility as a result of pain and joint destruction

rhodopsin pigment of the retina that is responsible for both the formation of the photoreceptor cells and the first events in the perception of light; also known as *visual purple*

rhythmicity property of all heart cells by which they can generate their own action potentials in a repetitive, uniform, or rhythmic manner. It naturally occurs in the nodal tissue of the healthy heart and is responsible for setting the intrinsic heart rate but can be manifested in ectopic foci and thus cause cardiac arrhythmias

rib cage 12 pairs of ribs that are hinged to the vertebral column

rickets disorder of normal bone ossification manifested by distorted bone movements during muscular action. Rickets is caused by vitamin D deficiency

right atrium upper chamber of the right heart. It receives the entire venous blood supply returning from all the systemic organs. It empties into the right ventricle through the tricuspid valve

right bundle branch block condition in which transmission of action potentials is blocked in the right branch of the bundle of His

right ventricle lower chamber of the right heart that receives blood from the right atrium through the tricuspid valve and pumps blood across the pulmonic valve into the pulmonary artery

right ventricular hypertrophy abnormally increased muscle mass in the right ventricle. It occurs in response to any factor that chronically increases wall stress on the right ventricle, such as pulmonary hypertension or a stenotic pulmonic valve

right-to-left shunt anatomical shunt in which blood goes directly from the pulmonary artery to the pulmonary vein and bypasses the alveoli. Shunted blood does not get oxygenated

rigor crossbridge myosin crossbridge strongly attached to actin and lacking adenosine triphosphate

rigor mortis postmortem muscle stiffness resulting from rigor crossbridges in the absence of adenosine triphosphate

rolling slow movement of leukocytes in blood vessels along the endothelium

rough endoplasmic reticulum (RER) eukaryotic organelle that forms an interconnected network of tubules, vesicles, and cisternae within cells; functions to synthesize proteins

rouleaux phenomenon stacking of erythrocytes like coins in blood

rubor redness associated with acute inflammation

rubrospinal tract motor control pathway for axons from the red nucleus to the spinal cord

rugae gastricae longitudinal folds in the corpus of the empty stomach

ryanodine receptor (RyR) controllable channel in the sarcoplasmic reticulum of muscle that allows calcium release

S cells type of endocrine cell in the intestinal mucosa that releases secretin induced by the entry of acidic chyme from the stomach into the small intestine during the intestinal phase of digestion

S-type enteric neurons synaptic-type enteric neurons; neurons with a single long process (axon) and so named because nicotinic fast excitatory postsynaptic potentials occur in virtually all of the S-type enteric neurons

S wave first downward (negative) deflection of the electrocardiogram following the R wave

SA node sinoatrial node; the normal "pacemaker" of the heart. It can generate its own action potentials in a repetitive manner and is responsible for setting the intrinsic heart rate. It is the first area of the heart to display electrical activity during the cardiac cycle

saccades rapid sideways movements of the eyeballs that change the direction of vision

safety factor at the myoneural junction, the amount by which the released transmitter substance exceeds that necessary for effective transmission

salivary glands heterogeneous group of exocrine glands that produces saliva; innervated by both the parasympathetic and sympathetic divisions of the autonomic nervous system. Parasympathetic stimulation of the salivary glands results in increased salivary secretion. The parotid, submandibular (submaxillary), and sublingual glands are the major salivary glands. They are drained by individual ducts into the mouth

salivon basic unit of the human submandibular gland, consisting of the acinus, the intercalated duct, the striated duct, and the excretory (collecting) duct—the network of acini and ducts that comprises the salivary glands

saltatory conduction process in a myelinated nerve whereby action potentials are successively generated at neighboring nodes of Ranvier and thus the action potential appears to jump from one node to the next

sarcolemma plasma membrane of a muscle cell, along with a fine fibrillar connective tissue network

sarcomere fundamental organized contractile unit of a skeletal or cardiac muscle; includes the overlapping myofilaments between two Z lines

sarcoplasmic reticulum elaboration of the endoplasmic reticulum of muscle that contains, releases, and takes up calcium during the control of contraction

satellite cell precursor to a muscle cell

satiety sensation of fullness or being well fed

scalene muscles three paired accessory skeletal muscles in the lateral neck that elevate the rib cage during forced inspiration

Scatchard plot graph used to find the equilibrium constant for a reaction such as [H] + [R] $\rightleftharpoons$ [HR] ([HR]/[H] versus [HR], or any functions proportional to these quantities); the magnitude of the slope of the graph is the equilibrium constant

schistocyte erythrocyte fragment generated as cell flows through damaged vessels

schizophrenia psychiatric illness marked by disordered thoughts, delusions (fixed false beliefs), inappropriate emotional response, auditory hallucinations, social withdrawal, and lethargy

scintigraphic techniques use of radioisotopes, either ingested with food or released from swallowed capsules, (in this case) to measure transit times through different regions of the gastrointestinal tract

scotopic (nighttime) vision noncolor vision that is served by the rod cells of the retina

scotopsin protein moiety in retinal rods that combines with 11-*cis*-retinal to form rhodopsin

scurvy disorder characterized by weakness, fatigue, anemia, and bleeding gums caused by vitamin C deficiency

seasonal affective disorder (SAD) mood disorder with a seasonal pattern, with episodes coming on in the autumn and/or winter with the depression resolving in the spring; also known as *winter depression* and *winter blues*

second heart sound second normally audible sound of contraction of the heart caused by vibrations set up in the ventricular chamber following abrupt closure of the aortic and pulmonic valves at the onset of ventricular relaxation (diastole); its intensity is proportional to the intensity of valve closure and is increased whenever aortic or pulmonary pressure is abnormally high; thus, it is taken as a clinical indicator of possible systemic or pulmonary hypertension, respectively

second messenger intermediary molecule that is generated as a consequence of the binding of a first messenger to a receptor. These intermediary molecules are generated in large quantities, and this amplifies the signaling initiated by the first messenger. Examples of second messengers are cyclic adenosine monophosphate, cyclic guanosine monophosphate, calcium, inositol 1,4,5-triphosphate, and diacylglycerol.

second polar body small haploid cell produced by unequal second meiotic division of an oocyte

secondary active transport mechanisms see *active transport*

secondary axons (type II) afferent axons from nuclear chair fibers of the muscle spindle, smaller in diameter than primary (type I) endings

secondary follicular stage stage during which several layers of granulosa cells and a single theca surround the oocyte

secondary lymphoid organs see *lymphoid organs*

secondary oocyte haploid oocyte produced by completion of the first meiotic division

secondary pacemaker ectopic foci that repetitively activates in the myocardium, thereby pacing the heart at a rate other than that set by the sinoatrial node

secondary peristaltic wave peristaltic propulsion in the esophagus initiated by an initial block of forward movement of a swallowed bolus or by the presence of acid in the esophagus

secondary sex characteristics external sex-typical male and female structures

secondary spermatocytes spermatocyte that divides to form the spermatids

second-degree atrioventricular block any condition in which some, but not all, action potentials from the atria reach the ventricles. It usually arises from blockage at the level of the atrioventricular node and results in an irregular heart rate

secretagogue substance stimulating secretion

secretin hormone secreted by the duodenal and jejunal mucosa when exposed to acid chyme that stimulates pancreatic secretion rich in bicarbonate ions. Secretin inhibits gastrin release

secretomotor neuron motor neurons in the submucous plexus, which innervate and evoke secretion from the intestinal glands

secretory phase phase of the endometrial cycle during which the endometrial glands become coiled, store glycogen, and secrete large amounts of carbohydrate-rich mucus. The stroma increases in vascularity and becomes edematous, and the spiral arteries become tortuous

segmentation specific pattern of small intestinal motility that starts with the ingestion of a meal and accomplishes mixing of the contents in the lumen; also called *digestive motility*

selective serotonin reuptake inhibitors (SSRIs) class of compounds typically used as antidepressants to treat depression, anxiety disorders, and some personality disorders

selenocysteine rare amino acid with properties that make it ideal for catalysis of oxidoreductive reactions

self-tolerance see *immune tolerance*

sella turcica depression in the sphenoid bone of the skull

semantic memory memory of facts

semen combination of sperm, seminal fluid, and other male reproductive secretions

semicircular canals three orthogonally oriented fluid-filled tubes of the vestibular apparatus

seminal emission movement of sperm and associated fluids from the cauda epididymis and vas deferens into the urethra

seminal plasma fluid portion of semen, secreted by the male sex accessory glands

seminiferous tubules tiny tubes in the testes where sperm cells are produced, grow, and mature

sensorineural deafness hearing impairment due to damage of the inner ear nerve endings, the cochlear portion of the eighth cranial nerve, the vestibulocochlear nerve, or the cortical hearing center

sensory modality specific type of sensory activity; the kind of sensation

sensory neuron neuron that conducts impulses arising in a sense organ or at sensory nerve endings

sensory receptor cell or part of a cell specialized and normally functioning to convert environmental stimuli into nerve impulses or some response, which, in turn, evokes nerve impulses; most sensory receptor cells are nerve cells, but some nonnervous cells can be receptors (e.g., intestinal enteroendocrine cells); one method of classifying receptors is by the form of adequate stimulus: chemoreceptors, osmoreceptors, and mechanoreceptors are normally stimulated by chemicals, osmotic pressure differences, and mechanical events, respectively

sensory transduction process whereby environmental signals (radiation, mechanical forces, or chemical quantities) are transformed into trains of nerve impulses

sepsis bloodstream infection causing excessive inflammation

septal nuclei group of cholinergic neurons located in the basal forebrain area with reciprocal connections with the hippocampus

septic shock decrease in blood pressure associated with overproduction of nitric oxide in infected tissue. It is characterized by high, unnecessary blood flow to the affected organs at the expense of flow to other organs in the body

serosa one of four layers of the stomach wall; lies over the muscularis externa, consisting of layers of connective tissue continuous with the peritoneum

serotonergic neuron type of neuron that uses serotonin as a neurotransmitter

serotonin mediator of the inflammatory response and a neurotransmitter used by central and peripheral neurons; also known as *5-hydroxytryptamine* or *5-HT*

serous cells cells that are found in the salivary glands, contain an abundance of rough endoplasmic reticulum, and secrete digestive enzymes

Sertoli cell cell, found in large numbers lining the semen-producing tubules of the testis, that provides support and nourishment for developing sperm

serum blood plasma without clotting factors

serum iron levels test measuring iron levels in blood as part of the iron profile

serum protein electrophoresis electrophoretic separation test of blood proteins

set point any one of a number of quantities or physiological variables (e.g., body weight, body temperature) that the body tries to keep at a particular value or in a state of dynamic constancy

sex hormone–binding globulin (SHBG) substance that binds sex steroids in the blood

SH2 domain structurally conserved protein domain contained in intracellular signal-transducing proteins that "helps" a protein find its way to another protein by recognizing phosphorylated tyrosine on the other protein

shallow-water blackout condition in which a person passes out underwater as a result of hypoxia to the brain

shear stress "rubbing" or frictional effect exerted along the walls of a cylinder as a result of fluid flowing in the cylinder. This force tends to strain or deform the inner portion of the wall in the direction of flow. In arteries, shear stress has an important influence on the function of the endothelium and influences the location of the development of atherosclerotic plaques

shivering bodily function in response to early hypothermia in warm-blooded animals. When the core body temperature drops, the shivering reflex is triggered. Muscle groups around the vital organs begin to shake in small movements in an attempt to create warmth by expending energy. Shivering can also be a response to a fever, as a person may feel cold, although the core temperature is already elevated

short hypophyseal portal vessels vessels that deliver blood into the sinusoids of the anterior pituitary

short-term memory newly acquired information that can be readily recalled for only a few minutes

shunt diversion of blood away from the alveoli

sickle cell anemia hereditary disorder that leads to an abnormal production of erythrocytes

sieving action selective gastric emptying of particles according to the particle size with the smallest particles emptying first

signal cascade series of biochemical reactions in which the products of one reaction are consumed in the next reaction

signal transduction process by which a cell converts one kind of signal or stimulus into another through a series of biochemical reactions that are often linked by second messengers, which serve to amplify the initial stimulus

silent nociceptor unresponsive sensory afferent nerve terminal that becomes responsive to mechanical stimulation during inflammatory states

simple acid–base disturbance one of the four primary acid–base disturbances

simple micelle micelle in which bile salts alone are present

single breath test test to determine oxygen uptake in a patient, using a single breath of a low concentration of carbon monoxide

single effect establishment of a modest osmotic gradient at any level of the loop of Henle. This gradient is developed into a larger gradient along the axis of the loop by countercurrent multiplication.

single nucleotide polymorphism (SNP) DNA sequence variation occurring when a single nucleotide in the genome differs between members of a species or paired chromosomes in an individual

sinus bradycardia decreased heart rate caused by normal physiological effects on the sinoatrial node

sinus tachycardia elevated heart rate caused by normal physiological effects on the sinoatrial node

sinusoidal endothelial cells cells that lack a basement membrane and have sievelike plates, thereby permitting the ready exchange of materials between the perisinusoidal space and the sinusoid

size principle pattern of muscle activation in which the smallest α motor neurons and associated motor units are activated first and then, as more power is needed, larger size motor neurons are activated

sleep apnea syndrome sleep disorder characterized by abnormal pauses in breathing that last a few seconds to minutes

sleep debt cumulative effect of not getting enough sleep; also called *sleep deficit*

sliding filament theory schema for muscle contraction involving the relative motion of actin and myosin filaments driven by the action of myosin crossbridges

slow adaptation process in static sensory receptors in which the output to the central nervous system is maintained (or falls slowly) while the applied stimulus is maintained in intensity

slow axoplasmic transport mechanism by which structural proteins, such as actin, neurofilaments, and microtubules, are transported from the cell body to the axon; occurs at 1 to 2 mm/day

slow excitatory postsynaptic potential a postsynaptic potential characterized by membrane depolarization and augmented excitability that lasts for extended periods of seconds or minutes

slow wave membrane electrical potential that waxes and wanes in a rhythmic fashion in the gastrointestinal tract; the slow wave determines the timing of a contraction but is not enough to produce a contraction in the absence of a burst of spike potentials triggered by the release of a neurotransmitter

slow-twitch fibers skeletal muscle fibers specialized for long-duration activity, using aerobic metabolism

slow-wave sleep (SWS) four stages of progressively deepening sleep during which the electroencephalogram becomes progressively slower in frequency and higher in amplitude

small-molecule transmitters amino acids and monoamines used as neurotransmitters

SNARES proteins found on both the synaptic vesicle and plasma membrane that cause the synaptic vesicle to dock and bind to the presynaptic terminal membrane

sodium appetite craving for salt

sodium-iodide symporter (NIS) substance located on the basal plasma membrane of the thyroid follicular cell that transports iodide into the cell

solvent drag one mechanism by which sodium is absorbed by the duodenum and jejunum

soma neuronal cell body; also called *perikaryon*

somatic pain sensation of pain arising from nonvisceral parts of the body

somatomedins class of small peptides responsible for the mediation of some growth hormone actions that increase the rate of cartilage synthesis by promoting the uptake of sulfate and the synthesis of collagen; also known as *insulin-like growth factors*

somatostatin 1. gastrointestinal hormone secreted by endocrine cells in the antrum of the stomach that inhibits the release of gastrin and consequently gastric acid secretion when the pH of the stomach lumen falls below 3; 2. polypeptide from the hypothalamus that inhibits secretion of growth hormone; also known as *somatotropin release–inhibiting factor (SRIF)*

somatotopic maps discrete groupings of motor or sensory neurons associated with functions of specific areas of the body

somatotrophs cells located in the anterior pituitary that secrete growth hormone

somatotropin release–inhibiting factor see *somatostatin*

space constant (λ) length along a membrane required for an electrotonic membrane potential to decay to 37% of its maximal value; also called *length constant*

spasticity feature of altered skeletal muscle performance occurring in disorders of the central nervous system impacting the upper motor neuron in the form of a lesion; when there is a loss of descending inhibition from the brain to the spinal cord, such that muscles become overactive, this loss of inhibitory control can cause an ongoing level of contraction, with decreased ability for the affected individual to volitionally control the muscle contraction, and increased resistance felt on passive stretch

spatial summation summation of two or more membrane potentials generated at different sites; also called *motor unit summation*

special senses (i.e., hearing, sight, smell, and taste) senses that have specialized organs devoted to them

specific compliance lung compliance divided by the functional residual capacity; used to correct for lung size

specific gravity measure of blood density

spectrin cytoskeletal protein that lines the intracellular side of the plasma membrane, forming a scaffold important role in maintenance of plasma membrane integrity and cytoskeletal structure

spermatids any of the four haploid cells formed by meiosis in a male organism that develop into spermatozoa without further division

spermatogenesis process of producing sperm, the male reproductive cells

spermatogonia any of the cells of the gonads in male organisms that are the progenitors of spermatocytes

spermatozoa mature sperm cells

spermiation release of mature spermatozoa from the surface of the Sertoli cell into the lumen of the seminiferous tubule

spermiogenesis process by which spermatids mature into sperm cells

spherocyte abnormal, spherical-shaped erythrocyte

sphincter of Oddi ring of smooth muscle surrounding the opening of bile ducts in the duodenum that regulates the flow of bile and pancreatic juice into the duodenum and prevents the reflux of intestinal contents into the pancreatic ducts

sphygmomanometer external, noninvasive device for measuring arterial pressure producing external pressure on a limb that is measured by a column of mercury or calibrated gage

spinal shock decrease in blood pressure resulting from physical trauma to the spinal cord

spinnbarkeit highly elastic cervical mucus

spinobulbar muscular atrophy neurodegenerative neuromuscular disease associated with mutation of the androgen receptor rarely affecting females; also known as *Kennedy disease*

spinocerebellar tracts afferent pathways bringing sensory axons from the spinal cord to the cerebellum or associated brainstem nuclei

spinocerebellum portion of the cerebellum concerned with coordination of axial and proximal limb motor function

spiral arteries specialized arteries that supply blood into the maternal placental sinus

spirogram recording from the spirometer, which is an instrument for recording respiratory movements

spirometry measurement of lung volumes by a spirometer

splanchnic nerves mixed nerves in the mesentery that have both sympathetic efferent and sensory afferent fibers

splay renal physiology, the rounding of the corner on the graph relating tubular reabsorption or secretion of a substance to its arterial plasma concentration, resulting in part from the fact that some nephrons reach their tubular maximum before others

splenomegaly enlargement of the spleen usually associated with hyperfunction of the organ SRY sex-determining region gene on the Y chromosome; required to produce male gonadal differentiation

SRY (sex-determining region, Y chromosome) found on the Y chromosome; the gene encodes a transcription factor required for the development of the testes

standard curve quantitative research tool and method of plotting assay data that is used to determine the concentration of a substance, particularly proteins and DNA

standard frontal lead system three frontal bipolar lead system developed by Einthoven. It consists of a lead I, in which the right arm has a negative polarity and the left arm is positive, a lead II, in which the right arm has a negative polarity and the left foot is positive, and a lead III, in which the left arm has a negative polarity and the left foot is positive

Starling–Landis equation illustrates the role of *hydrostatic* and *oncotic* forces in the movement of fluid across *capillary membranes*, which may occurs from diffusion, filtration, or pinocytosis: $J_V = K_h A\,[(P_c - P_t) - \sigma(COP_p - COP_t)]$

Starling forces all of the hydrostatic and osmotic forces within the capillary plasma and the interstitium that determine the direction and rate of bulk fluid transfer between the capillary blood and the interstitium

Starling law of the heart law that states that stroke volume increases when ventricular filling increases and decreases when ventricular filling decreases; also known as *Starling's Law of the Heart*

static compliance compliance curve generated with no airflow

statin drug used to lower plasma cholesterol level; inhibits HMG-CoA reductase, the rate-limiting enzyme of the mevalonate pathway of cholesterol synthesis

steady state state of a system in which the recently observed behavior of the system will continue into the future

steatorrhea excretion of unabsorbed lipids in the feces, caused by a pancreatic deficiency that significantly reduces the ability of the exocrine pancreas to produce digestive enzymes. Normally about 5 g/day of fat are excreted in human stool, but as much as 50 g/day can be excreted in the case of steatorrhea

stellate cells perisinusoidal cells containing lipid droplets in the cytoplasm that function in fat storage. During the inflammatory process, stellate cells transform into myofibroblasts that secrete collagen into the space of Disse, thereby regulating sinusoidal portal pressure by contracting or relaxing. Stellate cells may be involved in the pathologic fibrosis of the liver; also known as *Ito cells*

sternocleidomastoids paired muscles in the superficial layers of the anterior portion of the neck that, in addition to flexing and rotating the head, function as accessory muscles to elevate the rib cage during forced inspiration, along with the scalene muscles of the neck

stereocilia apical modifications of the cell characterized by their length and their lack of motility; found in the inner ear, the ductus deferens, and the epididymis

steroid hormone lipid hormones generally synthesized from cholesterol in the gonads and adrenal glands that can freely traverse the plasma membrane to enter the cytoplasm as well as membranes of other organelles such as the nucleus they are carried in the blood bound to a specific carrier protein

steroidogenesis process of steroid hormone production

steroidogenic acute regulatory (StAR) protein protein that facilitates transport of cholesterol into the mitochondria

stigma specific site where the follicular wall thins by cellular deterioration and bulges for rupture

stimulus input to a sensory receptor that causes it to produce its characteristic output

stress relaxation slow and time-dependent fall in force of a viscoelastic substance held at a stretched length

stria terminalis fiber bundle that interconnects the amygdala with the hypothalamus and basal forebrain

stria vascularis pseudostratified columnar vascularized epithelium containing melanocytes, which help maintain the high potassium concentration of the endolymph that is produced here

striae irregular areas of skin that appear as red or purple bands, stripes, or lines

striated duct duct lined with columnar cells that is part of the network of ducts that comprise the salivary glands, whose major function is to modify the ionic composition of the saliva

striatum caudate nucleus and putamen of the basal ganglia

stroke volume volume of blood ejected from the heart in one systole

stromal layer part of endometrium of the uterus; the closest layer to the myometrium

strong acid acid with a high dissociation constant, which consequently is highly ionized in solution

subcortical structures collections of neuronal cell bodies in the cerebrum that are located deep to the cerebral cortex

sublingual glands salivary glands that empty saliva into the mouth; located anterior to the submandibular gland under the tongue, beneath the mucous membrane of the floor of the mouth

submandibular ganglion location of synapse between pre–postganglionic parasympathetic neurons from the superior salivatory nucleus through cranial nerve VII, which are involved in control of the submandibular glands

submandibular glands (submaxillary glands) paired salivary glands located beneath the floor of the mouth, accounting for 70% of the salivary volume

submucosa layer of dense irregular connective tissue in the stomach wall (one of four layers) or loose connective tissue that supports the mucosa as well as joins the mucosa to the bulk of underlying smooth muscle (fibers running circularly within layer of longitudinal muscle)

submucosal plexus ganglionated plexus of the enteric nervous system situated between the mucosa and circular muscle coat of the small and large intestine; it consists of an inner ganglionated plexus called the *Meissner plexus* and an outer ganglionated plexus called the *Schabadasch plexus*

substance P neuropeptide used as a neurotransmitter by small dorsal root ganglion cells that signal pain

substantia nigra group of nerve cells in the upper brainstem, where dopaminergic neurons important in Parkinson disease are located

subthalamic nucleus group of nerve cells in the basal ganglia that receives input from the other basal ganglia nuclei and provides output to the globus pallidus internus

succinylcholine blocker of the myoneural junction that causes maintained depolarization of the endplate membrane

suckling reflex reflex in which the stimulation of sensory nerves in the breast by the infant initiates a neural signal to the hypothalamus, resulting in release of oxytocin to stimulate milk letdown

sudden infant death syndrome (SIDS) sudden death of an infant that is unexpected by history and cannot be explained following a thorough forensic autopsy; also known as *crib death*

sulcus valley on the surface of the brain

sulfonylureas any of a group of hypoglycemic drugs that act on the β cells of the pancreas to increase the secretion of insulin

superior cerebellar peduncles one of the connections of the cerebellum to the brainstem

superior cervical ganglion uppermost of the sympathetic ganglia located in the neck. Innervated by the sympathetic nervous system, sympathetic fibers arise in the upper thoracic segments of the spinal cord and synapse in the superior cervical ganglion. Postganglionic fibers leave the superior cervical ganglion and innervate the acini, ducts, and blood vessels, resulting in a short-lived and small increase in salivary secretion

superior colliculus layered structure in which the superficial layers are sensory related and receive input from the eyes as well as other sensory systems; the deep layers are motor related, capable of activating eye movements as well as other responses; and the intermediate layers have multisensory cells and motor properties

superior hypophyseal arteries arteries that give rise to a rich capillary network in the median eminence

superior mesenteric ganglion uppermost of the prevertebral sympathetic ganglia located in the abdomen

superior parietal lobe region of the cerebral cortex that integrates complex sensory information

superior salivatory nucleus region of the pons where parasympathetic preganglionic neurons involved in control of the lacrimal and nasal glands are located

superior vena cava (SVC) single major vein that collects blood from the brain, head, and upper extremities and empties into the right atrium

superoxide dismutase (SOD) antioxidant enzyme that neutralizes superoxide ions in cells by catalyzing the breakdown of superoxide to oxygen and hydrogen peroxide (H_2O_2)

superoxide ion (O_2•$^-$) free radical that is formed during oxygen use in the mitochondria. The free radical is an oxygen molecule with an unpaired electron.

supplementary motor area (SMA) cortical region on the medial aspect of the hemispheres that is active in motor control

suprachiasmatic nucleus (SCN) collection of neurons in the hypothalamus responsible for generating the body's circadian rhythms and entraining them to the day–night cycle

supraoptic nuclei portion of the hypothalamus in which magnocellular neuron cells bodies are found

supraventricular tachycardia abnormally high heart rate resulting from an ectopic foci in a region anatomically superior to the ventricles, usually the atria

surfactant surface-reducing material (lipoprotein) lining the alveoli

Sylvian fissure deep valley that separates the temporal from the frontal and parietal lobes of the brain

sympathetic division division of the autonomic nervous system, with its outflow from the central nervous system in the thoracolumbar segments of the spinal cord and having its ganglia in a pair of chains on either side of the spinal cord and in a grouping of three (coeliac, and superior and inferior mesenteric) in the abdomen

sympathetic tone concept that some level of neuronal activity is present in the sympathetic neurons, even in an apparent resting state

sympathy coordination of bodily activities by the nervous system

symport membrane transport process in which the driver ion (usually Na^+) and the solute move in the same direction

synapse functional (noncytoplasmic) connection between neurons consisting of a presynaptic site of chemical transmitter release and a postsynaptic site of action of the released transmitter

synaptic cleft extracellular space separating the presynaptic and postsynaptic membranes at a chemical synapse

synaptobrevin integral vesicular membrane SNARE protein

synaptotagmin integral vesicular membrane protein that acts as the Ca^{2+} sensor and binds proteins that cause the vesicles to dock and bind to the presynaptic terminal membrane

syncytiotrophoblast syncytial component of the trophoblast that lacks distinct cell boundaries

syndrome of inappropriate antidiuretic hormone (SIADH) disorder characterized by persistent hyponatremia

synergist muscle that acts in concert with the prime mover muscle

synthase enzyme that catalyzes synthesis of neurotransmitters in the enteric nervous system

systemic circulation circulation outside the thoracic cavity, not including the pulmonary circulation

systole period of time in which the heart is in its contraction phase

systolic pressure blood pressure in the ventricle or arterial system during the contraction of the heart, or systole

T cell type of lymphocyte with specific membrane receptors to recognize peptide antigens

T-cell receptor (TCR) recognition molecule of T cells for antigen–major histocompatibility complex protein complex

T wave represents repolarization of the ventricles in electrocardiography

tachyphylaxis rapid decrease in the response to a drug after repeated doses over a short period of time; increasing the dose of the drug will not increase the pharmacological response

tachypnea increased ventilation through rapid shallow breathing (i.e., increased breathing rate)

target cell 1. abnormal erythrocyte with central staining and a pale surrounding area; also known as *codocyte* or *leptocyte*; 2. cell that possesses specific receptors for a particular hormone

taste buds organized structures, located on the tongue and pharynx, that serve the gustatory sense

taste pore minute opening of a taste bud on the surface of the oral mucosa through which the gustatory hairs of the specialized neuroepithelial gustatory cells project

telencephalon region of the central nervous system; also called the *cerebrum* or *brain*

temporal lobe inferior and lateral region of the cerebrum

temporal summation time-wise addition of muscle force in response to repeated stimulation; the summation of two or more membrane potentials generated at the same location in quick succession

tension 1) *partial pressure* of dissolved gas in a liquid 2) force exerted on the wall of a cylindrical structure that would have the tendency to rip the wall as a result of a positive transmural pressure. It is directly proportional to both the transmural pressure and the radius of the cylinder, such that in two cylinders exposed to the same transmural pressure, the cylinder with the larger internal radius experiences a greater wall tension. This principle is applied to blood vessels and other hollow organs, such as the heart and urinary bladder.

terminal arterioles smallest arterioles; they lead directly into capillaries

terminal bronchioles smallest airways in the conducting zone

terminal cisternae lateral regions of the sarcoplasmic reticulum of striated muscle, from which calcium is released and where it is stored at rest; also called *lateral sacs*

terminal respiratory unit merging of alveolar ducts and their alveoli with adjacent pulmonary capillaries

tertiary follicular stage stage characterized by a large ovarian follicle containing an antral cavity

testicular feminization state of intersex in which the karyotype is male (i.e., 46,XY) and the gonads are testes (hence also male), but the body is completely unresponsive to testosterone and to its metabolite dihydrotestosterone. Thus, the body develops in the female way, with a normal vulva and vagina apparent at birth and with normal development of the breasts at puberty; because the testes still secrete antimüllerian hormone, there is no uterus.

testosterone main steroid producing the male secondary characteristics

tetanic fusion frequency stimulation rate beyond which individual muscle twitches are no longer discernible

tetanus maintained muscle contraction caused by restimulating a muscle before mechanical relaxation is complete

tetanus/twitch ratio comparison between the maximal tetanic contraction and that of a single twitch

tetrahydrocortisol glucuronide major excretable metabolite of cortisol degradation

tetraiodoacetic acid (tetrac) derivatives formed by degradation of thyroxine

thalamus collection of neuronal cell bodies deep in the cerebrum adjacent to the midline that relays information to and from the cortex

theca externa external dense vascular connective tissue layer of the theca

theca interna internal dense vascular connective tissue layer of the theca

theca lutein cells cells found in the corpus luteum that are formed from the theca cells

thelarche development of breast buds in girls during puberty

therapeutic hypothermia (TH) medical treatment that lowers a patient's body temperature order to help reduce the risk of the ischemic injury to tissue that follows a period of insufficient blood flow; also called *controlled hypothermia*

thermic effect of food increment in energy expenditure above resting metabolic rate resulting from the cost of processing food for storage and use; one of the components of total metabolism, along with the resting metabolic rate and the exercise component; also known as *specific dynamic action*

thermodilution in cardiovascular physiology, a technique used to measure cardiac output by the indicator dilution technique in which temperature is used as the indicator

thermogenic action action of thyroid hormones to regulate the basal rate of body heat production and of oxygen consumed

thermoneutrality range of ambient temperatures at which energy expenditure is lowest and no metabolic heat is required to maintain body temperature; thermal balance is maintained with neither shivering nor sweating.

thermoreceptors sensory receptors that respond to variation in environmental temperature; separate populations of receptors serve the sensations of warmth and coldness

theta waves electroencephalogram wave pattern with a rhythm ranging from 4 to 7 Hz, observed during sleep

thirst conscious sensation associated with craving for drink, ordinarily interpreted as a desire for water

thirst center located in the anterior hypothalamus, close to the neurons that produce and control arginine vasopressin release; this center relays impulses to the cerebral cortex, so that thirst becomes a conscious sensation

thoracic cavity cavity that houses the lung and is synonymous with the chest cavity

thoracolumbar division division analogous to the sympathetic division of the autonomic system; also called *sympathetic division*

threshold intensity level below which sensation is not possible. Accessory structures and processes may vary this level.

threshold concentration concentration that must be exceeded to begin producing a given effect or result or to elicit a response

thrombi fat globules or air bubbles that form in the bloodstream that act like clots and block vital blood flow

thrombin coagulation protein

thrombocyte cells that play a key role in blood clotting: in mammals, thrombocytes are anucleated cell fragments called *platelets*; thrombocytes of nonmammalian vertebrates have a nucleus, and thus, resemble B lymphocytes

thrombocytopenia condition characterized by a lower-than-normal number of circulating platelets

thrombopoietin liver/kidney hormone regulating production of thrombocytes

thymus lymphoid organ in which T lymphocytes mature

thyroglobulin (Tg) large protein that is a storage form of the thyroid hormones

thyroid hormone receptor (TR) receptor located in the nuclei of cells that binds thyroid hormones to facilitate their action on transcription

thyroid hormone response elements (TRE) sequence of DNA in a gene that binds the thyroid hormone receptor

thyroid peroxidase (TPO) substance that catalyzes the iodination of thyroglobulin

thyroid-stimulating hormone (TSH) hormone product of thyrotrophs located in the anterior pituitary

thyrotroph cells located in the anterior pituitary that secrete thyroid-stimulating hormone

thyrotropin-releasing hormone (TRH) polypeptide from the hypothalamus that stimulates secretion of thyroid-stimulating hormone

thyroxine (T_4) peptide hormone produced by the thyroid gland that is a precursor to triiodothyronine

thyroxine-binding globulin (TBG) glycoprotein synthesized and secreted by the liver that binds thyroid hormone in the blood

tidal volume volume of air inhaled with each breath

tight junction closely associated areas of two cells whose membranes join together forming a virtually impermeable barrier to fluid

time constant (τ) time required for the membrane potential to decay to 37% of its initial peak value following an influx of ions; a function of membrane resistance and capacitance

tissue growth factors chemical messengers that influence cell division, differentiation, and cell survival; they may exert effects in an autocrine, paracrine, or endocrine fashion

tissue thromboplastin (factor III) factor extrinsic to blood but released from injured tissue during hemostasis; also called *tissue factor*

tissue plasminogen activator (TPA) substance that has the ability to cleave plasminogen and convert it into plasmin, its active form

titin large filamentous muscle protein that may help prevent overextension of sarcomeres and maintain the central location of A bands

titratable acid acid involved in renal physiology; this term is used to explicitly exclude ammonium (NH_4^+) as a source of acid and is part of the calculation for net acid excretion

toll-like receptor transmembrane protein that recognizes pathogens, activates innate immune responses, and regulates adaptive immune responses

tonic contraction degree of tension, firmness, or maintained contraction in a muscle, in contrast to a phasic contraction

tonic–clonic seizures type of generalized seizure affecting the entire brain; convulsion, characterized by generalized muscle spasms and loss of consciousness, most commonly associated with epilepsy; formerly known as *grand mal seizures* or *gran mal seizures*

tonicity quality of a solution related to the concentrations of non-penetrating solutes, that is, solutes that do not cross the plasma membrane of a cell. Cells placed in a **hypotonic** solution will accumulate water and their volume will increase. Cells placed in a **hypertonic** solution will lose water and their volume will decrease. Cell volume will not change in an **isotonic** solution because there is no net movement of water into or out of the cell.

tonotopic organization place-specific frequency discrimination provided over the length of the basilar membrane in the organ of Corti

total bile acid pool total amount of bile acids in the body, primary or secondary, conjugated or free, at any time

total body water volume of water in the body

total iron-binding capacity (TIBC) test analyzing the amount of iron needed to bind to all transferrin

total lung capacity (TLC) total volume of air in the lungs at the end of maximum inspiration

total peripheral vascular resistance (TPR) resistance to flow created by all the systemic vascular segments in the cardiovascular system from the origin of the aorta to the entry of the right atrium. Quantitatively this value is dominated by the contribution made from the resistance of the arterial side of the circulation, which is much greater than the resistance to flow attributed to the veins. TPR is used in used in Poiseuille law to calculate mean arterial pressure or cardiac output in the cardiovascular system as MAP = CO × TPR. MAP, microtubule-associated protein; CO, carbon monoxide.

totipotency ability of a cell to divide and produce all the differentiated cells in an organism

toxoid inactivated toxin, used as vaccine against the toxin

trachea main airway of the lung through which respiratory air is brought into the lungs; it is held open by up to 20 C-shaped cartilage rings and prevents airway collapse during forced expiration; commonly known as *windpipe*

tractus solitarious brainstem structure that carries and receives visceral sensation and taste from the facial (VII), glossopharyngeal (IX), and vagus (X) cranial nerves

transairway pressure pressure difference across the airways (airway pressure–pleural pressure)

transcellular fluid specialized extracellular fluid that is separated from the blood plasma not only by a capillary endothelium but by a continuous layer of epithelial cells. It includes aqueous humor, cerebrospinal fluid, digestive secretions, sweat, synovial fluid, renal tubular fluid, and bladder urine.

transcellular transport transport of molecules through an epithelial cell, generally from the apical to basolateral side

transcobalamin protein to which vitamin B_{12} binds and by which it is transported in the portal blood

transducer any device that changes one form of energy (e.g., from the environment) to a signal for further signal processing (e.g., by the central nervous system)

transducin heterotrimeric G protein naturally expressed in vertebrate retina rods and cones by activation of the effector cGMP phosphodiesterase, which degrades cGMP to 5′GMP

ransferrin molecule that functions in the plasma transport of iron and in the maintenance of circulating iron level homeostasis. Plasma transferrin levels are inversely proportional to the iron ad of the body.

rrin saturation ratio of iron serum levels and total iron ng capacity

transient outward K^+ channels voltage-gated myocardial cell membrane channels that carry outward, hyperpolarizing K^+ current shortly after the initiation of a myocardial cell action potential. They are responsible for phase 1 of the action potential.

transient receptor potential (TRP) channels family of loosely related ion channels that are relatively nonselectively permeable to cations, including sodium, calcium, and magnesium

transit time time required for red blood cells to move through alveolar capillaries; under normal resting conditions, the transit time is 0.75 seconds

transmigration see *diapedesis*

transmitter-specific transporter integral membrane proteins that remove neurotransmitter from the synaptic cleft by high-affinity reuptake into the presynaptic terminal or perisynaptic astrocytes (in the central nervous system)

transmural pressure literally trans mural or "across the wall"; the pressure difference between the inside and outside of a hollow structure in the body

transneuronal transport transport of trophic (or other) substances from one cell to another in either the retrograde or anterograde direction

transpulmonary pressure pressure difference across the lung (i.e., alveolar pressure–pleural pressure)

transthyretin binds thyroxine and triiodothyronine in the blood

transverse tubules (T tubules) membrane-bound invaginations of the striated muscle surface membrane that conduct surface electrical activity inward as part of the excitation–contraction coupling process

triacylglycerols see *triglycerides*

triad structure in skeletal muscles, formed by a transverse (T) tubule surrounded by sarcoplasmic reticulum (including two terminal cisternae)

tricuspid valve a thin, flexible, trileaflet structure that separates the right atrium from the right ventricle. It opens into the right ventricle during diastole but closes off the right atrium when the right ventricle contracts

triglycerides most abundant dietary lipids, consisting of a glycerol backbone esterified in the three positions with fatty acids (more than 90% of the daily dietary lipid intake is in the form of triglycerides); also known as *triacylglycerols*

triiodothyronine (T_3) peptide hormone produced by the thyroid gland that regulates growth, development, and metabolism

trituration process of crushing and grinding ingested food by the gastric pump

trophic hormones hormones produced and secreted by the anterior pituitary that target endocrine glands

trophoblast single outer layer of the blastocyst consisting of extraembryonic ectodermal cells

tropomyosin fibrous protein constituent molecule of the actin filaments that participates in regulating striated muscle contraction

troponin protein complex in striated muscle thin filaments that binds calcium and, acting through tropomyosin, allows interaction between actin and myosin

Trousseau sign indication of latent tetany in which carpal spasm occurs when the upper arm is compressed, as by a tourniquet

trypsin one of three endopeptidases present in pancreatic juice that convert proenzymes to active enzymes. Trypsin splits off basic amino acids from the carboxyl terminal of a protein.

tuberoinfundibular system neurons located within the hypothalamus, with cell bodies in the arcuate nucleus and periventricular nuclei and terminals in the median eminence on the ventral surface of the hypothalamus; responsible for the secretion of hypothalamic releasing factors into a portal system that carries them through the pituitary stalk into the anterior pituitary lobe

tubular reabsorption transport of materials by the kidney tubule epithelium out of the tubular urine

tubular secretion transport of materials by the kidney tubule epithelium into the tubular urine

tubular transport maximum (Tm) maximum rate at which a particular substance is reabsorbed (e.g., glucose) or secreted (e.g., *p*-aminohippurate) by the kidney tubules

tubuloglomerular feedback mechanism blood flow control mechanism operating in the kidneys that limits changes in glomerular filtration rate

tubulovesicular membranes extensive smooth endoplasmic reticulum membrane system in the stomach

tumor swelling associated with acute inflammation

tumor necrosis factor (TNF) important cytokine (TNFα, cachexin, or cachectin) involved in systemic inflammation and the acute phase response; released by white blood cells, endothelium, and several other tissues in the course of damage, for example, by infection. Its release stimulates fever.

tumor-specific antigens molecules present only on tumor cell surfaces to be recognized by cytotoxic T cells or B cells

turbulence chaotic random motion of particles in a flow stream. It results when the kinetic component of flow exceeds the viscous forces holding fluid elements together. In the cardiovascular system, turbulence dissipates more pressure energy than does laminar flow of similar quantitative magnitude.

turbulent flow random movement of fluid in a tube (airway or blood vessel)

Turner syndrome congenital abnormality caused by a nondisjunction of one of the X chromosomes, resulting in a 45 X0 chromosomal karyotype

twitch single brief contraction of a muscle as the result of a single action potential

two-cell, two-gonadotropin hypothesis steroidogenesis completed by cooperation between granulosa and theca cells

type I collagen type of protein chemical substance that is the main support of skin, tendon, bone, cartilage, and connective tissue

type 1 diabetes (T1D) autoimmune disorder in which the body's immune system attacks the β cells in the islets of Langerhans, resulting in a decreased or complete absence of the production and secretion of insulin; formerly known as *childhood* or *juvenile diabetes* or *insulin-dependent diabetes*

type 1A diabetes subclass of type 1 diabetes characterized as immune-mediated

type 1B diabetes subclass of type 1 diabetes characterized as non–immune-mediated

type 2 diabetes (T2D) metabolic disorder characterized by insulin resistance, relative insulin deficiency, and hyperglycemia; formerly known as *diabetes mellitus type II*, *non–insulin-dependent diabetes (NIDDM)*, *obesity-related diabetes*, or *adult-onset diabetes*

type II pneumocytes granular, cuboidal cells typically found at the alveolar–septal junction that cover a much smaller surface area than *type I cells* (<5%), but are much more numerous; they are responsible for the production and secretion of surfactant (the majority of which are dipalmitoylphosphatidylcholine), a group of phospholipids that reduce the alveolar surface tension and are stored in type II pneumocytes in lamellar bodies (specialized vesicles unique to type II cells)

tyrosine kinase receptors family of receptors each having a tyrosine kinase domain (which phosphorylates proteins on tyrosine residues), a hormone binding domain, and a carboxyl terminal segment with multiple tyrosines for autophosphorylation

tyrosine hydroxylase rate-limiting enzyme in the synthesis of catecholamines that catalyzes the conversion of L-tyrosine to L-3,4-dihydroxyphenylalanine

UDP-glucose uridine diphosphate glucose; a molecule involved in the biosynthesis of glycoproteins or glycolipids

ultrafiltrate filtrate formed by ultrafiltration

ultrafiltration filtration through a selectively permeable membrane that allows passage of small molecules but restricts the passage of large molecules, such as proteins

umami one of five main taste sensations referring to that which is "meaty" or savory

umbilical arteries fetal blood vessels that carry blood containing wastes and CO_2 from the fetus into the fetal placenta

umbilical cord cord that connects the fetal circulation to the placenta

umbilical veins fetal blood vessels that pick up oxygen from the placental sinus and transport oxygenated blood to the fetus

uncoupling protein-1 mitochondrial protein that uncouples the membrane proton gradient from adenosine triphosphate synthesis

unitary-type smooth muscle predominant kind of smooth muscle in the gastrointestinal tract; it contracts spontaneously, contracts in response to stretch, and does not have structured neuromuscular junctions; contrast to *multiunit-type smooth muscle*

universal donor person with type O blood who can, therefore, donate to people with all blood groups

universal recipient person with AB blood group who can, therefore, receive blood from someone with any other blood type

unloading phase steep portion of the oxygen equilibrium curve that reflects the unloading of oxygen from hemoglobin

unsaturated iron–binding capacity (UIBC) calculated amount of transferrin not occupied by iron; UIBC equals total iron-binding capacity minus iron

unstirred water layer layer of poorly stirred fluid that coats the surface of the intestinal villi and that reduces the absorption of lipid digestion products

upregulation increase in the number of receptors on the surface of target cells, making the cells more sensitive to hormones or other agents

ureagenesis formation of urea, especially the metabolism of amino acids to urea

ureter tube that conducts the urine from the renal pelvis to the bladder; it consists of an abdominal part and a pelvic part, is lined with transitional epithelium surrounded by smooth muscle (both circular and longitudinal), and is covered externally by a tunica adventitia

urethra tube that carries urine from the urinary bladder to the outside of the body

urine titratable acid amount of acid excreted in the urine combined with buffers such as phosphate, determined by measuring the milliequivalents of strong base (NaOH) needed

urodilatin peptide produced by the kidneys that increases renal sodium excretion

uroguanylin polypeptide hormone produced by the small intestine that increases renal sodium excretion

uterus pear-shaped organ located in the abdomen of a female; the site of pregnancy

V_{max} hypothetical maximum velocity of shortening of an isotonic contraction of muscle; also the theoretically velocity obtained by muscle shortening with zero afterload; this value is essentially constant in a given skeletal muscle but can be altered by physiological and pathophysiological influences in cardiac and smooth muscle

V wave venous pressure wave that gradually increases during systole as blood continues to return to the heart while the atrioventricular valves are closed

V-type ATPase see *ATPase*

vaccine weakened or killed pathogen or piece of pathogen, administered to stimulate immunity

vagina tube extending from the uterine cervix to the vestibule

vagovagal reflexes pertaining to a process that uses both afferent and efferent vagal fibers: during the gastric phase of acid secretion, distention of the stomach stimulates mechanoreceptors, which stimulate the parietal cells directly through short local (enteric) reflexes and long vagovagal reflexes

vascular capacitance measure of the volume contained in a segment of blood vessel or vessels per unit of transmural pressure, measured in mL/mm Hg

vascular endothelial growth factor (VEGF) factor responsible for the growth of blood vessels (angiogenesis) at many sites in the body

vascular function curve relationship depicting how central venous pressure varies as a function of cardiac output

vascular tree branching of the pulmonary circulation, which parallels airway branching

vasoactive intestinal peptide (VIP) putative inhibitory neurotransmitter released by enteric musculomotor neurons to gastrointestinal muscles; also a putative excitatory neurotransmitter that is a potent stimulator of intestinal secretion by parasympathetic stimulation to increase blood flow to the salivary glands

vasovagal syncope loss of consciousness from profound fear or emotional stress; accompanying cardiovascular events include marked parasympathetic-induced bradycardia and depressed resting sympathetic vasoconstrictor tone, which cause a dramatic drop in heart rate, cardiac output, systemic vascular resistance, and thus, mean arterial pressure

vein any blood vessel that carries blood toward the heart. It is composed of three layers: an outermost elastic adventitia, a middle zone composed of circular layers of smooth muscle, and a single-cell lining of epithelial cells called the endothelium. Veins are thinner than arteries of similar outer diameter. They do not have structural rigidity and collapse without any positive transmural pressure in them

venipuncture process of obtaining intravenous access for the purpose of intravenous therapy or obtaining a sample of venous blood

venous admixture mixing of venous blood with oxygenated blood

venous–arteriolar response increase in arteriolar constriction in response to an increase in postcapillary venule pressure. It is a mechanism that helps prevent capillary hydrostatic pressure from becoming too high

ventilation movement of air between the mouth and alveoli

ventilation/perfusion ratio ratio of alveolar ventilation to simultaneous alveolar capillary blood flow in any part of the lung; because both ventilation and perfusion are expressed per unit volume of tissue and per unit time, which cancel, the units become liters of gas per liter of blood

ventral respiratory group (VRG) group of neuronal cells located in the ventral portion of the medulla that is active during inspiration and expiration

ventricular fibrillation random, chaotic electrical activation of all the ventricular muscle cells in which the heart no longer can effectively contract. It results in death of the person if not abolished by some intervention

ventricular tachycardia abnormally high heart rate resulting from an ectopic foci in any area of the ventricles

ventrolateral medullary area region of the medulla where neurons controlling cardiovascular and respiratory functions are located

ventromedial nuclei hypothalamic nuclei, which synthesize growth hormone–releasing hormone

venules small veins ~10 to 1,000 μm wide that drain capillary networks and pass blood to larger veins in the circulation. They are also involved with the control of intracapillary hydrostatic pressure and have some transport function between the bloodstream and the extracellular fluid surrounding all tissues, especially in the presence of histamine

vermis most medial portion of the cerebellar cortex

vertigo type of dizziness in which there is a feeling of motion even while stationary due to a dysfunction of the vestibular system in the inner ear; often associated with nausea and vomiting as well as difficulties standing or walking

very-low-density lipoproteins (VLDLs) one of two major classes of lipoproteins produced by the intestine; triglyceride-rich lipoproteins with densities less than 1.006 g/mL, made continuously by the small intestine during both fasting and feeding

vesicles small membrane-delimited spheres in nerve terminals filled with neurotransmitters whose contents are released into the synaptic cleft on stimulation of the nerve

vestibular apparatus sensory structures of the middle ear that are responsible for sensing body position and acceleration

vestibular nuclear complex groups of neurons in the pons and medulla that receive afferent axons from the vestibular portion of the inner ear

vestibulocerebellum medial portion of the cerebellum that is active in control of eye motions and equilibrium

vestibuloocular reflex (VOR) reflex eye movement that stabilizes images on the retina during head movement by producing an eye movement in the direction opposite to head movement, thus preserving the image on the center of the visual field

vestibulospinal tract axon pathway from the lateral vestibular nucleus to the spinal cord, which influences postural reflexes in the spinal cord

villi fingerlike projections of the mucosal surface that increase the surface area of the small intestine by about 30 times

virilization masculinizing effects of excessive adrenal androgen secretion on females

visceral hyperalgesia gut hypersensitivity; the appreciation, or exaggeration, of gut symptoms with stimuli (such as balloon distention of the gut) that would ordinarily not be noticed or considered noxious

 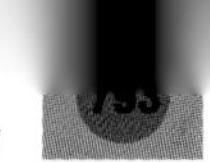

visceral pain sensation of pain arising from visceral parts (i.e., internal organs) of the body; often perceived as arising in somatic structures

viscoelastic materials substances whose mechanical reaction to applied forces is time-dependent

viscosity measure of the blood property that offers resistance to flow

vital capacity (VC) maximum volume of air that can be exhaled; the sum of expiratory reserve volume, tidal volume, and inspiratory reserve volume (note that this value is the same as *forced vital capacity*)

vitamin A lipid-soluble vitamin important in vision and in the normal growth of the skin

vitamin B1 (thiamine) water-soluble vitamin that plays an important role in carbohydrate metabolism; absorbed by the jejunum passively and by an active, carrier-mediated process

vitamin B2 (riboflavin) water-soluble vitamin that plays an important role in metabolism. It is absorbed by a specific, saturable, active transport system located in the proximal small intestine

vitamin B6 (pyridoxine) water-soluble vitamin involved in amino acid and carbohydrate metabolism; absorbed throughout the small intestine by simple diffusion

vitamin B12 (cobalamin) water-soluble, cobalt-containing vitamin that plays an important role in the production of red blood cells

vitamin C (ascorbic acid) water-soluble vitamin found in green vegetables and fruits that plays an important role in many oxidative processes by acting as a coenzyme or cofactor; absorbed mainly by active transport through the transporters in the ileum; uptake process is sodium dependent

vitamin D group of fat-soluble compounds collectively known as the *calciferols* that plays an important indirect role in the absorption of calcium by the gastrointestinal tract and is essential for normal development and the formation of bones and teeth; derived from the skin, which contains a rich source of 7-dehydrocholesterol that is rapidly converted to cholecalciferol when exposed to ultraviolet light and dietary vitamin D3; absorbed by the small intestine passively and incorporated into chylomicrons

vitamin D2 form of vitamin D generated by the irradiation of ergosterol; also known as *ergocalciferol*

vitamin D3 form of vitamin D normally manufactured in the skin, where ultraviolet light activates the compound 7-dehydrocholesterol

vitamin D–binding protein binding protein in plasma that transfers vitamin D3 during the metabolism of chylomicrons

vitamin E vitamin that is absorbed by the small intestine by passive diffusion and incorporated into chylomicrons, transported in the circulation associated with lipoproteins and erythrocytes, and is a potent antioxidant that prevents lipid peroxidation

vitamin K vitamin derived from green vegetables in the diet or the gut flora and incorporated into chylomicrons, where it is rapidly taken up by the liver and secreted together with very-low–density lipoproteins; essential for the synthesis of various clotting factors by the liver

vitreous humor thick clear gel that fills the bulk of the eyeball; provides support for the shape of the eyeball and its retinal structures; also known as *vitreous body*

voltage-dependent anion channel (VDAC) mitochondrial channel for metabolite diffusion across the mitochondrial outer membrane

voltage-gated ion channels ion-selective pores in the membrane of excitable cells that open (or close) in response to a change in membrane potential

vomeronasal organ (VNO) organ located at the base of the nasal cavity adjacent to the vomer bone in the nose (hence, its name); the sensory neurons of the VNO detect specific chemicals that give rise to animal scent

von Willebrand disease most common hereditary bleeding disorder; caused by deficiency of von Willebrand factor

von Willebrand factor a glycoprotein important for coagulation

wall stress force exerted on the wall of a cylindrical structure, as modified by the thickness of the wall of the structure, that would have the tendency to rip the wall apart as a result of a positive transmural pressure. It is directly proportional to both the transmural pressure and the radius of the cylinder but inversely proportional to the wall thickness. Wall stress is less in thick-walled vessels than in thin-walled vessels of the same internal radius when exposed to the same transmural pressure. This principle is applied to blood vessels and other hollow organs, such as the heart and urinary bladder

water diuresis excretion of a large volume of dilute urine

wavelength spatial period of the wave–the distance over which the wave's shape repeats

weak acid acid with a low dissociation constant, which consequently is poorly ionized in solution

Wernicke area area at the junction of the temporal and parietal lobes of the brain involved in sensory perception of language, usually on the left hemisphere

white blood cell (WBC) cell delivered by the blood to sites of infection or tissue disruption, where it defends the body against infecting organisms and foreign agents in conjunction with antibodies and protein cofactors in blood; five main types are neutrophils, eosinophils, basophils, lymphocytes, and monocytes; also known as *leukocyte*

white blood cell count measures the concentration of leukocytes in blood as an indicator of the body's immunologic stress state

white ramus branch of a spinal nerve that carries preganglionic sympathetic axons from the spinal cord to the sympathetic ganglia

whole-gut transit time time required for transit from the mouth to the anus

working memory mechanism whereby new experiences are processed, drawing on skills such as language and mathematical ability, to integrate new inputs in light of previously acquired learning

Wright-Giemsa common dual stain for peripheral blood and bone marrow smears

xanthine oxidase enzyme that catalyzes xanthine to urate. During ischemia-perfusion injury, hypoxanthine builds up and, in the presence of oxygen hypoxanthine, is converted to xanthine. Xanthine oxidase neutralizes xanthine by converting it to urate

xenobiotics hydrophobic compounds foreign to the body, some of which are toxic, that are metabolized by the liver. Xenobiotics are converted into hydrophilic compounds that are then absorbed and excreted by the kidneys

xenograft transplantation in which animal tissue is used to replace human organs

yellow marrow bone marrow of long bones consisting of fat and primitive blood cells

yolk sac extraembryonic membrane that contains a slight amount of yolk

Z line dark structure that crosses the center of the I band in muscle; sometimes termed a *Z disk* to emphasize its three-dimensional nature

zinc finger small globular protein structures that can coordinate zinc ions with cysteine residues to help stabilize their folds; they can be classified into families and typically function as interaction molecules that bind DNA, RNA, proteins, or small molecules

zona fasciculata inner zone of the adrenal cortex, which produces the glucocorticoid hormones: cortisol and corticosterone

zona glomerulosa outer zone of the adrenal cortex, which produces aldosterone

zona pellucida acellular membrane between the oocyte and membrana granulosa of an ovarian follicle, which surrounds the ovulated egg prior to implantation

zona reaction modification of the zona pellucida that blocks polyspermy; enzymes released by cortical granules digest sperm receptor protein so that they can no longer bind sperm

zona reticularis inner zone of the adrenal cortex, which produces the glucocorticoid hormones: cortisol and corticosterone

zones various regions of the lung (e.g., apex, base) through which blood flows, with respect to pulmonary arterial pressure, pulmonary venous pressure, and alveolar pressure

zygote fertilized egg

zymogen granules granules found in the apical region of acinar cells that store salivary amylase

Index

Note: Page numbers in *italics* indicate figures; those followed by "t" indicates table; those followed by "b" indicates boxes.

A

A band, of skeletal muscle, 141, *141*
α actinin, 141
α-melanocyte-stimulating hormone (α-MSH), 610
α motor neurons, 92–93
A wave, 252
abduction, defined, 91
abetalipoproteinemia, 526, 543
absence seizures, 127, *127*
absolute refractory period, 49
absorption
 of carbohydrates, 520–522, *520–522*
 disorders of
  abetalipoproteinemia, 526
  cystinuria, 528
  Hartnup disease, 528
  lactose intolerance, 522, *522*
 of electrolytes, 530–534
 of lipids, *523–525,* 523–526
 of minerals, 530–534
 of proteins, 526–528
 of vitamins, 528–530
 of water, 534
absorptive hyperemia, 302
absorptivity, defined, 556
accessory structures, in sensation, 62
acclimatization
 cardiovascular, 396
 cold, 568–570
 heat, 565–567, *566,* 568*t*
accommodation, in vision, 76
acetoacetic acid, 662
acetone, 662
acetylcholine (ACh)
 automatic nervous system, 109–110, *110*
 gastric secretion, 510
 neuromuscular junction, 143
 pancreatic secretion, 514
 psychiatric disorders, 133
acetylcholinesterase (AChE), 56, 143
achalasia, 490
acid
 defined, 451
 strong, 452
 weak, 452
acid–base balance
 acid production
  alkalinizing effect, 453–454
  nonvolatile, 453
  protein metabolism, 453
  source of carbon dioxide, 453
 balance disturbance
  clinical evaluation, 465t, 468–470, *469*
  compensation, 464
  metabolic acidosis, 466
  metabolic alkalosis, 467–468, 468t
  plasma anion gap, 466–467
  respiratory acidosis (*see* respiratory acidosis)
 blood pH regulation
  bicarbonate buffering system, 454, *454*
  chemical buffering (*see* chemical buffering)
  kidney, 457–458
  lungs, 457
 intracellular pH regulation, 462–463, *463*
 principle
  acid dissociation constant, 451–452
  conjugate base, 451
  Henderson–Hasselbalch equation, 452
  K_a logarithmic expression, 452
  pH stability, 452, *453*
acid dissociation constant, 451–452
acid tide, 514
acidemia, 364, 463
acidosis
 acid–base homeostasis, 464
 metabolic, 390, 466
 respiratory acidosis (*see* respiratory acidosis)
acinar cells, of pancreas, 512
acne vulgaris, 690
acoustic reflex, 78
acquired immune system. *see* adaptive immune system
acral regions, defined, 560
acromegaly, 616
acrosome, 685
acrosome reaction, 713
actin, 140
actin filaments, 80
actin-linked regulation, 147
action potential (s), 42
 cardiac, 228–229
 conduction velocity considerations, 50–51, *51*
 plateau region of, 229, *229*
 refractory periods, 49
 trigger zone, 47
action tremor, 106
activated partial thromboplastin time (aPTT), 185
activation gate, 47
active hyperemia, 289
active immunization, 197b
active membrane transport
 calcium pump, 33
 defined, 32
 proton pumps, 33
 sodium-potassium pump, *32,* 32–33
activin, 679, 694
activity front, migrating motor complex (MMC), 497
actomyosin ATPase activity, 157
acute mountain sickness, 395b
acute-phase blood proteins, 203
acute respiratory acidosis, 464–465, 468
acute respiratory distress syndrome (ARDS), 375
acyl-CoA cholesterol acyltransferase, 525, *525*
acyl-coenzyme A (acyl-CoA), 525
adaptation, 396
 sensation, 65, *65*
 skeletal muscle, 157–158
adaptive immune system, 188, 190
 antibodies of, 198–201
 cellular
  helper T cells, 197–198
  immunologic synapse, 194, *196*
  killer T cells, 197
  major histocompatibility complex, 194–195, *195, 196*
  memory T cells, 198
  T regulatory cells, 198
  T-cell differentiation, 195–197
  time delay in, 198
 clonal selection, 193
 defiined, 188, *190*
 diversity of, 193–194
 humoral
  antibody action, *200,* 200–201
  antibody classess, 198, 199t, 200
  antibody structure, 198, *199*
 memory of, 194, *194*
 specificity of, 193
adaptive relaxation, of gastric reservoir, 494
Addison disease, 438, 610, 640
adduction, defined, 91
adenohypophysis, 604–605, *605*
adenosine, 126
adenosine diphosphate (ADP), 147, 628
 oxidative phosphorylation of, 628
 thyroid hormone regulation of, 628
adenosine triphosphate (ATP), 139, 364b, 514, 610, 621, 653
adenylyl cyclase (AC), 12, 53, 164
adequate stimulus, 62
adipose tissue, lipogenesis in, 658
adipose triglyceride lipase (ATGL), 645
adrenal cortex
 cholesterol esters, 634
 dehydroepiandrosterone (DHEA), 634
 hormones, 634–647
  aldosterone (*see* aldosterone)
  cholesterol (*see* cholesterol)
  glucocorticoids (*see* glucocorticoids)
  steroids (*see* steroids)
 zona glomerulosa, 634
 zona reticularis, 634
adrenal gland
 disorders of
  Addison disease, 438
  congenital adrenal hyperplasia, 611, 640, 692, 729
  Cushing syndrome, 646b
 functional anatomy
  cortex, 633–634
  medulla, 633–634
adrenal hyperplasia, congenital, 611, 640, 692, 729

adrenal medulla
chromaffin cells, 634
hormones
catecholamines, 647–648, 647t
sympathetic postganglionic neurons, 647
adrenergic, defined, 109–110
adrenergic receptors, in autonomic nervous system, 57
adrenocorticotropic hormone (ACTH), 599, 605, 639
in cholesterol metabolism, 609
deficiency of, 611
proteolytic processing of, *609*
in regulation of adrenal cortex, 608–609
secretion of, 605
arginine vasopressin, 611
corticotropin-releasing hormone, 610, *610*
diurnal variation, 611–612, *612*
glucocorticoid hormones, 610–611, *611*
proopiomelanocortin, *609,* 609–610
stress-induced secretion, 611
adventitia, of blood vessels, 214
aerobic metabolism, muscle fibers and, 157
aerobic processes, 156
affect, 62, 132
affective disorders, 133–134
afterhyperpolarization, 49, 480
afterload
cardiac muscle contraction, 253, 254–356
defined, 150
age-related macular degeneration (AMD), 71b
aggression, limbic system mediation of, 132
aging
exercise, 586–587, *587*
influences on liver metabolism, 540
agonist muscle, 91
agonists, defined, 13
agranular leukocytes, 179
airflow
airway resistance, 349–350
lung volumes, 350, *350*
regional lung compliance, *344,* 344–345
airway
during forced expiration, 350–352, *351*
three-dimensional endoscope to scan, 380b
airway resistance
lung volumes, 350, *350*
major sites, *349*
overcoming, 354
smooth muscle tone effect, 350
airway tree, *327,* 327–328
albumin, 168, 544, 709
albuminuria, 410
aldosterone, 594, 609, 633
angiotensin II action, 643, *643*
angiotensin II formation, 643
synthesis, 643–644, *644*
alkalemia, acid–base homeostasis, 463
alkaline tide, 509
alkalinizing effect, acid–base homeostasis, 453–454
alkalosis
acid–base homeostasis, 464
defined, 390
metabolic, 467–468, 468t
respiratory, 390
allergies, 205, 207
allogenic, 181
allografts, 205
alpha cells, 594
alpha waves, 125
alveolar cells, 725
alveolar hypoxia, 373
alveolar pressure, 331
alveolar surface tension, 374, *374*
alveolar ventilation, 384–385
alveolar vessels, 373
alveoli
anatomic considerations, *327*
defined, 328
partial pressures, 330t
stability, 346, *346*
surface tension, 345–346
Alzheimer disease (AD), 21, 136
amacrine cells, for vision, 75
ambisexual stage, 722
amenorrhea
defined, 709
infertility, 709–710
lactational, 726
amine, 416
amino acids
liver metabolism of, *544,* 544–545
nonessential, 544–545
ammonia, 403, 417
amnesia
anterograde, 738
retrograde, 783
amniotic sac, 717
amphipathic, defined, 25
amphoteric, acid–base homeostasis, 455
ampulla, 713
amygdala, 118, 130
amylase, 505, 507
amylin, 651b
amylopectin, 521
amylose, 521
anaerobic metabolism
isometric exercise, 577–578, *578*
muscle blood flow, 304, 305
muscle fibers, 581
anaerobic threshold, 581
anal canal, 502
anal sphincter
external, 502
internal, 490
anamnestic response, 194
anaphylaxis, 207, 323
androgen-binding protein (ABP), 681
androgens, 594, 709
androstenedione, 699
anemia, 171, 366b
hemolytic, 177
iron-deficiency, 178
pernicious, 530
red blood cells (erythrocytes), 177, 178
sickle cell anemia, 171
aneurysms, 297
angina, 13b
exertional, 298
angiogenesis, 21, 186
angiotensin II formation, 643
angiotensin-converting enzyme (ACE), 369–370
angiotensinogen, 437
anion gap, metabolic acidosis, 468t
anions, composition in body fluids, 168
anisocytosis, 174
anorexia nervosa, 710
anorexigenic, 122
anovulation, 726
antagonistic pairs, 156
antagonists, 91
anterior lobe, 605
anterior periventricular region, 614
anterograde direction, 44
antibodies, 193, 198, *199,* 506, 507, 528, 717
classess, 198, 199t, 200
polyclonal, 193
structure, 198, *199*
anticoagulant, 167
antidiuretic hormone (ADH), 37, 432
antidromically, 116
antigen-presenting cells (APCs), 189
antigen-recognition molecules, 193
antigens, 189
epitopes, 193
immune system, 209
antioxidants, 21
antiport, 34
antral granulosa cells, 698
aorta, 213, 267–268, *268*
aortic baroreceptors, 313
aortic bodies, 389
aortic compliance, 267–268, *268*
aortic valve, 213
aphasias, 136
apical membrane, 36
apnea, 387
apo B proteins, 543
apoferritin, 533
apoproteins, 525
apoptosis, 22, *22,* 190
aquaporins, 37, 417
arachidonic acid, 60
arcuate nuclei, 614
arcuate nucleus (ARC), 122
areflexia, 98
areola, 725
arginine vasopressin (AVP), 37, 120, 311, 319, 432, 594, 605, 607, 611
proteolytic processing of, *607*
structural organization of, *607*
arrhythmias
compensatory pause, 243
complete atrioventricular block, 244–245
myocardial ischemia and, 231
premature atrial contractions (PACs), 243
respiratory sinus, 242, *244*
sinus bradycardia, 243
sinus tachycardia, 243
supraventricular tachycardia, 244
ventricular fibrillation, 244
ventricular tachycardia, 244
arterial baroreceptors, 313
arterial blood gases, 365t
arterial blood pressure
arterial compliance, 267–268, *268*
cardiac output and, 267, *267*
determinants of, 267–270, *267–270*
effects of exercise on, 269, *269*

factors affecting, 271, 272b
heart rate, 268–269, *269*
mean artrial pressure
cardiac output and, 267
defined, *267*
effects of exercise on, 269, *269*
equation for, 267
interactions of, 268–270, *269–270*
measurement of, 270–271, *271*
pulse pressure
arterial compliance and, 267–268, *268*
defined, 267
effects of exercise on, 269, *269*
interactions of, 268–270, *269–270*
stroke volume and, 267–268, *268*
systemic vascular resistance and, 269–270, *270*
arterial compliance
cardiac output and, 274–275
noninvasive assessment of, 274b
pulse pressure, 267–268, *268*
arteries, 213
blood circulation (*see* hemodynamics)
factors that affect muscular contraction, 215, 216t–217t
inferior hypophyseal, 606
pulmonary, 369, *370*
spiral, 307
umbilical, 306
arterioles, 213
arteriosclerosis, 274b
defined, 270b
systolic hypertension and, 270b
arteriovenous anastomoses, 305, *305*
arthritis
osteoarthritis, 584
rheumatoid arthritis (RA), 206, 584
ascending reticular activating system, 125
ascites, 539
association cortex, 129
asthma, 205
astigmatism, 70
ataxia, 106
atherosclerosis, 169
ATP-binding cassette (ABC) transporters, 33–34
atresia, 696
atrial fibrillation, 243, *244*
atrial natriuretic peptide (ANP), 311, 319–320, 438, 607
atrial septal defects, 309b
atrioventricular block, 244–245
atrioventricular (AV) node
functions of, 232–233
as secondary pacemaker, 245
as slow action potentials, 229, *229*
attention, 129
atypical antipsychotics, 134
auditory cortex, 82
auditory system
cochlea, 76
decibel (dB), 77
external ear functions, 77
hearing loss, 82
inner ear, 78–82
middle ear, 77–78, *78*
sound, 76–77
sound wave formations, *77*
Auerbach plexus, 479
augmented leads, 240
autograft, 205
autoimmune disease, 170, 171
thyroiditis, 630b
autoimmunity, 207
autologous, 181
automaticity, of cardiac cells, 228
autonomic nervous system (ANS)
efferent path of, 110, *110*
functions, 109
heart rate determination, 233
neurotransmitters, 110–111
overview of, 109–111
parasympathetic nervous system, 114–115
ganglia of, 115
sympathetic nervous system, 111–114 (*see also* sympathetic nervous system)
adrenal medulla interactions with, 114, *114*
ganglia of, 112, *113*
peripheral anatomy, *112*
autoregulation, 288–289, 407–408, *408*
axial streaming, 223–224
axon terminal, 51
axoneme, 685

B

band cells, 174–175
baroreceptor reflex
arterial pressure, 313
blood flow, 314–315
cardiac output and total peripheral resistance, 313–314
hormonal systems, 314
baroreceptors
aortic, 313
arterial, 313
cardiopulmonary, 313
carotid sinus, 313–314
Barr body, 722
Bartter syndrome, 424
basal forebrain nuclei, 135
basal ganglia, 128
disorders
Huntington disease, 104
Parkinson disease, 8
motor control, 103–104, 104*f*
basal metabolic rate (BMR), 121, 629
base, defined, 451
basement membrane, 409
basic metabolic panel (BMP), 169
basilar membrane, 79
basket cells, 105
basolateral membrane, 36
basophils, 179
B-cell receptors (BCR), 193
Becker muscular dystrophy (BMD), 142b
behavioral conditioning, cardiovascular responses, 317
benign paroxysmal positional vertigo (BPPV), 87b
benzodiazepines, 58
beriberi, 530
Bernoulli principle, 221
beta cells, 2, 288b
β-lipotropin, 609
beta waves, 125
bicarbonate, 352b, *362,* 362–363, 388–391, 532
bicarbonate buffering system, 454, *454*
bile acids, 516
bile canaliculi, 516
bile salts, 516
bile salt–sodium symport, 516
bilirubin, 177
biliverdin, 177
biologic clock, 123–124, *124*
biotin, 530
bipolar disorders, limbic system, 133
bitter receptors, 87–88
bladder, 448
blastocyst, 712
blastomeres, 715
bleeding time, 184
blind spot, 70
blood
circulation of (*see* hemodynamics)
clotting of, 182–186, *184* (*see also* hemostasis)
coagulation of
activated partial thromboplastin time, 185–186
coagulation cascade, 184, 185t
commom pathway in, 185
disorders of, 185–186
prothrombin time, 185
components of (*see also* blood cells)
plasma, 167
doping, erythopoietin and, 183b
functions, 166–167
indices for, 172
laboratory tests for
blood smears, 174–175, *175*
complete blood count, 172, 173t, 174
erythrocyte sedimentation rate (ESR), 171, *171*
hematocrit (Hct), 171
iron profile, 178, 178t
specific gravity, 170
pH regulation
bicarbonate buffering system, 454, *454*
chemical buffering (*see* chemical buffering)
kidney, 457–458
lungs, 457
viscosity, 171
volume of (*see* blood volume)
whole, 167
blood cells, 166, 167, 169, 170, *170,* 175–178
blood flow
anaerobic metabolism, 304, 305
baroreceptor reflex, 314–315
heart, 219–225
lungs, 376–378
skin in regulation of, 559–560
blood indices, 172
blood lipid profile, 169
blood pressure
diastole
defined, 216
factors affecting, 271, 272b
measurement of, 270–271, *271*
systolic
defined, 216

blood smear, 174
blood vessels
  factors that affect muscular contraction, 215, 216t–217t
  structure of, 213–214
blood volume
  central
    cardiac output and, 273–274
    central venous pressure and, 273
    defined, 272
  in circulation, 272
  extrathoracic, 272
  systemic, 272
  vein compliance, 272
  venous pressure in, 273
blood–brain barrier, 120, 388
blood–placental barrier, 717
body buffer, 455t
body fluids, 430t
body movement, 91–92
body plethysmograph, 335, *338*
body temperature, 551–574
  in cold acclimatization, 568
  in cold response, 567–570
  control of (*see* thermoregulation)
  core
    distribution of, 551, *551f*
    factors influencing, 563
    in heat acclimatization, 565–567, *566*
    measurement of, *552,* 552–553
    normal, 552
    set point and, 562–563
    thermal receptors lor, 562
  extreme, effects of, 567
  fever, 563
  in heat acclimatization, 565–567
  heat balance in, 553–556
  heat dissipation
    skin blood flow, 559–560
    sweating, 558
  heat exchange
    convection, 551–555, *555,* 559
    evaporation, *555,* 555–556, 558–559
    radiation, *553,* 553–556
    surface area and, *556,* 556–557
  heat production
    vs. heat loss, *553,* 553–554
    metabolism in, 554, 558
  heat storage, 554, 558
  heat stress
    disorders related to, 571–573
    effects of, 559–560
  heat transfer in, 551–553, 551*t*
  hypothermia, 568, 572b–574b
  nonthermal effects on, 563
  normal range, *551,* 569b
  shell, 551–553, 558, 559
    in cold reponse, 567–570
    conductance, 558–560, 562
  skin in regulation of
    blood flow and, 559–560
    convection, 552, 556, *557*
    evaporation, 557
    radiation, 556, 557, *557*
    surface area and, 556–558, *557*
  skin temperature
    in heat exchange, 552–553
    set point and, 562–563
    in thermoregulation, 561
body water
  extracellular fluid (ECF) in, 427
  in fluid compertments, 427–432
  intracellular fluid (ICF) in, 427
  total
    by age, 428
    distribution of, 428
    measurement of, 429
body weight, 122, 618b
Bohr effect, 360
bone
  disorders of
    osteogenesis imperfecta, 668, 669b
    osteoporosis, 673
    Paget disease, 675
    rickets, 529, 673–674, 674t
  exercise effects on, 582–584
  remodeling, 669
botulinum toxin, 144
Boyle law, 331
bradycardia, 243
brain
  functions of, 120t, 121–123, *122*
  split, 129b
brain natriuretic peptide (BNP), 438
brainstem
  motor control, 98–100, 99*f,* 100*f*
  parasympathetic nervous system, 115
brainstem neurons, parasympathetic nervous system, 115
breathing. *see also* ventilation
  airflow, 328–333
  changes in thoracic volume during, 328, *329*
  feedback control, 387–389
  medulla oblongata, 382–384
  pressures in, 328–333
    partial, 330, 330t
    pleural, 331
  during sleep
    arousal mechanisms, 394–395
    inspiratory flow rate, 392
    responsiveness to carbon dioxide, 392
    upper airway tone, 395
  unusual environments, 394–397
    acclimatization, 396
    cardiovascular acclimatization, 396
    diving reflex, 397
Broca area, 136, *136*
Brodmann cytoarchitectural map, 101*f*
bronchial circulation, 327, 380
buffer, chemical
  bicarbonate/carbon dioxide buffer
    acid–base physiology, 455
    Henderson–Hasselbalch equation, 455–456
    open system, *456,* 456–457
  changes, acid production, 457, *457*
  isohydric principle, 457
  metabolic acidosis, 466
  metabolic alkalosis, 468
  phosphate buffer, 454–455
  protein buffer, 455
buffy coat, 167
bulk flow, 356
bundle of His, 232

C

C-fiber endings, 387
C-peptide, 595b, 651
C wave, 252
Ca^{2+}-binding protein, 532
calbindin D (CaBP), 532
calcitonin, 622, *622,* 670
calcium, 532
  absorption in small intestine, 667
  in bone
    composition, 667–668
    neonatal development, 668
    osteoblasts, 668, *668*
    osteoclasts, 668, *668*
    osteocytes, 668, *668*
  composition in body fluids, *445,* 445–446
  distribution in bone cells, in healthy adult, 665, 665t
  inorganic constituents, 665, 665t
  kidneys, 667
  nerve and muscle function, 664
  plasma, 665
    calcitonin, 670
    exchangeable bone calcium pool, 669–670
    parathyroid hormone, 670, *670*
    protein-bound calcium, 669
    target tissues, 672–673
    vitamin D3 and 1,25-dihydroxycholecalciferol, *670,* 670–672, *671*
calcium channels
  in cardiac action potential(s), 228–229
  L-type, 230
calcium homeostasis, 665–666, *666*
calcium pumps, 33, 147
calcium release channel, 19, 146
calcium sparks, 249
calcium-induced calcium release, 19, 161, 249
calmodulin (CaM), 19, 161
calor, defined, 201
cAMP-dependent protein kinase, 16
canaliths, 87b
cancer immunotherapy, 209
capacitance, 46
capacitation, 713
capillaries, 213
  capillary distention, 371
  capillary filtration coefficient (CFC), 286
  diffusion, 358–359
  glomerular filtration
    capillary hydrostatic pressure, 412–413
    capillary pressure, 410
  microcirculation
    anatomic considerations, 280, *281*
    defined, 280
    endothelial vesicles in, 281, *281*
    pores in, 281
    porosity of, 281
  recruitment, 371
capillary distention, 371
capillary filtration coefficient (CFC), 286
capillary recruitment, 371
carbohydrates
  absorption, 520–522
  digestion, 520–522
  metabolism of, 540–541, *541*

carbon dioxide
chemoresponses
blood oxygen, 390–391
exercise-induced hyperpnea, 391–392
effects on oxyhemoglobin equilibrium curve, 361, *361*
carbon dioxide equilibrium curve, 363
carbonic acid, 362, 452–454, 467, 509, 532
carbonic anhydrase, acid–base homeostasis, 455
carboxypeptidase A, 528
carboxypeptidase B, 528
carboxypolpeptidase, 512
cardiac action potentials, 228–229
cardiac cells
automaticity of, 228
rhythmicity of, 228
cardiac cycle, *251,* 251–252
rapid ejection phase, 251–252, 252
reduced ejection phase, 251–252, 252
venous pressure waveforms, 252
ventricular diastole, 252
cardiac dipole, 236
cardiac function curve, 276, *276*
cardiac muscle
contraction
action potential and isometric twitch, *249*
afterload, 253, 254–356
calcium, 249–250, *250*
cardiac cycle, 251–252
ejection fraction, 258–260
excitation–contraction coupling, 249–251
force generation, 250
force–velocity curves, 255
isotonic, 254, *254*
length–tension curve, *254*
preload, 253, 254–356
pressure–volume loops, 256, *256*
stroke volume, 256–258
electrophysiology of
cardiac action potential, *228,* 228–229
depolarization, 229, *229*
pathological conditions, 234–235
repolarization, 230–231
resting membrane potentials, 229, *229,* 230t
voltage-gated sodium channels in, 229–230
vs. skeletal muscle, 227
cardiac output, 220
arterial compliance and, 274–275
arterial pressure and, 267, *267*
blood volume and, 273–274
cardiac function curve, 276, *276*
cardiac index, 260
effects on pulmonary vascular resistance, 371–372, *372, 373*
equilibriums in, *276,* 276–277
factors influencing, 260t
heart rate, 260–261
imaging technique, 263–265, *264*
indicator-dilution technique, *262,* 262–263
mean artrial pressure, 267, *267*
measurement, 262–265
vascular function curve, 476, *476*
venous filling pressure and, *275,* 275–276
stroke volume, 260–261
cardiac tamponade, 257
cardiopulmonary baroreceptors, 313
cardiovascular system
autonomic innervation, *312*
behavioral conditioning, 317
blood flow to organs, 215
blood vessel structure, 214–215
blood volume in, 218–219, *219*
circulation (*see* circulation)
dietary prevention, 324b
disorders of
arterial hypertension, 223b
atherosclerosis, 224b
estrogen benefits, 322b
functional organization of, 213–214, *215t, 216t*
hemodynamics, 217–225
nitric oxide (NO), 318b
output of right and left heart, 215
overview, 213–214, *214*
pressure profile across, 218–219, *219*
carotid bodies, 389
carotid sinus baroreceptors, 313–314
caspase, 22–23
catalase, 21
cataracts, 70
catechol-O-methyltransferase (COMT), 57
catecholamines, 555, 647–648, 647t
cations, 168
caudate nucleus, 103
caveolae, 25, 159
celiac ganglion, 112
celiac sprue, 521b
cell death, 22, *22*
cell model
HCO_3^-/CO_2 reabsorption, *460*
renal synthesis, *461*
titratable acid formation, *461*
cells
blood (*see* blood cells)
muscle (*see* muscle fibers)
nerve (*see* neurons)
cellular signaling
autocrine signaling, 7
cell-surface receptors
G protein–coupled receptors (GPCRs) in, 10–12, *11, 12*
ion channel-linked receptors in, 12
types of, 10
tyrosine-kinase receptors in, 12–14, *14*
first messengers, 9
gap junctions in, 6–7, *7*
hormone receptor signaling, 14–15, *16*
hormones in, 8–9
molecular basis of, 9–15
paracrine signaling, 7, *8*
second messengers in, 9, 15–21
signalling cascade in, 10
transducers, 9
central auditory pathways, 82
central autonomic network, 118, *118*
central blood volume, 272
central inspiratory activity (CIA) integrator neurons, 384
central pattern generator (CPG), 98
central sulcus, 128
central venous pressure, 273
centroacinar cells, 513
cephalic phase, 510
cerebellar peduncles, 105
cerebral circulations
autoregulation, *300*
blood pressures, 297–298
cerebral edema, 300
chronic hypertension, 300, *300*
sympathetic nerve activity, 299–300
vasodilation by CO^2 and H^+, 298–299
cerebral cortex, 128
cerebral edema, 300
cerebral peduncles, 103
cerebrocerebellum, 105
cervical ganglion
superior, 508
cervical mucus, 713
cGMP-dependent protein kinase (PKG), 7, 164
channelopathies, 49
Charcot-Marie-Tooth (CMT) disease, 59
Charles law, 331
chemical buffering
bicarbonate/carbon dioxide buffer
acid–base physiology, 455
Henderson-Hasselbalch equation, 455–456
open system, *456,* 456–457
changes, acid production, 457, *457*
isohydric principle, 457
metabolic acidosis, 466
metabolic alkalosis, 468
phosphate buffer, 454–455
protein buffer, 455
chemical synapse, 51
chemoattractants, 186
chemoreceptor, 62, 315–316, 387
chenodeoxycholic acid, 516
chest leads, 240
chest wall reflexes, 386–387
Cheyne-Stokes breathing, 392
chief cells, 508
chloride, 532
cholecystokinin (CCK), 122, 505
cholesterol, 25, 523
acyl-CoA, 636
adrenal steroid catabolism, *640,* 640–641
adrenal steroid transport, 640
apoprotein B100, 636
cholesterol acyl-transferase (ACAT, 636
cholesterol esterase, 635
genetic defects, 639–640
high- density lipoprotein (HDL), 636
3-hydroxy-3-methylglutaryl CoA reductase, 636
low-density lipoprotein, 636
oxidative cytochrome P450 enzymes
adrenarche, 637
17α-hydroxylase (CYP17), 637
17α-hydroxypregnenolone, 637
17α-hydroxyprogesterone, 637
aldosterone synthase (CYP11B2), 637
androstenedione, 637
11β-hydroxylase (CYP11B1), 637

- cholesterol (*Continued*)
    - 3β-hydroxysteroid dehydrogenase (3β-HSD II), 637
    - cholesterol side-chain cleavage enzyme, 636
    - 11-deoxycorticosterone (DOC), 637
    - 21-hydroxylase (CYP21A2), 637
    - isocaproic acid, 636
    - pregnenolon, 636
    - progesterone, 637
    - steroidogenic acute regulatory protein (StAR), 636
  - pregnenolone formation, 635, *636*
- cholesterol esterase, 524, 635
- cholesterol side-chain cleavage enzyme, 699
- cholic acid, 516
- choline acetyltransferase, 55
- cholinergic, 110
- cholymicrons
  - characteristics of, 543, 543t
  - defined, 543
  - functions of, 543
- chondrocytes, 668
- chorionic villi, 717
- choroid plexuses, 388
- choroidal neovascularization, 71b
- chromaffin cells, 113, 634
- chromatids, 700
- chromophore, 74
- chromosomal sex, 722
- chronic granulomatous disease, 179
- chronic myeloid leukemia (CML), 10
- chronic respiratory acidosis, 465, 468
- Chvostek sign, 664
- chylomicrons, 169, 525
- chyme, 508
- chymotrypsin, 512, 527, 528
- ciliary ganglion, 115
- cilium, 74
- cingulate gyrus, 130
- circadian rhythm, 123, 611
- circulation
  - bronchial, 380
  - cerebral, 297–300, *300*
  - coronary, 295–297, *297*
  - cutaneous, 305–306
  - fetal, 306–309
  - hepatic, 303–304
  - hormonal control, 317–321
    - arginine vasopressin, 319
    - atrial natriuretic peptide, 319–320
    - epinephrine, 317–318, *318*
    - renal hypoxia, 320
    - renin-angiotensin-aldosterone system, 318–319, *319*
    - short- and long-term blood pressure control, 320–321, *321*
  - neural control of, 311–317
    - arterial pressure, 313
    - brain and heart, 314–315
    - carotid sinus baroreceptors, 313–314
    - central blood volume, 315
    - chemoreceptor, 315–316
    - higher-order ANS responses, 316–317
    - hormonal systems, 314
    - pain and myocardial ischemia, 316
    - parasympathetic and sympathetic nerves, 312–313
    - pressure-and time-dependent, 315
  - placental, 306–309
  - shock (*see* circulatory shock)
  - skeletal muscle, 304–305
  - small intestine, *301,* 301–303
  - systemic
    - arterial pressures in, 267–271, *267–271*
    - blood volume, 271–274
    - cardiac function in, 274–277
- circulatory shock
  - anaphylaxis, 323
  - compensated, 321–322
  - etiology, 321
  - irreversible, 323
  - neurogenic, 323
  - progressive, 322–323
  - septic, 323
- circumventricular organs, 120
- climbing fibers, 106
- clomiphene, 691
- clonal selection, 193
- clonus, 98
- closure, 528
- clot retraction, 186
- coagulation cascade, 184
- coagulation factors, 184
- coated pits, 27
- cochlea, 79, *79*
- cochlear implants, 84b
- cold acclimatization, 568–570
- cold diuresis, 567
- cold pain, 68
- cold pressor response, 316
- cold stress, 562
- colipase, 524
- collagenase, 179, 701
- collateral ganglia, 112
- collateral vessels, coronary, 297
- colloid, in thyroid gland, 622, *622*
- colloid osmotic pressure, 285
- colloidal osmotic blood pressure, 168
- colon, 474, 500–502
- colostrum, 528, 725
- comfort, in thermoregulation
- commissures, 128
- compensations, acid–base homeostasis, 464
- compensatory pause, 243
- compensatory response, acid–base disturbances, 465t
- competitive binding assays, 598
- competitive exclusion, 191
- complement fixation, *200,* 200–201
- complement proteins activation, 190
- complete blood counts (CBCs), 168
- complete metabolic panel (CMP), 169
- complete proteins, 526–527
- compliance, 218
- compression, 67
- concentrations
  - in body fluid
  - measurement of
- concentric contraction, 91, 583
- conducting zone, 327
- conductive deafness, 82
- cone cells, 73
- congenital adrenal hyperplasia, 611, 640, 692, 729
- congenital nephrotic syndrome, 409
- conjugate base, 451
- connective tissue matrix, 160
- connexin, 6, *7*
- connexon, 6, *7*
- contraceptives, 727b
- contractility, 250
- convection
  - in body heat exchange, 551–555, *555,* 559
  - defined, 556
- cornea, 70
- corneal reflex, 76
- coronary artery disease
  - drug-eluting stents, 299b
  - exertional angina, 298b
- coronary circulation
  - autoregulation of, 297
  - blood pressure and, 295–296, *297*
  - collateral vessels, 297
  - oxygen demand and, 296
  - sympathetic nervous system in, 296–297
- coronary reserve, 295
- corpus callosum, 128
- corpus luteum, 696
- cortical granules, 715
- cortical motor area, 100
- cortical tissue, 722
- corticospinal tract, 101, 102–103, 103*f*
- corticosteroid-binding globulin (CBG), 596
- corticosteroid-binding protein (transcortin), 709
- corticosterone, 608, 633
- corticotropin, 605
- corticotropin-releasing hormone (CRH), 606, 610, *610*
  - actions of, 607t
  - structure of, 607t
- cortisol, 594, 608, 633
- cotyledons, 717
- coumarin, 185
- countercurrent heat exchange, 567
- coupling reaction, in thyroid hormone synthesis, 624
- craniosacral, 109
- creatine phosphate pool, 156
- creatinine, 404
- cremasteric muscle, 679
- critical micellar concentration, 524
- crossbridge cycle, 142, *142*
- crypts of Lieberkühn, 519
- cumulus granulosa cells, 698
- curare, 144
- Cushing reflex, 300
- Cushing syndrome, 646b
- cutaneous circulations, 305–306
- cyanosis, 176
- cyclic adenosine monophosphate, 595
- cyclic adenosine monophosphate (cAMP), 9, 164, 511, 610, 651
  - TSH effects on, 626
- cyclic guanosine monophosphate (cGMP), 7, 117, 682
- cyclooxygenase (COX), 438
- cystic fibrosis transmembrane conductance regulator, 514

cystinuria, 528
cytochrome P450 enzyme, 540
cytochrome P450 reductase, 540
cytokines, 18, 570
cytoskeleton, 26
cytotrophoblasts, 716

D

D cells, 509
D$_1$-like receptors, 57
D$_2$-like receptors, 57
Dalton law, 330, 356
dark adaptation, in vision, 75
dead space volume, 337
decerebrate rigidity, 100
2D-echocardiography, 265
decibel (dB), 77
decidua, 716, 717, 723
decidua basalis, 717
decidua capsularis, 717
decidua parietalis, 717
decidual reaction, 718
decidualization, 716
declarative memory, 135
decoding, in sensory transduction, 66
deep pain, 68, 316
defecation, 116, 471, 487, 499, 502–504
defensins, 179, 191
degranulation, 180
dehydration reaction, acid-base homeostasis, 455
dehydroalanine residue, 624
7-dehydrocholesterol, 671
dehydroepiandrosterone (DHEA), 608, 699
dehydroepiandrosterone sulfate (DHEAS), 720
deiodinase type 1, 624, *625*
deiodinase type 2, 624
deiodination
  5′, 625
  inner ring, 624
  outer ring, 624
  in thyroid hormone matabolism, *625,* 626
delayed afterdepolarizations (DADs), 235
delayed gastric emptying, 496b
delta 4 pathway, 699
delta 5 pathway, 699
delta waves, 125
delusions, 134
dementia, 135
dendritic cells, 192
denervation hypersensitivity, 318
dense bodies, 159
density, 170
dentate, 105
deoxycholic acid, 516
deoxyhemoglobin (Hb), 359
depolarization, in cardiac muscle, 229, *229*
depolarizing blockers, 145
desensitization, 602
diabetes insipidus, 37, 434, 607
diabetes mellitus, 407
  autoimmune disorder, 660–661
  obesity and insulin resistance, 661
  organ dysfunction
    diabetic neuropathy, 663
    metabolic and renal complications, 662
    microcirculatory complications, 662–663
  type 2 diabetes treatment, 661–662
diabetic ketoacidosis, 662
diabetic neuropathy, 663
diabetogenic action, 617
diacylglycerol (DAG), 611
dialysis, 401b
diapedesis, 202
diaphragm in breathing, 328–330
diastole, defined, 213
diastolic pressure, 216
diencephalon, 119
differential white blood count, 174
diffuse toxic goiter, 630
diffusing capacity, 358–359
diffusion
  alveolar-capillary membrane, 356–357
  blood hematocrit, 359–360
  in capillaries, 358–359
  pulmonary capillary blood flow, 357–358
diffusion-limited transport, 284
digestion
  carbohydrate, 520–522
  lipids, 523–526
  protein, 526–528, *527,* 527t
dihydropyridine receptors (DHPRs), 146
dihydrotestosterone (DHT), 709
1,25-dihydroxycholecalciferol, 529, 671
diiotyrosine (DIT), 624
dimerization, 13
dipeptidyl peptidase IV inhibitors, 655b
direct pathway, 104
distal nephron
  potassium secretion, 418–419, *419*
  proximal reabsorption, 418
diuresis, cold, 567
diurnal rhythm, 123
divergence, 112
diving reflex, 397
diving response, cardiovascular responses, 317
docking, 52
docosahexaenoic acid, 523
dolor, defined, 202
dominant follicle, 696
dopamine (DA), 8, 56, 606
doppler ultrasound, 265
dorsal motor nuclei, 115
dorsal motor nucleus, 476
dorsal respiratory group (DRG), in breathing rhythm, 383, *383*
dorsal vagal complex, 476
down-regulation, 602
drugs, metabolism of, 539–540, *540*
drusen, 71b
Duchenne muscular dystrophy (DMD), 142b
duct intercalated cells, *459*
ductus arteriosus, 308
ductus venosus, 308
dynamic exercise, 269, *269*
dysphagia, 491b
dyspnea, 581
dystonias
  treatment of, 145b
  types of, 145b
dystrophin, 141

E

early afterdepolarizations (EADs), 235
eccentric contractions, defined, 583
eccrine sweat glands, 559
echinocytes, 177
echocardiography, 265
ectopic foci, 234
ectopic pregnancy, 713
edema, 442
  myxedema, 631
Edinger-Westphal nucleus, 115
effective arterial blood volume (EABV), 433
effector, 3
effector caspases, 23
effector T cells, 194
eicosapentaenoic acid, 523
Einthoven triangle, 239, *240*
ejaculation, 683
ejection fraction, 258
elastase, 528
electrical synapses, 51
electrical syncytial property, 473
electrocardiogram (ECG)
  cardiac stress testing with, 246b
  dipole recording in, 236, *236,* 237–239, *238, 239*
  information from
    drug effects, 246
    mean QRS axis, 242, *243*
    metabolic conditions, 246, *246*
    ventricular myocardium, 245, *245*
  leads for, 239–241, *240, 241*
  normal waveforms, 236–237, *237*
electroencephalogram (EEG)
  abnormal brain waves, 127–128
  brainwave patterns on, 125–126, *126*
  defined, 125
  functions of, 125
  sleep stages, 126–127
electroencephalography (EEG), 125
electrogenic system, 36
electrolyte absorption, 530–534, *531, 533, 534*
electromechanical coupling, 473
electroneutral system, 36
electroneutrality, 168
electrotonic conduction, 46
emboli, 369
embryoblast, 716
embryonic hemoglobin, 176
emissivity defined, 556
emotional stress, cardiovascular responses, 316
emphysema, 582b
end-diastolic volume, 252
endocannabinoids, 60, 122
endocrine control, circulation, 317–321
endocrine system
  anatomic considerations, 589, *590*
  feedback regulation in, *591,* 591–592
  pleiotropic effects, 592
  signal amplification, 592
endogenous creatinine clearance, 404
endolymph, 79
endometrial cycle, 706
endometrial veins, 307
endopeptidase, 511, 527
endorphins, 210

endothelial vesicles, 281
endothelial-derived relaxing factor (EDRF), 17
endothelin, 291
endothelium, 214
endothelium and endothelial cells
  of liver, 536–537
  sinusoidal, 536–537, *537*
endothelium-derived relaxing factor (EDRF), 289
endotoxin, 202
endplate current, 143
endplate membrane, 143
endplate potential, 143
end-product inhibition, 3
end-systolic pressure-volume relationship, 256
end-systolic volume, 252
enkephalins, 210
enteric nervous system
  gastrointestinal musculature, 483–484
  inhibitory motor innervation, 486–487
  inhibitory musculomotor activity, 484–485
  intestinal circular muscle layer, 484
  musculomotor neurons, 484–485
  synaptic transmission
    ionotropic receptors, 481–482
    metabotropic receptors, 481
    presynaptic facilitation, 483
    presynaptic inhibition, 482–483
    slow excitatory postsynaptic potentials, 481
enteroceptors, 62
enterochromaffin cells, 472
enterochromaffin-like cells, 511
enterocytes, 505, 520
enteroendocrine cells, 472
enterogastrones, 511
enterohepatic circulation, 517
entero-oxyntin, 510
enteropeptidase, 527
enzyme-linked immunosorbent assay (ELISA), 600, *600*
enzymes
  in drug metabolism, 540, *540*
  of glycolytic pathway, 541
  pancreatic, 512–513
eosinophils, 179, 193
epidermal growth factor (EGF), 9, 186, 507
epididymis, 682
epilepsy, 125
epinephrine (EPI), 56, 112, 634
epiphyseal plate, 668
epithelial sodium channel, 86
epitopes, 193
equilibrium, fluid compartments in, 430–431
erectile dysfunction (ED), 13, 115, 682–683
erythroblasts, 181
erythrocyte sedimentation rate (ESR), 171, *171*
erythrocytes, 167
erythromelalgia, 56
erythropoiesis, 26, 181
erythropoietin, 181, 320, 396
esophageal motility, 490
esophageal motor disorders, 491b
essential fatty acids, 523
essential light chain, 140
estradiol, 698
estrogens, 594, 608, 718
estrone, 699
eunuchoidism, 691
eustachian tube, 78
euthyroid sick syndrome, 631b
evaporation
  in body heat exchange, *555,* 555–556, 558–559
  latent heat of, 556
  of sweat, 556
excitatory junction potentials, 484
excitatory postsynaptic potential (EPSP), 53
excretion, 403
  of electrolytes, 429
  renal clearance, 403
excretory (collecting) duct, 506
excretory duct, of salivary gland, 506
execution, 129
exercise
  aging, 586–587, *587*
  calcium homeostasis, 584, *584*
  caloric intake, 585, *585*
  cardiovascular response to, 316–317
    acute and chronic exercise, 578–579
    blood flow, 577–578, *578*
    cholesterol level, 579
  diabetes, 585–586, *586*
  dynamic, 269
    acute respiratory response to, 581t
    cholesterol, 579
  effects on arterial pressure, 269, *269*
  gastric emptying and intestinal absorption, 585
  immune response, 587
  ischemic heart, ECG, *579*
  isometric
    defined, 576
  lung dysfunction detection, 581–582
  obesity, 585
  oxygen uptake
    dynamic exercise, 575–576, *576*
    isometric exercise, 576
    maximal oxygen uptake, 576
    maximal voluntary contraction (MVC), 576
    oxygen debt, 575, *576*
  pregnancy and, 580
  respiratory response
    anaerobic threshold, 581
    effects of training, 581
    emphysema, 582b
    hypercapnia, 581
    minute ventilation, *580*
  skeletal muscle, 582–584
  thermoregulation during, 564–565, *565, 566*
  training, 575
    cardiovascular disease, 580
    endurance, 583, 583t
    respiratory system, 581
exertional angina, 298b
exocrine glands, 505
exocytosis, 27–28, 595
exopeptidases, 528
expiration, 384
expiratory reserve volume (ERV), 334
expired minute ventilation, 336
extension, 91
extensors, 156
exteroceptors, 62
extracellular fluid (ECF)
  defined, 430
  measurement of, 429
  osmolality of, 430
  osmotic equilibrium and, 430–431
  pH of, 446
  in potassium balance
  potassium effects on
  types of
extraction (E), 284
extraction ratio, 284
extrafusal muscle fibers, 92
extramedullary hematopoiesis, 181
extraocular muscles, 70
extrathoracic blood volume, 272
exudate, 202

F

fab region, of antibody, 198
farads, 46
fast axoplasmic transport, 44
fastigial, interposed, 105
fast-twitch fibers, 157
fat
  brown, 568
  dietary, 271, 508, 525
  soluble vitamins, liver storage of, 545, *545*
fatty acids, 542–543
fear, cardiovascular responses, 316
fecal continence, 502
feedforward control, *3,* 4
female athletic triad, 710
female pronucleus, 715
female reproductive system
  ampulla, 695, *695*
  cervix, *695,* 696
  cornification, *695,* 696
  corpora albicantia, 695, *695*
  corpora lutea, 695, *695*
  endometrium, *695,* 696
  estrogen, progestin, and androgen metabolism, 708–709
  ferning, *695,* 696
  fimbria, 695, *695*
  follicle selection and ovulation
    corpus luteum, 702–703
    dominant follicle, 701
    luteinizing hormone, 701–702
  follicles, 695
  follicular atresia, 700
  folliculogenesis, *696–697*
  graafian follicle, 697–698, *698*
  hormonal regulation
    germ cell, 693
    hypothalamic-pituitary, *694,* 694–695
  infertility (*see* infertility)
  infundibulum, 695, *695*
  isthmus, 695, *695*
  meiosis, 700–701
  menstrual cycle, 703–708
  myometrium, *695,* 696
  oogonia, 696
  oviduct, 695, *695*
  spinnbarkeit, *695,* 696

steroidogenesis, 698–700
stromal layer, *695,* 696
uterus, *695,* 696
vagina, *695,* 696
ferritin, 177, 533
ferroportin, 177
fertilization
implantation
acrosome reaction, *714,* 714–715
blastocyte, 715–716
decidualization, 716
embryoblast, 716
ovum transport, 713
sperm transport, 713
fetal development
of central nervous system (CNS), 628
endocrine system
growth, 721
hormones, 721
hormones, 722–723
parturition, 723–724
sex chromosomes, 722
thyroid hormones in, 628
fetal hemoglobin, 308, 717
fetus
circulation, 308
heart, 306
vasculature of, 306–307, *307*
fever
pyrogens in, 570
thermoregulation during, 564, *565*
fibrin, 169
fibrin stabilizing factor (factor XIII), 184
fibrinolysis, 186
Fick law, 356
fight-or-flight response, 132, 316
filtration fraction, 413
filtration pressure equilibrium, 411
filtration slit, 409
fimbria, 713
final common path, 92
first heart sound, 251
first messengers, 9
first polar body, 700
first-degree atrioventricular block, 245
flexion movement, 91
flexor withdrawal reflex, 97–98, 98*f*
flexors, 156
flow velocity, 220
flow-limited transport, 284
flow-mediated vasodilation (FMD), 290, 292b
fluid balance, composition in body fluids
extracellular fluid osmolality, 432, 432t
hyponatremia, 435b
plasma osmolality, 434
arginine vasopressin disorders, 434
habit and thirst, water intake, 434–435, *436*
water balance, 432
arginine vasopressin, 432–434
fluid compertments
body water in, 428, *428,* 428t
extracellular fluid (ECF) in, 430–432, *431*
intracellular fluid (ICF) in, 428–430, *431*
in osmotic equilibrum
size, measurement of, 427–428
fluid-phase endocytosis, 27, *27*
focal adhesions, 159
focal length (FL), 70
focal point, 70
folia, 105
folic acid, 530
follicle-stimulating hormone (FSH), 605, 693
follicular phase, 704
folliculogenesis, *696–697*
follistatin, 679, 694
foramen ovale, 308
fore brain, cerebral cortex, 128–130
fornix, 121
fresh frozen plasma (FFP), 167
frontal lobe, 128
F-type ATPases, 34
functio laesa, 202
functional residual capacity (FRC), 334, 373
functional syncytium, heart as, 228
fused tetanus, 148, *148*
fusimotor system, 94

G

G cells, 510
G protein–coupled receptors (GPCRs), 473, 484
G proteins
ras family of monomeric, 14
trimeric, 10–12
GABAergic neurons, 58
galactopoiesis, 724
gallstones, 519
gametes, 680
γ-aminobutyric acid (GABA), 58
gamma globulins, 169
γ motor neurons, 92
ganglia, cervical, 508
gap junctions, 51, 160, 472
gas exchange, pulmonary circulation, 369–370
gas tension, 330
gas transport
alveolar–arterial oxygen gradient, 363–364, *365*
bicarbonate, *362,* 362–363
carboxyhemoglobin, *361,* 361–362
of oxygen, 359–361
oxyhemoglobin-equilibrium curve, 360
gastric amylase, 511
gastric inhibitory peptide, 511
gastric lipase, 511
gastric motility, 490–495
excitatory musculomotor neurons, 493
gastric reservoir, 493–494
meal type, 494–495
motor behavior, antral pump, 490–493
relaxation, 494
gastric mucosa, 508
gastric phase, 510
gastric secretions
anatomic considerations, 508
chyme in, 508
endocrine control of, 511, *511*
gastric acid, 509, *509*
hormonal control of, 511, *511*
neural control of, 511, *511*
oxyntic glands in, 508–509, *509*
parietal cells in, 508–509, *509*
phases of digestion, 509–510, *510*
production of, 510, *510*
rate of, 509, *510*
gastrin, 509
gastrin-releasing peptide, 510
gastrointestinal motility patterns
emesis, 488–489
peristalsis, 487–488
peristaltic reflex circuit, 488
sphincters
ileocolonic, 490
internal anal, 490
lower esophageal, 489
pyloric, 489–490
sphincter of Oddi, 490
gastrointestinal secretions
biliary, 515–519
intestinal, 519, *519*
pancreatic, 511–515
saliva, 505–508
of stomach, 510
general gas law, 331
generator potentials, 63
genetic considerations, in liver metabolism, 540
genetic sex, 688, 720, 722
genital ridge, 722
gestational diabetes, 660, 721
GH-releasing hormone (GHRH), 606
ghrelin, 122
gigantism, 616
Gitelman syndrome, 424
glaucoma, 70
glial cells, 42
glicentin, 595
globulins, 169
globus pallidus (GP), 99
glomerular filtration
capillary hydrostatic pressure, 412–413
capillary pressure, 410
filtration coefficient, 413
glomerular ultrafiltration coefficient, 412
layers of, 409, *409*
osmotic pressure and hydrostatic pressure, 413
physical properties, 409–410, 409t, *410*
vascular resistance, 410–412, *412*
glomerular filtration rate (GFR), 403, 436
inulin clearance, 403, *404*
plasma creatinine, 404
plasma urea concentration, 405, *405*
renal blood flow, 407
serum creatinine concentration, 405, *405*
glomerulonephritis., 411b
glomerulus, 400, *400*
glucagon, 595
glucagon-like peptide-1, 595
glucagon-like peptide-2, 595
glucocorticoid hormones, 608, 610–611, *611*
glucocorticoids
fasting, *644,* 644–645
inflammation, and immunity, 645–646
negative-feedback control, 647
stress, 646
glucokinase, 541, 653
gluconeogenesis, 645
glucose
absorption of, 652
blood, 542
in glucagon secretion, 655
handling by kidneys, 406
glucose reabsorption, 406, *406*

glucose threshold, 406
glucose-galactose malabsorption (GGM), 34
glucosuria, 662
gluten-sensitive enteropathy, 521b
glycine (GLY), 58
glycinergic neurons, 58
glycogen phosphorylase, 658
glycogen synthesis, *657,* 657–658
glycogenesis
defined, 541
in metabolism of glucose, 541
glycogenolysis, 114, 541
glycolipids, 25
glycolysis, 156, 653
glycolytic pathway, 156
glycoproteins, 26
goblet cells, 519
goiter, diffuse toxic, 630
Goldman equation, 46
Golgi apparatus, 43
Golgi cells, 105
gonadal dysgenesis, 729
gonadal sex determination, 722
gonadarche, 703
gonadotrophs, 605
gonadotropin-releasing hormone (GnRH), 123, 606, 677, 693
high frequency, *678,* 678–679
low frequency, *678,* 678–679
in puberty sexual development, 729
gonadotropins, 608, 617–619
graded response, 63
graft-*versus*-host disease (GVHD), 181, 206
granular cells, 402
granular layer, 105
granule cells, 105
granulocytes, 179
granulosa cells
antral, 698
cumulus, 698
granulosa lutein cells, 702
granzyme, 193
Graves disease, 630, 630b, 670
gray ramus, 112
ground substance, 668
growth factors, 186
growth hormone (GH), 605
actions of, *614,* 617t
disorders of
Creutzfeldt-Jakob disease, 617b
deficiency, 616
excessive secretion, 616
expression of, thyroid hormones in, 628
inhibition of, 615, *615*
insulin-like growth factor I, 615
pulsatile hormone_secretion, 592b
pulsatile secretory pattern, 615–616, *616*
regulation of growth, 613–614, 616–617, 617t
regulation of metabolism, 616–617, 617t
secretion, 605, *614,* 614–615
synthesis, 614
growth hormone–releasing hormone (GHRH), 607t
actions of, 607t
structure of, 607t
guanosine diphosphate (GDP), 11
guanosine triphosphate (GTP), 11
guanylin, 439
guanylyl cyclase (GC), 7, 53
gustation, 85
gustatory system
chemoreception
bitter receptors, 87–88
salt receptors, 86
sour receptors, 86–87
sweet receptors, 88
umami receptors, 88
taste buds, 85–86, *86*
gustducin, 88
gynecomastia, 688
gyrus, 128

H

H band, 141
H_2 receptors, 511
H zone, of skeletal muscle, 141, *141*
habenula, 130
hair cells, 79–81
anatomic considerations, 80
inner, 79, 80
outer, 79–81
sound transduction in, 79, *80*
hair-follicle receptors, *67,* 113, 191
Haldane effect, 363
hallucinations, 134
hamartomas, 704b
haptens, 189
haptoglobin, 177
Hartnup disease, 528
Hashimoto disease, 630, 630b
haustration, 500
H_2CO_3. *see* carbonic acid
HCO_3. *see* bicarbonate
HCO_3-/CO_2 system, 456, *456*
hearing, 77, 125, 129, 134
hearing loss
cochlear implants for, 84b
conductive, 82
sensorineural, 82
heart, 213
blood flow
autoregulation of, *291,* 292
baroreceptor reflex, 313, 314
electrical activity of
cardiac action potential, *228,* 228–229
depolarization, 229, *229*
pathological conditions, 234–235
repolarization, 230–231
resting membrane potentials, 229, *229,* 230t
voltage-gated sodium channels in, 229–230
as functional syncytium, 228
output of right and left, 215
heart block, 234, 245, 253
heart failure
muscle mechanics, 259b
treatment, 261b
heart murmurs, 223
heart rate
acute exercise effects, 243
arterial blood pressure, 268–269, *269*
autonomic nervous system and, 249, 260
in heat acclimatization, 566
sinoatrial (SA) node, 260
heart sound, 251, 252
heart valves
anatomic considerations, 214–215
heat
as energy exchange, 553, 554
extreme, effects on body, 567
measurement of, 558
as by product of metabolism, 554–558
heat acclimatization, 565–567, *566,* 568*t*
heat balance, equation for, 553–554
heat dissipation
skin blood flow, 559–560
sweating, 558
heat exchange
conduction in, 551–554
convection in, 551–555, *555,* 559
countercurrent, 567
equation for, 556
evaporation in, *555,* 555–556, 558–559
heat dissipation
skin blood flow, 559–560
sweating, 558
heat production
vs heat loss, *553,* 553–554
at rest, 554–558, 554*t, 555*
radiation, *553,* 553–556
skin temperature, 552–553
surface area and, *556,* 556–557
total health flow in, 552
heat exhaustion, 571
heat flow, equations for, 560
heat injury, exertional, 572
heat storage, 558
heat stress
disorders related to, 571–573
effects of, 559–560
exercise and, 565
heat stroke, 572–573
heat syncope, 571
heat transfer. *see* heat exchange
heel stick, 168
helicotrema, 79, 81
helium-dilution technique, for residual volume, 335
hematocrit (Hct)
blood disorder, 171
blood flow velocity and, 222
defined, 171
hematologist, 167
hematopoiesis, 180–181
hematopoietic stem cells, 10, 181
hematopoietins, 181
heme oxygenase (HO), 18, 533, 546
hemifacial spasm, 145b
hemochromatosis, 547
hemocytometer method, 175
hemodialysis, 401b
hemodynamics, 217
axial streaming in, 224–225
Bernoulli principle, 222
defined, 217
flow velocity in, 221–222, *222*
fluid dynamic properties and, 223–224, *224*
physical determinants of, 219–225
Poiseuille law, 220–221

pressure in, 217–218, *218*
compliance, 218
defined, 217
effects of gravity, 217, *218*
profile across cardiovascular system, 218–219, *219*
pulmonary arteries, *222,* 222–223
transmural, 218
vascular capacitance, 218
Reynolds number, 223
shear stress in, 222
total resistance of multiple vascular segments, 221
turbulence in, 223
wall stress in, 219, *219*
hemoglobin (Hgb or Hb), 166
hemoglobinopathies, 176
hemolysis, 26, 167, 177
hemolytic anemia, 177
hemopexin, 177
hemophilia, 185
hemorrhage, 167
hemosiderin, 177
hemostasis
activation of, 185
angiogenesis, 186
blood coagulation in, 182–183
clot retraction in, 186
defined, 166–167
platelet plug formation in, 183–184
steps in, 182–186
vasoconstriction in, 182
Henderson–Hasselbalch equation, 452
Henry law, 356
heparin, 180, 193
heparin therapy, 185
hepatic arterial buffer response, 304
hepatic circulations, 303–304
hepatic portal vein, 538
hepatocytes, 515
hepcidin, 178
hereditary spherocytosis, 26
Hering–Breuer reflex, 386
hermaphroditism, 729
heterogametic, 722
heterologous desensitization, 602
hexokinase, 541, 658
hexoses, malabsorption of, 35b
hidromeiosis, 564, 566
high-density lipoproteins (HDLs), 34, 169, 518, 579
hilum, 399
hippocampus, 130
hirsutism, 690
histamine, 180, 193, 645
H^+/K^+-ATPase, 33, 509
homeostasis
coordinated body activity, 6
defined, 2
equilibrium *vs.* steady state, 4–6, *5*
forward control system in, *3,* 4
internal environment and, 2, *2*
intracellular, 2–3
negative feedback in, *3,* 3–4
positive feedback in, 4
homeotherm, 166, 550, 551
homogametic, 722
homologous desensitization, 602
hopping reaction, 102
horizontal cell, *73,* 75
hormonal mediators, 210
hormone control
of circulation, 317–321
arginine vasopressin (AVP), 319
atrial natriuretic peptide (ANP), 319–320
blood pressure, 319–320
of gastric secretion, 511, *511*
renal hypoxia stimulates, 319
renin–angiotensin–aldosterone system (RAAS), 318–319
hormone replacement therapy, 322b, 517, 630b, 672, 708, 708b
hormone response element (HRE), 15
hormones
amino acid and polypeptide hormone transport, 595
C-peptide, 595
defined, 589
degradation and excretion, 597–598
hormone receptor, 590–591
measurement of
bioassay measurements, 598
immunosorbent assay, 600, *600*
radioimmunoassay, 598–600, *599*
mechanisms of
dose-response relationship, 601–602, *602*
hormone-receptor binding kinetics, 600–601, *601*
pancreatic β-cell function, 595
peptide-derived hormones, *594,* 594–595, *595*
secretory vesicles and exocytosis, 595
peripheral transformation, 597
pleiotropic effects, 592
polypeptide hormone, 593, *593,* 594t
preprohormones, 594
prohormones, *594,* 594–595, *595*
receptors for, 590–591
releasing, 591
receptors for, 590–591
secretion pattern, 592
simple hormone, 593, *593*
steroid and thyroid hormone transport, *596,* 596–597, 596t
steroid hormones, *593,* 594
target cells, 590
tropic, 591
hormone-sensitive lipase (HSL), 616, 645
Horner syndrome, 114
human chorionic gonadotropin (hCG), 593, 681, 701, 712
human leukocyte antigen (HLA), 206
human placental lactogen (hPL), 718
humidity, defined, 558
humoral immune system, 197
Huntington disease, 104
hydration reaction, acid–base homeostasis, 455
hydraulic conductivity for water, 286
hydrochloric acid (HCl), 509
hydrogen, 175, 179, 304, 360, 440
hydrogen ion concentration, 3, 387, 452
hydrogen peroxide (H2O2), 20, 179
hydrophilic, defined, 25
hydrophobic, defined, 25
hydrostatic pressure, 285, 286, *286,* 287, 293, 302, 322, 375, 376
hydroxyapatite crystals, 667
hydroxyl radical, 20
5-hydroxytryptamine (5-HT), 58
hyperaldosteronism, acid–base homeostasis, 462
hypercalciuria, 445, *447*
hypercapnia, 581
hypercoagulation disorders, 186
hyperemia
absorptive, 302
active, 289, 292b
reactive, 292b, 293, 293b
hyperglycemia, 660
hyperkalemia, 215b, 227, 229, 246, *246,* 442, 443
hyperkinesis, 104
hypernatremia, 569
hyperopia, 70
hyperosmotic, defined, 37
hyperphagia, 585
hyperphosphatemia, 447
hyperpnea, 390
hyperpolarization, 49, 53, 58, 75, 81, 83, 164, *215, 216,* 231, 232
hyperprolactinemia, 692, 710
hypersensitivity reactions, 198, 207, 481, 528
hypertension
arterial, 222b, 272b, 309b
chronic, 222b, 302b, 309b
hypoxia-induced pulmonary, 375b
portal, 539b
hyperthermia
defined, 571
due to lesion, 571
malignant, 572–573
therapeutic, 573
hyperthyroidism, 630
hyperventilation, 390
hypervitaminosis A, 545
hypoaldosteronism, acid–base homeostasis, 462
hypocalcemia, 664
hypocalcemic tetany, 664
hypocapnia, 364, 390
hypochromic cells, 177
hypocretin, 124
hypogastric nerves, 448, 683
hypoglycemia, 609, 647–648
hypogonadism, 691
hypergonadotropic, 692
hypogonadotropic, 691–692
hypogonadotropism, 709
hypokalemia, 229, 231, 234, 246, *246,* 442, 443
hypokinesis, 104
hyponatremia, 434, 435b, 569b, 570b
hypophyseal arteries
inferior, 606
superior, 606
hypophyseal portal circulation, 606
hypophyseal portal vessels, 606, 612
hypophysiotropic hormones, 605
hypophysis. *see* pituitary gland
hypopituitarism, 677
hyposmotic, defined, 37
hypotension, 303

hypothalamic (or central) diabetes insipidus, 607
hypothalamic-pituitary axis, 604
hypothalamic-pituitary-adrenal axis, 610
hypothalamic-pituitary-GH axis, 615, *615*
hypothalamic-pituitary-thyroid axis, 612, 613, 630
hypothalamo-hypophyseal portal blood vessels, 119, *121*
hypothalamo-hypophyseal tract, 120
hypothalamus, 118
  components of, 119–120, *120*
  functions of
    biological clock, 123–124, *124*
    metabolism and eating behavior, 121–123, *122*
    reticular formation, 124–125, *125*
    sexual activity, 123
  hormones in, 120
  interactions with pituitary gland, 120
  limbic system, 121
  nuclei of, *121*
    paraventricular, 605
    suprachiasmatic, 121
    supraoptic, 605
  structure of, 119–120, *120*
hypothermia, 568, 572b–574b
  accident, 573
  classification of, 573b
  defined, 571, 573
  diagnosis of, 574b
  due to lesion, 571
  etiology of, 573b
  signs and symptoms of, 573
  treatment of, 574b
hypothyroidism, 630–631
hypotonia, 107
hypotonic, 38
hypoventilation, 341, 365, 366, 367, 380b, 390, *391*, 394, 456
hypovolemia, 433
hypoxemia, 363–368, 364, 373, 390
  diffusion block, 366–367
  generalized hypoventilation, 366
  pathophysiologic causes, 365, 365t
  respiratory dysfunction, 364–367
  shunt, 365, *367*
  venous admixture, 365, *367*
hypoxia, 373, 387, *388*, 394
hypoxia-induced pulmonary hypertension, 375b

I

I band, 141
I cells, 515
IgA antibody, 198
IgD antibody, 198
IgE antibody, 193
IgG antibody, 192, 198
IgM antibody, 198
ileocolonic sphincter, 490
ileus
  paralytic, 478, 489
  pathological, 478, 489
immune deficiency, 207
immune response, cell death and, 194
immune surveillance, 209
immune system, *189*
  activation of
    large complex molecules and proteins, 189
    severe injury and necrosis, 189–190
  adaptive, 190
    antibodies of, 198–201
    cellular, 194–198
    clonal selection, 193
    defined, 188, *190*
    diversity of, 193–194
    humoral, 198–201
    memory of, 194, *194*
    specificity of, 193
  antigens in, 209
  cells of, *189*
  defense mechanism
    physical barriers, 191
  disorders of
    allergies, 205
    asthma, 205
    cancer and, 207, 209
    etiology, 206–207
    rheumatoid arthritis, 206
    types of, 208t
  functions of, 188
  inflammation and
    acute inflammation, 201–204
    anti-inflammatory drugs, 204–205
    chronic inflammation, 204
  innate, 188, *190*
    cellular elements, 192t
    characteristics, 192t
    components of, 188–189, *189*
    defense mechanism, 191–193
    defined, 191
    humoral component of, 191
  neuroendoimmunology, *209*, 209–210
  organ transplantation, 205–206
  organs, *189*
  overview of, *190*
  self tolerance in, 191
  structures of, 198, *199*
immune therapies, 210
immune thrombocytopenia purpura, 184
immunity
  adaptive, 188
  humoral, defined, 194
  innate, 188, *190*
immunization. *see also* vaccinations
  active, 209
  passive, 197b
immunogens, 189
immunoglobulins. *see* antibodies
immunologic synapse, 194
immunosenescence, 210
immunosuppressive drugs, 209
immunosuppressive effects, 521
implantation, 712
implicit memory, 135
impotence, defined, 682, 690–692
impulse initiation region, 63–64
in vitro fertilization, 710, 714b
inactivation gate, 47, 48, 49, 145, 160
incisura, 252
incomplete proteins, 527
incretin effect, 652
incretin hormones, 653, 655
incus, 78
indicator-dilution technique, 262
indirect pathway, 104
inducible form of nitric oxide (NO) synthase in septic shock, 323
induction, of labor, 725b
infant respiratory distress syndrome (IRDS), 733
inferior cervical ganglia, 112
inferior hypophyseal arteries, 606
inferior mesenteric ganglion, 112
inferior olive, 106
inferior salivatory nuclei, 115
inferior vena cava (IVC), 214, 231, 272, 273, 308, 309, 538
infertility
  amenorrhea, 709–710
  treatment of, 710
inflammation
  acute
    biomarker, 202–203
    characterization, 201–202
    inflammatory mediators, 202
    self-terminates, 203–204
    vasodilation, 202
  chronic
    characterization, 204
    new treatments, 204
infundibular process, 605, *605*
infundibular stem, 605, *605*
inherited peripheral neuropathy, 59
inhibin, 694
inhibitors of apoptosis, 23
inhibitory junction potentials (IJPs), 484
inhibitory postsynaptic potential (IPSP), 53
innate immune system, 188, *190*
  cellular elements, 192t
  characteristics, 192t
  components of, 188–189, *189*
  defence mechanism
    nonphagocytic leukocytes, 192–193
    phagocytic leukocytes, 192, 192t
  defense mechanism, 191–193
  defined, 191
  humoral component of, 191
inositol trisphosphate (IP_3), 611
inotropic state, 250
inspiration, 384, *384*
insular and medial prefrontal areas, 118
insulin, 122, 595b
  carbohydrate and fat metabolism, 658
    GLUT4, 656–657
    glycogen synthesis, *657*, 657–658
    glycolysis, 658
    lipogenic and antilipolytic effects, 658, *658*
    pleiotropic action, 656, *657*
    protein synthesis and degradation, 658
  degradation of by liver, 548
insulin-like growth factor I (IGF-I), 615
insulin-like growth factor II (IGF-II), 615
insulin-like growth factors, 9
integral proteins, 26
integrins, 141
intention, 129
intercalated cells, acid–base homeostasis, 459
intercalated duct, 506
intercellular signaling, *6*

interferon-γ, 203
interferons, 191
interleukin-1, 18, 203
interleukin-6, 203
interleukin-8, 203
intermediate filaments, 159
intermediate lobe, 605
intermediate zones, 105
intermediate-density lipoproteins, 169
internal capsule, 103
international normalized ratio, 185
interstitial cells of Cajal (ICCs), 472
interstitial fluid, 430t
intestinal phase, 510
intestinal secretions
  crypts of Lieberkühn, 519, *519*
  goblet cells, 519
  mucoproteins, 519
  paneth cells, 519
  protective functions, 519
intrafusal muscle fibers, 92
intralaminar nuclei of the thalamus, 125
intrinsic factor, 509, 511, 530
inulin, 403
inverse myotatic reflex, 97, 97*f*
iodine, in thyroid hormone synthesis, 626
iodothyronine, 622, *622*
ion channels, 12, 228–229
ionization constant, acid–base homeostasis, 451–452
ionotropic, 53
iron, 533, 547, *547*
iron homeostasis, liver in, 546–547
iron profile, 178, 178t
iron-deficiency anemia, 178
irritant receptors, 387
ischemia
  myocardial
    arrhythmias, 231
    ectopic foci related to, 234
    reduced refractory period in, 231
  reperfusion, 364b
ischemia-reperfusion injury, 293b
ischemic phase, 706–707
islets of langerhans
  diabetes mellitus (*see* diabetes mellitus)
  functional units
    α cells, 650
    β cells, 650
    cell types, 650, *650*
    δ cells, 650–651
    F cells, 651
    vascular supply, 650
  glucagon
    cAMP-mediated, 658
    gluconeogenesis, 659, *659*
    glycogenolysis, 659, *659*
    ketogenesis, 659
    lipolysis, 659
    ureagenesis, 659
  hyperglycemia, 655–656
  hypoglycemia
    glucagon secretion, 655, 655t
    proglucagon synthesis, 655
  insulin (*see* insulin)
  insulin/glucagon ratio
    eating and fasting, 660
    type 1 diabetes, 660
  physiologic factor
    β-cell signaling, 652–654, *654*
    insulin-secretion regulation, 652, *653*
    proinsulin biosynthesis, 651–652, *652*
isoelectric potential, 236
isograft, 205
isohydric principle, 457
isometric contraction, 91, 150
isometric exercise, 576
isosmotic, defined, 37
isotonic, defined, 38
isotonic contraction, 91
isovolumic contraction, 251
isovolumic relaxation, 252
Ito cells, 538

J

J receptors, 387
jejunum. *see* small intestine
junctional complex, 146
juxtaglomerular apparatus, 402, *403*
juxtaglomerular nephron, 402
juxtapulmonary capillary receptors, 387

K

kallikrein, 507
Kallmann syndrome, 677
Kennedy disease, 139b
ketoacidosis, *457,* 468t, 544, 662
ketogenesis, glucagon in, 659
ketone bodies
  for energy, 542, 543
  during fasting, 645
ketone, production of, glucagon in, 544, 645
ketones, in diabetes mellitus complications, 225b
ketosis, 457, 544
17-ketosteroids, 641
kidney micropuncture technique, 413
kidney, pH regulation
  net renal acid excretion, 458
  new bicarbonate generation, 460–461, *461*
  reabsorption of filltered bicarbonate, 459–460, *460*
  renal excretion, 461–462, *462*
  titratable acid, 458
  urinary acidification, *458,* 458–459, *459*
kidney stones, 447b, 665
kidneys
  in acid–base balance, 427, 442–443, 457, 458
  in blood p^H^ regulation, 454–462
  blood supply to, 223b, 399–400
  in calcium balance, 445–446
  in calcium homeostasis, 665–666
  collecting duct system, 402, 413, 417
  distal nephron, 417–419
  functional anatomy, 321–622, 633–647
  glomerular filtration, 409–413
  glucose titration, 406, *406*
  inherited defects, 424–426, 425t
  innervation of, 113
  overview, 399–403
  in phosphate balance, 446–447
  proximal tubule (*see* proximal tubule)
  proximal tubule transport, 409–413
  structure and function
    blood supply, 399–400
    cortex, 399
    glomerulus, 400, *400*
    homeostatic functions, 400–401
    juxtaglomerular apparatus, 402, *403*
    medulla, 399
    nephron, 401–402, *402*
    ureters, 399
  transplantation of, 205, 401b
  urinary concentration and dilution (*see* urinary concentration and dilution)
  urine formation, 403–407
kidney tubule epithelium, *462*
kilocalorie, in heat measurement, 554
kinesin, 44
kininogen, 184, 306, 507
kinins, 645
kinocilium, 80, 83, 84
Klinefelter syndrome, 684, 692, 729
Korotkoff sounds, 271
Krebs cycle, 58, 156
Kupffer cells, 192, 537, *537,* 546, *546,* 547
Kussmaul respiration, 467b

L

labeled line, 63, 256
labor
  and delivery, 721, 723, 724
  hormones of, 721, 722, 727b
  induction of, 725b
laboratory tests, of blood, 175
  blood smears, 175
  complete blood count (CBC), 168, 172, 173
  erythrocyte sedimentation rate (ESR), 171, *171*
  plasma, 171
  serum tests, 171
  specific gravity, 171
lactase, *521,* 522
lactate, 156
lactate threshold, during exercise, 581
lactation, 608, 724
  lactogenesis, 724, 725
  milk ejection, 608, 724, 726
  oxytocin in, 607–608
lactational amenorrhea, 726
lactic acid, 156
lactiferous ducts, 725
lactoferrin, 507
lactogenesis, 724
  defined, 724
  process of, 725
lactose, 35b, 520, 521, *521,* 521b, 522, 725
lactose deficiency, *522*
lactose intolerance, 35b, 521b, 522
lactotrophs, 605, 607, 619, 710, 720
lacuna, 716, *716*
Lambert-Eaton myasthenic syndrome (LEMS), 49
lamellar inclusion bodies, 347–349
laminin, 141
laminin, in skeletal muscle, 141
langerhans cells, 192

language, association cortex in, 135–136
large intestinal motility
  ascending colon, 500
  defecation, 503–504
  fecal continence, 502
  power propulsion, 500–502
  somatosensory nerves, 502
  transverse colon, 500
large intestine. *see also* colon
  motility of, 489, 495–504
    anal canal, 502
    ascending colon, 499, 500
    sigmoid colon, 500–502, 504
    transverse colon, 500–502
  secretions of, 513
laser-coagulation, for macular degeneration, 71b
latch state, 163, *163*
latch state, of smooth muscle, 162–163
latency period, 96
lateral inhibition, of retina, 75
law of Laplace, 218, 280
leading contractions, of antral pump, 493
"leak" channels, 160
leak channels, in smooth muscle contraction, 160
learning, 134–135
  cholinergic innervation in, 135
  long-term potentiation in, 134
  studies of, 134
lecithin cholesterol acyltransferase (LCAT), 525
left atrium, 213, 214
left atrium, anatomical considerations, 213
left bundle branch block, 245
left ventricle, 213
left ventricle, anatomical considerations, 213
left ventricular end-diastolic volume (LVEDV), 255, *257*
left ventricular hypertrophy, 242
length-tension curve
  of cardiac muscle, 249, *254,* 256
  of skeletal muscle, 159
  of smooth muscle, 164
leptin, 122, 618b
leukemia, 207
leukocytes, 167
leukocytosis, 174
leukopenia, 174
leukotrienes, 645
lever system, 155–156, *156*
Lewis hunting response, 570
Leydig cells, 608, 677
LH-releasing hormone (LHRH), 606
Liddle syndrome, 424
ligand, 13
limbic system, 120, 128
  aggressive behavior, 132
  cerebral cortex, 128–130
  dopaminergic pathway, 131, *131*
  language and speech, 135–136, *136*
  learning and memory systems, 134–135
  limbic reward system, 132
  long-term and short-term memory, 135, *135*
  monoaminergic innervations, 130–131, *131*
  noradrenalin pathways, 131–132, *132*
  psychiatric disorders, 133–134
  serotonergic neurons, 132
  sexual arousal, 132–133
  structure of, 128, *128*
lipid microdomains, 25–26
lipid raft, 10, 25
lipids
  absorption, 523–526
  digestion, 523–526
  metabolism, liver, 540–544, *541*
lipocortins, 636
lipolysis, 114, 616
lipopolysaccharide (LPS), 194–195
Lipoprotein lipase, 543
lipoproteins
  characteristics of, 543, 543t
  high-density, 543, 543t
    characteristics of, 543, 543t
  liver metabolism of, 543
  liver synthesis of, 544
  low density
    characteristics of, 543, 543t
    liver metabolism and removal of, 543
  very-low-density
    characteristics of, 543, 543t
    functions of, 543t
    synthesis of, 543
lithium-induced polyuria, 37
lithocholic acid, 516
liver, 536–548
  anatomic considerations, *537*
  in blood glucose regulation, 541
  disorders of
    abetalipoproteinemia, 543
    ascites, 539
    blood glucose regulation and, 543
    cirrhosis, 539b
    portal hypertension, 538, 539b
    viral hepatitis, 538b
  endocrine functions of, 548
  functions of, 536
  hormone degradation of, 548
  in iron homeostasis, 546, *546*
  lymphatic functions of, 539
  metabolism in
    of ammonia, 544
    of carbohydrates, 540, *540*
    of cholesterol, 544
    of coagulation proteins, 547
    of drugs and xenobiotics, 540, *540*
    of glucose, 540, *541*
    influences on
    of lipids, 540
    of lipoprotein, 542, *542*
    of proteins, *547,* 547–548
    of urea, 544
  nutrient supply to, 541
  plasma proteins synthesized by, 544
  regeneration of tissue of, 539
  storage of fat-soluble vitamins, 545, *545*
  synthesis of blood-clotting proteins, 545, *546*
loading phase, 360
local axon reflex, 116
local excitatory current, 63
locus coeruleus, 131
long hypophyseal portal vessels, 606
long-term memory, 135
long-term potentiation, 134
loop of Henle
  ascending, water impermeable, 417
  sodium, 417
Lou Gehrig disease, 21
loudness, 82
low-density lipoproteins (LDLs), 169, 525, 579
  characteristics of, 543, 543t
  liver metabolism and removal of, 543
lung cancer, 368b
lung inflation reflex, 386
lung injury, free radical-induced, 364b
lung volumes
  effect on pulmonary vascular resistance, 372–373, *373*
  forced vital capacity, 334–335, *337*
  functional residual capacity, 334
  minute, 336–341
  residual lung volume, 335–336, *338*
  residual volume, 334
  spirometry, 333–336
  tidal volume, 334
lungs
  age-related changes, 328, 328t
  airway tree, *327,* 327–328
  alveolar pressure, 333
  blood flow, 376–378
    gravity, 376, *376*
    regional ventilation, 377, *378*
    ventilation/perfusion ratios, 378, *378*
    zones of, 376, *377*
  blood–gas interface, 328, *328*
  diffusing capacity, 358–359
  free radical-induced lung injury, 364b
  reflexes, 386–387
  spirometry, 333–336
  structural and functional relationships, 327–328
  ventilation/perfusion ratio, 364
luteal insufficiency, 702b
luteal phase, 706
luteinization, 608, 702
luteinizing hormone (LH), 605, 676, 693
luteinizing hormone–releasing hormone (LHRH)
  actions of, 607t
  structure of, 607t
lymph, 282
lymph nodes, 189
lymphagogic effect, 525
lymphatic bulbs, 282
lymphatic system, 539
lymphatic vessels, 189, 282, *282*
lymphoblasts, 181
lymphocytes, 174
lymphoid leukocytes, 179
lymphokine-activated killer (LAK) cells, 193
lymphokines, 210, 645
lysozyme, 179
lysyl-bradykinin (kallidin), 507

M

M line, of skeletal muscle, 141, *141*
macrocytes, 177
macrophages, 22, *22,* 169, 192, 525
macula densa, 402
maculae, 87b

macular degeneration, 71b
magnesium, 532
balance, 430t, 446
composition in body fluids, 446, *446*
magnetic resonance imaging (MRI)heart, 225b
magnocellular neurons, 605
major depression, 133
major histocompatibility complex (MHC), 190, 194
male infertility, 683b
male pronucleus, 715
male pseudohermaphrodites, 730
male reproductive system
androgen action
peripheral tissue, 688–689
puberty, 690
secondary sex characteristics and masculine phenotype, 690
sexual differentiation, brain, 690
disorders
erectile dysfunction, 114, 116b
hypogonadism, 691
hypogonadotropic hypogonadism, 691–692
prostate cancer, 689b
pseudohermaphroditism, 692
duct system, 682, *682*
endocrine glands of, 676–677, *677*
luteinizing hormone, 681–682
neural control, 682–683
sertoli cells, 680–681
spermatogenesis, 683–685
testicular function and regulation
GnRH, 677
luteinizing and follicle-stimulating hormones, 677–678
steroids and polypeptides, 679
testis, 679–680
malignant hyperthermia, 572–573
malleus (hammer), 78, *78*
maltose, 521
mammary glands, 724
fetal development of, 619
milk-secreting alveolar cells in, 725, *726*
mammillothalamic tract, 121
mammogenesis, 724
mania, limbic system role in, 133
manometric catheters, in esophagus motility disorders, 490, 491b, *492*
margination, 202
marrow stroma, 181
mast cells, 180, 193, 472
maternal recognition of pregnancy, 718
maximal oxygen uptake, 576
maximal response, 601
maximal voluntary contraction (MVC), 576
McCune-Albright syndrome, 704b
MDR-associated protein transporters, 34
mean artrial pressure
cardiac output and, 267
defined, *267*
effects of exercise on, 269, *269*
equation for, 267
interactions of, 268–270, *269–270*
mean circulatory filling pressure (MCP), 272
mean corpuscular (or cell) hemoglobin (MCH), 174
mean corpuscular (or cell) hemoglobin concentration (MCHC), 174
mean corpuscular (or cell) volume (MCV), 174
mean electrical QRS axis, 242
mean platelet volume (MPV), 174
mechanoreceptors, 62, 68
medial forebrain bundle, 121
median effective dose (ED50), 601
median eminence of the hypothalamus, 605, *605*
mediastinal membrane, 333, *333*
medium spiny neuron, 104
medulla
adrenal, *726*
renal
inner, 407, *407*
outer, 407, *407*
medulla oblongata, 313
in breathing rhythm, 382–384
ventrolateral, 313
medullary pyramid, 399, *400*
medullary reticulospinal tract, 99
medullary tissue, 722
megakaryocyte, 183
megaloblast, 177
megaloblastic anemia, 366b
meiosis, 683
Meissner corpuscles, 68
Meissner plexus, 479
melatonin, 124, 210
membrane attack complex (MAC), 201
membrane-associated dense bodies, 159
membranous labyrinth, *83*
memory
declarative, 135
long-term, 135
nondeclarative, 135
short-term, 135
memory cells, in immune system, 194, *194*
menaquinones, 530
menarche, 703
Ménière disease, 87b
menopause, 703, 707–708
hormone replacement therapy, 708b
osteoporosis and, 673
menses, 703
menstrual cycle
estradiol and progesterone
cervical mucus changes, 707
ischemic phase, 706–707
menstrual phase, 707
proliferative phase, 706
secretory phase, 706, *706*
menopause, 707–708
phases of
follicular phase, 704
luteal phase, 706
menstrual phase, 703–704
ovulatory phase, 704–706, *705*
puberty, 703
menstrual phase, 703–704
menstruation, 703–704
Merkel disks, 68
mesenchymal stem cells, 440b
mesenteric ganglia
inferior, 477, *477*
superior, 477, *477*
mesenteric receptors, in gastrointestinal motility control, 478
mesoderm, 716
mesolimbic/mesocortical system, 131
mesonephric (wolffian) ducts, 722
metabolic acidosis, 390, 466
anion gap in, 466
diabetes mellitus and, 466
vomiting in, 469b
metabolic alkalosis, 467–468, 468t
compensation in, 467
defined, 467
metabolic bone diseases, 673
metabolic clearance rate (MCR), 597
metabolic condition, electrocardiogram (ECG) in, *246,* 246–247
metabolic energy, 554
metabolic rate
in hypothermia, 573, 573b
in response cold, 568
metabolic syndrome, 663
metabolism
of cholesterol, 544
by liver
drugs and xenobiotics, 540, *540*
of lipoprotiens, 543
regulation of thyroid hormone in, *625,* 626
of xenobiotics, 540, *540*
metabotropic, 53
metabotropic glutamate receptors (mGluRs), 58
metabotropic receptor, 481
methemoglobin (metHb), 176
methylene blue, 175
MHC class I molecules, 195
MHC class II molecules, 194
micellar emulsification, 524
micelles, 515
microalbuminuria, 410
microcirculation, 279
arterial microvasculature, 279–280
capillaries in
anatomic considerations, 280, *281*
defined, 280
endothelial vesicles in, 281, *281*
pores in, 281
porosity of, 281
components, 279–282, *280, 281*
defined, 279
diabetes mellitus, 288b
lymphatic (*see* lymphatic system)
overview, *289*
regulation of
autoregulation, *291,* 291–293
endothelium cells, 289–291, *290*
myogenic, 288–289
organ blood flow, 289, *289*
reactive hyperemia, 293
sympathetic nervous system (SNS), 291
solute exchange in
capillary filtration coefficient, 286
capillary pressure, 285
colloid osmotic pressure, 285
diffusion, 284–285
diffusion-limited transport, 284–285
extraction ratio, 284
filtration, 286–287

microcirculation (*Continued*)
flow-limited transport, 284–285
hydrostatic pressure, *286,* 286–287
pathological conditions related to
permeability surface area coefficient, 284
Starling-Landis equation, 285–286
vascular resistance and, 279–280
venules in, 281–282
microcytes, 177
microfilaments, 43
microspherocytes, 26
microtubule-associated protein (MAP) kinesin, 44
microtubules, 44
microvilli, 520
micturition, 449
midbrain extrapyramidal area, 104
middle ganglia, 112
migrating motor complex (MMC)
activity front, *497*
phase of, *498*
milk ejection, 724
mineralocorticoid escape, 439
mineralocorticoids. *see also* aldosterone
action of various, 438
excess of, 438
minerals, absorption, 530–534, *531, 533, 534*
minute ventilation
alveolar ventilation, 338–340, 340t
dead space volume, 337
expired, 340–341, *340–341*
mitochondria, 139
mitogen-activated protein kinase (MAP kinase), 22
mitogenic agent, 615
mitogens, 186
mitosis, 683
mitral valve, 213, 214
mitral valve prolapse, 265
mittelschmerz, 705
mixed acid–base disturbance, 468–469, *469*
mixed contractions, 151
mixed micelle, 524
M-mode echocardiography, 265
molecular switches, 11
monoamine oxidase (MAO), 57
monoamine oxidase inhibitors (MAOIs), 132
monoaminergic neurons, 130–131
monoblasts, 181
monochromatic vision, 73
monoclonal antibodies, 180
monocyte-macrophage system, fixed, 547–548
monocytes, 179
monocytic leukocytes, 179
monoiodotyrosine (MIT), 624
mood, 132
mood disorders, 133–134
morula, 712
mossy fibers, 105
motion sickness, 84–85
motor control
basal ganglia, 103–104, *104*
brainstem in, 98–100, *99, 100*
cerebellum, *105,* 105–107, *106*
cerebral cortex role in, 100–103, *101, 103*
Golgi tendon organs in, 93, 94, *94*
motor cortex, primary, *101,* 101–102
motor neurons
alpha, 92–93
gamma, 92, 94–95, *95*
muscle fibers in
muscle spindles in, 93–94, *94*
rubrospinal tract, 99
sensory receptors in
spinal cord in
anatomical consideration, 96, *96*
basic locomotion, 98
reflexes, 96–98, *97, 98*
motor endplate, 143, *143*. *see also* neuromuscular junctions
motor neurons, in motor control
alpha, 92–93
gamma, 92, 94–95, *95*
motor reflex, *476,* 477
motor unit, 92, 149. *see also* motor neurons
motor unit summation, 149, *149*
mucins, 507
mucous cells, 505, 506
mucous gel layer, 509
mucous neck cells, 508
müllerian-inhibiting substance (MIS), 723
multidrug resistance (MDR), 34
multiple myeloma, 207
multiple sclerosis., 210
multipotent stem cells, 181
multiunit-type smooth muscle, 473
mural granulosa cells, 698
muramidase, 507
muscarinic receptor, 56, 110
muscle chemoreflex activity, 577
muscle fatigue, defined, 582
muscle fibers, 139
muscle function, 91–92
muscle(s)
blood flow during exercise, 552, 555–556
skeletal (*see* skeletal muscle)
smooth (*see* smooth muscle)
muscular dystrophy, 141
muscularis externa, 472, 508
myelin sheath, 42
myeloid leukocytes, 179
myenteric plexus, 479
myocardial infarction, 234
myocardial ischemia
arrhythmias, 231
defined, 231
reduced refractory period in, 231
myoclonus, 126
myoepithelial cells, 608, 725
myofibrils, 141
myogenic mechanism, 408
myogenic regulation, 288–289
myoglobin, 157
myoneural junction, 143, *143*
myopia, 70
myosin, 139
myosin light-chain kinase (MLCK), 161
myosin light-chain phosphatase (MLCP), 162
myosin-linked regulation, 161
myotatic (muscle stretch) reflex, 96–97, 97*f*
myxedema, 631

N

Na^+/K^+-ATPase, 507
NADPH-cytochrome P450 reductase, 540
narcolepsy, 124, 393b
natural killer (NK) cells, 180
nebulin, 141
necrosis, 23, 190
negative cooperativity, 601
negative feedback, *3,* 3–4
negative selection, 196
nephrolithiasis, 447
nephron, acidification, *458,* 458–459, *459*
nephrons, 401–402, *402*
nephrotic syndrome, 411b
Nernst equation, 46
nerve growth factors, 9
neural mediators, 210
neural presbycusis, 82
neural stem cells (NSC), 503b
neuroendocrine, 9
neuroendoimmunology, *209,* 209–210
neurofilaments, 44
neurogastroenterology and motility, 471
enteric nervous system (*see* enteric nervous system)
gastrointestinal tract, 472
musculature
action potentials, 473–474
afterhyperpolarization-type, 480
circular and longitudinal muscle, 472
electrical slow-wave frequencies, 474
electromechanical and pharmacomechanical coupling, 473
functional electrical syncytium., 473
interstitial cells of Cajal, 475
smooth muscles, 473
neural control
dorsal vagal complex, 476
enteric nervous system, 478–479
motor activity, 475–476
myenteric and submucosal plexuses, 479
parasympathetic neurons, 476
splanchnic nerves, 478
sympathetic nerves, 477–478
synaptic-type neurons, 480
vagovagal reflex, 476–477
neurogenic secretory diarrhea, 484
neurohypophysis, 604–605, *605*
neuroleptic drugs, 131
neuromuscular junctions (NMJ), *143,* 143–144, *144,* 482–483
neurons
action potential
conduction velocity considerations, 50–51, *51*
phases of, 46, *47*
refractory periods, 49
trigger zone
voltage-gated ion channels, 47, *48,* 49
anatomic features, 42–43, *43*
cholinergic, 55, *56*
conductance of, 46
cytoskeleton of, 43–44, *44*
electrical properties, 46
membrane potential, 44
protein synthesis, 43
resting membrane potential, 44–46, *45*

structure, *43*
synaptic transmission
chemical neurotransmitters, 51–53, *52*
excitatory postsynaptic potential, 53, *53*
inhibitory postsynaptic potential, 53, *53*
spatial summation, 53–54, *55*
temporal summation, 53, *54*
viability maintainance, 43
neuropeptide Y (NPY), 122
neuropeptides, 59–60, 59t
neurophysin, 607, *607*
neurosecretory cell, 9
neurotransmitters, 7
acetylcholine, 55–56, *56*
arachidonic acid, 60
catecholamines, 56–57, *57*
endocannabinoids, 60
GABA and glycine, 58–59
glutamate, 58
neuropeptides, 59–60, 59t
nitric oxide (NO), 60
serotonin, 58
neutralization, 200
neutropenia, 180
neutrophils, 174, 192
niacin, 530
nicotinamide adenine dinucleotide phosphate (NADPH) oxidases, 20
nicotinic ACh receptors, 143
nicotinic receptor, 55, 110
nigrostriatal system, 131
nipple, 725
nitric oxide (NO), 7, *8,* 60, 117, 289, 682
nitric oxide synthase (NOS), 7, *8*
NMDA receptor, 58
NO synthase, 289
nociceptors, 62
nodes of Ranvier, 43
nodose ganglia, 477
nondeclarative memory, 135
nonessential amino acids, 526
non-rapid eye movement (NREM) sleep, 126
nonvolatile acid-base homeostasis, 453
noradrenergic neurons, 131–132, *132*
norepinephrine (NE), 56, 111, *111,* 478, 593, *593*, 634
normoblast, 177
nuclear bag fibers, 93
nuclear chain fibers, 93
nucleus accumbens, 130
nucleus ambiguus, 115, 476
nucleus of the tractus solitarius, 118
nucleus reticularis gigantocellularis, 99
nucleus reticularis pontis caudalis, 99
nucleus reticularis pontis oralis, 99
nucleus tractus solitarius, 476
null cells, 180
nystagmus, 145b

O

obesity, 639b
occipital lobe, 76, 128
Ohm law, 46
olfactory system
mechanism of, 88–89
mitral cells, 89
olfactory bulb, 89
sensory cells, *89*
sneezing, 89
vomeronasal organ, 89–90
oncogenes, 10, 207
oncotic pressure, 168
oocytes, 676, 696
oogonia, 696
oolemma, 714
ooplasm, 714
opsonin, 179, 192
opsonization, 200
optic disc, 70
optic nerve, 70
optic radiations, 76
optic tract, 76
optimum length (L_o), 153
orexigenic, 122
orexin, 124
organ of Corti, 79, 81–82
organic anion transporting polypeptides (OATPs), 34
organification, 623
orthodromically, 116
orthophosphate, 665
osmolality, 38, 400
osmoregulation, 4
osmotic pressure, 37
osteoarthritis, 584
osteoblasts, 668, *668*
osteoclasts, 668, *668*
osteocytes, 668, *668*
osteogenesis imperfecta, 668, 669b
osteoid, 667
osteomalacia, 674
osteoporosis, 584, 673
otic ganglion, 115
oval window, 78
overshoot phase, 47
ovulation, 608
ovulatory phase, 704–706, *705*
oxidative burst, 179
oxidative phosphorylation, 156, 629
oxidative stress, 21
oxidoreductases, 20
oxygen
demands, small intestine, 301–302
partial pressure
measurement, 330–331
normal range, 330
at sea level, 330, 330t
oxygen carrying capacity (OCC), 174
oxygen content, 359–360
oxygen debt, 575, *576*
oxygen diffusion gradient, 356
oxygen uptake, 356
oxygen-carrying capacity, 359
oxyhemoglobin (HbO_2), 176, 359
oxyhemoglobin-equilibrium curve, 360
oxyntic glands, 508
oxytocin, 120, 605, 694

P

P wave, 236
pacemaker
for antral pump of stomach, *492,* 492–493
in gastrointestinal motility, 473
pacinian corpuscle, 68
Paget disease
defined, 675
etiology, 675
pain
cold, 68
deep, 68, 316
delayed, 68
initial, 68
referred, 67, 68, 115
somatic, 68
superficial, 68
visceral, 68, 386
pain sensory receptor, 62–63
palmitoyl transferase, 659
pampiniform plexus, 679
pancreatic amylase, 512, 521
pancreatic deficiency, 528
pancreatic lipase, 512, 524
pancreatic peptide
actions of, 649
F cell production of, 649
secreation of, 649
pancreatic proteases, 527
Paneth cells, 519
panhypopituitarism, 692
para-aminohippurate (PAH) clearance, 405–406
parabrachial nucleus, 118
paracrine mediators, 483
paracrine signaling, 7, *8*
paradoxical cold, 68
paradoxical sleep, 126
parafollicular cells, 622, *622*, 670
parahippocampal gyrus, 130
parallel fibers, 106
paralytic ileus, 478
paramesonephric (müllerian) ducts, 722
parasympathetic division, 109
parasympathetic nervous system (PNS)
brainstem neurons, 115
functions of, 111, 116t
ganglia of, 115
sacral neurons, 115
salivary secretion of, 507, *507*
spinal cord control of, 98
parathyroid hormone (PTH)
in calcium regulation, *445,* 445–446
in phosphate balance, 446–447, *448*
paraventricular nuclei, 605
paravertebral sympathetic ganglia, 111
parietal (oxyntic) cells, 508
parietal lobe, 128
parietal-temporal-occipital association cortex, 129
Parkinson disease (PD)
basal ganglia in, 104
dopamine and, 8, 8b
parotid glands, anatomic considerations, 505, *506*
pars distalis, 605, *605*
pars intermedia, 605
pars reticulata (SNr), 104
pars tuberalis, 605, *605*
partial pressure
of arterial carbon dioxide, 378, *378*
oxygen
of gas in liquid, 4
measurement of, 330–331

partial pressure (*Continued*)
normal range, 330
in respiratory gas exchange, 330, 330t
at sea level, 330, 330t
peripheral chemoreceptors, 389
in respiratory gas exchange, 330, 330t
partial seizures, *127,* 127–128
parturition, 712
parvicellular neurons, 610
passive force, 152
passive immunity, 528
passive transport, 26
patches, 159
pathogens
defined, 189
transmission of, 189
peak expiratory flow, during forced expiration, 336t, 350–352
pegaptanib sodium, for macular degeneration, 71b
pellagra, 530
pelvic nerves, 448, *448*
pendrin, 623, *623*
pepsin, 191, 511, 527
pepsinogen, 509, 511, 527
peptic ulcers, 513b
peptidases, 528
peptide hormones, 14
peptides
absorption of, 526–528
pancreatic, *527*
in protein digestion, 526–528
in regulation of insulin, 657
perception, 68
perforin, 193
perfusion pressure, 297
perfusion-limited gas uptake, 358
periaqueductal gray matter, 118
pericellular space, hepatic, 536, *537*
perikaryon, 42
perilymph, 79
peripheral lymphoid organs, 188
peripheral proteins, 26
peripheral sympathetic anatomy, *112*
perisinusoidal space, 536, *537*
perivitelline space, 714
permeability surface area coefficient, 284
pernicious anemia, 530
p-glycoprotein (MDRI), 33b
pH
buffer, 452
cerebrospinal fluid in ventilation, 388
defination, 452
of saliva, 507
phagocytosis, 27, 179, *179,* 191
phagosome, 179
pharmacomechanical coupling, 161, 473
phases of digestion
gastric secretions, 509–510, *510*
pheromones, 9, 89
phorbol esters, 18
phosphate
in bone, composition, 667–668
composition in body fluids, 446–447
distribution in bone cells, in healthy adult, 665, 665t
homeostatic pathways, 665–667
inorganic constituents, 665, 665t
kidneys, phosphaturia, 667
kidney stone disease, 447
pH buffering, 664–665
plasma, 665
phosphate buffer, acid-base homeostasis, 454–455
phosphatidylinositol 4,5-bisphosphate (PIP2), 611
phosphaturia, 667
phosphodiesterases (PDEs), 16, 74, 688
phosphoenolpyruvate carboxykinase, 645
phosphoenolpyruvate carboxylase, 658
phosphofructokinase, 658
phospholamban, 251
phospholipase, 636
phospholipase A2, 524
phospholipase C (PLC), 18, 53, 164, 611
phospholipids, 25
phosphorus homeostasis, *666,* 666–667
phosphorylation, oxidative
in skeletal muscle contraction, 162
in smooth muscle contraction, 161–162
thyroid hormones regulation of, 629
photons, 70
photopic (daytime) vision, 73
photoreceptors, 62
phylloquinones, 530
physiologic regulation, 1–6
physostigmine (eserine), 145
pigment epithelium, 69, 71b, 72
piloerection, 567
pineal gland, 124
pinna, 77, *77*
pinocytosis, 622
pitch discrimination, 82
pituitary dwarfism, 616
pituitary gland, 604
adenohypophysis, 604, *605*
anatomic considerations, 604–605, *605*
anterior hormones, 608–619
blood supply to, 605–606, *606*
disorders of
congenital adrenal hyperplasia, 611
tumors of, 609b
neurohypophysis, 604, *605*
posterior hormones
arginine vasopressin, 607
hypothalamic-releasing hormones, 607, 607t
oxytocin, 607–608
pituitary resistance to thyroid hormone, 627b
pituitary tumors, 609b
place theory, 81
placenta, 694
circulation of, 308
development of, 619
vasculature of, 306–307, *307*
placental villus, 306, *307*
placing reaction, 102
plasma, 167, 168t
electrolyte composition of, 430t
as extracellular fluid (ECF), 428
fresh frozen, 167
proteins in (*see* plasma proteins)
plasma cells, 180, 193–194
plasma lipids, 169
plasma membrane, 24
fluid mosaic model of, 25, *25*
memebrane transport, 26–37
structure, 24–26, *25*
lipid bilayer, 24, *25*
lipids, 25–26
proteins, 26
plasma membrane calcium ATPase (PMCA), 19, 20
plasma proteins
in hormone transport, 596, 596t
reflection coefficient, 429
synthesized by liver, 544
plasma volume
in circulatory shock, 323
effects of sweating on, 569
in heat acclimatization, 566
plasma water
electrolyte composition of, 430t
measurement of, 409t
plasmapheresis, 167
plasmin, 186, 701
plasminogen, 186
plasminogen activator, 680, 701
plate endings, 94–95
plateau region, of action potential, 229, *229*
platelet count (Plt), 174
platelet-derived growth factor (PDGF), 186
platelets, 167
plegia, 98
pleural pressure, 331
breathing rhythm, 327
effect on pulmonary vascular resistance, 371–374, *373*
electrical activity during breathing at rest, 382, *383*
in forced expiration, 329, 332
pneumothorax, 333
pleural space, 328
podocytes, 409
poikilocytes, 177
poikilotherm, defined, 550, 551
Poiseuille law, 267
polyclonal antibody, 193
polycyclic aromatic hydrocarbons, 540
polycystic kidney disease (PKD), 424b –425b
polycystic ovarian syndrome, 709–710
polycystin-1, 425b
polycystin-2, 425b
polycythemia, 171, 396
polyduction, 425b
polymorphonuclear leukocytes, 179
polyspermy, 714
polyuria, 37, 662
pontine respiratory group, 384
pontine reticulospinal tract, 99
pores, 281
portal hypertension, 538, 539b
portal vein, 214
positive feedback, 4
positive selection, 195
postcapillary venules, 280
posterior lobe, 605
postganglionic axon, 110
postganglionic sympathetic neurons, 477
postpartum and prepubertal periods
hypothalamic GnRH
childhood, 726–728
physical signs, puberty, 728
mammogenesis and lactogenesis, 724–725
suckling reflex, 725–726

postpartum thyroiditis, 630
postrema, 476
poststreptococcal glomerulonephritis, 411b
potassium, 532
  abnormal levels of
    hyperkalemia, 443–444
    hypokalemia, 443–444
    plasma, in pH regulation, 168
  acid-base balance, 442
  composition in body fluids
  counterbalancing effects, 444–445, *445*
  diarrhea, 443
  dietary potassium intake, changes, 444, *444*
  hyperkalemia, 442
  hypokalemia, 442
  metabolism, 442
  potassium influences, 442
  pseudohyperkalemia, 442
  transcellular fluids, 442–443, *443*
potassium channels
  in cardiac action potentials, 228–229
  inwardly rectifying, 231
  outwardly rectifying, 230–231
  transient outward, 230
potentiation, 511
PR interval, 236–237
prechondrocytes, 616
precocious puberty, 704b
precordial leads, 240
prefrontal association cortex, 129
prefrontal cortex, 130
preganglionic axon, 110
preganglionic sympathetic neurons, 477
pregnancy
  hormones secreted
    human chorionic gonadotropin, 718, *718*
    human placental lactogen, 718–719
  maternal metabolism and endocrine function change, 720
  placenta, 717
  postpartum period, 724–725
  postpartum thyroiditis, 630b
  steroids
    dehydroepiandrosterone sulfate, 719–720
    estriol, 720
pregnenolone
  in female steroidogenesis, 693
  in testosterone production, 685
pregranulosa cells, 696
premature atrial contractions (PACs), 243
premature ovarian failure, 709
premature ventricular contractions (PVCs), 235
premotor area, 130
presbyopia, 70
pressure, 216
pressure diuresis, 321, *321*
pretectal nuclei, 76
prevertebral ganglion, 112
primary adrenal insufficiency, 641b. *see also* Addison disease
primary follicular stage, 697
primary (or central) lymphoid organs, 188
primary motor cortex, 101–102, 101*f*
primary oocytes, 700
primary somatosensory cortex (S1), 102
primary spermatocytes, 684
prime mover, 91
primordial follicle, 696
procedural memory, 135
progesterone, 594, 608, 699, 718
progesterone block, 723
progestin, 709
programmed cell death, 22, *22*
prohormone, *594,* 594–595, *595*
prohormone convertase 1, 609
prohormone convertase 2, 609
prolactin (PRL), 605, 619, 688, 694
proliferative phase, 706
proopiomelanocortin (POMC), 122, 594, *609,* 609–610
proprio spinal cells, 96
proprioceptors, 62, 93
prostaglandins, 570, 645
prostate cancer, 689b
protein buffer, acid–base homeostasis, 455
protein inclusion bodies, 177
protein kinase A, 16
protein kinase A (PKA), 610
protein kinase C (PKC), 18, 611
protein kinase G (PKG), 7
protein S deficiency, 186
proteins
  absorption, 526–528, *527,* 527t
  aminoacids in, 527–528, 527t
  complete, 526–527
  digestion, 526–528, *527,* 527t
  genetic disorders, 528
  incomplete, 527
  metabolism of, *544,* 544–545
proteoglycans, 668
prothrombin, 546, *546*
prothrombin time (PT), 185
prothrombinase (factor X), 185
proton pumps, 33
proto-oncogenes, 10
proximal tubule
  isosmotic, 414
  sodium reabsorption, 414–415, *415*
  toxins and drugs, 415–417, *416–417,* 416t
  water reabsorption, 413–414
pseudohermaphrodite, 729
pseudohermaphroditism, 692
pseudohyponatremia, 435b
pterygopalatine ganglion, 115
ptosis, 114
P-type ATPase, 32–33
pubarche, 703
pubertal disorders, 730
pulmonary arteries, 369, *370*
pulmonary artery, 214
pulmonary capillaries
  anatomic considerations, 374
  fluid exchange, 374–376
  gravity effects on, 216
pulmonary circulation, 369–370, 370
  blood flow in lungs, 396
  in gas exchange, 369–370
  gravity effects on, 216
  hemodynamic features, 370–371, *371*
pulmonary edema, 375–376
pulmonary embolism, 215, 372b
pulmonary hypertension, hypoxia-induced, 375b
pulmonary stretch receptors, 386
pulmonary vascular resistance, 371–372
pulmonary vasoconstriction, 373–374
  hypoxemia, 373
  hypoxia-induced, 373
pulmonary wedge pressure, 370
pulmonic valve, 214
pulse pressure
  arterial compliance and, 267–268, *268*
  defined, 267
  effects of exercise on, 269, *269*
  interactions of, 268–270, *269–270*
  stroke volume and, 267–268, *268*
  systemic vascular resistance and, 269–270, *270*
pupillary light reflex, 76
purinergic neuron, 484
Purkinje cells, 105
Purkinje fibers, 228
putamen, 103
pyloric glands, 508
pyramidal tract, 102
pyrogens, role in fever, 570
pyruvic acid or pyruvate, 156

Q

Q_{10} effect, 573b
Q wave, 236
QRS wave, 236
QT interval, 237
quantum content, in synaptic transmission, 52
quantum, in synaptic transmission, 52, 143

R

R wave, 236
radial bundles, 80
radiation, in body heat exchange, *553,* 553–556
radioimmunoassay (RIA), 598–600, *599*
radioreceptor assay, 599–600
ranibizumab, for macular degeneration, 71b
rapid adapting receptors, 387
rapid ejection phase, 252
rapid eye movement (REM), 126–127, 392
rapid gastric emptying, 496b
rapid ventricular filling, 252
ras family, 14
Rathke pouch, 605
reactive hyperemia, 293
receptor adaptation, 315
receptor potential, 64
receptor(s), 9, 43
  acetylcholine, 55
  adrenergic, 113
  baroreceptors, 313–315
  in cellular signalling, 9
  chemoreceptors (*see* chemoreceptors)
  for hormones (*see* hormone, receptors for)
  mechanoreceptors (*see* mechanoreceptors)
  photoreceptors (*see* photoreceptors)
  sensory (*see* sensory receptors)
  thermal, 553, 562
  thyroid hormone, 624

receptor-mediated endocytosis, 27
reciprocal inhibition, 91–92
red blood cells (erythrocytes), 166, 175–178
  anemia and, 177, 178
  destruction of, 167, 177
  disorders of, 177
  erythrocyte sedimentation rate (ESR), 171, *171*
  formation of, 180–182, *182*
  hematocrit (Hct), 171
  hemoglobin in, 176
  iron recycling, 177–178, 178t
  life span of, 180
  mean corpuscular hemoglobin (MCH), 173t, 174
  mean corpuscular hemoglobin concentration (MCHC), 173t, 174
  mean corpuscular volume (MCV), 173t, 174
  morphology of, *176,* 177
  normal ranges for, 171
  in oxygen transport, *175,* 175–176
  specific gravity, 170
  structure and functions of, 175, *175,* 176, *176*
red cell distribution width (RDW), 174
red fiery feet, 56
red marrow, 181
red nucleus, 99
reduced ejection phase, 252
reduced ventricular filling, 252
reentry tachycardia, 234, 235
referred pain, 67
reflection coefficient, 286
reflective pain, 68
reflex sympathetic dystrophy (RSD), 118
reflex(es), 96
  lung
    irritant receptors, 387
    J receptors, 387
    mechanoreceptors, 386
    proprioceptors, 387
    pulmonary stretch receptors, 386
  lung inflation, 331
regional hypoventilation, 364
regulated exocytosis, 52
regulatory light chain, 140
regulatory volume decrease (RVD) mechanisms, 38
regulatory volume increase (RVI) mechanisms, 38
relative refractory period, 49
relaxin, 723
releasing factors, 120
releasing hormones, 605
renal blood flow (RBF), 405–406
  autoregulation, 407–408, *408*
  cortex, 407, *407*
  cortext, 407, *407*
  sympathetic stimulation and hormones, 408–409
renal compensation
  metabolic acidosis, 466
  metabolic alkalosis, 468
renal hypoxia, 320
renal response, acid–base homeostasis, 454
renin, 402
renin-angiotensin-aldosterone system, 318–319, *319*
repolarization phase, 47
residual volume (RV), 252, 334
resistance, to thyroid hormone, 627
respiration
  cellular, 326
  defined, 326
  gas exchange, 326
respiratory acidosis, 390
  carbon dioxide accumulation
    cell chemical buffering, 464–465
    renal compensation, 465
    ventilation stimulation, 465
  definition, 464
  loss of carbon dioxide
    alveolar hyperventilation, 465
    cell chemical buffering, 465–466
    kidney compensation, 466
respiratory alkalosis, 390
respiratory burst, 20, 179
respiratory compensation
  metabolic acidosis, 466
  metabolic alkalosis, 468
respiratory exchange ratio, 357, 553
respiratory quotient (RQ), 555
respiratory response, acid–base homeostasis, 454
respiratory sinus arrhythmia, 242, *244*
respiratory system compensation, metabolic alkalosis, 468
resting force, 152
resting membrane potential
  cardiac, 229, *229,* 230t
  electrochemical potential, 40
  Goldman equation, 40–41
  Nernst equation, 40
  steady state, 39, *39*
reticular formation, 99, 124–125
reticulocytes, 177
reticuloendothelial system (RES), 177, 192, 537
retina, 70
retinohypothalamic tract, 124
retinopathy, 663
retrograde direction, 44
Reynolds number, 222
rhesus factors (Rh), 172
rheumatoid arthritis (RA), 206, 584
rhodopsin, 74
rhythmicity, of cardiac cells, 228
rickets, 529, 673–674, 674t
right atrium, 214
right bundle branch block, 245
right ventricle, 214
right ventricular hypertrophy, 242
rigor crossbridge, 147
rod cell, 73
rods and cones, 70
rolling, 202
rough endoplasmic reticulum (RER), 43
rouleaux phenomenon, 171
rubor, defined, 201
rubrospinal tract, 99
ryanodine receptor (RyR), 19, 146

S

S wave, 236
saccule, 83, *83*
safety factor, 144
saliva
  electrolyte composition of, 506–507
  formation of, *507,* 507–508
  in immune response, 200
  osmolality, 506–507
  protective functions of, 507
  proteins in, 507
salivary glands, 505
  anatomic considerations, 505–506, *506*
  excretory duct of, 506
  primary secretion of, 506
salivary secretions
  anatomic considerations, 505, *506*
  in carbohydrate absorption, 113
  parasympathetic nervous system in, 507–508
  saliva, 507
  sympathetic nervous system in, 508
salivatory nuclei, superior, 114
salivon, 506, *506*
salt concentration, 3
salt receptors, 86
saltatory conduction, 51
salty taste, 86
sarcolemma, 145
sarcomere, *141,* 141–142
  defined, 141
  of skeletal muscle, 142
  structure of, 141
sarcoplasmic reticulum (SR), 145
satellite cells, 616
satiety, 121
saturation, of transport systems, 31
scala media, 79
scala tympani, 79
scala vestibuli, 79, *79*
scalene muscles, 329
scan airways, 380b
Scatchard plot, 601, *601*
schistocytes, 177
schizophrenia
  defined, 134
  limbic system, 134
Schwann cells, 143, *143*
sclera, 69
scotopic (nighttime) vision, 73
scotopsin, 74
scurvy, 530
seasonal affective disorder (SAD), 133–134
second heart sound, 252
second messengers, 9, 15–21
  1, 2-diacylglycerol (DAG), 18, *19*
  calcium, 18–20, *20*
  in cellular signaling, 9
  cGMP, 18–19
  cyclic adenosine monophosphate (cAMP), 9, 16
  inositol triphosphate (IP3), 18, *19*
  lipids, 18
  NO and CO, 18–19
  protein kinase A, 16–17, *17*
  reactive oxygen species, 20–21, *21*
  types of
second polar body, 701
secondary follicular stage, 697
secondary lymphoid organs, 188
secondary oocyte, 700
secondary pacemaker, atrioventricular (AV) node, 233

secondary sex characteristics, 703
secondary spermatocytes, 684
second-degree atrioventricular block, 245
secretagogues, 592, 653
secretin, 511, 514
  gastric secretion of, 511
  in pancreatic secretion, 514
secretomotor neurons, 475
secretory phase, 706, *706*
segmental arteries, 399
segmentation, of small intestine, 487
selective serotonin reuptake inhibitors (SSRIs), 133
selenocysteine, 624
self-tolerance, of immune system, 191
sella turcica, 604
semen, 683
semen secretions, 683
seminal emission, 683
seminal plasma, 713
seminal vesicles, secretion of, 592
seminiferous tubules, 679
senile plaques, 136b
sensation
  accessory structures for, 62
  adaptation in, 65
  hearing, 62
  sensory modality, 62
  smell sense, 88–90, *89*
  taste sense, 85–86, *86*
sensorineural deafness, 82
sensory information
  encoding and decoding in, 66–67
  transfer of, 66
sensory modality, 62
sensory receptors
  accessory structures for, 62
  accommodation in, 66
  action potential generation in, *63,* 63–64
  adaptation in, 65, *65*
  generator potential in, *63,* 63–64
  hearing
  smell sense, 88–90, *89*
  taste sense, 85–86, *86*
  touch sense, 67–68
sepsis, 204
septal nuclei, 130
septic shock, 323
serosa, 508
serotonergic neurons, 58, 132
serotonin, 58, 645
serous cells, 505, 506
Sertoli cells, 677
  functions of, 677
  Leydig cells interactions with, 677
  spermatogenesis in, 677
serum, 167
serum antibody test, 170
serum iron levels, 178
serum protein electrophoresis, 170
set point, in thermoregulation
  defined, 562–563
  factors influencing, 563
severe combined immunodefi ciency (SCID), 207
sex
  chromosomal, 722
  genetic, 722
  gonadal, 722
  secondary characteristics
    female, 123
    male, 123
sex chromosomes, 722
sex hormone–binding globulin (SHBG), 681, 709
sexual activity
  hypothalamus role in, 133
  limbic system and, 132
sexual development disorders
  gonadal dysgenesis, 729
  male pseudohermaphrodites, 730
  pubertal disorders, 730
  testicular feminization syndrome, 730
SH2 domain, 13
shallow-water blackout, 397
shear stress, 221
shell conductance, 558
shivering
  during cold stress, 562
  control of, 561
  primary motor center, 562
shock
  circulatory, *321,* 321–323
  neurogenic, 323
  septic, 323
  spinal, 98
short hypophyseal portal vessels, 606
short-term memory, 135
shunts, 378
  vs. dead spaces, 379, 380t
  hypoxemia related to, 365
  intrapulmonary, 365
  right-to-left, 378
sickle cell anemia, 171
sieving action, of stomach, 495
sigmoid colon, *500*
signal cascade, 10
signal transduction
  G protein-coupled receptors (GPCRs), 10–11
  ion channel-linked receptors, 12, *12*
  types of, 10
  tyrosine kinase receptors, 12–14, *14*
  visual, 71–75, *72–74*
signaling. *see* cellular signaling
signaling cascade, 13
simple acid–base disturbance, 464
simple micelle, 524
single-breath test, 358
sinoatrial (SA) node
  cardiac output, 233, 234
  functions of, 231–233
  heart rate and, 233
  as slow action potential, 229
sinus bradycardia, 243
sinus tachycardia, 243
sinusoid, hepatic, 536–539, *537*
sinusoidal endothelial cells, 536, *537*
size principle, 93
skeletal muscle
  adaptation of, 157–158
  antagonistic pairs of, 156
  blood circulation, 304–305
  vs. cardiac muscle, 227
  contraction of
    adenosine diphosphate (ADP), 147
    adenosine triphosphate, 156–157, *157*
    calcium, *146,* 146–147
    crossbridge cycle, 142, *142*
    excitation-contraction coupling, 143–148
    force-velocity curve, 152
    isometric, 150, *150,* 152–153
    isotonic, 150–152, *151, 152*
    length-tension curve, 152, *152*
    mixed, 151
    motor unit summation, *149,* 149–150
    neuronal control, *143,* 143–144
    sliding filament theory, 147–148
    twitch, *148,* 148–149
  disorders of
    muscular dystrophy, 142b
    myasthenia gravis, 145
  isotonic and isometric, 154–155, *155*
  lever system, 155–156, *156*
  mechanical factors, 150
  metabolic and structural adaptations, 157–158
  metabolic process of, 156, *157*
  neuromuscular junction (NMJ), *143,* 143–145, *144*
  red muscle fibers, 157
  sarcomere, 141, *141*
  structural muscle filaments, 141
  thick filaments, 139–140, *140*
  thin filaments, *140,* 140–141
  white muscle fibers, 157–158
  Z lines, 141
skeleton, 91, 92*f*
skin
  in body temperature regulation
    convection, 552, 556, *557*
    evaporation, 557
    radiation, 556, 557, *557*
    shell conductance, 558–560, 562
    surface area and, 556–558, *557*
  peripheral factors influencing, 553
  in response cold, 567–570
  set point and, 563
  sympathetic nervous system in, 560
  temperature
    in heat exchange, 552–553
    in thermoregulation, 561
  in thermoregulation
  wettedness, 557, 558
sleep
  apnea syndrome, 392b
  arousal from, 393
  debt, 127
  disorders of
    apnea, 392b
    narcolepsy, 124
  non-rapid eye movement (NREM), 126
  rapid eye movement (REM), 126
  slow-wave, 126
slow axoplasmic transport, 44
slow excitatory postsynaptic potentials, in gastrointestinal motility, 481
slow inhibitory postsynaptic potentials, in gastrointestinal motility, 481
slow-twitch fibers, 157
slow-wave sleep (SWS), 126, 392
small intestinal motility
  activity front, 497
  digestive state, 495
  interdigestive state, 495

small intestinal motility (*Continued*)
migrating motor, 498
power propulsion, 499
small intestine
absorption by, 302
anatomic consideration, *303*
bile salt recycling in, 517
blood flow
autoregulation, 301–302
nutrients absorption, 302
sympathetic nerve activity, 303
venous-arteriolar response, 302
during water absorption, 302–303
motility of
digestive state, 495
interdigestive state, 495
migrating motor complex (MMC), 495
power propulsion of, 495
vasculature of, 301
smell sense, 88–90, *89*
smooth muscle
adaptation of, 159
of bladder, 158
of blood vessels, 158, 215, 216t–217t
cell size, 163
contraction of
calcium, *160,* 160–161
cell-to-cell transmission, 159–160
control of, 158–159
latch state, 162–163, *163*
myosin filament proteins, phosphorylation, 161–162, *162*
relaxation, 164
electrical syncytial properties of, 473
gap junctions of, 473
gastrointestinal tract, 58
mechanical activity in, 163–164, *164*
structure of, 158
uterus, 696
SNARES, 52
sodium, *531,* 531–532
appetite, 438
balance, *441*
dietary intake, 441–442, *442*
in heat acclimatization, 566, 567
hormonal and neural factors
acute kidney injury, 440b
diuretic drugs, effect of, 441
estrogens hormones, effect of, 440
glomerular filtration rate, effect of, 436–437, 437t
hydrostatic pressure, effect of, 439
natriuretic hormones, effect of, *439,* 439–440
osmotic diuretics, effect of, 440–441
reabsorbed anions, effect of, 441
renin-angiotensin-aldosterone, effect of, 437–439, *438*
sympathetic stimulation, effect of, 440
kidneys excrete, 436, 436t, *437*
sodium channels, voltage-gated, 229–230, 231t
sodium-dependent glucose transporter-1 (SGLT1), 34
sodium-iodide symporter (NIS), 623
solute transport
active
calcium pumps, 33
defined, 27
proton pumps, 33
secondary, *34,* 34–36, *36*
sodium-potassium pumps, *32,* 32–33
endocytosis in, 27, *27*
in epithelial cell layers
exocytosis in, 27–28
passive
passive diffusion (*see* diffusion)
phagocytosis in, 27, *27*
water movement and
cell volume regulation, 38–39, *39*
osmotic pressure, 37–38
water channels, 39
solvent drag, 531
soma, 42
somatosensory system
nociceptors, 68–69
skin receptors, 67–68
thermoreceptors, 68
somatostatin, 509, 606, 607t
somatotopic maps, 101
somatotrophs, 605
somatotropin release–inhibiting factor (SRIF), 606, 607t
action of, 607t
structure of, 607t
sound
pressure, 76–77, *77*
transduction of, 77
wavelength, 77, *77*
sour receptors, 86–87
sour taste, 86
space constant, of a neuron, 46
spasmodic dysphonia, 145b
spasmodic torticollis, 145b
spasticity, 102
spatial summation, 53–54, 149, *149*
special senses, 67
specific gravity, 170
spermatogenesis, 608, 676, 683–685
meiosis, 684
mitosis, 684
phases of, 683–684, *684*
spermatozoon formation, 685
testosterone, 685
spermatogenic cycle, 684
spermatogonia, 680
spermatozoa, 676
spermiation, 680
spermiogenesis, 683
spherocytes, 177
sphincter of Oddi, 513
sphygmomanometry, 270–271, *271*
spinal cord motor neuron pools, 96, *96*
spinal shock, 98, 313
spinnbarkeit, 695, 696
spinobulbar muscular atrophy, 139b
spinocerebellar tracts, 94
spinocerebellum, 105
spiral arteries, 307
spiral ganglia, 80
spirogram, 334
spirometry, airflow, 334
splay, 406
spleen, functions of, 546
splenomegaly, 26, 26b
split brain, 129b
SRY gene, 722
standard curve, 599, *599*
standard frontal lead system, 240
starling forces, 286, 374
starling law of the heart, 256
starling-Landis equation, 286
steatorrhea, 526
stellate cells, 105, 538
stem cell hematotherapy, 181
stereocilia, 80, *80*
sternocleidomastoids, 329
steroid hormones, 14
steroidogenesis, 676
steroidogenic acute regulatory protein (StAR), 699
steroids
ACTH, 641
adrenocortical cell size, 642
cholesterol metabolism, 642
steroidogenic enzyme expression, 641–642, *642*
stigma, 701
stimuli, 61
stomach
anatomic considerations, 508
functions of, 508
mucosal lining of, 513
secretion of, 510
stress test, 579b
stria terminalis, 121
stria vascularis, 79
striated duct, 506
striatum, 103
stroke volume, 252
arterial pressure, 268–269, *269*
pulse pressure, 267–268, *268*
stroma, 160
strong acid, acid-base homeostasis, 452
subcortical structures, 128
sublingual glands, anatomic considerations, 505, *506*
submandibular ganglion, 115
submandibular glands, anatomic considerations, 505, *506*
submucosa, 508
submucosal plexus, 479
substantia nigra, 103
subthalamic nucleus, 103
succinylcholine, 144
sudden infant death syndrome (SIDS), 132
sulcus, 128
sulfonylureas, 653
superior cervical ganglion, 508
superior colliculus, 76
superior ganglia, 112
superior hypophyseal arteries, 606
superior mesenteric ganglion, 112
superior parietal lobe, 102
superior salivatory nuclei, 114, 115
superior vena cava (SVC), 214
superoxide, 20
superoxide dismutase (SOD), 20
superoxide ion, 179
supersonic shear imaging (SSI), 274b
supplementary motor area (SMA), 101, 102
suprachiasmatic nucleus, 76
suprachiasmatic nucleus [SCN], 121
supraoptic nuclei, 605
supraventricular tachycardia, 244
surfactant, 191

sweat glands
absence of, 561, 569b
apocrine, 559
eccrine, 559
evaporative heat loss from, 559
sweating
evaporation and, 559
in heat acclimatization, 559
set point and, 569b
skin temperature and, 563–564
skin wettedness and, 557, 558
in thermoregulation, 559, 561
water loss and, 569b
sweet receptors, 88
sympathetic division, 109
sympathetic nervous system (SNS), 111–114
adrenal medulla interactions with, 114, *114*
in circulation, skin, 560–561
fight-or-flight response, 114, *114*
functions of, 111–112, 116t
ganglia of, 112, *113*
neurotransmitters of, 112–114
peripheral anatomy, *112*
sympathetic tone, 112
sympathetic tone, 112
sympathy, process of, 109
symport, 34
synapse, 7, 43
synaptic cleft, 51
synaptic-type neurons, 480
synaptobrevin, 52
synaptotagmin, 52
syncytiotrophoblasts, 306, 716
syndrome of inappropriate ADH (SIADH), 434
synergist, 92
systemic circulation
arterial pressures in, 267–271, *267–271*
blood volume, 271–274
cardiac function in, 274–277
systemic vascular resistance
arterial pressure and, 269, *269*
interaction of, 268–270, *269–270*
mean artrial pressure, 269–270, *270*
in response to exercise, 269, *269*
systole, defined, 213
systolic hypertension, 270b
systolic pressure, 217

T

T cells, 180
cytotoxic CD8 +, 195
double negative, 195
double positive, 195
effector, 196t
helper CD4+, 197
memory, 197
specific antigen, 193
T helper 1 cells, 196, 197
T helper 2 cells, 197
T regulatory cells, 198
T wave, 236
tachycardia
reentry, 234, 235
sinus, 243, 244
supraventricular, 240, 260t
ventricular, 234, 235, 244
tachyphylaxis, 13
tachypnea, 392
target cells, 177
taste buds, 85–86, 506
taste pores, 85–86, *86*
taste sense, 86
T cell–mediated immunity, 192, 194–201
antigen presentation, *196*
clonal selection in, 193, *194*
cluster of differentiation in, 195, 196
immunological synapse in, 194
negative selection in, 196
positive selection in, 195
T-cell receptors (TCRs), 193
technetium-99 scans, of heart, 264
tectorial membrane, 80, 81
telencephalon, functions of, 103
temperature
body (*see* Body temperature)
environmental
during exercise, 564
in hypothermia, 573
temperature sensory receptors, 561
temporal lobe, 82, 128, *128*
temporal summation, 53, *54,* 148
tensions, 4, 218
tensor tympani muscle, 78, *78*
terminal arterioles, 280
terminal cisternae, 145
terminal intestinal peristalsis, 504
terminal respiratory unit, 328, 380
tertiary follicular stage, 697
testicular feminization, 692
testicular feminization syndrome, 730
testis, 679–680
luteinizing hormone, *687,* 687–688
steroidogenesis, Leydig cells, 685–687, *686*
testosterone, 594, 677
tetanic fusion frequency, 148
tetanus, 148, *148*
tetanus/twitch ratio, 148, *148*
tetany, hypocalcemic, 664, 672
tetrahydrocortisol glucuronide, 640
tetraiodoacetic acid, *625,* 626
thalamus, 121
theca externa, 697
theca interna, 697
theca lutein cells, 702
thelarche, 703
thermal receptors, 553, 561
thermodilution, in cardiac output measurement, 262
thermogenesis
nonshivering, 562, 568
thyroid hormone regulation of, 628–629
thermoneutrality, 559, 561
thermoreceptors, 62
thermoregulation, 561–574, 564–565. *see also* body temperature
behavioral, 561
body temperature, 561–564, 568, 570, 571, 573
central nervous system in, 553, 562
central thermoregulatory controller, 570
clinical aspects of, 570–574
cold acclimatization, 568, 570
cold response, 567–570
cold stress, 570
during exercise
cardiovascular system effects, 564, 573
compensatory responses, 565–566, *566*
vs. fever, 570–571
in heat, 564–565
during fever, 570–571
heat acclimatization, 565–567
heat stress
disorders related to, 571–573
effects of, 559–560
hypothermia, 568, 572b–574b
nonthermal effects on, 563
physiological, 561
set point in
defined, 562–562
factors influencing, 563
skin temperature, 561–564, 566
thermal stress, 554
thermoneutrality, 559, 561
theta waves, 125
thiamine, 529t, 530
thick filaments, of skeletal muscle, 147
thirst center, 435
thoracic cavity, components of, 328, 329
thoracolumbar division, 109
threshold, 47, 62
threshold concentration, 601
thrombi, 369
thrombin, 184
thrombocytopenia, 180
thrombopoietin, 183
thrombotic disorders, 186
thrombus, pulmonary circulation in protection against, 372
thymokine, 210
thymus, in immune system, 191, 195, 205
thyroglobulin (Tg), 622
thyroid gland, 621–632
anatomic considerations
blood supply, 621
changes during pregnancy, 628
disorders of
euthyroid sick syndrome, 631b
excess hormones, 630
hormone deficiency
resistance to thyroid hormone, 627b
follicles of, 622
functions of, 621–622
innervation of, 621
TSH in regulation of
thyroid gland innervation, 621
thyroid hormone receptors, 624
thyroid hormone response elements, 626
thyroid hormones. *see also* thyroxine, triiodothyroxine
actions of
in body growth, 628
in bone development
in CNS development, 626
in intermediary metabolism regulation, 629
in metabolism regulation, 626–627
calcitonin, 622, *622*
dietary iodide and, 626
mechanism of action, 626–627
metabolism of, 622–626
secretion of, 624

thyroid hormones (*Continued*)
self- regulation of
synthesis of, 622–626
thermogenic action of, 628–629
transport of, 624
TSH regulation of
thyroid hormones in regulation of, 629, 629t
thyroid peroxidase (TPO), 623
thyroiditis
autoimmune, 630
postpartum, 630b
thyroid-stimulating hormone (TSH), 210, 605
actions of
diurnal variation in, 613
effect of cold exposure on, 613
in regulation of thyroid gland, 612
secretion, 605
synthesis of, 612
in thyroid hormone regulation, 612–613, *613,* 626
TRH regulation of, 612, *613*
thyrotoxicosis, 630
thyrotrophs, 605
thyrotropin-releasing hormone (TRH), 593, 606
action of, 607t
structure of, 607t
thyroxine (T_4), 608
thyroxine, bile acids, and bilirubin, 34
thyroxine, triiodothyroxine, 593, *593*
metabolism of, 625
secretion of, 622, *622, 623,* 624
structure of, 622, *622*
synthesis of, 622, *622, 623*
transport of, 624
TSH regulation of
thyroxine-binding globulion (TBG), 624
tidal volume, 334
tight junctions, 36
time constant, 46
tissue growth factors, 9
tissue plasminogen activator (TPA), 186
tissue thromboplastin (factor III), 184
titin, 141
titratable acid, acid-base homeostasis, 458
toll-like receptors, 192
tonic-clonic seizures, 127, *127*
tonicity, 38
tonotopic map, 82
tonotopic organization, 81
total bile acid pool, 517
total body water, 427
total iron-binding capacity (TIBC), 178
total peripheral vascular resistance (TPR), 221
totipotency, 181
tractus solitarius, 116
transairway pressure, 331–333
transcellular transport, 36, *36,* 36–37
transcobalamin, 530
transducer, 63
transducers, 9
transducin, 74
transferrin, 177, 533
transferrin saturation, 178
transit time, 357
transmission, 67
transmural pressure, 218, 273, 331
transneuronal transport, 44
transplantation, 401b
transpulmonary pressure, 331
transthyretin, 624
transverse tubules (T tubules), 145, 146, 249
triad, 145
tricuspid valve, 214
triglycerides, 523
triiodothyronine (T_3), 608
metabolism of, 625
secretion of, 622, *622, 623*
structure of, 622, *622*
synthesis of, 622, *622, 623*
transport of, 624
TSH regulation of
trophoblast, 716
tropomyosin, 141
troponin, 141
Trousseau sign, 664
trypsin, 512, 527
t-SNARES, 52
tuberoinfundibular system, 131
tubular reabsorption, 403
tubular secretion, 403
tubuloglomerular feedback mechanism, 408, *408*
tumor, defined, 201
tumor necrosis factor (TNF), 18
tumor necrosis factor-α, 203
tumor-specific antigens, 209
turbulence, 223
Turner syndrome, 709
type 1 and type 2 Diabetes, 661b
type 1 diabetes, 586
type 2 diabetes (T2D), 586, 653
type I collagen, 667
tyrosine, in thyroid hormone synthesis, 622–626
tyrosine hydroxylase, 57
tyrosine kinase receptors, 13

U

UDP-glucose, 542
ultrafiltration, of plasma. *see* glomerular filtration
ultrasound, of heart, 265
umami receptors, 88
umbilical arteries, 306
umbilical vein, 306
uncoupling protein-1 (UCP-1), 629
underwater diving reflex, 397
unitary-type smooth muscle, 473
universal donor, 172
unsaturated iron-binding capacity (UIBC), 178
unstirred water layer, 524
upregulation, 602
urinary concentration and dilution
adaptation to life, 419
arginine vasopressin, 419–420, *420*
countercurrent exchange, 422
countercurrent multiplication, 420, *421*
hyperosmolarity, 422–423
plasma AVP, 424
urea, 423–424, *424*
urinary tract, composition in body fluids
bladder, 448, *448*
eliminating urine, 447–448
micturition, 449
ureters, 448
urethra, 449
urine formation, 403–407
excretion, 403
filtered load, 404
glomerular filtration, 403
glucose reabsorption, 406, *406*
inulin clearance, 403
plasma creatinine, 404
renal blood flow (RBF), 405–406
tubular reabsorption, 403
tubular reabsorption/secretion, 406
tubular secretion, 403
tubular transport maximum, 407, *407*
urobilinogen, 518
urodilatin, 439
uroguanylin, 439
uterus
anatomic considerations, 709
functions of, 159, 713
smooth muscle of, 158, 159, *159,* 307
utricle, 83, *85,* 87, 99

V

V wave, 252
vaccinations, 188, 197, 197b, 198
vagovagal reflexes, 510
vasconstriction
reflex, 560
of skin blood flow
during exercise, 564
in response to clod, 567–568
in thermoregulation, 571, 572
vascular capacitance, 217
vascular distensibility, 218
vascular endothelial growth factor (VEGF), 71b
vascular function curve, 276, *276*
vascular resistance, systemic
arterial pressure and, 269, *269*
interaction of, 268–270, *269–270*
mean artrial pressure, 269–270, *270*
in response to exercise, 269, *269*
vasoactive intestinal peptide (VIP), 111, 508
vasovagal syncope, 316
veins
compliance of, 272
defined, 213
factors that affect muscular contraction, 215, 216t–217t
structure of, 214–215
venipuncture, 168
venous admixture, 363
abnormal, 379, *379*
venous pressure
in blood volume, 273–274
central, 273
interaction with cardiac output, 273, *275,* 275–277
ventilation, 326–354. *see also* airflow; breathing
blood–brain barrier, 388–389
breathing pattern, 382–386
carbon dioxide, 387, *387*

cerebrospinal fluid pH, 388–389
metabolic demands, 384–385, *385, 386*
minute, 336–341
overview, 326–327
peripheral chemoreceptors, 389
phasic control, 382–386
ventilation/perfusion
functional organization, 369–370
hemodynamic features, 370–374
ventilation/perfusion ratio, 364
ventral respiratory group (VRG), in breathing rhythm, 383
ventricle
left, 213
right, 214
ventricular fibrillation, 244
ventricular tachycardia, 244
ventrolateral medullary area, 118
ventromedial nuclei, 614
venules, defined, 213
vermis, 105
vertebral sympathetic ganglia, 477
vertigo, 85, 87b
very low-density lipoproteins (VLDLs), 169, 525
characteristics, 543, 543t
functions of, 543
synthesis of, 542
vestibular nuclear complex, 99
vestibular system
dysfunction, 84–85
equilibrioception, 83
otoliths, 83, *83*
semicircular canal, 83
vestibular apparatus, 83, *83*
vestibulo-ocular reflex (VOR), 83–84
vestibulocerebellum, 105
virilization, 640
viscosity, 171
visual cortex, 76
visual system
age-related macular degeneration, 71b
depth perception, 76
eye layers, 69
focusing, 70
functional optics, 69–70
ganglion cell layer, 75
image crossover, 75–76
light regulation, 71
neural network layer, 75
phototransduction, 71–75
visual reflexes, 76
vitamin A (retinol), 529, 545, *545*
vitamin B1 (thiamine), 530
vitamin B2 (riboflavin), 530
vitamin B6 (pyridoxine), 530
vitamin B12 (cobalamin), 530
vitamin C (ascorbic acid), 530
vitamin D (cholecalciferol), 529, 545, *545*
vitamin D3 and 1,25-dihydroxycholecalciferol, *670,* 670–672, *671*
vitamin D–binding protein, 529
vitamin E, 529
vitamin K, 185, 529–530, 545
vitamins
absorption, 528–531
fat-soluble, 529–530, 529t, 545, *545*
functions, 528–529
water-soluble, 529t, 530
vitreous humor, 70
voltage-gated sodium channels, cardiac muscle, 229–230, 231t
vomeronasal organ (VNO), 89
von Willebrand factor, 183
v-SNARES, 52

W

wall stress, 218
water
absorption, 534, 534t
plasma
electrolyte composition of, 430t
measurement of, 430
water channels, 37, 417, 420
water diuresis, 433
wavelengths, 70
weak acid, acid–base homeostasis, 452
Wernicke area, 136, *136*
wettedness, skin, 557
white blood cells (leukocytes), 167, 174, 178–180
agranular, 179
basophils, 179, 180
developmental cell lines of, 179
eosinophils, 179–180
granulocytes, 179
life span of
lymphocytes, 180
monocytes, 180
mononuclear, 179
neutrophils, 179, *179*
normal ranges for
polymorphonuclear, 179
specific gravity, 170
specific gravity of
types of, 179
white ramus, 111
working memory, 135
Wright-Giemsa, 175

X

X chromosome, 722
xenobiotics
defined, 539
metabolism, 539–540, *540*
xenograft, 205

Y

Y chromosome, 722
Y4 receptor, 651
yellow marrow, 181
yolk sac, 717, *717*

Z

Z line, 141, 142b
zinc, 533
zinc finger, 14
zona fasciculata, 608
zona glomerulosa, 609
zona pellucida, 697, 713
zona reaction, 715
zona reticularis, 608
zonule fibers, 69
zygote, 676, 715
zymogen granules, 506